International Food Law and Policy

Gabriela Steier • Kiran K. Patel
Editors

International Food Law and Policy

Editors
Gabriela Steier
Food Law International, LLP
Boston, MA
USA

Kiran K. Patel
Food Law International, LLP
Washington, DC
USA

ISBN 978-3-319-07541-9   ISBN 978-3-319-07542-6 (eBook)
DOI 10.1007/978-3-319-07542-6

Library of Congress Control Number: 2017942795

© Springer International Publishing Switzerland 2016
This work is subject to copyright. All rights are reserved by the Publisher, whether the whole or part of the material is concerned, specifically the rights of translation, reprinting, reuse of illustrations, recitation, broadcasting, reproduction on microfilms or in any other physical way, and transmission or information storage and retrieval, electronic adaptation, computer software, or by similar or dissimilar methodology now known or hereafter developed.
The use of general descriptive names, registered names, trademarks, service marks, etc. in this publication does not imply, even in the absence of a specific statement, that such names are exempt from the relevant protective laws and regulations and therefore free for general use.
The publisher, the authors and the editors are safe to assume that the advice and information in this book are believed to be true and accurate at the date of publication. Neither the publisher nor the authors or the editors give a warranty, express or implied, with respect to the material contained herein or for any errors or omissions that may have been made. The publisher remains neutral with regard to jurisdictional claims in published maps and institutional affiliations.

The editors make no representations or warranties whatsoever as to the accuracy, completeness, or suitability for any purpose of the content, citations or attributions of this publication or any publications cited therein. Any opinions and views expressed in this publication are the opinions and views of the authors, and are neither the views of nor endorsed by the editors. All contributing authors agreed to the edits of the final version of this book and released the editors from any responsibility or liability directly or indirectly connected to this work, the supplements, errors, omissions, and works cited, quoted or otherwise used therein. This publication is intended for educational purposes only and any materials cited and statements made are not to be construed as or understood to be legal advice.

Printed on acid-free paper

This Springer imprint is published by Springer Nature
The registered company is Springer International Publishing AG
The registered company address is: Gewerbestrasse 11, 6330 Cham, Switzerland

This Book is Dedicated to
Morrice, Our Future
Liviu, Regina, Fany and Michael
Kanti, Damyanti and Pritesh
And
Prof. Dr. Kirk W. Junker
And
Dean and Prof. Jan M. Levine
And
Our Students

Foreword

As one who has been privileged to be involved both as an academic and legal practitioner in food law globally, it is an honor to be invited to write a foreword to a text that is much needed and that serves as a significant benchmark in the developing field of food law. This foreword will endeavor to promote the need for international food law as an academic discipline, frame international food law, and then describe how this text fleshes out this framework.

I. Academic Discipline

The modern global food system, which is governed in part by international law, is comprised of a series of systems on local, national, and international levels that produce, manufacture, and distribute foods in diverse and complicated ways and affects the lives of humans and animals in profound and unique ways. This author recalls when walking in the asparagus fields of Ica, Peru, being repeatedly asked by farm laborers, "what is Thanksgiving in America?," a question prompted by the extra long harvesting hours spent by the workers and their families in the fields during the late November holiday.

The complexities of these modern food systems have motivated countries in recent years to strengthen their legal regimes over food. Robust law development is aimed at regulating food safety, marketing (labeling and advertising), and nutrition. Examples include the US 2011 Food Safety Modernization Act[1] and China's 2015 Food Safety Law,[2] both of which mandate the use of innovative, preventive tools to ensure safe food produced and imported into these respective countries.

---

[1] Food Safety Modernization Act (FSMA), PL 111-353, § 106, 124 Stat. 3885, 3897 (2011) (codified in scattered sections of 21 U.S.C.).

[2] Food Safety Law (promulgated by the Standing Comm. of Nat'l People's Cong., Feb. 28, 2009, revised April 24, 2015, effective Oct. 1, 2015).

In addition to the regulation of food commerce, vast and profound consequences of the modern food systems have precipitated additional regulatory innovations. For example, to address concerns about nutrition and obesity, innovative regulations include the UK front-of-packing labeling regime,[3] Poland's innovative ban on both the advertising and sale of unhealthy foods in schools,[4] and Mexico's recent national tax on sugar-sweetened beverages.[5] In addition to public health regulation, laws that address environmental costs, food security, and food equity are promoted in numerous countries.

This quest to develop legal tools to address problems and challenges generated by the modern food systems has given rise to courses and programs in food law around the world.[6] While courses have been slow to incorporate international law into the food law curriculum, it appears that law programs are trending toward international analysis. For example, the esteemed Renmin University of China School of Law, which oversees the Center for Coordination and Innovation of Food Safety Governance, joined recently with the University of California-Los Angeles (UCLA) School of Law's Resnick Program for Food Law and Policy and East China University of Science and Technology (ECUST) in December 2015 to sponsor an Asia-Pacific Food Governance Symposium. Also, in October 2015, the University of Tuscia in Italy, which houses the European Food Law Center, hosted a conference addressing food law trends in the EU, the USA, and Asia. These conferences portend the development of international food law as a field of study and discipline, which is imperative if the legal academy is to help improve the modern global food system, from the farm to the table.

II. Framing International Food Law

The starting point in understanding international food law is to square it with the definition of international law: law that "consists of the rules and principles of general application dealing with the conduct of States and of international organizations in their international relations with one another and with private individual, minority groups and transactional companies."[7] The sources of international law

---

[3]See Martin Holle, Enrico Togni, Arianna Vettorel, *The Compatibility of National Interpretative Nutrition Labelling Schemes with European and International Law*, 9 Eur. Food & Feed L. Rev. 148 (2014).

[4]Sarah Lewis, Poland bans junk food ads and sales in schools, EU Food Law (Feb. 11, 2015).

[5]*See* Sarah Boseley, *Mexico to Tackle Obesity with Taxes on Junk Food and Sugary Drinks*, The Guardian, Nov. 1, 2013.

[6]See Baylen J. Linnekin & Emily M. Broad Leib, Food Law & Policy: The Fertile Field's Origins and First Decade, 2014 Wis. L. Rev. 557 (2014) (documents the history of the development of food law in American law schools).

[7]Robert Beckman and Dagmar Butte, Introduction to International Law, International Law Students Association, https://www.ilsa.org/jessup/intlawintro.pdf.

link to the International Court of Justice (ICJ), which is a permanent international court located in the Hague, the Netherlands, and the principal judicial organ of the UN.[8] Article 38(1) of the *Statute of the International Court of Justice* enumerates the sources of international law: (1) international conventions or treaties, (2) international custom, and (3) general principles of law.

There are numerous international agreements relevant to food. These agreements attempt to regulate how governments relate to one another on a host of issues, in which regulation ultimately affects the legal relationships between stakeholders in the global food supply chain and the food consumed by consumers. Trade agreements are among the most common types of international agreements that contribute to international food law and that enable a robust food commerce system.

The foundational trade agreement that impacts food is the General Agreement on Tariffs and Trade of 1947 (GATT). In addition to GATT, other WTO rules and agreements that today provide the legal framework for international trade for food were also produced during the Uruguay Round. These agreements include the Agreement on Sanitary and Phytosanitary Measures (SPS) and the Agreement on Technical Barriers to Trade (TB Agreement). The SPS and TBT agreements are intended to facilitate food trade. Additional significant trade agreements that impact food include the Agreement on Agriculture (AA) and the Agreement on Trade-Related Aspects of Intellectual Property Rights (TRIPS).

The World Trade Organization's (WTO) dispute resolution forum has been used to resolve high-profile food regulation disputes, including prohibitions of meat treated with growth hormones[9] and a moratorium on approving the import of GMO food products.[10] The current WTO trade negotiations (the Doha Round) commenced in November 2001, but have stalled, significantly reducing its chances of success and prompting countries throughout the world to pursue bilateral and regional trade agreements as an alternative, many of which resolve food regulatory issues.

International custom and general principles of law also have important roles as sources of international food law. For example, some scholars of international law advocate the right to food as customary international law, concluding that the right to food or the right to be free from hunger has achieved customary status.[11]

---

[8]To be clear, the ICJ has jurisdiction only over states that have consented to it. For example, the court cannot hear a dispute between two more state parties when one of the parties has not accepted its jurisdiction. Also, the ICJ does not have jurisdiction over disputes between individuals or entities that are not states. I.C.J. Stat. art. 34(1). ICJ also lacks jurisdiction over matters that are governed by domestic law instead of international law. Id. at art. 38(1).

[9]See Appellate Body Report, EC Measures Concerning Meat and Meat Products (Hormones), WT/DS26/AB/R, WT/DS48/AB/R (January 16, 1998), para. 181.

[10]Panel Report, EC—Measures Affecting the Approval and Marketing of Biotech Products, WT/DS291/R, WT/DS292/R, WT/DS293/R (September 9, 2006).

[11]Wieke Huizing Edinger, *Food, Safety and the Behavioural Factor of Risk*, 5 Eur. J. Risk Reg. 491, 494 (2014).

Alternatively, these scholars assert that general principles of law within the scope of Article 38 have extended to the right to food.[12]

An interesting and significant aspect of international food law is the harmonization of national food regulatory regimes. The SPS Agreement names the Codex Alimentarius Commission (Codex) specifically as the organization that is "to harmonize sanitary and phytosanitary measures on as wide a basis as possible"[13] by setting "international standards, guidelines or recommendations."[14] Although use of Codex standards creates favorable presumptions under the international legal instruments, states are not required to implement Codex standards into national regulation.[15] The differences that arise in the regulation of food in any particular country depend on a number of variables: the type of legal system (i.e., civil, common law, etc.), historical precedence in regulating food, cultural orientation to food, and available resources.

Comparative analysis of sovereign food regulatory systems is indispensable to the grasping of international food law. It is important to remember, however, that comparative law is more than just drawing comparisons. It is the study of the relationship of one legal system and its rules with another. The nature of any such relationship is discoverable only by a study of the history of the system or of the rules. It is also about the nature of law and the nature of legal development: otherwise known as jurisprudence. Hence, comparative law involves history and jurisprudence.[16] The value of comparative food law is to provide a deeper knowledge of food regulatory regimes and to provide impetus for improving and harmonizing these regimes.

III. Contributions of Text

This text assembles an impressive group of scholars that tackle complicated issues arising from the modern food systems by applying international law and norms. Such a text benefits scholars, teachers, students, practitioners, and policy makers.

This text addresses squarely the core international food law source—the GATT/SPS/TBT construct—by identifying and examining the legal barriers to trade and the major inconsistencies between the USA, EU, and Association of Southeast Asian Nations. This text also reviews TRIPS in the context of agriculture innovation. A novel examination of the normative capacities and activities of the World

---

[12]Dinah Shelton, *The Duty to Assist Famine Victims*, 70 Iowa L. Rev. 1309, 1313 (1985).

[13]Agreement on the Application of Sanitary and Phytosanitary Measures (SPS Agreement), Introduction (April 15, 1994), Marrakesh Agreement Establishing the World Trade Organization, Annex 1A, 1867 U.N.T.S. 493, 33 I.L.M. 1125.

[14]SPS Agreement, art. 12, para. 4.

[15]MICHAEL T. ROBERTS, FOOD LAW IN THE UNITED STATES, 27 (2015).

[16]MARY ANN GLENDON, ET AL., COMPARATIVE LEGAL TRADITIONS 2-8 (2007).

Health Organization (WHO) related to food is made by focusing on the WHO's articles of its constitution.

The text also moves beyond the GATT regime and addresses the application of international law and norms to the consequences of the global food system. For example, the text specifically addresses animal welfare, climate change, food waste, and aquaculture issues. The text also draws interesting and significant links between public health (including medicine and dental) and food policy, agriculture and environmental stewardship, marine conservation and food safety, and food and energy. These linkages provide important conceptual and legal frameworks from which to teach public health and environmental conservation. Woven into these linkages are interdisciplinary analyses from the fields of law, science, medicine, public policy, and public health.

The text provides a very helpful survey of food laws in countries around the world, including New Zealand, Australia, the EU, Sweden, Norway, the Americas, Canada, Brazil, Mexico, Africa, Asia, China, India, Russia, and Israel. More than merely stating the law in these countries, this text addresses cutting-edge food law and policy developments. While the text does not engage in comparative analysis (a single text can only cover so much), it sets the table for comparative scholarship. The text starts interesting comparative conversations about norms in food and agricultural policy by addressing ethical considerations in food production.

The organization of the text is conducive to learning. The text commences with two introductory sections that build on each other. Then, a section is devoted to each continent. Each section has an introduction to help orient students who will use this book as a textbook or to guide practitioners engaging in self-study. Within each section, chapters address emerging in food law and policy in those areas of the world. Some topics that are too new or undeveloped to merit a chapter are included in textboxes. Illustrations in several chapters help condense information and make it more accessible for students. Case studies are sprinkled throughout the text, including treatment of food law topics in Brazil, Nepal, and Senegal.

In short, the text is a novel approach to international food law and should serve as a useful start and framework for an introductory course on international food law. The text also will be a good guide to practitioners who wish to address a specific topic or who are simply looking for context in which to evaluate a particular issue or line of thought. Finally, the text will be an important resource and benchmark for scholars who wish to explore law and policy topics related to food and the problems of the modern global food system.

Michael Roberts

Preface

This book is our first and it is just the beginning. We set out with a table of contents as a list for the book that we wished we could have had in law school. Two years later, we have filled this table of contents with the wide array of topics and chapters covered in this first edition. It was a long process that we could only complete with the help of our families, friends, mentors, and editing interns.

We provide many different viewpoints, all of which are the authors' and not ours, but we selected these views to spark further research and scholarship, to provide an overview and an introduction into the intricacies of our field. Nobody else has attempted to put a similar book together because of the enormous amount of work and high level of professional difficulty that this interdisciplinary and international book encompasses both in scope as in complexity. Therefore, we consider this a start and by no means an all-encompassing version of what a book with this title could be. Instead, this is a comprehensive and carefully selected compilation of topics that are at the forefront in the cutting-edge field of food law and policy, which is still being developed through the law and remains an emerging field in many parts of the world. At best, this book is a global snapshot.

We hope that our work herein with will inspire many academics, practitioners, and students, that it will stimulate further scholarship, and that it will set the stage for sustainable, environmentally friendly, climate-resilient, diverse, kind, and conscientious law and policy development in this wonderful field of food law and policy. Onward!

Boston, MA, USA	Gabriela Steier
Washington, DC, USA	Kiran K. Patel

Acknowledgments

We thank the academic advisors, Prof. Dr. Kirk Junker, Prof. Jan M. Levine, Prof. Dr. Bernd van der Meulen, Mrs. Elizabeth Kucinich and Former Congressman Dennis Kucinich, Attorney Andrew Kimbrell, Attorney Alyssa Kaplan, Professor Tim Lang, Prof. Anne deLaire-Mulgrew, Prof. Isabella Perricone, Prof. Janice Freedman-Bellow, our colleagues and friends at the Center for Food Safety in Washington, D.C., and all of our accomplished colleagues who contributed to this breakthrough work of scholarship and who taught us how to make this book a reality. The authors who drafted chapters and textboxes for *International Food Law and Policy* are at the forefront of their field and helped us to create the book that we wished we could have had as students embarking upon the field of international food law. Gabriela Steier created the basis for this volume when she was still in law school. As one of the leading scholars in food law, she is contirbuting to this rich and varied field of studies with this seminal work. She thanks Kiran Patel for his support throughout the editing of this book and for his enthusiasm to embark upon this journey.

Our special thanks go out to our families, who supported and encouraged us during the years of planning, conceptualizing, and editing *International Food Law and Policy*. We especially thank Prof. Dr. Liviu Steier, Dr. Regina Steier, Mrs. Fany Sontag, Dr. Michael J. Nathenson, Mr. Kanti Nema Patel, Mrs. Damyanti Patel, Mr. Pritesh Patel, and our dear friend, Atty. Alyssa Kaplan. Many heartfelt thanks extend to little Morrice Nathenson Steier for his patience while Mami was working on editing this book.

This book would not have come together as quickly without the help of our editing interns. We thank all of these young professionals and our students for their help and support in creating the book that we all wish we had had for our studies (in alphabetical order): Alex Cherry, Christina Papillo, Daisy Zhang, E. Cody Samit-Shaw, Eva Saul Crawford, Ilana Madorsky, Kimia Shahi, Liza Howard, Marianne Flanagan, Mary Kate Engstrom, Nandini Shroff, Sabrina Rearick, and Tracy Bratt.

Many thanks go to Dr. Brigitte Reschke at Springer for believing in us and for her support and patience with this amazing piece of legal scholarship!

Contents

Part I Global Food Law and Policy

1 Information-Domination in the European Food Industry: Focus on Germany 3
Hans-Ulrich Grimm

2 Agriculture, Ethics, and Law 31
Bernard Rollin

3 The WHO in Global Food Safety Governance: A Preliminary Mapping of Its Normative Capacities and Activities 51
Ching-Fu Lin

4 The Right to Food in International Law with Case Studies from the Netherlands and Belgium 67
Bart Wernaart and Bernd van der Meulen

5 Intellectual Property and Food Labelling: Trademarks and Geographical Indications 101
Michael Blakeney

6 Agricultural Innovation: Patenting and Plant Variety Rights Protection 145
Michael Blakeney

7 Textbox: Cross-Contamination, Genetic Drift, and the Question of GMO Co-existence with Non-GM Crops 177
Gabriela Steier

8 Continuing Legal Barriers to International Food Trade 179
Peter Berry-Ottaway and Sam Jennings

Part II Interdisciplinary Facets of International Food Law

9   Food Policies' Roles on Nutrition Goals and Outcomes: Connecting of Food and Public Health Systems.............. 213
    Jessica Fanzo

10 Textbox: Planetary Boundaries in Food and Agriculture Law.... 253
    Gabriela Steier

11 Food and Nutrition in Cancer Prevention and Treatment........ 257
    Michael J. Nathenson

12 Textbox: Pesticides and Cancer in Conventionally-Grown Versus Organic Food... 285
    Gabriela Steier

13 Intersections of Dental Health and Food Law: The Conflict of Systemic Flouridation as a Public Health Instrument to Prevent Tooth Decay................................ 291
    Liviu Steier, Regina Steier, and Gabriela Steier

14 Textbox: The Hierarchy of Scientific Evidence................ 319
    Gabriela Steier and Liviu Steier

15 Food and Energy.. 321
    Stephen B. Balogh and Charles A.S. Hall

16 Textbox: Internalizing Externalities: Techniques to Reduce Ecological Impacts of Food Production.................... 359
    Randy Hayes

17 Textbox: Cooperatives' and Producer Organizations' Roles in Achieving Food Security............................... 365
    Denis Herbel, Mariagrazia Rocchigiani, and Nora Ourabah Haddad

18 Governing the Global Food System Towards the Sustainocene with Artificial Photosynthesis........................... 373
    Thomas Faunce and Alex Bruce

Part III European Food Law

19 Introduction to European Food Law and Regulation........... 409
    Gunnar Sachs

20 Food Safety and Policy in the European Union............... 451
    Giorgio Rusconi

21 European Food Trade.................................... 485
    Roland Kölcsey-Rieden and Lívia Hulló

22 European Food Labelling Law............................ 505
    Aude Mahy and Nicola Conte-Salinas

23	Consumer Protection Through Food Law in the European Union Raymond O'Rourke	541
24	The Problem of Food Waste: A Legal-Economic Analysis Harry Bremmers and Bernd van der Meulen	559
25	The Concepts of Transparency and Openness in European Food Law Nicola Conte-Salinas and Rochus Wallau	581
26	Food Law in Sweden Magnus Friberg	607
27	Food Law in Norway: Trade, Food Promotion, and Protection of Intellectual Property Within the Food Industry Marie Vaale-Hallberg and Nina Charlotte Lindbach	641
28	Food Law and Regulation in Germany Gunnar Sachs	671

Part IV The Americas

29	Introduction to Food Law and Policy in the United States and Canada Carly Dunster	695
30	Textbox: The U.S. Farm Bill and Textbox: The National School Lunch Program William S. Eubanks	713
31	Food Law in Canada: A Canvass of History, Extant Legislation and Policy Framework Ikechi Mgbeoji and Stan Benda	719
32	Brazilian Agriculture and Its Sustainability Luiz Antonio Martinelli, Luciana Della Coletta, Silvia Rafaela Machado Lins, Silvia Fernanda Mardegan, and Daniel de Castro Victoria	767
33	Food Law in Mexico: Regulatory Framework and Public Policy Strategies to Address the Obesity Crisis in Latin America Rebeca López-García	793
34	Global and US Water Law and Sustainability: The Tragedy of the Commons and the Public Trust Doctrine Zach Corrigan	823
35	Textbox: How the Public Trust Doctrine Supplements Existing Statutory Law Zach Corrigan	845

36	US and Global Regulation of Fisheries Beyond the Commons....	849
	Zach Corrigan	
37	Industrial Aquaculture: Human Intervention in Natural Law....	895
	Alfredo Quarto and Sara Lavenhar	
38	TEXTBOX: Creating Law and Policy for Resilient Urban Food Systems...	929
	Yassi Eskandari-Qajar and Janelle Orsi	

Part V Australia

39	Regulation of Food Labelling in Australia and New Zealand.....	939
	Michael Blakeney	
40	Present & Future Jurisprudence of Consumer Protection and Food Law in Australia................................	971
	Alex Bruce	

Part VI Africa

41	Food Law and Policy in Africa: Emerging Legal Framework, Key Issues, Major Gaps and Challenges.....................	1011
	Uche Ewelukwa Ofodile	
42	Disease Control, Public Health and Food Safety: Food Policy Lessons from Sub-Saharan Africa.........................	1061
	Kennedy Mwacalimba	
43	Regulatory Frameworks Affecting Seeds: Impacts on Subsistence Farmers in the Eastern and Southern African Region...........	1105
	Marcelin Tonye Mahop	
44	Textbox: Women in Agriculture and Labor Law: Cameroon's "Three C's" Legal Dilemma..............................	1131
	Wele Elangwe	
45	Intellectual Property Law and Food Security Polices in Ethiopia...	1137
	Getachew Mengistie Alemu	
46	Innovation and Development in Agricultural Biotechnology: Reflecting on Policy-Making Processes in Sub-Saharan Africa....	1181
	Julius Mugwagwa and Watu Wamae	
47	Food Law in South Africa: Towards a South African Food Security Framework Act..	1203
	Anél Gildenhuys	

Part VII Asia

48 Food Regulation and Policy Through the Association of Southeast Asian Nations (ASEAN) 1245
Jörn Westhoff

49 History of Asian Food Policy 1263
Peter Sousa Hoejskov

50 Branding, Regulation and Customs in Japan and Singapore 1295
Andreas Popper

51 Food Sustainability in the Context of Chinese Food Regulation ... 1327
Juanjuan Sun

52 Law and Regulation in India 1353
Dimpy Mohanty and Abhijeet Das

53 Food Regulation in the Customs Union of Belarus, Kazakhstan and Russia 1379
Alexey Petrenko, Anatoly Kutyshenko, and Victor Tutelyan

54 Israeli Regulation and Policy of GM Food and Crops 1409
Ronit Justo-Hanani

Editors and Contributors

About the Editors

Gabriela Steier, co-founded Food Law International, LLP, (FLI) in 2012. FLI gas grown into an LLC under Prof. Steier's guidance. She is Editor in Chief of the textbooks *International Food Law and Policy* (Springer 2017) and *International Farm Animal, Wildlife and Food Safety Law* (Springer 2016) and author of *Advancing Food Integrity: GMO Regulation, Agroecology, and Urban Agriculture* (forthcoming with Taylor and Francis).

She is an attorney licensed in the USA and focuses on food law, and policy, climate change, animal welfare, international trade, and GMO issues domestically and in the European Union. Prof. Steier has lectured on these topics and continues her research widely. She holds a B.A. (Cum Laude) from Tufts University, a J.D. from Duquesne University, and an LL.M. in Environmental Law focused on Food and Agriculture Law from the Vermont Law School and she has nearly completed a Doctorate (Dr. iur.) in Comparative Law at the University of Cologne in Germany. She speaks six languages.

Founding Partner of FLI, Prof. Steier continues her research in comparative law and focuses on food safety and sustainable food production. She worked as an LL.M. Fellow in Food and Agriculture Law at the Vermont Law School. Before joining the Vermont Law School, she completed a fellowship at the Center for Food Safety on Capitol Hill in Washington, DC, a national non-profit public interest and environmental advocacy organization working to protect human health and the environment by promoting organic and sustainable agriculture.

She is a distinguished legal scholar and has published widely on international food law, policy, and trade and has earned several awards for her work. Steier joined the Duquesne University School of Law in Pittsburgh, PA, as an Adjunct Professor teaching a breakthrough new course in "food law and policy" in 2015 and she also teaches "climate change law." As Visiting Professor at the University of Perugia, Italy, she also teaches EU-US comparative food law at the Department of Political Sciences. Additionally, as an experienced editor and with her numerous publications ranging from peer-reviewed articles in international medical journals to law reviews, Prof. Steier has gained widespread interdisciplinary interest and some of her articles have been on the Top Ten List on SSRN for several months and her work has been on Amazon.com's top lists in environmental law.

Kiran K. Patel, was Partner and Co-Founder of Food Law International, LLP, (FLI) until summer 2017 and served as co-editor of *International Food Law and Policy* and *International Farm Animal, Wildlife and Food Safety Law*. He is also a registered U.S. patent attorney (licensed to practice in New York and before the US Patent and Trademark Office) practicing as an Associate Attorney at Kramer & Amado, P.C. at, in the Washington, D.C. area, focusing on patent law, particularly the preparation and prosecution of patent applications in a wide range of technological fields, including the areas of electrical engineering and computer science. He also works in the chemical arts, biochemical compounds, mechanical devices and medical devices, including drug compositions and compounds. He joined Duquesne University School of Law as an adjunct professor teaching the intellectual property and environmental law section in the course "Food Law and Policy" in 2015. As a visiting professor at the University of Perugia, in Perugia, Italy, he also supports Prof. Steier by teaching about the intersections between food law and US intellectual property law.

Patel earned his bachelor of science (BS) in biology with *honors* from Drexel University in Philadelphia, PA. Subsequently, Patel earned his juris doctor (J.D.) at Duquesne University School of Law in Pittsburgh, PA, on a *McDonagh Scholarship* and an *Academic Award Scholarship* and his master of laws (LL.M.) in intellectual property law (specializing in patent law) as a *Dean's Merit Scholar* at the Benjamin N. Cardozo School of Law in New York, NY. Patel is currently pursuing his masters of science (M.S.) in electrical engineering and computer science at Johns Hopkins University School of Engineering.

Contributors

Getachew Mengistie Alemu is an intellectual property lawyer who drafted most of the present intellectual property laws of Ethiopia, headed the Ethiopian Intellectual Property Office as its first director general until August 2008, and taught intellectual property law to undergraduate and postgraduate students of the law school of Addis Ababa University for more than 12 years. After leaving government office, Mr. Getachew Mengistie Alemu has been serving as an IP consultant and attorney and involved in a number of projects supported by regional and international organizations aimed at strengthening the intellectual property systems and promoting the use of intellectual property as a tool for development in African countries. He had authored or coauthored a number of works, which are published in books, international journals, and workshop proceedings. He is a frequent speaker in regional and international intellectual property forums.

Stephen B. Balogh earned his doctorate degree from SUNY College of Environmental Science and Forestry (ESF). As a visiting assistant professor at ESF and adjunct at the Whitman School of Management at Syracuse University, he taught courses about energy systems, urban ecology, and sustainable enterprise. His research concentrates on food and energy flows in urban ecosystems and developing a social-ecological systems framework for sustainability and resiliency planning. His dissertation is entitled "Feeding and Fueling the Cities of the 21st Century." He has published nine papers and coauthored three book chapters on various topics related to energy, including the nexus of energy and the economy, social-ecological metabolism of cities, the importance of high-quality energy to societal development, the energetic efficiency of agricultural systems, and trade-offs involved in creating an urban green economy. He is currently employed as a postdoctoral researcher at US Environmental Protection Agency in the Atlantic Ecology Division.

Stan Benda is a graduate of the Royal Military College, and was a commissioned officer in the combat arms (armour corps). He was an exchange student at West Point. He was senior counsel with the Federal Department of Justice (of Canada). His primary duties were representing Agriculture Canada: the Intellectual Property Secretariat (technology transfer/licensing) as well as the International Science Section. In the latter instance, he acted before the UN FAO (Food and Agriculture Organization) on the International Treaty on Plant Genetic Resources for Food and Agriculture. He is an adjunct professor at Osgoode Hall Law School where he teaches LL.B. and LL.M. students. He is assistant professor at Ryerson University, Ted Rogers School of Management. There he teaches intellectual property law, information and technology law, business law and international trade law. Dr. Benda is also the part-time vice chair of the Ontario Agriculture, Food and Rural Affairs Appeal Tribunal. His LL.M. is in intellectual property. His Ph.D. (Osgoode) pertains to risk assessment, regulation and the labeling of genetically modified crops.

Michael Blakeney is Winthrop Professor of Law at the University of Western Australia and visiting professor of Intellectual Property and Agriculture at Queen Mary University of London. He has held academic positions at universities in Australia, Italy, Singapore, UK, and USA. He formerly worked in the Asia Pacific Bureau of the World Intellectual Property Organization. Professor Blakeney has advised the Asian Development Bank, Consulting Group for International Agricultural Research, European Commission (EC), European Patent Office, Food and Agricultural Organization, World Intellectual Property Organization, and a number of university and public research institutes on intellectual property management. He has directed EC projects concerned with the establishment of an IP infrastructure in a number of new EU Member States and EU Applicant States. He has also directed and been a consultant in a number of projects to assist developing countries to become members of the World Trade Organization.

Harry Bremmers studied business economics at Tilburg University (till 1977) and Dutch law at the Erasmus University in Rotterdam (till 1993). There he was appointed as assistant professor till 1993 at the economics and the law faculty. He graduated with a Ph.D. in 1994 on a legal-economic subject (effects of environmental law on content of financial reports). From 1993 to 2010, he occupied a full-time position at the business administration group of the Social Sciences Department of Wageningen University, the Netherlands. Since 2010, he is associate professor at the law and governance group of Wageningen University. He participates in national and international projects at the interplay of law and economics and has published on a broad range of legal and economic topics.

Alexander Bruce is an associate professor (reader) of law at the Australian National University. Alex is currently completing a DPhil in Comparative Theology at the University of Oxford. He is also a Buddhist monk ordained into the Tibetan tradition of His Holiness the Dalai Lama. Alex is one of Australia's leading authorities in competition and consumer law as well as the emerging field of animals and the law. He has published texts in all these areas of law. Before joining the ANU, Alex was a senior lawyer with the Australian Competition and Consumer Commission and has worked with the United Nations in capacity-building competition and consumer policies in several African countries. His research interests focus on the relationship of sentient beings to the economic structures of society (competition law), whether as consumers of goods or services (humans) or as resources (animals), and the regulation of corporate activity (through the Australian Competition and Consumer Commission), as well as the broader philosophical and religious assumptions associated with that relationship.

Luciana Della Coletta is an environmental engineer and earned her PhD in Sciences from University of São Paulo, Brazil, in 2015. Her research has focused in biogeochemistry cycling in terrestrial ecosystems. Currently, she is a lecturer at the Federal Institute of Education, Science and Technology, Brazil, and teaches classes in natural resources management for undergraduate students.

Nicola Conte-Salinas is a lawyer admitted to the Bar in Germany. She specializes in German and European Food Law, in particular in nutrition and health claims. She also advises and assists food companies in questions relating to the law against unfair competition, as well as in the area of contract law, in particular as regards general terms and conditions. She has been working in the food law practice of KWG from 2009 to 2015. Since September 2015, Nicola is the team leader for food law with one of the leading German retailers. Nicola is a regular speaker at food law seminars and conferences and author of various publications on food law issues. Nicola graduated from the University of Trier (Germany) and received a Diploma in Legal Studies (with distinction), as well as a Diploma in Higher Education from the University of Wales, Cardiff (UK). She advises and represents clients in German and English.

Zach Corrigan is the senior staff attorney for Food & Water Watch. He has been an advocate for sustainability and environmental conservation for more than a decade. He began working on seafood safety issues in 2002 when he served as a staff attorney for the US Public Interest Research Group. There, he was a lead advocate for limiting people's exposure to mercury from contaminated fish by fighting for more stringent EPA rules to curb mercury emissions from power plants. In 2004, Corrigan became a legislative representative for Public Citizen and, shortly thereafter, Food & Water Watch. In May 2006, he became the staff attorney for the organization, working with all of its teams to pursue litigation, regulatory, and legislative solutions for all of the issues on which Food & Water Watch works. He graduated from the University of Wisconsin and earned his J.D. from Northeastern University School of Law in Boston. He is a licensed attorney in the District of Columbia and Massachusetts.

Abhijeet Das (at the time of co-authoring the article, *Senior Associate at LexCounsel Law Offices*) has 7 years of experience and has been providing support and advice to a broad spectrum of domestic and foreign clients in their respective spear of activities. He has extensive experience on the Indian corporate and transactional laws and has represented many global and Indian entities in various transactions, *inter alia,* in relation to transactional documentation, advisory and coordination of transactions, including mergers, acquisitions, amalgamations and asset purchase arrangements. He has actively assisted multiple clients in due diligence exercises in private equity transactions as well as title verifications of real estate. Further, he has been active in the areas of intellectual property and provides transactional guidance vis-à-vis a diverse range of intellectual property assignment and licensing arrangements. In the recent years, he has also been engaged extensively in drafting various copyright development agreements, service level agreements, user agreements for website and mobile applications, etc. In the practice areas of labour and employment, Abhijeet has helped various entities in formulation/standardization of employment contracts, employee manuals and policies on standard employment practices, including leave policies, prohibition of harassment, employment benefits, discipline and grievance, confidentiality, intellectual property rights, anti-bribery obligations and social media. He has also dealt with pre and post termination restrictive covenants, concerning confidentiality, non-disclosure, non-solicitation and protection of trade secrets. Advisory on legal and regulatory issues pertaining to clinical trials, especially with reference to drafting, revising and negotiating clinical trial agreements is another aspect that he tackles on a regular basis. He has advised various pharmaceutical companies and institutions on various legal issues under the drugs regulatory regime in India.

Daniel de Castro Victoria is a researcher at Embrapa Agriculture Informatics, has an undergrad degree in Agronomy Engineering from University of São Paulo (2002), a master's degree from University of São Paulo (2004) and doctors degree from Center for Nuclear Energy in Agriculture (CENA—USP—2010). Research interests include geoprocessing, remote sensing, agrometeorology, land use, and cover change and hydrological modeling.

Carly Dunster is a sole practitioner and food lawyer at Carly Dunster Law, where she provides affordable and accessible legal and consulting services to those seeking to build more sustainable food systems. She has worked with many food and urban agriculture initiatives in Toronto, including the Toronto Underground Market, Food Truck Eats, the Food Constellation at the Centre for Social Innovation, and The Depanneur. She is a member of the Toronto Food Policy Council, as well as the Toronto Street Food Project, a group advocating for a friendlier regulatory structure for diverse street food in the city of Toronto. She has sat on the Metcalf Advisory Committee for their Environment Program's Sustainable Food Systems funding stream. She has spoken at a number of conferences on the practice of food law, including the Urban Agriculture Summit and the National Student Food Summit, and before students participating in the Laws in Action Within Schools Program, among others. She works as a capacity-builder in the small- and medium-sized food sector and also spends a significant amount of time convincing other lawyers to specialize in food.

Wele Elangwe is a legal specialist for the Maryland Agricultural Law Education Initiative (ALEI) and student services coordinator for the School of Graduate Studies at the University of Maryland Eastern Shore (UMES). She is an internationally trained lawyer with experience working with several law firms, the United Nations, and Catholic charities. She is knowledgeable in common law and civil law and procedure as well as business law, international law, and human rights law. She holds an LL.B. from the University of Buea and a maitrise en droit in business law from University of Yaoundé II in Cameroon and an LL.M. in international law from Indiana University and is pursuing a Ph.D. in organizational leadership at UMES. Wele has authored several reports submitted to the United Nations as well as published with the African Development Bank. Wele recently presented and won first place faculty oral presentation at the 5th annual UMES research symposium in April 2014.

Yassi Eskandari-Qajar directs policy at the Sustainable Economies Law Center (theselc.org). The Oakland-based Sustainable Economies Law Center aims to cultivate community resilience and grassroots economic empowerment through its unique approaches to legal education, research, advice, and advocacy. Yassi has co-authored several influential publications on local policy, including *Regulating Short-Term Rentals: A Guidebook for Equitable Policy* (2016) and *Policies for Shareable Cities* (2013). Yassi completed her legal education through apprenticeship in the California State Bar's Law Office Study Program, and received her undergraduate degree in Conservation & Resource Studies from the University of California, Berkeley.

Bill Eubanks is a partner at the public interest environmental law firm Meyer Glitzenstein & Crystal, where he litigates precedent-setting impact cases in federal appellate and trial courts. He specializes in litigation concerning public lands preservation, natural resource conservation, and wildlife and biodiversity protection. His notable cases include successfully challenging oil spill response strategies harmful to marine wildlife in the Gulf of Mexico after Deepwater Horizon, significantly curtailing environmentally damaging off-road vehicle use in Florida's Big Cypress National Preserve, prevailing in the nation's first federal lawsuit challenging an industrial wind energy project on environmental grounds, and coauthoring briefs in four recent US Supreme Court cases involving climate change, genetically engineered crops, naval sonar use, and logging road stormwater runoff. In addition to his litigation successes, Eubanks is a recognized legal scholar, having published numerous law review articles on diverse environmental law and policy topics and having recently coauthored and coedited a textbook titled *Food, Agriculture, and Environmental Law*. He also serves as an adjunct environmental law professor at American University's Washington College of Law, George Washington University Law School, and Vermont Law School.

Thomas Faunce has a joint appointment in the ANU College of Law and College of Medicine, Biology and the Environment. He was the recipient of an Australian Research Council Future Fellowship (2010–2013) investigating how nanotechnology could help resolve some of the great public health and environment problems. He has been awarded numerous competitive grants in the field of health technology regulation and has published over 100 refereed articles, over 25 book chapters, and 3 monographs in this field. His most recent research involves governance strategies for globalizing artificial photosynthesis—a field in which he has organized major international conferences and published with leading scientists in the field.

Jessica Fanzo, PhD is the Bloomberg Distinguished Associate Professor of Global Food and Agriculture Policy and Ethics at the Berman Institute of Bioethics, the Bloomberg School of Public Health, and the Nitze School of Advanced International Studies at the Johns Hopkins University. She also serves as the Director of the Global Food Ethics and Policy Program at Hopkins, and the Co-Chair of the Global Nutrition Report. Prior to joining Johns Hopkins, Jessica served as an Assistant Professor of Nutrition in the Institute of Human Nutrition and Department of Pediatrics and as the Senior Advisor of Nutrition Policy at the Center on Globalization and Sustainable Development within the Earth Institute at Columbia University. Prior to coming to academia, Jessica held positions in the United

Nations World Food Programme and Bioversity International, both in Rome, Italy. Prior to her time in Rome, she was the Senior Nutrition Advisor to the Millennium Development Goal Centre at the World Agroforestry Center in Kenya. Her area of expertise is on the linkages between agriculture, nutrition, health and the environment in the context of sustainable and equitable diets and livelihoods. She was the first laureate of the Carasso Foundation's Sustainable Diets Prize in 2012 for her work on sustainable food and diets for long-term human health. Jessica has a PhD in nutrition from University of Arizona.

Magnus Friberg is a lawyer in Sweden with Setterwalls Advokatbyrå and has some 20 years' experience in advising Swedish and international clients. His areas of expertise include consumer protection, marketing practice legislation, and industry codes. He also handles product regulatory legislation particularly in the field of life sciences legislation including food and pharmaceuticals. He provides advice to Swedish as well as foreign clients in these areas and also represents them before authorities and courts, and he is also a much appreciated lecturer. He earned his degrees from Lund University (LL.M.) 1990, University of Copenhagen, EC Law. He has been a practicing lawyer since 1990.

Anel Gildenhuys is a lecturer at the Faculty of Law, North-West University (Potchefstroom Campus), South Africa. Her research focus is on food security, the right to (have access to sufficient) food, human security, and framework legislation. In her doctoral thesis (2011) "Voedselsekerheid as Ontwikkelingsdoelwit in Suid-Afrikaanse Wetgewing: 'in Menseregte-gebaseerde Benadering" ("Food Security as Developmental Goal in South African Legislation: A Human Rights-Based Approach") (North-West University (South Africa)), she applied a human rights-based approach to discuss various means to accommodate food security, as a developmental goal, in South African national legislation.

Hans-Ulrich Grimm is Germany's foremost journalist covering food-related topics. He was born in the Allgäu and is a journalist in Stuttgart. He studied in Heidelberg history, German, and education and was a journalist and editor of the news magazine Der Spiegel. He has been a freelance writer since 1996. The total circulation of his books is one million copies. They have been translated into many languages. For his books, he researched worldwide such as in Europe, the USA, Japan, China, and the South Seas. He inspected food factories and interviewed researchers, politicians, and nutritionists and their patients. He is searching for good food, with dedicated farmers, good gardeners, and talented chefs.

Nora Ourabah Haddad has a background in economy and holds a degree in Business Administration from the High Business School of Grenoble, France (Ecole Supérieure de Commerce) and a Masters of Business and Administration in International Management from Laval University in Canada. From 2000 until 2010, Ms. Ourabah Haddad worked as Senior Policy Officer at the International Federation of Agricultural Producers (IFAP) on Policy analysis and formulation related to food security, sustainable agriculture, and rural development, as well as on developing the institutional, leadership and knowledge capacity building of agricultural and rural organizations. She worked on several projects aiming at strengthening farmers' organizations. One of her main responsibilities was to promote agricultural and rural development working closely with national and international policymakers (from both developed and developing countries) and to encourage them to develop investments in agriculture and rural areas. She developed several publications and policy papers reflecting the view of farmers' organizations worldwide. Prior to joining IFAP, Ms. Ourabah Haddad worked for a USDA funded program, the National Center for Food and Agricultural Policies (NCFAP). Ms. Ourabah Haddad joined FAO in 2010 as Rural Institutions Officer and soon became Team leader of rural institutions and people's empowerment. In 2013, she moved to the position of cooperatives and producer organizations coordinator within the division for communication, partnerships and advocacy. She coauthored several publications and articles on rural institutions, collective action, and rural development. Since May 1 2016, she was appointed by the Director General as FAO Representative to the Sultanate of Oman.

Charles A.S. Hall is a systems ecologist who received his Ph.D. under Howard T. Odum at the University of North Carolina at Chapel Hill. Dr. Hall is the author/editor of 13 books and some 300 scholarly articles and was awarded last year the distinguished Hubbert-Simmons Prize for Energy Education. He is best known for his development of the concept of EROI, or energy return on investment, which is an examination of how organisms, including humans, invest energy into obtaining additional energy to improve biotic or social fitness. He has applied these approaches to fish migrations, carbon balance, tropical land use change, and the extraction of petroleum and other fuels in both natural and human-dominated ecosystems. Presently he is developing a new field, biophysical economics, as a supplement or alternative to conventional neoclassical economics, while applying systems and EROI thinking to a broad series of resource and economic issues. In his latest book, Energy and the Wealth of Nations, Hall and coauthor Kent Klitgaard explore the relationship between energy and the wealth explosion of the twentieth century, the failure of markets to recognize or efficiently allocate diminishing resources, the economic consequences of peak oil, the EROI for finding and exploiting new oil fields, and whether alternative energy technologies such as wind and solar power meet the minimum EROI requirements needed to run our society as we know it.

Randy Hayes has been described in the *Wall Street Journal* as "an environmental pit bull." He is executive director at Foundation Earth, a new organization working to eradicate ecological externalities that threaten the biosphere's life support systems. Foundation Earth promotes a true cost economy. As a former filmmaker and Rainforest Action Network founder, he is a veteran of many high-visibility corporate accountability campaigns and has advocated for the rights of indigenous peoples. He served 7 years as president of the city of San Francisco's Commission on the Environment and as director of sustainability in the office of Oakland's mayor, Jerry Brown (now governor). As a wilderness lover, Hayes has explored the

High Sierras, the Canadian Rockies, and the rainforests of the Amazon, Central America, Congo, Southeast Asia, Borneo, and Australia.

Denis Herbel graduated with a degree in rural economics from Institut des Hautes Etudes de Droit Rural et d'Economie Agricole (Paris), in management at IAE (Rennes Business School), and in economics of commodity markets at Grenoble University (France). He has been working on development in Africa for 30 years. He specialized in agricultural cooperatives and producers' organizations (farmer collective action) in developing countries. He is the coauthor of different publications and papers on the subject. He has been working as advisor for producer organizations and cooperatives within the Partnerships, Advocacy and Capacity Development Office of FAO.

Peter Sousa Hoejskov is technical officer for food safety at the World Health Organization (WHO) Division of Health Security and Emergencies (DSE), Regional Office for the Western Pacific Region (WPRO) in Manila, Philippines. He assists countries in the Western Pacific Region in strengthening national food safety systems and their capacity to prepare for and respond to food safety incidents and emergencies. Mr. Hoejskov joined WHO in 2011, and before taking up his assignment in the Regional Office for the Western Pacific, he worked as a technical officer for food safety and noncommunicable diseases in the WHO Division of Pacific Technical Support (DPS) based in Suva, Fiji. Prior to joining WHO, Mr. Hoejskov worked as international technical advisor for the Food and Agriculture Organization of the United Nations (FAO) in Bangladesh (2010–2011) and as food quality and safety officer at the FAO Regional Office for Asia and the Pacific in Bangkok, Thailand (2006–2009). Mr. Hoejskov holds a master of science degree with specialization in food quality and safety, an AP degree in international trade and marketing with focus on international food laws and regulations, and a certificate in public health and nutrition promotion. Mr. Hoejskov is guest lecturer at the Food Law Internet Certificate Program offered by the Institute for Food Laws and Regulations and the College of Agriculture and Natural Resources at Michigan State University, USA.

Sam Jennings has a background in microbiology, and her work with the consultancy Berry Ottaway & Associates Ltd. spans almost 20 years. Her role covers scientific, technical, and regulatory aspects within the food industry, especially the dietary/food supplements and functional food sectors. She provides advice to industry and plays an active role in the development of European Community food legislation. She also liaises with governments across the world on topics particularly relating to dietary/food supplements. Sam is Technical Advisor to the Council for Responsible Nutrition UK (CRN UK) in the UK. She has been Chair of the International Technical Group (2012–2017), and is a member of the Scientific Council under the auspices of the International Alliance of Dietary/Food Supplement Associations (IADSA). Sam has co-authored chapters for three books, one of which has been translated and published in China and Russia.

Ronit Justo-Hanani is a visiting scholar at the University of California, Berkeley. She holds a bachelor's degree in law (LL.B) and master's degree in ecology and environmental studies (MSc) and has completed her Ph.D. requirements at the Faculty of Life Sciences affiliated with *Edmond J. Safra Center* for Ethics, Faculty of Law, Tel Aviv University. After her master's degree, she was awarded a UNESCO fellowship to study for a year at the Sheffield Institute of Biotechnology Law and Ethics, at the University of Sheffield, UK, where she studied GMO and agricultural biotechnology regulation with Prof. Kinderlelrer. She then moved back to Israel, where she wrote her Ph.D. dissertation on regulatory policies on nanotechnology risk in the EU and the USA. She also wrote several research reports on GMO regulation and policy for the Israeli Ministry of Agriculture and Rural Development. Her academic and policy work spans regulatory policies for biodiversity and ecosystems management, agricultural biotechnology, and environmental science policy.

Roland Kölcsey-Rieden was born in 1966 and is attorney-at-law (D, H) and mediator. He specializes in real estate law, corporation law, mergers and acquisitions, and food law. He attended the Faculty of Law at the University of Munster, Germany, and University of Bonn, Germany, and had his nostrification of German state examination and completion of Hungarian doctorate at the Peter Pazmany Catholic University, Budapest. Roland was an associate at the Oberlandesgericht in Cologne, had internships at Lewalder & Partner in Bonn and Krasznai & Partner in Budapest 1997, and is an attorney in the international partnership of Nörr Stiefenhofer Lutz in Munich and Budapest. From 2000, at Köves Clifford Chance Pünder in Budapest, he is a member of the real estate department and the head of the "German Desk." From 2003, he is a member of the Kölcsey-Rieden & Bánki partnership in cooperation with e/n/w/c Eiselsberg Natlacen Walderdorff Cancola Rechtsanwälte GmbH from Vienna. From April 1, 2006, he is a managing partner at BKRU. From June 3, 2008, he is a managing partner at Kölcsey-Rieden, Haslwanter & Partners Law Firm, a member of the COLAW International Legal Alliance. Roland speaks German, English, and Hungarian

Anatoly Kutyshenko graduated from the Department of Physics, Moscow State University, in 1996 and obtained a master's degree in biochemistry from McGill University in 2003. Prior to joining the private sector in 2004, he worked for McGill University and the National Research Council of Canada. Anatoly has previously worked at the various quality and regulatory management positions in Cargill and Mars and is currently heading the technical regulatory and quality department in Amway Russia and Kazakhstan.

Sara Lavenhar graduated in 2014 from Columbia University with a bachelor's degree in environmental science with a special concentration in sustainable development. While studying at Columbia, Sara began to conduct research into the food industry and became passionate about sustainable agriculture. She has had the opportunity to work with nonprofits like the Community Alliance for Global Justice and the Mangrove Action Project (MAP) to address challenges of food sovereignty and conservation. In a formal research setting, Sara investigated the microbiology of rooftop farms as a way to improve the efficiency and sustainability of urban agriculture for her senior thesis project. Sara is also an avid writer, cook, and archer. She intends to pursue a graduate degree and career in sustainable agriculture to promote and advocate for a healthy food system.

Ching-Fu Lin is Assistant Professor of Law at National Tsing Hua University (NTHU), where he teaches international health law, food law and policy, and international law and global governance. Professor Lin received his LL.M. and S.J.D. from Harvard Law School, with the honor of John Gallup Laylin Memorial Prize (best paper in public international law) and Yong K. Kim Memorial Prize (best paper in East Asian legal studies). He also holds a double degree in law (LL.B.) and chemical engineering (B.S.) from National Taiwan University. Before joining NTHU in 2015, Professor Lin served as Visiting Fellow at Graduate Institute of International and Development Studies in Geneva, Switzerland. He has also been Peter Barton Hutt Student Fellow at Petrie-Flom Center for Health Law Policy, Biotechnology, and Bioethics as well as Visiting Scholar Coordinator at East Asian Legal Studies Program, Harvard Law School. Active in legal academia, Professor Lin has been invited to lecture in many academic settings, such as Summer Academy in Global Food Law and Policy in Spain, Brescia University School of Law in Italy, and World Food Law Program and the United Nations Food and Agriculture Organization in the United States. His legal scholarship has appeared in numerous journals and edited collections, including Virginia Journal of International Law, Columbia Science and Technology Law Review, Food and Drug Law Journal, and Journal of World Trade.

Silvia Rafaela Machado Lins University of São Paulo, CENA, Piracicaba, SP, Brazil. She has a Bachelor in Biology from the Federal University of Alagoas, a Master and Doctor of Science in Applied Ecology at ESALQ/USP. She works with tropical ecosystems ecology, specifically with the nutrients' cycles in the Coastal Atlantic Forest, Brazil.

Nina Charlotte Lindbach is a senior associate at Haavind in the business area of technology, media, and IPR. Lindbach joined the firm in 2013 and primarily assists clients within the intellectual property sector. She graduated from the University of Bergen, Faculty of Law, in 2013.

Hulló Lívia was born in 1983 (attorney-at-law (H), LL.M. (master of laws)). Hullo specializes in consumer protection law, food law, competition law, European law, administrative law, and commercial law. She attended the Faculty of Law, University of Szeged, and she also studied, under Erasmus scholarship, at the Ludwig Maximilian University (LMU) of Munich, Munich, Germany. From 2004 to 2006, she pursued German and European commercial law studies at the University of Szeged and at the University of Potsdam. From 2007 to 2008, she studied comparative administrative law and European law studies (LL.M.) at the German-speaking Andrassy Gyula University, Budapest, Hungary. From 2008 to 2013, Hullo was a trainee at Kölcsey-Rieden, Haslwanter & Partners Law Firm and in 2013 became a partner at Kölcsey-Rieden, Haslwanter & Partners Law Firm, a member of the COLAW International Legal Alliance. Hullo speaks Hungarian, English, and German.

Rebecca Lopez-Garcia is the principal in Logre International Food Science Consulting, based in Mexico City. Dr. López-García works in the areas of food safety, toxicology, and regulatory compliance around the world. Rebecca has participated as a consultant in international cooperation projects with organizations such as the United Nations Food and Agriculture Organization (FAO) and the United States Agency for International Development (USAID). She is the lead instructor of Latin American food laws and regulations in the International Food Law Certificate Program of the Institute for Food Laws and Regulations (IFLR) at Michigan State University. Rebecca López-García holds a Ph.D. degree in food science/toxicology from Louisiana State University and a certificate in international food law from Michigan State University. She is a certified food scientist (CFS) by the Institute of Food Technologists and a certified quality auditor (CQA) by the American Society for Quality.

Marcelin Tonye Mahop is working as independent/freelance consultant based in London, UK, Dr. Mahop's interests extend to intellectual property policy and development (IP and agriculture, IP and pharmaceutical innovations, IP and biodiversity and traditional knowledge, management of IP) and to genetic resources policies including seed regulations and access to and utilization of biological and genetic resources and community rights. He earned the University of London Ph.D. in 2007. He has held research positions at Queen Mary University of London; Brunel University in London, UK; and Wageningen University in the Netherlands. He is published in international journals. Dr. Mahop is the author of *Intellectual Property, Community Rights and Human Rights: The Biological and Genetic Resources of Developing Countries*, published by Taylor & Francis in 2010, and coeditor of *Extending the Protection of Geographical Indications: Case Studies of Agricultural Products*, published in 2012 by Taylor & Francis. He is the lead author of a 2014 handbook of the African Agricultural Technology Foundation (AATF) which provides an overview of the plant variety protection regimes of 32 - sub-Saharan African countries and is the lead researcher and lead author of a major January 2013 report that explores seed systems and plant variety protection regimes in five sub-Saharan African countries.

Aude Mahy is a counsel and admitted to the Brussels bar and specialized on (international) commercial law and European and Belgian food law. Aude presides the Loyens & Loeff Food & Beverages team. In this capacity, she advises and assists food business operators on various matters such as nutrition and health claims, labeling issues, market practices and advertising, food safety (including additives and other issues with respect to the composition of foodstuffs), placement on the market, importation in the European Union, etc. This involves close collaboration with the relevant authorities and also includes dispute resolution before the courts. Aude is an active member of the European Food Law Association (EFLA) and of the Food Lawyers Network (FLN). She cooperates with the European Food and Feed Law Review (EFFL—Lexxion as the legal publisher) as a correspondent for Belgium and is a regular speaker at seminars and conferences on food law topics. She is mentioned as an expert in food law by the Legal 500 EMEA directory (editions 2014, 2015, 2016, and 2017). She graduated from the law faculty of the Université Libre de Bruxelles and completed a certificate in EU food law at the University of Michigan. Aude works in French, English, Dutch, and Spanish.

Silvia Fernanda Mardegan is assistant professor of the Federal University of Pará. She graduated as biologist at Londrina State University, and holds her masters in ecology from the National Institute for Research in the Amazon, and doctoral degree in applied ecology from the University of São Paulo. Her research interests are ecosystem processes related to nitrogen and carbon cycles in natural and disturbed areas.

Luiz Antonio Martinelli is full professor of the University of São Paulo. He graduated as an agronomist and earned his master and doctoral degrees also at the University of São Paulo. He is an expert in biogeochemistry of tropical forests and agriculture and development in developing countries.

Ikechi Mgbeoji is educated in Canada, Germany, and Nigeria, Prof. Dr. Ikechi Mgbeoji earned a doctorate degree in law from Dalhousie University, Halifax, Canada. His master's degree earned him the Governor General of Canada Academic Gold Medal. Ikechi is a full professor at Canada's Osgoode Hall Law School, York University, where he teaches patent law, international aspects of intellectual property law, and public international law. Ikechi is the author of four books and dozens of articles in reputable law journals.

Dimpy Mohanty is Partner at the Delhi Office of LexCounsel Law Offices and has over 18 years of experience advising and representing in the general corporate and commercial practice area as well as in the specialized practice areas *inter alia* of Labour & Employment and Life Sciences and Biotechnology. Her Life Sciences and Biotechnology practice includes advice on issues such as classification of BAFS, OTC, and herbal health products and cosmetic products, permitted, restricted, and prohibited ingredients, labeling and registration requirements, Indian food and drug laws including as applicable to companies engaged in the manufacture, distribution, import and export of ice creams, health supplements, cosmetics and medical devices and obtaining registration of products under the applicable laws. Dimpy is also active in the regulatory practice area and has experience in liaising with ministries including for invitation based comments/suggestions on the consolidated FDI Policy and participating in discussions held in this regard. She has been invited to speak at conferences and seminars included those hosted by the International Bar Association.

Julius T. Mugwagwa is a research fellow in the Development Policy and Practice (DPP) Unit and INNOGEN Institute in the UK. He received undergraduate and postgraduate training in biological sciences, biotechnology, and business administration in Zimbabwe before attaining a Ph.D. in science, technology, and innovation policy at the Open University in 2008. Julius has worked in a number of sectors including veterinary research, pharmaceutical production and quality assurance, medicine control, agricultural biotechnology research, and biotechnology/biosafety governance. He is a past recipient of a Leverhulme Early Career Fellowship (2009–2011) and currently (2013–2015) a holder of a research fellowship under the UK Economic and Social Research Council (ESRC) Future Research Leaders Scheme. Julius' ongoing research focuses on innovation systems in health and agricultural biotechnology, with a special interest in governance and adoption of technological and institutional innovations at sectorial, national, and cross-national levels, using concepts such as "multi-layered governance" and "policy gridlocks." His recent intellectual contributions include the concepts of "policy kinetics" and "innovative spending."

Kennedy Mwacalimba holds a Ph.D. in public health and policy from the London School of Hygiene and Tropical Medicine (2011). His core expertise is in the areas of policy analysis, epidemiology, public health, and risk analysis. His areas of interest include multi-sectorial risk management policy development, the sociology of risk, zoonosis risk assessment, cost-benefit analysis of zoonotic disease control, food safety, and international livestock and livestock product trade.

He is a former faculty member in public health at the University of Zambia, where he lectured in public health, epidemiology, livestock economics, environmental health, and food safety. He was also the course developer of the emerging and re-emerging diseases course and course leader of the health economics, policy, implementation, and evaluation course under the MSc in One Health Analytical Epidemiology hosted jointly by the Schools of Veterinary Medicine and Medicine of the University of Zambia. He is now an independent policy researcher based in Indianapolis, Indiana, where he conducts independent medical and pharmaceutical policy, livelihood impact, and One Health research. He has also worked as an independent policy researcher based in Indianapolis, Indiana, where he conducted independent medical and pharmaceutical policy, livelihood impact, and One Health research. He is now Associate Director, Outcomes Research for Zoetis (US Operations).

Michael J. Nathenson is Instructor in Medicine at the Dana Farber Cancer Institute at Harvard Medical School in Boston. He previously completed a sarcoma fellowship at the MD Anderson Cancer Center in Houston, Texas. Prior to that, he worked as a hematology oncology fellow at the University of Maryland Greenebaum Cancer Center in Baltimore, Maryland. He is a contributing author to the textbook *International Food Law and Policy*. He holds a BS from Brandeis University and an M.D. from Tufts University. He completed training in internal medicine at the University of Pittsburgh Medical Center and is boarded in internal medicine and in hematology/oncology.

Uche Ewelukwa Ofodile is a professor of law at the University of Arkansas School of Law in Fayetteville, Arkansas, where she teaches a broad range of courses including international trade law, international investment law, intellectual property law, the right to food, public international law, and business and human rights. Professor Ewelukwa also teaches in the law school's *LL.M. program in agriculture and food law* offering courses such as "the right to food," "intellectual property issues in food and agriculture," and "corporate social responsibility in the food and agricultural sector." Professor Ofodile is widely published. Her articles have appeared in many highly rated journals including *Michigan Journal of International Law*, *Minnesota Journal of International Law*, *Transnational Dispute Management*, *Vanderbilt Journal of Transnational Law*, *Yale Human Rights and Development Law Journal,* the *University of Miami Law Review*, and the *Cornell Journal of International Law* (forthcoming). Professor Ofodile is the recipient of several fellowships including a fellowship from the Carnegie Council on Ethics and International Affairs. In 2014, she received the *Outstanding Year-in-Review (YIR) Contribution Award* from the American Bar Association Section of International Law for the section on "Corporate Social Responsibility" that she edited and coauthored. Professor Ofodile was also the winner of the *2009 Human Rights Essay Award* from the Academy on Human Rights and Humanitarian Law, American University Washington College of Law. Professor Ofodile presently serves as the secretary general of the African Society of International Law and is on the advisory board of the *African Journal of Legal Studies* and the Africa International Legal Awareness, a nonprofit organization working to build capacity and promote sustainable economic development in African countries.

Raymond O'Rourke is a qualified Barrister and a specialist food regulatory and consumer affairs lawyer. He worked for many years in legal firms both in Brussels and Dublin and now has his own law practice. He has written two books European Food Law (3rd edition) (2005) Thomsen/Sweet & Maxwell and Food Safety & Product Liability (2000) and numerous articles on food and consumer protection law issues. He is presently Vice-Chair of the Management Board of the European Food Safety Authority (EFSA) and also is Chairman of the Consumers Association of Ireland (CAI).

Janelle Orsi is a lawyer, advocate, writer, and cartoonist focused on cooperatives, the sharing economy, urban agriculture, shared housing, local currencies, and community-supported enterprises. She is cofounder and executive director of the Sustainable Economies Law Center (SELC), which facilitates the growth of more sustainable and localized economies through education, research, and advocacy. Janelle has also worked in private law practice at the Law Office of Janelle Orsi, focusing on sharing economy law since 2008. Janelle is the author of *Practicing Law in the Sharing Economy: Helping People Build Cooperatives, Social Enterprise, and Local Sustainable Economies* (ABA Books 2012) and coauthor of *The Sharing Solution: How to Save Money, Simplify Your Life & Build Community* (Nolo Press 2009), a practical and legal guide to cooperating and sharing resources of all kinds. In 2014, Janelle was selected to be an Ashoka fellow, joining a robust cohort of social entrepreneurs who are recognized to have innovative solutions to social problems and the potential to change patterns across society. In 2010, Janelle was profiled by the American Bar Association as a legal rebel, an attorney who is "remaking the legal profession through the power of innovation." In 2012, Janelle was one of 100 people listed on The (En)Rich List, which names individuals "whose contributions enrich paths to sustainable futures."

Peter Berry-Ottaway is a food scientist and technologist with considerable experience in food law. During a career spanning over 50 years in the food industry, Peter has been involved in a very wide range of food sectors and products and accumulated considerable experience in the areas of food product development and food safety and control. In 1974, he formed a scientific consultancy specializing in food science, technology, nutrition, and food law. Peter has played an active role in the development of European Community food legislation and works with governments across the world. He is the author, editor, or contributor of numerous books and papers on food science, food technology, and food law and for 8 years was the editor in chief of the International Review of Food Science and Technology. Peter has been appointed chair of a number of expert committees and working groups. He has been involved with Codex Alimentarius Committees for nearly 30 years.

Alexey Petrenko is the general manager of EAS Strategic Advice, the global regulatory consultancy, in Russia and CIS. He is an expert in the Russian regulatory system and legislation covering food safety, sanitary requirements, and technical regulations. He also specializes in legislation of the customs union of Russia, Belarus, and Kazakhstan. Alexey had an extensive experience of working for and with large international companies and NGOs, e.g., the US-Russia Business Council and the American Chamber of Commerce, on a number of projects assisting US and EU businesses in Russia and beyond. Alexey holds a Ph.D. in organic chemistry from the University of Hull, UK.

Andreas Popper studied law at the Universidad "Pontificia de Salamanca" in Spain and at the Ludwig Maximilian University of Munich in Germany. He has earned his master's degree in intellectual property and transnational law at "Temple University" in Tokyo. After his time as assessor at the Higher Regional Court of Munich, he was an assistant to Prof. Dr. Kigawa in Tokyo and started his career at firms in Germany, Spain, the UK, and Belgium working mainly on EU foreign trade, anti-dumping, and regulatory matters. Since 1994, he spent an increasing amount of time in Japan, and in 2003, he became counselor for East–west relations at the Miyazaki (South Japan) local government. Subsequently, he worked in Tokyo at the law department of Sonderhoff & Einsel Law and Patent Office and as representative director of the Japanese subsidiaries of IQS Avantiq AG, a global branding specialist, and of Inovia Holdings Pty Ltd, a coordinator of PCT national stage applications. During his first years of work, he recognized the links between regulatory legislation and intellectual property law. In 2007, he cofounded the naexasCompass Group in Tokyo, to engage in the intersection between intellectual property law, product compliance, foreign trade, and import duty policies. Since then, he supported the strategic planning and implementation of global branding and market entry projects of medium- and large-sized enterprises. By 2014, his firm established offices in Tokyo, Shanghai, Singapore, the USA, and the EU. He is invited as expert speaker to conferences, seminars, and round tables of organizations such as the International Trademark Association (INTA), the Pharmaceutical Trade Marks Group (PTMG), and the International Bar Association (IBA).

Alfredo Quarto is the executive director and cofounder of the Mangrove Action Project (MAP), is a veteran campaigner with over 35 years of experience in organizing and writing on the environment and human rights issues. Formerly an aerospace engineer, his experiences range over many countries and several environmental organizations, with a long-term focus on forestry, indigenous cultures, and human rights. Prior to MAP, he was the executive director of the Ancient Forest Chautauqua, a multimedia traveling forum with events in 30 West Coast cities on behalf of old-growth forests and indigenous dwellers. Alfredo has published numerous popular articles, book chapters, and conference papers on mangrove forest ecology, community-managed sustainable development, and shrimp aquaculture. He lives on a small, organic farm in Port Angeles, Washington, and is conversant in Spanish.

Michael Roberts is the founding executive director of the newly established Resnick Program for Food Law and Policy at the UCLA School of Law. He is well versed in a broad range of legal and policy issues from farm to fork in local, national, and global food supply systems. He is a prolific author, having contributed to books and having written several articles on food regulation, trade, and policy issues. Roberts has published the first major treatise on food law, titled *Food Law in the United States*, Cambridge University Press. He is also coeditor of *Food Law & Policy*, a new casebook to be published by Wolters Kluwer. He has guest lectured on food law subjects at various law schools in the USA, Europe, and Asia, with frequent visits to the Renmin University School of Law (Beijing), East China University of Science and Technology (Shanghai), and the University of Tuscia European Food Law Center (Viterbo, Italy). He is a research fellow for Renmin University's Center for Coordination and Innovation for Food Safety. In 2000, He left his law practice and enrolled in the LL.M. program on agricultural law at the University of Arkansas School of Law, the only such program in the USA. Since then, Roberts has engaged in a variety of professional capacities related to food law and policy. A few years after completing the LL.M. program, he was invited to join the University of Arkansas School of Law as a research professor of law and as the director of the National Agricultural Law Center, where he taught food law and policy and founded the law school's *Journal of Food Law and Policy*. He is the former first chair of the Lex Mundi (world's largest association of private law firms) international agribusiness practice group. Roberts also was of counsel in Washington, DC, with Venable LLP, as a member of the firm's food and agricultural law practice group, and special counsel to the Roll Global farming and food companies headquartered in Los Angeles, where he was responsible for global food regulation, trade, and public policy. He was also a visiting scholar and consultant to the United Nation's Food and Agriculture Organization (FAO) in Rome. Roberts serves on the Los Angeles Food Policy Council's Leadership Board and chairs the Los Angeles Garden School Foundation. He also serves on the advisory board for the World Food Law Institute and serves on the editorial board for MDPI's *Laws*, an open access scholarly journal.

Mariagrazia Rocchigiani graduated in political science at the University of Rome La Sapienza, and she holds a master's degree in "sources, instruments and methods of social research." She has been working on development and food security topics for 11 years first at the UN World Food Programme (WFP) and then for the UN Food and Agriculture Organization (FAO). Since 2013, she has been working as a cooperatives officer within the Partnerships, Advocacy and Capacity Development Office of FAO where she focuses on organizational strengthening topics for producers organizations and cooperatives at global, regional, and national level.

Bernard Rollin's scholarly interests include both traditional philosophy and applied philosophy. In addition to numerous articles in the history of philosophy, philosophy of language, ethics, and bioethics, he is the author of *Natural and Conventional Meaning* (1976); *Animal Rights and Human Morality* (1981, 1993, & 2006); *The Unheeded Cry: Animal Consciousness, Animal Pain and Scientific Change* (1988 & 1998); *Farm Animal Welfare* (1995); *The Frankenstein Syndrome* (1995); Science *and Ethics* (2006); and *Putting the Horse Before Descartes* (2011). He has edited two volumes of *The Experimental Animal in Biomedical Research* (1989 & 1995). He is one of the leading scholars in animal rights and animal consciousness and has lectured over 1500 times all over the world. He is a weight lifter, horseman, and motorcyclist.

Giorgio Rusconi was born in Como, Italy, on May 31, 1971. He graduated from the State University of Milan (J.D.) in 1997 and was admitted in Italy in 2000. In 1996, thanks to a scholarship awarded by the European Commission, he attended the International Faculty of Comparative Law of Strasbourg, in France, obtaining a *doctorate* in comparative law. Giorgio has been assisting Italy-based multinational companies and inbound clients and dealing, inter alia, with food, drugs, cosmetics, and agricultural law with respect to a wide range of legal services toward food, drugs, and agricultural businesses including product liability, labeling and advertising, regulatory matters, and litigation. He has acquired significant experience in the field of food and drugs law as well as agriculture law, assisting his clients in connection with product liability matters, labeling and advertising, and applicable legislation, as well as in litigation, providing legal advice, among others, within advertising projects or for the creation of labels, both at the national and international level (for instance, verifying the presence of all information required by law, the choice of a trade name, or the wording adopted or ensuring that the label bears no misleading nutritional claim), providing assistance both out of court in drafting contracts for companies operating at all levels in food industry and in litigation before all authorities responsible for controlling food hygiene and safety. The approval of TV advertisements and marketing campaigns also constitutes work in this area. Storyboards are examined and clients assisted in their relationship with advertising agencies, as well as deal with legal issues concerning comparative advertising. Clients are represented in different legal proceedings concerning commercial disputes on labeling (unfair competition, industrial property), summary proceedings and injunction proceedings, actions for damages and criminal proceedings (e.g., defending company directors prosecuted for deception), and administrative proceedings. He is secretary general and founding member of FLN—*Food Lawyers Network Worldwide*—an international pool of professionals set up to the purpose of sharing opinions and enhancing exchange of views and experiences by lawyers with expertise on food law from all over the world as well as offering integrated legal services to multinationals operating in food industry.

Gunnar Sachs, Maître en droit (Paris) and Doctor of Law (Dr. iur), is partner in the Düsseldorf office of the global law firm Clifford Chance and Expert Lawyer for Intellectual Property Law, with long-standing experience in the food, healthcare, chemicals, and consumer goods sectors. He is an active member of the firm's worldwide industry groups "Consumer Goods & Retail" as well as "Healthcare, Life Sciences & Chemicals." He worked as a recognized EU expert in the EU-China Project as well as in the EU-Turkey Project on the Protection of Intellectual Property Rights and is a member of the German Association for Food Law and Food Science (BLL), the German Association for the Protection of Intellectual Property (GRUR) as well as of the German China Desk and Franco-German cross-border team of Clifford Chance. Gunnar Sachs represents national and international clients from regulated sectors, including the food industry. In addition to his daily work as a lawyer for Clifford Chance, he has also worked as external general counsel for and has assumed the direction of the German legal departments of one of the biggest US retail and multichannel companies and the world's leading TV shopping network, as well as of a leading US biotechnology company, which gave him further helpful insights into the relevant markets. During his legal clerkship, Gunnar Sachs has worked, inter alia, for the World Trade Organization in Geneva; for the French Ministry of Economics, Finance and Industry in Paris; for the French Ministry of Defense in Paris; for the Embassy of the Federal Republic of Germany in Paris; for the German Institution for Arbitration in Berlin/Bonn: as well as for the State Parliament of North Rhine-Westphalia in Düsseldorf. As a student and doctorate scholarship holder of the Konrad Adenauer Foundation, he has studied at the universities of Münster, Cologne (Dr. iur.), and Paris (Maître en droit). He is also a sought speaker at relevant food law events all over the world and has lectured at the University of Düsseldorf. Further, Gunnar Sachs is author of a handbook on "ethics for arbitrators" and coauthor of the WiKo legal commentary series on medical devices law, as well as coauthor of a handbook on medical devices law.

Liviu Steier is a specialist in prosthodontics and in endodontics. He has been appointed visiting professor at Tufts Dental School and the University of Florence and was clinical associate professor and the clinical and course director of the MSc in endodontics at Warwick Medical School, the University of Warwick. Steier holds various memberships and fellowships. He serves in the editorial board for the *Brazilian Dental Journal*, *Reality Esthetics*, and *Reality Endo*, in the scientific advisory board for the *Journal of Endodontics*, and as a reviewer for the *Journal of Photomedicine and Laser Surgery*. He has published commissioned chapters in three books and numerous articles and abstracts in international peer-reviewed dental journals. Steier maintains, together with his wife, a private practice in Mayen, Germany.

Regina Steier is a general dental practitioner in dental practice in Germany. She published together with her husband Prof. Dr. Liviu Steier and serves as clinical fellow teacher at Warwick Medical School—the University of Warwick. With her excellent clinical skills, she has gained widespread international acclaim in the field of dentistry.

Juanjuan Sun obtained her doctor of laws at Nantes University, France. Focusing on food law, the subject of her thesis is about the international harmonization of food safety regulation in light of the American, European, and Chinese Law. Currently, she is the postdoctor at China Renmin University as well as a researcher of the European food law program Lascaux, the Center of *Cooperative* Innovation for *Food Safety Governance, and* the China Food Safety Law Research Center.

Victor Tutelyan is the director general of the Institute of Nutrition of the Russian Academy of Sciences. He is a leading Russian expert in toxicology and nutrition and Russia's top regulator in safety and quality of foods. Tutelyan is an appointed expert of the World Health Organization (WHO) in food safety, a member of Codex Committee on Food Additives and Codex Committee on Contaminants in Food, the chair of WHO's collaborative center for monitoring food contamination, and a board member in the Institute of Life Sciences. He authored more than 500 papers including 8 monographs, 12 patents, and 60 regulations. He supervised 6 doctorships and 35 Ph.D.s.

Marie Vaale-Hallberg is an associated partner of Haavind law firm situated in Oslo, Norway. Vaale-Hallberg joined the firm in 2011 and primarily assists clients within the food and drink sector, but she also gives advice concerning intellectual property and contract law. Vaale-Hallberg's particular interest in the food industry awoke when working as in-house counsel at TINE SA, the biggest dairy company in Norway, owned by Norwegian farmers. In this period, Vaale-Hallberg dealt with, e.g., food law, regulatory issues, R&D, intellectual property, and marketing, but she also gained valuable commercial insight working as company secretary to the company's top management from 2009. She graduated from the University of Oslo, Faculty of Law, in 2003. Between her graduation and employment with TINE, Vaale-Hallberg worked as an associate within another Oslo-based law firm, Wiersholm. Marie Vaale-Hallberg is a member of the Food Lawyers Network and country correspondent for the EFFL (*European Food and Feed Law Review*).

Bernd van der Meulen is private consultant in food legal affairs and professor of Food Law at Wageningen University (the Netherlands). Wageningen University teaches an MSc specialization in Food Law and Regulatory Affairs which combines legal and scientific know-how. The food law curriculum includes courses in EU

Food Law; Food, Nutrition and Human Rights; Intellectual Property; US American Food Law, Chinese Law on Food and Agriculture and Comparative Food Law. Van der Meulen has a background in Public Law which comprises Administrative Law, Human Rights, European Law, International Law and Competition Law. His core competence is in the legal dimension of the Life Sciences in general and in Food Law in particular. For some of his books, see the European Institute for Food Law Series: http://www.wageningenacademic.com/series/227-1871-3483.

Rochus Wallau studied law, philosophy and Germanic studies at the University of Bonn. After several years as a criminal defense lawyer in the field of economic and environmental criminal law, Rochus gained experience as a consultant for food law in the Federation of Food Law and Food Science e. V. (BLL) in Berlin. Rochus has been working in the food law practice in the law firm of KWG from 2009 to 2014, consulting clients in the area of general national and European food law and specializing in the area of the law of consumer information, Food Criminal Law and the law governing administrative offences. Since August 2014, Rochus is the head of the department for food law and quality management with one of the leading German retailers. He is a lecturer for food law at the Geisenheim University. Rochus is a regular speaker at food law seminars and conferences. Rochus is the author of numerous publications on current food law, criminal law and criminal procedural issues. He advises clients in German and French.

Watu Wamae is an analyst in the innovation and technology policy team at RAND Europe since 2010. She is also a visiting fellow at the Open University and at the African Centre for Technology Studies. Prior to joining RAND, Watu worked both in academia and government and has significant experience informing innovation strategy and policy. In her previous appointments, she worked closely with governments at the national, regional, and pan-African levels. Watu continues to inform policy making in Africa and is a member of the Technical Experts Group for the African Observatory on Science, Technology and Innovation, which is a multidisciplinary team of five leading scholars and practitioners of science, technology, and innovation policy. She is also involved in a number of research networks and is a board member of the African Network for the Economics of Learning, Innovation, and Competence Building Systems whose main focus is to strengthen research capacity in Africa.

Bart Wernaart teaches law and ethics at Fontys University of Applied Sciences. He is also a professional musician (drums, mallets, and percussion), composer, and conductor. He earned his Ph.D. degree in 2013 at Wageningen University. Prof. Dr. B.M.J. van der Meulen is professor of law and governance at Wageningen University.

Jörn Westhoff was born in 1965 in Germany. He studied law, theology, and East Asian politics and culture at Ruhr-University in Bochum, Germany, as well as Japanese business and labor law at FernUniversität in Hagen, Germany. He also received legal training in St. Louis (Miss.), USA, and Tokyo, Japan. As a specialist for international business law, Jörn Westhoff worked as an assistant to Prof. Dr.

Toichiro Kigawa, Tokyo in 1996. In 1997, he took up a position within the Pricewaterhouse group in Düsseldorf, Germany, where in 1998 he joined a law firm specialized in international business law. In 2001, Jörn Westhoff was recruited as head of the legal department for a major German-Japanese law and patent office in Tokyo, Japan. In 2012, he returned to Germany and took his current position with Dr. Wehberg und Partner, a cooperation of internationally oriented lawyers, tax accountants and CPA in Hagen, Germany, where he is one of the seven senior partners. Additionally, in 2013 he was appointed professor for German and international business law at FOM University of Applied Science in Essen, Germany. In 2014, he was granted the title "Certified Specialized Lawyer for International Business Law" by the lawyers' chamber. Jörn Westhoff is an associate of the Chartered Institute of Arbitrators, London, a group a lawyers offering services as counsel or arbitrator in private international arbitration. Jörn Westhoff has published widely on Asian and German business law both in English, German and international journals and compilations.

Food Law International Interns

Daisy Zhang was an intern with Food Law International, LLP. She earned her Master's of Public Health (MPH) from CUNY School of Public Health, New York, and her Bachelor's of Science (BS) in Biological Sciences from University at Buffalo, NY.

Kimia Shahi earned her bachelor's degree from the University of California, Los Angeles in 2009. She studied History of Western and Near Eastern countries. In 2015, she successfully completed her master's degree in food studies at the New York University. Having lived in Iran and America, she researched production methods, national and international policies, and movement of food commodities across the globe. In particular, she focused on Iran under the imposed economic sanctions. Kimia is an up-and-coming law candidate and is expected to begin in fall of 2018. She aspires to use her knowledge, skills, and lived experience to eliminate regional instability, lift struggling societies out of poverty, and help the global community reach self-sufficiency.

Eva Saul Crawford grew up in Cleveland, Ohio. Her interest in sustainable food systems and international food regulation was sparked while working on a small, organic farm in Erfurt, Germany, during college. After receiving her B.A. in sociology from Beloit College, she moved to New York where she is currently working in communications at a nonprofit, intercultural exchange organization and running a local CSA. At the time of writing, Eva is pursuing her master's degree in food studies at New York University, with the goal of working in food policy.

Liza Howard is a master's candidate in the Food Studies Department at New York University. After completing her undergraduate degree in international development, she studied French in France before moving to the USA to continue her education. Her research interests include organic food policy, cultural constructions of taste, and food education for children.

Sabrina Rearick was an intern for Food Law International, LLP. She pursued her J.D. at Duquesne University School of Law in Pittsburgh, Pennsylvania. At Duquesne, she served as a member of law review and worked as a research assistant for the associate dean. In 2014–2015, she was the recipient of the Dean's Scholarship for Excellence for her academic performance during her first year. Sabrina graduated magna cum laude from Westminster College in 2014 as a part of the 3+3 early admissions program with a B.A. in political science. She is highly interested in sustainability, particularly in regard to community development and food.

Nandini Shroff, MPH currently works as a grants and research manager at the Institute for Family Health. She is primarily responsible for research and grant management. Her project portfolio ranges from women's health to chronic diseases research. Nandini's public health interests include health disparities and women's health with a particular emphasis on reproductive and sexual health. Nandini received her B.A. in history with a dual minor in anthropology and economics from the City College of New York. She earned her MPH at Hunter College in New York. She is currently a doctoral candidate at the CUNY Graduate School of Public Health and Health Policy.

Ilana Madorsky was an intern for Food Law International LLP. Her portfolio of research includes food access policy programs and the use of intermediated markets to support regional food systems. After living on a permaculture farm in Modi'in, Israel, she moved to New York where she is currently working at NYU School of Law. Ilana received her BS in Society and Environment, with honors, from UC Berkeley, and her MA in Food Studies from New York University.

Christina Papillo is a legal assistant at Katzke & Morgenbesser LLP in New York City and worked as an editorial intern at Food Law International, LLP. She works in executive compensation law and has experience in foreclosure litigation, commercial litigation, and intellectual property law. Christina is pursuing a juris doctor degree at Fordham University School of Law where she serves as an associate editor for the International Law Journal. She has a Master of Arts in Food Studies and a Bachelor of Science in Media, Culture and Communication, both from New York University. She focuses her education on food and agricultural law, with a specialization in international food systems and industrial farm animal production. She has performed field work in the United Kingdom, Hong Kong, and Ireland focusing on legal barriers to a nation's food system and its effects to food access. She plans to become a food and agriculture attorney, aspiring to provide guidance to small farmers and small food businesses on complex laws governing food production.

Victoria Yam holds a B.A. in French and German from the University of Rochester and an M.A. in German and Scandinavian studies from the University of Massachusetts, Amherst. She has lived in the cities of Berlin, Cologne, and

Heidelberg, Germany. Recently, she has taught as an adjunct instructor of both English and German at the University of Cologne, University of Massachusetts at Amherst, and Deutsches Haus at New York University. Victoria is the translator of the German-Mexican cookbook, *Authentic Mexican Cooking: 80 Delicious, Traditional Recipes for Tacos, Burritos, Tamales, and Much More*, coauthored by Scott Myers and by Gabriele Gugetzer, and has translated a variety of texts from German to English on a wide range of topics, including texts on psychoanalysis, excerpts from eighteenth- and nineteenth-century German literature, and personal biographies. In addition to translating in her spare time and working as a freelance instructor of German, Victoria's interests include international programs within study abroad and cultural sensitivity/minorities studying abroad. Victoria serves full-time as part of the administrative team at New York University's Department of German.

Alexander J. Cherry is an undergraduate Bachelor's candidate and Dean's List student of International Relations and Environmental Studies, with focuses on Global Health, Nutrition, the Environment, Food Systems, Sustainability, and Communications at Tufts University in Medford, Massachusetts. He has recently completed a semester-long program at the Umbra Institute, located in Perugia, Italy, studying Food and Sustainability. An avid traveler and lover of the outdoors, he can often be found working in a garden, rock climbing, hiking, or skiing.

List of Abbreviations

ACLs	Annual catch limits
ARIPO	African Regional Intellectual Property Organization
AU	Africa Union
BSE	Bovine spongiform encephalopathy or mad cow disease
CAADP	Comprehensive Africa Agriculture Development Program
CAC	Codex Alimentarius Commission
CANSEA	Conservation Agriculture Network in Southeast Asia
CAP	COMESA Agricultural Policy
CDQs	Community Development Quota
CEN-SAD	Community of Sahel-Saharan States
CEQA	California Environmental Quality Act
CETA	Comprehensive Economic and Trade Agreement
CETA	Comprehensive Economic and Trade Agreement
CJEU	Court of Justice of the European Union
COMESA	Common Market for Eastern and Southern Africa
CONSEA	Brazil's National Food and Nutrition Security Council
CoOL	Country of Origin Labeling
CWA	US Clean Water Act
DG	Directorate General
DG SANCO	EU Directorate General for Health and Consumer Affairs
DHHS	United States Department of Health and Human Services
DNA	Deoxyribonucleic acid
EAC	East African Community
ECCAS/CEEAC	Economic Community of Central African States
ECOSOC	Economic and Social Council of the United Nations
ECOWAS	Economic Community of West African States
EEA	European Economic Area
EEZ	Exclusive Economic Zone
EPA	Environmental Protection Agency
EPO	European Patent Office

List of Abbreviations

ESA	Endangered Species Act
EU	European Union
EurAsEC	Eurasian Economic Community
FAO	Food and Agriculture Organization of the United Nations
FDA	United States Food and Drug Administration
FIFRA	Federal Insecticide, Fungicide, and Rodenticide Act
FMPs	Fishery Management Plans
FOD	Department of Food Safety and Zoonoses
FSANZ	Food Standards Australia New Zealand
FSMA	Food Safety Modernization Act
GE	Genetic engineering *or* genetically engineered
GHG	Greenhouse gases
GM	Genetic modification *or* genetically modified
GMO	Genetically modified organism
GRAS	Generally recognized as safe
ICCPR	International Convention on Civil and Political Rights
ICCPR	International Covenant on Civil and Political Rights
ICESCR	International Convention on Economic, Social and Cultural Rights
ICESCR	International Covenant on Economic, Social and Cultural Rights
IFQs	Individual fishing quotas
IGAD	Intergovernmental Authority on Development
ILSI	International Life Sciences Institute
INFOSAN	International Food Safety Authorities Network
IPPC	International Plant Protection Convention
ITQs	Individual Transferable Quotas
KAFTA	Korea-Australia Free Trade Agreement
LAPPs	Limited Access Privilege Programs
MAFAP	Monitoring African Food and Agricultural Policies Project
MDGs	Millennium Development Goals
MSA	Magnuson-Stevens Fishery Conservation and Management Act
NAFTA	North American Free Trade Agreement
NEPA	National Environmental Policy Act
NEPAD	New Partnership for Africa's Development
NMFS	US National Marine Fisheries Service
OAPI	Organization Africaine de la Propriete Intellectuelle
OHIM	Office for Harmonization in the Internal Market
R&D	Research and development
RECs	Regional Economic Communities in Africa
SADC	Southern African Development Community
SFCA	Safe Food for Canadians Act
SPS	Agreement on the Application of Sanitary and Phytosanitary Measures
TAC	Total allowable catch

TBT	Agreement on Technical Barriers to Trade
TEFAP	Emergency Food Assistance Program
TFEU	Treaty on the Functioning of the European Union
TRIPS	Agreement on Trade-Related Aspects of Intellectual Property Rights
TTIP	Transatlantic Trade and Investment Partnership
TTMRA	Trans-Tasman Mutual Recognition Arrangement
UDHR	Universal Declaration of Human Rights
UMA	Arab Maghreb Union
UN	United Nations
UNCLOS	United Nations Conferences on the Law of the Seas
UPOV	International Convention for the Protection of New Varieties of Plants
USA	United States of America
USDA	United States Department of Agriculture
WHA	Health Assembly of the WHO
WHO	World Health Organization

Part I
Global Food Law and Policy

Gabriela Steier

This first section of the book lays the groundwork to ascertain the complexities of international food law and policy. It introduces the multitude of facets of the international regime governing food safety, security, regulation, and policy, while considering various perspectives of what is at stake, such as consumer protection, animal welfare, global governance, and democracy.

First, Dr. Hans-Ulrich Grimm, a German best-selling investigative journalist with deep insights into the depth of the system, explains the absurdities of the current food system and traces them back to their roots in industrialization, deregulation, and globalization. This chapter sets the stage for all the chapters that follow in this part and in Part II of this book by explaining how malnutrition, obesity, food insecurity, and fraud, relate back to law and policy.

Second, Prof. Bernard E. Rollin from Colorado State University, describes how animal welfare and ethics play a role in the current food system. He links industrial agriculture, animal husbandry, land stewardship, and the need for sustainability to the philosophical and ethical pillars upon which the food system depends if any positive changes are to be made possible in the future.

Third, Ching-Fu Lin, from the National Tsing Hua University, Institute of Law for Science and Technology, Harvard University, Petrie-Flom Center for Health Law Policy, Biotechnology, and Bioethics, and National Taiwan University, Asian Center for WTO & International Health Law and Policy, shares his expertise surrounding the global governance of food through the World Health Organization (WHO), Food and Agriculture Organization fo the United Nations (FAO), the Codex Alimentarius, and the World Trade Organization (WTO). In his chapter, he introduces the international cooperation in food safety around the world.

Fourth, Dr. Bart Wernaart and Prof. Dr. Bernd van der Meulen from Wageningen University, a leading academic institution in food law and policy from the Netherlands, give a lesson on the Right to Food. They explain the concept of adequate food, which encompasses food supplies, food safety, and cultural acceptability considerations. Based on case studies from the Netherlands and from Belgium, these concepts are then illustrated and contextualized.

In the fifth and sixth chapters, Professor Michael Blakeney, one of Australia's experts on intellectual property in food, teaches about the international intellectual property infrastructure and agricultural innovation. He describes how treaties, TRIPS and SPS, govern patenting of genetically modified crops (GMOs) and plant variety rights. His introduction of the International Convention for the Protection of New Varieties of Plants (UPOV) is a key prerequisite to understanding the critical evaluations of food system regulation and governance of GMOs in light of environmental protection, agrobiodiversity, and regulatory integrity. In Chap. 6, Professor Blakeney provides a detailed overview of the legal intricacies of bilateral and plurilateral agreements protecting trademarks and geographical indications. He explains how international intellectual property rights play into food trade and puts them in context with WIPO, GATT, and other leading treaties of global significance.

Finally, in Chap. 8, the esteemed and experienced Peter Berry Ottaway and Sam Jennings from England, analyse international barriers to food trade by contextualizing inconsistencies in food legislation, agricultural regulation, and food safety protection. This chapter focuses on some of the most prevalent aspects of international dissonance, thereby illustrating how the national laws come together when food becomes a commodity of global significance.

What all of these leading chapters have in common are astute introductions in the international framework of food law and policy and revelations as to the gaps of the food system. For the reason that what is being regulated is just as important as what is left out of legislative considerations, the shortcomings identified by thee chapters roll out a vast array of opportunities for stake-holders around the globe to contribute creative solutions to the problems that are currently unresolved.

Chapter 1
Information-Domination in the European Food Industry: Focus on Germany

Hans-Ulrich Grimm

Abstract Food safety is of vital importance for all people. Their lives and health depend on it. Food laws ought to serve to ensure food safety and the protection of consumer health, but currently they often fail to meet this standard. Presently, food laws do not meet many of the most significant threats, including Non Communicable Diseases (NCDs), such as heart disease, diabetes, Alzheimer's, and cancer. 35 million people worldwide will die from these types of diseases every year, according to the World Health Organization (WHO). In most cases, the causes of these diseases can be traced to modern, processed foods, fast food, take away food, and soft drinks. Conventional food laws treat these foods, however, as completely "safe". Food components associated with health risks in the long-term, such as sugar, salt, and some chemical additives, are not considered in the current conception of "food safety". The blame for these health consequences is taken by no one, and the responsibility for treating these diseases lies solely on the backs of the consumers. For these new realities, existing food laws and policies are inadequately prepared. Moreover, the manufacturers of these products are increasingly dominating the decision-making chain at both global and regional levels. The interests of consumers play only a minor role. Critics, therefore, are calling to bring the protection of consumer health to the fore again and to create the necessary legal conditions where significant changes can occur. The concept of food safety ought to be adapted to the new circumstances.

1.1 Introduction: What Is The Purpose Of Food Law?

Food safety is of vital importance for all people. Their lives and their health depend on it. Food laws ought to, therefore, serve to ensure food safety and the protection of public health, but currently they often fail to meet this standard. Current laws were designed in an era where hazardous conditions and threats in the overall food system were very different from those that exist today. For example, recent globalization

---

Translated by Victoria Yam, edited by Alex Cherry. Food Law and Food Safety. Who gets to decide what's healthy?The Influence of Big Food on Global Food Legislation.

H.-U. Grimm (✉)
Dr. Watson, Der Food Detektiv, Stuttgart, Germany
e-mail: grimm@food-detektiv.de

has increased the threat of people's exposure to common pathogens found in food, namely viruses and bacteria, resulting in increased disease transmissions. Furthermore, noncommunicable diseases (NCDs), the magnitude of the threat of which was not known until recently, continue to threaten lives. The WHO predicts that 35 million people each year will die of NCDs such as heart disease, diabetes, Alzheimer's disease, and cancer.[1] Rather than common pathogens, the so-called Western diet—the modern, processed diet consisting of fast food, frozen meals, and soft drinks—is considered to be the root cause of NCDs. Under the current system, the food products that fall into this category are fully considered "safe." However, within these products are certain components, such as sugar, salt, and some artificial chemicals, which are not currently considered within the concept of "food safety" in the technical sense but have been increasingly linked to the contraction of such diseases. The blame for these health consequences is taken by no one, and the responsibility for treating them lies solely on the backs of the consumers.

Law and politics are still inadequately prepared for the new food risks of the industrialized and highly-processed modern diets. Moreover, the manufacturers of these products are increasingly dominating global policy making and the chain of command of enforcement in food safety, security, and governance, on both a regional and a global level. In many regions of the world, democratic decision-making structures are being marginalized by industry lobbying and the dominant global regulatory entities. This is problematic because only specific, globally-active transnational corporate conglomerates are participating in the decision-making process and regulatory governance, with little leeway for any other significant influence. These political trends threaten the very nature of democracy itself.

Consumers themselves are the only ones concerned about the protection of their own health. The classic function of food law, consumer protection, steps into the background against the influence of powerful economic interests—more precisely, the particular interests of specific industry groups. Consequently, critics demand that the protection of consumer health be brought to the foreground again, and argue, therefore, that the creation of the necessary legal requirements to ensure these protections is paramount. Moreover, the concept of food safety should be readily adaptable to future circumstances. Current research maintains that our food can in fact be entirely safe. Nevertheless, every year, hundreds of thousands of people die from diet-related illnesses. Logically, this is an unsustainable situation.

1.2 Globalization and Food Safety: Increasing Risks, Missing Liability

In the system of global mass production, microsopic risk factors—pathogens, pollutants, and some additive chemicals—are becoming more widespread. When losses or damages occur, an investigation is prompted, but it is often impossible to

---

[1]World Health Organization (2010).

isolate the specific cause. This is because in massive transnational food corporations with global operations divided into highly specialized and hard-to-trace supply chains, identifying the point source of contamination can be a nearly unsolvable task.

Previously, health risks in food products were most often locally confined. However, with increased globalization what was once local can now be easily spread around the globe. According to the Swiss Federal Office of Public Health *(Bundesamt für Gesundheit, BAG)*, the risk of international outbreaks increases as a result of the globalized traffic of goods.[2] Consumers are confronted with a completely new scenario where a distinctive risk factor is created by industrial production. Previously, scientists in labs could determine pathogenic causes of illnesses with a microscope—now, researchers must take in the whole picture of the entire system. When viewed on this level, it becomes clear that the whole globalized food industry has become unhealthy. Unfortunately, this new world of industrially produced food is spreading all over the globe. And in its wake, new health risks are rising.

Food Quality News Industry Services has indicated that food safety fears are often exacerbated by globalization.[3] Microbiologist and professor Michael Doyle of the University of Georgia has warned that "sanitation practices for food production are not universally equivalent throughout the world. Importing foods can bring more diseases from areas where they are indigenous to locations where they are rare or do not exist" [4] (adapted).

Anxiety among consumers has also been growing in recent decades since China has been flooding the world with seemingly endless cheap goods and food products, including some risky products such as polluted shrimp, contaminated honey or sweets, and baby formula with blacklisted chemicals. In an international forum in Beijing in 2007, 600 delegates from 45 countries adopted a declaration, the Beijing Declaration on Food Safety, which championed global food transparency improvements and data-collecting programs on public consumption habits in order to assess and monitor the spread of food-borne illnesses. "This is not the first international agreement related to food safety... but it's the first time that we have countries getting together and saying, 'let's recognize that it's a joint responsibility and we should work together to improve it'," stated Jørgen Schlundt, Executive Director of the WHO's Food Safety Department.[5] Schlundt saw the collective declaration as "a significant step forward". Yet the concept of "food safety" still leaned toward the traditional health threats of food, and the declaration itself aimed at contamination

---

[2]Bundesamt für Gesundheit: Bulletin 32/08. Bern: Bundesamt für Gesundheit.
[3]http://www.foodqualitynews.com/Public-Concerns/Globalisation-raises-food-safety-fears-says-microbiologist Accessed on 5 May 2014.
[4]Vgl. Food microbiology: fundamentals and frontiers. Doyle, M. P.; Buchanan, R. L. (Editors): Food microbiology: fundamentals and frontiers 2013 http://www.cabdirect.org/abstracts/20143006614.html;jsessionid=4FF687D518DDDC09BA61869587D8A47E.
[5]http://www.chinadaily.com.cn/china/2007-11/28/content_6283416.htm.

reduction to protect consumers from "health risks posed by biological, chemical and physical hazards in food as well as by conditions of food".[6]

Illness caused by contamination was, until now, seemingly the biggest threat. However, over the past couple of decades an entirely new form of food supply has been established, for which this old concept of food safety is no longer adequate. "Through social and economical upheavals and technical innovations," claims an Austrian government study, "the kind of food supply and diet forms in the last 150 years has drastically changed."[7] Today, "consumers are transitioning more and more from raw material buyers to buyers of preprocessed, commercialized convenience products, right up to frozen foods." Previously food would come from local neighborhoods and be consumed shortly thereafter, but food products can now be brought into supermarkets from great distances and sit in stock or on the shelf for long periods of time before being brought to the fridge for further cooking, processing, and finally, consumption. "Generally speaking, prepared dishes are consumed in the household right after their preparation," the study notes. "On the contrary, with convenience products, . . . lies a more or less temporal and spatial tension between the processing or fermentation of the foods in manufacturing plants and the consumption by consumers." (adapted). Consequently, food must be adapted for this intermediate period, most notably through preservation. Chemical changes undergone during these food preservation processes create a new material composition that may affect the human body differently than the unprocessed version would have, as numerous studies on "nutrition transition" demonstrate with the evolution from traditional to modern, processed food.[8]

Because many ingredients now undergo preservation techniques for increased shelf life, a particular challenge for the human body is to process and adapt to the addition of these preservatives, flavor enhancers, artificial flavors, and artificial colors, as well as elevated levels of sugar and salt. These substances place the body under metabolic stress, which is unsurprising as they are not added for nutritional purposes, but for increasing the product's shelf life. In the Western diet, it is not only added pollution, corruption, and contamination that can lead to illnesses, but also the recipe of ingredients. In other words, the threat in food safety is now more than ever before a man-made one.

---

[6]Beijing Declaration on Food Safety. Adopted by consensus by the High Level International Food Safety Forum "Enhancing Food Safety in a Global Community" held in Beijing, People's Republic of China, 26 and 27 November 2007 http://www.who.int/foodsafety/fs_management/meetings/Beijing_decl.pdf.

[7]Bundesministerium für Gesundheit: Zusatzstoffe, Aromen und Enzyme in der Lebensmittelindustrie. Report zur Abschaätzung der Auswirkungen des FIAP auf Forschung, Entwicklung und Anwendung von Zusatzstoffen, Aromen und Enzymenin der Lebensmittelindustrie Institut für Lebensmitteltechnologie, Department für Lebensmittelwissenschaften und –technologie, Universität für Bodenkultur, Wien. Emmerich Berghofer http://www.bmg.gv.at/cms/home/attachments/5/8/9/CH1403/CMS1391427607538/forschungsbericht_zusatzstoffe_aromen_enzyme_2010.pdf.

[8]Popkin (1993), pp. 138–157.

In fact, beyond just the West, the world of soft drinks, snacks and sweets, powdered soups, fruit yogurts, hamburgers and frozen pizzas threatens serious risk for large parts of the global population as these products reach new markets. Allergies and food sensitivities are the most obvious problems associated with such industrial ingredients in food. But many additives have been linked to the development of health impairments in any consumer, not just those with preexisting risk factors. Flavor enhancers such as glutamate are listed as a suspicious trigger of weight gain. Additionally, hyperactivity (ADHD), learning disorders, and migraines can also be caused by food additives. Sugar, and even some artificial sweeteners, are suspected to be a carcinogen. Preservatives can damage the skin and upset the immune system. Additionally, a consumer who eats more fast food and frozen meals ages faster and has a higher probability of becoming sick from depression.[9] These risks constitute a huge burden for humanity, and according to international food experts are the biggest cause of this emerging conceptual parallel world of NCDs. According to the United Nations' calculations, 35 million people worldwide will die from these diseases with the majority stemming from food and 14 percent from tobacco consumption.[10]

Food's threat against health has acquired a new character and unprecedented prevalence. Yet the counter-measure protections and regulations of federal governments and other public authorities have also declined. Food supervision as a public operation is in its present form in no way proportional to the magnitude of the threat.

1.3 Strict Laws in the Middle Ages: Fraud Has Always Existed, but, Historically, It Used To Be Punished

In the past, consumer protection against the threat of inadequate food was of far higher priority. Producers who endangered consumer health with their products were prosecuted and severely punished. Swindling and forgery were pervasive in the European Middle Ages, but the authorities would intervene with all of their commanding power. Violations of food safety were not only viewed as fraud against citizens, but also as public health risks.[11]

In order to protect citizens from food-borne illness, an abundance of rules and regulations were established for bakers, butchers, innkeepers, and wine handlers.[12] For example, bakers tended to cheat a little with the sizes of their baked goods by

---

[9]Refer to: Grimm (2013).

[10]Persönliche Mitteilung Rüdiger Krech. Director, Department of Ethics, Equity, Trade and Human Rights World Health Organization, 22 May 2012.

[11]Refer to: Grimm (2014).

[12]Bitsch I: Gesundheitsschädigung und Täuschung im mittelalterlichen Lebensmittelverkehr. In: In: Irmgard Bitsch, Trude Ehlert und Xenia von Ertzdorff (Hgg.): Essen und Trinken in Mittelalter

adding plaster or clay to make their bread heavier. Since lighter-colored bread was considered more valuable, they would also lighten the color with limestone, ground bones, or even poisonous substances like lead and aluminium.

Furthermore, fraudulent bakers might regularly prepare two kinds of bread: one that complied exactly with the legal weight specifications and was examined by a controller, and the other that was sold to customers. The latter was somewhat lighter. Additionally, lighter bread was baked on holidays because the bread controllers were not on duty during these days. In those stricter times, it was even viewed as fraud when old baked rolls were sold as "fresh."

Another example is wine, which counted as a staple food, and was strictly controlled. In fact, the entire supply chain, from the wine growers and makers to the merchants and finally to the innkeepers who sold it, was monitored. In Cologne, for example, unannounced visits from inspectors called "wine observers" safeguarded against counterfeited products. At that time, claims Austrian historian Bettina Pferschy-Maleczek, "the law forbade wine adulteration" because it was "viewed... as a kind of wine falsification. This was considered fraud or theft, since it was essentially a misappropriation of buyers' money."[13]

Consequently, city or municipal authorities established long lists of forbidden substances for foods. For example, by the end of the fifteenth century in the southern German city of Ulm, limestone, mustard, and bacon, were blacklisted. Furthermore, clary sage, perry or apple cider were not allowed to be paired with white lead, mercury, impatiens, and sulphuric acid. Many of these additives were commonly steeped in the wine in a small pouch and later removed. Enforcing these bans was, of course, difficult for the lack of chemical methods in order to regulate the water content in the wine. Misconduct was often hard to prove, and thus usually went unpenalized. According to fraud researcher Pferschy-Maleczek, "one simply had to trust and hope in a godly justice that would discover the culprits and assign punishment in the afterlife."

Some transgressions, however, could be proven—and these were strictly punished. Fraudulent bakers had to endure a "baking baptism," also known as "bread christening" or "shed." For punishment and to the amusement of the public, they were shoved in a cage and submerged in a local river. In Alsatian Strasbourg, France, reports indicate bakers submerged in the River Ill. In Vienna, delinquents were plunged into the Danube, and in the year 1550, one of these punishments turned out to be fatal. In the so-called bakers' gallows, in which bread swindlers were chained and hung, some even faced unintended death.

Wine adulterators were also prosecuted and punished with fines, banishment from the city, closure of the cellar, and revocation of their liquor license. In the

---

und Neuzeit. Vorträge eines interdisziplinären Symposions vom 10. - 13. Juni 1987 an der Justus-Liebig-Universität Gießen. Sigmaringen 1987.

[13]Pferschy-Maleczek B: Weinfälschung und Weinbehandlung in Franken und Schwaben im Mittelalter. In: Schrenk C, Weckbach H (Hrsg.): Weinwirtschaft im Mittelalter. Verlag Stadt Heilbronn 1997.

1380s in Rothenburg, Bavaria, an adulteration conviction meant that the culprit would not only lose his adulterated wine product, but also the hand he used to make it. On occasion, wine adulterators feared even the death penalty, as proven by the public execution lists of several cities. In 1486, for example, in the upper Swabian city of Ravensburg, a man named Martin Geßmeister was beheaded because of wine adulteration.

These harsh punishments were not without consequences. Back then, wine was a basic food product in many places, as drinking water fountains were frequently polluted or infested with disease or pests. In these times, the demand for "natural" goods had already emerged. In 1419, Frankfurt am Main, for instance, created a provision that stated that no one should prepare wine in any way other than how God had intended for man to produce it. Strict penalties were developed for the protection of foods' integrity. For this purpose, food law is ultimately a matter of crucial significance. Protection against fraud is likewise the protection of health. Back then, these negative health consequences were usually directly apparent: stomach aches, cramps, indigestion, and even infertility and miscarriages.

Food fraud really took off, however, in the next few centuries. The technological innovations and new production methods of the industrial revolution made fraud and deception possible on a large scale. Aromas and flavor enhancers, for example, were used to imitate existing ingredients, as in ice cream or in industrial fruit yogurts. Similarly, food colorings could be used for deception even in high quality products such as whiskey or balsamic vinegar.

At the same time as the technical possibilities of deception were increasing, the propensity of governments to prosecute such fraudulences was decreasing. British historian Bee Wilson verifies in detail that the falsification of food, formerly strongly enforced, became increasingly tolerated during the industrial age, which was representative of the minimal governmental intervention typical of the time period.

In her book about food swindling in past centuries, *Swindled*, Wilson criticizes governments' tendencies to intervene less and less in the cases of food falsifications. She states that early on, governments have seen swindling as a threat to economic order and to their own authority. Adulteration is also a threat to civilized politics. Thus, governments have sought to police food fraud because permitting it would undermine their legitimacy and authority. A society in which swindling is permitted is one in which fundamental trust between buyers and sellers has broken down. It is therefore a vital political concern to stop it.

Nonetheless, for the past two hundred years many governments have allowed defrauders to get away with outrageous crimes. This is a story of corruption and greed, of the indifference some humans beings will exhibit towards their own actions that threaten others if it means that they are making money. According to Wilson, it is also a failure of politics; a story of the deep reluctance of post-industrial governments to interfere with the markets of food and drink products—something earlier governments were happy to do—to the extent that those markets have become dishonest and dangerous.[14]

---

[14]Wilson (2008), p. xiii.

1.4 The Dwindling State: A Ministerial Perspective on Industry Responsibility

Separate from the traditional system of locally produced food, of fresh carrots, potatoes, chickens, apples and pears, an industrial parallel system was established, with products flowing through a global supply chain, containing numerous chemicals, and producing different effects on the human body. This modern parallel world has not only substantially changed the nature of food products, but also the role of the state in its regulation. Previously, wherever there was pre-industrial food production, surveillance and control were essential tasks for the state. Ultimately, it was (and arguably still is) a question of life and death.[15] However, generally speaking, the more falsified food products grew, the larger the means of manipulation grew, the more the potential magnitude of health hazards grew, but the more countries' regulatory systems withdrew themselves and surrendered quality control to agents of the industry. The global corporations themselves are now largely responsible for the "safety" of their creations.[16] Along this process, the interests of consumers have by and large disappeared.

In the era of globalization, the small-time, locally-constrained frauds, the cheating of bakers, and the wine adulterations of vintners and innkeepers have been supplanted by large-scale manipulations, with high-tech methods and the support of researchers. The role of the State is being re-defined, beginning with lobbying, which allows the offenders to play a part in the legislative process and to contribute to the legalization of the most shrewd deception maneuvers in the food system. Additionally, the largest globally active corporations sell their products all over the world; they dominate global flavor development and ensure that children from all over the globe grow up accustomed to the same flavor chemicals. The power of nations over their own food supply is concurrently decreasing: they have forfeited their sovereignty when it comes to the nutrition of their population. In this industrial parallel world of food, the hazardous products of the food industry are dominating more and more in even the most remote territories.

In the Kingdom of Tonga, for example, a small island state in the South Pacific, there are still coconuts and bananas, carrots, and potatoes for sale in the public market hall of its capital Nuku'alofa. Yet in the stores, products are displayed from the parallel world: the 2-min noodles from Maggi (owned by Nestlé) in plastic packs, alongside other instant noodles from Indonesia and China. In the supermarkets, the South Pacific looks just like China or the U.S. or Europe. There are the Nestlé-noodles. There is the chocolate-spread Nutella. There are "Pringles" brand potato chips. And there are advertisements for Coca-Cola displayed at every bus stop[17].

---

[15]EuGH, Urteil vom 26. Oktober 1995 - C-51/94.
[16]Refer to: Grimm (2014).
[17]Grimm (2010).

The very real consequences of this universally available industrial food display can be seen in the nation's hospitals. The control over new food products is slipping away entirely from the authorities. In fact, in an orderly kingdom like Tonga, there are even exact numbers on these imports. For example, they have increased the amount of imported instant noodles from 271 to 664 tons from 1996 to 2006, and snacks, chips and the like have increased from 99 to 341 tons. "Previously, people died of tuberculosis, typhoid fever, and malnutrition," says Dr. Malakai Ake, who is a doctor in Tonga and cooperates closely with the World Heath Organization (WHO). "Today, we have created a new problem" he explains, where the four "top killers," i.e. the most important causes of death in Tonga, are heart diseases, hypertension, diabetes, and cancer. The climbing rate of these illnesses is directly correlated to the importing rates of these highly processed, unhealthy food products, the doctor claims. These diseases are hallmarks of the globalized industrial highly-processed food system that serves corporate interests before those of the consumer.

The Tongan royal administration clearly perpetuates this unfortunate system of priorities: "we don't know too much about additives," is the dismissive excuse of health officials responsible for food additives (see footnote 17).

Consumer protection should nevertheless be guaranteed. Ultimately, manufacturers are the first step in the chain of responsibility for the safety of their products. In this vein, the Beijing Declaration on Food Safety developed a set of principles for jurisdictions around the world to follow on the subject of food safety based on the idea that, "production of safe food is primarily the responsibility of the food industry."[18]

This position is, however, based on the traditional notion of food safety, which identifies safety risks only with illness pathogens and focuses on making producers ensure that their products are free from such unwanted risks. Yet this guideline only barely grasps what are the true triggers of modern illnesses—the intentionally added ingredients of normal industrial products such as sugar, salt, food additives. These ingredients are regarded as harmless—but only up to a certain amount. Accordingly, it is extremely important to have an overview of the scale of commonly consumed quantities in order to be able to assess the threat against the population. Food corporations of course cannot carry out this oversight. This is where the authorities ought to be responsible for food safety.

The Beijing Declaration on Food Safety thus formulated the state authorities' role in an attempt to "establish competent food safety authorities [to function] as independent and trusted public health bodies within a comprehensive production-to-consumption legislative framework." In a country like Tonga, this is not so easy because the small population is unable to control all of the chemicals that the industry uses to preserve their products and create artificial flavors and colors. Responsible authorities fit in a small "Ministry of Agriculture and Food, Forests and Fisheries - Food Division Vaololoa." Thus, the power of authorities to regulate food

---

[18]Beijing Declaration on Food Safety, a.a.O.

has noticeably diminished compared to the large industrial power of Big Food and Big Ag.

The European Union, by contrast, has a different attitude about the role of governmental institutions in monitoring and enforcing food safety. European food safety agencies have far more power to protect against risks and have even attempted to use their authority to gather data about consumer health risks from the overconsumption of "safe" food additives. In 1995, they mandated that member states implement a monitoring system which collected data about consumer quantities.[19] The European Commission's report states that the "European Parliament and Council Directives 94/35/EC, 94/36/EC and 95/2/EC require each Member State to monitor the consumption and usage of food additives. The Commission is required to submit a report on this monitoring exercise to the European Parliament and Council." Unfortunately, nothing came of this ordinance, although some countries did attempt to comply and engage in provisional inquiries. These inquiries yielded some partially significant evidence of transgressions over safe consumer quantities and portion sizes.

Other member states ignored the regulations—for years at that. The Federal Republic of Germany, for example, had indeed evaluated the consumption of a variety of food products in a so-called "national consumer study." Yet data about the consumption levels of food additives has not been collected, despite clear guidelines from the European parliament. For Professor Gerhard Rechkemmer, president of the Federal Research Institute for Nutrition and Food, the highest governmental nutrition research institution in Germany, and the man responsible for the "national consumer study,". He believes that ingredient lists are the intellectual property of the corporations that create them, and that the government does not have the right to demand this information, placing the burden of safety protection solely on these companies. He says, "The companies should have the information of how many food additives are present. No company can be forced to state their exclusive formulations for particular products. They ultimately have authorship of a certain composition. That is something upon which the manufacturer has a particular claim, if it has developed such products, not to be disclosed to the public."[20] Rechkemmer knows the wants and needs of the food industry very well, because he is not only an officer of the Ministry of Nutrition and as such, subordinate to the German government, but also a high-ranking official in the International Life Sciences Institute (ILSI), which with members such as Coca-Cola, Monsanto, Nestlé, and others, is the most important and influential lobby and policy think-tank associated with Big Food.

---

[19]The European Commission: Report from the Commission on Dietary Food Additive Intake in the European Union http://ec.europa.eu/food/fs/sfp/addit_flavor/flav15_en.pdf.
[20]Grimm (2012).

Rechkemmer is the highest governmental nutrition researcher in German for ILSI Europe, a member of the Board of Directors, and a member of the Board of Trustees, alongside leading managers from companies like Nestlé and Monsanto, for ILSI Global.[21] Thus the highest German governmental nutrition researcher is working two positions at once, one on behalf of the State, and the other in a position for the food industry. In his position working for the German state, he advocates with enormous vigor for the interests of the industry, and has represented the group in public events and conferences.[22]

By appointing industry leaders into high-level governmental positions meant to check the power of the same businesses from which they came, Big Food has fully taken over and weakened the food safety sector, under the radar of the global public. They are creating interest-focused arrangements between the industry lobby associations and the State's duties. This is surprisingly consistent throughout all types of political systems worldwide: the global food corporations have simply taken over.

Moreover, China's "Leading Food Safety Expert" is a high-ranking ILSI operative: Prof. Junshi Chen, senior research professor at the National Institute of Nutrition and Food Safety of the Chinese Center for Disease Control and Prevention.

He was the keynote speaker at a conference of the American Chamber of Commerce on May 30, 2013 at the Hotel Mandarin Oriental in Shanghai.[23]

The American Chamber of Commerce proudly pointed out the variety of activities of their keynote speaker: "Chen also serves as head of the Expert Advisory Committee on Food Safety of the Chinese Ministry of Public Health, is an member of the WHO Expert Advisory Panel on Food Safety, director of the International Life Science Institute (ILSI) Focal Point in China and adjunct professor in the Division of Nutritional Sciences at Cornell University in Ithaca, New York."[24]

Chen is also the presiding board member on the executive committee of food additives for Codex Alimentarius, a global institution established by the Food and Agriculture Organization within the United Nations, which determines internationally recognized food laws.[25]

The lobby association ILSI has become a top steering institution, through which the interests of the the most influential industrial factions can be found in the global regulation of food.

---

[21] http://www.mri.bund.de/de/personen/praesident.html Accessed on 4 April 2014.

[22] http://www.ilsi.org/Europe/Documents/FF%20Symposium%202011/Synopsis%20FF%20Sympo.pdf Accessed on 4 April 2014.

[23] http://www.amcham-shanghai.org/NR/rdonlyres/9DE2FF5F-DEF5-4054-94CB-C1517884129B/18409/2013ChinaFoodSafetyandSustainabilityConference_13M.pdf Accessed on 4 April 2014.

[24] http://www.amcham-shanghai.org/AmChamPortal/Committees/CommitteeNewsStory.aspx?ID=698&committeeid=32.

[25] http://www.codexalimentarius.org/contacts/exec/en/ Accessed on 15 April 2014.

1.5 The Power of Transnational Corporations: How Industry-Lobbying Is Taking Over Governmental Duties

Multinational food corporations have focused their interests in a globally-active lobbying institution, which is taking over governmental duties. The International Life Sciences Institute (ILSI), has gone widely unnoticed by consumers and the public but risen massively in importance in the global politics of food.

ILSI has abolished the borders between state and economy. It has substantial influence on laws and regulations, and even official consumption recommendations.

ILSI is widely unknown among the public. The transnational corporations themselves that compose the association are more prominent. These include[26]:

Ajinomoto
Barilla
BASF
Bayer
Cadbury
Campbell Soup
Cargill
Cereal Partners Worldwide
Clasado
Coca-Cola
Danisco
Danone/The Dannon Company
Dow
DSM
FrieslandCampina
Frutarom
Givaudan
General Mills Inc.
H.J. Heinz Company Herbalife International of
America, Inc.
H J Heinz
International Nutrition Company—INC
Kellogg Co.
Kraft Foods Group
Mars, Incorporated
McCain Foods
McDonald's
McNeil Nutritionals
Mead Johnson Nutrition

---

[26]http://www.ilsi.org/Documents/ILSI_2013_Member_List.pdf.

Mondelez International
Monsanto
Naturex
Nestlé
PepsiCo International
Premier Foods
Procter & Gamble
Red Bull
Roquette Frères
Sensus
Seven Seas
Solae Europe
Soremartec Italia—Ferrero Group
Südzucker/BENEO Group
Swiss Quality Testing Services
Syral
Tate & Lyle
Tetra Pak
Unilever
Valio
Wild Flavors
Wimm-Bill-Dann Foods
Yakult Europe

This is no ordinary lobbying group, but lobbying in a new dimension. ILSI is the most influential institution in the worldwide food industry, a globally operating network which creates the foundations for laws and regulations, as well as bans and recommendations. ILSI also sets standards for what people should eat, like necessary vitamin maximum and minimums consumption levels. It also sets standards for food safety.

ILSI is also not really representative of the vast majority of food manufacturers worldwide. Common bakers, gardeners and chefs are not members here. There are only the big transnational corporations. ILSI is an army of Big Food, a powerful association of the big boys, of those companies which dominate the global food supply with increasing power—especially throughout the so-called "Western diet."

ILSI is the most influential authority in the process of creating regulations for global food production, and it crafts them to serve the interests of its member corporations. These corporations are responsible for peddling the industrially manufactured products considered by independent researchers to be the leading causes of the "noncommunicable disease" epidemics worldwide, but especially in the Western world.

ILSI's influence extends across the globe. It is a world-spanning network operating on every continent. There are ILSI regional branches for Europe, North America, Argentina, Brazil, Mexico, South Africa, South Asia, Japan, Korea, China, and even the Northern and Southern Andes.

ILSI is the leading authority for centralized establishments all over the world. For example, ILSI Europe was given responsibility for creating the standardized nutritional recommendations (project abbreviation: EURRECA) for Europe, as well as the European Commission's concerted action efforts on food safety (Food Safety in Europe, FOSIE).

In a project called PASSCLAIM, ILSI developed the policies governing how the European food authority EFSA evaluates food manufacturers' health assertions.

ILSI was even financially supported to develop the official standards for food benefit and risk assessments (Benefit-Risk Analysis of Foods, abbr. BRAFO),[27] for the European Union.

Benefit-risk analysis of food products is incredibly important to accurately understanding and working to combat the threat of noncommunicable diseases.

The ILSI-Benefits-Risk-Project, however, focused far more research on oily fish, which are of relatively low-level importance as a global health risk, compared to the truly high-risk food products produced by its members.

The correlation between high consumption levels of processed foods such as soft drinks and the epidemic of noncommunicable diseases was completely ignored by this project.

This is because ILSI exists to protect the interests of its member corporations. For example, the leader in the world market for soft drinks, Coca-Cola, sat in the steering committee for BRAFO, along with the World Health Organization (WHO) and the European Food Safety Authority (EFSA).

This is the ILSI method: they set the topics and control the debates, and ensure that public-sector researchers at state universities and governmental organizations are following their lead. ISLI is gaining interpretational sovereignty with increasing frequency and is laying the groundwork for future decisions even more favorable to its member corporations.

ILSI cannot, of course, make any decisions administered by governments and parliaments. However, they write the texts and have infiltrated supposedly neutral institutions.

1.6 What Is ILSI Doing with EFSA? Examples of the Influence

The European Food Safety Authority (EFSA)'s deep ties to ILSI are often the center of critique. Many scientists who have acquired high level positions at EFSA are simultaneously working for the ILSI. Juliane Kleiner, for instance, leader of the Department of Scientific Strategy and Coordination, was a senior scientist for 7 years with ILSI Europe for Food Safety.[28]

---

[27]Hoekstra et al. (2012).

[28]http://www.efsa.europa.eu/en/staffdirectory/staff/julianekleiner.htm Accessed on 5 May 2015.

The most prominent example of personnel concurrence was Hungarian Diána Bánáti—she was temporarily the chairperson of the EFSA supervisory board and at the same time, a board member at ILSI. In May 2012, she was pressured into resignation from EFSA and fully switched to ILSI as their scientific director.[29]

These troubling industry connections have belonged to Project EFSA since its inception.

For years, Matthias Horst, the head lobbyist of the management board for the German food industry and chief executive officer of the National Association of the German Nutrition Industry (BVE) and of the Association for Food Law and Food Science (BLL), sat in on the EFSA council.

In 2014, yet another lobbyist, Jan Mousing, executive director of the Danish Agriculture & Food Council, an organization which, in his own words, "represents the farming and food industry of Denmark," was appointed to the management board.

Furthermore, the EFSA watchdogs are often closely related to industry. For instance, EFSA expert Professor Klaus-Dieter Jany, previously at the Karlsruhe Federal Research Institute for Nutrition (now the Max Rubner Institute), is regarded as an established friend of the genetics industry and had actively supported the GenTech corporation Monsanto. Numerous other EFSA scientists have already stepped in as consultants for food companies. More than half of the 209 EFSA experts have ties to the industry, as reported by the organization Corporate Europe Observatory (CEO).[30]

ILSI's influence on EFSA is traceable in its internal decisions. In the field of genetic engineering, consumers view feeding studies to be important for risk assessment. ILSI, however, is against feeding studies, so the EFSA is as well.

As the CEO report illustrates, some key phrases from ILSI papers can be found copied word-for-word in EFSA policy briefs.[31]

These economical expertise of these industry voices could arguably be an asset to the EFSA in its efforts to protect consumer health. However, as the EU-Act 178/2002 states, central, superior, and independent governmental government oversight is necessary for the serious protection of consumer health, as described below:[32]

> 1. This Regulation provides the basis for the assurance of a high level of protection of human health and consumers' interest in relation to food, taking into account in particular

---

[29]http://www.efsa.europa.eu/en/press/news/120509.htm Accessed on 22 April 2014.

[30]Corporate Europe Observatory (CEO): Unhappy meal. The European Food Safety Authority's independence problem. October 2013 http://corporateeurope.org/sites/default/files/attachments/unhappy_meal_report_23_10_2013.pdf Accessed on 5 May 2014.

[31]Corporate Europe Observatory (CEO): Conflicts on the menu. A decade of industry influence at the European Food Safety Authority (EFSA); February 2012 http://corporateeurope.org/sites/default/files/publications/conflicts_on_the_menu_final_0.pdf.

[32]Regulation (EC) No 178/2002 of the European Parliament and of the Council of 28 January 2002 http://eur-lex.europa.eu/legal-content/EN/TXT/?uri=CELEX:32002R0178 Accessed on 22 April 2014.

the diversity in the supply of food including traditional products, whilst ensuring the effective functioning of the internal market. It establishes common principles and responsibilities, the means to provide a strong science base, efficient organizational arrangements and procedures to underpin decision-making in matters of food and feed safety. The authorities, as Article (34) provides, "take on the role of an independent scientific point of reference in risk assessment and in so doing should assist in ensuring the smooth functioning of the internal market."

Thus the centralized regulatory bodies are obliged to set uniform safety standards on food, which will be used throughout the European community, while working to avoid economic ruin from overregulation providing too many obstacles to free market trade of food and feed.

The protection of health is thus placed in rhetorical opposition to economic stability in the internal food markets. When viewed in this way, it makes sense why the EFSA would choose to not actively proceed against problematic products.

1.7 The Aspartame Case Study: Controversial Artificial Sweetener Politics of EFSA

One example of EFSA's compromised ability to protect consumer health due to its deep industry ties is the case study of the controversial sweetener aspartame. The substance is 200 times sweeter than sugar and is particularly popular among figure-conscious people as well as parents who wish to safeguard their children from tooth decay caused by sugar.

Aspartame is lightly added to soft drinks like Cola, sugar-free gummies, and some dairy products such as milk drinks, yogurt, and quark. Fruit desserts, ice cream, canned goods and jam together with soups, snacks, and deli salads are also likely to contain the artificial sweetness of E951 (the EU codename for aspartame).

New scientific studies suggest, however, that the substance can have some serious side effects. Aspartame, like glutamate, can interfere with the control mechanisms of the brain. It can also cause acute disorders, headaches, dizziness, and even temporary amnesia. This sweetener is especially dangerous during pregnancy, as some scientists think, in some circumstances, that it can harm the brain of the developing child.[33]

In 1997 in the medicine journal "The Lancet," physician and author H.J. Roberts from West Palm Beach, Florida, summarized his research findings over the years thusly: "I believe that our society faces a preventable medical disaster if aspartame products are not promptly removed from public use."[34]

---

[33]Refer to Grimm (2003).

[34]Roberts (1997), p. 362.

Professor John Olney from Washington University in St. Louis in Missouri had also already previously cautioned against possible consequences, like the development of brain tumors.[35]

Digestion of aspartame releases aspartate, which, like its chemical relative glutamate, is a neurotransmitter. At a certain dosage, aspartate can be harmful to brain cells and also appears to be a neurotoxin. Therefore it is considered to be, like glutamate, a risk factor for diseases like Parkinson's, multiple sclerosis, as well as depression and epileptic seizures.

Just like glutamate, aspartate can additionally ease the crossing of aluminium through the blood-brain-barrier, which can increase susceptibility to Alzheimer's disease. Aspartate also impedes the access of glucose in the brain, which is problematic because it is the most important source of energy for brain activity.

Aspartame is especially risky for children, because their blood-brain-barriers are still not fully formed, and thus harmful substances can invade more easily than with adults.

During aspartame's first authorization period in the united States, numerous irregularities existed, according to an official U.S. government statement known as the Bressler Report.[36]

This report explicitly lists many abnormalities and inconsistencies throughout submitted investigations on the manufacturers' side.

In three manufacturer-established studies, the report found incredible flaws, sloppiness, and even falsifications. Additionally, FDA regulations were often ignored.

Here are some particular excerpts from the report:

> In one study, 98 of the 196 animals died but were not autopsied until as much as one year later. Because of the delay, much of the animal tissue could not be used and at least 20 animals had to be excluded from postmortem examinations.
> The original pathology sheets and the pathology sheets submitted to the FDA showed differences for 30 animals.
> One animal was reported alive at week 88, dead from week 92 through week 104, alive at week 108, and finally dead at week 112.
> An outbreak of an infectious disease was not reported to the FDA.
> Tissue from some animals were noted to be unavailable for analysis on the pathology sheets, yet results from an analysis of this "unavailable" tissue was submitted to the FDA.
> There was no documentation of the age or source of the test animals.
> There was no protocol established until one of the studies was well underway.
> Animals were not permanently tagged to prevent mix-ups.
> Some laboratory methods were changed during the study, but not documented.

Given such dubious methods, it should be no surprise that the responsible FDA expert committees refused for years to approve aspartame.

---

[35]Olney et al. (1970), pp. 609–610; Olney (1996), pp. 1115–1123.

[36]http://www.fda.gov/ohrms/dockets/dailys/03/Jan03/012203/02P-0317_emc-000202.txt accessed on 22 April 2014.

When Ronald Reagan became the U.S. president, he fired the former head of the FDA and appointed a man by the name of Arthur Hull Hayes. He determinedly worked towards the substance's approval. An FDA panel comprised of three experts had already delivered their verdict—and voted against aspartame. Hayes subsequently ordered an additional panel, increased the professional lineup to five people, but the approval was again denied three against two. At this point Hayes added another aspartame advocate to the panel, which then resulted in a deadlock where he had the possibility to cast the deciding vote in favor of its approval.

On the FDA websites, there is a detailed chronology of the background of the approval. There, it is noted that:

> On January 21, 1981, the day after Ronald Reagan takes office as U.S. President, G.D. Searle [a large pharmaceutical company] reapplied for the approval of aspartame. G.D. Searle submitted several new studies along with their application. It was believed that Reagan would certainly replace Jere Goyan, the FDA Commissioner. G.D. Searle president Donald Rumsfeld's connections to the Republican party were also thought to play a part in Searle's decision to reapply for aspartame's approval on the day after Ronald Reagan was inaugurated.

In 1981, aspartame would be approved for dry food, and in 1983 for drinks as well. Shortly after Hayes had enforced the approval, he left the FDA, and took on an assignment with the PR firm Burson-Marsteller, a public relations company that was, at the time, working for NutraSweet's parent company Monsanto.

Certain bitterness remained with officials of the supervisory authority.

The FDA toxicologist Adrian Gross spoke before the U.S. Congress that, first and foremost, the behavior of former aspartame manufacturer G.D. Searle during the procedure had absolutely not promoted his confidence in the safety of the product:

> They [G.D. Searle] lied and they didn't submit the real nature of their observations because, had they done so, it is more than likely that a great number of these studies would have been rejected simply for inadequacy. What Searle did, they took great pains to camouflage these shortcomings of the study. They would selectively filter and present to the FDA only what they wished it to know; they also used other questionable methods such as how when animals would develop tumors while they were under study, well, they would remove these tumors from the animals.

The FDA expert deemed the approval of the artificial sweetener via his administration as illegal—and it reminded him of the actual duty of the bureau—the protection of consumers. He said, "And if the FDA violates its own laws who are left to protect the health of the public?"

Furthermore, decades after the approval, there were new scientific reports about health risks.

Danish researchers from the State Serum Institute in Copenhagen have pointed out the increasing danger for preterm births in women who drank light lemonade. The baby might not only arrive early, but also enter the world with an increased risk of disease.[37]

---

[37]Englund-Ögge et al. (2012), pp. 552–559.

The Ramazzini Institute in the Italian city of Bologna has also pointed out the risks for various types of cancer scientifically linked to artificial sweeteners—leukemia, lung, lymph node, and liver cancers. These connections were noted with a partial daily dosage of 20 mg of aspartame per kilogram of body weight—only half of the acceptable limit for artificial sweeteners.[38]

Yet despite this knowledge, the European food authority EFSA sees no reason to take action against the widespread sweetener.

EFSA multiple times evaluated the results from the Ramazzini studies and still came to the conclusion that there was "no evidence" of cancer-causing or mutagenic effects from aspartame.

The EFSA experts did not doubt that lab animals suffered from various types of cancer in the course of their lives, but they linked them to other comorbidities—pneumonia, for example.

There was therefore no reason to reconsider this approval, according to EFSA in 2009. In 2013, the authority confirmed their previous stance once again: "The Panel concluded that aspartame was not of safety concern at the current aspartame exposure estimates or at the ADI of 40 mg/kg bw/day. Therefore, there was no reason to revise the ADI of aspartame."[39]

Critics view EFSA decisions to be strongly influenced by the numerous industrial affiliations of the EFSA experts who had participated in the aspartame assessment at various stages.

Some of the chairmen of the EFSA expert panels—such as the long-standing John Christian Larsen of ILSI—were actively involved in the aspartame decision of 2009 while concurrently serving for that industry lobby organization. Other members of the EFSA panels who had always argued in favor of the harmlessness of aspartame, such as the Dutch professor Ivonne Rietjens, the Frenchman Jean-Charles Leblanc, the Brit Susan Barlow, the chairwoman of an earlier aspartame assessment, the Italian Riccardo Crebelli, and the Swedish Kettil Svensson all also served ILSI.

The French professor Dominique Parent-Massin, who had also sat in on the panel of the 2013 decision, was representative of three active ILSI member corporations, the aspartame corporation Ajinomoto, the key aspartame customer Coca-Cola, and the Danish sweetener-suppliers Danisco.

Professor Karl-Heinz Engel had ties to the aspartame manufacturer Ajinomoto, while he played a part in an EFSA aspartame decision. He also holds the position of chair of general food technology at the Technical University of Munich, where from 2001 to 2004, he employed an Ajinomoto employee as a doctoral student in his laboratory whose research project was financed by Ajinomoto. Professor Engel himself had also already participated in workshops

---

[38]Soffritti et al. (2006), pp. 379–385. Soffritti et al. (2007), pp. 1293–1297.

[39]Scientific Opinion on the re-evaluation of aspartame (E 951) as a food additive EFSA Panel on Food Additives and Nutrient Sources added to Food (ANS) EFSA Journal 2013;11(12):3496. http://www.efsa.europa.eu/de/efsajournal/doc/3496.pdf.

co-financed by Ajinomoto and Monsanto concerning the topic of "Safety of DNA in Food" at ILSI Europe in Brussels in 2000.

The EFSA expert panel were aware of these relationships but judged there to be no conflict of interest, since Engel's ties to Ajinomoto, as well as the work of his doctoral student, had nothing to do with aspartame. "Because no other direct or indirect funding took place through Ajinomoto, the chairmen and the panels have agreed that this cannot be seen as a conflict of interest," according to the EFSA scientific panel on the assessment of the Ramazzini Foundation aspartame study.[40]

1.8 Codex Alimentarius: Democracy on Hold at the World Government of Food

The most important regulation entity in the world of globalized food supply is the Codex Alimentarius, an establishment of the Food and Agriculture Organization (FAO) and the World Health Organization (WHO). The Codex Alimentarius is, so to speak, the world government of food.

The rules that the Codex sets in respect to food are far more important than anything conceived in Vienna, Washington, Berlin, Bern, or Brussels on the subject. The rules of the Codex Alimentarius apply to 185 countries around the globe. The Codex sets standards for all types of food products trafficked all over the globe based on hygiene, additives, food labeling, and much more.

There are 16 subgroups known as "Codex Committees" for each of these categories of standards. The official seat of the Codex Alimentarius is in Rome, and sessions take place all over the world.[41] The Codex Alimentarius is also the most important organization for defining the global rules of food safety.

Thus the participants of the Beijing WHO conference on food safety promoted the international harmonization of these measures in consultation with the standards of the Codex Alimentarius (see footnote 6):

> Develop transparent regulation and other measures based on risk analysis to ensure safety of the food supply from production to consumption, harmonized with the guidance of the Codex Alimentarius Commission and other relevant international standards-setting bodies.

Despite its Roman-sounding Latin name, the Codex Alimentarius has only existed since 1962, and since then, the Codex committees have devoted themselves to setting universal standards for food, from principles on genetically-engineered food and organic goods to others on the quality of products such as fruit juices, margarine, soup, poultry, cornflakes, sugar, chocolate, and cheese, among several others. The Codex members issue hygiene guidelines, determine the limits for toxins in vegetables and for drug residues in meat, regulate the radioactive exposure

---

[40]Refer to: Grimm (2013).

[41]Refer to: Grimm (2014).

of seasonings, and investigate health risks that could spring from food, such as allergies. They also decide what goes on the label.

Previously, Codex decisions were non-binding. Yet since the liberalization of trade, they are actually in effect all around the globe. During this process the Codex Alimentarius Commission evolved into the informal world government in respect to food. The Codex Alimentarius is structured as an establishment of member states and only the official delegates of the nations are eligible to vote. However, representatives of globally active corporations virtually dominate this sphere.

During this process, the present industry representatives have played an essential role over the past few decades.

The statistics of an English consumer organization reveal the following: from 1989 to 1991, 2578 delegates participated in the expert panel sessions, from which merely 26 were dispatched from environmental or consumer organizations. Altogether, 105 countries were represented compared to 108 transnational companies. In the panel "Food Additives and Contaminants," as it was formerly called, the industry dispatched, for example, almost twice as many delegates as all of the EU governments put together (see footnote 41).

This was also the case in the time that followed. For instance, at a panel session from April 21–25, 2008 in Beijing, 262 delegates are listed, the majority of which were company and industrial association representatives.

In Germany's delegation, for example, there was a representative from Europe's biggest sugar corporation, Südzucker.

In the seven-person Swiss delegation, five industry representatives sat equally next to only two government officials. The industry representatives came from the food multinational Nestlé, the additive corporation Ajinomoto, the vitamin global leader DSM, the flavor corporation Givaudan, and a consulting firm for which various clients work.[42]

The industry has, of course, a huge vested interest in setting global policies preferable to its business capability. For example, for decades, rules and regulations for flavor have been determined with significant collaboration of the concerned lobby, which includes the flavor industry association (the International Organization of the Flavor Industry, abbr. IOFI), the individual flavor manufacturers, and also the general food industry lobby groups.

At the Codex panel session from November 5–11, 1985 in the Dutch capital The Hague, for example, national delegations were stuffed with industry representatives compared to those actually representing the nation's government.[43] In Germany's delegation, the government was hopelessly in the minority, supplied only three of

---

[42] Joint FAO/WHO Food Standards Programme Codex Alimentarius Commission: Report of the Fortieth Session Of The Codex Committee on Food Additives. Beijing, China 21-25 April 2008.

[43] Joint FAO/WHO Food Standards Programme Codex Alimentarius Commission: Report of the Fortieth Session Of The Codex Committee on Food Additives. The Hague, 5-11 November 1985.

twelve national delegates. The German flavor industry was represented by one representative, with another two from its lobby organization. There was a representative from the German food industry lobby organization (the German Federation of Food Law and Food Science, BLL), as well as the usual Südzucker; moreover, Coca-Cola was allowed to represent Germany, together with the chemical corporation BASF. The Unilever global corporation was also represented in the German delegation, with two representatives who also sat in at the table of the Dutch delegation.

In the Swiss delegation, moreover, the governmental representatives from Bern were similarly the minority compared to the seven industry representatives. Two of these were from flavor corporation Givaudan, two more from the pharmaceutical giant Roche, and one from the flavor company Firmenich. In the U.S. delegation, the flavor corporation IFF was present, along with Coca-Cola. Beside them was the aspartame manufacturer Searle, General Foods, and Pfizer. Governmental representatives were also a minority in that delegation.

The food industry was even represented in the three-person "technical administration" of the Codex committee by way of the British-Dutch food multinational Unilever. In Santiago in 2010, the subcommittee responsible for health food and diets ("Codex Committee on Nutrition and Foods for Special Dietary Uses", CCNFSDU) held a meeting, which dealt with health food products, vitamins, ready-made yogurts, and baby food.[44]

Nestlé was independently present at that session in Santiago and also represented in several other delegations, frequenting the Swiss, but also the Chilean and Mexican delegations with two representatives in each case.

Coca-Cola traveled with the American delegation, per usual, but also was present at the Chilean delegation, as well, as well as the German delegation, where a Südzucker manager was also present, just like in Beijing two years prior.

The pharmaceutical company Abbott was a member of the U.S. delegation, in which three representatives were from Mexico and from Chile.

Of course, Kellogg was also present (Mexico), as well as the Japanese additive corporation Ajinomoto (Switzerland), the Dutch vitamin world leader DSM (Chile/Switzerland), along with Herbalife (Chile), Dannon (France, the Netherlands), and Kraft Foods (Brazil).

In this system dominated by industry representatives, consumer interests play no noteworthy role in the decisions. At the Codex Committee panel session on food additives in Beijing 2008, there were only three consumer representatives among 262 delegates, for instance (see footnote 42).

---

[44]Codex Alimentarius Commission: Report Of The Thirty Second Session of The Codex Committee on Nutrition And Foods For Special Dietary Uses. Santiago, Chile, 1-5 November 2010.

1.9 The Predicament of the Consumer: Dwindling Influence, Increasing Personal Responsibility

The consumer, the citizen, who in democratic countries is supposedly sovereign, the subject of legal protection, is subject to double the pressure in the age of globalized food safety.

On the one hand, the consumer's influence on food is dwindling, and so is the possibility of their participation in establishing laws with global significance like the Codex Alimentarius. The consumers are underrepresented, and as a sovereign entity, they are also largely disempowered compared to the broad and impermeable scope of influence of the transnational corporations in the decision-making process.

Considering the complexity of modern processed food, and the broad, impenetrable supply chain, the law will continue to be limited, according to Frankfurt law Professor Wolf Paul's theory of "culinary self-determination." Paul stated that "the freedom of flavor and food is governed by one's own wishes and judgement." However, the "reign over food and drink" is now out of the hands of the individual consumer, and one's autonomy and self-determination is reduced to an unprecedentedly small size during times of industrial food production.[45]

On the other hand however, the consumers are carrying the full weight of responsibility for their hygienic well-being and health to a higher and higher extent. While countries have assumed a certain kind of monitoring and concern for classic illness pathogens, like viruses and bacteria, they reject to take on such a role where causes for "noncommunicable diseases" are concerned. It is solely up to the consumer to make that "healthy choice." Determining which food options are "safe" and choosing those over the widely-available alternatives is treated completely as the consumer's responsibility.

This means that if a consumer gets sick (from a non-communicable disease), or develops obesity, or dies prematurely due to their legally approved diet, it is the consumer's own fault. The manufacturer or authorities bear no responsibility. As actors in the vast realm of the consequences of these food products, they have withdrawn themselves.

Thus an entirely new situation prevails in the age of industrial food production: food products can now be legally considered entirely safe—and nevertheless, health is being unprecedentedly jeopardized.

Take soft drinks, for example: they are deemed to be completely safe, according to the current provisions on food safety, yet they are likely responsible for 180,000 deaths each year, according to calculations by the Harvard School of Public Health.[46] However, consumers themselves are the only ones legally responsible for their own protection against these health hazards.

---

[45]Paul W.: Die kulinarische Selbstbestimmung: eine menschenrechtliche Apologie des Feinschmeckers. In: Zur Autonomie des Individuums: liber amicorum Spiros Simitis; Simon D [Hrsg.] 2000.

[46]http://www.hsph.harvard.edu/news/hsph-in-the-news/roughly-180000-deaths-worldwide-linked-to-sugary-drink-consumption/.

Illustrating this point, the responsible state authority in Germany, the Federal Institute for Risk Assessment (BfR), officially believes the view that food today is "safer and obviously qualitatively better" than ever before.[47] If people have regarded "food as a potential threat against their health," then they are "profoundly wrong."

Instead, claims BfR department head Gaby-Fleur Böl, "the quality and safety of food is steadily increasing." She elaborates: "Although, from a natural scientific point of view, food is significantly safer and obviously qualitatively better nowadays in comparison to former times. This fact is diametrically opposed to the decreased trust in these goods."

The definition of "safe," from the regulator's perspective, solely means that food products "are free from disease-causing pathogens, bacteria, or viruses" as well as "chemical pollutants." This is what manufacturers are responsible for.

From governmental viewpoint, sugar or additives do not threaten safety, even if they are in excess very unhealthy. However, the burden of maintaining this safety is not the responsibility of the manufacturers who utilize these, but rather one of the consumers who consume these, according to risk manager Böl. "With each food product, I have to decide how much and how often I myself will consume this. Sugar can be harmful. We don't need to talk about that. The dosage determines toxicity. And through their own food habits, people can control how much they will gain from a single food product."[48]

The respective EU statutory regulation defines a food 'hazard' as[49]: "a biological, chemical or physical agent in, or condition of, food or feed with the potential to cause an adverse health effect."

In Article 14, "Food safety requirements" are listed.

Here, each common form risk is mentioned:

> 4. In determining whether any food is injurious to health, regard shall be had:
> (a) Not only to the probable immediate short-term and/or long-term effects of that food on the health of a person consuming it, but also on subsequent generations;
> (b) To the probable cumulative toxic effects.

Yet for the official safety assessment, the first considerations are clearly for the traditional criteria, i.e. those which can lead to short-term health hazards:

> 5. In determining whether any food is unfit for human consumption, regard shall be had to whether the food is unacceptable for human consumption according to its intended use, for reasons of contamination, whether by extraneous matter or otherwise, or through putrefaction, deterioration or decay.

---

[47]Böl, GFleur Lebensmittel als Sicherheitsrisiko? Von gefühlten und tatsächlichen Risiken. Vierteljahrshefte zur Wirtschaftsforschung: Vol. 81, (2012). Ernährungssicherung und Lebensmittelqualität: Herausforderungen für Agrarmärkte, pp. 183-198. http://ejournals.duncker-humblot.de/doi/abs/10.3790/vjh.81.4.183.

[48]Personal communication via E-mail on 25 April 2013.

[49]Regulation (EC) No 178/2002 Of The European Parliament And The Council of 28 January 2002 http://eurlex.europa.eu/LexUriServ/LexUriServ.do?uri=OJ:L:2002:031:0001:0024:EN:PDF Accessed on 6 May 2014.

This understanding of food safety corresponds to the reasoning about the risks of the past, and does not account for the new health hazards that originate from food products. This then implies: the concept of food safety itself is not a top priority. The actual legal concept of food safety does not account for new threats against life and health via these "noncommunicable diseases."

This is notable in Article 19 on the subject of "Responsibilities for food: food business operators," It dictates:

> 1. If a food business operator considers or has reason to believe that a food which it has imported, produced, processed, manufactured or distributed is not in compliance with the food safety requirements, it shall immediately initiate procedures to withdraw the food in question from the market where the food has left the immediate control of that initial food business operator and inform the competent authorities thereof. Where the product may have reached the consumer, the operator shall effectively and accurately inform the consumers of the reason for its withdrawal, and if necessary, recall from consumers products already supplied to them when other measures are not sufficient to achieve a high level of health protection.

With products of the "Western diet"—fast food, soft drinks, and candy—it is difficult to believe that after years of unobjectionable sale, manufacturers will suddenly nurture concerns about the safety of their products. Especially since officially, the components are considered to be harmless.

Take sugar, for example: since 1958, the GRAS status ("Generally Recognized as Safe") has been in place in the U.S., and in 1983, the FDA conferred both the GRAS predicate as well on high fructose corn syrup (HFCS). Despite increasing consumption quantities and newer insights about health risks, and despite the contribution of sweet substances to the obesity epidemic and to the so-called metabolic syndrome, the FDA still affirmed their safety assessment of these substances both in 1996 and in 2004.[50]

1.10 The Environment Must Change: Food Law and Food Safety in the Twenty-First Century

Critics are demanding that food elements with high hazard potential be more closely regulated, calling for a revision of the GRAS status.

Another suggestion would be regulations facilitating "healthy choices" on the part of the consumer with financial policy funds through taxes on unhealthy products like sugar.

In the New England Journal of Medicine, Yale Professor Kenny D. Brownell and Thomas Frieden, the director of the Center for Disease Control and Prevention (CDC), the highest U.S. health surveillance agency, proposed a tax on soft drinks.[51]

---

[50]Lustig (2014), p. 242.
[51]Brownell et al. (2009), pp. 1599–1605. Brownell and Frieden (2009), pp. 1805–1808.

They offer "approaches to designing a tax system that could promote good nutrition and help the nation recover health care costs associated with the consumption of sugar-sweetened beverages."

A penny per ounce or 11 cents for every 0.3L tax can reduce consumption by 13 % if it follows the cigarette tax model, and in 10 years, would save 50 billion dollars in healthcare costs. The state currently earns about 150 billion dollars per year.

"Health related food taxes could improve health", says Mike Rayner from the Department for Public Health at the elite University of Oxford. Together with his colleague Oliver Mytton, he had postulated in a sensational article, among other things, a sugar tax. The Oxford team argued that governmental interventions like taxation can be justified if the free market fails at its duty to promote good health and well-being.

> "Existing evidence suggests that taxes are likely to shift consumption in the desired direction." The tax would have to certainly be palpable, "at least 20% to have a significant effect on population health."[52]

Taxes like these have already been established in Denmark, Finland, France, and Hungary. Studies have also showed, for instance, that for every 10 % additional tax on soft drinks, it is estimated that there will be about an eleven percent decrease in consumption.

Moreover, the UNO special correspondent for diet and nutrition, Olivier de Schutter, wants to do away with the present-day form of subsidies for the agrifood industry because they were what had initially caused unhealthy foods and sweet soft drinks to become significantly cheaper than fresh products; and due to lack of money, those of a poorer background were then forced to eat unhealthy foods.

> The Special Rapporteur concludes that current food systems are deeply dysfunctional. The world is paying an exorbitant price for the failure to consider health impacts in designing food systems, and a change of course must be taken as a matter of urgency.

Tax payers were similarly asked to pay up in multiple ways: "Taxpayers pay for misguided subsidies that encourage the agrifood industry to sell heavily processed foods at the expense of making fruits and vegetables available at lower prices; they pay for the marketing efforts of the same industry to sell unhealthy foods, which are deducted from taxable profits; and they pay for health-care systems for which noncommunicable diseases today represent an unsustainable burden."

He therefore postulated a rerouting of these finance flows:

> Review the existing systems of agricultural subsidies, in order to take into account the public health impacts of current allocations, and use public procurement schemes for school-feeding programmes and for other public institutions to support the provision of locally sourced, nutritious foods, with particular attention to poor consumers.

---

[52]Mytton et al. (2012), p. e2931.

Furthermore, advertisements should be legally regulated, he stated in his report at the UN human rights committee. This is the most effective way to prohibit marketing strategies of unhealthy foods that are addressed towards children.[53]

"The environment must change," says Professor Kelly D. Brownell, director of the Center for Food Policy and Obesity at Yale University in the U.S. state of Connecticut.[54]

Brownell coined the concept of "toxic food environments". For him, there is a causal relationship that exists between the environment and rampant obesity. The only solution lies in the change of these environments, Brownell claims[55]: "As a society, we have two options," he says, "we can wait several thousands of years so that we will evolve to adapt to our toxic environment. Or, we can change our own environment so that that it no longer makes us sick."

References

Brownell KD, Frieden TR (2009) Ounces of prevention — the public policy case for taxes on sugared beverages. N Engl J Med 360:1805–1808. doi:10.1056/NEJMp0902392

Brownell KD, Horgen KB (2003) Food fight: the inside story of the food industry, America's obesity crisis, and what we can do about it. McGraw-Hill, Columbus, 352 p. hardcover, $24.95

Brownell KD et al (2009) The public health and economic benefits of taxing sugar-sweetened beverages. N Engl J Med 361(16):1599–1605, http://cspinet.org/new/pdf/soda_-_nejm2_-_9-09.pdf

Englund-Ögge L et al (2012) Association between intake of artificially sweetened and sugar-sweetened beverages and preterm delivery: a large prospective cohort study. Am J Clin Nutr 96 (3):552–559. doi:10.3945/ajcn.111.031567, Epub 2012 Aug 1

Grimm HU (2003) Die Ernährungslüge. Wie uns die Lebensmittelindustrie um den Verstand bringt. München 2011

Grimm HU (2010) Tödliche Hamburger. Hirzel, Stuttgart

Grimm HU (2012) Vom Verzehr wird abgeraten. Wie uns die Industrie mit Gesundheitsnahrung krank macht, München, Droemer

Grimm HU (2013) Chemie im Essen. Lebensmittel-Zusatzstoffe: Wie sie wirken, warum sie schaden. Droemer, München

Grimm HU (2014) Die Suppe lügt. Die schöne neue Welt des Essens. Komplett überarb., aktualisierte, erhebl. erw. Neuausg. Droemer, München

Hoekstra J et al (2012) BRAFO tiered approach for benefit-risk assessment of foods. Food Chem Toxicol 50(Suppl 4):S684–S698. doi:10.1016/j.fct.2010.05.049, Epub 2010 May 28

Lustig R (2014) Fat chance: the hidden truth about sugar, obesity and disease. Harper Collins, London, p 242

Mytton OT et al (2012) Taxing unhealthy food and drinks to improve health. BMJ 344:e2931. http://www.bmj.com/content/344/bmj.e2931?ijkey=anbZh0Tu5xuqTVc&keytype=ref

---

[53]United Nations General Assembly Human Rights Council. Nineteenth session Report submitted by the Special Rapporteur on the right to food, Olivier De Schutter A/HRC/19/59 http://www.ohchr.org/Documents/HRBodies/HRCouncil/RegularSession/Session19/A-HRC-19-59_en.pdf Abgerufen am 22. April 2014.

[54]Refer to: Brownell and Horgen (2003), 352 p.

[55]Der Spiegel 36/2012.

Olney JW (1996) Increasing brain tumor rates: is there a link to aspartame? J Neuropathol Exp Neurol 55(11):1115–1123

Olney JW et al (1970) Brain damage in infant mice following oral intake of glutamate, aspartate or cysteine. Nature 227:609–610

Popkin BM (1993) Nutritional patterns and transitions. Popul Dev Rev 19(1):138–157

Roberts HJ (1997) Aspartame and brain cancer. Lancet 349(9048):362

Soffritti M et al (2006) First experimental demonstration of the multipotential carcinogenic effects of aspartame administered in the feed to Sprague-Dawley rats. Environ Health Perspect 114 (3):379–385. doi:10.1289/ehp.8711.PMC1392232

Soffritti M et al (2007) Life-span exposure to low doses of aspartame beginning during prenatal life increases cancer effects in rats. Environ Health Perspect 115(9):1293–1297. doi:10.1289/ehp.10271

Wilson B (2008) Swindled: from poison sweets to counterfeit coffee - the dark history of the food cheats. John Murray (UK) and Princeton University Press (US), p xiii

World Health Organization (2010) Global status report on noncommunicable diseases. World Health Organization, Geneva, http://www.who.int/nmh/publications/ncd_report_full_en.pdf

Chapter 2
Agriculture, Ethics, and Law

Bernard Rollin

Abstract The development of human civilization was directly dependent on the creation of a secure and predictable food supply by way of plant and animal agriculture. Both forms of agriculture required sustainability. In animal agriculture, this was assured by *good husbandry*, i.e. respect for animals 'needs and natures'. The development of civilization and technology paradoxically led to the undoing of this ancient contract with animals, and of sustainability in plant agriculture. Industry and the search for profit and productivity supplanted husbandry and stability. Industrial agriculture for animals supplanted agriculture as a way of life. The loss of husbandry was pervasive throughout animal agriculture, and turned good husbandry and animal welfare into major moral issues, rather than a presupposition of raising animals. This in turn led to a societal demand for a new ethic for treatment of animals in agriculture. As Plato pointed out, new ethics evolve out of established ethics, so society turned to its ethic for humans, *mutatis mutandis*, as a basis for a new animal ethic and as a way of assuring respect for animal nature or *telos*. The chartering of the Pew Commission on Industrial Farm Animal Production created a vehicle for exposing the general public to the numerous problems growing out of industrial animal agriculture. These problems fell into five interconnected categories: antimicrobial resistance; environmental despoliation; rural sociology; human and animal health; and animal welfare. Though the Commission offered numerous recommendations for remedying these problems, none have as yet been legislatively mandated.

2.1 The Rise of Traditional Agriculture

The development of human civilization was directly dependent on the creation of a secure and predictable food supply. Such a food supply freed peoples from the uncertainties and vagaries of depending on hunting and gathering, and enabled the

---

B. Rollin (✉)
Colorado State University, Fort Collins, CO, USA
e-mail: Bernard.Rollin@colostate.edu

establishment of communities. Predictability regarding food was assured by the development of both plant and animal agriculture, which operated synergistically. Cultivation of crops and plants secured human ability to depend on (barring catastrophes of weather) foods of plant origin, and on a steady and local source of animal feed. Animal agriculture in turn provided a source of labor for crop production, as well as a predictable reservoir of animal protein for human consumption. The secure food supply ramified in the ability to develop manufacturing, trade, commerce and, in Hobbes's felicitous phrase, the "leisure that is the mother of philosophy," construed in the broadest sense as speculative thought, science, technological innovation, art and culture.

Presuppositional to the development of both agricultures was the concept of sustainability, i.e. assurance that the conditions and resources necessary to them were indefinitely renewable. As children, we might have learned about balanced aquariums. If we wished to keep a fish tank where the fish lived and we did not want to keep tinkering with it, we needed to assure that the system in question was as close to a "perpetual motion" machine as possible, a system that required little maintenance because all parts worked together. That meant including plants that produced oxygen and consumed carbon dioxide, enough light to nourish the plants, or rather plants that thrived in the available light source, water that was properly constituted chemically, scavengers to remove wastes. When such a system worked, it required minimal maintenance. If something were out of balance, plants and animals would die and require constant replacement. The fish tank is aimed at being a balanced ecosystem. Thus, it represents a model of traditional approaches to cultivation of land, wherein one sought to grow plants that could be grown with indefinitely available resources. This model conserved and maximized these resources, and which would not die out or require constant enrichment. Hence, the beauty of pastoral agriculture, where pasture nourished herbivores and herbivores provided us with milk, meat, and leather and their manure enriched the pastural land, in a renewable cycle.

Cultivation of land evolved locally with humans. If one did not attend to the constraints imposed by nature on what and how much can grow in a given region, the region would soon cease to yield its bounty, by virtue of salinization, depletion of nutrients, overgrazing, or insect infestation. Thus, over time, humans evolved to, as one book put it, "farm with nature," which became, like animal husbandry, both a rational necessity and an ethical imperative. Local knowledge, accumulated over a long period of trial and error, depicts how much irrigation was too much, what would not grow in given soils, what weeds left standing protected against insects, and where shade and windbreaks were necessary. Thus, accumulated wisdom was passed on—and augmented—from generation to generation, and was sustainable, i.e. required minimal tweaking or addition of resources. The genius of agriculture was to utilize what was there in a way that would endure. If the land did not thrive, you did not thrive. Traditional agriculture was inherently sustainable; through trial and error, over a long period, it evolved into a "balanced aquarium."

Not surprisingly, precisely isomorphic logic applied to sustainability in animal production. The maxim underlying continued success in rearing animals was *good husbandry*, which represented a unified synthesis of prudence and ethics. Husbandry meant, first of all, placing the animals into the optimal environment where they could maximally fulfill their physical and psychological needs and natures. The husbandman then augmented animals' ability to survive and thrive by watching over them, providing protection against predators, food during periods of famine, water during times of drought, shelter during extremes of climate, assistance in birthing, medical attention, and generally ministering to whatever needs the animals had. So powerfully ingrained was this imperative in the human psyche, that when the Psalmist searches for a metaphor for God's ideal relationship to human beings, he can do no better than seizing upon the conceit of the Good Shepherd.

We want no more from God than what the Good Shepherd provides for his flock.[1] As we know from other passages[2] in the Old Testament, a lamb on its own would live a miserable, nasty, and short life by virtue of the proliferation of predators—hyenas, raptors, wolves, bears, lions, foxes, jackals and numerous others. With the care and ministrations of the Shepherd, the animal lives well until such time as humans take its life, in the meantime supplying us with milk, wool, and, in the case of some domestic animals, the labor that became indispensable to the working of the land for crops.

The history of Western civilization cannot overestimate the power of this symbiotic image. In Christian iconography, for thousands of years, Jesus is depicted both as Shepherd and as the lamb, a duality built into the very foundations of human culture. The pastor, a word harking back *to pastoral*, tends to his flock; the members of his congregation are his sheep. Moreover, when Plato discusses the ideal political ruler in the *Republic*, he deploys the shepherd—sheep metaphor: The ruler is to his people as a shepherd is to his flock. Qua shepherd, he exists to protect, preserve and improve the sheep; any payment tendered to him is in his capacity as wage earner. So too the ruler; this again illustrates the power of the concept of husbandry on the human psyche.

Animal agriculture was indispensable to the subsequent development of society and culture. Husbandry agriculture is the *ancient contract* that was presuppositional to that entire evolutionary process. In one of the most momentous ironies in the history of civilization, this ancient contract with the animals, as well with the Earth, in terms of sustainability, contained within it the seeds of its own undoing. It was in virtue of a secure and predictable food supply that humans could proceed with the trade, manufacturing, invention and the general flourishing of culture.

By the late nineteenth century, industrial proliferation and innovation had reached a point where sustainability and good husbandry seemed to be no longer essential presuppositions of civilization. The ancient contract, which we may characterize as animal husbandry, and stewardship with regard to the land, was the presuppositional bedrock upon which economics, art, and culture rests. With the

---

[1] Rollin (2002).

[2] Cooper (n.d.).

profound *hubris* of Icarus, who challenged inherent human limitations, with blind and abiding faith in the humanly crafted tools, which repeatedly show themselves as impotent in the face of natural disaster, we thumbed our noses at both morality and prudence. As the ancients crafted the tower of Babel, we began to overreach the constraints imposed on us by the natural world. In both crop and animal agriculture, the ancient values of sustainability, stewardship, and husbandry inexorably gave way to modernist values of industrialization, productivity, and efficiency. The symbiotic partnership between humans and the Earth, and between humans and animals, rapidly transmutes into patent exploitation with no respect or attention to what priceless elements are lost.

Science and technology enabled extraction of greater crop yields than hitherto imagined. Instead of farming depending on idiopathic wisdom, local knowledge, know-how—passed from generation to generation—it is now conceived through technology, as applied science, as nomothetic,—law like—and in principle applicable to any locale. As in animal agriculture, capital and machinery supplanted knowledgeable labor. Farms got bigger and bigger; thus, food was plentiful and consequently, cheap.

With these gains came major costs, albeit costs that were not immediately obvious, and long term. If you forget about having a balanced aquarium, then you must pump resources in regularly to compensate for the loss of balance. *Sustainability* is abandoned. The new form of agriculture requires a great deal of fuel in order to run the machinery and make the chemicals. This new form requires massive amounts of water, as well. "Only 75 % of modern agriculture's consumption of water is replenished" (Dawkins).[3] These chemicals, fertilizers, pesticides, herbicides, and fuels left residue; they pollute both the air and water, which leads to the disturbance and death of fragile eco-systemic health. Growing crops requires land. As a result, forests had to be cleared, which uprooted and impoverished endogenous ecosystems, annihilated species, and lost wild plants.

Powerful and plentiful agricultural chemicals had negative health effect on workers and citizens. Indiscriminate use of pesticides predictably and inexorably led to inadvertent selective breeding of super-pests, highly resistant to these chemicals—even as the massive use of antibiotics in confinement animal agriculture both to promote growth and to mask the effects of bad husbandry led to the evolution of antibiotic-resistant pathogens. In addition, pesticides non-selectively killed off both desired pests, and their natural enemies.

As farmer debt load increased, increasing numbers of small farmers lost their farms, unable to afford the infrastructure required. As a result, they went out of business, and so died the little communities they inhabited, their culture, and their way of life.

---

[3]Dawkins and Bonney (2008).

2.2 Traditional Agriculture and Ethics

Was there a formalized ethic dictating how agriculturalists comported themselves in traditional agriculture? Self-interest is the most powerful motivator and sanction. Failure to practice good husbandry towards animals, and good stewardship towards the land would necessarily erode productivity, and thus harm producer self-interest. Hurting animals through any medium would diminish productivity in terms of weight gain, reproductive success, and ability to work, strength, endurance, coat quality, and every other expectation directed towards the animals. If self-interest did not move certain people, it is doubtful that enunciation of any ethic or laws mirroring that ethic would have any more success.

The only ethic in addition to the pragmatic teaching of husbandry in traditional agriculture was a prohibition against deliberate *cruelty*, aimed at sadists and psychopaths who were heedless of considerations of self-interest. Buttressing husbandry, the ancient biblical ethic even allowed for a violation of the Sabbath to help animals in distress.

Though no special additional ethic regarding animal treatment was historically necessitated, husbandry was widely taught in schools and by parents, both in actual lessons, and in behavior modeled for children by their parents and other elders. During this time, both "animal smarts" and "wisdom of the soil" were passed on through generations, in a ceremonious manner that we may call "the agricultural birthright." There was no shame, but rather pride in being a third or fourth or fifth or tenth generation agriculturalist, raising animals and/or crops on the same land multiple generations of your family had stewarded.

2.3 The End of Husbandry, Stewardship, and Sustainability

Industrialization of animal agriculture marked a genuine and major revolution in Western culture that has been insufficiently attended to by historians. First and foremost, the industrialization of animal agriculture was a complete transformation in values. Traditional agriculture has solid roots in husbandry of animals, good stewardship of the land, and was a sustainable way of life, more than it was a way of making a living. As industrial agriculture developed, the supreme values became efficiency and productivity. Traditional agriculture enabled people to live a life solidly enmeshed in, and extremely conscious of, one's dependence on nature. Industrialized agriculture is solidified in a hubristic rebellion against natural constraints. No longer was the agriculturalist forced to work with physical and animal nature. Instead of carefully and methodically putting square pegs in square holes, round pegs in round holes, while generating as little friction as possible, the industrial agriculturalist took it upon himself to force square pegs into round holes, round pegs into square holes by utilizing "technological sanders"[4] to grind the pegs in order to force fit them into contexts in which they did not naturally fit. Such sanders are epitomized by the liberal use of antibiotics that allow us to force animals into environments where they would sicken and die without the "technological fix" afforded by the drugs. Instead of seeking the healthiest possible environment for the animals, we worked our magic on highly pathogenic environments, and highly pathogenic conditions, such as crowding. The animals experienced major insults to their natural needs and natures and suffered in the sacrifice of good health and welfare, while at the same time the pathogens occupying their environmental niche mutated in the face of the selective pressures inherent in excessive antimicrobial use, generating new and dangerous drug-resistant pathogens wreaking havoc with both animal and human health.

Small, independent family farmers, whom Thomas Jefferson acknowledged as serving as the backbone of democracy by keeping control of the food supply away from governments and plutocrats, became slaves to large corporate entities, which were the only structures rich enough to capitalize agriculture according to the emerging mantra of "get big or get out." Capital in the form of machinery, huge buildings, and vast numbers of animals replaced "animal-smart" labor, and labor became unskilled manual labor, rather than applied husbandry knowledge. The number of small farms decreased exponentially. To take one depressing example, between 1980 and 1995, the U.S. lost 80 % of the small hog farmers who dominated the industry, only to have them replaced by five gargantuan mega-corporations who produced 95 % of U.S. pork. Citizens who complained about any of the issues mentioned were often shadowed and intimidated by company "security" police. Correlatively, thriving, rural communities based in small hog operations became ghost towns, as small operations were forced to sell to large corporate entities at pennies on the dollar.

As a result of the industrialization of animal agriculture, farm life went from being the stereotypical "healthy outdoor way of life" to a pathogenic one. The proliferation of animal wastes caused noxious air and water pollution, danger due to farm machinery accidents, and infections resulting from animal crowding and overuse of antibiotics. Animal wastes went from fertilizing pasture in a sustainable cycle in extensive agriculture to being a major source of water, ground, and air pollution in industrial agriculture.

The question naturally arises as to why, after thousands of years of successful animal agriculture based in husbandry, it was so quickly supplanted by industrial agriculture and its attendant problems. In fact, industrialization of animal agriculture in the U.S. occurred for a variety of understandable and even prima facie laudable reasons that are worth recounting.

1. When industrial agriculture began, in the early twentieth century, the U.S. was confronted with a variety of new challenges related to food. In the first place, the great economic Depression and Dust Bowl (severe drought) had soured many

---

[4]"Veterinary Care for Laboratory Animals – Animal Research in a ..." <http://www.ncbi.nlm.nih.gov/books/NBK91520/>.

people on farming, and even more dramatically, had raised the specter of starvation for the American public for the first time in U.S. history. Vivid images of bread-lines and soup kitchens drove the desire to assure a plentitude of cheap food. By the late 1960s and 1970s, the U.S. had large-scale industrialized animal agriculture with much bigger units compared to Europe.
2. Better jobs were to be found in cities, and rural people flocked to them in hopes of a better life, creating a potential shortage in agricultural labor.
3. Correlative with the growth of cities and suburbs came encroachment on agricultural land for various forms of development, raising land prices and moving acreage once available for agriculture out of that pool.
4. Many people who would otherwise have been happy with a slow, rural way of life were exposed to greater sophistication by virtue of military service in World Wars I and II, and thus were dissatisfied with an agrarian existence.
5. Demographers predicted a precipitous and dramatic increase in population, which turned out to be accurate.
6. With the success of industrialization in new areas, notably Henry Ford's application of the concept to the automobile, it was probably inevitable that the concepts of industrialization would be applied to agriculture. (Ford himself had already characterized slaughterhouses as "disassembly lines".)[5]

Thus was born an industrial approach to agriculture, with machines taking the place of labor. The traditional Departments of Animal Husbandry in agricultural schools symbolically marked this transition by changing their names to Departments of Animal Sciences, a field defined in textbooks as "application of industrial methods to the production of animals."

In this transition, as we have just seen, the traditional bedrock values of agriculture, husbandry, sustainability, agriculture as a way of life, not only a way of making a living, were transmuted into values of efficiency and productivity. With human labor replaced by machinery, in turn requiring large amounts of capital, farm units grew larger, eventuating in the mantra of the 1970s "get big or get out." Agricultural research stressed producing cheap and plentiful food, and moved in unprecedented directions. With animals confined for efficiency and away from forage, much research was directed towards finding cheap sources of nutrition, in turn leading to feeding such deviant items to animals as poultry and cattle manure, cement dust, newspaper, and most egregiously, bone meal to herbivores; the latter of which created BSE or "Mad Cow Disease."[6] Animals were kept under conditions alien to their natural needs for the sake of productivity.[7]

Industrial agriculture replaced good husbandry, which we have characterized as putting square pegs into square holes, round pegs into round holes, while producing

---

[5]"BONNIE BASH : 2-Day Festival in Costa Mesa Will Showcase ..." <http://articles.latimes.com/1991-05-23/news/ol-3377_1_costa-mesa>.

[6]"A Strategic Plan for Poverty Reduction, Food Security ..." <http://www.nast.ph/index.php?option=com_docman&task=doc_download&gid=142&Itemid=7>.

[7]"Improving Animal Welfare," written by Temple Gandhin, page 26 of ebook. Direct Quoting.

as little friction as possible, with technological "sanders" such as antibiotics, hormones, extreme genetic selection, air handling systems, artificial cooling systems, and artificial insemination designed to force animals into unnatural conditions while they nonetheless remained productive.

Consider, for example, the egg industry, one of the first areas of agriculture to experience industrialization. Traditionally, chickens ran free in barnyards, able to live off the land by foraging and express their natural behaviors of moving freely, nest-building, dust-bathing, escaping from more aggressive animals, defecating away from their nests, and, in general, fulfilling their natures as chickens. Industrialization of the egg industry, on the other hand, meant placing the chickens in small cages, in some systems with six birds in a tiny wire cage, so that one animal may stand on top of the others and none can perform any of their inherent behaviors, unable even to stretch their wings. In the absence of space to establish a dominance hierarchy or pecking order, they cannibalize each other, and must be "debeaked", producing painful neuromas since the beak is innervated. The animal is now an inexpensive cog in a machine, part of a factory, and the cheapest part at that, and thus totally expendable. If a nineteenth century farmer had attempted such a system, he would have gone bankrupt, with the animals dead of disease in a few weeks. Some genetic lines of pigs and chickens are so highly selected for egg and meat production that they have less disease resistance.

The steady state, enduring balance of humans, animals, and land is lost. Putting chickens in cages and cages in an environmentally controlled building requires large amounts of capital and energy and technological "fixes"; for example, to run the exhaust fans to prevent lethal build-up of ammonia. The value of each chicken is negligible, so one needs more chickens; chickens are cheap, cages are expensive, so one crowds as many chickens into cages as is physically possible. The vast concentration of chickens requires huge amounts of antibiotics and other drugs to prevent wildfire spread of disease in overcrowded conditions. Breeding of animals is oriented solely towards productivity, and genetic diversity—a safety net allowing response to unforeseen changes—is lost. Bill Muir, a genetics specialist at Purdue University, found that commercial lines of poultry have lost 90 % of their genetic diversity compared to noncommercial poultry. Dr. Muir is extremely concerned about the lack of genetic diversity (Lundeen 2008). Small poultry producers are lost, unable to afford the capital requirements; agriculture as a way of life as well as a way of making a living is lost; small farmers are superseded by large corporate aggregates. Giant corporate entities, vertically integrated, are favored. Manure becomes a problem for disposal, and a pollutant, instead of fertilizer for pastures. Local wisdom and know-how essential to husbandry is lost; what "intelligence" there is hard-wired into "the system". Food safety suffers from the proliferation of drugs and chemicals, and widespread use of antimicrobials to control pathogens in effect serves to breed—select for—antibiotic resistant pathogens as susceptible ones are killed off. Above all, the system is not balanced—not sustainable—constant inputs are needed to keep it running, and to manage the wastes it produces, and create the drugs and chemicals it consumes. And the animals live miserable lives, for productivity has been severed from well-being.

One encounters the same dismal situation for animals in all areas of industrialized animal agriculture. Consider, for example, the dairy industry, once viewed as the paradigm case of bucolic, sustainable animal agriculture, with animals grazing on pasture giving milk, and fertilizing the soil for continued pasture with their manure. Though the industry wishes consumers to believe that this situation still obtains—the California dairy industry ran advertisements proclaiming that California cheese comes from "happy cows", and showing the cows on pastures—the truth is radically different. The vast majority of California dairy cattle spend their lives on dirt and concrete and, in fact, never see a blade of pasture grass, let alone consume it. So outrageous is this duplicity that the dairy association was sued for false advertising and a friend of mine, a dairy practitioner for 35 years, was very outspoken against such an "outrageous lie".

In actuality, the life of dairy cattle is not a pleasant one. In a problem ubiquitous across contemporary agriculture, animals have been single-mindedly bred for productivity; in the case of dairy cattle, for milk production. Today's dairy cow produces 3–4 times more milk than 60 years ago. In 1957, the average dairy cow produced between 500 and 600 pounds of milk per lactation. Fifty years later, it is close to 20,000 pounds. (Colorado Dairy Facts 2005; NASS; Milk Production and Milk Cows 2006). From 1995 to 2004 alone, milk production per cow increased 16 %. The result is a milk bag on legs, and unstable legs at that. A high percentage of the U.S. dairy herd is chronically lame (some estimates range as high as 30 %), and these cows suffer serious reproductive problems. Whereas, in traditional agriculture, a milk cow could remain productive for 10, 15, or even 20 years, today's cow lasts slightly longer than two lactations, a result of a metabolic burnout and the quest for ever-increasingly productive animals, hastened in the U.S. by the use of the hormone BST to further increase production. Such unnaturally productive animals naturally suffer from mastitis, and the industry's response to mastitis in portions of the U.S. has created a new welfare problem by docking of cow tails without anesthesia in a futile effort to minimize teat contamination by manure. Still practiced, this procedure has been definitively demonstrated not to be relevant to mastitis control or lowering somatic cell count (Stull et al. 2002). (In my view, the stress and pain of tail amputation coupled with the concomitant inability to chase away flies, may well dispose to more mastitis.) Calves are removed from mothers shortly after birth, before receiving colostrum, creating significant distress in both mothers and infants. Bull calves may be shipped to slaughter or a feed lot immediately after birth, generating stress and fear.

The intensive swine industry, which through a handful of companies is responsible for 85 % of the pork produced in the U.S., is also responsible for significant suffering that did not affect husbandry-reared swine. Certainly the most egregious practice in confinement swine industry and possibly, given the intelligence of pigs, in all of animal agriculture, is the housing of pregnant sows in gestation crates or stalls—essentially small cages. The *recommended* size for such stalls, in which the sow spends her entire productive life of about 4 years, with a brief exception we will detail shortly, according to the industry is 0.9 m high by 0.64 m wide by 2.2 m long—this for an animal that may weigh 275 or more kilograms. In reality, many stalls are smaller. The sow cannot turn around, walk or even scratch her rump. In the case of large sows, she cannot even lie flat but must remain lying on her sternum. The exception alluded to is the period of farrowing—approximately 3 weeks—when she is transferred to a "farrowing crate" to give birth and nurse her piglets. The space for her is not greater, but there is a "creep rail" surrounding her so the piglets can nurse without being crushed by her postural adjustments.

Under extensive conditions, a sow will build a nest on a hillside so excrement runs off; forage an area covering 2 km a day; and take turns with other sows watching piglets and allowing all sows to forage. With the animal's nature thus aborted, she goes mad, exhibits bizarre and deviant behavior such as compulsively chewing the bars of the cage. She also endures foot and leg problems and lesions from lying on the concrete in her own excrement.

These examples are sufficient to illustrate the absence of good welfare in confinement. In general, all animals in confinement agriculture (with the exception of beef cattle who live most of their lives on pasture, and are "finished" on grain in dirt feed lots, where they can actualize much of their nature), suffer from the same generic set of affronts to their welfare absent in husbandry agriculture:

1. *Production Diseases*—By definition: a production disease is a disease that would not exist or would not be of serious epidemic import were it not for the method of production. Examples are liver and rumenal abscesses resulting from feeding cattle too much grain, rather than roughage. The animals that get sick are more than balanced out economically by the remaining animals' weight gain. Other examples are confinement-induced environmental mastitis in dairy cattle, weakness caused by beta-agonists to increase muscle mass in pigs and "shipping fever" in beef cattle. There are textbooks of production diseases, and one of my veterinarian colleagues calls such disease "the shame of veterinary medicine," because veterinary medicine should be working to eliminate such pathogenic conditions, rather than treating the symptoms.
2. *Loss of Workers who are "Animal Smart"*—In large industrial operations such as swine factories, the workers are minimum wage, sometimes illegal, often migratory workers with little animal knowledge. Confinement agriculturalists will boast that "the intelligence is in the system" and thus the historically collective wisdom of husbandry is lost, as is the concept of the historical shepherd, now transmuted into rote, cheap, labor.
3. *Lack of Individual Attention*—Under the husbandry systems, each animal is economically valuable. In intensive swine or poultry operations, the individuals are worth little. When this is coupled with the fact that workers are no longer caretakers, the result is obvious.
4. *The lack of attention to animal needs determined by their physiological and psychological natures*—As mentioned earlier, "technological sanders" allow us to keep animals under conditions violative of their natures, thus severing productivity from assured well-being.

2.4 The Demand for a New Ethic

The regnant ethic, born of necessity for the vast majority of agricultural history, was *good husbandry*. The advent of industrialized agriculture vitiated the relevance of husbandry to what agriculture has become. As confinement agriculture proliferated and came to dominate animal agriculture, the public began to realize that what they thought was "Old Macdonald's Farm," bucolic and pastoral, was no longer even related to that image. Public awareness began in Europe, specifically in Britain with the publication of journalist Ruth Harrison's seminal exposé of agricultural industrialization, *Animal Machines*. So powerful was the impact of this book during the mid 1960s, that the British government was compelled to charter the Brambell Commission, a group of ethnologists and biologists, charged with critically examining confinement agriculture (Brambell 1965). The Commission concluded that no system of production that failed to meet animals' basic needs and natures was morally acceptable. Though the Brambell Commission recommendations enjoyed no regulatory status, they served as a moral lighthouse for European social thought. In 1988, the Swedish Parliament passed, virtually unopposed, what the *New York Times* call a "Bill of Rights" for farm animals, abolishing in Sweden, in a series of timed steps, the confinement systems currently dominating North American agriculture (New York Times 1988). Much of northern Europe has followed suit, and the European Union moved in a similar direction with sow stalls for example, which were eliminated by 2011 (Rollin 2004).

People in the United States who found themselves morally concerned about the treatment of animals in confinement systems by industrialized agriculture found no moral recourse for effecting reform, and no legal grounds for challenging these systems. Attempts to utilize state anti-cruelty laws to attack modalities like veal crates and sow stalls had no traction. According to the legal history of anti-cruelty legislation in the United States, nothing accepted by an industry or "ministering to human needs" could ever count as cruel, with cruelty laws serving only to capture and punish sadistic, deviant, unnecessary practices. The anti-cruelty wrench could simply not fit the nut that needed to be turned.

Fueled by an ever-increasing social concern, a new societal ethic for animals began to emerge in the United States. In a study commissioned by USDA to answer why this took place, I distinguished a variety of social and conceptual reasons (Rollin 1995):

1. Changing demographics and consequent changes in the paradigm for animals:
   Whereas at the turn of the century, more than half the population was engaged in producing food for the rest, today only some 1.5 % of the U.S. public is engaged in production agriculture (AMC 2003). One hundred years ago, if one were to ask a person in the street, urban or rural, to state the words that come into their mind when one says "animal", the answer would doubtless have been "horse", "cow", "food", "work", etc. Today, however, for the majority of the population, the answer is "dog", "cat", "pet". Repeated studies show that almost 100 % of the pet-owning population views their animals as "members of the family" (The

Acorn 2002), and virtually no one views them as an income source. Divorce lawyers note that custody of the dog can be as thorny an issue as custody of the children!
2. We have lived through a long period of ethical soul-searching
   For almost 50 years society has turned its "ethical searchlight" on humans traditionally ignored or even oppressed by the consensus ethic—blacks, women, the handicapped, other minorities. The same ethical imperative has focused attention on our treatment of the non-human world—the environment and animals. Many leaders of the activist animal movement in fact have roots in earlier movements—civil rights, feminism, homosexual rights, children's rights and labor.
3. The media has discovered that "animals sell papers"
   One cannot channel-surf across normal television service without being bombarded with animal stories, real and fictional. (A *New York Times* reporter recently told me that more time on cable TV in New York City is devoted to animals than to any other subject.) Recall, for example, the extensive media coverage a number of decade ago of some whales trapped in an ice-floe, and freed by a Russian ice-breaker. This was hardly an overflowing of Russian compassion—an oxymoronic notion applied to a people who gave us pogroms, the Gulag, and Stalinism. Rather, someone in the Kremlin was bright enough to realize that liberating the whales was an extremely cheap way to score points with U.S. public opinion.
4. Strong and visible arguments have been advanced in favor of raising the status of animals by philosophers, scientists and celebrities (Singer 1975; Rollin 1981; Regan 1983; Sapontzis 1987).
5. Changes in the nature of animal use demanded new moral categories
   In my view, while all of the reasons listed above are relevant, they are nowhere near as important as the precipitous and dramatic changes in animal use that occurred after World War II. These changes were the huge conceptual changes in the nature of agriculture we have described, and second the rise of significant amounts of animal research and testing. The latter was also not conceptually able to be captured by the anti-cruelty laws. But ethical concepts do not arise *ex nihilo*.

Plato taught us a very valuable lesson about effecting ethical change. If one wishes to change another person's—or society's—ethical beliefs, it is much better to *remind* than to *teach* or, in my martial arts metaphor, to use *judo rather than sumo. In other words, if you and I disagree ethically on some matter, it is far better for me to show you that what I am trying to convince you of is already implicit— albeit unnoticed—in what you already believe.* Similarly, we cannot force others to believe as we do (*sumo*); we can, however, show them that their own assumptions, if thought through, lead to a conclusion different from what they currently entertain (*judo*). These points are well-exemplified in twentieth century U.S. history. Prohibition was *sumo*, not *judo*—an attempt to forcefully impose a new ethic about drinking on the majority by the minority. As such, it was doomed to fail, and in fact people drank *more* during Prohibition. Contrast this with Lyndon Johnson's civil rights legislation. As himself a Southerner, Johnson realized that even Southerners would acquiesce to the following two propositions:

All humans should be treated equally, and black people were human—they just had never bothered to draw the relevant conclusion.
If Johnson had been wrong about this point, if "writing this large" in the law had not "reminded" people, civil rights would have been as ineffective as Prohibition!

So, society was faced with the need for new moral categories and laws that reflect those categories in order to deal with animal use in science and agriculture and to limit the animal suffering with which it is increasingly concerned. At the same time, recall that western society has lived through almost fifty years of extending its moral categories for *humans* to people who were morally ignored or invisible—women, minorities, the handicapped, children, citizens of the third world. As we noted earlier, new and viable ethics do not emerge *ex nihilo*. A plausible and obvious move is for society to continue in its tendency and *attempt to extend the moral machinery it has developed for dealing with people, appropriately modified, to animals*. This is what has occurred. Society has taken elements of the moral categories it uses for assessing the treatment of people and is in the process of modifying these concepts to make them appropriate for dealing with new issues in the treatment of animals, especially their use in science and confinement agriculture.

What aspect of our ethic for people is being so extended? One that is, in fact, quite applicable to animal use, is the fundamental problem of weighing the interests of the individual against those of the general welfare. Different societies have provided different answers to this problem. Totalitarian societies opt to devote little concern to the individual, favoring instead the state, or whatever their version of the general welfare is. At the other extreme, anarchical groups such as communes give primacy to the individual and very little concern to the group—hence they tend to enjoy only transient existence. In our society, however, a balance is struck between these two extremes. Although most of our decisions are made to the benefit of the general welfare, fences are built around individuals to protect their fundamental interests from being sacrificed to the majority. Thus, we protect individuals from being silenced even if the majority disapproves of what they say; we protect individuals from having their property seized without compensation even if such seizure benefits the general welfare; we protect individuals from torture even if they have planted a bomb in an elementary school and refuse to divulge its location. We protect those interests of the individual that we consider essential to being human, to *human nature,* from being submerged, even by the common good. Those moral/legal fences that protect the individual humans are called *rights* and are based on plausible assumptions regarding what is essential to being human.

It is this notion to which society in general is looking in order to generate the new moral notions necessary to talk about the treatment of animals in today's world, where cruelty is not the major problem but where such laudable, general human welfare goals as efficiency, productivity, knowledge, medical progress, and product safety are responsible for the vast majority of animal suffering. People in society are seeking to "build fences" around animals to protect the animals and their interests and natures from being totally submerged for the sake of the general welfare, and are trying to accomplish this goal by going to the legislature. In husbandry, this occurred automatically; in industrialized agriculture, where it is no longer automatic, people wish to see it legislated.

As a mainstream movement, this new ethic does not try to give human rights to animals. Since animals do not have the same natures and interests flowing from these natures as humans do, human rights do not fit animals. Animals do not have basic natures that demand speech, religion, or property; thus according to them these rights would be absurd. On the other hand, animals have natures of their own and interests that flow from these natures, and the thwarting of these interests matters to animals as much as the thwarting of speech matters to humans. The agenda is not, for mainstream society, making animals have the same rights as people. It is rather preserving the common-sense insight that "fish gotta swim and birds gotta fly," and suffer if they don't.

This new ethic is *conservative*, not radical, harking back to the animal use that necessitated and thus entailed respect for the animals' natures. It is based on the insight that what we do to animals *matters* to them, just as what we do to humans matters to them, and that consequently we should respect that mattering in our treatment and use of animals as we do in our treatment and use of humans. *And since respect for animal nature is no longer automatic as it was in traditional husbandry agriculture, society is demanding that it be encoded in law.* Significantly, in 2004, no fewer than 2100 bills pertaining to animal welfare were proposed in U.S. state legislatures.

With regards to animal agriculture, the pastoral images of animals grazing on pasture and moving freely are iconic. As the 23rd Psalm indicates, people who consume animals wish to see the animals live decently, not live in pain, distress and frustration. It is for this reason, in part, that industrial agriculture conceals the reality of its practices from a naïve public—witness Perdue's advertisements about raising "happy chickens," or the California "happy cow" ads. As ordinary people discover the truth, they are shocked. When I served on the Pew Commission and other commissioners had their first view of sow stalls, many were in tears and all were outraged.

Just as our use of people is constrained by respect for the basic elements of human nature, people wish to see a similar notion applied to animals. Animals, too, have natures, what I call *telos* following Aristotle—the "pigness of the pig", the "cowness of a cow". Pigs are "designed" to move about on soft loam, not to be in gestation crates. If this no longer occurs naturally, as it did in husbandry, people wish to see it legislated. This is the mainstream sense of "animal rights".

As property, strictly speaking, animals cannot have legal rights. But a functional equivalent to rights can be achieved by limiting property rights. When I and others drafted the U.S. federal laws for laboratory animals, we did not deny that research animals were the property of researchers. We merely placed limits on their use of their property. I may own my car, but that does not mean I can drive it on the sidewalk or at any speed I choose. Similarly, our law states that if one hurts an animal in research, one must control pain and distress. Thus research animals can be said to have the *right* to have their pain controlled.

In the case of farm animals, people wish to see their basic needs and nature, *teloi*, respected in the systems that they are raised. Since this no longer occurs naturally, as it did in husbandry, it must be imposed by legislation or regulation. A Gallup poll conducted in 2003 shows that 75 % of the public wants legislated guarantees of farm animal welfare. This is what I call "animal rights as a mainstream phenomenon." Legal codification of rules of animal care respecting animal *telos* is thus the form animal welfare takes where husbandry has been abandoned.

Thus, in today's world, the ethical component of animal welfare prescribes that the way we raise and use animals must embody respect and provision for their physical and psychological needs and natures. It is therefore essential that industrial agriculture phase out those systems which cause animal suffering by violating animals' natures and replace them with systems respecting their natures. This would not be difficult to incorporate into a national law governing future constraints on animal agricultural systems, though no one has yet proposed such a radical and innovative set of changes to agriculture.

2.5 The Pew Commission on Industrial Farm Animal Production

While animal welfare is probably the best-known issue emerging from the industrialization of agriculture, numerous other profound problems for agriculture also emerged from this revolutionary approach to raising animals. The best and most probing account of these problems, including animal welfare, has been provided by the Pew Commission on Industrial Farm Animal Production, in a 2008 report entitled *Putting Meat on the Table* (The report may be found online at pcifap.org). This report received more than 800 positive editorials in the U.S. media. The rise of societal interest in where food comes from, expressed as the "slow food" movement, the organic movement, the rise of specialty groceries such as Whole Foods and Sunflower, and many restaurants focusing on local "natural" and humane food, coupled with burgeoning concern about animal welfare, led to the chartering of the first commission to systematically explore confinement agriculture under the Johns Hopkins University School of Public Health Center for a Livable Future, funded by The Pew Charitable Trusts. The commission can be seen as articulating society's nascent concerns about industrialized animal agriculture in a variety of areas.

The Pew Commission began when the Johns Hopkins School of Public Health, the best-funded school of public health in the U.S., garnering 25 % of federal research money in public health, was completing a study of water quality in the chicken industry in the Delaware-Maryland-Virginia area, home to a large segment of the poultry industry. Investigators from Hopkins were disturbed by much of what they found. Particularly disturbing was the presence in the water of cutting-edge human antibiotics, designed to be used as a last resort in human disease. They reported back to the Director of the Hopkins Center for a Livable Future, a unit concerned with health and sustainability, Dr. Robert Lawrence. He successfully petitioned the $6 billion Pew Charitable Trust to fund a study of industrial animal agriculture and issue a report.

The Chairman of the Commission was the former governor of Kansas, John Carlin, who was raised on a dairy farm and was a wise politician. The remaining commissioners, chosen for their knowledge in areas relevant to Commission concerns, were acknowledged as experts in their fields. This choice assured our credibility, which, undoubtedly, the industry would attack. Commission Members included: a former veterinary school dean and highest ranking veterinarian in the Public Health Service; a former executive director of the Catholic Rural Life Conference, an expert in rural sociology; founder and CEO of the first food service company to address issues of food ethics, catering 250 million meals a year; a former South Dakota state senator; a former U.S. Secretary of Agriculture; the Founding Director of the Center for Alternatives to Animal Testing at Johns Hopkins School of Public Health; a Professor Emeritus of Health Behavior and Health Education at the University of North Caroline School of Public Health; a rancher and former president of the Montana Stockgrowers Association; a Distinguished Fellow, Leopold Center for Sustainable Agriculture, Iowa State University; Dean of the University of Iowa, College of Public Health; Paulette Goddard Professor of Nutrition at NYU and best-selling author on food issues; the founder of Niman Ranch, a company supplied by 600 family-farmers producing humane meat; a University Distinguished Professor at Colorado State University; a leading expert in infectious disease at Harvard Medical School and Harvard School of Public Health; an expert on animal welfare in theory and practice, and a Senior Vice President at Cargill.

The Commission met for over 2 years on multiple occasions across the U.S., gave five congressional briefings, and released its final report in May of 2008. It was funded to hire whatever expert consultants were required, liberally used expert witnesses, and all of the final conclusions established by consensus, which was not easy, assured a united front.

In deliberations, five problematic interconnected areas associated with CAFOS (Confined Animal Feeding Operations) were the focus:

1. Antimicrobial resistance was very likely augmented by massive use of antibiotics in CAFOS to promote growth, prevent disease, and compensate for poor husbandry. As early as the mid-1940s, promotion of antibiotic resistant pathogens was foreseen as a Darwinian consequence of indiscriminate antibiotic use. An estimated 70 % of the antibiotics produced in the U.S. are used in CAFOS (Davies and Davies 2010).
2. Environmental despoliation and farm waste: CAFOS produce huge volumes of animal waste that often exceed the capacity of the land to absorb them,

especially in inappropriate areas such as flood plains. CAFOS also pollute air (as in large dairies), and contribute antibiotics, hormones, pesticides and heavy metals to water pollution. They utilize huge volumes of fossil fuel and water. Despite all this, they are not regulated as polluting industries.
3. Rural Sociology—CAFOS have replaced the independent, self-sufficient family farmer that Jefferson saw as the backbone of American democracy. In barely 40 years, the U.S. has lost over 80 % of its small hog farmers to a handful of huge corporate entities. This loss has increased rural poverty and degradation of small communities. Poor people often bear the brunt of CAFO pollution. Small farmers furthermore lack the resources to be able to compete with large corporations.
4. Other Public Health Issues—High confinement operations serve as incubators for pathogens, sources of antibiotic resistance that reduces the human armamentarium against infectious disease, and adversely affect the physical and mental health of people living near them. Workers in CAFOS suffer more health problems and can spread disease in communities.
5. Animal Welfare—CAFOS harm most relevant dimensions of animal welfare, from health of the animals to the ability to express their natural behaviors, including basic movements such as standing up and turning around or being with others of their own kind. The vast majority of farm animal diseases are "production diseases," i.e. diseases that would not be a major problem if animals were extensively raised (Pew Commission).

Obviously, all of these categories connect to the others (e.g. pollution and health). Through the research on the above issues the Pew Commission found that research on all of these issues is largely industry-sponsored, which requires getting results congenial to industry interests.

One of the most significant findings of the Commission was exploding the widespread belief that industrial agriculture is the source of cheap food. While creating animal products that are cheap at the cash register, the claim of cheapness excludes what economists call "externalization of costs"—or passing the hidden costs of production to the public. For example, pollution cleanup is passed to the public, as are the health costs of living near pathogenic CAFOS. As an additional example, every man, woman and child living near the mega-dairies in California's Central Valley spends $1500 more on health care than if the dairies were not there.

The Commission concluded with six basic recommendations, which a lobbyist for the industry who is a friend of mine called "a blueprint for the future of agriculture":

1. Phase out and ban the non-therapeutic use of antimicrobials
2. Improve disease tracking by a national animal identification system
3. Improve regulations of CAFO waste
4. Phase out intensive confinement of farm animals within 10 years (!)
5. Increase competition (reduce monopoly) in livestock production
6. Create publically funded research grants

The Pew Report demonstrated to the public the close connection between all of these issues. Now, environmentalists must be concerned about confinement of animals, and people concerned about rural life and public health must also see the relevance of animal welfare. As one member of the Commission said to me on our last day, "I used to think animal welfare was a fringe issue. Thank you for showing me the centrality of animal welfare to everything else." In the fate of the animals we raise is reflected our own fate—as I tell farmers, "the same forces that put animals in tiny boxes also put you in (financial) boxes."

2.6 Conclusion

Despite the extensive positive media coverage the Pew Commission and report received, as of November of 2013, not a single Commission recommendation had been legislatively instituted by the U.S. Congress, showing the extraordinary power of the agricultural industry to block reform in all of the areas discussed earlier. One can only hope that educated consumers can move the industry, as occurred with Smithfield and sow stalls.

Acknowledgements Notably, Prof. Rollin has been a long-standing thought leader and outstanding scholar in his field. The publisher acknowledges that some of the references cited merely used Prof. Rollin's original ideas, but the convention dictates that attributions must be made. The editors give Prof. Rollin credit for his own and original ideas, regardless of where those have been cited, quoted or printed.

References

Acorn, the, January 31, 2002. Survey says pets are members of the family
AMC: Agricultural Machinery Conference. May 5–7, 2003. http://www.amc-online.org/
Brambell FWR (1965) Report of the technical committee to enquire into the welfare of animals kept under intense livestock husbandry systems. HMSO, London
Colorado Dairy Facts, 2005
Cooper DL (n.d.) Biblical research studies group-the Gospel according to . . .. Biblical Research Studies Group. Web. 16 March 2014. http://www.biblicalresearch.info/page309.html
Davies J, Davies D (2010) Origin and evolution of antibiotic resistance. September 2010. http://www.ncbi.nlm.nih.gov/pmc/articles/PMC2937522/
Dawkins MS, Bonney R (2008) The future of animal farming: renewing the ancient contract. Blackwell, Malden
Gallup (2003) Available http://www.gallup.com
Lundeen T (2008) Poultry missing genetic diversity. Feedstufs, December 1, 11
New York Times (1988) Swedish farm animals get a bill of rights, p 1, October 25, 1988
Pew Commission. Putting Meat on the Table: Industrial Farm Animal Production in America. http://www.ncifap.org/_images/PCIFAPFin.pdf
Regan T (1983) The case for animal rights. University of California Press, Berkeley
Rollin B (1981, 1992, 2006) Animal rights and human morality. Prometheus Books, Buffalo

Rollin B (1995) Farm animal welfare: social, bioethical and research issues. Iowa State University Press, Ames

Rollin BE (2002) An ethicist's commentary on equating productivity and welfare. Can Vet J. Web. 16 March 2014. <http://www.ncbi.nlm.nih.gov/pmc/articles/PMC339160/>

Rollin B (2004) Annual keynote address: animal agriculture and emerging social ethics for animals. J Anim Sci. Web. 10 September 2014. <http://www.animal-science.org/content/82/3/955.full>

Sapontzis S (1987) Morals, reason and animals. Temple University Press, Philadelphia

Singer P (1975) Animal liberation. New York Review Press, New York

Stull CL, Payne MA, Berry SL, Hullinger PJ (2002) Evaluation of the scientific justification of tail docking in dairy cattle. J Am Vet Med Assoc 220:1298–1303

Chapter 3
The WHO in Global Food Safety Governance: A Preliminary Mapping of Its Normative Capacities and Activities

Ching-Fu Lin

Abstract The past decades have witnessed a surge of foodborne illnesses of diverse sources on every continent of the world, a testament of the complexity of the transformed supply chain of production, distribution, and consumption as well as the importance of global food safety governance. While the global health community continues to look to the WHO to address problems as such, the organization is arguably losing its institutional legitimacy for persistent governance inertia. The mandates and normative tools assigned by the WHO Constitution empower the WHO to actively engage and provide leadership in the governance of food safety, but the organization has comfortably nested in a soft approach, abstaining from assuming any international agreement for over 60 years. This chapter examines the WHO's normative capacities given by its Constitution and normative activities in practice to evaluate its influences in shaping and reshaping global food safety governance. While the mapping of this chapter resonates to the critiques that the WHO has not employed to the necessary extent the normative authority partly due to ossified bureaucracy, power politics, and budgetary weakness, this chapter further argues that the clear and enduring gap between the normative capacities and activities may ultimately render the organization's institutional legitimacy vulnerable and compromise its relevance as a whole. As a preliminary conclusion, this chapter offers a couple of recommendations for a stronger pivotal role in global food safety governance. It suggests that the WHO take active steps to build functional links with multiple stakeholders and partners, such as WTO and non-state actors, some of which have emerged as valid alternatives to the organization. The WHO needs to further solidify such functional links to coordinate efforts in global food safety, which may merit a more formal inter-institutional framework.

C.-F. Lin SJD, LLM (✉)
Institute of Law for Science and Technology, National Tsing Hua University, Hsinchu, Taiwan
e-mail: chingfulin@mx.nthu.edu.tw

3.1 Introduction

The past decades have witnessed a surge of foodborne illnesses of diverse sources, be it biological, chemical, or radiological, on every continent of the world. Documented outbreaks ranging from bovine spongiform encephalopathy (BSE) in beef, dioxin in pork, melamine-contaminated dairy products, and *E. coli* contaminated cucumbers have attested the increasing importance and complexity of global food safety governance and its public health, social, and economic implications. A 1999 World Health Organization (WHO) report suggests that every year, foodborne illnesses result in 1.5 billion cases of diarrhea in children and over 3 million premature deaths.[1] Contaminated food and water in developing countries are also responsible for 1.8 million children deaths.[2] Even so, according to a 2013 WHO report on advancing food safety initiatives, due to the fact that foodborne illness cases are frequently "under-reported," the total number and burden of the global food safety problem has yet to be quantified.[3]

The production, distribution, and consumption of food, moreover, have been transformed alongside the globalization of economic activities, advancements in food science, development of transportation technology, and integration and consolidation of agri-food industries, and the creation of the World Trade Organization (WTO).[4] Such transformed patterns have also posed new challenges to food safety, evidenced in the intensified scale, severity, frequency, and impact of foodborne illness outbreaks. Therefore, risks posed by unsafe food products can originate from a producer in one country and quickly spill over to many others, evolving from a local problem to a global concern within a short period of time. This necessitates effective and efficient international cooperation beyond unilateral efforts of individual countries—including standard setting, information sharing, technical and financial assistance, and collective responses to cross-border outbreaks.

The WHO is generally regarded as the first appropriate international body to play a crucial role in international cooperation of food safety governance, since such normative activity is within the ordinary understanding of the agency's public health authority and mandates.[5] WHO member states have recognized that ensuring food safety constitutes an essential and priority public health function.[6] The broad

---

[1] World Health Organization [hereinafter WHO], *Food Safety Programme, Food Safety: An Essential Public Health Issue for the New Millennium*, 9, WHO/SDE/PHE/FOS/99.4 (1999).

[2] Office for the South East Asia Region, WHO, *Health Situation in the South East Asia Region 1994–1997*, SEA/HS/209 (1999), pp. 213–214.

[3] WHO, *Advancing Food Safety Initiatives: Strategic Plan for Food Safety Including Foodborne Zoonoses 2013–2022*, p. 8 (2013).

[4] *See* World Economic Forum, Global Risk 2008: A Global Risk Network Report, http://www.weforum.org/pdf/globalrisk/report2008.pdf. Motarjemi et al. (2001), Käferstein et al. (1997), Käferstein and Abdussalam (1999).

[5] Constitution of the World Health Organization, Articles 2, and 19–23, July 22, 1946, 62 Stat. 2679, 14 U.N.T.S. 185 [hereinafter WHO Constitution].

[6] WHO, *supra* note 3, p. 11.

mandates and normative tools assigned to the WHO by its Constitution empower the WHO to actively engage in and provide leadership in global food safety governance. Part II of this chapter examines the normative functions of the WHO in ensuring food safety as an essential component in the broader context of achieving public health across the globe. After reviewing the normative functions of the WHO, Part III looks at the agency's normative activities on the ground, that is, what influences the WHO has in shaping and reshaping global food safety governance. This chapter points out the origins and repercussions of the clear and enduring gap between the normative capacities and activities of the WHO, and concludes by offering a few recommendations for a stronger pivotal role in global food safety governance.

3.2 Normative Capacities of the WHO

The WHO, headquartered in Geneva, has a decentralized structure with six regional offices and more than 190 member states. Established in 1948, the WHO is not only the principal institution but also the largest specialized agency and normative institution responsible for addressing global health issues.[7] The World Health Assembly (WHA), composed of all of the WHO member states, serves as the highest legislative organ that makes overall policy and adopts a variety of legal instruments including recommendations, regulations, and conventions. The WHO's central objective proclaimed by Article 1 of its Constitution is the "attainment by all peoples of the highest possible level of health."[8] The Constitution gives the WHO and WHA normative capacities to exercise and to fulfill this very objective. In particular, the authority derives from paragraphs (k), (o), (s), (t), and (u) of Article 2 along with Articles 19–23 of the Constitution. Specifically, the Constitution empowers the WHA to engage in three types of normative activities: Conventions and agreements (Article 19),[9] regulations (Article 21),[10] and nonbinding recommendations and standards (Article 23).[11]

---

[7]*Id.* Articles 18, 19, 21, and 23.
[8]*Id.* Article 1.
[9]*Id.* Article 19.
[10]*Id.* Article 21–22.
[11]*Id.* Article 23.

3.2.1 Adopting Conventions and Agreements

> Article 19
> The Health Assembly shall have authority to adopt conventions or agreements with respect to any matter within the competence of the Organization. A two-thirds vote of the Health Assembly shall be required for the adoption of such conventions or agreements, which shall come into force for each Member when accepted by it in accordance with its constitutional processes.

The WHO's normative capacity of treaty making is enshrined in Article 19 of the Constitution. As set by this provision, the WHA has the authority to adopt international conventions or agreements "with respect to any matter within the competence of the Organization" by a two-thirds vote. Such matters that fall within the agency's competence are set forth by Article 2, which provides over 20 functions of the WHO, ranging from facilitating "work to eradicate epidemic, endemic and other diseases" to setting and promoting "international standards with food, biological, pharmaceutical and similar products."[12] Article 19 stipulates the threshold (a two-thirds vote) for the WHA to pass any international conventions and agreements, yet such a procedural constraint leaves the WHA with relatively broad discretion in formulating substantive elements of a treaty. Examples of such broad discretion include the types of convention, structures and processes, and conditions for ratification, deposit, and entry into force.[13] While a treaty adopted under Article 19 of the Constitution does not bind WHO member states without their consent, Article 20 requires member states to affirmatively "take action relative to the acceptance of such convention or agreement" within 18 months after the WHA's adoption of the treaty.[14] More specifically, a WHO member state has to "notify the Director-General of action taken," and provide "a statement of the reasons" if it decides not to accept the convention or agreement.[15]

3.2.2 Promulgating Regulations

> Article 21
> The Health Assembly shall have authority to adopt regulations concerning:
> (a) sanitary and quarantine requirements and other procedures designed to prevent the international spread of disease;
> (b) nomenclatures with respect to diseases, causes of death and public health practices;
> (c) standards with respect to diagnostic procedures for international use;
> (d) standards with respect to the safety, purity and potency of biological, pharmaceutical and similar products moving in international commerce;
> (e) advertising and labelling of biological, pharmaceutical and similar products moving in international commerce.

---

[12]*Id.* Article 19.
[13]Burci and Vignes (2004), p. 124.
[14]WHO Constitution, *supra* note 5, Article 20.
[15]*Id.*

The WHO's normative capacity of promulgating regulations is found in Article 21 of the Constitution. Article 21 specifies five particular aspects where the WHO can exercise such normative capacity, ranging from sanitary and quarantine rules related to the prevention of the international spread of disease, disease nomenclatures, to standards regarding the safety of biological or pharmaceutical products in the flow of international trade.[16] In spite of the limited scope of authority specified above, Article 22 features an "opt-out" mechanism where regulations adopted by a simple majority vote under Article 21 automatically bind all WHO member states "except for such Members as may notify the Director-General of rejection or reservations within the period stated in the notice."[17] The authority to adopt binding international regulations with a simple majority vote and the unusual "opt-out" design together constitute the most impressive normative capacity of the WHO—a "quasi-legislative" power in international law.[18]

3.2.3 Making Non-Binding Instruments

Article 23
The Health Assembly shall have authority to make recommendations to Members with respect to any matter within the competence of the Organization.

The making of recommendations, guidelines, standards, or voluntary instruments under Article 23 of the Constitution is arguably the softest normative capacity of the WHO. Article 23 instruments are adopted by the WHO for pragmatic reasons, such as reduction of negotiation costs in sensitive issues, flexibility in varied national or regional circumstances, and adaptability to rapid scientific changes. In the context of public health, the need for reducing negotiation costs and increasing flexibility and adaptability appears manifest and renders such voluntary approach preferable. Article 23 is therefore the most frequently exercised normative power the WHO has employed to deal with global public health issues. All in all, while Article 23 has arguably the weakest legal binding force, it has been the most prolifically used of the three normative capacities.[19]

---

[16]WHO Constitution, *supra* note 5, Article 21.
[17]*Id.* Article 22.
[18]Fidler (1998) and Gostin (2007).
[19]Burci and Vignes (2004), p. 141.

3.3 Normative Activities of the WHO

With regard to normative activities in food safety, the WHO has relied on its Article 23 authority and refrained from adopting any legally binding instruments for over 65 years. This part of the chapter discusses the major normative activities the WHO has engaged in for promoting global food safety and its lack of success in addressing the problem.

The Department of Food Safety and Zoonoses (FOS) is in charge of leading and coordinating efforts in a large variety of areas. Its authority covers foodborne diseases, food hygiene, food technologies, microbiological risks, chemical risks, international food standards (Codex Alimentarius Commission, CAC), International Food Safety Authorities Network (INFOSAN), antimicrobial resistance, as well as zoonoses and intersectoral collaboration at animal-human-ecosystems interface.[20] As elaborated in the WHO's Strategic Plan for Food Safety Including Foodborne Zoonoses 2013–2022—which builds on the WHA's resolution WHA63.3 in 2010—the FOS should work in close collaboration with all the regional and country offices under an overarching framework composed of three essential strategic directions.[21] These three essential strategic directions adequately lay out the normative activities the WHO has engaged in on the ground.

3.3.1 WHO Scientific Advice and the Codex Alimentarius Commission

The WHO does not make international food safety standards, although it has the normative capacity, scientific expertise in the field, and the convening power over its 194 member states. Rather, the WHO participates *indirectly* in the CAC through ways of providing scientific advice and risk assessment.

The CAC is an international governmental body established under the two resolutions adopted by the Eleventh Session of the Food and Agriculture Organization of the United Nations (FAO) Conference in 1961 and the Sixteenth WHA in 1963.[22] The WHO and the FAO also adopted the Statutes and Rules of Procedure for the Commission,[23] which has 185 members (184 member countries and

---

[20]For further details, *see* World Health Organization, Department of Food Safety and Zoonoses (FOS) Areas of Work, at http://www.who.int/foodsafety/areas_work/en/.

[21]WHO, *supra* note 3.

[22]Food and Agriculture Organization of the United Nations (hereinafter FAO) and WHO, Understanding the Codex Alimentarius 13 (3d ed. 2006).

[23]The Statutes of the Codex Alimentarius Commission (Statutes) provide the legal basis for the Commission's work and formally reflect the concepts behind and reasons for its establishment. The Rules of Procedure of the Codex Alimentarius Commission (Rules of Procedure) describe and formalize working procedures appropriate for an intergovernmental body.

1 member organization as well as 24 active committees and task forces.[24] The mandate of the CAC is twofold: To develop international food standards, guidelines, and recommendations to "protect the health of consumers" as well as to "ensure fair practices in food trade."[25] It has formulated international standards for a wide range of food products and specific requirements covering pesticide residues, food additives, veterinary-drug residues, hygiene, food contaminants, and labeling and certification systems.

WHO (and FAO) scientific advice serves as the basis for CAC food standards and relevant measures taken by states along the food supply chain and international trade.[26] Scientific experts from diverse backgrounds convene in the standing institutions—the Joint FAO/WHO Expert Committee on Food Additives (JECFA), the Joint FAO/WHO Meeting on Pesticide Residues (JMPR) and the Joint FAO/WHO Expert Meeting on Microbiological Risk Assessment (JEMRA)—all of which are jointly administered by the WHO and the FAO, to prepare scientific evaluation and risk assessment.[27]

The WHO's scientific role in the CAC has been increasingly outweighed by that of the WTO, which explicitly refers to the CAC as the international standard-setter for disciplining WTO members' food safety regulatory measures in the international trade regime.[28] With the backing of the WTO's mandatory dispute settlement system, binding adjudicatory decisions, and retaliation mechanism,[29] the CAC is now commonly regarded as a quasi-legislator,[30] and its standards are *de facto* mandatory especially in WTO food safety disputes.[31] The CAC has therefore

---

[24]FAO/WHO, *Codex Members and Observers*. In addition to the Committee on General Principles and six regional coordinating committees (Africa, Asia, Europe, Latin America and the Caribbean, the Near East, North America and the Southwest Pacific), subsidiary bodies directly related to food safety include, *inter alia*, the Codex Committees on Contaminants in Foods, Food Additives, Food Hygiene, Pesticide Residues, Residues of Veterinary Drugs in Foods, Food Import and Export Inspection, and Certification Systems.

[25]FAO/WHO, *supra* note 22, p. 25.

[26]WHO, *supra* note 3, pp. 14–19.

[27]For example, the WHO calls for, selects, and enlists qualified experts to consider scientific evidence and perform toxicology risk assessments in the JECFA (and the FAO enlists scientists to evaluate residues). The JECFA evaluation reports, regarded as authoritative reviews of all available evidence and information concerning a given food safety risk, form the basis of CAC standards. *Ad hoc* expert meetings are convened when the WHO needs to assess issues in emergency or regards emerging and complex risks. WHO, *supra* note 3, pp. 14–19.

[28]World Trade Organization (hereinafter WTO), Agreement on the Application of Sanitary and Phytosanitary Measures, Annex 1A, The Legal Texts: The Results of the Uruguay Round of Multilateral Trade Negotiations 59 (2000), 1867 U.N.T.S. 493 (1994) [hereinafter SPS Agreement], Article 3.1 & Annex A.3. The SPS Agreement specifically refers to three international standard-setting bodies, now oft-called "Three Sisters:" The Codex Alimentarius Commission (CAC) dealing with food safety, the International Plant Protection Convention (IPPC) dealing with plant health, the World Organization for Animal Health (OIE) dealing with animal health.

[29]*See, e.g.*, Matsushita et al. (2006); WTO (2008), pp. 103–140.

[30]Trachtman (2006); Alemanno (2007), pp. 262–267; Silverglade (2000).

[31]Charnovitz (2002).

encountered controversial issues of "politicization of science"[32] where CAC member states have tended to evaluate proposed food safety standards for their potential impact on trade interests and act strategically,[33] and in some cases, have let trade interests overshadow food safety and public health. The CAC, despite being an international organization jointly established by the WHO and FAO, often becomes an extended WTO battlefield of public health versus international trade. As the international trade regime wields the most significant influence on CAC and food safety issues, the role of the WHO seems likely to fade.

3.3.2 International Cooperation in Information Exchange

The WHO facilitates international cooperation mostly in the area of information exchange and cross-sectoral coordination. After the serious BSE outbreak in the 1990s, the Fifty-Third WHA passed in 2010 Resolution WHA53.15 which urged member states to integrate food safety into their public health functions. The goal was to design systematic preventive measures that would reduce the occurrence of foodborne illnesses and support the development of science in the assessment of risks related to food.[34] Resolution 53.15 led to the WHO's later creation of INFOSAN, a voluntary international network of food safety authorities vis-à-vis food safety management. The FOS administrates the INFOSAN platform together with the FAO for rapid and timely exchange of food safety information and shares technical information on foodborne disease surveillance.[35] The FOS also endeavors to establish cross-sectoral linkages among relevant government agencies at the national level, such as the food safety, health, agriculture, and trade sectors to facilitate communication.[36]

Food safety risks and incidents in one country usually pose a public health concern to other countries because of the globalized pattern of food production and consumption. The WHA recognized the importance of having a mechanism by which states can exchange information on food safety issues, through Resolution 53.15. The Resolution requested that the Director-General "put in place a global strategy... for the efficient gathering and exchange of information in and between countries and regions."[37] In 2004, the WHO inaugurated INFOSAN, a global network of 102 national food safety authorities (at the time of creation), to promote the exchange of food safety information and to improve collaboration among food

---

[32]*See* Lin (2013).

[33]Veggeland and Borgen (2002), pp. 22–23.

[34]World Health Assembly (hereinafter WHA), Resolution WHA53.15, § 1(1)–(5).

[35]WHO, *supra* note 3, p. 23.

[36]WHO, *supra* note 3, p. 23.

[37]*Id.* § 2(4).

safety authorities at national and international levels.[38] As of today, the information-sharing system has increased to 181 country members.[39]

Under INFOSAN, member states are expected to designate one or more INFOSAN Focal Points and one INFOSAN Emergency Contact Point at the national level.[40] The role of the INFOSAN Focal Points is to receive notes and messages, WHO guidelines along with other important food safety information from INFOSAN, and to circulate the information to other government agencies, actors in the food industry, and nongovernmental organizations (NGOs).[41] The INFOSAN Contact Points serve to notify INFOSAN of food safety incidents, to respond to such notification by sending INFOSAN emergency alerts, and to exchange information during a food safety crisis.[42] At the international level, the WHO facilitates information exchange among members, verifies the extent of the problem, disseminates scientific knowledge, and provides food safety advice via INFOSAN during foodborne disease outbreaks or food contamination events.

While INFOSAN serves as a practical platform on which its members can cooperatively address food safety incidents through a wide variety of information-sharing strategies, there are two major weaknesses with the network. First, the voluntary nature of INFOSAN means possible non-adherence, which could in turn, bring about unexpected or unmanageable food safety crises. The lack of binding rules and a compliance mechanism renders the WHO powerless to strengthen coordination.[43] The INFOSAN framework imposes no legal obligations on WHO member states (and other members) to either promptly notify the network of urgent food safety incidents or to provide export and import data. Participating member states are merely expected to follow the relevant guidelines and principles, which are simply recommendations rather than requirements. While countries in general do not have difficulty designating Focal Points and Emergency Contact Points for INFOSAN, they have diverse concerns and broad discretion as to whether, when, and to what extent to provide information. In effect, political and economic considerations usually exacerbate the delay or concealment of internationally significant public-health information—as countries want to minimize the self-harm that would likely occur if they notify the world of their food safety outbreaks.[44] For instance, in the melamine-tainted milk incident, the Chinese authorities arguably delayed in notifying INFOSAN of key information which

---

[38] WHO (2004), pp. 1–4. For background information on INFOSAN, see WHO, *International Food Safety Authorities Network (INFOSAN)*.

[39] *Id.* There are also other non-country members within the INFOSAN framework, such as advisory group members, WHO regional food safety advisors; FAO regional food safety officers, or other regional food safety authorities.

[40] WHO (2006), pp. 1–6.

[41] *Id.*, p. 2.

[42] *Id.*, p. 4.

[43] Ruger (2014).

[44] *See e.g.* Forrest (2000).

aggravated the incident's scope and severity.⁴⁵ Despite the variety of considerations members have about information sharing, they face no legal responsibility when they decide not to notify or continue to share key information. To address this issue, 47 countries and regions, and 13 international organizations adopted, by consensus, the Beijing Declaration on Food Safety that urges all countries to actively notify the WHO of food safety incidents in a timely manner.⁴⁶

Second, the additional notification requirements of the International Health Regulations (IHR)⁴⁷ and other mechanisms such as the Global Early Warning System for Major Animal Disease, including Zoonoses (GLEWS) may weaken the effectiveness of INFOSAN and emergency responses. For example, when certain food safety outbreaks such as the avian influenza virus H5N1 may pose a serious international public health risk as per the IHR decision instrument (Annex 2), affected member states have a legal obligation to notify the IHR to facilitate information exchange so that the WHO is able to coordinate a global response (subject to a set of binding obligations, which however might not be relevant and tailored to food safety problems).⁴⁸ And such public health emergencies of international concern (PHEIC)⁴⁹ may, in principle, also fall within the scope of INFOSAN Emergencies, which operates pursuant to a different set of algorithms. Meanwhile, INFOSAN members are not legally obliged to notify.⁵⁰ Furthermore, food safety incidents that do not qualify as PHEIC are outside the IHR terrain but might be classified as INFOSAN Emergencies. Operating under the overarching IHR umbrella, these overlapping scopes, disparate algorithms, different criteria and requirements, and inconsistent memberships between INFOSAN and the IHR may cause fragmentation in food safety governance. Such fragmentation may render

---

⁴⁵According to the information provided to the INFOSAN by China on September 29, 2008, parents of the infants who consumed contaminated Sanlu formula filed complaints with the company as early as December 2007, and the company had detected melamine in its products in June 2008. The company only reported its findings to the local government in August 2008 and then a further delay prolonged the inadequate response to September 9 of that year, when the incident was reported to the provincial government. Some suggest that China was trying to cover up the incidents to protect its national image before the Olympic Games. *See e.g. China Milk Scandal Widens, More Recalls*, Canberra Times (Oct. 2, 2008).

⁴⁶High-Level International Food Safety Forum, *Report on High-Level International Food Safety Forum*, G/SPS/GEN/838 (Nov. 26–27, 2007).

⁴⁷WHO, International Health Regulations [hereinafter IHR] (2005). For a discussion of the 2005 IHR, see Baker and Fidler (2006), Gostin (2004a, b), Hardiman (2003).

⁴⁸The 2005 IHR is aimed at resolving problems of infectious diseases and therefore may be unsuitable for foodborne illness outbreaks. For example, Articles 21, 30, 31 and 32 deal with health measures and treatments to cross-border travelers, which is generally not applicable in the area of food safety. IHR, Articles 21, 30–32.

⁴⁹IHR, Article 1, Annex 2. "Public health emergency of international concern" means an extraordinary event that is determined, according to the decision instrument in Annex 2, to constitute a public health risk to other States through the international spread of disease and to potentially require a coordinated international response.

⁵⁰WHO (2006), p. 5.

cross-sectoral coordination at the national level extremely challenging and frustrate the effectiveness and the promptness of countries in risk management and risk communication as a whole.

Indeed, food safety governance requires multidisciplinary risk regulation, and may have implications for the international spread of human or animal diseases. The comprehensive information sharing and emergency response system then, suggests the inseparability of IHR, GLEWS, and INFOSAN to a certain extent. However, the balance between efficiency and effectiveness here has not been clarified by the WHO, the FAO, or the OIE, and a more streamlined approach throughout multiple interfaces is desirable.

3.3.3 Capacity Building Leadership and Assistance

Finally, the FOS helps the WHO provide leadership in assisting member states in building a risk-based and integrated food safety system at the national level.[51] Specifically, the FOS and the WHO regional offices help member states analyze their food systems and legal frameworks to build, update, or refine necessary legal and technical infrastructures and therefore improve member states' capacity to respond to food safety incidents.[52]

Capacity building serves as a fundamental and necessary supporting infrastructure for the first two normative activities. As pointed out in its report on advancing food safety initiatives, the WHO admits that "[m]any Member States still lack the necessary surveillance capacity for outbreak detection, assessment and response."[53] In particular, many developing countries are poorly equipped in terms of basic institutional framework, trained personnel, as well as technical and financial resources. In addition, many commentators have called for attention to developing countries' insufficient participation in the CAC international standard-setting process.[54] Multinational food companies have increasingly stretched their supply chains to many developing countries, such as Argentina, Brazil, China, India, and Indonesia, which together account for a major share of the global food supply.[55] Without an adequate core capacity in their food safety systems, these countries are unable to reasonably ensure the safety of food products that are exported and transported worldwide. Inadequate capacity translates into the inability of a country to control, monitor, and to notify the international community when a foodborne outbreak occurs.[56] Therefore, capacity building assistance to countries in need is of

---

[51] WHO, *supra* note 3, pp. 25–28.
[52] *Id.*
[53] *Id*, p. 21.
[54] *See generally* Livermore (2006).
[55] Keener (2010), pp. 139–40.
[56] Lin (2011).

crucial importance in the overall effectiveness of the WHO's normative activities in global food safety governance.[57]

The FOS leads capacity building programs mostly in collaboration with the WHO regional offices, aiming to promote participation in the CAC activities, to enhance surveillance and information-sharing capacity, and to help member states establish new or strengthen existing food safety regulatory systems based on risk, science, and relevant international standards.[58] In order to prepare member states' abilities to perform core functions and identify and solve problems in their food system, such as preventing, detecting, and managing food safety incidents, a set of modern food safety laws and a well-founded enforcement structure are needed. Subsequent core capacity development at the domestic level, such as training of inspection personnel, increasing laboratory competence, referencing to international standards and codes of good practices, and strengthening emergency response preparation,[59] empowers the states to collectively limit global foodborne illness burden.

3.4 The Gap Between Normative Capacity and Activity and the Marginalization of the WHO in Global Food Safety Governance

The role of the WHO in global food safety governance has been ancillary, for the organization has rather engaged in non-binding normative activities with a technical, scientific, and supplementary orientation at both national and international levels. First, the WHO has not employed to the necessary extent the normative authority given by the Constitution,[60] partly ensuing from the ossified bureaucracy, power politics, and budgetary weakness that make it a "compromised," if not an "irrelevant" organization.[61] The WHA, represented by all member states, has not exercised the Article 19 treaty-making capacity until the adoption of the Framework Convention on Tobacco Control (FCTC) in 2003.[62] As to its normative capacity to promulgate international regulations, the WHO has promulgated two international regulations under Articles 21(a) and 21(b)—the World Health

---

[57] As put by a leading scholar in global health law, "[c]apacity building must be a central focus of any effective global health governance regime." Gostin (2007), p. 378.
[58] WHO, *supra* note 3, pp. 25–28.
[59] Milen (2001).
[60] *See* Sridhar and Gostin (2011); Fidler (1998), p. 1079; Taylor (1996).
[61] Ruger (2014), p. 697. As criticized by Jack C. Chow, the WHO has become "outmoded, underfunded, and overly politicized. In a world of rapid technological change, travel, and trade, the WHO moves with a bureaucracy's speed. ... Taken together, these myriad dysfunctions are rendering the WHO closer and closer to irrelevancy in the world of global health." Chow (2010).
[62] WHO (2005).

Regulations in 1948 (the Nomenclature Regulations)[63] as well as the International Health Regulations (IHR) which was first adopted in 1951 as the International Sanitary Regulations (ISR), renamed in 1969, and significantly revised in 2005.[64]

Second, with regard to global food safety issues, the WHO has clearly preferred a flexible and non-binding approach through the provision of recommendation, guidelines, scientific advice, and information exchange, while abstaining from assuming any international agreement for over 65 years. There are no multilateral agreements under the WHO aimed at setting legal rights and obligations of member states regarding international cooperation, technical assistance, or risk management and communication so as to facilitate global food safety governance. In 2010, the WHO Executive Board finally suggested that the WHA adopt a resolution recognizing the need for an international agreement on global food safety.[65] In the same year, however, the sixty-third WHA did not accept such suggestions and refrained from assuming leadership in coordinating the various initiatives in global food safety governance. The Resolution WHA63.3 can thus be interpreted as a mere repetition of the existing INFOSAN mechanism. The Resolution again urged member states "to participate fully as members of the INFOSAN in its activities," and asked the Director-General to strengthen the existing strategy for food safety and the INFOSAN emergency function.[66] Nothing in the Resolution recognized the importance of an international agreement or an overarching framework dedicated to global food safety governance.

The WHO's role in global food safety governance seems to continue to be limited to the scientific and technical realm, as the organization is reluctant to assume leadership in this area of global health priority. The clear gap between the WHO's normative capacities given by its Constitution and normative activities on the ground may ultimately make the organization's institutional legitimacy vulnerable.[67] To be sure, the WHO's predilection for voluntary and supplementary approach may in the end provide a stronger moral basis for food safety governance as opposed to legal compulsion, but it may also make the organization ill-prepared for future international treaty administration and marginalize its role in coordinating global efforts. As the pivot of global health law and governance, the WHO needs first to take active steps to build functional links between itself and other stakeholders and partners, such as the WTO and multiple non-state actors (public health, trade, agriculture, science, and industry), some of which have emerged as

---

[63] See Doull and Kramer (1948), pp. 1379, 1400; Burci and Vignes (2004), p. 153.
[64] Gostin (2004a, b).
[65] WHO, *Advancing Food Safety Initiatives*, p. 2, EB126.R7 (Jan. 21, 2010).
[66] WHA, Resolution WHA63.3, WHA63/2010/REC/1 (May 21, 2010).
[67] The institutional legitimacy of an international organization justifies its position to exercise authority in a given field. The gap between the WHO's mandate and performance have a potential impact on the agency's institutional legitimacy and may in turn weaken its normative capacity to articulate in practice. For more discussion on institutional legitimacy in the global context, *see* Rocheleau (2011), pp. 562–564.

potential alternatives to the WHO in this issue area. The WHO needs to further solidify such functional links to coordinate efforts in global food safety, which may be challenging without an international agreement or an alternative inter-institutional framework especially when the WHO Constitutions has substantial procedural barriers to inter-institutional coordination.[68] Regrettably, the WHO has yet to show its political will and practical capability of fostering an overarching framework to coordinate future international cooperation efforts in global food safety governance.

References

Alemanno A (2007) Trade in food: regulatory and judicial approaches in the EC and the WTO. Cameron May, London
Baker MG, Fidler DP (2006) Global public health surveillance under new international health regulations. Emerg Infect Dis 12:1058
Burci GL, Vignes C-H (2004) World Health Organization. Kluwer Law International, The Hague
Charnovitz S (2002) Triangulating the World Trade Organization. Am J Int Law 96:28, 51
Chow JC (2010) Is the WHO becoming irrelevant? Why the world's premier public health organization must change or die. Foreign Policy
Doull JA, Kramer M (1948) Public Health Rep (1896–1970) 63:1379, 1400
Fidler DP (1998) The future of the world health organization: what role for international law? Vand J Transnat'l L 31:1079, 1088
Forrest M (2000) Using the power of the World Health Organization: the international health regulations and the future of international health law. Colum J Law Soc Probs 33:153, 166–167
Gostin LO (2004a) International infectious disease law: revision of the world health organization's international health regulations. JAMA 291:2623–2627
Gostin LO (2004b) International infectious disease law: revision of the World Health Organization's international health regulations. JAMA 291:2623–2626
Gostin LO (2007) Meeting basic survival needs of the world's least healthy people: toward a framework convention on global health. Geo Law J 96:331, 376
Hardiman M (2003) The revised international health regulations: a framework for global health security. Int J Antimicrob Agents 21:207
Käferstein FK, Abdussalam M (1999) Food safety in the 21st century. Bull World Health Org 77 (4):347–351
Käferstein FK et al (1997) Foodborne disease control: a transnational challenge. Emerg Infect Dis 3(4):503–510
Keener L (2010) Capacity building: harmonization and achieving food safety. In: Boisrobert CE et al (eds) Ensuring global food safety: exploring global harmonization. Academic, London
Lin C-F (2011) Global food safety: exploring key elements for an international regulatory strategy. Va J Int Law 51(3):637, 685–686
Lin C-F (2013) Scientification of politics or politicization of science: reassessing the limits of international food safety lawmaking. Columbia Sci Technol Law Rev 15:1

---

[68]WHO Constitution, Articles 69, 70, and 72; *See* Ruger (2014), p. 698. Ruger emphasizes "Despite its titular role as the coordinating health agency, the WHO constitution makes coordination arduous, requiring a two-thirds majority for approval of many WHO relations with other organizations," which "renders WHO efforts to coordinate and involve other organizations difficult, if not impossible." *Id.*

Livermore M (2006) Authority and legitimacy in global governance: deliberation, institutional differentiation, and the Codex Alimentarius. N Y Univ Law Rev 81:766

Matsushita M et al (2006) The world trade organization: law, practice, and policy, 2nd edn. Oxford University Press, New York

Milen A (2001) What do we know about capacity building? An overview of existing knowledge and good practice. WHO, Geneva

Motarjemi Y et al (2001) Future challenges in global harmonization of food safety legislation. Food Control 12(6):339, 340–341

Rocheleau J (2011) International institutional legitimacy. In: Chatterjee DK (ed) Encyclopedia of global justice. Springer, New York

Ruger JP (2014) International institutional legitimacy and the World Health Organization. J Epidemiol Community Health 68(8):697, 698

Silverglade B (2000) The WTO agreement on sanitary and phytosanitary measures: weakening food safety regulations to facilitate trade? Food Drug Law J 55:517, 518–524

Sridhar D, Gostin LO (2011) Reforming the World Health Organization. J Am Med Assoc 305 (15):1585–1586

Taylor AL (1996) An international regulatory strategy for global tobacco control. Yale J Int Law 21:257–304

Trachtman JP (2006) The world trading system, the international legal system and multilevel choice. Eur Law J 12:469, 480

Veggeland F, Borgen SO (2002) Changing the Codex: the role of international institutions, Working Paper 2002–16. Norwegian Agricultural Economics Research Institute

WHO (2004) Inauguration of the International Food Safety Authorities Network (INFOSAN). http://www.who.int/foodsafety/fs_management/infosan_1007_en.pdf

WHO (2005) WHO Framework Convention on Tobacco Control

WHO (2006) The International Food Safety Authorities (INFOSAN) users guide. http://www.docfoc.com/infosan-user-guide-final

WTO (2008) Understanding the WTO, 5th edn. WTO, Geneva

Chapter 4
The Right to Food in International Law with Case Studies from the Netherlands and Belgium

Bart Wernaart and Bernd van der Meulen

Abstract In this chapter, the enforceability of the right to adequate food is discussed in the context of industrialized countries. The right to food as a human right can be considered the fundament of food law. Human rights in themselves occupy a special position in the field of law. On the one hand they encompass rights of a high moral value which goes beyond the boundaries of a State or the consent of a State to be bound by it. On the other hand, human right agreements are put in the form of international treaties, whose effect is greatly depending on the willingness of its member States to act in compliance with their commitments. Therefore, enforcing an international human right in a domestic court, such as the right to adequate food, is not *per se* a matter of course. Two issues appear to be highly influential in determining whether an international human right can be effectively invoked in a domestic court. The first is the alleged difference between civil and political rights on the one hand, and economic, social and cultural rights on the other hand. Traditionally, it is assumed that the first type of rights require government abstaining and are therefore enforceable. The latter type implies government action and are not enforceable due to a margin of discretion the national governments enjoy in implementing these rights. However, there are sound arguments to oppose this traditional approach in human rights typology. These arguments are frequently pointed out in the context of the United Nation's specialized institutions as well as in literature. The second issue is the working of the domestic constitution that usually regulates the effect of international law in the domestic legal order. A case study of two industrialized countries who are favorable to human rights—the Netherlands and Belgium—was conducted. Where normally the right to food is addressed in the context of developing countries, poverty and large scale hunger, the selected countries do not suffer such constraints. Instead, the circumstances within these countries would allow an enforceable right to food to work. The case

B. Wernaart
Ethics and Law, Fontys University of Applied Sciences, Eindhoven, The Netherlands
e-mail: b.wernaart@fontys.nl

B. van der Meulen (✉)
Food Law, Wageningen University, Wageningen, The Netherlands
e-mail: bernd.vandermeulen@wur.nl

study reveals that the coincidental constitutional context of a country may be of greater influence to the enforceability of internationally recognized human rights, rather than the content of the rights in itself. In both countries the right to food can hardly be enforced through the domestic courts, in contrast to what these countries communicate in the international arena.

4.1 Introduction: The Right to Adequate Food Law Between the Markets and Human Rights

From this book, food law emerges as a building consisting of a wide variety of doors and windows, rooms and corridors. Can it be argued that this building of food law rest on a foundation? If so, what would this foundation be? Legal economists would be inclined to approach this question from the perspective of market failure. If the market fails to ensure values such as the safety of food, legislators need to take action. From a human rights perspective, however, this legal economic approach is flawed due to its Darwinistic nature. In Darwinism as in liberal economics, development hinges on a struggle to live with a survival of only the fittest. The notion of survival of the fittest implies the perishing of those who do not qualify as fittest. From the human rights perspective, it is not acceptable to simply give up on those who are unable to make their dollar count in the marketplace. If we want food law to protect each and every consumer not just the economic majority, the market economy cannot be the sole foundation of food law. In addition—or instead—human rights can be seen as the underlying notion of justice of any legal system and of food law as well. At the very least, human rights provide a yardstick to gauge when the effects of the market economy need to be re-considered because of the costs they inflict on the most vulnerable members of society.

This analysis explores to what extent the human right to adequate food can be viewed as, or can be developed into, the fundamental foundation of food law.

As will be elaborated below, the concept of an adequate food encompasses three elements: (1) food supply: there has to be a sustained quantity and quality of food that is sufficient to maintain a healthy and active life; (2) food safety: food should be free of adverse substances; (3) Cultural acceptability; food should be acceptable to the consumer, and adjusted to the consumers' way of life and beliefs, which includes foods that may be kosher, halal, GMO free, among other characteristics.

With regard to the sustained availability of food, three state obligations are distinguished: the obligation to respect the rights of people to feed themselves; the obligation to protect people in their exercise of this right from the interference by other people and finally; the obligation to provide food in situations where people through no fault of their own are unable to feed themselves.

From a legal perspective, it is important to distinguish state obligations regarding the population at large and from those regarding the individual. Ensuring food security for the population at large requires policies on agriculture and social security. However, it can only be maintained in any meaningful way that

Table 4.1 Aspects of the right to food

	Respect		Protect		Fulfill	
	Collective	Individual	Collective	Individual	Collective	Individual
Availability	Land Rights Policies	Legal Protection Against Expropriation			Agricultural Policy Social Security Policy	Justiciability of the Right to Food
Safety			Food Safety Law	Product Liability Law	Promote Healthy Eating Habits/ Education	
Acceptability	Consumer Autonomy and Freedom of Choice		Food Labeling Law/Misleading Advertising Directives		Culture Conscious Food Aid	

individuals have rights if the legal systems provides them with remedies that they can invoke in a court of law. The state's obligations regarding the food supply are represented in Table 4.1.

Some of the state's obligations have tentatively been filled in the table. Others have been left blank for the moment. The table can be seen as an invitation for further research on the interrelatedness of the human right to food and food law. As the table demonstrates, the food law can largely be understood as the legislator living up to the obligation to protect the population at large from the hazards of unsafe food. Apart from the fact that it is seldom framed as living up to human rights obligations, this obligation is relatively unproblematic.

In this chapter, we will first establish where the right to food fits within the human rights system. Then we will elaborate on the enforceability of the right to food, using two case studies. The case studies feature two Western, developed countries, generally favorable to human rights, and avoid the discussion in the context of underdevelopment, poverty and large-scale hunger. In such context, questions regarding the functioning of human rights may drown in economic adversity. In economically developed countries, the question regarding the human right to food can be approached as a more purely legal question. It will be seen that states tend to argue that the right to food needs to be realized through policy. This means that they place emphasis on collective obligations and deny accountability in the courts of law. In so far as this discussion focusses on enforceability, it places emphasis on the individual dimensions outlined in Table 4.1, and how they impact availability, safety, and acceptability.

4.2 Human Rights

In (international) law, human rights occupy a special position. On the one hand, they reflect fundamental values on which societies are built, and therefore have a high moral content that transcends the boundaries of the state's consent to be bound by them. Simply because a person is born, she/he has the right to live a life with human dignity, and therefore, the place where a person is born should be irrelevant. On the other hand, human rights are usually expressed in the form of international treaties. It is generally accepted in the law on treaties[1] that the sovereignty of the state is respected, and therefore, a state cannot be bound by international agreements without its consent. In terms of legal philosophy, human rights balance somewhere between natural law and legal positivism.[2]

Since the World War II, an impressive global system that aims to protect human rights has been adopted. This is done on a global level under supervision of the United Nations, but also on a regional level on most continents of the world, supervised by bodies installed for this purpose. On a global level, the Universal Declaration of Human Rights (UDHR), adopted by the UN General Assembly in 1948,[3] can be seen as a starting point for the development of internationally recognized human rights. In 1966, based on this declaration, two treaties were adopted: The International Convention on Civil and Political Rights (ICCPR), and the International Convention on Economic, Social and Cultural Rights (ICESCR).[4] Since then, also treaties that aim to protect particular groups of individuals were adopted, in which most human rights are specified to their particular situation or problem. The member states of the UN sign and ratify these treaties, and consequentially, human rights are mostly (but certainly not exclusively) formulated as obligations to these states, in varying degrees of compliance. To encourage states to implement the ratified human rights accordingly, reporting mechanism have been implemented in most human rights treaties. Such mechanisms oblige the ratifying states to periodically submit reports on the measures taken to implement the human rights at issue. Occasionally, but with considerably less enthusiasm, additional complaint procedures were adopted through which other states, and in some cases also individuals and/or NGO's, can file complaints against the ratifying state on the implementation of particular rights. An overseeing body (most already existing to monitor the periodic reports) is usually appointed to receive and comment on these complaints.

On the regional level, in Europe, Africa and the Americas, human rights treaties with equivalent human rights compared to the global documents have been adopted, but specified to the particularities of each continent. Europe and the Americas roughly adopted the same structure as the global human rights system, separately

---

[1] Vienna Convention on the Law of Treaties (1969), Articles 11–18.
[2] Mégret (2010), Chapter 6; McCrudden (2007).
[3] The universal declaration of human rights (1948).
[4] A/RES21/2200 (1966).

addressing civil and political rights on the one hand, and Economic, Social and Cultural (ECOSOC) rights on the other. In Africa, all human rights were adopted simultaneously. Also on the regional level, reporting and complaint procedures have been installed.

The right to adequate food is firmly embedded in the global and regional human rights system, mostly stipulated in the sphere of ECOSOC rights, and in specialized treaties aiming to protect particular groups of individuals. Indeed, from many wells the right to food thus streams throughout the international human rights system, with its final goal to benefit the final rights holder: the individual.[5] Notwithstanding the wide recognition of the right to food, and the many attempts to further clarify its content, this human right is perhaps more urgently recalled, and more often violated than any other human right.[6] It is not without reason thus, that the matter of state obligations and enforceability of human rights (more specifically, ECOSOC rights) is especially discussed in the context of the right to adequate food.[7]

4.3 The Enforceability of the Right to Adequate Food

The usual approach toward law, especially when adopted in a domestic parliament, is that violations of law can be brought before a court. The court then will have to decide whether the contested action, or lack of action, is in violation with the democratically established legal standard. Such proceedings can be found in the spectrum of private law, in which one individual sues the other individual, and in the spectrum of public law, in which legal conflicts between an individual (or a group of individuals) and a government can be brought before a court of law. Since human rights mostly address the relationship between the state and the individual, most case law and academic debate on the enforceability of human rights, concerns this relation. However, there is an increasing debate concerning the horizontal effect of human rights, between citizens and/or companies and citizens.[8]

Whether or not the right to food should be an enforceable right is a matter that is heavily debated in the international arena. On the one hand, specialized institutions such as the Committee on Economic, Social and Cultural Rights, and the special rapporteurs on the right to adequate food defend the position that the right to food

---

[5]Wernaart (2010).

[6]Alston and Tomasevski (1984), p. 9. Today, this assessment is still accurate, www.srfood.org, www.FAO.org.

[7]Kent (2008), Zeigler (2002), Eide (1999).

[8]OECD, *OECD Guidelines for Multinational Enterprises* (2011) Enterprises and Social Policy, adopted by the Governing Body of the International Labour Office at its 204th Session (Geneva, November 1977) as amended at its 279th (November 2000) and 295th Session (March 2006); and in more than one occasion in: *The voluntary guidelines to support the progressive realisation of the right to adequate food in the context of national food security*, adopted by the 127th session of the FAO Council, November.

should have at least, to some extent, an effect in the domestic courts. On the other hand, intergovernmental institutions seem to be reluctant in recognizing enforceability of internationally embedded ECOSOC standards fearing this would affect their sovereignty.

There are two major points of discussion regarding the matter. The first is related to the alleged distinction between civil and political rights on the one hand, and ECOSOC rights on the other hand, in line with the two UN treaties, the ICCPR and the ICESCR. A traditional approach toward human rights and state duties is that civil and political rights basically require governmental abstaining, and are enforceable through courts, whereas ECOSOC rights call for government action, which may not be enforced in court. This concerns both international and national standards. The matter will be discussed into more detail below. A second reason for debate is the fact that human rights are recognized in international legal texts. The constitutional system of a country plays a decisive role in determining how these rights have effect in the domestic legal order. It has always been a challenge for international human rights institutions to ensure that the international agreements on human rights are indeed complied with by their member states. Especially due to the fact that existing reporting and complaints procedures prove to be greatly dependent on the states' willingness to cooperate, and as a result are not as effective as originally foreseen. It is therefore no matter of course that individuals may have legal remedies against violations of the internationally embedded right to adequate food in either the national or the international arena (or both). This will be discussed in light of two case studies in Sect. 4.4.

4.4 The Right to Adequate Food in International Law

4.4.1 A Brief History of the Development of the Right to Food Since World War II

After the World War II, the United Nations was founded, with its main objective to maintain peace and security. Within this context, the right to food has been developed over decades. Three main drivers within the United Nations have been responsible for further clarifying the concept of the right to food.

The first is the treaty body of the ICESCR: the Committee on Economic, Social and Cultural rights. This Committee, in its capacity as overseeing body of the ICESCR reporting and complaints procedures, regularly adopts General Comments. These Comments are authoritative interpretations of the ICESCR Articles. Of particular importance are General Comment 3[9] and 9[10] on state obligations and

---

[9] E/1991/23, annex III at 86 (1991).
[10] E/C.12/1998/24 (1998).

domestic application of ICESCR Provisions, and general Comment 12[11] and 15,[12] on the right to food and the right to water. Furthermore, the Committee comments on the submitted periodic country reports.

The second driver is the work done within FAO context. Most famous are the adoption of the World Food Summit declarations[13] and the work that resulted from these summits, such as the adoption of the voluntary guidelines.[14] Also, much research has been conducted under the aegis of the FAO that is relevant for this topic.[15]

The third driver is the work done by the Special Rapporteurs. The Human Rights Council[16] and its predecessor, the Committee on Human Rights,[17] has regularly mandated special rapporteurs to investigate a particular human rights issue. In this light, Jean Ziegler has functioned as Special Rapporteur on the right to food in the period 2000–2008.[18] Since 2008, he is succeeded by Olivier De Schutter,[19] and another Special Rapporteur, Catarina de Albuquerque, was appointed to examine the right to safe drinking water and sanitation.[20] The Sub-Commission on the Promotion and Protection of Human Rights,[21] and its successor, the Advisory Council,[22] also installed mandates on the right to food and the right to water. Mr Asbjørn Eide[23] has functioned as Special Rapporteur on the right to food, and Mr El Hadji Guissé as Special Rapporteur on the right to drinking water.[24] The special rapporteurs report frequently on their research.

It is important to note here that in the sphere of the first and third driver, recommendations (or authoritative interpretations) are made by expert organizations, while in the sphere of the second driver, decisions are adopted in the context

---

[11] E/C.12/1999/5 (1999).
[12] E/C.12/2002/11 (2003).
[13] FAO Doc. WSFS 2009/2 (2009).
[14] FAO Council, 127th Session (2004).
[15] Knuth and Vidar (2011).
[16] A/res/60/251 (2006) *Human Rights Council*, Section 6.
[17] Economic and Social Council Resolution 5 (I), 16, (1946).
[18] Installed by Commission on Human Rights Resolution 2000/10, 17 April 2000, Section 10.
[19] Mr Olivier De Schutter was appointed Special Rapporteur on the right to food on March 26, 2008, by the Human Rights Council, his mandate was extended for another 3 years in: A/HRC/RES/13/4, 14 April 2010, Human Rights Council Resolution.
[20] Installed by Human Rights Council Resolution A/HRC/RES/7/22, 28 March 2008. The mandate was extended for another 3 years in: Human Rights Council Resolution A/HRC/RES/16/2, 8 April 2011.
[21] Before also named: *'Sub-Commission on the Prevention of Discrimination and Protection of Minorities.'*
[22] Human Rights Council Resolution A/HRC/RES/5/1, (2007, and A/HRC/RES/6/102, 27 September 2007).
[23] E/CN.4/Sub.2/1999/12 (1999).
[24] E/CN.4/Sub.2/1998/7, 10 June 1998; E/CN.4/Sub2/2004/20, 14 July 2004; E/CN.4/Sub.2/2005/25, 11 July 2005.

of intergovernmental relations, mostly based on voting procedures that require unanimous approval.

4.4.2 The Right to Food in UN Law, Regional Law, and Domestic Law

As stated above, the right to food is firmly embedded within the international and regional human rights system.

On a global level, within the context of the UN, the freedom from want is mentioned in the preamble of the most important general human right treaties.[25] Furthermore, the right to an adequate standard of living, including adequate food, is specifically stipulated in Article 25 UDHR,[26] and later as a treaty provision in Article 11 ICESCR. Also, the right to food is recognised as an independent right in documents that aim at the protection of a particular group of individuals. The right is stipulated in Article 27 of the International Covenant for the Rights of the Child,[27] Article 28 of the Convention on Persons with Disabilities.[28] In the particular context of healthcare and pregnancy, the right is recognized in Article 12 (2) of The Convention on the Elimination of All Forms of Discrimination against Women,[29] and Article 24 (2) of the International Covenant for the Rights of the Child.[30] Furthermore, the right to food is stipulated in several Articles of the Geneva Conventions and protocols[31] on humanitarian law that stipulate the right to food in the specific context of the beneficiaries of the treaties.[32]

---

[25] The Universal Declaration of Human Rights, the International Covenant on Civil and Political Rights and the International Covenant on Economic, Social and Cultural rights.

[26] 217 A (III) (1948) *The Universal Declaration of Human Rights.*

[27] A/RES/44/25 (1989) *Convention on the Rights of the Child,* Article 27.

[28] A/RES/61/106 (2007) *The Convention on the Rights of Persons with Disabilities,* Article 28.

[29] A/RES/34/180 (1979) *Convention on the Elimination of All Forms of Discrimination Against Women,* Article 12(2).

[30] A/RES/44/25 (1989) *Convention on the Rights of the Child,* Article 27.

[31] First Geneva Convention for the Amelioration of the Condition of the Wounded and Sick in Armed Forces in the Field (1949), Articles 32 (2) jo Article 27; Third Geneva Convention Relative to the Treatment of Prisoners of War (1949), Articles 20, 26, 28, 46, 51, 72; Fourth Geneva Convention relative to the Protection of Civilian Persons in Time of War (1949), Articles 15, 23, 49, 50, 55, 59, 76, 87, 89, 100, 108, 127; Protocol Additional to the Geneva Conventions (1949) and relating to the Protection of Victims of International Armed Conflicts (1977), Articles 54, 69, 70; Protocol Additional to the Geneva Conventions of 12 August 1949, and relating to the Protection of Victims of Non-International Armed Conflicts, 8 June 1977, Articles 5, 14, 18.

[32] In the four Geneva Conventions and their three protocols the human right to adequate food is recognized for the following groups of persons: medical personnel of a neutral country assisting one of the parties to a conflict, prisoners of war in general, prisoners of war who are being evacuated or transferred, civilians, detained civilians, and persons whose liberty is restricted. The starvation of civilians as means of pressure is forbidden in national and international armed

In line with the idea that human rights are "universal, indivisible and interdependent and interrelated,"[33] the right to food is often inextricably linked to other basic rights. There is an obvious link between the non-discrimination principle and the right to food. The Committee on Economic, Social and Cultural Rights underlined that "discrimination in access to food (...) constitutes a violation of the Covenant."[34] Such discrimination, especially in the sphere of the granting of social benefits, might be brought before a court more effectively than a direct violation of an ECSOC right.[35]

Another clear link exists between the right to food and the right to self-education. The right to food is frequently mentioned in a problem-specific context, such as land access in rural areas,[36] or access to resources in poor fishing communities,[37] indigenous people[38] and women.[39]

---

conflicts, as well as the deliberate destruction of foodstuffs and drinking water. Forced displacements of civilians leading to starvation are prohibited. There are also international rules concerning the protection of humanitarian assistance in occupied territories and during non-international armed conflicts. Also shipment/delivery of means of existence—including food—for prisoners of war or detained civilians should be allowed. In case of the establishment of a neutralized zone, the delivery of food supplies for (among others) the wounded and sick combatants or non-combatants and civilians should be agreed upon amongst the conflicting parties.

[33] A/CONF.157/23 (1993) the world conference on human rights, *Vienna declaration and programme of action*.

[34] E/C.12/1999/5 (1999), CESCR, *General Comment 12, Right to Adequate Food*, Sections 18 and 19, E/C.12/GC/20 (2009), CESCR, General Comment 20, *Non-Discrimination in Economic, Social and Cultural Rights (Art. 2, Section 2)*, Sections 6, 23, and 30.

[35] An interesting example is the extensive case law of the Belgian Constitutional Court that reviewed national legislation against the non-discrimination principle, in conjunction with *inter alia* the right to food.

[36] A/57/356 (2002), Jean Ziegler, *Report of the Special Rapporteur on the right to food to the General Assembly*, Chapter III.

[37] A/59/385 (2004), Jean Ziegler, *Report of the Special Rapporteur on the right to food to the General Assembly*, Chapter IV.

[38] A/RES/61/295 (2007) *United Nations Declaration on the Rights of Indigenous Peoples;* A/HRC/RES/7/14, 27 (2008) Human Rights Council Resolution, Section 12; A/RES/62/164 (2008) General Assembly Resolution, Section 12; A/CONF.151/26/Rev.1 (Vol. 1.), Rio de Janeiro (1992) *Agenda 21,* Chapter 26: *Recognising And Strengthening The Role Of Indigenous People And Their Communities; The voluntary guidelines to support the progressive realisation of the right to adequate food in the context of national food security*, adopted by the 127th session of the FAO Council, November 2004, preamble, Section 8.1; A/60/2005, 12 September 2005, Jean Ziegler, *Report of the Special Rapporteur on the Right to food to the General Assembly*, Chapter III. See also Lidija Knuth, *The right to food and indigenous people, how can the right to food help indigenous people?* Rome: FAO, 2009, especially Section 1.3.1.

[39] For instance: A/58/330, 28 August 2003, Jean Ziegler, *Report of the Special Rapporteur on the right to food to the General Assembly*, especially Section 22; A/CONF.177/20, Beijing, China, 4–15 September 1995, *Report of the fourth world conference of women*; A/HRC/RES/7/14, 27 March 2008, Human Rights Council Resolution, Sections 4–5.

The right to health(care) is often related to the right to adequate food. Unhealthy eating habits or malnutrition lead to bad health, and bad health may prevent an individual from consuming adequately.[40] In the context of pregnancy and health, the right to give (or receive) breastfeeding is stipulated in the ICRC, and the ESC.[41] The right to adequate food during pregnancy is stipulated in Article 12 (2).[42]

Enjoyment of the right to food could be considered a prerequisite to realise the right to life. Article 6 ICCPR implies, according to the Human Rights Committee, a duty for Member States to "take all possible measures to reduce infant mortality and to increase life expectancy, especially in adopting measures to eliminate malnutrition and epidemics."[43] Furthermore, to withhold (access to) food with the purpose to destroy life is in violation with Article II (c) of the Convention on the Prevention and Punishment of the Crime of Genocide.[44]

There is a strong interrelationship between the right to adequate food and the right to social security. This can be demonstrated by referring to two successive provisions of the ICRC. Article 26 ICRC stipulates that the State Parties "shall recognise for every child the right to benefit from social security (...) the benefits should (...) be granted, taking into account the resources and the circumstances of the child and persons having responsibility for the maintenance of the child (...)."[45] Article 27 ICRC states that "parents or other responsible for the child have the primary responsibility to secure, within their abilities and financial capacities, the conditions of living necessary for the child's development." The State Parties "shall take appropriate measures to assist parents and others responsible for the child to implement this right and shall in case of need provide material assistance and support programmes, particularly with regard to nutrition, clothing and housing."[46] Especially in industrialized countries, the right to food is both in case law and in country reports mostly discussed in light of the domestic system of social security. Furthermore, it is important to note here that in some international human rights

---

[40]See for instance: Office of the United Nations High Commissioner for Human Rights and the World Health Organisation, factsheet no. 31, *The right to health,* Geneva: UN, 2008.

[41]European Social Charter (revised), Strassbourg, 3.V.1996, Article 8; A/RES/44/25, 20 November 1989, *Convention on the Rights of the Child,* Article 24 (e).

[42]A/RES/34/180 (1979) *Convention on the Elimination of All Forms of Discrimination Against Women.*

[43]UN Human Rights Committee (1982), General Comment No. 6: *Article 6, Right to Life,* Section 5.

[44]A/RES/260 (III) (A) (1948) *International Convention on the Prevention and Punishment of the Crime of Genocide.*

[45]A/RES/44/25 (1989) *Convention on the Rights of the Child,* Art. 26.

[46]A/RES/44/25 (1989) *Convention on the Rights of the Child,* Art. 27.

treaties the choice was made not to include the right to food, but instead the right to social security.[47]

More than once, the interrelationship between the right to food and the right to education is stipulated. The combination of sound education and food are often considered to be key factors for successfully developing a country.[48] The CESCR emphasized that the right to education should imply, among others things, sanitation for both sexes and clean drinking water.[49]

Furthermore, elements of the right to food may relate to the right to freedom of thought, conscience and religion. In an interesting case, the European Court of Human Rights ruled that Article 9 European Convention on Human Rights, may under certain circumstances imply the obligation of a State to take into consideration the dietary wishes of a prisoner, coming forth from religious motives.[50]

On the regional level, the right to food is specifically stipulated in Africa, the Americas and the Islamic world.[51] In Africa, Article 15 of the additional protocol to the African Charter on Human and peoples' rights on the rights of women in Africa, stipulates the right to food security for women.[52] Also, the African Charter on the Rights and Welfare of the Child[53] stipulates the right to adequate food in Articles 14 and 20. Furthermore, in the African Union Convention for the Protection and Assistance of Internally Displaced Persons in Africa (Kampala Convention),[54] States Parties pledge themselves to provide internally displaced persons with adequate humanitarian assistance, including food and water,[55] and members of armed groups "shall be prohibited from denying internally displaced persons the right to live in satisfactory conditions of dignity, security, sanitation, food, water, health and shelter (...)"[56] In the Americas, the right to food is recognized in Article

---

[47] A/RES/2106 (XX) (1965) *The International Convention on the Elimination of All Forms of Racial Discrimination*, Article 5 (e) (iv); A/RES/34/180, 18 December 1979, *Convention on the Elimination of All Forms of Discrimination Against Women*, Article 11 (e); A/RES/429 (IV), 14 December 1950, *Draft Convention relating to the Status of Refugees*, Article 24 (1) (b); A/RES/45/158, 18 December 1990, *The International Convention on the Protection of the Rights of All Migrant Workers and Members of Their Families*, Article 27; European Social Charter (revised), 3 May 1996, Strassbourg, in: European Treaty Series 163, Article 12.

[48] Vivek (2008), Chapter 8.

[49] E/C.12/1999/10 (1999) CESCR, *General Comment 13, the Right to Education (Art. 13)*, Section 6.

[50] European Court of Human Rights (2013) case of Vartic v. Romania (no. 2, section 44–55).

[51] Asian Human Rights Charter (1998), Article 7.1.

[52] Additional Protocol to the African Charter on Human and Peoples' Rights on the Rights of Women in Africa, 11 July 2003, Article 15.

[53] OAU Doc. CAB/LEG/24.9/49 (1990) African Charter on the Rights and Welfare of the Child.

[54] African Union Convention for the Protection and Assistance of Internally Displaced Persons in Africa (Kampala Convention) (2009).

[55] African Union Convention for the Protection and Assistance of Internally Displaced Persons in Africa (Kampala Convention) (2009), Article 9 (2) (b).

[56] African Union Convention for the Protection and Assistance of Internally Displaced Persons in Africa (Kampala Convention) (2009), Article 7 (5) (c).

12 of The San Salvador Additional Protocol to the American Convention on Human Rights.[57] In the Islamic world, the right to food is stipulated in Article 17 of The Cairo Declaration on Human Rights in Islam,[58] and more in particular recognized in case of armed conflict (Article 3), and for children (Article 7). In Europe, the right to food is not explicitly recognized in a regional treaty. However, the European Social Charter stipulates ECOSOC rights, including most rights that closely relate to the right to food, including the right to social security (Article 12).[59]

In several countries, the right to food is recognized in the national constitutions. According to a FAO right to food study in 2011,[60] the right to food is recognized explicitly in the Constitution of 23 Countries. Furthermore, it is recognized implicitly in the constitutions of 33 countries, through a broader right or a directive principle. In addition, the right to food has effect in at least another 51 countries, due to the effect of international law in the domestic legal order. Considering all this, the right to food appears to be positive law in 107 countries. However, the fact that a right is part of positive law in a country does not necessarily mean that it is effectively implemented as such, or enforceable through the courts. To establish such things, a more in-depth analysis of the legal reality of states would be required, instead of a dogmatic comparison of constitutions.

4.4.3 *The Meaning of the Right to Adequate Food*

Article 11 ICESCR is often considered to be the earliest recognition of the right to food as a legally binding provision, and has the broadest scope, for the Covenant in which the right is stipulated addressed 'the right to everyone, and not a specific group'. Therefore, the content of the right to food is most often discussed in light of this Provision:

1. The States Parties to the present Covenant recognize the right of everyone to an adequate standard of living for himself and his family, including adequate food, clothing and housing, and to the continuous improvement of living conditions. The States Parties will take appropriate steps to ensure the realization of this right, recognizing to this effect the essential importance of international co-operation based on free consent.
2. The States Parties to the present Covenant, recognizing the fundamental right of everyone to be free from hunger, shall take, individually and through

---

[57]The San Salvador Additional Protocol to the American Convention on Human Rights (1988) adopted by the General Assembly of the Organization of American States, Article 19.

[58]Cairo Declaration on Human Rights in Islam (1990) adopted at the Nineteenth Islamic Conference of Foreign Ministers.

[59]European Social Charter (revised) (1996) Strassbourg, in: European Treaty Series 163.

[60]Knuth and Vidar (2011).

international co-operation, the measures, including specific programmes, which are needed:

(a) To improve methods of production, conservation and distribution of food by making full use of technical and scientific knowledge, by disseminating knowledge of the principles of nutrition and by developing or reforming agrarian systems in such a way as to achieve the most efficient development and utilization of natural resources;
(b) Taking into account the problems of both food-importing and food-exporting countries, to ensure an equitable distribution of world food supplies in relation to need.

Article 11 of the ICESCR was written in close cooperation with the FAO, which proposed Section (b), to provide for a legal basis for their 'freedom from hunger campaign' that was set up in 1960. It explains why sub (a) refers to 'the right to adequate food,' and sub (b) refers to 'freedom from hunger.'[61] The finally adopted version of Article 11 ICESCR was a compromise between countries who supported strong wordings that would include clear obligations and countries that preferred a larger margin of discretion for member states to implement the right in a way that suits best considering the particularities per country. Since the adoption of this provision, many attempts have been undertaken to further clarify the meaning of the right to adequate food. Especially because the wordings of Article 11 ICESCR, and other equivalent Articles, are perceived to be relatively vague by many countries, contributing to a denial of legal effect in these countries. Therefore, the Committee on Economic, Social and Cultural Rights stated in General Comment 12 that, "the right to adequate food is realised when every man, woman and child, alone or in community with others, has physical and economic access at all times to adequate food or means for its procurement."[62] In addition, the CESCR emphasized that the right to adequate food implies that Member States ensure: "The availability of food in a quantity and quality sufficient to satisfy the dietary needs of individuals, free from adverse substances, and acceptable within a given culture; and the accessibility of such food in ways that are sustainable and that do not interfere with the enjoyment of other human rights."[63] According to the CESCR, food should be adequate and sustainable, meaning that food is, "to a large extent determined by prevailing social, economic, cultural, climatic, ecological and other conditions," and implies, "long-term availability and accessibility."[64]

Furthermore, especially considering the second Section of Article 11 ICESCR, formulated in a negative understanding, the right to food has a certain minimum core that is the absence of hunger (insufficient quantity of food) and malnutrition

---

[61] Alston and Tomasevski (1984).
[62] E/C.12/1999/5 (1999) CESCR, *General Comment 12, Right to Adequate Food*, Section 6.
[63] E/C.12/1999/5 (1999) CESCR, *General Comment 12, Right to Adequate Food*, Section 8.
[64] E/C.12/1999/5 (1999) CESCR, *General Comment 12, Right to Adequate Food*, Section 7.

(insufficient quality of food).[65] Lastly, it is generally accepted that the right to adequate food includes the right to water,[66] and the right to (give or receive) breastfeeding.[67]

4.4.4 Enforcing the Right to Food: Human Rights and State Obligations

In general, human rights obligations are state obligations, and formulated as such. Of course, the realisation of human rights is not the sole responsibility of a government: all actors in society should contribute to the realisation of human rights,[68] with a primary responsibility for the individual who seeks to enjoy her/his human rights.[69] However, the international human rights system is shaped in a legal form, in which countries, through ratification of those documents, take the responsibility to implement human rights. In case of the right to food, Article 11 ICESCR specifies that this should be done by states individually and through international cooperation (the so called extra-territorial obligations[70]).

As already pointed out above, a traditional approach toward human rights implies that civil and political rights on the one hand bring mostly negative obligations for the state, while on the other hand ECOSOC rights imply positive obligations for the State. This distinction may be traced back to the choice made to split the rights stipulated in the UDHR in to two groups, separately recognized in UN treaties: the ICCPR and the ICESCR. While originally it was the intention to adopt one legally binding treaty stipulating the UDHR rights, the choice was made to draw two Covenants instead of one. This was mainly due to different perceptions of state duties between Eastern and Western countries and also between developing and developed countries.[71] This distinction has led to a widely accepted perception that civil and political rights, due to the fact that they imply government abstaining,

---

[65] www.wfp.org/hunger.

[66] The Commission on Human Rights requested the Special Rapporteur on the right to food in 2001 *'to pay attention to the issue of drinking water, taking into account the interdependence of this issue and the right to food'*. See: Commission on Human Rights Resolution 2001/25, 20 April 2001. See furthermore: A/56/210, 23 July 2001, Jean Ziegler, report of the Special Rapporteur on the right to food to the General Assembly, Chapter IV; E/CN.4/2003/54, 10 January 2003, Jean Ziegler, report of the Special Rapporteur on the right to food, Chapter II. See also: E/C.12/2002/11, 20 January 2003, CESCR, *General Comment 15, The Right to Water*. Furthermore, see the works of the Special Rapporteurs on the right to water: Mr El Adji Guissé for the sub-commission on human rights, and Mrs Catarina de Albuquerque for the Human Rights Commission.

[67] Article 24 (2) ICRC.

[68] Kent (2005), Chapter 6.

[69] Eide (2010).

[70] Skogly (2007).

[71] Ssenyonjo (2009).

are better suited for enforceability in the domestic courts, while ECOSOC rights, implying government action, are not, due to the margin of discretion of governments to determine what kind of action the particularities of their country requires. While this perception is (with some exceptions[72]) widely shared amongst the UN member states, the specialized institutions within the UN have tried to nuance this perception with compelling arguments. This is mostly done by introducing a typology of duties that can be applied to all human rights, without distinction. Inspired by various authors,[73] this typology was introduced in the UN human rights system and applied to the right to food by Asbjørn Eide in his capacity as Special Rapporteur.[74] In short, it is suggested that three different types of State Obligations are implied by all human rights: the duty to respect, the duty to protect and the duty to fulfill. In General Comment 12, the CESCR used this terminology to further clarify the State duties that come forth from ratifying the right to food:

> The obligation to respect existing access to adequate food requires States' parties not to take any measures that result in preventing such access. The obligation to protect requires measures by the state to ensure that enterprises or individuals do not deprive individuals of their access to adequate food. The obligation to fulfill (facilitate) means the state must pro-actively engage in activities intended to strengthen peoples' access to and utilization of resources and means to ensure their livelihood, including food security. Finally, whenever an individual or group is unable, for reasons beyond their control, to enjoy the right to adequate food by the means at their disposal, states have the obligation to fulfill (provide) that right directly. This obligation also applies for persons who are victims of natural or other disasters.[75]

While this typology is widely used amongst the UN institutions, especially when overseeing the various reporting mechanisms, it is mostly rejected by the member states.

---

[72]Constitutional Court of South Africa, *Government of the Republic of South Africa v. Irene Grootboom and others,* Case CCT 11/00, 4 October 2000. Swiss Federal Court, *V. v Resident Municipality X. and Bern Canton Government Council,* Case BGE/ATF 121 I 367, 27 October 1995. Supreme Court of India, *People's union for civil liberties v. Union of India and others,* case <!--Folio is a division of Open Market inc. Visit us at http://www.openmarket.com--> <!DOCTYPE HTML PUBLIC"-//W3C//DTD HTML 3.2//EN"> <!--Folio is a division of Open Market inc. Visit us at http://www.openmarket.com--> <!DOCTYPE HTML PUBLIC"-//W3C//DTD HTML 3.2//EN"> W.P(C) No. 196 of 2001, 23 July 2001. For the particularities of the later, see for more information: Human Rights Law Network (2009).

[73]Alston and Tomasevski (1984).

[74]E/CN.4/Sub.2/1999/12, 28 June 1999, *Updated study on the right to food, submitted by Mr. Asbjørn Eide in accordance with Sub-Commission decision 1998/106.*

[75]E/C.12/1999/5 (1999) CESCR, *General Comment 12, Right to Adequate Food,* Section 15.

4.5 The Enforceability of ECOSOC Rights

In a traditional approach, ECOSOC rights require government action, and human right articles stipulating these rights are usually not considered sufficiently specific to be enforced through domestic courts. This is due to the fact that such an article cannot take into account all particularities of each country around the globe, and therefore, governments need a margin of appreciation to implement the right to food, through government action, the way that best suits the needs of their specific country, and with the resources that are available. This is often taken as an argument in court to deny enforceability of ECOSOC rights when a claimant tries to seek legal remedies against an alleged violation of the right to food. This view is often substantiated by referring to Article 2 ICESCR, stipulating that,

> Each State Party to the present Covenant undertakes to take steps, individually and through international assistance and co-operation, especially economic and technical, to the maximum of its available resources, with a view to achieving progressively the full realisation of the rights recognised in the present Covenant by all appropriate means, including particularly the adoption of legislative measures.

In their General Comments 3 and 9, the CESCR oppose such a traditional approach towards the enforceability of ECOSOC rights. First of all, the CESCR defends the position that Article 2 ICESCR should not be thus understood that Member States are given some sort of carte blanche to implement ECOSOC rights as they see fit, without any consequences. Rather, the CESCR argues that Article 2 implies immediate core obligations concerning the ECOSOC rights stipulated in the ICESCR, requiring Member States "to ensure the satisfaction of, at the very least, minimum essential levels of each of the rights is incumbent upon every State Party."[76] Next to that, the obligation to progressive realisation should "not be misinterpreted as depriving the obligation of all meaningful content." This means that the States have an obligation of conduct, that is to "move as expeditiously and effectively as possible" towards full realisation of the ICESCR rights. This also means that "any deliberate retrogressive measures in that regard would require the most careful consideration and would need to be fully justified by reference to the totality of the rights provided for in the Covenant and in the context of the full use of the maximum available resources."[77] All this should contribute to the final goal, the realisation of ECOSOC rights: the obligation of result.

Second, the CESCR argued that phrase "all appropriate means" in Article 2 (1) ICESCR would imply "the Provision of judicial remedies with respect to

---

[76] E/1991/23, annex III at 86 (1991), 14 December 1990, CESCR, *General Comment 3: The Nature of States Parties' Obligations*, especially Section 10.

[77] E/1991/23, annex III at 86 (1991), 14 December 1990, CESCR, *General Comment 3: The Nature of States Parties' Obligations*, Section 9.

rights which may, in accordance with the national legal system, be considered justiciable."[78] The Committee underlined that:

> In relation to civil and political rights, it is generally taken for granted that judicial remedies for violations are essential. Regrettably, the contrary assumption is too often made in relation to economic, social and cultural rights. This discrepancy is not warranted either by the nature of the rights or by the relevant Covenant Provisions.

The CESCR added that, "there is no Covenant right which could not, in the great majority of systems, be considered to possess at least some significant justiciable dimensions."[79] The Committee even suggested that many ICESCR Provisions "are stated in terms which are at least as clear and specific as those in other human rights treaties, the Provisions of which are regularly deemed by Courts to be self-executing."[80] A provision is self-executing when it can be applied by a court without the need for any additional domestic legislation. The Committee listed 'by way of example',[81] some ICESCR Provisions that "would seem to be capable of immediate application by judicial and other organs in many national legal systems."[82] These Provisions are: 3, 7 (a) (i), 8, 10 (3), 13 (2) (a), (3) and (4), and 15 (3) ICESCR. The Committee underlined that it deemed arguments that would indicate that these provisions were not self-executing "difficult to sustain."[83] To conclude, the CESCR emphasizes the fact that there is no reason to assume that ECOSOC rights have no justiciable dimensions, and defends the position that at least some Articles are formulated sufficiently specific so that its content can be self-executing. As it seems, the CESCR uses the term self-executing to emphasize the possibility of direct application of the standard by a court, without the need of additional domestic legislation, while the term 'justiciability' is used in a broader context, underlining the possibility of any legal effect through the ECOSOC standard. The latter could also be indirect application, or the use of the standard as an interpretative standard.

---

[78] E/1991/23, annex III at 86 (1991), 14 December 1990, CESCR, *General Comment 3: The Nature of States Parties' Obligations,* Section 5.

[79] E/C.12/1998/24, 3 December 1998, CESCR, *General Comment 9, the domestic application of the Covenant,* Section 10.

[80] E/1991/23, annex III at 86 (1991), 14 December 1990, CESCR, *General Comment 3: The Nature of States Parties' Obligations,* Section 11.

[81] E/C.12/1998/24, 3 December 1998, CESCR, *General Comment 9, the domestic application of the Covenant,* Section 10.

[82] E/1991/23, annex III at 86 (1991), 14 December 1990, CESCR, *General Comment 3: The Nature of States Parties' Obligations,* Section 5.

[83] E/1991/23, annex III at 86 (1991), 14 December 1990, CESCR, *General Comment 3: The Nature of States Parties' Obligations,* Section 5.

4.6 The Enforceability of the Right to Adequate Food

In the list of self-executing provisions of the Committee, the right to adequate food was not included. However, the list was never meant to be exhaustive. It only addresses the possible self-executing nature of these provisions, and not necessarily any other legal effects that may fall under the scope of the broader concept of 'justiciability'. Within the three UN drivers as discussed above, there seem to be varying positions with regard to the enforceability of the human right to adequate food. On the one hand, the specialized institutions (labelled above as the first and third driver) seem to defend the position that the right to food should have some sort of legal effect in the domestic legal orders of the UN member states. On the other hand, the institutions in the second driver, usually taking decisions based on intergovernmental procedures, mostly requiring unanimous approval, seem to deny or even oppose such legal effect. The latter is no surprise, since in the second driver, the direct influence of UN member states, usually denying enforceability of ECOSOC rights in general, is most clearly reflected.

Although cautiously, the Committee in their General Comment 12, seems to defend the position that the right to food should have legal effect in the domestic legal order of the UN member states. The CESCR underlined that implementing the right to food, "can significantly enhance the scope and effectiveness of remedial measures (...)," and, "Courts would then be empowered to adjudicate violations of the core content of the right to food by direct reference to obligations under the Covenant."[84] However, a more straightforward approach can be noticed when the CESCR comments on the reports submitted by the member states. Especially during the sessions in which country delegations engage in a dialogue with members from the Committee and in their concluding observations, the CESCR is quite specific in their expectations: all ICESCR provisions, including the right to food, should be enforceable in the domestic courts.[85]

Also the rapporteurs on the right to food, perhaps in the strongest wordings, defend the position that the right to food should have legal effect in the member states. Jean Ziegler argued that "the right to food can be considered as justiciable by its very nature, and is therefore equal to civil and political rights."[86] Ziegler's successor, Olivier De Schutter, shares this view, and argues that states should adopt a framework legislation, "ensuring that the right to food is justiciable before national Courts or that other forms of redress are available (...)," This would

---

[84]E/C.12/1999/5, 12 May 1999, CESCR, *General Comment 12, Right to Adequate Food*, Section 33.

[85]Wernaart (2013).

[86]E/CN.4/2002/58, 10 January 2002, Jean Ziegler, report of the Special Rapporteur on the right to food, Section 49.

encourage, "Courts or other monitoring mechanisms, such as the human rights institutions (...) to contribute to ensure compliance with the right to food."[87]

However, especially in the context of the FAO hosted World Food summits, and the resulting Voluntary Guidelines, such enthusiasm towards legal effect of the right to food is not always shared. In the World Food Summit declarations, the matter of enforceability is simply ignored. The Voluntary Guidelines, finally adopted after stiff negotiations,[88] only, "invite" states "to consider (...) whether to include provisions in their domestic law (...) that facilitates the progressive realisation on the right to food in the context of national food security,"[89] or "to include provisions (...) to directly implement the progressive realisation of the right to food."[90] Regarding the enforceability of the right to food through juridical remedies, the guidelines merely underline that, "administrative, quasi-juridical and judicial mechanisms to provide adequate, effective and prompt remedies accessible, in particular, to members of vulnerable groups may be envisaged"[91] and, "States that have established a right to adequate food under their legal system should inform the general public of all available rights and remedies to which they are entitled."[92]

To conclude, the specialized UN institutions expect the right to food to be enforceable, while the member states seem to be reluctant in accepting this.

4.7 A Right to All and a Right to Each

In literature, but also within the framework of most human rights institutions, debates on the right to adequate food almost exclusively focus on developing countries. Understandable, and rightfully so, because it is in developing countries that peoples' right to adequate food most often fails to be realized. The right to food is then often discussed in light of feeding the population at large, i.e. a government

---

[87] A/HRC/9/23, 8 September 2008, Olivier de Schutter, report of the Special Rapporteur on the right to food to the Human Rights Council, Section 18.

[88] E/CN.4/2003/54, 10 January 2003, Jean Ziegler, report of the Special Rapporteur on the right to food, Chapter I, especially Section 24. See for a detailed analysis: See also Oshaug (2009).

[89] *The voluntary guidelines to support the progressive realisation of the right to adequate food in the context of national food security*, adopted by the 127th session of the FAO Council, November 2004, guideline 7.1.

[90] *The voluntary guidelines to support the progressive realisation of the right to adequate food in the context of national food security*, adopted by the 127th session of the FAO Council, November 2004, guideline 7.2.

[91] *The voluntary guidelines to support the progressive realisation of the right to adequate food in the context of national food security*, adopted by the 127th session of the FAO Council, November 2004, guideline 7.2.

[92] *The voluntary guidelines to support the progressive realisation of the right to adequate food in the context of national food security*, adopted by the 127th session of the FAO Council, November 2004, guideline 7.3.

obligation towards all, due to structural problems that surpass the matter of enforceability. However, in industrialized countries, it should—hypothetically—not be a problem to guarantee the right to food to each individual and to hold the government accountable in a court of law. It is to be expected that this right is invoked by the most vulnerable in society. The right to food should be a last resort when all existing social security legislation does not offer adequate protection to the individual. But it should also provide a yardstick to judge the adequacy and shortcomings of the social security system. In the first sense, the right to food functions as an individual right, in the second as a meta-law to which the lower law must conform. Here, the matter of enforceability plays a significant role. The concept of human rights was developed with the strong moral idea that due to the mere fact that a person is born, she/he has a right to lead a life in dignity.[93] Adequate food is a necessity to lead a dignified life. As discussed above, the specialized UN institutions strongly defend the position that the right to food implies that individuals should be empowered to bring violations of their rights before a court of law. This view is widely contested among the Member States.

4.7.1 Case Studies: The Netherlands and Belgium

In light of the discussion on the enforceability of the right to food, it may be enlightening to address, by way of example, the specific context of two industrialized countries and their approach to enforceability of the right to food: the Netherlands and Belgium. Both countries are relatively prosperous countries, have a stable democratic foundation, a functioning and accessible judiciary, and an extensive system of social security. Furthermore, both countries periodically report that they fully fulfill their duties that come forth from ratification of the right to food.[94] In these circumstances, one might expect that an individual would be able to claim her/his rights in court, when the individual falls through the cracks of the system of social security. The above mentioned FAO research suggests that in both countries, the right to food is indirectly embedded in the Constitution, and the constitutional mechanisms of both countries ensure that international law has supremacy over national law, and therefore, the right to food, as ratified by the Netherlands and Belgium is directly applicable in the national legal order of these countries.[95] However, as will be discussed in this section, the case law in both countries shows otherwise, and with some minor exceptions, the right to food turns out not

---

[93]The preamble of the UDHR stipulates that: *'All people are born free and equal in dignity and rights.'*

[94]See for the Netherlands for example: E/C.12/NLD/4-5, 17 July 2009, Sections 219–234; see for Belgium for example: E/C.12/BEL/3, 21 September 2006, Sections 418–558 (especially 541–558).

[95]Knuth and Vidar (2011), Chapter 6.

4.7.1.1 The Enforceability of the Right to Food in the Netherlands: Case Law Analysis

The Dutch Courts unanimously reject direct effect of Article 11 ICESCR. Although some variations to the theme can be found, there seem to be three main arguments why the courts consider the Article is not suitable for direct application. One relating to the treaty provision itself; one to Dutch constitutional law; and the last to the national interpretation of the provision at the time of ratification. First, the Article is, 'not binding on all persons,' as required by the Constitution for direct effect.[96] Second, the Article is not sufficiently precise for concrete use, and therefore further national legislation is required. Third, the legislator, in their ratification bill, considered that ICESCR rights in general would have no direct effect.[97] From this we can learn that the structure embedded in the Dutch Constitution[98] determines the scope of the competencies of the courts, and that the influence of the Legislature should not be underestimated. In general, the Courts state such arguments without taking into consideration the particularities of the case. Mostly, the cases concern people who support their claim to certain social benefits with a reference to the rights enshrined in Article 11 ICESCR. While there are also cases concerning people on low incomes, prisoners, elderly and disabled persons, a majority of the case law concerns the position of asylum seekers. In most cases, these asylum seekers reside illegally in Dutch territory, and for this reason are by Dutch legislation[99] excluded from all social benefits, except urgent medical care. The courts prefer to solve the matters by using domestic legislation. There are hardly any references to the authoritative interpretations of the CESCR, and when they are considered, they do not carry any weight of significance.[100]

---

[96]Kingdom of the Netherlands, Const. Art. 93, 94.

[97]Council of State of 19 April 2007, LJN BA4289; Central Court of Appeal 14 March 2011, NJB 2011, 755.

[98]Kingdom of the Netherlands, Const. Art. 93 94.

[99]The Dutch Linkage-Act. Wet van 26 maart 1998, *Stb*. 1998, 203, tot wijziging van de Vreemdelingenwet en enige andere wetten teneinde de aanspraak van vreemdelingen jegens bestuursorganen op verstrekkingen, voorzieningen, uitkeringen, ontheffingen en vergunningen te koppelen aan het rechtmatig verblijf van de vreemdeling in Nederland.

[100]Central Court of Appeal, 3 July 1986, TAR 1986, 215, and Central Court of Appeal, 22 December 2008, LJN: BG8789.

Somewhat different is the case law on Article 27 ICRC. While in general, also here the courts have the tendency to deny its direct application,[101] in some circumstances, the courts muster a milder approach, and use the article as an important interpretative norm, for instance in evaluating the margin of appreciation of the local administration in granting social assistance to children.[102] On several occasions, the Central Court of Appeal considered that in their decisions, the local administration had to take into account Article 27 ICRC when children were involved.[103] However, this approach towards Article 27 seems to be the exception to the rule, and is certainly not applied in case of illegally residing children.[104]

No case law of significance could be found on any other article stipulating the right to adequate food.

4.7.1.2 Dutch Qualified Monism

In essence, the Netherlands has a monistic system when it concerns the effect of international norms in the domestic legal order. In general, this means that international norms have direct effect in the domestic legal order, without the need for the adoption of further domestic legislation. However, the Dutch understanding of monism is limited to articles that are, 'binding on all persons.'[105] The courts then are obliged not to apply domestic law that is in violation with international provisions that are binding on all persons.[106] Oddly enough, it has never been clear who eventually determines when an international provision meets the requirements of being binding on all persons: the legislature or the judiciary. The official viewpoint of both the legislature and the judiciary is that it is the judiciary that has a final say in determining whether or not an international provision is binding on all persons. It is often portrayed that the court then evaluates this based on criteria that relate to the substance of the provision: its nature and content, as well as its wording. The opinion of the legislature could be used by the courts as a source of

---

[101] Mostly inspired by rulings of the Council of State, for instance: Council of State, 1 March 2005, JV 2005/176; 13 September 2005, JV 2005, 409; 15 February 2007, LJN AZ9524; 13 June 2007, www.rechtspraak.nl, 13 June 2007; 26 November 2007, www.rechtspraak.nl, 2 January 2008; 08 October 2010, LJN BO0685; 13 October 2010, LJN: BO0794; 22 February 2012, JV 2012, 200.

[102] The Dutch Work and Social Assistance Act. Article 16 (1) of this Act stipulates that *'to a person not entitled to assistance, the Mayor and Municipal Executive may, taking into consideration all circumstances, notwithstanding this Section, provide assistance if so required due to very urgent reasons.'* Original text in Dutch: *'Aan een persoon die geen recht op bijstand heeft, kan het college, gelet op alle omstandigheden, in afwijking van deze paragraaf, bijstand verlenen indien zeer dringende redenen daartoe noodzaken.'*

[103] 13 February 2007, LJN AZ8596; (2007) LJN BA6523 (regarding Dutch children); 6 October 2009, LJN BK0734; 20 July 2010, LJN BN3318 (regarding foreign children).

[104] Central Court of Appeal (2006) LJN AY9940; 7 April 2008, LJN BD0221; 14 July 2010, LJN BN1274.

[105] Const. Art. 93.

[106] Const. Art. 94.

inspiration, but the courts are autonomous in their decision on direct applicability. However, as can be observed in the case law on the enforceability of Article 11 ICESCR, the legal reality is somewhat different. The judiciary seems to greatly value the opinion of the legislature. From the parliamentary documents of the constitutional reforms and the ratification bills to human rights treaties, it can be deduced that it is the deliberate intention of the legislature to communicate its view on direct applicability of international standards, and also expects the courts to follow that view.[107] This has resulted in an established practice in which the Courts first considers the viewpoint of the legislature, and almost indiscriminately apply this to the case. Only when the legislature is unspecific or silent on the matter, the courts consider the substance of the provision. In case of the right to adequate food, as well as most ECOSOC rights, the legislature was very clear in its view: these rights are not binding on all persons, and therefore not suitable for direct application through the courts. Their main arguments pretty much resemble the usual concerns that are used in the global arena to deny enforceability of ECOSOC rights. That is mostly that most ECOSOC Provisions are not addressed to individuals but to the State,[108] and/or are focussed on the progressive realisation through implementation measures,[109] leaving a wide margin of discretion to the legislature. This was most clearly discussed during the parliamentary debates on the ratification bill of the ICESCR and ICRC. The Government argued that

> many provisions concerning substantive rights, embedded in Part III of the International Covenant on Civil and Political Rights, have, due to the content and wording of these provisions, and following the example of most provisions stipulating substantive rights of the European Convention, direct effect and can be applied by the Courts without the need for further legislation.' '(...) partly due to this, civil and political rights on the one hand, and economic, social and cultural rights on the other, are embedded in two separate documents, because in general the first group of rights is suitable for direct application, while the realisation of the second group often requires implementing measures.[110]

This approach of the Dutch Government, and taken into considering its apparent influence on case law, explains why the enforceability of the right to food—and ECOSOC rights in general—is consistently permanently denied by the courts, without any significant reference to the particularities of the case.

---

[107]Parliamentary Documents, II 1992–1993, (R1451), no. 3, p. 8.

[108]Explanatory Memorandum on the ratification Bill to the European Social Charter: Parliamentary Documents, II 1965–1966, 8606 (R 533), no. 6.

[109]Explanatory Memorandum on the ratification Bill to the ICESCR and ICRC: Parliamentary Documents, II 1975–1976, 13932 (R 1037), no. 3, pp. 12–13; Explanatory Memorandum on the ratification Bill to the CEDAW: Parliamentary Documents, II 1984–1985, 18950 (R 1281), no. 3, p. 7; Explanatory Memorandum on the ratification Bill to the revised version of the European Social Charter: Parliamentary Documents, II 2004–2005, 29941, no. 3, p. 4.

[110]Parliamentary Documents, II 1975–1976, 13932 (R 1037), no. 3, p. 13.

4.7.1.3 Dutch Reporting Behavior

Overall, the Netherlands have an extensive reporting duty, having ratified most global human rights instruments. In the UN arena, several issues can be distinguished that characterise the Dutch reporting behaviour towards enforceability of the right to adequate food. First of all, the Netherlands consistently report that they generously fulfill all the obligations that relate to the right to adequate food.[111] This is mostly substantiated by listing facts of the Dutch system of social security. Second, the Netherlands consistently report that in its view, ECOSOC rights are not suitable for enforceability. This has been a major cause for fierce debate between Dutch delegations and the Committee on Economic, Social and Cultural Rights. As also observed in the previous Section, the Dutch Government usually considers ECOSOC rights in general as non-enforceable rights. Therefore, especially during the reporting cycles on the implementation of the ICESCR Provisions, the matter of enforceability of ECOSOC rights is only discussed at the more general level, and not at the level of specific Provisions. The Dutch Government and Delegations mainly emphasize that there is a difference between on the one hand fulfilling the obligations that come from ratification of the rights enshrined in the treaty, and on the other hand, guaranteeing the enforceability of these rights.[112] The first would be a responsibility of the governments, while the latter would be up to the Judiciary to decide. A reasoning that was considered by especially the CESCR as incorrect, for it is a country that signs and ratifies a treaty that includes the Judiciary. Therefore, the State is responsible for ensuring that ECOSOC Provisions are enforceable.[113] The Dutch government repeatedly underscored that the Dutch Constitutional system is of a monistic nature, and as a result, the direct effect of 'eligible treaty Provisions is sufficiently guaranteed.'[114] However, the Dutch Government, despite criticism of various UN bodies in different degrees of urgency, also here persisted in its view that ECOSOC rights in general are not suitable for enforceability in the Courts. For instance, in their third periodic report on the implementation of the ICESCR Provisions, the Government argued that

> the nature and content of the Covenant, as well as the wordings of the Articles, indicate that it is aimed at the gradual implementation and increasing achievements of objectives by means of legislation and further implementation measures. As a result, most Provisions cannot be applied directly. All the more because, where further implementation laws are required, this implies a certain freedom of choice for the national Legislature regarding the way in which the rights to be guaranteed are given substance.[115]

Throughout the reporting procedures, although expressed in various ways, the essence of this argumentation remained the same.

---

[111] E/1994/104/Add.30 (2005), Section 329–343; E/C.12/NLD/4-5 (2009), Sections 219–232.
[112] E/C.12/NLD/Q/4-5/Add.1 (2010), Section 11.
[113] CESCR: E/C.12/NLD/CO/3 (2006), Section 19; E/C.12/NLD/CO/4-5 (2010), Section 6.
[114] E/1994/104/Add.30 (2005), Section 7.
[115] E/1994/104/Add.30 (2005), Section 8.

4.7.2 The Enforceability of the Right to Food in Belgium

4.7.2.1 Belgian trias politica

The Belgian constitutional context is characterized by a complex organization of the Legislature and Judiciary. As a result of profound differences amongst the Belgian people in language culture and economics, three main Legislators co-exist with equal and exclusive powers: the Federal Government, the centrally organized legislator representing all Belgian people, the Community Governments, representing the language groups,[116] and the Region Governments, representing the different economic regions in the country. The Belgian organization of the Judiciary is mostly based on functionality considerations. The Judiciary mainly consists of a subjective contentieux in which courts rule inter pares, and an objective contentieux, in which courts rule ergo omnes. However, throughout the years the rulings of both contentieux seem to converge. For instance, rulings in the objective contentieux (Council of State and Constitutional Court) result in effect for individuals, and in the subjective contentieux, especially in preliminary injunction procedures, courts rule on the basis of a balancing of interests rather than subjective rights.[117] While the results of rulings in both contentieux converge, the reasoning patterns are usually different, and these differences are highly relevant in light of the discussion on enforceability of the right to food.

In the objective contentieux, the constitutional court plays a rather distinctive role. Originally, the court was installed to rule in matters of conflict between the three highest legislative powers of Belgium as a court of arbitration. Throughout the years however, the competencies of this court were broadened gradually,[118] and as a result of several constitutional reforms, the court was authorized to review domestic legislation against the non-discrimination principle, and the right to equal treatment of different forms of education, as embedded in the Belgian Constitution.[119] The latter reflects the primary reason for this broadening of competencies; that is the protections of minorities in education, a rather sensitive topic in Belgian, considering the variety in language culture. Later, the competencies were broadened according to the already established legal practice, and the court was able to review national legislation directly against all rights stipulated in Chapter II of the Constitutional Act, in which most fundamental rights are enshrined. The Court established a legal practice in which national legislation was not only reviewed against constitutional provisions, but also indirectly against equivalent international human rights provisions. It is essentially in this light that the right to food plays a role in case law.

---

[116] Except the Brussels-Capital area, that falls under the combined authority of the Dutch and French Community.
[117] Maes (2003), no. 28–49.
[118] Alen (2005).
[119] Belgian Const. Art. 10, 11, 24.

In Belgium, it is generally accepted that the constitutional system is a monistic system, appearing from case law.[120] In the landmark Franco-Suisse Le Ski ruling, the Court of Cassation ruled that international standards that have direct effect prevail over contradicting national standards. Whereas only in the subjective contentieux individuals can invoke international standards, the courts in the objective contentieux do not engage in discussions on the functioning of international standards in the domestic legal order. As discussed above, the Constitutional Court basically reviews national standards indirectly against international standards in matters ergo omnes. This means that the debate on Belgian monism, and the criteria to determine whether or not an international Provision has direct effect exclusively plays a role in the sphere of the subjective contentieux. The courts within the subjective contentieux apply randomly either a subjective or an objective criterion to establish whether or not an invoked international standard has direct effect. When applying a subjective criterion, the court considers whether it was the intention of the State Parties to give such a direct effect to the provision. This can usually be deduced from the addressee of the provision: individuals or the State. In case of the first, the standard would have direct effect. In applying an objective criterion, the court considers whether the invoked standard is specific enough to deduce particular entitlements from its meaning. This means that the wider the margin of appreciation, the less likely it is that a court would rule that the Provision has direct effect.

4.7.2.2 Case Law Analysis[121]

In the subjective contentieux, the courts mostly deny direct effect to ECOSOC Provisions, without offering much of an explanation. The courts of last instance in this contentieux—the Court of Cassation and the Council of State—have the tendency to prefer to circumvent the issue of direct effect, and only rule on the matter when it is absolutely necessary. The general impression is that ECOSOC rights, but also specialized treaties such as the CEDAW and ICRC are entirely non enforceable. There are however some minor exemptions, in which the courts seem to apply international ECOSOC Provisions, or use such Articles as an interpretative standard.[122] However, due to the rather casuistic approach of the courts and the scarcity of such exemptions, it is hard to deduce a consistent approach towards the direct effect of ECOSOC rights.

---

[120]Van Eeckhoutte and Vandaele (2002).

[121]Wernaart (2013), Chapter 12.

[122]Court of Cassation (1996) Arresten van het Hof van Cassatie, 1996 (446) (Article 12 ICESCR); Council of State, 13 December 2000, case no. 91625; Council of State, 30 October 1995, case no. 56106 9Art. 13 ICESCR); Council of State, 22 March 1995, case no. 52424; Council of State, 3 December 2002, case no. 113168 (Art. 8 ICESCR); published at: www.raadvst-consetat.be.

Since the Constitutional Court, due to its function within the sphere of the objective contentieux is not restricted to matters of direct effect, it is able to review national legislation indirectly against international Provisions. The origin of this practice can be found in several cases in which the court reviewed several domestic standards against the discrimination Provisions in the Constitution, in conjunction with international Provisions that stipulate the right to education (such as Article 13 ICESCR).[123] Since then, the court reviewed against numerous international provisions, including those that stipulate the right to adequate food.

Most cases in which the right to food plays a role concern the status of aliens, mostly illegally residing foreigners. In these cases, mostly a particular domestic provision[124] was reviewed against Constitutional and international standards. The disputed domestic article excludes foreigners from social benefits—except urgent medical care—when they reside on Belgian territory illegally. In a landmark ruling,[125] the Court ruled that the national provision is not in violation with the invoked constitutional and international standards. In essence, the domestic Provision was considered to be a proportionate mean to realise the justified aim of restricting immigration. In a series of rulings, the Constitutional Court further nuanced this balancing of interests, in cases in which particular groups of individuals claimed they should be exempted from this general principle, for various reasons. The Court ruled inter alia that this principle should not be applied to illegally residing aliens who were incapable of leaving the territory due to medical impossibility,[126] to children of illegally residing aliens,[127] to their parents,[128] and to illegally residing parents whose child could not leave Belgian territory due to a medical impossibility.[129] In general, it can be observed that claims based on ICRC Articles seem to be more successful than claims based in ICESCR Provisions.

However, the effect of the international standard in itself should not be overestimated, for it seems that these provisions only play a minor role in the considerations of the court. The focus of the Constitutional Court is rather inwards, and the court prefers to solve the matters using domestic legislation instead of international.

4.7.2.3 Belgian Reporting Behaviour

The Belgian Government also has a very extensive reporting duty. With regard to the implementation of the right to food, and ECOSOC rights in general, the

---

[123] Constitutional Court, 33/92, 7 May 1992, in particular consideration B.8.2.
[124] Article 57 § 2 of the *'organic law of 8 July 1976 on public centres for social welfare'*.
[125] Constitutional Court, 51/94, 29 June 1994.
[126] Constitutional Court, 80/99, 30 June 1999.
[127] Constitutional Court, 106/2003, 22 July 2003.
[128] Constitutional Court, 131/2005, 19 July 2005.
[129] Constitutional Court, 194/2005, 21 December 2005.

following observations can be made. First, the Belgian reports primarily focus on all kinds of initiatives that were undertaken by the authorities to improve the human rights implementation embedded in the ratified treaties. The government seems to be very cautious in reporting on points for improvement. Second, Belgium hardly reports on the right to food as an individual right, and rather enlightens on the functioning on its system of social security. Third, an important issue that is frequently discussed during the reporting cycles seems to be the status of foreigners, in particular illegally residing aliens, and their access to social security. In fierce debates between the Belgian Delegations and the various UN committees the matter was discussed.[130] Especially the Committee on the Rights of the Child seems not convinced that the Belgian legislation on social security is in line with the ICRC.[131] Fourthly, it seems that the complex organization of the Belgian trias politica causes difficulties in the communication between the Belgian Government and the UN treaty bodies and leads occasionally to misunderstanding.[132] For example, the role of the Constitutional Court is not always clear to the committees, while the Belgian Government seems not to be bothered to nurture the impression that from the case law of the Constitutional Court it appears that some ECOSOC rights, including the right to food, have direct effect in the Belgian domestic order.[133]

Considering the function and the substance of the case law of this court, this impression is simply inaccurate. Furthermore, the Belgian Delegations seem to be not fully informed about the actual status of the domestic case law regarding the enforceability of ECOSOC rights. During the reporting cycles, the Delegation suggested that the CEDAW had direct effect in Belgium,[134] and did not correct the impression of the Committee on the Rights of the Child that the entire ICRC had direct effect in the Belgian legal order.[135] Finally, Belgium made an interpretative declaration to Article 2 ICRC,[136] in which they underlined that the non-discrimination principle would not guarantee foreigners similar rights compared to Belgian nationals. It is remarkable that while the Committee on the Rights

---

[130] E/C.12/BEL/CO/3 (2008).

[131] CRC/C/BEL/C/3-4, 18 June 2010, Sections 74–77.

[132] E/C.12/BEL/3, 21 September 2006, Section 7; E/C.12/BEL/CO/3, 4 January 2008, Section 11; Mr Citarella, CRC/C/SR.1523, 11 June 2011, Section 48; Mrs Kapalata, CEDAW/C/SR.559, 25 June 2005, Section 51.

[133] Mr Deneve, E/C/12/2000/SR.64, 27 November 2000, Section 38; E/C.12/BEL/Q/3/Add.1, 1 November 2007, Section 51.

[134] Mrs Paternottre, CEDAW/C/SR.559, 25 June 2005, Section 32.

[135] CRC/C/15/Add.38, 20 June 1995, Section 6; Mr Ahmed, E/C/12/2000/SR.64, 27 November 2000, Section 34; Mr Deneve, E/C/12/2000/SR.64, 27 November 2000, Section 38.

[136] With regard to Article 2, Section 11, according to the interpretation of the Belgian Government non-discrimination on grounds of national origins does not necessarily imply the obligation for States to automatically grant foreigners the same rights as their nationals. This concept should be understood as designed to rule all arbitrary conduct but not differences in treatment based on objective and reasonable considerations, in accordance with the principles prevailing in democratic societies.

of the Child does not necessarily oppose such a reasoning,[137] the Belgian Government nevertheless persists in maintaining the declaration.[138]

4.8 Conclusion

Human rights, and in particular the human right to adequate food, can be seen as a basis for food law, recognizing justice as its basic principle on which all other law is built. While the right to food is firmly embedded in the global human rights system, and is frequently—with utmost of urgency—emphasized, the meaning of this right in terms of state obligations, and in particular enforceability, is not a matter of course. There are many valid arguments to substantiate that a human right without individual entitlements, and thus legal means to bring violations of a right before a court, would be a hollow concept. However, with some exceptions, member states are reluctant in giving up part of their sovereignty for the sake of implementing internationally embedded ECOSOC rights through guaranteeing their direct effect in the courts. Also in industrialized countries in which theoretically the circumstances should allow each individual to enjoy the human right to adequate food, enforceability is not self-evident. The case study of the Netherlands and Belgium demonstrates that in both countries, it is the coincidental constitutional circumstances that greatly determine the possibility of enforceability of such rights, rather than its content or urgency. Not inconsequently, problems relating to mass immigration play a significant role in this viewpoint. It is remarkable that in the international arena, both countries seem to communicate that they generously fulfill the obligations coming forth from ratifying the right to adequate food, which seems to hardly be in line with the domestic legal reality. Where the normative value of the right to food and other ECOSOC rights seems to be uncontested, its legal effect is limited.

Only if food lawyers around the globe recognize and endorse in word and action that food law has a foundation in human rights is this state of affairs likely to ever change.

References

Vienna Convention on the Law of Treaties, 1969, Articles 11–18
Mégret F (2010) The nature of state obligations. In: Moeckli D, Sangeeta S, Sandesh S, Harris D (eds) International human rights law. Oxford University Press, Oxford, Chapter 6
McCrudden C (2007) Judicial comparativism and human rights. In: Őrűcű E, Nelken D (eds) Comparative law, a handbook. Hart Publishing, Portland, Chapter 16

---

[137]CRC/C/Q/BELG/2, 8 February 2002, part I, B.1.
[138]CRC/C/RESP/7, received on 3 May 2002, Section 1, B.1.

The universal declaration of human rights. Resolution 217 AIII (10 December 1948)
International Covenant on Economic, Social and Cultural Rights; International Covenant on Civil and Political Rights; Optional Protocol to the International Covenant on Civil and Political Rights A/RES21/2200 (16 December 1966)
Wernaart B (2010) The plural wells of the right to food and the enforceability of the human right to adequate food, a comparative study. In: Governing food security. Wageningen Academic Publishers, Wageningen, Chapter 3
www.srfood.org
www.FAO.org
Kent G (ed) (2008) Global obligations for the right to food. Rowman and Littlefield, Lanham
Ziegler J (2002) E/CN.4/2002/58. In: Special rapporteur on the right to food, Sections 32–34
Eide A (1999) E/CN.4/Sub.2/1999/12. In: Updated study on the right to food
OECD, *OECD Guidelines for Multinational Enterprises* (2011) OECD Publishing, especially part I, Section IV; E/CN.4/Sub.2/2003/12, 26 (2003) *Draft Standards on the Responsibilities of Transnational Corporations and Other Business Enterprises with Regard to Human Rights;* Tripartite Declaration of Principles concerning Multinational Enterprises and Social Policy, adopted by the Governing Body of the International Labour Office at its 204th Session (Geneva, November 1977) as amended at its 279th (November 2000) and 295th Session (March 2006); and in more than one occasion in: *The voluntary guidelines to support the progressive realisation of the right to adequate food in the context of national food security,* adopted by the 127th session of the FAO Council, November 2004
E/1991/23 (1991) annex III at 86 (1991), CESCR, *General Comment 3: The Nature of States Parties' Obligations*
E/C.12/1998/24 (1998), CESCR, *General Comment 9, the domestic application of the Covenant*
E/C.12/1999/5 (1999) CESCR, *General Comment 12, Right to Adequate Food*
E/C.12/2002/11 (2003) CESCR, *General Comment 15, The Right to Water*
FAO Doc, Rome Declaration on World Food Security (1996) WSFS 2009/2, Rome, 16–18 November 2009, *Declaration of the World Food Summit on Food Security;* FAO Doc. WSFS 2009/2, Rome, 16–18 November 2009, *Declaration of the World Food Summit on Food Security*
FAO Council (2004) *The voluntary guidelines to support the progressive realization of the right to adequate food in the context of national food security,* 127th session
Knuth L, Vidar M (2011) Constitutional and legal protection of the right to food around the world, right to food studies. FAO, Rome
A/res/60/251(2006) *Human Rights Council,* Section 6
Economic and Social Council Resolution 5 (I), (1946) Article 68 UN Charter
A/57/356 (2002) Ziegler J, *Report of the Special Rapporteur on the right to food to the General Assembly,* Chapter 3
A/59/385 (2004), Ziegler J, *Report of the Special Rapporteur on the right to food to the General Assembly,* Chapter 4
A/RES/61/295 (2007) *United Nations Declaration on the Rights of Indigenous Peoples;* A/HRC/RES/7/14, 27 March 2008, Human Rights Council Resolution, Section 12; A/RES/62/164, 13 March 2008, General Assembly Resolution, Section 12
A/CONF.151/26/Rev.1 (Vol. 1.), Rio de Janeiro, 3–14 June 1992, *Agenda 21,* Chapter 26: *Recognising And Strengthening The Role Of Indigenous People And Their Communities; The voluntary guidelines to support the progressive realisation of the right to adequate food in the context of national food security,* adopted by the 127th session of the FAO Council, November 2004, preamble, Section 8.1; A/60/2005, 12 September 2005, Jean Ziegler, *Report of the Special Rapporteur on the Right to food to the General Assembly,* Chapter III. See also Lidija Knuth, *The right to food and indigenous people, how can the right to food help indigenous people?* Rome: FAO, 2009, especially Section 1.3.1
A/58/330, 28 August 2003, Jean Ziegler, *Report of the Special Rapporteur on the right to food to the General Assembly,* especially, Section 22; A/CONF.177/20, Beijing, China,

4–15 September 1995, *Report of the fourth world conference of women*; A/HRC/RES/7/14, 27 March 2008, Human Rights Council Resolution, Sections 4–5

Office of the United Nations High Commissioner for Human Rights and the World Health Organisation, factsheet no. 31, *The right to health,* Geneva: UN, 2008

European Social Charter (revised), Strassbourg, 3.V.1996, Article 8; A/RES/44/25 (1989) *Convention on the Rights of the Child*, Article 24 (e)

A/RES/34/180 (1979) *Convention on the Elimination of All Forms of Discrimination Against Women*

UN Human Rights Committee (1982) General Comment No. 6: *Article 6, Right to Life*, Section 5

A/RES/260 (III) (A) (1948) *International Convention on the Prevention and Punishment of the Crime of Genocide*

Commission on Human Rights Resolution 2000/10 (2000) Section 10

Mr Olivier De Schutter was appointed Special Rapporteur on the right to food on March 26, 2008, by the Human Rights Council, his mandate was extended for another three years in: A/HRC/RES/13/4, 14 April 2010, Human Rights Council Resolution

Human Rights Council Resolution A/HRC/RES/7/22, 28 (2008) and Human Rights Council Resolution A/HRC/RES/16/2 (2011)

'Sub-Commission on the Prevention of Discrimination and Protection of Minorities.' Human Rights Council Resolution A/HRC/RES/5/1 (2007) and A/HRC/RES/6/102 (2007)

E/CN.4/Sub.2/1999/12 (1999) Updated study on the right to food, submitted by Mr Asbjørn Eide in accordance with Sub-Commission decision 1998/106

Guissé EH (1998) E/CN.4/Sub.2/1998/7, 10 June 1998; E/CN.4/Sub2/2004/20, 14 July 2004; E/CN.4/Sub.2/2005/25, 11 July 2005

217 A (III) (1948) *The Universal Declaration of Human Rights*

A/RES/44/25 (1989) *Convention on the Rights of the Child,* Article 27

A/RES/61/106 (2007) *The Convention on the Rights of Persons with Disabilities,* Article 28

A/RES/34/180 (1979) *Convention on the Elimination of All Forms of Discrimination Against Women*, Article 12(2)

First Geneva Convention for the Amelioration of the Condition of the Wounded and Sick in Armed Forces in the Field (1949) Articles 32 (2) jo Article 27

Third Geneva Convention Relative to the Treatment of Prisoners of War (1949) Articles 20, 26, 28, 46, 51, 72

Fourth Geneva Convention relative to the Protection of Civilian Persons in Time of War (1949) Articles 15, 23, 49, 50, 55, 59, 76, 87, 89, 100, 108, 127

Protocol Additional to the Geneva Conventions (1949) and relating to the Protection of Victims of International Armed Conflicts (1977) Articles 54, 69, 70

Protocol Additional to the Geneva Conventions (1949) and relating to the Protection of Victims of Non-International Armed Conflicts (1977) Articles 5, 14, 18

A/CONF.157/23 (1993) the world conference on human rights, *Vienna declaration and programme of action*

E/C.12/1999/5 (1999) CESCR, *General Comment 12, Right to Adequate Food*, Sections 18 and 19

E/C.12/GC/20 (2009) CESCR, General Comment 20, *Non-Discrimination in Economic, Social and Cultural Rights (Art. 2, Section 2),* in particular Sections 6, 23, and 30

A/RES/44/25 (1989) *Convention on the Rights of the Child*, Art. 26

A/RES/44/25 (1989) *Convention on the Rights of the Child*, Art. 27

A/RES/2106 (XX) (1965) *The International Convention on the Elimination of All Forms of Racial Discrimination,* Article 5 (e) (iv)

A/RES/34/180 (1979) *Convention on the Elimination of All Forms of Discrimination Against Women*, Article 11 (e)

A/RES/429 (IV) (1950) *Draft Convention relating to the Status of Refugees*, Article 24 (1) (b); A/RES/45/158, 18 December 1990, *The International Convention on the Protection of the Rights of All Migrant Workers and Members of Their Families*, Article 27; European Social Charter (revised), 3 May 1996, Strassbourg, in: European Treaty Series 163, Article 12

Vivek S (2008) Global support for school feeding. In: Kent G (ed) Global obligations for the right to food. Rowman and Littlefield, Lanham, Chapter 8
E/C.12/1999/10 (1999) CESCR, *General Comment 13, the Right to Education (Art. 13)*, section 6
European Court of Human Rights (2013) case of Vartic v. Romania (no. 2, section 44–55)
Asian Human Rights Charter (1998) South Korea article 7.1
African Charter on Human and Peoples' Rights on the Rights of Women in Africa (2003) Article 15
OAU Doc. CAB/LEG/24.9/49 (1990) African Charter on the Rights and Welfare of the Child
African Union Convention for the Protection and Assistance of Internally Displaced Persons in Africa (Kampala Convention), 22 October 2009
African Union Convention for the Protection and Assistance of Internally Displaced Persons in Africa (Kampala Convention), 22 October 2009, Article 9 (2) (b)
African Union Convention for the Protection and Assistance of Internally Displaced Persons in Africa (Kampala Convention). 22 October 2009, Article 7 (5) (c)
The San Salvador Additional Protocol to the American Convention on Human Rights, November 1988, adopted by the General Assembly of the Organization of American States, Article 19
Cairo Declaration on Human Rights in Islam, 5 August 1990, adopted at the Nineteenth Islamic Conference of Foreign Ministers
European Social Charter (revised), 3 May 1996, Strassbourg, in: European Treaty Series 163
E/C.12/1999/5 (1999) CESCR, *General Comment 12, Right to Adequate Food,* Section 6
E/C.12/1999/5 (1999) CESCR, *General Comment 12, Right to Adequate Food*, Section 8
E/C.12/1999/5 (1999) CESCR, *General Comment 12, Right to Adequate Food*, Section 7
www.wfp.org/hunger
Commission on Human Rights Resolution 2001/25 (2001) A/56/210, 23 July 2001, Jean Ziegler, report of the Special Rapporteur on the right to food to the General Assembly, Chapter IV; E/CN.4/2003/54, 10 January 2003, Jean Ziegler, report of the Special Rapporteur on the right to food, Chapter II. See also: E/C.12/2002/11, 20 January 2003, CESCR, *General Comment 15, The Right to Water*. Furthermore, see the works of the Special Rapporteurs on the right to water: Mr El Adji Guissé for the sub-commission on human rights, and Mrs Catarina de Albuquerque for the Human Rights Commission. How to handle multiple authors
Article 24 (2) ICRC
Kent G (2005) Freedom from want: the human rights to food. Georgetown University Press, Washington, Chapter 6
Eide A (2010) Adequate standard of living. In: Moeckli D, Shah S, Sivakumaran S (eds), Harris D (cons. ed) International human rights law. Oxford University Press, Oxford, Chapter 11
Skogly S (2007) Right to adequate food: national implementation and extraterritorial obligations. In: Von Bogdandy A, Wolfrum R (eds) Max Planck Yearbook of United Nations Law, vol 11. Koninklijke Brill N.V., Leiden, pp 339–358
Ssenyonjo M (2009) Economic, social and cultural rights in international law. Hart Publishing, Portland, Chapter 1
Constitutional Court of South Africa, *Government of the Republic of South Africa v. Irene Grootboom and others,* Case CCT 11/00 (2000) Swiss Federal Court, *V. v Resident Municipality X. and Bern Canton Government Council*, Case BGE/ATF 121 I 367, 27 October 1995. Supreme Court of India, *People's union for civil liberties v. Union of India and others*, case W. P(C) No. 196 of 2001, 23 July 2000 not sure how to handle
Human Rights Law Network (2009) Right to food, 4th edn. Human Rights Network, New Delhi
Alston P, Tomasevski K (eds) (1984) The right to food, International Studies in Human Rights. SIM, Utrecht, contributions of H. Shue: 'The interdependence of duties,' and G.J.H. van Hoof: 'The legal nature of economic, social and cultural rights: a rebuttal of some traditional views'
E/CN.4/Sub.2/1999/12 (1999) *Updated study on the right to food, submitted by Mr. Asbjørn Eide in accordance with Sub-Commission decision 1998/106*. Not sure how to handle
E/C.12/1999/5 (1999) CESCR, *General Comment 12, Right to Adequate Food,* Section 15
E/1991/23, annex III at 86 (1991), 14 December 1990, CESCR, *General Comment 3: The Nature of States Parties' Obligations*, especially Section 10 not sure how to handle two dates

E/1991/23, annex III at 86 (1991), 14 December 1990, CESCR, *General Comment 3: The Nature of States Parties' Obligations,* Section 9 not sure how to handle two dates

E/1991/23, annex III at 86 (1991), 14 December 1990, CESCR, *General Comment 3: The Nature of States Parties' Obligations,* Section 5

E/C.12/1998/24, 3 December 1998, CESCR, *General Comment 9, the domestic application of the Covenant,* Section 10

E/1991/23, annex III at 86 (1991), 14 December 1990, CESCR, *General Comment 3: The Nature of States Parties' Obligations,* Section 11

A/HRC/9/23 (2008) Olivier de Schutter, report of the Special Rapporteur on the right to food to the Human Rights Council, Section 18

E/CN.4/2003/54 (2003) Jean Ziegler, report of the Special Rapporteur on the right to food, Chapter 1 section 24

Oshaug A (2009) The Netherlands and the making of the voluntary guidelines on the right to food. In: Hospes O, van der Meulen B (eds) Fed up with the right to food. Wageningen Academic Publishers, Wageningen, Chapter 6

FAO Council (2004) 127th session, guideline 7.1

FAO Council (2004) 127th session guideline 7.2

FAO Council, (2004), 127th session guideline 7.3

Parliamentary Documents, II 1992–1993, (R1451), no. 3, p. 8

Ratification Bill to the European Social Charter: Parliamentary Documents, II 1965–1966, 8606 (R 533), no. 6

Ratification Bill to the ICESCR and ICRC: Parliamentary Documents, II 1975–1976, 13932 (R 1037), no. 3, pp. 12–13

Ratification Bill to the CEDAW: Parliamentary Documents, II 1984–1985, 18950 (R 1281), no. 3, p. 7; Explanatory Memorandum on the ratification Bill to the revised version of the European Social Charter: Parliamentary Documents, II 2004–2005, 29941, no. 3, p. 4

Parliamentary Documents, II 1975–1976, 13932 (R 1037), no. 3, p. 13

E/1994/104/Add.30 (2005), Section 329–343

E/C.12/NLD/4–5, 17 July 2009, Sections 219–232

E/C.12/NLD/Q/4-5/Add.1, (2010) Section 11

Maes G (2003) De afdwingbaarheid van sociale grondrechten, Oxford, Antwerp, Grondingen, Oxford: Intersentia, 2003, no. 28–49

Alen A (ed) (2005) Twintig Jaar Arbitragehof. Wolters Kluwer België, Mechelen

Belgian Const. Art 10, 11, 24

Van Eeckhoutte D, Vandaele A (2002) Doorwerking van internationale normen in de Belgische rechtsorde. Instituut voor Internationaal Recht K.U. Leuven, Working Paper No. 33

Wernaart B (2013) The enforceability of the human right to adequate food, a comparative study. Wageningen Academic Publishers, Wageningen, Chapter 12

For instance: Court of Cassation, 21 November 1996, Arresten van het Hof van Cassatie, 1996 (446) (Article 12 ICESCR); Council of State, 13 December 2000, case no. 91625; Council of State, 30 October 1995, case no. 56106 9Art. 13 ICESCR); Council of State, 22 March 1995, case no. 52424; Council of State, 3 December 2002, case no. 113168 (Art. 8 ICESCR); published at: www.raadvst-consetat.be

The earliest cases date back to 1992. See for instance: Constitutional Court, 33/92, 7 May 1992, in particular consideration B.8.2

Organic law of 8 July 1976 on public centres for social welfare' Article 57, sectuib 2 of the *July 8 1976*

Constitutional Court, 51/94 (June 29 1994)

Constitutional Court, 80/99, 30 June 1999

Constitutional Court, 106/2003, 22 July 2003

Constitutional Court, 131/2005, 19 July 2005

Constitutional Court, 194/2005, 21 December 2005

CRC/C/Q/BELG/2, 8 February 2002, part I, B.1

See for instance: CRC/C/RESP/7, received on 3 May 2002, Section 1, B.1

Chapter 5
Intellectual Property and Food Labelling: Trademarks and Geographical Indications

Michael Blakeney

Abstract This chapter looks at the international intellectual property regimes for the protection of trademarks and geographical indications (GIs), both of which play a role in food labelling. It is important for those involved in the marketing of food that they have to deal with harmonised trademark and GIs rules, so that they do not require a multiplicity of labels to comply with a multiplicity of rules. As this chapter explains a high level of harmonisation is achieved by the World Trade Organization (WTO) Agreement on Trade Related Aspects of Intellectual property Rights (TRIPS). As compliance with TRIPS is an obligation of all WTO members, this chapter focuses on the trademark and GIs provisions of TRIPS, while mentioning the other international instruments which deal with these subjects.

5.1 Introduction

Food Labelling typically involves either the use of trademarks to indicate the manufacturing origin of that food or the use of geographical indications to indicate the place of origin of that food. In both cases, the trademark or geographical indication will serve as a warranty of the quality of the designated food because the owner of those indications seeks repeat purchases by consumers. The global marketing of food that carries trademarks or geographical indications is facilitated by the international intellectual property system established by a number of international agreements and conventions. These agreements oblige countries to implement norms for protection set out in those instruments, thereby enabling marketers to deal with harmonised rules in relation to proprietary labelling.

The first of these international conventions was the 1883 Paris Convention for the Protection of Industrial Property, which contained norms dealing, inter alia, with trademarks and indications of product source. Supplementing the Paris Convention was the 1891 Madrid Agreement for the Repression of False or Deceptive

M. Blakeney (✉)
Faculty of Law University of Western Australia, Crawley, WA, Australia
e-mail: michael.blakeney@uwa.edu.au

Indications of Source of Goods[1] and the 1958 Lisbon Agreement on the Protection of Appellations of Origin and their International Registration.[2] The administration of these conventions was undertaken by an administrative agency of the Swiss Government: the Bureaux Internationaux Réunis pour la Protection de la Proprieté Intellectuelle (BIRPI). The BIRPI, however, was succeeded on July 14, 1967, with the creation of the World Intellectual Property Organization (WIPO) as a specialised agency of the United Nations.

A century after the promulgation of the Paris Convention, the United States sponsored the negotiations within the Uruguay Round of the General Agreement on Tariffs and Trade (GATT). The Uruguay Round of GATT commenced in 1986 and resulted in the Agreement on Trade-related Aspects of Intellectual Property Rights (TRIPS), which came into effect in 1995. The U.S. made the proposal that intellectual property rights (IPR) regulation be shifted to the GATT because of its disillusionment with WIPO as an effective custodian of the international IPR system.[3] The WIPO's inability to propose an effective means to deal with the apparently exponential growth in the global trade in counterfeit and pirated goods had caused the U.S. to turn to the GATT. In comparison with the WIPO, GATT was better equipped to handle the emerging problem with the ability to remove tariff preferences from developing countries involved in questionable trade practices and the ability to bring defaulting countries before the dispute resolution system.

The TRIPS Agreement was an annex to the Uruguay Round agreement of GATT which established the World Trade Organization (WTO). Members of the WTO are obliged to implement the IPR rules, which are set out in the TRIPS Agreement. Articles 15–20 of the TRIPS Agreement contain provisions dealing with trademarks. Articles 22–24 contain provisions dealing with geographical indications, and Articles 41–61 contain the machinery for the civil, administrative and criminal enforcement of IPRs.[4]

As the WTO had 159 members on March 2, 2013,[5] most countries have had to implement the rules regarding trademarks and geographical indications contained in the TRIPS Agreement. Therefore, the TRIPS agreement has effectively established global rules that have an impact upon the labelling of food products.

---

[1]Madrid Agreement for the Repression of False or Deceptive Indications of Source on Goods, Apr. 14, 1891, http://www.wipo.int/treaties/en/text.jsp?file_id=286779.

[2]Lisbon Agreement for the Protection of Appellations of Origin and their International Registration, October 31, 1958, http://www.wipo/int/lisbon/en/legal_texts/lisbon_agreement.html.

[3]Blakeney (1996).

[4]Agreement on Trade-Related Aspects of Intellectual Property Rights (TRIPS Agreement), *Annex 1C of the Marrakesh Agreement Establishing the World Trade Organization*, Apr. 15, 1994, available at www.wto.org/english/docs_e/legal_e/27-trips.pdf.

[5]World Trade Organization (2014) Members and Observers http://www.wto.org/english/thewto_e/whatis_e/tif_e/org6_e.htm.

5.2 Trademarks

5.2.1 Introduction

Trademark law developed from the common law action of passing off, which was an action to prevent the unfair competitive practice of misappropriating another's commercial reputation. A trademark was considered to be the quintessential symbol of a commercial reputation. Trademarks thus serve to protect commercial reputation. They are also used to facilitate advertising and product promotion through their ability to distinguish and identify goods and services. This is particularly important in markets where there is a proliferation of homogenous goods, as it allows purchasers to identify the goods of a particular trader. In the market for agricultural products, which tend to be homogenous and unpackaged, there has been some success in the marketing of products sold with an associated trademark such as "Chiquita" bananas and "Jaffa" oranges. In the market for packaged foods, branding plays a very important role in product differentiation.

A trademark is a concise way in which to refer to a product. Given the expense of advertising, trademarks serve a particularly important function in advertising and product promotion. The use of a trademark reduces the amount of information that needs to be communicated. The development of an advertised brand acts as a powerful incentive for the advertiser to offer goods of a consistently high standard because the brand will secure repeat purchases and cover the advertising spending. Further, consumers can use trademarks to identify goods, which will meet their needs, therefore creating an additional incentive for manufacturers and distributors to meet the reasonable expectations of consumers with regard to product quality. Accordingly, the use of trademarks tends to encourage trademark owners to maintain consistent standards of quality for goods and services offered under their marks.[6] Thus, a trademark serves as a form of "shorthand" upon which consumers can rely in making rational product selections. In jurisdictions where there is no consumer protection legislation or legislation regarding standards in relation to foodstuffs, the trademark performs a valuable function, by indicating quality, likely safety, and fitness for purpose.

The "goodwill" inherent in a trademark can be a valuable intangible property asset belonging to the trademark owner. The law recognizes this value and allows the trademark owner to prevent unauthorized uses of the trademark, which might tend to diminish the value of the mark. The trademark owner has the ability to protect its investment in creating the goodwill through infringement proceedings. The value of this goodwill can be used as security in raising new capital or in attracting further licensing.[7]

---

[6]Geographical Indications Trade, Nov. 2013, Free Essays- UK Law Essays http://www.ukessays.com/essays/law/geographical-indications-trade.php?cref=1.

[7]International Trademark Association Internet Subcommittee (Aug. 18, 1997) Inta "White Paper" The Intersection of Trademarks and Domain names. Available via www.ntia.doc.gov/legacy/ntiahome/domainname/not-emailed/INTA-whitepaper.htm.

For developing countries and least developed countries (LDCs), trademarks can be used first as a form of self-funded consumer protection since the trademark proprietor will be the person most vigilant in the policing of deceptive practices and in taking enforcement action against counterfeiters. They can also be used to facilitate the penetration of lucrative overseas markets. This will ultimately generate tax revenues, which can be used to underpin food purchases.

5.2.2 Registered Trademarks

Trademarks are typically protected by a system of registration. For the most part, trademark law is nationally based, but WTO Members are obligated to comply with the TRIPS Agreement. The TRIPS Agreement defines trademarks in Article 15 (1) as:

> Any sign, or any combination of signs, capable of distinguishing the goods or services of one undertaking from those of other undertakings, shall be capable of constituting a trademark. Such signs, in particular words including personal names, letters, numerals, figurative elements and combinations of colors as well as any combination of such signs, shall be eligible for registration as trademarks. Where signs are not inherently capable of distinguishing the relevant goods or services, Members may make registrability depend on distinctiveness acquired through use. Members may require, as a condition of registration, that signs be visually perceptible.[8]

Most trademark laws allow separate registrations for a mark in respect of each of the 45 categories of goods and services laid down in the International Classification of Goods and Services, which was established in accordance with the Nice Classification (NCL) established in the Nice Agreement of 1957[9] and its subsequent revisions. Registration of a mark may be permitted with the disclaimer of some elements of the mark. For example, in a word mark there may be a disclaimer of those words, which would be common to the relevant trade.

The application process usually requires an examination by the granting office to ensure compliance with the formal registration requirements, as well as with the substantive requirement of distinctiveness. There also has to be a check as to whether a mark is in conflict with prior rights. After the publication of an application, most countries provide for an opposition process whereby an interested third party may protest the registration of a mark, usually on the grounds of prior rights or deceptive similarity with another mark. Upon acceptance of a mark, registration is conferred for a term of between 10 and 20 years, with a possibility for renewal. A mark will expire if a renewal is not sought. Expungement of a mark may also be sought where its use becomes deceptive or where the mark becomes generic of

---

[8]TRIPS Agreement, 4, Art. 15(1).

[9]Nice Agreement Concerning the International Classification of Goods and Services for the Purposes of the Registration of Marks (June 15, 1957) www.wipo.int/classifications/nice/en/.

goods or services. For example, the marks "Vaseline" and "gramophone" are two examples of marks, which became generic descriptions of the type of goods to which they were appended.

A controversial requirement of some trademark laws is the requirement that registration of a trademark is contingent upon its use or a bona fide intention to use, upon or in close association with, the classes of goods or services in respect to which it is registered. A similar requirement provides for the removal of the registration after a prescribed period of non-use. Protection without registration may be extended to "well-known marks," i.e. those with a significant reputation in a country. Such marks invariably have a substantial international reputation through prior advertising and use.

Registration of a mark confers protection against emulation by traders using identical or substantially similar marks. Most systems of registration permit assignment or licensure. A system of registered users may be provided to record trademark licences. In the event of infringement of a registered mark, a trademark proprietor may seek relief in the form of injunction, compensation orders, and seizure of infringing goods.

5.2.3 Criterion of Distinctiveness

The touchstone for registration referred to in Article 15.1 of the TRIPS Agreement and under the Paris Convention is the criterion of distinctiveness. For example, Article 6 quinquies, of the Paris Convention allows the denial of registration to trademarks, which "are devoid of any distinctive character."[10] Most trademark laws have differentiated between marks, which are inherently distinctive, and those, which may subsequently become distinctive through use. This differentiation is recognized in Article 15.1 of the TRIPS Agreement, which provides that "where signs are not inherently capable of distinguishing the relevant goods or services, Members may make registrability depend on distinctiveness acquired through use."[11]

A trademark is generally understood as being inherently distinctive if it its association with the products in respect of which it is used is arbitrary or fanciful. Trademark laws based on statutes deriving their origin from the UK Trademarks Act 1875, list as inherently distinctive: word marks which comprise names represented in a special form; surnames or invented words; and marks which are otherwise inherently distinctive, such as device marks. As a general rule, a mark will be considered insufficiently distinctive if it is descriptive of the products or of

---

[10]Paris Convention for the Protection of Industrial Property (Paris Convention) (Mar. 20, 1883) Art. 6 *quinquies*, Sec. B(2) available at http://www.wipo.int/treaties/en/text.jsp?file_id=288514#P145_20374.
[11]TRIPS Agreement, 4, Art. 15(1).

the qualities of the products in respect of which it is used. It is possible for a descriptive mark of this nature to become distinctive where, through use, the description attracts the secondary signification of emanating from a particular enterprise. The Paris Convention disallows protection to marks, which consist of:

> [A] sign or indication, which may serve in the course of the relevant trade to designate the kind, quality, quantity, intended purpose, value, place of origin, or the time of production, or have become customary in the current language or in the bona fide and established practices of the trade of the country where protection is claimed.[12]

Therefore, a mark for a product containing some quality that is viewed as customary will be found insufficiently distinctive and will not be available for trademark registration.

5.2.4 Geographical Marks

A particular problem for food products is the tendency for traders to desire the use of geographical marks to indicate the origin of their products. As a general rule, trademark laws invariably refuse to allow geographical marks to be registered on the ground that they are insufficiently distinctive. Both the Council Regulation (EC) No 40/94 of December 20, 1993, on the Community trademark[13] (Community Trademark Regulation), and the First Council Directive 89/104/EEC of December 21, 1988, to approximate the laws of the Member States relating to trademarks[14] (Trademarks Directive), provide grounds for refusal of registration or invalidity of trademarks which:

(a) are devoid of any distinctive character;
(b) consist exclusively of signs or indications, which may serve, in trade, to designate the kind, quality, quantity, intended purpose, value, geographical origin, or the time of production of the goods or of rendering of the service, or other characteristics of the goods or service;
(c) consist exclusively of signs or indications, which have become customary in the current language or in the *bona fide* and established practices of the trade;
(d) are of such a nature as to deceive the public, for instance as to the nature, quality or geographical origin of the goods or service[15];

Additionally, both instruments limit the effects of a trademark by providing that the registered trademark shall not entitle the proprietor to prohibit a third party from using in the course of trade:

---

[12]Paris Convention, 10.

[13]Council Regulation No 40/94, Community Trademark Regulation of Dec. 20, 1993, O.J. (L 011) 14.1 (EC).

[14]First Council Directive 89/104/EEC, Trademarks Directive of Dec. 21, 1988, O.J. (L 040).

[15]Community Trademark Regulation, 13, Art. 7(1); Trademarks Directive, 14, Art. (3).

(a) his own name or address;
(b) indications concerning the kind, quality, quantity, intended purpose, value, geographical origin, the time of production of the goods or of rendering of the service, or other characteristics of the goods or service; ... provided he uses them in accordance with honest practices in industrial or commercial matters.[16]

Article 15(2) of the Trademarks Directive, under the heading "Special provisions in respect of collective marks, guarantee marks and certification marks," provides:

> By way of derogation from Article 3(1)(c), Member States may provide that signs or indications, which may serve, in trade, to designate the geographical origin of the goods or services may constitute collective, guarantee or certification marks. Such a mark does not entitle the proprietor to prohibit a third party from using in the course of trade such signs or indications, provided he uses them in accordance with honest practices in industrial or commercial matters; in particular, such a mark may not be invoked against a third party who is entitled to use a geographical name.[17]

The case law indicates that it must first be ascertained whether the relevant public as a geographical term knows the term. Secondly, the relevant public must understand it as a reference to the geographic place used in connection with the claimed goods or services. In other words, the geographic term must not be understood as having a mere suggestive or fanciful association with the relevant products, e.g. Mont Blanc would not be understood as a place of production for writing instruments. Thirdly, the geographical place must be currently associated, in the mind of the relevant public, with the category of goods in question, or the geographical name must be liable to be used in future by the undertakings concerned as an indication of the geographical origin of that category of goods.[18] In this assessment, regard must be given to the degree of familiarity amongst the public with the geographical name, the characteristics of the place designated by the name, and the category of the goods concerned.[19]

Registration is also excluded for those geographical names, which are liable to be used by undertakings.[20] However, the registration of geographical names is not excluded when the public is unlikely to believe that the category of goods concerned originate there.[21]

---

[16]Community Trademark Regulation, 13, Art. 12; Trademarks Directive, 14, Art. (3). Art. 6.
[17]Trademarks Directive, 14, Art. (3). Art. 15(2).
[18]*Windsurfing Chiemsee* [1999] [2000] 2 WLR 205, [1999] EUECJ C-108/97, [2000] Ch 523, [1999] ECR I-2779, paras 31, 37.
[19]Id. at para. 32.
[20]Case T-295/01, *Nordmilch eG v Office for Harmonisation in the Internal Market (Trademarks and Designs) (OHIM,)* EUECJ (Oct. 15, 2003) at para. 31.
[21]Id. at para. 33.

Community trademark law allows geographical marks to acquire the distinctiveness necessary for registration through use in the course of trade.[22] This means that registrable geographical marks will have attracted the secondary meaning of having the origin of the applicant undertaking.

5.2.5 Well-Known Marks

A special protective regime exists for marks, which are considered to be "well-known. This regime is created by Article 6bis(1) of the Paris Convention, which permits the countries of the Paris Union

> ...to refuse or to cancel the registration, and to prohibit the use of a trademark which constitutes a reproduction, an imitation, or a translation, liable to create confusion, of a mark considered by the competent authority of the country of registration or use to be well-known in that country as being already the mark of a person entitled to the benefits of this Convention and used for identical or similar goods.[23]

The Article also applies "when the essential part of a mark constitutes a reproduction of a well-known mark or an imitation likely to create confusion therewith." A period of 5 years from the date of registration is allowed by Art. 6bis(2) for the cancellation of unauthorised registrations of well-known marks,[24] but in relation to marks "registered or used in bad faith," Art. 6bis(3) provides that no time limit shall be fixed for requesting the cancellation or the prohibition of an unauthorized well-known mark.[25]

Article 16.2 of the TRIPS Agreement provides that "[i]n determining whether a trademark is well known, account shall be taken of the knowledge of the trademark in the relevant sector of the public...."[26] Market survey evidence is typically used to demonstrate the public repute of a brand.[27] The audience surveyed will be critical to a finding of repute. Article 16.2 also refers to "knowledge...obtained as a result of the promotion of the trademark," in determining whether a mark is well known. Thus, evidence of the amount of advertising spends and the extent of dissemination of advertising to the relevant target audience will have a bearing on the renown of a brand. The availability of widely distributed international journals and advertisements appearing in magazines distributed in airlines flying to the subject country are particularly useful in practice.[28]

---

[22]For example, the Community Trademark Regulation, Article 7, states: 3. Paragraph 1(b), (c) and (d) shall not apply if the trademark has become distinctive in relation to the goods or services for which registration is requested in consequence of the use which has been made of it.

[23]Paris Convention, 10, Article 6bis(1).

[24]Paris Convention, 10, Article 6bis(2).

[25]Paris Convention, 10, Article 6bis(3).

[26]TRIPS Agreement, 4, Art. 16.2.

[27]See Examples in Blakeney (1994), pp. 481–486.

[28]TRIPS Agreement, 26.

Other factors that will be taken into account in assessing the renown of a trademark are: (1) the registration history of a mark, including its date of first registration in the subject country; (2) the global extent of registration of a mark; (3) findings of repute by trademark offices in opposition hearings and by courts; and (4) the value of a mark as an asset on the books of a company.

5.2.6 Collective and Certification Marks

A special type of registered trademark is a collective mark, which may be registered by an association whose members may use it if they comply with the requirements fixed in the regulations concerning the use of the collective mark. Thus, the function of the collective mark is to inform the public about certain particular features of the product for which the collective mark is used. An enterprise entitled to use the collective mark may also use its own trademark in addition to the collective mark. In the U.S., collective marks are used by agricultural cooperatives of produce sellers. The collective mark owner is an organization, which does not sell its own goods, or renders services, but instead promotes the goods and services of its members.

Another type of registered trademark is a certification mark, which may only be used in accordance with defined standards. The main difference between collective marks and certification marks is that the former may be used only by particular enterprises, for example, members of the association, which owns the collective mark, while the latter may be used by anybody who complies with the defined standards. Another important requirement for the registration of a certification mark is that the entity which applies for registration is "competent to certify" the products concerned. Thus, the owner of a certification mark must be the representative for the products to which the certification mark applies.

In the U.S., agricultural producers typically use collective and certification marks in much the same way as geographical indications are used in Europe. U.S. State governments typically encourage the registration of certification marks to benefit agricultural producers. For example, the certification mark VIDALIA is owned by the State of Georgia's Department of Agriculture and is "intended to be used by persons authorized by certifier, and ... in connection with which it is used are yellow Ganex type onions and are grown by authorized growers within the Vidalia onion production area in Georgia as defined in the Georgia Vidalia Onion Act of 1986."[29] Similarly, FLORIDA CITRUS is owned by the State of Florida's Department of Citrus and certifies that goods bearing the mark "either consist of citrus fruit grown in the State of Florida, under specified standards, or are processed or manufactured wholly from such citrus fruit."[30] Non-U.S. agricultural producers

---

[29]U.S. Reg. No. 1,709,019.
[30]U.S. Reg. No. 1,559,414.

also have registered certification marks in the U.S. For example, the Thai Ministry of Commerce of Thailand has registered THAI HOM MALI RICE, "harvested in Thailand per the standards set by the Ministry of Commerce of Thailand in 'Regulations of the Department of Foreign Trade Re: Usage of the Certification Mark of Thai Hom Mali Rice.'"[31] Similarly, the Tea Board of India has registered DARJEELING to certify, "that the tea contains at least 100 % tea originating in the Darjeeling region of India and that the blend meets other specifications established by the certifier."[32]

The leading U.S. case involving the enforcement of a geographical indication as a certification mark is *Community of Roquefort v William Faehndrich, Inc.*[33] The case held that the designation "Roquefort" was not a generic designation of blue cheese and that the owner of the certification mark was entitled to prevent the use of the mark on all cheeses not made in the French city of Roquefort.

The system of registered certification marks is a departure from the trademark principle that no one can obtain an exclusive right in geographic names that other traders might legitimately wish to use. In Europe, the preference is for such marks to be registered as geographical indications.

5.3 Geographical Indications

5.3.1 *Introduction*

Marks indicating the geographical origins of goods were the earliest types of trademark. Until the commencement of the industrial revolution in the eighteenth century, the principal products entering international trade were agricultural products. In competition to earn revenues from the developing trade, it became apparent that the products of particular regions were more saleable than comparable products from other regions because of their superior quality. This superior quality resulted either from natural geographic advantages, such as climate and geology (e.g. Seville oranges, Kentish hops, Bresse poultry), or from recipes and food processing techniques local to a region (e.g. Roquefort cheese, Parma ham, Burgundy wine, Frankfurter sausages). In each case, the commercial attractiveness of these products was attributable to the traditional knowledge of the local communities. To protect the commercial reputation of these communities, local legislators passed laws to prevent the adulteration of local produce by the addition of inferior introduced goods or ingredients.

The legislation, which sought to protect the commercial reputation of traders in discrete geographical localities, evolved principally in Europe into systems for the

---

[31]U.S. Reg. No. 2,816,123.
[32]U.S. Reg. No. 2,685,923.
[33]*Community of Roquefort v. William Faehndrich, Inc.* 303 F. 2d 494 (Cir. 2,1962).

protection of geographical indications. As demonstrated below, these systems permit products emanating from the region to carry the geographic indication. Producer representatives from those regions police the use of geographic indications.

The evolution of the trademark system at the time of the Industrial Revolution did not result in the disappearance of geographic marks. Particularly in Europe, substantial processed foods markets and markets for alcoholic beverages remain dependent upon the continued recognition of geographical marks. These marks are protected typically within a *sui generis* system for the protection of geographical indications (GIs).

5.3.2 Modern GIs Protection

GIs may serve as indications of source, referring to the fact that a product originates in a specific geographical region. However, more frequently a GI is used as a sign that indicates that a product originates in a specific geographic region only when the characteristic qualities of the product are due to the geographical environment. The geographical environment includes both natural and human factors. A GI is a generic description, applicable to all traders in a particular geographic location referring to goods, which emanate from that location. Thus, a GI may be distinguished from a trademark, which is a sign that distinguishes the products of a specific trader from those of its competitors. In comparison to GIs, trademarks are not likely to be descriptive and cannot be generic.

The right to protect a geographical indication from wrongful appropriation is enjoyed by all traders from the particular geographical location, whereas a trademark is protected from wrongful appropriation only by the registered proprietor of that mark. Generally, GIs are monitored and protected by producer associations from the relevant region.

Unlike trademarks, geographical indications are not freely transferrable from one owner to another. This is because a user of a GI must have the appropriate association with the geographical region and must comply with the production practices of that region. GIs are obtained through registration. Typically, a specification is filed indicating the relevant geographical area and the product quality characteristics attributable to that area. A body representing the producers of that area usually files the application for registration. This body will also usually be responsible for bringing actions against wrongful users of the GI.

Geographical indications are becoming increasingly relevant for food marketing.[34] A study undertaken for the European Commission) estimated the worldwide sales value of products sold under geographical indications registered in the EU at

---

[34]Echols (2008); Anders and Caswell (2009), pp. 77–93; Bramley and Bienabe (2012), pp. 14–37.

€54.3 billion in 2010 and marked increases of 12 % between 2005 and 2010.[35] Some 43 developing countries and LDCs depend on exports of a single agricultural commodity for more than 20 % of their total revenues from merchandise exports. For example, Benin depends on cotton for over 80 % of its merchandise exports earnings. Ethiopia relies on coffee for over 70 % of agricultural exports.[36] The use of geographical indications, sometimes together with "fair trade" trademarking, could assist developing countries in their ability to market their food products in international trade and to support the sustainability of their agriculture.

A number of case studies of European food products have indicated that premium prices may be charged when a GI is used. For example, Bresse poultry in France receives quadruple the commodity price of poultry meat. Italian "Toscano" oil gains a 20 % premium above commodity oil; and milk supplied to produce French Comté cheese sells for a 10 % premium.[37] A case study of Comté cheese in France has indicated that French farmers receive an average of 14 % more for milk destined for Comté and that dairy farms in the Comté area have become more profitable since 1990, and now are 32 % more profitable than similar farms outside the Comté area. The retail price of Comté has risen by 2.5 % per year (against 0.5 % for Emmental), while the wholesale price has risen by 1.5 % a year (no change for Emmental).[38] In another example, the GI protection of 'Lentilles vertes du Puy' is said to have increased the production of lentils from 13,600 quintals in 1990, to 34,000 quintals in 1996, and 49,776 quintals in 2002 and reported that the number of producers has almost tripled from 395 in 1990, to 750 in 1996, and 1079 in 2002.[39]

However, some studies surrounding the designation of origin labelling have produced results indicating that consumers do not value the quality signal provided by the PDO label. For example, it was observed that at the same price, only a small proportion of consumers would prefer to buy a similar Camembert brand with a Protected Designation of Origin (PDO) label than without it. They noted that brand appeared to be more relevant information in the consumer's valuation of available products.[40]

There are fewer studies of premium prices for origin products outside Europe. However, some studies do exist, such as examples of the use of origin brands (certification marks in the Peoples Republic of China). The price of "Zhangqiu Scallion" per kilogram was raised from 0.2 to 0.6 yuan before the use of the

---

[35] Chever et al. (2012), available at http://ec.europa.edu/agriculture/external-studies/2012/value-gi/final-report_en.pdf.

[36] FAO Food Outlook (2005) Global Information and Early Warning System on Food and Agriculture (GIEWS)April 2005, No. 1, available at www.ftp.fao.org/docrep/fao/007/j5051e00.pdf.

[37] Babcock (2003), available at http://www.card.iastate.edu/iowa_ag_review/fall_03/article1.aspx.

[38] Gerz and Dupont (2006), pp. 75–87.

[39] O'Connor & Co. (2005), available at http://agritrade.cta.int/.

[40] Bonnet and Simioni (2001), pp. 433–449.

certification mark to 1.2–5 yuan in 2009. "Jianlian" lotus seed was registered as a GI in 2006, leading to a rise in price from 26–28 yuan per kilogram to 32–34 yuan per kilogram.[41]

5.4 Paris Convention for the Protection of Industrial Property, 1883

5.4.1 Scope

Article 2(3) of the Paris Convention provides that "industrial property shall be understood in the broadest sense and shall apply not only to industry and commerce proper, but likewise to agricultural and extractive industries and to all manufactured or natural products, for example, wines, grain, tobacco leaf, fruit, cattle, minerals, mineral waters, beer, flowers, and flour."[42]

5.4.2 Seizure of Goods Bearing a False Indication of Source

Article 9(1) of the Paris Convention provides for the seizure, upon importation, of all goods unlawfully bearing a legally protected "trademark or trade name". Article 9(3) provides that "seizure shall take place at the request of the public prosecutor, or any other competent authority, or any interested party, whether a natural person or a legal entity, in conformity with the domestic legislation of each country."

Article 10(1) provides for the application of the provisions of Article 9 "in cases of direct or indirect use of a false indication of the source of the goods or the identity of the producer, manufacturer or merchant." Unlike Article 9(1), which catches misleading indications, Article 10(1) requires the indications to be factually false and specifies that the indications may not be misleading. Although the provision only speaks of "indications of source," it is understood that it includes "appellations of origin," as referred to in Article 1(2).[43] As Article 10(1) refers to any direct or indirect use of a false identification, the false indication does not have to be expressed in words and appear on the product. Therefore, Article 10(1) includes the use of a false indication in advertising or on business documents.

The only sanction referred to in Article 10(1) is the seizure of the goods concerned, but no further civil or criminal sanctions are envisaged. Additionally, the obligation to seize goods on importation only applies to the extent that such a measure has been adopted under national law. Under Article 10(2), any

---

[41]Kireeva et al. (2009).

[42]Paris Convention, 10, Article 2(3).

[43]Pflüger (2011).

...producer, manufacturer, or merchant whether a natural person or legal entity, engaged in the production or manufacture of or trade in such goods and established either in the locality falsely indicated as the source, or in the region where such locality is situated, or in the country falsely indicated, or in the country where the false indication of source is used, shall in any case be deemed an interested party.

Therefore, any person engaged in the production, manufacturing, or trade of a protected good may bring suit to enforce Articles 9 and 10 of the Paris Convention.

5.5 Repression of Unfair Competition

Article 10*bis* also affords protection against false or misleading indications of source as a means of repressing unfair competition. Article 10*bis* (2) defined an act of unfair competition as "any act of competition contrary to honest practices in industrial or commercial matters."[44] In its various trademarks determinations, the European Court of Justice (ECJ) has observed that the requirement to act in accordance with honest practices in industrial or commercial matters "constitutes in substance the expression of a duty to act fairly in relation to the legitimate interests of the trademark proprietor".[45]

5.5.1 Madrid Agreement for the Repression of False or Deceptive Indications of Source of Goods, 1891

5.5.1.1 Seizure of Goods Bearing a False or Misleading Indication

The original form of Paris Convention prohibited the use of false geographical indications. However, a number of signatory nations proposed a more comprehensive form of regulation for false geographical indicators, as it was considered significant intellectual property abuse. Their eventual response was the 1891 Madrid Agreement concerning the protection of geographical indications. Article 1(1) of the Madrid Agreement provided that all goods "bearing a false or misleading indication" to signatory country, or to a place in that country "shall be seized on importation." Article 1(2) also provided for seizure "in the country where the false or deceptive indication of source has been applied, or into which the goods bearing the false or deceptive indication have been imported." Where the laws of a country

---

[44] Paris Convention, 10, Article 10*bis*.

[45] See Case C-63/97 *Bayerische Motorenwerke AG v Deenik* (1999) ECR I-905 at 61; Case C-100/02 *Gerolsteiner Brunnen GmbH & Co v Putsch GmbH* (2004) ECR I-691 at 24; Case C-245/02 *Anheuser-Busch Inc v Budejovicky Budvar np* (2004) I-10989 at 82, Case 228/03 *Gillette Co v LA-Laboratories Ltd Oy* (2005) ECR I-2337 at 41; and Case C-17/06 *Céline SARL v Céline SA* (2007) ECR I-7041 at 33.

do not permit seizure upon importation, Article 1(3) provides that such seizure shall be replaced by prohibition of importation. In the absence of any special sanctions ensuring the repression of false or deceptive indications of source, Article 1(5) provides that, "the sanctions provided by the corresponding provisions of the laws relating to marks or trade names shall be applicable."[46]

Article 2(1) provides that seizure shall take place at the insistence of the customs authorities. The customs authorities shall immediately inform the interested party, whether an individual person or a legal entity, in order that such desiring party may take appropriate steps in connection with the seizure effected as a conservatory measure. However, the public prosecutor or any other competent authority may demand seizure either at the request of the injured party or ex officio; the procedure shall then follow its normal course. Goods in transit are excluded from seizure by Article 2(2).[47]

5.5.1.2 Prohibited Use of Deceptive Indications in Advertising Etc

Article 3*bis* provides that signatory countries will undertake to prohibit the use, in connection with the sale or display or offering for sale of any goods, of all indications in the nature of publicity capable of deceiving the public as to the source of the goods. This includes indications and displays appearing on signs, advertisements, invoices, wine lists, business letters or papers, or any other commercial communication.[48]

5.5.1.3 Exception of Indications of Name and Address

Article 3 provides that the Madrid provisions shall not prevent the vendor from indicating his name or address upon goods coming from a country other than that in which the sale takes place. However, in such cases the address or the name must be accompanied in clear characters by an exact indication of the country or place of manufacture or production, or by some other indication sufficient to avoid any error as to the true source of the wares.[49]

---

[46]Madrid Agreement for the Repression of False or Deceptive Indications of Source on Goods (Madrid Agreement) (Apr. 14, 1891) Art. 1(1)-1(5). Available at www.wipo.int/treaties/en/text.jsp?file_id=286779.
[47]Madrid Agreement, 46, Art. 2(2).
[48]Madrid Agreement, 46, Art. 3bis.
[49]Madrid Agreement, 46, Art. 3.

5.5.1.4 Generic Indications

Article 4 permitted the courts of each signatory to decide what appellations do not fall within the provision of the Agreement on account of their generic character. However, Article 4 also excluded from reservation regional appellations concerning the source of products of the vine.[50] This provision has been noted as the explanation of why the Agreement failed to attract the accession of significant trading nations such as the U.S., Germany and Italy.

5.5.2 *International Convention on the Use of Appellations of Origin and Denominations of Cheeses ("Stresa Convention"), 1951*

The parties to the International Convention on the Use of Appellations of Origin and Denominations of Cheeses[51] (Stresa Convention) included the major cheese producing countries of Europe.[52] At the 1951 Stresa Convention, the countries "pledge[d] themselves to prohibit and repress within their respective territorial confines the use, in the language of the state or in a foreign language, of the 'appellations d'origine,' denominations and designations of cheeses contrary to the principles stated in Articles 2–9 inclusive." The Convention, which entered into force on September 1, 1953, applies to all specifications, which constitute false information as to the origin, variety, nature or specific qualities of cheeses, which are stated on products, which might be confused with cheese. The term "cheese," according to Article 2.1 of the Convention is reserved for "fresh and matured products obtained by draining after the coagulation of milk, cream, skimmed or partially skimmed milk or a combination of these," or by "products obtained by the partial concentration of whey, or of buttermilk, but excluding the addition of any fatty matter to milk."[53]

Article 3 provides that the appellations of origin of those cheeses "manufactured or matured in traditional regions, by virtue of local, loyal and uninterrupted usages" which are listed in Annex A are exclusively reserved to those cheeses. This is true regardless as to "whether they are used alone or accompanied by a qualifying or even corrective term such as 'type,' 'kind,' 'imitation,' or other term."[54] Annex A

---

[50]Madrid Agreement, 46, Art. 4.

[51]International Convention on the Use of Appellations of Origin and Denominations of Cheeses (Stresa Convention) (1951).

[52]The Stresa Convention was ratified by Austria (June 12,1953); Denmark (August 2, 1953); France (May 20, 1952); Netherlands (October 29, 1955); Norway (August 31, 1951); Sweden (January 27, 1951) and Switzerland (June 5, 1951).

[53]Stresa Convention, 51, Art. 2.

[54]Stresa Convention, 51, Art. 3.

lists: Gorgonzola, Parmigiana Romano, Pecorino Romano and Roquefort. Annex B lists a number of designations for cheese, which are prohibited by article 4.2 for products, which do not meet the requirements, provided by contracting parties in relation to "shape, weight, size, type and colour of the rind and curd, as well as the fat content of the cheese." Listed in Annex B are Asiago, Camembert, Cambozola, Danablu, Edam, Emmental, Esrom, Fiore Sardo, Fontina, Gruyére, Pinnzgauer Berkäse, Samsöe, and Svecia.[55]

The Stresa Convention came into force prior to the EEC Treaty and its regime providing for the free movement of goods.

5.5.3 Lisbon Agreement for the Protection of Appellations of Origin and their Registration, 1958

5.5.3.1 Introduction

The Lisbon Agreement established an international system of registration and protection of appellations of origin among members of the Lisbon Union that comprised signatory states. Article 1(2) obliged parties to the Lisbon Agreement to protect on their territories "the appellations of origin of products" of signatory countries, "recognized and protected as such in the country of origin" and registered at the International Bureau of WIPO.[56] Article 4 of the Agreement provides that the Agreement does not exclude the protection already granted to appellations of origin in each of the countries of the Lisbon Union by virtue of other international instruments, such as the Paris and the Madrid Agreement for the Repression of False or Deceptive Indications of Source on Goods, "or by virtue of national legislation or court decisions."[57]

The Lisbon Agreement failed to attract support from more than a few nations, receiving only 28 signatories by September 2013. An examination of all current appellations on the Lisbon register reveals that 11 countries hold 97.5 % of all entries, with the top 3 holding over 78 %. France alone holds 62.5 % (almost 90 % of which were for wines and spirits).[58] One problem lay in the fact that accession was confined to those nations, which protected appellations of origin "as such." Thus, states, which protected the appellation of origin form of intellectual property under trademark, unfair competition, or consumer protection laws, were locked out.

---

[55]Stresa Convention, 51, Annex A and B.

[56]Lisbon Agreement for the Protection of Appellations of Origin and their International Registration (Lisbon Agreement) (Oct. 31, 1958). Art. 1(2), available at www.wipo.int/treaties/en/text.jsp?file_id=285856.

[57]Lisbon Agreement, 56, Art. 4.

[58]Gervais (2010), pp. 67–126.

Also, the Agreement did not make an exception for geographic indications, which had already become generic in member states.

5.5.3.2 Protected Indications

Article 2(1) of the Agreement defined "appellation of origin" to mean "the geographical name of a country, region, or locality, which serves to designate a product originating therein, the quality and characteristics of which are due exclusively or essentially to the geographical environment, including natural and human factors."[59] The country of origin is defined in Art. 2(2) as "the country whose name, or the country in which is situated the region or locality whose name, constitutes the appellation of origin which has given the product its reputation."[60] Thus, the Agreement protects appellation of origin designations based on geographical indicators of quality and reputation.

5.5.3.3 Breadth of Protection

Article 3 of the Lisbon Agreement requires that "[p]rotection shall be ensured against any usurpation or imitation, even if the true origin of the product is indicated or if the appellation is used in translated form or accompanied by terms such as 'kind "type,' 'make,' 'imitation,' or the like." As will be seen below, this language was included in Article 23 of the TRIPS Agreement to provide for additional protection for wines and spirits. The Acts of the Lisbon Conference define usurpation as the "illicit adoption" or counterfeiting of an appellation.[61]

5.5.3.4 Registration

Article 5(1) provided for the registration of appellations of origin at the International Bureau of WIPO, at the request of the IP offices of the countries of the Lisbon Union, "in the name of any natural persons or legal entities, public or private, having, according to their national legislation, a right to use such appellations." Thus international protection is based upon the existence of a national registration. Article 5(2) requires the International Bureau, to notify, without delay, the relevant offices of the various countries of the Lisbon Union of such registrations and to be publishing such registrations in a periodical.[62]

---

[59]Lisbon Agreement, 57, Art. 2(1).
[60]Id. at Art. 2(2).
[61]*Actes De La Conference Reunie A Lisbonne Du 6 Au 31 Octobre 1958* BIRPI, Geneva, 1963.
[62]Lisbon Agreement, 57, Art. 5(1)-(2).

Article 5(3) provides for an IP office of a member country to "declare that it cannot ensure the protection of an appellation of origin whose registration has been notified to it." This notification must be made within a period of 1 year from the receipt of the notification or registration to the International Bureau of WIPO and must contain an indication of the grounds therefor. The declaration will be upheld "provided that such declaration is not detrimental, in the country concerned, to the other forms of protection of the appellation which the owner thereof may be entitled to claim under Article 4."[63] Article 5(4) provides that the Offices of the countries of the Union may not oppose such declaration after the expiration of the 1-year period from receipt of the notification.[64] Article 5(5) requires the International Bureau of WIPO to notify the office of the country of origin as soon as possible of any declaration made Article 5(3) by the office of another country. Article 5(5) provides that "the interested party," when informed by the national office of the declaration made by another country, "may resort, in that other country, to all the judicial and administrative remedies open to the nationals of that country."[65] The Lisbon Agreement does not define what is meant by "interested party," although Article 8 envisages that legal action required for ensuring the protection of appellations of origin may be taken in each of the countries of the Lisbon Union "by any interested party, whether a natural person or a legal entity, whether public or private."[66]

Where an appellation, which has been granted protection in a given country pursuant to notification of its international registration, has already been used by third parties in that country from a date prior to such notification, Article 5(6) provides that the competent office of that country "shall have the right to grant to such third parties a period not exceeding 2 years to terminate such use." The country granting such extended time must advise the International Bureau accordingly during the 3 months, following the expiration of the period of 1 year provided for in Art. 5(3).[67]

5.5.3.5 Duration of Protection

The Lisbon Agreement does not clearly define the duration of protection of a registered appellation of origin. Article 7, which is sub-headed "Period of Validity," provides "(1) Registration affected at the International Bureau in conformity with Article 5 shall ensure, without renewal, protection for the whole of the period referred to in the foregoing Article." Yet Article 5 makes no specific reference to a time period for protection. Its only reference to time periods relates to the process of declaring that certain appellations cannot be protected. However, since Article 7

---

[63] Id. at Art. 5(3).
[64] Id. at Art. 5(4).
[65] Id. at Art. 5(3)-(5).
[66] Id. at Art. 8.
[67] Id. at Art. 5(6).

(1) refers to an absence of renewals, it is assumed that an appellation is protected for as long as it remains an appellation in the relevant country of origin.[68]

5.5.3.6 Generic Appellations

Article 6 provides that an appellation which has been granted protection in one of the countries of the Lisbon Union, pursuant to the procedure under Article 5, cannot be deemed to have become generic in that country as long as it is protected as an appellation of origin in the country of origin.[69]

5.5.3.7 Enforcement

Article 8 of the Lisbon Agreement provides that legal action required for ensuring the protection of appellations of origin may be taken in each of the countries of the Lisbon Union under the provisions of national legislation:

1. At the instance of the competent Office or at the request of the public prosecutor; or
2. By any interested party, whether a natural person or a legal entity, whether public or private.[70]

5.6 The WTO TRIPS Agreement

The protection of geographic indications was a key demand of European negotiators at the Uruguay Round of the GATT. The competing positions were those of the EU and Switzerland, which proposed a French-style of protection, and those of the U.S., which favoured the protection of geographic indications through a certification mark system. As a result of the competing views, Section 3 of Part VII of the TRIPS Agreement covers four main topics: (a) protection of geographical indications (b) geographical indications and trademarks; (c) additional protection for geographical indications for wines and spirits; and (d) Review of Section 3. These topics are examined together below with an account of geographical indications disputes under the TRIPS Agreement.

---

[68] Id. at Art. 5 and 7.
[69] Id. at Art. 6.
[70] Id. at Art. 8

5.6.1 Definition

Article 22.1 of the TRIPS Agreement defines geographical indications for the purposes of the Agreement as "…indications which identify a good as originating in the territory of a Member, or a region or locality in that territory, where a given quality, reputation or other characteristic of the good is essentially attributable to its geographical origin."[71] This definition expands the Lisbon Agreement concept of appellation of origin to protect goods, which merely derive a reputation from their place of origin without possessing a given quality, or other characteristics, which is due to that place.[72]

In its only determinations to date on geographical indications under the TRIPS Agreement the WTO Dispute ruled that a "designation of origin" and "geographical indication," as defined in EC legislation in different terms, were a subset of geographical indications as defined in Article 22.1.[73] Thus, the TRIPS definition permits Members of the WTO to protect the geographical indications of goods where the quality, reputation. or other characteristic of goods is attributable to their geographical origin.

5.6.2 Permitted Methods for the Protection of Geographical Indications

Article 22.2 of the TRIPS Agreement requires that "in respect of geographical indications," Members of the WTO shall provide the "legal means" for "interested parties" to prevent:

(a) the use of any means in the designation or presentation of a good that indicates or suggests that the good in question originates in a geographical area other than the true Place of origin in a manner, which misleads the public as to the geographical origin of the good;
(b) any use which constitutes an act of unfair competition within the meaning of Article 10*bis* of the Paris Convention (1967).[74]

---

[71]TRIPS Agreement, 4, Art. 22.1.

[72]Blakeney, Michael. Geographical Indications and TRIPS. University of Western Australia – Faculty of Law Research Paper No. 2012-09.

[73]European Communities – Protection of Trademarks and Geographical Indications for Agricultural Products and Foodstuffs, Complaint by the United States, Report of the Panel ("*Panel Report, EC-Trademarks and Geographical Indications (US)*"), WT/DS174/R, March 15, 2005, para. 7.738; European Communities-Protection of Trademarks and Geographical Indications for Agricultural Products and Foodstuffs, Complaint by Australia, Report of the Panel (hereinafter "*Panel Report, EC-Trademarks and Geographical Indications (Australia)*"), WT/DS290/R March 15, 2005, para. 7.711.

[74]TRIPS Agreement, 10, Art. 22.2.

5.6.3 "Interested Parties"

In *EC-Trademarks and Geographical Indications (US)* the Panel explained that the obligation in Article 22.2 is to provide certain legal means to "interested parties" who are nationals of other Members in accordance with the criteria referred to in Article 1.3. The interested parties must qualify as "nationals of other Members," in accordance with the criteria referred to in Article 1.3. The Panel pointed out that these persons can be private parties, which is reflected in the fourth recital of the preamble to the agreement and reads "*[r]ecognizing* that intellectual property rights are private rights".[75]

Although the term "interested party" is also used in Article 10(2) of the Paris Convention (1967), as incorporated in the TRIPS Agreement, by Article 2(1) of the TRIPS Agreement, in *EC-Trademarks and Geographical Indications (US)*, the Panel observed that Article 10(2) of the Paris Convention (1967) did not set out a criterion of eligibility for protection for the purposes of the TRIPS Agreement. Article 10(2) of the Paris Convention did however provide guidance on the interpretation of Articles 22 and 23 of the TRIPS Agreement.[76]

5.6.4 Non-Diminution of Geographical Indications Protection

Article 24.3 of the TRIPS Agreement requires that in implementing the geographical indications provisions, a WTO Member shall not diminish the protection of geographical indications that existed in that Member immediately prior to the date of entry into force of the WTO Agreement.[77]

In *EC-Trademarks and Geographical Indications* the Panel found that the scope of Article 24.3 was limited to the implementation of Section 3 of Part II of the TRIPS Agreement on geographical indications, and did not apply to the implementation of Section 2 of Part II on trademarks.[78] The Panel interpreted the phrase "the protection of geographical indications that existed in that Member immediately prior to the date of entry into force of the WTO Agreement" to mean the state of protection of individual geographical indications immediately prior to January 1, 1995.[79]

---

[75]Panel Report, *EC-Trademarks and Geographical Indications (US)*, 73 at paras. 7.742–7.743.

[76]Id. at para. 7.170.

[77]TRIPS Agreement, 10, Art. 24.3.

[78]Panel Reports, *EC-Trademarks and Geographical Indications (US)*, 73 at paras. 7.631–7.632, and *(Australia)* paras. 7.631–7.632.

[79]Id. at para. 7.636, and *(Australia)*, para. 7.636.

5.6.5 Geographical Indications and Trademarks

The geographical indications provisions of the TRIPS Agreement sought to reconcile the two existing systems of protection for geographical indications: on one hand recognizing the novelty of geographical indications protections, and on the other the long-standing protection of registered trademarks. In *EC-Trademarks and Geographical Indications* the Panel recognized that the rights provided for in Article 22.2 and Article 16.1, concerned with trademark protection, could lead to a conflict between private parties. Notably, they considered that the treaty provisions themselves did not conflict.[80]

Article 22.3 of the TRIPS Agreement provides for *ex officio* action by a WTO Member. Additionally, if its legislation permits, *ex officio* action at the request of an interested party, to refuse or invalidate the registration of a trademark which contains or consists of a geographical indication with respect to goods not originating in the territory indicated, if use of the indication in the trademark for such goods in that Member is of such a nature as to mislead the public as to the true place of origin. In *EC-Trademarks and Geographical Indications* the Panel confined Article 22.3 to the resolution of conflicts between geographical indications and later trademarks, but not prior trademarks.[81]

Additionally, Article 22.4 provides that the protection under paragraphs 1, 2 and 3 of Article 22 shall be applicable against a geographical indication, which, although literally true as to the territory, region or locality in which the goods originate falsely represents to the public that the goods originate in another territory.[82]

Cognizant of the fact that for most countries the protection of geographical indications will be an innovation, Article 24.4 exempts from this form of protection trademarks which have been "applied for or registered in good faith," or where the rights to the trademark "have been acquired through use in good faith" either before the implementation of the TRIPS provisions, or before the geographical indication is protected in its country of origin.[83]

Article 24.5 provides that in a situation where a trademark has been applied for or registered in good faith, or where rights to a trademark have been acquired through use in good faith either: (a) before the date of application of these provisions in that Member as defined in Part VI; or (b) before the geographical indication is protected in its country of origin; measures adopted to implement the geographical indications provisions contained in Section 3 of the TRIPS Agreement shall not prejudice eligibility for or the validity of the registration of a trademark, or the right to use a trademark, "on the basis that such a trademark is identical with, or similar to, a geographical indication."[84]

---

[80]Id. at paras. 7.623–7.624, and *(Australia)*, paras. 7.623–7.624.
[81]Id. at para. 7.622, and *(Australia)*, para. 7.622.
[82]TRIPS Agreement, 10, Art. 22.4.
[83]Id. at Art. 24.4.
[84]Id. at Art. 24.5.

In *EC-Trademarks and Geographical Indications* the Panel interpreted Article 24.5 as an exception to geographical indications protection, and rejected arguments that it impliedly limited trademark rights or impliedly preserved any trademark rights that it does not specifically mention.[85]

Article 24.7 provides that a Member may provide that any request made under the section in connection with the use or registration of a trademark must be presented within 5 years after the adverse use of the protected indication has become generally known in that Member, or after the date of registration of that trademark, provided the registration has been published and "provided that the geographical indication is not used or registered in bad faith." Similar to the analogous provision in most trademark laws, Article 24.7 preserves "the right of a person to use, in the course of trade, that person's name or the name of that person's predecessor in business, except where such name is used in such a manner as to mislead the public."[86]

5.6.6 Use of Terms Common in the Trade

Article 24.6 provides that nothing contained in Section 3 of the TRIPS Agreement, containing the geographical indications provisions, shall require a WTO Member to apply its provisions in respect of a geographical indication to any other Member. The provision applies to goods or services for which the relevant indication is identical with the term customary in common language as the common name for such goods or services in the territory of that Member.[87]

5.6.7 Additional Protection for Geographical Indications for Wines and Spirits

In addition to the general protection for geographical indications for wines and spirits contained in Article 22, Article 23 accords further protection to geographical indications for wines and spirits. Article 2.1 provides that

> Each Member shall provide the legal means for interested parties to prevent use of a geographical indication identifying wines for wines not originating in the place indicated by the geographical indication in question or identifying spirits for spirits not originating in the place indicated by the geographical indication in question, even where the true origin of

---

[85]Panel Reports, *EC-Trademarks and Geographical Indications (US)*, 73 at para. 7.609 and *(Australia)* para. 7.609.

[86]TRIPS Agreement, 10, Art. 24.7.

[87]Id. at Art. 24.6.

the goods is indicated or the geographical indication is used in translation or accompanied by expressions such as "kind", "type", "style", "imitation" or the like.[88]

Article 23.2 provides that the registration of a trademark for wines or spirits which contains or consists of a geographical indication "shall be refused or invalidated, *ex officio* if a Member's legislation so permits or at the request of an interested party, with respect to such wines or spirits not having this origin."[89]

5.6.8 Multilateral System

Article 23.4 provides that to facilitate the protection of geographical indications for wines, negotiations shall be undertaken in the Council for TRIPS concerning the establishment of a multilateral system of notification and registration of geographical indications for wines eligible for protection in those Members participating in the system.[90] Preliminary work was initiated at the Council's meeting in February 1997.[91] Paragraph 18 of the Doha Ministerial Declaration, adopted on November 14, 2001, provided that

> [w]ith a view to completing the work started in the Council for Trade-Related Aspects of Intellectual Property Rights (Council for TRIPS) on the implementation of Article 23.4, we agree to negotiate the establishment of a multilateral system of notification and registration of geographical indications for wines and spirits by the Fifth Session of the Ministerial Conference.[92]

Article 24.1 of the TRIPS Agreement provides that WTO Members "agree to enter into negotiations aimed at increasing the protection of individual geographical indications under Article 23."[93] In addition, Article 24.2 contains a general obligation for the Council for TRIPS to keep under review the application of the provisions of Section 3 of the TRIPS Agreement, which contains its geographical indications chapter, and that "the first such review shall take place within 2 years of the entry into force of the WTO Agreement."[94]

As will be seen below, there has occasionally been confusion between the negotiations for a multilateral system for wines and spirits and between the general reviews of the geographical indications provisions. The general review of the provisions includes the possibility of extending the protection conferred upon wines and spirits to other products including food products.

---

[88]Id. at Art. 23.1.
[89]Id. at Art 23.2.
[90]Id. at Art. 23.4.
[91]Council for Trade-Related Aspects of Intellectual Property Rights, (TRIPS Council) Minutes of Meeting, Centre William Rappard, IP/C/M/12 (Feb. 27, 1997).
[92]Doha Ministerial Declaration, (Nov. 14, 2001) WT/MIN(01)/DEC/1, para. 18.
[93]TRIPS Agreement, 4, Art. 24.1.
[94]TRIPS Agreement, 4, Art. 24.2.

In June 2005, the EC submitted a proposal to amend the TRIPS Agreement to provide global protection for GIs in a multilateral system of registration. This proposal sought to bring international protection for GIs into conformity with the European Union. The EC submission set out provisions for a centralized register that would be compulsory and have legal effect.[95] The EC proposal aimed at preserving each WTO Member's prerogative to determine whether a certain sign, indication, or geographical name met the TRIPS definition of a geographical indication.[96] Opponents of the EC proposal included the U.S., Australia, Argentina, Australia, Canada, Chile, Ecuador, El Salvador and New Zealand. These countries opposed the extension of GIs protection, taking the position that the international protection of GIs was adequate as it stands, and such a drastic development would only serve to undermine future gains in market access for non-European food and agricultural products.[97] Concern has also been expressed about the additional costs and administrative burdens of implementing a distinct system of GI protection in addition to the TRIPS obligations. The opposing countries instead advocated for a system of voluntary notification and registration with no obligation to protect registered GI's.

5.7 The TRIPS Revision

Article 24.1 of the TRIPS Agreement provides that WTO Members "agree to enter into negotiations aimed at increasing the protection of individual geographical indications under Article 23." It also provides that the provisions of paragraphs four through eight of Article 24 "shall not be used by a Member to refuse to conduct negotiations or to conclude bilateral or multilateral agreements." It concludes with the observation that in the context of such negotiations, "Members shall be willing to consider the continued applicability of these provisions to individual geographical indications whose use was the subject of such negotiations."[98] Article 24.2 requires the Council for TRIPS to keep under review the application of the provisions of Section 3, containing the geographical indications provisions, and that the first such review shall take place within 2 years of the entry into force of the WTO Agreement.[99]

---

[95] Communication from the European Communities. The communication, dated, is being circulated to the General Council, to the TNC and to the Special Session of the Council for TRIPS at the request of the Delegation of the European Commission. (TN/IP/W/11) of June 13, 2005. This new proposal maintains the level of ambition of the EC as regards both "extension" and the multilateral register of GIs, as contained in its earlier proposals in documents IP/C/W/107/Rev.1 (on the GI register) and IP/C/W/353 (on "extension").

[96] Paragraph 3.2(a).

[97] See Communication from Argentina, Australia, Canada, Chile, Ecuador, El Salvador, New Zealand and the United States, TN/IP/W/9, 13 April 2004.

[98] TRIPS, Article 24.1.

[99] TRIPS, Article 24.2.

A submission by Turkey on July 9, 1999, prior to the Seattle Ministerial, proposed the extension of the multilateral register to products other than wines and spirits (see footnote 5). The African group of countries endorsed this proposal. On August 6, 1999, a document from Kenya filed on behalf of the African Group noted that at the Singapore Ministerial, Article 23.4 negotiations concerning a multilateral register for wines had been extended to include spirits, and that

> Considering that Ministers made no distinction between the two above-mentioned products, the African Group is of the view that the negotiations envisaged under Article 23.4 should be extended to other categories, and requests, in this regard, that the scope of the system of notification and registration be expanded to other products recognizable by their geographical origins (handicrafts, agro-food products).[100]

Paragraph 18 of the Doha Ministerial Declaration, adopted on November 14, 2001, noted "that issues related to the extension of the protection of geographical indications provided for in Article 23 to products other than wines and spirits will be addressed in the Council for TRIPS pursuant to paragraph 12 of this declaration."[101] In paragraph 39 of the Hong Kong Ministerial Declaration, Ministers *inter alia* "[took] note of the work undertaken by the Director-General in his consultative process on all outstanding implementation issues under paragraph 12 (b) of the Doha Ministerial Declaration, including on issues related to the extension of the protection of geographical indications provided for in Article 23 to products other than wines.[102]

On June 29, 2001, a joint communication was sent to the TRIPS Council by Argentina, Australia, Canada, Chile, Guatemala, New Zealand, Paraguay and the United States (Joint Communication) in opposition to the proposals for an extension for the protection of GIs for wines and spirits under TRIPS to all products.[103] The Communication argued that the advantages of Article 23 protection was overstated and, relevantly for the current project, that the proposals for the extension of the TRIPS wines and spirits provisions to all products had insufficiently addressed the costs and burdens of this extension. It stated that, "[t]hese new costs and burdens include administration costs, trade implications for producers, increased potential for consumer confusion, potential producer conflicts within the WTO Members and a heightened risk of WTO disputes."[104]

As was mentioned in the preceding section, in July 2008, a group of WTO members called for a "procedural decision" to negotiate the multilateral register and the extension of Article 23 in parallel, together with a proposal to require patent

---

[100]*Preparations for the 1999 Ministerial Conference the TRIPS Agreement Communication from Kenya on Behalf of the African Group*, WTO Doc. WT/GC/W/302, Aug. 6, 1999, paras 26-7.
[101]Doha Ministerial Declaration, 92.
[102]Hong Kong Ministerial Declaration (Dec. 18, 2005) WTO Doc. WT/MIN(05)/DEC.
[103]WTO Doc. IP/C/W/289.
[104]Id. Attachment at para. 13.

applicants to disclose the origin of genetic resources or traditional knowledge used in their inventions.[105] In relation to GI-Extension the proposed text was that

1. Members agree to the extension of the protection of Article 23 of the TRIPS Agreement to geographical indications for all products, including the extension of the Register.
2. Text based negotiations shall be undertaken, in Special Sessions of the TRIPS Council and as an integral part of the Single Undertaking, to amend the TRIPS Agreement in order to extend the protection of Article 23 of the TRIPS Agreement to geographical indications for all products as well as to apply to these the exceptions provided in Article 24 of the TRIPS Agreement *mutatis mutandis*.

On April 19, 2011, a Communication from Albania, China, Croatia, European Union, Georgia, Guinea, Jamaica, Kenya, Liechtenstein, Madagascar, Sri Lanka, Thailand, Turkey and Switzerland proposed that Section 3 of TRIPS be amended by removing the reference to Wines and Spirits in the heading of Article 23 and by deleting all the references to wines and spirits in that Article. The result would render Article 23 applicable to all goods.[106]

On April 21, 2011, WTO Director-General Pascal Lamy circulated a report on his consultations on geographical indications extension and proposals dealing with the relationship between the TRIPS Agreement and the Convention on Biological Diversity. In relation to extension he described the "state of play" as characterised by "divergent views" and "with no convergence evident on the specific question of extension of Article 23 coverage: some Members continued to argue for extension of Article 23 protection to all products; others maintained that this was undesirable and created unreasonable burdens."[107] The Director confirmed that trademark systems were legitimate forms of protecting geographical indications, and that they were in line with the general principle that Members are entitled to choose their own means of implementing their TRIPS obligations. On this basis, he reported that extension proponents sought guarantees that the trademark system could and would protect their GIs at the higher level for all goods.[108]

---

[105]TN/IP/W/8, 23 April 2003.

[106]'Draft Decision to Amend Section 3 of Part II of The TRIPS Agreement', WTO Doc. TN/C/W/6, 19 April 2011.

[107]'Issues related to the extension of the protection of geographical indications provided for in Article 23 of the TRIPS Agreement to products other than wines and spirits and those related to the relationship between the TRIPS Agreement and the Convention On Biological Diversity', Report by the Director-General, WTO DOC. WT/GC/W/633, TN/C/W/61, 21 April 2011, para. 17.

[108]Id.

5.8 The TRIPS GIs Disputes

A number of WTO Members had argued that the EU scheme for the protection of GIs was TRIPS-deficient in a number of areas. For example, the statement of the U.S. to the WTO on the WTO trade policy review of the EU expressed the concern that "foreign persons wishing to obtain protection for their GIs in the EU itself face a non-transparent process that appears to come into some conflict with the EU's TRIPS obligations," and that "EU rulemaking processes are often perceived by third countries as exclusionary, allowing no meaningful opportunity for non-EU parties to influence the outcome of regulatory decisions".[109] On June 1, 1999, the U.S. requested consultations with the European Communities pursuant to Article 4 of the *Understanding on Rules and Procedures Governing the Settlement of Disputes* (DSU) and Article 64 of the of the TRIPS Agreement regarding EC Council Regulation (EEC) No 2081/92 of July 14, 1992, on the protection of geographical indications and designations of origin for agricultural products and foodstuffs. The U.S. and the EC held consultations on July 9, 1999, and thereafter, but have thus far failed to resolve the dispute.

On the August 18, 2003, the U.S. and Australia requested the establishment of a WTO dispute settlement panel to review the consistency of the EU Regulation 2081/92 with the rules of the TRIPS and GATT Agreements. The U.S. and Australia argued that the EC Regulation was discriminatory and in violation of the national treatment obligations and the most-favoured-nation obligations in Articles 3 and 4 of the TRIPS Agreement and Articles 1 and 3 of the GATT 1994. The U.S. and Australia argued that: (1) Regulation 2081/92 did not provide the same treatment to other nationals and products originating outside the EC that it provided to the EC's own nationals and products; (2) the EU did not accord immediately and unconditionally to the nationals and products of each WTO Member any advantage, favour, privilege or immunity granted to the nationals and products of other WTO Members; (3) the EU diminished the legal protection for trademarks; (4) the EU did not provide legal means for interested parties to prevent the misleading use of a geographical indication; (5) it did not define a geographical indication in a manner that was consistent with the definition provided in the TRIPS Agreement; (6) the EU was not sufficiently transparent in its registration procedures; and (7) did not provide adequate enforcement procedures.[110]

The U.S. and Australia claimed that the EU Regulations imposed two requirements which contravened the national treatment principle contained in Article 2 (2) of the Paris Convention as incorporated by Article 2.1 of the TRIPS Agreement: (1) the requirement that enterprises seeking to register GIs possessed a commercial establishment in the EU; and (2) the requirement that GIs located in the territory of

---

[109]WTO Trade Policy Review of the European Union, Statement by the United States to the WTO, July 24, 2002.

[110]Evans and Blakeney (2006), available at www.gaileevans.com/EvansGIsAfterDohaJIEL06.pdf.

a WTO Member outside the EU could only be registered if that Member had adopted a system for GI protection that was equivalent to that in the European Communities and provided reciprocal protection to products from the European Communities.[111]

The Panel Report in the dispute was adopted at a meeting of the Dispute Settlement Body on April 20, 2005. Concerning the discriminatory conditions regarding the registration of foreign GIs and requirement for reciprocity of protection, the Panel decided in favour of the U.S. and Australia. Pursuant to Article 19.1 of the DSU, the Panel recommended that:

(a) The European Communities bring the Regulation into conformity with the TRIPS Agreement and GATT 1994.
(b) The European Communities could implement the above recommendation with respect to the equivalence and reciprocity conditions, by amending the Regulation so as for those conditions not to apply to the procedures for registration of GIs located in other WTO Members.

In an affirmation of the GI as intellectual property, the Panel endorsed the European principle of their coexistence with all but the most famous of prior trademarks. The Panel found that Article 14(2) of the Regulation was a "limited exception" permitted by Article 17 of TRIPS because it only allows use by those producers who are established in the geographical area of products that comply with the specification.

On the critical issue of whether the nationals of other WTO Members were accorded less favourable treatment than the European Communities' own nationals, the Panel ruled that the conditions in the Regulations modified the effective equality of opportunities to obtain protection with respect to intellectual property in two ways. First, GI protection was not available in respect of geographical areas located in third countries which the Commission had not recognized. It was confirmed that the European Commission had not recognized any third countries. Second, GI protection under the Regulation could become available if the third country in which the GI is located enters into an international agreement with the EU. For the Panel, both of those requirements represented a significant "extra hurdle" in obtaining GI protection, which did not apply to geographical areas located in the European Communities. The significance of the hurdle was reflected in the fact that currently, no third country had entered into such an agreement or satisfied those conditions (see footnote 54). Accordingly, the Panel found that the equivalence and reciprocity conditions modified the effective equality of opportunities with respect to the availability of protection to persons wishing to obtain GI protection under the EU legislation, to the detriment of those wishing to obtain protection in respect of

---

[111]European Communities-Protection of Trademarks and Geographical Indications for Agricultural Products and Foodstuffs. WT/DS174/R (Mar. 15, 2005), available at www.global-trade-law.com/WTO.Geographical%Indicators%20Case.Conclusions%20(March%202005).htm.

5 Intellectual Property and Food Labelling: Trademarks and Geographical... 131

geographical areas located in third countries, including WTO Members. This was held to be less favourable treatment.[112]

The Panel noted that, while the Regulation did not prevent a foreign national from producing goods within the territory of the European Communities, the different procedures, which applied to foreign nationals compared with those of the E, were perceived as disadvantageous to the nationals of other Members.

5.9 TRIPS Enforcement

5.9.1 Introduction

Article 41.1 of the TRIPS Agreement imposes upon Members of the WTO a general obligation to make available the enforcement procedures listed in the Agreement "so as to permit effective action against any act of infringement of intellectual property rights [covered by the Agreement]." These procedures are also required to include "expeditious remedies to prevent infringements and remedies which constitute a deterrent to further infringements." Consistent with the general trade liberalization objectives of the WTO, these procedures are required to be "applied in a manner as to avoid the creation of barriers to legitimate trade and to provide for safeguards against their abuse."

In amplification of the latter qualifications, Article 41.2 requires that "[p]rocedures concerning the enforcement of intellectual property rights shall be fair and equitable." More specifically, the paragraph requires that procedures "shall not be unnecessarily complicated or costly, or entail unreasonable time-limits or unwarranted delays." Article 41.3 requires that "[d]ecisions on the merits of a case shall preferably be in writing and reasoned, and they "shall be made available at least to the parties to the proceeding without undue delay." Due process is also required by the paragraph, which insists, "[d]ecisions on the merits of a case shall be based only on evidence in respect of which parties were offered the opportunity to be heard."

Article 41.4 requires an opportunity for judicial review of final administrative decisions and "the legal aspects of initial judicial decisions on the merits of a case". However, paragraph four provides that there is "no obligation to provide an opportunity for review of acquittals in criminal cases."[113]

---

[112]CITE.
[113]TRIPS Agreement, 4, Art. 41.1–41.4.

5.9.2 Civil Procedures

In relation to the intellectual property rights covered by the TRIPS Agreement, Article 42 requires Members to make available civil judicial procedures for the enforcement of those rights to rights holders, including federations and associations having legal standing to assert such rights. This will be important for those geographical indications, which are held or supervised by consortia. Article 42 also requires that these procedures are fair and equitable, in that defendants are entitled to "written notice which is timely and contains sufficient detail, including the basis of the claims."

Further, Article 42 also requires representation by independent legal counsel. All parties to such procedures "shall be duly entitled to substantiate their claim and to present all relevant evidence," without the procedures imposing "overly burdensome requirements concerning mandatory personal appearances." Finally, Article 42 provides that the procedure "shall provide a means to identify and protect confidential information, unless this would be contrary to existing constitutional requirements."[114]

5.9.3 Discovery and Interrogatories

As is conventional in civil proceedings in most jurisdictions, Article 43.1 provides for procedures in the nature of discovery and the administration of interrogatories, once a party has "presented reasonably available evidence to support its claims and has specified evidence relevant to substantiation of its claims which lies in the control of the opposing party." A particularly acute concern in patent actions is that these pre-trial procedures may result in trade secrets being revealed. Article 43.1 provides that the production of evidence may be compelled, "subject in appropriate cases to conditions which ensure the protection of confidential information."[115] In the UK, a plaintiff is required in these circumstances to show that there are "formidable grounds" for suspicion that the defendant is infringing a plaintiff's rights.[116] Where there are concerns about the disclosure of trade secrets to a commercial rival, the court may require that an independent expert conduct inspection of discovered evidence.

In the event that a party to a proceeding "voluntarily and without good reason refuses access to, or otherwise does not provide necessary information within a reasonable period, or significantly impedes a procedure relating to an enforcement action," Article 43.2 permits Members to accord the judicial authorities "the authority to make preliminary and final determinations, affirmative or negative on

---

[114]TRIPS Agreement, 4, Art. 42.
[115]TRIPS Agreement, 4, Art. 43.1.
[116]*Wahl & Anor. v. Buhler-Miag (England) Ltd.* (1979) FSR 183.

the basis of the information presented to them." This will include "the complaint or the allegation presented by the party adversely affected by the denial of access to information." Article 43.2 does, however, provide the opportunity for the parties to be heard on the allegations or evidence.[117]

5.9.4 Seizure Orders

Compelling a defendant to respond to interrogatories or requests for discovery presupposes the sort of defendant who may not be typical of the worst sort of infringer of intellectual property rights. For example, in cases where the defendant is conducting clandestine infringement activities on a large scale, the defendant will not usually remain available to answer interrogatories or to discover documents. Indeed, on detection, relevant evidence will immediately be removed or destroyed. To deal with this situation, the English Court of Appeal in *Anton Piller v Manufacturing Processes*[118] approved a procedure whereby on an *ex parte* application *in camera,* an order would be granted to an applicant. This order mandates that the defendant, advised by his legal representative, grant access to the applicant to inspect the defendant's premises to seize, copy, or photograph material, which may be used as evidence of the alleged infringement. The defendant may be obliged to provide any infringing goods and tooling, and may also be obliged to provide information about sources of supply and destination of infringing products.

Because of the exceptional nature of an *Anton Pillar* order and its impact upon an individual's civil rights, further safeguards have been imposed. After the demonstration that there is a very strong *prima* facie case of infringement, the courts have insisted upon proof that there is a strong possibility that evidence in the possession of a defendant is likely to be destroyed before an application *inter parts* can be made. Additionally, the British courts have insisted upon the safeguards of the attendance upon a search. According to these principles, searches must be conducted during business hours by both parties' legal representative, and sometimes, must be under the supervision of a neutral supervising solicitor who has experience in the execution of these orders. Refusal to comply with an *Anton Piller* order will result in a contempt of court. On the other hand, the use of the order for abusive purposes may result in the grant of substantial compensation to a defendant.

In the *Anton Piller* case itself, the Court of Appeal predicted that such orders would be extremely rare,[119] however, with the burgeoning of the large-scale copyright piracy and trademark counterfeiting, which precipitated the adoption of trade-related intellectual property rights as a matter for GATT, the use of this procedure has become increasingly common. Infringement of geographical

---

[117] TRIPS Agreement, 4, Art. 43.2.
[118] *Anton Piller v Manufacturing Processes* (1976) RPC 719.
[119] Id. at 725 per Ormrod, LJ.

indications rights have not yet attracted the use of *Anton Piller* orders, but with the increase in particular of wine label counterfeiting, such orders may provide a remedy in the future.

The *Anton Piller* order is adopted in the scheme provided in Article 50 of the TRIPS Agreement for the making of "provisional measures" by the judicial authorities.

Article 50.1(b) provides that the judicial authorities shall have the authority "to order prompt and effective provisional measures: to preserve relevant evidence in regard to the alleged infringement."[120] As with the *Anton Piller* order, Article 50.2 permits the judicial authorities "to adopt provisional measures *inaudita altera parte* where appropriate, where there is a demonstrable risk of evidence being destroyed."[121] Also the judicial authorities may have authority pursuant to Article 50.3 "to require the applicant to provide any reasonably available evidence in order to satisfy them with a sufficient degree of certainty that the applicant is the right holder," and that an infringement has occurred or is imminent.[122] Additionally, Article 50.5 provides that to assist the authority, which will enforce the provisional measure, "the applicant may be required to supply other information necessary for the identification of the goods concerned."[123]

As measures to prevent abuse and to protect a defendant's rights, Article 50.3 provides for an applicant to be ordered "to provide a security or equivalent assurance" and Article 50.4 provides that where provisional measures have been adopted *inaudita altera parte*, notice must be provided to the affected parties "without delay after the execution of the measures at the latest." Paragraph 4 also provides for "a review, including a right to be heard" upon the request of the defendant "with a view to deciding, within a reasonable period of notification of the measures" whether they should be "modified, revoked or confirmed." Additionally, if proceedings leading to a decision on the merits of the case have not been initiated within a reasonable period, Article 50.6 permits the defendant to request the revocation of the provisional measures or for a determination that they cease to have effect.

Similar to the safeguards which have been developed in relation to the *Anton Piller* procedure, Article 50.7 provides for the compensation of a defendant where "the provisional measures are revoked or where they lapse due to any act or omission by the applicant, or where it is found subsequently that there has been no infringement or threat of infringement of an intellectual property right."[124]

---

[120]TRIPS Agreement, 4, Art. 50.1(b).

[121]TRIPS Agreement, 4, Art. 50.2.

[122]TRIPS Agreement, 4, Art. 50.3.

[123]TRIPS Agreement, 4, Art. 50.5.

[124]TRIPS Agreement, 4, Arts. 50.3–50.4, 50.7.

5.9.5 Injunctions

Injunctive relief is an important civil remedy for the preservation of intellectual property rights. This is particularly true where infringement may damage or undermine the establishment of a commercial reputation immediately upon the launching of a new product. Similarly, it has marked importance where the widespread counterfeiting of a trademarked product may have the effect of destroying the distinctiveness of a proprietor's mark, thereby rendering the trademark registration voidable. Article 44 permits the conferral of power upon the judicial authorities "to order a party to desist from an infringement, *inter alia*, to prevent the entry into channels of commerce in their jurisdiction of imported goods that involve the infringement of intellectual property rights."

The injunctions which may be granted under Article 44 are grounded upon infringing conduct. Where proof of consumer deception is the central feature of the infringement, the remedy proffered by Article 44 may be rendered nugatory where a sufficient time is required to provide an opportunity for consumers to become deceived.[125] After this has occurred, it would be futile to hope that this deception can be undone. In this circumstance, the provision of interlocutory relief is essential.

5.9.6 Interlocutory Injunctions

Article 50.1 provides that the judicial authorities "shall have the authority to order prompt and effective provisional measures...(a) to prevent an infringement of any intellectual property right from occurring." The trade-related context of this remedy is emphasised by the supplementary particularization in sub-paragraph (a) that provisional measures may be taken to prevent the entry into the channels of commerce in their jurisdiction of goods including imported goods immediately after customs clearance.

As is mentioned above, a provisional order of particular utility in an intellectual property context is the grant of interlocutory injunctions for the purpose of freezing the status quo until a trial of the merits can take place.

As a matter of practice, the interlocutory injunction, although it is only intended to have a preservative effect, will actually be the basis of the final determination of parties' rights. It is very seldom that after the interlocutory hearing, the defeated party will proceed to the determination of final relief. If an appeal is to be taken, it will usually be on the issue of interlocutory relief. Provision is made in Article 50.6 for a defendant to request that provisional measures be revoked "if proceedings leading to a decision on the merits of the case are not initiated within a reasonable period, to be determined by the judicial authority. Where such a period is not

---

[125]TRIPS Agreement, 4, Art. 44.

determined, Article 50.6 prescribes 20 working days or 31 calendar days, whichever is the longer.[126]

A provisional order, interlocutory, or interim injunction is of particular utility in an IPR enforcement context to freeze the status quo until a trial of the merits can take place. As a matter of practice although these orders are intended to have a preservative effect, for the purpose of avoiding the infliction of uncompensable damage upon a right holder, until the merits can be decided, in most cases the provisional measures will actually be the basis of the final determination of parties' rights. In cases of egregious counterfeiting or piracy, the defendant is not likely to appear to oppose the grant of a provisional order. Even in contested IPR enforcement actions, it is very seldom that after the interlocutory hearing, the defeated party will proceed to the determination of final relief.

In the U.S., to obtain a preliminary injunction the applicant must establish

> (1) irreparable harm and (2) either (a) a likelihood of success on the merits, or (b) sufficiently serious questions going to the merits of its claims to make them fair ground for litigation, plus a balance of the hardships tipping decidedly in [its favor].[127]

The grant of all injunctive relief in common law countries is discretionary. The defendant's conduct is taken into account in disqualifying relief. Even a small delay, without reasonable grounds, in seeking the freezing of the status quo may debar an applicant from relief.

5.9.7 Final Injunctions

As is mentioned above, it will be usual that interlocutory relief will restrain allegedly infringing conduct. Although the courts are only obliged to ascertain whether there is a serious question to be tried, in practice these cases have begun to approximate final deliberations on the merits. Article 44 permits the judicial authorities 'to order a party to desist from infringement, *inter alia,* to prevent the entry into channels of commerce in their jurisdiction of imported goods that involve the infringement of an intellectual property right'.

The remedy of injunction is usually granted on a discretionary basis. Among the factors considered are whether: (a) damages provides an adequate remedy; (b) the order will require constant supervision by the court; (c) the applicant has engaged in some disentitling conduct, such as its own infringing activity; and (d) the applicant has delayed in seeking its remedy or has acquiesced in the respondent's conduct.

Another discretionary ground which is contained in Article 44 is that Members are not obliged to accord the remedy of injunction 'in respect of protected subject matter acquired or ordered by a person prior to knowing or having reasonable

---

[126]TRIPS Agreement, 4, Art. 50.6.

[127]*Monserrate v. N.Y. State Senate*, 599 F.3d 148, 154 (2d Cir. 2010). Applied Most Recently In *Louboutin v. Yves Saint Laurent America, Inc.*, 778 F.Supp.2d 445 (S.D.N.Y. 2011).

grounds to know that dealing in such subject matter would entail the infringement of an intellectual property right'. It is difficult to see the justification for this qualification and how it will operate in practice.[128] Article 50 permits the grant of provisional measures to prevent an infringement occurring on the application of a single party, where appropriate. A respondent may at that time discover that the products which it has purchased are infringing, but it cannot be enjoined from selling those products under Article 44, since it acquired the knowledge of infringement after the date of the contract of acquisition. Some sense may be made of this qualification by virtue of the fact that the respondent would still be liable to pay damages if it persisted in distributing infringing products.[129]

5.9.8 Damages

Article 45.1 provides that the judicial authorities shall have the authority to order 'the infringer to pay the rights holder damages adequate to compensate for the injury...suffered because of an infringement of that persons intellectual property right by an infringer who knowingly, or with reasonable grounds to know, engaged in infringing activity'.

Similarly to Article 13(1) of the Enforcement Directive,[130] Article 45.1 provides that the obligation to pay damages may be imposed only on infringers *"who knowingly or with reasonable grounds to know"* engaged in an infringing activity.[131]

5.9.9 Knowledge

Article 45.1 provides for compensation orders against infringers "who knowingly, or with reasonable grounds to know, engaged in infringing activity." A general standard of reasonableness is usually applied to the question of guilty knowledge. The courts have taken the view, for example that a person who copies a new product ought to have inquired whether it was patented.[132] Conventionally, the existence of relevant knowledge is sought to be established by the delivery of a cease and desist letter to an infringer. A continuation of infringing activity after receipt of such a letter is evidence of guilty knowledge.

---

[128] TRIPS Agreement, 4, Art. 44.
[129] TRIPS Agreement, 4, Art. 55.
[130] Directive 2004/48/EC of 29 April 2004 on the enforcement of intellectual property rights.
[131] TRIPS Agreement, 4, Art. 45.1.
[132] *Lancer Boss Ltd v. Henley Fork lift Co Ltd* (1975 RPC 301).

Article 45.2 permits Members to authorise the judicial authorities 'to order the recovery of profits and/or payment of pre-established damages even where the infringer did not knowingly, or with reasonable grounds to know, engage in infringing activity'. This sort of remedy is usually ordered in cases of unfair competition or passing off.[133]

5.9.10 Other Remedies

Article 46, under the justification of creating an effective deterrent to infringement, allows Members to empower the judicial authorities "to order that the goods which they have found to be infringing be, without compensation of any sort, disposed of outside the channels of commerce in such a manner as to avoid any harm caused to the rights holder." Alternatively, where existing constitutional requirements so permit, the infringing goods may be destroyed. A constitutional obstacle, which exists in some jurisdictions, is the obligation to provide "just terms" for any goods, which are compulsorily acquired.

A supplementary power, which is conferred upon the judicial authorities, is the power "to order that materials and implements, the predominant use of which has been in the creation of the infringing goods" is similarly disposed of outside the channels of commerce in such a manner as "to minimise the risks of further infringements."

In considering requests for orders to dispose of or destroy infringing goods and equipment used to produce such goods, the judicial authorities are required to take into account "the need for proportionality between the seriousness of the infringement and the remedies ordered as well as the interests of third parties." In the case of counterfeit trademark goods, Article 46 indicates, "the simple removal of the trademark unlawfully affixed shall not be sufficient, other than in exceptional cases, to permit the release of goods into the channels of commerce."[134]

5.9.11 Right of Information

A particularly useful innovation is the authority, which is conferred by Article 47 "to order the infringer to inform the right holder of the identity of third persons involved in the production and distribution of the infringing goods or services and of their channels of distribution." Article 47 counsels the exercise of this power where it is not "out of all proportion to the seriousness of the infringement." No guidance is provided as to how seriousness is to be evaluated or whether the

---

[133]TRIPS Agreement, 4, Arts. 45.1 and 45.2.
[134]TRIPS Agreement, 4, Art. 46.

touchstone of seriousness is damage to the party seeking the information, or whether from the perspective of the public interest in suppressing wrongful acts. For example, the large-scale counterfeiting of low quality trademarked goods may be of minimal concern to a trader producing high quality products, which are not likely to be confused with the counterfeiter's products. However there may be a public interest in the protection of consumers from the poorer quality goods. There may also be a more fundamental public interest in inculcating an ethos of commercial morality.

A limiting condition in Article 47 is that information will not be provided if it is out of proportion to the seriousness of the infringement, although no guidance is provided as to the test of proportionality.

TRIPS Article 47 provides that the information, which may be provided to a right holder, is "the identity of third persons involved in the production and distribution of the infringing goods or services and of their channels of distribution."[135] Article 8(1) of the EU's Enforcement Directive is even more broadly drawn. Although it commences with the qualification of justification and proportionality it provides that the competent judicial authorities may order that information on the origin and distribution networks of the goods or services which infringe an intellectual property right be provided not only by the infringer, but also by any other person who:

(a) was found in possession of the infringing goods on a commercial scale,
(b) was found to be using the infringing services on a commercial scale,
(c) was found to be providing on a commercial scale services used in infringing activities, or
(d) was indicated by the person referred to in point (a), (b) or (c) as being involved in the production, manufacture or distribution of the goods or the provision of the services.[136]

Article 8(2) of the Enforcement Directive then itemises the type of information, which may be provided, including:

(a) the names and addresses of the producers, manufacturers, distributors, suppliers and other previous holders of the goods or services, as well as the intended wholesalers and retailers,
(b) information on the quantities produced, manufactured, delivered, received or ordered, as well as the price obtained for the goods or services in question.[137]

The general qualification to the information, which may be provided under laws implementing the Directive is that, the information applies in respect of acts carried out on "a commercial scale". This term is not defined in the substantive part of the Directive but in Recital 14, which states that acts carried out on a commercial scale

---

[135] TRIPS Agreement, 4, Art. 47.
[136] EU Enforcement Directive 8(1).
[137] EU Enforcement Directive 8(2).

"are those carried out for direct or indirect economic or commercial advantage; this would normally exclude acts carried out by end consumers acting in good faith."[138]

5.9.12 Indemnification of the Defendant

Where "enforcement measures have been abused," Article 48.1 provides that the judicial authorities shall have the authority to order a party "at whose request enforcement measures were taken" to provide "adequate compensation for the injury suffered because of such abuse" to a person wrongfully enjoined or restrained. Article 48.1 also provides for the applicant to be ordered to pay the defendant's "appropriate attorney's fees."[139]

5.9.13 Criminal Sanctions

Article 61 provides that Members shall provide for criminal procedures and penalties "to be applied at least in cases of wilful trademark counterfeiting or copyright piracy on a commercial scale." The expression "at least" leaves it open for criminal penalties to be imposed in cases concerning other IPR offences, such as where a law might criminalise geographical indications infringements. Among the criminal sanctions, which are listed in the Article, are: "imprisonment, and/or monetary fines sufficient to provide a deterrent, consistently with the level of penalties applied for fines of a corresponding gravity." Also in appropriate cases, Article 61 provides for "the seizure, forfeiture and destruction of the infringing goods and any materials and implements the predominant use of which has been in the commission of the offence."

Article 61 also provides for criminal procedures and penalties to be applied in other cases of infringement of intellectual property rights, "in particular where they are committed wilfully and on a commercial scale."[140]

5.9.14 Border Measures

A key feature of the TRIPS Agreement was the obligation of Members to introduce border measures for the protection of intellectual property rights. It is obviously more effective to seize a single shipment of infringing products at the border rather

---

[138] EU Enforcement Directive, Recital 14.
[139] TRIPS Agreement, 4, Art. 48.1.
[140] TRIPS Agreement, 4, Art. 61.

than to await its distribution in the market. The stratagem of utilising border seizure to control the trade in infringing goods was foreshadowed in the Paris Convention, which in Article 9(1) provides that "all goods unlawfully bearing a trademark or trade name shall be seized on importation into those countries of the Union where such mark or trade name is entitled to protection." It was envisaged in Article 9(3) that this seizure would take place at the request of "the public prosecutor, or any other competent authority, or any interested party." The Paris Convention contains no provisions providing for the seizure upon importation of other intellectual property infringements.[141]

The key border control provision of the TRIPS Agreement is Article 51, which requires Members to:

> adopt procedures to enable a right holder, who has valid grounds for suspecting that the importation of counterfeit trademark or pirated copyright goods may take place, to lodge an application with competent authorities, administrative or judicial, for the suspension by the customs authorities of the release into free circulation of such goods.

As a footnote to this provision, the term "counterfeit trademark goods" is defined to mean:

> any goods, including packaging, bearing without authorization a trademark which is identical to the trademark validly registered in respect of such goods, or which cannot be distinguished in its essential aspects from such a trademark, and which thereby infringes the rights of the owner of the trademark in question under the law of the country of importation.

In addition to the suspension of release of goods involving a suspected counterfeit trademark, or which are pirated copyright goods, Article 51 also provides that an application for suspension may also be made in respect of other intellectual property rights infringements, such as carrying ornamentation which infringes a registered design or involving production in breach of a patented process.[142]

5.10 Protocol to the Lisbon Agreement

As was noted above, the Lisbon Agreement, 1958, failed to secure much support beyond the countries of the Mediterranean. The failure of the negotiations in the TRIPS Council to settle the operating principles as well as the details for a multilateral system for the registration of geographical indications, has led to an examination of the possibility that the Lisbon Agreement might be modified to become an acceptable registration option. At least one author has suggested a protocol to the Lisbon Agreement as a means of achieving this result. The simplest approach would be to "establish a new international register, possibly limited to wines and spirits, to be administered by WIPO, thus relying on the expertise of the

---

[141]Paris Convention, 10, Art. 9(1) and 9(3).
[142]TRIPS Agreement, 4, Art. 51.

Lisbon staff" and on WIPO's "experience in administering international intellectual property registration systems."[143] The protocol would mirror the current registration process but apply to GIs as defined in TRIPS definition with no substantive protection norms and leaving it to the WTO dispute system to deal with conflicts.[144] Alternatively, a protocol has been suggested that mirrors not just the administrative provisions of the current Lisbon system but also the substantive GIs provisions of TRIPS, including conflicts between GIs and trademarks.[145] To deal with the extension issue, the register might contain two distinct domains: one for wines and spirits for which TRIPS Article 23 protection would apply; and one for all other products, for which TRIPS Article 22 protection would apply.[146]

In September 2008, the Assembly of the Lisbon Union established a Working Group on the Development of the Lisbon System to explore possible improvements to the procedures under the Lisbon System to make it more attractive for users and prospective new members. Since 2009, the Working Group has engaged in a full review of the Lisbon International Registration System involving its possible extension to geographical indications in addition to appellations of origin. Various sessions of the Working Group considered drafts of proposed changes to the Lisbon system, culminating at its sixth session, in December 2012. In the sixth session, the Working Group on the Development of the Lisbon System (Appellations of Origin) requested the International Bureau of WIPO to prepare a Draft Revised Lisbon Agreement that would take the form of a single instrument covering both appellations of origin and geographical indications. The instrument would provide for a high and single level of protection for both, while maintaining separate definitions, on the understanding that the same substantive provisions would apply to both appellations of origin and geographical indications.[147] In advance of the seventh session of the Working Group, The Organization circulated a list of GIs of US products for an International Geographical Indications Network (oriGIn).[148] One function of this list may have been to indicate to US negotiators the significance of GIs for US industry. The list comprised various certification marks, which had been registered in the US for agricultural products.

A Draft Agreement and associated Regulations were presented to the Seventh Session of the Working Group, which was held April 29 to May 3, 2013.[149]

---

[143] Gervais, 58, at 121.

[144] Id. at 123.

[145] Id. at 124.

[146] Id. at 125.

[147] WIPO Secretariat, 'Draft Revised Lisbon Agreement on Appellations of Origin and Geographical Indications', WIPO Doc., LI/WG/DEV/6/2, Sep. 28, 2012.

[148] Mendelson and Wood (2013).

[149] WIPO Secretariat, 'Draft Revised Lisbon Agreement on Appellations of Origin and Geographical Indications', WIPO Doc., LI/WG/DEV/7/2, Mar 22, 2013.

5.11 Bilateral and Plurilateral Agreements

In the last decade it has become also a practice to incorporate provisions for the protection of specific geographical indications in some free trade agreements (FTAs), which contain provisions, modifying the TRIPS provisions dealing with trademarks and geographical indications. For example, the Trans-Pacific Partnership Agreement (TPPA) currently under negotiation between Australia, Brunei Darussalam, Canada, Chile, Japan, Malaysia, Mexico, Peru, New Zealand, Singapore, USA and Vietnam contains a number of trademarks and geographical indications provisions, which supplement those of the TRIPS Agreement.[150]

References

Anders S, Caswell JA (2009) The benefits and costs of proliferation of geographical labeling for developing countries. Estey Centre J Int Law Trade Policy 10(1):77–93
Babcock BA (2003) Geographical indications, property rights, and value-added agriculture. Iowa Ag Rev 9(4), available at http://www.card.iastate.edu/iowa_ag_review/fall_03/article1.aspx
Blakeney M (1994) 'Well-known' marks. EIPR 11:481–486
Blakeney M (1996) Trade related aspects of intellectual property rights. A concise guide to the TRIPS agreement. Sweet & Maxwell, London
Bonnet C, Simioni M (2001) Assessing consumer response to protected designation of origin labelling: a mixed multinomial logic approach. Euro Rev Agric Econ 28(4):433–449
Bramley C, Bienabe E (2012) Developments and considerations around geographical indications in the developing world. Queen Mary J Intellect Prop 2(1):14–37
Chever T, Renault C, Renault S, Romieu V (2012) Value of production of agricultural products and foodstuffs, wines, aromatised wines and spirits protected by a geographical indication (GI), Final report for European Commission, available at http://ec.europa.eu/agriculture/external-studies/2012/value-gi/final-report_en.pdf
Echols MA (2008) Geographical indications for food products: international legal and regulatory perspectives. Kluwer Law International, The Netherlands
Evans GE, Blakeney M (2006) The protection of geographical indications after DOHA: Quo Vadis? J Int Econ Law 1(40), available at www.gaileevans.com/EvansGIsAfterDohaJIEL06.pdf
Gervais DJ (2010) Reinventing Lisbon: the case for a protocol to the Lisbon Agreement (geographical indications). Chic J Int Law 11(1):67–126
Gerz A, Dupont F (2006) Comté Cheese in France: impact of a geographical indication on rural development. In: van de Kop P, Sautier D, Gerz A (eds) Origin-based products: lessons for pro-poor market development. KIT Publishers, Amsterdam, pp 75–87
Kireeva I, Xiaobing W, Yumin Z (2009) Comprehensive feasibility study for possible negotiations on a geographical indications agreement between China and the EU. EU-China IP2, Brussels
O'Connor & Co. (2005) Geographical indications and the challenges for ACP countries. Agritrade, CTA (paper available <http://agritrade.cta.int/>)
Mendelson R, Wood Z (2013) Geographical Indications in the United States: developing a preliminary list of qualifying product names. oriGIn paper, Geneva, OriGIn
Pflüger M (2011) Article 10. In: Cottier T, Véron P (eds) Concise International and European IP Law TRIPS, Paris Convention, European enforcement and transfer of technology. Wolters Kluwer, Alphen aan den Rijn, The Netherlands

---

[150] http://wikileaks.org/tpp/.

Chapter 6
Agricultural Innovation: Patenting and Plant Variety Rights Protection

Michael Blakeney

Abstract This chapter describes the international patent and plant variety rights systems which are regulated by the Trade Organization (WTO) Agreement on Trade Related Aspects of Intellectual property Rights (TRIPS) and the International Convention for the Protection of New Varieties of Plants (UPOV). It looks at the jurisprudence around plant and DNA patenting and plant breeding. The chapter also looks at the impact upon agricultural innovation of the International Treaty for Plant Genetic Resources for Food and Agriculture.

6.1 International Intellectual Property Infrastructure

The international intellectual property rights (IPR) regime based upon the World Trade Organization (WTO) Agreement on Trade-related Aspects of Intellectual Property Rights ("TRIPS Agreement") establishes global intellectual property norms. The TRIPS Agreement obliges the 159 member states of the WTO[1] to implement such norms by requiring the provision of legal protection for newly developed plant varieties and enabling the patenting and commodification of DNA. A key provision of the TRIPS Agreement in the context of food and agriculture is Article 27.1, which establishes a patenting regime extending to all WTO members. The provision provides that, "patents shall be available for any inventions, whether products or processes, in all fields of technology, provided that they are new, involve an inventive step and are capable of industrial application." The provision also requires that, "patents shall be available and patent rights enjoyable without discrimination as to the ... field of technology." Additionally, Article 27.3(b) of the TRIPS Agreement requires that WTO Members "shall provide for the protection of

---

Various parts of this chapter have previously been published by this author and are hereby impliedly cited.

[1] 159 Member States as of 13 March 2013. See http://www.wto.org/english/thewto_e/whatis_e/tif_e/org6_e.htm, accessed 31.10.2013.

M. Blakeney (✉)
Faculty of Law, The University of Western Australia, Crawley, WA, Australia
e-mail: michael.blakeney@uwa.edu.au

plant varieties either by patents or by an effective *sui generis* system or by any combination thereof."

Although the TRIPS Agreement does not prescribe a specific sui generis system for the protection of plant varieties, most countries have adopted the 1991 version of the International Convention for the Protection of New Varieties of Plants ("UPOV"). UPOV provides for the protection of new plant varieties which are "distinct", "uniform," and "stable."

Originally, under the 1978 version of UPOV, propagating material which had been harvested by farmers and retained for further planting or for sale was exempted from protection. Article 15 (2) of the 1991 version of UPOV Convention confined the farmer's seed saving exception to the use of saved material for propagating purposes on farmers' own holdings and in reasonable quantities. UPOV 1991 also contains a breeder's exception that permits the use of protected varieties for the purpose of breeding new varieties. As is indicated below, the seed saving and breeding exceptions become irrelevant where a new variety can be patented.[2]

The UPOV Convention has 71 signatories as of December 5, 2012.[3] Only a few countries have adopted alternatives to UPOV 1991 despite numerous commentaries and proposals for the adoption of alternative *sui generis* models.[4]

6.2 Patenting of DNA

The modern biotechnological revolution has enabled the engineering of desirable genetic traits from useful local species. These include: (1) pest control traits such as insect, virus and nematode resistance as well as herbicide tolerance; post-harvest traits such as delayed ripening of spoilage prone fruits; (2) agronomic traits such as nitrogen fixation and utilisation, restricted branching, environmental stress tolerance, male and/or seed sterility for hybrid systems; and (3) output traits such as plant colour and vitamin enrichment. The production of transgenic plants has become possible through the development of a number of enabling and transformation technologies.[5]

A key issue around the patenting of genetic resources was whether a DNA sequence could be characterised as an "invention." In the early history of patent law, an invention was thought to involve some kind of technical innovation, and a distinction was drawn between patentable inventions and non-patentable discoveries. In 1980 the United States Supreme Court ruled in a 4:3 majority decision that a bacterium genetically engineered to degrade crude oil was an invention in the

---

[2]Blakeney (2012b).
[3]http://upov.int/export/sites/upov/members/en/pdf/pub423.pdf accessed 31 10.2013.
[4]Eg. Leskien and Flitner (1997); Dhar (2002); Helfer (2002); Robinson (2007); Robinson (2008).
[5]Blakeney (2012b).

landmark case, *Diamond* v *Chakrabarty*.[6] This decision provided the legal underpinning for the U.S. biotechnology industry. The European Parliament's belated response. In 1998, The European Parliament responded with its belated Biotechnology Directive, providing in Article 3.2 that, "biological material which is isolated from its natural environment or produced by means of a technical process is deemed to be an invention even if this material previously occurred in nature.[7]"

The patentability of genetic materials and gene fragments, such as expressed sequence tags (ESTs) and single nucleotide polymorphisms (SNPs), as well as enabling gene-based technologies led to what has been described as a "genomic gold rush" in the 1990s as vast numbers of gene-based patent applications were filed, particularly in the USA[8] Significant misgivings were expressed by numerous commentators. Probably the most influential among these was the suggestion that genetic research tool patents could create a "tragedy of the anticommons" in which multiple patent owners would tie-up genetic materials in a thicket of IP patent rights.[9] The 'tragedy of the anticommons' was perceived to be a particular problem for the genetic improvement of crops since patenting is an incremental process and each new patent would constrain a researcher's "freedom to operate." The effect of the patentability of genetic materials and gene fragments therefore could particularly restrain public agricultural research institutes.[10]

Two recent US cases have questioned of the patentability of genetic material. In *Association for Molecular Pathology v. USPTO*,[11] a United States District Court Judge for the Southern District of New York delivered a summary judgement invalidating patents related to the BRCA 1 and 2 breast and ovarian cancer susceptibility genes, which had been patented by the company Myriad Genetics.[12] He ruled that DNA sequences in isolation were insufficiently distinct from naturally occurring genes in the body and were thus products of nature rather than inventions. He observed that DNA represents the physical embodiment of biological information, distinct in its essential characteristics from any other chemical found in nature and that DNA in an "isolated" form alters neither this fundamental quality as it exists in neither the body nor the information it encodes.[13]

---

[6]447 US 303 (1980).

[7]Directive 98/44/EC of the European Parliament and of the Council of the 6 July 1998 on the legal protection of biotechnological inventions, Official Journal L213, 30/07/1998 P.0013-0021.

[8]Joly (2003).

[9]Heller and Eisenberg (1998).

[10]See authorities referred to in Correa (2009).

[11]94 USPQ2d 1683 (S.D.N.Y. March 29, 2010).

[12]Association for Molecular Pathology et al v. United States Patent and Trademark Office et al, 09 Civ. 4515, March 29, 2010.

[13]Ibid at 121.

This decision was successfully appealed to the U.S. Court of Appeals for the Federal Circuit (CAFC) in Washington, D.C.[14] The Court of Appeals ruled that the District Court Judge had failed to consider whether the isolated DNAs were markedly different from naturally occurring DNAs, and improperly focused on whether the isolated DNAs had the same informational content as native DNA sequences. Nevertheless, the CAFC held that the District Court was correct in holding that Myriad's claims directed to comparing and analysing gene sequences were not patentable, as these claims contained no transformative steps and covered only patent ineligible abstract steps.

This reasoning was evaluated recently by the U.S. Supreme Court, in *Mayo Collaborative Services v. Prometheus Laboratories, Inc.*[15] The case concerned patents obtained by Prometheus which instructed doctors in the use of thiopurine drugs to treat autoimmune diseases. Mayo had developed its own diagnostic test, which Prometheus claimed infringed its patents. Justice Breyer, delivering the opinion of the Court, noted the long held view of the Supreme Court is such that laws of nature, natural phenomena, and abstract ideas are not patentable. He quoted from the Court's decision in *Diamond* v *Chakrabarty* that "a new mineral discovered in the earth or a new plant found in the wild is not patentable subject matter." The Court held that Prometheus' process was not patent eligible because the laws of nature recited by Prometheus' patent claims (i.e. the relationships between concentrations of certain metabolites in the blood and the likelihood that a thiopurine drug dosage will prove ineffective or cause harm) were not themselves patentable. The implications of the Supreme Court decision in *Prometheus Laboratories* regarding DNA patenting have yet to be worked out in the food and agriculture context, but as is explained below, DNA patenting has become an important feature of agricultural innovation.

6.3 DNA Patenting and Agriculture

The cultivation of GM crops has on occasion led to IPR liability for farmers, where genetically modified (GM) seed is patented and the cultivation of that seed by the patentee is unauthorized. The types of cases invoking IPR liability are divided between those where farmers knowingly cultivate patented GM seed and those where the cultivation of patented seed is apparently inadvertent, for example, where crops are pollinated by wind or insect-borne pollen.

An example of the first category of case in which a farmer knowingly cultivates patented GM seeds, is *Monsanto Co. v. Scruggs*.[16] The case concerned Monsanto's patented Roundup Ready ("RuR") glyphosate tolerant seeds. The RuR seeds were

---

[14]The Association of Molecular Pathology & Ors v. The USPTO and Myriad Genetics Inc, _F.3d_ (CAFC, 2011).

[15]132 S.Ct. 1289 (2012).

[16]342 F. Supp 2d 584 (2004).

licensed to seed companies, who were obliged to sell the seed only to growers who signed technology license agreements acknowledging Monsanto's patent and agreeing to the condition that thee seeds could only be used by growers for a single commercial crop, i.e. growers could not save seed produced from a harvested crop for replanting during the following growing season. Scruggs purchased a small quantity of RuR soybeans and cotton seeds for cultivation without signing a technology licensing agreement. After cultivation, Scruggs saved seed for further plantings. The Court decided that Monsanto's patent had been infringed by Scruggs. Scruggs raised the defence that neither Monsanto's biotechnology nor the plants in their fields were covered by the patent and that the first sale of the seed embodying the invention exhausted the patent rights of Monsanto. The Court noted that Monsanto did not make an unrestricted sale of its seed technology, but rather licensed its technology to seed companies with a proviso: subsequent sales of seed containing its transgenic trait must be limited to growers who obtained a license from Monsanto and for only a single growing season.[17]

A more recent variant of these facts occurred in *Monsanto Co v Bowman*,[18] where a farmer, Bowman, purchased commodity seeds from a local grain elevator which were not subject to a technology agreement. Bowman first applied glyphosate to the crops grown from these seeds. He then identified those which were glyphosate resistant and saved them for re-planting in subsequent years, enabling Bowman to utilize glyphosate-based herbicide. As a response, Monsanto filed a patent infringement claim against Bowman. In September 2009, the District Court in Indiana granted summary judgment on patent infringement for Monsanto and awarded damages of $84,456.

Bowman appealed to the Court of Appeals for the Federal Circuit. Bowman argued that Monsanto's patent rights were exhausted under the first sale doctrine in relation to all second-generation Roundup Ready soybean seeds that were present in the grain elevators. The Court of Appeals held that there would be no impact even if Monsanto's patent rights in the commodity seeds were exhausted, because once a grower, like Bowman, planted the commodity seeds containing Monsanto's RuR technology he created a newly infringing article upon development of the next generation of seed. It observed that "The fact that a patented technology can replicate itself does not give a purchaser the right to use replicated copies of the technology. Applying the first sale doctrine to subsequent generations of self-replicating technology would eviscerate the rights of the patent holder."[19]

The Supreme Court endorsed this reasoning, explaining that under the patent exhaustion doctrine, Bowman could resell the patented soybeans he purchased from the grain elevator, consume the beans himself, or feed them to his animals; however, "The exhaustion doctrine does not enable Bowman to make *additional* patented soybeans without Monsanto's permission (either express or implied)."[20]

---

[17]Ibid. at 591.
[18]569 U. S. ____ (2013).
[19]657 F. 3d, at 1348 (2010).
[20]569 U. S. ____ (2013) at 6.

Therefore, Bowman was denied protection under the exhaustion doctrine because he reproduced Monsanto's patented invention.

A case of apparently inadvertent infringement by a farmer is illustrated by the Canadian litigation between Monsanto Canada, Inc. and Percy Schmeiser. Schmeiser grew canola commercially in Saskatchewan. He had never purchased Monsanto's patented RuR Canola nor did he obtain a licence to plant it. Yet, in 1998, tests revealed that 95–98% of his 1000 acres of canola crop was made up of RuR plants. The origin of the RuR plants is unclear, but they may have been derived from RuR seed that blew onto or near Schmeiser's land. Regardless of the origin, Monsanto brought an action for patent infringement. In finding patent infringement, the trial judge ruled that the growth of the seed, the reproduction of the patented gene and cell, and the sale of the harvested crop constituted taking the essence of Monsanto's invention and using it without permission.

The Federal Court of Appeal ruled 5:4 that Schmeiser's saving and planting of seed, subsequently followed by the harvesting and selling of plants containing the patented cells and genes appeared to constitute "utilization" of the patented material for production and advantage within the meaning of s. 42 the Canadian *Patent Act*.[21] The argument that the infringing seed had merely grown as the result of wind pollination, or through the pollinating activities of birds and bees, was rejected by the majority Judges as denying "the realities of modern agriculture." Instead, the judges focused on the issue of sowing and cultivation, "which necessarily involves deliberate and careful activity on the part of the farmer". They noted that Schmeiser had actively cultivated RuR Canola as part of his business operations. Thus, in light of all of the relevant considerations, Schmeiser had used the patented genes and cells, and infringement was established.

6.4 Patenting of Stress-Tolerant Genes

DNA patenting has become of crucial significance for agriculture since the identification of stress tolerant genes which are of assistance in developing crops that are better able to tolerate climate change. In 2010, the Action Group on Erosion, Technology and Concentration ("ETC") conducted a study that, "examined patents containing claims concerned with abiotic stress tolerance (i.e. traits related to environmental stress, such as drought, salinity, heat, cold, chilling, freezing, nutrient levels, high light intensity, ozone and anaerobic stresses.[22]" The study noted "a dramatic upsurge in the number of patents published (both applications and issued patents) related to 'climate-ready' genetically engineered crops from June 30, 2008 to June 30, 2010." They identified 262 patent families and 1663 patent docu-

---

[21]*Monsanto Canada, Inc. v. Schmeiser.* [2004] 1 S.C.R. 902, 2004 SCC 34.
[22]ETC Group (2010).

ments.[23] The ETC report contrasted the ownership of patent families by public sector institutions, holding 9 % of the total, with the private sector, which holds 91 % of the total. The report pointed out that "just three companies – DuPont, BASF, Monsanto – account for two-thirds (173 [patents] or 66 %) of the total." This level of market concentration gives cause for concern in regards to the positive role of competition.[24]

Additionally, the market dominance of these private corporations has an important influence upon the sort of biotechnological research which is undertaken. For example, the dominance of private corporations in biomedical and agricultural research may direct research towards Northern concerns away from Southern food priorities.[25] Further, it has been estimated that only 1 % of research and development budgets of multinational corporations is spent on crops of interest that would be useful in the developing world.[26] Major corporations almost entirely neglect five of the most important crops essential to the poorest, arid countries, which include sorghum, millet, pigeon pea, chickpea and groundnut.[27]

6.5 The UPOV Convention

6.5.1 Introduction

The UPOV Convention defines "plant variety" in terms of a plant grouping within a single biological taxon of the lowest known rank, which grouping can be:

- defined by the expression of characteristics (such as shape, height, colour and habit) resulting from a given genotype or combination of genotypes;
- distinguished from any other plant grouping by the expression of at least one of the said characteristics; and
- considered as a unit with regard to its suitability from being propagated unchanged.

Generally, under plant variety rights legislation the plant breeder is conferred an exclusive right to do or to licence the following acts in relation to propagating material of the variety:

- produce or reproduce the material;
- condition the material for the purpose of propagation;
- offer the material for sale;

---

[23]Ibid, Appendix A.
[24]Eg see Lesser (1998).
[25]Alston et al. (1998).
[26]Pingali and Traxler (2002).
[27]Human Rights Council, Report of the Special Rapporteur on the Right to Food, Jean Ziegler, A/HRC/7/5, 10 January 2008, para 44.

- sell the material;
- import the material;
- export the material;
- stock the material for the purposes described above.

The protection under this legislation is afforded to a "breeder" or persons claiming through the breeder who is defined in Article 1 (iv) of the UPOV Convention as the person who bred, "or discovered or developed a variety." "Breeding" is generally defined as including the discovery of a plant together with its use in selective propagation so as to achieve a result.

The general duration of plant variety rights under legislation based on the UPOV Convention is 25 years in the case of trees and vines and 20 years for any other variety. During these periods, the breeder or other licensee or owner of the right is entitled to exclusivity in its exploitation and commercialisation.[28]

6.5.2 Criteria for Registrability

A plant variety is considered to be registrable if it has a breeder, is distinct, uniform, and stable, and has not been, or has only recently been, exploited. A plant variety is considered distinct if it is clearly distinguishable from any other variety whose existence is a matter of common knowledge. This issue was considered by the European Court of Justice of the First Instance in a case concerning a PVP application for a variety of the species *Plectranthus ornatus*.[29] The applicant's competitors successfully opposed the registration of the plant variety on the ground that it was not distinct from a wild variety originating in South Africa. This was regarded as a matter of common knowledge because the plant variety had been marketed for years in that country and was also found in private gardens.[30]

Additionally, some of the comments made by the court give insight to the questionable novelty of inventions derived from traditional knowledge. In deciding the question of common knowledge, the Court took into account the academic literature referring to the fact that *P.*, originally native in Ethiopia and Tanzania, was "cultivated and semi-naturalised" in South Africa.[31]

A variety is considered uniform if it is uniform in its relevant characteristics on propagation, subject to the variation which may be expected from the particular features of its propagation. A plant variety is stable if its relevant characteristics remain unchanged after repeated propagation. A plant variety is taken not to have

---

[28] Blakeney (2011a).
[29] Schrader v OCVV (SUMCOL 01) [2008] EUECJ T-187/06 (19 November 2008).
[30] Ibid.
[31] Codd (1975); confirmed by Andrew Hankey in Plantlife No 21, September 1999; Dr. H.F. Glen, "Cultivated Plants of Southern African names, common names, literature." 2002, p. 326.

been exploited if it or propagating material has not been sold to another person by or with the consent of the breeder.[32]

6.5.3 Farmer's Privilege ("Seed-Saving" Exception)

Seed saved by a farmer from harvested material and treated for the purpose of sowing a crop on that farmer's own land is usually excepted from plant variety rights. Article 15(2) of the UPOV Convention provides as an optional exception that "each Contracting Party may, within reasonable limits and subject to the safeguarding of the legitimate interests of the breeder, restrict the breeder's right in relation to any variety in order to permit farmers to use for propagating purposes, on their own holdings, the product of the harvest which they have obtained by planting, on their own holdings, the protected variety...." It should be noted that although this is colloquially defined as the "seed-saving" exception, Article 15(2) of UPOV refers to "the product of the harvest". The language of Article 15(2) can be contrasted with Article 9.3 of the International Treaty for the Protection of Plant Genetic Resources for Food and Agriculture, which provides that "Nothing in this Article shall be interpreted to limit any rights that farmers have to save, use, exchange and sell farm-saved seed/propagating material, subject to national law and as appropriate."

The use in Article 15(2) of the term "product of the harvest" instead of "seed" has raised questions about the proper scope of the exception. One of the aims of the farmer's privilege exemption was to achieve a balance between encouraging the development of new crops and avoiding the alienation of farming communities for whom saving seed was a traditional practice.[33] Under this rationale, it has been suggested that the exemption might not be appropriate for economies where farming is a quasi-industrial activity performed by a small minority of the population or where plant breeding has become an industrial activity. According to a former head of UPOV, the intention behind the exception was to provide UPOV member states with the opportunity to balance local interests and access to new crops and medicinal varieties against the interests of the breeder on a species by species basis.[34] At the time of the 1991 Diplomatic Conference, it was recommended that the provisions laid down in Article 15(2) "should not be read so as to be intended to open the possibility of extending the practice commonly called 'farmer's privilege' to sectors of the agricultural or horticultural production in which such a privilege is not a common practice on the territory of the Contracting Party concerned."[35]

---

[32]Blakeney (2011a).

[33]Llewelyn and Adcock (2006).

[34]Greengrass (1991).

[35]*Records of the Diplomatic Conference for the Revision of the International Convention for the Protection of New Varieties of Plants,* UPOV, Geneva, 1991, Recommendation Relating to Article 15(2), p. 63.

Some countries have clarified the scope of the exception in their national legislation. For example, Costa Rica[36] and the Dominican Republic[37] exclude "fruit, ornamental and forest species" from farmer's privilege "where planted for commercial ends". Mexico limits the exception to "grain for consumption or seed for sowing".[38]

An example of the farmer's privilege in an industrialised setting appears in Article 14 of the European Community Plant Variety Rights Regulation. Article 14 gives farmers the right to use farm saved seed without the consent of the owner (right holder) of the variety in question. However, the farmer, with the exception of small farmers, must pay the holder an equitable remuneration which shall be sensibly lower than the amount charged for the licensed product.[39] If the parties cannot agree upon the level of the remuneration, such remuneration should be 50 % of the amounts charged for the licensed production of propagating material.[40]

6.6 Critiques of the PVP System

Over the last two decades, commentators on the PVP system have begun to question to its relevance, raising the possibility that it might have become "the Neanderthal of intellectual property systems".[41] One reason for this critique is the impact of patents upon PVP, described above. At a more fundamental level it is observed that PVP in focussing upon a phenotypic paradigm, based upon "characteristics" and "features", has become outmoded as plant breeding moves towards a genotypic approach, based on genetic modification and molecular breeding techniques.[42] Mark Janis and Stephen Smith argue that plants should be reconceptualised as datasets that breeders manipulate to express particular characteristics, which could be better regulated by unfair competition laws rather than by a sui generis PVP scheme.[43]

It should be noted that critiques of PVP systems have tended to be intuitive, rather than empirically based, as there is an absence of data of the impacts of the genotypic approach upon PVP systems, as well as a general shortage of data of the impacts of PVP upon the development of new plant varieties. Some recent attempts have been made to redress the latter deficiency. In 2005 UPOV (2005) published a

---

[36]Law No. 8631 on the Protection of New Varieties of Plants, Article 23.

[37]Law on Protection of Breeder's Rights for Varieties of Plants, Article 18.

[38]Federal Law on Plant Varieties, Articles 2, 4, 5.

[39]Council Regulation (EC) No 2100/94 on Community plant variety rights, OJ No L 173/14, 25.7.95 CPVR, Article 14(3).

[40]Council Regulation (EC) No 1768/95 of 24 July 1995 implementing rules on the agricultural exemption provided for in Article 14(3) of CPVR.

[41]Fowler (1994).

[42]Janis and Smith (2007).

[43]Ibid at pp. 1607–14. See also Sanderson (2007).

report on the quantitative impacts of plant breeder's rights in Argentina, China, Kenya, Poland and the Republic of Korea.[44] The Report concluded that:

> Individual country reports have demonstrated increases in the overall numbers of varieties developed after the introduction of PVP. New, protected varieties have been developed for a wide range of crops including, for example, staple crops in the agricultural sector (e.g. barley, maize, rice, soybean, wheat), important horticultural crops (e.g. rose, Chinese cabbage, pear), traditional flowers (peony, magnolia, camellia in China) forest trees (e.g. poplar in China) and traditional crops (e.g. ginseng in the Republic of Korea.[45]

It has been generally assumed that the availability of PVPs has increased the number of varieties released and planted.[46] However, it is uncertain as to whether the availability of protection caused the increase in varietal release, as well as whether this is an economic good.[47]

Data measuring the impact of PVPs upon investment in plant breeding in developed countries suggests that this is uneven. For example, Lesser and Mutschler assessing US utility patent applications from 1970 to 2000, found that the overall increase in the use of utility patents in the mid-1990s could be attributed to two crops: corn and soybeans,[48] where patented GM technology has become important,[49] but insignificant, for example, in the case of wheat.[50] An Australian study of PVP applications in Australia from 1987 to 2007 observed a notable decrease in the number of applications for the period 2003 to 2007.[51] This was tentatively attributed to changing environmental conditions such as drought and increased salinity which effected plant breeding investment either by reducing the level of plant breeding, or by focusing breeding programs on developing particular traits (for example, drought resistance and salinity tolerance) in new plant varieties. In Australia the highest number of PVP applications (61 %) came from the nursery sector, which is perceived to be particularly vulnerable to changing climatic conditions.

6.7 Patenting of Plant Varieties

UPOV allows the protection of new varieties of plants which are distinct, uniform and stable. A variety is considered to be new if it has not been commercialized for more than 1 year in the country of protection. A variety is distinct if it differs from all other known varieties by one or more important botanical characteristics. A variety is uniform if the plant characteristics are consistent from plant to plant

---

[44]UPOV (2005).
[45]Ibid at p. 17.
[46]Lesser (1990).
[47]Rangnekar (2002).
[48]Lesser and Mutschler (2002).
[49]Dhar and Foltz (2007).
[50]Alston and Venner (2000). See also Rangnekar (2008).
[51]Sanderson and Adams (2008).

within the variety. A variety is stable if the plant characteristics are genetically fixed and therefore remain the same from generation to generation, or after a cycle of reproduction in the case of hybrid varieties.

The 1991 version of UPOV recognizes the right of breeders to use protected varieties to create new varieties. However, this exception is itself restricted to such new varieties as are not "essentially derived" from protected varieties. The drafters added this restriction to prevent second generation breeders from making merely cosmetic changes to existing varieties in order to claim protection for a new variety. The most contentious aspect of the 1991 Act is the limitation of the farmers' privilege to save seed for propagating the product of the harvest they obtained by planting a protected variety "on their own holdings," "within reasonable limits and subject to the safeguarding of the legitimate interests of the breeder." This is contrasted with earlier versions of UPOV which permitted farmers to sell or exchange seeds with other farmers for propagating purposes.

The seed saving privilege and the permitted development of non-essentially derived new varieties from protected material were compromises built in to the legislation to account for public policy concerns. One concern was that food security would be compromised if individuals were permitted to privatise food varieties by locking up breeding material and preventing farmers from seed saving for further harvests. However, from the perspective of plant breeders, any derivation of new varieties from their protected varieties, whether essential or non-essential, was inconvenient for them and any seed saving by farmers deprived them of new sales. Consequently, they looked to patent law, which does not contain these exceptions, to protect their new varieties.[52]

In the USA, plant varieties can be protected under a system of plant patents, or under a system of utility patents or under the Plant Variety Protection Act (PVPA).[53] The Plant Patent Act makes patent protection available to new varieties of asexually reproduced plants. Under this scheme, a plant variety must be novel and distinct and the invention, discovery, or reproduction of the plant variety must not be obvious. One of the disadvantages of the scheme is that only one claim covering the plant variety is permitted in each application.

The Federal Circuit Court of Appeals resolved any potential conflict between patent protection and protection under the Plant Variety Protection Act (PVPA) in its decision in *Pioneer Hi-Bred International Inc. v. J.E.M. Ag Supply Inc.*[54] Pioneer's patents covered the manufacture, use, sale, and offer for sale of the company's inbred and hybrid corn seed products as well as certificates of protection under the PVPA for the same seed-produced varieties of corn. The defendants argued that the enactment of the PVPA had removed seed-produced plants from the realm of patentable subject matter the Patents Act. The Federal Circuit rejected this

---

[52]WIPO/IP/UNI/DUB/04/10.

[53]Plant Patent Act, 35 U.S.C. §§ 161–164 (1994).

[54]200 F.3d 1374 (Fed. Cir. 2000), *cert. granted,* 148 L. Ed. 2d 954 (2001).

argument noting that the Supreme Court held that "when two statutes are capable of co-existence, it is the duty of the courts . . . to regard each as effective."[55]

The issue was further clarified in *Monsanto Co. V. McFarling*,[56] which concerned Monsanto's patent for glyphosate-tolerant plants, including the genetically modified seeds for such plants, the specific modified genes, and the method of producing the genetically modified plants. Monsanto required that sellers of the patented seeds obtained a "Technology Agreement," from purchasers in which they agreed that the seeds were to be used "for planting a commercial crop only in a single season," that the purchaser would not, "save any crop produced from this seed for replanting, or supply saved seeds to anyone for replanting." Mr. McFarling, a farmer in Mississippi, purchased Roundup Ready soybean seed in 1997 and again in 1998; he signed the Technology Agreement for both purchases. He saved 1500 bushels of the patented soybeans from his harvest during one season, and planted them as seed in the next season instead of selling these soybeans as crop. The saved seed retained the genetic modifications of the Roundup Ready seed. Mr. McFarling did not dispute that he violated the terms of the Technology Agreement but claimed that the contractual prohibition against using the patented seed to produce new seed for planting violated the seed saving provision of the PVPA as he had produced only enough new seed for his own use the following season. The Court declined to limit the patent law by reference to the PVPA, and Mr. McFarling was found to have infringed Monsanto's patent.

6.8 Patenting of Plant Breeding Methods

In addition to patenting the products of plant breeding, some patent laws allow for the patenting of plant breeding methods. For example, in the U.S. a patent has been obtained for the "selective increase of the anticarcinogenic glucosinolates in brassica species,"[57] and an application has been published concerning a "method for breeding tomatoes having reduced water content."[58] These demonstrate the possibility that methods of crop breeding to withstand climate stress can be privatized in the U.S., permitting so-called methods patents.

On the other hand, European patent legislation excludes "essentially biological processes for the production of plants or animals" defined in Article 2.2 of the EU Biotechnology Directive as consisting "entirely of natural phenomena such as crossing or selection." This exclusion resulted in the denial of patent protection for the same methods for breeding brassica and tomatoes that were able to be patented in the U.S.[59] The Board of Appeals of the European Patent Organization

---

[55]Ibid.
[56]302 F.3d 1291 (Fed. Cir. 2002).
[57]US Patent 6,340,784, January 22, 2002.
[58]US Patent Application 20,100,095,393, April 15, 2010.
[59]Blakeney (2012a).

(EBA) observed that with the creation of new plant varieties, for which a special property right was going to be introduced under the subsequent UPOV Convention in 1960, the legislative architects of the European Patent Convention were concerned with excluding from patentability the kind of plant breeding processes which were the conventional methods for the breeding of plant varieties of that time. These conventional methods predominantly included those based on the sexual crossing of plants deemed suitable for the purpose pursued and on the subsequent selection of the plants having the desired trait(s). These processes were characterised by the fact that the traits of the plants resulting from the crossing were determined by the underlying, natural phenomenon of meiosis. On the other hand, processes for changing the genome of plants by technical means, such as irradiation, were cited by the EBA as examples of patentable technical processes. The EBA pointed out that the provision of an explicit or implicit technical step in a process based on the sexual crossing of plants and on subsequent selection does not cause the claimed invention to escape the exclusion from patentability if that technical step only serves to perform the process steps of the breeding process.

A process leaves the realm of plant breeding and is not excluded from patentability in Europe if the process of sexual crossing and selection includes an additional step of a technical nature that by itself introduces a trait into the genome or modifies a trait in the genome of the plant produced so that the introduction or modification of that trait is not the result of the mixing of the genes of the plants chosen for sexual crossing. This principle applies only where the additional step is performed within the steps of sexual crossing and selection, independently from the number of repetitions Otherwise, the exclusion of sexual crossing and selection processes from patentability could be circumvented simply by adding steps which do not properly pertain to the crossing and selection process. This includes upstream steps dealing with the preparation of the plant(s) to be crossed or downstream steps dealing with the further treatment of the plant resulting from the crossing and selection process. The EBA noted that for the previous or subsequent steps *per se* patent protection was available. Patent Protection is also available for genetic engineering techniques applied to plants which differ from conventional breeding techniques, as they work primarily through the deliberate insertion and/or modification of one or more genes in a plant.[60]

6.9 Patenting of Genetic Resources (GRs)

One problem with determining the legal protection of genetic resources through IPRs or any other law, is the fact that scientific constructs do not always lend themselves to legal categorization. For example, Article 27.3(b) of the TRIPS Agreement provides that WTO Members may also exclude from patentability: "plants and animals other than micro-organisms, and essentially biological

---

[60]Blakeney (2012a).

processes for the production of plants or animals other than non-biological and microbiological processes." It has been noted that the division between plants and animals on the one hand and micro-organisms on the other is not as scientifically certain as the legal categories seem to suggest.[61]

Additionally, a number of international organizations, with varying levels of scientific competence, are now concerning themselves with IPRs and genetic and biological resources. At its sixteenth session, held from May 3 to 7, 2010, WIPO's Intergovernmental Committee on Intellectual Property and Genetic Resources, Traditional Knowledge and Folklore (IGC) Member States identified the need for a glossary to clarify the meanings of key terms related to genetic resources in order to facilitate negotiations of the Committee.[62] The Secretariat prepared a document drawing definitions from previous glossaries of the IGC and from existing United Nations and other international instruments, as well as taking into account definitions and glossaries which can be found in national and regional laws and draft laws, multilateral instruments, other organizations and processes and in dictionaries.[63]

6.10 Traditional Agricultural Knowledge and Farmers' Rights

The traditional knowledge of Indigenous peoples throughout the world has played an important role in identifying biological resources worthy of commercial exploitation. For example, the search for new pharmaceuticals derivable from naturally occurring biological material has been guided by ethno biological data.[64] Examples of traditional knowledge with an agricultural application include: "mental inventories of local biological resources, animal breeds, and local plant, crop, and tree species," as well as plants which are indicators of soil salinity, seed treatment, and storage methods and tools used for planting and harvesting.[65] Additionally, significant contributions have been made by the knowledge of indigenous peoples and farmers in the development of new crops types and biodiversity conservation, e.g. crops resistant to climate change.

The economic value of biological diversity conserved by traditional farmers for agriculture is difficult to quantify.[66] It has been suggested that "the value of farmers' varieties is not directly dependent on their current use in conventional

---

[61] Adcock and Llewelyn (2000).
[62] WIPO (2010a) Draft Report of the Sixteenth Session of the Committee (WIPO/GRTKF/IC/16/8 Prov. 2), para. 227.
[63] WIPO/GRTKF/IC/17/INF/13, October 4, 2010.
[64] Kate and Laird (2000).
[65] Hansen and van Fleet (2007).
[66] E.g. see Brush (1994).

breeding, since the gene flow from landraces to privately marketed cultivars of major crops is very modest" because "conventional breeding increasingly focuses on crosses among elite materials from the breeders own collections and advanced lines developed in public institutions.[67]" On the other hand, those collections and advanced breeding lines are often derived from germ plasm contributed by traditional groups. An increasingly significant economic value of biodiversity is the extent to which it provides a reservoir of species available for domestication, as well as genetic resources available for the enhancement of domestic species. The modern biotechnological revolution has enabled the engineering of desirable genetic traits from useful local species. It is estimated that about 6.5 % of all genetic research undertaken in agriculture is focussed upon germ plasm derived from wild species and land races.[68]

Traditional knowledge is particularly important in the development of farming systems adapted to the local conditions and farming practices. Development of this knowledge may enable the utilization of marginal lands, contribute to food security in enabling access to food in remote areas, and contribute to the management of the environment by preventing erosion, maintaining soil fertility, and agro biodiversity.

6.11 Traditional Knowledge and Prior Art

An alternative approach to the protection of traditional knowledge as a category of intellectual property, is its recognition as part of "prior art." Prior art calls into question the novelty and creativity of inventions which are the subject of patent applications. Patent examiners have practical difficulties in identifying relevant traditional knowledge as prior art, arising from the fact that they do not have access to traditional knowledge information in classified non-patent literature. Further, there are no effective search tools for the retrieval of such information.

The WIPO Intergovernmental Committee on Intellectual Property and Genetic Resources, Traditional Knowledge and Folklore (IGC) has begun to address practical measures to establish linkages between IP Offices and traditional knowledge documentation initiatives.[69] The draft Substantive Patent Law Treaty, which was submitted to the fifth session of the WIPO's Standing Committee on the Law of Patents (SCP), held in Geneva from May 14 to 19, 2001, contained two alternatives for a draft article on the definition of prior art. The draft provisions on the definition of prior art provide that any information made available to the public, anywhere in the world, in any form, including in written form, by oral communication, by display and through use, shall constitute prior art, if it has been made available to the public before the filing date, or, where applicable, the priority date.

---

[67]Correa (2000).
[68]McNeely (2001).
[69]WIPO Doc., WIPO/GRTKF/IC/2/6, July 1, 2001, para. 6.

6.12 Disclosure of the Source of GRs, Access and Benefit Sharing: Recent International Developments

One of the foundational tasks of the WIPO IGC has been the formulation of guidelines on the IP aspects of access and benefit-sharing in relation to GRs. A drafted set of guidelines was submitted to the seventh session of the IGC in November 2004, which sought to provide assistance in the negotiation of contracts for access to genetic resources and related information, including traditional knowledge, and for benefit-sharing arrangements.[70] This document has been through a number of drafts, the most recent of which was prepared for the third Intersessional Working Group, which met from February 28 to March 4, 2011.[71] This document, together with other documents prepared on the subjects of traditional knowledge and traditional cultural expressions, are to be taken into account in "text-based negotiations" by the IGC, ultimately with to the goal of formulating an international treaty.

At the Seventeenth Session of the IGC, which met in Geneva, December 6 to 10, 2010, the Secretariat identified the options which were then under consideration.[72] There were three categories of options: (1) those concerning the defensive protection of genetic resources; (2) those in relation to disclosure requirements; and (3) those concerning the IP aspects of access and benefit-sharing.

In relation to defensive protection, one category of options was the compilation of an inventory of existing periodicals, databases and other information resources which document disclosed genetic resources, with a view to discussing a possible recommendation that certain periodicals, databases and information resources may be considered by International Search Authorities for integration into the minimum documentation list under the Patent Co-operation Treaty. The second option concerned the extension of the Online Portal of Registries and Databases, established by the Committee at its third session, to include existing databases and information systems for access to information on disclosed genetic resources. A third option was for the formulation of recommendations or guidelines for search and examination procedures for patent applications to ensure that they better take into account disclosed genetic resources.

Options on disclosure requirements included the development of a mandatory disclosure requirement. Alternatively, it was proposed that the IGC could consider whether there is a need to develop appropriate (model) provisions for national or regional patent or other laws which would facilitate consistency and synergy between access and benefit-sharing measures for genetic resources and between national and international intellectual property law and practice. Another disclosure option was the development of guidelines or recommendations concerning the

---

[70] WIPO (2004) WIPO Doc., WIPO/GRTKF/IC/17/INF/10.

[71] WIPO (2011) WIPO Doc., WIPO/GRTKF/IWG/3/12.

[72] "Genetic Resources: Revised List of Options and Factual Update" WIPO/GRTKF/IC/17/6, September 15 (2010b).

interaction between patent disclosure and access and benefit-sharing frameworks for genetic resources.

On May 6, 2010, the delegations of Australia, Canada, New Zealand, Norway and the United States of America submitted a working document[73] on GR for the seventeenth session of the IGC held December 6 to 10, 2010. Comments on this document[74] were made by the Delegations of Chile, Colombia and the Russian Federation and a number of accredited observers, which resulted in a revised document identifying five objectives with underlying principles.[75] On December 8, 2010, the Delegation of Angola submitted the proposals of the African Group.[76] This suggested the commencement of negotiations on a mandatory disclosure requirement and an appropriate way to ensure prior informed consent and fair and equitable benefit sharing, in line with the Nagoya Protocol. The African proposal suggested that negotiations be based upon two current proposals on a mandatory disclosure requirement,[77] and the incorporation of the "internationally recognized certificate of compliance" as stipulated in the Nagoya Protocol, together with any other submission that may be tabled by member countries. In relation to the option for guidelines and recommendations on defensive protection, the African Group proposed consideration of the use of available databases on GR and/or associated Traditional Knowledge (TK).

The African Group proposed a number of amendments to the Submission made by Australia, Canada, New Zealand, Norway, and the United States of America. The common position between all groups of countries is that the objectives of the mandatory disclosure requirement should be: (1) the use of GRs and associated TK should be on the basis of benefit sharing; (2) patents should not be granted for inventions that are not novel or inventive in light of genetic resources and/or associated traditional knowledge; (3) patent offices should have available the information needed to make proper decisions on patent grant; (4) the principles developed should consistent with other international and regional instruments and processes; and (5) IP should maintain a role in promoting creativity and innovation. At the Third Intercessional Working Group of the IGC, which met from February 28 to March 4, 2011, a Working Group was appointed to review and rationalize the various Objectives and Principles which had been received by the IGC. The Working Group had a goal of clarifying the key and divergent policy positions

---

[73] WIPO/GRTKF/IC/16/7.

[74] WIPO/GRTKF/IC/17/INF/10.

[75] WIPO/GRTKF/IC/17/7.

[76] WIPO/GRTKF/IC/17/10.

[77] I.e. the "Declaration of the Source of Genetic Resources and Traditional Knowledge in Patent Applications: Proposal by Switzerland" (WIPO/GRTKF/IC/11/10) and EU Proposal "Disclosure of Origin or Source of Genetic Resource and Associated Traditional Knowledge in Patent Applications" (WIPO/GRTKF/IC/8/11) with a view to amending the Patent Cooperation Treaty (PCT) and the Patent Law Treaty (PLT) to reflect a mandatory disclosure requirement of the origin of the genetic resources.

6.13 Farmer's Rights Under the International Treaty on Plant Genetic Resources for Food and Agriculture (ITPGRFA)

The concept of Farmers' Rights was developed as "a counterbalance to intellectual property rights.[79]" This was a moral commitment by the industrialised commitment to reward "the past present and future contributions of farmers in conserving, improving and making available plant genetic resources particularly those in centres of origin/diversity". Farmers' rights were intended to promote a more equitable relation between the providers and users of germplasm by creating a basis for farmers to share in the benefits derived from the germplasm which they had developed and conserved over time.[80]

The first international enactment of Farmers' Rights occurred in the FAO International Treaty on PGRFA.[81] The preamble to the Treaty acknowledges that "the conservation, exploration, collection, characterization, evaluation and documentation of plant genetic resources for food and agriculture are essential in meeting the goals of the Rome Declaration on World Food Security and the World Food Summit Plan of Action and for sustainable agricultural development for this and future generations." It also acknowledges that PGFRA "are the raw material indispensable for crop genetic improvement" and affirms "that the past, present and future contributions of farmers in all regions of the world, particularly those in centres of origin and diversity, in conserving, improving and making available these resources, is the basis of Farmers' Rights."

The Preamble outlines that that "fundamental to the realization of Farmers' Rights, as well as the promotion of Farmers' Rights at national and international levels" are the rights "to save, use, exchange and sell farm-saved seed and other propagating material, and to participate in decision-making regarding, and in the fair and equitable sharing of the benefits arising from, the use of plant genetic resources for food and agriculture."

Under Art. 5.1 (c) the Contracting Parties agree, subject to national legislation, to promote or support, as appropriate, farmers and local communities' efforts to manage and conserve on-farm their plant genetic resources for food and agriculture

---

[78]'Draft Objectives and Principles Relating to Intellectual Property and Genetic Resources Prepared at IWG 3', WIPO/GRTKF/IWG/3/17, March 16, 2011.

[79]IWG (2011) Draft Objectives and Principles Relating to Intellectual Property and Genetic Resources Prepared at IWG 3, WIPO/GRTKF/IWG/3/17.

[80]Glowka (1998).

[81]International Treaty on PGRFA.

and in Art. 51 (d) to promote *in situ* conservation of wild crop relatives and wild plants for food production, by supporting, *inter alia*, the efforts of indigenous and local communities.

In Art. 9(1) of the Treaty the Contracting Parties "recognize the enormous contribution that the local and indigenous communities and farmers of all regions of the world, particularly those in the centres of origin and crop diversity, have made and will continue to make for the conservation and development of plant genetic resources which constitute the basis of food and agriculture production throughout the world."

Article 9.2 of the WTO International Treaty on PGRFA envisages that "the responsibility for realizing Farmers' Rights, as they relate to Plant Genetic Resources for Food and Agriculture, rests with national governments" and that national legislation should include measures relating to:

(a) protection of traditional knowledge relevant to plant genetic resources for food and agriculture;
(b) the right to equitably participate in sharing benefits arising from the utilization of plant genetic resources for food and agriculture;
(c) the right to participate in making decisions, at the national level, on matters related to the conservation and sustainable use of plant genetic resources for food and agriculture.

Finally, Article 9.3 provides that the Article shall not be interpreted "to limit any rights that farmers have to save, use, exchange and sell farm-saved seed/propagating material."

An assumption of Article 9 is that the landraces used by traditional farmers are a dynamic genetic reservoir for the development of new varieties and for the transmission of desirable genetic traits. The traditional knowledge of local and indigenous communities is similarly perceived. Farmers in subsistence systems have tended to utilize a diverse selection of crop species in order to assure their annual harvests and thus to guarantee a minimal level of production and to prevent food shortage. In many instances, seed production has involved the collection and domestication of locally known, wild varieties. Modern agricultural practices depend on crop species that promote productivity and resistance to disease that can only be maintained with the continuous input of new germ plasm. The diversity of landraces and the associated information on their specific qualities contribute invaluable information to formal breeding processes. It has been noted that the loss of biological diversity is paralleled by the loss of traditional knowledge. Where a plant variety becomes extinct, then the entire body of knowledge about its properties is condemned to irrelevancy.

As a means of remunerating these groups for their past contributions to the development of plant genetic resources for food and agriculture production, there can be little argument, except about the quantum and distribution of this remuneration. Inevitably, any calculation of the equitable share, which traditional farmers and indigenous communities might enjoy under a Farmers' Rights or Traditional Knowledge regime, will be arbitrary. However, the intellectual property system is no stranger to arbitrary calculations; thus, the 20 year length of a patent term is intended to provide an opportunity for the compensation of all inventors, whatever the area of technology. Similarly the 25 years exclusivity which the UPOV Convention provides for new varieties of trees and vines, takes no account of variations in R & D costs between the different varieties.

The principal ways in which plant genetic resources are translated into food and agriculture production is through plant breeding and plant patenting. Standing at the heart of a Farmers' Rights regime is the concept of the equitable benefit sharing of benefits with farmers for their contribution to innovations in plant breeding and plant patenting.[82]

Article 9.2 obliges the Contracting Parties to the Plant Genetic Resources Treaty "to take measures", subject to their national legislation to protect and promote Farmers' Rights. The content of these rights is defined in the balance of that provision and embraces the protection of traditional knowledge, equitable benefit sharing and the right to participate in decision making. The Treaty leaves open the legal context within which Farmers' Rights are to be enacted.

The only measure which has been implemented to provide for Farmers Rights is the International Fund for Plant Genetic Resources, which was envisaged in the Undertaking which preceded the Treaty. This Fund was to operate as a means of capacity building in the field of agricultural biotechnology in developing countries rather than as a reward to individual farmers or farming communities for their contribution to the development or improvement of plant varieties. However, to date this fund has not been established because funds were not made available by donor countries.

6.14 Recent Developments on Farmers' Rights

At its Third Session in Tunis in 2009, the Governing Body of the ITPGRFA adopted a resolution on Farmers' Rights (Resolution 6/2009), in which it requested the Secretariat to convene regional workshops on Farmers' Rights, subject to the agreed priorities of the Programme of Work and Budget and to the availability of financial resources. The aim of the workshops was to discuss national experiences on the implementation of Farmers' Rights as set out in Article 9 of the International Treaty, involving, as appropriate, farmers' organizations and other stakeholders.[83]

The fourth session of the Governing Body of the ITPGRFA held from March 14 to 18, 2011, in Bali, Indonesia adopted a resolution on Farmers' Rights that, *inter alia*:

---

[82]Blakeney (2007).

[83]IT/GB-4/11/Circ. 1 (2010) Fourth Session of the Governing Body. Global Consultations on Farmers' Rights in 2010.

- requests the Secretariat to convene regional workshops on Farmers' Rights, subject to availability of funding;
- encourages parties to submit views, experiences and best practices on the implementation of Farmers' Rights;
- invites parties to consider convening national and local consultations on Farmers' Rights with the participation of farmers and other stakeholders;
- requests the Secretariat to collect and submit these views, as well as reports from regional workshops to GB 5; and
- encourages parties to engage farmers' organizations and relevant stakeholders in matters related to the conservation and sustainable use of PGRFA, through awareness raising and capacity building.[84]

6.15 International Proposals for the Protection of Traditional Knowledge

The first international consideration of the protection of traditional knowledge (TK) occurred in a joint UNESCO/WIPO World Forum on the Protection of Folklore that convened in Phuket, in April 1997. The meeting was comprised of representatives of organizations of indigenous peoples calling for the promulgation of an international convention to protect Traditional Knowledge. WIPO responded in its 1998–1999 biennium by instituting a schedule of regional fact-finding missions "to identify and explore the intellectual property needs, rights and expectations of the holders of traditional knowledge and innovations, in order to promote the contribution of the intellectual property system to their social, cultural and economic development." Australia was chosen as the first port of call for such expert, fact-finding mission, which visited Darwin and Sydney from June 14 to 18, 1998. During 1998 and 1999, similar expert missions visited, Peru, South Africa, Thailand, and Trinidad and Tobago. Finally, in November 1999, WIPO convened a World Forum on Traditional Knowledge.[85]

Following the failure of the Seattle Ministerial in November 1999, WIPO became the focus of agitation for the inclusion of traditional knowledge within the international intellectual property regime. In a Note dated September 14, 2000, the Permanent Mission of the Dominican Republic to the United Nations in Geneva submitted two documents on behalf of the Group of Countries of Latin America and the Caribbean (GRULAC) as part of the debate on in the WIPO General Assembly on "Matters Concerning Intellectual Property and Genetic Resources, Traditional

---

[84]Earth Negotiations Bulletin (2011) Summary of the Fourth Session of the Governing Body of the International Treaty on Plant and Genetic Resources for Food and Agriculture: 14–18 March 2011. 9(550) http://www.iisd.ca/download/pdf/enb09550e.pdf.

[85]WIPO/IP/CAI/1/03/12 (2003) Intellectual Property, Traditional Knowledge, and Genetic Resources.

Knowledge and Folklore."[86] The central thrust of these documents was a request for the creation of a Standing Committee on access to the genetic resources and traditional knowledge of local and indigenous communities. "The work of that Standing Committee would have to be directed towards defining internationally recognized practical methods of securing adequate protection for the intellectual property rights in traditional knowledge."[87]

In order to clarify the future application of intellectual property to the use and exploitation of genetic resources and biodiversity, and also to the use of traditional knowledge, it was suggested that the Committee should clarify: (a) the notions of public domain and private domain; (b) the appropriateness and feasibility of recognizing rights in traditional works and knowledge currently in the public domain, and investigating machinery to limit and control certain kinds of unauthorized exploitation; (c) recognition of collective rights; (d) model provisions and model contracts with which to control the use and exploitation of genetic and biological resources, and machinery for the equitable distribution of profits in the event of a patentable product or process being developed from a given resource embodying the principles of prior informed consent and equitable distribution of profits in connection with the use, development and commercial exploitation of the material transferred and the inventions and technology resulting from it; (e) the protection of undisclosed traditional knowledge.[88]

6.16 WIPO Intergovernmental Committee

At the WIPO General Assembly, the Member States agreed on the establishment of an Intergovernmental Committee on Intellectual Property and Genetic Resources, Traditional Knowledge and Folklore. Three interrelated themes were identified to inform the deliberations of the Committee: intellectual property issues that arise in the context of (1) access to genetic resources and benefit sharing; (2) protection of traditional knowledge, whether or not associated with those resources; and (3) the protection of expressions of folklore.[89]

Despite this agreement, the work of the IGC has been very slow in practice. During the first 10 years of its existence, the IGC has concentrated on the formulation of "objectives" and "principles" which should animate the protection of

---

[86] WIPO Doc. WO/GA/26/9.
[87] *Ibid.*, Annex I, 10.
[88] WIPO/IP/UNI/DUB/04/10 Intellectual Property, Traditional Knowledge, and Genetic Resources: Policy, Law, and Current Trends.
[89] WIPO Matters Concerning Intellectual Property and Genetic Resources Traditional Knowledge and Folklore. WPO Doc, WO/GA/26/6, August 25, 2000.

TCEs and TK.[90] The African group of countries at WIPO were in the forefront of agitation to accelerate the international negotiations. A true reflection of their appreciation of the realistic likelihood of action was the promulgation by a diplomatic conference on August 9–10, 2010, in Swakopmund, Namibia, organized by the African Regional Intellectual Property Organization (ARIPO) of a Protocol on the Protection of Traditional Knowledge and Expressions of Folklore. The Protocol was meant to "protect creations derived from the exploitation of traditional knowledge in ARIPO member states against misappropriation and illicit use through bio-piracy." The protocol is also intended to prevent the "grant of patents in respect of inventions based on pirated traditional knowledge ... and to promote wider commercial use and recognition of that knowledge by the holders, while ensuring that collective custodianship and ownership are not undermined by the introduction of new regimes of private intellectual property rights."

A brief palpitation of enthusiasm on the international front was generated in October 2010, when the 17th session of the IGC, to be held December 6–10, 2010, was identified as the occasion for the first text-based discussion of the establishment an international TK and EC regime. However, the results of this session were not so exciting.[91] An "informal drafting group" was set up to provide a text on Traditional Cultural Expressions for the next meeting of the IGC on May 9–13, 2011. Further proposals for the protection of TK were made by a number of countries, and were considered by an Intercessional working group which met from February 21 to 25, 2011.

The slowness of the developments at WIPO reactivated Pacific considerations for a regional solution. The Pacific Island states made a decision to avail themselves of technical assistance being made available by the EU as part of the Partnership Agreement between the members of the African, Caribbean and Pacific (ACP) Group of States and the European Union (EU). The "Cotonou Agreement" was signed on June 23, 2000, and continued for a 20-year period from March 2000, to February 2020. Two EU projects were initiated under this Agreement. The first, entitled: "Technical Assistance to the Pacific Regional Action Plan for Traditional Knowledge Development,"[92] has a specific objective of providing technical assistance for the establishment of national systems of protection for TK in six of the member states of the Pacific Islands Forum (namely Cook Islands, Fiji, Kiribati, Palau, Papua New Guinea and Vanuatu). The second project aims to provide technical assistance to study the "Feasibility of a Reciprocal Recognition and Enforcement Mechanism," for TK between Fiji, PNG, Solomon Islands, the so called Melanesian Spearhead Group (MSG) countries.[93]

---

[90]The most recent contribution in this regard is a document of 7 June 2010 on the 'Protection of Traditional Cultural Expressions/Cultural expressions of Folklore: Revised Objectives and Principles', which was prepared for the 17th meeting of the IGC in December 2010, WIPO/GRTKF/IC/17/4Prov.

[91]See Decisions of the Seventeenth Session of the IGC, 10 December 2010 at http://www.wipo.int/meetings/en/details.jsp?meeting_id=20207.

[92]Project No: 9.ACP.RPR.007.

[93]Blakeney (2011b).

The Terms of Reference for the latter project recognized that a regional approach would operate as a parallel, viable, and faster alternative to the international developments. Additionally, any future collective arrangement would not preclude other countries from the wider Pacific region to participate in the system. These developments would instruct and inform global treaty making processes currently taking place in institutions such as WIPO and possibly lead to engagement with other like-minded regions given the slow impetus to conclude a global regime for TK at WIPO, the World Trade Organization (WTO), and the Convention on Biological Diversity (CBD).

Both projects have been productive. National mapping of TK and EC has been conducted in the target states, draft IP laws and policies have been formulated for Fiji, PNG and the Solomon Islands, and a collaboration treaty has been drafted for the MSG states. The Treaty was submitted to the 18th Melanesian Spearhead Group Leaders' Summit in Suva on March 31, 2011, which "agreed in principle pending decisions by members on the signing of the Treaty". The Government of Fiji proposes to sign the Treaty in May 2011, and the Governments of PNG and Vanuatu are currently undertaking in-country consultations on the Treaty before their Governments sign the Treaty.

6.17 Substantive Patent Law Treaty

In an endeavour to reach a consensus on substantive patent law issues, a Committee of Experts and WIPO's Standing Committee on Patents (SCP) considered a draft Patent Law Treaty (PLT) prepared by the International Bureau of WIPO. The draft PLT dealt with various procedural aspects of patenting. At the third session of the SCP in September 6–14, 1999, the delegation of Colombia proposed the introduction into the PLT as a means of achieving some global harmonization of patent registration procedures. The article provided that:

1. All industrial protection shall guarantee the protection of the country's biological and genetic heritage. Consequently, the grant of patents or registrations that relate to elements of that heritage shall be subject to their having been acquired made legally.
2. Every document shall specify the registration number of the contract affording access to genetic resources and a copy thereof whereby the products or processes for which protection is sought have been manufactured or developed from genetic resources, or products thereof, of which one of the member countries is the country of origin.

This proposal generated a heated debate about whether it raised a matter of procedural or substantive patent law. An agreement was eventually reached to defer consideration of this proposal to the occasion of the discussion of a proposed

Substantive Patent Law Treaty.[94] The SCP requested the International Bureau to include the issue of protection of biological and genetic resources on the agenda of a Working Group on Biotechnological Inventions, to be convened at WIPO in November 1999. The following month, the Working Group recommended the establishment of nine projects related to the protection of inventions in the field of biotechnology. The Working Group decided to establish a questionnaire for the purpose of gathering information about the protection of biotechnological inventions in the Member States of the WIPO, including certain aspects regarding intellectual property and genetic resources.[95]

An alternative approach to the protection of traditional knowledge is the recognition of such knowledge as part of "prior art." Prior art calls into question the novelty and inventive of inventions which are the subject of patent applications. The practical difficulty that patent examiners have in identifying relevant traditional knowledge as prior arteries from the fact that they do not have access to traditional knowledge information in classified non-patent literature. The problem is exasperated because there are no effective search tools for the retrieval of such information. The WIPO IGC has begun to address practical measures to establish linkages between IP Offices and traditional knowledge documentation initiatives. A number of the characteristics of traditional knowledge present difficulties in identifying the effect of prior art on technological information. These include:

(a) The transmission of traditional knowledge through oral communication. This requires the codification and fixation of traditional knowledge into what it is not.
(b) Traditional knowledge systems tend to dynamic evolution without necessarily being identified as "new".
(c) Traditional knowledge is expressed in local languages and its expression is contingent upon such languages.
(d) The transfer of knowledge from oral into written, printed, and electronic forms may involve a cultural, semantic, and symbolic transformation of the knowledge, which may affect the value of databases as a tool for the conservation of culture and knowledge.
(e) As knowledge must be in the public domain to be considered as prior art, this may provide some difficulties in those communities where knowledge is to be kept confidential.[96]

The draft Substantive Patent Law Treaty, which was submitted to the fifth session of the WIPO's Standing Committee on the Law of Patents (SCP), held in Geneva from May 14 to 19, 2001, contained two alternatives for a draft article on the definition of prior art. The draft provisions on the definition of prior art provide

---

[94]Blakeney, M. Proposals for the Disclosure of Origin of Genetic Resources in Patent Applications http://www.economia.uniroma2.it/conferenze/icabr2005/papers/Blakeney.pdf?origin=publication_detail.
[95]WIPO/ECTK/SOF/01/3.10 (2001).
[96]WIPO/IP/CAI/1/03/12 (2003).

that any information made available to the public, anywhere in the world, in any form, including in written form, by oral communication, by display and through use, shall constitute prior art, if it has been made available to the public before the filing date, or, where applicable, the priority date.[97]

6.18 TRIPS Agreement

One contemporary impetus for the formulation of international positions on the protection of traditional knowledge is the debate concerning review of Art. 27.3 (b) of the plant variety provision of the TRIPs Agreement.[98] Review of this provision was mandated by the TRIPs Agreement itself, and was to be completed by the end of 1999. Developing country participants in the review process have suggested. These provisions provide for equitable sharing of the benefits of the utilization of traditional medical knowledge with indigenous peoples.[99]

The African Group of countries proposed the inclusion of this issue in the Ministerial Conference to set the agenda for the Seattle Round of the WTO.[100] On July 25, 1999, a federation of Indigenous Peoples groups issued a statement for the purposes of the review, pleading for a legislative structure which, "Builds upon the indigenous methods and customary laws protecting knowledge and heritage and biological resources" and which prevents the appropriation of traditional knowledge and integrates, "the principle and practice of prior informed consent, of indigenous peoples as communities or as collectivities." The Statement concluded with an affirmation of the commitment of Indigenous Peoples "to sustain our struggle to have our rights to our intellectual and cultural heritage and our lands and resources promoted and protected."

On October, 4, 1999, Bolivia, Columbia, Ecuador, Nicaragua, and Peru specifically proposed that the Seattle Ministerial Conference establish a mandate within the framework of the Round:

(a) To carry out studies, in collaboration with other relevant international organizations in order to make recommendations on the most appropriate means of recognizing and protecting traditional knowledge as the subject matter of intellectual property rights.
(b) On the basis of the above-mentioned recommendations, initiate negotiations with a view to establishing a multilateral legal framework that will grant effective protection to the expressions and manifestations of traditional knowledge.

---

[97] WIPO Doc., WIPO/GRTKF/IC/2/6, July 1, 2001, para. 6.
[98] See Blakeney (1999).
[99] See Blakeney (1998/1999).
[100] *Communication to the WTO from Kenya, on behalf of the African Group,* WT/GC/W/3026, August 1999.

(c) To complete the legal framework envisaged in paragraph (b) above in time for it to be included as part of the results of this round of trade negotiations.[101]

A communication of August 6, 1999, from Venezuela proposed that the Seattle Ministerial should consider the establishment, "on a mandatory basis within the TRIPS Agreement a system for the protection of intellectual property, with an ethical and economic content, applicable to the traditional knowledge of local and indigenous communities, together with recognition of the need to define the rights of collective holders."[102]

A practical proposal for the integration of traditional knowledge with intellectual property rights can be found in India's suggestion that material transfer agreements be required where an inventor wishes to use biological material identified by traditional knowledge. That obligation would be incorporated through inclusion in Article 29 of the TRIPs Agreement, the requirement that the country of origin of source material be identified in patent applications.[103] Following the failure of the Seattle Ministerial, this agitation for the inclusion of traditional knowledge within the international intellectual property regime shifted to WIPO until it was picked up again at the Doha Ministerial.

Following the Doha approach, amendments have been proposed to the TRIPS Agreement (Art. 29*bis*) that would require WTO Members to oblige patent applicants to disclose the source of any traditional knowledge and to provide evidence of compliance with the source country's legal requirements of prior informed consent for access and fair and equitable benefit sharing arising from the utilization of the traditional knowledge. The African Group of Countries have proposed that as part of the review of Art. 27.3(b), TK should be protected as a "category of intellectual property rights".[104] The scheme of protection which they proposed would include the grant of rights to local or traditional communities concerning (1) respect for those communities on the commercialization of TK; (2) prior informed consent to the use of that TK; (3) full remuneration; and (4) the prevention of unauthorized third parties from utilizing that TK and incorporating that TK into any article or product.

Debate is still continuing within the TRIPS Council as to the whether it has a mandate to amend TRIPS by the inclusion of an Art. 29*bis* or whether that discussion is to be confined to the implementation of the existing text.

---

[101] WT/GC/W/362 12 October 1999.
[102] WT/GC/W/282.
[103] WT/GC/W/147.
[104] IP/C/W/404, 26 June 2003.

6.19 Convention on Biological Diversity

The Rio Declaration stated in Principle 22 that 'Indigenous peoples and their communities...have a vital role in environmental management and development because of their knowledge and traditional practices'. Chapter 26 of Agenda 21 detailed the relationship which conference participants recognised between indigenous peoples and their lands. The Agenda, at para. 26.3(a), required governments:

- to establish a process to empower indigenous peoples and their communities' through measures that include:
- recognition of their values, traditional knowledge and resource management practices with a view to promoting environmentally sound and sustainable development;
- enhancement of capacity—building for indigenous communities based on the adaptation and exchange of traditional experience, knowledge and resource-management practices, to ensure their sustainable development;
- establishment, where appropriate, of arrangements to strengthen the active participation of indigenous peoples and their communities in the national formulation of policies, laws and programs relating to resource management and other development processes that may affect them.

6.20 Conclusion

The application of IPRs to genetic resources has become a pronounced feature of agricultural innovation in the past decade. The Food and Agriculture Organization of the United Nations (FAO) Panel of Eminent Experts on Ethics in Food and Agriculture has observed that, "while most innovation for food and agriculture does not depend on IPRs, the acquisition and exercise of IPRs in this field raise a variety of ethical concerns.[105]" These concerns include the fact that "IPRs protection may just mean the lack of access to innovations for the poor" and concerns regarding the "patenting of merely isolated genes, the basic building blocks of life", which "are not invented, but are part of nature." Further, the ability of individuals and corporations to obtain proprietary rights over agricultural innovations has important implications for food security.[106] This is increasingly important as the expenses and general transactional costs of patenting have tended to concentrate such IPRs in a few hands. In effect, IPRs on genetic resources may impede the use of these resources for further research and breeding by third parties during the term of protection, and thereby inhibit the development of new products and the capacity

---

[105]FAO (2005).
[106]Blakeney (2009).

to address emerging problems, such as agricultural stresses caused by climate change.

The IPRs landscape confronting countries is dominated by the TRIPS Agreement. However, that Agreement contains a number of flexibilities. First, it allows the exclusion of plants and animals (whether genetically modified or not) from patent protection. Secondly, the criteria under which patents are granted, i.e. novelty, and inventive step and industrial applicability, may exclude materials identified through the application of traditional knowledge or genetic resources which exist in nature (even if isolated), as well as microorganisms. Article 30 of the TRIPS Agreement provides that, "Members may provide limited exceptions to the exclusive rights conferred by a patent, provided that such exceptions do not unreasonably conflict with a normal exploitation of the patent and do not unreasonably prejudice the legitimate interests of the patent owner, taking account of the legitimate interests of third parties." Thus, patent laws may allow third parties to undertake research and breeding during the patent term and farmers may be granted the right to save and re-use seeds where plant varieties, or certain components thereof, are subject to patent protection, in a way similar to the 'farmer's privilege' under PVP.

Over the last few years, there has been a significant amount of patenting in relation to genetic material which might be useful in permitting organisms to resist the stresses of climate change. This patenting mirrors the high market concentration levels which have already been observed in the seed industry and the control of patent thickets by a small number of companies. It should be noted in this regard, both in relation to patent rights and PVP that national laws may provide for compulsory licenses in situations of national emergency. There is also a possibility for the intervention of the competition authorities to remedy abuses in the exercise of patent rights.

The practical effects of the application of IPRs to genetic resources, is reflected in the actions which are brought for infringements of IPRs. To date, these actions have mainly been brought against farmers who have cultivated patented GM crops without the permission of the relevant rights holder, as well as actions against importers of products containing patented GM ingredients. Potential IPR liability lies against governments, research institutes (international and national) and seed breeders who supply or utilize proprietary technologies. The TRIPS Agreement establishes machinery to deal with international trade in infringing goods. This machinery is currently being supplemented by the proposed Anti-counterfeiting Trade Agreement (ACTA), of which the final text was settled on 3 December 2010.

References

Adcock M, Llewelyn M (2000) Micro-organisms, Definitions and Options under TRIPS. Quaker United Nations Office Programme, Occasional Paper 2

Alston J, Venner RJ (2000) The Effects of the U.S. Plant Variety Protection Act on Wheat Genetic Improvement. EPTD Discussion Paper No. 62

Alston J, Pardey G, Rosenboom J (1998) Financing agricultural research: international investment patterns and policy perspectives. World Dev 26(6):1057–1071

Blakeney M (1998/1999) Biotechnology, TRIPs and the convention on biological diversity. Bio-Sci Law Rev 4:144–150

Blakeney M (1999) International framework of access to plant genetic resources. In: Blakeney M (ed) Intellectual property aspects of ethnobiology. Sweet & Maxwell, London, p 1

Blakeney M (2007) Plant variety protection, International Agricultural Research, and exchange of Germplasm: legal aspects of sui generis and patent regimes. In: Krattiger A, Mahooney RT, Nelson L et al (eds) Intellectual property in Health and Agricultural Innovation: a handbook of best practices. MIHR/PIPRA, Oxford, Davis. available online at www.IPHandbook.org

Blakeney M (2009) Intellectual property rights and food security. Cab International, Wallingford

Blakeney M (2011a) Trends in intellectual property rights relating to genetic resources for food and agriculture. Background Study Paper no. 58, FAO Commission on Genetic Resources for Food and Agriculture, Rome

Blakeney M (2011b) Protecting traditional knowledge and expressions of culture in the Pacific. Queen Mary J Intellect Prop 1(1):80–89

Blakeney M (2012a) Patenting of plant varieties and plant breeding methods. J Exp Bot 63(3):1069–1074

Blakeney M (2012b) Climate change and gene patents. Queen Mary J Intellect Prop 2(1):2–13

Brush S (1994) Providing farmers' rights through in situ conservation of crop genetic resources. University of California, Berkeley

Codd LE (1975) Plectranthus (Labiateae) and allied genera in Southern Africa. Bothalia 11(4):371–442

Correa C (2000) Options for the implementation of farmers' rights at the national level. South Centre, Trade-Related Agenda, Development And Equity Working Papers, No. 8

Correa CM (2009) Trends in intellectual property rights relating to Genetic Resources for Food and Agriculture, Background Study Paper 49. Commission on Genetic Resources for Food and Agriculture, Rome

Dhar B (2002) Sui generis systems for plant variety protection. Options under TRIPS. A Discussion Paper. QUNO, Geneva

Dhar T, Foltz J (2007) The impact of intellectual property rights in the plant and seed industry. In: Kesan J (ed) Agricultural biotechnology and intellectual property: seeds of change. CABI, Wallingford

ETC (2010) Gene Giants Stockpile Patents on "Climate-ready" Crops in Bid to become "Biomassters" Patent Grab Threatens Biodiversity, Food Sovereignty, Issue no. 106, October 2010. Available at http://www.etcgroup.org/upload/publication/pdf_file/FINAL_climatereadyComm_106_2010.pdf

FAO (2005) Panel of Eminent Experts on Ethics in Food and Agriculture, 3rd Report. Available at http://www.fao.org/docrep/010/a0697e/a0697e00.htm

Fowler C (1994) Unnatural selection: technology, politics, and plant evolution. Gordon and Breach, Switzerland

Glowka L (1998) A guide to designing legal frameworks to determine access to genetic resources. IUCN, Gland

Greengrass B (1991) The 1991 act of the UPOV convention. Eur Intellect Prop Rev 13(12):466–472

Hansen SA, van Fleet JW (2007) Issues and options for traditional knowledge holders in protecting their intellectual property economies. In: Krattiger A, Mahoney RT, Nelsen L et al (eds) Intellectual property management in Health and Agricultural Innovation: a handbook of best practices. MIHR/PIPRA, Oxford/Davis

Helfer LR (2002) Intellectual property rights in plant varieties: an overview with options for national governments, FAO Legal Papers Online #31. FAO, Rome

Heller MA, Eisenberg RS (1998) Can patents deter innovation? The anticommons in biomedical research. Science 280:698–701
IWG (2011) Draft Objectives and Principles Relating to Intellectual Property and Genetic Resources Prepared at IWG 3, WIPO/GRTKF/IWG/3/17, March 16, 2011
Janis M, Smith S (2007) Technological change and the design of plant variety protection regimes. Chic Kent Law Rev 82:1557–1614
Joly Y (2003) Accès aux mèdicaments: le système international des brevets empêchera-t'il les pays du tiers monde de bénéficier des avantages de la pharmacogénomique? Les cahiers de Propriété Intellectuelle 16:131–185
Kate K, Laird SA (2000) The commercial use of biodiversity: access to genetic resources and benefit. Earthscan, London
Leskien D, Flitner M (1997) Intellectual property rights and plant genetic resources; options for a sui generis system, Issues in Genetic Resources No 8. International Plant Genetic Resource Institute, Rome
Lesser WH (1990) Sector issues II: seeds and plants. In: Siebeck WE, Evenson RE, Lesser, W, Primo Braga CA (eds) Strengthening protection of intellectual property in developing countries. The World Bank, Washington, pp 59–68
Lesser W (1998) Intellectual property rights and concentration in agricultural biotechnology. AgBioFOrum 1(2):56–61
Lesser W, Mutschler M (2002) Lessons from the patenting of plants. In: Rothschild MF, Newman S (eds) Intellectual property rights in animal breeding and genetics. CABI, Wallingford, pp 103–118
Llewelyn M, Adcock M (2006) European plant intellectual property. Hart Publishing, Oxford
McNeely R (2001) Biodiversity and agricultural development: the crucial institutional issues. In: Lee DR, Barrett CB (eds) Tradeoffs or synergies? Agricultural intensification, economic development and the environment. CABI, Wallingford, pp 399–408
Pingali PL, Traxler G (2002) Changing focus of agricultural research: will the poor benefit from biotechnology and privatization trends? Food Policy 27:223–238
Rangnekar D (2002) Access to genetic resources, gene-based inventions and agriculture – issues concerning the TRIPs agreement. Prepared for the UK Government Commission on Intellectual Property Rights, CIPR, London
Rangnekar D (2008) Is More Less? An evolutionary economics, critique of the economics of plant breeds rights. In: Gibson J (ed) Patenting lives: life patents, culture and development. Ashgate, London, pp 179–194
Robinson D (2007) Exploring components and elements of *Sui Generis* systems for plant variety protection and traditional knowledge in Asia. UNCTAD, IDRC and ICTSD, Geneva
Robinson D (2008) *Sui Generis* plant variety protection systems: liability rules and non-UPOV systems of protection. J Intellect Prop Law Pract 3(10):659–665
Sanderson J (2007) Back to the future: possible mechanisms for the management of plant varieties in Australia. Univ N S W Law J 30(3):686–712
Sanderson J, Adams K (2008) Are Plant Breeder's Rights Outdated? A Descriptive and Empirical Assessment of Plant Breeder's Rights in Australia, 1987–2007. Melb Univ Law Rev 32(3):980–1006
UPOV (2005) Report on the Impact of Plant Variety Protection (2005). UPOV Publication No. 353 (E). Accessible at: http://www.upov.int/en/publications/pdf/353_upov_report.pdf
WIPO (2004) WIPO Doc., WIPO/GRTKF/IC/17/INF/10
WIPO (2010) Draft Report of the Sixteenth Session of the Committee (WIPO/GRTKF/IC/16/8 Prov. 2), para. 227
WIPO (2010) Genetic Resources: Revised List of Options and Factual Update WIPO/GRTKF/IC/17/6, September 15, 2010
WIPO (2011) WIPO Doc, WIPO/GRTKF/IWG/3/12, January 10

Chapter 7
Textbox: Cross-Contamination, Genetic Drift, and the Question of GMO Co-existence with Non-GM Crops

Gabriela Steier

Abstract Genetically Modified Crops (GMOs or GE crops) cannot co-exists with organic and heirloom crops. GMOs decimate their organic ancestors at the expense of agrobiodiversity and with little regard for environmental consequences. The pollen of monoculture plants cross-pollinates plants of the same species that may be quite far away in a process called genetic drift. This would be natural and necessary if it were not for the unnatural and dangerous traits that are inserted into GMOs through human hands, thereby often recklessly infiltrating organic or heirloom plants with GMO traits.

The pollen of most agricultural crops, especially in monocultures on vast and flat areas, is light enough to travel many kilometers and, thereby, cross-pollinate plants of the same species that may be quite far away. This so-called cross-pollination is naturally necessary in many cases to combine genetic material and make the seeds of the respective plant species viable for recombination and proliferation of genetic material. Such traveling plant genetic material through the wind or via pollinators is called genetic drift. Nonetheless, where GM crops are planted, such as sugar beets, corn, wheat or rape, organic and non-GM crops are cross-contaminated by the pollen of GM plants, thereby unintentionally but often recklessly infiltrating plants with GMO traits.

Unfortunately, the dangerous cross-contamination of non-GM crops with GM pollen is inevitable anywhere around the world, because wind and pollinators, such as insects (e.g. bees and butterflies), birds and bats, carry seeds from GM field test sites or GM agricultural fields to organic and non-GM fields. Thus, co-existence is merely a way of masking the disastrous infection of organic and non-GM crops with unpredictably dangerous GMOs. The EU has buffer zones in place that strive to prevent the few GMOs that are grown in the European territory to cross-contaminate other crops. In the US, however, any preventive measures are likely to be less effective because there are massive field test sites, which often remain

---

G. Steier (✉)
Food Law International, LLP, Boston, MA, USA
e-mail: g.steier@foodlawinternational.com

© Springer International Publishing Switzerland 2016
G. Steier, K.K. Patel (eds.), *International Food Law and Policy*,
DOI 10.1007/978-3-319-07542-6_7

undisclosed and vastly underregulated, and because GMOs, such as corn, soy, wheat, and citrus are broadly planted across the nation. Some of the largest field test sites are in Hawaii, Florida and California. Even when buffer zones or isolated field test sites are maintained and when pollen-impermeable mesh bags are placed on blossoms and flowering crops, genetic drift is facilitated by nature's little helpers, namely by pollinators. This natural and crucial carrying of plant genetic material from one plant to another is one of evolution's ways to ensure biodiversity, plant resistance to climate change and survival of the uncountable organisms, including humans, dependent on these plants for food. However, the dangerous cross-contamination by GMOs which are engineered to withstand high levels of toxic pesticide and fertilizer applications threaten these very qualities that wind and insect pollination are naturally designed to accomplish.

In addition, the genetic engineering for pest or pesticide resistance of GMOs also threatens vital pollinators by facilitating monocultures and increased pesticide and fertilizer, mainly synthetic, uses where GMOs are cultivated. While pesticide-producers and monoculture advocates claim to increase crop production, the planting of industrial scale GMOs endangers food safety and sustainability around the world. In the case of insect pollination, for example, similar to bird and bat pollination, which plays a major role in global agriculture and shows how dependent our food production system is on these most useful and vital little creatures, GMOs, intensive agriculture and monocultures cause a decline in pollinator populations. The FAO, for example, estimates that at least one-third of the world's agricultural crops depend upon pollination provided by insects and other animals. In fact, bees, birds and bats increase outputs of 87 % of the leading food crops worldwide. Approximately 80 % of all flowering plant species are specialized for pollination by animals, mostly insects. Nonetheless, GMOs, monocultures, intensive agriculture and pesticide use endanger pollinators and cause devastating consequences for the global food supply system while claiming to improve crop yields. Such myths of improved crop yields, however, externalize the consequences of unsustainable agriculture focus on short-lived profits for BigFood and BigAg and fail to take the bigger picture into account: environmental conservation and social integrity.

New and emerging legislation should address this bigger picture around the world and support food safety for a greener future rather than maintain current short-sighted profit margins and trade balances at the cost of unpredictably great damage to the planet. Therefore, reducing, isolating and restricting the cultivation of GMOs would not only promote sustainable agriculture and conserve pollinators, it would also promote a safer food supply systems around the world. The better GM crop cultivation can be controlled and restricted, the safer our food supply system will become.

Co-existence of GM and non-GM crops is not possible due to genetic drift and cross-contamination. Stricter laws that regulate and ban GMOs from agricultural systems, however, may be the most powerful way to protect food safety for the future and to reshape the food industry through a top-down (government to producer) approach.

Chapter 8
Continuing Legal Barriers to International Food Trade

Peter Berry-Ottaway and Sam Jennings

Abstract International trade in food has been ongoing for millennia. However, despite the activities of certain global organisations whose aim is to harmonise food standards to ensure free trade, there are numerous barriers to trade that occur worldwide. These include inconsistencies in food legislation between countries or regions; inconsistencies in agricultural and production practices; barriers relating to the accepted composition of, and ingredients in, foods; differences in acceptance of new technologies, such as food irradiation, genetic modification and nanotechnology; and health risk scares owing to contamination, adulteration or communicable diseases. Whilst the need for a fully harmonised global trade in food has not been disputed, the reality is that it is not likely to be achieved within the next two decades.

8.1 Introduction

The international trade of foods and food components can be traced back over millennia. Even though most food then was produced, traded and consumed locally, historically, there has been an emphasis on global trade in foods and food ingredients, as can be seen with spices. By the second half of the nineteenth and the first half of the twentieth centuries, significant volumes of foods were being shipped across the world. Records show, in the early twentieth century, Australia exported around 684,000 tons of meat (including beef, mutton and lamb) of which about 50 % of this amount went to South Africa, 20 % to the Philippines and 15 % to the United Kingdom.[1] Since the late twentieth century, both developed and developing countries have had to place a greater reliance on the importation of food from other countries. This reliance was particularly important during and after the Second

---

[1] Australian Bureau of Statistics, *1301.0* (2000).

P. Berry-Ottaway (✉) • S. Jennings
Berry Ottaway & Associates Ltd, Hereford, UK
e-mail: boa@berryottaway.co.uk; spj@berryottaway.co.uk

© Springer International Publishing Switzerland 2016
G. Steier, K.K. Patel (eds.), *International Food Law and Policy*,
DOI 10.1007/978-3-319-07542-6_8

World War (1939–1945) when Europe had to become reliant on substantial food imports, primarily from North America.

A rapidly increasing world population, combined with increased urbanization and changes in agricultural policies led to a greater reliance on food importation in most parts of the world. Imports to be used in processing included commodities such as grains and meats, processed foods, specialist food ingredients and food additives. In the early twenty-first century, world food trade is estimated as being between US \$300 and \$400 billion per year.[2]

Food security, which is defined by the Food and Agriculture Organisation (FAO) as 'a condition that exists where all people, at all times, have physical and economic access to sufficient, safe and nutritious food to meet their dietary needs and food preferences for an active and healthy life,[3]' has become a key concern for governments. One commitment made by governments in the Rome Declaration on World Food Security in 1996 was, 'We will strive to ensure that food, agricultural trade and overall trade policies are conducive to fostering food security for all, through a fair and market-oriented world trade system'.[4]

The United Kingdom (UK) is an example of a country with a strong reliance on global food sourcing, and which is in the top ten of world economies. UK government data for 2011 shows that 48.25 % of the country's food supply was based on imports. Just over 20 % of the total food requirements came from five European Union countries (Netherlands, Spain, France, Irish Republic and Germany). Smaller, but economically significant, volumes came from a further 19 countries from across the globe, while food imports were recorded in smaller quantities from a further 144 countries. This data indicates that over 85 % of the countries in the world contributed to the food supply of the UK in 2011.[5]

8.2 Inconsistencies Causing Barriers to Trade

8.2.1 *Inconsistencies in Food Legislation*

Food legislation does not usually develop in a logical way. In most cases it is the result of a reactive response to circumstances, particularly those related to consumer protection, such as safety, adulteration or contamination.

Within most countries, food laws tend to develop around a cultural and parochial framework, and over the past century, little cognisance has been given to

---

[2]Codex and the International Food Trade, *FAO Corporate Document repository* available (2013).

[3]Codex and the International Food Trade, *FAO Corporate Document repository* available (2013).

[4]Food and Agriculture Organization of the United Nations, *World Food Summit – Rome Declaration on Food Security* (1996).

[5]Department of the Environment, *Food and Rural Affairs of the United Kingdom Food Statistics Year Book* (2011).

international trade. Globally, this has resulted in numerous disparate laws embracing the same areas of control.

This problem is well illustrated by the European Union (EU) where full harmonization of food legislation across the member states has not been achieved despite 45 years of intensive effort. Between 1965 and 2000, agreement in several legislative areas could not be reached when the European Community was relatively small, and today many of the same issues are still being discussed in the early twenty-first century.

For example, important legislation on Nutrition and Health Claims for Foods was finally adopted at the end of 2006 after 26 years in gestation, with many compromises and poorly constructed requirements. The result has been that more than 10 years after its adoption, it has not been possible for the legislation to be fully implemented.[6]

During the time the Nutrition and Health Claims legislation was being discussed in the EU, work was in progress in a number of countries across the globe to develop legislation in the same area. Work to develop guidelines on health claims was initiated in Codex Alimentarius, (adopted in 1997), in an attempt to aid global harmonisation in this area.[7]

Although the Codex Alimentarius Commission has been developing and adopting detailed technical Standards and Codes of Practice for over 50 years, there are still major inconsistencies in the technical food legislation between the major economic blocs such as the United States of America (USA), the European Union (EU) and the Association of South East Asian Nations (ASEAN). Many individual countries across the world also have legislation that is not compatible with the major blocs.

Multi-national food companies often have to modify the formula of processed foods to comply with the legal requirements of a particular market. Food additives and chemical contaminants provide numerous examples where differences in legislative requirements between the major economic blocs, such as the USA and the EU, can have a significant effect on product development.

Differences in the permitted substances and maximum levels of use are particularly common in the food additive categories of preservatives, antioxidants, colors, sweeteners and emulsifiers. Although the USA and EU are active participants in the Codex Alimentarius Committee on Food Additives (CCFA) and are collaborating on the development of the Codex General Standard for Food Additives (GSFA), which is intended to harmonize the use of food additives across the world, these differences are significant enough to result in barriers to trade. The GFSA contains a comprehensive list of additives that are toxicologically acceptable for use in foods, together with specific provisions for the use of each additive in each of the recognized food categories.[8]

---

[6]European Parliament and Council Regulation (EC) No 1924/2006 on nutrition and health claims made on food. Official Journal of the EU. L404 of December 30, 2006.
[7]Codex Alimentarius, *Guidelines for Use of Nutrition and health Claims*, CAC/GL 23 (1997).
[8]Codex General Standard for Food Additives (GSFA) Codex STAN 192-1995 as amended (2014).

In most cases, the provisions will include a maximum level of a specific food additive in a particular food category. The objective of the GFSA is to harmonize the use of food additives worldwide. However, due to the current disparate national approaches to the control of food additive usage, the achievement of the objective may still be a long way off.

8.2.2 Inconsistencies in Agricultural and Production Practices

Inconsistencies between major trading blocs in their approaches to agricultural and production practices can result in almost insurmountable barriers to international trade. This can be illustrated by the use of the hormone, recombinant bovine somatotropin (rBST). This substance is a veterinary medicine derived from a biotechnological process using recombinant DNA technology. The first patent for rBST was granted in the 1970s. Research going back to the late 1930s showed that the administration of rBST to lactating cows increased the milk yield by preventing mammary cell death. Until the mid-1980s, use in the dairy industry was limited, as the only source of the hormone was from dead cows. In the United States of America (USA), the Food and Drug Administration (FDA) had approved more than one source of rBST for use on dairy herds by 1994. The use in milk production in the USA was not without controversy, and there were a number of public campaigns to control the use of rBST. Some retailers made a point of marketing milk as being 'rBST free', although it was legal to use the hormone.[9]

The manufacturers of rBST then requested marketing authorization of their products as a veterinary medicine for use in the European Union (EU). These applications were originally referred to the EU Committee of Veterinary Medical Products (CVMP) in 1987, for assessment of the quality, safety and efficacy of the products.

In 1990, EU Regulation 2377/90 was adopted. This regulation introduced maximum residue limits (MRLs) for veterinary residues of active substances intended to be administered to 'food producing animals'. These residue limits applied not only to the milk but also to the meat and by-products of the animals.[10] Also in 1990, the European Council introduced a moratorium on the marketing of rBST, which was extended to the end of 1993.

Although the EU CVMP issued a positive opinion in 1993, which was in favor of the use of rBST, the European Commission proposed that the marketing of rBST should be prohibited. The moratorium was extended to the end of 1999. One of the

---

[9]*The Regulation of rBST: the European case*, AgBio Forum, 3(2 and 3), 164–172 (2000).

[10]European Council Regulation (EEC) No 2377/90 laying down a Community procedure for the establishment of maximum residue limits of veterinary medicinal products in foodstuffs of animal origin. Official Journal of the EU. L 224 of August 18, 1990.

reasons for extending the moratorium was political, based on the potential conflict of increasing milk production from the use of rBST with the strict EU-wide milk quotas on production.

In April 1999, the US FDA issued an analysis of the European report indicating that the findings did not appear to be consistent with the current state of scientific knowledge. The FDA maintained its determination that food products derived from cows treated with rBST were safe for consumers of all age groups, including infants. The Royal College of Physicians and Surgeons of Canada also concluded that there was no public health issue with the use of rBST on cows. Notwithstanding a number of reports from both national and international scientific committees, the Council of the European Union decided in December 1999 to an indefinite ban on the use of rBST in the European Union, effective beginning January 1, 2000. The reasons given in the ban were based on animal welfare, with the statement 'rBST is not used in cattle for therapeutic purposes, but only to increase milk production'. The European Commission concluded that scientific uncertainty remained over the use of rBST and further scientific studies were needed.[11]

The EU ban did not prohibit the production of rBST in the European Community provided that it was only exported to third countries. EU-based companies buying, producing or marketing rBST-based substances have to keep special registers, which must be made available to government authorities on request.

Curiously, the legislation imposing the EU ban did not prohibit the import into the European Community of meat or dairy products derived from rBST treated cows. This saga lasted for over 10 years and caused disagreements and barriers between the authorities in the USA and EU. It appears, in retrospect, to have had the science significantly influenced by politics, particularly in the EU. Issues used to justify the ban include internal agricultural policy on milk production, public health concerns, fears about a consumer backlash and finally, animal welfare concerns.

The long-running controversy has resulted in a situation where the USA and about 20 smaller countries have introduced legislation authorizing the use of rBST in cattle, while the use is not permitted in the EU, Canada, Japan, Australia and New Zealand. In addition, Codex Alimentarius has not yet approved the safety of rBST.

Another area of the food supply where there has been considerable controversy resulting in international barriers to trade is the genetic modification (GM) of foods and food ingredients. The World Health Organisation (WHO) defines genetic modification as:

> Genetically modified foods are foods derived from organisms whose genetic material has been modified in a way that does not occur naturally, e.g. through the introduction of a gene from a different organism.[12]

---

[11] European Commission, *Commission proposes ban on market release and use of BST in the Union as of 1st January 2000*, EC Press release No IP/99/758 (1999).
[12] World Health Organization, *Genetically Modified Food* (2014).

GM, which is also known as genetic engineering or recombinant-DNA technology, was first applied in the 1970s. GM techniques can include:

- transferring genes from one organism to another;
- moving, modifying, deleting or multiplying genes within a living organism;
- modifying existing genes or constructing new ones, and incorporating them into a new organism.

A considerable amount of research into GM technology was carried out in the 1970s and 1980s. Some of the early food applications were the development of GM microorganisms to increase various factors such as the production rate and yield in fermentation processes used to manufacture a range of food ingredients. This use went virtually unnoticed until serious health problems arose during the production in Japan of the amino acid, tryptophan, in 1992. A number of human deaths and permanent disabilities were eventually attributed to a small number of batches of tryptophan, which had been produced using a GM version of the fermentation organism. A detailed investigation carried out by the authorities in the USA and Japan eventually concluded that the health problems were caused by a metabolite formed during the fermentation, which had not been removed by the subsequent filtration and clean-up stages. It was also concluded that the clean-up failure was due to the increased production rate overloading the process.[13]

While the GM organism itself was exonerated, the fatal consequences of its use had an impact on the scientific community around the world, as many at the time had not realized the extent of the commercialization of GM technology in the food sector.

The first commercially grown GM food crop was a tomato created in California, USA, in the early 1990s. The technology produced a tomato that took longer to decompose after being picked and therefore had an extended shelf-life. Called the 'FlavrSavr' tomato, it was sold internationally, including into Europe where it could be found on supermarket shelves by the mid 1990s.[14]

The major turning point in the international acceptance of GM foods occurred in the second half of 1996. Prior to this period, much of the commercial focus of GM food technology in the USA had been to introduce pesticide and herbicide resistant varieties of the major food crops such as maize, soya and rape. These particular varieties had the advantages of both reducing the workload on the farms and increasing the yield of the crops, particularly in view of the huge areas of these crops that were planted each year.

The first large-scale pilot planting of the modified maize and soya was scheduled for the 1996 season. The agreement between the seed suppliers and the US FDA was that all seeds from these test crops were to be isolated from other crops until tested and approved.

---

[13]Institute of Food Science and Technology UK, *Addition of Micronutrients to Food* (1997).

[14]*The Case of the FlavrSavr tomato*. California Agriculture, 54, (4) 6–7 (2000).

Unfortunately, this did not happen. It was discovered that a proportion of the test crops of both maize and soya had been added indiscriminately to the bulk of the unmodified crop, to the extent that many thousands of tons of unmodified maize and soya had been contaminated with the GM varieties to some degree.

This problem was exacerbated by the discovery that a significant quantity of the GM contaminated crop was destined for the European Union for use as both human and animal food. When the European Commission received the information that some of the contaminated maize and soya was already in transit, the Commission refused entry into the EU until the safety had been assessed.

When the news broke in Europe, there was a widespread public and media reaction, mainly in a number of northern states of the EU. The European protests then spread to the United States and other countries and led the national farmers' associations in the USA to warn their members of the economic risks of planting genetically modified crops. Some American-based food companies such as Gerber, Heinz and Nestle banned the use of GM crops in their products.

The European Union first of all instituted an informal prohibition on the import of all food derived from GM technology, and then followed this with a series of laws introducing the requirements for scientific safety assessments and official authorization, and the labelling of foods containing approved GM sources.[15,16]

A number of countries outside the EU also instituted strict controls or outright bans on foods and food ingredients derived from GM technology.

Public opinion against GMOs (genetically modified organisms) gained momentum worldwide through the late 1990s and early 2000s. This public reaction was so strong that European food companies would not manufacture or import foods in the EU if there was any suspicion of the food or any of its ingredients being involved in GM technology. A large dichotomy emerged between public acceptance in the USA and the EU and many other countries, with American consumers in general, being unconcerned.

In 2002, Zambia shocked the world by prohibiting the entry of GM maize into the country during a regional food shortage. The Zambian government's reason was that the maize was unacceptable on two counts: the first was that the Zambian government scientists considered that GM maize was of unproven safety; and the second was that as the maize was in the form of viable seeds, they could potentially be planted, thus providing a threat to Zambian agriculture and public health. The issue escalated when three neighboring African countries, Malawi, Mozambique and Zimbabwe, also banned the import of food-relief GM maize. South Africa offered to mill the maize seed to prevent it being grown and to ship the maize-meal

---

[15] European Parliament and Council Regulation (EC) No 1829/2003 on genetically modified food and feed. Official Journal of the EU. L 268 of October 18, 2003.

[16] European Parliament and Council Regulation (EC) No 1830/2003 concerning the traceability and labelling of genetically modified organisms and the traceability of food and feed products produced from genetically modified organisms and amending Directive 2001/18/EC. Official Journal of the EU. L 268 of October 18, 2003.

to the four countries. Three accepted this route, but Zambia rejected the import of the maize in any form.[17,18]

Although the EU subsequently introduced an authorization and labelling procedure for GM crops and food ingredients, there has been a great reluctance amongst both food manufacturers and consumers to accept GM foods.

Food manufacturers in Europe and many other countries have spent a very considerable amount of money and human resources on the 'Identity Preservation' procedures to guarantee that all imported foods and ingredients have not been derived from GM technology.

In retrospect, it can now be considered that the accidental, or deliberate, act of adding the GM trial seeds to the bulk of the 1996 harvest, has set back the acceptance and consequently the research into the GM technology of foods a number of decades. The result is that the world is now split between those economic blocs that accept, and even encourage, the development and production of GM foods and those who ban or very strictly control their use.

The sad consequence of the barriers to international trade that have been built up, is that the potential benefits of GM technology in terms of sustainable food supply, in both over-populated and developing countries, have been suppressed.

The introduction of new technologies, as illustrated by both the use of Bovine Somatotropin and genetic modification, needs to be carefully managed in order to gain consumer acceptance. Decades of research can be negated overnight by the public misunderstanding of the intention of the technology and the creation of a real or imagined health issue.

What is emerging is that the acceptance of scientific developments appears to show national and cultural trends. For example, when comparing the attitudes of consumers in the USA and the EU, the Americans, in general, appear to embrace new technologies in the food sector more enthusiastically than their European counterparts. The Europeans have developed a consciousness of potential health concerns, as can be seen in the general food law legislation, which includes the Precautionary Principle, while Americans tend to have a greater acceptance of the potential benefits of a new technology, such as improved quality, increased shelf-life or reduced cost. Thus, the benefit:concern ratio can be quite different between the two continents.

An analysis of the success of the recent introductions of new food technologies has indicated that too little regard has been given to the national and cultural aspects of consumer acceptance. This has led to protective legislation, and ultimately to barriers in international trade.

---

[17]*Zambia bans GM Food Aid,* New Scientist, October 30, 2002.

[18]*Zambia and Genetically Modified Food Aid. Case study # 4-4 of the program: Food policy for developing countries: The role of the government in the global food system* (2007).

8.3 Compositional Barriers to International Food Trade

A surprisingly large number of trade barriers arising in processed foods can be attributed to the way different legal regimes regulate the composition of the food. These can be as simple as differences between countries in the approved coatings for the waxing of lemons to reduce dehydration, or as complex as a plethora of different issues that can arise in a compounded food containing a number of components.

8.3.1 Food Additives

As previously mentioned, inconsistencies between the food additive legislation of different economic blocs and individual countries has hindered the import and export of food products for many decades.

The problems arise for a number of reasons. One is from incompatible historic lists and maximum levels still used by countries. Probably the most important one is the national differences in their philosophical approaches to the use of food additives, which are mainly due to public health and safety concerns. Certain groups of additives have been particularly vulnerable to such concerns, such as synthetic colors, preservatives, antioxidants and intense sweeteners.

For example, the European Union has had a long-standing concern about a range of colors (mainly the azo dyes). Part of this concern relates to those colors, such as Sunset Yellow FCF, which may have an adverse effect on activity and attention in children.

The EU has for many years had strict controls on the use of the azo colors in foods. These controls include not only a restricted list of foods in which these colors can be used, but also relatively low maximum levels of use in these food categories. There are also restrictions on the total cumulative amounts of certain colors when used in combinations.

In 2008, following research suggesting a link between certain azo colors and 'Attention Deficit Hyperactivity Disorder' (ADHD) in children, a labelling requirement was placed on products intended for the European market that contained any of six colors. In 2008, EU legislation was amended to require the label statement '[name or E number of color] may have an adverse effect on the activity and attention in children'. This statement must appear on the label of all products intended for sale in the EU, including imported products (European Parliament and Council, 2008).[19]

The low levels and labelling requirements have generally not been adopted outside the EU.

---

[19]European Parliament and Council Regulation (EC) No 1333/2008 on food additives. Official Journal of the EU. L 354 of December 31, 2008.

A subsequent amendment removed Quinoline Yellow, Sunset Yellow FCF, Ponceau 4R and Cochineal Red A from the original table relating to maximum limits for combined colors. These four colors had specific maximum levels set for them in each relevant food category.

Another area of general inconsistency between countries in food additives legislation is that concerning intense sweeteners. Here an already confused situation was made worse with the introduction of a number of new sweeteners such as sucralose, stevia and trehalose.

The two earliest intense sweeteners, saccharine and cyclamate, became very controversial from the 1960s onwards. In 1969, following a single study showing that very high doses could cause bladder tumours in rats, cyclamates were prohibited in the USA and the UK.[20]

After several scientific reviews, the EU approved the use of cyclamates as sweeteners in foods in 1996, but with limitations on the foods in which they can be used and maximum levels of use in each category.[21] The USA continued with its ban, as did a number of other countries. The consequence was that in order to ensure trade across borders, large multinational companies were encouraged to search for intense sweeteners that would be more acceptable. Saccharine had already become controversial and another intense sweetener to enter the market was aspartame. Aspartame is a methyl ester of the aspartic acid/phenylalanine dipeptide and, as a low-calorie intense sweetener, is about 200 times sweeter than sugar, weight for weight.

Aspartame was approved by the FDA for use in dry food products in the USA in 1981, following a previous approval and stay of approval in 1974 and 1975, respectively.[22] It was approved in the UK and several other EU member states in 1982, and subsequently approved for use across the EU in 1994. During the period from the early 1980s to the mid-1990s, food products containing aspartame could not be traded across the EU, due to differences in controls. Current EU legislation restricts aspartame to specific product categories with maximum levels of use for each category[18].

As aspartame contains phenylalanine, it is contra-indicated for patients suffering from Phenylketonuria. This being so, EU legislation requires that food and drinks containing aspartame have the warning statement 'contains a source of phenylalanine' on the product label. This labelling requirement is not required in many other countries where the use of aspartame is approved. As a consequence, products imported into the EU from these countries require customized labels or label overstickers in order to comply.

---

[20]*Cyclamate sweeteners*, JAMA 25;236(17):1987–9 (1976).

[21]European Parliament and Council Directive 96/83/EC amending Directive 94/35/EC on sweeteners for use in foodstuffs. Official Journal of the EU. L 48 of February 19, 1997.

[22]United States General Accounting Office, *Food and Drug Administration Food Additive Approval Process Followed for Aspartame* (1987).

Since its FDA approval in 1981, aspartame has remained controversial. This controversy has continued into the twenty-first century and is largely centred on its safety due to several conspiracy theories relating to its approval by the FDA in the light of studies implying that high doses induced brain tumors in rats.

As a result of the controversy, there have been over 200 studies carried out on the sweetener, and a number of safety reviews carried out by both national and international expert committees.

An equally complex situation is emerging with a more recent intense sweetener, Stevia. Stevia is derived from the plant genus Stevia (family Asteraceae) which contains about 240 species. The plant of commercial concern is principally *Stevia rebaudiana*, extracts of which can have about 300 times the sweetness of sugar.

One of the first commercial cultivations of stevia took place in Japan following the controversies over saccharine and cyclamate. Steviosides were extracted from the plant leaves to produce the first stevia sweetener in Japan in 1971. It was widely used in Japan from the 1980s. Although used in Japan, China and some other Asian countries, stevia was not approved for use as a sweetener in other parts of the world for another two decades.

In the USA, the FDA accepted the addition of stevia leaves and extracts to dietary supplements from 1995, but only approved its use as a sweetener in foods (i.e. as a food additive) in 2008.[23] The EU permitted the use of steviol glycosides as a sweetener in specified categories of foods with specified maximum levels per category from 2011, but the general use of stevia leaves and extracts in foods is still prohibited. The steviol glycosides permitted as sweeteners in the EU must be in compliance with a detailed specification.[24,25]

Although stevia is currently permitted in a number of countries, there are significant inconsistencies in the permitted substances. The USA authorized Rebaudioside A as a sweetener; Brazil, stevioside extract; Russian Federation, stevioside; Mexico, mixed steviol glycoside extracts (not separate extracts); Australia and New Zealand, all steviol glycosides; and the EU, specified steviol glycosides. These differences have continued to result in barriers to international trade.

---

[23]*Ensuring the Safety of Sweeteners from Stevia*, Food Technology 65(4):42–49 (2011).

[24]European Commission Regulation (EU) No 1131/2011 amending Annex II to Regulation (EC) No 1333/2008 of the European Parliament and of the Council with regard to steviol glycosides. Official Journal of the EU. L 295 of November 12, 2011.

[25]European Commission Regulation (EU) No 231/2012 laying down specifications for food additives listed in Annexes II and III to Regulation (EC) No 1333/2008 of the European Parliament and of the Council. Official Journal of the EU. L 83 of March 22, 2012.

8.3.2 Novel Foods and Ingredients

A particularly difficult issue that can disrupt international food trade is that of novel foods and ingredients, including those resulting from novel technologies. These difficulties arise from the large variation in the levels of safety assessment required by countries before the new food or ingredient can be placed on the market. A consequence of these differences in requirements is that those countries that are less stringent tend to have the higher number of innovations.

Currently, one of the areas with the most stringent safety assessments is the EU, which has had legislation in place since 1997.[26]

In simple terms, the EU Novel Foods legislation prohibits any food, food ingredient or the product of a novel technology entering the EU market unless it has undergone a detailed safety assessment by the official EU scientific committee and has been approved by the European Commission. All approvals are published in the Official Journal of the EU, accompanied by any conditions of use or other specific requirements.

Only those foods or ingredients which can be demonstrated to have been sold in a significant amount in one or more EU member states before May 15, 1997 are exempt from the assessment. However, there is a provision in legislation adopted in 2006 for exempt foods or ingredients to be subject to this official assessment if a safety concern arises.

The data requirements for an application for approval of a novel food in the EU are considerable and include a full toxicological profile, detailed composition of the food or ingredient, confirmation of the consistency of the production process, and estimated intake/consumption data relevant to the EU member states. The cost of obtaining the data and the preparation and management of the application can be in excess of US $250,000. Although the time taken to assess the application is given in the law, the clock is stopped each time the reviewing committee calls for additional information and only restarts when the requested information is submitted by the applicant. Experience has shown that the time from initial application to publication in the Official Journal can be from 2 to 7 years. The food cannot be placed on the EU market until the application is approved. Due to the complexity of the process, the application normally has to be made by the manufacturer of the food or ingredient and not by companies or individuals wanting to import the food or ingredient into the EU.

The effect of the EU novel foods legislation is that since its introduction in 1997, it has effectively inhibited innovation by the smaller companies. In the 19 years after the law came into force, only a relatively small number of applications had been authorized.

In 1997 the United States FDA issued a proposed rule that would establish a notification procedure whereby any person could notify the FDA of a determination

---

[26]European Parliament and Council Regulation (EC) No 258/97 concerning novel foods and novel food ingredients. Official Journal of the EU. L 43 of February 14, 1997.

by that person that the particular use of a substance is Generally Recognized as Safe (GRAS). For a substance to fall under the classification of GRAS in the USA, the scientific information about the use of the substance must be widely known and there must be a consensus amongst qualified experts that the data and information given for the substance indicates that it is safe under the conditions of intended use.

The GRAS notification program established by the FDA provided a voluntary mechanism whereby a person may inform the FDA of a determination that the use of a substance is GRAS, rather than petition the FDA to affirm that the use of the substance is GRAS.[27] The FDA then evaluates the submission to ascertain whether the notification provides a sufficient basis for the GRAS determination. The FDA can query any of the detail and may consult with other agencies.

Following this evaluation, the FDA can respond in three ways. The first is a letter stating that the FDA does not question the basis for the notifier's determination of GRAS. This letter may also contain conditions of use or specific labelling requirements. The second type of response is when the FDA concludes that the notification does not provide a sufficient basis for a GRAS determination. For example, it does not include appropriate data or information, or because the available data and information supplied raises questions about the safety of the substance. A third response can be where the FDA, at the notifier's request, ceases the evaluation of the GRAS notification. The FDA maintains an inventory of the notifications and its responses. In the first 8 years from its introduction, over 200 GRAS notices were filed.

Australia and New Zealand have introduced a novel foods assessment procedure that is based on the EU law. In Australia and New Zealand, novel foods and ingredients cannot be placed on the market until assessed by a scientific committee using the official safety assessment guidelines. A novel food or ingredient cannot be sold by any form of retail sale in Australia and New Zealand, until approved and the approval is published by the government.[28]

Canada has also introduced a law that requires a formal assessment and approval of a novel food or ingredient. The Canadian law not only encompasses all foods and ingredients derived from genetic modification, but can also include new techniques for food preservation and processing. It is argued that new processing methods can alter the nutritional and toxic characteristics of a food.[29]

Other countries are also moving towards more control of new foods and ingredients. In October 2013, China's law on 'Administrative measures for safety of new food materials', came into force. The new system requires not only a safety

---

[27]How U.S. FDAs GRAS Notification Program Works, *Food Safety Magazine* (December 2005/ January 2006).

[28]Australia New Zealand Food Standards Code—Standard 1.5.1—Novel Foods. Federal Register of Legislative Instruments F2013C00142. Issue 139 (2013).

[29]Regulations Amending the Food and Drug Regulations (948—Novel Foods). Canada Gazette Part II, Vol. 133 No. 22 of October 27, 1999.

assessment by a panel of experts, but also includes public comment solicitation. A positive assessment can lead to approval and an official notification.[30]

In Japan, there is no term for or definition of a novel food, but there is a safety control built into the Japanese legislation on 'Foods for Specific Health Use (FOSHU)'.

The move towards controls on new foods, ingredients and technologies in the major economic blocs and major countries has led most other countries to assess their own legislative situation. The laws operating in this area in the EU are generally regarded as being the most stringent, and there is currently no mutual recognition of approved substances between the EU and other non-EU countries. For example, a number of novel foods and ingredients accepted in the USA as self-affirmed GRAS have not been allowed into the EU until the full EU assessment procedure has been carried out and the application approved. In many of these situations, the EU assessment has required a greater amount of detailed data than was presented for the US GRAS notification. Until some form of mutual recognition or agreement on minimal standards between countries is achieved, the problems that currently arise with trade in innovative foods and technologies will continue.

8.3.3 Addition of Micronutrients to Foods

Another area where there continues to be major inconsistencies between the national food legislation of countries is where micronutrients and some other substances are often added to foods.

The food categories most affected are fortified foods and drinks, food/dietary supplements, weight control products and some health foods.

The major problems preventing international trade in such products relate to both the chemical forms used to supply the micronutrients and the quantities added to the products. Developments through the second half of the twentieth century have resulted in a number of chemical forms being used to supply the 13 recognized vitamins. For example, over ten chemical forms of vitamin C (ascorbic acid) are in use worldwide, but far fewer are permitted in the legislation of many countries.

The situation with the chemical salts used to supply nutritional minerals and trace elements has become very complex. The EU permits over 20 forms of magnesium for the nutritional enhancement of foods. However, many more are in use in the USA and far fewer in some other countries. The consequence of these differences in the permitted lists is that one of two approaches are taken; either a nutritionally enhanced product is developed specifically for a particular market, or a 'common denominator' approach is applied, where multi-national companies use a

---

[30]Administrative Measures for Safety Review of New Food Materials. National Health and Family Planning Commission Decree No. 1. May 31, 2013.

significantly reduced list of those sources permitted in all the intended national markets.

Similarly, the maximum levels of micronutrients that can be used differ significantly between countries.

Within the EU, this issue has not yet been addressed in the harmonized food legislation, which means that national legislation still applies. This has resulted in many EU member states imposing low maximum levels of vitamins and minerals in products, many based on arbitrary limits based on the Recommended Daily Allowance (RDA). There is no consistency among the countries following this route, with some having a 1 × RDA limit for all the vitamins and minerals, while others can range between 1 and 3 × RDA depending on the micronutrient. Other countries, such as the UK, control these levels on the basis of a scientific assessment on consumer safety. The result has been that, within the EU, there are significant barriers to trade in food/dietary supplements, even those containing just vitamins and minerals. This problem is repeated worldwide and supplements, particularly, are one of the most difficult product categories to trade across borders.

In the context of fortified foods and supplements there is a surprising barrier to international trade, which is that the methods for calculating vitamin activity vary between countries. For decades, this has resulted in re-formulation or re-labelling of products traded between the USA and the EU.

While a number of vitamins are affected, the most extreme example is found in beta-carotene in its role as pro-vitamin A. Beta-carotene is a substance of plant origin that can be metabolized in the body to form vitamin A. It is particularly favored by vegetarians and vegans, as all other natural sources of vitamin A are of animal origin.

On a theoretical basis, the beta-carotene molecule contains two molecules of vitamin A (as retinol). In the USA, the convention is to take the conversion factor that 2 µg of beta-carotene is equivalent to 1 µg of retinol. However, in Europe the conversion factor is based on studies on biological activity, which indicates that the ratio is 6:1, or three times less than that calculated in the USA. These conversions obviously make a large difference in label claims between the two continents.

Other vitamins where there are significant differences in quantification between the USA and the EU are thiamin (vitamin B_1), where the EU calculation works out at 78 % of the American value and pyridoxine (vitamin B_6), where the EU reports only 82 % of the amount calculated by the US convention.

Another area relating to supplements and health foods and for which there is no agreement across the world is the addition of botanicals and botanical extracts to products sold under food law. In many countries, such products are subject to medicine law due to their traditional use in herbal remedies.

In the USA, the 'Dietary Supplement and Health Education Act' (DSHEA), adopted in 1994,[31] permitted botanicals as dietary supplement ingredients and,

---

[31]Dietary Supplement and Health & Education Act. Public Law 103-147; 108 stat. 4325-4335 103d Congress 2nd Session (October 15, 1994).

subsequently following the DSHEA, there was a rapid market growth in supplements containing botanicals, mainly botanical extracts.

In the EU and many other countries, the governments have taken a more cautious approach. Many still have a differentiation between supplements and traditional herbal medicines and tend to place products containing botanicals and their extracts in the latter category.

There have been moves in the EU at the national level to develop permitted lists of botanicals considered acceptable for use in supplements. The EU member states taking the lead in this area are Belgium, France and Italy. However, many other countries in Europe maintain strict national controls.

A problem that is challenging governments across the world is how to define and accommodate botanical extracts into food legislation.

In relation to the number of extracts in the commercial market, relatively few have an internationally agreed composition. Those that have an internationally agreed upon composition are normally well-established medicinal herbs used in traditional herbal remedies. For the majority of botanicals used in supplements and drinks, there is no standard extract, and for any particular botanical, a wide range of extraction ratios and extraction solvents can be found on the market. Many of these are produced for commercial advantage and may not have a well-established scientific basis. Analysis of a range of extracts, all purporting to represent the same botanical source, can have a disparity in chemical composition, and there can be no similarity between these extracts. This has been causing safety concerns among the authorities and in some cases has raised issues as to whether some of these extracts should be defined as novel foods.

8.4 Technological Barriers to International Trade

8.4.1 Food Irradiation

A food preservation technique that has been controversial for a number of decades is irradiation. Irradiation is the exposure of foods or food ingredients to ionising radiation from such sources as gamma rays from radionuclides or x-rays generated from machine sources.

The main commercial purpose of irradiation is to kill pathogenic bacteria in food. It has also been used commercially to delay the ripening of fruit and to delay sprouting in vegetables such as potatoes.

Although irradiation has been an established preservation method since the 1970s, there has, over the years, been a considerable amount of consumer opposition in some parts of the world. As a consequence, it has been banned in some countries, permitted with strict controls in some countries and accepted with few controls in others.

The reasons for consumer concerns have been found to fall into two categories. The first is a simplistic one, in that some people perceive a health risk associated with radioactivity and therefore wish to avoid all irradiated foods. The second also relates to a health concern, which is that an unscrupulous food business operator may use irradiation on microbiologically contaminated meats and vegetables that may then enter the market. The more sophisticated end of this concern is that any toxins produced by a high biological loading would remain in the food.

For some time, illegally irradiated foods and food ingredients have been one of the main causes of potential imports being rejected at EU borders. Much of the rejected product originated from the USA, Russia or China.

Since 1999, there has been EU-wide legislation controlling the use of irradiation on foods. While the EU permits the irradiation of specified groups of foods, there are strict conditions as to where and how the irradiation can carry out, together with a maximum dosage level for each of the permitted categories. There are controls on both the traceability of irradiated product and the documentation that is required to accompany the consignment. In addition, foods and ingredients that have been irradiated must have a statement on the label saying that they have been irradiated.[32]

One of the main criteria in the EU rules is that the irradiation can only be carried out in EU authorized premises. Some, but certainly not all, facilities in countries outside the EU have been approved. If the premises are not approved by the EU, any food processed in them is automatically illegal in the EU, even if it meets the other criteria.

The introduction of a simple, non-destructive and relatively cheap test for irradiation in food, the Photo Stimulated Luminescence (PSL) method has changed the dynamics of illegal irradiation. Prior to the widespread introductions of PSL, which is now used by both government enforcement agencies and industry, the testing of foods for suspected irradiation was both time consuming and costly. This was exploited by some companies deliberately supplying irradiated ingredients to food manufacturers. Owing to its simplicity in use, a large number of foods and ingredients can be tested for irradiation by PSL in a single day, giving both the authorities and manufacturers far greater control. The problems that still exist in international trade are mainly due to the differences in legislation, particularly between the USA, the EU and certain Asian countries, as to which foods can be irradiated.

---

[32]European Parliament and Council Directive 1999/2/EC on the approximation of the laws of the Member States concerning foods and food ingredients treated with ionising radiation. Official Journal of the EU. L 66 of March 13, 1999.

8.4.2 Genetic Modification Technology

The unfortunate effects of the mismanagement of the introduction of genetic modification (GM) into the global food sector has already been described. The after effects of the emotions generated in the late 1990s persisted into the second decade of the twenty-first century.

Notwithstanding these setbacks, GM technology has progressed, and a large number of foods and ingredients resulting from GM technology have become available. Much of the production of enzymes used for food processing uses GM microorganisms, as do the fermentation processes involved in the production of other food additives, such as amino acids and flavorings. Also, the number of plant and animal species for which there are GM forms is very considerable, with a number of vegetables, fruits and even fish with a GM counterpart, developed for either commercial or humanitarian reasons. An example of the latter is the development of 'Golden Rice', a rice variety enhanced with beta-carotene for use in regions where vitamin A intakes are well below the levels required for health. Unfortunately, in the commercial world in which we exist, there is far less funding for humanitarian projects than there is for the commercial exploitation of GM technology.

As discussed earlier, there is an imbalance in the levels of legislative control between countries on food and food ingredients derived from GM. This has meant that, for the foreseeable future, international trade in foods derived from GM technology will be seriously inhibited by the major differences in legislative approaches.

A major global initiative on genetically modified organisms (GMOs) came to fruition in 2000, with the adoption of the Cartagena Protocol by more than 130 countries, in Montreal, Canada.

The objective of this protocol was to ensure an adequate level of protection for the transfer, handling and use of GMOs that may have adverse effects on the environment and human health, specifically focusing on trans-boundary movements.

The Cartagena Protocol was the first agreement to mandate the need for consent of an importing country prior to trade in certain GMOs, to allow for assessment of potential risks posed by such an import on the biodiversity and human health in the importing country.[33]

---

[33]Cartagena Protocol on Biosafety to the Conventions on Biological Diversity. Secretariat of the Convention on Biological Diversity (2000).

8.4.3 Nanotechnology

A relatively recent technology with applications in the food sector is nanotechnology. Nanotechnology is described as the 'Engineering of functional systems at the molecular scale by controlling the shape and size of materials at the nanometre scale'. Nanoscience is defined as 'the study of phenomena and the manipulation of materials at the atomic, molecular and macromolecular scales, where the properties differ significantly from those of the material at a larger scale'.

The term 'nano' is derived from the Greek word for 'dwarf' and, to put the science into perspective, a nanometre (nm) is one-billionth of a metre, or approximately one hundred-thousandth the width of a human hair.

The electronics industry uses nanotechnology in its search for greater miniaturization of computer chips and enhanced data storage. Nanostructures are already being used in coatings, reinforced composites and certain polymers.

While some nanostructures are formed naturally in foods, such as in some food proteins, or are produced from conventional processing, such as the formation of foams and emulsions, there have been serious concerns over the relatively uncontrolled introduction of food ingredients deliberately engineered to produce nanoparticles for technological benefits.

One of the technological advantages of food ingredients reduced to nanoparticles is that, in powder form, they have a far greater collective surface area than their conventional counterparts.[34] This has a great advantage for certain applications such as the use of titanium dioxide as an opacifying food colorant or silicon dioxide as an anti-caking agent.

The introduction of deliberately engineered nanotechnology into the food industry in the first half of the first decade of the twenty-first century caused consternation within some governments. The UK government commissioned an independent expert scientific study into the current and future developments in nanoscience and nanotechnology. The report, published in 2004, addressed concerns about safety to humans of nanoparticles and the possible need for regulation and labelling of foods incorporating deliberately engineered nanotechnology. One of the main concerns was on the, then unknown, effects of the nanoparticles on the human gut. This was directly related to the very small particle size and the increased surface area of these particles.[35]

Concerns were also raised with the EU, and the European Commission commenced discussions on the regulation of nanotechnology for use in food, and the status of this technology in context of the EU Regulation on Novel Foods. An immediate problem faced by the EU authorities was to agree an accurate and workable definition of an engineered nanomaterial intended for use in food. This

---

[34]*Nanotechnology in Foods – A European Perspective*, International Review of Food Science and Technology, International Union of Food Science and Technology (2008).

[35]Nanoscience and nanotechnologies: opportunities and uncertainties. The Royal Society & The Royal Academy of Engineering (2004).

task became more complex than they first thought, particularly as other countries were also looking at the same problem.

Within the EU, it was considered that the control of nanotechnology impacted two legislative areas, the novel foods rules and those on the provision of food information to the consumer, i.e. food labelling. The culmination of these discussions is that an engineered nanomaterial is currently defined in European food law as:

> any intentionally produced material that has one or more dimensions of the order of 100 nm or less or that is composed of discrete functional parts, either internally or at the surface, many of which have one or more dimensions of the order of 100 nm or less, including structures, agglomerates or aggregates, which may have a size above the order of 100 nm but retain properties that are characteristic of the nanoscale.[36]

By derogation, natural substances in nanoform are excluded. Material intended for food use that fall into the above definitions are classified as novel foods in the EU, and are therefore required to undergo a safety assessment and authorization as required by the EU Regulation on Novel Foods.

The EU Regulation on Food Information for Consumers lays down specific labelling requirements for food derived from nanotechnology and which fall into the scope of the definition (see footnote 35). However, the accuracy and applicability of the definition of engineered nanomaterial is still being discussed by the EU authorities, and could be revised in the relatively near future.

While the EU definition has been written into EU food law, it is not necessarily the same as those definitions applying in other countries.

This problem has been further compounded by the term 'nano' becoming a marketing 'buzz-word' in the 2000s, particularly in the USA. This term was erroneously applied to foods and supplements where the particular materials did not fall into any definition of a nanomaterial. This caused unnecessary concern amongst the authorities globally, particularly those dealing with imported food products.

In the USA, the policy has been not to specifically legislate for nanotechnology used in food, but to issue a guidance document for manufacturers, suppliers and importers on the FDA's current thinking on whether FDA-regulated products contain nanomaterials or otherwise involve the application of nanotechnology.[37]

This policy obviously differs from the approach taken by the EU authorities, and in July 2013, the Transatlantic Trade and Investment Partnership (TTIP),[38] which took place in Washington D.C. in the USA, agreed to harmonize standards for food safety, nutrition and nanotechnology, which will be among the first of the

---

[36]European Parliament and Council Regulation (EU) No 1169/2011 on the provision of food information to consumers. Official Journal of the EU. L 304 of November 22, 2011.

[37]Considering whether an FDA-Regulated Product involves the Application of Nanotechnology—Guidance for Industry (2011).

[38]The Office of the United States Trade Representative, Transatlantic Trade and Investment Partnership (T-TIP) (2013).

collaborative research priorities to come out of the talks. Following these talks, the planned activities include access to each other's scientific information and exchange of experts.[39] It is hoped that this initiative will succeed in bringing two major economic blocs to a common view on the handling of nanotechnology.

8.4.4 Emerging Technologies

Recent history has shown that the way countries handle innovative technology in the food industry can seriously affect the viability of the technology. Innovators have failed to appreciate that there can be significant disparities in the way different nationalities embrace the concepts.

This has been illustrated by the introduction of GM technology, where American farmers and consumers were far more enthusiastic about GM foods and ingredients than their European counterparts. As a consequence, a large gulf has developed between the two continents, and there is very little trade between them in foods and ingredients derived from GM. Unless carefully managed, a similar situation could occur with nanotechnology and any other innovation.

8.5 Health Risk Barriers to International Trade

All governments will defend their food legislation on the basis that it is necessary for the protection of the health of their nation. However, there can be large differences in the approach to the risk assessment, risk management and risk communication.

The situation is never static and unexpected health risks will always emerge as scientific knowledge advances. For example, advances in analytical techniques and toxicology have relatively recently identified an increasing number of substances with carcinogenic potential that can be found in foods. Serious communicable diseases have been found to be carried in animals common in the human food chain, and there have been a number of problems associated with the deliberate or accidental adulteration of foods. Each of these introduce serious challenges to national authorities tasked with protecting the health of their citizens.

---

[39]European Commission, *The Transatlantic Trade and Investment Partnership* (2014).

8.5.1 Chemical Contaminants

The Codex Committee on Contaminants in Foods (CCCF) have already identified a large number of chemical contaminants in foods that could pose a health risk to the consumer unless their levels are controlled. Some of these substances have only recently been identified as toxicants.

Some, such as dioxins and polychlorinated biphenyls (PCBs) can be found in fats and oils from both animals and plants, and have to be regulated at picogram levels (that is, one trillionth of a gram). Unfortunately, there is not yet universal agreement between countries on the actual maximum levels that should be applied.

A group of toxic substances that are currently receiving attention are the mycotoxins. Mycotoxins are produced by moulds (fungi) when they proliferate on plants. Many cash crops are susceptible to mould contamination, particularly during harvesting and storage under certain climatic conditions. While growing on the plant matter, the moulds can produce toxic metabolites (mycotoxins), which can be carried through into animal feed or human food. A number of mycotoxins that are associated with food sources have already been identified, and controls on their levels have been instigated in a number of countries.

Another group of food contaminants of concern are the polycyclic aromatic hydrocarbons (PAHs). These are mainly found as by-products of combustion and are particularly prevalent in the atmosphere around areas of heavy industry. PAHs can be found in plant matter and plant oils and some have been found to be potentially carcinogenic.

Legislation on PAH levels in plant oils and plant foods varies widely across the world, with some countries imposing monitoring and controls, and with no current controls in others.

8.5.2 Communicable Diseases and International Trade

A major barrier to international trade is the transmission of a communicable disease, whether it be between plants used as food crops or between animal species.

The risk of plant diseases being rapidly transferred from country to country and continent to continent is very great, and many countries impose strict controls not only on bulk food plant sources, such as wheat, maize and soya, but also on the personal import of plant matter by individuals such as passengers on ships and aircrafts.

Similar restrictions apply to trade in animals, meat and ingredients of animal origin. The EU, for example, has introduced stringent controls on all products of animal origin. These rules cover a wide range of animal sources.

The severe effect an animal disease can have on international food trade can be illustrated by the case of Bovine Spongiform Encephalopathy (BSE), which became more popularly known as Mad Cow Disease. BSE is attributed to an aberrant protein known as a prion. It has been found that the prions are not destroyed when meat or material containing them are heat-treated, even at temperatures over 600 °C (about 1100 °F), and they have the ability to transfer the disease between animals. BSE is a type of transmissible spongiform encephalopathy (TSE), and transmission of the disease can occur when healthy animals come into contact with contaminated tissue from other animals with the disease.[40,41]

In BSE the disease affects the brain, spinal cord and digestive tract of the infected cattle and it can be transmitted to humans by eating these parts of the infected carcasses. In humans it became known as new-variant Creutzfeldt-Jakob disease (nvCJD). BSE in cattle was first identified in the UK in 1986. Laboratory tests confirmed the presence of the disease in 1987, and later that year the British Ministry of Agriculture, Fisheries and Food accepted that it had to deal with a new communicable disease.

It was later estimated that over 450,000 animals infected with BSE had entered the human food chain before the first controls on high-risk offal were in place in 1989. It has also been shown that around 160 people had died of nvCJD in the UK, and a further 44 elsewhere, in the 22 years between the first identification of the disease in 1987 and 2009.[42]

A British inquiry into the cause of BSE concluded that the cattle, which are naturally herbivores, had been fed parts of other cattle in the form of meat and bone meal added to their feed. There was also an indication that meat from sheep suffering from the disease scrapie could have also been involved.

After the cause of BSE, and subsequently nvCJD, had been identified, there were serious concerns about bovine by-products used in human foods and medicines. A widespread ingredient in the food chain was gelatine, and a large proportion of this originated from bovine sources. Similarly, a number of medicines from bovine sources, such as bovine insulin, were identified.

The EU introduced a ban on exports of British beef in 1996. This ban stayed in place for almost exactly 10 years and included trade to other EU countries. However, cases of BSE were eventually recorded in the majority of EU countries, although none were near the scale of that in the UK. The first reported case in North America was in Canada in 1993, and in the same year a case was found in the USA. A subsequent investigation confirmed that the cow was of Canadian origin. The first domestic case in the USA was in Texas in June 2005.[43,44]

---

[40] Animal/Plant Health Inspection Service, *Bovine and Spongiform Encephalopathy—An Overview*, United States Department of Agriculture (2006).

[41] *The origin of bovine spongiform encephalopathy: the human prion disease hypothesis*. Lancet 366 (9488) 856–61 (2005).

[42] *Estimation of epidemic size and incubation time based on age characteristics of vCJD in the United Kingdom*. Science 294 (5547), 1726–8 (2001).

[43] *Bovine Spongiform Encephalopathy and Canadian Beef Imports*. CRS (USA) Report for Congress No RL 32627, March 11, 2005.

[44] Center for Disease Control, *BSE Cases in North America, by Year and Country of Death, (1993–2008)*. Center for Disease Control and Prevention, Department of Health and Human Services (2008).

Japan, with 36 confirmed cases, was the only country outside of Europe and the Americas to have cases not related to imports. These cases resulted in a change to Japanese food safety policy and the establishment of the Food Safety Commission in 2003. As more and more countries revealed that they had confirmed cases of BSE, other countries introduced bans or placed severe restrictions on the import of beef. Japan halted beef imports from the USA after the discovery of the first case in the US. The same happened in Indonesia and eventually 65 countries had introduced full or partial restrictions on the imports of beef from the USA. Some of these restrictions were due to concerns that the US testing regimes for BSE were less rigorous than those used nationally.

8.5.3 Food Adulterants

Historically, there has always been the risk of deliberate adulteration of food. In most cases the reason foods are adulterated is economic and relates to the addition of cheaper 'fillers' to high value foods such as olive oils, fish oils, and spices. One of the first European food laws, created around the twelfth century, imposed severe penalties on those who added ground chalk to flour. Over 2500 years ago, a law in India prohibited the adulteration of grain, scents and medicine, and throughout the nineteenth century a number of countries introduced legislation to protect consumers from such practices. A centuries old global problem is that the adulteration tends to be ahead of the detection methods available to the authorities at that point in time.

Olive oil, due to its market value, has been a victim of adulteration and manipulation for many years, and there are now detailed specifications for olive oils obtained from different regions. Similarly, there have been cases of the adulteration of fish oils, mainly with less expensive vegetable oils.

A good example of a serious public health impact of food adulteration were the Chinese melamine problems, which started in China in 2007 and eventually affected consumers in many other countries.

Melamine is a synthetic substance, which in combination with formaldehyde can make a hard-wearing plastic that has been used to manufacture work-tops, white boards and durable tableware. Significant to the food adulteration, melamine has a very high nitrogen content (66.6 % w/w).

The problem first came to light in 2007, when there was an epidemic of kidney failure affecting thousands of cats and dogs in North America. Investigations showed that melamine and its analogue, cyanuric acid, were present in a rice protein concentrate and wheat gluten imported into the USA from China. These ingredients were widely used in pet foods for their binding and thickening properties.

A year later, in 2008, melamine was discovered in dairy products exported from China. Melamine was found in liquid milk in China and in dried milk, which was also being used to manufacture infant formula. The high levels of melamine in the milk, baby milk and other dairy-based products resulted in severe health effects and illness in nearly 300,000 Chinese infants and young children. The deaths of six children were attributed to melamine contaminated milk.

A Chinese government investigation indicated that the contamination was intentional and was probably occurring at the milk collection centers. It appears that the milk was being watered down and the melamine was being added to increase the milk's measured protein content. As stated earlier, melamine has a very high nitrogen content; the test for protein in milk is achieved by determining the nitrogen content and multiplying that value by a factor to give the protein content of the milk. Any increase in the nitrogen level would give a corresponding increase in the measured protein.[45]

Once the cause was identified, the Chinese government introduced emergency measures requiring all companies in the dairy industry to test their milk and dairy products for the presence of melamine.

However, by the time the problem was discovered, Chinese dairy products had been exported across the world and by the end of 2008, a number of countries had introduced controls on melamine limits for dairy products. These included the USA, the EU, Canada, Australia, New Zealand and many Asian countries. Many nations also implemented inspection and surveillance programs directed at products containing milk and milk containing ingredients exported from China.

It would appear from the investigations that the melamine problems were caused by a relatively small number of people for personal gain. However, the ramifications of their actions caused an international safety alert, which severely interrupted trade in dairy products.

A similar situation occurred with the illegal use of a broad-spectrum antibiotic known as Chloramphenicol in foods originating from Asia. Chloramphenicol is normally used as a topical antibiotic against both gram positive and gram negative bacteria. Unfortunately, it is widely available in Asia and has been found to be extensively used for both livestock and aquaculture. When ingested, chloramphenicol has been associated with a number of adverse conditions, including leukaemia, and its use with humans is strictly controlled in most countries. The illegal use of chloramphenicol in a wide range of foods and ingredients exported from Asia has been of international concern for a number of years. Many countries, including the EU, have had to introduce controls and surveillance for chloramphenicol residues in imports. Foods found to be contaminated include animal by-products, vitamin pre-mixes and enzyme preparations.[46]

As with the melamine issue, the problems with chloramphenicol have placed a major burden on countries around the world and have resulted in a high level of additional testing, and destruction of food products, with a consequential cost.

---

[45]*Background Paper on Occurrence of Melamine in Foods and Feed*. Prepared for World Health Organization Expert Meeting on Toxicological and Health Aspects of Melamine and Cyanuric Acid (2009).

[46]Enzyme Technical Association, *Statement on Chloramphenicol Contamination of Enzyme Preparations*, November 4, 2013.

8.5.4 European Union: Rapid Alert System for Food and Feed (RASFF)

As early as 1979, the European Commission introduced the Rapid Alert System for Food and Feed (RASFF) to provide the food control authorities in the member states with a procedure for the exchange of information and measures taken on food safety issues. Initially, this was an informal procedure.

In 2002, the RASFF was incorporated into law as part of the general food law requirements. This law requires that when a member state has any information about a serious health risk deriving from human food or animal feed, it must take the required action.[47]

The member states are required, in particular, to notify the European Commission if they identify a risk or potential risk, and to take such measures as instigating a withdrawal or recall of a food from the market to protect consumer health. This is particularly important when rapid action is required.

Member states are also to notify the Commission if they have agreed with a responsible food operator that a food or feed should not be placed on the market, if this measure is taken due to a serious risk to health. The Commission also requires notification of all rejections of food or feed at any border post in the EU, if there is an inherent health risk.

In 2006, animal health and environmental risks in relation to animal feed were added to the scope of the law. This also includes notifications about risks from pet food.

In order for the system to work effectively, each member state has a designated contact point that is responsible for sending that country's RASFF notifications to the Commission. The notifications summarize the issue and consolidate the information received from the national inspection agencies and laboratories. The Commission has templates to ensure the notifications from the member states are consistent in content.

The Commission is responsible for informing countries outside the EU, if a food which is the subject of the notification has been exported to, or imported from, that country.

Summaries of the notifications are regularly published by the European Commission and appear in the public domain. These reports are routinely accessed by the authorities in many countries outside the EU and it provides the basis of a global alert system.[48]

---

[47]European Parliament and Council Regulation (EC) No 178/2002 laying down the general principles and requirements of food law, establishing the European Food Safety Authority and laying down procedures in matters of food safety. Official Journal of the EU. L 31 of February 1, 2002.

[48]European Commission Rapid Alert System for Food and Feed (RASFF).

8.6 The Impact of Codex Alimentarius on Reducing Barriers to International Trade

In the late 1950s, the United Nations (UN), the World Health Organization (WHO) and the Food and Agriculture Organization (FAO) recognized the need to harmonize standards and controls internationally, to allow international trade, while maintaining the highest levels of food safety and consumer protection.

The deliberations resulted in the establishment of the Codex Alimentarius Commission (CAC) by WHO and FAO in 1963. The objectives of the CAC were to produce harmonized international food standards, guidelines and Codes of Practice to protect the health of the consumers and to ensure fair practices in the food trade. The Commission also had the responsibility for promoting the coordination of all food standards work undertaken by governmental and non-governmental organizations.

At the first session of the CAC in 1963, delegates from 30 countries and 16 international organizations were present at the meeting. By the 35th session, there were 145 member countries and 34 governmental and non-governmental organizations, and on its 50th anniversary in 2013 there were 185 member countries, plus the EU, and 221 observer organizations. It was estimated that Codex Alimentarius now covers 99 % of the world's population.[49]

The Codex Alimentarius organization plays a pivotal role in international food trade. Codex standards are taken into account in the development of national and regional food legislation, food products conforming to Codex standards are able to move more freely in international trade, and, since 1995, the standards form the basis for arbitration on food cases brought before the World Trade Organization (WTO).

The WTO was a consequence of the Uruguay Round Agreements that emerged from the General Agreement on Trade and Tariff (GATT) discussions held over 7 1/2 years between 1986 and 1994. There are two important agreements under the WTO that came out of the Uruguay Round: the Technical Barriers to Trade (TBT) agreement and the Sanitary and Phytosanitary (SPS) agreement.[50]

The TBT agreement seeks to ensure that technical regulations and standards (e.g. packaging, labelling and analytical procedures), do not create unnecessary barriers to trade.

Article 2 of the TBT Agreement reads:

> With a view to harmonising technical regulations on as wide a basis as possible, Member States shall play a full part, within the limits of their resources, in the preparation by appropriate international standardising bodies of international standards for products for which they have either adopted, or expect to adopt technical regulations.

The SPS agreement contains the following:

---

[49]Codex Alimentarius International Food Standards.
[50]World Trade Organization.

Article 2.2 of the SPS Agreement reads:

> Members shall ensure that any sanitary and phytosanitary measure is applied only to the extent necessary to protect human, animal or plant life or health, is based on scientific principles and is not maintained without sufficient scientific evidence.

Article 3.1 of the SPS Agreement reads:

> To harmonise sanitary and phytosanitary measures on as wide a basis as possible, members shall base their sanitary and phytosanitary measures on international standards, guidelines or recommendations, where they exist, except as otherwise provided for in this agreement.

Codex Alimentarius has 17 active technical committees, each dealing with a specific aspect of food trade. The 'horizontal' committees, such as the Codex Committee on Food Additives, the Codex Committee on Food Labelling and the Codex Committee on Contaminants in Food, deal with the issues within their remit affecting all categories of food. There are also commodity committees, which deal with specific food categories, such as the Codex Committee on Fish and Fish Products and the Codex Committee on Fats and Oils. Each Committee has specific tasks related to the development of Codex standards, guidance and Codes of Practice within their remit. This system ensures that the work is carried out by international experts in their respective fields.

The most important criterion for all Codex deliberations and decisions is that they must be based on an impartial assessment of all available scientific data. Decisions should not be influenced by national politics. Standards, Guidelines and Codes of Practice, are only adopted after consensus by the relevant Committee and endorsement by the Codex Alimentarius Commission.

While the Codex process can obtain international approval for the standards, which are then incorporated into the legislation of many countries, the procedural aspects of this work means that it can take a number of years between the initiation of the work on a standard and its adoption. Most committees only meet annually, and in a few cases biannually, and the procedure requires that the work progresses through a series of steps and consultations. The speed of progress can often be a disadvantage in a rapidly changing commercial environment, as some standards can take many years to come to fruition.

Although Codex Standards and Codes of Practice have no legal basis in international laws, an increasing number of countries world-wide are incorporating relevant standards into their national legislation. As a consequence, conformity to the standards are being used as the basis for accepting imports from other countries.

8.7 Conclusion

As can be seen from the range of issues discussed above, a number of factors can jeopardize the international trade in foods and food commodities.

Despite many efforts to harmonize food legislation both within and between economic blocs, free trade has not been achieved for many foods and trade barriers can be formed for a variety of reasons.

While the need for a fully harmonized global trade in food has not been disputed, the reality is that it is not likely to be achieved within the next two decades.

References

Administrative Measures for Safety Review of New Food Materials. National Health and Family Planning Commission Decree No. 1. May 31, 2013. Available at http://www.moh.gov.cn/fzs/s3576/201307/34de96581cfc4751be3bc83870360472.shtml. Unofficial translation by USDA Foreign Agricultural Service available at http://gain.fas.usda.gov/Recent%20GAIN%20Publications/Administrative%20Measures%20for%20Safety%20Review%20of%20New%20Food%20Materials_Beijing_China%20-%20Peoples%20Republic%20of_8-1-2013.pdf (last visited 28 February 2014)

Animal/Plant Health Inspection Service, Bovine and Spongiform Encephalopathy – An Overview, United States Department of Agriculture (2006). Archive available at http://www.aphis.usda.gov/publications/animal_health/content/printableversion/BSEbrochure12-2006 (last visited 25 January 2014)

Australia New Zealand Food Standards Code - Standard 1.5.1 - Novel Foods. Federal Register of Legislative Instruments F2013C00142. Issue 139 (2013)

Australian Bureau of Statistics (2000) 1301.0 – Year Book. Canberra, Australia

Becker GS (2005) Bovine Spongiform Encephalopathy and Canadian Beef Imports. CRS (USA) Report for Congress No RL 32627, 11 March 2005

Breuning G, Lyons JM (2000) The case of the Flavr Savr tomato. Calif Agric 54(4):6–7

Brinckman D (2000) The regulation of rBST: the European case. AgBio Forum 3(2 and 3):164–172. Available at http://www.agbioforum.org (last visited 25 February 2014)

Cartagena Protocol on Biosafety to the Conventions on Biological Diversity. Secretariat of the Convention on Biological Diversity, Montreal, Canada. ISBN 92-807-1924-6 (2000)

Center for Disease Control, BSE Cases in North America, by Year and Country of Death, (1993–2008). Center for Disease Control and Prevention, Department of Health and Human Services, USA (2008). Archive available at http://www.cdc.gov/ncidod/dvrd/bse/images/bse_cases_namerica_2008.gif (last visited 12 February 2014)

Codex Alimentarius (1997) Guidelines for use of nutrition and health claims, CAC/GL 23. Codex Alimentarius Commission, Rome

Codex Alimentarius International Food Standards. Available at http://www.codexalimentarius.org/ (last visited 15 April 2014)

Codex and the International Food Trade (2013) FAO Corporate Document repository available at http://www.fao.org/docrep/w9114e06.htm (last visited 20 February 2014)

Codex General Standard for Food Additives (GSFA) Online database. Codex STAN 192-1995 as amended (2014). Available at www.codexalimentarius.org/standards/gsfa/ (last visited 20 February 2014)

Colchester AC, Colchester NT (2005) The origin of bovine spongiform encephalopathy: the human prion disease hypothesis. Lancet 366(9488):856–861

Considering whether an FDA-Regulated Product involves the Application of Nanotechnology – Guidance for Industry. Draft guidance. Dept of Health and Human Services, Food and Drug Administration, Office of the Commissioner, USA (2011). Available at http://www.fda.gov/RegulatoryInformation/Guidances/ucm257698.htm (last visited 25 February 2014)

Department of the Environment, Food and Rural Affairs of the United Kingdom Food Statistics Year Book 2011. Web archive available at nationalarchives.gov.uk.defra/statistics-foodfarm-food-pocketbook (last visited 22 January 2014)

Dietary Supplement and Health & Education Act. Public Law 103-147; 108 stat. 4325-4335 103d Congress 2nd Session (October 15, 1994)

Enzyme Technical Association, Statement on Chloramphenicol Contamination of Enzyme Preparations, 4th November 2013. Available at http://www.enzymeassociation.org/?p=374 (last visited 22 January 2014)

European Commission Rapid Alert System for Food and Feed (RASFF). Available at http://ec.europa.eu/food/food/rapidalert/index_en.htm (last visited 15 April 2014)

European Commission Regulation (EU) No 1131/2011 amending Annex II to Regulation (EC) No 1333/2008 of the European Parliament and of the Council with regard to steviol glycosides. Official Journal of the EU. L 295 of 12 November 2011

European Commission Regulation (EU) No 231/2012 laying down specifications for food additives listed in Annexes II and III to Regulation (EC) No 1333/2008 of the European Parliament and of the Council. Official Journal of the EU. L 83 of 22 March 2012

European Commission, Commission proposes ban on market release and use of BST in the Union as of 1st January 2000, EC Press release No IP/99/758 (1999)

European Commission, The Transatlantic Trade and Investment Partnership (2014). Available at http://ec.europa.eu/trade/policy/in-focus/ttip/ (last visited 14 April 2014)

European Council Regulation (EEC) No 2377/90 laying down a Community procedure for the establishment of maximum residue limits of veterinary medicinal products in foodstuffs of animal origin. Official Journal of the EU. L 224 of 18 August 1990

European Parliament and Council Directive 96/83/EC amending Directive 94/35/EC on sweeteners for use in foodstuffs. Official Journal of the EU. L 48 of 19 February 1997

European Parliament and Council Regulation (EC) No 178/2002 laying down the general principles and requirements of food law, establishing the European Food Safety Authority and laying down procedures in matters of food safety. Official Journal of the EU. L 31 of 1 February 2002

European Parliament and Council Regulation (EC) No 258/97 concerning novel foods and novel food ingredients. Official Journal of the EU. L 43 of 14 February 1997

European Parliament and Council Regulation (EU) No 1169/2011 on the provision of food information to consumers. Official Journal of the EU. L 304 of 22 November 2011

European Parliament and Council Regulation (EC) No 1333/2008 on food additives. Official Journal of the EU. L 354 of 31 December 2008

European Parliament and Council Regulation (EC) No 1829/2003 on genetically modified food and feed. Official Journal of the EU. L 268 of 18 October 2003

European Parliament and Council Regulation (EC) No 1830/2003 concerning the traceability and labelling of genetically modified organisms and the traceability of food and feed products produced from genetically modified organisms and amending Directive 2001/18/EC. Official Journal of the EU. L 268 of 18 October 2003

European Parliament and Council Regulation (EC) No 1924/2006 on nutrition and health claims made on food. Official Journal of the EU. L404 of 30 December 2006

European Parliament and Council Directive 1999/2/EC on the approximation of the laws of the Member States concerning foods and food ingredients treated with ionising radiation. Official Journal of the EU. L 66 of 13 March 1999

Food and Agriculture Organization of the United Nations, World Food Summit – Rome Declaration on Food Security (1996). FAO Corporate Document Repository available at www.fao.org (last visited 20 February 2014)

Hilts C, Pelletier L (2009) Background paper on occurrence of melamine in foods and feed. Prepared for World Health Organisation Expert Meeting on Toxicological and Health Aspects of Melamine and Cyanuric Acid. Supported by Health Canada (2009). Available at http://www.who.int/foodsafety/fs_management/Melamine_3.pdf (last visited 15 April 2014)

How U.S. FDAs GRAS Notification Program Works, Food and Drug Administration, USA. Archive available at http://www.fda.gov/Food/IngredientsPackagingLabeling/GRAS (last visited 25 February 2014)

Institute of Food Science and Technology, Addition of Micronutrients to Food, London, UK (1997)

Jukes TH (1976) Cyclamate sweeteners. JAMA 236(17):1987–1989

Lewin AC (2007) Zambia and Genetically Modified Food Aid. Case study # 4-4 of the program: food policy for developing countries: the role of the government in the global food system. Cornell University, Ithaca

McQuate RS (2011) Ensuring the safety of sweeteners from Stevia. Food Technol 65(4):42–49

Morris VJ (2008) Nanotechnology in foods – a European perspective. International Review of Food Science and Technology, International Union of Food Science and Technology, Ontario, Canada

Nanoscience and nanotechnologies: opportunities and uncertainties. The Royal Society & The Royal Academy of Engineering, London, UK (2004). Available at http://www.nanotec.org.uk/finalReport.htm (last visited 3 March 2014)

Regulations Amending the Food and Drug Regulations (948 - Novel Foods). Canada Gazette Part II, Vol. 133 No. 22 of 27 October 1999

The Office of the United States Trade Representative, Transatlantic Trade and Investment Partnership (T-TIP) (2013). Available at http://www.ustr.gov/ttip (last visited 14 April 2014)

United States General Accounting Office (1987) Food and Drug Administration Food Additive Approval Process Followed for Aspartame. Report to the Honorable Howard M. Metzenbaum, U.S. Senate 133460. United States General Accounting Office, Washington. Available at http://archive.gao.gov/d28t5/133460.pdf (last visited 28 February 2014)

Valleron AJ, Boelle P-Y, Will R, Cesbron J-Y (2001) Estimation of epidemic size and incubation time based on age characteristics of vCJD in the United Kingdom. Science 294 (5547):1726–1728

Will Knight, Zambia bans GM Food Aid, New Scientist, 30 October 2002. Available at http://www.newscientist.com/article/dn2990-zambia-bans-gm-food-aid.html#.U16kHcfg-TQ (last visited 28 February 2014)

World Health Organisation (2014) Genetically modified food. Available at www.who.int/topics/food_genetically_modified/en (last visited 25 January 2014)

World Trade Organization. Available at http://www.wto.org/index.htm (last visited 15 April 2014)

Part II
Interdisciplinary Facets of International Food Law

Gabriela Steier

Food systems are far-reaching, global in scope, pervasive, and ubiquitous. Therefore, food law cannot be confined by statutes as it must cover the breadth of food systems. Food law, as Gabriela Steier is coining it in this work, is a dynamic, cross-disciplinary, colorful, complex, and complicated field that provides fascinating intricacies that challenge lawyers, policy makers and regulators to think critically and creatively and, most importantly, to collaborate with professionals from other disciplines and from all over the world. This section of *International Food Law and Policy* is instructive as to how these non-legal disciplines fit into food law.

Part I and this part build on each other. While Part I provides a general overview of the various consideration of international food law and policy, this part completes the broad picture by including some of the most important interdisciplinary aspects of the field and even some marginalized by many academics. The editors hope that, by including these chapters, scholars, students, practitioners, and anyone interested will be able to draw new connections and find inspiring solutions to the challenges of the current food system.

In Chap. 9, Professor Jessica Fanzo from Columbia University, New York, links nutrition to food law and policy. She highlights the role of food policies and nutrition goals and outcomes and links food policy, public health, and nutrition science. Her illustrations from around the world drive home the universality of importance of these links, and provide a solid basis to understand various later chapters.

Dr. Michael Nathenson, a brilliant oncologist devoted to the highest patient care, explains how food choices affect the number one leading cause of deaths, cancer. In Chap. 11, he explains which information is relevant, reliable, and instructive, and why. His description of various connections between cancer therapy and food choices bridge the topics of food policy, public health, medicine (oncology), and introduce scientific concepts into the broader food law field. In Chap. 12, a textbox, Gabriela Steier presents recent scientific studies that link agricultural pesticides and carcinogens.

Next, Prof. Dr. Liviu Steier, Dr. Regina Steier, and Gabriela Steier, compiled the first ever analysis of the legal implications of food systems, public health and dental care in Chap. 13. Professors and highly-acclaimed practitioners of dental medicine, Prof. Dr. Liviu Steier and Dr. Regina Steier, joined forces with their daughter, an attorney and editor of this book, to explain how and why mass-medication of caries through fluoridation is a food policy issue ripe for review.

In Chap. 15, Professors Charles Hall and Steve Balogh, breakthrough scholars and widely published and renowned systems ecologists with extensive academic experience, offer cutting edge insights into the connections of food law, energy and the environment. Their chapter rounds up (with a cherry on top!) the section about the interdisciplinary aspects of food law and policy around the world.

Of special note is Chap. 18, written by Professors Faunce and Bruce, explores whether global governance of new renewable energy and climate change mitigation technologies such as artificial photosynthesis may accelerate transition to such a reformed global food system in an epoch conveniently termed the Sustainocene. Here, readers are introduced to some aspects of the geoengineering debate.

Finally, this section also includes three textboxes about emerging topics that are new and important. In Chap. 17, Dr. Denis Herbel, Mariagrazia Rocchigiani and Nora Ourabah Haddad from the FAO introduce cooperatives' and producer organizations' roles in achieving food security by highlighting the importance of smallholder farming. Randy Hayes, a thought leader from the Center for Food Safety and Foundation Earth in Washington, D.C., explains in Chap. 16 the need to internalize ecologic externalities, which may affect food sticker prices in the future. To conclude the section, Gabriela Steier outlines the theory of planetary boundaries in food and agriculture law, which bring climate change issues into the realm of food law (Chap. 10).

Chapter 9
Food Policies' Roles on Nutrition Goals and Outcomes: Connecting of Food and Public Health Systems

Jessica Fanzo

Abstract Nutrition exists when food security is combined with a sanitary environment, adequate health services, and proper care and feeding practices to ensure a healthy life for all household members. Despite increased attention to undernutrition, it remains a devastating multi-faceted problem for infants, young children, and women around the world, resulting in increased morbidity, mortality, and long-term disability. Undernutrition can also lead to poor health into adulthood, which affects social and economic development of nations. On the other end of the malnutrition spectrum, overweight and obesity are growing problems, linked to changing diets and activity patterns, which also lead to serious health problems and impact the economies of nations. This chapter attempts to unpack the importance of food and agriculture policies on nutrition outcomes and why engagement of food and public health systems remain critically important. External pressures, such as climate variability and population growth, that tax these systems are discussed, as well as the globalization of our food system and why that has shifted dietary patterns and nutrition and health status trends. The multi-sectoral integration of food and health systems and its importance to improve nutrition is demonstrated through three models. Three very brief case studies are presented that help exemplify some of the food and health system trends that influence policy and ultimately, nutrition outcomes.

9.1 Introduction

Historically, malnutrition is broadly thought of as a lack of sufficient food. However, in light of rapidly changing economic landscapes in and between nations, multiple "burdens of malnutrition" contribute to poor health and development. For example, children who do not consume adequate calories and micronutrients over

J. Fanzo (✉)
Berman Institute of Bioethics and School of Advanced International Studies, Johns Hopkins University, Washington, 20036 DC, USA
e-mail: jfanzo1@jhu.edu

long periods—beginning in utero—do not achieve full genetic potential in cognitive, reproductive and immune development. The latest series on nutrition in the renowned peer-reviewed journal, *Lancet*, emphasized chronic malnutrition over acute malnutrition in terms of the overall detrimental effect on society. Micronutrient deficiencies of essential vitamins and minerals, such as iron, zinc, folic acid, Vitamin A and other, are also gaining importance as the scientific community proves links to disease and inhibited growth. It is, therefore, important to understand to connect food and public health systems to the role of food policies on nutrition goals and outcomes.

9.1.1 What Is the Triple Burden of Malnutrition?

Nutrition, by definition, exists when food security is combined with a sanitary environment, adequate health services, proper care and feeding practices to ensure a healthy life for all household members.[1] This definition may seem quite similar to the food security definition posed by the World Food Summit of 1996 that states: *Food security exists "when all people at all times have access to sufficient, safe, nutritious food to maintain a healthy and active life."*[2] However, there are definitions, on the one hand, and then there are the implications on how these definitions are translated into policies and programmatic implementation, on the other hand. Nutrition is often forgotten in the food security mandate. Thus, food and agriculture policies are generally less attuned to ensuring nutrition is central in their ultimate outcomes.

Not getting the right amount of food and nutrients or the right types of nutrients can lead to undernutrition or overweight, which, in turn, has serious deleterious effects on health, development, and productivity. Despite increased attention to undernutrition, it remains a devastating multi-faceted problem for infants, young children, and women around the world, resulting in increased morbidity, mortality, and long-term disability. Undernutrition can also lead to poor health into adulthood, which affects social and economic development of countries on large scales. Overweight and obesity are also growing problems, linked to changing diets and activity patterns, which can lead to serious health problems including increased risk of non-communicable diseases such as diabetes, cardiovascular disease and strokes.

The global burden of malnutrition can be described as a "triple" burden[3] in which countries, communities and households may be burdened with three manifestations of malnutrition: undernutrition (often in the form of chronic or acute), overnutrition (overweight and obesity) and/or micronutrient deficiencies. In FAO's State of Food and Agriculture 2013 report, countries were classified as having one,

---

[1] World Bank (2006).
[2] FAO (1996).
[3] FAO (2013).

two or three of these burdens and in different combinations. There are very few countries that do not have at least one malnutrition insufficiency or a combination.[4]

Global prevalence of stunting, which reflects chronic undernutrition during the early stages of life causing children to fail to grow to their full genetic potential, both mentally and physically, has declined to 35 % in children under 5 years of age since 1990, which is a reduction of 2.1 % per year.[5] Yet, there are still an estimated 150 million children who remain moderately or severely stunted.[6] Wasting, which reflects acute malnutrition and is a strong predictor of mortality among children, impacts 50 million children under 5 years of age, with the highest burden in South Asia. There has been an 11 % decrease since 1990.[7] On the other end of the malnutrition spectrum, an estimated 43 million children under 5 years of age are overweight, and two-thirds of those children reside in low- and middle-income countries.[8] Prevalence of low body mass index (BMI) in adult women has decreased in Africa and Asia in the last 4 decades, but still exceeds 10 % in the two regions. At the same time, prevalence of overweight and obesity has increased in all regions.[9] According to the world's leading medical journal, the Lancet, "[d]eficiencies of essential vitamins and minerals continue to be widespread and have significant adverse effects on child survival and development, as well as maternal health."[10] Thus, indicators of malnutrition have far-reaching effects world-wide.

9.1.2 Why Food and Agriculture Policies Should Have Nutrition Goals and Outcomes

> Where policies exist that support nutrition-sensitive approaches and where active government processes stimulate joint agriculture nutrition approaches, there is a relatively high likelihood of success in implementing such programmes and projects with the theoretical implication of improved, nutrient-rich and balanced diets, and eventually, improved health status of consumers. However, the sustainability of such initiatives relies heavily on sustained political will.[11]

The WHO maintains that "[a]griculture remains the largest employment sector in most developing countries and international agriculture agreements are crucial to

---

[4]FAO (2013).
[5]Black et al. (2013a) and UNICEF et al. (2013).
[6]Black et al. (2013a) and UNICEF et al. (2013).
[7]Black et al. (2013a).
[8]Black et al. (2013a) and UNICEF et al. (2013).
[9]Black et al. (2013a).
[10]Black et al. (2013b).
[11]Jaenicke and Virchow (2013).

a country's food security."[12] A recent report published by the UN Nutrition Sensitivity of Agriculture and Food Policies, however, concluded that the "complex role of how agricultural policies can effectively address nutrition is not yet well understood."[13] Additionally, the UN Standing Committee on Nutrition found that "[t]here is considerable conceptual knowledge on this topic, but little understanding of how to carry concepts and policy objectives into effective implementation and delivery of food-based approaches that impact nutritional status of populations."[14] By focusing on the entire human lifecycle, nutrition-sensitive agriculture and food policies should consider the nutrients and determinants that are important for development, growth, and maintenance of health at various stages of life. The approach should encompass the entire food system—a complete array of activities covering all stages of the food supply chain ranging from input distribution, on-farm production, marketing, processing, and storage. The goal should be to produce healthy and safe food containing essential micronutrients and to increase year-round, affordable access for both rural and urban communities, as the consumers.

Food policies and programs are relevant to ensure that food systems from production to processing to consumption are directed to ensuring improved dietary patterns and nutrition outcomes. Debate continues between those who argue that agricultural policy should play a large role in producing nutritious food and those who believe that it is more important for agricultural policy to focus on economic development and "feeding the planet" in the form of bulk calories.[15]

This chapter attempts to unpack the importance of food and agriculture policies on nutrition outcomes and why engagement of food and public health systems remain critically important. External pressures that tax these systems will be discussed, as well as the globalization of our food system and why that has shifted

---

[12] WHO, Trade, foreign policy, diplomacy and health: Food Security, available at http://www.who.int/trade/glossary/story028/en/ (last accessed May 2014).

[13] UN Standing Committee on Nutrition, The Nutrition Sensitivity of Agriculture and Food Policies 1-56, 6 (Nairobi, Kenya, 26–28 August 2013), available at http://unscn.org/files/Publications/Country_Case_Studies/UNSCN-Executive-Summary-Booklet-Country-Case-Studies-Nairobi-Meeting-Report.pdf (last accessed May 2014) (This UN report provides summaries of country case studies for Brazil, Malawi, Mozambique, Nepal, Senegal, Sierra Leone, South Africa, and Thailand).

[14] UN Standing Committee on Nutrition, The Nutrition Sensitivity of Agriculture and Food Policies 1-56, 6 (Nairobi, Kenya, 26–28 August 2013), available at http://unscn.org/files/Publications/Country_Case_Studies/UNSCN-Executive-Summary-Booklet-Country-Case-Studies-Nairobi-Meeting-Report.pdf (last accessed May 2014) (This UN report provides summaries of country case studies for Brazil, Malawi, Mozambique, Nepal, Senegal, Sierra Leone, South Africa, and Thailand).

[15] UN Standing Committee on Nutrition, The Nutrition Sensitivity of Agriculture and Food Policies 1-56, 6 (Nairobi, Kenya, 26–28 August 2013), available at http://unscn.org/files/Publications/Country_Case_Studies/UNSCN-Executive-Summary-Booklet-Country-Case-Studies-Nairobi-Meeting-Report.pdf (last accessed May 2014) (This UN report provides summaries of country case studies for Brazil, Malawi, Mozambique, Nepal, Senegal, Sierra Leone, South Africa, and Thailand).

dietary patterns and nutrition and health status trends. The multi-sectoral integration of food and health systems and its importance to improve nutrition will be demonstrated through three models. Finally, four case studies will be presented that help exemplify some of the agriculture, food and health system trends that influence policy and ultimately, nutrition outcomes.

9.2 External Pressures on Food and Public Health Systems

9.2.1 How Climate Variability Impacts Food and Public Health Systems

There is clear and convincing evidence for anthropogenic climate change.[16] The warming of the earth will have devastating consequences including more extreme weather conditions, rising seas leading to salination of agriculture and drinking water sources, and acidification of oceans, in which many people rely on for their diets and livelihoods. These effects will impact food security, nutritional outcomes and public health. The increased climate variability will intensify the severity and frequency of natural disasters. Both floods and droughts are and will continue to occur more frequently.[17] Predicting weather patterns will become much more difficult as the variability related to climate change increases.[18]

Climate change is having, and will have, very different effects depending on where people live and which resources are available to them. The poor will be impacted the most and suffer the greatest repercussions. Risks of food insecurity and health will also impact the poorest nations—ironically, those who had made the least anthropogenic contribution to climate change.[19] These changes are also likely to have the greatest impact in many low resource regions' agricultural outputs, reducing yields of crops, soil fertility, and forest and animal productivity, which may lower income, resiliency and subsequently, reduced access to sufficient, nutrient dense foods and impaired nutritional status of communities.[20] Unstable agriculture output could increase global prices for food and their volatility,[21] while urban areas will be especially vulnerable in accessing food.[22]

The poorest communities, especially female headed-households, will feel the consequences of rising food prices most strongly.[23] Increases in food costs force

---

[16] IPCC (2012).
[17] IPCC (2012).
[18] Hansen et al. (2007).
[19] Patz et al. (2005).
[20] Mason and Shrimpton (2010).
[21] FAO (2011) and World Bank (2013).
[22] Mason and Shrimpton (2010).
[23] Popkin et al. (2012).

people to reduce the quantity and nutrient-quality of food consumed, preferentially affecting those who are in need of nutrient dense foods like young children and pregnant or lactating women.[24] In addition, diminished biodiversity, which is an important source of diverse diets, may, "increase the risk of disease being transmitted to human beings, a phenomenon termed the dilution effect."[25] Thus, it has been estimated that 80 % of the burden of disease related to climate change will affect children and that by 2050, with a potential projection of a 20 % increase in malnutrition.[26] Climate variability, according to the UN Office for the Coordination of Humanitarian Affairs, "could eliminate much of the improvement in child malnourishment levels that would occur with no climate change."[27] In fact, some studies estimate an even greater impact with stunting increasing by as much as 30 % compared to a scenario in which climate is stable.[28]

Additionally, the disease burden will also be impacted with a warmer world, which will in turn, effect the public health system. As global temperature rises, health impacts will change. With increasing severity of heat waves, heat-related stress will increase short-term mortality from stroke, respiratory and cardiovascular incidents.[29] Rising temperatures will also increase the spread and transmission of vector and rodent-borne diseases as well as density, pathogen maturation and replication within mosquitos resulting in increases in infections of malaria, dengue fever and other vector born diseases.[30] It is estimated that malaria, diarrhea and protein-energy malnutrition together cause more than 3 million deaths each year.[31]

9.2.2 How Population Growth Will Impact Food and Public Health Systems

It is nearly certain that the world's population will continue to increase. Over the past decade, our global population has increased from 6 billion to 7 billion. Historically, most of the growth in population has occurred post industrial revolution following the boom in services and industry. It is estimated that by 2050, the world's population will continue to grow to 9 billion, which begs the question of

---

[24]Brinkman et al. (2010).
[25]Costello et al. (2009).
[26]IFPRI (2009) and Nelson (2010).
[27]UN Office for the Coordination of Humanitarian Affairs, IRIN Humanitarian News and Analysis: FOOD: The link between undernutrition and climate change, available at http://www.irinnews.org/report/87353/food-the-link-between-undernutrition-and-climate-change (last accessed May 2014).
[28]Lloyd et al. (2011).
[29]Robine et al. (2008), Kovats and Ebi (2006), and Husain and Chaudhary (2008).
[30]Costello et al. (2009).
[31]WHO (2004).

how much our planet and humanity can sustain and what the boundaries are of our food and health systems? (see Chap. 16).

The patterns of population growth are nuanced. Overall, fertility rates are declining and child survival is increasing. Explanations for these trends are analogous to the question of what came first, the chicken or the egg? Have declines in fertility resulted in increased child survival because families take care to invest in their few children, or has child survival, through improved primary care health systems led to declines in fertility because families are realizing that the children that they do have, not only survive, but also thrive? Although the trend is overall positive, the continued population growth and its resulting pressure will have impacts on our food and public health systems. Results will include pressures upon public health, sustainable food production, environmental conservation, and the prevention of climate change as a challenge of industrial food production.

As global populations expand, the health system will be increasingly strained. In high-income countries, the population is aging due to extended life expectancy. A UN Task Team, led by senior experts from over 50 UN entities and international organizations, the Department of Economic and Social Affairs, and the UN Development Programme, found that, "[t]he number and proportion of older persons aged 60 years or over are rising in all countries. Globally, the number of older persons aged 60 years or over is projected to increase from 810 million now to more than 2 billion in 2050."[32] With this longevity, comes increasing prevalence of expenses to treat chronic diseases in which the health sector must be in consistent demand.

At the same time, rapid economic growth and an emerging middle class in low- and middle-income classes will increase the demand for health services. Some countries will struggle to meet this demand.

Demographic changes in the past decades have led to the largest generation of youth in the world today. Globally there were 1.2 billion young people aged 15–24 in 2010. High-fertility countries in sub-Saharan Africa are projected to experience a rapid increase in the population aged 15–24, from 173 million at present to 362 million by mid-century.[33]

Thus, more of the world's population lives in urban areas than rural, resulting in a new set of challenges for governments and the health and food systems. In many low-income countries situated in sub-Saharan Africa, Asia, Latin America and even parts of the high-income world, like America and England, urbanization has resulted in slums with little access to health services, decent food shopping areas, or sanitation services. Data suggests that "[a]s urban populations swell, so do the incidence of illness such as hypertension, heart disease, obesity, diabetes and asthma."[34] As urban migration occurs, many cities are ill equipped to build a

---

[32]UNDESA/UNFPA (2012).

[33]UNDESA/UNFPA (2012).

[34]KPMG, Trends, risks and opportunities in healthcare, available at http://www.kpmg.com/Global/en/IssuesAndInsights/ArticlesPublications/care-in-a-changing-world/Pages/trends-risks-opportunities.aspx (last accessed May 2014).

healthy, functional food system. Some researchers examined food insecurity amongst slum dwellers in Nairobi and found that "[o]nly one household in five is food-secure, and nearly half of all households are categorized as food-insecure with both adult and child hunger."[35] Thus, current trends point toward increased food *insecurity*.

Food insecurity, however, is not only a conundrum of low-income countries. The effects of food insecurity are also felt in wealthy cities. In New York City, for example, one of the wealthiest cities in the world, more than 1.3 million residents, including one in five children and one in ten seniors (over the age of 60), live in households that lack sufficient food.[36] In the last 5 years, the number of city residents experiencing food insecurity has increased by more than 200,000 and the city's food pantries and soup kitchens reported a 10 % increase in demand, while 63 % reported food shortages in 2013.[37] A recent study has also shown that food insecurity was significantly associated with increasing body mass index in women not receiving food assistance in New York City.[38] This study further demonstrates that, "in an urban population, overweight and obesity are very common as is food insecurity"[39] and there is often a correlation between the two.[40]

With population pressures in the future will also come the necessity to ensure that enough food is produced and that the food produced is not wasted. An estimated 30–50 % of the world's food is never consumed, but, instead, wasted or disposed of somewhere along the value chain between the producer and the consumer.[41] In high-income countries, most of the food waste that occurs is with the consumer or the retailer, such as food markets, restaurants, convenience stores. In low-income countries, most of the food wasted is due to post harvest losses and a lack of infrastructure, transport and technology.

In order to ensure that enough food is available for the growing populations around the world and that food is not wasted, policy makers and consumers may need to better think about how food is grown, transported and distributed in an equitable and efficient way. In this process, nutrition is often forgotten or lost. Thus, changes in the types of food we eat, as well as shifts in our diet, will drive a new market of and demand for how food is grown, processed and consumed. While population is increasing, overall wealth amongst that growing population is also booming, particularly in places like India, China and Brazil. This diet will shift

---

[35]Faye et al. (2011).

[36]New York City Food Policy Center, New York City Food by the Numbers: Hunger, Food Insecurity and SNAP Enrollment, available at http://nycfoodpolicy.org/hunger-food-insecurity-snap-enrollment/ (last accessed May 2014).

[37]City Harvest 2014. New York City Food Policy Center, New York City Food by the Numbers: Hunger, Food Insecurity and SNAP Enrollment, available at http://nycfoodpolicy.org/hunger-food-insecurity-snap-enrollment/ (last accessed May 2014).

[38]Karnik et al. (2011).

[39]Karnik et al. (2011).

[40]Karnik et al. (2011).

[41]Stuart (2009).

towards more meat, dairy products, oils and processed food consumption and away from the more sustainable plant-based diets. Consequently, despite the need to grow more crops in a sustainable way, there is also a need to ensure that the food is available and accessible in a more equitable way. It follows that the food that is grown, provided and sold should be of better nutritional quality while consumers demand more nutritious foods.[42]

9.2.3 Potential Future Pandemics: Consequences for Food Supplies and Public Health

A pandemic is a global disease outbreak that represents a top global catastrophic risk [43]

Outbreaks are usually the result of rapidly transmitting pathogens. Many of these pathogens are still emerging and little is known about the potentiating implications, many of which originate from animals (termed zoonotic). Some concerning pathogens arose over the last decade—including severe acute respiratory syndrome (SARS), West Nile Virus and the H5N1 avian flu. According to the World Bank, "every year, 2.3 billion human infections occur in developing countries by zoonotic diseases, and 2.2 million" [44] people die as a result.[45] The food supply and the risks that come along with it are intricately linked to health outcomes that are also connected to the consumption of animal products.

Those that suffer the most are the poor and fragile communities, often because of the lack of public health services and their proximity to animals and livestock. The pathogens causing zoonoses result in diseases that can have profound impacts on food security, nutrition outcomes and livelihoods of these poor households. The World Bank states that "[p]andemic prevention requires robust public health systems (veterinary and human) to detect contagion early, ensure correct diagnoses, and respond rapidly to defend against contagion."[46] Prevention also requires oversight of the food system to do early screenings, oversight and governance of quality controls and assurances. To ensure better response systems, in 2008, the WHO, and the Food and Agriculture Organization (FAO), coordinated by the World Bank and UN System Influenza Coordinator, prepared a global strategy for using "One Health" approaches to reduce health risks at animal-human-environment

---

[42]FAO (2013).

[43]World Bank (2010). World Bank, Brief: Pandemic Risk and One Health (October 23, 2013), available at http://www.worldbank.org/en/topic/health/brief/pandemic-risk-one-health (last visited May 2014).

[44]World Bank, Brief: Pandemic Risk and One Health (October 23, 2013), available at http://www.worldbank.org/en/topic/health/brief/pandemic-risk-one-health (last visited May 2014).

[45]World Bank (2010).

[46]World Bank, Brief: Pandemic Risk and One Health (October 23, 2013), available at http://www.worldbank.org/en/topic/health/brief/pandemic-risk-one-health (last visited May 2014).

interfaces.[47] This multi-sectoral and disciplinary coordination is crucial to both food and health system integration to reduce pathogen risk.

9.3 Our Globalized Food System

9.3.1 The Impact of Food Policies on Diets and Nutrition Outcomes

Nutrition-sensitive agriculture involves the design and implementation of nutrition-based approaches within agriculture, sustainable farming, crop systems, value chains and market places.[48] Ultimately, nutrition-sensitive agriculture is aimed at improving the nutritional outcome of a population by maximizing the positive impact of food and agricultural systems and their value chains on nutrition while minimizing the potential for negative externalities on the sector's economic and production-driven goals.[49] It is agriculture with a nutrition *lens* and should not detract from the sector or consumer goals.[50] What is clear is that changing agricultural systems should not only meet basic caloric needs, but also the micronutrient and dietary quality needs of communities.[51]

There are five crucial entry points to improve nutrition-sensitive agriculture approaches[52] with two entry points in particular: (a) enabling policies and government structures expressing the political will to fight malnutrition and micronutrient deficiencies, and (b) appropriate mechanisms for inter-sectoral and inter-organizational collaboration within the countries. Studies have shown that for nutrition-sensitive agriculture programmes to be successful, partners from different sectors must be considered as active players.

Policies and processes of global market integration can influence long-term dietary change, but there is a need to look beyond the health sector.[53] Policies also need to focus on the promotion of healthy, high-quality diets over the long-term among populations living in lower income countries. Some researchers noted that there were few comprehensive sets of policies addressing obesity and diet-related chronic diseases in the developing world.[54] This remains true today. There is also very little that existing policies do to address the forces and institutions of the

---

[47]World Bank (2012a, b).
[48]Fanzo et al. (2013).
[49]Fanzo et al. (2013).
[50]Fanzo et al. (2013).
[51]Herforth et al. (2012) and FAO (2013b).
[52]Jaenicke and Virchow (2013).
[53]Hawkes (2006).
[54]Hawkes (2006).

global marketplace that can have more detrimental effects to the health of populations.[55]

Since the WHO adopted the Global Strategy in 2004, governments are increasingly beginning to implement food policies to encourage healthier eating. Although the main strategy has been to provide information for consumers, countries have made notable steps in reformulating food products, establishing school food standards, nutrition labeling, restricting food marketing to children, promoting fruits and vegetables, and more recently, implementation of food taxes. However, when it comes to national food and agricultural policies, the focus remains mainly on producers. The policies are also not designed with public health in mind. For example, the relationship between population nutrition and chronic disease risk is often ignored in most agriculture policies,[56] which was corroborated by other analyses. One particular supporting analysis points out that current food policies are largely incompatible with good public health, but interventions that include food producers, processors, and food providers can markedly improve a population's dietary health. Such an improvement would require cooperation between farming and commercial food producers in order to counter the current trends in food supplies and to reshape the nutrition transition.[57]

In the WHO nutrition policy review,[58] the most commonly reported policy activities in food security and agriculture were research (59 %) and supply of seeds (55 %), subsidized sales and construction of irrigation systems (48 %), construction of rural infrastructure (42 %), price control (41 %), international agreements to increase domestic food production (41 %), production credit from state-owned banks (39 %) and subsidized food for vulnerable groups (38 %).[59] The main policy goals of these programmes were to increase output and farm incomes, followed by improving the quality of the products. Few of the broad policy goals explicitly mentioned nutritional goals, such as combating undernutrition, reducing overweight or obesity, or promoting a healthy diet.

9.3.2 The Importance of Linking Food Policy to Public Health: Dietary Health Guidelines and Nutrient Needs

Increasingly, countries are developing and adopting dietary guidelines to promote healthy diets for their populations. These guidelines are often structured into select food groups and the recommended relative amounts to consume in order to obtain essential nutrients or to reduce risk of non-communicable diseases. Most guides

---

[55]Hawkes (2006) and Verstraeten et al. (2012).
[56]Nugent (2004).
[57]Wang and Lobstein (2006).
[58]WHO (2013).
[59]Id.

recommend a diet based on staples of cereals or starchy roots, combined with high vegetable and fruit consumption, moderate levels of animal and vegetable protein and small amounts of fat, salt and sugar. Many combine this dietary pattern type approach with physical activity recommendations.[60] Although many countries have developed their own set of guidelines, many countries have based their guidelines on those developed by the World Health Organization (WHO).

In the United States, dietary guidelines are updated every 5 years based on the latest science as well as some food policy. These guidelines are quite detailed and visual and, in their development, involve a public consultation. A study that compared the 2005 U.S. dietary guidelines to the food system supply that is available for the country, found that food supply does not, on average, equal the recommended dietary guidelines.[61] The volume supplied surpasses the volume required by some two-thirds in the case of grains, and by half for both fats and sugar, while the volume of protein supplied is more than twice as high as the amount required. Similarly, the amount of oil supplied is three times as high as the amount needed, while dairy supply exceeds requirements by one-fifth. Fruit supply stands at less than three quarters of the required amount. Only vegetables come close to the recommended levels. The authors of this study also found that this imbalance between supply and demand, or better, need, is even worse for the least developed countries.[62] In addition, there is an issue with what dietary guidelines recommend to what is possible within the planetary boundaries. For example, increasing consumption of fish rich in omega three fatty acids may quickly deplete the natural marine resources that contain these essential nutrients. Thus, changing diets may have massive implications on our environment and planet. The IOM has recently recommended that U.S. dietary guidelines be more thoughtfully based on environmental as well as nutritional considerations and consider more sustainable diet options (see Chap. 10).[63]

In middle- and high-income countries, attention and publicity is increasing to promote better quality of diets in terms of both nutrition and sustainability. However, it is clear that the cost of such diets is currently high[64] and even basic diets in much of the world remain costlier than daily wages due to increased food prices.[65] In South Africa, a middle-income country, a nutritious and healthy diet costs 69 % more than a typical South African diet.[66] Even in high-income countries, many people cannot meet the dietary guidelines because the cost is prohibitive or consumers do not know how to shift toward better diets. In particular, fresh fruit and vegetable recommendations are difficult to satisfy because fresh produce is more

---

[60] Keats and Wiggins (2014).

[61] Keats and Wiggins (2014).

[62] Keats and Wiggins (2014).

[63] IOM (2014).

[64] Keats and Wiggins (2014).

[65] Brinkman et al. (2010).

[66] Temple and Steyn (2011) and Keats and Wiggins (2014).

expensive across the globe. Conversely, the least expensive food alternatives are often also the least healthy and least sustainable—ultra-processed, high in sugar and fat, and high in energy cost for every dollar spent.[67] To some extent, this seemingly cheaper sticker price is due to the externalization of the food producers costs (see Chap. 16). Thus, sustainable and plant-based diets geared toward public health must become less costly so that consumers can have realistic alternatives to fast and processed foods.

9.3.3 Food Safety and Food Utilization

Food safety is of utmost concern as our global food system becomes more globalized, and the "movement" of foodstuffs is increasingly reaching most geographical pockets across the planet. Food pathogen outbreaks, for instance, that occur in one area can have devastating health impacts in another. Contaminated food is, therefore, a major cause of acute diarrhea, malnutrition and mortality in low- and middle-income settings, particularly among children who become vulnerable to diarrheal diseases when transitioning from breast milk to complementary foods.[68] Solving problems of contaminated food requires a multi-disciplinary approach involving experts in clean water resources, sanitation, public health, epidemiology, nutrition and of course, agriculture, as the major food source.

Some studies suggest that environmental enteropathy afflicts many children in the developing world.[69] Environmental enteropathy is a syndrome that causes changes in the small intestine of individuals who lack basic sanitary facilities and are chronically exposed to fecal contamination. This, in turn, decreases the ability of the intestinal tract to absorb critical nutrients necessary for optimum growth and development, leading to serious consequences in nutritional status. Environmental enteropathy is often seen in young children when complementary foods are introduced along with breastfeeding. Not only are children eating more solid foods, but they also have increased exposure to the outside environment itself, leading to an increased risk of consuming contaminated foods. The provision of toilets and community led sanitation, hygiene programs, and improved systems can reduce the incidence of enteropathy.[70] Therefore, robust community public health programs are critically important to nutrition improvements and consuming food is not the only pathway to better nutrition.

Notably, food toxins can be process-induced or naturally occurring, which is considered poisonous and can cause disease. There are many types of toxins but one in particular, aflatoxin, is receiving a significant amount of attention because of its

---

[67]Aggarwal et al. (2012) and Keats and Wiggins (2014).
[68]Motarjemi et al. (1993, 2012).
[69]Humphrey (2009).
[70]Guerrant et al. (2008) and Motarjemi (2000).

deleterious effects. Aflatoxins are fungal metabolites that contaminate staple food crops in many developing countries and have been loosely associated with growth impairments in children, i.e. stunting as a measure of chronic undernutrition. Foodborne aflatoxin exposure, for instance, in maize and groundnuts, is common in Africa and Asia.[71] More evidence is needed on how post-harvest storage and handling can control aflatoxin, which could indirectly have an impact on the nutritional status of households.[72] Thus, food toxins, whether process-induced or naturally occurring, should be addressed to improve nutrition outcomes as a whole.

This area of work is critically important in the infant and young child feeding interventions, and public health, education, and agriculture all play critical roles in ensuring food safety and nutrient availability. Most of the studies examine in this field, however, the contamination of foods with Escherichia coli. There are four major categories of diarrheagenic Escherichia coli: enterotoxigenic (a major cause of travelers' diarrhea and infant diarrhea in low-income countries), enteroinvasive (a cause of dysentery), enteropathogenic (an important cause of infant diarrhea), and enterohemorrhagic (a cause of hemorrhagic colitis and hemolytic uremic syndrome). Besides manifesting distinct clinical patterns, these categories of E. coli differ in their epidemiology and pathogenesis.[73] However, succinct outcomes need to be found as well as studies conducted to determine the effect of contaminated food directly on diarrhea incidence and indirectly on anthropometry measures. A lot of work still needs to be done in exploring connection, associations and correlations in young child feeding interventions, and public health, education, and agriculture but the evidence supports the need for overall improved food safety and nutrient availability.

9.4 Bridging Sectors and Disciplines

9.4.1 Integration of Sectors Across Food and Health Systems for Nutrition

The interactions between health, nutrition and agriculture are mutual: agriculture affects health and health affects agriculture—both positively and negatively.[74] Consequently, multi-sectoral approaches are vital to ensure improvements in food security and nutrition for individuals, households and communities.[75] However, integration across diverse sectors and distinct systems is complex. There are, at a minimum, three key sectors that need to engage, collaborate and contribute to

---

[71]Khlangwiset et al. (2011).
[72]Wild (2007) and Leroy (2013).
[73]Levine (1987).
[74]Hawkes and Ruel (2006).
[75]World Bank (2012a, b).

nutrition improvements: agriculture, health and environment sectors. These three sectors link nutrition with the functioning of and effectiveness of food, health, water and sanitation systems.[76]

A robust primary health care systems approach, complemented by improved water and sanitation, can improve nutritional status. The infectious disease burden, for example, impedes consumption and the body's ability to metabolize nutrients, resulting in poor nutrient absorption and nutritional deficiencies. Consequently, one of the most important premises to improve nutrition is to control and prevent the most common childhood infectious diseases by expanding immunization programs, providing diarrhea and malaria control, treating infected patients, and decreasing parasitic burden. The backbone of some of these programs is a clean water supply, hygiene and sanitation improvements at the household and school levels.[77] The importance of clean water must not be understated and plays a role across a variety of food law and policy aspects.

Synergistically, if food production systems are inadequate, there can be negative effects on health. Thus, sound food production systems can improve the overall health of communities. After all, health is considered a primary goal and endpoint of food systems.[78] Similarly, poor health and dysfunctional health systems can limit agricultural productivity. Conversely, good or improved health and nutrition allow for improvements in agriculture outputs.[79] With improvements in agricultural production, in turn, household income can potentially increase and with that, more can be spent on household healthcare and other goods. Additional income can also be used to purchase higher-quality food toward a more diverse diet. Changes in agricultural production can also result in the introduction of new foods into diets.[80] Thus, the feedback loop between food and public health affects multiple sectors. A multi-sectoral approach, therefore, is the foremost method to address these concerns because it brings together a coherent range of strategies with the aim of enhancing food and nutrition security. These necessarily include interventions in agriculture and business development, healthcare, clean water, hygiene and sanitation, basic infrastructure, gender equality, and education.

To strengthen food and health systems and to achieve integrated synergies, there needs to be thoughtful integration between interventions or approaches, especially when an already existing collection of distinct vertical programs exists.[81] "Every intervention, from the simplest to the most complex, has an effect on the overall system, and the overall system has an effect on every intervention."[82] Services,

---

[76]Field (1987).
[77]Fanzo et al. (2014).
[78]Pinstrup-Andersen (2009).
[79]Hawkes and Ruel (2006).
[80]Hoddinott (2012).
[81]Frenk (2009).
[82]de Savigny and Taghreed (2009).

interventions and solutions that are bundled or packaged across food, health and environmental systems can be more effective and advantageous.[83]

Recent calls for greater attention to hunger and under-nutrition highlight the importance of integrating technical interventions with broader approaches to address underlying causes of food insecurity while incorporating perspectives from agriculture, health, water and sanitation, infrastructure, gender and education.[84] Such an approach would inherently build on the knowledge and on the capacities of local communities to transform and improve the quality of diets for better child health and nutrition, as well. Moreover, such approaches also highlight the interdependence and the bidirectional relationships that exist between hunger and nutrition and a host of other health development challenges.[85] Recent research, for instance, has documented potential synergies between health and economic interventions, suggesting multi-sectoral approaches which may generate a wider range of benefits than single sector approaches acting alone.[86] Even while addressing broader determinants, a political priority is to address primary causes of hunger and under-nutrition in an equitable manner.

Evidence also suggests that increasing economic growth alone, while necessary and important, is unlikely to be sufficient to address hunger and under-nutrition. Food and nutrition security are complex and require efforts across a spectrum that include enhancing food production while simultaneously increasing access and utilization with substantive political commitment to address the most vulnerable populations with an equitable, basic human rights lens approach.[87] Therefore, addressing food and nutrition insecurity are inextricably linked to the wider progress towards the new Sustainable Development Goals. Durable gains will hinge on concurrent steps to reduce poverty, improve access to education, empower women and girls, and facilitate access to basic infrastructure including safe water and sanitation, energy, transport, and communication. Working on multiple fronts simultaneously has the potential to leverage synergies and catalyze gains that extend beyond those achieved through sector specific programs working in isolation.[88]

---

[83] Fanzo et al. (2014).

[84] Pinstrup-Andersen (2009) and Garrett and Natalicchio (2011).

[85] Fanzo and Pronyk (2011).

[86] UNMP (2005) and Kim and Abramsky (2009).

[87] Bloem (2013).

[88] Fanzo and Pronyk (2011).

9.4.2 Econutrition as an Integration Model

Ecologists, nutritionists and agronomists work in multi-dimensional systems, composed of organisms, energy, and the physical environment interacting at various spatial and temporal scales, which can be described in terms of composition, structure, functions, and resilience.[89] Though many ecologists have focused on the relationship between biodiversity and ecosystem functioning, there has been little focus on the role that ecosystems play in providing the essential elements of human diets. The same can be said for nutritionists and agronomists. Usually, experts stick to their own sectors without thinking of the role of their discipline on other sectors. More research is now in progress demonstrating how the combination of environment, communities, agriculture and human beings impact human nutrition and livelihoods.[90]

Econutrition is a discipline that integrates environmental health and human health, with a particular focus on the interactions among the fields of agriculture, ecology and human nutrition.

As humans modify their environment, they select and protect some species, crops and foods and exclude and eradicate others to achieve management goals and to maximize the provisioning of ecosystem services or optimize for nutrient diversity where increasing species richness increases the capacity of the agroecosystem to meet the entirety of human nutrition needs.[91]

The notion that nutrition, human and agricultural productivity, and environmental sustainability are interrelated has thus been coined as "econutrition."[92] It is the goal of econutrition to tackle malnutrition, where much can be gained by linking agriculture and ecology to human nutrition and health. Integrated and mixed agriculture systems such as the rice-fish aquaculture systems, poultry-orchard systems, livestock-cover crop systems all provide benefits for food production, ecological, diet and nutrition.[93]

Econutrition type approaches are being piloted that integrate ecosystems services, with food production and nutrition. For example, ecological complementarity that also results in net nutritional benefit comes from the Mesoamerican "three sisters"—the combination of corn (a grass), beans (a nitrogen-fixing legume), and squash (a low-lying).[94] Integrated and mixed agriculture systems such as the rice-fish aquaculture systems, poultry-orchard systems, and livestock-cover crop systems also provide food production, ecological, diet and nutrition benefits.[95]

---

[89] DeClerck et al. (2011).
[90] Remans et al (2011c).
[91] DeClerck et al. (2011).
[92] Deckelbaum et al. (2006).
[93] Fanzo and Hunter (2013).
[94] DeClerk et al. (2011).
[95] Fanzo and Hunter (2013).

One such example of mixed farming systems was implemented in Northern Malawi. Legumes were promoted in a maize-dominated farming society to improve soil health (nitrogen fixing from legumes) while improving child nutrition (consumption of legumes rich in protein and micronutrients). The agriculture intervention was paired with consistent community and household education and the delivery mechanism was done through the local primary health care system, i.e. the local hospital. After 6 years of intensive interventions, child growth improved with underweight and stunting levels decreasing. This project demonstrated the impact that using agriculture approaches, delivered through the health system, can impact nutritional outcomes.[96] Therefore, econutrition approaches have promising effects and great potential to reverse many of the food-related problems around the world.

9.4.3 Ecological Public Health

Building on the econutrition approach, Rayner and Lang (2012) developed an "ecological public health" model. This model addresses the divisions that often happen with human health and the integrity of the natural environment. Moreover, this model links disciplinary boundaries and "champions a rebalancing of what is meant by health activity"[97] with preventative approaches being central. Four dimensions are stressed. These are:

- **The Material**—physical and energetic infrastructure
- **The Biological**—bio-physical processes and elements
- **The Cultural**—how people think and how collective consciousness applies
- **The Social**—interactions between people and their mutual engagement as collectives

Time, human action and institutions shape these dimensions. The outcomes, ultimately, that this model is geared toward are sustainable health futures. Nutrition fits squarely in this model because it plays a significant role in altering public health as well as the ecosystems that support nutrition. Nutrition "refers to massive changes not just in what people eat but in how food systems operate: turning precious food into cheaper calorie commodities which can be marketed and sold." Nutrition is not just a health problem but also a larger systemic issue that demands action from many different sectors.[98]

---

[96]Kerr et al (2010).
[97]Rayner and Lang (2012).
[98]Rayner and Lang (2012).

9.4.4 Sustainable Diets

In recent years, a number of initiatives and studies have focused more directly on the question of diets and their impacts on human health, the environment, and food systems. In 2010, the FAO led an effort to develop the following consensus definition for Sustainable Diets: those diets with low environmental impacts, which contribute to food and nutrition security and to [a] healthy life for present and future generations. Sustainable diets are protective and respectful of biodiversity and ecosystems, culturally acceptable, accessible, economically fair and affordable; nutritionally adequate, safe and healthy; while optimizing natural and human resources.[99]

While elaborate, this definition reflects the important recognition that the health of human beings cannot be isolated from the health of ecosystems.[100] Thus, sustainable diets promote environmental and economic stability through low-impact and affordable, accessible foods, while supporting public health through adequate nutrition. Importantly, sustainable diets help promote sovereignty and preserve tradition involving culturally sensitive and acceptable foods.[101]

In order to advance commitments to sustainable diets as a central aspect to sustainable development, one must address the gaps in our understanding of what constitutes a sustainable diet for different populations and contexts. Additionally, it is important to understand how these diets can be assessed within the global food system and how environmental sustainability in global consumption patterns and dietary goals can be achieved. Finally, it is necessary to examine how (sustainable) diets can help transform the health of populations, while promoting economic development and the slowing of environmental degradation.[102] Although many of these processes are underway, they are not yet receiving due political attention or support. This lack of political attention is partly due to the complex web of interactions between food systems, manufacturers, the environment, public health, and consumer behavior and the challenges policy makers face in making appropriate choices whether they are in government, commerce or civil society.[103]

9.5 Case Studies

In this next section, four case studies are presented. In three studies, a political analysis was done in several middle- and low-income countries[104] in three major regions of the world, Africa, Asia and Latin America, to determine how nutrition is

---

[99] Burlingame and Dernini (2012).
[100] Johnston et al. (2014).
[101] Macdiarmid et al. (2011) and Burlingame and Dernini (2012).
[102] Johnston et al. (2014).
[103] Lang and Barling (2013).
[104] Fanzo et al. (2014).

integrated into food and agriculture policies, with particular focus on "nutrition sensitive" agriculture policies. These case studies demonstrate the opportunities and challenges of countries undergoing a triple burden of malnutrition, and how difficult it can be to integrate nutrition, public health and agriculture in policy and in practice.

The last case study highlights a particular project, the Millennium Villages Project in integrating not only just nutrition into public health and agriculture, but also, nutrition into a wider development framework to achieve the Millennium Development Goals (MDGs). This case study illustrates how impactful integration of multiple sectors can be to achieve results in the most difficult of settings.

9.5.1 Case Study Brazil: A Country in Transition

The major food commodities grown in Brazil include coffee, oranges, and soybeans. Notably, the Brazilian diet is shifting from traditional foods that are freshly prepared to ultra-processed foods.[105] Traditional foods include rice and beans as well as roots like cassava. These foods are being replaced with foods that are energy-dense and rich in salt, sugar, and fat. In fact, it has been observed that the reduction in consumption of traditional foods as well as fish, eggs, and vegetables has coincided with increased consumption of soft drinks, cookies, sausages, alcohol, and pre-made meals. Brazil has two agricultural models that define the food system. The agribusiness model focuses on large-scale monocultures grown primarily for export. Conversely, the family-farming model focuses on smallholders and domestic, diversified production. Although, the agribusiness model is the dominant model nationally, accounting for two thirds of agrarian production, the family-farm model is growing and is responsible for 70 % of the food consumed in Brazil.[106] Further and future-oriented support for family-farms is likely to be more beneficial for econutrition and food safety in Brazil than industrial agriculture.

Brazil is currently undergoing a nutrition transition. Within the past few decades, there has been a decrease in under-nutrition as well as large increases in overweight and obesity. Simultaneously, within Brazil's policies and programs, there is a strong emphasis on increased food production focusing on better food storage, concerns for vulnerable groups, empowerment of women, and explicit nutrition objectives and indicators. Some of these policies contain elements of commitments towards the over-nutrition agenda, including nutrition education and focus on nutrient-rich foods, increase of access to markets, and promotion of dietary diversification. The resulting dichotomy between the two main agricultural models parallels the nutrition challenges in Brazil. Logically, then, there is a dominance

---

[105]Monteiro et al. (2011).
[106]de Oliviera (2013).

of the agribusiness model, which reduces the ability of policies to target the most undernourished populations. It also is linked to the consumption of highly processed foods, which lead to obesity. Essentially, the policies that focus on family farming have more nutrition-sensitive interventions, but face challenges in implementation and sustainability, especially in competition with agribusiness approaches.[107] As part of the nutrition transition in brazil, a variety of challenges must, therefore, be overcome.

Brazil is an upper middle-income South American country with a population of 190 million and a rapidly growing economy. On average, per capita income increased by 22 % between 2004 and 2008 and roughly 30 million people have entered the middle class. Currently, Brazil is moving through the second stage of the nutrition transition. This means that most people have access to adequate calories, but not to adequate amounts of micronutrients. The typical diet in Brazil is unfortunately transitioning away from traditional and minimally processed foods—such as the combination of rice and beans—to energy-dense processed foods that are low in micronutrients.

Nevertheless, Brazil has made significant progress in combating chronic undernutrition. Between 1989 and 2006, the prevalence of childhood stunting under 5 years of age fell from 19.6 to 6.7 %. Similarly, the prevalence of underweight among children under 5 years of age decreased from 5.4 to 1.8 %. However, the prevalence of under-nutrition is higher among low-income groups and among traditional peoples and communities.

As is typical for a country moving through the second stage of the nutrition transition, the decrease in the prevalence of under-nutrition has been accompanied by an increase in the prevalence of overweight and obesity. Between 1989 and 2006, the prevalence of overweight adults and adolescent males increased from 29.9 to 50.1 %. Correspondingly, the prevalence of obesity increased from 5.4 to 12.4 %. Over the same time period, the prevalence of overweight females increased from 41.4 to 48.0 % and the prevalence of obesity increased from 13.2 to 16.9 %.

An UNSCN analysis of the nutrition sensitivity of food and agriculture policies and programs in Brazil reviewed nine policies, plans, and programs in various administrative sectors, which provide helpful examples in support of the aforementioned policies and food-related concerns outlined in this chapter.

9.5.1.1 Areas of Focus

There are two distinct and often contradictory models of agriculture in Brazil, namely the agribusiness model and the family-farm model. The agribusiness model accounts for 26.9 % of Brazil's GDP, and enjoys substantial political and financial support. The family farm model only accounts for 33 % of agrarian production, but it employs 74.4 % of rural workers and is responsible for 70 % of

---

[107] de Oliviera (2013).

the food consumed in Brazil. In general, policies and programs focused on the agribusiness model are much less nutrition sensitive than policies and programs focused on the family farm model.

Except for the Agriculture and Livestock Plans, all of the policies analyzed for this case study are associated with the family-farm model.

National Food and Nutrition Security Policy and Plan

The National Food and Nutrition Security Policy and the accompanying implementation plan take a comprehensive approach to improving food and nutrition Security in Brazil. The policy recognizes adequate food as a human right and seeks to sustainably improve access to food, especially among the most vulnerable members of society. Other policy objectives include: the creation of nutrition education processes, development and promotion of sustainable food systems, and increased integration of food and nutrition in all levels of health care. The plan also includes interventions targeting food production and supply, healthy eating education, and strengthening of family farming.

The Harvest Plan for Family Farming

The Harvest Plan for Family Farming is the overarching plan for the implementation of agricultural policy through the Ministry of Agriculture and Development. Its goal is to increase production, income, and use of technology within the family farming model. There are a number of nutritionally sensitive programs within the Harvest Plan for Family Farming, including the National Program for Strengthening Family Farming (PRONAF) and the Food Purchase Programme (PAA).

While all of the above programs meet many of the nutrition-sensitive criteria, the PAA, in particular, is an innovative program that simultaneously achieves improved food security for family farmer food producers and food insecure individuals while also securing additional funding for further food sensitive interventions. The PAA facilitates government purchase of nutritious food from family farms outside of the administrative procurement protocol typical for government purchases. This creates a source of dependable income for small-scale family farmers and improves food security through poverty reduction. The purchased food is then distributed to food insecure households and individuals as well as government institutions including hospitals, health care centers, and schools, where the food is used in the National School Feeding Programme. Finally, the program also stipulates that any income generated from the sale of food purchased through the PAA must be used solely for programs that combat hunger and improve FNS.

9.5.1.2 Successes, Challenges and the Way Forward in Brazil

Brazil has made great progress towards combating chronic malnutrition and promoting food and nutrition security. The right to adequate food is recognized in the Brazilian Constitution, and it is clear that there is substantial drive to see this right fulfilled both on the part of the government and that of civil society.

Nonetheless, the current policy in Brazil poses substantial challenges to the advancement of food and nutrition security and efforts to reduce the prevalence of overweight and obesity. Three such challenges are the apparent dichotomy between the agribusiness and family farming models, the struggle to achieve meaningful land reform, and the powerful lobby and legislative forces opposing some of the proposed food and nutrition initiatives. In the face of these challenges, however, the policies and programs analyzed are generally nutrition sensitive and provide a useful roadmap for improving nutrition in Brazil.

- *Agribusiness versus Family Farming*
  Stakeholders interviewed assert that the agribusiness model, which is dependent on monocultures and the extensive use of pesticides and genetically modified organism seeds, does not coexist harmoniously with the family-farm model. Agribusiness has had large positive effects on Brazil's economic growth and is powerful and well funded. Many stakeholders see the decision to prioritize agribusiness as diametrically opposed to promoting food and nutrition security.
- *Land Reform*
  Many stakeholders identified land concentration as a major hurdle to guaranteeing food and nutrition security among the most vulnerable groups in Brazil. While traditional and indigenous groups have access to social support programs, the right to land is critical for them to ensure long-term, sustainable food security. There has been progress toward family farming and settling of landless families since 2003, but the agribusiness model favors land concentration and, in recent years, land reform has been removed from the government's agenda.
- *Regulation and Legislative Challenges*
  Civil society plays a very important role in Brazil, and is in part responsible for many of the food and nutrition security advancements in the country, such as through the National Food and Nutrition Security Council (CONSEA). However, agriculture laws, as well as food and nutrition policies and regulations sometimes face powerful opposition from pro-agribusiness lobbyists and politicians in addition to wealthy national and transnational corporations (BigAg and BigFood). This is evidenced by the successful derailment of a regulation that intended to set limits on the advertisement of foods with low nutritional values to children.

9.5.1.3 Moving Toward Nutrition-Sensitive Agriculture in Brazil

All of the food and agriculture policies analyzed in this case study are nutrition sensitive to some degree. The plans that are related to promotion of the family farming model are especially nutrition sensitive. As a group, the policies and plans are strongest when it comes to taking a sustainable approach to improved food and nutrition security, increasing food production, targeting the most vulnerable, expanding access to markets, and improving food processing and storage. The current policies and plans, however, lack sufficient emphasis on increasing production of nutrient rich foods, improving processing to retain nutritional value, reducing post-harvest losses, and integrating nutrition education.

9.5.2 Case Study Nepal: A Landlocked, Post-Conflict Country with Great Momentum

In Nepal, the major crops are paddy (unmilled rice), wheat, maize, millet, barley, and legumes. Cereal crops are predominant in agricultural production, with 72 % of all agricultural households cultivating paddy, 64 % cultivating maize, and 57 % cultivating wheat. The Nepali diet varies depending on the landscape—the Terai (fertile lowland plains) population consumes rice and wheat, the Hill population consumes maize and millet, and the Mountain diet consists largely of millet, maize, and barley. The main Nepali cash crops grown are sugarcane, oilseeds, potatoes and pulses. Thus, the agricultural sector has diversified into fruits, vegetables, spices and condiments.

Overall, while the food system still focuses on the production of cereal crops, the percentage of households with holdings of fruits and vegetables is increasing. This trend has supplemented and changed farmer income, as well as increased the nutrition content of domestically produced foods and diets. However, this trend would have greater effect with an increase in the share of land used to cultivate the aforementioned crops.[108]

The government of Nepal has also a demonstrated commitment to improving food and nutrition security, which is evident in the planning structure and policies across sectors that emphasize the need for improved nutrition. tor instance, The country's programs and policies look toward diversification of production, increase of production of nutritious foods, improvements in post-harvest processes, and increases in women's income. These programs and policies could be further strengthened by focusing on additional nutrition education, managing natural resources, and empowering women through multiple channels, such as improving labor and time-saving technologies, access to extension services, and supporting their rights to land and employment.

---

[108]Fanzo and Andrews (2013).

However, there are also areas where nutrition-sensitive policies in Nepal are particularly weak. For example, there is little effort to assess the context and cause of malnutrition at the local level and to incorporate local insights and observations into sub-national planning. The Food and Nutrition Security Plan focuses on vulnerable groups, but there could be additional policy support to expand markets and market access to these groups. The plans could also elaborate on specific measures needed to increase equitable access, availability, and consumption of quality food, particularly in those areas that are geographically difficult to reach.[109]

In summary, Nepal is a low-income country with a population of about 27 million. It is located in southern Asia and is bordered by China on the north and India on the east, south, and west. Agriculture dominates Nepal's economy, accounting for 34 % of the GDP and employing 70 % of the workforce. Nepal is currently in the first stage of the nutrition transition, meaning that the typical diet is low in calories and micronutrients and under-nutrition is prevalent. Staple foods, such as grains including rice, wheat and maize, which are high in energy but low in micronutrients, account for 72 % of the caloric intake of the typical Nepalese diet.

Nepal has made significant strides in improving the nutrition situation over the past decade, reducing the prevalence of stunting for children under 5 years of age from 57 to 41 %. Moreover, the prevalence of underweight for children under 5 years of age fell from 43 to 29 %, and the prevalence of maternal anemia by 50–23 %. However, the Government of Nepal (GoN) recognizes that chronic malnutrition is still a serious problem. The major policies analyzed in the UNSCN case study seek to address this problem through a variety of nutrition-specific and nutrition-sensitive interventions.

A UNSCN analysis of the nutrition sensitivity of food and agriculture policies and programs in Nepal reviewed ten policies and programs in various administrative sectors. The main areas of focus are outlined in the following section.

9.5.2.1 Areas of Focus

The National Planning Commission (NPC), which is the advisory body for formulating development plans in Nepal, is responsible for leading the coordination of the three main plans: (1) the Multi-sectoral Nutrition Plan for Nepal (MSNP), (2) the Agriculture Development Strategy (ADS), and (3) the Food and Nutrition Security Plan of Action (FNSP).

Multi-Sectoral Nutrition Plan for Nepal (MSNP)

The MSNP sets specific reduction goals for the prevalence of stunting, underweight, and wasting among children under five and undernutrition among women

---

[109]Fanzo and Andrews (2013).

ages 15–49. In essence, the plan intends to accomplish these goals through interventions that focus on reducing diarrheal and other diseases that inhibit nutrition absorption, providing nutrition-focused maternal education, increasing the availability and consumption of nutrient-dense foods, and expanding capacity of national and local government to improve maternal and child nutrition among other factors. Under the MSNP, the Ministry of Agriculture and Development is responsible for increasing "consumption of diversified foods, especially animal source foods, particularly among pregnant women, adolescent girls, and young children".[110] They intend to increase production of foods rich in micronutrients, promote ideal Infant and Young Child Feeding (IYCF) practices, expand the percentage of children receiving immunizations and micronutrient supplements, and improve the distribution systems to reach subsistence farmers in rural areas.

Upon critical examination, the MSNP could be stronger, from an agricultural perspective, if it focused more on consumption and utilization activities of food security, indigenous food's role in improving nutrition, integration of food technology in improving IYCF practices through nutrient-dense complementary foods, and working toward food-based dietary guidelines and the introduction of a food labeling system.

Agricultural Development Strategy (ADS)

The ADS is long-term strategy to increase agricultural sector growth over the next 20 years. It focuses on four strategic components: governance, productivity, profitable commercialization, and competiveness. At the same time, ADS strives to promote inclusiveness, sustainability, multi-sectoral development, and market connectivity infrastructure. The ADS assessment report demonstrates a clear understanding of the difference between food sufficiency and food and nutrition security. Notably, nutrition remains 1 of the 12 thematic focuses of the ADS, while improving food and nutrition security are even included in the ADS vision statement. Thus, the four strategic components of the ADS are supposed to improve food and nutrition security both directly as well as indirectly through poverty reduction, agricultural trade surplus, and higher income for rural households. Nonetheless, the focus on profitable commercialization within the ADS remains cause for concern, as the commercialization of rice directly contradicts efforts within the FNSP to diversify diets. Additionally, the budget for agriculture in Nepal has historically been low and it is unclear if the plans can be accomplished without a significant increase in funding.

---

[110]MSNP (2012).

Food and Nutrition Security Plan of Action (FNSP)

The FNSP was developed as a collaborative effort between the GoN and the Food and Agriculture Organization to ensure that food and nutrition security was a part of the ADS. The FNSP is a 10-year plan that is intended to be the government's primary document for food security interventions. It will serve as a complement to the ADS and eventually become an entity of the ADS.

The FNSP seeks to reduce hunger and poverty by improving sustainable agriculture-based livelihoods, especially among Nepal's poorest households. Thus, the nine components of the FNSP focus mainly on increasing food availability and are:

(1) Agriculture Crops,
(2) Fisheries,
(3) Food Quality and Safety,
(4) Forestry,
(5) Gender Equity and Social Inclusion,
(6) Horticulture,
(7) Human Nutrition,
(8) Legislation, and
(9) Animal Health and Production.

The FNSP fills in gaps in the ADS by focusing on the most vulnerable and by promoting diversification of production systems. However, the FNSP would benefit from additional focus on access to and utilization of foods at the household level. It is the focus on production that ignores consumer-side factors that affect nutrition such as affordability, purchasing power, consumption and behavioral changes.

9.5.2.2 Successes, Challenges and the Way Forward in Nepal

The GoN has aggressively pursued ambitious policies to address chronic malnutrition in the country. Political will to improve the nutrition situation is critical to success, and the GoN has showed the commitment necessary to make a positive impact on chronic malnutrition. One demonstration of this success was the recently drafted Multi-sectoral Nutrition Plan for Nepal that truly represents multiple sectors of the GoN.

All three major policies have explicit nutrition objectives, nutritional impact measurements within the monitoring and evaluation systems, and opportunities for multi-sectoral collaboration. These plans all include activities or interventions that increase food access by diversifying production and income, increasing production of nutritious foods (with a focus on local foods rich in micronutrients and protein), improving processing and reducing post-harvest losses, increasing market access, and improving storage and preservation of food. The plans could be strengthened by a greater focus on incorporating nutrition education into interventions, long-term management of natural resources, and empowering women through increasing income, improving labor technologies and supporting their right to land, education, and employment. Finally, the plans are also weak in assessing the causes and context of malnutrition at the local level to maximize the effectiveness of interventions within the heterogeneous localities in Nepal and increasing equitable access to resources.

While the three major policies and plans analyzed in this case study have nutrition-sensitive elements, Nepal faces implementation challenges including a lack of capacity and insufficient coordination between plans and ministries. There is a lack of nutrition-related human resources at all levels of government in Nepal, which is a major obstacle to effective implementation of nutrition-sensitive interventions. The Government of Nepal is aware of this scarcity and is working with NGOs and the donor community to build the necessary capacity at the local and central levels. Even the MSNP and the ADS have built-in capacity objectives to help address this gap.

There is a large potential for collaboration within the plans, but it is not immediately evident that this potential is being fulfilled. Many stakeholders are unaware of their role in the MSNP and the ADS. The plans also do not take advantage of many opportunities for multi-sectoral coordination. In fact, the ministries are seen as secondary and are underutilized. For example, the Ministry of Education is not engaged in the nutrition education initiatives of the FNSP.

Toward Nutrition-Sensitive Agriculture in Nepal

All of the plans are ambitious with many outcome measures and target groups. To make substantive improvements, Nepal must focus on several key populations: children under two, pregnant and lactating women and the landless. If nutrition actions focused on these three populations within Nepal, coordination and impact of the plans would be optimized.

High-level government officials play a decisive role in the successful implementation of these plans. These official must coordinate all actions across ministries and government offices, channel donor and civil society efforts, and develop compelling narratives around nutrition as a poverty reduction priority. However, issues that repeatedly emerge include transient government and mandates which prove challenging for Nepal. Without a constitution, stable government and long-term positions in ministries, priorities shift. If Nepal can make a measurable impact in a short time with these new plans, it is in the best interest for Presidents to continue the work. It is also important for food and nutrition security to be embraced as a major objective of long-term national development strategies.

Finally, long lasting change takes time. Nepal's current food and agriculture plans are ambitious, and commendable. At the same time, Nepal is a young country, and faces a long path towards development and economic security. Under-nutrition reduction take time. With that said, nutrition goals and targets should be aggressive, but also realistic and achievable in the appropriate time scales.

9.5.3 Case Study Senegal: Improving Subsistence Agriculture Through Better Nutrition

As a largely rural country, Senegal relies predominantly on rain-fed, subsistence agriculture. Despite the fact that 75 % of the workforce is represented in the agricultural sector, Senegal remains a net food importer, particularly of rice. The main vegetable cash crops are green beans, tomatoes, melons, and mangoes. Other major commodities include peanuts, cotton, grains, and fish. In fact, fishing offers the biggest contribution to the Senegalese economy. Senegal has a traditionally diverse diet, including several forms of grains (millet, sorghum, rice), proteins (fish, goat, beef, ox), vegetables (carrots, lettuces, leaves), and starches (sweet and regular potatoes, cassava). Soil fertility and water issues are major barriers to agriculture in Senegal.

The Senegalese diet includes fruits, vegetables, meats, and grains. However, a primary constraint for Senegalese agriculture is water. The country depends on water as an agricultural input, but its proximity to the increasingly dry Sahel region makes the country subject to inconsistent rainfall and frequent droughts.[111] There are some approaches addressing the nutrition-vulnerable population and the use of agricultural programs as vehicles to deliver nutrition interventions but these are poorly developed in the agricultural sector.

Senegal's policies contain a number of objectives targeting vulnerable populations, empowering women, increasing production and diversification, improving the processing of agricultural products, and collaborating with other sectors. Policy commitments are nascent and clear nutrition objectives are absent in the national agricultural policy. Production of nutrient-rich foods, including nutrient value preservation, reduction in post-harvest losses, nutrition promotion and education, as well as market expansion and access carry the least focus in the analyzed policies. Coordinated action for nutrition has been high on the political agenda for the country for a long time, yet coordination is weak and delivery platforms are not used effectively.[112]

Thus, Senegal has followed many African neighbors by steadily improving life expectancy and health outcomes since 1995. Mortality for children under 5 years of age has decreased on average 6.4 % annually since 2000. Disparities exist particularly between rural and urban populations in Senegal. For instance, Senegalese children in rural areas face a 2.4-fold increased risk of dying compared to children who are living in an urban environment.

Malnutrition still underlies approximately one third of all child mortality. Overall, under-nutrition has decreased, from 28.5 % stunted and 16.4 % wasted in 1986 to 15.5 % stunted and 9.7 % wasted in 2012 in children under 5 years of age. Both wasting and stunting are much more prevalent in rural areas than near Dakar and

---

[111]Lachat (2013).
[112]Lachat (2013).

other cities, where overweight and obesity is rising. In fact, a study from a rural, central region showed that 15 % of surveyed children were severely deficient in iodine, showing a lack in coverage or consumption of iodized salt. Vitamin A coverage is very high across Senegal, reaching 97 % of children under 5 years of age in 2009. Anemia was estimated to be higher than 40 % in 2005 with more than 80 % of children under 5 years of age, almost 60 % of women and more than 70 % of pregnant women being affected. Anemia rates improved to 34 % of women in reproductive age in 2012.

A UNSCN supported analysis of the nutrition sensitivity of food and agriculture policies in Senegal reviewed 13 policies in various administrative sectors.

9.5.3.1 Areas of Focus

Senegalese agricultural policies and programs have a strong emphasis on food security. Nutrition objectives are largely absent in agricultural policies, even though there are several elements of the key recommendations on nutrition-sensitive agriculture built in to policy papers on food security. On average, the policies incorporated many of the key recommendations. The best covered item across policies is a sustainability approach, referring to the maintenance or improvement of the natural resource base, i.e. water, soil, air, climate and biodiversity. Conversely, the highest scoring policies are the Food Security Programs. For instance, the National Strategy and Priority Programmes for Food Security and the National Strategy for Food Security is rich in nutrition-sensitive approaches with a few exceptions, such as a focus on production of nutrient-rich foods and expanding markets, on the one hand, and access of nutrient rich foods, on the other hand.

9.5.3.2 Senegal's Agricultural Pastoral Orientation Law

The main agricultural program in Senegal is the Agricultural Pastoral Orientation Law. This law is by far the most robust national policy paper on Agriculture in Senegal, but it does not include key nutrition components or objectives. The main objective of the current agricultural programs is to ensure availability of food, which also aims to diversify food production in the country. Key informants noted that the current agricultural programs in which they were involved had no explicit nutritional goals. The main objective of the current agricultural programs is to ensure availability of food, with a first level of post harvest transformation. In addition to this, the programs aim to diversify food production in the country. Thus, regulators generally perceived food security or dietary diversity as the finality of their work, but essentially looked at this from an angle of food production and food availability.

The present administrative organization of the government of Senegal is not conducive to joint nutrition and agriculture programming and policy implementation. Agriculture, in the large sense, falls under the Ministries of Agriculture and Rural Equipment, Livestock and Fisheries and Maritime Affairs, while human nutrition is a matter for the Ministry of Health and Social Action. The current agricultural programs generally do not target the nutritional vulnerability or nutritional profile of communities.

A formal structure called the *Call Against Malnutrition (CLM)* was established in 2001 that reports directly to the Prime Minister's Office and was tasked with nutrition coordination at the national level. The CLM coordinates its activities with seven Ministries (Health, Education, Economy and Finance, Decentralization, Trade, Industry and Agriculture), National Association of Rural Advisors and the Civil Society. Senegal signed up to the *Scaling Up Nutrition Movement* in 2011 and aims to accelerate investment in nutrition, especially through the involvement of the Agricultural sectors. Senegal has begun to develop of a new national nutrition policy as of 2013.

In 2011, the government pledged to increase annual funding for nutrition to 2.8 billion FCFA per year in 2015. This direct investment will be strengthened to ensure full coverage of children and women in effective nutrition intervention programs. Following the *National Policy Paper on Nutrition* in 2001, the country is currently initiating the development of a multi-sectoral strategic plan for nutrition, called "*Lettre de Politique de Nutrition*" for 2013–2018. Some of the policies have specific monitoring and evaluation systems.

9.5.3.3 Successes, Challenges and the Way Forward in Senegal

There is a clear recognition from the highest levels of government in Senegal that nutrition is important for the development of a healthy nation. Stakeholders understand that nutritional goals can be built into agricultural plans at a national level, and they are willing to fund proven interventions. Overall, the nutrition sensitivity of the agricultural policy documents integrated several key recommendations but missed out on others. Targeting the vulnerable population groups, empowerment of women, increases of the production, diversification and improvement of processing of agricultural products, collaboration between sectors and sustainability approaches were all present in the large majority of the policy documents. Current agricultural programs are also reported to engage and target women in terms of wellbeing, empowerment and livelihoods.

Current agricultural programs, however, do not have explicit nutritional goals and are not monitored using nutritional indicators. Technical agencies collaborate typically at the implementation level and there is little joint thinking to share experiences and inform policy development upstream. Stakeholder interviews showed a misunderstanding of what nutrition is within the agricultural sector. Most of the respondents stated that they incorporated nutrition in their programs, as they (1) worked with food scientists for primary transformation of agriculture produce, (2) simply produced the food that people eat, or (3) looked at food safety, Such as postharvest reduction of aflatoxins in peanuts. Most of the programs barely considered how agricultural production was used in dietary intake. Although interviewees reported that data was collected on these concerns, the compilations were information on national food consumption levels, not individual food consumption data. In terms of dietary quality, the concerns of the respondents were mainly focused on ensuring enough dietary protein intake, dietary diversification, or increasing food availability.

Senegalese agricultural policies lacked aspects of incorporating nutrient-rich foods, nutrient value preservation, and preservation of nutritional quality of produce. Areas that were also weak within policies were reduction of post-harvest losses, nutrition education and promotion, food storage improvements, and market expansions and market access. All of these components were missing in more than half of the policies reviewed.

Various regions of the country suffer from persistent high rates of malnutrition despite a significant increase in agricultural productivity and income. Current agricultural programs insufficiently consider nutritional aspects and utilization of crops. Food availability at the macro level (regional—national level) has received the bulk of the attention of the agricultural sector but food availability at the individual level has received much less.

Moving Towards Nutrition-Sensitive Agriculture in Senegal

There are a number of experiences with value chain approaches within the food system in Senegal, such as fruit and vegetable value chain or innovation platforms for the incorporation of local cereals in bread. The choice selection of seed varieties, for example, is conducted on the basis of commercial, i.e. yield, pest resistance and appearance indicators only. Consequently, promoting crop varieties on the basis of micronutrient composition is considered a promising strategy to address micronutrient deficiencies and promote local foods.

Finally, there is a general willingness and enthusiasm to incorporate nutrition objectives into the overarching agricultural framework for Senegal. Nutrition can be built into the Agricultural Pastoral Orientation Law as a formative direction towards nutrition-sensitive agriculture. The initiative to develop a Policy Letter on Nutrition and the upcoming revision of the Orientation Law are opportunities to institutionalize nutrition-agricultural linkages in Senegal. Building nutrition capacity among government leaders in various sectors, particularly in agriculture, will address the knowledge gap and confusion that exists around nutrition-sensitive agriculture. Currently, nutrition is too poorly understood by the various professionals at the agricultural ministries to enable proactive dialogue.

9.5.4 Case Study Millennium Villages Project: Integration of Food and Health Systems for Nutrition

The Millennium Villages Project is attempting to address the root causes of extreme poverty by taking a holistic, community-led approach to sustainable development. In essence, the project focuses on the community level, through community-led development. With the help of new advances in science and technology, project personnel work with villages to create and facilitate sustainable, community-led action plans that are tailored to the villages' specific needs and designed to achieve the Millennium Development Goals (MDGs). The MDGs, in turn, reflect an understanding of the many interconnected factors that contribute to extreme poverty. Additionally, the MDGs include time-bound and measurable targets to address income poverty, hunger, disease, lack of adequate shelter and exclusion—while promoting gender equality, education and environmental sustainability.

The Millennium Villages (MVs) are demonstration and testing sites for the integrated delivery of science-based interventions in health, education, agriculture and infrastructure. By integrating these different sectors and interventions, it was originally thought that nutrition improvements would be seen. The aim of the Project is to accelerate progress towards the MDG targets, including MDG 1—to eradicate extreme poverty and hunger. The range of interventions adheres to a cost ceiling of $110 per capita sustained over a 5–10 year period, reflecting the full value of contributions from government, external donors, local communities, and the Project itself.[113]

Solutions like providing high-yield seeds, fertilizers, medicines, drinking wells, and materials to build school rooms and clinics and improving food and nutrition security. Improved science and technology such as agroforestry, insecticide-treated malaria bed nets, antiretroviral drugs, the Internet, remote sensing, and geographic information systems is included to enrich this progress. Over a 10-year period, community committees and local governments build capacity to continue these initiatives and develop a solid foundation for sustainable growth. To date, the Millennium Villages project has reached over 500,000 people in 79 villages. Clustered into 12 groups across 10 African countries (Ethiopia, Ghana, Kenya, Malawi, Mali, Nigeria, Rwanda, Senegal, Tanzania, and Uganda), the villages are located in different agro-ecological zones that reflect the range of farming, water, and disease challenges facing the continent.

The MVs are situated in 'hunger hotspots', where at least 20 % of children are malnourished and where severe poverty is endemic. The MVs were chosen to reflect a diversity of agro-ecological zones, representing the farming systems found in over 90 % of sub-Saharan Africa. Sites range from slash-and-burn in rainforest margins to pastoralism in deserts, reflecting varied levels of population density, soil conditions, climate instability, water access, disease profiles,

---

[113]Sanchez et al. (2007) and Remans et al (2011a).

environmental degradation, nutritional deficiencies and food availability, market access, education levels, cultural traditions and religious norms.[114]

The MVP model employs a three-pronged food and nutrition security approach. First, *clinical interventions* focus on persistent macro and micronutrient deficiencies in children, including vitamin A supplementation, treatment of severe acute malnutrition and regular growth monitoring. For cases of moderate malnutrition, families receive nutrient-rich flour and other food commodities. In addition, basic maternal health interventions such as basic antenatal care and institutional delivery are supported by efforts to promote adequate weight gain and improve coverage with iron and folic acid supplementation. These interventions are "core" to most nutrition programs in the developing world and for the most part, sit in the public health sector's responsibilities. However the MVP three-pronged nutrition approach called for more integration of other sectors to address the root causes of undernutrition.

This integration of other sectors was found in the second and third prongs. Second, *education and behavior-based interventions* include homegrown school meals programs, gardens and nutrition activities after school, along with de-worming and environmental enteropathy reduction campaigns. Balanced school meals have been demonstrated both to increase school attendance as well as improve learning outcomes. Food and nutrition education and increased knowledge for women is also a critical intervention addressed. As previously shown, such interventions are important to food safety and food utilization for optimal nutrition outcomes.

Finally, *household, community and livelihood-based interventions* engage longer-term realities of food and livelihood security. These consist of subsidized seed and fertilizer interventions to increase agricultural productivity; the introduction of high-value crops, and; agro-processing initiatives and microfinance programs to stimulate small-business development. Taken together, these efforts are an attempt to enhance nutritional intake and diet diversity, while affording households the additional income required addressing nutritional needs in a sustainable fashion. A community health worker program to promote exclusive breastfeeding and locally appropriate complementary feeding, home-based fortification, and proper food storage techniques complemented this approach. Without integrating the "food and agriculture" element, nutrition and food security achievements in these communities would be sub-par.

Three years after the start of the program in 2005–2006, consistent improvements were observed by researchers at Columbia University in household food security and diet diversity, whereas coverage with child care and disease-control interventions improved for most outcomes. The prevalence of stunting, or chronic undernutrition in children <2 years old at year 3 of the program was 43 % lower (adjusted OR: 0.57; 95 % CI: 0.38, 0.83) than at baseline. These findings provide encouraging evidence that a package of multi-sectoral interventions has the

---

[114]Sanchez et al. (2007) and Remans et al (2011a).

potential to produce reductions in childhood undernutrition.[115] What these data show is that multi-sectoral approaches are vital to ensure improvements in food security and nutrition for individuals, households and communities. By integrating different sectors—such as livelihood and agriculture based interventions, public health and education—major and rapid reductions in undernutrition can be made, in poor settings.

The Millennium Villages and other project that integrate sectors to improve nutrition have been important in informing larger scale initiatives such as the Millennium Development Goals Achievement Fund. The Millennium Development Goals Achievement Fund (MDG-F) was established in 2007 as a comprehensive development cooperative mechanism to help achieve the Millennium Development Goals (MDGs). The program was large—with 130 joint programs in 50 countries with the highest burdens of undernutrition—working with different UN agencies, governmental institutions, the private sector, communities and civil society entities. The Children, Food Security and Nutrition thematic area was the largest of the MDG-F and received over US$ 135 million to support 24 joint programs, implemented through the collaboration of several UN agencies.[116]

The joint programs of the MDG-F promoted multi-sectoral coordination to address food and nutrition security. Three case studies were analyzed for Peru, Brazil and Bangladesh by Levinson and Balarajan (2013). According to the authors, there were three major findings. First, convergence is important. Convergence that combines nutrition-specific and nutrition-sensitive interventions can be jointly targeted to vulnerable geographical areas and populations living in these areas. Second, the results-based incentives to sub-national governmental bodies and encouragement of more proactivity and accountability to demonstrate results related to reductions in undernutrition are critically important. Third, active and sustained civil society advocacy at the policy and programmatic levels is essential. At the policy level, this advocacy serves to ensure political and administrative commitment to nutrition and food security. At the programmatic level, advocacy helps to ensure adequate budgeting, well-designed and implemented programs that meet the needs of the population.[117]

9.6 Conclusion

The global community needs to better engage across the key sectors of agriculture, health and environment/ecology to improve nutrition. This engagement does not just need single, vertical interventions but bundled solutions that engage and revitalize food and health systems. More solutions are needed that ensure resilience

---

[115]Remans et al. (2011b).
[116]Levinson and Balarajan (2013).
[117]Levinson and Balarajan (2013).

of communities and sustainability of not only our food production and supplies, but also our diets, that take into account climate variability, food safety and population pressures.

New approaches and ways of thinking, that integrate sectors, such as the sustainable diets approach, econutrition and multi-sectoral integration could all be solutions to make nutrition improvements. The four case studies help illustrate how sectors can be integrated, albeit not without challenges. In the case of Brazil, Nepal and Senegal, there is commitment to integrate nutrition into agriculture and food policies but more needs to be done.

We are rapidly approaching 2015 and the shift to a post-2015 agenda. The Millennium Development Goals (MDGs) have brought much-needed attention to a number of priority areas in sustainable development policy however nutrition must be more central to the post-2015 goals and the strategies put forward to achieve food security. This will require countries to position nutrition objectives explicitly within their broader public health and agriculture agendas.

The case studies demonstrated that there is some level of commitment to achieving positive nutrition outcomes, as well as an understanding, to varying degrees, that the agricultural public health sectors have pivotal roles in achieving nutrition objectives. As we move forward into the post-2015 era, good practices and transferable lessons can be drawn from each country case study. The studies collectively highlight the importance of a supportive policy environment, well-developed human resources, and effective systems for planning, implementation, and monitoring impact for creating successful, nutrition-sensitive agriculture policies and programs.

References

Aggarwal A, Monsivais P, Drewnowski A (2012) Nutrient intakes linked to better health outcomes are associated with higher diet costs in the US. PLoS One 7(5). doi:10.1371/journal.pone.003753

Black RE, Victora CG, Walker SP et al (2013a) Maternal and child undernutrition and overweight in low-income and middle-income countries. Lancet 382:427–451

Black et al (2013b) Maternal and child undernutrition and overweight in low-and middle-income countries: prevalences and consequences, The Lancet Launch Symposium (June 6, 2013 London). Available at http://download.thelancet.com/flatcontentassets/pdfs/nutrition_2.pdf. Last accessed May 2014

Bloem M (2013) Preventing stunting: why it matters, what it takes. In: The road to good nutrition. DSM, Switzerland, Chapter 1

Brinkman HJ, de Pee S, Sanogo I, Subran L, Bloem MW (2010) High food prices and the global financial crisis have reduced access to nutritious food and worsened nutritional status and health. J Nutr 140(1):153S–161S

Burlingame B, Dernini S (eds) (2012) Sustainable diets and biodiversity—directions and solutions for policy, research and action. In: Proceedings of international scientific symposium on biodiversity and sustainable diets: united against Hunger. FAO, Rome, 3–5 November 2010

Costello A, Abbas M, Allen A, Ball S, Bell S, Bellamy R, Friel S, Groce N, Johnson A, Kett M, Lee M, Levy C, Maslin M, McCoy D, McGuire B, Montgomery H, Napier D, Pagel C, Patel J, de Oliveira J, Redclift N, Rees H, Rogger D, Scott J, Stephenson J, Twigg J, Wolff J, Patterson C (2009) Managing the health effects of climate change: lancet and University College London Institute for Global Health Commission. Lancet 373(9676):1693–1733

de Oliviera SI (2013) UNSCN Brazil Case Study. UNSCN, Geneva

de Savigny D, Taghreed A (eds) (2009) Systems thinking for health systems strengthening. Alliance for Health Policy and Systems Research, WHO, Geneva

Deckelbaum R, Palm C, Mutuo P, DeClerck F (2006) Econutrition, Implementation models from the Millennium Villages Project in Africa. Food Nutr Bull 27:335–343

DeClerck F, Fanzo JC, Remans R, Palm CA, Deckelbaum R (2011) Ecological approaches to human nutrition. Food Nutr Bull 32:41S–50S

Fanzo J, Andrews D (2013) UNSCN Nepal Case Study. UNSCN, Geneva

Fanzo J, Hunter D (eds) (2013) Diversifying food and diets, using agricultural biodiversity to improve nutrition and health. Earthscan as part of the series issues in agricultural biodiversity

Fanzo J, Pronyk PM (2011) A review of global progress toward the Millennium Development Goal 1 Hunger Target. Food Nutr Bull 32:144–158

Fanzo J, Marshall Q, Wong J, Merchan RI, Jaber MI, Souza A, Verjee N (2013) The integration of nutrition into extension and advisory services: a synthesis of experiences, lessons, and recommendations, Global Forum for Rural Advisory Services, Switzerland. Available at http://www.fsnnetwork.org/sites/default/files/gfras_nutrition_report.pdf. Last accessed May 2014

Fanzo J, Lachat C, Sparling T, Olds T (2014) The nutrition sensitivity of agriculture and food policies: a summary of eight country case studies. The UNSCN News 40: page 19, Geneva

FAO (1996) Food security definition at World Food Summit. FAO, Rome, Italy

FAO (2011) Resilient livelihoods – disaster risk reduction for food and nutrition security framework programme. Rome, Italy

FAO (2013) State of food and agriculture. Food and Agriculture Organization of the United Nations, Rome

Faye O, Baschieri A, Falkingham J, Muindi K (2011) Hunger and food insecurity in Nairobi's slums: an assessment using IRT models. J Urban Health 88(2):235–255

Field J (1987) Multisectoral nutrition planning, a postmortem. Food Policy 12:15–28

Frenk J (2009) Reinventing primary health care, the need for systems integration. Lancet 374:170–173

Garrett J, Natalicchio M (2011) Working multisectorally in nutrition : principles, practices, and case studies. IFPRI, Washington

Guerrant RL, Ori RB, Moore SR, Ori M, Lima A (2008) Malnutrition as an enteric infectious disease with long-term effects on child development. Nutr Rev 66:487–505

Hansen JW, Baethgen WE, Osgood DE, Ceccato PN, Ngugi RK (2007) Innovations in climate risk management: protecting and building rural livelihoods in a variable and changing climate. J Semi-Arid Trop Agric Res 4(1):1–38

Hawkes C (2006) Uneven dietary development: linking the policies and processes of globalization with the nutrition transition, obesity and diet-related chronic diseases. Glob Health 2:4. doi:10.1186/1744-8603-2-4

Hawkes C, Ruel M (2006) Overview, understanding the links between agriculture and health, in 2020 vision briefs. International Food Policy Research Institute IFPRI, Washington

Herforth A, Jones A, Pinstrup-Andersen P (2012) Prioritizing nutrition in agriculture and rural development: guiding principles for operational investments. The International Bank for Reconstruction and Development/The World Bank, Washington

Hoddinott J (2012) Agriculture, health and nutrition, toward conceptualizing the linkages. In: Fan S, Pandya-Lorch R (eds) Reshaping agriculture for nutrition and health. An IFPRI 2020 book. International Food Policy Research Institute, Washington, Chapter 2

Humphrey JH (2009) Child undernutrition, tropical enteropathy, toilets, and handwashing. Lancet 374(9694):1032–1035

Husain T, Chaudhary JR (2008) Human health risk assessment due to global warming—a case study of the gulf countries. Int J Environ Res Public Health 5:204–212

IFPRI (2009) Index insurance and climate risk: prospects for development and disaster management. In: Hellmuth ME, Osgood DE, Hess U, Moorhead A, Bhojwani H (eds) Climate and society No. 2. International Research Institute for Climate and Society (IRI), Columbia University, New York, USA

IOM (2014) Sustainable diets: food for healthy people and a healthy planet: workshop summary. The National Academies Press, Washington

IPCC (2012) Managing the risks of extreme events and disasters to advance climate change adaptation. In: Barros CBV, Stocker TF, Qin D, Dokken DJ, Ebi KL, Mastrandrea MD, Mach KJ, Plattner G-K, Allen SK, Tignor M, Midgley PM (eds) A Special Report of Working Groups I and II of the Intergovernmental Panel on Climate Change. Cambridge University Press, Cambridge, UK, and New York, NY, USA, 582 pp

Jaenicke H, Virchow D (2013) Entry points into a nutrition-sensitive agriculture. Food Secur 5:679–692

Johnston JL, Fanzo JC, Cogill B (2014) Understanding sustainable diets: a descriptive analysis of the determinants and processes that influence diets and their impact on health, food security, and environmental sustainability. Adv Nutrition 5(4):418–429

Karnik A, Foster BA, Mayer V, Pratomo V, McKee D, Maher S, Campos G, Anderson M (2011) Food insecurity and obesity in New York City primary care clinics. Med Care 49(7):658–661

Keats S, Wiggins S (2014) Future diets: implications for agriculture and food prices. ODI, London

Kerr RB, Berti PR, Shumba L (2010) Effects of a participatory agriculture and nutrition education project on child growth in northern Malawi. Public Health Nutr 14:1466–1472

Khlangwiset P, Shephard GS, Wu F (2011) Aflatoxins and growth impairment, a review. Crit Rev Toxicol 41:740–755

Kim JC, Abramsky T (2009) Assessing the incremental benefits of combining health and economic interventions: experience from the IMAGE Study in rural South Africa. Bull World Health Org 87:824–832

Kovats RS, Ebi KL (2006) Heat waves and public health in Europe. Eur J Public Health 16:592–599

Lachat C (2013) UNSCN Senegal Case Study. UNSCN, Geneva

Lang T, Barling D (2013) Nutrition and sustainability: an emerging food policy discourse. Proc Nutr Soc 72:1–12

Leroy J (2013) Child stunting and aflatoxin. In: Unnevehr L, Grace D (eds) Aflatoxins, finding solutions for improved food safety. IFPRI 2020 November 2013 brief. IFPRI, Washington

Levine MM (1987) Escherichia coli that cause diarrhea: enterotoxigenic, enteropathogenic, enteroinvasive, enterohemorrhagic, and enteroadherent. J Infect Diseases 155(3):377–389

Levinson JF, Balarajan Y (2013) Addressing malnutrition multisectorally: what have we learned from recent international experience? UNICEF Nutrition Working Paper, UNICEF and MDG Achievement Fund, New York

Lloyd SJ, Kovats RS, Chalabi Z (2011) Climate change, crop yields, and undernutrition: development of a model to quantify the impact of climate scenarios on child undernutrition. Environ Health Perspect 119(12):1817

Macdiarmid J, Kyle J, Horgan G, Loe J, Fyfe C, Johnstone A, McNeill G (2011) A balance of healthy and sustainable food choices: project report of Livewell. WWF-UK, London

Mason JB, Shrimpton R (2010) Sixth report on the world nutrition situation. UNSCN, Geneva

Monteiro CA, Levy RB, Claro RM, de Castro IRR, Cannon G (2011) Increasing consumption of ultraprocessed foods and likely impact on human health: evidence from Brazil. Public Health Nutr 14(01):5–13

Motarjemi Y (2000) Research priorities on safety of complementary feeding. Pediatrics 106 (suppl4):1304–1305

Motarjemi Y, Kaferstein F, Moy G, Quevedol F (1993) Contaminated weaning food: a major risk factor for diarrhoea and associated malnutrition. Bull World Health Org 71(1):79–92

Motarjemi Y, Steffen R, Binder HJ (2012) Preventive strategy against infectious diarrhea - a holistic approach. Gastroenterology 143(3):516–519

Nelson GC (2010) Food security, farming, and climate change to 2050: scenarios, results, policy options. IFPRI, Washington

Nepal Government. Multi-Sectoral Nutrition Plan. 2012. Ministry of Health, Government of Nepal

Nugent R (2004) Food and agriculture policy: issues related to prevention of non-communicable diseases. Food Nutr Bull 25(2):200–207

Patz JA, Campbell-Lendrum D, Holloway T, Foley JA (2005) Impact of regional climate change on human health. Nature 438:310–317

Pinstrup-Andersen P (2009) Food security: definition and measurement. Food Secur 1:5–7

Popkin BM, Adair LS, Ng SW (2012) Global nutrition transition and the pandemic of obesity in developing countries. Nutr Rev 70(1):3–21

Rayner G, Lang T (2012) Ecological public health: reshaping conditions for good health. Routledge/Taylor and Francis Group, London

Remans R, Fanzo J, Berg M, Atkins E, Mohammed A, Sachs JD (2011a) Climate change SCN News #38. UNSCN, Geneva, Switzerland

Remans R, Pronyk P, Fanzo J, Palm C, Chen J, Nemser B, Muniz M, Radunsky A, Abay A, Coulibaly M, Mensah J, Wagah M, Quintan E, Sachs SE, Sanchez P, McArthur J, Sachs JD (2011b) A multi-sector intervention to accelerate reductions in child stunting: an observational study from nine sub-Saharan African countries. Am J Clin Nutr 94(6):1632–1642

Remans R, Flynn D, Fanzo J, Declerck F, Lambrecht I, Sullivan C, Gaynor K, Siriri D, Mudiope J, Mutuo P, Nkhoma P, Palm C (2011c) Assessing nutritional diversity of cropping systems in African Villages. PLoS One 6:e21235. doi:10.1371/journal.pone.0021235

Robine JM, Cheung SLK, Le Roy S et al (2008) Death toll exceeded 70 000 in Europe during the summer of 2003. C R Biol 331:171–178

Sanchez P, Palm C, Sachs J, Denning G, Flor R, Harawa R (2007) The African millennium villages. Proc Natl Acad Sci 104(43):16775–16780

Stuart T (2009) Waste: uncovering the global food scandal. W. W. Norton & Company, London

Temple NJ, Steyn NP (2011) The cost of a healthy diet: a South African perspective. Nutrition 27:505–508

UNDESA, UNFPA (2012) UN system task team on the post 2015 UN development agenda. Population dynamics. Think piece. http://www.un.org/millenniumgoals/pdf/Think%20Pieces/15_population_dynamics.pdf

UNICEF, WHO, UNFPA, World Bank (2013) Levels and trends in child mortality: report. New York

UN Millennium Project (2005) Investing in development. A practical plan to achieve the millennium development goals. London

Verstraeten R, Roberfroid D, Lachat C et al (2012) Effectiveness of preventive school-based obesity interventions in low- and middle-income countries: a systematic review. Am J Clin Nutr 96:227–228

Wang Y, Lobstein T (2006) Worldwide trends in childhood overweight and obesity. Int J Pediatr Obes 1(1):11–25

WHO (2004) The world health report: changing history. Geneva, Switzerland

WHO (2013) Global nutrition policy review. Geneva, Switzerland

Wild CP (2007) Aflatoxin exposure in developing countries: the critical interface of agriculture and health. Food Nutr Bull 28(2 Suppl):S372–S380

World Bank (2006) Repositioning nutrition as central for development. Washington, DC

World Bank (2010) People, pathogens and our planet, volume 1: towards a one health approach for controlling zoonotic diseases

World Bank (2012a) Improving nutrition through multisectoral approaches. World Bank, Washington

World Bank (2012b) People, pathogens and our planet : the economics of one health. World Bank, Washington. https://openknowledge.worldbank.org/handle/10986/11892. License: CC BY 3.0 Unported

World Bank (2013) Food price watch. Washington

Chapter 10
Textbox: Planetary Boundaries in Food and Agriculture Law

Gabriela Steier

Abstract Visualizing the limits of sustainable food policy and agriculture law in light of how much pollution the planet can absorb, or how many resources the planet can recover, help in appreciating the multitude of factors that must be included in legislative proposals and legal frameworks around the globe. The planetary boundaries do just that. Taking the planetary boundaries into consideration in promoting sustainable food policy and shaping agriculture law toward more environmentally responsible practices will help avoid some of the climate change and ecological losses that current food systems often provoke. The benefits of at least steering clear of those points exceeding the planetary boundaries will inform positive ways to reshape diet, nutrition, public health, environmental integrity, and energy conservation.

Diet and nutrition are inextricably connected. Similarly, diet and public health are closely linked. In the present interdisciplinary study, it is also important to consider that environmental integrity, energy conservation, and climate change play vital roles in this network of concerns. All of these factors impact food policy and agriculture law—and should play even greater roles as food laws around the world develop.

Therefore, visualizing the limits of sustainable food policy and agriculture law in light of how much pollution the planet can absorb, or how many resources the planet can recover, help in appreciating the multitude of factors that must be included in legislative proposals and legal frameworks around the globe. The planetary boundaries according to Rockström, Steffen et al.,[1] thought-leaders in the field who have contributed greatly to the study of how to improve the planet's ecological resilience, are the foremost model to conceptualize at least some of the quantifiers of how much the planet can absorb and recover from. Figure 10.1 is a

---

[1] For the research and articles by Rockström and Steffen et al., please see the Stockholm Resilience Center at http://www.stockholmresilience.org/21/research/research-programmes/planetary-boundaries.html.

G. Steier (✉)
Food Law International, LLP, Boston, MA, USA
e-mail: g.steier@foodlawinternational.com

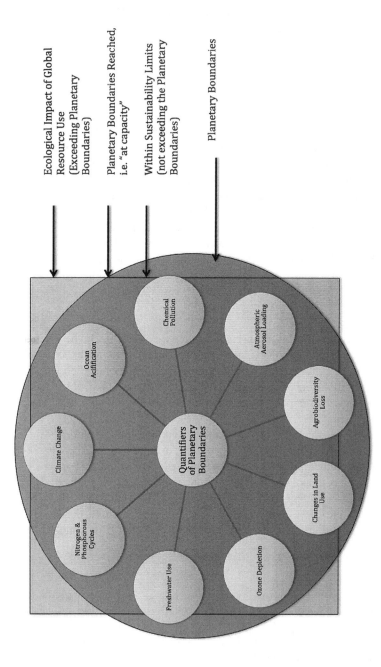

Fig. 10.1 The ecological impact model of the planetary boundaries. This figure illustrates the planetary boundaries (*large circle*) as they are partially exceeded by global resource use (*large square*). As the quantifiers of the planetary boundaries (*light circles*) are pushed past sustainable limits, the planet is "at capacity." Beyond this point, i.e. where the *square* extends beyond the *large circle*, resource use is no longer sustainable and the ecological footprint exceeds the rates at which resources could either be recovered, restored or the planet maintain ecological balance. Where the *square* is within the *large circle*, the planetary boundaries are within sustainable ranges. The closer to the center the *square* may be, the more sustainable the resource use, i.e. renewable resources. Although this simplified and generalized schematic view is not based on any data and does not reflect any specific levels of resource uses, the general trend has been that the planetary boundaries are increasingly exceeded because intensive agriculture is largely unsustainable and depletes the planet's resources on a large scale

simplified version of how these planetary boundaries may be visualized and focuses on some of the most pressing boundaries that damage the global ecosystems and environmental integrity. For the purpose of the present study on food law and policy, the list of the planetary boundaries is adapted to include only nine of these considerations, which each have a multitude of applications in food policy and agriculture law, as the following examples can merely begin to indicate:

1. **Agrobiodiversity Loss**—for example through the proliferation of GMO monocultures and the patenting of indigenous plants
2. **Changes in Land Use**—such as through the expansion of CAFOs
3. **Ozone Depletion**—as a consequence of processing and packaging fast food
4. **Freshwater Use**—for agricultural uses on monocultures
5. **Nitrogen and Phosphorous Cycles**—for instance in synthetic fertilizer and pesticide production and uses
6. **Climate Change**—as a result of industrial and centralized agriculture, unsustainable farming practices, and increased food mileage accumulation
7. **Ocean Acidification**—through industrial waste run-off and acid rain
8. **Chemical Pollution**—from food processing, transportation, and processing
9. **Atmospheric Aerosol Loading**—through greenhouse gas emissions

Although the planetary boundaries find application in various industries and across several fields of study, they are particularly useful in food law and policy because nearly every consideration surrounding diet, nutrition, public health, sustainable food production or energy use can be connected to the model of quantifying how resilient the planet is in the face of the challenges posed by these interdisciplinary issues. The dietary guidelines for Americans, for example, are a classic example of the compromises, greenwashing and industry lobby's domination over the recommendations of what consumers in the US should eat. In 2015, a revised version of the current myplate.gov recommendations is due and the advisory committee, composed of hopefully independent experts, proposed a shift away from processed and industrially produced food toward a more sustainable diet based on plants.[2] This shift, were the BigFood and Big Ag industry to let it pass, would turn away from animal-based protein and food sources, the production of which has especially damaging effects on the environment and wastes energy on multiple levels. Industrial animal production or CAFOs, in the US, for example, cause substantial damage to soil, pollute water, promote climate change, and yield unsustainable food products. In contrast, a shift toward a generally organic whole-foods plant-based diet would prevent some of this environmental damage and reduce a variety of public health complications and diseases, such as malnutrition, obesity and cancer.[3]

---

[2]See Chapter 4–6 in USDA and DHHS, Scientific Report of the 2015 Dietary Guidelines Advisory Committee (February 2015), http://www.health.gov/dietaryguidelines/2015-scientific-report/PDFs/Scientific-Report-of-the-2015-Dietary-Guidelines-Advisory-Committee.pdf.

[3]See Oxfam's GROW Report at https://www.oxfam.org/sites/www.oxfam.org/files/file_attachments/cr-growing-better-future-170611-en_0.pdf. See also Steier (2012).

Consequently, taking the planetary boundaries into consideration in promoting sustainable food policy and shaping agriculture law toward more environmentally responsible practices will help avoid some of the climate change and ecological losses that current food systems often provoke. The benefits of steering clear of the "at capacity" point (see Fig. 10.1) or from exceeding the planetary boundaries will inform positive ways to reshape diet, nutrition, public health, environmental integrity, and energy conservation.

Reference

Steier G (2012) Dead people don't eat: food governmentenomics and conflicts-of-interest in the USDA and FDA. Pittsbg J Environ Public Health Law 7(1)

Chapter 11
Food and Nutrition in Cancer Prevention and Treatment

Michael J. Nathenson

Abstract This chapter focuses on the interface of food, nutrition and oncology. Cancer is one of the major leading causes of death in the United States and throughout the world. The best treatment for cancer is still prevention. This chapter provides an introduction to cancer, clinical trials, and reviews the evidence that exists for the influence of physical activity, weight, and especially nutrition, including both macronutrients and micronutrients, on the risk of cancer development and on the success of cancer treatment. Additionally, this chapter discusses the difficulties and problems involved in developing and conducting large population based clinical trials examining the connection between nutrition and cancer. This, unfortunately, often results in conflicting results or results which are not clinically applicable. Overall, altering physical activity, weight, and diet is predicted to reduce the incidence of cancer by one third to one half. More support is needed to fund ongoing clinical trials, including trials that focus on the impact of GMO foods and pesticides on the risk of development of cancer. Finally, the consumers must become better educated on the relationship between exercise, weight, and diet on the development of cancer, in an effort to change population habits to reduce cancer incidences.

11.1 Introduction to Cancer

The two major leading causes of death in the US are cancer and cardiovascular disease, both of which are influenced by diet and lifestyle. The purpose of this chapter is to provide an introduction to cancer, clinical trials, and review the evidence that exists for the influence of physical activity, weight, and especially nutrition on the risk of cancer development and success of treatment.

M.J. Nathenson (✉)
Department of Medical Oncology, Dana-Farber Cancer Institute, Boston, MA, USA
e-mail: michaelj_nathenson@dfci.harvard.edu

11.1.1 Epidemiology and Overview of Cancer

Cancer is one of the major causes of death throughout the world. Worldwide there are 14.1 million new cases of cancer per year, with 8.2 million deaths per year. In addition there were at least 32.6 million people living with cancer in 2012.[1] In the United States alone there were an estimated 1,665,540 new cases of cancer with 585,720 cancer related deaths in 2014.[2] Cancer is the leading cause of death in people less than 85 years old in the United States. The lifetime risk of developing cancer is 1:3 for women and 1:2 for men in the United States.[3] This represents a huge burden of disease. To decrease the morbidity and mortality associated with cancer requires a multi-layer approach. This approach must include understanding cancer on a molecular and genetic level, improving screening for cancer, improving the local and systemic therapies for cancer, and improving our prevention of cancer through lifestyle modifications that can reduce the risk of developing cancer.

A tumor is a descriptive term for a growth or swelling, as is a neoplasm or new growth. These terms are not equivalent with cancer, as tumors and neoplasms can be benign or malignant. Only malignant tumors or neoplasms are considered cancer. Cancer is defined as a tumor or growth with malignant potential; the ability to spread beyond the primary site. Cancer is a heterogeneous disease with many different specific types. The two largest categories of cancer are carcinomas and sarcomas, the naming of which comes from the original ancient Greek description; karkinos- meaning crab, -oma meaning growth, and sarx- meaning flesh. Carcinomas are very infiltrating crab like tumors, and include the most common tumor types, including breast carcinomas, colon adenocarcinomas, prostate adenocarcinomas, lung adenocarcinomas, and lung squamous cell carcinomas. Sarcomas are much less common tumors of bone, nerve, muscle, fat, and connective tissue. Table 11.1 shows the ten leading types of cancer and Table 11.2 shows the ten leading causes of cancer related deaths in 2014.

Over the last 60 years, the field of oncology has moved from treating cancer with surgery alone, to the development of radiation therapy and chemotherapy, which has led to the ability to cure some cancers. For example cancers of the blood (leukemia and lymphoma) as well as testicular cancer can be cured with chemotherapy alone. Combined therapy with surgery, radiation and chemotherapy can lead to the cure of some early stage solid tumors, for example breast cancer. Unfortunately, for many other types of cancer, such as pancreatic and lung cancer, the existing therapies of surgery, radiation and chemotherapy have little impact on the overall mortality of patients suffering from these diseases (see footnote 3).

Patients with these types of cancers, more often than not, die of their disease. This death rate results in a huge impact on our society in terms of lives lost, emotional stress, loss productivity, cost of treatment and disease symptoms. The

---

[1] WHO, GLOBOCAN (2012).

[2] Siegel et al. (2014), p. 9.

[3] American Cancer Society, Cancer Facts & Figures (2014); Siegel et al. (2011); Siegel (2014).

Table 11.1 The ten leading causes of cancer

Male			Female		
Cancer	Estimated new cases	% of total	Cancer	Estimated new cases	% of total
Prostate	233,000	27%	Breast	232,670	29%
Lung	116,000	14%	Lung	108,210	13%
Colorectal	71,830	8%	Colorectal	65,000	8%
Bladder	56,390	7%	Uterine	52,630	6%
Melanoma	43,890	5%	Thyroid	47,790	6%
Renal	39,140	5%	Non-Hodgkin Lymphoma	32,530	4%
Non-Hodgkin Lymphoma	38,270	4%	Melanoma	32,210	4%
Head and neck	30,220	4%	Renal	24,780	3%
Leukemia	30,100	4%	Pancreas	22,890	3%
Liver and bile duct	24,600	3%	Leukemia	22,280	3%
All sites	855,220	100%	All sites	810,320	100%

This table shows the ten leading causes of cancer in the United States by gender in 2014 (Siegel et al. 2014, p. 12)

Table 11.2 The ten leading causes of cancer related deaths

Male			Female		
Cancer	Estimated deaths	% of total	Cancer	Estimated deaths	% of total
Lung	86,930	28%	Lung	72,330	26%
Prostate	29,480	10%	Breast	40,000	15%
Colorectal	26,270	8%	Colorectal	24,040	9%
Pancreas	20,170	7%	Pancreas	19,420	7%
Liver and bile duct	15,870	5%	Ovary	14,270	5%
Leukemia	14,040	5%	Leukemia	10,050	4%
Esophagus	12,450	4%	Uterine	8590	3%
Bladder	11,170	4%	Non-Hodgkin Lymphoma	8520	3%
Non-Hodgkin Lymphoma	10,470	3%	Liver and Bile Duct	7130	3%
Renal	8900	3%	Brain	6230	2%
All sites	310,010	100%	All sites	275,710	100%

This table shows the ten leading causes of cancer related deaths in the United States by gender in 2014 (Siegel et al. 2014, p. 12)

field of oncology still has a very long way to go in order to cure these patients. In fact, it is still quite true that the easiest cancer to treat is the cancer that never occurs at all. There are many causes of cancer including use of tobacco products, alcohol, infectious agents—particularly certain bacteria or viruses—sun exposure, exposure to other forms of radiation, air pollution, weight and obesity, physical activity, unsafe sex, chemical exposure, and exposure to carcinogens in food and beverages. An estimated 1/2 to 2/3 of cancer deaths are preventable by modification of these risk factors, and an approximate 1/3 of cancer deaths in the United States are preventable by modification of weight, physical activity and diet.[4] In addition it has been suggested that people with a body mass index (BMI) of less than 30, who have more than three and a half hours of physical activity weekly, have never smoked, and have a prudent diet, have a one third reduced risk of developing cancer.[5] Both the American Cancer Society (ACS) and the American Institute for Cancer Research (AICR) published guidelines for reducing the risk of cancer by modifying diet, weight and physical activity.[6] Table 11.3 shows a summary of the ACS and AICR guidelines. Results from the EPIC study, a large nutritional study, show a reduction in the risk of cancer when there was concordance with the AICR recommendations.[7] Unfortunately these recommendations are poorly implemented, and the rates of obesity and tobacco use in the US are still unacceptably high.[8]

A physician can have an impact on only one patient at time. In contrast, government regulation and public awareness campaigns can have a massive impact. For example, the decline in smoking combined with the increased tax on cigarettes has the potential to result in a decreased incidence of lung cancer. Public education with the goal of changing diet, increased physical activity, weight loss, use of sun protectants, but decreased the use of tobacco products and alcohol can greatly reduce the incidence of cancer in our society. Government regulation plays a key role as well. There are over 100 preventable human exposures associated with human cancers identified.[9] The government has an important role to protect the public from exposure to carcinogens in everyday products and foods, an area where there is much room for improvement.

11.1.2 Cancer Biology, Carcinogenesis and Carcinogens

Cancer is, in essence, a genetic disease that results from uncontrolled cell growth as the result of accumulating genetic mutations from exposure to intrinsic and environmental carcinogens. Historically, cancer was once thought to be caused by viruses. For, example the Rous sarcoma virus. Research into these viruses identified genes that could influence the development of cancer. These genes were found to be

---

[4]Marian and Robert (2010), Doll and Peto (1981), Ezzati et al. (2002), Danaei et al. (2005), Wolin et al. (2010), Jemal et al. (2011), and Brawley (2011).
[5]Ford et al. (2009).
[6]Kushi et al. (2012), American Cancer Society; American Institute for Cancer Research, Recommendations for Cancer Prevention.
[7]Romaguera et al. (2012).
[8]American Cancer Society, Cancer Prevention & Early Detection Facts & Figures (2014).
[9]Cogliano et al. (2011).

Table 11.3 ACS and AICR guidelines for weight, physical activity and diet

American Cancer Society Recommendations	American Institute of Cancer Research Recommendations
Be as lean as possible throughout life without being underwent, avoid excess weight at all ages	Be as lean as possible without becoming underweight
Get regular physical activity, get at least 150 min of moderate intensity or 75 min of vigorous intensity activity each week	Be physically active for at least 30 min a day
Limit intake of high-calorie foods and drinks	Avoid sugary drinks, limit consumption of energy-dense foods
Eat a healthy diet with an emphasis on plant foods, eat at least 2.5 cups of vegetables and fruits each day	Eat more vegetables, fruits, whole grains and legumes
Limit how much proceed meat and red meat you eat	Limit Consumption of red meats and avoid processed meats
Drink no more than 1 alcoholic drink per day for women, or 2–3 per day for men	Limit alcoholic drinks to 2 for men and 1 for women a day
	Limit consumption of salty foods and foods processed with salt
	Don't use supplements to protect against cancer

American Cancer Society, ACS Guidelines on Nutrition and Physical Activity for Cancer Prevention; American Institute for Cancer Research, Recommendations for Cancer Prevention

carried by viruses, but had originated from their host cells; this insight lead to the discovery of oncogenes and tumor suppressor genes. Oncogenes are genes, which require mutations to become active and induce the formation of cancer. Tumor suppressor genes, such as the well-known tp53 gene, are genes that must be inactivated, through mutation for the development of cancer.[10] It is now known that mutations can occur through a variety of mechanisms, from exposure to UV rays from the sun or other forms of ionizing radiation, to exposure to viruses, such as the human papilloma virus that causes cervical cancer, to inherited genetic disorders, to exposure to carcinogens in our environment. The development of cancer or carcinogenesis, is a multistep process over time, which requires the accumulation of multiple mutations of a period of months to years. This is best illustrated by the well-understood carcinogenesis of colon cancer, which requires multiple mutations over a period of several years. The first gene to be mutated is the APC gene leading to hyper-proliferation of colon cells, the mutations in K-ras and the DCC genes leading to the development of an adenoma, and final mutation in tp53 leading to the development of adenocarcinoma. In fact, some cancers may accumulate hundreds of mutations.

Cancer cells acquire, through these mutations, the ability to have self-sufficiency in growth signals, to have insensitivity to anti-growth signals, to have limitless replicative potential, to be able to invade tissues and metastasis, to cause sustained

---

[10]Cooper (1995).

angiogenesis, and to evade apoptosis, which includes evasion of the immune system. These are the six hallmarks for a cancer cell, which lead to their malignant potential.[11]

The development of mutations in a normal cell is a balance of the number of new mutations, the effectiveness of the DNA repair mechanisms, and pathways that are responsible for preserving the integrity of our genome. The baseline mutation rate that occurs with normal cell divisions, even without exposure to external carcinogens, is estimated to be approximately 1 in 9 billion. These mutations results from endogenous causes: oxidative stress, inflammation, and errors in DNA replication. There are multiple DNA repair mechanisms to correct mutations that occur, such as homologous recombination, non-homologous end joining, nucleotide excision repair, base excision repair, trans-lesion synthesis, and mis-match repair. The development of mutations can be increased by inheriting deficiencies in a DNA repair pathway or exposures carcinogens in the environment, such as UV light, ionizing radiation, smoking tobacco, infectious agents, medications, and diet.

The amount of carcinogens in our diet is very difficult to study and define. There are complex interactions between micronutrients, nutrients in our diet, with physical activity and cellular mechanisms and processes. There are carcinogens that naturally exist in our environment and food at low levels. Additionally, there are carcinogens that are formed during the cooking and preparation of meals. Finally, there are carcinogens that are introduced into the environment by humans. All ingested nutrients are filtered by the liver and many possible carcinogens are removed. This complex interaction makes it difficult to determine the exact cause of cancer in many cases, and to study the effect of diet or nutrition, physical activity and weight on the development of cancer. Well-designed clinical trials can help improve our understanding, of these complex interactions and start to determine the effect of nutrition, weight, and physical activity on the development of cancer.

11.1.3 Principles of Clinical Trials

Scientific evidence can be ranked by its level of quality (Fig. 14.1 in Chap. 14). The lowest quality evidence is expert opinion or practice guidelines which are inherently biased. Only slightly better are case reports, case series or retrospective reviews which are greatly limited by selection bias, small sample size, and lack of a control arm. The next level of evidence consists of case-control studies or prospective or retrospective cohort studies. Thus, the highest quality of evidence is meta-analysis, systematic reviews, and randomized controlled trials (RCTs). Systematic reviews and meta-analysis can be impaired by the quality of the smaller studies from which they draw. The gold standard, highest quality evidence, therefore comes from double blinded randomized controlled trials.

---

[11]Hanahan and Weinberg (2000) and Hanahan and Weinberg (2011).

These are large studies, usually of several hundred patients, powered to detect statistically significant differences between two populations. The study design is prospective with a control group, and group receiving the study intervention. Patients are randomly assigned to either the control or intervention group, thereby controlling the variances in the human population that could affect the results of the study. These studies are usually blinded so that the patients and treating physicians are not aware into which group patients have been randomly assigned, which helps to eliminate bias. With these controls in place RCTs are the least biased of the different clinical trial designs. If a randomized controlled trial does not meet the criteria above, then there are problems with its internal validity. External validity is also important, for example where randomized controlled studies can still be influenced by the patient selection. If a very specific patient population is selected then the results of the study are only applicable to that population.

Unfortunately RCTs are expensive to conduct, require a large support staff, years of follow-up, and, in terms of studies evaluating the effect of weight, physical activity, and nutrition on the development of cancer, it is very difficult to enforce a specific intervention over long periods of time. For example, in a RCT evaluating dietary fiber intake and the development of colon cancer, it is difficult to standardize the amount and frequency of fiber intake, as well as ensure patient compliance for 5 years. Finally, there are ethical considerations, conceivably asking a patient to undertake an intervention that could increase their risk of cancer is not morally viable.

Observational studies are another option, which take advantage of natural events, or exposures that occur, rather than imposing an intervention. Consequently the same ethical considerations as with RCTs do not apply. It is important to note that these types of studies help to show association, but do not prove causation, which is a very important distinction. Association does not prove causation. There are at least three types of non-causal associations. Random associations are associations that occur by chance when there is no actual link between exposure and outcome. Artificial associations occur due to a defect or error in the study design. Indirect associations happen when there is a third variable linking exposure and outcome. Once these non-causal associations have been ruled out, then the association is likely causal, though this is still not proof of causation. Epidemiologists have created a standard guideline for evaluating causal associations by a specific set of criteria. These criteria include the temporality, strength, dose response, replicability, and specificity of the association, as well as the influence of cessation of the exposure, biologic plausibility of the relationship, lack of alternative explanations, and consistency with other knowledge.[12] The more of these criteria are met, the more likely the association is truly causal for observational studies.

Types of observational studies include case reports, case series and cross-sectional studies, which are hindered by the lack of a control group and biases, as well as retrospective and prospective cohort studies and case control studies. In

---

[12]Gordis (2013), ch 14.

prospective cohort studies, the exposure or event has already occurred. However, in this type of studies, the outcome has not yet occurred. In retrospective cohort studies the exposure or event and outcome have already occurred. Conversely, in case-control studies, the key difference is the outcome that is identified first. The patients in this type of study are retrospectively examined for the exposure. Such a design is useful in a situation where the outcome is very rare.

Most studies examining the impact of diet, weight and physical activity on the prevention of cancer are observational, cross-sectional, retrospective cohort, or prospective cohort studies. There are specific difficulties associated with these studies, including multiple confounders, difficulty in quantifying carcinogen exposure, selection bias, reporting bias, poor patient compliance, insufficient follow-up, and difficulty in standardization of nutrients or vitamin dose. In addition, must studies focus on one nutrient rather than the inter play of an entire diet, as it naturally occurs in the human body. The human liver is a very good filter and eliminates many toxins and carcinogens that are ingested, but those substances can nonetheless reach the systemic circulation. So, a carcinogen seen in the lab (in vitro) by the standard ames test, may not be carcinogenic in the human body (in vivo). An additional difficulty, assuming the study is well designed, well conducted, and has statistically significant results, is the clinical significance of those results. For example, let's assume a study of 10,000 people is conducted, with 5000 people randomized to each arm. The intervention is eating one orange a day. The outcome under investigation is the prevention of scurvy, a disorder that results from low vitamin C levels.

The results show a 20% reduction in the risk of developing scurvy with consuming one orange a day, which is statistically significant. Based on these results, it is convincible to recommend that everyone eats one orange a day. However, the absolute risk reduction is important. If 5 out of 5000 people develop scurvy in the control group, and 4 out of 5000 people develop scurvy in the intervention group, this is a 20% relative risk reduction, but the absolute risk reduction is 1 person in 5000. This is not clinically significant, it makes no sense to have 5000 people eat one orange a day to prevent 1 incidence of scurvy. As a result, even if a study is well designed, well conducted and has statistically significant results, the results may not be important enough to recommend changes to people's behavior or clinical guidelines.

Overall, the world of clinical trials is complicated because it is difficult and expensive to conduct well-designed, ethical, and well-conducted studies, the results of which have to undergo vigorous evaluation before the results can be considered valid. There are numerous studies evaluating the effect of weight, diet, and physical activity on the development of cancer, which will be reviewed below. Following the hierarchy of credibility of clinical trials (Fig. 14.1 in Chap. 14), the highest quality evidence stems from RCTs or systemic reviews. Unfortunately, there are only a few RCTs evaluating the effect of weight, diet, or physical activity on the development of cancer, because they are too expensive, too large to conduct, too difficult to implement an intervention, and not ethical to conduct. Those RCTs that do exist evaluate the effect of a pill or vitamin supplement, which, as stated above, do not take into account the complicated workings and processing within the human body. Most studies about the effect of diet, physical activity and weight are observational and hindered by one or more of the difficulties discussed above. Finally, the large systematic review and meta-analysis, which are considered high quality evidence, draw mostly from the observational studies that exist and are therefore limited by the quality of those studies.[13] This is why the current body of literature regarding the influence of weight, physical activity and diet is at times conflicting. To help clarify these issues, larger populations, with better assessment strategies and more funding are needed. This can be accomplished through government support and regulation, and public awareness campaigns. The rest of this chapter will focus on describing and evaluating the existing evidence for the influence of weight, physical activity and nutrition on the development and treatment of cancer.

11.2 Weight, Physical Activity and Cancer

Obesity is a worldwide epidemic. In the US alone, the number of people with a Body Mass Index (BMI) of greater than 25, which is defined as overweight, is above 60 % in most states.[14] How much a human weighs is directly linked to their metabolism, and what, how much, and how often he or she eats. Overall reviews have suggested evidence for decreased risk and prevention of cancer with weight reduction.[15] As much as a one third risk reduction in the development of cancer has been reported with a BMI <30, never smoker, prudent diet, and >3.5 h a week of physical activity.[16] The American Institute for Cancer Research Expert Report[17] reported that concordance with these recommendations would lead to decreased cancer risk by one-third. Similar results were seen in the EPIC study.[18] The american cancer society guidelines on nutrition and physical activity in 2012 recommend at least 150 min of moderate intensity or 75 min of vigorous intensity activity each week, with limiting sedentary behavior, maintaining a healthy weight, and eating a healthy diet by choosing whole grains instead of refined grain, eating at least 2 and a half cups of vegetables and fruits each day, with limiting the amount of processed and red meat. Arguably these interventions can be considered insufficient; however it is important to note that even these recommendations are not followed by most of the American public. In fact, the implementation of cancer prevention guidelines is difficult to enact for a variety of factors, such as skepticism that cancer can be prevented, to disagreement among experts, to the interventions

---

[13]Colditz (2010).

[14]American Cancer Society, Cancer Prevention & Early Detection Facts & Figures (2014), p. 15.

[15]Wolin et al. (2010).

[16]Ford et al. (2009).

[17]Wiseman (2008).

[18]Romaguera et al. (2012).

deployed too late in life to show effectiveness, and the complexity of successful implementation.[19]

Weight and BMI have been directly linked to the increased incidence of specific cancers including: esophageal, thyroid, renal, endometrial, gallbladder, rectal, melanoma, pancreatic, leukemia, multiple myeloma, non-Hodgkin's lymphoma,[20] prostate cancer,[21] colon cancer,[22] and breast cancer.[23] This represents the majority of the most common cancers in the US, with the exception of lung cancer. Breast cancer, in particular, is a hormonally driven disease. Adipose or fat tissues have aromatase enzymes, which increase the peripheral production of estrogen and, therefore, increase the risk of the development of breast cancer. In fact, there are aromatase inhibitors that are currently used in the treatment of both localized and metastatic breast cancer.

Increased physical activity alone has been linked to the reduced risk of certain cancers[24] including breast cancer[25] and colon cancer.[26] For invasive and non-invasive breast cancers, the California teachers study showed a decreased risk with more than five hours versus less than one half hour per week of physical activity, in mostly estrogen receptor negative breast cancer.[27] Another prospective cohort study showed that more than seven hours versus less than one hour per week resulted in an 18 % decrease in risk of invasive breast cancer.[28] A third prospective cohort study showed a 23 % decrease in risk of breast cancer with increased physical activity in premenopausal women.[29] Meta-analyses of these and other prospective studies have confirmed these results.[30]

For colon cancer, a meta-analysis showed a 24 % decrease in risk of colon cancer with increased physical activity.[31] Similar results were seen in a second meta-analysis.[32] Another study showed that the risk of colon cancer decreases with increasing hours of physical activity.[33] Observational cohort studies have shown

---

[19]Colditz et al. (2012).
[20]Renehan et al. (2008), Wolin et al. (2010), and Calle et al. (2003).
[21]Freedland and Platz (2007).
[22]Campbell et al. (2007).
[23]Eliassen et al. (2006) and Kawai et al. (2010).
[24]Demark-Wahnefried (2006).
[25]Bernstein (2009).
[26]Colditz et al. (1997) and Barton (2013).
[27]Dallal et al. (2007).
[28]Rockhill et al. (1999).
[29]Maruti et al. (2008).
[30]Wu et al. (2013).
[31]Wolin et al. (2009).
[32]Boyle et al. (2012).
[33]Chao et al. (2004).

similar results.[34] Also, a Japanese cohort study reported a decreased risk of colon, liver, pancreatic and stomach cancers with increased physical activity.[35]

There even is evidence that physical activity reduces the risk for prostate cancer,[36] though this is indirect evidence and less robust than that reported for breast and colon cancer. There was a large cohort study showing a decreased risk in the rare subset of aggressive prostate cancer with increased physical activity.[37]

Thus, while some amount of physical activity has been associated with a reduction in risk for invasive cancer, with the best data for colon and breast cancer, the optimum duration, frequency, and intensity of the exercise or physical activity is as of yet unknown. In addition, the extent to which physical exercise at a young age versus older age impacts the risk of invasive cancer is unknown. Much research is still required, but it is plain that cancer can be reduced in the US and throughout the world by a decrease in weight and increase in physical activity.

Nutrition and diet are even harder to study than weight and physical activity. It is easy to measure weight and to assign physical activity into broad categories, so that general trends can be observed. However, it is much more difficult to study the specifics of individual diets, or to have patients keep to a diet for a study. The interactions among food as well as cellular processes within the human body, makes studying these factors much more complicated. The following sections focus on the conflicting role of food in cancer and the conflicting data that exists.

11.3 Effect of Food on Cancer

11.3.1 Introduction to Food, Diet, and Nutrition in Oncology

The human diet is highly complex and varied. It can be classified by food groups or by types of nutrients. In fact, the main food groups in most oncology-related studies are fruits and vegetables, grains, dairy, and meats. Meat can be divided into processed and non-processed meat or into fish, chicken, and red meat. Some fruits and vegetables are high in fiber. If classified by nutrients, food can be divided into macronutrients and micronutrients. Macronutrients break down into the categories of lipids (fat), protein (meats and legumes), and carbohydrates. Lipids include saturated and unsaturated fatty acids, which include olive oil and omega-3 fatty acids. Carbohydrates can be further sub-divided into foods with high glycemic load and low glycemic load. Micronutrients include Vitamins A (retinoids and beta-carotene), D (cholecalciferol and ergocalciferol), E (tocopherols), K (phylloquinones), C (ascorbic acid), B12 (cyanocobalamine), B1 (thiamine), B2 (riboflavin), B3 (niacin), B5 (pantothenic acid), B6 (pyridoxine), B7 (biotin), B9

---

[34] Martinez et al. (1997) and Giovannucci et al. (1995).
[35] Inoue et al. (2008).
[36] Antonelli et al. (2009).
[37] Patel et al. (2005).

(folic acid), calcium, iron, and selenium. There are other nutrients that have been studied in cancer prevention and treatment, which do not fit into specified categories, which include flavonoids and soy, lycopene and alpha linoleic acid. Specific herbal supplements, such as garlic, have been investigated as well as specific diets. Most studies, however, have been observational or meta-analyses. Nonetheless, some randomized controlled trials have been conducted using specific micronutrient supplementation. The difficulty with these randomized controlled trials is that there may be differences in the effect of micronutrients when absorbed in pill form versus absorbed in diet form with other micronutrients. Thus, the results of these studies can, at times be conflicting, with no clear answers. In the subsequent sections, macronutrients will be discussed first and then micronutrients.

11.3.2 Dietary Fat

The history of the development of recommendations for dietary fat in the prevention of cancer is instructive and revealing about the role that dietary fats play in cancer development. Studies in mice, for instance, lead to international observational and migration studies that suggested an increased risk of colon, breast and prostate cancer with dietary fat.[38] This led to a change in national recommendations, such as the recommendation of reduced dietary fat intake from the national cancer institute and the recommendation from the committee on diet, nutrition and cancer of the National Academy of Sciences to reduce dietary fat to 30 % of total daily calories. Unfortunately the previous studies were filled with confounders. For example, in breast cancer, the study populations were skewed by differences in physical activity, differences in diet other than fat intake, and differences in age at menarche, later age of first birth, lower parity, and higher post-menopausal body weight, all of which could potentially impact the risk of developing breast cancer.[39] Later prospective observational studies showed no effect on dietary fat and breast cancer.[40] This discrepancy led to the reversal of national guidelines. Consequently, dietary fat is not currently included in the ACS and AICR recommendations shown in Table 11.3.

As a result, there is currently no overall conclusive evidence for the increased risk of breast cancer from increased dietary fat intake. In addition to the prospective observational studies, the main RCT was negative,[41] as were three meta-analyses.[42] There was a second meta-analysis suggestive of a possible association of dietary fat and increased risk of breast cancer. This association, however, was certainly not

---

[38]Prentice et al. (1989).

[39]Willett (2001).

[40]Kushi and Giovannucci (2002), Kolonel (1996), and Clinton and Giovannucci (1998).

[41]Prentice et al. (2006a).

[42]Hunter et al. (1996), Smith-Warner et al. (2001b), and Alexander et al. (2010a).

Table 11.4 Summary of evidence for increased risk of prostate and colon cancer with increased dietary fat

Cancer	Prospective cohort studies	Randomized controlled trial	Critical appraisal	Pooled analysis of case-controlled studies
Prostate cancer	Two Positive[a] Two Negative[b]		One Negative[c]	
Colon cancer	Two Positive[d] One Negative[e]	One Negative[f]		One Negative[g]

[a]Giovannucci et al. (1993) and Veierod et al. (1997)
[b]Schuurman et al. (1999) and Severson et al. (1989)
[c]Fleshner et al. (2004)
[d]Slattery et al. (2001) and Willett et al. (1990)
[e]Stemmermann et al. (1984)
[f]Howe et al. (1997)
[g]Beresford et al. (2006)

overwhelming enough to change clinical recommendations, especially considering the other negative studies.[43]

Similarly, there has been negative data for the association of lung,[44] ovarian,[45] and endometrial[46] cancer with dietary fat intake. For colon and prostate cancer, the data is somewhat contradictory, but overall negative (see footnote 40). Table 11.4 summarizes the existing data for the role of dietary fat and the risk of prostate, and colon cancer. Thus, while initial associations suggested an increased risk of cancer with increased dietary fat, further work has indicated no conclusive evidence to support such an association. On further examination of these studies, one of the confounders may have been the increased intake of animal products, particularly red meat. People with a diet high in fat, also tended to have a diet high in red meat, thus the possible association seen with dietary fat and cancer was an indirect association, with animal products being the unknown third variable.

11.3.3 Red Meat

The American diet historically has been high in red meat and animal fat. Beef, a type of red meat, is an American staple food. Many American companies have been built on beef and red meat, including the numerous fast food chains serving hamburgers and other processed red meats. Recently there have been investigations, which suggested the increased risk of cancer with red meat. Particularly three prospective cohort studies showed an increased mortality in those with an increased consumption of red meat, including death from cancer and cardiovascular

---

[43]Boyd et al. (2003).
[44]Smith-Warner et al. (2002).
[45]Prentice et al. (2006b).
[46]Bandera et al. (2007).

disease.[47] Possible explanations for this increased risk include the production of carcinogens when meat is cooked, particularly heterocyclic amines or polycyclic aromatic hydrocarbons.[48] There are those that may have increased susceptibility to these carcinogens due to polymorphisms in the N-acetyltransferase (NAT2), an enzyme involved in the detoxification of many drugs and toxins.[49]

An alternate explanation was proposed by Dr. Harald zur Hausen at the American Society of Clinical Oncology meeting in 2014; he hypothesized the existence of an infectious agent, likely a virus, which has, specificity for American cattle and after ingestion by human's, resulted in an increased risk of colon cancer.[50] Neither hypothesis has been conclusively proven. Nonetheless, whatever the mechanism, numerous studies have shown an increase in the risk of colon cancer with increased red meat and processed meat.

There have been suggestions of increased risk for breast, pancreatic, and endometrial cancer as a result of red meat consumption, as well. For renal, prostate, stomach, and lung cancer the evidence is less convincing. Conversely there were several prospective cohort and meta-analyses that suggested no association between red meat and cancer. It should be noted that these studies were funded by National Cattlemen's Beef Association and the National Pork Board, so even though the evidence seems to be of high quality, the results are suspiciously biased. In addition an analysis of five prospective studies for mortality in vegetarians versus non-vegetarians showed a difference in mortality from cardiovascular disease, but no difference in mortality from cancer.[51] Unfortunately, such a study would fail to account for possible cofounders that result from other variations with the diet of vegetarians and non-vegetarians. Table 11.5 summarizes the evidence for the association of cancer and red meat. Presumably, vegetarian diets are higher in plants, fruits and vegetables. Another source of extensive investigation is the role of fruits and vegetable in the American diet and the risk of cancer.

11.3.4 Fruits and Vegetables

Over the last 30 years, research into the role of fruits and vegetables in decreasing the risk of cancer has produced conflicting reports. In some studies, there is, nonetheless, suggestion of minimal benefit to prevent cancer onset. Overall, there is no definitive evidence for the reduction of cancer risk. For example, a study comparing vegetarians to non-vegetarians showed no difference in cancer outcomes (see footnote 58). Some prospective cohort studies, such as the Greek cohort of the EPIC study, did show a risk reduction with increased vegetable intake in women.

---

[47]Pan et al. (2012) and Sinha et al. (2009).
[48]Zheng and Lee (2009) and Ferguson (2010).
[49]Ambrosone et al. (1998) and Deitz et al. (2000).
[50]Harald zur Hausen (May 2014).
[51]Key et al. (1999).

Table 11.5 Summary of the evidence for increased risk of cancer with increased intake of red and processed meat

Cancer	Prospective cohort or case-control	Meta-analysis
Prostate cancer	Five Positive[a]	Two negative[b]
Breast cancer	Four Positive[c]	One positive[d] Two negative[e]
Colon cancer	Seven positive[f] One negative[g]	One positive[h] One negative[i]
Pancreatic cancer	One positive[j] One negative[k]	
Renal cancer		One negative[l]
Endometrial cancer	One positive[m]	
Stomach cancer		One negative[n]
Lung cancer		One negative[n]

[a]Veierod et al. (1997), Giovannucci et al. (1993), Gann et al. (1994), Le Marchand et al. (1994), and Michaud et al. (2001)
[b]Key et al. (1999) and Alexander et al. (2010b)
[c]Toniolo et al. (1994), Fu et al. (2011), Zheng et al. (1998), and Steck et al. (2007)
[d]Boyd et al. (2003)
[e]Alexander et al. (2010c) and Missmer et al. (2002)
[f]Willett et al. (1990), Norat et al. (2005), Giovannucci et al. (1994), Chao et al. (2005), Goldbohm et al. (1994), Singh and Fraser (1998), and Hsing et al. (1998)
[g]Phillips and Snowdon (1985)
[h]Larsson and Wolk (2006)
[i]Alexander et al. (2011)
[j]Nothlings et al. (2005)
[k]Michaud et al. (2003)
[l]Alexander and Cushing (2009)
[m]Lonkhuijzen et al. (2010)
[n]Key et al. (1999)

However, the hazard ratio (HR) (an estimation of the strength of the association) was 0.91, indicating on a 9 % risk reduced with increased vegetable intake. A hazard ratio of one indicates no difference. When fruits and vegetable intakes were combined, the hazard ratio was 0.90 for women and 0.94 for men and women. Though, these hazard ratios barely met statically significance.[52]

The full European EPIC study showed similar results with a benefit seen in women with 100 g/day of vegetable intake, and a HR of 0.98, and benefit in men and women with a fruit and vegetable intake combined of 200 g/day and a HR of 0.96. Though again these barely met statically significance.[53] Also, a cohort study of the elderly showed a benefit in cancer risk reduction in elderly women, but not for men, with increased fruit and vegetable intake.[54] Conversely the NIH-AARP study showed no statistically significant benefit from fruit and vegetable intake

---

[52]Benetou et al. (2008).
[53]Boffetta et al. (2010).
[54]Shibata et al. (1992).

overall in men, women, or both, after adjustment for smoking intake.[55] The Nurse's health study showed no evidence of benefit for cancer risk reduction with increased fruit and vegetable intake.[56] The Japanese public health study showed no benefit to overall cancer reduction with fruits and vegetables either.[57]

With all of these studies, there are major limitations. First, there are geographical differences in diet. Different fruit and vegetables have varying amounts of micronutrients and could have different effects on the reduction of cancer. The types of fruits and vegetables consumed differ between the United States, Japan and even within Europe. The cited studies combining all types fruits and vegetables would not be able to detect this difference. Even if a difference is detected, however, it is unclear how much fruit and vegetables need to be eaten on a daily basis. Thus, each study has used different amounts.[58] In addition, none of these studies differentiated between organically grown fruits and vegetables versus those grown conventionally (with pesticides,) nor did they differentiate between genetically modified organisms (GMOs) and non-GMOs. So, again, to what extent these differences make an impact on cancer risk is unknown.

Furthermore, the risk of specific cancers may be affected differently by the consumption of fruit and vegetables. Table 11.6 summarizes the data for the decreased risk of development of certain cancers with increased fruit and vegetable intake; with the most evidence existing for the relationship of fruit and vegetable intake and the reduction of colon cancer risk.[59] In conclusion, much more research and funding are still needed to help answer questions about the possibility to prevent cancer through a plant-based organic diet because these studies are difficult to conduct on humans.

Another aspect that requires further research is the role of fiber. Fruits and vegetables are high in fiber, but evidence examining the role of fiber and the risk of cancer is currently limited to colon cancer. Early results and some prospective trials were negative.[60] However, a prospective study in the EPIC cohort showed an effect with 40 % reduction in the risk of colon cancer.[61] Then a pooled analysis showed an effect, but only in the univariate analysis.[62] A meta-analysis of prospective studies was positive as well,[63] though a Cochrane database review was

---

[55] George et al. (2009a).

[56] Hung et al. (2004).

[57] Takachi et al. (2007).

[58] Lee and Chan (2011).

[59] Slattery et al. (1999).

[60] Fuschs et al. (1999), Willett et al. (1990), and Giovanucci et al. (1994).

[61] Bingham et al. (2003).

[62] A univariate analysis is a form of quantitative analysis focused on the description of a single variable. By comparison, a multivariate analysis includes evaluation of multiple variables. *See* Park et al. (2005).

[63] Aune et al. (2011).

Table 11.6 Summary the data for the decreased risk of development of certain cancers with increased fruit and vegetable intake

Cancer	Prospective cohort study	Meta-analysis
Colon cancer	Three negative[a] Three positive[b]	Two positive[c]
Breast cancer	One negative[d] One positive, but only for estrogen negative breast cancer[e]	One negative[f]
Prostate cancer	One negative[g]	
Lung cancer	Two negative[h]	One positive[i]
Head and neck cancer	One positive[j] (Mainly in Legumes)	
Gastric cancer	Two negative[k]	

[a]Sato et al. (2005), Michels et al. (2000), and Steinmetz et al. (1994)
[b]van Duijnhoven et al. (2009), Terry et al. (2001), and Voorrips et al. (2000)
[c]Koushik et al. (2007) and Aune et al. (2011)
[d]Van Gils et al. (2005)
[e]Jung et al. (2013)
[f]Smith-Warner et al. (2001a)
[g]Key et al. (2004)
[h]Feskanich et al. (2000) and Liu et al. (2004)
[i]Smith-Warner et al. (2003)
[j]Freedman et al. (2008a)
[k]Freedman et al. (2008b) and Gonzalez et al. (2006, 2012)

negative.[64] Overall this evidence suggests that high fiber intake can decrease the risk of colon cancer.

A third aspect of fruit and vegetable intake, which also applies to any food high in carbohydrates, is the glycemic load and glycemic index of the food. The glycemic index (GI) is a ranking of carbohydrate rich food based on the rise in blood glucose levels after ingestion. The glycemic load (GL) is the glycemic index multiplied by the carbohydrate content of the food ingestion, which gives an estimation of the actual rise in blood glucose levels. The effects of GL and GI are controversial, with much of the evidence being contradictory, so no definitive recommendation can be made regarding the risk of cancer.[65] This is another area that should be further investigated, particularly since the American diet has a high incidence of high glycemic index foods.

---

[64]Asano and McLeod (2002).

[65]Shikany et al. (2011), Kabat et al. (2008), Hu et al. (2013), George et al. (2009b), Mulholland et al. (2009), and Romieu et al. (2012).

11.3.5 Dairy

The last major food group is dairy—although an argument against categorizing dairy as a food group can be made quite successfully. In terms of cancer prevention, unfortunately, most of the studies involving diary are inconclusive. Dairy products are usually high in calcium, which may be a confounder in these studies. As the role of calcium and vitamin D in the risk reduction of cancer is still being extensively studied, the overall the role of diary intake and cancer has been studied in ovarian cancer. Thus, in ovarian cancer, there were three meta-analyses that were negative.[66] In breast cancer a pooled analysis did not suggest an association.[67] Also, a meta-analysis with longer follow-up did not show any statistically significant results.[68] Nonetheless, there is a suggestion of a decreased risk of breast cancer in those premenopausal patients with increased calcium intake through low fat dairy products.[69] This suggests that more applicable results might be attained from examination of the effect of specific micronutrients, and other specific supplements or foods.

11.3.6 Other Nutrients

A variety of other nutrients have been studied within the context of their relation to cancer. Unfortunately the evidence is limited. There has been no overall benefit with omega-3 fatty acids, fish, garlic, linoleic acid or lycopene.[70] There is a suggestion of benefit with olive oil,[71] and a decreased risk of breast cancer with higher consumptions of flavonoids.[72] Also, there is suggestion of a decreased risk of lung cancer in female non-smokers with increased soy intake in the Shanghai Women's health study.[73] However, the baseline soy intake in the United States is very low compared to that in Asian countries, so it is difficult to compare across countries. A soy meta-analysis confirmed that the decreased risk of breast cancer in an asian population with high soy intake, but no effect in US population.[74] These studies are still preliminary and need to be further confirmed. Compared to the

---

[66]Larsson et al. (2006), Qin et al. (2005), and Genkinger et al. (2006).

[67]Missmer et al. (2002).

[68]Boyd et al. (2003).

[69]Shin et al. (2002).

[70]Kim and Kwon (2009), Wu et al. (2011), Zheng et al. (2013), Andreeva et al. (2012), Maclean et al. (2006), Zock and Katan (1998), Ilic et al. (2011), and Kavanaugh et al. (2007).

[71]Psaltopoulou et al. (2011).

[72]Fink et al. (2007).

[73]Yang et al. (2012).

[74]Wu et al. (2008).

studies of macronutrients, it is much easier to perform randomized controlled trials for micronutrients, because these can be included in a pill or supplement.

11.3.7 Micronutrients

Micronutrients intake has been studied in the prevention of cancer, and during the treatment of cancer. For the prevention of cancer, extensive studies, prospective and randomized controlled trials have been completed—some yielding negative results[75] and some positive results.[76] A systemic review by the National Institute of Health (NIH) suggested there was insufficient evidence to recommend daily multivitamin supplementation for cancer prevention.[77] A systematic evidence review by the U.S. Preventive Service Task Force reported similar results.[78] In regards to patients undergoing treatment for cancer, there is high use of vitamin supplements,[79] and there have been several reviews of cancer and vitamins.[80] Unfortunately, vitamin supplements may not be a good thing; there was increased cancer related mortality in male smokers with multivitamin use.[81] Also, a second study of patients with prostate cancer showed increased mortality in patients using multivitamins.[82] The role of food in cancer treatment is not limited to vitamin supplements.

11.4 Food as Medicine in Cancer Treatment: Food and Herb Uses During Cancer Treatment

While, multivitamin supplements cannot be recommended for patients undergoing cancer treatment, the significance of weight loss in patients with cancers is of paramount importance. In the majority of cancers, the amount of weight loss is of prognostic value. In other words the more weight loss a patient has, the shorter their survival is likely to be. Cachexia is the medical term for severe weight loss, from Greek kakos—meaning bad, and hexis—meaning condition. Nutritional interventions geared to maintaining or increase weight during the treatment of cancer can

---

[75]Neuhouser et al. (2009) and Park et al. (2011).
[76]Gaziano et al. (2012).
[77]Huang et al. (2006).
[78]Fortmann et al. (2013).
[79]Velicer and Ulrich (2008).
[80]Mamede et al. (2011) and National Institutes of Health (2007).
[81]Watkins et al. (2000).
[82]Stevens et al. (2005).

improve treatment outcomes.[83] In addition, there is a relationship between adequate nutrition and quality of life in cancer patients.[84] Specific diets have been tried in cancer patients as well as the Gerson, macrobiotic, and Kelley-Gonzales diets. Selected vegetable and herb mixes have been tried with soybeans, mushrooms, mung beans, red dates, scallions, garlic, lentils, leek, hawthorn fruit, onions, ginseng, angelica, dandelion, senegal root, licorice, ginger, olives, sesame seeds, and parsley. Though, none of these have been definitely proven to be of benefit. At the present time, just a balanced, nutritious diet is recommended for patients diagnosed with cancer.

In addition to diet, up to 90 % of patients will use one complementary and alternative therapies (CAM) for their cancer ranging from diet to vitamin supplements, as above, to herbal medications, to other therapies such as acupuncture, hypnotherapy, aromatherapy, mediation, Reiki (therapeutic touch), or support groups.[85] Although only one study showed shortened survival with CAMs, patients should be wary of fad treatments that have failed to show any benefit and have just shown harm, such as Koch therapies, Hoxsey, krebiozen, laetrile and Di bella.[86] In addition, herbal medications can have unwanted side effects and toxicities that can worsen a patient's quality of life. For examples St. John's Wort causes nausea, ginkgo can cause vomiting and headaches, while ginseng often results in diarrhea, headaches, and nausea. Herbal medications can interact with chemotherapy, as well, such as by increasing toxicity and interfering with the chemotherapeutic effect. For, example green tea inhibits cytochrome p450, an enzyme important in drug metabolism, and results in a decreased effect of bortezomib, an important treatment for multiple myeloma, a cancer of plasma cells. Green tea is the most widely used CAM,[87] however there is only limited data for its efficacy with both positive and negative studies.[88] St John's wort, a common herbal supplement, increases the activity of cytochrome p3A4 and therefore decreases the activity of irinotecan, taxanes, and imatinib, chemotherapeutic agents used in the treatment of many cancers. It is therefore recommended that herbal supplements and food such as black cohosh, echinacea, garlic, ginkgo, ginseng, green tea, grape seed, grapefruit, kava kava, milk thistle, soy, valerian, and St. John's Wort be avoided during cancer treatment.[89]

Conversely, some of the most effective chemotherapeutics have come from plants, such as taxols from the pacific yew plant, etoposide from mayapple, and vinca alkaloids from periwinkle. Also, traditional Chinese medicine such as, Huachansu, Fufangkushen and Shenqifuzheng can improve nausea, increase the white blood cell count while on chemotherapy, but have shown no evidence of

---

[83]Dewys et al. (1980), Tong et al. (2009), and Nitenberg and Raynard (2000).

[84]Nourissat et al. (2008).

[85]Kessler et al. (2001) and Gansler et al. (2008).

[86]Risberg et al. (2003).

[87]Bernstein and Grasso (2001).

[88]Jian et al. (2004) and Kikuchi et al. (2006).

[89]Meijerman et al. (2006) and Mathijssen et al. (2002).

activity against cancer.[90] Therefore, CAMs should be taken with great care and not without the treating physician's knowledge.

11.5 Conclusion and Future Directions

At the very least, the ACS and AICR recommendations should be followed to help decrease the risk of developing cancer by one third to one half. Though this is really a bare minimum of what should be done, not even these recommendations are met on a national basis. To further understand how the American diet should change and how it can prevent cancer, more research is needed. Though research alone is not sufficient, more support is needed for awareness campaigns to help educate Americans to change their diets toward an organic whole foods plant-based diet, low in saturated fats, simple sugars, and animal protein. In addition to Americans changing their diet, changes to food laws and regulation are needed. Federal Drug Administration (FDA) regulation of food and herbal supplements are quite inadequate. A new generation of medical practitioners, scientists, policy makers, and consumers needs to be trained to be aware of these problems and look for solutions.

References

Alexander D, Cushing C (2009) Quantitative assessment of red meat or processed meat consumption and kidney cancer. Cancer Detect Prev 32:340–351

Alexander D, Mink P, Cushing C et al (2010a) A review and meta-analysis of prospective studies of red and processed meat intake and prostate cancer. Nutr J 9:50–67

Alexander D, Morimoto L, Mink P et al (2010b) A review and meta-analysis of red and processed meat consumption and breast cancer. Nutr Res Rev 23:349–365

Alexander D, Morimoto L, Mink P et al (2010c) Summary and meta-analysis of prospective studies of animal fat intake and breast cancer. Nutr Res Rev 23:169–179

Alexander D, Weed D, Cushing C et al (2011) Meta-analysis of prospective studies of red meat consumption and colorectal cancer. Eur J Cancer Prev 20(4):293–307

Ambrosone C, Freudenheim J, Sinha R et al (1998) Breast cancer risk, meat consumption and N-acetyltransferase (NAT2) genetic polymorphisms. Int J Cancer 75:825–830

American Cancer Society: ACS Guidelines on Nutrition and Physical Activity for Cancer Prevention, available at http://www.cancer.org/healthy/eathealthygetactive/acsguidelinesonnutritionphysicalactivityforcancerprevention/index?ssSourceSiteId=null

American Cancer Society: Cancer Facts & Figures 2014, available at http://www.cancer.org/research/cancerfactsstatistics/cancerfactsfigures2014/index

American Cancer Society: Cancer Prevention & Early Detection Facts & Figures 2014, available at http://www.cancer.org/acs/groups/content/@research/documents/document/acspc-042924.pdf

---

[90]Mok et al. (2007), Guo et al. (2012), Yang et al. (2013), and Xu et al. (2009).

American Institute for Cancer Research: Recommendations for Cancer Prevention, available at http://www.aicr.org/reduce-your-cancer-risk/recommendations-for-cancer-prevention/

Andreeva V, Touvier M, Kesse-Guyot E et al (2012) B vitamin and/or omega-3 fatty acid supplementation and cancer. Arch Intern Med 172(7):540–547

Antonelli J, Jones L, Banez L et al (2009) Exercise and prostate cancer risk in a cohort of veterans undergoing prostate needle biopsy. J Urol 182:2226–2231

Asano T, McLeod R (2002) Dietary fibre for the prevention of colorectal adenomas and carcinomas (review). Cochrane Database Syst Rev 2

Aune D, Lau R, Chan D et al (2011) Nonlinear reduction in risk for colorectal cancer by fruit and vegetable intake based on meta-analysis of prospective studies. Gastroenterology 141:106–118

Bandera E, Kushi L, Moore D et al (2007) Dietary lipids and endometrial cancer: the current epidemiologic evidence. Cancer Causes Control 18:687–703

Barton M (2013) Higher levels of physical activity significantly increase survival in women with colorectal cancer. CA Cancer J Clin 63(2):83–84

Benetou V, Orfanos P, Lagiou P et al (2008) Vegetables and fruits in relation to cancer risk: evidence from the Greek EPIC cohort study. Cancer Epidemiol Biomarkers Prev 17:387–392

Beresford S, Johnson K, Ritenbaugh C et al (2006) Low-fat dietary pattern and risk of colorectal cancer the women's health initiative randomized controlled dietary modification trial. JAMA 295:643–654

Bernstein L (2009) Exercise and breast cancer prevention. Curr Oncol Rep 11:490–496

Bernstein B, Grasso T (2001) Prevalence of complementary and alternative medicine use in cancer patients. Oncology 15(10):1267–1272

Bingham S, Day N, Luben E et al (2003) Dietary fibre in food and protection against colorectal cancer in the European Prospective Investigation into Cancer and Nutrition (EPIC): an observational study. Lancet 361:1496–1501

Boffetta P, Couto E, Wichmann J et al (2010) Fruit and vegetable intake and overall cancer risk in the European Prospective Investigation into Cancer and Nutrition (EPIC). J Natl Cancer Inst 102:529–537

Boyd N, Stone J, Vogt K et al (2003) Dietary fat and breast cancer risk revisited: a meta-analysis of the published literature. Br J Cancer 89:1672–1685

Boyle T, Keegel T, Bull F et al (2012) Physical activity and risk of proximal and distal colon cancers: a systematic review and meta-analysis. J Natl Cancer Inst 104:1548–1561

Brawley O (2011) Avoidable cancer deaths globally. CA Cancer J Clin 61:67–68

Calle E, Rodriguez C, Waller-Thurmond K et al (2003) Overweight, obesity, and mortality from cancer in a prospectively studied cohort of U.S. adults. NEJM 348(17):1625–1638

Campbell P, Cotterchio M, Dicks E et al (2007) Excess body weight and colorectal cancer risk in Canada: associations in subgroups of clinically defined familial risk of cancer. Cancer Epidemiol Biomarkers Prev 16:1735–1744

Chao A, Connell C, Jacobs E (2004) Amount, type, and timing of recreational physical activity in relation to colon and rectal cancer in older adults: the Cancer Prevention Study II Nutrition cohort. Cancer Epidemiol Biomarkers Prev 13:2187–2195

Chao A, Thun M, Connell C et al (2005) Meat consumption and risk of colorectal cancer. JAMA 293:172–182

Clinton S, Giovannucci E (1998) Diet, nutrition, and prostate cancer. Annu Rev Nutr 18:413–440

Cogliano V, Bann R, Straif K et al (2011) Preventable exposures associated with human cancers. J Natl Cancer Inst 103:1827–1839

Colditz G (2010) Overview of the epidemiology methods and applications: strengths and limitations of observation study designs. Crit Rev Food Sci Nutr 50:10–12

Colditz G, Cannuscio C, Frazier A (1997) Physical activity and reduced risk of colon cancer: implications for prevention. Cancer Causes Control 8:649–667

Colditz G, Wolin K, Gehlert S (2012) Applying what we know to accelerate cancer prevention. Sci Transl Med 4(127):1–9

Cooper G (1995) Oncogenes. Jones and Bartlett, Boston

Dallal C, Sullivan-Halley J, Ross R (2007) Long-term recreational physical activity and risk of invasive and in situ breast cancer. Arch Intern Med 167:408–415

Danaei G, Hoorn S, Lopez A et al (2005) Causes of cancer in the world: comparative risk assessment of nine behavioural and environmental risk factors. Lancet 366:1784–1793

Deitz A, Zheng W, Leff M et al (2000) N-Acetyltransferase-2 genetic polymorphism, well-done meat intake and breast cancer risk among post-menopausal women. Cancer Epidemiol Biomarkes Prev 9:905–910

Demark-Wahnefried W (2006) Cancer survival: time to get moving? data accumulate suggesting a link between physical activity and cancer survival. J Clin Oncol 24(22):3517–3518

Dewys W, Begg C, Lavin P et al (1980) Prognostic effect of weight loss prior to chemotherapy in cancer patients. Eastern Cooperative Oncology Group. Am J Med 69(4):491–497

Doll R, Peto R (1981) The causes of cancer: quantitative estimates of avoidable risks of cancer in the United States today. J Natl Cancer Inst 66(6):1191–1308

Eliassen A, Colditz G, Rosner B et al (2006) Adult weight change and risk of postmenopausal breast cancer. JAMA 296:193–201

Ezzati M, Lopez A, Rodgers A et al (2002) Selected major risk factors and global and regional burden of disease. Lancet 360:1347–1360

Ferguson L (2010) Meat and cancer. Meat Sci 84:308–313

Feskanich D, Ziegler R, Michaud D et al (2000) Prospective study of fruit and vegetable consumption and risk of lung cancer among men and women. J Natl Cancer Inst 92:1812–1823

Fink B, Steck S, Wolff M et al (2007) Dietary flavonoid intake and breast cancer risk among women on long Island. Am J Epidemiol 165:514–523

Fleshner N, Bagnell P, Klotz L et al (2004) Dietary fat and prostate cancer. J Urol 171:19s–24s

Ford E, Bergman M, Kröger J et al (2009) Healthy living is the best revenge. Arch Intern Med 169(15):1355–1362

Fortmann S, Burda B, Senger C et al (2013) Vitamin, mineral, and multivitamin supplements for the primary prevention of cardiovascular disease and cancer: a systematic evidence review for the U.S. preventive services task force. Ann Intern Med 159(12):824–834

Freedland S, Platz E (2007) Obesity and prostate cancer: making sense out of apparently conflicting data. Epidemiol Rev 29:88–97

Freedman N, Park Y, Subar A et al (2008a) Fruit and vegetable intake and head and neck cancer risk in a large United States prospective cohort study. Int J Cancer 122:2330–2336

Freedman N, Subar A, Hollenbeck A et al (2008b) Fruit and vegetable intake and gastric cancer risk in a large United States prospective cohort study. Cancer Causes Control 19:459–467

Fu Z, Deming S, Fair A et al (2011) Well-Done meat intake and meat-derived mutagen exposures in relation to breast cancer risk: the Nashville Breast Health Study. Breast Cancer Res Treat 129:919–928

Fuschs C, Giovannucci E, Colditz G et al (1999) Dietary fiber and the risk of colorectal cancer and adenoma in women. N Engl J Med 340:169–176

Gann P, Hennekens C, Sacks F et al (1994) Prospective study of plasma fatty acids and risk of prostate cancer. J Natl Cancer Inst 86(4):281–286

Gansler T, Kaw C, Crammer C et al (2008) A population-based study of prevalence of complementary methods use by cancer survivors a report from the American Cancer Society's Studies of Cancer Survivors. Cancer 113:1048–1057

Gaziano J, Sesso H, Christen W et al (2012) Multivitamins in the prevention of cancer in men the physicians's Health Study II randomized controlled trial. JAMA 308(18):1871–1880

Genkinger J, Hunter D, Spiegelman D et al (2006) Dairy products and ovarian cancer: a pooled analysis of 12 cohort studies. Cancer Epidemiol Biomarkers Prev 15:364–372

George S, Mayne S, Leitzmann M et al (2009a) Dietary glycemic index, glycemic load, and risk of cancer: a prospective cohort study. Am J Epidemiol 169:462–472

George S, Park Y, Leitzmann M et al (2009b) Fruit and vegetable intake and risk of cancer: a prospective cohort study. Am J Clin Nutr 89:347–353

Giovannucci E, Rimm E, Colditz G et al (1993) A prospective study of dietary fat and risk of prostate cancer. J Natl Cancer Inst 85(19):1571–1579

Giovannucci E, Ascherio A, Rimm E et al (1995) Physical activity, obesity, and risk for colon cancer and adenoma in men. Ann Intern Med 122:327–334

Giovanucci E, Rimm E, Stampfer M et al (1994) Intake of fat, meat, and fiber in relation to risk of colon cancer in men. Cancer Res 54:2390–2397

Goldbohm R, van den Brandt P, van't Veer P et al (1994) A prospective cohort study on the relation between meat consumption and the risk of colon cancer. Cancer Res 54:718–723

Gonzalez C, Pera G, Agudo A et al (2006) Fruit and vegetable intake and the risk of stomach and oesophagus adenocarcinoma in the European Prospective Investigation into Cancer and Nutrition (EPIC-EURGAST). Int J Cancer 118:2559–2566

Gonzalez C, Lujan-Barroso L, Bueno-de-Mesquita H et al (2012) Fruit vegetable intake and the risk of gastric adenocarcinoma: a reanalysis of the European Prospective Investigation into Cancer and Nutrition (EPIC-EURGAST) study after a longer follow-up. Int J Cancer 131:2910–2919

Gordis L (2013) Epidemiology. Saunders, Philadelphia

Guo Z, Jia X, Liu J et al (2012) Herbal medicines for advanced colorectal cancer (review). Cochrane Libr 5

Hanahan D, Weinberg R (2000) The hallmarks of cancer. Cell 100:57–70

Hanahan D, Weinberg R (2011) Hallmarks of cancer: the next generation. Cell 144:646–674

Harald zur Hausen (2014) Do Some Human Cancers Originate from Infections Transmitted from Domestic Animals? American Society of Clinical Oncology Plenary Session

Howe G, Aronson K, Benito E et al (1997) The relationship between dietary fat intake and risk of colorectal cancer: evidence from the combined analysis of 13 case-control studies. Cancer Causes Control 8:215–228

Hsing A, McLaughlin J, Chow W et al (1998) Risk factors for colorectal cancer in a prospective study. Int J Cancer 77:549–553

Hu J, La Vecchia C, Augustin L et al (2013) Glycemic index, glycemic load and cancer risk. Ann Oncol 24:245–251

Huang H, Caballero B, Chang S et al (2006) The efficacy and safety of multivitamin and mineral supplement use to prevent cancer and chronic disease in adults: a systematic review for a National Institutes of Health state-of-the-science conference. Ann Intern Med 145:372–385

Hung H, Joshipura K, Jiang R et al (2004) Fruit and vegetable intake and risk of major chronic disease. J Natl Cancer Inst 96:1577–1584

Hunter D, Spiegelman D, Adami H et al (1996) Cohort studies of fat intake and the risk of breast cancer – a pooled analysis. N Engl J Med 334:356–361

Ilic D, Forbes K, Hassed C (2011) Lycopene for the prevention of prostate cancer (review). Cochrane Database Syst Rev 11

Inoue M, Yamamoto S, Kurahashi N et al (2008) Daily total physical activity level and total cancer risk in men and women: results from a large-scale population-based cohort study in Japan. Am J Epidemiol 168(4):391–403

Jemal A, Bray F, Center M et al (2011) Global cancer statistics. CA Cancer J Clin 61:69–90

Jian L, Xie L, Lee A et al (2004) Protective effect of green tea against prostate cancer: a case-control study in Southeast China. Int J Cancer 108:130–135

Jung S, Spiegelman D, Baglietto L et al (2013) Fruits and vegetable intake and risk of breast cancer by hormone receptor status. J Natl Cancer Inst 105:219–236

Kabat G, Shikany J, Beresford S et al (2008) Dietary carbohydrate, glycemic index, and glycemic load in relation to colorectal cancer risk in the Women's Health Initiative. Cancer Causes Control 19:1291–1298

Kavanaugh C, Trumbo P, Ellwood K et al (2007) The U.S. food and drug administration's evidence-based review for qualified health claims: tomatoes, lycopene, and cancer. J Natl Cancer Inst 99:1074–1085

Kawai M, Minami Y, Kuriyama S et al (2010) Adiposity, adult weight change and breast cancer risk in postmenopausal Japanese women: the Miyagi Cohort Study. Br J Cancer 103:1443–1447

Kessler R, Davis R, Foster D et al (2001) Long-term trends in the use of complementary and alternative medical therapies in the United States. Ann Intern Med 135:262–268

Key T, Fraser G, Thorogood M et al (1999) Mortality in vegetarians and non-vegetarians: detailed findings from a collaborative analysis of 5 prospective studies. Am J Clin Nutr 70:516s–524s

Key T, Allen N, Appleby P et al (2004) Fruits and vegetables and prostate cancer: no association among 1,104 cases in a prospective study of 130,544 men in the European Prospective Investigation into Cancer and Nutrition (EPIC). Int J Cancer 109:119–124

Kikuchi N, Ohmori K, Shimazu T et al (2006) No association between green tea and prostate cancer risk in Japanese men: The Ohsaki Cohort Study. Br J Cancer 95:371–373

Kim J, Kwon O (2009) Garlic intake and cancer risk: an analysis using the Food and Drug Administration's evidence-based review system for the scientific evaluation of health claims. Am J Clin Nutr 89:257–264

Kolonel L (1996) Nutrition and prostate cancer. Cancer Causes Control 7:83–94

Koushik A, Hunter D, Spiegelman D et al (2007) Fruits, vegetables, and colon cancer risk in a pooled analysis of 14 cohort studies. J Natl Cancer Inst 99:1471–1483

Kushi L, Giovannucci E (2002) Dietary fat and cancer. Am J Med 113:63s–70s

Kushi L, Doyle C, McCullough M et al (2012) American cancer society guidelines on nutrition and physical activity for cancer prevention reducing the risk of cancer with healthy food choices and physical activity. CA Cancer J Clin 62:30–67

Larsson S, Wolk A (2006) Meat Consumption and risk of colorectal cancer: a meta-analysis of prospective studies. Int J Cancer 119:2657–2664

Larsson S, Orsini N, Wolk A (2006) Milk, milk products and lactose intake and ovarian cancer risk: a meta-analysis of epidemiological studies. Int J Cancer 118:431–441

Le Marchand L, Colonel L, Wilkens L et al (1994) Animal fat consumption and prostate cancer: a prospective study in Hawaii. Epidemiology 5(3):276–282

Lee J, Chan A (2011) Fruit, vegetables, and folate: cultivating the evidence for cancer prevention. Gastroenterology 141:16–34

Liu Y, Sobue T, Otani T et al (2004) Vegetables, fruit consumption and risk of lung cancer among middle-aged Japanese men and women: JPHC study. Cancer Causes Control 15:349–357

Lonkhuijzen L, Kirsh V, Kreiger N et al (2010) Endometrial cancer and meat consumption: a case-cohort study. Eur J Cancer Prev 20:334–339

Maclean C, Newberry S, Mojica W et al (2006) Effects of omega-3 fatty acids on cancer risk a systematic review. JAMA 295:403–415

Mamede A, Tavares S, Abrantes A et al (2011) The role of vitamins in cancer: a review. Nutr Cancer 63(4):479–494

Marian M, Robert S (2010) Clinical nutrition for oncology patients. Jones and Bartlett, Boston

Martinez M, Giovannucci E, Spiegelman D et al (1997) Leisure-time physical activity, body size, and colon cancer in women. J Natl Cancer Inst 89:948–955

Maruti S, Willett W, Feskanich D et al (2008) A prospective study of age-specific physical activity and premenopausal breast cancer. J Natl Cancer Inst 100:728–737

Mathijssen R, Verweij J, de Bruijn P et al (2002) Effects of St. John's Wort on Irinotecan Metabolism. J Natl Cancer Inst 94:1247–1249

Meijerman I, Beijnen J, Schellens J (2006) Herb-drug interactions in oncology: focus on mechanism of induction. Oncologist 11:742–752

Michaud D, Augustsson K, Rimm E et al (2001) A Prospective study on intake of animal products and risk of prostate cancer. Cancer Causes Control 12(6):557–567

Michaud D, Giovannucci E, Willet W et al (2003) Dietary meat, dairy products, fat, and cholesterol and pancreatic risk in a prospective study. Am J Epidemiol 157:1115–1125

Michels K, Giovanucci E, Joshipura K et al (2000) Prospective study of fruit and vegetables consumption and incidence of colon and rectal cancers. J Natl Cancer Inst 92:1740–1752

Missmer S, Smith-Warner S, Spiegelman D et al (2002) Meat and dairy food consumption and breast cancer: a pooled analysis of cohort studies. Int J Epidemiol 31:78–85

Mok T, Yeo W, Johnson P et al (2007) A double-blind placebo-controlled randomized study of Chinese herbal medicine as complementary therapy for reduction of chemotherapy-induced toxicity. Ann Oncol 18:768–774

Mulholland H, Murray L, Cardwell C (2009) Glycemic index, glycemic load, and risk of digestive tract neoplasms: a systematic review and meta-analysis. Am J Clin Nutr 89:568–576

National Institutes of Health State-of-the-Science Conference Statement: Multivitamin/Mineral Supplements and Chronic Disease Prevention, NIH State-of-the Science Panel. Am J Clin Nutr 85:257s–264s (2007)

Neuhouser M, Wassertheil-Smoller S, Thomson C et al (2009) Multivitamin use and risk of cancer and cardiovascular disease in the Women's Health Initiative Cohorts. Arch Intern Med 169(3):294–304

Nitenberg G, Raynard B (2000) Nutritional support of the cancer patient: issues and dilemmas. Crit Rev Oncol Hematol 34:137–168

Norat T, Bingham S, Ferrari P et al (2005) Meat, fish, and colorectal cancer risk: the European Prospective Investigation into cancer and nutrition. J Natl Cancer Inst 97(12):906–916

Nothlings U, Wilkens L, Murphy S et al (2005) Meat and fat intake as risk factors for pancreatic cancer: The Multiethnic Cohort Study. J Natl Cancer Inst 97:1458–1465

Nourissat A, Vasson M, Merrouche Y et al (2008) Relationship between nutritional status and quality of life in patients with cancer. Eur J Cancer 44:1238–1242

Pan A, Sun Q, Bernstein A et al (2012) Red meat consumption and mortality results from 2 prospective cohort studies. Arch Intern Med 172(7):555–563

Park Y, Hunter D, Spiegelman D et al (2005) Dietary fiber intake and risk of colorectal cancer a pooled analysis of prospective cohort studies. JAMA 294:2849–2857

Park S, Murphy S, Wilkens L et al (2011) Multivitamin use and the risk of mortality and cancer incidence the multiethnic cohort study. Am J Epidemiol 173(8):906–914

Patel A, Rodriguez C, Jacobs E et al (2005) Recreational physical activity and risk of prostate cancer in a large cohort of U.S. men. Cancer Epidemiol Biomarkers Prev 14:275–279

Phillips R, Snowdon D (1985) Dietary relationships and fatal colorectal cancer among Seventh-Day Adventists. J Natl Cancer Inst 74(2):307–317

Prentice R, Pepe M, Self S (1989) Dietary fat and breast cancer: a quantitative assessment of the epidemiological literature and a discussion of methodological issues. Cancer Res 49:3147–3156

Prentice R, Caan B, Chlebowski R et al (2006a) Low-fat dietary pattern and risk of invasive breast cancer the Women's Health Initiative Randomized Controlled Dietary Modification Trial. JAMA 295:629–642

Prentice R, Thomson C, Caan B et al (2006b) Low-fat dietary pattern and cancer incidence in the Women's Health Initiative Dietary Modification Randomized Controlled Trial. J Natl Cancer Inst 99:1534–1543

Psaltopoulou T, Kosti R, Haidopoulos D et al (2011) Olive oil intake is inversely related to cancer prevalence: a systematic review and a meta-analysis of 13800 patients and 23340 controls in 19 observational studies. Lipids Health Dis 10:127

Qin L, Xu J, Wang P et al (2005) Milk/dairy products consumption, galactose metabolism and ovarian cancer: meta-analysis of epidemiological studies. Eur J Cancer Prev 14:13–19

Renehan A, Tyson M, Egger M et al (2008) Body-mass index and incidence of cancer: a systematic review and meta-analysis of prospective observational studies. Lancet 371:569–578

Risberg T, Vickers A, Bremnes R et al (2003) Does use of alternative medicine predict survival from cancer? Eur J Cancer 39:372–377

Rockhill B, Willett W, Hunter D et al (1999) A prospective study of recreational physical activity and breast cancer risk. Arch Intern Med 159:2290–2296

Romaguera D, Vergnaud A, Peeters P et al (2012) Is concordance with World Cancer Research Fund/American Institute for Cancer Research guidelines for cancer prevention related to subsequent risk of cancer? Results from the EPIC study. Am J Clin Nutr 96:150–163

Romieu I, Ferrari P, Rinaldi S et al (2012) Dietary glycemic index and glycemic load and breast cancer risk in the European Prospective Investigation into Cancer and Nutrition (EPIC). Am J Clin Nutr 96:345–355

Sato Y, Tsubono Y, Nakaya N et al (2005) Fruit and vegetable consumption and risk of colorectal cancer in Japan: The Miyagi Cohort Study. Public Health Nutr 8(3):309–314

Schuurman A, van del Brandt P, Dorant E et al (1999) Association of energy and fat intake with prostate carcinoma risk results from the Netherlands cohort study. Cancer 86:1019–1027

Severson R, Nomura A, Grove S et al (1989) A prospective study of demographics, diet, and prostate cancer among men of Japanese ancestry in Hawaii. Cancer Res 49:1857–1860

Shibata A, Paganini-Hill A, Ross R et al (1992) Intake of vegetables, fruits, beta-carotene, vitamin C and vitamin supplements and cancer incidence among the elderly: a prospective study. Br J Cancer 66:673–679

Shikany J, Flood A, Kitahara C et al (2011) Dietary carbohydrate, glycemic index, glycemic load, and risk or prostate cancer in the Prostate, Lung, Colorectal, and Ovarian Cancer Screening Trial (PLCO) cohort. Cancer Causes Control 22:955–1002

Shin M, Holmes M, Hankinson S et al (2002) Intake of dairy products, calcium, and vitamin D and risk of breast cancer. J Natl Cancer Inst 94:1301–1311

Siegel R, Ward E, Brawley O et al (2011) Cancer statistics, 2011 the impact of eliminating socioeconomic and racial disparities on premature cancer deaths. CA Cancer J Clin 61:212–236

Siegel R, Jiemin M, Zhaohu Z et al (2014) Cancer statistics, 2014. CA Cancer J Clin 64:9–29

Singh P, Fraser G (1998) Dietary risk factors for colon cancer in a low-risk population. Am J Epidemiol 148(8):761–774

Sinha R, Cross A, Graubard B et al (2009) Meat intake and mortality a prospective study of over half a million people. Arch Intern Med 169(6):562–571

Slattery M, Edwards S, Boucher K (1999) Lifestyle and colon cancer: an assessment of factors associated with risk. Am J Epidemiol 150:869–877

Slattery M, Benson J, Ma K et al (2001) Trans-fatty acids and colon cancer. Nutr Cancer 39 (2):170–175

Smith-Warner S, Spiefelman D, Yaun S et al (2001a) Intake of fruits and vegetables and risk of breast cancer a pooled analysis of cohort studies. JAMA 285:769–776

Smith-Warner S, Spiegelman D, Adami H et al (2001b) Types of dietary fat and breast cancer: a pooled analysis of cohort studies. Int J Cancer 92:767–774

Smith-Warner S, Ritz J, Hunter D et al (2002) Dietary fat and risk of lung cancer in a pooled analysis of prospective studies. Cancer Epidemiol Biomakers Prev 11:987–992

Smith-Warner S, Spiegelman D, Yaun S et al (2003) Fruits, vegetables and lung cancer: a pooled analysis of cohort studies. Int J Cancer 107:1001–1011

Steck S, Gaudet M, Eng S et al (2007) Cooked meat and risk of breast cancer-lifetime versus recent dietary intake. Epidemiology 18:373–382

Steinmetz K, Kushi L, Bostick R et al (1994) Vegetables, fruit, and colon cancer in the Iowa Women's Health Study. Am J Epidemiol 139(1):1–15

Stemmermann G, Nomura A, Heilbrun L (1984) Dietary fat and the risk of colorectal cancer. Cancer Res 44:4633–4637

Stevens V, McCullough M, Diver W et al (2005) Use of multivitamins and prostate cancer mortality in a large cohort of US men. Cancer Causes Control 16:643–650

Takachi R, Inoue M, Ishihara J et al (2007) Fruit and vegetable intake and risk of total cancer and cardiovascular disease Japan Public Health Center-based Prospective study. Am J Epidemiol 167:59–70

Terry P, Giovanucci E, Michels K et al (2001) Fruit, vegetables, dietary fiber, and risk of colorectal cancer. J Natl Cancer Inst 93:525–533

Tong H, Isenring E, Yates P (2009) The prevalence of nutrition impact symptoms and their relationship to quality of life and clinical outcomes in medical oncology patients. Support Care Cancer 17:83–90

Toniolo P, Riboli E, Shore R et al (1994) Consumption of meat, animal products, protein and fat and risk of breast cancer: a prospective cohort study in New York. Epidemiology 5:391–397

van Duijnhoven F, Bueno-de-Mesquita H, Ferrari P et al (2009) Fruit, vegetables, and colorectal cancer risk: the European Prospective Investigation into Cancer and Nutrition. Am J Clin Nutr 89:1441–1452

Van Gils C, Peeters P, Bueno-de-Mesquita H et al (2005) Consumption of vegetables and fruits and risk of breast cancer. JAMA 293:183–193

Veierod M, Laake P, Thelle D (1997) Dietary fat intake and risk of prostate cancer: a prospective study of 25,708 Norwegian men. Int J Cancer 73:634–638

Velicer C, Ulrich C (2008) Vitamin and mineral supplement use among US adults after cancer diagnosis: a systematic review. J Clin Oncol 26:665–673

Voorrips L, Goldbohm R, Poppel G et al (2000) Vegetable and fruit consumption and risks of colon and rectal cancer in a prospective cohort study. Am J Epidemiol 152:1081–1092

Watkins M, Erickson J, Thun M et al (2000) Multivitamin use and mortality in a large prospective study. Am J Epidemiol 152:149–162

WHO, GLOBOCAN 2012: Estimated Cancer Incidence Mortality and Prevalence Worldwide in 2012, Cancer Fact Sheets, available at http://globocan.iarc.fr/Pages/fact_sheets_cancer.aspx

Willett W (2001) Diet and cancer: one view at the start of the millennium. Cancer Epidemiol Biomarkers Prev 10:3–8

Willett W, Stampfer M, Colditz G et al (1990) Relation of meat, fat, and fiber intake to the risk of colon cancer in a prospective study among women. N Engl J Med 323:1664–1672

Wiseman M (2008) The Second World Cancer Research Fund/American Institute for Cancer Research expert report. Food, nutrition, physical activity, and the prevention of cancer: a global perspective. Proc Nutr Soc 67:253–256

Wolin K, Yan Y, Colditz G (2009) Physical activity and colon cancer prevention: a meta-analysis. Br J Cancer 100:611–616

Wolin K, Carson K, Colditz G (2010) Obesity and cancer. Oncologist 15:556–565

Wu A, Yu M, Tseng C et al (2008) Epidemiology of soy exposures and breast cancer risk. Br J Cancer 98:9–14

Wu S, Liang J, Zhang L et al (2011) Fish consumption and the risk of gastric cancer: systematic review and meta-analysis. BMC Cancer 11:26

Wu Y, Zhang D, Kang S (2013) Physical activity and risk of breast cancer: a meta-analysis of prospective studies. Breast Cancer Res Treat 137:869–882

Xu M, Deng P, Qi C et al (2009) Adjuvant phytotherapy in the treatment of cervical cancer: a systematic review and meta-analysis. J Altern Complement Med 15(12):1347–1353

Yang G, Shu X, Chow W et al (2012) Soy food intake and risk of lung cancer: evidence from the Shanghai Women's Health Study and a meta-analysis. Am J Epidemiol 176(10):846–855

Yang J, Zhu L, Wu Z et al (2013) Chinese herbal medicines for induction of remission in advanced or late gastric cancer (review). Cochrane Database Syst Rev 4:CD005096

Zheng W, Lee S (2009) Well-done meat intake, heterocyclic amine exposure, and cancer risk. Nutr Cancer 61(4):437–446

Zheng W, Gustafson D, Sinha R et al (1998) Well-done meat intake and the risk of breast cancer. J Natl Cancer Inst 90:1724–1729

Zheng J, Hu X, Zhao Y et al (2013) Intake of fish and marine n-3 polyunsaturated fatty acids and risk of breast cancer: meta-analysis of date from 21 independent prospective cohort studies. BMJ 346:f3706

Zock P, Katan M (1998) Linoleic acid intake and cancer risk: a review and meta-analysis. Am J Clin Nutr 68:142–153

Chapter 12
Textbox: Pesticides and Cancer in Conventionally-Grown Versus Organic Food

Gabriela Steier

Abstract Scientific publications have proven the links between pesticides in food and cancer prevalence. Conventionally-grown foods are often high in pesticides as compared to their organic counterparts. Pesticides are not only dangerous for the environment, but also for public health. One of the most dangerous diseases induced by pesticides are various types of cancers, such as: bladder cancer, bone cancer, brain cancer, cervical cancer, colorectal cancer, eye cancer, gallbladder cancer, kidney or renal cancer, larynx cancer, leukemia, lip cancer, liver or hepatic cancer, lung cancer, lymphoma, melanoma, mouth cancer, multiple myeloma, neuroblastoma, oesophageal cancer, ovarian cancer, pancreatic cancer, prostate cancer, soft tissue sarcoma, stomach cancer, sinonasal cancer, testicular cancer, thyroid cancer, and uterine cancer.

Conventionally-grown foods are often high in pesticides as compared to their organic counterparts. Pesticides are not only dangerous for the environment, but also for public health. One of the most dangerous diseases induced by pesticides are various types of cancers, such as: bladder cancer, bone cancer, brain cancer, cervical cancer, colorectal cancer, eye cancer, gallbladder cancer, kidney or renal cancer, larynx cancer, leukemia, lip cancer, liver or hepatic cancer, lung cancer, lymphoma, melanoma, mouth cancer, multiple myeloma, neuroblastoma, oesophageal cancer, ovarian cancer, pancreatic cancer, prostate cancer, soft tissue sarcoma, stomach cancer, sinonasal cancer, testicular cancer, thyroid cancer, and uterine cancer.[1]

---

[1] Beyond Pesticides, Pesticide-Induced Diseases: Cancer, http://www.beyondpesticides.org/health/cancer.php (*citing inter alia* Forastiere et al. (1993), Sharma et al. (2013), Koutros et al. (2009), Wesseling et al. (1999), Merietti et al. (2006), Carrozza et al. (2008), Holly et al. (1992), Moore et al. (2005), Nielsen et al. (2010), Rosso et al. (2008), Ruder et al. (2006); Van Wijngaarden et al. (2003); Efird et al. (2003), Holly et al. (1998), Pogoda and Preston-Martin (1997), Kristensen et al. (1996), Bunin et al. (1994), Cordier et al. (1994), Davis et al. (1993),

G. Steier (✉)
Food Law International, LLP, Boston, MA, USA
e-mail: g.steier@foodlawinternational.com

Although advocacy groups for a more sustainable, organic, plant-based whole-foods diet have long warned that conventional and intensive agriculture is high in carcinogenic pesticides, which persist in foods that people consume, a recent study finally offers conclusive and specific evidence. The renowned medical journal, *the Lancet Oncology,* published the findings of "17 experts from 11 countries [meeting] at the International Agency for Research on Cancer (IARC; Lyon, France) to assess the carcinogenicity of the organophosphate pesticides tetrachlorvinphos, parathion, malathion, diazinon, and glyphosate."[2] This group of organophosphate pesticides are highly toxic substances, similar and sometimes overlapping with those used in chemical warfare agents used in World War II.[3]

Some of the key findings include the following, which are directly cited from the study[4]:

- The insecticides tetrachlorvinphos and parathion were classified as "possibly carcinogenic to humans." ... Tetrachlorvinphos is banned in the European Union. In the USA, it continues to be used on animals, including in pet flea collars.
- For parathion, associations with cancer in several tissues were observed in occupational studies, but the evidence in humans remains sparse... Parathion is rapidly absorbed and distributed. Parathion metabolism to the bioactive metabolite, paraoxon, is similar across species. Although bacterial mutagenesis tests were negative, parathion induced DNA and chromosomal damage in human cells in vitro.

Wilkins and Sinks (1990), Wilkins and Koutras (1988), Gold et al. (1979), Samanic et al. (2008), Provost et al. (2007), Lee et al. (2005), Zheng et al. (2001), Viel et al. (1998), Mills (1998), Smith-Rooker et al. (1992), Musicco et al. (1988), Delzell and Grufferman (1985), Kross et al. (1996), Rodvall et al. (1996), Figa-Talamanca et al. (1993), Blair et al. (1983), Ortega Jacome et al. (2010), Shakeel et al. (2010), Teitelbaum et al. (2007), Charlier et al. (2003); Mills and Yang (2005), Brophy et al. (2002), Dolapsakis et al. (2001), Band et al. (2000), Duell et al. (2000), Fleming et al. (1999), Cerhan et al. (1998), Zhong and Rafnsson (1996), Lee et al. (2007), Lo et al. (2010), Samanic et al. (2006), Kang et al. (2008), van Bemmel et al. (2008), Giordano et al. (2006), Sharpe et al. (1995), Fear et al. (1998), Olshan et al. (1993), Tsai et al. (2006), Mellemgaard et al. (1994), Hu et al. (2002), Karami et al. (2008), Van Maele-Fabry et al. (2011), Turner et al. (2010), Rull et al. (2009), Soldin et al. (2009), Lafiura et al. (2007), Menegaux et al. (2006), Alderton et al. (2006), Reynolds et al. (2005), Ma et al. (2002), Alexander et al. (2001), Bonner et al. (2010), Van Maele-Fabry et al. (2007), Miligi et al. (2006), Beane Freeman et al. (2005), Chrisman et al. (2008), Yafune et al. (2013), Lee et al. (2004), Bumroongkit et al. (2008), Rusiecki et al. (2006), Chiu et al. (2006), McDuffie et al. (2001), Fortes et al. (2007), Tarvainen et al. (2008), Lope et al. (2008), Merhi et al. (2007), Landgren et al. (2009), Walker et al. (2007), Andreotti et al. (2009), Ragin et al. (2013), Svensson et al. (2013), Mullins and Loeb (2012), Budnik et al. (2012), Band et al. (2011), Cockburn et al. (2011), Tisch et al. (2002), Béranger et al. (2013), Leux and Guénel (2010).

[2]Guyton et al. (2015).

[3]Pesticide Action Network (PAN), Organophosphates, http://www.panna.org/resources/organophosphates

[4]Guyton et al. (2015), internal citations omitted.

- The insecticides malathion and diazinon were classified as "probably carcinogenic to humans." ... Malathion is used in agriculture, public health, and residential insect control. It continues to be produced in substantial volumes throughout the world. There is limited evidence in humans for the carcinogenicity of malathion. Case–control analyses of occupational exposures reported positive associations with non-Hodgkin lymphoma in the USA, Canada, and Sweden, although no increased risk of non-Hodgkin lymphoma was observed in the large Agricultural Health Study cohort (AHS). Occupational use was associated with an increased risk of prostate cancer in a Canadian case–control study and in the AHS, which reported a significant trend for aggressive cancers after adjustment for other pesticides. Diazinon has been applied in agriculture and for control of home and garden insects... Support for an increased risk of leukaemia in the AHS was strengthened by a monotonic increase in risk with cumulative diazinon exposure after adjustment for other pesticides. Multiple updates from the AHS consistently showed an increased risk of lung cancer with an exposure–response association that was not explained by confounding by other pesticides, smoking, or other established lung cancer risk factors.
- Glyphosate is a broad-spectrum herbicide, currently with the highest production volumes of all herbicides. It is used in more than 750 different products for agriculture, forestry, urban, and home applications. Its use has increased sharply with the development of genetically modified glyphosate-resistant crop varieties. Glyphosate has been detected in air during spraying, in water, and in food. There was limited evidence in humans for the carcinogenicity of glyphosate. Case–control studies of occupational exposure in the USA, Canada, and Sweden reported increased risks for non-Hodgkin lymphoma that persisted after adjustment for other pesticides... Glyphosate has been detected in the blood and urine of agricultural workers, indicating absorption. Soil microbes degrade glyphosate to aminomethylphosphoris acid (AMPA). Blood AMPA detection after poisonings suggests intestinal microbial metabolism in humans... The Working Group classified glyphosate as "probably carcinogenic in humans."[5]

All of this evidence supports the fact that pesticides used in conventional and intensive agriculture are not only carcinogenic for farm workers or people living in close proximity to farm land, but also linger in food and pose a cohort of dangers to public health. The Centers for Disease Control and Prevention, for example, reported that 29 different pesticides can be found in human bodies.[6] A Consumers Union paper cited USDA data that found a substantially higher accumulation of pesticides in conventionally grown produce:

> The USDA data showed that 73 percent of conventionally grown foods had at least one pesticide residue, while only 23 percent of organically grown samples of the same crops

---

[5]Guyton et al. (2015), *internal citations omitted*.
[6]Consumer Reports, Eat the Peach Not the Pesticides, http://www.consumerreports.org/cro/health/natural-health/pesticides/index.htm (citing the Centers for Disease Control and Prevention; internal citations omitted).

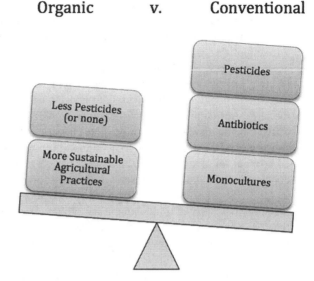

Fig. 12.1 Organic versus conventional agriculture

had any residues. More than 90 percent of the USDA's samples of conventionally-grown apples, peaches, pears, strawberries and celery had residues, and conventionally-grown crops were six times as likely as organic to contain multiple pesticide residues.[7]

Thus, buying organic may be the only tenable method to reduce one's exposure to pesticides, albeit not eradicating the dangers entirely (See the above figure provided by author).

The Consumer Report *From Crop to Table: Pesticide Use in Produce* reveals some of the most pressing risks of pesticide uses in food and described some of the US regulations—or lack thereof—under US law.

References

Alderton LE et al (2006) Am J Epidemiol 164(3):212–221
Alexander FE et al (2001) Cancer Res 61(6):2542–2546
Andreotti G et al (2009) Int J Cancer 124(10):2495–2500
Band PR et al (2000) J Occup Environ Med 42(3):284–310
Band PR, Abanto Z, Bert J et al (2011) Prostate 71(2):168–183
Beane Freeman LE et al (2005) Am J Epidemiol 162(11):1070–1079
Béranger R, Le Cornet C, Schüz J, Fervers B (2013) PLoS One 8(10), e77130
Blair A et al (1983) J Natl Cancer Inst 71(1):31–37
Bonner MR et al (2010) Cancer Causes Control 21(6):871–877

---

[7]Consumer's Union, Consumers Union Research Team Shows: Organic Foods Really DO Have Less Pesticides (May 2002), https://consumersunion.org/news/cu-research-team-shows-organic-foods-really-do-have-less-pesticides/

Brophy J et al (2002) Int J Occup Environ Health 8(4):346–353
Budnik LT, Kloth S, Velasco-Garrido M, Baur X (2012) Environ Health 11:5
Bumroongkit K et al (2008) Cancer Genet Cytogenet 185(1):20–27
Bunin GR et al (1994) Cancer Epidemiol Biomarkers Prev 3:197–204
Carrozza SE et al (2008) Environ Health Perspect 116(4):559–565
Cerhan JR et al (1998) Cancer Causes Control 9(3):311–319
Charlier C, Albert A, Herman P et al (2003) Occup Environ Med 60(5):348–351
Chiu B et al (2006) Blood 108(4):1363–1369
Chrisman JD et al (2008) Int J Hyg Environ Health 212(3):310–321
Cockburn M, Mills P, Zhang X et al (2011) Am J Epidemiol 173(11):1280–1288
Cordier S et al (1994) Int J Cancer 59(6):776–782
Davis J et al (1993) Family pesticide use and childhood brain cancer. Arch Environ Contam Toxicol 24:87–92
Delzell E, Grufferman S (1985) Am J Epidemiol 121(3):391–402
Dolapsakis G et al (2001) Eur J Cancer 37(12):1531–1536
Duell EJ et al (2000) Epidemiology 11(5):523–531
Efird JT et al (2003) Paediatr Perinat Epidemiol 17(2):201–211
Fear NT et al (1998) Br J Cancer 77(5):825–829
Figa-Talamanca I et al (1993) Int J Epidemiol 22(4):579–583
Fleming L et al (1999) J Occup Environ Med 41(4):279–288
Forastiere F et al (1993) Scand J Work Environ Health 19(6):382–389
Fortes C et al (2007) Eur J Cancer 43(6):1066–1075
Giordano F et al (2006) Int J Immunopathol Pharmacol 19(Suppl 4):61–65
Gold E et al (1979) Am J Epidemiol 109(3):309–319
Guyton KZ, Loomis D, Grosse Y, El Ghissassi F, Benbrahim-Tallaa L, Guha N et al (2015) Carcinogenicity of tetrachlorvinphos, parathion, malathion, diazinon, and glyphosate. Lancet Oncol 16:490–491. doi:10.1016/S1470-2045(15)70134-8
Holly EA et al (1992) Am J Epidemiol 135(2):122–129
Holly EA et al (1998) Cancer Epidemiol Biomarkers Prev 7(9):797–802
Hu J et al (2002) Occup Med 52(3):157–164
Kang D et al (2008) Environ Res 107(2):271–276
Karami S et al (2008) Carcinogenesis 29(8):1567–1571
Koutros S et al (2009) Int J Cancer 124(5):1206–1212
Kristensen P et al (1996) Int J Cancer 65(1):39–50
Kross BC et al (1996) Am J Ind Med 29(5):501–506
Lafiura KM et al (2007) Pediatr Blood Cancer 48(5):624–628
Landgren O et al (2009) Blood 113(25):6386–6391
Lee WJ et al (2004) J Natl Cancer Inst 96(23):1781–1789
Lee W et al (2005) Occup Environ Med 62(11):786–792
Lee WJ et al (2007) Int J Cancer 121(2):339–346
Leux C, Guénel P (2010) Rev Epidemiol Sante Publique 58(5):359–367
Lo AC et al (2010) Dis Colon Rectum 53(5):830–837
Lope V et al (2008) Cancer Epidemiol Biomarkers Prev 17(11):3123–3127
Ma X et al (2002) Critical windows of exposure to household pesticides and risk of childhood leukemia. Environ Health Perspect 110:955–960
McDuffie H et al (2001) Cancer Epidemiol Biomarkers Prev 10:1155–1163
Mellemgaard A et al (1994) Scand J Work Environ Health 20(3):160–165
Menegaux F et al (2006) Occup Environ Med 63(2):131–134
Merhi M et al (2007) Cancer Causes Control 18(10):1209–1226
Merietti F et al (2006) Int J Cancer 118(3):721–727
Miligi L et al (2006) Ann N Y Acad Sci 1076:366–377
Mills PK (1998) Arch Environ Health 53(6):410–413
Mills PK, Yang R (2005) Int J Occup Environ Health 11(2):123–131

Moore LE et al (2005) Int J Cancer 114(3):472–478
Mullins JK, Loeb S (2012) Urol Oncol 30(2):216–219
Musicco M et al (1988) Am J Epidemiol 128(4):778–785
Nielsen SS et al (2010) Childhood brain tumors, residential insecticide exposure, and pesticide metabolism genes. Environ Health Perspect 118(1):144–149
Olshan AF et al (1993) Cancer 72(3):938–944
Ortega Jacome GP et al (2010) J Toxicol Environ Health A 73(13–14):858–865
Pogoda JM, Preston-Martin S (1997) Environ Health Perspect 105:1214–1220
Provost D et al (2007) Brain tumours and exposure to pesticides: a case–control study in southwestern France. Occup Environ Med 64:509–514
Ragin C, Davis-Reyes B, Tadesse H et al (2013) Am J Mens Health 7(2):102–109
Reynolds P et al (2005) Epidemiology 16(1):93–100
Rodvall Y et al (1996) Occup Environ Med 53:526–532
Rosso AL et al (2008) Cancer Causes Control 19(10):1201–1207
Ruder AM et al (2006) J Agric Saf Health 12(4):255–274
Rull RP et al (2009) Environ Res 109(7):891–899
Rusiecki JA et al (2006) Int J Cancer 118(12):3118–3123
Samanic C et al (2006) Environ Health Perspect 114(10):1521–1526
Samanic CM et al (2008) Am J Epidemiol 167(8):976–985
Shakeel MK, George PS, Jose J, Jose J, Mathew A (2010) Asian Pac J Cancer Prev 11(1):173–180
Sharma T, Jain S, Verma A et al (2013) Cancer Biomark 13(4):243–251
Sharpe CR et al (1995) Am J Epidemiol 141(3):210–217
Smith-Rooker JL et al (1992) J Neurosci Nurs 24(5):260–264
Soldin OP et al (2009) Ther Drug Monit 31(4):495–501
Svensson RU, Bannick NL, Marin MJ et al (2013) J Environ Pathol Toxicol Oncol 32(1):29–39
Tarvainen L et al (2008) Int J Cancer 123(3):653–659
Teitelbaum SL et al (2007) Am J Epidemiol 165(6):643–651
Tisch M et al (2002) Eur Arch Otorhinolaryngol 259:150–153
Tsai J et al (2006) Int J Hyg Environ Health 209(1):57–64
Turner MC et al (2010) Environ Health Perspect 118(1):33–41
van Bemmel DM et al (2008) Environ Health Perspect 116(11):1541–1546
Van Maele-Fabry G et al (2007) Cancer Causes Control 18(5):457–478
Van Maele-Fabry G, Lantin AC, Hoet P, Lison D (2011) Environ Int 37(1):280–291
Van Wijngaarden E et al (2003) Am J Epidemiol 157(11):989–997
Viel JF et al (1998) Brain cancer mortality among French farmers: the vineyard pesticide hypothesis. Arch Environ Health 53(1):65–70
Walker KM et al (2007) J Agric Saf Health 13(1):9–24
Wesseling C et al (1999) Int J Epidemiol 28:365–374
Wilkins JR, Koutras RA (1988) Am J Ind Med 14(3):299–318
Wilkins JR, Sinks T (1990) Am J Epidemiol 132(2):275–292
Yafune A, Kawai M, Itahashi M et al (2013) Toxicol Lett 222(3):295–302
Zheng T et al (2001) J Occup Environ Med 43(4):333–340
Zhong Y, Rafnsson V (1996) Int J Epidemiol 25(6):1117–1124

Chapter 13
Intersections of Dental Health and Food Law: The Conflict of Systemic Flouridation as a Public Health Instrument to Prevent Tooth Decay

Liviu Steier, Regina Steier, and Gabriela Steier

Abstract Prevention of tooth decay via administration of flourides in drinking water, milk, and salt has been performed by governments around the world over the last decades. The literature was scrutinized to validate benefits and disadvantages of systematic flouridation for the individual as well to identify the legal background.

Regulation of maximal dose and most beneficial application methodology were identified. A great number of countries have interrupted systematic use in favor of local application due to genera health hazards generated while others continue despite seriously breaching valid law.

The most important finding of this chapter is that local application of flouride best contributes to decay prevention in conformity with biomedical ethics best complimenting general health.

13.1 Introduction

Dental caries is an infectious disease, which constitutes a continuously growing and global public health threat—some even call it a pandemic.[1] In a textbook on international food law and policy including public health, the laws and policies encompassing dental and oral health must not be overlooked. Although several

---

[1]Edelstein (2006).

L. Steier • R. Steier
Studienzentrum für Fortschrittliche Zahneilkunde, Berlin, Germany

Food Law International, LLP, Boston, MA, USA
e-mail: LSteier@gmail.com; Dr.R.Steier@gmail.com

G. Steier (✉)
Food Law International, LLP, Boston, MA, USA
e-mail: G.Steier@foodlawinternational.com

© Springer International Publishing Switzerland 2016
G. Steier, K.K. Patel (eds.), *International Food Law and Policy*,
DOI 10.1007/978-3-319-07542-6_13

chapters in this book focus on food production, consumption and trade, food intake starting at the mouth plays, an integral role in a complete interdisciplinary analysis. Eating and drinking, after all, bridge the production and consumption of food aspects. On the most basic level of logic, one cannot eat well with decayed teeth, so that nutrient intake is impaired by diseased mouths.[2] Those laws affecting what consumers eat and drink, consequently, should be rooted in scientifically sound data and analysis. Notwithstanding the firmly-established place of public health and nutrition policy within the food law framework, dental health should figure into any comprehensive analysis, especially in light of regulatory frameworks affecting oral health. The laws addressing dental caries as an infectious disease, therefore, use the scientific data to promulgate guidelines and treat dental decay as a pandemic.

One of the instruments to treat the pandemic of dental decay is fluoridation, i.e. the artificial addition of fluoride to water, milk, salt, tablets, toothpastes and other topical agents. Within the scope of this book are, of course, the fluoridation of water, milk and salt, all of which are undeniably ubiquitous sources of nutrition and hydration in every country and for every person around the world. Governments around the world have added fluoride to tap water and monitored fluoride levels in milk and salt with dental health in mind for over 70 years. Essentially, the addition of fluoride to water, milk and salt is a governmental instrument of public health intended to strengthen tooth enamel and to halt the pandemic of rampant caries spread. The corresponding goal is oral health improvement combined with topical flouride application and the protection of socio-economic interests. However, medical and scientific publications document the conflicts and the dangers of systemic fluoridation, where local application prevents tooth decay but ingestion fails to accomplish the same goal.

After decades of systemic water, milk and salt fluoridation in various countries around the world, the question of the efficacy of systemic fluoridation remains highly controversial and its efficacy has not been conclusively proven. For the reason that scientific data has not conclusively proven the efficacy of systemic fluoridation, the fact that governments continue to support this practice begs the question of whether this long-standing practice is truly in the public's best interest or whether it serves a different, possibly unintentional, purpose. This chapter analyzes and explains the conflict that permeates not dentists' and dental organizations understanding of the problem but also demonstrates how lay people aid in making executive decisions that affect public health in various ways. As food and water are inextricably intertwined, this chapter sheds light on a marginalized issue that warrants more attention than popular knowledge suggests.

This chapter sheds light on the premise that fluoridation is an effective public health measure, instrumentizing legal recommendations and regulations worldwide, to reduce oral diseases. Notwithstanding the intentions behind the schemes of mass-medicating entire countries through fluoridation of water, milk and salt, it remains a systematic medication. The successes and dangers associated with

---

[2]Touger-Decker and van Loveren (2003).

systemic versus topical fluoridation and the globally deep-rooted policy framework will be weighed against governmental authority and consumers' rights to chose not to be systematically medicated as part of a larger contagious pandemic.

13.1.1 Dental Caries and Systemic Fluoridation

The prevalence of dental caries as an omnipresent disease around the world validates its interdisciplinary relevance within public health, socio-economic concerns and food law. The US Surgeon General acknowledges dental caries as a severe national economic impact, thereby invoking aspects of food law because dental decay is a threat of substantial magnitude. Further statistical evidence supports the social and economic impact of tooth decay where a study published in the *American Journal of Dentistry* found that "[e]pidemiologic data from different countries show[s] an increase in dental caries prevalence affecting all ages."[3] As a result, the socio—economic impact of dental decay gave rise to the need of addressing caries disease as a high-priority national and international public health problem. Notably, a report of the US National Health Interview Survey in 1981, 4.87 million dental conditions caused 17.7 million days of restricted activity, 6.73 million days of bed disability, and 7.05 million days of work loss, according to the report.[4] Moreover, dental caries in pre-school children has an impact on psychological and social aspects of the their lives[5] while school children experienced dental pain, missed classes and performed poorly in coursework due to dental discomfort.[6] In the US, in the year 2000 alone, the national annual cost of dental treatment exceeded $60 billion according to the Health Care Financing Administration. The figure has not really changed over the past 15 years and, therefore, dental treatment costs permeate both social and economic avenues. These avenues, in turn, implicate food law and policy because they affect the mouths and chewing processes of consumers in the first step of food digestion, the ultimate goal of the entire food production chain (Fig. 13.1).

Thus, in a discourse "from farm to fork," as food law and policy often describe the food production chain, it is important to follow the path of food once it reaches the fork and, thereby, the mouth. This figurative speech is important but the connections have not been made until the writing of this chapter.

---

[3]Bagramian et al. (2009).
[4]Reisine (1985).
[5]Sheiham (2006).
[6]Jackson et al. (2011).

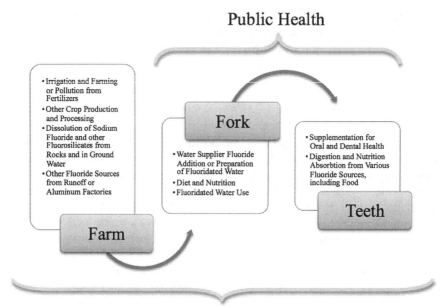

Fig. 13.1 Extension of fluoride accumulation in the farm to fork (to teeth) chain. Following the path from food production to consumption, a.k.a. from farm to fork, the pathway can be extended to digestion and nutrition and, thereby, to teeth. Through this extension of the fluoride dissolution, administration and distribution pathway, oral health becomes an integral step in the public health and regulatory understanding of fluoride accumulation

13.1.2 Tooth Decay and Predisposing Factors of Cavity Rates

The connections between food law, nutrition and dental decay lie in the food-related predispositions, such as a high-sugar and high-carbohydrate diet. Susceptibility to dental decay is directly related to an increased intake of carbohydrates. Although other factors also play important roles, such as the diminished salivary production associated with negligent oral hygiene and other major caries predisposing factors,[7] they are beyond the scope of this chapter.

First and foremost, dental caries is "an oral infectious disease of the teeth in which organic acid metabolites produced by oral microorganisms lead to demineralization and destruction of the tooth structures."[8] Even though a variety of terms have emerged over the years, such as tooth decay and cavities, the disease is a bacterially-induced destruction of enamel and dentin, the hard tissue of the tooth. The consumption of carbohydrates and sugars, which bacteria break down into

---

[7]Gupta et al. (2013); van Houte (1994), pp. 672–681.

[8]Touger-Decker and van Loveren (2003), pp. 881S–892S.

acids in a process called fermentation, generate a reduced pH level and facilitate the demineralization of protective healthy tooth tissue. The results are the common dental lesions, a.k.a. "holes in teeth."

Predisposing factors of dental decay have been widely studied. Scientific literature describes the major etiologic and predisposing factors involved in caries as follows: (a) the availability of cariogenic bacteria, (b) the intake of fermentable carbohydrates, (c) a susceptible tooth and host, (d) time[9] (e) individual, social, and community risk indicators.[10] The resulting socio economic burden on oral health imposes substantial complications for individual suffering from tooth decay. Notably, the World Health organization (WHO) finds that "the major risk factors relate to unhealthy lifestyles (i.e. poor diet, nutrition and oral hygiene and [the] use of tobacco and alcohol), ... limited availability and accessibility of oral health services" and "poor living conditions."[11] Therefore, dental decay is preventable to a large degree and falls squarely within the realm of public health and socio-economic regulatory decision-making.

The wide array of factors, including diet and nutrition, affect the severity of an individual's caries prevalence. In fact, the variables associated with dental caries are considered risk indicators, that likely predict the disease prevalence.[12] Figure 13.2 illustrates this Caries Balance and displays some of the most widely accepted factors. The scale tips toward caries rather easily. Public health and caries prevention measures, therefore, can directly impact the various factors by addressing the predictors listed in Fig. 13.2. Consequently, food policy can and should consider these factors in making dietary recommendations and setting nutrition goals.

13.1.3 Prevalence of Tooth Decay

The prevalence of tooth decay has been increasing. Especially developing countries[13] and lower-income populations have a higher prevalence of tooth decay but have limited, if any, dental care that can treat caries or halt the spread of infections.[14] From any age groups, elderly populations are most affected by tooth decay and loss.[15] The WHO estimates that, globally, about one third of people aged 65–74 have no natural teeth.[16] Of course, tooth loss correlates with poor nutrition choices

---

[9]Harris et al. (2004), pp. 71–85.
[10]Ismail et al. (2008), pp. 55–68.
[11]Petersen et al. (2005), pp. 661–669.
[12]da Silva Tagliaferro et al. (2008), pp. 408–413.
[13]Adiatman et al. (2013), pp. 262–269.
[14]Ismail et al. (2008), pp. 55–68.
[15]Ervin and Dye (2012).
[16]WHO, Oral Health Fact Sheet N°318 (2012), http://www.who.int/mediacentre/factsheets/fs318/en/

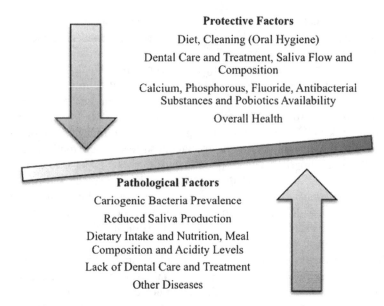

Fig. 13.2 The caries balance. Protective and pathological factors shift the balance between re- and demineralization of tooth enamel off the baseline and, ultimately, caries. This balance is based on Adrian Lussi, Elmar Hellwig, Joachim Klimek, Fluorides—Mode of Action and Recommendations for Use, 122 Schweiz. Monatsschr. Zahnmed. (Nov. 2012) (*citing* Featherstone J D B: The science and practice of caries prevention. 131 JADA 887–889 (2000))

or availability because of their deteriorated capability to chew solid foods, such as whole grains, fresh fruits and vegetables. Therefore, the prevention of caries impacts whether nutritional goals are met.

All age groups, however, are affected by caries. If left untreated, caries and tooth aches can even be fatal.[17] The WHO found that, globally, between 60 and 90 percent of school children and adults suffer from tooth decay.[18] In the US, about one half of all children between the ages of five and nine have at least one cavity and one filling of a primary or permanent tooth.[19] The caries prevalence increases to nearly 80 percent for teenagers and reaches almost 85 percent in adults.[20] Table 13.1 provides the caries prevalence in a selection of countries[21] and illustrates the wide reach of this infectious disease.

Caries, as the data shows, is an infectious disease that can lead to severe health detriments and must be prevented and treated. As a logic and immediate

---

[17] Owings (2007).

[18] Petersen and Esheng (1998).

[19] Oral Health in America: Report of the Surgeon General, http://www.nidr.nih.gov/sgr/sgrohweb/chap4.htm#dental_caries

[20] Id. (internal citations omitted).

[21] Bagramian et al. (2009), pp. 3–8.

Table 13.1 Selected rates of caries prevalence

Prevalence	Country	Year
50–78 %	USA	2004
97.1 %	Philippines	2006
100 %	China	2008
50 %	UK	2003
90.2 %	Mexico	2006
59.8 %	Norway	2006

Source: Bagramian et al. (2009)

consequence of tooth decay, the ability to eat and food choices work in conjunction with other factors to impact nutrition outcomes. The high prevalence of caries, therefore, justifies why governments all over the world address this public health threat—although it remains questionable whether aggressive water fluoridation is the best method to battle caries.

13.1.4 Fluoride: The Panacea

In 1901, Dr. Frederick S. McKay from Colorado Springs, identified the causal interrelation between water components (here fluorine)[22] and anomalies in tooth structure for the first time.[23] He distinguished fluorosis, the ingestion of excessive amounts of fluorine, in his research.[24] Then, in 1930, H. V. Churchill identified Fluoride as originating cause for fluorosis through spectrographic analysis.[25] Several years later, in 1945, scientists agreed on acceptable levels of fluoride implementation in water and the first prospective field studies using water fluoridation to reduce dental caries were initiated. Since then, water fluoridation has come a long way and has reached the peak focus of international attention and public health programs. Despite its limitations, current research, ranking high in scientific evidence, proves that fluoride exposure reduces dental caries between 25 and 27 % in children and as well as in adolescents.[26] Nearly 100 years of research yielded conclusive results that fluoridation may aid in the reduction of dental decay. Over

---

[22]To simplify, fluorite is the primary mineral source of fluorine, the noble gas F. Organic fluorides used in tooth pastes are mineral salts, which are not the same form of the cytotoxic F gas because salts are not as highly volatile. A lot more remains to be said about the various chemical forms and properties of F and how they affect the body, but such an analysis is outside the scope of this book.

[23]Centers for Disease Control and Prevention, Achievements in Public Health, http://www.cdc.gov/mmwr/preview/mmwrhtml/mm4841a1.htm

[24]Id.

[25]National Institute of Dental and Craniofacial Research, The Story of Fluoridation.

[26]Marinho et al. (2002a), CD002279; Marinho et al. (2002b), CD002280; Marinho et al. (2003a), CD002782; Marinho et al. (2003b), CD002284; Marinho et al. (2003c), pp. 448–458; Marinho et al. (2003d), CD002278; Marinho et al. (2004a), CD002781; Marinho et al. (2004b), CD002780; Griffin et al. (2007), pp. 410–415.

the decades, different patterns of fluoridation have been researched, systemic and topical ones. The crux consists in the proper selection of the best fluoridation approach with the highest decay prevention and the least adverse side effects for humans.

The American Academy of Pediatric Dentistry published guidelines on fluoride therapy. Through these guidelines, it is clear that fluoridation of water is a major strategy to address caries prevalence in children and adults. Thus, according to the American Academy of Pediatric Dentistry, the "[w]idespread use of fluoride has been a major factor in the decline in prevalence and severity of dental caries in the US and other economically developed countries. When used appropriately, fluoride is both safe and effective in preventing and controlling dental caries."[27] This is not to say, however, that there are no downsides to fluoridation of drinking water, which is merely one of many methods to make fluoride available for caries prevention. Consequently, the guidelines acknowledge that "[f]luoridation of community drinking water is the most equitable and cost-effective method of delivering fluoride to all members of most communities"[28] but it is not necessarily the safest or most effective method to contain the caries panacea. Of interest for this chapter, therefore, is the effect of cookie-cutter fluoride treatment as a public health and food safety concern.

13.2 The Safety of Fluoridation and the Avenues of Systemic Fluoridation

Understanding the protective mechanism of action of fluoride is a prerequisite to select the best fluoridation approach for public health. Government agents making executive decisions and even dental professionals often lose sight of the basic mechanism of action: Teeth have a hard and a soft tissue structure. The hard tissue component is build by the highly mineralized enamel and the organic matrix rich dentin and cementum. Calcium deficient material with different ions is more abundant in the hard structure building a less resistant apatite when compared to hydroxyapatite. Thus, local and targeted application of fluoride can strengthen the tooth structure. However, systemic application fails to protect tooth structure because the fluoride does not dissolve as needed.

Enamel, the outermost hard and mineral-based shell of the tooth, is characterized by its content of 3 % carbonate versus 5 % in dentin. Enamel accounts for its availability of flourhydroxyapatite where around 5 % of the OH groups are replaced by fluoride, stabilizing the apatite structure alongside calcium, phosphate, hydroxyl and fluoride ions. The flourhydroxyapatite concentration drops towards the inside of the tooth. Conversely, the presumed steady state of calcium ions made available

---

[27] American Academy of Pediatric Dentistry, Guidelines on Fluoride Treatment (2013).
[28] Id.

by saliva, enamel, and fluoride ions are responsible for the barrier against acid attack. Any imbalance of these minerals favors the dissolution of phosphate and hydroxyl ions, which then leads to hard tissue dissolution.

Demineralization through high sugar intake and insufficient oral hygiene are the decisive factors in decay propagation. Systemic fluoridation approaches, like water, milk and salt fluoridation favor the incorporation of fluoride into the mineral components with a minimal ability to dissolution when compared to topical approaches. Topical or local approaches grant a higher level of fluoride saturation around the enamel inhibiting effective demineralization. This has been scientifically proven. Thus, fluoride, on the most basic level, helps to prevent demineralization.

On a larger scale, the intention to combat tooth decay has freed simultaneous fluoridation paths with little consideration for the cumulative levels of fluoride in the body, which can add up quite significantly and lead to toxicity. Therefore, several countries subscribe to the concept of systemic water fluoridation parallel to the availability of fluoridated milk, salt, tablets and topic products. All of these avenues of fluoridation have cumulative effects on the body, especially in countries with high population density, such as India and China, as is described in greater detail below. To simplify, some countries add fluoride to their tap water and others remove fluoride. How these regulations of water fluoride levels play out is explained as follows.

Various countries regulate fluoride levels based on differing scientific data. For example, for tap water in the European Union, the upper levels of fluoride have been set at 1 mg/l. The European Food Safety Authority (EFSA) recommends,

> [a]n upper tolerable intake level (UL) of 0.1 mg/kg BW/day for fluoride has been derived by the EFSA Panel on Dietetic Products, Nutrition and Allergies (NDA) (EFSA 2005) based on a prevalence of less than 5% of moderate dental fluorosis in children up to the age of 8 years as the critical endpoint, i.e. 1.5 mg/day for children 1-3 years of age, and 2.5 mg/day for children aged 4-8 years. For adults, an UL of 0.12 mg/kg BW/day was based on a risk of bone fracture, which converts on a body weight basis into 7 mg/day for populations aged 15 years and older, and 5 mg/day for children 9-14 years of age.[29]

In contrast, in the US, upper levels have been set by the US Department of Health and Human Services at 0.7 mg/l. The Australian The Health (Fluoridation) Act 1973 set the levels at 1.5 mg/l, while in UK it is the Water Act of 2003 with set value at 1 mg/l. Such variations illustrate the inconclusive bases for government-regulated fluoridation of water. It follows that the regulations are unlikely at ideal levels and commutations can quickly result in overdoses of fluoride. The dangers of systemic fluoridation, then, are evident from the high variability and likelihood of accumulation.

---

[29]European Commission, Critical review of any new evidence on the hazard profile, health effects, and human exposure to fluoride and the fluoridating agents of drinking water, http://ec.europa.eu/health/scientific_committees/environmental_risks/docs/scher_o_139.pdf

Notably, the safety data of fluoridation is of utmost importance for food safety and public health because regulatory measures are based on this data. When fluoride levels from water, milk and salt accumulate in the bodies of consumers, the upper tolerance levels can be quickly reached and lead to toxic levels with severe side-effects. Therefore, regulators must be aware of the safety data and upper tolerance levels of fluoride before systemic fluoridation passes safety thresholds. Sodium fluoride (SF), the fluoride form most commonly occurring in food, is classified as a "hazardous product for the human health and the aquatic environment."[30] In its safety data sheet Solvay North America, a large salt producer, writes, "[s]odium fluoride is used for water fluoridation, as a metal surface treatment and cleaner, as a glass etchant, and for pH adjustment in industrial textile processing or laundries… Sodium fluoride is also used as a wood preservative and an insecticide."[31] This variety of fluoride uses, consequently, means that there are several ways for fluoride to come into the water supply, adding to the cumulative effects of water fluoride levels.

Three important interdisciplinary effects of fluoridation should be understood at the onset: the method of action, the dosage, and the safety data. These three considerations are important because the method of action determines the dosage and the dosage, in turn, determines the safety. The better the methodological application of fluoride, the lower the required dosage and the safer the fluoridation as a whole. It is, therefore, important to consider all three factors in determining the safety of fluoridation for public health and policy purposes.

First, the method of action of water fluoridation is directly on the tooth structure. Local replacement of the hydroxyl ion from the hydroxyapatite by Flouride generates the more stable and caries-resistant flour apatite of the tooth's natural crown, which is surrounded by enamel. This is significant for dental health because the crystalline composition of enamel is hydroxyapatite, a component prone to acidic attack generated by bacterial fermentation. Fluoridation, as a caries prevention treatment, wards off acid attack directly to the tooth but it is not a measure to improve overall nutrition and protect teeth as shown on the caries balance in Fig. 13.1.

Second, the upper levels of fluoride, though varying nationally, is usually limited to one source and fails to recognize other sources of fluoride that also add to the cumulative levels of a population's fluoride intake. Various countries regulate fluoride levels based on differing scientific data. Such variations illustrate the inconclusive bases for government-regulated fluoridation of water. It follows that the regulations are unlikely at ideal levels and accumulations can quickly result in overdoses of fluoride. The dangers of systemic fluoridation, then, are evident form the high variability and likelihood of accumulation.

---

[30]Sodium Fluoride Material Safety Data Sheet, http://www.nfc.umn.edu/assets/pdf/msds/sodium_fluoride.pdf

[31]Solvay, http://www.solvaynorthamerica.com/SiteCollectionDocuments/PDF/PS_SodiumFluoride Final.pdf

Third, the safety data of fluoridation is of utmost importance for food safety and public health. When fluoride levels from water, milk and salt cumulate in the bodies of consumers, the upper tolerance levels can be quickly reached and lead to toxic levels. Therefore, regulators must be aware of the safety data and upper tolerance levels of fluoride before systemic fluoridation passes safety thresholds. Sodium fluoride (SF), the fluoride form most commonly occurring in food, is classified as a "hazardous product for the human health and the aquatic environment."[32] In its safety data sheet Solvay North America, writes[33]: "Sodium fluoride is used for water fluoridation, as a metal surface treatment and cleaner, as a glass etchant, and for pH adjustment in industrial textile processing or laundries. ... Sodium fluoride is also used as a wood preservative and an insecticide." Multiple avenues and reasons of fluoride accumulation exist. Therefore, regulators and government agencies should exercise extreme caution in supporting mass-medication of entire states or regions through fluoridation of water.

13.2.1 Tablets

One way of administering fluoride is through tablets as nutrition supplements. Recently, an increased risk of mild fluorosis in anterior teeth has been identified as a result of fluoride tablets consumption.[34] The Cochrane systematic Review [35] concluded that, "[t]he effect of fluoride supplements was unclear on deciduous teeth. When compared with the administration of topical fluorides, no differential effect was observed." The WHO acknowledges the lack of scientific evidence in regards of the adequate quantity to be administrated.[36] In other words, taking fluoride supplements probably does little to prevent tooth decay.

13.2.2 Milk

It is widely—and erroneously—believed that milk is a source of fluoride for healthy teeth. In 1996, the WHO published a book entitled "Milk fluoridation for the prevention of dental caries" edited by Bánóczy et al.[37] based on studies performed

---

[32]Sodium Fluoride Material Safety Data Sheet.

[33]Solvay, http://www.solvaynorthamerica.com/SiteCollectionDocuments/PDF/PS_SodiumFluoride Final.pdf

[34]Meyer-Lueckel et al. (2010), pp. 315–323.

[35]Tubert-Jeannin et al. (2011), CD007592.

[36]WHO Expert Committee on Oral Health Status and Fluoride Use (1994), http://whqlibdoc.who.int/trs/WHO_TRS_846.pdf?ua=1

[37]Bánóczy et al. (2009).

in Bulgaria, Chile, China, Russia and the United Kingdom. Open questions conclude the publication and the authors could not reach a consensus, which illustrates the conflict of systemic fluoridation. Newer and higher level scientific research published in 2005[38] also failed to identify sufficient scientific support in favor of milk fluoridation. A request for labeling has been made by the WHO to label milk packages accordingly. Thus, there is no scientific support for the common belief that milk or fluoridation of milk contributes to dental health.

13.2.3 Salt

Another source of fluoride is table salt. In France and Germany, for example, 35–60 % of domestic salt is fluoridated.[39] However, to prevent undue accumulation of fluoride in the body, the WHO recommends utilization of either fluoridated water or fluoridated salt for caries prevention and warns of multiple concomitant sources of fluoridation.[40] The maximal dose should not exceed 200 mg F/kg salt. In response to these maximal dosages and warning levels, a request for regular homogeneity checks and concentration labeling on the package has been made by the WHO. This request further supports that mass-medication through fluoride addition to common food and drink items, such as water, milk and salt, remain doubtful in terms of safety and efficacy.

13.2.4 Ready-to-Feed (RTF) Infant Food and Drinks

An additional major source of fluoride are infant formulas and baby foods. A recently published study[41] reviewed 122 infant foods and 25 drinks for F concentration manufactured by 12 companies. The identified F concentration ranged up to 1.200 μg/g. The American Dental Association's Council on Scientific Affairs[42] recommends utilization of F free water for the restitution of RTF infant formula milks.[43] A request for labeling has been made by the WHO because the safety of these high and fluctuating concentrations of fluoride in infant formulas and baby foods are not yet well-understood and regulated.

---

[38]Yeung et al. (2005), CD003876.
[39]Marthaler and Petersen (2005), pp. 351–358.
[40]WHO Technical Report on Fluorides and Oral Health (1994), http://whqlibdoc.who.int/trs/WHO_TRS_846.pdf?ua=1
[41]Loganathan et al. (2012) pp. 26–36.
[42]Berg et al. (2011). pp. 79–87.
[43]Centers for Disease Control and Prevention, Overview: Infant Formula and Fluorosis, http://www.cdc.gov/fluoridation/safety/infant_formula.htm; Hujoel et al. (2009), pp. 841–854.

13.2.5 Other Food Sources of Fluoride

In addition to the aforementioned paths of fluoride accumulation in the body, there are several other avenues through water and food, but also through air, medicines, and cosmetics. Flouride is also naturally available in soil, even though it often does not leach into ground water reservoirs. Long-term applications of phosphorus fertilizers, however, accompanied by intensive irrigation likely raise the fluoride concentration in ground water[44] (see Fig. 13.1). Through fertilizer use and other farming practices, fluoride levels can also be higher in some foods and drinks than one may expect. In India, tea,[45] wheat,[46] spinach, cabbage, and carrots,[47] for example, showed elevated fluoride levels.

Fluoride pollution of the environment, such as through fertilizer use, cigarette smoking[48] or aluminum production, may interfere with and add to water fluoridation, which result in higher total fluoride intakes through food and drinks which far exceed safe limits. Modern industrialization contributes to Fluoride pollution via manufacture of semiconductors and integrated circuits,[49] production of hydrofluoric acid, the activities of phosphate fertilizer plants, textile dyeing, from plastic factories[50] and the production of enamel, glass, brick and tile works and several others. All of these add to fluoride pollution of water and accumulate in water supplies that reach consumers directly, thereby causing widespread fluorosis.

13.2.6 Water Fluoridation

One of the largest sources of fluoride is fluoridated water, the main subject of this chapter. Global trends indicate that the more impurities and pollution exist in water, the higher the fluoride levels. These higher fluoride levels are, however, not a result of intentional public health measures against dental decay. Systematic water fluoridation is, nonetheless, societies' response to the tremendous impact dental caries

---

[44]Loganathan et al. (2001), pp. 275–282; Ayoob and Gupta (2006a, b); Gilkes and Hughes (1994), pp. 755–766.

[45]Cao et al. (1998), pp. 1061–1063; Cao et al. (1997), pp. 827–833; Fraysse et al. (1989), pp. 39–46.

[46]Saini et al. (2013), pp. 2001–2008.

[47]Lakdawala and Punekar (1973), pp. 1679–1687.

[48]Okamura and Matsuhisa (1965), pp. 382–385.

[49]Deshmukh et al. (1995), pp. 1–20.

[50]Ayoob and Gupta (2006a, b), pp. 433–487.

has put on national economies around the globe because it increases healthcare costs.[51] In developing countries, for example, access to clean water is not guaranteed. Thus, it is estimated that out 200 million people from 25 nations are likely to be exposed to fluorosis, an excess of fluoride intake which likely correlates with high fluoride levels in drinking and tap water. In India alone 66 million people suffer from fluorosis although the fluoride levels are as low as 0.5 ppm,[52] numbers similar to one tenth of the Chinese population. A systematic review revealed that despite lower than standard level concentrations of fluoride in water, Iran also has high levels of fluorosis.[53] In Kenia, in turn, a daily fluoride intake of less than 0.03 mg F/kg body weight produces dental fluorosis.[54] Scientific studies demonstrated that in midlatitude areas, such as China, the Middle East, Africa and southern Asia, fluorosis could be endemic.[55] Consequently, fluoridation and endemic fluorisis are indicative of areas with low water quality, high levels of impurities and pollution, and not—as many people erroneously believe—of progressive public health measures against dental caries. Once again, the distinction between local application of fluoride must be contrasted with the systemic mass-medication of populations, whether it is intentional or not.

13.2.7 Topical Fluoride Use for Caries Prevention

Topical fluoride use through gels, liquids or toothpastes are direct application on the enamel, i.e. the outermost hard shell of the teeth. The goal is a targeted and controlled application to strengthen and remineralize enamel. According to some of the highest evidentiary level dental research, topical fluoride use is beneficial to prevent tooth decay.[56] Dental professionals use, recommend and prescribe topical fluorides for individuals at risk of developing dental caries. The current maximal strength guidelines are set based on scientific data.[57] Although the fluoride concentrations used in dentistry are much lower than those accumulated through a variety of sources from foods and the environment, there has not been any research that succeeded to summarize the total concentrations and intakes of fluoride from different sources and considered the overall and accumulated side effects.

---

[51]Centers for Disease Control and Prevention, Community Water Fluoridation: Benefits, http://www.cdc.gov/fluoridation/benefits/index.htm; Centers for Disease Control and Prevention, Community Water Fluoridation, http://www.cdc.gov/fluoridation/

[52]Ayoob and Gupta (2006a, b), pp. 433–487.

[53]Azami-Aghdash et al. (2013), pp. 1–7.

[54]Manji et al. (1986), pp. 371–380; Azami-Aghdash et al. (1987), pp. 452–456.

[55]Reardon and Wang (2000), pp. 3247–3253.

[56]Marinho (2009), pp. 183–191.

[57]Weyant et al. (2013), pp. 1279–1291.

13.3 Side Effects of Systemic Fluoridation

Systemic Fluoridation can and often does lead to a variety of side effects, such as dental fluorosis, higher risks of hip fractures and other chronic effects.[58] Nonetheless, there remains some controversy through differing interpretation of data[59] in regard to fluoridation side effects.

13.3.1 Dental Fluorosis

Dental fluorosis[60] is a disturbance of tooth formation with aesthetic consequences that become visible as white stains on the teeth. The excess of fluoride leads to hypomineralization of the enamel and developmental disruptions in the natural mineralization and remineralization processes.[61] Water fluoridation is a globally growing concern because of the raised fluoride levels that accumulate in the body. One of the consequences of these raised water fluoride levels is dental fluorosis.[62] Although "the predominant cariostatic effect of fluoride is not due to its uptake by the enamel during tooth development,"[63] excess levels of fluoride still damage teeth. Therefore, water fluoridation is among the identified risk factors[64] of fluorosis. A 2006 review by the US National Research Council revealed the occurrence of moderate dental flourisis at water level fluoridation levels of 0.7–1.2 mg/l.[65] Governments should act according to the scientific data that clearly outlines the risks of water fluoridation. The fluoridation recommendations of the US Centers for Disease Control and Prevention (CDC), for example, contradict the actual research in the field of fluorosis as a risk and consequence of water fluoridation[66] but the data, as mentioned above, remains controversial. The WHO states that "there are some undesirable side-effects with excessive fluoride intake. Experience has shown that it may not be possible to achieve effective fluoride-based caries prevention without some degree of dental fluorosis, regardless of which methods are chosen to maintain a low level of fluoride in the mouth."[67] Recent research, however,

---

[58] Chachra et al. (2008), pp. 183–223.

[59] Harrison (2005), pp. 1448–1456; http://www.york.ac.uk/inst/crd/CRD_Reports/crdreport18.pdf (last viewed on the 14 Feb 2014).

[60] Mascarenhas (2000), pp. 269–277.

[61] Fejerskov et al. (1990), pp. 692–700.

[62] Khan et al. (2005).

[63] Aoba and Fejerskov (2002), pp. 155–170.

[64] Mascarenhas (2000), pp. 269–277.

[65] Doull et al. (2006), (Council revealed the occurrence of moderate dental flourisis at water fluoridation levels of 0.7–1.2 mg/l).

[66] Horowitz (2003), pp. 3–8, discussion 9–10.

[67] WHO (2014).

emphasizes the need to review the available data to avoid a further spread of fluorosis.[68]

Flouride ingestion can result from any of the previously mentioned sources and government agencies as well as many public health professionals fail to act on the available data about the warning signs of fluoride overdoses. Based on the aforementioned, fluoride dietary supplements, such as those administered to small children, represent the most likely reason for the increased ingestion levels. Public health researchers acknowledge that it remains difficult to identify the exact amounts of fluoride that are ingested and to trace them back to their sources. Consequently, researchers have requested that public health authorities educate the public on fluoride sources and continuously revise the guidelines for fluoride supplementation based on the most scientifically sound data.[69]

13.3.2 Bone Weakening Through Fluorosis and Other Fluorosis Induced Diseases and Degeneration

High fluoride levels negatively impact human health beyond dental fluorosis. Fluoride ions can alter the mineral structure of the bone by substituting hydroxyl groups in the carbonate-apatite, thereby, creating fluorohydroxyapatite, which can possibly weaken the bone. Studies support the weakening effects that excessive fluoridation may have on bone. For example, studies performed in the US in the early and mid-90s demonstrated a small but significant increase in hip fractures within an elderly population.[70] A study performed in China "concluded that long-term fluoride exposure from drinking water containing more or equal to 4.32 ppm increases the risk of overall bone fractures as well as hip fractures."[71] In the US, water fluoridation levels starting at 1.5 mg/l showed correlation with bone fractures. Notably, in those areas where the correlation was evident, a daily intake of 1.4 mg/day of fluoride could be a reason for stage I skeletal fluorosis.[72] Additionally, data supports low risks of hip fractures in population exposed to fluoridated drinking water (concentrations around 1 ppm) in the UK.[73] Therefore, the risks associated with water fluoridation and its cumulative effects can be quite dangerous from a public health perspective.

---

[68] Ismail and Hasson (2008), pp. 1457–1468.

[69] Burt (1992), pp. 1228–1237.

[70] Danielson et al. (1992), pp. 746–748.

[71] Li et al. (2001), pp. 932–939.

[72] Doull et al. (2006).

[73] Hillier et al. (2000), pp. 265–269.

Fluoride intake can also affect the heart, liver, kidneys, gastrointestinal tract,[74] lungs, brain, blood, and hormones in potentially health threatening ways. In fact, fluoride has a potential neurotoxic[75] effect as well as teratogenic actions,[76] which disturb embryologic development.[77] A systematic review that investigated the possible association between the occurrence of Down's Syndrome and water fluoridation concluded that the available evidence was inconclusive[78] but did not rule the connection out.

13.3.3 Labeling Requirements

Due to its lack of efficacy and side effects fluoride supplements should not be prescribed or offered under the age of three, although the latest suggestions mark the age of six[79] unless prescribed by a dentist. The WHO, however, requests adequate labeling, child proof packaging, and indications of maximal doses per package not exceeding 120 mg.[80] For more information on nutrition and supplement labeling, see Chap. 16.

13.4 The Ethics of Water Fluoridation

At the core of the ethical controversy of water, salt, milk and other common types of fluoridation is the lack of consumers' consent to mass-medication. When governments and public health agencies make the decision for the general public, whether in the public's best interest or not, to fluoridate essential food and drink items, such as those mentioned in this chapter, ethical questions must be addressed. This is especially the case where these decisions to mass-medicate are not founded on sound data and science. Even worse, when such mass-medication as fluoridation occurs based on outdated sources or scientifically controverse principles, public health and environmental threats arise that can be difficult to control. As has been shown in this chapter, fluorosis and other negative health impacts arise from mass-medication with fluoride—a dangerous practice in the western world. In countries

---

[74]Spak et al. (1989), pp. 1686–1687; Spak et al. (1990), pp. 426–429; Susheela et al. (1993), pp. 97–104; Whitford and Pashley (1984)), pp. 302–307.

[75]Choi et al. (2012), pp. 1362–1368.

[76]Perumal et al. (2013), pp. 236–251; Takahashi et al. (2001), pp. 170–179; Tohyama (1996), pp. 84–91; Mastersa et al. (1999), pp. 435–449; Frenia (1994), pp. 109–121.

[77]Doull et al. (2006).

[78]Whiting et al. (2001).

[79]Wright et al. (2014), pp. 182–189.

[80]WHO Technical Report on Fluorides and Oral Health (1994).

with already high fluoride levels in drinking water, such as China and India, fluoride is a rather threatening problem for public health that remain inadequately addressed. Organizations such as the FDA and EFSA, for example, must review the regulations that are in place and update them to bring fluoridation back to safe levels and must stop propagating the myth that the fluoridation of drinking water decreases dental decay—the cost is simply too high.

The key considerations in systemic fluoridation are the negative influence upon tooth formation in children up to the age of 3–6 years in addition to the already mentioned side effects. There are different preconditions in the different countries of the world that make general recommendations for fluoride intake obsolete because some countries have high fluoride levels in drinking and tab water and others have low levels. One example of this controversy occurred in Australia, where the respective government agencies evaluated the fluoridation versus fluorosis evidence. These evaluations gave rise to the recently escalated political debate.[81] The goal of these evaluations was to investigate whether there is sufficient evidence to continue to support water fluoridation in the future. The researchers of the study suggested the application of proportionality "to resolve the conflict between the ethical principle of beneficence (prevention of dental caries) and ... non-maleficence ([to] reduce an increased risk of fluorosis and possibly hypothyroidism and bone fractures) in the water fluoridation controversy."[82] Efficiency in caries prevention by water fluoridation has to be bigger than possible harm generation. Thus, the researchers juxtaposed caries reduction worldwide induced by water fluoridation and topical application of fluorides. They identified a higher efficiency in the latter and concluded that the evidence is not sufficient to support systemic fluoridation. Thus, in Australia, the evaluation of the data resulted in conclusions against mass-medication through water fluoridation.

Similar debates evolved in other parts of the world. In Canada, for example, the debate about water fluoridation raised questions of basic human rights and freedom of choice.[83] The mass-medication through water-fluoridation is so concerning to a number of Canadians, that these ethical considerations surface, as mentioned above. Correspondingly, in the EU, surveys of European citizens' attitudes towards water fluoridation have been initiated in different population groups of 16 member countries.[84] The results of these surveys showed that a clear majority voted against systemic fluoridation. Therefore, pursuant to democratic principles, water fluoridation as mass-medication should not continue in the EU.

In the UK, a lawsuit shed further light on the ethics of water fluoridation. Where fluoride is used to prevent or to treat incipient disease, here dental caries, flouride has to be regarded as medicine under British law. Thus, the authority in charge, here the state, which mass-prescribes systematic water fluoridation has to obey to the

---

[81]Awofeso (2012), pp. 161–172.
[82]Id.
[83]Cohen and Locker (2001), pp. 578–580.
[84]Griffin et al. (2008), pp. 95–102.

same ethical rules applied to any health care worker. Therefore, the court in the aforementioned UK lawsuit[85] concluded that fluoridated water should be treated as medicine and falls squarely within the Medicines Act of 1968. Pursuant to this precedent, the state's mass-prescriptions of fluoride to the public should follow the same laws that other drugs must follow under the Medicines Act even though this is currently not the case yet.

One of the fundamental principles underlying the UK precedent and the debates in the EU, Canada and in Australia involve biomedical ethics. The fathers of biomedical ethics, Beauchamp and Childress, introduced four principles that apply to health care workers: (a) respect of patient autonomy, (b) beneficience, (c) nonmaleficience, and (d) justice. According to these well-established principles, patient can refuse medical treatment under the principal of informed consent, which is also supported by the Council of Europe Convention on Human Rights and Biomedicine 1997.[86] The problem here is that the cookie-cutter mass-medication of the public through water fluoridation violates these basic biomedical ethics principles. From an ethical point of view, therefore, systematic water fluoridation could be considered to oppose the all philosophical principles,[87] and various ethical rules that strip the public and individual consumers of basic rights.

13.5 Regulation of Water Fluoridation Around the World

13.5.1 The WHO Approach to Water Fluoridation

Despite the vast amount of scientific data bemoaning water fluoridation as an ineffective and dangerous public health measure, the WHO has supported water fluoridation for several decades. In fact, the "WHO established a guidance value for naturally occurring fluoride in drinking water of 1.5 mg/L based on a consumption of 2 L water/day, and recommended that artificial fluoridation of water supplies should not exceed the optimal fluoride levels of 1.0 mg/L (WHO 2006)."[88]In a coalition of large associations, "[t]he WHO Oral Health Programme, jointly with the FDI World Dental Federation (FDI) and the International Association for Dental Research (IADR), have embarked on an action plan for the promotion of using fluoride, particularly focusing on the disadvantaged and under-served population

---

[85]Cheng et al. (2007).

[86]Convention for the Protection of Human Rights and Dignity of the Human Being with regard to the Application of Biology and Medicine: Convention on Human Rights and Biomedicine, http://conventions.coe.int/Treaty/en/Treaties/Html/164.htm

[87]As stated by the "John Harris Centre for Social Ethics & Policy University of Manchester" in November 1989, http://www.bfsweb.org/facts/ethics/ethicsharris.htm

[88]European Commission, Critical review of any new evidence on the hazard profile, health effects, and human exposure to fluoride and the fluoridating agents of drinking water.

groups."[89] Although this program may appear to be a public health measure, it fails to take high-value scientific evidence and geographical differences into account, thereby, promoting a one-size-fits-all approach to caries prevention that has been shown to backfire through dangerous fluorosis levels around the world.

Contradicting its own program, the WHO already found that systemic water fluoridation should not follow a global cookie-cutter uniform set of fluoridation recommendations because fluoride exposure differs greatly around the world. These differing local availabilities of fluoride have been well-documented and support the position against the aforementioned mass-medication:

> [I]t is known that water is normally the major source of fluoride exposure, with exposure from diet and from burning high fluoride coal also major contributors in some settings. Fluoride occurs at elevated concentrations in many areas of the world including Africa, the Eastern Mediterranean and southern Asia. One of the best-known high fluoride areas extends from Turkey through Iraq, Iran, Afghanistan, India, northern Thailand and China. However, there are many other areas with water sources that contain high fluoride levels and which pose a risk to those drinking the water, notably parts of the rift valley in Africa. Many of these areas are arid and alternative sources of water are not available.[90]

Therefore, it is important to use both sound science as a basis for fluoridation recommendations—if any—and to adapt these recommendations to the localities around the world. As long as this is not the case, overexposure and fluorosis will continue to be likely and dangerous consequences of uncontrolled fluoridation around the world. A more streamlined approach should be instituted to evaluate the scientifically relevant and to use sound data to help redirect caries prevention efforts and steer them away from mass-medication through water fluoridation.

13.5.2 Water Fluoridation in the USA

Several agencies regulate water fluoridation in the USA but there is no single streamlined approach to what a safe level of fluoridation may be, especially because the current regulatory framework is based on several decade old data that has long needed an overhaul. The Centers for Disease Control and Prevention (CDC), a government agency, declares that water fluoridation "has been a safe and healthy way to effectively prevent tooth decay" for over 65 years and also recognized water fluoridation as "one of 10 great public health achievements for the 20th century"[91]—a rather sad prospect by comparison, especially in light of the dangers that water fluoridation has proved to pose. Official recommendations by the US

---

[89] WHO, Oral Health: Global consultation on oral health through fluoride, http://www.who.int/oral_health/events/Global_consultation/en/

[90] WHO, New WHO report tackles fluoride in drinking-water, http://www.who.int/mediacentre/news/new/2006/nw04/en/index.html

[91] Centers for Disease Control and Prevention, Community Water Fluoridation, http://www.cdc.gov/fluoridation/

Department of Health and Human Services (DHHS) have set the level of fluoride level in water at 0.7 mg/l,[92] which contrast other agency recommendations: "For values 0.1 mg/L above control range to 2.0 mg/L – Leave the fluoridation system on. For values 2.1 mg/L to 4.0 mg/L – Leave the fluoridation system on. For values 4.1 mg/L to 10.0 mg/L – Turn off the fluoridation system immediately."[93] Thus, there is clearly no true consensus and this chapter has demonstrated that water fluoridation should not be a one-size-fits-all approach.

The Environmental Protection Agency is empowered to regulate flouride in drinking water under the Safe Drinking Water Act. Adverse effects of water fluoridation, however, are proven to be suspected world wide and the EPA failed to act on the science supporting the dangers of water fluoridation. For example, in a 1998 case filed by EPA scientists against the agency itself, the petitioners called the "process by which EPA Arrived at the [recommendations] for fluoride ... irrational."[94] On the same note, a report by the National Research Council released in 2006, performed by request of the EPA, confirmed the need to lower the maximum level of safe fluoride to zero.[95] Nonetheless, drinking water continues to be fluoridated in the US.

13.5.3 Water Fluoridation in the European Community

The EU has issued a Directive on the quality of water intended for human consumption, namely Council Directive 98/83/EC (1998). In this Directive, the European Commission has included a list of chemical parameters that are safe in drinking water, where a fluoride level (both natural and as a result of fluoridation) for water intended for human consumption of less than 1.5 mg/l was determined.[96] Although this level is still high, it is substantially lower than the levels determined to be safe in other parts of the world.

---

[92]HHS Recommendation for Fluoride Concentration in Drinking Water for Prevention of Dental Caries, http://aspe.hhs.gov/oash/floridation.shtml; American Academy of Pediatric Dentistry, Guideline on Fluoride Therapy (revised 2014), http://www.aapd.org/media/Policies_Guidelines/G_fluoridetherapy.pdf

[93]US Department of Health and Human Services, Public Health Service Centers for Disease Control and Prevention, http://www.cdc.gov/mmwr/PDF/rr/rr4413.pdf (last visited on the 18 Feb 2014).

[94]*Natural Res. Def. Council, Inc. v. E.P.A.*, 812 F.2d 721 (D.C. Cir. 1987).

[95]Doull et al. (2006).

[96]http://ec.europa.eu/health/scientific_committees/environmental_risks/docs/scher_o_139.pdf (last visited on the 18 Feb 2014).

13.5.4 Water Fluoridation in Australia

In Australia, the Health (Fluoridation) Act of 1973 regulates the safe and effective addition of fluoride into drinking water supplies[97] and works in conjunction with the Australian drinking water guidelines and the Safe Drinking Water Act of 2003.[98] The dosage not allowed to be exceeded was set at 1.5 mg/l, the same level as in the EU.

13.5.5 Water Fluoridation in South Africa

Attributing water quality and health problems to water fluoridation, the South African Research Commission simply but most notably states that "fluoridation of groundwater is not advisable."[99] Thus, some governments are firmly opposed to water fluoridation despite the aforementioned contradictions within the WHO's recommendations.

13.5.6 Water Fluoridation in United Kingdom

Water fluoridation in the UK is regulated by paragraph 58 of the Water Act 2003. The Act limits the "the concentration of fluoride in the water supplied to premises in the specified area is maintained at the general target concentration of one milligram per litre." Moreover, individual (health care) authorities are empowered to request water fluoridation from water suppliers to further ascertain whether water fluoride levels are safe.[100]

13.5.7 Water Fluoridation in Israel

Similar to lawsuits in the US, UK and the EU, several lawsuits have been brought by concerned parties in an effort to update water fluoridation regulation. As a result of such litigation in Israel, the Supreme Court of Israel ruled against adding fluoride to drinking water in an opinion on July 29, 2013.[101] In 2014, the Minister of Health

---

[97]Health (Fluoridation) Act 1973, http://www.health.vic.gov.au/water/fluoridation/act.htm
[98]Id.
[99]Ncube and Schutte (2005), pp. 35–40, http://www.wrc.org.za/
[100]Water Act (2003), available at http://www.legislation.gov.uk/ukpga/2003/37/part/3/crossheading/water-fluoridation
[101]See translation of the Court's ruling at: http://www.fluoridealert.org/uploads/israel_supreme_ct_july2013.pdf

decided to stop water fluoridation because "fluoridation can cause harm to the health of the chronically ill," including "people who suffer from thyroid problems."[102] This breakthrough approach is evidence of progressive and scientifically-sound risk-benefit analysis, while simultaneously taking the ethics of water fluoridation into concern. Israel's cessation of water fluoridation sets a positive example for regulators around the world.

13.6 Conclusion

Many reasons speak for the inclusion of caries in any well-rounded discussion on public health and food policy. This chapter shows that the socio-economic impacts of caries prevalence affect individuals, public health, economic systems and international policies. One of the major public health measures taken against caries is water fluoridation, a government-regulated mass-medication following a one-size-fits-all approach. The various health concerns arising from this mass-medication include fluorosis and other health complications resulting from the systemic fluoridation of drinking water. While the US Center for Disease Control called the decrease of caries prevalence by half through fluoridation of water one of the greatest achievements since 1980,[103] other data tie caries into discussions of public health and food policy as socio-economic aspects and reveal well-founded and strong objections to water fluoridation because it poses a number of health threats.

In essence, this chapter analyzes the role of law and regulations in addressing public health and food policy concerns in light of fluoride deficiencies or accumulation through water and food fluoridation. The three fold problems, therefore, are the questionable outcome of water and food fluoridation, the associated health hazards resulting from fluoride accumulation, and the ethical implications of mass-medicating the public based on scientifically ambiguous data. As a result, this chapter presents the various measures taken around the world to address these three problems and illustrates how an interdisciplinary approach to food law, policy and regulation can have broad global application.

This chapter identifies that caries causes socio-economic costs and that the regulatory remedy in response to these rising costs around the world is fluoridation. There are, however, two major approaches to fluoridation: (1) Systemic fluoridation and (2) topical fluoridation. Systemic fluoridation is regulated through governments and international entities, such as the WHO. These governmental measures have both ethical and medical problems as consequences of mass-medication of the public. By comparison, the topical fluoridation approach remains an individualized

---

[102]Israel Will End Fluoridation in 2014, Citing Health Concerns, http://fluoridealert.org/articles/israel_fluoridation/

[103]Griffin et al. (2001), pp. 78–86, 79 (*citing* Ripa (1993), pp. 17–44; CDC, Fluoridation of public drinking water to prevent dental caries (1999), pp. 933–940).

and more targeted preventive measure against caries because it can be applied either by the dental professional or the individual consumer. However, obstacles to proper topical fluoridation treatments for public health remain both consumer or patient education and individual healthcare costs. At the same time, those individuals choosing topical fluoridation have little avenues to avoid being mass-medicated and fluoride accumulation by the government as long as water fluoridation remains a public health practice.

References

Adiatman M et al (2013) Functional tooth units and nutritional status of older people in care homes in Indonesia. Gerodontology 30(4):262–269

American Academy of Pediatric Dentistry, Guidelines on Fluoride Treatment (2013), http://www.aapd.org/media/Policies_Guidelines/G_fluoridetherapy.pdf (internal citations omitted).

American Academy of Pediatric Dentistry, Guideline on Fluoride Therapy (revised 2014), http://www.aapd.org/media/Policies_Guidelines/G_fluoridetherapy.pdf (last accessed Jan 2015)

Aoba T, Fejerskov O (2002) Dental fluorosis: chemistry and biology. Crit Rev Oral Biol Med 13(2):155–170

Awofeso N (2012) Ethics of artificial water fluoridation in Australia. Public Health Ethics 5(2):161–172

Ayoob S, Gupta AK (2006a) Fluoride in drinking water: a review on the status and stress effects. Crit Rev Environ Sci Technol 36:433–487

Ayoob S, Gupta AK (2006b) Fluoride in drinking water: a review on the status and stress effects. Crit Rev Environ Sci Technol 36:6

Azami-Aghdash S, Ghojazadeh M, Pournaghi Azar F, Naghavi-Behzad M, Mahmoudi M, Jamali Z, Baelum V, Fejerskov O, Manji F, Larsen MJ (1987) Daily dose of fluoride and dental fluorosis. Tandlaegebladet 91:452–456

Azami-Aghdash S, Ghojazadeh M, Pournaghi Azar F, Naghavi-Behzad M, Mahmoudi M, Jamali Z (2013) Fluoride concentration of drinking waters and prevalence of fluorosis in Iran: a systematic review. J Dent Res Dent Clin Dent Prospects 7(1):1–7

Bagramian RA, Garcia-Godoy F, Volpe AR (2009) The global increase in dental caries. A pending public health crisis. Am J Dent 22(1):3–8

Bánóczy et al., WHO: Milk fluoridation for the prevention of dental caries (2009), http://apps.who.int/iris/bitstream/10665/44152/1/9789241547758_eng.pdf?ua=1

Berg J, Gerweck C, Hujoel PP, King R, Krol DM, Kumar J, Levy S, Pollick H, Whitford GM, Strock S, Aravamudhan K, Frantsve-Hawley J, Meyer DM; American Dental Association Council on Scientific Affairs Expert Panel on Fluoride Intake From Infant Formula and Fluorosis (2011) Evidence-based clinical recommendations regarding fluoride intake from reconstituted infant formula and enamel fluorosis: a report of the American Dental Association Council on Scientific Affairs. J Am Dent Assoc 142(1):79–87

Burt BA (1992) The changing patterns of systemic flouride intake. J Dent Res 71(Spec Iss):1228–1237

Cao J, Zhao Y, Liu J (1997) Brick tea consumption as the cause of dental fluorosis among children from Mongol, Kazak and Yugu populations in China. Food Chem Toxicol 35(8):827–833

Cao J, Zhao Y, Liu JW (1998) Safety evaluation and fluorine concentration of pure brick tea and Bianxiao brick tea. Food Chem Toxicol 36(12):1061–1063

Centers for Disease Control and Prevention, Community Water Fluoridation: Benefits, http://www.cdc.gov/fluoridation/benefits/index.htm (last accessed Jan 2015)

Centers for Disease Control and Prevention, Community Water Fluoridation, http://www.cdc.gov/fluoridation/ (last accessed Jan 2015)

Centers for Disease Control and Prevention, Overview: Infant Formula and Fluorosis, http://www.cdc.gov/fluoridation/safety/infant_formula.htm (last accessed Jan 2015)

Centers for Disease Control and Prevention, Achievements in Public Health, 1900-1999: Fluoridation of Drinking Water to Prevent Dental Caries, http://www.cdc.gov/mmwr/preview/mmwrhtml/mm4841a1.htm (last accessed Jan 2015)

Chachra D, Vieira AP, Grynpas MD (2008) Fluoride and mineralized tissues. Crit Rev Biomed Eng 36(2-3):183–223

Cheng KK, Chalmers I, Sheldon TA (2007) Adding fluoride to water supplies. BMJ 335:699–702

Choi AL, Sun G, Zhang Y, Grandjean P (2012) Developmental fluoride neurotoxicity: a systematic review and meta-analysis. Environ Health Perspect 120(10):1362–1368

Cohen H, Locker D (2001) The science and ethics of water fluoridation. J Can Dent Assoc 67(10):578–580

da Silva Tagliaferro EP, de Castro Meneghim M, Ambrosano GMB, Pereira AC (2008) Risk indicators and risk predictors of dental caries in schoolchildren. J Appl Oral Sci 16(6):408–413

Danielson C, Lyon JL, Egger M, Goodenough GK (1992) Hip fractures and fluoridation in Utah's elderly population. JAMA 268(6):746–748

Deshmukh AN, Wadaskar PM, Malpe DB (1995) Fluorine in environment: a review. Gondwana Geol Mag 9:1–20

Doull J, Boekelheide K, Farishian BG, Isaacson RL, Klotz JB, Kumar JV, Limeback H, Poole C, Puzas JE, Reed N-MR, Thiessen KM, Webster TF (2006) Committee on Fluoride in Drinking Water, Board on Environmental Studies and Toxicology, Division on Earth and Life Studies, National Research Council of the National Academies, Fluoride in drinking water: a scientific review of EPA's standards, available for purchase online at: http://www.nap.edu) (last accessed Jan 2015)

Edelstein BL (2006) The dental caries pandemic and disparities problem. BMC Oral Health 15 (6 Suppl 1):S2

Ervin RB, Dye BA (2012) Number of natural and prosthetic teeth impact nutrient intakes of older adults in the United States. Gerodontology 20:e693–e702, e693

European Commission, Critical review of any new evidence on the hazard profile, health effects, and human exposure to fluoride and the fluoridating agents of drinking water

Fejerskov O, Manji F, Baelum V (1990) The nature and mechanism of dental fluorosis in man. J Dent Res 69(Spec Iss):692–700

Fraysse C, Bilbeissi MW, Mitre D, Kerebel B (1989) The role of tea consumption in dental fluorosis in Jordan. Bull Group Int Rech Sci Stomatol Odontol 32(1):39–46

Frenia SC (1994) Exposure to high fluoride concentrations in drinking water is associated with decreased birth rates. J Toxicol Environ Health 42(1):109–121

Gilkes RJ, Hughes JC (1994) Sodium-fluoride pH of South-Western Australian soils as an indicator of P-sorption. Aust J Soil Res 32(4):755–766

Griffin M, Shickle D, Moran N (2008) European citizens' opinions on water fluoridation. Community Dent Oral Epidemiol 36(2):95–102

Griffin O, Jones K, Tomar SL (2001) An economic evaluation of community water fluoridation. Journal of Public Health Dentistry 61(2):78–86, 79 (*citing* Ripa LW (1993) A half-century of community water fluoridation in the United States: review and commentary. J Publ Health Dentistry 53:17–44 ; CDC (1999) Fluoridation of public drinking water to prevent dental caries. Morbid Mortal Wkly Rep 48:933–940)

Griffin SO, Regnier E, Griffin PM, Huntley V (2007) Effectiveness of fluoride in preventing caries in adults. J Dent Res 86(5):410–415

Gupta P, Gupta N, Pawar AP, Birajdar SS, Natt AS, Singh HP (2013) Role of sugar and sugar substitutes in dental caries: a review. ISRN Dent 2013:519421, eCollection 2013

Harris R, Nicoll AD, Adair PM, Pine CM (2004) Risk factors for dental caries in young children: a systematic review of the literature. Community Dent Health 21(1 Suppl):71–85

Harrison PTC (2005) Fluoride in water: a UK perspective. J Fluor Chem 126(11–12):1448–1456

Health (Fluoridation) Act 1973, http://www.health.vic.gov.au/water/fluoridation/act.htm (last accessed Jan 2015)

HHS Recommendation for Fluoride Concentration in Drinking Water for Prevention of Dental Caries, http://aspe.hhs.gov/oash/floridation.shtml (last accessed Jan 2015)

Hillier S, Cooper C, Kellingray S, Russell G, Hughes H, Coggon D (2000) Fluoride in drinking water and risk of hip fracture in the UK: a case-control study. Lancet 355(9200):265–269

Horowitz HS (2003) The 2001 CDC recommendations for using fluoride to prevent and control dental caries in the United States. J Public Health Dent 63(1):3–8, discussion 9–10

http://www.york.ac.uk/inst/crd/CRD_Reports/crdreport18.pdf (last viewed on the 14 Feb 2014)

Hujoel PP, Zina LG, Moimaz SA, Cunha-Cruz J (2009) Infant formula and enamel fluorosis: a systematic review. J Am Dent Assoc 140(7):841–854

Ismail AI, Hasson H (2008) Fluoride supplements, dental caries and fluorosis: a systematic review. J Am Dent Assoc 139(11):1457–1468

Ismail AI, Sohn W, Tellez M, Willem JM, Betz J, Lepkowski J (2008) Risk indicators for dental caries using the International Caries Detection and Assessment System (ICDAS). Community Dent Oral Epidemiol 36:55–68

Israel Will End Fluoridation in 2014, Citing Health Concerns, http://fluoridealert.org/articles/israel_fluoridation/

Jackson SL, Vann WF Jr, Kotch JB, Pahel BT, Lee JY (2011) Impact of poor oral health on children's school attendance and performance. Am J Public Health 101(10):1900–1906

John Harris Centre for Social Ethics & Policy University of Manchester in November 1989, http://www.bfsweb.org/facts/ethics/ethicsharris.htm (last accessed Jan 2015)

Khan A, Moola MH, Cleaton-Jones P (2005) Global trends in dental fluorosis from 1980 to 2000: a systematic review. SADJ 60(10):418–421

Lakdawala DR, Punekar BD (1973) Fluoride content of water and commonly consumed foods in Bombay and a study of the dietary fluoride intake. Indian J Med Res 61:1679–1687

Li Y, Liang C, Slemenda CW, Ji R, Sun S, Cao J, Emsley CL, Ma F, Wu Y, Ying P, Zhang Y, Gao S, Zhang W, Katz BP, Niu S, Cao S, Johnston CC Jr (2001) Effect of long-term exposure to fluoride in drinking water on risks of bone fractures. J Bone Miner Res 16(5):932–939

Loganathan P, Hedley MJ, Wallace GC, Roberts AH, Maguire A, Omid N, Abuhaloob L, Moynihan PJ, Zohoori FV (2012) Fluoride content of Ready-to-Feed (RTF) infant food and drinks in the UK. Community Dent Oral Epidemiol 40:26–36

Loganathan P, Hedley MJ, Wallace GC, Roberts AH (2001) Fluoride accumulation in pasture forages and soils following long-term applications of phosphorus fertilisers. Environ Pollut 115(2):275–282

Manji F, Baelum V, Fejerskov O, Gemert W (1986) Enamel changes in two low-fluoride areas of Kenya. Caries Res 20:371–380

Marinho VC (2009) Cochrane reviews of randomized trials of fluoride therapies for preventing dental caries. Eur Arch Paediatr Dent 10(3):183–191

Marinho VC, Higgins JP, Logan S, Sheiham A (2002a) Fluoride varnishes for preventing dental caries in children and adolescents. Cochrane Database Syst Rev 3, CD002279

Marinho VC, Higgins JP, Logan S, Sheiham A (2002b) Fluoride gels for preventing dental caries in children and adolescents. Cochrane Database Syst Rev 2, CD002280

Marinho VC, Higgins JP, Logan S, Sheiham A (2003a) Topical fluoride (toothpastes, mouthrinses, gels or varnishes) for preventing dental caries in children and adolescents. Cochrane Database Syst Rev 4, CD002782

Marinho VC, Higgins JP, Logan S, Sheiham A (2003b) Fluoride mouthrinses for preventing dental caries in children and adolescents. Cochrane Database Syst Rev 3, CD002284

Marinho VC, Higgins JP, Logan S, Sheiham A (2003c) Systematic review of controlled trials on the effectiveness of fluoride gels for the prevention of dental caries in children. J Dent Educ 67(4):448–458

Marinho VC, Higgins JP, Sheiham A, Logan S (2003d) Fluoride toothpastes for preventing dental caries in children and adolescents. Cochrane Database Syst Rev 1, CD002278

Marinho VC, Higgins JP, Sheiham A, Logan S (2004a) Combinations of topical fluoride (toothpastes, mouthrinses, gels, varnishes) versus single topical fluoride for preventing dental caries in children and adolescents. Cochrane Database Syst Rev 1, CD002781

Marinho VC, Higgins JP, Sheiham A, Logan S (2004b) One topical fluoride (toothpastes, or mouthrinses, or gels, or varnishes) versus another for preventing dental caries in children and adolescents. Cochrane Database Syst Rev 1, CD002780

Marthaler TM, Petersen PE (2005) Salt fluoridation--an alternative in automatic prevention of dental caries. Int Dent J 55(6):351–358

Mascarenhas AK (2000) Risk factors for dental fluorosis: a review of the recent literature. Pediatr Dent 22(4):269–277

Mastersa RD, Myron J, Coplanb MJ (1999) Water treatment with silicofluorides and lead toxicity. Int J Environ Stud 56(4):435–449

Meyer-Lueckel H, Grundmann E, Stang A (2010) Effects of fluoride tablets on caries and fluorosis occurrence among 6- to 9-year olds using fluoridated salt. Community Dent Oral Epidemiol 38:315–323

National Institute of Dental and Craniofacial Research, The Story of Fluoridation, http://www.nidcr.nih.gov/oralhealth/Topics/Fluoride/TheStoryofFluoridation.htm (Last accessed Jan 2015)

Natural Res. Def. Council, Inc. v. E.P.A., 812 F.2d 721 (D.C. Cir. 1987)

Ncube EJ, Schutte CF (2005) The occurrence of fluoride in South African groundwater: a water quality and health problem. Water SA 31(1):35–40, http://www.wrc.org.za/

Okamura T, Matsuhisa T (1965) The content of fluorine in cigarettes. J Food Hyg Soc Jpn 6:382–385

Oral Health in America: Report of the Surgeon General, Chapter 4, http://www.nidr.nih.gov/sgr/sgrohweb/chap4.htm#dental_caries (last accessed Jan 2015)

Owings L (2007) Toothache Leads to Boy's Death, ABC News. Available at http://abcnews.go.com/Health/Dental/story?id=2925584 (last accessed 25 July 2014)

Perumal E, Paul V, Govindarajan V, Panneerselvam L (2013) A brief review on experimental fluorosis. Toxicol Lett 223(2):236–251

Petersen PE, Bourgeois D, Ogawa H, Estupinan-Day S, Ndiaye C (2005) The global burden of oral diseases and risks to oral health. Bull World Health Organ 83(9):661–669

Petersen PE, Esheng Z (1998) Dental caries and oral health behaviour situation of children, mothers and schoolteachers in Wuhan, People's Republic of China. Int Dent J 48:210–216, http://www.who.int/oral_health/media/en/orh_idj_210to216.pdf?ua=1

Reardon JE, Wang Y (2000) A limestone reactor for fluoride removal from wastewaters. Environ Sci Technol 34:3247–3253

Reisine ST (1985) Dental health and public policy: the social impact of dental disease. Am J Public Health 75(1):27–30

Saini P, Khan S, Baunthiyal M, Sharma V (2013) Mapping of fluoride endemic area and assessment of F(-1) accumulation in soil and vegetation. Environ Monit Assess 185(2):2001–2008

See translation of the Court's ruling at: http://www.fluoridealert.org/uploads/israel_supreme_ct_july2013.pdf

Sheiham A (2006) Dental caries affects body weight, growth and quality of life in pre-school children. Br Dent J 201:625–626

Sodium Fluoride Material Safety Data Sheet, http://www.nfc.umn.edu/assets/pdf/msds/sodium_fluoride.pdf

Solvay, http://www.solvaynorthamerica.com/SiteCollectionDocuments/PDF/PS_SodiumFluoride Final.pdf

Spak C, Sjostedt S, Eleborg L, Veress B, Perbeck L, Ekstrand J (1990) Studies of human gastric mucosa after application of 0.042 % fluoride gel. J Dent Res 69:426–429

Spak CJ, Sjostedt S, Eleborg L (1989) Tissue response of gastric mucosa after ingestion of fluoride. BMJ 298:1686–1687

Susheela AK, Kumar A, Bhatnagar M, Bahadur R (1993) Prevalence of endemic fluorosis with gastrointestinal manifestations in people living in some North-Indian villages. Fluoride 26:97–104

Takahashi K, Akiniwa K, Narita K (2001) Regression analysis of cancer incidence rates and water fluoride in the U.S.A. based on IACR/IARC (WHO) data (1978–1992), International Agency for Research on Cancer. J Epidemiol 11(4):170–179

The Health (Fluoridation) Act 1973, http://docs.health.vic.gov.au/docs/doc/Code-of-practice-for-fluoridation-of-drinking-water-supplies--Health-(Fluoridation)-Act-1973

Tohyama E (1996) Relationship between fluoride concentration in drinking water and mortality rate from uterine cancer in Okinawa prefecture, Japan. J Epidemiol 6(4):84–91

Touger-Decker R, van Loveren C (2003) Sugars and dental caries. Am J Clin Nutr 78(4):881S–892S

Tubert-Jeannin S, Auclair C, Amsallem E, Tramini P, Gerbaud L, Ruffieux C, Schulte AG, Koch MJ, Rège-Walther M, Ismail A (2011) Fluoride supplements (tablets, drops, lozenges or chewing gums) for preventing dental caries in children. Cochrane Database Syst Rev 12, CD007592

US Department of Health and Human Services, Public Health Service Centers for Disease Control and Prevention, http://www.cdc.gov/mmwr/PDF/rr/rr4413.pdf

van Houte J (1994) Role of micro-organisms in caries etiology. J Dent Res 73(3):672–681

Water Act (2003), available at http://www.legislation.gov.uk/ukpga/2003/37/part/3/crossheading/water-fluoridation

Weyant RJ, Tracy SL, Anselmo TT, Beltrán-Aguilar ED, Donly KJ, Frese WA, Hujoel PP, Iafolla T, Kohn W, Kumar J, Levy SM, Tinanoff N, Wright JT, Zero D, Aravamudhan K, Frantsve-Hawley J, Meyer DM (2013) Topical fluoride for caries prevention: executive summary of the updated clinical recommendations and supporting systematic review. J Am Dent Assoc 144(11):1279–1291

Whitford GM, Pashley DH (1984) Fluoride absorption: the influence of gastric acidity. Calcif Tissue Int 36:302–307

Whiting P, MacDonagh M, Kleijnen J (2001) Association of Down's syndrome and water fluoride level: a systematic review of the evidence. BMC Public Health 1:6

WHO, Global consultation on oral health through fluoride

WHO Expert Committee on Oral health Status and Fluoride Use (1994), http://whqlibdoc.who.int/trs/WHO_TRS_846.pdf?ua=1

WHO in collaboration with the World Dental Federation and the International Association for Dental Research (17 November 2006), http://www.who.int/oral_health/events/Global_consultation/en/ (last accessed Jan 2015)

WHO, Risks to oral health and intervention: Fluorides (2014), http://www.who.int/oral_health/action/risks/en/index1.html (last accessed Jan 2015)

WHO, New WHO report tackles fluoride in drinking-water, http://www.who.int/mediacentre/news/new/2006/nw04/en/index.html (last accessed Jan 2015)

WHO, Oral Health Fact Sheet N°318 (Apr 2012), available at http://www.who.int/mediacentre/factsheets/fs318/en/ (last accessed 25 July 2014)

WHO, Oral Health: Global consultation on oral health through fluoride, http://www.who.int/oral_health/events/Global_consultation/en/ (last accessed Jan 2015)

WHO Technical Report on Fluorides and Oral Health (1994), http://whqlibdoc.who.int/trs/WHO_TRS_846.pdf?ua=1

Wright JT, Hanson N, Ristic H, Whall CW, Estrich CG, Zentz RR (2014) Fluoride toothpaste efficacy and safety in children younger than 6 years: a systematic review. J Am Dent Assoc 145 (2):182–189

Yeung CA, Hitchings JL, Macfarlane TV, Threlfall AG, Tickle M, Glenny AM (2005) Fluoridated milk for preventing dental caries. Cochrane Database Syst Rev 3, CD003876

Chapter 14
Textbox: The Hierarchy of Scientific Evidence

Gabriela Steier and Liviu Steier

Abstract Not all studies are the same. Various types of publications and articles have different scientific weight. This short summary provides an overview of the scientific evidentiary value of various types of publications with the goal to create transparency in the level of bias of any given study so that readers may, in turn, determine the attention and importance they can give the study. Practically speaking, the more bias, the less evidentiary value a study or publication has and vice versa. A figure illustrates this concept and simplifies it.

In determining the validity of the scientific data, i.e. whether information in a medical publications and studies is scientifically sound, a variety of methods can be applied to grade publications and studies hierarchically. The goal is to create transparency in the level of bias of any given study so that readers may, in turn, determine the weight they can give the study. Practically speaking, the more bias, the less evidentiary value a study or publication has and vice versa. Figure 14.1 illustrates this relationship.

Figure 14.1 illustrates the five most common types of scientific evidence, systematic reviews, randomized clinical trials and observational studies, uncontrolled observational studies, case reports and case series, and expert opinions. With increasing bias, the scientific evidentiary value, i.e. the level of scientifically sound information, decreases. Conversely, with decreasing bias and controls, the evidentiary value increases. Understanding the hierarchy of scientific evidence is crucial to implementing the proper scientific data in policy and legislation based upon what should be scientifically sound.

G. Steier
Food Law International, LLP, Boston, MA, USA
e-mail: g.steier@foodlawinternational.com

L. Steier (✉)
Studienzentrum für Fortschrittliche Zahnheilkunde, Mayen, Germany
e-mail: lsteier@gmail.com

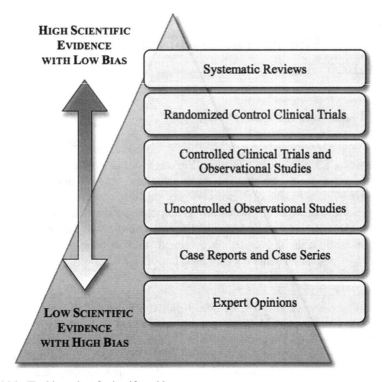

Fig. 14.1 The hierarchy of scientific evidence

Chapter 15
Food and Energy

Stephen B. Balogh and Charles A.S. Hall

Abstract There is nothing more critical to human existence than food. While a proper climate and water are arguably as important as food, they are usually present, but food shortages from population expansion, climatic extremes, conflict, and concentration of output in the hands of the powerful are a nearly constant characteristic of one part or another of the world's human population for as far back as we have records—and probably far before that. For example, huge famines occurred in China in the sixth and twentieth century AD, and many centuries in between, most of Europe in the fifteenth century, Ireland in the middle of the nineteenth century, Bosnia, Philippines, and Sudan during this past century and countless other locations all over the world. But we in most of the developed world live today in a situation of incredible food affluence, and famine seems to have left much of the world except for areas of political-military conflict. How has this come to be? The most general answer is the application of fossil fuel technology and its ancillary technologies, most notably the production of nitrogen fertilizer and substitution of mechanical work for human and draft animal labor, to food plant production. This has allowed an enormous expansion of food production and has allowed us to think about food from many other perspectives, including aesthetic, moral and political. We examine human food production over millennia with a particular focus on energy: the quantity and quality of the energy of the food and also of the energy required to produce it.

15.1 Introduction: Food as a Global Issue

For most of humanity's existence, extreme hunger and starvation have been constant companions. There are records of millions of people starving in China, India, Egypt, Russia, and elsewhere, as recently as the twentieth century. In many smaller

---

This paper derived in large part from: Hall CAS, Klitgaard K (2012) Energy and the Wealth of Nations: Understanding the Biophysical Economy. Springer, NY; Used with permission.

S.B. Balogh • C.A.S. Hall (✉)
State University of New York, College of Environmental Science and Forestry (SUNY-ESF), Syracuse, NY, USA
e-mail: sbbalogh@esf.edu; chall@esf.edu

nations and city-states, starvation has occurred periodically since human settlement began. Today, however, mass starvation is rare. There are two general ways to view this reduction in food shortage: from the "BigAg" perspective, which consists of industrial food producers, synthetic pesticide and fertilizer developers, and large food corporations, or, in the alternative, from the perspective of concerned consumers, environmental conservationists, and sustainability advocates.

From BigAg's one-sided point of view, the collective ability of today's farmers to meet—for the most part—the food demand of some seven billion people is allegedly a triumph of technological advances and human ingenuity. Artificial fertilizers, pesticides, novel cropping methods, and bioengineering have seemingly tripled yields of staple and commodity crops in the past half-century. This increase has increased per capita food availability by about 25 % from the 1960s to 2009[1]—despite the fact that these statistics play out differently around the world. Although agricultural output has increased during a period when the total area in farms has decreased[2] and per capita wheat, rice, and coarse grain production has risen over the last decade in all regions except Oceania (Australia and the South Pacific), food shortages continue to persist in many parts of the world. Despite this, the United Nations Millennium Development Goal to halve the proportion of the population that is chronically hungry and malnourished by 2015 may be within reach.[3] But the same advances that are publicized as agricultural breakthroughs are also the culprits of many challenges of the modern food system. The fact that anyone remains hungry in 2016 is more a consequence of resource distribution, politics, and a preference for higher animal protein diets and a high level of wastage in the food distribution, peparation and disposal systems, rather than intrinsic limits to agricultural production. Therefore, the alleged limits of agricultural production need not be resolved through biotechnology and synthetic pesticides or fertilizers alone. Instead, policy changes and a careful look at resource use and distribution may yield much more promising solutions.

From a biophysical and ecological perspective, however, modern farming has become increasingly unsustainable and is pushing the planetary boundaries. Perhaps most importantly, each kilogram of food produced through industrial agriculture is less and less a function of sunlight, soil, water, and labor inputs, as it should naturally be; instead, it is the product of fossil fuel inputs, chemicals, and biotechnology. These fossil fuel inputs can be either direct, in the form of diesel to run tractors, or electricity to pump irrigation water; or indirect, through petrochemical fertilizers and pesticides, as well as energy embodied in farm machinery and other infrastructure. Because of this dependence on petroleum, access to food markets, along with the price of food, is tied directly to changes in energy prices—especially oil.[4] Thus, the question "will we be able to produce enough food to meet the

---

[1] Food and Agriculture Organization (2013), p. 124.
[2] Food and Agriculture Organization (2013), p. 124.
[3] Food and Agriculture Organization (2013), p. 67.
[4] Baffes and Dennis (2013), p. 2.

demands of a growing and urbanizing population?" becomes, instead, a question of whether humanity will be able to continue to find and extract increasing amounts of oil and natural gas at a low cost.

The paradigm shift of asking another question entirely alters our view of the food production system. Furthermore, other issues associated with this heavy industrialization include growing pest resistance to pesticides, fertilizer saturation, soil depletion and toxification, and reduction of cultivar diversity and crop ecosystems. The collective long-term effects of the "new conventional" farming system, relying on oil, biotechnology, and industrialization of food production are difficult to predict but allude to a dangerous future. Rapid urbanization has changed how we produce and consume food. Since more than half of people around the world live in urban areas, humanity's dependence on petroleum for transporting food has increased. As cities grow, some of the most fertile land is being paved over with impermeable surfaces and is no longer available for food production. Moreover, as people move increasingly from rural areas to cities, their diets change. Greater income usually leads to higher protein demand and the transition to a more "global" diet that eschews locally produced food for cheaper, more processed imports and higher amounts of dairy, eggs, and meat. China, for example, has seen a four to tenfold increase in meat and milk consumption from 1980 through the 2000s.[5] The production of dairy and meat products is much less efficient than for staple grains such as rice and wheat. In fact, the former requires a much larger energy input for the amount of food produced, and this increases pressure to expand agricultural output. Consequently, rapid urbanization contributes to starvation and malnutrition when energy-intensive farming is the method of producing food.

15.1.1 Starvation and Malnutrition

The most important requirement for food production is to sustain our growing population. Large-scale starvation, while once common, is now relatively rare due to improved efficiencies in transporting food from areas of abundance to scarcity. This approach has historical precedence: for example, while starvation once occurred frequently in India, the completion of the national railroad system essentially eliminated the problem in the 1880s. Starvation still occurs in contemporary societies, but it tends to be linked to war, political instability, or strife rather than actual crop failure.[6] In other words, the planet can produce enough food to avoid human starvation, but it is not always distributed to where it is needed. This contradicting relationship is due to the industrialization of agriculture (and warfare) and increased urbanization.

---

[5]Food and Agriculture Organization (2009), p. 11.
[6]See Riley (1993); Devereux (2007).

Malnutrition, as opposed to starvation, is much more common and pervasive than starvation. Malnutrition, occurs when there is enough food, but not enough of the kinds that humans require to be healthy. Most fundamentally, malnutrition occurs when there are enough calories to survive, but not enough calories or nutrients for the body to do all the functions required for optimum well-being. A secondary issue to malnutrition is lack of protein: in general, plant—and especially animal—protein tends to be more expensive, both monetarily and energetically, than carbohydrates. This is because, ultimately, it is much more expensive for plants or humans to fix (take out of the air) nitrogen than carbon.[7] Paradoxically, a portion of the population in developed nations are over-fed and malnourished, with many people consuming up to 3700 kcal in protein-rich diets that lack sufficient potassium, calcium and vitamin D.[8] Thus, malnutrition can occur even when obesity becomes a problem. Thus, the sociological and environmental shifts must be considered in examining the problems of food security.

15.1.2 Food and the Environment

It is easy to look upon agroecosystems and see green, sustainable environments in harmony with nature; indeed, that is how they are often portrayed in the popular press. Yet at the most basic level, agriculture and the natural environment are intrinsically at odds. The very purpose of agriculture is to redirect the land's energy flow from diverse, sustainable ecosystems to simplified monocultures that require continual inputs of human and fossil energy to maintain their highly productive state. Natural ecosystems, on the other hand, usually maintain or build soils, are in carbon balance with the atmosphere, and retain nutrients. Agroecosystems, in contrast, typically lose soil and nutrients while adding carbon to the atmosphere. Depending upon local geography and cultural practices, agriculture also has a detrimental impact on water, air and soil quality: fresh and salt water bodies become eutrophied by excessive nutrients from farmland runoff; changes in land use increase greenhouse gas emissions; livestock and crop production contribute to global climate change; and tilling, fertilizers, and pest control lead to soil erosion and degradation.

Industrial agriculture also disrupts traditional farming systems that have developed over millennia and are tied to local environmental conditions and social structures. Perhaps most significantly, erosion from industrial agriculture is reducing our one-time allotment of arable soils.[9] Net soil losses under industrial farming are one to two orders of magnitude greater than soil production or erosion under

---

[7]On the most basic level, nitrogen fixation is required for the production of plant protein, while carbon fixation is part of the process of producing sugars and other carbohydrates.

[8]United States Department of Agriculture (2010), p. 8.

[9]Pimentel (2006).

native vegetation.[10] Excess fertilizers enter fresh water bodies and eventually end up in the ocean, leading to eutrophication and disruption of biogeochemical nutrient systems. These ecological impacts have energy costs, too, as nutrient-poor soils require increasing inputs of energy-intensive fertilizers. They also disrupt fresh and salt-water fisheries, leading trawlers to fish more intensively in order to maintain catch levels.

The water required to produce various food and forage crops ranges from 500 to 2000 liters (L) per kilogram of crop produced. A hectare of US corn, for instance, transpires more than 5 million L of water during the 3-month growing season. If irrigation is required, more than 10 million L of water must be applied to this crop.[11] It is possible that water, as opposed to energy, will limit agriculture in the future. Agriculture places enormous demands on global water resources, and climate change may further alter the distribution and amount of precipitation received in the current arable lands. Today, however, energy is still a limiting factor of production: water-pumping infrastructure and desalinization, for example, require significant energy inputs. Nonetheless, it is beyond the scope of this chapter to examine the potential impacts of climate change on agricultural production, other than to say that they are large, controversial, and act to increase some crops while decreasing others.

15.1.3   The Politics of Food

Since the agricultural revolution, higher incomes and social statuses have been linked to a higher quality and higher calorie diet. As nations become economically wealthier, their diets—and especially those of their richest populations—become much more protein-dense. Consequently, these diets became more energy intensive to produce. Still, while relatively affluent middle class Chinese urbanites dine on pork, or American workers lunch on hamburgers from the drive-thru window, countless poorer people around the world subsist on rice or sorghum mixed with a few scraps of vegetables. The evolution of agriculture has not brought positive impacts to the global population across the planet and continues to show discrepancies in terms of food safety and food security world-wide.

That this inequity in access to high-quality foods persists would likely surprise few readers. The world's poor, however, also face disproportionate environmental impacts from the shift to a "Western" diet by a growing middle class. Subsistence farmers in tropical nations, for example, have been forced from their traditional lands and their customary agricultural practices (such as *swidden*, or "slash and burn" farming). These techniques are replaced by large monocrop systems with outputs destined for export, such as soybean farming in Brazil.

---

[10]Montgomery (2007), p. 13268.
[11]Pimentel and Pimentel (2003), p. 660S.

The replacement of traditional century-old farming practices with intensive crop-monocultures is driven by economic incentives. During the 1980s, international financial institutions implemented neoliberal policies in the global south to provide development loans. However, their policies also eliminated trade barriers and flooded local markets with cheap, subsidized grains from the U.S. and Canada, which destabilized local production systems.[12] Many developing nations continue to rely on agricultural exports to generate income in order to run their economies and pay the interest on longstanding development loans. All of this has implications for each nation's food security and self-sufficiency.

Fad diets and changes in taste in Western countries also affect distant ecosystems and cultures. A recent example of this is the introduction of quinoa as a high-status food and healthy substitute for processed grains in the United States. Traditionally consumed as a staple food in Peru and Bolivia, Americans now import nearly 68 million pounds of quinoa per year.[13] Some scholars consider this an economic boon to poor farmers in the region who now have access to lucrative health food markets in the U.S. Yet others decry the disruption to local markets and the inability of the local poor to afford this once ubiquitous crop–the price of which has increased sevenfold from 2012 to 2014.[14] While the debate continues over costs and benefits to local farmers in Peru and Bolivia, it is clear that voluntary (or perhaps marketing-influenced) changes in Western diets have real implications for the diets and economies of far-flung cultures.

15.1.4 The Morality of Diet

What, then, should a moral person eat? Worldwide, an estimated two billion people live primarily on a meat-based diet, while an estimated four billion live primarily on a plant-based diet.[15] For some, this is a purely economic decision: they would consume meat if it were available and affordable. Yet, for others, the decision to eschew meat stems from religious, cultural, or moral beliefs. About one third or more of the 1.2 billion people living in India exist on a vegetarian diet. Another tenth of the population eats only grains, vegetables and some eggs. Certainly, there are also millions of relatively affluent people who are vegetarian for moral and environmental reasons, such as a desire to reduce resource consumption or concern about the welfare of animals raised for food. Other considerations for a moral diet include: the environmental impacts of food production (including, but not limited

---

[12]See, e.g. Costa Rica as described in Hall et al. (2000).
[13]Washington Post (2013).
[14]World Bank (2014).
[15]World Bank (2014).

to, greenhouse gas emissions, eutrophication, and pollution of water bodies); health impacts to humans, such as antibiotic resistance in bacteria; the often cruel treatment of livestock; the wages paid to farmers for their labor; and the protection of non-renewable and renewable resources, such as soil erosion.

The decision about what to eat is ultimately complex and nuanced; the growing global population, resource constraints—especially from peak petroleum production—and the increasing inability of the ecosphere to assimilate the wastes from our economy all complicate this process. Notably, however, the number of children one chooses to have, and at what age, may impact the food production system far more than what foods one decides to eat for a multitude of reasons explained hereinafter. Throughout this chapter, the goal is to educate the reader on the relation between fossil and natural energies and the contemporary food system, and provide insight for those looking to reduce the energy impact—and greenhouse gas emissions–of their diet. Ultimately, the morality of diet lies within one's food choices and the large-scale consequences of the accumulation of an individual's dietary decisions on the market and the planet. In other words, this chapter provides an outline that connects food choices with one's environmental footprint.

15.2 Food as Energy

Human bodies are, at their essence, biological machines. To operate well, these machines require daily inputs of water, fuel, and the essential chemical compounds that they cannot synthesize alone. Thus, whether one is rich or poor, the "work" one must do each day requires securing sufficient resources of fuel (food energy), water, and nutrients. Government agencies, intergovernmental organizations (such as, the United Nations), and non-governmental organizations (NGOs) research and provide suggestions about the amount and variety of food that humans should consume to meet dietary needs. These usually take the form of recommended daily allowances of calories,[16] along with macro- and micro-nutrients. They can also be presented as informal food-based dietary guidelines, such as "eat a variety of fruits and vegetables each day." It is telling that in our age of relative abundance, these guidelines often *warn* against overconsumption of calories,[17] rather than providing minimal daily energy consumption requirements. This oddity, especially in light of the problems associated with starvation and malnutrition, seem illogical. However, upon deeper reflection about the connections between the current food industry and the rising numbers of malnutrition, the links fall into place.

---

[16]Measured in kcal, where 1 kcal = 4.184 kJ.

[17]United States Department of Agriculture (2010), p. 8.

15.2.1 How Food Is Measured

Macronutrients provide the bulk energy content of food. These chemical compounds are grouped into three major classes: carbohydrates, proteins (and their amino acid components), and fats. The total combustible energy in food can be measured by using bomb calorimetry. Simply put, the food is combusted in a chamber full of oxygen submerged in a known volume of water, and the resulting increase in water temperature is measured. Although there are slight differences in the amount of energy per unit of mass for different compounds, it is generally accepted that 1 g of carbohydrates or protein contains about 4 kcal, while fats contain about 9 kcal of available energy. However, not all of the ingested chemical energy in food is available to the body. Insoluble fiber, for example, is combustible, but it passes through the digestive tract without being metabolized. More complex adjustments must be made to bomb calorimetry results to determine the energy available to humans in foodstuffs.[18]

The unit of energy typically used to measure the available energy in food is the kilocalorie. A calorie is defined as the amount of heat needed to raise 1 g of water by 1 °C at 15 °C. A kilocalorie is 1000 cal. One kilocalorie (kcal, dry weight) is equivalent to approximately four BTU, or 4.18 kJ (10^3 J—note Joules are the SI standards for energy and should be used for ALL energy calculations and representation, but calories are entrenched for food). A British Thermal Unit (BTU) is approximately equivalent to the energy found in the tip of a matchstick. Thus the digestion of each kcal of food liberates the energy contained in four matchsticks. There are usually about 4 to 9 kcal per (dry) gram of food, or 112–255 kcal per ounce, with the lower values characteristic of carbohydrates and proteins, and the higher values of fats. Using these metrics, a person needs roughly half to 1 kg of food per day, including food wasted. Thus, in a deeply simplified model, the planet must produce nearly 1 kg of food for per person per day. Some of this food will be wasted and some additional energy and water is needed to prepare and process the food. Thus, energy goes into producing food and food is, in turn, used to produce energy. The vast complexities within this seemingly simple equation, however, give rise to a universe of considerations that are crucial to understanding the modern food system.

15.2.2 The Fate of Ingested Food: Food as Physiological Energy

Due to incomplete digestion, not all of the gross energy available in food is available to the body. For every 100 units of gross energy ingested, approximately

---

[18]Food and Agriculture Organization (2003), p. 5.

three-quarters are assimilated into the blood in the form of simple sugars, amino acids, and fatty acids. The remainder of the energy is lost through egestion as feces and combustible gases. Other energy losses occur through the urine, as well as through the catabolism (breakdown) of protein and evaporation from the body's surface.[19] Metabolizable energy (ME) is the energy that remains after accounting for these important losses. Of this metabolizable energy, some must be used to run the processes of digestion, absorption, and intermediary metabolism, and is thus unavailable for other metabolic processes. The digestive tract is home to some 800 species of bacteria, which are important for the metabolism or generation of several vitamins, in addition to helping ferment indigestible carbohydrates. Additionally, the bacteria recirculate compounds excreted in bile from the liver.[20] The gut flora, in turn, uses some of the metabolizable energy in these processes. Correspondingly, the net metabolizable energy (NME) accounts for losses due to the aforementioned processes, while the remaining energy—that which passes from the gut into the bloodstream—becomes available for basal metabolism, active metabolism,[21] growth, and reproduction. About one percent of ingested energy is used for growth and reproduction.[22]

15.2.3 Human Energy Requirements

Today, humans require on average 2000–2500 kcal[23] for proper nutrition.[24] Depending upon their weight, sex, level of activity, and other factors, they can require as little as 1000 kcal (for children 2–3 years old) and up to 3200+ kcal (for active male adults) per day. This energy requirement is appreciably lower than the 3000 kcal/day estimated by some academics for modern and historical hunter-gather populations.[25] The notably higher levels of kcal for hunter-gatherers can be attributed to their greater levels of activity and higher resting metabolic rates. A modern human would have to walk nearly 19 km (12 miles) per day in addition to their current daily activities in order to expend the energy used by their typical !Kung or Ache counterpart.[26] The fact that the average human requires one-third less food energy than our Paleolithic ancestors (and modern hunter-gathers) can be attributed principally to our present use of fossil fuels for labor, transportation, and air-conditioning. Nonetheless, the variations in diet also change how much

---

[19]Food and Agriculture Organization (2003), p. 5.
[20]O'Keefe (2008) p. 51; Flint et al. (2012), p. 577.
[21]For example, locomotion and work.
[22]Hall et al. (1986), p. 12.
[23]Or approximately 8400–10,500 kJ.
[24]United States Department of Agriculture (2010), p. 8.
[25]United States Department of Agriculture (2010), p. 8.
[26]Kious (2002), p. 1.

nutrition a person can actually get from his or her diet. Arguably, an organic, locally-sourced, whole-foods plant based diet will provide more nutrients and more food for a certain amount of calories than a comparatively unsustainable processed-food animal-protein based mainstream meal that is not locally sourced. Thus, energy quality and nutrition also play a major role in the considerations at the heart of this chapter.

15.2.4 Energy Quality and Nutrition

Not only do humans need energy, they need the right kind of it. Thus, adequate nutrition depends on proper food safety and security—and vice versa. Centuries ago, when hunter-gatherers obtained their food, the hunters had to be good animal trackers, and the gatherers had to know where to search according to the season. Essentially, these populations got their food through their own energy investments. Similarly, early farmers had to understand many things about where and when to plant, cultivate and harvest. By contrast, with contemporary and intensive agriculture, food producers need not just human labor energy outputs and accumulated knowledge, but energy in the form of petroleum for tractors and transport, natural gas to fix nitrogen for fertilizer, and electricity for pumped irrigation. In addition, farmers seek affordable energy, i.e. energy that does not require a lot of energy to produce, such as in the form of fuels with a high-energy return on investment (EROI).[27]

The bulk of human food needs are met through the production and consumption of staple crops, mainly corn, wheat, and rice–all highly productive grasses. While cultivars vary by geographic region and culture, their structure and function—high carbohydrate (high energy) and high yield per effort (high efficiency)—vary little. Protein requires more energy per gram to produce, often substantially more. Vegetables, necessary for good nutrition, are more energy intensive than grains. Thus a complete diet requires a mix of both high efficiency staple crops, as well as higher quality foods that need a greater energy investment for production. If we were to eat the most energy efficient diet, it would probably still not be sufficiently nutritious. On the other hand, a modern diet too rich in proteins and fats is also energy intensive and early agriculture probably decreased the nutritional status of humans. Academic and scientific opinions vary greatly, and there is no consensus at this time about the optimal diet that combines good nutrition along with a low energy cost. Progressive thinkers suggest that a locally-sourced organic whole-foods plant-based diet may be the best approach for public health and environmental integrity[28] although further quantification of this is needed.

---

[27]Lambert et al. (2014), p. 153. For impact of early Agriculture see Angel (1975).
[28]T. Colin Campbell and Jacobson (2013).

Thus far, this chapter focused mostly on the nutritional requirements of the individual. Now, however, the focus shifts to a larger, more comprehensive, question: how should we look at food and energy for an entire country, or region, or the world? Nearly everyone asking this question goes back to the first important paper on this topic by Thomas Malthus.

15.2.5 Thomas Malthus's Question

A discussion of the resource versus population issue always starts with Thomas Malthus and his 1798 publication *First Essay on Population*:

> I think I may fairly make two postulata. First, that food is necessary to the existence of man. Secondly, that the passion between the sexes is necessary, and will remain nearly in its present state...., increases in a geometrical ratio. Subsistence increases only in an arithmetical ratio. Slight acquaintances with numbers will show the immensity of the first power in comparison of the second.[29]

Malthus continues with a very dismal assessment of the consequences of this situation for humans including even more disheartening and inhumane solutions that disadvantage poorer populations. Most people agree, however, that Malthus' premise has not held up between 1800 and the present, as the human population has expanded by about seven times along with concomitant increases in nutrition and general affluence—albeit the latter occurred only recently. In *The End of Food*, Paul Roberts (2008) reports that malnutrition was quite common throughout the nineteenth century. It was only in the twentieth century that cheap fossil energy allowed a sufficient level of agricultural productivity to avert famine. Many scholars have made this argument—that humans' exponential escalation in energy use, including that used in agriculture, is the principal reason that the food supply has grown parallel to the human population. Since Malthus' time, therefore, we have avoided wholesale famine for most of the Earth's people due to the expansion of fossil fuel use. This was something that Malthus could not have foreseen.

The first twentieth century scientists who argued consistently with Malthus' concern about population and resource distribution were ecologists Garrett Hardin and Paul Ehrlich. Hardin's essays in the 1960s on the impacts of overpopulation include the famous *Tragedy of the Commons*, in which he discusses how individuals tend to overuse common property to their own benefit even when it is disadvantageous to all parties involved.[30] Hardin wrote other essays on population, coining such phrases as "freedom to breed brings ruin to all" and "nobody ever dies of overpopulation," the latter implying that overcrowding is rarely a direct cause of

---

[29]Malthus (1798).
[30]Hardin (1968).

death, but rather it leads to disease or starvation or living in dangerous areas such as periodically hurricane-flooded deltas, which, in turn, kill people as a result of overpopulation. This idea is exemplified in an essay about the thousands of people in coastal Bangladesh who drowned in typhoons over the past centuries. Hardin argues that the residents knew this region would be inundated every few decades, but they lived there anyway because they had no other place to go in such a crowded country. The typhoon pattern recurred in 1991 and 2006, thus supporting Hardin's argument that overpopulation causes other problems, which then lead to death.

In *The Population Bomb* (1968), ecologist Paul Ehrlich argues that continued population growth will wreak havoc on food supplies, human health, and nature, and that Malthusian processes (such as war, famine, pestilence, and death) will sooner rather than later bring the human populations "under control" and down to the carrying capacity of the world. During the time of Ehrlich's work, agronomist David Pimentel and others,[31] ecologist Howard Odum, and environmental scientists John and Carol Steinhart quantified the energy dependence of modern agriculture and showed that technological development is almost always associated with increased use of fossil fuels. Other ecologists, including George Woodwell and Kenneth Watt, discuss in depth how people negatively impact ecosystems.[32] Kenneth Boulding,[33] Herman Daly and a few other economists begin to question the very foundations of economics,[34] including its dissociation from the biosphere necessary to support it and, especially, its focus on both growth and on infinite substitutability—the idea that something will always come along to replace a scarcer resource.[35] More recently, Lester Brown and others provide convincing evidence that food security is declining, partly because of distributional issues and partly because of declining soil fertility, desertification, and a decrease in the availability of fossil-fuel derived fertilizers.[36]

On another note, Jay Forrester is the developer of a series of interdisciplinary analyses and thought processes, which he calls system dynamics. He describes the impending difficulties posed by continuing human population growth in a world of finite resources. His analysis became known as the Limits to Growth model.[37] His computer models were refined and presented to the world by Forrester's students Donella Meadows, Dennis Meadows, and their colleagues in 1972.[38] They showed

---

[31] Pimentel et al. (1973, 2005).

[32] See Charles A.S. Hall, Kent A. Klitgaard, Energy and the Wealth of Nations: Understanding the Biophysical Economy (Springer 2012).

[33] See a list of Kenneth Boulding's work at http://www.nasonline.org/publications/biographical-memoirs/memoir-pdfs/boulding-kenneth-e.pdf.

[34] Hall and Day (2009).

[35] Hall and Day (2009).

[36] *See generally* Brown (2009a, b).

[37] Also known as the "Club of Rome" model, after the organization that commissioned the publication.

[38] Meadows et al. (1972)

that exponential population growth and resource use, in combination with finite resource and pollution assimilation, will lead to serious global economic instabilities, eventually resulting in a large decline in the material quality of life and the overall human population.[39] Around the same time as Forrester's writing, geologist M. King Hubbert predicted in 1956, and again in 1969, that oil production from the coterminous United States would peak in 1970 before declining. Although his predictions were dismissed at the time, U.S. oil production in fact peaked in 1970, and natural gas did so in 1973.[40] These predictions provided frameworks for an understanding of past and future food production challenges.

Before considering the present and future possibilities with respect to food, one must examine food production from the widest possible perspective. This analysis begs the questions: How has the present human food situation developed? Are Malthus' ideas still valid? Has the temporary availability of fossil fuels delayed the implementation of the "Malthusian dilemma," or have technological conditions changed the limits of food production? How can we understand the relation of human population and food over a long period of time? The following sections explore some of these issues.

15.3 History of Humans and Food

15.3.1 The Prehistory of Human Society: Living on Nature's Terms

Agriculture, by its definition, is a manipulation and cultivation of nature's abundance of foods. It is, therefore, important to understand the relationship between human evolution and food system evolution. In fact, people sufficiently similar to the modern human have been on Earth for roughly half a million years and have benefitted from nature's supply of food. Yet, scholars understand very little about how these people made their living, what they did day to day, or how they interacted with each other. The only existing evidence of their lives consists of human bones, the bones of their prey, and an occasional tool. Scientists are relatively certain that these early humans survived by hunting and gathering, i.e. by exploiting whatever food nature provided along with what could be obtained using relatively simple tools such as spears and baskets. Most of what we know about our hunter-gatherer ancestors is derived principally from anthropological studies of remaining hunter-gatherer cultures such as the !Kung, a group that still lives in the Kalahari desert of Southern Africa, as well as in towns and cities.[41] Nonetheless, all of those who examine what life must have been like for our ancestors, are indebted to the work of

---

[39]Hall and Day (2009), p. 220.
[40]Hall and Day (2009).
[41]Lee (1969)

Richard Lee, who studied the !Kung while they were relatively unaffected by modern civilization. Many academics believe that modern hunter-gatherers are the best mirror in which to see what life must have been like for our ancestors over the half million years between the evolution of our species and the development of agriculture.

Life for a hunter-gatherer is essentially about taking nature as it is found and finding ways to support oneself on those resources. Since most early human hunter-gatherers lived in tropical environments, the key issue was gaining needed energy from food. For the !Kung, this meant that women predominantly gathered mongongo nuts while men hunted. Mongongo nuts were a critical and abundant resource; today, they still provide the largest portion of energy and protein for the !Kung, in addition to nutrition from game. Life was good for the !Kung, at least before their major contact with outside civilizations.

According to Lee's studies, the !Kung spend far fewer hours working each day than most people living in industrial societies, and a lot of their time is spent in leisure activities. Life for the !Kung is not quite that simple however. Desert living is constrained by the need for water and food. In their homeland of Botswana, there are relatively few waterholes and it is essential to set up camp near one of these. As a result, the !Kung periodically exhaust the food resources near their present waterhole and must move to a new water source and establish a new camp. Mongongo trees are spread around part of the Kalahari desert and initially the !Kung have a relatively easy time obtaining the food they need from relatively short excursions from their camp. As time goes on, however, they deplete the nuts within easy reach so that each day they have to make a longer and longer trip to gather enough mongongo nuts to feed their families. At some point when they have gathered all the mongongo nuts within a day's hike, they have to make a much further overnight trip to get them. This has the effect of greatly increasing their energy expenditure and lowering their energy return on investment (EROI). Their energy investment is much greater because they need a lot of food both going and coming back, and may end up eating a substantial portion of the food they set out to gather. At this point, it is usually desirable to make the investment of moving to a new water hole.

It is becoming clear that our stone-age hunter-gatherer ancestors, just like hunter-gatherers today, were truly remarkable hunters. This had the net effect of drastically reducing the populations of the large birds and mammals of the earlier world. As humans spread about the world, they encountered, in each new place, large and presumably tasty herbivorous animals of the sort that no longer exist anywhere on Earth today. For example, the new arrivals to North America roughly 12,000 years ago found giant beavers, rhinoceros, two species of elephants, camels, and many other now unfamiliar creatures. Likewise, human arrivals in Australia found giant flightless birds, while the first humans in what is now contemporary Italy encountered enormous turtles. None of these large animals are there today. Furthermore, with the exception of those in Africa, there are few animals left larger than 100–200 kg—although such sizable animals were abundant prior to human contact.

There are two competing hypotheses for what caused the extinction of those large animals. First, since the climate was warming rapidly 10,000 years ago, it is possible that they succumbed to some effect of climate change. The second hypothesis is that humans hunted these animals to extinction. These large animals had no previous reason to be afraid of anything as small and seemingly weak as a human being. The first humans could simply walk up to these animals and stick a spear into their side. Africa still has many large herbivorous species, likely because these animals coevolved with humans as they became more proficient hunters with better weapons. Wherever humans migrated, most or all of the animals larger than 100–200 kg disappeared within 2000 years, lending support to the idea that *Homo sapiens* caused these animals' extinction.[42] In addition, the fact that these same animal species had survived many previous climate changes lends considerable— but not absolute—support to the human-caused extinction theory. Thus, significant environmental impact is hardly a new phenomenon of the human species, but rather something that has been occurring for millennia.

15.3.2 African Origin and Human Migrations

All available evidence suggests that humans and their predecessors evolved in Africa. It is the only place where scientists have found human fossils and evidence dating back to 1.7–1.8 million years ago. Take a mental time trip to East Africa about 2 or 2.5 million years ago: you will be at the epicenter of human evolution. What is remarkable, however, is that you will find not one, but perhaps half a dozen types of early humans (or hominids); each group as distinct from one another as chimpanzees are from gorillas. Most of these protohominids were found in small migratory bands more or less at the transition of forests to drier savannas. In the 1990s, scientists announced that they had found what appears to be the ancestor of humans; a being who lived some 4–6 million years ago. This discovery is cause for great excitement amongst those who are determining our lineage. The creature, named *Ardipithecus ramidus* (Ardi for short), walked more or less upright but still spent a significant portion of its time in trees, similar to chimpanzees.

The Ardis had several interesting characteristics. Recent research has found that a human uses only about one quarter the energy that a chimpanzee uses to walk 100 m., so there has clearly been a tradeoff of more energy-efficient walking for the ability to both walk and climb trees well. Probably most of the Ardis made, or at least used, tools of some sort. Studies show that even chimpanzees have a rather astonishing ability to make many different types of tools, including stone anvils. Most of the Ardis' tools were made from organic materials and were, therefore, not well preserved. Hence, scientists know little about the evolution of early protohominid tool-making. It seems clear for humans, however, that by about 2.5

---

[42]Sandom et al. (2014), p. 2, Martin (1973).

million years ago, they had developed skillful methods for making stone knives and spear points, such as by striking one rock on another in repeated, often sophisticated, patterns. There are even a number of ancient "industrial complexes" in, for example, Kenya's Olduvi gorge, which has become a rich hunting ground for information about our ancestors.

The development of tools is one of the factors setting Ardis and chimpanzees apart. Spear points and knife blades are energy-concentrating devices that allow the strength of a human arm to be multiplied many times. This, in turn, allowed humans to exploit many new animal resources, and eventually colonize cooler lands. Human ancestors were using stone tools for roughly two and a half million years, which is equivalent to about 100,000 human generations. By contrast, humans have been using metal tools for roughly 8000 years, or about 400 generations. Most of human history, therefore, has been without metal tools. Early copper and bronze tools were probably not much more effective than well-made rock or bone tools. In time, however, these tools became much more effective as their design and technology improved. An important reason behind the slow transition to metal tools is that stone tools could be made with a small energy investment (essentially human muscle power). Metal tools, on the other hand, required heat, which meant a much larger human investment of cutting trees, making charcoal, and finally making the tool itself.[43] Early smelting was probably technically inefficient, but it had the advantage, at least initially, of the availability of very high grades of ore. Thus, the development of tools became increasingly sophisticated. These stone spear points and knife blades were more or less the first in a long series of technological advances that helped increase the flow of energy to humans. The consequence of these tools is that they greatly expanded the ability of humans to exploit various plant and animal resources in their environment. They also diversified the climates in which humans could live by enabling them to kill large animals and use their skins for clothes.

Another important new energy technology was that of fire. While it naturally allowed people to stay warm in cooler climates, it more importantly increased the variability and utility of plant foods: cooking broke down the tough cell walls of plants, for example, and made them more digestible.[44] Following the discovery of this "technology" a little less than two million years ago, many humans left the relatively benign climate of Africa. Before long, the remains of both humans and their tools ended up in present day Middle East, Georgia and Indonesia. By one million years ago, human remains were common all throughout Asia. However, humans did not colonize Europe until roughly 500–800 thousand years ago. The first humanoid colonists of Europe are likely not our direct ancestors, for morphologically modern humans[45] appear to have left Africa in a separate migration only about 100,000 years ago. There are very strong debates in anthropological literature

---

[43]Perlin (1989), Ponting (1991).
[44]Wrangham (2009).
[45]Popularly known as "Cro-Magnons," and distinct from the earlier "Neanderthal" stocks.

as to whether all of these groups of people are the ancestors of modern humans or just the "Cro-Magnon" variety, but modern DNA analysis seems to favor the separate stock concept. For whatever reason—perhaps interracial warfare, climate change, or some indirect result of competition—the Neanderthal stocks were eliminated from Europe by 35–40,000 years ago, along with many other protohominid variations, leaving, it seems, a few of their genes with those of European stock. In sum, the evolution of humans is an important precursor to understanding how agriculture evolved.

15.3.3 The Dawn of Agriculture: Increasing the Displacement of Natural Flows of Energy

Some time roughly 10,000 years ago, in the vicinity of the Tigris and Euphrates valleys of present day Iraq, a momentous thing happened. Humans, previously completely constrained by their limited ability to exploit natural food chains (due to the low abundance of edible plants in natural systems), discovered that they could increase the flow of food energy to themselves and their families by investing some of the seeds that might otherwise be eaten into more food for the future. How this happened can never be known for certain.

The implications of agriculture development for humans were enormous. The first, seemingly counterintuitive, consequence of agriculture is that human nutrition declined. Studies of bones of people buried over the past 10 thousand years in Anatolia, which is the area roughly encompassing the border region of modern day Turkey and Greece, revealed the height and general physical condition, as well as their nutrition status of the people who used to live there. The data indicates that the people actually became shorter and smaller with the advent of agriculture, indicating a *decrease* in nutritional quality (see footnote 27). In fact, the people of that region did not regain the stature of their hunter-gatherer ancestors until about the 1950s. Therefore, although agriculture may have given the first agronomists an advantage in terms of their own energy budgets, that surplus energy was translated relatively quickly into more people with only an adequate level of nutrition as human populations expanded. Or perhaps, as outlined below, more of the farmers' net yield was diverted to artisans, priests, political leaders, and war, leaving less for the farmers themselves. One of the clear consequences of agriculture was that people could settle in one place, so that the previous pattern of human nomadism was no longer the norm. As humans occupied the same place for longer periods of time, it began to make sense to invest their own energy into relatively permanent dwellings, often made of stone and wood. This start of the construction of the durable human structures have left significant artifacts for today's archeologists and show some of the implications of agricultural development.

Another significant consequence of agriculture was the enormous increase in social stratification, which took place as economic specialization became more and more important.[46] For example, if one individual was particularly skilled at generating agricultural yield or understood the logic and mathematics (i.e. best planting dates) of successful farming, it made sense for the farmers of the village to trade with him some of their grain for knowledge, thereby initiating, or at least formalizing, the existence of markets. From an energy perspective, relatively low-skilled agricultural labor was being traded for the high-skilled labor of the specialist. The work of the specialist could be considered of higher quality in terms of its ability to generate greater agricultural yield per hour of labor. Considerable energy had to be invested in training that individual through schooling and apprenticeships. The apprentice had to be fed while he or she was relatively unproductive, in the anticipation of greater future returns. The energy return on investment (EROI) of the artisan was higher than that of the farmer (even if less direct), and as a result, so was his pay and status. Thus, social stratification was directly linked to agriculture and changed ancient societies tremendously.

Eventually, the concept of agriculture spread around Eurasia and Africa but resource depletion followed shortly thereafter. Another new phenomenon appeared with the development of agriculture: cities and other manifestations of urbanization. The first area this occurred appears to be in the Tigris Euphrates valleys in one of the first cities ever known, Ur.[47] This was roughly 4700 years ago, and there were many great cities in that region, including Girsu, Lagash, Larsa, Mari, Terqa, Ur and Uruk. These cities grew up in heavily forested land, as signified by the massive timbers in remaining ruins. Today we call that ancient civilization Sumeria and the people Sumerians but there are essentially no trees or cities left in that region. In fact, the forests were gone by 2400 BC, the harbors and irrigation systems were silted in, the soil became depleted and salinized, and barley yield dropped from about 2.5 tons per hectare to less than 1 ton. By 2000 BC, the Sumarian civilization was no longer extant. The world's first great urban civilization used up and destroyed its resource base and disappeared over a span of 1300 years.[48] Consequently, resource depletion turned out to be one of the reasons why entire civilizations and cities became extinct.

The interaction of people with cultivars,[49] in turn, greatly changed the plants themselves. Notably, all plants are in constant danger of being consumed by herbivores, ranging from bacteria, to insects, to large grazing or browsing mammals. In the planet's history, herbivorous dinosaurs predated today's mammals. Thus, the evolutionary response of plants to this grazing pressure was to develop various defenses, such as the physical protection of spines, which are especially abundant in desert plants. More common, however, was chemical protection in the form of alkaloids, terpenes, and tannins. These compounds place a heavy burden on

---

[46]Diamond (1999).

[47]The word "urban" is actually derived from the ancient city Ur.

[48]See Perlin (1989), Michener (1963), and Tainter (1988), who tell these stories in fascinating detail.

[49]Cultivars are plants that humans cultivate.

herbivores (or potential herbivores) by discouraging consumption or by requiring a high-energy cost to detoxify poisonous compounds. Humans do not like these frequently bitter, poisonous compounds either. For thousands of years, humans have been, therefore, preferentially saving and planting the seeds from plants that taste better or have other appealing characteristics. Partial exceptions include mustards, coffee, tea, cannabis, and other plants, that provide bitter alkaloids which would be poisonous if they were all that humans consumed, but present curious, interesting, or otherwise alluring dietary supplements in small doses to those humans who like them. Consequently, however, cultivars have poor defenses against insects and often require the use of external pesticides—a technology that has complex environmental and biological consequences. Many cultivars would not survive in the wild now, and have coevolved with humans into systems of mutual dependency. Meanwhile, all kinds of pests are themselves adapting to the concentration of humans, often with disastrous impacts to humanity. Humans have nevertheless survived, prospered, and multiplied, especially since the industrial revolution. Thus, the co-dependency of plants and humans is another aspect of the evolution and development of agriculture.

Other highly impactful energy-related events occurred during these prehistorical times. The domestication of animals may be one of the most significant developments. While some aspects of animal domestication predate agriculture, most domestication occurred more or less simultaneously with the inception of agriculture. Animal domestication and the increased sophistication of animal husbandry were critically important in increasing energy resources for humans in at least two important ways: First, for the reason that these animals ate plant material that humans did not eat, this greatly increased the amount of energy that humans could harvest from nature, especially in grasslands. Second, oxen and horses markedly increased the power output of a human.

The story of how the use of animal technology passed throughout Eurasia was critical in facilitating this transference. In fact, the majority of domestic animals came from Eurasia and could be moved East to West much more easily than North to South. Humans' most important animals included sheep, cows, horses, pigs and chicken. They were "corralled" in Eurasia by virtue of the area's geography, and consequently evolved into today's domestic animals. The increasing familiarity with beasts of burden, along with the development of roads and caravan technology, in turn, allowed for the expansion of long distance trade. Humans refined and passed on sailing and navigational skills, enriched agricultural knowledge and the biotic resources of many human groups.

As agriculture, settlement, and commerce expanded, a greater need for maintaining records arose. Some time around 3000 BC, humans developed formal writing, seemingly simultaneously in Egypt, Mesopotamia, and India. Writing had many societal implications, but perhaps most importantly, it allowed for agricultural and other technologies to be passed from one generation to another and transferred among cultures. These old records have also allowed scientists to estimate earlier patterns of human population changes and they suggest that the pattern of human population is hardly one of continuous regular growth; rather, it is one of periodic growth followed by decline. Sometimes this is manifest as a catastrophic drop in, and disappearance of, a particular population; or, more commonly, the demise of the political structure that once held them together. Edward Deevy suggests that there were three main historical increases in human populations: first, the corralling of animals; second, the development of agriculture; and third, the industrial revolution. We are still experiencing the last phase as global human population growth continues strongly, although at a lower rate than in earlier times. Commerce, nonetheless, continues to shape agriculture and international trade in an increasingly globalized world may have some of the greatest impacts on food law and policy.

15.3.4 Human Cultural Evolution as Energy Evolution

Most of the major changes in terms of humans' ability to exploit natural resources are associated with increased use of energy. Spear points and knives are, for example, energy concentrating devices; fire allows greater availability of plant energy to humans; agriculture significantly increases the productivity of land for human food; and so on. The evolution of humans' ability to control energy—such as through the harnessing of wind and water power—is best described in Fred Cottrell's book *Energy and Society*, which was published more than half a century ago.[50] Cottrell's focus was on the development of what he called "converters," i.e. specific technologies for exploiting new energy resources. As Cottrell shows, technological change is usually associated with an increase in the quantity or quality of exploited energy.

Cottrell's early chapters focus on herding as a means of exploiting biotic energy,[51] water power, and wind power. He shows the historical importance of situating cities downstream on a river so that the natural flow of the water allows citizens to easily exploit all upstream resources such as, timber, agricultural products, game, and ore. Through the use of barges, humans' carrying capacity upstream and downstream increased.[52] Likewise, the development of sailing ships increased the energy efficiency of a human porter enormously, and, according to Cottrell's calculations, the early sailing ships generally increased the load that a human could carry by a factor of 10; and by late Roman times it was as much as 100. The Romans needed to import large quantities of grain from Egypt,[53] in part because they had

---

[50]Cottrell (1955).

[51]Biotic means living parts of an ecosystem. In contrast, abiotic mean chemical or physical, non-living.

[52]The Nile is an exception, for the winds tend to blow north to south while the water flows south to north, so dhows could go both ways.

[53]Contrary to popular belief, Caesar and Mark Anthony were not in Egypt for Cleopatra—the real target was grain from continuously replenished flooding soil.

depleted their own soil. According to Cottrell, however, they were not the only ones who coveted grain and, initially, the Romans lost a lot of their supply to pirates. This required the Romans to transport the grain in heavily guarded narrow warships, while the soldiers on board consumed a significant portion. Therefore, another energy investment had to be made by the Romans, namely clearing the Mediterranean of pirates. With this accomplished, they adopted the use of wide-beamed merchant vessels, and Egypt became a significant net energy source for the Romans. Cottrell gives many other examples of the increasing use of energy by humans over time, including noteworthy chapters on the rise of industrial agriculture, steam power, and railroads in England. What all of these examples and developments have in common is humans' dependence on energy and the central role it plays in human development.

15.3.5 Industrial Agriculture

The next great leap forward in agriculture came in the twentieth century, when an increasing use of cheap fossil fuels, along with technological advancements, brought about a dramatic transformation of agriculture. It was the enormous surplus energy derived from fossil fuels (coal, oil, and natural gas), that made this development possible while 20–100 or more units of energy could be returned per unit invested. The high EROI (Energy Return on Investment) allowed surplus energy to be invested in agriculture and other industries, thus generating surplus wealth and boosting profit margins.

Between 1900 and 1970, the western world's shift from human and animal labor to predominantly mechanized labor changed the EROI of agriculture. In traditional cultures, 5–50 kcal of food were obtained for each kcal invested; by 1970 just one kcal of food was obtained for every 5–10 kcal of total invested energy (fossil and human labor), including transport and processing.[54] White hypothesized that the development of human societies is constrained ultimately by their ability to generate surplus energy, including food. This ability is a function of the quality of available energy and energy transformers (technology). Over the long run, the quality of available energy is determined by the amount of energy needed to return the next unit of energy. During this period, fossil-fuel-driven tractors and other machines replaced the labor of humans and draft animals. In the nineteenth century, up to 75 % of the U.S. labor force worked on farms. By the end of the twentieth century, it was less than 2 %. Astonishingly, far fewer Americans were working on farms in 2000 than in 1840, even though the American population is so much larger today. The increase in agricultural productivity is due primarily to the use of fertilizers and pesticides, along with the development of new varieties of crop plants. All of these shifts in energy use were made possible through the use of fossil fuels.

---

[54]Steinhart and Steinhart (1974), p. 307.

Perhaps the most important change in energy production is the development of the industrial Haber-Bosch process, which converts atmospheric nitrogen gas into ammonia. Up until 1908, crop plants were severely limited by the availability of nitrogen. This is true even though 80 % of the atmosphere is nitrogen (N_2). This nitrogen, however, is very difficult for most plants and humans to access due to the triple bonds in the di-nitrogen molecule ($N\equiv N$). Until Fritz Haber developed the Haber-Bosch process, only the tremendous energy of lightning or some very select algae and bacteria could break these bonds. Haber, in one of the most important scientific discoveries ever made, found that by heating and compressing air mixed with natural gas and using the right catalyst, the N_2 molecule could be split and turned into ammonia (NH_3). This, in turn, could be combined with nitrate (itself created by oxidizing ammonia) to generate ammonium nitrate, which is the basis for both gunpowder and fertilizer. The Haber-Bosch process requires significant energy input, but its development freed humans from the limits of natural processes such as manure fertilizing. Arguably, this freedom led to dangerous environmental consequences that may have been averted but the abuse of fertilizers through the industrial agricultural industry is beyond the scope of this chapter.

The Haber-Bosch process took off in 1946 at the end of World War II. As there was no further need for massive amounts of war explosives, the U.S. Federal Government asked whether there might be a different use for the weapons factories. The answer came from agricultural colleges: the technology could be used to significantly increase agricultural yield. This "industrialization of agriculture" freed food production from its former dependence upon manure fertilizers. With the concurrent development of machinery, far fewer Americans were needed to grow food. This shift in labor division created an exodus to the cities and led to the growing number of urban industrial jobs. Meanwhile, the increased use of oil, gas, and coal generated greater material wealth for workers. Thus, America changed from a relatively poor, agriculturally-based country into an increasingly industrialized and urban one while becoming enormously wealthy in the process. The net energy required for this economic work was increasing exponentially. The great increase in wealth prompted economists to develop theories to explain the economic forces behind this growth. And yet, interestingly, among those chronicling the process there is essentially no mention of energy as a catalyst for these changes.

The large agricultural yields generated by fossil-fuel-led agriculture allow a large surplus of energy, including food energy, to be delivered to society. In turn, this transfer allows most people and capital to be employed somewhere other than the energy industry. These energy surpluses, in other words, have helped to develop all aspects of our civilization—good and bad. The same can be said for technological advances that enabled the unearthing of phosphate rock deposits. Irrigation became widespread as a result of this industrialization, allowing crops to be grown in arid and semi-arid areas. The Central Valley of California is perhaps the best-known region where irrigation dramatically increased agricultural production. Worldwide, crops became more homogenized and food processing more widespread. All of these changes furnished the development of the globalized industrial food system that characterizes food production today.

15.4 Energy Cost of Food

15.4.1 Energy Production Efficiency for Agriculture in the United States

There are four relevant studies on the energy cost of food production in the United States. Each of these studies uses slightly different methodologies but express energy cost in terms of caloric output versus caloric input. Carol and John Steinhart,[55] for example, calculated energy use in the entire U.S. food system using data from governmental sources between 1940 and 1970. Output in the Steinhart and Steinhart study was based on the caloric requirements of the U.S. population, rather than actual crop production. The output amounts also excluded U.S. food production exports. Inputs included direct fuel and electricity use, energy used to create fertilizer, agricultural steel and farm machinery, and energy used to run irrigation systems. Steinhart and Steinhart concluded that U.S. agricultural energy efficiency declined by about threefold from 1940 through 1970, as tractors replaced animal power and farmers used more commercial fertilizers. In the end, agriculture was providing a return of less than one energy unit of food for one energy unit of fuel (even at the farm gate), and less than one unit of food for three units of fuel by the time the food reached the plate.

In another study, Cutler Cleveland examined the energy efficiency of food production in 1995.[56] Cleveland derived energy inputs and outputs from economic data and was thus able to make calculations from as far back as 1910. He determined the energy content of agricultural inputs by converting the dollar value of fossil fuel and electricity consumption, along with other farm input expenditures[57] to physical units at extant prices. Then, he converted these physical units to energy units using a dollar to energy conversion factor for the embodied energy in fuels using the energy intensities (kcal per dollar) derived by the energy research group at the University of Illinois.[58] Cleveland calculated agricultural output using two data sources: first, with the USDA index of total agricultural output, which includes dollar estimates of production of crops, fruits and vegetables, and animal products; and second, with the Gross Farm Product, or the value added in the farm sector in dollars. The results of Cleveland's research show that the energy efficiency of U.S. agriculture declined from about 5.5 calories of food energy output per one calorie of fuel in 1910 to a 1:1 ratio in 1980, leaving him in rough agreement with the Steinharts. Thus, the cost of energy to produce food could be estimated about 100 years ago.

---

[55] Steinhart and Steinhart (1974).
[56] Cleveland (1995), pp. 111–121.
[57] These include pesticides, fertilizers, machinery, energy used to generate electricity, and agricultural services.
[58] Herendeen and Bullard (1976), p. 383; Hannon et al. (1985).

The authors of this chapter have also summarized the energy it takes to grow food in the U.S. and Canada in recent decades using mainly physical data.[59] This study found that about two percent of all energy used in the United States goes towards growing crops. Pimentel and his colleagues[60] estimated that about 17 % of U.S. energy use goes into the entire food system, including growing, transporting, and preparing food, each sector consuming about one third of this total energy. We also found that the "Edible Energy Efficiency" (EEE) of U.S. agriculture has actually more than doubled from 0.8:1 in 1970 to 2.2:1 by 2000, followed by a slower increase to 2.3:1 by 2009. The energy efficiency of the agricultural sector in Canada has not changed appreciably since 1980, and has remained at about 2:1 from 1981 to 2009. The authors' study found that EEE improvements in the U.S. can be attributed not only to increased crop production *per* hectare and lower direct fuel consumption, but also to the increased use of less energy-intensive corn and changes to the diet of livestock.[61] Increases due to technological progress alone appear small for the last several decades, at less than 1 % a year.[62] In sum, although efficiency initially fell as agriculture was industrialized, technological advances in recent decades mean that efficiency has now stabilized or increased slightly.

Notably, there are several contributors to agricultural energy use. The production of N fertilizer contributes about 40 % of all energy used; on-farm fuel use requires 30 %, followed by K_2O (7 %), lime (6 %), the transportation of inputs (6 %), P_2O_5 (5 %), seed (5 %), herbicide (4 %), drying (2 %), and insecticide (1 %).[63] Energy use is lower for legume crops because they fix atmospheric N, and therefore do not require energy-expensive N fertilizer. How these contributors of energy use affect the food system is described in the following section.

15.4.2 The High (Energy) Cost of Meat, Dairy, and Processed Foods

The production of meat, dairy, animal products and processed foods is significantly more energy-intensive than plant-based food production. It is, therefore, important to appreciate the energy that is invested in producing the various types of foods. Grains, for example, are the most productive agricultural product.[64] Yields can be anywhere from 1 to 10 tons per hectare, and occasionally more. Temperate yields

---

[59]See Hamilton et al. (2013), p. 1764.
[60]Pimentel et al. (1989).
[61]For instance, increased use of meals and other by-products, which reduce the grain demand by livestock.
[62]See Hall et al. (2009a), pp. 25–47; Hamilton et al. (2013), pp. 1764–1793.
[63]Camargo et al. (2013) p. 263.
[64]This is due to their efficient photosynthetic pathways.

15 Food and Energy

tend to be higher than tropical, despite the longer growing season that tropical farmers enjoy. This is because the soil has lower nutrient levels and the longer nights consume more energy. Since people need a minimum of 1 kg of food per day,[65] anywhere from around 3 to 30 people can be supported by 1 ha[66] of land producing grains. The number of people fed per hectare of land will be significantly less if the crop is first fed to animals. This is a significant argument why vegetarianism, and even more specifically, veganism, makes for a more sustainable diet and leaves a lower environmental footprint than a diet rich in animal products.

In areas of the world with high human densities, such as India and China, the majority of people eat only grains, such as rice. Indeed, this grain-based diet is similar for poor people around the world. Energy yields per hectare of vegetables or animal products tend to be low; at best between one quarter and one half of energy invested, but more frequently the yields are as low as ten percent of the total invested energy per hectare. The conversion efficiency of plant to animal flesh is in fact only 10–20 %.[67] Despite this poor energy efficiency, animals can use lower quality, less productive land where it would otherwise be expensive or impossible to grow crops. This is readily observed in much of the world where wetter land is used for crops and drier land is used for pastures.

According to Pimentel and Pimentel,[68] farmers in the US raise and care for nine billion livestock in order to meet the animal protein demand by humans each year. Indeed, the total numbers of livestock are estimated to be some five times the U.S. human population. About 124 kg of meat is eaten per American per year. The average meat-eating diet consists of 35 % beef, 25 % pork, 39 % poultry, with the remainder made up of other meats. Americans also consume protein in the form of milk, eggs, and fish. In terms of the energy efficiency conversion, livestock must consume six units of plant protein for each unit of animal protein that they produce.[69] Thus, in terms of producing food for the world, the reliance on animal products reduces efficiency, productivity, and, as a side-note, causes substantial health concerns for human consumers and dangers for the environment.

15.4.3 Energy Distribution and Delivery: Food Miles

Researchers estimate that roughly equal amounts of energy are used in delivering food to the consumer as to grow it.[70] With the globalization of agriculture and other sectors of the economy, this quantity has almost certainly grown. A study at Iowa

---

[65] Or 365 kg—roughly one third of a ton—per year.
[66] 1 ha = 2.54 acres.
[67] The conversion is expressed in a "calories to calories" ratio—be careful not to equate this with weight because of varying moisture contents.
[68] Pimentel and Pimentel (2003).
[69] Pimentel and Pimentel (2003), p. 660S.
[70] Pimentel personal communication (on file with the author).

State University estimates that in the U.S. the average food item travels 1500 miles before it reaches the consumer.[71] There is also a growing movement toward locally-sourced food to reduce food miles, although the energy consequences of this movement have not been clearly evaluated yet. Syracuse, New York, for example, has a large and vibrant farmer's market. And yet, we observed that to deliver food locally to the farmer's market[72] would use as much delivery energy per kilogram of food as products delivered to grocery stores from 300 miles away in a full semi-truck[73] which can carry 50 times more food.[74] There are many good reasons to eat locally, but the extent to which transportation energy and overall food miles are actually saved in doing so need to be examined more carefully for conclusive evidence to support any one hypothesis of the sustainability of eating locally.

15.4.4 The Developing World

In the last decade, scholars have started to examine agricultural energy use in the developing world. For instance, Hamilton et al.[75] reviewed existing studies of agricultural energy efficiency[76] for developing nations. Cao et al.[77] found that the energy ratio for agriculture in China decreased by 25 % from 2:1 in 1978 to 1.5:1 in 2004, largely due to increases in fossil fuel use that outpaced food production. According to this study, for every two units of fossil fuel invested, there is a yield of approximately one unit of food energy although this ratio declined from 1990 to 2004. By contrast, Karkacier et al.[78] found a positive relationship between increasing an index of energy consumption and agricultural output in Turkey, with each additional ton of oil increasing an index of agricultural output by 0.167 units. Other studies looking at edible energy return on investment (EROI) have been conducted on national and international levels for specific crops such as rice. Pracha and Volk (2011) performed an analysis of the edible EROI for Pakistani rice and wheat from 1999 to 2009. The authors found that the average EROI was 2.9:1 for the edible portion of wheat, and 3.9:1 for rice. Going further, Mushtaq et al.[79] calculated EROI values for rice in eight nations and found that the EROI varied from 4:1 to 11:1 (which includes the energy stored in straw), and from 1.6:1 to 5:1 when including only the edible portion. Overall, it appears as though the efficiency of turning

---

[71]Pirog and Benjamin (2003), p. 1.
[72]"Local" meaning 30 miles away, and with a truck getting 15 miles per gallon.
[73]Which gets only 7 miles per gallon.
[74]Balogh et al. (2012).
[75]Hamilton et al. (2013).
[76]Kcal of food produced per kcal of input fossil energy.
[77]Cao et al. (2010).
[78]Karkacier et al. (2006).
[79]Mushtaq et al. (2009).

petroleum into food does not vary significantly between more and less developed nations, at least when production is dominated by the use of fertilizers and some machinery. The world is indeed globalized.

15.4.5 Waste

Although the efficiency of delivering food to the consumer has improved slightly in the U.S., waste remains prevalent throughout the food system. Some researchers[80] estimate that 27 % of food produced on American farms ends up as waste and is not consumed. More recent estimates, however, put food waste as high as 40 %, or some 1400 kcal per person per day.[81] Since agriculture requires such high quantities of water, this waste equates to one quarter of U.S. freshwater consumption and approximately 300 million barrels of oil.[82] While the reasons for food waste differ between high-income and low-income nations, post-consumer waste accounts for the bulk of waste in the U.S., while in lower income countries food waste/spoilage tends to occur before it is distributed to the end user.[83] Approximately 1.3 billion tons of food produced for human consumption is wasted globally each year.[84]

15.5 Challenges of Sustainable Agriculture in the US

It is no secret that the US has very high energy demands in its agricultural sector. Certainly, both Howard Odum's 1971 piece, *Potatoes Partly Made From Oil*,[85] and David Pimentel's early studies helped to raise awareness about this issue. Initially, most agronomists paid little attention and were generally dismissive that such problems were important. Nonetheless, there has been a large public response to the environmental concerns about agriculture, although these tend to focus on chemical threats to the health of humans and wildlife. Today, reducing energy use has become the goal of many agronomists.

While the general public response is much too broad to summarize in this chapter, it seems fair to say that the most typical responses are summarized with the word "sustainable." For instance, "sustainable health", "sustainable production systems", "sustainable farming or consuming cultures", "sustainable energy", and other terms, are at the centre of public discussions. Generally, as with most issues

---

[80]Kantor et al. (1997).

[81]Hall et al. (2009b), p. 2.

[82]Hall et al. (2009b), p. 2.

[83]Food and Agriculture Organization (2011), p. 10.

[84]Food and Agriculture Organization (2011), p. 10.

[85]See Howard T. Odum, *Environment, Power, and Society* (1971).

that appear in popular environmental literature, there is very little quantitative analysis. Certainly, Hamilton et al.'s findings[86] that U.S. agriculture is becoming somewhat *more* efficient, even as it is dominated by corporate and market forces, would likely surprise many people. Ironically, the largest single barrier to improved efficiency is the large U.S. governmental plan generating ethanol from corn as a replacement for gasoline. This program produces little, if any, net energy by the time the fuel goes into the vehicle,[87] and its existence is clearly based on a political strategies that have trumped current scientific data, which fail to support ethanol as an energy-efficient alternative to gasoline. Problems with the ethanol program include the removal of a substantial amount of food from a hungry world, enormous soil erosion, loss of wildlife habitat, and a net addition of carbon into the atmosphere. Since the land used for ethanol production tends to be the best corn land in the U.S., such as in Iowa, for example, the ethanol crops displace the remaining corn production to sub-optimal habits, such as in Minnesota and Texas. This replacement of crop land results in increased energy costs for American commodity production, such as Cornflakes and bacon. A recent program proposed by the U.S. Navy to fuel a large proportion of its ships and even airplanes with biofuels, nominally to improve energy security and efficiency, has completely flopped based on the enormous price and poor availability of such huge quantities of biofuels, in turn caused by their very low EROIs.

15.5.1 *Policy Constraints and Promotion of Sustainable Agriculture*

Sustainable agriculture movements, at least from the perspective of protecting soil and water resources, are becoming increasingly common in the US. This is exemplified by the rise of state and county soil conservation districts, agricultural colleges, and young people who are expressing an interest in the agricultural future of America. Probably the largest impact in terms of sustainability comes from the encouragement of no-till agriculture campaigns, whose goal is to disturb the soil cover as little as possible during planting and cultivation.

Certainly there is much popular and governmental lip service toward generating sustainable agriculture, and even some state and local policies that address it directly. And yet, most lawmakers are reluctant to act, resulting in the non-policy of leaving agricultural sustainability decisions to the market. What impact this may have on long-term environmental health is impossible to ascertain but gives many scholars reason to worry about the immediate and long-term consequences of allowing the industry-dominated status quo to continue. Leaving this issue to the market, which searches for the lowest production price as a matter of course, will argue against protecting the soil and continue to deplete the planet's resources.

---

[86]Hamilton et al. (2013).

[87]See, for example, Patzek (2004), Murphy et al. (2011), p. 179 and Conway (2007).

Techniques to prevent soil erosion, such as cover crops and erosion barriers, are crucial to maintaining soil health, the foundation of intact agriculture. However, these techniques require money to implement and stand in the way of short-term industry revenues that externalize the cost to the environment beyond market prices. (For more on the externalities of agriculture, see *Textbox: Internalizing Ecological Externalities*).

Another policy with significant environmental impact is the current production of huge amounts of corn-based ethanol. For the reason that corn is a highly soil erosive crop and ethanol production systems are characteristically placed on the best farmland, continued focus on ethanol production will have an enormous impact on the future of agriculture in the United States. How these larger national policies stack up against state and regional programs is impossible to calculate or predict with precision. It is imperative, however, to implement better policy to support sustainable agriculture, to protect the environment, to internalize the negative externalities of industrial agriculture, and to promote diets with lower environmental footprints. While there is much rhetoric on these issues we see precious little quantification of actual results, for example along the line of Cleveland (see footnote 54) and Hamilton et al. (see footnote 57).

15.5.2 Urban Agriculture

The common conception of agriculture is that of a rural enterprise. Indeed, the vast majority of food production comes from rural farms. There are exceptions, however, the most notable of which originated from Will Allen and his colleagues in the city of Milwaukee. Allen is a former professional basketball player from Milwaukee, who, at the end of his basketball career, returned to the city where he sought out his wife's parents' farm, which had been the site of many happy childhood memories. In the intervening years, the city had essentially expanded around the farm. Allen came up with an inspired idea: bringing more farms into the city.

Allen created a series of clever approaches, including making new soil, because the old soil in the city was polluted with industrial wastes, by combining old coffee grounds with city-generated wood wastes and discarded supermarket food. He placed this new soil mixture in plastic hoop houses to heat while it evolved into excellent compost; following that, it was used for growing crops on tables. This entire enterprise was coordinated with local residents including children, who were encouraged to grow and sell their own produce. As a result, vacant lots in the inner city started producing affordable and highly nutritious food. This entire effort was propelled by Allen's enormous charisma in a movement called "growing power", and it was eventually exported to Chicago and other cities around the U.S. Similar projects are springing up in many places. How much impact all this will have in the future remains to be seen, but it it certainly one of the most exciting new ideas we have seen. How many future farmers of America are now inner city kids? On a smaller scale we recommend Mel Bartholemew's (2006) book "Square foot gardening" (2nd edition) for another innovative way to personal low energy food production.

The full potential of urban agriculture remains to be seen, but it is certainly one of the most exciting new ideas on the rise. Cities, such as Baltimore, Washington, D.C., Pittsburgh, and New York are also starting to have more and more urban agriculture, including the White House bee hive.[88] Urban agriculture has the potential to create a sense of community, to raise awareness of sustainable agriculture, to reduce inner-city food deserts, to reduce food miles, and to feed populations. Although urban farms are limited in space, their potential exceeds the mere premise of food production and has potential to counter industrialized areas through some positive green space with tremendous potential to have positive impacts on agriculture.

15.5.3 Case Study: Syracuse and Onondaga County

Syracuse and Onondaga County provide an example of a geographic region that could decrease its environmental footprint by switching to a plant-based diet. A 2012 study by Balogh et al. quantified the food demand, production, and footprint for this small city and its surrounding county[89] over the past 100 years. Farms in this region have increased their caloric output since the 1930s despite a consistent decline in the overall area dedicated to farming. This can be attributed to increasing yields and the shift to more productive crops, as well as the increased inputs of fertilizers and energy more generally. They found that, from current farmland, the county could meet only 15 % of its food demand. Each year, the existing farms use energy equivalent to approximately 1.2 million barrels of oil. Furthermore, the county residents would require the equivalent of an 2.5 million barrels of oil per year to feed the local population solely from locally produced food. Transportation alone makes up 11 % of this annual energy demand. If the county were able to produce half of the food demanded by its residents, transportation energy could be reduced by 43 %. Larger reductions in energy consumption could be achieved by a shift to a low meat[90] or vegetarian diet. Two-thirds of county residents could be fed a vegetarian diet from the land that is currently under agricultural production. Despite this potential, it would require an area of farmland nearly twice the total size of the county, given the current meat consumption levels of the area's residents.

---

[88]See the White House Bee Hive video at https://www.whitehouse.gov/photos-and-video/video/inside-white-house-bees (last accessed April 7, 2015).

[89]Total population: 450,000. Based on Balogh et al. (2012).

[90]In this case, consuming meat once per week.

15.5.4 Case Study: Jevons' Paradox

Technology is usually seen as advancing and improving efficiency in different areas, such as food production or healthcare. In the past, most technological advances came from applying more fossil energy to the problem at hand, which ranged from fertilization of exploited soils, pesticide-treatment of GMO monocultures and similar unsustainable practices. Although, more recently makers of technology are attempting to use less energy, agricultural and biotechnology, however, is a double-edged sword, the benefits of which can be substantially blunted by Jevons' Paradox. This paradox centers on the idea that increases in efficiency often lead to lower prices, which, in turn, encourage greater use of the product in question.

Jevons[91] found that more efficient steam engines, which had been designed to use less coal, were cheaper to run and so that people used them more. A contemporary example is that more fuel-efficient automobiles tend to be driven more miles in a year and hence may consume greater energy resources overall. Likewise, Eva Alfredsson[92] found that, in Sweden, those who followed a lower-carbon, less energy-intensive diet saved money and tended to take vacations further away, often emitting as much or more carbon dioxide—essentially, using more energy—than they saved with their "green" diet.

Thus, without clear and distinct data to educate consumers and policy makers about what a truly sustainable diet or lifestyle represents and how it can be achieved, many practices intended to conserve resources will backfire and feed hypocritical greenwashing.

15.5.5 The Need for Quantitative Analysis

The current information gaps and lack of reliable predictions into the future of energy in the food system can be addressed through sound solution-oriented environmental science. There are countless examples of "greener" approaches to agriculture and food production, but on closer quantitative examination the benefits of these methods are ambiguous at best.[93] When it comes to "sustainability," there seems to be a deficit of hard, quantitative analysis in most U.S. national assessments, as well as in the research taking place at American colleges. Much of the authors' work has been to educate young people who think they already know the answers, when, most often, neither they nor their instructors actually have any of the answers. Many of students, for example, are anti-fracking, or anti-coal, or anti-nuclear or anti-something else. How we are going to balance the human

---

[91] Jevons (1865).
[92] Alfredsson (2004).
[93] Alfredsson (2004).

population's energy expectations remains to be seen. Nonetheless, in educational institutions, however, whatever the name of the program, it is crucial to teach "environmental science" and not just "environment" so as to properly educate students how to generate and test hypotheses, how to perform quantitative assessments, how the natural and social sciences are connected, and other important skills that will equip them to critically analyze how to best meet their country's future energy needs.

Government agencies and non-government organizations should strive to perform comprehensive systems-based analyses of current food production. These studies should include connections to the larger-scale global system as well as the impact on smaller-scale regional and individual agriculture production. Moreover, the studies should examine economic impacts, but should also be grounded in the biophysical reality in which agricultural systems exist—taking into consideration, for instance, a nation or region's stocks of freshwater and soil, as well as whether and how access to energy resources changes dynamically over time. It is important for the U.S. to have a well-funded National Institute of Agricultural Assessment where these issues can be studied thoroughly, independently, and objectively. Careful and thorough quantitative analyses are, in environmental science, the first step to creating policies that preserve the long-term health of the environment and ensure sustainable agricultural production. Legislation and policy can only yield effective solutions if it is grounded on sound environmental science.

15.5.6 Phosphorus: The Ultimate Limiter?

Plants need more than nitrogen fertilizer to survive and grow—especially in light of the often excessive use of artificial fertilizers in conventional agriculture. Phosphorous and potassium are critical too, as well as smaller quantities of sulfur, molybdenum, and perhaps a dozen other essential plant nutrients. When nuclear scientists Goeller and Weinberg examined the entire periodic table, they found that for all of the elements necessary to civilization, there is a substitute. For example, aluminum wires can substitute for copper; the Haber process can use energy to create substitutes for organic sources of nitrogen. Goeller and Weinberg found one exception, however: phosphorus.[94]

Phosphorus is essential for plant growth and life in general. However, it is somewhat rare and there is no substitute for it in plant metabolism. In the paraphrased words of geochemist Edward Deevey some five decades ago, "there is something peculiar about the geochemistry of the Earth today that life is so dependent upon phosphorus but it is now in such short supply."[95] In other words,

---

[94]Goeller and Weinberg (1976).
[95]Deevey (1960), p. 194.

life on earth may have initially evolved when phosphorus was more abundant. Today, most phosphorus comes from mines in Florida or Morocco, or is mined in the Western Sahara. Much of this phosphorous goes on a non-renewable trip from mine, to ship, to fertilizer bag, to crop application, to the crop itself, to animals, to humans, to toilets, to waterways, and finally to the ocean. The chemistry of phosphorus, therefore, is of great concern to modern economies because of its critical importance and non-substitutability for plant growth, indeed for all life. Another reason for its importance are that the main sources (in Florida and Morocco) are being increasingly depleted. Phosphorus now requires more energy to be produced and it also causes undesirable algae growths as a waste product in water bodies. With most phosphorus ending up diluted beyond recovery in the world's oceans, it is vital to invest in a better understanding our dependence on phosphorus and how to best conserve it. This means that the essential pathways of phosphorous use must not only be understood, but should also be explored in search of more sustainable and environmentally-friendly alternatives.

15.5.7 Continuing Population Growth

Thomas Malthus, who was mentioned earlier in this chapter, believed that humans would continue to have about the same number of children per female, and that this constant rate of increase would be applied to an increasing total number of families over time, thus leading to exponential growth. Malthus also believed, however, that food production would grow linearly, ultimately leading to starvation as the population outstripped food availability. Since Malthus' time, in fact, the human population and food production have both increased exponentially, with food production arguably increasing even somewhat more than the human population. The rise in food production tends to be attributed to technology, meaning plant breeding and better farm management, and especially an increased use of fertilizers and machinery. Arguably, the reasons for increased food production are indeed in large part due to industrial agriculture, but perhaps there could have been another way. Kimbrell has hypothesized that a more sustainable, less industrially intensive food production system could possibly bring about yields as great or greater.[96] In the meantime, we seem to be married to industrial agriculture by necessity, which is quite dangerous as human populations continue to grow and petroleum supplies seem less certain.

All of the energy inputs, ranging from water to chemicals, are based on an increasing use of petroleum in industrial agriculture and in the face of globalization. Until recently, petroleum production was also increasing exponentially—this is no longer the case and slowing growth in petroleum production and the substitution of ethanol for gasoline are causing a host of other environmental concerns for

---

[96]Kimbrell (2002), pp. 3–36.

modern agriculture. What Malthus' equations lacked, therefore, was a factor for the invention and enormous expansion of petroleum-based agriculture. Of course, if petroleum supplies becomes seriously constrained and good substitutes are not found, then, in the long run Malthus' predictions will prove to be correct. Thus, alternatives must be found that are less petroleum-dependent and more sustainable.

15.6 Conclusion

As long as conventional and intensive agriculture remain largely dependent on petroleum and as long as people fail to see the value in a locally-sourced organic whole-foods plant-based diet, true environmental sustainability cannot be achieved and agriculture will likely remain at odds with nature. In fact, many neoclassical economists, technology supporters, and empiricists argue that technological advancements will allow indefinite growth in agricultural productivity.[97] They postulate that new technologies, such as genetically modified organisms (GMOs) and better irrigation systems, will boost crop yields and crop efficiency. On the other hand, most economists believe that market incentives such as higher fuel prices will generate greater energy efficiency in agriculture through technical and managerial changes.[98] These changes could include reducing the amount of land in cultivation, thereby increasing the average quality of that land left in production, possibly increasing farm size, and reducing rates of energy use through technological improvements. Cleveland[99] concluded that from 1978 to 1990, U.S. agriculture made significant improvements in energy productivity through technical and managerial changes in response to higher fuel prices. By 1990, however, U.S. agricultural energy efficiency had returned to 1950s levels for a variety of reasons beyond the scope of this chapter. We need a much better assessement of energy and agriculture country by country.

Global energy resources face an uncertain future in the current post-peak and climate-challenged oil age.[100] Real crude oil prices have increased at least fourfold in recent decades.[101] As the US stands on the brink of what will undoubtedly be a significant change in how humans obtain and use energy, the uncertain future but certain price hikes (eventually) pose powerful and yet insufficient incentives for increasing energy efficiency. Therefore, it is important to determine the energy efficiency of agriculture using an energetic analysis, rather than a traditional economic cost-benefit analysis. An economically-focused cost-benefit analysis often ignores important factors, such as externalized costs to the environment and social integrity. The objective of a more complete energetic analysis should,

---

[97]Jorgensen (2011), p. 276; Minten and Barrett (2008), p. 797.
[98]Cleveland (1995), p. 111; USDA (2011), p. 87.
[99]Cleveland (1995).
[100]Hall and Ramirez-Pascualli (2012).
[101]United States Energy Information Administration (2014).

therefore, be to determine whether the energy efficiency in agriculture has increased substantially by region over the past several decades. Although this chapter focuses solely on human food energy produced by agriculture—as opposed to all energy produced by agriculture, which would include the energy implicit in inedible silage, fiber crops, animal bones and fuels, another objective remains to determine the amount of energy (in joules) used by each major agricultural input and to compare their individual efficiencies, to calculate the output percentage in the form of crops, meat, and livestock feed, to show the environmental impact of crops grown exclusively for biofuels, and to compare the results of this study against the results of two extant studies on the energy efficiency in the U.S.[102] and in other regions of the world. Such an analysis may help to determine the global energy resource availabilities, efficiencies, and vulnerabilities and whether all of the rhetoric about sustainability has obtained any real results that would compensate for depeltion of soils, fertilizers and petroleum and, especially, for increased wastage, affluence and human numbers.

References

Alfredsson E (2004) "Green" consumption - no solution for climate change. Energy 29 (4):513–524
Angel L (1975) Paleoecology, paleoecology and health. In: Polgar S (ed) Population, ecology and social evolution. Mouton, The Hague, pp 170–190
Baffes J, Dennis A (2013) Long term drivers of food prices. Policy Research Working Paper 6445. The World Bank Development Prospects Group and Poverty Reduction and Economic Management Network Trade Department. http://www-wds.worldbank.org/external/default/WDS$32#ContentServer/IW3P/IB/2013/05/21/000158349_20130521131725/Rendered/PDF/WPS6455.pdf. Accessed 25 June 2014
Balogh SB, Hall CAS, Guzman AM, Balcarce DE, Hamilton A (2012) The Potential of Onondaga County to Feed the Population of Syracuse New York: Past, Present and Future. In: Pimentel D (ed) Global Economic and Environmental Aspects of Biofuels. Taylor and Francis, Boca Raton
Bartholomew M (2006) All new square foot gardening: grow more in less space! Cool Springs Press, Nashville
Brown LR (2009a) Could food shortages bring down civilization? Sci Am 300(5):50–57
Brown LR (2009b) Outgrowing the Earth: The Food Security Challenge in an Age of Falling Water Tables and Rising Temperatures, http://www.earth-policy.org/books/out/out_table_of_contents
Camargo GGT, Ryan MR, Richard TL (2013) Energy use and greenhouse gas emissions from crop production using the farm energy analysis tool. BioScience 63(4):263–273
Cao S, Xie G, Zhen L (2010) Total embodied energy requirements and its decomposition in China's agricultural sector. Ecol Econ 69:1396–1404
Cleveland CJ (1995) The direct and indirect use of fossil fuels and electricity in USA agriculture 1900–1990. Agric Ecosyst Environ 55:111–121
Campbell TC, Jacobson H (2013) Whole: rethinking the science of nutrition. DanBella Books, Dallas, TX

---

[102]Hamilton et al. (2013), p. 1764 and Cleveland (1995).

Conway R (2007) The net energy balance of corn ethanol. In: Proceedings of intersection of energy & agriculture: implications of biofuels and the search for a fuel of the future, University of California: Berkley, CA, 4–5 October 2007

Cottrell F (1955) Energy and society. McGraw Hill, New York

Deevey E Jr (1960) The human population. Sci Am 203:194–204

Devereux S (2007) The new famines: why famines persist in an era of globalization. Routledge, London

Diamond J (1999) Guns, germs and steel: the fates of human societies. WW & Company. Norton

Ehrlich P (1968) The population bomb. Ballantine Books, New York

Flint HJ, Scott KP, Louis P, Duncan SH (2012) The role of the gut microbiota in nutrition and health. Nat Rev Gastroenterol Hepatol 9(10):577–589

Food and Agriculture Organization (2003) Food energy – methods of analysis and conversion factors. FAO Food and Nutrition Paper 77. ftp://ftp.fao.org/docrep/fao/006/y5022e/y5022e00.pdf. Accessed 25 June 2014

Food and Agriculture Organization (2009) Change in the livestock sector. In: The State of food and agriculture – livestock in the balance. http://www.fao.org/docrep/012/i0680e/i0680e02.pdf. Accessed 25 June 2014

Food and Agriculture Organization (2011) Global food losses and food waste – extent, causes and prevention. Rome http://www.fao.org/docrep/014/mb060e/mb060e00.htm. Accessed 29 June 2014

Food and Agriculture Organization (2013) FAO Statistical Yearbook 2013. Rome, Italy. http://www.fao.org/docrep/018/i3107e/i3107e00.htm. Accessed 29 June 2014

Goeller HE, Weinberg AM (1976) The age of substitutability. Science 191:683–689

Hall CAS, Day JW (2009) Revisiting the limits to growth after peak oil. Am Sci 97(3):230

Hall CAS, Day Jr JW (2009) Revisiting the limits to growth after peak oil. Am Sci 97:230–237

Hall CAS, Ramirez-Pascualli C (2012) The first half of the age of oil: an exploration of the work of Colin Campbell and Jean Laherrere. Springer, New York

Hall CAS, Cleveland CJ, Kaufmann R (1986) Energy and resource quality: the ecology of the economic process. Wiley, New York, p 12

Hall CAS, Perez CL, Leclerc G (2000) Quantifying sustainable development: the future of tropical economies. Academic, San Diego

Hall CAS, Balogh S, Murphy DJR (2009a) What is the minimum EROI that a sustainable society must have? Energies 2(1):25–47. doi:10.3390/en20100025

Hall KD, Guo J, Dore M, Chow CC (2009b) The progressive increase of food waste in America and its environmental impact. PLoS One 4(11):e7940

Hamilton A, Balogh SB, Maxwell A, Hall CAS (2013) Efficiency of edible agriculture in Canada and the U.S. over the past three and four decades. Energies 6(3):1764–1793. doi:10.3390/en6031764

Hannon BM, Casler SD, Blazeck T (1985) Energy intensities for the U.S. economy—1977. Energy Research Group: University of Illinois: Urbana, IL, USA, Document No. 326

Hardin G (1968) The tragedy of the commons. Science 162:1243–1248

Herendeen RA, Bullard CW (1976) Energy costs of goods and services. Energy Syst Policy 1(4):383–390

Hubbert MK (1956) Nuclear energy and the fossil fuels. Shell Development Company, Exploration and Production Research Division, Houston, TX, Publication No. 95. http://www.resilience.org/stories/2006-03-08/nuclear-energy-and-fossil-fuels. Accessed 29 June 2014

Hubbert MK (1969) Energy resources. In: The National Academy of Sciences–National Research Council, committee on resources and man: a study and recommendation. WH Freeman, San Francisco

Jevons WS (1865) The coal question: an inquiry concerning the progress of the nation, and the probable exhaustion of our coal-mines. Macmillan, London

Jorgensen D (2011) Innovation and productivity growth. Am J Agric Econ 93:276–296

Kantor LS, Lipton K, Manchester A, Oliveira V (1997) Estimating and addressing America's food losses. Food Rev 20:2–12

Karkacier O, Goktolga G, Cicek A (2006) A regression analysis of the effect of energy use in agriculture. Energy Policy 34:3796–3800

Kimbrell A (2002) Fatal harvest: the tragedy of industrial agriculture. Island Press, Washington

Kious BM (2002) Hunter-gatherer nutrition and its implications for modern societies. Nutr Noteworthy 5(1)

Lambert J, Hall CAS, Balogh SB, Gupta AJ, Arnold M (2014) Energy, EROI and quality of life. Energy Policy 64:153–167

Lee R (1969) Kung bushman subsistence: an input-output analysis. In: Vayda A (ed) Environment and cultural behavior. Natural History Press, New York, pp 47–79

Malthus TR (1798) First essay on population. Macmillan, London

Martin PS (1973) The discovery of America. Science 179:969–974

Meadows DH, Meadows DL, Rander J, Behrens WW III (1972) The limits to growth. Potomac Associates, Washington

Michener JA (1963) Caravans. Ballantine Books, New York

Minten B, Barrett CB (2008) Agricultural technology, productivity and poverty in Madagascar. World Dev 36:797–822

Montgomery DR (2007) Soil erosion and agricultural sustainability. Proc Natl Acad Sci 104(33):13268–13272

Murphy DJ, Hall CAS, Powers R (2011) New perspectives on the energy return on investment of corn based ethanol. Environ Dev Sustain 13:179–202

Mushtaq S, Maraseni TN, Maroulis J, Hafeez M (2009) Energy and water tradeoffs in enhancing food security: a selective international assessment. Energy Policy 37:3635–3644

O'Keefe SJ (2008) Nutrition and colonic health: the critical role of the microbiota. Curr Opin Gastroenterol 24(1):51–58

Odum HT (1971) Environment, power and society. Wiley Interscience, New York

Patzek TW (2004) Thermodynamics of the corn-ethanol biofuel cycle. Crit Rev Plant Sci 23:519–567

Perlin J (1989) A forest journey: the story of wood and civilization. W.W. Norton, New York

Pimentel D (2006) Soil erosion: a food and environmental threat. Environ Dev Sustain 8:119–137

Pimentel D, Pimentel M (2003) Sustainability of meat-based and plant-based diets and the environment. Am J Clin Nutr 78(3):660S–663S

Pimentel D, Hurd LE, Bellotti AC, Forster MJ, Oka IN, Sholes OD, Whitman RJ (1973) Food production and the energy crisis. Science 182:443–449

Pimentel D, Armstrong L, Flass C, Hopf F, Landy R, Pimentel M (1989) Interdependence of food and natural resources. In: Pimentel D, Hall C (eds) Food and natural resources. Academic, San Diego

Pimentel D, Hepperly P, Hanson J, Douds D, Seidel R (2005) Environmental, energetic, and economic comparisons of organic and conventional farming systems. BioScience 55:573–582

Pirog R, Benjamin A (2003) Checking the food odometer: comparing food miles for local versus conventional produce sales to Iowa institutions. Iowa State University; Leopold Center for Sustainable Agriculture, Ames. http://www.leopold.iastate.edu/sites/default/files/pubs-and-papers/2003-07-checking-food-odometer-comparing-food-miles-local-versus-conventional-produce-sales-iowa-institution.pdf

Ponting C (1991) A green history of the world: the environment and the collapse of great civilizations. Sinclair Stevenson, London

Pracha AS, Volk TA (2011) An edible energy return on investment (EEROI) analysis for wheat and rice in Pakistan. Sustainability 3:2358–2391

Riley SP (1993) War and famine in Africa. Research Institute for the Study of Conflict and Terrorism, London. http://catalog.hathitrust.org/api/volumes/oclc/30413941.html. Accessed 25 June 2014

Roberts P (2008) The end of food. Houghton Mifflin, New York

Sandom C, Faurby S, Sandel B, Svenning JC (2014) Global late quaternary megafauna extinctions linked to humans, not climate change. Proc R Soc B Biol Sci 281(1787):20133254

Smil V (1994) In energy in world history. Westview Press, Boulder

Steinhart JS, Steinhart C (1974) Energy use in the US food system. Science 184:307–316. http://www.sciencemag.org/site/feature/data/energy/pdf/se197400307.pdf

Tainter JA (1988) The collapse of complex societies. Cambridge University Press, Cambridge

United States Department of Agriculture (2010) Dietary Guidelines for Americans. http://www.health.gov/dietaryguidelines/dga2010/DietaryGuidelines2010.pdf. Accessed 25 June 2014

United States Department of Agriculture (2011) USDA Agriculture and Forestry Greenhouse Gas Inventory 1990–2008; Technical Bulletin 1930. GPO, Washington

United States Energy Information Administration (2014) Short Term Energy and Winter Fuels Outlook. http://www.eia.gov/forecasts/steo/realprices/. Accessed 29 June 2014

Washington Post, Quinoa should be taking over the world: This is why it isn't, July 11, 2013. http://www.washingtonpost.com/blogs/wonkblog/wp/2013/07/11/quinoa-should-be-taking-over-the-world-this-is-why-it-isnt/

White L (1943) Energy and the evolution of culture. Am Anthropol 45:335–356

World Bank (2014) Quinoa has reached NASA but is becoming inaccessible for Andean consumers. http://www.worldbank.org/en/news/feature/2014/01/06/quinua-llega-hasta-la-nasa-pero-se-aleja-de-los-consumidores-andinos. Accessed 30 June 2014

Wrangham RW (2009) Catching fire: how cooking made us human. Basic Books, New York

Chapter 16
Textbox: Internalizing Externalities: Techniques to Reduce Ecological Impacts of Food Production

Randy Hayes

Abstract Government regulation is only one technique to internalize a pollution impact (externality) from food production. When it leads to deregulation it typically does not solve the ecological impact or pollution problem.

In food production as in any other industry, externalities are market failures—"costs that are not borne by the actors in a particular transaction"[1]—which distort the perception of risk.[2] Externalities can be positive or negative. From an economist's view, "negative externalities are costs that are infeasible to charge to not provide."[3] This means that the market cannot sustain the externalization of costs without eventually failing. Although the government prefers the laissez-faire approach to a free market economy, it is a utopian ideal.[4] These governmental market failures generally fall into the six stages illustrated by Fig. 16.1.

The modern western food industry has proved to be such a failure because the costs of factory farming and industrial food production are externalized to the

---

[1] The editor has previously published excerpts of this introduction in Gabriela Steier, Externalities in Industrial Food Production: The Costs of Profit, 3 Dartmouth Law Journal 9, available at: http://works.bepress.com/gabriela_steier/1 and at http://www.dartmouthlawjournal.org/archives/9.3.6.pdf (citing Jan G. Laitos & Joseph P. Tomain, Energy and Natural Resources Law 21 (West 1992)).

[2] Steier, supra note i (citing Neal D. Fortin, *The Hang-Up with HACCP: The Resistance to Translating Science into Food Safety Law*, 58 Food & Drug L.J. 565 (2003)).

[3] Steier, supra note i (citing Bryan Caplan, *Externalities*. The Concise Encyclopedia of Economics (2008) Library of Economics and Liberty, http://www.econlib.org/library/Enc/Externalities.html (last visited Apr. 7, 2011)).

[4] Steier, supra note i (citing John Steven Kreis, *Lectures on Modern European Intellectual History, Lecture 21: The Utopian Socialists: Charles Fourier (1)*, http://www.historyguide.org/intellect/lecture21a.html (last visited Apr. 1, 2011)).

R. Hayes (✉)
Foundation Earth, Washington, DC, USA
e-mail: RHayes@FDNEarth.org

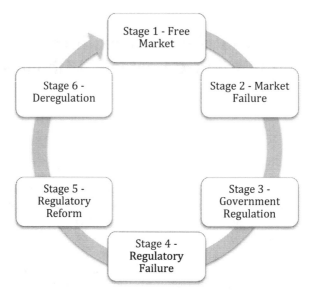

Fig. 16.1 The vicious cycle of market failures (This figure has previously been published in Gabriela Steier, Externalities in Industrial Food Production: The Costs of Profit, 3 Dartmouth Law Journal 9, available at: http://works.bepress.com/gabriela_steier/1 and at http://www.dartmouthlawjournal.org/archives/9.3.6.pdf (citing Jan G. Laitos & Joseph P. Tomain, Energy and Natural Resources Law 21 (West 1992))). Stage 1—Free Market, is the period when there is no government intervention in an industry or market... Stage 2 Market Failure are externalities, such as pollution. The existence and identification of market failure become the justification for government intervention into private enterprise that moves regulation from Stage 2 to Stage 3... The use of inadequate or incorrect regulation creates regulatory failure, or Stage 4... There are two reactions to regulatory failure. In the last two stages of the regulatory life cycle, either [the] government can respond by correcting the failure through regulatory reform (Stage 5), or, [the] government can extract itself from the market altogether by deregulation (Stage 6), ... thus reverting back to Stage 1

general public with far-reaching consequences for the future.[5] Air and water pollution created by many manufacturing operations are common examples of detrimental externalities ... If a product generates detrimental externalities, and the government does nothing, the market will cause more of the product to be produced than is optimal."[6] This happened with the those farms, which routinely increase in size and pollute the environment.

There are, however, several ways in which government regulation can reduce these detrimental externalities to consolidate the production and actual costs.[7] First,

---

[5]Gabriela Steier, Externalities in Industrial Food Production: The Costs of Profit, 3 Dartmouth Law Journal 9, available at: http://works.bepress.com/gabriela_steier/1 and at http://www.dartmouthlawjournal.org/archives/9.3.6.pdf.

[6]Steier, supra note i (citing Richard J. Pierce, JR. & Ernest Gellhorn, Regulated Industries 61 (West, 4th ed., 1999)).

[7]Steier, supra note 1.

the implementation of legislation to force the firms to internalize all of the production cost. Second, the imposition of taxes as high as the externalities produced by the firms would make up for the externalized costs. Third, pollution controls or emissions standards[8] that are environmentally sustainable. The logical result of such governmental action would be the inversion of the relationship between negative and positive externalities such that pollution could be reversed.[9] All of the measures within these three categories can be expanded further and should be explored in order to reduce the ecological footprint of food production world-wide.

On a broader scale, there are ten basic ways to internalize externalities:

1. **Eliminate Perverse Subsidies:** Eliminate existing subsidies, which contribute to social and environmental costs that have been externalized. This elimination can be accomplished by cutting government subsidies to polluting industries, such as oil, gas, industrial agriculture, nuclear or coal fired power plants, and, most recently, ethanol.
2. **Government Regulation Requiring *Internalization by Design*:** Government regulation could require all products to demonstrate a genuine **net** benefit to ecological and human well-being throughout the products' life cycle. This requirement may also extend to other goods and services within the supply chain. As *Internalization by Design* is implemented, industries can choose how to deal with their existing externalities. One approach would be to let the respective industries decide the means rather than over-regulating the entire sector. Another way to foster forethought or front-end analysis would be to incorporate the principles of ecological economics into social, economic, and environmental impact studies. This can be done at city, state or province and national levels with amendments to policies such as NEPA and California's CEQUA laws.
3. **Government Pollution Regulation:** Regulating pollution such as scrubbers on smoke stacks is a way to force internalization. Where externalization is not allowed in the first place, such regulation becomes unnecessary. Notably, the best approach may be to set strict standards and let businesses make their own decisions about how to meet those standards. Through natural competition, businesses may eventually reach the ideal levels.
4. **The "Polluter Pays" Model—Government Regulated Externalization Fee Assessment:** Assessing a "sin" tax on externalities at the front end of extraction or manufacturing could also internalize some of the costs that would otherwise be borne by the public and the environment. As long as the extraction of non-renewable resources is not banned, there should at least be an escalating sin-tax. The underlying "polluter pays" principle would also internalize

---

[8] Steier, supra note i (citing Jan G. Laitos & Joseph P. Tomain, Energy and Natural Resources Law 21 (West 1992)).
[9] Gabriela Steier, Externalities in Industrial Food Production: The Costs of Profit, 3 Dartmouth Law Journal 9, available at: http://works.bepress.com/gabriela_steier/1 and at http://www.dartmouthlawjournal.org/archives/9.3.6.pdf.

externalities from the onset and send the corrected price signal to the market. This corrected price would show that certain business activities are "ecological losers" over the medium and long term and are only focused on short-term profitability at the cost of negative externalities. In this way, fees could also be assessed for activities, which degrade or place burdens on ecosystems. Such fees could increase over time to allow companies time to adjust and to improve environmental protection while the aforementioned taxes may only be needed as an intermediate step when *Internalization by Design* is implemented. Ideally, however, most non-renewable resource extraction should be replaced by renewable substitutes, especially for the ones that are most crucial to our short term well-being. Conversely, by emphasizing recycling as a means of leaving virgin resources intact is an important way to circle back to internalize costs by design, which makes recycling, for example, more feasible.

5. **Valuing Priceless Ecosystem Services (or not):** This is controversial method, which is also called *ecosystem functions,* encourages payments for ecosystem functions (services). Essentially, it is a technique to give financial value to the otherwise priceless, such as watershed services, carbon sequestration, and fisheries stocks. The "polluter pays" principle, however, is much more desirable than putting a price on ecological externalities because it is better to commodify (pin a price to) pollution rather than nature.

6. **Unanticipated-Negative-Consequences Bond:** Another option to taxing pollution and a method to internalize negative externalities is to mandate that companies post a bond to help ensure that unanticipated negative consequences do not occur. This would help the company to follow the precautionary principle and to take it seriously because it has become financially significant for the company's bottom line.

7. **Certification Systems**: Independent certification systems such as the Forest Stewardship Council, organic agriculture, and fair trade strive to internalize some externalities and can be helpful, especially as monitors of success. The big change, however, has to come from something like the *Internalization by Design* system, that requires certification systems for products, goods and services that originate from ecologically responsible industries and satisfy the requirements to carry the respective label. Nonetheless, Post factum reviews are not desirable as they may discourage supply chain transparency and shroud other externalized ecological impacts in undeserved secrecy. Therefore, certification criteria should be integrated into the original designs, with high transparency and an adequate community review process and a local orientation included in the requirements.

8. **No Net-Loss Provisions**: Another way to internalize negative externalities with the goal to protect ecosystems is to ensure that all or more habitats are "replaced" when another area is degraded or destroyed. Under Canada's fisheries policy, there is a no net loss requirement, "whereby certain measures must be taken to replace the loss of fish habitat with newly created habitat or improve the productive capacity of some other natural habitat" if a project destroys fish habitat. A caveat to this approach is that the data must be studied because the

impact of such provisions remains questionable. It is generally not known how to "replace" habitat loss because not all the implications of the loss can be ascertained. Therefore, emphasizing the avoidance of the habitat the loss in the first place should be preferred along with preservation of such habitat, rather than "repair." Here, the analysis is one of tradeoff and should clearly favor preservation of habitat and species.

9. **Legal Structure:** Legal systems must allow for internalization mechanisms. An economic system needs to be embodied in a public governance system that allows citizens to mandate internalization. Such legal measures should include binding laws where citizens can reject companies, projects or products that are insufficiently internalized. This includes international law and WTO style rules. In a redesigned "True Cost Economy," prices would reflect externalities. In such an ideal economy, a WTO would become obsolete because trade agreements could no longer trump environmental treaties. Therefore, trade agreements should allow governments to impose tariffs on imports that do not meet required norms for resource stewardship, environmental protection, and labor practices.

   For example, trade sanctions against any nation that does not institute an *Internalization by Design* program would be a strong measure. What is needed is a new system of jurisprudence where nature is not just property, but has innate rights and where it becomes illegal to kill the planet's life support systems. Considering that corporate charters should only be awarded and renewed if certain environmental and social goals are met, could also require meeting certain conditions that justify a corporation of any given size. The burden should be on the applicant-company to justify the need other than large profit. At its heart, such laws would ask, "how does the company contribute to ecologic conservation and environmental protection in a way that no other smaller operation could?"

10. **Voluntary Measures**: Voluntary measures, such as a company's refraining from dumping chemicals into local rivers and, instead, investing in proper disposal or pollution reduction technologies, could be rewarded. Governments could support such voluntary measures by requiring disclosure of pollution emitted and resources consumed in annual reports filed with security exchanges. Currently, voluntary efforts are inadequate but often highlight what should be disclosed. In Japan, for example, practices that have been demonstrated by manufacturers as a "best practice" through voluntary means must later be incorporated in regulations as a minimum requirement, thereby consistently increasing the minimum standards. This approach would allow governments to monitor ecological performance of publicly trades companies and their entire supply chains.

Chapter 17
Textbox: Cooperatives' and Producer Organizations' Roles in Achieving Food Security

Denis Herbel, Mariagrazia Rocchigiani, and Nora Ourabah Haddad

Abstract Small-holder farming is the dominant form of agriculture in the world and it represents a significant potential for food security. However, in many developing countries, small-holder farmers face a number of constraints, some of them related to their size and geographical dispersion, which result in a lack of economies of scale and in a poor access to services. Efficient collective action in the form of producer organizations and cooperatives in particular, plays a key role in overcoming such challenges. However, organized collective action to stay effective need to "reinvent their business ideals" to adapt to a changing and competitive environment. They need to have committed leadership, a clear vision of finding solutions to the daily problems of their members as well as to build on strong endogenous processes. These processes rely on three interdependent types of relations: (1) bonding or intra-group relations among small farmers within the producer organization/cooperatives; (2) bridging or inter-group relations between producer organizations to create apex organizations; (3) linking or extra-group relations between producer organizations/cooperatives or agricultural cooperatives and economic actors and policy makers. Government and development practitioners can encourage the development of thriving producer organizations and cooperatives; however they need a shift in thinking and practice. They need to deeply understand their functioning, to recognize them as economic and political players, to adopt strategic capacity development approaches and finally to see such organizations as real partners, who can create the conditions to reduce poverty and achieve sustainable rural development.

Smallholder farming is the dominant form of agriculture in the world. Worldwide, farms of less than one hectare account for 72 % of all farms but control only 9 % of all agricultural land. They produce more than 80 % of the world's food in terms of

---

D. Herbel (✉) • M. Rocchigiani • N.O. Haddad
Office of Partnerships, Advocacy and Capacity Development, FAO, Rome, Italy
e-mail: Denis.Herbel@fao.org; MariaGrazia.Rocchigiani@fao.org;
Nora.OurabahHaddad@fao.org

© Springer International Publishing Switzerland 2016
G. Steier, K.K. Patel (eds.), *International Food Law and Policy*,
DOI 10.1007/978-3-319-07542-6_17

value.[1] Therefore, small farming represents a significant potential for food security. On the other hand the number of farm sizes in many lower-income countries continues to decrease. In fact, smallholder farmers face a number of constraints related to their size and geographical dispersion. Consequently, they lack economies of scale and access to services. These constraints limit their capacity to seize economic opportunities, or influence policies that affect them. They are often excluded from decision-making, whether in markets or in policy making processes. Addressing these constraints will unlock the significant potential of smallholders to contribute to food security.

In the meantime, efficient collective action in the form of producer organizations and cooperatives in particular, plays a key role in overcoming such challenges. Well-performing producer organizations and cooperatives can provide an array of services to enable small farmers to improve their productivity in a sustainable way by:

- accessing output and input markets;
- overcoming asymmetries of information and knowledge;
- providing access to financial services;
- managing natural resources sustainably;
- adapting to and mitigating climate change.[2]

Therefore, one overarching reason why farmers come into cooperatives, is the provision of this wide range of services.

17.1 How Do Small Farmers Benefit from Joining Cooperatives and Producer Organizations?

Producer Organizations and cooperatives can provide critical services in response to small holders' constraints. These services range from enhancing access to and management of natural resources, improving access to information and technologies, and facilitating participation in policy making. By providing an array of services, these organizations can increase farmers' incomes and improve their food security and livelihoods.

Private companies are selective when providing services to small farmers. Profitability drives their decisions according to the type of commodity, location and market contexts. These private companies normally direct their choices to high potential areas where market density is high and where they can secure their transactions in terms of accessing quantity and quality supply.[3] Instead, cooperatives and producer organizations can operate in low potential areas where markets

---

[1] FAO (2014).
[2] Herbel et al. (2011).
[3] Poulton (2014).

are "thin." They do so because their goals are different from those of for-profit organizations.[4] By forming cooperatives, farmers adopt an efficient strategy, which allows them to gain market power, thereby, rebalancing it to their advantage. Consequently, this enables farmers to mitigate the opportunistic behaviors of their trading partners in monopolistic and oligopolistic positions.

Joining cooperatives, which provide services to their members, also allow member farmers to achieve gains from economies of scale while reducing transaction costs that are lowered when farmers pool together their supply and demand. All these benefits derived from producer organizations and cooperatives enable farmers to minimize risk exposure. Indeed, building agricultural cooperatives and producer organizations is both an efficient way to share risks among their members and to minimize their exposure to contractual partners' opportunistic behaviors.[5]

On the other hand, evidence shows that producer organizations and cooperatives face significant challenges in performing their role as service providers. These challenges often translate into poor organizational performance such as "poor structure and weak governance, limited organizational capabilities, lack of financial capital, and difficulties in the institutional economic, and agro-ecological environment of small farms in poorest areas."[6] These challenges are often "exacerbated by ill-judged external support."[7] Therefore, in order to be efficient, they need to "reinvent their business ideals."[8]

17.2 What Are the Key Ingredients for Building Successful Producer Organizations and Cooperatives?

For this reinvention process to be effective there is a need to be pragmatic. Agricultural cooperatives and producer organizations can only succeed under certain conditions. It is not sufficient to have a strong board and enlightened management for a successful cooperative model.[9] Collective action—through producer organizations and cooperatives, needs to build on endogenous processes. And to develop, they rely on three interdependent types of relations:

- Bonding or intra-group relations among small farmers within the producer organization/cooperatives;

---

[4] Sexton and Iskow (1988).
[5] Cook (2012).
[6] Wanyama (2014).
[7] Wanyama (2014).
[8] Wanyama (2014).
[9] Cook (2012).

- Bridging or intergroup relations among producer organizations to create apex organizations;
- Linking or extra-group relations between producer organizations/cooperatives or agricultural cooperatives, and the economic actors and policy makers.

Through bonding relations, small farmers within producer organizations and cooperative organizations gain self-confidence and knowledge to analyse their own problems, make informed choices and act collectively. The close personal bonds of solidarity among small producers and their organizations, is the basis for their organizational development. By providing information and aligning member objectives, these first types of relations encourage members to cooperate towards the common goal: mitigating opportunistic behaviours. Bridging relations are intergroup relations, which, within a network, connect different organizations at the local, national, and regional levels. The creation of bridges among similar organizations by pooling their resources, allows them to better accomplish their mission, overcome constraints, communicate needs and offer a broader range of services. Just as the first two relations are established, farmers will have the ability to develop connections, which must be formed with influential economic and political actors. Subsequently, small producers will gain access to national and international markets and can voice their concerns.

17.3 Redefining the Role of Policy Makers and Development Practitioners for Thriving Producer Organizations and Cooperatives

Still, the roles of policy makers and development organizations should be redefined to facilitate the development of sustainable and thriving producer organizations and cooperatives. Reasons behind past failure of these organizations must be discussed and analyzed to pave the way for the future.

Until late 1980s, much had been said about the failure of cooperatives and producer organizations. Farmers often complained and expressed loss of faith in cooperatives and what they provided to their members. These organizations have frequently been considered "empty shells."

The period (1960s–1980s) is marked by state-led policies that led to the creation of organizations, i.e. the so-called cooperatives. Formation of these organizations took place under either the impulse of the state or the donor community, following a top-down approach. Specifically, the farmers, fisher folk, forest holders, and livestock keepers were not encouraged to form their own initiative. Consequently, the government gained control over decisions made by organization and its members. An immediate result of this top-down approach was demotivation from becoming a member, which translated (in some cases) into significant dropouts. Farmers considered their organizations as an extended arm of government.

Another important factor explaining the failure of these organizations is the nature and modalities of external support provided mainly by some donors and non-governmental organizations. Instead of offering support to organizations so that they become sustainable and autonomous, a dependency towards aid was developed. Indeed, this support often focused on the technical aspects of organizations instead of addressing long-term issues. Specifically, these issues are related to three dimensions of capacity development (strengthening individual, organization and institutional). Also, included is the so-called "soft skill development" such as: strengthening the effective participation of members, developing their managerial and financial skills, enhancing governance and equity aspects of organizations as well as looking at their external environment that impact them.[10]

Interventions meant to support their capacities contributed to disempowering farmers. It is the case of government direct interventions in the organization and the management of cooperatives—no respect of the principle of autonomy, or top-down donors and NGOs' interventions removing the responsibilities of farmers and creating financial dependency.[11] Rather than encouraging farmers to develop their capacities, they undermined both the farmers' incentives to acquire capacities and confidence to use them. Also, instead of looking at long-term issues and building on existing organizations and social dynamics, it adopted a narrow focus of technical results. Recognizing the capacity of producer organizations and cooperatives—individuals and as organizations, to gain ownership of their own change process is the key step to support their development path. Besides, support was often managed by donors along with the development community, and not by the organizations themselves. In short, these organizations failed because governments, the donor and development communities imposed their agenda, priorities and organizational models to further achieve their development policies. Indeed, this is certainly not sustainable.

In order to reverse this trend, governments and development practitioners need to understand and realize the function of producer organizations and cooperatives as both economic and political actors. This will enable the appropriate provision of support to facilitate the development of autonomous and sustainable organizations. The result is a need to build their capacities, which empowers autonomous service providers. Subsequently, the capacity development process requires a strategic and participatory approach—at three interrelated levels, including:

1. improved skills at the level of individual members;
2. strengthened organizational capacities of cooperatives and producer organizations; and
3. established and enablinging environments, in which organizations develop and thrive.

---

[10]The external environment encompasses: policies; the legal framework; incentives; and consultation platforms.
[11]See Develtere (2008).

At the individual level, it is important to develop "soft skills" such as: the strengthening of effective participation by members in policy for a and the decision making processes in addition to developing negotiation, mediation, and managerial skills.

At the organizational level, it is crucial to look at governance and equity aspects of such organizations for instance by:

- linking manager benefits to specific performance measure of the producer organization;
- attracting managers from the farming community in which the organization operates,
- professionalizing the Board of Directors through targeted coaching processes,
- engaging the organization in strategic planning processes,
- ensuring space to women and youth in the leadership and decision-making processes.

At the enabling environment level, it is important to understand the conditions of an enabling environment for such organizations, looking particularly at three dimensions: the regulatory framework, the investment climate, and the way in which the government and the producer organizations and cooperatives interact through existing policy consultation platforms.

In helping small farmers overcome the organizational challenges, this capacity development approach can contribute to effective governance and management practices, ensuring transparency and accountability. In this way, producer organizations and agricultural cooperatives can reach the "organizational maturation."[12] In this event, the process allows change in the nature and quality of relationships within the organization as well as among organizations contributing to the modernization of value chains and the changing food systems.

Therefore, governments and development practitioners play an important role in facilitating and supporting the development of autonomous and sustainable producer organizations and cooperatives. Hence, a shift in the thinking is formed based upon the transformation of relations between member-based organizations—cooperatives, producer organizations, and other actors such as the public sector, the development community (donors included). Relationships must be transformed into a "win-win partnership," replacing the member-based organizations as mere recipients and passive actors. In the same way, the "traditional beneficiaries" become partners on equal footing. After all, it is only through this transformed relationship that producer organizations and cooperatives can become agents of development alongside other powerful actors (including donors) and decision-makers.

Allowing small holders to actively participate in markets and the decision-making processes, the producer organizations and cooperatives will help create conditions to meet growing food demands, reduce poverty and achieve the sustainable development of rural areas.

---

[12]Badiane and Wouterse (2014).

References

Badiane O, Wouterse F (2014) Using ICT to promote social capital and build commercial and technical skills of producer. Collective Action among African Smallholders: Trends and Lessons for Future Development Strategies. Published by the West and Central Africa Office, Thematic Research Note 05. March 2014

Cook M (2012) Harnessing the cooperative advantage. The global forum and workshop, harnessing the cooperative advantage to build a better world. The United Nations Conference Centre (UNCC), Addis Ababa

Develtere P (2008) Cooperative development in Africa up to 1990s. In: Develtere P, Pollet I, Wanyama F (eds) Cooperating out of poverty—the renaissance of the African cooperative movement. International Labour Organization, Geneva

FAO (2014) The state of food and agriculture: innovation in family farming. Rome

Herbel D, Crowley E, Ourabah Haddad N, Lee M (2011) Good practices in building innovative rural institutions to increase food security. FAO, Rome

Poulton C (2014) The role of cooperatives in modernizing supply chains. In IFPRI–WCAO "Collective action among African smallholders". Thematic Research Note 05. March 2014

Sexton R, Iskow J (1988) Factors critical to the success or failure of emerging agricultural cooperatives. Department of Agricultural and Resource Economics, University of California, Davis. Giannini foundation, Information series #88-3

Wanyama F (2014) A history of cooperative development in Africa. In IFPRI–WCAO "Collective action among African smallholders". Thematic Research Note 05. March 2014

Chapter 18
Governing the Global Food System Towards the Sustainocene with Artificial Photosynthesis

Thomas Faunce and Alex Bruce

Abstract The development of the current global food production system has been predicated on multinational corporate market power coupled with the intensive use of pesticides and carbon-intensive fuels in mechanised, massive scale agricultural and slaughtered animal production and transportation (i.e., tractors, harvesters, trucks and container ships) but also in the production of fertilizer (particularly via the Haber-Bosch process for the conversion of atmospheric nitrogen to ammonia and in the phosphate industry). Our central hypothesis is that this corporatized global food system is fundamentally damaging to the sustainability of our environment and inhibitory of the growth of distributed or decentralised organic farming which is more likely to be the mainstay of food production in a future where humanity individually and collectively flourishes as stewards of resilient ecosystems. In the course of analysing that hypothesis we explore whether appropriate global governance of new renewable energy and climate change mitigation technologies such as artificial photosynthesis may accelerate transition to such a reformed global food system in an epoch conveniently termed the Sustainocene. In particular we examine whether and if so how international food law and policy can assist nanotechnology-based artificial photosynthesis become an 'off-grid' distributed family and community-based combined food, energy, water and climate change solution that establishes stable preconditions for humanity to realise its full potential as an ethical species.

T. Faunce (✉) • A. Bruce
ANU College of Law, Australian National University, Canberra, ACT, Australia
e-mail: thomas.faunce@anu.edu.au; alex.bruce@anu.edu.au

18.1 Introduction

18.1.1 Corporate Power and Carbon Intensive Energy Behind the Global Food System

The global food industry is dominated by a few multinational corporations, sometimes working in competition but also often in collusion and corrupt manipulation of the market and government regulation evidenced by large-scale antitrust litigation, especially in the United States. The industry inefficiently uses a large amount of the world's energy input to produce food for profit in selected markets. This energy takes the form of carbon-based fuels for mechanized production, transport in trucks, trains and ships. It is also required in huge amounts to make ammonia fertilizer via the Haber-Bosh process for fixing atmospheric nitrogen (Smil 2004). It is also used to mine and export phosphorus and, for example in the case of the Western Sahara, to import back this non-renewable resource once manufactured as fertilizer or even as a raw material from strategically protected sources in China, US and Morocco, it (Cordell et al. 2009). It is an indictment of this system that it not only perpetuates but tacitly encourages structural disparities in food access that see one billion people starving while being a direct cause of one billion people being obese. In addition to these structural disparities, corporate domination of the food production and distribution chains promotes severe information asymmetries where consumers are often unable to determine credence claims made by corporations about the food products being sold and consumed. It is in this context that policy makers are seeking for ways to shift this system to one which puts a greater premium on sustainability and consumer accountability, not just for human economies but for ecosystems (Sullivan 2012–2013).

18.1.2 Brief Overview and Background of the Corporate Global Food Industry

Since 1800 with the onset of the industrial revolution, the development of the capacity to fix atmospheric nitrogen as a fertilizer, improved sanitation, healthcare and transport, global human population and its impact have dramatically increased. Land ecosystems, for instance, were globally converted from mostly wild to mostly anthropogenic by the mid twentieth century (Steffen et al. 2011).

Using a new method for calculating undernourishment that began with the 2012 edition of the *State of Food Insecurity in the World* report, United Nations agencies estimate hunger in its most extreme form has decreased globally from over 1 billion in 1990–1992 (18.9 % of the world's population) to 842 million in 2011–2013 (12 % of the population) (FAO 2013). Yet these household figures, disturbing and revealing as they are, do not disclose short-term undernourishment, neglect inequalities in intra-household food distribution (particularly for children, women, frail and the elderly) and the necessity to perform physically demanding activities (Lappé et al. 2013). In fact the global food system at present perpetuates in developing nations lack of ready, safe access to clean water and diets with adequate micronutrients (i.e., iodine, of vitamin A or iron especially during pregnancy and in a child's first 2 years) while in middle and high-income countries the prevalence of obesity doubled between 1980 and 2008 creating heightened risk of type 2 diabetes, heart disease or gastrointestinal cancers (De Schutter 2014).

The current corporate-controlled global food system is characterized by use of monocultures of high-yielding plant varieties with increased irrigation, the mechanization of agricultural production and the use of nitrogen-based fertilizers and pesticides. Monsanto for instance has a monopoly for its genetically engineered-sterile corn and soy seeds that are now resistant to Monsanto's own brand of pesticide called Roundup (Patel 2008).

The acute structural imbalances of this system are buffered by large scale State subsidies and corporate lobbying and marketing, leading to surplus developed nation production of profitable maize, wheat and rice and soybean crops (Mellor 1988). These staple crop surpluses are then dumped with low prices on developing markets to function as a substitute for improved non-agricultural wages for workers in the non-agricultural sectors and for the establishment of adequate social infrastructure, but hindering local farmers' sales and making populations vulnerable to changes in price or supply of such food staple imports (Mellor 1988). One third of these cereals are used as food animal feed while poor people can not afford the resultant meat (De Schutter 2014).

This system has significant non-sustainable features. It results, for example, in a significant loss of agrobiodiversity, accelerated soil erosion, phosphate and nitrogen water pollution spurring algae growth that absorbs the dissolved oxygen required to sustain fish stocks (Chislock et al. 2013). It promotes unsustainable mass fishing practices such as those using drift and bottom dredging nets and supertrawlers (De Schutter 2014).

It is increasing becoming obvious that another of the most potentially devastating social and environmental impacts of this corporatized global food system in increased greenhouse gas emissions (De Schutter 2014). Corporatised mass-production techniques drive field-level practices causing 15 % of total human-made greenhouse gas emissions include nitrous oxide (N_2O) from the use of nitrogen fertilizers, methane (CH_4) from livestock, and carbon dioxide (CO_2) from the loss of soil organic carbon in croplands and (due to intensified grazing) on pastures (De Schutter 2014). The global agri-food business also causes significant carbon emissions through mass production of herbicides and pesticides, mechanized tillage, irrigation and fertilization, and the road, rail, sea and air transport, packaging and conservation of food (Campbell and Ingram 2012). By the early twenty-first century one third of total arable land was dedicated to crop production for animal feed or biofuels and this itself was becoming a major cause of deforestation and greenhouse gas emissions (FAO 2006) (Goodland and Anhang 2009).

It is likely that the technological boost to crop yields over past decades will soon be undone by what may be termed 'corporatogenic' climate change in many parts of the world (Lobell et al. 2011). Climate change is projected to progressively increase inter-annual variability of crop yields in the context of rapidly rising crop demand. All aspects of food security are potentially affected by climate change, including food access, utilization, and price stability (high confidence) (IPPC 2014). The fragility of global food systems is particularly concerning given predictions by the United Nations Population Division that the world's population will reach somewhere between 8 and 11 billion people.

In order to meet this expected demand for food generally and meat products particularly, the United Nations Food and Agricultural Organisation estimates that agricultural output will need to increase by 70 % but must do so in circumstances of a world-wide decline in agricultural land because of climate change, dwindling fossil fuel supplies and the general movement of people off the land and into cities, urban and sub-urban areas.

A significant feature of the global food system are ocean-going container ships. Each year 100 million containers, many carrying food, traverse the world's seas and thousands fall overboard. There are now approximately 15,000 pieces of plastic per square mile in many oceans and it takes about 500 years for such garbage to completely break down. Human waste which could be used to fertilise the soil if agriculture was more local, instead is piped or flushed during the increasing number of climate-change-driven fierce storms back into the ocean. Large trash vortices exist in the Atlantic and Pacific Ocean gyres leaching chemicals into the food chain (Moore et al. 2001; Barnes et al. 2009).

The global food system has been shaped to maximize efficiency gains and produce large volumes of a narrow range of profitable commodities for supermarkets. Indeed the corporate owners of supermarkets have become the 'masters' of the world food system, raising their profits by shaping through mass advertising the demand of citizens they rebadge as 'consumers' while restricting competition. Market concentration, information asymmetries, monopsony buying power and pricing discretion have converged in the form of global supermarket chain stores that exploit both up-stream (producers) and down-stream consumers. Supermarkets have had a major role in driving 'citizens' to become 'consumers' of processed foods rich in salt, sugar and saturated fats but with a conveniently long shelf life for sale (Baines 2014). Supermarkets often also use their tight control of the market to launch unrealistically cheap 'home' brands that force other food processors out of business (Baines 2014).

At the same time, World Trade Organisation (WTO) structural adjustment systems pushed upon developing and indebted nations, have had the effect of lowering prices therein for agricultural products while discouraging investors in local food infrastructure except for a narrow range of products destined for export markets (De Schutter 2014). The food supply of people in such countries has become dependent on cheap imports of staples and been compromised by a lack of governance support for community-focused agricultural production and food processing (De Schutter 2014).

In summary, a reasonable case can be made that the corporatized global food system has substituted chemicals for farm workers displacing them and their families and communities to cities where they became dependent on mass produced mono-cultured food produced by a few large corporations; this has not only deprived them of balanced nutrition and the flourishing involved in self-controlled work amidst nature but made them complicit in the savage exploitation of food animals (Patel 2008). What follows is an analysis of how international food law and policy can promote a technology that may help transform this system towards long-term sustainability.

It is part of our argument that if it is to maintain any claim to jurisprudential legitimacy, international food law and policy must drive fundamental changes in the global agri-food system including, where possible, harnessing domestic competition and consumer policies to mitigate market concentration, collusion and information asymmetries that exploit consumers. The first idea to be explored here is that this should involve support in academic writings, policy documents, legislation and international instruments for a jurisprudential shift in which traditional foundational social virtues such as justice, equity and respect for human dignity that have underpinned governance of the global food system are subsumed within the single foundational virtue of environmental sustainability. The second issue is whether this transition requires impetus from new technology and whether this is likely to involve globalization of artificial photosynthesis. Assuming that this approach is correct, the next issue is how governance of the global food system best be reframed so that it can most effectively and rapidly assist widespread equitable deployment of artificial photosynthetic technology. The final issue is whether this process can be continued so that international food law and policy include human aspirations towards sustainability that include shaping a billion-year period of human stewardship over a resilient earth ecology.

18.1.3 Importance of Topics for International Food Law and Policy

Our hypothesis is that international law and policy should embrace environmental sustainability as its foundational social virtue instead of the more anthropocentric concepts of justice or equity, or the more corporato-centric notions such as economic security or industry development. The second issue thus complements this by linking international food law and policy to prior revolutions in governance in which reform proposals supporting foundational social virtues have been driven by revolutions in technology that have remove the economic rationale for not developing virtues by consistently following universally applicable principles. The third issue overlaps as giving a practical dimension to how international food law and policy that best create the conditions responsive to the development of artificial photosynthesis. The final issue includes the others in a visionary approach of using this globally deployed renewable energy and food technology to restructure global governance towards sustainability over millennia.

18.2 Background and Contextualization

18.2.1 The Global Food System in the Corporatocene

It has been argued that this planet has passed from what is known as the Holocene geological epoch into what has been termed the Anthropocene period (Crutzen and Stoermer 2000). The term 'Anthropocene' was coined by Crutzen (2002). It refers to an epoch where human interference with earth systems (particularly in the form of influences on land use and land cover, coastal and maritime ecosystems, atmospheric composition, riverine flow, nitrogen, carbon and phosphorus cycles, physical climate, food chains, biological diversity and natural resources) have become so pervasive and profound that they are not only becoming the main drivers of natural processes on earth, but are threatening their capacity to sustain life (Steffen et al. 2007).

Five characteristic features of the Anthropocene epoch that tend to dominate policy debates over global food production include: population; poverty, preparation for war, profits and pollution (Furnass 2012). Salutary facts driving academic and policy interest in moving from the Anthropocene to a different type of human-controlled epoch are not only the greenhouse-gas driven increase in severe weather events, but the projected increase of global human population to around 10 billion by 2050 with associated energy consumption rising from ≈ 400 EJ/year to over 500 EJ/year beyond the capacity of existing fossil-fuel based power generation (Rogner 2004).

Yet is it really correct to term this period the 'Anthropocene?' Is it really the collective decisions of human beings, as this name implies, that are driving poverty, or a systemic perpetuation of preparation for war, profits and pollution. The thesis explored here is that it is really the activities of economically and politically powerful multinational corporations that primarily are driving these deleterious features of the present global food system. Governments have been lobbied by a carefully controlled and targeted approach from the corporate sector to embrace policies that privatise or sell core components of food infrastructure to the corporate sector, relax regulation that seeks to prevent corporate-initiated pollution, corruption and collusion or provide subsidies that distort markets. This is why governments frequently turn to domestic competition and consumer policies that directly target strategic corporate behaviour in searching for market-based solutions to industry concentration, collusion and consumer exploitation (Bruce 2013).

Critical analysis of the global food system and its corporate control in fact suggests that it is merely yet another deleterious outcome of corporate governance of the earth dedicated to maximising profit, an epoch that, with its philosophical foundations in neo-classical economics, may accurately be termed the 'Corporatocene.' The extent to which this is the case can be seen through analysis (in the next section) of some illustrative attempts made at the level of international food law and policy to govern aspects of this system that do not appear congruent with traditional and emerging foundational social virtues.

18.2.2 Attempts to Govern the Global Food Industry with International Law

In the outcome document of the United Nations *Rio+20 Conference on Sustainable Development*, entitled "The future we want", Heads of State and Government reaffirmed their "commitments regarding the right of everyone to have access to safe, sufficient and nutritious food, consistent with the right to adequate food and the fundamental right of everyone to be free from hunger." (UN Conference on Sustainable Development 2013) In its final report of May 2013, the High-level Panel of Eminent Persons on the Post-2015 Development Agenda proposed to include ensuring "food security and good nutrition" among the universal goals and targets to be agreed, with target 5 (a) referring to ending hunger and protecting the right of everyone to have access to sufficient, safe, affordable and nutritious food. Similar conclusions emerged from the *Madrid High-level Consultation on Hunger, Food Security and Nutrition* in the Post-2015 Development Framework, convened on 4 April 2013. At its fortieth plenary session, building on this emerging consensus, Committee on World Food Security (CFS) highlighted "the essential role of food security and nutrition and poverty eradication in the elaboration of the post-2015 development agenda", and it mandated its Bureau to ensure this key objective would be reflected in this agenda (CFS 2013). It is clear, however, even for the title of the conference that environmental sustainability was not viewed as the centralizing policy priority here. This is not to say, of course, that the stated goal are not important, but that without reference to environmental sustainability they risk losing the critical focus that makes their long-term achievement coherent.

Research underpinning the push to reorient , amongst other areas, international food law and policy towards environmental sustainability also emerges strongly from influential commentaries such as the Intergovernmental Panel on Climate Change (IPCC 2007) and the Stern Report (Stern 2007), as well as the United Nations *Millennium Development Goals* (UN MDGs).

The United Nations General Assembly in its 2012 *International Year of Sustainable Energy for All*, amongst its recommendations recognized that achieving equitable access to modern affordable energy services in developing countries was essential for transformation of global food production so that it became coherent with the achievement of the internationally agreed development goals, including the United Nations *Millennium Development Goals*, which would help to reduce poverty and to improve the conditions and standard of living for the majority of the world's population (UN Sustainable Energy for All 2012). Development of renewable energy for cooking appliances and fuels, for agricultural processes, for processing and transportation of food was emphasised, but in the context of enhanced 'business models' 'successfully engaging the private sector" "finance and risk management" "increasing private investment" and "targeted" use of public and philanthropic capital (UN Sustainable Energy for All 2012). Environmental sustainability did not appear as a fundamental organizing virtue behind the recommendations of this body, which indeed were so facile and nebulous as to create the impression that the whole process was an exercise in international level public relations for the directors of the major corporate organisations that were on the advisory committee.

A similar exercise with intersections for international food law and policy was the 2014 *International Year of Family Farming* (IYFF). The IYFF aimed to raise the public policy profile of family farming and smallholder farming in agricultural, forestry, fisheries, pastoral and aquaculture production by focusing world attention on its significant role in eradicating hunger and poverty, providing food security and nutrition, improving livelihoods, managing natural resources, protecting the environment, and achieving sustainable development, in particular in rural areas. Key issues included agro-ecological conditions and territorial characteristics; policy environment; access to markets; access to land and natural resources; access to technology and extension services; access to finance; demographic, economic and socio-cultural conditions and availability of specialized education (FAO 2014). One of the most relevant significant findings of the IYFF was the extent to which multinational corporations were having a deleterious impact on these objectives through promoting structural imbalances in the global food system (FAO 2014).

So, given the structural imbalances Corporatocene-style governance is causing with respect to traditional social virtues how does international food law and policy begin to build on initiatives such as those discussed here to move away from that destructive model?

18.2.3 *Artificial Photosynthesis for a Sustainable Global Agriculture*

There has been much policy interest in developing what is termed the 'hydrogen economy' in which hydrogen is used ubiquitously as a carbon-neutral energy vector (for example source of electricity via fuel cells or as a fuel itself when combined, for example, to form methanol) and source of small amounts of fresh water (when combusted) (Faunce 2012c). Major policy documents have outlined the case for such an economy (European Hydrogen and Fuel Cell Technology Platform 2011). (US DOE) (NSC) (E4 tech) Some of the challenges include the need to lower the cost of hydrogen fuel production to that of petrol, the difficulties in creating a sustainable and low carbon dioxide route for the mass production of hydrogen, the need to develop safe and more efficient storage (including the difficulties of compressing and cooling the hydrogen), the need to develop regulations and safety standards at national and international levels as well as the need to develop stable incentive systems for large scale investment in this area that will not fluctuate with oil prices.

Hydrogen (H_2) requires 3000 times more space for equivalent amount of energy, but on a weight basis H_2 has 3 times the energy content of gasoline. Liquifying H_2 requires complex and expensive process (pre-cooled with liquid ammonia to $-40\ °C$ then to $-196\ °C$ with liquid nitrogen, then helium in compression-expansion to get liquid H_2 at $-253\ °C$) 30–40 % of H_2 energy is lost in liquifaction and 1–5 % must be lost to atmosphere each day to avoid pressure build up and explosion. Compression similarly requires 10–15 % energy and requires cylindrical shape. One of the main problems at present with moving to a global hydrogen economy is the carbon-intensive energy required to produce hydrogen in large quantities by steam reformation of hydrocarbons, generally methane (Sartbaeva et al. 2008).

Yet a cheap and abundant source of H_2 is readily at hand as an output of technology that could engineer into buildings and roads the process of splitting water to get hydrogen using sunlight while absorbing atmospheric nitrogen and carbon dioxide. One of the reasons for focusing on this as the main energy supply of the Sustainocene is that more solar energy strikes the Earth's surface in 1 h of each day than the energy used by all human activities in 1 year. At present the average daily power consumption required to allow a citizen to flourish with a reasonable standard of living is about 125 kWh/day. Much of this power is devoted to transport (~40 kWh/day), heating (~40 kWh/day) and domestic electrical appliances (~18 kWh/day), with the remainder lost in electricity conversion and distribution (MacKay 2009). Global energy consumption is approximately 450 EJ/year, much less than the solar energy potentially usable at ~1.0 kW per square metre of the earth— 3.9×10^6 EJ/year even if we take into the earth's tilt, diurnal and atmospheric influences on solar intensity (Pittock 2009).

Photosynthesis as a natural process is equally important with DNA in the progress of humanity. Photosynthesis provides the fundamental origin of our oxygen, food and the majority of our fuels; it has been operating on earth for over two billion years. Photosynthesis can be considered as a process of planetary respiration: it creates a global annual CO_2 flux in from the atmosphere and an annual O_2 flux out to atmosphere. In its present nanotechnologically-unenhanced form, photosynthesis globally already traps around 4000 EJ/year solar energy in the form of biomass. The global biomass energy potential for human use from photosynthesis as it currently operates globally is approximately equal to human energy requirements (450 EJ/year).

It is a quite remarkable that contemporary energy policy analysts have failed to appreciate the significant implications for food and energy security that natural photosynthesis is capable of substantial improvement with nanotechnology. When photosynthesis is considered in this context it is usually as a source of biofuels, often with a deleterious cost to rainforests or agricultural lands. Further, even if

3000 m^2 per person is devoted to biomass production this will provide only fuel only 36 kWh/day per person (well short of the 125 kWh/day required for people to live comfortably) (MacKay 2009). Photovoltaic (PV) energy systems are improving their efficiencies towards 25 %, and the cost of the electricity they produce is nearing or has past grid parity in many nations. The development of "smart-grid" (allowing energy carrying capacity to fluctuate coherently in accord with input and output) and "pumped-hydro" (using diurnal PV electricity to pump water to high reservoirs so it can be run down through turbines at night) will assist the viability of this an a national energy source. But even large solar farms (for example taking up 200 m^2 per person with 10 %-efficient solar panels) could produce but ~50 kWh/day per person (MacKay 2009).

Many renewable energy researchers and policy makers are locked into the outdated view that only plants can do 'photosynthesis;' that the process is something that buildings, roads and vehicles are never destined to do unless we incorporate something biologic into those structures. Such thinking leads to a focus on genetically modifying plants and also using synthetic biology. Such researchers seek for example to genetically manipulate or even synthetically reproduce photosynthetic plants and bacteria to maximize their light capture and carbon reduction activities.

Yet, when we travel in aircraft across the world it is easy to see the extent to which human concrete and asphalt structures are proliferating and replacing photosynthetic plants across the face of the planet. Such human-made structures contribute little to the ecosystems around them. They do not enrich the soil or provide oxygen or absorb carbon dioxide. Yet we are almost at the point where nanotechnology and artificial photosynthesis can be engineered into such structures so they can be made to "pay their way" in an ecosystem sense.

Artificial photosynthesis can facilitate other energy options H$_2$—based fuels, the most promising perhaps by combining it with atmospheric nitrogen to make ammonia. Ammonia was used to power railcars in New Orleans in 1871 and in Brussels' buses in 1943 when the Nazis commandeered all the diesel for military purposes. Its combustion products are nitrogen and water, with an initial puff of small quantities from petroleum or diesel for ignition. Ammonia can be stored at around 130 psi and carried in cylinders in the boot of a motor vehicle (a trial run from Detroit to San Francisco recently requiring only one re-fill). NH3 is already is shipped, piped, and stored in large volumes in every industrial country around the world as an agricultural fertilizer. As a fuel, NH3 has been proven to work efficiently in a range of engine types, including internal combustion engines, combustion turbines, and direct ammonia fuel cells. Due to its high energy density and an extensive, existing NH3 delivery infrastructure, NH3 is ready for the market today as an alternative to gasoline (NH3 Fuel Association).

If such artificial photosynthetic technology is incorporated into every building, road and vehicle on the earth's surface than the positive outcome will be that humanity's structure will be producing abundant safe, low carbon fuels and fertilizers. In such a world it will be much more feasible for communities and families to support many of their basic food needs off–grid through organic farming rather than relying on distant sourced food provide by large corporate marketing chains. Dilution of market concentration will simultaneously reduce the potential for dominant corporations to act as monopsonists; dictating terms-of-trade with smaller upstream producers of food products. A market that is composed of a number of small-scale community food producers will enhance inter-brand competition for food products while also diminishing the pervasive information asymmetries that currently plague most consumers in the corporatised food market.

It also is likely that nanotechnology will play a significant role in the shift from the corporate globalisation model of food production and distribution. Nanotechnology is the science of making things from components that are not much bigger than a few atoms, less than 100 nm (a nanometer is a billionth of a metre). The chief policy interest to date with nanotechnology to date has been concerned with ensuring its safety (Faunce 2008). Interest is growing, however, in focusing nanotechnology on such problems. Experts have encouraged nanotechnology researchers to systematically contribute to achievement of the United Nations *Millennium Development Goals* particularly energy storage, production and conversion, agricultural productivity enhancement, water treatment and remediation (Salamanca-Buentello et al. 2005). In fact it has been argued by the first author that global artificial photosynthesis is the moral culmination of nanotechnology (Faunce 2012d).

18.2.4 Need to Globally Coordinate Artificial Photosynthesis Research

Artificial photosynthesis is the subject of intense and advanced research by large groups of scientists in all developed nations (Faunce et al. 2013a, b). A dozen European research partners, for example, form the Solar-H2 network, supported by the European Union and coordinated by Stenbjorn Styring at Uppsala University, Sweden. The US Dept. of Energy (DOE) *Joint Center for Artificial Photosynthesis* (JCAP) at the California Institute of Technology (Caltech) and Lawrence Berkeley National Laboratory coordinated by Nate Lewis has US$122 m over 5 years to build a solar fuel system. At Caltech Harry Grey coordinates a large National Science Foundation (NSF) grant to improve photon capture and catalyst efficiency and has initiated a Solar Army endeavour in which high school and college students are mentored to search the periodic table for suitable catalysts. In the US several Energy Frontier Research Centers including that of Mike Wasielewski at Northwestern funded by the US DOE are focused on scientific endeavours related to artificial photosynthesis (Faunce et al. 2013a). Other major solar fuels centres have been established in South Korea, The Netherlands, Germany (at the Max Planck institute) and Singapore and numerous other competitively funded research teams have dedicated artificial photosynthesis-related projects already underway (Faunce et al. 2013a, b).

A basic idea of solar fuels research amongst the large national and regional projects mentioned above is to develop solar fuel prototype devices that improve on how plants absorb sunlight and use it to create an electron flow and hydrogen by splitting water. The scientific challenges involved here are considerable.

Photosynthetic organisms absorb photons from a segment of the solar spectrum (~430–700 nm) by so-called 'antenna' chlorophyll molecules in thylakoid membranes, or chloroplasts. The absorbed photons' energy creates unstable spatially separated electron/hole pairs. The "holes" are captured by the oxygen-evolving complex (OEC) in photosystem II (PSII) to oxidise water (H_2O) to what can be termed a natural form of hydrogen (protons) and oxygen (O_2). This process can be written as the following chemical equation: $2H_2O \Rightarrow 4$ photons $\Rightarrow 4e^- + 4H^+ + O_2$. The protons released on water oxidation can be used to make hydrogen according to a chemical process recorded as: $2e^- + 2H^+ \Rightarrow H_2$. The electrons are subsequently captured in chemical bonds by photosystem I (PSI) to reduce NADP (nicotinamide adenine dinucleotide phosphate) to NADPH. Electro-chemical energy stored by the protons produces ATP (adenosine triphosphate). In the relatively less efficient "dark reaction", ATP and NADPH as well as carbon dioxide are used in the Calvin-Benson cycle to make a variety of energy rich chemicals, mainly sucrose and starch via the enzyme RuBisCO (Ribulose-1,5-bisphosphate carboxylase oxygenase) (Blankenship 2002). This capacity to store solar energy in chemical bonds is the feature that makes enhanced photosynthesis so intriguing as a form of renewable energy.

Some nanotechnological innovations for artificial photosynthesis focus on improved 'light capture.' This 'light capture' involves nanostructured materials or synthetic organisms absorbing photons from a much wider region of the solar spectrum (photon absorption by antenna chlorophyll molecules in thylakoid membranes of chloroplasts, for example, currently is restricted primarily to ~430–700 nm).

The next component involves creation of an equivalent catalytic system to the manganese-oxygen cluster called PSII. PSII in plants is a complex protein with 27 subunits and 32 co-factors involved in electron transfer and light harvesting. Researchers are working upon making a nanotechnological mimic of this protein (maquette) that is simpler (Koder et al. 2009) and incorporates designer molecules that prolong charge separation Carmieli et al. (2009). Nanotechnology is facilitating the construction of artificial photosynthetic electron pathways to this reaction centre that perform a single quantum computation, sensing many states simultaneously and so enhancing the efficiency of the energy capture and transfer at physiological temperatures (Lee et al. 2007; Ball 2010; Engel et al. 2007).

The most globally widespread water catalytic system will probably involve inexpensive and self-repairing components that operate at neutral pH with non-pure (salty or bacterially and chemically contaminated) water (Kanan and Nocera 2008) and be stable to a variety of exposure conditions in air, water and heat (Yin et al. 2010). A major scientific challenge will be to optimise the free energy required for the overall water splitting process. Multiwalled carbon nanotubes and singlewalled carbon nanotubes may produce the critical breakthrough here (Sgobba and Guidi 2009).

In the artificial photosynthesis version of the "dark reaction", ATP and NADPH as well as carbon dioxide (CO_2) will be used in an enhanced version of the Calvin-Benson cycle to make locally usable food or fuel (for domestic, heating, cooking, light and transport) in the form of carbohydrate via the enzyme RuBisCO (Gray 2009). Bio-inspired self-repair strategies will ensure that this aspect survives damage from repeated cycles of thermodynamically demanding reactions (Wasielewski 2006). New catalysts for H_2 production and methods for efficient H_2 usage (in a fuel cell to make electricity) or storage (as a fuel after cooling and concentrating) will need to be built (Magnuson et al. 2009).

Major publications in this area include those by Peidong Yang Sun et al. (2011). Dan Nocera's 'artificial leaf' configures a triple junction silicon photovoltaic cell with a cobalt catalyst for O_2 evolution and a ternary alloy (NiMoZn) as the H_2-evolving catalyst in a wireless configuration (Reece et al. 2011). Nobuo Kamiya of Osaka University, has encouraged the process of building mimics of the core part of the natural photosynthetic system with the publication by his team of a cubane configuration of the OEC in PSII to a level of 1.9A° (1.9 ångströms or 1.9×10^{-10} m) using an electron density map (Umena et al. 2011). Craig Hill of Emory University has developed a polyoxometalate water oxidation catalyst capable of strongly binding multiple transition metal centers proximal to one another, so facilitating multi-electron processes such as the 4-electron oxidation of H_2O to O_2 (Zhu et al. 2012). David Tiede of the Argonne National Laboratory has reported a new strategy for solar fuel production involving insertion of sustainable first-row transition metal molecular catalysts (cobaloxime) into Photosystem I (PSI) as a mechanism for H_2 production (Utschig et al. 2011). Gary Brudvig and Chris Moser are amongst the notable researchers who have focused on how insights from the natural photosynthetic system might develop bioinspired materials for photochemical water oxidation and fuel production (Blankenship et al. 2011).

An international conference coordinated by the first author at Lord Howe Island in August 2011 linked senior artificial photosynthesis and global governance experts purportedly as a precursor to a macroscience Global Artificial Photosynthesis (GAP) Project (GAP 2011). A second conference on the same theme sponsored by the UK Royal Society will establish a practical framework for a global project on artificial photosynthesis (GAP 2014).

18.3 Discussion and Analysis

18.3.1 *Ensuring Access to Food Resources*

The question for analysis then becomes how international food law and policy, can play a role in changing the global food system towards sustainability. The United

Nations Special Rapporteur on food recommended in 2014 that in a context in which commercial pressures on land are increasing, it is crucial that States strengthen the protection of land users (A/65/281) and implement the Voluntary Guidelines on Responsible Governance of Tenure of Land and other Natural Resources (De Schutter 2014). In relation to access to land in particular he recommended States should:

(a) Ensure security of tenure, by adopting anti-eviction laws and improving the regulatory framework concerning expropriation;
(b) Conduct decentralized mapping of various users' land rights and strengthen customary systems of tenure;
(c) Adopt tenancy laws to protect tenants from eviction and from excessive levels of rent;
(d) Respect the rights of special groups, such as indigenous peoples, fisherfolk, herders and pastoralists, for whom the protection of commons is vital;
(e) Prioritize development models that do not lead to evictions, disruptive shifts in land rights and increased land concentration, and ensure that all land investment projects are consistent with relevant obligations under international human rights law;
(f) Refrain from criminalizing the non-violent occupation of land by movements of landless people;
(g) Implement redistributive land reform where a high degree of land ownership concentration is combined with a significant level of rural poverty attributable to landlessness or to the cultivation of excessively small plots of land by smallholders, and supporting beneficiaries of land redistribution to ensure that they can make a productive use of their land; and
(h) Regulate land markets to prevent the impacts of speculation on land concentration and distress sales by indebted farmers (De Schutter 2014).

The Report therefore recommends both structural and behavioural changes to the ownership and stewardship of land resources. A common theme of these recommendations is the dilution of centralised corporate control and market concentration. If these legal and behavioural changes strengthened the security of access to land for communities then it would create the preconditions whereby artificial photosynthetic technology could assist in providing fuel and electricity to use farming machinery while also enriching the soil. These recommendations are unlikely to be supported by large corporations involved in land acquisition and exploitation.

18.3.2 Ensuring Access to Seeds and Genetic Diversity

In relation to guaranteeing food security and supporting crop genetic diversity including agrobiodiversity the United Nations Special Rapporteur on food noted that small-scale farmers in developing countries, who still overwhelmingly rely on seeds which they save from their own crops and which they donate, exchange or sell (De Schutter 2014). In order to ensure that the development of the intellectual property rights regime and the implementation of seed policies at the national level are compatible with the right to food, he recommended States should:

(a) Make swift progress towards the implementation of farmers' rights, as defined in article 9 of the International Treaty on Plant Genetic Resources for Food and Agriculture;
(b) Not allow patents on plants and establish research exemptions in legislation protecting plant breeders' rights;
(c) Ensure that their seed regulations (seed certification schemes) do not lead to an exclusion of farmers' varieties; and
(d) Support and scale up local seed exchange systems such as community seed banks and seed fairs, and community registers of peasant varieties.

Donors and international institutions should assist States in implementing the above recommendations, and, in particular:

(a) Support efforts by developing countries to establish a *sui generis* regime for the protection of intellectual property rights which suits their development needs and is based on human rights;
(b) Fund breeding projects on a large diversity of crops, including orphan crops, as well as on varieties for complex agroenvironments such as dry regions, and encourage participatory plant breeding;
(c) Channel an adequate proportion of funds towards research programmes and projects that aim at improving the whole agricultural system and not only the plant (agroforestry, better soil management techniques, composting, water management, good agronomic practices) (De Schutter 2014).

Once again changing international food law and policy to give small scale farmers better access to seeds would undoubtedly assist their nutrition once combined with greater legal security of a land and the cheap, non-polluting source of energy and fertilizer artificial photosynthesis could provide. The recommendations above would also clearly conflict in many instances with the financial plans of large corporations such as Monsanto who have invested in seed technology which requires repurchase by farmers every year.

18.3.3 Ensuring Access to Fisheries

In relation to fisheries the United Nations Special Rapporteur on the Right to Food food found it was urgent that States move towards sustainable resource use while ensuring that the rights and livelihoods of small-scale fishers and coastal communities are respected and that the food security of all groups depending on fish is improved (A/67/268). To reach this objective, he recommended States should take measures that in many cases would clearly cut across the interests of large corporations involved in the fishing industry:

(a) Respect the existing rights of artisanal and small-scale fishing communities;
(b) Refrain from taking measures, including large-scale development projects, that may adversely affect the livelihoods of inland and marine small scale fishers, their territories or access rights, unless their free, prior and informed consent is obtained;
(c) Strengthen access to fishery resources and improve the incomes of small-scale fishing communities by regulating the industrial fishing sector to protect the access rights of traditional fishing communities.
(d) Protect labour rights in the fishing industry;
(e) When engaging in fishing access agreements, agree to introduce provisions concerning conditions of work in the fishing industry and support efforts of coastal States to regulate the fishing practices of industrial vessels operating in exclusive economic zones.
(f) Implement their commitments under the Plan of Implementation of the World Summit on Sustainable Development, including to reduce their fishing capacity and to create marine protected areas;
(g) Implement the Agreement on Port State Measures to Prevent, Deter and Eliminate Illegal, Unreported and Unregulated Fishing; and
(h) Reduce the proportion of fish used for fishmeal purposes (De Schutter 2014).

18.3.4 *Governance Supporting Local Food Systems*

The United Special Rapporteur on the Right to Food found that reinvestment in agriculture and rural development should effectively contribute to the realization of the right to food (De Schutter 2014). In order to achieve this important goal, he recommended that the international community should:

18.3.5 *Prioritising of Public Goods*

(a) Channel adequate support to sustainable farming approaches that benefit the most vulnerable groups and that are resilient to climate change;
(b) Prioritize the provision of public goods, such as storage facilities, extension services, means of communications, access to credit and insurance, and agricultural research;
(c) In countries facing important levels of rural poverty and in the absence of employment opportunities in other sectors, establish and promote farming systems that are sufficiently labour-intensive to contribute to employment creation (A/HRC/13/33/Add.2); and

(d) Ensure that investment agreements contribute to reinforcing local livelihood options and to environmentally sustainable modes of agricultural production.

The system of corporate globalisation, through lobbying, mass advertising, collusion and corruption, has imposed on world wide food production an ideology postulating as the primary social virtue corporate profit maximisation and limitless economic growth, separating human beings from nature and community as it transforms the natural world into commodities. This ideology has its locus in the fundamental neo-classical idea of rational profit maximisation where the utility of nature and people are evaluated by reference to their economic usefulness. This is insidiously self-perpetuating since corporations churn out products that must be consumed while people consume ever more products, thus price-signalling endless demand for yet more products. Water, food, the human genome, cultural knowledge, biodiversity, education, justice, healthcare, even the capacity to live and die peacefully are no longer public goods. Worse still, because the value of an item is measured through the lens of profit-maximisation, traditional "public goods" that do not generate quantifiable profits are therefore considered inimical to future corporate investment, except where reluctantly forced by social policy legislation to do so. The system of corporate globalisation is backed by and is supported through taxpayer subsidies of a powerful military-industrial complex imposing control over natural resources and even thought and free association in the guise of national security. If public goods were prioritised as recommended by the United Nations Special Rapporteur on the right to food, then small-scale farmers would have better preconditions for secure living, particularly if their activities were backed by a local power and food source like artificial photosynthesis.

18.4 Agroecology

The United Nations Special Rapporteur on the right to food recommends:

Moving towards sustainable modes of agricultural production is vital for future food security and an essential component of the right to food. Agroecology has enormous potential in that regard (A/HRC/13/33/Add.2). States should support the adoption of agroecological practices by:

(a) Building on the complementary strengths of seeds-and-breeds and agroecological methods, allocating resources to both, and exploring the synergies, such as linking fertilizer subsidies directly to agroecological investments on the farm ("subsidy to sustainability");
(b) Supporting decentralized participatory research and the dissemination of knowledge by relying on existing farmers' organisations and networks.
(c) Increase the budget for agroecological research at the field level, farm and community levels, and national and sub-national levels; and
(d) Assess projects on the basis of a comprehensive set of performance criteria (impacts on incomes, resource efficiency, impacts on hunger and malnutrition,

empowerment of beneficiaries, etc.) in addition to classical agronomical measures.

The National Gardening Association associates the rise in food gardening to several reasons: An improving economy; strong national leadership, including the launch of the "Let's Move" initiative and White House Kitchen Garden by First Lady Michelle Obama during the time period; action by federal agencies such as the U.S. Department of Agriculture and the U.S. Department of Health and Human Services to increase awareness and educational efforts toward food gardening; and more engagement and public-private partnerships through organizations like the National Gardening Association, to promote and build food gardens in communities across the country.

18.4.1 Support Small-Holder Farmers

The United Nations Special Rapporteur on the right to food recommends:
The realization of the right to food for all will require proactively engaging in public policies aimed at expanding the choices of smallholders to sell their products at a decent price (A/HRC/13/33). To achieve this, States should:

(a) Strengthen local and national markets and support continued diversification of channels of trading and distribution;
(b) Support the establishment of farmers' cooperatives and other producer organizations (A/66/262);
(c) Establish or defend flexible and efficient producer marketing boards under government authority but with the strong participation of producers in their governance;
(d) Encourage preferential sourcing from small-scale farmers through fiscal incentives or by making access to public procurement schemes conditional on the bidders' compliance with certain sourcing requirements.

18.4.2 Contract Farming

The United Nations Special Rapporteur on the right to food recommends:
To ensure that contract farming and other business models support the right to food (A/66/262), Governments should ensure that regulatory oversight keeps pace with the level of the expansion and the complexity of business models. In particular, States should:

(a) Regulate key clauses of contracts, including those concerning price fixing, quality grading and the conditions under which inputs are provided, and the reservation of a portion of land for the production of food crops for self-consumption;

(b) Monitor labour conditions in contract farming;
(c) Link their support for contract farming to compliance with environmental conditions, such as reduced use of chemical fertilizers or the planting of trees, or to the adoption of a business plan that provides for a gradual shift to more sustainable types of farming; and
(d) Set up forums in which the fairness of food chains could be discussed among producers, processors, retailers and consumers to ensure that farmers are paid fair prices for the food they produce.

18.4.3 Agricultural Workers

The United Nations Special Rapporteur on the right to food recommends:
To guarantee that those working on farms can be guaranteed a living wage, adequate health and safe conditions of employment (A/HRC/13/33), States should:

(a) Improve the protection of agricultural workers by ratifying all ILO conventions relevant for the agrifood sector and ensuring that their legislation sets a minimum wage corresponding at least to a "living wage"; and
(b) Monitor compliance with labour legislation by devoting appropriate resources for an effective functioning of labour inspectorates in agriculture, and taking the requisite measures to reduce to the fullest extent possible the number of workers outside the formal economy to ensure that agricultural workers are progressively protected by the same social security schemes applicable to other industries.

Consistent with the mantra of profit-maximisation, the current corporate-dominated global food system promotes deforestation, monocrops, use of chemical fertilisers produced with high levels of carbon-intensive energy, industrial food processing marketing and transportation thousands of kilometres, destruction of aquifers, oceans (through fertilizer run-off, over-fishing and pollution) and ecosystems. It replaces in many instances farming land with biofuel monocultures and replaces sustainable organic farming methods with genetically modified organisms (GMOs) as a tool of corporations to control seeds and food globally.

Trade and investment agreements are a major factor in undermining local economies, food sovereignty, environmental, social and cultural rights and nutrition. Such agreements often also push monopolistic forms of intellectual property protection, denationalization and corporate exploitation of the land. In the intellectual property chapter of the Trans Pacific Partnership Agreement (The TPPA) provisions would require all involved countries to allow animal and plant life to be patented. This is not surprising given that the chief U.S. agricultural negotiator for the TPP is former Monsanto lobbyist, Islam Siddiqui. Monsanto, of course, is a company that has sued farmers over their genetically engineered (GE) seed patents lobbied vigorously over domestic GE labeling regulations.

The TPPA also includes provisions would include a NAFTA-like elimination of virtually all tariffs on U.S. agricultural products in the other signatory nations, which could lead to commodity dumping in those countries by US food corporations and subsequent dislocation of small farmers from their lands. Food safety rules—including rules about pesticide residue levels, labeling of GE ingredients, or limitations on additives—also could be challenged under the Investor-State Dispute Settlement (ISDS) provisions of the TPPA. ISDS is a provision in major free trade agreements that allows companies to sue governments when local laws disrupt trade and profit. Such rules under NAFTA.

Have permitted large US pesticide manufacturers to sue the Canadian government, for example for tightening environmental regulations or subsidizing local production. Such action have resulted in multimillion dollar damages claims and substantial legal costs.

18.5 National Governance Strategies on Food Security

The United Special Rapporteur recommended States should build national strategies for the realization of the right to adequate food, which should include mapping of the food—insecure, adoption of relevant legislation and policies with a right-to-food framework, establishment of mechanisms to ensure accountability, and the establishment of mechanisms and processes which ensure real participation of rights-holders, particularly the most vulnerable, in designing and monitoring such legislation and policies (De Schutter 2014).

18.5.1 Legal Status of Right to Food

For national strategies to be effective, the United Nations Special Rapporteur on the right to food recommended it be:

(a) Grounded in law, through the adoption of right to food/food and nutrition security framework laws and ideally through the inclusion of the right to food in national constitutions;
(b) Multisectoral and inclusive, ensuring the coordination amongst Government ministries and institutions and allowing for meaningful participation of civil society in their formulation and monitoring;
(c) Adequately funded.
(d) Monitored also by national courts and national human rights institutions as well as through social audits and community-based monitoring at the local level.

18.5.2 Human Rights Impact Assessments

The United Nations Special Rapporteur on the right to food recommended:

- To ensure consistency between domestic policies aimed at the full realization of the right to food and external policies in the areas of trade, investment, development and humanitarian aid, States should establish mechanisms that ensure that the right to food is fully taken into account in those policies. The Special Rapporteur has presented Guiding Principles on Human Rights Impact Assessments, based on a range of consultations with governmental and non-governmental actors, which provide guidance as to how to conduct such assessments, both ex-ante and ex-post (A/HRC/19/59/Add.5).

18.5.3 Women's Rights

The United Nations Special Rapporteur on the right to food recommended
In order to strengthen the protection of the right to food of women (A/HRC/22/50), States should:

(a) Remove all discriminatory provisions in the law, combat discrimination that has its source in social and cultural norms, and use temporary special measures to accelerate the achievement of gender equality;
(b) Recognize the need to accommodate the specific time and mobility constraints on women as a result of the existing gender roles, while at the same time redistributing the gender roles by a transformative approach to employment and social protection;
(c) Mainstream a concern for gender in all laws, policies and programs, where appropriate, by developing incentives that reward public administrations which make progress in setting and reaching targets in this regard;
(d) Adopt multisector and multi-year strategies that move towards full equality for women, under the supervision of an independent body to monitor progress, relying on gender-disaggregated data in all areas relating to the achievement of food security.

18.5.4 Social Protection

The United Nations Special Rapporteur on the right to food recommended:
The provision of social protection can substantially contribute to the realization of the right to food (A/68/268, A/HRC/12/31). States should:

(a) Guarantee the right to social security to all, without discrimination, through the establishment of standing social protection schemes;

(b) Ensure that, when targeted schemes are adopted, they are based on criteria that are fair, effective and transparent;
(c) Define benefits under national social protection systems as legal entitlements, so that individual beneficiaries are informed about their rights under social programs and have access to effective and independent grievance redressal mechanisms;
(d) Ensure that the design of social protection schemes is effectively transformative of existing gender roles (A/HRC/22/50); and
(e) Put in place a global reinsurance mechanism, creating an incentive for countries to set up robust social protection programmes for the benefit of their populations.

18.5.5 Nutrition

The United Nations Special Rapporteur on the right to food recommended:

To reshape food systems for the promotion of sustainable diets and effectively combat the different faces of malnutrition (A/HRC/19/59), States should:

(a) Adopt statutory regulation on the marketing of food products, as the most effective way to reduce marketing of foods high in saturated fats, trans-fatty acids, sodium and sugar (HFSS foods) to children, and restrict marketing of these foods to other groups;
(b) Impose taxes on soft drinks (sodas), and on HFSS foods, in order to subsidize access to fruits and vegetables and educational campaigns on healthy diets;
(c) Adopt a plan for the complete replacement of trans-fatty acids with polyunsaturated fats;
(d) Review the existing systems of agricultural subsidies, in order to take into account the public health impacts of current allocations, and use public procurement schemes for school-feeding programmes and for other public institutions to support the provision of locally sourced, nutritious foods; and
(e) Transpose into domestic legislation the International Code of Marketing of Breast-milk Substitutes and the WHO recommendations on the marketing of breast-milk substitutes and of foods and non-alcoholic beverages to children, and ensure their effective enforcement.

The private sector should:

(a) Comply fully with the International Code of Marketing of Breast-milk Substitutes, and comply with the WHO recommendations on the marketing of foods and non-alcoholic beverages to children, even where local enforcement is weak or non-existent;
(b) Abstain from imposing nutrition-based interventions where local ecosystems and resources are able to support sustainable diets, and systematically ensure that such interventions prioritize local solutions;

(c) Shift away from the supply of HFSS foods and towards healthier foods and phase out the use of trans-fatty acids in food processing.

18.6 Reshaping International Food Governance

The United Nations Special Rapporteur recommended the international community should find ways to better manage the risks associated with international trade and ensure that least-developed and net food-importing developing countries are better protected from the volatility of international market prices (De Schutter 2014).

18.6.1 Volatility on International Food Markets

To combat volatility on international markets, he recommended the international community should:

(a) Encourage the establishment of food reserves at the local, national or regional levels;
(b) Improve the management of grain stocks at the global level, including improved information about and coordination of global grain stocks to limit the attractiveness of speculation;
(c) Establish an emergency reserve that would allow the World Food Programme to meet humanitarian needs;
(d) Explore ways to combat unhealthy speculation on the futures markets of agricultural commodities through commodity index funds.

18.6.2 A New Framework for Trade and Investment in Agriculture

The United Nations Special Rapporteur on the right to food recommended:

The realization of the right to food requires designing trade rules that support the transition toward more sustainable agricultural practices. The multilateral trade regime as well as regional and bilateral trade agreements must allow countries to develop and implement ambitious food security policies including public food reserves, temporary import restrictions, active marketing boards, and safety net insurance schemes, in support of the progressive realization of the right to food (A/HRC/10/5/Add.2). In this regard, States should:

(a) Limit excessive reliance on international trade and build capacity to produce the food needed to meet consumption needs, with an emphasis on small-scale farmers;

(b) Maintain the necessary flexibilities and instruments, such as supply management schemes, to insulate domestic markets from the volatility of prices on international markets; and
(c) Encourage national parliaments to hold regular hearings about the positions adopted by the government in trade negotiations, and ensure that their undertakings under the WTO framework are fully compatible with the right to food;
(d) Fully implement the Marrakesh Ministerial Decision on Measures concerning the possible negative effects of the reform programme on least developed and net food-importing developing countries (NFIDCs) and, in order for it to be fully effective, ensure that it include a mechanism to systematically monitor the impact of the Agreement on Agriculture reform process on NFIDCs.

18.6.3 Regulating Agribusiness

The United Nations Special Rapporteur on the right to food recommended:

States should take steps towards the establishment of a multilateral framework regulating the activities of commodity buyers, processors, and retailers in the global food supply chain, including the setting of standards by these actors and their buying policies (A/HRC/13/33). In particular, States should use competition law in order to combat excessive concentration in the agribusiness sector. This requires having in place competition regimes sensitive to excessive buyer power in the agrifood sector, and devising competition authorities with mechanisms that allow for affected suppliers to bring complaints without fear of reprisal by dominant buyers.

Private actors of the agribusiness sector should refrain from practices that constitute an undue exercise of buyer power, as identified by the States in which they operate, and should:

(a) Seek to conclude international framework agreements with global unions;
(b) Consider unilateral undertakings to monitor compliance with ILO standards in the supply chain, while supporting their suppliers in achieving compliance;
(c) Engage in chain-wide learning to assure that participation in the chain is profitable for all involved, including small-scale producers;
(d) Involve smallholders in the elaboration of and compliance with food safety, labour or environmental standards; and
(e) Promote fair trade through increased shelf space and information campaigns.

18.6.4 Agrofuels

The United Nations Special Rapporteur on the right to food recommended:

The international community should reach a consensus on agrofuels, based not only on the need to avoid the negative impact of the development of agrofuels on the international price of staple food commodities, but also on the need to ensure that the production of agrofuels respects the full range of human rights and does not result in distorted development in producer countries. Public incentives for the production of crop-based biofuels must be reduced and eventually removed, while only those advanced biofuels that do not compete with food production for land or other resources should be incentivised.

18.6.5 Food Aid and Development Cooperation

The United Nations Special Rapporteur on the right to food recommended:

International aid remains an important component of the right to food (A/HRC/10/5). Donor States should:

(a) Maintain and increase levels of aid calculated as Official Development Assistance as a percentage of GDP;
(b) Provide food aid on the basis of an objective assessment of the identified needs in developing countries;
(c) Fully respect the principle of ownership in their development cooperation policies by aligning these policies with national strategies for the realization of the right to food;
(d) Promote the right to food as a priority for development cooperation.
(De Schutter 2014)

The United Nations Conference on Trade and Development (UNCTAD) report on reforming corporate sovereignty chapters stated:

> Concerns with the current ISDS system relate, among others things, to a perceived deficit of legitimacy and transparency; contradictions between arbitral awards; difficulties in correcting erroneous arbitral decisions; questions about the independence and impartiality of arbitrators, and concerns relating to the costs and time of arbitral procedures. (UNCTAD 2013)

18.7 Application Through Examples and Case-studies

18.7.1 Dow Agrosciences Investor-State Dispute Case

In 2009, the US corporation Dow Agrosciences sued the Canadian federal government using NAFTA's investor-state dispute process over Quebec's ban on the residential use of the dangerous herbicide 2,4-D, claimed $2 million in damages from the ban, which it called tantamount to expropriation. The Quebec government agreed to a statement that "products containing 2,4-D do not pose an unacceptable risk to human health or the environment, provided that the instructions on their label are followed" although the pesticide bans in Quebec and other jurisdictions will stay in place, Dow Agrosciences declared the settlement a victory, as commentators note the NAFTA case may discourage other Canadian governments at the local or provincial level from moving ahead with future public health and environmental regulation that may impede the profits of US corporations. The case showed how ineffective were public health or environment exceptions to investor rights in such agreements (Public Citizen 2010, 2013, 2014).

There are claims that recent changes to the wording of ISDS clauses in trade and investment agreements like the Korea-Australia Free Trade Agreement (KAFTA) are "safeguards" which will prevent foreign investors from suing governments over health, environment or other public interest legislation.

But the first "safeguard" sentence in the KAFTA reads: "except in rare circumstances non-discriminatory regulatory actions by a party that are designed and applied to protect legitimate public welfare objectives, such as public health, safety and the environment, do not constitute indirect expropriations" (KAFTA chapter 11, annex 2B). Many legal experts have pointed out that the phrase "except in rare circumstances" leaves a very big loophole, which recent cases have used to advantage. The second "safeguard" is a more limited definition of "fair and equitable treatment" for foreign investors (KAFTA chapter 11, clause 11.5.2 and Annex 2A). However tribunals have ignored these limitations and applied the previous higher standard. A third "safeguard" is a reference to the general protections for "human, animal or plant life" in article XX of the WTO General agreement on Tariffs and Trade (KAFTA Article 22.1). This article has only been successful in one out of 35 cases in the WTO which have attempted to use it to safeguard health and environmental legislation.

These same "safeguards" in recent trade agreements like the Central American Free Trade Agreement and the Peru-US Free Trade Agreement have not prevented foreign investors from launching cases against environmental legislation. For example, the US-based Renco Group is using ISDS in the Peru-US free Trade Agreement to contest a local court decision that it was responsible for pollution from its lead mine. Both cases are ongoing and may take several years (Public Citizen 2010, 2013, 2014).

This example shows government measures to promote or subsidise artificial photosynthetic technology could be blocked using ISDS mechanisms even though governments might claim there were scientifically sound public health and environmental reasons behind the implementation of the former strategies.

18.7.2 AbitibiBowater Investor-State Dispute Case

The Canadian federal government also had to pay US$130-million in damages under a NAFTA Chapter 11 ISDS claim to compensate US forestry corporation AbitibiBowater Inc. (incorporated in Delaware but with its head office in Montreal) for the actions of Canadian provincial governments Newfoundland and Labrador's over their alleged 'expropriation' of the corporation's assets (water and timber rights and hydroelectric assets) after the US company closed its mill in Grand Falls-Windsor, putting about 800 employees out of work. The Canadian government found it easier to pay the compensation rather than fight the case through the complex NAFTA ISDS process (Public Citizen 2010, 2013, 2014).

The Swedish energy company, Vattenfall, suing the German government over its decision to phase out nuclear energy. The Government of El Salvador has been sued by Pacific Rim Mining Corporation under the Central American Free Trade agreement, over a ban on mining to protect the nation's limited groundwater resources. US Lone Pine mining company suing the Québec provincial government of Canada over environmental regulation of shale gas mining (Public Citizen 2010, 2013, 2014).

This example also shows government measures to promote or subsidise artificial photosynthetic technology could be blocked using ISDS mechanisms even though governments might claim there were scientifically sound public health and environmental reasons behind the implementation of the former strategies.

18.7.3 Can ISDS Be Reformed to Not Limit the Capacity of Artificial Photosynthesis to Transform the Global Food Industry?

ISDS enables foreign investors to sue governments for compensation in an international tribunal if they can claim that a domestic law or policy "harms" their investment, not just by actual expropriation but by "indirect expropriation" of their investments defined in a very broad sense. The OECD estimates an average of $8 million per ISDS case, with some cases costing up to $30 million) and the compensation awarded to foreign investors, (often hundreds of millions and in some cases billions of dollars) can discourage governments from proceeding with legitimate domestic legislation. The highest compensation award so far is $1.8 billion against the government of Ecuador. The disputes are heard by international investment tribunals, which lack the safeguards of national legal systems, such as public disclosure, judicial independence a body of jurisprudence requiring the following of precedents encouraging certainty and predictability in the application of law, coherence with laws emerging from democratic processes such as legislation and referenda.

A new industry has arisen where corporate third parties invest in a claim in return for a portion of the damages, has been encouraged by large investment law firms which actively solicit business and encourage large claims.

Increasing numbers of governments are reviewing and terminating their involvement in ISDS. These include members of the European Union like France and Germany (European Parliamentary Research Service 2014) Brazil, Argentina and eight other countries in Latin America, India and South Africa. Indonesia has recently announced it will terminate all 67 of bilateral investment treaties (Bland

and Donnan 2014; Gaukrodger and Gordon, OECD, 2012, p. 7, European Parliamentary Research Service 2014. p. 2, Bland and Donnan 2014).

In June 2013 the United Nations Conference on Trade and Development (UNCTAD) issued a report on reforming investor-state dispute settlement which highlighted concerns including a "perceived deficit of legitimacy and transparency," "contradictions between arbitral awards," "difficulties in correcting erroneous arbitral decisions," "questions about the independence and impartiality of arbitrators," and "concerns relating to the costs and time of arbitral procedures" Recommendations for reform included (1) Promoting alternative dispute resolution (2) Tailoring the existing system through individual IIAs (3) Limiting investor access to ISDS (4) Introducing an appeals facility and (5) Creating a standing international investment court. These are all measures worth examining if ISDS is not to inhibit the capacity of globalized artificial photosynthesis to transform the global food industry.

18.8 Planetary Nanomedicine and the Sustainocene

18.8.1 Planetary Medicine

Planetary medicine is now a growing field in which the expertise of medical professionals in directed towards issues of global health and environmental protection, particularly including climate change (Vines et al. 2013; Faunce 2012a). A Global Artificial Photosynthesis (GAP) Project could well be promoted through domestic and international media as a defining symbolic endeavour of planetary nanomedicine (Faunce 2010a, b). One significance of this for artificial photosynthesis researchers is that funding agencies respond indirectly to public and governmental national interest concerns and nanotechnology, despite its great promise, still has a problematic place in the popular imagination owing to safety issues. A GAP Project therefore represents an excellent opportunity to create a high profile awareness of nanotechnology as a positive contributor to overcoming major contemporary public health and environmental problems.

The process of photosynthesis is as central to life on earth as DNA; thus there are likely to be similar major debates over whether patents should be allowed over any part of the photosynthetic process. Such a debate will be unlikely to inhibit patents being taken out over many aspects of a functional artificial photosynthetic process.

A larger issue for such governance approaches is that nanotechnology, despite its great scientific novelty and promise, still has a problematic place in the popular imagination owing to unresolved safety issues (Faunce 2008). A macroscience project to promote equitable global use of artificial photosynthesis therefore represents an excellent opportunity to create a high profile awareness of nanotechnology as a positive contributor to overcoming major contemporary public health and environmental problems. Provided an appropriate ethical regulatory structure was in place, such a project could well be promoted through domestic and international media as a defining symbolic endeavour of planetary nanomedicine.

18.8.2 The Sustainocene

One suggestion is that to achieve a transition from Corporatocene governance models, which conceptualises nature, people and animals as inputs to be exploited in the pursuit of profit maximisation, international food law and policy needs to become more coherent with a vision for humanity that is congruent with its most noble aspiration, resembling the process that occurred after WWII with the establishment of the international human rights regime. One such vision is encompassed by the term Sustainocene.

The term 'Sustainocene' was coined by the Canberra-based Australian physician Furnass (2012). It has been developed to encompass a billion year period (if humanity ultimately is to justify a characterisation as an ethical species and repay the evolutionary legacy of life on this planet) where governance structures and scientific endeavour coordinate to achieve the social virtues of ecological sustainability and environmental integrity (Faunce 2012c). It is a vision coherent with influential writings such as those of eco-economists such as the EF Schumacher (with his concept of 'small (and local) is beautiful') and Kenneth Boulding (with his idea of 'Spaceship Earth' as a closed economy requiring recycling of resources) as well as Herman Daly with his notion of 'steady state' economies drawing upon the laws of thermodynamics and the tendency of the universe to greater entropy (dispersal of energy) (Faunce 2012c, d).

One area of academic research and policy development that fits well with "Sustainocene' thinking is that centred on the idea that this planet should be treated not just as a distinct living entity (James Lovelock's Gaia Hypothesis), but as a patient (Lovelock 1991). 'Planetary medicine' as this field has become known has become a symbolic rubric focusing not just public and governmental attention on the interaction between human health, technological development and sustainability of the biosphere (Faunce 2012a). In this emerging discipline, characteristic features of the Corporatocene epoch such as climate change and environmental degradation, as well as gross societal imbalances in poverty as well as lack of necessary fuel, food, medicines, security and access to nature, are targeted as intrinsically global pathologies the resolution of which requires concerted efforts to implement a wide range of not just renewable energy technologies (such as nanotechnology-based artificial photosynthesis as will be discussed later) but ethical principles including those related to protecting the interests of future generations and preservation of biodiversity (Faunce 2012a).

18.9 Conclusion[1]

Moving into the twenty-first Century, humanity is faced with a stark choice in its ongoing governance of the environment. Since the Industrial Age, the dominant management model has been one of continuous exploitation where the environment, minerals, food products, animals and even people have been characterised and often used as mere inputs to profit. The gradual rise of corporations through the nineteenth century has effortlessly facilitated the pursuit of profit maximisation as the mantra of neo-classical economics. In setting the profit motive as the highest "good", corporations have thus set themselves against all those members of society that do not consume; the unemployed, the elderly, the animals and most particularly natural resources themselves.

These seeds of corporate design came to full flower in the twentieth century with the rise of global agribusinesses, State-corporate partnerships and market ideologies that permitted corporate concentration in key industries to generate anticompetitive and deceptive practices to the detriment of consumers, animals and the environment.

While States are attempting to mitigate the most egregious forms of market manipulation and consumer deception (Bruce 2013), a far more wide-ranging philosophical and culture shift in corporate governance is required. We believe a focus on global artificial photosynthetic technology can form a key element of the required shift.

However, no matter how significant the vision or advanced the science, the governance challenges of moving to a Global Artificial Photosynthesis (GAP) Project are considerable (Faunce 2012a). One model of a Sustainocene powered by solar fuels involves bio-mimetic polymer photovoltaic generators plugged in to the national electricity grid to power hydrogen fuel and waterless agriculture, chemical feedstocks and polymers for fibre production. This model has the advantage of the 'light' and 'dark' reactions being uncoupled in relation not only to energy/material flow balance, but also to the requirement to be co-located in space. Such an uncoupling will vastly extend the area for capturing light over otherwise barren land, and also allow the elimination or reduction of molecular oxygen in solar fuel reactions, enhancing longevity of the components. A model which the author favours emphasizes the greater potential for individual and community economic autonomy implicit in micro or local generation of fuel and food through solar fuel products installed as a policy priority on domestic dwellings and vehicles (Vines et al. 2013; Faunce 2012b). Large GSF facilities providing fuel for industry or backup supply can still be preferentially located under such a model near large sources of seawater, CO_2, waste heat, high solar irradiation and proximity to end use facilities. In the longer term every human engineered structure on the planet will have a built in artificial photosynthetic capacity allowing it to be a positive contributor to the biosphere—improving the atmosphere, providing fuel and basic

---

[1]The views and conclusions in this chapter are solely those of the authors, *not* the editors.

food and fertilizer. There is a simple public policy message at the core of a vision such as that of the Sustainocene. It involves telling people that nanotechnology will be used to make buildings function like trees. A device that can do this and is available to cheap purchase and installation, like the mobile phone or internet, could rapidly transform society to one that is more community and values-oriented.

References

Baines J (2014) Food price inflation as redistribution: towards a new analysis of corporate power in the world food system. New Polit Econ 19:79–112
Ball P (2010) Material witness: quantum leaves in fact and fiction. Nat Mater 9:614–621
Barnes DKA, Galgani F, Thompson RC, Barlaz M (2009) Accumulation and fragmentation of plastic debris in global environments. Philos Trans R Soc B Biol Sci 364(1526):1985–1998
Bland B, Donnan S (2014) Indonesia to terminate more than 60 bilateral investment treaties. Financial Times, March 27. http://www.ft.com/intl/cms/s/0/3755c1b2-b4e2-11e3-af92-00144feabdc0.html?ftcamp=published_links/rss/asiapacific/feed//product&siteedition=uk#axzz2x5CRbtpN
Blankenship RE (2002) Molecular mechanisms of photosynthesis. Blackwell Science, Oxford/Malden
Blankenship RE, Tiede DM, Barber J, Brudvig GW, Fleming G, Ghirardi M, Gunner MR, Junge W, Kramer DM, Melis A, Moore TA, Moser CC, Nocera DG, Nozik AJ, Ort DR, Parson WW, Prince RC, Sayre RT (2011) Science 332(6031):805. doi:10.1126/science.1200165
Bruce A (2013) Australian Competition Law, 2nd edn. Lexis Nexus, Sydney
Campbell B, Ingram J (2012) Climate change and food systems. Annu Rev Environ Res 37:195–222
Carmieli R, Mi Q, Ricks AB, Giacobbe EM, Mickley SM, Wasielewski MR (2009) Direct measurement of photoinduced charge separation distances in donor-acceptor system for artificial photosynthesis using OOP-ESEEM. J Am Chem Soc 131:8372–8378
CFS (2013) Committee on World Food Security, report on the fortieth session, Rome, 7–11 October 2013 (CFS 2013/40 REPORT), para. 81
Chislock MF et al (2013) Eutrophication: causes, consequences, and controls in aquatic ecosystems. Nat Edu Knowl 4(4):10
Cordell D, Drangert J-O, White S (2009) The story of phosphorus: global food security and food for thought. Glob Environ Chang 19:292–305
Crutzen PJ (2002) Geology of mankind: the anthropocene. Nature 415:23
Crutzen PJ, Stoermer EF (2000) The 'Anthropocene'. Glob Change Newslett 41:17–18
De Schutter O (2014) Report of the Special Rapporteur on the right to food, Final report: The transformative potential of the right to food United Nations General Assembly A/HRC/25/57 Jan (2014)
E4tech, Element Energy, Eoin Lees Energy, A Strategic Framework for Hydrogen Energy in the UK. http://www.berr.gov.uk/files/file26737.pdf
Engel GS, Calhoun TR, Read EL, Ahn T-K, Mancal T, Cheng Y-C, Blankenship RE, Fleming GR (2007) Evidence for wavelike energy transfer through quantum coherence in photosynthetic systems. Nature 446:782–793
European Hydrogen and Fuel Cell technology Platform. http://www.hfpeurope.org/hfp/keydocs
European Parliamentary Research Service (2014) Investor-State Dispute Settlement (ISDS): state of play and prospects for reform. European Parliamentary Research Briefing, Brussels. January. www.europarl.europa.eu/.../LDM_BRI(2014)130710_REV2_EN.pdf
FAO (2006) Livestock's long shadow: environmental issues and options. Rome

FAO (2013), International Fund for Agricultural Development and World Food Programme, The State of Food Insecurity in the World 2013: The Multiple Dimensions of Food Security. Rome, p. 8

FAO (2014) FAO Year of Family Farming. http://www.fao.org/family-farming-2014/en/

Faunce TA (2008) Toxicological and public good considerations for the regulation of nanomaterial-containing medical products. Expert Opin Drug Saf 7(2):103–106

Faunce TA (2010) 15th International Congress of Photosynthesis. Conference Proceedings ISPR, Beijing, August 2010

Faunce TA (2010) Nanotechnology for Sustainable Energy Conference sponsored by the European Science Foundation, July 2010. Obergurgl, Austria

Faunce TA (2012a) Governing planetary nanomedicine: environmental sustainability and a UNESCO universal declaration on the bioethics and human rights of natural and artificial photosynthesis (global solar fuels and foods). Nanoethics 6(1):15–27. doi:10.1007/s11569-012-0144-4

Faunce TA (2012b) Ch 21. Future perspectives on solar fuels. In: Wydrzynski T, Hillier W (eds) Molecular solar fuels book series: energy. Royal Society of Chemistry, Cambridge, pp 506–528

Faunce TA (2012c) Towards a global solar fuels project- artificial photosynthesis and the transition from anthropocene to sustainocene. Proc Eng 49:348–356

Faunce TA (2012d) Nanotechnology for a sustainable world. global artificial photosynthesis as the moral culmination of nanotechnology. Edward Elgar, Cheltenham

Faunce TA, Styring S, Wasielewski MR, Brudvig GW, Rutherford AW, Messinger J, Lee AF, Hill CL, deGroot H, Fontecave M, MacFarlane DR, Hankamer B, Nocera DG, Tiede DM, Dau H, Hillier W, Wang, Amal R (2013a) Artificial photosynthesis as a frontier technology for energy sustainability. Energy Environ Sci 6:1074–1076. doi:10.1039/C3EE40534F

Faunce TA, Lubitz W, Rutherford AW, MacFarlane D, Moore GF, Yang P, Nocera DG, Moore TA, Gregory DH, Fukuzumi S, Yoon KB, Armstrong FA, Wasielewski MR, Styring S (2013b) Energy and environment policy case for a global project on artificial photosynthesis. Energy Environ Sci 6(3):695–698. doi:10.1039/C3EE00063J

Furnass B (2012) From Anthropocene to Sustainocene. Challenges and Opportunities. Public Lecture. Australian National University, 21 March 2012

GAP (2011) Towards Global Artificial Photosynthesis: Energy, Nanochemistry and Governance. Lord Howe Island. Conference proceedings. http://law.anu.edu.au/coast/tgap/conf.htm

GAP (2014) UK Royal Society 2014 Chicheley Hall. https://royalsociety.org/events/2014/artificial-photosynthesis-global-project/

Goodland R, Anhang J (2009) Livestock and climate change: what if the key actors in climate change are cows, pigs, and chickens?. World Watch

Gray HB (2009) Powering the Planet with Solar Fuel Nature Chem 1(1):7–18

IPCC (2007) Climate Change 2007: Synthesis Report. Contribution of Working Groups I, II and III to the Fourth Assessment. Report of the Intergovernmental Panel on Climate Change In: Core Writing Team, Pachauri RK, Reisinger A (eds) IPCC, Geneva, Switzerland

IPPC (2014) Climate change 2014: impacts, adaptation, and vulnerability. IPCC WGII AR5 Summary for Policymakers, Geneva, Switzerland

Kanan MW, Nocera DG (2008) In situ formation of an oxygen-evolving catalyst in neutral water containing phosphate and carbon dioxide. Science 321:1072–1078

Koder RL et al (2009) Design and engineering of an O2 transport protein. Nature 458:305–311

Lappé FM et al (2013) How we count hunger matters. Ethics Int Aff 27.3

Lee H, Cheng Y-C, Fleming GR (2007) Coherence dynamics in photosynthesis: protein protection of excitonic coherence. Science 316:1462–1468

Lobell D, Schlenker W, Costa-Roberts J (2011) Climate trends and global crop production since 1980. Science 333(6042):616–620

Lovelock JE (1991) Gaia, the practical science of planetary medicine. Gaia Books, London

MacKay DJC (2009) Sustainable energy-without the hot air. UIT, Cambridge, pp 43–44

Magnuson A, Anderlund M, Johansson O, Lindblad P, Lomoth R, Polivka T, Ott S, Stensjö K, Styring S, Sundström V, Hammarström L (2009) Biomimetic and microbial approaches to solar fuel generation. Accounts Chem Res 42(12):1899–1918

Mellor JW (1988) Global food balances and food security. World Dev 16(9):997–1011

Moore CJ, Moore SL, Leecaster MK, Weisberg SB (2001) A comparison of plastic and plankton in the north pacific central gyre. Mar Pollut Bull 42(12):1297–1300. doi:10.1016/S0025-326X(01)00114-X

NH3 Fuel Association. http://nh3fuelassociation.org/

Patel R (2008) Stuffed and starved: markets, power and the hidden battle for the world food system. Portobello Books, London

Pittock AB (2009) Climate change. the science, impacts and solutions, 2nd edn. CSIRO Publishing, Collingwood, p 177

Public Citizen (2010) CAFTA Investor Rights Undermining Democracy and the Environment: Pacific Rim Mining Case, Washington found at http://www.citizen.org/documents/Pacific_Rim_Backgrounder1.pdf

Public Citizen (2013) Only One of 35 Attempts to Use the GATT Article XX/GATS Article XIV "General Exception" Has Ever Succeeded. https://www.citizen.org/documents/general-exception.pdf

Public Citizen (2014) Table of Foreign Investor-State Cases and Claims under NAFTA, August, Washington, found at http://www.citizen.org/documents/investor-state-chart.pdf

Reece SY, Hamel JA, Sung K, Jarvi TD, Esswein AJ, Pijpers JJH, Nocera DG (2011) Science 334:645. doi:10.1126/science.1209816

Rogner HH (2004) Ch 5 in United Nations Development World Energy Assessment. United Nations, Geneva, p 162

Salamanca-Buentello F et al (2005) Nanotechnology and the developing world. PloS Med 2:e97

Sartbaeva A, Kuznetsov VL, Wells SA, Edwards PP (2008) Hydrogen nexus in a sustainable energy future. Energy Environ Sci 1:79–85

Sgobba V, Guidi DM (2009) Carbon nanotubes-electronic/electrochemical properties and application for nanoelectronics and photonics. Chem Soc Rev 38:165–172

Smil V (2004) Enriching the earth: fritz haber, Carl Bosch, and the transformation of world food production. MIT Press, Cambridge

Steffen W, Crutzen PJ, McNeill JR (2007) The anthropocene: are humans now overwhelming the great forces of nature? AMBIO A J Hum Environ 36(8):614–621

Steffen W, Persson A, Deutsch L, Zalasiewicz J, Williams M, Richardson K et al (2011) The anthropocene: from global change to planetary stewardship. AMBIO A J Hum Environ 40(7):739–761

Stern N (2007) The economics of climate change: the stern review. Cabinet Office HM – Treasury. Cambridge University Press, Cambridge

Sun S, Liu Chong C, Yang P (2011) J Am Chem Soc 133(48):19306. doi:10.1021/ja2083398

Umena Y, Kawakami K, Shen J-R, Kamiya N (2011) Nature 473:55. doi:10.1038/nature09913

UNCTAD (2013) Reform of Investor-State Dispute Settlement UNCTAD June 2013. http://unctad.org/en/PublicationsLibrary/webdiaepcb2013d4_en.pdf

United Nations Committee on Trade and Development, (UNCTAD) (2013a) Recent Developments in Investor-State Dispute Settlement, IIA Issues Note, UNCTAD, May, found at http://unctad.org/en/PublicationsLibrary/webdiaepcb2013d3_en.pdf

United Nations Committee on Trade and Development, (UNCTAD) (2013b) Reform of investor state dispute settlement: in search of a roadmap, IIA Issues Note, UNCTAD, June, found at http://unctad.org/en/PublicationsLibrary/webdiaepcb2013d4_en.pdf

United Nations Conference on Sustainable Development (2013) "The future we want" (outcome document), para. 108

United Nations. Millennium Development Goals. http://www.un.org/millenniumgoals/

US Department of Energy: Hydrogen Posture Plan. http://www.hydrogen.energy.gov/

Utschig LM, Silver SC, Mulfort KL, Tiede DM (2011) J Am Chem Soc 133(41):16334. doi:10.1021/ja206012r

Vines T, Bruce A, Faunce TA (2013) Planetary medicine and the *Waitangi Tribunal Whanganui River Report*: global health law embracing ecosystems as patients. J Law Med 20:528–541

Wasielewski M (2006) Energy charge and spin transport in molecules and self-assembled nanostructures inspired by photosynthesis. J Org Chem 71:5051–5058

Yin Q, Tan JM, Besson C, Geletti YV, Musaev DG, Kuznetsov AE, Luo Z, Hardcastle KI, Hill CL (2010) A fast soluble carbon-free molecular water oxidation catalyst based on abundant. Metals Sci 328:342–349

Zhu G, Geletii YV, Kögerler P, Schilder H, Song J, Lense S, Zhao C, Hardcastle KI, Musaev DG, Hill CL (2012) Dalton Trans 41(7):2084. doi:10.1039/c1dt11211b

Part III
European Food Law

Raymond O'Rourke

This section is vital to understand the European precautionary principle, and especially to compare EU food law and regulation to the US and BRICS countries. The Treaty of Rome (1958) made no explicit mention of consumer protection or public health and these goals were only subsequently added as amendments to Article 3 by the Single European Act and the Maastricht Treaty. In that case, European Food Law developed piece-meal over a long period of time. In the early years, there was no central unifying text setting out the fundamental principles of European Food Law that would clearly define the obligations of those involved at every stage of the food production chain. The first EU food Directive, which was concerned with colours in foodstuffs, was adopted by the Council of Ministers in 1962. The Directive had no EEC number because at that time there was no formal system for the numbering of EU laws. This was not an auspicious start for the establishment of European Food Law as a major concern of EU legislators.

This part of the book demonstrates clearly that European Food Law has come a long way and is now a major facet of the EU's *acquis communitaire*. The initial approach of the European Commission in relation to European Food Law was to concentrate on the obligations enshrined in Article 3 of the Treaty of Rome to ensure the free movement of foodstuffs throughout the common market. The famous *Cassis de Dijon* ECJ case was the embodiment of this approach establishing the 'mutual recognition' principle whereby a foodstuff marketable in a Member State should be marketable in all other EU Member States. Trade is still an important aspect of European Food Law and this is covered by Dr. Roland Kölcsey and Dr. Livia Hullo in Chap. 21. This emphasis on trade was undermined by the BSE and Belgian Dioxin crisis. After much debate amongst Member States and the European Parliament the outcome was EU Regulation 178/2002 which established General Principles of European Food Law, re-orientated the Rapid Alert System and established the European Food Safety Authority (EFSA) in Parma. These major developments in European Food Law are covered by Dr. Gunnar Sachs and Avv. Giorgio Rusconi in Chaps. 19 and 20.

Other interesting aspects of European Food Law are covered in subsequent chapters. New issues that are being discussed in relation to European Food Law such as food waste and transparency issues are covered respectively by Prof. Dr. Mr. H.J. Bremmers, Prof. Dr. Bernd van der Meulen in Chap. 24 and by Atty. Nicola Conte-Salinas and Atty. Rochus Wallau in Chap. 25. No book dealing with European Food Law could neglect the issue of Food Labelling and this issue is covered by Atty Nicola Conte-Salinas and Avv. Aude Mahy in Chap. 22. In Chap. 23, Atty. Raymond O' Rourke discusses the role of the consumer in European Food Law and contrasts the maximum harmonisation approach of the EU in relation to Food Law to the minimum harmonisation approach to Consumer Law.

The last section of this part covers food laws in different European countries—some are EU Member States others are not. Adv Magnus Friberg covers Sweden in Chap. 26; Dr. Gunnar Sachs covers Germany within the wider European context in Chap. 19; Avv. Giorgio Rusconi covers Italy in the context of food safety in Chap. 20; and Atty. Marie Vaale-Hallberg and Atty. Nina Lindbach cover Norway in Chap. 27.

This section of the book demonstrates clearly that European Food Law has become in the wake of the BSE & Belgian Dioxin crisis a major tenet of EU Law. It is now on a par with EU Competition Law or EU Environmental Law. Food safety in the guise of consumer protection and public health has become the defining reason for any new prospective EU food regulation or directive. Trade has not been forgotten and still plays a major role in relation to European Food Law in particular in the international field. There have been important WTO cases taken against Europe over cattle growth promoting hormones, GMOs and chlorinated chicken.

These cases demonstrate a difference between a science-based food law model in the USA & Canada and a science-based plus consumer protection food law model in the European Union. This divergence is aptly demonstrated in the present on-going negotiations between the USA and EU over a Transatlantic Trade and Investment Partnership (TTIP).

Many in Europe in particular consumer and environmental NGOs and a large number of Members of the European Parliament and some Member States are concerned TTIP's "regulatory convergence" agenda will seek to bring EU standards on food safety closer to those of the USA. However, the USA does not demand mandatory labelling of foodstuffs containing genetically modified ingredients; has far laxer restrictions on the use of pesticides; permits the use of growth hormones in its beef which are banned in Europe—there is a concern that in an effort to enhance EU-US trade the emphasis on consumer protection and public health in European Food Law will be watered down. Without understanding the origin and basis of European Food Law as detailed in this part, you will not be able to understand why there is such a divergence between the USA and EU on this area of law—an area of law few would initially think could be so controversial.

Chapter 19
Introduction to European Food Law and Regulation

Gunnar Sachs

Abstract European food law provides for strict legal standards for foodstuffs, food ingredients and food packaging as well as for food labeling and advertising. Food manufacturers and distributors are therefore well advised to acquaint themselves with the applicable legal prerequisites before placing their products on the European market. Non-compliance with the pertinent requirements can have severe consequences, ranging from administrative fines via product withdrawals or recalls up to sensitive product liability. Relevant principles of European food law will be discussed and presented, including some fundamental rules on food labeling, packaging, hygiene, advertising and marketing, ingredients, required authorizations, health and nutrition claims, product liability and recalls.

19.1 Introduction

With more than 500 million citizens in 28 Member States, the European Union is one of the largest food markets in the world. According to the European Commission, the huge size of the European common market enables the food industry to constantly increase their productivity and to make more effective use of economies of scale.[1] Therefore, it is important to understand European food law, regulation and policy as part of the global food law system.

---

[1] Commission of the European Communities (2009).

G. Sachs (✉)
Attorney-at-law, Maître en droit (Paris), Clifford Chance, Düsseldorf, Germany
e-mail: Gunnar.Sachs@CliffordChance.com

19.1.1 The European Food Market: Facts and Figures

Today, the food industry is the second largest sector in the European manufacturing industry with a market share of more than 12 % and more than 14 % of the European manufacturing turnover. Germany, France, Italy, Spain and the UK account for 70 % of the food industry's total turnover. More than 310,000 companies belong to the food sector and provide jobs for more than 4 million people. The European food market had a stable growth in the last 15 years in both production and value added.[2]

The European food industry is the largest food exporter in the world with more than 20 % share in world exports and the second largest food importer with more than 18 % share in world imports. As to value added growth in food industries, the European food market ranks directly behind Brazil and China and before Australia, the United States of America, Canada and Japan.[3]

The most important food processing sub-sectors in Europe are, in order of their respective market shares, the meat industry followed by grain based, starch and further processed products, dairy produce, sugar and sugar based products, processed fruits and vegetables, oils and fats as well as fish and seafood. The European beverage industry is a global market leader with the largest production in the world and nearly 75 % share in world exports.[4]

19.1.2 Europe: A Highly Regulated Market

Europe is a highly regulated market. All consumer products including foodstuffs, food supplements, food contact materials, feedstuffs, pharmaceuticals, medical devices, cosmetics, biocides, textiles, electronic devices, toys, baby and children's articles are subject to harmonized rules providing for the same legal standards in all European Member States. As to foodstuffs and food supplements, there are harmonized requirements regarding, amongst others, product presentation and labeling, packaging, hygiene, advertising and marketing, ingredients, additives, flavorings and enzymes, authorization proceedings, good manufacturing practice, safety assessment, borderline products, functional food, dual use, health and nutrition claims, product risk management, consumer information as well as market surveillance.

There are four central principles of harmonized European food law. First, there is an assurance of a high level of protection of human health and consumers' interest in relation to food. Secondly, the variety in the supply of food, including

---

[2]Ibid and European Commission (2014).
[3]Commission of the European Communities (2009).
[4]Ibid.

traditional products is taken into account. Third, the law ensures the effective functioning of the internal market. Lastly, there is an approximation and resemblance of the laws, regulations and administrative provisions of the Member States.[5]

In view of these principles, European food law and policy provide for strict harmonized standards for foodstuffs and food supplements. Food manufacturers and distributors should, therefore, acquaint themselves with the applicable requirements before placing their products on the European common market. In the following sections, some major prerequisites of European food law and policy are discussed and presented, including rules on food hygiene, ingredients, labeling, packaging, authorization procedures, advertising and consumer rights.

19.2 Food Hygiene

In Europe, food manufacturers and distributors are subject to a strict regulatory regime of harmonized rules on food hygiene that are either directly applicable in the European Member States or have been transposed into their respective national laws. In 2004, the European legislator has fundamentally reformed the pan-European law on food hygiene by adopting three major laws. These laws are Regulation (EC) No. 852/2004 on the hygiene of foodstuffs; Regulation (EC) No. 853/2004 which sets specific hygiene rules for food of animal origin; and Regulation (EC) No. 854/2004 which states specific rules for the organization of official controls on products of animal origin intended for human consumption.

All three regulations are part of the so-called European "hygiene package" providing hygiene rules for foodstuffs. The "hygiene package" is supplemented by additional laws. The Regulation (EC) No. 178/2002 asserts the general principles of food law, establishing the European Food Safety Authority and laying down procedures in matters of food safety. The Regulation (EC) No. 882/2004 establishes the official controls performed to ensure the verification of compliance with feed and food law, animal health and animal welfare rules. The Directive 2002/99/EC describes the animal health rules governing the production, processing, distribution and introduction of products of animal origin for human consumption whereas the Commission Regulation (EC) No. 2073/2005 discusses the microbiological criteria for foodstuffs. Furthermore, the Commission Regulation (EC) No. 2074/2005 specifies implementing measures for certain products under Regulation (EC) No. 853/2004 and for the organization of official controls under Regulation (EC) No. 854/2004 and Regulation (EC) No. 882/2004. The Community legislation on food hygiene seeks to ensure the hygiene of foodstuffs at all stages of the production and

---

[5]Cf. Regulation (EC) No. 178 (2002).

supply chain by requiring food processors to follow the principles of Hazard Analysis and Critical Control Points (HACCP). HACCP was introduced by the Codex Alimentarius of the United Nations Food and Agriculture Organization that set international food standards.

19.2.1 General Requirements Regarding the Hygiene of Foodstuffs

According to the European Regulation (EC) No. 852/2004 on the hygiene of foodstuffs, all food business operators shall ensure that the processing and handling of foodstuffs is in accordance with the standards set by the Regulation at all stages in the production and supply chain for which the respective food business operators are responsible. Regulation (EC) No. 852/2004 provides for, inter alia, general hygiene requirements to be fulfilled by food business operators (other than at the level of primary production). These requirements include food premises, transport, equipment, packaging, heat treatment, food waste, water supply, personal hygiene of persons getting in contact with food and training of food workers. In particular, food business operators (other than at the level of primary production) shall apply, first of all, the principles of the HACCP.

19.2.2 Traceability

In accordance with the European Regulation (EC) No. 178/2002, which established the general principles of food law, food business operators shall develop traceability systems for foodstuffs, food ingredients and, where appropriate, for animals used for food processing. In case of any foodstuff presenting a serious risk to consumer health, the hazardous foodstuffs may immediately be withdrawn from the market and users or competent authorities may immediately be informed.

19.2.3 Approval and Registration

Pursuant to the European Regulation (EC) No. 853/2004, all food operators processing or handling foodstuffs of animal origin (except for those engaged only in primary production) have to be approved by the competent authorities of their respective Member States. After approval, the respective food companies shall be listed in the national register of approved food companies kept by their respective Member States. The companies are also given an approval number with additional codes indicating the type of products of animal origin manufactured. However, Regulation (EC) No. 853/2004 only applies to food production, processing, distribution, transport, and the other listed areas of products requiring temperature-controlled storage conditions and to those retail operations explicitly subject to this Regulation.

19.2.4 Specific Requirements for Foodstuff of Animal Origin

The European Regulation (EC) No. 853/2004 applies to unprocessed and processed foodstuff of animal origin but not to foods consisting partly of products of plant origin or, unless expressly indicated to the contrary, to the retail trade or primary production for private consumption for which the provisions of Regulation (EC) No. 852/2004 are sufficient.

Apart from establishing hygiene rules for food of animal origin, Regulation (EC) No. 853/2004 also specifies rules concerning microbiological criteria for foodstuffs, temperature control and compliance with the cold chain, sampling and analysis, where appropriate. Pursuant to this Regulation, for example, live bivalve mollusks and fishery products harvested from the wild and intended for human consumption must comply with high hygiene standards applicable at all stages of the production chain. Community legislation sets out specific hygiene standards regarding, inter alia, the transport, wrapping, labeling and storage, dispatch, purification and processing plants, essential equipment, freshness and viability of bivalve mollusks, microbiological criteria, presence of marine bio-toxins and harmful substances in relation to the permissible daily intake, health marking, conditions of hygiene during and after the landing of fishery products, protection against any form of contamination, fresh and frozen products, mechanically separated fish flesh, endo-parasites harmful to human health, cooked crustaceans and mollusks as well as processed fishery products.

Foodstuffs imported into the European Union must comply with the Community prerequisites for food hygiene or with equivalent standards. In order to guarantee a high level of food safety and public health, the European Commission keeps lists of non-EU member countries from which imports of products of animal origin are permitted. In addition, the Commission keeps up-to-date a list of establishments from which products of animal origin may be imported or dispatched. Foodstuffs of animal origin exported out of the Community shall at least comply with the requirements that would apply if they were marketed within the Community, as well as to any requirements that may be imposed by the importing country.

19.2.5 Specific Requirements for Organic Foodstuff

According to Regulation (EC) No. 834/2007, the production and distribution of organic products on the European market is subject to a strict certification process. Both farmers and processors of organic foodstuff are subject to inspections by the competent authorities to ensure their compliance with applicable organic food legislation. Before operators are granted organic certification and their goods can be labeled as organic, they have to undergo a conversion period of at least 2 years. Once the conversion process has been completed, farmers and food processors continue to be subjected to annual inspections. Each European Member State has established its own inspection scheme and designated specific public authorities and/or approved private inspection bodies to carry out the inspection and certification of organic production. In the case of operators not complying with the applicable legal requirements, their organic certification can be withdrawn and their right to market their products as organic be removed.

According to Regulation (EC) No. 834/2007 any terms such as "organic", "bio", "eco" and the like may not be used for non-organic products or products that contain genetically modified organisms (GMO). Since 1 July 2010, the use of the European organic logo is mandatory for pre-packaged food. The European logo may only be used for foodstuffs if (i) at least 95 % of the product's ingredients of agricultural origin have been organically produced, (ii) the product complies with the rules of the official inspection scheme, (iii) the product came directly from the producer or processor in a sealed package and (iv) the product bears the name of the producer, the preparer or vendor and the name or code of the inspection body.

19.2.6 Official Controls on Products of Animal Origin Intended for Human Consumption

The European Regulation (EC) No. 854/2004 lays down specific rules for the organization of official controls on products of animal origin intended for human consumption. The official controls include audits of good hygiene practices and HACCP principles as well as specific food sector-related controls (*e.g.* for milk and dairy products). Food business operators have to assist the competent authorities in their control by granting access to premises, documentation and records. Establishments that comply with the European regulations on food hygiene are formally approved by the respective authorities.

19.3 Food Ingredients

In 2008, the European Commission fundamentally reformed the pan-European legislation on food additives, flavorings and food enzymes by adopting the so-called "Food Improvement Agents Package" (FIAP). The FIAP includes Regulation (EC) No. 1331/2008 which institutes a common authorization procedure for food additives, enzymes and flavorings[6]; Regulation (EC) No. 1333/2008 on food additives[7]; Regulation (EC) No. 1334/2008 on food flavorings and certain food ingredients with flavoring properties for use in and on foods[8]; and Regulation (EC) No. 1332/2008 on food enzymes.[9] The FIAP Regulations are directly applicable in all European Member States and have successively repealed former European Directives on food additives, flavorings and enzymes (Directives 89/197/EEC,[10] 94/35/EC,[11] 94/36/EC[12] and 95/2/EC), respectively.[13] In addition, the FIAP Regulations provide for rules regarding the use, labeling[14] and maximum quantities of food additives, flavorings and enzymes as well as for reporting requirements. For instance, a producer or user of a food additive shall inform the Commission immediately of any new scientific or technical information which might affect the safety assessment of the respective food additive.[15]

According to the respective European legislation, food additives, flavorings and enzymes may only be placed on the market and added to food if they are previously approved by the European Commission and included in a Community list of authorized substances. The list will be regularly updated by the Commission and shall comprise of only those substances that comply with the respective sectoral food laws provided for in the Regulations on additives, flavorings and enzymes. In contrast, substances not allowed for use in foodstuffs shall be listed in separate "black lists", such as Annex III of Regulation (EC) No. 1334/2008.

---

[6]OJ L 354 (2008), pp. 1–6; cf. *Bertling, Streinz* (2009).
[7]Ibid.
[8]Ibid.
[9]Ibid.
[10]Directive 89/107/EEC (1988); cf. also OJ L 237 (1994), p. 1.
[11]OJ L 237 (1994), pp. 3–12.
[12]Ibid, pp. 13–29.
[13]OJ L 61 (1995), pp. 1–40.
[14]Cf. Hagenmeyer (2010), p. 3 et. seq.
[15]Cf. Regulation (EC) No. 1333 (2008), Art. 26 Para. 1.

19.3.1 The Regulation (EC) No. 1333/2008 on Food Additives

The new FIAP Regulation on food additives is applicable since 20 January 2010.[16] It covers various food additives including acids, acidity regulators, anti-caking agents, anti-foaming agents, antioxidants, bulking agents, carriers, colors, preservatives and sweeteners.[17] In contrast, processing aids, substances used for the protection of plants and plant products, nutrients added to food, substances used for the treatment of water and flavorings are only covered by the respective Regulation if they are used as food additives.[18]

Pursuant to Art. 3 Para. 2 lit. a) of the new Regulation, "food additive" means

> any substance not normally consumed as a food in itself and not normally used as a characteristic ingredient of food, whether or not it has nutritive value, the intentional addition of which to food for a technological purpose in the manufacture, processing, preparation, treatment, packaging, transport or storage of such food results, or may be reasonably expected to result, in it or its by-products becoming directly or indirectly a component of such foods.[19]

Art. 3 Para. 2 lit. a) contains, moreover, a list of certain substances that are not considered to be food additives (even if the criteria mentioned above should apply). This exclusion includes any substances used in covering or coating materials that do not form part of foods and are not intended to be consumed together with those foods. It also excludes any chewing gum bases and blood plasma, edible gelatin or protein hydrolysates and their salts, milk protein and gluten.[20]

According to Regulation (EC) No. 1333/2008, food additives may only be placed on the market and added to food if previously approved by the European Commission and included either in a Community list of authorized additives or in a list of food additives for use in other additives and food enzymes.[21] Food additives shall only be approved and included in such lists if they do not cause any safety concerns to consumer health, if their application is justified by a reasonable technological need that cannot be fulfilled in another economically and technologically feasible way, and if their use is not deceptive for the consumer.[22] Thus, before including food additives into said lists, the Commission had to re-examine all existing authorizations to ensure compliance with this criteria.[23] Food additives

---

[16] Regulation (EC) No. 1333 (2008), Art. 35.
[17] Ibid, Annex I; cf. *Bertling*, in: *Streinz*, l.c., no. 161d.
[18] Ibid, Art. 2 Para. 2.
[19] Id. Art. 3 Para. 2.
[20] For the complete list see ibid, Art. 3 Para. 2 lit.
[21] Ibid, Art. 4 Para. 1 and 2 (in connection with annexes I and II).
[22] Ibid, Art. 6 Para. 1 lit. a)–c); cf. *Bertling*, in: *Streinz*, l.c., no. 161 f.
[23] Regulation (EC) No. 1333 (2008), Art. 30 Para. 1.

that have been authorized before 20 January 2009 should also be assessed by the European Food Safety Authority (EFSA).[24]

The Community food additives list has been adopted by Regulation (EU) No. 1130/2001 of 11 November 2011 amending Annex III to Regulation (EC) No. 1333/2008 on food additives by establishing a Union list of food additives approved for use in food additives, food enzymes, food flavorings and nutrients. Until the Community list was accomplished, the Annexes to Directives 94/35/EC, 94/36/EC and 95/2/EC on colors, sweeteners and other food additives for use in foodstuffs as well as the respective national legislation based thereon remained in force and were regularly updated.[25] Moreover, the composition of all food additives placed on the market had to comply with the prerequisites of the consolidated Directives 2008/84/EC,[26] 2008/60/EC[27] and 2008/128/EC[28] which required specific purity criteria concerning sweeteners, colors and other food additives for use in foodstuffs.

Once a food additive has been included in a Community list, any considerable change in its raw materials or in its production methods means that the additive produced in this way is considered to be a new and different one.[29] Thus, before being placed on the market, such new additive must be submitted to EFSA for a separate health risk assessment.[30]

19.3.2 The Regulation (EC) No. 1334/2008 on Flavorings and Certain Food Ingredients with Flavoring Properties for Use in and on Foods

The FIAP Regulation on flavorings applies since 20 January 2011.[31] Some provisions, however, applied from a later date.[32] Foodstuffs placed on the market or labeled before 20 January 2011 which did not comply with the Regulation could be distributed until expiration of their date of minimum durability or use-by-date.[33]

The Regulation applies to all flavorings used to provide food with taste and/or odor such as flavorings which are used or intended to be used in or on foods, without prejudice to more specific provisions laid down in Regulation (EC) No. 2065/

---

[24]Regulation (EC) No. 1333 (2008) Art. 32 Para. 1.
[25]Regulation (EC) No. 1333 (2008), Art. 34.
[26]OJ L 253 (2008), pp. 1–175; OJ L 44 (2009), pp. 62–78.
[27]OJ L 158 (2008), pp. 17–40.
[28]OJ L 6 (2009), pp. 20–63.
[29]Regulation (EC) No. 1333 (2008), Art. 12.
[30]Ibid, Art. 12, 10.
[31]Ibid, Art. 30; cf. *Bertling*, in: *Streinz*, l.c., no. 166.
[32]Regulation (EC) No. 1334 (2008), Art. 30.
[33]Ibid.

2003.[34] Additionally, the Regulation is applicable to food ingredients with flavoring properties, food containing flavorings and/or food ingredients with flavoring properties; and any source materials for flavorings and/or source materials for food ingredients with flavoring properties.[35]

For the purposes of the Regulation, "flavorings" mean any product that is not intended to be consumed as such, added to food in order to impart or modify odor and/or taste and made or consisting of flavoring substances, flavoring preparations, thermal process flavorings, smoke flavorings, flavor precursors or other flavorings or mixtures thereof.[36] Substances which have exclusively a salty, sour or sweet taste as well as raw foods, smoke flavorings, mixtures of spices and/or fresh, dried or frozen herbs, tea mixtures and mixtures for infusion are only covered by the Regulation if they are used as food ingredients.[37]

The Regulation prohibits, in particular, any marketing or use of flavorings that do not comply with specific purity criteria and maximum amounts of dangerous or undesirable substances.[38] Moreover, according to the Regulation, an evaluation and approval is required for specific flavorings and source materials, for example, flavoring substances, several flavoring preparations of vegetable, animal or microbiological origin or even some thermal process flavorings obtained by heating ingredients.[39] These flavorings and source materials have to be approved by the Commission and included into a Community list and may only be placed on the market and used in or on food after such approval and in accordance with the conditions of use specified in such list.[40]

However, such restrictions applied only from 18 months after the date of application of the Community list.[41] Before including flavorings in the Community list, they were also subject to an EFSA assessment.[42] Some flavorings or food ingredients with flavoring properties may be used in or on food even without being subject to a previous assessment and authorization by EFSA and the Commission if they do not cause any risk for human health and if their use is not deceptive for the consumer.[43]

---

[34]OJ L 309 (2003), pp. 1–8.

[35]Regulation (EC) No. 1334 (2008), Art. 2 Para. 1.

[36]Ibid, Art. 3 Para. 2 lit.

[37]Ibid, Art. 2 Para. 2; cf. *Bertling*, in: *Streinz*, l.c., no. 162.

[38]Regulation (EC) No. 1334 (2008), Art. 5, 6 Para. 1 and 2, Annex 3.

[39]Ibid, Art. 9.

[40]Art. 10 Regulation (EC) No. 1334/2008; Natz et al. (2009), p. 184 et seq. (186).

[41]Regulation (EC) No. 1334 (2008), Art. 30.

[42]Ibid, Art. 12 Para. 1.

[43]Ibid, Art. 8.

19.3.3 The Regulation (EC) No. 1332/2008 on Food Enzymes

The Regulation on food enzymes was applied since 20 January 2009. Some provisions were applicable at a later date.[44] According to the Regulation, only food enzymes included in the Community list may be placed on the market and used in foods.[45] Moreover, no food enzyme nor any food in which a food enzyme has been used shall be placed on the market if the use of the food enzyme does not comply with the Regulation and its implementing measures.[46]

Until the application of the Community list, the national provisions regarding the marketing, distribution and use of food enzymes and food produced with food enzymes continued to be applicable in the Member States. The Regulation applies to all enzymes, including those used as processing aids that fulfill a technological function in foods, such as betaglucanase, invertase (E 1103), lysozyme (E 1105) and urease.[47] In contrast, the Regulation is not applicable to food enzymes used in the production of processing aids or of food additives already covered by the Regulation on food additives.[48]

The Community list is established on the basis of applications for authorization submitted to the European Commission and forwarded to EFSA. The applications for authorization had to be submitted to the Commission within a period of 24 months beginning with the enactment of the implementing measures for the Regulation (EC) No. 1331/2008 establishing a common authorization procedure for food additives, flavorings and enzymes.[49]

In addition, Regulation (EC) 1332/2008 sets forth detailed labeling requirements for food enzymes.[50] The labeling requirements differ depending on whether or not food enzymes are intended for sale to the final consumer. Food enzymes or food enzyme preparations not intended to be sold to the final consumer may be marketed only if their packaging contains the following information:

- the name laid down under the Regulation in respect of each food enzyme or a sales description which includes the name of each food enzyme or in the absence of such a name, the accepted name laid down in the nomenclature of the International Union of Biochemistry and Molecular Biology (IUBMB)[51];
- the statement "for food" or the statement "restricted use in food" or a more specific reference to the intended use;

---

[44]Regulation (EC) No. 1332 (2008), Art. 24 Para. 2 and 3.
[45]Ibid, Art. 4.
[46]Ibid, Art. 5.
[47]Ibid, Art. 18.
[48]Ibid, Art. 2 Para. 2.
[49]Ibid, Art. 7 Para. 3.
[50]Ibid, Art. 10–13; cf. Hagenmeyer (2010), p. 3 et. seq. (5, 7).
[51]Nomenclature Committee of the International Union of Biochemistry and Molecular Biology (1999), p. 264.

- the special conditions of storage and/or use (if necessary);
- a mark identifying the batch or lot;
- instructions for use if the omission thereof would preclude appropriate use of the food enzyme;
- the name or business name and address of the manufacturer, packager or seller;
- an indication of the maximum quantity of each component or group of components subject to quantitative limitation in food;
- the net quantity;
- the activity of the food enzyme(s); and
- the date of minimum durability or use-by-date.[52]

All above-mentioned information must be easily visible, clearly legible, indelible and shall be in a language easily understandable to purchasers.[53] When food enzymes and/or food enzyme preparations not intended for sale to the final consumer are sold mixed with each other and/or with other food ingredients, their packaging or containers shall bear a list of all ingredients in descending order of their percentage by weight of the total product.[54]

Food enzymes intended to be sold to the final consumer are subject to additional labeling requirements. Without prejudice to Directive 2000/13/EC,[55] Council Directive 89/396/EEC[56] and Regulation (EC) No. 1829/2003,[57] food enzymes and food enzyme preparations sold singly or mixed with each other and/or other food ingredients intended for sale to the final consumer may be marketed only if their packaging contains, besides the name of the enzyme or the accepted name such as registered in the nomenclature of the IUBMB, the specific function "for food", the statement "restricted use in food" or even a more specific reference to the intended food use.[58]

19.3.4 Additional Provisions on Food Labeling

Irrespective of the aforementioned Regulations (EC) 1331/2008, (EC) 1332/2008, (EC) 1333/2008 and (EC) 1334/2008, labeling of food additives, flavorings and enzymes must comply with the general labeling conditions provided in the national legislation implementing Directive 2000/13/EC. The labeling must include all information needed for the identification of the respective substances (i.e. name,

---

[52]Regulation (EC) No. 1332 (2008), Art. 11 Para. 1.

[53]Ibid, Art. 10 Para. 1.

[54]Ibid, Art. 11 Para. 2 lit. a)–k).

[55]OJ L 109 (2000), pp. 29–42.

[56]OJ L 186 (1989), pp. 21–22.

[57]OJ L 286 (2003), pp. 1–23.

[58]Regulation (EC) No. 1332 (2008), Art. 12 Para. 1; cf. also *Bertling*, in: *Streinz*, l.c., no. 161i.

lot, manufacturer etc.).[59] Moreover, labels of food flavorings must also include either the word "flavoring" or even a more specific name or description and mention either the specific statement "for food", the statement "restricted use in food" or even a more specific reference to the intended food use.[60] The term "natural" may only be used for substances or preparations derived directly from vegetable stuff or an animal.[61] All of this information must be easily visible, clearly legible, indelible and given in a language easily understood by the consumer.[62]

19.3.5 The Regulation (EC) No. 1331/2008 Establishing a Common Authorization Procedure for Food Additives, Enzymes and Flavorings

Finally, the Regulation (EC) No. 1331/2008 implements a common authorization procedure for food additives, food enzymes and food flavorings in the Community based on a previous risk assessment by EFSA. It also establishes a regularly updated Community list for each of the three categories of food substances. The European Commission adopted this Regulation after the passage of the FIAP Regulations on food additives, flavorings and enzymes. After the passage of each sectoral food law, EFSA presented a proposal on the data required for risk assessment of the substances concerned to the Commission.[63] Further, EFSA has approved respective guidance documents on food additives[64] and enzymes[65] and prepared a further draft scientific opinion on the data required for the risk assessment of flavorings.[66]

Any common updating of the Community list may be started either directly by the Commission or through an application submitted by a Member State or an interested party. The application must be submitted to the Commission, which will then be forwarded to EFSA for risk assessment. EFSA shall give an opinion on the application within a time limit of 9 months beginning with the receipt of the application.[67] Thus, the common procedure ends by updating the Community list within 9 months of receipt of EFSA's opinion. In case the Commission requests further information about risk management aspects, this period may be extended.[68]

---

[59]Cf. *Bertling*, in: *Streinz*, l.c., no. 161i.
[60]Ibid.
[61]Cf. *Bertling*, in: *Streinz*, l.c., no. 164.
[62]Directive 13/EC (2000), Art. 13 Para. 2, 16 Para. 1.
[63]Regulation (EC) No. 1331 (2008), Art. 9 Para. 2.
[64]EFSA (2009), pp. 1–7.
[65]EFSA (2009), pp. 1–26.
[66]EFSA (2010), p. 1623.
[67]Regulation (EC) No. 1331 (2008), Art. 5 Para. 1.
[68]Ibid, Art. 6 Para. 1.

19.4 Food Labeling

On 25 October 2011, the Regulation (EU) No. 1169/2011 of the European Parliament and of the Council on the provision of food information to consumers was adopted. The previous legislation on general food labeling dated back to 1978, and nutrition labeling rules were adopted in 1990. The new Regulation (EU) No. 1169/2011 combined existing rules on food labeling and nutritional information in one act. Regulation (EU) No. 1169/2011 introduced a set of new harmonized food labeling rules. Originally proposed in 2008, the Regulation was the subject of intense negotiations between members of the European Parliament and various European Member States. A compromise was finally brokered under the Hungarian presidency.

Several key aspects of the Commission's proposal were dropped in the negotiations (including placement of mandatory nutrition information on the front of packs, and exempting alcoholic beverages from the ingredient and nutrition labeling requirements). Calls by consumer organizations for a traffic light system giving consumers a visual warning for high fat, sugar or salt content were also dropped, with industry instead arguing in favor of a system of guideline daily amounts (GDA).

19.4.1 Mandatory Information

The following information is mandatory for all pre-packed food and must appear in a minimum font size of 1.2 mm (or 0.9 mm for small packets) [69] so that consumers will be able to read food labels "without having to use a magnifying glass" (in the words of former European Commissioner for Health and Consumer Protection, John Dalli):

- the name of the food and a list of its ingredients;
- allergens: allergenic substances (defined as having "a scientifically proven allergenic or intolerance effect") must be highlighted in the ingredient list; this also applies to non-packaged foods, for example food sold in restaurants or canteens; Member States can decide how to do this in practice;
- the quantity of certain ingredients;
- the net quantity of the food;
- the date of minimum durability or the "use by" date;
- any special storage conditions;
- the name and address of the business responsible for the food within the European Union;

---

[69]Regulation (EU) No. 1169 (2011), Art. 13 Para. 2 and 3.

- the country of origin must appear on a wide range of food (extended from beef, honey, olive oil, and fresh fruit and vegetables to fresh, chilled or frozen pork, lamb, poultry and goat meat) and where failure to indicate the country of origin might mislead the consumer; the Commission has introduced separate implementing rules for this purpose;
- instructions for use (where it would be difficult to make appropriate use of the food without them);
- alcoholic strength by volume for beverages with more than 1.2 % by volume of alcohol;
- a nutrition declaration, including energy content as well as fat, saturated fat, carbohydrate, sugar, protein and salt levels (per 100 g or 100 ml to allow comparisons between products, although the Commission may be given powers to adopt rules on the expression of the nutrition declaration per portion); this must be presented in a tabular format using the agreed minimum font size and following a template that is provided in Annex XV of the Regulation; the Commission may allow the use of pictograms and symbols if this conveys the same level of information to consumers as words and numbers.[70]

The obligation to provide information on food is imposed on the operator under whose name or business name the food is marketed or, if that operator is not established in the European Union, the importer into the European Union.[71]

Member States may require additional mandatory information on grounds of public health, consumer protection or the prevention of fraud. In those cases, they must notify the measures to the European Commission.[72] While many of the harmonized rules on food labeling have been agreed upon and included in the Regulation adopted by European Parliament, the Commission has been tasked with preparing additional measures. With the assistance of the Standing Committee on the Food Chain and Animal Health Committee composed of Member State representatives, the European Commission has done so in the form of delegated acts on the availability of certain mandatory particulars by means other than on the package or on the label such as pictograms or symbols; the list of foods not required to bear a list of ingredients; and the re-examination of the list of substances or products causing allergies, intolerances, or the list of nutrients that may be declared on a voluntary basis. It has also proposed implementing acts on the modalities of expression of one or more particulars by means of pictograms or symbols instead of words or numbers and the manner of indicating the date of minimum durability and the country of origin or place of provenance for meat. Other implementing acts are regarding the precision of the declared values for the nutrition declaration and the expression per portion or per consumption unit of the nutrition declaration.

---

[70] Ibid, Art. 9 Para. 1.
[71] Ibid, Art. 8 Para. 1.
[72] Ibid, Art. 39.

19.4.2 Package Labeling

The name under which the product is sold, the net quantity of pre-packaged foodstuffs, the date of minimum durability (or, in the case of foodstuffs which, from the microbiological point of view, are highly perishable, the use-by-date) as well as, if mandatory, the alcoholic strength by volume shall appear in the same field of vision. Where a product has been defrosted, this must be stated.[73] Food producers can also include GDAs or use the term "per portion", once the European Commission has defined portion sizes. The same compulsory information must be provided when food is sold over the Internet or through catalogues. It must be made available to the consumer before the purchase is concluded. This does not apply to food sold in automatic vending machines.[74] There are also other exemptions, including glass bottles intended for reuse which are indelibly marked and which, therefore, bear no label, ring or collar (they must only provide information on the name of the food, any allergens and the net quantity).[75] The exceptions also apply for packaging that is too small to accommodate the mandatory labeling requirements[76] and beverages with alcohol content of more than 1.2 % by volume for which no list of ingredients or nutrition declaration is required.[77]

When foodstuffs are pre-packaged, the mandatory labeling shall appear on the pre-packaging itself or at least on a label (e.g. a sticker) attached thereto. In all cases (including the sticker scenario), the respective labeling must be easily visible, clearly legible, indelible, and given in a language easily understood by the con sumer. They shall not in any way be hidden, obscured or interrupted by other written or pictorial matter. If a sticker is used, not only the text printed on it but the material as such should be indelible from the package, i.e. not easily removable from the package.

Where pre-packaged foodstuffs are intended for the final consumer but marketed at a stage prior to sale to the final consumer where sale to a mass caterer is not involved, or where pre-packaged foodstuffs are intended for supply to mass caterers for preparation, processing, splitting or cutting up, the mandatory labeling particulars may principally appear on the commercial documents if they either accompany the foodstuffs to which they refer or were sent before or at the same time as delivery.[78] However, the name under which the products are sold, the date of minimum durability (or, if need be, the use-by-date) as well as the name or business name and address of the manufacturer or packager, or of a seller established within the Community shall also appear on the external packaging in which the foodstuffs are presented for marketing. In the case of glass bottles intended for reuse which are

---

[73]Ibid, Annex VI Part A No. 2.
[74]Ibid, Art. 14.
[75]Ibid, Art. 16 Para. 1.
[76]Ibid, Art. 16 Para. 2.
[77]Ibid, Art. 16 Para. 4.
[78]Ibid, Art. 8 Para. 7.

indelibly marked and therefore bear no label, ring or collar and packaging or containers (the largest surface which has an area of less than 10 cm^2), only the name under which the product is sold, the net quantity and the date of minimum durability (or, if need be, use-by-date) shall be given.

Where foodstuffs are offered for sale to the final consumer or to mass caterers without pre-packaging, or where foodstuffs are packaged on the sales premises at the consumer's request or pre-packaged for direct sale, the Member States shall adopt separate rules concerning the manner in which mandatory labeling is to be shown. The Member States may decide not to require the provision of all or some of the respective particulars provided that the purchaser still receives sufficient information.

19.4.3 List and Quantity of Ingredients

In general, the quantity of ingredients to be indicated must be expressed as a percentage and correspond to the quantity of the ingredient or ingredients at the time of their use.[79] Only Community provisions may allow for derogations from this principle for certain foodstuffs. The quantity of certain ingredients or categories of ingredients shall only be mentioned in food labeling where the ingredient or category of ingredients concerned appears in the name under which the foodstuff is sold or is usually associated with that name by the consumer. Also if the ingredient or category of ingredients concerned is emphasized on the labeling in words, pictures or graphics or the ingredient or category of ingredients concerned is essential to characterize a foodstuff and to distinguish it from products with which it might be confused because of its name or appearance, then it must be included in the food labeling.[80]

The quantity indication shall appear either in or immediately next to the name under which the foodstuff is sold, or, alternatively, in the list of ingredients in connection with the ingredient or category of ingredients in question. However, the aforementioned labeling requirements for ingredients do not apply, inter alia, for ingredients or categories of ingredients where the drained net weight is duly indicated, or the quantities of which are already required to be given on the labeling under other Community provisions. It is also not applicable when such ingredients are used in small quantities for the purposes of flavoring, or which (while appearing in the name under which the food is sold) are not such as to govern the choice of the consumer because the variation in quantity is not essential to characterize the foodstuff or does not distinguish it from similar foods.[81] Given that the content of a number of nutrients varies over the shelf-life of a foodstuff, it is also essential to

---

[79]Ibid, Annex VIII No. 3 lit. a).
[80]Ibid, Art. 22.
[81]Ibid, Annex VIII No. 1.

establish to which point in time the labeled ingredient dosage refers. The decisive point in time could be, in particular, the production date (also referred to as time of use in the manufacture), the date at which the food supplement is put on the market, the end of the shelf-life, the use-by date, or any time until the end of the shelf-life or the use-by date.

Similar to Art. 7 Para. 4 of the former Directive 2000/13/EC of the European Parliament and of the Council of 20 March 2000 on the approximation of the laws of the Member States relating to the labeling, presentation and advertising of foodstuffs,[82] Annex VIII No. 3 lit. a of Regulation (EU) No. 1169/2011 also requires that the quantity of an ingredient indicated on the labeling of foodstuffs should be expressed as a percentage and correspond to the quantity of the ingredient or ingredients at the time of its/their use. The term "use" does not mean the time of consumption of the foodstuff or its ingredients but the processing of the respective ingredient into such foodstuff. Even though the English term "use" itself is rather vague, its denotation can be construed by referring to the corresponding terms in the German and French versions of the former Directive 2000/13/EC and of Annex VIII to Regulation (EU) 1169/2011 ("Verarbeitung"/"Verwendung" and "mise en œurvre").[83] This interpretation is also confirmed by Art. 6 Para. 5 of the former Directive 2000/13/EC according to which the list of ingredients should include all ingredients of the foodstuff, in descending order of weight, as recorded at the time of their use in the manufacture of the foodstuff. Pursuant to Art. 7 Para. 6 Directive 2000/13/EC, Art. 7 should apply without prejudice to Community rules on nutrition labeling for foodstuffs.

Consumers expect that food supplements will deliver the amounts of ingredients and nutrients mentioned on the product label at least for a reasonable time after purchasing the respective foodstuff. Whereas most minerals and trace elements are stable over time, the actual quantities of some nutrients (such as most vitamins) are not completely stable and may decline at different stages. Against this background, Art. 9 and 13 Directive 2002/46/EC require the European Commission, assisted by the Standing Committee on the Food Chain and Animal Health, to develop further rules with regard to differences between the values declared on the label and those established in the course of official checks.

Taking into account this situation, several non-governmental organizations have developed specific recommendations on acceptable tolerances for vitamins and minerals in order to improve consistency between declared values and quantities at the products' end of shelf-life. Some such non-governmental organizations are,

---

[82]OJ L 109 (2000), pp. 29–42.

[83]Cf. the respective wording in German and French: "Die als Prozentsatz anzugebende Menge entspricht der Menge der Zutat bzw. Zutaten zum Zeitpunkt ihrer Verarbeitung."/"Die Angabe der Menge einer Zutat oder Zutatenklasse erfolgt als Prozentsatz der Menge der Zutat bzw. Zutaten zum Zeitpunkt ihrer Verwendung" and "La quantité mentionnée, exprimée en pourcentage, correspond à la quantité du ou des ingrédients au moment de leur mise en oeuvre."/"L'indication de la quantité d'un ingrédient ou d'une catégorie d'ingrédients est exprimée en pourcentage et correspond à la quantité du ou des ingrédients au moment de leur mise en oeuvre".

e.g., the Association of the European Self-Medication Industry (AESGP),[84] the European Federation of Health Products Manufacturers Associations (EHPM)[85] and the European Responsible Nutrition Alliance (ERNA).[86] According to their recommendations, levels of variation between 80 and 180 %, shall be reasonable and a pragmatic target for manufacturers of foodstuffs, and should be established standards for circulation of food in the European Union.[87] Thus, ingredient quantities declared on the product label of foodstuffs should not fall outside the recommended ranges at the end of shelf-life. Regarding the time when the ingredient dosage of a foodstuff mentioned on its label shall be part of the product, the aforementioned recommendations differ from the wording of the European rules.

According to the recommendations of AEGSP, EHPM and ERNA, the values declared on the product label should be consistent with the quantities being part of the foodstuff and meet the recommended tolerance ranges at the product's end of shelf-life. In order to comply with the recommended tolerances at the product's end of shelf-life, the quantity of some nutrients that decline rather fast (e.g. most vitamins) often must be higher at the date when the foodstuff is put on the market. However, in no case, the nutrient dosages may exceed applicable maximum dosages (i.e. upper intake levels). Therefore, in case the original nutrient quantity being part of the food supplement at the date when it is put on the market might exceed such applicable maximum dosages, the end of shelf-life must be adjusted accordingly. However, the recommendations of AEGSP, EHPM and ERNA are not binding. Hence, there is sometimes still a divergence among the European Member States with their respective national recommendations on acceptable ranges of nutrients compared to declared values at the end of shelf-life.

19.4.4 Rules Against Misleading Labeling

In order to avoid any misleading of consumers, "imitation foods" (i.e. foods that look similar to other foods but are made of different ingredients, such as "cheese-like" foods made with vegetable products) must be clearly labeled on the front of the pack in a prominent font size and next to the brand name. Only mineral water and foods for particular nutritional uses may be claiming properties that prevent, treat or cure a disease.

---

[84] Association of the European Self-Medication Industry, 7 Avenue de Tervuren, B-1040 Brussels, Belgium (www.aesgp.be).

[85] European Federation of Associations of Health Product Manufacturers 50, Rue Jacques de Lalaing 4, B-1040 Brussels, Belgium (www.ehpm.org).

[86] European Responsible Nutrition Alliance, 50, Rue de l'Association, B-1000 Brussels, Belgium (www.erna.org).

[87] AESGP, EHPM and ERNA (2003), pp. 1–27.

19.5 Food Packaging

Materials and articles intended to come into contact with food and which, in particular, are used to package food, must comply with strict regulatory requirements. Food packaging, therefore, is subject, inter alia, to Regulation (EC) No. 1935/2004 of the European Parliament and of the Council of 27 October 2004 on materials and articles intended to come into contact with food[88] and Commission Regulation (EC) No. 2023/2006 of 22 December 2006 on good manufacturing practice for materials and articles intended to come into contact with food.[89] There are also diverse specific provisions applicable on plastic materials and articles intended to come into contact with foodstuffs, where, for example, the Commission Directive 2002/72/EC of 6 August 2002 relates to plastic materials and articles intended to come into contact with foodstuffs.[90] Rules on recycled plastic materials and articles intended to come into contact with foods are denoted in the Commission Regulation (EC) No 282/2008 of 27 March 2008.[91] The Council Directive 82/711/EEC of 18 October 1982 explains the basic rules necessary for testing migration of the constituents of plastic materials and articles intended to come into contact with foodstuffs.[92] The Commission Regulation (EC) No 1895/2005 of 18 November 2005 restricts the use of certain epoxy derivatives in materials and articles intended to come into contact with food.[93] The approximation of the laws of the Member States relating to materials and articles which contain vinyl chloride monomer and are intended to come into contact with foodstuffs is given under the Council Directive 78/142/EEC of 30 January 1978.[94] Lastly, the Council Directive 85/572/EEC of 19 December 1985 provides the list of simulants to be used for testing migration of constituents of plastic materials and articles intended to come into contact with foodstuffs.[95]

All aforementioned Regulations are directly applicable in the European Member States whereas the Directives mentioned above have been transposed into the respective national laws. The Regulations (EC) No. 1935/2004 and (EC) 2023/2006 provide for a general framework and common rules on good manufacturing practice (GMP) for materials and articles intended to come into contact directly or indirectly with food. In contrast, the Regulations (EC) No. 282/2008 and (EC) No. 1895/2005 as well as the Directives 2002/72/EC, 82/711/EEC, 78/142/EEC and 85/572/EEC specify purity requirements, maximum limits on migration, conditions

---

[88] OJ L 338 (2004), p. 1 et seq.; cf. *Rathke,* in: *Zipfel/Rathke,* Lebensmittelrecht, 154th delivery complement (2013), C Part 1, 102.

[89] OJ L 384 (2006), p. 1 et seq.

[90] OJ L 220 (2002), p. 1 et seq.; cf. *Bertling,* in: *Streinz,* l.c., III C., no. 187g.

[91] OJ L 86 (2008), p. 1 et seq.

[92] OJ L 297 (1982), p. 1 et seq.

[93] OJ L 302 (2005), p. 1 et seq.

[94] OJ L 44 (1978), p. 1 et seq.

[95] OJ L 372 (1985), p. 1 et seq.

of use, testing parameters and marketing prohibitions for plastic materials and articles intended to come into contact with foodstuffs. According to these Regulations and Directives, all respective materials and articles shall comply, in the general interest of food safety, with maximum limits on migration of constituents into or onto food and are subject to strict conditions of use. Besides these European provisions, all materials and articles intended to come into contact with food are also subject to several applicable national food laws.

19.5.1 Regulation (EC) No. 1935/2004 on Materials and Articles Intended to Come into Contact with Food

Regulation (EC) No. 1935/2004 shall apply to all materials and articles which, in their finished state, are intended to be brought into contact with food, are already in contact with food and were intended for that purpose, can reasonably be expected to be brought into contact with food, or can transfer their constituents to food under normal or foreseeable conditions of use.[96] Moreover, the Regulation introduces specific provisions concerning so-called "active" and "intelligent" food packaging (e.g., packaging that changes color when a foodstuff has expired). "Active" food packaging means materials and articles that are intended to extend the shelf-life or maintain or improve the condition of packaged food.[97] "Intelligent" food contact materials and articles mean materials and articles which monitor the condition of packaged food or the environment surrounding the food.[98] All materials and articles within the sense of Regulation (EC) No. 1935/2004, including active and intelligent packaging, shall be manufactured in compliance with GMP so that they do not transfer their constituents to food in quantities which could endanger human health, bring about an unacceptable change in the composition of the food or result in a deterioration of the organoleptic characteristics thereof.[99] Moreover, the labeling, advertising and presentation of any material or article must not mislead the consumer.[100]

Specific measures such as, criteria of purity, specific conditions of use, limits on the migration of certain constituents into or onto food or provisions for ensuring traceability may be adopted or amended for the groups of materials and articles listed in Annex I of the Regulation and, where appropriate, for all combinations of those materials and articles or recycled materials and articles used in the manufacture of those materials and articles.[101] The respective Annex I list includes, plastics,

---

[96] Regulation (EC) No. 1935 (2004), Art. 1 Para. 2.
[97] Ibid, Art. 2 Para. 2 lit. a.
[98] Ibid, Art. 2 Para. 2 lit. b.
[99] Ibid, Art. 3 Para. 1.
[100] Ibid, Art. 3 Para. 2.
[101] Ibid, Art. 5 Para. 1.

printing inks, varnishes, coatings and wood (see Annex I, no. 10, 11, 15, 17). In the absence of specific measures, the Regulation shall not prevent Member States from maintaining or adopting national provisions.[102] So far, specific measures have been adopted for active and intelligent materials and articles, printing inks, ceramics, regenerated cellulose, rubbers and, in particular, for plastics (see Directives 2002/72/EC and 85/572/EEC, for example). Moreover, without prejudice to the specific measures, materials and articles not yet in contact with food when placed on the market shall comply with specific labeling requirements and be accompanied by (a) the words "for food contact" or a specific indication as to their use, (b) special instructions to be observed for safe and appropriate use, if necessary; and (c) the name or trade name and, in either case, the address or registered office of the manufacturer, processor, or seller responsible for placing on the market established within the Community. The label must also (d) be adequate to ensure traceability of the material or article; and (e) in the case of active materials and articles contain information on the permitted use or uses and other relevant information such as the name and quantity of the substances released by the active component. This will enable food business operators who use these materials and articles to comply with any other relevant Community provisions or, in their absence, national provisions applicable to food, including the provisions on food labeling.[103]

All information required under Art. 15 Para. 1 shall be conspicuous, clearly legible and indelible.[104] However, the information provided for in lit. (a), (b) and (e) above shall be confined only to materials and articles which comply with the general requirements laid down in Art. 3 (GMP/non-misleading labeling, advertising and presentation), with the special requirements for active and intelligent materials,[105] if applicable, and finally, with the specific measures referred to in Art. 5 (for groups of materials and articles listed in Annex I) or, in the absence thereof, with any national provisions applicable to these materials and articles.[106] However, the information referred to in lit. (a) above shall not be obligatory for any materials and articles which, because of their characteristics, are clearly intended to come into contact with food.[107]

At the marketing stages other than the retail stage, the information required under Art. 15 Para. 1 shall be displayed either on the accompanying documents, on the labels or packaging, or on the materials and articles themselves.[108] Art. 15 Para. 4–7 provide for specific labeling requirements in case of retail trade to the end customer. The specific measures referred to in Art. 5 shall in any case require that materials and articles covered by those measures are accompanied by a written

---

[102]Ibid, Art. 6.
[103]Ibid, Art. 15 Para. 1.
[104]Ibid, Art. 15 Para. 3.
[105]Ibid, Art. 4.
[106]Ibid, Art. 15 Para. 9.
[107]Ibid, Art. 15 Para. 2.
[108]Ibid, Art. 15 Para. 8.

declaration of compliance stating that they comply with the rules applicable to them.[109] Further, appropriate documentation shall be available to demonstrate such compliance. Such documentation must be made available to the competent authorities at any time on demand.

Traceability of materials and articles must be ensured at all stages in order to facilitate the control, and recall of defective products, the consumer information and attribution of responsibility.[110] Thus, with due regard to technological feasibility, business operators shall have in place systems and procedures to allow identification of the businesses from which materials or articles and, where appropriate, substances or products covered by this Regulation and its implementing measures used in their manufacture are supplied. That information shall be made available to the competent authorities at any time on demand. The materials and articles placed on the European market shall be identifiable by an appropriate system which allows their traceability by means of labeling or relevant documentation or information. In order to enforce compliance with Regulation (EC) No. 1935/2004 in accordance with all other relevant provisions of European law, the Member States shall carry out official controls.[111]

19.5.2 Directive 2002/72/EC Relating to Plastic Materials and Articles Intended to Come into Contact with Foodstuffs

Directive 2002/72/EC as well as the respective national laws based thereon apply to all plastic materials, articles and parts thereof consisting exclusively of plastics or composed of two or more layers of materials, each consisting exclusively of plastics, bound together by means of adhesives or by any other means, which (in the finished product state) are intended to come into contact or are brought into contact with foodstuffs and are intended for that purpose.[112] The Directive provides for overall migration limits.[113] According to these limits, plastic materials and articles shall not transfer their constituents to foodstuffs in quantities exceeding 10 milligrams per square decimeter of surface area of material or article (mg/dm^2). However, the relevant limit shall be 60 milligrams of the constituents released per kilogram of foodstuff (mg/kg) in case of containers or articles which can be filled with a capacity of no less than 500 milliliters (ml) and not more than 10 liters (l), or which can be filled and for which it is impracticable to estimate the surface area in contact with foodstuffs.

---

[109]Ibid, Art. 16.
[110]Ibid, Art. 17.
[111]Ibid, Art. 24 Para. 1.
[112]Directive 72/EC (2002), Art. 1 Para. 2; cf. *Bertling,* in: *Streinz,* l.c., no. 187g.
[113]Directive 72/EC (2002), Art. 2.

Directive 2002/72/EC further lays down specific migration limits, restrictions and specifications for monomers, other starting substances and additives used for the manufacture of plastic materials and articles.[114] Verification of compliance with the specific migration limits shall be carried out in accordance with the testing parameters presented in Directives 82/711/EEC,[115] 85/572/EEC,[116] and further provisions in Annex I of Directive 2002/72/EC.[117] However, verification of compliance shall not be obligatory if it can be established that compliance with the overall migration limits specified in the Directive implies that the specific migration limits are not exceeded or by assuming complete migration of the residual substance in the respective material or article, it cannot exceed the specific limit of migration.

At the marketing stages other than the retail stages, the plastic materials and articles which are intended to be placed in contact with foodstuffs shall be accompanied by a written declaration of compliance with the rules applicable to them.[118] However, this does not apply to plastic materials and articles which, by their nature, are clearly intended to come into contact with foodstuffs.

19.5.3 Regulation (EC) No. 282/2008 on Recycled Plastic Materials and Articles Intended to Come into Contact with Foods

Regulation (EC) No. 282/2008 applies to recycled plastic materials, articles and parts thereof intended to come into contact with foodstuffs as referred to in Directive 2002/72/EC. This Regulation lays down specific authorization requirements and quality assurance standards for materials and articles containing recycled plastic. Recycled plastic materials and articles shall only be placed on the market if they contain recycled plastic obtained from a recycling process authorized by the Regulation.[119] The authorized recycling process shall be managed by an appropriate quality assurance system that ensures that the recycled plastic complies with the requirements of the authorization process. It follows that, the respective quality assurance system shall comply, in any case, with the detailed rules laid down in the relevant Annex to Regulation (EC) No. 2023/2006 (see below).

After authorization of a recycling process in accordance with this Regulation, the authorization holder or any other business operator using the authorized recycling process under license must comply with any conditions or restrictions

---

[114]Directive 72/EC (2002), Art. 3 et seq. together with Annexes II, III, IV and V.
[115]OJ L 297 (1982), p. 1 et seq.
[116]OJ L 372 (1985), p. 1 et seq.
[117]Directive 72/EC (2002), Art. 8.
[118]Ibid, Art. 9.
[119]Regulation (EC) No. 282 (2008), Art. 3.

attached to such authorization. Any converter using recycled plastic from the authorized recycling process or any business operator using materials or articles containing recycled plastic from the authorized recycling process must comply with any condition or restriction attached to such authorization. Moreover, the authorization holder or any other business operator using the authorized recycling process under license shall immediately inform the European Commission of any new scientific or technical information, which might affect the safety assessment of the recycling process in relation to human health.[120]

In addition to the requirements of Directive 2002/72/EC, the declaration of compliance of materials and articles containing recycled plastic shall confirm that only recycled plastic from an authorized recycling process has been used.[121] Moreover, this declaration shall contain the information that the recycling process has been authorized by listing the relevant EC register number of the authorized recycling process, attesting that the plastic input, the recycling process and the recycled plastic meet the specifications for which the authorization has been granted and, finally, that a quality assurance system according to Regulation (EC) No. 2023/2006 is in place.[122]

19.5.4 Regulation (EC) No. 2023/2006 on Good Manufacturing Practice for Materials and Articles Intended to Come into Contact with Food

Regulation (EC) No. 2023/2006 provides the rules on GMP for the groups of materials and articles intended to come into contact with food listed in Annex I to Regulation (EC) No. 1935/2004 and for combinations of those materials and articles or recycled materials and articles used in those materials and articles.[123] The Regulation applies to all sectors and to all stages of manufacture, processing and distribution of materials and articles, up to but excluding the production of starting substances.[124] Any business operator must ensure that manufacturing operations are carried out in accordance with the applicable rules on GMP.[125] The Regulation provides for specific quality assurance standards.[126] Pursuant to these standards, the business operator shall establish, implement and ensure compliance with an effective and documented quality assurance system necessary to certify that finished materials and articles comply with the rules applicable to them.

---

[120] Ibid, Art. 7.
[121] Ibid, Art. 12.
[122] OJ L 86 (2008), pp. 7, 9.
[123] Regulation (EC) No. 2023 (2006), Art. 1.
[124] Ibid, Art. 2 Para. 1.
[125] Ibid, Art. 4 Para. 1.
[126] Ibid, Art. 5.

Moreover, the quality assurance system shall ensure that starting materials are selected and comply with the rules applicable to them and that all operations are carried out in accordance with pre-established specifications, instructions and procedures.[127] This system is supposed to identify measures to correct any failure to achieve GMP. All corrective measures shall be implemented without delay and made available to the competent authorities for inspections.

The Regulation also requires the business operator to establish and maintain appropriate documentation in paper or electronic form with respect to specifications, manufacturing formulae and processing relevant to compliance and safety of the finished material or article. This includes to records covering the various manufacturing operations performed that are relevant to compliance and safety of the finished material or article, as well as the results of the quality control system.[128] Any documentation must be made available by the business operator to the competent authorities at their request at any time.[129]

19.5.5 Rules Against Misleading Packaging

In general, consumers expect that food packaging will be designed and filled in a way that does not pretend to contain more filling quantity than the actual quantity of the foodstuff contained in the respective packaging. According to this information, a deception of the consumer can only be avoided by a package design which (i) limits the remaining ullage to a necessary minimum or (ii) is transparent so that the actual net quantity can be clearly seen and easily evaluated from the outside or, finally, (iii) makes the actual quantity contained in the package otherwise clearly and easily perceptible from the outside (e.g. by using a striking filling mark or by graphically representing the filling level on the outside of the packaging). European food and food packaging law provides for limits of tolerance for so-called "negative quantity errors", i.e. for quantities by which the actual contents of a pre-package may be less than the nominal filling quantity. A product is considered to be pre-packed when it is placed in a package of whatever nature without the purchaser being present and the quantity of product contained in the package has a predetermined value and cannot be altered without the package either being opened or undergoing a perceptible modification.[130] In addition, European food and food packaging law prohibits any product packaging which is misleading and deceptive, particularly as to the quantity of the respective product. Pursuant to the pertinent provisions, pre-packaging must not give the consumer the false impression that the packaging contains more than the actual quantity of the respective foodstuff.

---

[127] OJ L 384 (2006), p. 2.
[128] Regulation (EC) No. 2023 (2006), Art. 7.
[129] OJ L 384 (2006), p. 2 et seq.
[130] Directive 76/211/EEC (1976), Art. 2 Para. 2.

Table 19.1 Directive 76/211/EEC negative error tolerance limits

Nominal quantity (Qn) in grams or milliliters	Tolerable negative errors	
	as of % of Qn	g or ml
Above 5 and less than 25	9	–
From 25 to 50	9	–
From 50 to 100	–	4.5
From 100 to 200	4.5	–
From 200 to 300	–	9
From 300 to 500	3	–
From 500 to 1000	–	15
From 1000 to 10,000	1.5	–

These numeric ranges determine limits of tolerances for so-called "negative errors" of prepackages containing liquid products, whereby the negative error means the quantity by which the actual contents of the pre-package is less than the nominal quantity

The Council Directive 76/211/EEC of 20 January 1976 on the approximation of the laws of the Member States relating to the making-up by weight or by volume of certain pre-packaged products only specifies negative variances. The Directive applies to pre-packages containing products intended for sale in constant unit nominal quantities which are equal to values predetermined by the packer and expressed in units of weight or volume not less than 5 g or 5 ml and not more than 10 kg or 10 L.[131] Directive 76/211/EEC further determines limits of tolerance for so-called "negative errors" of prepackages containing liquid products whereby negative error means the quantity by which the actual contents of the pre-package are less than the nominal quantity.[132] Such negative errors shall only be tolerable within the limits set out in Table 19.1.

19.6 Food Authorization Procedures

EFSA strives for the greatest possible transparency in the food authorization procedures it conducts. For this reason, it has a tendency to apply a restrictive interpretation to the relevant confidentiality regulations. This rings particularly true in authorization procedures for nutrition and health claims for foods pursuant to Regulation (EC) No. 1924/2006 of the European Parliament and of the Council of 20 December 2006. Whereas the drafters of the Regulation (EC) No. 1924/2006 had apparently forgotten to include a clear provision on confidentiality protection, the more recent European FIAP Regulations include such provisions.

With regard to the food authorization procedures of EFSA, many applicant food companies are learning by experience that the scientific data and other proprietary

---

[131] Ibid, Art. 1.
[132] Ibid, Sec. 2.3 et seq. of Annex I.

information they submit with their authorization applications are not being treated confidentially, but are, instead made available to the public. Often, this is due to the fact that the applicant food companies are not familiar with the relevant confidentiality standards and, as a result, do not claim the information and document protection for their authorization documents to which they are actually entitled. Interested companies should, therefore, find out for themselves at an early stage and to what extent, before the respective authorization procedures are launched, the relevant confidentiality standards apply to be able to claim the greatest possible protection for the information they file with their authorization applications.

19.6.1 Limited Data Protection Under the Regulation (EC) No. 1924/2006

The Regulation (EC) No. 1924/2006 merely contains a section on "data protection" against unauthorized use of the scientific data and other mandatory disclosures submitted by the applicant in the authorization application as part of the health claims authorization procedure by other applicants. Hence, the scientific data and other information may not be used for the benefit of a subsequent applicant for a period of 5 years from the date of authorization if (i) the scientific data and other information had been designated as proprietary by the prior applicant at the time the prior application was made, (ii) such prior applicant had an exclusive right of reference to the proprietary data at the time the prior application was made, and (iii) the health claim could not have been authorized without the submission of the proprietary data by the prior applicant.[133] The proprietary data may not be used by subsequent applicants unless such subsequent applicants have agreed with the prior applicant that such data may also be used for its benefit,[134] or the Commission takes a decision on whether a claim could or could not have been included in the list of authorized health claims even without the submission of data designated as proprietary by the prior applicant.[135]

---

[133] Regulation (EC) No. 1924 (2006), Art. 21.
[134] Ibid, Art. 21 Para. 1.
[135] Ibid, Art. 21 Para. 2.

19.6.2 More Extensive Confidentiality Protection Under the Regulations of the Food Improvement Agents Package

By contrast, the more recent FIAP Regulations states explicit provisions for comprehensive protection of confidential authorization documents. According to the Recitals of Regulation (EC) No. 1331/2008, the authorization procedure must be transparent and capable of review by the public, but at the same time preserve the confidentiality of certain information of the applicants so as not to impair their competitiveness. Only data relating to the safety of a substance, such as the findings of toxicological studies should, under no circumstances, be treated as confidential.[136] In this context, Regulation (EC) No. 1331/2008 further stipulates that at least those data in the application documents whose disclosure might significantly harm the applicant's competitive position may be treated as confidential.[137] All applicants, therefore, shall indicate, when submitting their authorization data, which information they wish to be kept confidential and furnish justification for this. The European Commission then consults with applicants about which information can actually remain confidential, and then notifies applicants and the Member States accordingly. After having been informed of the Commission's position, the applicants have 3 weeks to withdraw their application so as to preserve the confidentiality of the information provided which the Commission has not deemed proprietary. Up to the expiration of this period, the confidentiality of all data and information is preserved. If applicants finally withdraw their application, the Commission, EFSA and the Member States may no longer disclose to third parties the information deemed confidential.

The FIAP Regulation (EC) No. 1334/2008 also provides for special confidentiality standards for reporting by food business operators to the Commission. Producers or users of a flavoring substance, or the representative of such producer or user, shall, at the request of the Commission, inform it of the amount of the flavoring substance added to foods in the Community in a period of 12 months.[138] Information provided in this connection is also treated as confidential, provided that such information is not required for the assessment of its safety.

---

[136] Regulation (EC) No. 1331 (2008), Recitals 18 and 19.
[137] Ibid, Art. 12.
[138] Regulation (EC) No. 1334 (2008), Art. 19 Para. 1.

19.6.3 Restrictive Interpretation of Confidentiality Standards by EFSA

As mentioned above, EFSA tries to be as transparent as possible in the food authorization procedures, deferring in this regard to Art. 38 of Regulation (EC) No. 178/2002 of the European Parliament and of the Commission of 28 January 2002. Pursuant to Art. 38 Para. 1 sentence 1 of Regulation (EC) No. 178/2002, EFSA shall ensure that it carries out its activities with a high level of transparency. For this purpose, it must publish without delay, inter alia, all information on which its opinions are based.

The Regulation (EC) Nos. 178/2002 and 1049/2001 of the European Parliament and of the Council of 30 May 2001 regarding public access to European Parliament, Council and Commission documents, applies analogously to documents in EFSA's possession. Accordingly, all citizens of the European Union as well as any natural or legal person residing or having its registered office in a Member State has a right to access documents in EFSA's possession, subject to the principles, conditions and limits defined in Regulation (EC) No. 1049/2001.[139] Furthermore, pursuant to Regulation (EC) No. 178/2002 the conclusions of the scientific opinions delivered by EFSA relating to foreseeable health effects must never be kept confidential.[140] EFSA may only refrain from disclosing to third parties information for which confidential treatment has been requested and justified. However, disclosure of such information is permitted where such information has to be made public in order to protect public health.[141]

EFSA applies a restrictive interpretation to the confidentiality standards in food authorization procedures. In the Scientific and Technical Guidelines for the Preparation and Presentation of the Application for Authorization of a Health Claim published by EFSA, it is stated that the application is never treated as confidential and that passages which the applicant considers confidential should be kept to a minimum and clearly designated as such. Moreover, in its guidelines, EFSA expressly draws attention to the fact that it always publishes a summary of the authorization application and in its opinion deals with all documents submitted with the application, except, for such data and information, which EFSA itself considers confidential.[142]

In its restrictive interpretation of confidentiality protection, EFSA has gone so far as to repeatedly publish information in its scientific opinions in the health claims authorization procedures despite explicit prior requests by the concerned applicants for such information to be kept confidential. This especially raises hackles, for

---

[139] Regulation (EC) No. 1049 (2001), Art. 2 Para. 1.

[140] Regulation (EC) No. 178 (2002), Art. 39 Para. 3.

[141] Ibid, Art. 39 Para. 1.

[142] Cf. EFSA (2011), p. 8, margin no. 13: "[...] EFSA will also make public, once adopted, its scientific opinion on the data and information included in the application, excluding the information considered as confidential."

instance, in product-related studies, which the concerned applicants initially would like to keep confidential vis-à-vis their competitors or with a view to any applications for intellectual property rights that they might file.

The narrow interpretation of confidentiality protection applied by EFSA is highly questionable, not only with regard to Art. 39 of Regulation (EC) No. 178/2002 but also the confidentiality standards set by the recent FIAP Regulations. EFSA would appear to interpret the relevant provisions to mean that EFSA itself is authorized to decide on the (non-)confidentiality of the authorization data concerned. For the food industry, this restrictive approach is not only unacceptable but may also entail devastating competitive disadvantages for the affected applicants. Food companies are, therefore, well advised to familiarize themselves with the applicable confidentiality standards early on to ensure that they obtain the most extensive protection possible for the data and information submitted with their authorization applications.

19.7 Food Advertising

Food advertising in the European Union is subject to strict legal standards to prevent misleading and deceptive advertising and ensure, amongst others, scientific reliability of all health and nutrition related claims.

19.7.1 Rules Against Misleading Advertising

Pursuant to Directive 2006/114/EC of the European Parliament and of the Council of 12 December 2006 concerning misleading and comparative advertising,[143] misleading advertisements are prohibited. According to Directive 2006/114/EC, "advertising" means the making of a representation in any form in connection with a trade, business, craft or profession in order to promote the supply of goods or services, including immovable property, rights and obligations.[144] According to the same Directive, a "misleading advertising" is qualified as any advertising which, in any way, including its presentation, deceives or is likely to deceive the persons to whom it is addressed or whom it reaches and which, by reason of its deceptive nature, is likely to affect their economic behavior or which, for those reasons, injures or is likely to injure, a competitor.[145]

---

[143]OJ L 376 (2006), pp. 21–27; cf. *Köhler*, in: *Köhler/Bornkamm*, Gesetz gegen den unlauteren Wettbewerb, 32nd edition (2014), Introduction, no. 3.41 et seq.
[144]Directive 114/EC (2006), Art. 2 lit. (a).
[145]Ibid, Art. 2 lit. (b).

Also, Directive 2005/29/EC of the European Parliament and of the Council of 11 May 2015 concerning unfair business-to-consumer commercial practices in the internal market,[146] prohibits unfair and, in particular, misleading commercial practices. "Commercial practices" within the sense of Directive 2005/29/EC means any act, omission, course of conduct or representation, commercial communication, including advertising and marketing, by a trader, directly connected with the promotion, sale or supply of a product to consumers.[147] A commercial practice shall be regarded as misleading and, thus, illegal if it contains false information and is therefore untruthful or in any way, like overall presentation, deceives or is likely to deceive the average consumer (even if the information is factually correct) in relation to, e.g., the main characteristics of the product (such as its composition, ingredients, date of manufacture, fitness for purpose, quantity etc.), and if in either case causes or is likely to cause the consumer to take a transactional decision that he would not have taken otherwise.[148]

A commercial practice shall also be regarded as misleading if, in its factual context, taking account of all its features and circumstances and the limitations of the communication medium, it omits material information that the average consumer needs, depending on the context, to make an informed transactional decision and, thereby, causes or is likely to cause the respective consumer to make a decision that he would not have taken otherwise.[149] Moreover, it shall be regarded as a misleading omission when, taking account of the aforementioned matters, a trader hides or provides in an unclear, unintelligible, ambiguous or untimely manner such material information, and where this causes or is likely to cause the average consumer to make a transactional decision that he would not have taken otherwise.[150] In the case of an invitation to purchase, the main characteristics of the respective product as described above (e.g. composition, ingredients, date of manufacture, fitness for purpose, quantity etc.) shall also be regarded as material, if not already apparent from the context.[151] For both Directives 2006/114/EC and 2005/29/EC, the question of whether a certain practice is misleading shall always be assessed from the perspective of an average member of the group to whom the relevant practice is specifically aimed.[152]

---

[146]OJ L 149, 27/12/2006, pp. 22–39; cf. *Köhler*, in: *Köhler/Bornkamm*, l.c., no. 3.56 et seq.
[147]Directive 29/EC (2005), Art. 2 lit. (d).
[148]Ibid, Art. 6 Para. 1 lit. (b).
[149]Ibid, Art. 7 Para. 1.
[150]Ibid, Art. 7 Para. 2.
[151]Ibid, Art. 7 Para. 4.
[152]Cf. Ibid, Recitals 18, 19, Art. 5 Para. 3.

19.7.2 Nutrition and Health Claims on Foods

By adopting the Regulation (EC) No. 1924/2006 of the European Parliament and of the Council of 20 December 2006 on nutrition and health claims made on foods, the political decision-makers of the European Union were particularly interested in ensuring that any future nutritional and health claims on food would only be tolerated in the European Union when backed by clear scientific evidence. The European legislator wanted nutrition and health claims for foods to be truthful, clear, reliable and helpful for consumers trying to decide on a healthy diet. On food labels and in advertisements, consumers, henceforth, were to be provided with all relevant information without restriction so as to be protected against any misleading statements.[153]

To achieve this objective, the European legislator decided on a paradigm change. Prior to adoption of the Regulation (EC) No. 1924/2006, it was generally permitted to make nutrition and health claims in food labeling and advertising provided that their use was not expressly prohibited by certain legal provisions. The new approach was for everything to be prohibited unless explicitly allowed. Amongst others, nutrition and health claims shall be based on and substantiated by generally accepted scientific evidence. In addition, respective claims are only permitted if they are explicitly authorized by the European Commission. Further, the use of nutrition and health claims shall not be false, ambiguous or misleading, give rise to doubt about the safety and/or the nutritional adequacy of other foods or encourage or condone excess consumption of a food. Furthermore, all claims are forbidden to state, suggest or imply that a balanced and varied diet cannot provide appropriate quantities of nutrients in general or refer to changes in bodily functions that could give rise to or exploit fear in the consumer, either textually or through pictorial, graphic or symbolic representations.[154]

Along with the Regulation (EC) No. 1924/2006, the European Parliament and the Council first adopted a list of permitted nutrition claims together with specific requirements governing their use. For the most part, the list was based on the terms of use for nutrition claims that had been agreed internationally as well as defined in Community provisions (such as "sugar-free", "low-fat" "light" or "naturally") and was directly attached as an annex to the Regulation. Nutrition claims on food may only be used if they are listed in this annex and satisfy other conditions set in the Regulation.[155]

Health claims (such as "lowers cholesterol levels" or "promotes bone growth") are also prohibited unless they satisfy the requirements of the Regulation and are included in the list of permitted health claims within the meaning of Art. 13 and 14 of the Regulation.[156] Whereas Art. 14 specifically deals with reduction of

---

[153]Cf. Regulation (EC) No. 1924 (2006), Recitals 16 and 29.
[154]Ibid, Art. 3 Para. 2.
[155]Ibid, Art. 8 Para. 1.
[156]Ibid, Art. 10 Para. 1.

disease risk claims as well as statements on the development and health of children, Art. 13 applies to all other health claims relating to food. In view of the great number of possible statements in this area, the European legislator did not define the corresponding Community lists of permitted health claims at the same time as it adopted the Regulation but, instead, decided to adopt these within a reasonable period thereafter. In so doing, it left the food industry limited scope for participating in this process. Interested companies could apply to the competent national authorities in the Member States for any health claims they wanted to see included, which were then submitted to the EFSA. After consultation with EFSA, the European Commission would prepare the Community lists of permitted claims.

To give both consumers and the food industry sufficient legal certainty and clarity regarding the health claims to be permitted within a reasonable time after adoption of the Regulation (EC) No. 1924/2006, the Regulation provided for a stringent timetable in this regard. For example, where health claims within the meaning of Art. 13 Para. 1 of the Regulation (which neither relate to reduction of disease risk claims nor claims concerning the development or health of children), the Member States were to provide the Commission with the corresponding lists (along with the conditions applicable to them), as well as information on the corresponding scientific justification, by 31 January 2008. After hearing EFSA, the Commission was then to adopt a Community list of permitted claims within the meaning of Art. 13 Para. 1 of Regulation (EC) No. 1924/2006 by 31 January 2010.[157]

The competent authorities of the Member States have forwarded more than 44,000 applications for authorization of health claims to EFSA. The Member States and the Commission then compiled a list with a total of 4637 nutritional claims relating to food. The European legislator had no idea that such a huge number of applications would be made. The original deadline for approving the list of permitted health claims (31 January 2010) had already exceeded nearly a year (with still no definitive date in the offing for final adoption of the Community list). The European Commission at the end of September 2010 announced that it intended to make changes to the procedure of progressive approval of the list pursuant to Art. 13 Para. 1 of Regulation (EC) No. 1924/2006. The Community list was now to be drawn up in two stages. First, a list of health claims for all substances except "botanicals" would be adopted and the related opinions of EFSA were to be finalized by the end of June 2011. After that, food health claims relating to botanicals were to be assessed.

As the reasons for this change (which were also accompanied by the announcement of yet other significant delays in the authorization procedure), the European Commission referred, among other things, to alleged delays on the part of the respective applicants in submitting authorization applications for review by the Commission. It further stated that some applicants had raised concerns relating to potential market distortions and confronted the Commission with possible problems

---

[157]Ibid, Art. 13 Para. 2 and 3.

relating to differences in the treatment of botanical ingredients in the respective regulations relating to health claims and Traditional Herbal Medicinal Products (THMP). The Commission had, therefore, decided to take a "pragmatic" approach of adopting the Community list of permitted health claims in steps to protect consumers from unsubstantiated nutritional claims relating to foods. After adoption of the list, consumers should be assured that all health claims on the market had actually been substantiated by science.

By giving these reasons, the European Commission not only conceded that timely adoption of the Community lists of permitted health claims had failed, but at the same time put itself on a course of open confrontation with the food industry. Instead of admitting that it was simply too ambitious to implement the objectives relating to the adoption of the Regulation (EC) No. 1924/2006 within the planned timeframe (given the misjudgment of the relevant market by the European legislator) and, therefore, the prospects of this succeeding had been nil from the outset, the Commission put the blame for the delay on the food industry alone.

Finally, a first list of permitted health claims within the sense of Art. 13 Para. 1 of the Regulation (EC) No. 1924/2006 was adopted by Commission Regulation (EU) No. 432/2012 of 16 May 2012[158] establishing the permitted health claims made on foods, other than those referring to the reduction of disease risk and to children's development and health. The Regulation (EU) No. 432/2012 applied only as of 14 December 2012 giving food companies a brief transition period for adapting their product labeling and advertising. In addition, the European Commission published an online database of authorized and non-authorized claims. At that date, there were only 222 authorized health claims and 1719 non-authorized claims within the sense of Art. 13 Para. 1 of the Regulation (EC) No. 1924/2006.

Surprisingly, the food industry took a rather indulgent stance to these new advertising restrictions. Food business operators largely accepted the decisions taken by the European Commission and did not launch court challenges to the Commission's regulations on the non-authorization of health claims on foods. In principle, this was possible by lodging an action for annulment before the Court of Justice of the European Union (ECJ) pursuant to Art. 263 Para. 4, 256 of the Treaty on the Functioning of the European Union (TFEU). Based on the previous situation of legislation (Art. 230 Para. 4 EC Treaty, old version), legal acts of the applying European bodies generally could only be challenged by annulment if such acts were of direct and individual concern to such claimants. This legislative approach presented difficulties in the case of regulations which, although always applicable directly, do not as a general rule concern people individually because they broadly apply to everyone. Art. 263 Para. 4 of the TFEU provides for the possibility of lodging an action for annulment against "regulatory acts" which are of direct concern to the claimant and do not entail implementing measures. The decisive question in regard to the Commission's regulations on the non-authorization of health claims is whether these regulations are also covered by Art. 263 Para.

---

[158]OJ L 136 (2012), pp. 1–40.

4 TFEU. The ECJ has repeatedly ruled that the European treaties grant citizens and companies within the European Union comprehensive and effective legal remedies. This also applies to the legal remedies provided in Art. 263 Para. 4 TFEU. As a result, the specified "regulatory acts" must cover all acts of the European bodies claiming to be binding as well as generally and directly applicable in the Member States. Accordingly, all legal acts may be challenged by way of action for annulment within the sense of Art. 263 Para. 4 TFEU which (i) are aimed at the respective claimant directly, or (ii) are addressed to a third party but are of direct and individual concern to the claimant as a competitor or involved party, or (iii) apply generally and, because of special circumstances, are of similar individual concern to the claimant as those to whom the acts are actually addressed, or (iv) apply generally and are of direct concern to the claimant but whose application does not require any implementation measures.

In cases of doubt, these requirements are fulfilled through the regulations of the European Commission by which an authorization of health claims on food is refused. As expressly clarified by the recitals of the respective regulations, the regulations serve the purpose of deciding on individual and specific authorization applications for health claims on foods in accordance with the provisions of the Regulation (EC) No. 1924/2006. Pursuant to Art. 15 et seq. of Regulation (EC) No. 1924/2006, applicants are expressly granted own rights to participate and be heard in the authorization procedure. The regulations of the European Commission by which the authorization of health claims on foods are refused are, thus, of direct concern to the applicant within the meaning of Art. 263 Para. 4 TFEU and, therefore, must be capable of being challenged by way of annulment.

19.7.3 Specific Rules on Advertising for Alcoholic Beverages

According to the Regulation (EC) No. 1924/2006, beverages containing more than 1.2 % in volume of alcohol shall not bear any health claims.[159] As far as nutrition claims are concerned, only those referring to low alcohol levels, or the reduction of the alcohol content, or the energy content for beverages containing more than 1.2 % in volume of alcohol, shall be permitted. Further, in the absence of specific European Community rules regarding nutrition claims referring to the above parameters, relevant national rules may apply to comply with the provisions of the Treaty establishing the European Community.[160] In addition, there is a separate European Regulation (EC) No. 110/2008 of the European Parliament and of the Council of 15 January 2008[161] which sets forth specific provisions on the definition, description, presentation, labeling and the protection of spirit drinks. Both

---

[159] Regulation (EC) No. 1924 (2006), Art. 4 Para. 3.
[160] Ibid, Art. 4 Para. 4.
[161] OJ L 39 (2008), pp. 16–54.

Regulation (EC) Nos. 1924/2006 and 110/2008 are directly applicable in all European Member States.

19.8 Consumer Rights

On 25 October 2011, the European Parliament adopted the Directive 2011/83/EU on consumer rights,[162] which was originally proposed by the Commission in 2008. It replaced two Directives from 1985 and 1997, respectively, established contractual rights for consumers and concerned contracts negotiated away from business premises and distance contracts. This Directive 2011/83/EU takes a full harmonization approach, with a single framework regulating certain aspects of business-to-consumer contracts across the European Union. The rationale for the legislation is that domestic distance sales are growing significantly while the growth in cross-border distance sales has been limited. The Directive, therefore, pays particular attention to off-premises contracts (concluded in a place which is not the trader's business premises; for example the consumer's home or office) and distance contracts (concluded over the Internet, or by mail order, telephone or fax). The new rules apply to a wide variety of goods and services, but there are exemptions for sectors where separate legislation already exists. These sectors include electronic communications, energy markets, financial services, food labeling, medicine, medical devices and patients' rights in cross-border healthcare, and public transport.

19.8.1 Withdrawal

Under Directive 2011/83/EU, a harmonized 14 (calendar) days withdrawal period applies to all distance and off-premises contracts.[163] This is double the previously described 7 days. The withdrawal period starts from the moment the consumer receives the goods, rather than at the time of conclusion of the contract as was the case in the former European law. For service contracts, the withdrawal period will start on the day of the conclusion of the contract.[164] If a trader does not inform the consumer about the right of withdrawal, the withdrawal period is automatically extended to 12 months or to 14 days from the day the trader belatedly provides the information.[165] Consumers can withdraw without providing any reason and without incurring any costs (except in certain situations outlined below). Withdrawal must

---

[162] OJ L 304 (2011), pp. 64–88.
[163] Ibid, Art. 9.
[164] Ibid, Art. 9 Para. 2.
[165] Ibid, Art. 10.

be done by either using the standardized withdrawal form annexed to the Directive or a statement from the consumer.[166] Because the Directive is a full harmonization act, Member States may not alter the standard form in any way when transposing the legislation into their national laws.

The trader must reimburse the consumer for the goods or services using the same means of payment as the consumer used for the initial transaction. The trader must also reimburse the delivery costs within 14 days (except in cases where an express delivery was requested, in which case the consumer pays the difference between standard delivery and express delivery).[167] The trader may also be liable for the cost of the consumer returning the goods, unless the consumer has been clearly informed that he/she will bear this cost and has been provided with an estimate of the maximum costs of doing so. In the case of bulky goods, the costs could be significant.

There are many exceptions to the rules on withdrawal. These include items that may lose value, for example, wine due to market fluctuations or tailor-made items which only correspond to consumer's specifications. Contracts for services that have been performed are also included. Other exceptions apply to goods which deteriorate rapidly, sealed goods which are not suitable for return due to health protection or hygiene reasons once opened, and sealed audio and video recordings, and computer software that has been unsealed after delivery.[168] If an item has been used more than necessary, its relative decrease in value will be deducted from the amount reimbursed to the consumer.

19.8.2 Rules on Delivery

Rules on when delivery should occur will be determined at the national level but must not be more than 30 days from the day the contract is concluded. If the trader fails to meet that deadline, the consumer can ask for the delivery within a set period. If the trader still fails to deliver the goods within that time, the consumer can terminate the contract. Any delivery restrictions must be flagged at the beginning of the ordering process, as well as the accepted means of payment.[169]

19.8.3 Fees and Default Options

Businesses are prohibited from charging consumers' fees that exceed the cost borne by the trader for the use of a certain means of payment, for example, additional fees

---

[166] Ibid, Art. 11.
[167] Ibid, Art. 13.
[168] Ibid, Art. 16.
[169] Ibid, Art. 18.

for using a credit card.[170] Calls to a consumer must not be charged at more than the basic rate of a telephone call.[171] Default options, which the consumer is required to reject in order to avoid additional payment, are effectively banned. Online shoppers will be entitled to reimbursements since their express consent was not given. Under the harmonized rules, the final step in the order process must be clearly indicated and consumers must explicitly confirm that they understand that they have to pay a price by the statement "order with duty of payment" or some other language that indicates that proceeding with this transaction will entail an obligation to pay.

19.8.4 Information Requirements

All businesses, whether committing to a distance contract or not, must provide certain information to consumers. This mandatory information includes information about the characteristics of the good or service, identity of the trader, the trader's address and telephone number, and total price, including any taxes or delivery charges. Information about arrangements for payment, delivery and the trader's complaint handling procedure (where applicable), duration of the contract or conditions for terminating the contract, and special information requirements for digital content and the interoperability of digital content with hardware and software must also be provided.

For off-premises and distance contracts, there are additional information requirements related to contact details, the anticipated cost of returning goods (where this will be borne by the consumer) and the right of withdrawal. In order to boost cross-border online sales, Member States are not permitted to add any other requirements to those stipulated in the Directive 2011/83/EU.[172]

19.9 Conclusion

This chapter provides an introduction to European food law. As described above, this law provides for strict legal standards for foodstuffs, food ingredients and food packaging as well as for food labeling and advertising, as will be further discussed in the following chapters in this section. Food manufacturers and distributors are, therefore, well advised to acquaint themselves with the applicable legal prerequisites before placing their products on the European market. Non-compliance with the pertinent requirements can have severe consequences, ranging from administrative fines via product withdrawals or recalls to sensitive product liability.

---

[170]Ibid, Art. 19.
[171]Ibid, Art. 21.
[172]Ibid, Art. 5 and 6.

References

Amendment to Directive 2008/84/EC, Directive 2009/10/EC

Animal health rules governing the production, processing, distribution and introduction of products of animal origin for human consumption, Directive 2002/99/EC

Association of the European Self-Medication Industry. Available at www.aesgp.be (last accessed 7 Feb 2014)

Authorization procedure for additives, enzymes and flavorings, Regulation (EC) No. 1331/2008

Colors for use in foodstuffs, Directive 94/36/EC

Commission of the European Communities – Commission Staff Working Document (2009) European industry in a changing world. Available at http://ec.europa.eu/enterprise/policies/industrial-competitiveness/files/industry/doc/sec_2009_1111_en.pdf (last accessed 7 Feb 2014)

Consumer rights, Directive 2011/83/EU

Definition, description, presentation, labelling and the protection of geographical indications of spirit drinks and repealing Council Regulation (EEC) No. 1576/89, Regulation (EC) No. 110/2008

European Commission – SWD 14/3 (2014) State of the industry, sectoral overview and implementation of the EU industrial policy

European Federation of Associations of Health Product Manufacturers. Available at www.ehpm.org (last accessed 7 Feb 2014)

European Federation of Associations of Health Product Manufacturers (EHPM) (2003) EHPM response to commission consultation on nutrition labelling technical aspects. pp 1–27

EFSA (2009) Guidance of EFSA prepared by the Scientific Panel on Food Contact Materials, Enzymes, Flavorings and Processing Aids (CEF) on the submission of a dossier on food enzymes. EFSA J 1305:1–26

EFSA (2011) Scientific and technical guidelines for the preparation and presentation of the application for authorization of a health claim (revision 1). EFSA J 9(5):2170

EFSA (2010) Scientific Opinion by the EFSA Panel on Food Contact Materials, Enzymes, Flavorings and Processing Aids (CEF) on the data required for the risk assessment of flavorings. EFSA J 8(6):1623

European Food Safety Authority (EFSA) (2009) Scientific Statement of the Panel on Food Additives and Nutrient Sources added to Food: data requirements for the evaluation of food additives applications following a request from the European Commission. EFSA J 1188:1–7

European Responsible Nutrition Alliance. Available at www.erna.org (last accessed 7 Feb 2014)

Food additives authorized for use in foodstuffs intended for human consumption, Directive 89/107/EEC

Food additives other than colors and sweeteners, Directive 95/2/EC

Food additives, Regulation (EC) No. 1333/2008

Food and feed safety, Regulation (EC) No. 178/2002

Food enzymes, Regulation (EC) No. 1332/2008

Food flavorings and certain food ingredients with flavoring properties for use in and on foods, Regulation (EC) No. 1334/2008

Food supplements, Directive 2002/46/EC

Genetically modified food and feed, Regulation (EC) No. 1829/2003

Good manufacturing practice for materials and articles intended to come into contact with food, Regulation (EC) No. 2023/2006

Hagenmeyer M (2010) New labeling requirements for food additives, enzymes and flavorings – an overview of the labeling provisions of the "Food Improvement Agents Package". EFFL 1:3–9

Hygiene of foodstuffs, Regulation (EC) No. 852/2004

Implementing measures for certain products under Regulation (EC) No. 853/2004 and for the organization of official controls under Regulation (EC) No. 854/2004 and Regulation (EC) No. 882/2004, Regulation (EC) No. 2074/2005

Indications or marks identifying the lot to which a foodstuff belongs, Council Directive 89/396/EEC

Labelling, presentation and advertising of foodstuffs, Directive 2000/13/EC

List of permitted health claims made on foods, other than those referring to the reduction of disease risk and to children's development and health, Regulation (EU) No. 432/2012

List of simulants to be used for testing migration of constituents of plastic materials and articles intended to come into contact with foodstuffs, Directive 85/572/EEC

Materials and articles intended to come into contact with food and repealing Directives 80/590/EEC and 89/109/EEC, Regulation (EC) No. 1935/2004

Materials and articles which contain vinyl chloride monomer and are intended to come into contact with foodstuffs, Directive 78/142/EEC

Microbiological criteria for foodstuffs, Regulation (EC) No. 2073/2005

Misleading and comparative advertising, Directive 2006/114/EC

Nomenclature Committee of the International Union of Biochemistry and Molecular Biology (NC-IUBMB) (1999) Recommendations of the Nomenclature Committee of the International Union of Biochemistry and Molecular Biology on the Nomenclature and Classification of Enzymes by the Reactions they Catalyse. Eur J Biochem 264:610–650

Nutrition and health claims made on food, Regulation (EC) No. 1924/2006

Official food and feed controls, Regulation (EC) No. 882/2004

OJ L 6, 10/1/2009

OJ L 39, 13/02/2008

OJ L 44, 14/02/2009

OJ L 44, 15/02/1978

OJ L 86, 28/03/2008

OJ L 109, 6/5/2000

OJ L 136, 25/05/2012

OJ L 149, 27/12/2006

OJ L 158, 18/6/2008

OJ L 186, 30/6/1989

OJ L 220, 15/08/2002

OJ L 237, 10/9/1994

OJ L 253, 20/9/2008

OJ L 286, 18/10/2003

OJ L 297, 23/10/1982

OJ L 302, 19/11/2005

OJ L 304, 22/11/2011

OJ L 309, 26/11/2003

OJ L 338, 13/11/2004

OJ L 354, 31/12/2008

OJ L 376, 27/12/2006

OJ L 384, 29/12/2006

OJ L 372, 31/12/1985

Plastic materials and articles intended to come into contact with foodstuffs, Directive 2002/72/EC

Prepacked products, Directive 76/211/EEC

Provision of food information to consumers, Regulation (EU) No. 1169/2011

Public access to European Parliament, Council and Commission documents, Regulation (EC) No. 1049/2001

Recycled plastic materials and articles intended to come into contact with foods and amending Regulation (EC) No. 2023/2006, Regulation (EC) No 282/2008

Restriction of epoxy derivatives in food packaging, Regulation (EC) No. 1895/2005

Smoke flavorings used or intended for use in or on foods, Regulation (EC) No. 2065/2003

Specific criteria concerning colors for use in foodstuffs, Directive 2008/128/EC

Specific criteria concerning sweeteners for use in foodstuffs, Directive 2008/60/EC

Specific hygiene rules for food of animal origin, Regulation (EC) No. 853/2004
Specific purity criteria on food additives other than colors and sweeteners, Directive 2008/84/EC
Specific rules for the organization of official controls on products of animal origin intended for human consumption, Regulation (EC) No. 854/2004
Sweeteners for use on foodstuffs, Directive 94/35/EC
Testing migration of plastic materials in contact with foodstuffs, Directive 82/711/EEC
Union list of food additives approved for use in food additives, food enzymes, food flavorings and nutrients, Regulation (EU) No. 1130/2001
Unfair business-to-consumer commercial practices in the internal market, Directive 2005/29/EC
Bertling, in: *Streinz*, l.c., no. 161d
Bertling, in: *Streinz*, l.c., no. 161i
Bertling, in: *Streinz*, l.c., no. 162
Bertling, in: *Streinz,* l.c., III C., no. 187g
Natz A, Zumdick U, Heck M (2009) Bericht aus Brüssel. LMuR, p. 184 et seq. (186)
Rathke, in: *Zipfel/Rathke,* Lebensmittelrecht, 154th delivery complement (2013), C Part 1, 102
Köhler, in: *Köhler/Bornkamm*, Gesetz gegen den unlauteren Wettbewerb, 32nd edition (2014), Introduction, no. 3.41 et seq
Köhler, in: *Köhler/Bornkamm*, l.c., no. 3.56 et seq

Chapter 20
Food Safety and Policy in the European Union

Giorgio Rusconi

Abstract The chapter analyses the gradual evolution in EU food law, turning to food and feed safety protection following the serious food crises of the 1990s. The comprehensive, integrated principles underlying the 1993 Green Paper and 2000 White Paper became a reality with Regulation (EC) No. 178/2002, which ascribes a crucial role to consumer protection, setting forth the principles of transparency, traceability, and responsibility of business operators; establishing a rapid alert system and crisis management plan; and, most importantly, establishing the

---

Having worked as a food lawyer since the industry's infancy, Giorgio Rusconi has gained significant experience in the field of food law, assisting Italian and foreign clients with food hygiene, labelling, additives, organic farming, geographical indications/destinations of origin, packaging, and responsibilities regarding food products and within the industry.

Giorgio handles client contracts, including those for stakeholders within the food chain, ensuring regulatory compliance and advising on issues of Italian and European law.

Aware of the needs of multinational companies, Giorgio maintains close relationships with experts in different countries and provides assistance and advice on markets nationally and beyond. With relationships built over the years with these and other foreign colleagues, Giorgio helped establish FLN—Food Lawyers' Network Worldwide, in collaboration with Krell Weyland Grube, offering integrated legal services to multinationals in some 50 countries that are operating in the food industry.

To ensure that companies operating in the food industry have a wider range of services, Giorgio also works with a specialized laboratory capable of providing timely, individualized counter-analysis, as well as with a company specializing in the field of food audits.

He is the author, with Omar Cesana, of Guide to Food Law, which provides a broad and inclusive look at various aspects of food law in Italy.

In court, Giorgio handles all areas of civil litigation defense, centering on product liability and operators' obligations in the food sector. He helps clients protect their interests in disputes before authorities, and represents entrepreneurs and managers of food-related operations before the criminal court.

In 1996 Giorgio obtained a Higher Diploma in Comparative Law at the International Faculty of Comparative Law of Strasbourg (France) on a scholarship offered by the European Commission. He graduated with a degree in law from Università degli Studi di Milano, was admitted to the bar in 2000, and in 2014 he was admitted to practice before the Supreme Court of Cassation. He speaks Italian, English, and French.

G. Rusconi (✉)
Mondini Rusconi Studio Legale, Milan, Italy

FLN – Food Lawyers Network, Milan, Italy
e-mail: giorgio.rusconi@mondinirusconi.it

European Food Safety Authority (EFSA). EFSA provides scientific opinions and scientific and technical assistance, establishes surveillance procedures, promotes cooperation among food safety organisations, and plays a crucial role in risk assessment and communication. EFSA's first 10 years of activity has achieved a positive balance: despite the dramatic changes within the EU, it has successfully adapted itself. The need for the harmonisation of food hygiene requirements led to the adoption of the hygiene package and the HACCP system (Reg. 852/2004), as well as of specific hygiene rules applicable to products of animal origin (Reg. 853/2004). The conclusions drawn are encouraging: EU food law, by adopting harmonised vertical rules and regulations common to all operators, has committed to ensuring common, high quality standards. EU authorities' capacity to promptly and effectively react to significant food crises has led to one of the most developed systems worldwide, one that is functional for operators and safe for consumers. Yet the need for rapid, repeated actions, is a constant challenge, which certainly cannot be considered complete.

20.1 The Historical European Framework: From the Dawn of Common Policy to the 1997 Green Paper

20.1.1 The Social and Economic Frame of Reference

EC (now, EU) law has always played a crucial role in removing obstacles to the free movement of goods within the European single market, and this has been true for its food sector as well.

This role, however, also includes significant, practical assessment over the years of all food safety-related issues, and involves strategic action at the Community level, whether more or less binding on State Members, and influenced by high-profile episodes that have attracted the attention of legislators and the public alike.

Indeed, the greater attention paid to food safety within the single market has been the result of the serious food crises involving the European Union since the end of the 1980s.

The first and perhaps most notorious case was what has become commonly known as "mad cow disease." After reaching epic proportions in the United Kingdom, it was called to the attention of the relevant national and Community authorities due to a possible connection between the disease affecting the brain of cattle (bovine spongiform encephalopathy, BSE) and its human version.

Similarly scandalous was the case of chickens raised in Belgium on feed containing dioxin, residues of which were found in the animals' meat in excess of allowed limits under the legislation then in force.

Again great concern was caused by episodes of hoof-and-mouth disease, a highly infectious disease involving many ruminant species in Europe back in the 1990s.

Such episodes sparked a heated debate, and food safety has since become a matter of priority in domestic and Community (now, EU) policies committed to ensuring adequate safety standards and paying constant attention to consumers.

20.1.2 The Fundamental Aims of the Green Paper

Public concern thus led the EC Commission, back in the early 1990s, to ask three experts, namely Charles Castang, Amanda Clearly, and Dieter Eckert, to formulate a first regulatory project aimed at identifying the fundamental principles of future food law.

The experts' conclusions were presented in May 1993 at the European Institute of Florence and were then reported in an initial document, the Green Paper on "The General Principles of Food Law in the European Union".[1] The document, although of a merely programmatic nature, is the essence of what will be considered here in this chapter, and clarifies its aims and scope of analysis in its Summary. Indeed, in the General Background it specifies that *"A high level of security and effective public control is necessary to ensure that the food supply is safe and wholesome and to ensure the effective protection of the other interests of consumers."*

Based on this principle, six basic goals for Community food law have been identified:

1. to ensure a high level of protection of public health, safety, and consumers;
2. to ensure the free movement of goods within the European single market;
3. to ensure that law-making is based on scientific grounds and a correct risk assessment approach;
4. to ensure the competitiveness of European industry, also in promoting exports;
5. to place the primary responsibility for safe food on industry, producers, and suppliers, using systems backed by effective official control and enforcement measures; and
6. to ensure that legislation is coherent, rational, and user-friendly.

According to what was reported in the Green Paper, a consistent and systematic approach to this topic must begin from an analysis of the Community legislation then in force, a summary of which may be found in the first section of the paper. Within the economic framework described in Part I,[2] the European Community's role is clear in promoting a transparent and stable situation that may contribute to further stable development in a sector defined as "vital", in a competitive and pro-competition context.

Part II of the Green Paper identifies the process for the simplification and rationalisation of EC food law and highlights the need for further Community legislation in order to complete the internal market.

Part III re-examines the different measures that could be adopted in order to rationalise or simplify the Community legislation then in force. In this respect, the

---

[1]Document COM (97)176 final, of 30 April 1997.

[2]According to the Green Paper, within the Community every household then spent on average about 20 % of its disposable income on food and drink, with an estimated spending for 1996 for the consumption of food, drinks, and tobacco equal to approximately 500,000 million ECU, and an estimated production equal to 510,000 million ECU. Food and drink production and consumption between 1984 and 1992 had grown at a constant real rate equal to approximately 2–2.5 % a year.

Paper gives preference, where possible, to the use of Regulations rather than to Directives[3]; the need for updating the legislation in force by taking into consideration technical and scientific progress; and the rationalisation of the main definitions used in food law, starting from those of *"foodstuffs"*[4] and of *"placing on the market"*.[5]

Finally, the Green Paper stresses the need for stringent safeguards and protections for consumer health by introducing, among other things, a general obligation regarding the safety and wholesomeness of food involving the entire food chain at all stages from primary production to final sale to the consumer, and that takes into account the increasingly complex interactions among operators (Part IV of the Green Paper).

All of the above is predicated on the condition that efficient management of the internal market is ensured by supervising the implementation of EC Directives by Member States and ensuring that all EC rules are correctly applied (Part V of the Green Paper).

The Green Paper concludes with a final section that concerns the external dimension of the Community's food legislation, with specific reference to WTO Agreements and developments within *Codex Alimentarius*. In this respect, there is an increasing need for all adopted measures to be based on scientific evidence, and also to take into consideration the international context within which they are placed.

20.2 The 2000 White Paper on Food Safety

20.2.1 The Principles of Food Safety

The Green Paper on "The General Principles of Food Law in the European Union" was followed by another important publication. On 12 January 2000, the well-known "White Paper on Food Safety" was published, setting out and specifying the

---

[3]"[...] *it is suggested that consideration be given to greater use of Regulations in appropriate cases, both in primary and in secondary Community legislation. However, legislation which is limited in scope to the harmonisation of general principles and criteria, such as legislation on the official control of foodstuffs, would continue to be adopted by means of a directive.*"

[4]For which it proposes to examine the definition included in the *Codex Alimentarius*, pursuant to which *"foodstuff means any substance or product, whether processed, partially processed or unprocessed, intended to be ingested by humans, with the exception of tobacco as defined by Directive 89/662/EEC, medicinal products as defined by Directive 65/65/EEC, and narcotic or psychotropic substances controlled by Member States pursuant to the relevant international conventions"*.

[5]For which it proposes the following definition: *"any operation the purpose of which is to supply foodstuffs to a third party, including supply for sale or any other form of transfer against payment or free of charge to a third party and storage with a view to supply to a third party, with the exception of supply for the purposes of scientific research conducted under the supervision of the Member States"*.

priorities, aims, and projects for the new EU food law. Once again, the introduction outlines the goal of the project: *"The European Union's food policy must be built around high food safety standards, which serve to protect, and promote, the health of the consumer"*.

The guiding principle underlying the entire system proposed is of the "comprehensive, integrated" type. Namely, it proposes an analysis of the entire food chain (the "*farm to table*" concept is introduced), as well as of each sector involved in the food industry. At the same time, clear responsibilities are required of different players: feed manufacturers, farmers, and food industry operators in so far as primary responsibility is concerned; competent authorities for monitoring, enforcement, and surveillance purposes; and consumers regarding the proper storage, handling, and cooking of food.

Furthermore, in an effort to make food policy "more coherent, effective and dynamic", a fundamental role is ascribed to risk analysis (through a system ensuring risk assessment, management, and communication), the precautionary principle, and the necessary contribution of scientific advice in a sector constantly developing and evolving.

20.2.2 Essential Elements of the Project and Change Prospects

Based on the above principles, the White Paper sets the following goals:

1. creation of an independent Food Authority, acting as a reference scientific body at the European level whose duties would be preparing and giving scientific advice, gathering and analysing appropriate information to ensure adequate collaboration with EU legislative bodies, and monitoring the food sector (chapter 4);
2. creation of a new legal framework by EU bodies to develop a coherent and transparent set of rules in the different sectors of the food chain (chapter 5);
3. development of a Community framework of national control systems, ultimately providing for the improvement of administrative co-operation (chapter 6);
4. consumers' progressive involvement by consulting and discussing with the public, and by making available accurate, correct information aimed at allowing consumers to make informed choices, thus improving the labelling system currently in force (chapter 7);
5. constant attention to the Community's "international" dimension and compliance with the obligations arising from joining world Organisations.

In addition to the above goals, the White Paper also proposes a detailed action plan with respect to the measures to be taken.

20.3 Regulation (EC) No. 178/2002

20.3.1 The "Safety from Farm to Table" Strategy

The Papers described above became a regulatory reality with "Regulation (EC) No. 178/2002 of the European Parliament and of the Council of 28 January 2002 laying down the general principles and requirements of food law, establishing the European Food Safety Authority and laying down procedures in matters of food safety".

The title of the Regulation attests to the multiple purposes pursued by the new body of rules, ranging from the introduction of a European Food Safety Authority to the creation of a regulatory body of principles and grounds for legislation on the matter.

As a confirmation of the wide-ranging scope of the Regulation, it is sufficient to observe its "legal grounds", which do not only make reference to the provisions of the EC Treaty on agricultural products (Article 37) but also to Articles 95, 133, and 152.4(b).

The Regulation comprises sixty-six Whereas statements (with the last expressing the difficulty in intervening in matters the competences of which should be shared with the Member States[6]) and sixty-five Articles.

The Regulation implements and codifies the aforementioned "from farm to table" concept, which had first been introduced in the 2000 White Paper on Food Safety analysed above. The formula carries the spirit of the entire legislation on food law matters—that is, the commitment to and challenges of reaching an integrated approach covering the entire food chain, in order to ensure the movement of safe food and feed throughout the European market.

For a systematic analysis of the main issues addressed by the Regulation, it is worth examining several aspects in detail.

20.3.2 Risk Analysis

Risk analysis (Article 6 of Regulation (EC) No. 178/2002) is one of the priorities that national and EC legislators needed to set in order to protect the primary interest of human and animal safety. In line with the White Paper on Food Safety, "*risk analysis*" is defined here (Article 3.10) as the "*process consisting of three*

---

[6]Whereas 66: "*It is necessary and appropriate for the achievement of the basic objectives of this Regulation to provide for the approximation of the concepts, principles and procedures forming a common basis for food law in the Community and to establish a European Food Safety Authority. In accordance with the principle of proportionality as set out in Article 5 of the Treaty, this Regulation does not go beyond what is necessary in order to achieve the objectives pursued*".

interconnected components: risk assessment, risk management and risk communication". It is worth assessing each of these phases separately:

1. *"Risk assessment"* is defined in Article 3.11 as a *"scientifically based process consisting of four steps: hazard identification, hazard characterisation, exposure assessment and risk characterisation"*. As may easily be inferred from this definition, being a scientific process, it must be delegated to experts having adequate professional competence.[7] As stated by the Judgment of the Court of First Instance (Third Chamber) of 11 September 2002 in Case T-70/99 (Alpharma Inc. v Council of the European Union), where experts carry out a scientific risk assessment, the competent public authority must be given *"sufficiently reliable and cogent information to allow it to understand the ramifications of the scientific question raised and decide upon a policy in full knowledge of the facts. Consequently, if it is not to adopt arbitrary measures, which cannot in any circumstances be rendered legitimate by the precautionary principle, the competent public authority must ensure that any measures that it takes, even preventive measures, are based on as thorough a scientific risk assessment as possible, account being taken of the particular circumstances of the case at issue. Notwithstanding the existing scientific uncertainty, the scientific risk assessment must enable the competent public authority to ascertain, on the basis of the best available scientific data and the most recent results of international research, whether matters have gone beyond the level of risk that it deems acceptable for society [. . .]. That is the basis on which the authority must decide whether preventive measures are called for"*, or establish the expedient and necessary measures for eliminating the risk.
2. *"Risk management"* is the *"process, distinct from risk assessment, of weighing policy alternatives in consultation with interested parties, considering risk assessment and other legitimate factors, and, if need be, selecting appropriate prevention and control options"* (Article 3.12). The results of risk assessment and the European Food Safety Authority's opinions form the basis for appropriate planning of the risk management strategy providing for the choice of the most appropriate measures to prevent or control risk connected with the consumption of food or feed. Of course, the adoption of certain measures aimed at managing risk and providing for a limitation or restriction of the free movement

---

[7]*See* Judgment of the Court of First Instance (Third Chamber) of 11 September 2002—Pfizer Animal Health SA v Council of the European Union—Case T-13/99: "*A scientific risk assessment is commonly defined, at both the international level (see the provisional communication from the* Codex Alimentarius *Commission, cited at paragraph 147 above) and Community level (see the Communication on the Precautionary Principle, the Communication on Consumer Health and Food Safety and the green paper, cited at paragraphs 118 and 124 above), as a scientific process consisting of the identification and characterisation of a hazard, the assessment of exposure to the hazard and the characterisation of the risk. [. . .]A scientific risk assessment carried out as thoroughly as possible on the basis of scientific advice founded on the principles of excellence, transparency and independence is an important procedural guarantee whose purpose is to ensure the scientific objectivity of the measures adopted and preclude any arbitrary measures*".

of goods may be justified solely by real risks and must comply with the principles of non-discrimination and proportionality, while at the same time assessing the advantages and charges that each measure entails.
3. Finally, the entire risk analysis process concludes with the stage of *"risk communication"*, which is defined as *"the interactive exchange of information and opinions throughout the risk analysis process as regards hazards and risks, risk-related factors and risk perceptions, among risk assessors, risk managers, consumers, feed and food businesses, the academic community and other interested parties, including the explanation of risk assessment findings and the basis of risk management decisions"* (Article 3.13).

In conclusion, as stated in Whereas 17 of the Regulation, *"the three interconnected components of risk analysis—risk assessment, risk management, and risk communication—provide a systematic methodology for the determination of effective, proportionate and targeted measures or other actions to protect health"*.

20.3.3   The Precautionary Principle

Article 7 of Regulation No. 178/02/EC makes express reference to the precautionary principle, which is codified in the food sector for the very first time[8] via this regulation.

It is a general principle to be applied in the event of scientific uncertainty, where, following an objective assessment, it is impossible to exclude the existence of well-founded reasons for deeming that there might be harmful effects on human health or the environment.

The principle is recognised as significant in Whereas 21 of the Regulation, where it is stated that *"in those specific circumstances where a risk to life or health exists but scientific uncertainty persists, the precautionary principle provides a mechanism for determining risk management measures or other actions in order to ensure the high level of health protection chosen in the Community"*.

The intent, clearly, is of a mitigating character, taking into account the pre-eminence of the respect for and protection of human health in the presence of a situation of scientific uncertainty calling for the adoption of precautionary measures to prevent a risk to human health. In this respect, risk assessment, in practice, will necessarily have to be carried out on the basis of the most reliable scientific data available and the most recent international results rather than through a merely

---

[8]The precautionary principle arises within international environmental law with the World Charter for Nature adopted by the United Nations General Assembly in 1982. Within the European Union, the principle is first mentioned (with reference to environmental protection only, though) in Article 174.2 of the EC Treaty, which, however, fails to give any definition of the same.

hypothetical approach to risk based on simple assumptions without any scientific substantiation.[9]

As mentioned above, Article 7 sets forth the rules for applying the principle, providing that "*in specific circumstances where, following an assessment of available information, the possibility of harmful effects on health is identified but scientific uncertainty persists, provisional risk management measures necessary to ensure the high level of health protection chosen in the Community may be adopted, pending further scientific information for a more comprehensive risk assessment*" (Article 7.1).

Therefore, in the event that a product, a phenomenon, or a procedure may entail harmful effects on health, the competent institutions "may" decide whether to adopt "*provisional*" measures.

The provisional character of the measures adopted is specified in the last sentence of paragraph 2 of Article 7, where it is clarified that they "*shall be reviewed within a reasonable period of time, depending on the nature of the risk to life or health identified and the type of scientific information needed to clarify the scientific uncertainty and to conduct a more comprehensive risk assessment*". As to scientific uncertainty, it may concern both qualitative and quantitative aspects of the analysis.[10]

In any event, recourse to the precautionary principle does not allow the adoption of arbitrary measures or the waiver of the cornerstone principles of the legal system, such as the principles of good risk management, proportionality, non-discrimination, and coherence. Specifically, the measures adopted must comply with the principle of proportionality as described in the first part of paragraph 2 of Article 7, pursuant to which "*the measures adopted on the basis of paragraph 1 shall be proportionate and no more restrictive of trade than is required to achieve the high level of health protection chosen in the Community, regard being had to*

---

[9]*See* Judgment of the Court of 9 September 2003 Case C-236/01 (Monsanto Agricoltura Italia SpA and Others v Presidenza del Consiglio dei Ministri and Others): "[...] *protective measures may be taken pursuant to Article 12 of Regulation No. 258/97 interpreted in the light of the precautionary principle even if it proves impossible to carry out as full a risk assessment as possible in the particular circumstances of a given case because of the inadequate nature of the available scientific data (see to that effect Pfizer Animal Health v Council, cited above, paragraphs 160 and 162, and Alpharma v Council, cited above, paragraphs 173 and 175). Such measures presuppose, in particular, that the risk assessment available to the national authorities provides specific evidence which, without precluding scientific uncertainty, makes it possible reasonably to conclude on the basis of the most reliable scientific evidence available and the most recent results of international research that the implementation of those measures is necessary in order to avoid novel foods which pose potential risks to human health being offered on the market [...]*."

[10]For an application of the precautionary principle, although in connection with the cosmetic sector, see the Judgment of the Court of First Instance (Third Chamber) of 16 July 1998— Laboratoires pharmaceutiques Bergaderm SA and Jean-Jacques Goupil v Commission of the European Communities—Case T-199/96, pursuant to which "*...where there is uncertainty as to the existence or extent of risks to the health of consumers, the institutions may take protective measures without having to wait until the reality and the seriousness of those risks become fully apparent*".

technical and economic feasibility and other factors regarded as legitimate in the matter under consideration".

Finally, it is worth noting that there have been cases where, in the absence of scientific certainties, the application of the principle of "zero tolerance" by national legislation, totally prohibiting a certain substance in certain foods, has been deemed lawful.[11]

20.3.4 Protection of Consumers' Interests

Consumers' crucial role within the framework of Regulation (EC) No. 178/2002 may not only be inferred from the constantly referred to purpose of protection of consumer health, but also from the further provisions contained in Article 8 of the Regulation.

In particular, the provision aims to protect against:

1. "*fraudulent or deceptive practices*": the main reference is to the regulations on labelling and presentation of food and feed products;
2. "*the adulteration of food*" and any manipulation changing the composition of food more or less deeply significantly: see, in particular, Articles 14 and 15 of Regulation (EC) No. 178/2002 prohibiting the placing on the market of unsafe food insofar as it is injurious to human or animal health;
3. "*any other practices which may mislead the consumer*".

Obviously, a system for consumer protection cannot be implemented without an adequate official control structure and an effective sanctioning system.

---

[11]*See* Melkunie Case—Judgment of the European Court of Justice (Fifth Chamber) of 6 June 1984—Criminal proceedings against CMC Melkunie BV—C-97/83: "... *According to a consistent line of decisions of the Court, it follows from Article 36 that a national measure which has, or may have, a restrictive effect on trade is compatible with the Treaty only in so far as it is necessary for the purpose of effectively protecting human life and health. The proviso in Article 36 cannot therefore apply where human life and health can be protected just as effectively by measures less restrictive of intra-Community trade. [...] The data available at the present stage of scientific research do not make it possible to determine with certainty the precise number of nonpathogenic micro-organisms above which a pasteurised milk product becomes a source of danger to human health. In the absence of harmonisation in this field, it is for the member-States to determine, with due regard to the requirements of the free movement of goods, the level at which they wish to ensure that human life and health are protected. In those circumstances, national legislation seeking to ensure that at the time of consumption the milk product in question does not contain micro-organisms in a quantity which may constitute a risk merely to the health of some, particularly sensitive consumers, must be considered compatible with the requirements of Article 36*".

20.3.5 The Principle of Transparency

Section 2 of Chapter II of Regulation (EC) No. 178/2002 focuses on an issue to which predominant, and growing, attention is being paid within the European Union, namely, the principle of transparency.

More specifically, the above principle is defined in two different respects, both equally important and crucial to the matter considered here, as also mentioned in Whereas 9, which makes reference to the importance of a constructive relationship amongst all parties involved *("It is necessary that consumers, other stakeholders and trading partners have confidence in the decision-making processes underpinning food law, its scientific basis and the structures and independence of the institutions protecting health and other interests")*, namely:

1. Public consultation: according to Article 9 of the Regulation, *"there shall be open and transparent public consultation, directly or through representative bodies, during the preparation, evaluation and revision of food law, except where the urgency of the matter does not allow it"*. The Commission therefore must consult with the public and the parties concerned, also by organising committees and groups of undertakings with advisory purposes, and carry out specific (including online) consultations, both during the proposal for legislation stage and in the following phase of adoption of a measure; *and*
2. Public information: the goal of ensuring food safety is also achieved through the quick spreading of any information on the possible risks related to the use of certain products or feedstuffs. Therefore, where there are reasonable grounds to suspect that a food or feed may present a risk for human or animal health, then, pursuant to Article 10 of Regulation (EC) No. 178/2002, public authorities must take appropriate steps to inform the general public of the nature, seriousness, and extent of the risk to health, *"identifying to the fullest extent possible the food or feed, or type of food or feed, the risk that it may present, and the measures which are taken or about to be taken to prevent, reduce or eliminate that risk"*.

The European Food Safety Authority is also requested to play a fundamental role and comply with the principle of transparency, accomplishing its duties in abidance of said principle, as also laid down by Article 38 of Regulation (EC) No. 178/2002.

20.3.6 General Requirements of Legislation and the Principle of the Responsibility of Business Operators

In addition to the principles outlined above, Regulation No. 178/2002 also sets forth the "general requirements" with which the European and national laws and regulations need to harmonise.

In particular, Article 14 states that "***food shall not be placed on the market if it is unsafe***". Expressed in positive terms the rule included in the first paragraph of this Article states that exclusively safe food may be validly placed on the market and move freely therein. The EC legislator furthermore established that whenever food complies with specific Community provisions applicable to food safety, it must be deemed safe insofar as the aspects covered by the specific Community provisions are concerned.

Conversely, it is prohibited to place on the market any food deemed be "*unsafe*"—that is, considered to be "*injurious to health*" and "*unfit for human consumption*".

More specifically, to determine whether any food is injurious to health, reference must be made to a number of factors, namely: the use of the food by the consumer, the probable immediate and/or short-term and/or long-term effects of that food on the consumer, the probable cumulative toxic effects, and the particular health sensitivity of a specific category of consumers where the food is intended for that category (it is sufficient to think of the greater vulnerability of infants, the delicate character of follow-on formulae, or the specificity of dietetic foods). To determine whether any food is unfit for human consumption, consideration must be given to whether the food is unacceptable for human consumption according to its intended use, taking into account all useful elements (e.g., food contamination, deterioration, decay, etc.).

A fundamental role within food safety is furthermore ascribed to the **presentation, advertising, and labelling** of food or feed, which "*shall not mislead consumers*" (Article 16). This, obviously, is linked to the concept of providing correct information to consumers who, solely with a thorough, truthful presentation of food or feed, can make informed choices. For this purpose, reference is made to the content of recent Regulation (EU) No. 1169/2011 of the European Parliament and of the Council of 25 October 2011 on the provision of food information to consumers, which lays the basis to guarantee a high level of protection of consumers with respect to food information, and defines the related principles, requirements, and responsibilities.

An important aspect introduced by Article 18 of the Regulation is **traceability**, defined as "*the ability to trace and follow a food, feed, food-producing animal or substance intended to be, or expected to be incorporated into a food or feed, through all stages of production, processing and distribution*". It is a control system applicable to all food and feed business operators for all categories of food and feed (and, more generally, substances intended to be incorporated into a food or feed), thus allowing for the monitoring of the food chain as a whole, by collecting and spreading certain information allowing the tracing of a certain food or feed throughout the different stages of production, processing, storage, transport, and sale. The system makes it possible to identify the business operator at the source, from whom the product originates, and, downstream, the one to whom the company has supplied the food or feed. All that is required while carefully managing risks facilitating, if necessary, the procedures for action and market recall of goods unfit for distribution.

Also for the above purposes, Article 18 further provides that the food and feed intended to be placed on the market must be labelled or identified to *"facilitate its traceability, through relevant documentation or information in accordance with the relevant requirements of more specific provisions"*.

The system outlined by Regulation No. 178/2002 therefore introduces a cornerstone principle of the new food safety system, providing for the **responsibility of the food business operator** (defined as the *"natural or legal persons responsible for ensuring that the requirements of food law are met within the food business under their control"*), and the feed business operator. Said operators must ensure that the foods or feeds *"satisfy the requirements of food law"* (Article 17 of the Regulation).

As is clear, this rule does not only concern nor is it limited to the phase related to the placing on the market of the foods and feeds, but it also entails a series of further obligations, if a food business operator considers or has reason to believe that a food that it has imported, produced, processed, manufactured, or distributed, and no longer available or under its immediate control, *"is not in compliance with the food safety requirements"*. In such cases, Article 19 of the Regulation provides for a series of activities to be carried out by the food business operator, also in collaboration with the competent authorities. In particular, the operator in question must:

1. immediately initiate procedures to withdraw the food in question from the market—that is, take all measures aimed at preventing the distribution and exposure of an unsafe food as well as its offer to consumers;
2. inform the competent national authority, which must, if necessary, adopt all necessary measures to face the emergency;
3. effectively and accurately inform consumers of the reason for its withdrawal where the product may have reached consumers;
4. finally, if necessary, *"recall"* from consumers products already supplied to them, thus recovering the products already supplied or made available.

This type of approach, in any case, is flexible and does not specify how and by which means undertakings must manage any non-compliance of the products with food law, but leaves it to the operators to arrange their own organisation in the best and most congenial possible way to face any crisis.

20.3.7 Rapid Alert System and Crisis Management Plan

Regulation (EC) No. 178/2002, Chapter IV, introduces a further valuable instrument in food safety—the Rapid Alert System for Food and Feed, or RASFF.

This is a network linking Member States through their respective National Authorities, the Commission, and the European Food Safety Authority to implement a single system capable of identifying and rapidly (in real time) circulating all alerts concerning urgent food safety issues.

The system has a precedent, namely the rapid exchange of information system for all products intended for the end consumer, also included in foodstuffs (RAPEX), introduced by Council Directive 92/59/EEC on general product safety. However, this system failed to consider feedstuffs and, in light of the food crises involving the European scenario prior to the issue of Regulation No. 178/2002, it was apparent that a more effective and adequate structure was necessary to handle the existing situation.[12]

The parties involved in the new system are the Member States, the Commission, and the European Food Safety Authority, which are obligated to share any information of which they may become aware that may be deemed relevant for the system. They also have surveillance, monitoring, and control duties in connection with tasks related to food and feed business risks. Furthermore, the Commission is responsible for managing the network.

The information shared and circulated within the system concerns—with a deliberately wide and all-encompassing wording—any "*direct or indirect risk to human health deriving from food or feed*" (Article 50 of the Regulation).

This information is circulated through a system of different notifications, namely:

1. Alert (greatest level of hazard): there is a serious risk to health caused by a product already on the market, and it is necessary to take immediate action.
2. Information: the unsafe product has not reached the market or has expired in the meantime, and urgent measures are not necessary.
3. News: the information is of a general nature and concerns non-conformity ascertained in a Member State or in a non-EU Member State that may be useful for others to guide any and all official controls.
4. Border Rejection: information on the rejection of a batch at a border post for breach of EU rules and regulations.

The system allows the parties involved to forward and share information, thus encouraging an integrated use both in terms of monitoring and in order to react to emergencies rapidly, all with the aim of guaranteeing a high level of safeguards for human health.

In addition to a detailed explanation of the reasons for action, Member States must also provide notification of: (a) any measure they adopt aimed at restricting the placing on the market, or forcing the withdrawal from the market of, or the recall of, food or feed in order to protect human health and requiring rapid action;

---

[12]*See* Whereas 59: "*A system for rapid alert already exists in the framework of Council Directive 92/59/EEC of 29 June 1992 on general product safety. The scope of the existing system includes food and industrial products but not feed. Recent food crises have demonstrated the need to set up an improved and broadened rapid alert system covering food and feed. This revised system should be managed by the Commission and include as members of the network the Member States, the Commission and the Authority. The system should not cover the Community arrangements for the early exchange of information in the event of a radiological emergency as defined in Council Decision 87/600/Euratom.*"

(b) any recommendation or agreement with professional operators that is aimed, on a voluntary or obligatory basis, at preventing, limiting, or imposing specific conditions on the placing on the market or the use of food or feed on account of a serious risk to human health requiring rapid action; and (c) any rejection, related to a direct or indirect risk to human health, of a batch, container, or cargo of food or feed by a competent authority at a border post within the European Union.

It is worth noting that, in compliance with the principle of transparency generally laid down by the Regulation, Member States are bound to make all information on the risks to human or animal health caused by food or feed available to the public, subject to trade secrets or confidentiality duties.

In 2013 RASFF handled 3136 notifications, in 2012 3434 notifications, and in 2011 3721 notifications.

More specifically, 2649 notifications concerned food, 262 concerned feed (325 concerned feed in 2012), and 225 concerned the migration of materials intended to be in contact with food and feed.

In addition to the official controls made on the market, 4 % of notifications were implemented following consumer complaints, 13 % concerned the notifications of negative results made by way of self-control by companies, and 1 % were related to food contamination.[13]

20.4 The European Food Safety Authority

20.4.1 The Creation the Authority

The creation of the European Food Safety Authority is the core and immediately applicable part of Regulation (EC) No. 178/2002.

The creation of an Authority in the food sector has always been a priority, and it was first accomplished through the Scientific Food Committee set up by the Commission by Decision 74/234/EEC of 16 April 1974.[14] It was a committee mostly with advisory duties on issues related to human health and life, in connection with the consumption of food, in particular, in the light of issues concerning nutrition, hygiene, and toxicology. With Decision 76/791/EEC of the Commission

---

[13]Data drawn from the 2013 Annual Report on the Community Alert System of the Italian Ministry of Health, which has highlighted a drop of 8.7 % compared to the previous year of notifications forwarded through the alert system by Member States (the drop between 2012 and 2011 was 7.7 %). The analysis shows that Italy is the top Member State per number of notifications sent through the RASFF system, with 534 notifications. Regarding origin, Italy is the fourth Member State per number of received notifications, after Spain, Poland, and France.

[14]OJEC 1976 No. 136 p. 1, as amended.

of 24 September 1976[15] a Scientific Feed Committee was also set up, with similar advisory duties concerning the quality and salubrity of feed.[16]

Nonetheless, the significant food crises of the 1990s have made it clear that a reorganisation of the scientific and technical support system is necessary, something capable of contributing to providing adequate support for the new regulatory system that was taking shape. The different scientific Committees set up throughout time and initially working at the Directorate-General for the Internal Market or for the Directorate-General for Agriculture of the Commission were replaced by eight new committees by way of Decision 97/579/EC of the Commission of 23 July 1997 (then abrogated by Decision 2004/210/EC):

1. Scientific Committee on Food;
2. Scientific Committee on Feed;
3. Scientific Committee for Animal Health and Animal Welfare;
4. Scientific Committee on Veterinary Measures relating to public health;
5. Scientific Committee for Plants;
6. Scientific Committee on Cosmetic Products and Non-food Products Intended for Consumers;
7. Scientific Committee on Medicinal Products and Medical Devices;
8. Scientific Committee on Toxicity, Ecotoxicity and the Environment.

Nonetheless, only under the 2000 White Paper on Food Safety, that is, with the acknowledgement of the limits of the previous system,[17] was a proposal put forward to set up an independent European Food Authority with special responsibilities both in terms of risk assessment and risk communication, and adequate authority on food safety-related issues. The Authority furthermore acts following the rules set forth by the emerging single food legislation and, hence, pursuant to the principles of independence, excellence, and transparency.

---

[15]OJEC 1976 No. L 279 p. 35, as amended.

[16]Besides the committees described above, towards the end of the 1970s and in the early 1980s, other entities were set up, amongst which were the Scientific Committee on cosmetology (Decision 78/45/EEC of the Commission of 19 December 1977); the advisory Scientific Committee for the examination of toxicity and ecotoxicity of chemical compounds (Decision 78/618/EEC of the Commission of 28 June 1978); the Scientific Committee for pesticides (Decision 78/436/EEC of the Commission of 21 April 1978); and the veterinary Scientific Committee (Decision 81/651/EEC of the Commission of 30 July 1981).

[17]*See* White Paper on Food Safety, paragraph 25, where it is stated that "*It has become evident that the existing system is handicapped by a lack of capacity and has struggled to cope with the increase in the demands placed upon it. Furthermore, the recent dioxin crisis could only be managed by delaying works in other areas and has shown the need to have a system which is able to respond rapidly and flexibly. This lack of capacity has led to delays which have consequences both for the Commission's legislative programmes, and hence its ability to respond to consumer health problems, and for industry where commercial dossiers are involved. This situation will be exacerbated by the increased demands that will be placed on the scientific committees resulting, for example, from the proposed programme for reform of food legislation as set out later in this White Paper*".

As already noted, Regulation (EC) No. 178/2002 implements the provisions of the White Paper and sets up the European Food Safety Authority (EFSA), to which the duties of five of the eight former scientific Committees were transferred.

20.4.2 Tasks, Structure, and Operation

EFSA's main **task** is to ensure constant, competent, and independent technical advice and scientific support in the matter of food safety by collecting and analysing data either collected by it or brought to its attention. As specified in Whereas 35 of the Regulation *"The Authority should be an independent scientific source of advice, information and risk communication in order to improve consumer confidence".*

In performing its duties, the Authority must also: offer scientific advice, together with scientific and technical support on human nutrition; formulate scientific opinions on any issue related to the health and welfare of animals, and to the health of vegetables; and formulate scientific opinions on non-food and non-feed products that may be traced back to genetically modified organisms.

Therefore, it is a fundamental reference for the new system, capable of supporting the preparation, definition, and implementation of food legislation, also through a coordination of roles among the competent bodies of the different Member States and the Community bodies.

Article 23 of Regulation No. 178/2002 clarifies and lists the tasks of the Authority, by establishing that the latter must:

(a) provide scientific and technical assistance and advice;
(b) provide opinions on the subject matter;
(c) search for, collect, collate, analyse, and summarise scientific and technical data in the fields within its mission;
(d) contribute to collaboration amongst the different organisations operating in the fields within its mission and create a European network; and
(e) identify and communicate emerging risks.

The Authority is composed of four main **bodies**, to which specific tasks and duties are ascribed.

The Management Board, chaired by a Chairperson and numbering fifteen members, fourteen of whom are appointed by the Board, in consultation with the Parliament, plus a representative of the Commission, has the duty to ensure the operation of the Authority by preparing the work schedule and by reporting on the activity carried out by the Authority.

The Executive Director (who is also EFSA's legal representative) is appointed for a 5-year term by the Management Board and has, among other things, the following duties: draw up a proposal for the Authority's work programmes in consultation with the Commission; implement the work programmes and the decisions adopted by the Management Board; ensure the provision of appropriate scientific, technical, and administrative support for the Scientific Committee and the Scientific Panels; ensure that the Authority carries out its tasks in accordance with the requirements of its users, in particular with regard to the adequacy of the services provided and the time taken; prepare the statement of revenue and expenditure and the execution of the budget of the Authority; prepare all matters related to staff and annual accounts; develop and maintain contact with the European Parliament; and ensure a regular dialogue with its relevant committees.

The role of the Advisory Forum, which is composed of representatives from the competent bodies in the Member States that undertake tasks similar to those of the Authority, is to allow the exchange of information, avoid duplications of studies, and promote collaboration between the European Authority and the local authorities.

Finally, the Scientific Committee and permanent Scientific Panels are responsible for providing the scientific opinions of the Authority, each within their own spheres of competence.

The main instrument through which EFSA's **activity** is carried out is the formulation, by the Scientific Committee or permanent Scientific Panels, of "*scientific opinions*",[18] through which the core aim of the system (that is, risk assessment) is achieved. Opinions may be given on the Authority's own initiative or upon the request of the Commission, of the European Parliament, and of each single Member State (but not of consumers or of the economic operators or of their trade associations), as provided for under Regulation (EC) No. 1304/2003 of the Commission of 11 July 2003, which sets forth the procedure to be followed. The Commission must request a scientific opinion whenever Community legislation calls for consultation with EFSA. Requests for opinions must be accompanied by informative documentation on the scientific issue to be examined and the underlying Community interest. Should Community legislation fail to make reference to a deadline for giving the opinion, EFSA will give the opinions as per the relevant requests.

In case of "*divergence*" between the opinions given by EFSA and those given by other research institutes, Regulation No. 178/2002 provides specific procedures to evaluate and remedy it (Article 30) by establishing, first, that the Authority must refer to the body at issue in order to ascertain that all relevant information is shared. Should there be any substantive divergence over scientific issues, EFSA and the body with a non-aligned and divergent position (be it at a national or community level) must cooperate to rectify the divergence, in view of drawing up a joint document clarifying the contentious scientific issues and identifying the relevant uncertainties in the data.

Apart from the cases in which a scientific evaluation of the Scientific Committee or of the permanent Scientific Panels is necessary, the Authority may also be requested to provide "*scientific and technical assistance*" to the Commission in any field within its mission, both for the creation or assessment of technical criteria, and for the preparation of technical opinions, through scientific or technical work involving the application of well-established scientific or technical principles.

In fulfilling its own duties, EFSA may commission scientific studies and use the best independent resources available. It is also important that it searches for, collects, collates, analyses, and summarises the scientific data in the fields within

---

[18]Article 29 of Regulation No. 178/2002.

its mission so as to gather as wide as possible a range of useful information for assessing existing or potential risks. The above must be accomplished with specific regard for the consumption of food and the risks to which natural persons are exposed by consuming the food, the impact and spreading of biological risks, the contaminants in food and feed, and any residues.

As stated, EFSA's duties focus on risk assessment and, in this respect, not only is the reaction to any crisis important, but so is its prevention: indeed, an effective food safety policy entails different operational levels.

Precisely in this respect, Article 34 of Regulation No. 178/2002 introduces the concept of *"emerging risk"*[19]: it is the duty of the Authority to establish surveillance procedures for the systematic activity of searching, collecting, collating, and analysing information and data, in order to identify the emerging risks in the fields within its mission.

The second operational level entails the request by EFSA for further information to Member States, to other Community agencies, and to the Commission. Based on any such information, the Authority will identify, if necessary, an emerging risk and send the assessment and the information collected to the European Parliament, the Commission, and the Member States.

Finally, it is worth mentioning EFSA's duty to promote cooperation between organisations that are active at a European level in the fields of food safety, thus facilitating a scientific cooperation framework, *"by the coordination of activities, the exchange of information, the development and implementation of joint projects, the exchange of expertise and best practices in the fields within the Authority's mission"* (Article 36 of Regulation No. 178/2002).

Finally, just as risk assessment has a fundamental role from the point of view of the operation and tasks ascribed to EFSA, so does the *"risk communication"* activity. The Authority, of its own initiative, makes any disclosure that it may deem expedient in food safety matters, through its own competent offices.

20.4.3 The Role Played Throughout the Authority's First 10 Years

The balance of EFSA's first 10 years of activity has been positive.

Since it began operating, the Authority has published more than 2500 **scientific outputs** that have been used by the European Commission, Member States, and the European Parliament as the basis for risk management measures and policy initiatives.

---

[19]*See* Whereas 50 of Regulation No. 178/2002: *"Improved identification of emerging risks may in the long term be a major preventive instrument at the disposal of the Member States and the Community in the exercise of its policies. It is therefore necessary to assign to the Authority an anticipatory task of collecting information and exercising vigilance and providing evaluation of and information on emerging risks with a view to their prevention".*

EFSA has decisively contributed to the European Union's timeliness of action upon any and all food emergencies, such as in the case of dioxin contamination of pork in Ireland back in 2008 and the *E. coli* epidemic in Germany and France in 2011.

But the EU landscape has changed significantly since EFSA was established. The number of Member States has grown from 15 to 28 and the EU has become the largest importer and exporter of foodstuffs, especially processed goods, in the world. With the increase in population and territory, issues surrounding food production have become ever more complex given the emergence, for example, of new technologies such as "novel foods" and genetically modified organisms.[20]

In such a framework, EFSA has successfully adapted itself and turned into a complex organisation capable of considering the overall scenario without overlooking any detail.

And in the words of the Authority one may find its willingness to continue developing: "*As the organisation enters the next stage of its development it will continue to support the European food safety system with science-based risk assessments. As well as protecting consumers, EFSA aims to continue providing food producers, processors and distributors with a regulatory environment that is demanding but predictable. This will foster technological innovation in the economically important agrifood sector and support sustainable growth and development in the Europe of the future*".[21]

20.5 "Hygiene Package" Regulations and Implementing Regulations

Through the years, the European legislation on food safety has also been concerned with the harmonisation of food hygiene requirements. This was done first for specific categories of food[22] and, subsequently, through Directive No. 93/43/EEC of the Council of 14 June 1993 on the hygiene of foodstuffs, for all types of food.

---

[20] *See* EFSA Website (http://www.efsa.europa.eu/en/10thanniversary/achievements.htm), which reads: "*One area in which EFSA's work has changed significantly over the past 10 years is the evaluation of regulated products such as **food additives, GMOs, pesticides and health claims, which accounts for more than 60% of the Authority's outputs**. The resources committed to this work doubled between 2008 and 2010, from 20% to 40%*".

[21] *See* EFSA Website (http://www.efsa.europa.eu/en/10thanniversary/achievements.htm), where it also reads, "*As outlined in the Strategic Plan 2009-2013 and the Science Strategy 2012-2016 (adopted in 2011), the Authority increasingly will focus on **integrated multi-disciplinary advice** in areas such as meat inspection, nutrition and animal welfare. It will continue to ensure it performs to the highest standards through the development of state-of-the-art, **harmonised methodologies and the collection and analysis of quality data**"*.

[22] *See* by way of example, Directive 64/433/EEC of the Council of 26 June 1964 on health problems affecting intra-Community trade in fresh meat; Directive 89/362/EEC of the Commission of 26 May 1989 on general conditions of hygiene in milk production holdings; Directive

By introducing Regulation No. 178/2002, the need for integrating and consolidating the previous system through a new legal system in the matter of hygiene of foodstuffs in general became clear. Therefore, a package of Regulations (the so-called "hygiene package") was issued, setting forth harmonised rules within the field.

20.5.1 Regulation (EC) No. 852/2004 on the Hygiene of Foodstuffs and the HACCP System

The first regulation of the Hygiene Package is Regulation (EC) No. 852/2004 of the European Parliament and of the Council of 29 April 2004 on the hygiene of foodstuffs, which is aimed at implementing a global and integrated policy in the matter of food hygiene involving each stage of the food chain, save for primary production for private domestic use; for food intended for private domestic consumption; for domestic preparation, handling, or storage; and for the direct supply of small quantities of primary food from the manufacturer to the end consumer.

In this respect, Whereas 8 clarifies well that "*An integrated approach is necessary to ensure food safety from the place of primary production up to and including placing on the market or export. Every food business operator along the food chain should ensure that food safety is not compromised*".

From this standpoint, the core role of the food business operator is still the essence, as the latter has "*primary responsibility for food safety*" (Article 1(a)), as well as the duty to guarantee that all production, processing, and distribution phases of the food under their control meet the general hygiene requirements set forth by the Regulation.[23]

92/46/EEC of the Council of 16 June 1992, laying down the health rules for the production and placing on the market of raw milk, heat-treated milk, and milk-based products; Directive 89/437/EEC of the Council of 20 June 1989 on hygiene and health problems affecting the production and the placing on the market of egg products; Directive 91/492/EEC of the Council of 15 July 1991 laying down the health conditions for the production and the placing on the market of live bivalve molluscs; Directive 91/493/EEC of the Council of 22 July 1991 laying down the health conditions for the production and the placing on the market of fishery products; Directive 71/118/EEC of the Council of 15 February 1971 on health problems affecting trade in fresh poultry meat; Directive 91/495/EEC of the Council of 27 November 1990 concerning public health and animal health problems affecting the production and placing on the market of rabbit meat and farmed game meat; Directive 92/45/EEC of the Council of 16 June 1992 on public health and animal health problems relating to the killing of wild game and the placing on the market of wild-game meat; and Directive 94/65/EC of the Council of 14 December 1994 laying down the requirements for the production and placing on the market of minced meat and meat preparations.

[23] *See* Article 3 of Regulation No. 852/2004—"*General obligation – Food business operators shall ensure that all stages of production, processing and distribution of food under their control satisfy the relevant hygiene requirements laid down in this Regulation*".

Specifically, food business operators must implement and maintain one or more **permanent self-control company procedures, based on the HACCP** (*Hazard Analysis and Critical Control Points*) **principles system**, proving compliance with said obligation to the competent authorities, if requested.

It is the most significant and innovative aspect of the new approach introduced with Regulation No. 852/2004, since it replaces the previous system based on the control of the finished product, which was statistically unreliable and often retrospective, with a method allowing the identification, prevention, removal, or mitigation of any existing risks within the chain through a preventive analysis of each single production process.

The principles on which the HACCP system is based, as stated under Article 5 of Regulation No. 852/2004, consist of the following:

1. identifying any hazards that must be prevented, eliminated, or reduced to acceptable levels;
2. identifying the critical control points at the step or steps at which control is essential to prevent or eliminate a hazard or to reduce it to acceptable levels;
3. establishing critical limits at critical control points that separate acceptability from unacceptability for the prevention, elimination, or reduction of identified hazards;
4. establishing and implementing effective monitoring procedures at critical control points;
5. establishing corrective actions when monitoring indicates that a critical control point is not under control;
6. establishing procedures, to be carried out regularly, in order to verify that the measures outlined above are working effectively; establishing documents and records commensurate with the nature and site of the food business to demonstrate the effective application of the measures outlined above; and
7. when any modification is made in the product, process, or any step, review of the procedure and necessary changes are made to it.

The self-control system thus outlined requests the assistance of qualified experts, together with the installation of adequate equipment. Also, it must be specific for each single business reality, as well as flexible, and must be applied to all situations, also with respect to small businesses.[24]

If necessary, food business operators must also adopt, in addition to the self-control company procedures based on the HACCP system, further **specific hygiene**

---

[24]In this respect, see Whereas 15: *"The HACCP requirements should take account of the principles contained in the* Codex Alimentarius. *They should provide sufficient flexibility to be applicable in all situations, including in small businesses. In particular, it is necessary to recognise that, in certain food businesses, it is not possible to identify critical control points and that, in some cases, good hygienic practices can replace the monitoring of critical control points. Similarly, the requirement of establishing "critical limits" does not imply that it is necessary to fix a numerical limit in every case. In addition, the requirement of retaining documents needs to be flexible in order to avoid undue burdens for very small businesses"*.

measures related to compliance with the microbiological criteria of foodstuffs; achieve the aims fixed for accomplishing the purposes of Regulation No. 852/2004; comply with food temperature control requirements; keep the cold chain; and perform sampling and analyses.

Food business operators may also use the so-called "*Guides to good practice for hygiene and for the application of HACCP principles*" (Articles 7–9 of Regulation No. 852/2004). These guidelines, which are adopted voluntarily, are aimed at assisting companies and helping them in their duty to comply with the legislation on food hygiene and safety. Starting from the risk analysis referred to in the respective sector of reference, they put forward necessary and sufficient surveillance instruments in order to ensure compliance with the legislation at issue.

They are prepared by the food industry, in collaboration with the competent national bodies, and in the light of the relevant codes of practice included in the *Codex Alimentarius*.

Article 6 of Regulation No. 852/2004 then lays down a further obligation upon food business operators, who must **notify the appropriate competent authority, for registration purposes**, of each establishment under their control that carries out any of the stages of production, processing, and distribution of food, also ensuring that the competent authority always has up-to-date information on establishments, including by notification of any significant change in activities and any closure of an existing establishment.

Finally, the Annexes to the Regulation govern the general applicable food hygiene requirements in primary production and in any related operations (Annex I) and in the remaining phases (Annex II).

20.5.2 Regulation (EC) No. 853/2004 Laying Down Specific Hygiene Rules Applicable to Products of Animal Origin

Regulation (EC) No. 853/2004 of the European Parliament and of the Council of 29 April 2004, setting forth specific hygiene rules applicable to products of animal origin, was added to the general hygiene regulations because such products may present specific microbiological and chemical risks for human health, as was the case with recent food crises.

The Regulation is aimed at revising, consolidating, and replacing the different Directives intended to fix specific health rules in the sector of reference,[25] to obtain

---

[25]The pre-existing rules were specifically abrogated by Directive No. 2004/41/EC of the European Parliament and of the Council of 21 April 2004, which abrogates some Directives governing the hygiene of foodstuffs and the health rules for the production and sale of certain products of animal origin intended for human consumption, amending Directives 89/662/EEC of the Council and 92/118/EEC, together with Decision 95/408/EC of the Council.

further regulatory simplification common to all products of animal origin, and to "*secure a high level of consumer protection with regard to food safety, in particular by making food business operators throughout the Community subject to the same rules, and to ensure the proper functioning of the internal market in products of animal origin, thus contributing to the achievement of the objectives of the common agricultural policy*" (*see* Whereas 9).

The Regulation's recipients are both the food business operators carrying out the primary production and those acting in the following stages of production, processing, and distribution of food of animal origin, save for the primary production for private domestic use and for the domestic preparation, handling, and keeping of food intended for private domestic use,[26] for which the respective national laws and regulations remain in force.

The operators above must comply with the provisions under Annexes II and III for their own fields; register, and in some cases, have the establishments under their control approved (Article 4); meet the requirements for health and identification marking[27] (Article 5); and ensure that appropriate certificates or other documents accompany consignments of products, if requested (Article 7).

20.5.3 Official Controls

As analysed above (*see* Sect. 20.3.6), the system outlined under Regulation (EC) No. 178/2002 provides, as a fundamental condition for correctly managing all food safety related issues, not only for compliance by food business operators and/or feedstuffs with the food law provisions concerning their activities in all stages of production, transformation, and distribution, but also for an adequate official control system.[28]

---

[26]The Regulation does not apply (Article 1) to the direct supply, by the producer, of small quantities of primary products to the final consumer or to local retail establishments directly supplying the final consumer; the direct supply, by the producer, of small quantities of meat from poultry and lagomorphs slaughtered on a farm to the final consumer or to local retail establishments directly supplying such meat to the final consumer as fresh meat; or to hunters who supply small quantities of wild game or wild game meat directly to the final consumer or to local retail establishments directly supplying the final consumer. Save for special cases, the Regulation does not apply to retail, either.

[27]Thus implementing and abiding by the general traceability rules pursuant to Regulation (EC) No. 178/2002.

[28]*See* Article 17 Regulation (EC) No. 178/2002, paragraph 2: "*Member States shall enforce food law, and monitor and verify that the relevant requirements of food law are fulfilled by food and feed business operators at all stages of production, processing and distribution. For that purpose, they shall maintain a system of official controls and other activities as appropriate to the circumstances, including public communication on food and feed safety and risk, food and feed safety surveillance and other monitoring activities covering all stages of production, processing and distribution. Member States shall also lay down the rules on measures and penalties*

Regulation (EC) No. 882/2004 of the European Parliament and of the Council of 29 April 2004 therefore falls within the scope of the "hygiene package" on official controls performed to ensure the verification of compliance with feed and food law, and animal health and welfare rules.

The Regulation recipients are all Member States, which must appoint competent authorities responsible for achieving the goals and for all official controls. Such authorities, in their turn, must ensure that all of their staff performing official controls receive appropriate training, are kept informed of new developments, and have aptitude for multidisciplinary cooperation (Article 6 of Regulation No. 882/2004).

In so far as the uniform principles of general nature are concerned, Article 3 of the Regulation provides that all controls are carried out "*regularly, on a risk basis and with appropriate frequency*", if possible "*without prior warning*" and concerning "*any of the stages of production, processing and distribution of feed or food and of animals and animal products*".

In so far as the operational criteria are concerned, the national authorities must ensure that all official controls are carried out pursuant to the principles of effectiveness, fitness, impartiality, quality, and coherence at all levels, ensuring a high level of transparency (for instance, by promptly making available all relevant information to the public) as well as protecting in any event any and all information covered by confidentiality duties or trade secrets.

Regarding control methods and techniques, national competent authorities must use instruments deemed most appropriate, such as, for instance:

1. "*verification*", by examining and considering objective evidence, aimed at ascertaining whether specific requirements have been fulfilled;
2. "*audit*", that is, a systematic and independent examination to determine whether activities and related results comply with planned arrangements and whether these arrangements are implemented effectively and are suitable to achieve objectives;
3. "*inspection*", consisting of the examination of any aspect of feed, food, animal health, and animal welfare in order to verify that such aspect(s) comply with the law's provisions;
4. "*monitoring*", conducted through a planned sequence of observations or measurements in order to obtain an overview of the state of compliance with the applicable laws and regulations;
5. careful "*surveillance*" of one or more feed or food business operators or their activities;

---

applicable to infringements of food and feed law. The measures and penalties provided for shall be effective, proportionate and dissuasive". *See* also Whereas 6 of Regulation (EC) No. 882/2004: "*The Member States should enforce feed and food law, animal health and animal welfare rules and monitor and verify that the relevant requirements thereof are fulfilled by business operators at all stages of production, processing and distribution. Official controls should be organised for that purpose*".

6. "*sampling for analysis*", that is, the taking of samples—pursuant to the criteria set forth under Annex III of the Regulation and in compliance with the relevant Community legislation—of feed or food or any other substance (including from the environment) relevant to the production, processing, and distribution of feed or food, in order to verify compliance with the applicable law in force.

Specific rules are then set forth in connection with certain foodstuffs and feedstuffs. In this respect, **Regulation (EC) No. 854/2004** of the Parliament and of the Council of 29 April 2004 sets forth specific provisions for the arrangement of official controls on products of animal origin intended for human consumption and specifies the requirements and operational criteria that the national competent authorities need to comply with in carrying out the respective official controls.

20.5.4 The "Implementing" Regulations

Finally, a series of other "implementing" Regulations that are listed and summarised below, must be considered:

1. **Regulation (EC) No. 2073/2005** of the Commission of 15 November 2005 on microbiological criteria for foodstuffs sets forth the microbiological criteria (deemed as those criteria defining "*the acceptability of a product, a batch of foodstuffs or a process, based on the absence, presence or number of micro-organisms, and/or on the absence, presence or number of micro-organisms, and/or on the quantity of their toxins/metabolites, per unit(s) of mass, volume, area or batch*") that food business operators must comply with in the general and specific hygiene requirements under Article 4 of Regulation (EC) No. 852/2004;
2. **Regulation (EC) No. 2074/2005** of the Commission of 5 December 2005 laying down implementing measures for certain products under Regulation (EC) No. 853/2004 for the organisation of official controls under Regulation (EC) No. 854/2004 and Regulation (EC) No. 882/2004, derogating from Regulation (EC) No. 852/2004 and amending Regulations (EC) No. 853/2004 and (EC) No. 854/2004;
3. **Regulation (EC) No. 2075/2005** of the Commission of 5 December 2005 laying down specific rules on official controls for *Trichinella* in meat;
4. finally, **Regulation (EC) No. 2076/2005** of the Commission of 5 December 2005 laying down transitional arrangements for the implementation of Regulations (EC) No. 853/2004, (EC) No. 854/2004, and (EC) No. 882/2004, and amending Regulations (EC) No. 853/2004 and (EC) No. 854/2004.

20.5.5 A New Scenario for Official Controls: Regulation (EC) No. 625/2017

The importance of the subject has led the Commission to further refine Regulation 882/2004 in order to make up for the shortcomings of its drafting and implementation.

For that purpose the European Parliament and the Council adopted the new Official Controls Regulation 2017/625 on 15 March 2017. The Regulation was published in the Official Journal of the European Union on 7 April 2017 and entered into force on 27 April 2017. The new rules will gradually become applicable, with the main application date being 14 December 2019, fully replacing Regulation 882/2004 on official controls and other legislation that currently governs the control and enforcement of rules along the agri-food chain. In particular it will:

- Repeal: Regulations (EC) No 854/2004 and (EC) No 882/2004 of the European Parliament and of the Council, Council Directives 89/608/EEC, 89/662/EEC, 90/425/EEC, 91/496/EEC, 96/23/EC, 96/93/EC, and 97/78/EC, and Council Decision 92/438/EEC.
- Amend with respect to control rules: Regulations (EC) No 999/2001, (EC) No 396/2005, (EC) No 1069/2009, (EC) No 1107/2009, (EU) No 1151/2012, (EU) No 652/2014, (EU) 2016/429, and (EU) 2016/2031 of the European Parliament and of the Council, Council Regulations (EC) No 1/2005 and (EC) No 1099/2009, and Council Directives 98/58/EC, 1999/74/EC, 2007/43/EC, 2008/119/EC, and 2008/120/EC.

Extended Scope

The new Official Controls Regulation strengthens and enhances the EU system as an international reference for integrated rules covering the entire agri-food chain. Indeed, its scope has been extended, and will now also include official controls to verify compliance with plant health and animal by-product rules. As well, the regulation provides more specific rules for several areas already covered (e.g. animal health and animal welfare) and applies to all operators at all stages of production, processing, and distribution that handle animals, plants, food, feed, goods, substances, materials, or equipment. As such, businesses and enforcement authorities would benefit from a simplified framework that mainstreams official control rules into a single Regulation.

As per past official control regimes (i.e. Regulation 882/2004), the new Regulation does not deal with the verification of compliance with the rules on the common market organization of agricultural products already governed by established control systems, as it applies to those checks carried out under marketing standards rules (i.e. Article 89 of Regulation (EU) No 1306/2013) in order to identify potential instances of fraudulent or deceptive practices involving marketing standards of agricultural products.

Risk-Based Approach

Competent authorities shall perform controls without prior notice, unless this is necessary, and follow the same risk-based approach.

A new provision clarifies that official controls must be implemented in a manner that minimizes the burden on businesses. Thus, rigid, inflexible rules are withdrawn and control authorities would take advantage of a more integrated IT system that allows a more modern approach to tracking trade practices. This means that the frequency of controls will be linked to risks that a product presents with respect to fraud, health, safety, animal welfare, or the environment. The assessment of the risk also takes into consideration other factors, in particular the operator's past record of compliance and the likelihood that consumers are misled about properties, quality, composition, or country of provenance of the food.

Thus, this allows, on one side, the Commission to adjust control requirements to the specific enforcement needs of each sector (e.g. the establishment of minimum control frequencies where the risks warrant it) and on the other side national authorities to commit their resources where they are most needed.

Transparency

The new rules require increased transparency and greater accountability from Member State authorities, who are obligated to publish annual reports in order to enable consumers to obtain information about how agri-food chain rules are applied and enforced. These publications must specify the type and number of controls, the cases of non-compliance observed, and the cases where enforcement measures were taken and penalties were imposed. New rules allow Competent Authorities to make publicly available information about rating scheme systems for operators, based on the outcome of official controls. The rating criteria must be objective, fair, and transparent so that consumers are better informed about the level of compliance of businesses (such as manufacturers, retailers, restaurants, etc.).

In this regard, operators are required to assist and cooperate with the staff of competent authorities and to provide them updated details about their name, the places under their control, and the specific activities they carry out, including activities undertaken by means of distant communication (such as tele- and Internet sales). More specifically, to the extent necessary to perform official controls, operators must give access to their equipment, means of transport, premises, computers, documents, and relevant information, animals, and goods under their control.

If a list or register of operators already exists for other purposes, Competent Authorities may use the existing list or register. Furthermore, certain categories of operators for which registration would be disproportionate to the level of risk posed by the operator's activities, can be exempted via delegated act.

Delegation of Control Tasks

Rules on delegation of official control tasks remain largely the same, as provided for in Regulation 882/2004. However, new provisions specify that delegation can be made to delegated bodies or a natural person and cover the delegation of specific tasks regarding other official activities.

Moreover, the delegation of certain official control tasks shall meet precise conditions, including:

- the delegation must be in writing and shall contain a precise description of the tasks that the delegated body must perform;
- the delegated body must have the required expertise, equipment, suitable qualifications, and experienced staff to undertake the control;
- the delegated body must be impartial, free from conflict of interest, and accredited in accordance with standards relevant to the control tasks to perform;
- effective coordination between the delegating Competent Authority and the delegated body must be guaranteed.

Further to inspections or/and audits, if the Competent Authority finds that the delegated body or natural person is not meeting the necessary conditions, the delegation must be withdrawn fully or partially.

Responsibility for taking actions in the case of established non-compliance (i.e. enforcement measures) cannot be delegated: only the relevant Competent Authority may take such actions.

Official Laboratories

The Regulation establishes rules for the designation of the laboratories performing analyses, tests, or diagnoses for official controls and the conditions to be designated. Accreditation to EN ISO/IEC 17025 is still a mandatory condition for the designation of all official control laboratories.

Competent Authorities may introduce permanent derogation from mandatory accreditation of official control laboratories for laboratories with a limited scope of activities (e.g. Trichinella in meat).

Temporary derogation from mandatory accreditation may be introduced for methods of analysis, tests, or diagnoses that are not covered by accreditation. Although, this would be limited to specific conditions such as:

- emergency circumstances
- when the method is newly required by EU legislation.

Official laboratories, upon request of the Competent Authority, must make available to the public the names of the methods used for analyses, tests, or diagnoses.

The Commission has the power to establish EU Reference Laboratories (EURLs) in those sectors where there is a recognised need to develop uniform practices and reliable methods of analyses, tests, and diagnoses.

The decision to establish an EURL is taken by a delegated act involving the European Parliament and the actual designation is implemented by the Commission, in order to favor the openness and transparency of the selection process. The designation is valid for a minimum of 5 years.

Furthermore, within 1 year of the entry into force of the Regulation, new EU Reference Centres for animal welfare will be set up to provide coordinated assistance to Member States in the field of animal welfare. The tasks include:

- fulfilling scientific and technical studies;
- carrying out training courses and disseminating research findings and technical innovations;
- submitting scientific and technical advice for the development and application of animal welfare indicators.

Food Fraud

Fraud affects consumers' trust, undermines competition, and may compromise food and feed safety. Hence, the new rules require member States to perform regular, unannounced risk-based official controls to detect fraudulent or deceptive practices. This includes checking compliance against marketing standards for agricultural products. Financial penalties for fraud must reflect the expected economic gain or a percentage of the turnover made by the fraudulent operator. In addition, EU Reference Centers for the authenticity and integrity of the agri-food chain may be established. They would provide EU countries with up-to-date, reliable technical data and research findings to assist with the effective performance of their controls task.

Cascade of Methods

The new Regulation specifies that rules on methods of sampling, analysis, tests, and diagnosis are applicable to official controls as well as to other official activities in all sectors covered by the Regulation. The cascade of methods used for sampling, analysis, tests, and diagnosis appears as follows:

First, methods complying with the relevant Union rules must be considered
If there are no Union rules, the following methods must be used:

(a) Methods in compliance with internationally recognised rules or protocols, including those accepted by the European Committee for Standardisation (CEN), or
(b) Methods developed and recommended by European Union Reference Laboratories and validated in accordance with internationally accepted scientific protocols.

In the absence of the above rules or protocols, methods that comply with relevant rules established at a national level shall be used. If neither such rules are provided, those seeking compliance must use:

(a) Relevant methods developed or recommended by national reference laboratories and validated in accordance with internationally accepted scientific protocols; or
(b) other methods validated with intra-laboratory methods, in accordance with existing internationally accepted scientific protocols.

Products from Third Countries

The regulation establishes an integrated approach to import controls by eliminating the current fragmentation of requirements. A common set of rules applies to border controls carried out on animals, products of animal origin, plants, and other products and goods that pose a risk to health, safety, animal welfare, or in certain cases, the environment. All of these goods must be channeled through Border Control Posts (BCPs), which replace the various Border Inspection Posts (BIPs) and Designated Points of Entry (DPEs). All consignments to be presented at the border control posts shall undergo document checks, while identity and physical checks shall take place at a frequency depending on the risk linked to the specific animals or goods. The criteria to determine and modify the frequency of rates is established by the Commission. A Common Health Entry Document (CHED) is introduced for consignments from third countries and will be transmitted to the border control post through a new integrated computerised system for official controls (Integrated Management System for Official Controls, IMSOC).

Cooperation and Assistance Between EU Countries
The new regulation clarifies and strengthens rules on the cooperation and administrative assistance between EU countries to ensure cross-border enforcement of agri-food chain rules. Member States are obligated to facilitate the exchange of information between Competent Authorities and other enforcement authorities such as public prosecutors on possible cases of non-compliance. This allows a swifter and more efficient pursuit of non-compliance across borders.

Moreover, the Integrated Management System for Official Controls (IMSOC) integrates all existing and future computer systems (e.g. TRACES, RASFF, and Europhyt) to ensure optimal use of data, reduce burdens on businesses and national enforcers, and accelerate the exchange of information between Member States.

Financing of Official Controls and Other Official Activities
Mandatory fees are required of operators for certain official controls, and in particular those carried out:

(a) in slaughterhouses, cutting plants, on milk production and on the production and placing on the market of fishery and aquaculture products;
(b) at border control posts (or at control points other than border control posts where permitted) on animals, products of animal origin, germinal products, animal by-products, plants, and plants products;
(c) on goods originating from third countries on the basis of the risk posed;
(d) at borders on animals and goods subject to emergency measures or for which conditions or measures have been established to enter the Union;
(e) to verify the conditions for the approval of feed premises (feed mills);
(f) as necessary to follow up on non-compliance, but not originally planned.

As well, for the determination of the amount of fees for the controls at points (a) and (b) above, Competent Authorities have the option to choose one of the following methods:

1. fees set out at a flat-rate on the basis of the costs borne by the Competent Authorities over a given period of time, and applied to all operators irrespective of whether any official control activity is carried out at any particular operator's premises during the reference period. Member States will be required to take into account the impact of size and type of business concerned, and the relevant risk associated with these businesses;

2. fees calculated on the basis of the costs of each individual control and applied specifically to the operators subject to the control;
3. fees based on the amounts provided by the Regulation (detailed in the annex). Those amounts no longer constitute minimum fees as provided under the current regime of Regulation (EC) 882/2004.

For the official controls at points (c)–(f) above, the Competent Authorities will recover the actual cost of controls.

Moreover, Competent Authorities shall take into account the salary, social security, pension, and insurance costs of support and administrative staff (as well as of staff physically performing official controls); the same applies to the costs of services charged to Competent Authorities by delegated bodies for the official controls delegated to them.

Enforcement Actions and Measures

Competent Authorities shall consider the nature of the non-compliance and the operator's past record of compliance when deciding what action to take.

However, the non-exhaustive list of current enforcement actions has increased to reflect the broader scope of the new Regulation. For instance, the list of measures now includes:

- restriction or prohibition of movements of animals;
- the slaughter or killing of animals provided it is the most appropriate measure to safeguard public health and animal health and welfare;
- closing the website of the operator.

The Commission is authorized to adopt more suitable actions where there is evidence that the system of controls in a Member State faces serious disruption to the point that it constitutes a real risk to the agri-food chain, and the Member State has not addressed the shortcomings within a time limit set by the Commission. The Regulation also introduces more stringent rules for financial penalties imposed by Member States in order to deter fraudulent behaviours and foster fair competition among businesses. Those penalties shall reflect the economic advantage of the operator or a percentage of the operator's turnover.

E-commerce

Food law applies to food sold via the Internet. Consequently, e-commerce must be part of official controls. The new regulation confirms that Member states can, for control purposes, order products online without identifying themselves (i.e. mystery shopping), and use products purchased as official samples. Non-compliance can result in penalties regardless of the location of the operator.

20.6 Conclusions

The scenario outlined above shows the gradual evolution in food law, the fundamental cornerstone of which is the protection of food and feed safety.

This, as illustrated above, has been done by adopting vertical rules and regulations, harmonised throughout the entire territory and common to all operators, who will only then be able to achieve their goals and thus ensure high, common quality standards.

The significant food crises that have shaken the continent have been followed by serious and timely reaction by Community authorities, which have proven that they know how to react in relatively short periods of time, thus contributing to the creation of a system that is functional for operators and safe for consumers, and that is becoming one of the most developed systems in the world.

The need for rapid, repeated actions at the same time, however, is a constant challenge, one that certainly cannot be considered complete and that will continue testing not only the relevant national and European authorities but all stakeholders as well.

Chapter 21
European Food Trade

Roland Kölcsey-Rieden and Lívia Hulló

Abstract This chapter is a primer on EU market and food trade regulations. The internal market, food's free movement and the rules applied to agricultural product and food trade are reviewed. Economic integration and the harmonization of food trade are described with emphasis on the Treaty on the Functioning of the European Union to explain the customs union and the free movement fo goods in the EU. The second section of the chapter addresses the European Union's external market and world trade in light of the WTO and EFTA, EEA, and ACP countries.

21.1 Introduction to the Internal Market of the European Union

The European Union (EU) was first established in 1952 as the European Coal and Steel Community (ECSC) to economically and politically unite European countries in order to secure lasting peace.[1] In 1957, the six founding countries signed the Treaty of Rome, which extended the earlier cooperation within the ECSC and created the European Economic Community (EEC) or Common Market.[2] The EEC's objective was to bring about economic integration.[3] Member States the countries that comprise the EU remain independent sovereign nations, but they

---

[1] The ECSC was the first international organisation based on the principles of supranationalism. The Treaty came into force in 1958. European Union (2014).

[2] The six founders are Belgium, France, Germany, Italy, Luxembourg and the Netherlands. The Inner Six, or simply "The Six" was in contrast to the outer seven who formed the European Free Trade Association. The Treaty establishing the European Economic Community (TEEC) was signed on 25 March 1957. The EEC was also known as the Common Market and sometimes referred to as the European Community.

[3] Article 2 and 3 of the Treaty of Rome.

R. Kölcsey-Rieden • L. Hulló (✉)
Colaw - Kölcsey-Rieden & Partner Law Firm, Budapest, Hungary
e-mail: r.kolcsey-rieden@colaw.hu; l.hullo@colaw.hu

cede a part of their sovereignty in order to achieve a common market. Countries gain global influence as a Member State.

The Single European Act (SEA) represents the first major revision of the Treaty of Rome.[4] The SEA's chief objective was to complete the internal market.[5] The SEA transformed the Common Market into a single market on 1 January 1993 by creating new community competencies and reforming the institutions, such as collaborative legislative process.[6] The SEA opened the way to political integration and enshrined the economic and monetary union in the Treaty of Maastricht on the European Union.[7]

The Maastricht Treaty ("Treaty on European Union" or TEU) was signed on 7 February 1992 by the members of the European Community in Maastricht, Netherlands. It created the European Union on 1 November 1993.[8] By 1992, the Maastricht Treaty created the Economic and Monetary Union as the next stage of integration. The Maastricht Treaty was amended several times by the Treaties of Amsterdam, Nice, and Lisbon. The Treaty of Lisbon amended the two treaties Maastricht Treaty (1993) and Treaty of Rome (1952), which form the EU's constitutional basis. In this process, the Rome Treaty was renamed to the Treaty on the Functioning of the European Union (TFEU).

Primary law, secondary law and supplementary law are European Union law classifications. EU legislation divided into primary legislation embodied in the treaties, and secondary legislation in the form of regulations, directives and decisions used to implement the policies set out in the treaties.[9] Member States courts also apply the European Union law. Supplementary sources are elements of law not provided for by the Treaties.[10] The Court of Justice of the European Union (CJEU or "Court") is the highest court able to interpret European Union law.

21.1.1 Single Market of the European Union for Goods

The EEC's objectives were to develop a common market and a customs union among the Member States. The European Union's internal market

---

[4]The SEA was signed at Luxembourg on 17 February 1986 and at The Hague on 28 February 1986. It came into effect on 1 July 1987.

[5]Europea Summaries of EU legislation, The Single European Act.

[6]The European Parliament was given a say in legislating and was introduced with more majority voting in the Council of Ministers.

[7]See the Europea Summaries of EU legislation, The Single European Act.

[8]Article A TEU.

[9]European University Institute.

[10]This category includes Court of Justice case-law, international law and general principles of law.

(or single market) differs from other forms of integration.[11] It seeks to guarantee the so-called "four freedoms," the free movement of goods, capital, services, and people within the EU's 28 Member States. The internal market is open to three non-EU states.[12]

The fundamental idea behind the internal market is that the EU constitutes a single economic area that operates similarly to national markets. Accordingly, the free movement of goods is of primary importance for both the establishment of an internal market and for European integration as a whole. The free movement of goods, i.e. the free trade of goods within the EU, is secured by the customs union and by abolishing quantitative restrictions between the Member States.

21.1.2 Customs Union

TFEU Article 28 proclaims the EU Customs Union (EUCU) as a constituting element of the EU, which includes the entire trade of products between the Member States and with third countries.[13] Between Member States, the union prohibits (1) all customs duties on imports and exports, (2) all charges having equivalent effect to such duties, and (3) the adoption of a common customs tariff in their relations with third countries. Member States removed customs barriers between one another and introduced a common customs policy towards other countries. The overall purpose of the duties is "to ensure normal conditions of competition and to remove all restrictions of a fiscal nature capable of hindering the free movement of goods within the Common Market."[14] 28–30 TFEU regulates the customs union provision.

The Customs Union's integrative power enhanced trade between Member States from the EEC's beginning. The Customs Union aims to (1) eliminate customs duties between the Members States and (2) implement a common customs tariff policy with regard to third countries (the latter being the main point of difference between the Customs Union and a free trade zone). A Customs Union requires free movement of goods between Member States. The Member States acknowledge that they cannot restrict the movement of transit goods within the Union; they must adhere to the principle of the freedom of goods.

The TFEU does not define the term "goods." However, several European Union Court decisions define the concept as follows: "By goods, within the meaning of

---

[11] Article 26 states "The internal market shall comprise an area without internal frontiers in which the free movement of goods, persons, services and capital is ensured in accordance with the provisions of the Treaties."

[12] Agreement on the European Economic Area, entered into force on 1 January 1994, brings together the EU Member States and the three EEA EFTA States—Iceland, Liechtenstein and Norway—in a single market. EFTA agreement.

[13] Case C-173/05. Commission of the European Communities v Italian Republic.

[14] Case 27-67. Firma Fink-Frucht GmbH v Hauptzollamt München-Landsbergerstrasse.

Article 9 of the EEC Treaty [now Article 28], there must be understood products which can be valued in money and which are capable, as such, of forming the subject of commercial transactions."[15] It is important to draw a legal distinction between goods and services. The principle of the freedom to provide services enables an economic operator providing services in one Member State to offer services on a temporary basis in another Member State.[16] For example, while fish are considered goods, the free movement of goods principle does not cover fishing rights and angling permits. Fishing rights and angling permits constitute the "provision of a service" within the Treaty provisions relating to the freedom to provide services.[17]

The content behind the concept of "Community goods" as referred to in Article 28 TFEU is laid down in Article 4 (18) of Regulation (EC) No 450/2008. The Regulation defines "non-Community goods" as goods that comply with the import requirements of the state in question.[18] These requirements usually mean compliance with the Community legislation.

21.1.3 Customs Duties

Customs duties represent a charge imposed by the executive power in relation to the good's export, import, and transit.[19] TFEU article 30 prohibits Member States from levying any duties on goods crossing the border. Once a third country imports a good into the EU and the appropriate customs duty has been paid, then it is in free circulation between the Member States (TFEU article 29). There are not any systematic customs controls at the Members States' borders since the SEA.[20] Emphasis is on post-import audit controls and risk analysis. Physical controls of imports and exports now occur at traders' premises, rather than at the territorial borders.

---

[15] Case 7/68. Commission of the European Communities v Italian Republic.
[16] European Commission (EC) (2014f), Freedom to provides services/Freedom of establishment http://ec.europa.eu/internal_market/top_layer/living_working/services-establishment/index_en.htm and Article 56 TFEU (ex Article 49 TEC).
[17] Case C-97/98 Jägerskiöld [1999].
[18] Article 4 of Regulation No. (EC) 450/2008.
[19] Article 30 (ex Article 25 TEC).
[20] The Single European Act (SEA) was the first major revision of the 1957 Treaty of Rome. The Act set the European Community an objective of establishing a single market by 31 December 1992, and codified European Political Cooperation, the forerunner of the European Union's Common Foreign and Security Policy.

21.1.4 Charges Having Equivalent Effect to Customs Duties

TFEU article 30 prohibits "charges having equivalent effect." The European Court of Justice defined "a charge having equivalent effect" in Case 7/68:

> [A]ny pecuniary charge, however small and whatever its designation and mode of application, which is imposed unilaterally on domestic or foreign goods by reason of the fact that they cross a frontier, and which is not a customs duty in the strict sense, constitutes a charge having equivalent effect... even if it is not imposed for the benefit of the state, is not discriminatory or protective in effect and if the product on which the charge is imposed is not in competition with any domestic product.[21]

However, a charge escapes classification as "a charge having equivalent effect to a customs duty if it relates to a general system of internal dues applied systematically and in accordance with the same criteria to domestic products and imported or exported products alike."[22]

The meaning of "charges having equivalent effect to customs duties" as related to food products emerged in the Case *C-2-3/62 Commission v Belgium and Luxemburg*.[23] Belgium and Luxemburg levied tax on imported gingerbread to protect inland producers. They argued that the rey was more expensive in their countries than in the import countries. The Court of the European Union stated:

> A charge having equivalent effect within the meaning of Articles 9 and 12 of the EEC Treaty, whatever it is called and whatever its mode of application, may be regarded as a duty imposed unilaterally either at the time of importation or subsequently, and which, if imposed specifically upon a product imported from a Member State to the exclusion of a similar domestic product, has, by altering its price, the same effect on the free movement of products as a customs duty.

Pecuniary charge and fees charged by authorities are exceptions to the prohibition on charges imposed when goods or live animals cross the border, listed in Case 18/87 *Commission v Germany*.[24] For more information on charges and fees on live animals.

---

[21] Case 24/68 Commission v Italy.

[22] Case C-389/00 Commission v Germany.

[23] Joined cases 2/62 and 3/62, Commission of the European Economic Community v Grand Duchy of Luxembourg and Kingdom of Belgium.

[24] Case 18/87, Commission of the European Communities v Federal Republic of Germany, Charging of fees for inspections carried out during intra-Community transport of live animals.

21.2 The Treaty on European Union and the Treaty on the Functioning of the European Union (TFEU) and Food Regulation: Quantitative Restrictions and Measures Having Equivalent Effect

TFEU articles 34–35 address the general strategy concerning the free movements of goods.[25] TFEU articles 30–33 are the Customs Union's foundation by providing for the elimination of customs duties between the Member States and by establishing a Common Customs Tariff. Without these provisions, the Member States would be able to place quotas in the amount of goods that could be imported, and restrict the flow of goods by measures that have the equivalent effect to quotas.[26] The TFEU does not address what equivalent effect measures are and how they affect trade between Member States. The European Court of Justice, however, provides detailed case law interpreting article 34. Quantitative restrictions are measures that amount to a total or partial restraint on imports or goods in transit.[27] Examples include an outright ban or a quota system. However, this article only pertains to non-tariff quotas, since TFEU article 30 covers tariff quotas. A quantitative restriction is based on statutory provisions or may just be an administrative practice. Thus, TFEU article 34 can be applied to a covert or hidden quota system.

The conditions for TFEU article 34 application are (1) the case bearings pertain to the concept of goods crossing borders, (2) article 4 applied to import, and (3) the case bearings have cross-border elements. EU law does not prevent Member States from treating domestic products less favourably than imports through reverse discrimination, as seen in the need for a cross-border element. TFEU article 34 is, however, applicable when it is reimported, i.e. a domestic product is imported back into a Member State.[28] The Court mandates that the free movement of goods entail the existence of a general principle of free transit of goods within the EU.[29] All goods, irrespective of the manufacturing origin inside or outside the internal market, benefit from the principle of free movement in the internal market.[30]

While TFEU articles 34–36 are the groundwork for the general principle of the free movement of goods, they are not the only legal standard to check the compatibility of national measures with internal market rules. These Treaty articles do not apply when a product's free movement is fully harmonised by EU legislation, i.e. where a given product's specifications or conditions of sale are subject to harmonisation by EU directives or regulations. In other cases, more specific Treaty

---

[25] Quantitative restrictions on imports and all measures having equivalent effect shall be prohibited between Member States.
[26] Craig and de Búrca (2011).
[27] Case 2/73 Geddo.
[28] Case 78/70 Deutsche Grammophon v. Metro.
[29] Case C-320/03 Commission v. Austria.
[30] See European Commission (EC) (2014a), Free movement of goods.

rules, such as Article 110 TFEU on tax-related provisions that may hamper the internal market, prevail over the general provisions of Articles 34–36 TFEU. Therefore, any problem that is covered by harmonising legislation would be analysed through concrete terms and not according to the Treaty's broad principles.[31]

Food law harmonized areas address food labeling, food safety, organic products, and food quality standards.[32] Seconary law is applicable if areas are harmonized; the primarly law is applicable if harmonization is incomplete or there is no harmonization. By acknowledging that potential differences between the Member States' national rules could inhibit trade, the Court confirmed that TFEU article 34 could catch national measures applied equally to domestic and imported goods. In this case, Member States could derogate by having recourse not only to TFEU article 36 but also to the mandatory requirements. TFEU article 34 applies to national measures that discriminate against imported goods. The imported goods are required to comply with two sets of rules–rules by the Member State of manufacture, and rules by the Member State of importation.[33]

21.2.1 *Quantitative Restrictions*

Quantitative restrictions represent rough means of trade policy. Competitive products could not reach other Member States if quantiatitve restrictions are applied. Consequently, the free movement of goods on the internal market would not be realized if the Member States were allowed to impose quantitative restrictions against each other. The European Court of Justice determined "quantitative restriction" in the Geddo case as follows[34]:

> The prohibition, under Article 23 [now Article 34], of such a measure in the internal trade of the community is designed to ensure the free movement of goods within the community the prohibition of quantitative restrictions and measures having equivalent effect covers any total or partial prohibition on imports, exports or goods in transit and any encumbrance having the same effect.

---

[31] European Commission (EC) (2014a).

[32] Directive 2000/13/EC on the approximation of the laws of the Member States relating to the labelling, presentation and advertising of foodstuffs; regulation (EC) No 178/2002 of the European Parliament and of the Council of 28 January 2002 lays down the general principles and requirements of food law, establishing the European Food Safety Authority and lays down procedures in matters of food safety; Council Regulation (EC) No 834/2007 of 28 June 2007 on organic production and labelling of organic products and repealing Regulation (EEC) No 2092/91; Commission Implementing Regulation (EU) No 543/2011 of 7 June 2011 lays down detailed rules for the application of Council Regulation (EC) No 1234/2007 in respect of the fruit and vegetables and processed fruit and vegetables sectors.

[33] See European Commission (EC) (2014a), Free movement of goods.

[34] Case 2-73. Riseria Luigi Geddo v Ente Nazionale Risi European Court reports 1973 Page 00865.

A quantitative restriction constitutes restricting the import of an agricultural product from a specific country for a certain period.

21.2.2 Measuers of Equivalent Effect

The term "measure having equivalent effect" is broader in scope than a quantitative restriction. However, the rules apply similarly to quantitative restrictions and to equivalent effect measures. *Dassonville*, *Cassis de Dijon* and *Keck and Mithouard* represent milestone cases in the jurisdiction the "measures of equivalent effect."[35]

In *Dassonville,* the Court of Justice interpreted equivalent effect's meaning and scope of measures: "All trading rules enacted by Member States which are capable of hindering, directly or indirectly, actually or potentially, intra—Community trade are to be considered as measures having an effect equivalent to quantitative restrictions." The Court's case-law confirms the definition with minor variations. In the famous *Cassis de Dijon* case, German authorities did not allow the trading of French blackcurrant liqueur, claiming that its wine-spirit content did not reach the wine-spirit content provided by the German legislation on spirits. The Court ruled that the legislation is contrary to Community law. The Court of Justice stated that member states are obliged to recognise goods legally produced in another member state, according to the Union's fundamental principle of mutual recognition. However, the member state could justify the restriction by reference to a mandatory requirement.[36]

Technical regulations containing requirements for the presentation of goods (e.g. weight, composition presentation, form, size, packaging) are an example of typical measures of equivalent effects.[37] For example, margarine required to be packaged in cubes to distinguish it from butter is contrary to Article 34 TFEU, since there was not a harmonized law regarding margarine's packaging.[38] Other examples include national price controls and reimbursement, authorization procedure, advertising restrictions, indications of origin, quality marks, and incitement to buy national products.

Article 34 TFEU is characterised as a defense right invoked against national measures creating unjustified obstacles to cross-border trade. Accordingly, infringements of Article 34 TFEU presuppose state activity. The measures falling within the scope of Article 34 TFEU consist primarily of binding provisions of member states' legislation. However, non-binding measures can also constitute a breach of Article 34 TFEU. An administrative practice can amount to a prohibited

---

[35]Case 8-74 Procureur du Roi v Benoît and Gustave Dassonville [1974] ECR 937; Case C-120/78; Joined cases C-267/91 and C-268/91.

[36]Craig and de Búrca (2011).

[37]See European Commission (EC) (2014a), Free movement of goods.

[38]Case 261/81 Rau v De Smedt.

obstacle to the free movement of goods provided that this practice is, to some degree, of a consistent and general nature. In view of member states' obligations under Article 4(3) of the Treaty on the European Union (i.e. Article 10 EC), which require states to fulfill Treaty obligations and the *effet utile* of EU law, Article 34 TFEU may also be infringed by a Member State's inactivity. For example, a Member State refraining from adopting the required measures to address obstacles to the free movement of goods is considered a trade obstacle.[39]

Commission v. Ireland is an incitement Case to buy national products involved in a large-scale campaign encouraging the purchase of domestic goods rather than imported products. The Court decided that, as the campaign attempted to reduce the flow of imports, it infringed Article 34 TFEU. According to Community law, the national characteristic of goods cannot be emphasized, with the exception of cases where a product is bound to a certain region on the basis of its characteristics (e.g. Parma Ham Prosciutto) or where a product has been qualified as an "outstanding product."

Regulation (EC) No 2679/98 addresses the internal market's functioning and provides special procedures to cope with obstacles to the free movement of goods. There are different reasons behind obstacles to the free movement of goods; national authorities may be passive in the face of individual violent action or non-violent blockages of borders. Further, a member state's institutionalised boycott of imported products may also create obstacles. The regulation provides for an alert procedure and the information exchange between member states and the Commission. It reminds member states to adopt measures to ensure the free movement of goods and to inform the Commission. Lastly, it empowers the Commission to send notifications to the member states requesting that such measures are adopted within a deadline.

21.2.3 Exceptions to the Free Movement of Goods

A ban on a specific product's marketing is the most restrictive measure a member state can adopt. Foodstuff, including vitamins and supplements, are the majority of goods targeted by national bans.[40] Member States justify bans according to Article 36 TFEU for the protection of health, animals and plants, and the Court case law's mandatory requirements. The Member State imposing a national ban on a product/ substance must prove the measure is necessary. Further, the State must prove the marketing poses a public health risk and that the rules are in line with the principle of proportionality.[41] For example, when France banned caffeine's addition to

---

[39] See European Commission (EC) (2014a), Free movement of goods.
[40] See European Commission (EC) (2014a), Free movement of goods.
[41] So also European Commission (EC) (2014a), Free movement of goods. goods/files/goods/docs/art34-36/new_guide_en.pdf.

beverages above a certain limit, the Court held that "appropriate labelling, informing consumers about the nature, the ingredients and the characteristics of fortified products, can enable consumers who risk excessive consumption of a nutrient added to those products to decide for themselves whether to use them."[42] The caffeine ban above a certain limit was unnecessary in order to achieve consumer protection.[43]

The Danish government prohibited the enrichment of foodstuffs with vitamins and minerals if Denmark's population did not require the nutrients.[44] The Court agreed Denmark could decide on its intended level of protection. The Court remarked:

> Since Article 30 EC [now Article 36] provides for an exception, to be interpreted strictly, to the rule of free movement of goods within the Community, it is for the national authorities which invoke it to show in each case, in the light of national nutritional habits and in the light of the results of international scientific research, that their rules are necessary to give effective protection to the interests referred to in that provision and, in particular, that the marketing of the products in question poses a real risk to public health.

Member States can also require prior authorization for authorized substances from other Member States. In this case, Member States comply with EU law obligations if procedures are accessible, can be timely completed, and if the banning is challenged before the courts. This procedure must bind the national authorities. The Court established the "simplified procedure" characteristics in Case C-344/90.[45]

The *Comission v. Germany* case (C-178/84) prevented French manufacturers from entering the German market due to German legislation governing beer's manufacture (Biersteuergesetz). The Case is a quantitative restriction because of infringement and the causal relation between the measure and the manufacturers' loss. "Reverse discrimination" may also take place due to Community law's application; the national legislation may impose stricter regulations than the Community legislation, resulting in favored foreign producers.[46]

---

[42]Case C-24/00 Commission v France [2004].

[43]See European Commission (EC) (2014a), Free movement of goods.

[44]Case C-192/01 Commission v Denmark.

[45]Case C-344/90 Commission v France.

[46]In its judgment in Case C-366/98 *Geffroy*, the Court ruled that Article 34 TFEU "must be interpreted as precluding a national rule...from requiring the use of a specific language for the labelling of foodstuffs, without allowing for the possibility of using another language easily understood by purchasers or of ensuring that the purchaser is informed by other means."

21.2.4 Cases and Examples

21.2.4.1 Procureur du Roi v Benoît and Gustave Dassonville

There are exceptions of free movement of trade. Belgian Law provides that imported goods bearing a designation of origin must have a government certificate from the exporting country certifying its right to such designation. Dassonville imported Scotch whisky from France into Belgium without a British certificate. The certificate was difficult to obtain since the goods were already in free circulation in a third country. Belgium persecuted Dassonville and argued by wax of defence that Belgian rule constituted a measure having equivalent effect to a quantitative restriction.[47] This quantitative restriction entails trading rules enacted by member states.[48]

21.2.4.2 Case Cassis the Dijon

The French blackcurrant-based drink was at the heart of one of the European Court of Justice's most celebrated decisions. In 1979, Rewe-Zentral AG, one of Germany's largest food and drinks retailers, complained to the European Court of Justice that the German authorities made it difficult for the company to import and sell Cassis de Dijon. The Court ruled in the firm's favour and declared that if a company is allowed to make a product freely available for sale in one European Community country, then it must be allowed to do so in all Member States. As Cassis de Dijon was already available in France, the Court argued all other European citizens also had the right to purchase and drink it.[49]

21.3 Freedom to Provide Services and Taxation

The freedom to provide services (Article 56 TFEU) is closely related to the free movement of goods. Both freedoms relate to economic transactions of a commercial nature between Member States. Occasionally a specific national measure restricts both the circulation of goods (Article 34 TFEU) and the freedom to provide services. A given requirement relating to the distribution, wholesale or retail of goods may simultaneously restrict both the free movement of goods and the freedom to provide distributive trade services. As the Court recognized:

> The objective of retail trade is the sale of goods to consumers. That trade includes, in addition to the legal sales transaction, all activity carried out by the trader for the purpose of

---

[47]Craig and de Búrca (2011).
[48]Please note renumbering Article 36 is now Article 34.
[49]Case 120/78.

encouraging the conclusion of such a transaction. That activity consists, inter alia, in selecting an assortment of goods offered for sale and in offering a variety of services aimed at inducing the consumer to conclude the abovementioned transaction with the trader in question rather than with a competitor.[50]

Restrictions on advertising (e.g. alcohol advertisements) may affect the promotion sector as service providers. Such restrictions may create trade obstacles to specific goods and the market penetration possibilities.[51,52]

The Court argued that Article 57 TFEU does not establish any order of priority between the freedom to provide services and the other fundamental freedoms.[53] The Court usually examines national measures by focusing on only one fundamental freedom, even if the measure has the potential to affect more than one fundamental freedom. For this situation, the Court decides which of the fundamental freedoms prevails. Therefore, it is essential to identify the main focal point of the national measure: if it is goods-related, then Article 34 TFEU applies; if it is services-related, then Article 56 TFEU applies.[54]

Regarding taxation, Article 110 of the TFEU states:

> No Member State shall impose, directly or indirectly, on the products of other Member States any internal taxation of any kind in excess of that imposed directly or indirectly on similar domestic products. Furthermore, no Member State shall impose on the products of other Member States any internal taxation of such a nature as to afford indirect protection to other products.[55]

In the case Commission v UK, the low excise duty imposed on beer and the high excise duty imposed on wine resulted in the same bearings.[56] Wine discrimination is detected both in the products' similarity and the fact beer consumption and manufacturing is dominant. It can be difficult to determine whether a case falls under the scope of Article 30 TFEU governing charges equivalent to customs or Article 110 TFEU on discriminatory taxation. Regardless, the Court cannot invoke both Articles as the regulations provide for the clear categorization of each dispute.

---

[50]Case C-418/02 Praktiker Bau- und Heimwerkermärkte.
[51]Case C-405/98.
[52]See European Commission (EC) (2014a), Free movement of goods.
[53]Case C-452/04 Fidium Finanz.
[54]See European Commission (EC) (2014a), Free movement of goods.
[55]Article 110 TFEU has direct effect. Case 57-65. Alfons Lütticke GmbH v Hauptzollamt Sarrelouis.
[56]Case C-170/78.

21.4 Single Market for Food and Agricultural Products

Agriculture and fisheries have a special status within the European Union governed by regulations differing from the internal market rules. The TFEU exempts agriculture from the Common Market's general rules. According to Article 38 (2) TFEU the internal market rules are applied if the Treaty does not stipulate otherwise. The term "agricultural products" refers to stock farming and fisheries processing products.

The Court investigated first-stage processing in the König case.[57] The Court examined whether ethyl alcohol in general qualified as an agricultural product without regard to its alcoholic strengths:

> The concept of 'products of first-stage processing directly related' to the basic products, must be interpreted as implying a clear economic interdependence between basic products and products resulting from a productive process, irrespective of the number of operations involved therein. Processed products which have undergone a productive process, the cost of which is such that the price of the basic agricultural raw materials becomes a completely marginal cost, are therefore excluded.

Beyond the above definition of agricultural products, Article 38 (3) EEC also stipulates the products subject to the provisions of Articles 39–44 are listed in Annex I. The Court examined a similar problem in the Case 61/80 case and ruled that:

> The scope of regulation no 26 applying certain rules of competition to production of and trade in agricultural products was restricted by Article 1 thereof to the production of and trade in the products listed in annex ii to the treaty. that regulation may not therefore be applied to the manufacture of a product which does not come under annex ii even if it is a substance ancillary to the production of another product which itself comes under that annex.[58]

It follows from the above that concepts of foodstuffs and agricultural products are not identical. Should a product correspond to the concept of agricultural product, the special rules governing agricultural products must be applied to it. However, processed foodstuff is subject to the internal market rules.

21.5 The External Market of the European Union

As referred to in Chapter 2, the EU's common commercial policy is based on a set of uniform rules under the Customs Union and the Common Customs Tariff. The commercial policy governs the Member States' commercial relations with Non-EU

---

[57] Case C-185/73.
[58] Judgment of the Court of 25 March 1981—Coöperatieve Stremsel—en Kleurselfabriek v Commission of the European Communities.—Competition—Exclusive obligation to purchase rennet.—Case 61/80.

Member Countries. The tariff is common to EU members, but the rates of duty differ.[59,60] The rates depend on the product's economic sensitivity. The purpose of the instruments of trade defence and market access is mainly to protect European businesses from trade obstacles.

The EU evolved during globalisation by aiming for the harmonious development of world trade and fostering fairness and sustainability. It actively encourages open markets and trade development in the World Trade Organisation (WTO)'s multilateral framework. Simultaneously, it supports developing countries and regions through bilateral relations involving them in world trade using preferential measures.[61] However, the single market has several connections to these policies, including the underlying Internal Market principles. The Internal Market principles require consideration in the negotiation of international agreements. The Commission takes a position on Internal Market policies, whether bilateral or multilateral, in the context of enlargement and regulatory dialogues with third countries. In these negotiations, the services of the Commission responsible for the Internal Market cooperate with the services in charge of other policies to adequately represent and promote the European Internal Market's principles worldwide.[62]

21.5.1 Connection to the WTO

The World Trade Organisation (WTO) moderates world trade affairs and brings together 153 members representing more than 95 % of total world trade.[63] The WTO is a negotiating forum designed to liberalise world trade. Since the WTO's establishment in 1995, it has concluded a number of important agreements; for example, the Agreement on Agriculture came into being with the WTO's establishment on January 1, 1995.[64] The European Commission negotiates on behalf of all the member states in the WTO due to the common trade policy agreed on in Article 207 TFEU. To represent each EU country's interests, a special committee brings together officials from each of the 28 member countries. All aspects of international trade are addressed at these meetings, from WTO negotiations to export refunds.[65] The EU sets high safety standards on all its exports. The EU is

---

[59]Council Regulation (EEC) No 2658/87 of 23 July 1987 on the tariff and statistical nomenclature and on the Common Customs Tariff.

[60]European Commission (EC) (2014e), What is the Common Customs Tariff? http://ec.europa.eu/taxation_customs/customs/customs_duties/tariff_aspects/index_en.htm.

[61]See European Commission (EC) (2014e), What is the Common Customs Tariff?.

[62]European Commission (EC) (2014g), The external dimension, http://ec.europa.eu/internal_market/ext-dimension/index_en.htm.

[63]European Commission (EC) (2014c), External aspects of the EU's food industry, http://ec.europa.eu/enterprise/sectors/food/international-market/wto/index_en.htm.

[64]See European Commission (EC) (2014c), External aspects of the EU's food industry.

[65]See European Commission (EC) (2014c), External aspects of the EU's food industry.

as an international trade standard setter as the world's largest food exporter (20.8 %) and second largest food importer (18.1 %),[66] It has entered many bilateral and multilateral trade agreements with non-EU countries.

The EU and WTO negotiations aim to eradicate poverty by prioritizing trade development issues. A series of bilateral agreements governs the EU's trade in Processed Agricultural Products (PAP or non-annex I goods). [67] Two political entities sign the agreements and bind only the two territories concerned. Examples include free trade agreements, economic partnerships, association agreements, and stabilisation and association agreements. The foreign trade regime for the core Non-Annex I products consists of both import duties and export refunds. The import duties are based on an industrial and agricultural element. Export refunds are paid for through the basic products of milk, sugar, eggs, cereals and rice. Other products are also included, such as spirits, processed fruit and vegetables. Several particular regimes apply for different other products.[68]

Tariff and non-tariff barriers to third country market access have a negative impact on EU industrial competitiveness. Non-tariff barriers are restrictions to imports that do not take the usual form of a tariff. They are the result of food legislation discrepancies from one country to another. Legislation can diverge in areas such as labelling requirements, residue limits and test result recognition. To gain barrier-free market access to foreign countries, DG Enterprise and Industry encourages tariff liberalisation and makes use of the following:

- Regulatory measures for convergence with trade partners and international standardisation promotion (e.g. Codex Alimentarius Standards, Hazard Analysis and Critical Control Points). The measures help reduce the cost of complying with non-EU country regulations;
- Agreements on Sanitary and Phytosanitary Measures (SPS) and Technical Barriers to Trade (TBT) of the World Trade Organisation (WTO). These agreements are being developed to eliminate existing barriers and to prevent the emergence of new ones.
- In order to ensure a coherent food industry policy, the European Union has signed bilateral agreements with third countries.[69]

Moreover, the Commission encourages international business dialogue and makes funding available for measures in non-EU country markets that promote agricultural and food products from the EU.

---

[66]See European Commission (EC) (2014c), External aspects of the EU's food industry.

[67]The term "Annex I" refers to the Annex I of the Treaty on the EU's functioning. This Annex lists all agricultural products which could be subject to a Common Market Organization in the framework of the Common Agricultural Policy.

[68]European Commission (EC) (2014d), International Aspects, http://ec.europa.eu/enterprise/sectors/food/competitiveness/international-aspects/index_en.htm.

[69]See European Commission (EC) (2014d), International Aspects.

21.5.2 Connection to the EFTA, EEA, and ACP Countries

The European Economic Area (EEA) comprises three of four member states of the European Free Trade Association (EFTA) Iceland, Liechtenstein and Norway and 28 Member States of the EU.[70] The most important export destinations are USA, Japan, Switzerland, Russia, Canada and Norway.[71] The important suppliers of Processed Agricultural Products are USA, Switzerland, Ivory Coast, Bahamas, Malaysia and China.

The EU entered into bilateral agreements concerning trade in Processed Agricultural Products in addition to its international commitments under WTO.[72] The framework of bilateral agreements establishes specific preferential regimes for PAPs. The European Commission implements these preferential trade regimes on the EU's behalf. DG Enterprise and Industry participates in all bilateral trade negotiations for Processed Agricultural Products within a strict mandate defined by the Member States.

ACP countries must ensure their exports comply with changing EU standards (e.g. on food safety and animal welfare). For that reason, EPAs include technical support, training, and measures to promote knowledge transfer and strengthen public services. Examples include:

1. The EU pesticides programme for the horticulture sector and an EU fish health project
2. Training in food safety and quality control (PIP programme) for over 200,000 family-run fresh fruit and vegetable businesses.

21.5.3 EU Export and Import Conditions: Fish, Seafood and Aquaculture Products

The European Union is the world's biggest importer of fish, seafood and aquaculture products. The import rules for these products are harmonized; the same rules apply in all EU countries. For non-EU countries, the European Commission is the negotiating partner that defines import conditions and certification requirements.

The European Commission's Directorate-General for Health and Consumers (SANCO) is responsible for food safety in the European Union. The import rules for fishery products and shellfish (bivalve molluscs) seek to guarantee that all imports fulfil the same high standards as products from the EU Member States with respect to hygiene, consumer safety and, if relevant, to the animal health

---

[70]Switzerland, an EFTA member, has not joined the EEA.

[71]European Commission (EC) (2014b), Bilateral Trade Overview, http://ec.europa.eu/enterprise/sectors/food/international-market/bilateral-trade-overview/index_en.htm.

[72]Examples of agreements are free trade agreements, economic partnerships, association agreements, and stabilisation and association agreements.

status. The European Union's food law implements the principle of quality management and process-oriented controls throughout the food chain. Spot checks on the end product alone would not provide the same level of safety, quality and transparency to the consumer. To implement these harmonised principles, the Food and Veterinary Office of the European Commission is undertaking missions in all exporting countries.

21.5.3.1 General Rules for Fishery Products

Fishery products imported into the European Union are subject to official certification based on the recognition of the non-EU country's competent authority the European Commission. This formal recognition of the competent authority is a pre-requisite for the country to be eligible and authorized to export to the European Union. Public authorities with the necessary legal powers and resources must ensure credible inspection and controls throughout the production chain. The national competent authority must undertake all bilateral negotiations and other relevant dialogue concerning imports of fishery products. All other interested parties and private businesses wishing to export to the EU must contact their competent authority and communicate with the European Union via this channel.

21.5.4 EU Export and Import Conditions: Poultry and Poultry Products

The import rules in the European Union for poultry (including hatching eggs) and poultry products (including egg products) are fully harmonized. The European Commission acts as the competent authority on behalf of the 28 Member States. Member States' official veterinarians inspect animals and animal products entering the Community at a Border Inspection Post (BIP) to ensure they fulfil the EU requirements. Animals of a lower Community health status cannot transit the Community.

21.5.4.1 Import of Poultry and Poultry Products

The definition of poultry, hatching eggs, day old chicks, captive birds, and poultry products is laid down in legislation Council Directive 2009/158/EC. The legislation guides the animal health principles on which importation of live poultry is based, harmonises the rules, and establishes the general animal health conditions for live poultry imported into the EU. The objective is to verify that the Member States apply the same principles for poultry importation. Further, harmonization prevents poultry from entering EU territory carrying infectious diseases that are dangerous for livestock or humans. Council Directive 2002/99/EC are the animal health requirements for fresh meat. This Directive forms the legal basis for all animal health rules governing the production, processing, distribution and introduction of animal products for human consumption. Further, the Directive continues to provide harmonised rules and animal health guarantees for importing fresh meat into the EU.

21.5.4.2 Third Country Authorisation

Third countries of origin, i.e. countries that are not EU member states as trading partners, must be on a positive list of eligible countries. The eligibility criteria for the import of poultry and poultry meat are in Commission Regulation (EC) No. 798/2008. The criteria for the import of captive birds, for example, are in the Commission Implementing Decision (EU) No 139/2013.

The main criteria are:

- The exporting countries must have a competent veterinary authority which is responsible throughout the food chain;
- The country or region of origin must fulfil the relevant animal health standards;
- Adequate veterinary services must ensure effective enforcement of all necessary health controls;
- Imports are only authorised from approved establishments, such as slaughterhouses, cutting plants, game handling establishments, cold stores, meat processing plants, which have been inspected by the competent authority of the exporting country and found to meet EU requirements and the responsible agency provides the necessary guarantees and is obliged to carry out regular inspections;
- The veterinary authorities must have one or more laboratories that comply with certain minimum requirements, ensuring sufficient capability for disease diagnosis;
- The national authorities must also guarantee that the relevant hygiene and public health requirements are met. Specifically, the hygiene legislation contains specific requirements on the structure of establishments, equipment and operational processes for slaughter, cutting, storage and handling of meat. These provisions are aimed at ensuring high standards and at preventing any contamination of the product during processing. More information on the food hygiene legislation can be found on the webpages of the Directorate-General for Health and Consumers.[73]

---

[73]http://ec.europa.eu/dgs/health_food-safety/index_en.htm.

21.6 Conclusion

This chapter is a primer on EU market and food trade regulations. The internal market, food's free movement and the rules applied to agricultural product and food trade are reviewed. Economic integration and the harmonization of food trade are described with emphasis on the Treaty on the Functioning of the European Union to explain the customs union and the free movement fo goods in the EU. Notably, the European Union is the world's biggest importer of fish, seafood and aquaculture products. The import rules in the European Union for poultry (including hatching eggs) and poultry products (including egg products) are fully harmonized. The European Commission acts as the competent authority on behalf of the 28 Member States. The chapter concludes with an overview of the European Union's external market and world trade in light of the WTO and EFTA, EEA, and ACP countries.

References

Craig P, de Búrca G (2011) EU law text, cases, and materials, 5th edn. Oxford University Press, Oxford
EFTA agreement. Available at http://www.efta.int/eea/eea-agreement. (last accessed 25 Nov 2014)
Europea Summaries of EU legislation, The Single European Act. Available at http://europa.eu/legislation_summaries/institutional_affairs/treaties/treaties_singleact_en.htm. (last accessed 25 Nov 2014)
European Commission (EC) (2014a) Directorate C, Regulatory Policy of the Enterprise and Industry DG, Enterprise and Industry: Free movement of goods. Guide to the application of Treaty provisions governing the free movement of goods. http://ec.europa.eu/enterprise/policies/single-market-goods/files/goods/docs/art34-36/new_guide_en.pdf. (last accessed 25 Nov 2014)
European Commission (EC) (2014b) Enterprise and Industry: Bilateral Trade Overview. Available at http://ec.europa.eu/enterprise/sectors/food/international-market/bilateral-trade-overview/index_en.htm. (last accessed 25 Nov 2014)
European Commission (EC) (2014c) Enterprise and Industry: External aspects of the EU's food industry. http://ec.europa.eu/enterprise/sectors/food/international-market/wto/index_en.htm. (last accessed 29 May 2014)
European Commission (EC) (2014d) Enterprise and Industry: International Aspects. http://ec.europa.eu/enterprise/sectors/food/competitiveness/international-aspects/index_en.htm. (last accessed 25 Nov 2014)
European Commission (EC) (2014e) Taxation and Customs Union: What is the Common Customs Tariff?. http://ec.europa.eu/taxation_customs/customs/customs_duties/tariff_aspects/index_en.htm. (last accessed 29 May 2014)
European Commission (EC) (2014f) The EU Single Market: Freedom to provide services/Freedom of establishment. Available at http://ec.europa.eu/internal_market/top_layer/living_working/services-establishment/index_en.htm. (last accessed 29 May 2014)
European Commission (EC) (2014g) The EU Single Market: the external dimension, http://ec.europa.eu/internal_market/ext-dimension/index_en.htm. (last accessed 28 Nov 2014)
European Union (EU) (2014) The History of the European Union. Available at http://europa.eu/about-eu/eu-history/index_en.htm. (last accessed 25 Nov 2014)
European University Institute, EU Legislation. Available at http://www.eui.eu/Research/Library/ResearchGuides/EuropeanInformation/EULegislation.aspx. (last accessed 25 Nov 2014)

Chapter 22
European Food Labelling Law

Aude Mahy and Nicola Conte-Salinas

Abstract This chapter focuses on the regime of food information to consumers in Europe, covering labeling and advertisement rules. The difference between horizontal and vertical legislation is explained, and a good overview over both horizontal and vertical rules given. The chapter explains which information has to be provided to the consumer (mandatory information), and discusses certain mandatory elements in detail, such as the rules on the origin of food. This is complemented by rules for voluntary information, in particular the strict regime for nutrition and health claims. Further, the all-encompassing prohibition not to mislead the consumer is being discussed, illustrated by case-law.

22.1 European Union Food Labelling Laws

At the European Union (EU) level, several regulations and directives deal with the various aspects of food labelling. These legislative measures either apply to all foodstuffs alike, the so-called horizontal labelling rules, or apply to specific foods only, the so-called vertical labelling rules. These two types of rules either provide mandatory labelling requirements or they regulate information that should be voluntarily provided by the food business operators or manufacturers. Sections 22.1.1 and 22.2.1 introduce mandatory and voluntary labelling information.

A. Mahy (✉)
Department Litigation & Risk Management, Brussels, Belgium
e-mail: aude.mahy@loyensloeff.com

N. Conte-Salinas
Morsbach, Germany
e-mail: nicola_contesalinas@yahoo.com

22.1.1 Mandatory Information: FIC and Horizontal Labelling Rules for All Foodstuffs

EU legislation consists of regulations, which are directly applicable to all member states, and of directives, which require transposition and implementation into national legislations. According to the European Commission, "EU directives lay down certain end results that must be achieved in every Member State. National authorities have to adapt their laws to meet these goals, but are free to decide how to do so. Directives are used to bring different national laws into line with each other, and are particularly common in matters affecting the operation of the single market (e.g. product safety standards)."[1] On the other hand,

> Regulations are the most direct form of EU law - as soon as they are passed, they have binding legal force throughout every Member State, on a par with national laws. National governments do not have to take action themselves to implement EU regulations. They are different from directives, which are addressed to national authorities, who must then take action to make them part of national law, and decisions, which apply in specific cases only, involving particular authorities or individuals.[2]

Historically, rules governing mandatory labelling requirements for foodstuffs were contained in directives, giving the Member States of the European Union more leeway for developing their own legislation complementing national laws. One of the earliest principal instruments for the labelling of food was Directive 79/112/EC, then replaced by Directive 2000/13/EC,[3] relating to the labelling, presentation, and advertising of foodstuffs. Further examples of EU legislative directives on labelling involve rules on the indication of prices of certain foods,[4] rules concerning nominal quantities for pre-packed foods[5] and rules regarding the maximum tolerable deviation from the indicated net weight.[6] Eventually, harmonization of national food labelling laws was required to facilitate free trade within the Union. In an attempt to strengthen and enlarge the EU, Regulation (EU) No 1169/2011 (FIC)[7] was passed to consolidate and update existing labelling rules and

---

[1] See European Commission, Application of EU law: What are EU directives? Available at http://ec.europa.eu/eu_law/introduction/what_directive_en.htm (last accessed 7 Sept 2014).

[2] European Commission, Application of EU law: What are EU regulations?

[3] Directive 2000/13/EC (2000), p. 29.

[4] Directive 79/581/EEC (1979), p. 19; Directive 98/6/EC (1998), p. 27.

[5] Directive 2007/45/EC (2007) (repealing Directives 75/106/EEC and 80/232/EEC, and amending Council Directive 76/211/EEC).

[6] Directive 76/211/ EEC (1976), p. 1.

[7] Regulation (EU) No 1169/2011 of the European Parliament and of the Council of 25 October 2011 on the provision of food information to consumers, amending regulations (EC) No 1924/2006 and (EC) No 1925/2006 of the European parliament and of the council, and repealing commission directive 87/250/EEC, Council Directive 90/496/EEC, Commission Directive 1999/10/EC, Directive 2000/13/EC of the European Parliament and of the Council, Commission Directives 2002/67/EC and 2008/5/EC and Commission Regulation (EC) No 608/2004.

FIC repealed the preceding Directive 2000/13. As a regulation, FIC directly applies to all Member States of the EU, and achieves a high level of harmonization by leaving only limited leeway for the Member States to enact or maintain national rules concerning food information.[8]

The FIC Regulation covers labelling of all food information and sets forth mandatory labelling requirements. Food information is defined as information concerning a food and made available to the final consumer by means of a label, other accompanying material, or any other means including modern technology tools or verbal communication.[9] The regulation, therefore, covers information contained on a product label and information disseminated through advertising materials, such as websites, catalogues and flyers.

22.1.2 FIC Regulation: Mandatory Particulars

The FIC Regulation sets out requirements for mandatory information concerning all foods intended for the final consumer, including food delivered by and to mass caterers.[10] This important regulation first applies to all food business operators along the supply chain whose activities concern the provision of food information to final consumers.[11] Second, the FIC also applies to food business operators who deliver food not intended for the final consumer (e.g. ingredients, raw materials) to other food business operators. In these cases, information must be provided to ensure that food business operators are provided with sufficient information about the food so as to enable them to comply with their own obligations regarding food information.[12]

The elements of mandatory information required by the FIC Regulation are listed in Articles 9 and 10 of the regulation, and are further specified in the subsequent articles and annexes. Several labelling requirements exist, for example, for prepackaged foods.[13] Pursuant to the FIC Regulation, the following information is mandatory for prepackaged food:

- Name of food and particulars that shall accompany it[14];
- List of ingredients[15];

---

[8]Mainly as regards mandatory information requirements for non-prepacked food, Regulation (EU) No 1169/2011, *op cit.*, Art. 44; see also *infra*, national vertical labelling rules.
[9]Regulation (EU) No 1169/2011, *op cit.*, Art. 2 (2) (a).
[10]Regulation (EU) No 1169/2011, *op cit.*, Art. 1 (3).
[11]Regulation (EU) No 1169/2011, *op cit.*, Art. 1 (3).
[12]Regulation (EU) No 1169/2011, *op cit.*, Art. 8 (8).
[13]Regulation (EU) No 1169/2011, *op cit.*, Art. 2 (2) (e).
[14]Regulation (EU) No 1169/2011, *op cit.*, Art. 9 (1) (a), 17, Annex VI.
[15]Regulation (EU) No 1169/2011, *op cit.*, Art. 9 (1) (b), 18–20, Annex VII.

- Allergen labelling[16];
- Quantitative ingredient declaration[17];
- Net quantity[18];
- Date of minimum durability or "use by"-date[19];
- Storage conditions, conditions of use[20];
- Business name and address of the responsible food business operator[21];
- In certain cases: Country of origin/place of provenance[22];
- Instructions for use, if failure to do so would make appropriate use of the food difficult[23];
- Alcoholic strength (if greater than 1.2 %)[24];
- Nutrition declaration[25];
- Further specific mandatory particulars (e.g. for products with caffeine or which contain artificial sweeteners).[26]

For non-prepackaged food, only the provision of allergen information[27] is mandatory.[28] Member States are, however, entitled to enact additional requirements concerning information to consumers with respect to non-prepackaged food.[29]

The FIC regulation additionally sets standards for the presentation of mandatory particulars. First, the FIC regulation establishes a general[30] minimum font size requirement of 1.2 mm[31] for all mandatory particulars. Second, mandatory labelling requirements must be easy to understand and placed in a conspicuous place, so as to be easily visible, clearly legible, and indelible.[32] The mandatory requirements cannot be hidden, obscured, or interrupted by other intervening written or pictorial information, which are usually voluntary elements of labelling.[33] Third, mandatory

---

[16]Regulation (EU) No 1169/2011, *op cit.*, Art. 9 (1) (c), 21, Annex II.
[17]Regulation (EU) No 1169/2011, *op cit.*, Art. 9 (1) (d), 22, Annex VIII.
[18]Regulation (EU) No 1169/2011, *op cit.*, Art. 9 (1) (e), 23, Annex IX.
[19]Regulation (EU) No 1169/2011, *op cit.*, Art. 9 (1) (f), 24, Annex X.
[20]Regulation (EU) No 1169/2011, *op cit.*, Art. 9 (1) (g), 25.
[21]Regulation (EU) No 1169/2011, *op cit.*, Art. 9 (1) (h), 8.
[22]Regulation (EU) No 1169/2011, *op cit.*, Art. 9 (1) (i), 26, Annex XI.
[23]Regulation (EU) No 1169/2011, *op cit.*, Art. 9 (1) (j), 27.
[24]Regulation (EU) No 1169/2011, *op cit.*, Art. 9 (1) (k), 28, Annex XII.
[25]Regulation (EU) No 1169/2011, *op cit.*, Art. 9 (1) (l), 29–35, Annexes I, V, XIII-XV.
[26]Regulation (EU) No 1169/2011, *op cit.*, Art. 10, Annex III.
[27]Regulation (EU) No 1169/2011, *op cit.*, Art. 9 (1) (c), 21, Annex II.
[28]Regulation (EU) No 1169/2011, *op cit.*, Art. 44 (1) (a).
[29]Regulation (EU) No 1169/2011, *op cit.*, Art. 44 (1) (b), (2).
[30]For packages with the largest surface area of less than 80 cm^2, the minimum x-height is 0.9 mm instead of 1.2 mm.
[31]Regulation (EU) No 1169/2011, *op cit.*, Art. 12 (2), Annex IV.
[32]Regulation (EU) No 1169/2011, *op cit.*, Art. 12 (1).
[33]Regulation (EU) No 1169/2011, *op cit.*, Art. 12 (1).

particulars must be provided in a language easily understood by the consumers in the Member State(s) where the food is marketed and the Member States may stipulate which official language(s) have to be used.[34] Lastly, all mandatory particulars must be indicated with words and numbers.[35] Symbols or pictograms can be used to supplement words or numbers as long as they are not misleading. However, symbol or pictograms cannot be used to replace words or numbers regarding mandatory particulars.

22.1.3 Country of Origin Labelling

The place of provenance is defined as any place where a food's label indicates a country other than the country of origin,[36] such as a specific region or city. Country of origin—or place of provenance—labelling is currently mandatory for specific products, or when the absence of such information could mislead consumers as to the true origin of the product. Unless otherwise provided, the country of origin of a product is determined according to Regulation (EEC) No 2913/92[37] establishing the Community Customs Code (Articles 23–26). As a side-note, the name, business name or address of a food company or producer is not regarded as an indication of origin.[38]

Before the applicability of the FIC Regulation, country of origin labelling was only required for certain foods, such as beef—as a result of the BSE crisis in Europe.[39] The FIC Regulation now extends mandatory origin labelling for specific products to meat of swine, sheep or goats, and poultry.[40] However, the specific requirements for origin labelling of these products are laid down in an implementing regulation.[41]

Another important labelling rule in terms of the indication of origin is required where failure to indicate it might mislead the consumer as to the true place of origin of the food. This rule regulates the use of voluntary information indicating a certain

---

[34]Regulation (EU) No 1169/2011, *op cit.*, Art. 15 (1),15 (2).

[35]Regulation (EU) No 1169/2011, *op cit.*, Art. 9 (2).

[36]Regulation (EU) No 1169/2011, *op cit.*, Art. 2 (2) (g).

[37]For the full text of the Regulation, see http://eur-lex.europa.eu/LexUriServ/LexUriServ.do?uri=CELEX:31992R2913:en:HTML (last accessed 12 Sept 2014).

[38]Regulation (EU) No 1169/2011, *op cit.*, Art. 2 (2) (g).

[39]BSE is a fatal neurodegenerative disease. The BSE crisis first affected British cattle in the late 1990s and early 2000s leading to substantial media coverage of the epizootic. For more information, see USDA, A Focus on Bovine Spongiform Encephalopathy, available at http://web.archive.org/web/20080303135425/http://fsrio.nal.usda.gov/document_fsheet.php?product_id=169.

[40]Regulation (EU) No 1169/2011, *op cit.*, Art. 26 (2) (b), Annex XI.

[41]Regulation (EU) No 1337/2013 laying down rules for the application of Regulation (EU) No 1169/2011 regarding the indication of the country of origin or place of provenance for fresh, chilled and frozen meat of swine, sheep, goats and poultry, p. 19.

country of origin, particularly where other information provided for the food, such as through words, pictures, flags, and colour schemes, would imply that the food has a different country of origin or place of provenance.[42] For instance, country of origin labelling may be required in addition to the use of a voluntarily used Italian flag or a picture of the Eiffel Tower to allude to the place of origin. If voluntary information indicating a particular country of origin is used, but the food does not come from the country alluded to, it must be made clear that the product comes from another country. This can be done by a reference to the true country of origin, e.g. "produced in Germany." This reference must be clearly visible. Merely including the address of the national origin food business operator on the product label will not in itself be sufficient to prevent the consumer from being misled and the labelling to be made proper.

Additionally, the FIC Regulation requires the provision of mandatory country of origin information where the final product's origin is voluntarily indicated, but where it is not the same origin of that of its primary ingredients. "Primary ingredient" means one or more ingredient(s) of a food that represent over half of that food, or which is usually associated with the name of the food for which a quantitative indication is typically required.[43] A product may therefore contain several 'primary ingredients'. In these cases, the origin of the primary ingredient must be given or a statement added to the effect that the origin of the primary ingredient is different from the origin of the final product.[44] At the time of going to print, there was only a preliminary draft of the relevant implementing regulation.

22.1.4 Nutrition Declarations

The FIC Regulation further introduces a mandatory but general nutrition declaration[45] for all products save a limited number of exceptions.[46] It includes the following:

- Energy value; and
- The amounts of fat, saturated fats,
- Carbohydrates and sugars,
- Protein, and
- salt.[47]

---

[42]Regulation (EU) No 1169/2011, *op cit.*, Art. 26 (2) (a).
[43]Regulation (EU) No 1169/2011, *op cit.*, Art. 2 (2) (q).
[44]Regulation (EU) No 1169/2011, *op cit.*, Art. 26 (3).
[45]Exceptions are laid down in Regulation (EU) No 1169/2011, *op cit.*, Art. 16, Annex V.
[46]Regulation (EU) No 1169/2011, *op cit.*, Art. 9 (1) (l), 29-35, Annexes I, V, XIII-XV.
[47]Regulation (EU) No 1169/2011, *op cit.*, Art. 30 (1).

Correspondingly, the nutrition declaration must be made in the order listed above, and must be shown as a table whenever possible.[48] This mandatory nutrition declaration can be supplemented by one or more of the following:

- Mono-unsaturates;
- Polyunsaturates;
- Polyols;
- Starch;
- Fibre;
- Or any of the vitamins or minerals listed in the FIC Regulation (point 1 of Part A of Annex XIII), if they are present in significant amounts.[49]

The nutrition declaration must indicate the value per 100 mg or 100 ml respectively. Additionally, a nutrition declaration may indicate the nutrition values per portion or consumption unit as long as the quantification of the portion or unit on the label is in close proximity to the nutrition declaration and the statement of the number of portions or units contained in the package.[50] The declared values, in turn, shall be average values based on an analysis of the food, a calculation from the average values of the ingredients used, or a calculation from generally established and accepted data.[51]

Also for the first time, the Regulation outlines rules for the voluntary repetition of the nutrition declaration. A repetition is only allowed if a complete nutrition declaration is provided. The following information may be repeated: the energy value; or the energy value together with the amounts of fat, saturates, sugars, and salt.[52] The information must be repeated in the principal field of vision,[53] which generally will be the front of the package. When prepackaged food which are not under an obligation to label the nutrition declaration nevertheless label it on a voluntary basis, the declaration may be limited to the energy value, or the energy value together with the amounts of fat, saturates, sugars, and salt.[54]

Notably, the mandatory labelling rules on nutrition declaration do not apply to food supplements and natural mineral waters.[55] However, if a nutrition declaration is given on a voluntary basis, the requirements of the FIC Regulation must be met.

---

[48] See Regulation (EU) No 1169/2011, *op cit.*, Annex XV.

[49] As defined in Regulation (EU) No 1169/2011, *op cit.*, Point 2 of Part A of Annex XIII, Regulation (EU) No 1169/2011, *op cit.*, Art. 30 (2).

[50] Regulation (EU) No 1169/2011, *op cit.*, Art. 33.

[51] Regulation (EU) No 1169/2011, *op cit.*, Art. 31 (4).

[52] Regulation (EU) No 1169/2011, *op cit.*, Art. 30 (3).

[53] As defined in Regulation (EU) No 1169/2011, *op cit.*, Art. 2 (2) (l), Art. 34 (3) (a).

[54] Regulation (EU) No 1169/2011, *op cit.*, Art. 30 (5).

[55] Regulation (EU) No 1169/2011, *op cit.*, Art. 29 (1).

22.2 Vertical Labelling Rules for Specific Foodstuffs

Vertical legislation generally defines the specific products and assigns each one a legal name that must be used if the product meets the compositional definition. EU vertical legislation usually appears in harmonisation Directives to be implemented within the national law of each Member State, rather than in Regulations, which directly apply to the whole EU territory. Vertical EU legislation regulates products such as milk and milk products,[56] eggs and poultry,[57] spreadable fats,[58] olive oil,[59] honey,[60] chocolate products,[61] wine[62] and spirits,[63] fruit juices,[64] main fruits and vegetables,[65] sugar,[66] coffee and chicory extracts,[67] fruit jams and similar products, beef or veal,[68] and seafood.[69] There is also specific EU legislation involving

---

[56] Regulation (EC) No 1308/2013.

[57] Regulation (EC) No 1308/2013.

[58] Regulation (EC) No 1308/2013.

[59] Regulation (EU) No 29/2012 on marketing standards for olive oil, p. 14, as amended. Regulation (EC) No 1308/2013.

[60] Directive 2001/110/EC relating to honey, p. 47.

[61] Directive 2000/36/EC relating to cocoa and chocolate products intended for human consumption, p. 19, as amended.

[62] Regulation (EC) No 1308/2013.

[63] Regulation (EC) No 110/2008 on the definition, description, presentation, labelling and protection of geographical indications of spirit drinks and repealing Regulation (EEC) No 1576/89 (OJ L 39, 13.2.2008, p. 16), as amended.

[64] Directive 2001/112/EC relating to fruit juices and certain similar products intended for human consumption, OJ L 10, 12/01/2002, p. 58, as amended.

[65] Council Regulation (EC) No 2200/96 on the common organization of the market in fruit and vegetables, p. 1, as amended and Regulation (EC) No 1580/2007 laying down implementing rules of Regulations (EC) No 2200/96, (EC) No 2201/96 and (EC) No 1182/2007 in the fruit and vegetable sector.

[66] Directive 2001/111/EC relating to certain sugars intended for human consumption, p. 53.

[67] Directive 99/4/EC relating to coffee extracts and chicory extracts, p. 26, as amended.

[68] Regulation (EC) No 1760/2000 establishing a system for the identification and registration of bovine animals and regarding the labelling of beef and beef products and repealing Regulation (EC) No 820/97, p. 1, as amended, which sets out rules for compulsory and voluntary beef labelling and Regulation (EC) No 1825/2000 laying down detailed rules for the application of Regulation (EC) No 1760/2000 regarding the labelling of beef and beef products, p. 8, which provides detailed rules for the implementation of Regulation 1760/2000.

[69] Regulation (EC) No 104/2000 on the common organisation of the markets in fishery and aquaculture products, pp. 22–52, as amended and Regulation (EC) No 2065/2001 laying down detailed rules for the application of Regulation (EC) No 104/2000 regarding informing consumers about fishery and aquaculture products, p. 6, as amended: fish sold at retail in certain presentations (i.e. live, fresh, chilled or frozen fish, fresh, chilled or frozen fish fillets and other fish meat; smoked, dried, salted or brined fish; crustaceans and molluscs) must label their origin by area of catching.

particular labelling requirements for organic foods,[70] genetically modified ingredients,[71] and food supplements. Hence, parallel to the application of the FIC Regulation to all foodstuffs, there are additional vertical labelling rules, i.e. specific rules pertaining to particular categories of food or beverages.

Vertical directives often include requirements that go beyond the mere question of labelling and involve the manufacture and marketing of specific products on a global scale. Thus, if a particular matter is not harmonised, which means that it falls under standards common to all Member States at the EU level, Member States may adopt or retain specific national legislation. Here, the key requirement is that the vertical labelling rules must not impede the free movement of goods within the EU. Accordingly, if a particular matter is harmonised at the EU level, Member States may not enact legislation that prohibits the trade of a product which complies with the requirements of the EU. It follows that no national legislation may prohibit trade in a product which complies with the labelling requirements set out by the EU's FIC Regulation because food labelling is regarded as harmonised at EU level.

Importantly, Member States may impose national requirements for additional mandatory indications on the label of certain food stuffs, by way of exception, on the following grounds:

a) the protection of public health;
b) the protection of consumers;
c) the prevention of fraud;
d) the protection of industrial and commercial property rights, indications of provenance, registered designations of origin and the prevention of unfair competition.

Hence, when a Member State adopts measures concerning food labelling or presentation which are not properly justified on one of the aforementioned grounds, those listed measures will be rendered illegal and inapplicable, both in regards to imported foodstuffs and domestic foodstuffs. In practice, most national labelling measures are justified on the grounds of consumer protection. Examples of products most commonly subject to national requirements are cheeses, breads, yoghurts, and, essentially, products which are regarded as typical or traditional in the relevant Member State.

---

[70]Council Regulation (EC) No 834/2007 on organic production and labelling of organic products and repealing Regulation (EEC) No 2092/91, as amended. This Regulation lays lay down basic requirements with regard to production, labelling and control of organic products in the plant and livestock sector and Regulation 889/2008 laying down detailed rules for the implementation of Regulation 834/2007 provides among others for rules on production, processing, packaging, transport and storage of products and control requirements.

[71]Regulation (EC) No 1829/2003 on genetically modified food and feed, p. 1, and Regulation (EC) No 1830/2003 concerning the traceability and labelling of genetically modified organisms and the traceability of food and feed products produced from genetically modified organisms and amending Directive 2001/18/EC, p. 24.

22.2.1 Specific Rules for Labelling Food Supplements

Food supplements are concentrated sources of nutrients, i.e. vitamins and minerals, or other substances with a nutritional or physiological effect, which are intended to supplement a normal diet.[72] Usually, food supplements are marketed in "dose form," including capsules, pastilles, tablets, pills and other similar forms designed to be taken in a measurable small unit or quantity.[73] For the reason that food supplements are marketed in dose form, it can be difficult for consumers to distinguish them from medicinal products. They are however distinguished in terms of labelling laws.

In the EU, composition labelling pursuant to Annexes I and II of the Food Supplements Directive list vitamins and minerals, and their respective forms authorised for use in food supplements. Vitamins and minerals, or forms thereof, that are not listed in the Annexes cannot be legally used and distributed.

Food supplements can also contain "other substances," which are primarily so-called "botanicals," i.e. substances of plant origin. Since there is no EU harmonisation concerning the use of "other substances" in food supplements, national rules apply. However, if a product falls within the definition of "food supplement," then the term "food supplement" is to be used as the legal name of the product and cannot be replaced.[74]

For the labelling of food supplements and in addition to the general labelling requirements detailed above, the following information must be indicated on the label of a product defined as a food supplement[75]:

- the names of the categories of nutrients or substances that characterise the product or an indication of the nature of those nutrients or substances;
- the portion of the product recommended for daily consumption;
- a warning not to exceed the stated recommended daily dose;
- a statement to the effect that food supplements should not be used as a substitute for a varied diet;
- a statement to the effect that the products should be stored out of the reach of young children.

The amount of the nutrients or substances with a nutritional or physiological effect present in the product must be stated on the labelling in numerical form, per portion of the product as recommended for daily consumption.[76] Because of these specific detailed requirements for nutrition declarations for food supplements, the

---

[72]Directive 2002/46/EC, of the European Parliament and of the Council of 10 June 2002 on the approximation of the laws of the Member States relating to food supplements, OJ L 183, 12 July 2002, p. 51, Art. 2 (a).
[73]Directive 2002/46/EC, op cit., Art. 2 (a).
[74]Directive 2002/46/EC, *op cit.*, Art. 6 (1).
[75]Directive 2002/46/EC, *op cit.*, Art. 6 (3).
[76]Directive 2002/46/EC, *op cit.*, Art. 8.

horizontal, mandatory nutrition labelling rules under the FIC Regulation do not apply.

22.2.2 Distance Selling

The FIC Regulation also covers information requirements in business-to-consumer transactions in the case of distance selling.[77] Distance selling is defined to include "means of distance communication, including any means used for the formation of a contract between a supplier and the consumer without the simultaneous physical presence of the parties."[78] For practical reasons, vending machines or automated commercial premises are excluded from the scope of this article.[79]

To satisfy labelling requirements, any food supplied through distance selling must meet the same information requirements as food sold directly to the consumer in shops. All of the mandatory particulars according to Articles 9 and 10 of the FIC Regulation, and any other obligatory information according to EU law, must be made available before conclusion of the purchase.[80] This includes, among other things, the mandatory labelling requirements imposed by Regulation (EC) No 1924/2006 on nutrition and health claims made on foods (**HCR**).[81] Therefore, if a product contains a health claim, the information requirements under Art. 10 (2) HCR must be complied with before the contract is concluded, even if the web-shop does not repeat the health claim found on the labelling.[82] For the HCR labelling requirements, see infra, section on health and nutrition claims.

In distance selling, all of the mandatory information must also be made available "before the purchase is concluded." This language cannot be interpreted from a strict civil law perspective, since the rules about when a contract is formed may differ from one Member State to another. Generally, however, this language was enacted to ensure that the consumer has received the necessary information to make an informed decision about the impending purchase. Therefore, the consumer needs to have access to the mandatory information before he makes a purchase decision. To quote the Commission,

---

[77]Regulation (EU) No 1169/2011, op cit., Art. 14.
[78]Regulation (EU) No 1169/2011, op cit., Art. 2 (2) (u).
[79]Regulation (EU) No 1169/2011, op cit., Art. 14 (3).
[80]Regulation (EU) No 1169/2011, op cit., Art. 14 (1) (a).
[81]Regulation (EC) No 1924/2006 on nutrition and health claims made on foods, p. 3.
[82]Commission Implementing Decision of 24 January 2013 adopting guidelines for the implementation of specific health claims laid down in Article 10 of Regulation (EC) No 1924/2006, OJ L 22,251.2013, p. 25, point 2.1. (b).

[i]n the context of a product offer, when there is no possibility to buy the food directly, the information is not required; for example, in an advertising catalogue of products of a supermarket. However, when the catalogue is the only piece of available information before the delivery of products, the catalogue shall include all the mandatory particulars. Also, when there is the possibility to buy the product directly, for example, through a web-page on the internet, the mandatory information shall be available.[83]

However, the minimum durability date or best before-date does not have to be given before the time of purchase. This exception can be explained by the fact that it would be technically impossible, or at least impracticable, to indicate the minimum durability date on the website of an online store, since this information would have to be updated whenever the batch with the current best before date or use by-date is sold out. Nonetheless, at the moment of delivery of the product to the consumer, all mandatory information (including the best before date / use by date) must be made available.[84]

22.3 Voluntary Information

Voluntary information generally concerns marketing needs for purchasing decisions and subsequent use. There are some general rules regulating the presentation of voluntary information, so that the information provided must be accurate, verifiable, and not misleading. Additionally, there are specific and detailed rules regarding the standard formatting for nutrition and health claims, which are introduced hereinafter.

22.3.1 Misleading Labels

The ban on misleading advertising is a well-established prohibition and is embodied in numerous European legislative acts. Information that is misleading for consumers is also regarded as unfair vis-à-vis competitors. As far as foodstuffs are concerned, no less than six different sets of rules—general or food-related— may potentially apply,[85] a selection of which are summarised below.

The Unfair Commercial Practices Directive,[86] which is the main body of EU legislation regulating misleading advertising and other unfair practices in business-

---

[83]Working Document prepared by the Commission services, New Q&A items related to general labelling (February 21, 2014).

[84]Regulation (EU) No 1169/2011, *op cit.*, Art. 14 (1) (b).

[85]See Leible (2010), p. 322.

[86]Directive 2005/29/EC concerning unfair business-to-consumer commercial practices in the internal market and amending Council Directive 84/450/EEC, Directives 97/7/EC, 98/27/EC and 2002/65/EC, and Regulation (EC) No 2006/2004.

to-consumer applies to all sectors. It aims at ensuring that consumers are not misled or exposed to aggressive marketing and that any claim made by trading parties in the EU is clear, accurate, and substantiated, so as to enable consumers to make informed and meaningful purchase choices. Under the Unfair Commercial Practices Directive, the concept of "misleading commercial practice" is broadly defined to range from false information to misleading omissions, including deceptive practices, even if the information contained is factually correct.[87] In other words, for a commercial practice to be regarded as misleading for consumers, it must cause or be likely to cause the average consumer to take a transactional decision, i.e. to purchase goods that the consumer would not have otherwise bought if it were not for that misleading information. According to the European Commission, the Unfair Commercial Practices Directive "works as a safety net which fills the gaps which are not regulated by other EU sector-specific rules."[88] Therefore, this Directive, only applies to the extent that there are no specific EU provisions regulating specific aspects of unfair commercial practices, such as information requirements and rules on the way information is to be presented to the consumer.[89]

---

[87] According to Directive 2005/29, *op.cit.*, Art. 6 and 7, a commercial practice is misleading when:
  (1) contains false information and is therefore untruthful or if
  (2) in any way, including overall presentation, it deceives or is likely to deceive the average consumer, even if the information is factually correct, in relation to:
     a. the existence or nature of the product;
     b. the main characteristics of the product, such as its availability, benefits, risks, execution, composition, accessories, after-sale customer assistance and complaint handling, method and date of manufacture or provision, delivery, fitness for purpose, usage, quantity, specification, geographical or commercial origin or the results to be expected from its use, or the results and material features of tests or checks carried out on the product;
     c. the extent of the trader's commitments, the motives for the commercial practice and the nature of the sales process, any statement or symbol in relation to direct or indirect sponsorship or approval of the trader or the product;
     d. the price or the manner in which the price is calculated, or the existence of a specific price advantage;
     e. the need for a service, part, replacement or repair;
     f. the nature, attributes and rights of the trader or his agent, such as his identity and assets, his qualifications, status, approval, affiliation or connection and ownership of industrial, commercial or intellectual property rights or his awards and distinctions;
     g. the consumer's rights or the risks he may face.
  (3) the marketing of a product, including comparative advertising, creates confusion with any products, trademarks, trade names or other distinguishing marks of a competitor;
  (4) the trader does not comply with commitments contained in codes of conduct by which the trader is bound, where the trader indicates in a commercial practice that he is bound by the code.
  (5) the trader omits material information that the average consumer needs (or provides it in an unclear, unintelligible, ambiguous or untimely manner), according to the context, to take an informed transactional decision.
[88] Communication from the Commission of (March 14 2013) on the application of the Unfair Commercial Practices Directive.
[89] Directive 2005/29/EC, *op.cit.* Recital 10.

Consequently, the Directive complements the rules applying to specific sectors, including the food sector, and the latter will prevail in the event of any conflicts.

Corresponding to the Unfair Commercial Practices Directive for business-to-consumer relations, the Directive on Misleading and Comparative Advertising[90] provides rules concerning business-to-business relationships. Its purpose is to protect traders from misleading advertising and the unfair consequences thereof. This directive sets out the conditions under which comparative advertising is or is not permitted.[91] Under this set of rules, misleading advertising is regarded as behaviour which causes injury or is likely to cause injury to a competitor and is considered as an unfair practice towards that competitor. However, the practices prohibited under this Directive are minimum requirements, and Member States are, therefore, allowed to apply stricter rules, which provide extended protection with regard to misleading advertising for traders and competitors.[92]

Additional rules against misleading information specifically apply to the food sector. For example, the ban on misleading food information is covered by the General Food Law. This ban is widely drawn, prohibiting any misleading of consumers via labelling, advertising, and presentation of foodstuffs, including their shape, appearance or packaging, the packaging materials used, the manner in which they are arranged and the setting in which they are displayed, and the information which is made available about them through whatever medium.[93]

This general prohibition is nevertheless detailed in the FIC Regulation, which states that any food labelling, advertising, presentation of a foodstuff,[94] or any related method used must[95]:

(a) not mislead, particularly:

    i. as to the characteristics of the food and, specifically, as to its nature, identity, properties, composition, quantity, durability, country of origin or place of provenance, method of manufacture or production;

    ii. by attributing to the food effects or properties which it does not possess;

---

[90]Directive 2006/114/EC concerning misleading and comparative advertising, p. 21. The Directive also applies to products covered by the agricultural sphere, see Recital 12.

[91]See Art. 1. According to Directive 2006/114/EC, *op.cit.*, Art. 2 (b), "'misleading advertising' means 'any advertising which in any way, including its presentation, deceives or is likely to deceive the persons to whom it is addressed or whom it reaches and which, by reason of its deceptive nature, is likely to affect their economic behaviour or which, for those reasons, injures or is likely to injure a competitor.'"

[92]Directive 2006/114/EC, *op.cit.*, Art. 8.

[93]Regulation (EC) No 178/2002 laying down the general principles and requirements of food law, establishing the European Food Safety Authority and laying down procedures in matters of food safety, pp. 1–24, Art. 22.

[94]In particular, its shape, appearance or packaging, the packaging materials used, the way in which the food is arranged and the setting in which it is displayed.

[95]FIC Regulation, *op.cit.*, Art. 7 (Fair information practices).

iii. by suggesting that the food possesses special characteristics when in fact all similar foods have such characteristics, in particular by specifically emphasising the presence or absence of certain ingredients and/or nutrients;
iv. by suggesting, by means of the appearance, the description or pictorial representations, the presence of a particular food or an ingredient, while in reality a component naturally present or an ingredient normally used in that food has been substituted with a different component or a different ingredient;

(b) be accurate, clear and easy to understand for the consumer;
(c) not attribute to any food the property of preventing, treating or curing a human disease, nor refer to such properties (subject to specific derogations provided for by Union law applicable to natural mineral waters and foods for particular nutritional uses).[96]

In addition to general food-related regulations,[97] specific food law provisions are contained in other detailed regulations regarding misleading advertising. These include the Regulation on Nutrition & Health Claims made on food,[98] the Regulation on the Protection of Geographical indication of origin for agricultural products and foodstuffs,[99] and the Regulation on organic products[100].

22.3.2 Nutrition and Health Claims

The particulars of nutrition and health claims are brief and follow several of the principles outlined in this chapter. According to the Nutrition and Health Claims Regulation such claims must not be false, ambiguous or misleading, even if their use is authorised under the scope of that Regulation.

---

[96]FIC Regulation.
[97]See Regulation 1924/2006, *op.cit.*, Recital 3, Regulation No 1151/2012, *op.cit.*, Recital (8) and Art. 3, which stress the complementary nature of their provisions with respect to the Labelling Directive, whose rules are currently included in Regulation No 1169/2011, *op. cit.*
[98]Regulation (EC) No 1924/2006 on nutrition and health claims made on foods, p. 3, Art. 3(2)(a).
[99]Regulation (EU) No 1151/2012 on quality schemes for agricultural products and foodstuffs, p. 1, Art. 13 and 24.
[100]Regulation (EC) No 834/2007 on organic production and labelling of organic products, *op cit.*

22.3.3 Traditional Specialties and Products with a Geographical Indication of Origin

The European Regulation on food quality schemes protects foodstuffs that have a registered Protected Designation of Origin (PDO), a Protected Geographical Indication (PGI), and products that are recognised as Traditional Specialities Guaranteed (TSG).

PDO and PGI indications are protected against:

a. any direct or indirect commercial use of a registered name in respect of products not covered by the registration;
b. any misuse, imitation or evocation, even if the true origin of the products or services is indicated or if the protected name is translated or accompanied by an expression such as 'style', 'type', 'method', 'as produced in', 'imitation' or similar, including when those products are used as an ingredient;
c. any other false or misleading indication as to the provenance, origin, nature or essential qualities of the product, that appear on the inner or outer packaging, advertising material or documents relating to the product concerned, and the packing of the product in a container liable to convey a false impression as to its origin;
d. any other practice liable to mislead the consumer as to the true origin of the product.

TSGs are protected against any misuse, imitation or evocation, or against any other practice liable to mislead the consumer. Furthermore, Member States have to ensure that sales descriptions used at national level do not give rise to confusion with names that are registered.

22.3.4 Regulation of Organic Products

The European Regulation concerning organic products prohibits the use of the terms "organic," "eco," "bio," their derivatives or diminutives, alone or combined, for a product that does not satisfy the requirements set out under this Regulation. These terms may not be used for a product for which it has to be indicated in the labelling or advertising that it contains GMOs, consists of GMOs, or is produced from GMOs according to EU provisions and this prohibition applies in all languages used in the European Union. Furthermore, the prohibition applies to labelling, advertising and commercial documents.

As previously stated, a Member State may, however, further limit the free movement of foodstuffs on the basis of consumer protection, even if labelling requirements comply with EU law. The national measures adopted must be proportionate to the desired effect, which means that between various measures to attain the same objective, the one chosen must be the least restrictive on free trade.

Hence, ensuring that consumers are not misled by advertising or labelling is part of the standard of consumer protection that Member States must ensure.

The interpretation of whether and how a consumer could be misled will differ between Member States depending on the legal traditions of that Member State.[101] Although the European Court of Justice considers that the person deserving protection under the term "consumer" is "reasonably well-informed, reasonably observant and circumspect,[102]" the identification of such a consumer is a matter for national interpretation. Therefore, although the ban on misleading advertising is uniformly imposed at the European level, no standard European guidelines can be directly and automatically applied within the Member States,[103] a matter which presents a difficulty for food business operators, as there can be significant differences between the interpretations of different Member States.

National courts will make their assessments on a case-by-case basis whether an advertisement or label is misleading by referring to the group of consumers who are exposed to the labelling or advertising and who may or may not be misled under any given set of circumstances. On this note, the European Court of Justice has held that the courts may instruct a survey on the actual expectations of the consumer if they still have doubts as to the extent to which the statement at issue might be misleading.[104] In addition, the enforcement of food law is left to the national authorities. Member States have the duty of laying down appropriate rules and ensuring effective controls at all stages of the food chain on measures and penalties applicable to infringements of food law. They must apply effective, proportionate, and dissuasive penalties.[105] As a result, national differences are also apparent at the enforcement stage.

---

[101] See Judgment of the Court (Fifth Chamber) Case C-373/90 (January 16, 1992). Criminal proceedings against X (ECR., 1992 Page I-00131) ("*It is for the national court, however, to ascertain in the circumstances of the particular case and bearing in mind the consumers to which the advertising is addressed, whether the latter could be misleading...*"); see also CJEU, 16 July 1998, Case C-210/96 (Marketing standards for eggs), *op.cit.*

[102] CJEU, 6 July 1995, Case Mars, C-470/93, ECR 1995 Page I-01923, §24; CJEU, 16 July 1998, Case C-210/96 ('Marketing standards for eggs'), ECR, 1998 Page I-04657, §§ 29–37. See Meisterernst (2013), p. 91.

[103] For an analysis of the power limits of the authority control of the country of manufacturing and labelling, see Nilsson (2012), p. 22.

[104] CJEU, 16 July 1998, Case C-210/96 (Marketing standards for eggs), *op.cit.*

[105] Regulation No 178/2002, *op. cit.*, Art. 17(2),

> Member States shall enforce food law, and monitor and verify that the relevant requirements of food law are fulfilled by food and feed business operators at all stages of production, processing and distribution . . . . Member States shall also lay down the rules on measures and penalties applicable to infringements of food and feed law. The measures and penalties provided for shall be effective, proportionate and dissuasive.

> See also Regulation (EC) No 882/2004 on official controls performed to ensure the verification of compliance with feed and food law, animal health and animal welfare rules, p. 1, Art. 54 & 55; (Please note that the Regulation is currently being totally revamped). Directive 2005/29, *op.cit.* Art. 11–13.

22.3.5 Superfoods: Health and Nutrition Claim Regulation

In recent years, increasing numbers of superfoods, foods with special nutritive or health properties, have been introduced within and outside the European market. The 1997 Codex Alimentarius guidelines for the use of nutrition and health claims addressed the early food law concerns of such superfoods.[106] However, within the EU, the admissibility of and the requirements for nutrition and health claims on the labels of such superfoods were already regulated at the national level, a regulation which naturally resulted in differing approaches among the Member States. For example, a food labelled with a specific health claim might have been legally produced and marketed in one Member State, but might not have been legally marketed in another, thereby hindering the free trade of this food and, consequently, the free movement of goods. This hindrance created unequal conditions of competition,[107] which are not permissible within the EU.

In response to the aforementioned hindrances, Regulation (EC) No 1924/2006 (HCR) was adopted in 2006 to harmonise the law governing the use of nutrition and health claims and to ensure the effective functioning of the internal market while still providing a high level of consumer protection and has been applicable since July 2007.[108] The HCR covers the use of nutrition and health claims in commercial communications, including the labelling, presentation, or advertising of foods to be delivered as such to the final consumer.[109] Additionally, the HCR covers the use of trademarks and brand names, that qualify as or contain a nutrition or health claim. However, such claims do not have to be independently authorised so long as they are accompanied by an authorised nutrition or health claim.[110]

The HCR supplements the general labelling rules laid down in the FIC Regulation. In addition to general information rules, there are various legal instruments that cover the composition and, sometimes, labelling requirements of foods with specific nutritional or health benefits. These include: (i) the Regulation on the Fortification of Foods, (ii) the Food Supplements Directive,[111] and (iii) the Regulation on Dietetic Foods.[112]

Central to the HCR are the definitions of "nutrition claim"[113] and "health claim:"[114] A "nutrition claim" is a claim which states, suggests or implies that a

---

[106]Guidelines for Use of Nutrition and Health Claims, CAC/GL 23-1997.

[107]Regulation No 1924/2006 on nutrition and health claims made on food, p. 9, recital (2).

[108]Regulation (EC) No 1924/2006, *op.cit.*, Art. 1 (1).

[109]Regulation (EC) No 1924/2006, *op.cit.*, Art. 1 (2).

[110]Regulation (EC) No 1924/2006, *op.cit.*, Art. 1 (3).

[111]Directive 2002/46/EC *op.cit.*

[112]Regulation (EU) 609/2013 on food intended for infants and young children, food for special medical purposes, and total diet replacement for weight control and repealing Directive 92/52/EEC, Commission Directives 96/8/EC, 1999/21/EC, 2006/125/EC and 2006/141/EC, Directive 2009/39/EC and Regulations (EC) No 41/2009 and (EC) No 953/2009, p. 35.

[113]Regulation (EC) No 1924/2006, *op cit.*, Art. 2 (2) No 4.

[114]Regulation (EC) No 1924/2006, *op cit.*, Art. 2 (2) No 5.

food has particular beneficial nutritional properties. These beneficial nutritional properties must be due to the energy (calorific value) the food provides, does not provide, or provides at a reduced or increased rate, and/or due to the nutrients or other substances it contains, does not contain, or contains in reduced or increased proportions. A "health claim" is a claim that states, suggests, or implies that a relationship exists between a food category, a food or one of its constituents, and health.

The terms nutrition claim and health claim are to be broadly interpreted, as the mere suggestion or implication of a certain beneficial nutritional property or a health relationship falls within the definition of a nutrition or health claim. It can be difficult to establish whether beneficial nutritional properties are being referred to, so that the claim is specific enough to qualify as a nutrition claim. Alternatively, it can also be difficult to ascertain whether a reference is merely meant as the objective composition of a product. Likewise, it can be difficult to determine when a claim refers to a health relationship, which would qualify it as a health claim, or when it just contains a reference to the general, i.e. not health-related well-being.

The HCR operates by introducing a system of positive lists for both nutrition and for health claims. Such claims may, therefore, only be used if they are authorised, a so-called ban with permit reservation. Accordingly, claims not explicitly authorised are prohibited. In the field of nutrition and health claims, the HCR changes the general principle of EU food law in that food is rendered marketable if it is not specifically forbidden, a so-called permit with ban reservation. This effectually restricts the freedom of food business operators to advertise their products with references to nutritional properties and health effects, irrespective of whether these claims are scientifically correct or not.

For reference and administrative ease, the HCR established a Community Register[115] that includes all authorised nutrition and health claims. The Community Register has been published as an online document.[116] In addition to listing authorised claims, the register also includes a list of rejected health claims and the reason for their rejection, as well as those claims that are currently "on hold". Claims "on hold" indicate that final decisions have not yet been taken as to the admissibility of these claims, and for now, they can continue to be used provided the general requirements of the HCR are met. Prominent examples of "on hold claims" are claims concerning plants or plant parts (so-called "botanicals"). There are currently more than 2000 claims on hold.

---

[115]Regulation (EC) No 1924/2006, *op cit.*, Art. 20.

[116]EU Register of nutrition and health claims made on foods, http://ec.europa.eu/nuhclaims/ (last access April 9, 2014).

22.3.6 General Requirements for Health and Nutrition Claims

The HCR lays down general requirements that apply to nutrition and health claims alike.

Nutrition and health claims must not[117]:

- be false, ambiguous or misleading;
- give rise to doubt about the safety and/or nutritional adequacy of other foods;
- encourage or condone excess consumption of a food;
- state, suggest or imply that a varied diet cannot provide appropriate quantities of nutrients in general; or
- refer to changes in bodily functions which could give rise to or exploit fear in the consumer.

For the use of nutrition and health claims to be admissible, the following general requirements have to be met[118]:

- the presence, absence or reduced content in a food or category of food of a nutrient or other substance in respect of which the claim is made has been shown to have a beneficial nutritional or physiological effect, as established by generally accepted scientific data;
- the substance for which the claim is made is (not) contained in the final product in a significant quantity or in a quantity that will produce the nutritional or physiological effect;
- the substance is in a form that is available to be used by the body (bioavailable);
- the quantity of the product that can reasonably be expected to be consumed provides a significant quantity of the nutrient or other substance to which the claim relates.

If a nutrition claim is contained in the Annex of the HCR, food business operators do not have to provide further scientific evidence regarding nutritional benefits, since only those nutrition claims have been added that have beneficial nutritional properties deemed sufficiently scientifically proven. In regards to the claim "contains," however, scientific data has to be provided by the food business operator.

Many health claims are rejected because the scientific evidence filed is not considered sufficient. The European Food Safety Authority (EFSA) applies the highest scientific standard (the so-called, "gold standard"), that requires human intervention studies that are double-blind and placebo-controlled. If a claim has been authorised at EU level, Member States must accept the use of these claims and cannot request further scientific evidence.

---

[117]Regulation (EC) No 1924/2006, *op cit.*, Art. 3.
[118]Regulation (EC) No 1924/2006, *op cit.*, Art. 5.

22.3.6.1 Specific Requirements for Nutrition Claims

For a nutrition claim to be admissible, it must first be listed in the Annex of the HCR. The Annex contains claims referring to the reduction or absence of a substance, for example the claim "fat-free," as well as claims referring to the presence or increase of a substance, such as the claim "high protein." The Annex lists the claim "contains" followed by the name of the nutrient or other substance in question, thereby granting food business operators the possibility to highlight beneficial nutritional properties of substances other than those specifically named in the Annex.

Additionally, the specific requirements for the relevant nutrition claim must be met. In the case of "fat-free", the food must not contain more than 0.5 g of fat per 100 g or 100 ml respectively. The claim "X % fat-free" is not allowed.

22.3.6.2 Specific Requirements for Health Claims

The HCR distinguishes between various kinds of health claims:

- Claims referring to the role of a substance in growth, development and the functions of the body[119];
- Claims referring to psychological and behavioural functions[120];
- Claims referring to slimming or weight-control, reduction in the sense of hunger or increase in the sense of satiety or to the reduction of the available energy from the diet[121];
- Reduction of disease risk factor claims[122];
- Claims referring to children's development and health.[123]

The use of health claims further requires the labelling, or if no labelling exists, the presentation and advertising of foodstuffs, to contain the following information[124]:

- A statement indicating the importance of a varied and balanced diet and a healthy lifestyle;
- The quantity of the food and pattern of consumption required to obtain the claimed beneficial effect;
- Where appropriate, a statement addressed to persons who should avoid using the food; and
- An appropriate warning for products that are likely to present a health risk if consumed to excess.

---

[119]Regulation (EC) No 1924/2006, *op cit.*, Art. 13 (1) (a).
[120]Regulation (EC) No 1924/2006, *op cit.*, Art. 13 (1) (b).
[121]Regulation (EC) No 1924/2006, *op cit.*, Art. 13 (1) (c).
[122]Regulation (EC) No 1924/2006, *op cit.*, Art. 14 (1) (a).
[123]Regulation (EC) No 1924/2006, *op cit.*, Art. 14 (1) (b).
[124]Regulation (EC) No 1924/2006, *op cit.*, Art. 10 (2).

Health Claims may only be used if they are authorised. The HCR intended to create one Community list that contained all authorised general function health claims, but it was obvious that such agreement could not be reached, particularly concerning claims referring to plants and plant parts. Therefore, EU legislators decided to publish a first part of the list of authorised health claims.[125] In addition, risk reduction claims and claims referring to the development and health of children have been authorised in various separate regulations.[126]

22.3.7 Clean Labelling

Apart from the aforementioned nutrition and health claims, which are already highly regulated at the European level, food business operators are riding a wave of so-called "clean labelling" in conjunction with the current trend in consumer demand for quality and traditional products. Consequently, promotions of food products have increasingly placed emphasis on a foods' natural and healthful character. Common food product labels include: "additive-free," "100 % natural," "pure" or "home-made." The use of these terms has not been harmonised across the EU. Those aspects that have been harmonised, however, remain scattered among various bits of legislation and make an assessment of compliance difficult.

In short, the EU has laid down requirements for the use of the term "natural" with respect to:

- **Natural Flavourings**, which correspond to substances that are naturally present and have been identified in nature, i.e. flavouring substances obtained by appropriate physical, enzymatic or microbiological processes from material of vegetable, animal or microbiological origin either in the raw state or after processing for human consumption by one or more of the traditional food preparation processes[127];

---

[125] Regulation (EU) No 432/2012 establishing a list of permitted health claims made on foods, other than those referring to the reduction of disease risk and to children's development and health, p. 1, and its subsequent amendments.

[126] Among these are: Commission Regulation (EC) No 983/2009 on the authorisation and refusal of authorisation of certain health claims made on food and referring to the reduction of disease risk and to children's development and health, OJ L 277, p. 3; Commission Regulation (EC) No 1024/2009 on the authorisation and refusal of authorisation of certain health claims made on food and referring to the reduction of disease risk and to children's development and health, OJ L 283, p. 22. All admitted and rejected claims are listed in the Community register.

[127] Regulation (EC) No 1334/2008 on flavourings and certain food ingredients with flavouring properties for use in and on foods and amending Regulations (EC) No 1601/91, (EC) No 2232/96, (EC) No 110/2008, and Directive 2000/13/EC., Art. Art. 3(2)(c) and 22.

- **Nutrition Claims**, such as the terms "naturally" or "natural" may only be used as a prefix to a nutrition claim where the related food naturally meets the condition(s) laid down in by the EU Nutrition and Health Claims Regulation[128] for the use of these claims, e.g. "naturally rich in fibre," without the food's fortification with fibre.
- **"Mountain Product"** may only be used to describe agricultural foodstuffs in respect of which both the raw materials and the feedstuffs for farm animals come essentially from mountainous areas. In the case of processed products, the processing must also take place in mountain areas.[129]
- **"Traditional Specialities Guaranteed"** (TSG) are protected against any misuse, imitation or evocation, or against any other practice liable to mislead the consumer.[130]

To date, no further EU harmonisation exists. Therefore, the use of any other "clean" labelling must comply with the general requirement to provide the consumer with fair and accurate information. Additionally, national requirements may apply.

22.4 Advertisements in EU Food Law

The EU regulates advertisements as part of consumer law. Through provisions of consumer rights in the European Charter of Fundamental Rights and Article 12 of the European treaty, consumer guarantees include fairness (fair advertising), compliance with EU standards and a right of redress. As the following section shows, several of these principles are directly applicable to foodstuffs sold and marketed in the EU.

22.4.1 General Requirements of Food Advertisements

The obligation to provide the consumer with fair information (see supra on misleading labelling) applies equally to food advertising.[131] In addition to these rules, the EU legislative arsenal contains specific advertising restrictions when it comes to advertising alcoholic beverages through audio-visual communications, see Audio-visual Media Services Directive.[132] In this regard, television advertising

---

[128] Regulation (EC) No 1924/2006 on nutrition and health claims made on foods, *op cit.*

[129] Regulation (EU) No 1151/2012 on quality schemes for agricultural products and foodstuffs, p. 1, Art. 31.

[130] Regulation (EU) No 1151/2012, p. 1, Art. 24.

[131] Regulation No. 1169/2011, *op. cit.,* Art. 7(4).

[132] Directive 2010/13/EU on the coordination of certain provisions laid down by law, regulation or administrative action in Member States concerning the provision of audio-visual media services (Audio-visual Media Services Directive), Art. 9(1)(e) and 22. Please note that this Directive is currently open for review.

and teleshopping for alcoholic beverages must *at least*[133] comply with the following criteria, where the advertisement may not:

1. be aimed specifically at minors or, in particular, depict minors consuming these beverages;
2. link the consumption of alcohol to enhanced physical performance or to driving;
3. create the impression that the consumption of alcohol contributes towards social or sexual success;
4. claim that alcohol has therapeutic qualities or that it is a stimulant, a sedative or a means of resolving personal conflicts;
5. encourage immoderate consumption of alcohol or present abstinence or moderation in a negative light;
6. place emphasis on high alcoholic content as being a positive quality of the beverages.

Moreover, under the Audio-visual Media Services Directive, Member States are encouraged to co- or self-regulate at the national level, with respect to, inter alia, the advertisement of alcohol in audio-visual communication.[134]

Based on a further recommendation of the EU in 2001,[135] the Member States are also encouraged to establish effective mechanisms with the producers and the retailers of alcoholic beverages and relevant non-governmental organisations, in the fields of promotion, marketing and retailing, to ensure that regardless of the channel of communication used, producers do not produce alcoholic beverages specifically targeted at children and adolescents and that alcoholic beverages are not designed to appeal to children and adolescents. Particular attention should be paid to the use of styles, motifs and colours associated with "youth culture." The use of such images and the promotion of ideas associated with alcohol consumption through implications of social success, sexual or athletic prowess, features of children in spirits campaigns, and sponsoring of alcoholic drinks at sporting or musical events or on sport merchandising are prohibited.

22.4.2 Unhealthy Foods and Fair Advertising

Restricting the promotion of so-called "unhealthy food" is also one of the main objectives of fair advertising in the EU. Member States and the European Commission are required to encourage media service providers to develop codes of

---

[133]The Audio-visual Media Services Directive states that Member States remain free to impose more detailed or stricter rules provided that such rules are in compliance with Union law (Art 4 (1)).

[134]Audio-visual Media Services Directive, *op cit.*, Art 4(7).

[135]Council Recommendation 2001/458/EC of 5 June 2001 on the drinking of alcohol by young people, in particular children and adolescents.

conduct regarding inappropriate audio-visual commercial communication that accompanies or is included in children's programmes regarding foods and beverages containing nutrients and substances with a nutritional or physiological effect. Excessive intakes of fat, trans-fatty acids, salt/sodium, and sugars in the overall diet are particularly discouraged.[136]

Out of this set of precautions came the "EU Pledge."[137] The EU Pledge is an agreement among EU members that are major players in the manufacture of snacks and foodstuffs intended for children. Essentially, the EU Pledge aims to change food and beverage advertising in the EU directed at children under the age of 12 by:

- Not advertising food and beverage products to children under the age of 12 on TV, in print, and via the internet, except for products which fulfil specific nutritional criteria based on accepted scientific evidence or applicable national and international dietary guidelines;
- Not promoting products in primary schools, except where specifically requested by or agreed to with the school administration for educational purposes.

Member States may otherwise adopt legislation of their own choosing. It is therefore of primary importance for food business operators to be aware of the possible restrictions they may face when promoting their products internationally. In this regard, one should keep in mind that national legislation may not prohibit, impede or restrict the free movement of goods,[138] so there are limits to the permissible restrictions on advertisements.

22.5 EU Labelling Law: Cases and Examples

22.5.1 The Sauce Hollandaise Case: Replacement of Traditional Ingredients

An early, landmark decision[139] in the field of food labelling concerned an industrially produced food called "Sauce Hollandaise," see Case C-51/94 [1995].[140] As a traditional food, Sauce Hollandaise contains fresh eggs and butter, but the

---

[136] Audiovisual Media Services Directive, *op cit.*, Art. 9(1)(e).

[137] EU Pledge, http://www.eu-pledge.eu/.

[138] Joined Cases C-267/91 and C-268/91 Keck and Mithouard [1993] ECR I-6097, paragraph 17, "for national provisions restricting or prohibiting certain selling arrangements not to be caught by Article 28 EC, they must not be such as to prevent access to the market by products from another Member State or to impede access any more than it impedes access by domestic products"; See also Case C-239/02 Douwe Eghberts, ECR I-7037, § 51 and Case C-241/89 SARRP, ECR, 1990, I-04695, § 29.

[139] See http://eur-lex.europa.eu/legal-content/EN/TXT/PDF/?uri=CELEX:61994CJ0051&from=EN (last visited 12 Sept 2014).

[140] Joined cases C-10/97 to C-22/97, ECR, I-3617 (1998).

prepackaged Sauce Hollandaise in question contained plant fat instead of butter and eggs. Plant fat was properly labelled in the list of ingredients. However, the German authorities did not consider this labelling to be sufficient and requested that the departure from traditional Sauce Hollandaise be identified on the label in connection with the name of the food. The European Court of Justice decided that labelling the replacement ingredients in the list of ingredients would be sufficient since consumers interested in the composition of the food would check the list of ingredients and find all relevant information there. Additionally, consumers who are not interested in the composition of the food they buy, would likewise not be misled as to its composition. Therefore, the list of ingredients is generally considered the appropriate place for information regarding the composition of the product. Nonetheless, the Court did not rule out the possibility that in exceptional cases, the mere indication of replaced ingredients in the list of ingredients would not suffice, and accordingly, labelling in connection with the name of the food would be required.

The FIC Regulation overruled this decision by requiring a clear indication of the component or ingredient that has been used for the substitution in cases where an ingredient, that consumers expect to be normally used or to be naturally present in a food, has been replaced wholly or in part by another ingredient. Notably, the replacement ingredient or component must be clearly indicated in close proximity to the product name. Additionally, the labelling indication must be made using a font size of at least 75 % of the x-height of the name of the product (and, at least, of the prescribed minimum font size for mandatory particulars[141]).[142] Notably, there is an on-going debate as to whether the "product name" means the name of the food according to Article 17 of the FIC Regulation, or the brand—a.k.a. fancy name.

22.5.2 The Teekanne Case

As another example, the German Bundesgerichtshof (Federal High Court of Justice) initiated a preliminary ruling procedure before the CJEU in a case concerning a fruit infusion.[143] The product package showed raspberries, vanilla flowers and displayed the wording "all natural." However, the tea contained neither raspberries nor vanilla flowers, natural raspberry nor vanilla flavours, which would have been those produced from raspberries or vanilla flowers. Instead, the mixture contained other natural flavours that merely tasted like raspberry and vanilla. The Court held that this might be an example of "replaced ingredients" requiring an explanation in

---

[141]Regulation (EU) No 1169/2011, *op cit.*, Art. 12 (2), Annex IV.
[142]Regulation (EU) No 1169/2011, *op cit.*, Art. 9 (1) (a), 17, Annex VI Part A No 4.
[143]Decision of 26.02.2014, File no: I ZR 45/13, http://juris.bundesgerichtshof.de/cgi-bin/rechtsprechung/document.py?Gericht=bgh&Art=en&Datum=Aktuell&nr=66961&linked=pm.

connection with the product name, thereby, making the respective labelling mandatory.[144]

22.5.3 Misleading Labels: Cases

22.5.3.1 The Mars Case

The Mars case, Case C-470/93 – Mars +10 % [1995],[145] is an example of what types of labels must be circumspect and how far the prohibition of misleading information reach. Mars produced ice cream bars and presented them in wrappers marked "+10 %", in regards to the quantity of the product indeed being increased by 10 %. However, the package used different colouring and the coloured part, a band bearing the 10 % marking, occupied more than 10 % of the total surface. It was argued that consumers might therefore believe that the increase is larger than that represented. The European Court of Justice found that, "[r]easonably circumspect consumers may be deemed to know that there is not necessarily a link between the size of publicity markings relating to an increase in a product's quantity and the size of that increase."[146]

22.5.3.2 The Naturrein Case

The Naturrein case, Case C-465/98 – D'arbo Naturrein [2000],[147] dealt with a strawberry jam that contained the German word "naturrein" (naturally pure) on its label. It was established that the jam contained an additive gelling agent, pectin. Further analysis revealed that the jam also contained minimum levels of heavy metals and pesticides as traces or residues. The parties argued that these facts were incompatible with the claim "naturrein" (naturally pure), and would be misleading. The European Court of Justice, however, held that,

> As regards, first, pectin, it need merely be pointed out that its presence in d'arbo jam is indicated on the label on the packaging.[148]
>
> It is common ground that lead and cadmium are present in the natural environment as a result, in particular, of air pollution and pollution of the aquatic environment, as evidenced by several Community legislative instruments mentioned by the Advocate General in point 65 of his Opinion. Since garden fruit is grown in an environment of that kind, it is inevitably exposed to the pollutants present in it.

---

[144]CJUE, Case C – 195/14, Teekanne, 4 June 2015.
[145]Case C-470/93 – Mars +10 %, ECR, I-1936 (1995).
[146]Case C-470/93, *op cit.*, paragraph 24.
[147]Case C-465/98 – D'arbo Naturrein (2000), ECR I-2321.
[148]Case C-465/98, *op cit.*, paragraph 22.

The same conclusion is called for, thirdly, in relation to the presence of traces or residues of pesticides in d'arbo jam. As observed by the Advocate General in point 70 of his Opinion, the use of pesticides, even by private individuals, is one of the most usual means of combating the presence of harmful organisms on vegetables and agricultural products. Thus, that fact that garden strawberries are grown 'naturally' does not in any event mean that they are free of pesticide residues.

Lastly, it is necessary to verify whether the amounts of residues of lead, cadmium and pesticide measured in d'arbo jam render the presence of those substances incompatible with the description 'naturally pure' appearing on the label. Such a description might indeed be liable to mislead consumers if the foodstuff contained a high level of residues of toxic or polluting substances, even if they presented no risk to consumers' health.[149]

In those circumstances, it must be considered that, notwithstanding the presence of traces or residues of lead, cadmium and pesticides in d'arbo jam, the term "naturally pure" used on the label of the packaging of that foodstuff is not liable to mislead consumers as to its characteristics.[150]

These holdings illustrate the court's reasoning as to the alleged misrepresentation of "naturrein." Seeing how the rules outlined in this chapter have been applied is a helpful guide in learning how to read and interpret these labelling laws.

22.5.3.3 The Marketing Standards for Eggs Case

In Case C-210/96—Marketing Standards for Eggs [1998],[151] the German company "Gut Springenheide" marketed eggs ready-packed under the description "6-Korn — 10 frische Eier" (six-grain—ten fresh eggs). According to Springenheide, the six varieties of grains in question accounted for 60 % of the feed mix used to feed the hens, which were producing the eggs labelled as such. The disputes in this case gave rise to question whether the description on the egg packaging was likely to mislead a significant proportion of consumers into believing that the feed was made up exclusively of the six grains indicated, and that the eggs have particular characteristics. The German court referred the case to the European Court of Justice and asked how consumer expectation should be determined in this situation.

The European Court of Justice held that the question of whether the label misleads consumers is a matter for the national court, which "must take into account the presumed expectations which it evokes in an average consumer who is reasonably well-informed and reasonably observant and circumspect."[152] The Court did not decide if, in the case at hand, the consumer was misled or not but stated that consumer polls may help in ascertaining the level of likelihood that consumers are misled, "where the national court has particular difficulty in assessing the misleading nature of the statement or description in question, it may

---

[149]Case C-465/98, *op cit.*, paragraphs 28–30.

[150]Case C-465/98, *op cit.*, paragraph 33.

[151]Case C-210/96 ('Marketing standards for eggs'), *op.cit.*, http://curia.europa.eu/juris/liste.jsf?num=C-210/96.

[152]Case C-210/96, *op.cit.*, paragraph 37.

have recourse, under the conditions laid down by its own national law, to a consumer research poll, or an expert's report as guidance for its judgment."[153]

22.5.4 Food Advertisement Cases

22.5.4.1 The Douwe Egberts NV Case

In the Douwe Eghberts case (C-239/02 [2004]), the European Court of Justice ruled on an issue of national legislation[154] from Belgium, which imposed a general ban on the use of any reference to weight loss (slimming) in food advertising.

The Court stated that:

Foodstuffs lawfully manufactured and marketed in the other Member States in which particulars concerning health which are not misleading, may be mentioned under the provisions of Directive 2000/13, would be faced with restrictions on access to the Belgian market. In fact, the possibility cannot be ruled out that to compel a producer to discontinue an advertising scheme which he considers to be particularly effective, may constitute an obstacle to imports.[155]

Moreover, an absolute prohibition of advertising the characteristics of a product is liable to impede access to the market by products from other Member States more than it impedes access by domestic products, with which consumers are more familiar.[156]

The prohibition laid down by the national legislation, therefore, constitutes a fetter on intra-Community trade coming within the scope of Article 28 EC.

Such a fetter may be justified only by one of the public-interest grounds set out in Article 30 EC which include the protection of health and life of humans or by one of the overriding requirements ensuring, inter alia, consumer protection. It must also be appropriate for securing attainment of the objective which it pursues and must not go beyond what is necessary for attaining it.[157]

The grounds relied on to justify the aspects of the national legislation at issue relating to advertising are identical in scope to those relied on to justify the aspects of that legislation concerning labelling, namely the protection of the health of humans and prevention of fraud... those arguments cannot be upheld.

Nonetheless and unlike national legislation which in regard to labelling runs counter to Directive 2000/13 and cannot apply either to imported foodstuffs or to domestic foodstuffs, where national legislation on advertising is contrary to Articles 28 EC and 30 EC, application of that legislation is precluded only in regard to imported products and not domestic products.[158]

The Court finally concluded that:

---

[153]Case C-210/96, *op.cit.*, paragraph 37.

[154]Royal Decree of 17 April 1980 concerning the advertising of foodstuff, Official Belgian Gazette 6 May 1980.

[155]SARPP, paragraph 29.

[156]Case C-405/98 Gourmet International Products [2001] ECR I-1795, paragraph 21.

[157]Joined Cases C-34/95 to C-36/95 De Agostini and TV-Shop [1997] ECR I-3843, paragraph 45.

[158]SARPP, paragraph 16.

[g]iven that it is apparent from the order for reference that the dispute in the main proceedings does not concern imported foodstuffs, it is for the national court to ascertain to what extent national law requires a national producer to be allowed to enjoy the same rights as those which a producer of another Member State would derive from Community law in the same situation.[159]

Notably, however, that references to 'slimming' must now comply with the HCR (*supra*).

22.5.4.2 The Gourmet International Case

In Case C-405/98—Gourmet International Products [2001] (GIP), the company GIP published three pages of advertisements for alcoholic beverages a Swedish magazine, entitled "Gourmet." Those pages did not appear in the edition sold in shops. Ninety percent of the magazine's subscribers were traders, manufacturers, or retailers, while only 10 % of the subscribers were private individuals.

The relevant Swedish law prohibited any advertising for spirits, wines, or strong beers either in periodicals or in other publications subject to the Swedish Regulation on the Freedom of Press and comparable to periodicals by reason of their publication schedule. That prohibition did not, however, apply to publications distributed solely at the point of sale of such beverages in Swedish Law 1996:851 with title "Alkoholreklamlagen." The object of the Alkoholreklamlagen was that the prohibition on advertisements in periodicals only applies to marketing intended for consumers and does not apply to advertisements in the specialist press, i.e. the press targeting traders.

In this case, the European court of Justice held as follows:

> It is apparent that a prohibition on advertising, such as that at issue in the main proceedings, not only prohibits a form of marketing a product but in reality prohibits producers and importers from directing any advertising messages at consumers, with a few insignificant exceptions.
>
> Even without its being necessary to carry out a precise analysis of the facts characteristic of the Swedish situation, which it is for the national court to do, the Court is able to conclude that, in the case of products like alcoholic beverages, the consumption of which is linked to traditional social practices and to local habits and customs, a prohibition of all advertising directed at consumers in the form of advertisements in the press, on the radio and on television, the direct mailing of unsolicited material or the placing of posters on the public highway is liable to impede access to the market by products from other Member States more than it impedes access by domestic products, with which consumers are instantly more familiar.[160]

---

[159]C-239/02, (Douwe Egberts NV), §53–57 (emphasis added) (citing Case C-448/98 Guimont, paragraph 23).

[160]Case C-405/98, Gourmet International Products [2001] ECR I-1795, §§ 20 and 21.

Thus, impediments on the free movement of goods were struck down.

In a similar case invoking French national law, Case C-241/8912, SARPP [1990],[161] the European Court of Justice decided that a national legislation (in the case at hand: French legislation[162]) which prohibits any statement in the advertising of artificial sweeteners alluding to the word "sugar" or to the physical, chemical or nutritional properties of sugar that artificial sweeteners also possess, infringes the principle of free movements of goods.

22.5.5 Nutrition and Health Claim Cases and Examples

22.5.5.1 The Deutsches Weintor Case

Case C-544/10, Deutsches Weintor eG [2012] illustrates how wide the term "health claim" can be contrued. The case was referred to the European Court of Justice by the German Bundesverwaltungsgericht (Federal Administrative Court). A wine producer labelled certain wines with the description "bekömmlich" (easily digestible), and referenced the reduced acidity levels (gentle acidity) in these wines by including the following on the label as "ow[ing] its mildness to the application of ... [the] special 'LO3' protective process for the biological reduction of acidity." German authorities challenged the information as the use of an unauthorized health claim and found it inadmissible because since alcoholic beverages containing more than 1.2 % by volume of alcohol were not entitled to bear any health claims.[163] The German Federal Administrative Court questioned whether a description relating to the mere temporary maintenance of bodily functions would suffice to establish a link with health, or whether the description had to relate to the general health-related well-being of a person.

The European Court of Justice clarified what such a health claim should include:

> [i]n that regard it is apparent from the wording of Article 2(2)(5) of Regulation No. 1924/2006 that the starting-point for the definition of a 'health claim' within the meaning of that regulation is the relationship that must exist between a food or one of its constituents and health. That being the case, it must be noted that that definition provides no information as to whether that relationship must be direct or indirect, or as to its intensity or duration. In those circumstances, the term 'relationship' must be understood in a broad sense.
>
> Thus, the concept of a 'health claim' must cover not only a relationship implying an improvement in health as a result of the consumption of a food, but also any relationship which implies the absence or reduction of effects that are adverse or harmful to health and which would otherwise accompany or follow such consumption, and, therefore, the mere preservation of a good state of health despite that potentially harmful consumption.[164]

---

[161]Case C-241/89, SARPP [1990], ECR I-4714.
[162]French Law No 88-14 of 5 January, Article 10(1).
[163]Regulation (EC) No 1924/2006, *op cit.*, Art. 4 (3).
[164]Case C-544/10 *Deutsches Weintor* [2012], paragraphs 34–35 (emphasis added).

22.5.5.2 The Green Swan Case

Case C-299/12, Green Swan [2013] originated in the Czech Republic and concerned a food supplement marketed with the following statement: "The preparation also contains calcium and vitamin D3, which helps to reduce a risk factor in the development of osteoporosis and fractures." Among other questions raised, the referring Czech court explained that a health claim need not necessarily include the word "significantly" or a similar expression to be considered a "reduction of disease risk claim." Otherwise, the choice of a slightly different wording would lead to a circumvention of the requirements of the HCR. The was case brought before the European Court of Justice, which found as follows:

> Among health claims, Article 2(2)(6) of the regulation defines a 'reduction of disease risk claim' as 'any health claim that states, suggests, or implies that the consumption of a food category, a food, or one of its constituents significantly reduces a risk factor in the development of a human disease.'
>
> It follows from the use of the verbs 'suggests or implies' that classification as a 'reduction of disease risk claim,' within the meaning of that provision, does not require that such a claim expressly states that the consumption of a food significantly reduces a risk factor in the development of a human disease. It is sufficient that that claim may give the average consumer who is reasonably well informed and reasonably observant and circumspect the impression that the reduction of a risk factor is significant.[165]

Therefore, the Court held that "the answer to the first question is that Article 2(2)(6) of Regulation No 1924/2006 must be interpreted as meaning that, in order to be considered a 'reduction of disease risk claim' . . . a health claim need not necessarily expressly state that the consumption of a category of food, a food or one of its constituents 'significantly' reduces a risk factor in the development of a human disease."[166]

22.5.5.3 The Ehrmann Case

The Ehrman case, C-609/12, Ehrmann [2014], dealt with a milk curd that contained the label slogan "as important as the daily glass of milk." The product contained as much calcium as milk, but also contained 13 % sugar, whereas the sugar content of milk is 4.7 %. In the opinion of a German court, the slogan constitutes a health claim because the relevant public generally views milk as having positive effects on health due to its mineral content. According to the court, the slogan, therefore, expressed a positive effect of the curd in question by comparing it to a daily glass of milk.

Based on the belief that the reference was a health claim, the German court noted the absence of the additional mandatory labelling requirements, which would

---

[165]Case C-299/12, paragraphs 23–26.
[166]Case C-299/12, paragraphs 23–26.

usually be required when a health claim is made. The German court asked the European Court of Justice whether, according to Art. 10(2) HCR, these additional information requirements applied before the publication of the first part of the Article 13 listing all authorised health claims. The European Court of Justice decided that Art. 10(2) was applicable.[167]

22.5.6 Clean Labelling Examples

For the reason that clean labelling is essentially regulated at the national level, there are very few cases on point that have been decided by the European Court of Justice. Nonetheless, the following is an illustration of the practical implications of clean labelling:

22.5.6.1 The Pure Chocolate Case

In C-47/09, EC v. Italy (Pure Chocolate) [2010],[168] the European Court of Justice ruled on a provision of Italian law, which limited the use of the term "puro cioccolato" (pure chocolate) to chocolate products which did not contain any vegetable fat other than cocoa butter. According to the CJEU, such a limitation infringed upon the EU Chocolate Directive,[169] which the Court read in conjunction with the Labelling Directive.[170]

The EU Chocolate Directive authorises chocolate products to contain up to 5 % vegetable fats other than cocoa butter and also authorises supplementary information or descriptions relating to the quality criteria of the products in chocolate products which contain specific minimum characteristics, such as minimum content of cocoa solids and cocoa butter. Consequently, the specific requirements of Italian law regarding the right to use the description "pure chocolate," amounted to an infringement of the full harmonisation of the sales names of chocolate products introduced by the EU Chocolate Directive. Accordingly, the Court decided that such national legislation could not be justified on the ground of consumer protection.

---

[167]Case C-609/12 (2013).

[168]Case C-47/09, EC v. Italy ("Pure Chocolate") [2010], ECR, I-12083.

[169]Directive 2000/36/EC relating to cocoa and chocolate products intended for human consumption, p. 19, as amended.

[170]Directive 2000/13/EC on the approximation of the laws of the Member States relating to the labelling, presentation and advertising of foodstuffs, p. 29, as amended. This Directive is replaced by Regulation No 1169/2011, *op. cit.*

References

Meisterernst A (2013) A new benchmark for misleading advertising. Eur Food Feed Law Rev 2:91. Lexxion
Amended in 2001, 2008, 2009, 2010, 2011, 2012 and 2013. Annex adopted in 2009
Case C-210/96 Marketing standards for eggs, *op.cit.*, http://curia.europa.eu/juris/liste.jsf?num=C-210/96
Case C-239/02 Douwe Eghberts, ECR I-7037, § 51
Case C-241/89, SARPP [1990], ECR I-4714
Case C-241/8912, SARPP (1990)
Case C-405/98 Gourmet International Products [2001] ECR I-1795, paragraph 21
Case C-405/98, Gourmet International Products [2001] ECR I-1795, §§ 20 and 21
Case C-448/98 Guimont [2000] ECR I-10663
Case C-465/98 – D'arbo Naturrein (2000), ECR I-2321
Case C-470/93 – Mars +10%, ECR, I-1936 (1995)
Case C-47/09, EC v. Italy ("Pure Chocolate") [2010], ECR, I-12083
Case C-544/10 *Deutsches Weintor* [2012], paragraphs 34–35 (emphasis added)
CJEU, 6 July 1995, Case Mars, C-470/93, ECR 1995 Page I-01923, §24
CJEU, C-609/12, 10 April 2013
CJEU, case C-299/12, paragraphs 23–26
Codex Alimentarius *1-1985 (Rev. 1-1991)*, § 2, available at http://www.fao.org/docrep/005/y2770e/y2770e02.htm (last accessed 7 Sept 2014)
Commission Implementing Decision of 24 January 2013 adopting guidelines for the implementation of specific health claims laid down in Article 10 of Regulation (EC) No 1924/2006, OJ L 22,251.2013, p. 25, point 2.1. (b)
Commission Regulation (EC) No 1024/2009 on the authorisation and refusal of authorisation of certain health claims made on food and referring to the reduction of disease risk and to children's development and health, OJ L 283, p. 22
Commission Regulation (EC) No 983/2009 on the authorisation and refusal of authorisation of certain health claims made on food and referring to the reduction of disease risk and to children's development and health, OJ L 277, p. 3
Communication of 14 March 2013 from the Commission to the European Parliament, the Council and the European Economic and Social Committee on the application of the Unfair Commercial Practices Directive, COM(2013)138 final
Council Directive 76/211/EEC on the approximation of the laws of the Member States relating to the making-up by weight or by volume of certain prepackaged products, OJ L 046, 21.2.1976, p. 1
Council Recommendation 2001/458/EC of 5 June 2001 on the drinking of alcohol by young people, in particular children and adolescents [Official Journal L 161 of 22.06.2001]
Council Regulation (EC) No 2200/96 on the common organization of the market in fruit and vegetables, OJ L 297, 21.11.1996, p. 1, as amended and Regulation (EC) No 1580/2007 laying down implementing rules of Council Regulations (EC) No 2200/96, (EC) No 2201/96 and (EC) No 1182/2007 in the fruit and vegetable sector
Council Regulation (EC) No 834/2007 of 28 June 2007 on organic production and labelling of organic products and repealing Regulation (EEC) No 2092/91 (OJ L 189 of 20.7.2007), as amended
Decision of 26.02.2014, File no: I ZR 45/13, http://juris.bundesgerichtshof.de/cgi-bin/rechtsprechung/document.py?Gericht=bgh&Art=en&Datum=Aktuell&nr=66961&linked=pm
Directive 2000/13/EC of the European Parliament and of the Council of 20 March 2000 on the approximation of the laws of the Member States relating to the labelling, presentation and advertising of foodstuffs, OJ L 109, 6.5.2000, p. 29

Directive 2000/13/EC on the approximation of the laws of the Member States relating to the labelling, presentation and advertising of foodstuffs (OJ L 109, 06/05/2000, p. 29), as amended. This Directive is replaced by Regulation No 1169/2011, *op. cit.*
Directive 2000/36/EC relating to cocoa and chocolate products intended for human consumption (OJ L 197, 03/08/2000, p. 19), as amended
Directive 2001/110/EC relating to honey (OJ L 10, 12/01/2002, p. 47)
Directive 2001/111/EC of 20 December 2001 relating to certain sugars intended for human consumption (OJ L 10 12/01/2002, p. 53)
Directive 2001/112/EC relating to fruit juices and certain similar products intended for human consumption, OJ L 10, 12/01/2002
Directive 2002/46/EC
Directive 2005/29
Directive 2005/29/EC concerning unfair business-to-consumer commercial practices in the internal market and amending Council Directive 84/450/EEC, Directives 97/7/EC, 98/27/EC and 2002/65/EC of the European Parliament and of the Council and Regulation (EC) No 2006/2004
Directive 2006/114/EC concerning misleading and comparative advertising (OJ L, 376, 27.12.2006, p. 21)
Directive 2007/45/EC (2007) (repealing Directives 75/106/EEC and 80/232/EEC, and amending Council Directive 76/211/EEC)
Directive 2010/13/EU on the coordination of certain provisions laid down by law, regulation or administrative action in Member States concerning the provision of audio-visual media services (Audio-visual Media Services Directive)
Directive 79/581/EEC concerning the indication of prices of certain foodstuffs, OJ L 158, (1979) - Directive as last amended by Directive 95/58/EC, OJ L 299 (1995)
Directive 98/6/EC (1998) on consumer protection in the indication of the prices of products offered to consumers, OJ L 080
Directive 99/4/EC relating to coffee extracts and chicory extracts, OJ L 66, 13/03/1999, as amended
EU Pledge, http://www.eu-pledge.eu/
EU Register of nutrition and health claims made on foods, http://ec.europa.eu/nuhclaims/ (last accessed 9 April 2014)
European Commission, Application of EU law: What are EU directives? Available at http://ec.europa.eu/eu_law/introduction/what_directive_en.htm (last accessed 7 Sept 2014)
European Commission, Application of EU law: What are EU regulations? Available at http://ec.europa.eu/eu_law/introduction/what_regulation_en.htm (last accessed 7 Sept 2014)
FIC Regulation
French Law No 88-14 of 5 January, Article 10(1)
Guidelines for Use of Nutrition and Health Claims, CAC/GL 23-1997. Revised in 2004
Joined cases C-10/97 to C-22/97, ECR, 1995, I-3617 (22 October 1998), http://eur-lex.europa.eu/legal-content/EN/TXT/?uri=CELEX:61997CJ0010
Joined Cases C-267/91 and C-268/91 Keck and Mithouard [1993] ECR I-6097
Joined Cases C-34/95 to C-36/95 De Agostini and TV-Shop [1997] ECR I-3843, paragraph 45
Judgment of the Court (Fifth Chamber) Case C-373/90 (January 16, 1992). Criminal proceedings against X (ECR., 1992 Page I-00131)
Nilsson L (2012) Misleading? To whom? Eur Food Feed Law Rev 7(1):22–27
Nutrition and Health Claims, CAC/GL 23-1997
Regulation (EC) No 104/2000 on the common organisation of the markets in fishery and aquaculture products (OJ L 17, 21/01/2000, p. 22–52), as amended and Commission Regulation (EC) No 2065/2001 of 22 October 2001 laying down detailed rules for the application of Council Regulation (EC) No 104/2000 as regards informing consumers about fishery and aquaculture products (OJ L 278, 23.10.2001, p. 6), as amended
Regulation (EC) No 110/2008

Regulation (EC) No 1234/2007 of 22 October 2007 establishing a common organisation of agricultural markets and on specific provisions for certain agricultural products (Single CMO Regulation), 22.11.2007, OJ L 299/1

Regulation (EC) No 1234/2007, *op cit.*, Annex XV

Regulation (EC) No 1334/2008 on flavourings and certain food ingredients with flavouring properties for use in and on foods and amending Regulations (EC) No 1601/91, (EC) No 2232/96, (EC) No 110/2008, and Directive 2000/13/EC., Art. Art. 3(2)(c) and 22

Regulation (EC) No 1760/2000 establishing a system for the identification and registration of bovine animals and regarding the labelling of beef and beef products and repealing Council Regulation (EC) No 820/97 OJ L 204, (2000), as amended

Regulation (EC) No 178/2002 laying down the general principles and requirements of food law, establishing the European Food Safety Authority and laying down procedures in matters of food safety (OJ L 31 (2002)

Regulation (EC) No 1825/2000 laying down detailed rules for the application of Regulation (EC) No 1760/2000 as regards the labelling of beef and beef products OJ L 216/8 (2000) which provides detailed rules for the implementation of Regulation 1760/2000

Regulation (EC) No 1829/2003 on genetically modified food and feed OJ, L 168 (2003)

Regulation (EC) No 1830/2003 concerning the traceability and labelling of genetically modified organisms and the traceability of food and feed products produced from genetically modified organisms and amending Directive 2001/18/EC OJ L 268 (2003)

Regulation (EC) No 1924/2006 on nutrition and health claims made on foods OJ L 12 (2007)

Regulation (EC) No 2991/94 laying down standards for spreadable fats, OJ L 316 (1994).

Regulation (EC) No 834/2007 on organic production and labelling of organic products

Regulation (EC) No 882/2004 on official controls performed to ensure the verification of compliance with feed and food law, animal health and animal welfare rules OJ L 165 (2004)

Regulation (EEC) No 1906/90 (OJ L 173, 6.7.1990, p. 1)

Regulation (EU) No 1151/2012 on quality schemes for agricultural products and foodstuffs OJ L 343 (2012)

Regulation (EU) No 1169/2011

Regulation (EU) No 1337/2013 laying down rules for the application of Regulation (EU) No 1169/2011 on the indication of the country of origin or place of provenance for fresh, chilled and frozen meat of swine, sheep, goats and poultry, OJ L 335 (2013)

Regulation (EU) No 29/2012 on marketing standards for olive oil OJ L 12 (2012), as amended

Regulation (EU) No 432/2012 establishing a list of permitted health claims made on foods, other than those referring to the reduction of disease risk and to children's development and health OJ L 136 and its subsequent amendments

Regulation (EU) no 609/2013 on food intended for infants and young children, food for special medical purposes, and total diet replacement for weight control (repealing Directive 92/52/EEC, Commission Directives 96/8/EC, 1999/21/EC, 2006/125/EC and 2006/141/EC, Directive 2009/39/EC and Regulations (EC) No 41/2009 and (EC) No 953/2009, 29.6.2013, OJ L 181)

Regulation 1924/2006, *op.cit.*, Recital 3, Regulation No 1151/2012, *op.cit.*, Recital (8) and Art. 3, which stress the complementary nature of their provisions with respect to the Labelling Directive, whose rules are currently included in Regulation No 1169/2011

Regulation No 178/2002

Regulation No 1924/2006 on nutrition and health claims made on food, OJ L 404, 30.12.2006, p. 9, recital (2)

Regulation No. 1169/2011, *op. cit.*, Art. 7(4)

Royal Decree of 17 April 1980 concerning the advertising of foodstuff, Official Belgian Gazette 6 May 1980

Leible S (2010) Consumer information beyond food law. Eur Food Feed Law Rev 6:322. Lexxion

Sauce Hollandaise, see Case C-51/94 [1995]. See http://eur-lex.europa.eu/legal-content/EN/TXT/PDF/?uri=CELEX:61994CJ0051&from=EN (last visited 12 Sept 2014)

USDA, A focus on bovine spongiform encephalopathy, available at http://web.archive.org/web/20080303135425/http://fsrio.nal.usda.gov/document_fsheet.php?product_id=169 (last accessed 7 Sept 2014)

Chapter 23
Consumer Protection Through Food Law in the European Union

Raymond O'Rourke

Abstract In many countries lawyers would believe that food law is a component of consumer protection law. In the European Union, consumer protection law was not a priority at its establishment under the Treaty of Rome (1957), indeed consumer policy at that time was in its infancy both in Europe and the U.S.A. In the following years the EU attempted to agree laws to protect consumers but such laws were always to enhance the functioning of the Internal Market. This situation was highlighted by means of the famous *Cassis de Dijon* ECJ Court case. The EU introduced some minimal consumer protection legislation some of which covered issues like food labelling or food hygiene. Things changed dramatically after the BSE crisis after which a major body of EU Food Law has been introduced. These food laws have been introduced as maximum harmonisation measures unlike consumer laws which continue to be caught-up in a major struggle with Member States over whether they should be a minimum or maximum harmonisation measure.

23.1 Introduction

In many countries, lawyers believe that food law is a component of consumer protection law. In the European Union (EU), consumer protection law was not a priority at its establishment under the Treaty of Rome in 1957. In fact, consumer policy, at that time, was in its infancy both in Europe and the United States. In the years that followed, the EU attempted to agree on laws to protect consumers, but such laws were always to enhance the functioning of the so-called internal market. This situation was highlighted by means of the famous *Cassis de Dijon* case that was brought before the European Court of Justice (ECJ) where the EU introduced some minimal consumer protection legislation.[1] Some of this consumer protection

---

[1] Cassis de Dijon (1979) ECR 469.

R. O'Rourke (✉)
Food & Consumer Lawyer, Dublin, Ireland
e-mail: raymond.orourke7@gmail.com

legislation covered issues such as food labelling or food hygiene. After the notorious bovine spongiform encephalopathy crisis (also known as mad cow disease), however, European food law changed and the EU introduced a new body of food law in the form of maximum harmonization measures. Consumer laws, on the other hand, continue to be caught up in a major struggle with EU member states over whether they should be a minimum or maximum harmonization measure. This chapter provides an overview of European consumer protection laws through food law in the European Union and builds on the previous chapters to provide an advanced analysis of the relevant laws.

23.2 Establishment of European Food Law

The initial approach of the Commission in relation to EU food law was to concentrate on the obligations enshrined in Article 3 of the Treaty of Rome, so as to ensure the free movement of foodstuffs throughout the common market. For many years, EU food legislation pursued a path dictated by this approach using Article 94 of the EC Treaty as a basis for such legislation, thereby necessitating unanimity in the Council for adoption. The process of creating a body of EU food law was slow. It was only following a number of major court cases, including the Cassis de Dijon case in 1979 that the Commission attempted to reorient its approach to food law.

The Commission introduced its new approach in the 1985 Communication entitled "Completion of the Internal Market: Community Legislation on Foodstuffs."[2] It stated that Community legislation on foodstuffs should be limited to provisions justified by the need to: protect public health; provide consumers with information and protection in matters other than health and ensure fair trading; and provide for the adequate and necessary official controls of foodstuffs.

Shortly after the Commission released the above communication, it proposed to the Council a number of framework directives dealing with these essential requirements, which were subsequently adopted. These included framework directives on additives, labelling foods for particular nutritional needs, hygiene, and official controls.

Later, in 1989, the Commission published another communication, "The Free Movement of Foodstuffs within the Community,"[3] to provide further clarification on this subject. This communication was in response to the *Cassis de Dijon* court judgment, which established the principle that, in general, a food product lawfully produced and marketed in one member state should be allowed to be marketed in other member states—unless it could be proved that it was a threat to public health. The communication clarified the position of the trade description and the question

---

[2]Communication on completion of the internal market (1985) Legislation on foodstuffs (COM(85) 603).

[3]Communication on the free movement of foodstuffs in the community (O.J. 1989 C271/3).

of additives vis-à-vis the free movement of foodstuffs. Essentially, though, EU food law in the run-up to the 1992 establishment of the single internal market was still concentrated on questions of trade and free movement of goods. Consumer protection and public health aspects of food law were playing second fiddle to issues of trade. Indeed, during this time, there were often calls for the deregulation of EU food law back to member states. EU legislation relating to food was subject to critiques of over-regulation, incoherence, fragmentation, and lack of transparency and innovation.

The BSE crisis ensured that EU legislators once again had to go back and look at the way in which EU food law had developed. The European Commission, facing a motion of censure from the European Parliament for their handling of the BSE issue, agreed to reform the Commission's structures for preparing food legislation. The important outcome of these developments was that the Commission published a long-awaited Green Paper on Food Law in May 1997.[4]

23.3 EU Consumer Law

The EU's most recent Consumer Agenda from 2012 includes a strategic vision for EU consumer policy, and aims to maximise consumer participation and trust in the internal market. The Consumer Agenda is built around four main objectives:

1. increasing consumer confidence by reinforcing consumer safety;
2. enhancing consumer knowledge;
3. stepping up enforcement and redress; and
4. aligning consumer rights and policies to changes in society and the economy.

Unlike the Treaty of Rome, the EU's most recent Consumer Agenda directly references consumers. In the intervening years since it was enacted, the Treaty of Rome has been amended a number of times to make consumer protection a policy priority for the EU. Article 12 of the EC Treaty emphasizes the importance of consumer policy at the European level, which states "consumer protection requirements shall be taken into account in defining and implementing other Union policies and activities."[5]

Article 114 (1) addresses the harmonization and approximation of laws for the functioning of the internal market. As with Article 12, it emphasizes the need for a high level of consumer protection by ensuring that

> (t)he Commission, in its proposals envisioned in paragraph (1) concerning health, safety, environmental protection and consumer protection, will take high basic level of protection, taking into account in particular of any new development based on scientific facts. Within

---

[4]Green Paper (1997) The general principles of food law in the European Union. (COM(97)176) May 1997.

[5]Article 12. Consolidated version of the Treaty on the Functioning of the European Union.

their respective powers, the European Parliament and the Council will also seek to achieve this objective.[6]

Such measures must contribute significantly to promoting consumer interests while at the same time ensuring the smooth functioning of the internal market. Essentially, the EU must strike a balance between consumer protection and the internal market. As the *Cassis de Dijon* case highlighted, measures obstructing the movement of goods in the internal market may be introduced in order to protect consumer health and safety, but such measures must be adequately justified.

The European Charter of Fundamental Rights, which became primary law by means of the Treaty of Lisbon, states in Art. 38 that "union policies shall ensure a high level of consumer protection."[7] The use of the phrase "shall ensure" implies that this reference is more of a principle, rather than a hard-and-fast rule. It demonstrates once again an attempt to balance consumer protection rights with the pressures of the internal market.

As shown in the Unfair Commercial Practices directive, the aim and purpose of legislation in secondary law is "to contribute to the proper functioning of the internal market and achieve a high level of consumer protection by approximating the laws, regulations and administrative provisions of the Member States on unfair commercial practices harming consumers' economic interests."[8] Therefore, one of the main goals of the legislation, as specified in the Recitals, is to foster fair competition, provide for the smooth functioning of the internal market, and to do so with a high level of consumer protection. This will allow the consumer to make an informed and thus efficient choice. The directive highlights in Recital 18 that it:

> takes as a benchmark the average consumer, who is reasonably well informed and reasonably observant and circumspect, taking into account social, cultural and linguistic factors, as interpreted by the Court of Justice, but also contains provisions aimed at preventing the exploitation of consumers whose characteristics make them particularly vulnerable to unfair commercial practices.[9]

Many rulings discuss the question of the so-called "average consumer," and they all suggest that such a person is well informed—not an expert, but rather someone with general knowledge. These consumers will expand their knowledge when necessary and are able to find the information they require. In addition, they have the ability to correctly read labelling information.

It is useful to contrast the provisions in the Unfair Commercial Practices directive with those in the Food Information regulation, which deals with a consumer's right to adequate food labelling. In the past, the main principle for food labelling was that it did not mislead the consumer. Article 3(1) of the Regulation demonstrates the concept of food labels in assisting the consumer to make an informed choice. It is now included in EU food law that:

---

[6]Article 12. Consolidated version of the Treaty on the Functioning of the European Union.
[7]Article 38. European Charter of Fundamental Rights of the European Union.
[8]Article 1, Unfair commercial practices directive 2005/29 (O.J. L149 pp. 22–69).
[9]Recital 18—Unfair Commercial Practices Directive 2005/29 [O.J. L149 pp. 22–69].

The provision of food information shall pursue a high level of protection of consumers' health and interests by providing a basis for final consumers to make informed choices and to make safe use of food, with particular regard to health, economic, environmental, social and ethical considerations.[10]

In addition, food information shall facilitate the smooth functioning of the internal market, protect the interests of food producers, and promote the production of quality products.

There is one area, however, where there is still a major divergence between EU consumer protection laws and EU food laws—and that is in the degree of regulation harmonization between member states.

23.3.1 The Green Paper on Food Law

The Green Paper was a seminal publication because it generated an enormous debate about how the European Union should best legislate in the area of food law. The goals of the document included the need:

- to provide a high level of protection for consumer public health and safety;
- to ensure the free circulation of goods within the single market;
- for legislation to be based primarily on scientific evidence and risk assessment;
- to ensure the competitiveness of the European food industry;
- the need to place the primary responsibility for safe food with industry, including producers and suppliers, through self-checking provisions backed up by official controls and enforcement (so-called Hazard Analysis Critical Control Points systems, or HACCP systems); and,
- for legislation to be coherent, rational, consistent, simple, user-friendly, and developed in full consultation with all interested parties.

It was in this document that the Commission reaffirmed the fundamental goals mentioned above and achievements of EU food law, which were not to be questioned in any changes introduced following the debate on the Green Paper.

23.3.2 The White Paper on Food Safety

Although the Green Paper was a seminal work, it was not the final chapter in the re-orientation of EU food law that took place in the wake of the BSE crisis. The Commission was initially supposed to publish a communication subsequent to the Green Paper that would list new legislative proposals aimed at fulfilling the promises made to the European Parliament; namely, to establish an EU-wide

---

[10] Article 3(1)—EP & Council Regulation 1169/2011 on Food Information [O.J. L304 pp. 18–63].

food regulatory regime which placed consumer health and food safety at the top of the agenda.

A series of events conspired to change this plan, however, and, instead, in January 2000 the European Commission published a White Paper on Food Safety.[11] The Commission changed its approach due to the on-going fallout of the BSE crisis, continued consumer concerns over the safety of genetically modified (GM) foods, and the dioxin contamination scandal in Belgium. On 27 May 1999, Belgian authorities informed the European Commission about a heavy contamination of animal feed with dioxin. The Commission acted quickly and took protective EU-wide measures, removing all Belgian animal and poultry products, as well as those derived from these animals, from the market. Member states that had imported the contaminated animal feed were asked to trace the animal feed and foods produced from these farms and destroy them. Numerous food products were taken off the market and the EU had to work very hard to allay consumers' fears that items marketed throughout the internal market were unsafe. The European Commission activated its Rapid Alert System, whereby there is a co-ordinated exchange of information between the Commission and member states regarding potential threats to the health and safety of consumers.

In his first policy speech to the European Parliament in July 1999, the new Commission President, Romano Prodi, stated that enhancing food safety would be one of the main priorities of his tenure at the Commission. The Commissioner for Health and Consumer Protection, David Byrne, was given the task in the wake of these food crises to re-vamp and re-focus EU food law so that it would reflect a strong food safety/consumer-orientated perspective. The culmination of Byrne's work was the White Paper on Food Safety, published in January 2000.

The White Paper proposed a comprehensive range of over 80 new or amended items of European food law, including general food laws that outlined the responsibilities and obligations of all operators in the food chain. Other proposals involved a regulation on official food and animal feed controls, the re-casting of the EU hygiene rules, new labelling rules, limits on nutritional and health claims, a re-vamping of the Rapid Alert System, and a complete overhaul of EU legislation relating to genetically modified organisms (GMOs) and GM foods.

23.4 Consumer Protection and Food Law

23.4.1 Food Law and the Treaty of Rome

As is mentioned in other chapters in this section, the Treaty of Rome established the European Economic Community (EEC)—a precursor to the current European Union (EU)—and was signed into law in 1957 by France, Germany, Italy, and

---

[11]White Paper on Food Safety (COM(1999)719).

the Benelux countries. This EEC treaty gave rise to the European Economic Community and provides that the activities of the Community should include a variety of factors according to Article 3 in order to establish a common European market. Three factors stand out in particular as they relate to consumer protection and food law:

1. the elimination of customs duties and quantitative restrictions between member states in relation to the import and export of goods;
2. an internal market characterised by the abolition of obstacles to the free movement of goods, persons, services, and capital; and,
3. the approximation of the laws of member states to the extent required for the functioning of the common market.

The Treaty of Rome made no explicit mention of consumer protection or public health, and the Single European Act and the Maastricht Treaty only added these goals as amendments to Article 3. Thus, in contrast to legislation in most member states, EU food law developed gradually over a long period of time.

Establishing foods law in the EU came with distinct challenges. Initially, there was no central unifying text setting out the fundamental principles of EU food law to clearly define the obligations of those involved at every stage of the food production chain. The first EU food directive, which was concerned with coloring in foodstuffs, was adopted by the Council of Ministers in 1962. The directive had no EEC number because at that time, there was no formal system for numbering EU laws. This was not an auspicious start for the establishment of EU food law as a major concern of EU legislators. Ultimately, it is through the EEC treaty that food law in the EU began to follow harmonized principles. From this trend toward harmonization came the demand for free trade within the EEC member states, meaning the free movement of goods.

23.4.2 The Free Movement of Goods

One of the principal aims of the EU as stipulated in Article 3 of the EC Treaty is the elimination of all obstacles to the free flow of goods between member states. The most obvious obstacle is custom duties, but there are other ways nations may try to protect their domestic products from foreign competition. These ways include limiting the quantity of goods that may be imported (quantitative restrictions), as well as regulations and administrative practices that hinder the importation of specific products. Therefore, in order to provide for the free movement of goods, the EC Treaty states in Article 34 "quantitative restrictions on imports and all measures having equivalent effect shall be prohibited between Member States."[12]

---

[12] Article 34. Consolidated version of the treaty on the functioning of the European Union.

Article 35 of the EC Treaty requires that any restrictions on exports, or measures having an equivalent effect, should be abolished. The Treaty provides exemptions or derogations under which the basic rules on the free movement of goods may be avoided, as stipulated in Articles 34–35. Article 36 of the EC Treaty states:

> The provisions of Articles 34 and 35 shall not preclude prohibitions or restrictions on imports, exports or goods in transit justified on grounds of public morality, public policy or public security; the protection of health and life of humans, animals or plants; the protection of natural treasures possessing artistic, historical or archaeological value; or the protection of industrial and commercial property. Such prohibitions shall not, however, constitute a means of arbitrary discrimination or a disguised restriction on trade between Member States.[13]

23.4.3 Quantitative Restrictions and Article 28 of the EC Treaty

To return to Article 34, the term "quantitative restrictions" is defined as a measure that amounts to a total or partial restraint, according to the circumstances, of imports, exports, or goods in transit.[14] The term "measures having equivalent effect" creates much more difficulty with regard to interpretation. At first, the common reading was that Article 34 merely prohibited discrimination, meaning that it is a ban on measures by member states to hinder the free movement of goods. Barriers to trade arising from "differences" in national laws or regulations—for example, in compositional lists of ingredients for foodstuffs—were not initially seen as connected with Article 34's prohibition of quantitative restrictions. These barriers were considered to be a matter for EU harmonization of national laws.

23.4.4 Measures Having Equivalent Effect Defined: The Dassonville Case

It was not until the landmark decision in *Procureur du Roi v. Dassonville,* which is also discussed in earlier chapters in this section, that the European Court supported a "broad" view in relation to measures having equivalent effect. The Court defined measures of equivalent effect, in the context of Article 34 of the EC Treaty, as "all trading rules enacted by Member States, which are capable of hindering, directly or indirectly, actually or potentially, intra-Community trade."[15]

---

[13] Article 36. Consolidated version of the treaty on the functioning of the European Union.
[14] Geddo (1973).
[15] *Procurer du Roi v. Dassonville* (1974), ECR 837.

The case concerned Belgian legislation, which required a certificate of origin for imported goods bearing a designation of origin. Specifically, the legislation required all whisky sold in Belgium as "Scotch Whisky" to be accompanied by a certificate of origin, in addition to the existing designation of origin provided by the Scotch whisky manufacturers. The European Court ruled that this provision could not be applied to Scotch Whisky which had been lawfully imported and put into free circulation in France and then re-exported to Belgium. In its reasoning, it argued that it was much more difficult to obtain the certificate of origin for whisky imported into Belgium by way of a third country than for direct imports from Scotland. The Court ruled that the Belgian legislation acted as a hindrance to trade between member states.

Despite this ruling, the Court did note that, under certain circumstances, member states might reasonably require information about the place of origin of imported products. Thus, when the Belgian law was modified in relation to import certificates for Scotch Whisky, Belgium requested that manufacturers provide "certified" copies of the certificate of origin. The Court subsequently ruled that this new practice was lawful.

The main consequence of this European Court ruling was that literally, overnight, thousands of national laws and regulations relating to imported products suddenly fell within the scope of the provisions in Articles 34–36. A European Court ruling in *Rewe-Central AG v. Bundesmonopolverwaltung für Branntwein*, more commonly known as the *Cassis de Dijon* case, which also plays a role in other chapters of this section, however, soon modified this broad interpretation of Articles 34–36. The case involved a German law that prohibited the marketing of liqueurs with an alcoholic strength of less than 25 %. Rewe-Central requested authorization from the German authorities to import a liqueur from France, Cassis de Dijon, containing 15–20 % by volume of alcohol. The German authorities informed the company that Cassis de Dijon could not be sold in Germany, since under the German regulations only liqueurs of between 25 and 32 % alcohol by volume could be authorized. There were some exceptions to this rule, but regulators informed the company that Cassis de Dijon was not one of them.

The European Court ruled that this prohibition was discriminatory whether applied to imported or domestic products. The Court stated that: "the requirements relating to the minimum content of alcoholic beverages do not serve a purpose which is in the general interest and such as to take precedence over the requirements of the free movement of goods, which constitutes one of the fundamental rules of the Community."[16] It was the Court's judgment in Rewe-Central that these rules constituted an obstacle to trade that was incompatible to Article 34 of the Treaty. The Court believed that there was no valid justification for why alcoholic beverages should not be introduced into another member state, provided they had been lawfully produced and marketed in the EU, Consequently, this case marked the first time that the European Court articulated the principle of "mutual recognition,"

---

[16]Id.

according to which goods lawfully produced in one member state should be free to move throughout the rest of the EU. This principle has become one of the cornerstones of EU food law, as well as a fundamental part of the EU 1992 Single Market Program.

The Cassis de Dijon judgment also gave some clarification to the derogations coming within the scope of Article 34, as permitted under Article 30. The German government defended its position in the European Court by stating that these legal provisions were prompted by the wish to protect the consumer against the adverse effects of alcohol to his/her health. They took the view that a limitless authorization for liquors of different alcoholic strengths would be likely to lead to a rise in the consumption of alcohol as a whole, and would therefore increase the specific dangers of alcoholism. The Court ruled that although these arguments had some merit, they were not sufficient to permit Germany derogation from Article 34 obligations in line with the "public health" derogation in Article 36. Even so, by confirming the possibility of obtaining Article 36 derogation, the European Court "copper-fastened" those provisions into any future judgment on the possibility of quantitative restrictions existing in relation to free movement of certain foodstuffs throughout the EU.

On the basis of the case law of the ECJ, the following activities, rules, and practices constitute measures of equivalent effect to quantitative restrictions as defined in Article 34 of the EC Treaty.

23.5 Changes in the White Paper: EP and Council Regulation

The major changes envisioned in the White Paper were developed by means of EP and Council Regulation 178/2002.[17] Besides establishing the European Food Safety Authority and re-vamping the Rapid Alert System, it also outlined the general principles of EU food law. The regulation therefore harmonises at Community level many existing national requirements of food law, placing them in the European context so that there will be no ambiguity across the EU as to the basic philosophy and obligations flowing from EU food law.

23.5.1 Scientific Basis to Food Law

Depending on the nature of the measure, food law, and in particular measures relating to food safety, must be underpinned by strong science. Regulation 178/2002 establishes in EU law that the three inter-related components of risk

---

[17]EP & Council regulation 178/2002 (O.J. 2002 L31 p. 1).

analysis—risk assessment, risk management and risk communication—provide the basis for food law as appropriate to the measures under consideration. Not all food law has a strong scientific basis, however, and some measures do not need a scientific foundation. Exceptions include food law relating to consumer information and the prevention of misleading practices.

The new Regulation requires the scientific assessment of risk to be undertaken in an independent, objective, and transparent manner based on the best available science by means of the European Food Safety Authority (EFSA).

Risk management is the process of weighing policy alternatives in light of the results of a risk assessment and, if required, selecting the appropriate actions necessary to prevent, reduce, or eliminate the risk. The goal is to ensure a high level of health protection within the EU. In the risk management phase, the decision makers need to consider a range of information in addition to the scientific risk assessment. This includes, for example, the feasibility of controlling a risk, the most effective risk reduction actions depending on the part of the food supply chain where the problem occurs, the practical arrangements needed, the socio-economic effects, and the environmental impact. Accordingly, the regulation establishes the principle that risk management actions are not just based on scientific assessment of risk; rather, they must also take into consideration a wide range of other factors relevant to the matter at hand.

23.5.2 *Responsibilities of Food and Feed Business Operators*

The regulation establishes the basic principle that the primary responsibility for ensuring compliance with food law, and in particular the safety of foodstuffs, rests with the food business. Similarly, this principle is applied to feed businesses. To complement and support this principle, there must be adequate and effective controls organised by the competent authorities of the member states.

23.5.3 *Food Safety Requirements*

EP and Council Regulation 178/2002 establish a food safety requirement that comprises of two elements. In order for the food to be considered unsafe food should not be injurious to health, and food should not be unfit for human consumption. In considering whether a food is potentially injurious to health it is important to consider the use of the food, information provided with the food, and the processing or subsequent handling to which it will be subject. With regards to a food's effect on an individual, it is crucial to take into account possible long term, cumulative effects as well as short term, acute ones. Food unfit for human consumption is also considered to be unsafe in this regulation. Putrid food, for example, is unacceptable for human consumption and may be injurious to health. It may be almost impossible to prove injury or probable injury to health with such food, so this separate factor is included in relation to the overall food safety requirement. The regulation also makes it obligatory for food businesses to withdraw unsafe foods from the market, and to provide accurate information to consumers. It requires food safety to be considered along all stages of the food chain.

23.5.4 Traceability

The identification of the origin of feed, food, ingredients, and food sources is of prime importance for the protection of consumers—particularly when products are found to be unsafe. Traceability facilitates the withdrawal of foods from the market and provides consumers with targeted and accurate information concerning the implicated products.

EP and Council Regulation 178/2002 demands the traceability of all food and feed as they move between businesses, with information on the traceability of the food or feed to be made available to competent authorities if requested. Importers are similarly affected, as they are required to identify from whom the product was exported when that product is redistributed.

This traceability obligation ensures, at the very least, that businesses are able to identify the one step in the food supply above them and the one step below. In some cases, there may be specific provisions to provide further traceability.

23.6 Connections Between Consumer Law and Food Law

Food law is consumer-oriented and since the product is consumed, issues of safety are of paramount importance. On the other hand, much of EU consumer law aims to grant the consumer equal status contractually with the manufacturer or service provider.

In a special message to the U.S. Congress on March 15, 1962, President John F. Kennedy established the modern consumer movement by highlighting the need for legislators to take consumers into account when writing and adopting laws. He declared that the following consumer rights should be taken into account in future legislation. These rights are:

1. The right to safety—to be protected against the marketing of goods which are hazardous to health or life.
2. The right to be informed—to be protected against fraudulent, deceitful, or grossly misleading information, advertising, labelling, or other practices, and to be given the facts he needs to make an informed choice.
3. The right to choose—to be assured, wherever possible, access to a variety of products and services at competitive prices; and in those industries in which

competition is not workable and Government regulation is substituted, an assurance of satisfactory quality and service at fair prices.
4. The right to be heard—to be assured that consumer interests will receive full and sympathetic consideration in the formulation of Government policy, and fair and expeditious treatment in its administrative tribunals.[18]

The consumer rights highlighted by U.S. President Kennedy now form the basis of consumer policy and consumer law in the United States and throughout the world. These rights have also been the foundation of the EU's consumer protection policies since the 1980s.

It is evident, therefore, that EU Food Law must ensure the safety of foodstuffs while at the same time providing consumers with sufficient information so that they may make an informed choice about the items they purchase. In the same vein, if a consumer purchases a hair dryer, the law must ensure that the product is safe and the consumer understands the risks, if any, that may be associated with its use.

The extra dimension to consumer law—as compared to food law—comes in terms of providing the consumer with contractual powers so that they are placed on equal footing with the manufacturer or service provider. Consumer law, for instance, can outlaw unfair contract terms or unfair commercial practices. The economic interest of the consumer is therefore not as predominant in food law as it is in consumer law.

To highlight the differences in food law and consumer law, it is useful to look at how the two have been implemented by the EU.

23.7 Harmonization of Legislation with the Treaty of Rome and EC Treaty

The Treaty of Rome (1957) recognised that, in order to help the member states develop in a unified manner, the legislation in individual countries need to be harmonised (or, to use the wording of the Treaty, "approximated").

Article 114 EC Treaty allows the introduction of EU legislation for the purposes of the "establishment and functioning of the internal market" in line with Article 26 EC Treaty. Most aspects of food law are related to the internal market, and therefore Article 114 EC Treaty applies. The important point to note is that Article 114 EC Treaty provides for measures to be adopted under the co-decision procedure contained in Article 294 EC Treaty.

---

[18] JFK Special Message to the Congress on Protecting the Consumer Interest. March 15, 1962.

23.7.1 Legally Binding Acts: Regulations, Directives, and Decisions

The EC Treaty defines three types of legally binding acts: regulations, directives, and decisions.[19] Most food law is in the form of regulations or directives. There is a major difference between these two types of laws, which is important to understand. The regulation is 'directly applicable'—this means that it becomes the law of every member state from the time it is adopted in Brussels. Courts within the member states must use an EU Regulation as if it were national law. Food companies therefore have to comply with EU regulations even if there is no related national law, and even if national authorities have not informed them about the regulation.

The directive is binding upon the member states—it is actually an agreement by the member states to alter their own national legislation so as to implement the agreed provisions. It therefore only becomes law when it has been transposed into national legislation, either by an act of parliament or a statutory instrument. Most directives will have a date (12 months, 2 years, or often even longer) by which member states are required to have implemented the provisions of the directive into national law. The Council, the Commission, or the Council and the European Parliament acting jointly can all issue regulations and directives. The latter may occur when the Commission is given the power to take action on its own initiative, for instance in emergency situations such as with the BSE outbreak, dioxin contamination, FMD, and the Sudan Red crisis. The decision is binding upon those to whom it is addressed. A decision can be addressed to an individual member state, a company, or a private individual. In the food area, the most common decisions have been made in the case of emergency measures taken by the Commission, such as those just mentioned above.

There is another type of legal act permitted under the EC Treaty known as a recommendation. The recommendation has no legal binding force and is merely of persuasive authority. In the food area, recommendations might be adopted in relation to a member state government's policies to tackle obesity. The recommendation advocates certain actions, but it is up to each individual member state to decide whether or not to follow these recommendations.

23.7.2 Examining Regulations v. Directives

In the White Paper on Food Safety (2000),[20] the EU stated that in the future it would prefer to introduce major pieces of food legislation by means of regulations. Since a EU regulation becomes law throughout all member states once it is adopted in

---

[19][Article 288 EC Treaty].
[20]White Paper on Food Safety (COM (1999) 719).

Brussels, there is less chance of a difference in interpretation when the rule is enforced within each member state. In simple terms, the EU regulation offers the opportunity to establish a 'one size fits all' scenario for foodstuffs throughout the EU with no ambiguities. Indeed, in the wake of the White Paper, the EU adopted a major package of laws covering food hygiene by means of EU regulations. It has also done the same with the recent regulations on fortified foods, as well as on nutrition and health claims. The trend has been established: in the foreseeable future major pieces of EU horizontal food law are likely to be adopted by means of EU regulations, while directives will be the preferred format for vertical food laws.

23.7.3 EU Consumer Rights Directive

The European Commission has repeatedly pointed out that rules on consumer protection differ across the EU, and the level of consumer protection varies depending on the will of each member state to implement these rules. It has suggested that the fragmented nature of existing consumer protection legislation across EU member states has stunted the growth of cross border trade. Businesses seeking to export across Europe currently face the cost of interpreting up to 27 different regimes whilst consumers, unaware of their local rights within each member state, tend to stick with familiar domestic retailers and potentially lose out on the greater choice and lower prices that may exist in a more competitive pan-EU marketplace. The Commission has therefore seen minimum harmonization, on which directives regulating various aspects of consumer protection are based. Following a review of the Consumer Acquis,[21] the Commission presented a proposal for a directive on Consumer Rights in 2008. As a means of allaying this anomaly, the directive suggested maximum harmonization for consumer protection in four existing EU directives relating to unfair terms in consumer contracts, distance and doorstop selling, guarantees, and consumer sales. These trends are to be juxtaposed with the more sustainable local farming and shortened supply chain movements, but a detailed discussion thereof is beyond the scope of this chapter.

The ambitious scope of the proposal proved unacceptably wide, however. It encountered stiff opposition from many member states, particularly with regards to consumer sales, as maximum harmonization in this case would have actually reduced consumer protection in some states.[22] As a result, the scope of the directive was significantly reduced before being approved by the European Council in 2011. While the directive maintains maximum harmonization in order to standardize consumer protection regimes in all member states, it now only repeals two existing

---

[21]Green Paper on the Revision of the Consumer Acquis in 2007.

[22]Where the current minimum harmonization standard had been exceeded, a maximum approach would have led to an undesirably lower standard.

directives: the Distance Selling directive (97/7/EC), and the Doorstop Selling directive (85/577/EC).

At times, consumer legislation—such as the Unfair Commercial Practices directive—may be introduced as a maximum harmonization measure. As the proposed Consumer Rights directive demonstrates, however, more often than not the principle of shared competence between the Union and the member states applies in the area of consumer protection, so EU legislative measures need to be justified as necessary for the smooth functioning of the internal market.

Along similar lines, the Commission proposed an optional common European sales law to facilitate cross-border trade by offering a single set of rules for cross-border contracts. The proposal, however, was watered down in legislative discussions with the Council and the European Parliament. It has now been limited to distance contracts—notably online contracts—and, once again, issues over maximum and minimum harmonization were at the fore during the passage of this proposal through the EU legislative process.

23.8 EU Food Laws

No such debates over maximum v. minimum harmonization are heard when food legislative proposals are placed before the Council or Parliament. In addition, the number of legislative proposals adopted in the food area has increased immeasurably since the BSE crisis days in the late 1990s. In 1995, there were just five EU food laws; that number rose to 21 in 1996 and was at 117 in 2012. In 2013 alone, the EU adopted 130 food laws.

By contrast, there were only a few laws in the consumer protection area adopted in any 1 year from 1995 to 2013. This difference shows how EU food law has in many ways overtaken EU consumer law as the most important component of the EU's *acquis communitaire*.

Given the large number of EU food laws now being adopted, have member states or the European Parliament expressed unease about this increase? The answer appears to be no. In fact the only time member states have been concerned is when legislative proposals cover the official control of foodstuffs. Thus, the reason for such concerns is that member states are afraid that new EU provisions might force them to re-organize the nature and frequency of their official food controls. It follows that, the recent tranche of Commission proposals following the horsemeat scandal demonstrate this clearly: one proposal, for instance, is for a new regulation to replace EU Regulation 882/2004. This new regulation requires that official controls be performed to ensure compliance with feed and food law, as well as animal health and animal welfare rules. Despite the added administrative burden that such regulations may place on member states, they do not appear to be overly troubled by the rising adoption of food laws by the European Parliament.

Member states and the European Parliament were concerned about the inclusion of provisions to introduce mandatory fees for all official food controls and give the

Commission power to introduce delegated acts in numerous areas without needing the approval of the Council or European Parliament. These concerns were not about maximum harmonization in the food area; rather, they were adaptations to the way in which the Commission aimed to regulate the food sector. The member states were not arguing that the Commission had no role to play in overseeing their official food controls. Instead, they were arguing against the introduction of mandatory fees.

In relation to legislation, there is a similar dynamic between member states and the European Parliament. The issue was not over too much legislation or the Commission's use of the subsidiarity principle. Instead, member states and the European Parliament wanted more input and oversight of any new laws—a view that is entirely logical for a functioning democracy.

23.9 Conclusion

It is clear from the previous discussion that when assessing the differences between EU Consumer Law and EU Food Law, consumer protection was not a major policy issue at the time of the establishment of the European Union in 1957, initially consumer protection measures were only introduced if they could be seen to assist in the smooth functioning of the Internal Market. The Cassis de Dijon ECJ case was crucial in establishing the "mutual recognition" principle, i.e. that a product, in particular a foodstuff, if permitted to be marketed in one member state must be permitted to be marketed in any other member state. Then, the BSE crisis was the catalyst for the establishment of a major body of EU Food Laws including the introduction of General Principles of European Food Law. Furthermore, EU consumer law has become the *Cinderella* body of law to EU food law consisting of a few legislative proposals every year compared to approximately 150 food related laws in any 1 year. Finally, EU food laws are always introduced as maximum harmonization measures unlike EU consumer laws where there continues to be debates amongst member states as to whether they should be minimum or maximum harmonization measures.

References

Article 1, Unfair commercial practices directive 2005/29 (O.J. L149 pp. 22–69)
Article 3(1), EP & Council regulation 1169/2011 on food information (O.J. L304 pp. 18–63)
Article 12. Consolidated version of the treaty on the functioning of the European Union
Article 26, 114, & 294 Consolidated version of the treaty on the functioning of the European Union
Article 34. Consolidated version of the treaty on the functioning of the European Union
Article 36. Consolidated version of the treaty on the functioning of the European Union
Article 38. European charter of fundamental rights of the European Union
Article 114(3). Consolidated version of the treaty on the functioning of the European Union

Article 288. Consolidated version of the treaty on the functioning of the European Union
Cassis de Dijon (1979) ECR 469
Communication on completion of the internal market (1985) Legislation on foodstuffs (COM(85) 603)
Communication on the free movement of foodstuffs in the community (O.J. 1989 C271/3)
Directive http://ec.europa.eu/food/fs/sfp/addit_flavor/flav13_en.pdf
EP & Council regulation 178/2002 (O.J. 2002 L31 p. 1)
EP & Council regulation 1924/2006 on Nutrition & Health Claims (O.J. L12 pp. 3–18)
EU consumer rights directive proposal COM (2008)614. Consumer rights directive 2011/83 (O.J. L304 pp. 64–88)
Geddo (1973) ECR 865
Green Paper (1997) The general principles of food law in the European Union. (COM (97) 176) May 1997
JFK (1962) Special message to the congress on protecting the consumer interest. 15 March 1962
Procurer du Roi v. Dassonville (1974) ECR 837
Recital 18. Unfair commercial practices directive 2005/29 (O.J. L149 pp. 22–69)
White Paper on Food Safety (COM (1999) 719)

Chapter 24
The Problem of Food Waste: A Legal-Economic Analysis

Harry Bremmers and Bernd van der Meulen

Abstract This chapter reviews, building on legal-systematic and economic analysis, the origins for food waste vested in food law and states possible remedies. Several causes are identified: a policy of 'zero-tolerance', food information requirements, bans on use of hazardous materials, a policy of 'structural precaution' and strict top-down plant pest controls. In all of these, uncertainties as to how to behave and what the real risks are seem to play a key role in the early discard of consumable foodstuffs. Solutions can come from technical, legal as well as social sciences. In food law and policy, rule-makers should be more aware of the adverse effects of requirements on businesses that foster food safety. Technical sciences may provide solutions through nano- and it-innovations. From social sciences, it can be learned what factors induce humans to overemphasise risk exposure. Moral issues are connected to possible solutions, especially to certain potentially hazardous new techniques (like nanotechnology) and the acceptability to nudge, channel and restrict free human will and choice to reduce the waste stream.

Dr. H. J. Bremmers is associate professor at the Law & Governance group of the Social Sciences department of Wageningen University. He is lawyer and economist. He publishes and does research at the intersect of food law and economics. Prof. Dr. B.M.J. van der Meulen is professor of Law and Governance at Wageningen University. This contribution builds further on previous publications of the authors in official reports to the Dutch government on food waste, its causes and prevention. Especially in Annex 1 of *Verminderen van Voedselverspilling* (Transl.: *Reduction of Food Waste*) by Harry Bremmers, Report/LEI 2011-014, The Hague, the Netherlands, (2011); and *Houdbaarheidsdatum: Verspilde moeite* (transl. *Durability dating: a waste of effort?*), section 3.1, by Harry Bremmers and Bernd van der Meulen, Wageningen UR Food & Biobased Research, Wageningen, the Netherlands (2012).

H. Bremmers (✉) • B. van der Meulen
Law & Governance Group, Wageningen University and Research, Wageningen,
The Netherlands
e-mail: harry.bremmers@wur.nl; bernd.vandermeulen@wur.nl

24.1 Introduction

Food waste is a major policy concern in the European Union, in the USA, and internationally. The Food and Agriculture Organization of the United Nations (FAO) defined food waste in 1981 as "wholesome edible material intended for human consumption, arising at any point in the food supply chain that is instead discarded, lost, degraded or consumed by pests."[1] Delimitation in regional legal environments may be more or less narrow depending on the practical or legal context of the concept.[2] Food waste may occur anywhere in post-harvest food supply chains. In post-harvest stages, or prior to consumption, the wasted materials are called "food losses." Behaviour by retailers and consumers may result in "food waste." In the latter case, food waste has a narrow meaning as processors' activities are outside its scope.[3]

In this chapter the expression "food waste" depicts commodities that can be transformed or used to serve nutritional needs, but have lost this potential for economic, food safety or personal reasons. For the reason that processing is a key source of wasted food, such activities are viewed within the scope of the definition. The lack of a common delineation of the concept may limit the comparison of food waste over time and between regions. However, it only slightly hampers the search for factors in the legal-institutional environment that contributes to the early abandonment of foodstuffs. A common core in the concept is the image of the discarded food product in certain stages of food supply chains. Food waste may be given an inferior destination along the scale from food to feed, fertilizer, fuel, fire[4] and (land)fill.[5]

Why is food waste a problem in the first place? First, the occurrence of food waste threatens the realization of the "right to adequate food"[6] of humans, as nutrients are lost or simply not used for human consumption. The problem of securing food supply goes back to ancient times when there was ineffective storage, draughts, insect and rodent plagues, or other causes related to the governance of the food supply chain. However, the deliberate discarding of food must have been relatively rare due to the general food scarcity and substantive efforts to nourish growing populations.

---

[1] FAO (1981), cited in Parfitt et al. (2010), note 3. For the purpose of the EU waste Directive 2008/98/EC 'waste' has as definition: any substance or object which the holder discards or intends or is required to discard (Article 3 point 1. of the Directive).

[2] For instance including also edible material used as feed for animals or by-products that are not used as human food and/or over-nutrition.

[3] Parfitt et al. (2010), pp. 3065–3081.

[4] Under the 'f' of fuel, the biomass is used for energy. Fire, by contrast, is used to destroy.

[5] Vogdlander et al. (2001), pp. 344–355. These f's are based on the Order of Preferences of EoL Solutions in The Netherlands ('Ladder of Lansink'); Also: Article 4 of Directive 2008/98/EC of the European Parliament and of the Council of 19 November 2008 on waste.

[6] van der Meulen (2004); Hospes (2008), pp. 246–263; Wernaart (2010).

In contrast, nowadays the global population exceeds 7 billion and is estimated to grow to 9.6–11 billion by 2050.[7] It is estimated that globally yearly 1.3 billion tonnes of food destined for consumption by humans is lost at pre-marketing stages or wasted at retail or by the consumer. This food, if wasted food were included in the supply chain, would be enough to feed the world population.[8] It could also supply future generations with adequate food.

Second, from an economic perspective, wasted food represents financial losses. As losses occur in different phases of the supply chain—processing, retail, consumption—such losses are borne by multiple stakeholders. In a highly developed society, losses due to food waste are carried downstream. However, losses in developing countries occur primarily upstream of the supply chain for reasons.[9]

Third, for the reason that food production, of all human activities, has the biggest impact on the climate, next to transport and housing,[10] wasted food is causally connected to the depletion of natural resources and thus to long-term unsustainability. The prevention of waste is perfectly in line with EU goals as stated in the Europe 2020 strategy that fosters smart, sustainable and inclusive growth.[11]

24.1.1 Food Law as a Cause and Measure Against Food Waste

Notably, food law is an important *cause* for food waste. Precautionary measures are taken globally to avoid food risks. The USA and European Union are among the strongest risk avoiders.[12] In this context, the "wickedness" of food waste lies in the fact that such waste is considered socially and morally unacceptable, while the reduction of food waste might jeopardize economic[13] and food security goals. For instance, prevention of food waste, or its reuse, induces costs and will deplete other resources while the consumption of larger parts of the potential food stock, like animal by-products, might jeopardize food safety. The problem is apparently wicked[14] in the sense that more safety is inextricably connected to the increase of food waste; and thus the attainment of one goal is hampering the other.

---

[7]UN News Centre (2014).

[8]Future Directions International (2013).

[9]Gustavsson et al. (2011).

[10]Hall et al. (2009). The major part of freshwater is used to produce food; calculations show that about 25 % of freshwater is used inefficiently due to food waste; Kummu (2012), pp. 477–489; European Commission Joint Research Centre (2010).

[11]Communication from the Commission (2010).

[12]Wiener and Rogers (2002), pp. 317–349.

[13]Fischhoff et al. (1978), pp. 127–152.

[14]See on the origin of this concept: Skaburskis (2008), pp. 277–280.

Accordingly, the EU and the USA experience such controversial aims more than less developed regions. As indicated in a previous chapter, in affluent societies food waste occurs most in post-consumer stages.[15] To reduce food waste it may be necessary to nudge, channel or restrict the free choice of individuals and thus limit the exercise of fundamental rights. These opposite goals make the reduction of food waste even more of a "wicked problem."

The structure of food law is heavily influenced by regional and global food scares.[16] From an EU perspective, the BSE-crisis[17] is the most prominent. On a global level, the melamine scandal of 2008 had significant influence on policy choices, especially as to the traceability requirements and the protection against unsafe imported foods.[18] Food scares thus induce intensified actions to protect the consumer.[19] As a result of the BSE-crisis, the interests of the consumer were positioned in the center of food legislation, while the responsibility and burden to provide safe food is put on the shoulders of food businesses. To protect the consumer from unsafe food, safety enhancements may induce exaggerated risk avoidance measures and lead to food waste. Not only may the substantive content of rules cause food waste, but also their interpretation and imposed sanctions and liabilities. As to the content, earlier research has shown that food business operators in the EU perceive food law requirements to be clear, but the perceptions are not confirmed by practical evidence.[20] Moreover, international differences exist such as hygiene, labelling and processing requirements, which may not only impede international trade, but also lead to wasted food due to rejections at border control and removal of foods considered dangerous or unfit for human consumption.[21]

The threat of being sanctioned, held liable or publicly scrutinized induces business operators to take precautionary measures in their processing activities and follow or surpass publicly set standards (see below). This chapter addresses the major legal sources of food waste, the trade-off between food waste and safety from a legal-economic point of view, and states policy alternatives to mitigate the problem of food waste.

---

[15] Parfitt et al. (2010), pp. 3065–3081.

[16] In the EU, the BSE-crisis led to the issue of a White Paper on Food Safety by the European Commission in 2000, which initiated a cascade of European legislation to regain consumer confidence. The main EU law in this respect is the General Food Law (178/2002), which contains principles of food law and food safety, adopts risk analysis including assessment, and installs a rapid alert system for food and feed.

[17] Bovine Spongiform Encephalopathy.

[18] Petrum and Sellnow (2010).

[19] Knowles and Moody (2007), pp. 43–67 at p. 44.

[20] van der Meulen and Bremmers (2006), pp. 74–110; Report available at ec.europa.eu/enterprise/food/index_en.htm. While businesses stated to clearly know their legal obligations, interviews showed that many of them are not aware of the actual rules which apply to them.

[21] Clapp (2005), pp. 467–485.

24.2 Legal Triggers of Food Waste

Legal requirements induce food waste as they limit the attitudes and actions of actors in food supply chains. Almost all food legislation may cause food waste. However, some laws may have more severe impacts compared to others. In the next sections, several legal triggers for food waste are reviewed,[22] specifically limits as to contaminants and residues in food products, strict pre-market approval requirements, food information law, plant pest controls and the occurrence of animal by-products and legal barriers to their further use.

24.2.1 Contaminants and Microbiological Hazards

In the EU, contaminants[23] are considered substances that have not been intentionally added to food. Instead, they are present due to deliberate processing steps, like production, medical treatment of food producing animals, and/or environmental contamination. In the USA, the EU, as well as in the Codex Alimentarius, a fundamental principle is that food exceeding a level of contamination which is unacceptable to public health may not be placed on the market, or, if already marketed, must be removed immediately.

For example, in cases C-129/05 and C-130/05 of the Court of Justice of the European Union (CJEU) contaminated shrimp, rabbit and poultry originating from China were at issue after being detected at the border. The CJEU ruled that the lot had to be destroyed because of zero-tolerance, despite the fact that only low levels of contaminants had been found (parts per billion). The fact that the food is potentially hazardous was considered reason enough to destroy it instead of sending it back to its origin. This is questionable on social and economic grounds.[24]

Maximum tolerances, including analytical detection limits and limits for the same contaminant in different foods may be set, but regional differences are immanent. In other cases, good practices have to guarantee that contaminant levels are as low as reasonably achievable. Authorities may check agricultural imports from third world countries at border posts for salmonella, aflatoxin, or pesticide residue. In 2012, the EU rapid alert system[25] gave 1467 border control notifications resulting in 88.5 % of the cases in blocking the fruits and vegetables from entering the market.[26] Worldwide, 50 % of all fruits and vegetables are wasted.[27]

---

[22]Waarts, et al. (2011).
[23]The Council of European Communities (1993).
[24]Lelieveld and Kenner (2007), pp. S15–S19.
[25]This is basically a system that is designed and installed to share information between risk managers in the European Union; it was established in the General Food Law, 2002.
[26]Van Boxstael et al. (2013).
[27]FAO (2010).

Domestic products may also be taken off the market out of safety concerns, as for example the Dutch "potato peel"-case of 2004 shows. Products from dairy farms using factory potato peels as animal feed appeared to be contaminated with dioxin originating from clay that was used in the process to sort potatoes. Although they did not contain dangerous levels of dioxin, foods such as milk, and feed were destroyed.

For many hazardous substances, maximum levels have been set conservatively in the EU, compared to the Codex Alimentarius and in other countries. As already indicated, in some cases zero-tolerance is the standard.[28] Farms holding pigs that are treated with antibiotics before slaughter will have retain the animals for some weeks with higher feeding and housing costs as a consequence. Hence, businesses are confronted with a choice between delivering pigs with remnants of antibiotics and taking the risk to be fined and losing the license to produce, or to wait and accept losses. As long as the extra costs exceed the benefits, farms may be inclined to remove animals altogether, on pure economic grounds.

A case that illustrates the waste impact of contaminants in food is the European horsemeat fraud crisis. Apart from the fact that horsemeat had been marketed as beef in several European countries, in the Netherlands, two reasons were put forward to order the withdrawal of 50 million tonnes of meat product. The first was that some meat might have originated from horses treated with phenylbutazone (buta). Buta is a painkiller that in certain doses may threaten human health. However, the likelihood is very low.[29] The other reason was that the meat might be unsafe to consume because the traceability system had failed; tracing the meat back to animals approved for slaughter could not be assured due to fraudulent practices.

Some substances are considered unsafe by definition, like aflatoxin. Fungi may develop such toxic substances[30] during transport in the bulk of cargo ships, especially when they originate from warm or moist countries or countries with inferior hygiene systems. Border control agencies in the EU, US, and other parts of the world are well acquainted with this risk and examine shipped cargo. They are a major cause of food destruction. There are technical possibilities to reduce the aflatoxin content, but this brings about extra costs. Processing facilities may not be

---

[28] *See* Regulation (EC) 470/2009 Laying down Community procedures for the establishment of residue limits of pharmacologically active substances in foodstuffs of animal origin and the EU Commission's Implementing Regulation of 22 December 2009 on pharmacologically active substances and their classification regarding maximum residue limits in foodstuffs of animal origin, table 2.; van der Meulen (2009), specifically ch. 5.

[29] Buta was medicated for racehorses, which in the EU may enter the meat supply chain as horsemeat, contrary to the situation in the USA.

[30] Bennett and Klich (2003), pp. 497–516. This is a major source of border rejections in The Netherlands. The term was initiated in 1962 as a result of a veterinary crisis under turkeys in England, causing disease and death at thousands of animals. The 'mycotoxin family' includes a wide range of toxic fungi.

available in every country.[31] In 2013, the number of notifications in the European RASFF was 341 cases, with a total record amount of 642 notifications, due to pathogenic substances in meats and molluscs.[32]

24.2.2 Pre-Market Approval

The previous section brings us to the general question of whether food is considered unsafe, unless proven safe, or, in the alternative, safe, unless proven unsafe.[33] For a considerable number of foods the first is true in the European Union. Pre-market approval requirements are more common in the European Union compared to other countries, like the USA, for additives, enzymes, genetically modified (GM)-products and novel foods. In the USA, "GRAS"-exemptions[34] or notifications to the FDA are sufficient. While the general understanding is that the marketing of food is unbounded, a restrictive policy is in place in the EU. Pre-market restrictions may cause border rejections[35] and recalls of food for procedural reasons rather than for food safety concerns. In the EU and in a number of other countries, pre-market approval is mandatory for a range of innovative products such as food additives, GM foods and other novel foods.[36]

In the EU, foods are considered novel when new ingredients, such as fungi or algae, are brought to the market that have not been consumed in the EU to a significant degree[37] after the Novel Foods Regulation came into effect 15th May 1997. Not only the chemical composition of a food, but also a new production method that influences the structure of a food[38] can be reason enough to consider a food or ingredient novel. In the EU, foods that are considered novel or have novel ingredients are considered unsafe and banned from the market unless approved. This precautionary measure may be called 'structural precaution.'[39] This term underlines that precautionary measures are not taken temporarily to prevent hazards from becoming immanent during a period of scientific assessment of their nature, but are structural. Pre-market approval requirements close the border to all foods

---

[31] See the Guidance Document of the European Commission for Competent Authorities for the Control of Compliance with EU Legislation of Aflatoxin (2010).
[32] https://webgate.ec.europa.eu/rasff-window/portal/ accessed Sept. 2014; Whitworth (2014).
[33] van der Meulen et al. (2012), pp. 453–473.
[34] This stands for: Generally Recognized As Safe.
[35] Schoss (2011).
[36] In Regulation (EC) 258/97 concerning novel foods and novel food ingredients.
[37] See the guidance document on Human Consumption to a Significant Degree (n.d.).
[38] Like nanotechnology, see Van der Meulen et al. (2014).
[39] van der Meulen et al. (2012), pp. 453–473.

until risk assessment and authorization have taken place.[40] If foods are "substantially equivalent" to traditional products, a mild admittance procedure is applicable. However, in the EU unlike in the USA, "substantial equivalence" does not provide any exemption to GM-based foods. In the USA, "substantial equivalence" can be a reason for considering a food GRAS and granting it access to the market, whether it is a food derived with genetic modification or not. If a product containing GM material is substantially equivalent to existing products, measured by its chemical properties, pre-market permissions are not mandatory in the USA. In the EU, in case foods and also feed contain genetically modified organisms (GMOs), or have been made from GM-material, extensive and time-consuming authorization procedures are required. Extensive authorization requirements form a barrier to market entrance[41] and also to the development of new and better foods and more efficient production processes. Moreover, it prevents the improvement of plant materials and animal stock and so contributes to food losses in upper stages of food supply chains and food waste in the post-harvest stages.[42]

All in all, improvements of crops and products made thereof may contribute to better utilizing available resources. However, the more GM products circulate in international transport, the higher the probability that a contamination of non-gmo food or feed will take place. Contaminations will induce food losses at EU borders and on the EU market.

24.2.3 Food Information Requirements

In determining whether a food is unsafe, not only is its physical condition important but also its intended use. A human's physical condition may pose restrictions on food consumption. The information on food packaging provides clues on whether a food is safe to consume for certain individuals. Regional differences in mandatory food information may lead to barriers in international trade and induce border rejections. Comparing the European food information system with the USA and the Codex Alimentarius remarkable differences can be observed, which possibly lead to the removal of potential foods from supply chains and markets.

Products that carry invalid labels are taken off the market in the USA, where such products are considered misbranded, as well as in the EU, where wrong food information is considered misleading for the consumer. These foods are considered

---

[40]This can be exemplified with the long-grain rice dispute between the USA and the EU, the EU closing its borders for unauthorised genetically modified rice (2006 and beyond).

[41]Grossman (2009), pp. 257–304.

[42]Examples are the eradication of oomycetes (causing among others 'potato disease') and the improvement of the resilience of field crops against natural impact.

adulterated or regarded as genuinely unsafe. Unsafe foods may not be brought to the US or EU market.[43]

24.2.3.1 Allergens

Wrong or absent information of allergenic substances in foods is a major cause for considering a food unsafe. The Codex Alimentarius[44] requires revealing nine categories of allergens on the package or container of food. In the EU, the Food Information Regulation 1169/2011 in Annex II[45] requires highlighting the ingredient list for 14 categories. Beyond this extended list, industry standards may further add to regulation; for instance obliging to indicate possible cross-contamination during processing. In any case, foods with wrong or lacking allergen information may not enter or will be taken off the market. To reduce food-related risks, the industry may discard foods early depending on the probability of being held responsible or liable (see below).

24.2.3.2 Nutrition Labelling

Nutrition labelling requirements differ considerably across regions. It may be voluntary in one region and obligatory in another. Until 2016, nutrition labelling remains voluntary in the EU, while it has already been mandatory in the USA for years, as well as in a considerable number of other countries.[46] Comparing the USA with the EU, differences will remain even after 2016 as to fiber, cholesterol, the format and the scope. Such differences induce uncertainty for exporters and lead to refused imports at border posts. Consequently, this may induce destruction of foods if repatriation is too costly.

24.2.3.3 Expiration Dates

With the expression "expiry dates"[47] or "expiration dates," the indications on the package are meant as to the period in which consumption should preferably take

---

[43] Article 14 General Food Law, 178/2002.
[44] Codex General Standard for the Labelling of Pre-packaged Foods; Codex Stan 1-1985 (Rev. 1-1991), 4.2.1.4.: Cereals containing gluten; i.e., wheat, rye, barley, oats, spelt or their hybridized strains and products of these; crustaceans and products of these; Eggs and egg products; Fish and fish products; Peanuts, soybeans and products of these; Milk and milk products (lactose included); Tree nuts and nut products; and Sulphite in concentrations of 10 mg/kg or more.
[45] The European Parliament and of the Council (2011). Regulation (EU) 1169/2011 on the provision of food information to consumers.
[46] Hawkes (2004).
[47] Soethoudt et al. (2012), section 3.1, pp. 7–15.

place.[48] In the EU, it is mandatory to provide either a use-by date or a best-before date on pre-packaged foods.[49] Food has to be safe, and if the absence of expiry information puts the consumer at risk, a firm will be held responsible for the consequences in the EU as well as the USA.

The date of minimum durability, or the best before-date, is the date until which food remains marketable and retains specific characteristics such as taste, colour and nutritional properties when properly stored.[50] It is perceived as a quality indication, as it informs the buyer about the period in which they would maximally benefit from the product. The use-by date is a safety limit. Beyond the indicated time, a food is legally considered unsafe.[51] Surprisingly[52] in EU law, the two types of dates are alternative mandatory indications.[53]

The EU Regulation 1169/2011 regulating food information to consumers (FIC) is clear about the fact that the food business operator under whose name a food appears on the shelves[54] is the primary responsible actor for food information.[55] The manufacturer and/or the retailer that markets a food under a private label may experience uncertainty as to the appropriateness of the labelled expiry date. Uncertainties as to the durability of foodstuffs are a cause for food waste. Food waste causes follow three routes[56]: Businesses may avoid potential risks including private liability, public fines and/or scrutiny by setting and estimating expiry dates conservatively; Consumers may discard food early as they interpret expiry dates wrongly[57]; Public authorities contribute to uncertainties by requiring a choice between the use-by-date and the best-before-date as EU law exemplifies.

The latter obliges to make a wicked choice between the information about food safety and food quality. The business operator, most probably the manufacturer of

---

[48]Next to this, in certain cases EU law obliges to indicate the date of freezing, while in other jurisdictions (as delineated in the Codex) also a date of manufacture, of packaging and/or the sell-by-date may have to be indicated.

[49]In the EU, some exemptions apply; *See* Annex X to Regulation (EU) 1169/2011.

[50]Codex Stan 1-1985 (rev. 1-1991) point 2; similar in Article 2 (2) (r) of Regulation (EU) 1169/2011.

[51]In accordance with Article 14(2)-14(5) of Regulation 178/2002.

[52]As 'quality' and 'safety' are completely different categories and almost any food (with some exemptions, like sodium) becomes unsafe in due time.

[53]Article 24 of the FIC concerns the minimum durability date, 'use by' date and date of freezing. Under par. 1 it is stated: In the case of foods which, from a microbiological point of view, are highly perishable and are therefore likely after a short period to constitute an immediate danger to human health, the date of minimum durability shall be *replaced* by the 'use by' date (italics by authors).

[54]If the operator is not established in the Union, the importer into the Union market is responsible. 'Food business operator' has to be interpreted broadly, as defined in the General Food Law (EC) 178/2002. 'Food business' means any undertaking, whether for profit or not and whether public or private, carrying out any of the activities related to any stage of production, processing and distribution of food.

[55]Article 8 (1) of the FIC.

[56]Waarts et al. (2011).

[57]*See* Waarts et al. (2011) and Soethoudt et al. (2012) Annex 1 and section 3.1, respectively.

the product, which also packs it, not only has to choose the kind of date indication, but also the expiry time. They may be held liable if wrong dating induces damages to the consumer. In many jurisdictions the retailer of a foodstuff, next to the manufacturer, will be held liable in case of damages.

Setting expiry dates is a managerial act. Monetary considerations play a role as well. Liability-related and market-related factors influence responsibility. Short use-by dates reduce the probability of occurrence and effects of an incident, but also make products less valuable to market, as production batches are smaller and the shop display-time is shorter. While general court cases are limited in number,[58] collective non-legal actions, like naming and shaming, may negatively impact the reputation of the company and therefore indirectly bring about financial losses. The fear for loss of image is an important factor inducing conservative expiry dates and thus the early discarding of food.

On another note, conservative estimates of the date of minimum durability may lead to food waste, as wrong dates give ground for consumers to seek legal action. For example, using a best-before date where a use-by indication would have been appropriate. A product with a wrong date may be considered defective. In that case, product liability applies. Public authorities may impose penalties on improper or fraudulent acts. Strangely, despite the fact that most foods will perish gradually no matter the expiry date they carry, with a best-before date the responsibility to assess a food's "fitness for consumption" is transferred from the producer to the consumer. However, whoever carries the damages, the inherent uncertainties as to the microbiological status of foods will induce conservative strategies in marketing and/or consumption. Especially food products which by very nature have credence attributes[59] are subject to exaggerated risk avoidance.[60] A significant portion of food waste in western countries can be traced back to legal uncertainties as to the consequences of consuming foods that might be dangerous to health.

24.2.3.4 Claims

In accordance with the Codex in the Nutrition and Health Claims Regulation[61] (NHCR) of the EU, a "claim" means "any message or representation, which is not mandatory under [Union] or national legislation, including pictorial, graphic or symbolic representation, in any form, which states, suggests or implies that a food has particular characteristics." Just like in the USA,[62] a claim has a broad scope and encompasses information transferred via brand names, advertisements or online

---

[58]Holt (2008), pp. 1–20.
[59]Darby and Karni (1973), p. 69.
[60]Otsuki et al. (2001), p. 495.
[61]The European Parliament and of the Council Regulation (EC) 1924/2006 on nutrition and health claims made on foods.
[62]As regulated in the Nutrition Labelling and Education Act (1990).

webpages.[63] In the EU, all claims have to be authorized in advance before they can be used by businesses for marketing purposes. In the USA authorization is limited to health claims, i.e., disease risk reduction claims. In the EU, strict requirements, including substantiation with generally accepted scientific evidence,[64] have been put in place. The scientific assessment is carried out by an independent and centralized scientific entity (EFSA). In the USA, the FDA carries out this task.[65]

Businesses may easily make mistakes in applications for authorization of claims and, from the perspective of food waste, in the use of authorized claims for their marketed foods. Sources for errors with respect to claims are, for instance:

- Authorization has not been granted yet;
- Presenting legitimate claims in a way that is not allowed (for instance a suggestion is made that a disease is cured, while the claim has the reduction of a disease risk as an objective);
- Stating claims without the provision of the necessary additional information (for instance a disclaimer (USA), or information on the intake that is necessary to realise the promised effect);
- An allowed claim is used for a product that contains vitamins, minerals or other substances in amounts that cannot bring about the promised effect (Article 5 of the NHCR).

Where the requirement for authorization in the EU is "generally accepted scientific evidence"[66] the analog condition for authorization of a health claim in the USA is "significant scientific agreement" among experts. This more lenient position is further softened by US case law.[67] As a protection to the fundamental freedom of speech it is prohibited for the FDA to ban a potentially misleading health claim altogether if a disclaimer can compensate for the potentially misleading character of the claim.[68] The case law triggered a policy change of the FDA, considering credible evidence and a disclaimer sufficient for legitimate use of what we know now as a "qualified health claim." The risk of unlawful use of a claim may be bigger in the EU compared to the USA, but misrepresentations in both regions have as a consequence that a product should be taken off the market immediately, or even may not enter the market on beforehand, for instance is rejected at the border. Since repackaging or repatriation of products may be costly, wrong use of claims contributes to food waste.

---

[63]USFDA (2012).

[64]Article 6 of the NHCR.

[65]This is the European Food Safety Authority (EFSA), the tasks and competences of which have been delineated in the GFL.

[66]Article 6(1) of the NHCR.

[67]Pearson v Shalala, 164 F.3d 650.

[68]Lefevre ed. (2009) Appendix 5.

24.2.4 Phytosanitary Measures: Plant Pests and Pest Control

Anyone who has passed the Australian or US borders from Europe has possibly experienced the removal of foods to protect domestic plants, livestock and wildlife from being infected with alien organisms. Phytosanitary measures are taken to avoid plant diseases affecting crops or the local ecology. The growth of international trade and traffic increases the risk of spread of foreign hazardous species.[69] Plant pests are often hidden attributes, of which the seller may be better informed than the buyer. Information asymmetry and high transactional costs for bridging differences in levels of awareness[70] between public authorities, buyers and sellers of plants have stimulated the development of a strict top-down governance system.[71] However, the costs may outweigh the benefits of improved controls in terms of waste of potential foods and regulatory burdens.

The present plant protection regime, of which the foundation is laid down in the International Plant Protection Convention (IPPC), leaves little room for giving an alternative destination for infected plant material. It requires chain orchestration, as well as information-technological innovations, to create more transparency[72] and in doing so, to avoid destruction of potential foodstuffs. Improved technology such as tracking and tracing of individual plants or plant material, storage of information in ERP-systems, intelligent packaging and improved DNA-identification methods can improve the detection of plant pests. Then there would be potentially no need to destruct healthy plants, herbs, spices and other plant products, as a response to uncertainties with respect to the existence of plant pests.

24.2.5 Hazardous By-Products

By-products are foods or remnants thereof for which the production process is not intentionally set up, but which remain after processing or consumption of food. The recycling of by-products, especially of animal origin, is strictly regulated in the EU and elsewhere. The spark for this was the Bovine Spongiform Encephalopathy (BSE) crisis. Until the end of the last century, it was common to recycle animal by-products, such as sheep bones, swill, or brain tissue, into ruminant feed.[73] In the EU, business practice was governed by imposing restrictions on the reuse of animal protein, the most memorable of which is a ban of reuse of protein from ruminants in feed for animals.[74] Article 7(1) of Regulation 999/2001 establishes a prohibition of

---

[69]Dehnen-Schmutz et al. (2007), pp. 527–534.

[70]Lansink (2011), pp. 166–170.

[71]Henson and Traill (1993), pp. 152–162.

[72]van der Vorst et al. (2007), Hisiao et al. (2006).

[73]van Raamsdonk (n.d.).

[74]A prohibition of 'cannibalism'; see Article 11 of Regulation 1069/2009.

the feeding to ruminants of proteins derived from animals. Annex IV to this regulation aggravates the requirements, such as to feeding tri-calcium phosphate of animal origin to ruminants, or alleviates them, for instance with respect to fish feed used for farming non-ruminants. Specified risk-material of animal origin, like brain, eyes and spinal cord of bovine animals older than 12 months, originating from a country or region with possible BSE-risk must be removed in a prescribed way[75] and further kept out of the feed chain. Regulation 1069/2009[76] mentions categories of animal by-products that may not be consumed by humans due to control BSE-related risks and also because of dangerous levels of contaminants like dioxin or pesticide residues. It was not intended to expel all animal by-products from supply chains, but to guarantee a safe use in chemical, pharmaceutical, or feed applications and for the natural environment for humans and animals. In the EU, such waste streams are classified according to their risk profile. On the basis of this classification, some forms of food waste, like from catering services or households, may not be recycled as feed since it contains a mixture of hazardous and safe material. In the end, separating hazardous from safe material is a technological and behavioural problem. Solving this problem would reduce waste streams.

24.3 Legal-Economic Analysis

To keep a "license to produce," business operators will possibly show prudence to a level that surpasses publicly set precautionary measures. Uncertainty about the probability and legal consequences of bringing unsafe food to the market may induce the creation of multiple margins of safety at the importer's and/or farmer's level, the level of the manufacture, as well as at the level of the retailer and/or household. Figure 24.1 supports this argument.

Referring to Fig. 24.1, we will provide arguments that causes for food waste origin from excessive private and public safety margins, as a response to uncertainties on the required levels of food safety. We view food safety as a dimension of food quality. Costs of quality consist of costs of controls (prevention and appraisal costs) and failure costs (costs of non-compliance).[77]

Internal failure costs[78] as a result of product checks and other quality controls, can be ranked under prevention costs as they are intended to protect markets and consumers against foodborne diseases.[79] It is also realistic to suppose that in order

---

[75] Article 8 of this Regulation, Annex V.

[76] The European Parliament and of the Council Regulation (EC) 1069/2009 laying down health rules as regards animal by-products and derived products not intended for human consumption.

[77] See Feigenbaum (1991).

[78] Gyrna (1999), Williams et al. (1999).

[79] Tobers et al. (1996), pp. 1297–1301. It should be mentioned at this point that not only extra costs are made in manufacturing firms, but also for instance by retailers which lower the shelf time of products to assure safety (see Sect. 24.2).

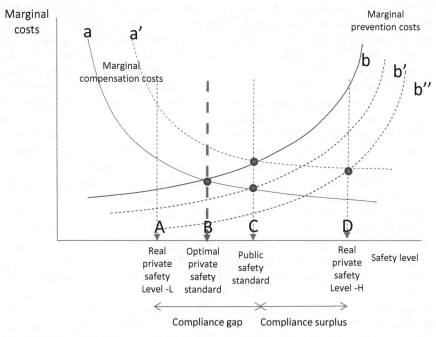

Fig. 24.1 The trade-of between marginal costs of prevention of safety risks and compensation once safety risks have manifested themselves in damages. Extra prevention costs (improved safety measures) induce lower levels of compensation (like financial compensation for injuries). *Dotted curves* represent higher marginal compensation costs (curve a′) and higher marginal prevention costs (curve b′ or b″) respectively, compared to the initial optimal level of aspired safety (B)

to reduce risks and hazards to higher levels, these costs increase progressively, as marginal improvements of food safety will be harder to realize at higher levels of food safety.

Suppose that private law instruments can successfully be used by consumers to seek compensation of damages due to unsafe food,[80] with the absence of fines for trespassing public safety standards. In that case, the optimal level of food safety that businesses aspire is where marginal private compensation costs (curve a) equal marginal private prevention costs (curve b); that is the level depicted as B. Total marginal prevention costs are extra internal failure costs, including waste after inspection, testing and other quality controls, and other future-oriented prevention costs, like innovation in better products and process improvements, or the installment of an ISO-system.[81]

---

[80]If consumers would have to bear damages themselves they would over-exaggerate the risk and therefore apply margins of safety leading to early discard or recycling of foodstuffs. The social effects would be similar to the ones here elaborated.

[81]International Standards Organization system.

In the short run, due to pressures from non-commercial stakeholder groups and as a result of policy priorities, public safety standards may be stricter than private commercial stakeholders would strive for using economic measures (level C). Attained safety levels might even surpass public standards, depending on the level of risk averseness of a business operator (level D). For instance, William Edwards Deming, the founder of the plan-do-check-act cycle and one of the initiators of total quality management.[82] Unintentionally, with his emphasis on superior quality as the dominant strategic goal, this may have induced business attitudes to reject products with minor deficiencies rather than bringing them to market at lower prices. Moreover, loss of image and/or brand value may be stronger motivations for risk avoidance than the fear of liability. Risk avoidance might boost aspired standards to higher levels than publicly required.[83] In many cases private standards for food quality and safety, like GlobalGAP,[84] may be stricter than, compete with, or even replace publicly established norms. The legitimacy of such non-governmental contracts may be questionable,[85] due to the limited possibilities of businesses to enter the system or opt out after committing to it.[86] It induces cautious behaviour and thus discarding of food products that do not completely meet customer-designed delivery and process specifications. For instance, the detection of hormone use in beef production in the EU[87] may not only lead to refusal of delivered animals by slaughterhouses, but also of future deliveries. Detection could thus result in the loss of a "license to deliver." Norms may vary across nations and sectors. For instance, in a similar situation, hormone use is considered safe and acceptable in the USA.[88]

Compliance surpluses due to risk avoidance will lead to premature removals of foods from supply chains. Compliance surpluses are not only induced by mandatory legal requirements, but also by private food quality standards. In the assessment of whether an inferior food should be transported back to its origin, repacked on the spot or degraded to a second-best destination. The transportation and operational costs of other options may be decisive.

Figure 24.1 indicates—with a shift of the curve (a) to 'a + marginal fines' (a')—that the threat of public fines in case of violation of safety standards could induce businesses to comply in any case. The risk of being fined can induce extra internal failure costs in terms of wasted food (i.e., point C in Fig. 24.1). It also shows that

---

[82]Hackman and Wageman (1995), pp. 309–342.

[83]Holt (2008), pp. 1–20.

[84]Global G.A.P. (2014).

[85]Busch (2011), pp. 51–75; Fuchs et al. (2011), pp. 353–367; Garcia-Martinez, et al. (2007), pp. 299–314; Havinga (2006), pp. 515–533.

[86]For instance as a result of asset-specific investments or binding delivery contracts.

[87]In the Netherlands this is governed by the semi-public IKB-scheme (Integrated Chain Governance).

[88]McNeil (1998–1999), pp. 90–112. This has led to international trade conflicts, as a ban of added hormones in specifically beef may be considered a barrier to trade.

regulatory threats of increased costs and responsive action in combination with structural preventive measures (like product or process innovation) may induce higher safety levels and reduce the failure costs in future (point D in Fig. 24.1).

The analysis gains further complexity if the expected costs show a probability distribution, or are uncertain altogether. In that case, risk may be over-emphasized even stronger.[89] For small and medium sized businesses (SME) relying on a limited product portfolio and scarcity in resources, it is not feasible to reduce waste through innovation. Worldwide, the majority of food and drink businesses are SME, that is with less than 250 employees according to European standards. These might keep extra margins of safety to meet the uncertainties vested in the limited predictability of risks, which they cannot reduce with technical measures or investments. The "cognitive closeness" of a potential hazard, because of a recent food safety crises, may further provoke economically unjustified risk controls by private and public actors. European experience shows that risk managers defend such measures with reference to the precautionary principle.[90]

24.4 Conclusion

Food waste has many origins, one of which is food law. Western countries tend to overemphasize the quest for safety and avoidance of risks,[91] but the exaggerated quest for safety comes at a price. One of the main drivers for exaggeration of risks is behavioural uncertainty. Risk perception propellers are connected to each of the legal sources of food waste.

One important factor is zero-tolerance with respect to contaminants and residues. Technological progress will improve the detection methods to a level that prohibited content is signalled immediately. In earlier times, when zero-tolerance was introduced, detection of minor remnants was difficult, if not impossible. Zero-tolerance may induce safety measures ad absurdum, to the expense of the consumer who will ultimately pay higher prices.

Another factor is the occurrence of by-products and limitations as to their re- or alternative use. No one will object to measures that protect the consumer against diseases like Kreuzfeldt Jacob. However, the number of lethal cases in humans has

---

[89]Kasperson et al. (1988), pp. 178–187. See Slovic's and others' concept of "social amplification of risk". Kahnemann (2011); Wiener and Rogers (2002), p. 328. In general and on an individual level, humans value potential losses higher than the same potential gains.

[90]Gollier et al. (2000), pp. 229–253. Precaution has found its place more in both jurisdictions, EU and American. Influences can be traced back to the Conference in Rio the Janeiro, the Rio Declaration: "where there are threats of serious and irreversible damage, lack of full scientific certainty shall not be used as a reason for postponing cost-effective measures to prevent environmental degradation". In EU law (Article 7 of the GFL), cost-effectiveness has been supplemented by proportionality.

[91]Walker Wilson (2011), p. 120.

been surprisingly low. The fear for catching a disease with a probability that is a fraction of a chance of dying from pneumonia may not be sufficient to stand in the way of legitimizing the recycling of food waste. For instance, restaurant waste and household garbage may be split into non-risk and risk material. It requires technological and social innovation in combination with improved waste management such as differentiation of waste streams in plant and animal waste, to extract valuable resources.

Next, the structural ban from the market of novel foods, including GM-foods, unless authorized or substantially equivalent to existing products, may lead to wasting valuable products, for instance after border controls. In some cases of contamination, like mycotoxin-infected nuts, it is a matter of applying existing processing techniques to avoid food waste or transform such products into feed. Additionally, food information requirements, like expiry dates, induce food waste. Neither the non-existence of a public information requirement, like in the USA, nor a mandatory choice between safety and quality, like in the EU, is efficient since uncertainty of the manufacturer, the retailer or the consumer are boosted.

The actors that cause most wasted food are consumers. A reason is the perceived uncertainty and lack of trust in the safety of the products they buy. Nowadays, extensive capabilities exist to detect and control risks. Food waste can be reduced by using newly developed techniques to signal decay, like intelligent packages. These can warn users that the content of a pre-packaged product is unsafe. But what if they exaggerate risks despite reliable information that is given? May we nudge consumers' behavior so that they actually consume what they buy? May we restrict their free choice to the benefit of the system as a whole? These considerations take us away from a legal-systematic review (Sect. 24.2) and an economic analysis (Sect. 24.3) and bring us to the moral level of food waste prevention. That is, the level at which a choice is due between the pursuit of competing aims: personal freedom, efficiency and sustainability.

References

Bennett JW, Klich M (2003) Mycotoxins. Clin Microbiol 16(3):497–516
Busch L (2011) Quasi states? The unexpected rise of private food law. In: van der Meulen B (ed) Private food law. Wageningen Academic Publishers, Wageningen, pp 51–75
Clapp J (2005) The political economy of food aid in an era of agricultural biotechnology. Glob Gov 11:467–485
Council regulation (EEC) No 315/93 of 8 February 1993 laying down Community Procedures for contaminants in food
Darby M, Karni E (1973) Free competition and the optimal amount of fraud. J Law Econ 16 (1):67–88 at p. 69
Dehnen-Schmutz K, Touza J, Perrigs C, Williamson M (2007) A century of the ornamental plant trade and its impact on invasion success. Divers Distrib 13(5):527–534
European Commission DG Joint Research Centre (EUR 22284EN) (2010) Environmental Impact of Products (EIPRO), 93
European Commission, Communication from the Commission "Europe 2020". Brussels, 2010

FAO (2010) Food loss prevention in perishable crops. FAO Agricultural Bulletin 43
Feigenbaum A (1991) Total quality control. McGraw Hill, New York
Fischhoff B, Slovic P, Lichenstein S (1978) How safe is safe enough? A psychometric study of attitudes towards technological risks and benefits. Policy Sci 9:127–152
Fuchs D, Kalfagianniand A, Havinga T (2011) Actors in private food governance: the legitimacy of retail standards and multi-stakeholder initiatives with civil society participation. Agric Hum Values 28:353–367
Future Directions International. Strategic Analysis Paper. 2013. http://www.futuredirections.org.au/files (accessed Aug 2014)
Garcia-Martinez M, Fearne J, Caswell JA, Henson S (2007) Co-regulation as a possible model for food safety governance: opportunities for public-private partnerships. Food Policy 32(3):299–314
Global G.A.P. (2014) http://www.globalgap.org/uk_en (accessed July 2014)
Gollier C, Jullien B, Treich N (2000) Scientific progress and irreversibility: an economic interpretation of the 'precautionary principle'. J Public Econ 75:229–253
Grossman M (2009) Protecting health, environment and agriculture: authorisation of genetically modified crops and food in the United States and the European Union. Deakin Law Rev 14(2):257–304
Guidance Document of the European Commission for Competent Authorities for the Control of Compliance with EU Legislation of Aflatoxin. 2010
Gustavsson J, Cederberg C, Sonesson U (2011) Global food losses and food waste – extent, causes and prevention. Save Food Congress, Dusseldorf (G)
Gyrna FM (1999) Quality and costs. In: Juran JM, Godfrey AB (eds) Quality handbook. McGraw Hill, New York, section 8.5
Hackman JR, Wageman R (1995) Total quality management: empirical, conceptual, and practical issues. Adm Sci Q 40(2):309–342
Hall KD, Guo J, Dore M, Chow C (2009) The progressive increase of food waste in America and its environmental impact. PLoS One 4(11):e7940, Accessed July 2014
Havinga T (2006) Private regulation of food safety by supermarkets. Law and Policy 28(4):515–533
Hawkes C (2004) Nutrition labels and health claims: the global regulatory environment. WHO, Geneva
Henson S, Traill B (1993) The demand for food safety: market imperfections and the role of government. Food Policy 18(2):152–162
Hisiao HI, van der Vorst J, Omta SWF (2006) Logistics outsourcing in food supply chain networks: theory and practices. In: Bijman J (ed) International agri-food chains and networks: management and organisation. Wageningen Academic Publishers, Wageningen, pp 135–150
Holt A (2008) Alternative liability theory: solving the mystery of who dunnit in foodborne illness cases. J Environ Public Health 1:1–20
Hospes O (2008) Overcoming barriers to the implementation of the right to food. Eur Food Feed Law Rev 4:246–263
Human Consumption to a Significant Degree. http://ec.europa.eu/food/food/biotechnology/novelfood/documents/substantial_equivalence_en.pdf (accessed Sept 2014)
Kahnemann D (2011) Thinking fast and slow. Farrer, Straus & Giroux, New York
Kasperson RE et al (1988) The social amplification of risk – a conceptual framework. Risk Anal 8(2):178–187
Knowles T, Moody R (2007) European food scares and their impact on EU food policy. Br Food J 109(1):43–67
Kummu M (2012) Lost food, wasted resources: global food supply chain losses and their impacts on freshwater, cropland, and fertilizer use. Sci Total Environ 438:477–489
Lansink AO (2011) Public and private roles in plant health management. Food Policy 36(2):166–170

Lefevre EC (2009) Food labelling – the FDA's role in the selection of healthy foods. Nova Publishers, New York

Lelieveld H, Kenner L (2007) Global harmonization of food regulations and legislation: the global harmonization initiative. Trends Food Sci Technol 18:S15–S19

McNeil DE (1998–1999) First case under WTO's sanitary and phytosanitary agreement: the European Union's hormone ban. Va J Int Law 39:90–112

Otsuki T, Wilson JS, Sewadeg M (2001) Saving two in a billion: quantifying the trade effect of European food safety standards on African exports. Food Policy 26:495–514

Parfitt J, Barthel M, McNaughton S (2010) Food waste within food supply chains: quantification and potential for change to 2050. Philos Trans R Soc 363:3065–3081

Petrum EL, Sellnow TL (2010) China's response to the melamine crisis: a case study in actional legitimacy. College of Communications and Information Studies, Lexington

Schoss S (2011) Food safety regulation and border rejections: what is the impact on food waste? Wageningen University, Wageningen

Skaburskis A (2008) The origin of "wicked problem". Plann Theory Pract 9(2):277–280

Soethoudt H, van der Sluis A, Waarts Y, Tromp S (2012) Expiry date: spilled effort?, Wageningen UR Food & Bio-based Research, pp 7–15, section 3.1 (Law and Guidances of Expiry Dates, Responsibilities at Surpassing Expiry Dates (by Harry Bremmers and Bernd van der Meulen)

Tobers T, Buzby JC, Ollinger M (1996) Using benefit and cost information to evaluate a food safety regulation. Am J Agric Econ 78:1297–1301

UN News Centre (2014) UN News Centre. http://www.un.org/apps/news/story.asp?NewsID=45165#.U9IPkPmSx8E (accessed July 2014)

USFDA (2012) Inspections, compliance, enforcement, and criminal investigations. http://www.fda.gov/ICECI/EnforcementActions/WarningLetters/2012/ucm340266.htm (accessed 2014)

Van Boxstael S, Jacxsens L, Uytendaele M (2013) Fresh produce rejections at EU border inspection posts. New Food Magazine 5. http://www.newfoodmagazine.com/12322/new-food-magazine/past-issues/issue-5-2013/fresh-produce-rejections-at-eu-border-inspection-posts/. Accessed Sept 2014

van der Meulen B (2004) The right to adequate food; food law between market and human rights. Elsevier Juridisch

van der Meulen B (2009) Reconciling food competitiveness. Wageningen Academic Publishers, Wageningen

van der Meulen B, Bremmers H (2006) The overhaul of EU food law as perceived by industry. In: Wijnands JHM, van der Meulen BMJ, Poppe KJ (eds) Competitiveness of the European food industry, LEI, The Hague, pp 74–110, Research report, Ch. 3

van der Meulen B, Bremmers H, Wijnands J, Poppe K (2012) Structural precaution: the application of premarket approval schemes in EU food legislation. Food Drug Law J 67(4):453–473

van der Meulen B, Bremmers H, Purnhagen K, Gupta N, Bouwmeester H, Geyer L (2014) Governing nano foods: principles-based responsive regulation. Elsevier Academic Press, Oxford

van der Vorst J, Duineveld MPJ, Scheer FP, Beulens AJM (2007) Towards logistics orchestration in the pot plant supply chain network. In: 14th annual international Euroma conference, Ankara

van Raamsdonk T (n.d.) ARIES: an expert system supporting legislative tasks: identifying materials using Linnaeus II software. In: Tools for identifying biodiversity: progress and problems, pp 145–150

Vogdlander J, Brezet H, Hendriks C (2001) Allocation in recycling systems – an integrated model for the analyses of environmental impact and market value. Int J Life Cycle Assess 6(6):344–355

Waarts Y, Eppink M, Oosterkamp E, Hiller S, van der Sluis A, Timmermans T (2011) Reduction of food waste: perceived barriers in legislation. LEI

Walker Wilson MJ (2011) Cultural understandings of risk and the tyranny of the experts. Oregon Law Rev 19:113–187

Wernaart B (2010) The plural wells of the right to food. In: Hospes O, Hadiprayitno I (eds) Governing food security: law, politics and the right to food. Wageningen Academic Publishers, The Netherlands, pp 43–80

Whitworth J (2014) Pathogenic microorganism notifications jump to an all-time high. http://www.foodqualitynews.com/Industry-news/RASFF-annual-report-reveals-EU-trends (accessed Sept 2014)

Wiener JB, Rogers MD (2002) Comparing precautions in the United States and Europe. J Risk Res 5(4):317–349

Williams ART, van der Wiele A, Dale BG (1999) Quality costing: a management review. Int J Manag Rev 1(4):441–446

Chapter 25
The Concepts of Transparency and Openness in European Food Law

Nicola Conte-Salinas and Rochus Wallau

Abstract This chapter focuses on the principles of transparency and openness, which are of increasing importance both at EU level and in national law and politics. It gives an overview of the areas of EU law that explicitly provide for transparency and openness, illustrated by case-law. In a second step, it explains how the principles of transparency and openness have been implemented in EU food law. Current transparency projects in Denmark, the UK and Germany are discussed in detail, in particular the practice of publishing information concerning food business operators, which is critically examined.

25.1 Introduction: The Concept of Transparency in EU Law

"With knowledge doubt increases"[1]—this thought-provoking quote from Johann Wolfgang von Goethe, one of the famous poets and authors in classic German literature, is incompatible with one of the most powerful current developments in European food law—the implementation of transparency. While the poet assumed that "more" knowledge would not lead to "more" orientation and security, the opposite is true today—even in the "post-wikileaks era,"[2] the possibility of information is used as a prescription against uncertainty and disorientation in view of the complexity of life.

---

[1] Johann Wolfgang von Goethe, Maximen und Reflexionen. The full quotation reads: "Eigentlich weiß man nur wenn man wenig weiß. Mit dem Wissen wächst der Zweifel." ("We know accurately only when we know little, with knowledge doubt increases."—Translation is ours).

[2] Cf. Compendium "Wikeleaks und die Folgen", edited by von Geiselberger, 2011.

N. Conte-Salinas (✉)
Morsbach, Germany
e-mail: nicola_contesalinas@yahoo.com

R. Wallau
Ingolstadt, Germany
e-mail: rwallau@gmx.de

There is a "trend" in secondary Union law to broaden the scope of rules concerning the publication of information. In particular, information known to authorities or institutes, will not anymore be used exclusively to avert existing dangers, but are increasingly used in situations where no security threats are present. Applied this way, transparency becomes an element of market regulation. Taking into account that the legislative process is characterized by consultation requirements intended to guarantee the involvement of an "organized civil society",[3] it becomes obvious that the concept of transparency encompasses the whole "regulation chain", starting from the legislative process to the implementation of law.

The societal background for the growing use of information can certainly be traced back (at least in part) to a growing mistrust of state institutions. In several areas, state institutions (or those perceived as such) have proven, at least partly, inefficient to prevent or deal with actual risks. There are two specific examples, the finance and food sectors. In her analysis of the social situation around 1870, historian Vera Hierholzer described conditions that led to the existence of national food law in the German Empire (Deutsches Kaiserreich). These findings, however, can be generalized and are still valid today.

> The increased feeling of insecurity was a logical consequence of the efforts for enlightenment. As key formative experiences, scandals attracted attention and seemed to prove that mistrust was justified. Growing knowledge and growing decision-making potentials of modern society did not only promote the feeling of being able to "make" the world, but also the awareness of risks, against which new risky constellations were identified constantly. The establishment of a critical public and an increasingly interconnected information society, reflected in the rapid rise of the free press, further strengthened this development.[4]

Positively phrased, Hierholzer outlines how transparency results in traceability, controllability and perhaps leads to acceptance of political or administrative decisions, thereby making it possible for many to directly participate, including, at any rate, those who have no official role in the decision-making process.

Transparency and openness can be applied to different categories of cases. One category concerns the activity of the EU (or national) institutions (e.g. granting access to documents of Council meetings). The other category, however, deals with the publication of information concerning private parties, for example, food business operators. An example of the latter would be the publication of results of hygiene inspections. Both categories are discussed in this chapter.

---

[3]Cf. European Governance—A White Paper, COM(2001) 428 final, OJ 287, 12.10.2001, p. 1; Communication from the Commission "Towards a reinforced culture of communication and dialogue – General principles and minimum standards for consultation of interested parties by the Commission" COM(2002) 704 final—not published in the Official Journal; and the Green Paper "European Transparency Initiative" COM(2006) 194 final—Official Journal C 151 of 29.6.2006.

[4]Hierholzer (2010) [translation is ours].

25.2 Applicable Laws with Respect to Transparency and Openness

With the aim to broaden the means of citizens' participation within the decision-making process and thereby enhancing legitimacy and accountability, the maxims openness and transparency[5] are laid down at the highest normative level in the EU, that is to say, in the Treaty on the European Union (TEU) and the Treaty on the Functioning of the European Union (TFEU). The principle was first established in Art. 255 (2) of the Treaty establishing the European Community, with its scope further broadened in Art. 15 (3) of the TFEU. Art. 15 (3) of the TFEU addresses all Union organs, in particular, the Commission, Council and Parliament, including their committees and working groups.

In the TEU, openness and transparency are established as democratic principles.[6] The TFEU obliges EU institutions, bodies and agencies to "conduct their work as openly as possible".[7] Therefore, the meetings of European Parliament are public, and so are Council meetings when discussions and voting on draft legislative acts is concerned.[8] EU citizens are granted a general right of access to documents of the EU institutions.[9] Secondary EU law gives shape to these principles of openness and transparency (see the following discussion).

25.2.1 Regulation (EC) No. 1049/2001[10]

Regulation (EC) No 1049/2001 further specifies the general right of access to documents, laid down in EU primary law.

The term access either means on-the-spot consultations, the provision of copies or of electronic versions for the benefit of the applicant. "Accessible" in this sense is any content, irrespective of the medium, concerning a matter relating to the policies, activities and decisions falling within the relevant institution's sphere of responsibility,[11] drawn up or received by the relevant institution and in its possession, in all areas of activity of the European Union.[12]

---

[5]For an analysis concerning the different content of transparency and openness, see Alemanno (2014).
[6]See Art. 1 (2), 10 (3) and Art. 11 TEU. Art. 10 and 11 are located under Title II—which is headed "provisions on democratic principles".
[7]Art. 15 (1) TFEU.
[8]Art. 15 (2) TFEU.
[9]Art. 15 (3) TFEU.
[10]Regulation (EC) No 1049/2001 of the European Parliament and of the Council of 30 May 2001 regarding public access to European Parliament, Council and Commission documents, Official Journal L 145, 31.05.2001 p. 43, as amended.
[11]Cf. Regulation (EC) No 1049/2001, *op. cit.*, Art 3 (a).
[12]Cf. Regulation (EC) No 1049/2001, *op. cit.*, Art 2 (3).

Recital 4 of said Regulation aims at a wide claim to access documents. According to the case law of the CJEU, it is therefore necessary to interpret and apply rules limiting this access strictly.[13] Exceptions to the right of access are laid down in Art. 4 Regulation (EC) 1049/2001: public interests of importance (e.g., as regards public security, defense and military matters), or international relations, the financial, monetary or economic policy of the Community or a Member State can give rise to a refusal of access, just as well as private interests of importance (e.g., privacy and the integrity of the individual, in particular, in accordance with Community legislation regarding the protection of personal data).[14]

The protection of commercial interests of a natural or legal person, including intellectual property, court proceedings and legal advice, as well as the purpose of inspections, investigations and audits lead to a refusal of access, unless there is an overriding public interest in disclosure.[15] In case of an overriding public interest, access is granted even to documents drawn up by an institution for internal use only or received by an institution, which relates to an ongoing matter where a decision has not yet been taken, and where disclosure of the document would seriously undermine the institution's decision-making process.[16] It is essential to note that the restriction of access to documents might only apply to a part or certain parts of a requested document,[17] and that a refusal of access can only be upheld for as long as the exception is justified, and generally for a maximum period of 30 years (with a counter exception for sensitive documents, protection of private and/or commercial interests).[18]

The application can be submitted in writing or electronic form. While it is not necessary to state reasons in the application for access to documents, it is, however, vital to be sufficiently precise in order to identify the relevant documents. The relevant organ generally has to respond to the application within 15 days, either by granting access or explaining the reasons for a full, or partial, refusal.[19]

A consultation on the spot is free of charge, just as well as the direct access in electronic form or through the register of documents and that the organs make available to the public.[20]

The general right of access to documents applies also to areas of law that have sector-specific rules, as long as these sector-specific rules do not provide for special rules concerning the access to European Parliament, Council and Commission documents.

---

[13]CJEU, 11.12.2008, C-524/07, 2008 I-00187.
[14]Regulation (EC) No 1049/2001, *op. cit.*, Art. 4 (1).
[15]Regulation (EC) No 1049/2001, *op. cit.*, Art. 4 (2).
[16]Regulation (EC) No 1049/2001, *op. cit.*, Art. 4 (3).
[17]Regulation (EC) No 1049/2001, *op. cit.*, Art. 4 (6).
[18]Regulation (EC) No 1049/2001, *op. cit.*, Art. 4 (7).
[19]Regulation (EC) No 1049/2001, *op. cit.*, Art. 7 (1).
[20]According to Art. 10 and 11 of Regulation (EC) 1049/2001, *op.cit.*, Art. 10 (1).

25.2.2 Sector-Specific Rules

There are many sector-specific rules that concern informing the public or public access to information. These can be divided into two groups: Environmental law, and other sector-specific rules that generally require risks or potential hazards, before information can or must be made public.

In this respect, Environmental law contains the most far-reaching rules. The rules foresee a right of access irrespective of potential hazards or any other risks involved. The concept behind those rules could be summarized as follows: By granting the citizen the possibility to access information, the citizen becomes—more or less—part of the monitoring system.[21]

As mentioned, other sector-specific rules generally require risks or potential hazards, before information can or must be made public. These rules tend to give (fundamental) rights of the people and companies concerned, more prominence.

25.2.3 Environmental Law

Environmental law can be regarded as the "pioneer" area of law with respect to transparency. Directive 2003/4/EC[22] introduces a right to access environmental information held by or for public authorities for everyone, without preconditions.[23] The purpose of this Directive is described as "a greater awareness of environmental matters, a free exchange of views, more effective participation by the public in environmental decision-making and, eventually, to a better environment."[24] Its predecessor, Directive 90/313/EEC of 7 June 1990 on the freedom of access to information on the environment[25] had already "initiated a process of change in the manner in which public authorities approach the issue of openness and

---

[21]Cf. Hansmann-Calliess, Grundzüge des Umweltrechts, 2012, p. 125: "Mit Blick auf das konstatierte Vollzugsdefizit setzen EuGH und Kommission schon lange auf den an der Wahrung seiner Rechte interessierten Bürger, der so zum Wächter der Einhaltung des Europäischen Umweltrechts instrumentalisiert wird." ("With respect to the recognised lack of enforcement, the CJEU and the Commission have long since counted on the citizen interested in protecting his rights, who thereby is used as the watchdog of the compliance with European Environmental Law"—translation is ours).

[22]Directive 2003/4/EC of the European Parliament and of the Council of 28 January 2003 on public access to environmental information and repealing Council Directive 90/313/EEC, OJ L 041, 14.02.2003, p. 26. Regulation (EC) No 1367/2006, of 6 September 2006, on the application of the provisions of the Aarhus Convention on Access to Information, Public Participation in Decision-making and Access to Justice in Environmental Matters to Community institutions and bodies, OJ L 264, 25.9.2006, p. 13 complements Regulation (EC) No 1049/2001.

[23]Directive 2003/4/EC, *op. cit.*, Art. 3 (1).

[24]Directive 2003/4/EC, *op. cit.*, recital 1.

[25]OJ L 158, 23.6.1990, p. 56.

transparency, establishing measures for the exercise of the right of public access to environmental information which should be developed and continued".[26]

However, the directive also foresees certain exceptions to the right to access environmental information.[27] These either take account of public (e.g. security interests, court proceedings) or private (e.g. privacy, commercial secrets) interests. These reasons, however, must not be used to refuse a request that relates to information on emissions into the environment.[28]

Member States are required to actively and systematically disseminate environmental information to the public.[29]

25.2.4 Consumer Products

The laws on consumer products do not know such extensive rights and obligations concerning access to information.

Directive 2001/95/EC foresees that "information available to the authorities of the Member States or the Commission relating to risks to consumer health and safety posed by products shall in general be available to the public, in accordance with the requirements of transparency and without prejudice to the restrictions required for monitoring and investigation activities. In particular the public shall have access to information on product identification, the nature of the risk and the measures taken."[30]

At the same time, however, it establishes that "Member States and the Commission shall take the steps necessary to ensure that their officials and agents are required not to disclose information obtained for the purposes of this Directive which, by its nature, is covered by professional secrecy in duly justified cases, except for information relating to the safety properties of products which must be made public if circumstances so require, in order to protect the health and safety of consumers."[31]

Rules with similar tendencies are also found in Regulation (EC) No. 765/2008 on market surveillance,[32] which shall ensure a harmonized level of market surveillance in the area of consumer products in the EU.

---

[26]Directive 2003/4/EC, *op. cit.*, recital 2.

[27]Directive 2003/4/EC, *op. cit.*, Art. 4.

[28]Directive 2003/4/EC, *op. cit.*, Art. 4 (2), subpara 2.

[29]Directive 2003/4/EC, *op. cit.*, Art. 7 (1).

[30]Directive 2001/95/EC of the European Parliament and of the Council of 3 December 2001 on general product safety, OJ L 011, 15.1.2002, p. 4, Art. 16 (1).

[31]Directive 2001/95/EC, *op. cit.*, Art. 16 (1).

[32]Regulation (EC) No 765/2008 of the European Parliament and of the Council of 9 July 2008 setting out the requirements for accreditation and market surveillance relating to the marketing of products and repealing Regulation (EEC) No 339/93, 13.8.2008, OJ L 218, p. 30, Art. 19.

25.2.5 Medicinal Products

Regulation (EC) No 726/2004[33] addresses the right to access documents concerning medicinal products. As a rule, the aforementioned general right of access to documents laid down in Regulation 1049/2001 applies to the European Medicines Agency (EMEA) as well,[34] but the Regulation also lays down sector-specific rules. Inter alia, EMEA is required to set up a database for medicinal products that contains information concerning, among others, assessment reports, summaries of product characteristics, and information on suspected adverse reactions.[35] In particular, assessments concerning scientific questions in connection with the assessment of medicinal products must be published,[36] as well as final decisions taken by EMEA concerning market-related measures.[37] If necessary to protect human health or the environment, a Member State may suspend the use of an authorized medicinal product in its territory, but must ensure that health-care professionals are rapidly informed of its actions and the reasons for the suspension.[38]

25.2.6 Chemicals

Regulation (EC) No. 1907/2006 concerning the registration, evaluation, authorization and restriction of chemicals[39] also addresses access to information concerning chemicals: "EU citizens should have access to information about chemicals to which they may be exposed, in order to allow them to make informed decisions about their use of chemicals."[40] Accordingly, certain information, such as the name of the substance, its classification and labeling, the result of each toxicological and eco-toxicological study, or the guidance on safe use shall be made

---

[33]Regulation (EC) No 726/2004 of the European Parliament and of the Council of 31 March 2004 laying down Community procedures for the authorisation and supervision of medicinal products for human and veterinary use and establishing a European Medicines Agency, Official Journal L 136, 30.04.2004, p. 1.

[34]Regulation (EC) No 726/2004, *op. cit.*, Art. 73.

[35]Regulation (EC) No 726/2004, *op. cit.*, Art. 57.

[36]Regulation (EC) No 726/2004, *op. cit.*, Art. 22.

[37]Regulation (EC) No 726/2004, *op. cit.*, Art. 20 (7).

[38]Regulation (EC) No 726/2004, *op. cit.*, Art. 20 (4), (5).

[39]Regulation (EC) No 1907/2006 of the European Parliament and of the Council of 18 December 2006 concerning the Registration, Evaluation, Authorisation and Restriction of Chemicals (REACH), establishing a European Chemicals Agency, amending Directive 1999/45/EC and repealing Council Regulation (EEC) No 793/93 and Commission Regulation (EC) No 1488/94 as well as Council Directive 76/769/EEC and Commission Directives 91/155/EEC, 93/67/EEC, 93/105/EC and 2000/21/EC, OJ L 396, 30.12.2006, p. 1.

[40]Regulation (EC) No 1907/2006, *op. cit.*, Recital 117.

publicly available, free of charge, over the Internet.[41] Further information, such as the trade name(s), shall also be published, free of charge, over the Internet, except when an objection against this publication (on the grounds that a publication would harm the commercial interests of the registrant or any other party concerned) has been accepted as valid by the European Chemicals Agency.[42]

Besides this right of access to information, the applicability and scope of Regulation (EC) No. 1049/2001 is massively limited in regards to information on chemicals, since it contains the refutable presumption that the right to access certain information (i.e., concerning the full composition of the substance or business relationships between manufacturers and its distributors) undermines the protection of the commercial interests of the person concerned (trade and business secrets).[43] A counter-exception exists, where urgent action is essential to protect human health, safety or the environment.[44]

25.2.7 Case Law

25.2.7.1 Case C-39/05 P [2008][45]

The Court had to decide whether a decision of the Council of the European Union to refuse access to an opinion of the Council's legal service concerning a proposal for a Council directive to a citizen (Mr. Turco), was in accordance with EU law.

The Court of Justice of the European Union (CJEU) stated that:

> Any exception to the right of access to the institutions' documents under Regulation (EC) No 1049/2001 must be interpreted and applied strictly.[46]
>
> ...the exception relating to legal advice laid down in the second indent of Article 4(2) of Regulation No 1049/2001 must be construed as aiming to protect an institution's interest in seeking legal advice and receiving frank, objective and comprehensive advice.
>
> The risk of that interest being undermined must, in order to be capable of being relied on, be reasonably foreseeable and not purely hypothetical.
>
> Third and last, if the Council takes the view that disclosure of a document would undermine the protection of legal advice as defined above, it is incumbent on the Council to ascertain whether there is any overriding public interest justifying disclosure despite the fact that its ability to seek legal advice and receive frank, objective and comprehensive advice would thereby be undermined.[47]

---

[41]Regulation (EC) No 1907/2006, *op. cit.*, Art. 119 (1).
[42]Regulation (EC) No 1907/2006, *op. cit.*, Art. 119 (2).
[43]Regulation (EC) No 1907/2006, *op. cit.*, Art. 118 (2).
[44]Regulation (EC) No 1907/2006, *op. cit.*, Art. 118 (2).
[45]CJEU, 1 July 2008, C-39/05 P, ECR 2008 I-04723.
[46]CJEU; C-39/05 P, *op. cit.*, Recital 36.
[47]CJEU; C-39/05 P, *op. cit.*, Recitals 42–44.

The Court further stated that:

> Openness in that respect contributes to strengthening democracy by allowing citizens to scrutinize all the information which has formed the basis of a legislative act. The possibility for citizens to find out the considerations underpinning legislative action is a precondition for the effective exercise of their democratic rights.[48]

25.2.7.2 Case T-233/09 [2011][49]

The case dealt with a request for access to a note concerning the proposal for a regulation of the European Parliament, Council and Commission documents. That document contained the proposals for amendments or re-drafting, submitted by several Member States at a meeting of the working party. The applicant was granted access, but the Member States, which had put those proposals forward, could not be identified. The Council argued that disclosure of the names would seriously undermine the decision-making process.

The CJEU held:

> If citizens are to be able to exercise their democratic rights, they must be in a position to follow in detail the decision-making process within the institutions taking part in the legislative procedures and to have access to all relevant information.[50]

25.2.7.3 Case C-92/09 [2010][51]

The case dealt with the validity of certain provisions of EU law that required the publication of information on the beneficiaries of funds deriving from the European Agricultural Guarantee Fund and the European Agricultural Fund for Rural Development.

The CJEU stated:

> The principle of transparency is stated in Articles 1 TEU and 10 TEU and in Article 15 TFEU. It enables citizens to participate more closely in the decision-making process and guarantees that the administration enjoys greater legitimacy and is more effective and more accountable to the citizen in a democratic system.[52]

On the other hand, the CJEU has pointed out

> it is necessary to bear in mind that the institutions are obliged to balance, before disclosing information relating to a natural person, the European Union's interest in guaranteeing the transparency of its actions and the infringement of the rights recognised by Articles 7 and

---

[48] CJEU; C-39/05 P, *op. cit.*, Recital 46.
[49] General Court, 22 March 2011, T-233/09, ECR 2011 II-01073. Judgment upheld: CJEU, 17 October 2013, C-280/11 P, ECLI:EU:C:2013:671.
[50] General Court, T-233/09, Recital 69.
[51] CJEU, 9 November 2010, C-92/09, ECR 2010 I-11063.
[52] CJEU, C-92/09, *op. cit.*, Recital 68, with further references.

8 of the Charter.[53] No automatic priority can be conferred on the objective of transparency over the right to protection of personal data (see, to that effect, Commission v Bavarian Lager, paragraphs 75 to 79), even if important economic interests are at stake.[54]

25.3 Transparency and EU Food Law

In the area of food law, the "transparency concept" is found on various levels of the food chain: It applies to the legislative procedure in Brussels, as well as to law enforcement in the Member States. For example, when carrying out risk assessments, the European Food Safety Authority (EFSA) is obliged to be open and transparent.

25.3.1 Regulation (EC) No 178/2002[55]

25.3.1.1 EFSA

The so-called General Food Law,[56] Regulation (EC) No. 178/2002, establishes the EFSA. It further explicitly states that EFSA "shall ensure that it carries out its activities with a high level of transparency"[57] and establishes specific publication requirements for, among others, agendas and minutes of the Scientific Committee and the Scientific Panels and its opinions.[58] It is important to note, however, that EFSA must not disclose information, for which confidential treatment has been requested and has been declared justified, with an exception for information that

---

[53]Art 7. Respect for private and family life
  Everyone has the right to respect for his or her private and family life, home and communications.

Art 8. Protection of personal data
  1. Everyone has the right to the protection of personal data concerning him or her.
  2. Such data must be processed fairly for specified purposes and on the basis of the consent of the person concerned or some other legitimate basis laid down by law.
  Everyone has the right of access to data which ahs been collected concerning him or her, and the right to have it rectified.
  3. Compliance with these rules shall be subject to control by an independent authority.

[54]CJEU, C-92/09, *op. cit.*, Recital 85.

[55]Regulation (EC) No 178/2002 Regulation (EC) No 178/2002 of the European Parliament and of the Council of 28 January 2002 laying down the general principles and requirements of food law, establishing the European Food Safety Authority and laying down procedures in matters of food safety, OJ L 031, 01.02.2002, p. 1, as amended.

[56]See *van der Meulen/van der Velde*, European Food Law Handbook, p. 253.

[57]Regulation (EC) No 178/2002, *op. cit.*, Art. 38.

[58]Regulation (EC) No 178/2002, *op. cit.*, Art. 38, 40.

must be made public in order to protect public health.[59] In the field within its mission, EFSA is obligated to inform the public "on its own initiative",[60] and to ensure that "the public and any interested parties are rapidly given objective, reliable and easily accessible information, in particular with regard to the results of its work."[61] Further, the right to access documents as laid down in Regulation (EC) Nr. 1049/2001 is also applicable to documents held by EFSA.[62]

25.3.1.2 Principles of Transparency

Chapter II Section 2 of Regulation (EC) No. 178/2002 is titled "Principles of transparency" and contains two articles. One addresses public consultation,[63] the other public information.[64]

During the legislative process, from preparation, to evaluation and revision of food law, the public shall be consulted[65] in an open and transparent manner.[66] An exception exists for cases of urgency that require immediate action and make such consultation impossible.[67]

The requirement of consulting the public during the legislative process serves a dual purpose: On the one hand, this ensures a maximum use of potential knowledge resources and information. On the other hand, the possibility to participate in the legislative process can increase the level of acceptance of the adopted rules. It has to be pointed out, however, that the term "public" primarily refers to consumer and trade associations.

The rule on public information reads as follows:

> ...where there are reasonable grounds to suspect that a food or feed may present a risk for human or animal health, then, depending on the nature, seriousness and extent of that risk, public authorities shall take appropriate steps to inform the general public of the nature of the risk to health, identifying to the fullest extent possible the food or feed, or type of food or feed, the risk that it may present, and the measures which are taken or about to be taken to prevent, reduce or eliminate that risk.[68]

As part of a regulation, this rule is directly applicable in all Member States.[69] As a consequence, Member States are obligated to inform the public in cases of a risk

---

[59] Regulation (EC) No 178/2002, *op. cit.*, Art. 39.
[60] Regulation (EC) No 178/2002, *op. cit.*, Art. 40.
[61] Regulation (EC) No 178/2002, *op. cit.*, Art. 40.
[62] Regulation (EC) No 178/2002, *op. cit.*, Art. 41.
[63] Regulation (EC) No 178/2002, *op. cit.*, Art. 9.
[64] Regulation (EC) No 178/2002, op. cit., Art. 10.
[65] Either directly or through representative bodies, Regulation (EC) No 178/2002, *op. cit.*, Art. 9.
[66] Regulation (EC) No 178/2002, *op. cit.*, Art. 9.
[67] Regulation (EC) No 178/2002, *op. cit.*, Art. 9.
[68] Regulation (EC) No 178/2002, *op. cit.*, Art. 10.
[69] TFEU, *op. cit.*, Art. 288.

for the health of humans or animals. Remarkably, the wording of the provision only covers those foodstuffs that are injurious to health,[70] but not those that are unsafe[71] because they are unfit for human consumption.[72] Accordingly, this rule follows the principle of proportionality when allowing or ordering the competent authorities to inform the public of injurious foods.

25.3.2 Regulation (EG) No 882/2004[73]

Regulation (EC) No. 882/2004 aims at ensuring a horizontally uniform level of control. According to its title, Art. 7 addresses the principles of "transparency and confidentiality" that authorities have to adhere to in their activities: The competent authorities shall ensure that they carry out their activities with a high level of transparency. For that purpose, relevant information held by them shall be made available to the public as soon as possible.[74] At the same time, the competent authority must take the necessary steps in order to ensure that members of their staff do not disclose information which is by its nature covered by professional secrecy in duly justified cases.[75] As a rule, however, protection of professional secrecy does not preclude the dissemination of information by the competent authorities according to Art. 10 Regulation (EC) No. 178/2002.[76] Some of the cases where information is covered by professional secrecy are listed in Art. 7 (3), among them the confidentiality of preliminary investigation proceedings and of current legal proceedings, but also cases of conflicting individual interests, such as the protection of personal data. Further examples are the right to a fair trial and the protection of trade and business secrets.[77] Art. 7 is worded as follows:

1. The competent authorities shall ensure that they carry out their activities with a high level of transparency. For that purpose, relevant information held by them shall be made available to the public as soon as possible.

   In general, the public shall have access to:

   (a) information on the control activities of the competent authorities and their effectiveness;
   and
   (b) information pursuant to Article 10 of Regulation (EC) No 178/2002.

---

[70]According to Regulation (EC) No 178/2002, *op. cit.*, Art. 14 (2) (a).

[71]For a detailed analysis of the concept of food safety, see Chap. 20.

[72]According to Regulation (EC) No 178/2002, *op. cit.*, Art. 14 (2) (b).

[73]Regulation (EC) No 882/2004 of the European Parliament and of the Council of 29 April 2004 on official controls performed to ensure the verification of compliance with feed and food law, animal health and animal welfare rules, OJ L 165, 30.4.2004, p. 1.

[74]Regulation (EC) No 882/2004, *op. cit.*, Art. 7 (1).

[75]Regulation (EC) No 882/2004, *op. cit.*, Art. 7 (2).

[76]Regulation (EC) No 882/2004, *op. cit.*, Art. 7 (2).

[77]Cf. *Grube/Immel/Wallau*, Verbraucherinformationsrecht, 2013, Part A, p. 32.

2. The competent authority shall take steps to ensure that members of their staff are required not to disclose information acquired when undertaking their official control duties which by its nature is covered by professional secrecy in duly justified cases. Protection of professional secrecy shall not prevent the dissemination by the competent authorities of information referred to in paragraph 1(b). The rules of Directive 95/46/EC of the European Parliament and of the Council of 24 October 1995 on the protection of individuals with regard to the processing of personal data and on the free movement of such data remain unaffected.

3. Information covered by professional secrecy includes in particular:
   – the confidentiality of preliminary investigation proceedings or of current legal proceedings,
   – personal data,
   – the documents covered by an exception in Regulation (EC) No 1049/2001 of the European Parliament and of the Council of 30 May 2001 regarding public access to European Parliament, Council and Commission documents,
   – information protected by national and Community legislation concerning in particular professional secrecy, the confidentiality of deliberations, international relations and national defence.

It is noteworthy that Regulation (EC) No. 882/2004 is being revised at the moment, and, in particular, the rules concerning openness and transparency. The Commission draft from 6 May 2013[78] introduces two new provisions that are worded as follows:

Article 7 Confidentiality obligations of the staff of the competent authorities

1. Competent authorities shall require members of their staff not to disclose information acquired when undertaking their duties in the context of official controls and other official activities which by its nature is covered by professional secrecy, subject to paragraph 2.
2. Unless there is an overriding public interest in its disclosure, information covered by professional secrecy as referred to in paragraph 1 shall include information whose disclosure would undermine:

   (a) the purpose of inspections, investigations or audits;
   (b) the protection of commercial interests of a natural or legal person;
   (c) the protection of court proceedings and legal advice.

3. Paragraphs 1 and 2 shall not prevent the competent authorities from publishing or making otherwise available to the public information about the outcome of official controls regarding individual operators, provided that the following conditions are met:

   (a) the operator concerned is given the opportunity to comment on the information that the competent authority intends to publish or make otherwise available to the public, prior to the publication or release;
   (b) the information which is published or made otherwise available to the public takes into account the comments expressed by the operator concerned or is published or released together with such comments.

---

[78]COM(2013) 265 final.

Article 10 Transparency of official controls

1. Competent authorities shall perform official controls with a high level of transparency and make available to the public relevant information concerning the organisation and the performance of official controls.

   They shall also ensure the regular and timely publication of information on the following:

   (a) the type, number and outcome of official controls;
   (b) the type and the number of non-compliances detected;
   (c) the cases where measures were taken by the competent authorities in accordance with Article 135;
   (d) the cases where the penalties referred to in Article 136 were imposed.

2. To ensure the uniform implementation of the rules provided for in paragraph 1 of this Article, the Commission shall, by means of implementing acts, lay down and update as necessary the format in which the information referred to in that paragraph shall be published. Those implementing acts shall be adopted in accordance with the examination procedure referred to in Article 141(2).

3. Competent authorities shall be entitled to publish or make otherwise available to the public information about the rating of individual operators based on the outcome of official controls, provided that the following conditions are met:

   (a) the rating criteria are objective, transparent and publicly available;
   (b) appropriate arrangements are in place to ensure the consistency and transparency of the rating process.

25.3.3 Case Law and Examples

25.3.3.1 German Court Decisions

There are several decisions by German administrative courts that deal with the interpretation and the scope of Art. 7 Regulation (EC) No. 178/2002. These decisions concern the scope of the term "preliminary investigation proceedings" and "current legal proceedings". According to German case law, the term "preliminary investigation proceedings" does not apply (any more) in the case an inspection has been carried out by the authorities and there has been a practical result.[79] If there has been an audit of a restaurant by the competent authority, for example, and the authority has come to the conclusion that the restaurant owner has violated food law (e.g. certain hygiene provisions), this is not a case of "preliminary investigation proceedings", irrespective of whether a decision to continue administrative proceedings and/or whether or not to impose a fine, has been taken.

The term "current legal proceedings" only applies in cases where opposition proceedings or court proceedings against the findings of an official control have been initiated.[80]

---

[79] OVG NRW (Higher Administrative Court of Northrhine-Westphalia), 27.08.2009, file no: 13a F 13/09.

[80] BayVGH (Higher Administrative Court of Bavaria), Decision of 22.12.2009, file no: G 09.1.

25.3.3.2 CJEU, Case C-636/11: Berger [2013][81]

In this case, the CJEU has clarified whether Art. 10 Regulation (EC) No. 178/2002 must be interpreted as precluding national legislation allowing the publication of information concerning a certain food, in a case where that food is unfit for human consumption, though not injurious to health. Until then, this question had been discussed very controversially. The decision was based on the following facts:

On 16 and 18 January 2006, the Passau Veterinary Office (Germany) carried out official inspections in several establishments of company B, which processes and markets game meat. The authorities found that the hygiene conditions were inadequate. On several dates, samples of the game meat concerned were taken and analyzed. Those analyses led to a finding that the food in question was unfit for human consumption and consequently was unsafe within the meaning of Regulation No 178/2002. On 23 January 2006, the competent authority informed company B of its intention to inform the public of their findings, that the listed foodstuffs were unfit for human consumption. Further, company B was informed that the authorities would not inform the public, if company B itself informed the public effectively and promptly.

As a result, company B prepared a consumer information that invited its customers to exchange certain products at their retail stores, due to possible sensory anomalies. There was, in its view, no risk to health. This information was declared insufficient by the competent authorities.

On 24 January 2006, the competent state authority published a press release, declaring that certain products marketed by company B were to be recalled. According to the press release, the relevant samples "gave off a rancid, nauseous, musty or acidic smell", in some samples, "the putrefaction process had already started". It was further stated that, during inspections of some establishments of company B, "revoltingly unhygienic conditions had been encountered", and that a temporary prohibition on company B from marketing products manufactured or processed by it in those establishments had been issued.[82] On 31 January 2006, company B declared itself insolvent. Company B brought an action for damages against the Freistaat Bayern (Free State of Bavaria) before the Landgericht München (Regional Court of Munich).

The CJEU had to decide, whether Art. 10 Regulation (EC) No 178/2002 precludes rules of national law allowing information to be issued to the public mentioning the name of a food and the name or trade name of the food manufacturer, processor or distributor, in the event that the food is not injurious to health but is unfit for human consumption, particularly food that is nauseating.

In a certain sense, the decision of the CJEU could be regarded as solomon-like, since the Court ruled:

---

[81]C-636/11, 11.04.2013.

[82]CJEU C-636/11, *op. cit.*, Recital 17–19.

Article 10 of Regulation No 178/2002 must be interpreted as not precluding national legislation allowing information to be issued to the public mentioning the name of a food and the name or trade name of the food manufacturer, processor or distributor, in a case where that food, though not injurious to health, is unfit for human consumption. The second subparagraph of Article 17(2) of that regulation must be interpreted as allowing, in circumstances such as those of the case in the main proceedings, national authorities to issue such information to the public in accordance with the requirements of Article 7 of Regulation No 882/2004.[83]

The question, in which specific cases a confidentiality obligation according to Art. 7 Regulation (EC) No 882/2004 applies, has not been answered by the CJEU in the case at hand. In particular, the Court did not take a position concerning the obvious frictions between Art. 10 Regulation (EC) No 178/2002 and Art. 7 Regulation (EC) No 882/2004. It is noteworthy in this respect, that Regulation (EC) No 882/2004 is being revised at the moment, and, in particular, the rules concerning openness and transparency (see supra). This planned modification could be interpreted as the legislative response to the decision of the CJEU.

25.3.4 Current Transparency Projects in Selected EU-Member States

Several Member States have introduced projects concerning the publication of official hygiene inspections. These programs were already in place before the revision process of Regulation (EC) No 882/2004 started, and are therefore based on existing rules. A comparative analysis concerning the consumer information law,[84] published in 2010 by a group of researchers of the University of Heidelberg, Germany, came to the following conclusion:

> The study revealed broad similarities among the examined jurisdictions with regard to legislation on consumer information: Hence, all surveyed jurisdictions know legal acts providing for the consumer's right to access certain public documents. Additionally, all examined jurisdictions impose a duty on businesses, though.
>
> However, despite these similarities on a general level, differences can be ascertained on a more specific level. None of the examined legal systems knows an act, which is comparable to the German Consumer Information Act. Meanwhile, consumers obtain the same information in the examined countries through Freedom of Information legislation. Additionally, we determined that especially Scandinavian countries have an extremely open culture when it comes to public documents. Even more, the public bodies of these countries tend to disseminate a wide range of different information automatically.

---

[83]CJEU C-636/11, *op. cit.*, Recital 37.

[84]"Rechtsvergleichende Untersuchung des Verbraucherinformationsrechts in Deutschland, Belgien, Dänemark, Frankreich, Großbritannien, Irland, Schweden und den Vereinigten Staaten von Amerika, Abschlussbericht vom 07.05.2010", Prof. Dr. Thomas Pfeiffer, Theresa Heinke, Philipp Portugall in Koorperation mit Prof. Evelyn Terryn, Prof. Peter Møgelvang-Hansen, Dr. Stephanie Rohlfing-Dijoux, Prof. Patrick Birkinshaw, Dr. Cliona Kelly, Dr. Jori Munukka und Prof. Anita Allen.

Especially the Danish legislator has chosen a very consumer protective information policy. There, all restaurants and other businesses handling food and beverages, such as supermarkets, have to display a ranking about their hygiene quality rendered by Danish public authorities at their entrance. A tendency towards the publication of public controls can also be found e. g. in Los Angeles County and Great Britain ("Scores on the doors").

25.3.5 The Danish Smiley-System[85]

As a pioneer in Europe, Denmark started its so-called Smiley-system in 2001. The system foresees the publication of the results of unannounced hygiene inspections. The results are depicted by using one of four different smileys. These symbolise that the official inspector had no remarks (happy smiley), has emphasized that certain rules must be obeyed (slightly smiling smiley), issued an injunction order or a prohibition (neutral smiley), or issued an administrative fine, reported the enterprise to the police or withdrew an approval (unhappy/sour smiley).

Not all rules and requirement are inspected each time. But at each inspection, several areas are checked. The areas may vary from inspection to inspection, so that after some time, all areas are being inspected. Each inspected area is assigned a result from 1 to 4. All results as well as the inspectors' remarks are published on the inspection report. The poorest result determines the smiley.

The inspection report contains the latest smiley, together with the rating of the previous three inspections. The reports must be displayed in the shop or restaurant in a way that enables the consumer to read them before entering the premises, but also on the homepage of the relevant company, where it must be easy to find. Finally, all inspection reports of the last four controls can be accessed via the website www.findsmiley.dk.

In order to reward food businesses with constantly excellent results, the so-called "elite-smiley" was introduced in 2008. It is awarded to those businesses that had only happy smileys on their last four reports and no remarks during the last year. It only applies to businesses that are checked at least once a year.

The enterprises are divided into risk groups, depending on their line of trade. For each group, there is a standard frequency per year, typically between one to three times a year. Enterprises that have received one of the three "not happy" smileys (i.e. that have not received an overall grade 1) will be re-inspected within a reasonable time period. These re-inspections have to be paid for by the enterprises concerned, whereas the regular inspections are paid for by the tax-payer.

The smiley system is intended to empower the consumer to make an informed choice, by using simple symbols that are easily understood by him or her. Thereby, the system is seen to introduce another powerful incentive for food businesses to secure a high level of food safety.

---

[85]Unless indicated otherwise, the information in this subheading was sourced from http://www.findsmiley.dk/en-US/Forside.htm (last accessed in May 2014).

25.3.6 The Food Hygiene Ratings in UK[86]

In England, Wales and Northern Ireland, consumers are informed about the results of hygiene inspections according to the so-called "Food Hygiene Rating Scheme". The system covers places where people can eat out, such as restaurants, pubs, cafés, but also schools and hospitals; but also covers places where food can be bought, such as supermarkets and bakeries. Certain businesses that only constitute a low health risk (e.g. because of the food they are selling, such as wrapped sweets or bottled drinks) are not rated.

The scheme is run by local authorities in England, Northern Ireland and Wales and in partnership with the Food Standards Agency. The "hygiene rating" results from the inspection by a food safety officer from the competent local authority. The safety officer checks: how hygienically the food is handled—how it is prepared, cooked, re-heated, cooled and stored, the condition of the structure of the buildings—the cleanliness, layout, lighting, ventilation and other facilities, and how the business manages and records what it does to make sure food is safe.

The ratings range from "0" (urgent improvement necessary) to "5" (very good hygiene standards). One of these six ratings is given at the end of the inspection. If improvement is necessary, the food safety officer will explain the required steps and measures needed to be taken in order to achieve the top rating.

The food safety officer's inspection report contains more detailed information on the hygiene standards of the inspected business. Unlike the Danish smiley system, the Food Hygiene Rating Scheme does not provide for a publication of this further information. However, this information can be accessed by consumers, if they make a 'Freedom of Information' request to the local authority that carried out the inspection.

With every inspection, the business is rated anew. The control frequency depends on the risk group the business belongs to. Businesses that prepare fresh food or sell unprocessed foods are controlled more frequently than businesses that offer only prepacked and refrigerated foods. The control frequency typically ranges between 6 months and 2 years.

Businesses that did not receive the top rate "5" and therefore had to take improvement measures, can ask to be inspected before the next planned inspection is due. In these cases, the food safety officer will check the improvements that have been made and see if a new rating should be given.

In cases where a new business has been set up, or an existing business had a change of ownership, the business will not have a food hygiene rating to begin with. Until the first inspection has taken place, at the end of which a rating will be

---

[86]Unless indicated otherwise, the information in this subheading was sourced from http://www.food.gov.uk/multimedia/hygiene-rating-schemes/ratings-find-out-more-en/#.U1-1pYF_uaW (last accessed in May 2014).

given, the business may display a sticker or certificate that says 'Awaiting Inspection'.[87]

Contrary to the Danish system, food businesses are not legally obliged to publish the sticker or certificate with the hygiene rating in England and Northern Ireland, although they are encouraged to do so. Since November 2013, the display of the sticker in a prominent place is mandatory in Wales, and food business operators have to provide information on their rating verbally, if asked. Further, the hygiene ratings can be accessed via the website http://ratings.food.gov.uk/.

Scotland runs a similar scheme, titled "Food Hygiene Information Scheme". This scheme only provides for two inspection results, i.e. "pass" and "improvement required". A "pass" means that the business has achieved an acceptable level of compliance with food hygiene law. The result "improvement required" means that the business has not achieved an acceptable level of compliance with food hygiene law.

If food business operators can demonstrate that their hygiene standards are better than those required by law, they can apply for and will receive the so-called "eat-safe-award". This award, however, is not part of the above-mentioned scheme.

25.3.7 Publication and Legislative Projects in Germany

25.3.7.1 Food Hygiene Rating Project

The debate concerning the publication of the results of hygiene inspections in Germany is dominated by legal arguments. Due to the federal structure of Germany, legislation and administration usually does not lie in the same hands. At the moment, there is no federal law that lays down the requirements for the publication of the results of hygiene inspections. Since the publication of these hygiene results is welcomed by the media audience, some Länder (states) have gone their own way to make these publications possible. The example of the Bezirksamt Pankow (local authority of Pankow) is worth mentioning: Pankow introduced a system of visualized hygiene results, based on the Danish smiley system. The results are classified into five categories from "sehr gut" (very good) to "nicht ausreichend" (not sufficient). Since 2011, the results can be accessed via a website.[88]

The system provides for the inspection of various different criteria, such as compliance with food law, traceability, staff training, effectiveness of self-monitoring, storage and refrigeration of foods, structural quality of the premises,

---

[87]In Wales, the sticker reads "rating awaited", http://www.food.gov.uk/multimedia/hygiene-rating-schemes/rating-schemes-faqs-en/fhrs/#.U1-2j4F_uaU (last accessed in May 2014).

[88]Please note that Pankow has preliminarily stopped the publication of the results, due to another judgment of a German administrative court, causing doubts as tot he legality of such a project, see https://www.berlin.de/ba-pankow/aktuelles/pressemitteilungen/2014/pressemitteilung.251465.php (in German).

cleaning and disinfection, and pest control. In case a post-control has already been carried out at the time of publication, this post-control and its date will be listed on the report, also whether corrective measures have been taken in between inspections. The results of the post-control do not influence the grading of the smiley, however.[89] This means, a sour smiley ("not sufficient") will still be displayed, even if deficiencies noted during the first inspection have been remedied in the meantime.

The Verbraucherzentrale NRW (consumer organisation of Northrhine-Westphalia), a non-governmental organization, has started their own "pilot project" named "appetitlich" (appetizing) in two German cities: They request information on the hygiene situation of food businesses in the relevant cities according to the German Verbraucherinformationsgesetz (Consumer Information Act). This information is then categorized and visualized by using a color bar. The color bar has three sections: green (requirements met), yellow (requirements partly met) and red (requirements not met). The color bar can be accessed via the internet[90] since December 2013, an app is also available.

25.3.7.2 Recent Legislative Developments

The German Lebens- und Futtermittelgesetzbuch (Food and Animal Feedstuff Code—hereinafter referred to as LFGB) contains a provision that obliges the competent authorities to inform the public also in cases where a health risk is not involved.[91] The provision reads as follows:

> The competent authority shall inform the public, stating the name of the food or feed, as well as that of the food or feed business operator under whose name or company the food or feed is produced or handled or placed on the market, if based on facts, in case of samples in accordance with Section 39 (1) Sentence 2, based on at least two independent studies from bodies according to Article 12 (2) of Regulation (EC) No 882/2004, there are sufficient grounds for suspecting that

---

[89]With its decision of 19.03.2014 (file no: 14 L 410.13), the Administrative Court of Berlin raised fundamental doubts concerning the lawfulness of such publication practice, and has preliminarily prohibited any further publication via the Internet by the authority concerned. The administrative court stated, in particular: "The publication of alleged hygiene deficiencies does not concern crisis management in unforeseen cases, but administrative measures in the area of health and consumer protection that aim at the handling of numerous specific cases and at remedying the resulting disadvantages. The publications have effects that are equivalent to those of a regulatory instrument: The food business operator concerned will be pilloried electronically, which is significantly more onerous than an administrative order to remedy the deficiencies found. [...] The publication on the internet undoubtedly infringes basic rights. The legality of the infringement has to be denied after summary examination. Administrative action by administrative information is irreversible, and in case of wrong information, this cannot be changed by the presentation of counter-arguments or other corrections, since the factual consequences of information, especially on the internet, regularly cannot be captured and completely erased." [translation is ours]

[90]http://www.verbraucherzentrale.nrw/kontrollbarometer.

[91]Section 40 (1a) LFGB [translation is ours].

1. permissible limits, maximum levels or limits laid down in provisions within the scope of this Code, have been exceeded, or
2. provisions within the scope of this Code, which serve the protection of consumers against health hazards, the prevention of fraud, or the compliance with hygiene requirements, have been violated to a non-insignificant extent or have been repeatedly violated, and the imposition of a fine of not less than three hundred and fifty Euros is expected.[92]

This provision has been controversially discussed, already during the legislative process. Within seven months after its entering into force, the provision was already subject of 20 administrative court proceedings (mainly concerning alternative No 2).

In the vast majority of cases (concerning alternative No 2), the courts requested a specific reference to a product, and did not authorize informing the public about general hygiene conditions in food businesses.[93] A majority of administrative courts further considered informing the public of deficiencies that had already been remedies to be disproportionate. They also requested a substantiated reasoning on the part of authorities as concerned the expected amount of the fine. These numerous decisions in the first instance already resulted in four decisions of appellate courts (Oberverwaltungsgerichte—Higher Regional Courts).

In three of the four decisions, the courts expressed doubts as to the constitutionality of the provision:[94] The courts voiced their concern in particular as regards the principle of legal certainty and clarity, as well as the proportionality principle: The principle of legal certainty and clarity might be infringed since there is no uniform catalogue of fines, and different authorities would handle comparative cases not in a uniform manner. It was further criticized that the provision did not provide rules on the duration of the publication of information, which, since it is essential for the extent of impairment of fundamental rights, should have been regulated by the legislator. It was not considered sufficient to regulate this issue only through ministerial decrees.

A violation of the proportionality principle was discussed, since an administrative fine of 350 Euros would only be imposed for minor offences, whereas a publication of violations of food law could lead to an extensive impairment of fundamental rights; further, the provision did not provide for exceptions or hardship clauses.

Publications according to alternative No. 1 have rarely been challenged before the courts. The Verwaltungsgericht Hannover (administrative court of Hannover) based its decision on the clear wording of the provision as well as the legislative

---

[92]Section 40 (1a) LFGB [translation is ours].

[93]VG Karlsruhe (administrative Court of Karlsruhe), 07.11.2012, file no: 2 K 2430/12; VG Berlin (administrative Court of Berlin), 28.11.2012, file no: 14 K 79.12.

[94]OVG Lüneburg (Higher Administrative Court of Lüneburg), 18.01.2013, file no: 13 ME 267/12; VGH Baden-Württemberg (Higher Administrative, Beschl. v. 28.01.2013, Az.: 9 S 2423/12; OVG Rheinland-Pfalz, Beschl v. 13.02.2013, Az.: 6 B 100035/13.OVG; Bayerischer VGH, Beschl. v. 18.03.2013, Az.: 9 CE 12.2755.

materials and decided that in case of samples two independent analyses from two different laboratories were needed. The court thereby rejected the common practice of authorities to have validation studies carried out in one and the same laboratory.[95]

The Bundesverfassungsgericht (Federal Constitutional Court) is currently assessing the constitutionality of the provision. A decision is expected soon, and it is hoped that the decision will set fundamental standards for national laws on transparency.

Just as well as the provision discussed supra, the German Verbraucherinformationsgesetz (Consumer Information Act—hereinafter referred to as VIG) has also undergone a revision.[96] The aim of the Act is to "improve market transparency, thereby enhancing consumer protection from food that is injurious to health or otherwise unsafe, and also from fraud as regards products and consumer products."[97] In order to achieve this aim, "within the framework of this law, everybody has free access to all data concerning (...) inadmissible infringements"[98] of food law requirements as established by the competent authorities.

An application to access information must be "sufficiently precise und must indicate which information it is directed at".[99] The authority that receives an application to access can grant the food business operator concerned the right to be heard.[100] In the interest of the applicant, the law foresees that a decision generally has to be taken within 1 month. The access to information can be granted orally, by access to files or by other means.[101] It is worth noting that the authority is not obliged to check the accuracy of the information it makes available to the public.[102]

At the time of writing, there was only one administrative court decision concerning the revised Consumer Information Act.[103] In this decision, the court does not address any of the varied legal concerns expressed by jurisprudence relating to the provisions of the Consumer Information Act.[104] It has to be pointed

---

[95]Beschl. v. 29.01.2013, Az.: 9 B 264/12.

[96]The revised Act entered into force on 01.09.2012.

[97]Section 1 VIG [translation is ours].

[98]Section 2 VIG [translation is ours].

[99]Section 3 VIG [translation is ours].

[100]Section 3 VIG [translation is ours].

[101]Section 6 VIG.

[102]Section 6 VIG.

[103]Verwaltungsgericht Oldenburg (Administrative Court of Oldenburg), 22.10.2013, file no: 7 A 1866/10 with a critical comment by Wallau/Theis in DVBL 2014, p. 330 ff. The Administrative Court has not even considered whether a violation of the right to a counter sample according to Art. 11 (5) and (6) of Regulation (EC) No 882/2004, *op. cit,* would preclude a publication of information. The CJEU had already decided that the right to a counter sample is essential (10. April 2003, Case C-276/01).

[104]See Becker ZLR 2011, p. 391 ff.

out that, in practice, mainly NGOs, journalists and the media in general have used the Consumer Information Act, whereas applications from consumers are rare.

25.4 Conclusion

The concepts of transparency and openness will lead to further changes in food law. The concept of a restricted public access to files ("beschränkte Aktenöffentlichkeit") and keeping the content of files secret ("Aktengeheimnis") traditionally predominant in German administrative law has come to an end, due to the development in and the primacy of EU law.

The revision of Regulation (EU) No 882/2004 clearly proves that on the EU level, the die is cast in favor of further developing transparency and openness in the food sector. This is consistent insofar as transparency and openness lead to a strengthening of the "organised civil society"; NGOs, for example, thereby are awarded the role of "information-watchdogs".[105]

It is questionable, however, whether the causes that initially led to an extension of transparency systems, can thereby be eliminated. The "problem of trust", it seems, will not be solved this way, but only be shifted: The citizen is generally not in a position to validate the published information; if he mistrusts the authorities, he is generally forced to trust the private institutions, such as NGOs, who filter and manage information. The "privatization" of information gathered during official inspections could lead to a decrease in legal safeguards for the food business operators concerned: The presumption of innocence, for example, only fully applies within the framework of sanction procedures, but only has limited effects with respect to media coverage.[106] It seems striking that through the publication of "negative" information via the Internet, an instrument from medieval times seems to return, albeit with a modern twist: The electronic pillory. The former vice-president of the German Bundesverfassungsgericht (Federal Constitutional Court), Winfried Hassemer, has pointed out a decisive difference between the medieval variant of the pillory and its modern form, in particular in the area of food law: "Also in our part of the world, the historical pillory was only used with relative restraint: Those, who were being pilloried, had at least been sentenced to this procedure by a final court judgment."[107]

---

[105] Cf. the proposal for a Directive of the European Parliament and of the Council on certain rules governing actions for damages under national law for infringements of the competition law provisions of the Member States and the European Union, COM(2013) 404 final, and *Mederer*, EuZW 2013, p. 847.

[106] Cf. further references at *Grube/Immel/Wallau*, Verbraucherinformationsrecht, 2013, Teil C Recital 17.

[107] Hassemer (2011), S. 107 ff [translation is ours].

References

Alemanno A (2014) Unpacking the principle of openness in EU law – transparency, participation and democracy. Eur Law Rev

Hassemer (2011) "Internetpranger": Kommunikative und rechtliche Vernunft. In: Schröder, Hellmann (eds) Festschrift für Hans Achenbach

Hierholzer V (2010) Nahrung nach Norm. Vandenhoeck & Ruprecht

Note of the European Parliament, Directorate-General for internal policies, Policy Department C, Citizens' rights and constitutional affairs; Openness, Transparency and access to documents and information in the EU; http://www.europarl.europa.eu/RegData/etudes/note/join/2013/493035/IPOL-LIBE_NT(2013)493035_EN.pdf

Transparency Portal of the EU: http://ec.europa.eu/transparency/index_en.htm

For an in-depth analysis of the principle of openness in EU law with further references:

Alberto Alemanno, Unpacking the Principle of Openness in EU Law – Transparency, Participation and Democracy, European Law Review 2014 (forthcoming); http://papers.ssrn.com/sol3/papers.cfm?abstract_id=2303644

Grube/Immel/Wallau, Verbraucherinformationsrecht, 2013, Part A, p. 11 et seq. (German)

All legislative acts can be accessed via the website http://eur-lex.europa.eu/advanced-search-form.html

Regulation (EC) No 178/2002 Regulation (EC) No 178/2002 of the European Parliament and of the Council of 28 January 2002 laying down the general principles and requirements of food law, establishing the European Food Safety Authority and laying down procedures in matters of food safety, OJ L 031, 01.02.2002, p. 1, as amended

Regulation (EC) No 1049/2001 of the European Parliament and of the Council of 30 May 2001 regarding public access to European Parliament, Council and Commission documents, Official Journal L 145, 31.05.2001 p. 43, as amended

European Governance – A White Paper, COM(2001) 428 final, OJ 287, 12.10.2001, p. 1

Communication from the Commission "Towards a reinforced culture of communication and dialogue – General principles and minimum standards for consultation of interested parties by the Commission" COM(2002) 704 final – not published in the Official Journal

Green Paper "European Transparency Initiative" COM(2006) 194 final - Official Journal C 151 of 29.6.2006

Sectoral Rules

Directive 2001/95/EC of the European Parliament and of the Council of 3 December 2001 on general product safety, OJ L 011, 15.1.2002, p. 4, Art. 16 (1)

Regulation (EC) No 765/2008 of the European Parliament and of the Council of 9 July 2008 setting out the requirements for accreditation and market surveillance relating to the marketing of products and repealing Regulation (EEC) No 339/93, 13.8.2008, OJ L 218, p. 30, Art. 25

Regulation (EC) No 726/2004 of the European Parliament and of the Council of 31 March 2004 laying down Community procedures for the authorisation and supervision of medicinal products for human and veterinary use and establishing a European Medicines Agency, Official Journal L 136, 30.04.2004, p. 1

Regulation (EC) No 1907/2006 of the European Parliament and of the Council of 18 December 2006 concerning the Registration, Evaluation, Authorisation and Restriction of Chemicals (REACH), establishing a European Chemicals Agency, amending Directive 1999/45/EC and repealing Council Regulation (EEC) No 793/93 and Commission Regulation (EC) No 1488/94 as well as Council Directive 76/769/EEC and Commission Directives 91/155/EEC, 93/67/EEC, 93/105/EC and 2000/21/EC, OJ L 396, 30.12.2006, p. 1

Regulation (EC) No 1367/2006, of 6 September 2006, on the application of the provisions of the Aarhus Convention on Access to Information, Public Participation in Decision-making and Access to Justice in Environmental Matters to Community institutions and bodies, OJ L 264, 25.9.2006, p. 13

Directive 2003/4/EC of the European Parliament and of the Council of 28 January 2003 on public access to environmental information and repealing Council Directive 90/313/EEC, OJ L 041, 14.02.2003, p. 26. complements Regulation (EC) No 1049/2001

Commission draft from 6 May 2013 COM(2013) 265 final: http://ec.europa.eu/dgs/health_consumer/pressroom/docs/proposal-regulation-ep-council_en.pdf

Case Law

a. CJEU, General Court

All judgments can be accessed via the website: http://curia.europa.eu
CJEU, Case C-64/05 P, 2007, ECR I-0000
CJEU, Case C-266/05 P, 2007, ECR I-1233
CJEU, Case C-139/07 P, 2010, ECR I-0000
CJEU, Case C-92/09 P, 9 November 2010, ECR 2010 I-11063
CJEU, Case C-28/08 P, ECR 2010 I-0000
CJEU, Case 41/00 P, ECR 2003 I-2125
CJEU, Case C-39/05 P, ECR 2008 I-04723
CJEU, Case C-524/07, 2008 I-00187
CJEU, Case C-636/11, not yet reported
General Court, Case T-2/03, ECR 2005 II-12121
General Court, Case T-166/05, 11 March 2009, not published in the ECR
General Court, Case T-20/99, ECR 2000 II-3011
General Court, Case T-191/99, 11 December 2001
General Court, Case T-309/97, ECR 1999 II-3217
General Court, Case T-233/09, ECR 2011 II-01073. Judgment upheld by:
CJEU, 17 October 2013, C-280/11 P [not yet reported]

b. German Courts

OVG NRW (Higher Administrative Court of Northrhine-Westphalia), 27.08.2009, file no: 13a F 13/09
BayVGH (Higher Administrative Court of Bavaria), Decision of 22.12.2009, file no: G 09.1
VG Berlin (Administrative Court of Berlin), Decision of 25.03.2014, file no: 14 L 410.13
VG Karlsruhe (Administrative Court of Karlsruhe), 07.11.2012, file no: 2 K 2430/12
VG Berlin (administrative Court of Berlin), 28.11.2012, file no: 14 K 79.12
OVG Lüneburg (Higher Administrative Court of Lüneburg), 18.01.2013, file no: 13 ME 267/12
VGH Baden-Württemberg (Higher Administrative Court of Baden-Württemberg), Decision of 28.01.2013, file no: 9 S 2423/12
OVG Rheinland-Pfalz (Higher Administrative Court of Rhineland-Pallatine), Decision of 13.02.2013, file no: 6 B 100035/13
BayVGH (Higher Administrative Court of Bavaria), Decision of 18.03.2013, file no: 9 CE 12.2755
Verwaltungsgericht Oldenburg (Administrative Court of Oldenburg), 22.10.2013, file no: 7 A 1866/10 with a critical comment by Wallau/Theis in DVBL 2014, p. 330 ff

Hygiene rating systems

http://www.findsmiley.dk/en-US/Forside.htm
http://www.food.gov.uk/multimedia/hygiene-rating-schemes/ratings-find-out-more-en/#.U1-1pYF_uaW
http://www.vz-nrw.de/kontrollbarometer

Chapter 26
Food Law in Sweden

Magnus Friberg

Abstract Sweden is a member of the European Union ("EU"). It has implemented all food legislation adopted by the EU. This chapter provides a brief background to the legal situation in Sweden including previous safety and labeling issues. It then continues with a description of applicable national legislation and "soft laws," the government agencies and the industry organizations and their self-regulatory measures. The chapter continues with a description of the supervisory procedure and sanctions. After this follows a brief presentation of the HACCP-rules and, in particular, the industry's guidelines that operators of a specific branch of the industry may follow. The chapter finishes with a review of labeling and marketing related issues.

26.1 Background

26.1.1 Food Law in Sweden

Sweden, a member of the EU since 1995, began adapting EU food legislation in 1992 as part of Sweden's dedication to the European Economic Area (EEA) agreement. Sweden implemented the EU *Acquis Communautaire* into its law as part of Sweden's and the other European Free Trade Association (EFTA) countries' commitments to the EEA agreement. At this time, the Swedish Food Act from 1970 and the Food Decree formed the core Swedish legislation; indeed, they still form the basis for all other law today. Additional legislation included the Veterinary import ordinance, Ordinance (1974:270) for the control on the importation of food, Ordinance (1974:271) on control of the export of food, and National Food Agency-adopted Regulations.

The Food Act gave the Swedish National Food Agency the parameters within which to adopt and implement detailed provisions in certain areas, a method still

M. Friberg (✉)
Advokatbyrån Gulliksson AB, Lund, Sweden
e-mail: Magnus.Friberg@gulliksson.se

used when implementing European Commission (EC) legislation. While regulations apply directly, directives are transposed into National Food Agency Regulations and published in the Agency's own Code of Statutes, LIVSFS (previously SLVFS). The authority to issue legislation is outlined in the Food Act and the Food Decree.

When assessing the need for revision and adaptation of the Swedish legislation in force in 1992, the Swedish government found that Swedish law in many respects harmonized with EU legislation. For example, the system of food control in Sweden prior to the implementation of EU law was such that it did not need any revisions or amendments. At this time, Sweden had also participated in an international context. Of note was Sweden's engagement in the Food and Agriculture Organization's (FAO) and the World Health Organization (WHO)'s Food Standards Program for the standardization of food products, the *Codex Alimentarius*, which Sweden had been engaged with for almost 30 years. Sweden's own Codex Contact point is the Swedish National Food Agency. Within the cooperation between the Nordic countries, there were also harmonization activities regarding food law. The provisions issued by the National Food Agency required adaptation, and in a few areas, there was a need for revisions of the applicable existing legislation. These were primarily labeling and foodstuffs for particular nutritional uses (PARNUTS).

Today, the Food legislation is in principle harmonized within the EU. The National Food Administration participates actively in the development of new legislation in collaboration with the other EU member states.[1] The Swedish Food Act not only complements the EC legislation, but also lists the food control authorities and contains stipulations on penalties and appeals.

26.1.2 Historic Safety Issues

The Swedish Board of Agriculture and the National Food Agency have long handled food safety issues. Historically, Sweden has been spared salmonella to a large extent. Salmonella and other forms of infectious diseases have been high on the agenda in order to maintain a high level of control and protection. This has also led to "skirmishes" between Sweden and neighboring countries over import restrictions, denigrating marketing, and other protective actions taken to ensure food safety. Other major issues include concern with anti-biotic resistance due to high levels in meat as well as pesticide residues in food. Animal welfare also ranks high on the agenda.

An example includes the criminal proceedings for the distribution of bovine semen from the Belgian Blue breed, and the subsequent insemination of cattle with this semen. Ultimately, the case was referred to the European Court of Justice. Sweden claimed protection of animal health, arguing that any breeding was liable to

---

[1] National Food Agency, *available at* http://www.slv.se/en-gb/.

entail suffering for animals or affect their behavior. The Court found that the Member State of import's national authorities were not entitled to reject the use of semen which was accepted for artificial insemination in another Member state on the grounds that it contained the muscular hypertrophy gene, that using the semen would likely entail suffering for the animals or affect their natural behavior, or that the national authorities regarded the breed as a carrier of genetic defects.[2] The case caused a stir in Sweden, as is generally the case with animal maltreatment or what is perceived as such. The Belgian Blue was not seen as natural, and this case sparked one of the first discussions about genetically modified (GM) foods and continued the ongoing discussion and dilemma of animal maltreatment.

26.1.3 Historic Labeling Issues

The Swedish agricultural community has remained a strong force on the Swedish market, traditionally for the protection of the farming community. This community had a particularly strong influence on protecting denominations for certain products—for example, milk, butter, and cream. Correspondingly, the community has weighed in on the topic of launching alternate products to these protected product names. Here we see a "battle" between consumer interests in alternative, cheaper products, and the interests of quality, prevention of misleading marketing, and the protection of the agricultural community's interests. Examples include the introduction of margarine, substitutions for single and double cream with vegetable oil based products, and the protection of names of cheeses.

These issues primarily address the use of signal color and product names. Double cream in Swedish, for instance, is ordinarily sold in red and white colored packaging single cream in green and white. Questions related to the introduction of substitute products based on vegetable oil included what color schemes could be used, how best to describe products and whether it was possible to mention cream on labels or product descriptions. The products were introduced under the names "Ädelvisp" and "Ädelkaffe" (cream is "grädde" in Swedish, double cream is "vispgrädde," and single cream is "kaffegrädde" in Swedish ("kaffe" is coffee). The products were introduced in red and white ("Ädelvisp") and green and white ("Ädelkaffe") color schemes. Other examples, however, further illustrate the issues that come in Swedish food law, as the following paragraphs show.

---

[2]Court of Justice of the European Union ("ECJ") Case C-162/05.

26.1.4 Health Claims: Originally a Self-Regulatory Measure

Since 1990, Sweden has allowed health claims. Swedish government agencies gave permission to the Industry to create and adopt self-regulatory measures for the use of health claims in the marketing of food. This was called "Health Claims in the Labeling and Marketing of Food" ("Hälsopåståenden vid märkning och marknadsföring av livsmedel"), generally referred to as the "Self-Regulatory Program" ("Egenåtgärdsprogrammet"). The Sweden Nutrition Foundation (SNF),[3] which administered these intended laws provided a platform and acted as a guarantor of a scientific level and objectivity. In 1990, Nils-George Asp, SNF's CEO and Professor of Applied Nutrition at Lund University, was appointed to participate in the writing of the law that the industry later agreed to. SNF's role also encompassed council and interpretation. Only claims that followed this code were permitted. Since the introduction of the code, SNF has taken on an advisory and coordinating role.

The first version, introduced in 1990, was based on general generic claims concerning the diminished risk for disease and the then eight established relationships between food and health. The claims were to be made in two steps—"X limits the risk for heart disease. Product Y contains X." By 2001, product specific physiological claims were allowed. SNF also introduced a special board for the assessment of claims on labels or in marketing. All products that wished to make these claims had to go through an independent scientific assessment. In 2003 and 2004, the terms and conditions for use of the generic claims were established. Additionally, nutrition claims were deemed as a form of generic claims. Beginning in 2008, the Industry appointed the Food Industry's Examiner ("Livsmedelsbranschens Granskningsman") to support and supervise the use of nutrition and health claims according to the Regulation 1924/2006.[4]

26.1.5 Current Issues and Problems

Issues regarding food safety continue to pervade Swedish food law and policy. Recent concerns include the re-packaging and relabeling of food products whose expiry date has passed, as well as the handling of fresh products such as minced meat in in-store packaging facilities. There have also been recalls due to salmonella and other contaminations; however, the number of incidents remains limited. Fresh drinking water has also become an issue where the drinking water in a number of municipalities has been found to be contaminated with bacteria or chemicals.

Concerning food labeling, the so-called horse meat scandal that included one of Sweden's leading food operators remains at the forefront of many consumers'

---

[3]Swedish Nutrition Foundation, *available at* http://snf.ideon.se/.
[4]*Id.*

minds. A number of similar incidents followed the scandal, including when colored filet of pork was sold as filet of beef. To some extent, this had religious implications, since certain religions do not permit the consumption of pork; undoubtedly, this caused some damage to the industry. In the case of the horse meat scandal, the operator, in whose business it all started in Sweden, launched a substantial and successful public relation campaign to turn the negative publicity around and save both the business and its reputation. Additionally, the operator made significant changes in its business implementing control functions, designed to avoid other incidents or similar cases from happening again. Also increasing are incidents and complaints concerning nutrition and health claims where the ongoing cases address the detailed interpretation of the Regulation 1924/2006/EG. There has also been a rising number of cases of food fraud for instance in the form of misleading origin labeling in recent years. The National Food Agency is addressing this situation and the industry organizations has also adopted a new program and licensing system to address origin fraud but also the promotion of food of Swedish origin called "Från Sverige" ("From Sweden").

26.2 Relevant Authorities, Legislation and "Soft Laws"

This section gives an introduction to some of the relevant legislation and soft laws applicable to the topics to be discussed for Sweden. Following is a description of the authorities and the industry's organizations and their roles on the market. The chapter concludes by discussing how the different parts work together towards a socioeconomically effective as well as ethical conduct on the market.

26.2.1 The National Food Agency and the Regional Authorities

The National Food Administration is responsible at the national level for enforcing the Food Act and regulations issued under the provisions thereof. The County Administrations have responsibility for coordinating food control at the regional level and the municipal Environment and Health Protection Committees have responsibility for food control at the local level. Most companies fall under the supervision of the County Administrations; only a few are supervised directly by the National Food Agency.

The Agencies' supervision primarily concerns food safety and labeling requirements. Advertising in other media rather than on labels is primarily handled by the Consumer Agency (see below).[5]

---

[5]The National Food Agency, What We Do, *available at* http://www.slv.se/en-gb/Group2/About-us/What-we-do/.

The National Food Agency has also been given the task of promoting healthy food consumption habits. The Agency issues, for example, advisory documents suggesting nutritious and healthy food for various consumer groups—children, pregnant women, and normal consumers. Together with its equivalent Scandinavian authorities, the National Food Agency publishes The Nordic Nutrition Recommendations, which form the basis of the national dietary recommendations in Nordic countries and are used academically in teaching nutritional science. An international collaborative effort, the Recommendations are the result of the work of more than 100 experts led by a working group under the Nordic Council of Ministers.[6]

The Keyhole symbol is another product of collaboration whose details will be discussed further in this chapter. The Keyhole symbol is intended to function as a guide for the consumer to find healthy food products, and operators may affix the symbol on their products that are in accordance with the requirements as to composition.

In 2004, the Government commissioned the National Food Agency and the National Institute of Public Health to produce a plan for healthy eating habits and increased physical activity in the Swedish population.[7] Not forgetting that environmental issues are often linked to food consumption, the National Food Agency is actively engaged in addressing sustainability concerns such as food waste; it is estimated, for example, that the average Swede throws away 25 kg of edible food per year.

26.2.2 The Swedish Consumer Agency

The Swedish Consumer Agency is a government agency whose purpose is to safeguard consumer interests. To achieve this goal, the Agency reviews product safety issues, advertising, consumer contract terms and conditions, environmental issues, and more. A Director General leads the Swedish Consumer Agency, and in addition to this role, is a Consumer Ombudsman (Konsumentombudsman, KO) who can represent consumer interests in relation to businesses and pursue legal action in the courts.

The Agency may act against all forms of marketing for food products, including labeling. Regarding the latter, any such action would primarily concern voluntary labeling—for instance, claims made for the product. It may also intervene in

---

[6]The Nordic Council, Nordic Nutrition Recommendations, *available at* http://www.slv.se/en-gb/Startpage-NNR/.

[7]The National Food Agency, Action Plan for Healthy Eating Habits, *available at* http://www.slv.se/en-gb/Group1/Food-and-Nutrition/Action-plan-for-healthy-eating-habits/.

product safety issues, and in these matters, the Agency works closely with the National Food Agency.[8]

26.2.3 The Medical Products Agency

Another government agency is the Medical Products Agency, responsible for regulating and supervising the development, manufacturing, and sale of drugs and other medicinal products. Its task is to ensure that both the individual patient and healthcare professionals have access to safe and effective medicinal products and that these are used in a rational and cost-effective manner.

The Medical Products Agency acts under the aegis of the Ministry of Health and Social Affairs, with many of its operations largely financed through fees. Approximately 750 people, mostly pharmacists and doctors, work at the Agency.

The Medical Products Agency may have cause to react to advertising for food or food supplements that make medical claims, or that contain substances which makes them medical products by function. This has on occasion led to a product being classified as a medicinal product and with a subsequent sales ban.[9]

26.2.4 Food Stuff Act and Agency Regulations

Sweden, as a member of the European Union ("EU"), follows EU-adopted legislation. If Sweden does not incorporate such legislation, the country can be brought before the EU Court of Justice. If a Swedish law conflicts with an EU law, the EU law takes priority. EU-legislation may apply directly, as it does with regulations, or may be implemented, which is the case with directives and decisions. In Sweden, most legislation concerning food is, or is based upon, EU-legislation.

The Swedish Food Act (SFS 2006:804) complements EU regulations and contains rules about control or supervision, governmental fees, penalties, and appeal of decisions. The Swedish Government has listed which EU regulations which wholly or partially supplemented by the Food Act.

The Swedish Food Ordinance (SFS 2006:813) regulates what authority shall supervise which facilities. Along with the Food Act, the Swedish Food Ordinance authorizes the National Food Agency to issue various regulations. Most EU-legislation and particular national regulations are adopted and implemented through the National Food Agency's regulations. These are published in the

---

[8]The Swedish Consumer Agency, About the Swedish Consumer Agency, *available at* http://www.konsumentverket.se/otherlanguages/English/About-the-Swedish-Consumer-Agency/.

[9]The Medical Products Agency, About the Medical Products Agency, *available at* http://www.lakemedelsverket.se/english/overview/About-MPA/.

National Food Agency's Code of Statutes, LIVSFS (previously SLVFS). These regulations are based on EU directives or complement the EU regulations. Only a few regulations are not derived from EU-legislation. All regulations are available on the agency's website in a consolidated form. ("Consolidated form" means that any changes made to a provision has been incorporated in the text).

Additionally, the National Food Agency issues guidelines to achieve uniform application of food legislation in Sweden. The purpose is to describe and interpret the content of the legislation. The guidelines also express the National Food Agency's position with regard to the supervision of the legislation. The guideline often contains additional information not expressed in the legislation, thereby increasing the understanding of the legislation and the National Food Agency's interpretation of it. Though not legally binding, the guidelines serve in the assessment and implementation of the legislation. The Courts may consider the guidelines in their application of the legislation and as an expression of good practice; the Courts may also disregard the guidelines and come to a different interpretation of the legislation. Naturally, court precedents from the Swedish national courts and the European Court of Justice amend or alter these guidelines.

26.2.5 Marketing Act

In addition to the Food legislation, the Marketing Act[10] applies to food labeling and marketing. The Marketing Act is the implementation of EU-directive 2005/29/EC concerning unfair business practices. In Sweden, the Marketing Practice Act of Sweden (Sw. *Marknadsföringslagen* (2008:486)) regulates general misleading advertising by implementing the Directive 2005/29/EC on unfair business practices. *Lex specialis* for food advertising is, of course, the Food Act and the specific regulations for e.g. labeling, nutrition and health claims and the prohibitions in those regulations against misleading labeling and advertising. Advertising in violation of these regulations is also considered a violation of the Marketing Act.

However, the Marketing Act is broad in its scope, applicable regardless of product, medium or intended target group. The Marketing Act is intended to protect both consumers and businesses and thus is applicable to both business-to-consumer advertising as well as business-to-business advertising. It also regulates practices which are not directly covered by the food legislation but which might still occur in food advertising such as misleading packaging size, passing-off (where a company imitates, for example, a product or its package design in a misleading way), comparative advertising, price claims and comparisons, denigration or ridiculing of competitors and/or their products where the denigration is not misleading.

---

[10]Marketing Act ("Marknadsföringslag") (2008:486), Swedish Code of Statutes, *available at* http://www.government.se/content/1/c6/05/03/14/6c7aa374.pdf.

In accordance with the Marketing Act, the assessment of a marketing measure could also differ from the assessment made according to the food legislation. One reason is that different courts handle cases according to the Marketing Act and the food legislation; another, that the Marketing Act requires not only an assessment of whether or not the advertising is misleading, but also if the advertisement is likely materially to distort the economic conduct of the average consumer.

Still, what is misleading food advertising under the food legislation is also, in general, misleading under the Marketing Act. The possibility of applying the Marketing Act to cases like this gives the operator another prosecutorial avenue in relation to a competitor's misleading advertising measures with different sanctions, including interim sanctions, as well as the possibility of demanding compensation for damages.

26.2.6 Industry Self-Regulatory Measures

The Swedish Food Federation (Livsmedelsföretagen)[11] represents a variety of companies in the food industry. It is a member of the Confederation of Swedish Enterprise (Svenskt Näringsliv) and FoodDrinkEurope. The Swedish Food Federation monitors and acts in the interest of its member companies. For instance, it monitors the application of the legislation in the different regions to ensure that all companies are treated equal according to the law. It maintains a continuous dialogue with the inspectors in each of the regional offices and pushes for further training. Additionally, the Swedish Food Federation provides guidance and advice regarding the interpretation and application of food legislation. It has, for example, issued guidance documents for its members, including a handbook on labeling.

Another group called Swedish Retail (Svensk Dagligvaruhandel)[12] organizes the Swedish retail industry and aims to satisfy consumers' need for reliable products. Among other activities, it issues guidelines on varying topics with relation to retail. It also represents its member companies in matters relevant to the Industry as a whole in negotiations before authorities and legislators.

Both the Swedish Food Federation and Swedish Retail sponsored the industry's previous support function for health claims.[13] This support was intended to enable the food industry to use nutrition and health claims in a responsible and balanced way, to accomplish fair competition among companies, and to maintain high confidence in consumers in the food they purchase and consume and the food industry at large.

The Sweden Nutrition Foundation ("SNF") coordinated support and advice, while a Food industry Examiner (LGM) engaged in market monitoring and cases

---

[11] The Swedish Food Federation, *available at* http://www.livsmedelsforetagen.se/.
[12] Swedish Retail, *available at* http://www.svenskdagligvaruhandel.se/.
[13] *Id.*

assessment. The Examiner continuously tracked the market and assessed claims in marketing in accordance with Regulation 1924/2006. The Examiner either tried cases on its initiative or upon notification from competitors or the general public. If the Examiner was of the opinion that an operator was in violation of the Regulation, the Examiner would approach the operator and invite them to respond to its assessment. If the assessment was maintained after the response and the company refused to change, the Examiner then issued a recommendation which was made public. The Examiner could also refer matters to the Food Agency or the Consumer Agency for further legal actions to have the labeling or marketing stopped. This support was discontinued in 2014.

Svensk Egenvård, or Swedish self-care,[14] organizes 75 % of the suppliers of plant-based pharmaceuticals, dietary supplements, weight loss products, sports nutrition, and "parnuts" (Foods for Particular Nutritional Use) on the Swedish market. To aid in examining product quality, Svensk Egenvård utilizes an external consultant whose monitoring consists of cluster sampling to ensure that the product content complies with the labeling and contains no unwanted substances. Svensk Egenvård created a Council for market supervision and also appointed an examiner to monitor advertising. This Board was discontinued in 2015. The Board collaborated with the appropriate government agencies, such as the National Food Agency, the Medical Products Agency, and the Consumer Agency. The Council's decisions were made public and available via Svensk Egenvård's website.

26.2.7 Other Private Initiatives

Numerous private organizations strive to improve food quality and encourage environmentally sustainable and ethical food production as well as ethical labeling and marketing of food. In particular, the environmentally sustainable and ethical food production organizations have gained considerable attention. An example of such organizations is KRAV, a key organization on the market for organic food. The KRAV-label has been on the market for many years and stands for a sound and natural environment, solid care for animals, good health, and social responsibility. Another organization started by two families is "Äkta vara" or "Real stuff," focused on labeling and food products without additives. Both organizations license the use of their labels and are registered trademarks.

Furthermore, there are a number of consumer organizations, including Konsumentföreningen Stockholm or "Stockholm Consumer Cooperative Society," that have profiled themselves with regard to food labeling, advertising, and safety.

---

[14]Swedish Retail, *available at* http://www.svenskegenvard.se.

26.2.8 Conclusions: The Players and Their Roles

Both the legislative and supervisory systems are based on three legs: the authorities, the industry organizations, and the operators themselves.

The commercial legislation for the health claim regulation, the Marketing Act, and the labeling regulations stems from industry self-regulatory measures where the embryos, later to be codified into legislation, were formed in industry's different branches. The industry's self-regulatory measures still plays an important part in forming new rules, codes of conduct, good practices. Such self-regulation also influences the legislation's application through the development of its own "case law" and co-operating closely, but also confronting, the authorities on their interpretation of the law. This is the case with the Marketing Act, its first version emerging in the early 1970s and based largely on the International Code Council ("ICC") Code of good marketing practice. Today, it follows the ICC Code for advertising and market communication, providing yet another example of an industry organization striving to maintain its regulatory system alongside the legislation.

The government agencies, the industry organizations, operators, and other initiatives on the market create "the law" in a broad sense, including actual legislation, codes of conducts or soft laws, and more. In turn, this contributes to the supervision of the market and in doing so aides in further development of "the law."

26.3 The Supervisory Procedure and Sanctions[15]

As previously stated under Sect. 26.2.1, the supervisory function with regard to the food legislation befalls on the National Food Agency and the municipal Environment and Health Protection Committees. This chapter will describe the food agencies' supervisory functions and the measures applied to food operators found transgressing the legislation. It will conclude with a discussion of the supervision and sanctions available according to the Marketing Act and some industry board cases.

---

[15] See the National Food Agency's guideline concerning sanctions in the food legislation "Vägledning – sanktioner i livsmedelslagstiftningen," *available (only in Swedish) at* http://www.slv.se/upload/dokument/livsmedelsforetag/vagledningar/Sanktioner_%20i_%20livsmedelslagstiftningen.pdf.

26.3.1 Food Agencies Inspections: Proportionate Measures

Primarily, municipal food inspectors conduct supervision through regular inspections, and in some cases, matters may be referred to the police and, potentially, for prosecution. When an inspector detects a deviation or a transgression, the first course of action is to urge the food operator to adhere to the set legislation and to inform the operator of the measures needed in order to comply. However, the instructions to the examiners from the National Food Administration for the initial contacts with the operator are meant to point out the deficiencies and to encourage the operator to comply with the law. If the demands raised by the examiner are perceived as a binding request for compliance, meaning that the authority may take legal action should the operator not comply with the law, such requests must be in the form of a decision so that the operator could appeal to a higher instance if necessary.

Suppose, for example, that an examiner detects serious deficiencies, or if, in a later follow-up inspection, he or she discovers that the operator never addressed the deficiency previously detected, or has only taken insufficient measures. The authority would consider using a suitable administrative penalty to achieve compliance. In assessing which measure should be taken, the authority would consider the nature and seriousness of the transgression, and whether or not the operator failed to comply with the legislation on previous occasions.

Whatever measure is taken, then, should be proportionate to the situation: tougher sanctions in relation to serious or repeated transgressions and failure or refusal to comply with the legislation, and less severe sanctions in cases of lesser transgressions and "first offences" apply. The principles of the supervision not only invite cooperation between the authority and the operator, but also put measures in place to achieve the goals of the legislation.

26.3.2 Communication During the Procedure

The food business operator must always have access to information relevant to the case and the opportunity to comment on the inspection report. The method used may vary and is decided by the authority. Normally, the operator is given a report in relation to the inspection, which may be sent by post or communicated orally; however, the latter is not recommend as it carries a risk of misunderstanding if the report is extensive and detailed. The communication should always be made to a person authorized to act for the operator.

In certain cases of a particularly serious nature, it may be necessary to decide on administrative measures directly and without communication. In such a case, the nature of the situation is such that a decision cannot be postponed—for example, if human life and health are in immediate jeopardy.

26.3.3 The Possibility to Appeal and the Court Procedure

The notification of the decision must contain the reasons for said decision and information of the possibility to appeal including the procedure for this. The County Administrative Board (Länsstyrelsen) takes the appeals in the county where the operator has its place of business. The Board's decision can be appealed to the Administrative Court, and that decision in turn can be appealed to the Administrative Court of Appeal, and, finally, to the Supreme Administrative Court.

The decision may have immediate effect. As a rule, the operator only has to abide by a decision that has come into force. If the finding is appealed, its effect will be postponed. However, in certain cases, the authority may decide that the decision should apply immediately and the operator must abide by the decision, even if it is appealed. This is only used in severe cases where there is an immediate danger to human life or health. Even in such cases, the authority must evaluate the implications for the operator—economically and otherwise. The authority should also consider whether or not the effects of the decision can be restored should the operator prevail on appeal and the decision be overturned. In such a case, the authority may have to compensate damages suffered by the operator.

In a case where a decision has an immediate effect, it is also possible for the operator to ask that the court grant inhibition of the decision—that is, a temporary stay in the enforcement. If granted, the operator may continue its business despite the decision for the duration of the court procedure. Inhibition, however, is rare and requires special circumstances in order to be granted. In making its decision, the court considers the implications of the decision on the operator, the nature of the transgression, and the likelihood of the decision being overturned.

A case before the Supreme Administrative Court concerned the repeal of authorization of a plant that handled the production and packaging of ostrich meat. The plant did not meet the requirements outlined in the applicable legislation. In order to meet these standards, the plant would have to make substantial changes in the business conducted there. The company appealed the decision and asked that the decision, which had an immediate effect, be subject to inhibition. The Supreme Administrative Court weighed both the interest of the company in having the decision lifted and the public interest in food safety. It had been established that there was no such risk, and the decision was based not on risk but rather on the assumption that the decision to repeal would not be effective. The operator also argued that its business was encompassed by an exemption according to an EU-directive and that most of the deficiencies that the authority had found had been remedied. The Court found that the outcome of the matter was not certain. Circumstances were such that public interest in food safety could not take precedence over the operator's interest in continuing its business during the process. The Court thus granted inhibition.[16]

---

[16]See Supreme Administrative Court, case no 3686-04.

26.3.4 Sanctions

In order to achieve compliance by the operator, the authority may issue sanctions. Among the most used sanctions are injunctions and prohibitive injunctions. The legal basis is found in both article 54 in the Regulation 882/2204/EC and section 22 in the Food Act.

Injunctions are used to get the operator to submit an action plan that the operator has not submitted or to get the operator to implement measures necessary to come into compliance with the legislation. Prohibitive injunctions are to compel the operator to refrain from a certain conduct. Which of these measures are used depend on the situation.

An injunction, prohibitive or other, must be precise in its scope and in the operator's requirements. The legal basis should also be clear on which rule in the legislation the decision is based.

As the supervision in effect concerns the operators' compliance programs the inspections should be directed primarily on detecting and remedying weaknesses in these programs. The measures taken should always address a specific problem. As the applicable regulations may be vague in their wording, it is important that the decisions issued are clear and reasoned.

26.3.5 Injunctions Conditioned on Penalties

In order to put further pressure on the operator, the injunctions are, as a rule, conditioned on a monetary penalty. If, for example, the operator does not comply with an injunction to implement a certain measure within a set time frame, a court may order the operator to pay a stipulated penalty. The amounts vary depending on the nature of the transgression, repeated transgressions, or unwillingness to collaborate. The penalty's purpose is to serve as an inducement to adhere to the decision. The decision must to be clear as to what the operator is required to perform; if unclear, the penalty cannot be ordered.

Though set in the decision, the penalty amount can be adjusted by the Court if it is found that ordering the full amount is deemed disproportionate to the nature of the operator's failure. This, however, is rare. The normal practice is that payment in the full amount should be ordered unless the incompliance can be justified as excusable due special circumstances, or that the penalty would affect the operator unreasonably hard. If the operator alleges any such circumstances, he or she has the burden of proving the fact.[17] The circumstances that the operator may call upon have to be of a severe nature, unforeseen and beyond the operator's control.[18] The circumstances, then, must be of an extraordinary nature. As an example, the failure

---

[17]Lavin, Lag om viten, En kommentar, 2010, p. 147.
[18]Government bill 1984/85: 96, p. 56 and Government bill no 57 år 1970 s. 92.

of legal counsel to inform his client of a prohibitive injunction was not considered sufficient grounds to adjust the amount of the penalty.[19]

A penalty can also be payable whenever a prohibition is violated. If this is the case there does not have to be issued a new decision after the penalty for a decision has been ordered. The injunction runs on and so does the penalty. Thus, the penalty may be order for each transgression or for each period that passes without the deficiency being remedied.

26.3.6 Repeal of Authorization and Temporary Shutdown

Authorization of an operator's place of business may be repealed pursuant to Article 31, paragraph 2 e) and Article 54 paragraph 2 f. of Regulation (EC) No 882/2004, and the National Food Agency's regulation (LIVSFS 2005:20) on food hygiene. This may be done temporarily or permanently, if the deficiencies in the operator's conduct are of a particularly serious nature or if the operator's business has been subjected to repeated shutdowns.

Primarily, the authority will choose to shut down the operation temporarily, but the authority may order a permanent shut down in cases where all attempts to rectify the deficiencies have been exhausted and the situation is such that public health and safety cannot be ensured.

The authorization can also be partially repealed, as is the case in the following examples:

- The use of a part of a store for a different purpose than that for which it is registered, thus constituting a significant change of the business in this part.
- When part of a place of business that, for hygienic reasons, cannot continue to be used—part of a restaurant, for example.
- Certain preparations that have been conditioned in the authorization, but where the conditions are not followed.

Repeals of authorization or the shutdown of operations are considered as actions with far-reaching consequences for the operator. The decision must be proportionate to the infringement of the law; a temporary shutdown, for example, should be taken when it would allow the operator to rectify the deficiency.[20]

---

[19] Stockholm Municipal Court's Decision Case No. B 7434-00.

[20] *See* Appellate Administrative Court in Stockholm, Case No. 4331-05. A local restaurant had been damaged by fire. The Court considered a temporary shut-down sufficient. A permanent repeal of the authorization was not warranted.

26.3.7 Seizure of Products

The authority may also seize products already on the market or intended to be put on the market. Such seizure may occur when the products: violate the requirements concerning the labeling and presentation of food (currently LIVSFS 2004:27); are handled or put on the market by an operator not authorized or registered, or which does not meet the requirements for authorization; contain banned substances; have concentrations of substances exceeding the limits defined by law or the EC Regulation; and are not considered safe according to Regulation (EC) No. 178/2002.

It is also possible to seize products covered by an injunction or prohibition if the operator does not follow the injunction or prohibition. The operator may remedy the deficiency in the products under the supervision of the inspection, or may use them for other purposes. In this case, it would not be an option for products to be destroyed at the operator's expense.

If a decision to seize products is made, and should the products be destroyed or damaged, the authority may be liable for damages.

26.3.8 Penal Provisions in the Food Act

The penal provision can be found in Section 29 of the Food Act. Violations of provisions in the Foods legislation, the Food Act, and regulations adopted in accordance with the Food Act may result in fines. (Until 2006, imprisonment was also a possible sanction. There is an ongoing discussion concerning the reinstatement of this sanction brought about by the recent rise in food fraud cases.) Offenders include food operators who do not meet their obligations to provide the authority with documents and other information upon request. Section 29 also encompasses violations of the applicable EU-legislation.

The penalty provision does not apply if the transgression concerns provisions on the authority's exercise of control—that is, by officers of the authority. Liability for such transgressions is regulated in the Penal Code. Minor offenses fall outside the scope of the penal provisions (Food Act Section 30).

Moreover, the penal provision in the Food Act does not apply to matters specifically regulated in the general Penal Code. Examples include: the spreading poison or infection (Penal Code 13:7), negligence with poison or disease agent (Penal Code 13:9), infliction of bodily harm or illness (Penal Code 3:8), and endangering of others (Penal Code 3:9).

The sanctions according to the Penal Code can only be directed to against natural persons. However, the Penal Code provides for corporate fines being issued against legal entities (Code of judicial procedure Ch 36 7-10a §§). This can be invoked if management has not been directly involved in the transgressions but where they were aware and omitted to prevent or address the violations.

The scope of the penal provision is substantial; it covers transgressions of both national and EU-legislation. According to the National Food Agency, the following transgressions are the most common, though the list is not exhaustive.

Examples of transgressions which may be pursued according to the penal provisions
Authorization/registration and ownership transfer   Food operators shall inform the competent authority of the intention to conduct a food business so that all facilities under its control can be registered or approved, if approval is required under national law or under Regulation (EC) No 853/2004.   Legislative foundation: Article 6 of Regulation (EC) No 852/2004, Article 4 of Regulation (EC) No 853/2004, 7 and Secs. 10 and 29 Food Act, and Secs. 12–13 Food Regulation, Secs. 4–13 in the National Food Agency's regulation on food hygiene (LIVSFS 2005:20).
Placing of food on the market in contravention of food law   Food operators placing food products on the market that are not considered safe should not be placing those products on the market.   Legislative foundation: Article 14 of Regulation (EC) No 178/2002 and other relevant EC regulations, 10 § and 29 § 1 and 2 of the Food Act.
Failure to report significant changes in the business to the authority   Legal foundation: Article 6 p 2 2 of Regulation (EC) No 852/2004 and 29 § 2 of the Food Act.
Violation of the hygiene regulations   Legal foundation: Article 4 and Annex II to Regulation (EC) No 852/2004 (general hygiene rules), Regulation (EC) No 853/2004 (specific hygiene rules for food of animal origin), the National Food Agency's regulations on food hygiene (LIVSFS 2005:20).
Violation of the personal hygiene regulations   Legal foundation: Regulation (EC) No 852/2004, Annex II, Chapter VIII, Sec. 6 of the Food Act, Sec. 8–11 Food Ordinance.
Violation of the labeling regulation, including misleading labeling and deficiencies in the obligatory labeling information   Legal foundation: Sec. 6 2, 10 1 and Sec. 29 1 in the Food Act and Sec. 7 in the Food Ordinance and applicable provisions in the NFA regulations on labeling and presentation of food, (currently) LIVSFS 2004:27.
Re-labeling with a later expiration date   Legal foundation: Sec. 15 of the National Food Agency's regulation 2005:20, reprinted 2007:6 on food hygiene and Sec 29 1 of the Food Act.
Failure to provide the authority requested documents and other information   Legal foundation: Sec. 20 1 and 29 1 of the Food Act.

26.3.9 Compensation Relating to Authority's Actions

The authority may be liable for damages in relation to its activities. Wrongful seizure of products or destruction in which the operator suffers loss is an example. Here, general tort law applies.

In extraordinary cases, a company may be eligible for governmental compensation due to seizures or orders for the destruction of food in order to prevent the spreading of disease.[21] One such case concerned a sales ban for beef contaminated with salmonella in which a company asked for compensation according to the law. The beef had been purchased from Romania, where inspectors had cleared the product free of salmonella. Upon entry into Sweden the veterinary took three samples but approved the beef for sale within the EU. Analyses of the samples showed salmonella contamination, leading the food authority to impart a sales ban. Subsequently, the company asked that the authority destroy the meat; it also asked for compensation from the government in the amount of SEK 536,446 (approximately $74,126.11 U.S. dollars). Though denied, the request was granted on appeal to the Appellate Administrative Court. The Supreme Administrative Court overturned this decision, finding that the conditions were not extraordinary; rather, this concerned control upon entry into Sweden, which is normal practice.[22]

Relevant national legislation and guidelines
Legislation   The Food Act (2006:804)   Food Ordinance (2006:813)   Act (1985:206) concerning penalties   Personal Data Act (1998:204)   Secrecy Act (1980:100)   Notice (2007:601) of the EC regulations complemented by the Food Act (2006:804)
Food Administration regulations   National Food Agency's provisions (LIVSFS 2005:20, reprinted LIVSFS 2007:6) on food hygiene   National Food Agency's provisions (LIVSFS 2005:21) on official controls   National Food Agency's provisions (LIVSFS 2004:27) on the labeling and presentation of food
Guidelines   National Food Agency's guidance on the approval and registration of food establishments   National Food Agency's guidance on hygiene   National Food Agency's guidance on the official control of foodstuffs   National Food Agency's guidance to introducing HACCP   National Food Agency's guidance on labeling and presentation of food

26.3.10 Hazard Analysis and Critical Control Points ("HACCP")

The food operator is primarily responsible for its risk management system. It is the operator's responsibility to put a control system into place in order to oversee the business. Specifically, the operator monitors the maintenance of an effective hazard

---

[21] Ordinance (1956:296) on state compensation, in certain cases, intervenes to prevent the spread of a contagious disease (Förordning (1956:296) om ersättning från staten i vissa fall vid ingripanden för att förhindra spridning av en smittsam sjukdom).

[22] Supreme Administrative Court, Case No. 5662-2000.

analysis and control of critical points in the business which could result in unsafe products.

Though the legislation is not regulated in detail, it is aimed at management through objectives. The legislation cannot regulate every form of food business down to the last detail. Therefore, the National Food Agency has asked the different industry organizations to produce national guidelines for their particular businesses. This constitutes the industry's own guide as to how the operators should achieve compliance with the hygiene legislation. These guidelines may contain definitions of otherwise vague expressions or terms in the law—for example, "where necessary," "adequate," and "sufficient."

26.3.11 The Marketing Act

The Marketing Act is applicable to all forms of marketing, regardless of products, medium used, or the target group for the marketing activity. Primarily, the Marketing Act is used in issues pertaining to food labeling and advertising.

The Consumer Agency and the Consumer Ombudsman are the government agencies designated to protect the consumers' interests. However, the Marketing Act gives the operators an effective avenue to legally pursue transgressing competitors' misleading labeling or advertising of their food products. Both the agencies and operators may file lawsuits against other operators.

Of all the available sanctions, the most used is a prohibitive injunction condition of a fine. This sanction aims to prevent the operator from continuing with the activity or launching the activity in the future. The prohibition not only covers the marketing activity as such, but also considers similar activities. Thus, it is not possible to make minor changes to circumvent the prohibition. In order to ensure adherence to the prohibition, it is standard practice that, should the operator re-engage in the prohibited activity, the Consumer Ombudsman or the competitor may ask a court to order payment of the fine. As previously stated concerning such fines, if the operator is found guilty of violating the prohibition, the court will order payment of the fine. The amounts are set in the interest of adherence; thus, it should not pay to violate the prohibition. SEK 1.000.000 is a common amount set. Only under exceptional circumstances will the amount be adjusted.

Furthermore, the operator may be ordered to pay a so-called Market disturbance fee, which is paid to the Swedish state in a set amount between SEK 5000 and 5,000,000.

The competitor or the consumers may be compensated for damages suffered, although such compensation may come from circumstances other than economic. Compensation can, depending on the nature of the transgression, be asked for directly, or be requested in a situation where the operator has violated a previous prohibition.

The operator may also be ordered depletion of the misleading materials if this is found to be a reasonable or necessary measure. Depletion can be ordered if it is the only way to ensure that the material will no longer mislead consumers. Although rare, such depletion could be used with regard to packaging or misleading products.

The current court system consists of two specialized courts: the Stockholm Municipal Court and the Market Court. The latter is a specialized court that handles cases according to the Competition Act, the Marketing Act, as well as other consumer and marketing law. The Market Court is also the highest court of appeal. If the plaintiff asks for a prohibitive injunction, the suit can be filed directly with the Market Court. In that instance, the case would only be heard by one court with no possibility to appeal. If the plaintiff requests the Market disturbance fee and/or damages, the suit must start in the Stockholm Municipal Court, with the possibility of appeal to the Market Court.

In addition to the operator, others can be held responsible and subject to sanctions. Employees, officers of the company and other parties, advertising agencies, media who have contributed significantly to the marketing activity can also be held responsible as complicit to the violation. This is often used when the advertiser is not established in Sweden, when it is important to prevent a certain medium (TV-channel or newspaper) from similar transgressions, or when an operator wants to safeguard against officers of the company starting a new business and continuing the transgressions. In such cases, it is important that the prohibition is personal, which would be a significant threat towards private individuals who would have to pay any penalties out of pocket.

26.3.12 Conclusions: How the System Works

The "Swedish approach" practiced by authorities is to resolve any conflicts or deficiencies through negotiations and convince the operator to address the deficiency on a voluntary basis. This means that authorities give the operator the opportunity to adhere to the authority's criticism and to make the necessary changes in its business before the authority proceeds with sanctions or lawsuits. For the most part, this has proven to be an effective method, since the number of cases actually brought to trial and the case law is limited. At first glance, it would appear that this method is economically effective as well as successful in maintaining adherence to the legislation since it enables cooperation rather than conflict between the operators and the authorities. However, the structure of the government authorities is such that it endangers a coherent application of the legislation in the different regions. This, in turn, may give certain companies a competitive advantage because of a regional authority's reluctance to intervene against a major employer on its "patch."

The industry's self-regulatory boards exercise supervision by intervening against operators, who are mainly members of their organization. The industry also provides advice and training for its members. In doing so, the organizations often apply a stricter interpretation of the applicable rules, thereby encouraging a higher ethical level for the operators on the market. Over time, this creates an expression of good practice applicable in the courts and thus works its way into the legislation.

Additionally, the operators contribute in their day-to-day "surveillance" of their competitors marketing measures, be it labeling, advertising, or other. The cases addressing misleading labeling or advertising seem primarily to be competitor driven, in which a competitor sues the food operator. However, there are cases where the authorities, the national and municipal food agencies, and the Consumer ombudsman have intervened and are the plaintiffs, examples of which will be discussed below. Such cases result from the refusal to comply with the authorities' demands, either due to a different interpretation of the legal situation or negligence. In some cases where there the authority sees a need for precedence, it may decide to file a lawsuit, regardless of the position of the operator.

Basing a claim on the Marketing Act is often also the most efficient avenue for an operator to proceed in seeking legal action against competitors' marketing measures. The alternative would be to refer the matter to the competent authority. By utilizing the former method, the operator can act swiftly against its competitor whose labeling or advertising the operator holds to be in violation of the food legislation; for example, the EU Regulation 1924/2006 on nutrition claims and health claims or the labeling regulation. The Court system provides a speedy and efficient option with a one-instance court system. These matters are referred to the Market Court, whose decisions cannot be appealed to a higher court. From the Market Court, an operator may achieve prohibitive injunction condition on a fine as well as preliminary injunctions.

Matters referred to the Market Court often have a faster procedure, particularly with issues of misleading labeling. It is possible to have a prohibitive injunction in one instance, and additionally there is the possibility of a preliminary injunction. The plaintiff must convince the Court that it is likely to prevail in the proceedings; it must also provide security for the potential loss suffered by the defendant should the final outcome go against the plaintiff, resulting in a lost or diminished effect of a prohibition. Time should also be considered in comparing the procedural system with letting the government agencies pursue the case solely according to the food legislation. On more than one occasion, there have been complaints that such procedures are too slow.

The procedural system is thus efficient, and its sanctions are similar to those available according to the food legislation. For an operator who wishes to put a stop to a competitor's activities, the available sanctions encompass the company and let the officers and employees of that company form an efficient and competitive tool. The losing side always pays the costs of the winner; this, of course, must be taken into consideration in any risk assessment.

26.4 Labeling

The so-called Information Regulation 1169/2011[23] applies in Sweden. Prior to the Regulation Directive 2000/13 on labeling, presentation, and advertising of foodstuffs,[24] was implemented into the Swedish law in the LIVSFS 2004:27. Swedish labeling law thus was harmonized with EU labeling legislation with the implementation of the directive. The Information Regulation applies in Sweden as in other EU-countries. The National Food Agency has adopted guidelines concerning the interpretation of the Regulation. The National Food Agency follows closely directives from the EU-Commission and cooperates with its neighboring Scandinavian countries on issues of interpretation. Sweden has also adopted legislation on the points where the Information Regulation leaves it to the Member States to regulate. This concerns provisions for non-packed foods (LIVSFS 2014:4) and NFA regulation (LIVSFS 2015:1) on the use of the so called Keyhole symbol which is used as an aid for consumers to find foods with a nutritious and healthy composition. The system has similarities to traffic light systems in other countries.

This section will, with the exception of misleading names of products, focus on misleading voluntary labeling. It will cite case law according the labeling regulations as well as cases that have been decided according to the Marketing Act.

26.4.1 Misleading According to the Food Act

The National Food Administration implements EC directives into its regulations and publishes them in the National Food Administration's Code of Statues ("LIVSFS").

Of the rules, first and foremost is that labeling cannot be misleading. Labeling must be clear and concise while explaining what a claim means. It is not sufficient to say that a claim is factually correct and substantiated. Rather, the claim must be presented in such a way that the consumers' impression and interpretation is not skewed by, for example, exaggerations, graphic depictions, and claims in large print and explanations (or attempts to explain) in smaller print. A food product must not be attributed with properties it does not have. The advertiser must be prepared to substantiate the claims made. Furthermore, is it not permissible to give the consumer the impression a property that is common for all or most products on the market is unique to that advertiser's product.

---

[23]Regulation (EU) No 1169/2011 of the European Parliament and of the Council of 25 October 2011 on the provision of food information to consumers.

[24]Directive 2000/13/EC of the European Parliament and of the Council of 20 March 2000 on the approximation of the laws of the Member States relating to the labelling, presentation, and advertising of foodstuffs.

In determining whether or not a claim is misleading, an overall assessment as well as a hypothetical assessment are made in understanding how the average consumer perceives the labeling.

It is permissible to make claims or pictures that go beyond the obligatory information, provided these elements are not misleading. This could include illustrations, pictures, and décor which might mislead the consumer as to product's content. For example, pictures of fruit on products that do not contain the fruit in question, or where the picture misleads as to the amount of fruit included in the products, would be considered misleading. Other examples are pictures or claims misleading the consumer as to the level of preparation.

In the National Food Agency's guideline,[25] the Agency defines and regulates the use of words of a more marketing nature, such as "traditional," "original," "natural," "natural ingredients," "genuine," "pure," "real," "fresh," and "luxury."[26]

If "traditional" is used for a product, it must be true, but the product should also state precisely what is traditional about the product.

"Original" refers to an original production method or recipe. If a product is labeled original, it should not have undergone any significant change in recipe or production.

In order for a food product to be considered "natural" or a "a natural product," it must exist in its present, natural condition, or have undergone minimal processing.

A composite food can never in itself be a "natural product." If the product claims it has "natural ingredients," all the ingredients must meet the requirement for "natural." To assess whether the term can be misleading, each case considers whether the expression's contrast can be used for other equivalent food.

In order to assess whether terms like "genuine," "pure," and "real" can be misleading, in each case it must be determined whether the expression's contrast can be used for other similar or equivalent food. "Genuine," "pure," "real," or comparable expressions can be deemed misleading if used on foods that have a reserved name (name protected).

Expressions such as "fresh" generally have no clear purpose and should be used restrictively. Their use and intension should be explained, for example, by specifying the meaning of "daily updated" combined with the current date. Additionally, there are foods where fresh is part of the name, such as fresh potatoes and fresh cheese (cottage cheese).

---

[25]Introduction to National Food Administration Regulations, http://www.slv.se/upload/dokument/livsmedelsforetag/vagledningar/vagledning_markning%26.pdf.

[26]Introduction to National Food Administration Regulations, *available at* http://www.slv.se/upload/dokument/livsmedelsforetag/vagledningar/vagledning_markning%26.pdf.

26.4.2 Supreme Administrative Court and Misleading Labeling[27]

Some Swedish companies sell the readily packed fish of the species Theagra chalcogramma (Swedish name "Alaska Pollock") under a trade name containing the word cod (Swedish name "torsk"). The trade name on the label "codfish filet" (Swedish tr. Torskfiskfilé) is in combination with the denomination "Alaska Pollock". The National Food Agency rendered a decision whereby the company was prohibited to use "cod" or combinations including "cod" in labeling Theagra fish. The grounds for the decision included misleading labeling and use of an illegal denomination, since cod is part of the species Gadus morhua.

The Supreme Administrative Court stated that one of the most important pieces of information on a label is the correct denomination for a product. The Court referred to, among other sources, an industry guideline that included a list of denominations of various species of fish. The Court ultimately concluded that the label was misleading to the consumer.

26.4.3 Public Prosecution Office v. Officers for a Grocery Store[28]

This case concerned the handling of ground meat in a grocery store. The case the authorities' attention after an investigative TV-program, using a hidden camera, showed employees of the store handling meat in a manner contrary to the food legislation. On more than one occasion, the employees had ground, packaged, and marked products with an expiration date. Employees had since taken back the mince to the cold store, removed the old packaging, put the mince in new packaging, marked it with a later expiration date, and then returned it to the store for sale.

The Court found that this procedure was contrary to the applicable food regulations; anyone intentionally or negligently violating the regulations shall be held liable under § 29 of the same Act.

The Court found that the officers charged had transgressed the food legislation as alleged and, subsequently, were fined.

---

[27]Supreme Administrative Court, case no 5675-1991 RÅ 1992 ref 95.
[28]Nacka Municipal Court, Case No. B 4330-08.

26.4.4 Findus v. the National Food Agency[29]

Findus, a frozen food brand, made the following claims on its packaging: "without unnecessary additives, without preservatives, without trans-fatty acids, only natural colors, without flavor enhancers." The National Food Agency ordered Findus to cease and desist from using these claims on its packaging "as soon as possible" (in Swedish "*snarast*"). It also issued an injunction against Findus, prohibiting the company from placing products on the market with these claims with a transitional period of 6 months after the decision. Findus appealed the decision, asking the Administrative Court in Uppsala Court to overturn the decision in its entirety.

The Court found that "without unnecessary additives" was misleading. No explanation was given concerning which relevant additives the product did not contain, or why the additives were considered unnecessary. The claim could, therefore, mislead the consumer about the product and its properties.

"Without preservatives" was also considered misleading. The products in question were frozen products, and freezing is considered a form of preservation. However, virtually no other products on the market contained preservatives. The claim was therefore irrelevant since it gave the consumer the impression that similar products on the market had properties that others did not.

"Without trans-fatty acids" likewise gave the consumer the impression that trans-fatty acids were common in this particular product category. The Court considered the claim in violation of Regulation 1924/2006/EC, as claims of trans-fatty acids do not appear in the appendix of permissible nutrition claims.

"Only natural colors" had not been used in the labeling or presentation of the products; rather, it had only appeared in a general description of Findus' ambitions. Thus, the labeling regulations did not apply and the Court overturned the decision concerning this claim.

"Without flavor enhancer" was considered relevant information because Findus had changed the recipe for its Indian Tikka Masala product. In such cases, such a claim is justifiable during a limited period following the change. The Court granted Findus a transitional period of six months following the Court's decision.

26.4.5 Nutrition and Health Claims

Regulation 1924/2006/EG applies in Sweden, and The National Food Agency has issued a guideline for the interpretation of said Regulation, which constitutes the

---

[29]Administrative Court in Uppsala case no 3453-10 and 3454-10 *Findus Sverige AB v. The National Food Agency*.

Swedish authority's interpretation of the Regulation.[30] The industry's self-regulatory body "Branschstödet"[31] has also published a handbook that closely follows the Agency's guideline.[32]

In addition to the above-mentioned sources, the Medical Products Agency has issued its guideline concerning the interface between health claims and medical claims and on the classification of medicinal products.[33]

The classifications are based on an overall assessment of each individual product. Although no medical claims are made in marketing or on the label of the product, it can be classified as medicinal due to content of a substance or function.

Assessing whether or not a product should be classified as a medicinal product is part of a larger, overall assessment. The Medical Products Agency generally refers a food product for which a medicinal claim is made either to the National Food Agency or the Consumer Agency. Due to their characteristics, food supplements, often similar to medicinal products, are more in danger of being classified, as medicinal products should a medicinal claim be used in the marketing or on the label.

26.4.6 The Keyhole Symbol

The National Food Agency created the Keyhole symbol,[34] which operators use in marketing healthier food products. The objective is to aide consumers in easily finding and choosing healthy foods. Specific rules regulate the use of the symbol for a number of food categories. The Keyhole symbol stands for less and/or healthier fat, less sugar, less salt and dietary fiber and whole grains.

Established in Sweden in 1989, the keyhole has become a well-known Nordic label for healthier food products in Denmark, Norway, and Sweden.

26.4.7 Fortification of Food Products

In relation to the description of the nutrition and health claim rule, it is worth mentioning that fortification today follows Regulation 1925/2006/EC. Sweden has

---

[30]National Food Agency, Guidance for Control Authorities, *available (in Swedish) at* http://www.slv.se/upload/dokument/livsmedelsforetag/vagledningar/vagledning_narings-_och_halsopastaenden.pdf.
[31]*Id.*
[32]*Id.*
[33]Gränsdragning mellan hälsopåståenden för livsmedel och medicinska påståenden, Medical Products Agency's guideline February 14, 2011.
[34]National Food Agency, The Keyhole Symbol, *available at* http://www.slv.se/en-gb/Group1/Food-labelling/Keyhole-symbol/.

abandoned the previous system of authorization regarding fortification. Certain products produced in Sweden must be fortified; such is the case with milk (with a maximum 1.5 % fat), margarine, and products of similar nature and use. These have to contain stipulated levels of vitamin D and A.[35] This fortification requirement does not apply to products produced in other EU Member states or Turkey.

Regarding enrichment, the list of plants and parts of plants deemed unsuitable for use in food including food supplements should be considered.[36]

26.4.8 Labeling Cases According to the Marketing Act

In *Svensk Mjölk v. Valio*,[37] the marketing of margarine with the claim "Soft butter using canola oil" was found to be in violation of the Food legislation (Regulation (EC) 1234/2007). Butter is a protected denomination and cannot be used for a product consisting of 67 % milk fat and 33 % canola oil.

The 2004 case *Vaasan & Vaasan Oy v. Fazer Bröd AB*[38] concerned the marketing of a particular type of thin crisp bread. Vaasan & Vaasan and Fazer had marketed this product for many years. Fazer created a new product in packaging that Vaasan & Vaasan found too similar to theirs. Primarily, this was a case of passing off, but two claims were also in question: "original" and "genuinely authentic." The market Court found that the claim "original" gave the impression that the product in some aspect was "original." Fazer had not substantiated in what sense the product was original, and the claim was, therefore, considered misleading. With regard to "genuinely authentic," the Court found that the claim gave the impression of quality; however, it could also be construed to mean that the crisp bread is based on a Finnish tradition which Fazer had argued. The Court did not find this claim misleading. Fazer was subjected to a prohibitive injunction conditioned on a fine of Swedish Krona ("SEK") 400,000.

26.4.9 Advertising Cases According to the Marketing Act

As has been said above misleading labeling and food advertising falls not just under food law, but is also actionable according to the Marketing Act. Marketing in violation of other legislation—in this case, food law—is simultaneously a violation of the Marketing Act. In fact, most cases concerning food advertising are handled

---

[35]National Food Agency's Regulation SLVFS 1983:2.
[36]National Food Agency, List of Plants and Plant Parts that Do Not Belong in Food, *available (in Swedish) at* http://www.slv.se/upload/dokument/risker/naturliga/vaxter/volm.pdf.
[37]Market Court's decision 2010:15 *Svensk Mjölk AB v. Valio Sverige AB*.
[38]Market Court's decision *Vaasan & Vaasan Oy v. Fazer Bröd AB*.

according to Marketing Act by the Consumer Agency of the operators sometimes with the assistance of the National Food Agency. This also reflects the division of responsibilities between the National Food Agency and the Consumer Agency.

The so-called Änglamark-case *Sveriges Spannmålsodlare (SpmO) AB v. Kooperativa Förbundet (KF)*[39] concerned environmental claims for a range of products sold under the trademark "Änglamark." That name first appeared in song lyrics, and Swedes strongly associated it with the preservation of nature and the environment. A TV-commercial for the products showed a restaurant scene with a couple that had just been served their food. A man with a pesticide sprayer on his back comes up to the table and starts spraying the food on the plates. When they ask what the man is spraying on their food, he answers, "Poison," and that there is no scientific research proving its danger.

"Everybody else is eating it." The commercial ended with "Änglamark – guaranteed unsprayed." The Court found the commercial to be misleading. First, it claimed that ecologically produced products were not sprayed with pesticides. This was not correct; the products were not sprayed with chemicals. Second, the commercial gave the impression that other food contained poison. Even if it had been shown that some food products contained residue from pesticides, the claim was too categorical and oversimplified. Additionally, the commercial discredited food producers who used pesticides in their production. Therefore, the Court considered the marketing misleading, and the Market Court, rendered a prohibitive injunction conditioned on a fine against the defendant, Kooperativa Förbundet.

In *Danske Slagterier, SA Bruxelles v. Scan Foods AB*,[40] the leading Swedish meat producer made the following claims: "Choose imported pork so you can get medicine in the bargain. That we cannot offer, because our meat is free of antibiotics," and "Christmas ham from the best pork in the world." The first claim was misleading because the advertiser could not prove that foreign-produced meat contained antibiotics while meat produced in Sweden did not. The second claim was also considered misleading as a general and unconditional quality assertion that should be understood as a claim that the advertiser's pork is truly superior to all other pork. The Court questioned if such a claim could ever be substantiated. Ultimately, the Court rendered a prohibitive injunction conditioned on a fine against the defendant Scan Foods AB.

The case *Institut National des Appelations d'Origine (INAO) et al. v. Arla Foods AB* involved a misleading geographical origin.[41] In 1999, Arla Foods launched two yogurt products: Yoggi Champagne and Yoggi Original Champagne. The marketing for these products also made reference to champagne. A press release contained the following text: "Champagne producers have busy days. Millennium approaches

---

[39]Market Court's decision 2002:8 *Sveriges Spannmålsodlare (SpmO) AB v. Kooperativa Förbundet (KF), ekonomisk förening.*

[40]Market Court's decision 2005:8 *Danske Slagterier, SA Bruxelles v. Scan Foods AB.*

[41]Market Court's decision 2002:20 *Institut National des Appelations d'Origine (INAO) et al. v Arla Foods AB.*

and hysteria increases. What to do? Go to a party, stay home, pretend like it's raining? While the darkness of winter settles over Sweden, what is needed is something that cheers you up. Something that adds flavor to life - Yoggi Champagne." In other marketing, the products were presented as yogurt with a champagne flavor. The advertising made reference to champagne with pictures of bubbles and people at parties with glasses of champagne.

The Market Court found that Arla exploited the goodwill of champagne in violation of the Marketing Act. The Court also took into consideration the design and color scheme of yogurt packaging and posters where glasses of champagne and champagne bottles were replaced with Arla yogurt packaging. In light of the foregoing, this was, in the Market Court's view, a clear case of a conscious passing-off.

The Court also found the marketing to be misleading. First, claiming a "champagne taste" in connection with the appearance of the packaging gave consumers the impression that the yogurt had been flavored with or tasted like champagne. The Court considered the list of ingredients on the pack insufficient to neutralize this impression. As for the taste of champagne, it was apparent from the evidence that the ingredients could not give this flavor. Arla was not able to show that the yogurt actually tasted of champagne, and thus, the company had misled consumers.

The Court rendered a prohibitive injunction against Arla with regard to these claims. However, the Court did not find that the claim "champagne flavor" mislead the consumer concerning the commercial or geographical origin. The Court reasoned that yogurt and champagne are separate products, and therefore the consumer was unlikely to be misled.

In the case Svensk Mjölk Ekonomisk förening ("Swedish Dairy") vs. Oatly AB (MD 2015:18), Swedish Dairy intervened against Oatly for their marketing of various oat-based products which can be used as substitutes for milk. Oatly used expressions like "No milk. No soy. No badness", "No milk, no soy, no nonsense", and "No milk, no soy, no craziness". There were claims used like "It's like milk but made for humans", "No milk, no soy, no badness". Swedish Dairy claimed that the expressions and claims were misleading and denigrating advertising. The Swedish Market Court delivered its decision finding in favour of Swedish Dairy with regard to the claims mentioned and others. The Court however did not find "No milk, no soy, no way" or "wherever and whenever you would find yourself drinking milk or using it in a recipe 'back in the day', you can use Oat Drink today", and "When should you use it? Whenever you would use old school milk from cows..." to be in violation of the Marketing Act. Swedish Dairy claimed that it was denigrating to indicate as some of these claims do that milk was a thing of the past.

In the matter Arla Foods AB vs. Unilever AB (MD 2105:20) Arla claimed that "Flora MED SMÖR Normalsaltat" ("Flora with butter normal salt") and "Flora MED SMÖR och Fint Havssalt" ("Flora with butter fine sea salt") was misleading and in violation of the Food Act for use of the protected denomination butter. Flora is a blended spread but containing butter. The Market Court did not find Unilever's labelling or marketing to be in violation of applicable EU-legislation, the Swedish Food Act of the Marketing Practices Act. The products were sold under their correct descriptive denomination.

26.4.10 Case Law Concerning Health Claims

The case *Bayer AB v. Bringwell Sverige AB and Bringwell AB*, which concerned the marketing of food supplements, was the first to apply the Regulation 1924/2006 on nutritional claims and health claims for food.[42] Bringwell, a company sells self-care items marketed a food supplement called Mivitotal. Bayer brought the case before the Market Court, which rendered a prohibitive injunction against no less than 74 claims used by Bringwell. The injunction was subject to a fine for each claim against the product in the amount of SEK 1,000,000 (approximately $138,615.50 U.S. dollars) in the event of non-compliance with the injunction. This demonstrated an exercise in the Regulation's articles on nutritional claims and health claims as well as in the application of the Medical Products Act on claims made for food products. Bayer's suit concerned general health claims, specific health claims, and medical claims. All prohibited claims were also considered misleading according to the Marketing Act. Below are a few examples of the claims that were prohibited.

The case demonstrates what happens when a company sues a competitor using the MPA with reference to the food legislation and the Medical Products Act. Not only does the company have to pay the claimant's costs, but the company will also suffer under an injunction which, if violated, will result in an order to pay a fine of SEK 1,000,000. Such is the unpleasant consequence for the defendant. Furthermore, it is possible for the claimant to ask for damages if the claims have been found misleading (see below).

26.5 Conclusions: How It Works

The current trend concerns health and nutritional claims regarding food marketing and labeling. Many issues remain unresolved, both at the EU and national levels. In accordance with the Marketing Act, the cases have primarily been handled between operators, or between the Consumer Ombudsman and an operator. However, there has been increased activity by the national and regional authorities in the past year.

There are cases for instance according to Regulation 1924/2006/EC that concerns the dairy manufacturer Arla, which launched a new product line called "Wellness," a probiotic dairy line, of yogurts and milk. A number of product variants have since been launched: "Balance," "Immune," "Protein control," and "Female." Here, the regional food agency in Stockholm intervened, leading to an ongoing procedure in the Administrative Court. Apart from the more practical issues, such as the placement of a claim on the pack, it is a case that demonstrates several interesting issues. It shows a weakness in the authorities' system for supervision. The product launch took advantage of the transitional rules with regard to health claims on the market, but which had not been approved by the

---

[42]Market Court's decision 2013:13 *Bayer AB v. Bringwell Sverige AB and Bringwell AB*.

EU-Commission. The regional agency demonstrated a reluctance to intervene with more drastic measures. Had Arla accepted the agencies' position, the Court would have given the company a transitional period to phase out the products on the market and replace them with packaging in accordance with the Regulation. As it currently stands, they have made some concessions and changes, but Arla was still able to make a forceful launch for the product, getting the message of the products' benefits across before having to get in compliance. This case also highlights the difficulties manufacturers of probiotic products face, since there are virtually no approved claims to make for the probiotic bacteria. Alternatively, manufacturers resort to making an approved claim; for example, an approved claim on the immune system based on the content of vitamin C or D, thus blurring the claim so it is perceived as concerning the probiotic as well.

The question remains: does legislation promote innovation or hamper it? Numerous resources and money go into the development of a probiotic, for example. Once the product gets on the market, there is no possibility to commercialize the innovation because there are simply no approved claims to be made. It is not only the industry's innovations that are thwarted; the initiative of new starts or even private individuals also suffers. Consider a group of students in Sweden who had an idea of creating an easy-to-read grading system that would make it easy for consumers to choose products graded on health, ethics, environment, and society. The products' scores on these qualities would appear on the label in a simple graphic. However, it is unclear whether such an idea would be possible to launch, taking into consideration the Regulation 1924/2006/EC.

Another case, which concerned the labeling regulation, concerned the use of "genuine," or in Swedish "äkta," in the context of an overall program to provide consumers information about the composition of food products. This was not tried in the Courts. The concept "äkta vara" or "real stuff" was launched by a private non-profit organization and addressed the use of additives in food. "Genuine" food was, according to this concept, food without additives. The organization registered a trademark and offered operators licenses to use the trademark, provided of course, that the products met the organization's stipulated requirements. Reactions came from both the food industry and from the National Food Agency, which called the trademark misleading unless the organization explained how the product carrying the trademark was "äkta." Several operators and a chain of grocery stores signed up as licensees and used the trademark in their businesses. The National Food Agency requested that the municipal agencies under which the operators in question were registered should take action. However, the agencies took no immediate actions. Simultaneously, the National Food Agency made public its position, making it difficult for the organization to attract new licensees. No licensee desired getting stuck with a license it may not be able to use. Specifically, National Food Agency reacted to the use of the word "äkta." The matter reached a resolution when the organization changed the trademark to just "Ä" with a reference to the organization's name, Äkta Vara.

Legal intervention is also a fickle character. The so-called horsemeat scandal, although considered a major labeling scandal, did not lead to prosecution in Sweden. With such an obvious fraud, it was expected that there would be legal consequences for the companies involved, including one of Sweden's oldest and most well-known food operators. Only one company filed a complaint with the local public prosecutions office, which dismissed it after examining the operator, in question. It was found to have its documents in order and therefore there was no longer suspected of any wrongdoing. Even though it should be said that the companies, which were most prominently exposed, took responsibility and adopted measures to avoid future scandals the total lack of prosecution might send the signal that food fraud is no big deal. Indeed the horsemeat scandal has been followed by other scandals; for example, pork meat sold as beef or lamb to kebab restaurants with obvious religious consequences.

There must be a balance in legislation and the application of the legislation. The responsibility for this falls on all the actors on the market, including the operators, authorities, industry organizations, and consumers.

A fine balance also exists between legislation and innovation. A topic not yet discussed in this chapter is new technology and the labeling requirements. It raises the question of whether or not legislation promotes or hampers innovation. New production techniques, perhaps necessary to help feed the world and to do it in an environmentally sustainable way, may be thwarted by a labeling requirement, forcing the operators to disclose the nature of the ingredients and how they have been produced. Consumers, advised by organizations' authority figures opposed to a product, may reject said product, making commercialization difficult,. This also affects global trade.

It is no secret that the view of genetically modified produce differs between the United States and the EU. In some Member States such as Sweden, the resistance is strong. A similar problem emerges with regards to nanotechnology, the cloning of animals and vegetables, and food produced in laboratories rather than on farms. On the other hand, it is in the consumers' interests to know the origin of food, which is used as an argument for disclosure on labels. If the operators or government agencies fail to convince the consumers that the product is safe, in the long run the problem may not be with labeling, but rather with communications.

Failure to disclose a product's genetically modified ingredients could result in the next "horsemeat scandal" in Europe, It is worth considering the precautionary principle as an unaffordable luxury, given that the world needs to be fed in a sustainable way from both an environmental and security viewpoint. These appear to be effects of the same problem. The further companies abuse nature and the environment, the more common the negative effects will be for food products and the more likely that the prediction that tomorrow's conflicts will concern access to fertile land. Consumers must decide if they are prepared to sacrifice their right to know what their food is made of, and to be able to choose whether to accept the government agencies and operators' assurances that the ingredients are safe for the public to eat.

References

National Food Agency
Court of Justice of the European Union ("ECJ") Case C-162/05
Swedish Nutrition Foundation
http://www.halsopastaenden.se/historik/
The Nordic Council, Nordic Nutrition Recommendations
The National Food Agency, Action Plan for Healthy Eating Habits
The Swedish Consumer Agency
The Medical Products Agency
Marketing Act ("Marknadsföringslag")
The Swedish Food Federation
Swedish Retail
National Food Agency's guideline concerning sanctions in the food legislation "Vägledning – sanktioner i livsmedelslagstiftningen"
Supreme Administrative Court, case no 3686-04
Lavin, Lag om viten, En kommentar, 2010, pp. 147
Government bill 1984/85: 96 p. 56 and Government bill no 57 år 1970 s. 92
Case no B 14536-11 Consumerombudsman v. L'Oreal
Stockholm Municipal Court's decision case no B 7434-00
Appellate Administrative Court in Stockholm, Case Number 4331-05
Ordinance (1956:296)
Supreme Administrative Court, case no 5662-2000
Regulation (EU) No 1169/2011 of the European Parliament and of the Council of 25 October 2011
Directive 2000/13/EC of the European Parliament and of the Council of 20 March 2000
Introduction to National Food Administration Regulations
Supreme Administrative Court, case no 5675-1991 RÅ 1992 ref 95
Nacka Municipal court, case no B 4330-08
Administrative Court in Uppsala case no 3453-10 and 3454-10 Findus Sverige AB v. The National Food Agency
National Food Agency, Guidance for Control Authorities
Gränsdragning mellan hälsopåståenden för livsmedel och medicinska påståenden, Medical Products Agency's guideline February 14, 2011
National Food Agency, The Keyhole Symbol
National Food Agency's Regulation SLVFS 1983:2
National Food Agency, List of Plants and Plant Parts that Do Not Belong in Food, available
Market Court's decision 2010:15 Svensk Mjölk AB v. Valio Sverige AB
Market Court's decision Vaasan & Vaasan Oy v. Fazer Bröd AB
Market Court's decision 2002:8 Sveriges Spannmålsodlare (SpmO) AB v. Kooperativa Förbundet (KF), ekonomisk förening
Market Court's decision 2005:8 Danske Slagterier, SA Bruxelles v. Scan Foods AB
Market Court's decision 2002:20 Institut National des Appelations d'Origine (INAO) et al. v Arla Foods AB
Market Court's decision 2013:13 Bayer AB v. Bringwell Sverige AB and Bringwell AB
Real Stuff (Akta Vara)
ECJ decision C-313/11 EU-Commission v. Poland, C-442/09 Karl Heinz Bablok et al. v Freistadt Bayern, C-36/11 Pioneer Hi Bred Italia SRL v Ministerio della politiche agricole alimentari e forestali

Chapter 27
Food Law in Norway: Trade, Food Promotion, and Protection of Intellectual Property Within the Food Industry

Marie Vaale-Hallberg and Nina Charlotte Lindbach

Abstract Being one of the northernmost countries in Europe, with an extensive coastline and with little opportunity for large-scale agriculture, facilitating for international trade and at the same time protecting Norwegian agriculture is of essential interest to Norway (This chapter was written in January 2015). Like in many other countries, the Norwegian customs tariffs function as effective barriers to trade, ensuring that imported products potentially threatening domestic agricultural products are not sold at lower prices than domestic products. International trade is however crucial as the Norwegian consumption presupposes extensive import of food. Instead of joining the EU, principally in order to protect agricultural and fishery resources, Norway has chosen to remain a member of the EFTA and a party to the EEA Agreement. As part of the internal market, a consequence of the EEA Agreement, many of the EU rules relating to foodstuffs are applicable, but also domestic rules with no equivalent EU legislation apply. Norway has traditionally been, and still is, one of the strictest countries in Europe when it comes to e.g. marketing towards children.

Investing in product development and innovation is important to most food suppliers. Protecting intellectual property and thus competitive advantages appear to be of growing interest. Up till now the legal battles between the market players have mainly been about packaging design and unfair business practices.

M. Vaale-Hallberg (✉)
Kvale Advokatfirma DA, Oslo, Norway
e-mail: mvh@kvale.no

N.C. Lindbach
Advokatfirmaet Haavind AS, Oslo, Norway
e-mail: n.lindbach@haavind.no

© Springer International Publishing Switzerland 2016
G. Steier, K.K. Patel (eds.), *International Food Law and Policy*,
DOI 10.1007/978-3-319-07542-6_27

27.1 Introduction

27.1.1 International Trade of Foodstuffs

Norway, a constitutional democracy, is Europe's northernmost country. It has a population slightly exceeding 5 million, mostly clustered in and around the bigger cities. Due to the offshore oil and gas deposits discovered in the 1960s, Norway enjoys one of the highest living standards in the world. While the general cost of living is rather high, the inhabitants only use approximately 11–12 % of their income on food, which is low compared to most other European countries.[1]

Norway stretches 2518 km from north to south and has a coastline of approximately 103 km. The topography and climate cause some obvious challenges when it comes to self-sufficiency of food. For example, large-scale farming is difficult and shipping food across long distances is costly. Consequently, food imports are necessary. Presently, Norway imports more than 50 % of the food consumed in the country, of which approximately 70 % is imported from the EU. However, access to the sea and other great fishery resources make Norway one of the largest seafood exporters in the world.

Balancing the interests between facilitating import and export on one hand, and supporting domestic agriculture on the other is difficult but inevitable. A country's desire to protect its own interests while at the same time allowing and facilitating increased sales from developing countries is often debated. This challenge is not new. Negotiations at WTO-level regarding an increase in international sales of agricultural products faced challenges during the DOHA negotiations in 2001, revealing the inevitable deadlocks created by the positions taken by the various countries. After more than a decade, the WTO members eventually managed to agree upon a text regarding agricultural products at meetings in Bali in December 2013.

As Norway is dependent on imports of food, it is essential for the country to form and build a strong cooperation with other countries. Thus, international trade has been important to Norway for centuries, and various agreements have been entered into to ensure that cooperation continues. One of the most important international collaborations regarding the trade of food began when Norway founded the European Free Trade Association (EFTA) in 1960, along with other European countries. The Convention was originally signed in Stockholm on January 4, 1960, and outlined the principles of the cooperation. It was later revised in the Vaduz Convention of June 21, 2001, with annexes, protocols and appendices. The EFTA is an intergovernmental organization promoting free trade and economic integration to the benefit of its members.[2] Since its foundation, it has managed to

---

[1] Eurostat Database, European Commission, http://epp.eurostat.ec.europa.eu/portal/page/portal/household_budget_surveys/Data/database.

[2] European Free Trade Association (EFTA) of 1960. Jan. 4, 1960, available at www.efta.int/sites/default/files/documents/legal-texts/efta-convention/Vaduz%Convention%20Agreement.pdf.

negotiate a range of well-functioning international free trade agreements, the most important being the agreement with the EU.

In the beginning, the EFTA formed a strong coalition and attracted countries such as the United Kingdom, Austria, Denmark, Portugal, Sweden, Switzerland and Finland. However, after the EU gained popularity and power, most of the members of EFTA shifted to the EU, thereby weakening the position of EFTA. Today, only Iceland, Liechtenstein, Norway, and Switzerland are members of EFTA.[3]

27.1.2 Norway and the EU

Despite Norway's failure to join the EU, the relationship between Norway and the EU remains extensive due to the European Economic Area agreement of 1992 (EEA Agreement) between EFTA and EU.[4] The EEA Agreement implies that Norway, Iceland, and Lichtenstein (EEA/EFTA Member States) will participate in the internal market established in the EU for all goods and services designated under the agreement. Under this arrangement, the fundamental principles of EU law apply to the EEA/EFTA Member States, such as the four freedoms of movement: goods, capital, services and people. The EEA Agreement also implies cooperation in other areas such as research and development, education, social policy, and environment. Moreover, the agreement is built upon the principles of equal rights and obligations within the internal market, both for citizens and economic operators in the EEA.

One of the core principles of EEA law lies in its dynamic character; development is both intended and wanted. EU Member States have passed on legislative power to EU bodies, whereas the legislative power of the EEA/EFTA Member States remains unaltered. By comparison, any alteration to the EEA Agreement is de facto a new international agreement, which by implication requires the consent of all EEA/EFTA Member States. In order to secure a more fluid decision-making process, the EEA Agreement lays down the principles of development and interaction between EU laws of relevance to the EEA Agreement and specifies how such legislation is to be adopted by the EFTA Member States. The EEA Joint Committee, a committee where both the European Union and the EEA/EFTA Member States are represented, makes the decision to incorporate new EU legislation into the EEA Agreement.[5] Accordingly, the EEA/EFTA Member States will implement the legislation into domestic law through legislative or regulatory decisions.

EU legislation is not the only factor that plays an important part in the harmonization process. The Norwegian Supreme Court has stated[6] that even though

---

[3]*Id.*, p. 2.
[4]Agreement on the European Economic Area (EEA Agreement), May 2, 1992.
[5]EEA Agreement, 4, Article 102.
[6]Rt. 2002, p. 391 (God Morgen), Rt. 1997, p. 1954.

relevant case law from the European Court of Justice (EJC) is not formally binding upon Norwegian law practitioners, such case law shall be taken into consideration in the interpretation of Norwegian law. Furthermore, practices from other important EU organs, like the Office for Harmonization in the Internal Market (OHIM) and the European Patent Office (EPO), will be taken into consideration, as these constitute relevant sources of law in Norway.

The EEA/EFTA Member States have no voting rights when it comes to the development of EU law, but they are virtually obligated to adopt the rules covered by the EEA Agreement. The co-operation does facilitate ways that the three EEA/EFTA Member States may influence the law making process, such as influence exerted through expert committees. Formally, the EEA/EFTA Member States may decline to incorporate new EU legislation into the EEA Agreement. In such cases, the affected part of the EEA Agreement will be suspended.[7]

Although there are no formal membership fees connected with the EEA, Norway is contributing financially to the EU and its Member States. In 2009–2014, contributions from the EEA/EFTA Member States amounted to €1.8 billion.[8] Whilst the contributions from Norway are extensive, Norwegian inhabitants and enterprises also benefit from the various EU programs.

Norway has formally applied for EU membership three times. The first application was sent in 1962, but was hindered by the veto laid down by France to stop the pending British application for EU membership. The second application was submitted in 1970, but was withdrawn after the national referendum showed that 53.5 % of the population (voter participation was 79.2 %) was in disfavour of membership. The third application was sent in 1992, but once again the national referendum proved that 52.2 % of the population was against membership. A voter participation rate of 88.6 % illustrates that the question of membership was both engaging and challenging. Some of the most disputed issues were, and still are, Norwegian agriculture, fishery resources, and sovereignty.

Accordingly, agricultural products are not a part of the EEA Agreement. However, Article 19 of the EEA Agreement aims at developing trade between the countries with regard to agricultural products and explicitly seeks liberalism within the trade of agricultural products. For instance, a new agreement between Norway and the EU entered into force January 2012, whereby the duty-free import quotas for cheese were considerably extended, from 4500 metric ton up to 7200 metric ton. Additionally, other duty-free import quotas were agreed upon, and some provisional arrangements were made with regard to products such as meat. Particular rules apply to manufactured agricultural products[9] and fish.[10]

---

[7]EEA Agreement, see *supra* n. 4, Article 102(5).
[8]"About the EEA and Norway Grants"("Hva er EØS-midlene?") http://www.regjeringen.no/nb/sub/europaportalen/eos-midlene.html?contentid=685567&id=684349.
[9]EEA Agreement, 4, Protocol 3.
[10]EEA Agreement, 4, Protocol 9.

27.1.3 Percentage-Based Tariffs

In January 2013, the Norwegian Government (Stoltenberg II) amended the customs tariffs and introduced a percentage-based customs duty on certain types of food. The main reason behind the change was an attempt to position Norwegian agriculture against future import. The percentage-based system would only apply to certain types of cheese, beef fillets, and lamb, all being goods particularly vulnerable to foreign competition. The various levels in the EU and EU Member States instantly expressed dissatisfaction, claiming that this new regime is not in line with the spirit of the EEA Agreement and the EEA Agreement Article 19. The new Government (Solberg I), which took over in the Fall of 2013, indicated that it would be taking measures to reverse the percentage-based tariffs but so far such reversal has not taken place.

27.2 Trading with Countries Outside Europe and the Transatlantic Trade and Investment Partnership (TTIP)

Through the EFTA-membership, Norway is a part of an extensive network of trade agreements and other partnership agreements. Currently, the EFTA Member States have 27 free trade agreements (covering 26 states) with countries such as Canada, Costa Rica and Panama, Chile, Colombia, Turkey, and Singapore.

Presently, there is no free trade agreement between Norway and the U.S., but there are on-going negotiations between the U.S. and the EU concerning a free trade agreement. If a free trade agreement between the U.S. and the EU is entered into, it would most likely affect Norway. The effect on Norway will occur partly because the agreement will likely render necessary changes in EU law relevant to the EEA Agreement, and partly because it will potentially pave the way for agreements between EFTA and the U.S. Undoubtedly, there will be issues that, for political reasons, are likely to be problematic for Norway, such as the position taken in U.S. regarding genetically modified food. The Transatlantic Trade and Investment Partnership (TTIP) is a feared and controversial source of friction that will hopefully not harm European food safety if it comes to pass.[11]

Norway has been a member of the World Trade Organization (WTO) since its establishment in 1995. The WTO collaboration covers several agreements, including: the General Agreement on Tariffs and Trade (GATT), the General Agreement on Trade in Services (GATS), and the Agreement on Trade-Related Aspects of Intellectual Property Rights (TRIPS).

---

[11] For more information on TTIP and the underlying concerns, see http://ec.europa.eu/trade/policy/in-focus/ttip/ and http://www.centerforfoodsafety.org/files/cfs_trade_matters_76070.pdf.

27.2.1 Restricting International Sales of Agricultural Products

In addition to national subsidies, customs barriers are generally seen as an essential measure to protect domestic agriculture. When imposing tariffs on imported products, the price naturally increases. Depending on the political motives, the retail price of the imported product may end up either equal to or considerably higher than domestically produced goods, thereby making imported goods less attractive from a price perspective. Most countries are in need of imported products due to fluctuations in internal production, and states prefer to have a flexible tariff and import regime that allows them to take into account the country's shifting needs.

The system selected in Norway is built upon methods of reducing tariffs and may be exemplified by the following overview:

- Certain products may, on a *general basis,* obtain a tariff reduction. For example, reductions will be imposed when the domestic production be too small and serious market disruption is likely without imports.[12]
- Certain products may obtain a tariff reduction on the basis of an *individual application.*[13] This reduction is relevant for types of foodstuffs that are not offered by domestic producers, e.g. canned mango, or products that include foodstuffs that are not offered by Norwegian producers. For such foodstuffs, there is no need to protect domestic agriculture.
- Certain products may be imported free of tariffs within certain quotas or otherwise enjoy a reduction of the tariffs within certain quotas.[14]
- Customs preferences for processed agricultural products exist for countries which Norway has entered into a free trade agreement with, e.g. the EEA Agreement.[15]
- The Generalized System of Preferences (GSP) of the EU allows tariff reductions for developing countries.[16] As Norway is an EFTA/EEA member state, this particular scheme applies.
- Particular uses of a certain product may substantiate a tariff free import.[17] For example, intermediate products to be used by the fishing industry are subject to tariff free importation.[18] One argument for this tariff reduction is that it helps to secure Norwegian aquaculture industry's competitiveness on the world market. The policy implies that Norwegians should be given equally favourable

---

[12]Regulation on Reduction of Customs Tariffs of 2005, Chapter 2. (*emphasis added*).

[13]Regulation on Reduction of Customs Tariffs of 2005, Chapter 3.

[14]Regulation on the Allocation of Tariff Quotas for Agricultural Products of 2008.

[15]Regulation on the Determination of Discounted Tariffs for Imports of Industrially Processed Agricultural Products, 2013.

[16]Generalized System of Preferences of 2008.

[17]Regulation on the Reduction of Customs Tariffs on Agricultural Products of 2005, Chapter 3.

[18]*Id.* at Section 15.

purchasing terms as their competitors so as not to be rendered less competitive by tariffs.

Surveying actual production, ensuring that the production meets the actual demand, and having measures to deal with under-coverage and excess production is handled in various ways. In Norway, such control and responsibility for certain product categories is regulated by market regulation; market regulation is one of the core principles of Norwegian agricultural policies.

27.2.2 Norwegian Market Regulation

The Norwegian market regulation schemes have been in place since the 1930s. The Act on Sales of Agricultural Products from 1936, and the preceding acts, came as a direct consequence of imbalances in the market that lead to dramatic price reductions and reduced income for farmers. The main purpose of the market regulation is to ensure that there is balance between demand and supply at all times. At present, there are market regulators for the following sectors: milk, red meat, corn, and eggs.[19]

The market regulation is financed through a fixed amount, payable per kg/litre of sales.[20] This sales tax is allocated to a fund that supports and funds immediate measures to stabilize the market or measures such as education and information activities. Various subsidies are also provided for farmers and producers in order to stimulate ecological production and even out transportation costs for various districts.

The main objective of the market regulator is to attain the maximum prices set out in the annual agreement between the farmers and the government, thus ensuring farmers their intended income. It follows that the market regulator is also intended as a means of securing product supply in line with actual demand, and helps organize the import of products should the domestic supply be insufficient. Another important obligation on the market regulator is the duty to take over products from all producers and the duty to supply members of the market regulation with products. As the market regulator is responsible for taking measures to ensure market stability, it is in the regulator's interest to ensure that the market is well taken care of. For the reason that it is the primary producers who pay the taxes which fund the market regulation, it is in their interest to limit over-production.

The appointed market regulators are strong cooperatives in charge of both collecting the food and processing and selling the food, principally under strong brand names. The market regulation has been criticized for having considerable financial interests in the sale of the products, as they are the lead players in both

---

[19]Regulation on the Market Regulation to Promote the Trade of Agricultural Products (2008).
[20]*Id.* at Sec. 3-3.

producing raw material and producing processed food. The Norwegian courts have assessed various aspects of the designed arrangements, but the systems have, so far, not been deemed invalid.[21]

In order to promote the trade of agricultural products, the Sales Council for Agricultural Products (the Council) was established in 1936.[22] The Council is an independent body, but its decisions may, to a limited extent, be re-examined by the Ministry.

The Solberg I government wants more competition within the food industry, and in March 2014 the Ministry of Agriculture and Food appointed a committee which shall evaluate the existing market regulation schemes. The result of this work is yet to be presented at the time this chapter is written.[23]

27.3 The Norwegian Legislation on Food and Beverages

27.3.1 Introduction to the Norwegian Legal Framework on Foodstuffs

The Norwegian legal framework on the import, production, handling, selling, and promotion of foodstuffs is found in the Act on Foodstuffs of 2003[24] (Food Act) and associated Regulations. Being a framework law, the Food Act establishes certain principles, rights, and obligations, all of which are further detailed in a range of regulations applicable to certain products or situations.

In retrospect, EU legislation has been the principal driver of changes within the legal framework applicable to foodstuffs. Most of the EU legislation on food is EEA relevant and thus, necessary for Norway to incorporate. In sum, Norwegian law is principally equivalent to EU law.

Despite this, there are currently still regulations of domestic origin in Norway, such as several "quality regulations", that are not based on equivalent EU law. These regulations pertain to products such as vegetable conserves (canned vegetables), milk, cream, cheese, edible ice, and butter.[25] The purpose of these quality regulations are, among other things, to ensure that manufacturers promote their

---

[21] Judgment of Borgarting Court of Appeal of 15 January 2007.

[22] Act on the Trade of Agricultural Products of 1936, Section 1.

[23] The committee delivered its report in June 2015. The committee unanimously agreed that the market regulation in place should not be abolished. However, the committee split into three factions that each recommended alternative models for the future which varied with regards to, amongst others, to what extent one should allow the strong cooperatives to take part in the regulation.

[24] Act on Foodstuffs of 2003.

[25] Regulation on the quality of milk and cream of July 17, 1953; Regulation relating to the manufacturing, labelling and sale of cheese of August 24, 1956; Regulation on the manufacturing, labelling and sale of butter and butter fat of November 16, 1962; Regulation on the manufacturing,

products in an honest manner and in a way that does not mislead consumers. They require the use of certain raw materials and production methods and set forth requirements regarding labelling that must be met in order for the manufacturer to be able to define his products in accordance with the definitions set forth in the relevant regulation.

It should be noted that the Food Safety Authority recently proposed a revision of these regulations, as they appear to be outdated to some extent.[26] Some regulations are therefore expected to be fully repealed in the future, and the remaining will be amended in order to line up with industrial developments.

Even though EEA/EFTA Member States strive at adopting EU legislation relevant to the EEA Agreement promptly, it sometimes takes considerable time due to matters concerning one or more of the EFTA Member States. This was the case when the EEA Joint Committee was to include Regulation (EC) No 178/2002 of the European Parliament and of the Council of January 28, 2002. Regulation (EC) No 178/2002 (the General Food Law Regulation) laid down the general principles and requirements of food law, established the European Food Safety Authority, and laid down procedures in matters of food safety.[27] Due to the delay on the EEA/EFTA-level, Norway initiated a revision of the legal framework applying to food and feed in June 2001, aiming to unite an extensive range of acts on food and feed. Taking into account development at EU level, the mandate was revised in 2002 to ensure that the new legal framework would be in compliance with the framework developed by the EU.

The purpose of the mandate was to ensure that no material changes to the Norwegian Food Act would be necessary on the day of formal incorporation of the General Food Law Regulation. The EEA Joint Committee considered and included the General Food Law Regulation into the EEA Agreement in October 2007, but because one of the EEA/EFTA Member States needed time to prepare for the implementation, the new rules were not effective in Norway until Spring 2010.

From a legal perspective, one should note that the wording used in the Food Act is not entirely similar to the General Food Law Regulation. This was done intentionally, partly to keep the Norwegian Food Act in line with Norwegian legal traditions, and partially because the contents of the General Food Law Regulation was found not to be easily accessible. These are valid arguments, but they also raise

---

labelling and marketing of edible ice of April 15, 1977; and Regulation on vegetable conserves of January 1, 2001.

[26]Cf. the request for comments of 1 November 2013 by the Norwegian Food Safety Authority (only available in Norwegian) which can be found here: http://www.mattilsynet.no/mat_og_vann/merking_av_mat/generelle_krav_til_merking_av_mat/horingsbrev.11601/binary/Høringsbrev. All of the before mentioned regulations was cancelled on 1 January 2016. The regulations relating to butter, edible ice and vegetable conserves have not been continued. The most important parts of the regulations on milk, cheese and butter have been continued in the Regulation on the quality of milk and milk products of 3 June 2015 (in effect from 1 January 2016).

[27]Regulation (EC) No. 178/2002 (The General Food Law Regulation) of Jan. 28, 2002, 2002 O.J. (L31/1).

some concerns: Norway has, from the time the General Food Law Regulation formally entered into force in Norway, to some extent, had two sets of rules on the same subject matter that are not worded the exact same way. This creates concern about the risk of conflict. EU sources of law, however, rank higher in cases of conflict of laws.[28]

Norwegian policies on food, including advertisements, are rather strict compared to other those of countries. For example, there is a full ban on the marketing of alcoholic beverages, justified by the need to protect public health.[29] The ban is further detailed in the Regulation of June 8, 2005, Chapter 14[30] which places restrictions on the use of trademarks and brand names of alcoholic beverages. Should a trademark or brand name be used for both non-alcoholic and alcoholic beverages, the advertisement of the non-alcoholic beverage is also restricted unless the alcoholic beverage uses its own distinct trademark. There are a few minor exceptions, such as advertisements in foreign magazines, giving information on point of sale and service, and showing information signs of small size in direct connection with the sale or serving.

Even though tobacco is not a food, the prohibition of tobacco advertisements also illustrates the strict Norwegian policy on advertising.[31] Even the mere display of tobacco in advertisements is illegal. This is significantly more progressive than most countries' counter-parts of legal regulation of tobacco advertisements and shows how well public health can be promoted in such regulation.

Except for the legal text itself, the sources of law regarding the application of the law in general are sparse. Few decisions made by the Food Safety Authority are tested by the courts; the Norwegian Supreme Court has only ruled in three cases involving the Food Act, the Court of Appeals appears to have ruled in 21 cases, and the District Court has ruled in only 17 cases made that have been made publicly available.[32]

27.3.2 Rules on the Promotion of Foodstuffs

Information requirements are extensive when it comes to selling food in Norway. Current legislation ensures that consumers take well-informed choices. Regulation (EU) No 1169/2011[33] of the European Parliament and of the Council on the provision of food information to consumers, which was implemented in Norway

---

[28] Act on Implementation of the main part of the Agreement on the European Economic Area (EEA) into Norwegian law, of 27 November 1992, Section 2.

[29] The Act on the Sale of Alcoholic Beverages of 2 June, 1989, Section 9-2.

[30] Regulation of June 8, 2005, Chapter 14.

[31] Act Relating to the Prevention of the Harmful Effects of Tobacco of March 9, 1973, Section 22.

[32] As of February 2nd 2015.

[33] Regulation (EU) No 1169/2011 of the European Parliament and of the Council of October 25, 2011,. 2011 O.J. (L304) 18.

on December 13, 2014, gives further details with regard to labelling requirements and rules on advertisements. The Regulation (EU) No 1169/2911 is binding law in Norway as it was incorporated into Norwegian law in the Regulation on the provision of food information of 28 November 2014. The regulation replaces the former Norwegian Regulation on Labelling of Food 1993. In the following paragraphs, the focus will be on the particular rules on promotion, thus the labelling requirements will not be discussed herein. (For labelling requirements in the EU, see Chap. 22.)

The general national legislation relating to the promotion of foodstuffs follows from the Food Act,[34] section 10, and is further detailed in Regulation on the provision of food information 2014. The core principles are that the advertisement shall be correct, adequate and not misleading, and, moreover, that the marking shall not mislead the buyer with respect to "the characteristics of the foodstuff, particularly with regard to its nature, identity, quality, composition, quantity, durability, origin or place of origin, the manufacturing or production means."[35]

The concept of misleading advertisements will be determined by an overall assessment of the advertisement in question. There are however, certain acts that always will be regarded as misleading, such as acts claiming that the foodstuff has certain effects or properties which are not correct, or giving the impression that the food product has special characteristics when all similar foodstuffs possess the same properties.

There is no case law illustrating misleading marketing of foodstuffs and no national guideline provided by the Food Safety Authority. There are, however, some trademark cases in which the courts assess whether a trademark has misleading effects (see below). In practice, some assistance and guidance is found in the guidelines available from Denmark and Sweden, both of which discuss sales-friendly terms such as "natural," "real," "home-made," etc. As these remain foreign guidelines, the Food Safety Authority will as such not be bound by them and may find reason not to emphasize the guidelines in a national case.

The Food Safety Authority did publish the report "Villedende merking – Kampanje 2013" (Misleading labelling – Campaign 2013) in the spring of 2013 after having conducted a survey of 195 foodstuffs, and checking whether the food was correctly labelled and not misleading. Out of 195 foodstuffs examined, the Food Safety Authority found that 94 of the foodstuff labels were not in compliance with the law.[36]

In particular, the Food Safety Authority found that the so-called Bread Scale could have misleading effects. There is no national definition of bread; however, NHO Mat og Drikke (Federation of Norwegian Food and Drink Industry) has developed a standard called the "Bread Scale, giving the consumers information on whether the bread is e.g. white or whole-grain". The Bread Scale may be used according to separate agreements with NHO Mat og Drikke.

---

[34]Food Act, Section 10.
[35]Food Act, Section 10.
[36]"Villedende merking – Kampanje 2013" (Misleading labelling – Campaign 2013).

With regard to the possible misleading effects, first the Food Safety Authority found that some of the breads labelled according to the Bread Scale were incorrectly labelled. Second, quite a few of the breads labelled as "whole grain" had just a little bit above 50 % whole grain. This is compliant with the Bread Scale, but the Food Safety Authority found this to be "bordering on misleading," as the consumer looking at the Bread Scale would believe that the bread contained 75 % whole grain.

This example illustrates some of the challenges faced when the industry develops their own standards. Even though intentions were good when developing the scale, aiming at giving customers an easy method to pick bread containing the desired amount of whole grain, the method chosen was nevertheless problematic. The Food Safety Authority's scepticism did pressure the industry to undertake measures to change the scale. One of the measures was to provide the actual percentage of wholegrain on the package, close to the picture of the scale.

The report from the Food Safety Authority raised a particular concern about the labelling of juice drinks such as smoothies, and more precisely, concerns about the use of images and pictures of fruit on such drinks. The manufacturer is obliged to ensure that illustrative photos and images are truthfully reflecting the product's actual content. For example, an image of a large passion fruit on the label of a juice drink that contained only 7 % passion fruit would be misleading.

27.3.3 The Applicability of the General Marketing Law

The food law is *lex specialis* to the general rules on marketing applying to all businesses. However, the general legislation on marketing establishes certain rules not found in the *lex specialis* food rules and so the food industry must abide by them. Some examples of general legislation on marketing rules that apply in the food context are presented below but the list is not exhaustive.

27.3.3.1 Marketing to Children: Progressive and Precautionary Marketing Control

First, marketing to children has been restricted for many years, both in the current Marketing Control Act of January 9, 2009, and the preceding acts.[37] Children are considered to be a particularly vulnerable group because of their innocence and gullibility. Thus, particular care must be taken when it comes to commercial practices affecting them. In essence, their lack of critical thinking skills makes them easy targets.

The particular rules protecting children set out in the Marketing Control Act, outline that such rules extend beyond promotional activities aimed at children and

---

[37]Marketing Control Act of January 9, 2009.

include all activities that may be seen or heard by children. Both promotional activities and all other activities seen or heard by children must take into account considerations that children are easily swayed, their lack of experience, and their natural credulity. Moreover, direct exhortation directed at children is prohibited, which includes messages that would inspire children to persuade their parents into buying the products in question.

27.3.3.2 Good Trade Business Practice

Second, section 25 of the Marketing Control Act[38] sets forth a general prohibition against acts in the course of trade that conflict with good business practice among traders. The rule is construed as a legal standard, meaning that its content may change over the course of time to be in line with social development. The prohibition is only applicable in the relationship between traders, and its area of application is thus delimitated from consumers.

The rule was established to protect businesses from unfair business practices. It is often applied as a secondary legal basis in cases involving imitations since it provides an opportunity for the Court to make an overall assessment of the facts of the case when deciding whether or not there has been a breach of law (see below).

Thirdly, based on section 26 of the Marketing Control Act,[39] the Ministry of Children, Equality, and Social Inclusion have developed the Regulation on Comparative Advertising (the Regulation); the Regulation implements Directive 97/55/EC, defining the several requirements to be met in order for a comparative advertisement to be legal. It follows from Section 3 of the Regulation that, amongst other restrictions, comparative advertising is only permitted: if it is not misleading; if it compares goods or services that meet the same needs or are intended for the same purpose; and if it objectively compares one or more of those goods and services' features that are specific, relevant, documentable, and representative.[40]

Thus, comparative advertising may be used as long as it is based on correct information and is carried out in a fair manner. As these practices regarding comparative advertising constitute good business practices, section 25 of the Marketing Control Act is often invoked when alleged section 26 violations.

Fourth, section 2 of the Marketing Control Act states, in general, that marketing shall not conflict with good marketing practice. This is further explained as an obligation for the marketer to ensure that the marketing does not violate gender equality, that it does not exploit the body of one of the sexes, or that it does not convey an offensive or derogatory appraisal of women or men.[41]

---

[38]Marketing Control Act, *supra* n. 37, Sec. 25.
[39]Marketing Control Act, *supra* n. 37, Sec. 26.
[40]Regulation on Comparative Advertising.
[41]Marketing Control Act, *supra* n. 37, Sec. 2.

27.4 Particular Protection of Children to Prevent Obesity

The WHO's Recommendations "Marketing of foods and non-alcoholic beverages to children" of 2010 have been implemented in Norway by way of a new self-regulatory regime introduced with full effect in January, 2014. The Norwegian Government first aimed at implementing the WHO Recommendations thorough a legally binding regulation, and a draft was introduced in June, 2012. The draft regulation met both acceptance and criticism, and particularly the latter as all marketing of particular food categories aimed at children below 18 years were suggested banned. An amended draft regulation was later launched. Having negotiated a self-regulatory regime in parallel, the Norwegian authorities chose to accept the regulatory regime suggested by the industry rather than bringing the second draft regulation into effect under the pressure that the self-regulatory regime would be evaluated within 2 years. Should the self-regulatory regime not function as intended, the regulation would be brought into effect.

The consequences of overweight and obesity are extensive. Not only is obesity the fifth leading risk for global deaths, but overweight and obesity may lead to cardiovascular disease, cancer, and type two diabetes. The socio-economic impact is not any less dramatic. Obesity has led to the persistent reduction of the working force and the growing need for health care and social security benefits. Additionally, concerns about the personal impact of obesity and the quality of life of those people suffering from obesity should not be disregarded.

Obesity has more than doubled among adults since 1980 and a similar trend has emerged in the average increased weight of children. This prevailing and growing health-problem causes major concerns globally. Therefore, finding good and appropriate measures to reduce and, in the end, stop this development are of the essence. The bigger picture and underlying reasoning is complex, but two main causes of the problem are nevertheless easy to detect: Less physical activity and large consumption of unhealthy food. Junk food and energy-dense food are to a large extent available day and night in most parts of the world.

The correlation between advertisement and obesity is not an unfamiliar theme for either authorities or the industry. For example, leading food and beverage companies created the EU pledge, which is a voluntary initiative, to change the way companies advertise to children. Additionally, the recommendations from WHO call for global action to reduce the impact that marketing of foods high in saturated fats, trans-fats, free sugars, and salt has on children. Industry guidelines with the purpose of reducing marketing pressure on children were in place and effective as of 2007, but the Norwegian authorities wanted to impose a stricter regime on the Norwegian industry. Consequently, in 2012, the Norwegian authorities proposed a ban on all marketing of unhealthy food aimed at children below 18 years of age. This proposal was massively criticized by the industry, alleging, inter alia, non-compliance with the EEA and its established principles of free movement of goods. A revised proposal was launched in 2013 that met only parts of the criticism raised and, again, questions arose regarding compliance with the

EEA Agreement. Further, EFTA Surveillance Authority (ESA) commented on the Revised Draft Regulation during the summer of 2013, underlining the need for proportionate measures.[42]

In parallel with the legislative works proposed by Norwegian authorities, the Ministry of Children, Equality and Social Inclusion, and the Ministry of Health and Care Services engaged in dialog with the industry to see whether the industry could come up with a better self-regulatory regime, thus replacing the need for new regulation. An amicable settlement between the industry and the authorities was made when the new industry guideline launched on June 5, 2013,[43] which imposed a self-regulatory regime. The new guideline was brought into effect on January 1, 2014. As the industry managed to develop a self-regulatory regime that was deemed appropriate by the authorities, the second legislative proposal has not been made effective yet. Whilst the self-regulatory regime does impose strict rules on marketing, it is considered that it may have less legal impact on the effectiveness of the implementation of the WHO Recommendations than a legally binding regulation.[44]

The industry was originally granted a 2-year trial period to prove the success of the industry guideline implementing a self-regulatory regime. Should the new regime not have the desired effect, the second proposal made by the Norwegian authorities was to be brought into effect. However, it remains to be seen what the outcome is after the self-regulatory regime is evaluated.

Essentially, the new industry guideline implies that that all food defined as "unhealthy" should not be marketed directly to children (below 13 years of age). Even if children are 13 years or older, it follows from the guideline that the industry is obliged to take into account both age and maturity.

The crux of the matter is to clarify which food falls within the term "unhealthy" and what marketing activities may be seen as aimed "directly at" children. First, unhealthy food is defined as certain energy-dense, salty, sweet, or nutrient-poor food, all exhaustively defined in the guideline. The category includes foods such as chocolate, biscuits, snacks, soda, and different kinds of fast food. In regards to the latter question, the guideline makes it clear that one must perform an overall assessment in order to decide whether the marketing is particularly aimed at children.

Notably, the guideline applies to all types of media sources and channels. Marketing is defined as any sales promotional act. However, the guideline clarifies that the following are not considered to be marketing: (a) The product itself, including the packaging; (b) General presentation of products at retail outlets;

---

[42]Letter from EFTA Surveillance Authority to the Ministry of Trade and Industry, "Comments by the EFTA Surveillance Authority to Norway concerning notification 2013/9005/N", dated 17 July 2013.

[43]Industry Guideline (June 5, 2013).

[44]For evaluation of self-regulatory regimes vs. legally binding regulations, see e.g. Garde and Bartlett (2013).

(c) Sponsorship which only involves the use of sponsors' name, the sponsor- or a product's trademark; or (d) Sampling of products if a parent or other responsible adult have consented to the sampling. Furthermore, television advertisements broadcast after 9:00 pm will not be regarded as marketing directed towards children.

If a product does not comply with the established guidelines, any entities, NGOs, or private persons may complain to the Complaint Commission of the Food industry' (Matvarebransjens Faglige Utvalg). The proceedings are confidential, but the decision will be made public. Moreover, the respondent will not be informed about whom the plaintiff is. Any complaint must give (1) information allowing identification of the plaintiff and the respondent and (2) a written presentation of the matter, including documentation. The respondent will be given a right to comment within 14 days. The complaint will thereafter be assessed by the commission.

The Guideline has been in effect for slightly more than a year when this is written. Thus, the Complaint Commission has so far given ten opinions related to the marketing ban in 2014.[45] The Complaint Commission concluded with illegal marketing in five of these. Finally, the repetitive question assessed in these cases has principally been whether the marketing was directed directly at children.[46]

27.5 Governmental Bodies and Actions

In case of non-compliance with the labelling or advertising requirements, actions may be taken by both the Consumer Ombudsman and the Food Safety Authority (Mattilsynet), the latter being the most likely. The Food Safe Authority is authorized to make any decision required in order to ensure compliance under the scope of the Food Act and the regulations attached thereto. Thus, depending on the violation's degree of severity, the Food Safety Authority will decide what sanction is deemed most appropriate and effective in each case.

The most common form of sanctioning is an injunction that requires amendment of the labelling or advertisement, or an injunction that requires that the foodstuff's content is changed so that it corresponds with the current labelling. The Food Safety Authority may also give compulsory fines for each day that the company does not comply with such injunctions. However, this happens rarely, and when it happens, the fines are normally low (between NOK 2000 and 20,000).

Furthermore, the authority may impose administrative fines, but this is not often used. Of practical interest is the fact that the Food Safety Authority may prohibit the sales or impose withdrawal of the foodstuffs in question. The Food Act, moreover, gives legal basis for criminal prosecution, which means criminal fines and

---

[45]The opinions are available in Norwegian on the Internet at http://mfu.as/39309-Aktuelt (last accessed 9 Jan 2015).

[46]Cf. the Guidelines clause 2, 7.

imprisonment. However, such reactions are reserved for the more serious cases that the Food Safety Authority is responsible for handling, e.g. cases of animal maltreatment. When a sanction is prepared by the authorities, the entity is normally entitled to comment. Correspondingly, when the sanction is officially made, there is a complaint procedure in place. The case may thereafter be brought before the courts.

Should the Consumer Ombudsman take action, the Ombudsman will primarily seek a solution based upon negotiation with the entity in question. In certain cases, the Ombudsman is authorized to make decisions involving prohibition, injunction, penalty and/or administrative fines. Normally, however, such decisions are taken by the Market Council, which is an administrative Council with certain parallels to the regular courts. Appeal is not possible, but the decisions may be brought before the regular courts.

A company may bring a case concerning its competitor(s) to court, so long as the company is considered to have legal standing to bring a case against the company in question according to the Dispute Act (Act No. 90 of June 17, 2005).[47] Company competitors may also bring notice to the authorities about the non-compliant activities of other companies. The industry has further established the Committee for Unfair Competition to assess whether activities are in compliance with the Marketing Control Act, Chapter 6, which concerns competition between entities rather than consumer perspective. The dispute system offers quick and low-cost proceedings, and even though these opinions are not legally binding, the parties will normally abide by them (see below).

27.5.1 Protecting Intellectual Property Rights in Norway

New products are launched into the Norwegian retail market three times per year. In 2013, approximately 1000 products were launched.[48] Although many of them will be short-lived and replaced by other products, the high number of new products illustrates the importance of innovation and product development within the sector. Innovation and product development are not just expected by the retailers, it is also cherished by the consumers. By implication, suppliers have strong incentives to innovate and develop new products. Building customer relations through branding is particularly present in the food and drink sector, and some of the world's most famous brands are represented here. Protecting the innovation and capital spent on developing new products is essential: It provides a legal arm's length to one's competitors. Consequently, seeking such protection is not just of interest to multinational companies.

---

[47]Dispute Act No. 90 of June 17, 2005.

[48]Nye produkter i butikken – noen fakta (2014). http://www.dagligvarehandelen.no/nye-produkter-butikken-noen-fakta/.

A food product may be protectable as intellectual property in various ways. To illustrate, a product name may be protected as a trademark, packaging may typically be protected as design, and the manufacturing process may involve use of patentable techniques or trade secrets. Product protection may also be found in the Market Control Act should there be a violation of the principle of good business practices.

International trade makes global cooperation within the intellectual property field necessary. Norway is a party to several important international agreements on intellectual property, amongst others, the Paris Convention for the Protection of Industrial Property, the TRIPS Agreement, the Berne Convention for the Protection of Literary and Artistic Works, and the European Patent Convention. Consequently, Norwegian legislation within the area of intellectual property is very much in line with parallel legislation in other countries.

Being a party to the EEA Agreement, Norway is obliged to accept EU legislation in the legal areas that are covered by this Agreement. As a result, Norwegian legislation on intellectual property rights is, for the most part, consistent with parallel EU legislation where the rules have been harmonized to ensure that intellectual property rights do not become an obstacle to trade. Procedural law is not a part of the EEA Agreement, and Norway is, therefore, not obligated to implement Directive 2004/48/EC on the Enforcement of Intellectual Property Rights OJ 2004 L 157/1. The Norwegian government decided that it was in the best interest of the industry that the rules were at least as favourable to Norwegian right holders as the applicable EU law and decided to adopt similar legislation. Thus, intellectual property right holders in Norway have at least as favourable law enforcement options as those under EU law. In some cases, the Norwegian enforcement provisions provide the intellectual property right holder with a stronger position than with what is provided by the minimum requirements within the EU.[49]

The Norwegian Industrial Property Office (NIPO) is the national centre for intellectual property rights. It has the authority to handle and decide on patents, trademarks, and design applications. Decisions from NIPO may be brought before the Norwegian Board of Appeal for Industrial Property Rights, which is an independent administrative board.

In the following, our focus will be on some of the most important ways for food suppliers to protect their products, and the presentation will not be exhaustive with regard to potential legal basis for claiming protection.[50]

---

[49]Prop. 81 L (2012–2013), pp. 1 and 42.

[50]Copyrights, business names, and domain names may also constitute potential legal basis for claiming protection for food suppliers. Copyrights and rights to business names are regulated in respectively the Norwegian Copyright Act of May 12, 1961, and the Norwegian Business Names Act of June 21, 1985. Domain names are protected by registration, for example, Norid (see www.norid.no).

27.5.2 Trademarks

Trademark protection is essential in the industry as some of the world's most famous brands are found within the food and beverages sector. Intense competition in the supply chain, including the growing presence of private labels, makes trademark protection important, not just for the multinational companies, but for any company desiring to sell products under its own name.

It follows from the Trademarks Act[51] that a trademark right has the effect that no one, without the consent of the proprietor, may in an industrial or commercial undertaking, use:

(1) any sign that is identical to the trademark on goods or services for which the trademark is protected, and
(2) any sign identical to or similar to the trademark on identical or similar goods or services if there exists a likelihood of confusion, such as if the use of the sign may give the impression that there is a link between the sign and the trademark.[52]

A trademark is any sign capable of distinguishing the goods or services of one undertaking from those of another. Potential signs include words and combination of words, including slogans, names, letters, numerals, figures and pictures, and also—if it is considered sufficiently distinctive—the shape of the goods or their packaging.[53]

Norway, not being a member of the EU, is not part of the established system of registration of EU-trademarks which implies a uniform effect all over the internal market.[54] Therefore, the registration of a Community Trademark through the Office for Harmonization in the Internal Market (OHIM) will not give the trademark protection in Norway. Instead, to achieve protection, it is necessary to apply for a national registration through the Norwegian Industrial Property Office (NIPO) or to designate Norway in an international trademark application via the Madrid Protocol.

There are two different ways of achieving trademark protection: registration or protection acquired through use.[55] Applicable in both alternatives is the requirement that the trademark must be distinctive to the goods or services that it relates.[56]

---

[51] Trademarks Act, Act No. 8 of March 26, 2010, Clause 4.
[52] *Id.*
[53] *Id.* at Clause 2; The Norwegian Trademarks Act of March 26, 2010 implements Directive 2008/95/EC.
[54] EU Council Regulation No 207/2009 of Feb. 26, 2009, OJ (L78) 24.3.2009.
[55] When protection is established by use, the possibility of acquiring a registration depends on whether the mark is well known as someone's brand within the relevant goods' and services' circle of trade in Norway, cf. the Trademarks Act, Clause 3.
[56] Trademarks Act, *supra* n. 47, Clause 14.

Protection acquired through registration is granted and valid in Norway. Protection acquired through use will only be valid within the geographic area (within Norway) where the trademark has been used and becomes well known within the circle of trade.[57]

A Norwegian trademark registration is valid for 10 years from the date of application. Registration may be subsequently renewed for 10 years at a time, thus making an indefinite protection theoretically possible.[58] However, in order to maintain protection, the trademark must be actively used. Otherwise the registration of the trademark may be deleted in full or in part based on the grounds of non-use.[59]

Several national landmark cases within trademark law concern trademarks used on food and drinks. In 1995, the Norwegian Supreme Court assessed the legality of the trademark "Mozell" used on soda in the so-called *Mozell-case*.[60] One of the claims made by the Deutscher Weinfonds, representing the wine producers of the Mosel-district in Germany, was that the word mark "Mozell" should be deleted as it was liable to mislead with regard to the geographical origin of the product it was registered for.

The Supreme Court ruled in favour of the trademark owner, based on the grounds that even though the trademark might give the customers certain associations to the wine district Mosel and Mosel-wine, the trademark was not misleading. It was emphasized that the trademark "Mozell" was only representing non-alcoholic beverages, and thus limiting the risk of deception as there was only a slight chance that consumers would believe that the soda in any way originated from the Mosel district. The court found that the trademark was a fantasy mark, rather than serving a descriptive purpose, and would thus not provoke a dilution of the origin. The trademark "Mozell" differed from Mosel, both in written format and pronunciation. Even though the word mark had "certain associations" to Mosel and Mosel wine, it was unlikely that consumers would believe that the product was Mosel wine.[61]

It follows that the most essential function of a trademark is its origin function: that the trademark identifies the products' and services' commercial source or origin. In order to ensure that a sign is capable of fulfilling this function, it must have the distinctive character of a sign for the relevant item or service it refers to.[62] Consequently, generic terms would normally not fulfil the fundamental requirements of a trademark. Moreover, there is a need to reserve the right to such use of a generic term for the competitors. The *Jo-Bolaget case*[63] illustrates a case on borderline distinctiveness. In 2002, the Supreme Court assessed whether a trademark

---

[57]*Id.* at Clause 3.
[58]*Id.* at Clause 32.
[59]*Id.* at Clause 37.
[60]Rt. 1995, s.1908.
[61]Id. (Rt. 1995, s.1908).
[62]The Trademarks Act, *supra* n. 47, Clause 14.
[63]*Jo-Bolaget Case*.

application consisting of words and figurative elements used on fruit juice could be registered due to the phrase "GOD MORGON" (good morning). NIPO rejected to register the mark due to the generic phrase used, and would only approve the application if Jo-Bolaget disclaimed the disallowed component of the mark. Effectively meaning they would not claim the exclusive right to use the phrase "GOD MORGON."

The Supreme Court found that the trademark could be registered, as the distinctiveness requirement was not to be interpreted too strictly. This was principally reasoned with reference to case law from The European Court of Justice (ECJ). Notably, having a low degree of distinctiveness impacts the scope of protection. Thus, the Supreme Court found that the trademark could only have a narrow scope of protection.[64]

As for the situation today, the Supreme Court might have ruled differently. Following the decision in the Jo-Bolaget case, the ECJ has emphasized the general interest in keeping certain signs freely available to all. Consequently, the ECJ seems to have applied a more restrictive norm for what constitutes distinctiveness.[65]

27.5.3 Patents

Protecting the manufacturing process and the technology used in the manufacturing process is done either by filing for patents or by keeping the information confidential (see below). Whereas food manufacturing processes are often based on common knowledge and, therefore, not possible to protect as patents, many companies are investing in research and development that would make the development of patentable inventions more likely.

In order for an invention to be patentable, it follows from the Norwegian Patents Act of December 15, 1967 (Patents Act) that the invention must be novel.[66] In short, this means that it must involve a characteristic which is not known prior to the application (prior art). The next requirement is that the invention must show an inventive step, which means that it must not be obvious to someone with knowledge and experience on the subject. Clearly, common knowledge is free for anyone to use, and not patentable. It logically follows that products that can easily be deducted from such knowledge also may not be patented. Patent rights are reserved for those inventions that bring technical development further and beyond the expected,

---

[64]The Oslo District Court ruling of 2005 (TOSLO-2003-18673) says that Jo-Bolagets trademark—even though registered—offers only weak protection, and thus the court did not consider a competitor's use of "God morgen" on their products as an infringement of Jo-Bolagets' exclusive rights.

[65]See Birger Stuevold Lassen and Are Stenvik, Kjennetegnsrett, 2011, page 70. For ECJ practice see e.g. Case C-53/01 Linde, and Case C-104/01 Libertel.

[66]Norwegian Patents Act (Patents Act) No. 9 of December 15, 1967, WIPO Lex No. NO056.

thereby illustrating that the act of inventing needs freedom to operate. The development and innovation processes should not be hindered by patents which involve solutions only within reach of technically skilled persons. Reserving patents to inventions having the "inventive steps" also gives strong incentives to seek extraordinary results. Patents and patent applications are official documents, and it is thought that access to such documents will trigger new developments since anyone, not just competitors, can learn from the techniques presented.

The invention must involve the solution to a problem that is of a technical character, that has a technical effect, and that may be re-produced. Consequently, ideas, scientific theories, mathematical methods, the mere discoveries of natural substances, commercial methods, and ingredients lists are generally not patentable.

When filing for a patent, it is principally possible to obtain 20 years of protection from the date of filing, provided that the annual renewal fees are paid.[67] If granted, the right holder obtains the exclusive right to exploit the invention commercially or operationally. A patent may, however, be deemed invalid, if the patent requirements are later found to be unfulfilled. In cases of alleged infringements, validity issues are often raised and invalidity proceedings initiated.[68]

Norway has ratified the European Patent Convention and has been a member of the European Patent Organization since January 1, 2008. European patents that have designated Norway will have the same validity as Norwegian patents, presupposed that the patent is translated into Norwegian and the annual fees are paid accordingly.[69] With only one application needed, this system provides the opportunity for a simple and cost-efficient way of obtaining a patent within several European countries at the same time. At the moment, there are no joint procedures on enforcements, and it is still the Norwegian courts that will have the competence to rule in matters revolving patents valid in Norway.[70]

In Norway, there is a particular Act on the Right to Inventions[71] made by Employees of September 1, 1970. The Act aims at balancing the interests between employers and employees and applies, unless the parties have previously agreed otherwise. Some clauses in the Act are mandatory and always apply in full, e.g. the right of the employee to have "reasonable payment." No other acts regarding intellectual property provide equivalent rules on the transferral of rights from employees to employers, except for the Norwegian Copyrights Act.[72] Such transferral of rights is principally regulated in the contracts or background law.

---

[67] *Id.* at Sec. 40.

[68] *Id.* at Sec. 52.

[69] *Id.* at Sec. 66c.

[70] For the time being there is on-going work within the EU relating to the creation of a unitary patent for Europe and a specialised patent court (UPC—"Unified Patent Court") that has exclusive jurisdiction for litigation relating to European patents and European patents with unitary effect. This work will not have any effects with regard to Norway.

[71] Act on the Right of Inventions (September 1, 1970), Section 4.

[72] Norwegian Copyrights Act, Section 39g.

27.5.4 Design

Whereas trademarks protect the ability to distinguish products and services from each other and patents protect technical features, a design registration will grant protection to the visual layout.[73] In Norway, design rights for packaging have been granted for foodstuffs such as yoghurt, which makes it possible to separate the yoghurt from its topping, and to special bottles and ornamentation.[74] Under the Norwegian Designs Act (Designs Act) of March 14, 2003, design is defined as the appearance of a product or of parts of a product.[75] Design rights, therefore, protect the overall visual appearance of a product or a part of a product. However, the protection does not cover the technical features of the product.

In order to obtain a design right, it is necessary to file for protection. The requirement for protection is that the design has to be new and have individual character, which is a requirement with resemblance to the patent criteria. This means that the design must differ significantly from previously known designs and cannot be made available to the public within the EEA before the date of filing or before the date of priority if priority is claimed.[76] The application procedure is simplified because NIPO does not conclude whether the requirements for registration are met, and issues on validity are raised at a later stage, e.g. in an administrative review initiated by a complaint or by the courts.

Protecting intellectual property gives no guarantee of success in the market. The strict novelty-requirement leaves designers, often with small budgets, no time to test the market's response to prior to application. Therefore, a grace period is granted, and the designer, or his successor in title, may disclose the design 12 months before the application for registration or before the date of priority without jeopardizing the novelty-aspect of the design.[77]

In contrast with trademarks, the established system of registration through OHIM of a Community Design, which implies uniform effect all over the internal market,[78] will not provide the owner of the design with protection in Norway. To obtain protection, it is necessary to either apply for a national registration through the NIPO, or to designate Norway in an international registration of design to the World Intellectual Property Organization (WIPO). The latter is done through the

---

[73] 3D trademarks will give protection to the visual layout. Practise illustrates that the threshold for 3D marks are strict as it is the design that needs to fulfil the principal requirement on distinguishing goods, but if obtained provides for a strong protection as the protection period may be indefinite, cf. e.g. Case C-468/01 Procter & Gamble and Case C-286/04 Eurocermex.

[74] *See* Design Registration No. 084229 and No. 083329.

[75] The Act implements Directive 98/71 EC on the legal protection of designs; Norwegian Designs Act (Designs Act) of Mar. 14, 2005.

[76] Designs Act, *supra* n. 67, Sec. 3; Priority may be claimed within six months after the first application was lodged, or after the exhibition of the design at an official or officially recognized international exhibition, cf. the Designs Act, Clause 16.

[77] Designs Act, 67, Sec. 6.

[78] Council Regulation (EC) No 6/2002 of Dec. 12, 2001.

Protocol Relating to the Madrid Agreement Concerning the International Registration of Marks (Madrid Protocol) of June 27, 1989, to which Norway is a member.[79]

A design registration is valid in Norway for at least one period of 5 years or more periods. A registration valid for a period shorter than 25 years may be renewed for further 5-year periods up to a total registration period of 25 years.[80]

27.5.5 The Marketing Control Act's Protection of Products

Unfair competition practices, such as deliberately taking advantage of others' brands, goodwill, or product developments, do not necessarily constitute a violation of registered rights. The Marketing Control Act is subsidiary to the acts on intellectual property. As the Act emphasizes aspects of unfair competition, it plays an important role in cases regarding infringements; in some cases the Act even supplies as the principal claim. The following paragraphs analyse how the Marketing Control Act supports registered rights holders by constituting an effective weapon against counterfeit products.

The starting point for assessing a specific business conduct is the principle that all acts must be in accordance with good business practices.[81] It is the clause that entails a legal standard which may be adjusted in time to reflect and adapt to changes in business practices relevant to the standard. In addition, the Marketing Control Act specifically addresses copy products and prohibits unreasonable exploitations of others' products when there is a risk of confusion.[82] Notably, if the copy is sold under another known brand (which often is the case when it comes to food and beverages), confusion is often not likely. Thus, in practice, the Marketing Control Act makes good business practices standards the most relevant legal basis as there is no need to establish confusion.[83]

Cases involving possible violations of the Marketing Control Act may be brought before the regular courts, claiming sales ban and compensation,[84] but the alternative is to request a statement from the Committee for Unfair Competition

---

[79]Madrid Agreement Concerning the International Registration of Marks (June 27, 1989).

[80]Designs Act, *supra* n. 67, Sec. 23.

[81]Marketing Control Act, *supra* n. 37, Sec. 25.

[82]*Id.* at Sec. 30.

[83]Even though there cannot be established a violation towards the involved intellectual property right, the Marketing Control Act may be used as a legal basis to prevent others from unreasonably exploiting your rights. However, the Supreme Court has stated (e.g. in the *Mozell-case*) that if section 30 is not applicable because confusion is not established, the use of section 25 presupposes that there are other facts to the case which goes beyond the exploitation as such. This because section 25 is a general prohibition contrary to section 30 which is a special provision, and therefore section 25 shall not be applied on exactly the same conditions as this would be contrary to the system of the Marketing Control Act.

[84]Marketing Control Act, *supra* n. 37, Chapter 9.

(Committee). The Committee is established by the industry and offers an inexpensive and efficient dispute system. It assesses whether activities are in compliance with the prohibitions of the Marketing Control Act concerning competition between business entities, and gives a statement which is not legally binding, however normally abided by.[85] The Committee does not grant any compensation or reimbursement, impose withdrawal, or impose the destruction of products. The Committee consists of a panel that does not include judges, but some of the members have legal expertise. The cases brought before the Committee provide concrete assessments of whether a counterfeit product is illegal. The rendered decisions are important even though they, although legal sources of law, are not as highly regarded as judgments from the courts.

An interesting aspect of the dispute resolution provided by this regime is that some of the cases are revolving branded goods and private labels. For the simple reason that private labels are offered by the retail in which branded goods are also sold, the supplier and the private label seller have a contractual relationship. Pursuing matters of infringements against a customer is obviously not preferred. Whereas bringing such a case before the regular courts normally results in a high conflict level, the dispute resolution offered by the Committee provides a smoother alternative and makes it easier for the supplier of the branded goods to claim infringement.[86]

The assessments concerning packaging design make it evident that the Committee normally emphasizes the extent of which the alleged infringing product is too close to the original, because in certain cases the freedom to make an alternative design is not fully exploited. Some examples may illustrate what factors are of relevance. For example, *The Ice-cream packaging case*[87] involved the two leading ice cream producers in Norway. The matter concerned whether the yoghurt ice cream package sold under the name "Please" was infringing the rights pertaining to the product sold as "Dream," illustrated in the picture below. Both products were sold in generic packaging, which by implication meant that the assessment would focus on the design, text, and color on the packaging.

Initially, the Committee underlined that the generic elements used on the packaging are free for anyone to use. The question was rather whether the competitor had used these generic elements in his own design in such a way as to conflict with good business practices. In a 5-2 decision, the majority stated that even though it was possible—when taking a closer look—to detect differences, the packaging gave an overall impression that was too similar to the product that was first on the market. The pictures used on the packages were of approximately the same size, the placement was similar, and the use of colours differed minimally. It was further highlighted that because the Norwegian ice cream market was a duopoly, one supplier should develop a design that differed more substantially than the other.

---

[85]*Id.* at Sec. 6.
[86]*See* e.g. Case No. 15/2010, Case No. 18/2011, and Case No. 01/2013.
[87]*The Ice-cream Packaging Case.* Case No. 13/2010.

The majority concluded that that the packaging violated section 25 of the Marketing Control Act.

To further illustrate, *the Pastilles case* concerns another packaging design-case, involving two of the leading producers of pastilles.[88] The Committee stated that both designs were "playful" and "fresh," and that even though it was possible that the latter producer could have been inspired by the claimed original, the producer had used the freedom of variations, where, for example, the colours used were principally linked to the flavours, the size of the packaging differed, and the material of the packaging differed. In sum, the Committee unanimously held that the packaging of the two products did not have the same overall impression.

27.5.6 Know-How and Trade Secrets

Food manufacturing is often based on processes that are considered common knowledge, but the methods are often refined by insight developed during the process which constitutes valuable know-how. In some cases the methods constitute trade secrets of food manufacturing. Patentable inventions are also made within this sector. Even though filing for a patent is often considered to be the best way of protecting an invention, there may be various reasons for keeping the invention confidential, for example, if the invention will be difficult to reverse-engineer.

Know-how may only be protected by way of contractual obligations, but trade secrets enjoy a certain protection under Norwegian law. According to the Marketing Control Act, it is illegal to exploit trade secrets in the course of trade.[89] The Marketing Control act does not define trade secret, but a definition has been established by way of case-law and also discussed in the preparatory works to the Act.[90] Decisive factors include: whether the information is considered crucial to the business' competitiveness, whether the information is kept secret, and what steps the business has made to ensure the secrecy of the information.[91]

For trade secrets that involve technical drawings, descriptions, formulas, models, or similar technical aids, there is a special provision in the Marketing Control Act that bans the unlawful use of such technical know-how in the course of trade.[92] Furthermore, unlawful use of trade secrets or unlawful revealing of trade secrets will imply risk of fines or imprisonment according to section 294 of the Norwegian General Civil Penal Code.[93]

---

[88]*The Pastilles Case*. Case No. 20/2011.
[89]Marketing Control Act, *supra* n. 37, Sec. 28.
[90](1971–1972), p. 24.
[91]The Norwegian Regulation on the Application of the Competition Act § 10 subsection 3 on categories of Technology Transfer Agreements, of 6th July 2006.
[92]Marketing Control Act, *supra* n. 37, Sec. 29.
[93]Norwegian General Civil Penal Code, Act No. 10 of May 22, 1902, WIPO Lex No: NO040. Sec. 294.

27.6 Protection of Specific Foodstuffs and Particular Labels Available for Foodstuffs

In Norway, there are various schemes created to promote and highlight certain characteristics of a foodstuff, such as traditional products and origin. The EU schemes on geographical indications and designations of origin are not part of the EEA Agreement. Thus, Norway has established its own scheme based on the EU model for the protection of designations of origins, geographical indications, and traditional specialities under the Regulation on the Protection of Designations of Origins, Geographical Indications and Designations of Traditional Character of Foodstuffs of July 5, 2002.[94]

At present, there are no other agreements between Norway/EFTA and the EU ensuring mutual and automatic acceptance and grant of protection for protected geographical origins, but such agreements should be expected in the future Norwegian protection of foreign geographical design of origins and geographical indications is however possible. One example is *Parmigiano Reggiano*, which enjoys protection as designation of origin under the Regulation on the protection of *Parmigiano Reggiano* of April 20, 2012.[95]

In addition, there are also other labelling schemes that are attractive to food suppliers. First, there is a voluntary labelling system that involves labelling foodstuff with the symbol of a keyhole to make it easier for the consumer to choose the healthiest alternative within certain types of food categories.[96] Products that are labelled with a keyhole contain less fat, salt, and sugar, and more fibre than other products within the same food category. Rules applying to this can be found in the Regulation of June 17, 2009, concerning the voluntary labelling of foods with the keyhole. This regulation is national, but it is based on collaboration between Norway, Sweden, and Denmark. The Food Safety Authority and the Directorate of Health are responsible for the keyhole labelling system.

Furthermore, DEBIO is an administrative body that controls and approves ecological food production, fishery, and aquaculture. Entities that apply for approval and comply with the applicable requirements are entitled to use the DEBIO label in its marketing. Imported products may use the label, providing certain requirements are met.[97]

---

[94]Regulation on the Protection of Designations of Origins, Geographical Indications and Designations of Traditional Character of Foodstuffs (July 5, 2002).

[95]Regulation on the Protection of Parmigiano reggiano (April 20, 2012).

[96]Regulation (Keyhole) of June 17, 2009.

[97]Regulation on Organic Production and Labelling of Organic Agricultural Products and Foodstuffs (Oct. 28, 2005) Section 4, referring to Council Regulation (EEC) No. 2092/91 on Organic production of agricultural products and indications referring thereto, on agricultural products and foodstuffs, OJ 1991 L 198/1.

Finally, the label "Nyt Norge" (Enjoy Norway) has been available since 2009. This label is a registered collective trademark[98] and is owned by Matmerk, which is a foundation established by the Ministry of Agriculture. The label is thus administered by Matmerk, and may only be placed on products made of Norwegian raw material and produced in Norway. As Norwegian topography and climate offer challenges with respect to the assortment of agricultural products and production timing, mixed products may contain foreign raw material, but must be less than 25 % of the product's contents.

References

About the EEA and Norway Grants ("Hva er EØS-midlene?") http://www.regjeringen.no/nb/sub/europaportalen/eos-midlene.html?contentid=685567&id=684349
Act on Foodstuffs (2003)
Act on Implementation of the main part of the Agreement on the European Economic Area (EEA) into Norwegian law (1992), sec. 2
Act Relating to the Prevention of the Harmful Effects of Tobacco (1973), sec. 22
Act on the Right of Inventions (September 1, 1970), sec. 4
Act on the Sale of Alcoholic Beverages (1989), sec. 9-2
Act on the Trade of Agricultural Products (1936), sec. 1
Agreement on the European Economic Area (EEA Agreement), (1992)
Agreement on the European Economic Area (EEA Agreement), 4, protocol 3
Agreement on the European Economic Area (EEA Agreement), 4, protocol 9
Agreement on the European Economic Area (EEA Agreement), *supra* n. 4, article 102
Agreement on the European Economic Area (EEA Agreement), *supra* n. 4, article 102 (5)
Allocation of Tariff Quotas for Agricultural Products, regulation (2008)
Application of the Competition Act § 10 subsection 3 on categories of Technology Transfer Agreements, regulation (2006)
Business Names Act (1985)
Case No. 15/2010
Case No. 18/2011
Case No. 01/2013
Case C-104/01 Libertel (ECJ practice)
Case C-53/01 Linde (ECJ practice)
Comparative Advertising, regulation
Copyright Act (1961), sec. 39g
Council Regulation (EC) No 6/2002 of Dec. 12, 2001
Dagligvarehandelen.no (2014). http://www.dagligvarehandelen.no/nye-produkter-butikken-noen-fakta/
Design Registration No. 084229 and No. 083329
Designs Act, 67, secs. 3, 6, 23
Designs Act, clause 16
Designs Act (2005) The Act implements Directive 98/71 EC on the legal protection of designs
Determination of Discounted Tariffs for Imports of Industrially Processed Agricultural Products, regulation (2013)
Dispute Act No. 90 (2005)

---

[98]Trademarks Act, *supra* n. 50, Sec. 2.

EC regulation No. 178/2002 (The General Food Law Regulation). 2002 O.J. (L31/1)
EC regulation No. 207/2009, OJ (L78) 24.3.2009
EC regulation No. 1169/2011 of the European Parliament and of the Council. 2011 O.J. (L304) 18
Eurocermex Case C-286/04
Eurostat Database, European Commission http://epp.eurostat.ec.europa.eu/portal/page/portal/household_budget_surveys/Data/database
European Free Trade Association (EFTA). Jan. 4, 1960, available at www.efta.int/sites/default/files/documents/legal-texts/efta-convention/Vaduz%Convention%20Agreement.pdf
Food Act, sec. 10
Garde A, Bartlett O (2013) Time to seize the (red) bull by the horns: the European Union's failure to protect children from alcohol and unhealthy food marketing. Eur Law Rev 38:498 et seq
General Civil Penal Code, Act No. 10 (1902), WIPO Lex No: NO040, sec. 294
Generalized System of Preferences (2008)
Guidelines clause 2, 7
Ice-cream Packaging Case. Case No. 13/2010
Industry Guideline (2013)
Jo-Bolaget Case. Oslo District Court ruling (2005) (TOSLO-2003-18673)
Judgment of Borgarting Court of Appeal of 15 January 2007
Keyhole, regulation (2009)
Labeling of Food, regulation (1993), sec. 5
Lassen, Birger Stuevold and Stenvik, Are, Kjennetegnsrett (2011) p 70
Letter from EFTA Surveillance Authority to the Ministry of Trade and Industry, "Comments by the EFTA Surveillance Authority to Norway concerning notification 2013/9005/N", dated 17 July 2013
Madrid Agreement on the International Registration of Marks (1989)
Marketing Control Act (2009)
Marketing Control Act, *supra* n. 37, sec. 2
Marketing Control Act, *supra* n. 37, sec. 25
Marketing Control Act, *supra* n. 37, sec. 26
Marketing Control Act, *supra* n. 37, sec. 28
Marketing Control Act, *supra* n. 37, sec. 29
Marketing Control Act, *supra* n. 37, ch. 9
Matbransjens Faglige Utvalg: Opinions http://mfu.as/39309-Aktuelt (last accessed 9 January 2015)
Norid.no
Norwegian Food Safety Authority (Mattilsynet) Request for comments of 1 November 2013 (only available in Norwegian) http://www.mattilsynet.no/mat_og_vann/merking_av_mat/generelle_krav_til_merking_av_mat/horingsbrev.11601/binary/Høringsbrev
Nye produkter i butikken – noen fakta
Oslo District Court ruling (2005) (TOSLO-2003-18673)
Pastilles Case. Case No. 20/2011
Patents Act No. 9 (1967), WIPO Lex No. NO056
Prop. 81 L (2012–2013) pp 1, 42
Regulation of June 8, 2005, ch. 14
Regulation on Organic Production and Labelling of Organic Agricultural Products and Foodstuffs (2005), sec. 4, referring to Council Regulation (EEC) No. 2092/91 on Organic production of agricultural products and indications referring thereto, on agricultural products and foodstuffs, OJ 1991 L 198/1
Manufacturing, labelling and marketing of edible ice, regulation (1977)
Manufacturing, labelling and sale of butter and butter fat, regulation (1962)
Manufacturing, labelling and sale of cheese, regulation (1956)
Market Regulation to Promote the Trade of Agricultural Products, regulation (2008)

Protection of Designations of Origins, Geographical Indications and Designations of Traditional Character of Foodstuffs, regulation (2002)
Protection of Parmigiano reggiano, regulation (2012)
Quality of milk and cream, regulation (1953)
Reduction of Customs Tariffs on Agricultural Products, regulation (2005), ch. 2 and 3
Rt. (1995) p 1908
Rt. (1997) p 1954
Rt. (2002) p 391 (God Morgen)
Procter & Gamble Case C-468/01
Trademarks Act, clause 3
Trademarks Act, Act No. 8 of March 26, 2010, clause 4
Trademarks Act, *supra* n. 47, clause 14
Trademarks Act, *supra* n. 50, sec. 2
Trademarks Act (2010) implements Directive 2008/95/EC
Vegetable conserves, regulation (2001)
"Villedende merking – Kampanje 2013" (Misleading labelling – Campaign 2013)

Chapter 28
Food Law and Regulation in Germany

Gunnar Sachs

Abstract In the Member States of the European Union, the European legislature has largely harmonized food law. Where Member States have still retained their own legislative power in the area of food law, they have implemented this in the form of national provisions. The German legislature has exercised its remaining legislative powers in the area of food law primarily by adopting a separate Code on Foods, Consumer Goods and Feedstuffs (*Lebensmittel-, Bedarfsgegenstände- and Futtermittelgesetzbuch*), in addition to various national ordinances and guidelines as well as enforcement provisions. In this chapter, relevant principles of German food law and regulation will be discussed and presented, including some fundamental rules on consumer and health protection, food ingredients, labeling, advertising and marketing as well as supervision, monitoring and enforcement of food law in Germany.

28.1 Introduction

In the Member States of the European Union, food regulations have been largely harmonized. For the purpose of achieving harmonization of food law the European legislature, invoking its power to approximate legislation,[1] has increasingly created secondary legislation. Whereas for the most part it initially adopted European directives whose requirements each of the Member States then still had to transpose into their national legislation, European food law in recent years has primarily been harmonized through European regulations which apply with immediate effect in all Member States and are thus designed to ensure a comprehensive approach ("from farm to fork") within the entire food sector. Some key examples of food legislation in the form of European regulations are:

- Regulation (EC) No 178/2002 of the European Parliament and of the Council of 28 January 2002 laying down the general principles and requirements of food

---

[1]Art. 114 of the Treaty on the Functioning of the European Union, OJ C 115, 09/05/2008, p. 47.

G. Sachs (✉)
Maître en droit (Paris), Intellectual Property Law, Clifford Chance Deutschland LLP, Düsseldorf, Germany
e-mail: gunnar.sachs@cliffordchance.com

law, establishing the European Food Safety Authority and laying down procedures in matters of food safety (known as the "basic Regulation")[2];
- Regulation (EC) No 882/2004 of the European Parliament and of the Council of 29 April 2004 on official controls performed to ensure the verification of compliance with feed and food law, animal health and animal welfare rules[3];
- Regulation (EC) No 852/2004 of the European Parliament and of the Council of 29 April 2004 on the hygiene of foodstuffs,[4] supplemented by Regulation (EC) No 853/2004 of the European Parliament and of the Council of 29 April 2004 laying down specific hygiene rules for food of animal origin[5] and Regulation (EC) No 854/2004 of the European Parliament and of the Council of 29 April 2004 laying down specific rules for the organization of official controls on products of animal origin intended for human consumption.[6]

Where harmonized European provisions exist, the regulatory areas covered by them are normally excluded from the legislative powers of the Member States. The remaining national rules on food law are understood as supplementing harmonized European food law in all areas in which the European legislature has not (yet) adopted any uniform rules.

28.2 The German Code on Foods, Consumer Goods and Feedstuffs

The German legislature has exercised its remaining legislative powers in the area of food law primarily by adopting a national Code on Foods, Consumer Goods and Feedstuffs (*Lebensmittel-, Bedarfsgegenstände- und Futtermittelgesetzbuch*, "**LFGB**")[7] as well as various guidelines, ordinances and enforcement provisions.

28.2.1 Purpose of the Code

The LFGB is understood as a national framework law supplementing the European basic Regulation (EC) No 178/2002. Having entered into force on 7 September 2005 and thus 3 years after the basic Regulation, the primary purpose of the LFGB,

---

[2] OJ L 31, 10/02/2002, p. 1.
[3] OJ L 191, 28/05/2004, p. 1.
[4] OJ L 226, 25/06/2004, p. 3.
[5] OJ L 226, 25/06/2004, p. 22.
[6] OJ L 226, 25/06/2004, p. 83.
[7] Foods and Feeds Code as published in the announcement of 3 June 2013 (Federal Law Gazette I, p. 1426), which was amended by Art. 4 Para. 20 of the Act of 7 August 2013 (Federal Law Gazette I p. 3154).

according to the official grounds stated for the draft bill,[8] is to adjust German national legislation to legal acts adopted in Community law. For example, it brought about an approximation in the term "food additive"[9] harmonized in European legislation, or the term "cosmetic products" also harmonized throughout Europe.[10]

In the interest of harmonization of food law at the Community level, the LFGB moreover combines numerous, previously separate national provisions. Whereas German food law previously had been enshrined in various different laws such as the Food and Consumer Goods Act (*Lebensmittel- und Bedarfsgegenständegesetz* (LMBG)), the Meat Hygiene Act (*Fleischhygienegesetz* (**FlHG**)) or the Poultry Meat Hygiene Act (*Geflügelfleischhygienegesetz* (**GFlHG**)), it underwent a consolidation with the LFGB in 2005, thus becoming significantly easier to apply. At the same time the "from farm to fork" approach in European law was also reflected through inclusion of feedstuffs law in the LFGB. Moreover, all relevant sanctioning rules for violations of provisions of food law were additionally consolidated under the LFBG.

28.2.2 Scope of Application

The LFBG supplements the European basic Regulation (EC) No 178/2002 in those areas in which the German legislature still has power to adopt legislation and considered it necessary to supplement European food law. Whereas the European basic Regulation (EC) No 178/2002 for example only applies to foods as well as feeds[11] and expressly excludes cosmetic products from its scope of application,[12] cosmetic products and consumer products as a general rule are included under national food law based on the traditional German understanding. Consequently, the provisions of the LFBG not only regulate food and feed within the narrower

---

[8]Bundestag printed matter 15/3657.

[9]Cf. Art. 3 Para. 2 lit. a) Regulation (EC) No 1333/2008 of the European Parliament and of the Council of 16 December 2008 on food additives; cf. also already Art. 1 Para. 2 Council Directive 89/107/EEC of 21 December 1988 on the approximation of the laws of the Member States concerning food additives authorized for use in foodstuffs intended for human consumption.

[10]Cf. Art. 2 Para. 1 lit. a) Regulation (EC) No 1223/2009 of the European Parliament and of the Council of 30 November 2009 on cosmetic products; cf. also already Art. 1 Council Directive 76/768/EEC of 27 July 1976 on the approximation of the laws of the Member States relating to cosmetic products.

[11]Art. 1 Regulation (EC) No 178/2002; cf. Art. 2 Para. 1 Regulation (EC) No 178/2002 for the definition of food; according to Art. 2 Para. 2 Regulation (EC) No 178/2002, food also includes drink, chewing gum and any substance, including water, intentionally incorporated into the food during its manufacture, preparation or treatment.

[12]Art. 2 Para. 3 lit. e) Regulation (EC) No 178/2002.

sense of the European basic Regulation (EC) No 178/2002 but *inter alia* also cosmetics and other consumer products.[13]

With regard to trade in foods, the LFGB provides among other things for health protection prohibitions, prohibitions and authorizations for food additives, radiation prohibitions and licensing authorizations, requirements for herbicide and other products as well as for pharmacologically active substances, regulations on preventing fraud, a prohibition on disease-related advertising as well as additional authorizations designed to protect health against fraud.[14] With regard to trade in feedstuffs, the LFGB by contrast provides *inter alia* for insurance requirements, feed bans and related authorizations, prohibitions for preventing fraud as well as a prohibition on disease-related advertisement and additional authorizations for protecting human and livestock health as well as promoting livestock production.[15] Moreover, the LFGB lays down sanctioning and fine regulations for violations of the aforementioned requirements and prohibitions.[16] An overview of some of the key regulations is provided in the following.

28.2.3   Health Protection Prohibitions

The LFGB first of all prohibits food being produced or treated for others in such a way as to make its consumption injurious to health.[17] The point in time when a food becomes injurious to health is in turn governed by the requirements harmonized under European legislation found in the basic Regulation (EC) No 178/2002. There it is stipulated that in deciding whether any food is injurious to health, regard shall be had not only to the probable immediate and/or short-term and/or long-term effects of that food on the health of a person consuming it but also on subsequent generations, to the probable cumulative toxic effects and where applicable also to the particular health sensitivities of a specific category of consumers where the food is intended for that category of consumers.[18] Moreover, the LFGB prohibits placing on the market as food any substances that are not food and whose consumption is injurious to health within the meaning of the European basic Regulation (EC) No 178/2002.[19] It is also prohibited to produce for others, treat or place on the market any products that might be confused with food.[20]

---

[13] Sec. 1 Para. 1 LFGB.
[14] Sec. 5 et seq. LFGB.
[15] Sec. 17 et seq. LFGB.
[16] Sec. 58 et seq. LFGB.
[17] Sec. 5 Para. 1 LFGB.
[18] Art. 14 Para. 4 Regulation (EC) No 178/2002.
[19] Sec. 5 Para. 2 no. 1 LFGB.
[20] Sec. 5 Para. 2 no. 2 LFGB.

Based on the comprehensive regulatory approach "from farm to fork" provided under European legislation, the LFGB additionally prohibits the production or treatment of feeds in such a way that, when fed as intended and objectively appropriate, the foodstuffs produced for others from food-producing animals can impair human health or are unsuitable for human consumption.[21] It is moreover prohibited to produce or treat feeds for others in such a way that, when used as intended and objectively appropriate, they are capable of harming livestock health.[22] Moreover, feeds may not be produced or treated in such a way that, when used as intended and objectively appropriate, they are capable of impairing the quality of the food or other products produced from farm animals or of posing a risk to the ecological balance as a result of undesirable substances found in animal excretions which in turn were already present in feed.[23] The LFGB also prohibits the placing on the market and feeding of feed capable of impairing the quality of foods or other products produced from farm animals or of posing a risk to the ecological balance as a result of undesirable substances found in animal excretions which in turn were already present in feed.[24] The LFGB further bans the feeding of certain fats.[25]

The provisions of the LFGB thus clarify the European basic Regulation (EC) No 178/2002, which merely contains a general prohibition of placing unsafe feeds on the market or feeding them to food-producing animals.[26] In this regard, feed, according to the basic Regulation, is deemed to be unsafe for its intended use if it is considered to be capable of having an adverse effect on human or animal health or of making the foods derived from food-producing animals unsafe for human consumption.[27]

28.2.4 Prohibitions for Certain Ingredients and Radiation Levels

For trade in food, the LFGB moreover provides for prohibitions for certain food additives, radiation levels, herbicides and other products, as well as for pharmacologically active substances.

For example, the LFGB *inter alia* also prohibits, with the exception of enzymes and microorganism cultures, using unauthorized food additives in unmixed form or in mixtures with other substances in the production or treatment of food intended to

---

[21] Sec. 17 Para. 1 sentence 1 LFGB.
[22] Sec. 17 Para. 2 no. 1 lit. a) LFGB.
[23] Sec. 17 Para. 2 no. 1 lit. b) LFGB.
[24] Sec. 17 Para. 2 no. 2 and 3 LFGB.
[25] Sec. 18 LFGB.
[26] Art. 15 Para. 1 Regulation (EC) No 178/2002.
[27] Art. 15 Para. 2 Regulation (EC) No 178/2002.

be placed on the market. Neither is it permitted to use ion exchangers in the production or treatment of foods where this results in unauthorized food additives making their way into the food. Also prohibited in the production or treatment of foods are methods used for the purpose of producing unauthorized food additives in the foodstuffs. Further, no food may be placed on the market which has been produced or treated in contravention to these prohibitions. The LFGB moreover prohibits the placing on the market of any food additives or ion exchangers not permitted for use in the production or treatment of food for the purpose of such use or for use in the production or treatment of food by consumers.[28]

It is moreover prohibited to use an unauthorized radiation with ultraviolet or ionizing rays in food or to place food on the market thus irradiated.[29] The LFGB also contains the prohibition on placing food on the market if certain herbicides, fertilizers, other plant or soil treatment products or biocidal products within the meaning of the German Chemicals Act (*Chemikaliengesetz*, "**ChemG**") are present in or on such food.[30] It is also prohibited by the LFGB to place on the market live animals or food originating from animals if certain pharmacologically active substances are present in or on them.[31]

In consultation with other ministries at the federal level, the competent federal ministry in each case is authorized to issue additional rules for the purpose of clarifying the respective provisions.

28.2.5 Prohibitions for Preventing Fraud

The LFGB contains separate provisions for preventing misleading labels, statements and presentations in food and feed.

Firstly, the Code prohibits the placing of foods on the market under a misleading label, statement or presentation or advertising for food, either generally or in the individual case, using misleading descriptions or other statements.[32] Impermissible misleading is deemed to exist if labels, statements, descriptions or other statements on properties capable of misleading consumers, particularly with regard to the nature, condition, composition, quantity, conservation, origin, provenance or type of manufacture or production, are used for a food. That furthermore means that it is misleading and impermissible to imply that a food has effects which it in fact does not have based on the current state of scientific knowledge, or which are not supported by sufficient scientific evidence. Impermissible misleading is also deemed to exist when it is suggested that the food possesses special characteristics when in fact all similar

---

[28] Sec. 6 Para. 1 LFGB.
[29] Sec. 8 Para. 1 LFGB.
[30] Sec. 9 LFGB.
[31] Sec. 10 LFGB.
[32] Sec. 11 Para. 1 LFGB.

foods possess the same characteristics, or if a food is ascribed the appearance of a pharmaceutical product. For the purpose of preventing fraud, the LFGB further prohibits imitations of food, as well as foods which, based on their characteristics, depart from the market perception generally associated with them and are thereby diminished to a not insignificant extent in their value, particularly in their nutritional or culinary value or their usefulness, or foods capable of giving the appearance of being better than their actual properties without the market being provided with sufficiently specific information in this regard.[33] Moreover, the LFGB prohibits placing on the market those foods that are not fit for human consumption and in this regard have not yet been prohibited by the European basic Regulation (EC) No 178/2002.[34]

The LFGB also prohibits placing feed on the market with misleading labeling or presentation or advertising for such feed either generally or in the individual case.[35]

In this regard also, the LFGB supplements the European basic Regulation (EC) No 178/2002 according to which in general the labeling, advertising and presentation of food or feed, including their shape, appearance or packaging, the packaging materials used, the manner in which they are arranged and the setting in which they are displayed, and the information which is made available about them through whatever medium, may not mislead consumers.[36] In connection with the provisions on prevention of fraud in feed, the LFGB moreover expressly refers to the requirements of Regulation (EC) No 767/2009 of the European Parliament and of the Council of 13 July 2009 on the placing on the market and use of feed.[37] Moreover, European legislation provides for further requirements for preventing fraud which are directly applicable in the Member States, such as in Regulation (EC) No 1924/2006 of the European Parliament and of the Council of 20 December 2006 on nutrition and health claims made on foods.[38]

28.2.6 Prohibition on Disease-Related Advertising

Whereas European Regulation (EC) No 1924/2006 creates framework conditions for the use of nutrition and health claims made on foods including reduction of disease risk claims, the German LFGB originally prohibited any kind of disease-related advertising for food outside specialist healthcare professional groups.

In particular, the LFGB prohibited making any claims to individuals outside the specialist community, in trading in food or in advertising for food, whether generally or in the individual case, which related to the removal, relief or prevention of

---

[33] Sec. 11 Para. 2 no. 2 LFGB.
[34] Sec. 11 Para. 2 no. 1 LFGB.
[35] Sec. 19 LFGB.
[36] Art. 16 Regulation (EC) No. 178/2002.
[37] OJ L 229, 01/09/2009, p. 1.
[38] OJ L 12, 18/01/2007, p. 3.

diseases.[39] Also prohibited were any references to recommendations made by physicians or expert opinions of physicians.[40] Medical histories or references to the same could not be used either.[41] Furthermore, the LFGB prohibited making reference to statements by third parties, in particular letters of appreciation, recognition or recommendation to the extent these referred to the removal or relief of diseases, as well as references to such statements.[42] It was also prohibited to present images of persons in occupational clothing or performing the work of members of the healthcare professions or healthcare industry or the pharmaceutical trade.[43] Statements capable of causing or exploiting anxiety as well as use of inscriptions or written statements inducing consumers to treat diseases with food were also prohibited.[44]

Today, the LFGB still prohibits as a general rule all claims, in trading in feed or in premixtures as well as in advertising for the same, whether generally or in the individual case, which relate to the removal or relief of diseases or the prevention of such diseases which are not the result of deficient nutrition.[45]

28.2.7 Supervision and Monitoring

By Regulation (EC) No 882/2004 of the European Parliament and of the Council of 29 April 2004 on official controls performed to ensure the verification of compliance with feed and food law, animal health and animal welfare rules,[46] the European legislature harmonized the key provisions on monitoring food and feed law in all Member States. The LFGB merely provides for supplementing provisions on competencies, duties, measures and exchange of mutual information amongst the national authorities, on the submission of data on Internet trade, on informing the general public, on measures in producer facilities, livestock trade and transport companies, on performance of monitoring and sampling, on duties of toleration, co-operation and submission on the part of the food and feed business operators concerned, on notification and submission requirements regarding study findings on substances undesirable from a health perspective, as well as on co-operation between the federal government and the federal states within the Federal Republic of Germany.[47] According to the LFBG, the authorities of the individual federal states in Germany are generally responsible for supervision measures. The

---

[39]Sec. 15 Para. 1 no. 1 LFGB (old version).
[40]Sec. 12 Para. 1 no. 2 LFGB (old version).
[41]Sec. 12 Para. 1 no. 3 LFGB (old version).
[42]Sec. 12 Para. 1 no. 4 LFGB (old version).
[43]Sec. 12 Para. 1 no. 5 LFGB (old version).
[44]Sec. 12 Para. 1 no. 6 and 7 LFGB (old version).
[45]Sec. 20 Para. 1 LFGB.
[46]OJ L 191, 28/05/2004, p. 1.
[47]Sect. 38 et seq. LFGB.

requirements that the competent authorities must satisfy are again set out in European Regulation (EC) No 882/2004.[48]

In addition, the LFGB provides for monitoring of food, food additives, feed, cosmetic products and consumer goods.[49] Monitoring in this regard means a system of repeated observations, measurements and assessments of levels of substances which are undesirable from a health perspective such as herbicides, pharmacologically active substances, heavy metals, mycotoxines and microorganisms in and on such products or livestock, which are performed for early detection of risks to human health using representative samples of individual products or animals, aggregate diet or another aggregate criterion of the same product.[50] The competent authorities of the Federal States submit the data gathered from performance of monitoring to the Federal Office of Consumer Protection and Food Safety (*Bundesamt für Verbraucherschutz und Lebensmittelsicherheit*, "**BVL**") for processing, aggregation, documentation and reporting. The BVL submits the data gathered from the monitoring performed to the Federal Institute for Risk Assessment (*Bundesinstitut für Risikobewertung*, "**BfR**") for assessment purposes. In this regard, personal data are not submitted. Instead, such data must be deleted to the extent not required for performing supervision or monitoring. The BVL annually publishes a report on monitoring results.[51]

28.2.8 Penalty and Fine Provisions

The LFGB provides for penalty and fine regulations for violations of food and feed law.

In trade in food and feed, violations subject to criminal sanctions include violations of the individual prohibitions for health protection and prevention of fraud, as well as violations of the regulations for certain ingredients, of the prohibition on radiation, of the requirements for the production and treatment of feed as well as violations of certain feeding prohibitions.[52]

Since the European legislature does not have power to adopt penalty and fine regulations, the LFGB at the same time provides for legal remedies for violations of regulations of European law directly applicable in Germany including certain requirements

- of the basic Regulation (EC) No 178/2002,
- of Regulation (EC) No 1332/2008 of the European Parliament and of the Council of 16 December 2008 on food enzymes,[53]

---

[48] Art. 4 Para. 2 et seq. Regulation (EC) No 882/2004.
[49] Sec. 50 et seq. LFGB.
[50] Sec. 50 LFGB.
[51] Sec. 51 Para. 5 LFGB.
[52] Sec. 58 and 59 LFGB.
[53] OJ L 354, 31/12/2008, p. 7.

- of Regulation (EC) No 1333/2008 of the European Parliament and of the Council of 16 December 2008 on food additives,[54]
- of Regulation (EC) No 1334/2008 of the European Parliament and of the Council of 16 December 2008 on flavorings and certain food ingredients with flavoring properties for use in and on foods,[55]
- of Commission Regulation (EC) No 124/2009 of 10 February 2009 setting maximum levels for the presence of coccidiostats or histomonostats in food resulting from the unavoidable carry-over of these substances in non-target feed,[56]
- of Commission Regulation (EU) No 10/2011 of 14 January 2011 on plastic materials and articles intended to come into contact with food,[57]
- of Regulation (EC) No 396/2005 of the European Parliament and of the Council of 23 February 2005 on maximum residue levels of pesticides in or on food and feed of plant and animal origin,[58]
- of Regulation (EC) No 1924/2006 of the European Parliament and of the Council of 20 December 2006 on nutrition and health claims made on foods,[59] and
- other directly applicable provisions of European legislation identical in content to the German requirements and prohibitions subject to criminal sanctions.[60]

Such violations are punishable in Germany by imprisonment of up to 3 years or by a fine. In particularly serious cases in which the health of a great number of persons is put at risk or which involve the risk of death or serious harm to health, or in which the perpetrators, acting out of gross self-interest, procure pecuniary gain for themselves or others on a large scale, prison sentences from 6 months to 5 years may be imposed.

Moreover, the LFGB provides for various fine regulations for other violations of requirements and prohibitions under European and national law, e.g. for violations of certain requirements relating to disease-related advertising for food or feed.[61] Depending on the provision violated, such violations may be punished in each case by fines of up to EUR 20,000, EUR 50,000 or EUR 100,000.

28.2.9 Other Provisions

The LFGB also contains other provisions dealing with matters of authorization, powers and co-operation as well as requirements relating to the import and export of food. The LFGB also provides for the compiling of guidelines on the production,

---

[54] OJ L 354, 31/12/2008, p. 16.
[55] OJ L 354, 31/12/2008, p. 34.
[56] OJ L 40, 11/02/2009, p. 7.
[57] OJ L 328, 11/12/2011, p. 22.
[58] OJ L 70, 16/03/2005, p. 1.
[59] OJ L 404, 30/12/2006, p. 9; OJ L 12, 18/01/2007, p. 3; OJ L 86, 2803/2008, p. 34.
[60] Sec. 58 and 59 LFGB.
[61] Sec. 60 LFGB.

quality and other properties of food of significance for their fitness for marketing in a "German Food Code" (*Deutsches Lebensmittelbuch*).[62]

The guidelines are to be adopted by a German Food Code Commission (*Deutsche Lebensmittelbuch-Kommission*) giving due regard to the international food standard recognized by the German government and published by the competent federal ministry. The German Food Code Commission is established within the competent federal ministry. The members of the Commission are appointed in equal numerical ratios from amongst science, food monitoring, consumer and food industry experts.[63]

Currently, guidelines have been published for meat and meat products, for delicatessen salads, for fish, crustaceans and mollusks and products made from them, for edible fats and edible oils, for bread and small bakery wares, for pastries, for pasta products, for vegetable products, for vegetable juice and vegetable nectar, for mushrooms and mushroom products, for potato products, for fruit juices, for fruit produce, for soft drinks, for honey, for oil seeds and for mixtures and sweets produced from them, for pudding powders and related products, for ice cream and semi-finished ice cream products, for tea, products similar to tea, their extracts and preparations as well as for spices and other seasoning products.

28.3 Additional Laws and Legal Ordinances

In addition to the LFGB, the German legislature has also issued various horizontal and vertical guidelines and legal ordinances clarifying national food law. Horizontal provisions apply to a great number of different products, whereas vertical provisions relate to only a certain category of products.

28.3.1 Horizontal Provisions

The most important horizontal provisions in Germany besides the LFGB are:

- the Metering and Calibration Act (*Mess- und Eichgesetz* (MessEG)) of 25 July 2013,[64]
- the Act on the Introduction and Use of Labeling for Organic Products (*Öko-Kennzeichengesetz* (ÖkoKennzG)) in the version as announced on 20 January 2009[65] together with the Ordinance on the Design and Use of the Organic Label (*Öko-Kennzeichenverordnung* (ÖkoKennzV)) of 6 February 2002,[66]

---

[62]Sec. 15 LFGB.
[63]Sec. 16 LFGB.
[64]Federal Law Gazette I, p. 2722.
[65]Federal Law Gazette I, p. 78.
[66]Federal Law Gazette I, p. 589.

- the Act Implementing the Legal Acts of the European Community or of the European Union on Certificates of Specific Characteristics of Agricultural Products and Foodstuffs (*Lebensmittelspezialitätengesetz* (LSpG)) dated 29 October 1993[67] together with the ordinance implementing the LSpG (*Lebensmittelspezialitätenverordnung* (LSpV)) of 21 December 1993,[68]
- the Ordinance on Hygiene Requirements for the Manufacture, Processing and Placing on the Market of Foods (*Lebensmittelhygiene-Verordnung* (LMHV)) in the version as announced on 21 June 2016,[69]
- the Ordinance on Hygiene Requirements for the Manufacture, Processing and Placing on the Market of Certain Foods of Animal Origin (*Tierische Lebensmittel-Hygieneverordnung* (Tier-LMHV)) of 8 August 2007,[70]
- the Ordinance on Food Labeling (*Lebensmittel-Kennzeichnungsverordnung* (LMKV)) in the version as announced on 15 December 1999,[71]
- the Ordinance on Nutritional Claims for Food and Nutritional Labeling of Food (*Nährwert-Kennzeichnungsverordnung* (NKV)) of 25 November 1994,[72]
- the Ordinance Implementing Provisions of Community Law on Novel Foods and Novel Food Ingredients (*Neuartige Lebensmittel- und Lebensmittelzutaten-Verordnung* (NLV)) in the version as announced on 14 February 2000,[73]
- the Ordinance on Food Supplements (*Nahrungsergänzungsmittelverordnung* (NemV)) of 24 May 2004,[74]
- the Ordinance on Dietary Foods (*Diätverordnung* (DiätV)) in the version as announced on 28 April 2005,[75]
- the Ordinance on Pre-Packed Products (*Fertigpackungsverordnung* (FertigPackV)) in the version as announced on 8 March 1994,[76]
- the Lot Labeling Ordinance (*Los-Kennzeichnungs-Verordnung* (LKV)) of 23 June 1993,[77]
- the Price Indication Ordinance (*Preisangabenverordnung* (PAngV)) in the version as announced on 18 October 2002,[78]
- the Ordinance on the Authorization of Food Additives for Technological Purposes (*Zusatzstoff-Zulassungsverordnung* (ZZulV)) of 29 January 1998,[79]

---

[67]Federal Law Gazette I, p. 1814.
[68]Federal Law Gazette I, p. 1996.
[69]Federal Law Gazette I, p. 1469.
[70]Federal Law Gazette I, pp. 1816, 1817.
[71]Federal Law Gazette I, p. 2464.
[72]Federal Law Gazette I, p. 3526.
[73]Federal Law Gazette I, p. 123.
[74]Federal Law Gazette I, p. 1011.
[75]Federal Law Gazette I, p. 1161.
[76]Federal Law Gazette I, pp. 451, 1307.
[77]Federal Law Gazette I, p. 1022.
[78]Federal Law Gazette I, p. 4197.
[79]Federal Law Gazette I, pp. 230, 231.

- the Ordinance on Requirements for Additives and the Placing on the Market of Additives for Technological Purposes (*Zusatzstoff-Verkehrsverordnung* (ZVerkV)) of 29 January 1998,[80]
- the Ordinance on Quick-Frozen Foodstuffs (*Verordnung über tiefgefrorene Lebensmittel* (TLMV)) in the version as announced on 22 February 2007,[81]
- the Ordinance on Vitamin-Enhanced Foods (*Verordnung über vitaminisierte Lebensmittel* (LMvitV)),[82]
- the Ordinance on the Limitation of Contaminants in Food (*Kontaminanten-Verordnung* (KmV)) of 19 March 2010,[83]
- the Ordinance on Maximum Amounts of Residues of Herbicides, Pesticides, Fertilizers and Other Products in or on Food (*Rückstands-Höchstmengenverordnung* (RHmV)) in the version as announced on 21 October 1999,[84]
- the Ordinance on the Use of Extraction Solvents and Other Technical Aids in the Production of Food (*Technische Hilfsstoff-Verordnung* (THV)) of 8 November 1991,[85]
- the Ordinance on Pharmacologically Active Substances (*Verordnung über Stoffe mit pharmakologischer Wirkung* (PharmStV)) in the version as announced on 8 July 2009,[86] and
- the Ordinance on the Performance of Veterinary Law Controls for the Import and Transit of Food of Animal Origin from Third Countries as well as on the Import of other Food from Third Countries (*Lebensmitteleinfuhr-Verordnung* (LMEV)) in the version as announced on 15 September 2011.[87]

Each of the aforementioned national acts and ordinances supplement the legal provisions relating to food as harmonized under European legislation in those areas in which the European legislature has not (yet) adopted any uniform rules.

Moreover, the German Act Against Unfair Competition (*Gesetz gegen den unlauteren Wettbewerb* (UWG)) in the version as announced on 3 March 2010[88] additionally lays down special rules on the protection of competition, consumers and other market participants against unfair business behavior including misleading advertising. The UWG applies to all goods and services including food and feed. The Act serves to implement various provisions harmonized under European

---

[80]Federal Law Gazette I, pp. 230, 269.
[81]Federal Law Gazette I, p. 258.
[82]In the revised version as published in the Federal Law Gazette Part III, No. 2125-4-23.
[83]Federal Law Gazette I, p. 278.
[84]Federal Law Gazette I, pp. 2082; 2002 I, p. 1004.
[85]Federal Law Gazette I, p. 2100.
[86]Federal Law Gazette I, p. 1768.
[87]Federal Law Gazette I, p. 278.
[88]Federal Law Gazette I, p. 254.

legislation[89] and thus supplements the advertising law requirements for food and feed under the LFGB.

28.3.2 Vertical Provisions

In addition there are numerous vertical provisions that apply only to certain categories of food.

28.3.2.1 Meat and Meat Products

These include for example the German Law on Meat (*Fleischgesetz* (FlG)) of 9 April 2008.[90] Moreover, the hygiene-related legal provisions for meat and meat products are governed by the "Community-law Hygiene Packages" (Regulations (EC) No 852/2004,[91] No 853/2004[92] and No 854/2004[93]) as well as the German LMHV and the Tier-LMHV. The production of meat and meat products is further subject in particular to the general requirements of the European basic Regulation (EC) No 178/2002 as well as European and national provisions on the use of additives.

---

[89]The German Act Unfair Competition (*Gesetz gegen den unlauteren Wettbewerb*, "**UWG**") serves to implement Directive 2005/29/EC of the European Parliament and of the Council of 11 May 2005 concerning unfair business-to-consumer commercial practices in the internal market and amending Council Directive 84/450/EEC, Directives 97/7/EC, 98/27/EC and 2002/65/EC of the European Parliament and of the Council and Regulation (EC) No 2006/2004 of the European Parliament and of the Council (OJ L 253, 11/06/2005, p. 22; OJ L 253, 25/09/2009, p. 18), furthermore to implement Directive 2006/114/EC of the European Parliament and of the Council of 12 December 2006 concerning misleading and comparative advertising (OJ L 376, 27/12/2006, p. 21) as well as to implement Art. 13 of Directive 2002/58/EC of the European Parliament and of the Council of 12 July 2002 concerning the processing of personal data and the protection of privacy in the electronic communications sector (OJ L 201, 31/07/2002, p. 37).

[90]Federal Law Gazette I, p. 714, 1025.

[91]Regulation (EC) No 852/2004 of the European Parliament and of the Council of 29 April 2004 on the hygiene of foodstuffs, OJ L 139, 30/04/2004, p. 1.

[92]Regulation (EC) No 853/2004 of the European Parliament and of the Council of 29 April 2004 laying down specific hygiene rules for on the hygiene of foodstuffs, OJ L 139, 30/04/2004, p. 55.

[93]Regulation (EC) No 854/2004 of the European Parliament and of the Council of 29 April 2004 laying down specific rules for the organization of official controls on products of animal origin intended for human consumption, OJ L 226, 25/06/2004, p. 83.

28.3.2.2 Fish, Crustaceans and Mollusks

For fish, crustaceans and mollusks, German law, in addition to the applicable horizontal provisions at the European level, provides for various vertical rules such as, for example, a national law for the implementation of the legal acts of the European Community regarding the labeling of fish and fishery products (*Fischetikettierungsgesetz* (FischEtikettG)) of 1 August 2002.[94] This serves for the implementation of particular European provisions on consumer information and labeling of fish and fishery products.

28.3.2.3 Milk and Milk Products

Milk and milk products are regulated in Germany vertically by the Act on Milk, Milk Products, Margarine Products and Similar Products (*Milch- und Margarinegesetz* (MilchMargG)) of 25 July 1990.[95] The Act contains provisions on the operation of dairy companies, on supervision and monitoring as well as provisions on penalties and fines. It moreover authorizes the adoption of standardization regulations. The Act is supplemented by an Ordinance on the Expertise for Operating a Milk Treating or Processing Business and a Commercial Dairy Business (*Milch-Sachkunde-Verordnung* (MilchSachkV))[96] of 22 December 1972, an Ordinance on Milk Products (*Milcherzeugnisverordnung* (MilchErzV)) of 15 July 1970,[97] an Ordinance on Butter and Other Dairy Spreads (*Butterverordnung* (ButtV)) of 3 February 1997,[98] a Cheese Ordinance (*Käseverordnung* (KäseV)) in the version as announced on 14 April 1986[99] and an Ordinance on the Quality Evaluation and Payment of Tanker Milk (*Milch-Güteverordnung* (MilchGüV)) of 9 July 1980.[100] Both the Butter and the Cheese Ordinances were primarily adopted for standardization purposes to promote the production, quality and sale of agricultural products and have a different scope compared with the Community-law provisions. For drinking milk there is moreover a vertical Ordinance on the Labeling of Heat-Treated Drinking Milk (*Konsummilch-Kennzeichnungs-Verordnung* (MilchKennzV)) of 19 June 1974.[101]

---

[94]Federal Law Gazette I, p. 2980.
[95]Federal Law Gazette I, p. 1471.
[96]Federal Law Gazette I, p. 2555.
[97]Federal Law Gazette I, p. 1150.
[98]Federal Law Gazette I, p. 144.
[99]Federal Law Gazette I, p. 412.
[100]Federal Law Gazette I, pp. 878, 1081.
[101]Federal Law Gazette I, p. 1301.

28.3.2.4 Cereals and Cereal Products

For cereals and cereal products, Germany has adopted separate implementing provisions for European legislation on the common organization of the market in agricultural products, such as for example a national law on the further development of the market organization in the agricultural sector (*Agrarmarktstrukturgesetz* (AgrarMSG)).

28.3.2.5 Fruit Juices, Fruit Nectar and Caffeinated Soft Drinks

Fruit juices, fruit nectar and caffeinated soft drinks are regulated in Germany vertically by the Ordinance on Fruit Juice, Certain Similar Products, Fruit Nectar and Caffeinated Soft Drinks (*Fruchtsaft- und Erfrischungsgetränkeverordnung* (FrSaftErfrischGetrV)) of 24 May 2004.[102] The purpose of the Ordinance is to implement Directive 2001/112/EC of the Council of 20 December 2001 relating to fruit juices and certain similar products intended for human consumption.[103]

28.3.2.6 Fruit Jams, Jellies, Marmalades and Sweetened Chestnut Purée

The vertical European law requirements of Directive 2001/113/EC of the Council of 20 December 2001 relating to fruit jams, jellies and marmalades and sweetened chestnut purée intended for human consumption[104] were transposed by the German legislature in the Ordinance on Fruit Jams and Similar Products (*Konfitürenverordnung* (KonfV)) of 23 October 2003.[105] The German Fruit Jam Ordinance exhaustively regulates the composition of those foodstuffs placed on the market in Germany as extra jam, jam, extra jelly, jelly, marmalade or sweetened chestnut purée.

28.3.2.7 Honey

For honey, German legislation provides for a vertical Honey Ordinance of 16 January 2004 (*Honigverordnung* (HonigV)).[106] The German Honey Ordinance

---

[102] Federal Law Gazette I, p. 1016.
[103] OJ L 10, 12/01/2002, p. 58.
[104] OJ L 10, 12/01/2002, p. 67.
[105] Federal Law Gazette I, p. 2151.
[106] Federal Law Gazette I, p. 92.

sets out the definitions and quality requirements as well as designation rules for honey and honey products.

28.3.2.8 Ice Cream

Between 1933 and 2007, there has also been a separate vertical ice cream ordinance in Germany.[107] This ordinance has been repealed step by step. At the end, the provisions still contained in the ordinance were primarily confined to authorized additives and the labeling of ice cream commercially marketed in bulk. Today, there still remains an ordinance on the professional education of qualified persons for ice cream dated 5 June 2014 (*Speiseeisfachkraftausbildungsverordnung* (EisAusbV)).[108]

28.3.2.9 Cocoa and Chocolate Products

The vertical European law requirements of Directive 2000/36/EC of the European Parliament and of the Council of 23 June 2000 relating to cocoa and chocolate products intended for human consumption[109] were transposed by the German legislature in the national Ordinance on Cocoa and Chocolate Products (*Kakaoverordnung* (KakaoV)) of 15 December 2003.[110] The German Cocoa Ordinance among other things stipulates requirements for the quality of cocoa and cocoa products, as well as labeling, penalty and fine regulations.

28.3.2.10 Coffee and Coffee Products

For coffee and coffee products, Germany's vertical Ordinance on Coffee, Coffee and Chicory Extracts (*Verordnung über Kaffee, Kaffee- und Zichorien-Extrakte* (KaffeeV)) of 15 November 2001[111] applies. This Ordinance sets out definitions, labeling regulations, specifications for analysis methods, market prohibitions as well as penalty and fine regulations.

---

[107]Ordinance on Ice Cream (*Verordnung über Speiseeis* ["**SpEisV**"] of 15 July 1933, RGBL I, p. 510.
[108]Federal Law Gazette I, p. 702.
[109]OJ L 197, 03/08/2000, p. 19.
[110]Federal Law Gazette I, p. 2738.
[111]Federal Law Gazette I, p. 3107.

28.3.2.11 Food Flavorings

To supplement Regulation (EC) No 1334/2008 of the European Parliament and of the Council of 16 December 2008 on flavorings and certain food ingredients with flavoring properties for use in and on foods,[112] Germany also has a vertical Ordinance on Flavorings (*Aromenverordnung* (AromV)) in the version as announced on 2 May 2006.[113] However, it only regulates the maximum levels of quinine for the placing of foods on the market, authorizing additives and identifying quinine in flavorings and non-alcoholic soft drinks as well as certain criminal and administrative offences.

28.3.2.12 Vinegar and Vinegar Essence

Vinegar and vinegar essence are regulated in Germany in a vertical national Ordinance on Trading in Vinegar and Vinegar Essence (*Verordnung über den Verkehr mit Essig und Essigessenz* (EssigV)) of 25 April 1972.[114]

28.3.2.13 Wine

By contrast, the requirements for wine in Europe have been largely stipulated directly and harmonized by Community law, such as Council Regulation (EC) No 1234/2007 of 22 October 2007 establishing a common organization of agricultural markets and on specific provisions for certain agricultural products (Single CMO Regulation).[115] Consequently, definitions, the delimitation of wine-growing zones and assignment of vineyards in the Community to the individual wine-growing zones, support measures, producer and industry organizations, oenological practices and treatments, designations of origin, geographical information and traditional definitions, trading with third countries and the production potential are regulated under European law. The Member States therefore have a legislating power only to the extent such power is expressly granted to them or a lacuna in the provisions exists. In addition, the German legislature has adopted a national Wine Act (*Weingesetz* (WeinG)) in the version as announced on 18 January 2011[116] as well as a Wine Ordinance (*Weinverordnung* (WeinV)) in the version as announced on 21 April 2009.[117] The Wine Act and the Wine Ordinance moreover stipulate *inter alia* additional national requirements for wine growing and processing as well

---

[112] OJ L 354, 31/12/2008, p. 34.
[113] Federal Law Gazette I, p. 1127.
[114] Federal Law Gazette I, p. 732.
[115] OJ L 299, 16/11/2007, p. 1.
[116] Federal Law Gazette I, p. 66.
[117] Federal Law Gazette I, p. 827.

as for geographical designations, labeling, monitoring, importing, sales promotion and special product specifications.

28.3.2.14 Beer

Beer is also subject to separate vertical regulation in Germany in the Beer Ordinance (*Bierverordnung* (BierV)) of 2 July 1990.[118] The Ordinance—in compliance with the judgment of the Court of Justice of the European Union of 12 March 1987[119]—regulates protection of the designation "beer". In that case the ECJ had objected that the Federal Republic of Germany had violated the Community-law prohibition on measures of equivalent effect to quantitative restrictions by having prohibited a beer, legally produced and placed on the market in another Member State, from being placed on the market which did not meet the requirements then in force in Germany for the production of beer and use of the designation "beer".

28.3.2.15 Spirits

To supplement Regulation (EC) No 110/2008 of the European Parliament and of the Council of 15 January 2008 on the definition, description, presentation, labeling and the protection of geographical indications for spirit drinks, Germany has adopted vertical quality requirements set out its Ordinance on Certain Alcoholic Beverages (*Alkoholhaltige Getränke-Verordnung* (AGeV)) in the version as announced on 30 June 2003.[120]

28.3.2.16 Drinking, Mineral and Table Water

There are also separate vertical regulations in Germany for drinking water. The German legislature has transposed the related requirements of the European Council Directive 98/83/EC of 3 November 1998 on the quality of water intended for human consumption[121] in the national Ordinance on the Quality of Water Intended for Human Consumption (*Trinkwasserverordnung* (TrinkwV)) in the version as announced on 10 March 2016.[122] There is also a separate vertical Ordinance on

---

[118]Federal Law Gazette I, p. 66.
[119]ECJ, judgment of 12 March 1987, Case 178/84, [1987] ECR, 1227.
[120]Federal Law Gazette I, p. 1255.
[121]OJ L 330, 05/12/1998, p. 32.
[122]Federal Law Gazette I, p. 459.

Natural Mineral Water, Spring Water and Table Water (*Mineral- und Tafelwasser-Verordnung* (Min/TafelWV)) of 1 August 1984.[123]

28.4 Enforcement of Food Law Provisions

As before, responsibility for enforcing Community law is incumbent on the Member States since there are no real enforcement bodies at the European level. The Member States are also responsible for enforcing the respective national regulations. As a result, power to decree enforcement regulations as a general rule is the prerogative of the Member States.

In this context the European basic Regulation (EC) No 178/2002 also stipulates that the Member States are to enforce as well as monitor and review food law to ensure that all relevant requirements of food law are fulfilled by the food and feed business operators at all stages of production, processing and distribution.[124] For that purpose, they are required to maintain a system of official controls and other activities as appropriate to the circumstances, including public communication on food and feed safety and risk, food and feed safety supervision and other monitoring activities covering all stages of production, processing and distribution. They are also required to lay down the rules on measures and penalties applicable to infringements of food and feed law. The measures and penalties themselves must be effective, proportionate and dissuasive.

In Germany, food law falls under a category of special safety law which, as a general rule, comes within the purview of the federal states.[125] Accordingly, separate enforcement acts are adopted at the level of federal state legislation in each case. To give just one example, the Act Implementing the Law on Food, Feed and Consumer Goods in North Rhine-Westphalia (*Gesetz über den Vollzug des Lebensmittel-, Futtermittel- und Bedarfsgegenständerechts in Nordrhein-Westfalen* (LFBRVG-NRW)) imposes the obligation on the competent district authorities (*Kreisordnungsbehörden*) to enforce, e.g., food, feed and consumer goods law and wine law as a mandatory duty to be fulfilled on instruction. The other federal states have similar provisions in place.

However, at least European law, by way of exception, is to be enforced by the bodies of the European Union. This applies particularly when differences in enforcement are to be addressed by a uniform decision.[126] Some examples that may be cited are the participation rights reserved to the European Commission in Regulation (EC) No 258/97 of the European Parliament and of the Council of 27 January 1997

---

[123]Federal Law Gazette I, p. 1036.
[124]Art. 17 Regulation (EC) No. 178/2002.
[125]*Zipfel/Rathke*, Lebensmittelrecht, 163rd edition 2016, introduction, no. 42e.
[126]Ibid, no. 42d.

concerning novel foods and novel food ingredients,[127] in Regulation (EC) No 1829/2003 of the European Parliament and of the Council of 22 September 2003 on genetically modified food and feed,[128] in Regulation (EC) No 1924/2006 of the European Parliament and of the Council of 20 December 2006 on nutrition and health claims made on foods[129] and in Regulation (EC) No 834/2007 of 28 June 2007 on organic production and labeling of organic products.[130]

---

[127] OJ L 43, 14/02/1997, p. 1.
[128] OJ L 268, 18/10/2013, p. 1.
[129] OJ L 12, 18/01/2007, p. 3.
[130] OJ L 189, 20/07/2007, p. 1.

Part IV
The Americas

Gabriela Steier

All people living on earth are trustees of the resources and environmental wonders available. Although, people are also consumers and need food, water, air and other vital substances from the world, there is no inherent right to capitalize upon these environmental withdrawals because the resources are nearing dangerous exploitation levels. Thus, in terms of food production, processing, use and trade, it is becoming increasingly important to engage in sustainable agricultural and fishing practices, environmentally responsible land and marine stewardship, and socially proactive strategies to conserve both food safety for consumers and environmental integrity for the future.

In the US, one of the leading food production and processing nations worldwide, sustainable and environmentally responsible approaches remain challenges. BigAg and BigFood companies are dominating the market, consumers are led to believe much of the "greenwashing" in popular media, and legislation often falls prey to the powerful industrial lobby. Nonetheless, the US also provides a set of highly regulated food safety and food policy aspects and a wide array of federal and state statutes, which provide legal tools that can be used to provoke progress toward more sustainable food systems.

The first chapter in this section, Chap. 29, provides an overview of the US and Canadian frameworks in the US. Then, the following two textboxes, on the US Farm Bill and the School Lunch Project illustrate two large federal aspects of US food law, that provide examples for readers interested in the larger picture of US domestic food policy (Chap. 30). Chapters 31–33 provide overviews of food law in Canada, Brazil, and Mexico, respectively, thereby contributing to the fundamentals of building an understanding of food law and policy in the Americas. These chapters are especially useful for a comparative approach and allow further analysis through the sources cited.

Chapters 34 and 36 introduce some of the most important legislative and regulatory frameworks of US and global waterbodies regulation. Through an initial description of the issues underlying the Tragedy of the Commons and the principles of the Public Trust Doctrine, Chap. 34 provides the basis to an understanding of water resource regulation, the foundation to appreciate how food law and policy can use existing law to tackle the problems of overfishing, marine pollution, and bycatch while aquacultures are becoming increasingly unsustainable. This understanding is further expanded by the subsequent Chap. 36, which goes beyond the commons and discusses global fisheries regulation.

Global fisheries regulation is of particular importance because it addresses a vast amount of food resources on a world-wide scale. Gaining an understanding of the foundations of the regulatory framework and the legal limitations provides legal tools to improve maritime conservation and fisheries management with the goal to further food safety and environmental protection. The US Magnuson Steven Fishery Management and Conservation Act (MSA), for example, allows coastal states to consider the effects on species that are not targeted, such as bycatch, through language that requires ecosystem management, where regulation is based on the biological relationships between species. Like the MSA, the Law of the Sea requires coastal states to base conservation and management measures on scientific support.

Thus, the principles of the Third United Nations Convention on the Law of the Sea, or UNCLOS III, and its purpose to establish a comprehensive set of rules governing the oceans are outlined in Chap. 36 and illustrated in Fig. 34.1, to provide a comprehensive foundation to understand maritime regulation. Finally, Chap. 37, outlines the effects of industrial aquaculture and the broad-sweeping impacts of human intervention in natural law.

Although this section on the Americas is not all-encompassing, the most relevant issues in current food law and policy are put into context and the following set of chapters provide a solid, albeit cursory, overview to understand how the law addresses concerns of sustainable and safe food production on the American continent. The chapters in this section are complimentary to one-another and have been written by practitioners, scholars and professors well-versed in their respective fields.

Chapter 29
Introduction to Food Law and Policy in the United States and Canada

Carly Dunster

Abstract This introductory chapter discusses the rise of the practice of food law in both the United States and Canada, and gives an overview of the legislative backdrop for the practice of food law. The chapter discusses the multi-layered landscape of food and agriculture laws and regulations in Canada and the United States, by outlining the key federal, provincial/state-wide, and regional/municipal legislative structures. The regulatory and legislative modernization efforts in both countries are assessed, with discussion on the factors that have driven their adoption by both countries at the federal level. The chapter goes on to discuss new legislative and policy developments in both jurisdictions, highlighting the issues and trends arising in the sector. These include new and more permissive legislative tools in the realm of urban agriculture, an increased legislative emphasis on local food and the role of procurement, the presence of cottage food laws, and the intertwined issues of the right to food and food sovereignty. In conclusion, the author discusses the proliferation of professional associations and corresponding opportunities for the practice of food law in both the United States and Canada, a development which demonstrates both the need for food lawyers and the legitimizing of the practice area on the whole.

29.1 Food Law on the Rise

Food law practitioners in the United States and Canada are a small but mighty bunch. Lawyers practicing in both countries have engaged with and provided advice on key federal, statewide or provincial legislation since their enactment. However, only recently can one enter the term "food lawyer" into a search engine and find professionals that define themselves as such. This proliferation in practices has been matched by a similar increase in food law and policy centers and clinics that are developing specializations and producing comprehensive reports on key

---

The author extends special thanks to Sara Zborovski, LL.B., and Glenford Jameson, LL.B., for their thoughtful contributions to the development of this chapter.

C. Dunster (✉)
Turnpenney Milne LLP, Toronto, ON, Canada
e-mail: carly@tmllp.ca

trends and issues within this growing field. There is also a corresponding rise in food law and policy educational courses and programs, both at law schools and as part of other related disciplines.

This growth in the realm of food law and policy can be explained in large part by two factors: (1) a heightened awareness on the part of consumers regarding their food and its journey from farm to fork, and (2) a realization on the part of practitioners that the patchwork nature of food regulations requires professional specialization to navigate. In both countries, the practice of food law demands a thorough understanding of how food is regulated on a federal, regional, and state- and province-wide level. Engaging with clients in the food sector often necessitates navigating between these layers of law. This chapter is intended to provide context for the subsequent chapters on the U.S. and Canada in this section of the book. First, this chapter will briefly review this multi-layered legislative and regulatory context within which food is subject to the law, and then introduce the reader to the significant modernization of food legislation that both Canada and the U.S. are undertaking. Finally, the chapter will review a selection of legislative developments that provide a window on to the current state of food law in these countries.

29.2 The Legislative Framework of Food Regulation

29.2.1 The Federal Level in the US

At the federal level, three executive branch agencies regulate food and agriculture among several other areas, such as drugs, cosmetics and pollution. First, the Food and Drug Administration (FDA) is an agency of the U.S. Department of Health and Human Services (DHHS) and acts to protect public health by assuring that foods within its purview are "safe, wholesome, sanitary and properly labeled."[1] In essence, the FDA asks whether the foods produced are safe to eat. While animal feed and drug regulation are also within the FDA's purview, the major legal instrument under which it operates is the Federal Food, Drugs and Cosmetics Act (FFDCA) enacted in 1938.[2] Food is regulated under chapter IV, 21 U.S.C. § 341 et seq. Specifically, it limits pesticide residues in food in interstate commerce, including imports. It also ensures the safety of any additives to food.

The second agency governing food regulation in the U.S. is the U.S. Department of Agriculture (USDA). It regulates livestock, meat and poultry products, as well as some egg products, with the mission to "provide leadership on food, agriculture, natural resources, rural development, nutrition, and related issues based on sound public policy, the best available science, and efficient management."[3] As the

---

[1] Food and Drug Administration (FDA) (2014), FDA Fundamentals.
[2] FDA (1938), Federal Food, Drugs, and Cosmetics Act (FFDCA).
[3] United States Department of Agriculture (USDA) (2014), Mission Statement.

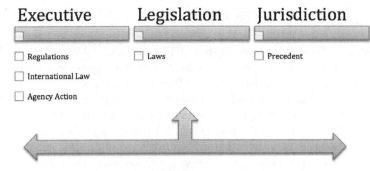

Fig. 29.1 Sources of federal-level food laws in the US

agency charged with agricultural oversight, the USDA is an agency for farmers, not consumers. The core question the USDA is concerned with is whether certain crops are safe to grow.

The third agency is the U.S. Environmental Protection Agency (EPA)[4] whose mission is to "protect human health and the environment." Agriculture and food production are inextricably linked to environmental concerns and pollution control. In terms of food regulation, the EPA regulates environmental values under the National Environmental Policy Act (NEPA),[5] biodiversity and threatened species under the Endangered Species Act (ESA),[6] pesticides under the Federal Insecticide, Fungicide, and Rodenticide Act (FIFRA),[7] and water and air pollution under the Clean Water Act[8] and the Clean Air Act,[9] respectively. With such a broad mandate, the EPA essentially overlooks agricultural and food practices and environmental safety (Fig. 29.1).

In addition to the aforementioned acts, the U.S. has begun a significant food and food safety legislative overhaul through the Food Safety Modernization Act (FSMA).[10] This legislation will shift the focus of food safety from response to prevention and give the FDA further tools to protect consumer food safety. A more detailed discussion on the FSMA continues below.

---

[4]Environmental Protection Agency (EPA) (2014), About EPA.
[5]EPA (2014c), National Environmental Policy Act (NEPA).
[6]National Oceanic and Atmospheric Administration (NOAA) (2014), Endangered Species Act (ESA).
[7]EPA (2014b) Federal Insecticide, Fungicide, and Rodenticide Act (FIFRA).
[8]EPA (2014a), Clean Water Act.
[9]EPA (2013), Clean Air Act.
[10]FDA (2014), Food Safety Modernization Act.

29.2.2 The Federal Level in Canada

The Canadian Food Inspection Agency (CFIA), created by the Canadian Food Inspection Agency Act of 1997,[11] is Canada's federal counterpart to the FDA. The CFIA reports to the Minister of Agriculture and Agri-Food. The mission of the CFIA is to safeguard food, animals, and plants and, through this pursuit, to enhance the wellbeing of Canadians, the Canadian economy and the environment.[12] It conducts activities such as the registering and monitoring of processing plants, inspection of foreign and domestic foods, certification of exports, and determination of quarantine necessity. Additionally, the CFIA oversees several legislative acts in the pursuit of its overall goals, including:

- **Canada Agricultural Products Act**,[13] which regulates the marketing of agricultural products in import, export and interprovincial trade and provides for national standards and grades of agricultural products, for their inspection and grading, for the registration of establishments and for standards governing establishments;
- **Consumer Packaging and Labeling Act**,[14] (with regard to food) which sets rules for packaging, labelling, sale, importation and advertising of prepackaged and certain other products;
- **Feeds Act**,[15] which controls and regulates the sale of feed;
- **Fertilizers Act**,[16] which is responsible for regulating agricultural fertilizers;
- **Fish Inspection Act**,[17] which sets rules inspection of fish and marine plants;
- **Food and Drugs Act**,[18] which regulates food, drugs, cosmetics and therapeutic devices;
- **Health of Animals Act**,[19] which considers diseases and toxic substances that may affect animals or that may be transmitted by animals to persons, and respecting the protection of animals;
- **Meat Inspection Act**,[20] which regulates the import and export of and interprovincial trade in meat products, the registration of establishments, the inspection of animals and meat products in registered establishments and the standards for those establishments and for animals slaughtered and meat products prepared in those establishments;

---

[11] Canadian Food Inspection Agency Act (1997), c. 6.
[12] Canadian Food Inspection Agency (2011), *Vision and Mission*.
[13] Canada Agricultural Products Act (1985), c. C-20.
[14] Consumer Packaging and Labeling Act (1985), c. C-38.
[15] Feeds Act (1985), c. F-9.
[16] Fertilizers Act (1985), c. F-10.
[17] Fish Inspection Act (1985), c. F-12.
[18] Food and Drugs Act (1985), c. F-27.
[19] Health of Animals Act (1990), c. 21.
[20] Meat Inspection Act (1985), c. 25.

- **Plant Breeders' Rights Act**,[21] which protects plant breeders' rights;
- **Plant Protection Act**,[22] which prevents the importation, exportation and spread of pests injurious to plants and provides for their control and eradication and for the certification of plants and other things and;
- **Seeds Act**,[23] which is in respect to the testing, inspection, quality and sale of seeds.

The CFIA also has ultimate oversight over the Safe Food for Canadians Act[24] (the SFCA), a new law that will be implemented in 2015. This legislation represents Canada's attempt to streamline, modernize, and in some cases, simplify its food laws.[25] Further discussion on the SFCA continues below.

Another agency, Health Canada,[26] is the federal department responsible for establishing safety and nutritional quality standards of all foods sold in Canada. This mandate is exercised in part via the Food and Drugs Act and its associated regulations.[27] The Food and Drugs Act is an integral part of the Canadian food law framework, as it addresses product safety and integrity and consumer protection from fraud regarding the sale and consumption of food.

29.2.3 Agricultural Law and Policy

One of the key U.S. food law and policy tools is the Farm Bill (the Bill),[28] a federal omnibus bill that is reviewed and renewed approximately every 5–7 years and is intended to provide some consistency in food and agricultural legislation and policy. Historically, it has supported staple farm commodities, such as corn, soybeans, and wheat. In recent years, however, its reach has grown and its contents have greatly diversified. The first "omnibus" Farm Bill was enacted in 1973[29] and is the resting place for many food and agriculture policies ranging from conservation of land, to trade and export programs, to crop insurance. The bill also provides authorization for both foreign food aid and domestic nutrition assistance, primarily via the Supplemental Nutrition Assistance Program (SNAP),[30] commonly known

---

[21] Plant Breeders' Rights Act (1990), c. 20.
[22] Plant Protection Act (1990), c. 28.
[23] Seeds Act (1985), c. S-8.
[24] Safe Food for Canadians Act (2012), c. 24.
[25] See generally the CFIA's "Action on Weatherill Report Recommendations to Strengthen the Food Safety System: Final Report to Canadians."
[26] Health Canada (2014), About Health Canada.
[27] Food and Drugs Act (1985), c. F-27.
[28] See the 2014 Farm Bill (P.L. 113-79): Summary and Side-by-Side by Chite R for an excellent introduction to the 2014 Farm Bill.
[29] Ibid, p. 3 Figure 1.
[30] USDA (2014), Supplemental Nutrition Assistance Program (SNAP).

as food stamps, and the Emergency Food Assistance Program (TEFAP).[31] Its most recent iteration is the Agricultural Act,[32] which was enacted into law in February 2014. Some key elements include increased support for access to crop insurance, fruits, vegetables, and organic programs, and diminished funding and subsidies for more traditional commodities. Money was also allotted to assist farms transitioning from traditional farming to organic practices.

This bill is a unique regulatory tool because its periodic review offers a perspective and a window into how food, agriculture and the law have been seen by legislators over time. For example, over the years, the bill has evolved to include titles that reflect key food and agriculture issues of the day, such as conservation laws in 1985 and local food systems in the 2008 farm bill.[33] In theory, it also reflects how consumers interact with food and agriculture.

29.2.4 Canadian Agricultural Law and Policy

Agriculture and Agri-Food Canada[34] is the Canadian federal department responsible for agricultural policy. However, there is no overarching legislative act akin to the U.S. Farm Bill. Instead, an Agricultural Policy Framework was developed by Agriculture Canada and serves as a comprehensive federal, provincial, and territorial document. Growing Forward 2[35] is the most current iteration of this policy document, and covers the period of 2013–2018. It highlights three key focus areas for agricultural policy and support: (1) encouragement of innovation, (2) international competitiveness, and (3) market development.

29.2.5 The Role of U.S. States and Cities and Canadian Provinces, Territories, and Municipalities

29.2.5.1 US State-Level Food Law

The regulatory behemoths, the FDA, EPA and USDA, are merely the tip of the food legislation iceberg in the U.S. These federal agencies have various subdivisions, such as the Animal and Plant Health Inspection Service (APHIS) within the USDA and several others. Additionally, there are state agencies, because states typically act as the implementers of federal legislation but they also have the liberty to develop their own food-related laws and policies within certain parameters. The

---

[31]USDA (1983), Emergency Food Assistance Program (TEFAP).

[32]United States Congress (2014), Agricultural Act of 2014—H.R.2642.

[33]Chite (2014), p. 4.

[34]Government of Canada (1868), Agriculture and Agri-Food Canada (AAFC).

[35]Agriculture and Agri-Food (2014), Growing Forward 2.

same is true at the local level. While it is rare for a local government to run explicitly afoul of any federal or state law, laws and policies can be created according to the needs of the region or the specific influences of that area. Some examples of state and city-specific laws are examined below.

29.2.5.2 Canada's Province-Level Food Law

The role of provinces, territories, and municipalities with respect to food law in Canada is similar to that of the U.S. Similarities include overarching federal laws that all jurisdictions must adhere to while maintaining some legislative freedom to tailor the legal framework to the needs of that jurisdiction. For example, although there are federal laws like the National Dairy Code[36] and the Dairy Products Marketing Regulations concerning the production and sale of dairy products,[37] almost every province or territory has their own dairy legislation specific to that region. Some examples of province and city-specific food laws are examined below.

29.3 Regulatory Modernization

The FSMA and SFCA represent a new era of regulatory modernization in both the U.S. and in Canada. In the U.S., federal food law has mostly remained the same for approximately 75 years.[38] In comparison, the SFCA compiles and streamlines distinct pieces of legislation and represents a robust overhaul of the piecemeal food safety laws that existed before. Both laws and their associated policies and programs demonstrate a clear overarching priority of federal governments in each jurisdiction: food safety.

29.3.1 United States Food Law

The U.S. has recently undergone a sizable legislative update with the introduction of the FSMA, which was signed into law in January of 2011. For clarity, the FSMA does not apply to meat, poultry, and some egg producers, which continue to be regulated by the USDA. The FSMA, nonetheless, empowers the FDA to regulate all levels of the food chain, at all points of food growth and production, and to work to prevent food borne illnesses rather than simply respond to an outbreak after it has occurred. This is a significant shift in approaching food safety. The FSMA's top

---

[36]Government of Canada (2013), National Dairy Code.
[37]Dairy Products Marketing Regulations (2009), (SOR/94-466).
[38]Strauss (2011), p. 355.

priority is to ensure food safety throughout all levels of domestic food production and distribution systems.

The FSMA has brought several changes in US food safety regulation. The FDA now has increased plant inspection powers and obligations and has the authority to initiate food recalls where necessary. It also increases the obligations of food facilities to develop food safety and contamination plans. To date, the producer of foodstuff has always borne the ultimate responsibility for contamination. The FSMA, however, requires producers to engage in preventative action and includes penalties for noncompliance. The law also levels the playing field for imported foods and their American-made counterparts. Under the FSMA, importers have an increased obligation to verify that foreign suppliers meet U.S. food safety standards, and the FDA can require credible third-party compliance certification as a condition of entry for high-risk foods.[39]

The rollout of the FSMA and its regulations and protocols remains in progress at the time this chapter was written. The act includes timelines within which the FDA must establish rules with respect to concepts such as mandatory produce safety standards and the implementation of written preventive control plans for food facilities.[40] Recently, the Center for Food Safety sued the FDA claiming that the FDA unlawfully withheld seven important regulations, which were to be implemented by July 2012, a deadline imposed by the FSMA. The U.S. District Court's recently released decision upheld that the FDA had not met the deadlines imposed by the FSMA and, therefore, was in violation of the Act. A consent decree released in February 2014 specified the deadlines for various rules contained in the Act.[41] The Court's decision emphasizes the shift in U.S. priorities and its uncompromising position to ensure and maximize food safety.

29.3.2 Canadian Food Law

Canada is also currently undergoing its own much-needed modernization of food laws, regulations, and policies. Legislative modernization has become, at least on the federal level, the dominant theme of food law discourse. The federal government has long held that the Food and Drugs Act[42] has been, from its inception in 1906, a law that prioritizes consumers. The new legislation, mainly in the form of the SFCA, demonstrates a heightened emphasis on consumer safety. However, increasingly important issues like the consumer's right to know all aspects of the food system via labeling, or the creation of a legislative infrastructure within which small producers can thrive, continue to be omitted from the federal legislative context. This omission prevents full disclosure for consumers and prevents them

---

[39]Department of Health and Human Services/FDA (2013), p. 3.
[40]Strauss (2011), p. 359.
[41]Case No.: 12-cv-04529-PJH (2014), Consent Decree para. 4.
[42]Food, Drug and Cosmetic Act (1906), 21 U.S.C. 1 et seq.

from fully exercising their rights to make informed and educated food choices. Furthermore, sustainability for small farmers is relegated to a lower priority when industrial farming is emphasized and protected by these laws. Note that this act should be effective as of the beginning of 2015.

The SFCA consolidates the authorities of the Fish Inspection Act, the Canada Agricultural Products Act, the Meat Inspection Act, and food related aspects of the Consumer Packaging and Labeling Act. The catalyst for the development of this law was a report by an independent investigator on the 2008 listeriosis outbreak,[43] where 22 deaths were all linked to listeriosis and were traceable to deli meats and related products processed by Maple Leaf Foods, emphasizing the lack of coordination between authorities and discordant laws. The Weatherill Report, released in 2009,[44] concluded that the regulation of food and consumer safety in Canada would greatly benefit from streamlining, modernizing, and in some cases, simplifying.

The SFCA focuses on three key areas of control—improved food safety oversight to better protect consumers, streamlined and strengthened legislative authorities, and enhanced international market opportunities for Canadian industry.[45] The law includes processes to ensure oversight of food commodities traded interprovincially or internationally, and is also meant to create improved opportunities for Canadian products in international markets. Moreover, it improves traceability of foods by providing the CFIA with the authority to develop regulations on the tracing and recalling of food. Through the act, inspection and enforcement powers are aligned to ensure more efficient inspections and increase compliance, ultimately resulting in more effective consumer food safety protections.

The FSMA and SFCA both modernize food safety regulations in their respective jurisdictions. It is hard to argue against the prioritization of food safety from a legislative focal point. Recent large-scale and high profile food contamination outbreaks in both the U.S. and Canada demand food safety to be a top priority. It should be noted, however, that there is very little contemplation at the federal level about why contamination cases have become so far-reaching, and why food safety is more difficult to establish and maintain. Both countries could benefit from a critical analysis of the scale of our food system, how it developed and what may need to be reassessed. This kind of critical analysis should work in concert with an updated food safety legislative framework; both elements are necessary to strengthen the integrity of food laws.

An overhaul of the principles involving consumer protection, food justice, farm animal welfare,[46] fair trade, and environmental protection would help to update some of the aforementioned laws. Additionally, an objective and unbiased evaluation of sound science—not only studies funded by the industry—could help to

---

[43] See generally the CFIA's "Action on Weatherill Report Recommendations to Strengthen the Food Safety System: Final Report to Canadians."

[44] Ibid.

[45] Safe Food for Canadians Act (2012), SFCA Overview.

[46] See Steier and Patel (Eds.), *International Farm Animal, Wildlife and Food Safety Law* (forthcoming with Springer 2016), http://www.springer.com/us/book/9783319180014.

revise the dietary recommendations and food safety standards, including clarifying labeling laws in favor of consumer protection, so that the public can make more educated and healthier food choices, thereby improving food safety from the public health perspective.

29.4 Legislative and Policy Developments

The issues and trends that food law practitioners should be cognizant of in the U.S. and Canada could fill a textbook in itself. For the sake of brevity, however, what follows is a discussion and analysis on a selection of legislative and policy developments on the federal, regional and provincial or statewide levels. While these developments address agriculture (both traditional and urban), issues of local food, cottage food laws, and the concept of food sovereignty, and other laws, regulations, and policies invoked by these developments provide a glimpse into innovation and updates by the U.S. and Canada, and also areas where more progressive policies can and should be developed in the future.

29.4.1 New Developments in U.S. and Canadian Food Law

29.4.1.1 Urban Agriculture

Urban agriculture is the act of growing food within a city's perimeters. It is often a smaller, more compact take on traditional agriculture, and can be done for one's own consumption or for use by a larger community. The sale of such crops has often been hampered by regulatory restrictions. Zoning by-laws and land use parameters also impact the growth of food within cities.

United States

Boston, Massachusetts, recently passed legislation that enables a more permissive regulatory regime for growing food within a city's limits than is commonly seen in the U.S. or Canada. Article 89[47] has been identified as the most comprehensive city re-zoning regulation of its kind and provides a transactional structure that will allow urban farmers to grow food within city limits and sell those items for a profit. This permissiveness distinguishes Article 89 from many other existing urban agriculture regulations in North America. As part of the regulation, planners have identified city-owned land parcels that are suitable for urban farming and offers them to

---

[47]Boston Redevelopment Authority Boston Redevelopment (2013a), Article 89, http://www.bostonredevelopmentauthority.org/getattachment/8405c72c-7520-43ad-a969-0e27dddae7a2.

potential farmers for $100/plot, with the requirement that the land be used for urban farming purposes for at least 50 years. Farm stands are also allowed on these plots. This used to be a major hurdle for urban growers, where people could grow produce but were not permitted to sell it on the same plot of land. The new Massachusetts law goes beyond simply articulating the new allowances. This progressive and modern law is accompanied by a User's Guide[48] that clearly and accessibly articulates the steps to conduct those activities for an urban farmer. This is notable because it shows that the new law is geared toward and vastly supportive of a more user-friendly urban agriculture. Creating such legislation and pairing it with resources like the User's Guide transforms an activity like urban farming away from a token nod for sustainable food to a very practical and accessible step towards food security in cities.

Canada

While not nearly as comprehensive as Boston's Article 89, the City of Vancouver, British Columbia, has taken the lead among Canadian municipalities on urban agriculture. The City has developed urban agricultural guidelines to assist developers with planning projects that incorporate those tenets. In 2008, the City approved a set of Urban Agriculture Guidelines for the Private Realm.[49] These guidelines are applied in conjunction with the Zoning and Development By-Law, and are to be consulted when an urban agriculture element is proposed in concert with a private development plan. In short, these guidelines provide a framework for those interested in urban farming.

Urban agriculture has inhabited the realm of the privileged in many regions of Canada and the U.S. This is because, in many cities, the space to conduct such activities is very limited and expensive, even where it is available. An urban farming activity, such as raising backyard chickens, requires a significant amount of time and resources. Legislation like Article 89 and the Urban Agricultural Guidelines build a framework that creates accessibility and eliminates some barriers usually inherent in urban agriculture.

29.4.2 *Local Food and Procurement*

29.4.2.1 United States

In 2010, the Illinois Food, Farms, and Jobs Act[50] was signed into law, and was in some ways a precursor to Ontario's legislation (see below). It was designed "to

---

[48]Boston Redevelopment Authority Boston Redevelopment (2013b), Article 89 Made Easy.
[49]City of Vancouver (2008), Urban Agriculture Design Guidelines for the Private Realm.
[50]Illinois General Assembly (2009), Local Food, Farms, and Jobs Act.

create, strengthen, and expand local farm and food economies throughout Illinois."[51] One of the tenets of this legislation was the establishment of procurements goals for state agencies. For instance, state agencies that provide food services need to increase the purchase of local food to 20 % by 2020. Comparatively, state-funded institutions have a goal of a ten percent increase by 2020.[52]

29.4.2.2 Canada

A notable development in food law at the provincial level in Canada is Ontario's Local Food Act,[53] which became law in November 2013. Ontario is the first Canadian province or territory to adopt a legislation that focuses specifically on local food. The purposes are to foster successful and resilient local food economies and systems throughout Ontario; increase awareness of local food in Ontario, including the diversity of local food; and encourage the development of new markets for local food.[54] It gives the Minister of Agriculture and Food discretionary power to encourage the use of local food by public sector organizations. In other words, this act could be used as a tool to provide more opportunities for local farmers and producers to compete for public sector procurement contracts.

These legislative acts are concrete evidence that legislators in some jurisdictions have begun to codify local food principles under law, giving discretionary powers to Ministers and legitimizing the role of local food. The advantage of promoting such local food movements is a reduction in food miles and greater potential for improved transparency in food origins, as well as improved sustainability if organic and responsible farming practices are encouraged and supported.

29.4.3 Local Procurement and International Trade Agreements

29.4.3.1 Canada

Local procurement policies may be directly impacted by international trade agreements. This is evidenced by local food policy activists' and members of the Toronto Food Policy Council's (TFPC) concerns with the Comprehensive Economic and Trade Agreement (CETA), a proposed free trade agreement between Canada and

---

[51]Illinois General Assembly (2009), Local Food, Farms, and Jobs Act, http://www.ilga.gov/legislation/publicacts/fulltext.asp?Name=096-0579.
[52]de Schutter (2014), Briefing Note.
[53]Legislative Assembly of Ontario (2013), An Act to enact the Local Food Act, 2013.
[54]Ibid, preamble.

the European Union.⁵⁵ The TFPC presented to the Executive Committee of Toronto's City Council in 2012, positing that CETA would have significant detrimental impacts on sub-national governments, primarily municipalities. The TFPC further argued that signing the agreement would prohibit municipal governments from favoring local, or even Canadian, goods or services, if those potential procurement contracts were valued over $340,000 CAD. The TFPC stated that CETA "would prohibit the City of Toronto from promoting food security and economic development through the local food procurement policy."⁵⁶ Toronto's City Council, subsequently, passed a motion to request that the Province of Ontario issue an exemption for Toronto. While it is highly unlikely that this exemption will be granted, it is an important symbolic gesture by the City, indicative, at least in part, of its awareness of the power of procurement with respect to local food. CETA has not yet been fully ratified and negotiations continued as of the summer of 2014.

29.4.4 Cottage Food Laws

29.4.4.1 United States

Cottage food laws are the colloquial name for laws that allow individuals to cook low-risk food items in their homes, then sell and distribute them to the public. These kinds of laws have been passed in 31 U.S. states, including Texas, Florida, and New York.⁵⁷ Nine states have pending cottage food laws at various stages of debate and approval. The impetus behind these cottage food laws is to remove a major stumbling block for start-up food entrepreneurs: access to a commercial kitchen. The laws stipulate that the approved foods that can be made in one's home must all pose only a minimal consumption risk. In terms of whom a producer can sell to, some states allow only direct-to-consumer distribution, while others allow an individual to conduct online sales, or sales to grocery stores, farmer's markets, and restaurants. Therefore, cottage food laws can provide viable economic opportunities for a growing community of food entrepreneurs, and the laws fostering this kind of growth in the U.S. are creating thriving businesses. In sum, cottage food laws support local businesses and promote a "back to the roots" approach of community-supported enterprises.

Canada has yet to pass any such laws, though there have been rumblings of advocacy campaigns amongst the growing community of small food entrepreneurs across the country.

---

⁵⁵Government of Canada (2014), Canada-European Union: Comprehensive Economic and Trade Agreement (CETA).
⁵⁶Toronto Food Policy Council (TFPC) (2012), Briefing Note—Procurement.
⁵⁷Forrager—Cottage Food Community (2014), Cottage Food Laws.

29.4.5 The Right to Food and Food Sovereignty

In conjunction with the right to food, food sovereignty is an important principle that deserves to be part of any discussion on food law in the U.S. and Canada. Food sovereignty is, essentially, the democratization of food and agriculture. Food sovereignty is a principle related to the right to food that is arising with some frequency in the U.S. and Canada. It is defined as the "right of peoples to healthy and culturally appropriate food produced through ecologically sound and sustainable methods. More importantly, it is the right to define and control our own food and agriculture systems, including markets, production modes, food cultures, and environments."[58] This term is sometimes used by and with respect to First Nations communities, who attach vast cultural significance to their interactions with food in all of its forms.

In 2009, the U.S. joined consensus on Resolution L.30, Rev. 1—the Right to Food,[59] a non-binding declaration that called for the human right to adequate food, for the first time. Joining consensus on this Resolution did not change international law or impose any obligations on the U.S. to feed its own citizens, or provide food aid, other than those that already existed. It was largely a symbolic gesture, albeit one that the U.S. had never agreed to before. Hence, there is also no legal or constitutional guarantee of the right to food in American law. In terms of the legal and policy framework, Canada is a signatory to the International Covenant on Economic, Social, and Cultural Rights.[60] While the Covenant requires countries to fulfill and protect the right to food for its citizens, there is no legal or constitutional guarantee of the right to food in Canadian law.

Many cities in Maine have introduced food sovereignty ordinances in the past 2 years, and other cities in Vermont, California, and Massachusetts have also passed similar legislation.[61] Critics deride the ordinances for violating constitutional principles, as well as jeopardizing public health by absolving those within the ordinances from complying with state and federally mandated regulatory oversight. Regardless, the ordinances illustrate existing support for a hyper-local food system on the part of at least a handful of jurisdictions in the U.S. Whether or not the principles of food sovereignty will have any lasting impact on, or contribution to, the field of food law remains to be seen[62]; but the concept is undoubtedly a powerful one for many communities in the U.S. and in Canada.

---

[58]Wittman and Desmarais (2012), Food Sovereignty in Canada.

[59]The Right to Food (2013), The Right to Food, A/C.3/68/L.60/Rev.1.

[60]United Nations Human Rights (1966), International Covenant on Economic, Social and Cultural Rights.

[61]See Town of Sedgwick (2011) Local Food and Community Self-Governance Ordinance and Vermont for Evolution (2012) The Vermont Resolution for Food Sovereignty.

[62]For food sovereignty and law, see Bellinger and Fakhri (2013).

29.4.6 Proliferation of Professional Associations in Food Law

29.4.6.1 United States

The number of high-profile law and policy centers and legal programs that focus on the study of food and its regulation and that have been launched in recent years is indicative of the increasing legitimacy of the practice of food law as a distinct discipline, particularly in the US. While certain stalwarts have existed for many years, entities like the Resnick Program for Food Law and Policy at the University of California, Los Angeles which was established in 2013, and the Harvard Law School's Food Law and Policy Clinic which opened in 2010, illustrate that established institutions now believe that the study of food law and policy merits greater attention. The University of Michigan continues to offer their Global Food Law Program, and the Vermont Law School opened the Centre for Agriculture and Food Systems in 2012. The editors of this book have started *Food Law International, LLP*,[63] an organization to promote legal scholarship in the area of sustainable food law and are teaching corresponding courses at law schools in the US and in Europe. The American Bar Association has conducted continuing legal education seminars on topics like providing counsel to the local food movement, food labeling laws, and a primer on the FSMA.

Nonetheless, vast differences exist within the budding field of law. While some lawyers call themselves "food lawyers," they are really not because their work mainly focuses on helping companies, such as the food industry's BigFood and BigAg, come into compliance with existing laws. Although this is relevant work, this area of law should not be counted as "food law" but rather as "resource law," the latter of which describes how coming into compliance with existing laws attains at most the lowest common denominator that was legally standardized. True food lawyers seek to use the law and precedent to raise the sustainability bar, to promote progressive action, to encourage farm animal welfare, environmental and social integrity in food law, and to use food policy to close the gaps that the globalized food system unveils. Colleagues, such as the legal teams at the Center for Food Safety,[64] Food and Water Watch,[65] Sierra Club,[66] Defenders of Wildlife,[67] the Environmental Integrity Project,[68] Beyond Pesticides,[69] the Center for Biological Diversity,[70] Xerces Society,[71] and many other national non-profit organizations

---

[63]See www.FoodLawInternational.com.

[64]http://www.centerforfoodsafety.org.

[65]http://www.foodandwaterwatch.org.

[66]http://www.sierraclub.org.

[67]http://www.defenders.org.

[68]http://environmentalintegrity.org.

[69]http://www.beyondpesticides.org.

[70]http://www.biologicaldiversity.org.

[71]http://www.xerces.org.

continue to make progress in food law that will greatly benefit the nation's wildlife, natural resources, consumers, and contribute to food safety on a multitude of levels and for many years to come.

29.4.6.2 Canada

In contrast, the overall academic infrastructure with respect to food law and policy is not as established in Canada. There is some movement with respect to the study of food sustainability, but there remains only a tenuous connection between the role of the law and this realm. However, the establishment of a Chair in the Sustainable Food Production at the University of Guelph in Ontario in 2011, the appointment of Canada's first "Food Laureate" by the University of Guelph in 2012, and Wilfred Laurier University's opening of the Centre for Sustainable Food Systems in 2013 are evidence of a growing awareness and need for study of our food systems. Bar Associations in Canada have yet to ensconce the practice of food law into their continuing legal education programs as many other associations have around the world.

There is a corresponding proliferation of food policy councils in the U.S. and Canada. These bodies play a meaningful role in the development of policy and practice regarding food in the municipal and sometimes provincial sphere, and are usually comprised of volunteers who represent a variety of stakeholders in the food system.

29.5 Conclusion

There is not only a need for legal advice and direction in this increasingly complex world of food and agriculture, but also a strong role for advocacy in these areas. Individuals and organizations need legal assistance, but consumers and citizens also need those with legal expertise to advocate for their rights on a policy level. The practice of food law can entail more than just assisting clients with the existing pieces of legislation and policy. It can also include the opportunity for legislative development and creative advocacy. Many are increasingly aware that food legislation was developed so long ago that it has been rendered archaic, or that it was developed with only large-scale manufacturers, developers, distributors, and creators in mind. The reach of the law can inhibit innovation in food and this limitation has led to an increase in awareness, involvement and activism. As the food sector diversifies and innovates, so must law professionals and legal frameworks in order to assist those endeavors.

References

Agriculture and Agri-Food Canada (2014) Growing Forward 2. Available at http://www.agr.gc.ca/eng/about-us/key-departmental-initiatives/growing-forward-2/?id=1294780620963 (last accessed 12 Jan 2014)

Bellinger N, Fakhri M (2013) The intersection between food sovereignty and law. Nat Resour Environ 28(2):45–48

Boston Redevelopment Authority (2013a) Article 89. Available at http://www.bostonredevelopmentauthority.org/getattachment/8405c72c-7520-43ad-a969-0e27dddae7a2 (last accessed 20 May 2014)

Boston Redevelopment Authority (2013b) Article 89 Made Easy: Urban Agriculture Zoning for the City of Boston. Available at http://www.bostonredevelopmentauthority.org/getattachment/4b74929b-920e-4984-b1cd-500ea06f1bc0 (last accessed 20 May 2014)

Canada Agricultural Products Act, R.S.C., 1985, c. C-20

Canada: Food and Drugs Act (1985), R.S.C. 1985, c. F-27

Canadian Food Inspection Agency (CFIA) (2011) Action on Weatherill Report Recommendations to Strengthen the Food Safety System: Final Report to Canadians. Available at http://www.inspection.gc.ca/food/information-for-consumers/food-safety-investigations/progress-on-food-safety/weatherill-report-recommendations/eng/1362425366007/1362425780005 (last accessed 12 Jan 2014)

Canadian Food Inspection Agency (CFIA) Act (1997), S.C. 1997, c.6

City of Vancouver (2008) Urban Agriculture Design Guidelines for the Private Realm. Available at http://vancouver.ca/files/cov/urban-agriculture-guidelines.pdf (last accessed 21 May 2014)

Center for Food Safety v. Margaret A. Hamburg, (2014) Case No.: 12-cv-04529-PJH

Consumer Packaging and Labeling Act, R.S.C., 1985, c. C-38

Dairy Products Marketing Regulations (2009), (SOR/94-466)

de Schutter O (2014) The Power of Procurement. Available at http://www.srfood.org/images/stories/pdf/otherdocuments/20140514_procurement_en.pdf (last accessed 21 May 2014)

Environmental Protection Agency (EPA) (2014) About Us. Available at http://www2.epa.gov/aboutepa (last accessed 22 May 2014)

EPA (2013) Clean Air Act (CAA). Available at http://www.epa.gov/air/caa/ (last accessed 22 May 2014)

EPA (2014a) Clean Water Act (CWA). Available at http://www.epa.gov/agriculture/lcwa.html (last accessed 22 May 2014)

EPA (2014b) Federal Insecticide, Fungicide, and Rodenticide Act (FIFRA). Available at http://www.epa.gov/agriculture/lfra.html (last accessed 22 May 2014)

EPA (2014c) National Environmental Policy Act (NEPA). Available at http://www.epa.gov/compliance/nepa/ (last accessed 22 May 2014)

Food, Drug and Cosmetic Act, 21 U.S.C. 1 et seq

Food and Drug Administration (FDA) (2014) FDA Fundamentals. Available at http://www.fda.gov/AboutFDA/Transparency/Basics/ucm192695.htm (last accessed 11 Jan 2014)

FDA (1938) Federal Food, Drugs, and Cosmetics Act (FFDCA). Available at http://www.fda.gov/RegulatoryInformation/Legislation/FederalFoodDrugandCosmeticActFDCAct/FDCActChaptersIandIIShortTitleandDefinitions/default.htm (last accessed 1 Nov 2014)

FDA (2014) Food Safety Modernization Act (FSMA). Available at http://www.fda.gov/Food/GuidanceRegulation/FSMA/default.htm (last accessed 25 May 2014)

Feeds Act, R.S.C., 1985, c. F-9

Fertilizers Act, R.S.C., 1985, c. F-10

Fish Inspection Act, R.S.C., 1985, c. F-12

Food and Drugs Act, R.S.C., 1985, c. F-27

Forrager – Cottage Food Community (2014) Cottage Food Laws. Available at http://forrager.com/laws/ (last accessed 23 May 2014)

Government of Canada (1868), Agriculture and Agri-Food Canada (AAFC). Available at http://www.agr.gc.ca/eng/about-us/what-we-do/?id=1360700688523 (last accessed 26 May 2014)

Government of Canada (2013) National Dairy Code. Available at http://www.dairyinfo.gc.ca/index_e.php?s1=dr-rl&s2=canada (last accessed 26 May 2014)

Government of Canada (2014) Canada-European Union: Comprehensive Economic and Trade Agreement (CETA). Available at http://actionplan.gc.ca/en/page/ceta-aecg/understanding-ceta (last accessed 21 May 2014)

Health Canada (2014) About Health Canada. Available at http://www.hc-sc.gc.ca/ahc-asc/index-eng.php (last accessed 27 May 2014)

Health of Animals Act, S.C. 1990, c. 21

Illinois General Assembly (2009) Local Food, Farms, and Jobs Act. Available at http://www.ilga.gov/legislation/ilcs/ilcs3.asp?ActID=3137&ChapterID=7 (last accessed 21 May 2014)

Legislative Assembly of Ontario (2013) An Act to enact the Local Food Act, 2013. Available at http://www.ontla.on.ca/bills/bills-files/40_Parliament/Session2/b036ra.pdf (last accessed 21 May 2014)

Meat Inspection Act, R.S.C., 1985, c. 25

National Oceanic and Atmospheric Administration (2014) Endangered Species Act (ESA). Available at http://www.nmfs.noaa.gov/pr/laws/esa/ (last accessed 22 May 2014)

Plant Breeders' Rights Act, S.C. 1990, c. 20

Plant Protection Act, S.C. 1990, c. 22

Safe Food for Canadians Act (2012), S.C. 2012, c. 24

Seeds Act, R.S.C., 1985, c. S-8

Strauss DM (2011) An analysis of the FDA food safety modernization act: protection for consumers and boon for business. Food Drug Law J 66(3):355

The Right to Food, A/C.3/68/L.60/Rev.1

Toronto Food Policy Council (2012) Letter to the City of Toronto Executive Committee re: CETA. Available at http://www.toronto.ca/legdocs/mmis/2012/ex/comm/communicationfile-28674.pdf (last accessed 22 May 2014)

Town of Sedgwick, ME (2011) Local Food and Community Self-Governance Ordinance. Available at http://www.sedgwickmaine.org/images/stories/local-food-ordinance.pdf (last accessed 22 May 2014)

United Nations Human Rights (1966) International Covenant on Economic, Social and Cultural Rights. Available at http://www.ohchr.org/EN/ProfessionalInterest/Pages/CESCR.aspx (last accessed 22 May 2014)

United States Congress (2014) Agricultural Act of 2014 - H. R. 2642. Available at http://www.gpo.gov/fdsys/pkg/BILLS-113hr2642enr/pdf/BILLS-113hr2642enr.pdf (last accessed 12 Jan 2014)

United States Department of Agriculture (USDA) (2014) Mission Statement. Available at http://www.usda.gov/wps/portal/usda/usdahome?navid=MISSION_STATEMENT (last accessed 11 Jan 2014)

USDA (1983) Emergency Food Assistance Program (TEFAP). Available at http://www.fns.usda.gov/tefap/emergency-food-assistance-program-tefap (last accessed 26 May 2014)

USDA (2014) Supplemental Nutrition Assistance Program (SNAP). Available at http://www.fns.usda.gov/snap/supplemental-nutrition-assistance-program-snap (last accessed 26 May 2014)

U.S. Department of Health and Human Services and the FDA (2013) Food Safety Modernization Act (FSMA) Facts. Available at (last accessed 12 Jan 2014)

Vermont for Evolution (2012) The Vermont Resolution for Food Sovereignty. Available at http://vermont4evolution.wordpress.com/human-rights/the-vermont-resolution-for-food-sovereignty/ (last accessed 22 May 2014)

White RM (2014) The 2014 Farm Bill (P.L. 113-79): Summary and Side-by-Side. Congressional Research Service. Available at http://www.farmland.org/programs/federal/documents/2014_0213_CRS_FarmBillSummary.pdf (last accessed 12 Jan 2014)

Wittman H, Desmarais AA (2012) Food Sovereignty in Canada. Canadian Centre for Policy Alternatives (CCPA). Available at https://www.policyalternatives.ca/publications/monitor/food-sovereignty-canada (last accessed 22 May 2014)

Chapter 30
Textbox: The U.S. Farm Bill and Textbox: The National School Lunch Program

William S. Eubanks

Abstract In the United States, Congress sets domestic legislative priorities and allocates funding for agricultural and child nutrition programs primarily through two omnibus bills that are each reauthorized approximately every five years—the U.S. Farm Bill and the Child Nutrition Reauthorization. This chapter discusses each of these laws, the histories behind their enactment, and current efforts in each bill designed to foster a sound agricultural economy, improved public health outcomes, and a more effective National School Lunch Program.

30.1 The U.S. Farm Bill

The Agricultural Adjustment Act was originally passed in 1933. This first "Farm Bill"—as it later came to be known—was viewed by Congress as a temporary solution to the struggles faced by farmers during this era. The Act imposed supply-side quotas, limits, and other restrictions on crop production, with the federal government stepping in to make up the difference where necessary so that farmers could obtain better market prices.[1] Although the federal government has long enacted measures to support and protect the agricultural sector, the first comprehensive legislation that aimed at providing family farms with a financial "safety net" came in 1933 in response to the Great Depression and the droughts of the Dust Bowl era, as part of President Franklin D. Roosevelt's New Deal.[2]

Notwithstanding the temporary nature of the first farm bill, Congress realized the importance of maintaining both supply controls and price controls in order to ensure

---

[1] *See id.* at 2–4.

[2] *See* William S. Eubanks II, *Chapter 1: A Brief History of U.S. Agricultural Policy and the Farm Bill* at 1–12, in Food, Agriculture, and Environmental Law (1st ed. 2013) (edited by Angelo, Czarnezki, and Eubanks) (published by Environmental Law Institute).

W.S. Eubanks (✉)
Meyer Glitzenstein & Eubanks LLP, Washington, DC, USA

Vermont Law School, Fort Collins, CO, USA
e-mail: BEubanks@MeyerGlitz.com

© Springer International Publishing Switzerland 2016
G. Steier, K.K. Patel (eds.), *International Food Law and Policy*,
DOI 10.1007/978-3-319-07542-6_30

as equitable an agricultural market as possible for farmers, and thus continued to reauthorize the farm bill approximately every 5 years subject to any revisions added by Congress.[3] Since its initial enactment 80 years ago, there have essentially been three distinct periods of farm bill implementation, each of which sought to achieve very different objectives—objectives that were often contrary to those embodied in previous farm bills. This disjointed evolution demonstrates that a farm bill is little more than a snapshot of legislative priorities vis-à-vis the agricultural sector at any given point in time.[4]

In the post-World War II era until the early 1980s, the farm bills enacted by Congress illustrated the United States' engagement in Cold War politics with the Soviet Union. These farm bills emphasized larger, more concentrated farming and food processing operations through subsidies and other financial incentives that were aimed at maximizing crop yields, above all other goals.[5] As a result, on-farm natural resource conservation was de-emphasized during this era while the predominant focus shifted to large-scale monoculture farming of commodity crops such as corn, wheat, rice, cotton, and soybean.[6] This era also coincided with the "Green Revolution"—the culmination of years of scientific experimentation with new plant breeding and hybridization techniques—and with the proliferation of new chemical pesticides, synthetic fertilizers, and mechanization. Cumulatively, these innovative agricultural inputs created the perfect storm for a highly efficient, albeit extremely resource-dependent economy in the agricultural sector.[7]

The post-World War II farm bill model collapsed when that system resulted in immense surpluses of certain commodities that significantly reduced the market prices farmers could yield, and when political tensions between the U.S. and the Soviet Union led to the suspension of certain farm exports to one of the US' largest trading partners.[8] This collapse began a tailspin into the 1980s farm crisis, which was—and remains—the worst financial period for farmers in American history outside of the Great Depression.[9] Recognizing the need to break away from the prevailing farm bill model, in 1985 Congress responded by creating a new framework that would survive for nearly 30 years. The reformed farm policy balanced the economic needs of farmers by continuing certain price support programs for covered commodity crops and by creating a bevy of new marketing loans to help farmers obtain much-needed capital. In exchange for the price support programs, farmers were required to meet their end of the social contract by satisfying certain environmental obligations such as conserving sensitive wetlands and preserving highly erodible soils.[10] These programs—colloquially known as *Swampbuster* and *Sodbuster*,

---

[3]*Id.* at 3–5.
[4]*Id.* at 4–10.
[5]Eubanks, *supra* note i, at 5–6.
[6]*Id.*
[7]*Id.* at 5.
[8]*Id.* at 7.
[9]*Id.*
[10]*Id.* at 7–8.

respectively—set the stage for the enactment of several dozen farm bill conservation programs over the next three decades, and ushered in an era in which Congress made clear that a resilient agricultural sector could ensure national food security in harmony with environmental laws designed to protect the country's natural resources.[11]

In 2014, Congress reauthorized the farm bill once again, which at this early stage appears to be the start of the third major epoch in farm bill history. Among other things, the 2014 farm bill drastically restructured certain farm-related payment programs, moving away from the subsidy system that had prevailed for decades and, instead, adopting a risk-management based approach. Within this framework, the current farm bill returns to its origins as a "safety net" by providing farmers with diverse crop and revenue insurance options that help protect farmers' crop investments from droughts, floods, predation, and other natural disasters that significantly affect farm income from year to year.[12] These new federally-backed insurance programs also create more equitable insurance packages for diversified operations—i.e., farms that produce more than just one or two staple crops—providing farmers with more flexibility in terms of deciding what to grow in a given season.[13] As for on-farm and off-farm resource protection, while the fiscal climate in the nation's capital led to some reduction in spending on conservation programs, Congress consolidated and improved several conservation programs and allocated billions of dollars to ensure that these programs operate efficiently and effectively when implemented by the U.S. Department of Agriculture—indicating Congress's ongoing commitment to natural resource protection.[14] Finally, compared to prior legislation, the 2014 farm bill provided significant funding opportunities for beginning farmers and ranchers, socially disadvantaged farmers and ranchers, organic producers, and local and regional food systems, indicating Congress's recognition that the face of farming in this country, as well as the composition of food marketplaces in the U.S., is changing rapidly. The U.S. must keep pace if it wants to maintain a vibrant, resilient, and environmentally protective food production system.[15]

Hence, as with farm bills that preceded it, the 2014 farm bill demonstrates Congress's current commitment to America's farmers, as well as to its citizens and the food system they covet. Although there certainly are challenges that Congress must address in future farm bills in order to avoid lapsing into the farm crises previously observed in this country,[16] for nearly a century, Congress has

---

[11] *Id.*; *see also* Mary Jane Angelo & Joanna Reilly-Brown, *Chapter 2: An Overview of the Modern Farm Bill* at 21–24, in Food, Agriculture, and Environmental Law (1st ed. 2013) (edited by Angelo, Czarnezki, and Eubanks) (published by Environmental Law Institute).

[12] Congressional Research Serv., *The 2014 Farm Bill (P.L. 113-79): Summary and Side-by-Side* at 2, 4, 17–18 (Feb. 12, 2014), http://www.farmland.org/programs/federal/documents/2014_0213_CRS_FarmBillSummary.pdf (last accessed July 2014).

[13] *Id.* at 17–18.

[14] *Id.* at 8–9.

[15] *Id.* at 17–19.

[16] For a list of concerns that should be addressed in future farm bills, *see* William S. Eubanks II, *Chapter 15: Achieving a Sustainable Farm Bill* at 270–79, in Food, Agriculture, and Environmental Law (Angelo, Czarnezki, and Eubanks, Eds.) (Environmental Law Institute, 1st ed. 2013).

proven its ability to respond with relative urgency to modify the "safety net" specifically created to protect the backbone of America—its farmers and ranchers—to address the challenges of the day.[17]

30.2 The National School Lunch Program

Government-assisted food service programs in educational settings have been in existence in Europe since the 1790s.[18] In the US, various efforts to involve the federal government in school meals began in the 1890s.[19] Despite these early efforts, however, the modern version of the National School Lunch Program (NSLP) was not created until 1946 with Congress's passage of the Richard B. Russell School Lunch Act.[20] In 1966, Congress significantly refined and expanded the reach of the NSLP by enacting the Child Nutrition Act, which, among other things, strengthened the USDA's authority to ensure that the NSLP would provide nutritional meals to students, and also extended the provision of school meals to include breakfast and snacks that occur during the school day.[21] Since that time, Congress has reauthorized the Child Nutrition Act approximately every 5 years, and in the process provided guidance to USDA on how to implement the NSLP to achieve the goals of the Act. The most recent reauthorization of the Act—which is discussed below—occurred in 2010, meaning that another reauthorization is expected in 2016.[22]

The current NSLP operates in more than 100,000 public and non-profit private schools throughout the nation. In 2012, the program served free and reduced-priced lunch to over 31 million American children each day during the school year.[23] Since 1998, the suite of school meal programs included within the NSLP umbrella has included not only traditional meals during the school day, i.e., breakfast and lunch, but also afterschool snacks.[24] Any student at a participating school may

---

[17]This piece only focuses on the farm-related provisions of the farm bill, rather than the nutrition provisions of the farm bill. The Supplemental Nutrition Assistance Program ("SNAP") is authorized via the farm bill, and approximately 70 % of farm bill funding is allocated for SNAP purposes. For more information about SNAP, *see* http://www.fns.usda.gov/snap/supplemental-nutrition-assistance-program-snap (last accessed July 2014).

[18]U.S. Dept. of Agric. (USDA), *National School Lunch Program: Background and Development in Europe*, http://www.fns.usda.gov/nslp/history_1 (last accessed July 2014).

[19]*Id.*

[20]USDA, *National School Lunch Program: National School Lunch Act*, http://www.fns.usda.gov/nslp/history_5 (last accessed July 2014).

[21]USDA, *National School Lunch Program: Child Nutrition Act of 1966*, http://www.fns.usda.gov/nslp/history_6.

[22]For a more detailed discussion of the historical origins of the NSLP, *see* Jason J. Czarnezki, *Chapter 13: The Food Statutes* at 233–40, in Food, Agriculture, and Environmental Law (Angelo, Czarnezki, and Eubanks, Eds.) (Environmental Law Institute, 1st ed. 2013).

[23]U.S. Dept. of Agric. (USDA), *NSLP Fact Sheet* at 1, http://www.fns.usda.gov/sites/default/files/NSLPFactSheet.pdf (last accessed July 2014).

[24]*Id.*

purchase a meal or snack through the NSLP. However, students satisfying certain family income eligibility requirements relative to the poverty level in a given year qualify for either free meals and snacks or reduced-price meals and snacks.[25] The total cost in federal resources per year to subsidize these free and reduced-price meals and snacks is approximately $11 billion.[26]

The bulk of federal funding received by schools participating in the NSLP "comes in the form of a cash reimbursement for each meal served,"[27] which amounts to $2.93 for free meals, $2.53 for reduced-price meals, and $0.28 for full-price meals to-date.[28] In addition, a school can receive a $0.06 bonus for each meal served as long as it can demonstrate compliance with the current USDA nutritional standards for school meals. These nutritional standards were established in the 2010 legislation and are now being implemented by USDA.[29] Finally, aside from cash reimbursements, participating schools receive "entitlement dollars" to use towards USDA-procured foods, a.k.a. USDA Foods, often comprised of surplus commodity products from that year's agricultural yields, such as grains, potatoes, frozen vegetables, canned fruits, pastas, or hamburger meat.[30] The percentage of a typical school lunch derived from USDA Foods obtained with entitlement dollars is 15–20 % of the volume of the lunch by weight.[31]

In light of the rising incidence of childhood obesity, type 2 diabetes, and other diet-related diseases, a critical focus of the NSLP in recent years has been to create health-based nutritional standards to which all NSLP meals and snacks must strive to adhere, with the goal being that the NSLP is not simply providing empty calories but rather giving students vital nutrients necessary for healthy bodies and strong minds in the classroom. In 2010, Congress reauthorized the Child Nutrition Act in legislation called the Healthy, Hunger-Free Kids Act, which specifically sought to transform the NSLP into a program emphasizing child nutrition.[32] Among other things, that Act: (1) provided the largest one-time increase per school meal served ($0.06 per meal) in the history of the Child Nutrition Act so long as a school is in compliance with all USDA nutritional standards, which allows school food administrators more flexibility in purchasing choices; (2) required that all NSLP-participating schools make potable water available anywhere that school meals/snacks are served so that less children rely solely on milk or soda; (3) mandated that USDA update its nutritional standards to ensure that meals are nutritionally

---

[25]*Id.* at 2.

[26]*Id.* at 3.

[27]Food Research and Action Center, National School Lunch Program, available at http://frac.org/federal-foodnutrition-programs/national-school-lunch-program/ (last accessed July 2014).

[28]USDA, *supra* note vi at 2.

[29]*Id.*

[30]*Id.* at 2–3.

[31]U.S. Dept. of Agric., *Food and Nutrition Service White Paper: USDA Foods in the National School Lunch Program* at 3, http://www.fns.usda.gov/sites/default/files/WhitePaper.pdf.

[32]Pub. L. No. 111-296 (2010).

balanced and contain few, if any, sugars, trans fats, and other low-nutrient ingredients; (4) placed limitations on the ability of external corporations to sell what are called "competitive foods," e.g., pizza, potato chips, and soda, which compete with NSLP meals for student attention and dollars; and (5) committed mandatory funding to a USDA Farm to School Grant Program since farm to school experiential education activities like school gardens, taste tests, and farm tours have proven effective in increasing children's consumption of fruits and vegetables while also supporting local and regional agriculture.[33]

In implementing the Healthy, Hunger-Free Kids Act, USDA has promulgated several regulations updating the NSLP nutrition standards to comport with current public health practices.[34] As these regulatory standards have taken effect, some food service administrators and politicians have raised concerns over the difficulty of compliance, children opting to avoid purchasing the more nutritional offerings, and increased food waste. However, careful scrutiny of the preliminary data indicates that those concerns are almost certainly outliers, as a recent peer-reviewed study of the NSLP's nationwide data suggest that more than 90 % of participating schools are, in fact, satisfying the updated nutrition standards; children are eating, on average, 23 % more fruit and 16 % more vegetables at lunch. Additionally, school lunch revenue at NSLP-participating schools is actually increasing because many regions are seeing much higher student NSLP participation than they previously experienced. Finally, food waste has not increased relative to the baseline.[35]

Accordingly, if current trends hold, the updated nutrition standards may prove to be a much-needed prescription for our nation's childhood obesity crisis. In light of the impending 2016 Child Nutrition Reauthorization, only time will tell whether Congress will recommit to these standards to give them sufficient time to bear fruit, or whether it will opt for a different path forward.

---

[33]U.S. Dept. of Agric., *Food and Nutrition Service: Summary of the Healthy, Hunger-Free Kids Act of 2010*, http://www.fns.usda.gov/sites/default/files/PL111-296_Summary.pdf; *see also* National Farm to School Network, *Fact Sheet: Benefits of Farm to School*, http://www.farmtoschool.org/Resources/BenefitsFactSheet.pdf.

[34]U.S. Dept. of Agric., Healthy, Hunger-Free Kids Act, http://www.fns.usda.gov/school-meals/healthy-hunger-free-kids-act.

[35]U.S. Dept. of Agric., *Fact Sheet: Healthy, Hunger-Free Kids Act Implementation*, http://www.fns.usda.gov/pressrelease/2014/009814.

Chapter 31
Food Law in Canada: A Canvass of History, Extant Legislation and Policy Framework

Ikechi Mgbeoji and Stan Benda

Abstract In this chapter, the structure, content and process of food regulation in Canada are examined. For ease of analysis, our analysis is structured into three main parts. Part 1 introduces the subject-matter of the history of Canadian food safety regime, and food labelling requirements. Part 2 explores extant federal and provincial regulatory frameworks with emphases on "product of Canada" requirements, organic food governance as well as public health issues in Canada. Part 3 deals with trade in food products, focusing on both import and export trade as well as on inter-provincial trade in food products. Part 3 concludes the chapter by examining international food trade regulatory instruments as they affect Canada.

31.1 Introduction

Wide ranges of legislation frame the legal regime of food in Canada. Historical accident, *ad hoc* reactions, and differing departmental mandates have created what Donald Buckingham—one leading commentator described as "complexity, opaqueness and difficulty in applying the law."[1] To undertake this overview of food regulation, one must first examine some basic concepts: what is food? Why regulate food and methods of food production? How do regulators choose? What is the legacy of progenitor legislation vis-à-vis food over the last 200 years in Canada? Having set that that table, we can look at the menu, namely the details of extant food regime in Canada. Next, the extant food regime of Canada will be discussed.

What is food? Food is philosophically, psychologically and culturally determined. The concept of "food" is fixed by the concepts of time and place. While food

---
[1] Buckingham (2005), p. 148.

I. Mgbeoji (✉)
Osgoode Hall Law School, Toronto, ON, Canada
e-mail: IkechiMgbeoji@osgoode.yorku.ca

S. Benda
Ryerson University, Toronto, ON, Canada
e-mail: Stan.benda@ryerson.ca

© Springer International Publishing Switzerland 2016
G. Steier, K.K. Patel (eds.), *International Food Law and Policy*,
DOI 10.1007/978-3-319-07542-6_31

is religion in India and Saudi Arabia, In China it is medicine.[2] Also, It may be considered taboo as is the case in East Africa.[3] Fourth century Christian taboos included blood, carrion, and things strangled.[4] Suggestion: Similar food taboos are manifested in Jewish, Muslim and Hindu strictures. The modern manifestations of food concepts are found in Kosher, Halal and Hindu strictures.

Similarly, food is psychologically determined. The concepts of rejection and acceptance will further explain this notion. Rejections range from distaste (spinach/chilli) through danger (allergens), inappropriate (sand) and disgust (insects/feces). Acceptance range from good taste (sweet, salty) through beneficial (honey) appropriate (ritual foods, e.g. bread and wine) and trans-valued (deity offerings).[5]

Genetics, age, gender, climate, or social status may also dictate what may be acceptable as food. For example, lactose intolerance, gluten sensitivity and bitterness may speak to genetic influences. The merits of the contradictions in determining what is acceptable as food is perhaps best exemplified in mushrooms. With the world history of the past 500 years characterized by global exploration, colonization, increased trade,[6] and massive movements and exchanges of people, goods, animals and food, the history of food is complex[7] and multifaceted. Medicine as food goes back to the Egyptians and honey. Coffee and chocolate came first as medicines to the European table. So did tea, garlic and sassafras.

Throughout time, different cultures of the world have influenced the definition of food, originally fixed by the climate, botany, riparian or littoral nature of the location. The Germans have the 1516 *Reinheitsgebot*, beer purity law. In India, food is tied to social status. In Japan, there is the tea ceremony.

Trade introduced new foods across the geographical regions of the world. Maize, rice, potatoes, avocado, tomato, peanuts, spices, citrus have all been spread far beyond their native centres of origin. The only crops indigenous to North America are blueberries, Jerusalem artichokes, sunflowers, squash, and Canadian fiddleheads (uncurled fern tops). The centres of origin of crop plants are the so-called *Vavilov Centres*,[8] named after the prominent Russian botanist who first discovered that plants had centres of origin that could be traced to specific geographic locals.

There are many gaps or fallacies in the consciousness about "old" food crops. Many crops are no longer associated with their place of origin. For example, potatoes are no longer associated with South America, ditto sugar cane and bananas with Papua New Guinea. Similarly, Cassava is not associated with South America, and neither is coffee with Ethiopia. The same applies to oranges with India, and apples with Kazakhstan.[9] Original crops have been improved and further

---

[2]Echols (2001), pp. 13–28.
[3]Id., p. 15.
[4]Grivetti (2000); Messer (2000), pp. 1495–1513.
[5]Rozin (2006); Messer (2000), pp. 1476–1485, 1478.
[6]For a history of each regions and foods introduced from where see: Civitello (2008), p. 83.
[7]For history of a particular plant food or spices see in general Kiple (2000).
[8]Janick (2000).
[9]Juniper and Mabberley (2006), p. 46.

differentiated over time from their parents by breeding, selection, adaption, and preference. Over half of all wild plants contain unattractive flavour, which is more or less toxic. Plants cannot run away, so their defenses against pests, fungus, and disease are chemicals and toxins. The preparation of such plants for food often requires complex processes of detoxification. In many instances, human response to plant toxicity has been to breed and reduce the level of toxins in plants by cooking and other means of food preparation.

Original crops were differentiated over time by breeding, selection, adaption, and preference. The Malus pumila, for example, became the sweet apple with the help of bear droppings and farmers. The Granny Smith was a mutation found on the farm in New South Wales on one Mrs. Smith in 1868. Entirely new species were bred that many assume always existed. In the mid-1700s two wild species of strawberry, from Virginia and Chile, were accidentally crossed in France to produce the modern "large" fruit strawberry. The progenitor for Kiwi is indigenous to China and is known as Chinese gooseberry. At the turn of the twentieth century in New Zealand this gooseberry was bred into the first Kiwi variety. Nectarine, a mutate peach is another crop that would not exist without human intervention.

While food is regulated as a product, the process begetting the product is sometimes regulated, especially in what is colloquially called GM or genetically modified foods. In order to understand the context and nuance of the applicable regulations, students of law must gain a better understanding of the science of breeding (Suggestion). With agriculture dating back about 10,000 years, early farmers probably separated seeds on the basis of yield, taste, and agronomic traits. Starting in the late 1800s this accelerated to the point that virtually everything in a supermarket was "improved," except for products such as North American wild blueberries. As it currently stands, some 75 %[10] of the food on the shelves in Canadian grocery stores is GM.

31.1.1 Food Safety Legislation and Labelling Requirements

Food regulation in Canada is derived from a melange of federal and provincial statutory provisions as well as general law provisions that affect producers, processors and distributors of food products. Ancient food regulations focused on *probity*. Ancient Egyptians had food labelling rules while the ancient Greeks and Romans had detailed laws on food adulteration (especially bread with chalk and wine with water).[11] Food safety legislation has an ancient and storied pedigree.[12]

Canadian food safety legislation is traceable to Britain[13] and the Bradford poisoning in 1858 is widely credited as the tipping point of effort to affect public

---

[10] Yaren (2001).
[11] Hutt (1984b).
[12] Hutt (1984a).
[13] Accum (1820).

food safety by way of legislative reform. In the Bradford scandal, arsenic was inadvertently substituted for plaster, which was the "usual" substitute for sugar, in the making of sweets.[14] Twenty died, mostly children, and hundreds became violently ill. The tragedy precipitated the *British Food and Drugs Act* of 1860.[15] The Act reoriented legislation from inspection for commercial purposes to the prevention of food adulteration.[16] The Act is reputed in the British Commonwealth to be the precursor of most food and drug legislations.[17]

Statutory measures to regulate agricultural activities were already well underway both before and after confederation in Canada. Indeed, both federal and provincial governments, since Confederation, have maintained regulatory mechanisms in agriculture. In 1869, Parliament enacted An Act *Respecting Contagious Diseases of Animals.*[18] The Act charged the Department of Agriculture with the task of controlling livestock diseases for draft animals and food animals and suppressing outbreaks of disease on farms, in places where animals were sold.

In the 1990s nutritional labelling of foods commenced in North America. As history suggests the main thrusts of regulatory food law have been safety and probity. The modern objectives of food labelling ensures fair competition; product marketability; increasing consumer access to information; protecting consumer's health and safety; and influencing consumer choices (e.g., nutritional labelling).[19]

In Canada, the earliest known legislation was *An Act of the Conseil Superieur de Quebec*, 1706, which set out regulations concerning the sale of bread.[20] After the union of Upper and Lower Canada in 1841, four new acts were promulgated: The *Act for the Inspection of Flour and Meal*[21]; *Act for the Inspection of Beef and Pork*[22]; *Act for the Inspection of Fish and Oil*[23]; and *Act for the Inspection of Hops.*[24] After

---

[14]London (2014).

[15]Id.

[16]Buckingham (2005).

[17]London (2014).

[18]1869, 32-33 Vict., c. 37.

[19]Mitchell et al. (2001).

[20]Buckingham (2005), p. 134.

[21]1841 (15 Victoria), c. 45 (Can.); replaced by Act for the Inspection of Four, Indian Meal and Oatmeal 1856, 19-20 Victoria, c. 87 (Can.) and amend by the Act for the Inspection of Flour and Meal, 1859 (22 Victoria) c. 48 (Can.).

[22]1859 (22 Victoria) c. 48 (Can.); in Consolidated Statutes of Canada and Upper Canada, Title 4 Trade and Commerce.

[23]1859 (22 Victoria) c. 50 (Can.); in Consolidated Statutes of Canada and Upper Canada, Title 4 Trade and Commerce.

[24]1859 (22 Victoria) c. 52 (Can.) in Consolidated Statutes of Canada and Upper Canada, Title 4 Trade and Commerce.

confederation the inspection acts were consolidated into the *General Inspection Act*.[25] Food labelling integrity was part of quality assurance.[26]

The next post-confederation legislative milestone was known by its abbreviated title as "The Inland Revenue Act of 1875".[27] In 1877, 51.7 % of 180 food samples were adulterated. In 1878, of 1500 food samples, 50.6 % were adulterated. In 1881 of 1500 food samples, 25 % were adulterated. Finally, in 1883 the number was 24.25.[28] The next key legislative step was the *Act to Amend and to Consolidate as Amended the Several Acts Representing the Adulteration of Foods and Drugs* (aka the *Adulteration Act*).[29] In 1890 the amendment regulatory standards under this Act were enabled, albeit not prepared until 1910.[30] The crux of the act was that food would be deemed adulterated "If its strength or purity falls below the standard, or its constituents are present in quantity not within the limits of variability fixed by the Governor in Council."

The *Food and Drugs Act* replaced the *Adulteration Act* in 1920.[31] Consequently, Canada became acquainted with the United States of America concept of misbranding,[32] where food is deemed misbranded if:

- Imitation of another article of food or drug without being plainly and conspicuously labelled so as to indicate true character[33];
- The label stated to be a product of a place or a country of which if was not[34];
- Food article sold by a name which belongs to another article[35];
- Food article coloured or coated or powdered to give the illusion of value[36];
- The label bore a statement which was false or misleading[37];
- The claims were false or exaggerated[38];

---

[25]S.C. 1874 (Can.).

[26]Buckingham (2005), p. 135. For a detailed discussion of food safety and regulation, see also *Halsbury's Laws of Canada Food (2014 Reissue);* Fuller and Buckingham (1999); Benson (1996) with particular reference to Saskatchewan; and Purich (1982).

[27]An Act to Impose Licence Duties on Compounders of Spirits; to Amend the Act Respect the Inland Revenue; and to Prevent the Adulteration of Food, Drink and Drugs. Proclaimed: January 1, 1875.

[28]Pugsley (1967).

[29]S.C. 1884 (47 Victoria), c. 34 (Can.).

[30]Buckingham (2005), pp. 134–148.

[31]S.C. 1920, c. 27 (Can.); R.S.C. 1927, c 76 (Can.); 1952, c. 123, 1952-53, c. 38, R.S.C. 1970, c. F-27 (Can.), and finally R.S.C. 1985, c. F-27 (Can.).

[32]Pugsley (1967), p. 449.

[33]S.C. 1920, c. 27 (Can.); R.S.C. 1927, c 76 (Can); 1952, c. 123, 1952-53, c. 38, R.S.C. 1970, c. F-27 (Can); and R.S.C. 1985, c. F-27; s. 5(a) (Can.).

[34]Id., s. 5(b).

[35]Id., s. 5 (c).

[36]Id., s. 5(d).

[37]Id., s. 5(h).

[38]Id., s. 5(e).

- The contents were not conspicuously and correctly stated (i.e., weight)[39];
- The product was sold as a mixture and not labelled as such.[40]

The courts found that misbranding *per se* was within the criminal orbit of the constitution irrespective of whether or not it was injurious to health.[41] The *Food and Drugs Act* of *1953* was the next milestone. *Inter alia*, this Act:

- Required the maintenance of books and records;
- Prohibited the sale, manufacture or storage of food made or stored in unsanitary conditions;
- Limited deceptive claims for offences only for food and drugs, not cosmetics;
- Shifted the conceptual lens to misbranding (a focus on specific prohibitions);
- Diminished the concept of adulteration; the issue now was non-compliance with an established standard.[42]

Even so, the notion of prior food inspection rather than prosecution was the underlying philosophy.[43] The subsequent amendments and consolidations achieved three things, namely:

- In 1949 regulations were consolidated into five parts with a common lexicon and organization that continues till today: to wit: Part A: General; Part B: Food; Part C: Drugs; Part D: Vitamins, and Part E: Cosmetics.
- The word « label » was defined in 1952 as « any legend, word, or mark attached to, included in, belonging to or accompanying any food, drug, cosmetic, device, or package».
- The 1934 and 1953 Acts created new food labelling offence provisions.

In 1960, the advent of food labeling required the disclosure of contents of the processed or manufactured food. Moreover, labels had to declare the net contents in proximity to the common name of the food.[44] Some products were given idiosyncratic treatment, either historically or presently. Canada in 1907 passed the *Meat and Canned Foods Act*[45] to assuage both Canadian and European importers.[46] Maple syrup, honey, and dairy all had special legislation.[47]

---

[39]Id., s. 5(f).

[40]Id., s. 5(g).

[41]Standard Sausage Co. v Lee (1934) 1 D.L.R. 706 (Can. B.C. S.C.).

[42]Pugsley (1967), pp. 424–425.

[43]Buckingham (2005), pp. 134–148.

[44]Pugsley (1967), pp. 387–449.

[45]S.C. 1907 (6-7 Edward VII) c. 27 (Can.); successor repealed 1985.

[46]Buckingham (2005), p. 134.

[47]Fruit, Vegetables and Honey Act, S.C. 1935, c. 672 (Can.); Natural Products Marketing Act, 1934, c. 57 (Can.); Canada Agricultural Products Standards Act, 1934, c. 57 (Can.) (progenitor of the Canada Agricultural Products Act (CAPA)); see also CRC c. 287, Honey Regulations under the CAPA.

The Department of Agriculture oversaw most of this legislation. In 1997, Agriculture was split into two organizations: The Department of Agriculture and Agri-Food Canada (AAFC) and the Canadian Food Inspection Agency (CFIA).[48] The CFIA was tasked with the regulation of food and food production at the production end of the chain for products destined for interprovincial and international trade. There were a number of failed legislative attempts to consolidate food regulations.[49]

A different legislative thread arose with the *Consumer Packaging and Labelling Act (CPLA)* in 1970.[50] The main thrust of CPLA regarded the importance of consumer information, so it addressed all manner of product, including pre-packaged foods. It also engaged false or misleading representation.[51] Lack of legislative co-ordination resulted in duplication and conflict between CPLA and the other food regulatory structures.[52]

At the provincial level, the following extant pieces of legislation deal with regulation of agriculture:

(AB) *Agricultural Operation Practices Act*, R.S.A. 2000, c. A-7
(BC) *Farm Practices Protection (Right to Farm) Act*, R.S.B.C. 1996, c. 131
(MB) *Farm Practices Protection Act*, C.C.S.M. c. F45
(NB) *Agricultural Operation Practices Act*, R.S.N.B. 2011, c. 107
(NL) *Farm Practices Protection Act*, S.N.L. 2001, c. F-4.1
(NS) *Farm Practices Act*, S.N.S. 2000, c. 3
(ON) *Farming and Food Production Protection Act, 1998*, S.O. 1998, c. 1
(PE) *Farm Practices Act*, R.S.P.E.I. 1988, c. F-4.1
(SK) *Agricultural Operations Act*, S.S. 1995, c. A-12.1.

31.2 Regulation Under Federal Jurisdiction

The legal regulation of agricultural activities is by extension, the regulation of food and the food industry. To this end, agricultural laws have four dimensions: (i) encouraging and assisting producers (farmers); (ii) promoting fair trade; (iii) ensuring food safety and quality; (iv) private law issues.[53]

In each province of Canada, it is constitutionally provided that each "legislature may make Laws in relation to Agriculture in the Province, and to Immigration, and it is hereby declared that the Parliament of Canada may from time to time make

---

[48]S.C. 1997. c. 6 (Can.).
[49]Canada Food Safety and Inspection Act (Bill C-80) died on the order paper.
[50]R.S.C. 1985, C-38 (Can.).
[51]Id. s. 4.
[52]Buckingham (2005).
[53]Halsbury's Laws of Canada, 1st ed., Agriculture, p. 125.

laws in relation to agriculture in all or any provinces, and to immigration into all or any of the provinces; and any law of the legislature of a province relative to agriculture or to immigration shall have effect in and for the province as long and as far as it is not repugnant to any Act of the Parliament of Canada."[54]

There are two schools of thought on this constitutional provision. The first school commencing with the advent of 1907s *Meat and Canned Foods Act* suggests that legislation could have an effect "beyond the farm gate."[55] The second school of thought, arising from a line of cases in 1922 suggests that legislation that went beyond the farm gate was invalid.[56] To this day the farm gate is the pivotal marker. Legislation pertaining to farming activities such as methods of production, inputs and animals is sustainable under s. 95 of the Constitution Act behind the farm gate, but marketing boards acting beyond the farm gate are not.[57]

Thus, under the constitutional laws of Canada, agriculture falls under both the federal and provincial jurisdiction.[58] Section 95 of the *Constitution Act, 1867* recognizes "Agriculture" as a subject matter of joint legislative power. As Cunningham rightly points out, "under Canadian constitutional law and the division of powers, exclusivity is the rule and concurrency is the exception. The scheme is opposite to that in the United States or Australia, for example, where exclusivity is the exception. Thus, the joint power over agriculture is unusual in its explicitness. Concurrency is usually determined by the courts when federal and provincial powers appear to, or implicitly, overlap."[59]

In sum, the preponderant view is "legislation with an impact on agriculture may seek its constitutional justification from s. 95 (joint agriculture power), from various subject-matters outlined in s. 91 (exclusive federal powers), or from various subject-matters outlined in s. 92 (exclusive provincial powers). Each subject-matter, or head of power, "clothes" the appropriate level of government with the competence to make legislation in relation to that subject-matter. Legislative activity outside that competence, by either the federal Parliament or a provincial Legislature, may be struck down by the courts."[60]

Beyond the farm gate a regulatory "grid" underpinned by the Federal jurisdiction over the criminal law enforces food safety.[61] Anything beyond food safety

---

[54]Note 52, Id.

[55]Halsbury's Laws of Canada, 1st ed., Agriculture, p. 125. See, *R v Manitoba Grain* C. [1922] M.J. No. 4, 66 D.L.R. 406 (Can. MAN).

[56]Halsbury's Laws of Canada, 1st ed., Agriculture, p. 125. See, King v Eastern Terminal Elevator Co., [1925] 3 D.L.R. 1 (S.C.C.) (Can.).

[57]For a detailed list of cases see Fuller and Buckingham (1999), pp. 142–144; and marketing boards history see *R v Bradford Fertilizer Co.*, [1971] O.J. No. 1763, [1972] 1 O.R. 229: 167–172. Fertilizers Are Behind the Gate; *R v Laboratoires Sagi Inc.*, [1985] C.S.P. 1073 (Can Que. C.S.P.). Animal Feed is Behind the Gate.

[58]Schedule B to the Canada Act 1982 (U.K.), 1982, c. 11 (Can.).

[59]Id. Note 53.

[60]Id. Note 55.

[61]s. 91(27), Schedule B to the *Canada Act* 1982 (U.K.), 1982, c. 11 (Can.); see also *R. v Wetmore*, S.C.J. No. 74, [1983] 2 S.C.R. 284 (S.C.C.) (Can.).

i.e. economic interests is probably unsupportable under criminal law.[62] At the provincial level, the various legislation dealing with agricultural operations include:

(AB) *Agricultural Operation Practices Act*, R.S.A. 2000, c. A-7, s. 1(b)
(MB) *Farm Practices Protection Act*, C.C.S.M. c. F45, s. 1
(NB) *Agricultural Operation Practices Act*, R.S.N.B. 2011, c. 107, s. 1
(NS) *Farm Practices Act*, S.N.S. 2000, c. 3, s. 3
(ON) *Farming and Food Production Protection Act, 1998*, S.O. 1998, c. 1, s. 1
(PE) *Farm Practices Act*, R.S.P.E.I. 1988, c. F-4.1, s. 1.1(a)
(QC) Regulation respecting the registration of agricultural operations and the payment of property taxes and compensations, CQLR c. M-14, r. 1, s. 1
(SK) *Agricultural Operations Act*, S.S. 1995, c. A-12.1, s. 2(a).

According to the provincial legislations listed above "agricultural operation" comprise operations carried out on farms: cultivating land, producing agricultural and horticultural crops, raising livestock, producing eggs and other animal products; operating farm machinery; distribution of goods produced; storing, handling and applying fertilizer, manure, organic wastes, soil amendments and pesticides.[63]

On the other hand, Federal law defines "agricultural product" as any animal, plant or any food or drink, wholly or partly derived from an animal or a plant.[64] The marketing of such post-farm products e.g. milk, margarine, et cetera may be deemed commercial activities beyond the joint jurisdiction conferred by s. 95 of the Constitution Act. Opined by one learned commentator, the "s. 95 might be the constitutional basis for compliance and enforcement legislation that imposes administrative rather than criminal law penalties for non-compliance by actors in the agriculture and agri-food system."[65]

The paramountcy of federal legislation in relation to agriculture remains undiminished. It has to be noted that in addition to s. 95, the extant provisions of s. 91 of the Constitution Act vests the federal Parliament with exclusive jurisdiction over subject-matters that relate to agricultural products and methods, processes, and commercialization of such products. The combined effect is to strengthen the federal Parliament in matters widely pertaining to agriculture and food regulation.

Presently, "food legislation in Canada is undergoing a significant transformation, not seen for almost 100 years, with the enactment of the *Safe Food for Canadians Act*, which received Royal Assent on November 22, 2012 and is expected to come into force in 2015. Once the Act is in force, two federal regimes will apply to food in Canada, instead of five, with the *Food and Drugs Act* and its Regulations applying to all foods sold in Canada, and the *Safe Food for Canadians*

---

[62]*Labatt Brewing co. v Canada (Attorney General)*, [1979] S.C.J. No. 134, [1980] 1 S.C.R. 914 (S.C.C.) (Can).
[63]Note 59, Id.
[64]*Canada Agricultural Products Act*, R.S.C. 1985, c. 20 (4th Supp.), s. 2. See, the Safe Food for Canadians Act, 2012.
[65]Note 66, infra.

Act and its Regulations applying to all food that is imported into Canada or is prepared in Canada for interprovincial or international trade. Simultaneous with the *Safe Food for Canadians Act* bringing into force an important overhaul of the federal food law regime, the *Food and Drugs Act* regime will also see the modernization of some of its provisions regulating food in Canada. This comprehensive overhaul of food legislation will take a number of years to complete, but the framework for such a transformation has already been set out in federal government documents."[66]

31.2.1 Public Health in Canada

The use of the term "regulatory grid" implies organization and underpinning strategy. The elements of this regulatory grid are:

- *Meat Inspection Act*[67];
- *Canada Grain Act*[68];
- *Canada Agricultural Products Act (CAPA)*[69];
- *Food and Drugs Act*[70]; and
- *Consumer Packaging and Labelling Act.*[71]

The first three acts prescribe standards; establish grades and grade names (the intellectual property in which vests in the federal Crown)[72] and establish necessary attendant elements of search and seizure.[73] In addition, The *Meat Act* addresses meat products and processes.[74] Similarly, the *Grain Act* requires the licensing of grain elevators, dockage, standards, shrinkage, infestation, et cetera.[75]

For its part, the *CAPA* sets grading standards for livestock and poultry; fresh fruit; vegetables; honey, maple products and processed foods. Section 2 of the Act defines an agricultural product as:

(a) An animal, a plant or product of an animal or plant,
(b) A product, either food or drink, wholly or partly derived from an animal or a plant, or
(c) A product prescribed for the purposes of this Act[76] Meat

---

[66]Halbury's Laws of Canada (2014) Reissue.
[67]R.S.C., 1985, c. 25 (1st Supp.) (Can.).
[68]R.S.C., 1985, c. G-10 (Can.).
[69]R.S.C., 1985, c. 20 (4th Supp.) (Can.).
[70]R.S.C., 1985, c. F-27 (Can.).
[71]R.S.C., 1985, c. C-38 (Can.).
[72]CAPA s.15 (Can.).
[73]For instance see CAPA, ss. 21–28 (Can.).
[74]Meat Inspection Regulations, 1990 (S.O.R. 90/288) (Can.).
[75]Canada Grain Regulations, C.R.C., c.889 (Can.).
[76]CAPA, s. 2 (Can.).

Fig. 31.1 Federal government meat stamp

In the same way, the *Meat Inspection Act*[77] prescribes registered establishments and key national trademarks. For instance there is the meat inspection legend stamped on the carcass by a federal inspector proving the fact of inspection in a registered premises and implicitly sanitary standard satisfaction (Fig. 31.1).[78]

This is required for all meat products destined inter-provincially or internationally.[79] A legend addresses food <u>safety</u>. And, Regulatory standards[80] encompass food standards or <u>quality</u>.

The *Meat Inspection Regulations*[81] further flesh this out:

"Adulterated" means, in respect of a meat product intended for sale, use or consumption as an edible meat product in Canada,

(a) Containing or having been treated with:

 (i) A pesticide, heavy metal, industrial pollutant, drug, medicament or any other substance in an amount that exceeds the maximum level of use prescribed by the *Food and Drug Regulations*,
 (ii) An ingredient, a food additive or any source of ionizing radiation not permitted by or in an amount exceeding the prescribed limits, by these Regulations or by the *Food and Drug Regulations*,
 (iii) Any poison, decomposed substance or visible contamination, or
 (iv) Any pathogenic microorganism in excess of levels published in the Manual of Procedures, or

(b) Failing to meet the standards set out in Part I;

The meat inspection regulations prescribe what common parlance names can be labelled to different cuts of meat. For example, section 94(9) provides that *"The

---

[77]R.S.C. 1985, c. 25 (1st Supp.) (Can.).
[78]Meat Inspection Regulations, S.O.R. 90/288, Schedule III (Can.).
[79]Id., s. 94.
[80]Id., Schedule I, Columns I - IV.
[81]S.O.R. 90/288 (Can.).

term "ham" shall not be used on the label of a meat product unless the meat product is derived from the hind leg of a dressed swine carcass above the tarsal joint."[82]

These regulations also address ante-mortem examination, the humane treatment and slaughter of food animals.[83]

Of relevance from a packaging and labelling perspective are the regulations that prescribe the minutiae of what goes into what size and type of package with what label including, production date coding/lot; best before dates; instructions; legend stamp references; cooking instructions, et cetera.[84] For instance, section 91(3) stipulates that:

> The label for any meat product identified as edible, used for medicinal purposes or animal feed shall indicate if the meat product should be kept refrigerated or kept frozen. Except if the meat product:
>
> (a) Is packaged in a hermetically sealed container and treated to achieve commercial sterility;
> (b) Is dried to attain a water activity of 0.85 or less;
> (c) Has a pH of 4.6 or lower;
> (d) Is packaged in salt or a saturated salt solution;
> (e) Is fermented and has a pH of 5.3 or less, and a water activity of 0.90 or less, at the end of the fermentation within the time set out in the Manual Procedures; or
> (f) Has been subjected to a treatment approved by the director, ensuring the stability of the meat product when it is stored at normal room temperature.

The regulations that address labelling issues such as net weight cross-reference the *Consumer Packaging and Labelling Act*.[85] Also, the regulations that address ingredients cross-reference the *Food and Drug Regulations*.[86] The minutiae of detail on labelling are found in Schedule IV.[87]

Surprisingly the grade names and standards—quality—are found in another set of regulations: *The Livestock and Poultry Carcass Grading Regulations*.[88] Section 29 speaks to beef grades.

There shall be 13 grades of beef carcasses with the grade names Canada A, Canada AA, Canada AAA, Canada Prime, Canada B1, Canada B2, Canada B3, Canada B4, Canada D1, Canada D2, Canada D3, Canada D4 and Canada E.

There are other regulations that identify grade standard elements e.g., maturity, muscling, tenderness, colour, et cetera. Reproduced in Fig. 31.2 are poultry marks showing respective grades.[89]

---

[82]Id., s. 94 (9).
[83]Id., s. 61 et al.
[84]Id., Schedule II, Column I - II.
[85]Id., s. 103(1).
[86]Id., s. 118.
[87]Id., Schedule IV, Columns I - II.
[88]Livestock and Poultry Carcass Grading Regulations, S.O.R./92-541 (Can.).
[89]Id., Schedule I under ss. 64–64.

Fig. 31.2 Federal government poultry grades

Still, the corresponding regulations for fish are found in the *Fish Inspection Act* and the *Fish Inspection Regulations*.[90]

The overall regulatory matrix identifies the names for the products, compositional standards; yield class; grading standards; authorized marks; authorized and licensed operations.

31.2.1.1 Grain

The bulk of the regulatory regime with respect to grains is a licensing and inspection scheme to ensure quality throughout the transportation chain: elevator, terminal, train and port terminal.[91] To this end, the Grain Commission is the responsible agency.[92] Grain and its various grades are defined in s. 5 of the regulations[93]:

(1) The following seeds are designated as grain for the purposes of the Act: barley, beans, buckwheat, canola, peas, corn, fava beans, flaxseed, lentils, mixed grain, mustard seed, oats, peas, rapeseed, rye, safflower seed, soybeans, sunflower seed, triticale and wheat.
(2) The grade names and the specifications for grades of each grain are set out in Schedule 3.

Grain safety is addressed in part by s. 61 and 64[94]:

(1) No hazardous substances, other than products used for fumigating infested grain, shall be used for the treatment of grain at an elevator.
(2) No hazardous substances shall be stored in an area of a licensed elevator or annex to it unless

    (a) there is no direct access from the area to any area used for the handling or storage of grain;
    (b) the operator of the elevator has specifically designated the area for the storage of hazardous substances; and
    (c) the storage of hazardous substances in the area is not a danger to the handling or storage of grain at the elevator.

(3) If grain that is being received into, stored in or discharged from a licensed elevator is found to be <u>contaminated</u>, the operator of the elevator shall without delay notify the Commission and shall dispose of the contaminated grain in accordance with directions given by the chief grain inspector for Canada under an order of the Commission under paragraph 118(d) of the Act. (Emphasis added).

---

[90] *Fish Inspection Act,* R.S.C. 1985, c. F-12; *Fish Inspection Regulations,* C.R.C. c. 802.
[91] *Canada Grain Regulations,* C.R.C., c. 889; *Canada Grain Act,* R.S.C. 1985, c. G-10.
[92] Canada Grain Act R.S.C., 1985, c. G-10. ss. 3–15 (Can.).
[93] Canada Grain Regulations, C.R.C. c. 889 s. 5(1)(2), Schedule 3 (Can.).
[94] Id.

In cases where grain stored in an elevator is found infested, the operator of the elevator shall,

(a) in the absence of Commission personnel at the elevator, without delay provide the Commission with full particulars of the nature and extent of the infestation;
(b) send to the Commission in a sealed container a 1 kg sample of the grain containing specimens of the pests with which the grain is infested;
(c) treat the infested grain in accordance with instructions issued by the Commission; (c.1) [Repealed, SOR/2005-361, s. 11] and
(d) clean out and treat any emptied annex or bin that contained infested grain and any equipment used in handling that grain in accordance with instructions issued by the Commission.

The seeming banality of these provisions belies the profundity of the problem. Moulds produce toxins called mycotoxins as by-products of their metabolism. Moulds on crops are especially dangerous since they are very resistant to high heat. Cooking is virtually ineffectual. Since mould cannot be removed, and cooking cannot neutralize mycotoxins, the entire crop, elevator and shipment must be destroyed.

The more infamous and destructive mycotoxin *aflatoxin*, from the family *Aspergillus flavus*, is commonly found on peanuts, spinach, and corn and a potent carcinogen with allowable rates measured in parts per billion. Also deadly are the mycotoxins of Fusarium and Claviceps that contaminate grains, especially rye and wheat.[95] Fusarium thrives on grains that became moist during storage. Consequently, it produces alimentary toxic aleukia, an often-fatal disease. Similarly, *Ergotism*, otherwise known as St. Anthony's fire, is a mould disease from *Claviceps purpurea*. Aflatoxins are particularly potent as they cause cirrhosis and liver cancer in animal species. And in infected human beings it causes fever, jaundice, edema, vomiting, enlarged livers, and death.[96] Ultimately, it is possible that hallucinations and tremors caused by ingesting grains infested with ergot precipitated the Salem witch trials.[97]

31.2.2 Canada's Agricultural Products Act

This Act is the foundational basis of a plethora of product specific regulations. Identified below are few of these regulations:

- *Dairy Product Regulations*[98];
- *Egg Regulations;*[99]

---

[95] Satin (1999), pp. 220–221.
[96] Id., p. 220.
[97] Caporeal (2002).
[98] S.O.R., 79 – 840 (Can.).
[99] C.R.C., c. 284 (Can.).

- Fresh Fruit and Vegetables Regulations[100];
- Honey Regulations[101];
- Livestock and Poultry Carcass Grading Regulations[102];
- Maple Products Regulations[103];
- Organic Products Regulations[104];
- Processed Egg Regulations[105]; and
- Processed Products Regulations.[106]

The *Dairy Product Regulations* mandate what is and is not a dairy product. Also, regulations define grades and standards for dairy products. For instance, regulations dictate when the word "whipped" and "cultured" may be used with butter products.[107] The crux of the regulations is found in s. 2.1, which states the following:

> Where a grade or standard is established under these Regulations for a dairy product, no person shall market any product in import, export or interprovincial trade in such a manner, which the product is likely to be mistaken for the dairy product.

The safety issue is identified in s. 2.2, thus:

> (1) Subject to subsections (2) and (3), no person shall market a dairy product in import, export or inter-provincial trade as food unless the dairy product
>
> (a) [Repealed, SOR/2004-80, s. 6]
> (b) is not contaminated;
> (c) is edible;
> (d) is prepared in a sanitary manner; and
> (e) meets all other requirements of the *Food and Drugs Act* and the *Food and Drug Regulations* with respect to the dairy product.

The same unadulterated food provision is found in the Egg Regulations, to wit:

> 6. Subject to section 6.1, no person shall market eggs in import, export or interprovincial trade as food unless the eggs:
>
> (a) are prepared in accordance with these Regulations;
> (b) [Repealed, SOR/2011-205, s. 3]
> (c) are not contaminated;
> (d) are edible; and
> (e) meet all other requirements of the *Food and Drugs Act* and the *Food and Drug Regulations*.

The *Processed Products Regulations* commences by saying:

---

[100]C.R.C., c. 285 (Can.).
[101]C.R.C., c. 287 (Can.).
[102]S.O.R./92-541 (Can.).
[103]C.R.C., c.289 (Can.).
[104]Organic Products Regulations, 2009, S.O.R./2009-176 (Can.).
[105]Processed Egg Regulations C.R.C., c. 290 (Can.).
[106]Processed Products Regulation C.R.C., c. 291 (Can.).
[107]Dairy Regulations, S.O.R. 79 - 840, s. 6(2)(a)(b) (Can.).

2.1 (1) Subject to subsections (2) and (3), no person shall market a food product in import, export or interprovincial trade as food unless the food product, including every substance used as a component or ingredient thereof,

(a) [Repealed, SOR/2011-205, s. 27]
(b) is not contaminated;
(c) is sound, wholesome and edible;
(d) is prepared in a sanitary manner;
(e) where irradiated, is irradiated in accordance with Division 26 of Part B of the Food and Drug Regulations; and
(f) meets all other requirements of the Food and Drugs Act and the Food and Drug Regulations with respect to the food product.

The bulk of the PPR is focused on grades, marks and standard containers. The container strictures prescribe size, headspace and net and drained weights.[108]

31.2.3 Organic Products Regulations

Organic is a word; a philosophy; ethic; an ideology—even a religion for some.[109] Organics is popularly perceived as an exemplar of safe, sustainable agriculture and healthy produce. And as such, the word 'organic' has an aura of inerrancy. Consumers invest into the term all manner of therapeutic and spiritual benefit with many inferring—rightly or wrongly, that the term means small agricultural undertakings or local produce. The consumer shibboleth is that the word means "pesticide free".

Despite public perception, all that can be legitimately promised is that no *synthetic* pesticides were *deliberately* added; organic pesticides or conventional pesticide drift are legitimate.[110] Multi-nationals and local farms both can use the label. There are no proven medical benefits to organic produce. And what is organic is one jurisdiction may not necessarily meet the criteria of another jurisdiction as demonstrated by differences in the United States of America, China and the European Union.[111]

The Canadian law defines organic as a method of food production. The food product is chemically indistinguishable from conventional food products. Consequently, certification and labelling to underpin the integrity of the food concept "organic" is geared towards the process by which the food is produced. This is reiterated by the Guidelines which explicitly provides thus: *Neither this standard nor organic products labelled in according with this standard represent specific claims about the health, safety and nutrition of such organic product.*[112]

---

[108]Processed Products Regulation C.R.C., c. 291, s. 26 (Can.).
[109]See Endres (2007), Conford (2001), Pollen (2001), Guthman (2004), Guthman (2003).
[110]Friedland (2005).
[111]Endres (2007).
[112]Note 113, infra.

Fig. 31.3 Federal trademark for organic

The regulatory scheme for the organic process has four parts:

- Regulations[113];
- Guidelines for Production[114];
- Permitted Substances[115]; and
- Quality Management System Manual.[116]

The latter three points focus on the process: techniques, soil management, authorized pesticides, et cetera. The Regulations stipulate the government trademark (called a logo and legend in the regulations) for organic products (Fig. 31.3).[117]

The logo cannot be affixed to a product that is less than 95 % organic. Therefore, an organic multi-ingredient food product can be up to 5 % non-organic.[118] Organic multi-ingredient food product that is less than 95 % organic can be labelled as "organic ingredients" immediately proceeded with the percentage rounded down to the nearest whole number and in the same size and prominence as the phrase "organic ingredients".[119]

As for international trade, in addition to stipulations mandating that products meet the standards of the Canadian regime, Canada has entered into agreements with other countries regarding exportation or importation and the requirement for certifications of imported/exported products.[120]

---

[113] Organic Products Regulations 2009; SOR/2009-176 (Can.). There is no Organic Act. These regulations are promulgated under the Canada Agricultural Products Act, (CAPA).

[114] "Organic Production Systems, General Systems and Management Systems"; (2006) CAN/CGSB-32-310-2006.

[115] Organic Production Systems Permitted Substances List, under Canada General Standards Safety Board.

[116] Canada Organic Regime Quality Management System Manual: online http://www.ota.com/standards/canadian.html.

[117] Organic Regulations, *op. cit.*, s. 22.

[118] Id., s. 30.

[119] Id., s. 24.

[120] Id., s. 27.

31.2.4 Food Safety, Health and Nutritional Information Requirements

The field of genetics and food is smudged with different lexicons, plagued by lobbies and infected with emotion. The first step in understanding food safety, health, nutritional regulation and food labelling is to understand plant breeding. Most are unaware that the plants that become food—root, leaves, stems or fruit, are the product of breeding techniques and genetic modification going back to X-rays and turn of the twentieth century chemistry.

Understanding the regulation of so-called genetically modified foods requires the understanding of planting methods. Unless a person has a grasp of the various scientific methods of plant breeding, the regulatory regime and its underlying approaches may elude easy comprehension.

31.2.4.1 Sexual

Some plants reproduce sexually, by the use of seeds. This is the oldest form of breeding and agriculture. Plant scientists estimate that wheat was bred 7000 years ago.[121] Bread wheat is widely believed to be a cross of tetraploid [durum] wheat with inedible goat grass.[122] It is also common knowledge that sexually reproduced plants have greater diversity. However, by definition, it is difficult to maintain seed vitality.

31.2.4.2 Asexual

There are also other plants such as strawberries, potatoes, perennial grasses, legumes, rootstocks, and bananas that can reproduce asexually. Asexual reproduction occurs by means of roots, tubers, and stem or leaf cuttings.

Scientists confirm that plants, which have been propagated vegetatively from a single plant are clones. Primary example of such clones includes plant cuttings. Clones have fixed traits, and may meet consumer demands, but they cannot adapt to environmental pressures since there is no genetic variation or evolution. Cloning techniques for plants are either vegetative (plant reproduction via rhizomes, bulbs, or root sprouts), or apomixes (seeds are produced asexually without the fusion of male and female gametes). The most obvious example of vegetative propagation is bananas. In such instances, they are all clones and all are seedless. Most fruit trees are clones. For example, all navel oranges in the United States are clones from 12 orange trees brought from Brazil in 1869.[123] To reproduce most tree fruits,

---

[121]Prakash (2001).

[122]Id. 81.

[123]Fedoroff and Brown (2004).

planters use grafting and not seed propagation. A planter of apple seeds would not harvest same fruits because the new trees from the seeds are new cultivars, not offspring true to the parental variety.[124]

31.2.4.3 Hybrid

In 1926, a major commercial breeding breakthrough occurred through the hybridization process.[125] Hybridization involves the crossing of any two parental lines to their parental lines. Often the interbreeding is of two distinct and genetically dissimilar parental lines.[126] More vigorous and higher yielding than either parent are the final progeny.[127] Plant scientists call the attendant vigor "heterosis."[128] The cross-pollination is usually between two parents from different lineages, but may refer to two different taxa. In hybridization, the offspring of the commercial crop reproduce the parental or grandparental lines of the commercial crop.[129] Hybridization is an expensive process—it takes a number of crosses and up to 12 years to develop market-ready seeds.[130]

Technically, the undesirable recessive genes combine, and their unwanted trait are expressed.[131] At first, hybridization process seemed counterintuitive. Farmers saw no reason to buy seed when they could replant their own. However, the agronomic qualities and yields changed that approach.[132] Hybridization was a foundation process for the seed industry, since no farmer could afford to spend years breeding a crop that did not reproduce true.[133] In other words, farmers' rights are irrelevant when dealing with hybrids.

31.2.4.4 Mutagenesis

Mutagenesis is the science and technology of inducing mutations (natural mutations are known as "sports"). It arose in the 1920s with X-ray-induced mutations,[134] and acquired momentum after 1955 due to the collaboration among UN, FAO, and the

---

[124] Id.
[125] The history of scientific crop breeding versus commercial crop breeding is in James (2005).
[126] Oczek (2000).
[127] Fernandez-Cornejo (2004).
[128] Fedoroff and Brown (2004).
[129] Oczek (2000).
[130] Fernandez-Cornejo (2004), p. 4.
[131] Oczek (2000).
[132] Fedoroff and Brown (2004), pp. 61–62.
[133] Murphy (2007), supra n. 26.
[134] Fedoroff and Brown (2004), p. 16.

Atomic Energy Commission.[135] The objective of mutagenesis is to cause lesions in the chromosomes of the plant, which causes chromosome breaks and rearrangements or deletions of the genes.[136] The resulting candidates are planted, and if agronomically successful, backcrossed into an elite cultivar, sometimes becoming a variety in their own right.

A more comprehensive definition of the technique would suggest damaging the DNA to generate agronomically useful mutants by any means—thermal neutrons, X-rays, γ-rays, gamma rays, fast neutrons, or chemicals such as ethyl methane sulphonate (a harsh carcinogenic).[137] The chemicals react directly with the DNA bases to modify their structure.[138] Breeders use a large number of seeds (about 100,000) to produce a second generation of 30,000–50,000 plants.[139] If successful, a tiny number of useful plants will be produced, with the desired as well as undesired traits. They will then be backcrossed for a number to generations to produce an elite mutant variety.[140] Gamma and X-rays are the most frequent mutagens.

Mutations can result at a number of levels.[141] The changes can be intragenic (within a gene), intergenic (between genes), or genomic (genome-wide). Genome changes can be polyploid (containing more than twice the number of chromosomes), haploid (containing half the number of chromosomes), or aneuploid (containing a complete set of chromosomes less one or more chromosomes). The chief practical difference between mutagenesis and what is commonly called genetically modified or GM—other than precision—is that the new manifesting trait tends to be recessive with mutation and dominant with GM.[142]

31.2.4.5 Interspecies Crossing

Interspecies crossing can occur between closely related species and more distant relatives. It is through non-rDNA cytogenetic manipulations,[143] in which portions of the chromosomes of involved species are recombined through natural processes (chromosomal translocations). To date, these have focused on pest and disease resistance. The technique has used chromosomes and not genes. Thus, large numbers of genes have also been transferred—be they useful, neutral, or deleterious. The technique is becoming more focused, limiting the amount of genetic

---

[135]Favret (1962).
[136]Eichelbaum et al. (2001).
[137]Federoff and Brown (2005), p. 24.
[138]Murphy (2007), supra n. 30.
[139]Id.
[140]Id.
[141]Jacobsen and Schouten (2007).
[142]Id. 2-3.
[143]National and Medicine 25.

material transferred. Corn, soybeans, rice, barley, and potatoes have been improved using this technique.[144]

31.2.4.6 Protoplast Fusion

Protoplast fusion, identified in 1909 and practiced since 1970s, is a technique that introduces novel genes into a crop genome from a donor with which the crop will not normally interbreed.[145] This process involves removing the cell walls of plant cells of two unrelated plant species (embryo rescue discussed below is ineffective),[146] suspending these protoplasts in a medium, and then adding a chemical or an electrical current that fuses the protoplasts together. These fused protoplasts regenerate a cell wall and begin to multiply when placed in a nutrient medium. The nuclei of the two different species also fuse to create a hybrid nucleus that contains both sets of parental genes.[147] From this cell culture, an undifferentiated mass of plant cells grows—a callus—which can be further manipulated to form "artificial" or "chimera" seeds. What results is a hybrid with DNA from both protoplasts. Sugar beets, potatoes, and oilseed rape were improved using this approach.[148]

31.2.4.7 Chromosome Doubling

When two unrelated plant species cross, the offspring is often sterile. Chromosome doubling is required for fertility. Their chromosomes are different and unable to form stable pairs during meiosis.[149] Culture or microspore culture is often used to generate haploid plants (containing only a single set of chromosomes), which too require chromosome doubling to be fertile. Through the application of the chemical compound colchicine—an extract from autumn crocuses discovered in 1937,[150] chromosomes could be doubled in a genome. Ethyl methane sulphonate or nitrogen mustard can be used instead of colchicine.

A particularly apt illustration of this technique is the food crop triticale—a forced mating between durum wheat and rye, two unrelated species.[151] It first appeared in 1876 as an infertile cross. Applying colchicine treatment to the hybrid

---

[144]Id. 26.

[145]Murphy (2007), 42; Prescott-Allen and Prescott-Allen (1988).

[146]Murphy (2007), supra n. 43.

[147]Id. There is an asymmetric version of the technique using a micro-dissection of a nucleus so as to only transfer a limited number of chromosomes from the donor, for instance a wild, unrelated species; Murphy (2007), 43.

[148]Custers supra n. 7; Murphy (2007), supra n. 43.

[149]Id. at 39.

[150]Fedoroff and Brown (2006). Colchicine is a powerful toxin and carcinogen. Prakash (2003).

[151]Fedoroff and Brown (2006).

enabled researchers to artificially double the number of chromosomes. The bulk of research work to make this new hybrid commercial has occurred since the 1970s.

31.2.4.8 Somaclonal Variation

Somaclonal variation exploits the fact that spontaneous mutations occur when plants cells are grown in vitro. In vitro multiplication of plants generates individuals with different phenotypes. These differences are thought to arise from genome instability during the in vitro phase, which may result in clones having different phenotypes than the parental clone.

Somaclonal variation is normally regarded as an undesirable by-product of the stresses imposed on a plant by subjecting it to tissue culture. Plant cells are passed through tissue culture, a process that may involve a callus phase (de-differentiating the cells into a mass of callus and then regenerating new plantlets or embryos), and they may undergo certain spontaneous genetic changes. Imposing physical stresses further controls the process: cold, drought, high salinity, excess or dearth of nutrients, chemical regulators, and challenges by pathogens. Depending on the type of plant, its age, the source of the tissue (e.g., root or leaf), its balance of growth hormones and nutrients, and length of time in culture, directed mutations can arise without radiation or chemicals. The objective is the detection of a single gene mutation or deletion; transposition of larger stretches of DNA; or duplication or loss of an entire chromosome.

31.2.4.9 Embryo Rescue

When making a wide genetic cross (e.g., wheat with rye), there may be cross-pollination, and the embryo may abort at some time during development. This differs from somatic hybridization, where an embryo cannot form since the two parental species are too distant.[152] In embryo rescue, the developing embryo is removed from the plant before abortion and cultured on an aseptic medium. The embryo will develop and germinate, growing into a plant. Generally, this approach is considered an intermediary technique—to transfer genes from incompatible species via intermediate, partially compatible relatives of the donor and recipient. It has been used to cross a cucumber, a melon, and triticale.

31.2.4.10 rDNA

Recombinant DNA molecules are DNA molecules formed by laboratory methods of genetic recombination to bring together genetic material from multiple sources,

---

[152]Id. 43; Prescott-Allen and Prescott-Allen.

creating sequences that would not otherwise be found in biological organisms. The developer would seek out the protein that manifests the trait, and the gene (called the target gene) that codes for the protein. The target gene is then decoded (base sequence order),[153] and a new copy (called the event) is synthesized. The developer identifies the appropriate promoter and regulator for the event[154] and chooses the most effective vector.[155] The developer inserts the cassette into the target genome and uses a gene marker to confirm incorporation.[156] This package is called the gene construct. If the event is not from the same species, one must use promoter and terminator genes that are harmonious within the target species genome.

For a breeder/developer, rDNA has several advantages. First, the genes may be identified in any source. The donor can be a plant, animal, or micro-organism.[157] rDNA is ruthlessly focused on chemistry. A corollary to this is that one can predict the presence and location of a gene in one species from what one knows of another. The conservation or consistency of gene content and gene order along chromosomes of different plant genomes is described by the term 'synteny'. Originally it was thought that each crop species had an idiosyncratic, if not unique, genome. Instead, a similarity in gene maps among different crop species has been found—the 12 chromosomes of rice align with the 10 chromosomes of maize and with the basic 7 chromosomes of wheat. Synteny allows one to pool knowledge of biochemistry, physiology, and genetics and transfer it between crops.

Secondly, one need not wait 10,000 years for the right mutation,[158] as rDNA allows for a relatively speedy production of new varieties,[159] as well as for extreme specificity in the "event" and associated genes incorporated by the recipient.[160]

With classical techniques of gene transfer, a variable number of genes can be transferred, the number depending on the mechanism of transfer; but predicting the precise number of the traits that have been transferred is difficult, and we cannot always predict the (characteristics) that will result. With organisms modified by

---

[153]Lewin (1997); McHughen (2000) pp. 24–32.

[154]Gupta and Ram (2004), p. 220; McHughen (2000), p. 31.

[155]Lewin (1997). The vector can be biological, physical or chemical. Kunich (2001). The biological vectors include *Agrobacterium* that has the natural capability to transfer DNA from its cells to plants cells during infection. Physical methods include particle bombardment, microinjection, sonicatin, silicon carbide (SiC) whisker treatment and electric current pulse. Gupta and Ram 221. Chemically based techniques include transfection using liposomes and polyethylene glycol. Gupta and Ram 221; Murphy (2007), 46.

[156]McHughen (2000). Since only some of the cells in the target organism are modified, it is necessary to destroy the non-modified cells. Here an event gene and a marker gene are regenerated in the presence of a selective agent (antibiotic) for which the marker and event gene have tolerance.

[157]Mandel (2004).

[158]Winn (1999).

[159]Id. 668.

[160]Mandel (2006), p. 85.

molecular methods, we are in a better, if not perfect, position to predict the (characteristics).[161]

31.2.4.11 Gene Silencing

Up to 95 % of a plant's genome comprise repetitive elements, heretofore pejoratively and incorrectly known as "junk" DNA. The mechanism that deals with this, "gene silencing,"[162] is a defence mechanism against viruses, transposons, and retroelements.[163] But gene mechanisms for silencing junk and viral infections also silence the construct. A number of techniques to overcome this problem exist (all of which are time-consuming). However, the science of gene silencing may also allow breeders to "un-silence" endogenous genes of merit without the need for a construct by inserting RNA or using antisense technology.[164] And it may also allow the silencing of key genes, such as those that code for ripening enzymes[165] or those that code for allergies.[166] The technique has been used to modify the phenotype, in particular to down-regulate expression of endogenous genes in the original plant, or confer resistance to infection by preventing expression of genes causing disease.[167]

31.2.4.12 Marker Assisted Selection

MAS is a method of performing conventional plant breeding in which researchers locate DNA sequences in a plant's genome that are consistently associated with desired trait(s) (selection).[168] Those sequences can then be used to screen for and predict the presence of those traits in the progeny of traditional crosses.[169] The developer does not breed plants to select traits; the developer's use of MAS identifies the plants manifesting the traits before they are grown to maturity. This method has proven particularly useful with plants that have long life cycles before they manifest the desired trait, e.g. tree crops such as oil palm, coconut, coffee, tea,

---

[161] Miller (2004).

[162] Ainley and Kumpatla (2004), p. 243.

[163] Id. 243, 51.

[164] Gao (2004), pp. 297–344. Parekh (2004), p. 305.

[165] Do GMOs Mean More Allergies? (2005), EU Commission. Online: http://www.gmo-compass.org/eng/safety/human_health/192.gmos_mean_more_allergies.html.

[166] Gao (2004), p. 305; Sir David King, *GM Science Review*.

[167] Sir David King, *GM Science Review* 54.

[168] Henry Miller *Rifkin Redux*, 2006, Food Safety Network (U of Guelph). Online: http://archives.foodsafetynetwork.ca/agnet/2006/7-2006/agnet_july_4.html, 4 Jul 2006; Murphy (2007), 51.

[169] Id.

cocoa, and mango.[170] The process is expensive, in that it requires mapping populations, assembling genomic markers, and creating genetic maps.[171]

Canadian biotechnology policy,[172] developed into the Canadian Federal Regulatory Framework for Biotechnology (1993), articulated the objective of realizing the benefits of biotechnology products and processes while protecting health, safety, and the environment. The principles include[173]: using existing laws and regulatory departments to avoid duplication; developing clear guidelines for evaluating biotechnology products that are in harmony with national priorities and international standards; providing a sound, scientific knowledge base on which to assess risk and evaluate products; ensuring that the development and enforcement of Canadian biotechnology regulations are open and include consultation; and fostering a favourable climate for investment, development, and innovation. This approach is consistent with the OECD Blue Book.[174]

In addition, the Canadian Food Inspection Agency's (CFIA) operating principles in evaluating and regulating include[175]: focusing on the product traits, not method of production; establishing safety levels and standards for each product, based on best scientific data and dealing with safety in the milieu of probability and magnitude of any adverse effects, rather than the absence of risk.

The Plants with Novel Traits (PNT) regime arose in Canada in the early 1990s when regulators faced the conundrum of the same trait created by different breeding methods.[176] The seed company Allelix[177] sought permission for confined trials of mutagenic, herbicide-tolerant canola. No regulatory requirements were suggested. Allelix subsequently identified the operative gene that bestowed the protein giving herbicide tolerance, and so used rDNA to insert that gene into its other canola lines. Field trials on these rDNA lines commenced. Then, the regulators interceded and demanded regulatory compliance. Also, Allelix complained that the same gene in the same crop species attracted two different regulatory responses depending on the breeding technique. It was certainly arbitrary if not unfair, given the costs. Then, the matter was referred to the higher echelons of government for direction.[178]

The resulting policy decision held that the threat to the environment from the plant traits arose irrespective of the breeding technique that introduced those

---

[170]Murphy (2007), 52.

[171]Id., pp. 52–53.

[172]Barrett, pp. 71–132.

[173]Rastogi (2005).

[174]*Recombinant DNA Safety Considerations, Safety Considerations for Industrial, Agricultural and Environmental Applications of Organisms Derived by Recombinant DNA Technologies* (OECD, 1986) 42, paragraphs 2 & 3.

[175]Prince, pp. 220–221; Rastogi (2005).

[176]Yarrow (2001), pp. 101–02.

[177]Allelix was later bought out by Pioneer.

[178]Yarrow (2001), p. 102. (Repeat of 176).

traits.[179] Henceforth the regulatory focus would be on the "novel" trait, and not on the breeding technique that begot or introduced that novel trait. Admittedly, it is easier in most instances to insert novel traits via rDNA techniques, but novelty can be had through other breeding techniques such as mutagenesis, somaclonal variation, and the dilemma of cisgenes and gene silencing.[180]

31.2.4.13 PNT: Field

The Canadian Food Inspection Agency (CFIA) conducts risk assessment. First, the CFIA seeks to determine if a risk assessment is necessary. If the plant is familiar, the trait is similar to one already approved. Meanwhile, the trait is derived by a technique that has been traditionally considered safe. Then a risk assessment may be omitted.[181] If not, a risk assessment commences to determine if the plant is substantially equivalent to an approved product. This entails both the genomics/proteomics and the effect on the environment.[182] If both familiar and substantially equivalent, then the CFIA assessment ceases. If not, the portion not substantially equivalent undergoes further risk assessment.[183] The crux of PNT, due to the novel trait from the introduced gene(s), is that element of the plant is not substantially equivalent to their progenitors:

> ... based on valid scientific rationale ... in terms of its specific use and safety both for the environment and for human health, to any characteristic of a distinct, stable population of cultivated seed of the same species in Canada, having regard to weediness potential, gene flow, plant pest potential, impact on non-target organisms and impact on biodiversity. [Emphasis added][184]

The policy interpretation of the regulation is that a plant contains a novel trait if the trait is either not present in plants of the same species already existing as stable, cultivated populations in Canada, or is present but at a level significantly outside the range of that traits in stable, cultivated populations of that plant in Canada. There are three contexts for interpreting novelty: trait introduction, trait modification, and incremental increase.[185] This means most PNTs have additional regulatory hurdles beyond the extensive strictures for a variety registration.

The nub of the regulation is that a PNT needs regulatory authorization for confined or unconfined releases. Extensive data required for justifying such authorization include all details about the donor organism, breeding method, trait, test

---

[179]Id.

[180]Roberts (2007), p. 7. See also Chapter 5, Section 16, Crop: Breeding & Misconceptions.

[181]Reimer and Schwartz (2001).

[182]Id.

[183]Id.

[184]Seeds Regulations C.R.C. c. 1400 s. 107(1) (Can.).

[185]Id. slide 14.

results, foreign filings, protocols,[186] weediness, outcrossing potential, ecology, potential interactions with other organisms, and impact on biodiversity, among other things.[187]

Every event is not subject to an extensive regulatory review; rather, every novel event creating a novel plant is subject to the review. Inserting the same event into the same crop species is known as re-transformation (e.g., canola variety C after insertion into canola variety A).[188] The same event can also be inserted into a new crop (e.g., tropical plant fungus resistance into a temperate crop, e.g., papaya into canola). The Seeds Regulations exempt seed that is derived from seed that was previously authorized for unconfined release that is "substantially equivalent, in terms of its specific use and safety both for the environment and for human health, to seed of the same species, having regard to weediness potential, gene flow, plant pest potential, impact on non-target organisms and impact on biodiversity."[189] In Canada in 2005 there were 403 varieties of soybeans of which 158 were PNTs. But these soybeans all had the same event, tolerance for the herbicide glyphosate.[190]

Some techniques require no regulation attachment until the final product exists and is submitted for registration. In contrast, rDNA use immediately fuses with the PNT scheme. A breeder can use any number of techniques—other than rDNA—to manifest a desired phenotype. Regulation attach only when the breeder has a stable plant manifesting the trait that fits the definition of PNT. And registration is sought only if the plant is a seed plant and variety. There are no regulatory strictures on laboratory safety concerning mutagenic or other processes that alter the genome at the chromosome or lower level. The average breeder of corn, soybeans, wheat, or potatoes may put 50,000 discrete new genetic variants a year into the field, not knowing what the genome will manifest.[191] To date, after decades of this practice, no human or environmental harm has transpired; and these plants are ploughed over.

Another economic reality is that those practising rDNA technologies labour under a heavy regulatory regime that by definition consumes prodigious amounts

---

[186] Seeds Regulations, C.R.C., c. 1400, s. 110(1)(2) (Can.). The requirements include: details of the donor organism; the methods of incorporation, if applicable, and details relating to expression of the novel trait, the stability of the incorporation, and a comparison of the characteristics of the plants derived from the modified seed with those derived from the unmodified host seed; all other information and test data that are relevant to identifying the risk to the environment, including the risk to human health; a list of other government agencies, either Canadian or foreign, that have been provided with information in respect of the PNT and the purpose for which the information was provided; a description of the analytical methodologies followed in generating any submitted data, including quality control and quality assurance procedures; the proposed starting date, completion date, and site of the confined release.

[187] Rastogi (2005), slide 25.

[188] Yarrow (2001), p. 107.

[189] Seeds Regulations C.R.C., c. 1400, s. 108 (Can.).

[190] Demek et al., p. 4.

[191] Miller (2003).

of time and money. This means that only large companies can truly operate in the field, given transaction costs of breeding, time, and professional services. Small companies, governments, and university researchers do not have that kind of wherewithal.

31.2.4.14 PNT: Fork

Having regulatory approval of a GMC does not equate approval for the GMF. GM corn, and GM cornflakes require different permits. Under the *Food and Drug Act Regulations*,[192] Health Canada continues with novel food, a concept that engages the concept of major change. "Major change" refers to change in the food, which based on the manufacturer's experience or generally accepted nutritional or food science theory, places the modified food outside the accepted limits of natural variations for that food with regard to:

(a) composition, structure or nutritional quality of the food or its generally recognized physiological effects;
(b) manner in which the food is metabolized in the body; or
(c) microbiological safety, the chemical safety or the safe use of the food.

Meanwhile, "novel food" means:

(a) a substance, including a microorganism, that does not have a history of safe use as a food;
(b) a food that has been manufactured, prepared, preserved or packaged by a process that
  (i) has not been previously applied to that food, and
  (ii) causes the food to undergo a *major change*; and
(c) a food that is derived from a plant, animal or microorganism that has been genetically modified such that
  (i) the plant, animal or microorganism exhibits characteristics that were not previously observed in that plant, animal or microorganism,
  (ii) the plant, animal or microorganism **no longer exhibits** characteristics that were previously observed in that plant, animal or microorganism, or
  (iii) one or more characteristics of the plant, animal or microorganism **no longer fall within the anticipated range** for that plant, animal or microorganism. [Emphasis added]

In short, a major change means the food is outside its natural limits in composition, metabolization, and safety. On the other hand, a novel food is complicated. This is the result of political compromise as much as by legal and scientific necessity. A food is novel if it has no history of safe use, or if it has a major change making it exhibit new characteristics or characteristics falling outside of the anticipated range, or fail to exhibit old characteristics.

---

[192]Food and Drug Regulations, C.R.C. c. 870 (Can.).

The effect of these regulations is to exclude GM foods, which are safely used in other countries or in Canada in a similar crop; Second, regulations may exclude minor food-processing changes—although most processed food contains GM (e.g., corn fructose, canola oil, soy protein).[193] True novelty, not GM per se, triggers these provisions. Regulators review field trials related to nutrition, toxicity, and allergenicity. If a PNT has a history of safe production and consumption in another country, regulators regard this as admissible data. However, this is not the case with PNT that has been regulated by the CFIA.[194]

According to regulations a novel food cannot be sold until the regulator is notified, provided prescribed data (nature of the trait, previous use, safety history), and an approval is issued. These provisions also speak to the parameters of not only what is approved, but also what might attract labelling, to wit, major compositional changes or health threats. The safety of a plant species as a food crop is established through familiarity and a record of use.[195]

For example, Canola—with risk assessment regime in place, would have been considered a PNT and a novel food given that its oil is edible but derived from rapeseed—while the oil is toxic.[196] Instead, canola was prosecuted through the *Seeds Act* system sans PNT, and into the marketplace without risk assessment—and to date without incident.[197]

31.2.4.15 PNT: Labels

Canada's approach to PNT, again a broader concept than the vernacular "GM" or genetically modified food, is premised on a major change in the food product. Only the nature of the product is pertinent. The product labelling approach in some literature is called the "need to know" (major changes, allergens etc.). The labelling approach as used in the EU is premised on the concept of consumers' inherent "right to know" how product are made. So rDNA breeding methods are captured for labelling purposes, but mutagenetic plants—irrespective of profound the breeding changes are not.[198]

There is probity in the PNT product labelling approach or "need to know." There is no distinction between *made with* a PNT and *made using* a PNT, as there is in EU regulations. There the former attracts a label, the latter does not. One neglected concept is whether the novel protein is or is not in the final product. If the novel protein is in the food product it must be labelled ("need to know"). Based on the EU

---

[193] Smyth et al., p. 27.

[194] Id. 29.

[195] Rasco (2008), p. 178.

[196] Gamma radiation was used to silence two genes—one produced a toxin, the other an anti-nutrient. Conko (2003).

[197] Barrett and Abergel, p. 6.

[198] Directive 2001/18, in particular annex 1A, & 1B.

regulation soils from genetically modified plants must be labelled, although the GM protein is not present in the oil, since the premise is process not product.

Nonetheless, there are voluntary guidelines in both Canada and the U.S. concerning GM *qua* GM labels.[199] First is the concept of a genetically engineered plant or food. Second, the guidelines state whether one cannot say "No GM" or "made without GMO" if there is no product on the market containing GM. For example, labeling "GM free bread" is considered breach of guidelines, since as of yet there is no GM wheat in commercial channels of trade in North America.[200] There can be no implication that GM is a threat or its absence is otherwise salutary.

Presently organic excludes genetically engineered (GE) or GM but not any other PNT or breeding technique. Organic requires segregated, traceable, and labelled channels of trade. On the whole, when process labelling comes to pass in Canada, then issues of segregation, traceability and labelling arise, and so will the attendant testing and enforcement.

31.2.5 Food and Drugs Act & Regulations (FDA&R)

The regulations have seven parts: (A) administration; (B) foods, (C) drugs, (E) vitamins, (F) Minerals and amino acids, (G) cyclamate and saccharine sweeteners, (H) controlled and restricted drugs. The provisions speak to the standards of composition, strength, potency, purity, quality, et cetera of the food or drug as applicable.

Part (B) the provisions on food, have 28 divisions, each division deals with a specific food or beverage type.[201]

Division 1 deals with labelling. Amendments made in 2002 and 2005 were geared to assist consumers in making informed choices. Amongst other things regulations require:

- the common name of the food, list of ingredients, place of business for whom the food was manufactured;
- Reference standards for a nutrient and serving size; and
- Prescribed data/format for the nutritional facts table.[202]

---

[199] Voluntary Labelling and Advertising of Foods that Are and Are Not Produces of Genetic Engineering, CAN/CBSB-32-3152004; Guidance for Industry: Voluntary Labelling Indicating Whether Foods Have or Have Not Been Developed Using Bioengineering: 656 Fed Reg (2001) 4,839–4,840.

[200] Canadian Guidelines, para. 6.1.4.

[201] Food and Drugs Act R.S.C. 1985 c. F-27 (Can.); Food and Drug Regulations C.R.C., c. 870 Div. 2-28 *op. cit.*

[202] Id., ss. B.01.450 - B.01.453.

The nutritional labelling requirements are being reviewed again and will be revised concerning the look, serving size, added sugar, et cetera.[203]

Amendments in 2012 respected allergy-labelling concerns with peanuts, sesame seeds, eggs, mustard seed, gluten and sulphites.[204]

As previously discussed, division 28 addresses novel foods, the products derived from PNTs.

Irradiated food under division 26 is illustrative of the role of legal definitions and thresholds. Division 26 authorizes:

- the irradiation by any of gamma rays, x-rays and electron bean;
- of no more than the prescribed dosage';
- only potatoes; onions; wheat; flour; whole wheat flour; whole or ground spices; and dehydrated seasonings;
- for an prescribed purpose (e.g. inhibit sprouting [potatoes]; control insect infestation [flour], reduce microbial load [spices]).[205]

Using the irradiation symbol will not be required in irradiation of previously mentioned above products. Additionally, only wholly irradiated foods must display the symbol on the basis that irradiated foods—although appearing the same as unirradiated, may undergo organoleptic changes.

In the same way, Division 15 on the topic of adulteration deals with arsenic (i.e. fruit juice); fluoride (ie. fish protein), lead (ie. tomato paste); tin (ie. canned foods) and pesticide residues beyond those authorized under the *Pest Controls Products Act*, ss. 9–10.[206]

31.3 Conclusion: Regulation and Agreements Affecting Food Import/Export—NAFTA, CETA, CANSEA

In Canada, food is regulated by both domestic laws and agreements at the provincial level (the collection of statutes and regulations discussed above) and at the international level through regional and multilateral agreements such as North American Free Trade Agreement (NAFTA), Comprehensive Economic and Trade Agreement (CETA) as well as global organizations such as the World Trade Organization (WTO). During the past several decades, regulation of food and related industries

---

[203] Proposed changes to the look of the Nutrition Facts Table and the List of Ingredients. Online: http://www.hc-sc.gc.ca/fn-an/alt_formats/pdf/label-etiquet/nutrition-facts-valeur-nutritive-fs-fr-eng.pdf (2014).

[204] SOR/2011 - 28, February 4, 2011, Enhanced Labelling for food Allergen and Gluten Sources and Added Sulphites.

[205] *FDA* Regs ss. B.26.003 - B.26.005.

[206] Id., ss.B.15.001 - B.15.003.

and commodities have increasingly been governed by laws and agreements made outside of the state-level (supra-nationally).

Such systems reflect a move towards a global governance model, which has arguably taken powers away from countries to regulate vast aspects of their economies. Globalization of trade has opened up Canada's national borders to international trade, flow of capital and food products. In fact, internationalism affects the shape and form of food policy in Canada. While NAFTA and CETA have emerged and are criticised as neoliberal institutions aimed at eroding state power at the behest of international capitalist interests. In contrast, CANSEA highlights the power states have managed to retain to influence trade to expand markets and national interests abroad.

Among emerging accords, the North American Free Trade Agreement (NAFTA) has perhaps had one of the largest impacts on the domestic food landscape. With its origin in the Canada-United States Free Trade Agreement, the addition of Mexico into the agreement On January 1st, 1994 saw the creation of a marketplace spanning the North American continent. According the Canadian government's own sources, the objective of NAFTA is "the elimination of most tariffs and reduction of non-tariff barriers, as well as comprehensive provisions on the conduct of business in the free trade area".[207] Since its inception, the push to create a tariff-free zone in North America has been largely successful. Official Canadian government sources indicate that since the singing of the agreement, "virtually all tariffs have been eliminated on Canadian agricultural exports to the U.S. (January 1, 1998) and Mexico (January 1, 2003). The only exceptions to this general tariff elimination are chicken, turkey and egg products, as well as refined sugar".[208] The provisions dealing with tariff issues are captured in Articles 701–708 of the Agreement, to wit:

Article 701: Scope and Coverage

1. This Section applies to measures adopted or maintained by a Party relating to agricultural trade.
2. In the event of any inconsistency between this Section and another provision of this Agreement, this Section shall prevail to the extent of the inconsistency.

Article 702: International Obligations

1. Annex 702.1 applies to the Parties specified in that Annex with respect to agricultural trade under certain agreements between them.
2. Prior to adopting pursuant to an intergovernmental commodity agreement, a measure that may affect trade in an agricultural good between the Parties, the Party proposing to adopt the measure shall consult with the other Parties with a view to avoiding nullification or impairment of a concession granted by that Party in its Schedule to Annex 302.2.
3. Annex 702.3 applies to the Parties specified in that Annex with respect to measures adopted or maintained pursuant to an intergovernmental coffee agreement.

---

[207]Schlachter (2007).
[208]Id.

Article 703: Market Access

1. The Parties shall work together to improve access to their respective markets through the reduction or elimination of import barriers to trade between them in agricultural goods.

Customs Duties, Quantitative Restrictions, and Agricultural Grading and Marketing Standards

2. Annex 703.2 applies to the Parties specified in that Annex with respect to customs duties and quantitative restrictions; trade in sugar and syrup goods; agricultural grading and marketing standards.

Special Safeguard Provisions

3. Each Party may, in accordance with its Schedule to Annex 302.2, adopt or maintain a special safeguard in the form of a tariff rate quota on an agricultural good listed in its Section of Annex 703.3. Notwithstanding Article 302.2, a Party may not apply an over-quota tariff rate under a special safeguard that exceeds the lesser of:

   a) the most-favoured-nation (MFN) rate as of July 1, 1991; and
   b) the prevailing MFN rate.

4. No Party may, with respect to the same good, the same country and at the same time:

   a) apply an over-quota tariff rate under paragraph 3; and
   b) take an emergency action covered by Chapter Eight (Emergency Action).

Article 704: Domestic Support

The Parties recognize that domestic support measures can be of crucial importance to their agricultural sectors. But they may also have trade distorting and production effects. Also, the domestic support reduction commitments may result from agricultural multilateral trade negotiations under the *General Agreement on Tariffs and Trade (GATT)*. Accordingly, a Party supporting its agricultural producers should endeavour to work toward domestic support measures that either:

a) have minimal or no trade distorting or production effects; or
b) are exempt from any applicable domestic support reduction commitments that may be negotiated under the GATT.

Moreover, Parties further recognize that a Party may change its domestic support measures, including those that may be subject to reduction commitments. This is done at the Party's discretion and subject to its rights and obligations under the GATT.

Article 705: Export Subsidies

1. The Parties share the objective of the multilateral elimination of export subsidies for agricultural goods and shall cooperate in an effort to achieve an agreement under the GATT to eliminate those subsidies.
2. The Parties recognize that export subsidies for agricultural goods may prejudice the interests of importing and exporting Parties and, in particular, may disrupt the markets of importing Parties. Accordingly, in addition to the rights and obligations of the Parties specified in Annex 702.1, the Parties affirm that it is inappropriate for a Party to provide an export subsidy for an agricultural good exported to the territory of another Party where there are no other subsidized imports of that good into the territory of that other Party.
3. Except as provided in Annex 702.1, where an exporting Party considers a non-Party exporting an agricultural good to the territory of another Party with the benefit of export subsidies, the importing Party shall, on written request of the exporting Party, consult

with the exporting Party with a view to agreeing on specific measures that the importing Party may adopt to counter the effect of any such subsidized imports. If the importing Party adopts the agreed-upon measures, the exporting Party shall refrain from applying, or immediately cease to apply, any export subsidy to exports of such good to the territory of the importing Party.
4. Except as provided in Annex 702.1, an exporting Party shall deliver written notice to the importing Party at least three days, excluding weekends, prior to adopting an export subsidy measure on an agricultural good exported to the territory of another Party. The exporting Party shall consult with the importing Party within 72 hours of receipt of the importing Party's written request, with a view to eliminating the subsidy or minimizing any adverse impact on the market of the importing Party for that good. The importing Party shall, while requesting consultations with the exporting Party, deliver written notice to a third Party of the request. A third Party may request to participate in such consultations.
5. Each Party shall take into account the interests of the other Parties in the use of export subsidies on an agricultural good, recognizing they may have prejudicial effects on the interests of other Parties.
6. The Parties hereby establish a Working Group on Agricultural Subsidies, comprising representatives of each Party. This group shall meet semi-annually-- or as the Parties may otherwise agree, to work toward elimination of all export subsidies affecting agricultural trade between the Parties. The functions of the Working Group shall include:

   a) monitoring the volume and price of imports into the territory of any Party of agricultural goods that have benefitted from export subsidies;
   b) providing a forum for the Parties to develop mutually acceptable criteria and procedures for reaching agreement on the limitation or elimination of export subsidies for imports of agricultural goods into the territories of the Parties; and
   c) reporting annually to the Committee on Agricultural Trade, established under Article 706, on the implementation of this Article.

7. Notwithstanding any other provision of this Article:

   a) if the importing and exporting Parties agree to an export subsidy for an agricultural good exported to the territory of the importing Party, the exporting Party or Parties may adopt or maintain such subsidy; and
   b) each Party retains its rights to apply countervailing duties to subsidized imports of agricultural goods from the territory of a Party or non-Party.

Article 706: Committee on Agricultural Trade

1. The Parties hereby establish a Committee on Agricultural Trade, comprising representatives of each Party.
2. The Committee's functions shall include:

   a) monitoring and promoting of cooperation on the implementation and administration of this Section;
   b) providing a forum for the Parties to consult on issues related to this Section at least semi-annually and as the Parties may otherwise agree; and
   c) reporting annually to the Commission on the implementation of this Section.

Article 707: Advisory Committee on Private Commercial Disputes regarding Agricultural Goods

The Committee shall establish an Advisory Committee on Private Commercial Disputes regarding Agricultural Goods, comprising persons with expertise or experience in the resolution of private commercial disputes in agricultural trade. The Advisory Committee shall report and provide recommendations to the Committee for the development of systems in the territory of each Party to achieve the prompt and effective resolution of such disputes, taking into account any special circumstance, including the perishability of certain agricultural goods.[209]

These provisions seek to remove distortions in trade. But they do little to whittle down state control over regulations on food safety. Indeed, Articles 716–718 of the Agreement reinforce state sovereignty over sanitary and phytosanitary measures related to agriculture and agricultural processes.

Article 716: Adaptation to Regional Conditions

These provisions stipulate:

1. Each Party shall adapt sanitary or phytosanitary measures relating to the introduction, establishment or spread of an animal or plant pest or disease, to the sanitary or phytosanitary characteristics of the area where a good subject to such a measure is produced and the area in its territory to which the good is destined, taking into account any relevant conditions, including those relating to transportation and handling, between those areas. In assessing characteristics of an area--whether an area is likely to remain a pest-free or disease-free area or low pest or disease prevalence, each Party shall take into account, among other factors:

    a) the prevalence of relevant pests or diseases in that area;
    b) the existence of eradication or control programs in that area; and
    c) any relevant international standard, guideline or recommendation.

2. Further to paragraph 1, each Party shall, in determining whether an area is a pest-free or disease-free area or an area of low pest or disease prevalence, base its determination on factors such as geography, ecosystems, epidemiological surveillance and the effectiveness of sanitary or phytosanitary controls in that area.

3. Each importing Party shall recognize that an area in the territory of the exporting Party is, and is likely to remain, a pest-free or disease-free area or an area of low pest or disease prevalence, where the exporting Party provides to the importing Party scientific evidence or other information sufficient to satisfy the importing Party. For this purpose, each exporting Party shall provide reasonable access in its territory to the importing Party for inspection, testing and other relevant procedures.

4. Each Party may, in accordance with this Section:

    a) adopt, maintain or apply a different risk assessment procedure for a pest-free or disease-free area than for an area of low pest or disease prevalence, or
    b) make a different final determination for the disposition of a good produced in a pest-free or disease-free area than for a good produced in an area of low pest or disease prevalence, taking into account any relevant conditions, including those relating to transportation and handling.

5. Each Party shall; in adopting; maintaining or applying a sanitary or phytosanitary measure relating to the introduction; establishment or spread of an animal or plant

---

[209] Id.

pest or disease, accord a good produced in a pest-free or disease-free area in the territory of another Party no less favourable treatment than it accords a good produced in a pest-free or disease-free area, in another country, that poses the same level of risk. The Party shall use equivalent risk assessment techniques to evaluate relevant conditions and controls in the pest-free or disease-free area and in the area surrounding that area and take into account relevant conditions, including those relating to transportation and handling.

6. Each importing Party shall pursue an agreement with an exporting Party--on request and on specific requirements, the fulfillment of which allows a good produced in an area of low pest or disease prevalence in the territory of an exporting Party to be imported into the territory of the importing Party and achieves the importing Party's appropriate level of protection.

Article 717: Control, Inspection and Approval Procedures

1. Each Party with respect to control and inspection procedure it conducts:

    (a) shall initiate and complete the procedure as expeditiously as possible and in no less favourable manner for a good of another Party or country.
    (b) shall publish the normal processing period for the procedure or communicate the anticipated processing period to the applicant on request;
    (c) shall ensure that the competent body

        (i) on receipt of an application, promptly examines the completeness of the documentation and informs the applicant in a precise and complete manner of any deficiency,
        (ii) transmits to the applicant as soon as possible the results of the procedure in a form that is precise and complete. Then the applicant may take any necessary corrective action.
        (iii) where the application is deficient, proceeds as far as practicable with the procedure if the applicant so requests, and
        (iv) informs the applicant, on request, of the status of the application and the reasons for any delay;

    (d) shall limit the information the applicant is required to supply necessary for conducting the procedure;
    (e) shall accord confidential or proprietary information arising from, or supplied in connection with the procedure conducted for a good of another Party

        (i) treatment no less favourable than for a good of the Party, and
        (ii) in any event, treatment that protects the applicant's legitimate commercial interests, to the extent provided under the Party's law;

    (f) shall limit any requirement regarding individual specimens or samples of a good, which is reasonable and necessary;
    (g) should not impose a fee for conducting the procedure that is higher for a good of another Party than is equitable in relation to any such fee it imposes for its like goods or for like goods of any other country, taking into account communication, transportation and other related costs;
    (h) should use criteria for selecting the location of facilities. The procedure conducted shall not cause unnecessary inconvenience to an applicant or its agent;
    (i) shall provide a mechanism to review complaints concerning the operation of the procedure and to take corrective action when a complaint is justified;
    (j) should use criteria for selecting samples of goods that do not cause unnecessary inconvenience to an applicant or its agent; and

(k) shall limit the procedure for a good modified. Resulting good shall fulfill the requirements of the applicable sanitary or phytosanitary measure necessary to fulfill the requirements of that measure.

2. Each Party shall apply—including necessary modifications, paragraphs 1(a) through (i) to its approval procedures.
3. Where an importing Party's sanitary or phytosanitary measure requires the conduct of a control or inspection procedure at the level of production, an exporting Party shall--on the request of the importing Party, take such reasonable measures available to facilitate access in its territory and to provide assistance necessary to facilitate the conduct of the importing Party's control or inspection procedure.
4. A Party maintaining an approval procedure may require its approval for the use of an additive, or its establishment of a tolerance for a contaminant, in a food, beverage or feedstuff, under that procedure prior to granting access to its domestic market for a food, beverage or feedstuff containing that additive or contaminant. Upon request, the party shall consider using a relevant international standard, guideline or recommendation as the basis for granting access until it completes the procedure.[210]

NAFTA has increased trade exponentially between the three signatories, e.g. agricultural trade from Canada to the United States since NAFTA has doubled,[211] while exports from Canada to Mexico "have more than tripled".[212]

Corresponding to the shift to free trade and open borders has been a movement away from national self-sufficiency in food and agricultural production. In this way, the North American food system has become more integrated and harmonized. Much scholarship exists exploring the effects of NAFTA in Mexico vis-à-vis the Mexican food sector and its relationship with its northern neighbours. While out of the scope of this chapter to delve into a detailed analysis of NAFTA's implications for Mexico, the empirical evidence largely suggests that Mexico has benefited the least of the three countries. One rationale often put forth to explain unequal benefits between the countries may be found in Mexico's status as a developing country with unequal bargaining power. Another rationale is that Mexican negotiators failed to closely examine the FTA between Canada and the U.S. that preceded NAFTA. Had they done so, they may have been more inclined to include provisions to exclude sensitive agricultural products.[213] Consequently, "high import quotas without tariffs were accepted for a broad range of products; there is no provision for the possibility of review, suspension, moratorium or the use of other instruments for protecting national production."[214] In addition to these stipulations, Mexico, (unlike its northern neighbours) was forced to put an end to the provision of agricultural subsidies. As Hellman writes:

> In the name of harmonization with its NAFTA partners, Mexico was required to withdraw almost all state support to peasant agriculture, including subsidies, low-interest or interest-

---

[210]Id.
[211]Id.
[212]Id.
[213]Hansen-Kuhn and Hellinger (2003), p. 52.
[214]Id.

free loans, the use of marketing boards to stabilize prices, and low-cost, state-produced, agricultural inputs like fertilizer and insecticides.[215]

In comparison to the situation in Mexico, both the American and Canadian agricultural sectors remain heavily subsidized.

Congruent with this process, many sources indicate that the decimation of Mexican agriculture through what has been referred to as 'the dumping' of cheaper North American imports has resulted in "Mexico losing its food sovereignty."[216] The inability of Mexican farmers to compete has been further compounded by the dismantling of the national food programs established during the more protectionist era of the early twentieth century. This has created an increased dependence on foreign imports as fewer Mexican farmers are now contributing to domestic food production. Findings in a report by the *Canadian Centre for Policy Alternatives* highlight this pattern: "in 1993 before NAFTA, Mexico imported 8.8 million metric tons of grains"[217] however, the think tank estimated that by 2002, Mexico was importing "more than 20 million metric tons or 2.3 times"[218] that amount. To bring this into perspective, with respect to national food sovereignty and food self-sufficiency, NAFTA has been credited with reversing Mexican national food self-sufficiency achieved in 1982.[219] While Mexico has been importing growing volumes of food, "the proportion of Mexicans earning a living from the soil [has been] cut in half."[220] Moreover, the factors cited above have contributed to the fact that Mexico, once self-sufficient in food now imports more than one-third of its food needs.[221] An inability to compete with agricultural imports has contributed to the Mexican countryside becoming unviable. Accordingly, this has prompted a massive immigration from rural Mexico, to the cities and across the border into the United States.

31.3.1 Canada & NAFTA

Canada and the United States have had quite different experiences with the NAFTA. One example is that Mexico now has less food sovereignty than it had prior to signing NAFTA. In contrast, the scholarship points to the fact that despite NAFTA, Canada remains "substantially self-sufficient" in food production.[222] Some scholars on the topic point out that "Mexico is far more food vulnerable

---

[215]Hellman (2008), p. 3.
[216]Hansen-Kuhn and Hellinger (2003), p. 56.
[217]Id.
[218]Id.
[219]Hellman (1988), p. 61.
[220]Id. 4.
[221]Barndt (2002), p. 175.
[222]Otero (2011).

than the other two NAFTA partners"[223] In fact, of the three NAFTA countries, only Mexico experienced significant increases in food prices during "the global food crisis, starting in late 2006".[224] Moreover, fluctuations in food prices during the food crisis of the mid to late 2000s did not have as much as an effect in Canada and the United States as they did in Mexico. The rationale for this can probably be found in Canada's more stable currency and high income status coupled with the fact that:

> [a]ccording to the United Nation's Food and Agriculture Organization (FAO) with Mexico being a "middle income" country, its households spend close to 35 % of their income on food. Still, any price increases have a much more serious impact than they do in either Canada or the United States where households spend on average 11 % to 12 % of their budgets on food.[225]

While there has been comparatively little written on the impacts of NAFTA in Canada, some trends have emerged in the general scholarship. First, as a free trade agreement—the main objective of which was to increase trade among member states—NAFTA has been wildly successful and Canada has in fact experienced an increase in trade exports and bilateral investment with the United States. For example,

> [t]wo-way investment has also increased markedly during the free trade era, both in terms of stock and flow of investment. The United States is the largest single investor in Canada with a stock of FDI into Canada reaching $351.5 billion in 2012, up from a stock of $69.9 billion in 1993.[226]

In addition, other benefits to the Canadian economy can be seen in that:

> Canadian FDI flows into the United States annually averaged $2.3 billion in five years prior to the FTA, and an annual average of $1.8 billion during the FTA years, but increased to an annual average of $9.9 billion from 1995 to 2012.[227]

While the apparent macro-economic gains are impressive and have been often lauded by both economists and politicians, other groups within Canadian society have been more critical of the NAFTA and in an effort to present a balanced analysis. Their views need be articulated. Perhaps no group has been more critical of NAFTA than groups representing agricultural and other food interests. One such group, the *National Farmers Union* (NFA) has in a number of policy briefs synthesized the sentiments of those at the frontlines of Canadian food production who are perhaps best placed to see the empirical consequences of the NAFTA. In a policy brief presented to Olivier De Shutter, United Nations Special Rapporteur on the Right to Food, the NFA cited the following statistics:

- Canadian agriculture exports have more than tripled from $10.9 billion in 1988 to over $35 billion in 2010 due to the "market access" these trade agreements bring

---

[223]Id.
[224]Id.
[225]Id.
[226]Angeles Villarreal and Fergusson (2014).
[227]Id.

- Net farm income has practically not changed: $3.9 billion in 1988 to just over $4 billion in 2010, although if you take 2007 as an example, there were $3.7 billion of taxpayer funded farm support payments masking a net income loss from the markets of $2.2 billion.
- Farm debt has tripled to over $65 billion in 2007 and the number of farmers in Canada has reduced by nearly 25 % from 293,089 farmers in 1988 to 229,373 in 2007.
- the number of farmers under the age of 35 has decreased by over 60 % in the 15 years leading to 2006 from 77,910 in 1991 to just under 30,000 in 2006. This is particularly worrying because if there are fewer young farmers starting to farm, there will be fewer people to carry on Canada's food production within a generation.[228]

In conjunction with these trends affecting farmers and their families, the NFA has also cited the following statistics with respect to general shifts in the Canadian control of food production since NAFTA:

- Before Canada signed on to the first free trade agreement, farmers owned 4 grain handling co-ops. Today there no farmer owned handling co-ops remaining.
- In 1988, 66 % of dairy was processed in farmer owned coops. This has dropped to 39 %.
- Canadian ownership in flour mills has dropped from 50 to 28 %.
- Beef packing plants were virtually 100 % Canadian owned in 1988, but Canadian
- ownership has since reduced to less than 30 %.
- In 1988, Canadians controlled 95 % of the brewing and malting capacity. By 2007 that had reduced to 10 %.
- The number of farm implement companies has reduced from 6 in 1988 to 3.[229]

Such statistics highlight that notwithstanding increased trade and foreign direct investment experienced since the signing of the FTA and NAFTA, the effects from a self-sufficiency perspective (and certainly from the perspective of Canadian agriculturalists) has not been entirely positive.

31.3.2 The Comprehensive Economic and Trade Agreement (CETA)

The discussion of the Comprehensive Economic and Trade Agreement (CETA), an economic and trade agreement between Canada and all 28 member states of the European Union in the food-policy context must like the preceding discussion of NAFTA, be couched within the currently reigning ideological and legal constructs.

---

[228] De Schutter (2011).
[229] Id.

This section argues these ideological, philosophical and legal (governing) parameters are neoliberal. Neoliberalism is at its core a set of economic principles, which advocate the supremacy of the market, less state regulation, deregulation, privatization and individualism. The neoliberal construct of our time is closely associated with the phenomenon of globalization, which has been characterized by increasing degrees of interconnectedness and interdependence between state actors and their citizens.

While Canada has generally been a strong supporter of neoliberalism and the opening up of its borders (and foreign borders) to trade, CETA represents an historic endeavour more ambitious and expansive than even NAFTA. At its core, CETA represents a FTA between Canada and EU—the world's largest market (consisting of over 500 million people[230]).

CETA is a new agreement. The CETA negotiations began in 2009 and since that time, Canada has been under considerable pressure by the EU to reconfigure its copyright and Intellectual Property Regime (IPR) legislation. It is important to note that food represents a very small portion of industry and trade sectors covered under the agreement. Notwithstanding this fact, the agreement once implemented will have broad implications for the food and agricultural industries of every province. While details of the agreement with respect to food only emerged in late 2013, and a leaked drafted surfaced in August 2014[231] the European objectives of the agreement were evident in earlier rounds of debates with respect to IPR (which could potentially impact food and agricultural policy). The EU objective, as Peter Smith has written was to "put pressure on Canada so that they take IPR issues seriously and remedy the many shortcomings of their IPR protection and enforcement regime".[232]

Additionally, surrounding the birth of CETA was contention and outrage over a general lack of transparency and the way in which the general public and Canadian civil society have been virtually excluded from the bargaining table. In this regard, negotiations around CETA have paralleled those of other multi-lateral trade agreements like the North American Free Trade Agreement (NAFTA) and the Anti-Counterfeiting Trade Agreement (ACTA) whose negotiations were shrouded in secrecy and characterized by their lack of civic dialogue, democratic input, parliamentary debate and popularity.

In response to the perceived lack of transparency in negotiation process of the agreements and in the wake of leaked contents of reports—as early as 2011, a coalition of over 70 Canadian and European civil society organizations joined forces in publishing a signed statement denouncing CETA. With respect to food, the group, which included *The Council of Canadians* noted that "[t]his agreement would reinforce intellectual property rights (IPR) at the expense of food

---

[230]Foreign Affairs, Trade and Development Canada. Online: http://www.international.gc.ca/trade-agreements-accords-commerciaux/agr-acc/ceta-aecg/index.aspx?lang=eng.

[231]Dupuis (2014).

[232]Peter (2010).

sovereignty and the right to health".[233] Critics of CETA generally point to it as the most recent manifestation of the neoliberal tradition of other institutions like the World Trade Organization (WTO), (for example) but as a multilateral trade agreement, lying firmly outside the institutional governance bodies that preceded it. For their part, the world's leading capitalist nations have been at the forefront of the push towards CETA (Canada and the EU).

An agreement on core political and policy issues was reached on 18 October 2013. Under the agreement, 94 % of agricultural products will be duty free compared to the current 24 %. European cheese makers will be allowed to sell Canada 29,000 tonnes of cheese, more than double the current quota. Naturally, Canadian farmers, especially, cheese makers are worried. The Canadian government has said it will provide compensation to cheese producers to address any adverse effects. While capital and corporations are infiltrating national borders with increasing ease, many observers claim that state power in the realm of international trade and capital flows has been largely curtailed by an international regulatory system designed to maximize international corporate profits to the detriment of national industries and economies. Herein can be found yet another critique of CETA. This criticism points to the fear that CETA may do more harm than its lauded benefits in that it will take away the power of every layer of the Canadian governance structure to regulate vast sectors of the economy—perhaps most notably the food sector. The publication of this book is especially timely in that CETA signatory states had until September 12th, 2014 to make any amendments and that the ratification process could take several years.

While it may be too early to forecast what the effects of the implementation of CETA will be, this has not stopped speculation on the part of observers. For his part, a recent article published by York University Professor, Rod MacRae in the Journal of *Canadian Food Studies* while acknowledging that relatively little has been published with respect to the agreements details, MacRae manages to highlight what CETA may entail for Canada's food landscape. Among other features, CETA:

- does have a sustainable development and trade section, a trade and environment section,
- and a trade and labour section, and CSOs are to participate in monitoring the provisions when worked out;
- calls for a ban on export subsidies pending tariff reductions;
- proposes collaborating on regulatory measures, including animal welfare, and technical standard equivalency through the Technical Barriers to Trade (TBT) process;
- sets out immediate EU tariff elimination on some foods, mostly processed;
- sets out quota duty free access for certain goods to each other's market;

---

[233]Online: http://www.tradejustice.ca/wpcontent/uploads/2013/08/EUCAN_DECLARATION_EN.pdf.

- has no changes to supply management, except increased cheese and milk proteins access for European producers.²³⁴

While some Canadian agribusiness stands to do well (most notably the beef industry), other sectors are not so optimistic. To provide a recent contextual example of how this aspect of CETA has been received by Canadian farmers—as leaked details of CETA emerged, Quebec and other provincial media outlets published articles highlighting the outpouring of discontent on the part of dairy farmers. Among their discontents are fears that CETA—an agreement seeking to roll back government protection through the elimination of export subsidies and tariff barriers of historically protected industries (the dairy and cheese industry is a prime example)—will have a detrimental effect on the Quebec dairy and cheese industry that sources cite will see cheese imports from Europe double.²³⁵ In the context of an agreement—shrouded in secrecy, it is easy to understand why such concern and fears of uncertainty have arisen among peoples whose livelihoods depend on the dairy industry. Such instances highlight that in any trade regime or new trade agreement, there are bound to be both winners and losers. As CETA has yet to be ratified, it can only be hoped that the benefits outweigh the risks to the Canadian agricultural and food industries. Again, like NAFTA, the leaked aspects of CETA pertaining to regulation of agriculture strongly suggest that domestic regulation of agriculture and agricultural products remains largely intact. The changed landscape is with respect to tariffs and subsidies.

31.3.3 CANSEA: Agri-Foods

In contrast to NAFTA and CETA which seek to create multi-lateral FTA's opening up national borders to increased trade through reduction in trade barriers, Canada-South East Asia (CANSEA), represents an attempt on the part of Canada's Department of Foreign Affairs Trade & Development Canada at actively promoting market expansion of Canadian agricultural products and brands within South East Asia.

To this end, a total of nine South East Asian countries are currently engaged in CANSEA. They are: Brunei, Indonesia, Malaysia, the Philippines, Singapore, Vietnam, Cambodia, Laos, and Burma (Myanmar). In this regard, Canada is not alone in using federal monies and manpower in the promotion of its brands outside of its boarders. Many countries including but not limited to the United States, France, Spain, etc. engage in such promotional activities.

An important component of CANSEA that can benefit Canadian agribusiness is its information gathering capacity. While out the scope of this chapter to delve into

---

[234]MacRae (2014).
[235]Fekete (2013).

every possible example, one recent attempt by the Canadian government at information gathering aimed at facilitating Canadian business expansion into the region can be seen in the example of Brunei. Brunei, a predominantly Muslim country holds countries exporting food products into the country to standards ensuring the products meet religious Islamic *Halal* requirements. To ensure that Canadian food producers are able to expand into this potentially lucrative market, the Canadian government through CANSEA has consulted with the Brunei government and clarified the Halal certification bodies in existence that will allow Canadian food producers to follow the requirements ensuring their food products are certified *Halal* and allowed into the country. To ensure wide dissemination of the information to Canadian Agri-business, the Department of Foreign Affairs & Trade publishes this information on its website[236] and publishes periodicals which are also available on its website.

References

1841 (15 Victoria), c. 45 (Can.); replaced by Act for the Inspection of Four, Indian Meal and Oatmeal 1856, 19-20 Victoria, c. 87 (Can.) and amend by the Act for the Inspection of Flour and Meal, 1859 (22 Victoria) c. 48 (Can.)
1859 (22 Victoria) c. 48 (Can.); in Consolidated Statutes of Canada and Upper Canada, Title 4 Trade and Commerce
1859 (22 Victoria) c. 50 (Can.); in Consolidated Statutes of Canada and Upper Canada, Title 4 Trade and Commerce
1859 (22 Victoria) c. 52 (Can.) in Consolidated Statutes of Canada and Upper Canada, Title 4 Trade and Commerce
1869, 32-33 Vict. c. 37
Accum F (1820) A treatise on adulteration of food and culinary poisons. USA 1(1):103–125
Angeles Villarreal M, Fergusson IF (2014) NAFTA at 20: overview and trade effects, congressional research service, Online: http://fas.org/sgp/crs/row/R42965.pdf. 18 August 2014
Ainley WM, Kumpatla S (2004) Gene silencing in plants: nature's defence. In: Parekh SR (ed) The GMO handbook, genetically modified animals, microbes, and plants in biotechnology. Humana Press Inc, Totowa
Barndt D (2002) Tangled roots: women, work and globalization on the tomato trail. Broadview Press, Peterborough
Benson ML (1996) Agricultural law in Canada 1867-1995. Canadian Institute of Resources Law, Calgary
Buckingham D (2005) Feeling the squeeze! National Food Labelling Legislation in a WTO world: case studies from France, Canada and Ghana. Dissertation, University of Ottawa & Montpellier
C.R.C., c. 284 (Can.)
C.R.C., c. 285 (Can.)
C.R.C., c. 287 (Can.)
C.R.C., c.289 (Can.)
Canada Agricultural Products Act, R.S.C. 1985, c. 20 (4th Supp.), s. 2
Canada Grain Act R.S.C., 1985, c. G-10. ss. 3 – 15 (Can.)

---

[236]Statistics and Market Information. CANSEA Agri-Food Update (2014) Online: http://www.atssea.agr.gc.ca/ase/5028-eng.htm#g.

Canada Grain Regulations, C.R.C. c. 889 s. 5(1)(2), Schedule 3 (Can.)
Canada Grain Regulations, C.R.C., c. 889; *Canada Grain Act*, R.S.C. 1985, c. G - 10
Canada Grain Regulations, C.R.C., c.889 (Can.)
CAPA s.15 (Can.).
CAPA, s. 2 (Can.).
"The Witches Curse" (2002), Director: Linda Caporeal, Public Broadcasting System (PBS), distributor; Series Director mark Lewis; Series Title: Secrets of the Dead; Producer Jenny Barraclough
Civitello L (2008) Cuisine and culture: a history of food and people. Wiley, Hoboken
Conko G (2003) Regulating genetically modified foods: is mandatory labeling the right answer. Rich J Law Tech X:15
Conford P (2001) The origins of the organic movement. Floris Books, Edinburgh
De Schutter O (2011) Canada & the Right to Food internationally: development cooperation, trade and investment. National Farmers Association. Online: http://foodsecurecanada.org/sites/foodsecurecanada.org/files/Canadian_trade_and_the_right_to_food_NFU_Brief.pdf. 24 August 2011
Dairy Regulations, S.O.R. 79 - 840, s. 6(2)(a)(b) (Can.)
Dupuis P (2014) CETA consolidated text. Commission DG Trade Dir. E.1. Online: http://www.tagesschau.de/wirtschaft/ceta-dokument-101.pdf
Echols M (2001) Food as culture. In: Food safety and the WTO: the interplay of culture, science and technology. Kluwer, Hague, pp 13–28
Eichelbaum T, Allan J, Fleming J et al (2001) Report of the New Zealand Royal Commission on Genetic Modification. Ministry for the Environment
Endres BA (2007) An awkward adolescence in the organics industry: coming to terms with big organics and other legal challenges for the industry's next ten years. Drake J Agric Law 12 (1):7–60
Favret EA (1962) Contributions of radio-genetics to plant breeding. Int J Appl Radiat Isot 13:445
Fedoroff N, Brown M (2004) Mendel in the kitchen: a scientist's view of genetically modified foods. Joseph Henry Press, Washington
Federoff N, Brown NM (2005) The story of wheat: ears of plenty, the story of man's staple food. Economist 24
Fedoroff N, Brown NM (2006) Mendel in the kitchen: a scientist's view of genetically modified food. Joseph Henry Press, Washington, DC
Fekete J (2013) Canada's dairy farmers 'angered and disappointed' by EU trade deal that would double cheese imports, The National Post, October 16, 2013. Online: http://news.nationalpost.com/2013/10/16/dairy-farmers-angered-by-reports-canada-close-to-eu-trade-deal-that-would-allow-more-cheese-imports. Last accessed on 16 Aug 2013
Fernandez-Cornejo J (2004) The U.S. seed industry in U.S. agriculture, an exploration of data and information on crop seed markets, regulation, industry structure, and research and development. Economic Research Service (USDA) 2. Washington, DC
Fish Inspection Act, R.S.C. 1985, c. F-12; *Fish Inspection Regulations*, C.R.C. c. 802
Food and Drug Regulations C.R.C., c. 870 Div. 2-28
Food and Drug Regulations, C.R.C. c. 870 (Can.)
Food and Drugs Act R.S.C. 1985 c. F-27 (Can.)
Foreign Affairs, Trade and Development Canada. Online: http://www.international.gc.ca/trade-agreements-accords-commerciaux/agr-acc/ceta-aecg/index.aspx?lang=eng
Friedland MT (2005) You call that organic? The USDA misleading food regulations. N Y Univ Environ Law J 13:379–440
Fruit, Vegetables and Honey Act, S.C. 1935, c. 672 (Can.); Natural Products Marketing Act, 1934, c. 57 (Can.); Canada Agricultural Products Standards Act, 1934, c. 57 (Can.)
Fuller RS, Buckingham D (1999) Agricultural law in Canada. Butterworth, Toronto

Gao Y (2004) Biosafety issues, assessment, and regulation of genetically modified food plans. In: GMO handbook: genetically modified animals, microbes, and plants in biotechnology. Humana Press, Totowa

Gupta M, Ram R (2004) Development of genetically modified agronomic crops

Guthman J (2004) Agrarian dreams, the paradox of organic farming. University of California Press, Berkeley

Guthman J (2003) Fast food/organic food reflexive tastes and the making of yuppie chow. Soc Cult Geogr 4(1):45–58

Grivetti LE (2000) Food prejudices and taboos. In: Kiple KF, Kriemhild Coneè O (eds) The Cambridge world history of food. Cambridge University Press, Cambridge

Halbury's Laws of Canada (2014) Reissue

Halsbury's Laws of Canada, 1st ed., Agriculture, p 125

Hansen-Kuhn K, Hellinger S (2003) Lessons from NAFTA: the high cost of free trade. CCPA, Ottawa

Hellman J (1988) Mexico in crisis. Holmes & Meier Publishers, New York

Hellman J (2008) The world of Mexican migrants: the rock & the hard place. The New Press, New York, p 3

Hutt PB (1984a) Government regulation of the integrity of the food supply. Annu Rev Nutr 4:1–21

Hutt PB (1984b) A history of government regulation of adulteration and misbranding of food. Food Drug Cosm Law J 39:2–73

James C (2005) Global Status of Commercialized Biotech/GM Crops 2005. International Service for the Acquisition of Agri-Biotech Applications (ISAAA), Ithaca, NY. Online: http://www.scientists-for-labour.org.uk. Excerpted from "Scientists for Labour" (UK) Policy Statement: "Science and the Development of Agriculture: GM Crops," 28 January 2003

Janick J (2000) Lecture 5 - Centers of origin of crop plants. Purdue University: Department of Horticulture and Landscape Architecture, West Lafayette, Indiana

Jacobsen E, Schouten HJ (2007) Cisgenics strongly improves introgression breeding and induced translocation breeding of plants. Trends Biotechnol 25(5):219–223.

Juniper B, Mabberley D (2006) The story of the apple. Timber Press, Portland

Kunich JC (2001) Mother Frankenstein, doctor nature and the environmental law of genetic engineering. South CA Law Rev 74:242–243

Labatt Brewing co. v Canada (Attorney General), [1979] S.C.J. No. 134, [1980] 1 S.C.R. 914 (S.C. C.) (Can)

Lewin B (1997) GenesVI. New York

Livestock and Poultry Carcass Grading Regulations, S.O.R. /92 -541 (Can.)

London J (2014) Tragedy, transformation and triumph: comparing the factors and forces that led to the adoption of the 1860 Adulteration Act in England and the 1906 Pure Food and Drug Act in the United States. Food Drug Law J 69(2):315–342, 326

MacRae R (2014) Do trade agreements substantially limit development of local/sustainable food systems in Canada? Can Food Stud :116

Mandel GN (2004) Gaps, inexperience, inconsistencies, and overlaps: crisis in the regulation of genetically modified plants and animals. William Mary Law Rev 45:267

Mandel GN (2006) The future of biotechnology litigation and adjudication. Pace Environ Law Rev 23:83–112

McHughen A (2000) Pandora's picnic basket, the potential and hazards of genetically modified food. Oxford University Press, Oxford

Messer E (2000) Culinary history. In: Kiple K (ed) The Cambridge world history of food. Bowling Green, Ohio

Meat Inspection Regulations, 1990 (S.O.R. 90/288) (Can.)

Meat Inspection Regulations, S.O.R. 90 / 288, Schedule III (Can.)

Miller HI (2004) How extremists are ruining earth day. Scripps Howard News Service 20 April 2004. Citing the U. S. National Research Council

Miller H (2003) The academy chokes on food biotech, public policy suffocates. Rich J Law Technol 10:1

Mitchell L, Kuchler F, Golan E et al (2001) Economics of food labelling. Economic Research Service: U.S. Department of Agriculture, Agriculture Economic Report 793. Washington, DC. Available via DIALOG. http://www.ers.usda.gov/media/532216/aer793.pdf. Accessed 11 Sept 2014

Murphy D (2007) Plant breedinhg and biotechnollogy: societal context and the future of agriculture. Cambridge UP, New York

Oczek J (2000) In the aftermath of the 'terminator' technology controversy: intellectual property protections for genetically engineered seeds and the right to save and replant seed. Boston Coll Law Rev 41:24–25

Organic Products Regulations 2009, SOR/2009-176 (Can.) Canada Organic Regime Quality Management System Manual: online http://www.ota.com/standards/canadian.html

Organic Products Regulations, 2009, S.O.R. /2009-176 (Can.)

Otero G (2011) Neoliberal globalization, NAFTA, and migration: Mexico's loss of food and labor sovereignty. J Poverty 15(4):384–402, 388

Parekh SR (ed) (2004) The GMO handbook, genetically modified animals, microbes, and plants in biotechnology. Humana Press, Totowa

Peter S (2010) The Anti-Counterfeiting Trade Agreement (ACTA) - Enclosing the Internet? Paper presented at the society for socialist studies, capital, connections, control, congress of humanities and social sciences, Concordia University, Montréal Québec 31 May–03 June 2010. http://auspace.athabascau.ca/handle/2149/2736. Last accessed on 29 Aug 2010

Prakash CS (2001) The genetically modified crop debate in the context of agricultural evolution. Am Soc Plant Physiol 126(1):8–15

Prakash CS (2003) Triticale gets the best of both worlds

Prescott-Allen R, Prescott-Allen C (1988) Genes from the wild: using wild genetic resources for food and raw materials. Earthscan, London

Pollen M (2001) Behind the organic-industrial complex. New York Sunday Times Magazine

Processed Egg Regulations C.R.C., c. 290 (Can.)

Processed Products Regulation C.R.C., c. 291 (Can.)

Processed Products Regulation C.R.C., c. 291, s. 26 (Can.)

Proposed changes to the look of the Nutrition Facts Table and the List of Ingredients. Online: http://www.hc-sc.gc.ca/fn-an/alt_formats/pdf/label-etiquet/nutrition-facts-valeur-nutritive-fs-fr-eng.pdf (2014)

Pugsley LI (1967) The administration of federal statutes of food and drugs in Canada. Med Serv J Can 23(3):387–449

Purich D (1982) Canadian farm law -- a guide for today's farmer. Weston Producer Prairie Books, Saskatoon

R.S.C. 1985, C-38 (Can.)

R.S.C. 1985, c. 25 (1st Supp.) (Can.)

R.S.C., 1985, c. 20 (4th Supp.) (Can.)

R.S.C., 1985, c. C-38 (Can.)

R.S.C., 1985, c. F-27 (Can.)

R.S.C., 1985, c. G-10 (Can.)

Rasco E Jr (2008) The unfolding gene revolution, ideology, science and regulation of plant biotechnology. International Service for the Acquisition of Agri-biotech Technologies, Manila

Rastogi T (2005) Government of Canada regulatory approach for plants with novel traits. In: 2nd Annual, Biotechnology, Biosafety and Trade Program. Canadian International Grains Institute, Winnipeg/Vancouver

Recombinant DNA Safety Considerations, Safety Considerations for Industrial, Agricultural and Environmental Applications of Organisms Derived by Recombinant DNA Technologies (OECD, 1986) 42, paragraphs 2 & 3.

Reimer P, Schwartz B (2001) Biotechnology: a Canadian perspective. Asper Rev Int Bus Trade Law 1: paragraph 17
Roberts MR (2007) Genetically modified organisms for agricultural food production: the extent of the art and the state of the science. In: Weirich P (ed) Labeling genetically modified food: the philosophical and legal debate. Oxford University Press, New York
Rozin P (2006) The integration of biological, social, cultural and psychological influences on food choice. In: Shepherd R, Raats M (eds) Impact of psychology of food choice. CABI, Oxfordshire
s. 91(27), Schedule B to the *Canada Act* 1982 (U.K.), 1982, c. 11 (Can.)
S.C. 1874 (Can.)
S.C. 1884 (47 Victoria), c. 34 (Can.)
S.C. 1907 (6-7 Edward VII) c. 27 (Can.); successor repealed 1985
S.C. 1920, c. 27 (Can.); R.S.C. 1927, c 76 (Can.); 1952, c. 123, 1952-53, c. 38, R.S.C. 1970, c. F-27 (Can.), and finally R.S.C. 1985, c. F-27 (Can.)
S.C. 1920, c. 27 (Can.); R.S.C. 1927, c 76 (Can); 1952, c. 123, 1952-53, c. 38, R.S.C. 1970, c. F-27 (Can); and R.S.C. 1985, c. F-27; s. 5(a) (Can.)
S.C. 1997. c. 6 (Can.)
S.O.R. /92- 541 (Can.)
S.O.R. 90 / 288 (Can.)
S.O.R., 79 – 840 (Can.)
Satin M (1999) Food alert! The ultimate sourcebook for food safety. Checkmark Books, New York
Schlachter A (2007) North American Free Trade Agreement (NAFTA) online: http://www.agr.gc.ca/eng/industry-markets-and-trade/agri-food-trade-policy/trade-agreements-in-force/north-american-free-trade-agreement-nafta/?id=1383938167884
Schedule B to the Canada Act 1982 (U.K.), 1982, c. 11 (Can.)
Seeds Regulations C.R.C. c. 1400 s. 107(1) (Can.)
Seeds Regulations C.R.C., c. 1400, s. 108 (Can.)
Seeds Regulations, C.R.C., c. 1400, s. 110(1)(2) (Can.)
Standard Sausage Co. v Lee (1934) 1 D.L.R. 706 (Can. B.C. S.C.)
Statistics and Market Information. CANSEA Agri-Food Update (2014) Online: http://www.atssea.agr.gc.ca/ase/5028-eng.htm#g
Winn LB (1999) Special labeling requirements for genetically engineered food: how sound are the analytical frameworks used by FDA and food producers. Food Drug Law J 34(4):667
Yaren K (2001) Frankenfears: a call for consistency. Asper Rev Int Bus Trade Law 1:150–151
Yarrow S (2001) Environmental assessment of the products of plant biotechnology in Canada. In: Gallaugher P, Wood L (eds) Food of the future: comparing conventional with genetically modified food crops: understanding and managing the risks. Vancouver

Chapter 32
Brazilian Agriculture and Its Sustainability

Luiz Antonio Martinelli, Luciana Della Coletta,
Silvia Rafaela Machado Lins, Silvia Fernanda Mardegan,
and Daniel de Castro Victoria

Abstract Brazil has emerged in this century as a powerhouse, developing a high productivity tropical agriculture, and today is one of the key players in the global food system. However, such increase of Brazilian agriculture was not without costs. One of the most important consequences was the loss of original vegetation and all the ecosystem services linked to this loss. Most of the Atlantic Forest was converted in urban or agricultural areas; approximately half of the Cerrado was also already converted, and more than 15 % of the Amazon forest was also lost. Coupled with loss of vegetation there is also environmental problems linked to agricultural practices such as: burning and heavy use of pesticides, and to a lesser extent of mineral fertilizers. However, the decoupling of agriculture production and deforestation observed in several regions of the country give us hope that in the future agriculture could advance without further vegetation loss. This mean that intensification will take place, and such has to be conducted under the umbrella of what is called "sustainable agriculture", which in turn is a series of practices aimed to give to an agroecosystem more complexity in order to mimic natural ecosystems. Among these practices several of them have already been adopted in large scale in the country, especially no-till, crop rotation, and lately crop-livestock systems. If Brazil succeeds in overcoming this challenge, it will not only benefit itself, but also other tropical countries that are pursuing such sustainability and, ultimately the entire world, given Brazil's importance in the global food system.

L.A. Martinelli (✉) • L.D. Coletta • S.R.M. Lins
CENA, University of São Paulo, Piracicaba, SP, Brazil
e-mail: martinelli@cena.usp.br

S.F. Mardegan
Federal University of Pará, Belém, PA, Brazil

D. de Castro Victoria
EMBRAPA Agriculture Informatics, Campinas, SP, Brazil

32.1 Introduction

During this century, demand for food will increase largely due to population growth and changes in nutritional habits, with many nations relying increasingly on animal protein as a significant fraction of daily food intake. Most of the expanding demand for food in recent decades has been in tropical regions of the world, and the loss of tropical forest is a direct consequence of agricultural expansion.[1] Future predictions are not very different from past one: most of the spread in agriculture will occur in the tropical areas of the globe.[2] Increases in food production can be achieved either by increasing productivity through intensification of inputs, or by expanding available farmland. While global warming raises concerns about such estimates, many researchers believe that increasing productivity remains possible. It seems, however, that intensification alone will not meet future demands for food, and consequently it will be necessary to expand agricultural areas.[3] Therefore, food production in tropical countries will have not only a regional impact, but a global one as well.

One country that is rising to meet the growing global food demand is Brazil, which is steadily becoming a powerhouse in the global food system.[4] Beginning in the 1990s, a series of macroeconomic interventions coupled with land availability, a favorable climate for agriculture, and abundance of inexpensive labor[5] set the stage for unprecedented development in the Brazilian agricultural sector.[6] The impressive figures related to this development can be seen in the high production rate of certain commodities in relation to the total world production.[7] For example, Brazil is responsible for about one-third of global production of coffee, soybean, and sugarcane.[8] Brazil is also responsible for approximately 15 % of beef and 8 % of the world production of maize. Often, tropical countries tend to dominate one specific commodity, as is the case of Cote d'Ivoire, which is the number one cocoa producer in the world. India, on the other hand, is the main producer of bananas, and Myanmar is the number one bean exporter. Yet even among these countries, Brazil still ranks highly in terms of output: currently it is seventh for cocoa production, sixth for banana production, and third in the world for bean production. The openness of Brazil's economy means that it is also a major exporter of soybean, sugar, coffee, beef, and chicken meat.

Based on these numbers, the authors find it is fair to conclude that Brazil plays a key role in the global food production system. Brazil also benefits from abundant

---

[1] See Gibbs et al. (2010).
[2] Ibid.
[3] See Rudel et al. (2009).
[4] See Martinelli (2012).
[5] See Gasques et al. (2004).
[6] See Chaddad and Jank (2006) and Barros (2008).
[7] See Martinelli et al. (2010) and Lapola et al. (2014).
[8] See FAO (2014).

natural resources: according to the 2010 Global Forests Resources Assessment, Brazil is the country with the greatest forest cover in the world, totaling approximately 5.9 million km^2, which corresponds to 60 % of the total country area.[9] Brazil harbors two of the most important forested biomes, the Amazon and the Atlantic Forest, as well as one of the most important tropical savannas of the world, the Cerrado. The Amazon biome is the largest continuous tropical forest and has some of the richest biodiversity on Earth. Moreover, the Amazon River is the largest river on the planet and its basin encompasses near 6.4 million km^2, with more than half of this area in Brazilian territory. The Amazon is home to several endemic and threatened species of plants and animals. For instance, scientists estimate that there are over 40,000 species of plants in the forest, and approximately 3000 species of fish in the rivers.[10] Due to its enormous breadth and vast number of trees, the Amazon region influences the global climate through enhanced evapotranspiration, which decreases air temperatures not only in the region but globally as well.[11] In a global warming scenario, the Amazon also serves as an important carbon reservoir, stocking carbon in plants and soil.[12] The Atlantic Forest another important environmental hotspot, and is the second most important forest cover in South America.[13,14] The forest originally covered 2 million km^2, but today only 15 % of the original biome remains, largely due to centuries of intense occupation and resource exploitation after Portuguese colonization.[15,16]

The Atlantic Forest harbors a great number of species of plants and animals, many of which are native. These species are distributed throughout a wide swathe of territory, from the Northeast to the Southern region of the country and grow under a series of edaphoclimatic conditions.[17] Endemism is high, for instance, among some groups such as plants (40 %), birds (16 %), mammals (27 %) and reptiles (31 %).[18] This diversity also extends to their physiognomies, and spreads across the coastal Atlantic Rain Forest (locally known as Mata Atlântica) and the Atlantic Semideciduous Forest.[19] The Atlantic Forest also harbors the rivers responsible for supplying freshwater to more than half of the Brazilian population,[20] and

---

[9] See Forest Resources Association—FRA (2010).

[10] See da Silva et al. (2005).

[11] See Rocha et al. (2004), (Bonan 2008), and Cox and Jeffery (2010).

[12] See Neill et al. (1996), Ometto et al. (2005), and Houghton et al. (2009).

[13] See Oliveira-Filho and Fontes (2000).

[14] See Myers et al. (2000) and Murray-Smith et al. (2009).

[15] See Ribeiro et al. (2009).

[16] See Morellato and Haddad (2000), Galindo-Leal and Câmara (2003), and Teixeira et al. (2009).

[17] See Murray-Smith et al. (2009).

[18] See Mittermeier et al. (2005).

[19] See Morellato and Haddad (2000) and Scarano (2002).

[20] See Kissinger (2014).

around 70 % of the country's population lives within its bounds, producing 80 % of its gross domestic product.[21]

Brazil, therefore, has a unique position because it is one of the few countries in the world with such mega biodiversity, and it is the first tropical country to develop an efficient export-oriented agriculture.[22] However, Brazil has not yet found a way to accommodate both. Some Brazilian scholars believe that it is impossible to reconcile environmental preservation and agriculture development, and it is difficult for groups on both sides to understand that agriculture is the most precious of all ecosystem services. Additionally, agriculture is fundamental to the Brazilian economy, not only for food supply but also for generating capital flows through commodity exports.[23] On the other hand, as an ecosystem service, agriculture depends upon complex interactions between the biological and physical aspects of an ecosystem. By oversimplifying ecosystems and transforming them from a multi-species to a mono-crop system, the basic ecosystem services provided by a multitude of species will collapse. Ultimately, Brazil's ability to produce food and maintain a sustainable agricultural system will suffer.[24] For instance, Brazil produces one-third of the world's coffee and is one of the main global exporters of coffee beans. Coffee depends upon insects to pollinate its flowers, and these insects depend on a minimally preserved ecosystem to survive. Researchers estimate that destruction of this unique environmental service would involve economic losses worth 12 trillion euros.[25]

So far, Brazilian agriculture has been progressing with few environmental concerns. Consequently, only a minor part of the Atlantic Forest has been fully protected, and it is in fragmented patches with low value in terms of biodiversity preservation and ecosystem services.[26] Half of the Cerrado has already been destroyed to make room for crops and livestock, and the Amazon region has already lost approximately 700,000 km^2 of forests. Recently, however, for the first time in several decades there has been a decline in deforestation rates in major Brazilian biomes.[27] An iconic example is the decrease of deforestation in the Amazon region that began roughly in 2004–2005, with only a small increase in last year's deforestation rate. At the same time, productivity of the main Brazilian crops has continuously increased in recent decades due to technological advances. These trends give us hope that it is possible to reconcile environmental preservation with agriculture development.

One of the main objectives of this chapter is to discuss how such a combination may be achieved. In order to achieve this, the chapter will be divided into four

---

[21] See Ministry of Environment (2013).

[22] See Martinelli et al. (2010).

[23] See Martinelli (2012).

[24] See Swinton et al. (2007) and Power (2010).

[25] See Gallai et al. (2009).

[26] See Ribeiro et al. (2009).

[27] See Lapola et al. (2014).

sections. First, we will provide an overview of Brazilian agriculture's expansion, focusing on the three major crops that cover the greatest land area: soybean, maize, and sugarcane. We will also look at the livestock sector, which also takes up significant land resources. Secondly, we will discuss the intensification of Brazilian agriculture and the environmental "price" of such intensification. Third, we will look at the environmental cost of such expansion and intensification, with emphasis on the Amazon region, the Atlantic Forest, and the Cerrado. Finally, we look to the future and propose potential ways of reconciling environmental preservation and agriculture development.

32.2 Brazilian Biomes

32.2.1 The Amazon

The Amazon region has several different boundaries and areas depending on the classification used. For instance, the Amazon Basin encompasses an area of approximately 6.4 million km^2 including the Tocantins Basin, while the Amazon biome is 7 million km^2, of which roughly 65 % is located in Brazilian territory. There is also the Legal Brazilian Amazon (LBA) that encompasses 5 million km^2, and was demarcated during the Brazilian dictatorship with government investments and tax exemptions in the area.

The main features of the Amazon region include the predominance of tropical rainforest vegetation intermingled with areas of cerrado-like savannas, along with extensive river networks that drain into the main Marañon-Solimões-Amazon River. Marañon is the name of the main river of the Amazon Basin before the river enters Brazilian territory, where its name changes to Solimões. It becomes the Amazon after meeting the Negro River leading up to the Atlantic Ocean. The Amazon River runs from West to East for more than 6600 km, making it one of the longest rivers on Earth.

In terms of biodiversity, the Amazon region has some of the richest on the planet.[28] It seems that the most plausible explanation for such mega-biodiversity is the combination of geological-geomorphologic events coupled with climate changes that created the perfect habitat for species of all kinds to flourish. For instance, tectonic events like the rising of the Andean Cordillera changed the river direction to its current west to east flow, thus shaping the Amazon biota.[29]

The Amazon's mega biodiversity began to be seriously threatened in the 1960s and 1970s. Before then, there had been limited human interventions in the Amazon region. During the Second Word War (WWII), for instance, people from the Northeast region migrated to the Amazon—mainly to the state of Acre—to extract

---

[28]See Hoorn et al. (2010) and Cheng et al. (2013).

[29]See Santos et al. (2009), Hoorn et al. (2010), and Cheng et al. (2013); Sacek (2014).

rubber in order to support the Allied effort. Those extracting the rubber were called "rubber soldiers," and they worked in slave-like conditions for their landlords in the middle of the jungle. After WWII was over, these "soldiers" were largely forgotten. The so-called 'development' of the region truly began during the military dictatorship of 1964–1985. The military generals in power established a system of tax exemptions and subsidies for large entrepreneurs in other parts of the country in order to bolster their support.[30] Their *modus operandi* was to purchase large tracts of land, deforest them, and replace the forest with cultivated pasture for livestock. The actual agricultural output of this enterprise was less important—it was land speculation on which the generals focused. With the opening of new roads, the price of the land increased exponentially and allowed entrepreneurs to sell their properties for enormous profits.

As a consequence of this rapid agricultural expansion, the Amazon region was left with a vast deforested area, an inefficient livestock system, and a booming market for land. Additionally, the military government also tried to alleviate population pressure in the impoverished Northeast region, which was suffering after several years of drought. With the opening of the Trans-Amazonian Highway, a series of "colonization" projects were implemented, re-settling small farmers along this road in the so-called *agrovilas*. Most of the colonies failed, mainly because there was no technology in place to cultivate the very poor soils of the Amazon. Some of the farmers went back to the Northeast region, but many remained in the Amazon region.

At the end of the military dictatorship in 1985, enormous swathes of the Amazon were transformed into pasture. The population increased substantially too, as the opening of roads along with the colonization projects attracted more people to the region. By the time the civilian government removed the tax exemptions and subsidies, the Amazon "gate" had already been opened for development. Even without government incentives, the livestock sector began to profit; this was primarily because the new roads dramatically decreased the cost of transporting goods to the main markets. At the same time, the Brazilian Agricultural Research Corporation (EMBRAPA) developed new varieties of *Brachiaria (Urochloa)* grasses, along with different types of soybean that could thrive in the Amazon region. This expanded the range of crops that farmers could grow and expedited the area's conversion to pastureland.

Currently in the Legal Brazilian Amazon, approximately 700,000 km^2 of original vegetation has been replaced with farmland and urban development (Fig. 32.1a). Since 1974, the Brazilian Institute of Space Research (INPE) has monitored and recorded the annual rates of deforestation (Fig. 32.1b). There have been three trends in different years which are worth noting: first, deforestation peaked in 1994–1995, following the election of the center-left-wing president Fernando Henrique Cardoso. Second, deforestation reached another peak in 2004–2005, right after the election of left-wing president Luiz Inácio Lula da

---

[30]See Hecht et al. (1988).

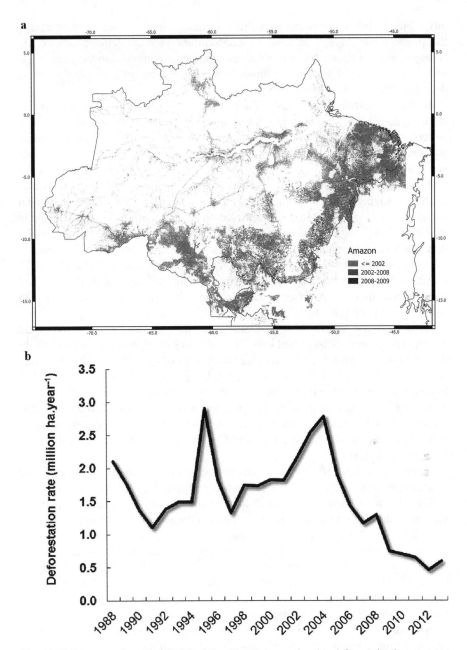

Fig. 32.1 *Upper panel*: map of the Legal Brazilian Amazon showing deforestation in gray tones prior to 2002, between 2002 and 2008, and between 2008 and 2009. *Lower panel*: deforestation rates in the Legal Brazilian Amazon. *Source*: PRODES-INPE

Silva (Fig. 32.1b). These peaks were probably caused by uncertainties surrounding the presidential change, as well as concomitant pent-up demand—especially because these presidents had a more leftist orientation compared to their predecessors. The third trend was in the opposite direction, seen by a steady decline in deforestation since the 2004–2005 peak.

In their research, Nepstad et al. summarize the causes of this decrease in deforestation.[31] They argue that deforestation declined due to a combination of factors, including an increase in governance in the municipalities where the deforestation rates were highest, coupled with a more rigorous system of surveillance and law enforcement. Another reason for the decrease in deforestation was the soybean and beef moratorium, which was an agreement between large companies not to buy soybean or beef from recently deforested areas. Lastly, after the global economic crisis of 2008 many companies could no longer afford to invest in land in the first place.[32] It is crucial to continue monitoring land-use changes in the Amazon region, especially because researchers have observed a small increase in the deforestation rate in 2013. In addition, 2014 was an election year and in 2015 and 2016 Brazil has been facing a profound economical and political crisis, and changes like that in government have historically led to strong variability in deforestation rates (Fig. 32.1b).

Land-use in the Amazon biome has shifted towards livestock grazing, and exotic African forage grasses have replaced most of the original Amazon vegetation. In the 1970s there were approximately three million heads of cattle, which is less than the number of cattle in the Cerrado and Atlantic Forest. In only four decades, however, the number of cattle in the Amazon has increased to more than 50 million, at an average annual increase of 125,000 heads per year. The Cerrado has also shown a rate of increase in recent decades. The total number of cattle in the Amazon is still lower than in the Atlantic Forest or Cerrado, but it is increasing by the same order of magnitude.

In the 1990s, maize was the most important crop in the Amazon biome, covering an area of 0.5 million ha. By 2012, the area of maize under cultivation has doubled to 1.0 million ha. However, since 2002 the Amazon has also seen a significant growth in the soybean crop, which currently occupies an area of approximately 1.6 million ha—larger than the area occupied by maize.

Sugarcane production has increased as well: in the beginning of the 1990s it occupied an incipient area in the Amazon, with just over 40,000 ha. By 2007–2008 it was at 100,000 ha. The government passed a law, however, that prohibited sugarcane to be cultivated on a large-scale in the Amazon biome, and as a result the sugarcane area did not change until 2012. As sugarcane has not traditionally been planted on a large-scale in the Amazon biome, its productivity has remained practically unchanged for the two last decades. During the 1990s, sugarcane productivity started at 36,000 kg ha^{-1} and increased to only 40,000 kg ha^{-1}. This

---

[31]See Nepstad et al. (2014).
[32]Ibid.

figure is well below the average of the Atlantic Forest biome, which is the main sugarcane producer in the country.

In contrast to sugarcane output over the 1990s, the productivity of soybean practically doubled in the same period. In the beginning of the decade its productivity was 1500 kg ha^{-1}, increasing to approximately 3000 kg ha^{-1}. Maize also increased in productivity, although at a lesser rate than the increase observed in soybean. From 1990 to 2012, maize productivity in the Amazon increased from 1000 kg ha^{-1} to 2000 kg ha^{-1}. However, while soybean productivity is not that different from other areas of the country, maize productivity is much lower, especially after taking into account the overall productivity of the Atlantic Forest biome, which totals 10,000 kg ha^{-1}.

32.2.2 The Atlantic Forest

The Atlantic Forest runs parallel to the Brazilian coast from north to south, encompassing an area of approximately 1.5 million km^2 (Fig. 32.2). According to Ribeiro et al.[33] there have thus far been more than 20,000 identified species of plants, and more than 1500 identified species of vertebrates. The Atlantic Forest has such rich biodiversity, largely due to its multitude of different landscapes.[34] It spans the eastern seaboard of the country longitudinally for almost 30°, which gives to this biome a high range of climatic variability. In terms of topography, it encompasses the full range of lowland sea level vegetation all the way up to mountain vegetation types at over 1500 m of altitude.

The Atlantic Forest biome is also the most densely populated biome of Brazil, since most of the human population is concentrated in the coastal areas of the country. The largest cities of Brazil—among them São Paulo and Rio de Janeiro, which have now been deemed mega-cities—are concentrated in this biome. Changes in land-use in this area are not new, however, and actually began in the 1500s with the arrival of Portuguese sailors. Since then, the Atlantic Forest has experienced a series of agricultural commodity cycles, beginning with sugar mills in the northeast region of the country. This was followed by the coffee cycle that expanded toward the southeast of the country up until the economic crisis of 1929. As a consequence of the urban and agriculture expansion that began more than 500 year ago, only 11–16 % of Brazil's original forest remains.[35] More importantly, most of this remaining cover is highly fragmented throughout the landscape. According to Ribeiro et al.,[36] the majority of the fragments (>80 %) are less than 50 ha, and the average distance between fragments is almost 1.5 km. Most of the

---

[33] See Ribeiro et al. (2009).
[34] See Murray-Smith et al. (2009).
[35] See Ribeiro et al. (2009).
[36] Ibid.

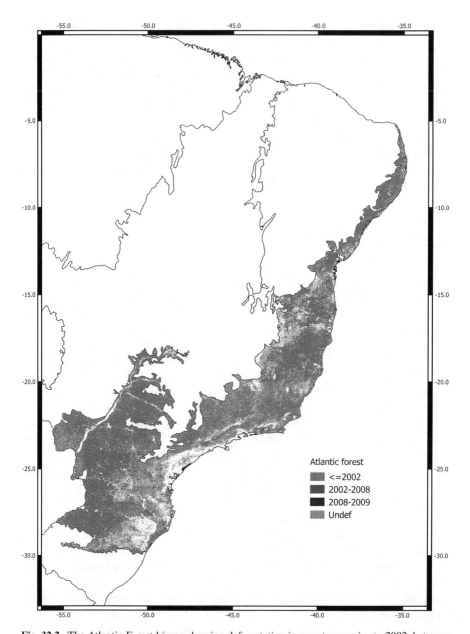

Fig. 32.2 The Atlantic Forest biome showing deforestation in gray tones prior to 2002, between 2002 and 2008, and between 2008 and 2009. Note that only 11–16 % of this biome's forests remain intact

largest preserved areas of the Atlantic Forest are in the scarps of the *Serra do Mar*. It's steep slopes, coupled with its designation in 1982 as a state park, helped to save large parts of the Atlantic Forest in São Paulo. Nevertheless, landscape fragmentation in the Atlantic Forest has a deleterious effect on the fauna and flora of this biome.[37]

The number of cattle in the Atlantic Forest has remained relatively constant over the past few decades. In the 1990s, there were more cattle in the Atlantic Forest than in the Cerrado, with approximately 45 million animals. Between 1994 and 1996, the number of cattle was almost the same in both biomes. Later, however, the number of cattle in the Atlantic Forest plateaued; currently this biome has approximately 50 million heads.

As with the Amazon, the main crop in the Atlantic Forest is soybean. Currently, there are approximately 10 million ha cultivated with this crop, up from roughly 5 million ha in the beginning of the 1990s, showing an average growth rate of 160,000 ha per year. It is important to note the large scale of soybean production in the Atlantic Forest, because it is commonly though that soybeans are grown almost entirely in the Cerrado and Amazon biomes—that is simply not true. However, the data show a significant increase in Atlantic Forest soybean production over the last two decades, especially in the states of Rio Grande do Sul, Paraná, and Mato Grosso do Sul.

The second largest crop in the region is maize, which covers an area of approximately 7 million ha. Most of the maize is planted in south Brazil, in the states of Rio Grande do Sul and Paraná. Unlike soybean cultivation, maize production has been relatively constant since the beginning of the 1990s. In recent years, however, there has been a sharp increase in the cultivated area of maize in the Mato Grosso do Sul state. In the search for renewable fuels there is a high demand for ethanol, which itself is derived from maize. In terms of sugarcane, the planted area in the Atlantic Forest has increased from 3.2 to 6.8 million ha in the last two decades, yielding a similar average growth rate of soybean (around 160,000 ha per year). Most of this growth took place in Brazil's southeast, and in the state of São Paulo especially.

In looking now at productivity rather than planted area, we can see that the productivity of the three major crops in the Atlantic Forest—soybean, maize, and sugarcane—has increased substantially in the last two decades. The sugarcane output, for instance, increased by 1.5 times in this period, while soybean and maize increased over 2.5 times. Soybean productivity is similar to that of other Brazilian biomes such as the Amazon and the Cerrado. The productivity of maize and sugarcane, by contrast, is highest in the Atlantic Forest. It is important to note, however, that the three crops have decreased in productivity in recent years. The decrease in sugarcane output could be explained by the economic crisis of the sugar-ethanol sector: due to the government's maintenance of an artificially low gasoline price, ethanol became a less competitive product, having a negative ripple

---

[37]See Ribeiro et al. (2009) and Tabarelli et al. (1999).

effect on the whole agricultural sector. Even today, sugarcane fields are still suffering from a lack of investment and a depressed market price for the crop.

32.2.3 The Cerrado

The Cerrado is the second largest biome in Brazil after the Amazon. It occupies approximately 2 million km^2, or 20 % of the country's total area. It stretches from south to north in the center of the country; encompassing 11 states in total (see Fig. 32.3). In terms of biodiversity, the Cerrado's is exceptionally high,[38] and many scientists consider it to be the richest tropical savanna of the world.[39] It was estimated that more than 10,000 species of plants are present in the Cerrado, with more than 40 % endemic to this region.[40] One of the unique features about the Cerrado is the widespread practice of slash-and-burn agriculture. There is also a dominant matrix of C_4 grasses intermingled with tree varieties that may have originated from the species-rich surrounding biomes like the Amazon and the Atlantic Forest.[41] Each of these factors helps to create an environment in which a multitude of plant and animal species can thrive.

The soils of the Cerrado are generally poor in nutrients. They also tend to have a low pH, and as a result aluminum toxicity levels are high.[42] Mainly because of these characteristics, the "opening" of the Cerrado as a new agriculture frontier began with livestock in the natural grassland fields. Later on, cultivated pastures, mostly of the genus *Brachiaria (Urochloa)*, replaced the native grasses.[43] In 1974, Brazil's center-west region, which encompasses a large area of Cerrado, was already home to 25 % of the total cattle in Brazil.[44] After the first phase of low-input livestock in the early 1970s the process known as the "Cerrado miracle" began. This "miracle" has been cited as one of the most important achievements of modern tropical agriculture. Today, much of the land in the Cerrado is used for pasture, covering 29 % of its total area. The number of cattle, at 80 million, makes up approximately one-third of the total number of animals in the country.[45] From an agricultural perspective, the Cerrado is highly productive.

---

[38] See Ratter et al. (1997).
[39] See Simon et al. (2009).
[40] See Mendonça et al. (2008).
[41] See Simon et al. (2009).
[42] See Sousa et al. (2001).
[43] See Sano et al. (2010).
[44] See Brazilian Institute of Geography and Statistics (2012).
[45] See Sano et al. (2010).

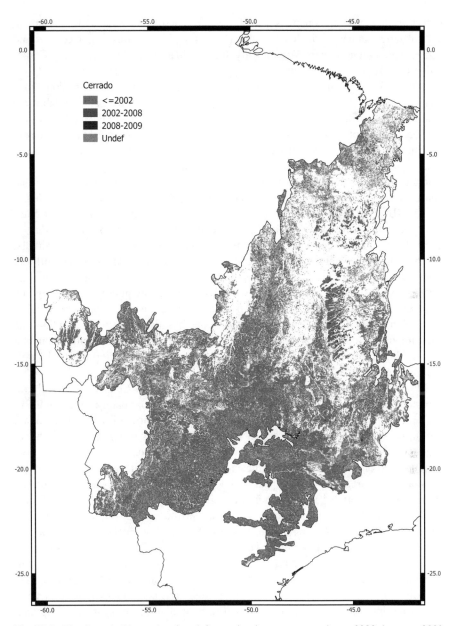

Fig. 32.3 The Cerrado biome showing deforestation in gray tones prior to 2002, between 2002 and 2008, and between 2008 and 2009. Note that almost half of this biome has already been altered by pasture and crop systems

The physical characteristics of the Cerrado soil make it well suited for agriculture. It is deep, well-drained, and the topography is mostly flat which facilitates mechanical harvesting. However, besides the negative soil characteristics cited in the previous paragraph, low calcium content is also observed.[46] These issues were solved by liming soils of agricultural use, which helped to raise the soil pH, increase calcium content, and decrease aluminum toxicity. Additionally, the Brazilian Agricultural Research Corporation (EMBRAPA) has specially designed soybean varieties that are adapted to the Cerrado soil conditions. Of the total area of the Cerrado biome (2 million km^2), approximately 60 % is still preserved (see Fig. 32.3).[47] Nevertheless, although more than half of the Cerrado is protected, it is important to note that in recent decades the conversion rates of the Cerrado have been similar to the rates observed in the Amazon region: the Cerrado has experienced a loss of about 16,000 to 17,000 $km^2 year^{-1}$. As in other Brazilian biomes, conversion rates in the Cerrado have decreased in recent years, reaching approximately 6000 km^2 in 2010.[48] These and other statistics indicate that conservation efforts are having the desired effect of protecting such valuable regions.

The IBGE-SIDRA databank is a useful source of information on land-use changes in the Cerrado.[49] Reviewing the data calls attention to the fact that in 1974 the number of cattle in this biome was approximately 30 million animals; 20 years later, the number had more than doubled to approximately 80 million. Therefore, the Cerrado continues to support a significant proportion of Brazilian cattle. If we assume that in Brazil an average of one hectare goes to supporting one head of cattle, the area occupied by livestock in the Cerrado would be roughly 80 million ha. The area occupied by major crops has also been increasing, and the most important today is soybean. At the beginning of the 1990s it occupied an area of only 4–5 million ha, but 20 years later it had jumped to an impressive 13 million ha—an increase of more than threefold.

Maize and sugarcane are also significant crops in the Cerrado's agricultural economy. Maize occupies the second largest area, at approximately 6 million ha. In the early 1990s, maize production was roughly one-third of what it is today. Similarly, sugarcane plantings began increasing in 2004 *in tandem* with ethanol production. As potential areas for the crop's expansion became increasingly scarce and more expensive in Brazil's southeast, the Cerrado was the next natural area for cultivating sugarcane. Over the past two decades, the area of sugarcane increased from approximately 0.6 million ha in the beginning of the 1990s to almost 2.5 million ha in 2012—an almost fourfold increase in the last 22 years. According to Ribeiro et al., until 2008 sugarcane was mostly replacing pasture areas (65 %), as opposed to other crops and natural vegetation (35 %).[50] Currently, there is no data available on the expansion of sugarcane in the last 5 years, and as such it is difficult to evaluate the effects of the recent crisis on the Brazilian sugar-ethanol sector. It is also difficult to measure the extent to which sugarcane is replacing other crops and expanding into natural vegetation areas. The combination of expanding cropland,

---

[46] See Sousa et al. (2001).

[47] See Sano et al. (2010).

[48] See Lapola et al. (2014).

[49] See Brazilian Institute of Geography and Statistics (2012).

[50] See Ribeiro et al. (2009).

increasing number of livestock, lack of preservation efforts, and the absence of a well-established and routinized deforestation surveillance program all suggest that the expansion of agriculture in the Cerrado is not yet over.[51]

32.3 Environmental Consequences of the Growth of Brazilian Agriculture

32.3.1 Environmental Issues with Agricultural Expansion

One consequence of Brazilian agriculture's rapid growth is the replacement of original vegetation with crops. As seen in the Amazon biome, there have been phases in history when the original vegetation was suppressed for land speculation purposes. The same is true for the Cerrado and especially for the Atlantic Forest, where land speculation for urban purposes has always been an important issue. Recently, however, the agricultural intensification in Brazil has not necessarily gone hand in hand with deforestation.[52] In each of the three major biomes, in fact, there has been a significant decrease in deforestation rates while the output of major commodities has remained unaffected.

Several authors have analyzed the reasons behind this reversal of the previous relationship between agricultural expansion and deforestation.[53] All agree that a series of conditions converged in order to decrease deforestation while allowing for agricultural growth: the creation of more protected areas; the presence of law enforcement in areas where illegal deforestation occurs; credit limits in municipalities with high rates of deforestation; and interventions in soybean and beef chains.[54] These measures, however, rely largely on the participation of corporations, and they could decide in the future that this deforestation moratorium is no longer worth upholding. With this in mind, government policies should rely less on punitive measures and instead place a greater emphasis on incentives that will encourage stakeholders to protect the natural vegetation.[55]

32.3.2 Vegetation Burning

One of the most destructive features of deforestation is the intentional burning of forest. In Brazil, as in most tropical countries, deforestation occurs by logging a few

---

[51]See Sano et al. (2010) and Lapola et al. (2014).
[52]See Lapola et al. (2014).
[53]See Nepstad et al. (2009, 2014), Barretto et al. (2013), and Lapola et al. (2014).
[54]See Nepstad et al. (2014).
[55]Ibid.

trees per hectare for timber, and then burning the rest of the vegetation. In addition to this practice, slash-and-burn farming methods are widely used in Brazil as a management tool to "clean" the pastureland for non-forage species, as well as in sugarcane fields to facilitate harvesting. In the Atlantic Forest biome, for instance, burnings of sugarcane fields are still common, although in some states (such as São Paulo), only 20 % of the sugarcane fields are burned—and these burnings could end by 2021.

The burning of vegetation has several environmental and public health consequences. Several studies link vegetation burning with various human diseases. In the southeast and Amazon regions, for example, burning practices have led to a sharp increase in respiratory problems.[56,57] Furthermore, the effects of burning on human health are not restricted to respiratory diseases: recently, a series of studies has shown a clear correlation between vegetation burnings and other health issues. Arbex et al., for instance, found that slash-and-burn agriculture increased cases of hypertension in regions where farmers practice these methods.[58] Barbosa et al. also reported an increase in cardiovascular disease in the areas that they studied.[59] Similarly, researchers found a higher incidence of cancer in areas with high vegetation burning, likely due to the emission of polycyclic aromatic hydrocarbon (PAHs).[60]

Along with human health, burning vegetation also affects the soil and atmosphere. One short-term benefit of vegetation burning in poor tropical soils is fertilization from the resulting ash. This effect is especially important for the many small family farms in the Amazon region that cannot afford to use mineral fertilizers. Despite this short-term boost, however, most other positive soil properties are reduced by the burnings. For instance, several authors found that sugarcane burning induces soil erosion by increasing soil compaction and decreasing the soil's water content.[61] Burning also diminishes the vegetation's ability to fix carbon into the soil. At the same time, carbon stocks in the soil are crucial to combatting current climate change because they help to reduce carbon emissions from the terrestrial system into the atmosphere. In their research, Galdos et al. showed that when sugarcane is burned during harvesting, the carbon soil stock is lower than when sugarcane is not burned for harvesting.[62] This could be an important advantage for the Brazilian ethanol industry in terms of fuel energy efficiency.[63]

---

[56] See Cançado et al. (2006), Goto et al. (2011), Riguera et al. (2011), Tsao et al. (2011), and Prado et al. (2012).
[57] See Jacobson et al. (2012).
[58] See Arbex et al. (2010).
[59] See Barbosa et al. (2012).
[60] See Magalhães et al. (2007) and Andrade et al. (2010).
[61] See Dourado-Neto et al. (1999), de Oliveira et al. (2000), and Tominaga et al. (2002).
[62] See Galdos et al. (2010).
[63] See Mello et al. (2014).

Effects of vegetation burning on the atmosphere became widely recognized after Crutzen and Andreae's seminal study in 1990.[64] Aerosol particles are of particular concern in areas of intense biomass burning, like in the Amazon biome.[65] These particles form part of the atmosphere's radiative budget over the Amazon,[66] and they affect the concentration of cloud condensation nuclei (CCNs). In turn, the concentration of CCN affects cloud albedo and rain droplet formation.[67] Studies show that aerosols produced in Amazon fires not only affect regional precipitation,[68] but also the precipitation distribution in other areas of the continent (for example, as aerosols from Amazonia shift around in air currents).[69] Just recently, researchers demonstrated that the La Plata Basin precipitation distribution is directly affected by the presence of Amazon aerosols in the atmosphere.[70] This finding is particularly important given that the La Plata Basin serves as a vital breadbasket for the whole South American region.[71]

Vegetation burning is also a serious environmental concern because soil erosion leads to soil degradation, which causes siltation in nearby water bodies and thus negative changes in aquatic systems.[72] Erosion in sugarcane fields in Brazil has been relatively well documented, and it is linked to the fact that most soil losses occur when the land is not covered by vegetation—such as after crops have been burned. Compounding the environmental issues is the fact that Brazil's sugarcane industry follows poor soil conservation practices and does little to prevent soil erosion from occurring in the first place.[73]

32.3.3 Pesticides and Fertilizers

Another environmental issue related to Brazil's agricultural expansion is that of chemical pesticides and fertilizers. With poor land care, soil particles transport pesticide run-off from fields to water bodies. This is of particular concern in a country like Brazil, which is one of the largest consumers of pesticides in the world. In just one decade, Brazil increased pesticide consumption four-fold, going from approximately 20,000 tons to almost 80,000 tons per year between 1990 and 2001. Despite this sharp increase in consumption, there are relatively few studies that look at pesticide contamination of soil and water bodies. This lack of information

---

[64] See Crutzen and Andreae (1990).
[65] See Artaxo et al. (2002) and Andreae et al. (2004).
[66] See Schafer (2002).
[67] See Roberts et al. (2001).
[68] See Martins et al. (2009).
[69] See Freitas et al. (2005).
[70] See Camponogara et al. (2014).
[71] See Martinelli (2012).
[72] See Martinelli and Filoso (2009).
[73] See Hartemink (2008) for a review.

precludes a full evaluation of the environment and human health risks involved in the use of these compounds. Undertaking such an evaluation, however, is vital considering that in 2009 there were 225 and 457 formulations registered for use in sugar cane and soybean cultivation, respectively.[74] More important, approximately 40 % of the formulas used on sugarcane and 50 % of those used on soybeans are considered extremely toxic to human health.[75] Not only that, but half of the formulas used for sugarcane and soybean production are considered highly dangerous to the environment.[76]

The use of fertilizers in agriculture has several effects on long-term environmental sustainability. Inputs of nitrogen and phosphorus impose a strong limitation on plants, as these nutrients dictate how big and productive they will become. This is particularly true for crops that were engineered to be highly productive, since alongside this trait comes their requirement for high levels of fertilizer. On the other hand, a lack of fertilizer—as is the case in some African countries—can lead not only to hunger, but to soil degradation as well. Crops extract nutrients from the soil, and over time they become less productive, decreasing the overall soil cover and leading to erosion that depletes the soil even further. In order to prevent this vicious cycle from occurring, farmers need to replenish soil nutrients—and fertilizers are often the most easily available option.

On the other hand, the excessive use of fertilizers leads to a surplus of nutrients in the soil, and these nutrients—especially nitrogen—become significant contaminants of the soil, atmosphere, and water bodies.[77] In Brazil, fertilizer consumption increased rapidly after the mid-1990s. This trend has continued relatively unchanged since, although from 2008 to 2009 there was a sharp decrease in consumption due to the global economic crisis. After 2009, however, fertilizer consumption continued to rise. Despite the demand for fertilizer in Brazil, its overall use still remains lower than that of the United States (US) and the European Union (EU). One of the metrics used to measure fertilizer consumption is to estimate the amount of N-fertilizer used in an area of arable land, including permanent crops. In Brazil, the N-fertilizer used in 2012 was approximately 50 kg ha^{-1}, which is lower than the 70 and 90 kg ha^{-1} used in the US and EU, respectively. Almost the same amount of phosphorus and potassium is applied per hectare in the country.

Even while Brazil consumes less fertilizer than other countries, it is important to note that no matter the total amount used in agriculture, plants cannot take up all of the fertilizer applied. Under good growing conditions, plants can take up to 40 % of the applied fertilizer, however most of the time this rate is less than 30 %. Therefore, there will always be fertilizer left in the field. Unlike phosphorus and potassium which can be trapped by the acidic tropical soil, nitrogen gets converted to N_2, which is inert, and N_2O, which is a potent greenhouse gas (GHG). While GHG emissions from fertilizer are notable, the Brazilian agricultural sector's main source

---

[74]See Schiesari and Grillitsch (2011).

[75]See Schiesari and Grillitsch (2011).

[76]See Schiesari and Grillitsch (2011).

[77]See Galloway et al. (2008).

of GHG emissions comes from enteric fermentation in cattle, along with the use of manure. Regardless, scientific research is increasingly calling attention the fact that GHG emissions are intimately linked with Brazil's agricultural development and have been increasing parallel to agricultural expansion.

With deforestation rates decreasing across the major biomes, there has been an important change in Brazil's overall GHG emissions. While previously emissions of GHG were primarily tied to land-use changes (deforestation), they now stem primarily from agricultural activities. In the 1990s, for instance, land-use GHG emissions were equivalent to 70 % of the agricultural emissions; by 2010, this proportion had decreased to only 30 %. Nevertheless, GHG emissions from agricultural land use in 2010 were still responsible for almost 60 % of *total* Brazilian GHG emissions.[78] Brazil's natural environment is certainly benefiting from the decrease in deforestation over recent years, and these GHG statistics demonstrate how land-use changes and the agriculture sector are possible contributors to global warming. This is the main reason that the country should adopt any possible measure to decrease its GHG emissions linked to these two sectors.

Deforestation is also relevant to the conversation about biodiversity, since destruction of habitat is responsible for most of the biodiversity loss on the planet, and the resultant loss of ecosystem services.[79] Biodiversity is not only important in terms of the ethical and aesthetic value of the flora and fauna; biodiversity itself helps to create the environmental conditions for a successful agriculture industry. For example, a good proportion of nitrogen input in Brazilian soybeans comes from biological nitrogen fixation (BNF) through a symbiosis process between the soybean plant and a *Rizhobium*-like bacterium.[80] Although genetic manipulation of the plant and bacterium has made them more efficient N-fixers, the process is still a natural one and has been a key component of the soybean expansion over such a large area in Brazil.[81] The pollination of coffee plants is also a valuable ecosystem service that depends on insects, which in turn depend on forests to survive. Therefore, the rich biodiversity of Brazil should be preserved at all costs not only for the sake of biodiversity itself, but also to ensure the functioning of the vital environmental services that make agriculture possible.

32.4 Sustainable Agriculture: A Win-Win Choice

By any metric, Brazil is exceptionally rich in its biodiversity, especially in the biomes of the Amazon, Atlantic Forest and the Cerrado. Likewise, by any metric, Brazil is also outstanding with regards to its development in tropical agriculture

---

[78]See Lapola et al. (2014).
[79]See Dirzo and Raven (2003).
[80]See Alves et al. (2003).
[81]See Alves et al. (2003).

over the last two decades. However, Brazilian agriculture growth is not only based on intensification (i.e., an increase in the productivity of the same land area and in the same location). This growth is also due to *extensification*, or the increase total agricultural area by expansion over natural vegetation. It is time to reconcile both preservation and agricultural growth. Agricultural sustainability in Brazil is still a long way off, and indeed it is much easier to *talk* about sustainability than to actually implement it. Nevertheless, implementing sustainable agriculture in Brazil is imperative, and not only because Brazil is one of the few countries in the world that still has most of its mega biodiversity preserved. It is also the first tropical country to develop a technically advanced agricultural sector. By making sustainable agriculture integral to agricultural expansion, Brazil can ensure the protection of its natural resources while providing nutritious and sufficient food for its growing population.

Sustainable agriculture recognizes the value of the landscape on which the agriculture sector depends, while supporting the preservation of valuable ecosystem services such as BNF, insect pollination, and natural pest control.[82] Sustainable agriculture also acknowledges that agro-ecosystems are simplifications of more complex, natural ecosystems. Therefore, one of the objectives of sustainable agriculture is to transform the simplistic agro-ecosystem into a more complex system in order to increase functional diversity.[83]

A series of specific agricultural techniques can aid in the development of this more complex system. Maybe the simplest one is no-till or minimum till farming, which consists of leaving crop residues from the previous harvest lying in the fields, thereby minimizing the exposure of bare soil. No-till methods build-up soil's organic matter and increase carbon stocks, which helps to mitigate the effects global warming.[84] The increase of soil organic matter content is followed by a general increase in nutrients, because organic matter enhances the soil's exchange capacity. Building up organic matter also protects against rain splash and compaction, preventing surface runoff and ultimately soil erosion.[85] Estimates of the area under no-till in Brazil vary substantially: SIDRA-IBGE reports that there are approximately 18 million ha under no-till,[86] while Boddey et al. estimate this area to be over 25 million ha.[87] In conjunction with no-till, crop rotation is another important technique as it keeps the soil permanently covered and prevents nutrient loss by maintaining a continuous cycling of nutrients in the soil-plant system.[88] Under proper crop rotation methods, it is common to replace a cash crop with a

---

[82] See Keating et al. (2010), Godfray et al. (2010), Power (2010), and Phalan et al. (2011).

[83] See Tilman et al. (1997) and Scherr and McNeely (2008).

[84] See Bayer et al. (2006) and Galdos et al. (2010).

[85] See Diekow et al. (2005), Hungria and Vargas (2000), Garcıa-Préchac et al. (2004), and Bernoux et al. (2006).

[86] See Brazilian Institute of Geography and Statistics (2012).

[87] See Boddey et al. (2010).

[88] See Zanatta et al. (2007) and Vieira et al. (2009).

nitrogen-fixing cover crop once the former is harvested.[89] Part of this extra nitrogen from the BNF is used by the next cash crop. Overall, this practice decreases the use of nitrogen fertilizers, prevents the nitrogen contamination of soil and water, decreases N_2O emissions, and saves farmers money by allowing them to spend less on mineral fertilizers.[90]

Another way to increase complexity in agro-ecosystems is by adding animals that will improve and increase nutrient cycling in soil-plant systems. These systems are not simple to manage; indeed, it is important to choose the appropriate stocking rates of animals to avoid excessive nutrient mining from the soil and ensure that the vegetation biomass of forage crops will be sufficient for mulching the next crop. In return, however, animals promote better economic yields for farmers, improve physical, biological, and chemical soil properties, and boost the productivity of subsequent crops.[91]

In Brazil, these complex agro-ecosystems are particularly developed in the south, where there are several well-established crop-livestock operations. On larger farms, for instance, mechanized soybean is planted in the summer while forage crops for beef cattle are planted in the winter. On smaller properties, maize, rice, beans, and other crops are combined with dairy cattle, sheep, and goats to create a diverse farm system.[92] In the Cerrado and Amazon biomes, these cropping methods were first implemented as a strategy for restoring degraded pastures.[93] More recently, farmers have adopted mixed crop-livestock systems as a long-term management practice for their farms.[94]

32.5 Conclusion

Brazil has still a long way to go towards agricultural sustainability; however, the impressive development of Brazilian agriculture in recent decades, coupled with improved dialogue between the agricultural and environmental sectors, gives us hope for the future. This dialogue needs to progress towards a common goal that will maintain ecosystem services while preserving them for the following generations. The only way to address this challenge is to end deforestation in Brazil, and at the same time make optimal use of the abundant cleared land that is already available. The Brazilian government must adopt regulations to facilitate the construction of more complex agri-scapes that go beyond monocultures. Together with agricultural stakeholders, the government must work towards creating a functional

---

[89] See Hungria and Vargas (2000).
[90] See Rosolem et al. (2004) and Jantalia et al. (2008).
[91] See Carvalho et al. (2009).
[92] See Balbinot-Junior et al. (2009).
[93] See Carvalho et al. (2010b).
[94] See Carvalho et al. (2010a).

landscape that will provide all citizens with an array of ecosystem services—the most important of which include the capacity to produce food, fiber, and fuel for the shared future.

References

Alves BJR, Boddey RM, Urquiaga S (2003) The success of BNF in soybean in Brazil. Plant Soil 252(1):1–9
Andrade TMB, Camargo PB, Silva DML, Piccolo MC et al (2010) Dynamics of dissolved forms of carbon and inorganic nitrogen in small watersheds of the coastal Atlantic Forest in Southeast Brazil. Water Air Soil Pollut 214(1–4):393–408. doi:10.1007/s11270-010-0431-z
Andreae MO, Rosenfeld D, Artaxo P, Costa AA et al (2004) Smoking rain clouds over the Amazon. Science 303(5662):1337–1342. doi:10.1126/science.1092779
Arbex MA, Saldiva PHN, Pereira LAA, Braga ALF (2010) Impact of outdoor biomass air pollution on hypertension hospital admissions. J Epidemiol Commun Health 64(7):573–579. doi:10.1136/jech.2009.094342
Artaxo P, Martins JV, Yamasoe MA, Procópio AS et al (2002) Physical and chemical properties of aerosols in the wet and dry seasons in Rondônia, Amazonia. J Geophys Res 107:1–14. doi:10.1029/2001JD000666
Balbinot-Junior AA, de Moraes A, da Veiga M, Pelissari A et al (2009) Integração lavoura-pecuária: intensificação de uso de áreas agrícolas. Cienc Rural 39(6):1925–1933. doi:10.1590/S0103-84782009005000107
Barbosa CMG, Terra-Filho M, de Albuquerque ALP, Di Giorgi D et al (2012) Burnt sugarcane harvesting – cardiovascular effects on a group of healthy workers, Brazil. PLoS One 7(9):158–170. doi:10.1371/journal.pone.0046142
Barretto AGOP, Berndes G, Sparovek G, Wirsenius S (2013) Agricultural intensification in Brazil and its effects on land-use patterns: an analysis of the 1975–2006 period. Glob Change Biol 19(6):1804–1815. doi:10.1111/gcb.12174
Barros G (2008) Brazil: the challenges in becoming an agricultural superpower. In: Brainard L, Martinez-Diaz L (eds) Brazil as an economic superpower? Understanding Brazil's changing role in the global economy. Brookings Institution Press, Washington, pp 2–35
Bayer C, Martin-Neto L, Mielniczuk J, Pavinato A et al (2006) Carbon sequestration in two Brazilian cerrado soils under no-till. Soil Tillage Res 86(2):237–245
Bernoux M, Cerri CC, Cerri CEP, Siqueira-Neto M et al (2006) Cropping systems, carbon sequestration and erosion in Brazil, a review. Agron Sustain Dev 26:1–8. doi:10.1051/agro
Boddey RM, Jantalia CP, Conceição PC, Zanatta JA et al (2010) Carbon accumulation at depth in ferralsols under zero-till subtropical agriculture. Glob Change Biol 16:784–795
Bonan GB (2008) Forests and climate change: forcings, feedbacks, and the climate benefits of forests. Science 320:1444–1449
Brazilian Institute of Geography and Statistics (IBGE) (2012) Sistema IBGE de Recuperação Automática. http://www.sidra.ibge.gov.br. Accessed 29 Sept 2014
Camponogara G, Silva Dias MAF, Carrió GG (2014) Relationship between Amazon biomass burning aerosols and rainfall over the La Plata Basin. Atmos Chem Phys 14:4397–4407
Cançado JED, Saldiva PHN, Pereira LAA, Lara LBLS et al (2006) The impact of sugar cane–burning emissions on the respiratory system of children and the elderly. Environ Health Perspect 114(5):725–729. doi:10.1289/ehp.8485
Carvalho FMV, de Marco P, Ferreira LG (2009) The Cerrado into-pieces: habitat fragmentation as a function of landscape use in the savannas of Central Brazil. Biol Conserv 142(7):1392–1403. doi:10.1016/j.biocon.2009.01.031

Carvalho JLN, Raucci GS, Cerri CEP, Bernoux M et al (2010a) Impact of pasture, agriculture and crop-livestock systems on soil C stocks in Brazil. Soil Tillage Res 110(1):175–186. doi:10.1016/j.still.2010.07.011

Carvalho PCF, Anghinoni I, de Moraes A, de Souza ED et al (2010b) Managing grazing animals to achieve nutrient cycling and soil improvement in no-till integrated systems. Nutr Cycl Agroecosyst 88(2):259–273. doi:10.1007/s10705-010-9360-x

Chaddad FR, Jank MS (2006) The evolution of agricultural policies and agribusiness development in Brazil. Choices 21(2):85–90

Cheng H, Sinha A, Cruz FW, Wang X, Edwards RL et al (2013) Climate change patterns in Amazonia and biodiversity. Nat Commun 4(1411):1–6. doi:10.1038/ncomms2415

Cox PM, Jeffery HA (2010) Methane radiative forcing controls the allowable CO_2 emissions for climate stabilization. Curr Opin Environ Sustain 2(5–6):404–408. doi:10.1016/j.cosust.2010.09.007

Crutzen PJ, Andreae MO (1990) Biomass burning in the tropics: impact on atmospheric chemistry and biogeochemical cycles. Science 250(4988):1669–1678. doi:10.1126/science.250.4988.1669

da Silva JMC, Rylands AB, Fonseca GAB (2005) The fate of the Amazonian areas of endemism. Conserv Biol 19(3):689–694. doi:10.1111/j.1523-1739.2005.00705.x

de Oliveira JCM, Reichardt K, Bacchi OOS, Timm LC et al (2000) Nitrogen dynamics in a soil-sugar cane system. Sci Agric 57(3):467–472. doi:10.1590/S0103-90162000000300015

Diekow J, Mielniczuk J, Knicker H, Bayer C et al (2005) Soil C and N stocks as affected by cropping systems and nitrogen fertilisation in a Southern Brazil acrisol managed under no-tillage for 17 years. Soil Tillage Res 81(1):87–95. doi:10.1016/j.still.2004.05.003

Dirzo R, Raven PH (2003) Global state of biodiversity and loss. Annu Rev Environ Resour 28 (1):137–167. doi:10.1146/annurev.energy.28.050302.105532

Dourado-Neto D, Timm LC, de Oliveira JCM, Reichardt K et al (1999) State-space approach for the analysis of soil water content and temperature in a sugarcane crop. Sci Agric 56 (4):1215–1221

Food and Agriculture Organization (FAO) (2014) http://faostat.fao.org/site/291/default.aspx. Accessed 9 Sept 2014

Freitas SR, Longo KM, Silva Dias MAF, Silva Dias PL et al (2005) Monitoring the transport of biomass burning emissions in South America. Environ Fluid Mech 5:135–167

Galdos MV, Cerri CC, Lal R, Bernoux M et al (2010) Net greenhouse gas fluxes in Brazilian ethanol production systems. Glob Change Biol Bioenergy 2(1):37–44. doi:10.1111/j.1757-1707.2010.01037.x

Galindo-Leal C, Câmara IG (2003) The Atlantic Forest of South America: biodiversity status, threats and outlook. Island Press, Washington

Gallai N, Salles J-M, Settele J, Vaissière BE (2009) Economic valuation of the vulnerability of world agriculture confronted with pollinator decline. Ecol Econ 68(3):810–821

Galloway JN, Cai Z, Townsend AR, Freney JR et al (2008) Transformation of the nitrogen cycle: recent trends, questions, and potential solutions. Science 320(5878):889–892

Garcıa-Préchac F, Ernst O, Siri-Prieto G, Terra JA (2004) Integrating no-till into crop–pasture rotations in Uruguay. Soil Tillage Res 77(1):1–13. doi:10.1016/j.still.2003.12.002

Gasques JG, Bastos ET, Bacchi MPR, Conceição JCPR (2004) Produtividade e fontes de crescimento da agricultura brasileira. Revista de Política Agrícola 3:73–90

Gibbs HK, Ruesch AS, Achard F, Clayton MK et al (2010) Tropical forests were the primary sources of new agricultural land in the 1980s and 1990s. Proc Natl Acad Sci U S A 107 (38):16732–16737

Global Forests Resources Assessment (2010) Main report. Food and Agriculture Organization of the United Nations, Rome

Godfray HCJ, Crute IR, Haddad L, Lawrence D et al (2010) The future of the global food system. Philos Trans R Soc B 365(1554):2769–2777. doi:10.1098/rstb.2010.0180

Goto DM, Lança M, Obuti CA, Barbosa CMG et al (2011) Effects of biomass burning on nasal mucociliary clearance and mucus properties after sugarcane harvesting. Environ Res 111(5):664–669

Hartemink AE (2008) Sugarcane for bioethanol: soil and environmental issues. Adv Agron 99:125–182. doi:10.1016/S0065-2113(08)00403-3

Hecht SB, Norgaard RB, Possio C (1988) The economics of cattle ranching in Eastern Amazonia. Interciencia 13(5):233–240

Hoorn C, Wesselingh FP, ter Steege H, Bermudez MA et al (2010) Amazonia through time: Andean uplift, climate change, landscape evolution, and biodiversity. Science 330:927–931. doi:10.1126/science.1194585

Houghton RA, Hall F, Goetz SJ (2009) Importance of biomass in the global carbon cycle. J Geophys Res 114(G00E03):1–13. doi:10.1029/2009JG000935

Hungria M, Vargas MAT (2000) Environmental factors affecting N_2 fixation in grain legumes in the tropics, with an emphasis on Brazil. Field Crop Res 65(2–3):151–164

Jacobson LSV, Hacon SS, de Castro HA, Ignotti E et al (2012) Association between fine particulate matter and the peak expiratory flow of schoolchildren in the Brazilian subequatorial Amazon: a panel study. Environ Res 117:27–35

Jantalia CP, dos Santos HP, Urquiaga S, Boddey RM et al (2008) Fluxes of nitrous oxide from soil under different crop rotations and tillage systems in the south of Brazil. Nutr Cycl Agroecosyst 82(2):161–173

Keating BA, Carberry PS, Bindraban PS, Asseng S et al (2010) Eco-efficient agriculture: concepts, challenges, and opportunities. Crop Sci 50(Supplement 1):S–109–S–119. doi:10.2135/cropsci2009.10.0594

Kissinger G (2014) Case study: Atlantic Forest, Brazil. In: Shames S (ed) Financing strategies for integrated landscape investment. Eco Agriculture partners, on behalf of the landscapes for people, food and nature initiative, Washington

Lapola DM, Martinelli LA, Peres CA, Ometto JPHB et al (2014) Pervasive transition of the Brazilian land-use system. Nat Clim Change 4(1):27–35. doi:10.1038/nclimate2056

Magalhães DR, Bruns E, Vasconcellos PC (2007) Polycyclic aromatic hydrocarbons as sugarcane burning tracers: a statistical approach. Quim Nova 30(3):577–581

Martinelli LA (2012) Ecosystem services and agricultural production in Latin America and Caribbean. Inter-American Development Bank, Washington

Martinelli LA, Filoso S (2009) Balance between food production, biodiversity and ecosystem services in Brazil: a challenge and an opportunity. Biota Neotropica 9:21–25. doi:10.1590/S1676-06032009000400001

Martinelli LA, Naylor R, Vitousek PM, Moutinho P (2010) Agriculture in Brazil: impacts, costs, and opportunities for a sustainable future. Curr Opin Environ Sustain 2:431–438

Martins JA, Silva Dias MAF, Gonçalves FLT (2009) Impact of biomass burning aerosols on precipitation in the Amazon: a modeling case study. J Geophys Res 114:1–19

Mello FFC, Cerri CEP, Davies CA, Holbrook NM et al (2014) Payback time for soil carbon and sugar-cane ethanol. Nat Clim Change 4(7):605–609. doi:10.1038/nclimate2239

Mendonça RC, Felfili JM, Walter BMT, Silva Júnior MC et al (2008) Vascular flora of the Cerrado biome: checklist with 12,356 species. In: Sano SM, Almeida SP, Ribeiro JF (eds) Cerrado: Ecology and Flora. Embrapa Cerrados/Embrapa Informação Tecnológica, Brasília, pp 421–1279

Ministry of Environment (MMA) (2013) Área da Mata Atlântica é habitada por 70 % da população brasileira. http://www.mma.gov.br/informma/item/9818-%C3%A1rea-da-mata-atl%C3%A2ntica-%C3%A9-habitada-por-70-da-popula%C3%A7%C3%A3o-brasileira. Accessed 01 Oct 2014

Mittermeier RA, Gil PR, Hoffmann M, Pilgrim J et al (2005) Hotspots revisited: Earth's biologically richest and most endangered terrestrial ecoregions. Cemex, Washington

Morellato LPC, Haddad CFB (2000) Introduction: the Brazilian Atlantic Forest introduction. Biotropica 32(4):786–792

Murray-Smith C, Brummitt NA, Oliveira-Filho AT, Bachman S et al (2009) Plant diversity hotspots in the Atlantic coastal forests of Brazil. Conserv Biol 23(1):151–163. doi:10.1111/j. 1523-1739.2008.01075.x

Myers N, Mittermeier RA, Mittermeier CG, Fonseca GAB et al (2000) Biodiversity hotspots for conservation priorities. Nature 403:853–858. doi:10.1038/35002501

Neill C, Fry B, Melillo JM, Steudler PA et al (1996) Forest- and pasture-derived carbon contributions to carbon stocks and microbial respiration of tropical pasture soils. Oecologia 107:113–119

Nepstad D, Soares-Filho BS, Merry F, Lima A et al (2009) The end of deforestation in the Brazilian Amazon. Science 326:1350–1351

Nepstad D, McGrath D, Stickler C, Alencar A et al (2014) Slowing Amazon deforestation through public policy and interventions in beef and soy supply chains. Science 344:1118–1123. doi:10.1126/science.1248525

Oliveira-Filho AT, Fontes MAL (2000) Patterns of floristic differentiation among Atlantic Forests in southeastern Brazil and the influence of climate. Biotropica 32:793–810

Ometto JPHB, Nobre AD, Rocha HR, Artaxo P et al (2005) Amazonia and the modern carbon cycle: lessons learned. Oecologia 143:483–500. doi:10.1007/s00442-005-0034-3

Phalan B, Balmford A, Green RE, Scharlemann JPW (2011) Minimizing the harm to biodiversity of producing more food globally. Food Policy 36:S62–S71. doi:10.1016/j.foodpol.2010.11.008

Power AG (2010) Ecosystem services and agriculture: tradeoffs and synergies. Philos Trans R Soc Lond [Biol] 365(1554):2959–2971. doi:10.1098/rstb.2010.0143

Prado GF, Zanetta DMT, Arbex MA, Braga AL et al (2012) Burnt sugarcane harvesting: particulate matter exposure and the effects on lung function, oxidative stress, and urinary 1-hydroxypyrene. Sci Total Environ 437:200–208

Ratter JA, Ribeiro JF, Bridgewater S (1997) The Brazilian Cerrado vegetation and threats to its biodiversity. Ann Bot 80:223–230

Ribeiro MC, Metzger JP, Martensen AC, Ponzoni FJ et al (2009) The Brazilian Atlantic Forest: how much is left, and how is the remaining forest distributed? Implications for conservation. Biol Conserv 142(6):1141–1153. doi:10.1016/j.biocon.2009.02.021

Riguera D, Andre PA, Zanetta DMT (2011) Sugar cane burning pollution and respiratory symptoms in schoolchildren in Monte Aprazivel, Southeastern Brazil. Rev Saude Publica 45(5):878–886

Roberts GC, Andreae MO, Zhou J, Artaxo P (2001) Cloud condensation nuclei in the Amazon Basin: 'marine' conditions over a continent? Geophys Res Lett 28(14):2807–2810. doi:10.1029/2000GL012585

Rocha HR, Goulden ML, Miller SD, Menton MC et al (2004) Seasonality of water and heat fluxes over a tropical forest in eastern Amazonia. Ecol Appl 14:22–32

Rosolem CA, Pace L, Crusciol CAC (2004) Nitrogen management in maize cover crop rotations. Plant Soil 264:261–271

Rudel TK, Schneider L, Uriarte M, Turner BL II et al (2009) Agricultural intensification and changes in cultivated areas, 1970–2005. Proc Natl Acad Sci U S A 106(49):20675–20680

Sacek V (2014) Drainage reversal of the Amazon River due to the coupling of surface and lithospheric processes. Earth Planet Sci Lett 401:301–312. doi:10.1016/j.epsl.2014.06.022

Sano EE, Rosa R, Brito JLS, Ferreira LG (2010) Land cover mapping of the tropical Savanna region in Brazil. Environ Monit Assess 166(1–4):113–131. doi:10.1007/s10661-009-0988-4

Santos JC, Coloma LA, Summers K, Caldwell JP et al (2009) Amazonian amphibian diversity is primarily derived from late Miocene Andean lineages. PLoS Biol 7(3):448–461. doi:10.1371/journal.pbio.1000056

Scarano FR (2002) Structure, function and floristic relationships of plant communities in stressful habitats marginal to the Brazilian Atlantic rainforest. Ann Bot 90(4):517–531. doi:10.1093/aob/mcf189

Schafer JS (2002) Observed reductions of total solar irradiance by biomass-burning aerosols in the Brazilian Amazon and Zambian Savanna. Geophys Res Lett 29(17):2–5. doi:10.1029/2001GL014309

Scherr SJ, McNeely JA (2008) Biodiversity conservation and agricultural sustainability: towards a new paradigm of 'ecoagriculture' landscapes. Philos Trans R Soc B 363(1491):477–494. doi:10.1098/rstb.2007.2165

Schiesari L, Grillitsch B (2011) Pesticides meet megadiversity in the expansion of biofuel crops. Front Ecol Environ 9(4):215–221. doi:10.1890/090139

Simon MF, Grether R, Queiroz LP, Skema C et al (2009) Recent assembly of the Cerrado, a neotropical plant diversity hotspot, by in situ evolution of adaptations to fire. Proc Natl Acad Sci U S A 106(48):20359–20364. doi:10.1073/pnas.0903410106

Sousa DMG, Vilela L, Lobato E, Soares WV (2001) Uso de gesso, calcário e adubos para pastagens no cerrado. EMBRAPA Cerrados, Circular Técnica, Planaltina

Swinton SM, Lupi F, Robertson GP, Hamilton SK (2007) Ecosystem services and agriculture: cultivating agricultural ecosystems for diverse benefits. Ecol Econ 64(2):245–252. doi:10.1016/j.ecolecon.2007.09.020

Tabarelli M, Mantovani W, Peres CA (1999) Effects of habitat fragmentation on plant guild structure in the Montane Atlantic Forest of Southeastern Brazil. Biol Conserv 91:119–127

Teixeira AMG, Soares-Filho BS, Freitas SR, Metzger JP (2009) Modeling landscape dynamics in an Atlantic Rainforest region: implications for conservation. Forest Ecol Manag 257(4):1219–1230. doi:10.1016/j.foreco.2008.10.011

Tilman D, Knops J, Wedin D, Reich P et al (1997) The Influence of functional diversity and composition on ecosystem processes. Science 277(5330):1300–1302. doi:10.1126/science.277.5330.1300

Tominaga TT, Cássaro FAM, Bacchi OOS, Reichardt K et al (2002) Variability of soil water content and bulk density in a sugarcane field. Aust J Soil Res 40(4):604–614

Tsao C-C, Campbell JE, Mena-Carrasco M, Spak SN et al (2011) Increased estimates of air-pollution emissions from Brazilian sugar-cane ethanol. Nat Clim Change 2(1):53–57. doi:10.1038/nclimate1325

Vieira FCB, Bayer C, Zanatta JA, Mielniczuk J et al (2009) Building up organic matter in a subtropical Paleudult under legume cover-crop-based rotations. Soil Sci Soc Am J 73(5):1699–1706. doi:10.2136/sssaj2008.0241

Zanatta JA, Bayer C, Dieckow J, Vieira FCB et al (2007) Soil organic carbon accumulation and carbon costs related to tillage, cropping systems and nitrogen fertilization in a subtropical Acrisol. Soil Tillage Res 94(2):510–519. doi:10.1016/j.still.2006.10.003

Chapter 33
Food Law in Mexico: Regulatory Framework and Public Policy Strategies to Address the Obesity Crisis in Latin America

Rebeca López-García

Abstract Mexico has a structured, institution-based food legislation system. The development of Food Law in Mexico has been an evolving process that has been molded and adapted by different challenges such as the opening of the Mexican market to numerous free trade agreements; the development of new technologies in agriculture and food processing; and the very complex health issues as related to food and nutrition, mainly matters related to the growing overweight and obesity problem and its impact on chronic disease. The system is founded on the Political Constitution of the United Mexican States where it is specifically affirmed that the protection of health is guaranteed. In 2011, the right to nutritious, sufficient and quality food was elevated to Constitutional right. Thus, all laws, regulations and norms associated with food including those developed to address the obesity crisis are based on Constitutional rights. Mexico, ranks among the highest in the list of most obese countries with alarming predicted obesity rates that will surmount the country's ability to face the expense associated with treating the diseases associated with the overweight and obese population. Thus, there has been a need to develop public policies that specifically address this crisis to try to slow down, or even better, reverse the trends. Mexico has adopted different public policies that follow the intervention strategies recommended by the Organization for Economic Cooperation and Development (OECD) that include: taxation of foods with high caloric density and sugary beverages, front of panel nutritional labeling, establishment of nutritional criteria for foods offered in schools and restricting advertisement of foods and beverages targeted to children. Other countries in Latin America are facing a similar crisis. Thus, different strategies implemented in Peru, Chile and Ecuador will also be discussed. The approaches taken in each country provide excellent examples of public policy and regulations that are used as an intervention of a system in order to solve a social problem. The only real measure of success will be evident if there is a positive impact in public health in the long term. Obviously, since this is a multifactorial problem, it will require multiple approaches. All strategies must have a solid scientific basis, have a proper regulatory basis for

R. López-García (✉)
Logre International Food Science Consulting, México, DF, Mexico
e-mail: rebecalg@prodigy.net.mx

implementation, be relevant to the population and provide useful and understandable information to ultimately have the desired positive impact.

33.1 Introduction

The development of food law in Mexico has been an evolving process that continues to address different challenges. These include the opening of the Mexican market to numerous free trade agreements; the development of new technologies in agriculture and food processing; and the very complex health issues related to food and nutrition, primarily obesity and its associated impact on chronic disease. The Mexican food industry has experienced a profound transformation since the late 1980s when Mexico joined the General Agreement on Tariffs and Trade (GATT, now the WTO) and more prominently since 1994–1995 when the North American Free Trade Agreement (NAFTA) finally opened Mexico's borders to free trade. Since Mexico used to be such a protected market, these events had a significant impact in many areas; food production and processing were not untouched. The food law and regulations arena rapidly evolved as laws, regulations and norms had to be updated or developed to adapt to the open market and the challenges presented by the new conditions in addition to harmonizing them with international regulations. In some cases, this has been challenging since Mexican cuisine is so varied and traditional products are so unique that some food categories that are important in Mexico do not match the *Codex alimentarius* food categories. Thus, harmonization of issues such as the approval of food additives has been difficult.

In Mexico, there has been a continuous evolution and improvement of the regulatory processes. During the 1980s, the regulation of several activities and the productive sector was excessive and in many cases, the implementation was not possible due to the prevalent economic conditions in the country. As implied above, joining the GATT led to an opening process that inevitably resulted in changes to the regulatory process and the development of public policies. In 1989, the federal government started a restructuring process in several areas of the economy to prepare the country for the transition from a low activity protected economy to international commerce. In 1995, a new program to improve regulation and simplify the regulatory process to promote economic activity was established. Finally, National Commission for Regulatory Improvement (COFEMER, for its acronym in Spanish) was created in 2000 and its activities have impacted all regulatory processes including developments that impact the food industry. According to the COFEMER, the regulatory improvement is part of a public policy whose objective is to promote transparency and the development and application of regulations that generate benefits that are greater than the costs and are advantageous to society.[1]

---

[1]Comisión Federal de Mejora Regulatoria (COFEMER), http://www.cofemer.gob.mx/contenido.aspx?contenido=86 consulted 4 Oct 2014.

33.1.1 Overview of Mexican Food Law, Regulation and Context of Legal Structure

Mexico has a structured, institution-based food legislation system. The Political Constitution of the United Mexican States is the fundamental law of the Mexican State in which Article 4 specifically guarantees the protection of health.[2] According to Mexico's National Standards Office (*Dirección General de Normas*—DGN), a division of the Secretariat of Economy *(Secretaría de Economía)*, formerly known as the Department of Commerce and Industrial Development (SECOFI), there are four basic principles in the legislative process: representation, consensus, modification and updating. According to these principles, the regulatory process is multidisciplinary and standards deriving from it are "living documents" that can be modified as new needs and information arises. In many cases, a particular product or service may not be covered by a specific norm even if it is covered by horizontal norms. If leaving this product or service unregulated has an impact on consumer health, the environment or the country's natural resources, then a mandatory standard or norm is proposed and goes through the whole approval process. All regulatory actions, regardless of the agency of origin, are published in the official gazette and a period for public comment is open before the law is finalized.

The law provides for two distinct types of standards. The first one is based on official, mandatory rules and is called Official Mexican Norms *(Normas Oficiales Mexicanas—NOMs)*. The second type is a set of voluntary references to determine the quality of goods and services. Compliance with NOMs is mandatory, and once a NOM has been published, it applies to all pertinent products in the market, including imports. NOMs are directly related to safety, public health, protection of natural resources or consumer protection. Thus, quality issues are excluded unless these have a direct effect on one or more of the issues aforementioned. NOMs can be of either a horizontal or vertical nature. An example of a horizontal standard is NOM-120-SSA1-1994, Goods and Services regarding hygienic and sanitary practices for food and beverage processing, which was published on August 28, 1995.[3] This standard is the equivalent to the Good Manufacturing Practices in the United States (21CFR110). An example of a vertical standard is NOM-035-SSA1-1993, Goods and Services, which was published on January 30, 1995 and provides sanitary specifications for whey cheeses.[4] This standard is very specific and deals with the sanitary specifications and standards for a specific type of cheese.

Article 39 of the Organic Law of the Federal Public Administration establishes the responsibilities of the Secretariat of Health *(Secretaría de Salud)* which include acting as the sanitary authority and exercising general health faculties as legally

---

[2]Constitución Política de los Estados Unidos Mexicanos (1917), Article 4.
[3]Secretaría de Salud (1995b), NOM-120-SSA1-1994.
[4]Secretaría de Salud (1995a), NOM-035-SSA1-1993.

conferred by the Federal Executive, as well as overseeing the implementation of the General Health Law *(Ley General de Salud)*, also known as, Article 17, its regulations and other applicable dispositions.[5] Mexico's food laws are part of and determined by its overall regulatory process. It should be noted that there is no special track for food laws by themselves. Most laws pertaining to food quality and composition regulation are based on the Federal Law of Metrology and Standardization *(Ley Federal Sobre Metrología y Normalización)*,[6] which was originally published in the official government gazette *(El Diario Oficial)* on July 1, 1992; the General Health Law originally published on Feb. 7, 1984[7]; and the Regulation on Sanitary Control of Products and Services *(Reglamento de Control Sanitario de Productos y Servicios)*.[8] The General Health Law provides the legal foundations for ensuring the right to health for all Mexicans.

The General Health Law (see footnote 3) establishes the duties of the Secretariat of Health, which are exercised through the Federal Commission for the Control of Sanitary Risks (COFEPRIS). These faculties concern the sanitary regulation and the control and promotion of: health establishments, disposal of organs, tissues, human cells and their components, disposal of blood, drugs, herbal remedies and other health related goods and services, food and dietary supplements, alcoholic and non-alcoholic beverages, cosmetics, products for beauty and grooming, tobacco, pesticides and fertilizers, vegetable nutrients, toxic substances or substances that represent a risk to health, essential chemicals, chemical precursors, narcotics and psychotropic drugs, biotechnological products, raw materials and additives as well as establishments which process or store these, sources of ionizing radiation for medical use, basic sanitation, imports and exports, publicity and promotion of activities, products and services as referred to by the law and other applicable dispositions, international sanitation or in general, the requirements for sanitary conditions that must be met by processes, products, methods, installations, services or activities related to the above-mentioned materials. Hence, COFEPRIS is responsible for identifying, analyzing, evaluating, regulating, promoting and disseminating the conditions and requirements for the prevention and handling of sanitary risks. COFEPRIS also issues official certificates for the sanitary condition of processes, products, methods, installations, services or activities related to materials as well as issuing, postponing or revoking sanitary authorizations under its jurisdiction. COFEPRIS exercises authority concerning sanitary regulation, control and promotion as established in or derived from the law and its NOMS, the current regulation and any other applicable dispositions. This agency is also responsible for evaluating and ensuring sanitary risks in conjunction or cooperation

---

[5]El Congreso General de los Estados Unidos Mexicanos (2013c), Ley Orgánica de la Administración Pública Federal.

[6]El Congreso General de los Estados Unidos Mexicanos (2014a), Ley Federal Sobre Metrología y Normalización.

[7]El Congreso General de los Estados Unidos Mexicanos (2014b), Ley General de Salud.

[8]COFEPRIS (2014), Reglamento de Control Sanitario de Productos y Servicios.

with other competent authorities, imposing administrative sanctions for non-compliance with the dispositions of the law, its regulations and other applicable ordinances. Furthermore, COFEPRIS determines the safety, preventive and corrective measures within the scope of its competencies, exercises corresponding acts of sanitary control, regulation and promotion in order to prevent and reduce the sanitary risks derived from the population's exposure to chemical, physical and biological factors.

Therefore, according to the General Health Law, the sanitary control of the processing, importation and exportation of all foods and beverages is the responsibility of the Secretariat of Health *(Secretaría de Salud)*. Additionally, DGN coordinates the regulatory process. Other federal agencies may promulgate regulations within their jurisdictions but they must work through the Secretariat of Economy. Some of the agencies involved in promulgating standards that affect agricultural and food products include the Secretariat of Agriculture, Livestock and Rural Development (SAGARPA); Secretariat of Natural Resources and Environment (SEMARNAP); and Secretariat of Health (SS).

One challenge in understanding the system is that most documents are "working documents" and the timing for updating may not be perfectly coordinated. For example, in the case of food and food products, information in each published document may not be consistent. Thus, understanding the hierarchy of the system is quite helpful to determine the most applicable disposition. For any issue, one must consult the appropriate framework in descending order of hierarchy: Constitution, law, regulation, norm, and agreement.

33.1.2 Food Labeling and Nutritional Claims

Food labeling has been a hot topic in Mexico for several years since NOMs were updated on several occasions. As new issues have been identified, food labeling has been acknowledged as a way to address these challenges and develop public policy. In general, food labeling is defined in Title XII, Chapter I, Articles 212, 213, and 214 of the General Health Law[9] and mandates that foods and non-alcoholic beverages labels must include nutritional information and have information that can be compared to the intake recommendations made by sanitary authorities. The law also requires the Secretary of Health to publish the appropriate Official Mexican NOMs in the gazette. The Regulation for Sanitary Control of Products and Services provides definitions and more specific information about the application of dispositions established in the Health Law. The regulation also refers to the NOMs that outline the requirements for labeling. Chapter II (Article 25)[10] of this

---

[9]Ley General de Salud (2014b), Título Decimosegundo, Capítulo I, Artículo 212 at p. 53.
[10]Reglamento de Control Sanitario de Productos y Servicios (2014), Capítulo II, Artículo 25 at p. 5.

regulation provides general guidance on the required information on a label. The label must not include any reference to illnesses, syndromes, signs or symptoms or any anatomic or physiologic data. There are two NOMs that are directly related to food labeling:

- Mexican Official Norm NOM-086-SSA1-1994, Goods and Services—Foods and Nonalcoholic Beverages with a Modified Composition under Nutritional Specifications[11] and its subsequent updates; and
- Mexican Official Norm NOM-051-SCFI/SSA1-2010, Goods and Services— General labeling specifications for pre-packaged food and non-alcoholic beverages—commercial and food safety information, published on April 5, 2010[12] and its subsequent updates.

These NOMs have been updated and adjusted to accommodate the new public policy measures that address obesity in Mexico, as described later in this chapter.

Although long overdue, NOM-086 has not been fully updated since its original publication, violating the official dispositions that mandate a full review and update every 5 years. This review is necessary to address issues pending with the authorization of health claims and develop specific definitions of permissible claims. During the last update of NOM-051, a section on health claims was included but it lacked the necessary details for proper control and guidance to obtain appropriate approval for a potential claim. Table 33.1 identifies and defines all general permissible claims in Mexico (mainly nutrient content and comparative claims as defined in the *Codex alimentarius*).[13] It is important to understand that, according to the most recent NOM-051-SCFI/SSA1-2010, the potential for making health claims is quite limited. No pre-packaged food or non-alcoholic beverages will be described or presented with false, wrong or misleading information, or in any other way that may mislead the consumer on the product's nature. Impermissible declarations are those that lead consumers to believe that a balanced diet based on ordinary foods cannot provide enough nutrients; cannot be proven; claim usefulness of the food or non-alcoholic beverage to prevent, treat or cure an illness, syndrome or physiologic condition; question the safety of an analog of the food or non-alcoholic beverages or may cause or provoke fear; and state that a particular food is an adequate source of all the essential nutrients. According to the same NOM, examples of misleading property declarations also include those that are unclear and contain comparisons and incomplete superlatives or refer to proper hygiene or trade such as genuine or healthy except of those indicated in other applicable law ordinances. Interestingly, according to these dispositions, labels should not contain any reference to

---

[11] Secretaría de Salud (1996), NOM-086-SSA1-1994. Bienes Y Servicios. Alimentos Y Bebidas No Alcohólicas Con Modificaciones en su Composición. Especificaciones Nutrimentales.

[12] Secretaría de Salud (2010), NOM-051-SCFI/SSA1-2010, Especificaciones Generales De Etiquetado Para Alimentos Y Bebidas No Alcohólicas Preenvasados-Información Comercial y Sanitaria.

[13] Codex alimentarius (1997), Guidelines for the Use of Nutrition Claims.

Table 33.1 Definitions of claims allowed in Mexico

Claim	Definition
Sodium free	Total sodium content is less than 5 mg/serving
Very low sodium	Total sodium content is less than or equal to 35 mg/serving. When a serving less than or equal to 30 g., sodium content must be less than or equal to 35 mg/50 g of product
Low sodium	Total sodium content is less than or equal to 140 mg/serving. When the serving is less than or equal to 30 g., sodium content must be less than or equal to 140 mg/50 g of product
Reduced sodium	Total sodium content is at least 25 % less than the content in the original product
Fat free	Fat content is less than 0.5 g/serving
Low fat	Fat content is less than or equal to 3 g/serving. When an individual serving is less than or equal to 30 g, total fat content must be less than or equal to 3 g/50 g of product
Reduced fat	Fat content is at least 25 % less than the content in the original product
Low saturated fat	Saturated fat content is less than or equal to 1 g/serving
Reduced saturated fat	Saturated fat content is at least 25 % less than the content in the original product
Cholesterol free	Cholesterol content is less than 2 mg/serving and the saturated fat content is less than or equal to 2 g/serving
Low cholesterol	Cholesterol content is less than or equal to 20 mg/serving. If an individual serving is less than or equal to 30 g., then the cholesterol content must be less than or equal to 20 mg/50 g of product
Reduced cholesterol	Cholesterol content is at least 25 % less than the content in the original product
No calories	Calorie content is less than 5 cal/serving
Low calorie	Calorie content is less than or equal to 40 cal/serving. When a serving is less than or equal to 30 g, calorie content must be less than or equal to 40 cal/50 g of product
Reduced calorie	Calorie content is at least 25 % less than the content in the original product
Gluten free	Total nitrogen content of the cereal used in the product must not exceed 0.05-g/100 g of dry matter. Vitamin and mineral content should not be less than that in the original product
Sugar free	Sugar content is less than 0.5 g/serving
Reduced sugar	Sugar content has been reduced at least 25 % from the original product

Source: NOM-086-SSA1-1994

certifications related to food safety since all food sold in the country should be processed under proper hygienic conditions.

Certain conditional property declarations are allowed according to NOM-051. For instance, a label can indicate that a food has acquired a special or superior nutritive value because of the addition of nutrients such as vitamins, minerals and amino acids but only if such addition has been made based on nutritional considerations according to the applicable law ordinance. Statements about the product's special nutritional qualities resulting from the reduction or lack of a nutrient can also be made based on the nutritional considerations but is subjected to the applicable law ordinance (see Table 33.1). Usage of terms such as "natural," "pure," "fresh," and "homemade," "organically grown" or "biologically grown" must adhere to the applicable law ordinance and comply with the prohibitions mentioned earlier. A food's ritual or religious preparation, for example Halal and Kosher, can be declared as long as the declaration is done according to the rules of the religious authorities or corresponding ritual. Declarations of properties that state that the food has special characteristics when all of the foods of the same kind have the same characteristics and is apparent in the declaration of properties are also allowed. Moreover, declarations could be made to highlight the lack or no-addition of particular substances in food as long as these are not misleading and the substance is not subject to special requirements in any norm; is not a substance that consumers expect to find normally in that food product; has not been replaced by another one providing equivalent characteristics, unless the nature of the replacement is explicitly declared with the same importance; or is the addition or the presence of a food ingredient that is allowed. Similarly, declarations that expose the absence or no-addition of one or more nutrients must be considered a nutritional property declaration and therefore, must comply with the mandatory nutritional declaration in the applicable law ordinance.

Health and nutritional declarations can be related to the caloric content, proteins, carbohydrates, fats (lipids) and derivatives, dietetic fiber, sodium, vitamins and inorganic nutrients (minerals) for which Reference Nutritional Values have been established. The following declarations of properties are conditionally permitted when:

- it is understood that declarations of properties related to nutrient content describes the content level of a particular nutrient, i.e., "source of calcium," "high fiber content," and "low fat" (see Table 33.1);
- it is understood that the nutrient properties comparison contrasts the nutrients and/or energetic value of two or more foods, i.e. "reduced," "less than," "increased," and "more than" (see Table 33.1);
- health-related properties are any representation that states, suggests or implies that there is a relation between a food or food component and health. See examples below.
  - a nutrient function claim describes a physiologic function of the nutrient in growth, development and normal body functions. For example, "Nutrient A (state a physiologic body function of nutrient A in relation to the maintenance of health, growth promotion and normal development). Food X is a source of ...high in nutrient A;" and
  - other nutrient function claims pertain to the beneficial effects of the consumption of a food or its components in the context of a total diet on normal biologic body activities or functions. Such property declarations are related to a positive contribution to health or the improvement of a function, modification or preservation of health. For example, "Substance A (meaning the

effects of A over the physiologic function or health association with the biologic activity). Food "Y" contains "X" grams of substance "A"".

However, declarations of properties will not be allowed when there is an intention to give the product characteristics that it does not contain or is associated with the reduction of illness or risk reduction.

Although this NOM and the definitions of declarations of properties were published in 2010, gaps still remain. For certain declarations, the applicable law ordinances mentioned in the NOM are incomplete and do not address many issues that are currently a challenge in the marketplace. Many questions remain open on how to get approval for specific claims. Basic nutrient content declarations or claims are common. However, more specific property claims are not common since, in practice, it is quite challenging to determine the specific requirements for approval. Thus, more specific rules are needed to level the playing field for all players.

33.2 The Overweight and Obesity Crisis and Public Policies as Potential Solutions

Obesity and overweight is quickly becoming a top health concern in Latin America. The entire Latin American region has exhibited a marked increase in the consumption of high energy density foods coupled with a decrease in physical activity. The region has gone through a difficult transition since social and economic progress has led to a decline in infectious diseases. However, the higher income has also fostered the consumption of meats, fats and oils, and sugar while reducing the consumption of grains and legumes. So, although there has been a gradual increase in the life expectancy at birth, there is also a greater burden of disease linked to obesity and other nutrition-related chronic diseases. The region is now facing the challenge of a double disease burden since many countries still experience the unresolved problem of malnutrition caused by nutritional deficits, while addressing the steady increase in chronic disease.[14]

Overweight and obesity has a strong impact on public wellbeing due to the high cost implications. Costs include medical attention, medicine, loss in productive hours and other costs associated with premature death. Government intervention is needed to reduce the impact of these externalities. On the other hand, access to public health is considered a common good and the problems associated with overweight and obesity restrict the resources available to address other illnesses that are not associated with nutrition, which negatively impacts the society.[15] Since obesity is a worldwide epidemic, there is much discussion about proper intervention

---

[14] Uauy and Monteiro (2004)), Food and Nutrition Bulletin.

[15] COFEMER (2012), Introduccion a la Regulacion Social.

strategies and initiatives to address this issue and help mitigate future impact. The development of public policy that has large impact on food choices is far from easy since individuals tend to make food choices that maximize utility. According to Fulponi, even if an individual knows what a healthy choice is, he may decide to choose an unhealthy option. There are many hypotheses behind this type of behavior. For example, changes in food preparation technology have relaxed time constraints on food preparation but may have increased the caloric content of food consumed. People also tend to heavily discount future events such as illness dependent on today's behavior.[16] According to a survey of policy initiatives on diet, health and nutrition performed by the Organization for Economic Cooperation and Development (OECD),[17] strategies implemented by different member countries focus mainly on two activities: increasing information on diet and health to consumers so they can make informed food choices and promoting increased consumption of fruit and vegetables, particularly amongst children. This report finds that the role of governments in establishing strategies to modify food choices is a delicate policy issue. While arguments like rising costs to the public purse can justify the need for interventions, there may be welfare losses if choices are restricted at the individual level. Consequently, many governments mainly opt to promote an environment conducive to healthy food choices through appropriate incentives and information provision. The most common initiatives are collaborations among different government agencies that focus on the provision of information through labeling and publicity campaigns, nutritional education programs for children and adults, promotion of fruits and vegetables, and partnerships with the food industry and producer groups.

In OECD countries, there is mounting evidence that school based programs are particularly effective; thus, efforts are increasingly focusing on school aged children. Interestingly, it was reported that while ministries of agriculture in most OECD countries do not play a major role in diet and nutrition issues, a growing number are becoming more involved through increased collaboration with public health agencies. The food industry, from producer groups to retailers, is also becoming more involved in campaigns that promote healthy eating by increasing fruit and vegetable consumption. According to the OECD survey, there are several studies around the world that concur that fruit and vegetable consumption confers a risk reduction for cardiovascular diseases. For example, Bazzano et al. found that consuming 3 servings or more a day of fruits and vegetables compared to 1 or less is associated with a 27 % reduction in stroke incidence, 42 % in lower stroke mortality and 24 % in lower ischemic heart disease mortality.[18] Likewise Joshipura et al. reported that for each 2 serving increase in intake of fruits or vegetables,

---

[16]Fulponi (2009), pp. 1–45.
[17]Ibid.
[18]Bazzano et al. (2002), pp. 93–99.

the risk of coronary heart disease decreased by 4 %.[19] High consumption of fruits and vegetables may also be important in reducing the risk of some cancers.

From a regulatory perspective, one of the underlying assumptions of this epidemic is that the consumer does not have enough information to make the optimal product choices. This illustrates that since the market is not self-regulated, a government intervention is justified.

Since consumers have asymmetrical information about the foods and beverages they are purchasing and consuming, measures should be taken to foster the provision of more information about them. One such measure is the regulation of food advertising by restricting the promotion of foods with low nutritional value during times when children may be exposed; or providing information so that the adults can promote an orderly consumption of those products. Another strategy is to promote better food labeling by ensuring information is clear and easily interpreted by consumers.

However, it is important to acknowledge that the problem is multifactorial and so potential solutions should also include multi-prong interventions. In order to have an adequate impact, international organizations like the OECD recommend the simultaneous application of three or more actions. The OECD specifically recommends intervention in four different areas: availability of healthy foods; access to healthy food consumption; knowledge of foods; and taxes. In his report for the OECD, Fulponi established that, in general, taxation, workplace interventions and food labeling have the most impact on the reduction of public health expenditure associated with overweight and obesity. Thus, multi-faceted interventions that utilize food labeling, taxes, self-regulation schemes for publicity and programs for health promotion and interventions at the school and workplace levels have been recommended.

33.2.1 Case Studies and Examples

Several countries in Latin America have started to adopt public policies to help mitigate the problems associated with overweight and obesity. These initiatives include one or more of the approaches recommended by the OECD. The following case studies will discuss the development of these policies and provide an insight into the approaches adopted by Mexico as well as other Latin American countries.

33.2.2 Mexico's Overweight and Obesity Crisis and Its Strategies

It is well known that Mexico is one of the most obese countries. Depending on the metrics used, Mexico ranks either the first or the second country in the world (after

---

[19]Joshipura et al. (1999), pp. 1233–1239.

the United States) with the highest number people who are obese or overweight. In Mexico, this trend was first noted towards the end of 1980s and it continues to grow. The increases in obesity and overweight affected the population in general, including both high and low-income population but with a higher prevalence in women. Although the rate of increase has slowed since 2006, 71 % of Mexicans are currently obese (32 %) or overweight (39 %) accounting for more than 48.6 million individuals in the country.[20]

Obesity is obviously a multifactorial problem that has evolved over decades. According to Rtveladze et al.,[21] Mexico has experienced a rapid increase in wealth in recent decades, bringing a significant shift in socio-economic status and geographical shift from rural to urban living settings. This has led to changes in diet and sedentary behavior coupled with an increased access to low-priced, high energy-dense foods (foods that are high in calories, fats and sugars) leading to a rapid growth in obesity and obesity-related non-communicable diseases (NCDs) prevalence. According to these authors, nutrition-related chronic diseases, such as type 2 diabetes and hypertension, are associated with increased obesity rates. Cardiovascular disease (CVD) and type 2 diabetes are now considered the main causes of adult mortality in Mexico. This is expected to impose a substantial burden on disease outcomes and health-care costs.

Based on the simulation model used by Rtyaladze et al. (see footnote 13), obesity is projected to increase across all age groups with particularly high levels among middle-aged men (50–59 years) and older females ($\geq$60 years). It is expected that the rate of obesity for the male population will increase from 68 % in 2010 to 88 % by 2050. Females do not fare any better; it is estimated that their rate will increase from 74 % in 2010 to about 91 % by 2050. The authors of this study concluded that the disease burden continues to increase every year and will result in twice the prevalence rates by 2050. In the same study, the authors show that the prevalence of normal weight individuals will decrease from 32 to 12 % in males and from 26 to 9 % in females and more people will be obese than overweight. Using these estimates, the projection for diabetes and CVD will raise alarmingly to 12 million diabetes cases and 8 million CVD cases in 2050 alone.

Based on information by the Organization for Economic Cooperation and Development (OECD) (see footnote 15), it is estimated that for every extra 15 kg, the probability of early death increases by 30 %. In 2008 alone, the loss of productivity due to premature death that is attributable to overweight and obesity in Mexico was estimated to be $1931 million USD. According to COFEPRIS,[22] the country spends more than the total federal budget for medicines to treat CVD and diabetes and is equivalent to 23 % of the total education expenditure in the country. For the 13 diseases that are considered related to overweight and obesity, costs will

---

[20]Astudillo (2014), pp. 15–16.
[21]Rtveladze et al. (2014), pp. 233–239.
[22]COFEPRIS (2013), Reformas al reglamento de la Ley General de Salud en material de publicidad.

skyrocket from $806 million USD in 2010 to $ 1.7 billion USD in 2050. Rtevaladze et al. report that even a 1 % reduction in mean body mass index (BMI) could save a total of $43 million USD in 2030 and $85 million USD in 2050. A more optimistic 5 % decrease would save $117 USD million in 2030 and $192 million USD in 2050. These figures show considerable savings in health care costs and future burden of diseases even without considering the social impact on the population if proper measures are not taken. Therefore, it is imperative that Mexico's public health community and policy makers develop and implement effective public health interventions to change the current trends.

In Mexico, the right to nutritious, sufficient and quality food was elevated to a Constitutional right by amending Article 4 of the Constitution, which was published in the official gazette on October 13, 2011.[23] Hence, all laws, regulations and norms that were developed to address issues associated with diet and nutrition are supported by this Constitutional right. Another amendment to Article 4 published in February 2012 also established that all individuals have the right to access, purify and dispose water for personal consumption and domestic use that is sufficient, healthy, acceptable and affordable (see footnote 23).

As stated by COFEPRIS' representatives, the main objective of the public health strategies is to stabilize and reduce the incidence of overweight and obesity in the medium to long range. In order to do this, the government has adopted several policies. In 2010, the Secretary of Health published the national strategy against overweight and obesity,[24] which establishes ten high priority objectives to achieve an effective national program. The objectives are to:

1. foster physical activity of the population in school, work, community and recreational settings through collaboration with public, private and social stakeholders;
2. increase the availability, access and consumption of drinking water;
3. decrease the intake of sugar and fats in beverages;
4. increase the daily consumption of fruits, vegetables, legumes, whole grains and fiber by increasing the availability and access and promoting the consumption of these products;
5. improve the population's informed decision making abilities regarding a healthy diet through useful and easy to understand labels that promote health and nutrition knowledge;
6. promote and protect exclusive breastfeeding until 6 months of age and favor adequate complementary nutrition after 6 months of age;
7. decrease the intake of sugars and other caloric sweeteners in foods by increasing the availability and access to foods that are reduced in sugar or without caloric sweeteners;
8. decrease the daily intake of saturated fats and reduce to a minimum industrial trans fats;

---

[23]Constitución Política de los Estados Unidos Mexicanos (1917), Article 4.

[24]Secretaría de Salud (2010a), Acuerdo Nacional para la Salud Alimentaria. Estrategia contra el sobrepeso y la obesidad.

9. provide guidelines to the population regarding serving size control when cooking homemade meals as well as promote the availability of reduced serving sizes in food service establishments; and
10. decrease the daily intake of sodium by reducing the amount of added sodium and increasing the availability and access to products that are low in or free of sodium.

The first six objectives depend on individuals and the surrounding environment that facilitate an increase in physical activity, intake of water and fruits and vegetables. The other four objectives rely on government regulation and participation of all stakeholders, including the processed food and food service industries.

In January of 2010, NOM-088-SSA3-2010 for the integrated treatment of overweight and obesity was published.[25] One of the very first steps taken towards achieving these objectives was to establish criteria for foods and beverages offered during school lunches.[26] These criteria became mandatory on January 1, 2011. The criteria were set for compliance on three stages. Table 33.2 presents the nutritional criteria for the last two stages of this strategy, which is being implemented in phases to allow for gradual changes in formulation. However, in practice, implementing the changes have been challenging since each reduction implies at least a change in label to display the new nutritional content and different formulation. On the other hand, there have been multiple controversies regarding specific definitions and terminology likes "without added sugars." Some reformulated products included ingredients such as concentrated fruit juice instead of added sucrose or corn syrup raising the question about the effectiveness of this strategy. After all, for practical purposes, concentrated fruit juice represents a load of fructose.

As the process evolved, a new norm, NOM-043-SSA2-2012, Basic Health Services—Promotion and education for health as related to nutrition,[27] was published to establish general criteria that unify the nutritional guidance strategies that are provided to the general population. This NOM provides information that is designed to promote the nutritional state and prevent nutrition-related illnesses. Also, it recommends appropriate food sources for nutrients for different ages and/or development stages. This NOM also promotes exclusive breastfeeding for infants under 6 months of age.

In 2013 as part of the tax reform, the Congress passed a tax bill that is popularly known as the "junk food and sugary beverage tax."[28] This tax was implemented on January 1, 2014 as part of an anti-obesity and revenue raising campaign. This

---

[25]Secretaría de Salud (2010b), NOM-008-SSA3-2010.

[26]Secretaría de Educación Pública y Secretaría de Salud (2010), ACUERDO mediante el cual se establecen los lineamientos generales para el expendio o distribución de alimentos y bebidas en los establecimientos de consumo escolar de los planteles de educación básica.

[27]Secretaría de Salud (2013), NOM-043-SSA2-2012.

[28]El Congreso General de los Estados Unidos Mexicanos (2013a), DECRETO por el que se expide la Ley de Ingresos de la Federación para el Ejercicio Fiscal de 2014 y se reforma el primer párrafo del artículo 2o. de la Ley de Ingresos de la Federación para el Ejercicio Fiscal de 2013.

Table 33.2 Nutritional requirements for products available at schools

Category	Nutritional criteria (per serving)[b]	Stage II (August 2011 To July 2012)	Stage III (to be implemented August 2012)
Prepared foods[a]	Energy (kcal)	180	180
	Protein (% kcal)	10	10
	Sugars and other caloric sweeteners	Without added sugars	Without added sugars
	Saturated fats (% kcal)	15	10
	Trans fatty acids (g/serving)	0.5	0.5
	Sodium (mg/serving)	230	220
	Whole grains (% of the products)	66	100
	Total fat	≤35 % of the total kcal	≤30 % of the total kcal
Beverages for elementary school[a,c,d]	The availability of potable drinking water must be guaranteed		
Beverages for middle school[a,c]	Serving (ml)	250	250
	Calories per serving (kcal maximum)	10	10
	Sodium (mg/serving)	250	250
	Non-caloric sweeteners (mg/100 ml)	45 (112.5 mg in 250 ml)	40 (100 mg in 20 ml)
Milk[a,e,f]	Serving (ml)	250	250
	Calories per 100 g (kcal)	50 (125 kcal in 250 ml)	50 (125 kcal in 250 ml)
	Total fat (in 100 g)	1.6 (4 g in 250 ml)	1.4 (3.5 g in 250 ml)
	Includes dairy formulas and combined dairy products; it does not consider dairy foods		
Yogurt and fermented dairy products[a,e,f]	Total fat (100 g or ml)	Solid = 2.5 (6.25 g in 150 g)	Solid = 2.5 (6.25 g in 150 g)
		Drinkable = 1.6 (4 g in 250 ml)	1.4 (3.5 g in 250 ml)
	Serving (g or ml)	Solid = 150	Solid = 150
		Drinkable = 200	Drinkable = 200
	Sugars (% of total calories of added sugars)	35	30
Fruit and vegetable juices	Total sugars (6/serving)	According to the dispositions of NOM-173-SCFI-2009	
	Serving (ml)	200	125
	Calories per serving (maximum)	110	70
Nectars[a,e]	Serving (ml)	200	125
	Calories per serving (maximum)	200	125

(continued)

Table 33.2 (continued)

Category	Nutritional criteria (per serving)[b]	Stage II (August 2011 To July 2012)	Stage III (to be implemented August 2012)
Liquid Soy beverages[a,e]	Serving (ml)	200	125
	Sodium (mg per 100 ml)	70	50
	Total fats (g per 100 ml). Saturated fats must not exceed 15 % of total fats	2.5 (5 g in 200 ml)	2.5 (5 g in 200 ml)
	Calories per serving (kcal, max)	100	60
	Protein (kcal per serving minimum)	6.5	6.5
Snacks[a,g]	Calories per serving (kcal)	130	130
	Total fat (% total calories)	40	35
	Saturated fat (% total calories)	25	15
	Trans fatty acids (g/serving)	0.5	0.5
	Added sugars (% total calories)	10	10
	Sodium (mg per serving)	200	180
Cookies, crackers, confectionery and desserts[h]	Energy per serving (kcal)	130	130
	Total fat (% total calories)	40	35
	Saturated fats (% total calories)	20	15
	Trans fatty acids (g per serving)	0.5	0.5
	Added sugars (% total calories)	25	20
	Sodium (mg/serving)	200	180

Source: ACUERDO mediante el cual se establecen los lineamientos generales para el expendio o distribución de alimentos y bebidas en los establecimientos de consumo escolar de los planteles de educación básica

[a]All packages must contain only one serving
[b]The parameters refer to less than or equal the amount
[c]In beverages, there are two proposals, elementary and middle school. Without caffeine or taurine
[d]These criteria incorporate industry's commitment to assist in the consumption of pure, potable water so that in a maximum period of two months after the beginning of the 2010–2011 school year all sugared beverages will be removed from schools
[e]Non caloric sweeteners in milk, nectars, yogurts, soy beverages may be allowed if these are approved in Codex Alimentarius for children
[f]The trans fatty acid restriction does not apply if it comes naturally with the food in question such as dairy products
[g]The oilseed group (for example, peanuts, nuts, almonds, pistachios, etc.) and dry legumes (for example dry fava beans) are not subject to the total fat content restriction due to their high nutritional value since despite their high fat content, their consumption in moderation has been associated with positive health effects as long as there is no additional added fat. The rest of the criteria for snacks apply
[h]The use of non-caloric sweeteners could be allowed in cookies, pastries and desserts as long as they are approved for their use in Children by Codex Alimentarius

proposal was the subject of heated public debate and was formally announced by the Finance and Public Credit Commission on September 8, 2013. Under this new taxation regimen, there is an increase of 8 % on the Special Tax over Products and

Services (IEPS) on processed foods that contain over 275 kcal per 100 g. This includes products such as snacks, confectionery, chocolate products, desserts, fruit jam and paste, peanut and hazelnut butter, ice cream and cereal based products. Chewing gum is also included in the tax reform as a product that has added sugar. However, based on the above definition, many products that are important in the diet are also now taxed. A clear example of this are nuts and seeds, which fall under the calorie definition even when they are sold in a natural state or dry-roasted without oil and/or added salt. Other products such as dried fruits, cocoa and gelatin also fall under the high calorie foods category. Products such as these that have great nutritional value and are traditionally consumed in the country are now too expensive to be part of a regular, healthy diet. So the question of the effectiveness of this strategy is also raised. Were the definitions provided in the regulation adequate? Should the authorities seek a different approach?

Additionally, a special tax was applied to "flavored beverages" that imposes payment of 1 Mexican peso per liter of processed beverages that contain sugar including powdered beverages. Again, this does not address beverages prepared at home, such as the traditional fruit juices made with water and sugar. Some organizations claim that taxing sugar would be a much better approach since it would address the global consumption instead of specific products. During the debate, many consumer organizations complained that obesity cannot be attributed to the consumption of specific products. It is estimated that this proposal will earn the government an estimated 39 billion pesos in revenue. Theoretically, the money is to be invested in prevention actions such as providing water fountains in schools and appropriate spaces to foster physical activity. The budgeted use of this additional income is counterintuitive since a successful taxation strategy would reduce the intake of these foods. The potential revenue decreases as the population stops or reduces the purchase and intake of these products. One question to think about is what measures would indicate the success of this strategy. Obviously, increased revenue, although useful, would indicate the failure of this strategy.

Apart from these policies, the Secretariat of Health specifically adopted two different approaches. The first approach was the adoption of a new "front" as part of label requirements and the development of a "nutrition seal of approval" that has been published throughout 2014. The second strategy is to develop standards for food advertising targeted towards children.

In summary, the front of label must include the following:

a. the declaration of caloric content (kcal), sugars (kcal), saturated fats (kcal), total fats (kcal) and sodium (mg) and the % of the recommended daily intake using the following values for the calculation of the % of the daily recommended value (Table 33.3);
b. the total caloric content of the product;
c. in family size packages, the energy content per serving and the number of total servings in the package; and
d. in family size packages of flavored beverages, snacks, confectionery products and chocolates, nutritional information of the total content of the package shall

Table 33.3 Nutrients to be included in the front of label declarations and the base to calculate the recommended intake

Source	Base for calculation of the Recommended Daily Intake (RDI)
Sugars	360 kcal
Saturated fats	200 kcal
Other fats	400 kcal
Sodium	2000 mg

Source: COFEPRIS (2013)

be included as well as the number of servings and caloric content of each serving.

The path to achieving legal approval of these new requirements has been rocky and has created a lot of confusion regarding compliance deadlines. In January 2013, a presidential decree was originally published to amend the Regulation of Sanitary Control of Products and Services.[29] This amendment was to prepare the legal roadway for the new labeling requirements. This publication established that compliance was mandatory effective the day after publication in the official gazette. However, this was physically impossible and required an update of the labeling norms. In April 2014, an agreement that established the guidelines for the implementation of the modifications to this regulation was published.[30] This agreement included the specific icons to be used for the front of label definitions and established the requirements for size and font of these icons.

The minimum size of the icons must occupy at least 0.5 % of the main display panel of the product and each icon must not be less than 0.6 cm in width and 0.9 cm in height. The width of the icon must be two thirds in relation to the height. The icons must always present the mandatory information in the following order from left to right: saturated fat (cal/kcal), other fats (cal/kcal), total sugars (cal/kcal), sodium (mg/g) and Energy (cal/kcal) with the % of the RDI in the lower part of the icons for saturated fat, other fats, total sugars and sodium. Figures 33.1 and 33.2 show examples of the icons to be used in the correct order from left to right.

These requirements will not apply to flavored beverages that have low energy content or those products in individual packages where the content of the package is less than the reference serving size published in the same agreement. In addition, when the caloric value for a nutrient equals zero, it shall be declared as zero "0" and

---

[29] El Congreso General de los Estados Unidos Mexicanos (2012), DECRETO por el que se reforman y adicionan diversas disposiciones del Reglamento de Control Sanitario de Productos y Servicios.

[30] Secretaría de Salud (2014), ACUERDO por el que se emiten los Lineamientos a que se refiere el artículo 25 del Reglamento de Control Sanitario de Productos y Servicios que deberán observar los productores de alimentos y bebidas no alcohólicas pre-envasadas para efectos de la información que deberán ostentar en el área frontal de exhibición, así como los criterios y las características para la obtención y uso del distintivo nutrimental a que se refiere el artículo 25 Bis del Reglamento de Control Sanitario de Productos y Servicios.

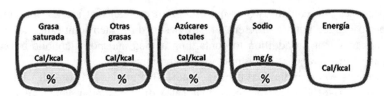

Fig. 33.1 Front of panel label requirements icons for Mexico for individual packages

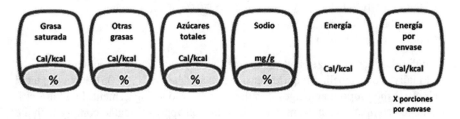

Fig. 33.2 Front of panel label requirements icons for Mexico for family size packages

this is also applicable to the percent value. Products with a value of less than 5 kcal shall be declared as zero "0." All values will be presented as round numbers using common rounding rules.

The color of the icons will be chosen by each producer but all icons must be the same color and font. Font will be presented in a contrasting color. When the main display panel is more than 60 cm^2, all icons shall be presented in this area. When the main display panel is between 20 and 60 cm^2, the energy content icon shall be displayed in the front, but the main display panel and other four icons can be presented in lateral panels or in the back panel if there are no lateral panels. When the main display panel is less than 20 cm^2, then only the energy content icon shall be displayed in the front panel. When the front panel is less than 10 cm^2, only the energy content icon will be required on the label. When the main display area is less than 5 cm^2, then no icons will be required.

The following products are exempted from the mandatory front of panel labeling:

1. herbs, spices and blends;
2. pure coffee extracts, whole coffee, ground coffee decaffeinated or not, soluble or not;
3. herbal infusions, decaffeinated tea or non-decaffeinated tea, instant or soluble that do not have added ingredients;
4. fermented vinegars or substitutes;
5. products and raw materials contained in packages that are destined exclusively for food service but the main display panel must show the statement "presentación institutcional" (institutional presentation);
6. foods and non-alcoholic beverages where each nutrient per serving represents an energy intake less than or equal to 1 % of the recommended daily nutrients;

7. packages that contain two or more units of product that are not individually pre-packaged but are different and are destined for their joint sale;
8. packages that, in addition to containing a food and non-alcoholic beverage, have a gift or decorative article as a promotional gift;
9. products where individual packages state "No etiquetado para su venta individual" (Not labeled for individual sale) or use similar language and are included in a collective or multiple package;
10. packages where the content corresponds to more than one type of product individually labeled as long as at least 70 % of the products in the package are labelled appropriately; and
11. products sold in bulk.

Individual serving sizes and requirements are found in this document.

Finally, since, labeling requirements are established in NOM 051, it was updated to include the proposed changes[31] in the aforementioned agreement. In addition, it mandated companies that want to use the nutrition approval seal to comply with the criteria established in the new agreement. Specific conditions have to be met to obtain approval for the use of this nutritional seal. These dispositions are effective as of June 30, 2015. Any company who has leftover labels or issues developing the new labels has the right to request a delay for up to one additional year prior to this date. Since NOM-051 rules foods and non-alcoholic beverage labelling, the established deadlines will have the higher hierarchy and therefore, these are the final dates and deadlines to comply.

These new front of label requirements have met a lot of criticism from different organizations. In addition to the many questions that still require clarification and guidance, many groups are against this new public policy. On one hand, consumer groups have asked to stop these requirements arguing that the values set for sugars are too high and were arbitrarily set. In addition, according to many groups, the information is unclear in the way it is asked to be presented since it misleads the consumer in thinking that they must get the recommended daily intake that is shown on the label, which is not the case nutrients such as total and saturated fats. There is also a lot of confusion regarding the declaration of sugars. Originally, the sugar label was for "added sugars" only. However, to address some of the issues discussed earlier, the sugar label now captures "total sugars." In addition to consumer groups, public health authorities are raising their voice against the labeling requirements. This is alarming since the National Institute for Public Health (INSP) along with other institutions are the pillars upon which nutrition

---

[31] Secretaría de Economía y Secretaría de Salud (2014), MODIFICACIÓN de la Norma Oficial Mexicana NOM-051-SCFI/SSA1-2010, Especificaciones generales de etiquetado para alimentos y bebidas no alcohólicas preenvasados-Información comercial y sanitaria, publicada el 5 de abril de 2010. Se adicionan los incisos 3.2; 3.5; 3.17; 3.18; 3.21; 3.40; 4.2.9 con sus subincisos y se ajusta la numeración subsecuente; 4.5 con sus subincisos y el Apéndice Normativo A. Se modifica el capítulo 2 Referencias, así como el literal b) del inciso 3.11; 3.15; 4.2.8.1. Se ajusta numeración del capítulo 3 Definiciones, símbolos y abreviaturas.

public policy should rest; yet they have complained about the lack of evidence for this strategy. The Director of Research and Nutrition Policies of the INSP sent a formal request to the Health authorities to abolish the new labeling requirements. One of his major concerns is the fact that the World Health Organization (WHO) recommends that an adult should limit sugar consumption to 50 g per day. However, in the current system, the amount of sugar allowed per day is 90 g. In addition, the clarity of the information presented on the labels has been questioned since surveys show that it is difficult to understand and make sound decisions based on the information provided. In spite of all the opposition and updates, the requirements are final and have been published in the official gazette. It is crucial to understand the implementation process and ultimately, identify indicators to evaluate and measure the actual impact on obesity, if any.

The second approach taken by authorities in 2014 was the regulation of food advertising. According to information published by the INSP, the publicity of food and beverages is one of the contributing factors for the change in food consumption patterns that results in overweight and obesity.[32] In Mexico, children are exposed to a large number of commercial spots for foods and beverages that can impact their food choices. Subsequently, a public policy strategy was implemented to reduce exposure to commercial spots for foods and beverages that are high in fats, sugar and salt. It is hoped to reduce the risk associated with this exposure. According to the INSP, an appropriate public policy for publicity include all advertising targeting children. All communication channels (radio, television, printed media, internet, etc.) are included. The policy also establishes mechanisms to impose fines when these dispositions are not met.

The updated Health Regulation for Publicity[33] clearly establishes through article 23 that foods and non-alcoholic beverages marketing shall include precautionary statements about the condition of the product as well as messages that promote a healthier diet. As of now, there will be no publicity of foods and beverages that do not comply with the nutritional criteria set by the Secretariat of Health from Monday through Friday from 14:40 to 19:30 and Saturday and Sunday from 7:00 to 19:30. There are some exceptions to this schedule. Flavored beverages, snacks, confectionery products and chocolates can be publicized during the transmission of the following types of shows: soap operas, news programs, sport events, series and movies with a classification B-15 and beyond. The deadline for these actions was January 01, 2015 for general categories and 90 days after the publication of the guidelines for flavored beverages, snacks, confectionery products and chocolates. Thus, at the time this chapter was written, the publicity rules for the special categories were already in place.

It is clear that Mexico is trying to implement public policies that address the obesity crisis. Many of these strategies are aligned with the recommendations

---

[32]Instituto Nacional de Salud Pública (2013), Evidencia para la política pública y prevención de obesidad. Publicidad en alimentos y bebidas.

[33]Secretaría de Salud (2006), Reglamento de la Ley General de Salud en Materia de Publicidad.

established by the OECD. However, education campaigns for the general population still seem to be underdeveloped. Although there are some campaigns to promote the consumption of fruits and vegetables such as the "5 a day" campaign, more efforts need to be focused on educating the general consumer. This was quite evident in the research performed by the INSP where the front of panel labels were still difficult to understand for consumers. This is an important part of the obesity crisis since most of these policies address processed foods; however little attention has been given to foods prepared at home. Mexico enjoys a rich culinary tradition and many traditionally cook foods that have high caloric density. While the taxes address processed foods, there is still an open gap regarding home-cooked meals. The same is true regarding processed foods in primary schools. Although some guidance was published, freshly prepared foods do not fall into any of the current control strategies. Several implementation challenges also still have to be conquered and in many cases, the policies have to be fine-tuned to address the concerns of different stakeholders.

33.3 Addressing Overweight and Obesity in Peru: Strategies Under Evolution

In Peru, a law that promoted healthy diet for boys, girls and teenagers was published on May 17, 2013.[34] According to Article 1 of this law, the objective is the promotion and effective protection of the right to public health; adequate growth and development through actions in the areas of education, physical activity, healthy lunches in elementary schools, and marketing and other promotional strategies related to foods and non-alcoholic beverages targeted to children and teenagers to reduce and eliminate illnesses associated with overweight and obesity. This law is comparable to Mexico's school lunch program; however, implementation is still pending. According to the dispositions stated in the law, the Ministry of Health (MINSA) would publish the exact parameters and definitions of healthy foods as well as the limits for sugar, fats, sodium and trans-fats present in the food and non-alcoholic beverages 60 days after the official publication. Interestingly, the Peruvian law accounts for trans-fatty acids unlike the Mexican policies that do not focus on these compounds. The law also prohibits advertising that encourages "immoderate consumption" of food and non-alcoholic beverages that contain trans-fats or high levels of sugar, salt and saturated fat. The law also established the timescales for the reduction of trans-fatty acids until full elimination is achieved. The reference values will be based on the recommendations published by the WHO and Panamerican Health Organization (PAHO).

However, MINSA has failed to meet these deadlines and without the applicable regulation in place, implementation of the law is not feasible. Additionally,

---

[34]El Congreso General de los Estados Unidos Mexicanos (2013b), Ley N°30021.

according to a report by Barbara Fraser,[35] Peru's law immediately drew criticism from legislators, food and beverage marketing industry and even the Catholic archbishop of Lima who said that shaping children's dietary habits was a job for parents, not the government. However, the efforts to redirect children's food choices are an important strategy.

33.4 Addressing Overweight and Obesity in Chile: The Warning Sign Approach

In Chile, labeling requirements are under development since 2012 when a new law on Nutritional Composition and Food Advertising (Law 20606) was published.[36] This law mandates food industry to declare the ingredients of their packaged products according to the criteria in Sanitary Regulation of Foods. All manufacturers, producers, distributors and importers of foods have to display the ingredients, including all additives, on their labels/packaging. The list must be in descending order and nutritional information must be expressed in percent composition, unit of weight or under the nomenclature established in current regulations. Through the Food Sanitary Regulation, the Ministry of Health will determine the shape, size, color, proportion, characteristics and content of the food nutritional labels with a particular emphasis on visibility and understandability by the population.

The Ministry of Health will identify foods that, by unit of weight or volume or by usual serving size, present in their nutritional composition, high contents of calories, fats, sugars, salt or other harmful ingredients. These foods will be labeled as "high in calories" or "high in salt" or with other equivalent denomination depending on the case. The content must be clearly declared on the package or label of all commercialized food products, if among the ingredients used in production are soy, dairy, peanuts, egg, shellfish, fish, gluten or tree nuts. Infant formula labels must not discourage breast feeding and include information relative to the superiority of breast feeding and state that the use of such substitutes must be under medical advice.

These foods cannot be sold, commercialized, promoted or publicized inside pre-school, elementary and middle school establishments. Energy dense, nutrient poor foods cannot be offered or given for free to children under 14 years of age. The Ministry of Health shall share with the Ministry of Education a mandatory monitoring system of nutrition education in pre-school, elementary and middle school where specialists will measure the body mass index and orient students to adopt healthy life styles. Regardless of educational level, elementary and middle schools must provide education and encourage the development of healthy eating habits and

---

[35]Fraser (2013), Latin American countries crack down on junk food, pp. 385–386.
[36]Ministerio de Salud (2012), Ley 20606.

warn against the negative effects of a diet with excess of fat, saturated fat, sugars, sodium and other nutrients where consumption in certain amounts or volumes can pose a health risk. They must also incorporate physical activity like sports to promote students to practice an active lifestyle.

Although the regulation was to be published a year after the promulgation of the law, it was finally published for public consultation in August 2014.[37] The dispositions were originally scheduled to be mandatory by June 2015. This new regulation is stricter than the preceding drafts since it now requires all products that contain a high amount of sodium, sugar or saturated fat to display a warning statement indicating that the product has "an excess of." Instead of the original plan to phase in the new dispositions, a single phase implementation was established for the new labeling requirements and ban from sale in schools. After critiques to the original law, the regulation was modified to redefine new maximum levels of fats, sugar, sodium and calories in order for these packaged products to be considered "healthy." If the limits on any of these key nutrients is exceeded, the product must comply with the precautionary statement on the label. The new regulation also changed the serving sizes from arbitrarily set values per product to a more universal unit of 100 g. Table 33.4 shows the limits for the critical nutrients that have been established.

Products that exceed these limits will be required to display a warning symbol on the main panel of the label (Fig. 33.2). For sodium, a gradual reduction has been established; there will be a progressive reduction of the limit by 15 % every 5 years until the value reaches a maximum of 150 mg/100 g. Figure 33.3 shows an example of the warning icons that must be included on labels of products that exceed set limits. In addition, when these icons are used, the label or advertising of such product cannot declare complementary nutritional information to make nutritional properties claims.

At the time of this publication, the period for public comment was still open and the regulation was under review. Based on the preliminary assessment, the regulation will be more flexible and implemented gradually instead of the single phase that was proposed in the update. Full implementation of the law will occur over 4 years instead of the original deadline of June 2015.

With this new phase-in process, the limits will decrease annually to finally achieve the desired limits by the fourth year (Table 33.5). This may be easier for companies trying to formulate new products; but may increase costs associated with annual label changes.

The new regulation also includes rules for advertising using any media. Any foods or food products that contain calories, sodium, sugars or saturated fat in amounts that exceed the limits (Table 33.4) cannot be advertised in any communication media directed to children under 14 years old. For example, these products cannot use cartoon characters, toys, music, animals, or characters that attract

---

[37]Ministerio de Salud (1996), Modifica Decreto supremo N° 977, de 1996, del Ministerio de Salud, Reglamento Sanitario de los Alimentos.

Table 33.4 Limits for the content of energy, sodium, total sugars and saturated fats

	Energy (kcal/100 g)	Sodium (mg/100 g)	Total sugars (g/100 g)	Saturated fats (g/100 g)
Limits for solid foods	275	400	10	4
	Energy (kcal/100 ml)	Sodium (mg/100 ml)	Total sugars (g/100 ml)	Saturated fats (g/100 ml)
Limits for beverages	70	100	5	3

Fig. 33.3 Warning icons that will be used in Chile

Table 33.5 New limits proposed by the Chilean Ministry of Health

Measured per 100 g or 100 ml				
Nutrient	First year	Second year	Third year	Fourth year
Saturated fat (g)	6	5.5	5	4
Sugar (g)	22.5	20	15	10
Sodium (mg)	800	700	500	400
Energy (kcal)	350	325	300	275

Source: Ministerio de Salud, Chile

children under 14 years old. The only exception to this disposition is the advertisement of foods where the content of energy, sugars, sodium or saturated fat is naturally present and is coherent with the values established in the Nutrition Guidelines published by the Ministry of Health. This is in direct contrast to Mexico's policy for "junk food" taxation where no exceptions are allowed. In addition, all publicity of foods and food products designed for mass communication must have a message that foster healthy living habits.

33.5 Addressing Overweight and Obesity in Ecuador: The Traffic Light Approach

Ecuador recently published the review of Ecuador's Technical Regulation PRTE INEN 022 (1R) modification.[38] The objective of this new technical regulation is to establish labeling guidelines to protect consumer health and prevent practices that mislead consumers. This is an update of a previous labeling regulation and establishes the requirements for using a traffic light approach to warn consumers about the total fat, sugars or sodium content. The limits are detailed in Table 33.6 and each concentration, low, medium or high will have a colored bar as well as text.

The label must have a colored system of horizontal bars to show the concentration of each component. The top or red bar include components with "high concentrations" and will have text showing that the product is "High in...." The middle or yellow bar display components with "medium concentrations" and will have text showing that the product is "Medium in...." The lower or green bar comprise components with "low concentrations" and will have text stating that the product is "Low in...." The font size and size of charts will be determined by the area of the main display panel of the food product and will always be located in the top left label corner. Products such as sugar, salt and animal fats (i.e., lard and

Table 33.6 Limits of components and concentrations permitted

Component/ level	"Low" concentration	"Medium" concentration	"High" concentration
Total fat	Less than or equal to 3 g/100 g	More than 3 and less than 20 g/100 g	More than or equal to 20 g/100 g
	Less than or equal to 1.5 g/100 ml	More than 1.5 and less than 10 g/ml	More than or equal to 10 g/100 ml
Sugars	Less than or equal to 5 g/100 g	More than 5 and less than 15 g/100 g	More than or equal to 15 g/100 g
	Less than or equal to 2.5 g/100 ml	More than 2.5 and less than 7.5 g in 100 ml	More than or equal to 7.5 g/100 ml
Sodium	Less than or equal to 120 mg/100 g	More than 130 and less than 600 mg/100 g	More than or equal to 600 mg/100 g
	Less than or equal to 120 mg/100 ml	More than 130 and less than 600 mg/100 ml	More than or equal to 600 mg/100 ml

Source: REGLAMENTO TÉCNICO ECUATORIANO PRTE INEN 022 (1R)

---

[38]Ministerio de Industrias y Productividad (2014), No. 14413.

butter) are exempted from the graphic icon but must use the following precautionary statement: "For your health, reduce the intake of this product."

In case of two or more messages, they should be illustrated together. Since these dispositions are included in the updated labeling regulation, no other strategy is defined. The deadline for these new mandatory requirements was August 29, 2014 for medium and large food companies. For small companies, the deadline is November 29, 2014. Companies are allowed to fix stickers that don't comply with these requirements. Since these changes were implemented, no other strategies have been formally announced.

33.6 Conclusion

Due to the alarming raise in overweight and obese people and the challenges associated with this epidemic, Mexico, like many other countries, have implemented various public policies to slow and reverse the trend. This has become a priority since the costs associated with treatment of nutrition-related, chronic diseases will be insurmountable in the near future. Strategies, specific to Mexico, involve taxes on energy dense foods and sugary beverages, nutritional labeling on front panel, advertising control of foods and beverages and promotion of proper nutrition and healthy lifestyles. These approaches still need to be refined as further gaps have been identified and in some cases, have met strong criticism even from public health authorities. The 2015 implementation of the new labeling requirements will be challenging and will still face strong criticism from various stakeholders. Peru, Chile and Ecuador have also adopted measures that use different approaches and are currently being implemented. Unfortunately, the criteria differ in each country; single labels for products sold throughout Latin America will no longer be possible since each country has adopted different systems. However, these strategies demonstrate that different approaches can be used to address the same challenge.

The approaches taken in each country are excellent examples of public policy, where regulations are used to intervene and solve a social problem. All strategies must have a solid scientific basis, have a proper regulatory basis for implementation, be relevant to the population of interest and provide useful and understandable information to be effective. One key factor that still needs to be urgently developed are proper indicators to show the success of the adopted strategies in the short term. Since most strategies only address processed foods, education will be critical to ensure that foods prepared at home and consumption of traditional foods are part of a balanced diet and healthy lifestyle. Change in consumer habits based on the information printed on the labels and its clarity and usefulness should also be evaluated. A positive impact in public health in the long term will be the true measure of success. Obviously, this is a multifactorial problem and will require multiple approaches and strategies to face the challenge.

References

Astudillo O (2014) Country in focus: Mexico's growing obesity problem. Lancet Diabetes Endocrinol 2(1):15–16

Bazzano LA, He J, Ogden LG, Loria CM, Vupputuri S, Myers L, Whelton PK (2002) Fruit and vegetable intake and risk of cardio vascular disease in US adults: the first National Health and Nutrition Examination Survey Epidemiologic Follow-up Study. Am Journal Clin Nutr 76:93–99

Chile Ministerio de Salud (1996) Modifica Decreto supremo No 977, de 1996, del Ministerio de Salud, Reglamento Sanitario de los Alimentos. Available at: http://www.dinta.cl/wp-dintacl/wp-content/uploads/DECRETO-modifica-RSA-reglamento-ley-20-606-18-06-2014-final.pdf (last accessed 4 Oct 2014)

Chile Ministerio de Salud (2012) Ley 20606 – Sobre Composición nutricional de los alimentos y su publicidad. Available at: http://www.leychile.cl/Navegar?idNorma=1041570 (last accessed 4 Oct 2014)

COFEPRIS (2014) Reglamento de Control Sanitario de Productos y Servicios. Available at: www.cofepris.gob.mx/MJ/Documents/Reglamentos/prodyser060409.pdf (last accessed 4 Oct 2014)

Comisión Federal de Mejora Regulatoria (COFEMER) (2012) Introducción a la Regulación Social. Available at http://www.cofemer.gob.mx/diplomadosv/Espanol/Diplomado/archivos/Caso%20Lectura%201%20M%C3%B3dulo%20III-%20Obesidad.pdf (last accessed 4 Oct 2014)

Ecuador: Ministerio de Industrias y Productividad (2014) No. 14413 – Reglamento Técnico Ecuatoriano Prte INEN 022 (1r) Rotulado De Productos Alimenticios Procesados, Envasados y Empaquetados. Available at: http://www.normalizacion.gob.ec/wp-content/uploads/downloads/2014/08/RTE-022-1R.pdf (last accessed 4 Oct 2014)

El Congreso General de los Estados Unidos Mexicanos (2012) DECRETO por el que se reforman y adicionan diversas disposiciones del Reglamento de Control Sanitario de Productos y Servicios. Available at: http://dof.gob.mx/nota_detalle.php?codigo=5332690&fecha=14/02/2014 (last accessed 4 Oct 2014)

El Congreso General de los Estados Unidos Mexicanos (2013a) DECRETO por el que se expide la Ley de Ingresos de la Federación para el Ejercicio Fiscal de 2014, y se reforma el primer párrafo del artículo 2o. de la Ley de Ingresos de la Federación para el Ejercicio Fiscal de 2013. Available at: http://www.dof.gob.mx/nota_detalle.php?codigo=5322823&fecha=20/11/2013 (last accessed 4 Oct 2014)

El Congreso General de los Estados Unidos Mexicanos (2013b) Ley N°30021 – Ley de promoción de la alimentación saludable para niños, niñas y adolescentes. Available at: http://www.leyes.congreso.gob.pe/Documentos/Leyes/30021.pdf (last accessed 4 Oct 2014)

El Congreso General de los Estados Unidos Mexicanos (2013c) Ley Orgánica de la Administración Pública Federal. Available at: http://www.normateca.gob.mx/Archivos/66_D_3632_22-01-2014.pdf (last accessed 4 Oct 2014)

El Congreso General de los Estados Unidos Mexicanos (2014a) Ley Federal Sobre Metrología y Normalización. Available at: http://www.diputados.gob.mx/LeyesBiblio/pdf/130_140714.pdf (last accessed 4 Oct 2014)

El Congreso General de los Estados Unidos Mexicanos (2014b) Ley General de Salud. Available at: http://www.diputados.gob.mx/LeyesBiblio/pdf/142_040614.pdf (last accessed 4 Oct 2014)

Federal Commission for the Control of Sanitary Risks (COFEPRIS) (2013) Reformas al reglamento de la Ley General de Salud en material de publicidad. Available at http://www.cofepris.gob.mx/AS/Paginas/Publicidad/Publicidad.aspx (last accessed 12 June 2014)

Food and Agriculture Organization (FAO) (1997) Codex alimentarius: guidelines for the use of nutrition claims. Available at www.codexalimentarius.net/input/download/standards/.../CXG_023e.pdf (last accessed 12 June 2014)

Fraser B (2013) Latin American countries crack down on junk food. The Lancet 382 (9890):385–386

Fulponi L (2009) Policy initiatives concerning diet, health and nutrition. OECD Food, Agriculture and Fisheries Working Papers, no. 14, pp. 1–45

Instituto Nacional de Salud Pública (2013) Evidencia para la política pública y prevención de obesidad – publicidad en alimentos y bebidas. Available at: http://www.insp.mx/epppo/blog/2984-publicidad-alimentos-bebidas.html (last accessed 4 Oct 2014)

Joshipura KJ, Ascherio A, Manson JE, Stampfer MJ, Rimm EB, Speizer FE, Hennekens CH, Spiefelman D, Willett W (1999) Fruit and vegetable intake in relation risk of ischemic stroke. J Am Med Assoc 282(13):1233–1239

Los Estados Unidos Mexicanos (1917) Political Constitution for the United Mexican States. Constitución Política de los Estados Unidos Mexicanos. Available at: http://www.diputados.gob.mx/LeyesBiblio/htm/1.htm (last accessed 4 Oct 2014)

Rtveladze K, Marsh T, Barquera S, Sanchez RLM, Levy D, Melendez G, Webber L, Kilpi F, McPherson K, Brown M (2014) Obesity prevalence in Mexico: impact on health and economic burden. Public Health Nutr 17(1):233–9

Secretaría de Salud (1995a) NOM-035-SSA1-1993: Bienes Y Servicios. Quesos De Suero. Especificaciones Sanitarias. Available at: http://www.salud.gob.mx/unidades/cdi/nom/035ssa13.html (last accessed 4 Oct 2014)

Secretaría de Salud (1995b) NOM-120-SSA1-1994: Bienes Y Servicios. Prácticas De Higiene Y Sanidad Para El Proceso De Alimentos, Bebidas No Alcohólicas Y Alcohólicas. Available at: http://www.salud.gob.mx/unidades/cdi/nom/120ssa14.html (last accessed 4 Oct 2014)

Secretaría de Salud (1996) NOM-086-SSA1-1994: Bienes Y Servicios. Alimentos Y Bebidas No Alcohólicas Con Modificaciones en su Composición. Especificaciones Nutrimentales. Available at: http://www.salud.gob.mx/unidades/cdi/nom/086ssa14.html (last accessed 4 Oct 2014)

Secretaría de Salud (2006) Reglamento de la Ley General de Salud en Materia de Publicidad. Available at: http://www.salud.gob.mx/unidades/cdi/nom/compi/rlgsmp.html (last accessed 4 Oct 2014)

Secretaría de Salud (2010a) Acuerdo Nacional para la Salud Alimentaria. Estrategia contra el sobrepeso y la obesidad. Available at: http://activate.gob.mx/Documentos/ACUERDO%20NACIONAL%20POR%20LA%20SALUD%20ALIMENTARIA.pdf (last accessed 4 Oct 2014)

Secretaría de Salud (2010b) NOM-008-SSA3-2010: Para el tratamiento integral del sobrepeso y la obesidad. Available at: http://dof.gob.mx/nota_detalle.php?codigo=5154226&fecha=04/08/2010 (last accessed 4 Oct 2014)

Secretaría de Salud (2013) NOM-043-SSA2-2012: Servicios básicos de salud. Promoción y educación para la salud en materia alimentaria. Criterios para brindar orientación. Available at: http://www.dof.gob.mx/nota_detalle.php?codigo=5285372&fecha=22/01/2013 (last accessed 4 Oct 2014)

Secretaría de Salud (2014) ACUERDO por el que se emiten los Lineamientos a que se refiere el artículo 25 del Reglamento de Control Sanitario de Productos y Servicios que deberán observar los productores de alimentos y bebidas no alcohólicas pre-envasadas para efectos de la información que deberán ostentar en el área frontal de exhibición, así como los criterios y las características para la obtención y uso del distintivo nutrimental a que se refiere el artículo 25 Bis del Reglamento de Control Sanitario de Productos y Servicios. Available at http://www.dof.gob.mx/nota_detalle.php?codigo=5340693&fecha=15/04/2014 (last accessed 12 June 2014)

Secretaría de Economía y Secretaría de Salud (2010) NOM-051-SCFI/SSA1-2010: Especificaciones Generales De Etiquetado Para Alimentos Y Bebidas No Alcohólicas Preenvasados-Información Comercial y Sanitaria. Available at: http://www.dof.gob.mx/nota_detalle.php?codigo=5137518&fecha=05/04/2010 (last accessed 4 Oct 2014)

Secretaría de Economía y Secretaría de Salud (2014) MODIFICACIÓN de la Norma Oficial Mexicana NOM-051-SCFI/SSA1-2010, Especificaciones generales de etiquetado para alimentos y bebidas no alcohólicas preenvasados-Información comercial y sanitaria, publicada el 5 de abril de 2010. Se adicionan los incisos 3.2; 3.5; 3.17; 3.18; 3.21; 3.40; 4.2.9 con sus subincisos y se ajusta la numeración subsecuente; 4.5 con sus subincisos y el Apéndice Normativo A. Se modifica el capítulo 2 Referencias, así como el literal b) del inciso 3.11; 3.15; 4.2.8.1. Se ajusta numeración del capítulo 3 Definiciones, símbolos y abreviaturas. Available at: http://www.dof.gob.mx/nota_detalle.php?codigo=5356328&fecha=14/08/2014 (last accessed 4 Oct 2014)

Secretaría de Educación Pública y Secretaría de Salud (2010) ACUERDO mediante el cual se establecen los lineamientos generales para el expendio o distribución de alimentos y bebidas en los establecimientos de consumo escolar de los planteles de educación básica. DOF: 16/05/2014. Available at http://www.sep.gob.mx/work/models/sep1/Resource/635/1/images/acuerdolineamientos_sept.pdf (last accessed 12 June 2014)

Uauy R, Monteiro CA (2004) Food and Nutr Bull 25(2). The United Nations University

Chapter 34
Global and US Water Law and Sustainability: The Tragedy of the Commons and the Public Trust Doctrine

Zach Corrigan

Abstract This chapter explores some of the theoretical foundations that underlie the governance of the production of seafood. It focuses on the theoretical problem at the heart of how to regulate oceans and other common areas: how to avoid the so-called "tragedy of the commons." It focuses on the two juxtaposing types of public and private property regimes that have been erected to prevent such a tragedy. The former embraces the concept known as the Public Trust Doctrine. The latter type, private-property or "market-based" regimes, are exemplified by Individual Fishing Quotas, (also known as Individual Trading Quotas) and aquaculture, both of which are explored more in-depth in other chapters. Not left unexplored is critical literature indicating that the "tragedy of the commons" may be far too simplistic a model on which to base regulatory regimes.

34.1 Introduction: Common Spaces, Sovereignty, Regulation, Privatization, and Its Relationship to Food

This chapter explores some of the theoretical foundations that underlie the governance of the production of seafood. While overlapping with the laws governing other areas of traditional food production, the laws governing seafood production are distinct and unique. In no small part, producing food from the planet's water bodies, whether from lakes, streams, estuaries, or oceans, has followed a "hunter-gatherer" model, where individuals hunt, kill, and bring in food from animals that are not raised within the confines of their private property. Therefore, the

---

Z. Corrigan (✉)
Food & Water Watch, Washington, DC, USA
e-mail: zcorrigan@fwwatch.org

© Springer International Publishing Switzerland 2016
G. Steier, K.K. Patel (eds.), *International Food Law and Policy*,
DOI 10.1007/978-3-319-07542-6_34

foundational legal theories underlying the regulation of this food source have focused on the common spaces where hunters and gatherers obtain the food. The regulation of food produced from the world's water bodies is thus intimately related to environmental and property law.

The following sections focus on the theoretical problem at the heart of how to regulate oceans and other common areas, i.e., how to avoid the so-called "tragedy of the commons." It focuses on the two juxtaposing types of public and private property regimes that have been erected to prevent such a tragedy. The former type embraces the concept known as the Public Trust Doctrine and is extensively discussed in this chapter. The latter type, private-property or "market-based" regimes, are exemplified by Individual Fishing Quotas, (also known as Individual Trading Quotas) and aquaculture, which, while briefly mentioned in this chapter, are explored more in-depth in Chap. 36. Not left unexplored is critical literature indicating that the "tragedy of the commons" may be far too simplistic a model on which to base regulatory regimes.

34.2 Problems of Regulating the Commons

34.2.1 Private Property Versus Communal Property and the Public Trust Doctrine

The seafood that finds its way on to people's plates typically does not come from someone else's private farm, unlike much of their other food. Even aquaculture, or fish farming, typically involves the raising of fish in areas that at some point in time were common to all and were only later appropriated for private use. So, many of the legal and policy issues related to seafood are about how *common* resources, specifically the fish and waterways shared by all, are governed and managed.

A basic appreciation of western theories of property and how they apply to common resources is necesssary to understand the national and international governance of seafood. Under western common law, property is the right to possess, use, and enjoy a determinant thing.[1] First-year U.S. law students learn the now-axiomatic saying that "property" is a "bundle of rights," or "bundle of sticks," which means that the property is defined by the ability of users do a number of things with it, including to possess, use, and transfer it, as well as exclude others from these activities.[2] Possessory rights, for example, allow the holder to both collect when property is damaged and be subject to liability caused by the property.[3] Use or "usufructuary" rights entitle the holder to use property for specific

---

[1]Black's (2009), p. 1232.
[2]For a discussion of the concept's disputed origins, *see* Duncan (2002), n. 1. *See* Osherenko (2006), p. 331, for a discussion of the four categories: possession, use, exclusion, and disposition.
[3]Osherenko (2006), pp. 332–333, citing Osherenko (1995), pp. 1086–1087.

purposes.[4] Conversely, exclusionary rights entitle the holder to exclude others from using or trespassing on it and set conditions for its use.[5] Disposition rights entitle the holder to alienate or sell the property.[6] This legal terminology essentially describes the rights that a person has over property when they say that they "own" it.

Property can also be classified by its "regime," which means the rules that govern it.[7] For example, under Roman law, property was classified into four regimes: *res publica*, *res communes*, *res nullius*, and res *privatae*.[8] *Res publica* is property held by the government for the use and benefit of the public.[9] It includes roads, ports, rivers, and public buildings.[10] *Res communes* is property that can be accessed by anybody but not exclusively acquired, such as air.[11] *Res nullius*, interpreted narrowly, is that which is susceptible to private appropriation but is presently without an owner.[12] Historically, unoccupied lands and wild animals were labeled as *res nullius*.[13] And *res privatae* is private property.[14]

Relevant to this chapter, the Romans classified the seas as common property.[15] The first recorded law of the seas was the Digest of Justinian, and it declared that the sea is available to all, including for fishing.[16] By natural reason, fishery resources were awarded to their captor. Notwithstanding this broad statement of "rule of capture" principles, however, it should be noted that even as early as Justinian,

---

[4]*Id.*

[5]*Id.*

[6]*Id.*

[7]Black's (2009), p. 1286.

[8]The first three are actually subcategories of a broader category of property, "*res extra patrimonium*," which are things owned by no individual in particular. Blumm and Ritchie (2005a), p. 677, citing Buckland (1963); Wise 503, 508 (1996).

[9]Buck (1998), p. 4.

[10]*Id.*; Blumm and Ritchie (2005a), p. 677, citing J. Inst. 2.1.2; Dig. 1.8.5; Sohm (1907), p. 303–304.

[11]Buck (1998), p. 4.

[12]Gorina-Ysern (2004), p. 665, citing Fenn (1926), pp. 47, 52–53. This is also the basis for "the capture doctrine" which U.S. first-year law students are often taught through the case *Pierson v. Post. See id.* Some have interpreted this concept and used the term more broadly to mean "incapable of appropriation in the form of private ownership," which is a definition that overlaps with property that is subject of appropriation by States such as *res publicae*, *res universalis*, and *res communes*. Gorina-Ysern (2004), p. 655.

[13]Blumm and Ritchie (2005a), p. 677, citing Wise (1996), p. 503, 508; Sohm (1907), pp. 303–304.

[14]Buck (1998), p. 4. Observers have pointed out the limitations of this categorical approach, as it does encapsulate the nature of the resource, the domain it is found, and how it is used. Other observers suggest that a better way to categorize the resources is by its multiple attributes. Buck (1998), p. 3. And some of these observers have noted that the most important characteristics for categorizing commons is the feasibility of excluding others from the resource and the degree to which one's use can diminish the amount left for others. *Id.*

[15]Buck (1998), p. 76.

[16]*Id.*

citizens' rights to take wildlife were not absolute, and the state maintained sovereignty over the harvest of animals.[17]

The fact that oceans and wild fisheries can have no singular owner means, by definition, that no individual acting alone can exclude others from using them. It follows that, if the resource is scarce, then each person, acting solely in his or her self-interest, has an incentive to use as much of the resource as quickly as possible; otherwise the amount of available resource will diminish because of others' use. When the resource is exhaustible, the result is overuse, making everybody worse off.[18] This challenge for common property was popularized in ecologist Garret Hardin's "Tragedy of the Commons."[19] Under the thought-experiment posed by Hardin, a rational herder whose animals graze on communal property and whose costs from overgrazing are delayed and shared among the other herders will continue to add animals to the property to increase personal gain until all are made worse off by the resulting overgrazing.[20]

While the name of the tragedy is associated with Hardin, he was not the first to recognize it, as more than a decade before, economist H. Scott Gordon outlined a similar theory with fisheries in his "An Economic Theory of Common Property Research. The Fishery."[21] Gordon pointed out that the value of fishery resources comes from its immediate use of such resources,[22] framing the discussion using the economic concept of "resource rent."[23] With other industries, if there is an excess amount of profit to be had due to too much supply (these extra-normal profits are known as "rents"), it will attract those who will seek these rents out, which will, in turn, optimally dissipate.

With certain natural resources such as fisheries, on the other hand, this optimal rent dissipation is not possible.[24] Attracted by excess profit, fishermen will continually enter the fishery. But, as they do, the biological availability of the resource will become threatened.[25] This means a greater expenditure on fishing effort (more boats or more gear, an inefficiency often called "overcapitalization") than is optimal.[26] As a result, there is an inefficient allocation of resources, as more and more of fishermen's resources are spent chasing fewer and fewer fish. In a regulated

---

[17]Blumm and Ritchie (2005a), p. 678, citing Wise, p. 503.

[18]Block-Lieb (1993).

[19]These two characteristics: (1) that it is difficult to exclude other potential users and (2) the resource is finite and extractable by one to the detriment of other potential users are what turns simple communal property into what is called a "common-pool resource" which is subject to the tragedy of the commons of which Hardin and Gordon speak. See Buck (1998), p. 4; National Research Council (1999), p. 39.

[20]Ostrom (1990), p. 2.

[21]Id. p. 3.

[22]id.

[23]Gordon (1954), p. 130.

[24]National Research Council (1999), p. 21.

[25]Id. p. 22.

[26]Gordon (1954), p. 131, 133, 141.

fishery, this means higher costs for fishermen or shorter seasons, which may additionally create dangerous fishing conditions because fishermen will fish for longer hours or in worse weather. Alternatively, in an uncontrolled fishery, you simply diminish or exhaust the resource, resulting in the tragedy described in Hardin's thought-experiment.

It has long been argued that this problem, the lack of regulation of fishermen who over-exploit exhaustible fish populations, is the primary threat to the world's fisheries. While there are other serious issues affecting fish populations, including agricultural and urban runoff, storm water and sewage overflows, discharges from wastewater treatment facilities, oil spills, atmospheric deposition of nutrients and toxic chemicals, global-warming-induced acidification (all discussed in Chap. 36), unsustainable fishing has long been pointed to as chiefly responsible for declining wild-fish populations.[27] The FAO has also indicated that the situation is more critical for some fishery resources that are exploited solely or partially in the high seas, which, as discussed in Chap. 36, is the area beyond any nations' Exclusive Economic Zone (EEZ).[28]

And while the pace of overexploitation of fisheries has slowed since 1990 and progress has been made in reducing exploitation rates and restoring overexploited fish stocks and marine ecosystems in some areas, the world's fisheries remain in bad shape. More than half of the world's fish populations are at or very close to their maximum sustainable production levels as of 2009.[29] Among the remaining stocks, close to 30 % were overexploited, producing lower yields than their biological and ecological potential.[30] Most of the fish populations that account for about 30 % of the world's marine fisheries production are fully exploited and, therefore, have no potential for increases in production.[31] The declining global marine catch over the last few years, the increased percentage of overexploited fish populations, and the decreased proportion of non-fully exploited fish populations has led the United Nation's Food and Agricultural Organization (FAO) to conclude that the state of the world's marine fisheries is worsening.[32]

The overexploitation of this exhaustible fish resource is not only harmful to the fish populations themselves and marine ecosystems, but it also means that there are fewer fish for people to consume, which is economically harmful for those that rely on fishing for income.[33] As a result of this imbalance, it can certainly be argued that

---

[27]Turnipseed et al. (2009), pp. 5–7.

[28]*Id.* p. 13.

[29]Food and Agricultural Organization (2012), pp. 11, 13, 53.

[30]*Id.*

[31]*Id.* p. 12.

[32]*Id.*

[33]*Id.*

the tragedy of the commons still exists for much of the world's fisheries, notwithstanding the significant legal developments in the governance of fisheries which are discussed in Chap. 36. Policy makers are still wrestling with how best to prevent the tragedy, either because the laws passed are insufficient or not adequately enforced.

The policy options that are often proposed to solve the so-called tragedy of the commons are bi-modal: either governmental regulation or private control of the common resources. Under the former model, the government assesses the carrying capacity of the resource, i.e., the amount each individual could consume that would benefit all, and establishes penalties for those that consume above these limits, thus ensuring that each user is only utilizing the amount that is in everybody's best interests.[34] Therefore, with fishing as an example, a government can set a limit on the total allowable amount that fishermen or any group of fishermen can catch that is best for the long-term health of the fishery. As discussed further in Chap. 36, the main law governing U.S. fisheries, the Magnuson-Stevens Fishery Conservation and Management Act (MSA), as it has been amended, requires the government to establish Annual Catch Limits (ACLs) for fishermen, which correspond to the annual amount of catch that would not result in overfishing, as reduced by scientific uncertainty.[35] Accountability measures, such as in-season fishing closures or measures to correct for when fishermen exceed these limits, are meant to prevent the catch from exceeding these ACLs or offset the overages.[36] These measures, established by the government, are an attempt to limit the overuse of the exhaustible fishery resource.

Under a privatization or unitization model (which are similar models but not necessarily the same), the resource is divided or parceled. The theory is that individuals who are allocated property interest have the exclusive right to the resource and the commons tragedy is averted because users no longer need to maximize the resource's value in the short-term. They can limit its use in order to maximize long-term gain.[37] The conversion of the resource from public to private property is also supposed to avoid economic waste, inefficiency, and internalize the benefits and costs of using the resource.[38] Resource owners incur the entire cost of overuse and thus carefully care for the resource.[39] Arguably, the most recent significant influence on fisheries policy has come from advocates of property-rights-based and free-market policies as a means to avert the tragedy of the commons.[40]

---

[34]Ostrom (1990), pp. 9–10.
[35]50 C.F.R. § 600.310(f)(1)-(7) (2014).
[36]*Id.* § 600.310(g)(1)-(3).
[37]*See* Ostrom (1990), pp. 12–13.
[38]Osherenko, p. 329, citing Demsetz (1967), p. 354.
[39]Thompson (2000), p. 244.
[40]Rieser (1999), p. 397.

In the United States, these two divergent models are perhaps best exemplified in two conceptual frameworks: On the one hand is the Public Trust Doctrine, traditionally a common law doctrine that some argue serves as a foundation for much of the MSA. The Public Trust Doctrine treats *res communes or res nullius* property, or property that has no owner, as *res publica* property, meaning property held by the government for the use and enjoyment of the public.[41] On the other hand, IFQs seek to make the property more closely resemble *res privatae*, or private property. Similarly, aquaculture, also known as fish farming, is a form of privatization or unitization, as it seeks to fence in the fisheries resource for exclusive use.

Neither IFQs or the Public Trust Doctrine, as they have been implemented in U.S. state and federal law, however, are perfect idealizations of either bi-modal solution—neither are a completely privately owned or government-managed scheme. Rather, the country has typically incorporated a blend of both into the management of its fisheries.[42]

Not only is there no such thing as a purely private or public regulatory regime for fisheries, the tragedy-of-the-commons-paradigm itself has come under sharp criticism for being unrealistic. Accordingly, the effectiveness of policy solutions that are supposedly dictated by the theories have been called into question. For example, Elinor Ostrom won the 2009 Nobel Prize in economics for her work demonstrating that communities can and have managed to avoid the tragedy of the commons without reliance on either a purely governmental or private alternatives.[43] She and others demonstrate that individuals can and do communicate and agree to rules of use that can improve their joint outcomes, and these critics believe that such rules can serve as design principles for institutional change.[44] Others have pointed out that Hardin was analyzing the tragedy of open access, or *res nullius* resource management regime, which is not an accurate representation of the present status of the oceans and fisheries resource, and thus does not form a good foundation for policy solutions to existing problems such as overfishing.[45] As U.S. professors Macinko and Bromley put it: "It seems useful to point out that the issue for fishery policy is certainly *not* a choice between some blissful state *with* property rights, and a state of nature in which anarchy is the order of the day. The issue is, rather, which

---

[41] As law professor Hirokawa puts it "By operating as a condition precedent to both ownership of land and its use, the public trust doctrine intervenes in capture scenarios to prevent common resources from consumption, transformation, or other capture-like consequences of privatization." Hirokawa (2010), p. 212.

[42] *See, e.g.*, Professor Rieser's discussion of IFQs in "The Ecosystem Approach: New Departures For Land And Water: Fisheries Management: Property Rights and Ecosystem Management in U.S. Fisheries: Contracting for the Commons?" Rieser (1997), pp. 818–820.

[43] Ostrom (1990), p. 14, 21; Rieser (1999), p. 400. For more on the Ostrom's Nobel Prize, *see* http://www.nobelprize.org/nobel_prizes/economic-sciences/laureates/2009/ostrom-facts.html, last visited Apr. 4, 2014.

[44] Rieser (1999), p. 402.

[45] Macinko and Bromley (2004), pp. 650–654; Rieser (1999), p. 399, citing Bromley (1991), p. 30.

property regime is best suited for particular settings and circumstances."[46] For example, at least since 1976, the MSA has regulated the fishery resources in the U.S. EEZ,[47] and the many refinements in the law have improved the status of many U.S. fisheries.

Notwithstanding these criticisms, the tragedy of the commons paradigm, along with its public and private policy solutions merit some understanding, as they underpin management of fisheries at both the national and international levels. Accordingly, the following section explores the Public Trust Doctrine in greater depth.

34.2.2 The Public Trust Doctrine

The Public Trust Doctrine is present in many countries' legal systems.[48] It perhaps finds its ancestry in the principles of common resources recognized by ancient dynasties in China, Islamic law, and various Native American cultures.[49] Scholars have most immediately traced the doctrine to the Roman Emperor Justinian, although some scholars criticize this historical analysis as flawed, arguing that the emperor's statements were those of ideals, not actual Roman legal doctrine.[50] Bracton incorporated the doctrine from Roman law into his writings on the English common law,[51] but it did not become a popular doctrine advanced to enhance environmental protection until much more recently, due to U.S. law professor Joseph Sax's 1970 seminal article "The Public Trust Doctrine in Natural Resource Law: Effective Judicial Intervention."[52] Since this time, the theory has become a significant component of the cultural and legal framework influencing all fisheries management in the United States.[53] As can be seen in the Chap. 36, it is a concept that also undergirds some international regulation of fisheries.

Broadly, this doctrine provides that the government holds certain public resources in trust for the benefit of the nation.[54] And as such, the state cannot abdicate direction and control of these resources by allowing them to be

---

[46]Macinko and Bromley (2004), pp. 650–654.

[47]*Id.*

[48]In their symposium article "Internationalizing the Public Trust Doctrine: Natural Law and Constitutional and Statutory Approaches to Fulfilling the Saxion Vision," Professor Michael C. Blumm and Rachel D. Guthrie outline the importance of the doctrine in the countries on four continents India, Pakistan, the Philippines, Uganda, Kenya, Nigeria, South Africa, Brazil, Ecuador, and Canada. Blumm and Guthrie (2012).

[49]Turnipseed et al. (2009), p. 10, citing Wilkinson (1989), pp. 428–430; Deveney (1976).

[50]Lazarus (1986), p. 634; *id.* citing Deveney (1976), p. 17, 29.

[51]Turnipseed et al. (2009), pp. 10–11.

[52]Blumm and Ritchie (2005b), p. 341, citing Sax (1970).

[53]National Research Council (1999), p. 49.

[54]*Id.*

permanently sold or transferred exclusively for private use without public benefit. This framing was put forward by the U.S. Supreme Court in the famous *Illinois Central Railroad* decision,[55] which Professor Sax seized upon, along with the decision's progeny, for developing the modern public trust doctrine thesis.[56] Under the doctrine, the government has authority to grant limited-use rights to this public property (including leases, easements, and permits) as long as it is short of total ownership.[57] Use of the property is limited to activities that do not harm or interfere with the public-trust purposes of the property.[58] The doctrine provides a theoretical foundation for the government's placing of limits on any individual's exploitation of scarce common resource, such as fisheries, as the resource is not solely to be utilized by the first captor. When the government manages a common resource in trust for the public, it is able to prevent self-interested users from overexploiting the resource to the detriment of all. Thus, the doctrine, at least in concept, has the effect of preventing the so-called tragedy of the commons.

While the doctrine seems straightforward, its scope and application, including to fisheries, has been the subject of much debate and confusion. The following sections outline the debate and analyze whether the doctrine even applies to fisheries. Questions have also been raised and will be discussed below about whether the doctrine is solely applicable to U.S. state resources or whether it applies where the states have no property interests, i.e., U.S. federal resources such as the EEZ. Finally, while there is no doubt that the Public Trust Doctrine undergirds U.S. fisheries law and serves as a foundation for a number of its provisions, there are questions about what the doctrine has to offer that existing statutory law does not. A detailed discussion of each of these controversies is presented in the following three sections of this chapter.

34.2.3 The Public Trust Doctrine's Application to Fisheries

The first question is whether the doctrine even applies to fisheries. This common-law doctrine has traditionally applied to the beds of waterways,[59] and, thus, confusion has arisen in relation to its applications to wild animals, also known as *feroe naturoe*. With wild animals such as fish, the concept of the state as the extant owner of the property runs headlong into a separate common-law doctrine that states that title for such resources can be acquired once the resource is reduced to possession, i.e., under "the rule of capture."

---

[55]*Ill. Cent. R.R. Co. v. Ill.*, 146 U.S. 387, 453–454 (1892).
[56]Lazarus (1986), pp. 641–642.
[57]Osherenko (2006), p. 362.
[58]Sax (1970), p. 477; *id.* p. 367.
[59]National Research Council (1999), p. 42.

This contradiction can be observed in the Supreme Court's eventual repudiation of its *Geer v. Connecticut* decision over the last century.[60] At issue in the *Geer* case was a state statute that prohibited the killing and possession of certain birds with the intent of bringing them out of state.[61] In evaluating whether this law violated the U.S. Constitution's Interstate Commerce Clause, the Supreme Court discussed the state's interest in regulating wild animals.[62] While recognizing that wild animals were common to all, and, therefore, all could acquire title by possession, this private right was still governed by the ultimate law-giving power of the state.[63] After quoting from scholars who concluded that this was the European legal tradition, the Court cited the seminal William Blackstone commentaries for the position that, while "every man from the prince to the peasant has an equal right of pursuing and taking to his own use all such creatures as are *feroe naturoe*, and, therefore, the property of nobody, . . . it follows . . . that this natural right . . . may be restrained by positive laws enacted for reasons of state or for the supposed benefit of the community."[64] The Court found that the state was therefore entitled to erect the statute prohibiting and this did not violate the Commerce Clause.[65]

Since this time, the U.S. Supreme Court has steadily interpreted the state's interests defined in *Geer* narrowly, suggesting that the government does not in fact "own" fisheries. The Court has held that a state's property interest in wild animals cannot invalidate a treaty under the U.S. Constitution's Tenth Amendment.[66] Such property interests cannot empower the state to impose far higher license fees on citizens of other states without violating the Privileges and Immunities Clause of the U.S. Constitution.[67] Nor are state statutes that prohibit federal licensees from catching fish in state waters permissible under the Constitution's Supremacy Clause, notwithstanding state claims that the state was granted ownership and title under the Submerged Lands Act.[68] Ultimately, the Supreme Court directly overruled *Geer* directly in a Commerce Clause challenge.[69] The Court, however, has never repudiated that the state has the authority to manage wildlife in trust for the public.[70] Rather, the Court has explained that the state's "ownership" of

---

[60]*Geer v. Connecticut*, 161 U.S. 519 (1895).

[61]*Id.* p. 521.

[62]*Id.* p. 522.

[63]*Id.*

[64]*Id.* p. 527 (quoting 2 Blackstone Commentaries 410).

[65]*Id.* p. 534.

[66]*See Mo. v. Holland*, 252 U.S. 416, 434 (1920) (calling the state's claim of title a "slender reed" because "[w]ild birds are not in the possession of anyone; and possession is the beginning of ownership").

[67]*See Toomer v. Witsell*, 334 U.S. 385, 401–402 (1948).

[68]*See Douglas v. Seacoast Prods., Inc.*, 431 U.S. 265, 283 (1977).

[69]*Hughes v. Oklahoma*, 441 U.S. 322, 335 (1979). For an in-depth treatment of these cases and the demise of *Geer*, see Macinko and Bromley (2004), pp. 631–634.

[70]National Research Council (1999), p. 42.

wild animals, as analyzed and portrayed in *Geer*, is not the same as title. Consequently, states exercise their duties with such property as part of its overall police power.[71]

Therefore, notwithstanding its character as *feroe naturoe*, the statements in *Geer* remains good law in the United States that the state has the power and authority to regulate wild fish populations as a public trust resource.[72]

As described below, there are other questions about whether the Public Trust Doctrine applies outside of state lands, whether it is a resource that is displaced by statutory law, and whether it is useful as a concept for the management of fisheries. Nonetheless, there is little strength in the argument that the Public Trust Doctrine does not apply to fish and fisheries solely because they are *feroe naturoe* and governed strictly by the rule of capture.

34.2.4 The Public Trust Doctrine's Application to Federal U.S. Resources

In order for the Public Trust Doctrine to have any protective value for resources such as federal fisheries, it must not simply remain a doctrine applicable to states. A related question is how, as a common law doctrine, it can coincide, if at all, with federal statutory law, or is it displaced by federal statutes like the MSA? There is no reason that the common law Public Trust Doctrine should have any importance to federal fisheries if it has been completely displaced by statutory law. The subsequent discussion sheds some light on these questions.

In the United States, the Public Trust Doctrine traditionally has been raised either strictly as a matter of state law or, as discussed above, as an (now, unavailing) argument for the state's interest to trump federal interests, such as those embodied in the U.S. Constitution's Treaty, Supremacy, or Commerce clauses. There has been little case law discussing its applicability to those areas where the federal government has asserted some sovereignty, such as the U.S. EEZ.[73] Indeed, the legal authority and responsibility of the U.S. government to protect public trust resources in the EEZ has never been fully and expressly established.[74] Neither courts nor Congress have ever directly stated that there is a common law or

---

[71]*See* Babcock (2007). And this concept has been bolstered by a number of state court decisions. National Research Council (1999), p. 42. Indeed, Justice Burger explained in his concurring opinion in *Baldwin v. Fish & Game Comm'n*: "A State does not 'own' wild birds and animals in the same way that it may own other natural resources such as land, oil, or timber. But, as . . . the doctrine is not completely obsolete. It manifests the State's special interest in regulating and preserving wildlife for the benefit of its citizens." 436 U.S. 371, 392 (1978).

[72]And, in fact, "absent federal-state conflict, it continues to endure today, as virtually all states claim ownership of wildlife in trust for their citizens." Blumm and Ritchie 2005a, p. 676.

[73]National Research Council (1999), p. 43.

[74]Turnipseed et al. (2009), p. 1.

statutory public-trust duty applicable to either federal lands or federal ocean waters, even though the principles underlying the doctrine appear in federal environmental statutory language, agency mission statements, and national policy recommendations.[75]

But the doctrine's development in state law in the United States strongly suggests that it applies to federal resources. In the latter half of the eighteenth and into the nineteenth centuries, the U.S. Supreme Court and various state courts described the public trust that pertains to state common resources as an essential attribute of sovereignty, first under the Crown, and then passing to the 13 American colonies when they gained independence.[76] Indeed, the common law "enhanced and extended" the public aspects of these resources.[77]

Under the English common law, a distinction was made between the tidal and non-tidal waterways. With the latter, the public had less of an interest. It only retained the right of passage; and title up to the center thread of the waters was typically held by the riparian landowners, which accorded them the exclusive right of fishery in the stream and entitled them to compensation for any impairment of their right to the enjoyment of their property.[78] In the United States, on the other hand, courts rejected the distinction between tidal and nontidal waters and concluded that states presumptively held title to navigable waterways.[79] Similarly, the common law in the United States did not recognize the sovereign leader's private property rights to lands under navigable waterways. "All these developments in American law are a natural outgrowth of the perceived public character of submerged lands, a perception which underlies and informs the principle that these lands are tied in a unique way to sovereignty."[80] It could therefore be argued that just like with state waterways, where the United States has asserted its sovereign authority over navigable waterways for the benefit of the public, the Public Trust Doctrine likewise attaches as a matter of U.S. federal law.[81]

One example of this exercise of sovereignty as detailed in Chap. 36, is President Reagan's 1983 declaration that the U.S. EEZ extends 200 nautical miles and within that zone:

> The United States has, to the extent permitted by international law . . . sovereign rights for the purpose of exploring, exploiting, conserving and managing natural resources, both living and non-living, of the seabed and subsoil and the superjacent waters and with regard

---

[75]*Id.* p. 8.

[76]*See, e.g., Martin v. Lessee of Waddell*, 41 U.S. 367, 416 (1842); *Arnold v. Mundy*, 6 N.J.L. 1, 71, 78 (Sup. Ct. 1821); *Shively v. Bowlby*, 152 U.S. 1, 57 (1894).

[77]*Idaho v. Coeur D'Alene Tribe*, 521 U.S. 285 (1997).

[78]*Id.*

[79]*Id.* p. 286.

[80]*Id.*

[81]*But see* Lynch (2007), arguing that unlike state waters, the EEZ has traditionally been treated as a global commons and thus should be treated as *res nullius*.

to other activities for the economic exploitation and exploration of the zone, such as the production of energy from the water, currents and winds; and jurisdiction with regard to ... the protection and preservation of the marine environment.[82]

The MSA likewise "exercis[es] ... sovereign rights for the purposes of exploring, exploiting, conserving, and managing all fish, within the [EEZ]."[83] Other observers have noted that the sovereignty (between 3 and 12 nautical miles) and sovereign rights (from 12 to 200 nautical miles) asserted by the United States in the EEZ is enough, not only for establishing the applicability of the Public Trust Doctrine, but also for the doctrine to be recognized under international law (Fig. 36.1).[84]

In addition to sovereignty-based arguments, some observers have also pointed to the broad language in several U.S. Supreme Court decisions such as *Illinois Central Railroad v. Illinois* as suggesting the doctrine's federal applicability.[85] *Shively v. Bowlby* also indicates that property acquired by the United States is a public trust resource.[86] Likewise, in settling the dispute between California and the federal government over oil leases within three miles of the coast, the Supreme Court refused to accept the state's argument that the federal government had waived its property interest, saying that "the Government, which holds its interests here as elsewhere in trust for all the people, is not to be deprived of those interests[,] ... [by] officers who have no authority at all to dispose of Government property...."[87] Thus, the federal government retains its property interests through the Public Trust Doctrine.

Several federal U.S. district court decisions support the view that there is a federal public trust. For example, in *In Re Steuart Transportation Co.*, an Eastern District of Virginia court held that both the federal and state governments could sue the owner of an oil transport vessel for injury to migratory wildfowl because "under the public trust doctrine, the State of Virginia *and the United States* have the right and duty to protect and preserve the public's interest in natural wildlife resources."[88] In *United States v. 1.58 Acres of Land*, a federal District Court in Massachusetts held that the United States could condemn state public trust property, but said that the federal government was restricted from "convey[ing] land below the low water mark to private individuals free of the sovereign's *jus publicum*,"[89] ostensibly referring to the government's public duty not to dispose of the property for private purposes. In the more recent *United States v. Ross* decision, a South Dakota district court found that the United States had sufficient interest in obtaining restitution from a person convicted in the aiding and abetting of

---

[82] 48 Fed. Reg. 10,605 (March 14, 1983).
[83] 16 U.S.C. § 1801(b)(1).
[84] *See* Turnipseed et al. (2009), pp. 34–40.
[85] National Research Council (1999), p. 42.
[86] *Shively v. Bowlby*, 152 U.S. at 57.
[87] *United States v. California*, 332 U.S. 19, 40 (1947).
[88] *In Re Steuart Transportation Co.*, 495 F. Supp. 38, 40 (E.D. Va. 1980) (emphasis added).
[89] *United States v. 1.58 Acres of Land*, 523 F. Supp. 120, 124 (D. Mass. 1981).

taking of migratory birds under the federal Victim and Witness Protection Act because "the Government surely has a legitimate and substantial interest in preserving and protecting hawks in its air space as part of the public's natural wildlife resources. This interest does not derive from ownership of the resources, but from the duty the government owes to the people."[90] These cases support the notion that the government can also recover for the federal trust, which extends to waters.

Notwithstanding, a recent Supreme Court decision has led other courts to conclude that the Public Trust Doctrine is strictly a state law doctrine and does not apply to federal resources. For instance, at dispute in *PPL Mont., LLC v. Montana* were the beds of three rivers on which a power company, PPL, had hydroelectric facilities.[91] In 2003, PPL was sued because the company had built its facilities on riverbeds that were state-owned and part of Montana's school trust lands.[92] The Court declared that the Montana Supreme Court erred is finding the river under state title. Most of the Court's decision relates to how to apply the test for "navigability" under the U.S. Constitution's "Equal Footing" doctrine.[93] By the end of the decision, however, the Court addressed Montana's argument that denying state title "will undermine the public trust doctrine, which concerns public access to the waters above those beds for purposes of navigation, fishing, and other recreational uses."[94] The Court concludes that,

> [u]nlike the equal-footing doctrine . . . the public trust doctrine remains a matter of state law . . . subject as well to the federal power to regulate vessels and navigation under the Commerce Clause and admiralty power. While equal footing cases have noted that the State takes title to the navigable waters and their beds in trust for the public, . . . the contours of that public trust do not depend upon the Constitution. Under accepted principles of federalism, the States retain residual power to determine the scope of the public trust over waters within their borders, while federal law determines riverbed title under the equal-footing doctrine.[95]

This language has been seized upon by at least two federal courts as an argument that the Public Trust Doctrine has no basis in federal law.[96] However, this is a mistaken interpretation of the *PPL* decision. The Public Trust Doctrine that

---

[90]*United States v. Ross*, No. CR. 11-30101-MAM, 2012 U.S. Dist. LEXIS 146285, at *5-6 (D.S.D. Oct. 10, 2012) (footnotes omitted). *See also United States v. Burlington N. R.R. Co.*, 710 F. Supp. 1286, 1287 (D. Neb. 1989) ("In view of this trust position, and its accompanying obligations, it appears that the United States . . . can maintain an action to recover for damages to its public lands and the natural resources on them, which in this action would encompass the destroyed wildlife.").

[91]*PPL Mont., LLC v. Montana*, 132 S. Ct. 1215, 1225 (2012).

[92]*Id.*

[93]*Id.* at 1228.

[94]*Id.* at 1234.

[95]*Id.* at 1235.

[96]*See, e.g., Alec L. v. Jackson*, 863 F. Supp. 2d 11, 15 (D.D.C. 2012), aff'd, 561 Fed. Appx. 7 (D.C. Cir. 2014). The courts found that the plaintiffs could not invoke federal subject matter jurisdiction for a claim against Defendant EPA for failing to reduce greenhouse gases in violation of the Public Trust Doctrine, as it was foreclosed by the *PPL* decision, which determined that it was a state law doctrine.

Montana was seeking to rely on was not one based in federal law. Rather, it was Montana's own state-law Public Trust Doctrine. In addition, while the Supreme Court used language that suggests the doctrine only pertains as a matter of state law, i.e., saying that the doctrine "remains a matter of state law . . .[,]" and that the contours of the doctrine "do not depend upon the Constitution[,]" the Court was contrasting this state doctrine relied upon by Montana, with the obviously supreme Equal Footing doctrine, which has its foundations in federal U.S. constitutional law.[97] The Supreme Court has not issued any interpretation about the existence of the doctrine as a matter of federal law. Instead, the court simply reached the rather mundane conclusion that under, the U.S. Constitution's Supremacy Clause, the state doctrine would not trump federal law.[98]

Regardless, the *PPL* decision's inartful language poses a new obstacle for advocates who argue that the Public Trust Doctrine is equally applicable to federal resources, such as fisheries in the U.S. EEZ, and that it is derived from the U.S. Constitution.[99] This, however, does not mean that the Public Trust Doctrine is not a legitimate source of common law derived from other positive sources of law, such as the MSA or the federal government's assertion of sovereign interest in the EEZ. As demonstrated later in this chapter, this use of principles to guide the MSA's interpretation may be where it is most useful for protecting fisheries.

34.2.5 The Public Trust Doctrine and the Magnuson Steven Fishery Management and Conservation Act (MSA)

Even if the Public Trust Doctrine does apply to federal resources, and particularly federal fishery resources, there remains an open question as to whether it is displaced by federal *statutory* law, such as by the MSA. When Congress addresses an issue governed under the federal common law, the federal statute displaces the common law.[100] The test is whether the statute directly addresses the particular question.[101] At least one federal court recently stated (in *dicta*) that the Public Trust

---

[97]*Id.*

[98]*Accord* Adler (2013), p. 1706 n. 14. As law professor Adler remarks "In *PPL Montana*, the Supreme Court reiterated earlier holdings that federal law controls the navigability for title test for purposes of the equal footing doctrine, but that state law governs the scope of the public trust doctrine within individual states."

[99]Since the U.S. Supreme Court's *Erie R.R. v. Tompkins*, 304 U.S. 64, 78 (1938) decision, it has been accepted that federal common law must be derived from a federal source of positive law, such as the Constitution or a statute. Chase at pp. 113, 138, 162 (2010). The *PPL* decision terminates the argument that the doctrine has a basis in the U.S. Constitution's Equal Footing Doctrine, as some say that the *Illinois Central* decision should be interpreted. *See, e.g., id.* pp. 140–142 (arguing that that the *Illinois Central* Court's articulation of the Public Trust Doctrine was derived from the U.S. Constitution's Equal Footing Doctrine).

[100]*Am. Elec. Power Co. v. Connecticut*, 131 S. Ct. 2527, 2537 (2011).

[101]*Id.*

Doctrine could not "supplant or supplement" the requirements established by the MSA for IFQ programs,[102] which simply begs the question: when has the statute or another source of law spoken so directly to an issue that it displaces the common law doctrine altogether?

Generally, if a statute does not "directly address" the question at issue, the U.S. Supreme Court has allowed the federal common law to supplement the statutory scheme. For example, where the Federal Water Pollution Control Act originally only provided broad authority to the states to enact pollution standards and gave the U.S. Attorney General the power to sue for abatement of pollution, the Court found that federal common law still could be applicable.[103] Where the Congress subsequently amended the act to create an all-encompassing program of water pollution regulation covering overflows from the source at issue, however, the court found that there was no interstice for the common law to fill.[104] In the latter case, "the relevant question for purposes of displacement is 'whether the field has been occupied, not whether it has been occupied in a particular manner.'"[105] It is not clear that the MSA has directly spoken to a number of issues that arise with federal fisheries management. For example, as discussed in Chap. 36, the issue of whether or not the act was designed to cover fish farming in federal waters is hotly disputed. To the extent that the act does not allow for federal regulation or provides for limited management of aquaculture in the EEZ, the state and federal public trust doctrines, just like other common law doctrines such as public and private nuisance, should be available to fill the gaps.[106]

Likewise, in terms of the IFQs, in the *Pac. Coast Fed'n of Fishermen's Ass'n* case mentioned above which seemed to curtail the applicability of the Public Trust Doctrine, the question was whether the federal government's National Marine

---

[102]*Pac. Coast Fed'n of Fishermen's Ass'n v. Locke*, No. C 10-04790 CRB, 2011 U.S. Dist. LEXIS 86662, at *52 (N.D. Cal. Aug. 5, 2011).

[103]*Ill. v. City of Milwaukee*, 406 U.S. 91, 102–103 (1972).

[104]*City of Milwaukee v. Ill.*, (*Milwaukee* II) 451 U.S. 304, 318–319, 23 (1981).

[105]*Am. Elec. Power Co. v. Connecticut*, 131 S. Ct. 2527, 2538 (2011) (citing *City of Milwaukee v. Ill.*, 451 U.S. at 324). For a discussion of the interplay between the two *City of Milwaukee* cases and the *American Electric* decision, *see* Winters (2011), p. 382.

[106]*See* Babcock (2007), pp. 60–61, who argues that federal laws involving the management of fish would not displace application of the common law to fish farming. *See also*, Turnipseed et al. (2009), pp. 48–49, which argues that at, with the EEZ, virtually all of the key prerequisites laid out by the Supreme Court in *Texas Indus., Inc. v. Radcliff Materials, Inc.*, 451 U.S. 630, 640 (1981) exist for when federal courts can create federal common law, except that Congress did not give this power expressly to the courts. *See also Michigan v. U.S. Army Corps of Eng'rs*, 667 F.3d 765, 769–771 (7th Cir. 2011) (denying that states would likely succeed on their common law nuisance claims against the U.S. Army Corps of Engineers, but assessing a greater likelihood than the District Court did because federal common law can apply where there is an overriding federal interest in the need for a uniform rule of decision, and nuisance law would cover the non-native species that were will migrating into Lake Michigan, even though they were not traditional pollutants.) *See United States v. Rainbow Family*, 695 F. Supp. 314, 327 (E.D. Tex. 1988) (allowing the federal government to pursue an injunction as a remedy for a common law nuisance claim in lieu of the applicability of its federal permitting scheme).

Fisheries Service's (NMFS) program had violated the Public Trust Doctrine because the agency allocated a "quasi-permanent harvest privilege" to certain individuals, without any pre-established criteria for the government's potential revocation, limitation, or modification of such privileges.[107] One of the *amicus curiae*, represented by this chapter's author, argued that the program violated the public-trust values embedded in the MSA by substantially impairing the public's ownership interest of the fishery resource because without such criteria, such privileges could not be terminated, modified, suspended, or even limited—or at least not in a timely manner—to address the public need.

Arguably, since Congress had not occupied the field by establishing the terms and criteria for revocation, modification, or suspension of such privileges in the MSA, the Public Trust Doctrine should be able to fill the gap and allow judicial precedent to guide agency action. The MSA, at least regarding this aspect of IFQs, might therefore be much more like the regulatory system of the Federal Water Pollution Control Act in the *Milwaukee I* case, mentioned above, which the Supreme Court found did not displace federal common law because it did not occupy the field. The district court in *Pac. Coast Fed'n of Fishermen's Ass'n* disagreed. But such a conclusion is only binding in the Northern District of California, and it may not foreclose the possibility of other common law challenges to IFQ programs under the MSA and Public Trust Doctrine. Therefore, it is at least arguable that the common law Public Trust Doctrine is applicable to provide judicially developed guiding principles that can fill certain gaps existing in the U.S. law governing fisheries. How this might strengthen the law is explored in Sect. 35.1. The next section, explores criticism of and limitations with the doctrine.

34.3 The Public Trust Doctrine's Limitations

Among the criticisms of the Public Trust Doctrine is that it is anachronistic and that it prevents holistic or ecological thinking.[108] Dissenters also warn that trying to protect natural resources by using private-property doctrines may render the resources more vulnerable to degradation.[109] They also challenge that the doctrine places emphasis on the judiciary and the common law rather than on more powerful and comprehensive legislative efforts to protect the environment.[110] Others challenge that it encourages judicial takings of private property without just compensation.[111]

---

[107] *See* 50 C.F.R. § 660.25(h)(2)(ii) (2013).
[108] Klass (2006), p. 699, citing Delgado (1991), pp. 1212–1218.
[109] Ryan (2001), p. 493, citing Lazarus (1986), p. 696.
[110] Klass (2006), p. 700, citing Lazarus (1986), pp. 656–715.
[111] *Id.* citing Huffman (1989), pp. 565–568, Smith and Sweeney (2006), pp. 322–341, Thompson Jr. (1990), p. 1449. *See also Stevens v. City of Cannon Beach*, 510 U.S. 1207, 1212 (1994) (Scalia, J., dissenting).

Nonetheless, it has been observed that some of these "green" dissenters, who believed that the doctrine would not best enhance environmental protection, have been less vociferous since the 1980s. At least on observer has questioned whether this may be because of hostility to protection of the environment expressed both by more recent U.S. presidential administrations and their judicial nominees, perhaps squelching any optimism that the Public Trust Doctrine is unnecessary in the presence of more democratic means of protecting the environment.[112] Additionally, court decisions have given substantial discretion to agency interpretations of their organic statutes and administrative records, while other judicial doctrines have curtailed citizen standing and limited the power of Congress and agencies to act without it amounting to a taking. These judicial decisions might have taken the wind out of the sails of Public-Trust-Doctrine critics, who thought that it was unnecessary for resource protection.[113]

And these recent developments may now be the inspiration for a rather recent revival of the doctrine.[114]

Notwithstanding its many controversies, the Public Trust Doctrine provides a conceptual framework for the regulation of certain scarce natural resources such as fisheries. The basic concept, that such resources are to be managed for the benefit of the public by the government, provides the theoretical basis that allows, and even obligates, the government to regulate the exploitation of such resources, thus avoiding the so-called tragedy of the commons. Despite its importance, and in no small part because of confusion surrounding the doctrine, it is currently a concept that, at best, remains in the background of the management of fisheries resources. Many advocates, however, hope that its recognition and development will better serve to protect fisheries and other U.S. federal resources, as it has done for individual states.

34.4 Conclusion

This chapter explores the "tragedy of the commons" as the theoretical catalyst underlying modern national and international fisheries regulation. The common-law Public Trust Doctrine represents a public model for tackling this problem. IFQs and aquaculture are attempts to solve this problem by privatizing the fisheries resources, and are discussed in later chapters. The management of fisheries has been done through a blending of these types of regimes, to limited success, as too much of the world's fisheries remain overexploited.

Because of the ongoing and worsening condition of fisheries, formal judicial and agency recognition of the Public Trust Doctrine in U.S. waters would enhance the

---

[112]Ryan (2001), p. 492.
[113]*Id.*
[114]*See, e.g.*, Turnipseed et al. (2009), pp. 34–40.

regulation of the fisheries. Not only would it ensure that regulators view fisheries with a proper long-term vision and serve as a model for the international regulation of fisheries—while limiting the discretion that unelected officials have to dispose of public property for purely unrecompensed private use—its proper development can fill gaps in the MSA and extraterritorially strengthen international law, provide the public with a means to protect the resource, and hold agency officials accountable when they abdicate their responsibilities. At least one group of legal scholars hopes that the U.S. President would officially recognize the Public Trust Doctrine in federal resources, but this has yet to happen.[115] With the Supreme Court's rather recent *PPL* decision, and subsequent courts interpreting it to preclude judicial recognition of a federal doctrine, it is more important than ever that the President or agencies recognize this doctrine, lest it may be forgotten or relegated to state law, where it can do little to protect the domestic and international fisheries that need the doctrine the most.

References

Adler RW (2013) The ancient mariner of constitutional law: the historical, yet declining role of navigability. Wash Univ Law Rev 90(14):1643–1706
Babcock HM (2007) Grotius, ocean fish ranching, and the public trust doctrine: ride 'Em Charlie Tuna'. Stan Environ Law J 26(5):60–61
Baer SD (1988) The Public Trust Doctrine – a tool to make federal administrative agencies increase protection of public land and its resources. Boston Coll Environ Aff Law Rev 15:385–433
Bevis KD (2005) Stopping the silver bullet: how recreational fishermen can use the Public Trust Doctrine to prevent the creation of marine reserves. Sea Environ Law J 13:171
Black's Law Dictionary 1232, 1286 (9th ed. 2009)
Block-Lieb S (1993) Fishing in Muddy Waters: clarifying the common pool analogy as applied to the standard for commencement of a bankruptcy case. Am Univ Law Rev 42:337–373
Blumm MC, Guthrie RD (2012) Internationalizing the Public Trust Doctrine: natural law and constitutional and statutory approaches to fulfilling the Saxion vision. UC Davis Law Rev 45:741, 745, 748, 765, 769, 780
Blumm MC, Ritchie L (2005a) The pioneer spirit and the Public Trust: The American Rule of capture and state ownership of wildlife. Environ Law 35:673, 675–679
Blumm MC, Ritchie L (2005b) Lucas's unlikely legacy: the rise of background principles as categorical takings defenses. Harv Environ Law Rev 29:321, 341
Bromley DW (1991) Environment and economy: property rights and public policy 22:30
Buck SJ (1998) A global commons, an introduction, vol 3–4. Island Press, Washington, DC, p 75
Buckland WW (1963) A text-book of roman law from Augustus to Justinian, 3rd edn. p 182
Chase CS (2010) The Illinois central Public Trust Doctrine and federal common law: an unconventional view. Hastings W -NW J Environ Law Policy 16:113, 138, 140–142, 162
Christie DR (2004) Marine reserves, the Public Trust Doctrine and intergenerational equity. J Land Use Environ Law 19:427–434

---

[115] *Id.*

Delgado R (1991) Our better natures: a revisionist view of Joseph Sax's Public Trust theory of environmental protection, and some dark thoughts on the possibility of law reform. Vand Law Rev 44:1209, 1212–1218

Demsetz H (1967) Toward a theory of property rights. Am Econ Rev 57:347–354

Deveney P (1976) Title, Jus Publicum, and the Public Trust: an historical analysis. Sea Grant Law J 1:13, 17, 29

Duncan M (2002) Essay: reconceiving the bundle of sticks: land as a community-based resource. Environ Law 32(1):773–807

Eichenberg T, Vestal B (1992) Improving the legal framework for marine aquaculture: the role of water quality laws and the Public Trust Doctrine. Terr Sea J 2:339–347

Fenn PT Jr (1926) The origin of the right of fishery in territorial waters, vol 47. Harvard University Press, Cambridge, pp 52–53

Food and Agriculture Organization of the United Nations (2012) The State of World Fisheries And Aquaculture 11, 13, 53. Available at http://www.fao.org/docrep/016/i2727e/i2727e.pdf (last visited April 10, 2014)

Gorina-Ysern M (2004) World Ocean Public Trust: high seas fisheries after Grotius – towards a New Ocean Ethos? Golden Gate U Law Rev 34:645, 655, 665

Hildreth RG (1993) The Public Trust Doctrine and coastal and ocean resources management. J Environ Law Litig 8:221–230

Hirokawa KH (2010) Property as capture and care. Alb Law Rev 74:175–212

Huffman JL (1989) A fish out of water: the Public Trust Doctrine in a constitutional democracy. Environ Law 527:565–568

Justinian Institutes 2.1.12. Available at http://www.fordham.edu/halsall/basis/535institutes.asp#I.%20Divisions%20of%20Things (last visited April 23, 2014)

Justinian Institutes 2.1.2, Available at http://www.fordham.edu/halsall/basis/535institutes.asp#I.%20Divisions%20of%20Things (last visited April 23, 2014)

Klass AB (2006) Modern Public Trust Principles: recognizing rights and integrating standards. Notre Dame Law Rev 82:699–700

Lazarus RJ (1986) Changing conceptions of property and sovereignty in natural resources: questioning the Public Trust Doctrine. Iowa Law Rev 71:631–634, 641–642, 654–715

Lynch KJ (2007) Application of the Public Trust Doctrine to modern fishery management regimes. New York Univ Environ Law J 15:285, 296, 307–310

Macinko S, Bromley DW (2004) Changing tides in ocean management: property and fisheries for the twenty-first century: seeking coherence from legal and economic doctrine. Vermont Law Rev 28:623, 625, 631–634, 650–654

National Research Council, Sharing the Fish Towards a National Policy on Individual Fishing Quotas (1999) National Academic Press, 21–22, 39, 42, 43, 49

Olson JM (1978–1979) Toward a public lands ethic: a crossroads in publicly owned natural resources law. Univ Det J Urban Law 56:739, 861–862

Osherenko G (1995) Property rights and transformation in Russia: Institutional change in the far North. Eur Asia Stud 47:1077–1108

Osherenko G (2006) New discourses on ocean governance: understanding property rights and the Public Trust. J Environ Law Litig. 21:317, 331–333, 362, 367, 369, 381

Ostrom E (1990) Governing the commons: the evolution of Institutions for collective action. Cambridge University Press, 2–3, 9–10, 12–14, 21

Rieser A (1997) The ecosystem approach: new departures for land and water: fisheries management: property rights and ecosystem management in U.S. fisheries: contracting for the commons? Ecol Law Q 24:813, 818–820

Rieser A (1999) Prescriptions for the commons: environmental scholarship and the fishing quotas debate. Harv Environ Law Rev 23:393, 397, 399, 400, 402

Ryan E (2001) Public Trust and Distrust: the theoretical implications of the Public Trust Doctrine for natural resource management. Environ Law 31:477

Sax JL (1970) The Public Trust Doctrine in natural resources law: effective judicial intervention. Mich Law Rev 68:471, 477

Scott Gordon H (1954) An economic theory of common property research. The fishery. J Polit Econ 64:130, 131, 133, 141

Smith GP II, Sweeney MW (2006) The Public Trust Doctrine and natural law: emanations within a Penumbra. Boston Coll Environ Aff Law Rev 33:307, 322–341

Sohm R (1907) The Institutes: a textbook of the history and system of roman private law (James C. Ledlie trans, 3rd edn), pp 303–304

Thompson BH Jr (1990) Judicial takings. Va Law Rev 76:1449

Thompson BH Jr (2000) Tragically difficult: the obstacles to governing the commons. Environ Law 30:241, 244

Turnipseed M, Roady SE, Sagarin R, Crowder LB (2009) The Silver Anniversary of the United States' exclusive economic zone: twenty-five years of ocean use and abuse, and the possibility of a blue water Public Trust Doctrine. Ecol Law Q 36:1, 5–8, 10–11, 18–19, 23, 34–40, 48–49, 55–57

Weiss EB (1990) Our rights and obligations to future generations for the environment. Am J Int Law 84:198

Wilder RJ (1992) The three mile territorial sea: its origins and implications for contemporary offshore federalism. Va J Int Law 32:681–689

Wilkinson CF (1989) The headwaters of the Public Trust: some thoughts on the source and scope of the traditional doctrine. Environ Law 19:425, 428–430

Winters P (2011) The erosion of federal common law: anticipatory delegation in American electric power company v. Connecticut. Buff Environ Law J 19:341–382

Wise SM (1996) The legal thinghood of nonhuman animals. Boston Coll Environ Aff Law Rev 23:471, 503, 508

World Comm'n on Env't & Dev., Our Common Future 8 (1987), available at http://www.un-documents.net/our-common-future.pdf (last visited April 23, 2014)

2 William Blackstone Commentaries *410-11. Available at http://avalon.law.yale.edu/18th_century/blackstone_bk2ch27.asp (last visited July 29, 2016)

Chapter 35
Textbox: How the Public Trust Doctrine Supplements Existing Statutory Law

Zach Corrigan

Abstract There is no doubt that the Public Trust Doctrine undergirds U.S. fisheries law and provides the basis for a number of its provisions (*See e.g.*, National Research Council (1999), p. 39, explaining this is true for IFQs). However, it has been argued that the doctrine can do more than simply provide the theoretical underpinnings for MSA, and the next question is what the doctrine offers that existing statutory law does not?

35.1 Procedural Constraint on the Alienation of Resources

The Public Trust Doctrine may serve as a procedural constraint on the alienation of resources.[1] An often overlooked provision of the *Illinois Central* public-trust decision is its holding that, to the extent that a public trust resource can be disposed of, it is only by way of a legislative grant that contains "express authority" to dispose of or divert the use of a public trust resource.[2] At least one observer has indicated that a strengthened public trust doctrine applicable to federal resources could serve as a rule of statutory construction to limit the deference usually afforded U.S. federal agencies in managing these resources.[3] Under such a view, and analogous to the doctrine's implementation in many states, the doctrine would require specific legislative sanction of an agency's decisions regarding federal trust resources, thus putting the power to control trust resources with those

---

[1] Olson, pp. 861–862 (1978–1979).
[2] *Id.* citing *Ill. Cent. R.R. Co. v. Ill.*, 146 U.S., at 452–454.
[3] Baer, p. 433 (1988).

Z. Corrigan (✉)
Food & Water Watch, Washington, DC, USA
e-mail: zcorrigan@fwwatch.org

who are democratically elected, rather than unelected administrative officials.[4] The downside of this approach, of course, is the potential ossification of federal environmental law and a lack of flexibility afforded to agencies to address environmental problems.[5]

35.2 Decreased Court Deference to Agency Action

Similarly, an adherence to the federal Public Trust Doctrine could yield less deference by courts to the agency in analyzing the factual record supporting agency actions. Typically, resource decisions, including those involving federal fisheries are reviewed under the federal Administrative Procedure Act, which provides that an agency's administrative record forms the evidentiary basis for any subsequent claim in court. One group of scholars has argued that recognition of the Public Trust Doctrine would bring such disputes under a breach of trust rubric, where claims do not require the courts to defer to agency factual determinations.[6]

35.3 Affirmative Duty to Protect Water and Fishery Resources

The Public Trust Doctrine may serve not simply as a negative restraint on the state's ability to alienate trust resources, but also as an affirmative duty "to protect the people's common heritage of streams, lakes, marshlands and tidelands, surrendering that right of protection only in rare cases when the abandonment of that right is consistent with the purposes of the trust[,]" as it has in the state of California.[7] Thus, under this doctrine, federal fisheries managers would be obligated to take a stronger role in conserving fishery resources.

35.4 Proper Frame of Reference

The doctrine may serve as a supplement to regulating fisheries, by giving managers a proper frame of reference. Some scholars contend that one of the largest shortcomings of the federal fishery management efforts is that they lack grounding in the Public Trust Doctrine. The Public Trust Doctrine would thus "provide[] an

---

[4]*Id.*, p. 426; Lazarus, pp. 654–655 (1986).
[5]*Id.*
[6]Turnipseed et al., pp. 56–57 (1999), citing Lum (2003).
[7]*Nat'l Audubon Soc'y v. Superior Court*, 658 P.2d 709, 724 (Cal. 1983).

organizing mission – to manage fisheries resources in the best interest of current and future citizens - as well as a valuable backstop when setting catch limits and other fishing regulations."[8]

35.5 Flexibility

As a potential common law doctrine, it entails a flexibility that statutory law might not have.[9] The traditional public trust purposes entailed navigation, commerce, and fishing, but it has been extended to recreational purposes (boating, fishing, and swimming) and to ecological and aesthetic purposes in some U.S. states, including preservation of lands in their natural state in California.[10] Notably and most importantly, in Hawaii, courts have interpreted the Public Trust Doctrine to embrace the precautionary principle in order to encourage resource protection in the absence of conclusive scientific proof.[11] For fisheries, the scope of the Public Trust Doctrine may be most relevant to protect ocean ecosystems to produce food, medicine, climate stabilization, recreation, aesthetic enjoyment, as well as navigation and commerce.[12]

35.6 Principle of Intergenerational Equity

Some observers consider the principle of intergenerational equity to be integral to the Public Trust Doctrine.[13] Drawing analogies to private trust law, this means that the trustee must ensure that the corpus of the trust is managed so that the needs of current beneficiaries are met without sacrificing the ability of future beneficiaries to meet their needs.[14] For example, certain fishery management measures such as catch limits and marine protected areas, while preventing some public access, might be justified for their ability to protect the resource.[15] Likewise, as asserted above

---

[8]Turnipseed et al., p. 55 (1999).

[9]*See id.*, p. 47.

[10]Osherenko, p. 367 (2006), citing *Marks v. Whitney*, 491 P.2d 374, 380 (Cal. 1971). The *Marks* decision recognized that one of the most important public uses of the tidelands is the "preservation of those lands in their natural state, so that they may serve as ecological units for scientific study, as open space, and as environments which provide food and habitat for birds and marine life, and which favorably affect the scenery and climate of the area." *Id.* at 380.

[11]Blumm & Guthrie, p. 748 (2012), citing In re Water Use Permit Applications, 9 P.3d 409, 467 (Haw. 2000).

[12]*See* Osherenko, p. 367 (2006).

[13]*Id.*, p. 369, citing Weiss (1990); World Comm'n on Env't & Dev. (1987).

[14]Turnipseed et al., p. 18 (1999).

[15]*Id.*; 21 J. Osherenko, pp. 369, 381 (2006), citing Christie at (2004a); *but see* Bevis, p. 171 (2005).

with IFQs and discussed more fully below, the doctrine may be employed to limit IFQ privileges from being treated as *de facto* private property.

35.7 Fill Regulatory or Legislative Gaps

As discussed, it may fill the gaps in existing federal law or as a state doctrine applying where the federal government has not regulated the field.[16] Similarly, as a state or federal doctrine, it may extend extraterritorially to regulate federal or international waters.[17] There is little reason that the public trust doctrine would not apply to state waters within the three nautical mile belt around the United States coastline.[18] Generally, courts have held that where there is no conflict with federal or international law, "'a state's interest in preserving nearby fisheries is sufficiently strong to permit such extra-territorial enforcement of its laws enacted for that purpose.'"[19] Thus, it follows that the doctrine could serve as a state or federal doctrine applying extraterritorially. The discussion of extraterritoriality is discussed in more detail in the next chapter.

35.8 Enhance Citizen Standing

It may provide as a doctrine that enhances citizen standing to protect resources. Some states have recognized that the doctrine provides both states and individuals the standing to enforce it.[20] In many other nations, citizens are likewise permitted to enforce the public trust doctrine, regardless of personal injury.[21]

---

[16]Babcock, pp. 5, 60–61 (2007).

[17]*See infra.*

[18]*Id.*, p. 64, citing Hildreth, p. 229 (1993); Eichenberg & Vestal, p. 347 (1992).

[19]*Id.*, p. 66, quoting *People v. Weeren*, 26 Cal.3d 654, 666 (1980). Professor Babcock talks about the application of the doctrine within the context of the Coastal Zone Management Act, but the Public Trust Doctrine might just as well stand on its own.

[20]Turnipseed et al., p. 19 (2009), citing *Center for Biological Diversity, Inc. v. FPL Group, Inc.*, 83 Cal. Rptr. 3d 588, 600 (Cal. Ct. App. 2008). This case notes that "'any member of the general public ... has standing to raise a claim of harm to the public trust.'" (quoting *Nat'l Audubon Soc'y v. Superior Court*, 658 P.2d 709, 716 n.11 (Cal. 1983)).

[21]Blumm & Guthrie, pp. 765, 769, 780 (2012).

Chapter 36
US and Global Regulation of Fisheries Beyond the Commons

Zach Corrigan

Abstract The regulation of fisheries touches upon issues of sovereignty, jurisdiction, and economics, all of which are the foundation for the actual specifics of fishery management laws. This chapter explores the history of international regulation of our world's oceans and fisheries, culminating in the Law of the Sea treaty. It then looks at the potential to extraterritorially apply nations' fisheries laws to cover gaps in the international regime. The chapter then explores how one particular nation's regime, the United States's Magnuson-Stevens Fisheries Conservation and Management Act, is structured to regulate U.S. domestic fisheries. It then explores two of the more recent trends in fisheries in depth: aquaculture, also known as fish farming, and Individual Fishing Quotas (IFQs), which are privatized or market-based means to manage fisheries. Finally, the chapter looks beyond the fisheries themselves and touches upon how people have attempted to regulate (or not regulate as the case may be) the waterways that are crucial for the survival of the world's fish populations. The chapter details how the movement to allocate resources that were once mistakenly thought to be inexhaustible has spurred significant theoretical and doctrinal developments, such as the movement away from treating such resources as open-to-all to a more communal concept of property. Practically, however, huge challenges for the conservation of the world's fisheries remain, whether it is from fish populations that continue to be overexploited; incomplete and disjointed management regimes; global warming; or developing industries like fish farming that seek to use ocean resources for their financial gain, perhaps to the detriment all.

36.1 Introduction

The regulation of fisheries touches upon issues of sovereignty, jurisdiction, and economics, all of which are the foundation for the actual specifics of fishery management laws. Chapter 34 looked at some of these issues, focusing on the

---

Z. Corrigan (✉)
Food & Water Watch, Washington, DC, USA
e-mail: zcorrigan@fwwatch.org

Public Trust Doctrine. This chapter explores the history of international regulation of our world's oceans and fisheries, culminating in the Law of the Sea treaty. It then looks at the potential to extraterritorially apply nations' fisheries laws to cover gaps in the international regime. Subsequently, this chapter explores how one particular nation's regime, the United States's Magnuson-Stevens Fisheries Conservation and Management Act (MSA), is structured to regulate U.S. domestic fisheries. The chapter then explores two of the more recent trends in fisheries in depth: aquaculture, also known as fish farming, and Individual Fishing Quotas (IFQs), which are privatized or market-based means to manage fisheries. IFQs provide fishermen a fixed portion of the amount of a fishery that they can fish, buy, sell, or trade to others.[1] Proponents argue that allocating a fixed share of the resource to fishermen limits their incentive to increase fishing effort to capture more fish than their competitors, thereby ending the "race to fish" or "derby fishing," as it is often called, and was conceptualized by Gordon and in Hardin's "Tragedy of the Commons," discussed in Chap. 34. The tradable nature of the fishing privileges is supposed to make the allocation of fishing privileges efficient. Fish farming, it is argued, will mitigate the so-called tragedy of the commons by increasing the supply, by allowing fish farmers to essentially build a fence around a particular "crop" of fish. Finally, the chapter looks beyond the fisheries themselves and touches upon how people have attempted to regulate (or not regulate as the case may be) the waterways that are crucial for the survival of the world's fish populations.

A review of fisheries and marine-pollution regulation over approximately the last 100 years shows a movement to allocate resources that were once thought to be inexhaustible. The debate about how public and private regimes—nationally, extraterritorially, and internationally—should best govern these resources has spurred significant theoretical and doctrinal developments, such as the movement away from treating such resources as open to all (as discussed in Chap. 34 on the theory underlying fisheries management), to more communal concept of property, as embodied by the 1982 Law of the Sea Convention. And even while nations have claimed more and more sovereignty over ocean waters and submerged lands, in no small part to accumulate greater economic wealth, such claims often have been couched in the language of conservation. In these respects, it might be argued that people have made some substantial strides in protecting common living marine resources such as fisheries.

Practically, however, huge challenges remain, whether it is from (1) fish populations that continue to be overexploited and regimes that are incomplete, ineffective, disjointed, or otherwise not up to the task of management; (2) relatively newer challenges such as global warming; and (3) developing industries like fish farming that seek to use ocean resources for their financial gain, perhaps to the detriment all. These challenges frame the national and international debates in law and policy governing the production of seafood.

---

[1] Rieser (1999), p. 407; National Research Council (1999), p. 33.

36.2 A Brief History of the Last 15 Centuries of Fisheries Protection as a Global Commons

The first question that arises over how fisheries are regulated internationally is who has the right to exploit, or, alternatively, conserve and manage them. The law has gradually changed over the last 15 centuries to define the areas that nation states can claim and what remains subject to open access to every person and nation. As will be seen, very little of the ocean's resources, even beyond territorial borders in the "open seas" are now completely free for unbridled exploitation, as national and international law has gradually expanded to protect a greater extent of these resources. As will also be seen in this chapter, however, there is great room to improve the management of fisheries in order to benefit all.

At least a cursory understanding of the history of the management of the oceans and its fisheries is needed to understand how human civilization has arrived at this point. This chapter explores this history from three important legal vantages: (1) nations' initial claims of sovereignty over the high seas prior to Grotius; (2) the enclosure movement; and (3) the law of the sea. The fourth section looks to the future, exploring whether the extraterritorial application of nations' domestic laws could serve to protect international fishery resources.

36.2.1 National Claims Over the High Seas

Under the Digest of Justinian, one of the first bodies of law on the world's oceans and fisheries that still influences the common law today, no state could extend its jurisdiction beyond the high water mark of the shore.[2] But as mercantile city-states grew more powerful during the Middle Ages, they began laying claims to various parts of the Mediterranean.[3] Other areas such as the Baltic Sea were also becoming subject to competing claims from other nations such as Denmark, Sweden, and Poland.[4]

The fifteenth century and the conquests for new lands brought even greater pressures on nations to control the high seas. Portugal and Spain in effect claimed the entire world's oceans for their control in the 1494 Treaty of Tordesillas.[5] Early in the sixteenth century, France, England, and Holland also expanded their ocean empires, rejecting these Spanish and Portuguese claims.[6] Denmark, Norway, and

---

[2]Buck (1998), p. 76.
[3]*Id.*, citing Swarztrauber (1972).
[4]*Id.*
[5]*Id.*, p. 77.
[6]*Id.*

Sweden's fishing interests propelled them into similar controversies in the sixteenth century.[7]

Perhaps the single largest watershed moment in this history came in 1602, when the Dutch East India Company seized a Portuguese ship in retaliation for resistance to Dutch trade in the East Indies. The company commissioned jurist Hugo Grotius to write a legal brief defending this seizure.[8] In his seminal work, Freedom of the Seas, written in 1604–1605, Grotius argued that the seas, like the air, cannot be appropriated to any nation, since the passage of a ship through these areas leaves no permanent trace.[9] He fathered the concept that nations could innocently pass through territorial seas, fostering commerce while preserving state sovereignty.[10] English jurist John Selden was commissioned by King James I of England to prepare the "Mare Clausum," a rebuttal in defense of British seizure of Dutch cargoes.[11] Under the Seldenian vision, the oceans were divided into jurisdictions for various nations.[12]

These two competing visions continue to frame the debate about how best to regulate our oceans to the present day. And while Grotius's view was ultimately accepted—as sovereignty is conceptualized more in terms of freedom of navigation and expanded commerce than national domination over the resource—nations have also advanced the Seldenian cause by dramatically expanding their claims to manage and control more and more of the high seas–first from a three to a twelve-nautical mile territorial sea, then to contiguous zones, the continental shelf, and finally to Exclusive Economic Zones (EEZs) mentioned later in this chapter.[13] This eventual expansion has become known as the enclosure movement.[14]

36.2.2 The Enclosure Movement

The international debates about how nations may assert control and sovereignty over the world's oceans initially focused on the high seas in the sixteenth century, but this area of the law would not begin to become settled until much later, after nations began establishing greater control over the ocean waters closest to their own shores. International custom has always provided that coastal nations were entitled to claim some part of their seas continuous to their shores, but it was disputed on

---

[7]*Id.* at 78.

[8]*Id.* at 79.

[9]*Id.*

[10]Osherenko (2006), p. 333 (citing Juda (1996), pp. 8–30).

[11]Buck (1998), p. 70.

[12]Osherenko (2006), p. 333 (citing Selden (1652)).

[13]*Id.*; Gorina-Ysern (2004), p. 659.

[14]Turnipseed et al. (2009), p. 28, citing Alexander (1983), p. 566.

how much they could claim.[15] One significant idea stemming from Grotius's work was the revival of an older concept that a nation could claim sovereignty over the amount that it could control.[16] Over the course of the sixteenth through eighteenth centuries, this concept of control became international custom, and the extent of this area was measured by the distance that could be claimed with a cannon shot,[17] evolving into the area of a marine league, or three nautical miles.[18] While other rules have been proposed, including one suggesting that the size should be determined by the nation's dependence on fishing, they have never been accepted.[19] The first treaty to recognize the three-mile limit was the Fishing Convention of 1818.[20] While this three-mile limit became largely the norm, it was often not adopted for nations when it was inconvenient, and such limits were often disputed.[21] (See Fig. 36.1 for a graphic illustration of the Maritime Zones under the Law of the Seas Convention.)

During the twentieth century, the United States and other coastal nations began asserting increasing authority over their adjacent ocean waters and seabed.[22] These nations sought to secure rights over the oil and gas, mineral, and fishery resources, as advancements in science and technology allowed them to utilize them.[23] In 1945, U.S. President Truman issued two presidential proclamations about the waters seaward of the U.S. three-nautical mile territorial seas. One declared exclusive jurisdiction and control over the oil, gas, and mineral resources of the continental shelf, and the other created U.S. federal authority to regulate fisheries in the waters above the shelf.[24]

Truman's unilateral and unprecedented actions gave rise to the enclosure movement, a frantic assertion of authority by other coastal nations.[25] Many nations' claims were aimed at excluding distant-water fishing fleets. Chile, Peru, and Ecuador were among the early states to assert 200-nautical-mile claims to protect their fishing interests for these reasons.[26] By the fall of 1977, 68 countries had claimed exclusive fishing zones beyond 12 miles, and these included 51 countries whose claims extended to 200 miles.[27]

---

[15]Buck (1998), pp. 76, 80.
[16]*Id.*, p. 80.
[17]*Id.*, pp. 80–81.
[18]*Id.*, p. 81.
[19]*Id.*, pp. 80–81 (citing Fulton (1911), p. 547).
[20]*Id.*, p. 81.
[21]*Id.*, pp. 81–84.
[22]Turnipseed et al. (2009), p. 27 (citing Alexander (1983), p. 561).
[23]*Id.*
[24]*Id.*, p. 28, citing Proclamations No. 2667, 10 Fed. Reg. 12,303, 12,303 (Oct. 2, 1945); No. 2668, 10 Fed. Reg. 12,304 (Oct. 2, 1945).
[25]*Id.* citing Alexander (1983), p. 566; Osherenko (2006), p. 347; citing Kalo et al., (2002), p. 311.
[26]Osherenko (2006), p. 348, citing Kalo et al.,(2002), pp. 312–313.
[27]*Id.* citing Eckert (1979), p. 129.

The Truman proclamation touched off a debate about the international legal validity of such claims, which would start to be settled over the course of the next 50 years, ending with the United Nations Convention on the Law of the Sea.[28]

36.2.3 The Law of the Sea

The last chapter in this legal history starts with the three United Nations Conferences on the Law of the Sea (UNCLOS), in 1958, 1960, and 1993, which sought to codify existing international law.[29] The products of UNCLOS I were four conventions, three of which codified generally accepted customary law of the sea.[30] First, the Convention on the Territorial Sea and the Contiguous Zone affirmed the sovereignty of coastal nations over internal waters and territorial sea, subject to the right of innocent passage for foreign-flag vessels.[31] It also provided that coastal states could exercise jurisdiction to implement and enforce customs, fiscal, immigration, and sanitary laws in a contiguous zone extending nine miles beyond the traditional three-mile territorial sea.[32] (*See* Fig. 36.1 for a graphic illustration of the Maritime Zones under the Law of the Seas Convention.)

The second convention, the Convention on the Continental Shelf confirmed coastal states' sovereign rights to explore and exploit the natural resources of the continental shelf.[33] Third, the Convention on the High Seas, codified the freedom of navigation, the freedom to fish, the freedom of overflight, and the freedom to lay cables and pipelines on the sea floor in the area beyond the territorial sea.[34] The fourth convention, the Convention on Fishing and Conservation of the Living Resources of the High Seas, allowed coastal nations to set nondiscriminatory conservation fishing rules for threatened stocks beyond their territorial seas.[35] Although this treaty was adopted and was ratified by a sufficient number of nations to enter into force, major nations with distant fishing fleets did not join or observe the regulations set by member countries, and it is thus largely discredited.[36]

---

[28]*Id.*, p. 348.

[29]Buck (1998), p. 83.

[30]Osherenko (2006), p. 336, citing Juda (1996), pp. 157–159.

[31]*Id.*

[32]*Id.*

[33]*Id.*, citing Geneva Convention on the Continental Shelf, Apr. 29, 1958, 419 U.N.T.S. 312, 312.

[34]*Id.*, citing Geneva Convention on the High Seas, Apr. 29, 1958, 450 U.N.T.S. 12.

[35]*Id.*, pp. 336–337, citing Juda (1996), pp. 159–160.

[36]*Id.*, p. 337, citing Juda, p. 150, 196 n.2. However, in what has been called a landmark case, in 1974 the International Court of Justice upheld Article 2 of this convention in the context of high-seas fishing and recognized that the freedom of the high seas must be exercised "with reasonable

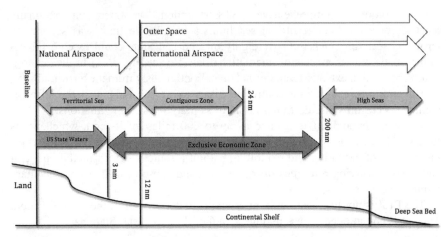

Fig. 36.1 Territorial Limits as Recognized by the Law of the Sea. As can be seen above, the Law of the Sea set a 12-mile limit for the territorial seas, to which the sovereignty of a coastal State extends. In the "contiguous zone," which extends 12 additional miles, a coastal State may exercise the control necessary to prevent and punish the infringement of its customs, fiscal, immigration or sanitary laws and regulations. The continental shelf is defined as up to 200 miles, or its end, whichever is further. In this area, coastal States exercises exclusive sovereign rights for the purpose of exploring and exploiting mineral and other non-living resources of the sea-bed and subsoil together and sedentary living organisms. In the EEZ, which is not to exceed 200 miles, a coastal State has sovereign rights for the purpose of exploring and exploiting, conserving, and managing the natural resources, sea-bed and its subsoil, and with regard to other activities, such as the production of energy from the water, currents and winds

None of these four 1958 conventions entailed the principle of abstention from fishing by nation states.[37] The movement for such a principle began long afterwards. Stemming from the realization that certain resources such as fisheries had reached exhaustion,[38] in 1967 the Maltese ambassador to the United Nations, Arvid Pardo, proposed the "common heritage of mankind" principle. It defined some resources as the property of the global human population and stated that these resources cannot be appropriated by any one individual or state.[39]

It was not until UNCLOS III conference that conservation of fisheries became an explicit concern.[40] After this conference, which began in 1973 and culminated in

---

regard to the interests of other States in their exercise of the freedom of the high seas." Gorina-Ysern (2004), p. 669, citing Fisheries Jurisdiction Case (U. K. v. Ice.) 1974 I.C.J. 3 (Feb. 2); Fisheries Jurisdiction Case (F.R.G. v. Ice.) 1974 I.C.J. Rep. 175 (Feb. 2). This finding validated the concept of extended fishery zones, which would be codified in 12 nautical-mile territorial sea limits and the adoption of 200 nautical-mile EEZs in the 1982 Law of the Sea Convention. *Id.*

[37] Gorina-Ysern (2004), pp. 668–669.
[38] *Id.*, p. 39.
[39] *Id.*, p. 28.
[40] Buck (1998), p. 92.

the adoption of the comprehensive 1982 Convention,[41] an agreement was finally reached for a range of territorial sea limits.[42] A 12-mile limit was set for the territorial seas, in which the sovereignty of a coastal State could extend, subject to the treaty and other rules of international law.[43] (See Fig. 36.1) In the "contiguous zone" which extends 12 additional miles contiguously from the territorial seas, coastal state may exercise the control necessary to prevent and punish the infringement of its customs, fiscal, immigration, or sanitary laws and regulations.[44]

The continental shelf is defined as up to 200 miles, or its end, whichever is further.[45] In the coastal zones, coastal states exercises exclusive sovereign rights for the purpose of exploring and exploiting natural resources, which include mineral and other non-living resources of the sea-bed and subsoil together and sedentary living organisms.[46]

The EEZ is not to exceed 200 miles and is under control of the coastal nation. Thus, it is not an open-access regime.[47] In the EEZ, a coastal State has

> [s]overeign rights for the purpose of exploring and exploiting, conserving, and managing the natural resources, whether living or non-living, of the waters superjacent to the sea-bed and of the sea-bed and its subsoil, and with regard to other activities for the economic exploitation and exploration of the zone, such as the production of energy from the water, currents and winds.[48]

One of the driving reasons for the expanded EEZ was to give coastal state jurisdictions the ability to manage their fisheries in these areas by allowing nation states to control the access to the area.[49] (See Fig. 36.1).

It is not simply the 1982 Law of the Sea Treaty's recognition of territorial limits that has enhanced ocean and fisheries management; the treaty also has a number of protective provisions. In fact, it includes 46 articles devoted to marine environmental protection, including the general obligation of States "to protect and preserve the

---

[41]Osherenko (2006), p. 339, citing Juda (1996), pp. 212–243. Little was accomplished at UNCLOS II. Buck (1998), p. 86. But for a discussion of the "cod war" between Norway and Britain that inspired it, *see id.* at 84–86 (1998). Buck argues that fisheries issues became dominant in the 1970s, thus inspiring the agreement at UNCLOS III. *Id.* at 86.

[42]Two implementation Agreements followed: Part XI (Seabed) in 1994 (the Agreement Relating to the Implementation of Part XI of the United Nations Convention on the Law of the Sea, July 28, 1994, 36 I.L.M. 1492) and Fish Stocks in 1995 (the Agreement for the Implementation of the Provisions of the United Nations Convention on the Law of the Sea, Relating to the Conservation and Management of Straddling Fish Stocks and Highly Migratory Fish Stocks, November, 1995 34 I.L.M. 1542) Osherenko (2006), p. 339 citing Juda (1996), pp. 256, 284.

[43]United Nations Convention on the Law of the Sea, Dec. 10, 1982, Article 2(1)-(3), 21 I.L.M. 1261.

[44]*Id.*, p. Article 33; Buck (1998), p. 94.

[45]*Id.*, p. Article 77.

[46]*Id.*

[47]Buck (1998), pp. 93–95.

[48]United Nations Convention on the Law of the Sea, Dec. 10, 1982, Article 56, 21 I.L.M. 1261.

[49]Christie (2004b), p. 2, quoting Christie (1999), p. 396 (quotation marks omitted).

marine environment."[50] For example, the treaty has requirements that EEZs have fishing limits based on Maximum Sustainable Yield, tracking the definition in the United States' MSA, which is discussed below.[51] Under Article 61(4), as is also true for the MSA, coastal States must consider the effects on species that are not targeted, also known as bycatch. This language may even be broad enough to require nation States to engage in ecosystem management, where regulation is based on the biological relationships between species.[52] Like under the MSA, the Law of the Sea requires coastal States to base conservation and management measures on the best science available.[53]

Perhaps the most significant development under the convention is its treatment of high seas, beyond the EEZs, which are treated as the "common heritage of mankind." States exercise the role as trustees for all the world's people as beneficiaries, including future generations.[54] Thus, the debate over whether the high seas is *res nullius* or *res communis*, as discussed extensively in Chap. 34, was ostensibly resolved in the 1982 Convention.[55] For example, the International Seabed Authority has the role of trustee over mineral resources of the seabed in areas not under any State's jurisdiction.[56] An exploiting State is required to pay a portion of the value of the production of its resources within this area to the authority to be distributed to the convention parties based on "equitable sharing criteria."[57]

In addition, while recognizing the right of all States to engage in fishing on the high seas, the 1982 Convention requires States to cooperate in the conservation of high seas resources and negotiate with others that exploit living resources. It calls for cooperation through the establishment of sub-regional, regional, and international organizations.[58] This has spurred several agreements including the U.N. Fish Stocks Agreement, (mentioned *supra*), and FAO Compliance Agreement,[59] which are both attempts to address the growing concerns about major problems facing international fisheries' management, including illegal, unreported, and unregulated fishing, the reflagging of vessels by nations that are unwilling to abide by fishing restrictions, the overcapitalization of fisheries (too many boats chasing too many

---

[50]Osherenko (2006), p. 342, citing Kalo et al., (2002), p. 421; Juda (1996), p. 235; United Nations Convention on the Law of the Sea, Dec. 10, 1982, Article 192, 21 I.L.M. 1261.

[51]*Id.* at Article 61.

[52]International Union for Conservation of Nature and Natural Resources at 23 (2009).

[53]United Nations Convention on the Law of the Sea, Dec. 10, 1982, Article 61(2).

[54]*Id.* at 372, citing Encyclopedia of Public International Law (1992).

[55]Gorina-Ysern (2004), p. 663.

[56]Osherenko (2006), pp. 372–373, citing United Nations Convention on the Law of the Sea, Agreement Relating to the Implementation of Part XI of the Convention, part. I, article 1, and part XI, section 2, article 137.

[57]Ashfaq (2010), p. 368, citing United Nations Convention on the Law of the Sea, Article 82.

[58]Gorina-Ysern (2004), p. 675, citing United Nations Convention on the Law of the Sea, Dec. 10, 1982, Articles 117 and 118.

[59]Agreement to Promote Compliance with International Conservation and Management Measures by Fishing Vessels on the High Seas, Nov. 20, 1993, 33 I.L.M. 968.

fish), and insufficient cooperation among States.[60] Despite some criticisms, others have praised the regional fishing organizations for being able to pressure nations to curtail unrestricted fishing in the high seas and instill measures including certain fisheries closures, compulsory registration, and catch limits.[61]

Thus, the 1982 Convention, including its adoption of the common-heritage-of-mankind principle, moved the international governance of the oceans towards the Public-Trust-Doctrine model, discussed in the previous Chap. 34, where the resource is governed by a trustee on behalf of all in order to prevent its over utilization to the detriment of all. While the convention was a step forward for ocean protection, as 90 % of the world's fisheries come from just three percent of the world's oceans, and the majority of these species are found within 200 miles of shore,[62] critics contend that the Law of the Sea remains insufficient. For example, one expert has remarked that while "[]the majority of rights vested on coastal and fishing nations under [the Law of the Sea] have crystallized as norms of customary international law[,] [t]he majority of duties... have not."[63] Critics point out that the near-complete discretion given to coastal states to interpret and implement their duties under the convention as the primary reason the decline in EEZ stock fish.[64] Additionally, over-exploitation in the EEZ is not prohibited unless it presents a danger to the maintenance of the living resources, and the law does not dictate what levels of fish populations should be maintained.[65] Other critics point to the convention's vague provisions, its lack of a global fisheries organization, and its lack of a compulsory EEZ-dispute-settlement mechanisms as other weaknesses.[66] Further, if coastal States and designated international organizations are to have public-trust duties, there must be some way to hold trustees accountable.[67] There does not exist the procedural and substantive mechanisms that would enable the beneficiaries to enforce the terms of the trust under the 1982 treaty.[68] This lack of comprehensive governance is perhaps one reason that the world's fish populations continue to remain at or very close to their maximum sustainable production levels,[69] with fishery resources from the high seas in the worst shape.[70]

---

[60]*Id.* at 680, citing Juda (2002), pp. 109–144.

[61]*Id.* at 683–685.

[62]Buck (1998), p. 94.

[63]Gorina-Ysern (2004), pp. 669–671, citing McLaughlin (2003).

[64]Christie (2004b), p. 3.

[65]International Union for Conservation of Nature and Natural Resources at 5 (2009), citing United Nations Convention on the Law of the Sea, Dec. 10, 1982, Article 61(2), 21 I.L.M. 1261.

[66]Gorina-Ysern (2004), pp. 671, 674 . For a comprehensive look at its conservation provisions applying in the continental shelf, EEZ, and the high seas, *see* Gorina-Ysern (2004), p. 673–675.

[67]Osherenko (2006), pp. 374–375.

[68]*Id.*

[69]Food and Agricultural Organization (2012), pp. 11, 13, 53.

[70]*Id.*, p. 13.

Another problem with the treaty has been the slow pace at which nations have implemented it. UNCLOS III came to a close in 1982, but the treaty did not go into effect until 1994, one year after it was ratified by 60 countries.[71] And there are still a number of notably absent parties. For example, even though it was instrumental in convening UNCLOS III, the United States, to date, has not ratified the treaty largely due to a few elected officials' arguments that the treaty infringes upon national sovereignty and that its deep-sea mining provisions limit free enterprise.[72]

Despite the slow pace of ratification, however, the UNCLOS has had a positive effect on domestic law, even for non-ratifying nations. U.S. President Reagan, for instance, acted on the treaty when he created the U.S. EEZ in 1983.[73] The United States secured "sovereign rights" and jurisdiction to the largest EEZ in the world, which stretches seaward out to 200 nautical miles from the U.S. mainland, Hawaii and Alaska, and U.S. island territories in the Atlantic and Pacific.[74] The U.S. EEZ covers 4.4 million square miles, larger than the combined area of the 50 states.[75] A 1988 proclamation extended the boundaries of the territorial sea from 3 to 12 nautical miles seaward of the coastlines of the United States and its territories, under which full sovereignty was claimed for the United States to the 12-nautical-mile territorial sea from the sub-surface seabed to the above airspace.[76] Finally, in 1999, U.S. President Clinton established the contiguous zone, which reaches from 12 to 24 nautical miles from U.S. and territorial coastlines.[77] (See Fig. 36.1).

---

[71]Turnipseed et al. (2009), p. 30 (citing Kalo et al., (2002), p. 388). A full 166 countries have ratified the treaty. The first 60 nations to do so were as follows (in alphabetical order): Angola (5 December 1990), Antigua and Barbuda (2 February 1989), Bahamas (29 July 1983), Bahrain (30 May 1985), Barbados (12 October 1993), Belize (13 August 1983), Bosnia and Herzegovina (12 January 1994), Botswana (2 May 1990), Brazil (22 December 1988), Cabo Verde (10 August 1987), Cameroon (19 November 1985), Costa Rica (21 September 1992), Côte d'Ivoire (26 March 1984), Cuba (15 August 1984), Cyprus (12 December 1988), Democratic Republic of the Congo (17 February 1989), Djibouti (8 October 1991), Dominica (24 October 1991), Egypt (26 August 1983), Fiji (10 December 1982), Gambia (22 May 1984), Ghana (7 June 1983), Grenada (25 April 1991), Guinea (6 September 1985), Guinea-Bissau (25 August 1986), Guyana (16 November 1993), Honduras (5 October 1993), Iceland (21 June 1985), Indonesia (3 February 1986), Iraq (30 July 1985), Jamaica (21 March 1983), Kenya (2 March 1989), Kuwait (2 May 1986), Mali (16 July 1985), Malta (20 May 1993), Marshall Islands (9 August 1991), Mexico (18 March 1983), Micronesia (Federated States of) (29 April 1991), Namibia (18 April 1983), Nigeria (14 August 1986), Oman (17 August 1989), Paraguay (26 September 1986), Philippines (8 May 1984), Saint Kitts and Nevis (7 January 1993), Saint Lucia (27 March 1985), Saint Vincent and the Grenadines (1 October 1993), Sao Tome and Principe (3 November 1987), Senegal (25 October 1984), Seychelles (16 September 1991), Somalia (24 July 1989), Sudan (23 January 1985), Togo (16 April 1985), Trinidad and Tobago (25 April 1986), Tunisia (24 April 1985), Uganda (9 November 1990), United Republic of Tanzania (30 September 1985), Uruguay (10 December 1992), Yemen (21 July 1987), Zambia (7 March 1983), Zimbabwe (24 February 1993).

[72]Id.; Ashfaq (2010), pp. 358–362.

[73]Id. p. 30, citing Proclamation No. 5030 (located at 48 Fed. Reg. 10,60 (March 14, 1983).

[74]Id.

[75]Id.

[76]Id. citing Proclamation No. 5928, 54 Fed. Reg. 777 (Jan. 9, 1989).

[77]Id., citing Proclamation No. 7219, 64 Fed. Reg. 48,701 (Aug. 2, 1999).

Each of the three proclamations was consistent with the 1982 Law of the Sea, and some scholars have argued that the United States has accepted the convention as a matter of international customary law,[78] despite its failure to ratify it.[79]

Thus, notwithstanding its limitations and the slow pace at which it has been implemented, the 1982 Law of the Sea Convention has dramatically changed the legal landscape governing the world's oceans and resources. It not only validated nations' moves to establish EEZs. But it also changed the dominant legal paradigm for these resources, moving it away from an open access or a *res nullius* resource to a public resource to be protected, akin to U.S. state waters under the Public Trust Doctrine. Moreover, the treaty, in conjunction with general international law and the treaty's implementing agreements, provides, as on expert has put it, a

> range of norms for national fisheries management, if carefully interpreted. Such legally binding norms include the coastal state's primary obligation to ensure that the maintenance of the living resources in its EEZ is not endangered by overexploitation; the duty to maintain or restore populations of target species at sustainable levels; the determination of catch limits for stocks actually or potentially affected by exploitation; the duty to apply the precautionary approach widely to conservation, management and exploitation of living marine resources; and duties to cooperate for the conservation and management of species not exclusively occurring within the coastal state's EEZ.[80]

36.2.4 Extraterritorial Jurisdiction for Protection of the Global Fisheries Commons

As mentioned in Chap. 34 and above, the United States has asserted jurisdiction over the fishery resources in the EEZ with the MSA, which codified the second

---

[78]The International Court of Justice has stated that for a treaty rule to acquire customary status, it must be:

(1) a fundamentally norm-creating character such as could be regarded as forming the basis of a general rule of law; (2) a very widespread and representative participation in the convention, including that of states whose interests were specially affected; (3) extensive and virtually uniform state practice, including that of states whose interests are specially affected; and (4) the passage of some time, short though it may be. *North Sea Continental Shelf*, Judgement, ICJ Rep. 3, at paras. 72–74 (1969).

[79]*Turnipseed et al.*, p. 70. n.169 (citing Restatement (Third) of Foreign Relations Law § 514 comment. a. (1987)). As evidence, another author points to the fact that the United States is a party to the 1964 Convention on the Territorial Sea and Contiguous Zone, which, like the Law of the Sea, precludes states' absolute claims to unlimited territorial seas and creates rules to restrict some forms of passage within their territorial seas. Ashfaq (2010), p. 364 The author also argues that the imposition of affirmative environmental and pollution-reducing obligations parallels the 1966 Convention on Fishing and Conservation of the Living Resources of the High Seas, to which the United States is also a party. *Id*. Finally, the author argues that the revenue-sharing provisions and dispute resolution mechanisms, which are the source of great controversy, were founded upon the "common heritage of mankind" principle are "customary law" supported in part by its widespread ratification. *Id*. Under general principles of international law, customary law is binding on all states, including the United States. *Id*.

[80]International Union for Conservation of Nature and Natural Resources (2009), p. 39.

1945 Truman Proclamation, and declared waters from three to 200 nautical miles off the shores of the United States and U.S. territorial possessions to be exclusively under federal jurisdiction. But what about areas beyond these limits? Might nations like the United States be able to use their extraterritorial jurisdiction to protect resources that exist beyond nation's boundaries? This is an important question, as nations have increasingly sought to enhance their ability to protect their fisheries from the fishing that happens beyond these areas, which, as discussed above, are governed by the less than completely comprehensive and binding Law of the Sea.[81]

In terms of "extraterritorial" jurisdiction, or the lawful control that a country can exercise beyond its borders, a nation generally can regulate the activities, interests, status, or relations of its own nationals, wherever they may be, as long as the exercise of jurisdiction is reasonable.[82] This is recognized, as an example, under the Law of the Sea under its "flag state jurisdiction" regime, which provides that the country in which a vessel is registered may take enforcement action against it.[83] Nations may also regulate conduct "that has or is intended to have substantial effect within its territory."[84] At least one scholar has argued that this latter, "substantial effects doctrine" is a clear basis for extending U.S. environmental laws extraterritorially, but there have been very few published court opinions that address the extraterritorial reach of such laws,[85] and even fewer applying federal fishing laws.

---

[81] Gorina-Ysern (2004), pp. 645–646.

[82] Showalter (2010), p. 229, citing Restatement (Third) of the Foreign Relations Law of the U.S., § 402(2). *See also* Restatement (Third) of the Foreign Relations Law of the U.S. § 403.

[83] For a criticism of the weaknesses in this regime, *see* Carr and Scheiber (2002). U.S. states have this same authority where there is no federal law to apply. *Skiriotes v. Florida*, 313 U.S. 69, 73 (1941); *Alaska v. Bundrant*, 546 P.2d 530, 555 (Alaska 1976). The extension of state law has been applied to fisheries based on the substantial effects doctrine, allowing the regulation of extraterritorial conduct, regardless of the citizenship of the target, so long as the exercise of jurisdiction does not conflict with federal law and is otherwise reasonable. *Alaska v. Jack*, 125 P.3d 311, 319 (Alaska 2005). Some U.S. state courts have recognized that a state's interests in preserving nearby fisheries are sufficiently strong to permit such extraterritorial enforcement of its laws enacted for that purpose. *California v. Weeren*, 607 P.2d 1279, 1285 (Cal. 1980). *But see North Alaska Salmon Co. v. Pillsbury*, 174 Cal. 1, 4 (1916) (stating that extraterritorial effect "will not be declared to exist unless such intention is clearly expressed or reasonably to be inferred from the language of the act or from its purpose, subject matter or history"); *Southeast Fisheries Assn. v. Dept. of Natural Res.*, 453 So. 2d 1351, 1355 (Fla. 1984) (finding that in order for state to apply outside of state waters, the legislature must have expressed a clear intent that the law is to apply extraterritorially). Where state law applies extraterritorially in the EEZ, it faces the additional issue of preemption. While there have been exceptions, especially where states have implemented outright bans on landings or landing limits that prevent fishermen from landing fish in amounts permitted under federal law, generally courts have allowed the extraterritorial application of state fishery laws where (1) the state law is not in conflict with a federal law or fishery management plan concerning the fishery resource at issue; (2) a vessel is registered under state law, as set forth under the MSA; and (3) a vessel that is not registered nevertheless submits itself to a state's jurisdiction by docking in state waters. *See* Mastry (2006), p. 235.

[84] *Id.*

[85] Nash (2010), pp. 998, 1003.

Extraterritorial jurisdiction, however, is not limitless and, in fact, is not fully recognized by some nations. For example, a U.S. legal canon of statutory interpretation known as the "presumption against extraterritorial application" limits the reach of U.S. federal and state law. It provides that a statute is presumed not to have extraterritorial reach unless it provides a clear indication that it was intended to do so. Essentially, it reflects the "'presumption that United States law governs domestically but does not rule the world.'"[86] It is meant to prevent U.S. courts from recognizing lawsuits without sufficiently definite norms of international law.[87] On the other hand, at least two U.S. Circuit Courts of Appeal have ruled that the presumption against extraterritoriality doctrine is inapplicable where the failure to extend a statute's reach to a foreign setting will result in adverse effects within the United States.[88]

For the reason that the federal fisheries law in the United States, the MSA, expressly provides a few situations where it applies beyond the EEZ, it suggests a limited extraterritorial reach for the statute, however, as it butts into the presumption against extraterritoriality.

Under the MSA, U.S. fisheries managers can manage anadromous species, which are fish that spawn in fresh or estuarine waters of the United States and which migrate to ocean waters[89] throughout their range except when they are found within foreign nations' waters.[90] The United States may also regulate continental shelf resources, including a number of coral, crab, crustacean, and sponge species, beyond the EEZ under the MSA.[91] Likewise, the MSA specifies that the United States is obliged to cooperate with nations involved in fisheries for highly migratory species to achieve optimum yield of such species throughout their range, both

---

[86]*Kiobel v. Royal Dutch Petro. Co.*, 133 S. Ct. 1659, 1664 (2013) (quoting *Microsoft Corp. v. AT&T Corp.*, 550 U.S. 437, 454 (2007)).

[87]*Id.* at 1664.

[88]*See Envtl. Def. Fund v. Massey*, 986 F.2d 528, 531 (D.C. Cir. 1993) (collecting cases); *H.K. & Shanghai Banking Corp. v. Simon (in Re Simon)*, 153 F.3d 991, 995 (9th Cir. 1998). It is unclear how the doctrine of substantial effects and the presumption against extraterritoriality relate to each other. The cited U.S. Court of Appeals decisions suggest that, as a threshold matter, the presumption does not apply when there are substantial domestic effects. Indeed, earlier courts indicate that when there are no substantial effects, then there is no question of extraterritoriality whatsoever. *See Laker Airways, Ltd. v. Sabena, Belgian World Airlines*, 731 F.2d 909, 923 (D.C. Cir. 1984) ("The territorial effects doctrine is *not an extraterritorial* assertion of jurisdiction.") Recently, in a Supreme Court decision of *Kiobel v. Royal Dutch Petro. Co.*, the Court indicated that claims that "touch and concern" the territory of the United States must do so with sufficient force to displace the presumption against extraterritorial application. 133 S. Ct. 1659, 1669 (2013). This suggests that courts will determine the domestic effects of statute-proscribing conduct as part of its determination of whether the presumption against extraterritoriality is rebutted. A concurring opinion by Justices Breyer, Ginsburg, Kagan, and Sotomayor argued the presumption against extraterritoriality is not even applicable when there are substantial effects on a domestic interest. 133 S. Ct. at 1671.

[89]16 U.S.C. § 1802 (2012).

[90]*Id.* §§ 1811(b), 1802 (2012).

[91]*Id.* § 1811(b) (2012).

within and beyond the EEZ.[92] In addition, it provides that the United States shall "promote" the provisions of the MSA where an international fishing organization does not have a process for developing a plan to rebuild depleted stock or a stock of fish is approaching a condition of overfishing.[93]

Thus, in order for the MSA to apply to other species not listed in the act, beyond the EEZ, and to non-citizens, the government would need to argue that, notwithstanding this express language, the basic common-law presumption against extraterritoriality is not applicable. For example, it might be argued that this doctrine is does not apply because of the substantial effects that fishing beyond the EEZ has on the United States' fishery resources. No court decision holds as much.[94] Therefore, while federal U.S. laws apply to some extent beyond the EEZ to regulate international fisheries, the reach of the MSA is perhaps limited by its express language.

36.3 U.S. Domestic Fisheries Regulation

Looking closer at domestic fisheries, this next subsection focuses on how one country, the United States, manages its fisheries, as a starting point for exploring the regulation of fisheries, more broadly. Covered by this discussion are both the MSA's more traditional government-based restrictions as embodied in National Standards and fishing limits, as well as the more recently employed market-based mechanisms such as IFQs for the management of fisheries. These programs are certainly not unique to the United States and are discussed primarily as an example of similar programs that exist elsewhere.[95] Another often-touted private property-

---

[92]*Id.*, § 1812(a) (2012).

[93]*Id.* § 1812(c).

[94]In the case of *Blue Water Fishermen's Ass'n v. Nat'l Marine Fisheries Serv.*, a U.S. District Court confronted the legality of the closing of a highly migratory species fishery for an area that extended beyond the EEZ. 158 F. Supp. 2d 118, 123 (D. Mass. 2001). The Court did not rule that the closure was permissible because of its effects in the EEZ, however. Rather, it found that another part of the MSA (16 U.S.C. § 1812) directed the United States to cooperate with other nations in managing highly migratory species, and the agency was simply implementing several treaties under provisions of the MSA, for which Congressional intent was clear extending jurisdiction extraterritorially. 158 F. Supp. 2d 118, 122–123, 123 n.19. Another U.S. District Court, faced with a lawsuit challenging that a possession ban on certain species was an impermissible extraterritorial application of the MSA, found that the statute was only being applied within the EEZ. While stating that the substantial effects doctrine can regulate conduct occurring outside its territory which causes harmful results within its territory, the court indicated it was not applying the doctrine, but rather was ruling that the presumption against extraterritoriality was inapposite because "applying only to United States vessels located within United States territory, the no-possession provision does not assert extraterritorial jurisdiction." *Nat'l Fisheries Inst., Inc. v. Mosbacher*, 732 F. Supp. 210, 215 (D.D.C. 1990).

[95]For a comprehensive look at IFQ programs internationally, at least as they existed in 1999, see National Research Council (1999).

based solution to reduce exploitation is aquaculture, discussed later in this chapter. Aquaculture seeks to provide private entities the exclusive dominion over a so-called particular "crop" of fish.

What will be seen with the MSA is a regime that, much like with international regulation, first sought to manage fisheries by establishing territorial limits to exclude foreign entities and then focused on perhaps the more difficult task of domestic conservation. While some gains have been made, fisheries management is still deficient in many respects. IFQs and aquaculture, often seen as silver bullets for solving fisheries management problems, remain controversial.

36.3.1 The Magnuson Steven Fishery Management and Conservation Act (MSA)

First passed in 1976, and amended several times since, the MSA[96] is the U.S. management regime for fisheries in its EEZ. It has been hailed as model for other countries,[97] but has also been criticized as being not adequate for the task.[98] It was initially passed to eliminate competition from foreign fishing operations, spurred by Congress's desire to return the economic benefits of fishing to domestic fishermen. Over time, however, it has been amended to establish a comprehensive system for the federal regulation of domestic fishing.

The act has a number of key provisions that are the bedrock of the regulatory system. As indicated, Title I of the 1976 act establishes jurisdiction in "a zone contiguous to the territorial sea of the United States," the inner boundary which was defined by the seaward boundary of each coastal state, and the outer boundary which was defined as 200 nautical miles out from these state waters.[99] (See Fig. 36.1). This language was amended in 1986 to reference the EEZ. Within the zone, the act establishes the United States' exclusive authority to manage its own fishery resources.[100]

Under the original act, foreign fishermen could only fish in what would become the EEZ if the fishing was governed by an international agreement and pursuant to a permit. They were also limited to the portion of fish not harvested by domestic fishing vessels.[101] Advocates hoped that this would boost the growth of the

---

[96] Its name has changed over time, but it will be referred to in this chapter as the MSA for simplicity's sake.

[97] See Daniel Pauly, Letter to the Editor, Apr. 17, 2011, available at http://www.nytimes.com/2011/04/21/opinion/l21fish.html?partner=rssnyt&emc=rss&_r=0, last accessed April 16, 2014.

[98] See Eagle et al. (2008), p. 649.

[99] 90 Stat. 336, Pub. L 94-265, Sec. 101 (April 13, 1976) (codified at 16 U.S.C. § 1811).

[100] Id. at sec. 102; (now codified at 16 U.S.C. § 1811)).

[101] 90 Stat. 337, Pub. L 94-265, Title II, Sec. 201 (April 13, 1976) (now codified at 16 U.S.C. § 1821)).

U.S. fishing industry. Nonetheless, in 1980, this provision had to be amended again to clarify that the only portion of the fisheries yield that could be allocated to foreign vessels was that which "cannot, or will not, be harvested by vessels of the United States."[102] Moreover, new criteria for determining allocations to foreign fishing nations were added in 1980 and 1986, including those related to whether the nation enacts trade barriers, whether the fish harvested are for the country's domestic consumption, and whether, and to what extent, the nation otherwise contributes to, or fosters the growth of, a sound and economic U.S. fishing industry.[103]

This new authority resulted in a near complete termination of foreign fishing in the U.S. EEZ, accomplishing the act's original goal.[104] But it would not be until 1996 that the act was amended to focus on conservation and reducing the threat of overfishing from domestic fishing, as discussed in the next few sections.

36.3.2 Fishery Management Councils and Fishery Management Plans Under the Magnuson Steven Fishery Management and Conservation Act (MSA)

The heart of the MSA provides a unique structure of shared governance between the federal and state governments. Federal fisheries are primarily managed pursuant to the advice that eight regional fishery management councils provide in their Fishery Management Plans (FMPs). The regional councils are composed of voting members, which include the head state fishery managers in each state and for each region; the regional director of the federal agency, the National Marine Fisheries Service (NMFS); state-governor-nominated and NMFS-approved regional experts representatives of commercial, recreational, charter fishing sectors, and Native tribes, depending on the region.[105] With this structure, Congress sought to "preserv[e] the states' ability to play a key [development] role" in fishery management programs,[106] according to the U.S. Court of Appeals in Washington, D.C.

Except in rare occasions, FMPs have not been interpreted to be stand-alone regulations, however.[107] Instead, under the MSA, after a regional council develops an FMP,[108] it must submit the plan to the Secretary of the Department of

---

[102]16 U.S.C. § 1821 (d) (2012).

[103]*See* 16 U.S.C. § 1821(e)(1)(E) (2012).

[104]Territo at 1369 (2000), citing Decker (1995), p. 335.

[105]16 U.S.C. § 1852(b).

[106]*C & W Fish Co. v. Fox*, 931 F.2d 1556, 1557 (D.C. Cir. 1991).

[107]*Gulf Restoration Network, Inc. v. Nat'l Marine Fisheries Serv.*, 730 F. Supp. 2d 157, 166 (D.D.C. 2010). *But see Am. Oceans Campaign v. Daley*, 183 F. Supp. 2d 1, 11 (D.D.C. 2000).

[108]The Secretary can also develop an FMP under specific circumstances. *See* 16 U.S.C. § 1854 (c) (2012).

Commerce, who must, after reviewing it for compliance with the various laws applicable to the fisheries, "approve, disapprove, or partially approve" it after providing for public comment.[109] Regional councils simultaneously submit FMP-implementing regulations for review.[110] The Secretary must review them for consistency with the governing FMP as well as with the MSA and other applicable law.[111] If the regulations are found to be inconsistent, they are returned to the council with proposed revisions.[112] Otherwise, the proposed regulations are published for public comment,[113] and after the public comment period, the Secretary promulgates them, consulting with the council on revisions and explaining the changes made.[114] All final regulations must be consistent with the FMP.[115]

Several aspects of this structure are important. For instance, the federal government is charged with the ultimate authority to regulate fisheries by approving or disapproving FMPs and developing regulations. But the states, through their representatives on the management councils, are able to use their regional expertise to advise and direct such management by writing FMPs. The federal government provides a national check on regional advice, so that any fishery management does not subvert the best interests of the nation as a whole.

Notwithstanding the express role of the states, however, one criticism of this structure has been that, in effect, the federal government, which staffs and funds the regional councils, still has an outsized role in pushing national policies on regional fishery management efforts, in turn, limiting the flexibility that should come from regional or localized management.

Another important structural element is the MSA's explicit involvement of commercial and recreational fishermen on the regional councils. Fishermen may know best how to manage the fisheries' resources, and they have a vested interest in doing so. After all, as was discussed in Chap. 34 on the tragedy of the commons, it is fishermen who will be harmed in the long-term by the over-exploitation of the resource. At the same time, having fishermen with dominant positions on the councils has led to the charge that other voices, such as those of consumers, conservationists, and other members of the public, are often under-represented. Fishermen are also said to have an inherent conflict of interest in regulating the fishery because they are charged with conserving a resource that they also have an interest in exploiting. The fact that NMFS, an agency within the U.S. Commerce Department, is ultimately responsible to implement the statute is another subject of

---

[109]16 U.S.C. § 1854(a), (a)(3) (2012).

[110]*Id.* § 1853(c) (2012).

[111]*Fishing Co. of Alaska v. Gutierrez*, 510 F.3d 328, 330 (D.C. Cir. 2007) (citing 16 U.S.C. § 1854 (b)(1)).

[112]*Id.* (citing § 1854(b)(1)(B)).

[113]*Id.* citing § 1854(b)(1)(A).

[114]*Id.* (citing § 1854(b)(3)).

[115]16 U.S.C. §§ 1854(b)(1)(B), (c)(7).

controversy, because it has been argued that the agency is limited in its ability to protect the resource because it also has a mission of promoting its development.

In addition to the act's structure, it is important to understand its management tools. The main documents that provide management tool are FMPs. FMPs contain the conservation and management measures that "are ... necessary and appropriate for the conservation and management of the fishery."[116] Additional provisions have been required of FMPs over time to strengthen how the agency and councils carry out this mandate. For instance, as part of a sweeping set of changes made to the act in 1996 and in response to the realization that fishery management efforts were not succeeding at reducing overfishing, the Sustainable Fish Act (SFA) amendments clarified that FMPs were to aim at "prevent[ing] overfishing and rebuild[ing] overfished stocks, and ... protect[ing], restor[ing], and promot[ing] the long-term health and stability of the fishery."[117] FMPs were also required to "describe and identify essential fish habitat for the fishery" and "minimize to the extent practicable the adverse effects on such habitat caused by fishing."[118] The plans must additionally establish a standardized reporting methodology to assess the amount and type of bycatch (which includes fish that are caught unintentionally), and include conservation and management measures that, to the extent practicable, minimize such bycatch and bycatch mortality.[119]

The act was also amended to require that when any species is found to be overfished, NMFS must approve a rebuilding plan that sets a time period for ending overfishing and rebuilding the fishery not to exceed 10 years, except in cases where the biology of the stock of fish, or other environmental conditions... dictate otherwise.[120] Under this provision, NMFS

> may consider the short-term economic needs of fishing communities in establishing rebuilding periods, but may not use those needs to go beyond the 10-year cap. To breach this cap, FMPs may only consider circumstances that 'dictate' doing so[,]... [including] an international agreement... [and] when the current number of fish in the fishery and the amount of time required for the species to regenerate make it impossible to rebuild the stock within 10 years ...[121]

---

[116] 90 Stat. 351, Pub. L 94-265, tit. III. sec. 303(April 13, 1976) (codified at 16 U.S.C. § 1811).

[117] 16 U.S.C. § 1853(a)(1)(A).

[118] *Id.*, § 1853(a)(7).

[119] *Id.* § 1853 (a)(11). Among other things, FMPs also are required to describe the fishery in detail, 16 U.S.C. § 1853 (a)(2) (2012); assess and specify the condition of, and the "maximum sustainable yield" and "optimum yield" from the fishery and include a summary of the information utilized in making such specification, *id.* § 1853 (a)(3); and assess and specify the capacity of the fishery, including the extent to which fishing vessels of the United States and foreign nations, can and will be able to annually harvest the fishery's optimum yield. *Id.* § 1853 (a)(4) (A)-(B). Finally, FMPs must include a fishery impact statement that analyzes the likely effects, if any, including the cumulative conservation, economic, and social impacts, of the conservation and management measures on fishermen and fishing communities. *Id.* § 1853 (a)(9).

[120] *Id.* § 1854(e)(4) (2012).

[121] *NRDC v. Nat'l Marine Fisheries Serv.*, 421 F.3d 872, 880 (9th Cir. 2005).

These measures, aimed at not only intentional fishing, but also unintentional damage to fish populations and habitat, reinforced that the MSA was to be aimed at overfishing. Nonetheless, the law did not live up to the task, and in 2007, the requirements for FMPs were amended again to mandate that they include Annual Catch Limits (ACLs), which correspond to the annual amount of catch that would not result in overfishing, as reduced by scientific uncertainty.[122] Accountability measures, such as in-season fishing closures or measures to correct for when fishermen exceed these limit were also added as requirements of FMPs to prevent the catch from exceeding these ACLs or to offset overages.[123] And accountability measures were added for all fisheries subject to overfishing.[124]

The various MSA revisions also demonstrate how the statute has changed from one that sought to manage fish exploitation by excluding foreign fishermen to one aimed at conserving fishery resources from domestic threats. These provisions remain controversial, as many fishermen have argued that the 10-year rebuilding requirement does not offer enough flexibility, especially where there is little information on the status of certain fish populations. Conservationists have argued that the calls for flexibility are simply an attempt to avoid regulation and that the act's overfishing and rebuilding provisions are precisely what is needed to conserve the fisheries.

36.3.3 National Standards Under the MSA

The National Standards are the criteria by which NMFS and courts evaluate the tools set out in the FMP documents. Notably, these very general standards have been the focus of numerous court decisions, as challengers have sought to invalidate various fishery management measures on the grounds that they did not comply with the standards.

36.3.3.1 National Standard 1

The lodestar is National Standard 1, which requires that "[c]onservation and management measures . . . prevent overfishing while achieving, on a continuing basis, the optimum yield from each fishery for the United States fishing industry." To break this down further, as amended in 1996, overfishing is the rate of fishing mortality that jeopardizes the capacity of a fishery to produce the maximum sustainable yield on a continuing basis. So under National Standard 1, conservation measures must reduce fishing mortality to the level that keeps fisheries at the

---

[122] 50 C.F.R. § 600.310(f)(1)-(7).
[123] *Id.* § 600.310(g)(1)-(3).
[124] *Hooks & Baylor* (2009), p. 194 (citing *Id.* § 1853 (a)(15)).

"Maximum Sustainable Yield," often referred to as MSY—language that is also present in the Law of the Sea.[125] The act does not define this language, but NMFS's guidelines provide that it is the scientific determination of "'the safe upper limit of harvest which can be taken consistently year after year without diminishing the stock so that it is truly inexhaustible and perpetually renewable.'"[126] Nonetheless, it is not enough under National Standard 1 that overfishing is prevented. Fishery measures must achieve "optimum yield," which is simply the MSY, reduced by any relevant social, economic, or ecological factor, and is the amount of fish that "will provide the greatest overall benefit to the Nation, particularly with respect to food production and recreational opportunities, and taking into account the protection of marine ecosystems." For overfished fisheries, it is the amount that provides for rebuilding to get to the MSY.

In short, under Nation Standard 1, FMPs are to include measures to prevent fishing mortality in a fishery that jeopardizes it as a renewable resource, and fishing may even be below this level, if needed to produce the greatest overall benefit to the nation. For those fisheries that are overfished, the standard requires that the fishing be at a level that allows for rebuilding. At least one U.S. federal Circuit Court has found that this means that the fishery management measure must have at least a 50 % chance of reaching the maximum sustainable yield, and in no way permits measures that only have as low as an 18 % chance of achieving this limit.[127]

36.3.4 National Standard 2

Perhaps ranking second in importance is National Standard 2, which requires that conservation and management measures be based on the "best scientific information available." The same language is used in the Law of the Sea.[128] Although no federal court of appeals has offered a definitive interpretation of this standard, it is clear that courts are "highly deferential" when determining whether NMFS and the regional councils meet the standard.[129] One district court has, for instance, stated that a complete failure to consider science and the introduction of better science would be needed for an FMP to be contrary to the standard.[130] The standard also does not mandate an affirmative obligation to collect data before establishing limits,

---

[125] *See* Article 61.

[126] *Maine v. Kreps*, 563 F.2d 1043, 1047 (1st Cir. Me. 1977) (quoting H.R. Rep. No. 445, 94th Cong., 1st Sess. 48 (1975)).

[127] *Natural Res. Def. Council, Inc. v. Daley*, 209 F.3d 747, 754 (D.C. Cir. 2000).

[128] Article 61(2).

[129] *See, e.g., Or. Trollers Ass'n v. Gutierrez*, 452 F.3d 1104, 1120 (9th Cir. 2006).

[130] *See Commonwealth v. Gutierrez*, 594 F. Supp. 2d 127, 132 (D. Mass. 2009) (collecting cases).

and the agency can act before information is even complete.[131] Notwithstanding the great discretion afforded to NMFS and the councils under National Standard 2, such discretion is not limitless. For example, it is not proper for NMFS to allocate fishing quota based on "pure political compromise" as opposed to "reasoned scientific endeavor."[132]

36.3.5 The Remaining Standards

The remaining standards[133] and case examples are listed below.

- **National Standard 3** requires stocks of fish to be managed as a unit throughout their range, and interrelated stocks of fish to be managed as a unit or in close coordination.
- **National Standard 4** prevents discrimination between residents of different States and says that any allocation of fishing privileges must be fair and equitable, reasonably calculated to promote conservation and ensure that no particular individual, corporation, or other entity acquires excessive shares.
  - At least one court has found that, under National Standard 4's fair-and-equitable requirements, measures must not "'impose a hardship on one group if its outweighed by the total benefits received by another group or groups.'"[134] But the same court has said that this standard is met, even if there is some discriminatory impact on fishermen, a long as the regulations are tailored to meet conservation purposes.[135]
- **National Standard 5** allows for fishery managers to consider efficiency, but no management measure can have economic allocation as its sole purpose.
  - Like standard 4, National Standard 5 prohibition on measures engendered by economic allocation is met as long as NMFS and the regional councils considered some justifications besides economic allocation.[136]
- **National Standard 6** requires conservation and management measures to take into account and allow for variations among, and contingencies in, fisheries, fishery resources, and catches.
- **National Standard 7** requires measures, where practicable, to minimize costs and avoid unnecessary duplication.

---

[131]*Mass. by Div. of Marine Fisheries v. Daley*, 10 F. Supp. 2d 74, 77 (D. Mass. 1998).

[132]*Midwater Trawlers Coop. v. DOC*, 282 F.3d 710, 720 (9th Cir. 2002).

[133]16 U.S.C. § 1851.

[134]*Fishermen's Finest Inc. v. Locke*, 593 F.3d 886, 895 (9th Cir. 2010) (quoting 50 C.F.R. § 600.325(c)(3)(i)(B)).

[135]*See Alaska Factory Trawler Ass'n v. Baldridge*, 831 F.2d 1456, 1464 (9th Cir. 1987).

[136]*Fishermen's Finest Inc.*, 593 F.3d at 895.

- It has been interpreted not to require a cost-benefit analysis, and it can be met with a limited study showing that the regulations will have positive environmental and sociological benefits.[137] Evidence showing that NMFS ignored a less costly, practicable approach, however, would render a plan in contravention of the standard.[138]
- **National Standard 8** requires management measures to take into account the importance of fishery resources to fishing communities by utilizing economic and social data in order to provide for the sustained participation of such communities, and minimize adverse economic impacts on such communities.
  - It has been interpreted solely to mean that measurement measures take into account the sustained participation of communities, not mandate a particular outcome for such communities.[139] As with all the National Standards, impacts to particular communities can be justified in the name of conservation.[140]
  - It has not been lost on the courts that there is a tension between Nationals Standards 8 and 1.[141] But the Courts have indicated NMFS "must give priority to conservation measures. It is only when two different plans achieve similar conservation measures that [NMFS] takes into consideration adverse economic consequences."[142]
- **National Standard 9** provides that conservation and management measures are to minimize bycatch and minimize the mortality of such bycatch.
  - National Standard 9 must also give way to competing conservation interests, but it is not permissible for the agency to adopt completely discretionary or voluntary measures to minimize bycatch.[143]
- **National Standard 10** requires conservation and management measures to the extent practicable, promote the safety of human life at sea.
  - Like Standard 9, National Standard 10 must be balanced in light of the other standards.[144]

---

[137] *Id.*

[138] *See Nat'l Coal. for Marine Conservation v. Evans*, 231 F. Supp. 2d 119, 133 (D.D.C. 2002).

[139] *See Pac. Coast Fedn. of Fishermen's Ass'ns v. Blank*, 693 F.3d 1084, 1093 (9th Cir. 2012).

[140] *See, e.g., Nat'l Coal. for Marine Conservation*, 231 F. Supp. 2d at 133.

[141] *See Natural Res. Def. Council, Inc.*, 209 F.3d at 753.

[142] *Id.*

[143] *See Ocean Conservancy v. Gutierrez*, 394 F. Supp. 2d 147, 159 (D.D.C. 2005); *Pac. Marine Conservation Council v. Evans*, 200 F. Supp. 2d 1194, 1201 (N.D. Cal. 2002).

[144] *Ocean Conservancy*, 394 F. Supp. 2d at 159. *Roche v. Evans*, 249 F. Supp. 2d 47, 57 (D. Mass. 2003); *Or. Trollers Ass'n v. Gutierrez*, No. Civil 05-6165-TC, 2005 U.S. Dist. LEXIS 34084, at *40 (D. Or. Sept. 8, 2005).

These are the touchstones of the act, and thus have been the focus of numerous court challenges by those seeking to invalidate fishery management measures on the grounds that they did not comport with the standards. Such cases were often brought by fishermen who were harmed by certain fishery regulations, or by conservation organizations that believed that the fishery restrictions were not strong enough.

The running theme through all of the court decisions is that National Standard 1 and its conservation purpose reign supreme. Moreover, while the National Standards provide a basic framework for managing fisheries, they also provide an enormous amount of flexibility and discretion for fisheries managers. While this discretion gives agencies the ability to implement very strong fishery management measures, it also allows them to implement measures that are ineffective. They can implement measures that are simply a product of the political power of the fishermen that have the most influence over the councils. Any such measures' opponents are left with a considerable challenge to make the case that managers have stepped too far in favoring one group of fishermen over another, designing measures that are aimed too much at economic allocation, or adopting measures not strong enough to reduce overfishing or bycatch or protect human safety.

Many of these precise complaints have been levied at what are perhaps the most recent tools used in fisheries management, IFQs and aquaculture, which are explored in depth in the next two subsections. IFQs, it is argued, are too focused on economic allocation because they are aimed at efficiency in the fishery and, thereby, do not adequately protect small, supposedly inefficient fishermen. Indeed, Congress passed specific measures aimed at addressing these issues in 1996. With fish farming, opponents claim that it is mainly aimed at displacing wild-catch fishermen and inappropriately aimed at economic allocation.

36.4 Individual Fishing Quotas (IFQs)

Perhaps the most significant current debate about how the United States currently manages its fisheries relates to Individual Fishing Quotas (IFQs). These are also often referred to as Individual Transferable Quotas (ITQs). The MSA classifies them in a broader category of Limited Access Privilege Programs (LAPPs).

IFQs are akin to "cap and trade" market-based regulatory regimes. The basic and primary objective of IFQs is to avert the so-called "tragedy of the commons," (discussed in Chap. 34), by providing fishermen a portion of the total fisheries harvest quota (this is the "cap;" it is called a Total Allowable Catch (TAC) in the United States. The theory is that when fishermen have their own share of the fishery that is exclusively theirs, they will not continue to invest money in boats and gear to utilize more of the diminishing, exhaustible resource. Another similar argument is that fisherman given an ownership interest in the fishery will trade off any short-term gain from fishing in order to reap any long-term gains. Fishermen will slow down, allowing fish to be caught all year. Proponents also argue that this makes fishing safer.[145]

Crucial to the concept of any IFQ program is that the fishing privilege—i.e., the shares or quota can be bought or sold. This provides the "trade" in the "cap and trade" scheme. Fishermen who want to exceed their cap are able to buy more quota from those fishermen who will not use theirs. If the purpose of establishing an IFQ program to reduce "overcapitalization," or the inefficient excess fishing boats and gear in the fishery, tradability allows purchasers of quota to finance those who sell their quota and thus remove their gear and boats. Quota is bought by those who can fish at the lowest costs or produce the most valuable product. Over the long-term, it is argued that transferability allows the most efficient operators to obtain the bulk of the fishing privileges, producing optimally sized fishing fleets.[146] In turn, those more efficient operators with lower costs are able to dedicate more money to invest in resource improvement.[147]

36.4.1 Criticisms of IFQs

In the Unitd States, IFQ programs have been controversial from the outset. Initially, the concern was that NMFS was implementing them without adequate consideration of the regional fishery management councils.[148] But another core concern by some conservation and consumer groups and fishermen was that the public resource would be privatized. In the run-up to approving the 1996 amendments to the MSA, U.S. Representative George Miller offered an amendment—which never passed—that would have ensured that IFQs could not be traded. Speaking of his concerns, he stated:

> [ITQs], which are allocated for free, can then be brought and sold, taking a public resource and turning it into a private commodity.
> . . .
> Proponents of quota systems tout their advantages. Allowing holders to fish when they want instead of in a derby fashion, they can produc[e] higher quality product, spread out their season, and stay at the dock when the weather is bad. All of these advantages will still hold true.
> But [this] does not merit nor does it require, the flatout giveaway of a public resource with no benefits to the taxpayers.
> . . .

---

[145] Carden et al. (2013), p. 51.

[146] National Research Council (1999), p. 169.

[147] Rieser (1997), p. 823.

[148] As the House Report for the 1996 SFA reports: "Recent efforts by [NMFS] to promote ... [ITQ] systems above any other type of limited access system concern the Committee and are inappropriate. ... Because ITQ systems have the potential to fundamentally alter fisheries management in the U.S., the Committee believes they must be used with great caution." H. Rept. No. 104-171 at 36 (1995).

> My amendment would ensure that the give-away of a public resource would be prevented; that big fishing corporations would not profit at the taxpayers expense; and the stewardship of our fisheries remains in the public trust where it belongs.[149]

Other concerns about IFQ programs include their distributional effects on smaller scale fishermen. Larger-scale fishermen end up purchasing smaller-scale fishermen's quota in order to reduce their costs and increase their ability to fish. Smaller fishermen, who cannot afford to fish at levels beneath their quota allocation—if they are allocated quota at all—and who cannot afford to purchase or lease more, often have to exit the fishery.[150] Smaller fishermen then must spend their money leasing access to the resource, akin to medieval serfs. Thus, it is argued, IFQs result in a radical redistribution in the fishery away from small independent fishermen to larger, sometimes international fishing corporations.

Other concerns, particularly leveled by consumer groups and academics, are related to the effects that IFQs have on competition and consolidation. The goal of IFQs is not to simply constrain fishing effort below the maximum sustainable yield, but to a point where the remaining fishermen use all of the entire fleet's "extra-normal profits" by excluding competitors, particularly smaller competitors.[151] This consolidation can be especially acute if quota is awarded based on fishermen's catch-history and for free, so that the largest fishermen start off with a windfall and a large capital advantage based on their historic fishing levels.[152] Additional concerns are related to whether the fishery will cause such concentration in the fishery that it will prevent new fishermen from entering,[153] and it is argued that this could harm consumers in the long run.

Other criticisms are that the efficiency aims of IFQ programs prevent adequate consideration of other equally beneficial non-market measures that may limit entry to the fishery, such as fishing-privilege leases, royalties, and auctions.[154] These measures unitize the common fisheries resource through privileges but either do not allocate the privileges for free based on catch histories or allow more of a government role in how privileges are allocated in the market, so as to keep the privileges more publicly controlled. True believers in IFQs, however, argue that these government interventions introduce inefficiencies in the market. For example, when carefully designed, it is argued, auctioning quota can serve the same conservation ends as allocating quota for free but at lower costs for smaller-scale fishermen, while compensating the public for the losses incurred from allocating fishing privileges for private use.[155] Auction opponents, on the other hand, argue that

---

[149] 141 Cong. Rec. H10, 223 (Oct. 18, 1995).

[150] *See* National Research Council (1999), pp. 173–174.

[151] Bromley (2008a), pp. 4–5.

[152] *See* National Research Council (1999), pp. 142–143; Bromley (2008b), p. 13.

[153] *Id.*

[154] *See generally* Bromley (2008b), pp. 13–15.

[155] *Id.*

auctions are unfair to bigger fishermen who have fished the most historically and should be awarded privileges based on this demonstration of efficiency.[156]

Further, some conservation and fishing groups and academics argue that that the conservation benefits of IFQs are myopic and that they may even hurt fisheries. The studies that are cited for the conservation benefits of IFQ program often fail to disaggregate their effects from other, non-market-based management measures implemented at the same time, such as the TACs alone.[157] Further, opponents argue that fisheries and habitats may be harmed if those who are granted the most catch shares are those that have been bad actors in the past and are those that use gear associated with higher levels of harm to habitat or bycatch.[158] Some critics believe that without increased spending on enforcement, IFQs incentivize cheating because the rewards of doing so are immediate, while the costs or losses are in the future and seemingly less certain.[159]

Fishermen and conservationist critics also argue that IFQs are incompatible with ecosystem management. Such regimes aim to only allow as much fishing in a fishery as is detrimental to the ecosystem, regardless of the fishing's effects on a particular target species, which has traditionally been how fisheries have been managed.[160] With IFQ programs, on the other hand, quota is allocated solely based up on the singular market value of a few species. There is a serious risk that other valuable components of the ecosystem will be ignored because they are not given a monetized value.[161]

Finally, a number of academic and fishermen critics of IFQs contend that the idea that fishermen will trade off short-term for long-term gains because they have a greater ownership interest in the fishery overlooks the inherently contingent and short-term nature of the limited access privileges allowed under the MSA.[162] In the United States, fisheries managers have not given LAPP privilege holders a permanent property interest—at least not a *de jure* interest. In other words, it is not an interest expressly recognized as such under the law. This may limit the conservation ethic, as the fisherman may fear that their share of fish could be subject to government modification or termination at any time.[163] Fishermen point out that fish mortality in any given year is not solely related to fishing, where, for example, some mortality may be due to factors like water pollution. This means that private price agreements between quota sellers and purchasers may not result in the long-

---

[156]*Id.*

[157]Bromley (2008a), p. 3; Food & Water Watch (2011), pp. 8–9.

[158]*See* National Research Council (1999), p. 177. As the National Research Council indicated: "[I]mplementing an IFQ regime may favor some technologies over others. If [such programs] typically involve more bycatch, bycatch rates can rise in the absence of enforcement." *Id.*

[159]Rieser (1997), p. 822, *citing* Congressional Research Service (1995), pp. 8–19.

[160]Pew Ocean Commission (2003), p. 44.

[161]Rieser (1999), p. 405.

[162]Bromley (2008a), p. 7.

[163]Rieser (1997), p. 822.

term allocation of privileges conducive to ending the race to fish and tragedy of the commons.[164] Thus, even with a fully functioning IFQ program, it is argued that there will still be overfishing, or the need for rebuilding and, consequently, the need for government intervention.

What resulted from the 1996 SFA amendments was a moratorium on IFQs so that a comprehensive study could be conducted by the National Academy of Sciences about the "controversial IFQ-related issues such as initial allocation, transferability, and foreign ownership."[165] The *de jure* moratorium lasted approximately 6 years. Then, in 2007, Congress ended the *de facto* moratorium by adopting some of the National Academies of Science's recommendations as criteria that would allow fisheries managers to "balance many of the concerns fishermen, crew, communities, conservation groups, and other interests have had over the potential impacts" of such programs.[166] These standards are discussed in the next subsection.

36.4.2 The MSA's Standards for IFQs and Other LAPPs

In the final version of the 2007 law, Congress amended the MSA to allow IFQ programs but also established a set of procedures and criteria for their adoption.[167] Some of these provisions were designed to incorporate fairness and equality into such programs.[168] Other parts of the law require consideration of the impacts of the programs on conservation, fishermen, and fishing communities.[169] Further, new provisions were aimed at giving fishermen in the New England and Gulf of Mexico regions the power to decide, through citizen referenda, whether they would want IFQ programs.[170]

Under the new provisions, LAPPs and IFQs are considered a permit, and they may be revoked, limited, or modified at any time, including if they jeopardize the sustainability of the stock or the safety of fishermen.[171] Such programs do not confer any right of compensation to the holder; they do not create, any right, title, or interest in or to any fish before the fish is harvested; and they are considered a grant of permission to engage in activities permitted by such limited access privilege or quota share.[172] The intent of these limitations was to prevent IFQ holders from

---

[164]*See* Bromley (2008a), p. 7.

[165]*See The Sustainable Fisheries Act*, Pub. L. No. 104-297, sec. 108(e), § 303(f), 110 Stat. 3576 (1996).

[166]S. Rept. No. 109-229 at 9 (2006).

[167]16 U.S.C. § 1802(26)(A)-(C) (2012).

[168]*See id.* § 1853a (2012).

[169]*See id.* § 1853a(c).

[170]*See id.* § 1853a(c)(6)(D).

[171]*Id.* § 1853a(b).

[172]*Id.*

developing investment-backed expectations that could require the government to compensate them for the elimination or modification of such rights.[173] IFQs, nevertheless, do convey at least a limited property interest, including the right to access and enjoy certain fish populations by fishing. They have two of the essential attributes of property: exclusivity and transferability.[174] Insofar as these property allocations are not done for the public interest and substantially impair what remains for others to access, one could argue that they would violate of the Public Trust Doctrine that inheres to the federal fish resources, as discussed in Chap. 34.[175]

Another key part of the 2007 provisions is the act's attempt to allow communities to hold quota.[176] Under Community Development Quota (CDQs), fishery managers grant quotas to the entire community, for the benefit of all. The benefits of this type of quota is its ability to subdivide the allocation of fishing privileges in a more refined manner than a regional council or individuals can in a free-market system.[177] CDQs may also be able to mitigate the inevitable social disruption caused by new IFQ programs.[178] Communities may also be able to enforce limits on fishing through informal norms and sanctions.[179] They also may better democratize the allocation of quota, which may decrease concerns about keeping the public out of the fishery.[180]

The MSA's community-based and other new LAPP provisions faced one of their first challenges in the case of *Pac. Coast Fedn. of Fishermen's Ass'ns v. Blank*, where a group of non-trawl (a type of fishing gear) fishermen challenged an IFQ program for the trawl-boat sector of the fishery off the west coast of the United States.[181] The challenging fishermen, who were concerned that they would be harmed by the consolidation impacts from the IFQ program, argued that NMFS had failed to adhere to new criteria established for LAPPS including that they (1) develop criteria for ensuring that quota shares are distributed to fishing communities, and (2) adopt other measures and policies to ensure the sustained participation of fishing communities.[182]

The court disagreed with the arguments made by the challengers of the IFQ program. It found that the statute did not require the agency to develop such criteria for ensuring community participation and that the sustained participation of

---

[173]Rieser (1997), p. 821.

[174]*Id.*, pp. 821–824, citing Scott at 35–36 (1996).

[175]*But see* Lynch (2007), pp. 307–310, arguing that the IFQ restrictions on public access meets the Public Trust Doctrine test laid out in the *Illinois Central* case.

[176]16 U.S.C. § 1853a(c)(3).

[177]Rieser (1997), p. 827.

[178]Lynch (2007), pp. 309–310.

[179]Rieser (1999), pp. 405–406.

[180]Lynch (2007), pp. 309–310.

[181]*Pac. Coast Fedn. of Fishermen's Ass'ns*, 693 F.3d at 1086.

[182]*Id.* at 1092.

communities only had to be considered by the agency, which it did, notwithstanding that communities could not participate in the programs by holding quota.[183] The regional fishery management council complied with the law by "survey[ing] the current status of fishing communities; ... describ[ing] the effects of quota programs and other management tools on those communities; and explain[ing] how communities participated in the Pacific Council's decisions."[184] In addition, the court credited several of the FMP measures aimed at the mitigation of the IFQ program's disruptive effects, including its adaptive management program that provided quota to be held and not completely allocated.[185] The quota could then be allocated later to communities and others, if NMFS determined that it was necessary to mitigate the harm caused by the initial allocations and the market. Finally, the other major issue in this case was whether quota holders had to be those that "substantially participate" in the fishery.[186] The court held that while those fishermen that substantially participate in the fishery had to receive quota, they did not have to be the only ones eligible.[187]

The implications of this decision are potentially far-reaching. In fact, the standards that Congress adopted in 2007 to prevent some of the negative impacts of IFQ systems, especially those aimed at promoting community involvement and sustained participation for smaller fishermen, were deemed discretionary. Certainly, many people's hopes were diminished that the new MSA's LAPP provisions would bring in a new era of CDQs. The decision also begs the question of whether courts will require IFQ programs to adopt any mitigation measures whatsoever in order to protect small fishermen and communities. Finally, the ruling means that quota can be almost completely privatized, notwithstanding the language of the MSA. Quota will be held by even those who have not substantially participated in the fishery, meaning that there is a strong potential that shares of the fishery will be acquired by those who have no connections to fishing whatsoever, including those who might seek to buy, sell, and trade their quota simply for financial gain, comparable to stocks on a stock market. Many consumer, conservation, and fishing organizations hope that this issue will be addressed in the MSA as it comes up for reauthorization in the next few years.

Thus, while the 2007 MSA amendments' termination of the *de jure* moratorium on IFQs was premised on the establishment of some standards to protect fishermen, the *Pac. Coast Fedn. of Fishermen's Ass'ns* decision perhaps demonstrates that these MSA standards were inadequate and that there is a need for more attention to these issues in the future.

---

[183]*Id.*

[184]*Id.* at 1093–1094.

[185]*Id.* at 1093.

[186]*See* 16 U.S.C. § 1853a(c)(5)(e).

[187]*Pac. Coast Fedn. of Fishermen's Ass'ns,* 693 F.3d at1095–1096.

36.5 Fish Farming

36.5.1 US Regulation of Aquaculture

Another attempt to unitize or privatize the common fisheries resource is with fish farming, also known as aquaculture. The U.S. federal government has defined it as the "propagating and raising of marine organisms."[188] Aquaculture is a very broad subject. It includes such diverse practices as the raising of bivalves, such as clams, molluscs, or scallops on strings or pots, raising alligators, inland aquarium-like recirculating structures, and the raising of finfish, such as salmon in pens made of nets or large spherical cages.[189] Aquaculture growth has increased rapidly in the last few decades and, in 2010, contributed to nearly half of the world's fish production for food.[190]

Fish farming, it is argued, will increase the supply of fish, thus preventing overfishing and preventing the so-called tragedy of the commons. While later chapters of this book will cover the harms of fish farming and their international regulation in depth, the issue about how best to regulate it in the United States, including under the MSA, has been the subject of much debate in the United States. Much of this debate has focused on whether it should be officially sanctioned as an allowed use of the EEZ, and whether NMFS should be allowed to regulate it under the MSA or some other statute aimed at targeting its unique potential environmental and public health impacts.

Fish raised on farms in rivers, streams, and within three miles of shore in the United States are governed by state law.[191] This includes most of the aquaculture production in the United States.[192] However, due to the crowded nature of state waterways, the non-uniformity of state laws, and advances in the technologies of raising fish, there has been an aggressive push by fish-farming advocates to move facilities beyond state waterways, in cages or netpens in open ocean waters beyond the three-nautical-mile state limit.[193]

In 1999, the U.S. Department of Commerce, which houses NMFS, called for the quintupling of the U.S. annual aquaculture production by the year 2025—from $900 million a year to $5 billion. The goal is to offset the so-called "seafood trade deficit," meaning that the United States exports more seafood than it imports, so as to create more jobs, and bring more high quality seafood to U.S. customers.[194] Part of this push is to increase the amount of fish farming in the EEZ.

---

[188]National Oceanic and Atmospheric Administration at 1 n.1 (2011).

[189]*See* Cicin-Sain et al. (2001), pp. 13–15.

[190]Food and Agricultural Organization at 26 (2012).

[191]*See id.* at 19.

[192]Government Accountability Office at 7 (2008).

[193]*See id.* at 13–21; Marine Aquaculture Task Force at 1. This argument is also being advanced internationally. *See* Kapetsky et al. at xiv (2013).

[194]U.S. Department of Commerce at 2 (1999).

Proponents of fish farming in the EEZ argue that the deep water and fast currents can minimize harmful environmental impacts from fish waste.[195] They also argue that offshore aquaculture would give operators more space to operate and minimize some of the conflicts between fish-farm facilities and fishermen.[196] Fish farms not only take up space that cannot be used by fishermen, they attract other fish, which can either harm or benefit fishermen depending on how much access fishermen are given to fish around the farms. It is also argued that with fish farms in areas where they can spread out, there will be fewer local impacts from excess fish feed and waste polluting the water and seafloor.

36.5.2 The Problems with Aquaculture

Critics of fish farming argue that the potential mitigation from allowing fish farms in deep waters will not sufficiently alleviate potential environmental, consumer, and socio-economic impacts. Concerns include the water quality and seafloor impacts from the untreated fish waste and excess feed waste from the farms.[197] In terms of water quality, the main concerns are nutrients, such as nitrogen and phosphorus. A recent technical memorandum for NMFS surveyed existing studies and concluded that increased nitrogen levels could cause potential localized water quality effects.[198] Further, it found that "[q]uestions remain about the cumulative impacts of [nitrogen] discharge from multiple, proximal farms, potentially leading to increased primary production and eutrophication at regional and far-field scales."[199] Eutrophication is the excessive richness of nutrients, which causes a dense growth of plant life and the death of animal life from a lack of oxygen. The one published study of offshore aquaculture in U.S. waters found that the farm's cages "grossly polluted" the sea floor and "severely depressed" marine life at some sampling sites very close to the fish cages, and, over the course of 23 months, these effects spread to sites up to 80 meters away.[200]

Other concerns are that escaped fish and disease from aquaculture cages in the ocean will harm surrounding wild-fish populations. Aquaculture cages can, for example, act as reservoirs for diseases. The two diseases that are often associated with fish farming are sea lice, a parasite that inflicts wild and farmed salmon alike, and Infectious Salmon Anemia. A 2012 study also found that 39 % of adult wild

---

[195]Government Accountability Office at 2 (2008).

[196]*See* Cicin-Sain et al. (2001), p. 18.

[197]Government Accountability Office at 9 (2008).

[198]Price and Morris (2013), p. 9.

[199]*Id.* at 46.

[200]Lee et al. (2006), pp. 175–185.

salmon in the Northeast Atlantic are killed by parasite infections and suggested that a large source of the sea lice are aquaculture facilities.[201]

Fish escapes are also a concern with open-water fish farms. Cages and net pens are not the natural habitat of fish, and so escapes are inevitable. When the fish do escape, farmed fish interbreed with wild fish. The offspring may have diminished survival skills, inherited from the farmed fish that may be genetically modified. The concern is that the result will be wild-fish populations that are genetically less fit and an overall lessening of biodiversity. A comprehensive review of the effects of escaped Atlantic salmon on wild populations conducted in 2008 concluded the outcome of escapee-wild fish interactions are frequently negative for wild salmon and can have devastating consequences for marine ecosystems.[202]

Another major issue is that the fish food that is used to feed farmed fish is often made from fishmeal and oil derived from small ocean forage fish such as sardines, anchovies, and herring, often caught in the Northeast Atlantic and off North and South America's Pacific coast.[203] Many such forage fish species are vulnerable to overfishing and have a high probability of population collapse.[204] Therefore, it is argued that there is a strong potential that an increase in fish farming could put further pressure on wild-fish populations, instead of alleviating overfishing, as is its promulgated goal.

Antibiotics, pesticides, and the other drugs or chemicals used in these operations can also be damaging to marine ecosystems and fish populations.[205] Studies have found antibiotics present in wild fish that were feeding on feces and food from fish farms, and thus may serve as a pathway for development of antibacterial resistance within wild populations.[206] Researchers believe that more needs to be understood about the accumulation of antibiotics from fish farms and human exposure,[207] but current trends offer cautionary tales.

Finally, there are concerns about the socioeconomic impacts to wild fishermen. Farmed fish can flood the market with product, thus decreasing prices that fishermen receive for their catch. For example, in the 1990s, commercial fishermen in British Columbia saw prices for salmon decrease by two thirds, in large part because of aquaculture's increasing of the salmon supply worldwide.[208] As prices declined, the ability of traditional fishermen to sustain a livelihood became more difficult.

Ultimately, the question is how to best prevent these impacts, especially because they are unique and quite different from many issues with wild-caught fishing.

---

[201] Krkosek et al. (2012), p. 5.

[202] Thorstad et al. (2008), p. 6.

[203] Food and Agricultural Organization (2011), p. 13.

[204] Lenfest Forage Fish Task Force (2012), p. 73.

[205] Marine Aquaculture Task Force (2007), pp. 74–76.

[206] Price and Morris (2013), p. 113.

[207] Id.

[208] Marshall (2003), p. 5.

Additionally, as mentioned in regards to the regulation of managing wild-fish populations, there are concerns that NMFS, whose mission is to increase the amount of fish produced from the seas, is not the proper agency to ensure that any growth in the industry is sustainable. The next section explores how this controversy has manifested itself with fish farming: whether and to what extent the agency has or should have the authority to regulate fish farming in the federal EEZ.

36.5.3 *Aquaculture Regulation Under the MSA*

In light of the controversies surrounding fish farming, the question arises as to whether fish farming should be conducted in the U.S. EEZ, and if so, how it should be regulated? Initially, proponents of more aquaculture development, including fish farmers and NMFS, pressed for new authorizing legislation from Congress, which, among other purposes, would provide fish farmers a property interest—a permit issued by NMFS—to begin operation in the EEZ.[209] Notably, the industry and NMFS were not the only advocates. Perhaps fearful that aquaculture operations would start in federal waters and not be comprehensively regulated, a few conservation organizations supported different legislation, the National Sustainable Offshore Aquaculture Act, first introduced in 2009 by Representative Lois Capps of California.[210] The legislation established a permit system for aquaculture facilities in federal waters, but also created environment standards for the permitting.[211] Other organizations argued that there was no need to provide a permit or any other property interest for aquaculture facilities in federal waters because this will simply provide the investment certainty needed to spur development. Without such certainty, aquaculture facilities are not likely to start operation in the EEZ.[212]

To date, these pieces of federal legislation have stalled. This has prompted NMFS to explore ways to authorize aquaculture operations in the U.S. EEZ under its existing authority under the MSA. In 1993, a legal opinion by then-Deputy General Counsel for the NMFS's parent agency, the National Oceanic Atmospheric Association, Jay S. Johnson, and NMFS's then-Assistant General Counsel, Margaret Hayes, asserted that NMFS and NOAA had the authority under the MSA to

---

[209]*See* National Offshore Aquaculture Act of 2007, H.R. 2010, 110th Cong. § 2(a)(3) (2007); S. 1609, 110th Cong. § 2(1)(c) (2007).

[210]H.R. 4363.

[211]After all, while there are plenty of laws that would apply to aquaculture facilities in the EEZ, including the Clean Water Act and Rivers and Harbors Act, some believe that, as Professor Hope Babcock has rightly noted, aquaculture in the EEZ is a classic example of what Professor William W. Buzbee has deemed a "regulatory commons." A regulatory commons exists when there is inaction or a lack of comprehensive oversight over problem due to shared disjointed jurisdiction between regulatory authorities. Babcock (2007), p. 69, citing William W. Buzbee (2003).

[212]*See* Cicin-Sain et al. (2001), p. 20.

regulate aquaculture in federal waters under the MSA's definition of "fishing." In particular, "fishing" as defined by the MSA is the "catching, taking, and harvesting of fish." Johnson argued that the use of the word "harvesting" in this definition connotes the gathering of a crop, so it could include fish farming.[213]

It is undisputed, however, that this opinion was simply Johnson's attempt to assert the agencies' limited authority to regulate fish farming in the face of a proposed salmon farm in federal waters in New England.[214] It was not until more recently, in 2009, that the Gulf of Mexico Regional Fishery Management Council (the Gulf Council) sought to extend this authority to establish a regional FMP for offshore fish farming. The plan was supposed to establish a permitting process that would allow an estimated 5 to 20 new aquaculture projects to the Gulf of Mexico portion of the EEZ over 10 years, collectively producing as much as an estimated 64 million pounds of fish annually. Operators would get 10-year permits with exclusive use of the area surrounding its operations.[215]

In 2011, after a number of public hearings and taking of public comment, the federal government also issued a new national policy with the chief aim to encourage and foster "sustainable aquaculture development that provides domestic jobs, products, and services and that is in harmony with healthy, productive, and resilient marine ecosystems, compatible with other uses of the marine environment, and consistent with [its] . . . National Oceans Policy."[216] Critics complained that the policy lacked tangible measures to prevent the harms from fish farming, and its perhaps-laudable goals were contradicted by other aims, such as "developing streamlined processes for authorizing aquaculture . . . in Federal waters[.]"[217] A month after the aquaculture policy was released, NMFS issued its first-ever permit to a commercial aquaculture facility in the EEZ off the coast of Hawaii. The permit was challenged in a lawsuit mounted by two non-profit organizations, with this chapter's author as one of its counsel. Among their claims, the plaintiffs argued that the agency did not have the authority under the regional FMP to issue permits for aquaculture. NMFS defended its permit issuance on the grounds that it had the broad authority under the MSA as it amounted to the "harvesting" of fish.

---

[213] Johnson (1993), p. 2.

[214] *Id.* at 1.

[215] Gulf of Mexico Regional Fishery Management Council at 1 (2009). In 2009 the FMP was finalized by the regional council, but the Secretary of Commerce under newly elected President Obama decided that he would neither approve nor disapprove the FMP and never issued any implementing regulations for the plan, instead indicating that it would develop a national policy by which to assess aquaculture. The final FMP was challenged in court, but a federal district court judge ruled that the lawsuit could not go forward until the agency finalized rules to implement it. *See Gulf Restoration Network, Inc. v. Nat'l Marine Fisheries Serv.*, 730 F. Supp. 2d 157, 162 (D.D.C. 2010).

[216] National Oceanic and Atmospheric Administration at 1–2.

[217] *Id.* The policy was issued on the same day that a new memo was issued from the NOAA General Counsel's office, citing the 1993 Johnson memo and stating that the MSA allows the agency to issue permits for aquaculture facilities in the EEZ.

In a 2012 decision, the District Court for the U.S. District of Hawaii found that because the MSA and its regulations do not define "harvesting," the agency was free to "look to the dictionary definition of 'harvest' as 'the act or process of gathering in a crop.'"[218] Likewise, the agency was reasonable in defining "'Crop,' . . . [as] 'a plant or animal . . . that can be grown and harvested extensively for profit or subsistence.'"[219] Thus, the court stated, "'Defendants' determination that [the aquaculture project] falls within the term 'harvesting' was reasonable. The project involves growing and gathering a 'crop' of [fish] to sell for human consumption."[220]

This decision was appealed. In a very brief 2013 unpublished decision, the U.S. Court of Appeals for the Ninth Circuit ruled that under the agency's regulations, "NMFS may review and issue special permits for proposals to fish 'with any gear not normally permitted,' and under NMFS's 'generally conferred authority'" under the MSA.[221] The decision leaves the issue of NMFS's authority to issue permits to aquaculture facilities under the MSA largely unresolved.

This decision is not binding precedent because it was not a published decision. It only serves as persuasive authority—even in the Ninth Circuit. Moreover, while the appellate court affirmed the lower court's decision, the former court did not adopt the latter's rationale and did not ratify the government's theory that aquaculture is "harvesting" under the MSA—failing to even use the words "aquaculture" in any part of its opinion. Opponents of extending NMFS's authority to issue permits for aquaculture facilities will certainly argue that the opinion suggests that the court was only affirming NMFS's permitting authority under the limited facts surrounding the particular permit.[222] Likewise, it is highly probably that aquaculture companies and NMFS will seize upon the Ninth Circuit's affirmance of the lower court's decision as meaning that NMFS has broad authority to issue permits to fish farms, especially with the implementing regulations for the long-stalled Gulf of Mexico aquaculture FMP.

Of further note about this case, as it ties together the role of common law and the Public Trust Doctrine discussed Chap. 34, is what NMFS argued in its brief and at oral arguments: If the agency did not have the authority to issue permits to

---

[218]*Kahea v. Nat'l Marine Fisheries Serv.*, No. 11-00474 SOM-KSC, 2012 U.S. Dist. LEXIS 59244, at *26–27 (D. Haw. Apr. 27, 2012). The decision quoted NMFS's filings and the "Harvest" definition found at Merriam-Webster, http://www.merriam-webster.com/dictionary/harvest.

[219]*Id.* The decision also quotes the "Crop" definition found at Merriam-Webster, http://www.merriam-webster.com/dictionary/crop (last visited Apr. 27, 2012).

[220]*Id.*

[221]*KAHEA v. Nat'l Marine Fisheries Serv.*, No. 12-16445, 2013 U.S. App. LEXIS 22046, at *3 (9th Cir. 2013) (quoting *United States v. Mead Corp.*, 533 U.S. 218, 229 (2001)).

[222]*See also Pazolt v. Director of the Div. of Marine Fisheries*, 631 N.E.2d 547, 551 (Mass. 1994) (finding aquaculture not within the boundaries of reserved right of public fishing, and saying that "aquaculture is a contemporary method of farming shellfish. . . . It is not incidental to or reasonably related to or a natural derivative of the public's right to fish.").

aquaculture facilities under the MSA, then there would be an unregulated industry operating in the EEZ. Plaintiffs mostly pointed to other MSA provisions that the agency could have relied upon for preventing the aquaculture project without actually having to provide a property interest to the facility with a permit. These provisions include ones that broadly allow FMPs to "include management measures in the plan to conserve target and non-target species and habitats."[223] Plaintiffs also pointed out that, even without statutory authority to regulate fish farms, NMFS could, if it chose to, prevent any aquaculture project's harms by using common law.[224] Thus, it may be that the common law can serve as a tool to prevent activities such as aquaculture that may harm EEZ resources, even if no statute provides such express authority. The U.S. government has sought and obtained such remedies with public lands, filing a common law public nuisance lawsuit in order to fill the gaps where statutory law did not apply.[225] While the U.S. Court of Appeals for the Ninth Circuit never reached this issue in its decision, in the future it might be argued that not only is NMFS fully authorized to prevent the harms from aquaculture facilities through common law actions such as trespass and nuisance, the agency has the obligation to do so under the common law Public Trust Doctrine, regardless of whether it has the authority to regulate aquaculture as "fishing" under the MSA.

Fish farming remains controversial in the United States. The authority of fisheries managers to regulate it and issue permits in the EEZ under the MSA is questionable and is certain to be tested in courts for years to come. The emergence of this industry as a means to meet a consumer demand for fish will be dependent on whether and how it is regulated with open questions about its environmental and socioeconomic costs.

---

[223] 16 U.S.C. § 1853(b)(12) (2012).

[224] The plaintiffs cited to the decision in *United States v. Ray*, for this authority. 423 F.2d 16 (5th Cir. 1970). In that decision, involving a number of defendants who had sought to establish an island nation on coral reefs beyond state jurisdiction off the coast of Florida, the U.S. government convinced the U.S. Court of Appeals for the 5th Circuit that it could issue an injunction, notwithstanding any express statutory authority to do so. The decision overturned the District Court's decision that the United States could not bring a trespass action under the common law because the government did not have sufficient property right in the reefs off the coast. The appeals court found that the reefs fell within the definition of seabed and subsoil in both the Outer Continental Shelf Lands Act and Geneva Convention on the Outer Continental Shelf, and long before President Reagan's 1983 proclamation about the EEZ. The convention provided that the United States had sovereign rights and the exclusiveness of those rights to explore the Shelf and exploit its natural resources. 423 F.2d at 21. The government's paramount rights established in international law and its vital interest "in preserving the reefs for public use and enjoyment[,]" were sufficient for it to obtain an injunction, even without fee simple interest in the property. *Id.* at 22.

[225] *See, e.g., United States v. Rainbow Family*, 695 F. Supp. 314, 327 (E.D. Tex. 1988). Indeed, the *Rainbow Family* decision cites *United States v. Ray*, and says that this power and duty to protect land come from its role as "conservator of the lands for the public interest[,]" *id.* at 326, using language strongly evoking the concept of the public trust doctrine.

36.6 Clean Water Regulation

Clean water regulation is integral to many of the subjects already discussed in this chapter, including international regulation of the global commons and aquaculture. This chapter's discussion, much like the discussion of fisheries thus far, is limited to marine issues and does not focus on pollution of freshwater bodies in lakes and streams that are only governed by domestic regulation. As detailed below, while there has been some improvement in the implementation of laws designed to prevent of marine pollution there still is much work to be done. Specifically, successful fisheries management in the United States will be dependent on reducing the aforementioned non-fishing threats to fish health and habitat.

Marine pollution can come from a number of different sources, including land-based runoff, vessel discharges, and oil drilling.[226] Discharges from land, for instance, include sewage, radioactive and industrial wastes, and agricultural run-off. A National Academy of Sciences study estimates that the oil that runs off U.S. streets and driveways into the oceans amounts to an *Exxon Valdez* oil spill every eight months.[227] Today, nonpoint sources pollution, or those that come from multiple sources on land like farms and urban runoff, presents perhaps the greatest pollution threat to oceans off the U.S. coasts.[228] In the United States, agriculture is one of, if not the most significant source of pollution because it is the source of nutrients such as nitrogen.[229] Nitrogen coming from animal waste that is applied to farmland as fertilizer is easily dissolved in water and is transported by rain into streams and rivers that eventually flow into the ocean. Further, tile drainage systems, which are constructed to collect and shuttle excess water from fields, act as an "expressway" for this nitrogen pollution.[230] For instance, the Mississippi River carries an estimated 1.5 million metric tons of nitrogen into the Gulf of Mexico each year.[231] Such nutrient pollution has been linked to harmful algal blooms and dead zones, including the Gulf of Mexico's dead zone that is more than 8000 square miles. Dead zones, which are areas of low oxygen, i.e. "hypoxic" conditions, kill large numbers of crabs, fish, and other species. In addition, this pollution results in the loss of seagrass and kelp beds, the destruction of coral reefs, and lower biodiversity in estuaries and coastal habitats.[232] These present a serious threat to fish habitat and consequentially to fish health.

---

[226]Buck (1998), p. 95.

[227]Pew (2003), p. 4, citing National Research Council (2002).

[228]*Id.* at 60.

[229]*Id.* at 60, citing National Research Council (2000).

[230]*Id.* at 62.

[231]*Id.* at 59, citing Goolsby et al. (1997).

[232]*Id.* at 62, citing Howarth et al. (2002).

Point sources of marine pollution also include animal waste overflowing from open lagoons at large industrial farms and oil spills[233] such as from the Deepwater Horizon explosion in the Gulf of Mexico, which released an estimated 210 million U.S. gallons of oil and dispersants.[234] Vessel discharges include contaminants from ballast tanks, sewage from cruise ships, and fuel spills.[235]

Global warming, caused by the burning of fossil fuels and other activities that release heat-trapping gases such as carbon dioxide, also poses a serious threat to ocean health. Recently, attention has focused on ocean acidification. The ocean absorbs about a quarter of the carbon dioxide released into the atmosphere every year. This has begun changing the chemistry of the seawater, including the amount of available calcium carbonate minerals, which serve as the building blocks for the skeletons and shells of many marine organisms, such as oysters, clams, and corals. In recent years, there have been near total failures in wild- and farmed-oyster production on the West Coast, which may be linked to global warming. When these shelled organisms are at risk, the entire food web is also at risk.[236]

These are just some of the marine-pollution threats that can affect the health of ecosystems and the size of fish populations, and which fishery managers must consider when deriving policy measures to curb fish mortality. In the United States and elsewhere, this becomes difficult when many of these threats are not directly under the jurisdiction of fishery management agencies. Much of this pollution, which comes from sources far upstream or even from air-pollution sources, is solely regulated domestically, and by other agencies. This creates a regulatory-commons problem already mentioned above.[237] Having multiple agencies exerting jurisdiction over different pieces of the problem limits the likelihood of comprehensive regulation.[238]

Much of this pollution from the United States into our oceans is simply a result of the lack of regulation. Only now has the country been moving to mandate national standards for greenhouse gas emissions from power plants and motor vehicles under the Clean Air Act, for example. As another example, much of the pollution from animal farms are excused from permitting requirements for manure discharges caused by precipitation.[239] Efforts to curb pollution from agricultural

---

[233]*Id.* at 63.

[234]Davies (2014).

[235]Buck (1998), pp. 96, 66.

[236]See http://www.pmel.noaa.gov/co2/story/What+is+Ocean+Acidification%3F, last accessed April 18, 2014.

[237]*See* Kundis Craig (2002), p. 664.

[238]As another example, Titles I and II of the Marine Protection, Research and Sanctuaries Act, 33 U.S.C. 1401–1445 (2012), also known as the Ocean Dumping Act, regulates ocean dumping and incineration at sea of materials other than vessel sewage waste. *Id.* (citing 33 U.S.C. 1402 (c)) Under this act, the U.S. Army Corps of Engineers issues permits for ocean dumping of dredged materials, while the EPA has permit authority for the dumping of all other materials. *Id.* (citing 33 U.S.C. 1412). The U.S. Coast Guard regulates garbage disposal of from vessels pursuant to the Marine Plastic Pollution Research and Control Act of 1987. *Id.* (citing 33 U.S.C. §1412).

[239]*See* 71 Fed. Reg. 37,744 (June 30, 2006).

sources have recently focused on pollution trading schemes, and have been challenged in court for not being authorized under the federal Clean Water Act (CWA).[240] As another example of failed regulation, the U.S. CWA does not cover untreated discharges of water from its largest source, cruise ships, anywhere in federal waters. Only in Alaskan state waters are cruise ships required to meet effluent standards, treat gray-water (waste water from sinks and showers that has not come into contact with feces) discharges, and monitor, record, and report discharges to state and federal authorities.[241]

Proponents of increasing the amount of aquaculture in the EEZ without further environmental standards point to the CWA as the chief tool to regulate pollution from such facilities, but critics argue that it is lacking in many respects. The primary tool under the CWA is its National Permit Discharge System (NPDES). The CWA makes it illegal to discharge any pollutant without a NPDES permit.[242] It defines a discharge as "any addition of any pollutant to the waters of the contiguous zone or the ocean from any point source other than a vessel or other floating craft." Such facilities are point sources under the CWA.[243]

To date, however, EPA regulates an aquaculture facility as a point source only if it qualifies as a "Concentrated Aquatic Animal Production Facility," which means that it discharges at least 30 days per year, and, for warm water facilities, produces 100,000 pounds of fish annually.[244] Critics contend that while these thresholds may prevent harm from aquaculture facilities operating on a commercial scale, they prevent the agency from collecting valuable information on smaller test projects.[245] Other limitations in existing CWA for aquaculture facilities in the EEZ is that the law is not clear whether escaped fish may also be considered a "pollutant" under the CWA.[246] Moreover, even though aquaculture facilities must comply with the act's technology- and water-quality-based effluent limitations, because EPA has set no water quality standards for ocean waters and its ocean-discharge criteria provide little guidance, the best tool that permit writers have for NPDES permits are the

---

[240]*Food & Water Watch v. United States EPA*, 2013 U.S. Dist. LEXIS 174430 (D.D.C. 2013). Available at http://switchboard.nrdc.org/blogs/aalexander/SJ%20decision.pdf, last accessed April 18, 2014. Non-profit organizations have also recently sued the U.S. Environmental Protection Agency for failing to implement water quality standards for nutrients in the Mississippi watershed in the face of states' failure to do so. See *Gulf Restoration Network et al v. Jackson et al*, 2:12-cv-00677-JCZ-DEK, Order and Reasons (Doc. 175) (September 20, 2013).

[241]Pew (2003), p. 66.

[242]33 U.S.C. § 1311(a) (2012).

[243]*Id.* § 1362(12)(B) (2012); see also Harvard Law School Emmett Environmental Law & Policy Clinic *et al*. at 4 (2012) (citing 40 C.F.R. § 122.2).

[244]Harvard Law School Emmett Environmental Law & Policy Clinic et al. at 5 (2012) citing 40 C.F.R. § 122.24(a). Facilities may also be designated as such on a case-by-case basis if they are deemed significant contributor of pollution to U.S. waters, but EPA has used this authority limitedly. *Id.*, citing 40 C.F.R. § 122.24(c).

[245]*Id.* at 6.

[246]*Id.* at 7–8.

EPA's effluent limitation guidelines, which do not set numerical limitations on pollution for aquaculture facilities, only establish best management practices.[247]

As can be seen, the United States has not been very aggressive in curbing marine pollution from all sources, including growing and novel offshore fish farming industry. International control has also been difficult.[248] Early conventions have failed.[249] A series of treaties have been ratified that target oil spills, but the costs of monitoring and compliance have given way to nations' shipping interests and the ratification of these treaties have been slow.[250] Dumping of wastes is also the subject of a number of treaties, but no treaty absolutely bans it. One treaty, the London Convention, regulates ocean dumping, prohibiting intentional dumping materials on the "black list," while allowing the dumping of less hazardous materials on the "grey list," if permitted by the International Maritime Organization.[251] In 1996, the parties to the London Convention produced the London Protocol, which entered into force in 2006[252] and further restricted intentional ocean dumping by banning it except for materials that are found on a "reverse list."[253] While monumental in its adoption of the "precautionary approach," where "appropriate preventative measures are taken when there is reason to believe that wastes or other matter introduced into the marine environment are likely to cause harm even when there is no conclusive evidence to prove a causal relation between inputs and their effects[,]"[254] the United States' failure to ratify this treaty has been criticized as sending a message of indifference about the global commons.[255] Moreover, critics have pointed to a lack of political leadership, legislative hurdles, insufficient resources, and pressure from regulated industries, as contributing to the weak implementation of the convention.[256]

Thus, as with fisheries regulation discussed earlier in the chapter, while some progress has been made limiting the harmful impacts of marine pollution in the United States and internationally, at least because a framework for its regulation has been established, there is much work that is needed to actually prevent the continued exploitation of the resources.

---

[247]*Id.* at 10–11.
[248]Buckat 97 (1998).
[249]*Id.*
[250]*Id.* at 98.
[251]Ghorbi (2012), p. 483, citing Hunter et al. (2002).
[252]http://www.imo.org/OurWork/Environment/LCLP/Pages/default.aspx, last visited April 19, 2014.
[253]Ghorbi (2012), p. 484 (citing 1996 Protocol to the Convention on the Prevention of Marine Pollution by Dumping of Wastes and Other Matter, Nov. 7, 1996, Article 4).
[254]http://www.imo.org/OurWork/Environment/PollutionPrevention/Pages/1996-Protocol-to-the-Convention-on-the-Prevention-of-Marine-Pollution-by-Dumping-of-Wastes-and-Other-Matter,-1972.aspx, last visited April 19, 2014.
[255]Sielen (2008), p. 52.
[256]*Id.*

36.7 Conclusion

Since ancient times, our oceans and its fisheries have been conceptualized as resources that are free for all to exploit. From Grotius and Seldenian to the enclosure movement and the Law of the Sea, however, there has been a dramatic change in how our world's oceans and fisheries resources are governed. Spurred by the realization that these resources are not inexhaustible, national, extraterritorial, and international legal management regimes have been erected in the last 70 years to prevent the so-called tragedy of the commons, discussed in the Chap. 34. The Law of the Sea is critical in this development. Its agreement of territorial limits and substantively protective articles has provided the means by which nations can assert the sovereignty necessary to protect the resources closest to their shores. These changes, and the treaty's adoption of the common heritage of mankind principle for the high seas, moved the international governance of the oceans and resources towards the Public Trust Doctrine model, also discussed in Chap. 34, where the resource is governed by a trustee on behalf of all in order to prevent its overutilization to the detriment of all.

While there have been significant developments in establishing legal regimes for fisheries and ocean-resource management in general, much more work has to be done. The Law of the Sea has a number of shortcomings, including its vague provisions, its lack of a global fisheries organization, its lack of a compulsory EEZ-dispute-settlement mechanism, and, notably, the fact that it does not mandate that nations adequately regulate their own EEZs. Domestic laws have also been ineffective to target overfishing. In the United States, for example, the MSA's shared governance between federal and state authorities and fishermen interests has been the source of division and regulatory stagnation instead of regulatory innovation. ACLs and 10-year rebuilding plans are some of the newest measures that seek to reduce overfishing, but they have been controversial. The MSA's National Standards—the MSA's "constitution"—has been interpreted by courts to maintain conservation as its primary focus, but they have also provided virtually unfettered discretion to fishery managers to cater to fishermen or special interest groups, further stymieing wise policy development. The result of all of these issues is that fisheries remain fully or over-exploited. Marine pollution regulation has also suffered as a result of a lack of comprehensive regulation, notwithstanding strides in regulating some of its sources.

IFQs and aquaculture, explored in detail in this chapter, are relatively recent and controversial attempts to solve the so-called tragedy of the commons. Nonetheless, attempts to implement privatized policies to manage fisheries and oceans cuts contrariwise to work that is needed to better manage fisheries. Supporters of privatization policies often neglect that the oceans are no longer a *res nullius* resource, as discussed in Chap. 34, and national and international management regimes do exist that are built on the premise that fisheries are public resources, to be managed for the benefit of all. Proponents of measures like IFQs and aquaculture overlook these foundations, treating the historical lack of *effective* management or *mismanagement* of the resources—often a product of a lack of political and the influence of fishing interests—as evidence that fishery management regimes *do not exist* or are broken beyond repair.

Instead of targeting these problems, the turn to privatization and market-based regulation, such as IFQs, weakens existing public management. For example, the premise that aquaculture can prevent overfishing by increasing supply neglects that this will come at the expense of displacing traditional fishermen and granting a private property interest that could spur development beyond the point that it will be effectively regulated. Perhaps, common law doctrines, such the Public Trust Doctrine or nuisance can fill in regulatory gaps, but this shift puts the onus on the public to protect the very resource for which they are supposed to be the beneficiaries. IFQs represent a similar shift, weakening a public system that has made some significant strides in limiting problems such as overfishing, and replacing it with an opaque system whose success is premised on granting a private-property interest in the fisheries for free, thus undermining the government's ability to control and manage the resource. Consequently, decision-makers should forgo such private regulatory measures and instead look to strengthen existing public management systems.

References

Alexander LM (1983) The ocean enclosure movement: inventory and prospect. San Diego Law Rev 20:561–566
Ashfaq S (2010) Something for everyone: why the United States should ratify the law of the sea treaty. J Transnatl Law Policy 19:357–62, 364, 368
Babcock HM (2007) Grotius, ocean fish ranching, and the Public Trust Doctrine: Ride 'Em Charlie Tuna. Standard Environ Law J 26:3, 69
Bromley DW (2008a) IFQs in the West Coast Groundfish Fishery: Economic Confusion and Bogus Reasons! Testimony before the Pacific Fishery Management Council 5–6 (Oct. 14, 2008a)
Bromley DW (2008b) The Crisis in Ocean Governance: Conceptual Confusion, Spurious Economics, Political Indifference, Maritime Studies 13 (MAST 2008, 6(2))
Buck SJ (1998) A global commons, an introduction. Island Press, pp. 70, 76, 80, 86, 92, 93–95
Buzbee WW (2003) Recognizing the regulatory commons: a theory of regulatory gaps. Iowa Law Rev 89:1
Carden K, White C, Gaines SD, Costello C, Anderson S (2013) Ecosystem service tradeoff analysis: quantifying the cost of a legal regime. Ariz J Environ Law Policy 4:39, 51
Carr CJ, Scheiber HN (2002) Dealing with a resource crisis: regulatory regimes for managing the world's marine fisheries. Stanford Environ Law J 21:45, 59
Christie DR (1999) The conservation and management of stocks located solely within the exclusive economic zone. In: Ellen Hey (ed) Developments in International Fisheries Law 395, 396
Christie DR (2004b) It don't come EEZ: the failure and future of coastal state fisheries management. J Transnatl Law Policy 14:1–3
Cicin-Sain B, Knecht RW, Rheault R, Bunsick SM, DeVoe R, Eichenberg T, Ewart J, Halvorson H (2001) Development of a Policy Framework for Offshore Marine Aquaculture

in the 3-200 Mile U.S. Ocean Zone 13-21 (July 2001), available at retiredsites/docaqua/reports_noaaresearch/sgeez1final.pdf (last visited 17 April 2014)

Congressional Research Service, Report for Congress, Environment and Natural Resources Policy Division, No. 95-849 ENR, Individual Transferable Quotas in Fishery Management 8-19 (1995)

Davies R (2014) Houston Oil Clean-Up on 25th Anniversary of Exxon Valdez Spill, ABC NEWS, Mar 24, 2014, available at http://abcnews.go.com/blogs/business/2014/03/houston-oil-clean-up-on-25th-anniversary-of-exxon-valdez-spill/ (last visited Apr. 17, 2014)

Decker CE (1995) Issues in the reauthorization of the Magnuson fishery conservation and management act. Ocean Coast Law J 1:323–335

Eagle J, Sanchirico JN, Thompson Jr BH (2008) Breaking the Logjam: environmental reform for the new congress and administration: protecting aquatic ecosystems: ocean zoning and spatial access privileges: rewriting the tragedy of the regulated ocean. New York Univ Environ Law J 17:646–649

Eckert RD (1979) The enclosure of ocean resources. Hoover Institute Press, p 129

1 Encyclopedia of Public International Law 692 (1992)

Food and Agriculture Organization of the United Nations (2011) Use of Wild Fish in Aquaculture. FAO Technical Guidelines for Responsible Fisheries 5(Suppl. 5):13

Food And Agriculture Organization of the United Nations, The State of World Fisheries And Aquaculture, 11, 13, 26, 53 (2012), available at http://www.fao.org/docrep/016/i2727e/i2727e.pdf, (last visited April 10, 2014)

Food & Water Watch, Fish Inc., 8-9 (2011), available at http://www.foodandwaterwatch.org/tools-and-resources/fish-inc/ (last visited May 1, 2014)

Fulton TW (1911) Sovereignty of the sea. Blackwood, p 547

Ghorbi D (2012) There's something in the water: the inadequacy of international anti-dumping laws as applied to the Fukushima Daiichi radioactive water discharge. Am Univ Int Law Rev 27:473, 483–484

Goolsby DA, Battaglin WA, Hooper RP (1997) U.S. Geological Survey, Sources and Transport of Nitrogen in the Mississippi River Basin, American Farm Bureau Federation Workshop, St. Louis, Missouri 14, July 15, 1997, available at http://wwwrcolka.cr.usgs.gov/midconherb/st.louis.hypoxia.html (last visited Apr. 10, 2014)

Gorina-Ysern M (2004) World ocean public trust: high seas fisheries after Grotius - towards a new ocean ethos?. Golden Gate Univ Law Rev 34:645, 645–646, 659, 668–671, 673–675, 680, 683–685

Government Accountability Office (2008) Offshore Marine Aquaculture Multiple Administrative and Environmental Issues Need to Be Addressed in Establishing a U.S. Regulatory Framework, GAO-08-594, 2, 7, 9

Gulf of Mexico Regional Fishery Management Council, Final Fishery Management Plan for Regulating Offshore Marine Aquaculture in the Gulf of Mexico, 1 (2009), available at http://www.gulfcouncil.org/Beta/GMFMCWeb/Aquaculture/Aquaculture%20FMP%20PEIS%20Final%202-24-09.pdf (last visited Apr. 17, 2014)

Harvard Law School Emmett Environmental Law & Policy Clinic, Environmental Law Institute, and The Ocean Foundation, Offshore Aquaculture Regulation Under the Clean Water Act, 4 (Dec 2012)

Hooks AM, Baylor E (2009) Recent developments: natural resources: fishery conservation and management after reauthorization of MSA. Texas Environ Law J 39:193–194

Howarth RW, Boyer EW, Pabich W, Galloway JN (2002) Nitrogen use in the United States from 1961–2000 and potential future trends. Ambio 31(2):88–96

Hunter D, Salzman JE, Zaelke D (2002) International Environmental Law and Policy, 2nd edn. Foundation Press, 735

International Union for Conservation of Nature and Natural Resources, Towards Sustainable Fisheries Law. A Comparative Analysis (Gerd Winter ed. 2009) 5, 39, available at /portals.iucn.org/library/efiles/edocs/EPLP-074.pdf, last visited May 20, 2014

Juda L (1996) International Law and Ocean Use Management: The Evolution of Ocean Governance. In: Smith HD (ed) pp 8–30, 157–160, 235

Juda L (2002) Rio plus ten: the evolution of international marine fisheries governance. Ocean Dev Int Law 33:109, 109–144

Kalo JJ, Hildreth RG, Reiser A, Christie, DR (2002) Coastal and Ocean Law 311–313, 388, 421 (W. Grp., 2nd edn)

Kapetsky JM, Aguilar-Manjarrez J, Jenness J (2013) Food And Agriculture Organization of the United Nations, A Global Assessment of Offshore Mariculture Potential From A Spatial Perspective, Fisheries and Aquaculture Technical Paper 549, xiv

Krkosek M, Revie CW, Gargan PG, Skilbrei OT, Finstad B, Todd CD (2012) Impact of parasites on salmon recruitment in the Northeast Atlantic Ocean. Proc R Soc B 5

Kundis Craig K (2002) Taking the long view of ocean ecosystems: historical science, marine restoration, and the oceans act of 2000. Ecol Law Q 29:649–664

Lee HW, Bailey-Brock JH, McGurr MM (2006) Temporal changes in the polychaete infaunal community surrounding a Hawaiian mariculture operation. Mar Ecol Prog Ser 307:175–185

Lenfest Forage Fish Task Force, Little Fish, Big Impact, 73 (2012)

Lynch KJ (2007) Application of the Public Trust Doctrine to modern fishery management regimes. New York Environ Law J 15:285, 296, 307–310

Marine Aquaculture Task Force (2007) Sustainable Marine Aquaculture: Fulfilling the Promise, 1, available at http://www.whoi.edu/cms/files/mcarlowicz/2007/1/Sustainable_Marine_Aquaculture_final_1_02_07_17244.pdf, (last visited at 17 Apr 2014)

Marshall D (2003) Fishy Business: The Economics of Salmon Farming in BC, 5 (July 2003)

Mastry M (2006) Extraterritorial application of state fishery management regulations under the Magnuson-Stevens fishery conservation and management act: have the courts missed the boat? UCLA J Environ Law Policy 25:225, 235

McLaughlin RJ (2003) Foreign access to shared marine genetic materials: management options for a quasi-fugacious resource. Ocean Dev Int Law 34:297

Memorandum from Jay S. Johnson, Deputy General Counsel, Nat'l Oceanic and Atmospheric Admin., and Margaret F. Hayes, Asst. General Counsel for Fisheries, Nat'l Oceanic and Atmospheric Admin., to James W. Brennan, Acting General Counsel, Nat'l Oceanic and Atmospheric Admin., 2 (Feb. 1, 1993) (on file with the author)

National Oceanic and Atmospheric Administration (2011) Marine Aquaculture Policy, 1(1), available at http://www.nmfs.noaa.gov/aquaculture/docs/policy/noaa_aquaculture_policy_2011.pdf, (last visited 17 Apr 2014)

Nash JR (2010) The curious legal landscape of the extraterritoriality of U.S. environmental laws. Va J Int Law 50:997, 998, 1003

National Research Council (1999) Sharing the fish towards a national policy on individual fishing quotas. National Academic Press, pp 33, 142–143, 169, 173–174, 177

National Research Council (2000) Clean coastal waters: understanding and reducing the effects of nutrient pollution. National Academic Press

National Research Council (2002) Oil in the Sea III: inputs, fates, and effects. National Academic Press

Osherenko G (2006) New discourses on ocean governance: understanding property rights and the Public Trust. J Environ Law Litig 21:317, 336, 339, 342, 347–348, 372–375

Pew Ocean Commission (2003) America's Living Oceans Charting A Course For Sea Change, 44, 59–63 (May 2003), available at http://www.pewtrusts.org/uploadedFiles/wwwpewtrustsorg/Reports/Protecting_ocean_life/env_pew_oceans_final_report.pdf, (last visited 17 Apr 2014)

Price CS, Morris Jr JA (2013) Marine cage culture and the environment: twenty-first century science informing a sustainable industry. NOAA Technical Memorandum NOS NCCOS 164, 9, 46

Rieser A (1997) The ecosystem approach: new departures for land and water: fisheries management: property rights and ecosystem management in U.S. fisheries: contracting for the commons? Ecol Law Q 24:813, 821–824, 827

Rieser A (1999) Prescriptions for the commons: environmental scholarship and the fishing quotas debate. Harv Environ Law Rev 23:393, 405–407

Scott AD (1996) The ITQ as a property right: where it came from, how it works, and where it is going. In: Crowley BL (ed) Taking ownership: property rights and fishery management on the Atlantic coast 31:35–36

Selden J (1652) Of the dominion, or, ownership of the sea. In: Leonard Silk advisory (ed) Arno Press, 1972

Showalter S (2010) Will California Law Apply to Hubbs-SeaWorld Research Institute's Offshore Aquaculture Demonstration Project? An analysis of the extraterritorial application of state aquaculture laws. Hast W-NW J Environ Law Policy 16:223, 229

Sielen AB (2008) An oceans manifesto: the present global crisis. Fletcher F World Aff 32:39, 52

Swarztrauber SA (1972) The three mile limit of territorial seas, vol 11. Naval Institute Press

Territo M (2000) The precautionary principle in marine fisheries conservation and the U.S. sustainable fisheries act of 1996. Vt Law Rev 24:1351, 1369

Thorstad EB, Fleming IA, McGinnity P, Soto D, Wennevik V, Whoriskey F (2008) Incidence and Impacts of Escaped Farmed Atlantic Salmon Salmo Salar in Nature. NINA Special Report 36, 6

Turnipseed M, Roady SE, Sagarin R, Crowder LB (2009) The silver anniversary of the United States' exclusive economic zone: twenty-five years of ocean use and abuse, and the possibility of a blue water Public Trust Doctrine. Ecology 36:27–28, 30

U.S. Dept. of Commerce, Aquaculture Policy, 2 (1999) available at www.nmfs.noaa.gov/aquaculture/docs/policy/doc_aq_policy_1999.pdf, (last visited 16 Apr 2014)

Chapter 37
Industrial Aquaculture: Human Intervention in Natural Law

Alfredo Quarto and Sara Lavenhar

Abstract The meteoric growth of both the shrimp and salmon farming industries has had noticeable adverse global environmental and social impacts. Shrimp aquaculture represents a powerful global industry that has an annual retail value of over $50–60 billion dollars. Meanwhile, vital coastal mangroves are being cleared to make way for expanding shrimp farming. Coastal poor fishing and farming communities are losing their once sustainable food sources as their traditional agriculture and fisheries are being steadily despoiled by the shrimp industry's operations, whose profits concentrate in the hands of wealthy investors. The shrimp produced have never become a food source for those who are truly hungry in the producer nations. The great majority of the farmed shrimp are exported to the wealthier nations. There is an urgent need to counter these market forces that are devastating mangrove forests and ruining the lives and livelihoods for tens of millions of indigenous peoples and traditional community residents who rely on healthy coastal environments for their lives and livelihoods. This chapter will focus mainly on shrimp aquaculture, but will also highlight some important and related aspects of salmon farming. Implications of the environmental, social and legal aspects of modern industrial aquaculture will be explored and the serious repercussions engendered by the present course of open, throughput systems of aquaculture will be presented. The power and effect of consumer choices and more sustainable, ecologically- and socially-friendly, closed-system aquaculture alternatives will be discussed.

A. Quarto (✉)
Mangrove Action Project, Port Angeles, WA, USA
e-mail: alfredo@mangroveactionproject.org

S. Lavenhar
Mangrove Action Project, Seattle, WA, USA

© Springer International Publishing Switzerland 2016
G. Steier, K.K. Patel (eds.), *International Food Law and Policy*,
DOI 10.1007/978-3-319-07542-6_37

37.1 Introduction: A Brief Background on the Advance of Aquaculture

Over half the world's human population is concentrated along coastal areas. These important zones also support a vast array of other life dependent upon healthy ocean ecosystems. Yet, today our oceans are beleaguered by over-fishing, pollution, and destruction of coastal resources via unsustainable forms of modern development. Serious declines in wild fish stocks amid increasing world consumer demands for seafood products have combined to present a dilemma on how best to meet these new challenges.

One proposed solution, aquaculture, is often highly lauded today by governments, inter-governmental agencies, world lending institutes, and industry. Many see it as the next logical step towards solving problems of sustainable seafood, and offering a revolution in modern fisheries, the "Blue Revolution." Following on the heels of agriculture's "Green Revolution," modern aquaculture promised to turn the tide on food production from the seas and waterways, delivering into the world's eager hands the key that unlocks the door to "farming the sea."

Aquaculture might be broadly defined as the establishment of man-made enclosures to raise aquatic life forms, such as shellfish, fish, and seaweeds for human consumption purposes. The aquaculture process itself is quite ancient, having appeared in traditional, less-intensive forms at least 2000 years ago in Asia and other parts of the world. The *gei wais* of Hong Kong or the *tambaks* of Indonesia, offer striking examples of traditionally derived forms of aquaculture, which still exist today.

Unfortunately, since the advent of more intensive modern industrial aquaculture, serious environmental and social issues have developed. Millions of indigenous and local coastal peoples are adversely affected, many losing their livelihoods, homes, and cultures to unsustainable aquaculture development. Meanwhile, in the cities and towns of the wealthy consumer nations, where imported seafood products are sold in great volumes, little is known of the great hardships created by these "revolutions" in farming the land and the sea. Few consumers of farm-raised seafood products are aware of the many serious problems caused by the incoming tide of the aquaculture industry, where ruin and riches run simultaneously like two parallel, but opposing sea currents (Fig. 37.1).

Modern industrial shrimp aquaculture is a case in point. In the last 30 years, the rapid and largely uncontrolled expansion of the shrimp aquaculture industry has led to immense environmental and social problems, which have only recently been brought to light. Among the most serious problems is the degradation and loss of natural coastal resources. Unsolved pollution and disease problems still plague the industry, despoiling once fecund waters of nearby estuaries and inshore coastal bays. Formerly rich fishing grounds are also impacted, and vital fish breeding and nursery habitat are being lost to the encroaching shrimp farms.

The construction and operation of industrial shrimp aquaculture is tremendously disruptive to the delicate and complex balance of coastal ecology. Vast stretches of

Fig. 37.1 Sketch of mangrove-shrimping overlap. This figure illustrates in how many areas of the world mangroves overlap with shrimp acquaculture. The *shaded areas* show where the most overlap occurs

invaluable mangrove forests are cleared to make way for shrimp ponds. Shrimp farms replace these diverse, multiple resource environments with large-scale monoculture operations. Worldwide, over a million hectares of valuable mangrove forests have been destroyed by shrimp farming alone—and this in only the last 2 decades!

Other important coastal habitats, such as mud flats, sea grass beds, and coral reefs have also been degraded or ruined by aquaculture. Once productive farmlands have been left fallow, and important waterways and underground aquifers have been dangerously contaminated. For many the shrimp industry has been aptly labeled a "slash and burn" enterprise, leaving in its wake both pain and loss.

In this chapter, we first give a brief background on the history of aquaculture from ancient times in Asia to modern times, exploring some of the issues afflicting modern, industrial-style aquaculture. We then detail some of the specific issues raised by shrimp and salmon aquaculture, and discuss the effects of shrimp and salmon farming on biodiversity and wild fisheries. In addition to the mentioned ecological problems manifest in the industry, aquaculture also affects essential food production processes. Both agriculture and fisheries are adversely affected. Salinization and pollution of land, waterways and aquifers by the shrimp farms ruins both fisheries and crop production. In later sections, we delve in the failures of current certification plans, which are engendering further reliance on failing systems; as well we explore food security, human rights and labor issues in the producer nations. A spotlight is also shown on a case of weakness of existing laws designed to regulate fishing and aquacultural activities, and suggest possible improvements. And finally, we explore alternatives to present-day industrial aquaculture that are more environmentally and socially responsible, and have much greater potential for being sustainable. Such systems will supplement, but not replace, conscientious management of wild fisheries production.

37.1.1 The Plight of the Mangroves

For many years, the mangrove forests were seen and actually often officially designated as wastelands, not fit for anything but mosquitoes and smelly swamps. Fortunately, this view of the tidal forests is changing, influenced by recent events and tragedies. Mangrove forests are one of the most productive and biologically diverse wetlands on earth. Growing in the intertidal areas and estuary mouths between land and sea, mangroves provide critical habitat for a diverse marine and terrestrial flora and fauna (Chapman 1976).

Healthy mangrove forests are key to a healthy marine ecology. Yet, these unique coastal tropical forests are among the most threatened habitats in the world. They may be disappearing more quickly than inland tropical rainforests, and so far, with little public notice (Valiela et al. 2001). In many areas of the world, mangrove deforestation is contributing to fisheries declines, degradation of clean water supplies and salinization of coastal soils, erosion, and land subsidence, as well as the release of carbon dioxide into the atmosphere (Pendleton et al. 2012; Alongi 2014).

In fact, mangrove forests fix more carbon dioxide per unit area than phytoplankton in tropical oceans, but when mangroves are cleared this stored carbon is released.

With an estimated original mangrove forest cover of 32 million hectares, mangroves once covered vast stretches of coastlines of tropical and sub-tropical countries. Today, less than 50 % of that mangrove forest area remains. Many factors contribute to mangrove forest loss, including the charcoal and timber industries, tourism, urban growth pressures, and mounting pollution problems (Valiela et al. 2001). However, one of the most significant causes of mangrove forest destruction in the past 3 decades has been the consumer demand for luxury shrimp, or "prawns," and the corresponding expansion of destructive production methods of export-oriented industrial shrimp aquaculture. Vast tracts of mangrove forests have been cleared to make way for the establishment of coastal shrimp farm facilities (Primavera 1997). The failure of national governments to adequately regulate the shrimp industry, and the headlong rush of multilateral lending agencies to fund aquaculture development without meeting their own stated ecological and social criteria, are other important pieces to this unfortunate puzzle.

37.1.2 The Fight for the Mangroves

The importance of mangroves has led to a growing number of organizations that are dedicated to mangrove conservation to stand firm in opposition to further expansion of the shrimp aquaculture industry. In Table 37.1, we list a few of those groups whose mission includes conscientious efforts to prevent and reverse the impacts of aquaculture.

Table 37.1 Organizations dedicated to mangrove conservation

Region	Name	Country
Latin America	The Committee for the Defense and Development of Flora and Fauna of the Gulf of Fonseca (CODDEFFAGOLF)	Honduras
	CCondem	Ecuador
	The Fundación de defensa ecológica (FUNDECOL)	Ecuador
	Redmanglar Internacional	Guatemala
Africa	African Mangrove Network	
Asia	The Coastal Poor Development Action Network (COPDANET)	India
	Nijera Kori	Bangladesh
	Yadfon Foundation	Thailand
	Kiara	Indonesia
Europe	Swedish Society for Nature Conservation	Sweden
	Forest Peoples Project (FPP)	UK
	Environmental Justice Fund (EJF)	UK
	PESCA/ Ecologistas en Accion	Spain
North America	Mangrove Action Project (MAP)	USA
	Food and Water Watch	USA

These are but a few of the organizations that are opposing destructive shrimp farm practices worldwide. One of the chief strategies employed by these groups and others is to help enhance and promote community-based resource conservation and management, which we will discuss later on in this chapter. Many groups fighting for mangroves and other coastal habitats, as well as coastal inhabitants, have engaged in legal action at the regional and national levels in opposition to industrial aquaculture.

37.2 The Seeds of the Blue Revolution

Beginning in the 1960s and 1970s, industrial processes were widely introduced into aquaculture to encourage commercial production. Then in the early 1980s, major improvements in hatchery production and feed processing allowed rapid advances in shrimp farming techniques, making it possible to dramatically increase yields (Primavera 2005).

This "Blue Revolution" in aquaculture has in many ways re-traced the steps of the "Green Revolution" in agriculture. The latter contributed to the growth of large-scale export-oriented agribusiness enterprises in developing nations, but it also generated widespread criticism for its environmental and social impacts. The new aquaculture techniques have resulted in an explosive expansion of coastal shrimp aquaculture throughout developing nations in Asia, Africa and Latin America (Baird and Quarto 1994). Over 80 % of worldwide-farmed shrimp is produced in Asia. Approximately two-thirds of it is exported to the United States, Japan and the EU, with the remainder divided among other foreign and luxury domestic markets (FAO 2010). Today, China is also stepping into the picture, emerging as a high-demand, shrimp-importing nation.

Though trawler-caught shrimp once dominated the world shrimp market, the rate of growth in farmed shrimp production allowed that sector to surpass the wild-caught production around 2006. Farmed shrimp production has truly skyrocketed, rising from just 26,000 metric tons of production in the 1970s to 100,000 metric tons in the early 1980s; by 2003, production reached over 1.6 million metric tons (mmt) (Barbier and Cox 2004). Global farmed shrimp production peaked in 2009 at 3.22 mmt, but has subsequently declined to 2.5 mmt in 2011 due to natural disasters, pollution, and disease (FAO 2012). The industry has recovered somewhat, and production supplies are expected to return to a trend of increase (FAO 2013a, b).

37.2.1 Bankrolling a Bankrupt System

Shrimp aquaculture has become a global industry that has an annual farm-gate value of over $12 billion dollars, and an annual retail value of around $50–60 billion dollars. It has great profit potential for the astute investor and entrepreneur. Spurred on by governments eager for increased export dollars, shrimp aquaculture development has been aided by generous support and incentives from international lending institutes, including the World Bank, the Asian Development Bank, and the Inter-American Development Bank (Lewis et al. 2003).

Throughout the 1980s and 1990s, shrimp farming itself was heavily supported by millions of dollars of World Bank loans, and by FAO research and development programs along the same coastlines where tsunamis had struck:

> The World Bank participated actively in the launching of the shrimp industry in Asia. Out of an investment of US$ 1.685 billion in 1992 for Indian agriculture and fisheries, the World Bank allocated US$ 425 million for aquaculture development (Mukherjee, 1994) A substantial part of this sum seems to be destined for intensification and expansion of shrimp ponds. The involvement of the World Bank in shrimp aquaculture, and the development of related hatcheries and other shrimp facilities, illustrates of the trends towards internationally organized vertical integration of this industry (Lewis et al. 2003).

This level of support by governmental and inter-governmental agencies led to rapid and uncontrolled expansion of the shrimp aquaculture industry in the developing world. The expansion of shrimp aquaculture impacts many fragile and essential coastal ecosystems, but has especially affected the mangrove forested regions.

Shrimp farming is considered to be the number one cause of mangrove deforestation by many researchers who have documented the rate of mangrove loss, which currently stands at around 1–2 % per year (FAO 2010). According to an eminent researcher, Dr. Jurgenne Primavera formerly of the Aquaculture Department, Southeast Asian Fisheries Development Center (SEAFDEC) in the Philippines:

Culture ponds for fish and shrimp account for the destruction of 20-50 % of mangroves worldwide in recent decades... Decimation of mangroves along the Philippine coastline accounts in part for the great losses to life and property inflicted by an average of 20 typhoons and tsunamis each year—around 3,000 deaths in Zamboanga province in 1976, 1,000 in Northern Panay in 1984, and 7,000 in Ormoc and other Leyte towns in 1991. In the Chokoria Sundarbans in Bangladesh, mangroves protected villagers from a 1960 tidal wave, but a similar one caused thousands of deaths in 1991, after the installation of shrimp farms. (Primavera 1997).

Around 3 million hectares of mangroves have been destroyed for shrimp farm development worldwide. The exact extent to which shrimp aquaculture has been responsible for mangrove loss is unclear, but evidence suggests shrimp farming has been a major contributor to global mangrove loss; it has been estimated that as much as 38 % of recent global mangrove loss may have been due to shrimp aquaculture, and that the major reasons for future destruction are unrestricted clearing of forests, overexploitation of fisheries, and aquaculture (Valiela et al. 2001; Alongi 2002). However, restoration and rehabilitation efforts are growing worldwide, revealing a glimmer of hope for mangroves as their value becomes globally and economically recognized.

The global economic figures and the allure of quick investment returns belie the fact that the shrimp aquaculture industry is a young adolescent giant with dramatic growing problems. The spread of deadly infectious viruses has ruined once thriving shrimp aquaculture industries in Taiwan, China, Vietnam, and Ecuador, causing hundreds of millions of dollars' worth of losses (Walker and Mohan 2008). In 2010, a new shrimp disease with far greater deadly consequences emerged—Early Mortality Syndrome ("EMS"), which has had devastating consequences for shrimp industries in China, Thailand, Malaysia, India and more recently Mexico (FAO 2013a, b).

After extensive investigation by a research team led by Donald Lightner at the University of Arizona, the pathogen causing early mortality syndrome (EMS), (technically known as acute hepatopancreatic necrosis syndrome ("AHPNS")), was identified. The researchers found that a bacterial agent, which is transmitted orally, colonizes the shrimp gastrointestinal tract, producing a toxin that causes tissue destruction and dysfunction of the shrimp digestive organ known as the hepatopancreas, causes EMS. It does not affect humans (Tran et al. 2013).

In response to this devastating new disease spread, several countries have implemented policies that restrict the importation of frozen shrimp or other products from EMS-affected countries. Since EMS was first reported in China in 2009, it has spread to Vietnam, Malaysia and Thailand, as well as Mexico, causing estimated annual losses of more than U.S. $1 billion. EMS outbreaks usually afflict shrimp ponds within the first 30 days after stocking, while mortality can exceed 70 % (Tran et al. 2013). With the emergence of EMS virus, there has actually been a shortage of shrimp for the export market, leaving many restaurants and retail stores scurrying to find new suppliers, as once quite dependable shipments from major producer nations such as Thailand have been hard hit by the disease outbreak. Meanwhile, some nations, such as India and Ecuador, are looking to greatly expand their productions to meet the demand left by the EMS affected nations, and taking advantage of the profit opportunities created by the disease plight of their competitor nations. Nevertheless, initiating these kinds of rapid expansions is a big gamble, because the disease EMS can spread rapidly and affect the new producers who may have invested heavily, risking everything in the process. Alternatively, should shrimp production recover from the impacts of EMS, surplus product may cause falling prices and negatively impact developing economies.

Despite these setbacks, the industry remains quite strong, and the market quite alluring, because of continuing demand for imported shrimp in the importing nations. Imported shrimp is not exclusively farmed, and may be derived from wild fisheries as well. In the United States, which contains the largest shrimp import market, shrimp consumption reached 4.2 pounds per capita in 2011 before dropping to 3.8 pounds the following year (NOAA 2012). Japan is the second largest importer of seafood in the world, capturing about 15 % of the market; 17 % of imports are fresh or processed shrimp (Popescu and Ogushi 2013). Domestic production in fisheries has been declining in Japan since 1980, while imports and aquacultural production have been on the rise to meet seafood demands.

This increase in farmed shrimp production has itself contributed disastrously to a decline of wild fisheries in the producer nations, thus creating a serious loss of food security for affected coastal communities in the Global South where food security is already a serious issue. Important factors contributing to local fisheries declines include the selective harvesting process to catch the wild shrimp larvae for stocking the ponds. Over the years, poor coastal villagers have carried out this task using fine mesh nets that catch all types of larval sea life. Those working the near shore waters in Latin America are called "larveros" and in Asia "larva catchers." They earn around $3 per day, and many suffer serious health issues from wading in chest deep in seawater for hours at a time.

Ill-managed shrimp farms, constructed without sustainable production in mind, do not tend to last many years. These farms are often constructed with quick profit in mind, and then abandoned when production declines. In Asia and Latin America, the average intensive farm has been found to survive only 3–5 years before serious pollution and disease problems cause early shrimp pond closures (Flaherty and Karnjanakesorn 1995; Dierberg and Kiattisimkul 1996). More recent data and studies on the average lifetime of shrimp ponds and other forms of aquaculture are limited, and this is an area of study that requires further attention. It is also difficult to determine the number of abandoned farms, but unofficial estimates suggest up to 70 % of intensive shrimp farms in Thailand are in disuse for environmental, financial, and political reasons (Stevenson 1997). Overstocking and indiscriminate use of feeds and water additives still are being widely practiced today. In Thailand, where nearly 85 % of the shrimp ponds are intensive systems, over half of the shrimp ponds have closed down in the first decade of Thailand's entry into the great race for world dominance in the shrimp export market. It is estimated that there are over 400,000 ha of abandoned shrimp farms around the world, (Lewis et al. 2003) and much of this area lies within former mangrove wetlands. In the process, vital coastal wetland habitats have been permanently lost for fish, mollusks, and crustaceans, as well as numerous birds, migratory species and endangered species.

37.2.2 Precautionary Principle Violations and Consumer Health Risks Mount

It is a tragic fact that industrial shrimp aquaculture has been practiced on a wide-scale production basis while still really in its research and development phase. It is still attempting to solve very grave and life threatening problems in the field, rather than in a closed test facility where failures will not be so ruinous. Indiscriminate expansion of the shrimp aquaculture industry might be likened to taking unsuspecting passengers on board a still untested proto-type commercial jet on its maiden flight. The reason test pilots are paid their high salaries is because of the uncertain risks they must take in putting their aircraft through its rigorous tests. However, with the rapid spread of shrimp farming, we are all being forced to fly this dangerous mission with great potential for an industry crash endangering all "passengers" and the very planet upon which we live.

Another tragic irony of industrial shrimp aquaculture is that the process requires clean water, yet it has become a source of severe water pollution, oftentimes fouling its own "nest" in its bid for higher shrimp production. As well, this spoiling of coastal waters has potentially serious consequences for human health. Even the shrimp product itself, though widely marketed and in popular demand in consumer nations, is questionable in regards to health risks. The often unrestricted use of chemical inputs, such as antibiotics, pesticides and water additives, when combined with the buildup on the pond bottoms of unused feeds and feces, has led to epidemic shrimp diseases and many early pond closures because of harmful accumulation of toxic effluents (Dierberg and Kiattisimkul 1996; Lewis et al. 2003; Primavera 2006).

Overuse of antibiotics in both shrimp and salmon farming is another reason for great concern, as this could cause the spread of antibiotic resistant bacteria. Antibiotic use in aquaculture is widespread, and unlike in the livestock industry it is required to make up for poor hygienic practices rather than as a growth promoter. The extensive use of antibiotics in industrial aquaculture is too often dismissed or overlooked; an excellent summary of the antibiotics used in aquaculture can be found in Burridge et al. (2010). Antibiotics kill bacteria which naturally exist in our body and play a role in our natural defenses against infection. These bacteria normally live in harmony with our bodies, protecting us from bacteria which we are exposed to and have the ability to cause disease. Once a bacterium develops a new resistance mechanism, it's only a matter of time before the resistant bacteria spread.

The antibiotics used in aquaculture are the same or closely related to those used in human medical treatment. Scientists have conducted studies to determine whether antibiotics used for shrimp and salmon production could result in bacterial resistance to these antibiotics in the humans who consume farmed shrimp. Presently, the question remains as to the development of resistant strains of human pathogens, but the risk has been identified as severe (Graslund et al. 2003). Studies have identified increased microbial resistance that may be the result of excessive antibiotic use in salmon aquaculture. Antibiotic use to control non-cholera *Vibrio* infections in shrimp farms may have contributed to cholera outbreaks in Ecuador, and shrimp have been tied to cholera cases in the United States (Blake et al. 1980; Weber et al. 1994). The state of antibiotic use is well-understood, but regulations for their use are highly varied from country to country. Due to escalating public concerns over health risks, Japan has identified over 20 antibiotics used in the farmed shrimp industry and has banned shrimp farmed with these antibiotics. Meanwhile, the United States Food and Drug Administration ("FDA") only looks for residues of 16 antibiotics compared to 57 in other countries, despite some of these drugs being unapproved for use in the United States. The FDA sampling of imported seafood is also limited and only 0.1 % of all imports was tested in 2009 (GAO 2011). As a result, it is likely that contaminated shrimp passes US inspections because of this serious lack of effective inspections.

Shrimp farming, along with other forms of aquaculture, also poses a real danger of genetic contamination and lowering of biodiversity. Accidental and incidental release of farm raised non-native shrimp or fish can have tremendous repercussions on the native species, which may come in contact with them. Competition for territory, genetic drift, disease spread, and excess demand on available resources are genuine concerns. Unfortunately, the number of escaped salmon is not well-known as escapes often go unreported, but their ecological and genetic impact is well documented (Thorstad 2008). Not much is known about the effect that accidental releases of shrimp are having on native wild species, but further study is urgently needed.

37.3 Farming Carnivores: Violating the Laws of the Sea

Rather than reducing the impact of seafood consumption on marine resources as many proponents of shrimp and salmon aquaculture purport, so-called "open-system aquaculture" adds new burdens. If you imagine a trophic pyramid, carnivores are typically at a higher trophic level, but the biomass of that trophic level is smaller than the one below it. By farming carnivorous fish intensively, their biomass increases, and the trophic pyramid is thrown out of balance. Changes at any given trophic level can cascade up and down, affecting primary consumers and top predators alike. Aquaculture supplies 46 % of the world's supply of food-fish, and is projected to exceed capture fishery production in the next few years; the FAO has cautioned that the growth of the industry is unsustainable, noting that current trends indicate a slight reduction in the annual rate of increase (FAO 2010). Without careful and effective management, the farming of carnivorous seafood will continue to contribute to widespread ecological impacts and may experience decline. Regardless, it is important to understand the direct and indirect consequences that can arise from unbridled aquacultural development.

Though shrimp farming is a major problem affecting many developing nations, there are many dangerous parallels between the culture of shrimp in countries such as Ecuador or Brazil and the rearing of salmon in Chile and Canada. When visiting Chile in 1994, 2 years after co-founding MAP, co-author Alfredo Quarto witnessed a very similar tragedy unfolding for fishing communities who were dependent upon a healthy world fishery for their livelihoods. A large salmon farm operation had been set up near the estuary around Cochamo Bay in southern Chile, and the fishing community was adversely affected by the salmon farm's operations which polluted the nearby estuary and greatly reduced the wild fishery upon which the local communities were dependent for their livelihoods. Fishermen complained that the salmon farm put them out of work, and the only jobs they could now find were part time with the salmon farm, and these were few, low paid and not very stable. The salmon farm was having disease problems at the time, and as a result many workers on the salmon farm were laid off. Conditions in the area were deteriorating, and local residents were very upset with the situation that befell them.

Similar problems exist in other areas where salmon farms have been placed, but the effect on residents of developing nations are especially deeply felt and more devastating for poorer fishing communities. The following is a short summary of similar problems plaguing both the shrimp and salmon farm industries.

- Use of public waterways for private industry gain, while badly affecting these same waterways, actually degrading such via pollution and aquaculture facilities construction.
- The introductions of non-native species into waters where frequent escapes can cause many problems, affecting biodiversity, spreading diseases, competing for local habitat, nesting sites and food sources. These escapes cannot be halted by present open system aquaculture facilities. The full range of serious consequences is hardly known, and the effects of shrimp and salmon culture cannot presently be fully understood, let alone controlled. Though the real threats to native species from these escapes are not yet understood, these accidental escapes may have disastrous consequences for already threatened wild species.
- Both shrimp and salmon farming utilize carnivorous species that consume large quantities of fish and fish oils. The industry is trying to reduce these protein needs, which effectively remove valuable local sources from the wild seas to feed farm raised species, but it still does not make sense to feed more weight of wild fish to raise less weight of farmed fish whose quality in taste and appearance is so much less than the wild caught fish.
- Our wild fisheries are not helped by either salmon or shrimp aquaculture, as too often the very habitat where wild fish thrive is adversely affected by industrial aquaculture. Nursery and breeding areas may be affected by the high levels of contaminants escaping these open systems which themselves are not sustainable

as they rely on good water quality to sustain themselves, but by their very nature in design pollute their own pure water sources. These so-called throughput systems are not an effective method of long-term, sustainable aquaculture.

By comparison, closed containment, land-based systems purify and recirculate water, deal effectively with effluent, leave no chance for escapes of farmed species into wild waterways, no chance of disease spread from farmed species to native wild species, and allow the farming operations to not interfere with the habitat and wildlife of our seas and other waterways. The aquaculture industry is undergoing changes as awareness of the social and environmental impacts are becoming better studied. For example, research into algae- and vegetable-based feeds as replacement for fishmeal appears promising and wild fish inputs to farmed fish outputs have dropped for some species (Naylor et al. 2009; Bendiksen et al. 2011). Although we focus below on shrimp farming, it is worth remembering that many of these same issues arise from other farm-raised seafood.

37.3.1 Shrimp Farming: Contribution to Habitat Loss, Wild Fisheries Decline, Endangered Species, and Overfishing

37.3.1.1 Habitat Loss

Multiple habitat types are impacted by industrial shrimp farming, but arguably the most significant degradation has been to mangrove forests. Since mangroves are nurseries for an immense diversity of sea life, their loss to shrimp farming and other types of unsustainable development spell disaster for our ocean's wild fisheries and biodiversity. This loss is further compounded by the fact that mangroves also filter out upland pollutants and prevent shoreline erosion and sedimentation of sea grasses and coral reefs, thus further protecting these nearby, inter-dependent ecosystems, which themselves play essential roles in supporting healthy marine environments essential for sea life.

This inter-connectedness, or inter-dependency, goes beyond just the mangroves, sea grass beds and coral reefs, but includes the whole array of coastal ecosystems from shore to uplands, and most definitely includes the variety of coastal wetlands, such as the mudflats, salt flats and salt marshes. Dr. Gilberto Cintron of the US Forest Service has described mangrove ecosystems as a dynamic mosaic, whereby seemingly discrete individual elements are actually connected by material flows, and remain stable despite potential changes to the state of a given element. Regardless of the changes, the whole mosaic is still a mangrove ecosystem. This holistic conceptual view constitutes what is called an ecotone—a region of transition between two biological communities. In this view of things, a mangrove forest is not an isolated, autonomous wetland, but is one part of a living and adaptive ecotone.

As Dr. Cintron goes on to clarify:

> Really, the problem is in our need to fragment things. We cannot study an ecosystem broken into components. It cannot be assembled into a whole. I call this the Humpty Dumpty Effect –'Humpty Dumpty had a great fall. All the king's horses and all the king's men couldn't put Humpty Dumpty together again'... This nursery rhyme enfolds a very important management lesson. We cannot manage successful ecosystems or landscapes in a fragmented way. Whereas scientists tend to like tearing things apart to see how they work, it is the manager's role to do the opposite. Managers must learn to integrate. The problem is that in nature the parts in isolation behave in different ways than when they are part of a whole. So it is not an easy job to go from parts to whole.

One approach is to identify the larger whole and manage it to conserve all of its parts. This is called the First Rule of Tinkering, as defined by Aldo Leopold: "Instead of learning more and more about less and less, we must learn more and more about the whole biotic landscape (Leopold 1953)."

Unfortunately, the mangrove forests are now viewed by many as somehow separate or isolated from their associate wetlands found on the tidal flats—the mud and salt flats, the salinas and salt marshes which are themselves really part of a greater, integrated tidal ecosystem. These are not really separate ecosystems, but are instead variations on a common theme—the tidal wetlands. Where there is now a mangrove forest, in the future there could be a salt marsh or salina, depending on changes in hydrology, sea level, or other factors. So, one must learn to view the entire functioning "whole" before assuming about the nature of its parts. In the case of mangroves, that whole is the tidal wetland that forms at the interface between the non-tidal upland to the lowest parts of the emergent sloping, tidally influenced platform. We call it mangrove but that is because sometimes it has the woody trees we call mangroves. Other times the trees are missing and we call it a mud flat. If the trees are missing but the surface is covered by herbaceous vegetation we call it a salt marsh. If the herbaceous vegetation is missing there will probably be an algal mat that dries out periodically. We then call it salt flat, but it is always the same wetland, capable of adapting to the local environmental conditions (Cintron GM, 2013, Personal communication).

This brings up the question of how we value the goods and services provided by an intact mangrove wetland vs. a shrimp farm. While the immediate profits from shrimp farming may satisfy a few wealthy investors, vast numbers of coastal residents, once dependent on healthy coastal ecosystems for fishing and farming, are being displaced and impoverished. Meanwhile, the greater long-term goods and services provided by healthy coastal wetland ecosystems are being sacrificed for short-term profits via unsustainable shrimp farm ventures. As well as robbing local communities of their traditional livelihoods and food security, the shrimp farms are leaving vast stretches of coast dangerously vulnerable to natural disasters brought on by climate change.

Dangerous sea level rise and more intense and frequent storms resulting from climate change are other factors wreaking havoc on coastal areas suffering from mangrove loss. Mangroves must no longer be misconstrued as worthless, muddy swamps or wastelands as they have been called in the past by those anxious to clear the forests to replace mangroves with shrimp ponds, tourist hotels and other developments. In an evaluation of estuarine and coastal ecosystems, it was found that coastal protection contributes about 80 % of total value of mangrove ecosystem services (estimated at $14,000–16,000); the remaining 20 % comes from erosion control, raw materials and food, fishery maintenance, and carbon sequestration (Barbier et al. 2011).

37.3.1.2 Wild Fisheries Decline

There can be little doubt that loss of mangrove to industrial shrimp farming has a direct correlation with decline in wild fisheries. Since the rapid and too often unregulated expansion of shrimp aquaculture has been the largest contributor to mangrove losses globally, a halt to further such expansion of the industry is called for if there is serious intent to revive our planet's now faltering wild fisheries. The encroachment of the shrimp farm industry along the coasts of Latin America, Asia and Africa has adversely affected local fisheries and those whose living is made by the sea (Barraclough and Finger-Stich 1997; Primavera 1997; Islam and Haque 2004). Many fishermen must venture further out to sea to find sufficient catch to earn their livelihoods and support their families. Fishers are often forced to take more risks in leaving the safety of the near shore fishing grounds once provided by the now-vanishing mangroves. They also must spend more time on the open waters away from their families, and often their daily catch falls far short of their earlier catches when they could fish near the mangroves.

While the world is grappling with global fisheries declines due to the rate of trawler by-catch and the number of commercial fishing vessels on the sea, the shrimp fry fishery for aquaculture has the highest by-catch rate in the world. Shrimp trawling may have a by-catch of 20 pounds of fish lost for every 1 pound of full grown shrimp caught (Eayrs 2007). Up to 99 % of total collection when capturing wild shrimp larvae may be non-target species, with a global average around 85 % capture of non-target species (Islam et al. 2004; Clucas 1997). Worse, the shrimp larvae by-catch consists mainly of other fish larvae that then never reach the reproductive stage. This contributes to declining wild fisheries, including decreases in wild stock of the very shrimp larvae required by the industry (Naylor et al. 2000). Efforts to decrease by-catch with specialized equipment are becoming more popular, especially in rating and certification systems. The fishmeal industry has also promoted the utilization of by-catch, rather than it being discarded or left to rot. This latter approach has an uncertain outcome and may limit the adoption of by-catch reduction practices (Naylor et al. 2000). Fishmeal production is a lucrative business, as we will discuss in depth later, and reductions in by-catch could mean losses for the industry. As such, shrimp trawlers may have little incentive to reduce by-catch in order to maintain profits. As well, the clearing of the mangrove forests and related degradation of the inter-tidal zones to establish the extensive areas of shrimp farms is having a deleterious effect on the wild fisheries. Mangroves are the marine nurseries for a myriad of marine life, and their loss is contributing to wild fisheries declines and food insecurity for the coastal poor (Islam and Haque 2004; Primavera 2006).

In Penang, Malaysia, a fishermen's cooperative sat idle because the loss of their mangroves to encroaching shrimp ponds, combined with trawlers and a nearby golf course, had reduced their once lucrative local fishery to near collapse, and they had to buy canned sardines if they now wanted to eat their preferred diet of fish. Countless fishers who once depended on fishing for their livelihoods could no longer make a living and moved to the cities in search of work as low-wage laborers, disrupting both their family and village life. With local fisheries threatened, many fishing communities face impoverishment with the loss of their traditional livelihoods, which have been callously usurped by manmade shrimp ponds where the rush for "pink gold" overrides both reason and law.

37.3.1.3 Endangered Species

Industrial shrimp aquaculture expansion has had a profound impact upon biodiversity; affecting marine life, migratory birds, and terrestrial life. Whole migratory flocks and endemic bird populations are being threatened by degradation and loss of vital habitat. A case in point is Brazil, where rapid expansion of the shrimp farm industry has had deleterious effects on the coastal wetlands, as well as the wide array of life these wetlands support. Brazil contains the second largest mangrove area in the world—more than 1 million hectares of mangrove forests are found along Brazil's long and curving coastline. In 2000, the Brazilian government released an ambitious 3-year plan to expand its area of shrimp aquaculture production sixfold—from 5000 to 30,000 ha. That expansion has since significantly surpassed this initial "modest" goal, threatening thousands of hectares of important coastal wetlands once too remote and obscure for most industrial development—areas that were once safe havens for immense assemblages of migratory shorebirds.

The north-central coast of Brazil is the most important wintering area in South America for Roseate Spoonbills, Black-bellied Plovers, Ruddy Turnstones, Whimbrels, and Willets, and is regionally important for Sanderlings, Semipalmated Sandpipers, Short-billed Dowitchers, and Red Knots. In the spring, some Red Knots likely use Brazilian coastlines as their final fueling stopover before departing for Delaware Bay. Because of its importance to migratory shorebird populations, coastlines of the Maranhão have been designated a Western Hemisphere Shorebird Reserve Network site of hemispheric importance—hundreds of thousands of shorebirds use the region each year.

Mounting pressure from environmentalists has forced the shrimp farm industry to pay heed to the negative image their industry evokes because of the rampant mangrove loss it has caused in the past. Unfortunately, the industry has increasingly taken the approach that the salt flats and salt marshes are not valuable coastal wetlands, and in countries such as Ecuador and Brazil, the shrimp farm industry is rapidly converting these wetlands to shrimp aquaculture ponds with impunity from laws meant to protect the mangrove forest zones. In Ecuador alone, nearly 80 % of the salt flat areas have been lost in the last 20 years. The justification for this, often supported by local "mangrove experts," is for the "protection of mangroves." However, this is clearly not the case, for as stated earlier mangroves are interdependent upon these related ecosystems.

Thus the threat from shrimp farm expansion goes beyond the continued loss of mangrove forests, which is quite serious in itself, but is also felt in the related loss of other associate tidal wetlands often targeted for shrimp culture as "safe substitutes" to the mangroves, such as the salt marshes. Nevertheless, these same wetlands are vital feeding grounds and resting places migratory shore birds. If these primary feeding and resting sites are lost to shrimp farms, whole species of migratory birds may be lost as well, and in such rapid fashion that little advance notice would be given in order to take remedial action to avoid such losses.

Although other development pressures, such as urban expansion, oil development and the tourism industry contribute to wetland loss, the burgeoning shrimp aquaculture industry is a major threat to Brazil's important coastal zones. The rapid, uncontrolled spread of shrimp culture could affect forever a delicate balance of nature that for too long has been taken for granted. That the future integrity of the Atlantic coastal migratory bird flyways may be lost to this unsustainable development is just one example of what is happening around the world in regards to mangrove loss and endangered species.

Other endangered species are being further threatened by shrimp farming, such as the Bengal tiger of the Sundarbans in Bangladesh and India, where massive mangrove clearing has eliminated key tiger habitat. The population of these apex predators has been steadily declining, along with the spotted deer, which is their main prey in the Sundarbans. Sundarbans means "the beautiful forests," and it surely lives up to its name in both benefits and productivity. Formed within the vast Ganges River Delta, these remote mangrove wetlands are composed of a complex network of meandering river tributaries coursing past innumerable small islands and forested lowlands, which taken together make the Sundarbans a perfect refuge for the tiger's last stand. The Sundarbans is also the single largest, contiguous mangrove forest in the world, extraordinarily rich in biodiversity and productivity. UNESCO has declared it a world heritage site. The Sundarbans has also been a huge natural safeguard against frequent cyclones, tsunamis, storm surges and other natural disasters (Rahman 2000). In every natural disaster, the Sundarbans has the potential to save the lives of hundreds of thousands of people while continuing to nurture a rich coastal ecosystem. The Sundarbans also vital to help counter climate change by sequestering CO_2 and storing carbon.

37.3.1.4 Overfishing

It might escape notice for many who study the multitudinous issues wrought by industrial shrimp aquaculture, but this destructive practice also contributes to overfishing. Though modern aquaculture is lauded as a way to decease pressures on wild fisheries, for many species it is actually achieving the opposite. We have already shown the harm to marine life caused by coastal wetland loss via conversion to shrimp ponds and supporting industry infrastructure. Nevertheless, there is a tremendous further draw down upon wild fisheries caused by the need to feed these small carnivores fishmeal and fish oils derived by trawlers that comb the sea floor. This highly destructive practice of bottom trawling has a terrible record of damaging ocean bottom habitat and causing massive losses to marine life. Industrial level farming of shrimp and salmon has been compared to raising tigers for popular consumption. The reasons are obvious: all involve the raising of carnivores, which require meat or seafood in their diet. For every one pound of farm raised shrimp, it might take on average 1.5–2 pounds of wild-caught fish and fish oils; for other species this ratio may be even higher (for salmon, it's 1 lb. of salmon raised for 3–5 lbs. fish feed) (Naylor and Goldburg 1998; Naylor et al. 2009).

At one time, shrimp and fishing bottom trawlers threw the non-target species, called "by-catch" or "trash fish," overboard, most of which died in the trauma of the catch and rough sorting on deck. Sea turtles, dolphins, octopus, starfish, small and large non-target fish species that were caught in the massive dragnets were killed in the non-discriminatory process. This itself was a terrible waste of marine life, contributing greatly to the decline of global wild fisheries. However, with the rising demand for fish feed and oils, trawlers are now seeing great profit in keeping this by-catch for later sale to the fish feed processors on shore where any and all marine life is ground up, pulverized and made into fish pellets and extracted oils for use by the shrimp aquaculture industry to feed its growing business of "farming tigers."

37.4 Global Food Security Issues: Robbing Rich Food from the Poor, Feeding Poor Food to the Rich

One high profile rationale used by international lending agencies to justify the investments in aquaculture has been its assumed importance as a tool to help augment food needs in developing countries, i.e. to "feed the poor." Ironically, the shrimp produced from these investments have been channeled mainly to luxury consumers in domestic and international markets, and have not become a serious food source for those who are truly hungry (Barraclough and Finger-Stich 1997; Primavera 1997; Stonich and Bailey 2000). Meanwhile, the coastal poor are being robbed of their once sustainable food sources as their traditional agriculture and fisheries are being steadily despoiled by the very nature of the shrimp aquaculture industry's operations (Deb 1998). Removing the mangroves, which are the marine nurseries, to emplace shrimp ponds means removing a sustaining wild fishery for the local communities. Removing the mangroves while siting shrimp farms on or near agricultural lands contaminates drinking water and aquifers with salt water intrusion and pond chemicals, thus ruining the ability to raise crops or farm animals (Primavera 2006).

In some villages, drinking water has to be hauled in by truck (or by foot) from many miles away because local wells are salinated by intruding seawaters that formerly were held back by the mangrove buffers (Dierberg and Kiattisimkul 1996). In some areas severe rice production losses have caused local agricultural economies to begin importation of what was once the region's staple food crop! It is the artificially created popular demand for cheap, farmed shrimp that has driven the industry in the Global South to so recklessly expand. The United States alone imports over a billion pounds of shrimp annually. However, this immense appetite for cheap shrimp in the wealthy importing nations comes at a very high price for the coastal communities in the producer nations.

> Once known as the Rice Bowl of the region, Nellore has become an industrial belt of aqua-factories, the land dug up and salinated for shrimp ponds. What was once lush and green, is now a concrete coast with sludge-filled reservoirs, canals for water supply, and huge jetties resembling highways that go right into the sea....high-security barbed wire fences and gates erected to protect the fortunes of the investors, who live in the comfort of the city. Meanwhile, the area is out of bounds for the coastal people who lived here for generations, kept their fishing boats along the coast, and fetched firewood and fruits from the mangrove forest. (Ahmed 1997).

And all of this to produce a luxury food aimed at an export market in the wealthier nations. Khushi Kabir of the NGO Nijera Kori in Bangladesh sums up well the terrible irony that is the shrimp aquaculture industry today:

> Producing luxury food in huge quantities, at the expense of our coastal poor, and making it affordable to overseas consumers — that doesn't make sense. Our priority is to produce food for our own people. (Ibid, p. 16)

To offer further insight into the grave issues raised by industrial shrimp aquaculture, a closer look at the problems engendered by the industry in Vietnam might help. In Vietnam's Ca Mau region in the late 1990s a terrible disease wiped out the industry that itself had wiped out the region's mangroves, wild fisheries and farm lands. Professor Hong, a renowned mangrove ecologist who had helped organize the planting of over 100,000 ha of mangroves after the US/Vietnam War, related how sad he was that the shrimp industry in Vietnam had so quickly destroyed an equal acreage of mangroves he had helped restore, and at that time of that shrimp epidemic he said that people were again facing hunger in the Ca Mau region because of this industry's sudden loss to disease (Rosenberry 2006).

A decade later, the same region was again hit by a disease epidemic of its farmed shrimp. Because shrimp farming is up to three or four times more profitable than rice farming, the industry has exploded in the Ca Mau District. The Aquaculture Development Programme for 1999–2010, Government Resolution No. 09 in 2001, and Decision 173 particularly promoted the conversion of rice to aquaculture (Lan 2011). Revenues in this area of Vietnam increased from 12 to 24 % between 2000 and 2005, dominating the local economy. Such profits may have predicted a rosy economic future for Ca Mau, but this shrimp boom could not last. Too soon, the boom became the classical bust, and the losses were terrible and irrevocable. The conversion of 60,000 ha of rice paddies and orchards to shrimp farms required farmers to break seawalls to allow for an influx of saltwater for the shrimp. As a result, rice paddy irrigation systems, based solely on freshwater, were destroyed; the salinization of coastal lands means agricultural production of rice will not be possible without costly interventions (Lan 2013). Shrimp farming has been recognized by local authorities in Ca Mau as unsustainable, but production continues.

In the face of hunger and degradation one must ask the FAO, World Bank, World Wildlife Fund (WWF) and other promoters of the shrimp aquaculture industry how they can continually advocate for the production and export of farmed shrimp from the poorer nations in the Global South to the more wealthy developed nations, and do so in the name of "food security." The consumers of farmed shrimp in wealthy nations do not usually require the additional protein source to supplement domestically produced food, and importing nations avoid the direct social and environmental drawbacks. Conversely, the entire production and export process in the Global South is clearly unsustainable, ecologically and socially unfriendly, and actually an infamous progenitor of food insecurity in the producer nations. The dichotomy is clear, and yet industrial shrimp production continues.

37.4.1 Human Rights Abuses Amid Loss of Resource Tenure Rights and Traditional Livelihoods

Any benefits from the accelerating expansion of the shrimp aquaculture industry in Asia, Latin America and Africa have come at a terrible price for countless local coastal communities and Indigenous Peoples whose surrounding environment, native cultures and traditional livelihoods have been adversely affected or ruined. The rapid clearing of vast tracts of mangrove forests has turned multiple-livelihood, multi-purpose natural resource systems into single sources of income and monoculture with no chance of sustainability. Many shrimp farms may last only 3–5 years before being closed down because of disease or pollution problems, oftentimes leaving behind wastelands where once productively rich wetlands stood (Barbier and Cox 2002).

The fishmeal industry that's sprung up in Peru and Chile to satisfy the voracious appetite for farmed seafood in the wealthier nations has raised a whole other set of issues involving food security. As mentioned above, intensified shrimp farming requires the addition of fishmeal to satisfy the growing needs of such densely packed stocks. Fishmeal is produced by cooking, drying and milling raw fish and trimmings, with the byproduct of fish oil. The massive demand from the aquaculture industry, for shrimp and other farmed fish, now fuels a $2.5 billion sector producing approximately 6 million tons of fishmeal and 1 million tons of fish oil annually. Advocates and activists in Peru and Chile claim that the fishmeal industry is rife with its own set of environmental and social costs, including pollution and overfishing. As aquaculture expands the demand for fishmeal rises, and so impacts from fishmeal and fish oil production also increases. It is perhaps unsurprising that the industry that feeds aquaculture is just as unsustainable, as the high demand for fishmeal and farmed fish alike drive each sector in a feedback loop with no clear end in sight.

37.4.2 Human Rights and Labor Abuses

It seems human rights abuses, including violence, child labor and human trafficking are common within the shrimp aquaculture industry in Asia. Over the years from the early industry expansions till today, along the coasts of Asia and Latin America, there have been numerous reports of threats and violence perpetrated by shrimp farm owners against the local community members who have protested against the rapid and destructive advance of the industry. Beatings, disappearances, rapes and even murders have been attributed to the shrimp aquaculture industry in producer nations like Thailand, Indonesia, India, Bangladesh, Malaysia, Honduras, Ecuador and Brazil.

Human rights abuses also permeate the labor force within the shrimp processing and feed industries themselves. For instance, in both Bangladesh and Thailand, a recent study by the American Center for International Labor Solidarity found major worker abuses occurring in the shrimp processing plants where armed guards watched over shrimp processing workers from guard watchtowers built as part of a barbed wire enclosure housing the shrimp processing facility, where many illegal immigrants from Burma and Cambodia worked long hours with little pay amid beatings and threats of violence if they complained or tried to leave. Often passports were taken away from the non-Thai workers so they could not then leave the country (Solidarity Center 2008).

As well, this human rights problem permeates the fish feed industry. Migrants working onboard Thai fishing boats suffer brutal exploitation. And the boats involved supply so-called 'trash fish' for use in feed given to farmed prawns and salmon exported globally. After the destruction of Typhoon Gay in 1989, combined with declining profits and rising education levels, Thai workers have been less willing to engage in the fishing industry. An industry that was once exclusively manned by Thai fishing crews is now primarily made up of Burmese and Cambodian migrants, most of whom are undocumented, but some come from as far as Malaysia, Vietnam, and Timor-Leste. The demands for labor remain high, however, and the shortage of workers has led to brutal exploitations, including human trafficking. Some fishers are outright kidnapped, but most voluntarily migrate in search of an escape from poverty (International Labour Organization 2013). A common pattern revealed from a recent investigation by the International Organization for Migration ("IOM") is as follows:

- A broker or agent will promise of work (not necessarily in the fishing industry) with decent pay to young men
- The costs of travel are prohibitive, so the worker agrees to pay the broker out of future wages
- The workers arrive at their destination, often without proper documentation and unable to speak Thai, and usually accept whatever is offered to them for work to avoid violence or arrest by Thai authorities

About 10–20 % of fishers reported experiencing "worse" or "much worse" conditions of wages, working hours, job function, and living conditions than they were told during recruitment. Ill-treatment, human trafficking, and forced labor are not uncommon, and this is in great part due to lax legal enforcements, especially around migrant workers. There are three main pieces of legislation that are important in regards to working in the fishing sector.

The first is The Labour Protection Act ("LPA"), enacted in 1998, which provides the basis for minimum wage, maximum working hours, and occupational health and safety. The law enables the Department of Labour Protection in Thailand to conduct inspections and impose penalties for violators. This law protects both documented and undocumented workers, but those working in the fishing industry are exempt (International Labour Organization 2013). The Ministerial Regulation No. 10 in particular places fisheries workers outside the protection of the LPA, and completely exempts boats with fewer than 20 employees or boats that spend a majority of their time operating outside of Thai waters. The Ministerial Regulation also stipulates that crew lists and payment of wages must be maintained, including a signed acknowledgement by the employee that wages are received. Workers are also supposed to be granted 30 days of sick leave and 30 days of paid holiday per year, and employment of children under the age of 15 is prohibited (unless the child's guardian works on the same boat or provides written consent). With the exception of the working age, few of these measures are effectively applied or enforced (International Organization for Migration 2011; International Labour Organization 2013). Reprehensible labor conditions on fishing boats are numerous, but often not reported for lack of easy scrutiny by third party observers who themselves have little institutional support or interest in enforcing the laws.

The second piece of legislation is The Recruitment and Job-Seekers Protection Act. This law was enacted in 1985, before the influx of migrant workers rose sharply to fill the gap in available labor, and is more focused on sending Thai workers to foreign nations for work. The Act lacks any specific protections for migrant workers entering Thailand, no regulation of brokers and other recruiting agencies, and no institutional mandate to protect incoming workers (International Labour Organization 2013). This shortfall in the Act is especially important in light of the patterns of abuse that arise specifically from the recruitment process of migrant workers.

The third law relates to human trafficking. The Anti-Trafficking in Persons Act was passed relatively recently, in 2008, and makes human trafficking a criminal act. The Act imposes severe penalties for trafficking, and expands Thailand's previous legal definition to include forced labor. This Act does cover the fishing industry, but enforcement has had mixed results. Corruption remains high among law enforcement officials, institutional support is limited, and few cases have been tried successfully. The legal process is slow and arduous, and many victims are discouraged from reporting and pursuing recompense for abuse.

37.5 Certification Schemes and Associated Laws/Policies

In recent years, there has been rising interest in certification, or ecolabeling, for food products of all varieties. Examples of certification schemes or systems that readers might be familiar with are Fair Trade, Certified Humane, and Rainforest Alliance. Certification standards for both wild and farmed seafood have emerged, both through private labels and sponsored by organizations like the World Wildlife Fund and Monterey Bay Aquarium (Vandergeest and Unno 2012). The demand for this kind of labeling arises largely from consumer concerns in the United States and the European Union over the environmental impact of the food they choose to eat, while coverage is lacking in most of Asia (Jonell et al. 2013). In this era of globalization, it is more and more difficult to determine where your food comes from and how it is produced; certification is offered as a way to answer some of these questions.

While it is a reasonable and important pursuit to develop criteria or guidelines to more effectively regulate industrial aquaculture, it is imperative that in our zeal to come up with solutions that we do not lose sight of the more complex social and environmental issues. While some of the concepts aiming to establish certain limited guidelines for better practices are commendable, certification is currently not comprehensive enough to justify further expansion of the aquaculture industry. Present schemes of certification of shrimp and other seafood are inadequate in many areas and potentially harmful to the cause of sustainability. Meanwhile, further studies of overall benefit and loss—particularly, who benefits and who loses—need to be undertaken, and further involvement and influence of indigenous and local communities must more assuredly permeate the makeup of such investigations.

37.5.1 *Impressive on Paper, Inadequate in Practice*

In reviewing numerous shrimp certification programs, it is possible the certifiers have the consumer's "best interest" in mind as they purport in their standards. However, one might justifiably ask, why are we so one-sidedly concerned about the expectations of the consumers? What about the expectations of the indigenous and local communities in the producer nations that are affected by these same shrimp or other aquaculture operations? And what of the local community members, who have no real, life-defining experience in shrimp farming, yet do have an immense stake in what happens to their surrounding coastal environment and its shared resources? Many developing countries lack the institutional resources to allow for a citizen voice against aquacultural production, or to promote sustainable production (Bush et al. 2013). This is complicated by land tenure issues, direct contradictions, and confusion between the standards set out by third-party, foreign certification schemes and the law of the land.

Certification guidelines often lay out a series of "best practices," but these best practices in one location may not serve well in another, and these depend as well on the numbers and proximity of producers. For instance, many certification standards emphasize environmental conservation and require that mangrove areas exceed 50 % of total farm area, even in areas where ponds have existed for 200–300 years and are not contributing to the more recent rise of mangrove deforestation (Hatanaka 2010). However, certification schemes are most often on a farm-by-farm basis and do not consider collective impact or compliance (Ha et al. 2012). This approach also defies the mosaic model of coastal wetlands, and many local farmers have opted out of certification because the standards do not reflect the realities on the ground. Certification for shrimp aquaculture is often cited as an effective way of drawing the line against "bad practices," such as mangrove clearing and misuse of chemicals in the shrimp ponds (Stonich and Bailey 2000). Nevertheless, certifications are designed for consumers in high-income countries, rather than those in production areas.

In considering the value of any given seafood certification label, it is worth examining whether the standards include or exclude the local communities and stakeholders most dependent upon the resources captured in aquaculture. A farmer complaint often heard by researchers is that the companies developing these certification schemes are basing them on a Western worldview, and that the producers have little or no input into the standards that are developed (Hatanaka 2010). The incentive for farmers to comply with standards is limited when they feel excluded from the process. Studies have also shown that community and local government regulation are more effective environmental regulatory networks than certification by transnational eco-labeling (Datta et al. 2012).

As well, too often in the producing countries, the laws are not adhered to in a dependable fashion, and enforcement is seriously lacking or biased in favor of "influential persons." This in itself gives little incentive to obeying the laws, just knowing how easily they can be broken. This is an internal problem involving governing infrastructure that will not be easily remedied by a simple code of conduct or best practice incentive. Shrimp farms that were constructed by illicit means, often associated with eviction of hundreds or thousands of people by violence and intimidation, may still be certified despite these injustices. Governments in support of shrimp farming sometimes turn a blind eye to companies illegally acquiring land, or in some cases even perpetrate the intrusion of shrimp ponds through the use of police or military action. Certification of the resulting ponds lends legitimacy to their presence and does not address the underlying sociopolitical problems or the disregard for domestic laws.

Governments in shrimp-producing nations often do not recognize the land title or tenure rights of indigenous and/or local communities, even though these said communities may have occupied those same waters and lands for many decades or centuries. In many cases, the expansion of the shrimp aquaculture industry follows the typical pattern of transformation of a commons, which includes the mangroves and other community resources into privately owned/operated lands, which in practice nearly always benefit the few and the rich rather than the many and the poor. In Mexico, for example, the Federation of Fishing Cooperatives of Southern Sinaloa presented a case against the construction of shrimp farms which would prevent seven of the cooperatives from fishing in areas granted to them; all complaints were ignored (Cruz-Torres 2000). There are many similar incidents in countries such as Colombia, Honduras, Ecuador, India, Vietnam, and Tanzania.

Remediation is another carrot offered by the eager certifiers, mimicking offset of traditional terrestrial forestry whereby new trees are planted for each that is cut down. Remediation ostensibly aims to set up mangrove tree plantations on mudflats or elsewhere to compensate for ecosystem loss due to shrimp farming. However, remediation invariably loses the original ecosystem value and doesn't reestablish the complex biodiversity that is being sacrificed. A healthy, functioning ecosystem, along with all of its robust productivity and functionality, cannot be simply planted elsewhere. Too often, valuable mangrove ecosystems are destroyed and a poor substitute is set up in their place. Most certification schemes do not include an audit to determine if replanting was successful (Jonell et al. 2013). These remediation schemes often fail miserably because they lack needed sensitivity and cognizance of the true ecology and inherent value of those ecosystems being remediated. Planting a new stand of mangrove or coastal wetland seedlings is no guarantee that it will grow and thrive to become a functioning ecosystem, although mass mangrove plantings are popular with authorities and celebrities touting such events as effective and significant rehabilitation methods for lost forest area.

Many remediation sites are also inappropriate for mangroves and other coastal wetland species. The reality remains that there is no other viable place to put in a new mangrove system. Mangrove systems, opportunistic and colonizing, already occupy the entire potential mangrove habitat. No mangrove remediation projects can recreate the habitat conditions required by mangroves, and usually an attempt to do so results in converting (and destroying) one valuable type of ecosystem, such as a salt marsh or mud flat, into a mangrove forest. The usual approach to restoration—hand planting—never really succeeds in recreating a viable, healthy mangrove ecosystem. In the Philippines, for example, most remediation efforts plant *Rhizophora* species, which has economic value as a forestry product, rather than the natural colonizers; mangroves are typically planted in tidal zones that are unsuitable habitats. The ideal locations are typically occupied by the brackish fishponds which demanded the mangrove relocation or remediation in the first place.

37.5.2 Aquaculture Certifiers

There are several third-party aquaculture certifiers currently in operation, including Naturland (organic certification), the Aquaculture Certification Council (which certifies the Global Aquaculture Alliance Best Aquaculture Practices), and the newly-established Aquaculture Stewardship Council (developed by the World Wildlife Fund). Other well-known certifiers are the International Food Standard (IFS) and GlobalGAP in Europe. The Aquaculture Certification Council (ACC) and the Aquaculture Stewardship Council (ASC) have both come under a great deal of criticism for ignoring local input and realities of human rights violation, displacement, and environmental degradation. On close review of the bulk of the World Wildlife Fund (WWF)/Aquaculture Stewardship Council (ASC) shrimp standards, there exist technical fixes for the on-site pond operations, but the social and environmental problems associated with industrial seafood farming are not eliminated even with these technical remedies set in place. The ASC standards represent the "best management practices" (BMP) side of the argument in favor of increasingly technological and intensive aquacultural systems (Stonich and Bailey 2000). What is put in place on paper is not so easy to accomplish in practice, especially when state actors regard transnational eco-certifiers with a great deal of suspicion (Vandergeest and Unno 2012).

Ensuring that these listed technical criteria are adhered to is not an easy matter, given the complexity and immensity of scale of these operations. Methods to ensure adherence to certification, such as auditing, are especially problematic as they do not yet seem able to guarantee validity and reliability (Albersmeier et al. 2009). Some certification schemes, such as Naturland, are also run using an Internal Control System (ICS), whereby a single local individual is responsible for corralling up to 100 others. Rather than directly investigating most of the farms, third-party certifiers instead audit the paperwork collected by that single representative; typically less than 10 % of farms are actually inspected by the certifiers (Ha et al. 2012). Studies have shown that ICS and similar systems allow for a significant level of fraud and adequate data collection for certain certification has been dismissed as unimportant even by Seafood Watch (Hatanaka 2010).

With or without certification, production of farmed shrimp continues to rise in volume each year to meet an artificially created demand for shrimp products in the wealthy consumer nations. There, shrimp is a luxury commodity, which recently made it to the top of the consumer hit record charts. This 2-decade surge in shrimp consumption in the high-income nations means that something is inherently off balance in the producer nations. Certifiers allay fears of consumers concerned about the environmental impact of farmed seafood, but few individuals actually understand the technical details of each certification scheme; this confusion is exacerbated by standards that are themselves ambiguous. Advocates of political ecology argue that certification will simply maintain the status quo of high consumption demand in the importing nations, while not addressing the overconsumption issues there nor the ongoing social and environmental issues in the producer nations

(Stonich and Bailey 2000). For instance, 90 % of shrimp sold in the U.S. comes from Asia and Latin America, where environmental regulations are lax and often not enforced.

The stated aim of WWF/ASC's standard setting process is to move industrial-style shrimp farming towards a more sustainable production system. This statement sounds good on paper, but is less simple to achieve in practice. How can such a resource dependent operation done on such a massive and growing scale be contained so as not to adversely impact the natural resource base upon which it depends for its continuing operations? This is a question that must be seriously considered as the industry moves forward. Meanwhile, this industry competes for the same resource base that the local populace ultimately depends upon for sustenance and traditional culture. The WWF is known worldwide, and is a well-respected organization. As with many such organizations, it is often assumed by the general public that their stated goals are the "right" direction. It is important to recognize that just because an organization is famous and its mission is admirable that its actions may reflect inertia in the system, maintaining status quo despite best intentions. It is the responsibility of concerned and educated global citizens to challenge these large institutions where needed and appropriate in order to promote a just and sustainable future.

One such researcher, Dr. Peter Vandergeest, labels this new move to certify shrimp as "the Ecocertification Empire," driven mainly by the interests of corporate buyers and environmental groups in Europe and North America. He highlights the fact that ecocertification is limited to species that are exported in great quantities to these regions in the Northern Hemisphere. In contrast, carp, which are the single most important aquaculture species for consumption in Asia, are entirely missing from the ASC dialogues presented by the WWF (Vandergeest and Unno 2012). The omission of species commercially important to the Global South from aquaculture dialogues reveals a serious flaw in certification as a means to achieve sustainability.

Certification covers very little of the seafood produced in aquaculture. Only around 4.2 % of production is certified in some way, and the majority is covered by GlobalGAP (Jonell et al. 2013). Full of loopholes, controversy, and sometimes trapped in a Westernized viewpoint, certification does not appear to be the most efficient method of achieving sustainable aquaculture. In contrast, engaging in community-based natural resource management that promotes stakeholder stewardship appears to be a more successful model (Vandergeest 2007). Although certification may help consumers make better choices about what they eat and the food system they engage in, progress has been slow and limited. Rather than attempting to impose standards and policies from the outside, promoting sustainable forms of aquaculture and environmental regulation may be better achieved by engaging with those it directly affects. Partnerships, rather than prescriptions, are an important piece of the puzzle that is a sustainable global food system.

37.5.3 Case Study: Aquaculture in Guatemala

Certification depends upon the certified producers following the laws of the producer nations that are meant to protect mangroves and other coastal resources. However, these laws too often are not enforced because there is little will or infrastructure to enforce them. Consequently, mangroves are cleared with growing impunity, as paper laws on the books do little to really protect them. Certified aquaculture represents a very small portion of the industry, but regulations to address aquacultural production are more widespread. Unfortunately, when countries begin to invest resources in aquaculture at the industrial scale, there are often growing pains when it comes to legislation and institutional support. The intentions for improvements in regulation and management are quite often genuine, but implementation is extremely complicated. As a result, there is often little visible change despite a law being on the books. Here, we provide a summary of a case study analyzing and comparing Guatemalan laws of aquaculture to international laws, providing a brief background of laws and policies, as well as suggestions for improvement. *The following is a summary adapted from Velascos (2009).*

Country: Guatemala.

Aquacultural activities: Marine shrimp farming (important for export), tilapia farming (important for internal consumption).

Relevant legislation: The *Ley General de Pesca y Acuicultura*, LGPA (Fishery and Aquaculture General Law) and its bylaw the *Reglamento de la Ley General de Pesca y Acuicultura*, RLGPA (Regulation of the Fishery and Aquaculture General Law). The law has been in place since 2002, and the regulation in force since 2005; prior to this time, Guatemala had no adequate laws and policies to govern aquaculture.

Background of aquacultural production: Aquaculture has been growing in Guatemala since 1950 largely due to increasing numbers of small tilapia farms. Tilapia are crucial to food security in Guatemala, and have remained important for rural income as prices for other agricultural products have dropped. Only recently has marine shrimp farming begun to expand. The growth of shrimp farming has been slow, but the FAO still expects greater future productivity based on increasing exports and a tendency for the industry to escalate quickly. Marine shrimp farming is essentially unlegislated at this time, creating serious complications for managing Guatemalan waters.

Current state of aquacultural law: The LGPA was developed when various fisheries and aquaculture stakeholders met and discussed the proposed law with MAGA (Guatemala's Ministry of Agriculture) and UNIPESCA (a MAGA agency). The LGPA is part of a larger effort in Guatemala to promote social and economic development and was a recognition that marine resources are part of the nation's wealth. Its primary purpose is to regulate fishing and aquaculture according to leading scientific principles in order to optimize the use of marine resources. The law was an enormous improvement over previous legislation. Some of these improvements are listed below:

- A glossary of scientific and technical terms;
- Categories and classifications of authorized fishing activities within Guatemala's jurisdiction;
- Defines concessions based on above classifications and categorization;
- Establishes bans, prohibitions, and sanctions; and
- Orders the creation of the National Registrar for Fishery and Aquaculture

The RGPA followed in 2005, despite the LGPA requiring it be created within 120 days of the 2002 publication. This bylaw or regulation contains details pertaining to fishery and aquaculture activities, such as:

- How to obtain concessions (divided into licenses and permits by activity);
- The equipment and methods allowed to ensure sustainable development;
- Defining areas where fishing activity is allowed in Guatemala; and
- Details of UNIPESCA's management

Although the letter of the law is much improved, the Guatemalan government has had trouble implementing the legislation. One of the greatest obstacles has been financial; for example, the National Registrar is barely functional for lack of staff and funding. Records of registration for boats, licenses, and permits are also woefully lacking. With some recent assistance from Spain, UNIPESCA began to issue identification cards to inshore fisherman as part of a census effort. After being registered in this census, many fishers requested formal permits. The identification cards demonstrated the importance of registering for permits without imposing burdens on the fishers. Such a solution may be useful in other countries where registration is hindered by limited institutional resources.

Crucially, aquaculture is not regulated in inland waters under either the LGPA or the RGPA. This is a serious omission because it means that farmers are not supported or protected by the government, but neither are they prevented from causing damage to the surrounding environment. The Guatemalan aquaculture industry has a unique opportunity to engage with UNIPESCA to develop regulation for sustainable development, according to the agency's mandate, but without enough institutional and financial support the industry may follow in the destructive footsteps of its Asian cousins.

Recommendations: Aquaculture is clearly an important resource in Guatemala, and the overall objective of the LGPA is to promote efficient use of fisheries while also maintaining them into the future. In order to realize this admirable goal, there are opportunities for further improvement in both the LGPA and the RGPA:

- Review and expand the glossary to cover gaps and explain confusing concepts that are poorly explained in the body of the law itself
- Include specific instructions for creating, revoking, and modifying fisheries and aquacultural management plans
- Allowing the use of management plans as ways to expand regulations for activities not currently managed under the law
- Clarify how regulations may be developed and implemented in order to enforce the law

- Expand and clarify how licenses and permits may be obtained
- Permit the addition of new prohibitions in the body of the law, such as illegal fishing gear or activities
- Provide protections for officials against abuse and ill-treatment
- Impose penalties for environmental destruction from aquacultural activity
- Define the power, duties, and eligibility requirements of officers, such as inspectors, in order to legitimize their activity
- Incorporate specific rules and regulations for the permitting and licensing of aquaculture
- Defining environmental regulations in accord with the latest scientific research.

These are only some of the improvements that the Guatemalan government might consider. However, it is important to keep in mind that UNIPESCA and Guatemala in general struggle to provide adequate institutional and financial support to implement the existing LGPA.

Conclusions: Laws are not always enough to ensure visions of sustainable development and sustainable aquaculture. Limited compliance with existing legislation may not be a conscious act of ill-will, but of ignorance or of limitation. It is important when examining these laws, policies, and certifications that we consider a holistic viewpoint. In concert with analyzing the legislation itself, we must also consider the social, economic, environmental, and political conditions where that legislation exists. Guatemala's aquaculture industry is a fine example of a socio-economic system, much like the ecotone of the mangroves, where there is significant intersection of human and environmental activity. There is no single solution to long term problems like sustainable aquaculture, although laws are certainly a critical part of addressing such a wicked problem, and Guatemalan officials are eager to embrace such solutions. Legislation of fisheries and aquaculture must keep in mind these complex realities in order to remain effective.

37.6 Conclusion: The Need to Redefine the Blue Revolution

The Blue Revolution, which praises industrial level aquaculture as a way to feed the world, has proven to be as flawed as its agricultural namesake, the Green Revolution. However, lessons may be learned from struggles and failures as well as the regions of success that aquaculture has achieved. Certification currently has its limits, and should not be considered the be-all and end-all of achieving sustainability in aquaculture. Studies have also demonstrated that aquacultural production of smaller seafood species may in fact be beneficial to the poor, but fish farming is rarely practiced in areas where it is most needed. A redefinition of the goals of the Blue Revolution is needed to address the challenges that have arisen from aquacultural production.

Perhaps closed containment, Recirculating Aquaculture System (RAS) is the "wave of the future" for attaining sustainable aquaculture, where no disease spread or escapes of these cultured species to the wild is possible, and production ponds or tank waters are cleaned and recirculated within the system itself. Closed containment systems can also be located outside the intertidal zones, which is vital in regard to protecting the coastal wetlands. Such closed-containment systems can also be integrated into aquaponics, allowing for the production of vegetables and herbs. This is not characteristic of the bulk of the industry today. One day these closed systems may replace the antiquated open systems. This is the industry's current challenge, but it will require a massive renovation of infrastructure for an already well-entrenched, unsustainable, "open system" industry, which existing certification programs, including the WWF/ASC scheme, still encourage and endorse.

Another noteworthy approach to present day aquaculture is to encourage more sustainable and less environmentally damaging forms of aquaculture that can be practiced to help the coastal communities supplement their incomes via small-scale, community-based aquaculture. One example is promoting a fish farmer field school as a bottom up approach intended to improve aquaculture practices. Rather than focusing on technical improvements, the fish farmer field school focuses on developing critical thinking skills and experimentation by the fish farmers themselves, attempting to democratize aquaculture, while also allowing aquaculture to provide a supplemental income. Aquaculture should supplement, not replace traditional fishing livelihoods.

Still we need to get beyond the consumer-and-producer-sidedness of the equation, to a more representative equation that gives more import to the values of indigenous and local communities, as well as a realistic view of what constitutes a healthy, biodiverse environment worthy of conservation. Although voluntary certification schemes with the best intentions can reward companies that meet its standards by giving them a "green seal" of approval, certification can do nothing to prevent the worst companies from continuing their destructive operations. Affecting consumer awareness to lower consumer demand can make a difference however.

Some possible goals for advocates, consumers, and legislators alike are:

- Raise public awareness of the environmental, social and health consequences of imported farmed shrimp and how that relates to growing demand for this product;
- Influence shrimp purchasing policies of institutional buyers, such as grocery stores, seafood markets, and restaurants, by providing guidance in sourcing environmentally and socially responsible shrimp;
- Contribute to the body of information on sustainable shrimp alternatives and further disseminate information on sustainable seafood programs and responsible consumer choices.

Even a relatively small-scale reduction in demand, of say 10 % or more, can have a decisive effect upon the rate of increase in farm area expansion. The majority of those buying, selling and consuming shrimp are not aware of the adverse ecological and social consequences of their demand. By initiating a combined consumer/markets campaign there is more certainty to affect the production side. Since the great majority of shrimp consumption occurs in restaurants, it also makes sense to target restaurateurs and chefs to reduce the import demand for warm water shrimp, thus affecting the market in a most sensitive "pocket."

It is important that aquacultural development does not eliminate or lessen the utility of long-term traditional livelihoods for coastal communities. Traditional communities may be "dollar poor," but culturally "wise and rich." Their livelihoods are often based upon a self-sufficient, small-scale, local economy that may not produce export dollars, but does produce a culture, traditions and skills that can be passed on from one generation to the next. This should not be endangered by new industries such as shrimp aquaculture that promise jobs that are inevitably low paid, unskilled and short-lived, while socially and environmentally bankrupt. Aquaculture may yet become a valuable, sustainable resource, but practitioners and researchers must be ever striving for better and wiser solutions to the problems we face in the global food system. There will never be just one solution, and even as we address one problem with aquaculture, another will arise. Being aware of the contexts and communities involved in the problems we face is critical to developing effective policies, especially when grappling with something as complex as the global food system.

References

Ahmed F (1997) In: Gillespie P, Rickman A, Ahmed F (eds) Defense of land and livelihood, vol 6. Sierra Club of Canada, Ottawa

Albersmeier F, Schulze H, Jahn G, Spiller A (2009) The reliability of third-party certification in the food chain: from checklists to risk-oriented auditing. Food Control 20(10):927–935. doi:10.1016/j.foodcont.2009.01.010

Alongi DA (2002) Present state and future of the world's mangrove forests. Environ Conserv 29(3):331–349. doi:10.1017/S0376892902000231

Alongi DM (2014) Carbon cycling and storage in mangrove forests. In: Carlson CA, Giovannoni SJ (eds) Annual review of marine science, vol 6, 2014. Annual Reviews, Palo Alto, pp 195–219. doi:10.1146/annurev-marine-010213-135020

Baird IG, Quarto A (1994) http://www.ecoemploy.com/. Nautilus Institute

Barbier E, Cox M (2002) Economic and demographic factors affecting mangrove loss in the coastal provinces of Thailand, 1979–1996. Ambio 31(4):351–357. doi:10.1639/0044-7447(2002)031[0351:eadfam]2.0.co;2

Barbier EB, Cox M (2004) An economic analysis of shrimp farm expansion and mangrove conversion in Thailand. Land Econ 80(3):389–407

Barbier EB, Hacker SD, Kennedy C, Koch EW, Stier AC, Silliman BR (2011) The value of estuarine and coastal ecosystem services. Ecol Monogr 81:169–193

Barraclough S, Finger-Stich A (1997) Some ecological and social implications of commercial shrimp farming in Asia. United Nations, United Nations Research Institute for Social Development, Geneva, Switzerland

Bendiksen EÅ, Johnsen CA, Olsen HJ, Jobling M (2011) Sustainable aquafeeds: progress towards reduced reliance upon marine ingredients in diets for farmed Atlantic salmon (Salmo salar L.). Aquaculture 314(1–4):132–139. doi:10.1016/j.aquaculture.2011.01.040

Blake PA, Allegra DT, Snyder JD, Barrett TJ, McFarland L, Caraway CT et al (1980) Cholera — a possible endemic focus in the United States. N Engl J Med 302(6):305–309. doi:10.1056/NEJM198002073020601

Burridge L, Weis JS, Cabello F, Pizarro J, Bostick K (2010) Chemical use in salmon aquaculture: a review of current practices and possible environmental effects. Aquaculture 306(1–4):7–23. doi:10.1016/j.aquaculture.2010.05.020

Bush SR, Belton B, Hall D, Vandergeest P, Murray FJ, Ponte S et al (2013) Certify sustainable agriculture? Science 342(6150):1067–1068

Chapman V (1976) Mangrove vegetation. Cramer, Vaduz

Clucas I (1997) A study of the options for utilization of bycatch and discards from marine capture fisheries. Food and Agriculture Organization of the United States, Rome

Cruz-Torres ML (2000) Pink gold rush: shrimp aquaculture, sustainable development, and the environment in Northwestern Mexico. J Polit Ecol 7:63–90

Datta D, Chattopadhyay RN, Guha P (2012) Community based mangrove management: a review on status and sustainability. J Environ Manag 107:84–95. doi:10.1016/j.jenvman.2012.04.013

Deb AK (1998) Fake blue revolution: environmental and socio-economic impacts of shrimp culture in the coastal areas of Bangladesh. Ocean Coast Manag 41(1):63–88. doi:10.1016/S0964-5691(98)00074-X

Dierberg FE, Kiattisimkul W (1996) Issues, impacts, and implications of shrimp aquaculture in Thailand. Environ Manag 20(5):649–666. doi:10.1007/bf01204137

Eayrs S (2007) A guide to bycatch reduction in tropical shrimp-trawl fisheries. Food and Agriculture Organization of the United Nations, Rome

FAO (2010) The state of world fisheries and aquaculture - 2010 (SOFIA). Food and Agriculture Organization, Rome. Retrieved from http://www.fao.org/docrep/013/i1820e/i1820e00.htm

FAO (2012) The food outlook: global market analysis. Food and Agriculture Organization, Trade and Market Division (ETS)

FAO (2013a) FAO/MARD technical workshop on early mortality syndrome (EMS) or acute hepatopancreatic necrosis syndrome (AHPNS) of cultured shrimp (under TCP/VIE/3304). Food and Agriculture Organization of the United Nations, FAO Fisheries and Aquaculture, Hanoi. Retrieved from http://www.fao.org/docrep/018/i3422e/i3422e.pdf

FAO (2013b) The food outlook: global market analysis. Food and Agriculture Organization, Trade and Markets Division (ETS)

Flaherty M, Karnjanakesorn C (1995) Marine shrimp aquaculture and natural-resource degradation in Thailand. Environ Manag 19(1):27–37. doi:10.1007/bf02472001

GAO (2011) FDA needs to improve oversight of imported seafood and better leverage limited resources. US Government Accountability Office

Graslund S, Holmstrom K, Wahlström A, Poungshompoo S, Bengtsson BE, Kautsky N (2003) Antibiotic use in shrimp farming and implications for environmental impacts and human health. Int J Food Sci Technol 38(3):255–266

Ha TT, Bush SR, Mol AP, van Dijk H (2012) Organic coasts? Regulatory challenges of certifying integrated shrimp–mangrove production systems in Vietnam. J Rural Stud 28(4):631–639. doi:10.1016/j.jrurstud.2012.07.001

Hatanaka M (2010) Governing sustainability: examining audits and compliance in a third-party-certified organic shrimp farming project in rural Indonesia. Local Environ 15(3):233–244

International Labour Organization (2013) Employment practices and working conditions in Thailand's fishing sector

International Organization for Migration (2011) Trafficking of fishermen in Thailand. International Organization for Migration (IOM)

Islam MS, Haque M (2004) The mangrove-based coastal and nearshore fisheries of Bangladesh: ecology, exploitation and management. Rev Fish Biol Fish 14(2):153–180. doi:10.1007/s11160-004-3769-8

Islam MS, Wahab MA, Tanaka M (2004) Seed supply for coastal brackishwater shrimp farming: environmental impacts and sustainability. Mar Pollut Bull 48:7–11. doi:10.1016/j.marpolbul.2003.11.006

Jonell M, Phillips M, Ronnback P, Troell M (2013) Eco-certification of farmed seafood: will it make a difference? Ambio 42(6):659–674. doi:10.1007/s13280-013-0409-3

Lan NT (2011) From rice to shrimp: ecological change and human adaptation in the Mekong Delta of Vietnam. In: Stewart MA, Coclanis PA (eds) Environmental change and agricultural sustainability in the Mekong Delta, vol 45. Springer, Netherlands, pp 271–285. doi:10.1007/978-94-007-0934-8_16

Lan NT (2013) Social and ecological challenges of market-oriented shrimp farming in Vietnam. Springerplus 2:675. doi:10.1186/2193-1801-2-675

Lewis RR, Phillips MJ, Clough B, Macintosh DJ (2003) Thematic review on coastal wetland habitats and shrimp aquaculture. World Bank, NACA, WWF and FAO. Consortium Program on Shrimp Farming and the Environment

Leopold A (1953) Round river: from the journals of Aldo Leopold. Oxford University Press, New York

Naylor R, Goldburg R (1998) Nature's subsidies to shrimp and salmon farming. Science 282:883

Naylor RL, Goldburg RJ, Primavera JH, Kautsky N, Beveridge MC, Clay J et al (2000) Effect of aquaculture on world fish supplies. Nature 405(6790):1017–1024. doi:10.1038/35016500

Naylor RL, Hardy RW, Bureaus DP, Chiu A, Elliott M, Farrell AP et al (2009) Feeding aquaculture in an era of finite resources. Proc Natl Acad Sci U S A 106(36):15103–15110. doi:10.2307/40484671

NOAA (2012) Fisheries of the United States. National Marine Fisheries Service/Office of Science and Technology Fisheries Statistics Division, Silver Spring

Pendleton L, Donato DC, Murray BC, Crooks S, Jenkins WA, Sifleet S et al (2012) Estimating global "blue carbon" emissions from conversion and degradation of vegetated coastal ecosystems. PLoS One 7(9), e43542. doi:10.1371/journal.pone.0043542

Popescu I, Ogushi T (2013) Fisheries in Japan. European Parliament, Policy Department

Primavera J (1997) Socio-economic impacts of shrimp culture. Aquac Res 28(10):815–827. doi:10.1046/j.1365-2109.1997.00946.x

Primavera JH (2005) Mangroves, fishponds, and the quest for sustainability. Science 310(5745):57–59. doi:10.1126/science.1115179

Primavera J (2006) Overcoming the impacts of aquaculture on the coastal zone. Ocean Coast Manag 49(9-10):531–545. doi:10.1016/j.ocecoaman.2006.06.018

Rahman LM (2000) The Sundarbans: a unique wilderness of the world. In: McCool SF, Cole DN, Borrie WT, Oloughlin J (eds) Wilderness science in a time of change conference, vol 2: wilderness within the context of larger systems. U.S. Department of Agriculture, Forest Service, Rocky Mountain Research Station, Ft. Collins, pp 143–148

Rosenberry B (2006) World shrimp farming. Shrimp News International, San Diego

Solidarity Center (2008) The true cost of shrimp: how shrimp industry workers in Bangladesh and Thailand pay the price for affordable shrimp. American Center for International Labor Solidarity

Stevenson NJ (1997) Disused shrimp ponds: options for redevelopment of mangroves. Coast Manag 25(4):425–435. doi:10.1080/08920759709362334

Stonich SC, Bailey C (2000) Resisting the blue revolution: contending coalitions surrounding industrial shrimp farming. Hum Organ 59(1):23–36

Thorstad EF (2008) Incidence and impacts of escaped farmed Atlantic salmon Salmo salar in nature. World Wildlife Fund

Tran L, Nunan L, Redman RM, Mohney LL, Pantoja CR, Fitzsimmons K, Lightner DV (2013) Determination of the infectious nature of the agent of acute hepatopancreatic necrosis syndrome affecting penaeid shrimp. Dis Aquat Org 105(1):45–55

Valiela I, Bowen JL, York JK (2001) Mangrove forests: one of the world's threatened major tropical environments. Bioscience 51(10):807–815. doi:10.1641/0006-3568(2001)051[0807: mfootw]2.0.co;2

Vandergeest P (2007) Certification and communities: alternatives for regulating the environmental and social impacts of shrimp farming. World Dev 35(7):1152–1171. doi:10.1016/j.worlddev. 2006.12.002

Vandergeest P, Unno A (2012) A new extraterritoriality? Aquaculture certification, sovereignty, and empire. Polit Geogr 31(6):358–367

Velascos BS (2009) The Guatemalan fishery and aquaculture general law versus international laws related to fishery and aquaculture. United Nations/Office of Legal Affairs Division for Ocean Affairs and the Law of the Sea, New York

Walker PJ, Mohan CV (2008) Viral disease emergence in shrimp aquaculture: origins, impact and the effectiveness of health management strategies. Rev Aquac 1(2):125–154. doi:10.1111/j. 1753-5131.2009.01007.x

Weber JT, Mintz ED, Cañizares R, Semiglia A, Gomez I, Sempértegui R et al (1994) Epidemic cholera in Ecuador: multidrug-resistance and transmission by water and seafood. Epidemiol Infect 112(1):1–11. doi:10.2307/3864427

Chapter 38
TEXTBOX: Creating Law and Policy for Resilient Urban Food Systems

Yassi Eskandari-Qajar and Janelle Orsi

Abstract In the field of ecology, resilience describes an ecosystem's ability to withstand, adapt to, and recover from disaster or change. Greater biodiversity brings greater ecosystem resilience, and, like ecosystems, urban and regional food systems are more resilient when a diverse network of producers, processors, distributors, retailers, and consumers interact within a robust local and regional economy. However, small-scale food businesses all along the chain of production face significant legal, regulatory, and financial barriers to their success. Policymakers have the opportunity to assess the range of roadblocks to local food, and cultivate a legal landscape for thriving urban food systems. Such policy changes will allow communities to rapidly transition towards greater food sovereignty, economic abundance, and resilience.

In the field of ecology, resilience describes an ecosystem's ability to withstand, adapt to, and recover from disaster or change. Greater biodiversity brings greater ecosystem resilience, and, like ecosystems, urban and regional food systems are more resilient when a diverse network of producers, processors, distributors, retailers, and consumers interact within a robust local and regional economy. Whereas unpredictable climate, resource insecurity, and changes in prevailing economic conditions may be catastrophic for rigid, industrial scale monocultures with faraway markets, resilient food systems are designed to endure. With a wide variety of crops and farming methods, and innovative agreements and enterprises connecting producers to local and regional markets, resilient food systems are capable of both stewarding the environment and feeding our communities for the long run.

A resilient urban food system, though focused on local benefits is nevertheless dependent on a robust regional network of small, sustainable farms. These urban, peri-urban, and rural farms are well connected to nearby markets, reducing reliance on long-distance transport and fossil fuels. Using a multiplicity of sustainable farming methods to cultivate a wide range of crop types in varying soils and

---

Y. Eskandari-Qajar (✉) • J. Orsi
Sustainable Economies Law Center, Oakland, CA, USA
e-mail: yassi@theselc.org; janelle@theselc.org

microclimates can lower the risk of widespread crop failure due to natural disaster, sudden resource limitation, pest invasions, and other symptoms of a changing climate. And whereas the large-scale globalized farm and food industry mass-produces food and concentrates ownership and profit in the hands of the few, food and farming enterprises in resilient urban food systems produce both environmentally- and culturally-appropriate food, and direct wealth to local food businesses and the stewards of agricultural lands. Simply put, resilient food systems are rooted in an understanding that the complexity and redundancy of small and micro farms and food enterprises are strengths, rather than opportunities for greater efficiency and agglomeration.

38.1 Benefits and Barriers

Just and resilient urban food systems can help cities achieve many goals, including increased community food security, economic opportunity, public health, education and skill-building opportunities, disaster preparedness, green spaces, and wildlife habitat. In addition, when farms and food businesses are both economically and geographically connected to their community through cooperative and community supported models, they are more likely to behave more responsibly overall.

However, small-scale food businesses all along the chain of production face significant legal, regulatory, and financial barriers to their success. Laws that were designed to temper the harms of large-scale industrial food systems are oftentimes improperly applied to small farms and food enterprises as well. For example, zoning laws that were intended to separate incompatible land uses in rapidly industrializing cities now separate the places where we work, live, and exchange—restricting positive and relatively low-impact activities such as home-grown produce sales and home-based microenterprise. Similarly, proponents of local food may find that zoning restrictions on the location and intensity of commercial activity also limit the expansion of farmers markets, mobile food vending, residential CSA distribution points, urban farms, and other key elements of a functional local food system.

In light of the tremendous public benefits of local food, policymakers have the opportunity to assess the range of roadblocks to local food, and cultivate a legal landscape for thriving urban food systems. Such policy changes will allow communities to rapidly transition towards greater food sovereignty, economic abundance, and resilience.

38.2 Policy Opportunities for Food Production and Access

38.2.1 Urban Farming Ordinances

Urban farmers must comply with a patchwork of regulations, including land use restrictions, building codes, health and safety regulations, and codes regarding the keeping of animals, landscaping, and commercial activities. Rather than leaving it to residents to find and interpret these rules, some cities, like San Francisco and Seattle, have created urban farming ordinances that provide clear, organized permissions and definitions for residential gardens, commercial urban farms, and community gardens. Other cities can follow their examples by revising planning codes for clarity and permitting farming in all zones.

38.2.2 Neighborhood Food Access Points

In order to increase local food production and access, cities should remove zoning barriers to the sale of produce grown in home gardens, community gardens, and vacant or underutilized lots. In addition, cities should follow the example of Portland, Oregon and allow residences to act as "food membership distribution points," where a farm can drop off produce boxes that customers will retrieve later in the day. To minimize concerns about associated traffic and nuisance, cities can establish rules about signage and hours and days of operation in residential neighborhoods. Policies should also exempt or reduce food safety regulatory burdens for CSAs in light of the direct relationship between producer and consumer.

38.2.3 Growing Food on Vacant or Underutilized Public Land

Access to arable land is an important part of community food security and self-sufficiency. Cities can rapidly increase local food access by making vacant and underutilized municipally owned lots available for food production. Local governments can conduct a vacant and underutilized public land inventory and layer data from reputable mapping tools like the United States Department of Agriculture's Food Environment Atlas and the California Environmental Protection Agency's CalEnviroScreen in order to identify the city's most food insecure and environmentally-burdened areas. By layering as much data as possible about the population and environmental factors, cities can ensure that support, incentives, and lease arrangements are targeted at residents and community based organizations in the areas that need it most.

38.2.4 Financial Incentives for Urban Agriculture on Privately-Owned Land

The 1965 California Williamson Act provides property tax reductions for farmed lands by assessing land based on its value for agricultural production, rather than the value of its "highest and best use." In 2013, a similar law, the California Urban Agriculture Incentive Zones Act (AB 551), was passed. Referred to as the "Urban Williamson Act," the law incentivizes food production in urban areas by authorizing cities and counties to offer lower property tax rates to owners of vacant or unimproved lots, so long as that land is used for small-scale agricultural production and animal husbandry for periods of at least five years. For local governments that cannot afford to provide such a financial incentive for food cultivation, an alternative strategy is to charge owners of vacant and abandoned lots a registry fee which could be lifted if the land was used for farming.

38.2.5 Preserving the Right to Save and Share Seeds

Saving and sharing seeds season after season helps communities cultivate and preserve a wide variety of crops that are well adapted to local conditions. These seeds, in turn, help to increase local biodiversity and climate resilience, as well as preserve local culture and natural history. Seed lending libraries – of which there are over 450 in the US and many more around the world – store local seed varieties and put them into circulation by allowing people to "borrow," plant, collect, and "return" them each season.

Unfortunately, seed libraries and the act of saving and sharing seeds in the US have come under attack in recent years. Government agencies are broadly interpreting laws that require commercial seed sellers to test, label, and register their seeds, and applying these expensive, burdensome, and unnecessary laws to seed libraries, threatening their ultimate existence. Governments should create clear language that exempts noncommercial seed sharing from the regulations imposed on commercial seed sales. Considering that four companies control 60% of the global seed market, and the UN Food and Agriculture Organization estimates that we have lost 75% of the world's plant genetic biodiversity in the last century, saving and sharing seed not only helps to increase local biodiversity, but can also improve global agricultural resiliency. After all, an heirloom seed variety cultivated and saved in one microclimate could very well help a community thousands of miles away bounce back from crop failure due to flood, drought, or pest invasion.

38.2.6 Incentives for Healthy Local Food Sales

Communities that lack access to a full-service grocery store often turn to corner markets for food purchases. However, corner stores tend to stock unhealthy processed foods, liquor, and sugary drinks with long shelf lives rather than fresh, local fruits, vegetables, and meats. Local governments could offer a tax credit or property tax reduction to those stores that sell a certain quantity of fresh foods that are locally or regionally sourced. Of course, reduced tax revenues may not be an option for some local governments, no matter how highly they value public health or local food economies. Funding from state or federal level programs could enable many more metropolitan areas to offer these healthy food incentives.

38.3 Policy Opportunities for Small and Micro Food Enterprises

38.3.1 Cottage Food Laws

Cottage food laws unlock the economic potential of home kitchens by permitting the sale of homemade foods such as jams, baked goods, and other food items deemed to carry a low-risk of causing food borne illness. A cottage food enterprise can be established with little or no start-up capital, which makes a cottage food law a powerful community economic development tool. Over thirty U.S. states currently have cottage food laws, and many are revising their laws to expand the list of permissible foods and methods of sale.

38.3.2 Mobile Food Vending and Produce Carts

Mobile food markets, food trucks, and produce carts create a low cost entry point into the food business, diversify local food economies, and increase fresh food access in the heart of food insecure neighborhoods. Though some are concerned that mobile food vendors can undercut storefront businesses in surrounding areas, adopting ordinances that prioritize mobile food permitting in economically-distressed districts may be way to both create new economic opportunities and increase access to fresh foods.

38.3.3 Subsidized Shared Commercial Kitchens

Most prepared foods intended for sale require access to a commercial kitchen. Because use of commercial kitchens can be a prohibitive cost for new food enterprises, cities can help incubate new food businesses by sponsoring local commercial kitchens. These kitchens, coupled with business support services, could have a particularly uplifting impact when located in or near low-income communities.

38.3.4 Local Food Procurement Policies

Contracting and purchasing preferences for sustainable, local food can even the playing field for these oftentimes small and struggling businesses. Government agencies and institutions such as schools, universities, and hospitals can provide large, stable markets for local food and could play a significant role in both increasing access to fresh food and supporting local food economies.

38.3.5 Land Use Policies that Promote Farmers' Markets

Farmers' markets create essential market access for local and regional farmers. However, establishing a new farmers' market requires licensing, permitting, and zoning compliance that ultimately limits the number, size, and activities of farmers' markets. Local governments can remove many of the planning and zoning barriers to new farmers' markets by declaring them an approved land use in more zones.

38.4 A Word on Liability and Waste

In nature, there is no such thing as "waste." Instead, diversity, complexity, and interconnectivity turn one system's outputs into another system's inputs. Resilient urban food systems revolve around a similar concept. Closed-loop farming processes and well-connected distribution networks limit opportunities for waste, and food gleaning programs, food pantries, and discount groceries for near-expiration foods combat hunger. Yet, even in a community with widespread sustainable farming practices, efficient food distribution networks, and hunger prevention programs, excess food may still be thrown away instead of being shared with others.

Fear of liability is a major cause of food waste in the US. Rather than sharing viable excess food with those in need, individuals, restaurants, and groceries may instead discard food in order to avoid potential liability should it cause illness to the consumer. The little-known federal Good Samaritan Food Donation Act limits civil and criminal liability in order to protect those who donate food to needy individuals with a good faith belief that it is safe to consume. However, several terms within the law call for further clarification, including a definition for the term "needy individuals," and clarifying whether food can be given directly to a needy individual or must first pass through a nonprofit organization. As is, the law provides important protections for food sharing, but if clarified and expanded to include liability protections for a broader range of food sharing activities, the law could play a central role in increasing community-level food security.

Fear of liability does not only cause food waste – it also causes land waste. Vacant and underutilized urban land, if shared for food production, could have a significant impact on neighborhood food access. However, landowners may be hesitant to share otherwise unused property for fear of liability should land users sustain injury while gardening. Consequently, many lots lay fallow rather than fruiting – and not just private land. Local public parks present an obvious opportunity for public food forests. Though local governments may recognize the benefits of neighborhood access to fresh fruits, cities, like land owners, may fear liability for potential injuries in instances where a park visitor was to slip on a fallen fruit. In the same ways that recreational use statutes can limit a land owner's liability when they allow access to their property for recreational use, and cities can be protected from liability for dog bites sustained in dog parks, new or expanded laws should protect private land owners and governments from liability when land is used for food production.

38.5 Food Systems in the Context of a Broader Shift Toward Just and Resilient Societies

Local and regional food systems are not insulated from the economic and political influences of states, nations, or international trade. In reality, the economic viability of robust regional food production is closely tied to land policy, government subsidies, and trade policies that affect competition, supply, and price. Similarly, the viability of small, local food enterprises is impacted by large-scale food industries that control the majority of food consumed in the US and influence both food prices and consumer choice. The above policy opportunities are not intended to fully address these larger issues or suggest new regulations on big food players. Rather, these policies help enable communities to remove many of the legal barriers to more just, resilient, and localized food economies and to support those activities and enterprises that confer the greatest amount of local health, economic, and environmental benefits – particularly for those most marginalized.

The many shining examples of, and opportunities for, equitable, community-centered food and farming practices present a sound vision for more ecologically-sound and socially just communities. Of course, achieving more just and resilient urban food systems will not necessarily lead into a broader societal shift toward economies that place people and planet first. However, these are important models to study and apply to other sectors as part of a transition toward community renewable energy, more equitable and democratic business entities, new models for the ownership and management of land and housing, and the humane and ecologically-sound stewardship of our common resources.

Resilience is a smart strategic approach to meeting all basic needs in uncertain times. Indeed, present-day economic, political, climate, and resource insecurities may be the major reason why the concept of resilience has found its way out of ecological circles and into some rather unlikely arenas. Even the Rockefeller Foundation – a foundation with a namesake made infamous through petroleum, banking, and real estate exploits – has begun to espouse resilience and fund resilience planning in cities all over the world. However, efforts to transition to resilience must go beyond simply security and climate adaptation goals. Resilience planning presents an opportunity to achieve a far more profound and lasting societal shift if plans are developed from the grassroots and are firmly rooted in the principles of equity, justice, democracy, and community empowerment.

Part V
Australia

Kiran K. Patel and Gabriela Steier

Australia and New Zealand have adopted the Food Standards Code, a governing body applicable to these two nations for the sole purpose for creating and accessing food standards. The Food Standards Australia and New Zealand (FSANZ) will be discussed in greater detail in Chap. 39 on Australia, however, it was introduced in 1991 to "develop, vary and review standards for food available in Australia and New Zealand and to "develop codes of conduct in collaboration with industry."[1] FSANZ was created under the Food Standards Treaty and works with a Board which has health ministers and other representatives. This section will also cover and review the food labeling laws and policy, including a brief introduction of these policies and how policy and politics drive these provisions. There will also be a discussion in public health and safety as it relates to nutrition. Additionally, the section will also discuss the government's response to the report of the review of food labeling laws and policy and recommendations that were made moving forward. Finally, a discussion regarding current amendments to the various numbers of food labeling laws and a discussion on cases in Australia regarding food standards and their respective outcomes and case precedents that were set as a direct result.

Notably, Chap. 39 includes an analysis of Australian labeling laws, country of origin labeling, nutrition and health claims, and the general food safety regulatory framework. A cutting-edge and provocative Chap. 40 pushes questions of sustainability through paradoxical legal constructs, such as animals as property but corporations as persons, to examine the Australian consumer protection framework.

---

[1]*Primary Production and Processing Standard for Dairy*, Food Standards Australia New Zealand (FSANZ), December 15, 2004.

Chapter 39
Regulation of Food Labelling in Australia and New Zealand

Michael Blakeney

Abstract This chapter describes the common arrangements between Australia, and New Zealand concerning food standards. The *Food Standards Australia New Zealand Act 1991* establishes the Food Standards Australia New Zealand (FSANZ) as an independent statutory authority with responsibility for developing food standards. Pursuant to this legislation a common "Food Standards Code" (the Code) has been agreed. Also described is the review of food labelling law and policy undertaken by the two countries in 2009–2010 which prioritised health concerns as the primary driver of food policy. The chapter concludes with a review of the principal food law cases which have been decided in Australia and New Zealand over the past 10 years.

39.1 Introduction

Australia and New Zealand share a common approach to food standards, the so-called *Food Standards Code* (the Code)[1] and a common body to determine food standards. These arrangements are part of a move towards a general uniformity of standards between the two countries supported by the Trans-Tasman Mutual Recognition Arrangement (TTMRA). The 1996 TTMRA was, "designed to remove regulatory barriers to the movement of goods and service providers between Australia and New Zealand and to ... facilitate trade between the two countries" through the implementation of mutual recognition principles for goods and occupations.[2] In 1996, the Government of Australia and New Zealand formalised the Joint Food Standards System via a Treaty. This treaty is called *the Agreement between the Government of Australia and the Government of*

Various parts of this chapter have previously been published by this author and are hereby impliedly cited.

[1] Food Standards Australia New Zealand, Standard 1.2.9 Legibility Requirements, issue 119, Australia New Zealand Food Standards Code, 2010, p. 1.
[2] Blewett et al. (2011), p. 24.

M. Blakeney (✉)
Faculty of Law, University of Western Australia, Crawley, WA, Australia
e-mail: michael.blakeney@uwa.edu.au

New Zealand establishing a System for the Development of Joint Food Standards and seeks to reduce unnecessary barriers to trade, to adopt a joint system of food standards, to provide for timely development, adoption and review of food standards and to facilitate sharing of information.[3] New Zealand joined this system under conditions that are set out in the 2002 *Agreement between the Government of Australia and the Government of New Zealand Concerning a Joint Food Standards System.*

Until September 2011, the food regulation system was overseen by the Australia and New Zealand Food Regulation Ministerial Council. This council had the responsibility for developing domestic food regulation policy and promoting a consistent approach to the implementation and enforcement of food standards.[4] In September 2011, the Council of Australian Governments (COAG) launched the *Legislative and Governance Forum on Food Regulation* ("the Forum"), which replaced the Australia and New Zealand Food Regulation Ministerial Council. The Forum is primarily responsible for the development of domestic food regulatory policy and the development of policy guidelines for setting domestic food standards.

Membership of the Forum comprises a Minister from New Zealand and the Health Ministers from Australian States and Territories, the Australian Government as well as other Ministers from related portfolios (Primary Industries, Consumer Affairs etc) where these have been nominated by their jurisdictions. The Australian Government Parliamentary Secretary for Health and Ageing is the Chair of the Forum. The Food Regulation Standing Committee (FRSC) is a sub-committee of the Forum. FRSC is responsible for coordinating policy advice to the Forum and ensuring a nationally consistent approach to the implementation and enforcement of food standards. It also advises the Forum on the initiation, review and development of FRSC activities.[5]

The *Food Standards Australia New Zealand Act 1991* establishes the Food Standards Australia New Zealand (FSANZ) as an independent statutory authority with responsibility for developing food standards. The agreed upon standards are placed in the Code. The Implementation Sub-Committee (ISC) of FRSC develops and oversees the consistent approach to implementation and enforcement of these standards.[6]

Under the TTMRA, New Zealand may opt out of any standard to be included in the code if they consider it inappropriate on prescribed grounds. In regards to labelling law, New Zealand exercised this option to opt out of mandatory country-of-origin requirements. In contrast, Australia has recently introduced legislation on mandatory country-of-origin requirements into its Federal Parliament.

---

[3]Blewett et al (2011), p. 24.
[4]*Id.*
[5]*Id.*
[6]Id., p. 24.

Each country has its own border control regime. In Australia, it is the Australian Quarantine Inspection Service (AQIS) and Australian Customs and Border Protection Service, and in New Zealand the border agencies include the Ministry of Agriculture and Forestry (MAF), Biosecurity New Zealand and the New Zealand Customs Service. Among the tasks of these bodies is responsibility for checking compliance of food with domestic standards. The TTMRA provides for the mutual recognition by one country of the standards applying in the other. This means that most goods imported from third countries that comply with standards applying in one partner country (and have cleared the border of that country) can be exported to the other partner country. One of the main reasons for this border control is the protection of domestic biodiversity and prevention of pest introduction and spread.

The major difference in the structure of government responsibility between the two countries can be attributed to the difference between the unitary nature of New Zealand on one hand and the federal nature of Australia on the other. Australia has a federal structure with legislation on food labelling emanating from the eight parliaments of the States and Territories in addition to that of the Federal Parliament. Therefore, there are eight separate state and territory food acts, and the associated instrumentalities operate in somewhat different ways in each of the eight jurisdictions. In contrast, New Zealand has a single Parliament and the enforcement of the Code is empowered under a single national act, the New Zealand *Food Act 1981*. This legislation operates across the whole country and is administered nationally by the New Zealand Food Safety Authority (NZFSA).[7]

In both countries, there is complementary support for food standards deriving from the general fair trading provisions relating to misleading or deceptive statements. In unitary New Zealand, consumer protections are based on national legislation (New Zealand *Fair Trading Act 1986*) and monitored and enforced by a single national body, the New Zealand Commerce Commission (NZCC). Australia also provides national consumer protection legislation (the 2010 *Competition and Consumer Act*) to be monitored by the Australian Competition and Consumer Commission (ACCC). However, there are additional state and territory consumer protection provisions monitored by particular state and territory consumer protection agencies. The requirements relating to misleading or deceptive statements operate independently from the Code and can impose different considerations in regards to accurate information, such as consumer values.[8]

Both countries are members of the World Trade Organization (WTO) and subject to the disciplines which it imposes through the Agreements on Technical Barriers to Trade (TBT), Sanitary and Phytosanitary Standards (SPS), and the Agreement on Trade Related Aspects of Intellectual Property Rights (TRIPS).

---

[7]Id., p. 26.
[8]Id., pp. 26–27.

39.2 Review of Food Labelling Law and Policy 2011

The Council of Australian Governments (COAG) requested The Australia and New Zealand Food Regulation Ministerial Council (Ministerial Council) to engage an independent panel of experts to undertake a comprehensive review of food labelling law and policy. In October 2009, the Ministerial Council released the terms of reference for the review. In Australia, this Review was jointly funded by the Australian Government and all the Australian States and Territories. In New Zealand, the New Zealand Government supported the consultations.

Under the terms of reference, the independent expert panel, chaired by former Australian Health Minister, Dr. Neal Blewett, was required to:

- Examine the policy drivers impacting on demands for food labelling;
- Consider what should be the role for government in the regulation of food labelling. What principles should guide decisions about government regulatory intervention?
- Consider what policies and mechanisms are needed to ensure that government plays its optimum role;
- Consider principles and approaches to achieve compliance with labelling requirements and appropriate and consistent enforcement;
- Evaluate current policies, standards and laws relevant to food labelling and existing work on health claims and front-of-pack labelling against terms of reference 1–4 above;
- Make recommendations to improve food labelling law and policy.[9]

Following the receipt of submissions from stakeholders and discussions held at a number of public fora, in January 2011 the panel issued its report: *Labelling Logic*. The report contained 61 recommendations on several key themes including: the policy drivers of food labelling; principles and criteria to guide government decision making on regulatory intervention; public health and safety; new technologies; consumer value issues; presentation; and compliance and enforcement.

39.2.1 Principles and Policy Drivers of Food Labelling

The panel reviewed the drivers for labelling policy by reviewing three major interest groups: consumers, industry, and government. Subsequently, the panel identified the drivers for labelling policy as "consumers' needs for information; industry's need for marketing flexibility and minimal regulatory burdens; and government's objectives in the area of individual and population health." Consideration of these policy drivers provides "a framework for deriving principles for regulatory intervention in order to steer the flow of labelling events." The panel

---

[9]Blewett et al. (2011) at Appendix C, p. 165.

further determined that exploration of these policy drivers "revealed the ubiquity and breadth of health concerns," including a growing acceptance of governments' preventative health role in reducing the risk of chronic diet-related disease.

Based on recognition of the prevalence of health concerns, the panel recommended that a comprehensive nutrition policy be developed that included a framework for the roles of the food label. This comprehensive Nutrition Policy, in turn, would inform the development or variation of labelling standards and establish principles that would guide decisions about government regulatory interventions in food labelling.[10]

Further, the panel formulated an issues hierarchy based upon risk which governs the initiation of regulatory action, the modes of intervention, and where rules and oversight should lie. The hierarchy in descending order is: food safety, preventative health, new technologies, and consumer values. The panel took the view that regulatory actions in relation to food safety, preventative health, and new technologies should be initiated primarily by government and referenced in the Code. Conversely, regulatory actions in relation to consumer values issues were to be initiated generally by industry. Generally, consumer values would rely on the 'misleading or deceptive' provisions in consumer protection legislation, not removing the possibility of some specific methods or processes of production being referenced in the Code.

There is little disagreement that the modes of intervention should be mandatory for food safety. However, for preventative health, the panel recommended a mixture of mandatory and co-regulation requirements, the choice to be dependent on government health priorities and the effectiveness or otherwise of co-regulatory measures. For new technologies the panel recommended mandated identification on the label of foods or ingredients treated or produced by such technologies for a period of 30 years after their introduction into the human food supply chain, at the end of which time the need for such identification should be reviewed. Finally, the panel proposed that the modes of intervention for consumer values issues should be self-regulatory but subject to more prescriptive forms of intervention in cases of market failure, such as in the case of country-of-origin issues created by New Zealand's exercise of the option to exclude mandatory country-of-origin labelling.[11]

39.2.2 Public Health and Food Safety

As mentioned above, the panel observed that public health and food safety were positioned at the top of the food labelling regulatory hierarchy. Food safety labelling issues relate to food choices that affect both the consumers' immediate

---

[10]Blewett et al. (2011), p. 1.

[11]Id., p. 2.

health and issues relating to impacts on consumers' long term health, such as preventative health measures. A broad array of government regulations, including the *Food Standards Code* (the Code), ensure the safety of the foods, as they can only be sold if certain strict requirements have been met.[12] The mandatory food label elements within the Code primarily require the provision of food safety information to the purchaser. This information includes warning statements, use and storage instructions, identification of allergens, date marking, batch code and contact details.

The Code outlines specific rules relating to the safe food handling of a product, primarily concerning date marking and use and storage instructions. The terms for date marking are specified in *Standard 1.2.5 Date Marking of Food* in the Code. Use and storage instructions are specified in *Standard 1.2.6 Directions for Use and Storage* in the Code. Use and storage instructions are applicable when, for reasons of health or safety, the consumer should be informed of specific use or storage requirements. Additional food labelling requirements relating to mandatory statements and declaration of allergens are listed in *Standard 1.2.3 Mandatory Warning and Advisory Statements and Declarations* in the Code. Directions are provided for the inclusion of warning statements, including their declarations on foods that are exempt from the requirement to bear a label and foods dispensed from vending machines.[13]

Concerns were expressed about the clarity of rules requiring allergen declarations and food components when purchasing unlabelled products, such as food purchased in restaurants and food outlets. While there is a requirement in the Code for the retailer/food outlet to declare this information on or in connection with the display of the food or to provide such information upon request, the high level of consumer dissatisfaction reported indicated that this responsibility was not well known in the food service sector. Based on this recognition, the panel recommended that there be more effective monitoring and enforcement of the existing requirements in the Code to provide mandatory warning and advisory statements and allergen declarations on all food products.[14]

39.3 Nutrition

The panel noted that Australia and New Zealand lack a comprehensive policy framework which would identify priority public health issues and specify strategies needed to address the issues. Such comprehensive policy would involve different agencies responsible for different food and nutrition policy areas, such as setting guidelines and public health goals, education strategies, primary and secondary

---

[12]Id., pp. 53–54.
[13]Id., p. 55.
[14]Id., p. 56, Recommendation 7.

prevention strategies, international food policies, monitoring and research. The need for a comprehensive was identified by the Preventative Health Taskforce in 2009,[15] but has not yet been implemented.

The panel recommended that a comprehensive Nutrition Policy be developed which includes a framework for the roles of the food label. Key aspects of the framework to be:

- the provision of food safety and nutrition information and education strategies to protect and promote the health of the population, including articulated roles for food label elements;
- the encouragement of the provision of healthy foods within the food supply to facilitate healthy diets;
- the setting and application of nutrient criteria and dietary guidance;
- the facilitation of social and other research to improve understanding of how label information is used and its impact on food selection, eating behaviours and the food supply;
- the establishment of monitoring and surveillance systems for dietary/nutrition practices that include the use and understanding of food labels.[16]

The Panel also recommended that the Food Standards New Zealand Act 1991 be amended to require Food Standards to give regard to the comprehensive Nutrition Policy when developing or reviewing labelling standards.[17]

39.3.1 Ingredient Lists

Ingredient lists serve to provide basic information in a standardized manner, enabling consumers to make decisions regarding the selection of foods to meet their own dietary needs. The primary role of ingredient lists is to reassure the purchaser that the food contains the ingredients expected to be present by providing a list of the components of the product, including the percentage of key or characterising ingredients. These lists are increasingly important for processed foods presented in packaging which make food selection using traditional methods of sight, smell, and comparison of like products difficult.

Standard 1.2.4 of the Code specifies that all ingredients must be listed in order of decreasing ingoing weight and food additives and colourings must be listed using their specific name or code number. *Standard 1.2.10* also requires that the percentage of characterising ingredients and components of certain foods is declared.[18] Currently, a compound ingredient should include its components in brackets unless

---

[15]Preventative Health Taskforce (PHT) (2009).
[16]Id., pp. 58–60, Recommendation 9.
[17]Id., p. 61, Recommendation 10.
[18]Id., p. 61.

it comprises less than 5 % of the food. In the case that it comprises less than 5 % of the food, there is only a requirement to identify any food additives in the compound ingredient that are providing a technological function, such as a preservative or stabilising agent.

The panel noted that the identification of food additives or flavourings in the ingredient list caused confusion, especially the combined use of scientific terms and code numbers. Consequently, the panel recommended that industry, in consultation with government, medical authorities and relevant consumer organisations, develop a voluntary code of practice and education initiatives to enable consumers to quickly identify label information relating to additives, colourings and flavourings that are of agreed medical priority for sensitive consumers.[19]

39.3.2 Nutrition Information Panels

Nutrition Information Panels (NIPs) provide specific, quantified information on major nutrients to inform consumers who have concerns regarding specific chronic illnesses or conditions. Standard 1.2.8 of the Code specifies the current requirements for NIPs appearing on packages. It requires the declaration of energy, carbohydrates and sugars, protein, fat, saturated fat, and sodium. It also requires an NIP if a nutrition claim is made regarding the food, even if the package is otherwise exempt. Certain conditions are included in the standard if particular nutrition claims are made. For example, claims related to polyunsaturated or monounsaturated fatty acids require a declaration in the NIP.[20]

NIP requirements are subject to change as the knowledge of the nutrition needs of the populations change over time. While regular review of the nutrients declared in the NIP will require regular review, the panel considered four specific issues and issued recommendations.[21] These issues included trans-fatty acids, fibre content, potassium levels, and sodium/salt content. The panel recommended that the NIP include a mandatory declaration of all trans-fatty acids above an agreed threshold, as well as of naturally occurring fibre content.[22] Noting concerns about potassium levels and their relationship with increased risk of high blood pressure, the panel recommended that a voluntary declaration of potassium content in the NIP be actively considered by industry. In the future, if nutritional policy guidance recommended the reduction in consumption of potassium for at-risk population groups, disclosure of potassium in the Nutrition Information Panel should become mandatory.[23]

---

[19]Id., p. 62, Recommendation 11.

[20]Id., p. 64.

[21]Id., p. 65.

[22]Id., pp. 66–67, Recommendations 13 and 14.

[23]Id., p. 68, Recommendation 15.

Because of the perceived confusion of consumers, the presentation of the NIP has also received considerable attention. Much of the confusion amounted to misleading information, particularly in relation to consumers' interpretations of statements about the amount of nutrients per serving. As a result of the serving sizes determined by the manufacturer, nominated serving size is not often consistent with how individuals would consume that food. Therefore, the panel recommended that the information of nutrients per serving be deleted from the NIP unless a daily intake claim is made.[24]

In the past decades, the Australian public increasingly consumed food outside the home obtained from chain food service outlets and vending machines. Chain food service outlets and vending machines often provide food purchased away from the home, have standardized recipes from a number of franchised outlets, and provide foods that are either not labelled or the label is not accessible prior to purchase. Noting the increase in consumption in Australia of food outside of the home and its association with poorer diet quality and greater risk for obesity in children, the panel recommended the mandatory declaration of energy content of standardised food items on menu boards or in close proximity and on vending machines.[25] These declarations were intended to provide more nutrition and ingredient information at the point of sale to assist consumers with their selections. While it is noted that some consumers intend to indulge, research has shown that consumers may change their orders after exposure to nutrition information at the point of sale, especially when ordering for children. The addition of nutritional values is thought to increase awareness and knowledge of the energy consumed at these food outlets.[26]

39.3.3 Health Claims on Food Packaging

The panel noted that health claims in food labelling had proliferated and that they could mislead and detract from public health messages. In particular, it referred to the use of such terms as "pure" and "natural" which may imply health benefits to consumers. FSANZ had devoted significant resources in an effort to develop a standard for nutrition, health and related claims. In 2008, FSANZ finalised a draft standard that was provided to the Ministerial Council. The Ministerial Council requested FSANZ to review the draft standard by October 2011. Accordingly, the panel recommended the finalisation of the standard and that it include the following:

---

[24]Id., p. 69, Recommendation 17.
[25]Id., pp. 70–71, Recommendation 18.
[26]Id., pp. 70–71.

- a hierarchy of substantiation of claims at the various levels, that would encompass use of defined nutrition words and terms, pre-approved relationships, authoritative sources, systematic review and pre-market assessment and approval;
- a requirement that all foods that carry a nutrition, health and related claim comply with an agreed nutrient profiling system;
- a requirement that the presence of a nutrition, health and related claim triggers relevant information disclosures in the Nutrition Information Panel or ingredients list; and
- a requirement that the presence of a general or high level claim triggers display of standardised front-of-pack label information.[27]

39.3.4 New Technologies: Irradiation, Genetic Engineering, and Nanotechnology

New technologies in food production present a new set of issues regarding safety assessments of the foods produced or processed by such technologies. The development of new technologies requires a cohesive approach is required to ensure consistent and effective food labelling regulation. Under the Panel's hierarchy of food labelling issues (See infra XX), there is strong rationale for time-limited, mandatory interventions by government regarding the labelling of foods produced with new technologies. Therefore, as a general principle, all foods or ingredients that have been process by new technologies be required to be labelled for 30 years from the time of their introduction into the human food chain: the application of this principle to be based on scientific evidence of direct impact on, or modification of, the food/ingredient to be consumed. At the expiration of that period the mandatory labelling should be reviewed.[28] The submissions to and consultations with the Panel raised specific concerns in regards to new technology primarily based on biotechnology, nanotechnology, and irradiation.[29]

Irradiation is a method of food preservation achieved through exposing certain types of food to a source of ionising energy. Currently, in Australia and New Zealand, irradiation is prohibited unless specific permission is granted. *Standard 1.5.3* sets out the permitted sources and levels of radiation and lists the foods permitted to be irradiated and the consequential labelling requirements. However, the panel referenced the 1997 report of a Study Group convened by the FAO, the International Atomic Energy Agency, and WHO concluded that "food irradiated to any dose appropriate to achieve the intended technological objective is both safe to

---

[27]Id., pp. 73–75, Recommendation 20.
[28]Id., p. 91, Recommendation 28.
[29]Id., p. 87.

consume and nutritionally adequate."[30] Since the conclusion of the Study Group, no contradictory conclusions have been presented in regards to further problems for humans occasioned by the consumption of foods treated with irradiation. Thus, the panel recommended that, on this basis, the mandatory labelling requirements prescribed for irradiated foods should be reviewed.[31]

Genetic Engineering produces genetically modified ingredients using gene technology, and is another new technology requiring legislative and regulatory frameworks to ensure food safety. Under the Code, genetically-engineered foods must be labelled. *Standard 1.5.2* of the Code, sets out the labelling requirements for foods produced using gene technology. This Standard requires that food be labelled genetically modified (GM) if novel deoxyribonucleic acid (DNA) or novel protein introduced by gene technology can be shown to be present in the final food or the food has altered characteristics as specified by the Code. However, genetic modification labelling is not required if GM ingredients or processing aids are used in the manufacturing process but there is no detectable residual genetic material or protein of the source in the final product and the food has no altered characteristics.[32]

There are a variety of additional exemptions in the labelling of GM foods. One such exemption is that flavours that contain GM material, but do not exceed a level of one part in a thousand in the final food, do not require genetic modification labelling. Another exemption is that if a food, ingredient, or processing aid includes unintentional traces of GM at one percent or less by weight per ingredient, it does not require genetic modification labelling. Finally, foods produced from animals fed GM products (i.e., animal foodstuffs) do not require genetic modification labelling.[33]

The panel noted that as there was no evidence that consumption of GM food produced any immediate detrimental effects in humans it recommended that only foods or ingredients that have altered characteristics or contain detectable novel DNA or protein be required to declare the presence of genetically modified material on the label.[34]

Nanotechnology refers to a technology that deals with microscopic particles sized 100 nanometres or less.[35] In regards to nanotechnology, the panel observed that the extent of use of nanomaterials in the Australian and New Zealand food supply appears at this stage to be minimal and adequate risk assessment techniques undeveloped. Nonetheless, it urged FSANZ and other relevant bodies develop a standard for regulating the presence of nanotechnology in the food production chain, consistent with the recommendations in this Report relating to new technologies.[36]

---

[30] Yan (2008).

[31] Id., p. 94, Recommendation 34.

[32] Id., p. 88.

[33] Id., p. 89.

[34] Id., p. 92, Recommendation 29.

[35] Id., p. 91.

[36] Id p. 94, Recommendation 35.

39.3.5 Consumer Values

Consumers' have an increasing desire to make food purchase decisions according to their personal values, their perceptions of the world, and their ethical convictions. These issues of consumer concern are frequently changing, and bring a new dimension to the food labelling debate. Generalised consumer values issues such as human rights, animal welfare, environmental sustainability and country-of-origin labelling were raised in a number of the submissions to the panel. In turn, the panel took the position that there is much to be said for self-regulation in the management of consumer values issues, particularly where "there are clearly defined problems but no high risk of serious or widespread harm to consumers."[37] That panel recommended that the value of industry-initiated self-regulatory intervention be recognised and that industry, in collaboration with special interest groups, further develop and apply a responsive and more structured self-regulatory approach to consumer values issues that incorporates:

- the role that voluntary codes of practice can play in relation to the evolution of standard definitions for values-based claims;
- the role that certification schemes can play in effectively communicating values-based messages; and
- the development of agreed standards through existing frameworks such as International Organization for Standardization, Standards Australia or Standards New Zealand.[38]

The panel proposed that a monitoring regime for self-regulatory measures be established. Further, they declared that evidence of systemic failure to provide accurate and consistent values-based information to enable consumers to make informed choices will require a more prescriptive mode of regulation. Marketplace failure will result in government intervention through government-mandated values claims, which have the advantage of being legally enforceable. In Australia, the only values-based label claim mandated is Country of Origin Labelling (CoOL).[39]

39.4 Country of Origin Labelling

Country of Origin Labelling (CoOL) was identified by the Panel as "a particularly contentious issue, much of this arising from the fact that country of origin is a generalised values-issue with ramifications far beyond food and where definitional precision is challenging." The divergence of approaches over CoOL represents a significant exception to the uniformity of the trans-Tasman food labelling regime.

---

[37]Taskforce on Industry Self-Regulation (2000), p. 42.
[38]Blewett et al. (2011), p. 106, Recommendation 38.
[39]Id., p. 106, Recommendation 39.

All packaged food sold in Australia must contain a statement on the package that identifies where the food was made or produced, identifies the country where the food was made, manufactured, or packaged for retail sale and that the food is constituted from imported or local ingredients. In contrast, New Zealand has no mandatory requirement for CoOL, apart from wine, although information can be voluntarily supplied.[40]

The Panel noted that the coverage of the Australian standard on CoOL coverage in Australia anomalously excepted beef, lamb and chicken and subsequently recommended that Australia's existing mandatory country-of-origin labelling requirements for food be maintained and be extended to cover all primary food products for retail sale.[41] Moreover, the panel noted that the consumer law provisions of the Federal Competition and Consumer Act 2010 offered a more comprehensive treatment of CoOL than did the Code, and consequently recommended that that CoOL regulation be dealt with under the Competition and Consumer Act 2010 rather than in the Food Standards Code.[42]

39.5 Presentation

Noting that the presentation of text and images influences consumers' ability to notice, locate, read and comprehend the information contained on a food label, the panel endorsed the view that food labelling should provide clear, simple and easy to interpret information that can be understood across demographic groups, particularly lower socio-economic groups. The current FSANZ Standard specifies, "Unless otherwise expressly permitted by this Code, each word, statement, expression or design prescribed to be contained, written or set out in a label must, wherever occurring, be so contained, written or set out legibly and prominently such as to afford a distinct contrast to the background and in the English language".[43] In addition, warning statements must be at least 3 mm in size (approximately equivalent to 8 point font) or 1.5 mm (4 point font) for small packages. Taking account of comments received on the legibility requirements, the panel recommended that a minimum font size of 3.5 mm in an open font style in mixed case be applied for mandated information, with the exception of small package sizes where the minimum font size should be 1.5 mm[44] and that a set of guidelines be developed in consultation with industry that includes reference to other presentation factors such as letter and line spacing, text justification, and stroke width.[45]

---

[40]Id., p. 106.

[41]Id., p. 108, Recommendation 40.

[42]Id., p. 108, Recommendation 41.

[43]Food Standards Australia New Zealand, *Standard 1.2.9 Legibility Requirements*, issue 119, - *Australia New Zealand Food Standards Code*, 2010, p. 1.

[44]Blewett et al. (2011), p. 116, Recommendation 44.

[45]Id., p. 116, Recommendation 45.

Some consumer submissions discussed difficulties associated with contrast, noting that particular combinations of foreground and background colour are especially problematic for legibility. The panel proposed a minimum contrast level of 70% for mandated information be stipulated in the Food Standards Code.[46] It also proposed that the warning and advisory statements be emboldened and allergens emboldened both in the ingredients list and in a separate list.[47]

Finally, based on the FSA's recommendation that all mandatory information should be positioned on a face of the pack within defined boarders, the panel recommended that industry be encouraged to develop a set of guidelines relating to the co-location of mandatory health information presented in a standardised manner on the label.[48] Further, it was recommended that an automated label assessment tool be investigated that can gauge a label's compliance with mandated legibility requirements and those stipulated in relevant voluntary codes.[49]

39.6 Images

Interpretive symbols, claims, and endorsements on labels have the potential to convey important nutritional information when included and have been termed front-of-pack labelling (FoPL). Acknowledging complaints that consumers lack adequate time to read food labels while shopping, the panel recommended that an interpretative FoPL system be developed that is reflective of a comprehensive Nutrition Policy and agreed public health priorities.[50]

Currently, there is no single consensus on the best form of FoPL labelling. However, of the labelling systems which had been developed and tested, the panel preferred the colour-coded multiple traffic lights (MTL) system of warnings and recommended that a MTL FoPL system be introduced on a voluntary basis. In addition, such labelling should be mandatory where general or high level health claims are made, or where equivalent endorsements/trade names/marks appear on the label.[51] The MTL system comprises coloured lights for key nutrients, "a green light signifies a healthy choice, an amber/orange light is an 'okay' choice and a red light a less health or unhealthy choice." Additional forms of the MTL system include the use of the words low/medium/high or numbers (e.g. nutrient amounts per serving) associated with each light.[52]

---

[46] Id., p. 116, Recommendation 46.
[47] Id., p. 117, Recommendation 47.
[48] Id., p. 118, Recommendation 48.
[49] Id., p. 118, Recommendation 49.
[50] Id., p. 121, Recommendation 50.
[51] Id., p. 124, Recommendation 51.
[52] Id., p. 123.

The panel also proposed that government advice and support be provided to producers adopting the MTL system and that its introduction be accompanied by comprehensive consumer education to explain and support the system.[53] This would be backed-up by monitoring and evaluation of the MTL system to assess industry compliance and the effectiveness of the system in improving the food supply and influencing consumers' food choices. Chain food service outlets were to be encouraged to display the MTL system on menus/menu boards, but on a mandatory basis where general or high level health claims are made or equivalent endorsements/trade names/marks are used.[54]

39.7 Compliance and Enforcement

Food labelling statements and claims must be actively enforced to be effective. In Australia and New Zealand, food labelling requirements are enforced through the specific labelling requirements of the Code and through the general provisions of the Australian and New Zealand consumer laws relating to misleading or deceptive conduct. These two systems operate independently from one another, as a claim can be in breach of consumer protection laws but not in breach of the code; another can be in breach of the Code but not in breach of consumer protection laws. Although FSANZ developed the standards, they have no role in the enforcement under either system.

In Australia, non-compliance with outlined labelling requirements risks prosecution under the food legislation of the States and Territories.[55] New Zealand has imported the Code into its *Food Act 1981*. In Australia, action can be taken on conduct involving misleading or deceptive conduct that breaches of the Code under the Australian Consumer Law provisions of the Federal *Competition and Consumer Act 2010*. In New Zealand consumers are protected from misleading or deceptive terms under the *Fair Trading Act 1986*.[56]

The panel reported many submissions from both industry and consumer organisations which were critical of levels of enforcement for food labels, because of low levels of funding and because of inconsistencies in the interpretation of the requirements across the ten jurisdictions responsible for interpreting the Code.[57] Reference was made to critical comments in a 2008 NSW Supreme Court case[58] which

---

[53]Id., p. 125, Recommendation 52.

[54]Id., p. 126, Recommendations 54 and 55.

[55]*Food Act 2003* (NSW), ss 18, 21; *Food Act 1984* (Vic), ss13, 16; *Food Act 2006* (Qld), ss37, 39; *Food Act 2001* (SA), ss 18, 21; *Food Act 2008* (WA), ss 19, 22; *Food Act 2003* (Tas), s 18, 21; *Food Act 2004* (NT), ss 17, 20; *Food Act 2001* (ACT), ss 24, 27.

[56]Blewett et al. (2011), p. 130.

[57]Id., p. 131.

[58]*Tumney (NSW Food Authority) v Nutricia Australia Pty Ltd*; [2008] NSWSC 1382.

highlighted a number of problems with the Code, including lack of definitions, its "piecemeal nature", duplication, and because the document was not prepared by specialist drafters. Consequently, the panel recommended that food standards always be drafted with the understanding that they are intended to be enforceable legal documents and that where current deficiencies in the labelling requirements have been identified; standards should be re-drafted to make the obligations clear.[59]

Under the current legislation, formal enforcement actions for breaches of food labelling are mainly limited to prosecutions, which depend upon prosecutorial discretion and resources, as well as being time-consuming to mount. The panel recommended a more versatile range of enforcement provisions, such as the power to make orders or require user-paid compliance testing consequent on a breach or impose enforceable undertakings in relation to non-compliant labelling.[60]

The problem of inconsistency is exacerbated by the many agencies involved. For example, across Australia there are as many as 29 authorities and agencies in some way responsible for the regulation of food and its labelling. Each of the States and Territories has a principal agency and in many cases a number of primary production authorities are also involved The Federal Government is involved through its quarantine service and the Australian Customs and Border Protection Service in the case of imported food. In New Zealand, there is the NZFSA as the principal agency with the New Zealand Customs Service and MAF Biosecurity New Zealand also having responsibility for imported foods at the border.[61] The panel concluded that: If food labelling is to be taken seriously by governments, a new entity, which for the purposes of this Review is called the Food Labelling Bureau (the Bureau), should be established to advise Australian and New Zealand ministers on all aspects of labelling policy. Resources for this Bureau must reflect the high profile of food labelling as the most public face of food policies, standards and laws.[62]

The following specific functions were suggested for the Bureau; it should:

- be the primary source of food labelling advice to the Ministerial Council, industry, government and the community in relation to the operation of the Code and existing and emerging food labelling issues and technologies;
- undertake or commission research relating to new and existing issues in food labelling;
- educate and inform consumers and industry about labelling requirements and other nutrition and public health initiatives relevant to labelling; assist regulators with compliance; and assist FSANZ with the development and review of labelling standards as necessary;

---

[59] Blewett et al. (2011), p. 138.
[60] Id., p. 134, Recommendation 58.
[61] Id., p. 135, Para 8.14.
[62] Id., p. 139, Para 8.26.

- provide information and guidelines that will assist industry to comply with current requirements and support the development and operation of compliance tools (such as computer-generated labels or pre-approvals);
- be a clearing house for complaints made to it, facilitating their resolution where possible and referring matters to the appropriate jurisdiction for formal enforcement where necessary;
- monitor and report on food labelling compliance across jurisdictions (e.g., for nutrition, health and related claims, compliance with NIP and FoPL requirements); and oversee self- and co-regulatory arrangements; and
- monitor consumer values issues claims on food labels and provide a point of contact between the ACCC in Australia and the NZCC and other relevant agencies in relation to food labels that are potentially misleading or deceptive under consumer protection laws.[63]

Government Response to Report of the Review of Food Labelling Law and Policy The government released its response to the Report of the Review of Food Labelling Law and Policy on 9 December 2011.[64] The Forum considered that a number of the labelling recommendations of the review panel had been anticipated by industry action and that it should have a watching brief to see how these initiatives developed. The principal responses concerning positive action in response to the review report are outlined below.

39.8 Role of FSANZ

In response to the review panel's report of ambiguity regarding the role of Food Standards Australia New Zealand (FSANZ) in developing and reviewing food standards, particularly whether it had a role in relation to broader public health issues, the Legislative and Governance Forum ("the Forum") made several proposals. First, a Ministerial Policy Guideline will be developed detailing the expectations of FSANZ in relation to the role of food standards in supporting public health objectives. The aim of the Ministerial Policy Guideline is to decrease ambiguity regarding the role of FSANZ in developing and reviewing food standards. Additionally, it would require that FSANZ take into consideration both long-term health impacts and immediate health risks in the development of food standards while not changing the overall function and role of FSANZ. The Forum proposed that in 2 years the Ministerial Council Guideline would be evaluated to determine whether it has been effective in addressing the issue or whether there is still a need to include a definition of 'public health' in the FSANZ Act.[65]

---

[63]Id., p. 141, Recommendation 61.

[64]Legislative and Governance Forum on Food Regulation on Food Regulation (Convening as the Australia and New Zealand Food Regulation Ministerial Council) 2011.

[65]Legislative and Governance Forum on Food Regulation, pp. 9–10.

39.8.1 Food Labelling Principles

The Forum endorsed the review's proposals in relation to the clarity of food labelling and stated that "it will develop an overarching policy statement on food labelling that supports the principle that information on food labels be presented in a clear and comprehensible manner to enhance understanding. The policy statement will guide decisions and actions by both government and industry." There is widespread recognition of the importance of optimal comprehension and readability of information that appears on food labels. However, the Forum noted that there are some design and presentational challenges in developing label information that is understood "across all levels of the population."[66]

39.8.2 Food Safety Labelling

The Forum indicated that it would request that FSANZ undertake a technical evaluation to ensure consumers' ability to access relevant information. The review would aim to maximize the effectiveness of food safety communication and provide advice on the food safety elements on food labels.[67]

39.8.3 Nutrition Policy

The Forum sought the advice of the Standing Council on Health which supports the development of a comprehensive national nutrition policy and has referred the development of this policy to the Australian Health Ministers' Advisory Council (AHMAC).[68]

39.8.4 Ingredient List and Nutrition Information Panel

The Forum indicated that it would request FSANZ "to undertake a technical evaluation and provide advice on the proposed changes to the ingredient listing and Nutrition Information Panel." This advice would assist the Forum in its consideration of the expected benefits and cumulative impacts of possible changes to labelling requirements prior to considering any amendments to the Food Standards Code.[69]

---

[66]Id., p. 14.
[67]Id., p. 16.
[68]Id., p. 19.
[69]Id., p. 21.

39.8.5 Mandatory Health Messages

The Standing Council on Health advised the Forum that the majority of Health Ministers supported mandatory preventive health messages on food labels being instigated by government. However, some Health Ministers would prefer to address health messages on labels in the context of specific proposals rather than through a generic approach.[70]

39.8.6 New Technologies

The Forum agreed not to pursue recommendation 28 of the review panel which concerned mandatory labelling where new technologies were involved in food production because it took the view that this requirement for new technologies constituted a technical regulation under the World Trade Organization Agreement on Technical Barriers to Trade (TBT Agreement). It preferred FSANZ to continue to apply a case-by-case approach to the labelling requirements for new technologies.[71] The Forum undertook to develop a conceptual framework and issues hierarchy that will include a Ministerial Policy Guideline for the case-by-case consideration of regulatory (i.e. labelling) and non-regulatory measures applying to food produced using a new technology requiring pre-market safety assessment.

39.8.7 Presentation

The Forum accepted the recommendation of the review panel and indicated that it would request FSANZ to undertake a technical evaluation and provide advice on the application of the Perceptible Information Principle to the presentational aspects of food labels, as well as whether the Perceptible Information Principle as a tool to aid food label design has benefits over other tools.[72]

In relation to the recommendations concerning proposed font sizes and a minimum contrast level of 70 % the Forum took the view that these should not be taken up as highly prescriptive requirements may place considerable design limitations on industry, and may lead to an information density that reduces readability for consumers.[73]

---

[70] Id., p. 27.
[71] Id., p. 34.
[72] Id., p. 46.
[73] Id., p. 47.

39.8.8 Compliance and Enforcement

The Forum undertook to explore a range of policy options for increasing the capacity of the food regulatory system to monitor and enforce food laws in a way that met stakeholders' expectations. It expressed it support for the existing risk-based approach to monitoring and enforcement and indicated that it would request the Food Regulation Standing Committee (FRSC) to consider the work already completed by the ISC and the range of potential enforcement tools that could be available to address noncompliant labelling and make recommendations to the Forum regarding amendments to the *Model Food Provisions* to be adopted by all jurisdictions.[74]

39.8.9 Clarity of Standards

The Forum agreed with the review panel that the Food Standards Code should be clear, unambiguous and legally enforceable. However, it declined to initiate new action on this subject as issues with the wording and enforcement of the Food Standards Code had been previously recognised by the then Ministerial Council and action commenced in a number of areas. An audit of the Food Standards Code was conducted by the Australian Government's Office of Legislative Drafting and Publishing in 2010. An implementation plan to address the identified deficiencies had been developed by FSANZ.

The newly established Food Standards Code Interpretation Service (CIS) within FSANZ commenced operation on 1 July 2011. The CIS will provide public interpretive advice for chapters 1 and 2 of the Food Standards Code. The interpretive advice issued by the CIS is to be adopted and applied by relevant State and Territory food regulatory enforcement agencies in the course of their monitoring and enforcement activities. This interpretive advice is to be issued as guidelines to assist the interpretation of the Food Standards Code.[75]

39.8.10 Food Labelling Bureau

The Forum declined to accept the recommendation of the review panel that a trans-Tasman Food Labelling Bureau be established under the *Food Standards Australia New Zealand Act 1991*. The Forum took the view that this bureau would establish another bureaucratic layer to the food regulation system, without providing any

---

[74]Id., pp. 57–58.
[75]Legislative and Governance Forum on Food Regulation, p. 60.

additional capacity for enforcement.[76] The Forum explained that the recently established Food Standards Code Interpretation Service (CIS), operated by FSANZ, was designed to address the inconsistent interpretation and enforcement of labelling standards between jurisdictions. A key aim of this service is to facilitate consistent implementation of food standards by relevant regulatory agencies and industry across jurisdictions and local governments. The effectiveness of the CIS was set to be reviewed after 2 years of operation.

The Forum also cited to the work of the ISC in promoting a consistent approach to the implementation of food regulation across the jurisdictions. As part of this role, ISC has developed a framework entitled the *Strategy for the consistent implementation of food regulation in Australia*, which has facilitated the drafting of food standards that are legally enforceable in both jurisdictions.[77]

39.9 Country of Origin Rules in Australia

Country of origin requirements for food labels in Australia are contained in the *Food Standards Code* ("the Code"), the *Commerce (Trade Descriptions) Act 1905*, the *Commerce (Imports) Regulations 1940* (the Regulations), the *Australian Consumer Law* (ACL) and schedule 2 of the *Competition and Consumer Act 2010* (CCA). The Code and the Regulations require packaged and particular unpackaged foods to have labels that denote country of origin.

The ACL does not specifically include country of origin obligations for food labels. However, it prohibits a person from engaging in misleading or deceptive conduct[78]; making false or misleading representations that a good has had a particular history[79]; or making false or misleading representations concerning the place of origin of goods.[80] Section 255 of the ACL provides a "safe harbour" for producers by providing that a representation as to the country of origin of goods (such as "Made in Australia") will not be considered misleading so long as the goods have undergone a "substantial transformation" in that country. A substantial transformation is said to occur when the goods undergo a "fundamental change in form, appearance or nature," so that the goods after the transformation are new and different to those existing before; and 50 % or more of the cost of production is attributable to processes that occurred in that country.[81] The safe harbour provisions also impose stricter requirements for "Product of..." and "Grown in..." claims, and allow claims about the origin of particular ingredients. Confusion was identified by the Blewett review arising from "Made in..." claims where the food

---

[76]Legislative and Governance Forum on Food Regulation, p. 61.
[77]Legislative and Governance Forum on Food Regulation, pp. 61–62.
[78]Australian Consumer Law Section 18.
[79]ACL Section 29(a).
[80]ACL Section 29(k).
[81]The method of calculating the "cost of production" is set out in s 256 of the ACL.

comes from outside the country, but the safe harbour requirements are met in Australia.

The Blewett review recommended amending the country of origin rules by moving the country of origin rules from the Code to the CCA[82] and basing the "safe harbour" for country of origin claims on the weight of goods (excluding water) rather than costs of production.[83] This would focus the safe harbour on where the food itself came from, compared with where it was packed or prepared. The panel believed that this would lead to country of origin information that would be more in line with consumer expectations.

39.9.1 Senate Select Committee on Australia's Food Processing Sector, 2011–2012

On March 24, 2011, the Senate established the Select Committee on Australia's Food Processing Sector (Select Committee) to investigate possible policy responses to the challenges and pressures within the broader economy that threaten the ongoing viability and competitiveness of food processing in Australia. The committee was also tasked with examining certain broader areas of government policy to assess the appropriateness of the overall regulatory environment Australia's food processing industry operates in. Following public hearings and site visits, the select committee reported to the Senate in August 2012.

In view of the recent publication of the report of the panel chaired by Dr. Blewett,[84] the Select Committee confined its remarks on the subject of food labelling mainly to country of origin labelling requirements.[85]

39.9.2 Country of Origin Labelling Regime

The Senate Select Committee's report noted that, essentially, the Blewett Review had recommended creating a specific regime regulating food product country of origin labelling claims that would operate separately from the general laws governing country of origin claims for other products, such as manufactured goods. The government's response explicitly rejected the idea of creating a separate regime governing country of origin claims for food, largely because of increased expenses for industry in implementing these changes.[86]

---

[82]Blewett et al. (2011), p. 109, Recommendation 41.
[83]Id., p. 111, Recommendation 42.
[84]Blewett et al. (2011).
[85]Senate Report, p. 63.
[86]Senate Select Committee (2012) at paras 4.26-32.

The Select Committee disagreed with the government's rejection of the Blewett proposal, and posited the rejection as a misunderstanding of consumer and industry expectations regarding the purpose and clarity of country of labelling laws with respect to food. Therefore, the select committee supported the case for simplifying and clarifying these laws.[87]

On the basis of the evidence presented to it, the select committee observed that "that food labelling issues, particularly to do with country of origin requirements, loom large in the minds of many industry participants,"[88] but that the current laws "are not at all transparent and potentially mislead consumers."[89] The Select Committee considered several options for reform, including adjusting the existing labelling regime to make labels less confusing or to provide greater detail, educating consumers, and using technology to better connect consumers with the food they were purchasing.[90] The Senate Select Committee concluded that

> ...there would be merit to reforming the current country of origin labelling laws to make them more transparent. ... Industry must do more to understand consumer preferences and behaviour. Government can assist this by providing a strong and clear country of origin labelling regime upon which processors can more confidently base their claims.[91]

The Committee recommended "that the government expand the application of food labelling requirements to require all primary food products for retail sale to display their country of origin, in accordance with recommendation 40 of the Blewett Review."[92] They further required that "the government reform country of origin labelling requirements for food so that these requirements are clearer, more transparent and focus on the consumer's understanding."[93] Finally, the select committee recommended that "the government move mandatory country of origin labelling requirements for food to a specific consumer product information standard under the *Competition and Consumer Act 2010*, consistent with recommendation 41 of the Blewett Review."[94]

Essentially, the Select Committee endorsed the proposals of the Blewett report, recognizing the need for clear and detailed country of origin information.

---

[87]Id. at para 4.34.
[88]Id. at para 4.55.
[89]Id. at para 4.59.
[90]Id. at para 4.62-.83.
[91]Id. at para 4.84.
[92]Id. at para 4.88.
[93]Id. at para 4.97.
[94]Id. at para 4.111, Recommendation 12.

39.9.3 The Competition and Consumer Amendment (Australian Country of Origin Food Labelling) Bill 2013 (Cth)

Following the rejection by the Senate Standing Committee of the Blewett committee's recommendations for reform of the Australian country of origin regime, bills were introduced into the Australian Senate and the House of Representatives, by the minority Australian Greens Party attempting to reinstate Recommendations 41 and 42 of the Blewett Committee.[95] Those bills were subsequently abandoned, and on May 16, 2013, the leader of the Australian Greens Party, Senator Milne, introduced the Competition and Consumer Amendment (Australian Country of Origin Food Labelling) Bill 2013 (Cth) (the Bill).[96] The Bill includes the new country of origin requirements in the CCA which were contained in Recommendation 41 of the Blewett Review, but it dropped the "weight of the goods" rule contained in Recommendation 42.

The Bill outlines a new Part 3-4A in the CCA for food-specific country of origin claims, which overrides overlapping requirements in the Code. Under Part 3-4A, one of the following country of origin claims must be used for packaged and certain unpackaged foods:

- "Product/Produce of Australia:" a premium claim for packaged food wholly manufactured or processed in Australia, where all the significant ingredients are grown in Australia;
- "Australian Manufactured/Manufactured in Australia:" for packaged food "substantially transformed" in Australia, where at least 50 % of the total cost of processing is incurred in Australia (instead of "Made in Australia" as the safe harbour allows);
- "Packaged in Australia:" for packaged food (but not, for example, "Packaged in Australia from local and imported ingredients").

Similar to the s 255 ACL safe harbour, the Bill allows significant ingredients to be labelled with an Australian origin, and would also allow "Grown in Australia" to be used in connection with certain unpackaged foods. Finally, the Bill creates an offence for failing to comply with the labelling requirements, with criminal penalties of $250,000 for corporate bodies and $50,000 for individuals.

The Bill failed to obtain parliamentary approval prior to the Australian Parliament being dissolved for the September 2013 parliamentary election, but it is an indication of likely future legislative developments on the subject of country of origin labelling requirements.

---

[95]Competition and Consumer Amendment (Australian Food Labelling) Bill 2012 (Cth) and Competition and Consumer Amendment (Australian Food Labelling) Bill 2012 (No 2) (Cth).

[96]Competition and Consumer Amendment (Australian Food Labelling) Bill 2013 (Cth).

39.10 Food Law Jurisprudence

Outlined below are the principal food law cases which have been decided in Australia and New Zealand over the past 10 years.

39.10.1 Interpretation of the Food Standards Code

Tumney (NSW Food Authority) v Nutricia Australia Pty Ltd[97] concerned the consideration by the Supreme Court of New South Wales of 92 criminal charges in relation to alleged breaches of the Food Act 2003 (NSW) by Nutricia Australia Pty Ltd ("Nutricia") and its principal officers. Nutricia was an Australian company that manufactured and supplied infant baby milk formula products. Two components of the formula in question were fructo-oligosaccharides and galacto-oligosaccharides. The Prosecutor argued that the products were packaged and labelled in such a way as to represent that they were of a particular nature or substance, and that this amounted to a false description of the food.

Section 18(3) of the Food Act provided that "a person must not, in the course of carrying on a food business, sell food that is packaged or labelled in a way that falsely describes the food." Section 22 (a) gave as an example of food which was falsely described "food ... represented as being of a particular nature or substance for which there is a prescribed standard under the Food Standards Code and the food does not comply with that prescribed standard ..."

Section 21 of the Food Act provided:

1. A person must comply with any requirement imposed on the person by a provision of the Food Standards Code in relation to the conduct of a food business or to food intended for sale or food for sale.
2. A person must not sell any food that does not comply with a requirement of the Food Standards Code that relates to the food.
3. A person must not sell or advertise for sale any food that is packaged or labelled in a manner that contravenes a provision of the Food Standards Code.

The Court pointed out that the "*Food Standards Code*" referred to in the legislation was the *Australia New Zealand Food Standards Code*. "The Code is a national Code promulgating detailed standards in respect of the production, sale and importation of food, devised under a co-operative arrangement between the governments of Australia and New Zealand and, in NSW, given the force of law by the *Food Act* section 21."[98]

The trial Judge pointed out that the Code was "an unusual breed of instrument, something of a hybrid".[99]

---

[97] *Tumney (NSW Food Authority) v Nutricia Australia Pty Ltd* (2008) NSWSC 1382.
[98] Id. at para 48.
[99] Id. at para. 69.

It was developed and maintained by "Food Standards Australia New Zealand", which, as its name proclaims, is a joint instrumentality of Australia and New Zealand. So far as the Australian contribution is concerned, it is authorised by the Food Standards Australia New Zealand Act 1991 (Cth). But enforcement and policing of food standards are within the constitutional realm of the states. For that reason, the Code is given the force of law by the Food Act of NSW.

Accordingly, in my view … construction of the Code falls to be determined by reference to NSW interpretation law. That is because it is given the force of law by a state Act, and this is a prosecution brought under state law, governed by rules of evidence and interpretation of NSW law.

In construing the Code, it seems to me to be appropriate to bear in mind that, while it has legal force, it is a document drafted, apparently, by non-lawyers. That being so, it may be appropriate to take a more liberal, or more purposive, approach to its construction.[100]

Noting that, in a criminal prosecution, any ambiguity must be resolved in favour of the defendants and against the Prosecutor, the Judge dismissed that charges because of the imprecise and ambiguous wording of the relevant provisions of the Code.

Axiome Pty Ltd on behalf of Cognis GmbH v Food Standards Australia New Zealand[101] concerned a review of a decision of a delegate of Food Standards Australia New Zealand (FSANZ). The delegate rejected an application to amend Standard 1.5.1 of the Australia New Zealand Food Standards Code (the Code) to approve the use of a chemically defined mixture of conjugated linoleic acid triglycerides (CLA), Tonalin CLA, as a novel food in Australia. Standard 1.5.1 of the Code regulates the sale of 'novel foods' and novel ingredients and prohibits the sale of these foods unless they are listed in the Table to clause 2 of Standard 1.5.1 and comply with any special conditions of use. Failure to comply with Standard 1.5.1 results in significant criminal offences and fines being imposed under the various State and Territory *Food Acts*.

The application stated that Tonalin CLA "meets the definition of a non-traditional food and a *Novel Food*," and that its purpose was as "a useful adjunct in weight control programmes and diets" (original emphasis). Examples of its potential food applications "include milk products, soy beverages, fruit based beverages, yoghurt and yoghurt products, nutrition bars, and table spreads." On May 13, 2011, the application was rejected by FSANZ pursuant to s 30(1)(b) of the *Food Standards Australia New Zealand Act 1991* (Cth). Among the stated reasons for the rejection were that the overall evidence base was not sufficient to demonstrate the safety of Tonalin CLA at the recommended intake of 4.5 g/day.

On June 6, 2011, the Applicant applied to the Administrative Appeals Tribunal for a review of this decision. The Tribunal has jurisdiction to review such decisions pursuant to s 143(1) of the FSANZ Act. The Tribunal identified the issues as, first, whether Tonalin CLA was a "novel food" as defined in clause 1 of Standard 1.5.1 of

---

[100] Id. at para. 69–71.

[101] *Axiome Pty Ltd on behalf of Cognis GmbH v Food Standards Australia New Zealand* (2012) AATA 551.

the Code and, second, whether FSANZ's decision to reject the application for variation of Standard 1.5.1 was the correct or preferable decision. On the first issue, the Tribunal ruled that there was insufficient evidence to find that Tonalin CLA was not a non-traditional food.

On the second issue, whether the decision of FSANZ that it would not vary Standard 1.5.1 is the correct and preferable decision, the Tribunal observed that FSANZ must give regard to the object of the FSANZ Act stated in Section 3 "to ensure a high standard of public health protection." This goal is specifically reinforced by Section 18(1), which sets out the objectives of FSANZ in developing or reviewing food regulatory measures and variations of such measures, the most important being "the protection of public health and safety." In performing such functions, pursuant to Section 18(2), FSANZ must have regard, amongst other matters, to "(a) the need for standards to be based on risk analysis using the best available scientific evidence," and "(e) any written policy guidelines formulated by the Council for the purposes of this paragraph and notified to the Authority."

In assessing an application, FSANZ is also required to have regard to the matters set out in Section 29(2), including whether the costs that would arise relevantly from a variation of a food regulatory measure outweigh the direct and indirect benefits to the community.

The Tribunal ruled that, having regard to the protection of public health and safety, the test to be applied was whether the best available scientific evidence satisfied that Tonalin CLA was safe for human consumption as an additive to food. In this regard, the Tribunal found the weight of evidence suggested that Tonalin CLA required further assessment of the public health and safety considerations, having regard to the potential for adverse effects in humans.

Consequently, the Tribunal affirmed the decision to reject application for the amendment of Standard 1.5.1 of the Code in order to approve the use of Tonalin CLA as a novel food in Australia.

39.10.2 Misleading Conduct in Relation to Food Advertising

Australian Competition and Consumer Commission v Nudie Foods Australia Pty Ltd[102] concerned the labelling of fruit juice products described as cranberry juice through the use of pictures and text. The Federal Court of Australia found this to be misleading because less than 20 % of the drink was cranberry juice. The balance of the drink was composed of apple juice. The Court ruled that Sections 52 and 53 of the Trade Practices Act had been breached.[103] The Court issued an injunction

---

[102]*Australian Competition and Consumer Commission v. Nudie Foods Australia Pty Ltd* (2008) FCA 943.

[103]Now ss. 18 and 29 of the *Australian Consumer Law* which appears as Schedule 2 to the *Competition and Consumer Act 2010*.

restraining the defendant for a period of 3 years from supplying, or marketing, or causing to be supplied or marketed, any products bearing labels or packaging that use a product name or description that includes the name of a fruit but does not in the product name or description either:

(a) identify all of the fruits used in making the product; or
(b) alert the consumer to the fact that fruit or fruits other than those identified in the product name or description were used in making the product.

The Court further ordered corrective advertising and required the defendants' establishment of an education training and trade practices compliance program designed to minimize the risk of future consumer protection infringements.

A similar case was *Australian Competition & Consumer Commission v Cadbury Schweppes Pty Ltd*[104] which involved the public sale of a flavoured cordial with a label containing the words "banana mango flavoured cordial." This phrase was coupled with pictorial representations of bananas and mangoes and a logo containing a caricature of a monkey with a half-peeled banana and an unpeeled banana. However, the cordial did not contain bananas or mangoes. The pictorial representations were found to be likely to mislead or deceive, contrary to Section 52(1) of the Trade Practices Act 1974 (Cth). The false representation that the cordial was of a particular composition was found contrary to Section 53(a) of the Act. Finally, engaging in conduct that was liable to mislead the public as to the nature or characteristics of the cordial was found contrary to Section 55 of the Act. Conclusion.

Australian Competition and Consumer Commission v Harvey Fresh (1994) Limited[105] concerned representations on packaged cheese that indicated it was produced in Western Australia. In reality, the cheese was produced in the State of Victoria. This was found to be misleading conduct in breach of Section 52 of the Trade Practices Act 1974(Cth). It was also found to be a false representation as to the origins of the cheese in breach of Section 53(eb) of the Act. The respondent was enjoined from further advertising of this nature and it was required to publish corrective advertisements. Its staff was also required to undergo compliance training.

---

[104]*Australian Competition & Consumer Commission v Cadbury Schweppes Pty Ltd* (2004) FCA 516.

[105]*Australian Competition and Consumer Commission v Harvey Fresh (1994) Limited* (2009) FCA 853.

39.10.3 Sanctions for False and Misleading Statements Used in Food Advertising

In *Australian Competition and Consumer Commission v Turi Foods Pty Ltd (No 2)*[106] the Federal Court found that advertising materials for the 'La Ionica' brand of chickens falsely represented the conditions in which the chickens were raised. Over a period spanning between 2004 and 2010, Turi Foods published various advertising posters and painted murals on the side of its delivery trucks, both of which represented that the "La Ionica" chickens were raised in barns where they had substantial space to roam around freely. The artwork included statements such as "free roaming" and "no cages." In reality, Turi Foods' production barns held up to 18 chickens per square metre (reducing to 12 chickens per square metre as they grew in size). The ACCC argued that the representations of the conditions in its barns were false, misleading, or deceptive in breach of sections 52 and 53 of the Trade Practices Act 1974 (TPA) in respect of conduct which had taken place prior to 1 January 2011 and sections 18 and 29(1)(a) of the *Australian Consumer Law* (ACL) which appears as Schedule 2 to the *Competition and Consumer Act 2010* (Cth) (CCA). Turi Foods accepted the ACCC's argument. The CCA replaced the Trade Practices Act 1974 on 1 January 2011. The Court ordered that Turi Foods pay a fine of $100,000, publish corrective advertisements, and implement a compliance training program.

Australian Competition and Consumer Commission v Turi Foods (No 4)[107] concerned similar findings against other defendants concerning false and misleading advertising that chickens were "free to roam" in barns or sheds during their growth cycle. The defendants appealed against a number of the penalties imposed by the Court, including declarations under Section 21 of the *Federal Court of Australia Act 1976* (Cth), injunctions under Section 232 of the ACL; pecuniary penalties under Section 76E of the TPA and Section 224 of the ACL; orders for publication and disclosure under Sections 86C and 86D of the TPA and under Sections 246 and 247 of the ACL; orders for implementation of compliance programmes under Section 86C of the TPA and Section 246 of the ACL. The Court affirmed the various remedies which had been ordered by the court below.

Australian Competition and Consumer Commission v Turi Foods Pty Ltd (No 5)[108] addressed the principles that courts apply in quantifying penalties, including the necessity for deterrence. These principles include:

- the size of the contravening company;
- the deliberateness of the contravention and the period over which it extended;

---

[106] *Australian Competition and Consumer Commission v Turi Foods Pty Ltd (No 2)* (2012) FCA 19.
[107] *Australian Competition and Consumer Commission v Turi Foods (No 4)* (2013) FCA 665.
[108] *Australian Competition and Consumer Commission v Turi Foods Pty Ltd (No 5)* (2013) FCA 1109.

- whether the contravention arose out of the conduct of senior management of the contravener or at some lower level;
- whether the contravener has a corporate culture conducive to compliance with [the ACL] as evidenced by educational programmes and disciplinary or other corrective measures in response to an acknowledged contravention;
- whether the contravener has shown a disposition to cooperate with the authorities responsible for the enforcement of [the ACL] in relation to the contravention;
- whether the contravener has engaged in similar conduct in the past;
- the financial position of the contravener; and
- whether the contravening conduct was systematic, deliberate or covert.[109]

Australian Competition and Consumer Commission v Pepe's Ducks Ltd[110] provided a recent illustration of the sanctions which courts impose in relation to false or misleading advertising in breach of the federal Australian consumer protection laws. The Respondent (Pepe's Ducks) acknowledged its liability for falsely advertising its duck meat products as "Grown Nature's Way" or an Outdoor "Barn Raised" as well as packaging indicating that the ducks were raised in other than factory conditions. It was ordered at its own expense to:

(a) establish and maintain for 3 years a Trade Practices Compliance Program which met set out in an annexure to the court order;
(b) within 14 days of the date of the court order to send a notice in the form of annexed to the court order to each customer that it is able to identify from its commercial records as being a customer to whom it has supplied Duck Meat Products during the 2 year period prior to the date of this order;
(c) for a period of 90 days, publish on its Australian website homepage and on each of the internal webpages accessible by following the 'click-through' icons which appear on the Pepe's Ducks Homepage a notice in the terms and form annexed to the court order;
(d) within 14 days of the date of the order display a notice in the form prescribed by the Court at the front of each of its business premises which shall be viewable to the public, and keep that notice on display for at least 90 days.

Additionally a pecuniary penalty of $375,000 was imposed.

References

Blewett N et al (2011) Labelling logic. Review of food labelling law and policy. Commonwealth of Australia, Canberra

---

[109] Applying *Australian Competition and Consumer Commission v Singtel Optus Pty Ltd (No 4)* (2011) FCA 761; (2011) 282 ALR 246 at 11.

[110] *Australian Competition and Consumer Commission v Pepe's Ducks Ltd* (2013) FCA 570.

Legislative and Governance Forum on Food Regulation (convening as the Australia and New Zealand Food Regulation Ministerial Council) (2011) 'Response to the Recommendations of Labelling Logic: Review of Food Labelling Law and Policy (2011)' available at http://www.foodlabellingreview.gov.au/internet/foodlabelling/publishing.nsf/Content/ADC308D3982EBB24CA2576D20078EB41/$File/FoFR%20response%20to%20the%20Food%20Labelling%20Law%20and%20Policy%20Review%209%20December%202011.pdf

Preventative Health Taskforce (PHT) (2009) Australia: The Healthiest Country by 2020 – National Preventative Health Strategy – the roadmap for action. Commonwealth of Australia, Canberra

Senate Select Committee on Australia's Food Processing Sector (2012), Australia's food processing sector, Report, 16 August Commonwealth of Australia, Canberra

Taskforce on Industry Self-Regulation (2000) Industry self-regulation in consumer markets. Commonwealth of Australia, Canberra

Yan W (2008) International peer review of FSANZ GM food safety assessment process. FSANZ, Canberra. <http://www.foodstandards.gov.au/_srcfiles/GM%20Peer%20Review%20Report.pdf>

Chapter 40
Present & Future Jurisprudence of Consumer Protection and Food Law in Australia

Alex Bruce

Abstract In this chapter, I explore how the new *Australian Consumer Law* ("the ACL") embedded in the *Competition and Consumer Act 2010* (Cth) is intended to protect consumers from misleading, deceptive and false credence claims associated with food products made by large corporate food producers and distributors. By "credence claims" I mean representations made by corporations in marketing food products that convey to consumers an impression that the product possesses some added quality that similar products may not possess. Credence claims are often associated with values-choices made by consumers such as "gluten free", "not tested on animals" or "free range". The differentiating quality may also appeal to the consumer's ethical or religious values associated with food production and consumption.

40.1 Introduction

In this chapter, I explore how the new *Australian Consumer Law* ("the ACL") embedded in the *Competition and Consumer Act 2010* (Cth) is intended to protect consumers from misleading, deceptive and false credence claims associated with food products made by large corporate food producers and distributors. By "credence claims" I mean representations made by corporations in marketing food products that convey to consumers an impression that the product possesses some added quality that similar products may not possess. Credence claims are often associated with values-choices made by consumers such as "gluten free", "not tested on animals" or "free range". The differentiating quality may also appeal to the consumer's ethical or religious values associated with food production and consumption.

A. Bruce (✉)
Australian National University, College of Law, Canberra, ACT, Australia
e-mail: alex.bruce@anu.edu.au

The interest in but also potential vulnerability of consumers to misleading credence claims associated with the production and marketing of food products has assumed a high level of importance in Australia since 2011. There are two principal reasons for this. First, a significant Commonwealth (Federal) government review into Food Labelling Regulation recommended that "values claims" (credence claims) associated with food products should be regulated by the *Competition and Consumer Act 2010* (Cth) ("the CCA") and, in particular the new *Australian Consumer Law* ("the ACL") that sits within the CCA. Instead of direct regulation of food marking practices through specific legislation, the Commonwealth government intends the CCA generally, and the ACL specifically to assume a much more important profile in ensuring the fair, competitive and informed marketing of food products.

Secondly, and related to the first, the Commonwealth regulator of the CCA, the *Australian Competition and Consumer Commission* ("the ACCC") has specifically made credence claims a National Enforcement Priority. Since 2011, the ACCC has instituted legal proceedings against several major food producers alleging breaches of the ACL in relation to credence claims associated with food products made by those producers.

Throughout this chapter, I will use examples of credence claims associated with the production and marketing of food animal products, eggs particularly, to demonstrate how the ACL has assumed a greater role in the regulation of food production and marketing in Australia. The chapter analyses several important decisions of the Federal Court of Australia concerning credence claims associated with food products. In doing so, the chapter identifies both the strengths and weaknesses of the ACL as the principal mechanism intended to regulate food production and marketing claims.

I argue that although this litigation does suggest the ACL functions well in capturing misleading or deceptive food product credence claims, it is only addressing the symptoms of a larger underlying dysfunction associated with the world-wide food production industry. While large agribusinesses dominate markets for food products generally and food animal products particularly, they will continue to face an intractable dilemma.

The nature of this underlying dilemma is this: On the one hand, agribusinesses have become reliant on large-scale, intensive farming and animal husbandry practices to meet demand for their products. However, on the other, those same agribusinesses are increasingly relying on clever advertising and marketing involving credence claims to exploit the growing interest shown by consumers and for whom such claims are important.

These strategies are largely antithetical. It is almost impossible for a large agribusiness processing chickens and eggs (for example) using intensive husbandry practices to simultaneously market chicken meat and eggs with "free range" or "free to roam" credence claims. As recent and frequent Australian litigation demonstrates, credence claims of this nature are simply not sustainable within the context of current corporate agricultural and animal husbandry practices.

In avoiding this dilemma, and associated economic externalities such as litigation, damages, corrective advertising and court-sanctioned compliance programs, agribusinesses need to explore alternative ways of food production that are not fundamentally damaging to the sustainability of our environment and to the human and animals inhabiting this fragile ecosystem. My chapter therefore concludes that in the future, litigation under the ACL involving credence claims may be avoided if agribusinesses implement alternative food production practices that emphasise nanotechnology-based artificial photosynthesis technologies forming the basis for a transition from anthropocentric-oriented food production systems to "sustainocene – oriented" food production systems.

This chapter is intentionally written for well-informed readers who are not necessarily familiar with the Australian legal system, the *Competition and Consumer Act 2010* (Cth), the *Australian Consumer Law* or the *Australian Competition and Consumer Commission*. Accordingly, the chapter gently introduces readers to these important pieces of legislation before embarking on more substantive analysis exploring how the CCA and the ACL have been utilised by the ACCC and then applied by the Courts in addressing misleading or deceptive credence claims associated with food products.

My discussion in this chapter proceeds in the following manner. Following this Introduction, Sect. 40.2 introduces the economic and regulatory context to food regulation and consumer protection in Australia. It identifies the market forces that have resulted in corporate domination of the Australian food industry and explains how those corporations are employing credence claims associated with food products generally and food animal products particularly, to capture market share. In Sect. 40.3 it discusses growing consumer preferences for food product credence claims and demonstrates how and why Australian consumers are vulnerable to exploitation by large corporate producers and suppliers of food products.

Importantly, this discussion underscores the intractable dilemma faced by food producers and suppliers in marketing credence claims discussed in Sect. 40.1; that it is almost impossible for a large agribusiness processing chickens and eggs (for example) using intensive husbandry practices to simultaneously market chicken meat and eggs with "free range" or "free to roam" credence claims. As recent and frequent Australian litigation demonstrates, credence claims of this nature are simply not sustainable within the context of current corporate agricultural and animal husbandry practices.

In Sect. 40.4, the chapter introduces the *Labelling Logic Report* and the Commonwealth government's response to the *Report*, giving the *Australian Consumer Law* a greater role in addressing corporate exploitation of consumer preferences for credence claims associated with food products. The nature, role and method by which the *Competition and Consumer Act 2010* (Cth) and the *Australian Consumer Law* have assumed a greater role in regulating food product claims is then set out in Sect. 40.5 and explains the mechanism by which 18 of the ACL prohibits misleading and deceptive conduct.

Having established the regulatory, economic and legal framework, Sect. 40.6 turns to an analysis of recent Federal Court litigation instituted by the *Australian*

Competition and Consumer Commission against food producers and suppliers. This litigation focuses on misleading and deceptive conduct associated with credence claims made about food animal products, particularly eggs. It demonstrates how the *Australian Consumer Law* can be deployed by the ACCC to address misleading or false food credence claims.

Section 40.7 concludes with a discussion of alternative agribusiness technology in the form of artificial photosynthesis. Employing sunlight to catalyse water into hydrogen and oxygen, as well as capturing atmospheric carbon-dioxide has the potential to revolutionise global agribusiness. Artificial photosynthesis processes would also enable hydrogen to be combined with nitrogen to produce ammonia, a staple input in agrifood production. In this way, artificial photosynthesis technology has the potential to go some way to resolving the dilemma facing agribusiness discussed earlier.

40.2 Economic and Regulatory Context

40.2.1 The Patchwork Nature of Food Law & Regulation in Australia

Regulating and managing a safe, competitive and informed relationship between Australian consumers on the one hand, and food retailers on the other, is problematic. With no express Constitutional power concerning food production, the creation of food standards, food product labelling or food product marketing, the Australian Commonwealth (Federal) government instead relies on a complex network of *Intergovernmental Agreements, Codes, Legislative Instruments, Standards* and other policy documents co-operatively enforced by the Commonwealth government, State and Territory governments and the government of New Zealand.

The starting point is the *Australia New Zealand Food Standards Code* ('the Food Code') administered through *Food Standards Australia New Zealand* ('FSANZ'). The Food Code itself is not a legislative document. However the Food Code includes certain *Food Standards* ('the Standards') that are given legal status as legislative instruments under the *Legislative Instruments Act 2003* (Cth). Industry compliance with the Standards is therefore mandatory.

In Australia, the Standards are then reinforced through the *Food Standards Australia New Zealand Act 1991* (Cth) ('the FSANZ Act') intended to prevent misleading and deceptive conduct by ensuring that consumers have adequate information to make informed food choices.[1] Actual day-to-day responsibility for the Standards is shared across the Commonwealth, State and Territory governments through a series of *Intergovernmental Agreements* assigning responsibility between the Commonwealth *Department of Agriculture, Forest and Fisheries*, the

---

[1] *Food Standards Australia New Zealand Act 1991* (Cth), s 3(c).

Australian Quarantine Inspection Service and State and Territory *Primary Industries* or *Health Departments*.

Importantly, *enforcement* of the Standards is the joint responsibility of the *Australian Competition and Consumer Commission* ('the ACCC') under the ACL and the *Australian Quarantine Inspection Service* ('AQIS').[2]

In this way, the Commonwealth government anticipates misleading or deceptive conduct associated with food labelling to be dealt with at the Commonwealth level by the ACCC and at State and Territory levels by relevant Fair Trading Departments through the ACL as it applies in their jurisdiction.[3]

40.2.2 Corporate Dominance of Food Production & Distribution

Food distribution in Australia is characterised by near duopoly, with Coles Supermarkets Pty Ltd and Woolworths Limited responsible for approximately 70 % of packaged grocery sales and 50 % of fresh food product sales in Australia.[4] This concentration of market power in the hands of two significant vertically integrated retailers enables them to exercise considerable up-stream and down-stream market power.[5]

In addition, most of the animal meat produced in Australia for both domestic consumption and export is processed by a few dominant corporations. The Australian chicken meat industry is a virtual duopoly. According to the Australian Chicken Meat Federation ('ACMF'): 'the two largest (companies) Baiada Poultry and Inghams Enterprises, supply more than 80 % of Australia's chicken meat'.[6]

The beef industry is dominated by four producers. Swift Australia, Cargill Australia, Teys Brothers and Nippon Meats supply almost 50 % of meat products in Australia.[7] And in 2011, the Australian Competition and Consumer Commission

---

[2]*Issues Consultation Paper: Food Labelling Law and Policy Review*, 5 March 2010, Food Labelling Secretariat, Canberra, Part 1, 1.3.

[3]Food Labelling Law and Policy Review Panel, *Labelling Logic: Review of Food Labelling Law and Policy*, 27 January 2011, Commonwealth of Australia, 97, [6.3].

[4]Australian Competition and Consumer Commission, *Report of the ACCC Inquiry into the Competitiveness of Retail Prices for Standard Groceries*, July 2008, Commonwealth of Australia at xv.

[5]Australian Competition and Consumer Commission, *Report of the ACCC Inquiry into the Competitiveness of Retail Prices for Standard Groceries*, July 2008, Commonwealth of Australia at xv.

[6]Australian Chicken Meat Federation Inc, *The Australian Chicken Meat Industry: An Industry in Profile*, 2012 at 13. http://www.chicken.org.au/industryprofile/. Accessed on 29 March 2012.

[7]*Top 25 Red Meat Processors*, 'Feedback', Meat & Livestock Industry Journal Supplement, October 2005.

('the ACCC') cleared a proposed acquisition of Teys Brothers by Cargill Beef Australia; an acquisition that permitted the merger of Australia's second and fourth largest beef processors leading to a further concentration of corporate production of animal food products.[8]

40.2.3 The Scale of Food Animal Production and Consumption

The quantities of chickens, cattle and pigs slaughtered and processed by these corporations each year is enormous. In 2010–2011, almost 500 million chickens were slaughtered[9] and by 2015, cattle processing is expected to reach 2.4 million tonnes.[10] To give these figures some perspective, and in relation to the 2010 figures, Meat and Livestock Australia report:[11]

> Over the 12 months to September 2010, fresh meat purchases increased 3 % to about 133 million serves/week. Contributing to the trend was a rise in beef (by 4 %), lamb (up 2 %) and chicken purchases (up 6 %) to 52 million serves/week, 22 million serves/week and 38 million serves/week, respectively.[12]

Australia also exports a significant amount of cattle and sheep to Muslim countries for slaughter according to religious 'halal' procedures. According to Animals Australia, some 22 million sheep and cattle have been exported to Kuwait alone over the past 20 years.[13]

Australia's exports of beef and sheep generally continued to grow throughout 2011. Beef exports during 2011 increased by 22 % to 165,000 tonnes when compared to the previous quarter with sheep exports increasing by 2 % to 644,000 tonnes.[14]

Most farmed animals in Australia for both domestic consumption and export are confined in Concentrated Animal Feedlot Operations ("CAFO's") described as:

---

[8]*ACCC will not Oppose Teys Bros and Cargill Beef Australia Proposed Merger*, ACCC Media Release dated 6 July 2011.

[9]2010 Australian Chicken Meat Federation: *Industry Facts and Figures*: http://www.chicken.org.au/page.php?id=4#Production.

[10]*Farm Facts*, 2011, National Farmers Federation, Canberra, Australia, 10.

[11]Meat and Livestock Australia (MLA) is a corporation whose members are Australian cattle producers. MLA is the corporate entity that acts as the cattle farmer's advocate in the development of Commonwealth primary industry policies. It also provides marketing and research on behalf of its member cattle farmers.

[12]Meat and Livestock Australia, *Australian Fresh Meat Consumption Increases*, 3 December 2010, on www.mla.com.au at *Prices & Markets*, then *Market News* at *Dec 2010* (cited 21 August 2011).

[13]Animals Australia, *Eye on Live Export*, on www.animalsaustralia.org (cited 21 August 2011).

[14]Australian Bureau of Statistics, 7215.0—Livestock Products, Australia, Dec 2011 at http://www.abs.gov.au/ausstats/abs@.nsf/lookup/7215.0main+features4Dec+2011 (cited 27 March 2012).

... a system of raising animals using intensive production line methods that maximise the amount of meat produced while minimising costs. Industrial animal agriculture is characterised by high stocking densities and/or close confinement, forced growth rates, high mechanisation and low labour requirements.[15]

Australian Animal Advocacy Group 'Voiceless' states that 'more than 5 million pigs, 13 million hens and 420 million meat or 'broiler' chickens are raised for food production in Australia every year. Most of these animals spend their lives crammed together in giant factory farms'.[16]

By 2050, the United Nations Population Division predicts that the world's population will reach somewhere between 8 and 11 billion people.[17] Much of this population growth will occur in developing countries where a growing middle class, with more disposable income is expected to generate substantial demand for meat products as part of their diet.[18] This is particularly so in China and India where demand for meat products is quickly growing.[19]

In order to meet this expected demand for food generally and meat products particularly, the United Nations Food and Agricultural Organisation estimates that agricultural output will need to increase by 70 % but must do so in circumstances of a world-wide decline in agricultural land because of climate change, dwindling fossil fuel supplies and the general movement of people off the land and into cities, urban and sub-urban areas.[20]

Most suggestions for meeting these challenges involve increasing the output of CAFO's through more efficient breeding and production techniques rather than advocating plant-based diets or even artificially grown meat products.[21] In these circumstances, the challenge for most Western countries will be to increase the efficiency of existing CAFO's in order to produce sufficient meat products for domestic consumption and emerging foreign demand for meat products.

For example, the National Farmers Federation ('the NFF'), the peak industry representative body for farmers in Australia has specifically noted the strategic advantages available to Australian meat and grain producers in satisfying future demand from developing countries. In its *NFF Farm Facts: 2012* Report, the NFF observes:

> The prospects for agriculture are huge, with the need to feed, clothe and house a booming world population. Expanding Asian societies need food and fibre like never before and, due to their growing affluence, are demanding produce of the highest quality. The challenge for Australian agriculture and our famers will be in meeting this booming need for food and

---

[15] Sharman and Kossew (2008), p. 9.

[16] *Lifting the Veil of Secrecy: The Animal Behind your Food*, Voiceless, May 2007.

[17] United Nations Department of Economic and Social Affairs; *World Population Prospects—The 2010 Revision* http://www.un.org/popin/ (accessed 29 March 2012).

[18] Thornton (2010), pp. 2854–2855.

[19] Hocquette and Chatellier (2011), p. 20.

[20] Hume et al. (2011), p. 2.

[21] Galyean et al. (2011), pp. 29–32.

fibre through increasing production. Agriculture has an enormous uptake of new technology.[22]

Australia, North America and other Western countries are therefore proposing to meet the expected increase in world demand for food animal products generally, and meat products particularly, by increasing the efficiency and productivity of agricultural practices generally and CAFOs specifically particularly through technology.[23]

In the process, attention is being drawn to the suffering that food animals inevitably experience as a result of the growth in corporate exploitation of more efficient and productive intensive animal husbandry practices.[24] And the role of corporations in the pursuit of profits through efficiency is central to this enterprise.

40.2.4 Animals as Property, Corporations as People

These industries are made possible in Australia by a legal system that characterises sentient, feeling animals as property without enforceable rights, and non-sentient corporations as legal persons with rights to own and exploit property. Section 124 (1) of the *Corporations Act 2001* (Cth) provides that corporations have the legal capacity and powers of an individual and of a body corporate, including the power to own property.

Darian Ibrahim explains the consequences of this characterisation: 'corporate personhood and animal thing-hood allow for the corporate ownership of animals. Corporate ownership of animals exists wherever animal use has been institutionalised, but it figures most prominently in animal agriculture'.[25] Beginning in the second half of the twentieth century, corporations began owning and exploiting large numbers of animals to produce food for people at profit. In this process, the primary responsibility of corporation is to shareholders, manifesting as duties to trade profitably.[26]

40.2.5 The Privileging of Profit Over Welfare

In these circumstances, the welfare of animals exploited by corporations is therefore a relative consideration necessarily balanced against the imperative of the

---

[22]National Farmers Federation, *NFF Farm Facts: 2012* at 3. http://www.nff.org.au/farm-facts.html accessed on 29 March 2012.
[23]Vinnari and Tapio (2009), p. 269.
[24]Winders and Nibert (2004), p. 76.
[25]Ibrahim (2007), pp. 89–93.
[26]*Greenhalgh v Ardene Cinemas* [1951] Ch 286; Horrigan (2003), p. 321.

profit motive. In *Department of Local Government & Regional Development* v *Emanuel Exports Pty Ltd*,[27] Magistrate C.P. Crawford noted that the relationship between corporate profit motives and suffering of animals was a balancing act; 'in the context of this case, that commercial gain has to be balanced against the likelihood of pain, injury or death to relevant sheep shipped in the second half of the year'.[28]

Achieving this balancing act may be difficult when the profit motive is pressing. In his text *Diet For a New America*, John Robbins quotes livestock auctioneer Henry Pace:

> We're no different from any other business. These animal rights people like to accuse us of mistreating our stock, but we believe we can be most efficient by not being emotional. We are a business, not a humane society, and our job is to sell merchandise at a profit. It's no different from selling paper-clips, or refrigerators.[29]

And so consumers are not readily placed to think about the often difficult animal husbandry practices associated with efficiently producing animal meat. One reason for this is the significant lack of connection between the way animals are portrayed in promotional material and the clinical, plastic-wrapped meat products purchased from the supermarket. Consumers don't make the explicit connection between the slaughter of another sentient being—an animal—and the meat that is then consumed. It's a form of consumer 'cellophane fallacy' that disassociates the consumer from the method of production of the consumer's food.[30]

While consumers might prefer to think of animals as living carefree lives on verdant open-air farms, the reality is very different. In fact, the number of animal farmers in Australia has declined dramatically. For example, between 1980 and 2007, the number of Australian pig-farmers declined by 90 %. However, during that same period, the average pig herd increased by 900 % while pig-meat production almost doubled.[31]

If the number of pigs being harvested for their meat has increased so dramatically, while the number of farms has decreased by 90 %, then where are the pigs being raised and slaughtered? The reality is that large corporations are now involved in industrial-scale intensive animal 'farming'. These corporations do not

---

[27]*Department of Local Government & Regional Development* v *Emanuel Exports Pty Ltd* (Unreported decision of 8 February 2003, WA Magistrates Court).

[28]*Ibid* at 99.

[29]Robbins (1987), p. 104.

[30]The 'cellophane fallacy' is a term used in competition law (antitrust) that refers to an error in market definition. Named after the decision of the United States Supreme Court in *United States v El Du Pont de Nemours & Co* 351 US 377 (1956), the cellophane fallacy is incurred if the price that is employed in the 'hypothetical monopolist test' in market definition is taken to be the current monopoly price and not the *competitive price*. Employing the monopoly price results in an overbroad market definition and thus erroneously gives the impression of less market power than might actually exist. See Corones (2010), p. 68.

[31]Voiceless, *Pig Factories*, consumer action sheet, November 2009, on www.voiceless.org.au at *Resources* (cited 21 August 2011).

raise pigs on lush, green pastures. Instead, they employ intensive farming techniques where the animals are removed from their natural environment and confined or caged in great numbers to live under controlled conditions.

Producing animal meat or harvesting eggs using these intensive production methods is legal in Australia. Commonwealth *Model Codes of Practice* ('MCOPs') relating to beef cattle, poultry and pigs permits the industrial processing of animals for human consumption. These MCOPs were issued by the Primary Industries Ministerial Council ('PIMC'), now the *Standing Committee on Primary Industries* ("SCoPI") whose stated objective is 'to develop and promote sustainable, innovative and profitable agriculture, fisheries/aquaculture and food and forestry industries'.[32]

In other words, the goal of the Commonwealth Authority responsible for creating welfare standards for food animals in Australia is to create profitable and exploitable food animal and animal product industries. When the primary purpose is the creation and maintenance of profitable industries, animal welfare is subordinated to economic efficiency.

40.3 Food Product Credence Claims & Consumer Vulnerability

40.3.1 Consumer Demand for Credence Claims

In 2006, then Commissioner John Martin of the Australian Competition and Consumer Commission ('ACCC') launched the first edition of the ACCC's *Food and Beverage Labelling Guidelines*.[33] The *Guidelines* are intended to assist food and beverage providers in understanding the implications of the law relating to misleading or deceptive conduct.

Commissioner Martin made two important points during his presentation that are relevant to the discussion in Part 3 of this thesis. First, Commissioner Martin noted that 'consumers are becoming increasingly sophisticated and discerning. They are demanding products that offer health benefits, are fresher or are Australian produced'.[34]

Second, he noted that 'products that can highlight such benefits have a better chance of standing out from the pack and grabbing the attention of shoppers on crowded shelves. But this creates temptation for producers and their marketers to "push the envelope" and in some cases break the law in an effort to gain an edge over the competition'.[35]

---

[32]http://www.mincos.gov.au/about_pimc.
[33]Martin (2006).
[34]Ibid 3.
[35]Ibid.

Over the last 5 years, Australian consumers *have* become more discerning about the way in which food animals are treated. In Australia, the InterContinental Hotels Group, which owns the Crowne Plaza Canberra, the National Convention Centre and Parliament House Catering Services, announced in 2008 that it would alter its purchasing decisions to buy eggs pursuant to the *Choose Wisely* Campaign.[36]

The same month, the Australian Capital Territory Government announced that by May 2009, 'all ACT Government agencies including our hospitals, correctional facilities, CIT campuses and schools, will use barn laid or free-range eggs' pursuant to the RSPCA's *Choose Wisely* Campaign.[37] And in its 2012–2013 Budget, the Tasmanian government introduced the *Intensive Animal Farming Development Program* under which $2.5 million will be spent over 2 years in phasing out battery-hen farms and the use of sow stalls.[38] In introducing these food animal welfare initiatives, the Tasmanian Treasurer specifically noted that 'changes in market and consumer demand' motivated the Budget initiatives'.[39]

Evidence from the United Kingdom, the European Union and the United States reflects evidence from Australia that consumers are becoming increasingly concerned about credence claims associated with food products generally and food animal products particularly.[40]

This willingness on the part of consumers to pay a premium for food products that bear credence claims about quality, content and creation provenance has been noted by the Federal Court. False representations concerning free-range eggs were described by the Court in, *ACCC v Turi Foods (No 4) Ltd* as amounting "to a cruel deception on consumers who mostly seek out free range eggs as a matter of principle, hoping to advance the cause of animal welfare by so doing."[41]

Conversely, food producers and suppliers are also beginning to recognise consumers' concerns and are attempting to differentiate their food animal products on the basis of animal welfare.

---

[36]*Government Welcomes Action Against Battery Eggs*. http://www.chiefminister.acy.gov.au/media.php?v=7457.

[37]Ibid.

[38]Treasurer Lara Giddings MP, *2012–2013 Budget Speech 'Strong Decisions. Better Future'*, 17 May 2012, delivered on the Second Reading of the *Consolidated Fund Appropriation Bill (No 1) 2012*.

http://www.treasury.tas.gov.au/domino/dtf/dtf.nsf/v-budget-budget-papers/0 (Accessed 21 May 2012).

[39]Ibid 12.

[40]Napolitano et al. (2010), pp. 537–538.

[41][31].

40.3.2 Consumer Vulnerability

These market concentrations and practices potentially expose both food producers (farmers) and food consumers to both price and information asymmetries and exploitation through deceptive marketing practices.[42] Indeed, 2009 OECD figures indicated that domestic food prices in Australia had increased by 40 % over the decade 2000–2009.[43] The situation in Australian simply reflects the wider shift in agricultural practices world-wide, characterised by increased corporate concentration at all stages of the supply chain.[44]

Unfortunately, the decisions of the Federal Court in *Australian Competition and Consumer Commission v C.I. & Co Pty Ltd*,[45] and *Australian Competition and Consumer Commission v Turi Foods Pty Ltd (No 2)*[46] confirm Commissioner Martin's fears that in doing so, some producers will attempt to take advantage of these concerns by labelling food animal products in ways that deceive consumers about welfare issues.

Why are consumers vulnerable to price exploitation and misleading marketing practices as a result of information asymmetries? Consumers are rarely in a position to evaluate credence claims associated with the production of food products. This is particularly so in considering whether and to what extent welfare-friendly animal husbandry methods have been employed in the production of animal based food products. In such circumstances information asymmetries create opportunities for producers and retailers to exploit consumer demand for credence attributes associated with food products.

How do these information asymmetries arise? In centuries past, consumers may have been able to purchase their food products directly from the farmer/producer at village markets thereby satisfying themselves of the provenance of the food being purchased. While this did not completely immunise the consumer from exploitation, that possibility was at least minimised.[47]

---

[42] *ACCC v Turi Foods (No 4) Ltd* (2013) ATPR 42-448.

[43] Cited in Richards et al. (2012), p. 250.

[44] United Nations Conference on Trade and Development, *Trade and Environment Review 2013* at 285 (Commentary IV: Getting Farmers off the Treadmill), United Nations.

[45] *Australian Competition and Consumer Commission v C.I. & Co Pty Ltd* [2010] FCA 1511 (23 December 2010).

[46] *Australian Competition and Consumer Commission v Turi Foods Pty Ltd (No 2)* [1012] FCA 19 (2 December 2011).

[47] Bruce (2013), p. 4. In the Middle Ages, most trade in domestic goods and services was conducted at markets, held once each week. The owners of the markets or fairs charged the merchants who were selling goods at their market a fee called a 'stallage' for the space. Unscrupulous traders would often attempt to prevent other traders from entering the market. This had the effect of preventing price competition for various goods as well as preventing the market owner from collecting the stallage fee. This practice was known as 'forestalling'. When it was practiced during times of poor harvests or famine, it enabled some traders to manipulate the market. The absence of outside competition allowed the traders to extract supra-competitive prices for their goods.

However, in the twenty-first century the corporatisation of food production displaces the consumer in time and space from the primary producer of the food products being consumed. In contemporary Western societies, consumers will generally purchase their food from corporate-owned supermarket chains. In Australia, the average family spends approximately 12–14 % of their after-tax income on standard groceries at supermarkets.[48] In making these spending decision, labels on food packaging provide valuable information to consumers at point of sale.[49]

In these circumstances, significant information asymmetries are created; the actual producer possesses almost all relevant information concerning credence claims associated with their food products, while the consumer is usually unable to verify the truth of the credence claim at point of sale.[50] For example, once eggs are placed in a carton, it is impossible for consumers to discern whether the eggs were free-range or cage.[51]

This 'information asymmetry' creates an imbalance of power between producers and suppliers on one hand, and consumers on the other.[52] In circumstances of such information asymmetry, where consumers are unable to verify the truth of credence claims, they are then unable to make informed purchases of products.[53]

The most intuitive response to information asymmetries associated with credence claims is for governments to initiate food labelling reform. It is therefore important to explore the Commonwealth government's response to these difficulties.

40.4 Australian Commonwealth Government Labelling Initiatives

40.4.1 *The* Labelling Logic Report

In Australia a major Commonwealth Government Review into food labelling laws was recently concluded. On 23 October 2009, the Council of Australian Governments (COAG) and the Australia and New Zealand Food Regulation Ministerial Council (Ministerial Council) agreed to undertake a comprehensive review of food

---

[48]Australian Competition and Consumer Commission, *Report of the ACCC Inquiry into the Competitiveness of Retail Prices for Standard Groceries*, July 2008, Commonwealth of Australia at xiii.

[49]Mhurchu and Gorton (2007), pp. 105, 105.

[50]Kehlbacher et al. (2012), pp. 627, 628.

[51]*ACCC v C I & Co Pty Ltd* [2010] FCA 1511, [31]; *ACCC v Bruhn* [2012] FCA 959, [50].

[52]Lee (2013), p. 42.

[53]Ibid.

labelling law and policy. This is the review that resulted in the *Labelling Logic Report*.

After the first round of consultations during which it received over 6000 public submissions, the Review Panel issued its *Issues Consultation Paper* on 5 March 2010 ('the *Consultation Paper*') and invited further submissions.[54] Question 17 of the *Consultation Paper* asked whether 'there is a need to establish agreed definitions of terms such as 'natural', 'lite', 'organic', 'free range', 'virgin' (as regards olive oil), 'kosher' or 'halal'? If so, should these definitions be included or referenced in the Food Standards Code?'[55]

However, in relation to consumer values issues relating to specific food production methods the *Labelling Logic Report* did recommend the Commonwealth government adopt specific values-based definitions in the *Food Standards Code* in order to achieve consistency of definitions.[56]

This recommendation was rejected by the Commonwealth government in its December 2011 Response. Instead, the Commonwealth stated that where regulation concerning labelling representations was needed, the mechanisms in the *Competition and Consumer Act 2010* (Cth) were more appropriate to the task.[57] The decision to employ the ACL in this way is consistent with earlier Commonwealth regulatory initiatives such as the *Food Standards Australia New Zealand Act 1991* (Cth) intended to prevent misleading and deceptive conduct by ensuring that consumers have adequate information to make informed food choices.[58]

40.4.2 Greater Role of the Australian Consumer Law

What this means is that Instead of simply legislating to prohibit certain animal husbandry practices, the Commonwealth government is intending market forces in the form of consumer demand exerting up-stream market pressure on primary industry producers to implement food animal welfare initiatives.

In attempting to satisfy this consumer demand, food animal products accentuating animal welfare will be subject to careful scrutiny under the misleading or

---

[54]*Issues Consultation Paper: Food Labelling Law and Policy Review*, 5 March 2010, Food Labelling Law Secretariat, Canberra, Australia.

[55]Ibid 6.

[56]Food Labelling Law and Policy Review Panel, *Labelling Logic: Review of Food Labelling Law and Policy*, 27 January 2011, Commonwealth of Australia, Recommendation 36, 12.

[57]*Response to the Recommendations of Labelling Logic: Review of Food Labelling Law and Policy (2011)*, Commonwealth Government, December 2012, 40.

[58]*Food Standards Australia New Zealand Act 1991* (Cth), s 3(c).

deceptive conduct provisions of the ACL. Product differentiation based on food animal welfare claims requires careful substantiation.[59]

However, the unstated assumption behind this policy of preventing deception associated with food labels involves the effective operation of market forces of supply and demand. It assumes that market dynamics will facilitate consumers' desires for accurate information about welfare friendly food animal products. In an increasingly competitive market for food products, it is anticipated that consumer demand for ethically produced animal products will signal producers to implement food animal welfare practices such as free-range farms.[60]

40.5 The Nature and Function of the *Australian Consumer Law*

40.5.1 Australia's New Consumer Protection Regime

Until 2011, consumer protection and product liability in Australia was regulated by an often confusing patchwork of Commonwealth, State and Territory legislation, regulations and subordinate legislation. At the Commonwealth (Commonwealth) level and since 1974, the *Trade Practices Act 1974* (Cth) ('the TPA') provided Australia's principal source of consumer protection legislation. Complimenting and in many places duplicating the TPA were individual State and Territory *Fair Trading* legislation.

In May 2008, some 2 years after it had formally commenced its inquiry, the Australian Productivity Commission's (PC) *Review of Australia's Consumer Policy Framework* was tabled in Federal Parliament. That review recommended the implementation of a single, national Australian consumer law to replace the consumer protection regimes in both the TPA and the various State and Territory Fair Trading Acts.

Pursuant to this Review and on 1 January 2011, this relatively fragmented landscape of consumer protection and product liability law in Australia fundamentally changed. The *Trade Practices Amendment (Australian Consumer Law) Act (No 2) 2010* ('the ACL Act') completed a process of reform that had been gaining momentum since the early 2000s and that culminated in the creation of a single, nation-wide consumer protection and product liability regime in the form of the *Australian Consumer Law*.

In the process, the *Australian Consumer Law* replaced 17 generic consumer protection laws that existed across States and Territories with a single national Consumer

---

[59]In its *Food Labelling Guide*, the ACCC warns that it has 'become increasingly concerned about representation on the labels, packaging and advertisements of food and beverage products.' ACCC *Food Labelling Guide*, 2009, Canberra, Australia, 3.

[60]Food Labelling Law and Policy Review Panel, *Labelling Logic: Review of Food Labelling Law and Policy*, 27 January 2011, Commonwealth of Australia, paragraph 3.20 at p 47.

Law found in Schedule 2 to the CCA and implemented as a law of the Commonwealth in Part XI of the CCA and as an 'applied law' of the States and Territories in Part XIAA of CCA and for the most part, embedded in *Fair Trading* legislation.

It is the largest reform of Australian consumer protection laws ever undertaken. I have explained in more detail elsewhere the policy and constitutional background to, and the mechanics of the national implementation of the ACL.[61]

40.5.2 The ACL & Food Product Labelling

At first glance, the ACL appears to have little to do with food product labelling generally and food animal product labelling specifically. Sitting as it does within the *Competition and Consumer Act 2010* (Cth) ('the CCA') the ACL is intended to facilitate the larger objective of the CCA in enhancing the welfare of Australians through the promotion of competition, fair trading and consumer protection.[62]

Despite its principally economic focus, the CCA and the ACL *does* have a significant role to play in protecting consumers from misleading or deceptive credence claims associated with food products. This is because s 18 of the ACL prohibits conduct that is "misleading or deceptive or likely to mislead or deceive". Almost all of the recent litigation associated with credence claims has involved the *Australian Competition and Consumer Commission* instituting proceedings against a corporation alleging a contravention of s 18 of the ACL in making misleading credence claims. This litigation is discussed below.

Before exploring this litigation, it is important to understand how s 18 of the ACL is structured and has been interpreted by the Courts.

40.5.3 ACL s 18: The Prohibition Against Misleading or Deceptive Conduct

The text of s 18 of the *Australian Consumer Law* is relatively straightforward; there are 3 elements that must be satisfied in order to establish a contravention:

1. A corporation or person engages in conduct;
2. In trade or commerce; that is
3. Misleading or deceptive or likely to mislead or deceive.

At this point, it should be noted that although s 52 of the TPA is now s 18 of the *Australian Consumer Law*, all of the case-law relating to the interpretation of the

---

[61]Bruce (2011a), Chapters 1 and 2, pp. 1–50.

[62]*Competition and Consumer Act 2010* (Cth) s 2.

former s 52 will continue to guide the Courts in evaluating conduct alleged to breach s 18 of the *Australian Consumer Law*.

The *Explanatory Memorandum* to the *Trade Practices Amendment (Australian Consumer Law) Act (No 2) 2010* Cth, specifically states:

> Section 18 of the ACL replaces the repealed Section 52 of the TP Act. The substance of the drafting of the prohibition has not been changed, other than changing the reference to 'a corporation' to 'a person'. Accordingly, the well-developed jurisprudence relating to s 52 of the TP Act is relevant to the interpretation or understanding of the meaning and application of Section 18 of the ACL.[63]

Most of the case law discussed in this chapter involves conduct evaluated under the former s 52 of the TPA. However, for the sake of clarity, I will substitute '[s.18]' for 's.52' in relevant case-law extracts. There is no material difference in the sections and the substitution is intended for ease of conceptualising the argument as involving ACL s 18 rather than s 52 of the TPA.

While it is true that the elements of ACL s 18 are clear, the interpretation and application of those elements has not been straightforward. The High Court in *Parkdale Custom Built Furniture Proprietary Limited v Puxu Proprietary Limited* admitted:

> The words of [s.18] have been said to be clear and unambiguous....Nevertheless they are productive of considerable difficulty when it becomes necessary to apply them to the facts of particular cases. Like most general precepts framed in abstract terms, the Section affords little practical guidance to those who seek to arrange their activities so that they will not offend against its provisions.[64]

These difficulties often arise in satisfying threshold requirements for successfully establishing a contravention of ACL s 18. First, the conduct of the person or corporation must be assessable under the *Competition and Consumer Act 2010* (Cth).

Second, the conduct must be 'in trade or commerce'; third, there must be an identifiable 'class of consumers' who are alleged to have been misled; and fourth, the conduct must have in fact caused the misled state of mind, and to succeed in recovering damages, the loss or damage must have been caused by the allegedly misleading or deceptive conduct.

In working through these requirements the Courts have developed a *methodology* in relation to the former s 52 of the *Trade Practices Act 1974* (Cth) that enables conduct to be evaluated. There are certain elements built into that methodology that must be examined in interpreting ACL s 18.

---

[63]*Explanatory Memorandum* to the *Trade Practices Amendment (Australian Consumer Law) Act (No 2) 2010* (Cth), 37, [3.11].

[64]*Parkdale Custom Built Furniture Proprietary Limited v Puxu Proprietary Limited* (1988–1982) 149 CLR 191, 197 per Gibbs CJ.

40.5.4 A Norm of Conduct Not an Imposition of Liability

Section 18 of the *Australian Consumer Law* does not actually create a cause of action. It simply establishes a standard of conduct; a person must not in trade or commerce, engage in conduct that is misleading or deceptive or likely to mislead or deceive. The Court in *Brown v Jam Factory Pty Ltd* noted:

> Section [18] does not purport to create liability at all; rather it establishes a norm of conduct, failure to observe which has consequences provided for elsewhere in the same statute or under the general law.[65]

When the Court mentioned 'consequences provided for elsewhere in the same statute' it was referring to the remedies and orders available under the then *Trade Practices Act 1974* (Cth) for a contravention of Part V of the TPA that included s 52.

Likewise, the 'norm of conduct' provided for by ACL s 18 does not of itself establish the consequences for a breach. The remedial provisions for a contravention of ACL s 18 are found in other parts of the ACL; principally in Chapter 5 and include injunctive relief[66] and damages.[67]

40.5.5 Conduct 'In Trade or Commerce'

For the purposes of this article, and consistent with the authoritative interpretation of that term by the High Court in *Concrete Constructions (NSW) Pty Limited v Nelson*,[68] the sale for profit of grocery and other food animal products is conduct that is in trade or commerce for the purposes of the ACL. Consistent with the decision in *Concrete Constructions*, the retail sale of products is an activity that of itself bears a trading or commercial character.

40.5.6 When Is Conduct 'Misleading or Deceptive'?

Conduct is misleading or deceptive when it 'leads into error'. The High Court in *Parkdale Custom Built Furniture Proprietary Limited v Puxu Proprietary Limited* explained:

> The words of [s.18] require the Court to consider the nature of the conduct of the corporation against which proceedings are brought and to decide whether that conduct

---

[65]*Brown v Jam Factory Pty Ltd* (1981) 53 FLR 340, 348.
[66]ACL s 232.
[67]ACL s 236.
[68]*Concrete Constructions (NSW) Pty Limited v Nelson* (1990) 169 CLR 594.

was, within the meaning of that section, misleading or deceptive or likely to mislead or deceive...One meaning which the words 'mislead' and 'deceive' share in common is 'to lead into error'.[69]

The Full Federal Court in *Astrazeneca Pty Ltd v Glaxosmithkline Australia Pty Ltd* formulated the requirement as follows:

> In order to determine whether there has been any contravention of [s.18] of the Act, it is necessary to determine whether or not the conduct complained of amounted to a representation which has or would be likely to lead to a misconception arising in the minds of that section of the public to whom the conduct (which may include refraining from doing an act) has been directed.[70]

Whether evaluating the conduct of corporations or of persons, there are three threshold issues that need to be addressed before the prohibition in ACL s 18 can be established.

1. Whether conduct is 'in trade or commerce';
2. The *Taco Bell*[71] methodology for evaluating misleading or deceptive conduct; and
3. The *Campomar*[72] methodology employed for evaluating the relevant 'class of consumers' alleged to have been misled.

These foundational methods or principles influence whether ACL s 18 even applies (because the conduct in question might not be 'in trade or commerce') and if it does, who might have been misled (identifying the 'class' of consumers through the *Campomar* methodology) and then whether that conduct is misleading or deceptive in breach of s 18 of the *Australian Consumer Law* (the *Taco Bell* methodology).

40.5.7 *The* Taco Bell *Methodology*

Conduct is misleading or deceptive if it 'leads into error'. But what is the method by which conduct is considered to have led into error and therefore breached ACL s 18? What processes does the Court undertake in making its assessment? In *Apotex Pty Ltd v Les Laboratoires Servier (No 2)*, the Court stated:

> In *Taco Co of Australia Inc v Taco Bell Pty Ltd* (1982) 24 ALR 177 at 202 Deane and Fitzgerald JJ outlined a series of propositions to be considered in assessing whether conduct is misleading or deceptive under [s 18] of the Act:

---

[69] *Parkdale Custom Built Furniture Proprietary Limited v Puxu Proprietary Limited* (1981–1982) 149 CLR 191, 198.
[70] *Astrazeneca Pty Ltd v Glaxosmithkline Australia Pty Ltd* (2006) ATPR 42-106, 44,890.
[71] *Taco Co of Australia Inc v Taco Bell Pty Ltd* (1982) 24 ALR 177.
[72] *Campomar Sociedad v Nike International* (2000) 202 CLR 45.

- It is necessary to identify the relevant section(s) of the public by reference to whom the question of whether conduct is or is likely to be misleading or deceptive falls to be tested;
- Once the relevant section of the public is established, the matter is to be considered by reference to all who come within it, including the astute or the gullible, the intelligent or not so intelligent, educated or not educated and men and women of various ages and vocations;
- Evidence that some individual has in fact formed an erroneous conclusion is admissible and may be persuasive but is not essential. Regardless, such evidence does not of itself conclusively establish the conduct to be misleading or deceptive, the test is objective and the Court must determine for itself;
- It is necessary to inquire why any proven misconception has arisen. It is only by this investigation that the evidence of those who are shown to have been led into error can be evaluated and it can be determined whether they are confused because of misleading or deceptive conduct on the part of the respondent.[73]

40.5.7.1 Application of the Taco Bell Methodology

This process of evaluation, sometimes referred to as the 'Taco Bell Steps' has been adopted and elaborated upon, either explicitly or implicitly by almost all decisions in which the Court is required to evaluate whether conduct is misleading or deceptive or likely to mislead or deceive. For example, in *Domain Names Australia Pty Ltd v .au Domain Administration Ltd*, the Full Court stated:

It has long been established that:

- When the question is whether conduct has been likely to mislead or deceive, it is unnecessary to prove anyone was actually misled or deceived...
- Evidence of actual misleading or deception is admissible and may be persuasive but is not essential...
- The test is objective and the Court must determine the question for itself...
- Conduct is likely to mislead or deceive if that is a real or not remote possibility, regardless of whether it is less or more than 50 %.[74]

See also *AMI Australia Holdings Pty Ltd v Bade Medical Institute (Australia) Pty Ltd*.[75]

In other cases, Courts have not *explicitly* set out the *Taco Bell* steps, but have *implicitly* adopted them. For example, in *Astrazeneca Pty Ltd v Glaxosmithkline*

---

[73]*Apotex Pty Ltd v Les Laboratoires Servier (No 2)* (2008) ATPR 42-235, 49,206.

[74]*Domain Names Australia Pty Ltd v .au Domain Administration Ltd* (2004) 139 FCR 215, [17]–[18].

[75]*AMI Australia Holdings Pty Ltd v Bade Medical Institute (Australia) Pty Ltd.* (2009) 262 ALR 458, 472.

Australia Pty Ltd, the Full Court referred to *Taco Bell* in explaining its approach to the evaluation of the conduct under challenge in that case.

The Full Court stated:

> For [s.18] of the Act to be enlivened it is sufficient that the conduct complained of, in all the circumstances, answers the statutory description, that is to say, that it is misleading or deceptive or likely to mislead or deceive. It is unnecessary to go further and establish that any actual or potential consumer has taken or is likely to take any positive step in consequence of the misleading or deception. That is not to say that evidence of actual misleading or deception and of steps taken in consequence thereof is not likely to be both relevant and important on the question whether the relevant conduct in fact answers the statutory description.[76]

A similar re-formulation of the *Taco Bell* steps was adopted by the Court in *Johnson & Johnson Pacific Pty Limited v Unilever Australia Limited (No 2)*.[77] The Court in *Unilever Australia Limited v Goodman Fielder Consumer Foods Pty Ltd*[78] also adopted the *Taco Bell* steps without referring to the case itself. See also *Australian Competition and Consumer Commission v Australian Dreamtime Creations Pty Ltd*.[79]

40.5.8 A Consistent Approach to Principles: The Basic Evaluative Framework

The consistent approval of the *Taco Bell* steps enables a summary of the basic principles employed by the Courts in evaluating whether conduct is misleading or deceptive or likely to mislead or deceive.[80]

The principles set out below have been extracted from a number of recent decisions of the Federal Court including *ACCC v Clarion Marketing Pty Ltd*[81]; *Butcher v Lachlan Elder Realty Pty Limited*[82] and *ACCC v Australian Dreamtime Creations Pty Ltd*.[83]

There are many other cases in which these principles have been stated (in various ways) and elaborated upon, and they form a basic conceptual framework against which allegations of misleading or deceptive conduct may be evaluated.

---

[76]*Astrazeneca Pty Ltd v Glaxosmithkline Australia Pty Ltd* (2006) ATPR 42-106, 44, 890.

[77]*Johnson & Johnson Pacific Pty Limited v Unilever Australia Limited (No 2)* (2007) ATPR 42-136, 46,618 - 46,619 and 46,625.

[78]*Unilever Australia Limited v Goodman Fielder Consumer Foods Pty Ltd* (2009) ATPR 42-305, [18]–[23].

[79]*Australian Competition and Consumer Commission v Australian Dreamtime Creations Pty Ltd* (2009) 263 ALR 487, 493.

[80]Bruce (2011b), pp. 51–73.

[81]*ACCC v Clarion Marketing Pty Ltd* [2009] FCA 1441, 8–9.

[82]*Butcher v Lachlan Elder Realty Pty Limited* (2004) 218 CLR 592, 621–626.

[83]*ACCC v Australian Dreamtime Creations Pty Ltd* (2009) 263 ALR 487, 493–494.

The following ten principles form that basic evaluative framework:

1. Whether a representation is likely to mislead or deceive is an objective question of fact, to be determined having regard to all the circumstances of the conduct and not just some isolated aspect of that conduct;
2. Conduct is misleading or deceptive if leads into error. It is likely to be misleading or deceptive if there is a real chance that the conduct or representations will mislead or deceive;
3. It is necessary to identify some conduct, whether in the form of a representation, an omission or some other form, that led the consumer(s) into error;
4. It is necessary to identify the class of consumers toward whom the allegedly misleading conduct was directed;
5. Having identified the relevant class of consumers, the test to be applied is objective, that is, whether a ordinary and reasonable person from the class is likely to have been misled or deceived;
6. The process involved in identifying the 'ordinary and reasonable' person from the class differs depending on whether the allegedly misleading or deceptive conduct was directed toward specific and identified individuals or to a large class;
7. Actual intention to mislead or deceive is not necessary to establish a breach of s 18 of the ACL, but if intention is present, a Court may be more likely to find that the conduct complained of was misleading;
8. Conduct may be misleading or deceptive if it induces error, but it is not sufficient merely to show that it may have led to confusion or caused people to wonder;
9. Actual evidence that some people may have been misled is not essential but is admissible and may be persuasive if given;
10. A corporation does not avoid liability for breach of s 18 because a person who has been the subject of misleading or deceptive conduct could have discovered the misleading or deceptive conduct by proper inquiries;

40.6 Enforcement of the ACL and Food Product Credence Claims

Consistent with the intention of the *Labelling Logic Report*, credence claims associated with food products are now principally regulated in Australia through the lens of the *Australian Consumer Law* that is enforced by the *Australian Competition and Consumer Commission*. Accordingly, credence claims are evaluated in terms of whether they have the potential to mislead or deceive consumers in breach of the ACL, principally ACL s 18 discussed above.

40.6.1 Credence Claims Now an ACCC Focus of Litigation

This closer relationship between food product credence claims and the ACL is evidenced by the February 2013 decision of the ACCC to designate credence claims, particularly in the food industry as well as competition and consumer issues in the supermarket industry as an explicit enforcement priority.[84]

This scrutiny paid by the ACCC is reflected in a spate of litigation instituted in the Federal Court of Australia[85] since 2010 against food producers and retailers, principally involving credence claims associated with eggs, chicken and meat products. The decisions in *Australian Competition and Consumer Commission v C.I & Co Pty Ltd*,[86] *Australian Competition and Consumer Commission v Turi Foods Pty Ltd*,[87] *Australian Competition and Consumer Commission v Bruhn* [2012] FCA 959, *Australian Competition and Consumer Commission v Pepe's Ducks Ltd* [2013] FCA 570 and *Australian Competition and Consumer Commission v Luv-a-Duck Pty Ltd* (2013) FCA 1136, provide useful guidance about the relationship between product labelling and how the misleading or deceptive conduct provisions of the ACL can be deployed to ensure food animal products accurately reflect antecedent animal husbandry practices.

40.6.2 When Free Range Is Not "Free to Range"

Three of these five decisions concern credence claims associated with eggs and this chapter will focus on those cases.[88]

In *Australian Competition and Consumer Commission v C.I & Co Pty Ltd*, a Western Australian based family owned company, C.I & Co Pty Ltd ('CI') acquired eggs from egg farms and supplied them to a number of retailers, cafes and restaurants. Between June 2008 and April 2010, CI acquired over a *million* dozen eggs produced by battery cage hens and 12,000 dozen free-range eggs.

However, in that period, CI supplied nearly 900,000 dozen eggs to customers that it had labelled 'free range', conduct described by the Court as involving 'a high

---

[84]ACCC, *ACCC Compliance and Enforcement Policy*, Commonwealth of Australia, February 2013.

[85]Part XI, s138(1) of the *Competition and Consumer Act 2010* (Cth) confers jurisdiction on the Federal Court of Australia in relation 'to any matter arising under this Part or the Australian Consumer Law'.

[86]*Australian Competition and Consumer Commission v C.I & Co Pty Ltd* [2010] FCA 1511.

[87]*Australian Competition and Consumer Commission v Turi Foods Pty Ltd (No 2)* [2012] FCA 19.

[88]The decision in *Australian Competition and Consumer Commission v Pepe's Ducks Ltd* [2013] FCA 570 concerned misleading claims that duck meat was sourced from ducks that spent a substantial time outdoors when in fact the ducks were entirely barn raised and never permitted out-doors. The decision in *Australian Competition and Consumer Commission v Luv-a-Duck Pty Ltd* (2013) FCA 1136 concerned similar allegations.

level of dishonesty'.[89] In doing so, CI and its directors earned a significant amount of revenue they would not otherwise have earned if the eggs had been truthfully labelled as 'cage eggs'. For example, the Court noted that in a 2-week representative period between 15 and 30 April 2010, CI and its directors earned between $5744 and $9008 in revenue 'which they would not have derived had the eggs been labelled clearly as 'cage eggs''.[90]

Following an ACCC investigation, CI and its directors admitted the deception and that they had contravened Sections 52, 53(a) and 55 of the *Trade Practices Act 1974* (Cth).[91] Both CI and the directors consented to certain orders being made against them, including declarations, injunctive relief and corrective advertising.

Because CI and its directors had admitted the contraventions and consented to orders being made, the Court was not required to establish CI's liability through the *Taco Bell* and *Campomar* methodologies discussed above. It has become increasingly common for respondents to agree to consent orders and making joint submissions on penalties with the ACCC thereby avoiding a substantive trial on the issues.

Nevertheless, the Court cannot simply make orders and impose penalties just because parties consent to them. Before it does so, the Court must be satisfied that the facts before it actually do disclose a breach of the CCA or ACL.[92] In this case, the Court accepted that the relevant sections of the TPA (now ACL) had been breached 'following many years of unlawful conduct which must have yielded considerable undeserved profit'.[93]

40.6.3 Relationship Between Consumer Interests and Misleading Labelling

Significantly, the Court clearly explained the relationship between the misleading labelling and consumer interest in food animal welfare. According to the Court, the misleading labels 'amounted to a cruel deception on consumers who mostly seek out free range eggs as a matter of principle, hoping to advance the cause of animal welfare by so doing.'[94]

This is a simple and direct statement of the intended use of the ACL anticipated by the Commonwealth *Labelling Logic Report*.

---

[89] *Australian Competition and Consumer Commission v C.I & Co Pty Ltd* [2010] FCA 1511, [31].
[90] Ibid [14].
[91] These are now Sections 18, 29(1)(a) and 33 respectively, of the ACL.
[92] *Australian Competition and Consumer Commission v SIP Australia Pty Ltd* (1999) ATPR 41-702, 43,000.
[93] *Australian Competition and Consumer Commission v C.I & Co Pty Ltd* [2010] FCA 1511, [21].
[94] Ibid [31].

Growing consumer demand for welfare friendly animal products is unfortunately reflected in the continued willingness of some suppliers to mislead consumers. For example, in *Australian Competition and Consumer Commission v Bruhn* [2012] FCA 959 the ACCC instituted proceedings against Ms Rosemary Bruhn, trading as "Rosie's Free Range Eggs" in South Australia. In instituting proceedings, the ACCC 'allege(d) that from March 2007 to October 2010, Ms Bruhn represented that eggs she supplied to business customers including 117 customers in South Australia such as retail outlets, bakeries, cafes and restaurants, were free range eggs when a substantial proportion of the eggs were not free range but cage eggs'.[95]

Between January 2009 and October 2010, the company earned AUD $186, 978 as a result of selling cage-eggs deliberately misrepresented as free-range eggs. The Court imposed a pecuniary penalty of AUD $50,000 and required Ms Bruhn to pay AUD $15,000 toward the ACCC's legal costs.

40.6.4 Eggs Before Chickens

Similar to the *Bruhn* case, the ACCC instituted proceedings in September 2011 against Turi Foods Pty Ltd, Baiada Poultry Pty Ltd, Bartter Enterprises Pty Limited and the Australian Chicken Meat Federation Inc ('the ACMF') also alleging misleading or deceptive conduct in relation to the supply of chicken mean products.

Baiada Poultry and Bartter Enterprises supply chickens throughout Australia under the well-known 'Steggles' brand name while Turi Foods supplies 'La Ionica' brand chickens in New South Wales and Victoria. The ACCC alleged that these corporations engaged in misleading or deceptive conduct in breach of both the TPA and the ACL in making certain representations associated with the chicken meat products they supplied.[96]

The ACCC alleged that 'Baiada Poultry and Bartter Enterprises made false or misleading claims in print advertising and product packaging, that Steggles meat chickens are raised in barns with substantial space available allowing them to roam freely' when this was not the case at all'.[97]

Similar allegations were made against Turi Foods where it was alleged that 'Turi Foods made false or misleading representations through in-store displays and advertising on delivery trucks. La Ionica brand meat chickens were claimed to be

---

[95] *ACCC Takes Court Action Against SA Egg Supplier*, ACCC Media Release 8 March 2012.
http://www.accc.gov.au/content/index.phtml/itemId/1037910 (Accessed 12 May 2012).

[96] *ACCC Takes Action over 'Free to Roam' Chicken Claims*, ACCC Media Release 7 September 2011.
http://www.accc.gov.au/content/index.phtml/itemId/1006465/fromItemId/966100 (Accessed 12 May 2012).

[97] Ibid.

able to roam freely in barns with substantial space and in conditions equivalent to a free range system.'[98]

This litigation has been bitterly contested. In December 2011, the Australian Chicken Meat Federation sought interlocutory orders dismissing the ACCC's proceedings either on the basis that no reasonable cause of action was disclosed[99] or that the ACCC had no reasonable prospect of successfully prosecuting its claims.[100]

The actual evidence indicated that the average space available to each chicken was about 500 square centimetres.[101] To provide some perspective, an A4 sheet of paper has an area of 625 square centimetres. A standard laying hen is at least 40-cm high when she stands erect and is approximately 45-cm long and 18-cm wide, without her wings extended. Her body space takes therefore takes up an area of about 810 square centimetres.

Despite the mathematical bleakness of the evidence, the ACMF claimed that there were simply no grounds for alleging that chickens were not 'free to roam' as they had represented.[102] Perhaps not surprisingly, Tracey J refused to strike out the ACCC's action, observing that 'five hundred centimetres squared is a remarkably small space. In order for any one chicken to have a larger area of movement, others would have to be confined within an even smaller space.'[103]

By January 2012, Turi Foods Pty Ltd had decided to conclude the proceedings against it by admitting the contraventions and submitting to consent orders.[104] After reviewing the evidence, Tracey J concluded:

> the stock densities, which La Ionica has admitted are to be found in the barns in which its chickens are raised, are maintained at such a level that the chickens have severe restrictions placed on their capacity to roam, if, indeed any such capacity exists. [105]

These conclusions would later return to haunt Tracey J. The three other respondents continued to fight the ACCC's allegations and[106] at least two of the respondents decided to personally attack the Judge.

---

[98]Ibid.

[99]Rule 26.01(1)(c) *Federal Court Rules 2011*.

[100]Rule 26.01(1)(a) *Federal Court Rules 2011*.

[101]*Australian Competition and Consumer Commission v Turi Foods Pty Ltd* [2011] FCA 1382, [14] (Unreported decision of Tracey J dated 2 December 2011).

[102]Ibid [10].

[103]Ibid [13].

[104]*Australian Competition and Consumer Commission v Turi Foods Pty Ltd (No 2)* [2012] FCA 19 (unreported decision of Tracey J dated 23 January 2012).

[105]Ibid [23].

[106]It is usual for food animal suppliers to vigorously litigate against persons who threaten to expose their treatment of animals—see *Takhar & Anor v South Australian Telecasters Ltd* (BC 9702320, Unreported decision of Perry J of the Supreme Court of South Australia, 1997) involving an application for an injunction to restrain a current affairs program from airing footage of a battery hen egg farm operated by a supplier falsely selling eggs as 'free range'; *ABC v Lenah Game Meats Pty Ltd* (2001) 185 ALR 1 in which a supplier of possum meat sought an injunction to restrain display of footage taken of the plant's processing practices.

In February 2012 Baiada Poultry Pty Ltd and Bartter Enterprises attempted to have Tracey J disqualify himself from hearing the case on the grounds of apprehended bias.[107] Both Baiada Poultry and Bartter Enterprises owned and operated chicken growing sheds in the same way as Turi Foods, including equivalent stocking densities.

Accordingly, they alleged that Tracey J's conclusion about stocking densities made in concluding the proceedings against Turi Foods 'travelled beyond the agreed facts...and constituted findings independently made by me.'[108] His Honour rejected the application and, in 2012, pecuniary penalties of $100,000 AUD were imposed upon the company.

40.6.5 Ongoing ACCC Litigation and Labelling Logic Report

These decisions demonstrate that the misleading or deceptive conduct provisions in the ACL can be effectively deployed to prevent suppliers of food animal products from deceiving consumers about value issues such as animal welfare conditions. They also reflect an awareness by the Courts of the relationship between consumer concern for food animal welfare and the information provided by suppliers on labels and advertising material.

At least in relation to positive representation such as 'free-range' or 'free-to-roam', the policy behind the *Labelling Logic Report* is likely to be well served by the effective enforcement of the ACL. By preventing suppliers from making misleading or false claims associated with food animal products, consumers will be provided with sufficient and accurate information enabling them to make informed choices about their purchases. This is especially important when, as North J in *Australian Competition and Consumer Commission v C.I & Co Pty Ltd* noted, consumers 'seek out free range eggs as a matter of principle, hoping to advance the cause of animal welfare by so doing.'[109]

At the time of writing, the ACCC has also instituted proceedings against Snowdale Holdings Pty Ltd and Pirovic Enterprises Pty Ltd alleging that those corporations made false, misleading or deceptive representations that the eggs supplied and labelled as 'free range' were produced 'by hens that were farmed in

---

[107] Apprehended bias exists where 'a fair-minded lay observer might reasonably apprehend that the judge might not bring an impartial and unprejudiced mind to the resolution of the question the judge is required to decide.' *Michael Wilson & Partners v Nicholls* (2011) 282 ALR 685, 692.

[108] *Australian Competition and Consumer Commission v Turi Foods Pty Ltd (No 2)* [2012] FCA 19 (unreported decision of Tracey J dated 23 January 2012) [18].

[109] *Australian Competition and Consumer Commission v C.I & Co Pty Ltd* [2010] FCA 1511, [31].

conditions so that the laying hens were able to move about freely on an open range every day'.[110]

The ACCC alleges that in fact, the eggs were produced by hens that were not able to move about freely on an open range every day due to several factors, including the stocking density of the barns, physical openings of the barns, conditions of the outdoor range, and the manner in which the hens were trained.

40.7 Artificial Photosynthesis as Alternative Agribusiness Technology

40.7.1 Revisiting the Dilemma

Earlier in this chapter, I suggested that while large agribusinesses dominate markets for food products generally and food animal products particularly, they will continue to face an intractable dilemma. I suggested that on the one hand, agribusinesses have become reliant on large-scale, intensive farming and animal husbandry practices to meet demand for their products.

However, on the other, those same agribusinesses are increasingly relying on clever advertising and marketing involving credence claims to exploit the growing interest shown by consumers and for whom such claims are important.

These strategies are largely antithetical. It is almost impossible for a large agribusiness processing chickens and eggs (for example) using intensive husbandry practices to simultaneously market chicken meat and eggs with "free range" or "free to roam" credence claims. Much of the litigation associated with credence claims discussed in this chapter are examples of this dilemma, resolving against the interests of both consumers and food suppliers.

However, what if alternative agribusiness technologies in the form of artificial photosynthesis could be realistically developed and implemented? Could artificial photosynthesis provide a "circuit breaker" to this dilemma?

40.7.2 Artificial Photosynthesis

More solar energy strikes the Earth's surface in 1 h of each day than the energy used by all human activities in one year. At present the average daily power consumption required to allow a citizen to flourish with a reasonable standard of living is about 125 kWh/day. Much of this power is devoted to transport (~40 kWh/day), heating (~40 kWh/day) and domestic electrical appliances (~18 kWh/day), with the remainder lost in electricity conversion and distribution (MacKay 2009). Global energy consumption is approximately 450 EJ/year, much less than the solar energy

---

[110]ACCC Media Release, 10 December 2013 at http://www.accc.gov.au/media-release/accc-institutes-proceedings-against-free-range-egg-producers.

potentially usable at ~1.0 kW per square metre of the earth—3.9×10^6 EJ/year even if we take into the earth's tilt, diurnal and atmospheric influences on solar intensity (Pittock 2009).

Photosynthesis as a natural process is equally important with DNA in the progress of humanity. Photosynthesis provides the fundamental origin of our oxygen, food and the majority of our fuels; it has been operating on earth for over two billion years. Photosynthesis can be considered as a process of planetary respiration: it creates a global annual CO_2 flux in from the atmosphere and an annual O_2 flux out to atmosphere. In its present nanotechnologically-unenhanced form, photosynthesis globally already traps around 4000 EJ/year solar energy in the form of biomass (Kumar, Jones and Hann 2009). The global biomass energy potential for human use from photosynthesis as it currently operates globally is approximately equal to human energy requirements (450 EJ/year) (Hoogwijk, Faaij, van den Broek, Berndes, Gielen, Turkenburg 2003).

Artificial photosynthesis can facilitate other energy options H_2–based fuels, the most promising perhaps by combining it with atmospheric nitrogen to make ammonia. Ammonia is already is shipped, piped, and stored in large volumes in every industrial country around the world as an agricultural fertilizer. As a fuel, ammonia has been proven to work efficiently in a range of engine types, including internal combustion engines, combustion turbines, and direct ammonia fuel cells. Due to its high energy density and an extensive, existing ammonia delivery infrastructure, ammonia is ready for the market today as an alternative to gasoline.

If such artificial photosynthetic technology is incorporated into every building, road and vehicle on the earth's surface than the positive outcome will be that humanity's structure will be producing abundant safe, low carbon fuels and fertilizers. In such a world it will be much more feasible for communities and families to support many of their basic food needs off–grid through organic farming rather than relying on distant sourced food provide by large corporate marketing chains.

References

Bruce A (2011a) Consumer protection law in Australia. LexisNexis Butterworths, Chatswood, Chapters 1 and 2, pp 1–50
Bruce (2011b) Introduction to misleading or deceptive conduct, Chapter 3 in consumer protection law in Australia. LexisNexis Butterworths, Chatswood, pp 51–73
Bruce A (2013) Australian competition law, 2nd edn. LexisNexis Butterworths, Chatswood, p 4
Corones S (2010) Competition law in Australia, 5th edn. Thompson Reuters, Sydney, p 68
Galyean M, Ponce C, Schultz J (2011) The future of beef production in North America. Anim Front 1(2):29–32
Hocquette J-F, Chatellier V (2011) Prospects for the European beef sector over the next 30 years. Anim Front 1:20
Hoogwijk M, Faaij A, van den Broek R, Berndes G, Gielen D, Turkenburg W (2003) Biomass Bioenergy 25:119
Horrigan B (2003) Adventures in law and justice: exploring big legal questions in everyday life. UNSW Press, Kensington, p 321

Hume DA, Whitelaw CBA, Archibald AL (2011) The future of animal production: improving productivity and sustainability. J Agric Sci 1:2

Ibrahim D (2007) A return to descartes: property, profit and the corporate ownership of animals. Law Contemp Probl 70:89–93

Kehlbacher A, Bennet R, Balcombe K (2012) Measuring the consumer benefits of improving farm animal welfare to inform welfare labelling. Food Policy 37(627):628

Kumar A, Jones DD, Hann MA (2009) Energies 2:556

Lee F (2013) False or misleading credence claims: what's the harm? Compet Consum Law News 29:42

MacKay DJC (2009) Sustainable energy-without the hot air. UIT, Cambridge, p 204

Martin J (2006) Misleading claims and the trade practices act. Presentation to the 8th annual food regulation and labelling standards conference, Sydney, Australia, 23 November 2006

Mhurchu CN, Gorton D (2007) Nutrition labels and claims in New Zealand and Australia: a review of use and understanding. Aust N Z J Public Health 31(105):105

Napolitano F, Girolami A, Braghieri A (2010) Consumer liking and willingness to pay for high welfare animal-based products. Trends Food Sci Technol 21:537–538

Pittock AB (2009) Climate change. The science, impacts and solutions, 2nd edn. CSIRO Publishing, Collingwood, p 177

Richards C, Lawrence G, Loong M, Burch D (2012) A Toothless Chihuahua? The Australian competition and consumer commission, neoliberalism and supermarket power in Australia. Rural Soc 21(3):250

Robbins J (1987) Diet for a New America. H.J. Kramer Publishers, Tiburon, p 104

Sharman K, Kossew S (2008) From nest to nugget: an expose of Australia's chicken factories. Voiceless, Paddington, p 9

Thornton P (2010) Livestock production: recent trends, future prospects. Philos Trans R Soc Lond B Biol Sci 368:2854–2855

Vinnari M, Tapio P (2009) Future images of meat consumption in 2030. Futures 41:269

Winders B, Nibert D (2004) Consuming the surplus: expanding 'Meat' consumption and animal oppression. J Soc Policy 24(9):76

Part VI
Africa

The views expressed are those of the author and not necessarily those of the institutions and associations that she is associated with.

Food Law and Policy in Africa: An Introduction

The Africa Union (AU), the successor of the Organization of African Unity (OAU),[1] declared 2014 *the Year of Agriculture and Food Security in Africa.*[2] The Sub-themes of the Year—Increased agriculture production, productivity and value addition; Functioning agricultural markets (regional markets & trade); Increased investment financing (public & private) along the agriculture value chains; Towards ending hunger in Africa by 2025; and Building resilience to vulnerability and Managing Risks—lend insight into the current priorities of countries in Africa as far as food law and policy is concerned.

Food law is conceptualized differently in Africa than is the case in the West. Food law in Africa encompasses issues integral to food production and distribution (e.g. food safety) as well as issues relating to human rights and governance (freedom from eviction, right to take part in public affairs, and freedom from discrimination.). While food safety issues appear to dominate food law and policy in many developed countries, the situation is different in most countries in Africa is somewhat different. In Africa, food insecurity, food trade, and foreign investment in food and agriculture, rather than food safety, currently dominate law and policy. Food law in Africa is primarily organized at the country level and wide variations exist in terms of national approaches to food regulation and agricultural policies. However, with the push towards regional integration in the continent, laws and

---

[1] Africa Union, AU in a Nutshell, http://www.au.int/en/about/nutshell. The Durban Summit (2002) launched the AU and convened the 1st Assembly of the Heads of States of the African Union.

[2] Briefing Note: on 2014 Year of Agriculture and Food Security in Africa, Marking 10th Anniversary of the Adoption of CAADP. (2014).

policies at the continental level and sub-regional level are becoming increasingly important and highly relevant.[3]

Food law as a distinct legal discipline is at its nascent stage in most countries in Africa. The AU does not recognize food law as a distinct legal discipline, and there is no Africa-wide agency charged with developing food law and policy for the region or harmonizing the laws of the different countries in the region. Nevertheless, a growing number of continent-wide conventions and declarations are pertinent to food and agriculture in Africa and will undoubtedly shape food law and policy in the continent in the future. At the continental level, key instruments and initiatives include: the Constitutive Act of the African Union (2000); the New Partnership for Africa's Development (NEPAD) (2001); the Maputo Declaration on Agriculture and Food Security in Africa (2003); the AU Framework and Guidelines on Land Policy in Africa (2009), the African Model Law on Biosafety; the African Model Legislation for the Protection of the Rights of Local Communities, Farmers and Breeders, and for the Regulation of Access to Biological Resources (2000); the Draft Policy Framework for Investment in Agriculture (NEPAD-OECD); the Monitoring African Food and Agricultural Policies (MAFAP) Project; the African Water Vision for 2025; the African Water Facility; and the Comprehensive Africa Agriculture Development Program (2003). Other pertinent instruments and initiatives include: AU Africa Agenda 2063, "A shared Strategic Framework for inclusive Growth and Sustainable Development';" the Science, Technology and Innovation Strategy for Africa-2024 (STISA-2024)[4]; the Framework Work Programme on Climate Change Action in Africa[5]; and the 1 July 2013 Declaration to End Hunger in Africa by 2025.[6]

The Constitutive Act of the Africa Union ("Constitutive Act") provides a framework for the development of robust food law regimes in Africa.[7] According to the Constitutive Act, the objectives of the AU includes *inter alia* to: "Promote and defend African common positions on issues of interest to the continent and its peoples;" "Promote peace, security, and stability on the continent;"[8] "Promote democratic principles and institutions, popular participation and good

---

[3]The Treaty establishing the African Economic Community (AEC)—1991: commonly known as the Abuja Treaty.

[4]Adopted at the 23rd Ordinary Session of the Summit of the African Union concluded in Malabo, Equatorial Guinea on 27 June 2014.

[5]At the 23rd Ordinary Session of the Summit of the Africa Union, Member States endorsed this Framework as a continental framework that will guide the African Union, its Member States and the RECs in addressing climate change in the near future.

[6]The declaration was adopted during the High Level Meeting of African and International Leaders on a Renewed Partnership for a Unified Approach to end Hunger in Africa by 2025 within the CAADP Framework, initiated and organized by the African Union Commission, FAO, and the Lula Institute.

[7]Organization of African Unity (OAU), *Constitutive Act of the African Union*, 1 July 2000, available at: http://www.refworld.org/docid/4937e0142.html [accessed 5 November 2014].

[8]Id., Article 3(f).

governance;"[9] "Promote and protect human and peoples' rights in accordance with the African Charter on Human and Peoples' Rights and other relevant human rights instruments;"[10] "Establish the necessary conditions which enable the continent to play its rightful role in the global economy and in international negotiations;"[11] "Promote sustainable development at the economic, social and cultural levels as well as the integration of African economies;"[12] "Advance the development of the continent by promoting research in all fields, in particular in science and technology;"[13] and "Work with relevant international partners in the eradication of preventable diseases and the promotion of good health on the continent."[14]

Pursuant to Declaration 1 (XXXVII) of 2001, NEPAD was adopted as the integrated and comprehensive socio-economic development programme to accelerate Africa's renewal. NEPAD manages a number of programmes and projects in six theme areas that include "Agriculture and Food Security," "Climate Change and Natural Resource Management," and "Human Development."[15] Endorsed in 2003, the Comprehensive Africa Agriculture Development Program (CAADP) drives the agricultural program of NEPAD. The overall goal of CAADP is to "eliminate hunger and reduce poverty by improving agriculture across the African continent." CAADP is premised on four pillars: (i) Sustainable land and reliable water control systems; (ii) Private sector development, rural infrastructure, improved trade and market access; (iii) Increasing food supply and reducing hunger; and (iv) Agricultural research and dissemination of agricultural technology.[16] In furtherance of the goals of CAADP, in 2003, African governments committed to two "targets." First, governments made a commitment to achieving an annual agricultural growth target of 6 percent by 2015.[17] Second, governments pledged to allocate 10 per cent of their national budget to agriculture by 2008.[18] Governments have the option of signing the CAADP compact and incorporating the CAADP Compact into their agricultural agenda. Governments who sign the CAADP Compact are subjected to an independent review process. As of June 2012, about 30 have signed CAADP compacts and about 40 have engaged the CAADP process.[19]

Regional Economic Communities (RECs) in Africa are playing a significant role in terms of laying the foundation for robust food law regimes to emerge in the

---

[9] Id., Article 3(g).
[10] Id., Article 3(h).
[11] Id., Article 3(i).
[12] Id., Article 3(j).
[13] Id., Article 3(k).
[14] Id., Article 3(n).
[15] NEPAD, About. http://www.nepad.org/about.
[16] NEPAD, Overview, http://www.nepad.org/foodsecurity.
[17] http://www.nepad.org/foodsecurity/agriculture/about.
[18] Id.
[19] Mwangi Kimenyi, Brandon Routman, Andrew Westbury, CAADP at 10: Progress Towards Agricultural Prosperity (2012) (hereinafter "CAADP at 10").

continent. Africa currently boasts of multiple RECs, some with overlapping memberships. Eight RECs are considered the building blocks of the African Economic Community: (i) Economic Community of West African States (ECOWAS); (ii) East African Community (EAC); (iii) Southern African Development Community (SADC); (iv) Common Market for Eastern and Southern Africa (COMESA); (v) Intergovernmental Authority on Development (IGAD); (vi) Economic Community of Central African States (ECCAS/CEEAC); (vii) Community of Sahel-Saharan States (CEN-SAD); and (viii) Arab Maghreb Union (UMA). Important policy instruments emanating from the RECS include: Agriculture and Rural Development Strategy for the East African Community (2005–2030); the EAC's Agriculture and Rural Development Policy; the Framework for the West African Agricultural Policy-ECOWAP (July 2004), SADC's Regional Agricultural Policy; the SADC Protocol on Wildlife Conservation and Law Enforcement; the SADC Protocol on Fisheries; and the COMESA Agricultural Policy (CAP).

Because of grinding poverty and risks of famine and food insufficiency, food security concerns have thus far dwarfed concerns about food safety in Africa. Not surprising, the food law architecture in Africa is presently dominated by laws and policies pertaining to food security. At the continental level, the *Framework for African Food Security System* (FAFS) and the *Action Plan for Africa of the Global Strategy for Improving Statistics for Food Security, Sustainable Agriculture and Rural Development (2011–2015)*, articulate Africa's strategy for combating food insecurity. At the domestic level, laws such as the Zanzibar Food Security and Nutrition Act of 2011 and policy documents such as Angola's National Food and Nutrition Security Strategy for 2010–2015 are emerging.

Regarding climate change, relevant laws and policies are just now emerging with a good number of countries still in the process of developing and implementing National Adaptation Programmes of Action (NAPAs) and National Climate Change Response Strategies. Examples include: *National Policy on Climate Change*—Namibia; *National Climate Change Response Strategy and Policy*—Zambia; *National Climate Change Response Policy White Paper*—South Africa; *Programme of Adaptation to Climate Change*—Ethiopia; Ethiopia's *Climate-Resilient Green Economy (CRGE)* initiative; *National Strategy on Climate Change and Low Carbon Development* –Rwanda; and *National Coastal Adaptation Law*—Gabon.

Biopiracy is an issue of grave concern for indigenous communities in Africa and for governments in the region. One significant legal response to date is the *Swakopmund Protocol on the Protection of Traditional Knowledge and Expressions of Folklore* that was adopted in 2010 within the framework of the 17-member African Regional Intellectual Property Organization (ARIPO).[20] In the Protocol, Member States expressed concern "at the gradual disappearance, erosion, misuse, unlawful exploitation and misappropriation of traditional knowledge and

---

[20] ARIPO, Swakopmund Protocol on the Protection of Traditional Knowledge and Expressions of Folklore (2010). http://www.wipo.int/wipolex/en/treaties/text.jsp?file_id=201022.

expressions of folklore."[21] One of the goals of the Protocol is to "protect traditional knowledge holders against any infringement of their rights as recognized."[22] The Protocol defines, protection criteria for traditional knowledge (Section 4), the formalities relating to protection of traditional knowledge (Section 5), the beneficiaries of protection of traditional knowledge (Section 6), and the rights conferred to holders of traditional knowledge (Section 7). The Protocol also addresses issues such as assignment and licensing (Section 8), equitable benefit-sharing (Section 9), exceptions and limitations applicable to protection of traditional knowledge (Section 11), compulsory licence (Section 12), duration of protection of traditional knowledge (Section 13), administration and enforcement of protection of traditional knowledge (Section 14) and access to traditional knowledge associated with genetic resources (Section 15). WIPO Director General Francis Gurry described the adoption of the Protocol as "a significant milestone in the evolution of intellectual property."[23] The Organization africaine de la propriete intellectuelle (OAPI), also adopted an instrument similar to the Protocol in July 2007. Despite winning the praise of organizations like WIPO, the Swakopmund Protocol is yet to gain sufficient ratification to enter into force.[24] Only a few countries have so far implemented the terms of the Swakopmund Protocol in their domestic legislation.

Courts and judicial tribunals—at the national, regional and continental levels—are increasingly shaping food law and policy in Africa and are increasingly being called upon to decide thorny issues pertaining to access to and right over land and water resources. In *The Animal Network for Animal Welfare v. The Attorney General of the United Republic of Tanzania* decided on 20 June 2014, the East African Court of Justice issued a decision barring the United Republic of Tanzania from constructing and maintaining a road across the northern wilderness of the Serengeti National Park on the grounds that such a move would violate Tanzania's obligation under the East African Community (EAC) Treaty to conserve and protect the environment. The tension between agribusiness and smallholder farmers is also looming large in Africa as exemplified by cases such as *West Coast Rock Lobster Association v The Minister of Environmental Affairs and Tourism*.[25] In *West Coast Rock Lobster*, appellants (commercial fishing entities), unsuccessfully sought an order nullifying a government decision granting subsistence fishers the right to catch and sell west coast rock lobster. In *West Coast Rock Lobster*, the Supreme Court of South Africa aptly noted that "For a fortunate few, rock lobsters conjure up images of exotic cuisine. For others, like communities

---

[21] Id., Preamble.

[22] Id.

[23] WIPO Director General Welcomes Moves to Enhance Protection of Traditional Knowledge & Folklore in Africa, Geneva, August 31, 2010, PR/2010/654.

[24] Id.

[25] *West Coast Rock Lobster Association v The Minister of Environmental Affairs and Tourism* (532/09) [2010] ZASCA 114 (22 September 2010).

who engage in subsistence fishing, they are a means of survival and a modest source of income."[26]

Although Africa has seen growing political commitment to promote food security, transforming political commitments into concrete results remains a challenge in most countries in the continent. Episodes of food shortages and food famines are still common in Africa.[27] Despite the growing number of legal instruments and policy documents that address food security, countries in Africa fall short on relevant indicators. Of any region, Sub-Saharan Africa has the highest prevalence of undernourishment (23.8%).[28] Between 1990–1992 and 2012–2014, the prevalence of undernourished Sub-Saharan Africa fell from 33.3 % to 23.8 %. Despite the decline in the prevalence of undernourished people in Africa, the absolute number of people who are undernourished in the region remains disturbingly very high (214.1 million up from 176.0 million in 1990–1992). According to *The State of Food Insecurity in the World 2014*:

> In general, in Africa, there has been insufficient progress towards international hunger targets, especially in the sub- Saharan region, where more than one in four people remain undernourished—the highest prevalence of any region in the world.[29]

While the foundation for robust food law regimes are in place in Africa, food law as a distinct legal discipline is yet to emerge in Africa and food law and policy is yet to receive adequate attention in the continent. There are major loopholes and weaknesses in the food law architecture of many countries in Africa. Some of the problems include:

- Reliance on outdated laws, most dating back to the colonial period;
- Complex regulatory and institutional environment dominated by overlapping and often inconsistent pieces of legislation that are managed by agencies with overlapping and competing mandates. Doting Kenya's food and agricultural law landscape are some 131 separate pieces of legislation, 41 implementing bodies, and about 1186 by-laws.[30]
- Inadequate or minimal enforcement of existing laws and policies;
- Policy incoherence and major inconsistencies in existing laws;
- Externally-driven legal framework and failure by governments to promote inclusive and participatory food laws and policies;

---

[26]Id.

[27]Somalia famine 'killed 260,000 people', BBC News, 2 May 2013. http://www.bbc.com/news/world-africa-22380352 (discussing the famine that hit Somalia from 2010 to 2012 killing nearly 260,000 people). See also Suzanne Goldenberg, Africa famine: soaring food prices intensifying crisis, report warns, The Guardian, 16 August 2011. http://www.theguardian.com/environment/2011/aug/16/africa-famine-food-prices-world-bank (discussing the food crisis in the Horn of Africa in the 2011 and tracing it to volatile global food supply).

[28]Food and Agricultural organization, *The State of Food Insecurity in the World 2014* 8 (2014).

[29]Id., at 9.

[30]Dr. John Omiti, CAADP and Agricultural Sector Governance Reform Processes in Kenya, in CAADP at 10, *supra* note 19, at 17.

- Lack of transparency in the development and implementation of food laws and regulations;
- Top-down approach to the development and implementation of food and agricultural law in the continent contrary to the promise made in instruments such as the African (Banjul) Charter on Human and Peoples' Rights (1981), the African Charter for Popular Participation in Development (1990), and the *Maputo Declaration on Agriculture and Food Security in Africa* (2003).
- Weak institutions and poor governance mechanisms;
- Prevalence of laws and policies that are nor informed by science and sound agricultural statistics;
- Laws and policies that are reactive rather than proactive;
- Lack of coordination at the national and regional level between different agencies responsible for food law administration;
- Confusion arising from lack of harmonization of laws and policies at the national and sub-regional level; and
- Lack of sensitivity in the development and implementation of laws, to the needs and problems of vulnerable groups and communities including women, persons with disabilities, and indigenous groups.

All six pillars identified in the *Common African Position on the Post-2015 Development Agenda* adopted on 31 January 2014, have implications for food law and policy in Africa.[31] The six pillars are: (i) structural economic transformation and inclusive growth; (ii) science, technology and innovation; (iii) people-centred development; (iv) environmental sustainability natural resources management, and disaster risk management; (v) peace and security; and (vi) finance and partnerships.[32] Within the rubric of "structural economic transformation and inclusive growth," African governments explicitly commit to promote "sustainable development, food self-sufficiency and nutrition" and resolve to:

- "Enhance the production, storage, transportation, availability, accessibility, utilization, safety and quality of food."[33]
- "Improve the productivity of smallholder agriculture and livestock through extension of technological support, small-scale irrigation schemes, rural infrastructure, credit and social services."[34]
- "Support modernization and diversification of agricultural sectors through: private sector participation in agriculture; agri-business development; improved agroindustry linkages; providing special support to integrate women into agri-business value chains; equitable access to land; and sustainable land

---

[31] African Union, Common African Position (CAP) on the Post-2015 Development Agenda (2014).
[32] Id., Article 16.
[33] Id., Article 22 (i).
[34] Id., Article 22 (ii).

management practices, including on our arable lands, for present and future generations."[35]
- "Promote agricultural marketing and information flows by establishing national and regional information centres and cooperation mechanisms in agriculture, food and nutrition security."[36]
- "Adopt sustainable agricultural, ocean and freshwater fishery practices and rebuild depleted fish stocks to sustainable levels."[37]
- "Strengthen resilience to external and climate shocks, such as droughts, floods, commodity price volatility, food shortages and export restrictions, particularly on staple foods."[38]
- "Urgently call for multilateral partnerships aimed at food loss reduction, resilience to commodity price fluctuations, and addressing food shortages and export restrictions during crises."[39]

Clearly, the foundations for the emergence of very robust food law regimes in Africa are already in place. What is needed is purposeful action directed at translating lofty commitments into tangible reality for millions in Africa.[40] Legal and regulatory reform and sound agricultural policy and food policy are priority areas in Africa. Without the necessary policy, legal and regulatory reform, Africa's aspirations for 2063 will remain but a distant dream. In *Agenda 2063: The Africa We Want*, among the vision espoused is that of Africa as a prosperous continent where: "[m]odern agriculture for increased production, productivity and value addition contribute to farmer and national prosperity and Africa's collective food security,"[41] "[t]he environment and ecosystems are healthy and preserved, and with climate resilient economies and communities,"[42] and Africa's agriculture is "modern and productive, using science, technology, innovation and indigenous knowledge."[43]

Strengthening the enabling legal and regulatory environment needed to increase agricultural productivity, improve food security and upgrade food governance

---

[35] Id., Article 22 (iii).

[36] Id., Article 22 (iv).

[37] Id., Article 22 (v).

[38] Id., Article 22 (vi).

[39] Id., Article 22 (vii).

[40] In addition to the instruments already mentioned, other notable instruments are: the African (Banjul) Charter on Human and Peoples' Rights (1981); the African Charter for Popular Participation in Development (1990); the African Charter on the Rights and Welfare of the Child (1990); the 1993 Cairo Declaration Establishing the Mechanism for Conflict Prevention, Management and Resolution; the Protocol on the Establishment of an African Court on Human and Peoples' Rights (1998); the 1999 Grand Bay (Mauritius) Declaration and Plan of Action for the Promotion and Protection of Human Rights; and the Conference on Security, Stability, Development and Cooperation (CSSDCA) Solemn Declaration (2000).

[41] Africa Union, Agenda 2063: The Africa We Want, Article 10 (2014).

[42] Id.

[43] Id., Article 13.

apparatus in countries in the Africa will not be easy. A host of thorny and complex issues will need to be addressed including, how to entrench gender equality particularly with regards to ownership and control of land and other productive assets; how to strike the proper balance between the goal of improving the productivity of smallholder agriculture and the goal of modernizing and diversifying the agricultural sectors through industrialization and agri-business development; how to protect traditional knowledge and stimulate home-grown innovations in the agricultural sector while respecting international commitments to protect intellectual property rights; and how to manage the continent's external relations and make relationships with traditional partners in the West and new partners in emerging markets such as China and Brazil work to Africa's advantage.

In conclusion, robust, pro-active, participatory and transparent food law architecture is urgently needed in Africa at the national, sub-regional and continental level. The necessary foundation has been laid. Across the board, countries have strengthened their political commitment to food security and nutrition. The time has come for purposeful action. Delay is costly and is no longer an option.

Chapter 41
Food Law and Policy in Africa: Emerging Legal Framework, Key Issues, Major Gaps and Challenges

Uche Ewelukwa Ofodile

Abstract This chapter offers a rare and penetrating insight into the legal and institutional framework of food law and policy in Africa. The chapter analyzes continental, regional and national laws and judicial decisions that are shaping food law policy in Africa and identifies key actors and institutions. How food law is conceptualized in Africa, the legal instruments (treaties, resolutions, declarations, statutes, policy papers, etc.) that underpin and inform decisions about food regulations in Africa, and current struggles over food and land in the continent are some of the issues addressed in this chapter. The chapter is the four parts. Part Two provides a survey of Africa Union (AU) instruments pertaining to food law. The section also offers an overview of legal and policy developments at the sub-regional and national levels. In all, legal instruments pertinent to food and agriculture in four regional economic communities in Africa—the East African Community (EAC), the Economic Community of West African States (ECOWAS), the Common Market on Eastern and Southern Africa (COMESA) and the South African Development Community (SADC)—are examined in this section. Part Three examines key issues now emerging in food law and policy discussions and debates in Africa as well as issues that are likely to shape discussions about food law and policy in the continent in the coming years. In all, seven issues are examined: Food Security; Food Safety; Women and Agriculture in Africa; Climate Change; Indigenous Peoples and Food in Africa; Agribusiness/Foreign Investment in Land; and Food Trade. Part Three, "*Towards a Coherent and Functioning Food Law Regime in Africa,*" identifies existing gaps in the food law and policy framework of countries in Africa, draws attention to the to the development of robust food law regimes in Africa, discusses the potential role of the

---

U.E. Ofodile (✉)
University of Arkansas School of Law, Fayetteville, AR, USA

Arkansas Bar Foundation, Little Rock, AR, USA

African Society of International Law (AFSIL), Midrand, South Africa

Committee on Corporate Social Responsibility, American Bar Association Section of International Law (ABA-SIL), Chicago, IL, USA

International Investment and Development Committee, ABA-SIL, Chicago, IL, USA
e-mail: uchee@uark.edu

legal profession in Africa in the development of food law in the continent, and offers some concluding thoughts about the future direction of food law and policy in Africa. Overall, answers to several questions are explored in this chapter. For example, how is food law conceptualized in Africa? Is food law conceptualized differently in Africa than is the case in the West? Is food law recognized as a distinct legal field in countries in Africa? What laws, policies and institutions inform food regulation in Africa today? What areas of food law and policy in Africa call for more rigorous analysis and urgent action? Is food law and policy in Africa proactively addressing emergent issues and challenges such as climate change? What local, regional and global issues are likely to shape food law in Africa in the coming years? The conclusion reached is that food law is a new and underdeveloped field of law in Africa, although most countries in the continent have laws that deal with various aspects of food law. To the extent that it is treated as a distinct legal field, food law is conceptualized somewhat differently in Africa than is the case in the West. Food law in Africa encompasses issues integral to food production and processing (e.g. food safety), issues pertinent to food distribution (e.g. regional and global food trade), as well as issues relating to human rights and governance (access to land, freedom from eviction, right over traditional knowledge, and freedom from discrimination.). However conceptualized, food law is still in its infancy in Africa. While the foundations for the emergence of strong food law regimes in Africa appear to be very strong, robust and credible food law frameworks are not yet in place at the national, sub-regional or continental level.

41.1 Introduction

How is food law conceptualized in Africa? Is food law conceptualized differently in Africa than is the case in United States, the European Union and other Western Nations? Is food law recognized as a distinct legal field in countries in Africa? What laws, policies and institutions inform and implement food law in Africa today? When it comes to food law and policy in Africa, what issues presently dominate and who are the key actors? What areas of food law and policy in Africa call for more analysis and action? What is the place of food law in Africa's regulatory framework? Is food law and policy in Africa proactively addressing emergent issues and challenges such as climate change? What local, regional and global issues are likely to shape food law in Africa in the coming years? These questions and more are timely because although food and agriculture is very important to livelihood, survival and economic growth in Africa, very little is known about food law in the continent. Food law will be important in Africa in the coming years given increased pressure on land and water resources in the continent, growing conflicts over land and other natural resources, old and new threats to food security, and increased push by formerly disenfranchised groups for equity and justice as regards access to and control of land and other natural resources.

The conclusion reached is that food law is a new and underdeveloped field of law in Africa, although most countries in the continent have laws that deal with various aspects of food law. To the extent that it is treated as a distinct legal field, food law is conceptualized somewhat differently in Africa than is the case in the West. Food law in Africa encompasses issues integral to food production and processing (e.g. food safety), issues pertinent to food distribution (e.g. regional and global food trade), as well as issues relating to human rights and governance (access to land, freedom from eviction, right over traditional knowledge, and freedom from discrimination.).

However conceptualized, food law is still in its infancy in Africa. While the foundations for the emergence of strong food law regimes in Africa appear to be very strong, robust and credible food law frameworks are not yet in place at the national, sub-regional or continental level. Forming part of the framework of continental policies pertaining to food and agriculture in Africa are a host of instruments including:

- the 1968 African Convention for the Conservation of Nature and Natural Resources (revised in 2003);
- the African Model Legislation for the Protection of the Rights of Local Communities, Farmers and Breeders, and for the Regulation of Access to Biological Resources (2000)
- the African Model Law on Safety in Biotechnology (first finalized in 2001);
- the Maputo Declaration on Agriculture and Food Security in Africa (2003);
- the Declaration on Climate Change and Development (2007);
- the Sharm El-Sheik Declaration on Water and Sanitation (2007);
- the Resolution of the Abuja Food Security Summit (2006);
- the Decision and Declaration of the Africa Union on Climate Change and Development in Africa (2007);
- the Algiers Declaration on Climate Change (2008);
- the Swakopmund Protocol on the Protection of Traditional Knowledge and Expressions of Folklore within the Framework of the African Regional Intellectual Property (2010);
- the Common African Position on the Post-2015 Development Agenda (2014); and
- the Africa Agenda 2063 (2014).

Although there are many continent-wide policy statements, declarations resolutions, and initiatives pertaining to food, policies have not been thoroughly and consistently written into national legislation. Poor policy alignment and lack of harmonization is a problem. Food law is not harmonized in Africa. Presently, the African continent lacks an overarching food law legislation or policy or anything comparable to Regulation (EC) No 178/2002 of the European Parliament and of the Council of 28 January 2002 which laid down the general principles and requirements of food law, established the European Food Safety Authority and laid down procedures in matters of food safety. Poverty, poor governance mechanisms, and lack of political will appear to be major factors that have thus far undermined the development of food law in Africa.[1] Because the primary producers of food in the

---

[1] "Poverty and the International Covenant on Economic, Social and Cultural Rights" (E/C.12/2001/10).

continent are poor, disenfranchised and marginalized rural dwellers, little attention has, to date, been given to the role of law and institutions in food production, processing, and distribution. Key stakeholders have typically been excluded when food law and policies are developed and implemented in Africa. Thus, although integral to the food system in Africa, small holders, pastoralists, fisherfolks, forest users, indigenous people, landless workers such as agricultural laborers and sharecroppers, and women are rarely invited to participate in food governance.

There is an urgent need for strong food law frameworks in Africa at the national, sub-regional, and continental level. Urgent attention needs to be paid to designing and implementing appropriate and effective food laws and policies in Africa. The 2014 Ebola outbreak in West Africa underscored the need for strong food law regimes in Africa—regimes strong enough to promptly identify and effectively manage the complex linkages between food and public health. Clearly, food can be a vehicle for tropical disease. Conversely, public health emergencies can undermine, overwhelm and compromise local food systems with devastating consequences.[2]

While painting a general picture of the laws, policies and institutions that structure and shape food law in Africa today, this chapter focuses on seven key issues: (i) food security; (ii) food safety, including biosafety; (iii) the gender dimension of food law in Africa; (iv) climate change; (v) food law and indigenous groups in Africa; (vi) agribusiness, including foreign investment in agricultural land in Africa; and (vii) the intersection of food law and global trade rules. Section 41.2 offers an overview of the legal framework for food law in Africa at the continental and sub-regional levels. Section 41.3 examines in some detail, the seven issues identified above. The chapter will conclude in Sect. 41.4 with a look at the challenges to the development of robust food law frameworks in Africa and the potential role the legal profession in Africa can play to encourage development in this field.

41.2 Food Law and Policy in Africa: The Emerging Legal Framework

An analysis of food law architecture in Africa should rightly start with an assessment of the policy initiatives emanating from the Regional Economic Communities (RECs) in the continent as well as the activities of the principle organization of African States—the African Union (previously Organization of Africa Unity). This section offers a survey of Africa Union (AU) instruments relevant to food law. The section also offers an overview of legal and policy developments at the sub-regional and national levels. Africa currently boasts of multiple RECs, some with overlapping memberships. Eight RECs are considered the building blocks of the

---

[2]DiLorenzo (2014).

African Economic Community: (i) Economic Community of West African States (ECOWAS); (ii) East African Community (EAC); (iii) Southern African Development Community (SADC); (iv) Common Market for Eastern and Southern Africa (COMESA); (v) Intergovernmental Authority on Development (IGAD); (vi) Economic Community of Central African States (ECCAS/CEEAC); (vii) Community of Sahel-Saharan States (CEN-SAD); (viii) Arab Maghreb Union (UMA). In addition to AU instruments and policies, instruments pertinent to food and agriculture in four RECS—EAC, ECOWAS, COMESA and SADC—will be examined in this section.

41.2.1 Food, Law, and the Africa Union

The AU does not have an overarching food law treaty. Food law is not harmonized in Africa and there is no Africa-wide agency presently tasked with coordinating the food law legislation of the member states of the AU. Nevertheless, food and agriculture features prominently in *the New Partnership for Africa's Development* (NEPAD) adopted as a Programme of the AU in 2001. In 2003, African Heads of State and Government endorsed the *"Maputo Declaration on Agriculture and Food Security in Africa"* (Maputo Declaration).[3] In the Maputo Declaration, African Heads of State made several commitments in the area of food and agriculture. First, they resolved to revitalize the agricultural sector "through special policies and strategies targeted at small scale and traditional farmers in rural areas and the creation of enabling conditions for private sector participation."[4] Second, they resolved to implement the Comprehensive Africa Agriculture Development Program (CAADP). Third, they agreed to adopt sound policies for agricultural and rural development, and to allocate at least 10 % of national budgetary resources for their implementation within 5 years.[5] In the Maputo Declaration, African Heads of State also resolved to ensure the establishment of regional food reserve systems, and to ensure the development of policies and strategies to fight hunger and poverty in Africa.[6] Established in 2003, the CAADP is the agricultural programme of NEPAD and has as its goal the elimination of hunger and reduction of poverty in Africa through agriculture. CAADP aims "to help African countries reach a higher path of economic growth through agriculture-led development."[7] In 2014, the AU Joint Conference of Ministers of Agriculture, Rural Development, Fisheries and Aquaculture adopted seven Africa Accelerated Agricultural Growth and

---

[3] Maputo Declaration on Agriculture and Food Security in Africa, Assembly/AU/Decl. 7(II).
[4] Id., Para. 1.
[5] Id., Para. 2.
[6] Id., Para. 6.
[7] NEPAD, About CAADP, http://www.nepad-caadp.net/about-caadp.php.

Transformation Goals (3AGTGs) for 2025.[8] Food and agriculture also feature strongly in the *Common African Position on the Post-2015 Development Agenda* that was adopted on 31 January 2014 and the AU Africa Agenda 2063, "A shared Strategic Framework for inclusive Growth and Sustainable Development" that was adopted in 2014.

Although not directly related to food or agriculture, a number of conventions and declarations in the field of human rights and environmental law are shaping food law and policy in Africa and are likely to remain very relevant in the coming years. These include: the African (Banjul) Charter on Human and Peoples' Rights (1981),[9] the African Charter on the Rights and Welfare of the Child (1990),[10] and the Protocol to the African Charter on Human and Peoples' Rights on the Rights of Women in Africa (2003); the African Charter for Popular Participation in Development (1990); the 1999 Grand Bay (Mauritius) Declaration and Plan of Action for the Promotion and Protection of Human Rights; and the Protocol on the Establishment of an African Court on Human and Peoples' Rights (1998).

41.2.2 Food Law and the Regional Economic Communities in Africa

41.2.2.1 The East African Community

The EAC is a regional intergovernmental organization made up of five countries—the Republics of Burundi, Kenya, Rwanda, the United Republic of Tanzania, and the Republic of Uganda.[11] While the EAC does not have an overarching food law treaty and there is no agency charged with coordinating the food law legislation of Member States, Chapter 18 of the Treaty Establishing the East African Community (EAC Treaty) deals with 'Agriculture and Food Security.'[12] In the EAC Treaty, EAC Member States agree to ensure a common agricultural policy: food sufficiency within the Community, an increase in agricultural production, and post-harvest preservation and conservation and improved food processing.[13] In the EAC Treaty, Member States also agree to cooperate in specific fields of agriculture, including, *inter alia* "the development of food security within the Partner States and the Community as a whole, through the production and supply of foodstuffs"[14] as

---

[8]Meeting, held in Addis Ababa, Ethiopia, from 1st to 2nd May 2014.

[9]Adopted 27 June 1981, OAU Doc. CAB/LEG/67/3 rev. 5, 21 I.L.M. 58 (1982), entered into force 21 October 1986.

[10]OAU Doc. CAB/LEG/24.9/49 (1990), *entered into force* Nov. 29, 1999.

[11]http://www.eac.int/, http://www.eac.int/.

[12]Treaty For the Establishment of the East African Community (As amended on 14th December, 2006 and 20th August, 2007) (hereinafter "EAC Treaty."). Section 2(2) of the EAC Treaty provides that "the Contracting Parties shall establish an East African Customs Union and a Common Market as transitional stages to and integral parts of the Community."

[13]Id., at Article 105(2).

[14]Id., at 105(2)(b).

well as joint actions in combating drought and desertification.[15] Presently, the *Agriculture and Rural Development Strategy for the East African Community (2005–2030)* is guiding policy and programming in the EAC region.

41.2.2.2 The Economic Community of West African States

Pursuant to Article 3(2)(a) of the Treaty of ECOWAS, ECOWAS shall, by stages, ensure; "the harmonization and coordination of national policies and the promotion of integration programmes, projects and activities, particularly in food, agriculture and natural resources, industry, transport and communications, energy, trade, money and finance, taxation, economic reform policies, human resources, education, information, culture, science, technology, services, health, tourism, legal matters."[16] Chapter IV of the Treaty of ECOWAS is titled "Cooperation in Food and Agriculture" and calls on Member States to "co-operate in the development of agriculture, forestry, livestock and fisheries."[17]

ECOWAS is responsible for implementing CAADP in West Africa and has been organizing policies in line with this mandate. Legal and policy developments to date include: *Framework for the West African Agricultural Policy* adopted in July 2004; Decision A/DEC.11/01/05 adopting an agricultural policy for the Economic Community of the West African States—ECOWAP and annex to the decision (2005)[18]; Regional action plan for the implementation of the ECOWAS agricultural policy and CAADP/NEPAD in West Africa between 2006 and 2009 (June 2005); Memorandum on the increase in food prices: the situation, outlook, strategies and recommended measures (May 2008); the Regional Initiative for Food Production and the Fight Against Hunger (adopted in 2008); and the establishment of a Regional Fund for Agriculture and Food. ECOWAS has also endorsed a Regional Agricultural Investment Programme (RAIP).[19]

41.2.2.3 The Southern Africa Development Community

Food law is not recognized as a distinct field within SADC. However, food and agriculture are specifically mentioned in the SADC Treaty of 1992.[20] In the SADC Treaty, Member States agree to cooperate in a number of areas including, *inter alia*

---

[15]Id., at 105(2)(h).
[16]Treaty of ECOWAS. http://www.comm.ecowas.int/sec/?id=treaty&lang=en.
[17]Id., at Article 25(1).
[18]On the 19th January 2005, the Heads of State and Government Conference of West Africa adopted the Agricultural Policy of the Economic Community of West African States.
[19]http://www.aidfortrade.ecowas.int/programmes/raip.
[20]SADC Treaty 1992. http://www.sadc.int/documents-publications/sadc-treaty/.

"food security, land and agriculture;"[21] and "natural resources development and environment"[22] Pursuant to Article 22 of the SADC Treaty, which calls on Member States to conclude "such Protocols as may be necessary in each area of cooperation," SADC Member States have concluded a number of Protocols relevant to food and agriculture including the Protocol on Wildlife Conservation and Law Enforcement[23] and the Protocol on Fisheries.[24]

41.2.2.4 Common Market for Eastern and Southern Africa (COMESA)

COMESA does not have an overarching food law legislation or policy. However, food and agriculture are mentioned in the 1993 COMESA Treaty. In the COMESA Treaty, Member States specifically agree to: cooperate in agricultural development; adopt a common agricultural policy; enhance regional food sufficiency; cooperate in the export of agricultural commodities; coordinate their policies regarding the establishment of agro-industries; cooperate in agricultural research and extension; and enhance rural development.[25] Chapter 15 of the COMESA Treaty is titled "Standardization and Quality Assurance," Chapter Sixteen is titled "Co-operation in the Development of Natural Resources, Environment and Wildlife," while Chapter Eighteen is titled "Co-operation in Agriculture and Rural Development." Chapter Eighteen lays down the framework for cooperation in a number of areas including: Supply of Staple Foods,[26] the Export of Agricultural Commodities,[27] Agricultural Research and Extension,[28] Drought and Desertification Management,[29] Rural Development, as well as Strengthening Farmers Participation in Agricultural development.[30]

COMESA has developed a number of initiatives relevant to food law and policy in the region. These include: the COMESA Agricultural Strategy, the COMESA Agricultural Policy, and the *Declaration of the Second Meeting of the Ministers of Agriculture on "Expanding Opportunities for Agricultural Production, Enhanced Regional Food Security, Increased Regional Trade and Expanded Agro-exports through Research, Value Addition and Trade Facilitation"*. Two major objectives of COMESA's agricultural programs are sustainable regional food security and enhanced regional integration. COMESA's Food and Agricultural Marketing

---

[21] Id., Article 21(3)9a).
[22] Id., Article 21(3)(e).
[23] Adopted 18th August 1999 and entered into force on November 30, 2003.
[24] Adopted on 14 August 2001 and entered into force 8 August 2003.
[25] COMESA Treaty, Article 4(5).
[26] Id., at Article 131.
[27] Id., at Article 132.
[28] Id., at Article 134.
[29] Id., at Article 135.
[30] Id., at Article 27.

Information System—a web portal that provides up-to-date information on food-related initiatives (regional food security, market information system, and sanitary and phytosanitary measures) in the COMESA region is now up and running.[31]

41.3 Food Law in Africa: Emerging Issues

This section focuses on key issues emerging in the food law architecture of Africa as well as issues that will shape discussions about food law and policy in the continent in the coming years. The decision about focus areas is based on an assessment of important policy documents at national and regional levels in Africa. An assessment of these documents suggest that in Africa a number of issues are addressed under the rubric of food law, including: food security; right to food; food safety; women's access to land; climate change; the plight of indigenous peoples; agribusiness; and food trade.

41.3.1 Food Security in Africa

Food security is said to exist, "when all people, at all times, have physical, social and economic access to sufficient, safe and nutritious food that meets their dietary needs and food preferences for an active and healthy life."[32] The *Plan of Action and the Declaration of the World Food Summit* states that food security means "the access of all people to sufficient, safe and nutritious food to meet their dietary needs and food preferences for an active and healthy life." This is not yet the situation in Africa. Although Africa has the potential to eliminate hunger, ensure sustainable food security and produce enough food for export, poverty, malnutrition, and food insecurity remains prevalent in the region. At least 30 % of the population of Africa is chronically and severely undernourished.[33] 214.1 million people in Africa are considered undernourished according to *The State of Food Insecurity in the World* 2014.

Constraints to achieving food security in Africa are legion and include, *inter alia*: inadequate laws and institutions; poor infrastructure; low levels of public and private investment in agricultural production in Africa; political instability and social unrest; extreme weather patterns due to climate change; lack of support for small-scale farmers; poor or limited access to affordable agricultural credit for small-scale farmers; limited attention to the development of value-added agro-

---

[31]COMESA, Food and Agricultural Marketing Information System. http://famis.comesa.int/com/option.com_news/yid.32/Itemid.120/pillar.foodsecurity/lang.en/sectionid.COMESA-Policy-Reform/.
[32]FAO (2001).
[33]Maputo Declaration, *supra* note 3.

processing industries in Africa; limited attention to the risks in agriculture (production risks, market risks, input cost risks, transaction risks as well as food safety risks); poor and inadequate farmer's institutions; prevalence of major human and animal transboundary diseases; increased pressure on land and natural resources as a result of other development imperatives; social unrests and civil strife that disrupt food production and food distribution; entrenched discrimination against vulnerable groups such as women, indigenous groups, and people with disabilities with respect to access to and control of land and other productive resources; and post-harvest losses, a result of inadequate storage facilities and limited opportunities for food processing. Poor statistics and information also contribute significantly to food insecurity in Africa.

41.3.1.1 The Legal Framework

Food insecurity is a major concern in Africa and is one of the primary focus of legislative and policy development in the region. At the continental level, initiatives that have been adopted to address food insecurity include the *Framework for African Food Security System* (FAFS) and the *Action Plan for Africa of the Global Strategy for Improving Statistics for Food Security, Sustainable Agriculture and Rural Development (2011–2015)*.[34] The African Development Bank Group has also launched a number of initiatives specific to the agricultural sector and is involved in implementing *the Action Plan (2011–2015) on Improving Statistics for Food Security, Sustainable Agriculture, and Rural Development*.[35]

Several RECs in Africa have adopted specific measures to address food insecurity. In the EAC region, measures to address food insecurity include the East African Community Food Security Action Plan (2011–2015)[36] and the EAC Climate Change Policy.[37] Priority areas for the EAC Food Security Action Plan include: provision of enabling policy, legal and institutional framework; increasing food availability; improving access to food; improving stability of food supply and improving capacity for emergency preparedness and response; and enhancing the efficiency of food utilization, nutrition and safety. Food Security is specifically addressed in Article 110 of the EAC Treaty with Members committing to, *inter*

---

[34] The plan has been endorsed by at least four institutions: The UN Economic Commission for Africa (ECA); the Food and Agricultural Organization (FAO), the African Union Commission; and the African Development Bank.

[35] African Development Bank, The Bank Group's Participation in Implementing The Action Plan (2011–2015) on Improving Statistics for Food Security, Sustainable Agriculture, and Rural Development (June 2013).

[36] EAC Secretariat, East African Community Food Security Action Plan 2011–2015, (February 2011).

[37] Id.

alia: "harmonise food supply, nutrition and food security policies and strategies"[38] and "initiate and maintain strategic food reserves".[39]

Article 25 of the ECOWAS Treaty stipulates that "Member States shall co-operate in the development of agriculture, forestry, livestock and fisheries in order to: a) ensure food security; b) increase production and productivity in agriculture, livestock, fisheries and forestry, and improve conditions of work and generate employment opportunities in rural areas; c) enhance agricultural production through processing locally, animal and plant products.; and d) protect the prices of export commodities on the international market."

Ensuring regional food security and sufficiency is explicitly listed as one of the objectives of cooperation in agricultural development in the COMESA Treaty.[40] The COMESA Secretariat has also implemented a number of initiatives including: Agricultural Market Promotion and Regional Integration Project, Regional Food Security/Food Reserve Initiative among member states, the Food Security Policy and Vulnerability Reduction Program, and Coordinated Agricultural Research and Technology Interventions. Food security is a topic addressed in COMESA's FAMIS web portal. The web portal provides information on: COMESA Policy Reform, Food Security Alerts, Annual Food Balance Sheet, and Food Security Updates.[41]

In the *Dar-es-Salam Declaration on Agriculture & Food Security in the SADC Region* (2004), Member States of SADC agreed to implement short-, medium-, and long-term objectives to advance the state of agriculture and food security in Southern Africa. Other initiatives of SADC include the *SADC Multi-Country Agricultural Productivity Programme* (MAPP),[42] SADC Protocol on Fisheries (2001), and the Regional Indicative Strategic Development Plan (RISDP).[43] "Promot[ing] and enhanc[ing] food security and human health" as well as "safeguard[ing] the livelihoods of fishing communities" are among the expressed objectives of the SADC Fisheries Protocol.[44]

At the national level, countries are at different stages in terms of formulating and/or implementing national food security policies and establishing the necessary governance structure. Examples of national commitment to the food and nutrition security agenda include: Government Resolution No. 6/2004 of 18 February 2004 on the National Food Security Sustainable Strategy (Cape Verde)[45]; National Food

---

[38]EAC Treaty, *supra* note 12 Id., Article 110 (d).

[39]Id., Article 110 (e).

[40]COMESA Treaty, *supra* note 25, at Article 129.

[41]http://famis.comesa.int/com/option.com_news/yid.32/Itemid.120/pillar.foodsecurity/lang.en/sectionid.COMESA-Policy-Reform/.

[42]http://www.sadc.int/documents-publications/show/SADC%20Multi-country%20Agricultural%20Productivity%20Programme%20(MAPP)%20Document.

[43]http://www.sadc.int/about-sadc/overview/strategic-pl/regional-indicative-strategic-development-plan/.

[44]SADC Fisheries Protocol, Article 3(a).

[45]Creating the National Food Security Strategy (ENSA) and the corresponding National Food Security Plan (PNSA) for the period 2003–2015.

and Nutrition Security Strategy (Mozambique)[46]; National Food and Nutrition Security Plan (Mozambique); and the National Food and Nutrition Security Strategy for 2010–2015 (Angola).

Signed into law on 6 July 2011, the Zanzibar Food Security and Nutrition Act (Zanzibar Food Act) is intended to cover all issues related to food availability, food accessibility, food utilization, food stability and the realization of the right to adequate food in Zanzibar.[47] The Zanzibar Food Act establishes the National Food Security and Nutrition Council and vests it with the responsibility of overseeing the implementing Zanzibar's Food Security and Nutrition Policy and Programme.[48] Additionally, the Zanzibar Food Act establishes the Zanzibar Food Reserve, the purpose of which is to ensure a reliable supply of food in Zanzibar, meet local shortfalls in the supply of food, meet any other food emergencies, and solve problems relating to the supply of food in Zanzibar.[49] The Act calls for the establishment of the Food Insecurity and Vulnerability Information Mapping System, which is intended to provide the information needed to galvanize and strengthen the capacity to respond to food emergency and food aids.[50]

Countries are also establishing inter-sectoral governance structures to support the implementation of their food security laws. Examples include: the Technical Secretariat for Food and Nutrition Security (Mozambique); the National Food Security Council (Cape Verde) and the National Food Security Network (Cape Verde); and the National Food and Nutrition Security Council (Angola). As part of the Community of Portuguese Language Countries (CPLP), the Portuguese-speaking African Countries (PALOP)—Angola, Mozambique, Cape Verde, Guinea-Bissau and Sao Tome and Principe—are also addressing food insecurity.[51] CPLP is working on establishing a Working Group on Food Security and Agriculture and adopting an Action Plan for Food Security. In the Dili Declaration, adopted in July 2014 at the Tenth Summit of Heads of State and Government of CPLP, Member States placed food and nutrition security as a permanent issue on the organization's agenda until 2025.[52]

41.3.1.2 Food Security Litigation in Africa

According to the Office of the High Commissioner for Human Rights, "the concept of food security itself is not a legal concept per se and does not impose obligations

---

[46]Resolution No. 56/2007 of 16 October 2007 reviewed the 1998 strategy and ushered in ESAN II for the period 2008–2015.

[47]The Zanzibar Food Security and Nutrition Act, 2011, Article 3(2).

[48]Id., Article 5 and 6.

[49]Id., Article 25(1) and (2).

[50]Id., Article 26(1).

[51]Members of CPLP are: Angola, Brazil, Cape Verde, Guinea-Bissau, Mozambique, Portugal, Sao Tome & Principe, Timor-Leste and Equatorial Guinea.

[52]Food and Agricultural Organization, FAO welcomes commitment to fight hunger from the Community of Portuguese Language Countries, 23 July 2014.

on stakeholders nor does it provide entitlements to them."[53] There are few, if any, lawsuits challenging the food security initiatives that the different governments in Africa have adopted or pressing for their full implementation. However, a growing number of cases that have implication for the food security of particular groups have been decided in a number of countries in Africa and by some regional courts. In *George v. Ministers of Environmental Affairs and Tourism Appeal,* respondents, representing 5000 artisanal fishers in South Africa, lodged applications in both the high court and the Equality Court asserting that the Minister of Environmental Affairs and Tourism failed to provide fair and just access to fishing rights and demanding an order giving them equitable access to fishing. In *Commercial Farmers Union v Minister of Lands,*[54] the Supreme Court of Zimbabwe, while addressing the issue of land expropriation, discussed the history of land injustice in Zimbabwe and appeared to acknowledge the need for a land reform program that is guided by the rule of law.

At the regional level, the African Commission on Human and Peoples' Right (African Commission) has ruled on the right to food and related rights in two important cases: *The Social and Economic Rights Action Center and the Center for Economic and Social Rights v. Nigeria* (SERAC)[55] and the *Centre for Minority Rights Development (Kenya) and Minority Rights Group International on behalf of Endorois Welfare Council v Kenya (Endorois).*[56] In *Endorois,* a case involving the forced eviction of the an indigenous group in Kenya from their ancestral land, the African Commission on Human Rights concluded that the community's right to culture protected under Article 17(2) and (3) of the African Charter had been violated. According to the African Commission: *"By forcing the community to live on semi-arid lands without access to medicinal salt licks and other vital resources for the health of their livestock, the Respondent State have created a major threat to the Endorois pastoralist way of life."*[57] In *SERAC,* complainants alleged that the oil exploration and extraction activities of oil companies operating in the Niger Delta region of Nigeria resulted in the contamination of water, soil and air and caused serious short and long-term health impacts, including skin infections, gastrointestinal and respiratory ailments, increased risk of cancers, and neurological and reproductive problems. Although the African Charter does not explicitly provide for the right to food, the African Commission concluded that African Charter implicitly guaranteed the right to food and that the minimum core of the right to food "requires that the Nigerian Government should not destroy or contaminate food sources"[58] and "should not allow private parties to

---

[53]Id., p. 4.

[54]2001 (2) SA 925 (ZSC).

[55]*Social and Economic Rights Action Center & the Center for Economic and Social Rights v. Nigeria.* Cited as: Communication No. 155/96.

[56]*Centre for Minority Rights Development (Kenya) and Minority Rights Group International on behalf of Endorois Welfare Council v Kenya.* 276/2003 (hereinafter "Endorois").

[57]Id. at § 251.

[58]Communication No. 155/96.

destroy or contaminate food sources, and prevent peoples' efforts to feed themselves."[59]

41.3.2 Food Safety and Food Law in Africa

Most countries in Africa have a host of laws and regulations pertaining to food safety. However, because of grinding poverty and recurring episodes of famine and food crisis, food security concerns have thus far dwarfed food safety concerns in Africa.[60] The result is that although most countries have laws on the books that address various aspects of food safety, most of the laws are outdated, are not based on evidence, and are rarely, if ever, implemented. Most countries in the region lack analytical capacity for food-borne disease surveillance and research. Linkages, cooperation, collaboration and coordination among national food safety agencies is extremely limited. Furthermore, only a few countries include food safety in their national development plans and national health policies. Finally, although most countries in the region have taken a stand against products containing genetically modified organisms, most do not have the legal and institutional framework necessary to ensure that such products do not enter their border.

There is presently no continent-wide food safety law or system, no comprehensive or integrated approach to food safety in Africa, no comprehensive system of traceability within food and feed businesses in the continent, and no generally-agreed principles upon which food and feed may be traded throughout the continent. There is also no continent-wide accepted definition of food, no continent-wide requirement for feed, and no measure in place at the continental level to guarantee that unsafe foods do not enter the stream of commerce or to address food safety emergencies. Finally, there is no agency comparable to the European Food Safety Authority tasked with providing scientific advice and scientific and technical support for continent-wide legislation and policies that have direct or indirect impact on food safety. At the level of the regional economic communities, there are plans on-going to harmonize food safety laws and policies. However despite political commitments to harmonize rules at the sub-regional level, little progress has been made on this front.

Risks associated with food in Africa are microbiological, chemical, physical, radioactive, and biotechnological. The major causative agents of foodborne diseases in the African region according to the World Health Organization (WHO) are bacteria, parasites and viruses. Other concerns include foodborne zoonotic diseases and chemical contamination of food from pesticides and veterinary drug residues. Several countries in Africa have experienced outbreaks of food-borne diseases such

---

[59]Id.

[60]AFR/RC57/4 1 (30 August 2007). ("Because of insufficient food to meet demand on the African continent, the majority of people are only concerned with satisfying hunger and do not give due attention to the safety of food.").

as salmonellosis, entero-haemorrhagic *Escherichia coli* (EHEC), hepatitis A, cholera, and acute aflatoxicosis.

41.3.2.1 Food Safety Laws in Africa: A Survey

There is much diversity in the food safety law situation of countries in Africa. In the absence of harmonization, countries are taking varied and different steps to strengthen the food safety laws, policies and institutions.

Food Safety Laws at the Sub-Regional Level

There are plans to harmonize food safety laws along sub-regional lines. Under the EAC Treaty, Members States agree to: "harmonise policies, legislation and regulations for enforcement of pests and disease control;" "harmonise and strengthen regulatory institutions;" "harmonise and strengthen zoo–sanitary and phyto-sanitary services inspection and certification;" "establish regional zoo–sanitary and phyto-sanitary laboratories to deal with diagnosis and identification of pests and diseases;" "adopt common mechanism to ensure safety, efficacy and potency of agricultural inputs including chemicals, drugs and vaccines;" and "co-operate in surveillance, diagnosis and control strategies of transboundary pests and animal diseases".[61] In the ECOWAS Treaty, Member States agree to cooperate in "the adoption of a common agricultural policy especially in the fields of research, training, production, preservation, processing and marketing of the products of agriculture, forestry, livestock and fisheries."[62]

The EAC, ECOWAS, COMESA and SADC all have in place protocols or draft protocols on sanitary and phyto-sanitary measures.[63] For instance, the *COMESA Regulations on the Application of Sanitary and Phytosanitary Measures* (COMESA SPS Regulation) was adopted in 2009 with the goal of setting out principles and creating mechanisms for cooperation in the implementation of SPS measures by Member States.[64] The COMESA SPS Regulation established a COMESA Green Pass program—"a commodity-specific SPS certification scheme and authority for movement of food and agricultural products within the Common Market, issued by a National Green Pass Authority."[65] Approved in 2009, the EAC Protocol on Sanitary and Phytosanitary Measures remains in a draft form awaiting approval from the *EAC Sectoral Council on Legal and Judicial Affairs*. The East African

---

[61]EAC Treaty, supra note 12., at 108.
[62]ECOWAS Treaty, supra note 16, at 25(2)h).
[63]See e.g. Sanitary and Phytosanitary (SPS) Annex to the SADC Protocol on Trade (Approved by the SADC Committee of Ministers of Trade on 12 July 2008, Lusaka, Zambia. See also,
[64]The COMESA Regulations on the Application of Sanitary and Phytosanitary Measures (2009). Article 2(a). http://famis.comesa.int/popups/articleswindow.php?id=220&print=print.
[65]7(1).

Community Sanitary Measures on Fish and Fisheries Products was launched recently in March 2014.[66] Also existing within the EAC are: Harmonized Phytosanitary Measures and Procedures for Plants Vol. I; Harmonized Sanitary Measures and Procedures for Mammals, Birds and Bees Vol. II; Harmonized Sanitary Measures and Procedures for Fish and Fishery Products Vol. III; and Harmonized Measures for Food Safety Vol. IV. Altogether the four volumes specify rules that apply to the importation and exportation of specified products. The EAC Sanitary Measures on Fish and Fisheries products was launched on 12 March 2014, and consists of documents on Sanitary and Phytosanitary (SPS), Manual of Standard Operating Procedures (MSOP) and the Inspector's Guide.[67]

Food Safety: National-level Laws and Policies

Most countries in Africa have laws that address some aspects of food safety.[68] Some laws address public health generally, some laws address food safety very broadly, and other laws address specific issues such as pests, food inspection, or use of pesticide. Examples of general public health legislations include: Public Health Act Cap 242—Kenya; Public Health Act No. 36 of 1919—Namibia; Decree-law No. 1/16 of May 1982 regarding a code on public health—Burundi); Law No. 64/LF/123 of November 1964 regarding the public health protection—Cameroon). Examples of general food safety laws in Africa include: Law 84–009, Basic law governing the control of staple food (Benin Republic); Food Control Act, 1993 (No. 11 of 1993)—Botswana; Food Act 1987 (Act No. 14 of 1987)—Seychelles; Standards Act No. 29 of 1993—South Africa; the Agricultural Products Standards Act, 1990 (Act No. 119 of 1990—South Africa; and Food and Drugs (Amendment) Decree 1999 (No. 21 of 1999)—Nigeria. In addition to general food safety legislation, many countries have laws addressing specific products. Meat appears to be heavily regulated in many countries.[69] A growing number of countries have laws that address specific food safety concern such as pesticide,[70] pests,[71] labeling,[72]

---

[66]NEPAD, East African Community (EAC) sanitary measures for fish and fisheries products launched in Mombasa , http://www.africanfisheries.org/news-tags/sanitary-and-phytosanitary-sps.

[67]East African Community, EAC Sanitary Measures for Fish and Fisheries Products Launched in Mombasa, http://www.eac.int/index.php?option=com_content&view=article&id=1515:eac-sanitary-measures-for-fish-and-fisheries-products-launched-in-mombasa&catid=146:press-releases&Itemid=194.

[68]See generally, Yankey (2004).

[69]Meat and Meat Products Act—Malawi; Meat Inspection Regulations—Malawi; Meat Marketing Regulations—Malawi; Meat Act 3/665 November 1974—Mauritius; Meat Safety Act, 2000—South Africa; Livestock and Meat Industries Act (No. 32 of 1962)—Botswana.

[70]Control of Pesticide Law 041-96—Burkina Faso and Pesticides Act 528, 1997—Ghana.

[71]See e.g.: Disease and Pest Act (Botswana); Pest Control Products Act Cap 346—Kenya.

[72]See e.g. Regulations on food imports and exports, Street food, food hygiene and labelling—Benin; General Labelling Rules, 1992 (L.I. No. 1541, 1992)—Ghana.

counterfeiting,[73] and street food vending. Also common are laws regulating the export and/or import of food,[74] laws regulating medicines and related substances,[75] and laws addressing food adulteration.[76]

Some but not all countries in Africa have functioning agencies tasked with establishing and elaborating on food standards. National agencies tasked with food safety regulation in Africa include: South Africa Bureau of Standards (South Africa), Agence National de Contrôle Sanitaire et Environmental des Produits (Algeria), Food Control Authority (Mali, Morocco, Zimbabwe); Tanzania Bureau of Standards (Tanzania), Kenya Bureau of Standards, Rwanda Bureau of Standards, Food and Drug Laboratory, Ministry of Health (Zambia), Public Health and Marketing, Ministry of Agriculture, Animal Husbandry and Fisheries (Uganda). Lesotho has a Director of Standards within the Ministry of Trade and Industry but also has a Food Hygiene and Safety Programme Manager within the Ministry of Health and Social Welfare.[77] Although legally mandated to develop food safety standards, some of domestic agencies are yet to establish functioning and accessible systems for the development and review of food standards. Finally, more and more countries are passing laws to establish national Codex Alimentarius Committees and National Codex Contact Points. Examples include: Decree n° 834/PR/MAEDR in Tunisia establishing a national Codex Alimentarius Committee and Decree n° 2005-388 amending decree n° 2000-2574 on the establishment of "Codex Alimentarius" Committee, its ruling, organization and composition.

Consumer right is increasingly enshrined in national constitutions and may form the basis for stronger food safety system in the future. Article 46(1) of the Kenyan Constitution declares that "Consumers have the right—(a) to goods and services of reasonable quality; (b) to the information necessary for them to gain full benefit from goods and services; (c) to the protection of their health, safety, and economic interests; and (d) to compensation for loss or injury arising from defects in goods or services." Pursuant to Article 46(2), "Parliament shall enact legislation to provide for consumer protection and for fair, honest and decent advertising."

Some countries are stepping up laws relating to food labelling and advertising. On 29 May 2014, South Africa's Department of Health published The draft Regulations Relating to the Labelling and Advertising of Foods (R. 429), and

---

[73]Counterfeit and Fake Drugs and Unwholesome Processed Food Decree, (Act No. 25 of 1999)—Nigeria.

[74]Decree-Law No. 100/92 of 17 August 1992 regulating the export of bananas (Cape Verde) and Fishery Products Importation and Exportation Regulations (L.N. No. 69 of 2003)—Eritrea.

[75]The Medicines and Related Substances Act, 1965 (Act 101 of 1965)—South Africa.

[76]Sale of Adulterated Food Act No. 25, 1968—Swaziland; Adulteration of Produce Decree (Cap. 109)—Tanzania.

[77]Other examples include: Quality and Standards Authority—Ethiopia; Senegal Standardization Institute—Senegal; and National Center for Food and Hygiene—Mauritania.

supporting guidelines.[78] R. 429 is based on Foodstuffs, Cosmetics and Disinfectants Act, 1972 (Act No. 54 of 1972) and addresses the commercial marketing of foods and non-alcoholic beverages to children of school-going age (up to Grade 12). R. 429 affects persons who manufacture, import, sell, donate or offer for sale any pre-packed food. R. 429 joins existing laws in the area including: R.246 of 11 February 1994: *Regulations governing the maximum limits for pesticide residues that may be present in foods;* R.1809 of 3 July 1992: *Regulations governing the maximum limits for veterinary medicine and stock remedy residues that may be present in foods*; R.500 of 30 April 2004: *Regulations relating to Maximum Levels for Metals in Foods*; R.491 of 27 May 2005: *Regulations relating to Marine Biotoxins*; R.1145 of 8 October 2004: *Regulations governing tolerances for fungus produced toxins in foods* (mycotoxins); and R.911 of 28 September 2001: *Regulations governing certain solvents in foods* (benzene and methanol).

The legal and regulatory response to biotechnology and biosafety issues in Africa remains extremely weak despite continental and regional activities in this area. An Africa-wide biosafety system is not yet in place. Efforts to implement an Africa-wide biosafety system as well as an Africa-wide capacity building program in biosafety are ongoing, however. Most countries in Africa have ratified the Cartagena Protocol on Biosafety. In its Decision EX/CL/Dec.26 (111) adopted in 2003, the Executive Council of the AU "stressed the need for Member States to equip themselves with the necessary human and institutional capacities to deal with Biosafety issues and implicitly the need for a common African approach to addressing issues pertaining to modern biotechnology and biosafety."

Positive developments at the continental level include: the development of the African Model Law on Safety in Biotechnology first finalized in 2001; the implementation of the AU-Biosafety Project—a project on capacity building for an Africa-wide biosafety system developed between the AU and German Technical Cooperation (GTZ); the development of an African Strategy on Biosafety; the establishment of an AU Biosafety Unit under the AU's Department of Human Resources Science and Technology; the establishment of the High-level African Panel on Biotechnology (APB); and the launch of the African Biosafety Network of Expertise (ABNE). Developments at the sub-regional level include the establishment of a SADC Advisory Committee on Biotechnology and Bio-safety (SACBB) and the development of a SADC Guidelines on GMOs, Biotechnology and Biosafety.

41.3.2.2 Gaps in Africa's Food Safety Law Apparatus

Despite the impressive array of laws and regulation that address food safety in countries in Africa, major gaps in the legal framework for food safety persists. First, although many countries have laws on the books addressing different aspects of food safety, "[t]he traditional food control systems in most African countries do not

---

[78]Republic of South Africa, Regulation Gazette No. 37695 (29 May 2014).

provide the concerned agencies with a clear mandate and authority to prevent food safety problems."[79] Second, existing food legislation is often not in line with international requirements.[80] Third, for most countries, existing laws are outdated, inadequate, fragmented, and confusing.[81] Fourth, although most countries have agencies charged with some aspect of food safety management, the scope of the mandate of the agencies, the capacity of the agencies to fulfill their mandate, the accountability of these agencies, and the level of coordination and cooperation among the different agencies tasked with food safety management are often not clear. Fifth, many countries in Africa still lack effective, functioning and comprehensive food inspection mechanisms; while some countries have some sort of system in place for inspecting food imports and export, mechanisms directed at general food inspection are lacking in most countries. Sixth, food safety laboratory support services is absent in many countries. Overall, most countries in the region lack a well-conceived food safety policy, do not have an effective national food control systems, have no system in place for dealing with emerging food safety challenges, have not included food safety in their national development goals and strategies, lack the technical capacity to develop and implement of food safety policies and legislation, and lack the capacity to monitor food imports even though food import has increased significantly in Africa.

There is an urgent need for increased standardization of the African food safety system. In 2005, the Food and Agriculture Organization of the United Nations (FAO) and the World Health Organization (WHO) jointly convened the FAO/WHO Regional Conference on Food Safety for Africa. Conference participants unanimously adopted a resolution recommending a nine-point *Five-year Strategic Plan for Food Safety in Africa* for adoption by FAO and WHO, along with the African Union.[82] At its 57th session in 2007, the WHO Regional Committee for Africa endorsed the Regional Strategy on Food Safety and Health.[83] In 2006, the Summit of the Heads of States of the AU adopted the *African Union Regional Nutrition Strategy 2005–2015*. Following African Union Commission's (AUC) call for a pan-African food safety body, discussions are now underway for an African Food Safety Authority.[84] There are also

---

[79]CAF 05/2.

[80]Id.

[81]Id.

[82]Nine elements: (i) Food safety policies and programmes; (ii) Legislative and institutional aspects; (iii) Standards and regulations; (iv) Food inspection programmes and techniques; (v) Food analysis and food safety testing laboratories; (vi) Monitoring food-borne diseases and the safety of foods on the market; (vii) Participation in Codex; (viii) Communication and stakeholder involvement (including industry officials and consumers); (ix) National, regional and international cooperation.

[83]AFR/RC57/4 (53.92 kB) Food Strategy and Health: A Strategy for the WHO African Region.

[84]Gretchen Goetz, African Union Makes Plans for Food Safety Authority, November 2, 2012 http://www.foodsafetynews.com/2012/11/african-union-makes-plans-for-food-safety-authority/#. U8aVB7bn_rc.

on-going talks about establishing an African Rapid Alert System for Food and Feed (RASFF).

Despite efforts to implement an Africa-wide biosafety system, major gaps remain in the legal and regulatory response to biosafety concerns. First and foremost, most countries in Africa do not have strong biosafety regulations that are in line with internationally-agreed rules and guidelines. Additionally, existing regulations often do not strike appropriate balance between promoting innovation and safeguarding human health and the environment. When it comes to biotechnology and biosafety, the legal terrain in Africa is uneven. There are wide variations in biosafety policies and regulations of countries in the region. It is not clear if and to what extent the Model Law on Biotechnology has influenced the drafting of national safety acts and bills. It is also not clear how many countries in Africa have implemented the call, in the Model Law on Biotechnology, for countries to establish National Focal Points, National Biosafety Committees, as well as Institutional Biosafety Committees. The level of public participation in African Biosafety regulations and policies is extremely low and deserve closer scrutiny, as does the extent to which emerging biosafety regulations in the continent address issues such as liability, redress, compliance, and dispute settlement mechanisms for biosafety. Presently, GMO laws and regulations are not harmonized. Efforts to harmonize GMOs may yet emerge at the sub-regional level.[85] The intersection of biosafety and food aid, biosafety and trade rules, biosafety and intellectual property rights are issues of concern to countries in Africa and deserve closer analysis.

41.3.2.3 Conclusions

Food security concerns have typically taken precedence over food safety issues in Africa. However, in the light of the 2014 Ebola crisis and other public health crises, food safety is becoming a matter of grave concern in the continent. The pressure on countries to address gaps in their food safety system is growing. Pressure for change is coming from the general consuming public, non-governmental organizations, professional associations, the media, international trading partners and trade associations, and from international organizations such as the WHO. With the influx of international fast-food franchises into Africa and noticeable changes in eating habits of ordinary Africans—from eating home-prepared food to consumption of ready-to-eat foods—food safety concerns is also growing. Functioning food standards systems, effective inspection mechanisms, and functioning laboratory support services are all needed. Effective national food control systems will ensure that consumers are protected and can contribute to efforts to address food insecurity in Africa. Effective national food control systems will help boost agricultural export from Africa and allow countries in Africa take full advantage of international food

---

[85]East African Community Protocol on Environment and Natural Resource Management, Article 27 (calling for a common approach to biosafety.).

trade opportunities. There are important economic gains to be had as well. Addressing food safety allows countries to control economic consequences due to absenteeism, hospital fees and international trade losses.

To create a credible and effective food safety system, adequate policies, programs and laws developed in an open and transparent manner and with full participation of relevant stakeholders are necessary. Such laws must of necessity address all aspects of the food production, processing, and distribution chain, lay down basic food safety requirements, and require that food safety measures be based on risk analysis including risk assessment, risk management and risk communication. Such laws must also require that risk assessment be undertaken in an independent, objective and transparent manner. Finally, such laws ought to be administered in a transparent manner and should prioritize the principles of public consultation, public participation, and public information. Adequate standards and regulations are necessary and ought to be communicated to all stakeholders in a timely fashion. Capacity building must be an important component of food safety laws and policies in Africa. While some governments have focused attention on developing the capacity of the export food industry, few have paid attention to addressing capacity constraints of enterprises that produce food for domestic consumption. Effective monitoring and enforcement mechanisms are also urgently needed as are functioning food safety testing laboratories.

Work is also needed at the regional and continental level. The absence of a coherent regime at the continental level undermines regional trade in food. With the goal of ensuring continuous monitoring of food safety, a networking of laboratories at the regional and continental level is important. The establishment of an African Food Safety Authority is an idea worth pursuing. As is the case with the European Food Safety Authority, such a body can take on the role of an independent scientific point of reference in risk assessment. Also necessary is the development of a pan-African food safety standard based on Codex standards. The Ebola crisis in West Africa and past food incidents in Africa demonstrate the need for effective emergency food safety measures at the national, regional and continental levels.

At the global level, African countries must strive to contribute meaningfully to the development of international standards that underpin food law, including sanitary and phytosanitary standards.[86] Increasingly, agreements on recognition of the equivalence of specific food measures are becoming important. Individually and collectively, countries in Africa must push to contribute to the development of equivalence standards. Increased participation in the relevant committees of the Codex Alimentarius Commission is also necessary. Given limited resources, there is an obvious need for countries in Africa to strengthen joint efforts in capacity-

---

[86]Several RECs in Africa have observer status in the SPS Committee of the World Trade Organization including: SADC, COMESA, ECOWAS, Intergovernmental Authority on Development (IGAD), West African Economic and Monetary Union (WAEMU), Economic Community of Central African States (ECCAS), and the Community of Sahel Saharan States (CEN SAD).

building, international standard setting, information sharing and food contamination monitoring.[87]

Establishing functioning food law systems in Africa will not be easy. Challenges to improving food safety in Africa are legion and include: (i) inadequate commitment to a transparent, participatory, science-based food safety system; (ii) outdated food laws regulations; (iii) weak law enforcement; (iv) inadequate capacity for food safety surveillance; (v) lack of coordination and communication among the different agencies responsible for food safety; (vi) lack of adequate research on food safety; (vii) weak private sector and inability of small- and medium-scale producers to produce safe food; (ix) inadequate cooperation among stakeholders; (x) ignorance; (xi) non-functional laboratories; (xii) long and very porous national borders; and (xi) persistence of some cultural beliefs, attitudes and practices that obstruct food safety goals. African nations need increased capacity in almost every component of the food control system. Food safety capacity building is urgently needed in a number of areas including: strengthening national and regional laboratories, strengthening consumer associations, developing effective legislations, standards and guidelines, as well as educating all stakeholders in the food chain, particularly the general public and small business owners.

41.3.3 The Gender Dimension of Food Law in Africa

Women in Africa are heavily involved in agriculture and food processing, and are highly dependent on agriculture for their livelihood. However, women experience significant inequality and injustice in the food and agricultural sector. Gender is yet to be fully and holistically integrated into the food and agricultural strategies of countries in Africa. Women face significant economic risks, labor risks, social risks, and health and safety risks in the agricultural sector.[88] As food producers and caregivers, African women are also disproportionately affected when large-scale development projects displace poor peasant communities from their land, destroy local means of livelihood, or trigger land-related conflicts and violence. One of the greatest problems women face in Africa is lack of access to and ownership of land and other agricultural assets, inputs, and services such as credit, technology, information and extension services, and natural resources. With growing pressure on land, increased commercialization of land, and increases in land-related disputes and conflicts in the continent, the land-related problems African women face are likely to increase with resulting consequences for food law and policy in the continent.

---

[87]International, regional, sub-regional and national cooperation in food safety (CAF 05/6).

[88]Fair Labor Association, Assessing Women's Role in Nestle's Ivory Coast Cocoa Supply Chain (2014).

The land question in Africa is complicated by a number of factors, including: the multiplicity of claims to land; the plurality of legal systems (customary, religious and secular) and property systems; the dominance of communal land tenure system in most parts of the continent; the enduring influence of customary laws and practices; and increased pressure on land as a result of globalization of agricultural trade and agricultural investment. While many countries have adopted new laws and policy aimed at addressing gender inequality in respect of access to land, land reforms often do not go far enough in terms of redressing imbalance, and sometimes create problems of their own. Moreover, implementing and enforcing laws and policies designed to strengthen women's land rights remains a major challenge in most countries in Africa as a result of corruption, lack of information about land reform and land rights, and limited access to judicial and non-judicial grievance mechanisms for addressing land-related claims. Addressing and strengthening women's land rights can contribute to poverty reduction efforts in the continent, enhance food security in the region, and boost agricultural productivity in the continent. Additionally, addressing women's land rights also has implications for women's right to health. Organizations like the International Centre for Research on Women (ICRW) have established strong linkages between women's property and inheritance rights and HIV/AIDS in Africa.[89]

41.3.3.1 Women and Access to Land: Present Problems and Challenges

Lack of access to and control over land is a major problem women face in Africa, although the level and severity of the problem varies from country to country and from one ethnic group to another. In many countries, customary law still precludes women from owning and inheriting land. In other countries, women can legally inherit but relatives prevent them from accessing and using the land they have inherited. Problems range from customary restrictions on land acquisition and ownership, lack of awareness about land rights, failure to operationalize and enforce equal-rights legislation on the books, to lack of access to legal representation. The result is that in many parts of Africa, customary norms continue to limit women's ownership and control of land. Through relationships with male relatives—father, brother, son, husband—many women enjoy access to land which they lose whenever the underlying relationship is disrupted. Even when women have access to land to cultivate, their rights to such land is tenuous and vulnerable at best and is usually lost through widowhood or divorce.[90]

---

[89]Strickland (2004).
[90]Rural Development Institute, Women's Land Rights in Rwanda 6 (2006).

41.3.3.2 Women and Land in Africa: The Legal Framework

Article 4(L) of the Constitutive Act of the African Union specifically provides that the African Union "shall function in accordance with the promotion of gender equality." Policy makers in Africa have made numerous commitments to gender equality and women's empowerment. Documents encapsulating this commitment include: the African Union Gender Policy 2009; AU Assembly Decision AU/Dec. 134-164 (IIIV) on the Establishment of the African Women Trust Fund; the Protocol to the African Charter on Human and People's Rights on the Rights of Women in Africa, and the Solemn Declaration on Gender Equality in Africa (SDGEA).[91] Also suggesting a renewed commitment to gender equality is the fact that on 18 October 2010—the International Day for Rural Women—the AU officially launched the African Women's Decade (AWD) (2010–2020) under the theme "Grassroots Approach to Gender Equality and Women's Empowerment." Agriculture and Food Security, Environment and Climate Change, Peace and Security, and Violence Against Women are some of the sub-themes of the Decade.

Binding commitments in sub-regional protocols and declarations offer another layer of protection for women. Examples include the SADC Protocol on Gender and Article 3(e) of the EAC Treaty 3(e) which provides that the Community shall ensure the mainstreaming of gender in all its endeavors and the enhancement of the role of women in cultural, social, political, economic and technological development. Many governments in Africa have ratified the Convention on the Elimination of All Forms of Discrimination against Women (CEDAW) as well as ILO Conventions Nos. 100 and 111. Although many countries in Africa have ratified CEDAW, in many countries in Africa, the treaty has yet to be incorporated into national and state laws.[92]

Gender features strongly in most national constitutions with most explicitly prohibiting sex discrimination. Gender equality and non-discrimination feature in the 2010 Constitution of Kenya under Article 10 (National values and principles of governance), Article 19 (Rights and fundamental freedoms) and Article 27 (Equality and freedom from discrimination). Article 27(3) declares "Women and men have the right to equal treatment, including the right to equal opportunities in political, economic, cultural and social spheres." Article 27(6) goes a step further and requires the State to "take legislative and other measures, including affirmative action programmes and policies designed to redress any disadvantage suffered by individuals or groups because of past discrimination."

Some national constitutions in Africa address the land question, as exemplified by Article 25 of the South African constitution. The Kenyan Constitution addresses

---

[91] Adopted by AU Heads of State and Government at their July 2004 Summit.

[92] For countries whose constitution provide for direct applicability of treaties, international treaties protecting the right of women such as the CEDAW Convention are directly applicable and enforceable. Article 2(5) of the Constitution of Kenya states that "The general rules of international law shall form part of the law of Kenya." Article 2(6) further provide: "Any treaty or convention ratified by Kenya shall form part of the law of Kenya under this Constitution." See The Constitution of Kenya 2010.

the thorny issue of inheritance, land rights, and right to property. Indeed, Chapter 5 of the Kenyan Constitution is titled "Land and Environment." Article 40(1) of the Constitution that "Subject to Article 65, every person has the right, either individually or in association with others, to acquire and own property—(a) of any description; and (b) in any part of Kenya." Article 60 (1) states that "Land in Kenya shall be held, used and managed in a manner that is equitable, efficient, productive and sustainable." Article 60(1) lists several principles of land policy including: equitable access to land[93]; security of land rights[94]; and "elimination of gender discrimination in law, customs and practices related to land and property in land."[95] Pursuant to Article 68, Parliament shall enact legislation "to protect the dependants of deceased persons holding interests in any land, including the interests of spouses in actual occupation of land."[96] Beyond the constitution, a good number of countries are enacting new laws aimed at addressing discrimination against women in land matters. An example of this is the Devolution of Estate Act of 2007 in Sierra Leone.

41.3.3.3 Legal Battle for Women's Right in Africa

In Africa, women are waging the battle for equality on the legislative front as well as on the judicial front. While some court decisions have been favorable to women, others have merely affirmed entrenched customary laws and practices.

In *Ephraim v. Pastory*[97] the issue was whether a customary law that barred women (not men) from selling clan land violated the Tanzanian Constitution and international Human Rights Law. In the case, the respondent, Holaria d/o Pastory, sold clan land that she inherited from her father to a non-clan member. The appellant, Bernardo s/o Ephrahim, the respondent's nephew by marriage, filed a suit at Kashasha Primary Court in Muleba District, Kagera Region, praying for a declaration that the sale of the clan land by the respondent was void as females under Haya Customary Law have no power to sell clan land. The Primary Court agreed with the appellant, voided the sale and ordered the respondent to refund the money she received on the land. The customary law on the issue stated thus: "Women can inherit, except for clan land, which they may receive in usufruct but may not sell. However, if there is no male of that clan, women may inherit such land in full ownership."[98] Citing the Bill of Rights, which was incorporated in the 1977 Constitution as well as international human rights instrument, Mwalusanya J. of the High Court of Tanzania at Mwanza observed that the customary law in question

---

[93]Id., Article 60(1)(a).
[94]Id., Article 60(1)(b).
[95]Id., Article 60(1)(f).
[96]Id., Article 68(c)(vi).
[97](2001) AHRLR 236 (TzHC 1990).
[98]Id., p. 2.

"flies in the face of [Tanzania's] Bill of Rights as well as the international conventions"[99] and noted that "Courts are not impotent to invalidate laws which are discriminatory and unconstitutional."[100] The Judge ultimately found the law in question to be discriminatory and inconsistent with article 13(4) of the Bill of Rights of Tanzania's Constitution which bars discrimination on account of sex.[101]

Legal victories for women in Africa are few and far in between. Even when the constitution prohibits gender discrimination, claw-back clauses in the constitution sometimes render the non-discriminatory provisions useless. Section 23 of the Constitution of Zimbabwe prohibits gender discrimination. However, Section 23 (3) of the Constitution states: 'Nothing contained in any law shall be held to be in contravention of subsection 1(a) to the extent that the law in question relates to any of the following matters—(a) ... devolution of property on death or other matters of personal law; (b) the application of African customary law in any case involving Africans ...'. The plaintiff in the Zimbabwean case of *Magaya v Magaya*[102] had to contend with Section 23(3). In *Magaya v Magaya*, the Supreme Court of Zimbabwe ruled that a woman could not inherit property under customary law in any case occurring prior to the passing of the Administration of Estates Amendment Act No. 6 of 1997. Section 23(3)(a) and (b) effectively immunized laws relating to devolution of property on death and to the application of customary law from the reach of the provisions of the constitution prohibiting gender discrimination.

41.3.3.4 Conclusions

Emerging food law in Africa must grapple with the gender question because African women contribute about 70 % of the food produced in the continent, account for nearly half of all farm labor, and account for 80–90 % of food processing, storage, and transport, as well as hoeing and weeding done in the continent.[103] Beyond the question of ownership of and access to land, the law must address broader issues including: the impact of mining and oil exploration on women's access to land and food; the link between land rights and violence against women; land governance issues; and water grabbing and its impact on women and local food systems.

There is need to enhance women's participation in decision-making processes pertaining to food production and distribution, encourage and support women's entrepreneurship in food production and distribution, and address cultural impediments to women's ownership of resources such as land. In the 2004 Sirte

---

[99]Id., § 10.
[100]Id.
[101]Id.
[102]*Magaya v Magaya* [1999] ICHRL 14 (16 February 1999).
[103]See more at: http://www.un.org/africarenewal/magazine/special-edition-women-2012/women-struggle-secure-land-rights#sthash.nOct7LZu.dpuf.

Declaration, African Heads of State and Government pledged to ensure gender balance in access to training, education, land, natural resources, loans and development programs. A comprehensive and holistic approach is called for—one that addresses entrenched legal, institutional, cultural, and practical obstacles to the realization of women's land rights in Africa.

41.3.4 Indigenous Peoples and Food Law in Africa

Food law in Africa is incomplete without an examination of the role of indigenous peoples in food production and processing in Africa, as well as the present threats to indigenous food systems in the continent.[104] Despite their contribution to environmental conservation and food production, indigenous people in Africa are rarely consulted or informed when food-related laws and policies are designed and implemented. Worse, indigenous peoples face many hardships in food-related matters and their right to food is under considerable attack.[105] In an increasing number of countries in Africa, indigenous groups are threatened with extinction. Obstacles to indigenous peoples' food security and sovereignty in Africa and globally are legion and include: the domination of globalization and free trade; the imposition of industrial models by the government; increased bio-piracy and the illicit appropriation of the biological diversity and traditional knowledge of indigenous people; militarization and repression in indigenous territories; the policies and demands of international financial institutions; and the imposition of unsustainable development projects by governments and private companies in territories belonging to indigenous peoples without consultation or prior informed consent, and without taking into account the rights and values of the indigenous peoples affected.[106] In Africa, indigenous groups have lost land through colonization, nationalization and privatization.

Development projects designed and implemented without adequate consultation with affected local communities are a big problem for indigenous communities in Africa. The African Commission's Working Group on Indigenous Populations/Committees (WGIP) noted in a 2005 report that:

> Dispossession of land and natural resources is a major human rights problem for indigenous peoples. They have in so many cases been pushed out of their traditional areas to give way for the economic interests of other more dominant groups and large-scale development

---

[104] Indigenous groups in Africa identified as hunter-gatherer include: the Pygmies of the Great Lakes Region, the San of Southern Africa, the Hadzabe of Tanzania and the Ogiek of Kenya. Those identified as pastoralists and agro-pastoralists include the *Pokot* of Kenya and Uganda, the *Barabaig* of Tanzania, the *Maasai* of Kenya and Tanzania, the *Samburu, Turkana, Rendille, Orma* and *Borana* of Kenya and Ethiopia, the *Karamojong* of Uganda.
[105] According to the 2002 *Declaration of Atitlàn*, indigenous people "face a higher risk of suffering the consequences of Food Insecurity."
[106] Id. The *Declaration of Atitlàn*.

initiatives that tend to destroy their lives and cultures rather than improve their situation. The establishment of protected areas and national parks has impoverished indigenous pastoralist and hunter-gatherer communities, made them vulnerable and unable to cope with environmental uncertainty and, in many cases, even displaced them. Large-scale extraction of natural resources such as logging, mining, dam construction, oil drilling and pipeline construction have had very negative impacts on the livelihoods of indigenous pastoralist and hunter-gatherer communities in Africa. So has the widespread expansion of areas under crop production. They have all resulted in loss of access to fundamental natural resources that are critical for the survival of both pastoral and hunter-gatherer communities such as grazing areas, permanent water sources and forest products.[107]

With respect to Kenya, the former UN Special Rapporteur on the Situation of Human Rights and Fundamental Freedoms of Indigenous People, Rodolfo Stavenhagen observed, in a 2007 report, that "[m]ost of the human rights violations experienced by pastoralists and hunter-gatherers in Kenya are related to their access to and control over land and natural resources"[108] and that "[t]he land question is one of the most pressing issues on the public agenda." Stavenhagen added that "[h]istorical injustices derived from colonial times, linked to conflicting laws and lack of clear policies, mismanagement and land grabbing, have led to the present crisis of the country's land tenure system" and that inappropriate development and conservationist policies have aggravated the violation of the economic, social and cultural rights of indigenous communities in Kenya in recent decades.[109]

The problems of indigenous people in Africa are compounded when development projects that threaten their livelihoods receive financial support and the blessing of international financial institutions such as the World Bank and the International Monetary Fund. The good news is that increasingly, indigenous people in Africa are fighting back and are using mechanisms available at the national, sub-regional, regional and international level to defend their right to land and to food.

41.3.4.1 Case Study 1: Sengwer Forest Dwellers of Kenya

In Kenya, indigenous communities in Embobut in Elgeyo Marakwet County, Western Kenya (Sengwer forest dwellers) are victims of forced eviction from the Embobut Forest that they claim as their ancestral land. On 12 December 2013, the Kenya Forest Service (KFS) issued a 21-day eviction notice to the Sengwer forest dwellers. The KFS proceeded with the eviction on or about 5 January 2014 with about 150 police and forest guards including 30 General Service Unit riot

---

[107]Report of the African Commission's Working Group on Indigenous Populations/Committees 20 (2005).

[108]See Report of the Special Rapporteur on the Situation of Human Rights and Fundamental Freedoms of Indigenous People, Rodolfo Stavenhagen, on "Implementation of General Assembly Resolution 60/251 of 15 March 2006, A/HRC/4/32/Add.3, 26 February 2007: "Mission to Kenya" from 4 to 14 December 2006 (hereinafter "Special Rapporteur on Kenya")", p. 25.

[109]Id.

police.[110] Hundreds of Sengwer homes were burnt in January 2014 alone.[111] The KFS proceeded with the planned eviction despite urgent appeals from the international community and despite an order from a court in Kenya enjoining the eviction.[112] Implicated in the Sengwer forest saga is the World Bank-funded Natural Resource Management Project (NRMP) with the Government of Kenya.[113] Kenya has reportedly received over $600 million from the World Bank to protect the forests.

The Kenyan Government claimed that the presence of the Sengwers threatens the forest's biodiversity as well as urban water supplies. According to the KFS, the eviction was necessary "in order to pave way for the Government to restore the Cherangani forests, a major national water tower serving more than 2 million Kenyans through the Nzoia and Turkwel rivers."[114] The Government allegedly offered Sengwer households $4600 compensation for moving, although thousands claim that they did not receive any compensation. In a 5 January 2014 letter addressed to the President of Kenya, Sengwer forest dwellers stated that "[a]t no stage of the current process have the Embobut's indigenous residents been meaningfully consulted in relation to this resettlement, nor has their free, prior and informed consent been sought and obtained" and that "[w]hen the Sengwer have been consulted in the past they have refused to move, and they have made clear ... that they refuse to move."[115] The letter further stated that "[t]here has ... been no legitimate consultation, compensation or legally valid approval agreed with the indigenous communities over their ancestral lands."[116] Addressing the Government's claim that compensation was paid to evictees, the Forest Peoples Program acknowledged that on 15 November 2013 the President, Deputy President and County Senator (Senator Kipchumba) visited Embobut and the President promised 400,000 Kenyan shillings per family to the 'Evictees' but are quick to point out that "the 400,000 Kenyan shillings would buy the equivalent of 4 cows or one or two

---

[110]Forest Peoples Program, Press Release: Forced eviction by Kenya threatens indigenous communities' human rights and ancestral forests, 6 January 2014.

[111]Forest Peoples Programme, Kenyan Government torches hundreds of Sengwer homes in the forest glades in Embobut, 20 January 2014. http://www.forestpeoples.org/topics/legal-human-rights/news/2014/01/kenyan-government-torches-hundreds-sengwer-homes-forest-glade.

[112]Forest Peoples Programme: http://www.forestpeoples.org/topics/rights-land-natural-resources/news/2013/12/urgent-appeal-against-forced-eviction-sengwerchera%20p.

[113]Global Information Network, World Bank Linked to "Cultural Genocide" of Kenya's Sengwer People, 30 September 2014. http://www.ipsnews.net/2014/09/world-bank-linked-to-cultural-genocideof-kenyas-sengwer-people/. Vidal (2014).

[114]Kenya Forest Service, Clarifying Information on the Embobut Forest: The NRMP Perspective. http://www.kenyaforestservice.org/index.php?option=com_content&view=article&id=568:clarifying-information-on-the-embobut-forest-the-nrmp-perspective&catid=228:press-releases&Itemid=168.

[115]*Dear President Uhuru Kenyatta, Deputy President William Ruto, Senator Kipchumba Murkoman, Legal Advisor Korir Sing'Oei and other Kenyan Government Authorities and Parliamentarians*, 5 January 2014.

[116]Id.

acres of land in Trans Nzoia District," and "is therefore ... completely inadequate for enabling families to secure their livelihoods."

41.3.4.2 Case Study 2: The Bushmen of Botswana

The Basarwa minority group in Botswana (also known as Bushmen or San) has experienced several rounds of forced eviction from their ancestral land in the Central Kalahari Game Reserve (CKGR). Since 1997, the Bushmen have been the target of at least three separate eviction exercise, in 1997, 2002 and 2005. The Botswana Government takes the position that the presence of the Basarwa and other indigenous communities on the land is incompatible with the reserve's conservation objectives and status. Accusations of arbitrary arrests, police harassment and brutality, and intimidation are rife.[117] In 2012, paramilitary police allegedly beat two Bushmen severely and subsequently buried one in a shallow grave, after accusing them of hunting without permits in CKGR.[118] In 2013, three Bushmen children were arrested for being in possession of antelope meat in the CKGR. Also in 2013, two Bushmen were reportedly arrested and tortured for killing an antelope, and were fined US $190 each.[119] Following a 2006 court decision, Bushmen can only hunt in the CKCR with licenses.[120] Today, many of the Bushmen live in resettlement camps and are unable to hunt or gather. Furthermore, many "are now gripped by alcoholism, depression, and illnesses such as TB and HIV/AIDS" according to Survival International.[121] One Bushman told Survival International, 'The Bushmen are being hunted and their rights are being denied because of tourism (....) Police are given guns to go out and hunt and arrest Bushmen gathering bush food. The Bushmen of the CKGR cannot eat, cannot drink. How will they survive without food?'[122] Another Bushman told Survival International "We depend on the natural resources of the CKGR for our food. How are we expected to survive if we cannot hunt?"[123]

The Bushmen face an even greater threat—lack of access to water. A law in Botswana prohibits the Basarwa from using existing boreholes in the CKGR and prevents them from drilling new ones. In 2002, in a bid to get the Bushmen off the land, the Government sealed and capped a borehole in the reserve. Critics believe that the forced eviction of the Bushman has nothing to do with conservation but is

---

[117]Survival International, Bushman children arrested under renewed government repression 15 January 2013. http://www.survivalinternational.org/news/8919.

[118]http://www.survivalinternational.org/news/8883.

[119]Id.

[120]Survival International. Bushmen beaten, suffocated and buried alive for killing an antelope 13 December 2012. http://www.survivalinternational.org/news/8883.

[121]http://www.survivalinternational.org/tribes/bushmen/courtcase#main.

[122]http://www.survivalinternational.org/news/8919.

[123]http://www.survivalinternational.org/news/8883.

about diamonds. Fortunately or unfortunately for the Bushmen, CKGR "lies right in the middle of the world's richest diamond-producing area."[124] The diamond deposit at Gope, in the centre of the reserve, is reportedly valued at $3.3 billion. To Professor James Anaya, the UN Special Rapporteur on the Right of Indigenous Peoples:

> the Government's position that habitation of the reserve by the Basarwa [Bushmen] and Bakgalagadi communities is incompatible with the reserve's conservation objectives and status appears to be inconsistent with its decision to permit Gem Diamonds/Gope Exploration Company (Pty) Ltd. to conduct mining activities within the reserve, an operation that is planned to last several decades and could involve an influx of 500-1200 people to the site, according to the mining company.

The forced eviction of indigenous people from their ancestral land is a food law issue. Force eviction threatens and compromises indigenous food systems. Forced eviction also leads to the destruction of biodiversity and traditional knowledge. Even when eviction is accompanied by resettlement, oftentimes resettlement camps are not adequate. In the case of the Basarwa, the resettlement camps have been described as "disgusting places where rape, prostitution and drunkenness [are] rampant."[125] According to Maude Barlow, Senior Advisor on Water to the 63rd President of the United Nations General Assembly, "It's hard to imagine a more cruel and inhuman way to treat people. One can only conclude Botswana's authorities view Bushmen as less important than wildlife."[126]

41.3.4.3 Forced Eviction, Biopiracy and Indigenous Groups in Africa: The Legal Framework

In 2007, three countries in Africa—Burundi, Kenya, and Nigeria—abstained from the United Nation's vote adopting the Declaration on the Rights of Indigenous People.[127] Also, many countries in Africa have yet to ratify or accede to the International Labour Organization Indigenous and Tribal Peoples Convention, 1989 (No. 169).[128] Nevertheless, most countries have ratified treaties that protect the rights of minorities as well as treaties that enshrine freedom from discrimination and the right to equality and equal protection, including the African Cultural Charter (1976), the African Charter on Human and Peoples' Rights, and the Convention on the Elimination of Racial Discrimination. Many countries in Africa have also ratified the Convention on Biological Diversity, the Cartagena Protocol on Biosafety, as well as its Nagoya Protocol on Access to Genetic Resources, and the Fair and Equitable Sharing of Benefits Arising from their Utilization (Nagoya

---

[124]Simpson (2011).
[125]Id.
[126]http://www.survivalinternational.org/news/6307.
[127]Declaration on the Rights of Indigenous Peoples, E/CN.4/Sub.2/1994/2/Add.1 (1994).
[128]United Nations, *Treaty Series*, vol. 1650, No. 28383.

Protocol). Furthermore, the constitution of most countries in Africa incorporate the right to equality and freedom from discrimination. In 2001, the African Commission on Human Rights established a Working Group on the Rights of Indigenous Populations/Communities. Fundamentally, most governments in Africa are reluctant to recognize any particular ethnic group as indigenous and prefer to use terms such as "minorities," "vulnerable communities," or "marginalized communities." The term "indigenous" is contested in a continent where almost all Africans living in the continent are descendants of the original inhabitants.

Although Africa is rich in generic resources and traditional knowledge, and although many indigenous groups in the continent are custodians of this knowledge, a complete inventory of the continent's rich resources is not available. The direct and indirect cost of loss of genetic resource and traditional knowledge is astronomical. African governments are beginning to take steps to protect the continent's genetic resources and traditional knowledge, these efforts are undermined by poor institutional framework, weak and poorly enforced laws, corruption, and limitations imposed by global trade rules, particularly rules encapsulated in the Agreement on Trade-related Aspects of Intellectual Property Rights (TRIPS Agreement). Although many countries in Africa have ratified the Convention on Biological Diversity, the Cartagena Protocol on Biosafety, as well as its Nagoya Protocol on Access to Genetic Resources, and the Fair and Equitable Sharing of Benefits Arising from their Utilization (Nagoya Protocol), few have enacted domestic legislation to implement these treaties. At the continental level, relevant instruments include: the 1968 African Convention for the Conservation of Nature and Natural Resources (revised in 2003), the African Model Legislation for the Protection of the Rights of Local Communities, Farmers and Breeders, and for the Regulation of Access to Biological Resources (African Model Legislation on Access to Biological Resources),[129] and the Swakopmund Protocol on the Protection of Traditional Knowledge and Expressions of Folklore within the Framework of the African Regional Intellectual Property (the Swakopmund Protocol).[130] The period 2001–2010 was declared "the Decade for African Traditional Medicine."

Ultimately, few countries in Africa have developed national laws for the protection of traditional knowledge and cultural expressions. The problem is that many countries already model their domestic laws after the intellectual property laws in place western countries with its emphasis on exclusivity and private ownership. South Africa is one of the few countries in Africa that is making a serious effort to find a solution to this dilemma. The *Intellectual Property Laws Amendment Act 2013* (Act No. 28 of 2013) was approved by the President of South Africa and published in the Government Gazette on December 10, 2013. With the goal of

---

[129] OAU Model Law, Algeria, 2000—Rights of Communities, Farmers, Breeders, and Access to Biological Resources.

[130] Adopted by the Diplomatic Conference of ARIPO at Swakopmund (Namibia) on August 9, 2010. The protocol was initially signed by 9 of the 17 Member States that are members of ARIPO.

ensuring effective protection mechanisms for indigenous knowledge (IK) as a form of intellectual property in South Africa, the 2013 Acts amended several previous legislations, including the Performers' Protection Act 1967 (Act No. 11 of 1967), the Copyright Act 1978 (Act No. 98 of 1978), and the Trade Marks Act 1993 (Act No. 194 of 1993).

While there are many steps that governments in Africa can take, through legislation, to ensure maximum protection for the continent's genetic resources and traditional knowledge, most nations are not taking these steps. Broad debates and discussions about the future of intellectual property rights and law in Africa and the appropriate role for intellectual property rights in national development are needed. Another necessary topic of discussion is the potential role of traditional medicine in Africa's technological progress and overall economic development. South Africa is taking the lead on this with the publication, for public commentary, of the country's first Draft National Policy on Intellectual Property, 2013.[131] Chapter Three of the Draft National Policy focuses on agriculture and genetic resources. Chapter Four addresses intellectual property and indigenous knowledge. In line with the 2001 *Decision on Intellectual Property, Genetic and Biological Resources, Traditional Knowledge and Folklore in Africa*, governments in Africa need to examine ways and means of raising awareness about the protection of genetic resources and indigenous knowledge and identify, catalogue, record, and document the genetic and biological resources and traditional knowledge held by their communities. An Africa-wide database would be a good starting point.

41.3.4.4 Legal Battle for the Right of Indigenous People in Africa

Increasingly, indigenous peoples in Africa are fighting for their survival and for their ancestral land through the courts and other judicial and quasi-judicial tribunals. Indigenous peoples in Africa are also making their voice heard in international fora such as the UN Permanent Forum on Indigenous Issues[132] and are using available UN human rights complaints mechanisms such as Urgent Action/Early Warning Procedure of the UN Committee charged with implementing the UN Convention on the Elimination of all forms of Racial Discrimination (UN CERD). In a letter dated 30 August 2013, the CERD Committee reiterated its previous calls for state parties (including Kenya):

> to recognize and protect the rights of indigenous peoples to own, develop, control and use their communal lands, territories and resources and, where they have been deprived of their lands and territories traditionally owned or otherwise inhabited or used without their free and informed consent, to take steps to return those lands and territories. Only when this is

---

[131]Department of Trade and Industry South Africa, Draft National Policy on Intellectual Property, 2013, Notice No. 918 of 2013.

[132]Africa Caucus statement to the 13th Session of UNPFII, delivered by Mr Kanyinke Sena, May, 2014. Available at: http://natural-justice.blogspot.co.uk/2014/05/13th-session-of-unpfii-african-caucus.html.

for factual reasons not possible, the right to restitution should be substituted by the right to just, fair and prompt compensation. Such compensation should as far as possible take the form of lands and territories.

Legal battles are also occurring in domestic courts. In Botswana, the Bushmen initiated a case challenging their illegal eviction from their ancestral land in 2004. In a 2006 decision, the Court declared the evictions illegal and unconstitutional.[133] The Court also recognized the Bushmen's right to live, and hunt on their ancestral land and ordered the Government to issue hunting permits to the Bushmen. In 2010, the Bushmen commenced another lawsuit directed specifically at access to water in the CKGR. The Question before the court was simple: can the Bushmen use existing boreholes and can they sink new boreholes in the CKGR? The High Court sided with the Government but the Court of Appeal reversed. The Court of Appeal called the case "a harrowing story of human suffering and despair." The Botswana Government remains defiant and is yet to comply with the decision of the Court of Appeal. Finally, indigenous people in Africa are also increasingly using regional and sub-regional judicial and quasi-judicial mechanisms to safeguard their interest.[134]

41.3.4.5 Conclusion

Forced eviction from ancestral lands has a devastating impact on indigenous peoples' right to food, particularly the availability of culturally appropriate foods. Forced eviction make it impossible for indigenous groups to engage in traditional hunting and gathering activities, lead to the destruction of biodiversity, and compromise traditional knowledge. Clearly, the principle of Free, Prior and Informed Consent enshrined in key international instruments is yet to be fully implemented in most countries in Africa.[135] Whether to recognize certain groups as indigenous and which groups to recognize is a thorny question which governments in Africa must address in line with the *Advisory Opinion of the African Commission on Hu Human and Peoples' Rights on the United Nations Declaration on the Rights of Indigenous Peoples*.[136] Governments also need to establish fair, transparent mechanisms for

---

[133] *Sesana and others v. The Attorney-General* (2006) (2) BLR 633 (HC).

[134] ACHPR 2009, Decision on Communication 276/2003, Centre for Minority Rights Development (CEMIRIDE) and Minority Rights Group International (MRG) on behalf of Endorois Welfare Council v Kenya.

[135] United Nations Declaration on the Rights of Indigenous Peoples (A/RES/61/295) (2007) (hereinafter "the Declaration"). See also General Recommendation XXIII (51) concerning Indigenous Peoples. Adopted at the Committee's 1235th meeting, 18 August 1997. UN Doc. CERD/C/51/Misc.13/Rev.4 requires that 'ensure that members of indigenous peoples have equal rights in respect of effective participation in public life, and that no decisions directly relating to their rights and interests are taken without their informed consent.'

[136] Adopted by the African Commission on Human and Peoples' Rights at Its 41st Ordinary Session Held in May 2007 in Accra, Ghana (2007).

sharing benefits that accrue from the use of resources located on land belonging to indigenous groups. Additionally, laws and policies addressing the very vulnerable situation of indigenous women are needed.

41.3.5 Agribusiness/Foreign Investment in Land: A Growing Controversy

On 28 October 2014, the Wall Street Journal reported that the Democratic Republic of Congo planed to lease a quarter of the country (an area larger than France) for agricultural purposes, in a bid to attract technology and foreign capital.[137] Congo's first agribusiness parks are under construction—a result of a partnership between the Congolese government and Africom Commodities (Pty) Ltd. A looming battle in Africa's food system is that between smallholder farmers and giant food and agricultural companies. The race appears to be underway to control the future of food consumption in Africa. Implicated in this looming battle are hundreds of bilateral investment treaties that countries in Africa have concluded with foreign governments as well as thousands of investment agreements that these countries have concluded with foreign investors.

International agricultural companies, international retailers, international goods distributor, and even fast food companies are jumping into the race. Upscale food supermarkets are appearing across the continent.[138] With a growing middle class, steady urbanization, and rising incomes, there is definitely a market for big retailers and specialised food chains in Africa. Proponents argue that "creating a well-functioning African food industry will have a far bigger impact on the local economy as it will allow farmers to reduce their reliance on export markets, and will allow consumers to buy better quality locally produced products."[139] As a sector, the food industry in Africa is performing very well and is reportedly "amongst the best performing sectors in Africa and has shown an average annualized growth of 20 percent in the last 4 years."[140] In 2013, Carrefour—the world's second largest retail group—entered Africa in a joint venture partnership with CFAO, a distributor of manufactured goods. The two companies are reportedly eyeing eight countries in the continent.[141]

---

[137]Clark (2014).
[138]Walmart acquired South African-based Massmart in 2011.
[139]Zin Bekkali, Why food could be Africa's sweet spot.
[140]Id.
[141]Green (2013a).

41.3.5.1 The Case for Agribusiness in Africa

Although pledging commitment to smallholder farmers, many countries in Africa are aggressively seeking investors for new agricultural processing zones. Nigeria's new 'Staple Crop Processing Zones' (SCPZs) offers land, infrastructure, and tax incentives to private investors; the country plans to create about 20 SCPZs by 2015. The government believes that the successful implementation of the SCPZ's will add somewhere between N660 billion and N1.4 trillion to the Nigerian economy and create about 250,000 jobs.[142] In April 2014, the Nigerian government upgraded Olam Farms to a Staple Crop Processing Zone (SCPZ) bringing the number of such zones in the country to 15.

International organizations, including the World Bank, are making a strong push for Africa to be opened to agribusiness.[143] The case for agribusiness in Africa is based largely on evidence of declining agricultural productivity over the past 50 years. Countries that were at one time net food exporters are now dependent on imported food for survival. Nigeria is a case in point. In the 1960s, long before the oil boon, agriculture contributed over 60 % of Nigeria's GDP and 70 % of exports. Nigeria was a dominant producer of products like palm oil (exceeding Malaysia and Indonesia) and groundnuts (exceeding the US and Argentina). Today, Nigeria spends around $11 billion a year on food imports.[144]

Agribusiness and foreign direct investment (FDI) in land can attract top farming talent and much-needed capital to Africa's agricultural sector, allowing the continent to address declining exports and providing an opportunity for Africa to move up the value chain into businesses related to production, including storage, milling and transportation. Presently, Thailand exports more agricultural products and earns more from agricultural exports that the whole of sub-Saharan Africa combined. Agribusiness can also help hydrocarbon-dependent economies in Africa to diversify and move away from oil and gas dependence. Furthermore, foreign investment in land and agricultural markets in Africa can help countries meet food security targets and reduce foreign exchange spending. Marc Engel, chief procurement officer at Unilever, Africa, said recently, "We have 13 factories in Africa that use products like soft oils, tomatoes or starch-based compounds on a daily basis, but much of this is imported, wasting foreign exchange and increasing our carbon footprint."[145] According to the EAC Food Security Action Plan (2011–2015), agro-industries "create forward and backward linkages, leading to significant multiplier effects, generating demand for agricultural produce ... creating on- and off- farm employment, enhancing incomes and contributing to value

---

[142]http://www.thisisafricaonline.com/Business/Nigeria-seeks-investors-for-agri-processing-zones.
[143]World Bank (2013).
[144]Green (2013b).
[145]Green (2013c).

addition and increased public sector revenues."¹⁴⁶ The EAC Food Security Action Plan (2011–2015) also take the position that expanded agro-industries "will contribute to poverty reduction through combined effects of employment gains, income enhancement, inclusiveness and food security"¹⁴⁷ and that this could potentially benefit smallholder producers, promote their inclusion into the value chain, and facilitate their access to market, finance and technical assistance.¹⁴⁸

To the World Bank, agriculture and agribusiness should be at the top of the agenda for economic transformation and development in Africa. The World Bank projects that by 2030, agriculture and agribusiness will be a US $1 trillion industry in sub-Saharan Africa (up from US $313 billion in 2010). To the World Bank, in Africa, agribusiness "can play a critical role in jump-starting economic transformation through the development of agro-based industries that bring much-needed jobs and incomes."¹⁴⁹ To the World Bank, the current attention on agriculture in Africa, which focuses primarily on production agriculture "will not achieve its development goals in isolation from agribusinesses."¹⁵⁰

Also making the case for agribusiness in Africa is Nigeria's former Minister of Agriculture, Akinwumi Adesina, who has lamented the fact that Nigeria is known for nothing else besides oil. According to Adesina, this is a sad state of affairs because Nigeria never used to have oil, and before the oil boom of the 1970s relied almost entirely on agriculture.¹⁵¹ To Adesina, "[a]griculture is the future of Nigeria," And not just any type of agriculture but "agriculture that is modernised, that is productive, that is competitive." Nigeria is reportedly seeking to add 20 million metric tons to the domestic food supply by 2015, and is seeking to create 3.5 million jobs through agriculture. Nigeria's error, according to Adesina, was that the country "[was] not looking at agriculture through the right lens." Nigeria, Adesina claims, "[was] looking at agriculture as a developmental activity, like a social sector in which you manage poor people in rural areas." To Adsesina, "agriculture is not a social sector. Agriculture is a business. Seed is a business, fertiliser is a business, storage, value added, logistics and transport - it is all about business."

41.3.5.2 The Case Against Agribusiness in Africa

Organizations like ActionAid, OXFAM, and War on Want, challenge the fundamental assumption that the principal solution to food insecurity in Africa is to simply producing more food,¹⁵² and are of the view that small farmers are the key to

---

¹⁴⁶The EAC Food Security Action Plan (2011-2015).
¹⁴⁷Id.
¹⁴⁸Id.
¹⁴⁹Id., p. xiv.
¹⁵⁰Id.
¹⁵¹Green (2013d).
¹⁵²*An October 2013 report from ActionAid (*http://www.actionaidusa.org/publications/feeding-world-2050*) challenges the assumption that the principal solution to food security is simply producing more food. "Rising to the Challenge: Changing Course to Feed the World in 2050".*

reducing poverty and food insecurity in the region.[153] In his report on Benin, the UN Special Rapporteur on the Right to Food alludeed to the potential risks of pursuing "green revolution" based on the model followed in Latin America and Asia.[154] Opponents also argue that agribusiness has a price and is being developed in Africa at the expense of the environment, small-scale farmers, and indigenous as well as rural communities. To opponents, the promised gains from agribusiness are not materializing. Presently, governments do not appear to be striking a balance between encouraging foreign investment in land and protecting small-scale farmers and local communities. For the latter, the price of agribusiness and land concession agreements between governments and foreign investors are numerous, and include: the loss of land to businesses,[155] total disregard for communities' right to free, no prior and informed consent in land acquisition, the destruction of crops and water sources, as well as intimidation, arrests, and harassment of community leaders.

The development of palm oil plantations in Liberia is coming in at a steep price.[156] The 16 August 2010 concession agreement between the Liberian Government and the Golden Veroleum Liberia (GVL) is creating a great deal of trouble in the country. The agreement provides for a 65-year lease for 220,000 ha of land to GVL in five counties in Liberia: Sinoe, Grand Kru, Maryland, River Cess and River Gee and also a new port with 100 ha of adjacent land. Under the lease agreement, GVL has the option to extend the lease for an additional 33 years. Affected communities are fighting back. On 1 October 2012, a complaint was filed with the RSPO.[157] In December 2012, the RSPO called for a freeze in plantation development until the complaint was resolved.[158] A report by The Tropical Forest Trust (TFT), a third party that GVL contracted to independently assess the

---

[153]International Fund for Agricultural Development, Africa's Small Farmers Key to Reducing Poverty, Increasing Food Security (press release), 15 May 2014.

[154]A/HRC/13/33/Add.3.

[155]Vermeulen and Cotula (2010).

[156]Global Witness. "UK's Equatorial Palm Oil accused of human rights abuses in Liberia." December 20, 2013. Accessed on June 15, 2014. http://www.globalwitness.org/Liberia/EPO. See also Sustainable Development Institute, "Community Complaint against Equatorial Palm Oil." Accessed on June 15, 2014. http://www.forestpeoples.org/sites/fpp/files/news/2013/10/Community%20Complain_ LiberiaGrandBassaCounty_Oct2013.pdf.

[157]Forest Peoples Programme (nd) Letter of complaint to Round Table on Sustainable Palm Oil (RSPO) from indigenous Butaw Kru tribes and inhabitants from several local communities within the proposed Golden Veroleum 220,000 ha oil palm concession in Liberia, October 2012. All sources pertaining to the complaint to the RSPO available from http://www.forestpeoples.org/topics/palm-oil-rspo/news/2012/10/letter-complaint-round-table-   sustainable-palm-oil-rspo-indigenous.

See http://www.forestpeoples.org/sites/fpp/files/news/2012/10/Final%20complaint%20 to%20%20RSPO%20on%20Golden%20 Veroleum-%20Butaw-sinoe%20county%20(2).pdf. See also http://www.forestpeoples.org/topics/palm-oil-rspo/news/2012/10/letter- complaint-round-table-sustainable-palm-oil- rspo-indigenous.

[158]RSPO 2012 Letter from RSPO to Golden Veroleum (Liberia) Inc. 13th December 2012. Available at http://www.forestpeoples.org/sites/fpp/files/news/2012/10/RSPOLetter_GVL_ PremFindings_ December2012.pdf.

company's operations with particular reference to FPIC, was very critical of the company's activities.[159] The GVL case highlights the dangers that come when governments promote an export-led developmental model without addressing the competing interest of all stakeholders. According to the Forest Peoples Programme:

> Although agreed to by the government and ratified by the Liberian legislature, GVL's concession contract itself and the contracting process leading up to and including the conclusion of the contract have been subject to criticism on grounds that they violate national law (including the constitution), as well as Liberia's international human rights law commitments.[160]

A human rights-based analysis of the agricultural concession agreements that many governments in Africa are concluding suggest that the agreements are not designed with the goal of securing maximum benefits for local communities.[161] Although in some cases domestic courts have stepped in to address forced eviction, in many cases, displaced communities are denied justice. In Uganda, the Government's sale of a land to foreign investors for a coffee plantation led to forced eviction of over 2000 individuals. Kaweri Coffee Plantation Ltd. is wholly-owned by the Neumann Gruppe, a German coffee producer. In a lawsuit filed against Kaweri and the Ugandan Government, plaintiffs alleged violent forced eviction from their land, mistreatment and destruction of their houses and other property. In a 28 March 2013 decision, a High Court in Uganda at Kampala found that the plaintiffs were illegally evicted from their land and ordered that compensation be given to the evictees.[162] James Nangwala and Alex Rezida, advocates for Kaweri, were ordered to pay 37 billion shillings for defrauding the government and the tenants, and for dishonesty, negligence, theft and misappropriation of client monies, violation of the Land Act, Bias, Eviction, Tenant's compensation, advocates, and Account Rules. The judge specifically found that the plaintiffs, who were tenants on the land, "were violently evicted without any relocation or compensation" and that "[t]he officers of the German Investor's company ... were active participants at the meetings on eviction of the tenants." Appeal is currently pending.

With the expansion of agribusiness in Africa, the issue of workers' rights is coming to the forefront and protests by farm workers is growing and escalating in many countries. Abuse has been cited in a number of industries, including the fruit

---

[159]Tropical Forest Trust 2013 Independent assessment of free, prior & informed consent process—Golden Veroleum (Liberia) Inc. February 2013. Final report. Available at http://www.forestpeoples.org/sites/fpp/files/news/2012/10/TFT_GVL_Liberia_FPIC_ Report_Final_Eng_low%20res.pdf.
[160]Kenrick and Lomax (2013).
[161]Lomax (2012).
[162]"Baleke Kayira & 4 Ors v Attorney General & 2 Ors", 28 Mar 2013. [2013] UGHC 47.

and wine industry,[163] the cocoa sector,[164] the flower industry,[165] the tea industry,[166] the fruit industry,[167] and the palm oil sector.

41.3.5.3 Agribusiness in Africa: A Confused Legal Framework

African governments are committing to promote agribusiness at the same time that they commit to protect the interest of small scale farmers. What is lacking are the laws, regulations, policies and institutions necessary to ensure that the interests of different stakeholders are adequately protected.

Policy incoherence—vertical and well as horizontal—is a major problem in Africa. Regarding "vertical" incoherence, most governments in Africa have taken on human rights commitments at the continental and international levels but are not taking effective steps to implement these obligations and to write the obligations into national legislation. Horizontal incoherence refers to weak policy alignment, in most countries in Africa, between human rights law and other areas of law including the tax law, investment law, and corporate law. Whether the bilateral investment treaties that countries in Africa are concluding are designed to achieve sustainable development objectives is a question that is receiving increased scrutiny today. As noted in the 2008 *Report of the Special Representative of the Secretary-General on the issue of human rights and transnational corporations and other business enterprises*,

> To attract foreign investment, host States offer protection through bilateral investment treaties and host government agreements. They promise to treat investors fairly, equitably, and without discrimination, and to make no unilateral changes to investment conditions. But investor protections have expanded with little regard to States' duties to protect, skewing the balance between the two. Consequently, host States can find it difficult to strengthen domestic social and environmental standards, including those related to human rights, without fear of foreign investor challenge, which can take place under binding international arbitration.[168]

African countries have concluded a considerable number of BITs, but critics argue that the BITs are not designed with sustainable development in mind.[169] In a

---

[163]Human Rights Watch, Ripe with Abuse: Human Rights Conditions in South Africa's Fruit and Wine Industries (2014).

[164]Griek et al. (2010); See also, Orla Ryab, Labouring for Chocolate, BBC News, 27 April 2007.

[165]Kenya Human Rights Commission, 'Wilting in Bloom'—The Irony of Women Labour Rights in the Cut-Flower Sector in Kenya (2010).

[166]Kenya Human Rights Commission, A comparative Study of the Tea Sector in Kenya (2008).

[167]Kenya Human Rights Commission, Exposing the Soft Belly of the Multinational Beast: The Struggle for Workers' Rights at Del Monte Kenya (2002).

[168]*Report of the Special Representative of the Secretary-General on the issue of human rights and transnational corporations and other business enterprises*, A/HRC/8/5 (7 April 2008).

[169]Peterson and Gray (2003); Mann (2011). See also Shemberg (2008).

recent article that examined the BITs that the Ethiopian Government has concluded, this author observed that Ethiopia's BITs:

> [D]o not: (i) specifically reference human rights, environmental protection, or other social issues; (ii) avoid provisions that could constrain the ability of a host government to regulate in the public interest; (iii) impose binding obligations on investors; (iv) establish clear mechanisms for monitoring compliance or enforcing human rights or environmental rights claims; (v) incorporate, directly or indirectly, specific human rights treaties or environmental protection treaties; (vi) confer on investment tribunals the jurisdiction to consider human rights norms and principles when assessing a State's liability under a BIT; and (vii) do not condition the availability of investor rights on the observance of international law by the investors.[170]

For agribusiness development to contribute to sustainable development in Africa, policy coherence is important. Organizations like the United Nations Conference on Trade and Development now call for a new generation of BITs[171] and a reform of the Investor-State dispute settlement system.[172] Overall, there is a growing call for BITs that "accommodate the home state's interests in conserving regulatory space by introducing provisions that avoid liability for treaty violations by identifying circumstances in which a state may regulate foreign investment."[173]

41.3.5.4 Conclusion

Domestic and foreign investment in land and in agricultural market in Africa is growing. On the one hand, large scale land acquisition could mean the infusion of much needed agricultural capital and innovative technology to the agricultural sector in Africa. On the other, there is a growing recognition that the development benefits of large-scale land acquisition are not automatic and that such investments can have negative implications for affected communities. As the former UN Special Rapporteur on the right to food, Olivier De Schutter, put it:

> Investment is flooding into the continent's land and agricultural markets, but question marks remain about how this will be turned to the benefit of the 250 million Africans suffering from food insecurity. Are small farmers – themselves often food insecure – gaining new income opportunities? Are the customary rights of herders being respected? Is enough being done to ensure that adequate food is affordable and accessible to poor urban communities?

How to manage the new agricultural investors, including agribusinesses, private equity groups, international pension funds, sovereign wealth funds, foreign

---

[170] Professor Uche Ewelukwa Ofodile, Foreign Investment in Land in Africa: Mapping the Role of International Investment Contracts and Bilateral Investment Treaties, Law and Development Review Issue No. 2 (2014).

[171] UNCTAD, Towards a New Generation of International Investment Policies: UNCTAD's Fresh Approach to Multilateral Investment Policy-Making, IIA Issues Note No. 5 (2013)(hereinafter "A New Generation of International Investment Policies").

[172] UNCTAD, Reform of Investor-State Dispute Settlement: In Search of a Roadmap, IIA Issue Note No. 2 (June 2013).

[173] Sornarajah (2011).

governments, and domestic investors is a growing challenge for governments in Africa. From a food law perspective, attention must be paid to a host of policy issues that impact smallholders farmers and local communities. How to integrate women into agri-business value chains, ensure equitable access to land, establish an enabling legal framework for sustainable land management practices must all be part of the discussion.

Expansion of agribusiness in Africa must be matched by expansion in the legal and institutional framework necessary to prevent negative environmental and social impacts. Laws must be put into place to address issues such as agro-chemicals and pollution, pesticide use, soil deterioration, land grabbing, labor rights violations, inhumane working conditions, and other community impacts.[174] This calls for a fresh approach to corporate governance in Africa. Countries must find a way to ensure that agribusiness is not advanced at the expense of agricultural workers, small farmers, and consumers.

Expansion of agribusiness must also be matched by expaned the provision of grievance mechanisms for those affected by the activities of corporations. Presently, there are very few examples of instances where the courts in Africa have held enterprises in the food and agricultural sector in violation of their legal obligations.[175] Cases such as *Haribo Mohammed Fukisha v. Redland Redroses Limited* (Kenya) and *Baleke Kayira & 4 Ors v Attorney General & 2 Ors* (Uganda) involving law suits against agribusinesses are still rare in Africa.

How to balance the goal of encouraging industrial and commercial model of agriculture with its emphasis on large producers and private investors against the goals of protecting and promoting family agriculture and small farmers is one issue that will shape food law in Africa in the coming years. According to Neil Crowder, managing partner at Chayton Capital, "a sustainable business model for farming in Africa must not only take advantage of the continent's impressive agricultural potential, but also implement strategies that will make Africans successful farmers in the years ahead."[176] To Crowder, "[m]odels that combine a commercial farming approach, underpinned by significant capital investment, with local small-scale farming have the potential to develop the African agricultural opportunity in a sustainable and profitable way." Crowder calls for a community-based approach to developing agricultural potential in Africa. As he puts it:

---

[174]See also International Labor Rights Forum, The Sour Taste of Pineapple: How an Expanding Export Industry Undermines Workers and Their Communities (October 20, 2008) (discussing the working conditions and labor rights violations in the pineapple fields in Costa Rica and the Philippines and finding that massive infestation of insects came with the expansion of large-scale pineapple production and had a dire effect on local ranchers' ability to maintain livestock.).

[175]See e.g. *Haribo Mohammed Fukisha v. Redland Redroses Limited* [2006] eKLR (applying Kenya's Workmen's Compensation Act (Cap. 236) and hold that an obligation falls on the defendant to compensate the plaintiff who was injured in a mishap involving potentially dangerous substances used in the course of employment.).

[176]Crowder (2010).

it is our view that large-scale agricultural investment must be made in conjunction with efforts to improve returns for small farmers. Efforts should be made to improve regional access to storage and transportation. Education and skill transfer can have economic as well as social benefits within the local community, while smaller farmers can take advantage of improved local and regional infrastructure, including access to storage and transportation.[177]

As a matter of urgency, governments in Africa must implement the *UN Guiding Principles on Business and Human Rights: Implementing the United Nations 'Protect, Respect and Remedy' Framework (the Guiding Principles)*,[178] and create an enabling environment for corporate social responsibility to thrive in the continent. Businesses such as Nestlé, Coca-Cola, and Pepsi, are beginning to make public commitments on land and land rights. In July 2014, Nestlé made a public commitment to protect land rights in agricultural supply chains.[179] Food giants like General Mills and Nestlé have reportedly initiated projects aimed at improving productivity and providing farmers with greater market access, training and technology.[180] Whether these commitments will yield tangible benefits for smallholder producers, poor rural families and local communities in Africa remains to be seen.

41.3.6 Food Law and Food Trade in Africa

Africa's food system is largely shaped by the global trade in food and by the rules of the multilateral trading system. Consequently, how trade liberalization in agricultural products and global trade rules affect food production, food export, and food security in Africa are proper concerns for food law and food law regimes in Africa. Of the 54 countries on the African continent, 42 are Members of the World Trade Organization (WTO). In addition, nine countries in Africa are in the process of acceding to the WTO.[181] Of particular relevance are: agricultural subsidies in developed countries; the impact of unrestricted food import on local producers and local production; and limited market access in developed countries as a result of imposition of high levels of protection and domestic support.

---

[177]Id.

[178]Special Representative of the Secretary-General on the Issue of Human Rights and Transnational Corporations and Other Business Enterprises, *Guiding Principles on Business and Human Rights: Implementing the United Nations "Protect, Respect and Remedy" Framework*, U.N. Doc. A/HRC/17/31 (Mar. 21, 2011) (by John Ruggie) [hereinafter *Guiding Principles*].

[179]Nestlé, Nestlé Commitment on Land & Land Rights in Agricultural Supply Chains (July 2014). The new policy is an Appendix to The Nestlé Policy on Environmental Sustainability.

[180]Wendy Atkins. 28 May 2013. http://www.thisisafricaonline.com/Perspectives/Professionalising-smallholder-organisation-in-Africa. Professionalising smallholder organisation in Africa.

[181]Algeria, Comoros, Equatorial Guinea, Ethiopia, Liberia, Libya, Sudan, Seychelles and Sao Tome and Principe, are in the process of accession to the WTO.

Individually and/or with other Members, African countries have made contributions to the negotiations on agriculture at the WTO. To the WTO Africa Group, any reform of trade rules relating to agricultural products should "Strengthen the rules and disciplines governing trade in agriculture to promote development;" "Ensure that trade liberalisation takes into account non-trade concerns such as food security, sustainable rural development and poverty alleviation;" "Ensure commercially viable market access for all agricultural products originating in developing countries including those at the higher end of the processing chain;" and "Level the playing field in the international trading system, taking into account different structural constraints among countries."[182]

A complete evaluation of how global trade rules affect food systems in Africa and ultimately shape food law in the continent is beyond the scope of this paper. A host of issues are cause for concern, including: market access issues; export competition issues (export subsidies and export support); domestic support issues; special and differential treatment; food insecurity; the impact of food aid on the agricultural production systems of countries in Africa; arbitrary imposition of sanitary and phytosanitary measures by some governments; policy incoherence; and the fate of net food importing developing countries.

41.4 Conclusion: Towards a Coherent and Functioning Food Law Regime in Africa

The goal of this part of the paper is to: identify gaps in the food law and policy of countries in Africa, draw attention to challenges to the development of robust food law regimes in Africa, briefly discuss the potential role of the legal profession in Africa in the development of food law in the continent, and offer some concluding thoughts about the future direction of food law and policy in Africa.

41.4.1 Food Law in Africa: Gaps

There are noticeable gaps in Africa's food law and policy architecture. Water, REDD "Reducing Emissions from Deforestation and Degradation," Carbon Trading, agricultural bankruptcy, agricultural insurance, water resources, water rights, and conflicts over water are issues that have not been properly addressed in the legal framework of many countries. Although a few regional and sub-regional resolutions and declarations address water, many issues related to water rights and water

---

[182]World Trade Organization, WTO Africa Group: Joint Proposal on the Negotiations on Agriculture, G/AG/NG/W/142 (23 March 2001).

use are yet to be addressed.[183] Many countries do not address in a holistic and comprehensive fashion issues such as how to implement sustainable ocean and freshwater fishery practices and how to strengthen resilience to climate shocks associated with droughts and floods. The land question, including questions of ownership, access and control of farmland, rangeland, forest, fishery, wetlands, pasture, and hunting territories is also yet to be fully and effectively addressed in the land law, policy, and programming of most countries in Africa despite the adoption in 2009 of the AU *Framework and Guidelines on Land Policy in Africa*.

To date, limited attention has been paid to ensuring the availability and accessibility of grievance mechanisms in Africa's food and agricultural sector. Very few farm-level grievance mechanisms exist in Africa even though an overwhelming majority of Africans live and work on farms. There is a need for a thorough mapping of available grievance mechanisms and a need to create more State-based and non-State-based grievance mechanisms accessible to farmers and farm workers in Africa. There is also a need to assess the effectiveness of existing grievance mechanisms. In light of the recent push to develop agribusiness in Africa, designing and operationalizing appropriate grievance mechanisms must be on the agenda of governments and businesses in the continent. According to Principle 26 of the United Nations Guiding Principles on Business and Human Rights, "States should take appropriate steps to ensure the effectiveness of domestic judicial mechanisms when addressing business-related human rights abuses, including considering ways to reduce legal, practical and other relevant barriers that could lead to a denial of access to remedy."[184] Principle 29 adds: "To make it possible for grievances to be addressed early and remediated directly, business enterprises should establish or participate in effective operational-level grievance mechanisms for individuals and communities who may be adversely impacted."[185]

In some countries, national human rights institutions are beginning to receive and respond to complaints from farmers and farm workers. The Kenya National Commission on Human Rights, a body established by an Act of Parliament in 2002, has in the past responded to allegations of abuses in Kenya's tea industry and flower industry.[186] In addition to its role of investigating complaints regarding human rights violation, the Kenya National Commission also independently carries out human rights studies and has in the past focused attention on businesses in the food and agricultural sector. More studies are needed to understand the potential role of

---

[183] Sirte Declaration on the Challenges of Implementing Integrated and Sustainable Development on Agriculture and Water in Africa (2004). Ex/Assembly/AU/Decl. 1 (II).

[184] Special Representative of the Secretary-General on the Issue of Human Rights and Transnational Corporations and Other Business Enterprises, *Guiding Principles on Business and Human Rights: Implementing the United Nations "Protect, Respect and Remedy" Framework*, U.N. Doc. A/HRC/17/31 (Mar. 21, 2011) (by John Ruggie) [hereinafter *Guiding Principles*].

[185] Id.

[186] The Kenya National Commission on Human Rights Act, 2002. Another example is the Commission for Conciliation, Mediation and Arbitration (CCMA) in South Africa—a body charged with resolving disputes in labor relations. www.ccma.org.za.

national human rights institutions in promoting respect for human rights in the food and agricultural sector in Africa.

41.4.2 Food Law in Africa: Challenges

Lack of political will as well as weak policy alignment and harmonization are major barriers to the development of food law and food governance in Africa. The result is outdated laws, major gaps in the food law architecture of most countries in Africa, and lack of enforcement of existing laws. Another challenge is poor agricultural statistics and low priority accorded to agricultural research in Africa. According to Judith Francis, senior programme coordinator, Science and Technology Policy, at the ACP-EU Technical Centre for Agricultural and Rural Cooperation (CTA), "Agricultural research and development projects in sub-Saharan Africa are still heavily dependent on donor funding and agricultural, science, technology and innovation systems are weak."[187]

Limited public participation and lack of transparency in legislative processes pose another set of challenges to the development and maturation of food law in Africa. African governments so far have not kept to the promises made in the Maputo Declaration to promote inclusive and participatory food and agricultural regimes in the continent. In Paragraph 4 of the Maputo Declaration, African Heads of State resolved to: "**ENGAGE** in consultations at national and regional levels with civil society organizations and other key stakeholders, including the small-scale and traditional farmers, private sector, women and youth associations, etc., aimed at promoting their active participation in all aspects of agricultural and food production."[188]

A final set of challenges arise out of the fact that much of Africa's food policy is externally driven, shaped by the agenda of donor governments and donor agencies rather than by local concerns and domestic constituencies.

41.4.3 The Legal Profession in Africa & Food Law

Food law is not taught in most law schools in Africa. Lawyers, law firms, and bar associations in Africa have historically not catered to the needs of the main stakeholders in Africa's food and agriculture sector—the small-scale farmers. Law firms with food law practice are rare in Africa, and cater largely to the needs of corporate clients. There is an urgent need to explore how lawyers and bar associations in Africa can add value to the continent's food and agricultural sector. Law firms in Africa must explore ways to develop expertise specific to the food and

---

[187]http://www.thisisafricaonline.com/Analysis/Raising-the-bar-on-agricultural-innovation.

[188]Maputo Declaration, supra note 3 (emphasis in the original).

agriculture industry, make connections with stakeholders in the industry, and serve the particular needs of food producers and processors in Africa. The bar associations in the different countries in Africa can take a leading role in exploring and explaining to farmers in the continent the implications for food producers and processors of various areas of law, particularly Administrative Law, Labor & Employment Law, Environmental Law, Cooperative Law, Energy Law, Corporate/Commercial Law, as well as Immigration Law. The bar associations can also play a key role in bringing together stakeholders in the industry and updating the industry on current legal issues. The law firms and law societies in the continent can explore partnership opportunities with relevant agencies, provide increased training opportunities for food producers/processors in Africa, and promote laws and policies beneficial to the industry. Finally, law firms and bar associations in Africa can foster understanding on the linkages between agribusiness and human rights in Africa in line with the *Business and Human Rights Guidance for Bar Associations* that the International Bar Association released in 2014.[189]

41.4.4 Conclusions

Food law in Africa is in its nascent stages. There is as yet no coherent or holistic approach to food law on the continent. Moreover, in the nascent food law regime in Africa, legacies of colonialism are evident. Food law in Africa is likely to be the site for much debate in the coming years. In almost every country in Africa, there is a noticeable change from a prior emphasis on government intervention in agricultural production and marketing to a shift towards liberalized markets, deregulation, and private sector-led agricultural development strategies.[190] Many countries have adopted or are in the process of adopting national strategies for revitalizing agriculture.[191] Laws and regulation that countries adopt in the coming years to implement their respective agricultural strategies are likely to define the contours of food law in the continent. The food law regimes in Africa will be tested in part by the extent to which they empower key stakeholders, including small scale farmers, consumers and private investors. In many countries, the food and agricultural sector is presently bogged down by the prevalence of old, outdated and sometimes draconian pieces of legislation. Countries will need capacity to revise and update their food and agricultural law, and capacity to address major inconsistencies in these laws.

---

[189]International Bar Association, Business and Human Rights Guidance for Bar Associations (With Commentaries) (2014).
[190]See e.g. Kenya's Crop Act, 2012 (No. 16 of 2013).
[191]See e.g., The Strategy for Revitalising Agriculture (Kenya) and Agricultural Sector Development Strategy 2010–2020 (Kenya).

Many factors shape and are likely to continue to shape food law and policy in Africa including international trade rules and practices, international investment law and the investor-state dispute settlement system, the international aid system, and climate change. Food law in Africa will also be shaped by laws and policies designed to promote and manage intra-African trade in agricultural and fishery products. In 2004, African Heads of State and Government established the African Common Market for agricultural products, according to the Lusaka Summit decision.[192] In the 2004 Sirte Declaration, African Governments pledged to promote "intra-African trade in agricultural and fishery products in order to correct discrepancies in food balances at both national and regional levels."[193]

In many respects, food law and policy in Africa is still externally driven. Decisions about food volumes, prices, and qualities are largely driven by actors outside Africa. Developments in the global arena, including the continued liberalization of agriculture and agricultural markets, steady rises in foreign investment in agricultural land and agribusiness, growth in South-South trade and investment, the growth in the number of voluntary food safety standards and other corporate social responsibility initiatives are all likely to shape food policy in Africa in the coming years.

There are many challenges and obstacles to the development and maturation of food law as a discipline in Africa. Some of these include the general disregard for the rule of law in some countries, the lack of capacity needed to develop and implement new laws, and the fact that very few universities and law schools have food law as a distinct and separate discipline and academic offering. The University of Pretoria's Institute for Food Nutrition and Well-being (IFNuW), established in 2011 and officially launched in May 2012, is presently one of the few academic centers dedicated to food law.[194] Academic collaboration between scholars in Africa and those in countries where food law is more developed will be crucial to efforts to develop food law as a distinct legal discipline in Africa. Food law academies and centers committed to promoting and fostering legal and policy developments in the area of food law are thus urgently needed in the continent. Overall, although the future looks bright for the development of robust food law and

---

[192] Sirte Declaration on the Challenges of Implementing Integrated and Sustainable Development on Agriculture and Water in Africa (2004). Ex/Assembly/AU/Decl. 1 (II).

[193] Id., para. 14.

[194] The Institute involves five faculties: Natural and Agricultural Sciences, Health Sciences, Education, Law and Veterinary Science. The activities of the Institute for Food, Nutrition and Well-being (IFNuW) are organised around five research areas focusing on: (1) Sustainable animal- and plant-based food production in a resource-constrained environment; (2) Food safety, biosecurity, public health and regulatory control; (3) Identifying and promoting beneficial compounds in foods to promote health and address diseases of lifestyle; (4) Facilitating behaviour change for improved health and well-being; and (5) The food security and nutrition impacts of policies and programmes. See http://web.up.ac.za/default.asp?ipkCategoryID=17839&subid=17839.

policy regimes in Africa, the road is likely to be tortious, controversial and very challenging.

References

Clark S (2014) Congo seeks investors for farmland bigger than France. Wall Street J
Crowder N (2010) Capturing Africa's agricultural potential. This is Africa. 11 January 2010
DiLorenzo S (2014) Thousands break Ebola quarantine to find food, Yahoo News. 4 November 2014. http://news.yahoo.com/thousands-break-ebola-quarantine-food-124818527.html
FAO (2001) The state of food insecurity in the world 2001. Rome
Green A (2013a). Brains plus brawn: big food retailers seeking local partners. This is Africa. 21 August 2013. http://www.thisisafricaonline.com/Business/Brains-plus-brawn-Big-food-retailers-seeking-local-partners
Green A (2013b) Africa's food imports on the rise. This is Africa. 16 May 2013
Green AR (2013c) The value addition imperative in agriculture. 19 August 2013. This is Africa. http://www.thisisafricaonline.com/News/The-value-addition-imperative-in-agriculture
Green A (2013d) Interview: Akinwumi Adesina, Minister of Agriculture, Nigeria. This is Africa, 30 July 2013
Griek L, Penikett J, Hougee E (2010) Bitter harvest: child labour in the cocoa supply chain
Kenrick J, Lomax T (2013) Summary case study on the situation of Golden Veroleum Liberia's oil palm concession. In: Conflict or consent? The oil palm sector at a crossroads
Lomax T (2012) Human rights-based analysis of the agricultural concession agreements between Sime Darby and Golden Veroleum and the Government of Liberia. Forest Peoples Programme, Moreton-in-Marsh. Available at http://www.forestpeoples.org/sites/fpp/files/publication/2012/12/liberiacontractanalysisfinaldec2012_0.pdf
Mann H (2011) Stabilization in investment contracts: rethinking the context, reformulating the result. Investment Treaty News
Peterson L, Gray K (2003) International Bilateral Investment Treaties and in Investment Treaty Arbitration
Shemberg A (2008) Stabilization clauses and human rights, 11 March 2008. http://www.ifc.org/ifcext/enviro.nsf/AttachmentsByTitle/p_StabilizationClausesandHumanRights/$FILE/Stabilization+Paper.pdf
Simpson J (2011) The Kalahari Bushmen are home again. The Guardian. 13 December 2011
Sornarajah M (2011), Mutations of neo-liberalism in international investment law. Trade Law Dev 3(1):203, 228
Strickland RS (2004) To have and to hold: women's property and inheritance rights in the context of HIV/AIDS in Sub-Saharan Africa, ICRW Working Paper, June 2004
Vermeulen S, Cotula L (2010) Over the heads of local people: consultation, consent and recompense in large-scale land deals for biofuel products in Africa. J Peasant Stud 34:899, 904
Vidal J (2014) World Bank accuses itself of failing to protect Kenya forest dwellers. The Guardian, 29 September 2014
World Bank (2013) Growing Africa: unlocking the potential of agribusiness
Yankey LE (2004) FAO Background paper for Global Forum

Chapter 42
Disease Control, Public Health and Food Safety: Food Policy Lessons from Sub-Saharan Africa

Kennedy Mwacalimba

Abstract This chapter reviews the agro-economic environment in Sub-Saharan Africa as it relates to animal production, public health, and disease control to contextualize the concept of risk and food safety. Drawing mostly from the experience of Zambia, it analyzes food safety actors and interests in Sub-Saharan Africa, and provides an outline of the general regulatory framework that is in place on the continent, to explain how food safety governance is impacted by different interest groups and agendas. Two case studies are provided, zoonotic tuberculosis and avian influenza. The chapter demonstrates how the two zoonoses, both important food safety concerns, have been prioritized differently in the case of Zambia, as a result of multiple sociopolitical and economic factors. The chapter concludes that, in order to be useful, a definition of food safety risks should include multiple contextual issues and stakeholders along the food supply chain. It is important to keep in mind what national food safety governance actors perceive the risks to be, and how their definitions fit into the broader picture of food safety in general. Food safety governance regulatory processes should take into consideration local realities, local food supply chains and local food safety threats to ensure the appropriateness and sustainability of any and all disease control measures instituted. Context will always matter, and therefore, local ecological, biological and policy considerations should be given primacy.

42.1 A Risk Management Approach to Examining Food Safety, Disease Control and Public Health

Taking an African perspective on food law and food safety, this chapter tackles the concept of risk as it relates to food derived from animals. To bring the various issues that impact risk and food safety in Sub-Saharan Africa into focus and highlight the pertinent concerns for disease control, it is necessary to provide both a descriptive

K. Mwacalimba, BVM, MSc, DLSHTM, PhD (✉)
Outcomes Research, US Operations, Zoetis LLC, Parsippany, NJ, USA
e-mail: kennedy.mwacalimba@zoetis.com

© Springer International Publishing Switzerland 2016
G. Steier, K.K. Patel (eds.), *International Food Law and Policy*,
DOI 10.1007/978-3-319-07542-6_42

and an explanatory analysis of the issues surrounding food safety on the continent. This chapter's explanations are rooted in the analysis of the narratives of food safety actors and interests, and includes some detail on the general regulatory framework that is in place in the region, how it is influenced by different interest groups and agendas, and the politics of the policy process. These provide important lenses for understanding food policy in Africa. In order to deepen this discussion, some of the perceptions of risk and policy issues concerning the African agroecosystem in general are also highlighted. This chapter does not purport to provide a complete analysis. It merely provides contextual depth to case studies from a perspective of animal disease control, public health and food safety to illustrate important food policy lessons for the African region.

Food-borne diseases remain a significant problem for public health around the world. An estimated 70 % of diarrheal incidents across the globe are due to biological or chemical contamination of food.[1] The burden of food-borne illness is borne by both developed and developing countries. However, the literature suggests that the incidence is highest in the African region.[2] Over the last 15 years, according to the World Health Organization Regional Office for Africa (WHO-AFRO), the African region has suffered several major food-borne disease outbreaks, including aflatoxicosis in Kenya, anthrax in Zimbabwe, bromide poisoning in Angola and chemical intoxication in Nigeria.[3] WHO-AFRO suggests that the vast majority of food-borne incidents in the African region are unreported. Therefore, the true extent of this public health problem is unknown.

In Africa, an appropriate establishment and consistent maintenance of adequate food safety infrastructure would go a long way in reducing the burden of the public health threat of food-borne illnesses. But multipartite food supply chains make food safety a complex policy issue to unpack, particularly in the era of globalization. Therefore, the evolving issues around food safety and the complexity of global food supply chains[4] require the development of context-appropriate food safety and disease control structures and policies that appropriately mitigate the myriad threats to public health presented by food and food trade. This is because, the various stakeholders along the food chain—producers, retailers, consumers and regulators—all play a role in assuring that food is safe, sound and wholesome.

Food of animal origin carries multiple risks. In the case of food-borne illnesses, these risks can generally be grouped under intoxications or infections. The livestock production related threats to public health include acaricides used for tick control, antibiotic residues in meat and milk, infectious diseases, and pollutants from

---

[1] Buzby and Roberts (2009), pp. 1851–1862.
[2] Dewaal et al. (2010), pp. 483–490; WHO-AFRO (2012).
[3] *ibid.*
[4] see Kimball (2006).

agricultural runoff. Other important public health threats are zoonotic diseases. Rudolf Virchow, the pioneer of the concept of One Health,[5] first coined the term zoonosis in 1855.[6] Originally, zoonoses were defined according to the direction of disease transmission. Additional important terminology included zooanthroponosis, infections humans could acquire from animals, and anthropozoonosis, diseases that humans could transmit to animals.[7] This conceptualization of zoonoses, however, failed because the two terms were used indiscriminately, leading to an expert committee decision to abandon them. Instead, the committee recommended that the term zoonoses should be defined holistically as diseases and infections naturally transmitted between vertebrate animals and humans.[8]

Animal health is only one link in a long food production chain that contributes to the final quality of food. The promotion of animal health is vital to the enhancement of the quality and quantity of products derived from this source. This is especially true for food supply chains with global dimensions. Chemical and biological contamination can occur at any point from production to consumption. Through their associated impacts within the food production chain, animal breeding, feed, fertilizer and pesticide use, producers, processors, and retailers all add to or subtract value from the final product.

Globalization, economic development, expansion and diversification of agricultural food trade form causal links that increase public health risks, food safety hazards, and the spread of diseases.[9] Globalization transcends the nation state,[10] and brings with it social phenomena such as power and politics. Furthermore, in most low-income countries in the African region, economic considerations[11] are given primacy over health concerns.[12] This is reinforced in global context, where trade considerations run ahead of the implementation of measures that protect health.[13] Additionally, due to the process of globalization, policy-makers have seen a decline in their ability to control the determinants of health. Power relationships in many low-income countries are complicated by external relationships with advisors, experts, aid donors and financial institutions and internal institutional

---

[5] The American Veterinary Medical Association has defined One Health as "the collaborative effort of multiple disciplines — working locally, nationally, and globally — to attain optimal health for people, animals and the environment." American Veterinary Medical Association, One Health (2008), available at https://www.avma.org/KB/Resources/Reports/Documents/onehealth_final.pdf (last accessed Jan 2015).

[6] Kahn et al. (2007), pp. 5–19.

[7] Krauss et al. (2003).

[8] Hubálek (2003), pp. 403–404; Mwacalimba (2013).

[9] Slingenbergh (2004).

[10] Lee et al. (2002).

[11] Walt and Gilson (1994), pp. 353–370.

[12] Lee and Koivusalo (2005), p. e8.

[13] *ibid.*

relationships characterized by large power gaps between actors,[14] with power and politics playing a role in shaping the process.[15]

This chapter examines zoonoses through multiple and layered lenses, the first of which is the concept of "the human-animal interface,"[16] with the goal of better capturing the broad socio-economic and political landscape of food safety. The understanding of the human-animal interface, as it relates to infectious disease governance, is important from two standpoints: First, the human-animal interface facilitates the examination of the relevant public health risks related to animals and their products in different contexts. Second, through food governance, and its relationship to risk enabling policy activities, the human-animal interface further adds to the understanding of the complexities associated with risk management. The corresponding risk enabling activities include the governance of land use, wildlife use and livestock production and chosen routes for economic growth and trade promotion. These activities both foster and enhance disease transmission.[17] Specific examples are provided through case studies of two zoonoses, one of which is a neglected disease, bovine tuberculosis (BTB), and the other an emerging disease, highly pathogenic avian influenza (HPAI, a.k.a. bird flu).

42.1.1 Overview of the Epidemiological Framework for Disease Control and Public Health in Food Safety

Epidemiology is defined as the study of the distribution and determinants of disease in defined populations. Disease determinants include risk factors for emergence which are both multifactorial and highly contextual. While epidemiology studies the dynamics of disease in defined populations, it also seeks solutions to disease problems to mitigate their impact on individual and public health. This chapter focuses on food safety within a disease prevention framework.

In a disease prevention framework, three levels of prevention are important: primary, secondary and tertiary prevention.[18] In primary prevention, the focus lies on averting incidents of disease in the first place. Secondary prevention strives to avert clinical manifestation of the disease state. Tertiary prevention tries to avert the complications of a disease state, such as extended morbidy, secondary infection or death.[19] These levels are also dependent on access to interventions, i.e. health system or government capacity and response, and do not factor in the role that industry, for instance, plays in disease prevention.

---

[14]Walt and Gilson (1994).
[15]Navarro (1998) pp. 742–743.
[16]Greger (2007), pp. 243–299.
[17]Kimball (2006); Greger (2007).
[18]Kimball (2006).
[19]*Id.*, pp. 13–14.

The three levels of prevention are, of course, permeated by food safety governance, which reinforces the importance of this discussion within this book. For instance, primary prevention in food safety would involve the removal of chemical, physical or biological hazards from food before it is consumed. This includes the establishment of herd health programs for food animals, provision of safe feed, monitoring antibiotic use in animals, physical control mechanisms such as abattoir inspections and milk pasteurization, and the development and enforcement of food safety standards. Secondary prevention may include mechanisms for early health system response to food-borne disease outbreaks, institutional capacity to conduct timely food-borne illness outbreak investigations, food product recalls, destruction of contaminated produce or quarantine of implicated markets, farms, or animals to contain outbreaks. Tertiary prevention includes the appropriate treatment of food-borne illnesses, such as administering the correct medication to treat zoonotic tuberculosis. This prevention framework does not easily accommodate the roles that multiple stakeholders play in disease prevention, particularly those stakeholders that lie outside traditional food safety and health systems, but still impact the social determinants of health.[20] Logically, this begs the question; how can multiple stakeholder interests and influences be incorporated in this basic prevention framework?

One way of looking at disease prevention and management is to utilize a risk-based epidemiological model. In the epidemiological framework for disease control and public health in food safety, a risk based framework considers multiple risk factors along the food supply chain, from farm to fork, and the holistic impact of stakeholders on food safety and food-borne disease management. However, the concept of risk and its understanding is highly contextual, as our case studies will demonstrate. Risk is socially constructed. Therefore, risk identification and assessment are both innately human and dependent upon social activities that generate meaning. In other words, risk perception is contingent upon a shared understanding of reality.[21]

Risk is also "politicized" through several social processes, which separate risk from the actual dangers presented by various hazards.[22] But, the use of scientific information to inform policy is difficult. Paradoxically, one of the reasons for the difficulty to use science to inform policy on risk is scientific limitation to only the objective assessments of risk.[23] As Stirling and Scoones[24] contend in their 2009 paper on risk assessment and knowledge mapping, scientific assessments of risk often attempt to aggregate complex social and biological phenomena into a set of probabilities and outcomes, thereby often structuring the phenomena to become

---

[20]The social determinants of human health are the conditions of the environment where people live and work (Exworthy 2008).
[21]Horlick-Jones (1998), pp. 64–67.
[22]Douglas and Wildavsky (1982).
[23]Mwacalimba (2012), pp. 391–405.
[24]Stirling and Scoones (2009).

policy, rather than inform it. This also applies to risk assessment as applied in food safety and food policy.

Food-borne disease risks and their implications for holistic management require a multi-sectoral or shared understanding of Africa's food safety problems with a goal of a pragmatic development of solutions. Conceptually, the integration of the socio-political structures of disease risk and risk management should provide a useful framework for understanding food safety, disease prevention and public health in the African region. In other words, to be useful, an epidemiological framework for disease control and public health in food safety needs to be inclusive and open to multiple perspectives of African food safety problems. In order to succeed, it is important to properly frame food safety problems, and identify the different stakeholders involved along the food chain. Additionally, it is crucial to recognize the various stakeholders' roles and capacity for risk facilitation or management, as influenced by their institutional norms, priorities and ideas. The goal is to develop a consensual view of risk. In reality, however, the various stakeholders force aggregates of complex social and biological phenomena into risk metrics to which food governance systems are expected to somehow respond.

To create a usable disease control, response coordination and risk management framework, a proper understanding of both institutional power and the food policy framework are critical. Using this simplified risk-based epidemiological model, this chapter explains the various issues affecting food-borne diseases, their impact on public health and the policy implications of food safety governance. Fragmentation unfortunately is the complicated reality of many legal and policy issues at local, national and international levels. In applying this framework to the African region's multi-sectoral context, the boundaries between sectoral or "silo" responses to food safety will blur. Nonetheless, it should be clear by the end of the chapter that a one-cap-fits-all solution to food safety problems in the African region is unlikely effective.

42.1.2 Food Policy Regulatory Framework: Actors, Interests and Conflicts

The African continent's position on the development continuum puts it in the precarious state of having to deal with a double burden of disease,[25] i.e. both chronic, or diseases of affluence and infectious diseases.[26] Within this spectrum of disease threats, are food-borne illnesses. In turn, food-borne diseases remain a

---

[25] A double burden of disease is a state in which the prevalence of risk factors for chronic diseases (diabetes, heart diseases and cancers) increase at the same time that traditional health problems such as maternal and child deaths caused by infectious diseases are still major public health threats for the majority of the population.

[26] Green (1999).

significant problem for the region, given its high levels of poverty and urgent nutritional needs for its most vulnerable citizens. As highlighted in the introductory section, food-borne diseases are frequently reported in Africa, with high incidence. Most cases of food-borne diseases remain unreported. There are several factors responsible for this. In many countries in the region, there is the existence of poor food safety surveillance systems and weak obligatory reporting mechanisms for food-borne disease outbreaks, outside high profile diseases such as cholera. Reporting is also inaccurate in some cases, with only a handful of countries reporting incidence of food-borne illnesses. These knowledge gaps mean that an accurate picture of incidence is nonexistent, which limits our understanding of the public health impact of food-borne diseases in the region.

There are multiple stakeholders in food safety governance in the African region. These include both national and international actors. From the multilateral trade environment, characterized by overlapping regional trading blocs at continent level to professional rivalries at national and local levels, the food safety governance environment is fairly complex. Furthermore, the continent's myriad problems, make prioritizing food safety difficult. For instance, the WHO asserted that food-borne diseases and food safety do not feature very highly on national agendas although both have a major public health impact throughout the region. This assertion is based on a weighting of the paucity of resources directed at the issue. In fact, it is the political impetus to focus on food-borne diseases in the African region that is often absent.[27]

Fundamentally, the management of public and environmental health risks in the African region is characterized by fragmentation and the existence of conflicting national food safety standards.[28] The mandate to assure the safety of food of animal origin falls under various agencies, which can result in professional rivalries that impact negatively on food safety governance.[29] These mandates, enshrined in law and legal documents such as statutory instruments,[30] compound the problem of the policy disconnectedness surrounding food safety in Africa (see Table 42.1).

Understanding the role that external influences play in the policy processes for food safety and management in the African region is equally important. In many developing country settings, policy actors are forced to balance the external desires of funding agencies and international bodies as well as contend with internal power struggles.[31] Therefore, a hindrance to the management of risk in food safety in the African region is its dependence on development partners. External partners play a key role in defining national development agendas, by focusing their support on water delivery, primary healthcare, or particular infectious diseases such as tuberculosis, HIV and malaria. The dependence on development partners and the politics

---

[27] WHO-AFRO 2012.
[28] Wilson and Otsuki (2001).
[29] Muma et al. (2014).
[30] Statutory Instruments are a means of creating delegated or secondary legislation.
[31] Walt and Gilson (1994).

Table 42.1 African regional and subregional economic partnerships

Countries	Subregional	Regional
Algeria Morocco Libya Tunisia Mauritania	**AMU** *Arab Maghreb Union*	
Ghana Nigeria Cape Verde Gambia		**ECOWAS** *Economic Community Of West African States*
Benin Togo Niger Mali Côte d'Ivoire Burkina Faso	**WAEMU** *West African Economic and Monetary Union*	
Senegal Guinea-Bissau Liberia Guinea Sierra Leone	**MRU** *Mano River Union*	
Chad Cameroon Central African Rep. Gabon Equat. Guinea Rep. Congo	**CEMAC** *Communauté Économique et Monétaire de l'Afrique Centrale (Central African Economic and Monetary Community)*	**ECCAS** *Economic Community of Central African States*
Rep. Congo Burundi Rwanda	**ECGLC/CEPGL** *The Economic Community of the Great Lakes Countries/ Communauté Économique des Pays des Grand Lacs*	
Angola		
South Africa Botswana Lesotho Namibia Swaziland	**SACU** *Southern African Customs Union*	**SADC** *Southern African Development Community*
Angola Malawi Zambia Zimbabwe Mauritius Seychelles Mozambique		
Somalia Ethiopia Eritrea Sudan Kenya Uganda	**IGAD** *Inter-Governmental Authority on Development*	**COMESA** *Common Market for Eastern & Southern Africa*
Angola Egypt Burundi Rwanda Comoros Madagascar Malawi Zambia Zimbabwe Mauritius Seychelles		
Burkina Faso Chad Libya Mali Niger Sudan Central African Republic Eritrea Djibouti Gambia Senegal Egypt Morocco Nigeria Somalia Tunisia Benin Togo Ivory Coast Guinea-Bissau Liberia Ghana Sierra Leone Comoros Guinea Kenya São Tomé and Príncipe Equatorial Guinea	**CEN-SAD** *The Community of Sahel-Saharan States*	
Egypt Sudan Ethiopia Uganda Kenya Tanzania Burundi Rwanda the Rep. Congo Eritrea	**NBI** *Nile Basin Initiative*	
Tanzania Kenya Uganda	**EAC** *East African Community*	

The table shows the overlap of international economic partnerships and trading blocs in Africa. This multilateral trade environment, with its various agreements, provisos and foci, is an amorphous set of issues that, while not immediately obvious, have implications for disease spread and control. Typically, in such an environment, trade issues take primacy over health concerns, with profound implications for public health. The left hand side of the table lists the countries and the right shows the acronyms of economic partnerships they belong to (bold and italic terms)

of aid has implications for the capacity of African countries to control the focus of funding.[32] The assistance provided by donors towards food safety may not foster ownership among local stakeholders, given that the focus is donor driven and not aligned with existing needs and local realities.[33] In addition, the WHO states that development partners are not always willing to commit to sustaining the strengthening of food safety management systems.

42.1.3 Key Stakeholders and Mandates for Risk Management and Food Safety in Africa

There are contextual differences in food control system implementation across the African continent. In general, however, the existing food safety and control systems do not provide the policy coherence necessary for stakeholder agencies to synergistically prevent food safety problems. International guidelines such as the *Codex Alimentarius*[34] do not have the supporting legislative framework in many African countries. Many of the laws governing food safety are outdated, inadequate or fragmented.[35] The provisions relating to animal health, for instance, can be found in multiple statutes, codes and standards (both legal and voluntary), which spurs institutional rivalries and blurs the boundaries of responsibility. There is inadequate protection of consumers from contaminated food products, fraudulent practices and the importation of substandard food for domestic markets.[36] Furthermore, enforcement of food law remains an important concern. The absence of a coherent policy framework for food safety has created an environment in which these government agencies operate according to their own institutional perspectives on food safety, often leading to effort duplication.[37]

The African Regional Office of the WHO highlights the inadequate address of food safety concerns in national policies in the African region and recommends national food safety policy coherence to be the foundation for effective food safety management systems.[38] While linked, these are, in reality, two different issues. The letter of food safety policies, i.e. policy content, are based on regulation inherited from Africa's former colonial masters. Food safety governance in the region, like many institutions on the continent, cannot be easily separated from the legacy of colonialism.

---

[32]Mwacalimba (2013).

[33]WHO-AFRO (2012).

[34]The Codex Alimentarius Commission was established jointly by the United Nations' (UN) Food and Agriculture Organization (FAO), and the World Health Organization (WHO).

[35]*Ibid.*

[36]*Ibid.*

[37]FAO/WHO (2005).

[38]WHO-AFRO (2012).

Many existing laws are rooted in legislation that was created during the colonial period and are, therefore, not properly oriented towards contemporary realities. So, while secondary legislation through statutory instruments may help address some concerns, it is the underlying definitions, institutions and authorities that form the basis of the foundational legislation that needs to be reformulated. Therefore, while it is true that the inadequate address of food safety concerns in national food safety regulations hampers the effectiveness of government responses to the issue, the lack of movement may also be due to the myriad of issues governments have to deal with using limited resources.

These issues are also symptomatic of a fragmented polity. Some researchers[39] explain that fragmentation is the norm in contemporary policy, where different sectors of government work as so-called silos, pursuing individual sector interests and mandates. As would be expected in a fragmented polity, the key institutions involved in food safety governance and the implementation of risk management for the protection of public, animal and environmental health, all operate with different priorities, agendas and mandates. Each institution along the supply chain has a different value system. How various stakeholders perceive food safety, i.e., as either "low politics" or "high politics,"[40] also affects overall policy coherence. Others[41] suggest that while issues of "high politics" or those of macro or systemic importance may be formulated and imposed by a narrow group of elites, those of "low politics" are subject to the influence of many different groups. Certainly, the interaction of low and high politics amongst the various international and national stakeholders involved in food safety governance in the African region provide an interesting dynamic to the management of food-borne risks to public health.

The food safety governance boundaries in the African region are also shaped through legislation and international standards adopted within the respective mandates of national and international institutions. The confusion only increases when regional and global actors in food safety governance come into play. A WHO study examining the status of food law in Africa suggested that in most African countries, there existed a discord between national food law and international requirements, such as the *Codex Alimentarius*. The study further states that this discord has led to the rejection of food exports from the region.[42] Such complex phenomena, however, cannot be treated as mere events, but should be considered institutionally embedded processes with distinct histories that need to be uncovered to help illuminate more general problems,[43] especially where many countries in the region face several non-tariff barriers that limit their ability to export their produce. These include Sanitary and Phytosanitary (SPS) requirements, technical barriers to trade, quotas, and market standards, restrictive rules of origin and complex tariff

---

[39]Kingdon (2003).

[40]Walt and Gilson (1994), Buse et al. (2005).

[41]Walt (1994).

[42]WHO-AFRO (2012).

[43]Omamo and Farrington (2004).

structures and import requirements.[44] The enforcement of import controls and inspections is also problematic for most governments, which presents the risk of importing unsafe and substandard food and food products.[45] A report[46] on Zambia, for example, argues that European Union (EU) and United States of America SPS standards are "dynamic" and have resulted in the rejection of Zambian goods at ports of entry. Others[47] state that EU standards for food safety are high to meet the perceived requirements of its affluent consumers. The following will shed further light on the case of Zambia.

42.2 Legislative Overlap in Food Law Governance

42.2.1 The Case of Zambia

First, the key ministry in charge of food safety in Zambia is the Ministry of Health, which is responsible for the review of law and policies, as well as the mobilization of resources to monitor and evaluate the quality of the health care delivery system. Enshrined in law (The Public Health Act Cap 295[48] and The Food and Drugs Act, Cap 303[49] of the Laws of Zambia), the Ministry is charged with protecting the

---

[44]Ndulo (2006).

[45]WHO-AFRO (2012).

[46]Mudenda (2005).

[47]Barling and Lang (2004).

[48]The date of the original text for this law is April 11th, 1930. It was last consolidated in 2006. The Act 'makes provision with respect to matters affecting public health in Zambia including prevention and suppression of infectious diseases including diseases communicable from animal to man, sanitation, protection of food, supply of water, protection from mosquitoes and pollution in general.

The Minister is granted certain regulation-making powers in respect of infectious diseases. Importation of animals may be restricted. The Act also prohibits the sale of unwholesome food and grants in general regulation-making powers to the Minister especially for the control of quality and hygiene of food. Water shall be kept in such a manner so as to avoid stagnant water. Local authorities shall take all possible measure for the prevention of the pollution of water and to purify any polluted water supply. The Minister may make, on the recommendation of the Central Board of Health, certain Orders for the protection of milk.

Descriptors (Livestock): animal health; pests/diseases; data collection/reporting

Descriptors (Food): food quality control/food safety; hygiene/sanitary procedures; milk/dairy products

Descriptors (Water): water supply; freshwater quality/freshwater pollution

Descriptors (Waste & hazardous substances): pollution control; waste disposal' Cap 295 of the Laws of Zambia.

[49]This act originated as S.I. No. 244 of 1972 as at 2006. The Food and Drugs regulations 'prescribe that no manufacturer or distributor of, or dealer in, any article shall sell such article to a vendor unless he gives to the vendor a warranty in a form set out in the Schedule and applicable to such sale. "Article" in the Act means any food, drug, cosmetic or device and any labelling or advertising

public from food hazards, be they chemical, physical or chemical agents. To achieve this, the Ministry monitors food quality and safety along the production chain using set standards, usually developed by the Zambia Bureau of Standards (ZBS). Interestingly, while the health ministry operates under compulsory legal guidelines such as the Public Health Act and the Food and Drugs Act, the standards created by the ZBS are voluntary.[50] Consequently, industry is under no obligation to comply with ZBS standards.

Second, Zambia's Agricultural Ministry's mandate focuses on animal and plant health. Its responsibility is animal disease control and the prevention of novel plant pest incursions into the country. The veterinary department, under the Ministry of Agriculture, is responsible for controlling hazards that may enter the food chain through food of animal origin. The Ministry of Agriculture draws on the Stock Diseases Act Cap 252 of 2010, which repealed the Stock Diseases Act of 1961,[51] and the Control of Goods Act Cap 421. Local government is responsible for meat inspection in abattoirs, the setting up of appropriate structures for animal slaughter and municipal waste management under Cap 281[52] of the Laws of Zambia. These legal mandates all empower these arms of government in food safety governance. It should be easy to see that there is considerable overlap when it comes to inspection of food of animal origin. Furthermore, the Public Health Act in Zambia does not cover many modern public health concerns in food processing and the Food and Drugs Act is superseded by international standards or country of export legal provisos, in cases of multilateral trade. This structure is fairly typical of many African countries.

In countries like Zambia, the informal sector, which plays a key role in food supply to the general population, is mostly considered to be outside the purview of official control mechanisms, except municipal authorities.[53] The sector is thus

---

materials in respect thereof or anything used for the preparation, preservation, packing or storing of any food, drug, cosmetic or device.

Descriptors (Livestock): animal health; drugs

Descriptors (Food): food quality control/food safety' Cap 303 of the Laws of Zambia.' Source: http://faolex.fao.org/.

[50]FAO (2005).

[51]The year this law was repelled is revealing. Zambia only became independent on October 24th, 1964.

[52]In some by-laws of this act, animal health and food safety is addressed. For instance, concern the slaughtering of animals and sale of meat in the area under the jurisdiction of the Katete District Council. They also provide for the control of stray animals.

'Butcheries shall be approved by the District Council. A person shall not expose, offer, deposit, accept or have in his or her possession for resale any meat unless such meat has been examined and passed by the Meat Inspector as fit for human consumption and stamped and marked accordingly.

Descriptors (Livestock): grazing/transhumance; slaughtering

Descriptors (Food): food quality control/food safety; meat; slaughtering; inspection' Source: http://faolex.fao.org/.

[53]WHO-AFRO (2012).

usually beyond the purview of official control and falls prey to substandard practices in the marketing of food products. In some cases, foods are processed and sold in unhygienic environments with little regard for cold chain requirements and pest control, for instance in the case of the street vending of raw and cooked food, a significant risk for food safety. Given the complicated mix of actors affecting the regulatory environment for food safety in Africa, in addition to the continent's rampant and myriad disease problems, inadequate or outdated legislation, solutions to the continent's problems in food safety are difficult to find.

42.2.2 The World Organization for Animal Health and Food Safety

An important international body in food safety regulation with regard to food of animal origin is the World Organization for Animal Health (OIE).[54] It was established on January 25, 1924, and is headquartered in Paris, France. It is not a United Nations body, like the Food and Agricultural Organization (FAO) and WHO, but the World Trade Organization (WTO) recognizes the OIE as a reference organization.[55] Thus the OIE has adopted an active pro-trade stance in their address of issues surrounding trade, health protection and food safety. In its Terrestrial Animal Health Code, the OIE lists international animal health standards that are the basis for facilitating safe trade in animals and animal products, standards that are recognized under the Sanitary and Phytosanitary (SPS) Agreement of the WTO.[56] Countries that are involved in animal and animal product trade are expected to comply with the SPS Agreement in order to reap the full benefits of international trade.[57] The SPS Agreement states that public health measures to ensure food safety and to control plant or animal diseases should be based, as far as appropriate, on international standards.[58] In addition, the SPS Agreement sets forth that measures to protect public health, animal health and plant health should only minimally interfere with trade. On issues of food safety, however, it must be noted that produce from developing countries, particularly those from the African region, cannot easily enter the more lucrative Western markets. Furthermore, both weak national food governance legislation and the facilitative intent of the SPS Agreement biased towards trade, may compound the problem of the importation of substandard food and food products. Therefore, the view that measures to protect health should only minimally interfere with trade remains problematic.

---

[54]This is the French acronym (Office International des Epizooties—OIE).
[55]Thiermann (2005), pp. 101–108.
[56]Bruckner (2009), pp. 141–146; OIE (2010); OIE (2004).
[57]Thiermann (2005).
[58]Zepeda et al. (2005), pp. 125–140.

The OIE's Terrestrial Animal Health Code prescribes the role that national veterinary services of member countries should play in food safety governance. It is the OIE's position that a veterinarian's background and training places him or her in a unique position as far as the assurance of food safety of foods of animal origin is concerned, emphasizing the proper training of veterinarians to meet the challenges in food safety. It also provides guidelines for evaluating national veterinary services. Interestingly, the OIE's veterinary services evaluation process sets independence from political influence as a primary benchmark.[59] The OIE's aim of separating science from politics is impractical. In fact, it has been suggested that the veterinary profession is likely limited by its dependence on scientific or authoritative opinion and its exclusion of political and social phenomena.[60] For food safety in particular, influencing policymaker perceptions on the human risks of diseases from livestock, cannot only be an exercise in science, it must also recognize politics.[61]

42.3 The African Agroecosystem: International Legal Perspectives of Epidemiology and Disease Control

Now we examine the international policy scenario as it relates to the African agroecosystem and disease control. Globalization under the current multilateral trading system has created vast inequalities between the world's richest and poorest nations.[62] Even the OIE's perspective on global issues, particularly its views on trade, has significant impacts on developing countries such as those found in the African region. Before the OIE disease lists were revised, of the 15 "List A Diseases" considered to be transboundary in nature and prioritized as threats to global animal health in the trade of livestock products, 12 were endemic to sub-Saharan Africa.[63] The presence of transboundary animal diseases in the African region and their unlikely eradication in the foreseeable future means that under WTO global trade rules, these countries will continue to be excluded from involvement in international trade.[64] Many countries in the African region have to deal with a range of animal diseases simultaneously, and are likely to continue doing so. This makes technical considerations and regulation of trade extremely difficult, even within a facilitative global trade environment.[65] Some publications,[66] for

---

[59]e.g. Vallat and Pastoret (2009), pp. 503–510.
[60]Hueston (2003), pp. 3–12.
[61]Green (2012), pp. 377–381; Mwacalimba and Green (2014).
[62]Stiglitz (2009), pp. 363–365.
[63]Thomson et al. (2004), pp. 429–433.
[64]*Ibid.*
[65]Upton and Otte (2004).
[66]Rweyemamu and Astudillo (2002), pp. 765–773.

example, suggest that the global distribution of Foot and Mouth Disease (FMD) interestingly mirrors the global economic make-up with industrialized countries generally being free of the disease while developing countries are endemic. Furthermore, reviews of WTO agreements and their effect on livestock production and trade in Africa, highlight the lack of transparency and equality among negotiating countries that has excluded many developing countries.[67] Sub-Saharan Africa has given out more concessions on tariff reduction than what it received from its trading partners.[68] This state pushes trade in a North–south direction and, unfortunately, international standards have been used to reinforced this direction of trade, primarily on health grounds.[69]

As was the case with food safety monitoring mechanisms, developing countries such as those found in the African region have serious problems in their surveillance systems and veterinary infrastructure.[70] With these problems, Africa's trading partners automatically assume that products exported from the continent are risky.[71] Focusing on development, some researchers[72] have argued that WTO SPS measures marginalize the world's poor producers the most and may even contribute to global poverty and disease. A review of the history of free trade[73] argues that current approaches to global trade deny poorer countries the opportunity to implement policies that fostered the development of the world's wealthy economies. These arguments suggest that developing countries are not only purposefully restricted from participating in global trade in reciprocal ways, their development opportunities are also restricted by these multilateral systems.

The dominant international perspective is that the African region and similar developing world contexts, pose the greatest risk as sources of infectious diseases.[74] Some[75] state, for instance, that the FAO's philosophy is to control these diseases at their developing country source. This view that resource-constrained countries are the biggest sources of infectious disease risks for the rest of the globe also suggests that disease control efforts would focus on the "global impacting" disease problems. It further implies a fostering of particular methods of control that may not be appropriate for different settings, which could actually harm local livelihoods or worse, encourage further disease spread.[76] The issues

---

[67]Tambi and Bessin (2006).
[68]see also MacDonald and Horton (2009) pp. 273–274.
[69]Mwacalimba (2013).
[70]Stärk et al. (2006), p. 20; Zepeda et al. (2005).
[71]Tambi and Bessin (2006).
[72]Hall et al. (2004), pp. 425–444.
[73]Chang (2003).
[74]Hampson (1997), pp. S8–S13; Domenech et al. (2006), pp. 90–107; Kruk (2008), pp. 529–534.
[75]Domenech et al. (2006).
[76]see Scoones (2010).

that lie at the confluence of health and trade also impact food safety and food safety governance.

The resulting "controlling-risk-at-source" narrative obscures the cultural, disease management and stigmatization challenges that the African region faces when its member states strive to embrace these global perspectives on trade and infectious disease control. For instance, livestock in many countries in the African region are not just kept as articles of commerce, but have cultural significance as well, what some[77] term multifunctionality; where they serve such functions as assuring domestic food security, provide access to nutrition for the less privileged and play key roles in the maintenance of distinctive rural cultures and ways of life.[78] Furthermore, compared to more industrialized countries where there are mechanisms for farmer compensation following livestock culling, this is not usually possible in resource-limited countries because of a dependency on livestock for rural livelihoods and difficulties in obtaining replacement stock.[79] In addition, although the two systems are sometimes loosely integrated, it must be understood that both traditional and commercial production practices co-exist in many of these contexts. Finally, it is possible that the diseases and risks prioritized in the global West are not necessarily the ones of most significance in these contexts. Therefore, part of the problem with this global aversion to infectious disease risk is that there is little effort made to understand the context in which developing countries attempt to negotiate global imperatives, be they public health, animal health or trade concerns. In the context of food safety in particular, not much has been done to investigate context-relevant ways of addressing these problems.[80]

42.3.1 Zoonotic Tuberculosis and Food Safety in Africa

To explain the links between zoonotic tuberculosis and food safety in Africa, it is important to first define and describe the wildlife-livestock interface. The wildlife-livestock interface can be simply defined as an area in which both wildlife and livestock commonly reside.[81] This definition, however, does not adequately capture the dynamic nature of this interface. Its nature is best defined by the context in which it exists. Kock,[82] provides several interesting contextual descriptions of what constitutes a wildlife-livestock interface, including migratory bird contact with intensive pig operations in North America, China and Europe, and pastoral cattle

---

[77]Smith et al. (2002).

[78]Mwacalimba et al. (2013), pp. 274–279.

[79]Zinsstag et al. (2007), pp. 527–531.

[80]Mwacalimba (2013).

[81]Grootenhuis and Olubayo (1993), pp. 55–59.

[82]Kock (2003), ftp://ftp.fao.org/docrep/nonfao/LEAD/x6198e/x6198e00.pdf.

foraging in African wildlife sanctuaries. The key descriptors of the interface include health, conservation, culture and economics. In the next segment, this chapter will focus on cattle production, the transmission dynamics of bovine tuberculosis (BTB) between cattle and wildlife reservoirs, the pertinent food safety concerns in the agroeconomy of wildlife-livestock interface and their implications for public health. It will also attempt to highlight key cultural and economic details of the interaction of these myriad facets that are important for food safety governance.

42.3.2 Epidemiology of Zoonotic Tuberculosis

Mycobacterium bovis is a member of the mycobacterium tuberculosis complex.[83] The bacterium causes Bovine tuberculosis (BTB), a zoonosis characterized by the development of specific granulomatous lesions in the lung, lymph nodes and other tissues.[84] Comparative genomics suggest that *M. bovis* evolved from *Mycobacterium tuberculosis*, the primary cause of human TB, but has since developed a capacity for infecting a large host range.[85] BTB's susceptible species range spans domestic and wild animal species, and man.[86] Although the specifics of the evolutionary biology of the two infectious agents remain controversial,[87] one school of thought is that the deletions in the *M. bovis* genome that occurred from its evolution from *M. tuberculosis* increased its host range.[88] *Mycobacterium bovis* is hardy, and can survive outside the host depending on environmental conditions. It can survive for up to 2 years in the environment, 1 year within dung pats, and between 5 and 7 months in manure, slurry or water.[89]

In Africa, approximately 85 % of cattle and 82 % of humans live in areas where BTB is either partly controlled or not controlled at all.[90] Like many of the African region's disease problems, a clear picture of the extent of the problem is yet to be developed. There is a paucity of data on the prevalence of BTB in cattle, a lack of species differentiation of human TB isolates and the presence of significant wildlife reservoirs.[91] The epidemiology of the condition in cattle is as follows: A vast majority of cattle excrete *M. bovis* almost from the inception of a lesion.[92] However, it is still not fully understood to what extent tuberculosis latency exists in

---

[83] O'Reilly and Daborn (1995), pp. 1–46.
[84] Ayele et al. (2004), pp. 924–937.
[85] Colston (2001); Gibson et al. (2004), pp. 431–434.
[86] Niemann et al. (2000), pp. 152–157.
[87] Brosch et al. (2002), pp. 3684–3689.
[88] Colston (2001).
[89] Hancox (2000), pp. 87–93.
[90] Cosivi et al. (1998).
[91] Ayele et al. (2004), pp. 924–937.
[92] Menzies and Neill (2000), pp. 92–106.

cattle and the associated problems of this potential source of infection. BTB will persist in cattle as long as the bovine host lives.[93] A single bacillus that becomes aerosolized is sufficient to establish infection, droplet size being more important than number of bacilli.[94] With droplet nuclei as a vector, animals do not need to be in close contact with a tuberculosis disseminator to become infected.[95] We must cross reference this scientific fact with the social role that cattle play in African rural culture and livelihood i.e. not as a source of food, but a status symbol. This is important for understanding disease transmission in the wildlife-livestock interface.

In humans, tuberculosis caused by *M. bovis* is similar to *M. tuberculosis*.[96] Infection occurs via aerosol inhalation due to close contact with infected cattle,[97] oral consumption of infected food, and through skin wounds.[98] In a broad sense, these routes of infection are all important, but it must be understood that BTB is predominantly a milk-borne zoonosis.[99] During the pre-eradication period in western countries, milk was the main source of infection for human beings, especially children.[100] Studies have been done where *M. bovis* has been isolated from milk in Africa.[101] In cases were the bacillus enters a human being orally, i.e. through the ingestion of unpasteurized and BTB contaminated milk or milk products, disease manifestation is mainly extra pulmonary with abdominal, bone and joint forms, as well as infection of cervical and mesenteric lymph nodes.[102]

Humans in close contact with infected cattle may acquire *M. bovis* via the respiratory route.[103] Gibson et al.[104] discuss a case from Gloucester in which a 20 year-old male became infected with *M. bovis* following inhalation of infectious aerosols. He was frequently sprayed with nasal mucus from cattle. He is thought to have later infected his sister who was diabetic and pregnant, thus immunocompromised. Similar transmission may occur in pastoral communities in individuals who frequently handle their livestock or infected wildlife meat and secretions.

Meat from infected animals may also contain viable *M. bovis*[105] that could pose a risk of infection for humans. Specific tissues such as liver, spleen, kidney, mammary glands and lymph nodes may contain sufficient organisms detectable by culture or guinea pig inoculation, even though not showing evidence of infection

---

[93]Hancox (2000).
[94]Menzies and Neill (2000).
[95]*Ibid.*
[96]O'Reilly and Daborn (1995); Cosivi et al. (1998).
[97]Moda et al. (1996), pp. 103–108; Grange (2001); Mfinanga et al. (2003a, b), pp. 933–941.
[98]Wilkins (2000).
[99]Unger et al. (2003).
[100]O'Reilly and Daborn (1995); Grange (2001).
[101]Ameni et al. (2003).
[102]Unger et al. (2003).
[103]Moda et al. (1996); Grange (2001); Mfinanga et al. (2003a, b), pp. 695–704.
[104]Gibson et al. (2004).
[105]Aranaz et al. (2004), pp. 2602–2608.

at necropsy.[106] Haematogenous spread is believed to be responsible for secondary lesions.[107] Thus meat and blood from infected animals pose potential risks of zoonotic infection if not properly prepared or if consumed raw.[108]

In Africa, the relative proportion of human tuberculosis cases caused by *M. bovis* is unknown because differentiation from infections caused by *M. tuberculosis* is not commonly performed.[109] However, significant risk factors for the transmission of *M. bovis* occur in communities where there is close human-to-livestock contact.[110] Some[111] estimate that in countries where pasteurization of milk is rare and bovine tuberculosis is common,[112] 10–15 % of human cases are caused by *M. bovis*. In many rural parts of the African region, BTB control is nonexistent and poor food hygiene practices, husbandry methods, and consumption of raw milk still present risks of human infection.[113] Infection in these areas is by ingestion of unpasteurized milk, poorly cooked meat and close contact with infected animals, tissues or secretions.[114] In pastoralist communities of Northern and Southern zones of Tanzania, for instance, 16 % of culture-positive mycobacteria isolates from human cases were *M. bovis*.[115] Kazwala[116] states that there was a definite link between the number of cattle owned and the incidence of non-pulmonary tuberculosis in the human population. Similarly, the ownership of reactor cattle herds in Zambia was shown to be statistically associated with human tuberculosis cases,[117] although in other countries no statistically significant association between reactor cattle and associated households was found, even though there were human tuberculosis cases in households that owned reactor cattle.[118]

Grange[119] cites other studies that exemplify human cases of *M. bovis* in Africa. This includes surveys where 0.4, 5.4 and 6.4 % of pulmonary tuberculosis cases respectively were due to *M. bovis* in Egypt, while 3.9 % of 102 isolates of pulmonary tuberculosis was due *M. bovis* in Nigeria. Prior to effective tuberculosis control in cattle in developing countries, the positive correlation between human prevalence and infection was well recognized. Up to 6 % of human pulmonary

---

[106]FSAI Scientific Committee (2003), http://www.fsai.ie/publications/other/zoonotic_tuberculosis.pdf.
[107]Neill et al. (2001).
[108]Moda et al. (1996).
[109]Moda et al. (1996); Kazwala et al. (2001), pp. 87–91.
[110]Mfinanga et al. (2003a, b).
[111]cited by Unger et al. (2003).
[112]Tamiru et al. (2013), pp. 288–295.
[113]Ameni et al. (2003).
[114]Ayele et al. (2004).
[115]Kazwala et al. (2001).
[116]*Ibid.*
[117]Cook et al. (1996).
[118]Ameni et al. (2003).
[119]Grange (2001).

tuberculosis cases were attributed to infection of bovine origin in the south of England between 1931 and 1937.[120] In one review,[121] pulmonary disease was found to be more common in rural areas of England and Wales in the early part of the twentieth century, possibly due to aerogenous infection from cattle. The current situation in Africa is thought to be similar to the pre-eradication era in Europe.[122]

Finally, although human-to-human transmission is considered rare,[123] other humans with *M. bovis* infection may be potential sources of disease. The highest risk groups for acquiring *M. bovis* are individuals with concomitant HIV/AIDS infection.[124] Thus the high prevalence of HIV/AIDS in sub-Saharan Africa is an added risk. Transmission among HIV infected individuals may be particularly high because immunosuppression increases susceptibility to infection.[125] Some researchers[126] suggest that most TB cases in African HIV/AIDS patients are due to exogenous re-infection rather than reactivation of endogenous *M. tuberculosis* and could have a similar risk of exogenous disease upon exposure to *M. bovis*. In such individuals, it may be difficult to determine if the zoonosis is a reactivation or a new infection.[127]

42.4 Malevolence or Benevolence? Issues at the Nexus of Food Safety and Food Security in the Wildlife-Livestock Interface of Southern Zambia

In Zambia the incidence of HIV has been steadily declining since the 1990s. However, according to country statistics, the HIV prevalence still stands at 12.7 % while human TB is at 433 per 100,000 of the population.[128] The reduction in HIV incidence is a result of control strategies instituted by the government, with the support of donor agencies such as the Global Fund to Fight AIDS, Tuberculosis and Malaria. In this country, research has demonstrated that BTB may be an important neglected disease whose impact on public health is underestimated because of the high prevalence of human TB coupled with the ongoing HIV/AIDS pandemic.[129]

One area where BTB may be significant is in a wildlife-livestock interface area in the south of Zambia, in the flood plains of the 6500 km^2 Kafue Flats. Lochinvar and Blue Lagoon national parks are two contiguous Game Management Areas in the catchment of the Kafue Flats in which human settlement, small-scale

---

[120]O'Reilly and Daborn (1995).
[121]Grange (2001).
[122]Ayele et al. (2004).
[123]O'Reilly and Daborn (1995); Grange (2001).
[124]Ayele et al. (2004).
[125]*Ibid.*
[126]cited by Ayele et al. (2004).
[127]Zumla et al. (2000), pp. 259–268.
[128]UNGASS (2012).
[129]Mwacalimba et al. (2013).

agriculture, livestock production and fishing are permitted. This area brings the nexus of food safety, disease control and risk into sharp focus. The risk factors for the zoonotic transmission of the disease to humans in this area include livestock management methods, food preparation and hygiene methods, and socio-economic and health status.[130]

The Tonga and Ila tribes constitute the largest ethnic groups living in this area. Their main economy is livestock production.[131] Their cattle are kept for prestige, milk, draft power, dowry, savings and to offset crop failure[132] and are rarely slaughtered except during ceremonies.[133] Cattle rearing in this area is predominantly pastoral with grazing based on the cycle of flooding in the floodplains, which provides year-round pasture. Nearly three quarters of the area's cattle graze in the floodplains for 6 months out of every year.[134] There are three contiguous herding systems practiced in this area. In village resident herding, cattle are reared in and around villages. Transhumant grazing involves the trekking of cattle into the floodplains during the dry season. These herds return to the villages when the rains start and pasture becomes abundant closer to the villages. The last herding system is interface herding. These are large herds of cattle whose numbers cannot be supported by pasture around the villages.[135] Transhumant and interface cattle interact freely with wildlife such as the Kafue lechwe (*Kobus leche kafuensis*).[136] This is a highly sociable semi-aquatic marsh antelope that can only be found in the floodplains in and around Lochinvar and Blue Lagoon National Parks.[137] It is[138] estimated that the herd level prevalence of BTB in cattle from this area at 49.8 %, while the individual cattle prevalence was estimated around 6.8 %.

The first zoonotic TB food-borne risk comes from cattle as a milk source for local communities and human contact with infected herds. As stated earlier, BTB is primarily a milk-borne zoonosis. Some[139] estimate that 50 % of milk derived from these cattle is consumed locally. Pasteurization is able to kill *M. bovis* in milk[140] but where this is not done, human infection is likely to occur. The milk consumed in the wildlife-livestock interface is rarely pasteurized and is sometimes consumed as curdled or soured milk, which is considered a delicacy. The local breeds of cattle are not bred for milk production, and, hence, have low outputs. Thus, milk pooling is a common practice, which increases the risk of humans acquiring BTB from

---

[130]Munyeme et al. (2010a, b).
[131]Mumba (2004).
[132]Mwacalimba et al. (2013).
[133]Cook et al. (1996).
[134]Chabwela and Mumba (1998).
[135]Munyeme et al. (2008).
[136]*Ibid*.
[137]Mumba (2004); see also Phiri et al. (2011), pp. 20–27.
[138]Munyeme et al. (2008).
[139]Mumba et al. (2011), p. 137.
[140]Kells and Lear (1959).

milk. Although only 1 % of cattle with BTB excrete *M. bovis* in their udders, infected milk from a single cow can contain enough viable bacilli to contaminate milk from up to 100 cows if pooled.[141] Furthermore, *M. bovis* has been shown to survive in soured milk for up to 14 days.[142]

The second zoonotic food-borne risk comes from wildlife. The predominant species and primary wildlife maintenance host for BTB is the Kafue lechwe.[143] There is a single population of lechwe in the floodplain coexisting with humans and livestock herds.[144] Dated reports on BTB burden in lechwe are varied, estimating the prevalence around 14 and 30 %.[145] In a 2010 study,[146] Munyeme and his colleagues estimated the prevalence of BTB in hunter-harvested lechwe to be around 24.3 %. These findings represent the risk to public health for the communities living in and around the wildlife-livestock interface.

The food safety significance of lechwe lies in the fact that, of all the wild animal species in Zambia, this small antelope is the most sought after species for game meat.[147] The legal off-take of lechwe amounts to around 800 lechwe a year.[148] An estimated 47.7 tonnes of lechwe meat is produced annually and consumed by about 39,780 people.[149] Poaching, of course, remains a significant problem for this and other species of wildlife. However, the concern here is that even meat obtained legally (through the official quota utilization system) by members of the community is not subject to food safety enforcement mechanisms such as meat inspection. In Africa, abattoir inspected meat is usually consumed in urban areas[150] while rural communities do not routinely submit their animals for meat inspection. For Zambia, wildlife conservation usually occurs in areas remote to veterinary services.[151] The result is that game continues to be consumed, and wildlife trophies handled, without veterinary clearance. The implication of this is that, while providing local communities access to meat from the Kafue lechwe offers an important source of supplementary protein, the food derived from this source also carries the risk of transmission of zoonotic tuberculosis.

Of course, there are other routes available for the zoonotic transmission of BTB, as the aforementioned epidemiological review of *M. bovis* has tried to demonstrate. Therefore the food safety and infectious disease dangers to the communities living in and around the wildlife-livestock interface are very real. Clearly, BTB may play

---

[141] Ameni et al. (2003).
[142] Kazwala et al. (2001); Ayele et al. (2004).
[143] Munyeme and Munang'andu (2011).
[144] Jeffery et al. (1991), cited by Kock et al. (2002), pp. 482–484.
[145] Cook et al. (1996); Cosivi et al. (1998); Pandey (2004), pp. 17–20.
[146] Munyeme et al. (2010a, b), pp. 305–308.
[147] Siamudaala et al. (2003).
[148] Simasiku et al. (2008).
[149] Siamudaala et al. (2003).
[150] Ayele et al. (2004).
[151] Siamudaala (2004), pp. 48–52.

a role in the epidemiology of human TB, particularly in the context of HIV and AIDS. However it is rarely, if ever, linked to the policy narratives that focus on the big three; tuberculosis, HIV and malaria. Furthermore, while the scientific literature has linked wildlife management, community nutrition and livestock husbandry to BTB epidemiology and risk, the relevant policy actors in this triad have not been appropriately engaged and thus BTB remains a neglected disease with real consequences. Based on data extrapolated from other countries and the prevalence of human tuberculosis in the region, it was estimated that over a 10 year period, around $1.5 million (US) are costs attributable to the treatment of zoonotic tuberculosis in this area.[152]

42.4.1 Avian Influenza: The Zambian Experience of Pandemics and Food

42.4.1.1 The Epidemiology of Highly Pathogenic Avian Influenza H5N1 Spread and the International Response

The infamous H5N1 highly pathogenic avian influenza (HPAI)[153] was first identified at a goose farm in Guangdong Province, southern China in 1996.[154] Subsequently, high H5N1-related mortalities were reported on three chicken farms in Hong Kong, just adjacent to Guangdong Province, between March and early May, 1997.[155] In May of the same year, a child died of viral pneumonia, the first reported case of zoonotic H5N1 influenza.[156] Following the identification of 17 more human infections that resulted in five deaths between November and December of 1997,[157] H5N1 became recognized as a zoonosis of possible public health concern. As a result, in December 1997, total and rapid depopulation of all poultry in markets and chicken farms in Hong Kong was carried out to control the outbreak, a move that both policy and virology experts believed had averted a potential human pandemic.[158] Arguably, live poultry markets were important in the transmission of the H5N1 virus to other avian species and humans during these outbreaks.[159] The control measures instituted, i.e. the total culling of all farmed chickens and all poultry in markets in Hong Kong, appeared effective, as the responsible genotype

---

[152]Mwacalimba et al. (2013).

[153]Avian influenza exists in two forms, highly pathogenic avian influenza (HPAI) and low pathogenic avian influenza (LPAI). Continuous existence of LPAI virus in avian populations may provide chances for the virus to undergo mutation and convert to a highly pathogenic form. Highly pathogenic avian influenza, especially of the H5 and H7 subtypes, has the potential to infect human beings.

[154]Xu et al. (1999), pp. 15–19; Webster et al. (2002), pp. 118–126.

[155]Shortridge et al. (1998), pp. 331–342.

[156]Ibid.

[157]Ibid.

[158]Fidler (2004b), pp. 799–804; WHO (2005b); Webster and Hulse (2005), pp. 415–416.

[159]Shortridge et al. (1998).

of H5N1 (A/goose/Guangdong/1/96) has not been reported since the execution of these controls.[160] However in February 2003, during the SARS epidemic, and after a 6-year hiatus, three more human H5N1 infections with two fatalities were identified in China. This suggested viral persistence, despite the control measures that had been instituted in 1997.[161] An epidemiological review[162] states that outbreaks had continued to occur in poultry in Hong Kong from 2001 to early 2002, caused by a different H5N1 lineage. While there is some suggestion that the H5N1 problem had been subdued in 1997,[163] it was, in fact, entrenching itself in the poultry systems of Hong Kong, and likely elsewhere in Southeast Asia, between 1997 and 2003.

Between December 2003 and February 2004, the first wave of an H5N1 panzootic in poultry was reported nearly simultaneously in eight countries in South and Southeast Asia, most of which occurred in commercial poultry establishments. This was followed by a second wave of spread from July 2004.[164] The WHO states that the second wave was associated with more rural settings.[165] The countries initially affected were China, Indonesia, Cambodia, Japan, Laos, Korea, Thailand and Vietnam, with a ninth country, Malaysia, joining the list in August 2004.[166] The pro-poor advocacy group, GRAIN, states that the initial outbreaks in Vietnam, Thailand, Cambodia, Laos and Indonesia all occurred in closed, intensive factory farms.[167] During the first wave, millions of poultry either died or were culled in an effort to control the disease.[168] Human infections were then reported in Hanoi, Vietnam, in January, 2004, a few days prior to a report of massive H5N1-related poultry mortalities in two poultry farms in the south of the country.[169] Vietnam had initially experienced an H5N1 outbreak in 2001.[170] In early 2004, during the first wave of the panzootic, the WHO declared the outbreak an unprecedented catastrophe for agriculture in Asia and a "global threat to human health."[171]

Coinciding with the second wave of the panzootic, the period between August and October 2004 saw eight more human deaths in Thailand and Vietnam.[172] The third wave began in December 2004, involving new poultry outbreaks in Indonesia, Thailand, Vietnam, Cambodia, Malaysia and Laos.[173] Fresh human cases were

---

[160] Sims et al. (2005), pp. 159–164.
[161] WHO (2005b).
[162] Sims et al. (2005).
[163] WHO (2005b).
[164] Alexander (2007); Paul et al. (2010).
[165] WHO (2005b).
[166] Sims et al. (2005).
[167] GRAIN (2007).
[168] WHO (2004), at http://www.who.int/mediacentre/releases/2004/pr7/en/.
[169] WHO (2005b).
[170] Sims et al. (2005); Sims and Narrod (2008).
[171] WHO (2004).
[172] WHO (2005b).
[173] WHO-AFRO (2005).

reported in Vietnam, Thailand and Cambodia.[174] At this point, after reviewing the unfolding situation, a writing committee of the WHO consultation on human influenza established that Vietnam led the human death toll.[175] According to a WHO pandemic threat report,[176] by 2005, H5N1 had crossed the species barrier three times, namely in 1997, 2003, and the period between 2004 and early 2005, which recorded the largest occurrence of human H5N1 cases of the period in question. With the report of migratory birds being affected with H5N1 in Mongolia and China, particularly at Lake Qinghai in China in April 2005, concern grew that this posed a potential risk of southward and westward spread of the virus in poultry.[177] Around 6345 birds of different species died in the weeks following the Qinghai outbreak.[178] This is probably the single most important event linking H5N1 to migratory bird spread. This outbreak singularly raised the profile of the role of migratory birds in the global spread of H5N1.

H5N1 had spread through the diverse market and poultry production systems of Southeast Asia. There is much debate around the primary causes and drivers of the H5N1 problem, revolving around poultry production and marketing practices. An important factor in the Asian panzootic is that ducks appeared to have played a key role in the maintenance of the virus, primarily as silent carriers of H5N1. While outbreaks in poultry were still possible, this suggests that in areas where duck production was less significant, the chances of endemicity could be lower. By 2005, H5N1 had become endemic in the duck population of poultry, providing a reservoir of the virus for other poultry species as asymptomatic shedders of H5N1 influenza.[179]

As of November 11, 2010, H5N1 HPAI had claimed a cumulative 508 confirmed cases and 304 human deaths.[180] The panzootic cost the global poultry industry well over $10 billion (US) in losses and continued to persist in poultry populations of parts of Europe, Southeast Asia, Egypt and Nigeria.[181] The primary public health concern had been H5N1's likely candidacy for the next human influenza pandemic, which many experts believed was overdue.[182] Interestingly, it was the rapid spread, public health and economic ramifications of the SARS outbreak in 2003 that appear to have alerted the global health community to the conceivable need for pandemic

---

[174]*Ibid.*

[175]Beigel et al. (2005), pp. 1374–1385.

[176]WHO (2005a), http://www.who.int/csr/disease/influenza/H5N1-9reduit.pdf.

[177]Chen et al. (2005), pp. 191–192; Webster and Govorkova (2006), pp. 2174–2177; Alexander (2007), pp. 5637–5644; Cattoli et al. (2009), p. e4842.

[178]WHO (2005c).

[179]Webster and Hulse (2005); Sims et al. (2005); Sims and Narrod (2008), www.fao.org/avianflu.

[180]WHO (2010), http://www.who.int/csr/disease/avian_influenza/country/cases_table_2010_11_19/en/index.html.

[181]WHO (2005a, b); GRAIN (2006); Eurosurveillance (2006) E061221.1.; Kilpatrick at al. (2006), pp. 19368–19373; FAO (2010), www.fao.org/avianflu/en/maps.html.

[182]Conly and Johnston (2004), pp. 252–254; Kilpatrick et al. (2006); Bartlett (2006), pp. 141–144.

preparedness.[183] In the wake of SARS, H5N1 presented an unprecedented challenge to the animal health, public health and trade policy communities, identified as a threat to the poultry industry, a pharmaceutical interest, a trade-related epidemic, public health threat, and a human pandemic concern.[184]

For avian influenza, one[185] explanation is that public health experts and epidemiologists did not know whether an H5N1 human pandemic was actually imminent, only that it was plausible. Adding to the complexity was the need to determine how exactly to respond to a potential H5N1 pandemic. Reviewing available surveillance data on past human influenza pandemics, there was no pattern to the epidemiology of occurrence,[186] or standard manifestation of pandemics, including which segments of the population would be affected the most.[187] This suggests that there is no clear precedent for definitively predicting how the next pandemic will behave.

Despite these uncertainties, a multi-sectoral approach to H5N1 management and pandemic preparedness across the policy sectors affected was advocated at national, regional and international levels. The purpose of such an approach was to foster a coherent response to H5N1.[188] Correspondingly, the main thrust of the H5N1 avian and pandemic influenza response was the coordination of public health and animal health agencies at national and international levels with the goal of developing preparedness interventions for those areas that had not yet been affected. An alternative goal was to reinforce control measures in locations where the disease had become endemic, where, based on WHO pandemic preparedness guidelines, OIE recommended control measures and FAO devised surveillance strategies.[189] However, H5N1 presented unique challenges for pandemic planning, involving the weighing of sector-specific risk against wider ecological and socio-economic interests that challenged the traditional public health and animal health based interventions of the pre-SARS era. Global infectious disease governance post-SARS dictated, at least in the context of avian and pandemic influenza preparedness, the attempt to balance the interests of pharmaceutical, conservationist, transnational business and commercial poultry as well as bridge the previously growing divide between public and animal health.[190]

In the policy domain, many of the concerns over a pandemic began to sound apocalyptic. The often-cited comparator was the 1918 Spanish flu pandemic which

---

[183]Scoones and Forster (2008a).

[184]WHO (2006), www.who.int/csr/disease/avian_influenza/avian_faqs/en/index.html; ALive (2006); Karesh et al. (2005), pp. 1000–1002; Ong et al. (2008); Scoones and Forster (2008a).

[185]Osterholm (2005), pp. 1839–1842.

[186]Monto et al. (2006), pp. S92–S97.

[187]Nicoll (2005), pp. 210–211.

[188]Ong et al. (2008); UN (2010), http://www.un-influenza.org/node/4040.

[189]WHO (2005a); Webster and Hulse (2005).

[190]Fidler (2004a); WHO/DFID-AHP (2005); Fidler (2008), pp. 88–94; Scoones and Forster (2008b), http://www.steps-centre.org/PDFs/Avian%20flu%20final%20w%20cover.pdf; Rabinowitz et al. (2008), pp. 224–229.

one health policy scholar suggests killed over 50 million people.[191] This particular pandemic was said to have its origins in Kansas military camps and was spread to Europe by US troops during the war in 1918.[192] With contemporary concerns focusing on human population growth, increased intensity of production systems and the unprecedented nature of globalization, which many have argued allow for the faster and further transmission of infectious disease threats,[193] there was a fear that a pandemic in modern times could kill millions. Some researchers,[194] on the other hand, held the view that a pandemic now is likely to result in considerably lower deaths than the one that occurred in 1918. The WHO[195] estimated that a pandemic arising from H5N1 could result in two-seven million deaths at the minimum. Other commentators were skeptical[196] and dismissed the avian influenza issue, and SARS before it, as elaborate political conspiracies of corporate and pharmaceutical interests disguised as national security threats and pandemic concerns. There was, therefore, a lot of politics surrounding the issue of avian and pandemic influenza.

In the debates on global responses to avian influenza, some researchers[197] mapped recurring themes and summarize the four themes characterizing the core political issues: (1) risk and uncertainty, (2) economy and livelihood impacts, (3) effects on health and extent of disease, and (4) effects on food and farming. Drawing from these four issues, there are six linked debates identifiable in the international policy discourse concerning avian and pandemic influenza. The first debate involved the scientific uncertainty of the likelihood of the occurrence of a pandemic caused by H5N1. As mentioned above, some authors[198] state that public health experts were uncertain of its likelihood. The concern over a possible pandemic resulted in called to focus control on the likely source of this risk, such as Southeast Asia, where most of the impact of H5N1 had been felt. In fact, some authors have referred to Southeast Asia as an "influenza epicenter."[199] Of course, a complex interplay of cultural factors and production practices led to the exposure and eventual succumbing of humans in this region to H5N1.[200] These factors have been identified and reviewed in various context-specific network analyses,[201] HPAI risk

---

[191] Osterholm (2005).
[192] Webster (1997), pp. S14–S19; Hollenbeck (2005), pp. 87–90.
[193] Käferstein et al. (1997), pp. 503–510; Hampson (1997), pp. S8–S13; Kimball et al. (2005), p. 3; Kimball (2006).
[194] Morens and Fauci (2007), pp. 1018–1028.
[195] WHO (2006).
[196] Horowitz (2005).
[197] Scoones and Forster (2008a).
[198] Osterholm (2005).
[199] see Hampson (1997).
[200] see Webster (1997); Osterholm (2005).
[201] Van Kerkhove et al. (2009), pp. 6345–6352; Soares Magalhaes et al. (2010), p. 10.

mapping studies,[202] risk factor studies[203] and risk analysis.[204] However, in some reviews of infectious diseases risk management and governance of global risks,[205] it was argued that although surveillance had focused on H5N1, there was still a lot of uncertainty about both its evolution as a zoonosis and its effects on public health.

The second debate involved linking poultry production practices, HPAI epidemiology and disease spread through trade, poultry and poultry product and migratory bird movement. According to the epidemiological reviews by Capua and Alexander[206] and Alexander,[207] recent increases in intensive poultry production practices were responsible for the increasing incidence of highly pathogenic influenza in the world. It was stated by van den Berg[208] that all parts of the world were at risk of H5N1 incursions as a result of the globalization of trade. Some authors took the view that it was migratory birds that would spread H5N1 across the globe,[209] yet others claimed that wild birds were only capable of short range spread.[210]

The third debate involved the 'One Health' approach response to mitigate the pandemic threat. This was characterized by calls to strengthen veterinary control systems in addition to human pandemic preparedness, addressing the pandemic risk at-source but involving other sectors to mitigate the risk.[211] A key question here was how do countries incorporate other policy sectors in risk mitigation? As one study demonstrated, each sector, and indeed each country, would view the HPAI problem differently.[212] In addition, while the international community recommended 'at-source' controls, the 'standardized' approaches adopted worked in some areas and failed in others.[213] In their examination of the epidemiology of H5N1, Yee, Carpenter and Cardona,[214] state that control measures such as culling, disinfection and stamping out had been successful in controlling H5N1 outbreaks in Europe, but were not as effective in Southeast Asia.

The fourth debate involved the potential effects of a human pandemic on the global economy. This resulted in HPAI risk mitigation responses perceived to largely affect only the livelihoods of those in outbreak areas.[215] The brunt of these control efforts was largely felt by poor farmers, impacting on food security,

---

[202] Gilbert et al. (2008), pp. 4769–4774.

[203] Yupiana et al. (2010), pp. e800–e805.

[204] e.g. Kasemsuwan et al. (2009).

[205] Pitrelli and Sturloni (2007), pp. 336–343.

[206] Capua and Alexander (2004), pp. 393–404.

[207] Alexander (2007).

[208] van den Berg (2009), pp. 93–111.

[209] Normile (2006), p. 1225; Chen et al. (2005).

[210] See e.g. Webe and Stirlianakis (2007), pp. 1139–1143.

[211] FAO (2004); WHO (2005a).

[212] Mwacalimba and Green (2014).

[213] Scoones and Forster (2008b).

[214] Yee et al. (2009), pp. 325–340.

[215] Scoones and Forster (2008a).

livelihoods and farming. Stirling and Scoones,[216] for example, estimated that over 2 billion birds were slaughtered with the greatest losses suffered by the poor. Nicoll[217] states that the effect of H5N1 was mostly felt in the social sphere, particularly in Southeast Asia, where several countries (e.g. Thailand) had their poultry exports prejudiced and rural livelihoods affected by control interventions. This has links to contentions between business and livelihood interests and controversies over the role of intensive vs. backyard farming in disease spread.[218]

The fifth debate involved pharmaceutical interests, covering influenza virus sharing and concerns that genetic sequence information collected from outbreak areas would be used to create vaccines for market that would not be distributed equitably in case of a pandemic.[219] The policy response involved Western countries scrambling to stockpile antiviral drugs and vaccines for 'high level pandemic preparedness efforts', the vaccines of whose production depended on H5N1 virus strains recovered from outbreak centers in developing countries.[220] In an effort to globalize this policy response, there were calls for affected countries to either develop pharmaceutical capacity or consider non-pharmaceutical interventions.

Linked to this was the sixth debate, involving the 'securitization' framing of the avian and pandemic influenza issue, which, Elbe[221] argued, contributed to, and caused difficulty in resolving, the controversy over influenza virus sharing. In implementing this 'securitization' approach, Western countries spent massively on pandemic preparedness. Burgos and Otte's (2008) study[222] citing Jonas's study[223] state that the US and European countries had spent approximately US$2.8 billion 'at home' versus US$950 million 'abroad' for disease control 'at-source' by the end of 2008. This forms the background against which developing countries generated their avian and pandemic influenza intervention policies guided by the WHO global pandemic preparedness plan.[224]

The African response was coordinated by the WHO African Regional Office (WHO-AFRO), the African Union Inter-African Bureau for Animal Resources (AU-IBAR) and some regional trading blocs such as the Southern African Development Community (SADC), with funding from the African Union and the World Bank.[225] This was under the global coordination of United Nations System Influenza Coordinator (UNSIC), with the main participants being WHO, OIE and FAO.[226] These global and regional actors set out a framework to guide the

---

[216]Stirling and Scoones (2009); Scoones and Forster (2008b).
[217]Nicoll (2005).
[218]GRAIN (2006); GRAIN (2007).
[219]Garrett and Fidler (2007), p. e330; Fidler (2008).
[220]Elbe (2010), pp. 476–485.
[221]Id.
[222]Burgos and Otte (2008).
[223]Jonas (2008) cited by Burgos and Otte (2008).
[224]WHO (2005a); ALive (2006).
[225]WHO-AFRO (2005); ALive (2006); UNSIC and World Bank (2008).
[226]UNSIC (2006a, b); Scoones and Foster (2008a).

development of national avian and human influenza prevention and control responses. Among these guidelines was a recommendation for multi-sectoral integration.[227] By 2007, response plans on the African continent were at different stages of development with most aimed at containment of avian influenza in poultry to the neglect of pandemic preparedness.[228]

42.4.2 Chicken: A Cheap Source of Protein or a Pandemic Threat?

At the height of the global avian influenza crises, a WHO AFRO risk assessment made comparisons between Asian and African poultry production systems to justify the continent's risk of an incursion as well as recommend similar control measures to those used in Southeast Asia.[229] The statement read in part, 'Though the densities of human and poultry populations are generally lower in Africa than in south-east Asia, the poultry production systems have many similarities which could create multiple opportunities for human exposure, if outbreaks occur in African poultry'.[230] Despite the different contextual realities, such as the role that ducks, mixed farming, and wet markets played in the evolutionary epidemiology of avian influenza,[231] or the fact that Southeast Asia is considered to be a viral mixing pot most likely to be the epicenter for the emergence of novel viruses such as H5N1,[232] the unexamined underlying assumptions of this statement are what formed the mould for preparedness efforts in Africa.

In Zambia, the international call for pandemic preparedness was met first by local media reports of possible avian influenza outbreaks in Zambian poultry. The result was a nearly $7 million (US) loss to the poultry industry over a 3 month period as production scaled down.[233] Producers reduced their production, consumers feared the consequences of eating infected chicken and this had knock-on effects on the feed industry, the veterinary pharmaceutical industry and poultry breeders.[234] The reports of outbreaks in poultry turned out to be false. Under WHO and FAO oversight, the Zambian government commissioned a 20-person multi-sectoral task force on avian influenza at the end of October 2005, to be the nation's eyes and ears concerning avian and human influenza and prepare for what they perceived to be an inevitable incursion of H5N1 in the country.[235]

---

[227]WHO-AFRO (2005); UNSIC (2006a, b).
[228]Ortu et al. (2008), pp. 161–169; Ortu et al. (2007).
[229]WHO-AFRO (2005).
[230]*Id.*, p. 7.
[231]Mwacalimba (2012), pp. 391–405.
[232]Hampson (1997).
[233]Mwacalimba (2012).
[234]*Ibid.*
[235]GRZ (2006).

Early in the evolution of Zambia's response to avian and pandemic influenza, the threat of an incursion of H5N1 was successfully presented as an imminent threat. Between 2005 and 2009, this framing of the H5N1 threat as an emergency led to both government and academic scientists in the country searching for the elusive virus in both traditional poultry and wildlife.[236] The public response to media reports suggests a perception of a food safety concern. The perception of policy makers, however, reflected their institutional standpoints. For example, stakeholders from the Ministry of Health viewed it as a pandemic concern and a potential triage strain for the health system should human infections occur. The Agricultural Ministry, represented by the veterinary department, viewed it as an exotic poultry problem with the potential for zoonotic transmission. It was also viewed as likely to emerge from poor poultry producers with no knowledge of biosecurity. The Trade Ministry and poultry industry viewed it as a threat to both Zambia's poultry sector as a whole and Zambia's international trade opportunities.[237]

Around 64 % of Zambia's households keep chickens.[238] It provides an affordable source of protein for many of the country's citizens. Production is primarily traditional, based on the rearing of indigenous breeds with around 10–15 of chickens per household. The commercial sector in the country is a mix of backyard producers, emergent broiler and layer farmers supplied by locally produced feed and imported breeding stock.[239] The production capacity of the Zambian commercial sector in 2010 was estimated at 30 million broiler birds annually and 6 million eggs monthly.[240] Production is nowhere near the levels found in Southeast Asia. Furthermore, the agricultural ministry was oriented to focus on cattle and cattle diseases and not poultry and poultry problems.

External partners had a disproportionate role in influencing the pandemic preparedness policy process in Zambia. International finance, evidence and prescriptions provided a preconceived view of how pandemic preparedness should be pursued over what should have been a response based on a contextualized understanding of Zambia's policy structure, priorities, and material needs. The result was a one-size-fits-all policy that reflected global narratives that were at odds with Zambia's needs. In retrospect, the confluence of interests surrounding pandemic preparedness and economic development in Zambia presented unique challenges which required careful weighing in the financing and development of the country's avian and pandemic influenza prevention and control policy. Even at the stage where it became evident that H5N1 was unlikely to affect Zambia, international influences continued to emphasize avian influenza as a poultry problem and immi-

---

[236]Mwacalimba (2012).

[237]Id.

[238]CSO (2004).

[239]DVLD (2009).

[240]Munang'andu et al. (2012).

nent threat, pushing agriculture to the fore and inadvertently underplaying what the issue was truly about: the pandemic concern.

42.5 Policy Issues for Food Policy and Risk Management

Our case studies presented two very different food safety risks. In the case of bovine tuberculosis, the issues highlighted a problem obscured by bigger health concerns, i.e. HIV/AIDS and human tuberculosis. The second case was that of an indeterminate risk of incursion of a disease alien to Zambia. Both presented interesting cases for food safety. The BTB case study presented a twofold risk to communities living in and around the livestock interface. First, there is a public health risk from the consumption of uninspected meat from lechwe, a wildlife species known to be a maintenance host for BTB, and second, there is the major risk from consumption of unpasteurized milk derived from tuberculosis positive herds. Milk, as previously stated, is known to be the primary means of zoonotic conveyance to humans. Of course there are other risks of acquiring BTB in this area, including aerosols from cattle with active tuberculosis lesions in their lungs, during the evisceration of infected cattle or lechwe carcasses and during the consumption of undercooked meat contaminated with *Mycobacterium bovis*. However, the two risks highlighted, and the related food processing and handling practices of the area are pertinent concerns for food safety and zoonotic risk management.

Because lechwe meat provides communities with a supplementary source of protein, lechwe that are culled for consumption within wildlife-livestock communities actually bolster local food security. This service, however, is perhaps being provided at the expense of public health. The provision of meat inspection services would help address this concern, but requires a concerted response by local health, wildlife and veterinary officials. Although meat inspection is not very sensitive and up to 60 % of discrete tuberculosis lesions may go undetected,[241] it is still necessary to mitigate the food safety risks to public health in rural livestock keeping communities such as those found in Zambia's wildlife-livestock interface. The difficulty in controlling the condition in lechwe means that the disease is unlikely to be eliminated in cattle. Food hygiene thus seems to be the primary prevention mechanism available for food governance stakeholders in the area. A long term solution would require exploring control mechanisms in the lechwe population, in addition to control of the disease in cattle.[242]

On the other hand, avian influenza presents a food safety issue that occupies a different level of importance, particularly for the international donor community, when compared to BTB. An important risk question that was not asked in rolling out the avian and pandemic influenza response in Zambia was how the country was

---

[241]FSAI Scientific Committee (2003).
[242]Mwacalimba et al. (2013).

linked to the global poultry industry. The focus of control instead was on the Zambia's traditional and backyard production systems, a response that addressed a risk scenario mirroring the Southeast Asian H5N1 experience where backyard production was strongly implicated in the maintenance and spread of H5N1. A few key elements where missing from Zambia, the mixed farming systems, use of wet markets and duck production.[243]

Fortunately, an outbreak of zoonotic avian influenza H5N1 never occurred in Zambia. The abatement of the threat, however, served to highlight some significant policy conflicts. Poultry and poultry production were low priorities for the Zambian Agricultural Ministry, whose veterinary department had a long list of diseases of national economic importance to tackle. The man-hours spent in pursuit of the elusive H5N1 were thus viewed as wasteful. In short the donor-driven response was not properly aligned with local economic realities.[244]

These two case studies demonstrate the complexity of issues surrounding food governance. In the case of bovine tuberculosis, there is an apparent lack of understanding for the need for its control in rural populations, despite the myriad studies demonstrating the risk to humans living in contact with infected wildlife and livestock. Some of the reasons, presumably, are based on larger health problems (human tuberculosis and HIV/AIDS), an absence of veterinary support, and a wildlife management system that is incognizant of the risks to public health. In the case of H5N1, there was no real local risk, but international interest and finance pushed its importance up the government agenda. Outside Egypt and Nigeria, the African countries that had been most impacted by H5N1, other countries in the region responded to the avian and pandemic influenza more or less because of international finance. In areas where H5N1 was a problem, there were notable food security repercussions. For example, in Nigeria, a ban in poultry and poultry product movement resulted in those regions with low poultry production unable to obtain poultry and poultry meat from the high poultry producing areas. The impact was the reduced availability of the cheapest and commonest source of protein for low-income consumers.[245]

42.6 Key Learning Points for Public Health and Food Policy

This chapter sought to describe and explain the confluence of complex issues surrounding the issue of animal health, public health and food safety governance in the African region. The background sections served to demonstrate that politics is rife around food governance and food law in the African region. Using a risk-

---

[243]Mwacalimba (2012).
[244]Mwacalimba and Green (2014).
[245]Muma et al. (2014).

based epidemiological model, the chapter highlighted the various problems African countries face in addressing food safety concerns, including resource constraints, legislative overlap and redundancy, donor dependency, trade rule complexity, different perceptions on livestock use, problem overload and globalization.

Because food-borne diseases and food safety are important concerns in the African region, there remains an urgent need to understand both the public health and economic impact of food-borne diseases on the continent in general and in its various countries in particular. This knowledge, however, is not a guarantee for the adequate address of food safety on national governmental agendas. It must be understood that food governance policy coherence is not a state, but a process. It is a bargained 'collective' construct, requiring a level of oversight, coordination and consensus among actors for whom coherence is relevant.[246] Therefore, considerable advocacy and policy entrepreneurship is required to garner the necessary support for food safety reform both in the African region and with the international donor community. Furthermore, it is difficult to develop a coordinated and sustainable approach to the holistic management of food safety in the African region, especially when the impact of food safety on public health and the general economy have not been adequately assessed. Therefore, while important, getting the various stakeholder institutions to understand the public health benefits of coordinated and improved food safety mechanisms remains a fundamental challenge. This is because, as alluded to earlier, modern polity is fragmented as a matter of necessity. Viewing government as a unitary body that must generate knowledge on the economic and public health impacts of food safety and develop coherent national and international food safety policies in consultation with all stakeholders along the food supply chain is difficult.

Fundamentally, a flexible structuring of food safety risk in ways that is stakeholder inclusive is what is required. It is also important to include local ecological, biological and policy considerations. A clear picture of key stakeholders needs to be developed for effective food-borne disease risk management. This could include wide stakeholder consultation in understanding food safety risks and their assessment.[247] In the case of food safety governance in the African region, there is a need to have a multi-actor view of the food supply chain, and to develop context and time specific definitions of food safety problems that would help inform food safety agendas for both national governments and the global public health community.

The Bovine Spongiform Encephalopathy (BSE) crisis in the United Kingdom provides interesting lessons for understanding the governance needs of food-borne disease risk management, particularly as it relates to zoonoses.[248] In the context of the BSE/Creutzfeldt-Jakob Disease crisis in Europe, it was found useful to assess

---

[246] Ashoff (2005); Blouin (2007), pp. 169–173.
[247] Stirling and Scoones (2009), http://www.ecologyandsociety.org/vol14/iss2/art14/.
[248] See Dora (2006).

public perceptions through the lens of lay epidemiology[249] where the understanding of risk problems mirrored expert knowledge.[250] For zoonosis control, such a conceptualization of risk can be extended to accommodate multiple decision-makers along the food supply chain. That is, each decision-making body, with its institutional norms and ideas, can contribute their expertise and understanding of food-borne disease risk and understanding specific to the role they play along the food chain. Of course, this requires a deliberative approach emphasizing dialogue, particularly in defining the problems and analyzing and evaluating food safety risk issues.

42.7 Conclusion

In conclusion this chapter has hinted at the fact that defining food safety risks and, indeed the process of problem identification, could be developed by considering multiple contextual issues and stakeholders along the food supply chain. It is also important to keep in mind what they perceive the risks to be, and how their definitions fit into the broad picture of food safety in general. Certainly, the application and utility of the disease prevention framework presented in the conceptual section of this paper would be greatly advanced. However, because disease control is highly politicized, a more inclusive approach is required to use evidence to support responses to global disease concerns aligned with local priorities and realities.[251] As in the case of general disease control, food safety governance requires the right questions to be raised to foster the socio-political and economic change that the international community expects from the African region.[252] As one study argues,[253] the process of instituting change cannot remain the "purview of the global North, especially when the questions asked, and the responses advocated, favor a Northern perspective of globalized risk over the 'real' needs of the global South." Context will always matter and therefore, local ecological, biological and policy considerations should be given primacy. In conclusion, food safety governance regulatory processes should take into consideration local realities, local food supply chains and local food safety threats to ensure appropriateness and sustainability of any and all disease control measures instituted.

---

[249]This describes the processes through which health risks are understood and interpreted by laypeople. Allmark and Tod (2006).
[250]Dowler et al. (2006).
[251]Mwacalimba (2012).
[252]Colvin (2011), pp. 253–256.
[253]Mwacalimba (2012).

References

Alexander DJ (2007) An overview of the epidemiology of avian influenza. Vaccine 25:5637–5644
ALive (2006) Avian influenza prevention and control and human influenza pandemic preparedness in Africa: assessment of financial needs and gaps. In: Fourth international conference on avian influenza, Bamako, Mali, Dec 2006
Allmark P, Tod A (2006) How should public health professionals engage with lay epidemiology? J Med Ethics 32(8):460–463. doi:10.1136/jme.2005.014035
Ameni G, Bonnet P, Tibbo M (2003) A cross-sectional study if bovine tuberculosis in selected dairy farms in Ethiopia. Int J Appl Res Vet Med 1(4)
American Veterinary Medical Association (2008) One health: a new professional imperative, Final Report of the One Health Initiative Task Force (July 15, 2008)
Aranaz A et al (2004) Bovine tuberculosis (*mycobacterium bovis*) in wildlife in Spain. J Clin Microbiol 42(6):2602–2608
Ashoff G (2005) Enhancing policy coherence for development: justification, recognition and approaches to achievement. German Development Institute, Bonn
Ayele WY, Neill SD, Zinsstag J, Pavlik I (2004) Bovine tuberculosis: an old disease but a new threat to Africa. Int J Tuberc Lung Dis 8(8):924–937
Barling D, Lang T (2004) Trading on health: cross-continental production and consumption tensions and the governance of international food standards
Bartlett JG (2006) Planning for avian influenza. Ann Intern Med 145:141–144
Beigel JH, Farrar J, Han AM, Hayden FG, Hyer R, de Jong MD, Lochindarat S, Nguyen TK, Nguyen TH, Tran TH, Nicoll A, Touch S, Yuen KY (2005) Avian influenza A (H5N1) infection in humans. N Engl J Med 353:1374–1385
Blouin C (2007) Trade policy and health: from conflicting interests to policy coherence. Bull World Health Organ 85:169–173
Brosch R, Gordon SV, Marmiesse M, Brodin P, Buchrieser C, Eiglmeier K, Garnier T, Gutierrez C, Hewinson G, Kremer K, Parsons LM, Pym AS, Samper S, van Soolingen D, Cole ST (2002) A new evolutionary scenario for the Mycobacterium tuberculosis complex. Proc Natl Acad Sci U S A 99(6):3684–3689, Epub 2002 Mar 12
Bruckner GK (2009) The role of the World Organisation for Animal Health (OIE) to facilitate the international trade in animals and animal products. Onderstepoort J Vet Res 76:141–146
Burgos S, Otte J (2008) Animal health in the 21st century: challenges and opportunities: Pro-Poor Livestock Policy Initiative, Research report 09-06
Buse K, Mays N, Walt G (2005) Making health policy. Open University Press
Buzby JC, Roberts T (2009) The economics of enteric infections: human foodborne disease costs. Gastroenterology 136(6):1851–1862. doi:10.1053/j.gastro.2009.01.074
Capua I, Alexander DJ (2004) Avian influenza: recent developments. Avian Pathol 33:393–404
Cattoli G, Monne I, Fusaro A, Joannis TM, Lombin LH, Aly MM, Arafa AS, Sturm-Ramirez KM, Couacy-Hymann E, Awuni JA, Batawui KB, Awoume KA, Aplogan GL, Sow A, Ngangnou AC, El Nasri Hamza IM, Gamatie D, Dauphin G, Domenech JM, Capua I (2009) Highly pathogenic avian influenza virus subtype H5N1 in Africa: a comprehensive phylogenetic analysis and molecular characterization of isolates. PLoS One 4, e4842
Chabwela HNW, Mumba W (1998) Integrating water conservation and population strategies on the Kafue Flats. In: de Sherbinin A, Dompka V (eds) Water and population dynamics. American Association for the Advancement of Science, Washington
Chang HJ (2003) Kicking away the ladder: the "real" history of free trade: FPIF Special Report "Globalization and the Myth of Free Trade", April 18, 2003, The New School University in New York City
Chen H, Smith GJ, Zhang SY, Qin K, Wang J, Li KS, Webster RG, Peiris JS, Guan Y (2005) Avian flu: H5N1 virus outbreak in migratory waterfowl. Nature 436:191–192
Colston MJ (2001) Mycobacterial infection from the cellular point of view. Acta Cient Venez 52 (Suppl 1):13–15

Colvin CJ (2011) Think locally, act globally: developing a critical public health in the global South. Crit Public Health 21(3):253–256

Conly JM, Johnston BL (2004) Avian influenza - The next pandemic? Can J Infect Dis Med Microbiol 15:252–254

Cook AJC, Tuchili LM, Buve A, Foster SD, Godfrey-Faussett P, Panday GS, McAdam KPWJ (1996) Human and bovine tuberculosis in the Monze district of Zambia-a cross-sectional study. Br Vet J 152:37–46

Cosivi O, Grange JM, Daborn CJ, Raviglione MC, Fujikura T, Cousins D, Rabinson RA, Huchzermeyer HFAK, de Kantor I, Meslin FX (1998) Zoonotic tuberculosis due to Mycobacterium bovis in developing countries. Emerg Infect Dis 4(1):59–70

CSO (2004) Living conditions monitoring survey. Central Statistics Office, Lusaka

Dewaal CS, Robert N, Witmer J, Tian XA (2010) A comparison of the burden of foodborne and waterborne diseases in three world regions, 2008. Food Protect Trends 30(8):483–490

Domenech J, Lubroth J, Eddi C, Martin V, Roger F (2006) Regional and international approaches on prevention and control of animal transboundary and emerging diseases. Ann N Y Acad Sci 1081:90–107

Dora C (2006) Seeking lessons from BSE/CJD for communication strategies on health and risk: In: Dora C (ed) Health, hazards and public debate: lessons from risk communication from the BSE/CJD saga. WHO

Douglas M, Wildavsky A (1982) Risk and culture. University of California Press, Berkeley

Dowler E, Green J, Bauer M, Gasperoni G (2006) Assessing public perception: issues and methods: In: Dora C (ed) Health, hazards and public debate: Lessons from risk communication from the BSE/CJD saga. WHO

DVLD (2009) Zambia poultry sector study: risk mapping activity – ILRI EDRS-AIA project: early detection, reporting and surveillance –Avian Influenza in Africa. C/09/079:MK04 STA USA064 102

Elbe S (2010) Haggling over viruses: the downside risks of securitizing infectious disease. Health Policy Plan 25:476–485

Eurosurveillance (2006) Highly pathogenic avian influenza A/H5N1 – update and overview of 2006: Eurosurveillance weekly release. Euro Surveill 11(12):E061221.1

FAO (2004) Poultry production sectors: Food and Agriculture Organization of the United Nations. Available at http://www.fao.org/docs/eims/upload//224897/factsheet_productionsectors_en.pdf

FAO (2005) Analysis of the food safety situation in Zambia. In: FAO/WHO regional conference on food safety for Africa, Harare, Zimbabwe, 3–6 October 2005

FAO (2010) Avian influenza outbreaks: Food and Agriculture Organisation of the United Nations. Available at http://www.fao.org/avianflu/en/maps.html

Fidler DP (2004a) Global outbreak of Avian Influenza A (H5N1) and international law. American Society of international law 8(1). January 25, 2004

Fidler DP (2004b) Germs, governance, and global public health in the wake of SARS. J Clin Invest 113:799–804

Fidler DP (2008) Influenza virus samples, international law, and global health diplomacy. Emerg Infect Dis 14:88–94

FSAI Scientific Committee (2003) Zoonotic tuberculosis and food safety. Zoonotic tuberculosis-Final report, Food Safety Authority of Ireland Scientific Committee, July 2003 http://www.fsai.ie/publications/other/zoonotic_tuberculosis.pdf

Garrett L, Fidler DP (2007) Sharing H5N1 viruses to stop a global influenza pandemic. PLoS Med 4, e330

Gibson AL, Hewinson G, Goodchild T, Watt B, Story A, Inwald J, Drobniewski FA (2004) Molecular epidemiology of disease due to Mycobacterium bovis in humans in the United Kingdom. J Clin Microbiol 42(1):431–434

Gilbert M, Xiao X, Pfeiffer DU, Epprecht M, Boles S, Czarnecki C, Chaitaweesub P, Kalpravidh W, Minh PQ, Otte MJ, Martin V, Slingenbergh J (2008) Mapping H5N1 highly pathogenic avian influenza risk in Southeast Asia. Proc Natl Acad Sci U S A 105:4769–4774

GRAIN (2006) The top-down global response to bird-flu: against the GRAIN. http://www.grain.org/articles/?id=12

GRAIN (2007) Bird flu: a bonanza for 'Big Chicken.' Against the Grain. http://www.grain.org/articles/?id=22

Grange JM (2001) Mycobacterium bovis infection in human beings. Tuberculosis 81(1/2):71–77

Green A (1999) An introduction to health planning in developing countries, 2nd edn. Oxford Medical Publications/Oxford University Press, Oxford

Green J (2012) 'One health, one medicine' and critical public health. Crit Public Health 22:377–381

Greger M (2007) The human/animal interface: emergence and resurgence of zoonotic infectious diseases. Crit Rev Microbiol 33:243–299

Grootenhuis JG, Olubayo RO (1993) Disease research in the wildlife-livestock interface in Kenya. Vet Q 15(2):55–59

GRZ (2006) National task force on avian influenza: operational guidelines 2006. Lusaka Government of Zambia

Hall DC, Ehui S, Delgado C (2004) The livestock revolution, food safety, and small-scale farmers: why they matter to us all. J Agric Environ Ethics 17:425–444

Hampson AW (1997) Surveillance for pandemic influenza. J Infect Dis 176(Suppl 1):S8–S13

Hancox M (2000) Letter to the editors: cattle tuberculosis schemes: control or eradication? The society for applied microbiology. Lett Appl Microbiol 31:87–93

Hollenbeck JE (2005) An avian connection as a catalyst to the 1918-1919 influenza pandemic. Int J Med Sci 2:87–90

Horlick-Jones T (1998) Social theory and the politics of risk. J Conting Crisis Manag 6:64–67

Horowitz L (2005) The Avian flu fright: politically timed for global "Iatrogenocide." Global Research, October 12, 2005

Hubálek Z (2003) Emerging human infectious diseases: anthroponoses, zoonoses, and sapronoses. Emerg Infect Dis 9(3):403–404. doi:10.3201/eid0903.020208

Hueston WD (2003) Science, politics and animal health policy: epidemiology in action. Prev Vet Med 60:3–12

Jeffery RCV, Malambo CH and Nefdt R (1991) Wild mammal surveys of the Kafue flats. A Report to the Director, National Parks andWildlife Service, Chilanga, Zambia

Käferstein FK, Motarjemi Y, Bettcher DW (1997) Foodborne disease control: a transnational challenge. Emerg Infect Dis 3:503–510

Kahn LH, Kaplan B, Steele JH (2007) Confronting zoonoses through closer collaboration between medicine and veterinary medicine (as 'one medicine'). Vet Ital 43:5–19

Karesh WB, Cook RA, Bennett EL, Newcomb J (2005) Wildlife trade and global disease emergence. Emerg Infect Dis 11:1000–1002

Kasemsuwan S, Poolkhet C, Patanasatienkul T, Buameetoop N, Watanakul M, Chanachai K, Wongsathapornchai K, Métras R, Marcé C, Prakarnkamanant A, Otte J, Pfeiffer D (2009) Qualitative risk assessment of the risk of introduction and transmission of H5N1 HPAI virus for 1-km buffer zones surrounding compartmentalised poultry farms in Thailand Mekong Team Working Paper No. 7

Kazwala RR, Daborn CJ, Sharp JM, Kambarage DM, Jiwa SFH, Mbembati NA (2001) Isolation of Mycobacterium bovis from human cases of cervical adenitis in Tanzania: a cause for concern? Int J Tuberc Lung Dis 5(1):87–91

Kells HR, Lear SA (1959) Thermal death time curve of Mycobacterium tuberculosis var. bovis in artificially infected milk. Paper of the journal series, New Jersey agricultural experiment station, Rutgers, the State University, Department of Dairy Science, New Brunswick, New Jersey, vol 8

Kilpatrick AM, Chmura AA, Gibbons DW, Fleischer RC, Marra PP, Daszak P (2006) Predicting the global spread of H5N1 avian influenza. Proc Natl Acad Sci U S A 103:19368–19373

Kimball AM (2006) Risky trade: infectious diseases in the era of global trade. Ashgate Publishing, London

Kimball AM, Arima Y, Hodges JR (2005) Trade related infections: farther, faster, quieter. Glob Health 1:3

Kingdon JW (2003) Agendas, alternatives and public policies, 2nd edn. Addison-Wesley Educational Publishers Inc, New York

Kock RA (2003) What is this infamous "wildlife/livestock disease v. interface?" A review of current knowledge for the African Continent. In: Osofsky SA (ed) Conservation and development interventions at the wildlife/livestock interface implications for wildlife, livestock and human health

Kock ND, Kampamba G, Mukaratirwa S, Du Toit J (2002) Disease investigation into free-ranging Kafue lechwe (Kobus leche kafuensis) on the Kafue Flats in Zambia. Vet Rec 151:482–484

Krauss H, Weber A, Appel M, Enders B, Isenberg HD, Schiefer HG, Slenczka W, von Graevenitz A, Zahner H (2003) Zoonoses: infectious diseases transmissible from animals to humans, 3rd edn. ASM press, Washington

Kruk ME (2008) Emergency preparedness and public health systems lessons for developing countries. Am J Prev Med 34:529–534

Lee K, Koivusalo M (2005) Trade and health: is the health community ready for action? PLoS Med 2, e8

Lee K, Fustukian S, Buse K (2002) An introduction to global health policy. In: Lee K, Buse K, Fustukian S (eds) Health policy in a globalising world. Cambridge University Press, Cambridge

MacDonald R, Horton R (2009) Trade and health: time for the health sector to get involved. Lancet 373:273–274

Menzies FD, Neill SD (2000) Cattle-to-cattle transmission of bovine tuberculosis. Vet J 160:92–106. Harcourt Publishers Ltd

Mfinanga SG, Mørkve O, Kazwala RR, Cleaveland S, Sharp JM, Shirima G, Nilsen R (2003a) The role of livestock keeping in tuberculosis trends in Arusha, Tanzania. Int J Tuberc Lung Dis 7 (7):695–704

Mfinanga SG, Mørkve O, Kazwala RR, Cleaveland S, Sharp JM, Shirima G, Nilsen R (2003b) Tribal differences in perception of tuberculosis: a possible role in tuberculosis control in Arusha, Tanzania. Int J Tuberc Lung Dis 7(10):933–941

Moda G, Daborn CJ, Grange JM, Cosivi O (1996) The zoonotic importance of Mycobacterium bovis. Tuber Lung Dis 77:103–108

Monto AS, Comanor L, Shay DK, Thompson WW (2006) Epidemiology of pandemic influenza: use of surveillance and modeling for pandemic preparedness. J Infect Dis 194(Suppl 2):S92–S97

Morens DM, Fauci AS (2007) The 1918 influenza pandemic: insights for the 21st century. J Infect Dis 195:1018–1028

Mudenda D (2005) Zambia's trade situation: implications for debt and poverty reduction. A report on Zambia's trade situation by the Jesuit Centre for Theological Reflection

Muma JB, Mwacalimba KK, Munang'andu HM, Matope G, Jenkins A, Siamudaala V, Mweene AS, Marcotty T (2014) The contribution of veterinary medicine to public health and poverty reduction in developing countries. Vet Ital 50:117–129

Mumba M (2004) Biodiversity challenges for invaded wetland ecosystems in Africa: the case of the Kafue Flats floodplain system in southern Zambia. In: Proceedings of a global synthesis workshop on 'Biodiversity loss and species extinctions: managing risk in a changing world' Sub theme: invasive alien species-coping with aliens

Mumba C, Samui KL, Pandey GS, Hang'ombe BM, Simuunza M, Tembo G, Muliokela SW (2011) Economic analysis of the viability of smallholder dairy farming in Zambia. Livest Res Rural Dev 23:137

Munang'andu HM, Kabilika SH, Chibomba O, Munyeme M, Muuka GM (2012) Bacteria isolations from broiler and layer chicks in Zambia. J Pathog 2012, 520564

Munyeme M, Munang'andu HM (2011) A review of bovine tuberculosis in the kafue basin ecosystem. Vet Med Int 2011:918743

Munyeme M, Muma JB, Samui KL, Skjerve E, Nambota AM, Phiri IGK, Rigouts L, Tryland M (2008) Prevalence of bovine tuberculosis and animal level risk factors for indigenous cattle under different grazing strategies in the livestock/wildlife interface areas of Zambia. Trop Anim Health Prod 41:345–352

Munyeme M, Muma JB, Siamudaala VM, Skjerve E, Munang'andu HM, Tryland M (2010a) Tuberculosis in Kafue lechwe antelopes (*Kobus leche Kafuensis*) of the Kafue Basin in Zambia. Prev Vet Med 95(3–4):305–308

Munyeme M, Muma JB, Munang'andu HM, Kankya C, Skjerve E, Tryland M (2010b) Cattle owners' awareness of bovine tuberculosis in high and low prevalence settings of the wildlife-livestock interface areas in Zambia. BMC Vet Res 6:21

Mwacalimba KK (2012) Globalised disease control and response distortion: a case study of avian influenza pandemic preparedness in Zambia. Crit Public Health 22:391–405

Mwacalimba K (2013) Pandemic preparedness and multi-sectoral zoonosis risk management: implications for risk assessment of avian influenza in Zambian trade, health and agriculture. LAP LAMBERT Academic Publishing

Mwacalimba KK, Green J (2014) 'One health' and development priorities in resource constrained countries: policy lessons from avian and pandemic influenza preparedness in Zambia. Health Policy Plan 30:215–222. doi:10.1093/heapol/czu001

Mwacalimba KK, Mumba C, Munyeme M (2013) Cost benefit analysis of tuberculosis control in wildlife–livestock interface areas of Southern Zambia. Prev Vet Med 110(2):274–279

Navarro V (1998) Comment: whose globalization? Am J Public Health 88:742–743

Ndulo M (2006) Zambia and the multilateral trading system: the impact of WTO agreements, negotiations and implementation. United Nations Conference on Trade and Development. UNCTAD/DITC/TNCD/2005/16

Neill S, Bryson D, Pollock J (2001) Pathogenesis of tuberculosis in cattle. Tuberculosis 81:79–86

Nicoll A (2005) Avian and pandemic influenza--five questions for 2006. Euro Surveill 10:210–211

Niemann S, Richter E, Rüsch-Gerdes S (2000) Differentiation among members of the Mycobacterium tuberculosis complex by molecular and biochemical features: evidence for two pyrazinamide-susceptible subtypes of M. bovis. J Clin Microbiol 38:152–157

Normile D (2006) Avian influenza. Evidence points to migratory birds in H5N1 spread. Science 311:1225

O'Reilly LM, Daborn CJ (1995) The epidemiology of Mycobacterium bovis infections in animals and man: a review. Tuber Lung Dis 76(Suppl 1):1–46

OIE (2004) Handbook on import risk analysis for animals and animal products. Vol 1 Introduction and qualitative risk analysis. World organisation for animal health, Paris, pp 1–46

OIE (2010) Manual of diagnostic tests and vaccines for terrestrial animals

Ong A, Kindhauser M, Smith I, Chan M (2008) A global perspective on avian influenza. Ann Acad Med Singapore 37:477–481

Omamo SW, Farrington J (2004) Policy research and African agriculture: time for a dose of reality? ODI. Natural resource perspectives no. 90

Ortu G, Mounier-Jack S, Coker R (2007) Pandemic influenza preparedness in the African continent: analysis of national strategic plans. London School of Hygiene and Tropical Medicine

Ortu G, Mounier-Jack S, Coker R (2008) Pandemic influenza preparedness in Africa is a profound challenge for an already distressed region: analysis of national preparedness plans. Health Policy Plan 23:161–169

Osterholm MT (2005) Preparing for the next pandemic. N Engl J Med 352:1839–1842

Pandey GS (2004) Tuberculosis in the Kafue lechwe (Kobus leche Kafuensis) and its public health significance particularly game meat utilization in Zambia. In: Proceedings at the Commonwealth Veterinary Association/Veterinary Association of Zambia joint Regional Conference for Eastern, Central and Southern Africa. Special Issue, pp 17–20

Paul M, Tavornpanich S, Abrial D, Gasqui P, Charras-Garrido M, Thanapongtharm W, Xiao X, Gilbert M, Roger F, Ducrot C (2010) Anthropogenic factors and the risk of highly pathogenic avian influenza H5N1: prospects from a spatial-based model. Vet Res 41:28

Phiri AM et al (2011) Helminth parasites of the Kafue lechwe antelope (Kobus leche kafuensis): a potential source of infection to domestic animals in the Kafue wetlands of Zambia. J Helminthol 85(1):20–27. doi:10.1017/S0022149X10000192, Epub 2010 Apr 14

Pitrelli N, Sturloni G (2007) Infectious diseases and governance of global risks through public communication and participation. Ann Ist Super Sanita 43:336–343

Rabinowitz PM, Odofin L, Dein FJ (2008) From "us vs. them" to "shared risk": can animals help link environmental factors to human health? Ecohealth 5:224–229

Rweyemamu MM, Astudillo VM (2002) Global perspective for foot and mouth disease control. Rev Sci Tech 21:765–773

Scoones I (2010) The international response to avian influenza: science, policy and politics. In: Scoones I (ed) Avian influenza: science, policy and politics, Pathways to sustainability series. Earthscan, London

Scoones I, Forster P (2008a) HPAI and International policy processes – a scoping study: Research Report. Controlling Avian Flu and Protecting livelihoods, A DFID-funded collaborative research project. STEPS Centre

Scoones I, Forster P (2008b) The international response to highly pathogenic avian influenza: science, policy and politics steps centre. http://www.steps-centre.org/PDFs/Avian%20flu%20final%20w%20cover.pdf

Shortridge KF, Zhou NN, Guan Y, Gao P, Ito T, Kawaoka Y, Kodihalli S, Krauss S, Markwell D, Murti KG, Norwood M, Senne D, Sims L, Takada A, Webster RG (1998) Characterization of avian H5N1 influenza viruses from poultry in Hong Kong. Virology 252:331–342

Siamudaala VM (2004) The role of wildlife in poverty alleviation. Zambian J Vet Sci (Special Issue on the Proceedings of the CVA/VAZ Joint Regional Conference for Eastern, Central and Southern Africa):48–52

Siamudaala VM, Muma JB, Munang'andu HM, Mulumba M (2003) Disease challenges on the conservation and utilisation of the Kafue lechwe (Kobus leche Kafuensis) in Zambia. In: Conservation and development innovations at the wildlife/livestock interface: implications for wildlife, livestock and human health

Simasiku P, Simwanza H, Tembo G, Bandyopadhyay S, Pavy J (2008) The impact of wildlife management policies on communities and conservation in game management areas in Zambia. Natural Resources Consultative Forum, Zambia

Sims L, Narrod C (2008) Understanding avian influenza – a review of the emergence, spread, control, prevention and effects of Asian-lineage H5N1 highly pathogenic viruses: United Nations Food and Agriculture Organization (FAO). www.fao.org/avianflu

Sims LD, Domenech J, Benigno C, Kahn S, Kamata A, Lubroth J, Martin V, Roeder P (2005) Origin and evolution of highly pathogenic H5N1 avian influenza in Asia. Vet Rec 157:159–164

Slingenbergh J (2004) Environmental, climatic risk factors: abstracts of keynote speeches. Report of the WHO/FAO/OIE joint consultation on emerging zoonotic diseases, 3–5 May 2004

Smith VH, Sumner DA, Rosson CP (2002) Bilateral and multilateral trade agreements. In: Outlaw JL, Edward G (eds) The 2002 farm bill: policy options and consequences, Smith Publication No. 2001-01. Farm Foundation, Oak Brook, IL, September 2001

Soares Magalhaes RJ, Ortiz-Pelaez A, Thi KL, Dinh QH, Otte J, Pfeiffer DU (2010) Associations between attributes of live poultry trade and HPAI H5N1 outbreaks: a descriptive and network analysis study in northern Vietnam. BMC Vet Res 6:10

Stärk KD, Regula G, Hernandez J, Knopf L, Fuchs K, Morris RS, Davies P (2006) Concepts for risk-based surveillance in the field of veterinary medicine and veterinary public health: review of current approaches. BMC Health Serv Res 6:20

Stiglitz JE (2009) Trade agreements and health in developing countries. Lancet 373:363–365

Stirling AC, Scoones I (2009) From risk assessment to knowledge mapping: science, precaution and participation in disease ecology. Ecol Soc 14. http://www.ecologyandsociety.org/vol14/iss2/art14/

Tambi E, Bessin R (2006) The WTO agreement on agriculture: effects on livestock production and trade in Africa African Union/Interafrican Bureau for Animal Resources Pan African programme for the Control of Epizootics European Commission, 16 January 2006

Tamiru F, Hailemariam M, Ter W (2013) Preliminary study on prevalence of bovine tuberculosis in cattle owned by tuberculosis positive and negative farmers and assessment of zoonotic awareness in Ambo and Toke Kutaye districts, Ethiopia. J Vet Med Anim Health 5 (10):288–295

Thiermann AB (2005) Globalization, international trade and animal health: the new roles of OIE. Prev Vet Med 67:101–108

Thomson GR, Tambi EN, Hargreaves SK, Leyland TJ, Catley AP, van't Klooster GG, Penrith ML (2004) International trade in livestock and livestock products: the need for a commodity-based approach. Vet Rec 155:429–433

UN (2010) Hanoi declaration at the international ministerial conference: "Animal and pandemic influenza: the way forward". Hanoi, Vietnam, 19–21 April 2010. IMCAPI Hanoi 2010: http://www.un-influenza.org/node/4040

UNGASS (2012) Monitoring the declaration of commitment on HIV and AIDS and the universal access. Zambia Country report 2012: UNGASS 2012 Country reports, UNAIDS

Unger F, Münstermann S, Goumou A, Apia CN, Konte M (2003) Risk associated with Mycobacterium bovis infections detected in selected study herds and slaughter cattle in 4 countries of West Africa. Animal Health Working Paper 1. ITC (International Trypanotolerance Centre), Banjul, The Gambia, p 25.

UNSIC (2006) Avian and human influenza: UN system contributions and requirements. A strategic approach: Office of the United Nations System Influenza Coordinator. UNSIC/Strategy/Final

UNSIC (2006) Pandemic planning and preparedness guidelines for the United Nations system

UNSIC and World Bank (2008) Responses to avian influenza and state of pandemic readiness. Fourth Global Progress Report

Upton M, Otte J (2004) The impact of trade agreements on livestock producers pro-poor livestock policy initiative. Research Report. RR Nr. 04-01

Vallat B, Pastoret P-P (2009) The role and mandate of the World Organisation for Animal Health in veterinary education. Rev Sci Tech 28(2):503–510

van den Berg T (2009) The role of the legal and illegal trade of live birds and avian products in the spread of avian influenza. Rev Sci Tech 28:93–111

Van Kerkhove MD, Vong S, Guitian J, Holl D, Mangtani P, San S, Ghani AC (2009) Poultry movement networks in Cambodia: implications for surveillance and control of highly pathogenic avian influenza (HPAI/H5N1). Vaccine 27:6345–6352

Walt G (1994) Health policy: an introduction to process and power. Zed Books, London

Walt G, Gilson L (1994) Reforming the health sector in developing countries: the central role of policy analysis. Health Policy Plan 9:353–370

Weber TP, Stilianakis NI (2007) Ecologic immunology of avian influenza (H5N1) in migratory birds. Emerg Infect Dis 13:1139–1143

Webster RG (1997) Predictions for future human influenza pandemics. J Infect Dis 176(Suppl 1): S14–S19

Webster RG, Govorkova EA (2006) H5N1 influenza — continuing evolution and spread. N Engl J Med 355:2174–2177

Webster R, Hulse D (2005) Controlling avian flu at the source. Nature 435:415–416

Webster RG, Guan Y, Peiris M, Walker D, Krauss S, Zhou NN, Govorkova EA, Ellis TM, Dyrting KC, Sit T, Perez DR, Shortridge KF (2002) Characterization of H5N1 influenza viruses that continue to circulate in geese in southeastern China. J Virol 76:118–126

WHO (2004) Press release: unprecedented spread of avian influenza requires broad collaboration-FAO/OIE/WHO call for international assistance: January 27, 2004, at http://www.who.int/mediacentre/releases/2004/pr7/en/

WHO (2005a) Avian influenza: assessing the pandemic threat: http://www.who.int/csr/disease/influenza/H5N1-9reduit.pdf

WHO (2005b) WHO global influenza preparedness plan: the role of WHO and recommendations for national measures before and during pandemics: World Health Organization, WHO/CDS/CSR/GIP/2005.5

WHO (2005c) Avian influenza and the pandemic threat in Africa: risk assessment for Africa World Health Organization. http://www.who.int/csr/disease/avian_influenza/riskassessment Africa/en/index.html. Accessed 10 June 2010

WHO (2006) Avian influenza frequently asked questions: WHO. Available at: http://www.who.int/csr/disease/avian_influenza/avian_faqs/en/index.html

WHO (2010) Cumulative number of confirmed human cases of avian influenza A/(H5N1) Reported to WHO: WHO. Available at: http://www.who.int/csr/disease/avian_influenza/country/cases_table_2010_11_19/en/index.html

WHO/DFID-AHP (2005) Meeting on control of zoonotic diseases: a route to poverty alleviation among livestock-keeping communities. WHO headquarters, Geneva (salle d). 20 and 21 September 2005

WHO-AFRO (2005) Avian influenza and the pandemic threat in Africa: risk assessment for Africa World Health Organization. http://www.who.int/csr/disease/avian_influenza/riskassessment Africa/en/index.html

WHO-AFRO (2012) Manual for integrated food-borne disease surveillance in the WHO African region. The World Health Organisation African Regional Office (WHO-AFRO)

Wilkins M (2000) Assessing the Risk of Human Health from bovine tuberculosis in Michigan. Epi Insight, Michigan Department of Community Health. DCH-0709 (Rev. 12/00)

Wilson JS, Otsuki T (2001) Global trade and food safety: winners and losers in a fragmented system. Policy Working Group Research Paper. The World Bank

Xu X, Subbarao, Cox NJ, Guo Y (1999) Genetic characterization of the pathogenic influenza A/Goose/Guangdong/1/96 (H5N1) virus: similarity of its hemagglutinin gene to those of H5N1 viruses from the 1997 outbreaks in Hong Kong. Virology 261:15–9

Yee KS, Carpenter TE, Cardona CJ (2009) Epidemiology of H5N1 avian influenza. Comp Immunol Microbiol Infect Dis 32:325–340

Yupiana Y, de Vlas SJ, Adnan NM, Richardus JH (2010) Risk factors of poultry outbreaks and human cases of H5N1 avian influenza virus infection in West Java Province, Indonesia. Int J Infect Dis 14:e800–e805

Zepeda C, Salman M, Thiermann A, Kellar J, Rojas H, Willeberg P (2005) The role of veterinary epidemiology and veterinary services in complying with the World Trade Organization SPS agreement. Prev Vet Med 67:125–140

Zinsstag J, Schelling E, Roth F, Bonfoh B, de Savigny D, Tanner M (2007) Human benefits of animal interventions for zoonosis control. Emerg Infect Dis 13:527–531

Zumla A, Malon P, Henderson J, Grange J (2000) Impact of HIV infection on tuberculosis. Postgrad Med J 76(895):259–268. doi:10.1136/pmj.76.895.259

Chapter 43
Regulatory Frameworks Affecting Seeds: Impacts on Subsistence Farmers in the Eastern and Southern African Region

Marcelin Tonye Mahop

Abstract With the current growth of the population in sub Saharan Africa, the need to produce and make available sufficient food for the needs of this population put agriculture one of the prominent development concerns in this part of the world. Using their traditional farming practice, small holder farmers currently play a significant role in food production. It is suggested that one critical factor which help this category of farmers to play this role is their ability to exercise their rights to handle seeds as they have been doing before the advent of the seed industry. Therefore to address food insecurity in Sub Saharan Africa, seeds related programmes, initiatives and regulatory framework need not undermine farmers' rights. Using the Eastern and Southern African regions, the chapter explores the extent to which current programmes (of pan African and sub-regional scope) and regulatory frameworks impact on smallholder farmers' rights and suggest the way forward.

43.1 Introduction

Food insecurity is a global challenge that is garnering the attention and efforts of a wide array of actors and agencies the world over. With the world population now surpassing seven billion people, the search has intensified for ideas and solutions that can contribute to the provision of affordable and sufficient food. Despite the efforts made in addressing this struggle, the reality remains: there is a lack of sufficient food to feed the world. As such, food security remains at the top of the international development agenda.[1] The simplest description of what amounts to

---

[1] One of the focus areas for the sustainable development goals which are being discussed under the umbrella of the United Nations Open Working Group on Sustainable Development Goals,

M.T. Mahop (✉)
United Nations Environment Programme, Nairobi, Kenya

United Nations Environment Programme, London, UK
e-mail: tonye2169@aol.com

food security is when every individual has, at all times, both physical and economic access to sufficient food to meet their dietary needs for a productive and healthy life (FAO 1996, 2009). Based on this description, there are a number of requirements that must be addressed if food security is to be realised. These include but are not limited to: sufficient quantities of appropriate foods being consistently available; individuals having adequate incomes or other resources to purchase or barter for food; food being properly processed and stored; and individuals were having enough nutritional and family care knowledge to attain adequate health and sanitation practices (Bremner 2012; Ecker and Breisinger 2012). These critical requirements for food security are applicable in both the developed and developing worlds.

While the lack of sufficient food is a global issue, the situation is particularly dire in Sub-Saharan Africa. The African population is growing rapidly, with some estimates suggesting that by 2050 the population will have reached 2 billion people—more than double the current estimate of roughly 850 million.[2] Out of the current 850 million people, more than 240 million lack adequate food for a productive and healthy life. According to its own statistics, the New Partnership for Africa's Development estimates that around 97 % of the continent's population that is considered food insecure lives in Sub-Saharan Africa, and out of the continent's total food insecure population, 34 % is described as undernourished.[3] This means that not only does this category of people not have sufficient food, but the food they do have is lacking nutritional value.

Agriculture is critical to addressing the food insecurity problem in Africa, evidenced in the design of the Comprehensive Africa Agricultural Development Programme ("CAADP").[4] CAADP is a component of the New Partnership for Africa's Development ("NEPAD"), and was adopted by the AU heads of States in 2003. CAADP recognizes that, among other things, low agricultural production is one of the main causes of the increasing issue of food scarcity in Sub-Saharan Africa. The CAADP proposes that steps be taken to improve research and innovations in agriculture and to encourage the use of improved technologies across the continent (NEPAD 2003). As such, embedded in the CAADP is the idea that intensification of agricultural practices—the cultivation of farmland with high amounts of inputs such as fertilizers, pesticides, and others—is the way forward and may represent an effective means for boosting agricultural production. Supporting this approach is the deployment of initiatives such as the Alliance for Green Revolution in Africa. This initiative is focused on the utilization of improved

---

established in January 2013 pursuant to the recommendation of the outcomes of the 2012 Rio+20 conference, is focus area 2: Food Security and Nutrition, http://sustainabledevelopment.un.org/focussdgs.html.

[2]*Id.*

[3]The Comprehensive Africa Agriculture Development Programme (CAADP): http://www.nepad-caadp.net/.

[4]*Id.*

seeds that are protected by plant breeders' rights and are regulated through seed regulations (AGRA 2013).

Equally, the CAADP acknowledges the importance of smallholder farmers and their farming practices in the production and supply of food in Sub-Saharan Africa.[5] Not only does subsistence agriculture practiced by smallholder farmers provide a major source of food at an estimated 75 % of total food produced,[6] it also plays a critical role in employment and income generation at the community level. Thus, while the use of new technologies like improved seeds is likely to influence regulatory frameworks pertaining to agriculture in SSA, such frameworks must not undermine the contribution of smallholder farmers in food production.

This chapter will examine the utilisation of seeds in the eastern and southern African regions, exploring the extent to which relevant regulatory frameworks have an impact on smallholder farmers' practices, as well as the implications they have on food production. Among other inputs, seeds are probably the most critical in farming as they are the key determinant of yields. However, to harness their yield potential, seeds are aided by other inputs including fertilizers and pesticides. The extent to which farmers employ the most closely-adapted agronomics and cropping techniques to their own agro-ecological conditions is undoubtedly another determinant of agricultural production; indeed, seeds do not do it alone.

Historically, seeds have been produced by the farmers' themselves[7] and farmers have used traditional techniques to process them for conservation and stockade. From the saved grains, they have selected which ones will be used for either consumption or planting the following season. Today, farmers continue to exchange seeds among themselves, and have shared within their communities their knowledge about how to use the seeds they exchange. However, since the nineteenth century, seed production has increasingly been separated from agriculture, becoming a specialist area in its own right. This shift has encouraged the emergence of the seed industry in Europe and the United States.[8] The rediscovery of the Mendelian science of heredity and its mastering by professionals involved in seed production has contributed to not only the emergence of a "breeder" profession, but also the development of conventional breeding techniques of crossing and selecting in varietal production. Over the past decades, advances in molecular biology and genetic engineering have provided breeders with additional knowledge and tools in varietal production.[9] Today, seed production involves a range of activities including varietal development through conventional or biotech-based breeding techniques,[10] seed multiplication, testing and certification, quality control, distribution/commercialisation, and, lastly, utilisation by farmers.

---

[5]*Id.* at 4.
[6]Livingston et al. (2011).
[7]Dutfield (2008).
[8]Poonia (2013).
[9]Stafford (2009).
[10]Hansen (2000).

There are therefore two platforms for seed production: traditional and modern. The traditional platform revolves around farmers, (who, in Sub-Saharan Africa, are largely smallholder farmers who are still developing plant varieties that are adapted to their agro-ecological conditions and farming practices). The varieties they produce are mostly excluded from plant variety protection regimes and seed regulations because they do not fulfil the requirements set by these regulatory frameworks. This traditional platform for seed production, also called the informal seed system, is known to supply about 80 % of the seeds used in farming in Sub-Saharan Africa, and produces about 70 % of the food.[11] The regulatory frameworks that undermine this platform of seed production are likely to have a negative impact on food security in SSA, as these smallholder farmers' practices are useful for the conservation of agro biodiversity and are actually advocated for at the international level.[12] The second platform is the modern platform, or formal seed system. This platform involves many actors, including scientific breeders, research organisations, multinational corporations, and seed merchants. Activities in this platform are strongly regulated, e.g. the rights of breeders of new varieties of plants are protected, and only seeds produced according to relevant regulations may be legally made available on the market for (commercial) farmers. A recent assessment of the plant variety protection regimes of more than 20 countries in Sub Saharan Africa reveals a considerable imbalance in the treatment given to breeders' interests and commercial farming as compared to the very moderate consideration given to farmers' rights and smallholder farming.[13]

The chapter begins with an explanation of the realities of subsistence agriculture, addressing its fundamentals and main actors with a focus on seed use. It will then explore the key regulatory instruments affecting seeds in the eastern and southern African regions, focusing on the rights of smallholder farmers to pursue their seed-handling practices. These ancient practices of saving and exchanging seeds are proven to be instrumental in their farming systems and agricultural production. Then, the chapter will investigate the extent to which some programs and initiatives have either failed or succeeded in accommodating farmers' seed-handling practices, focusing on both programs with a pan-African scope that are spearheaded by the African Union as well as those contained within a single nation. The final section will consider how to give priority to smallholder farmers' concerns within the mainstream regulatory frameworks of the region. We will see the impact of the formal support provided for famers' rights in the 2001 International Treaty on Plant

---

[11]*Id.* at 2.

[12]The International Treaty on Plant Genetic Resources for Food and Agriculture of the United Nations Food and Agricultural Organisation (FAO) provides for farmers' rights (article 9) to save, use, exchange and sell farm saved seeds and advocates the involvement of farmers in breeding undertakings as very critical in the conservation (article 6 of the plant treaty) of agro biodiversity which is very important to smallholder farmers.

[13]Mahop et al. (2013).

Genetic Resources for Food and Agriculture ("ITPGRFA")[14] of the Food and Agriculture Organisation of the United Nations. Although the current trend in the region is to adopt seeds-related regulations that are more supportive to commercial farming, there is a need to consider creating a legal space for smallholder farmers' interests in these regimes. Because these farmers are still playing an instrumental role in food production, their interests must be protected.

43.2 Understanding Subsistence Agriculture: Fundamentals and Actors

The vast majority of people living in rural Sub-Saharan Africa rely on agriculture for many reasons. As the main economic activity providing employment to the deprived people of the region,[15] it is critical for providing both income and food. Estimates suggest that three out of every four people among the 240 million people of SSA live in rural areas and depend on small-scale subsistence agriculture for their livelihood.[16] The focus in these parts is on smallholder and subsistence agriculture because it is best suited to these poor populations.

Smallholder and subsistence agriculture is often described in contrast to large-scale and intensive agriculture. The latter involves considerable financial investment, requires changes to the infrastructure, uses large farms, and engages skilled experts such as agronomic engineers, soil scientists, and plant breeders. Intensive agriculture is generally characterised by its use of technological inputs such as seeds that have been developed and multiplied by professional plant breeders. Generally speaking, breeders develop improved varieties that bear specific desirable traits. Such traits are identified prior to the process and include, but are not limited to, resilience to drought and excess water, resistance to pests and diseases, agronomic characteristics such as early maturity, and other nutritional or food-processing features for the food industry.[17] Intensive agriculture is also characterised by its considerable use of fertilizers in tandem with these improved seed varieties. An interconnected web of actors is responsible for the development of new plant varieties, including the production and quality control of seeds (sometimes through a certification scheme), and the multiplication, packaging and commercialisation of these seeds.[18] It is through this formal sector that farmers involved in commercial and intensive agriculture source their seeds. The varieties

---

[14]The international Treaty on Plant Genetic Resources for Food and Agriculture: http://www.planttreaty.org/.
[15]United Nations Development Programme (2012).
[16]*Id.* at 2.
[17]Vose (1983).
[18]Minot et al. (2007).

and seeds produced and exchanged in this context are subject to stringent regulatory schemes that include plant breeders' rights[19] and other seed regulations.[20] To harness the science and technology embedded in these inputs, effective irrigation techniques and cropping systems adapted to the specific features of the soil must be applied in the course of farming. Together, these parameters make intensive agriculture an expensive undertaking, generally well beyond the financial and technical reach of poor rural dwellers.

In contrast, subsistence farming is mainly practised by poor farmers on relatively small tracts of land.[21] About 80 % of farms in Sub-Saharan Africa are less than 2 ha, and with the rural population still expected to grow,[22] farm size is expected to shrink even further due to a lack of available land.[23] Smallholder farming is considered sustainable because it uses little to no chemical inputs, and these inputs, specifically fertilizers and pesticides, cause harmful effects on the environment.[24] Rather, in the place of chemical inputs, smallholder farmers tend to use organic manure they produce themselves on their own farms.[25] Lacking the financial means to install sophisticated irrigation systems, subsistence farmers lucky to have nearby water sources rely on very rudimentary methods like watering their farms using watering cans. Most, however, rely mainly on rainfall as their main source of water. Another sustainable approach to water management used by smallholder farmers involves protecting watersheds and reinforcing their capacity to hold water and bring it to those most in need.[26] These principal features of smallholder agriculture illuminate the constraints faced by farmers: access to land, access to credit, and access to inputs and output markets. Other notable constraints include the policy and regulatory constraints that dictate access to such facilities as agricultural extension advisory services.[27]

The informal seed system is the main channel through which seeds are supplied, accessed and exchanged among subsistence farmers.[28] A key feature is that smallholder farmers are both breeders and farmers. They tend to select the best grains

---

[19]Plant breeders' rights are the category of intellectual property right designed purposely for the protection of new varieties of plants and to entitle specific rights to breeders of such varieties. The rights of breeders are generally meant to prevent anyone from producing, selling, importing and exporting plants and seeds protected by plant breeders' rights.

[20]Broadly defined, seed regulations refer to the rules and procedures guiding the development and release of a new variety of plant, the production of seeds based on a newly developed variety, and the release and delivery of seeds including quality control measures.

[21]*Id.* at 2.

[22]*Id.* at 7.

[23]*Id.* at 2.

[24]ActionAid (2011).

[25]Svotwa et al. (2009).

[26]Munang and Andrews (2014).

[27]Selami et al. (2010).

[28]*Id.* at 19.

from their current harvest and save them for use as seeds in the following planting season. Season after season, traditional farmers can observe and select varieties that are the best adapted to the agro-ecological conditions in their area, are most resistant to pests and diseases prevalent in the region, and boast other useful traits such as grain colour. Traditional farmers are also known to have the ability to further improve and adapt seeds that were purchased in the commercial sector to their specific needs.[29] A core characteristic of small-scale farming is that farmers generally exchange seeds among themselves. Through such interactions, a transfer of knowledge also occurs.[30] Despite similarities to their commercial counterparts, seeds developed using these methods are largely unregulated and are not eligible for protection within the existing frameworks of plant variety protection laws. It is with this understanding that the next section will discuss the impact of regulatory frameworks.

43.3 Regional Regulatory Frameworks

43.3.1 Harmonisation of Plant Variety Protection (PVP) Regimes in the SADC Region

The Southern African Development Community ("SADC")[31] was formed in 1992 with the signing of the SADC Treaty. It aims to achieve economic growth, alleviate poverty, and enhance the quality of life of the peoples of Southern Africa through regional integration.[32]

The SADC Food, Agriculture and Natural Resources Directorate have been working on a Draft Protocol for the Protection of New Varieties of Plants in the SADC region, the latest version of which was circulated in November, 2012 (SADC 2012). The Draft Protocol is modelled after the 1991 UPOV (International Union for the Protection of New Varieties of Plants) Act, and will, if enacted, establishes a regional plant breeders' rights ("PBR") office that shall grant or reject a PBR application on behalf of all member states. Like UPOV'91, the Protocol applies to all genera and species of plants, providing 25 years of protection for trees and vines and 20 years for all other genera or species.[33] It also covers the same

---

[29] Santilli (2012).
[30] Swiderska et al. (2011).
[31] The 15 SADC members are Angola, Botswana, DR Congo, Lesotho, Madagascar, Malawi, Mauritius, Mozambique, Namibia, Seychelles, South Africa, Swaziland, Tanzania, Zambia and Zimbabwe: http://www.sadc.int/.
[32] Article 5, Treaty of the Southern African Development Community (SADC), Windhoek, 17 August 1992.
[33] Articles 3 and 25 of the Draft 2012 Protocol on PBR of SADC.

activities that require authorization of the right holder, and includes the UPOV 1991 provisions on Essentially Derived Varieties ("EDV") (UPOV 2009).[34]

The closest the Draft Protocol comes to accommodating smallholder farmers is by providing an important exception to the rights of plant breeders in the form of farmers' privileges. These are formulated in the Draft Protocol as "acts done by subsistence farmers for the use for propagating purposes, on their own holdings, the product of the harvest which they have obtained by planting, on their own holdings the protected variety..."[35] An optional farmers' privilege in UPOV 1991 that allows farmers on their own farm to save and replant seed of protected varieties "within reasonable limits and subject to the safeguarding of the legitimate interests of the breeder"[36] is thus included in the Draft Protocol, but is only applicable to subsistence farmers.

The Draft ARIPO Legal Framework for the Protection of New Varieties, (another regulatory framework which will be discussed in further detail later in the chapter), takes a different approach to farmers' privileges. The ARIPO Draft grants farmers the privilege of specific agricultural crops to be listed by the Administrative Council. It further states that the different levels of remuneration to be paid by small and large farmers shall be stipulated in the implementing regulations.[37]

Both the SADC and ARIPO drafts thus allow for very restrictive farmer privileges and are strongly criticized by civil society organizations in the region. These organisations point out that smallholder farmers in both the ARIPO and SADC countries heavily rely on saved seeds and seed exchanges with relatives and neighbours.[38] They find both draft legislations to be inflexible and restrictive, imposing a "one-size-fits-all" system that limits the ability of individual member countries to design a PBR system appropriate to their diverse agricultural needs and priorities, and to balance such a system with the protection of farmers' rights.[39]

As is noted by the organisations leading communications on these issues, the proposed PBR systems may have the opposite effect of what they are intending to accomplish. While they are expected to encourage plant breeding and facilitate agricultural development by allowing farmers to access "a wide range of improved varieties to contribute to the attainment of the regional goal of economic development and food security",[40] they will likely fall short; while these policies may be justified for commercial farming, there is a high potential of detrimental effects to

---

[34] Article 26 of the draft SADC protocol on PBR. According to the Union for the Protection of New Variety of Plants, EDV refer to varieties developed by a breeder and which is essentially derived from a protected variety, the production of which requires a repeated use of a protected variety.

[35] Article 27.d, the 1992 SADC Draft Protocol on PBR.

[36] Article 15.2 of the 1991 Act of UPOV.

[37] Article 22 of the ARIPO Draft Legal Framework on Plant Breeder Rights.

[38] African Center for Biosafety (2012).

[39] Civil society concerned with the Draft Protocol Draft Protocol for the Protection of New Varieties of Plants (Plant Breeders' Rights) in the Southern African Development Community Region (SADC), 2 April 2013. Accessible via: http://www.ip-watch.org/2013/04/05/african-regional-plant-variety-protection-draft-legislation-raises-protest/ (last visited 20 January 2014).

[40] Preamble of SADC Treaty 1992; Introduction of 2012 Draft ARIPO Legal Framework on PBR.

the many farmers that simply cannot afford to adopt formal seed system practices and high-input farming.

43.3.2 Harmonization of Seed Policies in the Common Market for Eastern and Southern Africa (COMESA) Region

The Common Market for Eastern and Southern Africa ("COMESA"), a regional trading bloc of nineteen countries, is currently spearheading a seed policy harmonisation process.[41] The core objective of COMESA is to create a common market that will enhance regional trade amongst member countries. Through COMESA, various cooperative programmes are being developed to enhance integration through the removal of all physical, technical, fiscal and monetary barriers to intra-regional trade and commercial exchanges. One such programme focuses on the harmonisation of seed trade regulation between member countries.

A critical piece in the history of the harmonisation process is the 2008 Declaration by COMESA Ministers of Agriculture under the theme, "Consolidating Regional Economic Integration through Value Addition, Trade and Food Security."[42] Under this declaration, member states committed themselves "to harmonizing within two years, seed trade regulations in the region and to finalize a regional protocol for the protection of new varieties of plants within the same period."[43] A specialized agency of COMESA, the Alliance for Commodity Trade in Eastern and Southern Africa ("ACTESA"), is spearheading the regional seed trade harmonisation efforts and has published a draft COMESA Seed Trade Harmonization Regulations, 2013 ("draft Seed Trade Regulations"). The core objective of the draft Seed Trade Regulations is to facilitate seed trade among all member states; as such, once adopted by the COMESA Council of Ministers, these regulations will be binding for all member states.[44] Secondary objectives of this draft instrument, such as the harmonisation of phytosanitary measures for seed in the region and the establishment of the COMESA seed certification and variety release system, are meant to support the seed trade objective.[45]

---

[41]The Treaty Establishing the Common Market for Eastern and Southern Africa (COMESA) was signed on 05 November 1993 in Kampala, Uganda and ratified on 08 December 1994 in Lilongwe, Malawi. COMESA is an international organization whose member states are Burundi, Comoros, DR Congo, Djibouti, Egypt, Eritrea, Ethiopia, Kenya, Libya, Seychelles, Madagascar, Malawi, Mauritius, Rwanda, Sudan, Swaziland, Uganda, Zambia and Zimbabwe. More details are available at: http://about.comesa.int/index.php?option=com_content&view=article&id=95&Itemid=117 (last visited on 22 January 2014).

[42]Victoria Declaration of the 5th meeting of the COMESA Ministers of Agriculture held in Victoria, Mahe, Seychelles, 14–15 March 2008.

[43]Ibid, paragraph 10.

[44]Article 10, 1993 Treaty establishing the Common Market for Eastern and Southern Africa.

[45]Rule 3, draft COMESA Seed Trade Harmonisation Regulations, 2013.

The COMESA seed certification system establishes four seed classes: pre-basic; basic; first generation certified; and second generation certified.[46] There are also colour labels for each class,[47] intended to aid in recognition of the different classes of seed trading within the COMESA market. A seed certification system is also established with the Council of Ministers, who are mandated to adopt specific rules dealing with field and laboratory certification standards, seed testing methodologies based on International Seed Testing Association (ISTA) rules, post-control tests, and accreditation of laboratories, among others.[48]

With respect to variety release, preliminary tests are required for Distinctiveness, Uniformity and Stability (DUS) and Value for Cultivation (VCU) or National Performance Trials (NPT).[49] DUS tests are required to be carried out in accordance with the UPOV guidelines. VCU tests are implemented based on performance data typically established through multi-locational testing to show that a new variety has value to be released for cultivation. National authorities bear the responsibility of ensuring that released varieties have passed the aforementioned tests before being entered into the COMESA Variety catalogue.[50] The 2013 Draft Seed Harmonisation Regulations of COMESA also have rules on quarantine and phytosanitary measures and standardised testing procedures for the control of pest and diseases[51] in relation to the import and export of seeds. These rules are largely based on the procedures established by the ISTA.[52]

The scope of the draft Seed Trade Regulations is twofold. In regards to certification standards, there are only 12 crops regulated for certification of basic and certified seed.[53] These are provided in Schedule D to the Regulations, and include beans, maize (open pollinated varieties and hybrids), rice, groundnut, cotton, wheat, sunflower, sorghum (open pollinated and hybrid), soybean, pearl millet, cassava and Irish potato. The second level relates to variety release and plant quarantine, which applies to all crops.

The draft Seed Trade Regulations has received several criticisms. The process through which these regulations have been developed has been seen as non-inclusive, as some farmers feel they have not been consulted.[54] To the extent that the draft regulations only facilitate certification and release of those varieties meeting the DUS criteria, these regulations could appear to discriminate against trade and dissemination among farmers themselves. The regulations have also been

---

[46]Ibid, Rule 14.
[47]Ibid, Rule 15.
[48]Ibid, Rule 13.
[49]Ibid, Rule 20.
[50]IbId.
[51]Ibid, Rule 40.
[52]Ibid, Chapter 5.
[53]Ibid, Rule 18.
[54]*Id.* at 40.

criticized for not providing safeguards for smallholder farmers saving their own seed or exchanging it with their neighbours, a common practice in these countries.[55] Lastly, while established as a multi-stakeholder institution, the COMESA seed committee does not actually include any farmers, despite the fact that it is meant to provide technical support.[56]

43.3.3 The African Regional Intellectual Property Organisation (ARIPO) Draft Legal Framework for the Protection of New Varieties of Plants

Discussed briefly above, the African Regional Intellectual Property Organisation ("ARIPO")[57] is the successor to the English Speaking African Regional Industrial Property Organisation ("ESARIPO"), created by the 1976 Lusaka Agreement. ARIPO administers the Harare, Banjul and Swakopmund Protocols that were adopted respectively in 1982, 1997 and 2010. The purpose of the Harare and the Banjul Protocols is mainly to streamline the processes of registration, filing, processing and granting of patents, utility model, industrial design and trademark applications. These two protocols are the two major regional intellectual property (IP) systems administered by ARIPO. The Swakopmund Protocol is the most recent, and it seeks to provide a regional system for protection of traditional knowledge and expressions of folklore. Looking at the possibility of linking these three regional IP protocols, ARIPO is discussing the regional framework for the protection of new varieties of plants from which a regional PVP protocol could originate.

Concerns about the ARIPO draft pertain to its development process and potential impact on farmers' rights and food security. With regards to the development process, civil society organisations are concerned that the instrument has been developed without consultation with member state actors; as such, they contend that the PVP legal framework does not reflect the realities of plant breeding and seed systems of member states. Rather, it is more a reflection of UPOV 1991, and therefore not supportive of farmers' rights and farmers seed systems in ARIPO member states.[58]

---

[55] See civil society & smallholder farmer statement at the awareness creation on COMESA seed trade harmonisation regulations for the COMESA region 27–28th march 2013, Lusaka, http://www.twnside.org.sg/title2/susagri/2013/susagri255.htm.

[56] Rule, 11 and 12 draft COMESA Seed Trade Harmonisation Regulations, 2013.

[57] ARIPO member states include: Botswana, Gambia, Ghana, Kenya, Lesotho, Liberia, Malawi, Mozambique, Namibia, Rwanda, Sierra Leone, Somalia, Sudan, Swaziland, Tanzania, Uganda, Zambia, Zimbabwe: http://www.aripo.org/.

[58] *Id.* at 59.

The draft ARIPO regional policy also proposes the development of agricultural innovation systems, taking into account the phenomenon of climate change, the development of the seed industry, the role of biotechnology in exploring and protecting agricultural genetic resources, and the establishment of an effective plant variety protection system at the national and regional levels.[59] With regard to this last point, the draft regional policy appears to have made a choice for an effective *sui generis* system for the protection of plant varieties based on the 1991 Act of the UPOV Convention (ARIPO 2012). The basis for this choice, according to ARIPO, is that adopting such a legal framework will facilitate ARIPO application to join UPOV as a full member in the future. Therefore, based in the UPOV 1991 Act, the draft legal framework has provisions for the protection of new plant varieties and measures for conducting examination of varieties. The exceptions to plant breeders' rights include a provision on the so-called "farmers' privilege," which also is attuned to the UPOV 1991 Act. However, article 22 of the ARIPO draft legal framework narrows the extent of farmers' privilege to a specific list of crops that shall be defined by the administrative council of ARIPO but shall not include fruits, ornamentals, vegetables and forest trees. Thus, while the ARIPO draft does not embrace farmers' rights specifically, (for example, the right to exchange and sell seeds obtained from protected varieties amongst themselves), even the limited privilege that is provided by the draft framework is contained to a specific category of crops determined by the administrative council.

The road map for the continued development of this framework will take into account the concerns of CSOs regarding consultations with member states as it plans meetings from January to August, 2013. By November 2013, the final text and a decision on possible diplomatic conference will hopefully be adopted by the 14th session of the ARIPO Council of Ministers. The road map proposes the formulation of the legal framework into the draft between December 2013 and January 2014.[60] There is, however, no indication that the road map defined within ARIPO is being implemented. The outcome document of the fourteenth session of the ARIPO council of ministers, held in Kampala, Uganda on 28–29 November 2013, does not specifically mention any amendment to the farmers' privilege provisions that would appear to be responsive or sympathetic to the concerns of CSOs (ARIPO 2013). As such, it is not clear whether or not they actually adopted the draft legal framework.[61]

In light of the lack of proper incorporation of farmers' rights in seeds-related regulatory frameworks, the following section explores how broader programmes and initiatives in Africa have attempted to accommodate the interests of smallholder farmers.

---

[59]*Id.* at 59.
[60]Kabare (2013).
[61]*Id.* at 59.

43.4 Africa-Wide Programmes and National Initiatives: Impacts on Smallholder Farmers

Smallholder farmers' interests and practices in Sub-Saharan Africa are supported by various programmes and initiatives at both the continental and national level. The effectiveness of any support mechanism depends on how it addresses the smallholder farming issues described in the previous section.

43.4.1 The Comprehensive Africa Agriculture Development Programme (CAADP) and African Seed and Biotechnology Programme (ASBP)

At the continental level, the African Union adopted the Comprehensive Africa Agriculture Development Programme ("CAADP") in July 2003, whose principle aim is to eliminate hunger and reduce poverty through the development of the agricultural sector in Africa. The programme's approach is to increase investment on four mutually-reinforcing pillars, which include: (1) extending the area under sustainable land management and reliable water control systems; (2) improving rural infrastructure and trade-related capacities for market access; (3) increasing food supply and reducing hunger; and (4) promoting agricultural research and technology dissemination and adoption (NEPAD 2003). The Comprehensive Africa Agriculture Development Programme is Africa's agricultural project and vision of the New Partnership for Africa's Development (NEPAD). In setting up the CAADP, Africa is working toward the Millennium Development Goals (MDGs) agreed upon in 2000 by the international community. The CAADP particularly addresses Goal 1: the eradication of extreme poverty and hunger.[62] Achieving the four pillars identified by CAADP should assist NEPAD in the pursuit of its overall vision for Africa's agriculture. Specifically, by 2015, the continent of Africa should:

- Attain food security (availability, affordability, and accessibility to adequate food and nutrition);
- Improve agricultural productivity to attain an average annual growth rate of 6 %; with particular attention to small-scale farmers, especially women;
- Develop dynamic agricultural markets between nations and regions;
- Integrate farmers into the market economy, including improved access to agricultural export markets;
- Achieve a more equitable distribution of wealth;
- Become a strategic player in agricultural science and technology development; and,

---

[62]MDG 1: http://www.undp.org/content/undp/en/home/mdgoverview/mdg_goals/mdg1/.

- Practice environmentally-sound production methods and develop a culture of sustainable management of the natural resource base, including biological resources for food and agriculture.[63]

A noteworthy element of the NEPAD vision is the 6 % increase in productivity, which specifically notes the contribution of small-scale farmers and women. What could be problematic is that the CAADP does not appear inclined to supporting their farming practices, especially as they relate to accessing and utilising seeds at the local level. In fact, an in-depth assessment of the CAADP does not reveal any mention of support to farmers' rights to save, exchange, re-sow, or sell seeds, including those derived from the cultivation of proprietary seeds. Instead, the CAADP approach is designed to build agricultural productivity and economic growth through better access to agricultural inputs.[64]

Undoubtedly, small-scale farmers' access to agricultural inputs such as improved and protected plant varieties is an important factor in reaching improved agricultural productivity. However, one could argue that the CAADP should consider the constraints that smallholder farmers face and adapt to their specific agro-ecological context and farming practices. NGOs have voiced their concerns over the impacts of PVP and seed laws on food security based on the manner in which these laws affect farmers' rights and farming systems (ACB publications on COMESA AND SADC PVP ARIPO). Arguably, considering its weakness on farmers' rights and emphasis on modernizing agricultural practices, the CAADP is overlooking the constraints that smallholder farmers face and only advocating for intensive agriculture.

Charged with implementing the fourth pillar of the CAADP, the Forum for Agricultural Research in Africa ("FARA") developed the Framework for Africa's Agricultural Productivity (FAAP)[65] as a tool for stakeholders working on agricultural research and development. One of the identified objectives of the fourth pillar is '*to empower farmers, livestock producers and their organisations by investing in their acquisition and absorption of the outputs of research and innovations in the agricultural sector.*' The way FAAP intends to achieve this is by ensuring that there is an integrated approach that brings together researchers, extension services, smallholder farmers, pastoralists, the private sector and NGOs.[66] While there is no denying that agricultural technology is one important long-term answer to low agricultural productivity, it bears of the risks of not always answering the critical needs of farmers as they relate to their specific conditions. Smallholder farmers are largely known to cultivate a variety of crops adapted to their agro-ecological conditions; unless the integrated approach contemplated under FAAP involves cooperation between smallholder farmers and breeders in defining the objectives

---

[63]*Id.* at 4.

[64]*Id.* at 4.

[65]Framework for Africa's Agricultural Productivity (FAAP): http://www.fara-africa.org/media/uploads/File/FARA%20Publications/FAAP_English.pdf.

[66]*Id.* at 67.

of breeding programmes and their crop selections, there is a risk of farmers being forced to adopt varieties of crops that are not necessarily of interest or utility to them. Recognition of the importance of smallholder farmers' practices in food production should thus be central to the integrated approach. The manner in which smallholder farmers access and utilise seeds is a critical area for policy support, which requires targeted investment in its own right.

Another important initiative at the continental level that came out of the CAADP is the African Seed and Biotechnology Programme ("ASBP"). The ASBP is coordinated by the African Union, and was adopted during the eighth ordinary session of the African Union Assembly in January 2007 in Addis Ababa, Ethiopia. Its primary pursuit is to strengthen the development of the seed sector in Africa, taking advantage of new developments in plant breeding. The overall goals of the programme are the realisation of food security in Africa, improved nutrition and poverty alleviation through the establishment of effective and efficient seed systems, and enhanced application of biotechnologies and methodologies within the seed sector. Thus, the programme shows a clear signal that its emphasis is on the inclusion of biotechnology in strengthening the seed sector in Africa. One can assume that the application of biotechnologies and other modern methodologies will be aimed at all facets and actors involved in the development of the seed sector, including small-scale farmers. The ASBP does, however, have one objective designed specifically for smallholder farmers, which is to strengthen the connection between the informal and formal seed sectors to better respond to farmers' needs.

The ASBP programme comprises 20 interrelated components, which are sub-divided into achievable outputs at the continental level, regional level, and national level. One of the national level components (component 2), which focuses on policy-setting in relation to the seed sector, sets out *"to improve seed quality and supply of crops of national importance including minor crops"* (see footnote 37). Output 1 for this component proposes that national seed policies, systems and activities should be analysed, reviewed and endorsed. To achieve this output, activities will ensure the establishment of integrated seed development policies, spanning germplasm conservation, characterization, utilization and improvement, application of biotechnologies, variety release and seed production and distribution. It is our view that, in the interest of smallholder farmers, these activities should offer the platform for stakeholders to take into consideration the manner in which this category of farmers handles seeds. Seed-handling includes, but is not limited to, systems of production and improvement of seed cultivation, taking into account the utilization of proprietary seeds and the knowledge and practices to improve or adapt them at the local level. The proposed reviews of seed laws and policies and the update to the national seed compendium should therefore provide recognition of farmer-produced seeds in the interest of agriculture at the national level. But, beyond these pan African programmes/initiatives that have minimal direct impact on promoting smallholder farming practices, there are other nationally- and locally-adapted initiatives which provide direct support to smallholder farmers, as discussed hereunder.

43.4.2 Rwandan Initiatives

In addition to continental-level programmes there are several national endeavours supporting smallholder farmers that have resulted in increased agricultural productivity and availability of food both in terms of quantity and diversity. One prominent example is Rwanda, where policy makers identified smallholder farmers as the most practical target for addressing food shortage and tackling poverty reduction.[67] The Rwandan government instituted a number of changes, most notably the increase of budgetary allocation to agriculture, which moved from 4.2 % in 2008 to about 11 % in 2011.[68] This budget increase complies with the recommendation set out in the CAADP, to which Rwanda was the first country to sign up in March 2007. By undertaking these efforts, the government of Rwanda is demonstrating a desire for the intensification of agriculture, shifting from subsistence-based farming to productive, high value and market-oriented farming (MINAGRI 2009). However, two major policies have been developed to these ends which still have elements of support to smallholder farmers: (1) the Vision 2020 and the Economic Development and Poverty Reduction Strategy ("EDPRS") 2007–2012, and; (2) the Strategic Plan for the Transformation of Agriculture ("PSTA II") 2008–2012 and the Crop Intensification Programme ("CIP") 2008–2012. As the principal leader in the implementation of these policies, the ministry of agriculture prioritises investments that can enhance access to, and use by, smallholder farmers of such inputs as fertilisers and 'improved' seeds. These inputs are, it must be stressed, largely imported from the neighbouring Kenya and Tanzania in the case of maize and wheat, while the National Institute of Agricultural Research ("ISAR")[69] provides some pest resistant varieties of cassava.

As a result of these budget increases and implementation of the above-mentioned policies, there has been a purported increase in the production/outputs in key crops from 2007 to 2010, such as maize increasing by 322 %, wheat by 213 %, and cassava by 206 %. Interestingly, these increases in outputs are not based on increased land use, but in the increased yields.[70] As attractive as these results are, one key question remains unanswered regarding smallholder farmers' rights to recycle the improved and proprietary seeds that they use. Rwanda at the moment lacks a plant variety protection law (Mahop et al. 2013, 2014), so perhaps the issue of infringement of PBR is of less concern to smallholder farmers in this country

---

[67]Willoughby and Forsythe (2011).

[68]World Bank (2011).

[69]*Id.* at 69.

[70]It must be noted that access to inputs such as improved seeds and fertilisers were not the only tenets of the policies developed and implemented in support to smallholder farmers. Other critical tenets of the broader goal intensification of agriculture included training in sustainable production systems such as irrigation, the professionalization of smallholder farmers through provision of extension advice in cropping systems and the promotion of commodity chains and agribusiness development in areas of post-harvest handling and processing.

than it is in neighbouring countries such as Kenya and Tanzania where PBR regimes are in force. Considering that an LDC like Rwanda is currently not under any legal obligation to promulgate a PBRs law at the domestic level,[71] the policy changes are arguably beneficial to Rwanda's smallholder farmers because they can maintain their age-old practices of seeds saving, exchanging and selling. One element that has not been tested is the extent to which the withdrawal or reduction in government support to farmers' access to agricultural inputs will affect productivity. Farmers' rights to domesticate improved seeds and better adapt them to their agro-ecological conditions is likely critical to sustaining these increased yields.

43.5 Giving Priority to Seed-Related Concerns of Subsistence Farmers in Ongoing Processes

There is a clear trend of plant variety protection regimes being developed alongside market-oriented seed regulations at the domestic and regional levels in Sub-Saharan Africa.[72] However, opinions about their suitability for the region are divided. On the one hand, UPOV-based sui generis plant variety protection regimes and seed regulations—both of which are based on international standards[73] will contribute to boosting national plant breeding activities for crops of national or regional importance and will assist farmers in accessing new and improved varieties from other countries, thereby contributing to domestic diversity.[74] Additionally, there is an opportunity for external investments in the seed sector, given the increased assurance that investors' interests would be protected.[75] Therefore, some believe that countries and regions that uphold international standards on seeds-related policies are more likely to increase international trade through commercialisation of both

---

[71]On 11 June 2013, The TRIPS Council, through its decision IP/C/64, extended the general TRIPS compliance transition period for LDC Members for all obligations under the TRIPS Agreement, other than Articles 3, 4 and 5, until 1 July 2021 or until such date on which a Member ceases to be an LDC, whichever date is earlier. This means that LDCs are not obliged to develop national laws for the protection of new variety of plant and the rights of breeders of such varieties. Least developed countries still in considerable needs for technical and financial capacities to fully implement TRIPS have been calling for such extension as evidenced through the various communications from Uganda and Tanzania at the last symposium on LDC priority needs for technical and financial cooperation that was held at the WTO in Geneva from 31st October to 2nd November 2012.

[72]*Id.* at 73.

[73]By international standards here, we refer to the seed laws that comply with the standards of seed testing and release developed by the International Seed Testing Association (ISTA). ISTA develop and publish standard procedures for field seed testing that should be applied by member states. The principal mission of ISTA is to have uniform seed testing evaluation procedures worldwide, http://www.seedtest.org/en/home.html.

[74]Alliance for Commodity Trade in Eastern and Southern Africa (2013).

[75]*Id.* at 77.

seeds and agricultural products, thus bringing in the much needed export revenues for the countries' or regions' economies.

On the other side of the argument, there are those who question the rationale for developing and implementing stronger PVP and seed regulations in sub Saharan countries where agriculture is currently dominated by small-scale farmers and undertaken largely for subsistence purposes.[76] Those opposed to stronger laws at the domestic or regional level argue that there is little evidence that stronger regimes have contributed to more plant breeding activities. Where there is evidence of any impact, it has not been on food crops of national importance, but largely on crops that are commercially valuable to seed companies.[77] The passage of stronger plant variety protection and seed laws could lead to breeding research priorities being focused on a narrow set of homogenous, commercially valuable crops for the industry, much to the detriment of the diversity of food security crops that are used by small-scale farmers.[78] Back in 1999, an IPGRI[79] report questioning the move by many developing and least developed countries to implement Article 27.3.b of TRIPS to promulgate UPOV-based *sui generis* plant variety protection laws regardless of their UPOV membership, stated: *'Countries whose agricultural economy is mainly geared towards domestic markets and which depend largely on traditional varieties cultivated by small-scale and subsistence farmers will have less to gain from the introduction of stronger PVP regimes.*[80] This is because in the context of subsistence agriculture, small-scale farmers use traditional varieties that contain a lot of genetic diversity. Introducing stronger PVP will likely affect this agro biodiversity leading to the exacerbation of food insecurity in these countries. As such, NGOs and other actors view smallholder and subsistence farmers' rights, as outlined by the 2001 FAO International Treaty on Plant Genetic Resources for Food and Agriculture, as a serious factor in the pursuit of food security and poverty reduction strategies.

It is absolutely reasonable that sub Saharan countries would wish to take advantage of recent developments in biotechnology in designing programmes after the CAADP and ASBP models. Appropriate laws and policies are required, however, to regulate the development and commercialisation of improved varieties bearing such features as higher yields, pest resistance, lesser vulnerability to climatic changes, and industry requirements for nutrition and food processing. Seeds-related regulatory regimes must be able to protect the rights and rewards

---

[76]Singh (2002).

[77]Tripp et al. (2007).

[78]Dutfield (2008).

[79]The International Plant Genetic Resources Institute (IPGRI) is the successor to the International Board for Plant Genetic Resources (IBPGR) which was established in 1974 by the Consultative Group on International Agricultural Research (CGIAR). IBPGR became IPGRI in 1991. In 1994, IPGRI took over the governance of the International Network for the Improvement of Banana and Plantain (INIBAP). In 2006 IPGRI and INIBAP were merged into a single organization called Bioversity International.

[80]International Plant Genetic Resources Institute (1999).

of those actors who utilise their genius to develop plant varieties with these features.

Because stronger PVP and seed regulations are focused on commercial farming, they are cause for concern in the immediate to medium term because they do not take into account the interests of small-scale farmers whose subsistence farming is the current response to food needs in Sub-Saharan Africa. Some argue that smallholder farmers will not be affected by the Draft ARIPO plant variety protection legal framework and the COMESA harmonised seed regulations as long as they continue to use their traditional varieties in the context of their agricultural practices. However, this is not entirely true. Should small-scale farmers want to commercialise their traditional varieties in the formal seed markets, they will not be able to unless otherwise permitted by the laws under strict conditions. For example, in Tanzania, the Seed Act of 2003 has regulated the production and local commercialisation of Quality Declared Seeds ("QDS"), which are seeds produced by and for smallholder farmers who do not have access (physical and financial) to formal/certified seeds.[81] In any case, small-scale and subsistence farmers who maintain using traditional varieties are generally not eligible for government support in terms of access to credit, agricultural inputs, or advisory support provided by the extension services.

In sum, to sustain or bolster the contribution of subsistence agriculture to food production in Sub-Saharan Africa, this chapter suggests that on-going initiatives and processes at the continental, regional and national levels should refocus to give priority to the concerns of smallholder farmers. This can be chiefly achieved by pursuing full implementation of farmers' rights in the plant variety protection and seed regulations of the Eastern and Southern African regions. The realisation of farmers' rights is a core tenet of the 2001 International Treaty on Plant Genetic Resources ("ITPGR") for Food and Agriculture to which most Sub-Saharan African countries are members. Article 9.1 of the International Treaty recognises the enormous contribution that indigenous communities and farmers of all regions of the world have made and will continue to make to the conservation and development of Plant Genetic Resources for Food and Agriculture ("PGRFA"). Furthermore, protection of traditional knowledge relevant to plant genetic resources is part of the fulfilment of Farmers' Rights,[82] and the rights of farmers to save, use, exchange and sell farm-saved seed and propagating material are not limited by

---

[81]The production and commercialisation of QDS does not follow the stricter quality control and release procedures that are designed for certified seeds meaning that, because they are produced by registered smallholder farmers, the fields were QDS are produced are not subject to the same number of visits of the inspectors employed by Tanzanian Official Seed Certification Institute (TOSCI). Seed processing procedures, including storage and labelling for the purpose of commercialisation are less stringent for QDS as compared to certified seeds.

[82]Article 9.2 of the International Treaty on Plant Genetic Resources for Food and Agriculture 2001.

the Treaty.[83] As encapsulated in article 9.1 of the International Treaty, farmers' rights are more elaborate than the very narrow concept of farmers' privilege which is provided by the 1991 Act of UPOV, designed purposely to promote and protect the rights of breeders, not those of smallholder farmers. The common denominator of both farmers' rights and farmers' privilege is that they are implemented at the national and/or regional level through domestic and regional policy making, unlike the plant breeders' rights, which have an international dimension promoted by the UPOV convention.

At the international level, the governing body of the international treaty is indeed encouraging national and regional agricultural policy processes for the realisation of farmers' rights. In 2009, the Governing Body initiated activities aimed at collecting views on the implementation of Farmers' Rights.[84] In its resolution 6/2011, the Governing Body encouraged Parties to submit further views, experiences and best practices, and requested the Secretariat to the Seeds Treaty to compile these views and disseminate the same to Parties.[85] African countries have been at the forefront of not only submitting views on farmers' rights, but also in encouraging that farmers' rights become a permanent agenda item for discussion by the Governing Body. Work on the farmers' rights provisions of the treaty has continued, leading to the recent adoption by the Governing Body in September 2013 of resolution 8/2013 on the Implementation of Farmers' Rights by the Fifth session of the Governing Body of the Treaty. This Resolution called on contracting parties to engage farmers' organizations and relevant stakeholders in matters related to the conservation and sustainable use of plant genetic resources for food and agriculture, and will consider their contributions in awareness-raising and capacity-building efforts.[86] Furthermore, the resolution invites parties—previously involved or otherwise—to consider reviewing and, if necessary, adjusting its national measures affecting Farmers' Rights.[87] Reviewing and/or designing national regulatory measures will provide legal safeguards to the seed handling practices that are deeply rooted in smallholder farming systems and are very instrumental in smallholder farmers' contribution to food production and food security in Sub-Saharan Africa. The realisation of these rights has direct bearing on another critical pillar of smallholder farming, which is the maintenance of agro-biodiversity. Agro-biodiversity can be achieved through conservation and sustainable use of plant genetic resources for food and agriculture by smallholder farmers.[88] Conflict lies in the fact that the ultimate impact of seed regulations and

---

[83] Article 9.3 of the International Treaty on Plant Genetic Resources for Food and Agriculture 2001.

[84] Resolution 6/2009 and subsequent reports on Global Consultations on Farmers' right in 2010, IT/GB-4/11/Circ.1.

[85] IT/GB-4/11/Report.

[86] Clause 4 of Resolution 8/2013 of the Fifth Session of the Governing Body of the International Treaty on Plant Genetic Resources for Food and Agriculture, Muscat, Oman, 24–28 September 2013.

[87] Clause 6, Ibid.

[88] Santilli (2012).

plant variety protection laws is the establishment of uniformity and homogeneity across agricultural systems, which would lead to the displacement of agro biodiversity. Homogenous or uniform crops are generally criticised for their vulnerability to pest and diseases. For smallholder farmers, agro biodiversity is a far-stronger shield to such vulnerability. The role that subsistence farming plays in food supply in Sub-Saharan Africa should be a strong incentive to frame the conservation of agro-biodiversity as a key tenet of plant variety protection and seed laws in the region. This would greatly assist countries fulfilling their obligations under the ITPGR.

Additionally, some feel that the current focus on phenotypic aspects of new plant protection, which is based on external features of protectable varieties and related crops, is outdated.[89] They argue that plant breeding activities are increasingly—and perhaps irreversibly—moving towards a focus on the genotypic features of plants, given the current progress in molecular biology and genetic engineering. In this regard, it has been proposed that new varieties of plants be rather considered as datasets that the breeder can manipulate in order to express a specific feature in the crops, such as a stress tolerant feature. If new varieties of plants are considered to be datasets that can be modified through biotechnology-based breeding techniques, competition laws—rather than *sui generis* plant variety protection laws—may be the best tool to ensure their legal protection. In practice, despite the advances in biotechnology-based breeding techniques, conventional methods do remain at the core of plant breeding activities. Biotechnology-based breeding is used as complementary to, not instead of, conventional breeding.[90] Taking advantage of the possibilities offered by molecular biology and genetic engineering in the manipulation of the genetic make-up of plant varieties, it is understandably possible in the context of biotechnology based breeding activities to minimise homogeneity and uniformity and maintain the crop diversity that is central to subsistence farmers' agricultural systems. Ensuring access by smallholder farmers to these biotechnology-engineered diverse and proprietary new varieties of plants is an issue for policy makers to clarify at the domestic level.

One approach that works towards the conservation and sustainable use of agro biodiversity also has undoubtable implications on the realisation of farmer's rights. This approach is the creation of a legal space for participatory breeding in the regulatory instruments pertaining to seeds developed in the region. Participatory breeding is noticeably absent in the regional seed harmonisation regime underway in the COMESA; therefore, there is an opportunity for revision and reorientation to include it. Indeed, article 6 of the International Treaty on Plant Genetic Resources for Food and Agriculture, which deals with sustainable use of plant genetic resources, encapsulates conservation and participatory breeding under its coverage. To achieve conservation of agro-biodiversity, article 6.2.b provides for the strengthening of research that enhances and conserves biological diversity by maximizing intra- and inter-specific variation for the benefit of farmers. The article

---

[89]Blakeney (2011).
[90]*Id.* at 92.

specifically mentioning those farmers who generate and use their own seed varieties and apply ecological principles in maintaining soil fertility and combating diseases, weeds and pests. This special mention implies a need for strong involvement of such farmers in the breeding process. Thus, article 6.2.c of the International Treaty calls for promoting plant breeding efforts that strengthen the capacity to develop varieties particularly adapted to social, economic and ecological conditions in all areas, and specifically calls for the participation of farmers—particularly those in developing countries.

According to the International Center for Tropical Agriculture ("CIAT"), a research institution under the Consultative Group on International Agricultural Research ("CGIAR"), participatory breeding is the systematic and regular involvement of farmers as decision makers in all stages of the breeding program. From the perspective of CIAT, farmer involvement in a breeding program can take many forms, including: defining breeding goals and priorities; selecting or providing germplasm sources; hosting trials; selecting lines for further crossing; evaluating results; planning for the following year's activities; suggesting methodological changes; multiplying seeds; and, commercialising seeds of selected lines.[91] Participatory breeding differs from conventional breeding in that farmers' participation is very limited in conventional breeding. If it occurs at all, it may be restricted to the evaluation of materials produced by breeders. In participatory breeding, farmers apply their traditional knowledge and skills to crop selection. Ultimately this process provides farmers with the best-adapted seeds for their own agro-ecological and accounts for their individual and communal nutritional needs. Because smallholder farmers are involved at all stages of the process—including the beginning when program objectives and priorities are defined—they generally select a wide variety of crops and avoid the narrow, commercially-focused approach of conventional breeding.

One major constraint to participatory breeding is its financing. The private sector generally invests in breeding programs for commercially valuable crops, and can claim intellectual property protection over new varieties in order to recoup the investment and make a profit. Because of this interest, it is important that participatory breeding programs are financed and implemented by the public sector. One objective of the publicly-funded participatory program can be to ensure that smallholder farmers use and exchange the seeds developed through the program freely, with no restriction on the extent to which they can exchange, plant or sell the product of their harvest as seeds to fellow farmers. This objective is achievable if there is a restriction on the acquisition of plant breeders' rights by the publicly-funded research center, and there are safeguards for the improved varieties to remain in the public domain.[92]

---

[91]*Id.* at 91.
[92]*Id.* at 92.

43.6 Conclusions

The seed and plant variety protection regulations currently being developed in the Eastern and Southern African regions are pursuing an agenda of modernisation, resulting in a shift from subsistence to commercial agriculture. On-going regulatory processes are driven by continent-wide programmes initiated by the African Union like the CAADP and the ASBP. Programs such as these have a strong focus on boosting agricultural productivity through research and development and the dissemination and adoption of agricultural technologies by farmers, including smallholder farmers. These approaches are not necessarily or inherently bad for Africa, especially in the medium- to long-term. As smallholder farmers acquire the financial and technical capacity to integrate new innovations, there could be some benefit. However, in the short term, these regulatory frameworks and programmes at the continental and regional levels are not being observed to properly uphold farmers' rights. Furthermore, their lack of integration of sustainable objectives of agro biodiversity can seriously affect smallholder farmers' contributions to food production and supply and thus have a profound impact on food security. Despite the preferential treatment that the FAO Plant Treaty has over other existing international treaties (such as the WTO TRIPS agreement and the UPOV) with regards to the protection of farmers' rights, on-going seeds-related regulatory processes do not encapsulate the Plant Treaty's provisions on farmers' rights in the same manner that they provide for strong protection of breeder's rights. A number of research questions therefore arise which are worth exploring, including: how can regulatory processes be rebalanced to emphasise FAO Plant Treaty farmer's rights and agro-biodiversity conservation principles? Are there any major constraints countries in these regions will face in their efforts to crystallise such plant treaty principles in their regional or domestic seeds-related regulatory frameworks? If any constraints are envisaged or encountered in the course of the rebalancing process, what would be the best workable approach to overcome them? For Sub-Saharan Africa, where smallholder farmers are still playing a very instrumental role in food production, this chapter advocates the development of seeds-related laws and policies that accommodate farmers' rights and include conservation and sustainable use objectives of agro-biodiversity. This approach is recommended in order to avoid a disruption in agricultural production that would affect the food supply at the local, community, national and regional levels.

References

ActionAid (2011) Smallholder-led Sustainable Agriculture. ActionAid International Briefing, 2011

African Center for Biosafety (2013) Submission on the COMESA Draft Harmonisation of Seed Trade Regulations to Dr John Makuka of the Alliance for Commodity Trade in Eastern and

Southern Africa (ACTESA). http://www.acbio.org.za/images/stories/dmdocuments/ACB-COMESA-seed-Trade-Regs-submission.pdf. Accessed 23 Jan 2014

African Center for Biosafety (2012) Harmonisation of Africa's Seed Laws: a Recipe for Disaster, Players Motives and Dynamics. http://www.acbio.org.za/index.php/media/64-media-releases/409-aripos-pvp-law-undermines-farmers-rights-a-food-security-in-africa. Accessed 22 Jan 2014

African Regional Intellectual Property Organisation (2012) The Draft ARIPO Legal Framework for the Protection of New Varieties of Plants. http://www.aripo.org/index.php/resources/laws-and-protocols/finish/13-laws-protocols/77-draft-aripo-legal-framework-for-the-protection-of-new-varieties-of-plants. Accessed 24 Jan 2014

African Regional Intellectual Property Organisation (2013) Consideration of the Revised ARIPO Legal Framework for Plant Variety Protection. Council of Ministers Fourteenth Session, Kampala, Uganda, 28–29 November 2013

Alliance for a Green Revolution in Africa (2013) Africa Agriculture Status Report: Focus on Staple Crops, Nairobi

Alliance for Commodity Trade in Eastern and Southern Africa (2013) Status of the COMESA Seed Trade Harmonisation Regulation and its Implementation. Presented in Port Louis, Mauritius by the Alliance for Commodity Trade in Eastern and Southern Africa (ACTESA), March 2013

Blakeney M (2011) Trends in Intellectual Property Rights Relating to Genetic Resources for Food and Agriculture. Commission on Genetic Resources for Food and Agriculture, Food and Agriculture Organisation of the United Nations, Background Study Paper No 58, July 2011

Bremner J (2012) Population and Food Security: Africa's Challenge. Population Reference Bureau, Policy Brief, February 2012

Dutfield G (2008) Turning Plant Varieties into Intellectual Property: the UPOV Convention. The Future Control of Food: A Guide to International Negotiations and Rules on Intellectual Property, Biodiversity and Food Security, London, pp 27–47

Ecker O, Breisinger C (2012) The Food Security System: a conceptual framework. Discussion Paper No 01166, International Food Policy Research Institute (IFPRI), Washington

Food and Agriculture Organisation of the United Nations (1996) The State of Food and Agriculture

Food and Agriculture Organisation of the United Nations (2009) The special challenge for Sub-Saharan Africa. Presented at the high level expert forum 'How to Feed the World 2050,' Food and Agriculture Organisation of the United Nations, Rome, October 2009

Hansen MK (2000) Genetic Engineering is not an Extension of Conventional Plant Breeding: How Genetic Engineering differs from Conventional Plant Breeding, hybridisation, wide Crosses and horizontal Gene Transfer Consumer Policy Institute/Consumers Union

International Plant Genetic Resources Institute (1999) The Agreement on Trade-Related Aspects of Intellectual Property Rights - A Decision Check List, Rome, Italy

International Union for the Protection of New Varieties of Plants (2009) Explanatory notes on essentially derived varieties under the 1991 Act of the UPOV convention. Geneva, UPOV/EXN/EDV/1

Kabare JN (2013) Latest developments in the African Regional Intellectual Property Organisation (ARIPO): Draft ARIPO regional framework on plant variety protection. Presented at the AFSTA annual congress, Mauritius, 3–6 March 2013

Livingston G, Schonberger S, Delaney S (2011) Sub-Saharan Africa: the state of smallholders in agriculture. Conference on new directions for smallholder agriculture. International Fund for Agriculture Development (IFAD), Rome, 24–25 January 2011

Mahop MT, De Jonge B, Munyi P (2013) Plant variety protection regimes and seed regulations in Sub-Saharan Africa: current trends and implications. BioSci Law Rev 13(3):101–111

Mahop MT, Tejan-Cole A, Shikoli AM (2014) The protection of new varieties of plants in Sub-Saharan Africa: review of existing intellectual property regimes. African Agricultural Technology Foundation (AATF), Nairobi

Ministry of Agriculture and Animal Resources of Rwanda (2009) Strategic Plan for the Transformation of Agriculture in Rwanda – Phase II (PSTA-II), Rwanda, February 2009

Minot M, Smale M, Eicher C, Jayne T, Kling J, Horna D, Myers R (2007) Seed development programmes in Sub-Saharan Africa: a review of experiences. Rockefeller Foundation, Nairobi

Munang R, Andrews J (2014) Despite Climate Change Africa can feed Africa. Africa Renewal Special Edition on Agriculture, 6. http://www.un.org/africarenewal/magazine/special-edition-agriculture-2014/despite-climate-change-africa-can-feed-africa. Accessed 27 Feb 2014

New Partnership for Africa's Development (2003) Comprehensive Africa Agriculture Development Programme (CAADP). July 2003

Poonia TC (2013) History of seed production and its key issues. Int J Food Agric Vet Sci 3(1):148–154

Santilli J (2012) Agrobiodiversity and the law: regulating genetic resources, food security and cultural diversity. Earthscan from Routledge, New York and Oxford

Selami A, Kamara AB, Brixiova Z (2010) Smallholder Agriculture in East Africa: trends, constraints and opportunities. African Development Bank Group Working Paper 105, April 2010

Singh H (2002) Emerging plant variety legislations and their implications for developing countries: experiences from India and Africa. Paper presented in the National Conference on TRIPS – Next Agenda for Developing Countries, Shyamprasad Institute for Social Service, Hyderabad, 11–12 October 2002

Southern African Development Community (2012) Draft Protocol for the Protection of New Varieties of Plants (Plant Breeders' Rights) in the Southern African Development Community (SADC). http://www.ip-watch.org/2013/04/05/african-regional-plant-variety-protection-draft-legislation-raises-protest/. Accessed 10 Feb 2014

Stafford W (2009) Marker assisted selection (MAS): key issues for Africa. African Centre for Biosafety (ACB), Johannesburg

Svotwa E, Baipai R, Jiyane J (2009) Organic farming in the smallholder farming sector in Zimbabwe. J Org Syst 4(1):8–14

Swiderska K, Song Y, Li J, Reid H, Mutta D (2011) Adapting agriculture with traditional knowledge. International Institute for Environment and Development (IIED) Briefing, London, October 2011

Tripp R, Louwaars N, Eaton D (2007) Plant variety protection in developing countries. A report from the field. Food Policy 32:354–371

United Nations Development Programme (2012) Africa human development report: towards a food secure future. Regional Bureau for Africa, New York

Vose PB (1983) Rationale of selection for specific nutritional characters in crop improvement with *Phaseolus vulgaris*, L as a case study. Plant Soil 22(2–3):351–364

Willoughby R, Forsythe L (2011) Farming for impact: a case study for smallholder agriculture in Rwanda. Concern Worldwide/University of Greenwich Natural Resources Institute, London

World Bank (2011) Rwanda economic outlook: seeds of change. Kigali

Chapter 44
Textbox: Women in Agriculture and Labor Law: Cameroon's "Three C's" Legal Dilemma

Wele Elangwe

Abstract This textbox discusses the impact of the three c's in the empowerment of women in the agricultural sector of Cameroon's economy. The 3 c's refer to Cameron's use of the common law, civil law and customary law systems developed out of country's complex colonial past. Customary law in Cameroon evolves from the indigenous patriarchal culture which promotes male dominance over females. This law is practiced in most rural Cameroon where agriculture is the mainstay of economic and social life. Although efforts have been made in the harmonization of the three legal systems, current agricultural policies emanating from the customary law are an impediment to the growth and development of women as independent economic actors in the agricultural field. With the globalization of the agricultural market, the importance of women as key economic players in Cameroon's agricultural sector cannot be overemphasized. The textbox thus concludes by recommending that the laws be aligned in a manner that promotes the empowerment of women as equal players in the agricultural industry in order to boost the economy of Cameroon.

In various countries around the world, women contribute substantially to the agricultural output both directly and indirectly. However, some developing countries fail to acknowledge the important role of women in agriculture and are lacking in providing legal rights to women. Consequently, an enormous loss of potential, that of women as property owners, entrepreneurs, and equal citizens, makes it difficult for these countries to compete in the global market and stifles various aspects of the countries' food and agricultural laws and policies. One such example is Cameroon. This example sheds some light of how these legal imbalances affect women and ties in concerns affecting the agricultural sector. Although labor law is essentially beyond the scope of this book, it is important to consider the enormous effects that legal protections have on the market strengths of any given country.

W. Elangwe, LL.M (✉)
University of Maryland Eastern Shore, Princess Anne, MD, USA
e-mail: welangwe@umes.edu

Cameroon is a country with a rather unique legal system due to its unusual colonial past. After gaining independence from Britain and France in 1960 and 1961 respectively,[1] Cameroon inherited Common Law from Britain, Civil Law from France, both of which came in addition to existing Customary Law from African tradition and customs.[2] These three systems of law combine to form "The Three Cs," Common Law, Civil Law and Customary Law.

Common Law is applicable in the two English-speaking regions of Cameroon and Civil Law in the eight French-speaking regions.[3] Considering that Common Law and Civil Law function independently in their countries of origin, an amalgamation of the two systems in Cameroon has proven to be rather complex, resulting in a hybrid system of law that is deserving of its own name. As a result of this complexity, the different areas of law, such as labor, land, family, and commercial law have taken new forms and approaches under these very unusual circumstances in Cameroon.

Labor and land law, however, happen to be two of the few areas of law that have been 'harmonized' in Cameroon through the 1992 Labor Code[4] and the 1974 Land Tenure Ordinance.[5] Business law, too, has been harmonized through the OHADA law (Organization for the Harmonization of Business Law in Africa),[6] a regional treaty with 17 member countries.[7] However, despite these harmonized laws, implementation of these laws is still a challenge because these laws remain dominated by old habits that die hard. Thus, the influence of culture through traditions and customs, on the one hand, and Civil and Common Law, on the other hand, from which these new harmonized laws were inspired, continue to strongly affect the implementation of laws in Cameroon. The resulting implementation challenges are felt strongly, especially by women who, due to long-standing cultural patriarchy, have long been considered subordinate to men.

It is extremely difficult to provide an accurate appraisal of women's contribution to agriculture due to the lack of documented statistics. However, it has been estimated that about 90 % of Cameroon's agricultural production can be attributed to the hard work of its rural women.[8] Even in the cash-crop sector, which is male dominated, women equally participate, contributing about six to eight hours a day during high-growing seasons.[9] In general, it is reported that rural women in Cameroon work one-and-a-half to three times longer than men.[10]

---

[1]Manga (1997), pp. 209–211.
[2]Id.
[3]USAID (2012).
[4]Labor Code, Law No. 92/007 (1992) http://www.ilo.org/dyn/natlex/docs/WEBTEXT/31629/64867/E92CMR01.html.
[5]Land Tenure Ordinance No. 74-1 (1974).
[6]OHADA Uniform Acts (2014).
[7]http://www.ohada.com/traite.html.
[8]See FAO, Role of women in Agriculture (2014).
[9]Id.
[10]Id.

In spite of these significant contributions made by women to the agricultural sector; customary law excludes them from the right to land ownership - the same land they tirelessly toil every day. Nonetheless, if there is any uniform custom in the regions of Cameroon, it pertains to the fact that women cannot own property because they are considered property themselves.[11] In fact, a woman is by the very nature of a dowry regarded as property[12] and proponents of customary law and tradition purport that this dowry payment confers to a man proprietary rights over his bride. Therefore, in a bride's new capacity as "property", she cannot own property because property cannot own property. Although section 70(1) of the Civil Status Registration Ordinance of 1981[13] provides that the dowry shall not affect the validity of a marriage, this is not the case in reality.

Interestingly, the decision in *Zamcho Florence Lum v. Chibikom Peter Fru and others*[14] brought about an entirely different trend in legal reasoning under customary law. Inspired by the provisions of Section 27(1) of the Southern Cameroon High Court Law 1955, which overrides any law that is repugnant to natural justice, equity and good conscience,[15] this decision set a legal precedent in English-speaking Cameroon. Preventing a woman from owning or inheriting property was held to be contrary to public order and the constitution.[16] Although this judicial decision could be seen as a great victory for women from the stifling dictates of customary law, the freedom and rights it promises are still nothing more than a dream for most rural women who continue to battle with the oppressive forces of customs and traditions pertaining to land ownership.

While a female child's right to succession is not recognized under customary law in the English-speaking regions in Cameroon, Common Law (applicable predominantly in the urban areas in the English-speaking regions), offers a little reprieve from the constraints of customary law. This is particularly by virtue of Section 15 of the Southern Cameroon High Court Law of 1955, which sanctions the application of post-1900 English Statutes in matters of Probate, Divorce and Matrimonial Causes whenever there is a lacuna in the local laws.[17] Thus, the 1925 Administration of the Estate Act,[18] which is applicable in the English-speaking regions of

---

[11]Ngwafor (1993).

[12]Inglis J. in Achu v. Achu, Appeal No BCA/62/86, unreported.

[13]Civil Status Ordinance no. 81-02 (1981) http://www.cameroonhighcomabuja.com/Republic%20of%20Cameroon%20Civil%20Status%20Registration.pdf. Accessed Nov 6 2014. Hereinafter Civil Status Ordinance, 1981.

[14]Supreme Court of Cameroon Judgment No 14/L (1993).

[15]Manga (2009).

[16]This view was also upheld in French-speaking Cameroon in Affaire *Puete Jacqueline v. Ngouoko Joseph*. Supreme Court Arrêt No.15/L of 12 April, as cited by Nzalie J (2008) The Structure of Succession Law in Cameroon: Finding a Balance between the Needs and Interests of Different Family Members. http://etheses.bham.ac.uk/300/1/NzalieEbi09PhD_A1a.pdf. Accessed Nov 6, 2014. Hereinafter, Nzalie 2008.

[17]*Id.*

[18]*Id.*

Cameroon confers survivor and spousal right of succession and makes it possible for a female or a male child to succeed and inherit property.

However, the case is a little different in the French speaking regions of Cameroon. Generally, the application of customary law also reigns in rural areas and civil law in urban areas. Noteworthy is the fact that customary courts have jurisdiction to handle civil matters that have not been expressly reserved for the formal courts.[19] Because civil law is heavily reliant on specific codes,[20] where there is a gap in the law, the courts resort to international law[21] or customary law.

Conversely, under Civil Law in Cameroon, married women are subject to some restrictive rules of employment. Civil Law provides that a married woman can engage in professional activity separate from that of her husband.[22] The husband, however, may object in the interest of the household and the children, unless discharged by the court in case of unjustified opposition.[23] This view originates from Section 1421 of the 1804 Civil Code,[24] which states that the husband alone shall administer the joint estate of the family as head of the family. He may mortgage the same without the consent of his wife.[25] These limitations to a woman's right to work and total dominance provided to men by the law in the administration of the joint estate of the family only further restricts women's ability to play an active role in the nation's economy.

Notwithstanding all of these laws, according to a 2012 report,[26] women in Cameroon produce 80 % of the country's food needs, yet own only two percent of the land. With an increased demand from agricultural enterprises in industrialized countries to set up production in many developing countries, better laws that protect women who dominate the labor force in the agricultural sector are urgently needed in order for Cameroon to explore the potential of collaboration with such industrialized countries. In doing so, the efforts, which in the past have been short-lived because of the absence of or inadequate environmental protection legislation, labor laws and labeling and food safety standards, will have the potential to be successful and thrive. Thus, the fragmentation of the legal system through the "Three C's" in Cameroon, illustrates how the law affects women, how women, in turn, affect agriculture, and how agriculture, in turn, affects the success of a nation.

---

[19]*Supra* note Manga (2009). See also Global Environmental Facility (2006). Submission to the World Bank for the Cameroon sustainable agro-pastoral and land management promotion. http://www.thegef.org/gef/sites/thegef.org/files/repository/Cameroon_Sustainable_AgroPastoral_LM_Promotion_PNDP.pdf. Accessed Nov 6, 2014.

[20]See Berkeley.edu, The common law and civil law traditions.

[21]Cameroon.1996 Consti. Art 45.

[22]Art. 74 (1) of the Civil Status Ordinance, 1981.

[23]Art. 74 (2) of the Civil Status Ordinance, 1981.

[24]See French Civil Code (1804).

[25]*Id.*

[26]Ngala (2012).

References

Manga FC (1997) An experiment in legal pluralism: the Cameroonian Bi-Jural/Uni-Jural Imbroglio. Univ Tasmania Law Rev 16:209–211
Manga FC (2009) Update: researching Cameroonian law. http://www.nyulawglobal.org/Globalex/Cameroon1.htm. Accessed 6 Nov 2014
Ngala K (2012) Giving Women Land, Giving them a Future. http://www.ipsnews.net/2012/10/giving-women-land-giving-them-a-future/. Accessed 6 Nov 2014
Ngwafor EN (1993) Family law in Anglophone Cameroon. University of Regina Computer Services, Regina-Saskatchewan
USAID (2012) Land Tenure and Property Rights Portal, Strengthening Women's Property Rights in Cameroon. http://usaidlandtenure.net/commentary/2012/11/strengthening-womens-property-rights-in-cameroon. Accessed 6 Nov 2014

Chapter 45
Intellectual Property Law and Food Security Polices in Ethiopia

Getachew Mengistie Alemu

Abstract Ethiopia has abundant resources that may enhance agricultural output but could not be food self sufficient. Successive regimes in the past four and half decades took various measures to deal with the problem of food shortage but attaining food security has been very difficult. The effort to ensure food security at a household level may be affected by a number of factors including policies, laws and intellectual property (IP) protection. This chapter aims at highlighting the food security situation in the country, reviewing relevant policies, strategies, plans and laws dealing with or affecting food security; exploring the relationship between intellectual property and food security; examining the extent to which the existing intellectual property laws in particular the patent and plant variety protection laws may help in dealing with food security issues and concerns; identifying challenges and gaps; and recommending measures.

45.1 Introduction

Ethiopia has abundant resources that can enhance agricultural output, yet is not food self-sufficient. Food insecurity within Ethiopia is widespread. As one of the least developed countries in the world, ensuring food security has been a major challenge for Ethiopia for years. Successive regimes in the past four and half decades took various measures to deal with this problem but attaining food security has been very difficult. There are a number of factors, such as natural resource degradation, increased population growth, irregular rainfall and recurrent drought, behind the persistent food shortage. The main cause for food insecurity, however, is the poverty of the people and the low level of socio-economic development of the country. The effort to ensure food security at the household level is also affected by a number of factors including policies, laws and intellectual property ("IP") protection. This chapter examines existing relevant policies, strategies, plans and laws

G.M. Alemu (✉)
Getachew & Associates Law Office, New Haven, CT, USA
e-mail: getachewal@gmail.com

dealing with or affecting food security, explores the relationship between IP and food security, identifies challenges and gaps; and recommends measures.

To provide a framework and demonstrate Ethiopia's potential in attaining food security, the chapter first focuses on its geography and socio-economic conditions. Second, it examines the situation of food insecurity in the country and identifies factors behind the persistent food shortage, such as continued dependence on food aid and imports. It also discusses the reasons behind food deficit using low agricultural productivity and international trade as examples, highlights the impact of programs that were designed and implemented with the assistance of the donor community, and reflects on the challenges of attaining food security. The relationship between IP, IP protection and food security concerns of developing countries is discussed afterwards. Moreover, the chapter concentrates on IP policies and laws of Ethiopia, examines the extent to which the existing IP laws can contribute to food security issues, and identifies gaps and challenges. Enhancement of food security requires clear policies and implementation instruments, such as laws and institutions. Policies and laws that deal with or support food security are also examined. An overview of the relevant policies, strategies and plans in this section is limited to those that emphasize agriculture, target IP issues and can be supported by IP. Gaps and weaknesses of the policy instruments are identified, and the extent to which the existing IP system can complement policy goals and support implementing strategies is also examined. The last section is the conclusion, which includes recommendations to strengthen the policy and legal framework as well as enhance effective use of suitable policy instruments to support food security at the household level.

Much has been written on food security in Ethiopia. However, little or no study has been conducted to examine relevant laws and policies and IP issues. This chapter will fill this gap as well as serve as a basis for future studies in the field.

45.2 Geographical Location and Socio-Economic Context of the Country

Ethiopia is a landlocked country located on the Eastern Horn of Africa between latitudes 3° and 15° North and Longitude 33° and 48° East. The country shares borders with Eritrea in the north, Djibouti and Somalia in the east and southeast, Kenya in the south, Sudan and Southern Sudan in the west. The total area of the country is 1.13 million sq. km.[1] The country has a ragged topography featuring the vast central highlands separated from the eastern highlands by the Great Rift Valley running from the northeast to the southwest. Altitudes range from 148 m below sea

---

[1]*See* Food & Agriculture Organization ("FAO") & African Union ("AU") (2012), Chanyalew et al. (2010).

level at Dallol depression[2] to over 4600 m above sea level at Ras Dashen Mountain.[3]

Ethiopia had a population of 73,750,932 in 2007, of which 50.5 % were males and 49.5 % were females.[4] The population density in the same year was estimated to be 65.4 per sq. km.[5] According to the demographic statistical report by the Central Statistical Authority of Ethiopia, the population is expected to grow annually at about 2.7 %.[6] The population size has increased since the 2007 census. In 2012, the population was estimated to be more than 84 million,[7] making Ethiopia the second most populous country in Africa after Nigeria. Most of the population lives in rural areas. According to the 2007 Census report, more than 83 % of the population lives in rural areas compared to only 16.1 % that lives in urban centers.

Ethiopia has registered impressive economic growth rate averaging 11.0 % per year since 2003 and the current 5-year Growth and Transformation Plan aims to sustain the growth rate at a minimum of 10 % per annum.[8] The continued economic growth was possible due to the good performance of agricultural production, significant contribution of manufacturing and services, and the expansion of the construction sector (mainly housing, roads and hydroelectric dams).[9] Agriculture is the backbone of the Ethiopian economy and engine of long-term economic growth and development. The sector contributes nearly 42 % to the GDP, 90 % of the country's export earning, 85 % of employment opportunities and 70 % of the raw material requirements of local industries. Nevertheless, the country is still at the lowest level of development and is one of the 48 least developed countries in the world,[10] characterized by low gross national income, low level human capital index and high economic vulnerability index.[11] Looking at the 2013 UNDP human development index, which ranked Ethiopia 173rd out of 186 countries, the low level of development is evident.[12]

The country is endowed with rich biological and natural resources that can strengthen the economy and specifically, the agricultural sector. Ethiopia is known as a major center of origin and diversity of important crops. It is the sole

---

[2]This place is known as the lowest depression in the world.

[3]Ras Dashen is the highest mountain in the country situated in North Gondar Zone, north western part of Ethiopia.

[4]Central Statistical Agency (2007).

[5]*Id*.

[6]FAO & AU (2012). The population is projected to reach 129.1 million by the year 2030.

[7]World Bank (2012).

[8]Chanyalew et al. (2010).

[9]*Id*.

[10]For a list of LDCS, *see* http://unctad.org/en/pages/aldc/Least%20Developed%20Countries/UN-list-of-Least-Developed-Countries.aspx.

[11]For details on the criteria, see the Committee for Development Policy ("CDP") by the United Nations ("UN") Economic and Social Council at http://www.un.org/en/development/desa/policy/cdp/ldc/ldc_criteria.shtml.

[12]UNDP (2013).

or the most important center of genetic diversity of crops, such as Arabica coffee, *teff* (*eragrotis tef*), *enset* (*enset ventricosa*) and *anchote* (*Coccinia abyssinica*).[13] It is also one of the main centers of diversity for sorghum, finger millet, field pea, cowpea, perennial cotton, safflower, castor bean and sesame.[14] Additionally, because of genetic erosion in other parts of the world, it is now the most important center of genetic diversity for durum wheat, barley and linseed.[15] Ethiopia has the largest livestock resource in Africa, including 49 million cattle, 47 million small ruminants, nearly 1 million camels, 4.5 million equines and 45 million chickens.[16] This resource is vital to the livelihood of an estimated 80 % of the rural population, provides nutritious food, additional emergency and cash income, transportation, draught power (over 11 million oxen are utilized for plowing fields and other farming activities), fuels for cooking, and helps cope with shocks by accumulating wealth and serving as a store of value in the absence of formal financial institutions and other missing markets.[17]

The country has huge arable land that can be used to expand agriculture. It was noted that out of the total area of 1.13 million hectares of land, 69 % is suitable for crop and livestock production, but only 14 million hectares or 17 % is cultivated.[18] Ethiopia has vast surface and underground water resources[19] that can also strengthen the agricultural sector and ensure food security. However, people in many regions of the country face water shortages.[20] Little effort is made to use these water resources for irrigation to ensure all-year-round crop cultivation in areas where farming depends on rain.[21] Moreover, water is not used to meet the needs of people involved in livestock, which would reduce movement of both people and livestock searching for water.[22]

In spite of the fact that Ethiopia has abundant resources that can enhance agricultural output, it is not food self-sufficient. The agricultural sector is unable to feed the country's ever-increasing human population and obtain sufficient foreign exchange to purchase agricultural and industrial inputs to enhance food production. The sector is dominated by subsistence small-scale farmers, who contribute to 95 % of agricultural production[23] using low input and low output traditional technologies and depending mainly on rain.[24] As a result, a large number

---

[13]FDRE (1997).
[14]Id.
[15]Edwards (1991).
[16]FAO/WFP (2012).
[17]Id.
[18]FAO & AU (2012), p. 14.
[19]FDRE (2003).
[20]Id.
[21]Id.
[22]Id.
[23]Feyissa (2006), p. 1.
[24]This requires getting enough rain regularly at the required time and amount. However, the rainfall pattern is irregular causing drought or flood resulting in crop failure.

of households face persistent food shortage, are vulnerable to shocks that will result in loss of assets, hunger and famine, and suffer from food insecurity.

45.3 Food Security Situation in Ethiopia

Food security is defined differently depending on the purpose and the context in which the concept is used. For the purpose of this chapter, food security refers to a situation when all people, at all times have physical and economic access to sufficient, safe and nutritious food to meet their dietary needs and food preferences for an active and healthy life.[25] Food insecurity, which can either be transitory or chronic,[26] is a persistent problem in Ethiopia. The country is known as one of the most famine-prone countries with a long history of famines and food shortages, which cost the lives of about a million people.[27]

Food shortage is a serious problem that has affected millions of Ethiopians under different regimes in the last 4 decades. A report indicated that 1.5 million or 5 % of the population during the imperial regime in mid-1974, 7 million or 17.4 % of the population during the military regime[28] and by the end of 2003, under the current regime, about 14.5 million or 22 % of the total population was unable to feed themselves during periods of drought.[29] Although life loss as a result of famine and food shortage is no longer reported, an increased number of people depend on food aid and handouts under the present regime. See Fig. 45.1 for the number of people who received emergency food assistance from 2000 to 2010.[30] The people who are most vulnerable to food insecurity are the poor living primarily in rural areas. The incidence of food poverty at the national level is estimated at 50 %, of which 52 % and 37 % is in rural and urban areas, respectively.[31]

Ethiopia is one of the countries listed by the Food and Agriculture Organization ("FAO") as being in a protracted food crisis, which is characterized by long-lasting, or recurring crises often with limited or little capacity to respond, and exacerbates food insecurity problems.[32] The severity of the food insecurity and shortage problem that caused a national disaster and impacts the lives of millions in the

---

[25] FAO (1996).

[26] Transitory food insecurity is temporary in that it occurs due to shocks such as drought, flood, crop or livestock disease and frost while in chronic food insecurity households suffer food shortage in any given year.

[27] Van der Veen and Tagel (2011).

[28] The Socialist Military Government, which is also known as the "Derg," ruled the country from 1974 to 1991.

[29] Belete (2011), p. 24.

[30] Id.

[31] FDRE (2002).

[32] FAO (2010).

Fig. 45.1 People in need of food aid

Year	Number of People in million
2000	7.7
2001	6.5
2002	5.2
2003	12.2
2004	7.2
2005	3.8
2006	2.6
2007	2.3
2008	2.29
2009	6.3
2010	5.2

country at present is well articulated by one of the renowned researchers and scientists in this field in Ethiopia as follows:

> Food insecurity and famine are deeply rooted in rural Ethiopia: they have brought hardships to rural people for countless generations, shaped livelihood strategies and social relationships, conditioned attitudes to the land and the environment, and regulated the rhythm of production and consumption. The country has archived the dubious distinction of being the epic center of human disaster since at least the 1970s: food emergencies have occurred here with greater frequency than in any other country in the world in recent history. The high profile disasters and emergencies, which have attracted worldwide attention, should not, however, obscure the grim day-to-day reality of persistent hunger and malnutrition, which is part of the lives of millions of peasants and pastoralists and which in the end, provide the fuel for the large-scale catastrophes.[33]

The food deficit in the country is met by food aid and imports.[34] In addition to receiving food aid, the size of which fluctuates from year to year depending on the domestic production, Ethiopia has been importing food crops. The value of commercial food import in any given year ranges from 5 to 7 % of total imports.[35] This figure is higher when parts of the country suffer from drought and is affected by famine. In 2002/03, for example, Ethiopia spent 231.7 million United States Dollars (USD) or 7.38 % of total imports to import food.[36]

There are a number of factors that affect food security in Ethiopia and explain continued dependence on food aid and food imports. These include low agricultural productivity, severe natural resource degradation, recurrent drought, population growth, inadequate transport and market infrastructure, inadequate access to

---

[33]Rahmato (2013), p. 114.
[34]Häberli (2012).
[35]Belete (2011), p. 25.
[36]*Id.*, p. 23.

finance and credit facilities, lack of alternative income sources outside of agriculture and problems related to international trade. Since the purpose of this section is not to identify and discuss all factors that impact food security in Ethiopia but to provide an overview, the discussion will be limited to highlighting the problem of food insecurity and shortage by using low agricultural productivity, including factors that affect agricultural production and productivity, and international trade challenges as examples.

Agricultural productivity is extremely low. Small-scale farmers that contribute to 95 % of agricultural production use outdated agricultural implements, such as animal driven ploughshare[37] and hand hoes, and depend mainly on rain.[38] Currently, only 1 % of arable land is irrigated in the country.[39] The availability, accessibility and use of inputs, like improved varieties of seed, is limited. There are public research and higher learning institutions involved in the generation, development and supply of new and improved high yielding varieties. Moreover, the Ethiopian Seed Enterprise (a public-owned enterprise), multinational seed companies,[40] regional seed enterprises, cooperative unions and approximately 35 private producers have been engaged in the production and distribution of improved seed in the country.[41] However, the seed supplied by these entities is inadequate to meet demand. The total annual seed requirement by the agricultural sector in Ethiopia is estimated at about 700,000 t of which only 15 % was made available in 2011 by these entities in the formal seed system.[42] The remaining 85 % of the seed requirement is met by an informal seed supply system involving exchange and selling of farm saved seeds and landraces.[43]

The rapid population growth[44] and other undesirable characteristics, such as fragmentation of land and environmental degradation, have affected agricultural productivity. The average land holding in the rural areas decreased to less than a hectare. It is estimated that without expansion of cultivated land and given forecasted population growth, the average land holding size in highland areas will be reduced to 0.7 ha by 2020.[45] Increase in population is accompanied by a decline in food production. When Ethiopia's population increased from 23.5 million in 1960/61–48.6 million in 1989/90, the per capita food output declined from 240.2 to

---

[37]These technologies have been known for thousands of years and are passed from generation to generation with little improvement.

[38]Lack of or erratic rainfall and flooding have affected agricultural production.

[39]FAO & AU (2012), p. 14.

[40]The Major Seed Company is Pioneer Hi-bred Ethiopia that deals mainly with hi-bred maize. It began operations in Ethiopia beginning 1990.

[41]*See*, FAO/WFP (2012), p. 14.

[42]*Id*.

[43]Merso (2011), p. 127.

[44]The size of the population increased from 11.8 million in 1900 to 23.6 million in 1960, doubling in 60 years but took only 28 years to double to 47.3 million in 1988. See Minas (2008), p. 23.

[45]Bill & Melinda Gates Foundation (2010).

141.7 kg in the same period.[46] It was noted that high population growth rate undermines Ethiopia's ability to be food secure and provide effective education, health and other essential social and economic services.[47]

Increased deforestation, expansion of desertification and land degradation has also contributed to low levels of agricultural productivity and will continue to have an impact on food production. The rate of deforestation is estimated to be between 140,000 and 200,000 ha per annum and the forest coverage is limited to about 3.6 % of the country's landmass.[48] This figure is alarming when compared to the 40 % of forest cover before the turn of the twentieth century.[49] Ethiopia is also affected by the expansion of desertification in the Sahelian region of Africa. A large part of the country, approximately 75 % of the landmass, is projected to be threatened by desertification.[50] Land degradation, which includes loss of soil fertility, loss of biodiversity and soil acidification, occurs due to improper land use and poor land management practices, population pressure, overgrazing, deforestation and the use of crop residues and dung for fuel in rural areas.[51] Ethiopia has one of the highest rates of soil nutrient depletion in sub-Saharan Africa and the annual phosphorus and nitrogen loss nationwide from use of dung for fuel is estimated to be equivalent to the total amount of commercial fertilizer applied.[52]

The second crucial factor that impacts food security is international trade. Measures taken by trading partners and fluctuation of agricultural commodity prices on the international market contribute to food insecurity. Ethiopia depends on few agricultural products[53] to generate export revenue. Often, the export of Ethiopian agricultural products is affected by measures taken by major importing countries. For instance, in February 1998 and April 1999, Saudi Arabia decided to ban the import of meat and livestock from Ethiopia and other Horn of Africa countries. This decreased the livestock prices by around 30 %, affecting the majority of people in Eastern and North Eastern Ethiopia, who depend on livestock production and export as their main source of income.[54] The price of agricultural commodities has fallen for the past 30 years[55] and there is no price stability of agricultural products on the international market. The buyers and primary product importing countries often determine the price in which producers has no role. They are mere price takers affecting countries such as Ethiopia and other agricultural producers that depend on the export of primary products. In Ethiopia, coffee is a major export product that contributes to more than

---

[46]Minas (2008), p. 24.
[47]FDRE (2002).
[48]FAO & AU (2012).
[49]Minas (2008), p. 25.
[50]FAO & AU (2012).
[51]*Id.*
[52]Chanyalew et al. (2010).
[53]These include coffee, cereals, flowers and pulses.
[54]Belete (2011), p. 29.
[55]Light Years IP (2008), p. 9.

one-third of the foreign exchange earnings and is the source of livelihood for one-fourth of the population.[56] The collapse of the coffee price in 2001 threatened the livelihood of coffee growers [57] and reduced the foreign exchange earning that affect, amongst others, investment in agriculture and food import.

Ethiopia has designed and implemented programs that address food production and environmental protection, water management and irrigation, employment creation, resettlement and credit provision with the support of the international donor community to achieve food security since the 1970s.[58] It should, however, be noted that extensive measures have only been taken and comprehensive programs have been designed and implemented with the assistance of the international donor community since the current government took power in 1991. There are encouraging developments resulting from economic growth, increased agricultural production, and the investment made and measures taken by the government and the international donor community. The economy has registered a double-digit agricultural growth continuously for several years. The average growth of agricultural output has been about 10 and 13 % in 1996/7 and 2004/05, respectively. The government allots nearly 16 % of its budget to agriculture, the bulk of which goes to food security programs.[59] Food poverty headcount decreased from 44 % in 1999/00 to 38 % in 2005/06.[60] There is also reduction in the number of people vulnerable to food insecurity and improvement in the position of Ethiopia on the Global Hunger Index, which captures three dimensions of hunger—insufficient availability of food, shortfalls in the national status of children and child mortality. The position of Ethiopia on the Global Hunger Index fell from 43.5 % in 1990 to 30.8 % in 2009.[61] The percentage of vulnerable rural population has decreased despite the growth in population size. In 2008, 23.1 % of 63.5 million people were vulnerable and this figure decreased to 22.3 % and 19.5 % in 2009 and 2010, respectively,[62] in spite of a population increase to 65.1 million in 2009 and 66.8 million in 2010. It should be noted that of these, 7.2 million in 2009 and 7.8 million in 2010 were beneficiaries of the productive safety net program[63] launched by the government in cooperation with the donor community to reduce dependence on food aid and help the population to stand on its own. The government has also taken a number of other complementary measures, including the establishment of the Agricultural Transformation Agency. This agency is mandated to modernize agriculture and address bottlenecks and set up commodity exchange (the first in Sub-Saharan Africa to

---

[56]Mengistie (2010a, b, c), p. 1.
[57]Belete (2011).
[58]Pankhurst and Rahmato (2013).
[59]*Id.*
[60]Chanyalew et al. (2010).
[61]IFIPRI (2009).
[62]Id., p. 116.
[63]Id., p. 113.

provide for transparent and predictable market), deal with the abuse of the middlemen and increase the benefits for farmers.[64]

Enhancing agricultural production and ensuring food security is a priority on the government's agenda as clearly evident by relevant policies, strategies, plans and laws discussed under Sect. 45.5 of this chapter. However, a lot remains to be done to ensure food security at the household and national level in Ethiopia. According to Dessalegne Rahamato, ensuring food security for a sizable population of rural (and increasingly urban) households remains an elusive goal.[65] It is a tremendous challenge to attain food security while coping with a growing size of the population, increased environmental degradation and impact of climate change.

45.4 Intellectual Property and Food Security

Intellectual property (IP) refers to exclusive rights conferred by law over the creations of the human mind for a prescribed period.[66] Third parties may not make, produce and exploit the work, in any other form[67] without securing authorization from the right holder, except under circumstances provided by law. IP consists of three major elements—copyright and related rights; industrial property, which includes patents, industrial designs, trademarks, geographical indications, trade secrets; and plant variety protection (PVP).

Various IP titles such as patents, copyright, industrial designs, plant variety protection, trade secrets, trademarks and geographical indications are used to protect and market agricultural innovations, such as new and improved plant varieties,[68] that may enhance agricultural productivity and ensure food security. In discussing the relationship between IP and food security, the focus of this section will be limited to patents and PVP. A patent is a legally enforceable right granted by law to a person to exclude others from certain acts related to a described product or process invention that meets the requirements set by patent law[69] for a limited period of time. The patent system was initially established to cater for new inventions related to non-living things but could not accommodate the need to stimulate research and development activities in the field of plant varieties and

---

[64]For details of these measures, *see* von Uffelen (2013), pp. 2–13.

[65]Rahmato (2013).

[66]Please note that trademarks, geographical indications and trade secrets may be protected for an indefinite period of time provided the conditions for protection remains intact.

[67]An example is the use of a patented invention for the purpose of scientific research and experimentation, which is allowed in patent laws.

[68]Blakeney (2009), pp. 22–56; Tansey (2008); Dutfield (2008).

[69]Article 27(1) of the TRIPS agreement provides that patents will be granted to inventions in any field of technology. National patent laws define the requirements that should be met in order to grant a patent. Moreover, the laws exclude subject maters that may not be protected by a patent due to the nature of the subject matter itself or based on policy considerations.

reward results thereof. As a result, beginning from the 1930s, several countries introduced legislation that gradually evolved to a special (*sui generis*) system of protection of breeders' rights that is distinct from the patent system.[70] PVP, commonly known as "plant breeders' right," is a form of protection that evolved and was developed to protect new plant varieties. Plant variety laws allow breeders to make use of the protected variety in developing and disseminating improved varieties. However, this privilege is restricted by the 1991 International Union for the Protection of New Varieties of Plants (UPOV) convention.[71] Varieties that are "essentially derived" from a "protected variety" cannot be exploited without the authorization of the breeder or breeders of the prior protected variety.

Patents and PVP can stimulate agricultural innovation and help meet food security needs. However, there is no common understanding on the role and contribution of these IP tools to address food insecurity problem of mainly low income developing and least developed countries. There are arguments in favor and against IP protection of biotechnological inventions,[72] new plant varieties, plants and animals. Those who are in favor support their argument based on the objective and role of IP protection. IP is meant to stimulate creative activities. Since biotechnology is a result of such an activity, it should not be denied IP protection. IP protection encourages inventive and innovative activities, facilitates the disclosure and availability of technological information and stimulates the transfer of technologies, thereby meeting the needs of society and fostering socio-economic development. Others argue against IP protection of creations and innovations involving life forms on a number of grounds—moral or ethical, technical and economic. Each of the grounds is extensive and beyond the scope and purpose of this chapter. The discussion in this section will thus be limited to patents and PVP and food security concerns related to developing and least developed countries.

Patents and PVP for creations and innovations involving life forms such as plants is justified based on the need to increase agricultural productivity by stimulating research and development (R&D) activities and encouraging investment in the generation of new and improved technologies, which involves considerable money, time and labor. In the absence of patents or PVP, persons who have done very little will be in a position to make use of these agricultural innovations with no or little investment. This will discourage R&D activities and investment affecting agricultural productivity and food security. This argument looks sound if not seen

---

[70]Blakeney (2009), pp. 80–81.

[71]Developing countries that are not members of this convention or are members of the 1978 UPOV Convention may not have this problem. Under this Convention, a protected variety can be slightly been modified and exploited without requiring the authorization of the earlier breeders so long as the modified variety is distinct from the protected variety.

[72]Article 2 of the Convention on Biological Diversity defines biotechnology as "any technical applications that use biological systems, living organism, or derivatives thereof, to make or modify products or processes." Biotechnology consists of a set of techniques, such as genetic engineering, cell fusion, tissue culture, in-vitro fertilization and the use of recombinant DNA that use, make or modify life forms for a defined purpose.

within the context of developing and least developed countries. In these countries, there is little to no conclusive evidence to show an increase in R&D as a result of a patent and PVP system.[73] The R&D effort of the private sector, which contributes to about a third of global agricultural research[74] and mainly done in developed countries, focuses not on minor food crops that small, resource-poor farmers grow in developing countries to meet nutritional needs but major crops, such as wheat, rice and maize, that have demand in large markets to ensure return on research investment.[75] The major beneficiaries of IP protection system in developing countries are foreigners. For example, in Kenya, the foreign-owned companies engaged mainly in exporting flowers and vegetables use variety protection system.[76] The PVP system in Kenya facilitates the availability of a new plant variety that contributes to the expansion of export industries and commercial agriculture, but contributes very little in stimulating local research that meets the needs of poor farmers.[77]

The search for high yield, disease resistant, etc. plant varieties to increase agricultural productivity requires the creation and expansion of agricultural research development capacity. This search is weakened as a result of IP protection, which further strengthens the position and dominance of big transnational corporations in the domestic market. IP has facilitated an increase of R&D investment by the private sector, mainly big northern companies, and enabled private agro industrial enterprises to assume a dominant position in agricultural research.[78] Similarly, IP is used as a means to control and ward off the local R&D institutions from the domestic market. In fact, it is very difficult for the public agricultural research institutions to produce varieties that will compete in the market with those developed by transnational corporations. There are a few public agricultural R&D and higher learning institutions involved in developing and least developed countries such as Ethiopia, whose main objectives include the development and dissemination of new or improved crop varieties. However, the R&D capacity of these institutions is weak and contributes very little to the country's objectives. Public research organizations may, for example, use a patented technology for the purpose of scientific research,[79] but may not be in a position to develop and disseminate the research result. The patent owner can prohibit the use, production, sale or offer for sale of any research output that contains the patented subject matter.[80] The weakening of these public institutions and the penetration of the northern private R&D institutions into the domestic market also changes the purpose and goal of research

---

[73]Commission on Intellectual Property Rights (2002); Blakeney (2009).
[74]Blakeney (2009), p. 97.
[75]*Id.* and see also Dutfield (2008), p. 41.
[76]Commission on Intellectual Property Rights (2002), p. 62.
[77]*Id.*
[78]Blakeney (2009), p. 97.
[79]Patent laws often provide for such an exception.
[80]Correa (2012).

outputs as well as affect its use for small farmers. For example, these private enterprises now focus on innovations that have commercial potential and are not necessarily congruent with the interest of small farmers in developing countries, particularly in relation to the food crops.[81]

IP rights are also private rights that facilitate misappropriation of genetic resources and associated traditional knowledge with no or little benefit to the country of origin and local communities that have maintained and developed these resources for generations.[82] For example, a cereal called *teff* (*Eragrostis tef*) from Ethiopia has been accessed and exploited by foreigners without any or insignificant sharing of benefit. *Teff*, which is believed to be native to Ethiopia, is an important food crop. It has been conserved and developed for generations by Ethiopian farmers. It is a staple food for millions of Ethiopians and is the largest crop grown in large hectares of land by majority of Ethiopian farmers.[83] *Teff* was accessed from the country, protected and owned by foreign companies outside of Ethiopia. These incidents include two cases that occurred before and after the country joined the Convention on Biodiversity and adopted a national biodiversity policy and set up a government body, the Institute of Biodiversity, to issue permits to access genetic resources to ensure equitable sharing of benefits. The first case relates to one of the *teff* varieties, grown in the Wollo province. The genetic resource was taken out of the country, and an application for plant variety protection was submitted to the US department of Agriculture by the applicant, a US company called the "*Teff* Company." The variety was named "Dessie."[84] The office granted a plant breeders right certificate no. 8900033 on November 1996, which conferred an exclusive right valid for 20 years from the date of this certificate. The company had the right to exclude others from selling, producing, importing or exporting the variety for propagation or making it for any of the above purposes or using it in producing a hybrid or a different variety to the extent provided by the US PVP law.[85] The second case concerns the *teff* genetic resource accessed after the adoption of biodiversity policy and the establishment of the Institute of Biodiversity. A Dutch company called Health and Performance Food International accessed *teff* varieties, undertook research and produced beverages and food products that were gluten free.[86] In 2004, the company filed an application for a patent, EP 1 646 287 B1, for processing of *teff* flour, which was granted by the European Patent Office on 10 January 2007.[87] The patent, which is valid in six countries: the

---

[81]*Id. See also* Tansey (1999), p. 20.

[82]Correa (2009), p. 6.

[83]In 1980, 50 % of the total area used for cereal cultivation was devoted to *teff*. *See* Andersin and Winge (2012), Grains (1996), p. 1.

[84]Dessie is the name of the capital city of the then Wollo province from which the genetic resource was taken and the present Eastern Wollo zone in north eastern Ethiopia.

[85]Mengistie (2001).

[86]The products have a demand on the international market because of a growing number of people that are allergic to gluten products. Moreover, there is a growing demand for *teff* products in Europe and USA for its nutritional value.

[87]Andersin and Winge (2012), p. 50.

Netherlands, Germany, France, Spain, Great Britain and Turkey, covers *teff* grain and processes, which are known and practiced in Ethiopia, such as seed storing after harvesting. This process improves the baking quality of the flour and the making of dough as well as a range of non-traditional products and food products made from *teff* flour including bread, pancakes, shortcakes, cookies and cakes of various kinds.[88] The company has also applied for patent protection in USA and Japan. The above IP titles are unprotected in Ethiopia and will not affect the production and sale of *teff* products as well as products made using the patented or similar processes in the country. However, they will affect the country and stakeholders regarding the export of *teff* products, generating foreign exchange and meeting food security and other related needs. It may also limit the country's ability to provide access to *teff's* genetic resource to other companies that may be involved in the use of genetic resources in developing and marketing similar products using similar processes. The title holders have the right to prevent export of *teff* products or products processed using *teff* from Ethiopia as well as other companies that would like to develop products, process, and manufacture and sell similar products using *teff* in countries where the rights are validly protected. There is a belief that the above titles may be challenged and revoked. However, this will require expertise and resources, which the country does not have.

IP rights protection encourages medium and large-scale farming and results in the displacement of small-scale farmers, as noted below:

> Patents and plant variety protection will facilitate the commercialization of farming systems along the lines of farming systems in the industrialized countries and so rapidly undermine the whole base of small-scale subsistence and local market-based production systems. If R&D produces varieties and methods most suitable for medium and large-scale farmers rather than products and methods geared to small farmers need, many small farmers will be squeezed out. Such a result would probably greatly increase population movements to urban centers.[89]

The displacement of small farmers and their migration to urban centers will worsen socio-economic problems such as unemployment and increase the number of food insecure in developing and least developed countries such as Ethiopia.

The attainment of food security by increasing agricultural productivity, among others, requires access to and use of new and improved crop varieties. However, farmers are not in a position to make use of such varieties when they are subject to IP rights. Traditionally, farmers used to save seeds, replant, exchange, or sell farm-saved seeds. This is not possible or is restricted when seeds are protected. Farmers may not save a patented seed that they collect from their farmland, sell and replant it unless authorized by the right holder.[90] Such authorization may not always be obtained. Even when secured, it is often subject to compliance with stringent terms and conditions. One of these requirements is the obligation to pay royalty

---

[88]*Id.*, pp. 50–51.
[89]Tansey (1999), p. 25.
[90]Commission on Intellectual Property Rights (2002), p. 63.

for reuse of a seed from a previous harvest, which is beyond the means of most small farmers in developing countries. Small farmers are allowed to save and reuse seed protected by PVP laws.[91] However, this privilege does not extend to exchange of seeds, which is a common practice in developing countries since there is no well-developed formal seed supply system, or procuring seeds from sources other than the right holder or sources designated by him.[92] This has resulted in the privatization of agricultural innovation, erosion of traditional farmers' rights and privileges, and changed the position of farmers from 'seed owners' to mere 'licensees' of a patented product.[93] New and improved protected seeds often require additional inputs, such as herbicides and pesticides, which are very costly.[94]

This is worsened as a result of improved marketing position of big companies resulting from IP protection and vertical and horizontal integration of global seed by agricultural input industries to maximize return on investment by controlling the distribution channels.[95] The excessive cost of acquiring and using improved seeds would, therefore, be beyond the financial means of most farmers in primarily low-income and developing countries. This makes access to new and improved agricultural innovations impossible and attainment of food security difficult.

IP protection of plant varieties results in monoculture and loss of food crop varieties. Graham Dutfield noted that "plant variety protection may contribute to a trend whereby traditional diverse agro ecosystems, containing a wide range of traditional crop varieties, are replaced with monocultures of single agrochemical-dependent varieties, with the result that the range of nutritious foods available in local markets becomes narrower."[96] The erosion of biological diversity and displacement of landraces may cause serious food crises when the variety is lost due to new diseases or climate change.

Some studies conducted on the impact of IP on developing countries concluded that certain forms of protection should not be made and recommended measures that cater to their interests and address concerns. For example, a study conducted by the United Kingdom Commission on Intellectual Property Rights (CIPR or the commission) noted that patent protection of biotechnological related inventions is not in the interest of developing countries with little or no capacity in this technology.[97] The commission proposed that developing countries should generally not provide patent protection for plants and animals because of the restrictions it imposes on farmers in reusing, exchanging and selling seed as well as researchers

---

[91] This privilege was automatic under the 1978 International Plant variety Protection Convention (UPOV) system but made conditional under the 1991 Convention, including payment of some form of remuneration.

[92] Blakeney (2009); Dutfield (2008), p. 41.

[93] Tansey (1999), p. 9.

[94] Rajotte (2008), p. 162.

[95] Commission on Intellectual Property Rights (2002), p. 65.

[96] Dutfield (2008), p. 41.

[97] Commission on Intellectual Property Rights (2002).

in developing and disseminating research results using and involving patented material.[98] However, these may be difficult to implement as the policy space was eroded by the Trade Related Aspects of Intellectual Property (TRIPS) and bilateral agreements that developing countries are party to. The TRIPS agreement, which was concluded on April 15, 1994 as part of the multilateral trade agreements administered by the World Trade Organization, has narrowed the policy space that member countries used to have prior to the conclusion of this agreement. The agreement defines minimum requirements that all member states must comply with. One such requirement is the obligation to accord protection for plant varieties either by patent or by an effective *sui generis* system or by any combination thereof.[99] "Effective *sui generis* system" is not defined. To some, it means a PVP scheme, which is already in place in compliance with the UPOV convention.[100] Moreover, there are people who argue that developing countries can use this requirement as a basis to provide protection for community achievements and address issues of interest. Blakeney, for example, argues that "the TRIPS agreement does not prohibit the development of additional protection systems. Nor does it prohibit the protection of additional subject matter to safeguard local knowledge systems or informal innovations as well as to prevent their illegal appropriation."[101] This can be possible only if the protection system is accepted as effective. The TRIPS agreement does not provide guidance to determine whether a *sui generis* system is effective or not. What an "effective sui generis system" means will, therefore, remain to be seen. Nevertheless, developing countries should use this policy space or flexibility in designing a system that will enable them to comply with the requirements of the TRIPS agreement and address their concerns and meet their needs. An example of such a PVP system can include the farmers and breeders privilege recognized under the 1978 UPOV convention. However, it should be noted that the limited policy space under the TRIPS agreement was further eroded by bilateral and regional free trade agreements (FTAs). For instance, the FTAs concluded between USA and Lebanon, Morocco, Tunisia, Jordan, Peru and Central America (under CAFTA) require these countries to join the 1991 UPOV convention.[102] Moreover, the FTAs that the USA has entered into with Jordan, Mongolia, Nicaragua, Sri Lanka and Vietnam mandate the provision of patent protection for plants and animals.[103]

---

[98]Id., pp. 66–68. See also Tansey (1999), p. 14.

[99]Article 27(3) (b) of the Agreement on Trade Related Aspects of Intellectual Property (TRIPS) agreement.

[100]GRAIN (1997).

[101]Blakeney (2009), p. 88.

[102]Rajotte (2008), p. 142.

[103]Id.

45.5 Intellectual Property Policies and Laws of Ethiopia and Food Security Issues

45.5.1 National IP Policy

The various policies issued by the government clearly recognize the importance and need for IP protection, promotion of local creative, inventive and innovative activities as well as facilitating the acquisition and exploitation of foreign technology. These include the 2012 Science, Technology and Innovation Policy,[104] the 1992 Seed Policy,[105] the 1997 Cultural Policy[106] and the 5 years Growth and Transformation Plan ("GTP").[107] In addition to policies that recognize conventional IP, the 1997 Environment Policy envisages the development of a scheme to protect community achievements and IP. However, Ethiopia does not yet have a consolidated national IP policy. The government realized the need to fill this gap and solicited the support of the World Intellectual Property Organization (WIPO). An IP assessment was carried out in 2005 and a draft national IP policy and strategy was prepared in 2013 with the support of WIPO.

The main policy objectives of the draft IP policy and strategy are to:

a. Stimulate and foster the generation, protection and commercialization of IP assets in Ethiopia and facilitate transfer and acquisition of foreign technology;
b. Facilitate integration of IP into national and sectorial development policies, strategies and plans and ensure its contribution in the realization of their development goals;
c. Provide for the development and strengthening of IP protection, administration and enforcement of legal and institutional framework;
d. Build capacity including development of human resources needed for effective protection, management and enforcement of IP and the use of IP as a tool for development;
e. Promote greater awareness of IP by potential users, government officials and the general public and improve the use of IP in meeting technological, social, cultural and economic needs; and

---

[104]The Science, Technology and Innovation Policy was issued in 2012 replacing the first National Science and Technology Policy that was adopted in 1993. The policy aims to build and promote technology transfer capacity that will meaningfully support the socio-economic development endeavors of Ethiopia; Identifies IP as one of the critical issues; recognizes IP as an element of the national innovation system and enumerates a number of strategies to use IP as a tool in promoting technology transfer.

[105]The policy requires protection of breeder's rights in Ethiopia.

[106]The policy provides that existing laws of the country related to copyright and other related rights shall be amended and new laws pursuant to the advanced technology shall be affected and the right of ownership of the people concerned shall be protected while traditional fine arts of the different nations and nationalities are put to use.

[107]*See* Sect. 45.6.3 of this chapter for additional details on the Growth and Transformation Plan.

f. Strengthen linkage between the national and the international IP system, facilitate membership of Ethiopia to international IP agreements and maximize benefits from the opportunities offered.

Once the national government approves the draft IP policy and strategy, the implementation of the policy is expected to ensure effective use of IP in fostering accelerated national socio-economic development of Ethiopia and improvement in the welfare of Ethiopians.

45.5.2 Intellectual Property Laws of Ethiopia

The protection of IP is *recognized* by the 1994 Constitution of the Federal Democratic Republic of Ethiopia (FDRE). Article 40 *recognizes* the right to property over intangible assets. Moreover, Articles 51 (19) and 77 (6) expressly require the Federal Government to protect patents and copyrights. Since then, the following IP laws were enacted:

a) Inventions, Minor Inventions and Industrial Designs Proclamation No. 123/1995;
b) Inventions, Minor Inventions and Industrial Designs Regulations No. 12/1997;
c) Copyright and Neighboring Rights Proclamation No. 410/2004;
d) Trademark Registration and Protection Proclamation No. 501/2006;
e) Trademark Protection and Registration Regulation No. 273/2012;
f) Plant Breeders' Right Proclamation No. 418/2006; and
g) Trade Practice and Consumers' Protection Proclamation No. 685/2010.

Moreover, there are other laws that complement each of the above IP laws. These include the 1960 Civil Code and the 2004 Criminal Code and the Customs Proclamation No. 622/2009, which are used to enforce IP rights against infringers. Discussing the IP laws of Ethiopia is beyond the scope and purpose of this chapter and is thus limited to the patent and PVP laws, particularly the provisions that impact food security.

There was no specific legislation that dealt with patents, utility models and industrial designs until May 1995. The first industrial property law Ethiopia, the Proclamation Concerning Inventions, Minor Inventions and Industrial Designs (referred to as "patent law"), and the implementing regulations were issued on May 10, 1995 and March 1997, respectively. The law states that a patent can be granted to a product or process invention.[108] The law does not exclude inventions based on the type of technology and extends patent protection to any invention in any field of technology.[109] In order to be patented, an invention must meet the

---

[108] Article 2(3) of the Patent law defines an 'invention' as an Idea of an inventor, which permits in practice the solution to a specific problem in the field of technology.

[109] This is in line with the requirement of the Article 27(1) of the TRIPS agreement in spite of the fact that country has not yet acceded to the agreement.

criteria of patentability: novelty, inventive step, and industrial applicability.[110] An invention is considered new if it does not form part of the state of art, which is the sum total of knowledge made available to the public, in any form or in any part of the world before the filing date or where appropriate prior to the priority date of the patent application.[111] An invention is inventive when it advances prior technological knowledge. An invention is presumed to be industrially applicable if it can be made or used in handicraft, agriculture, fishery, social service or any other sector.[112] The industrial applicability requirement refers to the fact that an invention must be more than an abstract theory and that it can be put into practice. Meeting these three criteria of patentability is still not enough to obtain a patent. The invention should also not fall under the category of excluded subject matters. These include plant or animal varieties or essentially biological processes for the production of plants or animals.[113] Plant or animal varieties are thus not patentable.[114] Moreover, inventions involving essentially biological process such as conventional breeding of animals and plants are excluded from being patented. Once a patent is granted, it is for an initial period of 15 years beginning from the date of filing of the application provided that annual fees are paid.[115] This period may be extended for an additional period of 5 years where proof is furnished that the invention is being properly utilized in Ethiopia.[116] A patent may be terminated or invalidated prior to the expiry of its duration when the patentee fails to pay the annual or maintenance fee during the prescribed period of time or when the invention is found not to be patentable or the description does not disclose the invention in a clear and complete manner for it to be carried out by a person skilled in the art.[117]

The exclusion of plant or animal varieties from patentability is justified based on technical and policy consideration. Plants or animals may not be patented in that they are found in nature and do not meet the requirements for patent protection such as novelty of inventions. The exclusion is also justified on the basis of public policy consideration that may include protecting plant varieties in a separate legal regime with distinct requirements and limitations, such as those that cater to the interest of small farmers.[118] The Issuance of the Plant Breeders' Right Proclamation No. 418/2006 discussed below reflects this public policy consideration.

---

[110] Article 3(1) of the patent law.

[111] *Id.*, Article 3(2).

[112] *Id.*, Article 3(5).

[113] *Id.*, Article 4(1) (b).

[114] It should be noted that the law does not exclude all inventions involving life forms from being patented. Microorganisms are not excluded. Biological materials that consist of microorganisms or resulted from non-essential biological processes are patentable.

[115] *Id.*, Articles 16 and 17(1).

[116] *Id.*, Article 16.

[117] *Id.*, Articles 17, 34 and 36.

[118] Mengistie (2013).

In order to address the potential adverse impact of these exemptions, the law provides various limitations and safeguards. Unlike other national patent laws and the TRIPS agreement,[119] the Ethiopian patent law does not recognize the right to prevent import of a patented product or a product made using a patented process as part of the exclusive rights conferred to the patentee.[120] This exclusion prevents abuse of a patentee and encourages the exploitation and use of a patented invention in Ethiopia.[121] This is in response to the fear that a patentee will use the monopoly power to control the domestic market and exploit a patented invention by importing the patented product or the product made using a patented process into the country.[122] The patentee also cannot prevent the use of a patented invention solely for the purposes of scientific research and experimentation.[123] This enables researchers in Ethiopia to use patented inventions for research purposes. However, this limitation forbids exploitation if the research result contains or falls into the claims made by the patent. The safeguards built into the law to prevent potential abuse by the right holder include issuance of a license authorizing the exploitation of a patented invention without the consent of the right holder. The patent office may initiate such an authorization or upon request of an interested party. The office may authorize a government agency or a third person to exploit a patented invention without the consent of a patentee where the public interest, in particular, national security, nutrition, health or the development of other vital sectors or the national economy, so requires. [124] The areas where public interest is presumed to exist are not exhaustive. There is thus a possibility of granting a compulsory license by the office in any other area or sector where public interest is deemed to exist. A compulsory license may also be granted to an interested party in the case of dependent inventions and when a patentee fails to exploit a patented invention without justifiable reason.[125] When a compulsory license is issued by the initiative of the patent office or upon request of an interested person, the patentee will be entitled to payment of an equitable remuneration that may be set by the office or the parties.[126] If the patentee is unhappy with the amount of remuneration, he has the right to appeal against the decision to the Federal High Court.[127]

PVP is a recent phenomenon in Ethiopia. There was no law governing the protection of plant varieties prior to 2006. The first law, which accords protection to new plant varieties, is the Plant Breeders' Right Proclamation, Proclamation

---

[119] Article 28 of the TRIPS agreement provides that the exclusive right conferred by a patent includes importing a patented product or a product made using a patented process.

[120] Article 22(2) of the patent law.

[121] Draft patent legislation explanatory note.

[122] *Id.*

[123] Article 25(1)(b) of the Patent law.

[124] *Id.*, Article 25(2).

[125] For details of the requirements, *see* Id., Articles 29–33.

[126] Id., Articles 25(2) and 33.

[127] Id., Articles 25(2) and 54.

No. 418/2006 (hereafter referred to as "plant variety protection or plant breeder's right law") that was enacted on 27 February 2006. Protection of a breeder right (PBR) is not available to all plant genera and species but to specified varieties,[128] that is new, distinct, uniform and stable and must not fall outside the list of protectable plant varieties. A plant variety is considered new if the variety is not commercialized prior to the date of application for a grant of a plant breeder's right. Unlike the requirement of novelty for patentable inventions, non-commercial use of the variety and disclosure of the variety in any media prior to filing of the application will not defeat the novelty of the plant variety.[129] This makes the novelty requirement less stringent than a patent. For example, the description of the variety made available to the public through publication prior to the date of application will not result in loss of novelty of the plant variety, unlike in the case of a patent. A variety distinct when it is clearly distinguishable by one or more important characteristics from any other variety known at the time of filing of an application for protection.[130] A variety is uniform in that it must be homogeneous regarding the particular features of its sexual reproduction or vegetative propagation. Hence, the degree of uniformity is determined by taking into account the particular features of the variety's propagation.[131] A variety shall be stable in the sense that its reproduction should remain unchanged in its essential hereditary characteristics.[132] The four criteria are similar to those provided under foreign PVP laws and the International Convention for Protection of New Plant varieties.[133] However, the fulfillment of these four requirements is not enough to obtain a plant breeder's right. The plant variety should not fall outside of the list of plant varieties genera and species determined by a directive issued by the Ministry of Agriculture.[134] Moreover, proof must be produced that the breeder has obtained the genetic resource used to develop the variety in accordance with the law regulating access to genetic resources.[135] When these two additional requirements are met, a plant breeder right is granted. The holder of such a right has the exclusive right to sell or license other persons to sell the plants or propagating material of the protected variety and produce or license other persons to produce propagating material of the protected variety for sale.[136] A plant breeder's right is validly protected for a period

---

[128] Article 3(1) of the plant variety protection law applies to new plant varieties of the genera and species determined by a directive that is issued by the Ministry of Agriculture and Rural Development (referred to as 'Ministry of Agriculture'). However, such a directive is not yet put in place.

[129] *Id.*, Article 2(5).

[130] *Id.*

[131] *Id.*

[132] *Id.*

[133] *See*, for example, Article 5 of the 1991 Act of the International Convention for Protection of New Plant varieties, commonly known or referred to as UPOV 1991.

[134] The Ministry of Agriculture has not yet issued the directive that was envisaged under Article 3 (1) of the plant variety protection law.

[135] Id., Article 14 (3).

[136] Id., Article 5(1).

of 20 years in the case of annual crops and 25 years in the case of trees, vines and other perennial trees beginning from the date of acceptance of an application.[137] Such a right may, however, be revoked or terminated prior to the expiry of its duration for not meeting substantive requirements of protection or due to failure of the holder to discharge his obligations.[138]

The rights conferred by the law are exclusive in that no other person is entitled to perform any of these acts without the authorization of the right holder.[139] Unauthorized selling or production of the seed or propagating material of a protected plant variety constitutes infringement of the right and entails legal liability. It should be noted that the exclusive right conferred under the Ethiopian law is narrower than those provided n some national laws[140] and UPOV 1991.[141] The rights over a protected plant variety are limited to defined activities; namely, the sale of the plant or propagating material of a protected variety or production of a propagating material for sale. The right does not extend to other activities such as importation, storage or exportation of a propagating material. Moreover, there are exemptions from and restrictions imposed on the exclusive rights conferred under the PVP law. The exemptions entitle any person or farmer community to propagate, grow and use a protected variety for non-commercial purposes and to sell plants or propagating material of the protected variety for use as food or for any other use that does not involve growing the plant or the propagating material of the protected variety. It also allows selling plants or propagating material of a protected variety as long as they are within a farm or any other place where plants of the variety are grown and use plants or propagating material of a protected variety as an initial source of variation for the purpose of developing another new plant variety, except where the person makes repeated use of plants or propagating material of the variety for the commercial production of another variety and sprout a protected variety for use as food for home consumption or for the market. Lastly, it allows the use of a protected variety in further breeding, research or teaching and to obtain, with the conditions of utilization, a protected variety from gene banks or plant genetic centers.[142] Thus, the PVP law recognizes farmers' rights that stem from their contribution to the conservation and sustainable use of plant genetic resources and entitles them to save, use, multiply, process and sell farm-saved seed of protected varieties.[143] However, such a right does not extend to selling farm

---

[137]Id., Article 9.

[138]Id., Article 22.

[139]Id., Article 5(2).

[140]See, for example, the 1972 Seeds and Plant Variety Act, as amended in 2002, of Kenya. Section 20(1) provides that the exclusive right extends to produce reproductive material of the variety for commercial purposes, to commercialize it, to offer it for sale, to export it, to stock it for any of these.

[141]Article 14(1) of UPOV 1991.

[142]Id., Article 6(1).

[143]Id., Articles 27, 28 (1) (c) and 6 (1) (c). The farmer privilege is broader than what is recognized under the 1991 UPOV condition. It includes the right to sell farm saved seeds, which is not recognized under UPOV.

saved seed or propagating material of a protected variety in the seed industry as a certified seed[144] or in the seeds industry on a commercial scale.[145]

In addition to the above exemptions, the law restricts exclusive rights of a PBR holder upon consideration of public interest.[146] In other words, the holder of a PBR may not exercise his right to prevent the sale or production of the propagating material of a PVP without his consent when measure is taken to restrict the right. The Ministry of Agriculture also has the power to grant a compulsory license to safeguard public interest, upon application of an interested person, under specified conditions.[147] A compulsory license can only be granted when the holder is not producing and selling the propagating material of the PVP in sufficient amount to meet the needs of the general public and has refused to license other persons to produce and sell the propagating material of the protected variety; or unwilling to give such license under reasonable terms; or there is no condition under which the holder can be expected to give a permit to use his protected variety.[148] A person who is granted a compulsory license has an obligation to pay remuneration to the right holder. The Ministry of Agriculture will determine the amount of compensation, duration and other conditions of using the license.[149] However, this PVP law has not yet been implemented due to the lack of implementing regulations and directives envisaged by the law. Efforts are being made by the Ministry of Agriculture to amend the law and develop implementing regulations. One of the reasons behind the revision of the existing law is the complaint that the emerging flower and horticulture industry in the country is unable to access improved varieties from aboard, which affects its competitiveness in the export market.

The patent and PVP laws have provisions that can address food security related issues. Some of the provisions are broader or contrary to the requirements of relevant international treaties. This is possible as a result of Ethiopia's non-membership to a number of international IP agreements. At present, Ethiopia is a member of the 1981 Nairobi Treaty on the Protection of the Olympic Symbol, which it joined in 1982, and the Convention establishing the World Intellectual Property Organization that the country acceded to in 1998. This will soon change. Ethiopia is in the process of acceding to the WTO, which will require revising the laws to comply with the requirements of the TRIPs agreement. Moreover, there is a

---

[144]Id., Article 28(2).

[145]Id., Article 6(2).

[146]Article 7(1) of the Plant Variety Protection law provides that the Ministry of Agriculture has the power to restrict the exercise of a PBR where problems arise due to anti-competitive practices of the right holder; food security, nutritional or health needs or biological diversity are adversely affected; a high proportion of a protected variety offered for sale is being imported; and the requirements of the farming community for propagating material of a particular protected variety are not met; and where it is considered important to promote public interest for socio-economic reasons and for developing indigenous and other technologies.

[147]Id., Article 8(1).

[148]Id., Article 8(2).

[149]Id., Article 8(3).

growing recognition of the need to join relevant international IP agreements and efforts are being made by the Ethiopian Intellectual Property Office to develop accession proposals that will be submitted to the government for approval.[150] The economic partnership agreement that the country is negotiating with Eastern, Central and Southern Africa countries, and the European Union will include a chapter on IP that may erode the flexibilities available in international IP treaties such as the TRIPS Agreement and will require making changes to the existing law.[151] The changes due to these new developments and their implication for food security concerns remain to be seen.

45.6 Food Security Policies and Laws of Ethiopia and IP Issues

45.6.1 Relevant Policies

Enhancement of food security in least developed countries such as Ethiopia requires a package of policies that address the production and marketing of agricultural products.[152] This is well recognized in Ethiopia. There are relevant policies, strategies and plans issued by the current government that aim at realizing or supporting food security in the country.[153] These include the:

a) Agriculture Led Industrialization Strategy;
b) Growth and Transformation Plan;
c) Rural Development Policy and Strategies;
d) Food Security Strategy;

---

[150] Draft National IP Policy and Strategy.

[151] The EU has concluded similar agreements with other developing countries that belong to the African, Caribbean and Pacific (ACP) groups such as CARIFORUM consisting of Antigua and Barbuda, the Bahamas, Barbados, Belize, Dominica, the Dominican Republic, Grenada, Guyana, Haiti, Jamaica, Saint Lucia, St Kitts and Nevis, St Vincent and the Grenadines, Suriname and TrinIdad and Tobago. See Malcolm Spence, "Negotiating Trade, Innovation & Intellectual Property: Lessons from CARIFORUM EPA — Experience from a negotiator's Perspective," UNCTAD-ICTSD Project on IPRs and Sustainable Development, Policy Brief No. 4 (September 2009).

[152] Blakeney (2009), p. 88.

[153] It should, however, be noted that the previous governments, namely the Imperial regime (1930 to 1974) and the Military government (1974 to 1991) had taken similar measures to enhance agricultural productivity and deal with the problem of food insecurity. However, the measures were not as extensive and comprehensive as the present government. In addition to the federal government, regional governments have also issued policies and strategies to foster sustainable development and support the goal of attainment of food security at the house hold level. The discussion in this section, however, is limited to federal policies, strategies and plans of the present regime.

e) National Seed Industry Policy;
f) Science, Technology and Innovation Policy[154];
g) Agricultural Research policy;
h) National Population Policy (NPP)[155];
i) Conservation Strategy of Ethiopia ("CSE")[156];
j) Environment Policy; and
k) Biodiversity Policy.

The brief review of the relevant policies, strategies and plans in this section is limited to those that give emphasis to agriculture, involve IP issues and can be supported by IP. Gaps and weaknesses of the policy instruments are identified and the extent to which the existing IP system may complement identified policy goals and support implementing strategies is examined below.

45.6.2 Agriculture Led Industrialization Strategy

Ethiopia's economic growth and development is intrinsically linked to the development of its agriculture sector. The country is highly dependent on this sector for income, employment and export earnings. As a result, the government adopted the Agricultural Development Led Industrialization Strategy (ADLI) in 1993[157] as a long term development strategy for sustainable economic development in which agriculture is its driving force. ADLI aims to increase agricultural production and productivity in order to improve the living conditions of majority of its population and enhance export earnings to finance investment in the industrial sector and support other development efforts of the country. The strategy focuses mainly on

---

[154]The Science, Technology and Innovation Policy of the Federal Democratic Republic of Ethiopia, which was issued in 2012 replacing the national science and technology policy that was adopted in 1993, aims to build and promote technology transfer capacity that will meaningfully support the socio-economic development endeavors of Ethiopia. The policy is relevant to food security in that the development of indigenous technological capability will help generate new and improved technologies that may enhance agricultural productivity and facilitate effective transfer of technology.

[155]The national population policy of Ethiopia that was issued in April 1993 is relevant to food security in that it aims to deal with the problem of increased population that had an adverse impact on agricultural production and natural resources. The policy aims to harmonize population growth rate and the capacity of the country for development and rational utilization of natural resources to maximize the level of welfare of the population over time. For details of the policy, its impact and the challenges faced, see Minas (2008), pp. 23–45.

[156]The Conservation Strategy of Ethiopia, which was issued in 1997, is an umbrella strategy that considers all sectors of human activity and provides a comprehensive cross-sectorial analysis of conservation and resource management issues. The strategy will help to arrest environmental degradation, which affects agricultural productivity and food security.

[157]The strategy was adopted by the transitional government of Ethiopia and continues to guide the development policies and strategies of the present government.

rural development[158] and industries that will optimize use of abundant resources such as labor and land against the scarce resource-capital. The strategy envisages that improved farm income will generate sufficient demand for the industrial sector instigating dynamism and inter sectorial linkages.

A number of sectorial policies and strategies, including the food security strategy, reflect and complement the goals and policy directions of ADLI. IP can support ADLI by providing incentives that will stimulate inventive and innovative activities, encourage investment and facilitate transfer of technology.

45.6.3 Growth and Transformation Plan

The Federal Democratic Republic of Ethiopia has a 5 year Growth and Transformation Plan (GTP) covering the period 2010/11 to 2014/15, which was adopted by the Council of Ministers and House of People's Representatives to carry forward the important strategic directions pursued in the Plan for Accelerated and Sustained Development to end poverty (PASDEP).[159] The GTP outlines the areas of emphasis and measures that will be taken to enhance agricultural production and productivity, manage natural resources and the environment, develop livestock resources of the country, and put in place a transparent, efficient and effective marketing system in order to ensure food security. In particular, the section related to increasing agricultural production and productivity focuses on strengthening agricultural research, encouraging the adoption, use and transfer of foreign technologies, attracting investment, and improving competitiveness of agricultural export products in the international market. The plan's target is to decrease the number of households in safety net program from 7 million to 1.3 million and increase the country's food reserve from 0.41 million tons to 3 million tons at the end of the plan period.

The strategies and measures set forth by the GTP can be meaningfully supported by IP. The GTP acknowledges the significance of IP and enumerates the measures that will be taken to support Ethiopia's development efforts. The plan prioritizes IP rights to encourage local innovators to build local technological capacity; strengthens the use of IP as a tool for economic development; and recognizes the significance of technological information in supporting indigenous innovative activities and meeting technology needs. The latter requires that information from 5 million patents be used for technology transfer and adaptation during the plan period.

---

[158]The strategy emphasizes increasing the productivity of small farmers and promotes expansion of large-scale private commercial farms.

[159]PASDEP was the second 5 year plan that covered the period 2005–2010 following the Sustainable Development Plan to Reduce Poverty (SDPRP) that was implemented during 2000/1 to 2004/05. Each of these plans, including GTP, aimed at reducing poverty and enhancing socio-economic development within the framework of ADLI.

45.6.4 Rural Development Policy and Strategies

The Rural Development Policy and Strategies was issued in April 2003 to guide rural and agricultural development to ensure rapid economic growth, enhance benefits for the people, eliminate the country's dependence on foreign aid, such as food assistance, and promote the development of a market oriented economy. The policy is articulated based on the recognition of factor endowments of the country. These include the scarcity of capital and abundance of labor and land resources, the contribution of agriculture in accelerating economic growth, and the linkage between the agricultural sector growth to the development of other sectors, like trade and industry.

The policy defines strategies aimed at fostering rural and agricultural development in general and addressing specific food security issues. These strategies include increasing agricultural productivity, improving natural resources management and development, producing marketable output of acceptable quality, improving the marketing system, facilitating access to finance, developing rural infrastructure, encouraging private sector participation and attracting foreign investment. IP can meaningfully support some of these strategies by providing an environment that encourages the generation of new and improved agricultural technologies and facilitates the transfer and use of foreign technologies.

45.6.5 Food Security Strategy

Ethiopia issued the first and current food security strategies in 1996 and 2002, respectively, to ensure food security at the household level in food insecure rural areas.[160] The current food security strategy divides food insecurity as chronic[161] and acute[162] but recognizes the interrelationship between the two.[163] Unlike the 1996 food security strategy, the present strategy recognizes the need to maintain,

---

[160] Many criticize the limitation on the scope of this strategy since there are food insecure households in rural areas that are considered food secure as well as in urban areas. See, for example, Negatu (2008).

[161] Chronic food insecurity reflects the situation where households face food shortage because of overwhelming poverty and lack of assets. The shortage is permanent since food insecure people require food assistance in any given year even where there are no natural or manmade shocks. Chronic food insecure households live mainly in drought prone, moisture deficit areas and peripheral pastoral areas that do not produce enough food and do not have enough income to purchase food from the market.

[162] This is a situation where food shortage is caused as a result of natural or manmade shocks such as drought and war. The impact of the shock is temporary and the food shortage can be met with short term assistance.

[163] The division is theoretical in that unpredictable shocks do not suddenly lead to acute food insecurity unless people are already very poor, as in the case of the chronically food insecure.

develop and sustain the environment necessary to meet food security. This strategy is based on three pillars: increasing the availability of food through increased domestic production, ensuring access to food for food deficit households and strengthening emergency response capabilities. A number of programs were developed and implemented to realize the goals of this strategy.[164] The productive safety net program (PSNP), implemented in 2005 targeted approximately 5 million chronic food insecure people.[165] It aimed to prevent asset depletion at the household level and create assets at the community level by providing employment opportunities in public works to those who are able and giving cash or food to those unable to work in chronically food insecure areas.[166]

IP can support the food security strategy by providing incentives to persons that engage in the generation of new or improved agricultural technologies such as agricultural implements and disease and drought resistance and high yielding varieties that help increase agricultural production and enhance agricultural productivity.

45.6.6 National Seed Industry Policy

The National Seed Industry Policy was approved in October 1992. The policy is very broad since it is not confined to seed improvement, development and utilization as one may expect from its name. In fact, it deals with a number of broad biodiversity and R&D issues that should have been left to separate policies.

One of the major objectives of the national seed industry policy is to develop an effective system to produce and supply high quality seeds of important crops to satisfy national seed requirements. The generation of improved seeds that enhance agricultural productivity and meet the need for food security involves research that requires investment, time and effort. Such an investment may not be possible in the absence of IP protection. This is recognized by the policy itself when it envisaged the enactment of a law that will protect Plant Breeders' Rights. The issuance of the Plant Breeders Proclamation in 2006 is believed to emanate from the policy direction set under the national seed industry policy.

---

[164]These include the productive safety net program ("PSNP"), voluntary resettlement, Household Asset Building Program (HABP), which aims to build assets for food insecurity vulnerable households, and a Complementary Community Investment facility (CCI) meant to Identify and fund small-scale community infrastructure to establish essential community assets. See Rahmato (2013), p. 117.

[165]This figure has increased to 8 million in 2008. See Negatu (2008).

[166]Id., pp. 12–13.

45.6.7 Agricultural Research Policy

The Agricultural Research Policy, issued in 1993, has broad and specific policy objectives and policy directions relevant to food security. The overall objective of the policy is to enhance the use of science and technology in agriculture and to give clear direction and guidance for agricultural research and application of research results in agricultural development. The specific policy objectives include target oriented, coordinated and integrated research in various agricultural fields to solve major agricultural problems in Ethiopia while simultaneously focusing on the protection and development of the environment. A number of policy directives and strategies are specified to realize these objectives. These include undertaking research to generate and select technologies that will enhance agricultural productivity, help overcome environmental problems such as erosion, land degradation and deforestation and enable sustainable use of natural resources.

IP can be used to support agricultural research endeavors and protect research results. The technological information contained in patent documents can be used to reproduce technologies that are already available but not validly protected in Ethiopia, enabling researchers to keep abreast of current developments and reorient research endeavors.[167] Moreover, the IP system can be used to protect and exploit agricultural research results. Public research and higher learning institutions in Ethiopia mainly do agricultural research. However, there is no clear policy direction on the ownership of research results made using public resources. This gap is well recognized. Efforts are being made to develop and implement institutional IP policies by public higher learning and R&D institutions such as the Addis Ababa University and the Ethiopian Institute of Agricultural Research.[168]

45.6.8 Environment Policy

The Environment Policy was adopted in April 1997 to ensure a sustainable socio-economic development of the country through sustainable management and utilization of natural, manmade and cultural resources and the environment. The environmental policy gives due attention to the conservation and utilization of biodiversity resources that can support food security. The policy acknowledges community IP rights and explicitly states the need to create a system for the protection of community IP rights. The concept of community rights is a new concept. It is very difficult to define the subject matter of protection, who the holders of such a right are, and how the right will be acquired, exercised and

---

[167] For details of the role of the patent system as a source of valuable technological information that can be used to strengthen and support research and innovative endeavors, *see* Mengistie (2010b), *parts I and II*. Int. T.L.R.

[168] Background study for the draft National IP policy.

enforced. Moreover, the concept of property in its conventional sense is difficult to apply for community achievements that are not considered a commodity and are freely exchangeable. It may be due to this problem that the African model legislation for the recognition and protection of local communities, farmers and breeders[169] deal with community rights and community intellectual rights without referring to property.[170] Moreover, the concept of community IP is different from the conventional intellectual property rights (IPRs), which are private rights. The current IP regimes are designed to protect readily identifiable and new contributions to existing knowledge. Community knowledge is gradually built over decades or centuries and often lacks novelty in the sense required by the existing IPR regime.

The community IP rights envisaged in the environmental policy requires the development of a *sui generis* system that defines and safeguards such rights. However, such a *sui generis* system is not yet in place. The Access to Genetic Resources and Community Knowledge, and Community Rights Proclamation No. 482/2006 is devoted to community rights, such as community knowledge associated to genetic resources but not community IPRs.

45.6.9 Biodiversity Policy

On April 23, 1998, the Ethiopian government adopted the National Policy on Biodiversity Conservation and Research, which explicitly targets the conservation, development and sustainable use of biological resources as well as equitable sharing of benefits. The rationale behind the adoption of the policy is stated in the preamble as follows:

> Lack of comprehensive guidelines, awareness and appreciation of plant, animal and microbial genetic resources importation, and exportation and exchange activities has resulted in the movement of genetic resources into and out of the country in unregulated and uncoordinated manner. This has resulted in the loss of benefits from valuable indigenous genetic resources and in the introduction of unverified genetic materials, diseases, pests and weeds in the country. The National Policy on Biodiversity Conservation and Development is formulated based on the rational that the conservation of biodiversity is one of the conditions of the overall socio-economic development and sustainable environmental management goals. Hence, because of its vital importance in the socioeconomic well-being of the Ethiopian people, the conservation, proper management and the use of biodiversity need to be supported by policy, legislation and national capacity building.

The policy is tailored to a number of provisions of the Convention on Biological Diversity (CBD), which Ethiopia had joined in 1994. The national biodiversity policy recognizes the sovereign right of Ethiopia over its biological resources and provides guidance on the conservation and sustainable utilization of biological

---

[169]The OAU head of summit adopted a model law on protection of community rights in Ouagadougou, Burkina Faso in 1998.

[170]Ekpere (2001).

resources as well as regulation of access to biological resources and benefit sharing in compliance with the CBD requirements. The policy supports food security by facilitating the conservation, development and use of genetic resources that are essential for agricultural production. Moreover, the sharing of monetary and non-monetary benefits with local communities supports food security not only by providing income-generating opportunities but also by making new and improved technologies accessible and building capacity.

Access to genetic resources and benefit sharing raise IP issues that can be addressed by IP laws. Moreover, IP laws can be a tool to complement the objectives of this policy by facilitating sharing of equitable benefits by requiring disclosure of the origin of these genetic resources. The extent to which the existing IP laws address IP issues and complement objectives of the biodiversity policy and the access and benefit sharing law issued under the policy will be discussed under the subsequent section where laws relevant to food security are examined.

45.7 Relevant Laws for Food Security

Instruments such as laws and institutional arrangements should support the implementation of various policies and strategies related to food security. There are a number of relevant laws and institutions that can facilitate implementation and contribute to food security at the household level. These include:

a) Constitution of the Federal Democratic Republic of Ethiopia, Proclamation No. 1/1995 (here in after referred to as "constitution");
b) Federal Democratic Republic of Ethiopia Rural Land Administration and Land Use Proclamation No. 456/2005;
c) Access to Genetic Resources and Community Knowledge, and Community Rights Proclamation No. 482/2006;
d) Access to Genetic Resources and Community Knowledge, and Community Rights Council of Ministers regulation No. 169/2009;
e) Ethiopia Commodity Exchange Proclamation No. 550/2007;
f) Coffee Quality Control and Marketing Proclamation No. 602/2008,
g) Coffee Quality Control and transaction Council of Ministers Regulation No. 161/2009;
h) Sesame and White Pea beans, council of Ministers regulation no. 178/2010;
i) Definition of Powers of the Executive Organs of the Federal Democratic Republic of Ethiopia Proclamation No. 691/2010[171];

---

[171] Article 19 of this law defines the powers and duties of the Ministry of Agriculture, which include the power and duty to formulate and facilitate the implementation of a strategy for natural resources protection and development through sustainable agricultural development; build capacity for supplying, distributing and marketing of agricultural inputs; ensure the supply of inputs; undertake disaster prevention and preparedness activities and ensure proper implementation of

j) Forest Development, Conservation and Utilization Proclamation No. 542/2007;
k) Agricultural Transformation Council and Agency Establishment Council of Ministers Regulation No. 198/2010[172];
l) Emergency Food Security Council of Ministers Regulations No. 67/2000;
m) Seed Proclamation No. 20612000;
n) Ethiopian Agricultural Research Organization Establishment Proclamation No. 79/97 as amended by Proclamation No. 328/2004;
o) Biosafety Proclamation No. 655/2009;
p) Environmental Impact Assessment Proclamation No. 299/2002;
q) Environment Protection Organs Establishment Proclamation No. 295/2002; and
r) Institute of Biodiversity Conservation and Research Establishment Proclamation No. 120/1998 as amended by Proclamation No. 381 /2004.

It will be difficult to discuss each of the above laws due to space and time constraints. As a result, an attempt is made to cite relevant provisions of some laws mentioned above in foot notes when the law's title is not self-explanatory to its link to food security. Discussion is limited to selected laws that focus on land tenure, access and benefit sharing resulting from use of genetic resources, and marketing of agricultural products. Each of these laws can support agricultural production and marketing of agricultural products and serve as examples to highlight linkage with food security and explore relevant issues.

45.7.1 Land Law

An appropriate land tenure system can ensure or foster food security. Such a system creates a sense of stability and security as well as encourages landholders to conserve and utilize the land and resources thereon on a long-term basis. The Ethiopian Constitution declares that the right to ownership of rural land as well as of all natural resources is exclusively vested in the state and in the peoples of

---

food security programs; ensure creation of an enabling environment for the provision of credit facilities to farmers and pastoralists; monitor events affecting agricultural production and set up an early warning system; establish a system whereby stakeholders of agricultural research coordinate their activities and work in collaboration; and expand small-scale irrigation schemes to enhance agricultural development.

[172]The Council, which is chaired by the Prime Minster and consists of Ministers from relevant Ministries and the Director General of the Agency, has powers and duties that include providing leadership in Identifying, designing and effectively implementing solutions to basic hurdles of agricultural development and providing policy directions and leadership to ensure that effective coordination is realized among different actors involved in agricultural development. (See Articles 4 (1) and 5(1)). The Agricultural Transformation Agency, accountable to the Ministry of Agriculture and established under this law, Identifies systemic constraints of agricultural development by conducting studies, recommends solutions to ensure sustainability and structural transformation, and supports the establishment of strong linkages among agricultural and related institutions and projects to ensure the effectiveness of agricultural development activities. (See Articles 7 and 9).

Ethiopia[173] and that the federal government has the duty to enact laws for the utilization and conservation of land and other natural resources.[174] There are federal and regional land laws that define rights and obligations related to land in conformity with the Federal Constitution. The federal law that governs rural land use and administration is the Federal Democratic Republic of Ethiopia Rural Land Administration and Land Use Proclamation No. 456/2005 (herein after referred to as the "federal land law"). Regional land use and administration laws are enacted by regional governments based on the federal land law,[175] which specifies the basic principles of rural land distribution and utilization and defines the ownership and land use rights.

The federal land law states that ownership of land belongs to the state and peoples of Ethiopia. Individuals have a land use right called "holding right." This right refers to the right of any peasant farmer or semi-pastoralist or pastoralist to use rural land for the purposes of agriculture and natural resource development, lease and bequeath it to members of his family or other lawful heirs, the right to acquire property produced on his land thereon by his labor or capital, and to sale, exchange and bequeath same.[176] A holding right may or may not be limited in time depending on who has the holding right. The rural land use right of peasant farmers, semi-pastoralists and pastoralists is not limited in time while the duration of the right of others, such as private investors, is determined by the rural land administration laws of regions.[177] In order to ensure security of land tenure, the law provides for registration and issuance of land holding certificate[178] and prohibits eviction or displacement from the land unless for the purpose of public use. A person who is evicted from his land is entitled to compensation proportional to the development he has made on the land and the property acquired, or will be given substitute land.[179]

Some argue that the present land tenure system hinders an increase to agricultural productivity and achievement of food security at the house hold level. Dessalegne, for example, argues that the current land regime is inflexible and a barrier to enterprising endeavors, inhibiting peasant initiative and increased effort.[180] Since land is state owned, farmers do not have the right to sell or mortgage land. Moreover, the government has the power to distribute land,[181] creating fear

---

[173] Article 40(3) of the Constitution of the Federal Democratic Republic of Ethiopia, Proclamation No. 1/1995.

[174] Id., Article 51(5).

[175] Article 17(1) of the federal land law provides that each regional council shall enact rural land administration and land use law consisting of detailed provisions necessary to implement the federal land use and administration proclamation.

[176] Id., Article 2(4).

[177] Id., Article 7.

[178] Id., Article 6.

[179] Id., Article 7(3).

[180] Rahmato (2013), p. 128.

[181] Article 9 of the federal land law.

and insecurity of land use. The law recognizes that the holder of rural land use right has a right to transfer his right. However, such a right is highly restricted. Transfer of land through inheritance is limited to family members who permanently live with the holder of holding right and shares the livelihood of the latter.[182] Moreover, use of land through rental is subject to a number of conditions including the size of land to be rented and the duration of land rent. A holder of rural land use right can only rent out a portion of his land,[183] which in some regions cannot be more than half of his plot of land.[184] Short-term rents also cannot exceed 2 years and long-term rent cannot be more than 25 years.[185] The land lease agreement is valid only if it secures the consent of all the members who have the right to use the land and is approved and registered by the competent authority.[186]

45.7.2 Access and Benefit Sharing Law

Ethiopia's genetic resources have been accessed freely and in an unregulated manner for years due to the attitude that prevailed not only in Ethiopia, but also elsewhere, before the adoption of the Convention on Biological Diversity ("CBD").[187] Biological resources were considered as the "common heritage of mankind." The CBD, for the first time, recognized sovereign rights of states to regulate access to genetic resources within their jurisdictions.[188] However, such a right is accompanied by an obligation to enable access to genetic resources. The convention requires member states to create conditions to facilitate access to genetic resources for environmentally sound uses and not impose restrictions that run counter to CBD's objectives.[189]

CBD is a framework agreement, which merely prescribes principles, overall goals and basic rules. Thus, a national legal framework should be put in place to complement the convention. The convention itself calls on the contracting parties to take legal, policy and administrative measures that facilitates access to genetic resources by other parties and ensures fair and equitable sharing of benefits.[190]

---

[182] Id., Article 2(5).

[183] Article 8(1) of the federal land law provides that peasant fanners, semi pastoralist and pastoralist who are given holding certificates can lease to other farmers or investors land from their holding of a size sufficient for the intended development in a manner that shall not displace them.

[184] Rahmato (2013), p. 127.

[185] Id.

[186] Article 8(2) of the federal land law.

[187] The convention was adopted in 1992 at the UN conference on environment and development and entered into force on December 29, 1993. Ethiopia signed the Convention in 1992 and ratified it on 31 May 1994.

[188] Article 15(1) of the Convention on Biodiversity.

[189] Id., Article 15(2).

[190] Id., Article 15(7).

Accordingly, Ethiopia adopted the national biodiversity conservation policy that affirms their sovereign right over its biological resources, established the Institute of Biodiversity,[191] and enacted the Access to Genetic Resources and Community Knowledge, and Community Rights Proclamation, Proclamation No. 482/2006 (referred to as the "Access and Benefit Sharing ("ABS")" law),[192] which requires a permit to access genetic resources. This law was established to conserve biodiversity and ensure sustainable use of resources for the benefit and development of the people, recognize the contribution of local communities in the conservation, development and sustainable utilization of biodiversity resources, and comply with CBD requirements.[193] The law aims to ensure that the country and its communities obtain fair and equitable share of benefits arising from the use of genetic resources and to promote the conservation and sustainable utilization of the country's biodiversity resources.[194]

The ABS law defines the preconditions, requirements and procedures to obtain access permit and the obligations of an access permit holder. One of the preconditions is that the state and the concerned communities shall obtain fair and equitable share of benefits arising from the utilization of genetic resources (GRs) and community knowledge accessed.[195] The list of monetary and nonmonetary benefits includes license fee, upfront payment, milestone payment, royalty, research funding, joint ownership of IP rights, employment opportunity, participation of Ethiopian nationals in research, priority to supply raw materials of genetic resources for producing products from there, access to products and technologies developed through the use of the GRs or community knowledge, training at both institutional and local levels, provision of equipment, infrastructure and technology support and other benefits as appropriate.[196] The list of benefits is not exhaustive but illustrative. This allows room for negotiation and provision for other forms of benefits that were not envisaged during the enactment of the law. The obligations of access permit holder include negotiating new agreement with the Institute of

---

[191]The Institute, established in 1998 by Institute of Biodiversity Conservation and Research Establishment Proclamation No. 120/98, was called the Institute of Biodiversity and Research. Articles 6(2), 12 and 13 of this law respectively states that genetic resources cannot be accessed without permit. Permit is required for the collection, dispatching, importing or exporting of any biological specimen or sample. Engaging in any of these activities without securing a permit constitutes a criminal offence punishable with 5–10 years of imprisonment and a fine of fifteen to twenty thousand Birr.

[192]The law was enacted in February 2006, nearly 13 and 9 years after the country joined the CBD and adopted a national biodiversity policy, respectively.

[193]See preamble of the ABS law.

[194]Id., Article 3.

[195]Id., Article 12(3).

[196]Id., Article 19.

Biodiversity[197] based on relevant Ethiopian laws when he seeks to acquire IP rights over the genetic resources.[198] Moreover, the access permit holder has a duty to recognize the locality from where the genetic resource was accessed in the application for IP protection of the product developed using the genetic resource.[199] Such a requirement helps prevent bio-piracy and facilitates equitable sharing of benefits arising from the use of genetic resources. However, existing IP laws do not have similar provisions to complement this requirement. Both plant breeders' and patent laws do not require applicants to indicate the origin of genetic resources used to develop the product for which a patent or plant breeder's right is sought for. This gap is recognized and the draft national IP policy requires revision of the IP laws to complement the ABS law and contribute to the realization of national biodiversity policy objectives.

The ABS law supports the policies and strategies related to food security and will contribute to furthering food security at the household level. For example, the requirement of benefit sharing with communities[200] can serve as a source of off-farm income, meet the need for access to improved agricultural technologies and seeds, and build the capacity of farmers, thereby improving agricultural productivity through non-monetary forms of benefit sharing.[201] It is easier to state the potential contribution of the ABS regime based on its relevant provisions than corroborating it by tangible evidences in Ethiopia.

The number of access agreements is very few and each of them could not live up to its expectations.[202] An example is the *teff* agreement concluded between the Ethiopian Institute of Biodiversity Conservation and a Netherlands Company, Health and Performance Food International (referred to as "the genetic resource recipient" or the "company"), to enable the genetic resource recipient to access and use specified *teff* genetic resources to develop food and beverage products specified in the agreement. These products were considered to have considerable marketing potential in Europe and the USA because *teff* is gluten-free and high in nutritional

---

[197] The Institute is the competent government authority entrusted with the implementation of the law and issuance of permit to access genetic resources. See Articles 12 and 13 of the Institute of Biodiversity Conservation and Research Establishment Proclamation, Proclamation No. 120/1998 and Article 6 (12) of Institute of Biodiversity Conservation and Research Establishment (Amendment) Proclamation. No. 381/2004.

[198] Article 17(12) of the ABS law.

[199] Id., Article 17(14). Please note that there is a difference in the Amharic (the national working language) and English versions of the provision. The Amharic version refers to "intellectual property application" while the English version refers to "commercial property application". The problem seems to be poor translation. Where there is nonconformity between the Amharic and the English versions, the former prevails over the latter. The Amharic version is always the governing version.

[200] Article 9(2) of the ABS law provides that local communities have the right to obtain 50 % of the benefit shared by the state in the form of money and such money shall be used to the common advantage of local communities.

[201] The rest of non-monetary benefits listed by the ABS law can also contribute and support the endeavor to ensure food security at the household level.

[202] Reported access and benefit sharing agreements in addition to *teff* include the Vernonia agreement, which has also failed. See Andersin and Winge (2012).

value.²⁰³ The agreement offers various forms of benefit sharing, namely lump sum upfront payment, annual royalty payment, license fees, research collaboration, sharing of research results, develop *teff* business in Ethiopia, and an annual contribution of 5 % of the company's net profit, which should not be less than 20,000 Euros per year, to a fund used to improve living conditions of local communities.²⁰⁴ The agreement was lauded as the most advanced of its time and expectations were high.²⁰⁵ However, little of the expected benefits were gained. The company was declared bankrupt in 2009 and the only benefit that Ethiopia received was a payment of 4000 Euros and a small research project that was discontinued in less than a year.²⁰⁶ The genetic resource recipient transferred values to new companies that were established before its bankruptcy and the new companies continued with the production and sale of *teff* products.²⁰⁷ Ethiopia had no recourse against the new companies as they are not party to the agreement. The failure of the *teff* agreement had an adverse impact as the Ethiopian authorities have now become reluctant to enter into agreements.²⁰⁸

45.7.3 Law Governing Marketing of Agricultural Products

One of the factors that affect food security at the household level relates to market value of agricultural products. Farmers lack capacity to store agricultural products, bigger markets are inaccessible, a lack of market information is profound, and abusive market power of intermediaries is high. The majority of small farmers do not have capacity to store their grain and have no income outside of agriculture that will enable them to wait for a season when the prices of agricultural products are relatively higher. It was noted that 79 % of grain sales occur during the primary harvest season (January–March) owing to farmers' fear of storage loss and urgent need for cash.²⁰⁹ The sale of grains often takes place in small markets that are close to farmers due to a lack of modern transport that can help them reach bigger markets.²¹⁰ The market position of farmers is weak since they do not have information on current market price and often take the price offered by buyers.

Cognizant of this problem, the government enacted the Ethiopia Commodity Exchange, Proclamation No. 550/2007. The objectives of this law include creating an efficient, transparent, and orderly marketing system that serves the needs of

---

²⁰³Id.

²⁰⁴Section 8 on the agreement on access to and benefit sharing from *teff* Genetic Resources.

²⁰⁵Andersin and Winge (2012).

²⁰⁶Id.

²⁰⁷Id.

²⁰⁸Id.

²⁰⁹von Braun and Olofinbiyi (2007).

²¹⁰Id.

buyers, sellers and intermediaries, promotes increased market participation of small scale Ethiopian producers, provides a centralized trading mechanism in which bids are coordinated through a physical trading floor with open bidding or an electronic order matching system or both and provides timely market information to the public.[211] The Ethiopian Commodity Exchange (ECX) started operation as a coffee transaction center in November 2008 and later included sesame and white pea beans. It currently serves as a marketplace for these three agricultural products where transactions are conducted in a transparent manner, as required by the exchange and relevant laws dealing with quality control and marketing of each product.

The Coffee Quality Control and Marketing Proclamation, Proclamation No. 602/2008 and the Sesame and White Pea Beans Council of Ministers, Regulation No. 178/2010[212] define where and how the product transactions will be made as well as outline the obligations of the persons involved in coffee, sesame and white pea beans transactions. The coffee quality control and market law requires that coffee transaction take place at designated primary transaction centers and[213] the auction centers or local markets of the Exchange that were established by the Ministry of Agriculture & Rural Development or an appropriate regional body.[214] The laws define the obligations of persons involved in the transaction and services of each product.[215] Failure to comply with the requirements and obligations of the laws constitutes a punishable criminal offence.[216] The penalty provided under the Coffee Quality Control and Marketing Proclamation include a fine of Birr 20,000 (about $1000 USD at the current exchange rate) and imprisonment of at least 1 year but not exceeding 3 years.[217] The sesame and white pea beans regulation does not specify the penalty and leaves for the punishment to be determined in accordance with the relevant provisions of the criminal code.[218]

The establishment of the ECX, which is the first of its kind in Sub-Saharan Africa and hailed as a big step forward in the fight for transparency and against

---

[211] Article 6 on Ethiopia Commodity Exchange, Proclamation No. 550/2007.

[212] The regulation was issued pursuant to Article 31(2) of the Ethiopian Commodity Exchange, Proclamation No. 550/2007 as amended.

[213] This is a wholly state owned market institution established by law. See Ethiopia Commodity Exchange, Proclamation No. 550/200 on 4 September 2007.

[214] Articles 2(2), 5 and 6 of the Ethiopia Commodity Exchange, Proclamation No. 550/2007 and Articles 3, 4 and 13 of the Sesame and White Pea Beans Council of Ministers, Regulation No. 178/2010.

[215] Articles 7–12 of the Coffee Quality Control & Marketing Proclamation that define the obligations of coffee suppliers, coffee exporters, domestic coffee whole sellers, coffee roasters, coffee producers and service providers engaged in facilitating the marketing of coffee and Articles 13–18 on the obligation of persons involved in sesame or white pea beans transactions or services including suppliers, exporters, processors, service providers and producers.

[216] Article 12 on the Coffee Quality Control and Marketing Proclamation and Article 24 on the Sesame and White Pea Beans Council of Ministers, Regulation no. 178/2010.

[217] Article 15(2) for the Coffee Quality Control and Marketing Proclamation.

[218] Article 26 of the Sesame and White Pea Beans Council of Ministers, Regulation No. 178/2010.

market power abuse by the intermediaries, enables farmers to take their crops to a nearby warehouse, agree with the manager on the quality grading and present their minimum offering price, which can be modified within a month without a penalty or after a month with penalty.[219] The open market system, supported by electronic means, enables farmers to follow market price information for each agricultural product through their cell phones.[220] Such transparent marketing system has improved the farmers' negotiating position and income. According to a study by ECX, farmers now receive, on average, 65 % of the trading price.[221] It is important to note that this trading arrangement does not include all or bulk of the agricultural products of small farmers; it is limited only to three of the cash crops.[222] Nevertheless, this positive development can improve the income of small farmers and help ensure food security at the household level.

45.8 Conclusion and Recommendations

Since the main purpose of this chapter is examining relevant policies, strategies and laws relevant to or impacting food security and IP, the recommendations below are limited to measures that can be taken to strengthen existing policy and legal framework and ensure the effective use of these instruments to support food security at the household level.

The formulation and adoption of policies, strategies, and the enactment of laws is a very good beginning but not an end in itself. The policies and strategies can be well articulated on paper but are meaningless if not implemented. The population policy that can complement the food security strategy, for example, was inadequately implemented due to the absence of a detailed action plan and a legal framework.[223] Moreover, the 2006 PVP law aimed to encourage the development and facilitate the acquisition of new and improved varieties could not be implemented due to a lack of regulation and directive. Implementation of policies, strategies and laws presupposes the existence of institutional capacity. However, such capacity is absent.[224] The government should develop detailed programs and action plans, enact laws and implement regulations as well as design and establish capacity building programs for adequate implementation of policies and laws to achieve intended goals and objectives.

---

[219] Häberli (2012), p. 6.
[220] Id.
[221] Id.
[222] Id.
[223] Minas (2008), pp. 39–40.
[224] A number of policies and laws are not adequately implemented. This may be attributed to inadequate implementation capacities that include inadequate human resource, facilities and systems.

Policies, laws and institutions are instruments to achieve an end. Consequently, they should be dynamic and revised to accommodate new needs and developments. However, there are no current mechanisms to ensure dynamism of these instruments. The average age of these instruments is more than a decade. During this time, few revisions were made to the majority of these policies, strategies and laws. Furthermore, the impact of these policy instruments should be assessed on a periodic basis. However, no or little policy impact assessment related to IP and food security was conducted in Ethiopia. The government should develop mechanisms to follow up, evaluate and study the impact of these policy instruments. This will help ensure periodic revision of policies, strategies and laws to take into account the findings from these impact studies and any other new developments and needs.

IP can contribute to food security by supporting agricultural R&D and stimulating the generation of new and improved technologies and varieties that enhance agricultural productivity. The technical information contained in patent documents, for example, can be used to strengthen agricultural research. The Ethiopian Intellectual Property Office claims that it has a collection of more than 30 million patent documents consisting of technical information for inventions made since 1790. This includes inventions that can technologically advance the agricultural sector, such as improve traditional agricultural technologies used by small farmers for generations. However, this information is inadequately utilized and leveraged primarily due to a lack of awareness of the significance of technical details contained in patent documents.[225] Agricultural research results generated by public research organizations are not protected mainly due to a lack of policy direction on the ownership of research results made using public resources.[226] Thus, there is a need to encourage exploitation of patent documents to support agricultural research as well as to expedite the process of determining institutional IP policies and establishment of technology transfer offices in the public higher learning and agricultural research institutions.

Protection of community achievements and knowledge can also support food security by providing the opportunities for monetary and non-monetary benefits. The legal framework for existing IP law is inadequate to protect community achievements and knowledge. Hence, there is a need to develop a *sui genris* system of protection based on the interests and needs of local communities, as well as learning from experiences of other countries. The establishment of a national *sui genris* system for protection of community achievements and knowledge may be inadequate if not supported or complemented by an international system. Ethiopia should, thus, take part in the ongoing endeavors to develop international legal instruments for protection of traditional knowledge and traditional expressions

---

[225]See background study for the 2013 draft national intellectual property policy and Mengistie (2006).

[226]See background study for the 2013 draft national intellectual property policy.

under the auspices of the WIPO intergovernmental committee on IP, genetic resources, traditional knowledge and traditional cultural expressions.[227]

Ethiopia is in the process of acceding to the World Trade Organization and negotiating economic partnership agreement with the European Union, which may result in changes to existing IP laws. Efforts are also being made to revise the PVP law due to pressure from the emerging horticulture and floriculture industry. The country should take advantage of the flexibilities available in international IP treaties, such as the TRIPS agreement, and incorporate safeguard mechanisms and tools when amending existing and enacting new laws that address food security issues and concerns.

This chapter illustrates that Ethiopia has a clearly defined goal of ensuring food security at the household level. There are a number of policies, strategies, plans and laws that support and can contribute this goal. The commitment of the government is evident in its allotment of 16 % of its budget to strengthen the agricultural sector, the various measures taken to enhance agricultural productivity and address problems related to marketing, and the positive achievements made with the support of the international donor community are commendable. However, a lot remains to be done. Recommendations noted above can help Ethiopia achieve the desired goal of food security at the household level.

References

Access to Genetic Resources and Community Knowledge, and Community Rights, Proclamation No. 482/2006
Access to Genetic Resources and Community Knowledge, and Community Rights Council of Ministers, Regulation No. 169/2009
Act of International Convention for Protection of New Varieties of Plants (UPOV) (1991)
Agricultural Transformation Council and Agency Establishment Council of Ministers, Regulation No. 198/2010
Andersin R, Winge T (2012) The access and benefit sharing agreement on teff genetic resources: facts and lessons. Fridtjof Nansen Institute. Available at http://www.fni.no/doc&pdf/FNI-R0612.pdf (last accessed 6 Jan 6 2014)
Belete M (2011) Consequences of agricultural trade liberalization: a food insecure country's perspective. In: Krajewski M, Merso F (eds) Acceding to the WTO from a least developed country perspective. Nomos, Baden-Baden, pp 21–46

---

[227]The WIPO Intergovernmental Committee on Intellectual Property and Genetic Resources, Traditional Knowledge and Folklore was established in 2000 and entrusted with the mandate to develop an international legal instrument by the Assemblies of WIPO in 2009. The WIPO assembly renewed this mandate in its recent session held from September 23 to October 2, 2013. The decisions of the Assembles include that "the Committee will, during the next budgetary biennium 2014/2015, and without prejudice to the work pursued in other fora, continue to expedite its work with open and full engagement, on text-based negotiations with the objective of reaching an agreement on a text(s) of an international legal instrument(s) which will ensure the effective protection of GRs, TK and TCEs." See http://www.wipo.int/export/sites/www/about-wipo/en/assemblies/pdf/synthesis_2013.pdf (last accessed on 10 Feb 2014).

Bill & Melinda Gates Foundation (2010) Accelerating Ethiopian agriculture development for growth, food security, and equity: synthesis of findings and recommendations for the implementation of diagnostic studies in extension, irrigation, soil health/fertilizer, rural finance, seed systems, and output markets (maize, pulses, and livestock). Available at http://www.ata.gov.et/wp-content/uploads/Ethiopia-Agriculture-Diagnositc-Integrated-Report-July-2010.pdf (last accessed 10 Jan 2014)

Biosafety, Proclamation No. 655/2009

Biothai, GRAIN (1997) Sign posts to sui generis rights: resource materials from the international seminar on sui generis right. Biothai, Bangkok

Blakeney M (2009) Intellectual property rights and food security. CABI, Oxfordshire

Board on Science and Technology for International Development, National Research Council (1996) Lost crops of Africa: volume 1, Grains. Available at http://www.nap.edu/download.php?record_id=2305 (last accessed 14 Mar 2014)

Central Statistical Agency (2007) Summary of census report. Available at http://www.csa.gov.et/index.php?option=com_rubberdoc&view=category&id=62&Itemid=70 (last accessed on 26 Jun 2012)

Chanyalew D, Adenew B, Mellor J (2010) Ethiopia's Agricultural Sector Policy and Investment Framework (PIF) 2010–2020, Draft Final Report. Federal Democratic Republic of Ethiopia. Ministry of Agriculture and Rural Development, Addis Ababa

Coffee Quality Control and Marketing, Proclamation No. 602/2008

Coffee Quality Control and Transaction, Council of Ministers, Regulation No. 161/2009

Commission on Intellectual Property Rights (2002) Integrating intellectual property rights and development policy. CIPR, London

Constitution of the Federal Democratic Republic of Ethiopia, Proclamation No. 1/1995

Copyright and Neighboring Rights, Proclamation No. 410/2004

Correa C (2009) Trends in intellectual property rights relating to genetic resources for food and agriculture. Available at ftp://ftp.fao.org/docrep/fao/meeting/017/k533e.pdf (last accessed 17 Feb 2014)

Correa C (2012) TRIPS-related patent flexibilities and food security: options for developing countries, policy guide. QUNO-ICTSD, Geneva

Definition of Powers of the Executive Organs of the Federal Democratic Republic of Ethiopia, Proclamation No. 691/2010

Dutfield G (2008) Turning plant varieties into intellectual property: the UPOV convention. In: Tansey G, Rajotte T (eds) The future control of food: a guide to international negotiations and rules on intellectual property, biodiversity and food security. Earthscan, Abingdon, pp 27–47

Edwards S (1991) Crops with wild relatives found in Ethiopia. In: Engles JMM, Hawkes JG, Worede M (eds) Plant genetic resources of Ethiopia. Cambridge University Press, Cambridge

Ekpere JA (2001) The African model law for the protection of the rights of local communities, farmers and breeders, and for the regulation of access to biological resources. Available at http://www.wipo.int/edocs/lexdocs/laws/en/oau/oau001en.pdf (last accessed 10 Jan 2014)

Emergency Food Security, Council of Ministers, Regulations No. 67/2000

Environment Protection Organs Establishment, Proclamation No. 295/2002

Environmental Impact Assessment, Proclamation No. 299/2002

Ethiopia Commodity Exchange, Proclamation No. 550/2007

Ethiopian Agricultural Research Organization Establishment, Proclamation No. 79/97 as amended by Proclamation No. 328/2004

FAO (1996) Rome Declaration on World Food Security and World Food Summit Plan of Action. World Food Summit. 13–17 November 1996. Available at http://www.fao.org/docrep/003/w3613e/w3613e00.HTM (last accessed 6 Jan 2014)

FAO (2010) The state of food insecurity in the world: addressing food insecurity in protracted crisis. Available at http://www.fao.org/docrep/013/i1683e/i1683e.pdf (last accessed 6 Jan 2014)

FAO/WFP (2012) Crop & food security assessment mission to Ethiopia: Special Report. Available at http://www.fao.org/docrep/015/al987e/al987e00.pdf (last accessed 6 Jan 2014)
FDRE, Rural Land Administration and Land Use, Proclamation No. 456/2005
FDRE, Environment Policy (1997)
FDRE, Food Security Strategy (1996)
FDRE, Food Security Strategy (2002)
FDRE, Food Security Program 2004–2009 (2004)
FDRE, Growth and Transformation Plan (GTP) 2010/11-2014/15
FDRE, Science, Technology and Innovation Policy (2012)
FDRE, National Policy on Biodiversity Conservation and Research (1998)
FDRE, Rural Development Policy and Strategies (2003)
Feyissa R (2006) Farmer's Rights in Ethiopia: a case study. The Fridtjof Nansen Institute. Available at http://www.fni.no/doc&pdf/FNI-R0706.pdf (last accessed 7 Jan 2014)
Food & Agriculture Organization (FAO) & African Union (AU) (2012) National strategy and action plan for the implementation of the Great Green Wall initiative in Ethiopia. Available at http://www.fao.org/fileadmin/templates/great_green_wall/docs/FINAL_National_strategy_and_plan_of_implementation_Ethiopia_FV.pdf (last accessed 6 Jan 2014)
Forest Development, Conservation and Utilization, Proclamation No. 542/2007
Häberli C (2012) Ethiopia' food reserve policies and practice. Available at http://www.wti.org/fileadmin/user_upload/nccr-trade.ch/wp4/publications/NCCR_WP_2013_2_Ethiopia.pdf (last accessed 7 Jan 2006)
Institute of Biodiversity Conservation and Research Establishment, Proclamation No. 120/1998 as amended by Proclamation No. 381/2004
International Food Policy Research Institute (2009) Global hunger index 2009. The challenge of hunger: focus on financial crisis and gender equality. Available at http://www.ifpri.org/sites/default/files/publications/ib62.pdf (last accessed on 7 Jan 2014)
Inventions, Minor Inventions and Industrial Designs, Proclamation No. 123/1995
Inventions, Minor Inventions and Industrial Designs, Regulations No. 12/1997
Light Years IP (2008) Distinctive values in African export products: how intellectual property can raise export income and alleviate poverty. Available at http://www.lightyearsip.net/downloads/Distinctive_values_in_African_exports.pdf (last accessed 6 Jan 2014)
Mengistie G (2001) Bio prospecting in Ethiopia: enhancing scientific and technological capacity. ACTS Press, Nairobi
Mengistie G (2006) Intellectual property assessment in Ethiopia. Ethiopian Intellectual Property Office, Addis Ababa
Mengistie G (2010a) Intellectual property as a tool for development: the Ethiopian fine coffee designations trade marking and licensing experience. Int TLR 16(1):1–24
Mengistie G (2010b) The patent system in Africa: it's contribution and potential in stimulating innovation, technology transfer and fostering science and technology (part 1). Int TLR (5):138–152
Mengistie G (2010c) The patent system in Africa: it's contribution and potential in stimulating innovation, technology transfer and fostering science and technology (part 2). Int TLR (6):175–189
Mengistie G (2013) Intellectual property legal framework of Ethiopia, international encyclopedia of laws. Kluwer International, Netherlands
Merso F (2011) A critical reflection on the legal framework providing protection for plant varieties in Ethiopia. JEL 25(1):113–158
Minas G (2008) A review of the national population of Ethiopia. In: Assefa T (ed) Digest of Ethiopia's national policies, strategies and programs. Forum for Social Studies, Addis Ababa, pp 23–45
Ministry of Finance and Economic Development (MoFED), A plan for accelerated and sustained development to end poverty (PASDEP) (2006)

Negatu W (2008) Food security strategy and productive safety net program. In: Assefa T (ed) Digest of Ethiopia's national policies, strategies and programs. Forum for Social Studies, Addis Ababa, pp 1–22

Pankhurst A, Rahmato D (2013) Introduction. In: Rahmato D, Pankhurst, van Uffelen J (eds) Food security, safety nets and social protection in Ethiopia. Forum for Social Studies, Addis Ababa, pp xxv–xlv

Plant Breeders' Right, Proclamation No. 418/2006

Rahmato D (2013) Food security and safety nets: assessment and challenges. In: Rahmato D, Pankhurst, van Uffelen J (eds) Food security, safety nets and social protection in Ethiopia. Forum for Social Studies, Addis Ababa, pp 113–146

Rajotte T (2008) The negotiations web: complex connections. In: Tansey G, Rajotte T (eds) The future control of food: a guide to international negotiations and rules on intellectual property, biodiversity and food security. Earthscan, London, pp 141–170

Seed, Proclamation No. 20612000

Sesame and White Pea Beans, Council of Ministers, Regulation No. 178/2010

Tansey G (1999) Intellectual property, food and biodiversity, key issues and options for the 1999 review of Article 27 (3) (b) of the TRIPS Agreement. Available at http://www.sristi.org/material/mdpipr2003/MDPIPR2003CD/M13%20trade%20ip%20food.pdf (last accessed 6 Jan 2014)

Tansey G (2008) Food, farming and global rules. In: Tansey G, Rajotte T (eds) The future control of food: a guide to international negotiations and rules on intellectual property, biodiversity and food security. Earthscan, London, pp 3–26

TGE, Agricultural Research Policy (1993)

TGE, National Population Policy of Ethiopia (1993)

TGE, National Seed Industry Policy and Strategy (1992)

Trade Practice and Consumers' Protection, Proclamation No. 685/2010

Trademark Protection and Registration, Regulation No. 273/2012

Trademark Registration and Protection, Proclamation No. 501/2006

Transitional Government of Ethiopia (TGE), Agricultural Development Led Industrialization Strategy (ADLI) (1993)

United Nations Development Program (2013) Human development report: the rise of the south human progress in a diverse world. UNDP, New York

United Nations, Convention on Biological Diversity (1992)

Van der Veen A, Tagel G (2011) Effect of policy interventions on food security in Tigray, Northern Ethiopia. Ecol Soc 16(1):18

von Braun J, Olofinbiyi T (2007) Famine & food insecurity in Ethiopia - Case Study #7-4 of the program: food policy for developing countries: the role of government in the global food system. Available at http://faculty.apec.umn.edu/kolson/documents/4103_cases/case_7-4.pdf (last accessed 7 Jan 2014)

von Uffelen JG (2013) Social protection in situations of chronic food insecurity and poverty: lessons from different models and implications for Ethiopia. In: Rahmato D, Pankhurst, van Uffelen J (eds) Food security, safety nets and social protection in Ethiopia. Forum for Social Studies, Addis Ababa, pp 5–40

World Bank (2012) Summary of ease of doing business in Ethiopia Report. Available at http://www.doingbusiness.org/data/exploreeconomies/ethiopia (last accessed 26 Jun 2012)

Chapter 46
Innovation and Development in Agricultural Biotechnology: Reflecting on Policy-Making Processes in Sub-Saharan Africa

Julius Mugwagwa and Watu Wamae

Abstract The aim of this chapter is to improve our understanding of underlying policy processes, conflicts and contradictions in agricultural biotechnology in Sub-Saharan Africa (SSA). The chapter examines the nature of relationships that emerge in policy making processes and how these influence attainment of social imperatives such as food security. The significance of broad-based public participation approaches and processes is recognised as one important aspect of policy processes if harnessing and sustainable deployment of innovations is to emerge in the agriculture and food security sector.

46.1 Introduction

New technologies play an important role in addressing global food security challenges especially in Sub-Saharan Africa ("SSA").[1] However, efforts to effectively access and exploit technological knowledge are a persistent challenge shaped by various economic, political, social and cultural realities. In SSA countries, where the vast majority of the population relies on farming for their livelihood, it is important to consider the impact of policy-making processes on food security and the socio-economic status of poor farmers.

The mechanisms developed for technology governance reflect a myriad of complex realities. Regulation as an instrument of technology governance provides the norms and standards for quality, safety, effectiveness, environmental protection

---

Material for this chapter was first presented in the Second Science with Africa Conference, Addis Ababa, Ethiopia (2010) and appears in non-referred conference proceedings of the same conference.

[1] FAO (2004).

J. Mugwagwa (✉) • W. Wamae
Development Policy and Practice Group, The Open University, Walton Hall, Milton Keynes MK7 6AA, United Kingdom
e-mail: Julius.Mugwagwa@open.ac.uk

and intellectual property protection. Regarding modern biotechnology in agriculture, developing countries recognize the importance of effective regulatory systems.[2] The ways in which biotechnology[3] is governed not only determines its ability to achieve socially desired aims, but also gives important signals about the direction of technological development.[4] The two elements are important for the credibility and legitimacy of new technologies, which to some extent, incorporating public consensus in policy may strengthen processes, technology use and development.[5] Tait notes that "increasingly, the actual outcomes of huge public and private investments in basic science are moderated by the attitudes, values and interests of a wide range of citizens and their representative groups."[6]

Demand is a key factor in the selection and success of a technology and must be factored into policy. As final consumers of biotechnology products, the public influences the orientation of biotechnology research.[7] Cohen notes, "the public controls the fate of biotechnology in its willingness or refusal to accept products produced ... communication efforts should recognize that the public is a full partner in deciding if, when, and how the technology is to be used..."[8] However, intermediary consumers of biotechnology and, in particular, private investors in the biotechnology industry also play a critical role in stimulating demand and consequently orienting the innovation trajectory.

The decisions of private investors are constantly influenced by both the dynamic nature of demand and technology. They play a key role in matching the demand within a specific context to specific technologies, such as biotechnology, and create supply incentives for technology development and commercialization. The uncertainty that surrounds potential risks of biotechnology coupled with resulting tensions may encourage the perception that SSA offers less resistance to testing the capacity for success of biotech innovations before launching them into other regions.[9] The nature of demand articulated by private investors may differ from that of poor consumers in SSA.[10]

The extent to which the final consumers, particularly low-income earners in SSA, can influence demand for biotechnology may be relatively limited in comparison to the demand articulated by intermediary consumers that target secondary

---

[2]Persely (1999).
[3]Defined as a continuum of traditional and modern biological techniques.
[4]Cohen and Paarlberg (2004), pp. 1563–1577.
[5]Jaffe (2004), pp. 5–19.
[6]Tait (2009).
[7]Rayner (2003), pp. 163–170.
[8]Cohen (2001).
[9]SSA markets are characterized by low-income consumers and are generally of peripheral interest to the dominant innovators (western multinationals).
[10]DPP (2009).

markets.[11,12] It is not presumptuous to suggest that the degree to which the concerns of the poor in SSA can be incorporated into policy frameworks is fairly limited. Moreover, other actors who have considerably more influence in shaping such frameworks, including scientists engaged in biotechnology research, government agencies that seek to coordinate and balance priorities across sectors, pursue specific interests that could inadvertently consign food security concerns to the background. The range of interests suggests that striking a balance in the framing of policy is complex.

This chapter illuminates latent aspects of policy-making processes and examines the potential for institutional change. It addresses two questions: What is the nature of interactions amongst actors in biotechnology decision-making processes in SSA? What are the implications of this for the design of effective innovation policies for food security? The chapter seeks to build a case for recognition of the importance of holistic processes in addressing concerns of poor farmers in general and food security in particular.

46.2 Framework for Analysis

This section frames the two core issues—power and scientific expertise in biotechnology—that will guide the analysis of decision-making processes. The ability of SSA countries to effectively address the issue of food security depends not only on the capacity to access and exploit modern agricultural biotechnology, but also on the ability to develop institutions that reflect the concerns of the population, which relies on traditional agricultural practices.[13] The underlying elements that define the nature of governance are critical in orienting innovation activities into specific trajectories that may not necessarily enhance the socio-economic welfare of poor farmers and consumers.

Knowledge and power are two central elements to policy-making processes. Biotechnology is knowledge intensive and problems around it, including social ones, are generally couched in scientific terms. Nevertheless, scientific considerations and political power are intertwined in decision-making processes. Shore and Wright highlight that "policies are most obviously political phenomena... political technologies advance by taking what is essentially a political problem, removing it from the realm of political discourse, and recasting it in the neutral language of

---

[11]Clark et al. (2002).

[12]Pressure from industry to speed up the approval of regulatory frameworks may limit opportunities to sufficiently engage other actors and incorporate their concerns. However, this is not unique to developing countries (Newell 2002).

[13]It is difficult to argue that access to new knowledge or modern biotechnology, in the case of agriculture, is the only constraint to food security.

science.[14] Central to this process is the use of expert knowledge in the design of institutional procedures".[15] The relationships that are dominant in framing regulations are based on a dynamic and complex interplay between knowledge and power.

Scientific evidence plays a dominant role in the decision-making process of biotechnology. However, experts in science do not necessarily override in all instances of policy-making. The uncertainty surrounding biotechnology provides a possibility to supersede scientific evidence and apply the precautionary principle. Hajer explains that:

> The precautionary principle is not a natural scientific concept but a policy principle, which is meant to illuminate the credibility of the idea of anticipatory policy and to create new coalitions. In that context the precautionary principle holds that policymakers will sometimes have to decide on action even if there is no scientific evidence of a causal link...[16]

This could mean that if science does not lead to a favorable decision in accordance with the political power that the latter can safely disregard it on the basis of the uncertainty of potential risks and evoke the precautionary principle.[17] In the case of SSA countries, the need to attract foreign investment, develop economic competitiveness or access specific export markets may outweigh other local priorities inadvertently strengthening the influence of the biotechnology industry.[18] This implies that questions relating to food security for the poor or safeguarding their socio-economic status, though in theory important, may remain peripheral to considerations that shape regulatory frameworks.

Various efforts have defined specific strategies and undertaken action to reduce hunger, which afflicts large populations in developing countries. For example, the Millennium Development Goals set the target of reducing hunger in half by 2015. These efforts are in tandem with recent views on improving the livelihoods of marginalized populations in developing regions that are articulated in development theory. For example, Sen views development as social transformation, which relates to issues such as hunger, environment and basic needs. He recognizes capabilities that include the ability to avoid starvation and undernourishment as the fundamental factor for development.[19,20] The capability to avoid hunger relates to various complementary aspects; for example, the ability of those afflicted by hunger and deprivation to influence policy-making processes that may consist in reconfiguring participation cultures. The relationships that are established in policy processes as a result are more inclusive. The ways in which actors with varying interests interact and the nature of relationships they form in decision-making

---

[14] Quoted from Dreyfus and Rabinow (1982:196) by Shore and Wright (1997).

[15] Shore and Wright (1995).

[16] Hajer (1995).

[17] See the case of Zambia's refusal to accept milled GM maize during the 2002–2003 food crisis (Clark et al. 2007).

[18] Newell (2002).

[19] Sen (1993).

[20] Sen (1999).

processes are perceived to occur within a dynamic context and offer potential for institutional change.

The importance of an integrated approach to development in developing countries has already been underscored within literature a few decades earlier.[21] Development economics in the 1960s stressed the significance of a holistic approach that integrates economic growth to socio-political change in addressing development challenges. For example, Adelman and Morris emphasised that an interdisciplinary approach is important for integrating political, social and institutional aspects to economic growth in developing countries. However, these perspectives were waylaid in the 1970s and 1980s during the economic booms and oil crises that eventually led to stringent restrictions on public expenditure following the Washington consensus. The resurgence of integrated development approaches over the last two decades has aggressively attempted to influence policy perspectives. In particular, the innovation systems (IS) perspective calls attention to the systemic nature of innovation, which is seen as a salient feature of development.[22] Within the IS framework, the underlying complex dynamics are determined by the nature of interactions amongst actors in defining and responding to institutions as well as triggering institutional change.[23]

The acknowledgement that innovation occurs within a system underlies the recognition that it is not only shaped by structural factors, but also institutional and social factors that determine social transformation. In addition, the IS framework views innovation as a cumulative process that is path-dependent, but open to change and therefore characterized by uncertainty and selectivity resulting in specific innovation trajectories. The formulation of biotechnology policy is framed around the uncertainty surrounding potential environmental and health risks, but at the same time attempts to define a trajectory that engenders uncertainties around broader socio-economic impacts. Furthermore, cumulativeness and path dependency are intrinsic features of decision-making processes. Perceptions that develop in the interactive environment in which relationships are formed are not static; rather they are subject to continuous review and reconfiguration. The ability to adequately develop practical problem-solving mechanisms for socio-economic challenges, including agriculture, requires a systems approach to innovation.[24]

---

[21]Hoselitz (1957).

[22]The IS perspective is based on three major components—organizations i.e. actors (firms, universities, research organization, policy agencies etc.), institutions i.e. the rules of the game that influence how organizations undertake innovative activities (such biotechnology regulations) and relations i.e. the linkages that determine the nature of interactions between and among organizations and institutions.

[23]Despite the wide acceptance of the IS perspective in the intellectual spheres, policy prescriptions still tend to reflect the science and technology approach, which is based on a linearity assumption of neoclassical growth theory.

[24]Hall (2005), pp. 611–630.

This chapter acknowledges that the relevance of IS framework of developing countries continues to attract debate.[25] Nevertheless, the framework provides scope for analysing the nature of relationships that shape technology governance including biotechnology. The nature of relationships in decision-making processes influences the ways in which actors organize themselves in achieving goals.[26,27,28] Understanding the context-specific differences that influence the nature of interactions, and the relationships that determine them within the decision-making processes, is an important step towards providing clarity on linkages as a major component of the IS perspectives.

46.3 Framing the Discussion

The complexity of adopting a broad framework for biotechnology decision-making processes suggests that different networks aligned to specific types of interests emerge. Understanding how actors group surface provides insights into how the relationships that form within policy-making processes determines regulatory frameworks.[29] Based on the notion of systems within the IS perspective, this chapter provides understandings into the emergence and nature of the mechanisms that have led to existing dominant networks within biotechnology decision-making processes in SSA. It also examines how the dominant networks interact with other networks to shape decision-making processes. The implications of these interactions are also discussed to understand the extent to which the dominant networks are tenacious in shaping decision-making processes and what this means within the broader framework of biotechnology innovation trajectories and food security in SSA.

There are a number of theoretical approaches that attempt to clarify how relationships in decision-making processes are formed and how the nature of those relationships shapes debates among policymakers. The analysis in this chapter draws insights from theoretical approaches linked to three aspects of biotechnology regulations: risk—a major issue relating to the uncertainty and management of biotechnology; the dynamic nature of discourses around regulation design; and

---

[25]Some of the main aspects that surround this debate relate to the fact that the concept originated in industrialized countries, where there is relatively significant innovation at the technology frontier, interactions amongst actors appear to be much tighter and the general innovation systems have relatively well established organizations and institutions.

[26]Nelson and Sampat (2001), pp. 31–54.

[27]Chataway et al. (2009).

[28]Nelson and Sampat (2001) and Chataway et al. (2009) discuss the importance of creating organizational structures and cultures, which are key in developing and distributing technologies. They refer to them as social technologies. An analysis of social technologies lies beyond the scope of this paper and is therefore not undertaken.

[29]Scoones (2002).

the participation of actors in the formulation of regulatory frameworks, which is key in achieving credibility and legitimacy of biotechnology.

Risk approaches are essential in understanding how sound science determinately guides the formulation of biotechnology regulation in SSA. While this is not a peculiarity of SSA countries, the relationships that emerge may take different forms owing to context-specific realities. It is posited that the focus on sound science has contributed to shifting the focus of the debate from principles to principals,[30] which leads to differences in the analysis of decision-making processes. Discourse analysis shows how biotechnology is framed and evolves in an environment that consists of groups of actors with different perspectives, which may be contradictory and fragmented. It is argued that while specific social perceptions, attributing to their malleability, permeate contexts with different realities, the core content may remain unchanged.

Approaches that relate to participation in policy formulation processes are key in understanding how public participation is conceived and the forms it assumes in SSA. Deliberative and performative participation are two forms of participation that are commonly discussed within literature on innovation and development. The former is based on discursive consultation and negotiation in decision-making processes while the latter is based on modes of actions such as farmer field schools. The analysis of decision-making processes in this chapter draws on deliberative participation. The discussion contends that public participation is crucial. Constant changes in decision-making processes could both provide and limit opportunities for influencing regulatory frameworks. However, decision-making processes are viewed as evolutionary and therefore the existence or absence of opportunities for public participation in one period does not create a deterministic path for the future.

46.4 Power Dispersal Issues Within Policy Processes of Agricultural Biotechnology

Knowledge and power are two guiding principles to policy-making processes. A discussion of these two core issues will guide the analysis of decision-making processes in SSA countries. The ability of SSA countries to effectively address the pressing issue of food security depends not only on the capacity to access and exploit modern agricultural biotechnology, but also on the ability to develop institutions that reflect the concerns of the majority of the population.[31] The

---

[30]In this discussion, the term principle refers to the core elements of the biotechnology debate. This relates to the uncertainty surrounding potential environmental and health risks, but also the definition of a biotechnology innovation trajectory that engenders uncertainties around broader socio-economic and ethical issues. Principal refers to biotechnology experts (champions) who are generally an integral part of the biotechnology decision-making processes.

[31]It is difficult to argue that access to new knowledge or modern biotechnology is the only constraint to food security and the improvement of most of SSA population, which relies on traditional agricultural practices.

underlying principles that define the nature of governance relationships that emerge in decision-making processes are critical in aligning innovation activities into specific trajectories. These may not necessarily enhance the socio-economic welfare of poor farmers and consumers.[32]

Complex forms of interactions amongst actors drive the processes through which regulatory systems emerge. To a large extent, the interactions are underpinned by the nature of knowledge for which the regulatory mechanisms are developed as well as the power dispersal, which influences the orientation of the regulations. The knowledge intensive nature of biotechnology and the high uncertainty surrounding possible human and environmental risks tend to overwhelmingly favor scientific and technical expertise in decision-making processes. Other actors that may have a relatively strong influence in shaping biotechnology regulations are private investors, although their influence closely depends on the government's position with regard to strategic economic priorities.

In the case of SSA countries, the need to attract foreign investments, develop economic competitiveness or access specific export markets may outweigh local priorities. The influence of the biotechnology industry may be inadvertently strengthened.[33] This implies that questions relating to food security for the poor or safeguarding their socio-economic status, though in theory important, may remain peripheral to considerations that shape regulatory frameworks. Decision-making processes are "inherently political processes, rather than simply the instrumental execution of rational decisions."[34]

It is possible to draw similar parallels with regard to other countries including developed countries, where risk management and commercial interests may override societal choices notwithstanding credibility and legitimacy problems from the public. Public participation exercises in Europe intended to address societal conflict in terms of desired technologies have not altered the framework, which essentially consists of minimizing health and environmental risks and maximizing benefits for the agricultural biotechnology industry. The social desirability of technologies in Europe is related to the joy of living that is derived from cultural landscapes and foods. These issues lead to public contestation of biotechnology because the existing framework renders them incompatible.[35] The framing of biotechnology regulation as a technical problem reduces the ability to include public concerns about the desirability of technologies; participatory exercises have 'biotechnologies democracy.'[36] Nevertheless, there are differences between Europe and SSA

---

[32] As will become evident in the discussion, these two core principles are important as they tend to overrule public concerns about the desirability of technologies despite their importance in achieving socially oriented imperatives.

[33] Newell (2002).

[34] Keeley and Scoones (2003).

[35] Rayner (2003), pp. 163–170.

[36] For example, although public participation in the UK GM debate did shape the government's policy on biotechnology and public participation is likely to continue to be seen as necessary, this does not foreclose the possibility of achieving contrasting outcomes in the future.

countries, including the perception of desirable technologies but it does not necessarily suggest radical variations in the importance of public participation.

The power balance between consumers and producers directly influences the institutional environment in which decisions are made. The ramifications of the nature of relationships conjured up in the process of establishing regulatory requirements have important implications, including the food security issues of the poor. Van Zwanenberg explains, "regulations inevitably privilege some classes of technology producers/users over others whether deliberately or through insufficient reflexive framing of issues and problems those regulations are designed to address".[37]

46.5 Scientific and Technical Considerations of Agricultural Biotechnology

Agricultural biotechnology is a knowledge intensive form of modern technology. It requires a strong scientific knowledge base of core competencies. This can facilitate adequate assessment of potential risks and benefits as well as provide sufficient flexibility for incorporating the emerging scientific evidence and shifting boundaries of social and ethical debates that raise fresh challenges to the credibility of biotechnology. Suitable scientific and technical competences in risk assessment are limited in many SSA countries. This is further compounded by the fact that other forms of competencies for articulating socio-ethical concerns tend to be scarce in SSA. For example, competencies for information communication and management, which are critical in developing an official and effective information strategy, are aimed at providing actors with sufficient transparency to allow for better articulation of specific interests. An adequate information strategy has multiple roles including educating the public, countering extremist views and gauging public attitudes, which is important in guiding institutional changes that are necessary in facilitating biotechnology innovation strategies.

The scenario in SSA could imply that those with scientific and technical competencies drive decision-making processes based on the deficit model of knowledge production.[38] SSA countries for the most part rely on bilateral and international assistance as well as multinational companies to develop scientific and technical competencies and formulate regulatory frameworks.[39] However, it is increasingly evident that public distrust in SSA, particularly with regard to the plight of poor farmers, is challenging the adequacy of institutional capacity to address public

---

[37] STEPS Centre (2008).

[38] The deficit model of knowledge production assumes that the public is not able to make informed decisions in policy-making processes because they have limited knowledge and therefore need to be educated (Bryant 2009).

[39] Cohen and Paarlberg (2004), pp. 1563–1577.

concerns. The scientists are at the center of distrust with regard to public interest. It is not their competencies, but rather their ability to make impartial decisions with respect to donors and multinationals that raises concerns.

Scoones notes that a key assumption of pro-poor agricultural biotechnology advocates is that "regulatory issues will be dealt with throughout the world by international 'capacity building' efforts in developing standardised, harmonised regulations for the agricultural biotechnology sector. With new regulations in place these will be enforced consistently and effectively throughout the developing world". Such efforts of capacity building and harmonization of regulation may provide simple and standardized regulatory procedures that encourage investment in the biotechnology industry or facilitate trade, but may not substantially promote the achievement of context specific moral imperatives. There are numerous aspects that limit the ability to effectively incorporate the concerns of the poor in biotechnology regulatory frameworks.[40,41]

The science-based approach' to problem framing is intricately and determinately tied to the knowledge intensive nature of biotechnology and appears to be core in determining the extent to which objectives are achieved. Attempts to incorporate the concerns of poor farmers and consumers in SSA are generally undertaken by champions who mainly operate within non-governmental organizations (NGOs).[42] NGOs are therefore thought to have strong links with the poor and are in a better position to articulate socio-ethical concerns in the highly scientific problem framing contexts of biotechnology regulation. Rayner points out that "NGOs tend to explain their motives for supporting public participation in terms of extending democratic control".[43] However, in SSA champions, or the NGOs within which the champions operate, who wield the strongest influence on decision-making processes cannot be assumed to hold interests that are entirely compatible with extending democratic control to the public.[44] The formulation of a regulatory framework is further complicated by the continuously changing regulatory environment and therefore offers no standard approach of reflecting the heterogeneity of the complex contextual realities. This raises critical questions in terms of incorporating context-specific realities of SSA, such as the challenge of food insecurity and the socio-economic status of poor farmers.

---

[40]Newell (2002).

[41]Glover (2003).

[42]NGOs are the main vehicle through which champions operate. However, it is the champions rather than the NGOs that are the subject matter of this discussion. As further discussed, the position held by champions may vary between pro and anti biotechnology and this may be related to champions moving from one NGO to another.

[43]Rayner (2003), pp. 163–170.

[44]Harsh (2009).

46.6 The Nexus Between External and Local Interest in Biotechnology Innovation: How the Debate Shifted from Principles to Principals

In SSA countries, research activities are undertaken in the public domain; 60–70 % of funding is absorbed by public research institutes compared to 30–40 % in the private sector.[45,46] In agriculture it was estimated that the public sector accounted for 96 % of the research expenditure in Kenya in 1996 and the corresponding proportion for Zimbabwe in 1998 was 84 %.[47] Most of the population in SSA derive their livelihood from agriculture and it is often the largest contributor to income and employment. There is a strong case for public investment in agricultural research, particularly with regard to the poor.[48,49] On the whole, however, the debate around improving the capacity to harness new knowledge, including biotechnology, is viewed as an issue of strengthening science-related disciplines within education systems and scientific capacity in public research institutes. Extreme academic conservatism generally characterizes higher education and public research institutes. Within a linear framework, it is difficult to engage in problem solving activities based on the trans-disciplinary approach that is better adapted to issues of food security and the problems faced by poor farmers.

Public research institutes are not only the purveyors and developers of knowledge in SSA, but are also the intermediate organizations. Although a number of successes have been reported, particularly in agriculture and health, the positive impact of public research institutes as intermediate organizations on social imperatives is less obvious. Other intermediate organizations including those that cut across disciplinary lines and particularly those that deal with socio-ethical issues are either absent or weak. In particular, channels for ensuring that the needs of poor consumers and farmers are represented in decision-making process are wanting.[50] The contribution of public research institutes in policy-making processes tends to focus on more funds for basic and applied research couched in publically palatable language of need to address developmental challenges.

The need to create framework for coordinating scientific efforts across Africa emerged as far back as the late 1920s and led to the African Survey in 1936.[51] The

---

[45] Arnold and Bell (2001).

[46] Although the structure of developed countries varies significantly from that of developing countries, it is important to note that in industrialized countries, private firms generate the bulk of new knowledge in-house.

[47] Cohen (2001).

[48] Clark (2002), pp. 353–368.

[49] For example, basic research in certain areas of agriculture where vast amounts of knowledge do not already exist, due to the intricate connection between agriculture and the context of biological systems, may be critical to addressing specific problems peculiar to the African context.

[50] Cohen and Paarlberg (2004), pp. 1563–1577.

[51] Gruhn (1971), pp. 459–469.

organizations that emerged out of these efforts primarily aimed at establishing a common communications network that could enhance utilization of African resources.[52] They were not intended to create a vehicle to integrate scientific and technical matters within the continent to address the socio-economic issues of indigenous populations. Latter efforts to harness scientific knowledge for the benefit of African economies are inclined towards the consolidation of infrastructure and emphasize knowledge through research and development.[53] The relationship between science and development, including perceptions of how public research institutes contribute to policy-making processes, has firm roots in mode one.[54] The top-down approach views innovation as a process driven by big-pushers through research and development. Policy-making is essentially the purview of scientists, although their interests may not necessarily prioritize or adequately articulate the concerns of food security and socio-economic welfare of poor farmers. This approach has been variously criticized in literature—policy that is based on this perception tends to be counter-productive. According to Arnold and Bell, "the high (almost religious) status, which the basic science establishment has managed to achieve, has made it hard to question the allocation of resources to it."[55]

Increasingly, public research institutes obtain funding through public support from external sources. Based on the general principle of limiting benefits that may accrue to the donor while maximizing those intended for the beneficiary, donors are naturally most confident in supporting public sector R&D. Historically, the public sector has been the main beneficiary of donor assistance and efforts have been made over time by changing the nature of relationships between donors and beneficiary. For example, there has been radical change from tied aid to more collaborative assistance. In practice, however, it may be argued that other forms of misalignments have emerged, or reinforced, and the principle may not render donor assistance significantly more successful in strengthening the delivery of knowledge assets for the socio-economic benefit of the vast majority of the populations in developing countries. As suggested by Hall and Dijkman, "instead of incentives for developing effective alliances for co-development, development assistance is a professional environment where succeeding (in reducing poverty) means working your way out of a job."[56]

---

[52]These efforts led to the creation of the Scientific Council for Africa South of the Sahara (CSA) in 1949 and the Commission for Technical Co-operation in Africa and Scientific Council for Africa South of the Sahara (CCTA) in 1950.

[53]The creation of African research centers has benefited from various sources including the United Nations Economic Commission for Africa (ECA) that has made notable efforts on shaping policy and has made attempts to incorporate STI directly into the development agenda of African countries through collaboration with NEPAD. Other initiatives have taken place in the form of conferences and workshops include CASTAFRICA I (1974), Lagos Plan of Action (1980), and CASTAFRICA II (1987).

[54]Chataway (2005), pp. 597–610.

[55]Arnold and Bell (2001).

[56]Hall and Dijkman (2008).

Private investment in SSA agricultural innovation also tends to collaborate closely with public research institutes. Various reasons could be advanced to support this form of partnership, including the latent weakness of African innovation systems; scientific research capabilities, though limited, are generally concentrated in public research institutes. However, broader reasons related to the credibility that accrues from such collaborations cannot be underestimated. Nevertheless, the relationships that form are not devoid of bias. Bryant carried out a study in Mali,[57] which indicated that the local scientists' views on the benefits of genetically modified (GM) crops for the country were essentially based on the linear model of more science leading to more innovation, and invariably providing solutions to socio-economic problems.

Clark[58] observes that some donor-funded structures may not be abreast with the knowledge-related restructuring that is occurring in developing countries, hence the dismal outcome observed over the last few decades. In particular, the validation processes that led to the introduction and use of modern technology is strictly confined to scientists and technology pushers who provide the necessary resources. The underlying assumption is that demand, a key factor in influencing the selection and success of a technology, is sufficiently addressed through considerations on the supply side. Little attention is paid to the final consumers of biotechnology products even though it is widely acknowledged that they influence the orientation of biotechnology research. Private investors in the biotechnology industry,[59] the public sector and donors are seen as adequate in articulating and addressing the concerns of final consumers. In SSA they are viewed as unable to make informed contributions to the definition of the biotechnology innovation trajectory.

Ayele et al., demonstrates that "partnership efforts tend not to be end-user oriented, but rather supply driven ... partnership projects tend to be limited in scope and not cast within holistic efforts to innovate in ways that affect food production and hunger".[60] Legitimacy and credibility issues are increasingly plaguing decision-making processes that relate to biotechnology in SSA. The public does not directly question their scientific and technical competencies, but rather it is their impartiality in dealing with donors and multinational firms that is treated with caution and is perhaps at the center of distrust with regard to public interest.[61] While the public recognizes that there are internal contradictions and tensions that prohibit their governments from adequately responding to social challenges, they

---

[57]Bryant (2009).

[58]Clark (2008).

[59]Private investors in the biotechnology industry have the twin role of intermediary consumers of biotechnology on the one hand and technology providers through investment in research and commercialization on the hand.

[60]Ayele et al. (2006).

[61]With the exception of a few countries, such as Zambia, scientists in most SSA countries and particularly those with a significant private sector in the biotechnology industry tend to be less critical of private sector interests. The private sector invests in biotech research and innovation in collaboration with local public research institutes (see Clark et al. 2007, p. 104).

do not assume that external pressure has no adverse effect on their governments' politics and policies.[62] The extent to which the local public sector is able to collaborate with donors and multinationals in ways that do not compromise the concerns of the poor and assert national sovereignty is fraught with mistrust.[63]

It cannot be assumed that the gaps that have emerged from this scenario have not been anticipated, albeit within the science push model. It would be presumptuous to suggest that the emergence of champions, particularly local ones within decision-making processes, is coincidental. In SSA economies the debate in many areas is shifting from the top-down approach, which has received incessant criticism in recent times, to a bottom-up approach. However, the bottom-up approach in not evident in the biotechnology decision-making processes. The techno-savvy champions are a direct product of the pre-existing structure. They generally have a history within public research institutes and have been entrusted with representing the concerns of the poor in biotechnology decision-making process, but not because of their apparent advisory independence. They are thought to have strong links with the poor and therefore well imbued with their concerns and, at the same time, to be in a strong position for independently articulating the socio-ethical concerns in the highly scientific problem framing contexts of biotechnology policy.

The champions operate within networks and play an important role of mobilizing and engaging a wide range of actors, whose views it is assumed are represented in policy-making processes. The extent to which such activities revolve around knowledge exchange rather than persuasion is debatable. It is difficult to deny that a form of cumulative and path-dependent learning is occurring within the linear model owing to its ability to recreate itself and remain tenacious. However, it is scarcely possible to suggest that any form of meaningful reflexivity with some potential for institutional change that reflects the need of poor consumers and farmers is taking place. The diversity of sites and the types of knowledge that contribute to decision-making processes are limited. The next section looks at how the focus on principals rather than principles is restructuring the relationships that shape decision making processes in SSA.

46.7 The Impact of the Shift on Biotechnology Decision-Making Processes

The creation of champions responds to cultural differences in SSA in that a certain sense of divinity is attached to science. Furthermore, scientists and science tend to be compounded. It would be foolhardy to suggest that SSA has a long tradition of engaging in heated criticism of science. The creation of local champions serves to palliate the public mistrust with regard to the government's ability to counter

---

[62]Smith (2003).
[63]Clark et al. (2002).

external pressure from donors and multinationals by providing a semblance of local advisory independence. At least two problems emerge from the unfolding situation in SSA that occurs within the broader non-static context of biotechnology. Firstly, the focus on principals further confuses or camouflages any cracks that may emerge around the divinity of science. It diverts public attention from questioning science as the only solution to local socio-economic challenges. Furthermore, other critical issues are pushed into the periphery of the debate. Literature on food insecurity has pointed out that food shortages with regard to the poor are often not a question of whether sufficient amounts of food exist, but rather whether the poor access it.[64,65] The question of access is intricately related to distribution of resources, including land tenure systems and infrastructure, which is determined by power dispersal.

Secondly, the demystification of scientists is easier to achieve through interactive processes compared to science itself, which is highly abstract.[66] For example, the stance held by a specific champion may oscillate between advocating and campaigning against biotechnology depending on changes in perceptions of the individual or changes in funding sources that require a change in perceptions. This demystification may lead to a blanket rejection of certain forms of biotechnology that are irrationally perceived as having nothing positive to offer. Such perceptions find a comfortable outlet in existing polemics within and outside a specific context. This situation has virulent repercussions. It may transpire as a direct affront to the champions who, in a desperate attempt to reassert themselves, may engage in counter extremist views. It is not unusual to come across blunt statements by champions in the media. The acrimonious debates that ensue suggest that there are no quick and simple strategies for constructive public engagement. These debates cannot be further from ensuring that concerns of poor farmers and consumers are addressed.[67]

While it is obvious that issues of safety and environment dominate the ways in which biotechnology innovation trajectory is evolving in SSA,[68] it is difficult to

---

[64]Cohen (2001).

[65]Smith (2003).

[66]The demystification of scientists in SSA is not restricted to champions in the biotechnology debate. In Kenya for example, the need to elect highly qualified professionals with doctoral and professorial credentials as MPs rather than candidates with minimal education was highly contested in the media in early 2007. The debate focused on the ability to deliver social services, particularly highly visible ones for the improvement the livelihoods of the poor. It was largely argued that the Nobel Prize holder lost her seat in the previous elections because of her inability to deliver development enhancing services. The sanctioning of champions takes different and subtler forms given that ascendancy to influential positions does not depend on the ballot.

[67]Cohen (2001).

[68]For example, in Ethiopia where there is wide ecological diversity and rich biological resources, the leading authority in the formulation of a biosafety policy is reported to have adopted protective principles that limit the use and development of biotechnology; it weakens the purported argument of focusing on small farmers and seems strongly guided by concerns of bio-piracy. In a different scenario, it has been suggested that SSA serves as a battle ground for EU-US disagreements on GM trade. The US aggressive approach confronts the EU cautious approach to GM crops by

argue that these aspects dominate the public debates relate to policy-making processes at the local level. Contestations of biotechnology are closely tied to the apprehension that stems from the power imbalance in externally triggered processes.

There are clear but insufficient attempts to define risk more broadly and to include the relationship between biotechnology innovation and food security.[69] Nevertheless, this does not interrogate the ability of sound science to tackle the broader socio-economic challenges of the poor. Furthermore, sound science is viewed as a magnet for pursuing the harmonization of regulatory systems in the region. Numerous discussions about the importance of pooling resources in the region encourage countries without sufficient scientific and technical capabilities or funds to engage in harmonization processes because missing such opportunities is viewed as a risk in its own right.[70] It is also not uncommon for SSA countries to borrow key elements of bio-policy from the few that already have one without questioning core elements of innovation trajectories. Wafula and Clark report, "Uganda was open in terms of borrowing key elements and tenets of a biotechnology policy from countries such as Kenya, Zimbabwe, South Africa, Namibia and the European Union."[71]

The biotechnology discourse is enshrined in champions whose framings of the connection between biotechnology innovation and food insecurity are shaped by contradictory and fragmented perspectives. Zambia is perhaps an exception as demonstrated by the 2002–2003 food crises. It is noted that the decision-making process on whether to accept GM food prominently featured the Zambian scientific community that was able to focus 'not only on purely scientific evidence, but also on the potential political and economic impacts of allowing GM food into the country'.[72,73] Nevertheless, Glover notes that:

> In countries which lack the capacity to compete in biotechnology, or where the degree of vested interests or the intensity of controversy is low, it is more likely that participation will be feasible and that public concerns will be allowed to frame the issues under consideration, as well as shape the decisions to be made...[74]

---

expanding biotechnologies in SSA—a source of EU agricultural products—and in so doing seeks to encourage the EU to develop more accommodating biotechnology policies (Clark et al. 2007, p. 101). The financial resources accruing to SSA from such disagreements cannot be overlooked as an important element drives local debates.

[69]The extent to which such efforts influence the outcome of policy-making processes is a different matter.

[70]Mugwagwa (2010), pp. 352–366.

[71]Wafula and Clark (2005), pp. 679–694.

[72]Clark et al. (2002).

[73]This may have been and perhaps paradoxically facilitated by the absence of a vibrant biotechnology industry. Zambia is yet to draw up a biosafety policy.

[74]Glover (2003).

On the whole, however, narrow perspectives of biotechnology in most SSA countries continue to have a disproportionate role in influencing policy. Sound science guides the formulation of biotechnology regulation and innovation in SSA. While this is not a peculiarity of SSA countries, the emergence of champions influences the relationships in decision-making processes specifically. For example, the existence of champions takes the public away from questioning the extent to which science can adequately take into account the challenge of food insecurity and the socio-economic status of poor farmers.

The scope of issue framing within the technology arena also influences how policy processes are shaped. Schattschneider (1960) discusses policy entrepreneurs engaging in 'venue shopping', i.e. searching for arenas from which to frame policy problems, and that the policy entrepreneurs may 'limit the venues in which they set their feet'.[75] For example, taking science only as a policy venue, the leeway for venue shopping is likely to vary across countries, across other sub-national arrangements and among policy actors.[76] Different actors seek access to different types of venues,[77] and this illuminates how public problems are a result successful imposition of problem definitions by one group on others.[78] In SSA, for many policy actors, biosafety is about safe application or use of products of modern biotechnology, while to others, it is about ensuring safety of all biological processes and products.[79] Science and scientific knowledge are key venues in both cases, but the extent to which these are explored and incorporated in the science-policy nexus differs because of the different levels of focus on science. These different framings result in what Schattschneider referred to as issues being 'organized into or out' of politics. In the final analysis, this has a bearing on the policy processes and the impact on food security.

It would be disingenuous to suggest that the divinity attached to science is unshakable in SSA, particularly owing to the rapid changes occurring in the wider spectrum, with regard to information and communication technologies. Nevertheless, public participation remains encapsulated in champions and is about persuasion rather than consultation.[80] The adherence of champions to an elite biotechnology community gives them considerable influence in decision-making processes, particularly owing to the uncertainty that surrounds biotechnology.[81] It is difficult to argue that the biotechnology debate in decision-making processes is evolving towards more mature debates in the north. However, decision-making processes are intrinsically evolutionary and the existence or absence of

---

[75] Schattschneider (1960).
[76] Renn (1995).
[77] Pralle (2003), pp. 233–260.
[78] Hajer (1995).
[79] Kelemu et al. (2003).
[80] Harsh (2009).
[81] Keeley and Scoones (2003).

opportunities for effective public participation in one period does not preclude variations in the future.

46.8 Implications for Food Insecurity and the Concerns of Poor Farmers

The inclusion of public participation in biotechnology decision-making processes is important and will continue to be necessary even if it is difficult to argue that it has had a positive and significant impact in aligning biotechnology innovation to social needs in SSA. In some countries, decisions about biotechnology innovation trajectories are driven by external efforts, such as international research institutes, donors and non-governmental organizations. Public participation has been facilitated informally through public-private partnerships. The role of the government has been mainly reactive and its ability to guide biotechnology innovation has been minimal.[82] Nevertheless it is important to emphasize that public participation is not the panacea for addressing the needs and ecological environments of poor farmers and consumers. Poor farmers and consumers in SSA face a wide range of challenges, including access to food and farm inputs because of market failure or distributional networks, agricultural employment, better wages and access to markets for farm outputs that cannot be entirely resolved by public participation. Moreover, improved public participation does not imply that the concerns of the public will automatically be reflected in the outcome of policy-making process and that as a result the problem food insecurity will become history.[83,84] Public participation may be viewed as a public relations exercise rather than a genuine form of consultation that is expected to shape the biotechnology innovation trajectory.

Thus the introduction of champions, who supposedly serve as knowledge brokers and help to strengthen interactions across a diverse range of actors through multiple activities are not likely to alter the core principles of biotechnology innovation. The networking role of the champions gives them significant leverage in bringing key individuals and institutions together and consolidating their influence over policy-making processes. The definition of key individuals and institutions may be based on a pre-determined type of contribution to decision-making that may not prioritize food security and the livelihood of poor farmers.[85] In the bio-policy formulation process in Uganda, key institutions, such as the national consumer organizations and the agricultural research organizations, were excluded from the process.[86] Public participation that is confounded with champions casts

---

[82]Harsh (2005), pp. 661–677.

[83]Rayner (2003), pp. 163–170.

[84]Ayele et al. (2006).

[85]Harsh (2009).

[86]Wafula and Clark (2005), pp. 679–694.

serious doubts on the degree to which the concerns of the poor can be constructively addressed.

The ability to adequately factor practical problem-solving mechanisms for socio-economic issues into innovation processes requires a systems approach to innovation.[87] Even within a holistic approach public participation is not the final solution to agricultural constraints. One of the shortcomings of the systems approach in SSA relates to the central question of demand and innovation. Whilst demand dynamics are viewed as critical in stimulating and defining technology paths that are congruent to the consumer and may better reflect the preferences and concerns of the public, SSA does not provide a strong case in support of this argument in agricultural biotechnology. As mentioned previously, intermediary consumers of biotechnology, particularly private investors in the biotechnology industry that target secondary markets, play a critical role in stimulating demand and orienting the innovation trajectory.

In the case of industrialized countries it may be argued that, "[O]ur consumption decisions are likely to have a greater impact in shaping our lives than our ballots. Thus, popular choices about governance seem to be increasingly made in the marketplace rather than in the legislature."[88] Neither ballots nor consumption decisions in SSA seem to offer much in terms of shaping governance. However, a systemic approach provides a premise for addressing the shortcoming of agriculture in an integrated way by identifying gaps and providing solutions from holistic perspectives. Other strands of literature, such as the value chain analysis, also provide avenues to identify broader and complementary channels for responding to the concerns of poor farmers and consumers in SSA.[89] It would be useful to discuss some broader prospects and processes for addressing the needs of the poor in agriculture. However, these remain beyond the scope of this paper.

46.9 Conclusion

This chapter sought to illuminate the nature of relationships that exist in policy processes in SSA. It demonstrated that effective public participation, specifically adequate representation of the socio-economic realities facing poor farmers and consumers in biotechnology decision-making processes, is limited despite its importance. However, the existence or absence of opportunities for effective public participation in one period does not preclude variations in the future. Decision-making is a continuous process involving negotiations that evolve with changing needs in a path-dependent manner and are principally based on continuous learning. Nonetheless, while it is not expected that improved representation would be

---

[87]Hall (2005), pp. 611–630.
[88]Rayner (2003), pp. 163–170.
[89]Kaplinsky (2005).

sufficient in addressing the challenge of food insecurity and the plight of poor farmers in SSA, it can make an important contribution to attempts for aligning biotechnology to the socio-economic welfare of the poor. Additionally, the significance of broader prospects and processes must also be taken into account.

References

Arnold E, Bell M (2001) Some good ideas for research and development. In: Partnerships at the leading edge: a Danish vision for knowledge, research and development, report of the commission on development-related research. Copenhagen

Ayele S, Chataway J, Wield D (2006) Partnerships in African crap biotech. Nat Biotechnol 24 (6):619–621

Bryant (2009) Deliberative governance; political fad or a vision of empowerment? In: Lyall C, Papaionnou T, Smith J (eds) The limits to governance; the challenge of policy-making for the new life sciences. Ashgate, England

Chataway J (2005) Introduction: is it possible to create pro-poor agriculture-related biotechnology? J Int Dev 17:597–610

Chataway J, Hanlin R, Muraguri L, Wamae W (2009) PDPs as social technology innovators in global health: operating above and below the radar. In: Penea O (ed) Innovating for the Health of All, vol 6, Global Forum update on research for health. Global Forum for Health Research, Geneva, pp 123–126

Clark N (2002) Innovation systems, institutional change and the new knowledge market: implication for third world agriculture development. Econ Innov New Technol 11(4–5):353–368

Clark N (2008) Science and technology for developing countries: the 'Sussex Manifesto' Revisited. In: Learning innovation knowledge. United Nations University

Clark N, Hall AJ, Reddy P (2007) Client-driven biotechnology research for poor farmers: a case study from India. Int J Technol Manag Sustain Dev 5(2):125–145

Clark N, Mugabe J, Smith J (2002) Biotechnology in Africa. African Centre for Technology Studies. ACTS Press, Nairobi

Cohen J (2001) Harnessing biotechnology for the poor: challenges ahead for capacity, safety and public investment. J Hum Dev 2(2):239–263

Cohen J, Paarlberg R (2004) Unlocking crop biotechnology in developing countries: a report from the field. World Dev 32(9):1563–1577

DPP (2009) Below the Radar: what does innovation in the Asian driver economics have to offer other low income economies? In: Development policy and practice. The Open University, Milton Keynes

Dreyfus HL, Rabinow P (1982) Michel Foucault: beyond structuralism and hermeneutics. Harvester, Brighton, Chapter 5

FAO (2004) The state of food security in the world. Food and Agriculture Organization of the United Nations, Rome

Glover D (2003) Public participation in national biotechnology policy and biosafety regulation. Institute of Development Studies

Gruhn I (1971) The commission for technical co-operation in Africa, 1950–65. J Mod Af Stud 9 (3):459–469

Hajer MA (1995) The politics of environmental discourse: ecological modernization and the policy process. Clarendon, Oxford

Hall A (2005) Capacity development for agriculture biotechnology in developing countries: an innovation systems view of what it is and how to develop it. J Int Dev 17:611–630

Hall A, Dijkman J (2008) New global alliances: the end of developmental assistance? In: Learning innovation knowledge. United Nations University

Harsh M (2005) Formal and informal governance of agricultural biotechnology in Kenya; biotechnology accountability in controversy surrounding the draft biosafety bill. J Int Dev 17:661–677

Harsh M (2009) Non-governmental limits: governing biotechnology from Europe to Africa. In: Lyall C, Papaionnou T, Smith J (eds) The limits of governance: the challenge of policy-making for the new life sciences. Ashgate, England

Hoselitz BF (1957) Feudalism in history. World Polit 267–279

Jaffe G (2004) Regulating transgenic crops: a comparative analysis of different regulatory processes. Transgenic Res 13:5–19

Kaplinsky R (2005) Globalisation, poverty and inequality. Polity Press, Cambridge

Keeley J, Scoones I (2003) Understanding environmental policy processes: a review. Institute of Developmental Studies

Kelemu S, Mahuku G, Fregene M, Pachico D, Johnson N, Calver L, Rao I, Buruchara R, Amede T, Kimani P, Kirkby R, Kaaria S, Ampofo K (2003) Harmonization the agricultural biotechnology debate for the benefit of African farmers. Afr J Biotechnol 2(11):394–416

Mugwagwa JT (2010) Alone or together? Can cross-national convergence of biosafety systems contribute to food security in Sub-Saharan Africa? J Int Dev 22:352–366

Nelson RR, Sampat BN (2001) Making sense of institutions as a factor shaping economic performance. J Econ Behav Organ 44:31–54

Newell P (2002) Biotechnology and the politics of regulation. Institute of Development

Persely GJ (1999) Agricultural Biotechnology and the Poor: Promethean Science. http://www.cgiar.org/biotech/rep0100/persley.pdf

Pralle SB (2003) Venue shopping, political strategy, and policy change: the internationalization of Canadian forest advocacy. J Public Policy 23(3):233–260

Rayner S (2003) Democracy in the age of assessment: reflections on the roles of expertise and democracy in public-sector decision making. Sci Public Policy 30(3):163–170

Renn O (1995) Style of using scientific expertise: a comparative framework. Sci Public Policy 22 (3):147–156

Schattschneider EE (1960) The semisovereign people. Holt, Reinhadt and Winston, New York

Scoones I (2002) Science, policy and regulation: challenges for agricultural biotechnology in developing countries. IDS working paper 147, University of Sussex, UK

Sen AK (1993) Capability and well-being. In: Nussbaum MC, Sen AK (eds) The quality of life. Clarendon Press, Oxford, pp 30–53

Sen AK (1999) Development as freedom. Oxford University Press, Oxford

Shore C, Wright S (1995) Policy: a new field of anthropology. In: Anthropology of policy: critical perspectives on governance and power. Routledge, London

Shore C, Wright S (eds) (1997) Anthropology of policy: critical perspectives on governance and power. Routledge, New York

Smith J (2003) Poverty, power and resistance: food security and sovereignty in Southern Africa. Centre for Social and Economic Research on Innovation in Genomics

Tait J (2009) Forward. In: Lyall C, Papaionnou T, Smith J (eds) The limits of governance: the challenge of policy-making for the new life sciences. Ashgate, England

Wafula D, Clark N (2005) Science and governance of modern biotechnology in Sub-Saharan Africa: the case of Uganda. J Int Dev 17:679–694

Chapter 47
Food Law in South Africa: Towards a South African Food Security Framework Act

Anél Gildenhuys

Abstract The right to have access to sufficient food is enshrined in section 27(1)(b) of the *Constitution of the Republic of South Africa, 1996*. Section 27(2) mandates the South African state to take reasonable legislative and other measures, within available resources, to progressively realize this right. This chapter accordingly sets out the current (this chapter reflects the South African legal framework until September 2014) food security legal framework by outlining both legislative and other measures (referring to relevant policies, strategies and programs) as required by section 27(2). The United Nations Food and Agricultural Organization indicated in 2009 that legislative measures can take the form of: (a) constitutional inclusion; (b) a food security framework act; and (c) inclusion in sectoral legislation. The mentioned three legislative levels are outlined within the South African food security context. Specific attention is given to the enactment of a South African food security framework act in order to address various challenges faced within the regulatory framework, specifically the current fragmented approach. In addition to food security specific policies, strategies and programs, food security as a development priority is also outlined within government strategic plans and programs. Due to the fact that food insecurity is especially high in rural South Africa, special attention is given to relevant rural development and agrarian reform measures. Attention is also given to farmers' rights with reference to biopiracy and plant breeders' rights.

47.1 Introduction

The right to have access to sufficient food is constitutionally enshrined in section 27 (1)(b) of the Constitution of the Republic of South Africa, 1996 (the Constitution).[1] Importantly, for purposes of this chapter, are the duties that are placed on the

---

[1] Constitution of the Republic of South Africa (1996a).

A. Gildenhuys (✉)
Faculty of Law, North-West University, Potchefstroom, South Africa
e-mail: Anel.Gildenhuys@nwu.ac.za

South African government in realizing the right to have access to sufficient food.[2] Section 27(2) of the Constitution obliges the state to take reasonable legislative and other measures, within its available resources, to achieve the progressive realization of, inter alia, the right to have access to sufficient food.[3] The South African government's commitment to food security[4] currently manifests in primarily other related policies, strategies and programs.

However, despite the prioritization of food security as a developmental priority, abovementioned measures, and the fact that South Africa is currently food self-sufficient,[5] on-going food shortages remain a daily reality for approximately 13.8 % individuals.[6] According to the Integrated Food Security Strategy for South Africa (the IFSS),[7] various food security challenges exist in South Africa, inter alia: inadequate safety nets,[8] weak institutional support networks and disaster management systems, inadequate and unstable household food production, lack of purchasing power and poor nutritional status. Further challenges that were identified by the Department of Agriculture, Forestry and Fisheries (DAFF) and the Department of Social Development (DSD) in the subsequent National Policy on Food and Nutrition[9] include inadequate access to knowledge and resources as well as the underutilization of productive land. In some jurisdictions, yet other factors influence the availability and accessibility of food as well as the stability of food supply.[10] These factors include infrastructure,[11] food prices, international trade,

---

[2] Ibid, sections 27(1)(a-c), which includes other socio-economic rights entrenched in the Constitution, including (but not confined to) the right to have access to health care services, the right to have access to sufficient water, and the right to have access to social security.

[3] Ibid, section 7(2), which asserts that the "state must respect, protect, promote and fulfil the rights in the Bill of Rights." The Bill of Rights refers to chapter 2 of the Constitution, where various human rights and fundamental freedoms are enshrined.

[4] "Food security" is defined in the South African context as "access to and control over the physical, social and economic means to ensure sufficient, safe and nutritious food at all times, for all South Africans, in order to meet the dietary requirements for a healthy life." See DAFF & DSD (2013), p. 8.

[5] Ibid, p. 3. The National Policy for Food and Nutrition Security was subsequently published as Government Notice 637 in the South African Government Gazette (No. 37915) of 22 August 2014.

[6] Ibid, p. 6; The Government of South Africa (2013), Millennium Development Goals: Country Report 2013d; StatsSA (2011), GHS Series Volume IV.

[7] National Department of Agriculture (2002), pp. 25–27.

[8] DAFF & DSD (2013), p. 4.

[9] Ibid, p. 4.

[10] National Department of Agriculture (2002), pp. 8–9, 19.

[11] Ibid, Addendum A, which states that "the physical accessibility of food in many rural areas presents problems because of the lack of infrastructure, such as roads, electricity and trading facilities. In the context of a stagnant economy and rapid population growth, the emphasis on commercialization and ineffective support programs for small-scale farmers had a negative impact on the food security of many rural households, even though national food security was enhanced." Poor storage and distribution of food also threatens household food security as mentioned in DAFF & DSD (2013), p. 3.

land reform,[12] environmental aspects like climate change,[13] and poor storage facilities.[14]

Fragmentation is a major criticism of the current South African food security legal framework.[15] Various governmental departments, including the DAFF, DSD, Department of Health, Department of Rural Development and Land Reform (DRDLR), Department of Environmental Affairs, Department of Basic Education, and the Department of Public Works, are responsible for addressing the abovementioned challenges and factors. Further fragmentation also occurs between the different levels of government: national, provincial and local.[16]

Despite section 27(2) of the Constitution's mandate to take reasonable legislative (and other)[17] measures for the progressive realization of the right to have access to sufficient food, the South African government has yet to adopt an integrated and comprehensive food security act, in other words a "framework act," regarding this right.[18] Consequences related to the absence of a national overarching and integrated framework law and the current fragmented approach include "no assured way of measuring progress in and monitoring the implementation of the right."[19] Khoza[20] is similarly of the opinion that "maintaining a fragmented, and often weak and inadequate legislative system relating to the implementation of the right to sufficient food [is] at the expense of people's basic needs, human dignity and life."

---

[12]In 1994, ownership of agricultural land was racially skewed as a result of the segregation policy of the apartheid government as well as a consequence of colonization. The majority of agricultural land was owned by the minority whites. As such, a land reform program was formulated and implemented in order to address: (a) land restitution; (b) land redistribution; and (c) land tenure reform in accordance with the mandate in section 25 of the Constitution. For more information on the various pieces of legislation, policies, strategies and programs within the land reform program (and it subsequent development). For more information see Pienaar (2014).

[13]DAFF & DSD (2013), p. 3.

[14]Ibid.

[15]Brand (2005a), p. 181; Khoza (2004), p. 664.

[16]National Department of Agriculture (2002), pp. 5–6 illustrates this statement with the following example: "Despite Government's intentions to promote food and agricultural production, various deficits in service delivery still exist. For example, the current extension and advisory services are inadequate due to a number of reasons including the fragmentation of service delivery within the three tiers of government The provision of post settlement support to farmers who benefit from land reform's restitution, redistribution and tenure reforms requires better coordination primarily between the Department of Agriculture and Land Affairs, between the National and Provincial, and local authorities and farmers organizations and business. The roles and responsibilities of the various departments therefore need to be clarified and the necessary institutional mechanisms for project implementation and monitoring should be in place. Furthermore, this disjuncture reflects the broader challenge government has in identifying the most appropriate mechanisms for ensuring the resourcing of interdepartmental programs."

[17]"Other" programs, strategies and policies will be outlined later in this chapter.

[18]Terblanche and Pienaar (2012), pp. 229, 232.

[19]Khoza (2005), p. 197.

[20]Ibid, p. 197.

The Food and Agricultural Organization (FAO)[21] of the United Nations recently indicated that constitutional provisions alone are often not enough to "ensure concerted action for the realization of the right to food. For this there is a need for implementing legislation, such as framework laws on food security and nutrition and sectorial legislation that advances the right to food, as well as adequate programs that support its realization for all."

The current chapter, firstly, will give an exposition[22] of the current[23] South African food security legal framework in light of the constitutional mandate in section 27 (2) as well as the FAO's encouragement above. This will include an exposition of the current constitutional and sectorial legislative measures, other measures in the form of policies, strategies and programs, and selected international and regional obligations. Secondly, the chapter will argue that despite the fact that various food security measures are in place and despite the fact that progress is being made in addressing food insecurity,[24] drafting and enactment of a South African food security framework act could address the various shortcomings in the current legal framework, including the current fragmentation. Because food insecurity mostly occurs in rural South Africa,[25] special attention will be given to rural development measures, including agrarian and land reform measures. Furthermore, farmers' rights with reference to biopiracy and plant breeders' rights will also be discussed briefly.

47.2 The Current South African Food Security Legal Framework

47.2.1 Legislative Measures

The FAO[26] indicated in the Guide on Legislating for the Right to Food, that the right to food[27] could be legislated on three different legislative levels: by inclusion of the right (or implied in other rights) in constitutions, specific framework legislation, and sectoral legislation.[28] It was previously mentioned that the South African government has failed (to date) to adopt a framework food security

---

[21]FAO (2014), Legal Developments in the Progressive Realization of the Right to Adequate Food: Working Paper.

[22]Please take note that due to the nature and extent of this chapter, detailed discussions of the various measures are not necessarily possible.

[23]This chapter reflects the South African legal framework up until 30 September 2014.

[24]President Jacob Zuma recently indicated that "the overall food security figure is declining due to government programmes," specifically "the percentage of households that were vulnerable to hunger declined from 29,3 % in 2002 to 12,6 % in 2012." Zuma (2013).

[25]Du Toit (2011).

[26]FAO (2009).

[27]Including food security measures.

[28]FAO (2009).

act leaving a void in the current legal framework. While the importance and benefits of a food security act will be discussed in Sect. 44.3 of this chapter, current legislative measures, including constitutional provisions and sectoral legislation will be explained in this section.

47.2.2 Constitutional Provisions

The right to have access to sufficient food is specifically enshrined in section 27(1)(b) of the South African Constitution, providing everyone the right to have access to sufficient food. Subsequent section 27(2) of the Constitution dictates the state to take reasonable legislative and other measures, within its available resources, to achieve the progressive realization of this right.[29] It must be highlighted that the right to have *access* to sufficient food is classified in the South African law as a "qualified socio-economic right," meaning that a person is not entitled to be provided with food, but rather that *access* to food must be facilitated by the state.[30] The duty to progressively realize *access* to food is twofold: firstly, facilitate and subsequently, to provide. The duty to facilitate refers to the state's duty to create an environment that will ensure access to sufficient food for all people at all community levels.[31] In other words, it implies "the provision of a framework of laws and policies that assist individuals and groups to enjoy the right."[32] A number of sectors are involved in the creation of such an "enabling environment" since food security is a complex issue that covers a range of related economic, social and political factors,[33] and has interdisciplinary and cross-sectorial application and implications. Areas that accordingly need to be addressed (by means of legislative and other measures) include (but are not confined to)[34] food systems, economy and trade, environmental matters such as disaster management, land and agrarian matters, labor matters like employment strategies, and measures related to the realization of related human rights (for example access to medical care services, education, water and sanitation, land, social assistance). On the other hand, the duty to provide refers to situations where individuals or groups are unable to access food on their own with the means at their disposal, despite the creation of an enabling environment.[35] In this case, the state has an obligation to

---

[29]Constitution of the Republic of South Africa (1996a), section 7(2).
[30]Brand (2005a), p. 3.
[31]Liebenberg (2010).
[32]Ibid, p. 85.
[33]Clover (2003), pp. 5, 7.
[34]See in this regard CESCR (1985), paragraph 25.
[35]Liebenberg (2010), p. 85.

provide by means of (amongst other measures) food banks, school feeding programs,[36] social assistance, etc.[37]

It must be noted that till date of finalization of this chapter[38] the right to have access to sufficient food has not been the subject of direct constitutional litigation.[39] Other food related rights enshrined in the Constitution, include every child's right to basic nutrition, shelter, basic health care services and social services (section 28 (1)(c)) and detainees' (including sentenced prisoners) rights to provision of nutrition at state expense (section 35(2)(e)).

47.2.3 Sectorial Legislation

In the outline of section 27(1) (b) and (2) of the Constitution above, it was mentioned that the state must facilitate access through legislation and other measures. Despite the absence of a specific food security framework act, these legislative measures also include sectorial legislation. Sectorial legislation accordingly implies legislation that has an influence on the normative contents of the right to have access to sufficient food within the broader context of food security.[40] Stated differently,[41] sectorial legislation refers to legislation that have an influence on the accessibility, availability (including stability and sustainability of food supplies), sufficiency, safety and utilization of food.[42] Such legislation includes laws relating to[43]: (a) land matters (including access to land)[44]; (b) labor matters[45]; (c) social

---

[36]See, for example, the National School Nutrition Programme, which is described later in this chapter.

[37]Liebenberg (2010), p. 85.

[38]July 2014.

[39]For a detailed discussion on the right to have access to sufficient food, the impact of the general limitation clause (section 36) and the test for reasonableness, see Brand (2005a), pp. 153–295.

[40]Terblanche (2011), p. 309.

[41]Ibid.

[42]For example, see FAO (2009), p. 6, where it contends "sectoral laws are important to the progressive realization of the right to food, since the right depends upon many factors and actors. Legislation concerning access to and management of land and natural resources can partly determine whether rural people are able to feed themselves and produce a surplus to feed urban dwellers. Trade legislation influences affordability of food as well as the ability of farmers to compete. Agricultural laws frame conditions for food production. Labour laws have an impact on whether or not wage labourers can earn enough to buy the food they need, and social protection supports food purchasing."

[43]Ibid, p. 191.

[44]For example, the Restitution of Land Rights Amendment Act 15 of 2014, Provision of Land and Assistance Act 126 of 1993, Extension of Security of Tenure Act 62 of 1997b, and Prevention of Illegal Eviction from and Unlawful Occupation of Land Act 19 of 1998e.

[45]For example, the Employment Equity Act 55 of 1998; Skills Development Act 97 of 1998f; and the Basic Conditions of Employment Act 75 of 1997a.

assistance[46]; (d) agriculture[47]; (e) environmental matters[48]; (f) production and marketing[49]; (g) consumer protection[50]; (h) health matters[51]; and (i) trade and foreign investments.[52] A detailed account of all applicable sectorial laws is beyond the nature and scope of this chapter.

47.2.4 Other Measures: Programs, Policies and Strategies

Since 1994,[53] food security, on national and/or household level, has received attention as a development priority in various government policies, strategies, and programs, etc. For example, the Reconstruction and Development Program of 1994[54] (the RDP)[55] sought to "ensure that as soon as possible, and certainly within 3 years, every person in South Africa can get their basic nutritional requirement each day and that they no longer live in fear of going hungry."[56] According to the RDP,[57] the following measures were key in achieving food security: (a) the provision of productive employment opportunities through land reform, job programs and the reorganization of the economy; (b) nutrition education; (c) stable, low-cost supply of staple foods combined with targeted income transfers and food subsidies; (d) the exemption of basic foodstuffs from value added taxes; (e) improved social security payments; (f) price control on standard bread; (g) enhanced marketing

---

[46]See, for example, the Social Assistance Act 13 of 2004.

[47]See, for example, the Agricultural Product Standards Act 119 of 1990 and the Genetically Modified Organisms Act 15 of 1997c.

[48]For example, see the National Environmental Management Act 107 of 1998c and the National Water Act 36 of 1998d.

[49]See, for example, the Foodstuffs, Cosmetics and Disinfectants Act 54 of 1972, Marketing of Agricultural Products Act 47 of 1996b, Standards Act 8 of 2008b, and the Genetically Modified Organisms Act 15 of 1997.

[50]For example, the Consumer Protection Act 68 of 2008a.

[51]See the National Health Act 61 of 2003.

[52]See, for example, the Competition Act 89 of 1998a and the International Trade Administration Act 71 of 2002. Various sectoral bills have been developed between 2013 and 2014 that must still follow the legislative process, but which may (when enacted) have an impact on food security, including the Spatial Planning and Land Use Act 16 of 2013e (not yet commenced), Draft Extension of Security of Tenure Amendment Bill of 2013c, Land Protection Bill of 2014a, Draft Expropriation Bill of 2013b, and Protection and Development of Agricultural Land Framework Bill of 2014b.

[53]The year when the first democratic elections were held.

[54]ANC (1994), Reconstruction and Development Programme.

[55]The RDP is a socio-economic policy framework adopted after the 1994 elections by the elected African National Congress. For more information on the RDP, see ANC (1994), Policy Documents.

[56]Ibid, para 2.11.2.

[57]Ibid, para. 2.11.2–2.11.6; 4.5.2.2–4.5.2.3.

(by curbing the powers of marketing boards and monopolies and the revision of the effects of tariffs); (h) a structured agricultural sector (with a spread ownership base, encouragement of small-scale agriculture, development of the commercial sector and increased production and employment); and (i) the removal of unnecessary controls, levies and unsustainable subsidies in the commercial agricultural sector. These aspects were to be provided for in subsequent measures and only selected measures are explained in this section.

More recently food security has been identified as a priority area in; inter alia, the War on Poverty Campaign of 2008,[58] the New Growth Path of 2010,[59] and more importantly, the Medium Term Strategic Frameworks of 2009–2014 (the MTSF 2009–2014) and 2014–2019 (the MTSF 2014–2019)[60] and the National Development Plan of 2011.[61] The MTSF 2009–2014[62] follows the 2009 election manifesto of the African National Congress (the ANC) wherein various challenges such as crime, poverty, unemployment and the high cost of living were identified.[63] To address these challenges, priority areas were identified and incorporated in the MTSF 2009–2014 of President Zuma's 2009 elected administration, including Strategic Priority 3—"A Comprehensive Rural Development Strategy linked to Land and Agrarian Reform and Food Security."[64] Strategic Priority 3 called for an[65] (a) aggressive implementation of land reform policies; (b) stimulation of agricultural production[66] "with a view to contributing to food security"; (c) improvement of rural livelihoods and food security[67]; (d) improvement of service delivery to ensure quality of life; (e) implementation of a development program for rural transport; (f) skills development; (g) revitalization of rural towns; (h) exploration of and support for non-farm economic activities; (i) institutional

---

[58]WOP (2009), War on Poverty Campaign.

[59]The Government of South Africa (2011), The New Growth Path.

[60]The Presidency – Republic of South Africa (2009), Medium Term Strategic Framework: A Framework to Guide Government's Programme in the Electoral Mandate Period 2009–2014; The Presidency – Republic of South Africa (2014), Medium Term Strategic Framework (MTSF) 2014–2019.

[61]National Planning Commission (2011), National Development Plan.

[62]The MTSF 2009–2014 was meant to guide the government's planning and resource allocation between 2009 and 2014, following the ANC government's 4th consecutive victory in the 2009 elections. For more information on the MTSF 2009–2014 see The Presidency – Republic of South Africa (2009). The MTSF 2009–2014 was recently followed up by the MTSF 2014–2019 following the ANC government's 5th consecutive victory at the 2014 elections.

[63]ANC (2009), Election Manifesto.

[64]The Presidency – Republic of South Africa (2009), pp. 18–22.

[65]Ibid, pp. 19–22.

[66]Amongst other measures, the "government has pledged over R2.6 billion in conditional grants to provinces for agricultural infrastructure, training and advisory services and marketing, and for upgrading agricultural colleges." See ibid.

[67]Amongst other measures, agricultural starter packs were to be provided to 140,000 households per annum under the Ilima/Letsema Campaign. The Ilima/Lesema Campaign is outlined later in this chapter.

capacity development; and (j) cooperative government.[68] As with the RDP, these aspects were to be provided for in subsequent measures. Selected measures in this regard will receive attention later in this section.

The National Development Plan of 2011 follows the RDP and focuses on, amongst others things,[69] job creation, infrastructure improvement, and an inclusive and integrated rural community. Food and nutrition security is seen in the National Development Plan as "both a consequence of poverty and inequality as well as a cause."[70] A number of steps are identified in the National Development Plan to address food security, including the expanded use of irrigation, security of land tenure[71] and the promotion of nutrition education.[72] As with the RDP and the MTSF, these aspects were to be provided for in subsequent measures and selected measures are discussed later in this section.

Food security is also listed in the recent strategic plan of the DRDLR[73] as well as the DAFF.[74] In the 2014–2019 strategic plan of the DRDLR (in Program 3: Rural Development), a baseline of 7800 households in rural communities is targeted for the provision of support to help improve their livelihoods by 2019.[75] The strategic objective[76] is to offer "support to rural communities to produce their own food in all rural districts."[77] Other objectives include[78] the provision of quality infrastructure[79]; facilitation of the establishment of rural enterprises and industries[80]; and the creation of jobs and skills development in rural areas.[81] Accordingly, "[s]pending

---

[68]The subsequent MTSF 2014–2019 reaffirms the government's commitment to the establishment/maintenance of vibrant, equitable and sustainable rural communities that contributes towards food security for all. It is envisioned in the MTSF 2014–2019 that, by 2013, the rural economy (with specific reference to agriculture) should "create close to 1 million new jobs, contributing significantly to reducing overall unemployment." Republic of South Africa. Policy imperatives mentioned in the MTSF 2014–2019 include (a) improved land administration and spatial planning, (b) sustainable land reform for agrarian transformation, (c) development and support for smallholder farmers, (d) increased access to quality basic infrastructure and services (with specific mention of education, healthcare and public transport in rural areas), (e) support for sustainable rural enterprises/industries (with rural-urban linkages), and (f) increased investment in agro-processing, trade development, and improved access to markets and financial services. See The Presidency – Republic of South Africa (2014).
[69]National Planning Commission (2011), pp. 10–16.
[70]DAFF & DSD (2013), p. 5.
[71]Especially for women.
[72]DAFF & DSD (2013).
[73]DRDLR (2014), Strategic Plan: Rural Development and Land Reform 2014–2019.
[74]DAFF (2012a), Strategic Plan 2012–2017.
[75]DRDLR (2014).
[76]Ibid, Objective 3.1.
[77]Ibid, p. 39.
[78]Ibid.
[79]Ibid, Strategic objective 3.2.
[80]Ibid, Strategic objective 3.3.
[81]Ibid, Strategic objective 3.4.

over the medium term will focus on implementing rural livelihood strategy, providing technical support to municipalities, coordinating and facilitating infrastructure projects, supporting irrigation schemes, developing and implementing a rural enterprises and industrial development strategy, skills and youth development and job creation."[82] Food security as a key priority also features in the DAFF's strategic plan[83] for 2012–2017.[84] While food safety receives attention under Program 2—Agricultural Production, Health and Food Safety,[85] food security features as a sub-program in Program 3—Food Security and Agrarian Reform,[86] with an emphasis on the promotion of household food security and agrarian reform programs and initiatives targeting subsistence and smallholder producers.[87,88] Sector capacity development[89] and national extension support services[90] are two other sub-programs within Program 3.[91] As such, one of the strategic goals of the DAFF[92] is to increase the profitability of food production and fiber and timber products by all producer categories.[93] Consequently, a coordinated government food security initiative is identified as an objective of Program 3. This initiative should "improve support mechanisms for food security production for subsistence and smallholder producers"[94] as well as "220,000 existing and 80,000 new

---

[82]Ibid, pp. 39–40.

[83]DAFF (2012b), pp. vii–ix.

[84]Ibid, p. ix: "Against the background of an increasing global population, the gradual decrease of natural resources and the effects of climate change, food security is a vital focus area for the department." More specifically it is stated in DAFF, at xi that it is a strategic priority to "[i]mprove the food security initiative by coordinating production systems to increase the profitable production, handling and processing of food, fibre and timber products by all categories of producers," referring to both commercial and emerging farmers.

[85]Ibid, pp. 31–35.

[86]Ibid, pp. 35–38.

[87]Ibid, p. 35.

[88]See the outline of initiatives focused on subsistence and smallholder farmers as well as agriculture in general later in this chapter.

[89]The Sector Capacity Development sub-program is aimed at "... the provision of agriculture, forestry and fisheries and training in support of sustainable growth and equitable participation in the sector. This will be achieved by facilitating and supporting education and training skills, promoting the development of centres of excellence on skills training and developing, managing and coordinating the sector transformation policy and strategy in line with the government objectives for the departments." See DAFF (2012b), p. 35.

[90]The National Extension Support Services sub-program provides "national extension policies, norms and standards on the transfer of technology. The sub-program will provide strategic leadership and guidance for the planning, coordination and implementation of extension and advisory services in the sector. It will also provide leadership and strategic support in the implementation of norms and standards for extension." See ibid.

[91]Ibid.

[92]Ibid, p. 36.

[93]As previously mentioned "all categories of producers" includes commercial and emerging farmers.

[94]DAFF (2012b), p. 36.

smallholder producers receiving comprehensive support."[95] A Food Security Bill is identified as a strategic intervention for the 2012–2017 period[96] and coordinates "comprehensive support to smallholder producers."[97] It was envisioned that the number of smallholder producers will increase from 200,000 to 500,000 by 2020.[98] An "agrarian beneficiary assistance policy" was accordingly also envisioned.[99] Emphasis was also placed on a (still then proposed) Food Security Policy as well as on the enforcement of the Zero Hunger Campaign.[100] The Food Security Policy was approved by the DAFF and DSD in 2013 as the National Food and Nutrition Policy and will be discussed together with the Zero Hunger Campaign later in this section. This food security initiative includes the "department partnership model (sustainable farming model) pilot project" that:

> [p]rovides comprehensive support to targeted smallholders to enable them to produce crops, in collaboration with agribusiness[101] ... The project assisted 15 farmers in the North West and Free State provinces to grow bitter sorghum, maize, sunflower and beans. The market of these grains was secured through Grain South Africa and Noord-Wes Koöperasie. Altogether 650 ha of grain crops were planted. Over the medium term, the focus will be on implementing the approved Zero Hunger Campaign and implementing the policy on mechanization support. Altogether 15 000 producers will receive comprehensive production support.[102]

The strategic objectives of the DAFF and DRDLR were to be addressed in subsequent measures. Selected measures, specifically regarding rural and agrarian reform relating to subsistence and smallholder farmers, is covered later in this section.

47.2.5 Selected Agricultural and Rural Development Programs, Policies and Strategies[103]

Building upon the reforms envisioned in the RDP and the strategic objectives of several of the abovementioned documents, food security measures have since been incorporated in the agricultural and rural development, including land reform and

---

[95]Ibid.

[96]It was initially planned that a Green Paper on Food Security (namely a discussion document) was to be published in the South African Government Gazette in 2012/2013, while the actual Food Security Bill would be developed in 2013/2014. See ibid, p. 36. However, no Food Security Bill has been published to date. The benefits of enacting a framework act will be discussed later in this chapter.

[97]Ibid, p. 36.

[98]Ibid, p. 37.

[99]Ibid, p. 38.

[100]Ibid.

[101]Agribusinesses include Grain South Africa, Agri South Africa, Noord-Wes Koöperasie Beperk. See ibid, p. 38.

[102]Ibid, p. 38.

[103]Note that due to the nature and scope of this chapter, only those programs, policies and strategies that directly influence food security (on national and/or household level) were selected.

legal frameworks. Small-holder farming receives particular attention as a strategy to increase household food security. For example, the 1995 White Paper on Agriculture[104] stated that while the South African government should support the "full spectrum of production systems and practices,"[105] new farming systems and appropriate technology must be developed through a "program of integrated research and technology development" to meet the needs of small-scale farmers for household food security purposes.[106] It was also stated that incentives should be created for the public and private sector to support such development.[107]

The Land Redistribution for Agricultural Development (hereafter LRAD) program of 2001,[108] accordingly made it possible for beneficiaries[109] like emergent small-scale farmers to purchase public or private land for an agricultural use of their choosing. These uses include "improve[ing] food production to improve household consumption, grazing, production for markets, and other agricultural activities."[110] Under the LRAD program, a range of grants (from R20,000 to R100,000 or approximately $1175 to 8871) are available depending on the amount of beneficiaries' own contribution (either in kind, labor and/or cash).[111] The grants may be used to "acquire land for food crop and/or livestock production to improve household food security."[112] Furthermore, eligible people that already have access to land may apply for assistance in order to make productive investments in, inter alia, infrastructure or land improvements, on their land.[113]

Launched in 2004, the Comprehensive Agricultural Support Program (the CASP),[114] aims[115] to "enhance the provision of support services to promote and facilitate

---

[104]National Department of Agriculture (1995), White Paper on Agriculture.

[105]The range of this "spectrum" refers to "urban food gardens and small scale production for household income and food security to large-scale production systems which can add considerably to national food security." See ibid, para 2.2. Previously, it was believed that the development of small-scale farming was detrimental to South Africa's food self-sufficiency. Ibid, Preamble. See also ibid, Addendum A, which affirms that "at national level food security will be enhanced by promoting the realisation of agriculture's potential within the constraints of comparative and competitive advantages, and of other resources that may be required for sustainable agricultural development. While acknowledging that large-scale commercial farms will still make a valuable contribution to national food security and that the policy environment must support them, small farms may be of increasing importance for improving both national and household food security."

[106]Ibid, para 8.9.

[107]Ibid.

[108]Ministry for Agriculture and Land Affairs (2001), Land Redistribution for Agricultural Development.

[109]For the qualifying criteria, see ibid, p. 8.

[110]Ibid, p. 5.

[111]Ibid, p. 1.

[112]Ibid.

[113]Ibid, p. 2.

[114]National Department of Agriculture (2004b), Comprehensive Agricultural Support Programme.

[115]The aims of CASP are to "reduce poverty and inequalities in land and enterprise ownership, improved farming efficiency, improved national and household food security, stable and safe rural

agricultural development targeting the beneficiaries of the land and agrarian reforms."[116] Four categories of beneficiaries have been identified[117]: the hungry and vulnerable, household food producers,[118] beneficiaries of land and agrarian reform programs,[119] and those operating within the macro-economic environment. Six areas of support are[120] information and knowledge management, technical and advisory assistance and regulatory services, training and capacity building, marketing and business development, on-farm and off-farm infrastructure and production inputs, and financial assistance.

Additional relevant rural development and land reform initiatives include the Comprehensive Rural Development Program of 2009 (the CRDP)[121] and the Green Paper on Land Reform of 2011.[122] The CRDP flows from Strategic Priority 3 of the MTSF 2009–2014 and encourages the establishment of a comprehensive rural development strategy that is directly linked to land and agricultural reform and food security. Similarly, the CRDP aims to "create vibrant, equitable and sustainable rural communities."[123] The CRDP makes provision for agrarian transformation,[124] land reform[125] and rural development like "capacity building initiatives, where rural communities are trained in technical skills, combining them with indigenous knowledge to mitigate community vulnerability to, especially climate change, soil erosion, adverse weather conditions and natural disasters, hunger and food insecurity."[126] Following the moderate success of the government's land reform program, the Green Paper on Land Reform of 2011 envisions a "re-configured single, coherent four-tier system of land tenure, which ensures that all South Africans, particularly rural blacks, have reasonable access to land with secure rights in order to fulfill their basic needs for housing and productive livelihoods."[127] The principles underpinning land reform are

---

communities, reduced levels of crime and violence, and increased creation of wealth in agriculture and rural areas and finally increased pride and dignity in agriculture as an occupation and sector." See ibid, p. 2.

[116]Ibid, p. 1.

[117]Ibid, p. 8.

[118]These categories of beneficiaries are provided agricultural starter packs. See ibid, p. 8.

[119]Supported Funded through farm level support. See National Department of Agriculture, p. 8.

[120]Ibid, p. 8.

[121]Ministry of Rural Development and Land Reform (2009), The Comprehensive Rural Development Programme Framework.

[122]DRDLR (2011), Green Paper on Land Reform.

[123]"The CRDP is aimed at being an effective response against poverty and food insecurity by maximizing the use and management of natural resources to create vibrant, equitable and sustainable rural communities." See Ministry of Rural Development and Land Reform (2009), p. 3.

[124]Ibid, pp. 4, 13–14.

[125]Ibid, pp. 5, 16–20.

[126]Ibid, pp. 5, 14–16.

[127]DRDLR (2011), para 3.1.

threefold, one being "a sustained production discipline for food security."[128] New land reform measures[129] must "improve on past and current land reform perspectives, without significantly disrupting agricultural production and food security."[130]

Building upon the agrarian transformation envisioned in the *CRDP* and following the South African government's "undertaking to review all land reform policies as enunciated in the 2011 Green Paper on Land Reform" as well as in line with the MTSF 2009–2014's vision for vibrant, equitable and sustainable rural communities and food security for all communities, the Policy for the Recapitalization and Development Program of 2013[131] (the PRDP) of the DRDLR "seeks to provide black emerging farmers with the social and economic infrastructure and basic resources required to run successful agricultural business."[132] The rationale for the PRDP is based on an evaluation of the land reform programs,[133] which found that the majority of land reform projects were unsuccessful mainly due to a "lack of adequate and appropriate post-settlement support."[134] As a result, the lands that were transferred through the land reform programs are in distress or lying fallow[135] or are on the verge/in the process of being auctioned or had already been sold.[136] For the purposes of this PRDP, recapitalization refers to "the capital renewal or restructuring of poor and previously disadvantaged and under-producing agricultural enterprises of Emerging Black farmers who are beneficiaries of the State's land reform program. Development here refers to support of human (capacity development), infrastructural development and operational inputs on other newly acquired properties."[137] This will be possible through mentorship-programs,[138] co-management,[139] share-equity arrangements,[140] and contract farming and

---

[128]The other principles are "de-racialising the rural economy" and "democratic and equitable land allocation and use across race, gender and class." See ibid, para 4.1.

[129]Recent developments include the Agricultural Landholdings Policy Framework of 2013a for the establishment of land ceilings and the re-opening of land restitution cases by means of the Restitution of Land Rights Amendment Act 15 of 2014c.

[130]DRDLR (2011), para 6.1.

[131]DRDLR (2013b), Policy on the Recapitalisation and Development Programme.

[132]Ibid, p. 10.

[133]Therefore, as early as 1991, but specifically after the first democratic elections held in 1994.

[134]DRDLR (2013a), p. 11.

[135]Ibid.

[136]Ibid, p. 10.

[137]Ibid, p. 5.

[138]These programmes include not only free support from neighboring or local farmers but also aligned remuneration or reimbursement packages. See Ibid, p. 12.

[139]Co-management is "an arrangement where two or more parties define and guarantee amongst themselves a fair sharing of the management functions, entitlements and responsibilities for a given territory or set of natural resources." See Ibid.

[140]Share-equity arrangements are defined as "[p]artners [who] acquire shares in an existing agricultural farm or other enterprises across the value chain with farmers or entrepreneurs." See ibid, p. 14.

concessions.[141] The PRDP will apply to the following categories or properties[142]: selected distressed land reform properties, properties selected by District Land Reform Committees, sited within the former homelands[143] and other communal areas, and farms acquired by individuals or collectives from historically disadvantaged communities that require strategic support. The PRDP aims to contribute to "the transformation of the rural economy through the establishment or enterprise and industrial development in the various agricultural value chains and other industries in order to ensure national and household food security, sovereignty and job creation."[144]

Various policies are in the process of being developed or finalized following the 2014 elections.[145] These are the[146] Agricultural Land Holdings Policy Framework,[147] State Assets Acquisition and Lease Disposal Policy, Policy on the Establishment of a Rural Development Agency, Policy on the Establishment of a Rural Investment and Development Financing Facility, and Protection and Development of Agricultural Land Framework Policy.

47.2.6 Food Security Specific Strategies, Policies and Programs

The Integrated Food Security Strategy (IFSS) was adopted in 2002[148] and identified various strategic goals, including[149] the promotion of increased household food production and food trade, improving income and job creating opportunities, improving food and nutritional security, ensuring an increase in safety nets and food emergency managements systems, improving analytical and information

---

[141] This is "an agreement between farmers (generally small-scale) and processors or marketing firms, the basis of which is 'a commitment on the part of the farmer to provide a specific commodity in quantities and at quality standards determined by the purchaser and a commitment on the part of the company to support the farmer's production and to purchase the commodity.'" See ibid.

[142] Ibid, p. 17.

[143] See Apartheid Museum, The Homelands: "The policy of separate development sought to assign every black African to a 'homeland' according to their ethnic identity. Ten homelands were created to rid South Africa of its black citizens, opening the way for massed forced removals. In the 1970s, the government granted sham independence to South Africa's black homelands. This served as an excuse to deny all Africans political rights in South Africa."

[144] DRDLR (2013a), p. 12.

[145] Held on 7 May 2014. The ANC was again democratically elected for another term (2014–2019).

[146] Hendriks et al. (2014).

[147] This was already approved by the Minister of the DRDLR in 2013. See DRDLR (2013a), Agricultural Land Holdings Policy Framework.

[148] National Department of Agriculture (2002).

[149] Ibid, pp. 6–7.

management systems, provision of capacity building, and holding of discussions with stakeholders. These strategic goals function within different programs like the[150] Special Program for Food Security, Integrated Nutrition and Food Safety Program, Safety Program, Comprehensive Social Security System and Disaster Management Program, Food Security Capacity Building Program, and Food Security Stakeholder Dialogue Program.

Eleven years after the adoption of the IFSS, the DAFF and DSD finally approved a food security policy, the National Policy on Food and Nutrition Security.[151] The policy identifies five pillars upon which existing[152] and future food and nutrition initiatives must be based upon. These pillars are[153] effective food assistance networks through the availability of improved safety nets, improved nutrition education, the alignment of investments in agriculture towards local economic development,[154] improved market participation of the emerging agricultural sector by leveraging government food procurement to support community-based food production initiatives and smallholders,[155] and food and nutrition security risk management. This policy is underpinned by the interdepartmental programs, Household Food and Nutrition Security Strategy and the Fetsa Tlala Food Production Intervention. Hence, the focus is on the challenges related to both household food shortages as well as increased food production.[156]

The Household Food and Nutrition Security Strategy, under the auspices of the DSD and Departments of Health and Basic Education, focuses on providing social assistance safety nets, such as food parcels, food banks and soup kitchens.[157] It further provides nutritious meals to children between the ages of 0–4 years,[158] in addition to more than eight million children in schools.[159] The Fetsa Tlala Food Production Intervention aims to put all underutilized agricultural land under

---

[150] Ibid, p. 7.

[151] DAFF & DSD (2013).

[152] The policy aims to "build on existing initiatives and systems, and to put in place mechanisms that ensure stricter alignment, better coordination, and stronger oversight." Ibid, p. 6.

[153] Ibid, pp. 6, 19.

[154] Investments include the "revitalisation of irrigation schemes, and the development of production, storage and distribution of food." Ibid, p. 19.

[155] This will be achieved through "public-private partnerships, including off-take and other agreements, a government food purchase programme that supports smallholder farmers, as well as through the implementation of the Agri-BEE Charter, which requires agro-processing industries to broaden their supply bases to include the emerging agricultural sector." See ibid. Regarding the government food purchase program, see also the Zero Hunger Program mentioned later in this chapter. Agri-BEE stands for Broad-Based Black Economic Empowerment in the agricultural sector. See in this regard the Department of Agriculture (2004a), Agri-BEE, Broad-Based Black Economic Empowerment Framework for Agriculture.

[156] Agripen (2013), South Africa Moves to Improve Food Security.

[157] Zuma (2013).

[158] Ibid.

[159] Ibid.

production.[160] More specifically, the intervention intends to "put one million hectares of land under production of maize, beans and potatoes. There is a significant amount of land that still lies fallow especially in rural areas and some of the land that has been acquired through land reform."[161] Furthermore, smallholder farmers, communities and households will be "assisted through the provision of mechanization support and distribution of production inputs and technical services."[162] In this regard, the former Minister of Agriculture, Forestry and Fisheries, Joemat-Pettersson, was quoted saying,[163] "We will work with communities to ensure that we use every bit of land to produce food. Through the intervention, we have brought 200 000 ha of land under production in seven provinces. Our targeted goal is one million hectares in the next 5 years."

Some programs and campaigns that have launched include the Ilima/Letsema Campaign, the National School Nutrition Program and the Zero Hunger Program. The Ilima/Letsema Campaign was launched to "stimulate food production to ensure that the vulnerable groups in our Society becomes food secure".[164] Other objectives include the stimulation of economic activities of vulnerable groups in society and the transfer of technical skills to food insecure households.[165] This is made possible through various interventions, including the provision of production inputs like seeds, fertilizers, tools, feed, chickens etc., as well as monitoring and evaluation of the implemented projects.[166] The National School Nutrition Program[167] was launched in 1994 through the Department of Basic Education and strives to provide one nutritious meal to all learners in primary and secondary schools[168] in order to "enhance learning capacity."[169] The program also teaches "learners and parents on ways of living a healthy lifestyle" and promotes the development of school vegetable gardens.[170] In addition to the use of vegetables grown in the school vegetable garden, the program also encourages schools to buy their supplies from local suppliers.[171] In conjunction

---

[160]Ibid.

[161]Ibid.

[162]Ibid.

[163]Agripen (2013).

[164]Department of Agriculture (2008), Ilima/Letsema Campaign.

[165]Ibid.

[166]Ibid.

[167]Department of Basic Education (1994), National School Nutrition Programme.

[168]Specifically in Quintile 1–3 schools. South African schools are classified into quintiles for resource allocation. Quintile one schools are regarded as the poorest schools and quintile 1–3 schools are declared "no-fee" schools. The poverty levels of the surrounding communities are (inter alia) taken into account when ranking the schools. See Western Cape Government (2013), Media Release: Background to the National Quintile System. See also Department of Basic Education (1994).

[169]Ibid.

[170]Department of Basic Education (2013), Q & A for NSNP Schools.

[171]Department of Basic Education (1994).

with the National School Nutrition Program, the Zero Hunger Program[172] (launched during 2011) aims to:

> ... link subsistence producers and smallholder producers/producers to government institutions such as government schools (i.e. to supply the School Nutrition Program), public hospitals and prisons, and in the medium term also be a conduit through which food produced by smallholders can be used to meet the nutritional needs of low-income individuals and households in communities at large. As such, the Zero Hunger Program seeks to provide a boost to existing smallholder producers/producers, and an opportunity through which subsistence producers can start generating a sustainable income through farming, and thereby become smallholder producers in their own right.[173]

47.2.7 Selected International and Regional Obligations

South Africa is a signatory to several fundamental international instruments related to food security and the right to food, including the International Covenant on Economic, Social and Cultural Rights of 1966[174] (ICESCR). Article 11 is of import to this chapter since it provides the right to an adequate standard of living, including sufficient food. South Africa signed the ICESCR on 3 October 1994[175] but till date has failed to ratify it. Despite the failure to ratify the ICESCR, section 39(1) (b) of the South African Constitution refers to the non-binding international law, such as treaties and conventions to which South Africa is not a member state, when interpreting the rights enshrined in the Bill of Rights. Furthermore, Brand[176] suggests that the socio-economic rights enshrined in the Bill of Rights, including the right to have access to food, are based on the ICESCR. Therefore, the ICESCR, as an interpretative source, is of great importance for the analysis of the socio-economic rights enshrined in the South African Constitution. Similarly, the normative contents in article 11 of the ICESCR as well as the exposition of state duties in the realization of this right by General Comment 12: The Right to Adequate Food[177] will hold great value for the interpretation of section 21(1) (b) of the South African Constitution if this right becomes subject to constitutional litigation.[178]

South Africa is also a signatory to the Millennium Development Goals (hereafter the MDGs)[179] and the MDGs, including goal 1, which aims to eradicate extreme poverty and hunger, are addressed in various government programs and

---

[172]DAFF (2012b), p. 3.

[173]The Zero Hunger Program is also linked to CASP.

[174]CESCR (1996), International Covenant on Economic, Social and Cultural Rights.

[175]After the first democratic elections in April 1994.

[176]Brand (2005b), p. 7.

[177]CESCR (1999), General Comment 12: The Right to Adequate Food.

[178]See also the recent United Nations General Assembly (2012), Resolution 67/174: The Right to Food.

[179]United Nations (2000), Millennium Development Goals.

strategies.[180] The South African government also incurred food related obligations under various other international instruments, including the Convention on the Discrimination of All Forms of Discrimination against Women (CEDAW)[181] and Geneva Conventions and protocols.[182] It must be noted that the right to food, the right to be free from hunger and food security measures are often the subject of various declarations and plans of action adopted at international conferences and summits. Of particular importance is the Rome Declaration on Food Security of 1996[183] and the subsequent World Food Summit Plan of Action[184] that was adopted by 112 heads of states, including the (then) Deputy President of South Africa, Thabo Mbeki. Cohen and Brown[185] show that although these declarations[186] do not create binding international obligations, its importance lies in the fact that it represents international consensus and offers guidance for appropriate action.

Regionally, South Africa is a member of the African Union (the AU) and sub-regionally, it is a member of the Southern Africa Development Community[187] (the SADC). Of regional importance is the Protocol to the African Charter on Human and Peoples' Rights on the Rights of Women in Africa, which was ratified by South Africa on 17 December 2004.[188] The right to food security or Article 15 of the protocol obliges member states to "provide women with access to clean drinking

---

[180] For the various programs and strategies, see Republic of South Africa (2013), Millennium Development Goals: Country Report 18–19. Goal 1 is addressed in, inter alia, the *RDP* program as well as in the National Planning Commission (2011), National Development Plan. Also, outcome 7 of the *MTFS* or the establishment of "vibrant, equitable, and sustainable rural communities with food security for all" is mapped to goal 1 of the MDGs. See Republic of South Africa (2013), Millennium Development Goals: Country Report 19 and 21–26 for South Africa's progress regarding goal 1 as at 2013.

[181] Article 12 of CEDAW, for example, states that "notwithstanding the provisions of par. I of this article, States Parties shall ensure to women appropriate services in connection with pregnancy, confinement and the post-natal period, granting free services where necessary, as well as adequate nutrition during pregnancy and lactation." See United Nations Human Rights—Office of the High Commission for Human Rights (1979), CEDAW.

[182] The *Geneva Conventions* consist of four treaties and three additional protocols that set international standards for the humanitarian treatment of war victims. See ICRC (1949), The Geneva Conventions. Various articles in the four conventions are applicable to food issues related to war victims.

[183] FAO (1996), Rome Declaration on Food Security.

[184] FAO (1997), World Food Summit Plan of Action.

[185] Cohen and Bown (2006), p. 225.

[186] As well as other declarations adopted at the subsequent FAO World Food Summits.

[187] "The main objectives of Southern African Development Community (SADC) are to achieve economic development, peace and security, and growth, alleviate poverty, enhance the standard and quality of life of the peoples of Southern Africa, and support the socially disadvantaged through Regional Integration. These objectives are to be achieved through increased Regional Integration, built on democratic principles, and equitable and sustainable development." SADC (1992), SADC Objectives.

[188] African Commission on Human and Peoples' Rights (2003), Protocol to the African Charter on Human and Peoples' Rights on the Rights of Women in Africa.

water, sources of domestic fuel, land, and the means of producing nutritious food and establish adequate systems of supply and storage to ensure food security."[189] Of sub-regional importance is the South Africa's involvement in the initiatives of the SADC Food, Agriculture and Natural Resources Directorate[190] as well as the Dar-Es-Salaam Declaration on Agriculture and Food Security in the SADC Region of 2004.[191] The Dar-Es-Salaam Declaration seeks to develop a competitive agricultural sector in the SADC region through[192] (a) improved access to agricultural inputs, (b) promotion of draught power and equipment for tillage, (c) disease control as well as improved crop storage and handling, (d) the development of drought tolerant crops, and (e) improved fish stock management, processing and handling.

47.2.8 Plant Breeders' Rights, Farmers' Rights and Biopiracy

Another concern within the food security debate is the protection of (commercial) plant breeders' rights at the cost of farmers' rights. Hence, the development of plant breeders' rights and its infringement will be discussed briefly by (a) first, defining biopiracy in reference to the development of plant breeders' rights in international instruments; (b) outlining the South African regulatory framework for plant breeders' rights; and (c) highlighting the plant breeders' rights versus farmers' rights debate with special reference to the acknowledgement of traditional knowledge.

The development and subsequent legal protection of plant breeders rights by means of intellectual property rights has led to a situation where the harvesting of patented seeds by farmers without the patent holder's permission (in other words piracy of seeds or biopiracy), is punishable by law.[193] Borowiak[194] indicates that this development is despite the fact that:

> [f]or virtually the entire history of agricultural production, up until the twentieth century, seed collection and distribution resided in the hands of farmers. Farmers collected the seeds from their fields after harvest and then used them for the next crop, for feed, for exchange, and for the breeding of new varieties of crops.

---

[189]On AU regional level, the New Partnership for Africa's Development (NEPAD) aims to "address critical challenges facing Africa, namely poverty, development and Africa's marginalisation internationally." (See NEPAD, About) The accompanying Comprehensive Africa Agriculture Development Programme (CAADP) is also a regional initiative. (See NEPAD (2003), Comprehensive Africa Agriculture Development Programme, which focus on the improvement and promotion of agriculture in Africa.

[190]SADC (2012a), Food, Agriculture and Natural Resources.

[191]SADC (2004), Dar-Es-Salaam Declaration on Agriculture and Food Security in the SADC Region.

[192]SADC (2012b), Food Security.

[193]Borowiak (2004), pp. 511, 519.

[194]Ibid, p. 513.

Consequently, farmer-breeders "did not need to keep purchasing seeds but rather collect, use, and exchange the seeds they harvested from their own fields."[195] This state of affairs was permitted until the early twentieth century[196] since intellectual property rights protection did not apply to plants.[197,198] Borowiak,[199] however, shows that plant breeding became:

[i]ncreasingly specialized and "scientific," and, as it did so, the site of breeding began to shift from farmers in their fields to scientist in laboratories. Correspondingly, plant breeding began to resemble modes of industrial innovation and production: plants began to resemble inventions.

"Plant variety protection" or "breeders' rights" received international recognition in the conventions of the Union for the Protection of New Varieties of Plants (UPOV) of 1961 and as amended in 1972, 1978 and 1991.[200] Subsequently, section 27 of the Agreement on Trade-Related Aspects of Intellectual Property Rights (TRIPS)[201] of the World Trade Organization (WTO) recognized plant variety as patentable subject matter. Section 27(3b) urges member states to "provide for the protection of plant varieties either by patents or by an effective *sui generis* system or by any combination thereof." An outcome of plant variety protection is formulated as follows: "Even though it has been a tradition in most countries that a farmer can save seed from his own crop, it is under the changing circumstances not equitable that a farmer can use this seed and grow a commercial crop out of it without payment of a royalty...."[202]

South Africa became a member of the WTO on 1 January 1995 and subsequently, a member to TRIPS (with inclusion of article 27(3b)), but plant breeders' rights have been legislatively protected in South Africa since 1977 through the Plant Breeders' Rights Act. Shortly after the enactment of the Plant Breeders' Rights Act, South Africa was accepted as the tenth member of the International Union for the Protection of New Varieties of Plants (UPOV),[203] and subsequently became a party to the 1978 International Convention for the Protection of New Varieties of Plants[204] on 8 November 1981.[205] Even though South Africa is not

---

[195]Ibid, p. 514.

[196]Ibid.

[197]Ibid.

[198]For a discussion on why intellectual property rights did not initially apply to plants, see ibid, p. 514.

[199]Ibid, p. 514.

[200]South Africa is a signatory to the 1978 International Convention for the Protection of New Varieties of Plants.

[201]WTO (1994), Agreement on Trade-Related Aspects of Intellectual Property Rights.

[202]Quoted in Sivia, *Captive Minds, Captive Lies* p. 107 as quoted in Borowiak (2004), p. 518.

[203]Department of Agriculture (1976).

[204]UPOV (1978), International Convention for the Protection of New Varieties of Plants.

[205]UPOV (1991b), Members of the International Union for the Protection of New Varieties of Plants.

party to the 1991 iteration of this convention,[206] South Africa reported to the FAO in 1999, that its Plant Breeders' Protection Act 15 of 1976 adheres to the (more restrictive) 1991 International Convention for the Protection of New Varieties of Plants.[207]

According to South Africa's Plant Breeders' Rights Act,[208] a breeder is defined as the (a) person who bred, or discovered and developed the variety; (b) employer of the person mentioned above if that person bred or discovered and developed the variety in the performance of his duties to such an employer; or (c) the successor in title of the person referred in (a) above or the employer referred to (b) above. According to the South African Department of Agriculture,[209] a plant breeder's right provides protection of the registered plant variety against "exploitation without the consent (permission) of the holder of the right, allowing him to obtain royalties as remuneration."[210] Key provisions of the Plant Breeders' Rights Act are highlighted below:

(a) Application: According to section 2(1), the Plant Breeders' Rights Act is applicable to every variety of any kind of plant if it is new,[211] distinct,[212] uniform[213] and stable.[214]
(b) Registration of plant breeders' rights: All granted plant breeders' rights must be entered in a register maintained by the Registrar of Plant Breeders' Rights.[215]

---

[206]UPOV (1991a), International Convention for the Protection of New Varieties of Plants.

[207]FAO (2004), Part III: Options Available to National Governments under existing International IPR Agreements Protecting Plant Varieties and Plant Breeders' Rights.

[208]Department of Agriculture (1976), Plant Breeders' Rights—Section 1.

[209]Ibid.

[210]Furthermore, "a plant breeder's right therefore provides the holder with a means of gaining financial remuneration for his efforts, encouraging breeders to continue with the breeding of new and better varieties, a process that is very time-consuming and expensive." Ibid.

[211]A variety is regarded as "new" if, according to section 2(2)(a), "propagating material or harvested material thereof has not been sold or otherwise disposed of by, or with the consent of, the breeder for purposes of exploitation of the variety- (i) in the Republic, not more than one year; and (ii) in a convention country or an agreement country, in the case of- *(aa)* varieties of vines and trees, not more than six years; or *(bb)* other varieties, not more than four years, prior to the date of filing of the application for a plant breeder's right."

[212]A variety is regarded as "distinct" if, according to section 2(2)(b), "at the date of filing of the application for a plant breeder's right, it is clearly distinguishable from any other variety of the same kind of plant of which the existence on that date is a matter of common knowledge."

[213]A variety is regarded as "uniform" if, according to section 2(2)(c), it is "subject to the variation that may be expected from the particular features of the propagation thereof, it is sufficiently uniform with regard to the characteristics of the variety in question."

[214]A variety is regarded as "stable" if, according to section 2(2)(d), "the characteristics thereof remain unchanged after repeated propagation or, in the case of a particular cycle of propagation, at the end of each such cycle."

[215]Department of Agriculture (1976), Section 4.

(c) Application for and granting of plant breeders' rights: The criteria for persons who may apply for plant breeders' rights are listed under section 6[216]; the application process is provided for under section 7; consideration and examination of applications are provided for under section 19, while the actual granting of plant breeders' rights is contained under section 20.[217]
(d) Period of plant breeder's right: Section 21 states that a plant breeder's right shall be granted for a period of 25 years in the case of vines and trees and 20 years in all other cases.

However, more importantly for the purposes of this chapter are the rights that are granted to the successful plant breeder applicant under section 23 as well as possible methods of infringement as identified under section 23A of the Plant Breeders' Rights Act. According to section 23(1), the rights give the plant breeder control of the (a) production or reproduction; (b) conditioning for the purpose of propagation; (c) sale/marketing; (d) exporting; (e) importing; and (f) stocking of the concerned plant and, under certain circumstances,[218] the plant variety.[219] According to section 23A, a plant breeder's right is infringed if a person who:

> *(a)* not being the holder of the plant breeder's right, performs, or causes to be performed, an act contemplated in section 23 (1) without a license obtained under section 25 or 27; *(b)* has obtained a license under section 25 or 27 but fails to comply with any term or condition thereof; *(c)* uses the approved denomination of a protected variety in relation to plants or propagating material of any other variety for any purpose whatsoever; and *(d)* sells plants or propagating material of a protected variety under any other denomination than the approved denomination of that variety.[220]

---

[216]Including South African citizens or persons who are domiciled in South Africa or a convention country or an agreement country and a juristic person that has a registered office in the Republic of South Africa or a convention country or an agreement country. Ibid, Sections 1, 6(2)(a-b), 7(2).

[217]If a plant breeder's right is granted, the register must: (a) issue a certificate of registration (section 20(2)(a)); (b) enter the details in the register (section 20(2)(b)); and (c) publish the particulars in the *Government Gazette* (as may be described)(section 20(2)(c)).

[218]See ibid, section 23(4)(a): "The provisions of subsections (1), (2) and (3) shall also apply to varieties which are essentially derived from the protected variety, where the protected variety is not itself an essentially derived variety; not distinguishable from the protected variety as contemplated in section 2(2)(b); or the production of which requires the repeated use of the protected variety." See also ibid, section 23(4)(b): "For the purposes of paragraph (a) (i) a variety shall be deemed to be essentially derived from another variety if - (i) it is predominantly derived from that other variety, or from a variety that is itself predominantly derived from that other variety, while retaining the essential characteristics of that other variety; and (ii) it is clearly distinguishable from that other variety; and (iii) except for the differences which result from the process of derivation, it conforms to that other variety in respect of the essential characteristics."

[219]"Variety" refers to "plant grouping within a single botanical taxon of the lowest known classification, which grouping, irrespective of whether or not the conditions for the grant of a plant breeder's right are fully met, can be—(a) defined by the expression of the characteristics resulting from a given genotype or combination of genotypes; (b) distinguished from any other plant grouping by the expression of at least one of the said characteristics; and (c) considered as a unit with regard to its suitability for being propagated unchanged." See ibid, section 1.

[220]Appropriate compensation in respect to infringement of a plant breeder's right is provided for in ibid, section 47.

The DAFF[221] affirms that traditional farmer practices "such as exchange and saving protected varieties for re-sowing" may be constituted as an infringement of plant breeders' rights, but section 23(6)(f) allows farmers to re-sow protected material on his/her own holding:

> a person who procured any propagating material of a variety in a *legitimate manner*[222] shall not infringe the plant breeder's right in respect of the variety if he or she ... *(f)* is a farmer who on land occupied by him or her uses harvested material obtained on such land from that propagating material for purposes of propagation: Provided that harvested material obtained from the replanted propagating material shall not be used for purposes of propagation by any person other than that farmer.

Section 23(6) (f) has been labeled the "Farmers' Privilege Provision,"[223] which allows farmers to re-sow seed harvested from protected varieties for non-commercial use.[224] The DAFF[225] has, however, indicated that this provision is often abused since no clear definition of "farmer" is provided in the Plant Breeders' Rights Act and "no scale of production and the scope of varieties" are stipulated. Consequently, the DAFF[226] announced in the Plant Breeders' Rights Policy of 2011 that norms and standards, including the consideration of the farm size and its enterprises and the category of crops, will be developed. In section 9(1) (d) of the Plant Breeders' Rights Bill, that was tabled during 2013 and has not yet been approved, it is stated that a plant breeder's right with respect to a variety obtained in a legitimate manner does not extend to:

> a farmer who - (i) within reasonable limits; and (ii) subject to the safeguarding of the legitimate interests of the breeder, uses the protected variety concerned as prescribed in accordance with subsection 2(a). (2) The Minister must, in respect of subsection 1*(d)*, prescribe the - *(a)* category or categories of farmers; *(b)* category or categories of plants; *(c)* uses to which such a protected variety may be put; and *(d)* where applicable, the - *(i)* conditions for payment of royalties; and *(ii)* labeling requirements.

While plant variety protection was welcomed by the plant breeder industry, including the seed industry, various concerns arose for farmers, specifically smallholder farmers in developing countries like South Africa. These included the fact that plant breeders' rights pose a threat to their cultural autonomy[227]; the fear of dependence on multinational agribusinesses[228]; and the "end of locally controlled food production."[229] In South Africa, farmers and NGOs' displeasure with the protection of plant breeders' rights was formulated in the Valley of 1000 Hills

---

[221]DAFF (2011), p. 9.
[222]Own emphasize.
[223]DAFF (2011), p. 13.
[224]Ibid, p. 23.
[225]Ibid, p. 13.
[226]Ibid, p. 19.
[227]Borowiak (2004), pp. 512, 520.
[228]Ibid, p. 520.
[229]Ibid.

Declaration[230] and the Johannesburg Declaration on Biopiracy, Biodiversity and Community Rights.[231] A major criticism against the protection of plant breeders' rights is that "as multinational seed companies reap great rewards from their innovations, many farmers believe that their and their communities' historical contributions to biodiversity and seed development[232] are going largely unrecognized."[233] It must be mentioned that the unauthorized use of patented seeds by farmers is not the only form of biopiracy; biopiracy could also be committed by plant breeders themselves. In the course of their bioprospecting activities, they often exploit local biodiversity resources and/or traditional knowledge.[234] It must also be noted that there is a South African legal framework for the sustainable use of biodiversity. In the National Environmental Management Biodiversity Act 10 of 2004a (Biodiversity Act), "bioprospecting" is defined as "any research on, or development of application of, indigenous biological resources for commercial or industrial exploitation..."[235] This act regulates bioprospecting of genetic material[236] as well as the sustainable use of indigenous biological resources.[237] It further provides for a fair and equitable sharing of benefits arising

---

[230]In Motion (2003), Valley of 1000 Hills Declaration.

[231]Anon (2002), The Johannesburg Declaration on Biopiracy, Biodiversity and Community Rights.

[232]Regarding the traditional role of farmers as stewards of agro-biodiversity, see DAFF (2011), p. 8. "Farmers' Rights consist of the customary rights that farmers have had as stewards of agro-biodiversity since the dawn of agriculture to save, grow, share, develop and maintain plant varieties; their legitimate right to be rewarded and supported for their contribution to the global pool of genetic resources as well as to the development of commercial varieties of plants; and their rights to participate in decision making on issues that may affect these rights."

[233]Borowiak (2004), p. 512.

[234]DAFF (2011), pp. 9–10.

[235]See the Government of South Africa (2004), Biodiversity Act—section 1. Various examples of bioprospecting are included such as "(a) the systematic search, collection or gathering of such resources or making extractions from such resources for purposes of such research, development or application; (b) the utilization for purposes of such research or development of any information regarding any traditional uses of indigenous biological resources by indigenous communities; (c) research on, or the application, development or modification of, any such traditional uses, for commercial or industrial exploitation; or (d) the trading in and exporting of indigenous biological resources in order to develop and produce products, such as drugs, industrial enzymes, food flavors, fragrances, cosmetics, emulsifiers, oleoresins, colors, extracts and essential oils."

[236]Genetic material is defined as "material of animal, plant, microbial or other biological origin containing functional units of heredity." See ibid.

[237]See ibid, section 80(2) For bioprospecting purposes, biological resources are defined as "(i) any indigenous biological resources as defined in paragraph (b) of the definition of 'indigenous biological resource' in section 1, whether gathered from the wild or accessed from any other source, including any animals, plants or other organisms of an indigenous species cultivated, bred or kept in captivity or cultivated or altered in any way by means of biotechnology; (ii) any cultivar, variety, strain, derivative, hybrid or fertile version of any indigenous species or of any animals, plants or other organisms referred to in subparagraph (i); and (iii) any exotic animals, plants or other organisms, whether gathered from the wild or accessed from any other source which, through the use of biotechnology, have been altered with any genetic material or chemical compound

from bioprospecting involving such indigenous biological resources,[238] specifically when "traditional knowledge accompanies the genetic resource."[239] Baker[240] highlights the relevant provisions in the Biodiversity Act below:

> [b]efore a party may legally bioprospect in South Africa, in terms of the Biodiversity Act, that party must obtain a permit from the South African government,[241] and this permit will only be granted if the community possessing the indigenous resource has consented to the terms of a benefit-sharing agreement.[242] In addition to this agreement, a material transfer agreement must be entered into with the holders of the traditional knowledge, if such is associated with the genetic material.[243]

In addition to the Biodiversity Act, the Patents Act 57 of 1978 (Patents Act)[244] requires every applicant who lodges an application for a patent in South Africa to attest in a statement if the invention for which protection is claimed is based on or derived from an indigenous biological resource,[245] genetic resource,[246] or traditional knowledge or use.[247] Baker[248] argues that both the Biodiversity Act and the Patents Act show South Africa's commitment to comply with its international obligations like the Convention on Biological Diversity,[249] which was ratified by South Africa.[250]

In terms of the plant genetic resources for food and agriculture (PGRFA), South Africa is not a contracting party to the FAO's International Treaty on Plant

---

found in any indigenous species or any animals, plants or other organisms referred to in subparagraph (i) or (ii)." Section 1 includes: "when used in relation to any other matter, means any resource consisting of- (i) any living or dead animal, plant or other organism of an indigenous species; (ii) any derivative of such animal, plant or other organism; or (iii) any genetic material of such animal, plant or other organism".

[238] Ibid, preamble.

[239] Baker (2010), Forget the piracy off the Somali Coast; South Africa has some piracy of its own – Biopiracy.

[240] Ibid.

[241] The Government of South Africa (2004), section 81.

[242] Ibid, section 83.

[243] Ibid, section 84.

[244] Ibid, section 30(3A).

[245] Refers to an "indigenous biological resource" as defined in ibid, section 1. See ibid, patent act section.

[246] Genetic resource is any indigenous genetic material or the genetic potential or characteristics of any indigenous species. See ibid, section 1.

[247] "Traditional knowledge" means the knowledge that an indigenous community has regarding the use of an indigenous biological or a genetic resource. "Traditional use" means the way in which or the purpose for which an indigenous community has used an indigenous biological resource or a genetic resource. See ibid.

[248] Baker (2010).

[249] Convention on Biological Diversity (1993), Introduction.

[250] See also the Cartagena Protocol (United Nations Convention on Biological Diversity (2001), Cartagena Protocol on Biosafety) that South Africa ascended to as well as the Nagoya Protocol (United Nations Convention on Biological Diversity (2010), Nagoya Protocol) which South Africa ratified.

Genetic Resources for Food and Agriculture[251] (ITPGRFA). Even though South Africa is not a signatory, the DAFF[252] recently referred to the protection of farmers' rights in its Plant Breeders' Rights Policy of 2011 as reflective of article 9 in the ITPGRFA. These rights protect traditional knowledge relevant to PGRFA, ensure equitable participation in benefit-sharing arising from the utilization of PGRFA, and give the right to participate in decision-making.[253] The ITPGRFA rests the responsibility for realizing farmers' rights relating to PGRFA on the national governments. The DAFF[254] expressed that the use of PGRFA is important in the development of new plant varieties; however, neither the Plant Breeders' Rights Act nor the UPOV Convention "offers protection for these varieties due to their lower level of distinctness, uniformity and stability.[255] It is, therefore, not possible to use these systems as benefit-sharing mechanisms in recognition of farmer's contributions as stewards of agro-biodiversity."

Apart from the more specified provisions regarding farmers' privilege in the 2013 Plant Breeders' Rights Bill and section 42(1)(e) for the appointment of one person to represent indigenous interests in respect of new plant varieties,[256] the 2013 Bill does not make provision for the protection of farmers' rights. It is recommended that the provisions of the Biodiversity Act, Patents Act, Conservation of Agricultural Resources Act 43 of 1983 and the Plant Breeders' Rights Act be harmonized in order to address the current fragmented approach to conservation, farmer varieties, agricultural biodiversity and traditional knowledge.[257] Additionally, better protection must be afforded to traditional farmers' rights particularly by the signing and ratification of the ITPGRFA.[258]

---

[251]ITPGRFA was implemented on 29 June 2004. See FAO (2001), The International Treaty on Plant Genetic Resources for Food and Agriculture.

[252]DAFF (2011), pp. 8–9.

[253]Refers to decision-making at the national level.

[254]DAFF (2011), p. 9.

[255]These are the requirements for protection of plant variety as listed in Department of Agriculture (1976), section 2(2)(a)-(c).

[256]As well as the source, use and impact for of new plant varieties.

[257]Wynberg et al. (2012), p. 1; DAFF (2011), p. 14.

[258]For further recommendations see Wynberg et al. (2012), pp. 1–2.

47.3 Towards a South African Food Security Framework Act[259]

From the exposition of the South African legal framework (both legislative and other) for food security, it is clear that food security measures are in place. The DRDLR[260] recently attested that "trends recorded in indicators measuring food security interventions at household level show that although progress has been slow, there is a definite increase in the number of households assisted to produce their own food; this has had a positive impact on food security at household level." The DAFF also[261] recently disclosed that:

> Some successes have been recorded in different priority areas of the Strategy, and South Africa is presently able to boast national food sufficiency through a combination of own production and food imports. The General Household Survey (GHS) has also indicated that the food access index has been improving, and the incidence of hunger is declining.[262]

However, the National Policy for Food and Nutrition Security[263] warns that "secure access to food by all is still not guaranteed" and without coordinated interventions "increasing numbers of the population may experience inadequate access to food and many more will fail to benefit from proper nutrition." As mentioned in the introduction to this chapter a major criticism of the current South African food security legal framework is its fragmentation due to the discordant responsibilities of various government departments that address various aspects of food security and related measures. It was also mentioned that the South African government has failed to yet adopt an integrated and comprehensive food security framework act for the realization of the right to have access to food.[264] Although a draft food security framework bill was prepared during 2002, further attempts at a framework act have been postponed because of DAFF's priority to first establish a strategic and policy framework for food security.[265] As previously mentioned, the IFSS was adopted in 2002 and the National Policy on Food and Nutrition Security was approved in 2013, thereby establishing the strategic and policy framework. Thus, the drafting and eventual enactment of a food security framework law should be the next logical step in the regulatory process:

---

[259]This section is partially based on chapter 6 of the author's doctoral thesis. For a detailed discussion of a proposed South African food security act, see Terblanche and Pienaar (2012).

[260]DRDLR (2014), p. 23.

[261]DAFF & DSD (2013), p. 3.

[262]This corroborates with President Zuma's indication that "the overall food security figure is declining due to government programmes" namely that "the percentage of households that were vulnerable to hunger declined from 29,3 % in 2002 to 12,6 % in 2012." Zuma (2013).

[263]DAFF & DSD (2013), p. 3.

[264]Terblanche and Pienaar (2012), p. 232.

[265]Terblanche (2011), pp. 199–203.

In line with the directive of the Constitution and conforming to its international obligations, South Africa has to consider the proposal of the Food and Agriculture Organization of the United Nations (FAO),[266] that member states should consider the enactment of legislation on the right to access to food. The approval of the National Food Security policy[267] will be an initial step towards a Food Security Act for South Africa...Currently there is no legislation regulating food security and its co-ordination in South Africa...South Africa needs a Food Security Act which will enforce the public and private sector to promote the non-violation of the basic human's right of having access to food and water as prescribed by our Constitution.[268,269]

Framework legislation for the realization of constitutionally enshrined qualified socio-economic rights is not a strange phenomenon in the South African law.[270] Examples of framework legislation include the National Health Act 61 of 2003,[271] National Environmental Management Act 107 of 1998c,[272] National Water Act 36 of 1998d,[273] National Housing Act 107 of 1997,[274] and Social Assistance Act 13 of 2004b.[275] In Government of the Republic of South Africa v Grootboom[276] (the Grootboom-case), the Constitutional Court addressed the necessity of framework legislation in the realization of the right of everyone to have adequate housing access.[277] The Constitutional Court indicated that the constitutional requirement of

---

[266]For example, see FAO (2009).

[267]The National Food Security Policy has since been approved.

[268]DAFF (2012b), pp. 5, 40.

[269]See also DAFF & DSD (2013), p. 18, which states that "in line with its international obligations, South Africa has to consider the recommendation of the Food and Agriculture Organisation of the United Nations (FAO) that Member States should consider the enactment of legislation on the right to access to food. The approval of this National Food and Nutrition Security policy could be an initial step towards a Food and Nutrition Security Act for South Africa, which would give statutory force to such structures. A Green and White Paper process is envisaged to prepare for this."

[270]Terblanche and Pienaar (2012), pp. 232–233.

[271]Enacted per section 27(2) of the Constitution as a legislative measure for the realization of the right to have access to health care services, including reproductive health as enshrined in section 27(1)(a) of the Constitution.

[272]Enacted per section 24(b) of the Constitution as a legislative measure for the realization of everyone's right to an environment that is not harmful to their health or well-being as enshrined in section 24(a) of the Constitution.

[273]Enacted via section 27(2) of the Constitution as a legislative measure for the realization of the right to have access to sufficient water as enshrined in section 27(1)(b) of the Constitution.

[274]Enacted via section 26(2) of the Constitution as a legislative measure for the realization of the right to have access to adequate housing as enshrined in section 26(1) of the Constitution.

[275]Enacted per section 27(2) of the Constitution as a legislative measure for the realization of the right to have access to social security (including, if they are unable to support themselves and their dependents, appropriate social assistance) as enshrined in section 27(1)(c) of the Constitution.

[276]Government of the Republic of South Africa v Grootboom (2001), 1 SA 46 (CC).

[277]As enshrined in the South African Constitution. See section 26(1-2) of.

legislative measures may necessitate framework legislation at national level by[278] arguing that such act can serve to guide the state's duty to ensure that legislation, policies, programs and strategies are sufficient to meet the state's obligations.[279] Although section 27(1)(b), which gives the right to have access to sufficient food, has not been judicially decided and interpreted, the textual similarities between the state's obligations imposed in sections 26(2)[280] and 27(2)[281] of the Constitution (with reference to the Grootboom-case) justifies the enactment of a nationally coordinated and overarching piece of legislation within this context. Khoza[282] even contends that the government's failure to enact such legislative measure is indicative of the state not fulfilling its obligations and such failure is open to judicial testing.

The possibility of countries enacting framework legislation as a means to implement national strategies for the realization of the right to food was also suggested by the United Nations Committee on Economic, Social and Cultural Rights (CESCR) in General Comment 12: The Right to Adequate Food[283] in 1999.[284] Such encouragement was later repeated by the FAO in its 2009 Guide on Legislating for the Right to Food[285] and again, in 2014, in its[286] working paper Legal Developments in the Progressive Realization of the Right to Adequate Food.[287] Framework legislation is regarded as a component of a "contemporary

---

[278]However, the Constitutional Court did not need to elaborate on the enactment of a framework law for housing since the Housing Act 107 of 1997 already existed at the time of the hearing of the *Grootboom*-case. See the *Grootboom*-case (2001), para. 40.

[279]Ibid with reference to the right to have access to adequate housing.

[280]The section states that "the state must take reasonable legislative and other measures, within its available resources, to achieve the progressive realization of this right." This refers to section 26 (1) of the Constitution, which provides for the right to have access to adequate housing. See ibid.

[281]The section obliges the state to "take reasonable legislative and other measures, within its available resources, to achieve the progressive realization of each of these rights." This refers to section 27(1) of the Constitution, including section 27(1)(b) which ensures the right to have access to sufficient food. See ibid.

[282]Khoza (2005), p. 192.

[283]CESCR (1999).

[284]Ibid, para. 29 states that: "in implementing the country-specific strategies referred to above, States should set verifiable benchmarks for subsequent national and international monitoring. In this connection, States should consider the adoption of a framework law as a major instrument in the implementation of the national strategy concerning the right to food. The framework law should include provisions on its purpose; the targets or goals to be achieved and the time frame to be set for the achievement of those targets; the means by which the purpose could be achieved described in broad terms, in particular the intended collaboration with civil society and the private sector and with international organizations; institutional responsibility for the process; and the national mechanisms for its monitoring, as well as possible recourse procedures. In developing the benchmarks and framework legislation, States parties should actively involve civil society organizations."

[285]FAO (2009).

[286]The FAO indicated that an increasing number of countries have enacted food security framework laws over the last 10 years, including a few African countries like Angola, Mozambique and Zanzibar. For the complete list see ibid.

[287]Ibid.

legislative process."[288] This means that the relevant piece of legislation sets broad legal and operational principles without necessarily giving content thereto.[289] According to the FAO,[290] contents are provided by ensuing measures like subsequent legislation, regulations, administrative decisions, policies and financial measures.[291]

The primary objective of a framework act is to operate as a coordinated instrument for the implementation of national strategies and policies.[292] Subsequent objectives, as emphasized by Khoza[293] and others,[294] include (a) identification of further development of specific policies and further legislation where necessary,[295] (b) harmonization and integration of certain measures where necessary, (c) creation of structures in order to facilitate co-ordination between and accountability of relevant organs of state, departments and non-governmental role players,[296] (d) provision of operational instruments such as benchmarks, norms, targets and objectives[297] as well as monitoring mechanisms,[298] and (e) the provision of a set of guiding principles to which the development of policy and legal reform must conform to.[299] The FAO[300] emphasizes the importance of a framework act as "useful as they articulate the normative content of the right to food and provide various means of enforcement at the administrative, judicial and quasi-judicial levels." It "seeks to get a systematically defined and complex process of implementation of the right started by, first and foremost, identifying duplications, gaps and obstacles encountered in implementing the current legislative and policy measures."[301] Khoza[302] asserts that a

---

[288] Garrett (2005), pp. 717–718; Coomans and Yakpo (2004), pp. 17, 20.

[289] Health Systems Trust (2005), National Health Act Proclaimed by the President; FAO (2009), p. 57.

[290] Ibid.

[291] See FAO (2009), p. 57. The general nature of framework legislation is summarized as "establish[ing] a general frame for action, framework legislation does not regulate the areas it covers in detail. Instead, it lays down general principles and obligations but leaves it to implementing legislation and other authorities to determine specific measures to be taken to realize such obligations, possibly within a given time frame."

[292] Khoza (2004), p. 672; CESCR (1999), p. 29; Terblanche (2011), pp. 194–196.

[293] Khoza (2004), p. 672.

[294] See also Coomans and Yakpo (2004), p. 20 and Garrett (2005), p. 741.

[295] Ibid, p. 733 indicates that a framework act may indicate the legislator's decision to solve or address a certain problem or matter, can lay down neutral laws, and facilitate coordinated action.

[296] Vapnek and Spreij (2005), pp. 155–158, 168–173, 196–198.

[297] Or at least identify the relevant body/official responsible for establishing the abovementioned operational instruments (usually within a given timeframe). See Khoza (2005), p. 197 and CESCR (1999), para. 29.

[298] Ibid.

[299] Vapnek and Spreij (2005), pp. 158–160.

[300] FAO (2009), pp. 4–5.

[301] Khoza (2004), p. 672.

[302] Ibid, p. 669.

framework act is part of the rule of law and thereby strengthens the idea that society must primarily be governed by legislation and secondarily by policy. Framework legislation must direct policies, which must conform and comply with the principles stated in said legislation.[303] Policy measures allow too much flexibility and uncertainty unless it is directed by legislation.[304] Khoza[305] further argues that a policy can be changed at any moment by an executive authority and without the participation of or consultation with the legislative authority.

Due to the possible inclusion of operational instruments such as norms, indicators, time-defined goals and objectives, framework legislation is also often considered a "transformed policy framework" since such operational instruments are usually found in policy measures rather than in statutory provisions.[306] Even if the proposed framework act makes provision for the mentioned operational instruments, the implementation, application and enforcement thereof will be determined by subsequent policy and/or strategy measures.[307] The importance of incorporating such operational instruments in the proposed framework act lies in the fact that if the legislatively required monitoring, for example, does not take place, such failure is not merely considered non-compliance with policy or strategy measures, but rather as a non-compliance with a legislative provision.[308]

In addition to the general primary and subsequent aims of framework legislation, other benefits of a food security framework act includes:

(a) providing normative contents to the right to have access to sufficient food and related principles[309];
(b) strengthening the justiciability of section 27(1)(b) of the Constitution by making provision for concrete remedies in the case of infringement[310];
(c) enforcing human right norms and constitutional principles[311];
(d) emphasizing the interdependence of rights[312];
(e) serving as a starting point for the identification of loopholes and gaps in the current legal framework[313];
(f) providing for better coordination and enhanced accountability by awarding specific responsibilities to the different spheres of government,[314] government

---

[303]Ibid, pp. 669–670.
[304]Ibid.
[305]Ibid.
[306]Ibid, p. 197.
[307]Ibid, pp. 187–204.
[308]Ibid.
[309]Ibid; Coomans and Yakpo (2004), p. 22; FAO (2009), p. 4.
[310]Khoza (2004), p. 677.
[311]For example, transparency, participation and accountability. See ibid, p. 194.
[312]Coomans and Yakpo (2004), p. 21.
[313]Khoza (2005), p. 194; FAO (2009), p. 4.
[314]Namely on national, provincial and local level.

departments, organs of state, government bodies and non-governmental stakeholders[315];

(g) enhancing integrated food security management by making provision for comprehensive and coordinated implementation of national strategies and policies as well as sectorial legislation relating the right to have access to sufficient food[316]; and

(h) serving as a means for measuring and monitoring progress in the realization of socio-economic rights by providing norms and time-defined goals and objectives.[317]

Ergo, a food security act will serve as a confirmation of the South African government's commitment to the realization of the right to have access to sufficient food as enshrined in the Constitution.[318]

47.4 Conclusion and Way Forward

It is clear from the above account of the South African food security legal framework that various legislative and other measures are in place to address various food related challenges. Although coordinated by different departments and within different budget allocations, various measures often seek the same outcomes, demonstrating the fragmentation of the food security legal framework. The National Food and Nutrition Security Framework[319] accurately summarize:

> Food and nutrition security is a multifaceted and multidimensional issue which will not be attained through a single approach – be it in the form of social relief or agricultural production. Food and nutrition security requires well-managed inter-sectorial co-ordination, and the genuine integration of existing policies and programs in health, education, and environmental protection, as well as in agrarian reform and agricultural development.

Furthermore, the National Food and Nutrition Security[320] calls for mechanisms that ensures stricter alignment, better coordination, and stronger oversight. While the general benefits of enacting a proposed food security act were listed earlier in this chapter, it is also suggested that the proposed framework act should make provision for the establishment of a national coordinating body, such as a National Food and Nutrition Security Authority, to achieve the goals of stricter alignment, better coordination and stronger oversight. Such a coordinating body could be

---

[315]Khoza (2004), p. 677; Khoza (2005), pp. 194, 196; FAO (2009), pp. 4, 54; Coomans and Yakpo (2004), p. 22.
[316]Khoza (2005), p. 194; FAO (2009), p. 4.
[317]Khoza (2004), p. 677; Coomans and Yakpo (2004), p. 22.
[318]Khoza (2005), p. 197.
[319]DAFF & DSD (2013), p. 6.
[320]Ibid.

composed of representatives from relevant government departments[321] as well as other stakeholders such as the South African Human Rights Commission, research and statistic institutes[322] like the South African Human Science Research Council, the private sector and academia[323] non-governmental organizations, and representatives of civil society.[324] The composition of the proposed national food and nutrition coordinating body should, in essence, reflect the multi-sectorial nature of the right to food and food security.[325]

According to the FAO,[326] the proposed framework act could mandate the national coordinating body to (a) advise the government and coordinate the activities and actors involved at national, regional and local level; (b) formulate, negotiate, adopt and review food security measures; (c) determine "appropriate benchmarks for measuring progress in the implementation of the framework law and the realization of the right to food;" (d) collect information and ensure the dissemination of information among all relevant actors; (e) provide advice on the harmonization of relevant sectorial policies; (f) make recommendations for change, amendments to a law, regulation or policy, as well as the adoption of new measures; (g) set priorities and coordinate the allocation of resources "according to priorities;" and (h) report to Parliament "on the state of implementation of the right to food and the framework law itself."[327]

Further research into the structure, composition, representation and general mandate for the proposed food and nutrition national coordinating body is encouraged as well as for further provisions of the proposed food security framework act. This chapter concludes with reference to the words of the late Nelson Mandela.[328]

We know it well that none of us acting alone can achieve success. We must therefore act together as a united people, for national reconciliation, for nation building, for the birth of a new world. Let there be justice for all. Let there be peace for all. *Let there be work, bread, water and salt for all.*[329] Let each know that for each the body, the mind and the soul have been freed to fulfill themselves.

---

[321]These include the national departments of Basic Education; Higher Education and Training; Health; Agriculture, Forestry and Fisheries; Environmental Affairs; Social Development and the Department of Women, Children en People with Disabilities. See FAO (2009), pp. 139–142, 144–145 for possible structure and composition options.

[322]Ibid, p. 144.

[323]Ibid.

[324]See also ibid, p. 6, which states that "the participation of stakeholders during the drafting stages will reveal the scope of interests and concerns pertaining to the realization of the right to food. As such, framework laws provide improved policy coherence and can ensure that the right to food is central to a country's development strategies. In addition, the institutional framework established by the framework law for coordination and consultation purposes should include provisions on participation of non-governmental stakeholders, as most of them do."

[325]Ibid, p. 144.

[326]Ibid, pp. 142–144.

[327]Ibid.

[328]Mandela (1994), Statement of Nelson Mandela at his Inauguration as President.

[329]Own emphasize.

References

African Commission on Human and Peoples' Rights (2003) Protocol to the African Charter on Human and Peoples' Rights on the Rights of Women in Africa. Available at http://www.achpr.org/instruments/women-protocol/ (last accessed 13 July 2014)

African National Congress (ANC) (1994) Policy Documents, A Basic Guide to the Reconstruction and Development Programme. Available at http://www.anc.org.za/show.php?id=234 (last accessed 13 July 2014)

Agripen (2013) South Africa Moves to Improve Food Security. Available at http://www.agriculturalwriterssa.co.za/agripen_1309.html (last accessed 13 July 2014)

ANC (2009) Election Manifesto. Available at http://www.anc.org.za/elections/2009/manifesto/manifesto.html (last accessed 10 July 2014)

Anon (2002) The Johannesburg Declaration on Biopiracy, Biodiversity and Community Rights. Available at http://www.iatp.org/files/Johannesburg_Declaration_on_Biopiracy_Biodiver.htm (last accessed 20 July 2014)

Apartheid Museum (YEAR) The Homelands. Available at http://www.apartheidmuseum.org/homelands (last accessed 3 July 2014)

Baker A (2010) Forget the piracy off the Somali Coast; South Africa has some piracy of its own – Biopiracy. Available at http://www.adamsadams.com/index.php/media_centre/news/article/Forget_the_piracy_off_the_Somali_Coast_South_Africa_has_some_piracy_of_its/ (last accessed 20 July 2014)

Convention on Biological Diversity (1993) Introduction. Available at http://www.cbd.int/convention/ (last accessed 20 July 2014)

Borowiak C (2004) Farmer's rights: intellectual property regimes and the struggle over seeds. Polit Soc 32(4):511–543

Brand D (2005a) Introduction to socio-economic rights. In: Brand D, Christof HC (eds) Socio-economic rights in South Africa. PULP, Pretoria, pp 1–56

Brand D (2005b) The right to food. In: Brand D, Christof HC (eds) Socio-economic rights in South Africa. PULP, Pretoria, pp 153–295

CESCR (1996) International Covenant on Economic, Social and Cultural Rights. Available at http://www.ohchr.org/EN/ProfessionalInterest/Pages/CESCR.aspx (last accessed 13 July 2014)

Clover J (2003) Food security in Sub-Saharan Africa. Afr Secur Rev 12(1):5–15

Cohen MJ, Bown MA (2006) The right to adequate food. In: Borghi M, Blommestein LP (eds) The right to adequate food and access to justice. Schulthess, Geneva, pp 219–252

Committee on Economic, Social and Cultural Rights (CESCR) (1999) General Comment 12: The Right to Adequate Food. Available at http://www.refworld.org/docid/4538838c11.html (last accessed 13 July 2014)

Coomans F, Yakpo K (2004) A framework law on the right to food – an international and South African perspective. Afr Hum Rights Law J 4(1):17–33

DAFF (2011) Plant Breeders' Rights Policy

DAFF (2012a) Strategic Plan for the Department of Agriculture, Forestry and Fisheries 2012/3–2016/7. Available at http://www.nda.agric.za/doaDev/topMenu/StratPlan201213-201617.pdf (last accessed 10 July 2014)

DAFF (2012b) Zero Hunger Program. Available at http://www.namc.co.za/upload/agricultural_industry_trusts_workshop/DAFF%20Food%20Security%20Policy%20-%20Zero%20Hunger%20&%20Masibambisane.pdf (last accessed 10 July 2014)

Department of Agriculture (1976) Plant Breeders' Rights. Available at http://www.nda.agric.za/doaDev/sideMenu/geneticResources/docs/undrestandingPlantBreedersRights.pdf (last accessed 20 July 2014)

Department of Agriculture (1995) White Paper on Agriculture. Available at http://www.nda.agric.za/docs/Policy/WHITEPAPER.htm (last accessed 15 July 2014)

Department of Agriculture (2002) Integrated Food Security Strategy. Available at http://www.nda.agric.za/docs/Policy/FoodSecurityStrat.pdf (last accessed 15 July 2014)

Department of Agriculture (2004a) Agri-BEE, Broad-Based Black Economic Empowerment Framework for Agriculture. Available at http://www.nda.agric.za/docs/agribee/agriBEE.htm (last accessed 13 July 2014)

Department of Agriculture (2004b) Comprehensive Agricultural Support Programme (CASP). Available at http://www.nda.agric.za/docs/CASP/casp.htm (last accessed 20 July 2014)

Department of Agriculture, Forestry and Fisheries (DAFF) & Department of Social Development (DSD) (2013) National Policy for Food and Nutrition Security. Available at http://www.nda.agric.za/docs/media/NATIONAL%20POLICYon%20food%20and%20nutririon%20security.pdf (last accessed 20 July 2014)

Department of Agriculture, Ilima/Letsema Campaign (2008)

Department of Basic Education (1994) National School Nutrition Programme. Available at http://www.education.gov.za/Programmes/NationalSchoolNutritionProgramme/tabid/440/Default.aspx (last accessed 13 July 2014)

Department of Basic Education (2013) Q & A for NSNP Schools. Available at http://www.education.gov.za/LinkClick.aspx?fileticket=mQuw0D0RzsU%3d&tabid=440&mid=1911 (last accessed 13 July 2014)

Department of Rural Development and Land Reform (DRDLR) (2011) Green Paper on Land Reform. Available at http://www.gov.za/sites/www.gov.za/files/land_reform_green_paper.pdf (last accessed 20 July 2014)

DRDLR (2013a) Agricultural Land Holdings Policy Framework. Available at http://www.ruraldevelopment.gov.za/legislation-and-policies/file/2052 (last accessed 10 July 2014)

DRDLR (2013b) Policy on the Recapitalisation and Development Programme. Available at http://www.dla.gov.za/phocadownload/Policies/rdp_23july2013.pdf (last accessed 20 July 2014)

DRDLR (2014) Strategic Plan: Rural Development and Land Reform 2014–2019. Available at http://www.ruraldevelopment.gov.za/publications/strategic-plans (last accessed 10 July 2014)

Du Toit DC (2011) Food Security. Available at http://www.nda.agric.za/docs/genreports/foodsecurity.pdf (last accessed 01 July 2014)

FAO (1996) Rome Declaration on Food Security. Available at http://www.fao.org/docrep/003/w3613e/w3613e00.HTM (last accessed 13 July 2014)

FAO (1997) World Food Summit Plan of Action. Available at http://www.fao.org/docrep/003/w3613e/w3613e00.HTM (last accessed 13 July 2014)

FAO (2001) The International Treaty on Plant Genetic Resources for Food and Agriculture. Available at http://www.planttreaty.org/content/article-xiv (last accessed 20 July 2014)

FAO (2004) Part III: Options Available to National Governments under existing International IPR Agreements Protecting Plant Varieties and Plant Breeders' Rights. Available at http://www.fao.org/docrep/007/y5714e/y5714e04.htm (last accessed 20 July 2014)

FAO (2009) Guide on Legislating for the Right to Food. Available at http://www.fao.org/fileadmin/templates/righttofood/documents/RTF_publications/EN/1_toolbox_Guide_on_Legislating.pdf (last accessed 01 July 2014)

FAO (2014) Legal Developments in the Progressive Realization of the Right to Adequate Food: Working Paper. Available at http://www.fao.org/fsnforum/righttofood/sites/default/files/files/3_Legal%20development%20RTF.pdf (last accessed 08 July 2014)

Garrett E (2005) The purpose of framework legislation. J Contemp Leg Issues 4(3):717–766

Health Systems Trust (2005) National Health Act Proclaimed by the President. Available at http://www.hst.org.za/news/national-health-act-proclaimed-president (last accessed 13 July 2014)

Hendriks S, Olivier NJJ, Pienaar CJ, Olivier N, Williams C (2014) Current regulatory framework relating to, and impacting on, food security. Paper presented at the KAS/NWU conference on land reform and food security, Hakunamatata, Muldersdrift, 19–20 June 2014

In Motion Magazine (2003) Valley of 1000 Hills Declaration. Available at http://www.inmotionmagazine.com/global/hills.html (last accessed 20 July 2014)

International Committee of the Red Cross (ICRC) (1949) The Geneva Conventions of August 12 1949. Available at http://www.icrc.org/eng/resources/documents/publication/p0173.htm (last accessed 13 July 2014)

Khoza S (2004) Realising the right to food in South Africa: not by policy alone – a need for framework legislation. S Afr J Hum Rights 20:664–683

Khoza S (2005) The role of framework legislation. In: Eide WB, Kracht U (eds) Food and human rights in development: volume I legal and institutional dimensions and selected topics. Intersentia, Oxford, pp 187–204

Liebenberg S (2010) Socio-economic rights: adjudication under a transformative constitution. Juta, Claremont, p 541

Mandela N (1994) Statement of Nelson Mandela at his Inauguration as President. Available at http://www.anc.org.za/show.php?id=3132 (last accessed 14 July 2014)

Ministry for Agriculture and Land Affairs (2001) Land Redistribution for Agricultural Development. Available at http://www.nda.agric.za/docs/Policy/redistribution.htm (last accessed 10 July 2014)

Ministry of Rural Development and Land Reform (2009) The Comprehensive Rural Development Programme Framework. Available at http://www.ruraldevelopment.gov.za/publications/file/670-the-comprehensive-rural-development-programme-framework (last accessed 20 July 2014)

National Planning Commission (2011) National Development Plan. Available at http://www.npconline.co.za/medialib/downloads/home/NPC%20National%20Development%20Plan%20Vision%202030%20-lo-res.pdf (last accessed 20 July 2014)

NEPAD (2003) Comprehensive Africa Agriculture Development Programme (CAADP). Available at http://www.nepad.org/foodsecurity/agriculture/about (last accessed 13 July 2014)

New Partnership for Africa's Development (NEPAD) (2002) About. Available at http://www.nepad.org/about (last accessed 13 July 2014)

Pienaar JM (2014) Land Reform: Reflections and Dimensions [COULD NOT FIND PUBLISHING INFO]

SADC (1992) SADC Objectives. Available at http://www.sadc.int/about-sadc/overview/sadc-objectiv/ (last accessed 13 July 2014)

SADC (2004) Dar-Es-Salaam Declaration on Agriculture and Food Security in the SADC Region. Available at http://www.sadc.int/files/6913/5292/8377/Declaration_on_Agriculture__Food_Security_2004.pdf (last accessed 13 July 2014)

SADC (2012a) Food, Agriculture and Natural Resources. Available at http://www.sadc.int/sadc-secretariat/directorates/office-deputy-executive-secretary-regional-integration/food-agriculture-natural-resources/ (last accessed 13 July 2014)

SADC (2012b) Food Security. Available at http://www.sadc.int/themes/agriculture-food-security/food-security (last accessed 13 July 2014)

Statistics South Africa (2011) Food security and agriculture 2002–2011: In-depth analysis of the General Household Survey data. GHS Series Volume IV. Available at http://www.statssa.gov.za/publications2/Report-03-18-03/Report-03-18-032011.pdf (last accessed 27 June 2014)

Terblanche A (2011) Voedselsekerheid as Ontwikkelingsdoelwit in Suid-Afrikaanse Wetgewing: 'n Menseregte-gebaseerde Benadering. North-West University, South Africa

Terblanche A, Pienaar G (2012) Framework legislation for the realisation of the right to have access to sufficient food. Potchefstroom Electron Law J 15(5):185–307

The Government of the Republic of South Africa (1972) Foodstuffs, Cosmetics and Disinfectants Act

The Government of the Republic of South Africa (1990) Agricultural Product Standards Act

The Government of the Republic of South Africa (1993) Provision of Land and Assistance Act

The Government of the Republic of South Africa (1996a) Constitution of the Republic of South Africa

The Government of the Republic of South Africa (1996b) Marketing of Agricultural Products Act

The Government of the Republic of South Africa (1997a) Basic Conditions of Employment Act

The Government of the Republic of South Africa (1997b) Extension of Security of Tenure Act
The Government of the Republic of South Africa (1997c) Genetically Modified Organisms Act
The Government of the Republic of South Africa (1998a) Competition Act
The Government of the Republic of South Africa (1998b) Employment Equity Act
The Government of the Republic of South Africa (1998c) National Environmental Management Act
The Government of the Republic of South Africa (1998d) National Water Act
The Government of the Republic of South Africa (1998e) Prevention of Illegal Eviction from and Unlawful Occupation of Land Act
The Government of the Republic of South Africa (1998f) Skills Development Act
The Government of the Republic of South Africa (2002) International Trade Administration Act
The Government of the Republic of South Africa (2003) National Health Act
The Government of the Republic of South Africa (2004a) National Environmental Management: Biodiversity Act 10 of 2004 (Biodiversity Act). Available at http://www.saflii.org/za/legis/consol_act/nemba2004476.pdf (last accessed 20 July 2014)
The Government of the Republic of South Africa (2004b) Social Assistance Act
The Government of the Republic of South Africa (2008a) Consumer Protection Act
The Government of the Republic of South Africa (2008b) Standards Act
The Government of the Republic of South Africa (2009) War on Poverty Campaign. Available at http://www.waronpoverty.gov.za/WOP_PHP_JOOMLA/index.php?option=com_frontpage&Itemid=1 (last accessed 10 July 2014)
The Government of the Republic of South Africa (2011) The New Growth Path. Available at http://www.gov.za/documents/download.php?f=135748 (last accessed 10 July 2014)
The Government of the Republic of South Africa (2013a) Agricultural Landholdings Policy Framework
The Government of the Republic of South Africa (2013b) Draft Expropriation Bill
The Government of the Republic of South Africa (2013c) Draft Extension of Security of Tenure Amendment Bill
The Government of the Republic of South Africa (2013d) Millennium Development Goals – County Report 2013. Available at http://www.za.undp.org/content/dam/south_africa/docs/Reports/The_Report/MDG_October-2013.pdf (last accessed 10 July 2014)
The Government of the Republic of South Africa (2013e) Spatial Planning and Land Use Act
The Government of the Republic of South Africa (2014a) Land Protection Bill
The Government of the Republic of South Africa (2014b) Protection and Development of Agricultural Land Framework Bill
The Government of the Republic of South Africa (2014c) Restitution of Land Rights Amendment Act
The Government of the Republic of South Africa and Others v Grootboom and Others (2001) Case No. CCT11/00 ZACC 19; 1 SA 46; 11 BCLR 1169. Available at http://www.saflii.org/za/cases/ZACC/2000/19.html (last accessed 20 July 2014)
The Presidency – Republic of South Africa (2009) Medium Term Strategic Framework: A Framework to Guide Government's Programme in the Electoral Mandate Period 2009–2014. Available at http://www.thepresidency.gov.za/docs/pcsa/planning/mtsf_july09.pdf (last accessed 10 July 2014)
The Presidency – Republic of South Africa (2013) Address by His Excellency Presidency Jacob Zuma at the launch of Fetsa Tlala Integrated Food Production Initiative, Kuruman, Northern Cape Province. Available at http://www.thepresidency.gov.za/pebble.asp?relid=16329&t=79 (last accessed July 08, 2014)
The Presidency – Republic of South Africa (2014) Medium Term Strategic Framework (MTSF) 2014–2019. Available at file:///C:/Users/nandini/Downloads/MTSF%202014-2019.pdf (last accessed 10 July 2014)
United Nations (2000) Millennium Development Goals. Available at http://www.un.org/millenniumgoals/ (last accessed 13 July 2014)

United Nations Convention on Biological Diversity (2001) Cartagena Protocol on Biosafety. Available at http://www.unep.org/dewa/Africa/publications/AEO-2/content/163.htm (last accessed 20 July 2014)

United Nations Convention on Biological Diversity (2010) Nagoya Protocol. Available at http://www.cbd.int/abs/nagoya-protocol/signatories/ (last accessed 20 July 2014)

United Nations General Assembly (2012) Resolution 67/174: The Right to Food. Available at http://www.refworld.org/docid/51e64c4d4.html (last accessed 13 July 2014)

United Nations Human Rights (1985) Committee on Economic, Social and Cultural Rights (1985)

United Nations Human Rights - Office of the High Commission for Human Rights (1979) Convention on the Elimination of All Forms of Discrimination against Women New York. Available at http://www.ohchr.org/EN/ProfessionalInterest/Pages/CEDAW.aspx (last accessed 13 July 2014)

UPOV (1978) International Convention for the Protection of New Varieties of Plants. Available at http://www.upov.int/export/sites/upov/upovlex/en/conventions/1978/pdf/act1978.pdf (last accessed 20 July 2014)

UPOV (1991a) International Convention for the Protection of New Varieties of Plants. Available at http://www.upov.int/en/publications/conventions/1991/act1991.htm (last accessed 20 July 2014)

UPOV (1991b) Members of the International Union for the Protection of New Varieties of Plants. Available at http://www.upov.int/export/sites/upov/members/en/pdf/pub423.pdf (last accessed 20 July 2014)

Vapnek J, Spreij M (2005) FAO Legislative Study 87: Perspectives and Guidelines on Food Legislation, with a New Model Food Law. Available at http://www.fao.org/fileadmin/user_upload/legal/docs/ls87-e.pdf (last accessed 13 July 2014)

Western Cape Government (2013) Media Release: Minister of Education Donald Grant Western Cape - Background to the National Quintile System. Available at http://wced.pgwc.gov.za/comms/press/2013/74_14oct.html (last accessed 13 July 2014)

World Trade Organization (WTO) (1994) Agreement on Trade-Related Aspects of Intellectual Property Rights. Available at http://www.wto.org/english/tratop_e/trips_e/t_agm0_e.htm (last accessed 20 July 2014)

Wynberg R, van Niekerk J, Williams R, Mkhaliphi L (2012) Policy Brief: Securing Farmers' Rights and Seed Sovereignty in South Africa. Available at http://agrobiodiversityplatform.org/par/2012/07/03/policy-brief-securing-farmers-rights-and-seed-sovereignty-in-south-africa/ (last accessed 20 July 2014)

Part VII
Asia

This part begins with chapters by Prof. Dr. Jörn Westhoff (Chap. 48), an international business law specialist from Germany, and by Prof. Peter Sousa Hoejskov (Chap. 49) from the World Health Organization. These two chapters discus food regulation under ASEAN, the Association of Southeast Asian Nations. ASEAN will be introduced in great depth as it largely affects countries that have a variety of different issues surrounding food laws as many countries are included under ASEAN.

Subsequently, in Chap. 50, Andreas Popper provides an important introduction to Branding, Regulation and Customs in Japan and Singapore, thereby linking food regulation and intellectual property law to business aspects.

This will be followed by an in-depth discussion of food law in China in Chap. 51. Dr. Juanjuan Sun describes how food safety, food quality, and sustainability are regulated while China's agricultural policy continues to address challenging and fascinating questions.

No food law and policy textbook would include a complete section on Asia without a discussion on food law in India. Thus, for Chap. 52, Dimpy Mohanty summarized how India's law and policy are reshaped while the country is advancing its food regulatory system. In light of rich cultural and dietary variations, India has begun to unify previously multifaceted and cumbersome regulatory regimes, with respect to standards of food and its manufacture, sale and import, to ensure availability of safe and wholesome food for human consumption.

In Chap. 53, an outstanding and comprehensive discussion on food law in Russia follows with an introduction into various food safety and trade regulatory frameworks and a comparison between the Eurasian Economic Community and the European Union. Three distinguished authors, Dr. Anatoly Kutyshenko, Dr. Alexey Petrenko and Dr. Victor Tutelyan, combined their efforts to provide an unprecedented overview of one of the largest economic zones in the world and emphasized the food law and policy aspects with exceptional elegance.

As a final but lasting issue of importance, Chap. 54 sets out to describe and explain the Israeli regulatory policy for GM crop and food safety. By linking food law issues from economics, science and technology, Dr. Justo-Hanani explains institutional structures and cultural factors that affect Israel's GMOs policy. Finally, the chapter demonstrated an increase in academic awareness and public activities, which resemble those activities of global policy trends more closely, and may profoundly impact policy in the future.

Chapter 48
Food Regulation and Policy Through the Association of Southeast Asian Nations (ASEAN)

Jörn Westhoff

Abstract The Association of Southeast Asian Nations (ASEAN) is a supranational organization comprising of ten rather inhomogeneous countries striving to develop into a common single market with legal and social harmonization. Food security is a major concern of many of them, since the region is exposed to severe hazards to its environment and agriculture, caused by natural disasters as well as man-made threats. ASEAN with the help of other countries has managed to organize a food safety system ensuring the supply of rice, a staple food of predominant importance in the region. Food security is still not harmonized on an ASEAN level and remains within in the responsibility of each member states. All ASEAN member states, though, are members of the Codex Alimentarius, yet implementation of the codex standards differs widely among ASEAN member states.

48.1 Introduction to General ASEAN Policy

48.1.1 ASEAN: History and Organization

The Association of Southeast Asian Nations (ASEAN) was founded on August 8, 1967 in Bangkok with the signing of the ASEAN Declaration[1]—also referred to as "Bangkok Declaration"—in the aftermath of de-colonization following World War II. Stemming from smaller, yet not very efficient regional agreements (like ASA and MAPHILINDO), ASEAN was established under the impression of the Vietnam war and with the declared goal to foster peace in the region by common efforts to stabilize the then beginning economic growth, to pursue social advance

---

[1]English text at: http://www.asean.org/news/item/the-asean-declaration-bangkok-declaration (last visited April 21, 2014).

J. Westhoff (✉)
Dr. Wehberg and Partner GbR, Feithstr. 177, D-58097 Hagen, Germany

German and International Business Law, FOM University, 45141 Essen, Germany
e-mail: joernwesthoff@hotmail.com

and to support cultural development. Founding members were Thailand, Malaysia, Indonesia, the Philippines and Singapore, later joined by Brunei Darussalam (1984), Viet Nam (1995), Myanmar and The People's Democratic Republic of Lao (1997) and finally Cambodia (1999) to form a group of ten nations, with a population of approximately 575 million, which is 8 % of the world population and more or less as much persons as are living in the European Union. Papua New Guinea and East-Timor have a status as observers. Through progressing economic integration, ASEAN developed into an ASEAN Economic Community (AEC), obviously modeled after the European Union, as a region of equitable economic development joined in a single market and politically based on three pillars, namely the ASEAN Political-Security Community, the ASEAN Economic Community and the ASEAN Socio-Cultural Community. Legal status and institutional framework of ASEAN, its norms, rules and values as well as clear targets are now set out in the ASEAN Charter which entered into force on December 15, 2008.

ASEAN is administered by the ASEAN Secretariat in Jakarta, Indonesia, headed by a Secretary General, who coordinates the work of ASEAN. Political decisions are made by the ASEAN Summit of heads of state or government of the member states and the ASEAN Coordinating Council, which meets three to five times a year and consists of the hosting country's foreign minister and the ambassadors of the other states. Resolutions on specialized matters are passed in ministerial conferences on industry, mining and energy, trade and tourism, nutrition, agriculture and forestry, banking and finance, transport and communication. The foreign ministers of the ASEAN member states meet annually in the Council of Ministers passing resolutions on the further development structure and related guidelines. Every 3 years, the summit of heads of state and governments gives ASEAN new impetus and encourages new developments deemed desirable.

48.1.2 ASEAN: Not a Homogenous Group of Countries

Quite different from the European Union (as an obvious comparator), ASEAN by no means is a group of countries with more or less equal economic strength (at least this was true for the EU of 16 countries). ASEAN member states differ very much in terms of population, size of the agricultural sector and other economic variables, etc.

The following is a short outline on the history as well as the political and legal system of the ten ASEAN member states.

48.1.2.1 Brunei Darussalam

The Sultanate of Brunei, a major regional force between the fifteenth and seventeenth centuries lost influence over internal dynastic conflicts, European colonization, and piracy. A British protectorate since 1888, Brunei gained independence in 1984 and is ruled by an absolute monarch in a dynasty stemming from 600 years ago. Brunei wealth roots in his petroleum and natural gas fields and has the highest per capita GDP in ASEAN. The legal system is based on English common law with influence by Islamic sharia in some fields.

48.1.2.2 Cambodia

Cambodia looks back on the great history of the Angkor empire which in the tenth through thirteenth century extended over big parts of Southeast Asia. The country came under French protection in 1863, was made part of French Indochina and in World War II was occupied by Japanese troops. Independence was gained in 1953, and over time Cambodia evolved into a constitutional monarchy with democratic elections to parliament. Cambodia has the mixture law system typical for a country under varying foreign influence and with several internal law sources. It is now a conglomerate of French-influenced statutes, royal decrees, parliamentary legislation, often referencing customary law, and influence from common law principles.

48.1.2.3 Indonesia

Indonesia, a Dutch colony for centuries and experiencing Japanese occupation in World War II, gained independence from the Netherlands in 1949. It is now one of the world's largest democracies, with a population widely spread over about 6,000 densely inhabited islands (of a total of 17,508, most of which are wilderness), with hundreds of distinct ethnic groups and as many languages and dialects. Being Javanese and/or Muslim gives a shared identity to a large group, but by far not to all of the population, and "Bhinneka Tunggal Ika" ("many, yet one") is the Indonesian motto that describes the situation and binds together the Indonesian people. The Indonesian legal system is based on Dutch law, influenced by indigenous concepts.

48.1.2.4 Lao PDR

The ancient kingdom of Lan Xang ruled the region from fourteenth through seventeenth century, but came under Siamese (now Thailand) influence in the eighteenth century and was colonized by France in the late nineteenth century as part of French Indochina. Having experienced Japanese colonization, Laos declared independence in 1945, but had to fight French re-colonization, and only in July 1954 France gave up its claims to French Indochina and accepted Lao's full sovereignty. In the aftermath of Vietnam war Lao came under communist influence and Lao PDR was declared in 1975. Lao PDR's economy largely depends on the agricultural sector, employing about 80 % of the work force. The legal system is based on French influenced statutes, mixed with traditional customary law.

48.1.2.5 Malaysia

The area of what is now Malaysia was a conglomerate of British colonies and protectorates from late eighteenth through nineteenth century. Freed from World War II Japanese occupation, the former British territories on the Malay Peninsula formed the Federation of Malaya and as such gained independence in 1957. The state of Malaysia today is under the rule of the Yang di-Pertuan Agong, a constitutional monarch, elected for a 5-year term from the heads of the nine sultanates which together with four other federal states and three federal territories form Malaysia. Major ethnic groups in Malaysia are Malayans and Chinese, official language is Bahasa Malaysia, but English as language i.a. for official documents and several Chinese languages are of great importance as well. The legal system is based on British common law with some influence of Islamic law in family and religious matters of the Muslim population.

48.1.2.6 Myanmar

The Union of Myanmar was—as Burma—a British colony from 1824 and became part of the British-Indian Empire. Occupied by Japan in World War II, Myanmar was re-colonized by Great Britain and gained independence in 1948. Ruled by military forces, Myanmar has only recently started a democratization process. The population consists of 135 distinct ethnic groups with indigenous languages. The biggest ethnic group (70 % of the population) is the Bamar. More than two thirds of the population work in the agricultural sector. The legal system is based on British common law.

48.1.2.7 Philippines

The Philippine islands were fragmented in various kingdoms, city states and sultanates when in the sixteenth century Magellan claimed the islands for Spain, which united the islands under colonial rule as the "Spanish East Indies" or the "Viceroyalty of New Spain". The name of the islands stems from this time, when the islands of Leyte and Samar were named "Filipinas" after the Prince of Asturias, who later became King Philip II of Spain. 20 million dollars bought the islands for the United States of America in the 1898 Treaty of Paris. While still struggling for independence with the United States, the Philippines were occupied by Japan during World War II. Independence and sovereignty was recognized by the United States in 1946. The Philippines are a multi-ethnic states, with appr. 28 % of the population the Tagalog are the biggest ethnic group. 171 languages are spoken in the Philippines, with Filipino and English as the official languages, and 19 regional languages as auxiliary official languages. The legal system of the Philippines is based on Spanish law and heavily influenced by American common law.

48.1.2.8 Singapore

Singapore was a British trading outpost established in 1819 within the Sultanate of Johor. Later under the jurisdiction of British India, Singapore was occupied by Japanese troops during World War II and reclaimed by Great Britain after the Japanese surrender. Singapore gained independence in 1963, joined the Federation of Malaya, but left it 2 years later. 40 % of the inhabitants of Singapore are foreigners, underlining the importance of the state as center of trade for Southeast Asia. English, Malay, Mandarin and Tamil are the four official languages of Singapore, with English dominating business, government and education. The Singapore legal system bases on British common law.

48.1.2.9 Thailand

Thailand was known as the Kingdom of Siam from fourteenth century until 1939. It has never been colonized by European countries, yet had several severe conflicts with its neighboring states. From 1932, Thailand developed into a constitutional monarchy, ruled by the Chakri dynasty since general Chao Phraya Chakri became King Rama I in 1782. Thailand was invaded by Japan in 1941. 75 % of the population are Thai, the second largest ethnic group (15 %) is Chinese. 73 languages are used in Thailand, with Thai as the official language. Traditionally, Thailand's economy is agriculture-oriented, with 39 % of the work-force employed in this sector, but the so-called "dual-track" economy is aimed to support a process of industrialization. Thailand has a civil law system.

48.1.2.10 Viet Nam

Viet Nam was a part of Han-China until 938, when the Trieu-Dynasty was able to gain sovereignty. This lasted until the nineteenth century, when Viet Nam became part of French Indochina. Having been occupied by Japan during World War II, Viet Nam fought against French re-colonization, in the Indochina-War. In 1954, Viet Nam was divided into a Northern and a Southern part, both sovereign states, which were united in 1976, following the Viet Nam War, as the Socialist Republic of Viet Nam. Viet Nam's economy is agriculture-oriented, with 65 % of employees working in this sector. The Viet Nam population is relatively homogeneous, with 88 % ethnic Vietnamese, yet 53 ethnic minority groups, among which the Chinese are the biggest group. Viet Nam's legal system is based on French law, influenced by socialist legal theory.

48.1.3 Rule of Law in ASEAN

From the very beginning, ASEAN in its internal policy has relied more on diplomacy rather than law.[2] Consultation and consensus were the means of managing political relations, and only few legaly binding treaties were put into force. Yet, the founding members had been able to create among them an organization aimed to foster peaceful relations with each other and to put up principles for their cooperation which were laid down in Art. 2 of the Treaty of Amity and Cooperation in 1979, namely:

- Mutual respect for the independence, sovereignty, equality, territorial integrity and national identity of all nations
- The right of every State to lead its national existence free from external interference, subversion or coercion;
- Non-interference in the internal affairs of one another;
- Settlement of differences or disputes by peaceful means;
- Renunciation of the threat or use of force;
- Effective cooperation among themselves.

These principles describe "the ASEAN Way" of refraining from interference with the internal affairs of other member and a—very Asian—way of resolving disputes and eventually organize cooperation, "based on realism, the self-interest of each state or at best functionalism based on the common interests of each state."

Things changed when in 2007, ASEAN member states signed the ASEAN Charter to establish a legal and institutional framework for ASEAN, and blueprints for each of the three pillars named above were formulated and adopted together with new treaties and protocols often depicting detailed obligations and related procedures for dispute settlement, all with the declared goal to resolve disputes peacefully, forge closer economic integration and be bound by a common regional identity.[3] Dynamic and rapid though the recent developments may seem, it has to be noted that, other than with the European Union, integration within ASEAN is not promoted by some states of major influence in the group. The position which France and Germany have assumed in the European Union as the two states with the biggest population and highest economic strength cannot be claimed by any of the ASEAN member states. Whereas Indonesia is by far the biggest population, it is also one of the poorest countries of the group, whereas wealthy nations like Brunei Darussalam and Singapore are the smallest states. Furthermore, other than in Europe, not all member states of ASEAN have democratic political systems, but some are military dominated like Myanmar, claim to be socialist states like Vietnam or Laos, and some are even absolute monarchies, namely Brunei Darussalam. With the lack of dominant states, the establishment of ASEAN intergovernmental or supranational institutions and their ability to influence politics and

---

[2]Ewing-Chow and Hsien-Li (2013), p. 1.
[3]AESAN Vision 2020, available at: http://wwww.asean.org/1814.htm (last visited April 21, 2014).

legislation in the member states is of decisive importance. Art. 11 (2) of the ASEAN Charter does bestow upon the ASEAN Secretariat a substantive role in facilitating and monitoring the implementation in the member states of ASEAN legal acts, resolved in particular by the ASEAN Summit, and in pursuing the goal of ASEAN integration. But, as a matter of fact, the Summit depends on unanimous resolutions and the Secretariat's role is still weak, the details of its competences are not clear, and the funding for ASEAN institutions in general is difficult, their budget low. Courts are of no help either in implementing ASEAN legal acts. Unlike the European Union with the European Court of Justice (ECJ) and the system of preliminary ruling which enables all domestic courts of European member states to directly apply European legislation authentically interpreted by the ECJ, ASEAN neither has a supranational court nor a system to enforce ASEAN law, not to speak of harmonized interpretation.

Nevertheless, ASEAN legislation as such does exist. It is laid down in numerous agreements, declarations, MOU, policies and framework, of which the binding character often remains unclear. It is with a political approach rather than a legal one that those instruments are drafted, and the wording is a result of compromises sought in order to reach consensus among the member states as a general principle. This fact, however, is quite in line with the perception of law in Asia in general. As has often been emphasized, Asian culture is influenced by principles of group orientation, harmony and informality rather than individualism, contradiction and formality which are perceived as "Western" principles, particularly with regard to legal culture. Hence, it has to be noted that ASEAN as an institution cannot, as to now, in the same way as the European Union harmonize member states' policies and laws in any field. This has to be born in mind when examining the policy and legal acts of ASEAN on food safety and food security and the implementation thereof in the ASEAN member states.

48.1.4 Food Law and Food Policy: A Central Issue in ASEAN Economy and Development

With food production as a major industry in ASEAN member states, food policy is a central issue in ASEAN's strives to build the AEC as a means not only of mutual commercial benefit but of peace and prosperity as well. However, integration of the member states' economies is often hindered by the disparity of the member states themselves, with a multitude of ethnic and religious groups and occasional armed conflicts among member states. Often is it bemoaned that ASEAN member states tend to put national interests before ASEAN group solidarity. One of the more recent examples was the ASEAN member states reaction to the 2007/2008 food price crisis. With the ASEAN member states having diverse food production capacities, soaring prices for basic staple foods, particularly rise, during the crisis had a major negative impact on the consumer price index in food importing countries and even put into question whether net rice buyers would be able to secure enough food supplies for their populations, while rice exporting countries benefitted from the situation. Among them, ASEAN member states Thailand and Viet Nam disregarded the needs of fellow member states, in particular the Philippines, possibly in breach of the 1979 Agreement on the ASEAN Food Security Reserve,[4] banned exports and thus forced the Philippines, the world's major rice importer, to even increase its imports to secure the supply of rice for their population.

In systematic terms, the political conflicts which came up among ASEAN member states could be described as a tension between food self-sufficiency and the food self-reliance principle: While food self-sufficiency means to rely on domestic producers, not to import food to a major extend and, thus, support local food production and diets based on what can be grown on domestic soil, the food self-reliance principle aims on the availability of a variety of foodstuffs for domestic consumption, yet includes international trade and exchange as an important source to provide such variety. Consequently, in countries following the self-reliance principle, the protection of domestic production focuses on large scale producers able to compete and succeed in international trade. In all ASEAN countries rice is a domestic product and an important component of regular local diets. Hence, where the self-sufficiency and the self-reliance principle relate to the very same product, conflicts between countries with a different approach are quite viral, and the product itself must be in the focus of any policies striving to foster cooperation and harmonization. It goes without saying, therefore, that ASEAN food law and policy to a major extend is virtually a rice law and policy.

48.2 Implementation of Food Laws and Policies by ASEAN Member States

48.2.1 Food Security: The ASEAN Plus Three Emergency Rice Reserve

As the Asian Development Bank (ADB) has pointed out in a 2012 survey on "Food Security and Poverty in Asia and the Pacific",[5] "poverty is the single most common cause of food insecurity." Food security is defined as the situation when "all people, at all times, have physical, social and economic access to sufficient, safe and nutritions food that meets their dietary needs and food preferences for an active an healthy life".[6] This does not only refer to the supply of food in sufficient quantity and quality, i.e. the question whether there is enough good food to feed everyone,

---

[4]English text available at: http://www.jus.uio.no/english/services/library/treaties/13/13-02/asean_food_security.xml (last visited March 25, 2014).
[5]Asian Development Bank (2012).
[6]Food and Agriculture Organization (2002).

but also to availability of food for every individual in terms of affordability, utilization and stable supply, i.e. the question whether a single person can be sure about when he will have the next meal, where it will come from and whether it will be enough to nourish him in the way he needs. Very unfortunately, this question cannot be answered affirmatively for all nations in South East Asia. South Asia as a whole, ADB states, has the highest rate and largest number of undernourished children in the world,[7] and much of this situation, according to the ADB, is caused by a vicious circle in which "poverty deprives people of access to adequate, good quality food ..., (m)alnutrition undermines productivity, keeps incomes low, and traps people in poverty."[8] ADB, therefore, advocates global, regional, and domestic policies of simultaneously fighting against poverty and for food security.

ASEAN, while it does not expressively follow such a simultaneous approach, has undertaken to organize food security on a large scale. In 1979, ASEAN member states signed the Agreement on the ASEAN Food Security Reserve (ASFR), thereby establishing the ASEAN Emergency Rice Reserve (AERR). AERR consists of domestic food security stocks which are designated as such by the member states on a voluntary basis and were planned to be released by the member states to a fellow member state in a state of emergency, following bilateral negotiations. As a matter of fact, however, no releases ever were made from the AERR, due to, according to an ADB survey, (1) too small quantity of reserves, (2) cross-border transactions for food becoming more complex by bilateral negotiations, and (3) the inability of the AFSR Board to operate the AERR because of lacking funds for the secretariat.[9] However, as ASEAN member states continuously supported the idea of a regional food reserve scheme, in 2001 the strengthening of the AERR was put on its way in a special workshop on Food Security Cooperation and Rice Reserve Management System in East Asia, held in 2001 in Nakhon Pathom, Thailand, starting with a review of the scheme. The review's outcome showed the need of emergency mechanisms for effective immediate relief. Food emergency is often caused, as was pointed out, by the high volatility of the price for rice, partly due to environmental disasters or at least unforeseeable climate variations which render the region's agricultural sector particularly vulnerable.

Together with The People's Republic of China, Japan, and the Republic of Korea, the ASEAN member states formed the "ASEAN Plus Three"-group which put up as a pilot project from 2003 to 2010 the East Asia Emergency Rice Reserve (EAERR), with funding from Japan and contribution in kind from the other states, mounting up to a reserve of 787,000 tons of rice.[10] Stocks can be withdrawn from the EAERR as an emergency loan or grant or on the basis of a commercial contract. The latter transaction was the only one ever performed, when in March 2010 the Philippines acquired 10,000 tons of rice from Viet Nam. In October 2011, an

---

[7] Asian Development Bank (2012), p. 8.
[8] Asian Development Bank (2012), p. 22.
[9] Briones (2011), p. VIII.
[10] Briones (2011), p. IX.

agreement was signed among the members of the ASEAN Plus Three group to establish the "ASEAN Plus Three Emergency Rice Reserve" (APTERR) scheme as permanent institution which obliges the member states to cooperate whenever food-related emergency situations need response. The APTERR Agreement entered into force on 12 July 2012, and the first meeting of the APTERR Council was held on 28 March 2013 in Bangkok, Thailand. The APTERR Secretariat was officially launched on 29 March 2013.

APTERR provides two kind of stock, namely "earmarked emergency rice reserves", which are specific quantities of milled rice which remain owned and/or controlled by the government of the earmarking country for the purpose of meeting emergency requirements of one or more APTERR member countries, and "stockpiled emergency rice reserves", which is rice voluntarily donated to APTERR in the form of cash and/or in kind and which are owned collectively by the APTERR member countries and managed by the APTERR Secretariat under the supervision of the APTERR Council.[11] Three programs regulate the release of rice from the APTERR stock:

"Tier 1 – involves the release of earmarked stocks under a pre-arranged scheme to address problems of food availability. This program is designed for anticipated emergencies. The pre-arranged release of rice reserves under Tier 1 is formalized as a forward contract, stating the specific quantity and grade of rice, pricing method, terms of payment and delivery, and other requirements between a supplying country and a recipient country. Delivery of rice from the supplying country will be made in the event of an emergency in the recipient country, with payment based on the prevailing international market price. The amount of rice under a forward contract is based on an estimate of rice shortfall in the event of an emergency over the medium term.

Tier 2 – involves the release of earmarked stocks for emergencies not addressed by Tier 1. This program is designed for unanticipated emergencies. The release of rice reserves under Tier 2 is made available to an APTERR member country to meet an emergency requirement of rice under other arrangements. This program provides for the release of rice reserves beyond what is already arranged under Tier 1. Delivery follows an on-the-spot agreement between a supplying country and a recipient country. Pricing is similar to Tier 1: payment can be made in cash or through long-term loan or grant, based on mutual agreement of the countries involved.

Tier 3 – involves the release of stockpiled emergency rice reserves to address problems of food accessibility. This program is designed for acute emergencies and for other humanitarian responses to food insecurity. The release of rice reserves under Tier 3 is a donation of rice as humanitarian assistance to a recipient country affected by calamity upon their request in response to acute emergency. In special cases, rice distribution can be fast-tracked under

---

[11] http://www.apterr.org/about-us/how-we-work/apterr-mechanism (last visited April 21, 2014).

automatic trigger. Moreover, rice stocks may be also released for poverty alleviation and malnourishment eradication programs to address other humanitarian purposes."[12]

APTERR on its website lists ten "accomplishments", i.e. cases in which rice was distributed from the rice reserve[13]:

- From December 2004–June 2005, 87 households and students in Vientiane province in Lao PDR were provided with a total quantity of 13.37 tons of rice under a poverty alleviation program.
- From November 2005–November 2006, 9,992 people in Indonesias Sampang district and 22,825 people in Jember district received 100 tons of rice under a relief program to help people affected by flood.
- From July 2006–December 2006, 154,500 households in Leyte, Cebu, Davao and Manila City in the Philippines were provided with 930.24 tons of rice after a volcanic eruption and typhoons.
- From July 2007–January 2008, in Cambodia, under a relief program to help people affected by flood and under a poverty alleviation program, 379.76 tons of rice were distributed to 11,798 households in Kampong Thom, Ratanakiri, Kandal, Kompong Chhnang and Takeo provinces.
- From March 2008–May 2009, 18,182 households affected by flood in Central Java and East Java, Indonesia, received a total of 186.5 tons from the APTERR.
- From November 2008–January 2009, under a rehabilitation program to help people affected by cyclone Nargis in Myanmar, 13,120 people in Laputta and Bogalay townships where helped with 164 tons of rice.
- From November 2008–January 2009, a rehabilitation program to help people affected by typhoon Ketsana and flashfloods in the Philippines provided 520 tons of rice to 7,137 households in Metro Manila and Ifugao provinces.
- From July 2010–October 2010, 9,207 villages in Saravan and Attapeu provinces in Lao PDR, affected by typhoon Ketsana, received 347 tons of rice.
- From November 2011–December 2011, 50 tons of milled rice and 31,000 cans of cooked rice were distributed from the stock under an emergency relief program to help 8,100 households affected by flood in Central region of Thailand.
- From October 2012–December 2012, 200 tons of rice were distributes to 20,000 Indonesian households in Yogyakarta, Central Java, Banten and East Java provinces as food assistance for a poverty alleviation and malnutrition eradication program.

---

[12]http://www.apterr.org/about-us/how-we-work/apterr-mechanism (last visited April 21, 2014).
[13]http://www.apterr.org/about-us/how-we-work/apterr-accomplishment (last visited April 21, 2014).

48.2.2 Food Safety

48.2.2.1 General Outline

Food safety here is understood as "the assurance that food will not cause harm to the consumer when it is prepared and/or eaten according to its intended use."[14] ASEAN and its member states have made various efforts to implement regulatory frameworks on food safety both on a national and a regional level. The ASEAN Common Food Control Requirements (ACFCR) were set up as a guideline for national food control systems in the member states, promoting, inter alia, a set of five components for a food control system, namely:

- Food legislation,
- Food control management,
- Inspection activities,
- Laboratory service,
- Information, education, communication and training.

These five points, of which legislation will be in the focus of this paper, meet the basic requirements commonly enumerated as the necessary parts of an effective food safety regime. It is, however, by no means an easy task for some ASEAN members to fulfil those requirements. Partly so because of lack of awareness, funds and efficiency, as *Othman* states[15]:

> National food safety programs in Southeast Asia generally lack the following critical elements, namely: an appreciation of the nature and extent of national food safety problems, an awareness of the consequences of contaminated food on the nation's health status and economic development, and a sense of urgency for the need to investigate and do research. There is a shortage of sound, cost-effective methods for identifying specific food safety problems. The responsibility for ensuring food safety is based on a multi-agency approach due to historical or political reasons, and there is lack of coordination among agencies. In addition, specific food safety policies are either nonexistent, inadequate or of low priority in most of these countries. This situation is further compounded by the presence of other areas of concern which compete for the limited resources.

48.2.2.2 Food Safety Legislation

Other than food security, food safety cannot be assured just by piling on each other the efforts of all ASEAN member states (and some others), as was done in the case of APTERR. Food safety is about control on a large and on a small scale, starting where food is produced and proceeded and following the distribution chain to the end user. For efficient control, standards need to be set by legislation,. Those standards must not by all means meet the standards agreed upon internationally,

---

[14]Othman (2006), p. 83.
[15]Othman (2006), p. 84.

but for developing countries with an export-oriented food industry, like some of the ASEAN member states, adhering to internationally accepted standards is a question of economic success and, thus, part of the fight against poverty. Countries which depend on import to supply their population's nutrition needs, on the other hand, profit from having their domestic standards harmonized on an international level as then the standards themselves will not stand in the way of the import of food. Setting standards only to protect the domestic food industry might make sense for exporting states, it does not help anybody in importing states.

The probably most influential internationally accepted standard for the production of food is the "codex alimentarius" (Latin for "food code", hereinafter: the "codex"), which was established in 1963 by the Food and Agriculture Organization of the United Nations (FAO) and the World Health Organisation (WHO). Art. 3 of the World Trade Organization's (WTO) Agreement on Sanitary and Phytosanitary measures (SPS Agreement) references the codex, which comprises of several regulations for harmonized international food standards, guidelines and codes of practice to protect the health of consumers and ensure fair food trade. While the regulations are mere recommendations for voluntary application by the codex members, global coordination of food standards is a major concern of the Codex Alimentarius Commission and the committees organized to administer the codex. In addition to the SPS Agreement, the Agreement on Technical Barriers to Trade (TBT Agreement) does not directly refer to the production or procession of food, but covers all technical regulations on traditional quality factors, fraudulent practices, packaging, labeling, etc.

Whereas all ASEAN member states are codex members, ASEAN as an organization is not (other than the European Union, which joined in 2003). Implementation of the codex, therefore, is a domestic task for ASEAN member states. However, ASEAN has created several bodies dealing with food safety in the region: The ASEAN Expert Group on Food Safety (AEGFS) monitors, facilitates and coordinates food safety activities of ASEAN member states; the ASEAN Food Safety Improvement Plan (AFSIP), developed by AEGFS, provides the related policy outlines.[16] So far, AEGFS identified ten areas which need improvement, among them, legislation is one of the five named top priority issues. Single ASEAN member states have been entrusted as lead countries for each of the five top priority areas, the country in charge of legislation is the Philippines.

In the following text, a brief outline on each state will give an overview on the current state of food safety activities in ASEAN member states, mainly summarizing the results of a survey[17] made in the context of the FAO regional project "Support to Capacity Building and Implementation of International Food Safety Standards in ASEAN Countries, launched in December 2011, aiming on better

---

[16]For detailed information see: http://aegfs.aseanfoodsafetynetwork.net/ (last visited April 21, 2014).

[17]FAO (2012).

implementation of codex standards, and information provided in the internet by the "Asian Food Regulation Information Service".[18]

Brunei Darussalam

Two major legal acts form the body of law dealing with food safety in Brunei Darussalam, namely the Public Health (Food) Act, which provides for regulation of public health in respect of food in general, and the Halal Meat Act, which regulates the supply and importation of Halal Meat. The setting and enforcement of standards for Halal Meat lies in the responsibility of the Ministry of Religious Affairs and several other institutions like the Majis Ugama Islam (Islamic Religious Council), while the ministry of health takes care of the setting, enforcement and inspection of standards for food in general. In 2013, Brunei announced a plan to set up a National Standards Committee on Food,[19] but as of yet, national standards for food in particular seem to refer to Halal standards only, which deal with the production, preparation, handling, distribution and storage of Halal food.[20]

Cambodia

In Cambodia, the Law on the Management of Quality and Safety of Products and Services serves as an umbrella law covering, in a general manner, all regulated products including food, giving power to regulatory agencies to set up technical regulations for food. Cambodia's food control system follows a farm to table approach, setting up a multiple agency system on several levels of production, each level under the supervision of another national ministry. Food standards on a national level are included in the Public Health (Food) Act, with an extra set of standards existing for Halal. Under the regime of the Public Health (Food) Act, Cambodia currently has 18 voluntary food commodity standards and recommended 5 general and 24 codex commodity standards for adoption as technical regulations. Basically, Cambodia follows a policy to adopt codex norms for regulatory.

---

[18]http://www.asianfoodreg.com (last visited April 21, 2014).

[19]The Brunei Times, June 25, 2013, available at: http://dns.bt.com.bn/2013/06/25/prepare-national-food-rules-f-b-firms-urged (last visited April 21, 2014).

[20]Halal Food Standard, available under: http://wwww.asianfoodreg.com (last visited April 21, 2014).

Indonesia

Indonesia has the following legislation on food safety: Law No 7 of 1996 on Food; Government Regulation No. 69 of 1999 on Food Labeling and Advertisement; Government Regulation No. 28 of 2004 on Food Safety, Quality and Nutrition. Food control is under the responsibility of the Ministry of Agriculture, the Ministry of Marine Affairs and Fisheries and the National Agency for Drugs and Food Control. The Ministry of Trade and the Ministry of Industry share in controlling special standards together with local governments. National standards are developed by the technical committee for standard development under the supervision of the National Standardization Agency of Indonesia (NSA), which is also responsible to coordinate the implementation of international food standards, in particular the codex. The named authorities have agreed on the establishment of a National Codex Alimentarius (Codex Indonesia).

Lao PDR

Lao has announced a National Food Safety Policy as a reference scheme for the implementation of the control and management of quality and safety of food. There are several regulations, rules and codes of practice related to food, and according to the survey, they have been revised or drafted to meet codex standard. A National codex Committee was established in 1998, and 76 national standards are registered and adopted by the Ministry of Science and technology and the Food and Drug Department of the Ministry of Health.

Malaysia

The main law for protecting the Malaysian public against health hazards caused by food is the Food Act 1983 and related Food Regulations 1985 and 2009, the Food Hygiene Regulations 2009, The food Irradiation Regulations 2011 and the Food Analyst Act 2011. Regulatory rules are set by the Ministry of Health and its Food Safety and Quality Division is in charge of control along the food supply chain. There are mandatory standards covering all major aspects of food safety, however, codex provisions are mostly referred to in voluntary standards. Those standards and related practices are in line with codes guidelines and standards. A National Codex Committee (NCC) was established in 1985, convening government, industry, consumers, academia and professional bodies.

Myanmar

Myanmar has a set of National Food Law, Pesticide Law, Fisheries Law and Animal Health and Development Law, administered by several ministries on a multi-level scope following the food supply chain. Standards are developed by the Food and Drug Supervisory Committee and the Food and Drug Administration, supported by various committees and sub-committees.

Philippines

In the Philippines, the Food, Drug and Cosmetics Act was set into force in 1963 and amended in 1987 and 2009, a Sanitation Code was enacted in 1975. Food control is conducted by a multi-agency system of several ministries on a national level, and standards are developed and released as Administrative Orders by the Bureau of Agriculture and Fisheries Product Standards, the Food and Drug Administration as well as several other agencies under the supervision of the Department of Agriculture. 110 standards and codes of practice cover all major areas of food supply and food safety. A National Codex Organization was established in 2005, which defines its own role as an advisor to the government "on the implications of various food standardization and food control issues which have arisen and are related to the work undertaken by the Codex Alimentarius Commission (CAC). Such a consultative group provides important benefits for the government so as to assist in ensuring a safe supply of food to consumers while at the same time maximizing the opportunities for industry development and expansion of international trade."[21] The influence of the National Codex Organization on the enforcement of codex standards in national Philippine legislation remains unclear, adherence to codex stipulations by the government and legislative body of the Philippines is be a matter of further detailed research.

Singapore

Singapore enacted a Sale of Food Act in 1973 and amended it in 2002, Food regulations of 1998 were last revised in 2005 and cover more than 200 food standards. Codex standards are adopted as national standards as need is assessed by local risk examination.[22]

Thailand

Thailand has a Food Act of 1979 regulating the mandatory control on specific foods and imported foods, food labeling and food advertising. The Agriculture Standards Act of 2008 establishes standards for producers, exporters and importers alike of

---

[21]http://www.fdc.net.ph/index.php?id1=23.

[22]For more detailed information particularly to labeling standards in Singapore, see Chap. 50 of this volume.

agricultural commodities and with binding effect to certification bodies. 181 Thai Agricultural Standards have been elaborated, setting up commodity standards, production system standards and standards for general requirements such as maximum pesticide residue limits. Conformity with international standards is pursued on a political level by the National Agricultural Standard Committee, which, however, cannot set standards itself.

Viet Nam

Vietnam passed a Food Safety Law in 2010, specifying responsibilities for safety of food to various ministries. Standards are development following a system established under the Law on Standards and Technical Regulations. Viet Nam follows a multiple agency approach for food control. 46 national technical regulations, dealing, e.g., with heavy metals contamination are set up as mandatory regulations by the Ministry of Health, and 752 standards issued by the Ministry of Science and Technology appeal to be followed on a voluntary basis. The Viet Nam codex Committee, established in 1994, coordinates codex activities among its members, who are leaders from the relevant ministries, government agencies, food businesses and academia.

48.3 Suggestions for Further Research

- In many ASEAN member states, access to safe drinking water is a major concern of many, and a need not provided to all. The problem is often linked to hygiene and sanitation, but water can also be polluted with pesticides or herbicides. Find out how ASEAN and its member states deal with this particular problem.
- Food safety can only be effectively implemented if producers have access to comprehensive explanation and advice. One means of giving such help can be a "codex country manual". Find out in which of the Asian member states such a manual exists and whether it is available via internet in all major languages spoken in that country.
- Pick a particular aspect of food safety, e.g. contamination by intoxicating substances, check out the relevant regulations of each or selected ASEAN member states at http://www.asianfoodreg.com and other internet sources and compare the contents with each other and the Codex Alimentarius. Put up a synopsis and discuss the reasons for differences.
- Many of ASEAN member states experienced Japanese occupation in World War II. Examine how Japan today helps ASEAN members with their food security and food safety policy and how Japanese and ASEAN food markets work with each other.

References

Asian Development Bank (2012) Food security and poverty in Asia and the Pacific: key challenges and policy issues. Mandaluyong City

Briones RM (2011) Regional Cooperation for Food Security: the case of emergency rice reserves in the ASEAN Plus Three. ADB Sustainable Development Working Paper Series No 18, November 2011, p VIII

Ewing-Chow M, Hsien-Li T (2013) The role of the rule of law in ASEAN integration. EUI Working Papers of the Robert Schuman Center for Advances Studies, RSCAS 1013/16, p 1

FAO (2012) Status of national codex activities in ASEAN countries. Bangkok

Food and Agriculture Organization (2002) The State of Food insecurity 2001. Rome

Othman NM (2006) Food safety in Southeast Asia: challenges facing the region. Asian J Agric Dev 4(2):83

Chapter 49
History of Asian Food Policy

Peter Sousa Hoejskov

Abstract Asia is a diverse region consisting of over 48 countries and is home to about 4.4 billion people. Trade of food and agricultural products in Asia is crucial for the development of the region and is contributing to the Sustainable Development Goals (SDGs) by eradication of food insecurity and rural poverty. With increasing trade in food and agricultural products and with the establishment of the Sanitary and Phytosanitary (SPS) and Technical Barriers to Trade (TBT) Agreements of The World Trade Organization (WTO), food safety issues are receiving increased attention in Asia. Within Asia two groups of countries (ASEAN and SAARC) have strengthened their collaboration in terms of economic, political and social development and is striving towards harmonizing their food legislative frameworks as a means to facilitate trade with food and agricultural products within the regions. This chapter explores the legal foundations and important issues surrounding all of these aspects of food safety regulation in Asia.

49.1 Introduction

Asia is a diverse region consisting of over 48 countries and is home to about 4.4 billion people. Some of the world's richest as well as poorest countries are found in Asia and within the last 20–25 years, the region has been subject to impressive economic as well as social and political development.

Over the past two decades, trade in food and agricultural products within and beyond Asia has been increasing, making the food and agriculture sector one of the most international sectors.

Trade of food and agricultural products in Asia is crucial for the development of the region and is contributing to the Sustainable Development Goals (SDGs) by eradication of food insecurity and rural poverty. However, increasing diversity and volume of trade in food and agricultural products, together with changing

---

P.S. Hoejskov (✉)
Food Safety, World Health Organization (WHO) Regional Office for the Western Pacific, Manila, Philippines
e-mail: hoejskovp@who.int

© Springer International Publishing Switzerland 2016
G. Steier, K.K. Patel (eds.), *International Food Law and Policy*,
DOI 10.1007/978-3-319-07542-6_49

agricultural practices, human ecology and behavior and new technologies, is a key contributor in the cross-border spread of food safety hazards.

With increasing trade in food and agricultural products and with the establishment of the Sanitary and Phytosanitary (SPS) and Technical Barriers to Trade (TBT) Agreements of The World Trade Organization (WTO), food safety issues are receiving increased attention in Asia. Food safety has become a shared concern among all countries in the region and in response to increasing political and social demands on government regulatory agencies, many countries have allocated substantial resources into applying new methods for regulating the food industry and updating, revising and harmonizing national regulatory systems in line with international requirements, standards and recommendations.

The restructuring of the national food regulatory systems has taken place along with enhanced regional collaboration in the area of food safety, food legislation and regulation, inspection and certification as well as coordination of capacity building activities.

Within Asia two groups of countries (ASEAN and SAARC) have strengthened their collaboration in terms of economic, political and social development and is striving towards harmonizing their food legislative frameworks as a means to facilitate trade with food and agricultural products within the regions.

This paper describes food laws and regulations in Asia with main focus on ASEAN. The purpose of the paper is to present the overall framework for food legislation in the region and at the same time highlight some of the main issues with regard to food quality and safety and how these issues impact on the development of food laws and regulations in the region. The purpose of the paper is not to give a detailed insight into food legislation development and implementation in all Asian countries, but rather to give an overview of the regional context and some of the mechanisms for collaboration between countries in the region.

49.2 What Is ASEAN and Its Legal Foundation?

ASEAN was established in August 1967 by the five countries Indonesia, Malaysia, Philippines, Singapore, and Thailand. Brunei Darussalam joined in 1984, Vietnam in 1995, Lao PDR and Myanmar in 1997 and Cambodia in 1999.

The ASEAN region has a population of about 500 million people, a total area of 4.5 million square kilometers, a combined gross domestic product of almost US$ 700 billion and a total trade of about US$ 850 billion.

The ASEAN Declaration (http://www.aseansec.org/1212.htm) states that the aims and purposes of the Association are: (1) to accelerate economic growth, social progress and cultural development in the region and (2) to promote regional peace and stability through abiding respect for justice and the rule of law in the relationship among countries in the region and adherence to the principles of the United Nations Charter.

The ASEAN Vision 2020 (http://www.aseansec.org/1814.htm), adopted by the ASEAN Leaders in 1997, agreed on a shared vision of ASEAN as a concert of

Southeast Asian nations, outward looking, living in peace, stability and prosperity, bonded together in partnership in dynamic development and in a community of caring societies.

In 2003, the ASEAN Leaders resolved that an ASEAN Community should be established by 2015 comprising three pillars, namely:

- ASEAN Security Community
- ASEAN Economic Community
- ASEAN Socio-Cultural Community

A year later ASEAN established the Vientiane Action Programme (http://www.aseansec.org/VAP-10th%20ASEAN%20Summit.pdf) to realize this goal and in 2009, with the Cha-am Hua Hin Declaration on the Roadmap for the ASEAN Community (2009–2015) (http://www.aseansec.org/publications/RoadmapASEANCommunity.pdf), the way towards an ASEAN Community was defined.

The process of ASEAN Community building is a result of the considerable change in the association's mission in the recent two decades. The end of the Cold War, the advance of globalization, the rise of China and India in economic size and political influence as well as the Asian financial crisis have forced ASEAN to shift from its original preventive diplomacy of maintaining peace and harmony among its members and in the region to the constructive diplomacy of community building to cope with increasing political and economic competition in a globalised world.

The Treaty of Amity and Cooperation (http://www.aseansec.org/1217.htm) in Southeast Asia (TAC) spell out the following fundamental principles for the ASEAN countries in relation to each other:

- Mutual respect for the independence, sovereignty, equality, territorial integrity, and national identity of all nations
- The right of every State to lead its national existence free from external interference, subversion or coercion
- Non-interference in the internal affairs of one another
- Settlement of differences or disputes by peaceful manner
- Renunciation of the threat or use of force
- Effective cooperation among themselves.

ASEAN did not conclude a legally binding treaty until the first ASEAN Summit in Bali in 1976—the Treaty of Amity and Cooperation in Southeast Asia—was held. This was followed by the 1977 Preferential Trading Arrangement which carried some measures of legal obligation. In 1987 ASEAN concluded, at the third ASEAN Summit, the Agreement for the Promotion and Protection of Investments and the agreement on the standstill and rollback of non-tariff barriers, which conferred legal rights and obligations upon their signatories.

In the intervening period, three industrial cooperation schemes were agreed upon—the ASEAN Industrial Projects (1980), the ASEAN Industrial Complementation (1981), the ASEAN Industrial Joint Ventures (1983)—and the Brand-to-Brand Complementation Scheme (1988).

The development towards economic integration of ASEAN was strengthened in 1992 with the agreement on the Common Effective Preferential Tariff for the

ASEAN Free Trade Area (AFTA) which the ASEAN countries agreed upon during the ASEAN Summit in 1992.

Under this agreement, the first six signatories to the AFTA treaty—that is, the first six members of ASEAN—were legally committed to reducing tariffs on their trade with one another, with a few exceptions, to a range of zero to five percent by the beginning of 2002 or, in some cases, the beginning of 2003. The newer members were given a little more time.

The tariff reductions are to be carried out through national legal enactments by each party to the agreement in accordance with an agreed schedule. The ASEAN countries have also formally agreed on an arrangement to govern delays in the inclusion of products in the AFTA scheme or suspensions of AFTA concessions under very stringent conditions.

In legally committing ASEAN's members to reduce and eventually remove tariff barriers between them, AFTA is the first substantial step toward integrating the ASEAN market that ASEAN countries have recognized as essential for making their production and commercial processes more efficient, bringing down costs, encouraging investments, and, in general, strengthening their economic competitiveness.

ASEAN has extended the scope of the AFTA from trade in goods to the equally important realm of trade in services like transportation, telecommunications, financial services, construction and tourism. At the summit of December 1995, ASEAN concluded the Framework Agreement on Services. The agreement is to be given strengths through sector-by-sector negotiations. The resulting agreements are legally binding upon the parties.

At the same summit of 1995, ASEAN also entered into an agreement committing its members to undertake national measures to protect intellectual property rights, encourage investments as well as encourage industrial and scientific innovation.

In October 1998, ASEAN decided to allow the freer flow of investments through the ASEAN Investment Area agreement, under which each country legally undertakes to open up its industrial sector to investments from other ASEAN countries and accord national treatment to such investors.

As a means to facilitate trade and encourage investments in the ASEAN region, ASEAN Member countries agreed, at the summit of December 1998, on a framework agreement on mutual recognition arrangements for goods-in-transit. This framework agreement is to be carried out by agreed arrangements on specific product groups, under which ASEAN countries are obligated to recognize results of conformity assessments issued by any one of them, such as test reports, product certifications or registration approvals.

Two things are to be noted about the legally binding agreements that ASEAN has concluded so far. The first is that they are overwhelmingly economic in nature. The second is that, since the conclusion of the AFTA agreement in 1992, ASEAN has entered into such agreements with increasing frequency. This may be an indication of ASEAN's growing realization that closer regional economic integration requires that it is based on binding legal foundations if integration and that it is to be stable, credible and effective. The commitments undertaken must be clear, firm and enforceable.

As ASEAN moves into further integration, the number of binding undertakings can be expected to increase. The e-ASEAN framework agreement, which the ASEAN leaders signed in November 2000, may require legally binding agreements on such things as the authentication of signatures, the use of electronic documents in business transactions, privacy and confidentiality, and so on. The further deepening of ASEAN economic integration, involving tariff nomenclatures, product standards, policy coordination, banking and finance, transportation and telecommunications, would surely need clear and enforceable agreements in these areas.

More broadly and fundamentally, ASEAN countries are working towards harmonization of domestic laws and regulations that govern trade and investment. This is to ensure that differences in domestic laws and regulations on "fair trade," competition policy, government procurement and product standards are not used to frustrate the purposes of AFTA and the benefits of an integrated market. It is to provide the harmonized regional investment regime that investors increasingly require.

This developing rules-based economic regime will most likely gradually extend to other areas of ASEAN cooperation. After all, ASEAN is more than an economic association. ASEAN is working towards initiatives that undertake legal obligations related to transnational regional problems as the marine environment, the preservation of biodiversity, money-laundering, trafficking in human beings, drug-trafficking and piracy.

ASEAN is evolving into a more rules-based association. However, the success of regional agreements undertaken by ASEAN depends on effective national legislation and enforcement mechanisms to carry them out. This would also help strengthening the national legal systems of the member-states as well as the rule of law in the region as a whole.

49.3 Regional Coordination and Collaboration in the Area of Food Safety and Standards Setting Within ASEAN

The goals of the ASEAN countries are to produce and supply foods that are safe and meet requirements of importing countries as well as international standards. Strengthening Member Countries' participation and implementation of international food safety standards, especially Codex standards setting process, is also the main goal of the region. However, due to the different stages of development of food safety and quality standards in Member Countries, the regional approach has been concentrated on harmonization and networking of food safety and food standards among member countries. For this purpose different working groups and task forces have been established

- **ASEAN Task Force on Codex (ATFC)**
  The ASEAN Task Force on Codex has been established to serve as a forum for Member Countries to discuss Codex issues of common interest and to possibly identify common positions on Codex issues of importance to Member Countries. It is also a forum to harmonize standards and regulations in ASEAN by using Codex

standards as a reference. Four meetings of the Task Force have taken place in Kuala Lumpur, Bali, Bangkok and Cebu. Many joint ASEAN positions on Codex issues were discussed and agreed. Activities on the strengthening and improving participation of ASEAN countries in Codex, for which a Member Country was nominated to be a focal point for each Codex Committee, have commenced.

- **Expert Working Group on the Harmonization of Maximum Residue Limits (MRLs) of Pesticide among ASEAN Member Countries**

  The main objectives of the programme of this Expert Working Group are to harmonize MRLs among ASEAN countries in order to protect consumer's health, the harmonization of standards on agricultural and food commodities among ASEAN Member Countries and to provide Member Countries with a means for coordination and information sharing to reach international standards. The total number of harmonized MRLs of pesticides which have been endorsed by the ASEAN Ministers on Agriculture and Forestry (AMAF) is 369, involving a total of 28 pesticides. The number of draft harmonized MRLs in the process of consideration is 258. Even though most of the harmonized MRLs are based on Codex MRLs, more focus is on the harmonization of MRLs for minor crops important to the region for which no Codex MRLs are available. Issues on regional collaboration in the generation of residue data for harmonization and the principles and criteria on harmonization of MRLs are being considered in the Expert Working Group.

- **ASEAN Food Safety Network**

  The 25th meeting of the ASEAN Ministers on Agriculture and Forestry which was held on 21 August 2003 in Malaysia expressed full support to the initiative on establishing an ASEAN Food Safety Network http://www.aseanfoodsafetynetwork.net. The Meeting noted that the proposed establishment of the Network would provide cohesive direction for the ASEAN working groups to help resolve recurring problems of non-tariff barriers encountered in the trade of food and agricultural products for ASEAN regarding food safety.

  The network provides a communication tool towards a cohesive approach for the relevant ASEAN bodies in addressing the issues of food safety encountering the region's agricultural trade. Thailand is the coordinator of the ASEAN Food Safety Network.

  An electronic coordination has been set up with the aim of coordinating, networking, information sharing and providing early warning among various national authorities and also working groups/task forces in ASEAN. This is also a forum for on-line bilateral or multilateral discussions and information sharing with regard to capacity building activities.

- **ASEAN Expert Group on Food Safety (AEGFS)**

  ASEAN Expert Group on Food Safety (AEGFS) is a subsidiary body under Senior Officials Meeting on Health Development (SOMHD). The main objectives of AEGFS are not only to improve food safety of ASEAN countries, but also to facilitate food trade and formulate a strategic plan to address important food safety issues for mutual benefits. Besides those responsibilities, AEGFS also provides assistance to ASEAN Governments to develop and strengthen food safety infrastructures and programmes which support them to deal with their

new obligations and rights related to the safety and quality of food in both regional and international trade.

To achieve its objectives, AEGFS developed ASEAN Food Safety Improvement Plan (AFSIP) covering activities in ten areas which Information Sharing Programme is one of the priority areas. According to the agreement of the 4th AEGFS in Laos PDR, ASEAN Food Safety Network (AFSN) which has already served as a platform for coordinating and exchanging information on non-tariff barrier for ASEAN member countries and ASEAN bodies related to food safety, was assigned to serve as a coordinating forum for Information Sharing for AEGFS to cooperate and deal with information on food safety.

In the ASEAN Post-2015 Health Development Agenda and in the ASEAN Socio-Cultural Community (ASCC) Blueprint, food security and food safety is included as a cross-cutting issue within the human development directorate and the sustainable development directorate of the ASCC department. Under the ASEAN Post-2015 Development Agenda, Cluster 4 focuses on 'Ensuring Food Safety'. In 2016, the work programme for 2016–2020 was developed and presented at the 11th Senior Officials Meeting on Health Development (SOMHD). From 2016 onwards, Cluster 4 'Ensuring Food Safety' will replace the ASEAN Expert Group on Food Safety as the ASEAN coordination body for food safety matters.

- **ASEAN Consultative Committee on Standards and Quality (ACCSQ)— Prepared Foodstuff Product Working Group (PFPWG)**

  The purpose of the PFPWG is to exchange information on existing standards, technical regulations and conformity assessment procedures, develop mutual recognition agreements for prepared foodstuff as well as common food control requirements for the ASEAN countries. The PFPWG has also been working on harmonizing HACCP and GMP requirements, requirements for import and export inspection and certification systems and establishment of a Rapid Alert System for Food and Feed in the region. One of the recent initiatives by the PFPWG has been to establish an ASEAN coordinating committee on food safety which is aimed at coordinating national as well as international initiatives, assistance and support in the area of food safety.

49.4 AFSIP: A Framework for Food Laws and Regulations in ASEAN

Food safety has since the foundation of ASEAN been one of the most important areas of collaboration. In 2000 this resulted in the development of the ASEAN Food Safety Improvement Plan (AFSIP) pursuant to a directive of the Declaration of the 5th. ASEAN Health Ministers Meeting on Healthy ASEAN 2020 held in Jakarta in 2000, to formulate an ASEAN Food Safety Policy and an ASEAN Framework on Food Policy, as part of a comprehensive programme of action to address the impact of globalization and trade liberalization in the health sector. The implementation of AFSIP contributes towards the realization of the ASEAN Health Ministers' vision of Healthy ASEAN 2020 as well as addresses a priority of the Declaration of ASEAN Concord II adopted by the 9th. ASEAN Summit in 2003, to intensify cooperation in the area of public health.

The development of AFSIP recognizes that food safety is a shared concern and responsibility of all stakeholders along the food supply chain including food industry, governments, consumers and academics and that it has to be controlled by using a "farm-to-table" approach. In this context the industry is responsible for producing safe foods, the government is responsible for providing a conductive and enabling environment for the production of safe foods and ensure that food safety requirements are met by enforcement, monitoring and other means, the consumers are responsible for making informed choices and safe handling practices and the academics are responsible for providing scientific data and risk assessments that can be used by food safety regulators to mange food safety risks.

The development of AFSIP has been guided by a scientific and risk-based approach which is the modern preventive and process based approach to control food safety. The plan recognizes the need to harmonize with international standards, and in particular Codex standards, guidelines and recommendations. The overall goal and expected outcome of the AFSIP is significantly improvements in the level of food safety in ASEAN Member Countries as reflected by (1) strengthen protection of consumers' health and (2) enhance competitiveness of the ASEAN food industry and improve food export potential globally. The plan aims at:

1. Strengthening the food control system from farm to table with the involvement of relevant stakeholders
2. Increasing the level of credibility and competency of regulatory authorities
3. Enhancing the industry and consumer awareness and participation in food safety
4. Coordinating ASEAN common positions in international fora, as and when appropriate
5. Facilitating the alignment of member countries' food safety policies with obligations under the WTO SPS and TBT agreements
6. Facilitating the harmonization of national regulatory standards with Codex standards.

The AFSIP recognizes that food safety activities in ASEAN are carried out by several agencies and bodies at the national level. Thus it aims at contributing towards an integration and coordination of these activities and include activities to support capacity building in food safety at the national level and implement joint regional activities in food safety at the regional level.

The activities necessary to improve food safety within the ASEAN region by cooperative efforts are classified under the following programme headings:

- Legislation
- Laboratories
- Monitoring and surveillance
- Implementation of food safety systems
- Food inspection and certification

- Education and training
- Information sharing
- Research and development
- International participation in food safety standard setting including Codex
- Consumer participation in food safety

49.4.1 Common Principles for Food Control Systems in ASEAN

Under the AEGFS, the ASEAN countries have prepared some common principles for food control systems. http://www.aseansec.org/21915.pdf The principles recognize that effective national food control systems are essential to protect the health of consumers from foodborne illnesses and to facilitate trade within and beyond the ASEAN region. Development of national food control systems should take into account the obligations under the WTO's SPS and TBT agreements which have great significance for international trade with food and agricultural products.

An important component of the principles for food control systems in ASEAN is the establishment of a food control and regulatory framework including food legislation in the ASEAN countries.

The ASEAN Member Countries have recognized that development of relevant and enforceable food laws and regulations is an essential component of the national food control systems and that food legislation should provide a high level of consumer's health protection and provide mechanisms to facilitate food recalls in case of non-compliance and food safety emergencies. In addition to the necessary legal powers to ensure food safety, food laws should also allow food control agencies to build preventive approaches in to the food control system. There should also be a mechanism for reviewing and revising food laws and regulations. In preparing or updating food regulations and standards, ASEAN countries should take full advantage of Codex standards, guidelines and recommendations.

49.4.2 Requirements for the Labeling of Prepackaged Foods

In 2007, a draft of ASEAN common principles and minimum requirements for the labelling of prepackaged foods were developed. The purpose of the principles and guidelines are to harmonize and mainstream the requirements for labelling of prepackaged foods in the ASEAN countries and thereby facilitate trade between the countries. The common principles are based on Codex guidelines on food labelling and say that "Prepackaged food shall not be described or presented on any label or in any labelling in a manner that is false, misleading or deceptive or is likely to create an erroneous impression regarding its character in any respect" and

"Prepackaged food shall not be described or presented on any label or in any labelling by words, pictorial or other devices which refer to or are suggestive either directly or indirectly, of any other product with which such food might be confused, or in such a manner as to lead the purchaser or consumer to suppose that the food is connected with such other products".

The common principles consist of a number of mandatory requirements including information about e.g. manufacturer, expiry date, instructions for use, nutrition declaration, storage instruction. Moreover, there are some additional labelling requirements that countries may require based on local needs. This can be declaration of ingredients like alcohol, ingredients of animal origin etc. The principles also include some optional labelling requirements such as claims or recognized logos on religion or ritual preparation such as halal or kosher. Along with the mandatory and optional labelling requirements, the common principles and minimum requirements for the labelling of prepackaged foods also include some prohibited claims. This include claims on the suitability of a food for use in the prevention, alleviation, treatment or cure of a disease, disorder or other physiological condition, claims stating that any given food will provide an adequate source of all essential nutrients, claims implying that a balanced diet or ordinary foods cannot supply adequate amount of all nutrients, claims which could give rise to doubt about the safety of similar food or which could arouse or exploit fear in the consumer, claims which highlight the absence or addition of any food additive or nutrient supplement, the addition of which is prohibited and claims on the absence of beef or pork or its derivatives, or lard or added alcohol if the food does not contain such ingredient. http://www.aseansec.org/21915.pdf (page 10)

49.4.3 Food Legislative Frameworks in ASEAN

49.4.3.1 Malaysia

The Food Act 1983 and the Food Regulations 1985 are the legislative documents that form the backbone of food laws and regulations in Malaysia. The Food Act 1983 prescribes the powers of the Minister of Health, enforcement officers and analyst, whereas, the Food Regulations 1985 specifies the standards for various types of food, the use of food additives, labeling and packaging requirements. It is the minimal mandatory standards which must be complied with in order for food to be sold in the country. The Food Act is enforced by the Ministry of Health and the Local Authorities. The legislation, applicable to all foods sold in the country either locally produced or imported, covers a broad spectrum from compositional standards to food additives, nutrient supplements, contaminants, packages and containers, food labelling, procedure for taking samples, food irradiation, provision for food not specified in the regulations and penalty. Efforts in harmonization of legislation with international standards are actively undertaken. Codex Alimentarius is the benchmark for food safety legislation.

In order to ensure overall coordination and exchange of information between ministries and agencies involved in food control activities, Ministry of Health in 2001 established the National Food Safety and Nutrition Council. The Council provides a platform in enhancing national coherence of food control activities. The Council is chaired by the Minister of Health and includes all relevant Government agencies, industry and consumer representatives as well as other stakeholders from farm to table. In 2002, the Council took the lead in developing the National Food Safety Policy which provides direction for implementation of food safety measures. The implementation of the Policy is outlined in the National Plan of Action on Food Safety which clearly defines the enforcement role of each concerned agency.

49.4.3.2 Philippines

The two main agencies tasked with developing and enforcing food laws, regulations and standards in Philippines are the Food and Drug Administration (FDA) (previously known as Bureau of Food and Drugs (BFAD)) under the Department of Health (DOH), and the Bureau of Agriculture and Fisheries Product Standards (BAFPS) of the Department of Agriculture (DA). FDA is responsible for the safety of processed food and enforces the Food, Drug, Devices & Cosmetics Act, 1963 (Amended in 1987) as well as a number of other Acts and Orders pertaining to food quality. BFAD and the Food Development Center (FDC) also certifies food establishment according to national GMP and HACCP standards.

BAFPS is accountable for fresh and primary agricultural and fisheries products and the enforcement of the Agriculture and Fisheries Modernization Act (AFMA). Other Acts and Codes implemented by BAFPS include the Code on Sanitation of the Philippines, 1975; the National Meat Code, 2004; the Philippine Clean Water Act, 2004; the Agriculture and Fisheries Modernization Act, 1997; the Meat Inspection Code of the Philippines, 2004; the Philippine Fisheries Code, 1998; the Plant Quarantine Decree, 1978; the National Dairy Development Act, 1995. The Bureau of Fisheries and Aquatic Resources (BFAR) of the DA is the competent authority for GMP and HACCP accreditation for fish exports to EU.

49.4.3.3 Thailand

In Thailand, the Food Act of B.E.2522 (1979) is the main law aimed at protecting and preventing consumers from health hazards occurring from consumption of food. The Act prohibits unsafe food from being produced, imported and distributed in Thailand. The Act divides unsafe food into four categories based on the causes that make the food unsafe to consumers: (1) impure food, (2) adulterated food, (3) substandard food and (4) other foods which prescribed by the minister. Other areas covered by the Act include control of packaging and labelling as well as restrictions on advertisement.

Table 49.1 Acts in the area of food safety in Thailand

Act	Responsible body	Scope of the Act
Public Health Act 1992	Department of Health, Ministry of Public Health	To control the hygiene practices of markets, restaurants (including food stalls) and food storage facilities
Communicable Diseases Act 1980	Department of Disease Control, Ministry of Public Health	To protect and control communicable diseases
Plant Quarantine Act 1999	Department of Agriculture and Agricultural Extension, Ministry of Agriculture and Cooperatives	To prevent and control plant diseases, insects and pests as well as the import and export of plants and plant products, including GM products
Fisheries Act 1947	Department of Fisheries, Ministry of Agriculture and Cooperatives	To control fishery catching and hatching methods as well as the import and export of aquatic animals and products
Animal Epidemic Act 1999	Department of Livestock Development, Ministry of Agriculture and Cooperatives	To effectively prevent and control epidemics
Feed Control Act 1999	Department of Livestock Development, Ministry of Agriculture and Cooperatives	To control the quality of feed, including raw materials, feed processing methods, production equipments, storage and packaging facilities
Animal Slaughter Control and Sale of Meat Act 1992	Department of Livestock Development, Ministry of Agriculture and Cooperatives	To protect the meat consumed within the country from contamination by setting standards for slaughterhouse facilities and slaughtering processes as well as livestock farms
Goods Import and Export Control Act 1979	Department of Foreign Trade, Ministry of Commerce	To control the import and export of food and food products
Hazardous Substance Act 1992	Department of Industrial Works, Ministry of Industry	To control production, import, export and possession of hazardous substances for the use in agricultural production, food industry and by consumers
Sugarcane and sugar Act 1984	Thai Industrial Standards Institute, Ministry of Industry	To control the production, import and export of sugarcanes and sugar
Consumer Protection Act 1979	Office of the Consumer Protection Board, Prime Minister Office	To protect the consumer's rights to safe food. To control advertisements and labelling of food products
National Health Act 2007	The National Health Commission Office, Prime Minister's Office	To set up guidelines for strategically monitor the status of public health
The Draft of Agricultural Commodities Standards Act	National Bureau of Agricultural Commodity and Food Standards, Ministry of Agriculture and Cooperatives	To set food safety and quality standards as national references for production and trade of agricultural commodities

The Food and Drug Administration under Ministry of Public Health is responsible for the execution of the Act. The Act empowers the Ministry of Public Health to promulgate ministerial regulations, to appoint the Food Committee and competent officers, and to set up other activities in order to carry out the provisions of the Act. Along with the Food Act of B.E. 2522 (1979), the main Acts in Thailand protecting consumers from unsafe food are listed in Table 49.1.

The ministerial regulations describe the procedures for applications of manufacturing licenses, import licenses, and registration including the rates of fees, the identification card of the competent officers and the labelling of food products for exports. The Food Act B.E. 2522 (1979) classifies foods into three main categories as follows:

- **Specially Controlled Foods**—Registrations are required for foods in this category. Legal provisions are related to food standard quality, specifications, packaging, and labelling requirements, as well as other aspects of good manufacturing practices. The Food Committee may make recommendations to the Minister of Public Health specifying specially controlled foods.
- **Standardized Foods**—Standard foods do not require registration but their quality and labelling have to meet the standard requirements as specified in the Notification of the Ministry of Public Health.
- **Other Foods**—Foods, raw or cooked, preserved or non-preserved, processed or non-processed, if not listed under category 1 or 2 will be considered as general foods. Although registrations are not required, general food products are controlled and monitored with regard to hygiene, safety, labelling and advertisement. Foods in this category may be subdivided into (a) foods that must bear standard labels and (b) other general foods.

The control measures for each category of foods differ; with measures for "Specially Controlled Foods" being strictly controlled. Before producing or importing such foods, the application for product registration is required. For standardized foods, the application for such permission is not required, but they must be produced up to the prescribed quality or standard. For labelled foods, however, the main objective is to control the labelling in order to avoid misleading or cheating of consumers.

The Food and Drug Administration (FDA) of the Ministry of Public Health and the Provincial Offices of Public Health is responsible for the enforcement of the Food Act as well as legal food control operations with the support of food analytical services of the Department of Medical Sciences, Ministry of Public Health. The FDA also has the responsibility to ensure that the development of food safety control measures corresponds with international standards.

49.4.3.4 Vietnam

The national legislation in Vietnam is based on a complex array of laws, ordinances, decrees, decisions, instructions, circulars and regulations. A diverse set of legislative documents addressing food safety and quality has been drafted in recent years. The National Assembly is empowered to develop the country's laws. Permanent Committees of the National Assembly establish ordinances, decrees and resolutions. Each Ministry may also prepare and issue decisions, instructions, circulars, regulations and standards.

In 2011, the Government of Vietnam released a draft implementing regulation, intended to implement several articles of the Law on Food Safety (No.55/QH12/2010), which the National Assembly approved in June 2010, and which entered into force on 1 July 2011. While the Law is a broad, overarching legislative document that attempts to outline all aspects of Vietnam's food safety system, the draft implementing regulation is comparatively sparse, dealing with a relatively small number of provisions.

When all provisions of the Law on Food Safety are implemented, the Law will replace the Ordinance on Food Safety and Hygiene which has been in place since 2003. The Ordinance aims at ensuring the safety and hygiene of food during the process of their manufacture and trading and the prevention and remedy of poisonous food. The Ordinance specifies requirements for manufacturing and selling fresh and raw food, processing food, storing and transporting food and importing and exporting food in Vietnam. For "high-risk" foods, state certification of satisfaction of business conditions is required. The Ordinance also regulates the proclamation of food standards and the advertising and labelling of food.

Prior to 1990, Vietnamese food safety standards were primarily based on standards from the former Soviet Union. In the beginning of the 1990s, Vietnam started to develop its own standards through the establishment of technical advisory committees and sub-committees. A total of 17 committees have been established with participation from more than 100 senior experts representing governmental agencies, research institutions, universities, food enterprises and consumer groups. Vietnam became a member of the Codex Alimentarius commission in 1994 and started the process of harmonizing and developing national standards in accordance with Codex standards and recommendations.

49.4.3.5 Indonesia

The main legislative documents pertaining to food quality and safety in Indonesia are the Food Act, 1996; the Food Safety, Quality and Nutrition Regulation, 2004; and the Food Labelling and Advertisement Regulation, 1999.

The Food Act includes provisions for food safety including issues related to food sanitation; food quality and nutrition; food label and advertisement; food import and export; responsibilities of the food industry; food resilience; participation of the community; supervision and implementation of the Act; as well as criminal provisions (National Agency of Drug and Food Control 2011). The Food Safety, Quality and Nutrition Regulations specify the provisions of the Act and set detailed requirements for food safety, quality and nutrition including packaging and labelling; sanitation; food additives; quality assurance and testing; and contaminants in food. The Regulations also specify the legal requirements for food import and export inspection and certification; and control and supervision (National Agency of Drug and Food Control 2011).

National food control in Indonesia is managed by the national Integrated Food Safety System (IFFS). The IFFS is a national program shared by all key stakeholders involved in food safety from farm to table. It combines the skills and experience of governments, industry, academia and consumers to synergistically address the emerging challenges influencing the food supply.

The IFSS model was created to achieve equivalence of food safety and laboratory standards internationally. IFSS provides the national framework for food safety—it transcends government, academia, industry and consumers and enables them to work together to maximise resources and improve food safety in Indonesia.

Within the IFFS framework, three functional stakeholder groups (networks) have been organized to reflect the risk analysis principles. The networks enable improved communication between stakeholders, provide for greater knowledge sharing and build food safety capital at local, regional and national levels (National Agency of Drug and Food Control 2011).

49.4.3.6 Myanmar

Based on the Public Health Law, 1972, Myanmar enacted its National Food Law in 1997. The Law was developed with guidance from World Health Organization (WHO) and is in line with international recommendations on food legislation including Codex standards and guidelines. The Law was enacted to enable the public to consume food of genuine quality, free from danger, to prevent public from consuming food that may cause danger or are injurious to health, to supervise production of controlled food and to control and regulate the production, import, export, storage, distribution and sale of food (WHO 2008).

Under the National Food Law, 1997 the Food and Drug Board of Authority was established as an entity responsible for coordination of food control activities. Enforcement and implementation of the Food Law is carried out by the Food and Drug Administration (FDA) which was established in 1995 under the National Drug Law, 1992. The FDA enforces mandatory requirements for Good Manufacturing Practices in food processing industries and encourages all food businesses to apply risk-based approaches to food quality and safety assurance including HACCP.

There are no food regulations or standards in Myanmar.

49.4.3.7 Brunei Darussalam

In Brunei Darussalam food quality and safety is regulated by the Public Health Food Act, 2000 and a number of subordinate regulations. The Public Health Food Act makes specific provisions for public health measures including measures in respect of food. It specifies the role and power of enforcement officers including Directors and inspectors in direct contact with food businesses. It also defines the concept "halal" can defines when it can be used, offences, sale of prohibited items and other overall provisions for food quality and safety. The Public Health Food Regulations, 2001 specifies some of the requirements of the Act including measures for the use of food additives; irradiated Food; labeling as well as standards and particular labelling requirements for the following food items: flour, bakery and cereal products; aerating ingredients; meat and meat products; fish and fish products; edible fats and oils; milk and milk products; sauces and vinegars; sugar and sugar products; nuts and nut products; tea, coffee and cocoa; fruit and fruit products; jams; non-alcoholic drinks; alcoholic drinks; salts; and spices and condiments.

Other legislation related to food quality and safety includes the Halal Meat Act, 2000 which regulates the supply and importation of halal meat and related matters. The Act specifies the requirements for halal import permits; inspection reports from abattoirs; halal certificates; halal labels; slaughtering requirements and procedures; procedures for obtaining a licence and certificate for supplying local halal meat; storage and packaging of halal meat; and fees. The Fisheries Act, 1973 form the basis for food safety and quality requirements for the catch, processing and handling of fish and fisheries products while the Agricultural Pests and Noxious Plants Act, 1971 aims at providing protection of plants and cultivated products from diseases and provides for the destruction of pests and noxious plants. The Miscellaneous Licenses Act, 1983 makes provisions for licensing, regulation and control of certain commercial places and activities including licenses for food establishments, manufacturers, hawkers etc.

49.4.3.8 Lao PDR

Lao PDR has enacted a Food Law, 2004. The Law defines principles, rules, methods and measures on the administration and inspection of activities relating to food, with the aims to control the quality and standard, in particular the safety, of food; to ensure the consumers' nutrition and health; to promote production and business relating to food; and to guide the citizens to be aware, to understand and to adapt to the consumption of food that is safe, hygienic and that contains nutrients for physical strength, including to make them understand the usefulness and danger of food and to know how to care for their health, in order to contribute to national protection and development (Lao PDR National Assembly 2004).

There are no food regulations in Lao PDR but standards exist for selected food items including drinking water, tomato sauce, mineral water, roasted coffee, rice, sugar and fish sauce. Additionally, there are and no specific requirements for registration of food businesses; food inspection and enforcements; use of additives; and control of agricultural inputs.

49.4.3.9 Cambodia

Cambodia has four laws related to food safety. Law on the management of quality and safety of products and services (2000), Law on Fishery (2006), Law on Cambodia Standards (2007), and Law on the management of pesticides and fertilizers (2012). In addition, a number of voluntary and mandatory standards are in place and managed by the Institute of Standards of Cambodia (ISC).

In 2015, a draft Food Law was developed with assistance from FAO. The aim of the Law is to establish a food safety system for the protection of human and consumer health by preventing, controlling and eliminating food contamination and hazards to consumers. The Law covers the entire food chain. These include products standards as well as standards for food hygiene and labelling. For products where no national standard has been developed, Codex standards apply as mandatory food standards. Additionally, under the Food Safety and Quality Law a number of sub-decrees have been established addressing food safety and quality. This include requirements for good manufacturing practices (GMP); food hygiene; phytosanitary inspection. The Cambodia National Codex Committee (CNCC) is hosted by the Ministry of Commerce (MOC) and Camcontrol (MOC) serves as the national Codex Secretariat. The CNCC considers matters related to food policies and measures for safety and quality of products and services necessary for consumer protection, fair trade and coordinated food control action by relevant ministries.

Camcontrol under MOC serves as the enforcement agency for food quality and safety. The agency aims at preventing the production and distribution of unsafe, poor quality, adulterated, misbranded or contamination products including food items. Camcontrol inspects domestically produced food products as well as imported and exported foods. Camcontrols activities are supported by enforcement activities implemented by the Ministry of Industry (MOI). MOI controls the quality of industrial products including food items. It also inspects samples of processed foods and undertakes microbiological and/or chemical analysis on a diversity of products e.g. bottled water, beers, wines, fish and soy bean sauce and vinegar (Soeun 2004).

49.4.3.10 Singapore

In Singapore, the main food legislation include the Sale of Food Act, 2002 which aims at securing the wholesomeness and purity of food; providing a framework for developing food standards; preventing the sale or other disposition or the use of articles dangerous or injurious to health; providing regulations for food establishments. The Sale of Food Regulations, 2005 have been developed as subsidiary legislation to the Act. The Regulations include provisions for food labeling; health claims; date marking; marketing and advertisement; irradiated food; approved food additives; products standards; as well as chemical residues in food (AVA 2011).

Food control and implementation of food and agriculture related legislation falls under the responsibility of Agri-Food and Veterinary Authority (AVA) which was established in 2001 by the Agri-Food and Veterinary Authority Act, 2000. The role of AVA is to regulate the safety and wholesomeness of food for supply to Singapore; promote and regulate animal and fish health, animal welfare and plant health; promote, facilitate and regulate the production, processing and trade of food and products related to or connected with the agri-food and veterinary sectors; develop, manage and regulate any agri-food and veterinary centre or establishment; promote the development of the agri-food and veterinary sectors; advise and make recommendations to the Government on matters, measures and regulations related to or connected with the agri-food and veterinary sectors; and represent the Government internationally on matters related to or connected with the agri-food and veterinary sectors (AVA 2011).

49.5 SAARC

The South Asian Association for Regional Cooperation (SAARC) is an organization of South Asian nations, founded in December 1985. SAARC consists of eight member countries which include Afghanistan, Bangladesh, Bhutan, India, the Maldives, Nepal, Pakistan and Sri Lanka. The Headquarters of the SAARC Secretariat is located in Kathmandu, Nepal.

SAARC is the home of 1.567 billion people which constitutes about 23.77% of the global population. Based upon the data available with Asian Development Bank, about 451 million people live below poverty line. The reason attributed to poverty is the low per capita income and inequitable distribution of income in the region, which can be judged by the fact that 23.7% of the global population has only 2.6% share in the global income.

The majority of the population in SAARC region lives in rural areas and agriculture is the mainstay of the economies of the South Asian countries and is still the main source of livelihood of the majority of the population. Agriculture sector is facing stiff competition in the region, which threatens its growth and sustainability. On one hand, food security challenges have raised alarms and on the other food safety concerns have poised problems for the policy makers, planners and economic masters.

49.5.1 The Objectives of SAARC

SAARC aims at achieving enhanced economic growth through regional cooperation.

The objectives of SAARC are:

(a) to promote the welfare of the peoples of SOUTH ASIA and to improve their quality of life;
(b) to accelerate economic growth, social progress and cultural development in the region and to provide all individuals the opportunity to live in dignity and to realise their full potentials;
(c) to promote and strengthen collective self-reliance among the countries of SOUTH ASIA;
(d) to contribute to mutual trust, understanding and appreciation of one another's problems;
(e) to promote active collaboration and mutual assistance in the economic, social, cultural, technical and scientific fields;
(f) to strengthen cooperation with other developing countries;
(g) to strengthen cooperation among themselves in international forums on matters of common interests; and
(h) to cooperate with international and regional organisations with similar aims and purposes (SAARC Secretariat 2011).

49.5.2 Cooperation in the Area of Food Quality and Safety

SAARC has 16 stated areas of cooperation. The areas of cooperation are managed by a number of Technical Committees and Groups consisting of technical experts from SAARC member countries. The Technical Committees and Groups are responsible for the implementation, coordination and monitoring of the programmes in their respective areas of cooperation. This includes formulation of programmes and preparation of projects as well as implementation and coordination of sector programmes.

The stage of development of national food control systems in SAARC and the implementation of food quality and safety assurance systems varies greatly between the eight SAARC countries. Some SAARC countries have invested in developing their food control capacity in terms of new food control structures and management systems while others are yet to update and implement food safety laws and regulations as well as other measures to protect the health of consumers and facilitate trade with food and agricultural products.

The Technical Committee on "Agriculture and Rural" and the Technical Committee on "Economic Cooperation" as well as the "Standing Group on Standards" and "SAARC Standards Coordination Board" are the Committees and Groups that are directly involved in work related to development and harmonization of food laws and regulations.

- **Technical Committee on Agriculture and Rural Development (TCARD)**
  TCARD has contributed to identifying numerous concrete areas for pursuing regional actions and projects in the area of agriculture and rural development.

An important output from the Technical Committee was the **SAARC Agricultural Perspective/Vision 2020** which was finalized in 2008. The document articulated the long-term regional challenges and priority measures *inter alia* in production augmentation, natural resource management, bio-safety and bio-security, technology development and dissemination, seed and other inputs, food safety/standards, climate change adaptation and risk mitigation, and livelihoods of small and marginal farmers in farming and non-farm activities. It also brings forth the recent challenges e.g. Avian Influenza that appears to threaten much of the gains achieved in rural South Asia over the past years (SAARC Secretariat 2011).

During 2007–2008, as a SAARC-FAO collaborative effort, the **SAARC Regional Strategy and Regional Programme for Food Security** was drawn up. The strategy and programme identifies a number of priority projects under four broad clusters. The Cluster "Ensuring Bio-Security" includes priority projects on development/updating of national SPS measures in line with Codex Alimentarius and other internationally recognized standards; support/assistance in capacity-building in the areas of food safety, quality and standards; and development/up-gradation of accredited laboratories in SAARC. These are important activities that will strengthen the regional framework for food safety and quality and contribute to facilitating trade with food and agriculture products between countries in the region (FAO 2010).

- **Standing Group on Standards and Quality Control and Measurement (SGSQCM)**
  The SAARC Standing Group on Standards, Quality Control and Measurement (SGSQCM) focuses on developing regional standards for products in the interest of intra-SAARC trade facilitation as well as trade beyond the region. The Standing Group has agreed on developing mutual recognition agreement to facilitate trade and works towards harmonization of standards; conformity assessment procedures; testing and metrology as well as certification and accreditation. The group has developed a regional action plan to enable the SAARC Region to take a collective position at the international standardization for a like the Codex Alimentarius Commission (CAC) and International Standardization Organization (ISO) (www.kuenselonline.com). The Group is supported by the SAARC Standards Coordination Board which aims at strengthen regional cooperation in the area of standardization and harmonization of standards between SAARC countries (SAARC Secretariat 2011).

49.5.3 Legislative Framework for Food Quality and Safety in SAARC

The legislative framework for food quality and safety in SAARC varies greatly between the SAARC member countries. The region's biggest country India has within recent years invested considerable resources in developing and updating its food quality and safety legislation and in building capacity among enforcement officers. The smaller and less developed countries have outdated and fragmented legislation that is not based on risk and contemporary approaches to food quality and safety. In many cases, regulations and standards are not in line with Codex Alimentarius standards, guidelines and recommendation and other internationally recognized food quality and safety measures.

The implementation and enforcement of food quality and safety legislation varies between the SAARC countries as well. Generally, SAARC countries have limited resources in terms of staff (food inspectors, public health officers etc.) and equipment (tools, guidelines, vehicles etc.) to carry out food inspection and enforcement activities. The resource constraints and different levels of technical expertise result in inconsistently implemented and weakly coordinated inspection and enforcement activities.

In most SAARC countries, the organization of food safety management structures is divided between multiple ministries and agencies as well as units working at different administrative levels (federal, state, district, sub-district etc). India has a federal as well as state level structure for food safety. Both Bangladesh and Pakistan have weak federal structure with emphasis on provincial level (districts and sub-districts) enforcement activities. Sri Lanka has also limited central structures. Nepal has a central level operational structure with five regional units to implement food legislation.

In most SAARC countries, the roles and responsibilities between enforcement agencies are not clearly defined and communication, coordination and exchange of information between the different agencies is in most cases very limited. With the enforcement agencies often having mandates that go beyond food control and include vector control, water safety, sanitation etc, food control activities in many cases do not receive the necessary attention to address the growing demand of sophistication in managing food safety risks and prevent foodborne disease outbreaks.

49.5.4 Food Legislative Frameworks in SAARC

In the SAARC region food legislation was formulated over a period of four to five decades. The overall thrust of the legislation is primarily based on end product testing and heavily relies on penalisation in cases of food safety violations. After the emergence of the WTO, the existing legislative measures implemented in many SAARC countries do not comply with international requirements to protect the health of consumers and facilitate trade with food and agriculture products. Several SAARC countries are in the process of revising and updating their legislative frameworks for food quality and safety and strengthen the implementation and enforcement structures. In countries like Bangladesh and Maldives have primarily been focusing on updating legislation and enforcement services for food exports (fish and fisheries products) while the legislation covering food sold on the domestic markets in most cases remain weak.

49.5.4.1 Bangladesh

The Pure Food Rules, 1967; the Pure Food Ordinance 1959 and the Food Safety Amendment Act, 2005 make the basic framework of food legislation in Bangladesh. The Pure Food Ordinance, 1959 was established to provide better control of the manufacture and sale of food for human consumption. This Ordinance was revised and reissued on 22 September 2005 as Act No. 27 of 2005. The revised Act provided for the constituting of a National Food Safety Advisory Council (NFSAC) headed by the Minister of Ministry of Local Government Rural Development and Cooperatives (MOLGRDC), and the establishment of Pure Food Courts, with power and jurisdiction under the Act. The NFSAC comprises 16 Ministries and agencies involved in food safety and is responsible for advising Government on all matters related to food safety and administration of the Act; standards and quality control for food in terms of ensuring purity, safety and nutritional value; technical matters; development of man-power and facilities required for ensuring safety, and policies and strategies related to food safety and quality control. The Amendment Act prohibits the sale or use of poisonous or dangerous chemicals, including intoxicating food colours; sets standards of purity for milk, butter, ghee, wheat flour, mustard and rape seed oil and other food articles; prohibits the sale of diseased animals and unwholesome food; prohibits the use of false labels; sets requirements for analysis of food and certificates of analysis; addresses the inspection and seizure of food; and sets offences and penalties.

The Directorate General of Health Services (DGHS) is the main implementing agency responsible for food safety throughout Bangladesh. DGHS implements the Pure Food Ordinance; the Pure Food Rules and the Food Safety Amendment Act. Enforcement activities are carried out by sanitary inspectors based at Divisional, District and Upazila levels. City Corporations and Municipalities under Ministry of Local Government, Rural Development and Cooperative (MLGRDC) implement the same legal documents in their respective administrative areas. The City Corporation has health officers, public analysts, chemists and inspectors to carry out food safety activities in areas under their command. Each City Corporation has a food-testing laboratory. The main task of enforcement agencies is to check for food adulteration, food hygiene and sanitation.

Other agencies involved in food inspection include Department of Fisheries and Department of Livestock Services under Ministry of Fisheries and Livestock; Bangladesh Standard and Testing Institution (BSTI) under Ministry of Industries; Ministry of Food and Disaster Management (MFDM); and Department of Agricultural Extension under ministry of Agriculture. BSTI is the Codex contact point for Bangladesh and develops food standards in line with Codex standards and guidelines. BSTI certifies processed products for which the Institute has developed mandatory standards. Currently, there are 53 mandatory food standards in

Bangladesh. MFDM exercises the power to assess the quality and safety of imported and domestically produced food grains and cereals like rice, wheat and pulse. The Ministry of Agriculture is responsible for assuring the quality and safety of other agricultural produce and for certifying exportable agricultural products.

49.5.4.2 Bhutan

Bhutan has a Food Act, 2005 and Food Rules and Regulations, 2007 that cover issues related to food safety. The Food Act, 2005 created the Food Quality and Safety Commission which is a multi-sectoral mechanism that (1) formulates policies to maximize industry development, protect consumers and foster trade; (2) addresses any matter related to the enhancement of food control activities including enforcement structures; (3) coordinates national responses to food related emergencies; (4) review and approve the work of the National Codex Committee; (5) makes recommendations to the Minister on the level of any fees and penalties to be assessed under the Act; (6) recommends educational activities to inform food businesses and the people of Bhutan of the importance of food safety; and (7) coordinates responses to the media with regard to all food control issues in Bhutan. The Act also outlines the mandate of the Bhutan Agriculture and Food Regulatory Authority (BAFRA) and the mandate and functioning of the National Codex Committee. Other issues covered by the Food Act include appointment and responsibilities of food inspectors and analysts; operation of food businesses; import and export of food; and enforcement. The Food Rules and Regulations, 2007 specifies the mandates of food control authorities and outlines the procedures for food control action (Ministry of Agriculture Bhutan 2005).

49.5.4.3 India

Food quality and safety in India is regulated by the Food safety and Standards Act, 2006. The Act consolidates the laws relating to food and regulates the manufacture, storage, distribution, sale and import of food. The Act provides for the establishment of the Food Safety and Standards Authority (FSSA) in order to regulate the food sector and ensure safe and wholesome food. The Authority is aided by several scientific panels and a central advisory committee to lay down standards for food safety which include specifications for ingredients, contaminants, pesticide residue, biological hazards and labels. Every entity in the food sector shall be required to get a license or a registration, which will be issued by local authorities. State Commissioners of Food Safety and other local level officials are responsible for the enforcement of the Act. The Act further provides for the following matters: recognition and accreditation of laboratories, research institutions, food analysts; offences and penalties; finances, accounts audits and reports of the Food Authority; and provisions of miscellaneous nature.

The implementation of the Food Safety and Standards Act, 2006 formally repeals the regulatory framework established by the Prevention of Food Adulteration Act (PFA), 1954, the Fruit Products Order, 1955, the Meat Food Products Order, 1973, the Vegetable Oil Products (Regulation) Order, 1998, the Edible Oils Packaging (Regulation) Order 1988, the Solvent Extracted Oil, De-Oiled Meal and Edible Flour (Control) Order, 1967, Milk and Milk Products Order, 1992, and the Essential Commodities Act, 1955 (USDA 2011).

In August 2011, the Food Safety and Standards Regulations, 2011 came into force under the Ministry of Health and Family Welfare. The Food Safety and Standards Regulations, 2011 contain labeling requirements and standards for packaged food, permitted food additives, colors, microbiological requirements, etc. These regulations officially transpose the Prevention of Food Adulteration Rules, 1954 as amended (USDA 2011).

49.5.4.4 Maldives

Maldives has a Food Act and some general food regulations specifying requirements for food service establishment; food processing businesses; food exports and imports; and frozen foods. Apart from fish product, there are very few guideline; standards and codes of practices for food products and processes in the country. Within recent years some effort have been put into developing guidelines for food processing units; food service establishment with main focus on resorts, hotels and guest houses, food import; food transportation; and healthy market places, but they do not seem to cover the entire food chain continuum (Moosa 2009).

The Consumer Protection Act 1996 does cover areas related to food quality and safety. These include provisions for labelling such as name of the product; ingredients; weight/quantity; production date; information on the use of products; use before expiry date etc. Public Health Department (PHD) under the Ministry of Health carries out quality monitoring and inspection activities. The PHD further issues certificate or license for the sale of locally manufactured foods. It monitors foods prepared in restaurants, catering and processing plants. The frequency of inspection is relatively high in Male but low in other parts of the country. Public Health Officials (PHO) examine and inspect bakeries, cafeterias, and restaurants from hygiene point of view.

49.5.4.5 Pakistan

Pakistan does not have an integrated legal framework but has a set of laws, which deals with various aspects of food safety. The Pure Food Ordinance 1960 and Pure Food Rules 1965 form the legislative framework of food safety in Pakistan. The rules give authority to provincial governments to appoint public analysts for the investigation of quality and safety of food. There is no federal structure of food safety programme in Pakistan. The Pure Food Rules are enforced through health service delivery channels of the provincial government. The District Health Officer and Deputy Health Officer function as food inspector for sampling and inspection. Municipality Corporation may also appoint food inspectors and sanitary inspectors for sampling purposes. Any other public servant can also be appointed as inspector and can execute the power of food inspector.

The Pakistan Standard Institute (PSI) with its Food and Agriculture Division develops voluntary standard for food products. Mandatory food standards are established in the Pakistan Pure Food Laws (PFL) of 1963. The PFL is basis of the existing trade-related food quality and safety legislative framework. It covers 104 food items falling under nine broad categories: milk and milk products; edible oils and fat products; beverages; food grains and cereals; starchy food; spices and condiments; sweetening agents; fruits and vegetables; and miscellaneous food products. These regulations address purity issues in raw food and deal with additives; food preservatives; food and synthetic colors; antioxidants; and heavy metals.

Pakistan's Hotels and Restaurant Act of 1976 applies to all hotels and restaurants in Pakistan and seeks to control and regulate the rates and standard of service by hotels and restaurants. The Pakistan Standards and Quality Control Authority Act of 1996 provides for the establishment of Pakistan Standards and Quality Control Authority (PSQCA), which is the apex body to formulate or adopt international standards. The PSQCA is also responsible for enforcement of standards in the whole of Pakistan and has the mandate to inspect and test products and services, including food items, for their quality, specification and characteristics during use, and for import and export purposes.

The federal government generally applies Codex standards and guidelines in its regulation of imported food products. For animal products including dairy, halal certification is required. Importation of food products containing pork or pork products is prohibited. The Customs Department and Plant Protection and Quarantine (PPQ) are the two main agencies involved in regulating food imports. The Customs Department's primary functions are: ensure that imported foods meet Pakistan's labeling and shelf-life requirements prevent imports on the list of banned items, and assess appropriate import tariffs. PPQ also ensures that shipment of bulk commodities and live animal shipments meet phytosanitary requirements.

49.5.4.6 Nepal

The Food Act 1966 and Food Regulation 1970 are the basic legislative frameworks for ensuring safe food supplies in Nepal. With Nepal's accession to WTO amendments to the food legislative framework have been initiated and training and capacity building of enforcement officers has taken place. Draft Food Regulations have been prepared, but is still subject to endorsement.

The Food Legislation of Nepal (Food Act, 1966 and Food Rules (1970) dates back to more than three decades. The Act has been amended a few times, but still it does adopt a risk-based approach to food control. The Act and Rules were promulgated for the welfare of the people with the objective of protecting people from health hazards and commercial fraud. The salient feature of the Food Act 1966 is definition of food, contaminated food and substandard food (WHO 2008).

The Department of Food Technology and Quality Control (DFTQC) is the focal agency responsible for conducting inspections of industries, import/export and market. The DFTQC with its five Regional Food Laboratories are assigned tasks to conduct inspections and laboratory investigations in their respective areas. The District Attorney Officers and Chief District Officers are responsible for administering food legislation in the country.

Food Inspectors carry out inspections of food processing industries, customs points, warehouses, markets, (wholesale/retail) and collect suspicious samples in duplicate. One part of the sample is sent to the public analysts and the other part to the Chief District Officer (CDO) for further verification if needs be. Public Analysts send the report to the concerned Food Inspector. In case the report does not comply with the set standard, the food inspector files a case against the owner of the sample in the office of CDO. If the concerned party is not satisfied with the decision, he may request the Director General of DFTQC through the CDO for re-analysis.

The Food Standardization Committee (FSC) is responsible body to recommend appropriate standards and to advise the Government on all matters relating to food safety. The FSC is chaired by the secretary, Ministry of Agriculture and Co-operatives, with representatives from Ministry of Law and Justice, Ministry of Industry, Commerce and Supplies, Ministry of Home, Representatives of Kathmandu Municipal Corporation, Food Industry, Federation of Nepalese Chambers of Commerce and Industry (FNCCI), Consumers, and the Director General of DFTQC as the Member Secretary. Nepal Quality Standard (Quality Mark) Act 1980 was implemented to fix the quality standards for Nepal Council of Standards under the Act, which has been empowered to make standards and regulations for ensuring quality marks. While food standard under the Food Act 1966 is mandatory, Nepal Standards under the Nepal Quality Standard Act is voluntary in nature and, this operates as a third party certifying authority for the quality assurance of the food products.

49.5.4.7 Sri Lanka

The main legislative document for food quality and safety in Sri Lanka is the Food Act, 1980 (amended in 1991). The Act contains elements such as manufacture, sale, distribution, and import, and seizure of food. It also lays down standards for natural or added substances, which may pose potential health risks. The Act and its amendments prohibit the importation, manufacture for commercial purposes, transportation, storage, distribution, sale, or offer for sale of any food, raw or processed, or any ingredient of food or food additive that has been subjected to genetic modification using DNA recombination technology or any food that contains one or more ingredient or additive that has been subjected to genetic manipulation. In 2009, a regulation implemented under the Act which implements mandatory standards for 158 food items came into force.

The food control infrastructure is comprised of high-level and multi-sectoral Food Advisory Committee (FAC) at the apex level that also functions as an advisory body to the Ministry of Health (MoH) on all issues relating to food safety. FAC has three sub-committees: technical sub-committees for standards and regulations, national codex committee, and a food control co-ordination committee. The Central Administration operates in the Ministry of Health with the Food and Drug Administration. The import and export inspection is carried out by the food and drug Inspectorate of the MoH in close co-ordination with customs authorities. Sri Lanka Standard Institute (SLSI) controls imports of some food items such as canned fish, condensed milk, fruit squashes, fruit syrup, and fruit cordials, synthetic/artificial cordials, fruit concentrates, ready to serve fruit drinks, brown sugar, soybean oil, peanut oil, sunflower seed oil, palm oil, and coconut oil.

49.6 Food Laws and Regulations in North East Asia

Within recent years a number of food safety incidents and related trade implications have occurred between the three main countries in North East Asia: China, Japan and South Korea. Most issues occurred from Chinese exports of contaminated food products which have led to closer collaboration and cooperation between the countries in the area of food safety.

As an output of the closer relations between the countries, the three nations in 2009 signed a food safety memorandum where they agreed to notify each other immediately if a food safety problem surfaces and to clarify the process of investigation. Japan and China have also worked towards a joint food safety law that could allow Japanese authorities to conduct safety inspections in cases of tainted Chinese exports. This development towards closer collaboration between China and Japan comes after several cases of Japanese consumers getting sick after consuming food imported from China. The latest incident was in 2008 when 10 people in Japan fell ill from consuming Chinese-made dumplings laced with the insecticide methamidophos. The closer collaboration is expected to have a positive impact on the Chinese inspection processes and on preventing similar incidents from happening in the future (www.2point6billion.com).

49.6.1 Food Legislation in North East Asia

Food trade between the North East Asian countries China, Japan and Korea is increased extensively over the last 20 to 30 years. The increasing trade in food and agricultural products has increased the need for closer collaboration on food law and regulation development and implementation.

In 2011, China and Japan have held talks on a new draft food safety law that could allow Japanese authorities to conduct safety inspections in cases of tainted

Chinese exports. The new law is in response to a 2008 incident that led to 10 people in Japan to fall ill from Chinese-made dumplings laced with the insecticide methamidophos (www.2point6billion.com). The development of this new Law follows a food safety pact that was signed between China, Korea and Japan in 2009. Under the pact, the three nations have agreed to notify each other immediately if a food safety problem surfaces and to clarify the process of investigation.

49.6.1.1 Japan

The food law system in Japan is modeled after European civil law system with English-American influence. In the mid-nineteenth century to early twentieth century, German and French codes were introduced, and served as a model for the major Japanese food codes. After World War II, under Allied occupation, some laws were amended or replaced on the basis of American law.

Within the last 10 years, there has been a growing concern and distrust of food safety among the Japanese consumers, triggered by various food safety scandals. Among these the occurrence of BSE in 2001, pesticide contaminated Chinese dumplings in 2008 and most recently the risk of nuclear contaminated domestically produced food products from areas around the damaged Fukushima nuclear power plant in North Japan.

These food safety incidents have led to changes in the regulatory system. In the early 2000s discussions focused on the significant overlap of responsibilities of the two lead ministries for food safety: Ministry of Agriculture, Forestry and Fisheries (MAFF) and Ministry of Health, Labor and Welfare (MHLW). It was proposed to combine these two regulatory agencies into one new organization, but this proposal was rejected and replaced by another proposal to establish an independent advisory committee under the Prime Minister's Office: the Food Safety Commission. The Food Sanitation Law was amended, and the new Food Safety Basic Law, under which the Food Safety Commission was established, took effect in 2003.

The role of the Food Safety Commission is to undertake risk assessment in relation to the health impact of food consumption. The Commission works independently from the enforcement agencies under MAFF and MHLW as well as independent from consumer and private sector related organizations. The primary goals of the Commission are to:

- Present opinions to the Prime Minister
- Conduct scientific and independent risk assessments
- Make recommendations and provide scientific advice to the government (through the Prime Minister
- Monitor the implementation of food safety policies
- Examine and deliberate on important matters regarding food safety policy implementation and provide technical advice
- Plan and implement risk communication activities targeting consumers, food businesses and food safety regulators

The Food Safety Basic Law, 2003 was developed together with a number of related laws and decrees to ensure food safety of domestically produced and imported food products. The legislative framework for food quality and safety is built around the risk analysis principles and the different pieces of legislation target different components of the risk analysis framework. The Food Safety Basic Law, 2003 establishes basic principles and direction for policy formulation and clarifies the responsibilities of the state, local governments, food business operators and consumers.

The Food Sanitation Law is very comprehensive law which covers all types of foods. The law regulates food labelling, food additives, residue of pesticides, veterinary drugs and additives to feed, food import quarantine, inspection, GMO, food with health claims, HACCP, and business licenses. However, nutrition labeling; nutrition and health claims; and food for specified use such as food for infants, food for lactating mothers, and food for pregnant women are regulated separately under the Health Promotion Law, 2002. Food labeling however, is also covered the JAS Law which is under the responsibility of MAFF.

Within the last 20 years, a variety of health foods began to appear in the Japanese market as the dietary habit changed and consumers' concern about nutrition and health increased. To prevent misleading labelling on health food, the regulation of "Food for Specified Health Use (FOSHU)", which is officially allowed to have indication of health function claims, was created in 1991. The system of "Food with Health Claims", which is composed of FOSHU and Nutrition Functional Food was established in 2001.

MHLW is responsible for enforcement of the Food Safety Basic Law, 2003 and the Food Sanitation Law, 2006. Enforcement at the local level however, is carried out by prefectures. In principle, prefectures are responsible for monitoring and inspection (except import inspection), appointment of monitoring officer and order of withdrawal of violating products from the market, as well as licensing of businesses. To perform these duties, each prefecture has public health centers. Import quarantine and related inspection at the point of entry are performed by the state.

Food quality related issues are regulated by the Law Concerning Standardization and Proper Labelling of Agricultural and Forestry Products (JAS Law) which is being implemented by MAFF. The JAS Law was established in 1995. This law succeeded the standards of foods under the Government Inspection Law for Agricultural and Forestry Products in association with the supply and demand adjustment policy including the government procurement and ration system. The initial objective of the JAS Law was to prevent distribution of low quality food in the market by establishing common standards for various foods. The Law covers quality certified products such as organic and other specially grown products, special distributed products, specified regional food, food representing environmental protection and sustainable agriculture as well as compositional quality standards for fisheries products, rice and products containing GMOs. The law also includes standards for labeling of the food covered by the Law.

49.6.1.2 China

Following a number of food safety incidents and emergencies in China, food safety has become a priority issue for the Government. Within recent years, a number of new food laws and statutes have come into place and a number of plans and notices to strengthen the legal framework for food safety and rectify the implementation of food laws and regulations.

The Food Safety Law of the People's Republic of China came into effect in 2009. The Law is based on the risk analysis principles and integrates all existing food safety standards and regulations which facilitates implementation by the authorities. The Law protects consumers from sub-standard products and give them right to claim compensation from food producers in case a product does not comply with specified requirements.

The Food Safety Law is divided into 104 articles that cover the following issues: Surveillance and Assessment of Food Safety Risks; Food Safety Standards; Food Production and Trade; Inspection and Testing of Food; Food Import and Export; Response to Food Safety Incidents; Supervision and Administration; and Legal Liabilities.

The Law highlights that the responsibility for assuring the safety of food lies with the food producers and not the government. For issues where no national standard exists local governments are entitled to establish local legal requirements that are valid for food businesses in the particular geographic area.

The Law also stipulates that imported food, food additives and food-related products must comply with China's national food safety standards, including labeling, records and inspection. The government will develop a coordinated plan to manage food safety. Activities include inspect food businesses, test samples, inspect records, detain unsafe food, and terminate business operations. In addition, government will formulate a food safety emergency plan; respond to alerts and information about unsafe food and establish an information release system.

In April 2015, a modern amended food safety law was adopted and went into force on 1 October 2015. The revised food safety law of 2015 is divided in 10 chapters: General Principles; Food Safety Risk Surveillance and Assessment; Food Safety Standards; Food Production and Trading; Food Testing; Food Import and Export; Handling of Food Safety Incidents; Regulatory Work; Legal Liabilities; and Supplementary Provisions. Further regulatory details will be specified in the Implementation Regulations which are currently being developed.

49.6.1.3 Primary Legislation Sources

A total of 12 food and food related laws constitute the frame of China's food hygiene and safety system.

The Food Hygiene law, 1995 stipulates the hygiene standard and safety for food, food containers, food additives, and packaging materials of food, the equipment and place for food production. The food quality is also stipulated by the Standardization Law, 1988; Product Quality Law, 1993; and Entry and Exit Commodity Inspection Law, 1989. The Fishery Law, 1986; Agriculture Law, 1993; the Entry and Exit Animal-Plant Quarantine Law, 1991; Animal Epidemic Prevention Law, 1997; the Frontier Hygiene and Quarantine Law, 1987; and Seed Law, 2000 cover aspects related to primary production. The Trademark Law, 2001 protects the food brands and labels. The Consumer Protection Law, 1993 ensures that the consumers enjoy quality food.

49.6.1.4 Secondary Legislation

A total of 13 statutes are classified as administrative statutes and statute documents under the primary legislation.

Implementation Detailed Rules of Fishery Law, 1987; Implementation Detailed Rules of Entry and Exit Commodity Inspection Law, 1992; Administration Ordinance of Iodization of Salt to Eliminate the Endangerment of Iodine Deficiency, 1994; Implementation Detailed Rules of Plant Entry and Exit Quarantine Law, 1996; Pig Slaughter Administration Ordinance, 1997; Safety Administration Ordinance of Agricultural Genetically Modified Organism, 2001; and the Public Hygiene Emergency Ordinance, 2003.

1. The State Council Office Issue on Food and Nutrition Development Program国务院办公厅关于印发中国食物及营养发展纲要2001–2010 (11-3-2001)
2. The State Council on Strengthening the New Stage of the "Vegetable Basket" of Notice国务院关于加强新阶段"菜篮子"工作的通知 (8-3-2002)
3. The State Council Office On the Implementation of Food and Drug Safety Project Notice 国务院办公厅关于实施食品药品放心工程的通知 (7-16-2003)
4. The State Council Office Issue on Food Safety Special Punishing Program of Work Notice 国务院办公厅关于印发食品安全专项整治工作方案的通知 (5-17-2004)
5. The State Council Office on Further Food Safety Decisions国务院办公厅关于进一步食品安全工作的决定 (9-1-2004)
6. The State Council Office Issue on National Food and Drug Special Rectification Work Notice国务院办公厅关于印发2005年全国食品药品专项整治工作安排的通知 (3-30-2005)
7. The State Council on Strengthening the Food Products Safety Supervision and Management of the Special Provisions国务院关于加强食品等产品安全监督管理的特别规定 (7-27-2007).

Currently, the administration system of food hygiene and safety is constituted by the State Council, the State Food and Drug Administration, the Ministry of Health, the Ministry of Agriculture, General Administration of State Quality Supervision, Inspection and Quarantine, the Ministry of Commerce, the State Industry and Commerce Administration, and the Ministry of Science and Technology. In addition to the above, Ministries and Administration extension agencies in each of the provinces, counties and cities form a nation-wide administration and monitoring network.

References

AVA (2011). http://www.ava.gov.sg/docs/default-source/legislation/sale-of-food-act/51web_saleoffoodact1
FAO (2010). http://www.fao.org/docrep/meeting/019/k8774e.pdf
Lao PDR National Assembly (2004). https://www.wto.org/english/thewto_e/acc_e/lao_e/WTACCLAO17A1_LEG_2.pdf
Ministry of Agriculture Bhutan (2005). http://www.wipo.int/edocs/lexdocs/laws/en/bt/bt014en.pdf
Moosa (2009). http://www.ilsi-india.org/PDF/international_conference_on_infrastructure_needs_for_food_control_system_roadmap_for_regional_harmonization/Ms.%20Satheesh%20Moosa,%20Microbiologist,%20Maldives%20Food%20and%20Drug%20Authority,%20Ministry%20of%20Health,%20Maldives.pdf
National Agency of Drug and Food Control (2011). http://faolex.fao.org/docs/pdf/ins9666.pdf
SAARC Secretariat (2011). http://www.saarc-sec.org/SAARC-Charter/5/
Soeun (2004). http://www.chinaaseansps.com/upload/2012-07/12072018579293.pdf
USDA (2011). https://gain.fas.usda.gov/Recent%20GAIN%20Publications/India%20Enforces%20the%20New%20Food%20Safety%20Law_New%20Delhi_India_8-10-2011.pdf
WHO (2008). http://www.searo.who.int/myanmar/documents/HealthinMyanmar_2012_5_policylegislationplans.pdf

Chapter 50
Branding, Regulation and Customs in Japan and Singapore

Andreas Popper

Abstract This section introduces the regulatory systems on food safety and the trademark systems of Japan and Singapore. Both countries are strong economies and faced with extremely low food self-sufficiency rates. Despite these similarities, they arrive at very different conclusions as to how to secure food safety. The Madrid Protocol serves the purpose of harmonizing trademark laws and facilitating the registration of trademark rights. Both countries, Japan and Singapore, joined the Protocol, but, while the registration of a trademark has become a very straight forward matter in Singapore, the handling of trademark matters was barely simplified in Japan. The following three sections will give insight into the logic of two very different Asian systems and expose the links between their regulatory and trademark laws.

50.1 Introduction

Section 50.2 introduces the regulatory system on food safety in Japan, a country that stands as one of the world's largest economies and yet, is struggling with a wide range of problems. Japan is faced with an extremely low and continuously falling food self-sufficiency rate of only about 39 %.[1] The local agricultural industry is economically barely viable against foreign competition and is therefore dependent on governmental support and protection in the form of subsidies and import restrictions. Serious incidents during the last 2 decades related to food imports made food safety become a high priority issue to the legislator.

Answering these problems by applying restraints on international trade has become increasingly problematic in view of the formation of numerous free trade

---

[1] Ministry of Agriculture, Forestry and Fishery (in Japanese) (http://www.maff.go.jp/j/zyukyu/zikyu_ritu/012.html), figures 2013 (http://www.maff.go.jp/j/zyukyu/zikyu_ritu/pdf/himoku.pdf), graph 2013 (http://www.maff.go.jp/j/zyukyu/zikyu_ritu/pdf/25mekuji.pdf).

A. Popper (✉)
NaexasCompass Group, Tokyo, Japan
e-mail: a.popper@naexascompass.asia

zones worldwide. Trade restraints are objective economic measures directed against foreign economies. The application of trade restraints by one triggers counter measures of other governments and leads to economic conflicts between them. Regulatory legislation, however, relates to subjective health, safety, and welfare measures. The protection of health, welfare, and safety of people is a common interest of governments, and the rightfulness of the application of protective regulatory measures is unquestionable. Therefore, a global harmonization of regulatory legislation relating to food safety should basically be a logical outcome.

Section 50.3 introduces the regulatory system on food safety in Singapore, a country that is almost completely dependent on imports to satisfy the local demand for food, beverages and water. Despite this similarity to Japan, the approach of the Singapore governments to achieve food safety is very different from that of the Japanese. It is hard to explain why Singapore and Japan arrive at very different conclusions as to how to secure food safety, unless one seeks for reasons lying in the differences between the economic interests of the two countries. While the Japanese economy strongly depends on in- and exports, Singapore's economy is built on trading. Whether or not there is any relation between trade restraints and regulatory legislation, both are applied by the governments to protect not only their consumers, but also their domestic markets, industries and economic interests. Therefore, a basic understanding of regulatory legislation has become indispensable for many experts, for example those engaged in trading, marketing, product development as well as intellectual property law.

Section 50.4 introduces the links between regulatory legislation and trademark law. No business can prosper without a strong brand. Due to globalization, companies engaging in international business need viable regional or even global brands in order to secure their chance of steady growth. As brands open doors into foreign markets, the securement of vast portfolios of intellectual property rights by the local industry is in great and, beyond any doubt, the rightful interest of any government. Hence, as in the case of regulatory legislation, a global harmonization of trademark law should therefore basically be a logical outcome. However, despite having international systems[2] established to harmonize trademark laws among countries, governments are also designing and applying them in a way that they protect domestic markets from foreign competition. This similarity in attitude inevitably causes similarities between trademark law and regulatory legislation.

Both countries, Japan and Singapore joined the Madrid Protocol.[3] The establishment of international systems, such as the Madrid Protocol, serves the purpose of harmonizing trademark laws and procedures globally to facilitate the securement

---

[2]For example the Madrid Union: http://www.wipo.int/madrid/en/members/.

[3]The Madrid Protocol is an international system that was established in order to facilitate the registration of trademarks in multiple jurisdictions around the world. By filing a trademark application to the local patent/trademark office, applicants are enabled to designate simultaneously any member state of the Madrid Protocol. The designation of a member state is comparable with the filing of a national trademark directly in the member states. (http://www.wipo.int/madrid/en/) Japan and Singapore joined the Madrid Protocol in 2000: http://www.wipo.int/about-wipo/en/

of trademark rights. In Singapore, the registration of a trademark has become a very straight forward matter. In contrast, the Madrid Protocol has barely simplified the handling of trademark matters in Japan to foreigners. It has opened doors for initiating trademark applications from abroad, but hurdles such as language barrier,[4] a unique local similarity code system,[5] and quirky, inconsistent procedural rules[6] result in the majority of foreign applications undergoing costly examinations or failing them altogether. The dramatic differences between the trademark laws of these two countries that are members of the same international system is hard to explain unless one seeks, as mentioned in connection with regulatory legislation, for reasons lying in the differences between the economic interest of the two countries.

In Asia, the characteristics of the local languages and scripts enable local governments in particular to apply trademark law in a way that it protects the local markets. Misleading connotations, exaggerated meanings, concepts and appearances play a great role in both product presentation—which is a part of regulatory legislation—and trademark law. These marketing elements are all very strongly influenced by the languages and local scripts. Accordingly, there can only be limited consensus among the different countries. To experts working on both fields, the links are apparent: Securing a product name as a trademark without any foresight of regulatory compliance, or preparing a product for import under a product name that is not registrable as a trademark in the target market, may lead to a dead-end road. The following three sections will give insight into the logic of two very different Asian systems and expose the links between regulatory and trademark law.

These three sections do not seek to present a complete overview of Japanese or Singapore Food Laws or of Trademark Law. Rather, they aim to provide a basic

---

offices/japan/wipo_japan.html, http://www.wipo.int/treaties/en/notifications/madridp-gp/treaty_madridp_gp_128.html.

[4]In Japan, the similarity of trademarks is determined mainly based on their Kana transliterations, regardless of the question whether they have been filed in Kana or Latin. Therefore, conflicts with prior registrations are unpredictable to a foreign applicant. The framework for internationality is completely non-existent in Japanese trademark law, despite being a member of the Madrid Protocol.

[5]Japan has reorganized the goods and services according to the international system, but the determination of similarity between goods and services is still based on the old national system. One of the main purposes of organizing goods into classes in trademark law would however be to determine similarities.

[6]Japan, is one of the only three countries (the other two being Ghana and Cuba) among the 92 member countries (http://www.wipo.int/export/sites/www/treaties/en/documents/pdf/madrid_marks.pdf) where the official fees are to be paid in two parts: http://www.wipo.int/madrid/en/faq/fees.html#P3_4; http://www.wipo.int/edocs/madrdocs/en/2002/madrid_2002_48.pdf. This is a rather odd and unpredictable requirement. Another example, in Japan, the "letters of consent system" is not accepted, but trademarks may be assigned regardless of the question whether the assignee has marks that are confusingly similar to the assigned trademark. This rule is inconsistent.

understanding for how food safety and trademark issues are being dealt with in parallel in these jurisdictions.

50.2 Japan

This Sect. 50.2 focuses on the Japanese regulatory system on food safety. In Japan, food production is governed by the Food Sanitation Law.[7] This law was drafted in light of the consideration that the traditional legal system did not sufficiently address the fast technological development and the resulting complexity of food production and distribution processes. At its core, this law aimed to change policies to become more consumer-oriented rather than industry-oriented. Additionally, in order to take food safety issues and the globalization of food trade better into consideration, the Japanese Food Sanitation Law sets forth the general principles of "protection of consumers", "measures based on science" and "from farms to tables." These principles were developed, because traditional corrective food safety measures which were only applied to the final products became insufficient. The goal of the Food Sanitation Law, therefore, is to improve the realization of these principles.

One of the most important principles are precautionary measures in the food production system. Preventive or precautionary measures refer to decisions or actions that aim to prevent potential risks to human health, and which may be based on available information and do not necessarily have to be supported abundantly by scientific evidence. Henceforth, the new "preventive or precautionary" measures under the new Food Sanitation Law were designed to apply in all stages of production and distribution of food and agricultural materials including feed, pesticides and veterinary drugs.

The following paragraphs show how the fast technological development and the resulting complexity of food production and distribution processes are addressed by the precautionary approach to food sanitation legislation under Japanese law.

50.2.1 The Food Sanitation Law

The Food Sanitation Law applies to all foods and drinks, food additives and natural flavoring agents manufactured in or imported into Japan.[8] It also applies to tableware, kitchen utensils, machines and other articles which are used for handling, manufacturing, processing, preparing, storing, and transporting, displaying,

---

[7]The Japanese Food Sanitation Law, *Food Sanitation Law in Japan*, is available at http://www.itp.gob.pe/normatividad/demos/doc/Normas%20Internacionales/OTROS/Japn/RegulaciJapn.pdf.

[8]Art. 4 Food Sanitation Law.

delivering or consuming food. Generally, anything that comes into direct contact with food, for example parts of water filters, is subjected to this law. Moreover, the Food Sanitation Law regulates containers and packaging, and also all hygiene related issues in connection with human consumption of food. However, food, drinks, additives and flavoring agents that are categorized as "quasi-drugs" according to in the Pharmaceutical Affairs Law are not subject to the Food Sanitation Law. The same applies to products that have effects on human health or contain ingredients that are categorized to be quasi-drug ingredients or pharmaceutical ingredients.

Article 2 of the Food Sanitation Law puts the state, prefectures, cities and special wards[9] in charge with the execution of the regulations of the Food Sanitation Law through their public health centers and officers. They monitor, inspect and detain products that are not in compliance with the law, and control whether activities related to food are covered by the approved related business licenses. The state is in charge with the establishment of systems concerning food sanitation examination of imported food. This means that the state must establish and maintain import inspection and quarantines at the harbors and airports. The implementation of these duties by the state is not necessarily homogeneous as each port has its own circumstances. Importing food through Nagoya is, according to experience, much more time consuming and troublesome than through Osaka, Yokohama or Tokyo. Inspectors in Osaka have a much larger work-load and consequently more experience than inspectors in Nagoya. This reflects in the speed of handling imports. More about this is written under the section Import Procedures and Inspection.

Structurally, the first three Articles of the Food Sanitation Law state its purpose, and define the duties of public and private bodies and incorporations dealing with food and related matters. Article 4 contains definitions for important terms used in this law. The following articles contain the main regulations of the Food Sanitation Law.

- Art. 5 and 6 stipulate that the "dealing with" (handling, manufacturing, importing, processing, using, preparing, storing or displaying for sale, distributing and related activities) food and food additives is only allowed under clean and sanitary conditions such that no harm may be caused to the human health.
- Art. 6–18 further specify prohibitions, such as, for example dealing with newly developed food, unregistered food additives, polluted or contaminated food products, as well as with toxic, injurious and polluted apparatus, containers and packaging.
- Art. 11, 12, 13 and 18 authorize the Ministry of Health, Labour and Welfare (MHLW) to establish, in cooperation with the Ministry of Agriculture Forestry and Fisheries, specifications and standards that have to be complied with when dealing with food and food additives and their packaging. The MHLW also

---

[9]There are multiple ways to translate the word "city" into Japanese, depending on the number of habitants. The size of some wards in Tokyo and Osaka is comparable with that of major European cities. Accordingly they are put in charge with duties that are taken care typically by cities.

decides whether certain food manufacturing and manufacturing plants are in compliance with regulatory legislation.
- Art. 19 and 20 authorize the MHLW to establish standards for product labelling, prescribe that product presentation must comply with established standards and prohibit exaggerating and misleading product presentation.
- Art. 21 authorizes the MHLW to establish and manage standards and specifications for food additives.
- Art. 22 and 23 put the MHLW in charge with the establishment of procedures and guidelines for the inspection of food, food additives, apparatus and packaging, generally and specifically with respect to imported food.
- Art. 24 addresses governors and mayors to establish health centers and elaborate procedures and guidelines for the inspection of food sanitation within their regional responsibility.
- Art. 26–30 prescribe that food, food additives, apparatus and packages shall undergo an examination by the MHLW, regulate procedures in connection with the examination and stipulate that incurred expenses shall be borne by the applicant. They oblige importers to follow the orders of the MHLW with respect to import notifications, reports, on-site inspections and the collection of samples. Furthermore, they regulate the establishment of examination facilities ant the appointment of inspectors.
- Chapter 8 of the law, consisting of Articles 31–47, specifies in detail all matters related to the establishment of examination facilities, such as for example qualification, registration, validity, work regulations, compliance, reporting and discontinuation.
- Chapter 9 of the law, consisting of Articles 48–56, addresses the obligation of dairy products and food additives manufacturers to appoint food sanitation supervisors and the training of these supervisors. They assign the right to the MHLW to establish standards for the manufacturing of food and food additives and authorize the prefectures to establish standards for public health measures, generally and specifically for each type of business. These Articles oblige business owners to obtain respective business licenses from the governor of the prefecture, who regulates the inheritance, violation, revocation, prohibition or suspension of such licenses.
- Chapter 10, consisting of Articles 57–70, deals with miscellaneous issues, such as the liability of public bodies for costs, notifications with respect to poisoned persons, investigations, autopsies of corpses, hearings, appeals, black-listings of violators, and announcements to inform the public about risks.
- Chapter 11, consisting of Articles 71–79, comprises provisions imposing fines or imprisonment for violations of this law.

This paragraph gave an overview of the structure of the Food Sanitation Law, the most important regulations on food safety in Japan. The following four articles address selected issues within or related to the Food Sanitation Law, such as regulations on pesticides, on food additives on import and inspection procedures and the Hazard Analysis and Critical Control Points (HACCP) system.

50.2.2 Pesticides: Regulations on Agricultural Chemicals

Japanese law regulates pesticides mainly to protect consumers from potential hazards to human health. A side-effect of this law is also the protection of the environment. The agricultural industry is admonished to use less and only registered pesticides that meet highest industrial standards, because otherwise, they infringe the law and will not be able to sell their products. Food manufacturers are urged to use in their products only those ingredients that comply with the regulations on agricultural chemicals. They refrain from buying raw materials that contain non-compliant amounts of pesticide residues or forbidden agricultural chemicals, or else, they infringe the law and bear the risk of not being able to sell their products. In order to make sure that the raw materials they buy comply with regulations, they will conduct laboratory test on each lot of products they purchase originating from the same source. Traders and importers suffer substantial damages if their merchandise does not pass the inspections by the authorities. Every food product of a new source is subject of laboratory testing at the customs. Products that originate from a source that is known from a past import record are being re-tested every year at least once. Moreover, in view of the vast volume of food imports to Japan, this law has an impact on food manufacturing and trading not only in Japan, but also in other countries producing for the Japanese market.

In its unrevised form, the Agricultural Chemicals Regulation Law of 1948[10] prohibited the commercial "import and sales" of non-registered pesticides. However, it did not apply to "private imports" and the "use" of non-registered pesticides. After a number of serious incidents[11] with such chemicals, in 2002 the law was revised, and the prohibition was extended to the "use" of non-registered pesticides.[12] In order to specify this new regulation, the Ministry of Welfare and Health (MHLW) compiled a list of substances allowed for import, distribution and use. This list is called the "Positive List System of Agricultural Chemicals". This list covers pesticides, feed additives and veterinary drugs. In addition, each substance in this list is connected with a certain limit, called the "Maximum Residue Limits (MRLs)", which expresses how much of a certain substance a food product may contain to be compliant with the law. This "Positive List" and the "Maximum Residue Limits" have to be seen as inseparably linked to each other.

The Ministry of Welfare and Health (MHLW) introduced the Positive List System in 2006.[13] It prohibits the manufacturing, import, processing, use, sale

---

[10] The Agricultural Chemicals Regulation Law of 1948, Agricultural Chemicals Regulation Law (Provisional Translation) (Law No. 82 of July 1, 1948, last amended on 30 March 2007), is available at: http://www.env.go.jp/en/chemi/pops/Appendix/05-Laws/agri-chem-laws.pdf.

[11] One example is the import and distribution of unauthorized pesticides (Difolatan and Plictran), in the Yamagata Prefecture in 2002 (http://www.thefreelibrary.com/Man+held+on+charge+of+selling+unauthorized+farm+chemicals.-a091757985).

[12] Articles 9, 11 and 12 Food Sanitation Law.

[13] Introduction of the Positive List System for Agricultural Chemical Residues in Foods: http://www.mhlw.go.jp/english/topics/foodsafety/positivelist060228/introduction.html.

and distribution of food that contains agricultural chemicals, pesticides, feed additives or veterinary drugs that are not listed. The Positive List comprises two kinds of substances. There are substances that had been regulated by the MHLW already before the establishment of the Positive List System. Many Maximum Residue Limits (MRLs) for substances contained in food had been set by the MHLW already by 1959. The limits were set based on experience with these substances and on scientific evaluation. These limits were revised during the years 2003–2005.[14] There are also substances cited in the Positive List that the MHLW did not have sufficient experience with, but with respect to which the MHLW had reasonable grounds to rely on Codex standards.[15] For such substances without existing MRLs, "provisional MRLs" were set taking into consideration Codex standards, the Agricultural Chemicals Regulation Law, the Pharmaceutical Affairs Law, the Law for Safety Assurance and Quality Improvement of Animal Feed, and Standards established by Australia, Canada, the EU, New Zealand and the USA based scientific studies. Food containing chemicals above the MRLs are subject to the mentioned prohibition. For agricultural chemicals, for which provisional MRLs could not be determined[16] yet, MHLW decided to set a uniform limit at 0.01 ppm.[17] This uniform limit has been calculated such that, taking the food consumption of an average Japanese into account, the estimated intake of the respective agricultural chemicals does not exceed the toxicological threshold of 1.5 μg/day. The source of this threshold is findings of the Joint FAO/WHO Expert Committee on Food Additives (JECFA), the US Food and Drug Administration (FDA), the EU and different organizations. Food containing chemicals above the uniform limit and where MRLs are not set, are also subject to the mentioned prohibition. Finally, there are substances that are exempted from the prohibition set by the law. These are agricultural chemicals and their decomposition products which are determined not to pose adverse health effects, specified agricultural chemicals shown in the Agricultural Chemicals Regulation Law,[18] chemicals for which registration withholding limits are not established and which are determined not to pose adverse health effects even if crops exposed to these chemicals are consumed,[19] and agricultural chemicals which are determined not to require any MRL in foreign countries and whose uses are not restricted.

---

[14]The maximum residue limits of substances used as ingredient of agricultural chemicals in food: http://www.mhlw.go.jp/english/topics/foodsafety/positivelist060228/dl/index-1b.pdf; see also: http://www.m5.ws001.squarestart.ne.jp/foundation/search.html.

[15]http://www.mhlw.go.jp/english/topics/foodsafety/positivelist060228/introduction.html.

[16]The maximum residue limits of substances used as ingredient of agricultural chemicals in food: http://www.mhlw.go.jp/english/topics/foodsafety/positivelist060228/dl/index-1b.pdf.

[17]Positive List System—Uniform Limit at 0.01 ppm (http://www.ffcr.or.jp/zaidan/FFCRHOME.nsf/pages/MRLs-p-UL).

[18]AGRICULTURAL CHEMICALS REGULATION LAW: http://www.env.go.jp/en/chemi/pops/Appendix/05-Laws/agri-chem-laws.pdf.

[19]List of substances that have no potential to cause damage to human health: http://www.mhlw.go.jp/english/topics/foodsafety/positivelist060228/dl/n02.pdf.

50.2.3 Food Additives

The Food Sanitation Law governs food additives. Food additives are defined as substances that are added, mixed or infiltrated into food or by other methods in the process of producing food or for the purpose of processing or preserving food.[20] This means that food additives are not only substances that remain in the final product (colorants and preservatives), but also that have been applied during the food manufacturing process and do not remain in the final products (microorganisms, control agents, filtration aids). Only additives, preparations and food containing additives that have been designated by the MHLW as posing no risk to human health may be imported, produced, processed, sold, used, stored, or displayed for the purpose of marketing.

The MHLW differentiates among "Designated Additives", "Existing Food Additives", "Food Additive with Use Standards", "Natural Food Additives" and "Ordinary Foods Used As Food Additives". The structure of legislation on food additives, having beside the "Designated Additives" and a number of subcategories focusing on exceptions from the main category implies the adaption of a positive list system. As of August 2014, the MHLW has designated 443 substances as food additives. These "Designated Additives"[21] are substances that are unlikely to harm human health.[22] Food additives that have been served for human consumption and that are used as additives as listed in the list of "Existing Food Additives",[23] are not subject to the designation system of food additives. Consequently, they may be freely produced and used. The MHLW is continuously reviewing the additives in this list and removing items that are assessed to create potential hazards to human health. The number of listed items was reduced in the time from 2005 to 2014 from 489 to 365. The MHLW has set standards (i.e., target foods and maximum use limits/residue limits) for a number of additives.[24] Food preparations containing these "Food Additive with Use Standards" must comply with the established standards. The same applies when such additive is used for the manufacturing of the food preparation. "Natural Flavoring Agents" are natural products that are obtained from animals and plants and used for flavoring food. (e.g. vanilla flavoring and crab flavoring in small amounts). Such substances and mixtures as listed in the List of plant or animal sources of natural flavorings[25] are also not subject to the

---

[20]Article 4.2 of the Food Sanitation Law defines "food additives" that are governed by the Food Sanitation Law.

[21]List of Designated Additives: http://www.ffcr.or.jp/zaidan/FFCRHOME.nsf/pages/list-desin.add-x.

[22]Article 10 of the Food Sanitary Law.

[23]List of Existing Food Additives is available at http://www.ffcr.or.jp/zaidan/FFCRHOME.nsf/pages/list-exst.add.

[24]http://www.ffcr.or.jp/zaidan/FFCRHOME.nsf/pages/stanrd.use.

[25]List of plant or animal sources of natural flavorings is available at http://www.ffcr.or.jp/zaidan/FFCRHOME.nsf/pages/list-nat.flavors.

designation system of food additives. Ordinary Foods Used as Additives are substances that are generally provided for eating or drinking as food and also used as food additives (e.g., blueberry juice used as colorant). These substances, as listed in the List of food additives generally provided for eating or drinking as foods and which are used as food additives,[26] are also are exempted from the designation system of food additives.

Manufacturers, importers and distributors may file a petition with the MHLW and apply to add a specific food additive to the list. The application must be supported by documents validating the safety of the additive. The Pharmaceutical Affairs and FSC examine and decide on such petitions, and may establish standards for manufacturing, processing, and use. The documents required for risk assessment by the FSC are specified and published in the Guideline for Assessment of the Effect of Food on Human Health Regarding Food Additives[27] by the MHLW.[28] The petitioner must bear all costs connected with this procedure. Many food additives which have been approved to be safe in other countries are yet prohibited in Japan. Also the MHLW may proceed ex officio and add new items to the list of "Designated Additives". However, prior to this the Pharmaceutical Affairs and Food Sanitation Council (FSC) shall be heard in order to determine which additives may be designated.

Besides the specific rules, generally, food additives shall not be distributed or used if they are decomposed or immature, which contain toxic or injurious substances or which are suspected to contain such substances, which are contaminated or suspected to be contaminated with pathogenic microorganisms, and which could be injurious to human health due to uncleanness.[29]

50.2.4 Import Procedures and Inspection

There is no other way to ensure the safety of imported food and its compliance with local regulations but by inspections. The same product might be categorized differently in different jurisdictions. An oil blend that is meant to help against anxiety, emotional distress or hyperactivity might be categorized depending on the application method in one jurisdiction as quasi-drug or health care product, in another as cosmetic, and in a third as a product *sui generis* that is under the control of none of the regulatory authorities. Therefore, the inspectors cannot rely on

---

[26]List of food additives generally provided for eating or drinking as foods and which are used as food additives is available at http://www.ffcr.or.jp/zaidan/FFCRHOME.nsf/pages/list-general.provd.add.

[27]Guideline for Assessment of the Effect of Food on Human Health Regarding Food Additives is available at http://www.fsc.go.jp/english/standardsforriskassessment/guideline_assessment_foodadditives_e2.pdf.

[28]http://www.mhlw.go.jp/english/topics/foodsafety/foodadditives/.

[29]Art 6 Food Sanitation Law.

product labels or import documents. The laboratory testing method and the interpretation of formula and of test results can be in every jurisdiction different. Therefore, not even a laboratory test result can fully be relied on as objective source of information. Only a thorough product review followed by a local laboratory test will confirm the correct categorization of a product. The correct categorization is necessary to determine the applicable laws.

The state is in charge with the establishment of systems concerning food sanitation examination of imported food.[30] Prior to each import of food, food additives, and any apparatus, container and package that is subject to the Food Sanitation Law, the submission of import notifications to the Quarantine Station of the MHLW is required.[31] Each submission must comprise the notification form and be supported by documentation disclosing all specifications of the respective product that are necessary for the evaluation of the product. The Quarantine Station may request for any document that supports the evaluation, for example in formation to the source of ingredients, place of manufacturing, etc. The first evaluation as to which category the product falls into is based on the review of the submitted documentation. Then the inspectors review whether the used ingredients are allowed for use in Japan, the limits are within the allowed range, and the sources of ingredients fall under any restrictions or regulations. The manufacturing flow might reveal the application of forbidden recipients, filtering or other steps of the manufacturing process. The import of a number of food items such as meat, meat products and shellfish must also be supported by a "sanitary certificates" issued by a governmental organization of the exporting country. Applications may be submitted by post, e-mail or upon registration with the MHLW through an electronic information processing system. The applications must pass a thorough examined by food sanitation inspectors at the quarantine station. There, the inspectors evaluate the compliance of the import with the Food Sanitation Law by focusing on the provided information, such as the exact product category, the ingredients, the applied manufacturing method, the additives, chemicals and materials applied during the manufacturing, the country of export, the manufacturer and the place of manufacture. Based on the above-mentioned information, available track record of the place of manufacturing or any record of incompliance of the respective product category, the inspector also evaluates the risk that the products might be contaminated with poisonous or hazardous substance, or have some issues with sanitation.

It is up to the discretion of the inspectors to grant the import of a product or issue an inspection order. An inspection order means the suspension of the import procedure until the compliance of the food concerned is confirmed. Such a confirmation may be obtained following an on-site inspection of the goods, or by a laboratory test. This procedure may take only a week, or it could take months, in

---

[30] Article 2 Food Sanitation Law.
[31] Article 27 Food Sanitation Law.

case the specifications of the product are complex. The importer must bear the costs of the inspection.

Food products that are imported to Japan for the first time are typically subjected to thorough inspection along with laboratory testing. New and complex products often get stuck at the customs because the inspectors need time to get familiar with the formulation and manufacturing of the products, and then decide on the scope of the necessary laboratory testing. In order to decrease the work load of the Quarantine Station, the inspectors may decide to apply "monitoring inspections" on products that are unlikely to be incompliant with the Food Sanitation Law. Products that may be subject to monitoring inspection are designated based on the annual import amount and the record of compliance for each item. Designated products may be processed without any delay caused by waiting for the inspection results.

50.2.5 HACCP

The Hazard Analysis and Critical Control Points or HACCP is a system that aims to prevent biological, chemical, and physical hazards in any production process that might result in the finished product to be defective or unsafe. Traditional quality control methods, which focus on sample inspection of finished products and often conducted by the purchasers of products, have become insufficient and ineffective in view of recent advanced and complex processing technologies and in the case of mass production. The HACCP is borne out of an awareness that "preventive or precautionary" measures must be developed and applied during the manufacturing process to recognize and avoid defects in the finished product. Accordingly, the system provides scientific measurements that help identify such risks and reduce them to a safe level. One additional advantage of the HACCP system is that it allows not only the industry but also the regulatory authorities to establish and audit safe production practices.

The HACCP was not developed specifically for the food industry. It is believed to originate from a monitoring system developed for the manufacturing process of artillery shells during the Second World War (MIL-STD-105 official record), to avoid duds or misfiring. The HACCP has been, in the meantime, also applied in other industries, such as the manufacturing of cosmetics and pharmaceuticals. HACCP was applied in the food industry for the first time in the 1960s as a logical tool for adapting traditional inspection methods to a modern, science-based, food safety system. It was when General Mills, Inc. started designing and manufacturing the first food for space flights for the NASA. The HACCP was introduced in 1973 as Good Manufacturing Practices for low-level acidity canned food in 1973, and provided solutions in the 1980s in a number of food poisoning cases caused by O-157, salmonella and lead contamination. In 1985, the Science Academy recommended the USDA (US Department of Agriculture) to introduce HACCP in the manufacturing of meat. The US regulatory authorities introduced compulsory HACCP in 1995 for large-size fish products manufacturing facilities, in 1998 for meat and poultry manufacturing, and in 2001 for fruit and vegetable juice manufacturing facilities (http://haccpalliance.org/alliance/HACCPall.pdf).

Although HACCP focuses only on the safety and not on the quality of products, it has become a basis for many food quality and safety assurance systems, for example, Title 21 parts 120, 123 of the Code of Federal Regulations in the USA, and *FAO/WHO guidance to governments on the application of HACCP in small and/or less-developed food businesses*.[32] In the meantime, the HACCP has been recognized internationally as a system that helps ensuring the safety of food for human consumption internationally.[33] The European Union (EU) ordered member countries to oblige all facilities to adopt the HACCP principles by the *Council Directive 93/43/EEC of 14 June 1993 on the hygiene of foodstuffs*.[34] Therefore, the HACCP principles are regarded as a standard to be observed by any facility in EU countries. In light of such development of HACCP in the USA and the EU, the Food and Agriculture Organization (FAO) Codex Commission drafted principles and guidelines for the application of HACCP system.[35]

Japan started studying the HACCP system, when the import of certain Japanese fish products into the EU was rejected for not complying with the EU Directive on hygiene of food stuffs. In 1995, the Food Sanitation Law was amended and the HACCP system was introduced under the name "Integrated Sanitation Management of Process of Manufacturing". Articles 13 and 14 of the Food Sanitation Law in connection with The Food Sanitation Law Enforcement Ordinances authorize the MHLW to decide whether the manufacturing of milk, milk products, meat products, fish paste products, high pressured, heated, sterilized food in containers or packages, and soft drinks is in compliance with legislation and standards, and also to grant related approvals (to manufacturing and facilities). To ensure the effectiveness of HACCP, the validity of approvals is limited to 3 years.

However, while the EU and the USA have introduced HACCP as an obligatory management system for food safety, and along with it minimum standards to be respected by all manufacturers, in Japan it is not compulsory for manufacturers to apply for an HACCP approval with the MHLW. Manufacturers who have received approval do not have any significant benefits; their products cannot sell at higher prices within Japan. On the other side achieving an HACCP approval is connected with substantial investments. As the number of approved facilities remained small, the HACCP Promotion Law was created, which sought to promote the HACCP system by supporting the grant of low-interest-rate loans and tax reliefs to HACCP-

---

[32]*FAO/WHO guidance to governments on the application of HACCP in small and/or less-developed food businesses* is available at ftp://ftp.fao.org/docrep/fao/009/a0799e/a0799e00.pdf.

[33]http://www.fao.org/docrep/v9723t/v9723t0e.htm.

[34]*Council Directive 93/43/EEC of 14 June 1993 on the hygiene of foodstuffs* is available at http://ec.europa.eu/food/food/biosafety/salmonella/mr06_en.pdf.

[35]http://www.google.com.sg/url?sa=t&rct=j&q=&esrc=s&source=web&cd=1&ved=0CBwQFjAA&url=http%3A%2F%2Fwww.codexalimentarius.org%2Finput%2Fdownload%2Fstandards%2F23%2FCXP_001e.pdf&ei=cVf8U7_RIpSiugTIj4HABw&usg=AFQjCNGx1qgI0C4LDmiOfbZubFHjIbk2ng.

approved facilities. Nevertheless, the number of approved facilities barely increases in Japan.

Presently, the HACCP system is typically adopted by large-size manufactures and by exporters to demonstrate their general corporate performance in food safety and that their products have been manufactured in compliance with the standards of the HACCP system to avoid problem with EU and US customs.

50.3 Singapore

This section focuses on the Singapore regulatory system on food safety. The Singapore Food Regulations are derived from scientific knowledge and take the fast technological development, the complexity of food production and distribution processes into account. At the same time, it considers that Singapore is almost completely dependent on imports to satisfy the local demand for food, beverages and water. The Singapore's Office of Agriculture (AVA) regulates food, from its production to its retail. It takes a science-based risk analysis approach and applies international standards to evaluate and ensure food safety. AVA is responsible for the enforcement of the Sale of Food Act with the objective of setting standards and ensuring food quality and reducing risks to the health of humans.

50.3.1 The Agri-Food and Veterinary Authority (AVA)

The Agri-Food and Veterinary Authority (AVA) is a statutory board in charge with food safety in Singapore. The AVA evolved from the Primary Production Department (PPD), which was formed in 1959 by merging the agriculture, co-operatives, fisheries, rural and veterinary authorities to ensure a coordinated development and regulation of the local farming and fishing industry.[36] As over the decades Singapore's farming activities diminished and the trade with food expanded, the PPD was put in charge with food safety, testing and facilitation of trading, and in 2000 it was restructured into the AVA.

The Food Control Division of the AVA supervises the safety of local and imported raw and processed food, develops, implements and enforces food safety policies, operates an inspection program and monitors health threats. Apart from the accuracy and compliance of shipping documentation, the AVA inspects each batch of food import for absence of contamination, disease, damage food-borne pathogens and parasites and banned chemical preservatives. The investigations are supported by laboratory tests and microbiological examinations. Without

---

[36]History on PPD is available at *HistorySG* http://eresources.nlb.gov.sg/history/events/14f36ce1-a54c-4bb3-8993-91c57fee7c50 and AVA website http://www.ava.gov.sg/AboutAVA/History.

exception, importers are obliged to refrain from the distribution of meat and poultry products, oysters, clams, mussels, scallops and cooked crabmeat until the completion of the laboratory testing. Shellfish products may only enter Singapore if they are supported by health certificates and from sources with acceptable sanitation programs. Fruit and vegetable imports are spot-checked for pesticide residues. Shipments that fail to meet the Singapore safety standards due to the presence of forbidden ingredients, preservatives, additives or colorings are rejected and destroyed under AVA's supervision.

50.3.2 Legislation

Raw and processed food, beverages, and ingredients, whether manufactured in or imported to Singapore, must comply with the Food Regulations and related guidelines. Part III of the Food Regulations comprises of two main sections. The first section lists provisions that set general rules with respect to:

- Labeling (Art. 5) and exemptions (Art. 6)
- Containers (Art. 7, 37)
- Hampers (Art. 8)
- Nutrition information panel (Art. 8A)
- False or Misleading statements (Art. 9)
- Date marking (Art. 10)
- Claims as to presence of vitamins or minerals (Art. 11)
- Misleading statements in advertisements (Art. 12)
- Food and appliances offered ad prizes (Art. 13)
- Registration of imported food (Art. 14)

The second section deals with:

- Food additives (Art. 15–28)
- Incidental constituents (Art. 29–35)
- Mineral Hydrocarbons (Art. 36)
- Irradiated food (Art. 38)

Part IV of the Food Regulations sets specific standards and requirements for labeling and naming with respect to a catalog of pre-packed food and beverage products:

- Flour, bakery and cereals (Art. 39–56)
- Aerating ingredients (Art. 57–58)
- Met and meat products (Art. 59–70)
- Fish and Fish products (Art. 71–77)
- Edible fats and oils (Art. 78–92)
- Milk and milk products (Art. 93–125)
- Ice-cream, frozen confections and related products (Art. 126–129)

- Sauces, vinegars and relishes (Art. 130–142)
- Sugar and sugar products (Art. 132–152)
- Tea, coffee and cocoa (Art. 153–170)
- Fruit juices and fruit cordials (Art. 171–175)
- Jams (Art. 176–179)
- Non-alcoholic drinks (Art. 180–184)
- Alcoholic drinks (Art. 185–210)
- Salts (Art. 211–212)
- Spices and condiments (Art. 231–236)
- Flavoring essences or extracts (Art. 237–245)
- Flavor enhancers (Art. 246)
- Special purpose foods (Art. 247–254)
- Miscellaneous foods (Art. 255–259)
- Rice (Art. 260)

Furthermore, the Food Regulations comprise also 12 schedules that support the interpretation of the general end specific provisions. They provide specifications, set limits, and deal with penalties.

The following passages address the most salient provisions of the Singapore Food Regulations by focusing on the compulsory content of food product labels. The Food Regulations stipulate that pre-packed food and beverages for sale must be labeled and that the labels must provide basic product information in English.[37] The basic labeling requirements apply to samples, rewards in the form of food products, and to containers holding pre-packed food, too. The labeling requirements do not apply to food that is un-packed or weighed, counted or measured in the presence of the purchaser. It does not apply to sugar confectionery, chocolate and chocolate confectionery. Labels of intoxicating liquors do not have to state ingredients. As an exception of these exceptions, on the labels of food that contains synthetic coloring, the name and address of manufacturer, importer, packer or distributor must be displayed.

The label requirements must not be applied in a misleading manner, for example to food that is not destined to be, or that does not qualify for human consumption.[38] The name and description of the product reflecting its true nature must be displayed on the label. The name and description must not be misleading or implying forbidden claims, and it must be in compliance with the specific provisions of Part IV of the Trademark Regulations.[39] A complete list of ingredients and additives (statement of ingredients) should be declared on food labels. The ingredients

---

[37]This is remarkable, because Singapore has four official languages, which are English, Chinese, Malay and Tamil. Although Singapore was part of the British Empire until 1963, in the meantime, nearly 75 % of the population is Chinese, about 13 % Malay, and about 9 % Indians. The economic focus of the country on trading provides a more plausible reason for the mentioned regulation.

[38]http://www.ava.gov.sg/NR/rdonlyres/B96B0EC2-1D1E-4448-9C25-ABD8470D2BF4/26937/AGuidetoFoodLabellingandAdvertisementsVersionSepte.pdf.

[39]Article 5 Food Regulations.

have to be displayed in descending order with respect to the weight proportions with their registered generic names as listed in the first schedule, Permitted Use of General Terms in the Declaration of Ingredients.[40] The International Numbering System (INS) and E numbers may be used for the declaration of food additives. Ingredients known to cause hypersensitivity, such as:

- Cereals containing gluten
- Crustacean and
- Crustacean products
- Eggs and egg products
- Fish and fish products.
- Peanuts, soybeans and their products
- Milk and milk products (including lactose).
- Tree nuts and nut products.
- Sulphites in concentrates of 10 mg/kg or more.

...must be declared as ingredient, food additive or component of a compound.[41] If a food product contains aspartame, a respective indication must be displayed on its label.

Based on the minimum or the average quantity system, the Net Content, or quantity of food must be displayed on the label of packed food. The quantity must be expressed using net weights (grams, kilograms) for solid food, and volumetric measures (milliliters, liters) for liquid food. The label of solid food filled in water, aqueous solutions of sugar and salt, fruit and vegetables juices or vinegar must display net and drained weight. The font size of information displayed on labels must be at least 1.5 mm in height. The name and address of manufacturer, packager or distributor must be displayed on the labels of food manufactured in Singapore in order to enable to identify the responsible person. The name and actual (not virtual) local address of importer, distributor or agent must be displayed on the labels of imported food. The country of origin of the product must be displayed on food product labels. Recipes, serving suggestions with respect to packed food in written or in pictures must be indicated as such and the size of the printed letters must be at least 1.5 mm.

Nutritional labeling is a mandatory component of product labels, where (permitted) explicit or implied nutrition, vitamins, minerals or health claims are made. Nutrition claims are allowed in Singapore as long as they comply with the requirements of the Food Regulations and the nutrient claims guidelines published in "A Handbook on Nutrition Labelling".[42] Nutrition claims are defined as claims that suggest or imply that a certain food has nutritive properties, or as comparison of the nutritive property of a certain product in terms of energy, salt (sodium or potassium), amino acids, carbohydrates, cholesterol, fats, fatty acids, fibre, protein,

---

[40] Article 5 (4) (b) Food Regulation.
[41] Article 5(4)(ea) Food Regulation.
[42] http://www.ava.gov.sg/FoodSector/FoodLabelingAdvertisement.

starch or sugars, vitamins or minerals, or any other nutrients. Examples of nutrition claims are "Low in calorie", "Sugar free" and "Reduced sodium". Nutrition claims on pre-packed food have to be provided in an adequate information panel, as for example[43]:

....

Servings per package (here insert number of servings)*
Serving size: (here insert the serving size)*

	Per Serving* or	Per 100 g (or 100 ml)
Energy	kcal, kJ or both	kcal, kJ or both
Protein		
Fat	g	g
Carbohydrate	g	g
(here insert the nutrients for which nutrition claims are made, or any other nutrients to be declared)**	g	g
	g	g

*Applicable only if the nutrients are declared on a per serving basis.
**Amounts of sodium, potassium and cholesterol are to be declared in mg.

....

There are some exceptions of this rule that apply to where the claims on a label only refer to salt, sodium or potassium and where the total surface area of the label is less than 100 square centimeters.

Additional labeling requirements apply where the claim that a food product is a source of energy or protein is made. Such claims must be supported by an adequate information panel stating the quantity that needs to be consumed within one day such that it provides at least 300 kcal to the consumer. A product only qualifies as a source of protein, if at least 12 % of the total calorie contained in it is derived from protein. It qualifies as an excellent source of protein if at least 20 % of the total calorie is derived from protein, and if the stated amount of one-day consumption contains at least 10 g of protein. The daily recommendation statement may be put into words as "Recommended daily intake: 3 servings" or "Add 20 g powder in 200 ml water. Drink 2 times daily."

The Food Regulations prohibit[44] statements to particulars of products that are false or misleading with respect to the nature, stability, quantity, strength, purity, composition, weight, origin, age, effects, or proportion of the food or any ingredients. They forbid suggestions that a product has therapeutic, prophylactic medical or health effects. It only allows the use of the attribute "pure" if the food is free from other added substances and if it is of a quality as required under the Food Regulations.

---

[43] Food Regulations—A Guide to Food Labelling and Advertisements p. 17: http://www.ava.gov.sg/NR/rdonlyres/B96B0EC2-1D1E-4448-9C25-ABD8470D2BF4/26937/AGuidetoFoodLabellingandAdvertisementsVersionSepte.pdf.

[44] Article 9 Food Regulations.

The law sets strict requirements[45] to the display of vitamin and mineral claims on product labels. The claim to contain vitamins and minerals may only be displayed on the labels of products that contain at least one-sixth of the daily allowance of vitamins and minerals as stated in Tables I and II. Such claims must be supported by statements to the quantity of the food that contains the claimed quantity of the respective vitamin or mineral or comparable information. The law stipulates further limitations to the language of the claim and to the maximum amount of vitamin A content (750 mcg of retinol activity) per reference quantity (Table II).

The food regulations stipulate additional labeling requirements for a number of products, for example Irradiated Food, Wholegrain, Bakery Products, Edible fats and oils, Milk, Coffee, Fruit Juice, Natural mineral water, Fruit wine, Compounded liquor, Infant Formula, Rice.[46]

The Food Regulations stipulate the display of warning statements for products containing certain ingredients, such as for example aspartame or royal jelly.

The expiry date information must be marked or embossed on all product packaging, and printed in letters of not less than 3 mm height. The day of the month must be expressed in figures, where the figure is a single digit it should be preceded by a zero. The month of the year may be expressed in figures or in words and may be abbreviated (first three letters). The year must be expressed in figures in full or by the last two figures of the year. Pre-packed food listed in the second schedule of part V of the Food Regulations must be labeled with their expiry dates:

- Cream, reduced cream, light cream, whipped cream and sour cream excluding sterilised canned cream.
- Cultured milk and cultured milk drink.
- Pasteurised milk and pasteurised milk drink.
- Yoghurt, low-fat yoghurt, fat-reduced yoghurt, non-fat yoghurt and yoghurt products.
- Pasteurised fruit juice and pasteurised fruit juice drink.
- Pasteurised vegetable juice and pasteurised vegetable juice drink.
- Tofu, "tauhu" or "doufu", a soya beancurd product made of basically soya beans, water and a coagulant, including "egg tofu", "taukua" or "dougan", and the soft soya beancurd desert known as "tauhui", "tofa" or "douhua", but excluding the oil fried tofu in the form of a pouch known as "taupok", and the dried beancurd stick.
- Food which is stored or required to be stored at a chilling temperature to maintain or prolong its durable life, but excluding raw fruits and vegetables.
- Vitaminised fruit juice and vitaminised fruit juice drink.
- Vitaminised vegetable juice and vitaminised vegetable juice drink.

---

[45] Article 11 Food Regulations.
[46] Articles 38, 40A, 53, 79, 109, 158, 159, 161, 171, 183A, 195, 210, 254, 260 Food Regulations.

- Liquid milk and liquid milk products excluding condensed milk, sweetened condensed milk, evaporated milk and canned sterilised milk and milk products.
- Flour.
- Salad dressing.
- Mayonnaise.
- Raisins and sultanas.
- Chocolate, milk chocolate and chocolate confectionery in which the characteristic ingredient is chocolate or cocoa, with or without the addition of fruits and nuts.
- Breakfast-cereal with or without fruits and nuts except cereal in cans.
- Infants' food.
- Edible cooking oils.

The display must comprise any of the expressions:

- "USE BY (here insert the day, month and year)";
- "SELL BY (here insert the day, month and year)";
- "EXPIRY DATE (here insert the day, month and year)"; or
- "BEST BEFORE (here insert the day, month and year)" or other words of similar meaning.

Where the way of storage might influence the validity, storage directions must be provided on the product label, for example: "BEST BEFORE: JAN 30 2012. Store in a cool, dry place." The display of the date of packing, as:

- "PACKING DATE (here insert the day, month and year)";
- "PACKED ON (here insert the day, month and year)"; or
- "PKD (here insert the day, month and year)

...is sufficient for raw produce. Raw produce must also comprise the reference "raw".

The Singapore Ministry of Trade and Industry put the Genetic Modification Advisory Committee (GMAC)[47] in charge with the exploitation of products derived from genetic manipulated organisms. The GMAC has four subcommittees specializing in release of agriculture-related GMOs, research on GMOs, labeling and public. The GMAC advises with respect to all related matters, such as approvals, research and development, production, use, reviews, monitoring. The GMAC informs the public on planned releases, facilitates the exchange of information with foreign agencies and the harmonization of guidelines with international authorities, and monitors international developments on the labeling of GM products to see how these may be of relevance to Singapore. Singapore does not have any legislation or guideline for the labeling of GM food, yet. However, genetically modified food is subject to special declaration, review, inspection and testing procedures. The GMAC will consider a new agriculture-related genetically

---

[47] http://www.gmac.gov.sg/.

manipulated organism to be safe and recommend its admission, if it is found to be substantially equivalent to an existing food or food component.

The labels of "Special Purpose Food", such as of diabetic food, low sodium food, gluten-free food, low protein food, carbohydrate-modified food, low-calorie energy food, infant formula and formulated food, must state the special suitability of the food, and comprise a nutrition information. Special purpose food may only be labeled as "sugar-free" if they contain equal or less than 0.5 g sugar per 100 g or 100 ml.[48] Special purpose food may only be labeled as "low-calorie food" (Art. 149 Food Regulations) if they contain equal or less than 8 kcal per 100 ml in case of beverages, 100 kcal/100 g in case of bread spreads including jam substitutes or 50 kcal/100 g in case of all other kind of food. Diabetic food must be labeled as such and the label must contain a nutrition information panel (twelfth schedule of the Food Regulations) as well as a statement as to nature of the carbohydrates present in the food.[49] Infant food and infant formula is food for persons not older than 12 months of age. The law[50] sets strict and very specific requirements for the content of infant food and specific labeling requirements for packaging of infant formula, such as:

- directions as to the method of preparing the food;
- the amount of energy and the number of grams of protein, fat and carbohydrate per 100 ml or other equivalents of formula prepared in accordance with the directions on the label;
- the total quantity of each vitamin and mineral per 100 ml or other equivalents of formula prepared in accordance with the directions on the label;
- a statement suggesting the amount of the prepared food to be given each time, and the number of times such amount is to be given per day; such statement shall be given for each month of the infants' age up to 6 months;
- directions for storage and information regarding its keeping qualities before and after the container has been opened; and
- information that infants over the age of 6 months should start to receive supplemental foods in addition to the formula.

Forbids the use of mineral hydrocarbons is forbidden in the composition or preparation of any food and the sale of food that contains mineral hydrocarbons, unless the product belongs to a catalog of products listed under this Article, and unless the residues of mineral hydrocarbons does not exceed the limits as set in Article 26 Food Regulations. Except under specific license, the law[51] prohibits the import or sale of irradiated food, or food that has been exposed to ionizing radiation. Ionizing radiation only falls under the exception of this Article, if it has been conducted in accordance with or meets the standard of the Codex Recommended

---

[48] Article 248 Food Regulations.
[49] Article 250 Food Regulations.
[50] Articles 252–254 Food Regulations.
[51] Article 38 Food Regulations.

International Code of Practice for the Operation of Radiation Facilities Used for Treatment of Food. This Article further proscribes the obligatory display of the words: "Treated with ionizing irradiation" or "Irradiated (name of food)" in letters not smaller than 3 mm of height.

50.3.3 Provisions Regulating Ingredients, Contaminants and Pesticides

Raw and processed food, beverages, and ingredients, whether manufactured in or imported into Singapore, must comply with the Food Regulations and its guidelines. Food and beverages that are manufactured, imported and sold in Singapore may strictly only contain additives that comply with the specifications and limitations (purity, amount), as stated in the Food Regulations.[52] These articles set official definitions, specify permitted uses and proportions and establish specific labeling requirements for Anti-caking agents, Anti-foaming agents, Anti-oxidants, Sweetening agents, Chemical preservatives, Coloring matter, Emulsifiers and stabilizers, Flavoring agents, Flavor enhancers, Humectants, Nutrient supplements, Sequestrants, Gaseous packaging agents, General purpose food additives.[53] Contaminants, incidental constituents, toxic substances, pesticide residues, heavy metals, antibiotic residues, oestrogen residues, mycotoxins and microbiological contamination are forbidden in Singapore, others may not exceed given limits in Food.[54] The Food Regulations comprise nine detailed schedules supporting the interpretation of the law[55]:

- The third schedule: oxidants
- The fourth schedule: specific permitted chemical preservatives
- The fifth schedule: coloring matters; synthetic organic and other colors
- The sixth schedule: emulsifiers and stabilizers
- The seventh schedule: nutrient supplements
- The eighth schedule: general purpose food additives
- The ninth schedule: maximum amounts of pesticides
- The tenth schedule: permitted maximum amount of arsenic, lead and
- The eleventh schedule: microbiological standards for milk powder, buttermilk powder, pasteurized milk, ice cream, cooked crab meat, prawns and shrimps, mollusk ready for consumption, edible gelatin, fish ready for consumption, pastry, meat ready for consumption and any solid or liquid food ready for consumption

---

[52] Articles 16–28 Food Regulations.
[53] Articles 16, 16A, 17–28 Food Regulations.
[54] Articles 29–35 Food Regulations.
[55] Article 16–35 Food Regulations.

These schedules are continuously updated by the AVA. In the absence of specific local legislation, the recommendations of the Joint Food and Agriculture Organization of the United Nations[56] and World Health Organization (FAO/WHO) Expert serve as guidance. With respect to pesticides (ninths schedule) the AVA might also consult the Codex Alimentarius Commission.[57]

The import, sale, consignment, delivery or use of containers that contain more than 0.05 ppm vinyl chloride monomer or compounds known to be carcinogenic, mutagenic, teratogenic, poisonous or injurious is forbidden. The same applies to the import, sale, consignment, delivery or use of appliances, containers or vessels that are intended to be used in the storage, preparation or cooking of food, if it is capable of imparting lead, antimony, arsenic, cadmium or any other toxic substance to the food stored, prepared or cooked in it. The import, sale, consignment, delivery or use of ceramic food ware that contains lead beyond limits as stated in the law is restricted.[58]

50.3.4 Business Licenses and Control Programs

Singapore is running a number of food control programs. All meat, fish and egg processing establishments, warehouses and slaughter-houses require specific business licenses to be issued by the AVA to carry out food processing or storage for wholesale and distribution. They are subjected to regular and unannounced inspections by AVA. The AVA categorizes businesses (A for Excellent, B for Good, C for Average and D for Pass) based on their food hygiene and food safety standards, determines the frequency of such investigation, the need for the collection of samples and their laboratory analysis, and provides businesses with consultation on Good Manufacturing Practices (GMPs) and safety programs such as the Hazard Analysis and Critical Control Point (HACCP).

50.3.5 Import Requirements

Importers must ensure that their products comply fully with the requirements of the Food Regulations, meet the import requirements and answer the inspection and sampling requests for laboratory tests of the AVA. Meeting the specified import requirements does not exempt the consignments from inspection and sampling for laboratory tests by the AVA. However, it helps avoiding unnecessary delays and storage of goods at the customs. All products are subjected to document and

---

[56]http://www.fao.org/home/en/.
[57]http://www.codexalimentarius.org/.
[58]Article 37 Food Regulations.

physical inspection and sampling for laboratory analysis. The AVA updates regularly and publishes specific import requirements for selected products.[59]

- Dairy Products from Foot-and-Mouth Disease (FMD) affected Countries.
- Pasteurized Liquid Milk from FMD-free countries
- Infant Formula (Age 0–12 Months)
- Infant Cereals
- Traditional Cakes & Nasi Lemak
- Coconut Products—Coconut Milk, Grated Coconut, Jelly Coconut, Shelled Coconut
- Ready to Eat Minimally Processed Cut/Peeled Fruits and Vegetables from Malaysia
- Minimally Processed Cut/Peeled Fruits and Vegetables
- Minimally Processed Cut Sugar Cane
- Mooncakes
- Processed Land Snail and Snail Caviar
- Beef extract and any food products containing beef extract
- Packaged Mineral and Drinking Water
- Ice
- Soy Sauce & Oyster Sauce
- Absinthe
- Other Imported Food Products

There are also a number of regulations that regulate the import of food products, such as for example the Wholesale Meat and Fish Act, the Control of Plants Act and the Animals and Birds Act.

The Wholesale Meat and Fish Act[60] regulates the import of meat and poultry, fish, their parts and products, no matter whether they are chilled, frozen, processed or canned form. Meat products may only be imported from approved sources (countries and establishments).[61] Such restriction does not apply to fish products, with exception of a few problematic species (live oysters, chilled crab meat etc.), and such listed in the Convention of International Trade in Endangered Species (CITES).[62] Every consignment of meat, oyster, blood cockle meat, prawns and crab meat product must be shipped directly from the country of export and be supported by bills of lading, airway bills and invoices also by a health certificated issued by a veterinary authority of the exporting country that confirms the compliance of the

---

[59]http://www.ava.gov.sg/NR/rdonlyres/2BA0A4AA-05D8-4E3C-A8F9-60F26F90EA76/27312/ImportRequirementsofSpecificFoodProducts.pdf.

[60]http://www.ava.gov.sg/NR/rdonlyres/F35FFB95-C706-49CD-B9E4-E402788BAF78/13756/Attach65_legislation_WMFAct.pdf.

[61]http://www.ava.gov.sg/AVA/Templates/AVA-GenericContentTemplate.aspx?NRMODE=Published&NRNODEGUID=%7b2CFF1FE3-8D20-4E7A-BC89-6368076F427B%7d&NRORIGINALURL=%2fFoodSector%2fImportExportTransOfFood%2fAccredOfOverseasMeatEgg%2f&NRCACHEHINT=Guest#List.

[62]http://www.cites.org/.

products with the Singapore's animal health and food safety requirements. The label of every unit of meat and meat product imported must contain:

- A description of the meat product;
- The country from which the meat product originates;
- The brand name of the meat product, if any;
- The name and designation number of the processing establishment in which, and the date on which, the meat product was processed, if applicable;
- In the case of a processed meat product, the name and designation number of the slaughter-house in which the animals used in the production of such meat product were slaughtered and the date of the slaughter
- The name and designation number of the establishment in which, and the date on which, the meat product was packed;
- The batch number and, where the meat product is canned, the canning code.
- The net weight of meat product as contained in each basic packaging and outer carton.

The Control of Plants Act[63] (Import & Transshipment of Fresh Fruits & Vegetables) regulates the import of raw (unprocessed) fruits and vegetables and makes reference to the ninth schedule of the Food Regulations. The Animals and Birds Act,[64] in particular the regulations to "Veterinary Conditions for the Importation of Table Eggs" regulate the import of fresh hen eggs. Fresh and processed eggs may only be imported from approved establishments, which can be found on the website of the AVA.[65] Fresh egg consignment must be from the same source (one specific farm), be supported by an import license and a Veterinary Health Certificate, and dated within 7 days of the import. Imports of processed food must be supported by documentary proof that the production took place in a regulated establishment under proper supervision of the competent food authority of the exporting country. The use of artificial sweetening agents is strictly regulated and limited. Artificial sweetening agents may be used only in special dietary foods which are formulated for specific consumer groups, to meet their dietary needs (thirteenth schedule in the Singapore Food Regulations). Their import, sale, manufacture and use must be supported by an artificial sweetening agent license (Art. 18 Food Regulations). The product labels must comply with specific labeling requirements and comprise the information that the product contains artificial sweetening agents stating the exact name of the used sweetener.

Only companies or businesses registered with the Accounting and Corporate Regulatory Authority (ACRA) holding a Unique Entity Number (UEN) and

---

[63] http://www.ava.gov.sg/NR/rdonlyres/2BA0A4AA-05D8-4E3C-A8F9-60F26F90EA76/13760/Attach27_legislation_COP_ImportandTranship_rules.pdf.

[64] http://statutes.agc.gov.sg/aol/search/display/view.w3p;page=0;query=DocId%3A%22f0719c63-6c52-4222-b991-3804d749ea36%22%20Status%3Ainforce%20Depth%3A0;rec=0.

[65] http://www.ava.gov.sg/FoodSector/ImportExportTransOfFood/AccredOfOverseasMeatEgg/List+of+overseas+meat+and+egg+processing+establishments.htm.

registered with the Singapore Customs (SC) are qualified to register products with the AVA. The registration of processed food with the AVA and the acquisition of a registration number from the Quarantine & Inspection Department—QID are prerequisites to any application for import permits. The registration of each product is connected with the payment of official fees and is valid for 1 year. Any change in the particulars of the product makes a new application necessary. Each food item must be declared accurately with Correct HS code, product description, quantity and unit of measurement, brand name, and country of origin.

The Food Control Division (FCD) of the AVA is in charge with the control of food safety, licensing and inspection of food and controls whether processed food product imports comply with the Sale of Food Act (SFA), the Food Regulations (FR) and all labeling requirements. AVA has adopted a risk-based approach and places food products that have been identified through trend studies or by precedent cases to be of high potential health risk, under strict import control (mineral water, coconut milk, infant formula, ready-to-eat fruits and vegetables, etc.). These products require pre-import inspection such as the submission of health certificates or laboratory reports to certify the safety of the products. Prior to releasing them for sale the FCD conducts document and physical inspection (eventual sample taking and laboratory testing) on all products that are under intensive surveillance. Other products are being released for sale before they are investigated (post import inspection and eventual sampling and laboratory testing).

50.4 Links Between Regulatory Legislation, Marketing and Trademark Law

Although the category of a product is primarily determined by its composition and the applied manufacturing process, the outcome may be influenced by the intended use, the applied marketing method, or the chosen product name. An oil blend that is meant to help against emotional distress might be categorized, conditional to the application method (consumption or external application on the human skin or inhalation upon vaporization), in the same jurisdiction differently. The same product with the same application method might be categorized in Japan and Singapore based on the applicable regulations differently. Depending on the hurdles set by regulations of his target markets, a manufacturer or trader might choose to import the same product in one jurisdiction as quasi-drug or health care product, in another as food, and in a third as general merchandize.

The marketing method chosen for a product might have influence on the product categorization. The decision whether or not to make a claim for a product may shift it from one into another category. If the manufacturer decides to claim for the above mentioned oil blend that "its consumption will provide relieve from emotional distress", or "help relaxing the muscles" or even "solve insomnia", it is likely that the product will be classified as health product or quasi drug, while without such claim, the same product might be categorized as food additive, food or even general merchandize. For the import and distribution as a quasi-drug, the product would have to pass a complicated and costly product registration procedure. Moreover, the distributor would have to obtain specific import and distribution licenses. For the import and distribution as general merchandize, no specific requirements have to be complied with.

The same applies to the choice of the product name accordingly. A product name might imply connotations that are equivalent to claims. Names like "AntInsomnia" or "StressRelieve" or "ImmunBallance" imply effects that food products are not supposed to have. The regulatory authorities might object to such product names during the product registration or import procedures or order a recall of products labeled with such product names. Product names are typically created by marketing experts or name creation companies (marketing companies) who are trying to use descriptive product names with strong connotations and implied claims. The name candidates created by marketing experts are then reviewed by regulatory experts who try to avoid using product names that imply claims and risky connotations. Regulatory experts would prefer using product names that fit to the category of the product. The trademark experts evaluate the candidates that have "survived" the review by the regulatory experts, and eliminate those that are descriptive or generic and that are not available, because they have been already taken by competitors. It is easy to imagine, that some name creation projects end up without any surviving name candidates. Each of these experts is looking at very different objectives. The subject of greatest controversy is the product name that needs to be registered with the regulatory authorities and that also needs to be secured as trademark.

A trademark may consist of or comprise letters, numerals, devices, colors, aspects of packaging and any combination of these. In some countries, colors, sounds and slogans may also be registered as trademarks. A trademark is registrable if it is distinctive with respect to the goods or services it is used or intended to be used for and available because not identical or similar to prior registered marks. A trademark registration is valid for 10 years from the date of application and is renewable without limitation. Marks that are descriptive or misleading as to the nature of the goods they are intended to be used for are not registrable. The same applies to marks that are "common to the trade" or that are in contrary to public policy or morality.

Similar to product categories in regulatory legislation, the trademark law arranges products and services in classes. According to the Nice Classification,[66] they are arranged in 45 classes, of which the first 34 contain goods and the remaining 11 classes contain services. Trademarks must be designated for specific goods or/and services. In other words, the scope of a trademark registration is determined by the goods or services the trade mark is registered for. The trademark does not necessarily have to reflect the claimed class, but it must not be misleading or descriptive as to the goods it is intended to be used for. Hence, the product name

---

[66]http://www.wipo.int/classifications/nice/en/.

Table 50.1 Same terms, different definitions in trademark law, regulatory and marketing

	Trademark law	Regulatory	Marketing
Linguistics	Connotations	Acceptable and appropriate	Brand strategy
Classification	Nice or nat. classification	HS	Possibly expensive
Availability	Registrable, useable	Risk of confusion	Image
Distinctiveness	Inherently registrable	Not misleading as to the product category	Outstanding

"water" cannot be registered for water, as such trademark would give its owner the right to exclusive use of the word water for water, but it might also be difficult to register it for lemon juice, which would be misleading. However, it would not be a problem to register it for example for a computer, or a chair.

With respect to product names, trademark law and regulatory legislation often overlap. A product name, for example "DToX", implies the health claim of being able to "detoxify". According to regulatory legislation, such a product name may only be used for quasi drugs or health care products, but not for food or beverages. According to trademark law, "DToX" is a coined word and its mentioned connotation is not necessarily strong enough to deem it to be descriptive of misleading when used for food or beverages. Registering the product name "DToX" as trademark in class 29 or 30 (food and beverage products) would therefore be possible, but it would make little sense, because according to regulatory legislation the names could not be used for food and beverage products.

The links between trademark law and regulatory legislation become even more complex when a reference to a specific place of manufacturing is embedded in the product name.[67] The place of manufacturing indicates whether certain specific regulations apply to the product (BSE, CITES). Wrong or misleading geographical indications as part of the product name may lead to the non-compliance of the product with regulatory legislation.[68] The product name "Norwegian King Salmon Fish Oil" would be misleading for "fish oil", but it would be registrable as trademark for "fish oil from Norway".[69] When the product is manufactured in Malaysia, regulatory legislation requires the display "Made in Malaysia". From trademark law point of view, this display stands in clear contradiction to the registered scope of protection, which is "fish oil from Norway". The trademark might become vulnerable to cancellation suits based on non-use in Japan after 3, in Singapore after 5 years. This specific example should demonstrate how difficult it

---

[67]This may happen if the marketing department sees a sales point in such approach.

[68]In modern trade, the source of the product is often in one, but its place of manufacturing in another jurisdiction.

[69]Another problem, which is that such trademark would be descriptive, could be solved by registering the trademark along with a distinctive device. The limitation of the scope of protection of such trademark to "fish oil from Norway" would still be necessary to avoid misleadingness.

sometimes is to comply with both the demands of trademark law and regulatory legislation.

When dealing with trademarks, it is unwise to fade-out regulatory aspects or the efforts of marketing. Finding a product name that best reflects what makes the product special, thereby descriptive, maybe even exaggerating and misleading, is the goal of marketing while those in charge with regulatory compliance try to avoid them, and trademark experts just do their best to avoid conflict with prior rights of competitors. Interestingly, all three fields are making use of the same terms to express different things (see Table 50.1).

The following overview gives insight into the meaning of the used terms depending on the context.

Linguistics
– Trademark Law: Does the mark have any negative connotations? Is it unique and memorable? Is it relevant for the target consumer?
– Regulatory: Does the mark have any connotations that are misleading as to the products for which it is going to be used? Can these connotations lead to a rejection of the product name by the authorities?
– Marketing: Does the mark have connotations that do not fit to the marketing strategy of the client? Is the mark future-proof?

Classification
– Trademark Law: Which is the correct Nice class? Do the relevant goods/services have any similarity to such in other classes? Which is the acceptable wording of the specification?
– Regulatory: Which HS refers to the goods of the client? According to the design, formulation and manufacturing of the products, which categories do they belong to? According to the applicable regulations, are there any needed business licenses for the import and distribution of the products, any claimable effects and quality, any limitations to the choice of the product name, and is there a need for clinical trials and laboratory testing?
– Marketing: Can the product be sold above its actual value? What is the highest rank and quality claim that can be believably made for the product?

Availability
– Trademark Law: Evaluation of risks of (1) a rejection of the registration of the mark and of (2) oppositions against the registration of the mark based on relative grounds. Evaluation of risks resulting from the use of the mark without a registration.
– Regulatory: Is the mark already being used in the target market for a similar or related product (danger for consumer) or for a product or service the authorities will not want the product to be related to (misleading or confusing for the consumer)?
– Marketing: Is the mark already being used in the target market for a similar or related product or for a product or service the client does not want the product to

be related to? Can the product name be secured in the widest range possible as a trademark, a domain, and by copyright or patent law? Is the brand future-proof for the life of the brand?

Distinctiveness
– Trademark Law: Is the mark inherently registrable?
– Regulatory: Is the mark misleading as to any qualities, functions, capabilities or categorization of the product?
– Marketing: The best possible way to emphasize that the product is unique, outstanding, the best available, and legendary in its class.

50.4.1 Search and Appraisal of Compliance of Product Names

In most countries, the procedure to clear a food product with the regulatory authorities and the procedure to register a product name as a trademark are legally and administratively not related to each other. Traders will typically not wait with the import of a product until the completion of the registration of the respective product name as trademarks.[70] Therefore, the rejection of the registration of the product name as trademark usually happens at a very bad time, when products have already been labeled, registered, shipped imported and distributed. The distribution of a product with a product name that cannot be registered as trademark because of a conflict with earlier trademarks of competitors can result in the infringement of these prior trademark rights. The importer may be sued for injunction and damages. The importer would have to collect the distributed products, rename them and begin with the product clearance from zero. In conclusion, passing a product name through the stringent regulatory examination procedure, without securing it as trademark is imprudent. The only way to avoid the described scenario is using product names that are regulatory compliant and registrable in the first place. Product names must be created, walked through strict, selective screening procedures, tested for regulatory compliance, distinctiveness, negative connotations and availability, before the begin of import procedures. Public and commercial trademark data bases must be reviewed for potential conflicts of the chosen product name with prior trademark registrations. The searcher[71] must evaluate the chosen product name on compliance with the legislations of the target market. Registering a product name in one market before moving to the next, just to recognize that the name that passed in the first failed in the second, results in that the same product will need different names in every jurisdiction. The creation of regional or global

---

[70]The registration of a product name as trademark takes typically much longer than the clearance of a food product for import.

[71]A trademark expert with experience and language skills in the target markets.

Table 50.2 Different forms of similarities in Japanese and Chinese

Japanese	
Phonetic similarity	*Examples*
Latin : Latin	Releus : Liliasu
Latin : Japanese Kana	Flower : フラワー (furawā)
Latin : Japanese Kanji	Tōki : 東機 (tōki)
Japanese : Japanese	山陽 : 三洋 (sanyō)
Visual similarity	*Examples*
Latin : Latin	OTI : QT1
Japanese : Japanese	全力 : 金刀
Similarity in meaning and concept	*Examples*
Latin : Japanese Kana	flower : フラワー (frawaa)
Latin : Latin	petite : petty
Latin : Japanese Kanji	porcelain : 陶器 (procelain)
Japanese : Japanese	観念 : 意味 (meaning)
Chinese	
Phonetic similarity	*Examples*
Latin vs. Latin	neksas : nexas
Latin vs. Chinese pinyin	naexas : nakesasi
Chinese vs. Chinese (same intonation)	极祥 (jí xiáng) : 级详 (jí xiáng)
Chinese vs. Chinese (different intonation)	极地 (jí dì) : 际迪 (jì dí)
Latin vs. Chinese transliterations	apollo : 阿波罗 (ā bō luó)
Visual similarity	*Examples*
Latin : Chinese	naexas : naexiasi
Chinese : Chinese (same reading)	极祥 (jí xiáng) : 级详 (jí xiáng)
Chinese : Chinese (different reading)	兔七 (tù qī) : 免匕 (miǎn bǐ)
Latin : Latin	CTI : GTI
Similarity in meaning and concept	*Examples*
Latin : Chinese	deep blue : 深蓝 (deep blue)
Chinese : Chinese	玫瑰 (rose) : 玫瑰花 (rose flower)
Latin : Latin	petite : petty

brands means coordinated trademark searching in all jurisdictions simultaneously. Such multinational and global trademark searches, involve substantial costs, and are conducted as a preliminary step to trademark and product registrations.

The regulatory compliance check of the product name goes beyond a similarity and risk-of-confusion evaluation that is typical in the trademark law. It takes into account meaning, connotations and the use of the product name in context with the product representation, while keeping the product formulation, manufacturing process and design as well as the form of application in mind.

The meaning, connotation and appearance of the product name plays a crucial role for both the evaluation of the product name on regisrability as trademark and its evaluation on compliance with regulatory legislation. Inevitably, in multi-national projects, language issues have to be dealt with. The phonetic diversity of languages and the question whether alphabets, syllabaries or pictorial characters are in use has a major impact on meaning connotations and appearance of product names. Product names must be registered with local regulatory authorities in the local script; hence, they must be transliterated or translated. Transliteration is the transfer of a sound from a source language into a target language. The differences in the phonetic ranges of the source and the target languages and their overlap determine whether a word can be transliterated along with its original sound successfully. Often, the sound of the product name or trademark in the original language is "squeezed" into the phonetic limits of the target language. A transliteration from English (Latin script with about 44 sounds) into Japanese (Kana with about only 24 sounds) transforms the sound of a mark completely. The transliteration of a trademark from a Latin based source language into Japanese, Korean or Chinese is like a metamorphosis. In addition, the understanding of similarity is very different in each language. The overview in Table 50.2 shows different forms of similarities in Japanese and Chinese.

The transliteration shifts or changes completely the sound, appearance, connotations and meaning of a mark.

In conclusion, the review of product names requires expertise in many fields. Marketing, regulatory and trademark experts can only come to feasible solutions by combining their expertise.

Chapter 51
Food Sustainability in the Context of Chinese Food Regulation

Juanjuan Sun

Abstract Facing the challenges like limited natural resources, increasing population and climate changes, it is still an urgent call to produce food in the harmony with the needs of sustainability. Therefore, having a perspective of food policy and law, this chapter is dedicated to elaborate a concept of food sustainability by taking into account food related concepts, including food security, food quality and food safety. Based on this, the case of China is further provided to state how policies and laws concerning agriculture, quality labels and food safety put into practice, with a view to make sure the harmony between food supply and sustainability.

51.1 Food Sustainability

Food is important for both human and national security. For the former, it is quite literally impossible to have a life without food. As a basic human need, there is a right to adequate food. Accordingly, every man, woman and child, alone or in communities with others, should, at all times, have physical and economic access at all times to adequate food or means for its procurement.[1] For the latter, since ancient times, food gained power as a weapon has been acknowledged in which controlling food in a food-scarce society can be powerful and a key to victory in at times of war. In this sense, it is recognized in China that only when a country is basically self-sufficient in its food production, can it take the initiative in food security and grasp the overall situation for economic and social growth.[2] Additionally, it is important to note that the risk of tampering or other malicious, criminal or terrorist actions regarding food may also threaten the national security. Therefore, food defense programs are being developed in the United States and support the

---

[1] Committee on Economic, Social and Culture Rights (1999).
[2] Shanghai Daily (2013).

J. Sun (✉)
Law school of Renmin University of China, Beijing, China
e-mail: sunjuanjuan_1984@163.com

weight of the field of priority due to food law policy, with consideration of elements of food sustainability, including food security, food quality and food safety.

Facing the food shortages due to the World War II, the guarantee of an adequate food supply was at the core of the food policy and law. Yet, the evolution of food policy and law witnessed a shift from a consideration of quantity to an emphasis on quality. In this regard, the EU is a good example and has fifty years of experience in the Common Agriculture Policy (CAP)[3] and food safety regulation.[4] In 1962, CAP was launched with the primary objective of meeting food security. Under the framework of CAP and with its financial underpinnings, food supply was boosted, changing what was a food shortage into a food surplus. However, without considering the market needs, the food surplus became a new issue of its own in the 1980s due to high costs of food storage and disposition. In response to this issue, CAP reform has placed more emphasis on quality versus quantity by taking into account food diversity, environmental concerns and rural development. Meanwhile, the outbreak of Bovine Spongiform Encephalopathy (BSE, also known as mad cow disease) has brought the food safety issue into the forefront of public concern. To restore consumer confidence in the food safety guarantee, a series of law reforms has been introduced, in particular, the establishment of General Food Law.[5] In view of this, food policy and law now takes into account different food issues, including food security, food quality, and food safety.

For the world as a whole, progress has also been made in the food supply, with a near balance in the growth of population and per capita food production reached in the 1980s.[6] However, challenges arise from inequitable development between north and south and from shortsightedness in the pursuit of economic growth. Compared to the people in more developed countries, the poor people in developing countries in Asia and Africa still suffer from hunger or cannot afford to buy food. Although those countries, especially many on the African continent, make strenuous efforts to attract foreign investment in the agricultural sector, the investment in terms of access to natural resources, in particular land and water, is still at the cost of environmental deterioration as well as the small farmers' ability to self-supply.[7] Putting economic profit ahead of safety concerns, the outbreak of food safety issues such as the BSE crisis in the EU have raised consumers' interests in food safety and food quality as well.

It is important to point out the fact that it is not the lack of resources, but the way in which the food is produced that has led to such unsustainability. For example, overuse of chemicals can increase the food production quickly in the short run, yet, the pollution of both soil and water undermines long-term agricultural

---

[3]European Union, The Common Agricultural Policy (2012).

[4]European Union 50 Years of Food Safety in the European Union (2007).

[5]European Parliament and of the Council Regulation (EC) No 178/2002 of 28 January 2002 laying down the general principles and requirements of food law, establishing the European Food Safety Authority and laying down procedures in matters of food safety, Official Journal L 31, 1 February 2002.

[6]Pinstrup-Andersen (1994).

[7]Food and Agriculture Organization (2013), p. 3.

development, putting both farmers and the land being farmed under pressure, resulting from such unsustainable development. In addition, the polluted environment can introduce health-threatening substances into the food chain. As a result, the costs can include environmental decline, lower standard of living for farmers, and be a detriment to consumers' health. Therefore, there remains an urgent call to support sustainable food production. To understand how this is possible, it is important to review the definition of sustainability.

A famous Chinese leader, Deng Xiaoping, emphasised that development is a top priority, especially, economic development. However, emphasizing the economic development at the cost of the environment became the root of environmental problems, which, in turn, impeded further economic development. Additionally, economic growth was believed to solve the problem of poverty. But the reality is that the unbalanced development between regions (north and south), and between sectors (agriculture and industry) has given rise to the issue of inequitable distribution of economic benefit, in particular, the inability of the poorest people to access any share in such economic growth.

As a reaction to the challenges of ensuring fair and equitable economic development, the concept of sustainable development is raised in the report of the World Commission on Environment and Development (known as Our Common Future or the Brundtland report),[8] that encourages the concept that the exploitation of resources, the direction of investments, the orientation of technological development, and institutional change can be made consistent with future as well as present needs. This concept of sustainable development stresses the reconciliation of the needs of development with the protection of the environment since an environment without equality can still deprive poor people of their opportunity to share benefits resulting from economic growth. This reasoning requires that economic growth shall be at the same time environmentally and socially sustainable. That is to say, to put sustainable development into practice, three pillars are interdependent and mutually supportive: namely, economic growth, environmental protection and social progress.

When it comes to food, a variety of in-depth changes in the food supply are required for this practice to be effective. Food is originally a gift of nature that could be identifiable with one agricultural input.[9] Yet, while modernization and with industrialization, food becomes the outcome of food science which has undergone substantial transformation and cannot be identified with one principal agricultural input. Also, the food supply chain has been extended from a local supply to a long-distance supply, especially as global food supply is on the rise with the development of international food trade. Regrettably, the positive results of being able to transport food away from its original local source are accompanied by negative

---

[8]Brundtland Report (1987) Our common future. Report of the World Commission on Environment and Development.

[9]Bunte F, Dagevos H (eds) (2009) The food economy: global issues and challenges. Wageningen Academic Publishers, pp 48–49.

effects. Scientific and technical breakthroughs have improved food supply by revolutionizing agriculture; yet, it also degrades the environment as well as opens up a wide gap between the rich north and the poorer south. In addition, food trade can enable a great increase in the generation of wealth, but has not been able to reduce hunger and poverty while preserving the resources.[10] Therefore, in order to move food in the direction of sustainability, the purpose is to promote non-market oriented, environmental, and social and culture values.[11] To this end, the policy and law formulation in relation to food should take into account the perspective of sustainable development, making sure that not only the economic growth but also the environmental and social needs are considered by politicians or legislators. As far as food is concerned, there are three major subject matters that specifically address food: food security, food quality, and food safety.

51.1.1 Food Security

From the perspective of quantity, a healthy food supply is aimed at providing sufficient food for individual survival. However, the challenges to ensuring such an adequate food supply are multiple. For example, the adverse impact of weather as well as a chain reaction from the rising oil prices towards food prices caused the food crisis in 1973.[12] More recently, factors including water scarcity, climate change, and rising food prices continue negatively affect people's access to food, especially the most poor and vulnerable people. As a result, a concept of food security has come into being, defined roughly as: all people, at all times, have physical, social, and economic access to sufficient, safe, and nutritious food that meets their dietary needs and food preferences for an active and healthy life.[13] As far as the evolution of this definition is concerned, for one thing, it has extended the requirements specifically addressing food from sufficiency alone to safety and nutrition. That is to say, people should be secured not only from hunger but also from food contamination and malnutrition. It also includes values such as fairness and dignity in the concept, while remaining in harmony with the needs of sustainability. For instance, food security does not exist if vulnerable people have no access to food or if people cannot have food in a way that is culturally acceptable.

While the major function of agriculture is to enhance food security, it seems that local agriculture is becoming less important. In fact, food imports have become an alternative to local food supply; with nearly 70–80 % of food sold to consumers a produced or processed product.[14] Yet, this reliance on food imports is not necessarily

---

[10]Collart Dutilleul (2012).
[11]Collart Dutilleul (2009).
[12]Timmer (2010), pp. 1–11.
[13]Food and Agriculture Organization (2003).
[14]Valceschini (2013), p. 49.

less important. For one thing, given the volatility of food prices at the international level, true national security is still heavily reliant on self-supply. For another thing, the role of agriculture is always essential to provide product as raw material for the food industry. Therefore, it is always a necessity to consider the function of agriculture in the guarantee of food security. To fully realize the implication of food security, sustainable agriculture and rural development (SARD) is designed to ensure food security by taking into account the environmental protection, natural resource management, alleviation of poverty, and to supply nutritional food to vulnerable groups.[15] For this purpose, the agricultural policy and law should integrate the objective of food security with sustainable development.

51.1.2 Food Quality

From the perspective of quality, food quality has been a long-term consideration in the food supply. Although it is a subject of debate as to what food quality is, one common agreement is that food quality is characterized by multidimensional attributes. For example, quality attributes can be classified as safety attributes, nutrition attributes, value attributes and package attributes (Hooker and Caswell 1996). Hygiene, nutrition, enjoyment and use have been proposed as the four key components in defining quality (Chiaradia Bousquet 1995). Indeed, with endless food safety issues, the attribute of food safety has been separated from quality and becomes an independent target for regulation.[16]

As far as quality is concerned, there is a quality specification that can distinguish food in terms of features such as organoleptic qualities, taste, material, source of origin, etc. There is also a classification in the direction of upgrading, taking the quality as a reflection of value and the ability of a product to satisfy the users.[17] Therefore, the quality attributes can be used to differentiate a food product and meet consumers' needs. Consumers are increasingly willing to pay more for a value-added food (Caswell and Siny 2007). Meanwhile, a differentiated quality attribute can be a comparative advantage for food in both the competitive domestic and international markets. Therefore, when safety becomes a legally provided requirement for entering into market, a variety of other attributes can differentiate food and may influence consumer choice. These attributes include origin, nutrition, environmental consideration, animal protection, etc. While information, including a quality assurance label or health claim can be helpful to indicate quality attributes, both public and private parties are engaged to ensure that consumers are not misled.

Different from food safety standards, the public intervention in the standardization of other food quality attributes could impede the diversity in the provision of

---

[15]United Nations (1992).
[16]Sun (2013).
[17]M'hatef (2007).

value-added food. This is why the early composition requirement (also known as recipe legislations) in a given food product was not suited to quick development in the chemical industry. Although private food operators have the freedom to promote value-added food as long as they respect food safety standards, it may not be easy to evaluate certain quality attributes for consumers. The quality attributes in forms of credence regarding production methods, environmental and social orientation cannot be verified by the consumer due to lack of technical expertise or practical possibilities. Therefore, they suggested that the information communication of these quality attributes usually is in the form of labels.[18] From public policy perspective, public intervention in support of quality assurance labels could help private food operators to establish a niche market on the basis of a given quality attribute.[19] As they asserted, to transfer the societal concerns of environment and animal welfare into obligatory food standards can also be a market advantage as long as they are widely accepted by both producers and consumers.[20] In the interest of consumers, the public intervention can provide certification and allow the use of a corresponding quality assurance label, such as the labels for geographic indications, traditional specialties or organic food. In addition, it can also define a condition under which a quality related term can be used; for example, what "fat free" actually means as a nutrition claim.

In general, private parties are more sensitive in their desire and ability to meet the consumer's needs for safer and higher quality food products by applying private food standards. However, the private food standards may be inconsistent and there is currently no mechanism to ensure its compliance, especially for long-distance cooperation. Significantly third-party certification has provided an objective or impartial technical tool to guarantee its conformity.[21] Similarly, the skepticism on the reliability of private certification has further required accreditation, (independent evaluation of conformity against recognized standards to ensure their impartiality and competence). The imposition of private food standards and requirements on the third-party certification as well as accreditation has led to a kind of private control in the form of Tripartite Standards Regime.[22]

51.1.3 Food Safety

For the sake of public health, only the food presenting no health risk for human consumption can be placed on the market. In this sense, there exists a "quality threshold" or "objective quality" from the perspective of food safety concern. For

---

[18] Fernqvist and Ekelund (2014), pp. 340–353.
[19] Chrysochou et al. (2012), pp. 156–162.
[20] European Commission (2008).
[21] Hatanaka et al. (2005), p. 355.
[22] Busch (2011), p. 59.

the former, the quality threshold is defined by sanitary requirements, ensuring the cleanness of food. Based on this, there are quality specifications (Food and Agriculture Organization 1995). For the latter, the objective quality is defined as the technical and measurable criteria to meet sanitary requirements.[23] By contrast, the subjective quality refers to the different perception of quality depending on the consumers' needs. As mentioned earlier, such safety attributes have been separated from quality and become an independently regulated target.

Since food safety can be measured by technical standards, a great number of food safety standards have been developed. For example, the Codex Alimentarius Commission (CAC) was created in 1963 by the Food and Agriculture Organization (FAO) and World Health Organization (WHO) with the aim of protecting consumer health and ensuring fair trade in the food domain. Since then, it has become a significant international reference point for developing food-associated standards. In addition, the Agreement on the Application of Sanitary and Phytosanitary Measures (SPS Agreement) has provided the international standards, guidelines or recommendations established by the CAC as legal points of reference. While the SPS Agreement is designed to promote international harmonized food safety standards, there remains a dispute concerning what constitutes food safety. Why does this argument still emerge from time to time?

To answer this question, it is important to note the so-called risk society.[24] Life is full of risks, and risk is not a new invention. Nevertheless, modernity, characterized by the striking advancement of science and technology, has redefined the risk. Accordingly, having their basis in the modernization, risks come as by-products of science and technology. The costs present in the form of the technological risk, the probabilities of physical, chemical, or biological harm due to a given technological or other processes. Different from those in the past, modern technological risks have many characteristics, including:

Threat From risk-taking to risk-avoiding, early risks carry a note of bravery and adventure and seem to be seen as an attempt to a new discovery or development. On the contrary, technological risks are those having something to do with the destruction that has not yet happened, but is threatening, which means they should be managed in an anticipatory way. These widespread technology-originated risks can put public health in danger and thereby people would do everything possible to avoid them.[25]

Ubiquity From personal risk to global danger, the spread of technological risks is no longer limited to their place of origin. For example, travelling along the food supply chain from local to nationwide, and then to a global level, a food safety issue in one location can endanger all forms of life on the planet. In addition to the spread in terms of physical space, there is a generational concern. Not only does the current

---

[23]M'hatef (2007).
[24]Beck (1992).
[25]Steele (2004), p. 29.

generation, but also future generations face the risk of long-term incubation of a given toxin.

"democracy" From poor people to rich people, democracy means that everyone is equally confronted with technological risks. Certainly, the rich man can have more opportunities and better capabilities to avoid technological risks by living in a more sustainable environment or eating healthier food, but he still ultimately will suffer from if everything turns into a hazard.

Circularity Described as the boomerang effect, circularity refers to the result of a person engaging in some risk-taking activities in order to profit regardless of harm to others; he will, sooner or later, become a victim as well. As with the example of the overuse of chemicals, eventually the deterioration of the environment will circle around and undermine production.

Therefore, safety has become a counter-concept to the risk.[26] In reality, there is no risk-free behavior and absolute safety cannot be achieved. When it comes to food, the human diet is, in fact, loaded with toxins of all kinds in the foods we eat regularly, including mushrooms and peanut butter.[27] This means it is hard to find a diet that would support life and at the same time impose no risk on the consumers. Considering this concept, safe food usually means food that is safe enough. What's more, since risk can be managed but not eradicated, concerns in the face of risk uncertainty does not consider whether or not a risk exists, but the adverse consequences inevitable risks bring about. Therefore, from a risk perspective, the significance of safety lies in the measurement of the acceptable level of risk (adverse consequence). Then, the question arises as to what can be used to decide such acceptability?

With the introduction of risk assessment, the management of food-related risk is established on a scientific basis. For example, quantitative standards can be formulated to prevent physical, chemical, biological and nutritional hazards. As a result, a scientific judgment of food safety can be defined as the state of being certain that harmful agents under defined conditions will not cause adverse effects. However, such certainty in decision-making is only based on the sound, naturally scientific truth, as a scientific principle in the SPS Agreement against the failure of risk perception in the legal reasoning.[28] As a matter of fact, the role of social science, especially, the research on public perception, can also contribute to a broad acceptability since acceptability not only involves the science but also perceptions, opinions, and values.[29] Accordingly, both natural science and social science play essential roles in the risk regulation: scientific rationality without social rationality remains empty, but social rationality without scientific rationality remains blind.[30]

---

[26]Luhmann (2008), p. 19.
[27]Ruckelshaus (1984), pp. 161–162.
[28]Alemanno (2012), p. 2.
[29]Nestle (2003a), p. 16.
[30]Beck (1992).

Therefore, to consider food safety as scientific judgment, the comprehensiveness of scientific certainty contains both scientific rationale and social rationale.

Equally important regarding risk, all risks are uncertain to some extent because we can never know the future with complete certainty.[31] Current risks spread across national boundaries and over times, threatening people thousands of miles away and even future generation without compensating many of those affected by them. Despite the contribution of science to risk prevention, there may be still scientific uncertainty in identifying and characterizing risks and hazards. Such uncertainty is influenced by both insufficient information and scientific controversy. Therefore, precaution can be seen as a proactive way to manage globalization and a so-called risk society, especially that of potentially irreversible harm.[32] What's more, when the trust in the government and industry has declined due to the malpractice in the food domain caused by prioritizing economic growth over safety, the precautionary principle can be invoked to replace a non-democratic technocracy with a more humanistic and community-orientated approach to decision-making.[33] In this context, originating in the field of environmental protection but now widely applied in the regulation of risk, the precautionary principle has been an important but disputed instrument in the risk regulation. In spite of the existing arguments, the precautionary principle has been extended into food safety regulation in Europe, in order to prevent decision makers from using "lack of full scientific certainty" as a reason for not taking precautionary action where there are threats of health harm.

What's more, food safety is not only a scientific judgment but also a value judgment. The purpose of the law is to accomplish certain social ends based on a value judgment, in other words, on an appreciation of values.[34] Legal norms, therefore, are provided to attain the chosen goal. The subjectivity of a value may be decided by a particular historical period or may depend on a particular society while their influence in the legal decision-making is inevitable.[35] What's more, when it comes to prioritizing which value is the highest goal in the legal order—freedom or security for example—the conflicts of values are unavoidable. Therefore, instead of a right-or-wrong choice among the conflicting values, the decision-maker is supposed to choose a preferred value.[36] About food, research demonstrates that values in relation to food are multiple, such as safety, nutrition, fairness, tradition, economic interest, etc.[37] Among them, the safety on average is the most important. Although each state may have its own social reality and different food preference, the prioritization of safety as a value in relation to food is a common priority. However, such a realization is only resulted from the lessons that point out

---

[31]Wiener (2002), p. 319.
[32]Feintuck (2005), pp. 371–372.
[33]Goldstein and Carruth (2003), p. 249.
[34]Bodenheimer (1970), p. 349.
[35]Freeman (2001), pp. 50–51.
[36]Ruckelshaus (1984), pp. 161–162.
[37]Lusk and Briggeman (2009), p. 191.

the ad-hoc based and economic oriented food safety legislation are failed to address the protection of public health.[38]

More recently, the tendency in shaping the legislative framework concerning food safety is to provide a basic food law, which lays down a legislative foundation for the provision of safe food, in order to prioritize the protection of public health.[39] Based on such basic food law, the universal subject matter in the guarantee of food safety can be generalized in three aspects: including substance, process and information.

Substance Generally speaking, food is composed of substances. When it comes to substance, Paracelsus says that poison is in everything, and it is the dosage that makes it either a poison or a remedy. Therefore, when the recipe legislation in the chemical composition is inconsistent with the chemical development, the general rule has been introduced that the safety of a given substance can only proved by scientific assessment: can it be used in food or during food production? Based on this, the different legal instruments have been applied to the substances in line with their intended use. For example, substances used in a food can be categorized to different groups, such as food additive, and pesticide. Accordingly, the use of food additives is authorized and must follow the quantity limitations and use of condition while pesticide use must comply with the positive list and maximum residue limit.

Process Initially, safety considerations from the perspective of process focused on the sanitary conditions of the production environment. With the improvement of sanitary conditions in modernized food technology, food contamination, in particular biological contamination has become a new concern during the food production. What's more, as far as the food supply chain is concerned, heavy metal environmental pollution and pesticide residues also contribute to the food safety issue. Therefore, risk-targeted and scientific-based systems have been introduced to control the production process. Currently, a Hazard Analysis and Critical Control Point (HACCP) system has been developed to ensure food safety during the food production. In addition, the advancement of food science also contributes to inventing new methods of food production. As a result, it is arguable that food may be substantially different due to the new scientifically invented production methods. This is a concept of substantial equivalence, which stresses that assessment of a novel food, in particular one that is genetically modified, should demonstrate that the food is as safe as its traditional counterpart (Organization for Economic Cooperation and Development 2000). Food products can be different depending on what they contain, but also depending on where they are produced or how they are produced. For example, it is the way of production that differentiates genetically modified crops from the organic crops.

---

[38]Vos (2000), pp. 227–255.
[39]Vapnek and Spreij (2005).

Information Food information may refer to the information concerning a food and make available to the final consumer by means of a label, other accompanying material, or any other means including modern technology tools or verbal communication. In comparison to the rigid recipe legislation, a well-developed and clear system of labelling, presentation and advertising can help to make sure the consumers know their food product well (European Commission 1985). As far as food safety is concerned, the necessity of mandatory prohibition or provision of food information is threefold. First, the prohibition of misbranded food can protect consumers from mislabelled food. Second, the provision of nutritional information and harmonization of terminology in nutritional and health claims, can help to prevent the nutrition-related food-borne diseases, especially the chronic diseases. Third, the saying "one man's meat is another man's poison" describes vividly a fact that for certain groups of people, they can have an abnormal physiological response to a particular food. For instance, food allergies can be life threatening and have a larger impact on certain population groups.

Although it is a trend to harmonize food regulation at the international level, using tools such as an official standardization concerning on food safety under the framework of CAC, private standardization regarding food safety and quality initiated by food operators, the policy and law for food regulation still differs from nation to nation given the fact that each country has its own priority regarding food and its own legal and institutional system to address it.

51.2 Chinese Legal Reform in the Realization of Food Sustainability

It is difficult to talk about China without reference to food. First, due to the pressure of the world's largest population, food security is always a concern of priority. Although there has been great achievement, there are still many issues about food supplies. The challenges in this regard include an on-going increase in population, the reduction of arable land due to urbanization, and climate changes (Li, 2013; Guo and Brandt 1998). Second, well-publicized food safety issues, on the one hand, have created both external and internal pressures to modernize the food safety regulation while, on the other hand, raising consumers' interests in food quality. As a result, food policies and laws are established or updated to ensure food security, food safety and food quality, respectively.

Although the political and legal systems are beyond the scope of this chapter, it is an essential backdrop for food policy and law. Though the People's Republic of China has been established since 1949, the current legal system was developed only 30 years after the implementation of reform and more open policies since 1978. According to the Law on Legislation of the People's Republic of China, the statutes take many forms including the laws from legislative bodies and regulations or rules from the executive body.

Law (Fa lü, 法律) Exercising the legislative power of the State, the National People's Congress enacts and amends basic laws (Ji Ben Fa lü, 基本法律) governing criminal offences, civil affairs, the state organs and other matters while the Standing Committee of the National People's Congress enacts and amends laws other than the ones to be enacted by the National People's Congress, and when the National People's Congress is not in session, partially supplements and amends laws enacted by the National People's Congress, but not in contradiction to the basic principles of such laws.

Administrative regulation (Xing Zheng Fa Gui, 行政法规). In accordance with the Constitution and the laws, the State Council formulates administrative regulation. Provided that such regulations do not contradict the Constitution, the laws and the administrative regulations, the people's congresses or their standing committees of the provinces, autonomous regions and municipalities directly under the Central Government may, in light of the specific conditions and actual needs of their respective administrative areas, formulate local regulations (Di Fang Xing Fa Gui, 地方性法规). Similarly, there are also Autonomous Regulations (Zi Zhi Tiao Li, 自治条例) and separate regulation (Dan Xing Tiao Li, 单行条例) in the national autonomous areas.

Administrative rule (Bu Men Gui Zhang, 部门规章). In accordance with the laws as well as the administrative regulations, decisions and orders of the State Council and within the limits of their power, the ministries and commissions of the State Council, the People's Bank of China, the State Audit Administration as well as the other organs endowed with administrative functions directly under the State Council may formulate rules.

Against this context, it is interesting to note that the strong involvement of government in the law making by means of administrative rules. For example, due to the involvement of more than five departments in the food safety regulation, one of long-existing problems in the official control is the fragmentation in both law making and law enforcement. There are three reasons for such a powerful government. First, China is a highly centralized country in which there are no checks or balances between the legislative power and the executive power. Since the government can also propose a bill, they can more easily to turn their own will rather than the public interests to legally protected interests. Second, the long-term legal tradition still exerts its influence in the legal practice, which may conflict with the current legal principles, as the struggle between rule of law and rule of man. As a result, it seems that although a western legal system has been planted in China, however, the practice is often still guided by the rules we have practiced for over two thousand years.[40] Third, even the critics have been raised on the old way of regulation; the public still put their confidence in the government to enforce regulations rather than on the private parties' capacities in self-regulation because

---

[40]Jiao (2007), pp. 88–89.

as an important characteristic of China's political culture is that the public has regarded the government as a parent to the public. As a result, it is difficult, if not impossible for the public to judge or even raise questions concerning government decisions.

51.2.1 Agriculture Policy and Law

In 1995, Lester raised the question: "Who will feed China? (Lester 1995)" Food insecurity in China is potentially a great threat at both the national and international level due to population pressures. The best argument regarding this is the fact that the supply and demand of food in China has been balanced while there has been a surplus each harvest year since the 1990s.[41]

In analysis of the achievements of the last 30 years researchers concluded that the key, but most difficult point, lies in the modernization of agriculture.[42] As a starting point, the land reform in the application of the household contract responsibility system since 1982 has contributed considerably to boosting the food productivity. In China, land in both rural and suburban areas is mainly owned by collectives. With the household contract responsibility system, individual farmers on behalf of family members can rent the land from collective organization by signing a contract. In this case, they enjoy right of use of the land, which allows the self-determination in the management of contracted land. The right to use land for long periods of time can encourage the use of land-saving investments.[43] Therefore, to stabilize the contractual relationship, the Law on Land Contract in Rural Areas has provided legal protection for land tenure, such as a 30-years period for arable land.

When it comes to the major function of agriculture, namely the food production, officials from the Ministry of Agriculture (MoA) stress that the bottom line for food security is always the self-sufficiency. To ensure this, the development of agriculture will always rest on political will. Each year, the No. 1 Central Document of the central government puts emphasis on the modernization of agriculture since 2004. This means a great number of investments would be granted to this domain. In addition to improving the agricultural input, including through finance and technology, this kind of policy has also paid attention to the environmental issue since the agricultural development would fail to enhance productivity if there is a decline of arable land and other natural resources in the long term. As stressed by the latest No. 1 Central Document in 2014, the agricultural development shall be advanced in an environmentally friendly way, taking into account the protection of arable land, water resources and agro-ecology.

---

[41]China (2008).
[42]Gong (2008), p. 230.
[43]Guo and Brandt (1998), p. 69.

Agriculture is not just about food; it also concerns the rural environment and the people who live in them. In this aspect, it is interesting to mention a special phenomenon of "abandoned arable land" in China. Whereas the reduction of arable land is a threat to ensuring food security, there is still much arable land abandoned in the rural areas since many young people prefer working and living in cities. Consequently, what is lacking in the rural areas is a workforce, rather than arable land. At this point, an "agriculture/countryside/farmers" policy has been developed to address these interrelated issues in a systematic and comprehensive way. For these three pillars, the major concerns are the low efficiency of agriculture; the slow development of the countryside; and the poor incomes of the farmers. With regard to agriculture, the high-speed economic growth in China has resulted in low efficiency in agriculture, even though the objective of agricultural modernization was raised in the 1950s. Following industrialization, greater attention has been paid to the urbanization with more significant investments in education, social security and public service than in the rural area. Notably, to control the population in an urban area, a family register system has been introduced to avoid the circulation of people from the countryside to cities. As a result, the farmers' standard of living is relatively poor in comparison to the citizens of urban areas.

To keep development balanced between agriculture and industry, the rural and urban areas, farmers and citizens, the "agriculture/countryside/farmer" policy has integrated the industrialization of agriculture, rural development and an increasing standard of living for a farmer. In addition, advancements have also been made in the direction of sustainable development. As an instrument in implementing this policy, building a new socialist countryside is designed to enhance production, improve the standard of living of farmers, promote a polite way of life in the countryside, keep the cleanliness of the rural environment, and manage the countryside in a democratic way.

In addition, agriculture, countryside, and farmers also are significant targets for the agriculture law. According to the amended Agriculture Law in 2002, one of newly added provision to move agriculture development toward sustainability way by protecting the natural resources such as soil and water as well as enhancing farmers' economic and social interests by protecting their right to education and social security. As far as food security is concerned, this law has stressed the realization of food security in many aspects, including, the maintenance of arable land; reinforcement of agricultural investments in major grain production areas supportive of stable purchasing price in the case of low market price; creation and maintenance of an alert system in relation to the grain storage, and; creation of a risk fund to guarantee food security and to fight against food waste.

Additionally, a special law concerning on food security, the Grain Law, is under the deliberation, in order to provide a legal basis for the production, circulation, and consumption of grain. Interestingly, this drafted law has provided that no unit or person can apply the transgenic technology to this staple food without approval. During public consultation, critics have been raised the concern that such a provision is still vague in the government's attitude toward genetically modified crops. When reading the text literally, the so-called "without approval" can mean that the application of transgenic technology to rice production can be possible as long as the approval is granted. In this case, it is still uncertain if the final attitude of the government in regards to genetically modified food in this concern is "for" or "against."

51.2.2 Certification and Quality Assurance Labels

As mentioned above, the environmental consideration is to be integrated into food production, as the new orientation stated in the No. 1 Central Document of 2014. As a result, it is said that both quantity and quality have been prioritized in the national food policy. When it comes to the food quality, especially, agricultural food, a vertical classification of quality attributes regarding substance use has been provided, including the hazard-free agricultural food, green food and organic food. The difference between them is the degree of strict oversight of the use of chemical substances during the food production. In addition, to ensure food is free from hazardous substances, this kind of quality promotion is also contributing to sustainable agriculture by protecting the environment. Also in the interest of humanity and sustainability, a ChinaGAP is developed specifically in the model of GlobalGAP, with the purpose of promoting good agriculture practices.

Protection of geographical indication was introduced into China in the 1980s, with the purpose of satisfying the obligation under the Paris Convention on the Protection of Industrial Property.[44] This is also an important instrument to promote food quality for China given its diversity in the natural resource and food culture. For example, a "10 plus 10" project between Europe and China is completed, in order to protect 10 famous EU food names in China. Reciprocally, 10 Chinese food names have been registered for the European market. Regrettably, the legal protection of this quality attribute is only at the level of administrative rules, such as a General rule on the protection of geographical indication issued by the General Administration of Quality Supervision, Inspection and Quarantine (AQSIQ) in 2005, or a special rule on the protection of geographical indication for agricultural food. In view of this, there have been calls for a special law to lay down the legal basis for protection of geographical indication, in order to make full use of this tool to promote Chinese food, especially on the international market.

Confronted with the high aspiration toward safer and higher quality food arising from consumers, the private efforts take many forms in order to gain and retain consumers' confidence.

Private Food Standard With their reputation, applying internationally recognized private food standards is a shortcut to assure consumers' confidence and enter into an international market for Chinese food operators. For example, the

---

[44]Huang (2008), p. 18.

certification by the British Retail Consortium (BRC) standards and the International Food Standard (IFS) can facilitate the entry into European markets while the European standard is considered as a representative of stricter standards for Chinese people.

International Cooperation For food operators, the advanced experience of management or raw material of higher quality can be obtained through the cooperation with foreigner parties. As a key target for food regulation, milk companies engage actively in international cooperation for raw milk of better quality. For example, as China's second-large diary producer, Inner Mongolia Yili Industrial Group has carried out cooperation with Dairy Farmers of America Inc. and Synutra International, Inc. has invested in France to construct a milk factory.

51.2.3 Food Safety Regulations

Food safety regulations in China started as early as 1953 with the rules on fresh food only, with an emphasis on the sanitary condition. However, the emphasis on food safety started with the revision of Food Hygiene Law in 1995. There are three primary reasons proposed for this: first, as far as China is concerned, food security has always been a priority due to the large population, in particular for a new established country after the internal war in 1940s. Second, there were few food safety issues during 1950s and 1960s because, at that time, food production factories were state-owned and controlled, and the standards were simple to follow and traditional methods used few chemical substances. Third, with the introduction of the reform and more open policies in the late 1970s, the economic development has become the priority, and the emergence of food safety issues relates to a background of profound changes resulting from economic development.[45]

While economic development did bring about the progress in the food domain, it also gave rise to many food safety issues. During the last 30 years, there are two breakthroughs with great influence on the food domain: one is the market-orientated economic development after 1992, and the other is the entry into World Trade Organization (WTO) in 2001, after which food trade developed quickly. For the former, the controlling power of government has been loosened due to the transition to a market economy. As a result, counterfeit products are manufactured and distributed through a largely unregulated and chaotic distribution system.[46] For the latter, China should follow the international rules to regulate its food safety as required by the WTO rules, such as laying down the scientific basis for food safety regulation. In addition, food safety issues around "Made in China"

---

[45]Bian (2006), p. 169.
[46]Thompson and Ying (2007), p. 4.

also imposed high pressure on China to handle its food safety issues, such as the melamine-tainted pet food issue in the United States in 2007.

Against this context, the number of food related laws, regulation or rules (including more than 40 repealed ones) has reached 832 since December of 1978. Despite those regulatory efforts, a systematic legal framework in the food domain has not been established.[47] When the melamine scandal occurred, far-reaching influence in the politic and economy has brought about reform in the food safety legislation. In the shadow of melamine-tainted milk scandal, the first Chinese Food Safety Law came into effect in 1st June of 2009, in order to protect the public health and life through the food safety guarantee.

The newly established basic law for food safety has provided a science based preventive system by means of risk assessment. To carry out risk assessment, the responsible competent authority-Ministry of Health (MoH) has set up the first Expert Committee on Risk Assessment, consisting of 42 experts from the domains of medicine, agriculture, food, nutrition, etc. With this practice, the organizational arrangement for risk assessment has been further developed by the establishment of the China National Center for Food Safety Risk Assessment. As a newly established public authority subordinated to MoH, the Center plays a central role in organizing the risk monitoring, risk assessment and risk communication in the food domain other than agricultural food. The cooperation between the China National Center for Food Safety Risk Assessment and the Expert Committee on Risk Assessment is that when the former prepares the assessment result, it is the latter that evaluates and makes the final decision to formulate a scientific opinion.

As far as agricultural food is concerned, the risk management of agricultural food is separated from other foods and the legal basis is laid down by the Quality and Safety Law on Agricultural products, so is the risk assessment in this field. Although the risk management of other foods follows the requirements established by Food Safety Law, the function for official control is also divided among several departments. As a result, the official control for food safety is characterized by a sector-based regulatory system. Historically, there are major departments involved in food safety regulation. Along the food supply chain, MoA is responsible for the primary sector, AQSIQ for the production sector as well as exportation and importation, and the State Administration for Industry and Commerce (SAIC) for the circulation sector, and the State Food and Drug Administration (SFDA) for the food catering service.

MoA With the engagement of MoA as the official control for agricultural food, it can be said that the double-system separating the agricultural food and other food products exists in both risk assessment and risk management, which allows the MoA to formulate of rules, and agricultural food standards as well as to enforce these rules.

---

[47]Gangjian (2008), pp. 54–55.

AQSIQ AQSIQ takes the responsibility for supervision and management of food production on the one side and exported and imported food on the other side. For this purpose, it is interesting to note the organizational characteristic of this department. With the so-called "One Father Two Sons" characteristic, it has set up the Bureaus of Quality and Technical Supervision (QTS) at the provincial level for the function of quality and technical supervision. In addition, it also establishes in total of 35 Entry-Exit Inspection and Quarantine Bureaus (CIQ) in China's 31 provinces, in order to perform the function of entry-exit inspection and quarantine. Against this context, QTS is responsible for the domestic food while CIQ is responsible for imported and exported food.

SAIC SAIC is responsible for the food circulation. Initiatively, it was the QTS that took charge of the inspection for processed food in accordance with the Product Quality Law. Yet, after the reorganization of the SAIC in 2001, it has obtained the authority for inspection of the product quality in the sector of food circulation.

SFDA The purpose of establishing the SFDA in 2003 was to coordinate the food safety regulation among the multiple departments. Due to the failure of food safety regulation, SFDA has been integrated into MoH during the reform of administrative system in 2008, taking responsibility for food catering service.

Due to the involvement of multiple agencies, the overlaps and gaps in the official control were contributing considerably to the failure in food safety guarantee. To try to solve this issue, the Food Safety Law has further clarified the function and responsibility among the involved departments.

To this purpose, the Food Safety Law has emphasized the coordination and cooperation through strengthening the regulatory power of MoH and organizing a committee for food safety at State Council level. For the former, the MoH has the responsibility for the comprehensive coordination of food safety matters, food safety risk assessment, the formulation of food safety standards, the release of food safety information, the accreditation of food inspection and testing agencies and the formulation of inspection and testing standards. For the latter, a Food Safety Committee was set up in 2010. Represented by a State Office on Food Safety for daily work, its major functions include organization of investigation into serious food safety events, provision advice for food policy, promotion of a joint-mechanism in official control, and review of the effectiveness of implementation of both food safety legislations and policies, etc. Since the co-existence of the MoA and the Food Safety Committee can give rise to the duplication in the work of coordination and cooperation, reorganization between them has been further addressed with the transition of three functions of MoH into the State Office on Food Safety, including the comprehensive cooperation of food safety, origination of investigation during the food safety incidences and uniform distribution of food safety information.

As mentioned earlier, the SFDA has been restructured for several times in order to coordinate and co-operate the official controls against the sector-based multiple agency system. Implied by its name, the initiative of this department is to copy the

American Food and Drug Administration (FDA), with the purpose of reinforcing the food and drug regulation. However, this vice-ministry-level organization failed to meet its mission since it lacked adequate power to command the ministry-level organizations to follow its suggestions in reality (like the MoA or the MoH for example). After integrating into the MoH as an internal body for official control at the stage of catering, the new round of administrative system reform in 2013 has tried again to reorganize this "China's FDA" as a ministry-level department, in order to integrate the official control including:

- The work of coordination owned by the State Office on Food Safety;
- The official control of food safety at the stage of production owned by the AQSIQ;
- The official control of food safety at the stage of circulation owned by the SAIC;
- The official control of food safety at the stage of catering owned by the original SFDA.

With the above-mentioned arrangement, this newly organized China's FDA is responsible for the unified official control for food safety, covering the food supply chain from production to circulation, and to the final stage of consumption. In addition, it is also responsible for the cooperation and coordination. After being responsible for overall food safety regulation, China's FDA has placed emphasis on the self-regulation in food safety assurance. To this purpose, the way of assuming primary responsibility on the part of individual food operators has been elaborated on in that a food enterprise assigns a director specifically responsible for food safety matters over the whole process from buying raw material from upper suppliers to release the product after testing. As a pilot project, it is supposed to solve the dilemma of "everyone's responsibility is no one's responsibility".

As a reminder, it is important to note that this so-called one-department-led system is not absolute since other departments also are involved, including the MoA being responsible for the official control of agricultural food, the MoH for the risk assessment and food safety standard settings, and the AQSIQ for the official control of imported and exported food. Last but not least, it is important to address that the above-mentioned Food Safety Law has been revised and put into effect since 1st October 2015, in order to put the emphasis on the prevention first, risks management, farm-to-fork control, and co-governance by involving all the stakeholders and the public in the food safety guarantee. To this purpose, the China'FDA led regulatory system has been strengthened by its empowerment to take part in the standard-setting, to carry out the official control of agro-food when it place into market, to supervise the responsibility for food safety of the thirdparty platform that engaging in providing online food trade, etc. What's more, the way of food safety regulation also becomes more and smarter with the introduction of risk raking based management, risk communication, as well as increasing application of information announcement in the form of black or red list and credit system. In addition to the update of official control, one of the most striking futures of this revised Food Safety Law is to encourage all kinds of stakeholders as well as the public to fight against food safety issues by means of co-governance, such as the flexibility in the local official control which allows the local counterparts taking into account the specialty in a given region, the monetary incentive for whistle-blower who turns in the food crime, the involvement of experts as well as the mass media in the risk communication. In April and May 2016, a nationwide evaluation of law enforcement has been carried out in the field of food safety regulation, which has pointed out that more efforts should be addressed to the re-organization of regulation system at the local level as well as the training for capacity building at both food industry and competent authorities. In view of this, this so-called strictest Food Safety Law in the history has been placed with high expectation to improve the public confidence in the food safety regulation in China.

51.2.4 An Integrated Approach

Depending on the regulatory purpose, food legislation have a wide range of topics, including agriculture law designed to safeguard food security, legislations regarding quality indicators such as the protection of geographic indication, or food safety standards for food additives or pesticides. With the above-mentioned analytical framework, it seems that these subject matters are separated from each other. Actually, they are in fact closely linked and mutually supportive, although sometimes in conflict with each other. For instance, safety requirements are a precondition for ensuring food security and satisfying different consumers' needs from the perspective of quality. Nevertheless, safety considerations can conflict with the economic interests causing food regulations to swing like a pendulum between them. The BSE crisis is a valuable lesson.

Generally speaking, the BSE and CJD crisis chronology goes as follows. The first diagnosed BSE in the cattle was found in 1986 in the UK, and was considered a "prion" disease causing the fatal brain damage, without knowing whether it was transmissible or not at that time. However, it was confirmed as a zoonosis passing from animals to humans in 1988. To protect the human health, protective measures such as monitoring slaughterhouses and banning milk production from the suspect cows were adopted by the UK government. However, while the government believed that there was no risk for human health by claiming the safety of British beef in 1990, the outbreak of the BSE cases reached a peak in 1992 at which time it was discovered that three cows in every 1000 in Britain had the disease. Then, the first victim of a new variant of Creutzfeld Jakobs Disease (vCJD) was discovered in 1995 while the government scientists rejected a connection between vCJD and BSE. In the following year, however, the link between vCJD and BSE was confirmed and tightened measures were adopted through a 30-month slaughter scheme, which meant all cows over the age of 30 months at the time of slaughter were prohibited from entering into the human food or animal feed chain.

The European Commission adopted a set of measures more aggressive than those in the UK against BSE. These included the ban of sending live cattle born before 18 July 1988 or born to females in which BSE was suspected or had been officially confirmed in an effort to prevent BSE from spreading to other Member States for the sake of animal health in 1989 (Commission Decision 1989), the ban of the tissues and organs including brains, spinal cord, thymus, tonsils, spleen, and intestines from bovine animals aged more than six months at slaughter in 1990, the ban of feeding of protein derived from mammalian issues to ruminant species given such protein was the only significant potential source of spongiform encephalopathy agents, the ban of the bovine animals and bovine products from the UK to other Member States or their countries considering the possibility of transmission of BSE to human, After the UK took these strict official controls against the BSE crisis, such as a bovine passport system, the agreement banned these exports of certain bovine products.

To reveal the events and decision that led to the spread of vCJD and BSE, the BSE inquiry was carried out at both the UK and EU levels.

As far as Britain's BSE inquiry is concerned, there are two kinds of hazards involved in the BSE crisis, one is the known hazard to animal health, and the other is an unknown hazard to human health. Although the protective measures had been taken for the sake of animal health, (Phillips 2001), found that the measures for human health were neither timely nor adequate since they were affected by the belief of many prior to early 1996 that BSE was not a potential threat to human life. Notably, being afraid of the economic losses and the public's overreaction due to the BSE disease, the UK government had repeatedly gave assurance that the British beef was safe rather than giving the public factual information, in particular the uncertainty around the BSE crisis with its transmission. As a result, critics claimed that the government misled the public.[48]

When it comes to the BSE inquiry at the EU level (European Parliament 1997), the Council should bear the responsibility on the ground that it took no effective steps to monitor the enforcement of the protective measures by the Commission and put the economic interests of the meat industry over health protection. Similarly, the Commission was responsible since it gave the priority to the management of the market rather than the possible human health risks in the light of scientific uncertainties concerning the BSE, and thus took no preventive action. Ironically, no checks of the British inspections at the EU level were carried out by the Commission, and when Member States such as France took safeguard measures against the British beef exports, however, the Commissioner opposed by threatening an initiative of infringement proceedings and even a lawsuit, on the other side.

As a valuable lesson, the BSE crisis has illustrated that public confidence in public administrations as well as in the food industry would be devastated if economic concerns override safety on matters of health. Therefore, to structure food law in an integrated way, the basic food safety law is not only a legal foundation for all food safety-related laws but also for laws regarding other matters. In this sense, such basic law plays the role of a constitution in the domain of food legislation, in order to prioritize the protection of public health.

---

[48]Phillips (2001), pp. 3–4.

In practice, the regulation of agricultural food usually is separated from other foods because the inspection of those foods of animal origin requires veterinary specialties that could recognize sick animals and keep them out of the food supply, for example the Quality and Safety Law on Agricultural Food in parallel to Food Safety Law in China (Nestle 2003b). The disadvantage of such separation is the difficulty in avoiding conflict of jurisdiction. Currently, food has been defined, as any substance, whether processed, semi-processed or raw, which is intended for human consumption, and includes drink, chewing gum and any substance which has been used in the manufacture, preparation, or treatment of food but does not include cosmetics or tobacco or substances used only as drugs (Codex Alimentarius Commission 2013). In view of this, the scope of the basic law can cover the whole food supply chain from farm to fork, with the purpose of having food safety responsibility shared by all involved food operators. Notably, the emphasis on taking food as a whole from the perspective of the food supply chain does not conflict with the specialization since functional separation can be further realized in the regulatory rules, such as the administrative rules established by the involved competent authorities including AQSIQ and China' FDA to implement the Food Safety Law.

Based on the basic food law, the food laws can be established respectively depending on the subject matter. Yet, a cross-reference can also be arranged. Taking the EU as example, the CAP was initiated to increase agricultural productivity, ensure a fair standard of living for the agricultural community, and promote reasonable prices of the suppliers when they reach the consumers. Since the General Food Law places the requirements on the public, animal, and plant health in the primary sector, the mechanism of Cross Compliance has been introduced under the CAP in 2003. Depending on the severity of the infringement on the Statutory Management Requirements relating to public, animal and plant health, to animal welfare and to the Good Agricultural and Environmental Conditions, farmers who break the law can be sanctioned and the EU may decrease the support it provides them with. In addition, the new direction of CAP after 2013 is to respond to the economic, environmental and territorial challenges. In the face of fierce competition on the international market, adjustments also can improve the comparative advantage of agricultural food from the perspective of food quality.

In addition to the internal harmony, the realization of the external harmony in terms of sustainability moves in two directions. As mentioned earlier, to make food production in a sustainable way, the food policy and law must take into account the objective of sustainable development, and vice versa. That is to say, the general legislation for sustainable development also provides a legal ground to sustain food production. As far as sustainable development is concerned, human law must be reformulated to recognize and respect the reciprocal rights and responsibility of individuals and states regarding sustainable development (United Nations 1987). For instance, the right to an adequate environment correspond to the state's responsibility for protecting the environment. However, it is arguable how best legalize sustainable development. Comparatively, Chen (2002) found that it is rather narrow to understand it as the same as environmental law or only as a part of economic law since the sustainability lies in the economic promotion, social progress and environmental protection.[49] Therefore, to consider sustainable development law as an independent legal branch, the framework shall contain the legislations regarding these three pillars. Accordingly, the Chinese legislative body and government pay special attention to crafting sustainable development as a purpose and intent of legislation. For example, 28 laws in the field of resources and environment have been established, including the Land Administration Law, Forest Law and Water Law (China 2012).[50] Undoubtedly, these laws in the support of sustainable development can also provide a sound legal background to enable food production in the harmony with sustainability.

References

Alemanno A (2012) Public perception of risk under WTO law: a normative perspective. Available via Social Science Research Network. http://papers.ssrn.com/sol3/papers.cfm?abstract_id=2018212, p 2. Accessed 28 Sept 2014

Beck U (1992) Risk society, towards a new modernity (trans Ritter M.). SAGE Publication, London

Bian Y (2006) Current Chinese law on food safety: an overview. In: Mahiou A, Snyder F (eds) La sécurité alimentaire/food security and food safety. Académie De Droit International de La Haye/Hague Academy of International Law, Netherlands, p 169

Bodenheimer E (1970) Jurisprudence, 3rd edn. Harvard University Press, p 339

Bunte F (2009) The food economy of today and tomorrow. In: Bunte F, Dagevos H (eds) The food economy: global issues and challenges. Wageningen Academic Publisher, Netherlands, pp 48–49

Busch L (2011) Quasi-state? The unexpected risk of private food law. In: van der Meulen B (ed) Private Food Law, governing food chains through contract law, self-regulation, private standards, audits and certification schemes. Wageningen Academic Publishers, Netherlands, p 59

Caswell JA, Siny J (2007) Consumer demand for quality: major determinant for agricultural and food trade in the future? Working paper No. 2007-4, Department of Resource Economics, University of Massachusetts Amherst

Chen Q (2002) A tentative exploration of sustainable development law. Modern Law Sci 24 (5):108–110

Chiaradia Bousquet J-P (1995) Legislation governing food control and quality certification. Food and Agriculture Organization Legal Office, p 4

China (2008) The National Framework for Medium-to-Long-Term Food Security (2008–2020). http://www.gov.cn/jrzg/2008-11/13/content_1148414.htm. Accessed 25 April 2014

China (2012) The national report on the sustainable development. http://www.china-un.org/eng/zt/sdreng/P020120608816288649663.pdf. Accessed 25 April 2014

Chrysochou P, Krystallis A, Georges G (2012) Quality assurance labels as drivers of customer loyalty in the case of traditional food products. Food Q Pref 25:156–162

Codex Alimentarius Commission (2013) Procedural manual, 21st edn. p 22

---

[49]Quansheng (2002), pp. 108–110.
[50]China, The national report on the sustainable development (2012).

Collart Dutilleul F (2009) Analysis and assessment of the new European agri-food law in the contexts of food safety, sustainable development and internationaltrade.http://www.droit-aliments-terre.eu/documents/sources_lascaux/projet_lascaux/projet_lascaux_EN.pdf. Accessed 25 April 2014

Collart Dutilleul D (2012) The law pertaining to food issues and natural resources exploitation and trade. Agric Food Secur 1:6. doi:10.1186/2048-7010-1-6

Du G (2008) Certain issues for establishment of food safety law. Pac J 2(52–65):54–55

European Parliament (1997) EP BSE inquiry report, A4-0020/97/A. http://www.mad-cow.org/final_EU.html

European Union (2007) 50 years of food safety in the European Union. Office for Official Publications of the European Communities, available at http://ec.europa.eu/food/food/docs/50years_foodsafety_en.pdf

European Union (2012) The Common Agricultural Policy A story to be continued. doi:10.2762/35894

European Commission (1985) The completion of the internal market: community legislation on foodstuffs, COM(85) 603 final

European Commission (2008) Green Paper on agricultural product quality: product standards, farming requirements and quality schemes, COM (2008) 641 final

Feintuck M (2005) Precautionary maybe, but what's the principle? The precautionary principle, the regulation of risk and the public domain. J Law Soc 32(2):371–398

Fernqvist F, Ekelund L (2014) Credence and the effect on consumer liking of food-a review. Food Q Pref 32:340–353

FAO (1995) Legislation governing food control and quality certification

Food and Agriculture Organization (2003) Trade reforms and food security. http://www.fao.org/docrep/005/y4671e/y4671e00.htm#Contents. Accessed 28 Sept 2014

Food and Agriculture Organization (2013) Trends and impacts of foreign investment in developing country agriculture, evidence from case studies, p 3

Freeman M (2001) Lloyd's introduction to jurisprudence, 7th edn. Sweet & Maxwell Limited, London, pp 50–51

Goldstein B, Carruth RS (2003) Implications of the precautionary principle for environmental regulation in the United States: examples from the control of hazardous air pollutants in the 1990 Clean Air Act Amendments. Law Contemp Probl 66:247–261

Gong J (2008) From the household contract responsibility system towards a new socialist countryside-30 years' review and outlook. Jiangxi Soc Sci 5:229–238

Guo R, Brandt L (1998) Tenure, land rights, and farmer investment incentives in China. Agric Econ 19:63–71

Hatanaka M, Bain C, Busch L (2005) Third-party certification in the global agrifood system. Food Policy 30:354–369

Hooker NH, Caswell JA (1996) Trends in food quality regulation: implications for processed food trade and foreign direct investment. J Agribus 12(5):411–419

Huang L (2008) On the legal protection of geographical indication in China: history, status quo and future. Law Econ 12:17–18

Jiao L (2007) Tradition and future of Chinese law of administration. J Shanghai Adm Inst 8(1):84–90

Lester RB (1995) Who will feed China, 1st edn. W.W. Norton & Company, Inc

Li E (2013) China. In: Collart DF, Bugnicourt J-P (eds) Legal dictionary of food security in the world. Larcier, p 87

Luhmann N (2008) Risk a sociological theory (trans: Barrett Rhodes), 4th edn. Piscataway, Aldine Translation, p 19

Lusk JL, Briggeman B (2009) Food value. Am J Agric Econ 91(1):184–196

M'hatef M (2007) Gestion de la qualité des aliments. http://www.umc.edu.dz/buc/theses/agronomie/MHA5362.pdf. Accessed 28 Sept 2014

Nestle M (2003a) Food politics. University of California Press, Berkeley, p 16

Nestle M (2003b) Safe food: bacteria, biotechnology, and bioterrorism. University of California Press, p 52
Organization for Economic Cooperation and Development (2000) Agriculture policies in OECD countries: monitoring and evaluation 2000: glossary of agricultural policy terms
Pinstrup-Andersen P (1994) World food trends and future food security, Food Policy Statement, Number 18. International Food Policy Research Institute. http://www.ifpri.org/sites/default/files/publications/ps18.pdf. Accessed 28 Sept 2014
Phillips L (2001) Lessons from the BSE inquiry. J Found Sci Technol 17(2):3–4
Quansheng C (2002) A tentative exploration of sustainable development law. Modern Law Science 24(5):108–110
Ruckelshaus WD (1984) Risk in a free society. Risk Anal 4(3):157–162
Shanghai Daily (2013) China to stick to policy of food self-sufficiency. http://www.shanghaidaily.com/national/China-to-stick-to-policy-of-food-selfsufficiency/shdaily.shtml. Accessed 23 April 2014
Steele J (2004) Risks and legal theory. Hart Publishing, Oxford, p 29
Sun J (2013) The international harmonization of food safety regulation in light of the American, European and Chinese law. Thesis in the Nantes University, France
Thompson D, Ying H (2007) Food safety in China: new strategies. Glob Health Gov 1(2):4
Timmer PC (2010) Reflections on food crises past. Food Policy 35(1):1–11
United Nations (1987) Our common future. Report of the World Commission on Environment and Development, also known as Brundtland report
United Nations (1992) 21 Agenda, Promoting sustainable agriculture and rural development. http://www.fao.org/sd/erp/toolkit/Books/SARDLEARNING/CD-SL/Sources/Agenda%2021-chapter%2014.html. Accessed 28 Sept 2014
Valceschini E (2013) Agri-food activities. In: Collart DF, Bugnicourt J-P (eds) Legal dictionary of food security in the world. Larcier, Belgium, p 39
Vapnek J, Spreij M (2005) Perspectives and guidelines on food legislation, with a new model food law. The Development Law Service, Food and Agriculture Organization Legal Office
Vos E (2000) EU food safety regulation in the aftermath of the BSE crisis. J Consum Policy 23:227–255
Wiener JB (2002) Comparing precaution in the United States and Europe. J Risk Res 5(4):317–349

Chapter 52
Law and Regulation in India

Dimpy Mohanty and Abhijeet Das

Abstract Indian laws concerning standards of food and its manufacture, sale and import were overhauled in the year 2006. Encompassed primarily in the Food Safety and Standards Act, 2006 (FSS Act) and related Rules and Regulations, the laws were introduced stage wise from the year 2007 onwards. The proponents of the FSS Act referred to international trends and international legislations including instrumentalities and Codex Alimentarius Commission. They intended for the FSS Act to take care of international practices and envisaged a policy framework designed to eliminate multi departmental control and guide and regulate persons engaged in manufacture, marketing, processing, handling, transportation, import and sale of food. With the objective of consolidating previous multiple laws and regulatory bodies overseeing the laws, the FSS Act repealed the Prevention of Food Adulteration Act, 1954 and six other laws/orders related to specific food products and established the Food, Safety and Standards Authority of India (Food Authority).

52.1 Introduction

52.1.1 General Overview of Food Law

For a long time, India had lagged behind in reconciling its food laws with contemporary needs. As a result, India overhauled its food regulatory regime (including standards applicable to food itself, and to the manufacture, sale, and import of food) in 2006. These changes were primarily contained in the Food Safety and Standards Act ("FSS Act" or "FSS") and related rules and regulations. These changes have been implemented in stages beginning in 2007.

Research Assistants:
(i) Mr. Tanmay Mohanty, 3rd year student of B.A.LLB, Amity University, Noida, Uttar Pradesh, India; and
(ii) Ms. Shubhangi Mehrish, 2nd year student of B.A.LLB, University School of Law and Legal Studies, G.G.S.I.P.U, Delhi India.

D. Mohanty (✉) • A. Das
LexCounsel, Law Offices, New Delhi, India
e-mail: dmohanty@lexcounsel.in

The proponents of the FSS Act were guided by international trends and international law the Codex Alimentarius Commission. FSS proponents intended for the FSS Act to adhere to international practices and envisaged a single policy framework designed to guide the manufacture, processing, handling, transport, import, and sale of food, and to eliminate multi departmental control of these processes.

In keeping with the objective of consolidating India's food regulatory regime, the FSS Act repealed the Prevention of Food Adulteration Act, 1954 and six other laws/orders related to specific food products and established the Food, Safety and Standards Authority of India (Food Authority).

52.1.2 Food Law as Applied to Manufacturers, Sellers and Consumers: Licensing, Product Approval, Ingredients and Labelling Requirements and Some Cultural Drivers Behind India's Food Regulatory Regime

This chapter summarizes India's food licensing, product approval, ingredient and labelling requirements. This chapter will also illustrate some cultural drivers that have influenced India's food regulatory regime. Further, this chapter will note where India may face challenges in implementing the FSS, or in introducing globally accepted products to India. Finally, this chapter provides a brief overview of the requirements for importing food into India and India's foreign direct investment policy concerning food industry.

52.1.3 Discussion and Analysis of Specific Aspects of Non Vegetarian Foods and Genetically Modified Foods

This chapter also provides an analysis of how India's food law has dealt with controversial non-vegetarian and genetically modified foods, and the impact that other areas of legislation have had on India's food sector.

52.2 Food Laws: Multiple Regulations; Key Definitions and Authorities

52.2.1 Act, Rules and Regulations

The FSS Act is read along with a host of regulations as well as the rules framed thereunder, and essentially regulates a whole spectrum of activities, including, manufacture, import and retail, vis-à-vis food in India. The Food Safety and Standards Rules, 2011 ("FSS Rules"), provides for enforcement structure and procedures, including the seizure of articles of foods and sampling and analysis by Food Safety Officers (at the State level).

The regulations framed under the FSS Act apply to various facets of the food industry. For example: the FSS Licensing and Registration of Food Businesses Regulations, 2011 (Licensing Regulations), provides for licensing requirements not only for manufacturers and retailers of all foods (including packaged and tinned foods) but also for street food vendors and those who distribute foods on religious and/or social occasions.

Further, the FSS Food Products Standards and Food Additives Regulations, 2011 ("Food Standards Regulations"), provides quality standards for food products and their ingredients, and enumerates those foods or ingredients whose use are prohibited or restricted. Similarly, the FSS Contaminants, Toxins and Residues Regulations, 2011 ("Food Toxin Regulations"), provides certain restrictions on the use of metals such as lead, copper, etc., and also regulates the use of fumigants/insecticides in foods. These regulations are often read along with other regulations, such as those prepared by the Nutritional Advisory Committee and the Nutrition Expert Group of the Indian Council of Medical Research (which prescribes the desirable daily dose of certain metals, etc.). Furthermore, the FSS Prohibition and Restrictions on Sales Regulations 2011 ("Sales Regulations") compliments the Food Standard Regulations vis-à-vis specified products including admixtures.

Additionally, the FSS Packaging and Labelling Regulations, 2011 ("Packaging and Labelling Regulations") provides for the packaging and labelling requirements of food articles, which is typically read together with the Legal Metrology Act and Legal Metrology Packaged Commodities Rules, 2011 (LMPC Rules), which also prescribe labelling requirements.

52.2.2 Key Definitions[1]

Certain key definitions prescribed under the FSS Act and the rules and regulations framed thereunder are:

---

[1] For ease of reference, the definitions have been edited for brevity, while keeping the essence thereof.

(i) 'Food' means any substance which is intended for human consumption and includes genetically modified or engineered food/ingredients, infant food, packaged drinking water, alcoholic drinks, and chewing gum, but does not include drugs and medicinal products[2];

(ii) 'Food Business' means any undertaking carrying out any activity related to the manufacture, processing, packaging, storage, transportation, distribution, or import of food, and includes food services, catering services, and the sale of food or food ingredients[3];

(iii) 'Manufacturer' includes a packer who packs and labels articles of food or a person who obtains articles of foods and labels them[4];

(iv) 'Misbranded Food' includes foods that are offered/promoted for sale with false and misleading claims. Such false claims may relate to the type of food, the manufacturer thereof, or to the contents of the food, including the absence or presence of additives, nutrients, etc.[5];

(v) 'Package' includes case, pouch, receptacle and wrappers[6];

(vi) 'Principal Display Panel' means that part of the package/container which is intended or likely to be displayed or presented or shown or examined by the customer under normal and customary conditions of display, sale or purchase of the commodity contained therein[7]; and

(vii) 'Unsafe food' are foods which are injurious to health and which could be unsafe by reference to a number of factors including unhygienic processing, addition of prohibited ingredients, misbranding, etc.[8]

52.2.3 Food Authorities

At the Central (federal) level, the Food Authority is comprised of senior government officials, including those from the Departments of Agriculture, Food Processing, Health, Consumer Affairs, and Commerce. The Food Authority is also comprised of representatives of the food industry, consumer organisations, food technologists/scientists, farmer organisations, and retailer organisations.[9] The Food Authority is supported by a Central Advisory Committee, various scientific

---

[2] See section 3 (j) of the FSS Act.
[3] See section 3 (n) of the FSS Act.
[4] See section 3 (zd) of the FSS Act.
[5] See section 3 (zf) of the FSS Act.
[6] See section 3 (zh) of the FSS Act.
[7] See regulation 1.2.1 (9) of the Packaging and Labelling Regulations.
[8] See section 3 (zz) of the FSS Act.
[9] See section 5 of the FSS Act.

panels and a Scientific Committee.[10] The day to day activities of the Food Authority are carried out by a CEO and other officers and employees.[11]

In addition to the authorities at the Federal level, each State has a Commissioner of Food Safety/Health and Family Welfare and multiple officers known as "Designated Officers" who, *inter alia,* are responsible for processing applications for and issuing licenses. Each State also employs "Food Safety Officers" who oversee the active sampling and seizure procedure, and forward the same to Food Analysts, also appointed under the terms of the FSS Act.[12] The Food Analysts use recognised and accredited food laboratories for discharge of their duties.

52.3 Law Applicable to Manufacturers, Sellers and Consumers

52.3.1 Licensing and Registration

Registration is required for Petty Food Business Operators, a category constituted, *inter alia,* of food stall operators (even stalls of a temporary nature), itinerant vendors, those working from home based kitchens, road-side tea vendors, etc.[13]

The implementation of the registration requirement for Petty Food Business Operators is an enormous practical challenge. Authorities are attempting to regulate those who do not even have a temporary place of business but who may move around on bicycles vending milk, or with their daily output of food produced in their home kitchens and moving on foot. A vast majority of these vendors are illiterate, unaware of the requirements of law, and have no access to newspapers or other media where the Food Authority regularly advertises various facets of law for the awareness of the public at large.

Interestingly, the Licensing Regulations also seek to bring within its purview arrangements made for religious gatherings.[14] It is not unusual in India to celebrate religious, or even personal, occasions by distributing food and/or beverages to the less privileged. Nothing in the Licensing Regulations suggests that the registration requirement is limited to places of worship which distribute food to the needy on a daily basis. While the effort to implement the registration requirement is laudable from the point of view of controlling sanitary and hygiene conditions, even in the case of free distribution of food to the poor, it remains to be seen if such a requirement can be strictly implemented by its officials against individuals setting out to feed the poor.

---

[10]*See* section 12 of the FSS Act.
[11]*See* section 10 of the FSS Act.
[12]*See* sections 30 and 31 of the FSS Act.
[13]*See* regulation 2.1.1 read with regulation 1.2.1 (4) of the Licensing Regulations.
[14]*See* regulation 1.2.1 (4) of the Licensing Regulations.

Separate from the registration process is the license process which is applicable to those who are not Petty Business Operators (with reference to the scale of operations).[15] While a license for a food business is primarily obtained at the State/local level where the business is operational, for specified businesses the license is granted by at the Federal level—including for importers and food business operators operating in two or more States.

Food business operators who manufacture any article of food containing ingredients or substances or using technologies or processes, or a combination thereof, whose safety has not been established through FSS Act and related regulations, or which do not have a history of safe use, or food containing ingredients being introduced for the first time into the country need to apply for product approval separately before applying for license (discussed in some detail in Sect. 52.3.2 below).

The Food Authority has clarified the circumstances when a single or multiple licenses would be required for the same premises.[16] If the products being handled at a single premises can be treated as connected then only one license will be required. According to the Food Authority, if processing of meat, milk, fruits and vegetables is being undertaken in the same premises, separate licenses must be obtained as each activity will be done in separate portions and the businesses are not connected. However, where the premises have facilities for fruit and vegetable processing as well as a store along with it, since these are connected activities, only one license would be required.

Indian food laws do not specify requirements for separate facilities for nuts and/or soy or other ingredients which are known internationally to cause allergies, or labelling requirement concerning traces of such ingredients—this may be due to a general perception that such allergies are uncommon in the Indian population and/or the fact that there may be inadequate data concerning the existence of such allergies in the Indian population.

52.3.2 Product Approval

As stated in Sect. 52.3.1 above, foods for which standards are not prescribed under the FSS Act and related regulations or which contain ingredients or substances whose safety has not been established, or which do not have a history of use in India, need to apply for product approval separately before, applying for license.[17] The product approval requirements are either relatively relaxed or more arduous depending on the category into which the food products fit.

For instance, if the products do not contain plants or botanicals or animal origin substances and the safety of the ingredients are known and permitted under the applicable FSS regulations/Codex and other regulatory bodies like the EU/USFDA/

---

[15]*See* regulation 2.1.2 of the Licensing Regulations.

[16]*See* clarification dated January 24, 2012 issued by the Food Authority.

[17]*See* 'Procedure Regarding New Product Approval' dated December 10, 2012 issued by the Food Authority.

FSANZ the products would be granted approval as a matter of formality with relative ease.

However, if the products contain plants or botanicals or animal origin substances irrespective of the safety of the ingredients being known and permitted under the applicable FSS regulations/Codex and other regulatory bodies like the EU/USFDA/FSANZ, the application form for such products is not only more detailed requiring exhaustive information and documents, the timelines typically are extended before the approval or No Objection Certificate (NOC) is granted. The NOC serves as an interim permission to place the food in the market for a year during which time the product is assessed by the Scientific Panel for safety. If the safety assessment is prolonged beyond a year, the NOC is extended.

There is a separate identified category of foods referred to as "traditional/ethnic" food which is treated in a more relaxed fashion and is exempt from the requirement of product approval—these foods range from traditional savouries such as spice fried dals (lentils/pulses) to indigenous sweets such as gulab jamuns.

While the FSS Act is intended to take care of international practices and the Food Authority, when dealing with product approval, refers to acceptability under international regulatory bodies such as EU/USFDA/FSANZ, there are certain challenges in maintaining complete uniformity for well-established international products. For example, if nutritional supplements are to be introduced into India, their acceptability in other regions for years does not necessarily mean an automatic acknowledgment of safety and hence approval. This is especially so in the case of added minerals and vitamins where the ICMR guidelines stipulating Recommended Dietary Allowance (Indian RDA) are the sole acceptable standard in India, and any deviations from the Indian RDA are liable to be rejected.

No doubt, the Indian RDA being based on Indian population should take precedence over any other country's RDA. However, even when one accepts the supremacy of the Indian RDA for the Indian population, there are certain challenges in such universal implementation which may have been unforeseen by the Food Authority.

The Food Authority does not clarify the treatment of products aimed at those who are outside the mean population that the Indian RDA is based on. Any nutritional supplement for, say, the elderly (whose nutritional requirement may merit additional vitamins or minerals) will fall afoul of the Indian RDA (which are calculated on the basis of "men between 18–29 years of age and weighing 60 kg with a height of 1.73 m with a BMI of 20.3 and who are free from any disease and physically fit for active work"). There is no denying that with increases in age the requirement for minerals and nutrients in the body change due to various factors such as loss of appetite, difficulty in absorption due to medicine intakes, etc., and as a result there may be a requirement for increased nutrient intake.

Some examples of the differences between the Indian RDA and the international RDA requirements follow:

(i) For Vitamin C the Indian RDA limit is **40 mg**, while the limit prescribed in the European Union Directive 90/496/EEC on nutrition labelling for foodstuffs as

regards recommended daily allowances, energy conversion factors, and definitions is **80 mg**;
(ii) For Vitamin B12 the Indian RDA limit is **1 µg**, while the limit prescribed in the European Union Directive 90/496/EEC on nutrition labelling for foodstuffs as regards recommended daily allowances, energy conversion factors and definitions is **2.5 µg**;
(iii) For Selenium the Indian RDA limit is **50 µg**, while the limit prescribed in the European Union Directive 90/496/EEC on nutrition labelling for foodstuffs as regards recommended daily allowances, energy conversion factors, and definitions is **55 µg**.

Would a product aimed at the elderly pass muster because it is within the limits prescribed by the EU? Say, for Vitamin C, even in the Indian scenario an amount over 40 mg but below 80 mg could be considered as normal variation from Indian RDA given the target consumer for such a product would be older people than those considered by Indian RDA.

Additionally, there have been queries raised regarding vitamins and/or minerals for which no Indian RDA has been prescribed. For example, questions have been raised concerning quantities of Beta Carotene in a product. Beta Carotene is a pigment/carotenoid—an abundant source of Vitamin A but not by itself a nutrient per se. Accordingly, no RDA limit is prescribed for Beta-Carotene, though one is prescribed for Vitamin A.

It may be more appropriate for the Food Authority to summarily reject products containing vitamins and minerals beyond the Indian RDA limits ordinarily but make allowance for variation on account of the target consumer. The reference to international guidelines would have then perhaps be appropriate, relevant, and real, taking into account products aimed at a specialised population in countries where the safe consumption of the products have been established.

52.3.3 Food Standards and Ingredients

Standards including permitted ingredients and quantities for ingredients are prescribed for about 13 broad categories of foods (e.g.: dairy products, fats and oils, fruit and vegetable products, cereals and cereal products, meat and meat products, fish and fish products, spices and condiments, sweets and confectionary, sweetening agents, beverages, etc.[18]).

Each category comprehensively covers specific food items totalling about 150 identified food and beverage items ranging from globally identifiable items such as different types of cheese, chewing gums, and chocolates to more locally relevant products, such as ghee (in the milk products category) and jaggery

---

[18]*See* regulation 2 of the Food Standards Regulations.

(categorized under sweets and confectionary). As discussed in Sect. 52.3.2 above there is also the identified category of foods referred to as "traditional/ethnic" food.

Additionally, standards are prescribed for food additives such as permitted colours, artificial sweeteners, preservatives, anti-oxidants, emulsifying agents and anticaking agents. For example, only lecithin, ascorbic acid, and tocopherols are generally permitted for use as anti-oxidants. Further, colouring agents are classified into the following sub-categories: (i) Natural Colouring Matters such as beta carotene, chlorophyll, saffron, curcumin, caramel etc. are generally permitted to be used; (ii) Inorganic Colouring Matters and pigments are generally not permitted for use unless otherwise provided in for the Food Additive Regulations; and (iii) select Synthetic Food Colours such as ponceau, carnosine etc. are permitted to be used only in specified food products.

Further, the Food Standard Regulations classifies 'Preservatives' into two classes as follows: (i) Class I preservatives such as common salt, sugar, spices, vinegar, honey and edible vegetable oils, etc., which can generally be used; and (ii) Class II preservatives which are restricted for use in specified foods to specified limits. There is also a restriction on using more than one Class II preservative even in the foods where individual Class II preservatives are permitted.[19]

Furthermore, Flavouring Agents and Related Substances are permitted to be used subject to appropriate labelling and are classified as follows: (i) Natural Flavours and Natural Flavouring substances; (ii) Nature-Identical Flavouring Substances; and (iii) Artificial Flavouring Substances—covering those substances which have not been identified in natural products intended for human consumption, either processed or not.[20]

At the opposite end of the requirements of specific standards is the category of Proprietary and Novel Foods. This category covers all food for which standards have not specifically been prescribed. However, regulatory restrictions concerning permissible ingredients and limits on the use of certain other ingredients all apply to Proprietary Foods.[21]

The Food Standard Regulations also identify the foods which can undergo irradiation, the minimum and maximum doses of irradiation, and the facilities at which foods can undergo irradiation.[22]

52.3.4 Labelling

The food laws related to labelling are primarily contained in the Packaging and Labelling Regulations which are read along with the LMPC Rules. The LMPC

---

[19] *See* regulation 3.1.4 of the Food Standards Regulations.

[20] *See* regulation 3.1.10 of the Food Standards Regulations.

[21] *See* regulation 2.12 of the Food Standards Regulations.

[22] *See* regulation 2.12 of the Food Standards Regulations.

Rules provide for the manner of packaging of commodities and the declarations to be made on packed commodities. The LMPC Rules expressly provide that no person shall pre-pack, or cause or permit to be pre-packed, any commodity for sale, distribution, or delivery unless the label is securely affixed and the requisite declarations have been made on the package in which the commodity is pre-packed.[23]

The Packaging and Labelling Regulations stipulate certain mandatory information, declarations, and specifications to be provided on the labels of every pre-packaged food proposed to be imported, stored, and/or sold in India.[24]

As per LPMC Rules, separate labelling requirement apply to wholesale packages which consist of multiple packages which can then be retailed individually. Since the labelling requirement for retail packages are extensive, the declarations for wholesale packages are more relaxed, requiring only the name and the address of the manufacturer, importer (and in some instances the packer), the identity of the commodity contained, and the total number or weight of the retail package (s) contained in the wholesale package.[25]

52.3.4.1 "Vegetarian" or "Non-Vegetarian" Food Labelling

While the general food labelling requirements discussed in Sect. 52.3.4.2 below echo labelling requirements in various other countries, a more distinctive labelling characteristic in India is the compulsory requirement to label each food as either "vegetarian" or "non-vegetarian".[26] This requirement reflects the vast population of Indians who are vegetarian for religious reasons.

It is not only the use of meat products per se that require a food to be labelled "non-vegetarian", but even ingredients such as gelatine (which traditionally has been sourced from animal products) would lead to a food item being labelled "non-vegetarian". Gelatine itself is a standardised ingredient: i.e. standards are prescribed for it and special labelling requirement (other than "non-vegetarian") applies to the use of gelatine.

The declaration to the effect that a food is non-vegetarian or vegetarian is made by a symbol with the former being identified by a brown colour filled circle inside a square with brown outline and the latter by similar symbol in green colour (see Fig. 52.1).

From a consumer information standpoint, both the Packaging and Labelling Regulations as well as the LMPC Rules specify details such as, *inter alia*, the quantity declaration being free from printed information on all its sides by specified space, the heights of numerals made in the declarations, and the label's proportion

---

[23] *See* rule 4 of the LMPC Rules.
[24] *See* rule 6 of the LMPC Rules.
[25] *See* rule 24 of the LMPC Rules.
[26] *See* regulation 2.2.2 (4) of the Packaging and Labelling Regulations.

Fig. 52.1 Non-vegetarian (brown)/Vegetarian (*green*) declaration

"*Non-Vegetarian Declaration*" (brown)   "*Vegetarian Declaration*" (green)

in comparison to the principal display panel (with the principal display panel being pretty much the entire package/container of food).

There are further requirements concerning the non-vegetarian or vegetarian symbols, including the diameter of the sides of the square and size of the symbol in comparison to the principal display panel, and for the symbol to be prominently displayed on the labels having a contrasting background on the principal display panel, and being placed in close proximity to the name or brand name of the products.

52.3.4.2 Other Information, Declarations and Specifications Required to be Stated

In addition to the non-vegetarian or vegetarian declaration, some more India specific labelling requirements include: (i) the labelling language must be either English or Hindi[27]; (ii) the format for declaration of maximum price at which the product may be sold to the end consumer viz. 'MRP Rs......incl. of all taxes' or Maximum or Max. retail price...... inclusive of all taxes' or in the form.[28]

The Packaging and Labelling Regulations mandate the use of the International System of units in declaring the net quantity of the commodity on the labels including kilogram and grams.[29]

Other information, declarations and specifications required to be stated on the labels of foods include[30]:

(i) Name/description of the Food Product.
(ii) List of Ingredients—in the mode and manner prescribed including, but not limited to: listing the name of the ingredients in descending order of their composition by weight or volume, as the case may be; declaration of constituents of compound ingredients, etc.
(iii) Nutritional Information per 100 g or per serving of the product with specifics of kcal, protein, carbohydrates from sugar, and fat.
In cases where a nutrition or health claim is made, depending on the kind of claim, a declaration may need to be made concerning the amount of fatty

---

[27] *See* regulation 2.2.1 (2) of the Packaging and Labelling Regulations and rule 9 (4) of the LMPC Rules.
[28] *See* rule 2 (m) of the LMPC Rules.
[29] *See* regulation 2.2.2 (5) of the Packaging and Labelling Regulations.
[30] *See* regulation 2.2.2 of the Packaging and Labelling Regulations.

acids, monounsaturated fatty acids, and polyunsaturated fatty acids and/or cholesterol and/or trans fatty acids.

Further standards are specified as to the amount of trans fats or saturated fats that may be present in a product for a health claim of 'trans fat free' or 'saturated fat free'.

(iv) Declaration Regarding Food Additives—under the following class titles together with the specific names or recognised International Numerical Standards (**INS**):

Acidity Regulator, Acids, Anticaking Agent, Antifoaming Agents, Antioxidant, Bulking Agent, Colour, Colour retention Agent, Emulsifier, Emulsifying Salt, Firming Agent, Flour Treatment Agent, Flavour Enhancer, Foaming Agent, Gelling Agent, Glazing Agent, Humectant, Preservative, Propellant, Raising Agent, Stabilizer, Sweetener, Thickener.

There are specific recommended declarations for the extraneous addition of colouring matters related to the sub-category of the colouring matter used. If the statement is displayed along with the name or INS number of the food colour, the colour used need not be declared in the list of ingredients.

Similarly there are specific recommended declarations for the extraneous addition of flavouring agents and also separately for cases where both colour and flavour are used.

(v) Instructions for use including for reconstitution, if necessary, to ensure correct utilization of the product.

(vi) Name and Complete Address of the Manufacturer and Packer.

If the address of the manufacturer and the manufacturing unit are different, both have to be specified. The name and address of the packer is required when the manufacturer is not the packer.

In cases of contract manufacturing for some other manufacturer or company under the latter's brand name, the label will need to state the name and complete address of the manufacturing or packing unit, as the case may be, and also the name and complete address of the brand name owner for, and on whose behalf, the product is manufactured or packed.

(vii) Contact Details of the person who can be contacted in case of consumer complaints.

(viii) Lot/Code/Batch Identification, enabling tracking in manufacture and identification after distribution.

(ix) The date, month and year of manufacture, packing or pre-packing of the product.

(x) Best Before and Use By Date in any one of the three specified formats.

(xi) The net weight or number or measure of volume of contents which would need to exclude the weight of wrappers and materials of the package.

Additional and specified declarations need to be made for certain foods such as for infant foods, irradiated foods, artificial sweeteners, food colours, edible oils, drinking water, foods containing monosodium glutamate, etc.

The LMPC Rules and the Packaging and Labelling Regulations mandate clear and unambiguous declarations including specifying instances where declarations must be printed in a colour which contrasts the background.

The LMPC Rules and the Packaging and Labelling Regulations proscribe any description or presentation or labelling that is false, misleading or deceptive or is likely to create an erroneous impression regarding the product's character in any respect. The aforesaid rules and regulations stipulate against use of words such as "recommended by the medical profession" or any words which imply or suggest such recommendation. They direct against declarations of quantity with words, or expressions which may lead to exaggerated, misleading or inadequate impressions, such as the use of the following words or phrases: 'minimum', 'not less than', 'average', 'about', 'approximately', etc.

The required declarations need to be printed on each label, as LMPC Rules do not permit affixation of individual stickers on the package for altering or making declarations. The only exception is that a sticker may be used for reducing the retail sale price of a commodity, as long as such a sticker does not cover the original retail sale price stated on the package.

52.3.4.3 Exemptions from Labelling Requirements: By Legal Provisions and by Practice

Packaging and Labelling Regulations prescribe, *inter alia*, limited and specific exemptions from labelling requirements for packaged products if the surface area of the package is not more than 100 square centimetres. There is additional relaxation for packages whose surface area is less than 30 square centimetres.

Certain exemptions also apply to a package which qualifies as a 'multi piece package', i.e., it is a package containing two or more individually packaged or labelled pieces of the same commodity of identical quantity, intended for retail either in individual pieces or as a whole.

While the aforesaid labelling exemptions are provided in law, certain labelling requirements are routinely not enforced in practice, leading to a perception that exemptions have been extended by the authorities. The risk of relying on such practice is that at any point in time, the Food Authority, and hence other government departments, may decide otherwise. This sudden decision could be the result of economic/foreign policies which cannot be foreseen.

For example, the requirement to label beverages, especially imported alcoholic beverages had not been enforced by the Food Authority for some time. However, as per recent reports there are containers of imported Scotch whisky and wines pending clearance at customs since the Food Authority had taken a decision to enforce labelling provisions for such beverages. While European manufacturers and exporters have reportedly dug in their heels concerning labelling, the recent stringent application may also be a direct response to the EU banning import into Europe of Indian mangoes and other produce products.

The Indian authorities stance, that it does not matter what the US and EU prescribe since Indian laws provide otherwise, while technically sound, does seem contrary to one of the objects and reasons for the new Indian food laws—moving towards international practices. If world over the consumer can be expected to know what Scotch whisky and wine are comprised of, or at least not be the worse for wear without that knowledge, then the Indian labelling requirements being made applicable to imported wines and whiskeys seems excessive.

52.3.5 Trademarks Protection

Products of various food brands from across the world have become a common sight in the marketplaces and dining rooms in India. The effective branding strategies for these products are essential ensuring their distinctiveness and making them easily recognisable to consumers and to help in achieving and sustaining a competitive market share. Given the appeal of these global/Indian brands in the food sector, trademark protection has become a growing priority for the major players in the food sector.

Trademarks can include words, logos, shapes (of the food product itself or its packaging), slogans, specific colours and sounds. Some of the more well know food brands in India include McDonalds, Coca Cola, Maggi, Dominos, Cadbury, and many others. In terms of the Trade Marks Act, 1999, proprietors of these brands are afforded adequate protection against the infringement of the rights in their respective brands. Below are some recent examples of protection by the courts in India, of some well-known food brands:

(i) *Cadbury Ltd and Ors. v. Tims Foods Private Ltd. and Anr.*, 2009 (39) PTC 544 (Del),
This suit was filed by Cadbury against Tims Food, seeking permanent injunction against Tims Food from infringing Cadbury Dairy Milk 'eclairs' labels/wrappers. The Court ruled that the trade dress/wrapper adopted by Tims Foods was deceptively similar to that of Cadbury and concluded that there had been undue enrichment to Tims Foods to the detriment of Cadbury. Injunction was granted to *Cadbury* and it was awarded damages amounting to Rs. 500,000/-.
(ii) *The Coca-Cola Company & Anr. vs. K.M. Salim*, MIPR 2014 (1) 217
The High Court of Delhi held, that when a trademark is identical to a registered trademark and goods/services for which it is used are also identical to goods or services for which registration has been granted, then the Court shall presume that it is likely to cause confusion on part of public and therefore granted injunction to the Coca-Cola Company.
(iii) *Heinz Italia and Anr. v. Dabur India Ltd.*, 2007 (7) SCALE 608
Injunction was granted against marketing of a product under the brand 'Glucose-D', as the brand was deceptively similar to 'Glucon-D', a registered trade mark of Heinz Italia.

52.3.6 Import Related Provisions

Pursuant to obtaining an Import–export Code (IEC) under the Export–import Policy of India from the office of the Director General of Foreign Trade, importers may import food for which there are few additional requirements as well as some relaxation of existing requirements.

52.3.6.1 Licenses

As discussed above, all importers would require a license from the Food Authority at the Federal level.

52.3.6.2 Ingredients

The provisions of the FSS Act apply in totality with reference to the use of ingredients, and restrictions for all imported food items.

52.3.6.3 Labelling Requirements

A food item's country of origin needs to be stated on the labels of imported foods. If a food product has undergone processing in a second country which changes the nature of the food product, then the country in which the processing is performed is considered to be the country of origin for the purposes of labelling. In the case of imported food the name and address of the importer in India is also required. Additionally the discussion at Sect. 52.4.1 below, concerning non-vegetarian foods, may be relevant for the import of meat products.

52.3.6.4 Foreign Direct Investment Policy as Applicable to the Food Industry

The Government of India in general permits investments in the food processing sector under the "automatic route", i.e., without the requirement for further approvals for investment. However, there are certain categories of food products which are reserved for the micro and small scale sector where foreign direct investment is eligible for the automatic route only for up to twenty four per cent (24 %) foreign direct investment.

52.4 Discussion and Analysis of Specific Aspects of Non-Vegetarian Foods and GM Foods

52.4.1 Non Vegetarian Foods: The Controversy Around Beef and Whether Food Laws Permit the Sale of Beef

The discussion here revolves around the permissibility of selling and consuming beef products, as the issue appears to flow from food laws, and the barriers which surround the possibility of such sale and consumption arise from the food laws read with other laws.

52.4.1.1 Food Laws

The Food Standard Regulations, in its treatment of various meat products, appears to permit import and sale of beef products along with other non-vegetarian products.[31] At the same time most Indian states restrict the slaughter of cows. To a lesser extent, most States also prohibit slaughter of other bovine animals for the manufacture of packaged foods in India. Below is an attempt to break down the complex treatment of meat products, especially beef, as per the food laws.

The definition of food includes animals, prepared and processed. As discussed above, standards are prescribed for about 13 broad categories of foods which include the category of meat and meat products. Meat products, as described in the Food Standard Regulations, includes bovine meat product.

Standards are prescribed for, *inter alia*, "canned corned beef". Canned corned beef is defined as product prepared from the boneless meat of the carcasses of bovine animals including buffalo meat. While the definition does not restrict canned corned beef to buffalo meat, from a harmonious construction of the import policy (discussed below) and food laws it would appear to be limited to buffalo meat. The definition of canned corned beef (as also the definition of canned cooked ham) does not refer to slaughter house/abattoir which is a reference point for almost all other standardised meat products, whether bovine, caprine, ovine, suilline, or poultry.

The other categories of standardised bovine meat products are "Canned Luncheon Meat" and "Canned Chopped Meat". Both of the aforesaid categories refers to product prepared from the meat of mammalian animal (i.e. including bovine) slaughtered in an abattoir. A third category titled "Frozen Mutton, Chicken, Goat and Buffalo meat" appears to be broader than the title suggests as it refers to products from the meat of animals specified under the Food Standard Regulations, including buffalo meat slaughtered in an abattoir.

---

[31]*See* regulation 2.5.2 of the Food Standard Regulations.

In the Food Standard Regulations the term "slaughter house" has been defined as a building licensed by the local municipal authority for slaughter.[32] The term has, however, not been used, with the term "abattoir" having been used interchangeably instead.

The relevance of licensed abattoir and conditions concerning such slaughter cannot be ignored—from an import policy view point these conditions have to be satisfied for imported products. More importantly, the licensed abattoir will operate within the four corners of other laws. Since a majority of States have passed legislation banning the slaughter of specified cattle and provided conditions under which slaughter can be done, whether beef and beef products can be sold would also be impacted by these legislations.

52.4.1.2 Preservation of Cattle Legislation

While legislation which regulates the slaughter of cattle is nominally for the protection of agricultural cattle as mandated by the Directive Principles contained in the Constitution of India, there is no denying that a part of the purpose sought to be achieved by certain States by such legislation is to take into account the sensitivity attached to cow slaughter. A vast number of the Indian population considers the cow an animal not to be slaughtered for food due to religious and cultural beliefs. For some of those who hold such a belief, their belief is restricted to cows and not to the slaughter of buffaloes. The sensitivities of the Indian people against the slaughter of cows can be traced back to the Indian rebellion of 1857 against British Colonisation, which can be principally attributed to the use of ammunition for the '1853 Enfield Rifle' which was greased with cow fat. This greatly outraged the Indian sepoys, eventually leading to the mutiny. Further, in 2001, the rumours of use of beef extracts in the 'French Fries' served in McDonalds restaurants was met with great outrage in India, eventually leading to McDonalds India issuing a statement dismissing such rumours.

Broadly, these pieces of State level legislation can be divided into three categories, as follows: (i) States that completely ban the slaughter of all bovine animals, i.e., Chhattisgarh, Jammu and Kashmir, Madhya Pradesh and Rajasthan; (ii) States that ban cow slaughter and permit, either expressly or impliedly, the slaughter of bull and bullock and/or other bovine animals, i.e., Andhra Pradesh, Bihar, Daman and Diu, Delhi, Goa, Gujarat, Haryana, Himachal Pradesh, Karnataka, Maharashtra, Manipur, Orissa, Pondicherry, Punjab, Uttar Pradesh; and (iii) States that permit cow slaughter either by express legislation or by lack of it, i.e., Assam, Kerala, Tamil Nadu and West Bengal in the former category, and Meghalaya and Nagaland in the latter.

Most legislation expressly allowing slaughter, whether of cows or other bovine animal, stipulates that the animals may only be slaughtered if they are over a certain

---

[32]*See* regulation 2.5.1 of the Food Standard Regulations.

specified age, usually 14 or 15 years of age (however in some cases lesser and some as high as 25 years) and/or upon production of a "fit-for-slaughter" certificate. The certificate is indicative of the animal not being economical for the purpose of breeding or agriculture.

Leaving aside the States which permit cow slaughter, it is an enormous task to attempt to unravel the definitions, the restrictions and the permissions provided in the legislations of the other States. For example, it appears that the State of Andhra Pradesh limits itself to prohibition of slaughter and does not concern itself with the sale of beef products. Further, the State of Bihar bans the slaughter and export of certain bovine animals but does not concern itself with sale of beef products. On the other hand, in the National Capital Territory of Delhi, the States of Madhya Pradesh, and Jammu and Kashmir, the possession of flesh of agricultural cattle has been categorised as an offence. Similarly, in the State of Chhattisgarh, beef is referred to as being the flesh of agricultural cattle and possession of such slaughtered agricultural cattle, in contravention of the applicable legislation, is illegal. In the Union territory of Daman and Diu, the States of Goa, Punjab, Haryana, Himachal Pradesh and Pondicherry, cow slaughter is prohibited, and while beef is defined as the flesh of a cow in any form, any flesh brought into the States in sealed containers is outside the purview of this definition.

Further, these States (and Union Territories), other than Goa, further prohibit the sale of beef except for prescribed medical purposes. The pertinent legislations for the States of Punjab, Haryana, Himachal Pradesh, define the term "beef products" as including beef extracts and prohibits the sale of beef products as well (except for medicinal purposes). Goa however permits the sale of beef and beef products being derived from animals other than cows, or imported from other States with the requisite certificate. It also requires the registration of vendors of such beef.

The State of Uttar Pradesh also prohibits cow slaughter, defines beef as the flesh of a cow, and takes outside the purview of the definition any flesh brought into the States in sealed containers. Interestingly while prohibiting the sale or transport of beef or beef-products in any form except for such medicinal purposes as may be prescribed, it permits the sale of severed beef or beef-products for consumption by a bona fide passenger in an aircraft or railway train.

Furthermore, the State of Gujarat prohibits the slaughter of all bovines, viz., bulls, bullocks, cows, calves, male and female buffaloes and buffalo-calves, without a certificate, and provides for a certificate to be issued under strict condition only with respect to buffaloes. Transport of the aforesaid specified animals for purposes of slaughter is also an offence in the State of Gujarat and there is an inherent presumption that such transport is for slaughter except where proven otherwise. The sale and purchase of beef and beef products is illegal in the State of Gujarat, where beef is defined as flesh of animals specified.

In the State of Karnataka, the slaughter of cows and calves, including the calves of buffaloes is prohibited. The transport, purchase, and sale of the aforesaid animals are also restricted. Karnataka permits the slaughter of other bovines with appropriate certificates related to age and incapacity to breed and milk. Similarly, in the State of Maharashtra, the slaughter of cows and requires a certificate not just for bovines but also for ovines and caprines (sheep and goats) and is otherwise prohibited. The State of Orissa prohibits cow and bull slaughter but does not provide for any restriction on transport, sale and possession beyond the prevention of slaughter. Slaughter of specified animals is allowed under specified conditions.

The extent of prohibition on slaughter of bovine animals has been the subject of court decisions including decisions by the Supreme Court of India (Supreme Court). The position established by the Supreme Court in *State of Gujarat Vs. Mirzapur Moti Kureshi Kassab Jammat and Ors.*, AIR 2006 SC 212 (Mirzapur Case) is that States can prohibit the slaughter of animals, even those animals which are not fit for breeding and or providing milk. However, as noted by the Supreme Court in *Akhil Bharat Gosewa Sangh Vs. State of A.P. and Ors.*, 2006 (4) SCC 162, the decision of the Supreme Court in Mirzapur Case does not warrant a conclusion that permitting slaughter of bovine cattle is itself unconstitutional.

In view of the cultural sensitivity, the lack of clarity in some of the legislation could present issues for the food industry and/or the consumer. The issues which appear open to debate are:

(i) In States like Goa, Punjab, Haryana, Himachal Pradesh, the Union Territories of Daman and Diu, and Pondicherry, which define beef as the flesh of a cow in any form but take outside the purview of the definition any flesh brought into the States in sealed containers, would processed food imported in cans from other States be treated as "flesh of cow in any form"?
If yes, since these States, other than Goa, further prohibit the sale of beef except for medicinal purposes, it may be impractical to import products since their sale comes with attached condition of medicinal purpose.
It is, however, possible that the legislation on its face permits the sale of beef which is from bovine animals other than cows and/or brought from other States, be it for sale as food products or to enable processing for other purposes/industries.
(ii) The State of Gujarat prohibits the sale of beef and makes the purchase of beef and beef products illegal. It also defines beef as the flesh of animals specified. However, there no further definition of beef products.
The State of Madhya Pradesh prohibits the slaughter of cows, calves and female buffaloes and further prohibits the sale or purchase of such animals for slaughter. The States of Madhya Pradesh, Karnataka and the National Capital Territory of Delhi make the possession of the flesh of such slaughtered animals in contravention of the applicable Act illegal.
In the above cases, the interpretation of beef products should rationally relate to the intent of the legislation which is to preserve cattle in the State.

Similarly, it would be practical to say that beef products sourced from outside the State are not the flesh of slaughtered agricultural animals per se and are not a violation of the legislation concerning slaughter and/or tantamount to possession of flesh of animals slaughtered in contravention of the State legislation. However, such arguments have not been tested in practice.

52.4.1.3 Import Laws

While the import duties specified for bovine animals, at first appearance, may indicate permission to import beef, there is a prohibition on the import of beef. As per the import policy, the import of beef, in any form, is prohibited and all consignments of edible oils and processed food products imported in bulk have to carry a declaration of the exporter in the shipping documents to the effect that *"the consignment does not contain beef in any form"*. Additionally, all consignments of edible products, imported in consumer packs, are compulsorily required to declare on the label of the package *"the product does not contain beef in any form."*

Thus, it appears that bovine animals would refer to bovine animals other than cows. This interpretation is supported by the export policy which also prohibits the export of beef which is defined as including the meat of cows. However, since India is the world's largest exporter of beef, it would appear that as per the Export Import Policy of India, "beef" refers to products derived from cows only.

Additionally in terms of the Import Policy, the import of meat and meat products, including poultry products, are subject to compliance with conditions regarding manufacture, slaughter, packing, labelling and quality conditions as laid down in the FSS Act and related rules as well as labelling requirements prescribed under the LMPC Rules. Manufacturers of meat/poultry products exporting goods to India are required to meet the sanitary and hygienic requirements stipulated in the FSS Act.

Separately, the import of all other meat and meat products (which includes milk and milk products and egg and egg products) are mandatorily subject to a sanitary import permit. From time to time to check the spread of *avian* influenza or other like diseases, the import policy may also restrict import of specified meat and from specified countries.

52.4.1.4 Conclusion

If one connects the prohibition of slaughter of animal legislation (with no definition of beef and/or blanket ban on *sale* of beef), import laws and the food laws, it is apparent that even import, and possession of, "Canned Corned Beef" which appears at first glance to include beef derived from cow would be illegal. This, when the product is not derived from slaughter of animals in India, would be in consonance with Indian cultural norms and not food laws.

While imported canned beef products have been intermittently visible on the shelves of shops in Delhi, it is to be understood that such products are derived from bovine animals other than cows.

Further, insofar as standardised products (Canned Luncheon Meat, Canned Chopped Meat and Frozen Mutton, Chicken, Goat and Buffalo meat, etc.) are concerned, the same would, by reference to a licensed abattoir, in a narrow sense refer to products not imported but procured in India in the States where such slaughter is permitted. However, in view of the import policy as long as the imported products are processed as per the guidelines of the FSS Act, the import and sale of such products would be allowed—however, caution should be exercised with respect to which States completely ban possession of bovine meat!

52.4.2 Genetically Modified Food: The Legal Position

Food has been defined to include genetically modified or engineered food. The FSS Act provides for the constitution of a Scientific Panel for genetically modified organisms and foods. It further stipulates that no person can manufacture, distribute, sell or import genetically modified organisms and foods except as provided under the FSS Act or related regulations. Under India's food laws there are no further regulations concerning genetically modified foods including for labelling of such foods.

While food laws are seemingly open about allowing genetically modified foods on the shelves, there has been an extremely strong reaction to genetically modified produce/crops in India. India permitted trials for BT cotton in the early 2000. However, following an outcry from farmers and their interest groups, the government banned farming of the first GM food crop (BT brinjal/eggplant). Other than brinjal/eggplant some trials for BT okra were also conducted. There is a petition against GM crops pending in the Supreme Court. The Supreme Court had, in the course of the hearing of the petition, set up a technical expert committee which had recommended a moratorium until regulatory process were refined. In view of the recommended moratorium, until recently, only trials of BT cotton were permitted and there was a moratorium on field trials for GM seeds in food crops. However, in March 2014 the Government cleared field trials in GM food crops with the stance that the Supreme Court had not laid down an embargo against such trials and that regulatory processes governing trials were adequate.

Without prejudice to the erstwhile ban on BT food crop trials, food laws have provided for the introduction of GM foods without any reference to any specific labelling or other regulatory requirements.

However it appears from the Food Authority's paper concerning "Operationalising the Regulation of Genetically Modified Foods in India" (GM Paper), that the Food Authority is considering the mode and manner of implementing regulations for GM Food.

According to the GM Paper, the Food Authority will develop guidelines to describe the regulatory framework for GM food including the interim process for regulation of GM food until formal regulations under the FSS Act are released. The GM Paper delineates the likely organisational structure to be set up including a scientific panel for GM Foods.

The GM Paper signifies the Food Authority's intent to adopt guidelines concerning safety assessment of GM foods and protocols approved by the Genetic Engineering Approval Committee (GEAC) and the Review Committee on Genetic

Manipulation under the Environment Ministry, which are stated to be consistent with standards developed by Codex Alimentarius.

The Food Authority, in terms of the GM Paper, contemplates taking over the responsibility for food safety assessment of viable and processed GM foods and approval for commercial release of processed GM foods while letting the GEAC maintain control of approval for commercial release of viable GM foods/Living Modified Organisms (LMOs).

Since the publication of the GM Paper, it is noteworthy that no guidelines or regulations have been developed. Hence, GM foods as per the Food Authority's own admission in the GM Paper continue to be governed as per the Rules for the Manufacture/Use/import/Export and Storage of Hazardous Microorganisms, Genetically Engineered Organisms, or Cells, 1989 under the provisions of the Environment Protection Act (MOEF GM Rules).

In terms of Rule 11 of the MOEF GM Rules, food stuffs, ingredients in food stuffs and additives including processing and containing or consisting of genetically engineered organisms or cells, cannot be produced, sold, imported or used except with the approval of the GEAC.

Thus, while food laws do not presently provide for any regulatory coverage of GM foods other than identifying the category, the Ministry of Environment and Forests (MOEF) is technically the regulating authority (although as per its own claims its role is limited and other departments would need to be involved—refer discussion below concerning Doritos and averment of the GEAC in its meeting).

The Import Policy makes the import of GM items, including GM food intended for use as food, subject to the provisions of the Environment Protection Act and Rules and subject to approval of the Genetic Engineering Approval Committee (GAEC). The Import Policy also requires a compulsory declaration to the effect that any GM food is genetically modified.

As per information in the public domain, no entity has sought approval of the GEAC except when third parties have raised the issue. However, according to anti-GM activists, GM foods are in the Indian market.

As per the minutes of the meeting of the GEAC in 2008, Greenpeace had filed a complaint concerning import of Doritos' Cool Ranch Corn chips which were alleged to be containing GM ingredients. GEAC requested the sample stated to have been collected by Greenpeace. The minutes also reflect the stance of the GEAC that it would not, as a matter of course, test food products for GM ingredients, and appeared to reflect that it would be proactive only on the basis of a complaint—in this case the complaint of Greenpeace.

In the instant case the manufacturers of Doritos, Pepsico, had submitted that while Doritos was their brand they were neither manufacturing it in India nor importing it themselves or through authorised third parties, and averred that they did not use GM ingredients and were in compliance with all Indian laws relating to GM organisms.

Further, after the above meeting of the GEAC, there was another meeting in January 2009 wherein the GEAC disclaimed direct involvement in the import, sale or distribution of any GM commodity including GM food products. The GEAC stated that the matter had been taken up by the Ministry of Commerce and Industries and the Directorate General of Foreign Trade (DGFT) in view of the restriction concerning GM food items intimating them of the lack of prior approval of the GEAC in the case.

Finally, in May 2009, the GEAC considered a NOC request from an Indian entity to import Doritos chips which were claimed to be made from corn produced in Taiwan. The items had been detained at Mumbai customs in view of the communication issued by the MOEF and DGFT concerning the illegal import of Doritos chips from the USA, which was based on a representation that had been earlier received from Greenpeace.

The GEAC, while noting and relying on the submissions of the importer that (a) the products were produced in Taiwan,(b) the product is made from corn indigenously produced in Taiwan which was not genetically modified corn, and (c) Taiwan does not produce any GM food crop, decided to convey 'no objection' for release of the consignment.

Possibly in view of the declaration of the importer that Taiwan imports GM soybeans and corn from USA and that Taiwan's labelling policy is applicable beyond a threshold of 5 % of GM content, or otherwise, the GEAC made the release subject to certain conditions—that if the products were later found to contain genetically modified material, the importer would be liable for prosecution and advised that the customs officials be directed to retain random samples from the import consignment for further verification, if necessary.

If all of the aforesaid facts are taken at face value then it would appear that third parties may have been importing food products with GM ingredients without complying with Indian law. With no authority ensuring active compliance this is evidence of the gap between the law as it exists on paper and as it is actually implemented. It would require an extremely interested and motivated entity (as was evident in the Doritos case) to actually prevent the entry of GM foods into India, law notwithstanding.

Part of this conclusion is bolstered by the allegedly conflicting responses given by the Food Authority. Anti GM activists have used the Right to Information Act to assail Food Authority's response disclaiming responsibility for labelling of GM foods.

However, where the Food Authority has not been proactive, the Legal Metrology Department has taken steps to the extent that was within its powers to provide for consumer awareness concerning GM foods. Pursuant to an amendment to the LMPC Rules, effective January 2013, there is now a requirement to include a declaration on every package of genetically modified food at the top of the label using the words "GM". While anti-GM activists consider this a positive step, they are of the view that more needs to be done. Their view is that the requirement of the Legal Metrology department in the first place may not add to consumer awareness and may instead lead to confusion in the face of common usage of the metric unit 'gm' indicating gram. Secondly, in the absence of any clarity on the authority which will test the presence of GM content and potential labelling violations (the Legal

Metrology department not being equipped for such an exercise) the law, for now, lacks teeth.

The law has drawn the attention of the All India Food Processors' Association which has appealed for a suspension of the requirement for declaration. There has been no action in their favour and the declaration requirement remains in force.

52.5 Conclusion

With the advent of the FSS Act, India has taken a welcome step in unifying previously multifaceted and cumbersome regulatory regime, with respect to standards of food and its manufacture, sale and import, to ensure availability of safe and wholesome food for human consumption. The FSS Act emphasizes consolidating the laws related to food and establishment of the Food Authority, laying down scientific standards for and regulating articles of food. With the harmonization of laws relating to food quality and standards with established international norms, the FSS Act aims at scientifically developing the food industry. Thus, while the food processing industry may ultimately see the FSS Act as a mixed blessing, the practical application of this legislation, being at its nascent stage, will require some time to come into full force. There are certain elements which must be further tweaked to bring Indian food law completely in line with the international standards, while carefully balancing the interests of the food industry as well as the consumer.

Even with the aforesaid effort of unification, India still has a lot of laws that apply to the food sector with implications on entities ranging from the corner 'chai wala' to the 'McDonalds' of this world. While compliance with these requirements may not be strictly followed or enforced by the regulatory authorities at the moment, that doesn't guarantee, that contraventions will fly under the radar for the indefinite future.

References

Department of Industrial Policy & Promotion. Consolidated FDI policy circular of 2014. http://dipp.nic.in/English/Policies/FDI_Circular_2014.pdf. Accessed 26 June 2014

Food Safety and Standards Authority of India, FSSI interim regulations on 'Operationalizing the regulation of genetically modified foods in India'. http://www.fssai.gov.in/Portals/0/Pdf/fssa_interim_regulation_on_Operatonalising_GM_Food_regulation_in_India.pdf. Accessed 26 June 2014

Food Safety and Standards Authority of India. http://www.fssai.gov.in/. Accessed 26 June 2014

Guidelines stipulating recommended dietary allowance. Indian Council of Medical Research. http://www.icmr.nic.in/final/RDA-2010.pdf. Accessed 26 June 2014

Indian Ministry of Agriculture Department of Animal Husbandry, Dairying and Fisheries. Executive summary of the report of the National Commission on Cattle. http://www.dahd.

nic.in/dahd/reports/report-of-the-national-commission-on-cattle/chapter-iii/annex-iii-1.aspx. Accessed 26 June 2014

Indian Ministry of Agriculture Department of Animal Husbandry, Dairying and Fisheries. Report of the National Commission on Cattle. http://www.dahd.nic.in/dahd/reports/report-of-the-national-commission-on-cattle/chapter-iii/annex-iii-1.aspx. Accessed 26 June 2014

Indian Ministry of Environment and Forests. Decisions taken by the Genetic Engineering Approval Committee in its 87th meeting on July 9, 2008. http://www.envfor.nic.in/divisions/csurv/geac/decision-jul-87.pdf. Accessed 26 June 2014

McClain S (25 June 2014) India faces dry spell as customs blocks booze. The Wall Street Journal, India. http://blogs.wsj.com/indiarealtime/2014/06/25/india-faces-dry-spell-as-customs-blocks-booze/. Accessed 26 June 2014

Ministry of Environment and Forests. Decisions taken by the Genetic Engineering Approval Committee in its 91st meeting on January 1, 2009. http://www.envfor.nic.in/divisions/csurv/geac/decision-jan-91.pdf. Accessed 26 June 2014

Ministry of Environment and Forests. Decisions taken by the Genetic Engineering Approval Committee in its 93rd meeting on May 13, 2009. http://www.envfor.nic.in/divisions/csurv/geac/decision-may-93.pdf. Accessed 26 June 2014

The Indian mutiny, in: The British empire. http://www.britishempire.co.uk/forces/armycampaigns/indiancampaigns/mutiny/mutiny.htm. Accessed 26 June 2014

"No beef in MDonald's fries" (4 May 2011) BBC News. http://news.bbc.co.uk/2/hi/south_asia/1312774.stm. Accessed 26 June 2014

Chapter 53
Food Regulation in the Customs Union of Belarus, Kazakhstan and Russia

Alexey Petrenko, Anatoly Kutyshenko, and Victor Tutelyan

Abstract The chapter discusses the regulation of food markets in the Eurasian Economic Union of Armenia, Belarus, Kazakhstan, Kyrgyzstan and the Russian Federation (the Union), five countries that lead regional integration in the region of the Former Soviet Union. The first part presents a brief overview of the regional co-operation in the post-Soviet world and highlights basic principles of the legislative and regulatory harmonization that are essential for understanding the concept of the regional food law. The contribution of the region to the globe agriculture is also discussed. The second part lays out general approach to technical regulation and sanitary measures adopted in the Union while the third details essential food regulations and explains basic principles of their interpretation. In the final part, two practical cases on food labeling and registration of foods for special dietary use are discussed to provide examples of practical application of technical regulations in the Union.

53.1 Economic Integration in the Commonwhealth of Independent States

53.1.1 CIS Free Trade Zone and Its Evolution Towards a Single Economic Space

The 15 independent states that emerged in the wake of the subsidence of the Soviet Union possess close to 14 % of the world's agricultural resources. According to the

---

A. Petrenko, PhD (✉)
EAS Strategies, Moscow, Russian Federation
e-mail: alexeypetrenko@eas-strategies.com

A. Kutyshenko
Amway Corporation, Sushevskiy Val 18, Moscow 123018, Russian Federation

V. Tutelyan, Sc.D, MD, PhD
Federal Nutrition, Biotechnology and Food Safety Research Center, Ustyinskiy proezd 2/14, Moscow 109240, Russian Federation

FAO report[2], the region has almost 1 ha of arable land per capita—one of the highest rates in the world—and yields a per capita daily output of over 8000 kcal of food.

In 2013, the former Soviet Union (FSU) countries contributed 5 % to the global agricultural output.[1] Together they produced 178 million tons of wheat (8 % of global output), of which 52 million were exported to the rest of the world. At the same time the region has become an important producer of meat, delivering 8,173,000 t or 3 per cent of the global output in 2013.[2]

The high standing of the FSU countries in the global agricultural output chart has been partly rooted in the once prominent farming industry of the USSR, their parent state which made up a substantial part of the global food industry and was one of world's leading agro-exporters in the 1970s and 1980s. In 1988, the USSR gross agricultural product reached 465.8 billion Rubles (116 billion dollars), being comparable to 136 billion dollars of US output.[3] Even though there was a significant decline in agricultural production and efficiency in the late 1980s (close to 25 % of agricultural products failed to reach the market), the Union was exporting over 20 billion USD worth of wheat during the final years of its existence.[4]

Nevertheless, the benefits of the wealthy Soviet heritage were offset by the harsh economic crisis that gripped the FSU in the 1990s immediately after the single market and farming industry broke up. Some countries had relied heavily on supplies from other neighbors, which were now no longer available. Six out of fifteen Soviet republics, for example, were not capable of producing their own sugar or vegetable oil. Other countries that had been big contributors to Soviet agriculture, such as Russia and Ukraine,[5] failed to adapt their old-style infrastructure to the new market economy and were not able to compete in the open market.

The economic hardship and overall impact of the market reforms pushed the FSU countries to attempt a revival of their economic cooperation, albeit somewhat unsuccessfully at first. In December 1991 three FSU countries—Belarus, Kazakhstan and the Russian Federation—established the Commonwealth of the Independent States (CIS) to promote closer political and economic integration. In a short while, the CIS was joined by all[6] but the three Baltic States, which sought closer ties with the European Union as soon as they became independent. From the start, the CIS had became a stage for political disputes and negotiations rather

---

[1]FAOSTAT (2016).
[2]CIS Executive Council (2013).
[3]Goskomstat (1990).
[4]FAO (1989).
[5]Russia and Ukraine contributed 47 % and 22 % respectively in 1988 (Goskomstat, 1990).
[6]Georgia was the last to join in 1993 and the first to leave the CIS in 2008.

than an efficient platform for economic partnership. The first agreement on the economic front came in 1994 when the CIS members agreed to create a free-trade zone.[7]

The following year, the three leading CIS members—the Russian Federation, Kazakhstan and Belarus—announced they were planning to form a customs union, thereby completely removing tariff trade barriers in the mutual trade and setting up a uniform tariff schedule in trade with other countries.[8] As with most other declarations and agreements, however, the plan remained unfulfilled and was postponed due to the CIS structural complexity and bureaucracy. The management structure of the CIS included the heads of states council, the prime ministers council, the executive council, the economy council, the parliamentary assembly, in addition to four other bodies. The councils often confused their responsibilities and authority, and commonly made decisions without any follow-up action.

The integration also stumbled as member states were clearly at different stages of economic development and had different priorities. At the same time, most of them had to deal with severe internal economic crises, and the CIS integration was not at the top of their agenda. In order to speed up regional integration, in 2000 Belarus, Kazakhstan, Kyrgyzstan, Russia, and Tajikistan formed the Eurasian Economic Community (EurAsEC).[9] Three other CIS members, Ukraine, Moldova, and Armenia, joined the new organization as observers. The EurAsEC was designed to accelerate the launch of the customs union and a single capital market. The members also set an ambitious goal to introduce a single currency and monetary union.

The EurAsEC has become a grand-scale regional structure stretching from the eastern borders of the European Union (EU) to the Pacific Ocean, and uniting a population of 214 million people. Economically, it was much smaller than the neighboring EU with total GDP being 2.2 times less than that of Europe (see comparison in Fig. 53.1). Nevertheless, the Community possessed an unprecedented growth potential and was one of the EU largest trading partners. In 2012, trade with the EU alone made up 44 % of EurAsEC's exports and 38 % of its imports.

No doubt, the formation of the EurAsEC accelerated economic integration in the CIS by enabling fast progression towards the objectives set when the CIS was first created. The CIS free-trade zone began operating in 2003 and in 2007 the Customs Union of Belarus, Kazakhstan, and the Russian Federation was finally launched.[10]

---

[7]CIS (1994).
[8]CIS (1995).
[9]EurAsEC (2000).
[10]CU Commission (2007).

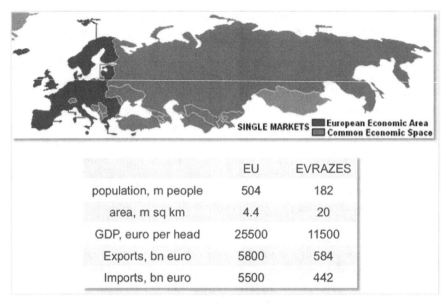

Fig. 53.1 Comparison of EU and EurAsEC geography and economics. Data by Erostat (2016) and EurAsES Secretariat (2014).

53.1.2 Customs Union of Belarus, Kazakhstan and Russia

The primary objective of the Customs Union of the Republic of Belarus, Republic of Kazakhstan, and the Russian Federation (the CU) was to completely remove tariff barriers that were hampering mutual trade in the free-trade zone. The three countries agreed on and implemented a unified code of tariffs for trade with the third countries, thereby creating a single customs territory. This essentially led to the removal of interstate customs borders with goods imported to one country being automatically cleared for circulation in the two other Union members.[11] As of today, around 370 customs points operate at the Union borders following the same customs clearance rules and procedures.

The CU was administered by the Customs Union Commission, which was established as a chief executive authority after having assumed some functions and responsibilities of national authorities.[12] The Commission's key function was to supervise member state compliance with the Union treaties and agreements.

---

[11]Isakova et al. (2012).
[12]CU Commission (2007).

53.1.3 Eurasian Economic Commission

As the CU progressed and internal trade grew, the member states agreed to further strengthen their economic integration by transforming the customs union into a single economic space (SES). The intention was to create single goods market to begin with strategic commodities, such as mineral resources and energy, and later expand to all other markets. At the heart of the SES was the principle of four freedoms that envisaged the free flow of goods, services, money, and labor within the single territory. The SES required member states to pursue coordinated policies in macroeconomics, the financial sector, transportation, energy, trade, and industry. Furthermore, the SES entailed that member states should share a uniform tax code though a single taxation regime still remains a long way from its implementation. For example, there is a difference in rate of the value added tax. Russia continues to apply 18 % rate for most goods and services offering discount rates for selected products (e.g. baby foods) while both Belarus and Kazakhstan use flat rate of 12 %.

In 2011, the Eurasian Economic Commission was established to supervise and drive the process of the CU transformation into the single economic space.[13] At the top of Commission's structure was the Supreme Council, represented by heads of the CU member states. Under the Supreme Council, the two collective bodies—the College of nine commission ministers,[14] and the Commission Council represented by national deputy prime ministers—were set to execute the Supreme Council's decisions. The College was authorized to make rulings with direct enforcement in member states in the following areas:

- customs regulation,
- technical regulations,
- sanitary, veterinary, and phytosanitary measures,
- macroeconomics policies,
- subsidies,
- energy,
- intellectual property,
- migration and financial market.

The Commission Council of deputy prime ministers serves as an advisory board for the Supreme Council and also retains the right to veto the College's decisions and rulings.

---

[13]CU Commission (2011).

[14]The ministers' responsibilities are assigned in the areas of integration and macroeconomics, economics and financial policies, industry and agriculture, trade, technical regulation, customs, energy and infrastructure and antitrust regulation.

53.1.4 Eurasian Economic Union

Most recently, the single economic space has been transformed to the economic union. On May 29, 2014, the Republic of Belarus, Republic of Kazakhstan and the Russian Federation signed a treaty of the Eurasian Economic Union (the Union) which has been designed to enhance the single economic space launched by the three countries in 2012.

The treaty represents by far the largest and the most detailed piece of the international legislation within the CIS to date. Treaty's 28 sections, 118 articles and 33 annexes occupy 1014 pages and address almost all spheres and sectors of economy.

As the closest form of the regional economic integration, the Union has been widely viewed as a step closer to a political union between the three countries (and possibly some other CIS members) on analogy with the European Union.

As of 2016, the Union has five members: Armenia, Belarus, Kazakhstan, Kyrgyzstan and the Russian Federation.

In a broad sense, the treaty re-enforces the existing laws and practices that have been developed by the member states part of the Customs Union and the single economic space.

53.2 Common Technical Regulation and SPS Measures

53.2.1 Agreement on Common Sanitary and Phytosanitary Measures

The Union harmonization of food and agriculture-related legislation is centered on common sanitary and phytosanitary (SPS) codes, in addition to veterinary measures, as described by the three basic CU agreements between the Republic of Belarus, the Republic of Kazakhstan and the Russian Federation. These agreements, which in many aspects echo the WTO SPS agreement,[15] include the agreement on sanitary measures, the agreement on veterinary sanitary measures, and the agreement on quarantine of plants adopted in 2009.[16] The common principle of the Union SPS agreements states that an SPS measure should be scientifically substantiated and dedicated primarily to the safety and health of humans and animals. At the same time, the agreements proclaim the commitment to the mutual recognition of SPS measures introduced by third countries, as envisaged by the WTO.[17]

---

[15] WTO (1994). Currently, only the Russian Federation is a member of the WTO. Both Belarus and Kazakhstan are seeking the WTO membership, and the CU legislation in most cases is drafted in compliance with WTO agreements.

[16] The Customs Union Commission (2009a, b, c).

[17] WTO (1994).

The SPS measures applicable to foods are based on a risk analysis of any adverse effects associated with a particular food additive, contamination, toxin, or microorganism present in food, drink, or animal feed. The member states are restricted to the measures described in the agreements, although they are allowed to introduce their own interim SPS measures provided that:

- there is a substantial deterioration of the sanitary and epidemiological safety;
- there is trustworthy information on existing threats to sanitary and epidemiological safety, or an additional SPS measure was introduced by a non-member state;
- There is a documented proof of unsafe goods circulated in the territory of the member state.

Additional SPS measures should be transparent, with all other member states and third countries appropriately informed on new procedures and inspections.

The agreement on sanitary measures reads that member states share a single list of goods subject to the sanitary and epidemiological control.[18] Products on the list must conform to the Uniform Sanitary Requirements[27], which describe detailed hygienic requirements to goods on the market (foods are covered in Chapter 2). The requirements contain two chapters and 23 sections that list sanitary criteria for various categories of goods. In particular, section 1 of chapter 2 describes requirements for the safety and nutritional value for foods, and contains over 400 pages of specific safety criteria applicable to foods and food raw materials. Sections 22 and 23 list safety parameters for food additives and technological aids. Harmonizing the SPS and veterinary measures was an important step in reducing trade barriers.

53.2.2 Common Principles of Technical Regulation

The Union system of technical regulation was designed with the purpose of removing non-tariff and technical barriers to trade between member states, and offsetting trade protectionism policies to which the member states often resorted during trade disputes. The system is based on five interstate agreements (Table 53.1) that harmonized national technical regulations and defined common principles shared by the member states, such as maintaining the safety and conformity of goods, along with processes of manufacturing, transportation, storage, sales, and disposal.

The system operates through special legislative acts called "technical regulations" that make up an integral part of the Union legislation and, as such, have direct enforcement in the member states. In other words, technical regulations adopted by the Eurasian Commission do not require further adoption or approval by national parliaments or governments. A technical regulation is a document that describes harmonized mandatory safety requirements for goods and processes. The member

---

[18]CU Commission (2010a).

Table 53.1 Legal acts adopted in the Union being part of the technical regulation system

Document	Key provisions/Stipulations
Agreement on circulation of goods subject to mandatory conformity assessment.[a]	• Shared database of certification authorities. • Uniform template for conformity certificates and declarations. • Single list of goods subject of conformity assessment.
Agreement on uniform principles of technical regulation in Republic of Belarus, Republic of Kazakhstan, and the Russian Federation.[b]	• Enforcement of technical regulations in the status of law. • Details of technical regulation drafting, discussion, adoption, and application.
Agreement on policy in technical regulation and sanitary and phytosanitary (SPS) measures of the Eurasian Economic Community (EurAsEC).[c]	• Mutual recognition of national standards (if in agreement with international norms). • Mutual recognition of national conformity assessments and state inspections. • Technical regulations in status of the law.
Agreement on the use of the Conformity symbol for EurAsEC member states market.[d]	• Introduction of the symbol of conformity for goods compliant with CU technical regulations.
Agreement on establishing the common information system in the sphere of technical regulation and SPS measures.[e]	• Shared databases of accredited certification authorities; certificates and declarations of conformity; records of dangerous goods, penalties, and violations.

[a]CU Commission (2009d)
[b]CU Commission (2010b)
[c]CU Commission (2008b)
[d]CU Commission (2006)
[e]CU Commission (2008a)

states have identified the single list of 66 goods categories[19] for which mandatory safety requirements must be described by technical regulations. These categories include oil and gas, machinery, railways, roads, cars, ships, lifts; food products, animal feed, and grain.

Goods included in the single list have to pass the conformity assessment, which is a pre-market approval process of evaluating product characteristics against the mandatory requirements described in the technical regulations. The assessment should take place *before* goods are imported and/or marketed in the Union. The conformity is formally endorsed by a declaration of conformity issued by a goods manufacturer/owner, or by a conformity certificate produced by an accredited certification authority.

The declaration of conformity is a statement by the manufacturer or importer that the goods conform to the mandatory requirements outlined in the technical regulations. All foods require a declaration of conformity (except those subject to the state registration). Declarations have to be registered in the Union electronic

---

[19]Eurasian Commission (2012b).

Fig. 53.2 Eurasian conformity symbol

database, as implied by the agreement on establishing the common information system in the sphere of technical regulation and SPS measures.[20]

The agreements on common technical regulatory principles and policies hold that a food operator, manufacturer, or importer bears full responsibility for the safety and quality of goods that they bring to the market.[21] At the same time, certification authorities that issue conformity certificates or register manufacturer's (importer's) declarations are responsible for the authenticity of test results and the entirety of the procedure needed to run the conformity assessment. Goods that pass the conformity assessment and are certified or declared as compliant should be labeled with the Eurasian conformity symbol (Fig. 53.2). The symbol is designed to inform consumers that a product has passed all necessary conformity procedures as prescribed by relevant technical regulations.[22] By legislating conformity between the products that pass across Union internal borders, member states understand that they importing a safe and consistent product.

53.2.3 Technical Regulations and Interstate Standards

The Union agreement on uniform principles of technical regulation implies that Union technical regulations are always given priority over the national safety norms and standards of the member states.[23] Furthermore, member states have committed to harmonizing their respective national standards and norms with the regulations in the case they are different or contradictory. The agreement states that the primary purpose of technical regulation is to protect the health and life of consumers, as well as to ensure safety of the environment. The regulation should also be designed to prevent misinforming or misleading consumers.

In general, technical regulations are drafted by an appointed member state which has to follow a uniform drafting and submission procedure approved by the Council of the Eurasian Commission.[24] A technical regulation should clearly specify: the goods that are covered, key safety requirements (including sanitary, hygienic, and

---

[20]CU Commission (2008a).
[21]CU Commission (2008b, 2010b).
[22]CU Commission (2006).
[23]CU Commission (2010c).
[24]Eurasian Commission (2012a).

veterinary criteria), rules to be followed in product identification and conformity assessment, basic principles of the in-market surveillance, and responsibilities of government authorities.

Once the regulation has been drafted and submitted to the Eurasian Commission by the appointed member state authority, the Commission publishes the text for public discussion and accepts comments and proposals from the public, industry groups, and non-government organizations. The text is then revised to account for all comments and proposals received during the hearing before being forwarded to the member states for interstate approval. The final step is a Commission ruling to formally adopt the regulation.

The first technical regulations adopted were constructed to account for as much detail as possible. The goal was to make them a single point of reference that could address or resolve any technical aspect of product circulation and safety. Thus, the technical regulation TR TS 021/2011 On Safety of Foods, adopted in 2011, detailed the specific food safety criteria for numerous categories of foods including meat, milk, sea foods, oil and fats, confectionaries, specialized foods, and dietary supplements. This approach was justified at the earlier stages of the Union as the regulator wanted to introduce fast and universal regulations that could immediately cover safety issues for the most vulnerable and high-risk industries, with foods being top priority. These universal (or horizontal), technical regulations were designed to cover a broad range of goods. Today, however, specialized (or vertical) regulations, which cover only a specific group within the category of goods, are most preffered.

The CU has adopted 35 technical regulations, which cover just over half of the 66 categories of goods that are considered to pose a risk to the human health and wellbeing.[25] The Eurasian Commission aims to adopt any missing technical regulations within the next several years. In the food sector ten regulations have been adopted to date, including the horizontal regulation TR TS 021/2011 On Safety of Foods, and TR TS 021/2011 Labeling of Foods (Fig. 53.2).

Technical regulations describe product or process safety requirements including safety-related technical specifications, and codes of practice. Technical regulations are always accompanied by a list of international or interstate standards that must be used in conformity assessment of goods to the mandatory safety requirements. They may also be supplemented by technical guidance that outlines how to comply with the requirements of the regulation, also called a 'deemed-to-satisfy' provision. It is important to note that if a product falls under the scope of more than one technical regulation, that product should conform to the requirements listed in all applicable regulations.

In conclusion, it should be emphasized that while Union technical regulations describe general provisions and criteria of product safety and conformity, in specific measurements and detail they rely on interstate standards adopted by the CIS Interstate Standardization Council, or on international standards issued by a

---

[25]Koreshkov (2014).

recognized international authority.[26] The standards are more detailed and technically comprehensive than the regulations—they typically describe a test method, a specific product characteristic, or a particular handling procedure. A recent report by the Russian Technical Regulation Agency shows that despite government effort and member state commitment, there is still a big gap between the number of standards that have to be developed in support of the adopted technical regulations, and those that are actually enacted (or taken at least to the development stage).[27] As of 2014 there are over 2400 standards awaiting development in order to provide support for adopted and drafted technical regulations. The technical regulations are designed to protect consumers, and as implied by the agreement on uniform principles, they take precedence over individual state standards.

53.3 Food Technical Regulations

In this section, we look in detail at the Union food regulations and analyze mandatory safety requirements. This includes an examination of the fundamental principles and important regulatory concepts of the horizontal food regulation TR TS 021/2011 On Safety of Foods. Special attention is given to the use of the food additives and technological aids.

We also discuss the subject of the pre-market approval—a safety concept that envisages the formal approval of foods before they enter the market. Understanding this concept is of paramount importance for ensuring undisturbed food imports to the Union member states. A separate section describes basic import requirements for foods. The national authorities responsible for food market surveillance are briefly mentioned in the final part of the section.

53.3.1 Regulatory Framework and General Safety Requirements for Foods

As of 2016, the Union has adopted ten technical regulations related to the circulation of foods, four of which are horizontal and cover safety aspects for all foods:

1. TR TS 021/2011 On Safety of Foods,
2. TR TS 022/2011 Labeling of Foods,
3. TR TS 029/2012 Safety Requirements for Food Additives, Flavorings and Technological Aids
4. TR TS 005/2011 On Safety of Packaging

---

[26]For Example, the International Organization of Standardization (ISO).
[27]Zazhigalkin (2014).

The horizontal regulations describe basic principles of food safety, list hygiene and microbiology criteria per food category, and establish key general requirements for food labeling and food packaging.

The Eurasian Economic Commission has also adopted six specialized vertical regulations that cover meat, milk, grain, fruit and vegetable juices, specialized foods (medical and preventive), and oil and fats. For food categories not listed and not covered by respective technical regulations, general provisions of the TR TS 021/2011 and TR TS 022/2011 are applied together with respective sections of the Uniform Sanitary Requirements. Since both technical regulations and Uniform Requirements address foods safety, they are closely related and often have overlapping safety standards, as seen for example for the microbiology and hygiene criteria for foods.

Figure 53.3 schematically represents the relationship between the Uniform Sanitary Requirements, horizontal technical regulations, and vertical technical regulations.

The fundamental objective of the TR TS 021/2011 On Safety of Foods regulation is to achieve the highest level of protection for consumer health. All raw materials and components of food products, including additives, are treated the same way as consumer foods and are subject to the same safety requirements and procedures. For food handling processes such as manufacturing, transportation, storage, sales, and disposal, the regulation introduces risk management recommendations based on the hazard analysis of critical control points (HACCP).[28] Implementation of HACCP principles is mandatory in food manufacturing.

The regulation TR TS 021/2011 defines foodstuffs as products of animal, plant, microbiological, mineral, artificial, or biotechnological origin in their natural,

Fig. 53.3 Schematic representation of the food regulation structure in the Union. The sanitary requirements are designed to address mostly hygienic aspects of food safety however they often overlap and duplicate safety requirements listed in the horizontal and vertical technical regulations

---

[28]Fortin (2009).

treated, or processed form. Foodstuffs are intended to be consumed by humans as food. In addition to traditional foods, the definition of foodstuffs covers the following categories:

- foods for special use,
- drinking water filled in containers and drinking mineral water,
- alcohol (including beer and beer drinks),
- non-alcoholic drinks,
- bioactive (food) supplements,
- chewing gum,
- probiotic cultures and/or microorganisms,
- food additives and flavors,
- and any raw materials used in food manufacturing.

There is a key difference between this definition that cover foods only, and the concept used in the EU food law. The latter additionally includes feed produced for, or fed to, animals raised for food. Thus, the supply chain that falls under the scope of the Union food regulation is one step shorter than that of the EU, where it starts from unprocessed or untreated food raw materials used in animal feed manufacturing.

The TR TS 021/2011 offers an extensive and risk-based safety criterion that includes:

- microbiology criteria and hygienic requirements of consumer foods (Annexes 1–3),
- irradiation tolerances (Annex 4),
- requirements for unprocessed raw materials of animal origin (Annex 5),
- and microbiology criteria for fish and seafood (Annex 6).

There is also a separate list of pesticides that are not allowed in manufacturing of raw materials of baby foods.

The TR TS 021/2011 has introduced three new concepts that cover foods with modified composition, novel foods, and foods with intended health effects. The first concept defines 'foods for special use' as foods with a tailored composition: those that possess a claimed effect or benefit for human health and are designed for consumption by a targeted population group. The second concept covers novel foods that do not have a prior history of safe use in the Union. In particular, novel foods include foods with altered molecular structure; foods produced from, or consisting of, microorganisms, micro fungi and/or algae, plants, and animals; and foods sourced from genetically modified organisms (GMOs) or nano-structures.

The third concept deals with regulating enriched or fortified foods. The term 'enriched or fortified foods' refers to those that contain nutrients, probiotics, or bioactive substances that have been added during manufacturing or processing. In most cases, the amount of added substance should be above 15% of a recommended daily allowance value.

The TR TS 021/2011 implies that industrial facilities intended for the processing or handling food raw materials of animal origin be authorized—that is, registered with state authorities—before being operated. This requirement is mandatory for slaughterhouses, milk and egg farms, and fish and seafood farms. The registration process includes submitting an application for the state permit and, once the permit is issued, registering the facility in the electronic database. Other types of manufacturing facilities that handle foods or food components are not required to register, although their owners must notify authorities on the nature and the size of the operation.

In 2012, the Eurasian Economic Commission adopted horizontal regulation TR TS 029/2012 On Safety of Food Additives, which summarized complete requirements for the use of food additives, flavorings, and technological aids across all product categories that fall under the definition of foods. In addition, the law deals with foods that contain additives, bioactive substances in the form of flavoring and color,[29] and technological aides. It also describes the safety requirements for related processes of manufacturing, storage, transportation, and disposal of the additives.

The regulation lists all additives that are permitted in foods in clases as defined by the International Numbering System for Food Additives.[30] In general, the Union positive list of additives is similar to the Codex Alimentarius General Standard of Food Additives and European list approved by EU regulation 1131/2008.[31] At the same time, TR TS 029/2012 lists as approved some plant-origin additives that are not on the EU positive list, such as RED RICE color, extract of licorice, extract of Acantophyllum root, and the leaf powder and syrup of Stevia Rebaudiana Bertoni. Overall, the TR TS 029/2012 regulation represents a comprehensive guide for food manufacturers on what additives can be used in what foods, and in what amounts.

Annexes 3–8 of the regulation describe maximum limits of food additives as defined per food category. Annex 10 specifies a list of foods for which only selected colors are allowed, and Annex 18 describes foods for which only selected additives may be used.

53.3.2 Pre-Market Approval of Foods

The technical regulation TR TS 021/2011 On Safety of Foods reads that food manufacturers and importers are allowed to take their products to the market only after passing a pre-market assessment of conformity to the mandatory requirements. Essentially, this process is an assessment that goods meet requirements of technical regulations, which is run by national authorities responsible for the safety of all

---

[29]Augustin and Sanguansri (2012).
[30]Codex Alimentarius (1989).
[31]EU Parliament (2008).

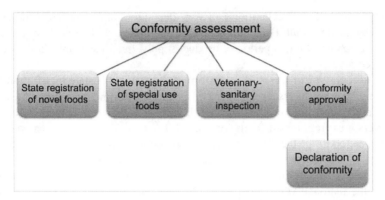

Fig. 53.4 Forms of conformity assessment adopted in the Eurasian Economic Union

goods imported or manufactured in the Union territory. Depending on the category of the product, the pre-market approval of foods could take the form of:

- a declaration of conformity;
- state registration (including state registration of novel foods, bioactive supplements, and GMO foods), and/or
- veterinary-sanitary expertise (applicable only to goods subject to the veterinary control, mostly materials of animals and maritime origin) (Figure 53.4).

The Annex 9 of the Union Treaty clarifies the difference between terms conformity approval and conformity assessment which have often been confused by regulators and the industry alike. The conformity approval is defined as a documental prove of conformity with mandatory requirements of technical regulations (e.g. certificate, declaration, permission, etc). The conformity assessment is a direct or indirect examination of product properties and characteristics.

The declaration of conformity is issued based on evidence collected by a declarant (a manufacturer or an importer) or a third party (an authorized certification body). The evidence should clearly demonstrate that the product being declared was produced under strict manufacturing control and, as a result, complies with the safety criteria listed in technical regulations or the interstate standards referred to in the regulations.

The declaration of conformity is the simplest form of the pre-market approval. It differs, too, from the notification widely used in the US and the EU, as the declaration must be supported by reliable evidence of compliance with the regulations. In the Union, all foods, except those being subject to state registration (see below), require declaration of conformity before being imported or marketed in member states. The declaration must be issued in the standard form approved by the Eurasian Economic Commission.

Before being officially issued, the declaration is submitted to a certification authority, which evaluates the body of evidence presented in support of the declaration. On the authority's approval, the declaration is recorded in the Union database of conformity declarations and henceforth regarded as released. The validity period of the declaration can vary depending on the approval procedure chosen by the declarant. For example, the declaration could be issued for a specific product batch or a limited period of product supply. On no accounts, however, is it to exceed 5 years.

The second type of pre-market conformity assessment is state registration, seen by many manufacturers and importers as the most rigid and time-consuming procedure. According to the TR TS 021/2011 regulation, manufacturers and importers must register all foods for special use, including bioactive supplements and novel foods. In the Russian Federation foods are registered by the Federal Service on Customer's Rights Protection and Human Wellbeing Surveillance (Rospotrebnadzor). In the Republics of Belarus and Kazakhstan, registrations are run by the national Ministries of Health. The registration consists of an expert review of the product dossier, as well as laboratory safety tests and analyses. The official length of time for processing a complete product dossier and completing the laboratory analyses is 40 days from the submission date. In practice, however, the procedure could take several months depending on the quality of submitted documentation and the scope of analytical work required.

As mentioned above, conformity assessments are mutually recognised by the member states/The accreditation of laboratories, certification bodies and assessment centres is also guided by the principle of mutual recognition which implies that all accredited bodies in one member state are recognised in two others. However, a laboratory or centre that resides in one member state may apply for accreditation in another member state only if its national accreditation authority does not cover a particular area of accreditation (goods tested, tests performed etc).

53.3.3 Import Requirements

The Eurasian Union Treaty implies that the member states share a single customs territory.[32] Therefore, the customs authorities of the five member states use a uniform customs code and a harmonized procedure of customs clearance for imports from third countries. This procedure requires all food shipments that are imported to the Union territory to be accompanied by documents of conformity approval as assigned by the TR TS 021/2011 technical regulation.

It is essential to consult the TR TS 021/2011 on what types of conformity assessment to follow for specific foods before importing to the Union. As described in the previous section, all foods need to pass a pre-market assessment that generates either the declaration of conformity or the product registration certificate. In addition, some foods will need to pass sanitary, veterinary and/or phytosanitary inspections. The Eurasian Economic Commission maintains three lists of goods, grouped in accordance with international harmonization codes, which require safety assessment

---

[32]EurAsEC (2007).

at the CU customs border. The first list defines goods that are subject to sanitary-epidemiological inspection at the border—the inspection is run to evaluate the hygiene safety of the shipment and review documents that testify the goods' safety.

The second list describes goods that are subject to veterinary inspection. The clearance of these goods will require submission of a veterinary certificate issued by the country of origin, and/or an import permit issued by a member state authority. The third list of goods requiring customs clearance is for products containing potentially contaminated units. These require the submission of a phytosanitary certificate that has been issued by the country of origin.

53.3.4 In-Market Surveillance Authorities

In the Union, the in-market surveillance of foods' conformity to technical, sanitary, veterinary, and phytosanitary regulations remains the responsibility of the national authorities.

In the Russian Federation, the Rospotrebnadzor was appointed to supervise compliance with technical regulations in consumer markets (including food). The Rospotrebnadzor functions as a consumer market watchdog: it runs the state registration of novel foods, foods for special use, and bioactive supplements. It also processes notifications from manufacturers on new food products. In contrast to processed products, the safety and conformity of raw food materials and unprocessed/untreated foods is controlled by the Federal Service of Veterinary and Phytosanitary Surveillance (Rosselkhoznadzor) under the Ministry of Agriculture. Broad supervisory functions of the Rosselkhoznadzor cover veterinary safety, plant quarantine, circulation of pesticides and agrochemicals, and good land cultivation practices. The service also supervises the circulation of grains and cereals, two categories that remain one of the largest and strategic agricultural export commodities for Russia and other Union member states.

In Republic of Belarus, the state authority on food markets is shared between the Ministry of Health and the Ministry of Agriculture. The Belarus Ministry of Health runs consumer market surveillance and sanitary monitoring, while the Ministry of Agriculture is responsible for veterinary and phytosanitary measures. In Kazakhstan, the recently established State Agency of the Consumer Rights Protection has been put in charge of food safety. This new body ensures product compliance to the requirements of the Union technical regulations, a responsibility that was formerly under the Ministry of Industry and New Technologies. It also monitors sanitary-epidemiological safety, a task that was previously observed by the Ministry of Health. In addition, the Agency is responsible for registering foods for special use with the state.[33]

---

[33]Foods for special use make up an important category of products designed for consumption by specific population groups, and for specific purposes.

53.4 Case Studies

Two case studies in this section discuss labeling requirements, product claims, and food advertising in the Union, highlighting practical aspects of the technical regulations.

The first case presents a detailed analysis of two technical regulations, TR TS 022/2011 and TR TS 005/2011, which cover food labeling and packaging requirements. It is useful in demonstrating practical use of provisions provided in the technical regulations. The second case covers the specifics of the dossier submission for state registration of foods for special use.

53.4.1 Case Study I: Consumer Information on Food Products

53.4.1.1 Labeling Requirements

Providing consumers with full and trustworthy information is one of the most important responsibilities of food manufacturers and importers. The Union law implies that conformity approval for foods extends beyond the product itself and covers its packaging and labeling as well. According to TR TS 022/2011 Labeling of Foods, the information on the label of any food product must conform to a number of mandatory requirements. The most essential of these include:

- product name,
- composition (the list of ingredients),
- net quantity,
- date of manufacture,
- manufacturer and importer details,
- nutritional value,
- GMO information, and
- Eurasian conformity symbol (figure 53.1).

In what follows below, each requirement is described in detail. The name of a food product should reflect the product's essence and allow for easy product identification and differentiation. The name should not directly or indirectly refer to components that are not present in the product.

The composition (list of ingredients) should be listed on the product label in order of decreasing weight percentage. Food additives may be referred to by their chemical names and/or INS number according to the International Numbering System. In this context, some specific requirements for information on food additives need to be mentioned. Foods containing polyol sweeteners should be labeled with a warning statement that the product contains sweeteners and may cause diarrhea if consumed irresponsibly. For foods with added so-called Southampton colors, Carmoisine E122, Tartrazine E102, Quinolone Yellow E104, Sunset Yellow E110, Allura Red E129, and Ponceau 4R E124, there must be a warning statement that the components could adversely affect children's activity and concentration.[34] Finally, the TR TS 022/2011 regulation contains an extensive list of 15 categories of allergens and substances that should be avoided by people with certain diseases. Examples include broad food categories such as nuts, eggs, seafood, and milk, as well as specific substances like aspartame and sulphur dioxide.

The net quantity of foods must be quoted in a weight or volume measure, or in the number of product units. TR TS 022/2011 also allows the use of two measuring units at the same time—for instance, mass/volume. It is *not* permissible to indicate an approximate amount of product or give a range of quantities such as "around 100 g" or "100-150 grams".

It is also essential to include the date of manufacture on the packaging. The format of the manufacture date depends on the product's shelf life. For shelf lives of less than 72 h, the date of manufacture should include the hour, day, and month. Shelf lives up to 3 months require indication the day, month, and year of manufacture. Anything longer than 3 months only requires listing the month and year.

Foods must contain manufacturer and importer details. Products that were manufactured at one facility and then packed in another have to be labeled with the information of both facilities. The same approach needs to be used for foods that were produced by more than one manufacturer, provided the information allows a consumer to clearly identify the manufacturer of the packaged product.

Information on the nutritional value should reflect the energy value (in calories), as well as the amount of protein, fat, and carbohydrates in the product. This requirement is not mandatory for foods that contain less than 2% of the recommended daily allowances for energy and macronutrient intake, as defined in Annex 2 of regulation TR TS 022/2011. The amounts of vitamins and minerals that are added to foods during manufacturing must be included in the nutritional value information. Vitamins and minerals may also be quoted as part of the nutritional value if their amounts per 100 g or 100 ml of the food product exceed 5% of the recommended daily allowance (listed in Annex 2 of regulation TR TS 022/2011).

Manufacturers must also disclose the use of any genetically modified organisms (GMOs) on the label, if present in amounts over 0.9%. This should be in the form of a "genetically modified product" statement on the packaging. The regulation does not require a GMO statement for foods that contain GMOs as "technically unavoidable residue" in amounts less than 0.9%.

Finally, foods that receive conformity approval need to be labeled with the Eurasian Conformity symbol (as described in Sect. 53.2.2). The labeling should also provide information about the product's shelf life, storage conditions, and instructions for use or cooking instructions.

---

[34] Abbey et al. (2014).

Fig. 53.5 Symbols of food-contacting and recyclable packaging to be used in food labeling, according to the TR TS 005/2011 technical regulation

As of 2016, the CU has not yet adopted a specific regulation that would cover food-contacting materials.[35] These issues are currently regulated by the horizontal technical regulation TR TS 005/2011 On Safety of Packaging which regulates safety of all packaging and packaging materials. In particular, Annex 1 of the regulation lists sanitary and hygienic criteria for food packaging and food-contacting materials, as well as test methods and testing models. The regulation additionally requires that food-contacting materials be labeled with a special "suitable for foods" symbol. Recyclable packaging should be marked with the Mebius loop (see Fig. 53.5).

The packaging materials, along with the packaging itself, need to have declarations of conformity issued before being used for food products and released on the market. As with food items, packaging that passes conformity approval needs to be marked with the Eurasian conformity symbol (see Sect. 53.2.2).

53.4.2 Advertising

In 2013, food was the second largest category of goods advertised on Russian TV, trailing only behind pharmaceuticals.[36] Despite the significance of advertising in the marketing of food products, advertising regulations are not yet harmonized across Union member states which often apply rather different approaches in regulating advertising. In the Republic of Kazakhstan, for example, Law No. 508-II "On Advertising" does not contain any references to foods or food products at all. By contrast, the Republic of Belarus and the Russian Federation have tighter policies, as explained below.

In the Republic of Belarus, Law No. 225-Z "On Advertising," adopted in 2007, was recently amended with Article 15–1 "Advertising for bioactive (food) supplements." The amendment introduced the following provisions relating to bioactive

---

[35]In 2013, the Russian Federation initiated work on drafting new CU technical regulation dedicated exclusively to food-contacting materials. It is not yet clear how the new law would interact with the existing regulation TR TS 005/2011 as both would cover the same safety aspects of food-contacting materials.

[36]TNS Global (2014).

(food) supplements, which comprise one category of Foods for Special Use. Article 15–1 states that:

1. Advertising for food supplements is permitted only with approval from the Belarusian Health Ministry. In other words, adverts must be pre-approved before going public.
2. Advertising for food supplements that have not passed through the state registration process is prohibited.
3. Advertising for food supplements must contain: the name of the supplement producer (manufacturer); information indicating that the product is a food supplement, and not a medicine nor intended for treating diseases; and information advising consumers to read the instructions carefully.
4. Advertising for food supplements must *not* contain: information directly appealing to the underage; statements or suggestions implying that the target audience has certain conditions that require the use of the advertised product; statements which may give healthy people the impression that they need to consume the advertised product; suggestions of possible financial reward in exchange for purchasing the advertised product; and information which does not correspond to the information mentioned on the food supplement retail label.

In the Russian Federation, Federal Law No. 38-FZ "On Advertising" contains special provisions related to the advertising of baby foods, food additives, and bioactive (food) supplements. In relation to baby foods, for instance, the Law forbids positioning infant formulas as a replacement to breastfeeding or to state that infant formulas are advantageous for child development. In addition, Article 25 of the Law reads that advertising of baby foods, food additives, and bioactive (food) supplements should not:

- form the impression that additives and supplements are medicines and/or have therapeutic properties;
- contain references to concrete instances of people recovering or experiencing a health improvement as a result of using such products;
- contain praises of dietary supplements from individual persons;
- impel the consumer to give up healthy nutrition; and
- suggest certain advantages of a supplement or an additive by making references to the results of the mandatory testing undertaken for the pre-market approval process (state registration), or the results of any other tests that have been performed with the product.

53.4.3 Product Claims

Until recently, claims for food products remained a largely unregulated area of food labeling and marketing. Currently, there is no harmonized procedure for approving claims for food products in the Union. Manufacturers have no clear guidance what evidence to submit and what criteria are used in consideration of claim applications. Instead, the consumer market authorities in member states make individual ad-hoc judgments on product labeling and consumer information during the pre-market approval process.

The technical regulation TR TS 022/2011 on Labeling of Foods made the very first attempt to introduce the term "nutrition claims" and define a criterion for its use in labeling across all food product categories. The regulation describes specific nutrition claims and the conditions for their use (see Table 53.2).

The regulation reads that claims are part of the general description of a product's discernible characteristic, which manufacturers may voluntarily include to inform consumers about their products. The claims, however, may only be used if the "conditions for use" terms are met. Claims must be accompanied by scientific evidence. This evidence may be produced independently, by the applicant, or by a third party. An organization or an individual responsible for releasing the food product into the Union market should keep the evidence and supporting documents, and they must be made available on request of national or Union authorities.

A nutrition claim is, in fact, the only type of claim about a food product that is permitted and regulated in the Union. The current legislation lacks the whole set of definitions adopted in the USA and the EU that address health claims, functional

Table 53.2 Selected examples of nutrition claims that can be used as described in Annex 5 of the TR TS 022/2011

Claim	Condition of use
Low in calories	Energy value is below 40 kcal per 100 g of solids or below 20 kcal for liquids. For sweeteners consumed as a sugar replacement, the value is below 4 kcal (sweetening value is equal to 6 g of sucrose).
Source of protein	The amount of protein contributes at least 12 % of the product's energy value, provided that this amount is at least 5 % of the recommended daily allowance (RDA) of protein per 100 g of solids or 100 ml of liquids.
Low fat	Fats amount to less than 3 g per 100 g of solids, and less than 1.5 g per 100 ml of liquids.
Source of fiber	Product contains at least 3 g of fiber per 100 g of solids, or at least 1.5 g per 100 ml of liquids.
High levels of vitamins and minerals	Vitamins and minerals constitute at least 30 % of RDA per 100 g of solids, per 100 ml of liquids, or per portion.
Low level of sodium (table salt, sodium chloride)	Level of sodium (or its sodium chloride equivalent) is below 0.04 g per 100 g of solids or 100 ml of liquids.
No cholesterol	Amount of cholesterol is below 0.005 g per 100 g for solids or 100 ml of liquids, under condition that the product contains less than 1.5 g of saturated fatty acids per 100 g of solids or less than 0.75 g per 100 ml of liquids.

claims, qualified claims, and others. Gaps in legislation do not help food marketing and hinder food trade with the rest of the world.

It should be noted that there is no uniform Union-wide procedure in place for conducting an assessment (expert evaluation) of claims for market approval. National authorities deal with existing market practices and new products on a case-by-case basis, randomly allowing some claims and repelling the others.

53.5 Case Study II: Registration of Foods for Special Use

53.5.1 Classification of Foods for Special Use

TR TS 021/2011 On Safety of Foods introduces a new classification of products for special use. Under this criterion, they can be categorized as:

- Baby foods, including drinking water (split into three age categories: babies of 0–3 years, pre-school age of 3–6 years, and school age of 6 years and older).
- Dietetic medical foods—defined as foods with controlled nutrition and energy values, and physical and organoleptic properties. They are to be used in medical diets.
- Dietetic foods for disease prevention (prophylactic) purposes—defined as those used for correcting carbohydrate, fat, or protein metabolism; those with a modified ratio of ingredients, those containing foreign substances (components); or those designed to reduce the risk of disease.
- Mineral water with mineralization of at least 1 g/L, or containing bioactive substances.
- Sport foods—defined as foods of specific chemical composition, designed to have increased nutritive value and/or a tailored effect to improve the adaptive human response to physical and emotional stress.
- Foods for pregnant and breast-feeding women.
- Bioactive (food) supplements—defined as natural (or identical to natural) biologically active substances, including pro-biotic microorganisms, which are designed to be taken with food or made part of food products.

As has already been noted throughout this text, Foods for Special Use must pass a pre-market approval in the form of state registration before they can be marketed and sold.

53.5.2 Registration Dossier

All required documents in the registration dossier have to be issued in their original version, officially translated and approved by a notary. A typical dossier should include the following documents:

1. A formal registration request and a filled-in application form.
2. Power of attorney authorizing the applicant to apply for product registration, unless the product manufacturer or importer applies directly.
3. Documents issued by the applicant or by authorities in the product's country of origin testifying that the product is food and is safe for human consumption. These documents include:

   - a hygienic certificate, health certificate, or safety certificate;
   - a declaration that the product does not contain nano-materials, GMOs, hormones, or pesticides;
   - a free-sale certificate;
   - a certificate on the legal status of the manufacturing facility in the country of origin;
   - information about the country of origin authority that is responsible for legal compliance of the product.

4. A full ingredients list, including data from the quantitative and qualitative analysis, signed and stamped by the manufacturer.
5. A short summary of the manufacturing process, signed and stamped by the manufacturer.
6. An explanatory note on the product's special use, including:

   - indication of the recommended daily dose;
   - recommended duration of intake;
   - contraindications, if any;
   - support of any health benefit claims by clinical study reports or other scientific data;
   - documents containing a toxicological and hygienic assessment of the product;
   - restrictions of use, if any; and
   - efficacy.

7. Foods that contain herbal ingredients additionally require:

   - manufacturer's declaration that the product does not contain narcotic and psychotropic substances;
   - information on the herb's scientific botanical name and the mode of preparation (e.g. extract—1:4, solution—1–10, etc.).

8. Foods that contain live microorganisms additionally require:

   - information on the category and type of microorganism (name should be given in Latin);
   - information of the name of the strain and registration information (such as passport, certificate, or similar).

9. Foods that contain GMOs additionally require a manufacturer's statement on the ingredients obtained from genetically modified sources,
10. Sports foods require additional documents for dossier submission, including:

    - manufacturer's statement that the product does not contain substances classified as doping.

11. Mock-up of a product label and a sample of the original label both sealed with the manufacturer's company seal.
12. 10 product samples.
13. Completed form of the random sampling report sealed with the manufacturer's company seal, specifying time and place of sample selection, the quantity and names of samples, manufacturer's address, manufacturing date, and names and positions of staff members who collected the samples.
14. Notarized approval of the translation for all documents listed above.

53.6 Conclusion

The economic and political integration in the CIS area has been gathering momentum after the CIS free-trade zone, established in the early 2000s, was transformed into a fully functioning Customs Union and most recently into the Eurasian Economic Union. The single consumer market has also been developing with food and agriculture sectors at the forefront of legislative, regulatory, and supply chain unifications.

The Union member states still need to resolve disagreements on the regulatory front, as they have become stumbling blocks on the path to closer integration. At the top of the issue list is inconsistency between sanitary and technical regulations. In addition, there continues to be in some cases a straight conflict between Union technical regulations and member states' national standards and norms.

On the one hand, in Kazakhstan, national food sanitary norms (SanPiN) have been left completely unrevised to account for newly adopted technical regulations in the Union, thereby creating conflicts with the national legislation.[37] On the other hand, the Republic of Belarus continues with its state program of expanding the national system of technical standards. The Belarusian government regularly introduces new classifications of foods and imposes excessive safety requirements on the top of what has been written in the Union technical regulations.

Another challenge for Union members has to do with the absence of interstate technical standards which are urgently required to support the enforcement of the technical regulations. Instead, market operators are left with the responsibility of declaring the conformity of their own goods, but they lack a clearly defined criteria and testing method for doing so.

---

[37]World Bank (2012).

It also needs to be noted that the Treaty of the Eurasian Economic Union introduces two important principles still to be fully realised. First, it declares that a member state authority responsible for the market surveillance cannot simultaneously be in charge of the conformity assessment. In the Russian Federation, for example, the market watchdog Rospotrebnadzor currently combines the functions of market surveillance with the responsibility of running the product registration of Foods for Special Use. Second, the treaty emphasizes that national law and safety requirements are no longer applicable if there is a respective Union technical regulation that covers safety of goods and processes in question.

Republics of Armenia and Kyrgystan have become most recent members of the Treaty and more countries are considering joining, including Tajikistan and Uzbekistan. The Union is now driven by mostly transparent and WTO-compliant legislation. Since Russia's and Kazakhstan's WTO accessions, the Union legislation has been developing in line with Russia's WTO agreement, regardless of the fact that Kyrgyzstan is not a WTO member. For instance, Russia systematically pushes amendments to the Union tariff code along with its WTO commitments. In the same manner, the WTO sanitary and phytosanitary measures agreement is taken into account when new technical regulations are drafted or sanitary measures are applied.

Russia has been engaged in close discussion with the Organization for Economic Co-operation and Development (OECD) on membership, which would trigger the application of the OECD principles in accreditation and analytical services in the Union. In particular, good laboratory practices[38] and a mutual recognition of laboratory results and data between OECD members would greatly facilitate the conformity approval of imported products in the Union.

At the same time, however, food markets could face more rigid state regulation considering growing concerns over the safety of food supply chains both regionally and globally. With so many international operators involved in the manufacturing and marketing of a single food product, the Commission and national market authorities find it increasingly difficult to control the quality and safety of foods at different stages of a product's circulation in the market. Thus, we expect that pre-market approval (declarations, product registrations) will remain an integral part of the regulatory system for the foreseeable future.

Finally, the member states are increasingly concerned that they continue to top global charts of non-communicable disease rates, while the scientific community unanimously agrees that unhealthy food choices and poor dietary habits represent a serious health risk. In both Kazakhstan and Russia, the state policies for national healthcare development seek to promote healthy nutrition and strengthen the local production of foods for special use, enriched foods, and bioactive supplements.[39] It has been frequently repeated that a consolidated effort from the government and the industry is required to reduce vitamin and nutrient deficiencies amongst the

---

[38] OECD (2014).

[39] The Government of the Russian Federation (2010).

general population. In the next decade, therefore, we could see more regulations aimed at facilitating the manufacture and distribution of "healthy" foods, including foods for special use.

References

Abbey J, Fields B, O'Mullane M (2014) Food additives: colorants. Encycl Food Saf 2:459–365
Augustin MA, Sanguansri L (2012) Challenges in developing delivery systems for food additives, nutraceuticals and dietary supplements. Nutrition 19–48
CIS Executive Council (2013) Prognoz proizvodstva, potrebleniya, vvoza i vyvoza vayzhneyshikh vidiv produktsii gosudarstv-uchastnikov CNG (Forecast of manufacturing, consumption, import and export of most essential commodities by CIS countries in 2014), http://www.e-cis.info/index.php?id=323. Accessed 1 Mar 2014
CIS - The Commonwealth of Independent States (1994) Dogovor o zone svobodnoy torgovli (The agreement on a Free Trade Zone). http://cis.minsk.by/reestr/ru/index.html#reestr/view/text?doc=3183. Accessed 19 Mar 2014
CIS - The Commonwealth of Independent States (1995) Soglasheniye o tamozhennom soyuze (The agreement of the Russian Federation, Republic of Belarus and Republic of Kazakhstan on the Customs Union). http://www.eurasiancommission.org/docs/. Accessed 1 Mar 2014
Codex Alimentarius (1989) Class names and the international numbering system for food additives, CAC/GL 36-1989, 1–57
Erostat (2016) Database. http://ec.europa.eu/eurostat/data/database. Accessed 1 May 2016
EurAsEC (2000) Dogovor ob uchrezhdeniii Evraziyskogo ekonomicheskogo soobshestva (The Treaty on establishing the Eurasian economic community). http://www.evrazes.com/docs/view/3. Accessed 1 Mar 2014
EurAsEC (2007) Dogovor o sozdanii yedinnoy tamozhenoy territory i formirovanii Tamozhennogo soyuza (The agreement on formation of a common customs territory and the customs union of Belarus, Kazakhstan and Russia). http://www.eurasiancommission.org/docs/Download.aspx?IsDlg=0&ID=3039&print=1. Accessed 1 Mar 2014
EurAsEC secretariat (2014) EurAsEC Today 2013. http://www.evrazes.com/i/data/item7621-1.pdf. Accessed 19 Mar 2014
FAO - Food and Agriculture Organisation of the United Nations (1989) Commodity review and outlook 1988–89. Rome
FAO Food and Agriculture Organization of the United Nations (2013) FAO statistical year book. http://www.fao.org/docrep/018/i3107e/i3107e00.htm. Accessed 19 Mar 2014
FAOSTAT (2016) Production domain. http://faostat3.fao.org. Accessed 1 May 2016
Fortin ND (2009) Food regulation - law, science, policy and practice. Wiley, Hoboken, pp 240–245
Goskomstat USSR (1990) Razvitiye selskokhozyaystvennoy promyshlennosti (Development of the agricultural industry). Concise Statistical Book. Goskomstat, Moscow
Isakova A, Koczan Z, Plekhanov A (2012) How much do tariffs matter? Evidence from the customs union of Belarus, Kazakhstan and Russia. European Bank for Reconstruction and Development, http://www.ebrd.com/downloads/research/economics/workingpapers/wp0154.pdf. Accessed on 19 Mar 2014
Koreshkov VN (2014) Creating technical regulatory system go the Customs Union. In-market control and ensuring conformity with technical regulations. Paper presented at the international conference "Russia - Customs Union - European Union. Technical regulation. Standardization. Conformity assessment. Achievements. Hurdles. Perspectives", Moscow, 13 February 2014

OECD - Organization for Economic Co-operation and Development (2014) OECD series on principles of Good Laboratory Practice (GLP) and compliance monitoring. http://www.oecd.org/. Accessed on 19 Mar 2014

The Customs Union Commission (2006) Soglasheniye o preminenii yedinogo znaka obrasheniya produkstsii na rynke gosudarstv-clenov Evraziyskogo ekonomicheskogo prostranstva (The Agreement on the use of Conformity Symbol for products in the market of member states of the Eurasian economic space). http://www.eurasiancommission.org/docs/Download.aspx?IsDlg=0&ID=3104&print=1. Accessed 19 Mar 2014

The Customs Union Commission (2007) Dogovor o komissii tamozhennogo soyuza (The Treaty on the Customs Union Commission), http://www.eurasiancommission.org/docs/. Accessed 1 Mar 2014

The Customs Union Commission (2008a) Soglasheniye o provedenii soglasovannoy politiki v oblate tekhnicheskogo regulirovaniya, sanitarnykh i phytosanitarnykh mer (Agreement on concurred policy in technical regulation, sanitary and phytosanitary measures). http://www.tsouz.ru/DOCS/INTAGRMNTS/Pages/Sogl_teh_san_1part.aspx. Accessed 1 Mar 2014

The Customs Union Commission (2008b) Soglasheniye o sozdanii informatsionnoy sistemy Evraziyskogo ekonomicheskogo soobshestva v oblate tekhnicheskogo regulirovaniya, sanitarnykh i phyotsanitarnykh mer (Agreement on establishing the common information system of the Eurasian Economic Community in the sphere of technical regulation, sanitary and phytosanitary measures). http://www.tsouz.ru/Docs/IntAgrmnts/Pages/evrazes_12122008_2.aspx. Accessed 1 Mar 2014

The Customs Union Commission (2009a) Soglasheniye tamozhennogo soyuza po sanitarnym meram (The Customs Union Agreement on Sanitary Measures). http://www.eurasiancommission.org/docs/Download.aspx?IsDlg=0&ID=5238&print=1. Accessed on 1 Mar 2014

The Customs Union Commission (2009b) Soglasheniye tamozhennogo soyuza po vetirenarno-sanitarnym meram (The Customs Union Agreement on Veterinary and Sanitary Measures). http://www.eurasiancommission.org/docs/Download.aspx?IsDlg=0&ID=5239&print=1. Accessed 1 Mar 2014.

The Customs Union Commission (2009c) Soglasheniye tamozhennogo soyuza o quarantine rasteniy (The Customs Union Agreement on Quarantine of Plants). http://www.eurasiancommission.org/docs/Download.aspx?IsDlg=0&ID=5240&print=1. Accessed 1 Mar 2014

The Customs Union Commission (2009d) Soglasheniye ob obrashenii produktsii, podlezhayshey obyazatelnoy otsenke sootvetstviya, na tamozhennoy territory Tamozhennogo soyuza (Agreement on circulation of goods subject to mandatory conformity assessment). http://www.tsouz.ru/MGS/mgs-11-12-09/Pages/mgs25-27-pril1.aspx. Accessed 1 Mar 2014

The Customs Union Commission (2010a) Edinyy perechen tovarov, podlezhashikh sanitarno-epidemiologicheskomu nadzoru na granites tamozhennoy territory Tamozhennogo soyuza (The Single List of Goods Subject to Sanitary and Epidemiological Supervision (Control) at the Customs Border and Customs Territory of the Customs Union). http://www.eurasiancommission.org/ru/act/texnreg/depsanmer/regulation/Documents/perechen-73.pdf. Accessed 19 Mar 2014

The Customs Union Commission (2010b) Soglashenie o yedinykh printsipakh tekhnicheskogo regulirovaniya v Respublike Belarus, Respublike Kazakhstan and Rossiyskoy Federatsii (Agreement on uniform principles and rules of technical regulation in Republic of Belarus, Republic of Kazakhstan and the Russian Federation). http://www.eurasiancommission.org/ru/act/texnreg/deptexreg/normbaza/Documents/Soglachenie%20o%20principah.pdf. Accessed on 19 Mar 2014

The Customs Union Commission (2010c) Uniform sanitary and epidemiological and hygienic requirements for products subject to sanitary and epidemiological supervision (control). http://ec.europa.eu/food/safety/docs/ia_euru_sps-req_req_san-epi_chap-2_1_en.pdf. Accessed on 1 May 2016

The Customs Union Commission (2011) Dogovor o Evraziyskoy ekonomicheskoy komissii (The Treaty on the Eurasian Economic Commission). http://sudevrazes.org/main.aspx?guid=19461. Accessed 1 Mar 2014

The Eurasian Economic Commission (2012a) Polozheniye o poryadke razrabotki, prinyatiya, vneseniya izmeneniy i otmeny tekhnicheskogo reglamenta Tamozhennogo soyuza (Guidelines on drafting, adoption and amending and suspending a technical regulation of the Customs Union). http://www.tsouz.ru/eek/RSEEK/RSEEK/SEEK6/Documents/P_48.pdf. Accessed on 19 Mar 2014

The Eurasian Economic Commission (2012b) Yedinyy perechen produktsii, v otnoshenii kotoroy ustanavlivayutsya obyazatelnyye trebovaniya v ramkakh Tamozhennogo soyuza (The single list of goods subject to mandatory requirements in the Customs Union). http://www.eurasiancommission.org/ru/act/texnreg/deptexreg/tr/Documents/Ed%20perech%20new.pdf. Accessed on 19 Mar 2014

The European Parliament and the Council (2008) Regulation (EC) No 1331/2008 of The European Parliament and of the Council of 16 December 2008 establishing a common authorization procedure for food additives, food enzymes and food flavorings. Official Journal of the European Union L354:1–6

The Government of the Russian Federation (2010) Resolution 1873-r. The policy of healthy nutrition of the Russian Federation until 2020

The World Trade Organization (1994) The Uruguay Round Agreement, Agreement On Application of Sanitary and Phytosanitary Measures. http://www.wto.org/english/docs_e/legal_e/15sps_01_e.htm. Accessed on 19 Mar 2014

TNS Global (2014) Monitoring data report: TV advertising in 4 quarter of 2013. http://tns-global.ru/services/monitoring/advertising/description/. Accessed on 24 Mar 2014

World Bank (2012) Report 65977-KZ assessment of costs and benefits of the Customs Union for Kazakhstan, New York

Zazhigalkin AV (2014) Standards as a body of evidence in technical regulations. Paper presented at the international conference on the industry role in technical regulation system of the single economic space, Russian Business Week, Moscow, 19 March 2014

Chapter 54
Israeli Regulation and Policy of GM Food and Crops

Ronit Justo-Hanani

Abstract Israel's policy on GMOs during the past decades focused primarily on managing its agricultural research and development to promote national economics. Similar to the European Union's policy on the same issue, one of the critical and most important activities for Israel are GM crops, over which the government has asserted formal authority. *Israeli law permits the development and growth of genetically modified organisms (GMOs) for research purposes in accordance with requirements established by subsidiary legislation.* This chapter explores Israel's environmental and safety policy on GM food and crops and places it in an explanatory perspective. First, this chapter briefly outlines the Israeli GMO policy, and the differences in the development of GM regulations for different sectors. Second, the chapter continues to analyze the governmental, legal and regulatory frameworks for sectoral activity on GMOs. Third, the chapter follows with an examination of the roles of economic interests, institutional structure and cultural factors in explaining Israel's GMO policy. Fourth, the discussion points to contemporary developments and their implications for sustainability. Finally, the chapter concludes by emphasizing the importance of sustainability in providing a basis for the development of future policy.

54.1 Introduction

This chapter explores Israel's environmental and safety policy on GM food and crops and places it in an explanatory perspective. First, this chapter briefly outlines the Israeli GMO policy, and the differences in the development of GM regulations for different sectors. Second, the chapter continues to analyze the governmental, legal and regulatory frameworks for sectoral activity on GMOs. The evaluated data suggests that Israeli policy places a higher priority on establishing a policy for Agro-technology

---

R. Justo-Hanani, Adv., LL.B, M.Sc, PhD (✉)
Faculty of Life Sciences, Tel Aviv University, Tel Aviv, Israel
e-mail: ronitjus@post.tau.ac.il

© Springer International Publishing Switzerland 2016
G. Steier, K.K. Patel (eds.), *International Food Law and Policy*,
DOI 10.1007/978-3-319-07542-6_54

and economic activities, compared to environmental, health and safety-related issues. Third, the chapter follows with an examination of the roles of economic interests, institutional structure and cultural factors in explaining Israel's GMO policy. Fourth, the discussion points to contemporary developments and their implications for sustainability. Finally, the chapter concludes by emphasizing the importance of sustainability in providing a basis for the development of future policy.

54.2 A Brief Overview of Israeli GMO Policy

Israel's policy on GMOs during the past decades focused primarily on managing its agricultural research and development to promote national economics. Similar to the European Union's policy on the same issue, one of the critical and most important activities for Israel are GM crops, over which the government has asserted formal authority. In 2005, the Ministry of Agriculture and Rural Development (MOARD) established regulation that effectively placed all research and development activity on GM crops and seeds into government scrutiny, including microorganisms related to plants over their lifecycle, such as pollinators and pathogens.[1] No research and development could be used for research purposes without a government permit and all research and development activity has come under government regulation and oversight.

Israeli law permits the development and growth of genetically modified organisms (GMOs) for research purposes in accordance with requirements established by subsidiary legislation. The national GM crop management was placed in the hands of The National Committee for Transgenic Plants (NCTP) under the Plant Protection and Inspection Service (PPIS), which is responsible for determining who receives experiment permits, for which crop or microorganism, and under which conditions or circumstances. The committee also controls the import, marketing and export licensing policies of plant propagating materials, such as seeds. Israel's policy on GM crops corresponds with its highest priority for agricultural biotechnology.

Notably, Israel's agricultural research is highly developed. Universities are widely involved in GM research projects financed by Israeli or international governments and foundations.[2] The results of the research, however, cannot be tested on a large scale or implemented in Israel due to local restrictions on GMO crops. They inform tests abroad.

Israel also established a formal authorization system for novel foods or ingredients through the Public Health Ordinance on Food of 1983. This ordinance grants the Ministry of Health and the National Food Control Services (FCS) the extensive

---

[1] The Seed Regulations (Genetically Modified Plants and Organisms), 2005 (herein: the GM plants regulations).
[2] OECD (2010).

power over food import, marketing and production. It is this legislation, and its subordinate procedures, which require national food authorities to conduct safety assessments for food products in general as well as the authorization of new foods. In Israel, however, the pressure for economic development and the lack of strong, visible public awareness means that the sale and use of GM products are practically permitted, although GM crops cannot be commercially grown in the country. Thus, GMOs are widely used in Israel's pharmaceutical industry and food products include a variety of GM elements.[3]

Other governmental bodies in Israel, whose policies have a major impact on the country's GMO policy, have generally delayed environmental, health and safety related policies. For example, the Ministry of Health has not yet enforced specific restrictions on GM food products. Nor has the Ministry required the mandatory labeling of GMOs, even though many GMOs have been imported to Israel and sold in local food chains. MOARD has also failed to encourage the establishment of regulatory oversight over transgenic animals, despite the research on transgenic animals that has been conducted in the country, such as on mice, cattle or chickens.[4] Thus, even though this research on transgenic animals should fall under government oversight, no regulations are specifically targeting such research yet.

Israel's highest priority for agricultural development is to promote the national agro-biotech industry and biotechnology research. As Shimon Peres, the former, ninth President of Israel, put it, "in twenty five years, Israel increased its agricultural yields seventeen times. This is amazing... agriculture is ninety-five percent science, five percent work."[5] Unlike in many other countries, throughout the last 2 decades in Israel, pressures for economic development meant that agreements for research and development with the seed giant Monsanto were approved, despite a great deal of controversy worldwide.[6] In recent years, Monsanto has been involved in strategic cooperation agreements and investments in Israeli crop bioengineering companies, such as Rosetta Green, as well as acquisitions of their activity.[7] Biotechnology advocates point to the implication of these activities' marking the beginning of a new tech era for Israel, while genetically modified food becomes more prominent. These advocates push toward additions of agricultural technology to Israel's expertise in networking, social media, and mobile technology,[8] a strategy that could backfire as long broader topics of agricultural sustainability, public health and environmental conservation continue to be neglected.

---

[3]Israel Ministry of Health (2014), FAO (2014), Grunpeter (2013).
[4]See also Israel Ministry of Economy (2014).
[5]Cited in Senor and Singer (2009).
[6]For example, a general cooperation agreement was signed by the Ministry of Economy with Monsanto. Recently, the Minister of Economy, Naftali Bennett, was called for removing confidentiality from this agreement as it became clear that confidentiality was imposed in practice, due to the Corporation requirement. For discussions within Israeli social networks see Hildesheim (2014).
[7]**Winrav (2013).**
[8]Shama (2013).

54.3 The Israeli Regulatory Framework of GMO Policy

54.3.1 Research on GM Crops

Guidelines and subsidiary legislation were introduced in Israel, particularly to manage research on GM crops and seeds. The GM plant regulations, for example, apply to transgenic plants and organisms that affect a plant's life cycle. MOARD passed these GM plant regulations as a subsidiary legislation in 2005, after long discussions since the early 1990s, and based the regulation on general authorities provided under the Seeds Law of 1956 and the Plant Protection Law of 1956.[9] These regulations define a GMO as "[a]n organism, including a microorganism, virus, viroid, and any single-celled or multi-celled entity, that has undergone a modification by genetic engineering and is involved with plants in any way during its life cycle." The goals are aimed at the protection from gene leakage and contamination by GMOs. Although the regulation initially applied primarily to seeds and vegetative propagation material, it also applies to the marketing of flowers.[10]

Nonetheless, GMOs are not yet widely cultivated in Israel. GM crops were released into the environment in Israel only as part of controlled field experiments, such as in planting tomato, potato, strawberry, banana, corn, cotton, flowers and eucalyptus plants. According to data provided by MOARD, the size of the study areas ranges from 5 square feet to 10 acres. So far, commercial use of any transgenic plants is not permitted in Israel's agricultural systems.[11]

Israel's main export market for agricultural products is Europe, where consumers' concern over GMO safety is considerable. Accordingly, Israeli importers require an exporter's declaration that the product is GMOs free.[12] Given that most agricultural production is exported to Europe where strict legislations exist there is no commercial cultivation of GM plants in Israel.[13] Thus, MOARD also established a formal authorization and enforcement system for GM crops to ensure that future trade with the EU may continue.

The Plant Protection and Inspection Services (PPIS) at the MOARD is a regulatory authority that enforces regulation through laboratory services and inspectors. Correspondingly, the regulations established a statutory committee, the NCTP, with the authority to oversee and establish procedures for laboratory, greenhouse and field tests. Nonetheless, the potential environmental effect of allowing future commercial activity has not been assessed.[14]

---

[9]Maoz (1996).

[10]Due to commercial intention regarding flowers with a longer shelf life developed in Israel through genetic engineering. See also Blizovsky (2003).

[11]Israel Ministry of Environmental Protection (2010), p. 104.

[12]13 Commercial Service (2013).

[13]OECD (2010), FAO (2014).

[14]Supra note 6.

Unlike in the case of GM plants regulations, there is no legally binding instrument with regard to transgenic and cloned animals in Israel. Thus, transgenic and cloned animals, as such, are not specifically regulated under the Israeli law. GM animals and their reproductive materials, i.e. semen, ova and embryos, nonetheless, are regulated under general animal health and welfare, zootechnical and veterinary legislation and rules. These rules, however, do not specifically address GMOs' environmental and safety-related issues.[15]

54.3.2 GM Food Products Labeling in Israel

The Israeli Ministry of Health reports that Israel imports food products containing genetically modified ingredients. Additionally, the Israeli food industry uses genetically modified raw materials imported from the United States and other countries, such as corn, soybeans and canola.[16]

Israel has chosen to use the trigger of "novel food" for the regulation of GM food. Practically speaking, it is the novel food as defined in internal registration procedures from 2006 that triggers the regulatory process. According to information provided by the Ministry of Health, all new food, including food that was genetically engineered, goes through a risk assessment process before being approved. Such a risk assessment, conducted by the FCS, includes an evaluation of aspects related to its safety, nutrition, and consumption.[17]

In the early 2000s, a discussion on labeling GM food products was held in a Knesset (the Israeli parliament) committee, that was, among others, sparked by the Remedia baby formula scandal.[18] In the Remedia scandal, where the kosher formula made by Remedia was linked to the deaths of three infants, the formula did not contain Vitamin B1, although the package claimed it was in the formula. The resulting crisis renewed the call for examining Israel's food labeling policy, including GM food. As a reaction to the crisis, the Israeli Ministry of Health established an internal committee to examine the issue of GM food. This committee recommended a set of mandatory labeling requirements. However, recommendations were not adopted due to political resistance and pressure exerted by the food industry.[19] To date, legislation specifically regulating the labeling of GMO components in food have not been passed. In addition, governmental regulation has not established a threshold for foods that may be labeled "non-GMO." Consequently, Israel has no streamlined government policy on GM food regulation. Regulations

---

[15]For general discussion on GM animals' regulation in Israel, see Justo-Hanani and Dayan (2011).

[16]Israel Ministry of Health (2014); For information on Israel's labeling, importation and marketing requirements on GM food products, see also the US Commercial Service (2013).

[17]Supra note 3.

[18]See discussion on the Knesset Science and Technology Committee (2003).

[19]Lavie (2013).

are, nonetheless, being prepared during the last decade which will require positive labeling when a product unlisted in the approved GMO list can legally enter the food market and food supply. Manufacturers or importers are, therefore, required under the draft regulations to submit an application for the approval of novel foods to the Novel Food Commission in the FCS at the Ministry of Health, including applications for GMOs, which are not listed in the approval list.[20]

54.3.3 GMO Threats to Biodiversity

GM foods and crops may pose wide-reaching, and often underestimated risks to biodiversity and ecosystems functioning that remain *not* well-understood. The adoption of biotech-agriculture, for example, may damage biodiversity by promoting greater use of pesticides associated with GM-crops, which are especially toxic to other species as well as by introducing exotic genes and organisms into the environment, disrupting ecosystems and natural biotic communities.[21]

In recent years, threats to biodiversity began to occupy a more prominent place on the agro-policy agenda but Israel has not caught up yet. This policy change was, in part, prompted by the rapid expansion in the commercial use of GMOs among highly-industrialized countries. Regulatory agencies in several countries, for example, have started to show awareness for GMOs risks and have increased environmental and food safety regulations. Similar awareness is emerging in Israel, although to a somewhat lesser extent, which is partially due to Israel's current pre-commercialization phases of GMOs. Israel has, overall, been exposed to only a minor threat to biodiversity as compared to other relatively affluent countries, such as the USA.

Notably, Israel, geographically located at the junction of three continents, actually represents a unique habitat for species within a small heterogeneous landscape with an extraordinarily rich biodiversity at the genetic, species, and ecosystem levels.[22] With this vast biodiversity in mind, Israel has also made progress in complying with the provisions of the Convention on Biological Diversity (CBD), which was signed in 1992 and ratified in 1995.

The Ministry of Environmental Protection and the Israel Nature and Parks Authority created a national implementation plan according to the CBD guidelines; the "National Plan for Biodiversity in Israel" was published in January 2010. This National Plan identified and recognized GMOs as a priority for future research. According to the National Plan,

---

[20]Rabinovich (2010), FAO (2014). An updated version of the draft regulation was posted in The Manufacturer Association of Israel website.
[21]Snow et al. (2005).
[22]See also Israel Ministry of Environmental Protection (2009).

[e]ngineered organisms are already in widespread commercial use in some countries; there is a need in learning the potential impact on biodiversity from the accumulated experience and to examine whether it is appropriate to use GMOs for agricultural purposes in light of the composition and the distribution of Israel's biodiversity. Study on the impact of GMOs introduction should be extensive, encompassing not only the bio-physical aspects, but also the cultural, economic and social processes which produce and nourish these threats.[23]

Israel has, however, neither signed nor ratified the 2000 Cartagena Protocol on Biosafety, which was designed to protect biological diversity from undesirable effects of modern biotechnology products.[24] The National Plan for Biodiversity recommends that Israel should join the Protocol and further stresses that

Israel is obligated to obey the Protocol in exporting GMOs products to other nations that signed the protocol. Nevertheless, other nations can export to Israel such products without early notice. This situation is likely to threaten Israel's biodiversity and therefore joining the Protocol is necessary and urgent.[25]

It follows that Israel's policy on GMO regulation still needs to be refined and adapted to the modern challenges to biodiversity and disastrous threats caused by biotechnology.

54.4 Insights into Israel's GMO Regulation and Policy

54.4.1 Economic and International Considerations of GMO Trade

Two structural economic factors can be viewed as motivating the Israeli GM food and crops policy. First, Israel's agricultural biotechnology and research and development funding increased technology progress through research and bio-entrepreneurships, which rapidly changed the agricultural sector and made Israel a world leader in agricultural technologies, particularly in farming under arid conditions.[26] There is a lot of research on the subject and very advanced genetic engineering in plants. The search for new cultivars and the development of new verities involves the application of GM methods. GM crops and seeds are developed mostly by Israel's private sector seeds companies, and also by the agricultural research institutes. Basic and applied research on GMOs is conducted at various sites including seeds companies' research stations, the Agricultural Research Organization (ARO) at the MOARD, the Weizmann Institute of Science, the Hebrew University of Jerusalem, Ben-Gurion University of the Negev,

---

[23]The Ministry of Environmental Protection (2010), pp. 210–211.
[24]See OECD (2011).
[25]Supra note 6, p. 268.
[26]OECD (2010).

Tel-Aviv University, and Bar-Ilan University.[27] One the one hand, the investment horizons of Israel in technology infrastructure, such as advanced facilities and resources for the research community, and, on the other hand, private investments, are expected to grow, as well as the volume and heterogeneity of research conducted on GMOs. Increasing regulatory clarity that affects the aggregated behavior of the economy is, therefore, essential to ensure that technology and research and development spending will bring innovation and competitiveness to Israel. The MOARD, for example, triggered the improvement of risk management strategies, i.e. risk assessment, testing guidelines, accidental contamination with GM crops, and further improved the reliability of future research and services on GM plants. These emerging issues promoted the pressure for appropriate and timely GM crop and seed regulation in the country.[28]

Structural economic factors in Israel are becoming evident while a global trend of agricultural and food production is emerging. Consideration of consumer preferences in export markets, especially in the EU, has, for example, limited the growth and adoption of GMOs in Israeli agriculture[29] because the EU resists GM crop imports. As previously noted, Europe is a main target market for Israel's agricultural production and Israel remains mindful of the EU market. This dependency created a need for parallel regulatory instruments in Israel, starting with risk assessment procedures and safety testing, both of which are essential to introducing Israel into the European market. Aside from ensuring a high-level regulatory environmental and safety oversight and consistency with the Cartagena protocol, regulatory oversight in Israel is expected to ensure that trade agreements and relationships with the single European market remain intact. Israel must, therefore, avoid market distortion which may emanate from GMO contamination or different national treatment of GM crops. Accordingly, there is no commercial cultivation of GM crops in Israel and any experiments in cultivating GM crops are strictly supervised by the NCTP.[30]

The impact of the aforementioned economic growth factors, however, is mixed. On the one hand, the MOARD restrictions on GMOs activity were established after the EU GM crops policy. Accordingly, GM crops are not grown or cultivated in Israel for commercial use to minimize the risk of GMO contamination. On the other hand, GMOs ingredients and products may be imported, sold, and used in the production of food in Israel, and are not required to be labeled in a way that identifies any GMO components.[31] Moreover, Israel, which is a signatory to the World Trade Organization agreements (including the SPS, and TBT agreements),

---

[27] See, for example, MOARD (2012).

[28] Personal interviews with Dr. Edna levy, former head of the NCTP, and Dr. Arie Maoz, former member of the NCTP and Chief Scientist assistant. See also Maoz (1996).

[29] See also Prainsack and Firestine (2006).

[30] As recently reported, however, this situation might be changed in the future, as the MOARD discusses commercial production of biotech crops and seeds, and as the private industry puts pressure to allow their production in Israel. See US & Foreign Commercial Service and US Department of State (2014).

[31] See discussion, Sect. 54.3, above.

maintains relatively few restrictions on agricultural imports.[32] From this perspective, Israel might be closer to the United States GMO policy and not to Europe's, where the US has regulated GM food issues in a rather lax manner. Notably, in this context, the funding for GMO research and testing in Israel derives from Israeli and foreign sources, including the US.[33]

Overall, economic growth and competition poses two major challenges to GMO policy in Israel: First, the volume of agricultural research and technological innovation activities are sources of constant and rapid development. Therefore, there is a need for appropriate regulation, and environmental and safety oversight as part of economic interests. Second, global markets and trade relationships for agricultural crops (export) and food products (import) with the EU and the US remain influential in "shaping" domestic sectoral regulatory policies and oversight priorities in Israel.

54.4.2 Institutional Structure in GMO Regulation: Public Policy Divided

Another explanation for the nature of Israeli GMOs risk policy looks directly at the policy making pattern and institutional structure. Public authority on GMOs in Israel is rather fragmented due to differentiation in authority structures, and the country's policy on GM food and crops mirrors this structure. Each Ministry, as well as regulatory agencies, such as the NCTP or the FCS, operates relatively independently, with their own resources, policy objectives and priorities. Authority for approvals for various uses of GMOs is divided. Lab and field trials on GM crops and recombinant DNA used in agro-biotechnology research go through the MOARD. GM uses in the process of food production go through the Ministry of Health. Livestock biotechnology (such as breeding and genetic manipulation for improved growth rate, milk and egg production) first goes through the MOARD. GM uses in the process of trade and commercialization also go through the Ministry of Economy.

Compared to other industrial nations, the power to influence GMOs policy of Israel's Ministry of Environmental Protection is limited due to this political-authority structure. The GMOs regulation and policy conforms to the preferences of the MOARD, the Ministry of Health and the Ministry of Economy in their respective policy domain. So far, the Ministry of Environmental Protection and Israel Nature and Parks Authority did not play a pivotal role in their regulatory policies on GMOs. In addition, green positions, either "greener" Knesset or those taken by green political parties, did not play a crucial role in driving legislative measures on risks associated with GM food and crops.[34] Recently, however, Sinaia

---

[32]For the Israeli report on food and agricultural import regulation and standards see USDA (2013).
[33]US Congress (2014).
[34]for recent policy initiatives and their results, see discussion, Sect. 54.5, below.

Netanyahu, the Ministry of Environmental Protection's Chief Scientist, initiated a discussion forum on GMO risk policy in Israel, which included representatives from MOARD and Israeli experts, such as ecologists and molecular biologists. This initiative will hopefully pave the way for more streamlined and sustainability-oriented policy in Israel, in line with policy trends and biodiversity-oriented regulation worldwide.

54.4.3 Cultural and Social Values

Cultural and social factors were identified as crucial for public policymaking on GMOs. Accordingly, many scholars believe that religious constrains, social narratives and public endorsement or disapproval of genetic engineering technologies might have a significant effect on the formation of national GMO risk regulation.[35] Social and cultural values permeate the GMO food safety and regulatory discourse in both Israel and abroad. Jewish communities around the world raised concerns about whether products that include GMO components are kosher and thus fulfill strict Jewish dietary standards.[36] This determination has been contested by some Jewish groups in Israel and the United States. On this note, a newspaper has reported that:

> [t]he religious kashrut authority [which certifies products as Kosher] in Israel had ruled that genetic engineering "does not affect kosher status" because genetic material is "microscopic." But there are Jewish groups that dispute this decision and consider GMOs a violation of the biblical prohibition against "kilayim," mixed breeding both in crops and in livestock. Those believing GM products cannot be labeled kosher quote the well-respected 13th century Kabbalist Rabbi Moshe ben Nachman (known as [Nahmanides or] "the Ramban"), who said mankind should not disturb the fundamental nature of creation.[37]

The Ministry of Health recognized the question of religious and social views on GM food. According to information posted on its website,

> [t]here is a possible contradiction between ethical values, such as kashrut, vegetarianism and veganism, as a result of consuming genetically modified food. For example...the consumption of tomatoes with genes that were added from animal is a problem for vegetarians and vegans - people whose faith is opposed to the consumption of meat? Such problems can be solved by clear labeling on the packaging of genetically modified food.[38]

Based on this paragraph, the GMO issue seems to be one of labeling. The National Food Service Authority, however, has not yet taken an official position

---

[35] For the prominent role of cultural and social explanations in current global discussions over GMOs regulation and governance, see for example Jasanoff (2005); Stephan (2012).

[36] US Congress (2014), Aikhenbaum (2013).

[37] Grunpeter (2013).

[38] Supra note 3.

on this issue, or on the validity of the kashrut argument to justify GM-food labeling. In a meeting of the Knesset committee for the Labor, Welfare and Health in July 2013, Yifat Kariv, a Knesset member from the major party Yesh Atid, pressed toward the inclusion of specific labeling requirements for GM food products within draft amendments to Public Health Regulations for food (Nutritional).[39] But legislation specifically regulating the labeling of GMO components in food does not appear to have been passed to date. This GM food policy raises another question which further complicates the debate: what actually determines the Ministry of Health GM food policy preferences?[40] Obviously, the Israeli policy on GM food is not rooted in public concern or strong pressure for stricter and clearer regulatory statements. Thus, while public demands for GM food labeling increased significantly worldwide, setting a path for stricter government oversight, a similar but less strong and intense pressure emerged in Israel.[41]

Specific political and social narratives are also important factors. Such explanations, for example, stress that research area priorities and risk policies must not disregard social values and public concerns. Many scholarly papers have been written about the Israeli ethos of being a knowledge base society, rooted in innovation and technology entrepreneurship and in different context, such as medicine, space, and hi-tech.[42] Some researchers point to positive and uncritical attitudes towards genetic technologies, including controversial ones. [43] Moreover, the Israeli government embraces science and technology as crucial tools to promote innovation and growth, maintaining its global position as a key research and development center, which also comprises academic and research collaborations.[44] More work to promote environmental conservation, preserve biodiversity and support sustainable and organic agriculture is urgently needed from Israel's talented and progress-driven research sector.

Yair Shamir, former Minister of Agriculture and Rural Development, Knesset member, said that

> the future is in GMO... we are just doing it for research purposes and its only restricted to the labs,...in order to prepare ourselves for the future, we examine the possibility of what can be done because when the time comes and people decide that it is time to go for the new way of doing things, we will have been ready for it.[45]

---

[39]Knesset Labor, Welfare and Health Committee (2013).

[40]For an extended discussion on the role of public pressure and risk perception in triggering stricter regulatory policies see Vogel (2003, 2012).

[41]For scholars and ethicists view on the pressing need for a public debate in Israel on GM food issues, see Efron and Ravitsky (2003). See also discussion on contemporary development, Sect. 54.5, below.

[42]See for example, Trajtenberg (2001); Barok (2013).

[43]Cited in Mozersky (2013). See also Prainsack and Firestine (2005), and Tirosh-Samuelson (2009).

[44]USISTF (2014); see also Cohen and Scheer (2013) for general information on Life Sciences research growth in Israel.

[45]Quoted in OKine and Acheapong (2014).

In this respect, the policy path taken on GM plant research clearly fits into Israel's ethos of creativity and as a knowledge society, attributing the foremost priority for science and technology. In fact, the Ministry of Economy' Chief Scientist, Avi Hason, explained the decision to join with Monsanto saying that it will provide Israeli companies opportunity to demonstrate their capacity to respond with innovative research and development.[46] Nonetheless, it remains vital for Israeli research and development to take sustainable and organic agricultural alternatives into greater consideration for the sake of protecting biodiversity and the promote sustainable agriculture rather than to help optimize the dangerous GMO agriculture.

Similarly, in the absence of a persistent, active public demand for GMO scrutiny, the Israeli discourse over GMO issues mainly takes place within the academic community. The interest of Israeli academy in GMO issues, however, primarily pertains to the development of GM crop and weed technologies—not in studying their ecological implications. In fact, the research interests of most Israeli biotechnologists do not even include an environmental risk component. To-date, ecological impacts of GMO remain underexplored in Israel. In the absence of complete data, academic and policy discussions in Israel have mainly taken place with reference to ecological and biodiversity research abroad, through reviews of literature and existing common practices.[47]

54.5 Current Developments of Israeli GMO Policy and Their Implications for Sustainability

Over the past decade, Israeli academic awareness of the risks of GMOs and the concern for improved sustainability have increased[48] even though there is still further space for improvement toward environmental responsibility and protection of biodiversity. Recent discussions demonstrate a growing interest in ecological implications and consumers' choices, especially regarding food labeling. For example, the first academic conference on Agricultural biotechnology safety and GM-food in Israel took place in 2003. It was founded by two academic scholars—Noach Efron and Vardit Ravitsky, and with the support of the Porter School for Environmental Studies at Tel-Aviv University and the Heschel Center for Sustainability, a Non-Governmental Organization (NGO). The conference's aim was to promote a public discourse on responsible development of agricultural biotechnology and GM-food in Israel. Participants of the conference included representatives from Israeli academy, public authorities, industry and NGOs. According to the conference organizers, the hope was "to facilitate an open discussions between

---

[46]Available at: http://www.holesinthenet.co.il/holesinthenet-media-story-34213.
[47]See for example, Dayan (2003, 2005).
[48]See for example, Ulanovsky and Sapir (2013).

Israeli individuals and groups that...rarely meet with each other, in a way that assist [s] in redesign[ing] political decisions, as well as social or industrial perceptions."[49] Thus, much of the conference's focus was on consumer choices, rights and ethical issues,[50] a positive step that set a path for further development in the same direction.

Two years after the first conference, a second academic conference was organized in collaboration with international and Israeli experts. This conference focused on risk assessment and public policy issues.[51] A few years later, in the 36th Annual conference of the Israel Society of Ecology and Environmental Sciences, a session on GMOs risk regulation and governance featured a discussion of science policy in Israel.[52] Much of its focus, in turn, was on designing more environmentally and consumer-oriented regulatory oversight, in line with global policy trends.

The academic discourse about the risks of GMOs increased the pressure for further discussions within the MOARD and the Ministry of Environmental Protection Chief Scientists forums on the need to promote regulatory oversight over GMO issues in Israel so that this oversight may comply, in line with global policy trends. The immediate reaction was a heightened sensitivity to the ambiguity over the essential components needed for an Israeli GMO risk regulation. Accordingly, in 2008, the MOARD Chief Scientist fund offered a financial grant to Israeli academic scholars to explore legal and regulatory frameworks for environmental, health and safety issues of GMOs. About US$ 80,000 have been allocated for a 3-year research period.[53] In 2012 and 2013, the research, among others, led to further discussions about risk regulation of GM crops in an internal symposium with the Chief Scientists of the Ministry of Environmental Protection and MOARD.[54] These important discussions, however, have not yet been translated into further policy statements or further institutional steps.

Recently, there has also been a growth in environmentalists' urgent calls for action on GM food and agricultural biotechnology regulation in Israel. Between 2005 and 2014, environmental activists in Israel waged a number of discussions in the Knesset interior committees and a public campaign. These activists attempt to prevent the commercial cultivation of GM crops, to increase transparency on GM ingredient use in food products, to address environmental and social impacts in public R&D activity, and to ensure regulatory oversight on GMOs with the overall goal of improved food safety and security.

The general context of these activities is the growing recognition of the limits, or even the unwillingness, of the Israeli government to address the potential risks of emerging technologies, which also offer enormous economic benefits. In essence,

---

[49]Supra note 17.
[50]Efron and Ravitsky (2003).
[51]The PSES (2005).
[52]The ISEE (2008).
[53]Supra note 14.
[54]See discussion, Sect. 54.4, above.

the conflict underlying the government's inaction is an apparent short-term economic gain versus long-term food safety and environmental integrity. For example, during a discussion in the Knesset Science and Technology Committee on July 12, 2005, the Commissioner of Future Generations and environmental activists called for the adoption of a precautionary approach to emerging technologies, by engaging environmental and safety considerations in national research and development policies.[55] Other examples include *Greenpeace's* recent call for action to the Israeli government to provide transparency for consumers. *Greenpeace* is an international organization for the protection of wildlife and the environment. In the same *Greenpeace* action, a unique coalition of members from the Israeli Bio-Organic Agriculture Association, food allergy association and *Greenpeace Israel and Mediterranean* have joined forces to initiate a discussion in the Knesset in an attempt to balance the discourse around agricultural biotechnology in Israel. This imbalanced discourse was ripe for change because it had been dominated by strong technology and biotechnology supporters.[56] *Greenpeace* also invited Israeli citizens to sign a petition on labeling GM ingredients in food products.[57] The results of these campaigns and calls for action were significant. Over the past 10 years, the Knesset devoted a number of interior committee discussions to GMO issues. The increase in NGOs concerns has also led to some important lobbying activity in the Knesset.[58]

The aforementioned increase in NGOs' activities, and academic awareness, however, has not yet led to the much-needed and important policy changes. Since 2006, the Israeli Ministry of Health has only formulated regulations for Novel foods, including GM food labeling, but none of these regulations have come into force yet.[59] Although the MOARD has addressed GM crops and seeds, the issue of GM animal use, either for experiments or future commercial activity, remains unregulated. Moreover, the MOARD has not yet ruled out the possibility of commercial cultivation of GM crops in the future.

Thus, while these growths in awareness are important, they represent a far cry from the "greening" of Israel's GMOs policy. In marked contrast to many other heads of state, no Israeli Prime Minister has ever publicly expressed either any interest in overarching policy on EHS aspects of GMOs, nor concerns for the future of Israeli moves.

---

[55]Knesset Science and Technology Committee (2005).

[56]Aikhenbaum (2014).

[57]Greenpeace Israel (2013).

[58]For example, Knesset members Michal Rosin and Dov Hanin from the social and environmental lobby alliance submitted a Knesset query to Minister of Economy Naftali Bennett on the topic of Monsanto agreement. This query has been addressed in June 2014.

[59]See discussion, Sect. 54.3, above.

54.6 Conclusion

This chapter sets out to describe and explain the Israeli regulatory policy for GM crop and food safety. The aforementioned shows that Israeli GMO policy has evolved differently in the agricultural as opposed to the food sector, and that a higher priority was given to the promotion of economic and science and technology development, as compared to environment and consumers choice issues. It has been shown that the Israeli GMO policy path cannot be attributed to a single factor. Instead, this chapter explores the role of international trade and economic interests, institutional structure and cultural factors in explaining Israel's GMOs policy. Finally, the chapter demonstrated an increase in academic awareness and public activities, which resemble those activities of global policy trends more closely, and may profoundly impact policy in the future.

Acknowledgements The author would like to thank Prof. David Vogel for helpful discussions and generous hospitality in UC Berkeley. The author would also like to thank Prof. Julian Kinderlerer for sharing expertise and invaluable guidance during her year as a visiting scholar at Sheffield Institute for Biotechnology Law and Ethic (UK). She also extends her special appreciation and thanks to her PhD advisor Prof. Tamar Dayan, who provided her with scientific guidance, encouragement and support for this research. The material is based upon research supported by the Israel's Ministry of Agriculture and Rural Development (MOARD) Chief Scientists Fund (grant no. 891-0210-08, P.I., with Prof. Dayan). Finally, thanks to the editors of this book for their helpful review. The author is solely responsible for the chapter's content.

References

Aikhenbaum J (2013) GM plant commercialization: Jewish aspects, questions and answers (In Hebrew). Available at: http://www.tevaivri.org.il/Resources/handasa_genetit.pdf. Accessed 27 Sept 2014

Aikhenbaum J (2014) GMOs agriculture or sustainable agriculture? A position paper. Greenpeace Mediterranean (In Hebrew)

Barok D (2013) Cooperation in Space between Europe and Israel in light of the recent ESA-ISA agreement. In: Hulsroj P, Pagkratis S, Baranes B (eds) Yearbook on space policy 2010/2011: the forward look. Springer, Wien

Blizovsky A (2003) Engineered plants – fears and necessity. Ha-Yad'an (In Hebrew). Available at: http://www.hayadan.org.il/gal60gm-101003. Accessed 29 Dec 2014

Cohen T, Scheer S (2013) After tech success, Israel seeks life sciences growth. Available at: http://www.reuters.com/article/2013/06/06/us-israel-biomed-idUSBRE9550IU20130606. Accessed 27 Sept 2014

Dayan T (2003) In towards public discourse on agricultural biotechnology and genetically modified food in Israel. In: Proceedings of the 1st porter conference for sustainability, Tel-Aviv University, Israel, 29–30 May 2003 (In Hebrew)

Dayan T (2005) In Brave New World Technologies GMO/NANO: freedom of scientific inquiry and society's right to protect the environment and public health. Invited presentation, Tel Aviv University, 15–17 May 2005

Efron N, Ravitsky V (2003) Towards public discourse on agricultural biotechnology and genetically modified food in Israel. In: Proceedings of the 1st porter conference for sustainability, Tel-Aviv University, Israel, 29–30 May 2003 (In Hebrew)
Food and Agriculture Organization of the United Nations (FAO) (2014) GM food platform: Israel. Available at: http://www.fao.org/food/food-safety-quality/gm-foods-platform/browse-information-by/country/country-page/en/?cty=ISR. Accessed 27 Sept 2014
Greenpeace Israel (2013) A petition (In Hebrew). Available at: http://www.greenpeace.org/israel/he/getinvolved/GMO/. Accessed 27 Sept 2014
Grunpeter MA (2013) GMOs, A global debate: Israel a center for study, Kosher concerns. Epoch Times, 5 August 2013
Hildesheim E (2014) What are they afraid of? Israel imposed confidentiality on the agreement with Monsanto (In Hebrew). Available at: http://kalkala-amitit.blogspot.com/2014/06/. Accessed 29 Dec 2014
Israel Ministry of Environmental Protection (2009) Fourth Country Report to the United Nations convention on biological diversity. Available at: http://www.cbd.int/doc/world/il/il-nr-04-en.pdf
Israel Ministry of Environmental Protection (2010) The national plan for biodiversity (In Hebrew). Available at: http://www.sviva.gov.il/InfoServices/ReservoirInfo/DocLib2/Publications/P0501-P0600/P0540.pdf. Accessed 27 Sept 2014
Israel Ministry of Agriculture and Rural Development (MOARD) (2012) Israel's agriculture (In Hebrew). Available at: http://www.moag.gov.il/agri/files/Israel%27s_Agriculture_Booklet.pdf. Accessed 29 Dec 2014
Israel Ministry of Health (2014) Genetically modified food (In Hebrew). Available at: http://www.health.gov.il/unitsoffice/hd/ph/fcs/novelfood/pages/engfood.aspx. Accessed 27 Sept 2014
Israel Ministry of Economy (2014) Technological Incubators Program. CEO Provisions No. 8.3. Available at: http://www.moital.gov.il/NR/exeres/427E7C02-4026-4F7C-B897-9797E0EF5562.htm. Accessed 27 Sept 2014
Israel Society of Ecology and Environmental Sciences (2008) GMOs science policy. In: The 36th annual conference, 17–18 June 2008
Jasanoff S (2005) Designs on nature: science and democracy in Europe and the United States. Princeton University Press, Princeton
Justo-Hanani R, Dayan T (2011) Regulating risks of agricultural biotechnology: research on and use of modern biotechnology in animal agriculture in Israel – regulatory aspects. Internal report submitted for the MOARD Chief Scientists Fund, 13 January 2011 (In Hebrew)
Knesset Science and Technology Committee (2003) Genetically engineered food. Protocol No. 33, Sixteen Knesset, Second Session, 18 November 2003 (In Hebrew). Available at: http://www.knesset.gov.il/protocols/data/html/mada/2003-11-18.html. Accessed 27 Sept 2014
Knesset Science and Technology Committee (2005) Chances and risks of breakthrough technologies – nano and genetic engineering and how they impact on human and its surroundings. Protocol No. 134, Sixteen Knesset, Third Session, 12 July 2005 (In Hebrew). Available at: http://www.knesset.gov.il/protocols/data/html/mada/2005-07-12.html. Accessed 27 Sept 2014
Knesset Labor, Welfare and Health Committee (2013) Protocol No. 47, Nineteenth Knesset, First Session, 3 July 2013 (In Hebrew)
Lavie A (2013) The Ministry of Health will require GM products labeling, Israel News, 22 September 2013 (In Hebrew). Available at: http://www.nrg.co.il/online/1/ART2/508/322.html. Accessed 27 Sept 2014
Maoz A (1996) GM agriculture goes into the environment. Synthesis 12:43 (In Hebrew). Available at: http://telem.openu.ac.il/courses/c20237/gntengagr-s.htm. Accessed 27 Sept 2014
Mozersky J (2013) Risky genes: genetics, breast cancer and Jewish identity. Routledge, London
OKine CB, Acheapong J (2014) The future for agriculture is GMOs. The Times of Israel, 2 April 2014
OECD (2010) OECD review of agricultural policies: Israel 2010. Available at: http://www.oecd-ilibrary.org/agriculture-and-food/oecd-review-of-agricultural-policies-israel-2010_9789264079397-en. Accessed 27 Sept 2014

OECD (2011) Environmental performance reviews. Israel 2011. Available at: http://www.oecd-ilibrary.org/environment/oecd-environmental-performance-reviews-israel-2011_9789264117563-en. Accessed 27 Sept 2014

Prainsack B, Firestine O (2005) Genetically modified survival: red and green biotechnology in Israel. Sci Cult 14(4):355–372

Prainsack B, Firestine O (2006) Science for survival. Biotechnology regulation in Israel. Sci Public Policy 33(1):33–46

Rabinovich M (2010) Food products labeling: a comparative survey. The Knesset Research and Information Center. 8 February 2010 (In Hebrew). Available at: http://www.knesset.gov.il/mmm/data/pdf/m02463.pdf. Accessed 27 Sept 2014

Senor D, Singer S (2009) Start-up nation: the story of Israel's economic miracle. Grand Central Publishing, New York

Shama D (2013) Israeli agritech IPO could be first of a controversial wave. The Times of Israel, 21 November 2013. Available at: http://www.timesofisrael.com/israeli-agritech-ipo-could-be-first-of-a-controversial-wave/. Accessed 27 Sept 2014

Snow AA, Andow DA, Gepts P, Hallerman EM, Power A, Tiedje GM, Wolfenbarger AA (2005) Genetically modified organisms and the environment: current status and recommendations. Ecol Appl 15:377–404

Stephan HR (2012) Revisiting the transatlantic divergence over GMOs: toward a cultural-political analysis. Glob Environ Polit 12(4):104–124

Tirosh-Samuelson H (2009) Jewish philosophy, human dignity and the new genetics. In: Sutton SD (ed) Biotechnology: our future as human beings and citizens. State University of New York, New York

The Porter school for Environmental Studies (2005) "Brave New World" Technologies GMOs/NANO: freedom of scientific inquiry and society's right to protect the environment and public health. Tel-Aviv University, 15–17 May 2005

Trajtenberg M (2001) Innovation in Israel 1968–97: a comparative analysis using patent data. Res Policy 30:363–389

US Congress (2014) Restriction on genetically modified organisms: Israel. Report. The Law Library of Congress. Available at: http://www.loc.gov/law/help/restrictions-on-gmos/israel.php. Accessed 27 Sept 2014

US Commercial Service (2013) Israel: trade regulations, customs and standards. Available at: http://www.export.gov/israel/marketresearchonisrael/countrycommercialguide/eg_il_026160.asp. Accessed 27 Sept 2014

USDA (2013) Israel: food and agricultural import regulations and standards – narrative. Available at: http://gain.fas.usda.gov/Recent%20GAIN%20Publications/Food%20and%20Agricultural%20Import%20Regulations%20and%20Standards%20-%20Narrative_Tel%20Aviv_Israel_12-24-2013.pdf. Accessed 29 Dec 2014

US & Foreign Commercial Service and US Department of State (2014) Doing business in Israel. 2014 country commercial guide for U.S. companies. Available at: http://www.export.gov/israel/eg_il_076545.asp. Accessed 29 Dec 2014

Ulanovsky H, Sapir Y (2013) Environmental risks and benefits of GM plants used in agriculture in Israel (In Hebrew). Ecol Environ 4(4):340–342

US ISRAEL Science and technology Foundation (2014) Israel 2028: vision and strategy. Available at: http://www.usistf.org/wp-content/uploads/2014/03/U.S.-Israel-Science-and-Technology-Collaboration-2028-Israel-2028-Appendix-2.pdf. Accessed 27 Sept 2014

Vogel D (2003) The Hare and the Tortoise revisited: the new politics of consumer and environmental regulation in Europe. Br J Polit Sci 33(4):557–580

Vogel D (2012) The politics of precaution: regulating health, safety, and environmental risks in Europe and the United States. Princeton University Press, Princeton

Winrav G (2013) Monsanto acquires Rosetta Green activity. February 2013. Available at: http://www.globes.co.il/en/article-1000818988. Accessed 27 Sept 2014